Encyclopedia of Earthquake Engineering

Michael Beer
Ioannis A. Kougioumtzoglou
Edoardo Patelli • Siu-Kui Au
Editors

Encyclopedia of Earthquake Engineering

Volume 2

F–P

With 2431 Figures and 278 Tables

Editors
Michael Beer
Institute for Computer Science in
Civil Engineering
Gottfried Wilhelm Leibniz
University Hannover
Hannover, Germany

Edoardo Patelli
Institute for Risk & Uncertainty
and Centre for Engineering
Sustainability
Liverpool, UK

Ioannis A. Kougioumtzoglou
Department of Civil Engineering &
Engineering Mechanics
Columbia University
New York, NY, USA

Siu-Kui Au
Institute for Risk & Uncertainty
and Centre for Engineering
Dynamics
Liverpool, UK

ISBN 978-3-642-35343-7   ISBN 978-3-642-35344-4 (eBook)
ISBN 978-3-642-35345-1 (print and electronic bundle)
DOI 10.1007/978-3-642-35344-4

Library of Congress Control Number: 2015946601

Springer Heidelberg New York Dordrecht London
© Springer-Verlag Berlin Heidelberg 2015
This work is subject to copyright. All rights are reserved by the Publisher, whether the whole or part of the material is concerned, specifically the rights of translation, reprinting, reuse of illustrations, recitation, broadcasting, reproduction on microfilms or in any other physical way, and transmission or information storage and retrieval, electronic adaptation, computer software, or by similar or dissimilar methodology now known or hereafter developed.
The use of general descriptive names, registered names, trademarks, service marks, etc. in this publication does not imply, even in the absence of a specific statement, that such names are exempt from the relevant protective laws and regulations and therefore free for general use.
The publisher, the authors and the editors are safe to assume that the advice and information in this book are believed to be true and accurate at the date of publication. Neither the publisher nor the authors or the editors give a warranty, express or implied, with respect to the material contained herein or for any errors or omissions that may have been made.

Printed on acid-free paper

Springer-Verlag GmbH Berlin Heidelberg is part of Springer Science+Business Media (www.springer.com)

Preface

The scope of the *Encyclopedia of Earthquake Engineering* covers the interaction between earthquake events and our engineering installations and infrastructures. It is expected to range over buildings, foundations, underground constructions, lifelines and bridges, roads, embankments, and slopes. Although a plethora of references exist in the context of treating/addressing individual earthquake engineering topics, there is no literature dealing with earthquake engineering in a comprehensive, versatile, and unified manner. In this regard, the extreme event of an earthquake has a multifaceted impact on a variety of activities. These include day-to-day operations of public and private services which have greatly increased their exposure to risk.

The Encyclopedia is designed to inform technically inclined readers about the ways in which earthquakes can affect engineering installations and infrastructures and how engineers would go about designing against, mitigating, and remediating these effects. It is also designed to provide cross-disciplinary and cross-domain information to domain experts. Specifically, the proposed work introduces a coupling between traditional topics of earthquake engineering, such as geotechnical/structural engineering, topics of broader interest such as geophysics, and topics of current and emerging value to industrial applications, such as risk management. In this regard, risk management is included in a comprehensive manner, which addresses the dimension and complexity of earthquake hazards. This elucidates the vital connection of the technical contents to the societal context. The main benefit of the Encyclopedia is its breadth of coverage to provide quick information on a substantial level to virtually all groups of readers from academia, industry, and the general population who would like to find out more in the area of earthquake engineering.

Overall, this work is a concerted effort to provide with a holistic perspective on earthquake engineering–related issues of recent currency. Its innovative, modular, and continuously updated form facilitates a potent, self-contained, and readily accessible exposition of multi/interdisciplinary elements in the broad field of earthquake engineering, thus enabling the researcher/practitioner/designer to identify links and potential future research themes in an efficient and timely manner.

Acknowledgments

The Editors greatly appreciate the effort made by the section editors, the contributions of the authors related to work of high standards, as well as the determination of the reviewers to preserve high quality levels. Further, the Editors would like to thank Springer for its guidance and leadership on making this great project a reality.

Michael Beer, Dr.-Ing.
Professor and Head
Institute for Computer Science in Civil Engineering
Gottfried Wilhelm Leibniz University Hannover
Hannover, Germany

Ioannis A. Kougioumtzoglou, Ph.D.
Assistant Professor
Department of Civil Engineering & Engineering Mechanics
Columbia University
New York, NY, USA

Edoardo Patelli, Ph.D.
Lecturer in Uncertainty and Engineering
Institute for Risk & Uncertainty and Centre for Engineering Sustainability
Liverpool, UK

Siu-Kui Au, Ph.D.
Ir Professor of Uncertainty, Reliability and Risk
Institute for Risk & Uncertainty and Centre for Engineering Dynamics
Liverpool, UK

About the Editors

Michael Beer Institute for Computer Science in Civil Engineering, Gottfried Wilhelm Leibniz University Hannover, Hannover, Germany

Michael Beer, Dr.-Ing., obtained his degrees in civil engineering from the Technical University of Dresden, Germany. He worked at Rice University as a fellow of the Alexander von Humboldt-Foundation and held a faculty position at National University of Singapore in the Department of Civil and Environmental Engineering. Dr. Beer was a professor and the founding director of the Institute for Risk and Uncertainty in the University of Liverpool. He is now professor and head of the Institute for Computer Science in Civil Engineering at Gottfried Wilhelm Leibniz University Hannover, Germany, as well as a guest professor in Tongji University, Shanghai, China and a part-time professor in the Institute for Risk and Uncertainty, University of Liverpool.

Ioannis A. Kougioumtzoglou Department of Civil Engineering & Engineering Mechanics, Columbia University, New York, NY, USA

Ioannis Kougioumtzoglou, Ph.D., obtained his M.Sc. and Ph.D. degrees from the civil and environmental engineering department of Rice University, Texas, USA. He also holds a five-year Diploma in Civil Engineering from the National Technical University of Athens (NTUA), Greece, and is a professional/chartered civil engineer in Greece. He is currently an assistant professor at the Department of Civil Engineering and Engineering Mechanics at Columbia University, New York, USA.

Edoardo Patelli Institute for Risk & Uncertainty and Centre for Engineering Sustainability, Liverpool, UK

Edoardo Patelli, Ph.D., obtained his degrees in nuclear engineering and Ph.D. in radiation science and technology from the Politecnico di Milano, Italy. He worked at the Institute for Engineering Mechanics, University of Innsbruck, Austria. He is currently a lecturer in uncertainty and engineering at the Institute for Risk and Uncertainty at the University of Liverpool, UK, and the head of computational technology for the EPSRC Centre for Doctoral Training in Risk and Uncertainty Quantification. Patelli is also an honorary member of the National Tsing Hua University, Taiwan.

Siu-Kui Au Institute for Risk & Uncertainty and Centre for Engineering Dynamics, Liverpool, UK

Siu Kui Au, Ph.D., is a chartered civil engineer in Hong Kong and obtained his B.Eng. and M.Phil. from the Hong Kong University of Science and Technology and Ph.D. from the California Institute of Technology (USA). He is currently a professor of uncertainty, reliability and risk at the Institute for Risk and Uncertainty. Before joining Liverpool he has held faculty positions at the Nanyang Technological University (Singapore) and the City University of Hong Kong. He was a visiting professor at the Tokyo City University (Japan) and Wuhan University (China).

Section Editors

Section: Artificial and Other Sources and Mechanisms

Hans Thybo Department of Geography and Geology, University of Copenhagen, København K, Denmark

Section: Aseismic Design

Dimitrios G. Lignos Civil Engineering, McGill University, Montreal, QC, Canada

Section: Case Histories

Rafael Riddell Departamento Ingeniería Estructural y Geotécnica, Pontificia Universidad Católica de Chile, Santiago, Chile

Section: Computational and Sensing

Muneo Hori Earthquake Research Institute, University of Tokyo, Bunkyo, Tokyo, Japan

Section: Computational Rock Mechanics

Jui-Pin Wang Department of Civil and Environmental Engineering, The Hong Kong University of Science and Technology, Kowloon, Hong Kong

Section: Construction Techniques

Polat Gulkan Civil Engineering Department, Cankaya University, Ankara, Turkey

Section: Disaster Recovery and Reconstruction and Loss Modeling

Erica Seville Resilient Organisations, University of Canterbury, Sheffield, New Zealand

John Vargo Department of Accounting and Information Systems, University of Canterbury, Christchurch, New Zealand

Section: Geotechnical Engineering

Kok-Kwang Phoon Department of Civil and Environmental Engineering, National University of Singapore, Singapore, Singapore

Hesham El Naggar Geotechnical Research Centre, Department of Civil and Environmental Engineering, The University of Western Ontario, London, ON, Canada

Dimitrios Zekkos Department of Civil and Environmental Engineering, University of Michigan, Ann Arbor, MI, USA

Section: Mathematical Tools

Sondipon Adhikari College of Engineering, Swansea University, Swansea, UK

Subhamoy Bhattacharya Chair in Geomechanics, Faculty of Engineering and Physical Sciences, University of Surrey, Guildford, UK

Section: Paleoseismology

Sarah J. Boulton Centre for Research in Earth Sciences, Fitzroy 115, Plymouth University, Plymouth, Devon, UK

Iain Stewart School of Geography, Earth and Environmental Sciences, Plymouth University, Plymouth, Devon, UK

Section: Random Vibration

Pol D. Spanos Department of Mechanical Engineering and Materials Science, Rice University, Houston, TX, USA

Antonina Pirrotta Dipartimento di Ingegneria Civile, Ambientale, Aerospaziale, dei Materiali, Università di Palermo, Palermo, Italy

Section: Reliability and Robustness

Hector A. Jensen Department of Civil Engineering, Santa Maria University, Casilla, Valparaiso, Chile

Section: Remote Sensing

Salvatore Stramondo Istituto Nazionale di Geofisica e Vulcanologia, Rome, Italy

Section: Retrofitting and Strengthening

Agathoklis Giaralis City University London, Room: C174, School of Engineering and Mathematical Sciences, London, UK

Andreas Kappos School of Engineering and Mathematical Sciences, City University London, London, UK

Section: Risk Management: Decision Analysis

Ayhan Irfanoglu School of Civil Engineering, Purdue University, West Lafayette, IN, USA

Section: Risk Mitigation Policies and Approaches

Bijan Khazai Center for Disaster Management and Risk Reduction Technology (CEDIM), Karlsruhe Institute of Technology, Karlsruhe, Germany

Section: Seismic Risk Assessment

Fatemeh Jalayer Department of Structures for Engineering and Architecture, University of Naples "Federico II", Naples, Italy

Carmine Galasso Department of Civil, Environmental and Geomatic Engineering and Institute for Risk and Disaster Reduction University College London, University College London, London, UK

Section: Sensors and Sensor Systems

Jens Havskov Department of Earth Science, University of Bergen, Bergen, Norway

Gerardo Alguacil Física Teórica y del Cosmos, Universidad de Granada, Beiro, Granada, Spain

Section: Structural Analysis for Earthquake-Resistant Design

Charis J. Gantes Institute of Steel Structures, School of Civil Engineering, National Technical University of Athens, Athens, Greece

Section: Structural Health Monitoring

Eleni Chatzi Institute of Structural Engineering ETH Zürich HIL E, Zürich, CH, Switzerland

Costas Papadimitriou Department of Mechanical Engineering, University of Thessaly, Volos, Greece

Geert Lombaert Department of Civil Engineering, KU Leuven, Leuven, Belgium

Section: Tectonic Sources and Mechanisms

Jan Sileny Institute of Geophysics, Academy of Sciences, Praha, Czech Republic

Section: Volcanic Seismology

Silvio De Angelis Earth, Ocean and Ecological Sciences, School of Environmental Sciences, University of Liverpool, Liverpool, UK

Contributors

S. Abhinav Department of Civil Engineering, Indian Institute of Science, Bangalore, Karnataka, India

Nick Ackerley Nanometrics, Inc, Kanata, Ottawa, ON, Canada

Christoph Adam Department of Engineering Science, Unit of Applied Mechanics, University of Innsbruck, Innsbruck, Austria

Sondipon Adhikari College of Engineering, Swansea University, Swansea, UK

Vadim M. Agafonov Center for Molecular Electronics, Moscow Institute of Physics and Technology, Moscow, Russia

John J. Sanchez Aguilar Departamento de Geociencias, Universidad Nacional de Colombia, Bogotá, Colombia

Nopdanai Ajavakom Department of Mechanical Engineering, Chulalongkorn University, Bangkok, Thailand

H. Serdar Akyüz İstanbul Teknik Üniversitesi, Maden Fakültesi, Jeoloji Müh. Bölümü, Ayazağa, İstanbul, Turkey

David Alexander University College London, London, England

N. A. Alexander Civil Engineering Department, University of Bristol, Bristol, UK

Gerardo Alguacil Instituto Andaluz de Geofísica, University of Granada, Granada, Spain

Clive Allen Faculty of Engineering and Built Environment, The University of Newcastle, Callaghan, NSW, Australia

Hernán G. Alvarado Departamento de Geociencias, Universidad Nacional de Colombia-Sede Bogotá, Bogotá, Colombia

Sotiris Argyroudis Department of Civil Engineering, Aristotle University, Thessaloniki, Greece

Aysegul Askan Civil Engineering Department, Middle East Technical University, Ankara, Turkey

Domenico Asprone Department of Structures for Engineering and Architecture, University of Naples "Federico II", Naples, Italy

Kuvvet Atakan Department of Earth Science, University of Bergen, Bergen, Norway

Siu-Kui Au Institute for Risk & Uncertainty and Centre for Engineering Dynamics, Liverpool, UK

Luis David Avendaño-Valencia Department of Mechanical & Aeronautical Engineering, Stochastic Mechanical Systems and Automation (SMSA) Laboratory, University of Patras, Patras, Greece

Jack Baker Stanford University, Stanford, CA, USA

Richard J. Bathurst GeoEngineering Centre at Queen's-RMC Civil Engineering Department, Royal Military College of Canada, Kingston, ON, Canada

Josep Batlló Instituto Dom Luiz (IDL), Faculdade de Ciências da Univ. de Lisboa, Lisbon, Portugal

Sarah Beaven Department of Geological Sciences, University of Canterbury, Christchurch, New Zealand and Natural Hazards Research Group, University of Canterbury, Christchurch, New Zealand

André T. Beck Department of Structural Engineering, São Carlos School of Engineering, University of São Paulo, São Carlos, SP, Brazil

Julia Becker GNS Science, Lower Hutt, New Zealand

Danilo Beli Department of Computational Mechanics, Faculty of Mechanical Engineering, UNICAMP, Campinas, SP, Brazil

Andrew F. Bell School of GeoSciences, University of Edinburgh, Edinburgh, UK

Amadeo Benavent-Climent Department of Structural Mechanics and Industrial Constructions, Polytechnic University of Madrid, Madrid, Spain

Fouad Bendimerad Earthquakes and Megacities Initiative, Quezon City, Philippines

Ninfa L. Bennington Department of Geoscience, University of Wisconsin-Madison, Madison, WI, USA

Djillali Benouar University of Bab Ezzouar, Algiers, Algeria

Mounir Khaled Berrah Ecole Nationale Polytechnique, Algiers, Algeria

Sanjaya Bhatia UNISDR Recovery Platform, Kobe, Japan

Subhamoy Bhattacharya University of Surrey, Guildford, UK

Christian Bignami Istituto Nazionale di Geofisica e Vulcanologia, National Earthquake Center, Rome, Italy

Mustafa Bilal Civil Engineering Department, Middle East Technical University, Ankara, Turkey

Daniel Binder Zentralanstalt für Meteorologie und Geodynamik, Vienna, Austria

Gian Maria Bocchini Institute of Geodynamics, National Observatory of Athens, Athens, Greece

Teddy Boen PT Teddy Boen Konsultan, Jakarta, Indonesia

I. Bondár Research Centre for Astronomy and Earth Sciences of the Hungarian Academy of Sciences, Budapest, Hungary

Franco Bontempi Department of Structural and Geotechnical Engineering, Sapienza University of Rome, Rome, Italy

Moses Kent Borinaga Earthquakes and Megacities Initiative, Quezon City, Philippines

Peter Bormann Formerly GFZ German Research Center for Geosciences, Potsdam, Germany

Jitendra Kumar Bothara NSET, Kathmandu, Nepal and Miyamoto Impact, Christchurch, New Zealand

Sarah J. Boulton Plymouth University, Centre for Research in Earth Sciences, Plymouth University, Devon, UK

Nouredine Bourahla Civil Engineering Department, University Saâd Dahlab, Blida, Algeria

Francesca Bovolo Fondazione Bruno Kessler (FBK), Povo, Trento, Italy

Enrico Brandmayr Department of Mathematics and Geosciences, University of Trieste, Trieste, Italy and The Abdus Salam International Centre for Theoretical Physics, SAND Group, Trieste, Italy

Jochen Braunmiller University of South Florida, Tampa, FL, USA

Ewald Brückl Department of Geodesy and Geoinformation, TU Wien, Vienna, Austria

Marco Breccolotti Department of Civil and Environmental Engineering, University of Perugia, Perugia, Italy

Florent Brenguier Institut des Sciences de la Terre, University of Grenoble, Grenoble, France

Christopher Bronk Ramsey Oxford Radiocarbon Accelerator Unit, University of Oxford, Oxford, UK

Charlotte Brown Christchurch Polytechnic Institute of Technology, Christchurch, New Zealand

David Brunsdon Kestrel Group Ltd., Wellington, New Zealand

Lorenzo Bruzzone Department of Information Engineering and Computer Science, University of Trento, Povo, Trento, Italy

Christian Bucher Institute of Building Construction and Technology, Vienna University of Technology, Wien, Austria

Pierfrancesco Cacciola School of Environment and Technology, University of Brighton, Brighton, UK

Luigi Callisto Department of Structural and Geotechnical Engineering, Sapienza Università di Roma, Rome, Italy

Luigi Carassale Department of Civil, Chemical and Environmental Engineering, University of Genova, Genoa, Italy

Omar-Dario Cardona Universidad Nacional de Colombia, Manizales, Colombia

Mehmet Çelebi Earthquake Science Center, US Geological Survey, Menlo Park, CA, USA

Yunbyeong Chae Department of Civil and Environmental Engineering, Old Dominion University, Norfolk, VA, USA

Chung-Han Chan Department of Geosciences, National Taiwan University, Taipei, Taiwan, ROC and Earth Observatory of Singapore, Nanyang Technological University, Singapore, Singapore

Ya-Ting Chan Department of Geosciences, National Taiwan University, Taipei, Taiwan

A. A. Chanerley School of Architecture, Computing and Engineering (ACE), University of East London, London, UK and The University of Auckland, Auckland, New Zealand

Alice Yan Chang-Richards The University of Auckland, Auckland, New Zealand

Eleni N. Chatzi Department of Civil, Environmental and Geomatic Engineering, ETH Zurich, Institute of Structural Engineering, Zurich, CH, Switzerland

Da-Yi Chen Central Weather Bureau, Taipei, Taiwan

Jianbing Chen School of Civil Engineering & State Key Laboratory for Disaster Reduction in Civil Engineering, Tongji University, Shanghai, China

Chin-Tung Cheng Disaster Prevention Technology Research Center, Sinotech Engineering Consultants, Inc., Taipei, Taiwan, ROC

Tai-Lin Chin Department of Computer Science and Information Engineering, National Taiwan University of Science and Technology, Taipei, Taiwan

Marco Chini Luxembourg Institute of Science and Technology (LIST), Environmental Research and Innovation Department (ERIN), Belvaux, Luxembourg

Christis Z. Chrysostomou Department of Civil Engineering and Geomatics, Cyprus University of Technology, Limassol, Cyprus

Federico Cluni Department of Civil and Environmental Engineering, University of Perugia, Perugia, Italy

Simona Colombelli Department of Physics, University of Naples Federico II, Federico II – AMRA S.c.ar.l, Napoli, Italy

Joel P. Conte Department of Structural Engineering, University of California at San Diego, CA, USA

Vincenzo Convertito Istituto Nazionale di Geofisica e Vulcanologia, Osservatorio Vesuviano, Napoli, Italy

R. Corotis Department of Civil, Environmental and Architectural Engineering, University of Colorado, Boulder, CO, USA

Wayne C. Crawford Institut de Physique du Globe de Paris, Sorbonne Paris Cité, Univ Paris Diderot, UMR 7154 CNRS, Paris, France

Michele Crosetto CTTC Division of Geomatics, Av. C.F. Gauss, Castelldefels, Spain

Laura D'Amico School of Environment and Technology, University of Brighton, Brighton, UK

James Edward Daniell Geophysical Institute, Karlsruhe Institute of Technology, Karlsruhe, Germany

Shideh Dashti Department of Civil, Environmental, and Architectural Engineering, University of Colorado at Boulder, Boulder, CO, USA

Craig A. Davis Waterworks Engineer, Los Angeles Department of Water and Power, Los Angeles, CA, USA

Dina D'Ayala Department of Civil, Environmental and Geomatic Engineering, University College London, London, UK

Silvio De Angelis Earth, Ocean and Ecological Sciences, School of Environmental Sciences, University of Liverpool, Liverpool, UK

José Roberto de França Arruda Department of Computational Mechanics, Faculty of Mechanical Engineering, UNICAMP, Campinas, SP, Brazil

Flavia De Luca Department of Civil Engineering, University of Bristol, Bristol, UK

Raffaele De Risi Department of Structures for Engineering and Architecture, University of Naples Federico II, Naples, Italy

Guido De Roeck Department of Civil Engineering, KU Leuven, Leuven, Belgium

A. Deraemaeker FNRS Research Associate, Building Architecture and Town Planning (BATir), Brussels, Belgium

Mario Di Paola Dipartimento di Ingegneria Civile, Ambientale, Aerospaziale, dei Materiali (DICAM), Universitá di Palermo, Palermo, Italy

Amod Mani Dixit NSET, Kathmandu, Nepal

Mustapha Djafour RIsam Laboratory, Faculty of Technology, University Abou Bakr Belkaïd, Tlemcen, Algeria

Matjaž Dolšek Faculty of Civil and Geodetic Engineering, University of Ljubljana, Ljubljana, Slovenia

Baiping Dong Department of Civil and Environmental Engineering, Lehigh University, Bethlehem, PA, USA

Doug Dreger Department of Earth and Planetary Science, College of Letters and Science, University of California, Berkeley, CA, USA

Hossein Ebrahimian Department of Structures for Engineering and Architecture, University of Naples Federico II, Naples, Italy

Páll Einarsson Institute of Earth Sciences, University of Iceland, Reykjavík, Iceland

Göran Ekström Department of Earth and Environmental Sciences, Lamont-Doherty Earth Observatory, Columbia University, Palisades, NY, USA

Hany El Naggar Department of Civil and Resource Engineering, Dalhousie University, Halifax, NS, Canada

Gaetano Elia School of Civil Engineering and Geosciences, Newcastle University, Newcastle Upon Tyne, UK

E. Elwood Department of Civil, Environmental and Architectural Engineering, University of Colorado, Boulder, CO, USA

Antonio Emolo Department of Physics, University of Naples Federico II, Federico II – AMRA S.c.ar.l, Napoli, Italy

E. R. Engdahl Department of Physics, University of Colorado, Boulder, CO, USA

Murat Altug Erberik Civil Engineering Department, Middle East Technical University, Ankara, Turkey

Niki Evelpidou Faculty of Geology and Geoenvironment/National and Kapodistrian University of Athens, Athens, Greece

Licia Faenza Istituto Nazionale di Geofisica e Vulcanologia, Centro Nazionale Terremoti, Bologna, Italy

Giuseppe Failla Dipartimento di Ingegneria Civile, dell'Energia, dell'Ambiente e dei Materiali, (DICEAM), University of Reggio Calabria, Reggio Calabria, Italy

Michael N. Fardis Department of Civil Engineering, School of Engineering, University of Patras, Patras, Greece

Spilios D. Fassois Department of Mechanical & Aeronautical Engineering, Stochastic Mechanical Systems and Automation (SMSA) Laboratory, University of Patras, Patras, Greece

Gaetano Festa Department of Physics, University of Naples Federico II, Federico II – AMRA S.c.ar.l, Napoli, Italy

Christopher Corey Fischer Wright State University, Dayton, OH, USA

Tomáš Fischer Faculty of Science, Charles University in Prague, Prague, Czech Republic

Steven L. Forman Department of Geology, Baylor University, Waco, TX, USA

G. Fornaro Institute for Electromagnetic Sensing of the Environment, National Research Council, Naples, Italy

G. R. Foulger Department of Geological Sciences, Durham University, Durham, UK

Michalis Fragiadakis School of Civil Engineering, Laboratory for Earthquake Engineering, National Technical University of Athens (N.T.U.A.), Athens, Greece

Tinu Rose Francis Department of Civil and Environmental Engineering, The University of Auckland, Auckland, New Zealand

Guillermo Franco Head of Catastrophe Risk Research – EMEA, Guy Carpenter & Company Ltd., London, UK

Fabio Del Frate Department of Civil Engineering and Computer Science Engineering, University of Tor Vergata, Rome, Italy

Kohei Fujita Department of Architecture and Architectural Engineering, Graduate School of Engineering, Kyoto University, Kyoto, Japan and RIKEN Advanced Institute for Computational Science, Kobe, Japan

N. Ganesh Department of Applied Mechanics, Indian Institute of Technology Madras, Chennai, Tamil Nadu, India

Charis J. Gantes Institute of Steel Structures, School of Civil Engineering, National Technical University of Athens, Zografou Campus, Athens, Greece

Alexander Garcia-Aristizabal Center for the Analysis and Monitoring of Environmental Risk (AMRA), Naples, Italy

Paolo Gasparini Center for the Analysis and Monitoring of Environmental Risk (AMRA), Naples, Italy

Vincenzo Gattulli DICEAA - Dipartimento di Ingegneria Civile, Edile-Architettura, Ambientale, CERFIS - Centro di Ricerca e Formazione in Ingegneria Sismica, University of L'Aquila, L'Aquila, Italy

Lind S. Gee Albuquerque Seismological Laboratory, U.S. Geological Survey, Albuquerque, NM, USA

Eric L. Geist U.S. Geological Survey, Menlo Park, CA, USA

Carmelo Gentile Department ABC, Politecnico di Milano, Milan, Italy

Debraj Ghosh Department of Civil Engineering, Indian Institute of Science, Bangalore, Karnataka, India

Sonia Giovinazzi Department of Civil and Natural Resources Engineering, University of Canterbury, Christchurch, New Zealand

Konstantinos Gkoumas Department of Structural and Geotechnical Engineering, Sapienza University of Rome, Rome, Italy

Katsuichiro Goda Department of Civil Engineering, University of Bristol, Bristol, UK

Tatiana Goded GNS Science, Lower Hutt, New Zealand

Marco Götz Institute for Structural Analysis, Technische Universität Dresden, Dresden, Germany

Wolfgang Graf Institute for Structural Analysis, Technische Universität Dresden, Dresden, Germany

Ramana V. Grandhi 210 Russ Engineering Center, Wright State University, Dayton, OH, USA

Damian N. Grant Arup Advanced Technology and Research, London, UK

Michael Gray Cast Connex Corporation, Toronto, ON, Canada

Rebekah Green University of Western Washington, Bellingham, WA, USA

Mircea Grigoriu Civil and Environmental Engineering, School of Civil & Environmental Engineering, Cornell University, Ithaca, NY, USA

Luca Guerrieri Department of Geological Survey, ISPRA, Istituto Superiore per la Protezione e la Ricerca Ambientale, Rome, Italy

P. Gülkan Department of Civil Engineering, Çankaya University, Ankara, Turkey

Manu Gupta SEEDS of India, Delhi, India

Sayan Gupta Department of Applied Mechanics, Indian Institute of Technology Madras, Chennai, Tamil Nadu, India

Vittorio Gusella Department of Civil and Environmental Engineering, University of Perugia, Perugia, Italy

Amir M. Halabian Department of Civil Engineering, Isfahan University of Technology, Isfahan, Iran

Achintya Haldar Department of Civil Engineering and Engineering Mechanics, University of Arizona, Tucson, AZ, USA

Matthew M. Haney U.S. Geological Survey, Alaska Volcano Observatory, Anchorage, AK, USA

Youssef M. A. Hashash Department of Civil and Environmental Engineering, 2230c Newmark Civil Engineering Laboratory, University of Illinois at Urbana-Champaign, Urbana, IL, USA

Jens Havskov Department of Earth Science, University of Bergen, Bergen, Norway

Mark B. Hayman Nanometrics Inc., Ottawa, Canada

Eric M. Hernandez Department of Civil and Environmental Engineering, College of Engineering and Mathematical Sciences, University of Vermont, Burlington, VT, USA

William T. Holmes Rutherford + Chekene Engineers, San Francisco, CA, USA

Andrew Hooper School of Earth and Environment, The University of Leeds, Maths/Earth and Environment Building, Leeds, UK

Josef Horálek Institute of Geophysics, Academy of Sciences of the Czech Republic, Prague, Czech Republic

Muneo Hori Earthquake Research Institute, University of Tokyo, Bunkyo, Tokyo, Japan

Shigeki Horiuchi COE, Home Seismometer Corporation, Shirakawa, Fukushima, Japan

Alicia J. Hotovec-Ellis Department of Earth and Space Sciences, University of Washington, Seattle, WA, USA

Roman Hryciw Department of Civil and Environmental Engineering, University of Michigan, Ann Arbor, MI, USA

Nai-Chi Hsiao Central Weather Bureau, Taipei, Taiwan

Pao-Shan Hsieh Disaster Prevention Technology Research Center, Sinotech Engineering Consultants, Inc., Taipei, Taiwan, ROC

Charles Robert Hutt Albuquerque Seismological Laboratory, U.S. Geological Survey, Albuquerque, NM, USA

Luis F. Ibarra Department of Civil and Environmental Engineering, University of Utah, Salt Lake City, USA

Tsuyoshi Ichimura Earthquake Research Institute, University of Tokyo, Bunkyo, Tokyo, Japan

Aristidis Iliopoulos Peikko Greece SA, Marousi, Athens, Greece

Ioanna Ioannou Department of Civil, Environmental and Geomatic Engineering, University College London, London, UK

Antonio Iodice Dipartimento di Ingegneria Elettrica e delle Tecnologie dell'Informazione, Università degli Studi di Napoli Federico II, Napoli, Italy

Tatjana Isakovic University of Ljubljana, Ljubljana, Slovenia

Radoslaw Iwankiewicz Institute of Mechanics and Ocean Engineering, Hamburg University of Technology, Hamburg, Germany

Steinunn S. Jakobsdóttir Faculty of Earth Science, University of Iceland, Reykjavík, Iceland

Li-ju Jang Department of Social Work, National Pingtung University of Science and Technology, Pingtung, Taiwan

Hector A. Jensen Department of Civil Engineering, Santa Maria University, Casilla, Valparaiso, Chile

Randall W. Jibson U.S. Geological Survey, Golden, CO, USA

Sarb Johal Joint Centre for Disaster Research, GNS Science/Massey University, Wellington, New Zealand

Jessica H. Johnson School of Earth Sciences, University of Bristol Wills Memorial Building, Bristol, UK and School of Environmental Sciences, University of East Anglia, Norwich, UK

David Johnston Joint Centre for Disaster Research, GNS Science/Massey University, Wellington, New Zealand and Risk and Society, GNS Science, Lower Hutt, New Zealand

Lucy Johnston University of Canterbury, Christchurch, New Zealand

Bruce R. Julian Department of Geological Sciences, Durham University, Durham, UK

Ioannis Kalpakidis KBR, Houston, TX, USA

George S. Kamaris School of Engineering, University of Warwick, Coventry, UK

Viswanath Kammula Structural Engineer-in-Training, Soscia Engineering Limited, Toronto, Canada

Ram Chandra Kandel Toronto, ON, Canada

Andreas Kappos Department of Civil Engineering, City University London, London, UK

Volkan Karabacak Eskişehir Osmangazi Üniversitesi, Mühendislik-Mimarlık Fakültesi, Jeoloji Müh. Bölümü, Meşelik, Eskişehir, Turkey

Dimitris L. Karabalis Department of Civil Engineering, University of Patras, Patras, Greece

Spyros A. Karamanos Department of Mechanical Engineering, University of Thessaly, Volos, Greece

Theodore L. Karavasilis School of Engineering, University of Warwick, Coventry, UK

Toshihide Kashima Building Research Institute, Tsukuba, Japan

Tsuneo Katayama Professor emeritus of the University of Tokyo, Bunkyo-ku, Tokyo, Japan

Edward Kavazanjian, Jr. School of Sustainable Engineering and the Built Environment, Arizona State University, Tempe, AZ, USA

Miklós Kázmér Department of Palaeontology, Eötvös University, Budapest, Hungary

Ilan Kelman Institute for Risk and Disaster Reduction and Institute for Global Health, University College London, London, England and Norwegian Institute of International Affairs, Oslo, Norway

Bijan Khazai Karlsruhe Institute of Technology, Center for Disaster Management and Risk Reduction Technology (CEDIM), Karlsruhe, Germany

Christopher R. J. Kilburn Department of Earth Sciences, Aon Benfield UCL Hazard Centre, University College London, London, UK

Anastasia A. Kiratzi Department of Geophysics, Aristotle University of Thessaloniki, Thessaloniki, Greece

Ulrich Klapp AREVA GmbH, Erlangen, Germany

Takaji Kokusho Department of Civil & Environmental Engineering, Chuo University, Tokyo, Japan

Petros Komodromos Department of Civil and Environmental Engineering, University of Cyprus, Nicosia, Cyprus

Indranil Kongar EPICentre, Department of Civil, Environmental and Geomatic Engineering, University College London, London, UK

Ivan Koulakov Trofimuk Institute of Petroleum Geology and Geophysics, SB RAS, Novosibirsk, Russia and Novosibirsk State University, Novosibirsk, Russia

George P. Kouretzis Faculty of Engineering and Built Environment, The University of Newcastle, Callaghan, NSW, Australia

Ioannis Koutromanos Department of Civil and Environmental Engineering, Virginia Polytechnic Institute and State University, Blacksburg, VA, USA

D. S. Kusanovic Department of Civil Engineering, Santa Maria University, Casilla, Valparaiso, Chile

Simon Laflamme Department of Civil, Construction, and Environmental Engineering, Iowa State University, Ames, IA, USA

Nikos D. Lagaros Institute of Structural Analysis & Antiseismic Research, Department of Structural Engineering, School of Civil Engineering, National Technical University of Athens, Athens, Greece

Andreas P. Lampropoulos School of Environment and Technology, University of Brighton, Brighton, UK

Roberto Leon The Charles Edward Via, Jr. Department of Civil and Environmental Engineering, Virginia Tech, Blacksburg, VA, USA

Jie Li School of Civil Engineering & State Key Laboratory for Disaster Reduction in Civil Engineering, Tongji University, Shanghai, China

Giorgio Antonino Licciardi GIPSA-Lab – INP Grenoble, Grenoble Institute of Technology, Saint Martin d'Hères, France

Po-Shen Lin Disaster Prevention Technology Research Center, Sinotech Engineering Consultants, Inc., Taipei, Taiwan, ROC

Ting Lin Department of Civil, Construction and Environmental Engineering, Marquette University, Milwaukee, WI, USA

Wen Liu Department of Urban Environment Systems, Chiba University, Chiba, Japan

Geert Lombaert Department of Civil Engineering, KU Leuven, Leuven, Belgium

Rafael H. Lopez Civil Engineering Department, Federal University of Santa Catarina, Florianópolis, SC, Brazil

Santiago López Departamento de Geociencias, Universidad Nacional de Colombia-Sede Bogotá, Bogotá, Colombia

Paulo B. Lourenço Department of Civil Engineering, University of Minho, ISISE, Guimarães, Portugal

Xiao Lu School of Civil Engineering, Beijing Jiaotong University, Beijing, People's Republic of China

Xinzheng Lu Key Laboratory of Civil Engineering Safety and Durability of China Education Ministry, Department of Civil Engineering, Tsinghua University, Beijing, People's Republic of China

Björn Lund Department of Earth Sciences, Uppsala University, Uppsala, Sweden

Guido Luzi CTTC Division of Geomatics, Av. C.F. Gauss, Castelldefels, Spain

Lucia Luzi Istituto Nazionale di Geofisica e Vulcanologia, Milan, Italy

Fai Ma Department of Mechanical Engineering, University of California, Berkeley, Berkeley, CA, USA

John Hugh George Macdonald Civil Engineering Department, University of Bristol, Bristol, UK

Kristof Maes Department of Civil Engineering, KU Leuven, Leuven, Belgium

Andrea Magrin Department of Mathematics and Geosciences, University of Trieste, Trieste, Italy

Sankaran Mahadevan Department of Civil and Environmental Engineering, Vanderbilt University, Nashville, TN, USA

Ian G. Main School of GeoSciences, University of Edinburgh, Edinburgh, UK

Ljubica Mamula-Seadon Centre for Infrastructure Research at the Department of Civil and Environmental Engineering, Faculty of Engineering, The University of Auckland, Auckland, New Zealand

Gaetano Manfredi Department of Structures for Engineering and Architecture, University of Naples "Federico II", Naples, Italy

Sandeeka Mannakkara Department of Civil and Environmental Engineering, The University of Auckland, Auckland, New Zealand

C. S. Manohar Department of Civil Engineering, Indian Institute of Science, Bangalore, Karnataka, India

Carlo Marin Department of Information Engineering and Computer Science, University of Trento, Povo, Trento, Italy

Rui Marques Engineering Department, Civil Engineering Section, Pontifical Catholic University of Peru, Lima, Peru

Justin D. Marshall Auburn University, Auburn, AL, USA

Mark J. Masia Faculty of Engineering and Built Environment, The University of Newcastle, Callaghan, NSW, Australia

Neven Matasovic Geosyntec Consultants, Huntington Beach, CA, USA

Annibale Luigi Materazzi Department of Civil and Environmental Engineering, University of Perugia, Perugia, Italy

George P. Mavroeidis Department of Civil and Environmental Engineering and Earth Sciences, University of Notre Dame, Notre Dame, IN, USA

Garry McDonald Market Economics Ltd, Takapuna, New Zealand

Hamish McLean School of Humanities, Griffith University, Nathan, QLD, Australia

Nuno Mendes ISISE, University of Minho, Guimarães, Portugal

Stefan Mertl Mertl Research GmbH, Vienna, Austria

Alessandro Maria Michetti Dipartimento di Scienza e Alta Tecnologia, Università dell'Insubria, Como, Italy

Danielle Hutchings Mieler Earthquakes and Hazards Resilience Program, Association of Bay Area Governments, Oakland, CA, USA

Leandro F. F. Miguel Civil Engineering Department, Federal University of Santa Catarina, Florianópolis, SC, Brazil

Jon Mitchell Jon Mitchell Emergency Management Ltd., Queenstown, New Zealand

Chara Ch. Mitropoulou Institute of Structural Analysis & Antiseismic Research, Department of Structural Engineering, School of Civil Engineering, National Technical University of Athens, Athens, Greece

Babak Moaveni Department of Civil and Environmental Engineering, Tufts University, Medford, MA, USA

Tracy Monk Families for School Seismic Safety, Vancouver, Canada

Troy A. Morgan Exponent, New York, USA

Lalliana Mualchin Headquarters: Studio legale Avv. Wania Della Vigna, International Seismic Safety Organization (ISSO), Arsita (TE), Italy

Catherine Murray Market Economics Ltd, Takapuna, New Zealand

Giuseppe Muscolino Dipartimento di Ingegnria Civile, Informatica, Edile, Ambientale e Matematica Applicata, Università degli Studi di Messina, Messina, Italy

S. C. Myers Lawrence Livermore National Laboratory, Livermore, CA, USA

Satish Nagarajaiah Department of Civil and Environmental Engineering, Rice University, Houston, TX, USA and Department of Mechanical Engineering, Rice University, Houston, TX, USA

Hesham El Naggar Geotechnical Research Centre, Department of Civil and Environmental Engineering, The University of Western Ontario, London, ON, Canada

Alexander V. Neeshpapa R-sensors LLC, Dolgoprudny, Russia

Meredith Nettles Department of Earth and Environmental Sciences, Lamont-Doherty Earth Observatory, Columbia University, Palisades, NY, USA

Ching Hang Ng US Nuclear Regulatory Commission, Office of Nuclear Reactor Regulation, Washington, DC, USA

Nikolaos Nikitas School of Civil Engineering, University of Leeds, Leeds, UK

Nicola Nisticò Dipartimento di Ingegneria Strutturale e Geotecnica, Università La Sapienza, Rome, Italy

Viviana Novelli Department of Civil, Environmental and Geomatic Engineering, University College London, London, UK

Ilan Noy School of Economics and Finance, Victoria University, Wellington, New Zealand

David D. Oglesby Department of Earth Sciences, University of California, Riverside, Riverside, CA, USA

Kenji Oguni Department of System Design Engineering, Keio University, Yokohama, Japan

Norio Okada Graduate School of Science and Engineering, Kumamoto University, Kumamoto City, Japan

Izuru Okawa Building Research Institute, Tsukuba, Japan

Scott M. Olson Department of Civil and Environmental Engineering, University of Illinois at Urbana-Champaign, Urbana, IL, USA

Rolando P. Orense Department of Civil and Environmental Engineering, University of Auckland, Auckland, New Zealand

Lars Ottemöller Department of Earth Science, University of Bergen, Bergen, Norway

Daria Ottonelli Department of Civil, Chemical and Environmental Engineering, University of Genoa, Genoa, Italy

Francesca Pacor Istituto Nazionale di Geofisica e Vulcanologia, Milan, Italy

Sara Paganoni Ziegert|Roswag|Seiler Architekten Ingenieure, Berlin, Germany

Alessandro Palmeri School of Civil and Building Engineering, Loughborough University, Loughborough, Leicestershire, UK

Bishnu Pandey University of British Columbia, Vancouver, Canada

Chris P. Pantelides Department of Civil and Environmental Engineering, University of Utah, Salt Lake City, UT, USA

Giuliano F. Panza Department of Mathematics and Geosciences, University of Trieste, Trieste, Italy and The Abdus Salam International Centre for Theoretical Physics, SAND Group, Trieste, Italy and Institute of Geophysics, China Earthquake Administration, Beijing, China and International Seismic Safety Organization (ISSO), Arsita, Italy

C. Papadimitriou Department of Mechanical Engineering, University of Thessaly, Volos, Greece

Vissarion Papadopoulos Institute of Structural Analysis and Seismic Research, School of Civil Engineering, National Technical University of Athens (N.T.U.A.), Athens, Greece

Manolis Papadrakakis Institute of Structural Analysis & Antiseismic Research, Department of Structural Engineering, School of Civil Engineering, National Technical University of Athens, Athens, Greece

Thanasis Papageorgiou Department of Structural Engineering, School of Civil Engineering, National Technical University of Athens (N.T.U.A.), Athens, Greece

Ioanna Papayianni Laboratory of Building Materials, Department of Civil Engineering, Aristotle University of Thessaloniki, Thessaloniki, Greece

Byeongjin Park Department of Civil and Environmental Engineering, Korea Advanced Institute of Science and Technology (KAIST), Yuseong-Gu, Daejeon, Republic of Korea

Douglas Paton School of Psychology, University of Tasmania, Launceston, TAS, Australia

Manuel Pellissetti AREVA GmbH, Erlangen, Germany

George Gr. Penelis Civil Engineering Department, Aristotle University of Thessaloniki, Thessaloniki, Greece

Marla Petal Risk RED (Risk Reduction Education for Disasters), Los Angeles, CA, USA

Francesco Petrini Department of Structural and Geotechnical Engineering, Sapienza University of Rome, Rome, Italy

Matteo Picchiani Department of Civil Engineering and Computer Science Engineering, University of Tor Vergata, Rome, Italy

Paolo Pirazzoli Laboratoire de Géographie Physique, Paris, France

Kyriazis Pitilakis Department of Civil Engineering, Aristotle University, Thessaloniki, Greece

Stephen Platt Cambridge Architectural Research Ltd, Cambridge, UK

Nikos Pnevmatikos Department of Civil Engineering, Surveying and Geoinformatics, Technological Educational Institute of Athens, Egaleo-Athens, Greece

Panayiotis C. Polycarpou Department of Civil and Environmental Engineering, University of Cyprus, Nicosia, Cyprus

Silvia Pondrelli Istituto Nazionale di Geofisica e Vulcanologia, Sezione di Bologna, Italy

Keith Porter Civil, Environmental, and Architectural Engineering, University of Colorado, Boulder and SPA Risk LLC, Denver, CO, USA

Regan Potangaroa Unitec School of Architecture, Auckland, New Zealand

Francesco Potenza DICEAA - Dipartimento di Ingegneria Civile, Edile-Architettura, Ambientale, CERFIS - Centro di Ricerca e Formazione in Ingegneria Sismica, University of L'Aquila, L'Aquila, Italy

Alessandro Proia Dipartimento di Ingegneria Strutturale e Geotecnica, Università La Sapienza, Rome, Italy

Carsten Proppe Institut für Technische Mechanik, Karlsruhe Institute of Technology, Karlsruhe, Germany

Ioannis N. Psycharis School of Civil Engineering, Department of Structural Engineering, National Technical University of Athens (N.T.U.A.), Athens, Greece

Wolfgang Rabbel Institute of Geosciences, University of Kiel, Kiel, Germany

Alin C. Radu Civil and Environmental Engineering, School of Civil & Environmental Engineering, Cornell University, Ithaca, NY, USA

Mircea Radulian National Institute for Earth Physics, Măgurele, Romania

D. Reale Institute for Electromagnetic Sensing of the Environment, National Research Council, Naples, Italy

Klaus Reicherter Institute of Neotectonics and Natural Hazards, Department of Geosciences and Geography, RWTH Aachen University, Aachen, Germany

Robert K. Reitherman Consortium of Universities for Research in Earthquake Engineering, Richmond, CA, USA

Edwin Reynders Department of Civil Engineering, KU Leuven, Leuven, Belgium

Sanaz Rezaeian U.S. Geological Survey, Golden, CO, USA

Daniele Riccio Dipartimento di Ingegneria Elettrica e delle Tecnologie dell'Informazione, Università degli Studi di Napoli Federico II, Napoli, Italy

James M. Ricles Bruce G. Johnston Professor of Structural Engineering, Department of Civil and Environmental Engineering, Lehigh University, Bethlehem, PA, USA

Adam T. Ringler Albuquerque Seismological Laboratory, U.S. Geological Survey, Albuquerque, NM, USA

Janise Rodgers GeoHazards International, Menlo Park, USA

Maria Ines Romero-Arduz Department of Civil and Environmental Engineering, University of Illinois at Urbana-Champaign, Urbana, IL, USA

Tiziana Rossetto Department of Civil, Environmental and Geomatic Engineering, University College London, London, UK

Daniel Roten San Diego Supercomputer Center, University of California, San Diego, La Jolla, CA, USA

James Olabode Bamidele Rotimi Auckland University of Technology, Auckland, New Zealand

Theodoros Rousakis Department of Civil Engineering, School of Engineering, Democritus University of Thrace, Xanthi, Greece

Masayuki Saeki Department of Civil Engineering, Tokyo University of Science, Shinjuku, Japan

David Saftner Department of Civil Engineering, University of Minnesota Duluth, Duluth, MN, USA

Antonella Saisi Department ABC, Politecnico di Milano, Milan, Italy

Simone Salimbeni Istituto Nazionale di Geofisica e Vulcanologia, Sezione di Bologna, Italy

Zeynep Türkmen Sanduvaç Risk RED, Istanbul, Turkey

Shankar Sankararaman Intelligent Systems Division, SGT Inc., NASA Ames Research Center, Moffett Field, CA, USA

Evangelos Sapountzakis Department of Structural Engineering, School of Civil Engineering, National Technical University of Athens (N.T.U.A.), Athens, Greece

Wendy Saunders GNS Science, Lower Hutt, New Zealand

Richard Sause Joseph T. Stuart Professor of Structural Engineering, Department of Civil and Environmental Engineering, Lehigh University, Bethlehem, PA, USA

Mechita C. Schmidt-Aursch Alfred-Wegener-Institut, Helmholtz-Zentrum für Polar- und Meeresforschung, Bremerhaven, Germany

Johannes Schweitzer NORSAR, Kjeller, Norway

Erica Seville Resilient Organisations, Sheffield, New Zealand

Anastasios G. Sextos Division of Structural Engineering, Department of Civil Engineering, Aristotle University of Thessaloniki, Thessaloniki, Greece

Anna S. Shabalina Center for Molecular Electronics, Moscow Institute of Physics and Technology, Moscow, Russia

Ayman A. Shama Ammann & Whitney, New York, NY, USA

Nikolay Shapiro Institut de Physique du Globe de Paris Laboratoire de Sismologie, Paris, France

Rajib Shaw University of Kyoto, Kyoto, Japan

P. Benson Shing Department of Structural Engineering, University of California, San Diego, La Jolla, CA, USA

Nilesh Shome Model Development, Risk Management Solutions, Newark, CA, USA

Pablo G. Silva Departamento de Geología, Universidad de Salamanca, Escuela Politécnica Superior de Ávila, Avila, Spain

Priscilla Brandão Silva Department of Computational Mechanics, Faculty of Mechanical Engineering, UNICAMP, Campinas, SP, Brazil

E. Simoen Department of Civil Engineering, KU Leuven, Leuven, Belgium

Manuel Sintubin Department of Earth and Environmental Sciences, Geodynamics and Geofluids Research Group, KU Leuven, Leuven, Belgium

Reinoud Sleeman Seismology Division, Royal Netherlands Meteorological Institute (KNMI), De Bilt, Netherlands

Nicola Smith Market Economics Ltd, Takapuna, New Zealand

Patrick Smith Montserrat Volcano Observatory, Flemmings, Montserrat and Seismic Research Centre, University of the West Indies, Trinidad and Tobago, West Indies

Richard Smith Science and Education, Earthquake Commission, Wellington, New Zealand

Andrew W. Smyth Department of Civil Engineering and Engineering Mechanics, Columbia University, New York, NY, USA

Emily So Department of Architecture, University of Cambridge, Cambridge, UK

Hoon Sohn Department of Civil and Environmental Engineering, Korea Advanced Institute of Science and Technology (KAIST), Yuseong-Gu, Daejeon, Republic of Korea

Minas D. Spiridonakos Department of Civil, Environmental and Geomatic Engineering, ETH Zurich, Institute of Structural Engineering, Zurich, CH, Switzerland

Srinivas Sriramula Lloyd's Register Foundation (LRF) Centre for Safety and Reliability Engineering, School of Engineering, University of Aberdeen, Fraser Noble Building, Aberdeen, UK

George Stefanou Institute of Structural Analysis & Antiseismic Research, School of Civil Engineering, National Technical University of Athens, Athens, Greece and Institute of Structural Analysis & Dynamics of Structures, Department of Civil Engineering, Aristotle University of Thessaloniki, Thessaloniki, Greece

Mark Stirling GNS Science, Lower Hutt, New Zealand

Dieter Stoll Lennartz Electronic GmbH, Tübingen, Germany

Salvatore Stramondo National Earthquake Center, Remote Sensing Lab, Istituto Nazionale di Geofisica e Vulcanologia, Rome, Italy

Kosmas-Athanasios Stylianidis Civil Engineering Department, Aristotle University of Thessaloniki, Thessaloniki, Greece

Xiaodan Sun Southwest Jiaotong University, Chengdu, China

V. S. Sundar Department of Civil Engineering, Indian Institute of Science, Bangalore, Karnataka, India

Ricardo Taborda Department of Civil Engineering, and Center for Earthquake Research and Information, University of Memphis, Memphis, TN, USA

Izuru Takewaki Department of Architecture and Architectural Engineering, Graduate School of Engineering, Kyoto University, Kyoto, Japan

T. P. Tassios Department of Structural Engineering – Reinforced Concrete Laboratory, National Technical University of Athens (N.T.U.A.), Zografou, Athens, Greece

Georgia E. Thermou Department of Civil Engineering, Aristotle University of Thessaloniki, Thessaloniki, Greece

Geoff Thomas School of Architecture, Victoria University of Wellington, Wellington, New Zealand

Glenn Thompson School of Geosciences, University of South Florida, Tampa, FL, USA

Clifford Thurber Department of Geoscience, University of Wisconsin-Madison, Madison, WI, USA

Lucia Tirca Department of Building, Civil and Environmental Engineering, Concordia University, Montreal, QC, Canada

G. Tondreau Postdoctoral Researcher, Building Architecture and Town Planning (BATir), Brussels, Belgium

Ikuo Towhata Department of Civil Engineering, University of Tokyo, Bunkyo-ku, Tokyo, Japan

Bruce Townsend Nanometrics, Inc., Kanata, Ontario, Canada

Konstantinos Daniel Tsavdaridis School of Civil Engineering, University of Leeds, Leeds, UK

Yiannis Tsompanakis Computational Dynamics Research Group, School of Environmental Engineering, Technical University of Crete, Chania, Crete, Greece

Angelos S. Tzimas School of Engineering, University of Warwick, Coventry, UK

Marcos Valdebenito Department of Civil Engineering, Santa Maria University, Casilla, Valparaiso, Chile

Suzanne Vallance Lincoln University, Christchurch, New Zealand

Dimitrios Vamvatsikos School of Civil Engineering, National Technical University of Athens (N.T.U.A.), Athens, Greece

John Vargo Resilient Organisations, University of Canterbury, Christchurch, New Zealand

Nick Varley Facultad de Ciencias, Centre of Exchange and Research in Volcanology, Universidad de Colima, Colima, Mexico

Graça Vasconcelos Department of Civil Engineering, ISISE, University of Minho, Guimarães, Portugal

George Vasdravellis Institute for Infrastructure and Environment, Heriot-Watt University, Edinburgh, UK

Marcello Vasta INGEO, Engineering and Geology Department, University of Chieti-Pescara "G. D'Annunzio", Pescara, Italy

Maria Vathi Department of Mechanical Engineering, University of Thessaly, Volos, Greece

Václav Vavryčuk Institute of Geophysics, Czech Academy of Sciences, Prague, Czech Republic

Ioannis Vayas Traffic Engineering Laboratory, National Technical University of Athens (N.T.U.A.), Athens, Greece

Gerardo M. Verderame Department of Structures for Engineering and Architecture (DiSt), University of Naples Federico II, Naples, Italy

Elizabeth Vintzileou Department of Structural Engineering, Faculty of Civil Engineering, National Technical University of Athens, Athens, Greece

Peter H. Voss Geological Survey of Denmark and Greenland – GEUS, Copenhagen K, Denmark

Christos Vrettos Division of Soil Mechanics and Foundation Engineering, Technical University of Kaiserslautern, Kaiserslautern, Germany

Gregory P. Waite Department of Geological and Mining Engineering and Sciences, Michigan Technological University, Houghton, MI, USA

Jui-Pin Wang Department of Civil and Environmental Engineering, The Hong Kong University of Science and Technology, Kowloon, Hong Kong

Kai-Shyr Wang Ministry of Science and Technology, Taipei, Taiwan

Friedemann Wenzel Geophysical Institute, Karlsruhe Institute of Technology, Karlsruhe, Germany

Lalith Wijerathne Earthquake Research Institute, University of Tokyo, Tokyo, Japan

Dennis Wilken Institute of Geosciences, University of Kiel, Kiel, Germany

Suzanne Wilkinson Department of Civil and Environmental Engineering, The University of Auckland, Auckland, New Zealand

Thomas Wilson Department of Geological Sciences, University of Canterbury, Christchurch, New Zealand and Natural Hazards Research Group, University of Canterbury, Christchurch, New Zealand

Ben Wisner University College London, London, England and Oberlin College, Oberlin, OH, USA

Guoxi Wu Wutec Geotechnical International, New Westminster, Metro Vancouver, BC, Canada

Yih-Min Wu Department of Geosciences, National Taiwan University, Taipei, Taiwan

Zhen Xu Key Laboratory of Civil Engineering Safety and Durability of China Education Ministry, Department of Civil Engineering, Tsinghua University, Beijing, People's Republic of China

Fumio Yamazaki Department of Urban Environment Systems, Chiba University, Chiba, Japan

Yongchao Yang Department of Civil and Environmental Engineering, Rice University, Houston, TX, USA

Yin-Tung Yen Disaster Prevention Technology Research Center, Sinotech Engineering Consultants, Inc., Taipei, Taiwan, ROC

Ka-Veng Yuen Faculty of Science and Technology, University of Macau, Macau, China

Daniil Yurchenko Institute of Mechanical, Process and Energy Engineering, Heriot-Watt University, Edinburgh, UK

Cengiz Zabcı İstanbul Teknik Üniversitesi, Maden Fakültesi, Jeoloji Müh. Bölümü, Ayazağa, İstanbul, Turkey

Jerome Zayas Earthquakes and Megacities Initiative, Quezon City, Philippines

Djawad Zendagui Risam Laboratory, Faculty of Technology, University Abou Bakr Belkaïd, Tlemcen, Algeria

Christos A. Zeris Department of Structural Engineering, Faculty of Civil Engineering, National Technical University of Athens, Athens, Greece

Enrico Zio European Foundation for New Energy–Electricité de France, Ecole Centrale Paris and Supelec, Paris, France and Energy Department, Politecnico di Milano, Milan, Italy

Aldo Zollo Department of Physics, University of Naples Federico II, Federico II – AMRA S.c.ar.l, Napoli, Italy

Konstantin M. Zuev Department of Computing and Mathematical Sciences, California Institute of Technology, USA

Features Extraction from Satellite Data

Fabio Del Frate and Matteo Picchiani
Department of Civil Engineering and Computer Science Engineering, University of Tor Vergata, Rome, Italy

Synonyms

Dimensionality reduction; Image classification; Image processing; Parameter retrieval

Introduction

Earth Observation (EO) images can be very useful in a wide range of human activities, from the understanding of global phenomena to decision-making processes. Mostly, the exploitation of the acquired images in the area and period of interest is still performed by visual interpretation. In this way, experts in EO sensing and application domains (e.g., seismology, agriculture, meteorology, forestry, etc.) add their knowledge after their analysis. When the experts' interpretation is concluded, only a minimal part of the data, the portion containing the useful information, is kept, while the remaining part can be discarded. However, this process is highly demanding in terms of costs, human effort, and time. Therefore, it cannot be performed on a regular basis, which involves the loss of information potentially useful for research activities, service providers, and final users. The consequences of such a lack of information flow might be very significant as it may prevent operators who rely on EO data from making the right decisions, and leave relevant phenomena undetected or to be discovered too late. In this context, important support can be provided by feature extraction methods, more automated and direct procedures for the extraction of information from images. Relying on "intelligent" automatic techniques, these methods support the image interpreters in discovering the essential elements of information in an image. Sometimes the extracted feature is already the final object of interest, for example, an oil spill in the ocean or a thermal anomaly. In other cases, it consists of intermediate processing aiming at generating a new representation of the image data where particular statistical or geometrical characteristics are highlighted.

Pixel averaging is one of the simplest techniques used when extracting features in image processing. This means replacing a box of pixels with a single pixel. In particular, the value of the pixel will be obtained by computing the average of the values of all the pixels belonging to the considered box. We can say that the new pixels' values are features of the original image, representing its information content in a more compact way. From the example given, it may be understood that feature extraction can also play another crucial role in the context of image processing and, more in general, of pattern recognition: dimensionality reduction. In fact, dimensionality reduction can be a fundamental step in the implementation of many classification or retrieval problems. If we are forced to work with a limited quantity of data, as we often are, then increasing the dimensionality of the space rapidly leads to the point at which the data is very sparse, therefore providing a very poor representation of the mapping. To describe the extraordinarily rapid growth in the difficulty of problems as the number of variables (or dimension) increases, the phrase "the curse of dimensionality" has been coined by Richard Bellman (1957). Reducing the number of input variables can lead to improved performance for a given data set, even though information is being discarded. The mapping is better specified in the lower-dimensional space, which more than compensates the loss of information.

An Overview of Feature Extraction Techniques

According to the type of satellite data to be analyzed (SAR, optical/multispectral, time series) and the considered application, different approaches can be used for feature extraction:

Spatial and Contextual Features

Spatial and contextual features try to synthesize the information contained in neighboring pixels with respect to a pixel under examination. Different contextual descriptors can be defined. One of the earliest and most important techniques used to express textural and tonal variations in a remotely sensed image data is the one based on the gray-level co-occurrence matrix (GLCM) (Haralick et al. 1973). The GLCM is the core of a mathematical framework capable of representing the distance and angular relationships among the pixels over an image subregion of a specified size. Each element of the GLCM is a measurement of the probability of occurrence of two gray scale values separated by a given distance in a given direction. Several quantities can be computed from the elements of the GLCM which describe quantitatively the textural behavior of the subregion considered. Another approach for textural characterization relies on probabilistic modeling. Markov random field (MRF) models fall under this category and can be quite successful for texture modeling (Cross and Jain 1983). They capture local characteristics of an image considering regional patterns and by assuming a local conditional probability distribution. Spatial feature extraction can be based on multichannel filtering as well. The image is decomposed into a set of sub-bands which are calculated by convolving it with a bank of linear filters, nearly uniformly covering the spatial-frequency domain, followed by some nonlinear procedures. Gabor filters can be an example for this type of techniques (Jain and Farrokhnia 1991). Mathematical morphological filters should also be mentioned for the detection of textural features. In this case, the extraction is obtained by performing shape-oriented operations, like openings and closures, based on set theory (Soille 2003).

Transform Domain Features

An image transform is a mathematical operation that re-expresses in a different, and possibly more meaningful, form all, or part of, the information content of a multiband or gray scale image. One of the most used transform methods for feature extraction is the principal component analysis (PCA) (Singh and Harrison 1985) where a multispectral set of m images is re-projected in terms of a set of m principal components. Two particular properties exist: the components are extracted for decreasing variance, and the principal components are statistically unrelated or orthogonal. For multiband images, the first PCA components can represent effective features to be used in the operational scenarios to be considered, as they reveal the structure behind the correlation of the variables. Indeed no linear combination (i.e., weighted sum) of the original bands can contain more information than is present in the first principal components.

An alternative approach considers harmonic analysis, a field of mathematics that studies the

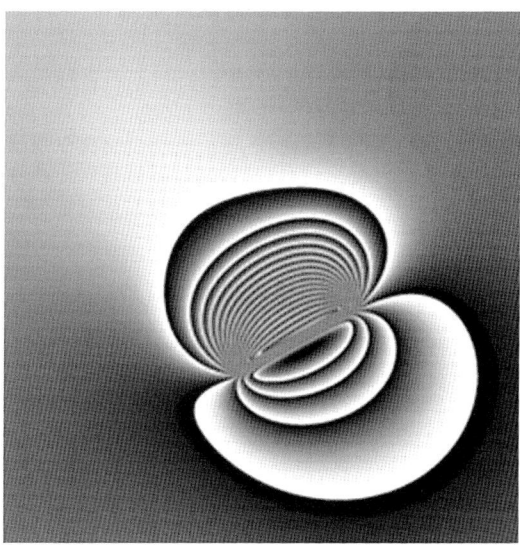 

Features Extraction from Satellite Data, Fig. 1 1,500 × 1,500 pixels synthetic interferogram's example (*left*) and reconstructed one by means of first 500 DWT components (*right*)

representation of functions as overlapping of fundamental waveforms. It is known that a multivariate function f can be well approximated by the linear combination of the elements of a given basis. This type of operation is called harmonic analysis. When harmonic analysis is applied to image data, the discrete image can be seen as a two-dimensional signal, and it is possible to consider a set of mathematical tools that perform a transformation suitable to extract some features otherwise difficult to be identified. This can be done by means of particular functional operators like the DFT (discrete Fourier transform) and the DWT (discrete wavelet transform). Both the transforms express the signal as coefficients in a function space spanned by a set of basis functions. The basis of the DFT is complex exponential functions, representing sinusoid functions in the real domain, and the multiplying coefficients are also complex numbers. The basis of the DWT is scaled and shifted versions of a mother wavelet real-valued function. In this case, the coefficients have real values. Low pass filtering of the DFT and DWT transformed images can be considered for the extraction of low spatial-frequency features (Picchiani et al. 2012). See an example of application in Fig. 1.

Advanced Nonlinear Techniques

Linear techniques such as PCA may be unable to capture nonlinear correlations so they may fail when dealing with highly complex data. One way of performing a nonlinear extension is to lift the input space to a higher dimensional feature space by a nonlinear feature mapping and then to find a linear dimension reduction in the feature space (Scholkopf et al. 1999). Kernel techniques allow such nonlinear extensions without explicitly forming a nonlinear mapping or a feature space, as long as the problem formulation only involves the inner products between the data points and never the data points themselves.

Neural networks can also be applied to effectively extract nonlinear features from data. For example, this can be obtained by using Autoassociative Neural Networks (AANN) (Bishop 1995). These are multilayer perceptron (MLP) networks which are trained to the purpose of reproducing the inputs in the output layer (Fig. 2). This means that the targets used in the training phase are simply the input vectors themselves. Hence, the net is said to perform autoassociative mapping (autoencoder). One of the hidden layers of the AANN, known as the bottleneck layer, is characterized by a number of units significantly lower than the number of

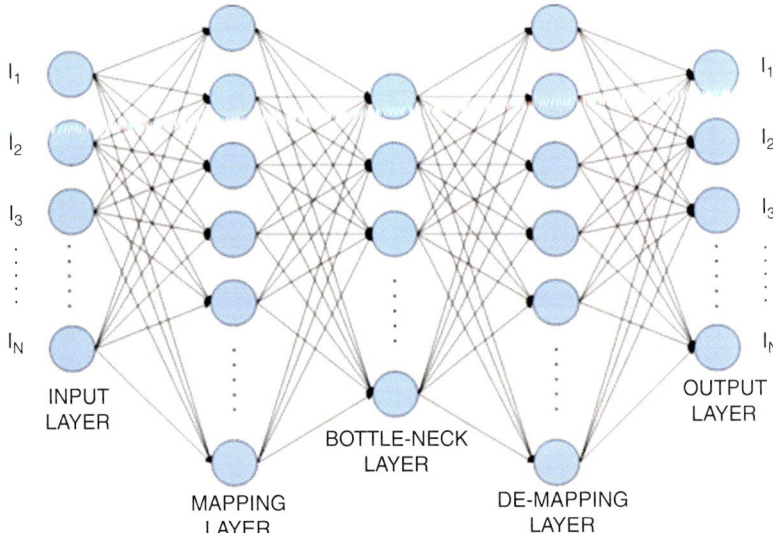

Features Extraction from Satellite Data, Fig. 2 Autoassociative Neural Network topology

inputs/outputs. If during the training phase the actual outputs of the network are capable of resembling those desired with a satisfactory level of error, this implies that the bottleneck layer is capable of providing, in a lower dimension, good representation of the input. In Fig. 3, two nonlinear principal components calculated from a Landsat image are shown. SOM (Self-Organizing Maps) represent another possibility of performing feature extraction using neural networks. The SOM map or Kohonen map is a two-layered network. The first layer of the network is the input layer. The second, called the competitive layer, is usually organized as a two-dimensional grid. All interconnections go from the first to the second layer. In the training phase, the data are presented in a random order to the network. All the nodes in the competitive layer compare the inputs with their weights and compete with each other to become the winning unit with the lowest difference. After that, the weights of the winning neuron and of its neighbors are adjusted towards the input vector. The magnitude of change decreases with time and with distance from the winning neuron. In this way, a SOM can automatically form 1D or multidimensional maps of the intrinsic features of the input data and reorganize them in several output clusters/classes.

Feature Extraction in Seismological Applications

There are several features of interest for seismology that can be extracted from Earth Observation data. A major category regards the thermal anomalies. Due to pressure increase and subsurface degassing activities, a sudden rise in land surface temperature (LST) can characterize the area of interest prior to an earthquake. The detection of such thermal changes can be very important in an effort to predict earthquake activities and provide time to give suitable warnings. Appropriate satellite data with thermal bands can be used to extract this kind of feature from multi-temporal images. Moreover, besides land surface, due to the release of gas emissions causing a local greenhouse effect, anomalies can also be detected in the concentration of some gases in the atmosphere. Another important application is focused on retrieving information about the surface deformation once earthquake occurs. Earthquakes produce static displacements of the ground that can be observed near the fault that slipped during a seismic rupture. The measurement of such displacements is a crucial issue in seismotectonics, as it gives insight into the geometry of the ruptured fault and the energy released by the earthquake. Such measurements are generally performed using geodetic techniques that can

Features Extraction from Satellite Data, Fig. 3 Original Landsat image (*left*) and two computed nonlinear principal components (*center* and *right*)

only provide a sparse covering, because the areas of coseismic ground displacement are not known a priori. Satellite imagery is naturally suited for such measurements, because it regularly provides comprehensive and detailed images of the ground with high radiometric and geometric quality. The displacement field can be measured by the comparison of images acquired before and after the event. Both InSAR (interferometric SAR) and optical data can be exploited for this type of investigation. Other effective uses of Earth Observation data for seismology are connected with disaster management. In the management of a seismic event to immediately draw up a draft, damage map of the stricken urban areas is of great importance in order to organize civil protection interventions. Remote sensing can provide valuable items of information in this respect, thanks to its synoptic capability, in particular when the seismic event is located in remote regions or a failure affected the main communication systems. Both optical and radar sensors can be exploited for the extraction of this type of feature. Despite the great potential provided by Earth Observation data for supporting operational activities in the field seismology, it has been pointed out that the level of automatic data processing is mostly inadequate with "still too much manual labour and author arbitrariness" (Tronin 2010). This is not the case when considering the estimation of tectonic parameters from SAR interferometry where feature extraction, according to innovative technologies based on neural networks, can be fully automated.

Automatic Tectonics Parameter Retrieval from InSAR Data

Until SAR interferometry (InSAR) broke out in the beginning of the 1990s as an effective technique for analyzing a seismic event, most of the measurements used for this kind of study were based on the ground recording of the waves generated by the earthquake. In fact, such data are still very helpful in providing information on the location or the magnitude of the seismic event. However, these measurements may not be sufficient if a more accurate estimation of the earthquake parameters, such as the fault depth, is desired.

The first application of the InSAR technique for the investigation of the characteristics of a seismic event regarded the Landers Earthquake (Massonet et al. 1993). The innovative method was capable of providing effective information on the generated displacement field. After that, a few studies using the same approach were conducted on various both strong (Kobe, Japan, 1997; Izmit, Turkey, 1999; Hector Mine, CA, USA, 2000; Denali, AK, USA, 2002; Bam, Iran, 2003; Sichuan, China, 2008) and moderate (Colfiorito, Italy, 1997; Abruzzi, Italy, 2009) earthquakes. It was discovered that, thanks to its ability to generate a distributed map of the total strain release, in many cases the InSAR technique was the only one capable of providing the investigators with a higher level of information.

In fact, by comparing pre- and post-event SAR data and calculating the interferometric phase, InSAR can reconstruct the coseismic surface displacement field with a centimetric accuracy. The potential of InSAR data can also be combined with GPS measurements and with more standard seismological and geophysical data, such as strong motion records, for the assessment of normal fault models, generally stemming from the Okada formulation (Okada 1985). In particular, for modeling radar interferometric data, specific software packages can be used to calculate the displacement components (Feigl and Duprè 1999) which, in an elastic half space, are the expression of the seismic source at the surface.

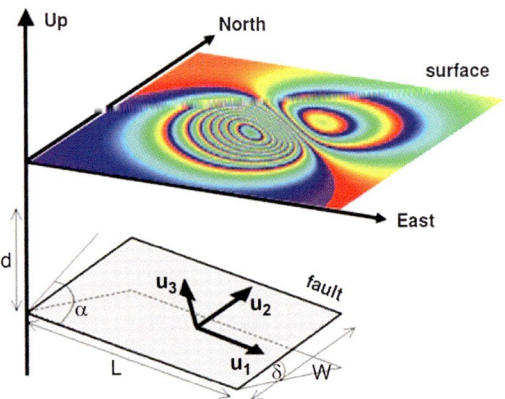

Features Extraction from Satellite Data, Fig. 4 Fault geometry and symbols

In fact, the inverse problem is of great interest. This means the attempt to retrieve, also in a quantitative way, the source parameters starting from the knowledge of the InSAR surface displacement field. In particular, some useful information to define the fault geometry (dip and strike angle; width and length), the extension of the rupture, and the slip distribution on the fault plane can be obtained.

For the estimation of the desired parameters, two main steps have to be carried out after the computation of the interferogram. The first one regards the dimensionality reduction of the data. The second one is the design of a suitable retrieval algorithm. With reference to Fig. 4, besides the length and the depth of the fault, tectonic parameters that can be estimated are the following: the slip vector $U = [U1, U2, U3]$, which represents the movement of the hanging wall with respect to the foot wall signed such that $U1$ is positive for left lateral strike slip, $U2$ is positive for thrusting dip slip, and $U3$ is tensile slip; the strike (α), which is the clockwise angle between the direction of the fault trace and the North–South one; and the dip (δ), that is, the angle between the fault plane and the surface. Another possibility, optionally complementary to parameter estimation, is the classification of the fault. The main classes of normal fault, strip-like fault and reverse fault (thrust), are typically considered.

Dimensionality Reduction of the Data

Data dimensionality reduction is one of the key issues in retrieval and classification problems. In this specific case, it is certainly not appropriate to feed a classificator or a parameter estimator with all the DN numbers forming the interferometric image. In such a context, a standard technique for performing dimensionality reduction is to sample the interferogram with an adequate frequency. A better performance can be obtained considering the combination of two techniques for feature extraction. In particular, the first processing step relies on the harmonic analysis, and in the second, the AANN is applied. Such an approach leads to the determination of a considerably reduced number of components for the interferogram representation without involving the training of huge AANN topologies. An interferogram with a dimension of 1,500 × 1,500 pixels can effectively be represented by a vector of 50 components if the chaining of the AANN and the DWT is applied (Picchiani et al. 2012).

Classification and Retrieval Using Neural Networks

Once the data dimensionality reduction has been performed, and the interferogram is represented by means of a vector expressing a limited number of significant feature values, an algorithm is necessary to transform such a concentrated information content into a set of values to be associated with the desired quantities. Such desired quantities may either be values of tectonic parameters or probabilities of belonging to a specific fault class. In both cases, the problem is rather complex due to its inherent nonlinearities. For this reason, the use of neural networks can be considered as they are recognized as a powerful tool for inversion procedure in remote sensing applications (Stramondo et al. 2011).

Among various topologies, multilayer perceptrons (MLPs) have been found to have the best suited topology for classification and inversion problems and can also be used in this case. These are feedforward networks where the input only flows in one direction, towards the output, and each neuron of a layer is connected to all neurons of the successive layer but has no feedback to neurons in the previous layers.

Compared to other methods, the use of NNs is often effective because they can simultaneously address nonlinear dependences and complex physical behavior with reduced computational efforts and without requiring any a priori information (Bishop 1995). Moreover, NNs learn the input–output relationships exclusively from the training patterns; therefore, no explicit expression addressing the interactions between the parameters of interest is needed with such an approach.

In the training phase, the weights (internal parameters of the network) are progressively adjusted to enable the approximation of the needed functional relationships. The performance of the algorithm is monitored by an error (or cost) function which has a nonlinear dependence on weights. Once the error is minimized, the trained MLP performs the desired mapping of input space onto output space.

As far as the source mechanism classification is concerned, the training and validation sets can be created through the generation of synthetic differential interferograms. Such a generation can be performed using a forward model computing the surface deformation (the displacement vector) associated to specific values of the given fault parameters. For this purpose, the Okada formulation can be considered as it explains, in a close analytic form, the surface deformation due to a seismic event by a dislocation model in an elastic half space. The data set is obtained by running the forward model varying the values of its parameters appropriately all along a certain range (see Fig. 5). In the classification case, the output layer of the network consists of three units each associated with one of the three considered fault mechanisms. The network is trained in order to give the value one to the unit corresponding to the desired type of mechanism and the value zero to the other two units. In the parameter retrieval case, the output of the network is formed by the desired quantities. Once the network is trained, the technique performs in a fully automatic mode, providing the fault classification and/or the estimation of the fault parameters in real time.

Features Extraction from Satellite Data, Fig. 5 Different examples for the three considered types of faults of differential interferograms extracted from the simulated data set generated by means of RNGCHN software. *Top*: normal fault; *middle*: strike slip fault; *bottom*: reverse fault

Summary

The contribution introduces the feature extraction techniques and explains the rationale for their use especially in the field of image processing. The main feature extraction approaches are briefly reviewed and an insight of the most important applications for the analysis of earthquakes with Earth Observation data is given. The complete methodology addressing the problem of the estimation of tectonic parameters from SAR interferometric data is described.

Cross-References

▶ Building Damage from Multi-resolution, Object-Based, Classification Techniques
▶ Damage Detection in Built-up Areas Using SAR Images

- Earthquake Damage Assessment from VHR Data: Case Studies
- Earthquake Magnitude Estimation
- Earthquake Mechanism Description and Inversion
- Earthquake Mechanisms and Tectonics
- Estimation of Potential Seismic Damage in Urban Areas
- Hyperspectral Data in Urban Areas
- InSAR and A-InSAR: Theory
- Land Use Planning Following an Earthquake Disaster
- Noise-Based Seismic Imaging and Monitoring of Volcanoes
- Remote Sensing in Seismology: An Overview
- SAR Images, Interpretation of
- Seismic Event Detection
- Spatial Variability of Ground Motion: Seismic Analysis
- Urban Change Monitoring: Multi-temporal SAR Images

References

Bellman R (1957) Dymanic programming. Princetown University Press, Princetown

Bishop C (1995) Neural networks for pattern recognition. Oxford University Press, New York

Cross GR, Jain AK (1983) Markov random field texture models. IEEE Trans Pattern Anal Mach Intell PAMI-5(1):25–39

Feigl KL, Duprè E (1999) RNGCHN: a program to calculate displacement components from dislocations in an elastic half-space with applications for modeling geodetic measurements of crustal deformation. Comput Geosci 25(6):695–704

Haralick RM, Shanmugam K, Dinstein I'H (1973) Textural features for image classification. IEEE Trans Syst Man Cybern SMC 3(6):610–621

Jain AK, Farrokhnia F (1991) Unsupervised texture segmentation using Gabor filters. Pattern Recogn 24(12):1167–1186

Massonnet D, Rossi M, Carmona C, Adragna F, Peltzer G, Feigl K, Rabaute T (1993) The displacement field of the Landers earthquake mapped by radar interferometry. Nature 364(6433):138–142

Okada Y (1985) Surface deformation due to shear and tensile faults in a half space. Bull Seismol Soc Am 75(4):1135–1154

Picchiani M, Del Frate F, Schiavon G, Stramondo S (2012) Features extraction from SAR interferograms for tectonic applications. EURASIP J Adv Signal Process 2012:155

Schölkopf B, Mika S, Burges CJC, Knirsch P, Muller K, Ratsch G, Smola AJ (1999) Input space versus feature space in kernel-based methods. IEEE Trans Neural Netw 10(5):1000–1017

Singh A, Harrison A (1985) Standardized principal components. Int J Remote Sens 6(6):883–896

Soille P (2003) Morphological image analysis: principles and applications. Springer, New York

Stramondo S, Del Frate F, Picchiani P, Schiavon G (2011) Seismic source quantitative parameters retrieval from InSAR data and neural networks. IEEE Trans Geosci Remote Sens 49(1):96–104

Tronin AA (2010) Satellite remote sensing in seismology. A review. Remote Sens 2010(2):124–150

Frequency-Magnitude Distribution of Seismicity in Volcanic Regions

John J. Sanchez Aguilar[1], Hernán G. Alvarado[2] and Santiago López[2]
[1]Departamento de Geociencias, Universidad Nacional de Colombia, Bogotá, Colombia
[2]Departamento de Geociencias, Universidad Nacional de Colombia-Sede Bogotá, Bogotá, Colombia

Synonyms

b-value anomalies; b-values; Earthquake size distribution; Seismic b-values around volcanoes; Volcanic seismicity; Volcano-tectonic earthquakes

Introduction

The spatial and temporal distribution of earthquakes sizes near and around volcanoes can be characterized through the well-known frequency-magnitude distribution:

$$\log N = a - bM \qquad (1)$$

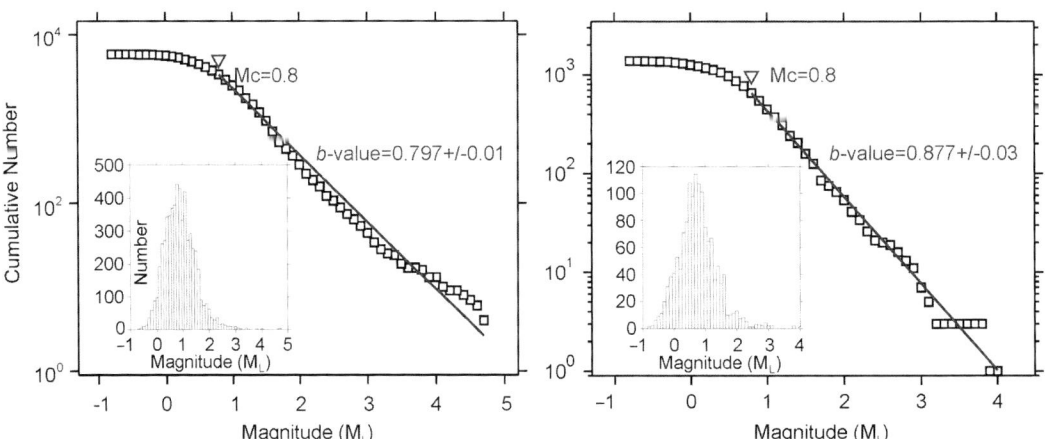

Frequency-Magnitude Distribution of Seismicity in Volcanic Regions, Fig. 1 Average magnitude of completeness and b-value for the data compiled by Observatorio Vulcanológico y Sismológico de Pasto (OVSP, Servicio Geológico Colombiano) at Galeras Volcano, Colombia. *Left*: Data for the period 2004–2013. Total number of earthquakes used is 5,858. The *solid line* shows a least squares best fit for the linear part of the frequency-magnitude distribution (FMD). The *insets* show histograms of magnitudes, with bins equal to 0.1. *Right*: Data for the period 2008.3–2010.6. Total number of earthquakes is 1,384. Conventions as in *left* panel

where N is the cumulative number of earthquakes with magnitude $\geq M$ and a and b are constants representing the seismic productivity of a region and the slope of the linear part for the distribution, respectively. This is a well-known power law (Gutenberg and Richter 1944) that is applied herein.

Earthquake catalogs compiled by volcano observatories around the world are commonly used to estimate b values and map their spatial distribution. From Eq. 1, it is evident that earthquake magnitude plays a fundamental role and is important to mention that its computation may vary from one catalog to another. Firstly, different magnitudes have been traditionally calculated at different observatories (duration magnitudes, M_D, and local magnitudes, M_L, are most common) and secondly, detection capabilities vary because of the different network configurations and equipment. The magnitude of complete detection or magnitude of completeness (Mc) is the value above which the catalog is considered to be approximately complete (Fig. 1).

In a practical way Mc can be estimated by visual inspection of any of the forms of the FMD. For example, in Fig. 1 the FMD for earthquakes at Galeras Volcano, Colombia, is shown. The histogram of the number of events versus magnitudes shows a slightly skewed distribution with a right tail and a maximum that occurs in the range $0.7 \leq M_L \leq 0.9$, a first indication of the probable value of M_C in the catalog. Another estimate of completeness may be obtained by finding the maximum curvature of the FMD using a combination of 90–95 % confidence in the curvature (derivative) estimation. Indeed, this procedure gives $Mc = 0.8$ for the whole catalog implying that earthquakes with $M_L < 0.8$ are probably not being always detected. Inspection of a curve of cumulative number of earthquakes versus time revealed that during 2004–2013 rates of seismicity at Galeras varied, most likely because of occurrence of swarms, groups of earthquakes with similar magnitudes that occur closely separated in time and not associated to a mainshock. The period 2008.3–2010.6, however, exhibited a fairly stable behavior and the corresponding FMD did not vary (Fig. 1, right panel). The data for both time periods follows fairly well a Gutenberg-Richter law and the difference in average b-values between both datasets is small. Earthquake catalogs usually are culled at the Mc value to ensure that the dataset is mostly complete. The linear part of the FMD can be fit by a least squares procedure to obtain a straight line with negative slope.

The slope is called the *b*-value and it represents the average size distribution of earthquakes.

Because the *b*-value is related to the mean magnitude in a given region, it is important to spatially map its distribution in both volcanic and tectonic environments because of the possible relationships with the state of stress in the crust. When calculating the spatial distribution of *b*, one must consider (1) the calculation method, which can either be maximum likelihood or weighted least squares, and (2) the parameters of the gridding scheme and the number of events to sample around each node of the grid (Wiemer and McNutt 1997).

To calculate the *b*-values and estimate their spatial distribution, it is necessary to create a grid in two or three dimensions. As a general rule the node separation should be proportional to the formal errors in the spatial parameters of the earthquake locations. Regarding the sampling mode around each node, there are two possibilities: either selecting a constant number of events within circles of variable radii or selecting a variable number of events inside circles of constant radii. In both methods the sampling volume has the shape of a cylinder, with its base centered on a particular node of the grid. If the calculation takes place in map view, the cylinder will be vertical, but if the calculation is on a vertical cross section, the cylinder will be horizontal, with variable height and radius depending on the sampling method. In this way the cylinders can be coin shaped or pipe shaped. If the grid is for a 3D region, the sampling volumes will be spheres.

Within each cylinder the *b*-value is calculated from the FMD obtained from the sample and in this case one can either assume a constant Mc (usually equal to the minimum magnitude within the sample) or calculate the FMD locally. It is usually a good practice to calculate *b*-values using both methods and trying several sampling schemes to test the stability of the results.

The estimation of *b*-values has been an active research area over the last two decades and here a description of results for volcanoes around the world is presented, then the possible implications for magma processes are summarized and the body of accepted knowledge on seismic *b*-values near volcanoes is presented. Finally, a short case study on one of the most active volcanoes is shown to exemplify applications of the technique and to describe some possible implications for volcano and earthquake hazards and risk.

Seismic *b*-Values at Volcanoes Worldwide

The distributions of seismic *b*-values at depth have been studied (by spatial mapping mostly) at a number of volcanoes around the world using a variety of catalogs. A compilation of results of 2D and 3D mapping at 16 volcanoes is presented (Table 1).

Table 1 highlights that *b*-value anomalies are common regardless of the type of volcano, configuration of the seismograph network, magnitude scale used, or amount of earthquakes in the dataset. The pattern of the anomalies varies from place to place, however (Wyss et al. 1997), and it is established that, with no exception, volcanic areas exhibit irregularly shaped volumes of anomalously high *b*-values surrounded by regions of normal-to-low *b*-values (Wiemer and McNutt 1997; Wiemer and Wyss 2002). From Table 1, it is observed that there is notable spatial variability of *b*, with several volcanoes having more than one region of increased *b*-values. There is variability in the computed *b*-values as well, with *b*-anomalies in the range $1.0 < b < 3.0$, and although it would seem appealing to compare these values among volcanoes, conclusions drawn from such task would not be solid, because of the different magnitude scales used. For instance, the *b*-values of ~1.55 reported for shallow anomalies at Mt. St. Helens and Nevado del Ruiz are not necessarily comparable.

It is reasonable to assume that mineral stabilization (formation of crystals) in magmas takes place at depth within the magma chamber and that exolution of gas and vesiculation might occur at shallower levels. The mineral and volatile species and types of magmas studied vary and thus the calculated depths for stabilization of minerals or gas exolution and formation of

Frequency-Magnitude Distribution of Seismicity in Volcanic Regions, Table 1 Main features of b-value anomalies and data on mineral stabilization and gas processes at volcanoes around the world. Reported and calculated stabilization pressure values from various authors are converted to pascals (Pa, $N/m^2 = m^{-1} \cdot kg \cdot s^{-2}$) or megapascals (MPa $= 10^6$ Pa) as this is the SI derived unit for pressure or stress

Volcano	Type	Number of earthquakes used	Magnitude scale	Mc	b-value (anomaly)[a]	Anomaly centroid depth (km)	References	Pressure (MPa)	References
St. Helens (USA)	Stratovolcano	1,674	M_L	0.4	1.5 ± 0.3	3.25 ± 0.55	Wiemer and McNutt (1997)	175	Pallister et al. (2008)
Mt. Spurr (USA)	Stratovolcano	643	M_L	0.1	1.6	8.35 ± 1.65	Wiemer and McNutt (1997)		
					1.7	3.4 ± 1.1			
					1.2	>10.0			
Mammoth Mtn. (Long Valley Caldera)	Caldera dome		M_L	1.3	1.8	5.0	Wiemer et al. (1998)	20	Birkett (2007)
					1.73 ± 0.06	8.0			
Resurgent Dome (Long Valley Caldera)	Active dome		M_L	1.3	1.9?	4.0			
					1.36 ± 0.05	10.0			
Teishi Knoll (Izu-Tobu group), located off Ito (Japan)	Submarine Crater, within pyroclastic cones field	10,000	M_L?	1.3	1.00 ± 0.03	8.5	Wyss et al. (1997)	800	Kushiro (1991)
					1.54 ± 0.05	13.0			
Soufriere Hills (West Indies)	Stratovolcano	1,900	M_D	1.7	3.07 ± 0.07	2.0	Power et al. (1998)	150	Barclay et al. (1998)
Mt. Etna (Italy)	Stratovolcano	450	M_D	2.5	3.0 ± 0.5	7.5 ± 2.0	Murru et al. (1999)	155	Trigila et al. (1990)
Martin (Alaska, USA)	Stratovolcano	2,500	M_L	0.7	1.6 ± 0.05[b]	14.5 ± 2.5	Jolly and McNutt (1999)	62.5	Coombs and Gardner (2001)
						6.0 ± 1.0			
Mageik (Alaska, USA)	Stratovolcano		M_L	0.7	1.8 ± 0.24[c]	6.0 ± 1.5			

Frequency-Magnitude Distribution of Seismicity in Volcanic Regions

Volcano	Type	N	Mag.	Mc	b	a	Reference	
Katmai (Alaska, USA)	Stratovolcano, Caldera		M_L	0.7	1.30 ± 0.25	8.0 ± 1.0		37.5	Coombs and Gardner (2001)
Kilauea-East Rift Zone	Shield Volcano	16,963	M_p[d]	1.5	1.73	12.0 ± 3.0 5 ± 2.0	Wyss et al. (2001)	
Mt. Redoubt (USA)	Stratovolcano				1.7 ± 0.1	4.0	Wiemer and Wyss (2002)	180.9	Gerlach et al. (1994)
Nevado del Ruiz (Colombia)	Stratovolcano	2,769	M_D	0.7	2.2 ± 0.2 1.55 ± 0.25 1.65 ± 0.05 1.55 ± 0.15 1.55 ± 0.15	9.0 1.0 1.0 ± 0.5 5.35 ± 0.35 2.0 ± 0.2	Rodríguez (2003)	50	Stix et al. (2003)
Pinatubo (Philippines)	Stratovolcano	1,406	M_D	0.73	1.7 ± 0.1	2.0 ± 1.0 11.0 ± 1.0	Sánchez et al. (2004)	220	Bernard et al. (1996)
Galeras (Colombia)	Stratovolcano	1,918	M_L[e]	1.2	1.3 ± 0.05	2.5 ± 0.5	Sánchez et al. (2005)	60	Stix et al. (1997)
Miyake-jima-Tokushima (Izu Islands, Japan)	Stratovolcano	10,000	M_L?	2.5	1.86 ± 0.1 1.3 ± 0.1 1.3 ± 0.1	8.0 ± 0.5 12.5 ± 2.5 19 ± 2.0	Rierola (2005)	224	Amma-Miyasaka et al. (2003)
Makushin (Alaska, USA)	Stratovolcano	491	M_L	1.25	1.9 ± 0.1	4.0 ± 1.7	Bridges and Gao (2006)[f]	169	Goldberg (2010)

[a] Values for b-anomalies and their locations are taken from results of 3D mapping and vertical cross sections, where reported

[b] During July 27, 1995, to December 31, 1997, the Alaska Volcano Observatory reported over 2,500 earthquakes located in the general area Martin-Mageik-Trident-Katmai Caldera. The b-values reported in the table were estimated from Fig. 6 (Jolly and McNutt 1999)

[c] b-values at Martin-Mageik region may include data from swarms in Sept. 1996–April 1997. Values calculated pre- and post swarm were in the range $0.92 < b < 0.98$

[d] Mp is a "preferred magnitude" chosen hierarchically and defined in the data description prepared by F. Klein for distribution of the earthquake catalog at Hawaii. Mostly equal to M_D and in few cases equal to M_L

[e] Magnitudes were recalculated by regression equation from M_D to M_L using a sample of earthquakes simultaneously recorded by broad band and short period subnetworks

[f] Bridges and Gao report a second, deeper b-value anomaly ($b \sim 1.6$) detected at roughly 12 km depth southeast of Makushin, but it was deemed statistically insignificant

Frequency-Magnitude Distribution of Seismicity in Volcanic Regions,
Fig. 2 Histogram of b-value anomaly centroid depths beneath volcanoes

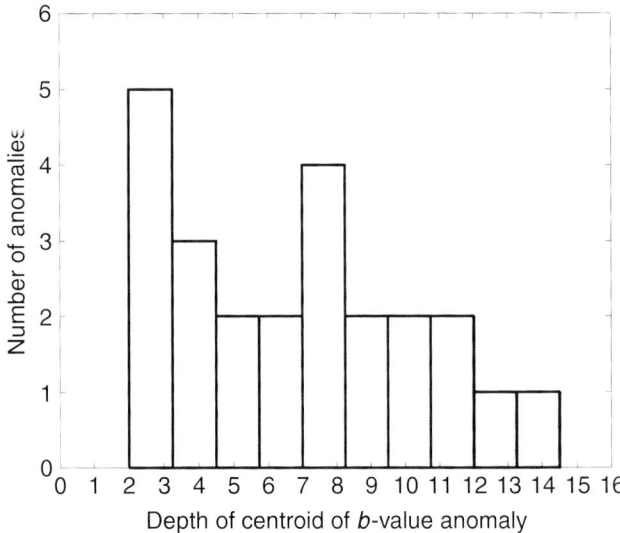

vesicles: For Mt. Etna, Italy, Trigila et al. (1990) report a range of 85–225 Mpa pressure range for Plagioclase stabilization in basaltic lavas and it has been suggested that the feeding system for magmas at Etna may be composed of different reservoirs; at the Katmai cluster, USA, Coombs and Gardner (2001) estimated a range of 40–100 Mpa for Plagioclase stabilization in andesite-dacite-rhyolite magmas erupted during the 1912 eruption from Novarupta; at Redoubt, USA, Gerlach et al. (1994) suggested that SO_2 is stored as vapor at 6–10 km depth in the magma before the 1990 eruptions and perhaps at shallower levels (4–6 km) prior to the 2009 eruptions (Coombs and Bacon 2012); at Nevado del Ruiz, Colombia (Stix et al. 2003), it has been suggested that ascent of volatile-bearing magma took place from 9 to 15 km depth and remained at shallow residence depth of <3 km; at Mt. Pinatubo, Philippines, estimation of volatile contents in the range 5.1–6.4 % suggested a vesiculation depth of 6 km for pre-eruptive magmas and stabilization of olivine at 2.2 kbar or a depth of magma storage in the range 7–11 km was also inferred (Bernard et al. 1996); at Galeras, Colombia (Stix et al. 1997), gas exolution might have taken place at 20–100 Mpa or 0.8–3.9 km depth; and at Makushin, USA, scoriae within mafic ignimbrite layers were analyzed (Goldberg 2010) and a stabilization pressure of pyroxenes in a range of 1.26–2.12 kbar has been suggested. Evidence from Geology, Petrology, and Geochemistry thus indicates that several processes take place at different depths beneath volcanoes.

From Table 1 it is also evident that seismicity studies have also revealed that beneath many volcanoes, b is anomalously high at different depths. All volcanoes listed (17) exhibit b-value anomalies embedded in an otherwise region of normal-to-low b-values. With respect to the distribution of b-values at depth, most volcanoes studied (16, b-values at Yellowstone were not mapped at depth) have at least one anomaly, and 11 out of 16 volcanoes exhibit more than one anomalous region. The centroid depths for b-value anomalies vary in the range 2–19 km and their distribution are shown in Fig. 2.

From Fig. 2 it is evident that, in many cases, volcanoes exhibit two anomalies at different depths: one shallow anomaly is common in the range 2–4 km and a deeper one appears in the range 7–8 km. The first anomaly range includes not only the depth at which volatiles in magma with 4 % weight fraction may start to exsolve but also the region where the existence of open cracks is plausible. At depths similar to the range 7–8 km, alternate geophysical evidence and geology studies support the nearby presence of magma bodies. At Galeras Volcano, Colombia, the shallow plumbing magmatic system has been conceived as a vertically elongated feature such as a reservoir extending down to 5–6 km depth or a fractured plug beneath de active cone that overpressurizes and acts as the source of material during Vulcanian explosions (Stix et al. 2003) and a deeper magma reservoir below 4 km depth.

Regarding the cause of anomalous b-values, there is general agreement that several factors may be involved; these are summarized as follows:

Accepted Knowledge on Seismic b-Values at Volcanoes

- That the frequency-magnitude distribution is related to the heterogeneity (degree of cracking) in the upper crust and can be appropriately measured by a power law expressed in the Gutenberg-Richter law. The parameters of this law are understood as the seismic productivity of the volume ("a-value") and the mean magnitude of the linear part ("b-value").
- Estimating the b-value is equivalent to estimating the mean magnitude and it is reasonable to assume the mean magnitude as proportional to mean crack length.
- Heterogeneity in volcanic areas can be pervasive and may be related not only to the nature and amount of cracking that the magma system or volcano edifice may have but also to the nature of rocks and deposits that form volcanoes, usually a combination of hydrothermally altered lava flows and pyroclastic deposits.
- Although volcanic areas were usually reported as having higher-than-normal b-values, with respect to a "global" average of $b = 1.0$, it is now known that instead, volcanoes exhibit regions of anomalously high b-values within a "normal" crust.
- That magma (molten rock inside the earth) cannot support shear stresses and therefore earthquakes cannot occur there. b values give information on faults and fractures in the vicinities of magma chambers. The resolution of earthquake locations may not be still enough to reveal details on the configuration of magma systems, although the concept of complex configurations in the magma and plumbing systems beneath volcanoes is generally accepted.
- That factors that may contribute to variations in b are:
  (a) To increase b:
   1. Increase in material heterogeneity. At volcanoes this can be associated to increase in cracks, caused by vesiculation and fragmentation of ascending magma, processes that likely take place at relatively shallow levels (roughly 4 km depth for magmas with 4 wt. % H_2O).
   2. Increase in pore pressure caused by magma nearing groundwater (which would imply decreasing the effective stress). This process may also likely take place at relatively shallow levels.
   3. Increase in the thermal gradient. Quite possible near magma systems or intrusions, because magmas may reach temperatures near 1,200 °C.
  (b) To decrease b:
   1. Increase in applied shear stresses. Difficult to model this near volcanoes, but plausible along the faults that are common in volcanic areas. Zones of lower-than-normal b-values have been associated to regions of high stress load and therefore areas where large earthquake nucleation could take place. In this perspective, seeking for regions of lower-than-normal b-values is just as important and may have hazard implications.
   2. Increase in effective pressure or decrease in pore pressure. Could be conceivable during or after volcanic eruptions.
  (c) Either to increase or decrease b:
   1. Changes in the configuration of the detection network or changes in analyses procedures. These may introduce artifacts in the catalog. For example, a change in the way magnitudes are determined may cause a "stretching" of the magnitude scale, offsetting b toward lower values.

It has been suggested that processes such as vesiculation, fragmentation, and magma intrusions can affect the crack distribution beneath a volcano. Gas exolution and subsequent bubble nucleation and growth inside magmas take place under colossal pressure conditions. Bubbles can increase in numbers and accumulate near the top of magma reservoirs increasing pressure on the host rock and facilitating cracking. Also, magmas intrude through fractures and weak zones of the crust and this process may cause new cracks and fracture propagation thus changing the homogeneity of the rock masses.

From Table 1 a scatter plot of the centroid depths for b-value anomalies versus the pressure at which mineral stabilization and/or gas exolution might take place or the estimated depth for magma storage (from Petrology, Geochemistry, and Geology data) was built. Errors in the centroids of b-value anomalies were taken as the maximum and minimum observable extents at depth from published and authoritative material. Results are shown in Fig. 3.

Based on Fig. 3 (right panel), a positive relation between the depth of deeper b-value anomalies and the magma process beneath volcanoes is mildly revealed for a group of eight volcanoes: Mammoth Mt. (Long Valley Caldera), Nevado del Ruiz, Mt. St. Helens, Katmai-Novarupta, Redoubt, Mt. Pinatubo, Mt. Etna, and Miyakejima. These include diverse-sized volcanic systems from a variety of geological environments: six continental arc volcanoes (formed above oceanic-continental subduction zones), a hot spot volcano, and an island arc volcano (formed above oceanic-oceanic subduction zone).

b-Values at Galeras Volcano, Colombia, and Implications for Earthquake Hazard and Risk

Galeras Volcano, one of the most active volcanoes in Colombia, looms over the city of San Juan de Pasto (pop. 400,000) which lies just a few kilometers east of the active crater. Not only have repeated eruptions of Galeras affected life and property over the years and volcanic processes do represent a significant hazard for the region (INGEOMINAS 1997), but also earthquakes possibly associated to volcanic-tectonic interactions have caused damage and fatalities in recent times. Thus, seismically characterizing the region near Galeras is of relevance and mapping of b-values offers the opportunity to identify zones of smaller-than-average earthquakes as well as regions where relatively large earthquakes are typically produced.

Using preliminary earthquake locations between September 1995 and June 2002, a first approach to spatially map the earthquake size distribution around Galeras was taken (Sánchez et al. 2005). Their results of 3D mapping readily allowed to identify a prominent high b-value anomaly beneath the active crater and vertically elongated down to 5 km depth which was inferred to be the region adjacent either to the volcano's conduit, the shallow magma reservoir, or alternatively the remnants of a semi-brittle intrusion. In map view, this anomalous feature was basically surrounded by regions of normal-to-low b-values. Of particular interest here is the cluster of earthquakes located northeast of Galeras, where the source of three swarms that included damaging earthquakes ($M_L \leq 4.7$) was identified during 1993–1995 (April 1993, November–December 1993, and February 1995). This source falls within a region of low b-values detected preliminarily in the early 2005 study (Fig. 4).

Recent data for Galeras has been analyzed and the catalog of earthquakes for the period 2004–2013 was used to spatially map b-values using both the entire catalog data and a subset of it (2008–2010), during which seismicity rates were observed to be fairly stable. Results are presented in Fig. 5. In map view the region north-northeast of Galeras still appears as a region of low b-values and remains as a source of moderate-sized earthquakes (a sequence of four $4.6 < M_L < 4.7$ earthquakes occurred there during mid 2005). The active crater region remains highly productive, but it does not show as a zone of smaller-than-average earthquakes, on the contrary, falls within an area of low b-values and was the source of two separate $4.6 < M_L < 4.7$ earthquakes in 2006 and 2010.

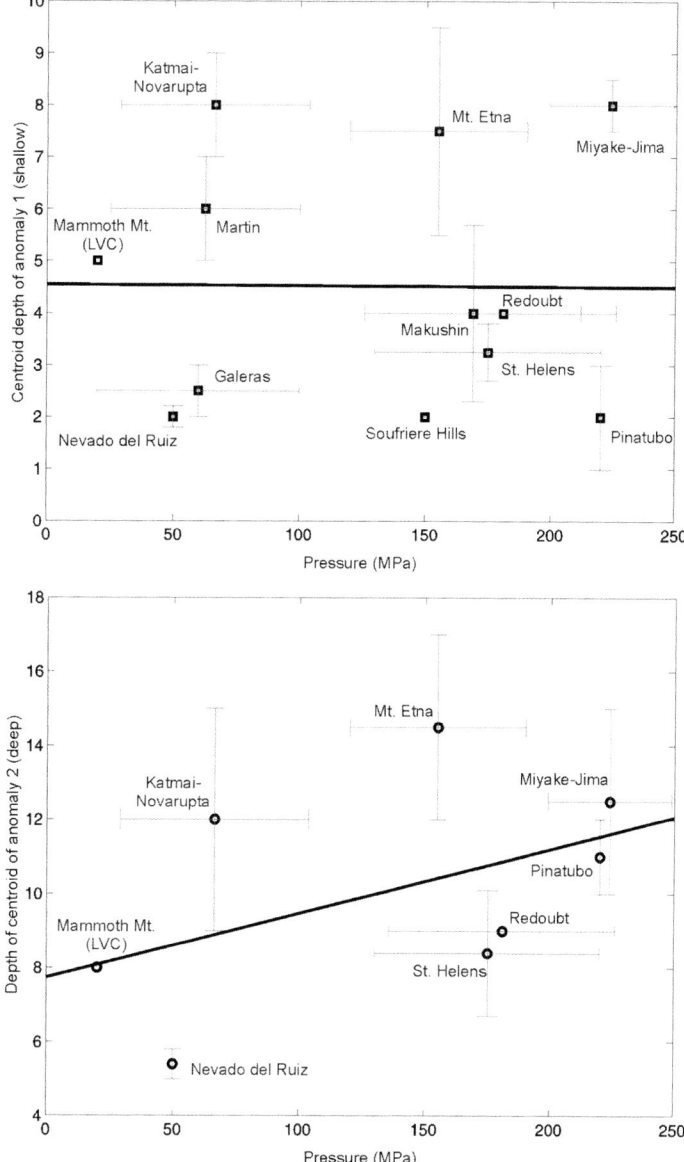

Frequency-Magnitude Distribution of Seismicity in Volcanic Regions, Fig. 3 Scatter plots of the depths of b-value anomalies observed at volcanoes versus pressure conditions calculated for gas exolution or mineral stabilization (from Geology, Petrology and Geochemistry data). Many volcanoes from Table 1 exhibit more than one b-value anomaly usually at different depths. *Left*: Centroid depth of "shallower" anomalies. *Right*: Centroid depth for "deeper" anomalies. Thick *black lines* in both panels show a basic linear fit (R = 0.0 for the data on the *left* panel, R = 0.5 for data on the *right* panel)

It is conceivable that once a volcano or a region of the crust has acquired a given crack distribution, this is difficult to change and thus the distribution of b-values would be fairly constant through time. At volcanoes, however, multiple intrusions are common and therefore a much fractured host rock may be plausible. For example, the well-recorded earthquake swarms beneath the Teishi Knoll Volcano (off Ito, Japan) were associated to magma intrusions and the detected change of b-value with time was attributed to new cracks (Wyss et al. 1997). Thus, changing the heterogeneity of the shallow crust near volcanoes, or the difficulty in causing changes in such systems, could be viewed as a matter of time scale. The results described for Galeras could signal a change in conditions of the shallow crust near the active crater.

As it can be seen in Figs. 4 and 5, the Galeras region is crossed by several faults in a general southwest-northeast direction and the volcano

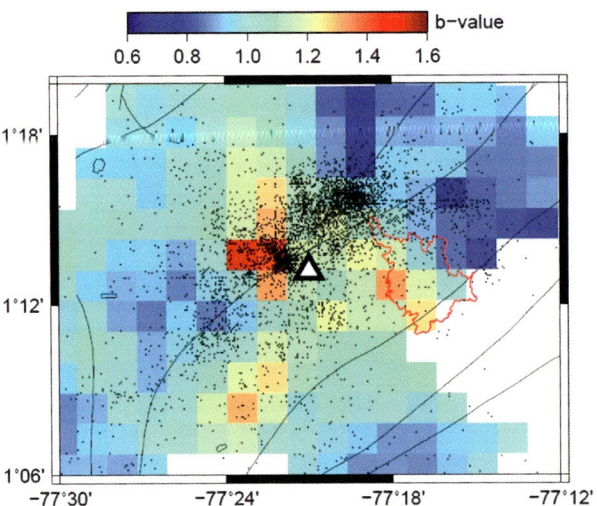

Frequency-Magnitude Distribution of Seismicity in Volcanic Regions, Fig. 4 Map of *b*-values in the Galeras region for the period 1995–2002. High *b*-values are mapped as regions of hot colors (*red-orange*), regions of low *b*-vales appear as areas of cold colors (*blue*). Dots: earthquake epicenters. Triangle: active vent. Continuous *black lines*: faults. Thick *red* contour: limits of San Juan de Pasto. Swarms of damaging earthquakes were produced between 1993 and 1995 in the area of low *b*-values northeast of the volcano (Modified from Sánchez et al. 2005)

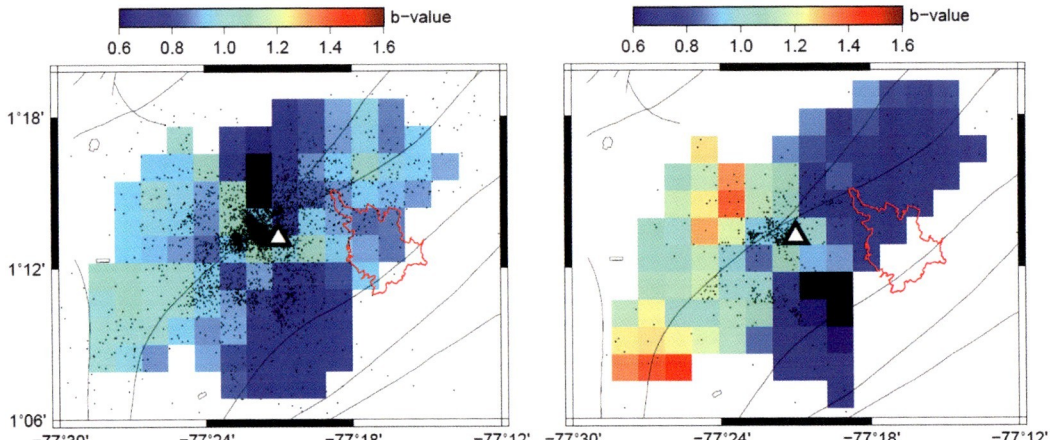

Frequency-Magnitude Distribution of Seismicity in Volcanic Regions, Fig. 5 Maps of *b*-values in the Galeras region. *Left*: period 2004–2013, using all earthquakes with $M_L > 0.8$. Relatively large earthquakes ($4.6 < M_L < 4.7$) occurred in 2006.16 (near the crater), 2010.64 (just east of the crater), and during 2005.64–2005.65 (north-northeast of the crater, four earthquakes). *Right*: period 2008–2010, during which the seismicity rate was fairly constant. The different *b*-values in the western portions of Galeras between the different datasets may be an artifact caused by the different sampling around each node, necessary because of the *lower* number of earthquakes in the shorter period 2008–2010. The region of interest, northeast of the active crater remains mostly unchanged. Conventions as in Fig. 4

itself is spatially related to these structures, particularly to the splay of the two faults near the active crater. It is conceivable that magma has used faults as pathways to the surface, making this area a particularly complex one in terms of stresses and earthquakes.

Interestingly, the area southeast of the volcano which shows persistently as a region of very low *b*-values has not produced relatively large earthquakes damaging to San Juan de Pasto.

The concept that faults include both locked and unlocked portions, the former conceived as asperities that resist sliding and the latter known as creeping segments, implies that asperities are regions that accumulate stress whereas in creeping segments stresses are constantly relieved. The distribution of *b*-values near active faults has been successfully used to mark the sites of asperities and creeping segments along faults (Wiemer and Wyss 2002), with areas of low *b*-values signaling asperities and areas of higher-than-normal *b*-values representing creeping segments. Volcanic regions are always affected by active faults, and considering that moderate-sized earthquakes have been detected near the volcano, it is reasonable to assume that near Galeras Volcano the distribution of *b*-values may help in identifying areas of relatively high stress load, and from Figs. 4 and 5, it may be advisable to use a higher-quality dataset and map *b*-values along the faults that have been documented there. Such analyses might have implications for hazards and risk not only for the city of San Juan de Pasto but also for other towns surrounding the volcano.

Summary

The *b*-value (slope of the frequency-magnitude distribution of earthquakes) around active volcanoes has been mapped at a number of volcanoes around the world and, with information from published sources on types of magmas (Geology data) and pressure conditions inferred for gas exolution, vesiculation, and mineral stabilization within magma systems (Petrology and Geochemistry data), a possible relation between the depth of *b*-value anomalies with magma processes is presented. The body of generally accepted knowledge on *b*-values at volcanoes is summarized and recent data for *b*-values at Galeras Volcano, Colombia, are analyzed in terms of volcano and earthquake hazards and risk.

Cross-References

▶ Earthquake Location
▶ Earthquake Magnitude Estimation
▶ Seismic Anisotropy in Volcanic Regions
▶ Seismic Monitoring of Volcanoes
▶ Seismic Tomography of Volcanoes
▶ Volcanic Eruptions, Real-Time Forecasting of
▶ Volcano-Tectonic Seismicity of Soufriere Hills Volcano, Montserrat

References

Amma-Miyasaka M, Nakagawa M (2003) Evolution of Deeper Basaltic and Shallower Andesitic Magmas during the AD 1469–1983 Eruptions of Miyake-Jima Volcano, Izu-Mariana Arc: Inferences from Temporal Variations of Mineral Compositions in Crystal-Clots, J. Petrology 44:2113–2138.

Barclay J, Rutherford MJ, Carroll MR, Murphy MD, Devine JD, Gardner J, Sparks RSJ (1998) Experimental phase equilibria constraints on pre-eruptive storage conditions of the Soufrière Hills magma. Geophys Res Lett 25:3437–3440

Bernard A, Knittel U, Weber B, Weis D, Albrecht A, Hattori K, Klein J, Oles D (1996) Petrology and geochemistry of the 1991 eruption products of Mount Pinatubo (Luzon, Philippines). In: Newhall CG, Punongbayan RS (eds) Fire and mud: eruptions and lahars of Mount Pinatubo. Philippine Institute of Volcanology and Seismology/University of Washington Press, Quezon City/Seattle, pp 767–798

Bridges DL, Gao SS (2006) Spatial variation of seismic *b*-values beneath Makushin Volcano, Unalaska Island, Alaska. Earth Planet Sci Lett 245:408–415

Burkett SM (2007) Geomorphic mapping and petrography of mammoth mountain, California. MS thesis, University of New York at Buffalo, Buffalo, I14 pp

Coombs ML, Bacon CR (2012) Using rocks to reveal the inner workings of Magma Chambers below volcanoes in Alaska's National Parks. In: Shah M (ed) Alaska Park science-volcanoes of Katmai and the Alaska Peninsula. Alaska Geographic, Anchorage, pp 27–33

Coombs ML, Gardner JE (2001) Shallow-storage conditions for the rhyolite of the 1912 eruption at Novarupta, Alaska. Geology 29:775–778

Gerlach TM, Westrich HR, Casadevall TJ, Finnegan DL (1994) Vapor saturation and accumulation in magmas of the 1989–1990 eruption of Redoubt Volcano, Alaska. J Volcanol Geotherm Res 62:317–337

Goldberg AR (2010) Petrologic and volcanic history of Point Tebenkof ignimbrite, Unalaska, Alaska. In: de Wet AP (ed) Proceedings of the twenty-third annual Keck research symposium in geology. Keck Geology Consortium, Houston, pp 335–340

Gutenberg B, Richter CF (1944) Frequency of earthquakes in California. Bull Seismol Soc Am 34:185–188

INGEOMINAS (1997) Mapa de amenazas volcánicas del Galeras-tercera versión. Instituto de Investigaciones en Geociencias, Minería y Química. http://www.ingeominas.gov.co/Pasto/Volcanes/Volcan-Galeras/Mapa-de-amenazas.aspx. Accessed 4 June 2013

Jolly AD, McNutt SR (1999) Seismicity at the volcanoes of Katmai National Park, Alaska; July 1995–December 1997. J Volcanol Geotherm Res 93:173–190

Kushiro I (1991) Origin of volcanic rocks in Japanese island arcs. Episodes 14:258–263

Murru M, Montuori C, Wyss M, Privitera E (1999) The location of magma chambers at Mt. oEtna, Italy, mapped by b-values. Geophys Res Lett 26:2553–2556

Pallister JS, Thornber CR, Cashman KV, Clynne MA, Lowers HA, Mandeville CW, Brownfield IK, Meeker GP (2008) Petrology of the 2004–2006 Mount St. Helens lava dome – implications for magmatic plumbing and eruption triggering. In: Sherrod DR, Scott WE, Stauffer PH (eds) A volcano rekindled: the renewed eruption of Mount St. Helens, 2004–2006, U.S. Geological Survey professional paper 2008-1750. U.S. Geological Survey, Vancouver, pp 647–702

Power JA, Wyss M, Latchman JL (1998) Spatial variations in the frequency-magnitude distribution of earthquakes at Soufriere Hills Volcano, Montserrat, West Indies. Geophys Res Lett 25:3653–3656

Rierola M (2005) Temporal and spatial transients in b-values beneath volcanoes. Diploma thesis, ETH Zurich, Institute of Geophysics, Zurich, 77 pp

Rodríguez SP (2003) Mapeo tridimensional del valor b en el Volcán Nevado del Ruiz. Trabajo de Grado, Universidad de Caldas, Manizales, Colombia, 72 pp

Sánchez JJ, McNutt SR, Power JA, Wyss M (2004) Spatial variations in the frequency-magnitude distribution of earthquakes at Mount Pinatubo Volcano. Bull Seismol Soc Am 94:430–438

Sánchez JJ, Gómez DM, Torres RA, Calvache ML, Ortega A, Ponce AP, Acevedo AP, Gil-Cruz F, Londoño JM, Rodríguez SP, Patiño J de J, Bohórquez OP (2005) Spatial mapping of the b-value at Galeras Volcano, Colombia, using earthquakes recorded from 1995 to 2002. Earth Sci Res J 9:30–36

Stix J, Torres RA, Narváez L, Cortés GP, Raigosa J, Gómez DM, Castonguay R (1997) A model of vulcanian eruptions at Galeras volcano, Colombia. J Volcanol Geotherm Res 77:285–303

Stix J, Layne GD, Williams SN (2003) Mechanisms of degassing at Nevado del Ruiz Volcano, Colombia. J Geol Soc 160:507–521

Trigila R, Spera FJ, Aurisicchio C (1990) The 1983 Mount Etna eruption: thermochemical and dynamical inferences. Contrib Miner Petrol 104:594–608

Wiemer S, McNutt SR (1997) Variations in the frequency-magnitude distribution with depth in two volcanic areas: Mount St. Helens, Washington, and Mt. Spurr, Alaska. Geophys Res Lett 24:189–192

Wiemer S, Wyss M (2002) Mapping spatial variability of the frequency–magnitude distribution of earthquakes. Adv Geophys 45:259–302

Wiemer S, McNutt SR, Wyss M (1998) Temporal and three-dimensional spatial analyses of the frequency-magnitude distribution near Long Valley Caldera, California. Geophys J Int 134:409–421

Wyss M, Shimazaki K, Wiemer S (1997) Mapping active magma chambers by b values beneath the off-Ito volcano, Japan. J Geophys Res 102:20,413–20,422

Wyss M, Klein F, Nagamine K, Wiemer S (2001) Anomalously high b-values in the south flank of Kilauea Volcano, Hawaii: evidence for the distribution of magma below Kilauea's east rift zone. J Volcanol Geoth Res 106:23–37

Friction Dampers for Seismic Protections of Steel Buildings Subjected to Earthquakes: Emphasis on Structural Design

Lucia Tirca
Department of Building, Civil and Environmental Engineering, Concordia University, Montreal, QC, Canada

Synonyms

Damping; Earthquake; Energy dissipation; Friction dampers; Seismic design; Steel buildings

Introduction

To reduce the seismic demand, researchers have proposed to incorporate supplemental energy dissipation devices into the structural system of buildings. According to the primary dissipation mechanism, supplemental energy dissipation devices are grouped into two categories: hysteretic and viscoelastic. Hysteretic devices rely on the relative displacements of components within the device and are typically based on either metallic yielding or frictional sliding, while viscoelastic devices are velocity dependent. More specifically, friction devices dissipate energy through the relative sliding developed between two solid interfaces. Depending on the type of friction devices, they could be installed in line with single-diagonal or chevron steel braces, at the intersection of X-bracing system, and in parallel with the beam located at the top of chevron bracing system. The activation of slip forces that characterize the designed friction dampers occur simultaneously with the maximum internal forces allowable to develop in the system during the ground motion excitation. The building reaches the peak interstorey drift when the available slip distance provided by friction damper devices was consumed. "The forces generated by these devices installed in the structural members are usually in phase with the internal forces resulting from ground motion shaking" (Christopoullos and Filiatrault 2006).

The total input energy, E_I, induced by a seismic event into a structural system can be expressed as a summation of kinetic energy, E_k, cumulative strain energy, E_S, inherent damping, E_D, and the hysteretic damping E_h of the seismic force resisting system (SFRS). In this study, E_h is the damping induced by friction devices. The energy balance equation is:

$$E_I = E_k + E_S + E_D + E_h \quad (1)$$

The kinetic and cumulative strain energy are accumulated into the primary structural system and rely on structural damage (Akiyama 2000; Tirca 2009), while the system is damped by both E_D and E_h, which are amplitude-dependent. In general, the contribution of E_D and E_h is related to the amount of post-yielding response and Eq. 1 can be rearranged as follows:

$$E_k + E_S = E_I - (E_D + E_h) \quad (2)$$

In Eq. 1, the term $(E_k + E_S)$ expresses the vibrational energy, E_v, or more specifically the potential damage energy, while $(E_D + E_h)$ is the energy dissipated by viscous damping in the structural members and supplemental devices. Thus, by adding damping into a structural system the elastic vibration energy diminishes, while structural

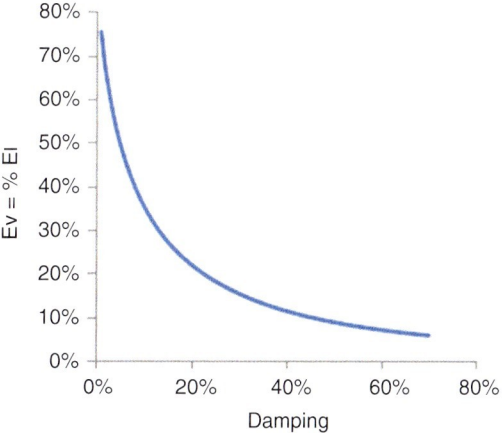

Friction Dampers for Seismic Protections of Steel Buildings Subjected to Earthquakes: Emphasis on Structural Design, Fig. 1 Vibrational energy versus supplemental damping

members are protected from damage associated with permanent deformations. In this light, the main design objective is to minimize the difference between the seismic input energy and that dissipated by the dampers (Christopoullos and Filiatrault 2006). The variation of E_v as a function of supplemental damping is depicted in Fig. 1.

Pioneering work on friction devices was conducted by Pall (1979) and Pall and Marsh (1981). Since then, several types of friction dampers have been developed and studied in the literature. These devices differ in their mechanical shape and materials used for the sliding surfaces.

Pall friction dampers dissipate energy through friction developed by the relative sliding within two surfaces in contact which are clamped by posttensioned bolts. In order to obtain stable rectangular hysteresis loops similar to that of Coulomb friction, different types of surface treatment and lining material were studied experimentally. After investigating the response of slip bolted joints under monotonic loading, Pall reported that the most stable behavior was obtained when brake lining pads in contact with mill scale surface on plate was chosen (Fig. 2a). Nevertheless, minor differences between the static and dynamic friction coefficient were observed. However,

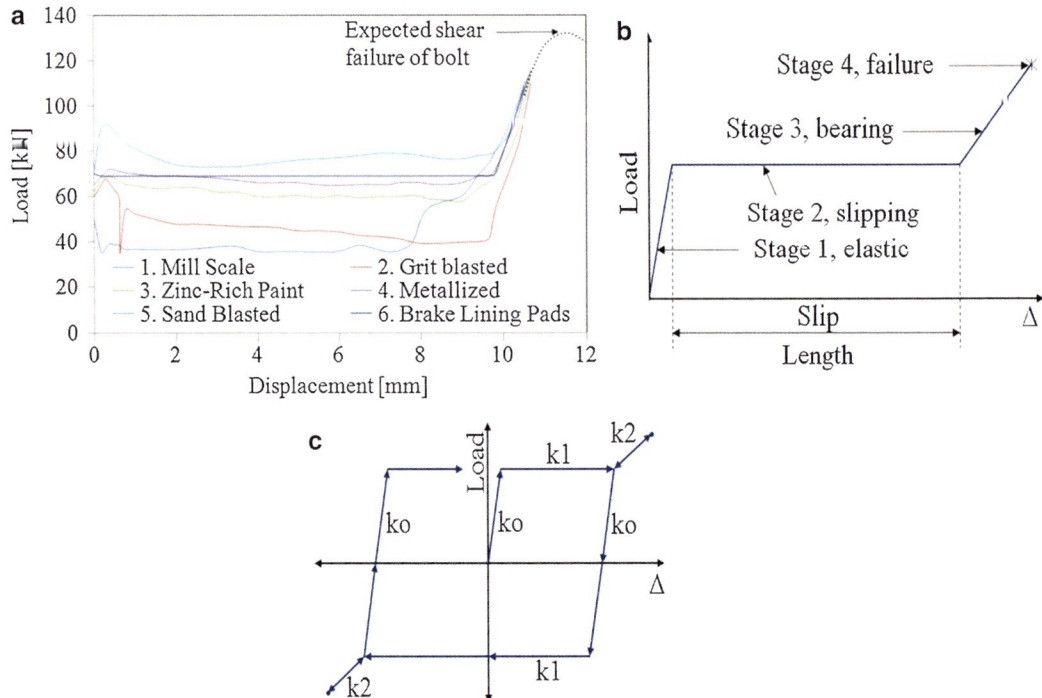

Friction Dampers for Seismic Protections of Steel Buildings Subjected to Earthquakes: Emphasis on Structural Design, Fig. 2 Response of slip-bolted joint: (**a**) monotonic test, (**b**) back-bone curve, and (**c**) hysteretic behavior (After Pall 1979)

under large seismic excitations, the posttensioned bolts of friction dampers may impact into the end of slotted hole and undergo bearing or even bolt shear failure. The hysteretic behavior of friction damper shown in Fig. 2b is similar to that of an elasto-perfectly plastic system. Herein, the backbone curve is composed of four segments: elastic, slipping, bearing, and bolt shear failure. When the demand is higher than the available slip length which is equal to the length of slotted hole, a sudden increment in storey shear forces accompanied by decreasing of Coulomb damping is encountered. The hysteresis behavior depicted in Fig. 2c in terms of slip load versus the slip length, Δ, shows rectangular symmetrical loops which are largely influenced by the fluctuation of friction coefficient during the slipping stage. The elastic stiffness, k_o, shown in Fig. 2c, is the stiffness of the attached brace member.

Due to their efficiency (Pall and Pall 2004), Pall friction dampers were employed in more than 40 new and retrofit buildings in Canada, United States, and India (Christopoullos and Filiatrault 2006). Examples of Pall friction damper installed in an X-bracing system and single-diagonal brace are showed in Fig. 3.

In the last two decades, the following types of friction devices have been developed: slotted bolted connections (Fitzgerald et al. 1989; Grigorian et al. 1993; Tremblay 1993), Sumitomo devices developed by Sumitomo Metal Industries Ltd. of Japan and reported by Aiken and Kelly (1990), energy dissipating restraint damper (EDR) developed by Flour Daniel Inc. (Nims et al. 1993), friction variable damper developed based on the EDR damper (Zhou and Peng 2009), friction damper developed by Mualla (2000), etc.

The energy dissipated by the Sumitomo friction damper is due to friction generated when friction pads, made of cooper alloy with graphite plug inserts, slide directly on the inner surface of the outer cylinder. The precompressed internal spring, incorporated into the device, applies a

Friction Dampers for Seismic Protections of Steel Buildings Subjected to Earthquakes: Emphasis on Structural Design, Fig. 3 Pall friction dampers manufactured by Pall Dynamics Lmt. Montreal: (**a**) installed in X-bracing at Concordia Library, Montreal, and (**b**) installed in single-diagonal brace at Boeing Commercial Airplane Factory Everrett, USA (Courtesy Dr. Pall)

force that is converted through the action of inner and outer wedges into a normal force on the friction pads. Sumitomo device can be installed in parallel to the beam located on top of chevron bracing system or in-line with steel braces. As reported by Aiken and Kelly (1990), Sumitomo dampers were used in two high-rise buildings in Japan in order to resist small intensity earthquakes and ground floor vibrations.

The EDR damper is similar to Sumitomo device. It includes the following components: an internal spring, steel compression wedges, bronze friction wedges, stops at both spring's ends, and the outer steel cylinder. Its functioning comprising the generating slip force depends on the length of the internal spring. In addition, the normal force acting on the cylinder wall and the friction force developed between the bronze friction wedges and the inner surface of the outer cylinder determine the slip force in the device. As noted by Nims et al. (1993), the EDR device can produce a wide range of hysteretic behavior such as double-flag shape and triangular lobed shape. Both types of hysteresis loops are self-centering. By using the same principles, Zhou and Peng (2009) proposed a new friction variable damper where a sliding shaft and a friction ring were used to replace the spring and wedges. In addition, two zones with high and low friction coefficient have been defined in the internal walls of the outer cylinder.

The friction damper proposed by Mualla (2000) was added to connect the top part of chevron braces to the mid-span of a moment resisting frame beam. This friction damper, patented by DAMPTECH Denmark, consists of three steel plates that are able to rotate against each other around a pre-stressed bolt which passes through. Two circular friction pad discs inserted between the aforementioned steel plates provide dry friction lubrication and ensure stable friction forces. As reported by Liao et al. (2004), a full scale test conducted on a three-storey moment resisting frame with added chevron bracing system and Mualla's dampers was investigated. The total weight of the frame was about 34 t. The bracing system consisted of bar members that were pre-tensioned in order to avoid buckling. Due to the application of lower scaling simulated ground motions, no damage of structural members was reported during the conducted experimental tests.

Design Provisions

Since 1980, seismic response control techniques have been used as complementary solutions to the existing SFRSs. Despite of their wide-spread applications, design guidelines for structures equipped with supplemental energy dissipation devices (hysteretic and viscoelastic) are still in evolving phases. This study refers to the design philosophy of structures with incorporated friction damper devices (FDD). As mentioned above, FDDs are hysteretic devices. For example, buckling-restrained braces are also known as hysteretic damper devices and are designed to dissipate part of vibration energy through yielding of the core plate.

In general, hysteretic devices are installed in braces and are able to undergo large inelastic response, while they dissipate most of the hysteretic energy. The purpose of installing these devices into the structural system is to maintain the main structure either elastic or within low inelastic deformations. Both, the main frame and the supplemental energy dissipation system, share the same deformation, which in turn is that of the entire system. It is important to assure stable response of these devices under dynamic loading. Thus, the added dampers installed in new or retrofit buildings should yield or slip before the shear resistance of the main structure is reached.

In North America, the first design guidelines addressing provisions for passive energy dissipated devices were introduced in FEMA 356 (2000) and FEMA 450 (2003) that refer to the seismic rehabilitation of buildings and seismic design for new buildings, respectively. Later on, FEMA 450 was replaced by FEMA-P 750 (2009). The aforementioned design guidelines were incorporated in Chapter 14 of ASCE/SEI 41-13 (Seismic evaluation and retrofit of existing buildings), as well as in Chapter 18 of ASCE/SEI 7-10 standard. It is noted that *Commentary J* of NBCC 2010 (National Building Code of Canada) includes two clauses in regard to seismic design with supplemental energy dissipated devices. However, both CSA/S16-2009 and ANSI/AISC 360-2010 standard provide design regulations for the "ductile buckling-restrained braced frames" system.

In Europe, the Eurocode 8 part 1 (CEN 2004) contains Chapter 10 entitled *Base isolation*, but this chapter does not cover regulations for passive energy dissipation devices installed into several storeys of building's structure.

The aforementioned guidelines specify general requirements, analysis procedures, required testing program, etc. A brief review of these guidelines with highlights on friction damper devices design is presented below.

FEMA P-750 and ASCE/SEI 7-10 Provisions

Design Philosophy
General requirements:

– Structures with a damping system have a seismic force resisting system that provides a complete load path. The base shear used to design the employed SFRS shall not be less than V_{min}, where V_{min} is the grater of:

$$V_{min} = V/B_{v+1} \text{ and } V_{min} = 0.75 \, V.$$

Herein, V is the seismic base shear in the direction of calculation determined in accordance with the code procedure corresponding to the selected SFRS and B_{v+1} is the numerical coefficient set for effective damping reduction factor that is equal to the damping provided by supplemental dampers in addition to the inherent damping (2 % for steel structures). For irregular building structures and for damping system that has less than four damper devices per floor disposed in the direction of loading, the minimum base shear will not be less than $1.0 \, V$.

– Damping system may be used in addition to the SFRS in order to meet interstorey drift limits and it may be located external or internal to the structure, while it shares or not the members of SFRS. In Fig. 4 is illustrated a SFRS consisting of one bay moment resisting frame (MRF) and a damping system composed of diagonal braces with installed FDs. The FD system is located independently to the MRF system.

– Energy dissipating devices must be tested before installation for displacement and slip force corresponding to seismic demand that is required by code.

– Linear static and response spectrum analysis methods for design are accepted for regular

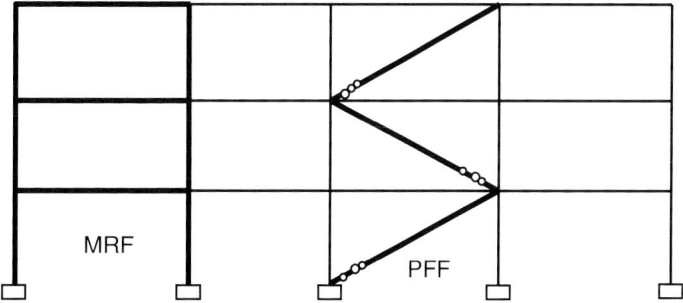

Friction Dampers for Seismic Protections of Steel Buildings Subjected to Earthquakes: Emphasis on Structural Design, Fig. 4 Configuration of SFRS and damping system (no shared elements)

structures that have at least four damper devices per each floor located in one principal direction of the building. Like MRF or braced frame structures, damped structures yield during the design ground motions, while they need to behave elastically under the wind loads.
- For detailed design and seismic response investigations, nonlinear dynamic time-history analyses are recommended.
- Dampers are modeled to perform in the plastic range. The computer model should be calibrated based on experimental test results. For designs in which the SFRS is expected to yield, the postyield behavior of the structural elements must be modeled explicitly.

In the ASCE/SEI 7-10 provisions, damping system is defined as being "the collection of structural elements that includes all the individual damping devices, all structural elements or bracing required to transfer forces from damping devices to the base of the structure, and the structural elements required to transfer forces from damping devices to the seismic force-resisting system." For details of design procedure, the reader is referred to the ASCE/SEI 7-10 document. It is noted that other design methodologies can be applied to design structures equipped with friction dampers.

Design of Structures Equipped With Pall Friction Dampers

As noted above, metallic dampers and friction dampers are hysteretic dampers. The design of structures equipped with metallic dampers or friction dampers is similar. The difference consists of changing the yield load corresponding to metallic dampers with the slip load of friction dampers. As mentioned above, friction dampers are installed in braces and are designed not to slip under the wind forces. On the other hand, in order to increase the building performance under earthquake loads, systems composed of friction-damped braces are added to the MRF structures, while the MRF system should posses enough strength, stiffness, and ductility. Furthermore, all members of the SFRS and those attached to FDs should be designed to carry the internal forces developed when slip forces are activated. Thus, the application of capacity design checks to structural members is required. According to Christopoullos and Filiatrault (2006) the design of structures equipped with friction dampers can be divided in four steps:

- Calculate the demanded slip force and size the in-line brace by employing simplified methods (e.g., the equivalent static force procedure). Then, the demanded peak interstorey drift (e.g. linear dynamic analysis methods) should be evaluated.
- Optimize the design of friction dampers and adjacent members.
- Apply the capacity design and size all structural members to carry the designed slip forces.
- Use the nonlinear time-history dynamic analysis to check the design performance of building structure under an ensemble of selected ground motions scaled to match the design response spectrum corresponding to 2 % probability of exceedance in 50 years (NBCC 2010).

A review of different design procedures from the literature is discussed below. However, in all design procedures, the building structure is composed of a bare frame system and a supplemental damping system, whilst its response is analyzed before and after the friction dampers are activated.

Design Procedures from the Literature

The first design procedure for MRF system equipped with friction damped braces was proposed by Baktash and Marsh (1986). They considered a single storey MRF structure with friction damped braced bays. Based on this procedure, the slip force is activated when yielding in the moment resisting frame brace is reached. For a single bay MRF frame that was designed based on the principle "strong column weak beam", the shear force resisted by the frame action, V_s is related to the plastic moment capacity of the MRF beam, M_p. Herein, $M_p = V_s h/2$, where h is the storey height. Thus, the proposed design method is based on the assumption that the lateral shear force leading the damper in-line with brace to slip must be equal to the shear force causing the MRF to yield and is expressed by the following equation:

$$V_s = 2M_p/h \quad (3)$$

The corresponding shear deflection of the storey, Δ_s is:

$$\Delta_s = (V_t - V_{br})/k_u \quad (4)$$

where V_t is the total shear force exerted by the frame and braces, V_{br} is the total shear force exerted by braces alone, and k_u is the lateral stiffness of the bare MRF.

The hysteretic energy dissipated by friction devices E_f is:

$$E_f = V_{br}\Delta_s = V_{br}(V_t - V_{br})/k_u \quad (5)$$

By differentiating Eq. 5 with respect to V_{br} and setting it equal to zero, the maximum energy dissipated by friction devices is obtained:

$$\partial E_f/\partial V_{br} = V_t/k_u - 2V_{br}/k_u = 0 \quad (6)$$

From Eq. 6, the shear force exerted by braces is:

$$V_{br} = 0.5 V_t \quad (7)$$

Thus, to maximize the energy dissipated by friction devices, the total shear force must be equally shared between friction devices and the bare frame (MRF). In addition, it was reported that the optimum slip force of friction devices installed in structure depends on the structural characteristics only and not on the ground motion intensity (Baktash and Marsh 1987).

Filiatrault and Cherry (1988) proposed to determine the optimum activation of friction devices based on minimizing the *Relative Performance Index*, RPI, devised from the application of the energy concept.

$$RPI = \tfrac{1}{2}[SEA/SEA_o + U_{\max}/U_{\max 0}] \quad (8)$$

where SEA is the strain energy area of all structural members of a friction damped system, SEA_0 is the strain energy area corresponding to a zero activation force, $U_{\max}$ is the maximum strain energy stored in all structural members of a friction damped system, and $U_{\max 0}$ is the maximum strain energy for a zero slip force. The values resulted for the RPI yield to three cases: (i) RPI $=$ 1 corresponds to the response of a bare frame structure; (ii) RPI $<$ 1 corresponds to the response of a damped structure which is smaller than the response of the bare frame structure; (iii) RPI $>$ 1 corresponds to the response of a damped structure which is larger than the response of the bare frame structure. Authors recommended the selection of diagonal braces to comply to $T_b/T_u < 0.4$, where T_b and T_u are the fundamental period of the braced frame structure and bare frame structure, respectively. Herein, the fundamental periods T_b and T_u can be expressed as $T_b = 2\pi/\omega_b$ and $T_u = 2\pi/\omega_u$, where ω_b is the natural circular frequency of the fully braced frame before dampers are activated, $\omega_b = (k_b/m)^{0.5}$ and ω_u is the natural circular frequency of bare frame, $\omega_u = (k_u/m)^{0.5}$. In addition, k_b and k_u are the lateral stiffness of the braced frame structure and bare frame structure, respectively. In these expressions, m is the total seismic mass of the system. Later on, Filiatrault and Cherry (1990) proposed the following equation to calculate the total shear force, V_o that is required to activate all friction damper devices in a structure:

$$V_o/W = (a_g/g) \times Q(T_b/T_g, T_b/T_u, N_f) \quad (9)$$

where N_f is the number of floors, a_g is the design peak ground acceleration, g is the gravity acceleration, T_g is the predominant period of the selected design ground motion, and Q is an unknown single valued function that is given in Eq. 10a for $0 \leq T_g/T_u \leq 1$ and Eq. 10b for $T_g/T_u > 1$.

$$Q = (T_g/T_u) \times [(-1.24N_f - 0.31)T_b/T_u + 1.04N_f + 0.43] \quad (10a)$$

$$Q = (T_b/T_u) \times [(0.01N_f + 0.02)T_g/T_u - 1.25N_f - 0.32] + (T_g/T_u) \times (0.002 - 0.002N_f) + 1.04N_f + 0.42 \quad (10b)$$

Equations 9 and 10a or 10b can be used directly to calculate the total base shear force V_O. It is noted that for a single diagonal brace located at floor i, the slip force is P_i and the activation shear at that floor is $V_i = P_i \cos\theta_i$, where θ_i is the angle of brace inclination reported to a horizontal line.

In the meantime, studies on hysteretic bracing systems were carried out by Ciampi et al. (1990, 1992, 1995, etc.) who considered that both MRF and bracing system are arranged in parallel. They concluded that the activation of friction damper devices should occur before yielding of moment resisting frame. In addition, Ciampi et al. (1995) referred to four key parameters such as:

- T_f, which is the fundamental period of the bare moment resisting frame. A deductable parameter is $\alpha = T_b/T_f$ where T_b is the fundamental period of the bracing system.
- $\lambda = k_b/k_f$ which is the ratio between the stiffness of bracing system and the stiffness of the bare moment resisting frame.
- $\beta = \delta_{by}/\delta_{fy}$ which is the ratio between the displacement that cause yielding of bracing system and that that causes yielding of bare MRF system.
- $\eta_f = F_{fy}/ma_g$ which is the yield strength of the frame normalized to the mass of the structure multiplied by the design peak ground acceleration. Similarly, it can be defined $\eta_b = F_{by}/ma_g$, where F_{by} is the yield force of bracing system. The total normalized force is $\eta_{II} = \eta_f + \eta_b$.

The selection of the above parameters suggests the variation effect of β and λ on the building response. When $\beta = 1$, bracing system and MRF system yield for the same lateral displacement. When $\beta = 0$, the structural system corresponds to the MRF system only. They recommended an optimal value of β around 0.5.

The application of this design methodology to MDOF systems requires uniform distribution of stiffness and slip forces along the building height, aiming at uniform activation of friction devices in order to avoid damage concentration within a specific floor. In addition, they proposed to maintain at each storey the stiffness of braces proportional to the bare frame stiffness. Also, they recommended keeping proportionality between the horizontal projection of slip forces per floor and the corresponding static or dynamic distributed storey shear over the building height. In addition, two design assumptions were made: (i) the MRF system is designed to respond elastically while the friction damper devices behave in the non-linear range and (ii) the MRF system may yield when δ_{fy} was reached, therefore after friction damper devices were activated. The tri-linear force-displacement constitutive low is depicted in Fig. 5.

Fu and Cherry (1999, 2000) proposed a simplified design method that leads to a code compatible procedure for a *friction-damped steel braced frame system*. First, they developed the method for a SDOF system based on establishing an equivalent ductility-related force modification

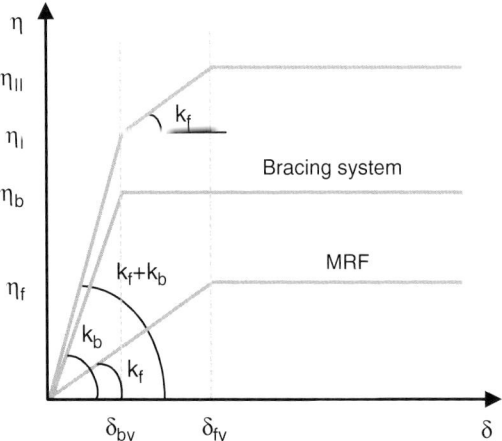

Friction Dampers for Seismic Protections of Steel Buildings Subjected to Earthquakes: Emphasis on Structural Design, Fig. 5 Trilinear force-displacement constitutive low (After Ciampi et al. 1995)

factor R, which reflects the capacity of the system to dissipate energy. They have developed a trilinear model similar with that illustrated in Fig. 5, where η parameter was replaced by the restoring force of a SDOF system, $f(t)$. In addition, η_I and η_{II} parameters were replaced by f_s and f_y, where f_s is the slip restoring force (after dampers were activated) and f_y is the yielding restoring force developed when $f(t)$ exceeds f_y and the frame members yield. Thus, to design the friction-damped steel braced frame system, the design base shear is first calculated based on the code procedure (e.g. equivalent static force procedure). In this case, the value assigned to the R factor (R_d in the current NBCC code) is evaluated based on the selected system parameters. Then, all members are designed based on forces resulted from the distribution of design base shear over the building height, and the brace stiffness and slip forces are evaluated at each floor. In this study, friction-damped braces were installed in a ductile steel MRF system. The response of building designed based on the employed R factor should be verified such that the demanded lateral frame deflection under the wind and earthquake load to be maintained below the code limits. For example, according to the NBCC 2010 requirements, the system must respond elastically under the wind load, while the interstorey drift should not exceed $h_s/500$. Under earthquake loads, the maximum interstorey drift should be less than $2.5\%h_s$ for a building classed under the normal category. Herein, h_s is the storey's height.

Tirca (2009) proposed a design methodology to upgrade the seismic resistance of an existing steel moment frame building designed before 1970. The method complies with FEMA 356 (2000) provisions. First, the available elastic base shear, V_f, provided by the existing steel moment frame system was evaluated by using modal response spectrum method and a three-dimensional building model. Then, the required base shear carried by friction-damped braces V_{br} was computed as being the difference between the code demand, V, and V_f. Herein, V is the minimum lateral earthquake force computed based on NBCC provisions by setting $R_d = 1.0$. The effect of torsion was included. By considering the nonlinear behavior of friction dampers, the resulted base shear assigned to friction-damped braces, V_{br} was divided by B_{v+1} as per FEMA 356 and then was distributed over the structure height. According to FEMA 356 recommendation, minimum four friction-damped braces displaced in the direction of loading were installed in each floor. The slip force of each device located at storey i was computed by divided the corresponding storey shear force resulted from a dynamic analysis to the number of friction-damped braces and to the $cos\theta_i$, where θ_i was defined above. All steel braces were designed to behave elastically while sustaining the nonlinear response of attached friction damper. Thus, braces were sized such that 130% C_r to be larger than the slip force, whereas C_r is the brace compressive resistance. To preserve the existing columns of MRF system in elastic range, it was proposed to stagger the installation of friction-damped braces such that to minimize torsion. To prevent the concentration of damage at specific storeys, it was proposed to maintain a constant ratio between the horizontal projection of slip forces and dynamic distributed storey shear over the building height. However, the available slip length should be recommended in the design phase in order to avoid sudden failure. The displacement at yield, Δ_y, depends on the

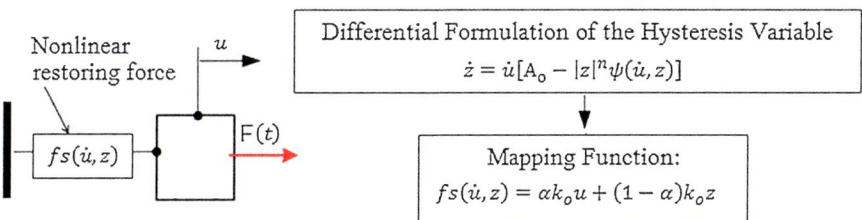

Friction Dampers for Seismic Protections of Steel Buildings Subjected to Earthquakes: Emphasis on Structural Design, Fig. 6 Mechanical model of a SDOF system

stiffness of friction-damped braces, k_{bd}, as shown in Fig. 2c. As noted above, the stiffness of the frame and the stiffness of friction-damped braces are in parallel ($k_f + k_{bd}$), whereas $\Delta_y = F_s/(k_f + k_{bd})$ and F_s is the slip restoring force. In this study, the seismic response of retrofitted building equipped with friction-damped braces was verified to remain within the interstorey drift code limits and forces developed in structural members to be below the members' resistance. It was concluded that friction-damped braces installed in moment frame buildings are able to control the interstorey drift and floor acceleration.

Numerical Modeling of Friction-Damped Braces

In general, researchers select the bilinear force-deformation model for simulating the behavior of friction damper devices. When performing nonlinear time-history analyses, convergence problems may arise due to the sharp transitions from the elastic to inelastic stages during the loading, unloading, and reloading cycles. Due to the large number of friction devices installed in a building structure, the bilinear model "can become computationally inefficient" when these devices are in different phases of stiffness transition (Moreschi and Singh 2003). To overpass this modeling drawback, the Bouc-Wen model is recommended. In addition, in the case of Bouc-Wen model, it is the same differential equation that governs its response (Eq. 11) and more specifically its behavior during different transition stages. Therefore, the Bouc-Wen model is able to simulate the highly nonlinear Coulomb friction and has the ability to represent different hysteresis shapes according to the values of the parameters involved (Morales Ramirez 2011). Since the desired shape of the Coulomb dry friction law is symmetric and strength and stiffness degradation is neglected, the Bouc-Wen model is reduced to a nonlinear restoring force (Eq. 11) of a SDOF system shown in Fig. 6. The evolutionary variable z, given in Eq. 11, is defined in Eq. 12.

$$f_s(\dot{u}, z) = \alpha k_o u + (1-\alpha) k_o z \quad (11)$$

$$\dot{z} = \dot{u}\left\{\frac{A - |z|^n[\gamma + \beta sgn(\dot{u}z)]v}{\eta}\right\} \quad (12)$$

In the above equations, α is the participation ratio of the initial stiffness in the nonlinear response, k_o is the initial stiffness of the system, u is the displacement of the SDOF system, and z is the hysteresis variable. In Eq. 12, γ and β are parameters controlling the shape of the hysteresis cycle and the exponent n influences the sharpness of the model in the transition zones. The remaining parameters A, v, and η control the degradation process in stiffness and strength. When the degradation process is neglected, the aforementioned parameters are $A = A_o$, $v = 1$, and $\eta = 1$. The considered SDOF system is characterized by the restoring force $f_s(du/dt, z)$ that has a linear and a nonlinear component, as defined in Eq. 11. Using $n = 1$ might yield to a flexible behavior, while increasing the period of vibration and reducing the inertial forces. The smooth transition toward a rectangular hysteresis shape is obtained for a large value of variable n which might become computationally expensive, because the transient

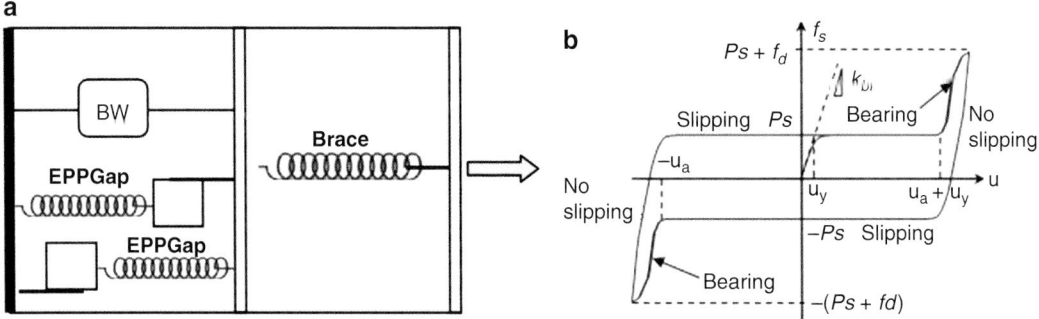

Friction Dampers for Seismic Protections of Steel Buildings Subjected to Earthquakes: Emphasis on Structural Design, Fig. 7 Friction-damped brace: (**a**) Schematic model and (**b**) Hysteresis model plus the bearing state

analysis requires the calculation of the evolutionary variable z at each step of time for a single element. By using $n = 10$, it gives an acceptable level of prediction because the difference is quickly reduced throughout the evolution of the slipping stage.

To simulate the slip and slip-lock phases of friction-damped brace device in OpenSees (McKenna and Fenves 2004), Morales Ramirez (2011) proposed a novel equivalent uniaxial material composed of BoucWen material (BW) in parallel with gap springs made of Elastic-Perfectly Plastic Gap material (EPPGap), which are already available in the OpenSees library (Mazzoni et al. 2006). The friction damper was arranged in series with the brace member made of steel material. The brace was modeled with a truss element. The schematic representation of friction-damped brace device is shown in Fig. 7a. In the illustrated model, *BoucWen* material was selected to replicate the smooth hysteresis behavior of friction damper (stick–slip and slip phase), which was activated when the axial stress σ_s generated by the slip force P_s was reached ($\sigma_s = P_s/A_{br}$ and A_{br} is the gross area of the attached brace). Each friction damper is characterized by its slip force and slip length that are resulted from calculation. However, under strong ground motions, the provided slip length, u_a, may be limited and the pretensioned edge bolt may drive the friction damper into the slip-lock phase as depicted in Fig. 7b (Morales Ramirez and Tirca 2012). Herein, u_y is the elastic axial deformation of the brace. To simulate the slip-lock phase, one equivalent EPPGap spring made of three bilinear gap springs arrange in parallel each other are defined to act in tension and another set in compression. Each uniaxial *ElasticPPGap* material (OpenSees notation) has a defined stress–strain or force-deformation relationship either in tension or in compression. Thus, one equivalent EPPGap spring is activated when the displacement demand exceeds the available slip length, $+u_a$, and the other is activated when the $-u_a$ slip length is reached. Once activated, this component is able to limit the displacement and to increase the force experienced by the friction damper. The threshold force of these gap elements is related to the maximum force that the device is able to withstand ($P_s + f_d$) without reaching failure (Fig. 7b). To control the failure phase, the *MinMax* material was set to decouple the friction device when the strain of the *ElasticPPGap* material exceeds the predefined bounds either in tension or compression. In addition, when the *MinMax* material is activated (e.g., $t = t_i$), the device is decoupled from $t = t_i$ until the end of analysis.

Detailed design examples of 4-, 8-, and 12-storey MRF building equipped with Pall friction dampers installed in-line with braces are given in Morales Ramirez (2011) reference. The building is located in Montreal, Canada, and was subjected to ten simulated and historical ground motions scaled to match the design response spectrum over the period of interest $0.2\ T_1$–$1.5\ T_1$, where T_1 is the fundamental period of the building. All analyses were performed using OpenSees.

To evaluate the seismic response of structures equipped with friction-damped braces, the following parameters should be investigated: interstorey drift, lateral deflection, and residual interstorey drift.

For an accurate OpenSees model of friction-damped brace, it is required to calibrate the model against experimental test results. Presently, experimental tests conducted on Pall friction dampers installed in-line with braces are carried out and the obtained results will be disseminated. Furthermore, the development of a code-based design method for structures equipped with friction-damped braces is required.

Summary

Since 1980, seismic response control techniques have been used as complementary solutions to the existing seismic force resisting systems and several types of friction damper devices were developed. Among them, Pall friction damper devices are the most popular. However, despite of their wide-spread applications, design guidelines for structures equipped with supplemental energy dissipation devices are still in an evolving phase. This study refers to the design philosophy of structures with incorporated friction dampers that in most cases are installed in-line with steel braces. A brief review of design guidelines and other seismic design provisions that emphasize on the design of friction-damped braces is presented. In addition, a brief discussion on the numerical modeling of friction-damped braces was conducted.

Cross-References

▶ Nonlinear Dynamic Seismic Analysis

References

Aiken ID, Kelly JM (1990) Earthquake simulator testing and analytical studies of two energy-absorbing systems for multistory structures. Report No UBC/EERC-90/03, Earthquake Engineering Rresearch Center, University of California, Berkeley

Akiyama H (2000) Evaluation of fractural mode of failure in steel structures following Kobe lessons. J Construct Steel Res 55:211–227

American Society of Civil Engineers (2010) Minimum design loads for buildings and other structures ASCE Reston. SEI 7–10

American Society of Civil Engineers (2013) Seismic evaluation and retrofit of existing buildings ASCE Reston. SEI 41–13

Baktash P, Marsh C (1986) Seismic behavior of friction damped braced frames. In: Proceedings of the 3rd U.S. National Conference on Earthquake Engineering, Charleston, pp 1099–1105

Baktash P, Marsh C (1987) Damped moment-resistant braced frames: A comparative study. Can J Civil Eng 14:342–346

CEN (2004) Eurocode 8: Design provisions for earthquake resistance structures. ENV 1998–2. Brussels, Begium

Christopoullos C, Filiatrault A (2006) Passive supplemental damping and seismic isolation. Instituto Universitario di Studi Superiori di Pavia, IUSS Press, Pavia

Ciampi V, Ferretti A (1990) Energy dissipation in buildings using special bracing systems. In: Proceedings of the 9th European conference on earthquake engineering, vol 3. Moscow, pp 9–18

Ciampi V, Paolone A, De Angelis M (1992) On the seismic design of dissipative bracing. In: 10th world conference on earthquake engineering, vol 7. Madrid, pp 4133–4138

Ciampi V, De Angelis M, Paolone F (1995) Design of yielding or friction-based dissipative bracing for seismic protection of buildings. Eng Struct 17:381–391

Federal Emergency Management Agency (FEMA) (2000) Prestandard and commentary for the seismic rehabilitation of buildings. FEMA-356, Washington, D.C.

Federal Emergency Management Agency (FEMA) (2003) NEHRP Recommended seismic provisions for new buildings and other structures. FEMA-450, Washington, DC

Federal Emergency Management Agency (FEMA) (2009) NEHRP recommended seismic provisions for new buildings and other structures. FEMA-P-750, Washington, DC

Filiatrault A, Cherry S (1988) Seismic design of friction damped braced steel plane frames by energy methods. Earthquake Engineering Research Laboratory report UBC-EERL-88-01. Department of Civil Engineering, University of British Columbia, Vancouver

Filiatrault A, Cherry S (1990) Seismic design spectra for friction damped structures. ASCE J Struct Eng 116(5):1334–1355

Fitzgerald TF, Anagnos T, Goodson M, Zsutty T (1989) Slotted bolted connections in a seismic design of concentrically braced connections. Earthquake Spectra, EERI 5(2):383–91

Fu Y, Cherry S (1999) Simplified seismic code design procedure for friction damped steel frames. Canadian J Civil Eng 26:55–71

Fu Y, Cherry S (2000) Design of friction damped structures using lateral force procedure. Earthq Eng Struct Dyn 29:989–1010

Grigorian CE, Yang TS, Popov EP (1993) Slotted bolted connection energy dissipators. Earthq Spectra 9(3):491–504

Liao WI, Mualla I, Loh CH (2004) Shaking table test of a friction-damped frame structure in. Struct Des Tall Spec 13:45–54

Mazzoni S, McKenna F, Scott M, Fenves G (2006) Opensees command language manual. PEER, University of California, Berkeley

McKenna F, Fenves GL (2004) Open system for earthquake engineering simulation (OpenSees), PEER, University of California, Berkeley

Morales Ramirez JD (2011) Numerical simulations of steel frames equipped with friction-damped diagonal-bracing devices. MASc thesis, Building, Civil and Environmental Engineering, Concordia University, Montreal

Morales Ramirez JD, Tirca L (2012) Numerical simulation and design of friction-damped steel frame structures. In: The 15th world conference in earthquake engineering, Lisbon, paper #2538

Moreschi LM, Singh MP (2003) Design of yielding metallic and friction dampers for optimal seismic performance. Earthq Eng Struct Dyn 32:1291–1311

Mualla IH (2000) Experimental and computational evaluation of a novel friction damper device. PhD thesis, Department of Structural Engineering and Materials, Technical University of Denmark

National Research Council of Canada (2010) National building code of Canada. National Research Council of Canada, Ottawa

Nims D, Ritcher R, Bachman R (1993) The use of the energy dissipating restraint for seismic hazard mitigation. Earthq Spectra 9(3):467–489

Pall A (1979) Limited slip bolted joints, a device to control the seismic response of large panel structures. PhD thesis, Building, Civil and Environmental Engineering, Concordia University, Montreal

Pall A, Marsh C (1981) Response of friction damped braced frames. J Struct Div ASCE 108(ST6):1313–1323

Pall A, Pall R (2004) Performance based design using Pall friction dampers – An economical design solution. In: 13th world conference on earthquake engineering, Vancouver, paper #1955

Tirca L (2009) A simple approach for the seismic retrofit of buildings with staggered friction dampers. In: Mazzolani FM (ed) Proceedings of the first international conference on: protection of historical buildings. CRC Press, Italy, pp 761–768

Tremblay R (1993) Seismic behavior and design of friction concentrically braced frames for steel buildings. PhD thesis, University of British Columbia, Vancouver

Zhou X, Peng L (2009) A new type of damper with friction. Earthq Eng Eng Vibrat 8(4):507–520

Geotechnical Earthquake Engineering: Damage Mechanism Observed

Ikuo Towhata
Department of Civil Engineering, University of Tokyo, Bunkyo-ku, Tokyo, Japan

Synonyms

Damage; Deformation; Earthquake; Fault; Liquefaction; Shaking

Introduction

In the modern community, there are a variety of problems caused by earthquakes, including such traditional ones as collapse of buildings and slope failures as well as lifeline problems and economic damages which became important in the recent times. Thus, new types of problem have been recognized as our culture and civilization change with time. It is also important to know that the safety demand is rising continuously with time.

Earthquake engineering addresses a large number of issues that are related with the past experiences of damage, their causative mechanisms and mitigations. Because of the complexity of the problem, the issues to be discussed range widely from geological and mechanical topics to societal and legal issues. It is obviously impossible for one author to deal with everything. The present author was requested to write general remarks on geotechnical issues and there are many things to be picked up. Because of the page limitation and expectation that other authors will write more about technical issues, it was decided that this entry should address geotechnical damages so far experienced and their causative mechanisms with some suggestions on future directions. Because the human civilization is changing continuously, it cannot be stressed too much that more new problems will occur at any time and that all the readers of this encyclopedia make efforts to develop new solutions. To achieve this goal, it is important to visit the real earthquake damage sites and draw lessons from these real case studies.

Damage Caused by Ground Shaking

Intensity of Seismic Motion and Subsoil Conditions

Collapse of houses has been and still is the most important kind of seismic damage. Figure 1 illustrates a building crash. While this building was subjected to a complicated effect of ground shaking, it is generally agreed that the horizontal shaking is more responsible for the damage than the vertical component. The reason for this is that any structure has a safety margin under the gravitational action, and the vertical seismic action is

Geotechnical Earthquake Engineering: Damage Mechanism Observed, Fig. 1 Seismic collapse of RC building (2001 Gujarat earthquake of Mw = 7.7 in India)

resisted to a reasonable extent by this margin, whereas there is no significant safety margin against the horizontal action produced by earthquake excitations. Consequently, many efforts have been made in the recent times to improve horizontal resistance in structures.

It has been widely recognized that earthquake damage is more substantial upon soft soil deposits than on firm soil or rock outcrop. This belief is supported by an experience during the 1923 Kanto earthquake of Mw = 7.9 in Tokyo. Figure 2a illustrates the geomorphology of the present Tokyo at the end of the sixteenth century when the urban development was started. The eastern part of the area used to be a mouth of a big river surrounded by swamps and rice paddies underlain by soft and loose soils. In contrast, the western part of the area is composed of higher terraces covered by volcanic ash materials, associated with occasional small valleys that were produced by erosion and filled with soft soils. Figure 2b illustrates the distribution of earthquake damage percentage of traditional wooden houses. Note that this damage implies the damage by shaking and does not include those by big fires that finally destroyed most parts of the city. In this figure, a good consistency can be found between the distribution of soft soils in the eastern part and in valleys and the location of high damage rate.

Amplification of Ground Motion in Soft Soil

The intensity of ground motion is increased by the surface soft soil deposits and this phenomenon is called amplification. It is known that the extent of amplification changes basically with the thickness of the surface soil (H) and the velocity of S-wave propagation (V_s). The S(econdary)-wave means a propagation of shear deformation and the theory of elasticity defines its velocity by

$$V_s = \sqrt{\frac{G}{\rho}} \qquad (1)$$

in which G and ρ stand for the shear modulus and mass density of soil, respectively. More precisely, the extent of amplification is a function of $2\pi f H/V_s$ where f designates the number of shaking cycles per second (frequency). The most significant amplification, which is called resonance in subsoil, occurs when $2\pi f H/V_s = \pi/2$, and therefore the frequency is equal to $V_s/(4H)$. This implies that the surface ground motion is dominated by a component of this special "resonant" frequency although the motion in the base rock at a deep elevation includes many frequency components in a random manner.

Resonance occurs also in structures upon the ground surface. Generally, soft wooden houses with heavy roofs have low resonant frequency.

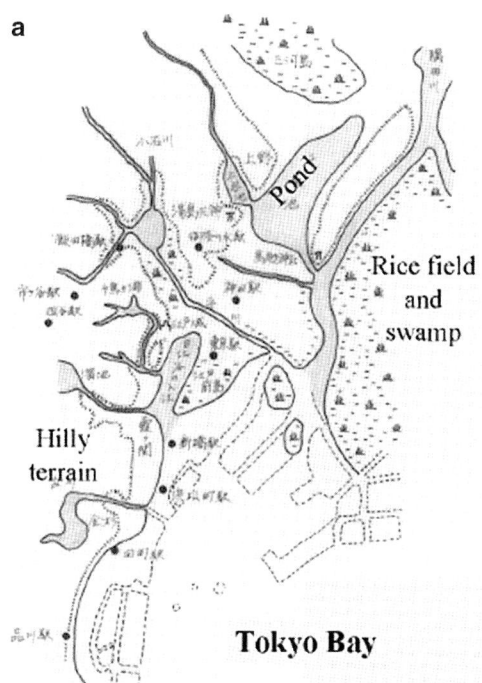 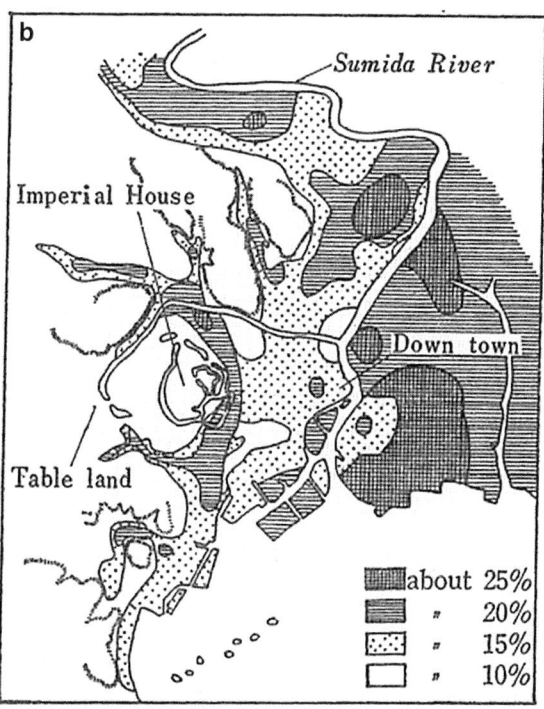

Geotechnical Earthquake Engineering: Damage Mechanism Observed, Fig. 2 Correlation between subsoil conditions and earthquake damage of wooden houses in Tokyo during the 1923 Kanto earthquake (Drawn after Murai 1994). (**a**) Original geomorphology of Tokyo (Drawn after Murai 1994). (**b**) Distribution of house damage in 1923 (Okamoto 1973)

Therefore, if the surface ground motion is rich with (dominated by) low-frequency components, the seismic response of such house is significant and heavy damage is likely. In contrast, rigid structures have higher resonant frequencies and are prone to damage if located upon stiff soil where H is small, V_s is high, and $V_s/(4H)$ is close to the structure's resonant frequency. Figure 3a, b illustrates two wooden houses, one crushed house on soft alluvium and survival of the other on a firm gravelly fun deposit. On the contrary, Fig. 3c shows a rigid warehouse that was located on a firm gravelly fun and suffered a significant damage during the earthquake.

Seismic Inertial Force

The earthquake resistant design in the modern times was established by the use of seismic inertial force. Theoretically, this force is related with the maximum acceleration (*PGA*) by

$$\text{Seismic inertial force for design} = (\text{Weight of structure}) \times PGA/g \quad (2)$$

where g stands for the gravitational acceleration (=9.8 m/s^2) and *PGA* is affected by the aforementioned amplification of motion. Noteworthy is that this inertial force is considered to be a static force that lasts permanently despite that the real earthquake action occurs for a much shorter time. Hence, Eq. 2 overestimates the real seismic effect. To compensate for this, *PGA* in Eq. 2 has to be made smaller than the real maximum acceleration.

Equation 2 implies that the maximum acceleration or its equivalent inertial force controls the seismic failure of structures. This type of failure is the case of a brittle structure which collapses if the external load exceeds the resistance; see Fig. 4. Because traditional masonry buildings and wooden houses were of brittle nature, seismic design by means of this static seismic design

Geotechnical Earthquake Engineering: Damage Mechanism Observed, Fig. 3 Different damage extents of houses because of different resonant frequency and ground conditions (after the 1995 Kobe earthquake of $M_{jma} = 7.3$). (**a**) Soft house on soft soil. (**b**) Soft house on firm soil. (**c**) Rigid house on firm soil

Ductile and brittle behaviors of structure

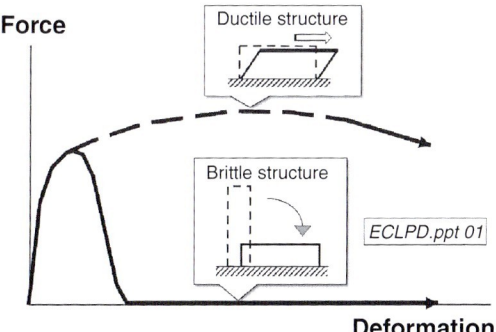

Geotechnical Earthquake Engineering: Damage Mechanism Observed, Fig. 4 Schematic illustration of ductile and brittle behaviors

force achieved significant success in the first half of the twentieth century. Noteworthy is that the second half of the century developed ductile structures (Fig. 4) of which seismic deformation plays major roles in their seismic performance-based design. Moreover, underground lifelines are more affected by ground deformation than the inertial force. Therefore, PGA is not so important today, and the maximum velocity (PGV), response spectrum, duration time of shaking, and the number of loading cycles are often referred to in seismic risk assessment. Note that Mexico City experienced significantly intensified ground motion and elongated duration time of shaking due to its basin structure and soft soil deposits during the 1985 earthquake of Mw = 8.3.

Seismic Soil-Structure Interaction

Structures in contact with soil are always subject to the soil-structure interaction. Mononobe-Okabe theory of seismic earth pressure (Mononobe and Matsuo 1929; Okabe 1924) addresses the interaction of a rigid retaining wall and backfill soil which has yielded during shaking and exerts pressure on a wall. Note that the theory hypothesized that the wall has translated substantially to produce the active earth pressure in the backfill. Thus, the Mononobe-Okabe earth pressure is not the pressure during earthquakes but the active pressure after large deformation. Because of the long history of

Geotechnical Earthquake Engineering: Damage Mechanism Observed, Fig. 5 Failure of retaining wall due to significant seismic earth pressure from backfill soil (near Hanshin Ishiyagawa Station during the 1995 Kobe earthquake; $M_{jma} = 7.3$)

practice, the Mononobe-Okabe theory has successfully reduced the number of failure of retaining walls. An exceptional case is shown in Fig. 5. More recent experiences suggest that attention should be paid to the foundation of a retaining wall. For example, Fig. 6 shows a case in which a concrete wall resting on unstable fill failed and translated substantially.

Pile foundation is subject to soil-structure interaction during earthquakes as well. As illustrated in Fig. 7, piles are fist subjected to the horizontal seismic load that occurs in the superstructure (inertial effect). Second, piles are affected by the differential movement between pile and soil (kinematic pile-soil interaction). In case that soil is very soft, soil translates more than piles and hence piles are loaded in the same direction as the seismic force in the superstructure (Fig. 7a). It seems that such a difficult situation was hardly considered in the past; soil was supposed to be stable and resisted against the pile deformation (Fig. 7b). When the horizontal capacity of pile foundation is not sufficient, one

Geotechnical Earthquake Engineering: Damage Mechanism Observed, Fig. 6 Translation of retaining wall triggered by failure of unstable fill in its foundation (in Takamachi residential area, Nagaoka, during the 2004 Niigata-Chuetsu earthquake of Mw = 6.6)

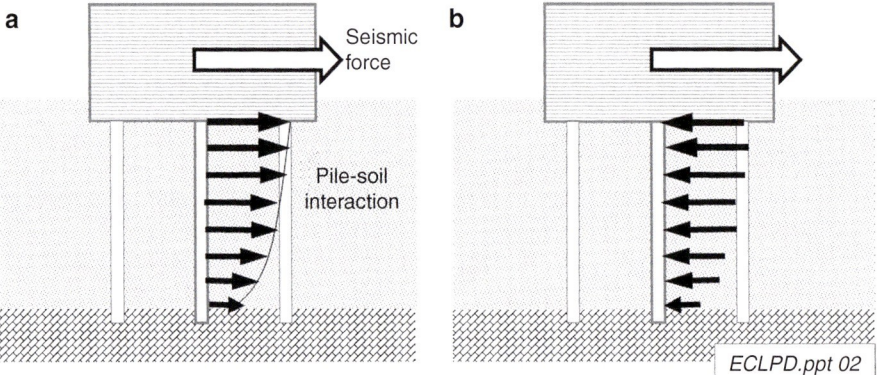

Geotechnical Earthquake Engineering: Damage Mechanism Observed, Fig. 7 Schematic illustration of pile-soil interaction during earthquakes. (a) Soil displacement > pile displacement. (b) Soil displacement < pile displacement

mitigation is to enlarge the size of foundation and increase the number of piles (Fig. 8).

Underground Structure and Earthquake Motion

The deformability of subsoil as illustrated in Fig. 7a is important because this is one of the main reasons why underground structures are destroyed during strong earthquakes. Figure 9 illustrates depression of ground surface that was caused by the collapse of a metro station in Kobe in 1995. This station collapsed because the surrounding backfill soil was not sufficiently firm and exerted load on the station. Central columns in the station were destroyed, the ceiling fell down, and the ground surface caved in. Noteworthy is that soil becomes softer during strong shaking because shear strain in soil during strong shaking is greater and shear modulus of soil decreases as shear strain amplitude increases. Such a situation was not fully considered in the past when subsoil was optimistically assumed to reduce the deformation of underground structures. After this lesson was learned, central columns in many subway stations were reinforced to avoid future problems (Fig. 10). However, the

Geotechnical Earthquake Engineering: Damage Mechanism Observed, Fig. 8 Increasing the number of piles to resist stronger earthquake force (Kobe)

Geotechnical Earthquake Engineering: Damage Mechanism Observed, Fig. 9 Depression at ground surface above the collapsed subway station (Kobe 1995, Photo by Prof. Nozomu Yoshida)

success of reinforcement depends on the intensity of earthquake shaking during future earthquakes.

Similar situation is found in the significant deformation of a tunnel in "soft" rock (Fig. 11). Note that tunnels used to be considered stable during earthquakes because surrounding "hard" rock was expected to reduce the deformation of tunnel during earthquakes.

Geotechnical Earthquake Engineering: Damage Mechanism Observed, Fig. 10 Seismic retrofitting of central columns in Oshiage subway station in Tokyo

Geotechnical Earthquake Engineering: Damage Mechanism Observed, Fig. 11 Significant seismic deformation of tunnel in soft rock (Haguro Tunnel, 2004 Niigata-Chuetsu earthquake of Mw = 6.6) (this tunnel did not have an invert at the bottom). (**a**) Bending and compression in the top of tunnel (*crown*). (**b**) Uplift at the bottom of tunnel

Slope Instability

Gigantic Failure of Slopes

In addition to collapse of houses, slope failure caused by earthquake shaking is another traditional and important type of seismic disaster. If its size is huge, slope failure may kill thousands of people. For example, the failure of the slope of Mt. Huascarán in Peru triggered by the 1970 Ancash earthquake of Mw = 7.9 caused a huge avalanche flow that overtopped a natural protection hill, buried Yungay City (Fig. 12), and instantaneously killed 20,000 residents (Plafker et al. 1971). After this tragedy, Yungay moved to a safer place.

The 1964 Alaska earthquake of Mw = 9.2 caused a submarine landslide in the Port of Valdez. The abrupt and significant change of the submarine topography triggered a huge tsunami that washed the Valdez township and claimed more than 30 lives. Figure 13 shows an aerial view of the site of former Valdez. It seems that the failed submarine slope consisted of fine nonplastic sediments that were transported

Geotechnical Earthquake Engineering: Damage Mechanism Observed, **Fig. 12** Former site of Yungay City (see a hill in the back side and Mt. Huascarán with snow)

Geotechnical Earthquake Engineering: Damage Mechanism Observed, **Fig. 13** Site of former Valdez, Alaska (Photograph taken by Dr. K. Horikoshi in 2000)

from the glacier through the river in the figure into the sea bottom. The fine grain size of the material made the sedimentation rate very low in the seawater, resulting in loose deposit, while the lack of cohesion made the material highly prone to liquefaction. Even worse was that the fine grain size reduced the permeability and hence the dissipation of excess pore water pressure during shaking was very slow. Consequently, this submarine slope easily liquefied and failed during the gigantic earthquake and triggered a tsunami.

Natural Dam

The 1999 Chi-Chi earthquake in Taiwan registered Mw = 7.7. The rock mass in the geologically young Taiwan island was not fully lithified. The complicated tectonic environment around

Geotechnical Earthquake Engineering: Damage Mechanism Observed, Fig. 14 Tsao-Ling slope failure in Taiwan caused by the 1999 Chi-Chi earthquake

the Taiwan island disturbed the rock mass. The high extent of annual rainfall made steep slopes in the mountainous regions. These situations were combined to develop many slope failures during the earthquake. Among them the biggest slope failure occurred in Tsao-Ling (Fig. 14), and 120 million m³ of materials flowed down into the river to form a natural dam (Hung 2000). This slope had failed several times in the past because of heavy rains or earthquakes in 1862, 1941, 1942, 1979, and 1999 (Hung 2000). Noteworthy was that the 1999 event produced a big natural dam, similar to the previous event in 1941 (Kawata 1943), and a reservoir which was later filled with river sediments.

The 2008 Wenchuan earthquake of Mw = 7.9 in the Sichuan Province of China affected mountainous area where body of mountains has been deteriorated by the tectonic action. Slope failures produced natural dams (Fig. 15). The problem of the natural dams is the possibility of breaching and flooding in the downstream area. Therefore, it is important after big earthquakes in mountain regions to find the formation of natural dams and if a dam is subject to overtopping, erosion, and breaching to drain water as soon as possible. Empirically it is known that seepage and piping failure is rather rare in natural dams (Tabata et al. 2002).

The travel distance of a failed landslide mass (runoff distance) is an important issue because the longer runoff results in damage in a bigger area. Hsü (1975) proposed to study the H/L ratio of the soil flow geometry in which H stands for the height of fall of the soil mass, while L the horizontal travel distance, H/L suggesting *tan (friction angle)* where the friction angle is in terms of the total stress. Figure 16 was drawn by collecting data from many literatures. It is important herein, as has been pointed out by many authors, that the greater volume of soil mass is associated with the smaller H/L and, hence, the friction angle. Submarine slope failures are accompanied by even smaller friction angle.

Instability of Slope Surface

It is often the case that a mountain slope is covered by a weathered unstable material. This weak surface material may fall down during strong shaking. Although the size is small, this type of slope failure occurs at many places and often close mountain roads, making emergency activities very difficult.

Figure 17 reveals a typical seismic failure of a mountain slope. While rainfall-induced slope failures often start from the shoulder of a slope, earthquake-induced failures start from the mountaintop. It should be stressed that slopes of

Geotechnical Earthquake Engineering: Damage Mechanism Observed, Fig. 15 Site of natural dam of Qing zhu Jiang River near Qingchuan, Sichuan Province, China

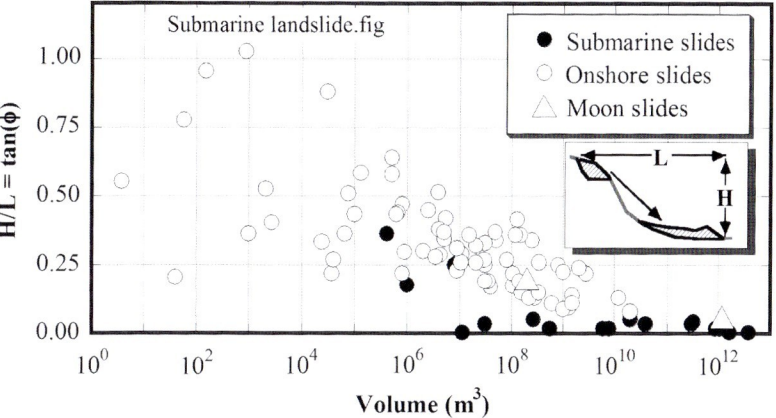

Geotechnical Earthquake Engineering: Damage Mechanism Observed,

Fig. 16 Empirical relationship between height of fall (H) and horizontal travel distance (L) of landslide mass (Reproduced from p. 309 of Towhata (2008))

volcanoes are more vulnerable to failure because, first, generally the volcanic slopes are made of ashes and pumices without rigid rock mass and, second, these materials are pervious and include a big amount of water. The volcanic slope in Fig. 18 started its failure from the top.

Another problem of slope instability has emerged in residential land developments in hilly areas where cut and fill is a common way of construction. Being different from public and industrial projects, individuals cannot always afford reliable but expensive safety measures. As a business practice, therefore, residential lands have been constructed in a less costly manner. As a consequence, unfortunately, fill parts of land became unstable during strong earthquakes. See Fig. 19. Because residents do not have engineering knowledge, quality assessment of residential lands is recommended as an urgent necessity.

Long-Term Effects of Earthquake Shaking on Slope Instability

It is noteworthy that strong earthquake shaking in mountainous regions may initiate long-term instability of slopes and debris-flow hazards.

Geotechnical Earthquake Engineering: Damage Mechanism Observed, Fig. 17 Seismic failure in the surface of mountain slope (Sichuan Province in China)

Geotechnical Earthquake Engineering: Damage Mechanism Observed, Fig. 18 Failure of slope of volcanic deposits, Costa Rica, 2008, after the Cinchona earthquake of Mw = 6.1

The slope in Fig. 20 seismically collapsed in 1707 (120 million m³) and, thereinafter, has produced debris flows during many heavy rainfalls. It seems that the strong shaking disturbed the entire mountain body and the rock mass became subject to disintegration more easily after a huge mass of rock fell down and stress was relieved. Similarly, the slope behind the city of Muzaffarabad in northern Pakistan became unstable after the 2005 earthquake of Mw = 7.6; see Fig. 21.

The second type of long-term effect is caused by a huge deposit of debris in mountain valleys. Figure 22 shows a huge mass of debris in a valley.

Geotechnical Earthquake Engineering: Damage Mechanism Observed, Fig. 19 Seismic instability of residential land in Sendai City after the 2011 gigantic earthquake of Mw = 9 in Japan

Geotechnical Earthquake Engineering: Damage Mechanism Observed, Fig. 20 Ohya slide site in Shizuoka, Japan, which became unstable after gigantic collapse caused by the 1707 earthquake of magnitude = 8 or more

This debris deposit was produced by slope failures during the 2008 Wenchuan earthquake, China. Since then, the risk of debris flow and avalanche has increased significantly.

Fault-Induced Problems

Fault movement or fault rupture is an important causative mechanism of earthquake shaking and

Geotechnical Earthquake Engineering: Damage Mechanism Observed, Fig. 21 Seismically disturbed slope behind Muzaffarabad, Pakistan

Geotechnical Earthquake Engineering: Damage Mechanism Observed, Fig. 22 Huge debris deposit in a valley as a consequence of slope failures during an earthquake in Sichuan Province of China

is classified into normal, reverse, and strike-slip types. Those types of fault depend on the local or regional tectonic stress conditions. Traditionally fault has been a target of scientific study and has not been investigated for engineering purposes. Recently, however, it attracts more engineering concern because of the increasing demands for safety.

Damage Examples

Figure 23 illustrates the destroyed shape of a concrete gravity dam in Taiwan. The energy and power of fault action overwhelmed the resistance of a massive concrete structure. As such, fault is generally perceived as a formidable natural disaster. In contrast to this idea, Fig. 24 shows the situation around the fault displacement in Taiwan as well. Noteworthy is that the damage was limited within the area of ground distortion, and the houses out of this distortion were not affected. This implies that the essence of fault-induced damage lies in the substantial distortion of ground, while the intensity of shaking is not so significant as may be imagined. Further, Fig. 25

Geotechnical Earthquake Engineering: Damage Mechanism Observed, **Fig. 23** Shikang Dam in Taiwan after the 1999 Chi-Chi earthquake

indicates the ground distortion caused by an underlying strike-slip fault. The indicated distortion in the surface sandy soil was a consequence and was not the fault action. Therefore, the power of the ground distortion was not significant, and the massive concrete foundation of the house was able to resist the stress exerted by the surface soft soil.

Mitigation of Fault-Induced Damage

One of the measures to avoid the effects of fault displacement is a land-use control by which construction on a known fault line is prohibited. Thus, the area in Fig. 26 remains vacant as a park. The range of control is typically within 15 m from the fault line, as practiced in California and New Zealand. Lifelines, however, cannot sometimes avoid a known fault and may be damaged (Fig. 27). It should be recalled that the Trans-Alaska Pipeline successfully survived the effects of strike-slip action of a fault that it crossed during the 2002 Denali earthquake of Mw = 7.9. This success was brought about by the extremely flexible structure that absorbed the fault-induced ground distortion (USGS 2003). Thus, there are two kinds of mitigation, rigid

Geotechnical Earthquake Engineering: Damage Mechanism Observed, Fig. 24 Limited damage in a fault area (1999 Chi-Chi earthquake)

Geotechnical Earthquake Engineering: Damage Mechanism Observed, Fig. 25 Survival of rigid house resting on strike-slip type of ground distortion (1995 Kobe earthquake)

Geotechnical Earthquake Engineering: Damage Mechanism Observed, Fig. 26 Prohibition of land use along fault (Yokosuka, Japan)

foundation in Fig. 5 and this soft structure. The shield tunnel for the subway of Osaka City is made of more ductile iron segments at the crossing of the Uemachi fault (Azetori et al. 2006). The Shin-Kobe Station is situated on a normal fault, and the station building is divided into three parts without structural connection among them so that a possible fault movement may not destroy the building. The gravity-type Clyde Dam in New Zealand is divided into two parts across the fault line, and a concrete key block is installed between them in order to prevent possible leakage of water after fault rupture (Hatton et al. 1991).

Coseismic Subsidence and Uplift

Because earthquakes are induced by rupture and distortion in the earth crust, vertical displacement often remains and affects the society. Figure 28 is the result of the subsidence of the earth crust (coseismic subsidence) after the 2011 gigantic earthquake in Japan.

The lowered ground level leads to difficulty in water drainage and protection from high sea waves. The lowered ground level may or may not be recovered within years. On the contrary, the 1804 Kisagata earthquake of the magnitude being around 7.0 caused uplift and drainage of water from a formerly famous beautiful lagoon. Today, former small islands remain as small mounds in the rice field (Fig. 29).

Liquefaction of Sandy Ground

Soil liquefaction is a phenomenon in which cyclic shear deformation of soil causes high pore water pressure and reduces dramatically the shear rigidity of soil. Liquefied subsoil deforms profoundly and the function of the affected structure is lost. Liquefaction is likely to occur in loose, young, cohesionless, and water-saturated soil that is subjected to strong earthquake shaking. The high pore water pressure induces water flow toward the ground surface and this water flow transports soils. As a consequence, sandy deposits of crater shape remain (Fig. 30).

Geotechnical Earthquake Engineering: Damage Mechanism Observed, Fig. 27 Fault-induced buckling of embedded pipeline (1990 Manjil earthquake of Mw = 7.4, Iran)

Geotechnical Earthquake Engineering: Damage Mechanism Observed, Fig. 28 Coseismic subsidence in Mangoku-Ura, Japan

Geotechnical Earthquake Engineering: Damage Mechanism Observed, Fig. 29 Coseismic uplift in Kisagata, Japan (former lagoon is rice field today)

Geotechnical Earthquake Engineering: Damage Mechanism Observed, Fig. 30 Deposit of sand ejecta (Urayasu City on March 13)

Typical Liquefaction Damage

The 1964 Niigata earthquake of Mw = 7.6 triggered liquefaction-induced damage at many places. Although many liquefaction had occurred in the human history prior to this event, they did not attract engineering concern because they occurred in rice fields, abandoned river channels, etc. Liquefaction became an important problem in the recent time because human settlement started to expand into areas of poor soil conditions.

Liquefaction causes large deformation of ground and structures. Figure 31 shows subsidence and tilting of a building. Note that the eccentricity of the gravity force (shape of the building) induced tilting. Because consolidation settlement is quick and small in sandy ground, no pile foundation was installed under many

Geotechnical Earthquake Engineering: Damage Mechanism Observed, Fig. 31 Subsidence and tilting of building with eccentric center of gravity (Niigata, 1964)

Geotechnical Earthquake Engineering: Damage Mechanism Observed, Fig. 32 Falling of Showa Bridge in Niigata, 1964

buildings, and hence subsoil liquefaction easily induced this damage. Figure 32 depicts the falling down of Showa Bridge. Although the exact cause of this damage is not yet known, liquefaction in the riverbed somehow affected the stability of the bridge foundation. After 1964, much effort was made to understand the mechanism of liquefaction damage and to develop preventive measures as well as assessment of liquefaction risk. By 1980, it had become possible to protect important structures from liquefaction problems by densification, installing drains and grouting.

Earthfill dams with insufficient compaction are subject to liquefaction as exemplified by the collapse of the Lower San Fernando Dam in California in 1971 caused by the San Fernando earthquake of Mw = 6.6 (Seed et al. 1975); see Fig. 33. Moreover, tailings, which are a waste from mining industries, are prone to liquefaction because it is disposed into reservoir water, its

Geotechnical Earthquake Engineering: Damage Mechanism Observed, Fig. 33 Liquefaction-induced slope failure of Lower San Fernando Dam, California, USA, in 1971

Geotechnical Earthquake Engineering: Damage Mechanism Observed, Fig. 34 Liquefaction of mine tailings in Mochikoshi, Japan, caused by the 1978 Izu-Ohshima-Kinkai earthquake of $M_{JMA} = 7.0$

grain size is small, being of silty size, and it deposits softly in water to form loose water-saturated cohesionless subsoil. Figure 34 shows an example of liquefaction in a tailings dam.

Although end-bearing pile is an effective protection measure of buildings from subsurface liquefaction, it is not perfect. Because water and sand are ejected after liquefaction (Fig. 30), the volume of subsoil contracts, leading to consolidation settlement in the ground around a pile-supported building (Fig. 35). Accordingly, lifeline connections are damaged.

Horizontal Displacement of Liquefied Subsoil

Those preventive measures that were developed before 1980 were suitable for important structures for which soil improvement was financially feasible. Since the 1980s, liquefaction problem of

Geotechnical Earthquake Engineering: Damage Mechanism Observed, Fig. 35 Elevation difference between pile-supported building and surrounding liquefied ground (Urayasu 2011)

Geotechnical Earthquake Engineering: Damage Mechanism Observed, Fig. 36 Buckling and floating of water pipeline in Dagupan, the Philippines

less expensive structures has attracted concern. For example, a lifeline network cannot improve soil conditions although its possible liquefaction damage affects the entire community significantly. Accordingly, an idea emerged in which the liquefaction-induced damage or deformation should be reduced to an acceptable extent. This idea has developed to the seismic performance-based design in more recent times.

Figure 36 indicates a water pipeline after extensive liquefaction caused by the 1990 Luzon earthquake of surface-wave magnitude = 7.8. The buckling of this pipeline was caused by compressional deformation of the local subsoil. Whether compression or extension, the horizontal deformation and displacement of liquefied ground exert serious effects on underground structures. Hamada et al. (1986) found a

Geotechnical Earthquake Engineering: Damage Mechanism Observed, Fig. 37 Devastated gravity quay wall in Kobe, 1995

Geotechnical Earthquake Engineering: Damage Mechanism Observed, Fig. 38 Cracks parallel to quay wall in Nishinomiya Harbor after 1995 Kobe earthquake

liquefied gentle slope to have moved downward and destroyed gas pipelines. Kawamura et al. (1985) reported bending failure of a concrete pile foundation in Niigata where subsoil translated toward the river and the associating lateral displacement of a surface building generated significant bending moment in the pile. Furthermore, in 1995, gravity quay walls moved toward the sea in Kobe and other harbors where backfill soils liquefied (Figs. 37 and 38). The lateral expansion of backfill is evidenced by many cracks parallel to the quay wall line in Fig. 38.

Liquefaction Problems in the Twenty-First Century

The gigantic earthquake in 2011 in Japan (Mw = 9.0) and the sequence of seismic events in New Zealand in 2010 and 2011 caused many liquefaction problems. They are characterized by damage in relatively inexpensive structures such as personal houses, lifelines, and river levees. It is the

Geotechnical Earthquake Engineering: Damage Mechanism Observed, Fig. 39 House damage in Christchurch in 2011

case therein that the presently available mitigation technologies have reduced the liquefaction problems of important structures for which financial resource is available for soil improvement. In contrast, individuals cannot afford the cost of perfect soil improvement, which may be too conservative for a small house. Moreover, lifelines and river levees do not have sufficient financial resources either for soil improvement over their long length. Thus, liquefaction problem is still important today.

Figure 39 illustrates a house that tilted because of subsoil liquefaction during the 2011 Christchurch earthquake of $Mw = 6.2$ in New Zealand. Although the damage may not appear serious, minor tilting can cause headache and dizziness to residents. The Christchurch officials declared liquefaction-prone area where all the residents were advised to move to safer places. Figure 40 shows liquefaction-induced tilting and subsidence of a low building resting on a shallow foundation without pile. Figure 41 demonstrates liquefaction in a recent artificial land in Itako, Japan, in 2011. Being different from Christchurch, residents in Japan wish to restore their houses in order to continue to live, while improving the subsoil condition. Apparently, soil improvement under an existing structure is expensive and time-consuming.

River levee is prone to liquefaction because it is often situated upon water-saturated loose sand. Formerly it was supposed that earthquake resistant design is not always necessary in levees and that seismic damage of levees should be restored within a short time prior to the next flooding. Probabilistically earthquake and flooding are unlikely to occur at the same time. Importantly, levees allow a certain extent of seismic deformation and do not require a strict control of safety. Such an idea started to change in the 1990s. Figure 42 illustrates the damaged shape of Yodo River levee in Osaka City after the 1995 Kobe earthquake. See sand ejecta on the ground surface as an evidence of liquefaction. Because this site is close to the sea where high tide occurs twice a day, a significant subsidence of this levee could have led to overtopping of water and "flooding" in the back area that was lower than the average sea level. Since then, the importance of seismic design of river levees has been recognized under certain circumstances.

During the gigantic earthquake in 2011 in Japan, liquefaction damage occurred in many river levees that were located on clayey subsoil

Geotechnical Earthquake Engineering: Damage Mechanism Observed, Fig. 40 Liquefaction in building foundation in Christchurch

Geotechnical Earthquake Engineering: Damage Mechanism Observed, Fig. 41 Liquefaction in residential area of Itako City

which was unlikely to liquefy (Sasaki et al. 2012). It was interpreted that the sandy body of the levee caused consolidation settlement in the clayey subsoil, got saturated with water, and developed liquefaction (Fig. 43). Figure 44 illustrates an example of this type of liquefaction in which the original ground surface has no liquefaction in contrast to the significant distortion of the levee. From the engineering viewpoint, this kind of liquefaction revealed such problems as identification of liquefaction-prone sites, improvement of vulnerable soils, and assessment of seismic performance (assessment of seismically induced deformation), all at reasonable costs.

Buried lifelines are subject to liquefaction problems as well. In particular, liquefaction in

**Geotechnical Earthquake Engineering: Damage Mechanism Observed,
Fig. 42** Liquefaction effects on Yodo river levee at Torishima of Osaka, 1995

**Geotechnical Earthquake Engineering: Damage Mechanism Observed,
Fig. 43** Schematic illustration of mechanism of liquefaction in the body of river levee

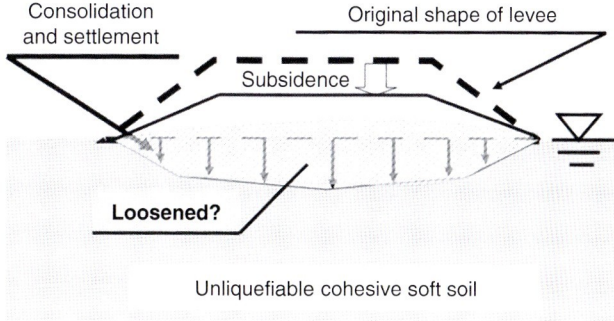

Geotechnical Earthquake Engineering: Damage Mechanism Observed, Fig. 44 Example of liquefaction damage in a river levee resting on clayey subsoil (Naruse River, Japan, after the 2011 earthquake)

Geotechnical Earthquake Engineering: Damage Mechanism Observed, Fig. 45 Floating of water pipe caused by liquefaction (Itako, Japan, 2011; by Prof. J. Koseki)

Geotechnical Earthquake Engineering: Damage Mechanism Observed, Fig. 46 Two-meter floating of manhole caused by liquefaction (Urayasu City, 2012)

the backfill soil frequently occurs because compaction of the backfill soil around a pipe in a small trench is not easy. Figures 45 and 46 show floating of a pipe and a manhole. Liquefaction problem of sewerage pipes is more important than that of water supply pipes because the former is embedded at deeper elevation below the groundwater table than the latter and also because

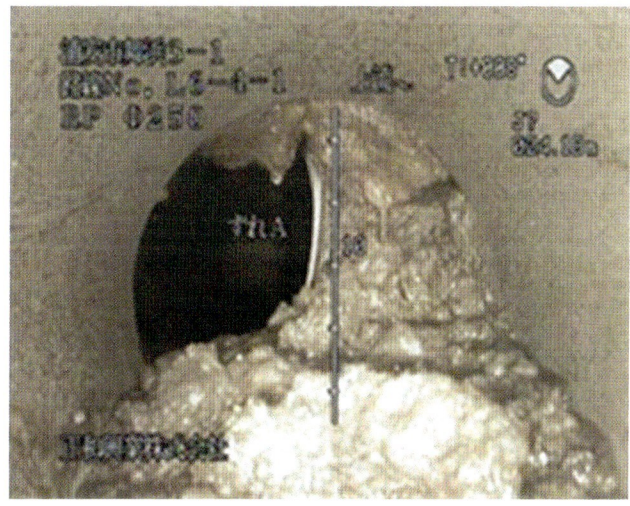

Geotechnical Earthquake Engineering: Damage Mechanism Observed, Fig. 47 Disconnection at sewerage pipe joint and flow-in of liquefied sand (Urayasu City, 2011)

the sewerage water is not pressurized and flows in accordance with the carefully designed slope of pipes; minor change of pipe slope makes the flow difficult. Figure 47 demonstrates one type of liquefaction damage in which a pipe joint was disconnected by large deformation, and the invasion of liquefied sand clogged the water flow.

Summary

This section reviewed the history of earthquake-induced damage in ground and geotechnical structures, focusing on local soil conditions, intensity of ground shaking, seismic earth pressure, underground structures, slope failures, fault effects, tectonic subsidence, and liquefaction. It is therein seen that the types of damage have been changing with the change of our culture and lifestyles. In particular, liquefaction and lifeline problems were not important 50 years ago and fault problems are now seen as concerns. Another example is the ductile behavior of structures for which the conventional seismic inertial force for design is not good enough. It is therefore important for engineers to pay attention to the changing situation and, when a big earthquake occurs, conduct damage reconnaissance survey in order to detect new types of damage and develop new mitigation measures. Recent earthquakes revealed that earthquake engineering has not yet been helpful for such relatively inexpensive structures as lifelines, river levees, and residential houses. Because people are demanding a higher level of seismic safety, engineers have to develop methodologies by which their lives and properties are protected from earthquake effects at affordable costs.

Cross-References

▶ Analysis and Design Issues of Geotechnical Systems: Rigid Walls
▶ Liquefaction: Performance of Building Foundation Systems
▶ Seismic Actions Due to Near-Fault Ground Motion

References

Azetori R, Nagataki M, Izumiotani T, Kitada N (2006) Protection of railway shield tunnel from fault displacement. Proc 61st Annu Conv JSCE 3:187–188 (in Japanese)

Hamada M, Yasuda S, Isoyama R, Emoto K (1986) Generation of permanent ground displacements induced by soil liquefaction. Proc JSCE 376(III-6): 211–220

Hatton JW, Foster PF, Thomson R (1991) The influence of foundation conditions on the design of Clyde Dam. Transactions of 17th ICOLD, Question 66, pp 157–178

Hsü KJ (1975) Catastrophic debris streams (Sturzstroms) generated by rockfalls. Geol Soc Am Bull 86:129–140

Hung J-J (2000) Chi-Chi earthquake induced landslides in Taiwan. Earthq Eng Eng Seismol 2(2):25–33

Kawamura S, Nishizawa T, Wada K (1985) Seismic damage in piles detected by excavation 20 years after earthquake. Nikkei Architecture, Tokyo, July 29th Issue, pp 130–134 (in Japanese)

Kawata S (1943) Study of new lake created by the earthquake in 1941 in Taiwan. Bull Earthq Res Inst Univ Tokyo 21:317–325 (in Japanese)

Mononobe N, Matsuo H (1929) On the determination of earth pressure during earthquakes. In: Proceedings world engineering conference, Tokyo, vol 9. pp 177–185

Murai M (1994) Edo castle and life of shogun family. Chuko Shinsho 45:5 (in Japanese)

Okabe S (1924) General theory on earth pressure and seismic stability of retaining wall and dam. Proc JSCE 10(6):1277–1330 (in Japanese)

Okamoto S (1973) Introduction of earthquake engineering. University of Tokyo Press, Tokyo, p 62

Plafker G, Ericksen GE, Fernández Concha J (1971) Geological aspects of the May 31, 1970, Peru earthquake. Bull Seism Soc Am 61(3):543–578

Sasaki Y, Towhata I, Miyamoto K, Shirato M, Narita A, Sasaki T, Sako S (2012) Reconnaissance report on damage in and around river levees caused by the 2011 off the Pacific coast of Tohoku Earthquake. Soils Found 52(5):1016–1032

Seed HB, Lee KL, Idriss IM, Makdisi FI (1975) The slides in the San Fernando Dams during the Earthquake of February 9, 1971. J Geotech Eng Div ASCE 101(GT7):51–688

Tabata S, Mizuyama T, Inoue K (2002) Natural dam and disasters. Kokin Shoin, Tokyo, ISBN4-7722-5065-4 C3051, pp 50–53 (in Japanese)

Towhata I (2008) Geotechnical earthquake engineering. Springer-Verlag Berlin Heidelberg

USGS (2003) Rupture in South-Central Alaska – the Denali Fault Earthquake of 2002, USGS Fact Sheet, 014-03

Global Navigation Satellite System (GNSS) in Earthquake Engineering, Usage of

Masayuki Saeki
Department of Civil Engineering, Tokyo University of Science, Shinjuku, Japan

Synonyms

Displacement monitoring; GPS (Global Positioning System)

Overview

GNSS (Global Navigation Satellite System) is a satellite based navigation system. Since it provides accurate position and precise time, the GNSS receiver is employed as a time synchronization device and/or a displacement monitoring sensor in earthquake engineering.

The United States' GPS (Global Positioning System) is one of the operational GNSSs. The constellation of GPS satellites consists of $24 + \alpha$ operational satellites which are deployed in six orbital planes. At least four satellites are arranged in a plane, and several spare satellites are also operated for malfunction. With this constellation, four to eight satellites will be always visible with the elevation mask of 15°. The positioning accuracy reaches to a few meters with point positioning, a few centimeters with kinematic positioning, and a few millimeters with static positioning. Since GPS was at first organized for military purpose, GPS was not perfectly accessible to civil and commercial users. However, SA (Selective Availability) was discontinued in May 2000 to promote its civil use.

In addition to GPS, other GNSS systems are in use or under development. The Russian GLONASS (Global Navigation Satellite System) is also operational GNSS. European Union Galileo positioning system, Chinese Compass navigation system, Indian Regional Navigation Satellite System, and Japanese Quasi-Zenith Satellites are under development as of April 2013.

In the following sections, at first, the concept of GNSS positioning analysis is mentioned. And after that, examples of usage of GNSSs in earthquake engineering are described, such as time synchronization, observation of quasi-static displacements of the Earth's surface, and measurement of the dynamic displacement.

Concept of GNSS Positioning Analysis

Point Positioning and Time Synchronization

A GNSS receiver collects the signals broadcast from the GNSS satellites, which include satellite navigation information and the precise time when the GNSS satellite emits the signals. As receiving the signals, the receiver estimates the travel time by subtracting the broadcast time from the received time and calculates the range between the receiver and satellite by multiplying the travel time by the propagation velocity. Since an inexpensive crystal clock is employed in the receiver for decreasing its cost, the travel time is contaminated with receiver's clock error. Therefore, the range is called "pseudorange" instead of true range. So, in the point positioning, the receiver's position (x, y, z) and the receiver's clock error dt are treated as unknown parameters.

The receiver calculates the satellite position from the satellite navigation information. Thus, the receiver has the information about the satellites' positions and the ranges from the satellites. So the receiver is able to estimate its position and clock error with the least squares method if more than four GPS satellites are available. It is said that the accuracy of estimated position is about a few to several meters.

Since the velocity of light is about 3.0×10^{-8} m/s, for instance, 3 m error corresponds to 10 ns. So, in this case, the receiver is able to synchronize to the GPS system time with an accuracy of about 10 ns. The GPS system time is derived from the atomic clocks which are implemented in the GPS ground control stations and the GPS satellites. The GPS Navigation Message included in the GPS signals contains parameters that allow the receiver to calculate the correction time from the GPS system time to the UTC (Universal Time Coordinated) with the accuracy of sub-microsecond order (Dana 1997). Therefore, the GPS receiver for time synchronization provides the precise UTC time.

Interferometry Positioning (Static and Kinematic Positioning)

In order to achieve an accuracy of a few or sub-centimeters, interferometry positioning technique is needed. In the interferometry positioning, phase of the carrier wave is analyzed instead of the code pseudorange. The carrier phase (expressed in cycles) multiplied by the wavelength gives the information associated with the distance between the receiver and satellite. Since the wavelength is about 19.0 cm in case of L1 carrier wave and the phase can be determined with an accuracy of 1–0.1 %, the range can be estimated less than 2 mm accuracy. This phase pseudorange enables us to estimate the receiver position with an accuracy of a few millimeters. However, how many cycles exist in the range is unknown. So there is an ambiguity in phase pseudorange. This ambiguity is called "phase ambiguity" or "integer ambiguity" and treated as unknown parameters should be estimated as an integer value.

The phase pseudorange between receiver i and satellite k is generally described as the following equation:

$$\phi_i^k = \rho_i^k + \lambda N_i^k + c\Delta t_i + c\Delta t^k + \Delta_{trop,i}^k - \Delta_{ion,i}^k + \Delta_{ant,i}^k + \varepsilon_i^k$$

where ϕ_i^k is the phase in meters, ρ_i^k is the true range, λ is wavelength, N_i^k is the phase ambiguity (or called integer ambiguity), Δt_i is clock error of receiver, Δt^k is clock error of satellite, $\Delta_{trop,i}^k$ is tropospheric delay, $\Delta_{ion,i}^k$ is ionospheric delay, and $\Delta_{ant,i}^k$ is error related to the antenna including antenna phase center variation and multipath noise. The true range ρ_i^k is expressed as

$$\rho_i^k = \|\mathbf{x}_i - \mathbf{X}^k\| = \sqrt{(x_i - X^k)^2 + (y_i - Y^k)^2 + (z_i - Z^k)^2}$$

where $\mathbf{x}_i = (x_i, y_i, z_i)$ is the GPS antenna position and $\mathbf{X}^k = (X^k, Y^k, Z^k)$ is the satellite position.

To eliminate the clock errors related to the receiver and satellite, the double-difference operation is used in the relative positioning. In this method, the relative position from a base station is accurately estimated. The DD (double-differenced) phase pseudorange is expressed as the following equation:

$$\phi_{ij}^{kl} = \rho_{ij}^{kl} + \lambda N_{ij}^{kl} + \Delta_{trop,ij}^{kl} - \Delta_{ion,ij}^{kl} + \Delta_{ant,ij}^{kl} + \varepsilon_{ij}^{kl}$$

where

$$*_{ij}^{kl} = *_i^k - *_j^k - \left(*_i^l - *_j^l\right)$$

If the baseline is short enough to assume that the measured phases are equally affected by the troposphere and ionosphere at both receivers, $\Delta_{trop,\ ij}^{kl}$ and $\Delta_{ion,\ ij}^{kl}$ are also canceled through the double-difference operation. However, such assumption is generally incorrect in case of determining the crustal deformation, because the baseline is generally over a few to tens of kilometers.

Δ_{trop} is estimated with a mathematical model such as Hopfield model, modified Hopfield models, and Saastamoinen model. In these models, the tropospheric delay in zenith direction is estimated first and is multiplied by the mapping function which represents the effect of elevation angle. For more precise application, Δ_{trop} is treated as an unknown parameter. The ionospheric delay Δ_{ion} also can be calculated with a mathematical model, but it is efficiently eliminated using the combination of L1 and L2 frequency measurements (Hofmann-Wellenhof et al. 1994). Therefore, L1/L2 GPS receiver is commonly used for determining the crustal deformation. L1 GPS receiver is used only in the case of short baseline.

In the DD phase pseudorange equation, the receiver position $\mathbf{x}_i$ and DD integer ambiguities N_{ij}^{kl} are unknown. If n satellites are available, $3+(n-1) = n+2$ unknowns should be solved although only $n-1$ equations are there at a single epoch. Therefore, in general, equations obtained at different epochs are solved together with the least squares method or Kalman filter. The solutions are called "float solution," and the integer ambiguity should be resolved as an integer value. LAMBDA method is the most famous method for determining integer ambiguities (Teunissen et al. 1997).

In the relative positioning, relative position from a base station is estimated. Therefore, if the base station and observation point are close to each other, these points are equally displaced due to earthquake and the relative position cannot be observed. So, to measure the displacement due to earthquake, the base station is selected enough far from the observation points. It is said that the accuracy is correlated with the baseline length and amounts to 1–0.1 ppm for baselines up to some 100 km and even better for longer baselines in case of static positioning (Hofmann-Wellenhof et al. 1994).

The static positioning is generally employed when coseismic, postseismic, and interseismic displacement is estimated. In this analysis, the unknown parameters (x, y, z) are treated as constant value and several hours measurements are analyzed together in a single batch. Since the effects of various noises are averaged, the accuracy reaches to a few millimeters. On the other hand, if the seismogram or dynamic motion is needed, kinematic positioning is applied. In this case, the unknown parameters (x, y, z) are treated as time-dependent variables and generally estimated with Kalman filter. The accuracy is said to be a few centimeters or worse. For the rapid use such as earthquake early warning, RTK (Real Time Kinematic) positioning is used. In the kinematic positioning, since the analysis is carried out in the postprocessing manner, relatively high accurate satellites' orbit or additional constraints can be involved in the analysis. However, in the RTK GPS, the kinematic positioning is timely carried out following an earthquake event. As a result, the accuracy is relatively worse than that of the kinematic positioning.

Example of Time Synchronization

As mentioned in the previous section, the GPS receiver for time synchronization is able to output timing precisely synchronized to UTC. This enables us to observe seismogram with synchronized clocks at many stations extremely far from each other.

The National Research Institute for Earth Science and Disaster Prevention (NIED) has developed and maintained the nation-wide high-sensitivity seismograph network called "Hi-net" (High Sensitivity Seismograph Network in Japan). In this system, the seismic data originally sampled at 1 or 2 kHz is decimated to 100 Hz and created a data packet of 1 s length to which a GPS time stamp is added (Okada et al. 2004).

Examples of Static Displacement Determination with GNSS

Displacements of the Earth's surface are monitored with permanent GNSS networks. The determined displacement vectors are used for studying crustal deformation associated with earthquakes, determining the geometry of fault plane, and estimating strain accumulation.

Permanent GNSS Networks

IGS (the International GNSS Service) operates an international network of over 350 continuously operating dual-frequency GPS stations and collects GPS and GLONASS observation data sets. These data sets are analyzed and combined to form the IGS products. The primary IGS products are the GPS satellites' IGS final orbit and clock corrections. These accuracies are within 5 cm and 0.1 ns, respectively (Dow et al. 2009). Any users are freely accessible to the orbit data on line. Geocentric coordinates of IGS Tracking Stations are weekly updated and also provided (IGS HP 2013).

GEONET (GNSS Earth Observation Network System) is the largest continuous GNSS network that is operated by GSI (the Geospatial Information Authority of Japan). The GSI operates GNSS-based control stations that cover Japanese country with over 1,200 stations at an average interval of about 20 km. The GPS and GLONASS observation data are sampled at 1 s interval. The 1 Hz sampling data sets and the data sets resampled at 30 s interval are both available online for actual survey work in Japan as well as studies of earthquakes and volcanic activities (GSI HP 2013).

Coseismic, Postseismic, and Interseismic Displacement

It is considered that there are spatial variations in the degree of coupling on the fault plane (asperity model). This means that some parts on the fault plane are strongly coupled and the other parts slowly slip or rupture at relatively small earthquakes. As the plates are moved, the crustal deformation occurs and strains are accumulated around the strongly coupled zone. The accumulated strain will be released at a large earthquake. Therefore, strain accumulation monitoring is very important to assess earthquake potential. The strain accumulation and its release are estimated by analyzing the data measured by seismograph, tiltmeter, leveling, and the permanent GPS networks.

Coseismic displacements have been measured for studying geometry of fault surface. The surface displacement of the Earth is associated with the slip on the fault surface through mathematical models. The radiated seismic wave field is dependent on the slip amplitude, rupture velocity, and source time function, whereas the quasi-static displacements depend only on the final slip amplitude. Therefore, the displacements measured with GNSS are complementary constraints on the geometry of fault surface (Segall and Davis 1997).

Hudnut et al. (2002) observe the postseismic displacements associated with the 1999 Hector Mine earthquake using the rapidly deployed continuous GPS network and SCIGN (Southern California Integrated GPS Network) stations and analyze the temporal character and spatial pattern of the postseismic transients. They report that the displacements measured at some sites display statistically significant time variation in their velocity.

Ozawa et al. (2011) investigates the spatial distribution of the coseismic and postseismic slips associated with the 2011 Tohoku earthquake (Mw9.0, March 11) by analyzing the ground displacement detected using the GEONET and reports that the detected coseismic slip area matches the area of the pre-seismic locked zone.

At subduction zones, slow-slip events have been detected by geodetic measurements in interseismic period. Hirose and Obara (2005) report the repeating occurrence of short-term and long-term slow-slip events which are accompanied by deep tremor activity around the Bungo Channel region, southwest Japan. They detect these activities using NIED Hi-net equipped with a tiltmeter and a high-sensitivity seismograph and GPS stations of GEONET. The GPS is only used to detect the long-term slow-slip events because the short-term slow-slip event is too small in magnitude to be detected by GNSS.

Examples of Dynamic Displacement Determination with GNSS

The permanent GNSS network is used to measure dynamic displacements as well as static displacements.

Strong Motion

Strong motion associated with a large earthquake is generally observed with seismograph that is inertial sensor and sensitive to acceleration. Therefore, higher frequency components can be detected accurately with the sensor, but the lower frequency components are not detectable and largely contaminated by random noise. Furthermore, if the seismograph is tilted during a large earthquake, the observed acceleration has offset due to the gravitational force. Therefore, the dynamic displacement estimated with the double integration of acceleration diverges proportional to the square of time. Sensor rotation also distorts the measurements. So it is very difficult to detect long-term displacements with inertial sensor measurements. On the other hand, GNSS sensor directly detects the displacement itself. Besides, some GPS receiver is able to sample the data up to 20–50 Hz. From these reasons, GPS sensor is also used for monitoring dynamic displacements. The accuracy (standard deviation) is said to be a few centimeters in horizontal direction and about two times larger in the vertical direction.

Nikolaidis (2001) et al. report that the dynamic ground displacement caused by the Hector Mine earthquake in southern California (Mw 7.1, October 16, 1999) is determined by analyzing the GPS data collected with SCIGN continuous GPS network. The baseline distances are from 53 to 205 km. In this research, the raw GPS data is analyzed with instantaneous positioning method to obtain displacement time series, while kinematic positioning is generally used. In the instantaneous positioning method, integer ambiguities are resolved independently for each epoch. Hence, the analysis is not affected by the cycle slips of phase measurements.

Dynamic displacements obtained from GPS network are used in the inversion for rupture processes. The dynamic displacements include coseismic displacement as well as lower frequency components. So it is expected to merge the seismic wave inversion and geodetic observation inversion. Yue and Lay (2011) estimate the rupture process of the 2011 Tohoku earthquake (Mw 9.0, March 11) using 1 Hz sampling GPS data measured with GEONET.

Tsunami

Kato et al. (2000) develop a tsunami observation system using RTK GPS to detect a tsunami before it reaches the coast. This system consists of the Support-buoy, the Sensor-buoy, and the base station. The Sensor-buoy is designed not to react to wind wave motion and is equipped with only GPS antenna. The antenna cable is connected to the GPS receiver mounted on the Support-buoy. The Support-buoy is designed to move with wind waves and is equipped with GPS receivers and antenna, a number of sealed lead batteries, a wind generator, solar panels for power supply, and a pair of radio receiver and transmitter devices. Their experimental results show that the developed system is able to track the sea surface quite well.

Examples of Disaster Mitigation with GNSS

Early Warning System

Allen and Ziv (2011) investigate the possibility of application of RTK GPS to earthquake early warning using the data collected in the El Mayor-Cucapah earthquake in northern Baja California (Mw 7.2, 2010). Earthquake early warning is the rapid detection of a large earthquake and prediction of the expected ground shaking within seconds so that a warning can be broadcast to the nation. In general, earthquake early warning system estimates the magnitude of an earthquake with seismic waves observed around the epicenter. The problem of how accurate the magnitude is estimated with only seismic waves for a large earthquake is considered. The unique information provided by the GNSS-based displacement time series is the coseismic displacement that is hardly obtained from acceleration time series. This information is expected to provide effective constraints to the magnitude estimation.

A GPS-based tsunami early warning component, developed by the German Research Centre for Geosciences (GFZ) within the German-Indonesian Tsunami Early Warning System (GITEWS) project, was installed in Indonesia after the disastrous Indian Ocean tsunami (December 26, 2004). This component consists of four subcomponents; the GPS real-time reference stations (GPS RTR) for ground motion detection and reference station for buoy GPS receivers; GPS sensor stations at tide gauges for ground motion detection and tide gauge data correction; GPS sensors on buoys for sea level measurements and direct tsunami detection; and external GPS stations for providing external reference frame. At a large earthquake, GPS data is transmitted from GPS stations in near real time, and displacements are determined. The ground displacements are used to select the most probable tsunami scenario from some thousands of pre-calculated scenarios, based on a matching process. The sea level height is determined by analyzing the GPS data from offshore buoys to detect tsunami directly (Falck et al. 2010).

Postseismic Damage Detection

Application of the GPS wireless sensor network (GWSN) to postseismic building-wise damage detection is studied. The details of GWSN are described in the later section.

The GWSN consists of a central server and many sensor nodes. The sensor node has a wireless communication module, microcontroller, small battery, and inexpensive L1 GPS module connected to a small patch antenna. At a large earthquake, the sensor nodes collect raw GPS data according to the command from a central server and share the collected data between the neighboring sensor nodes. After that, the sensor nodes determine the relative position from the neighboring sensor nodes onboard and transmit the estimated relative positions to the central server through wireless communication network. The central server combines the relative positions to organize the sensor mesh and detects the displacements of each sensor. From the estimated displacements, the useful information about the building collapses and road closures due to the debris is investigated for the early stage of the rescue action after a large earthquake (Oguni et al. 2011).

Summary

GNSS is a very useful system in earthquake engineering because it enables us to determine the displacements due to earthquake at any point wherever the antenna and receiver are properly deployed. The estimated coseismic, postseismic, and interseismic displacements are used for determining the geometry of fault plane and rupture process, estimating the magnitude of earthquake, and estimating the future earthquake potential. This helps our understanding of earthquake. GNSS is also used as a time synchronization system for seismograph network.

GNSS receiver is employed as a displacement sensor in various systems for disaster mitigation. A GPS-based tsunami early warning system is installed in Indonesia. And other tsunami early warning systems are studied. Earthquake early warning system with RTK GPS and postseismic building-wise damage detection system are also studied.

As of April 2013, only the United States' GPS and Russian GLONASS are fully operational GNSSs. In the near future, European Union Galileo positioning system, Chinese Compass navigation system, Indian Regional Navigation Satellite System, and Japanese Quasi-Zenith Satellites will join in the fully operational GNSSs. Therefore, the accuracy and robustness of the system with GNSS are expected to be improved further.

GPS Wireless Sensor Network

System Overview

Wireless sensor network is a system in which many wireless sensor nodes automatically organize and maintain a wireless network and transmit their data or information to each other. The sensor node consists of a microcontroller, physical sensor, low power radio module, and small battery. In the GPS wireless sensor network (GWSN), an inexpensive L1 GPS receiver connected to a small patch antenna is employed as the physical sensor (Saeki et al. 2006). Figure 1 shows a prototype of the sensor node.

This prototype employs a wireless communication module having a 32 bit microcomputer, 128 kbyte RAM, several serial communication ports, as well as an RFIC (Radio Frequency Integrated Circuit) based on IEEE802.15.4.

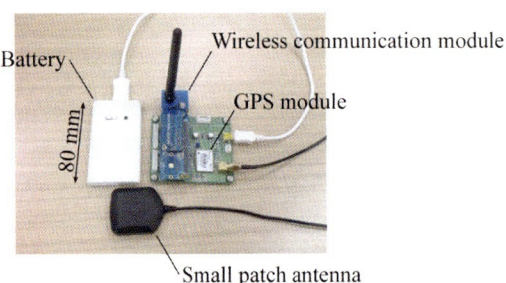

Global Navigation Satellite System (GNSS) in Earthquake Engineering, Usage of, Fig. 1 Wireless sensor node equipped with an inexpensive L1 GPS module connected to a small patch antenna

The electric current is 14.6 mA (3.0 V) at receiving data and 17.4 mA at transmitting data. This kind of wireless communication module has some sleep modes depending on usable functions, and the electric current reaches at a few μA in the deep sleep mode.

An inexpensive L1 GPS module is connected to the wireless sensor module through a serial communication. The electric current is about 50 mA (3 V) which is highest among the electric components of the prototype. In the prototype, a small patch antenna is employed, which is generally never used for the precise displacement determination because such a small patch antenna has weak directivity and then the data is largely affected by the multipath noises. This noise contamination makes it difficult to resolve the integer ambiguities.

A schematic view of GWSN is shown in Fig. 2. This system consists of many sensor nodes and a central server. The sensor nodes are ordinarily in sleep mode for saving their battery energy. They wake up, for example, every 1 min to ask their task to the server. If the task is "sleep," they go into sleep mode again. If the task is "observation," they turn the GPS module on and wait a trigger signal. After receiving the

Global Navigation Satellite System (GNSS) in Earthquake Engineering, Usage of, Fig. 2 A schematic view of the GWSN (GPS wireless sensor network)

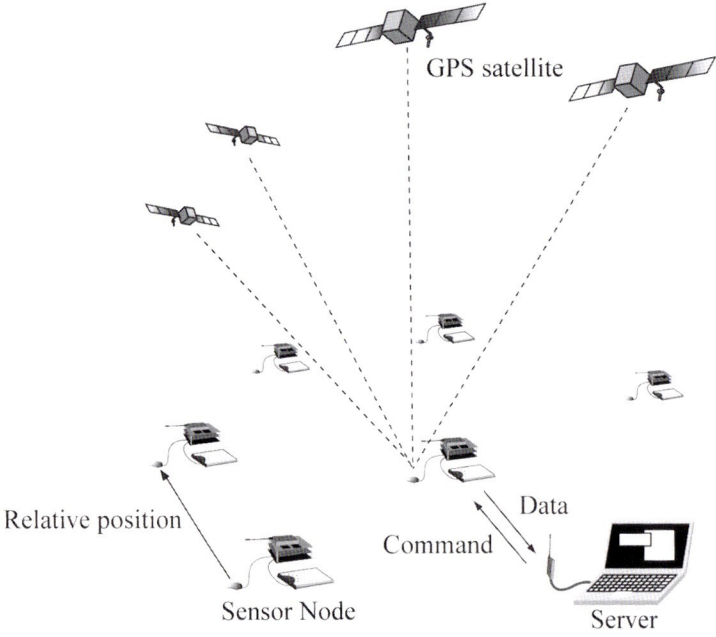

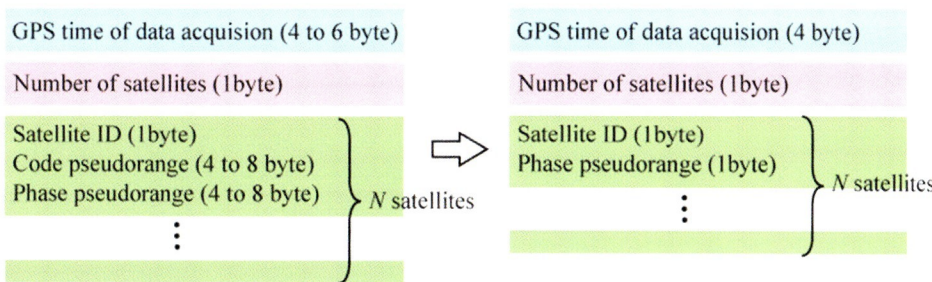

Global Navigation Satellite System (GNSS) in Earthquake Engineering, Usage of, Fig. 3 A format of compressed data of a single epoch in case of static relative positioning with short baseline

trigger signal, the sensor nodes save the required raw GPS data on their RAM during a specified time interval. After finishing data acquisition, they turn the GPS module off and return back to sleep mode. If the received task is "data collection," the sensor node sends the saved GPS data to the server through wireless communication. The sampling interval and data length can be changed according to the command from server.

Data Compression Technique

Data compression is a very important technology for wireless sensor network, because a large amount of data requires long communication time and high energy consumption which results in short battery lifetime. For example, if a GPS module outputs about 300 bytes per epoch (practically the data size of original message depends on the type of GPS module) and a sensor node saves the whole data for 10 min, the amount of data reaches to 180 kbyte. It takes about 18 s for a sensor node to transmit the data to the server with a transmit rate of 100 kbps. If a single server covers 100 sensor nodes, it needs at least 30 min (=1,800 s) to collect data from all sensor nodes. (GWSN for the application of building-wise damage detection is specially customized so that the sensor nodes locally share their data between neighboring sensor nodes, estimate their relative position onboard, and transmit only the estimated results to the server to avoid transmitting large amount of data.) Since the original raw GPS data is large in amount, truly required data should be extracted. Figure 3 shows a format of compressed data packet.

In the GWSN, quasi-static condition is assumed, which means a relative position of sensor nodes is invariant during the short observation period (e.g., 10 min). The quasi-static assumption and dense sensor deployment enable us to omit the data of code pseudorange and integer part of phase pseudorange. As a result, the amount of data can be drastically decreased to 28 byte per epoch (Saeki et al. 2006).

The integer part of phase pseudorange is very important information especially for kinematic positioning. As mentioned above, the phase pseudorange includes phase ambiguity as unknown value. The value of phase ambiguity is constant as long as the GPS module continuously tracks the satellite's carrier waves. If the integer part of phase pseudorange is missed, the phase ambiguity becomes different value and should be determined again. This phenomenon is called "cycle slip" and recognized as an important problem to be solved.

On the other hand, in case of static positioning, the missing integer part can be fixed by shifting the corrected DD phase pseudorange $\hat{\phi}_{ij}^{kl}(t)$ (appears in the following section) by integer value so that the difference between the successive two values is minimized. This is because the corrected DD phase pseudorange can be approximated as a linear function of time with a small slope in case that the relative position of sensor nodes is invariant and the baseline is short enough. Figure 4 shows an example of the corrected DD phase pseudorange. The fluctuation is less than 0.2 cycles after repairing the cycle slips.

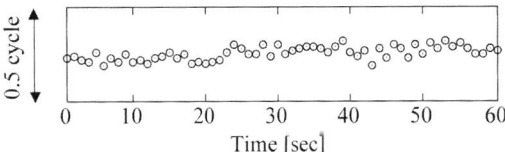

Global Navigation Satellite System (GNSS) in Earthquake Engineering, Usage of, Fig. 4 An example of the corrected DD (double-differenced) phase pseudorange in case of short baseline and static condition

Static Positioning Using Linear Approximation Method

Since energy consumption of GPS module is highest among the electric components of the sensor node, it is effective to shorten the acquisition time of GPS for saving battery energy. However, in general, short data length makes it difficult to successfully determine phase ambiguities. The wrong phase ambiguities largely deteriorate the accuracy of positioning. Therefore, an algorithm improving the success rate of phase ambiguity determination with short data length is required to solve the trade-off relationship between the energy consumption and accuracy.

At first, a conventional approach is described here. The observation equation of DD phase pseudorange can be linearized by substituting $\mathbf{x}_i = \hat{\mathbf{x}}_i + \Delta \mathbf{x}$ into the equation of ϕ_{ij}^{kl}.

$$\phi_{ij}^{kl} = \hat{\rho}_{ij}^{kl} + \nabla \rho_{ij}^{kl} \cdot \Delta \mathbf{x} + \lambda N_{ij}^{kl} + e_{ij}^{kl}$$

where $\hat{\rho}_{ij}^{kl}$ is the calculated DD range between the approximate position of GPS antenna $\hat{\mathbf{x}}_i$ and the GPS satellite position and e_{ij}^{kl} is the noise including residuals of DD ionospheric and tropospheric delay, antenna noise, and random noise. The equations obtained for available satellites provide the following simultaneous equation:

$$\mathbf{U}(t) = A(t)\Delta \mathbf{x} + \lambda \mathbf{N} + \mathbf{e}(t)$$

where $\mathbf{U}(t)$ is the vector of corrected DD phase pseudorange defined as $\hat{\phi}_{ij}^{kl} = \phi_{ij}^{kl} - \hat{\rho}_{ij}^{kl}$, $A(t)$ is the design matrix, $\mathbf{N}$ is the vector of DD phase ambiguities N_{ij}^{kl}, and $\mathbf{e}(t)$ is the vector of noise $e_{ij}^{kl}(t)$.

The above simultaneous equation is generally solved by using Kalman filter. The solution is called "float solution" because the phase ambiguities are determined as float values. Accurate position can be estimated after resolving the phase ambiguities as integer values. The resolved integer values are called "fixed solution," which is searched so that the following objective function J is minimized:

$$J = (\mathbf{N} - \hat{\mathbf{N}})^T Q_{\hat{N}}^{-1} (\mathbf{N} - \hat{\mathbf{N}})$$

where $\hat{\mathbf{N}}$ is the float solution of phase ambiguity vector and $Q_{\hat{N}}$ is the variance-covariance matrix of the float solution. The fixed solution is validated by checking the ratio J_2/J_1 where J_1 and J_2 are the minimum and the second minimum residuals, respectively. It is said that the integer ambiguity may be successfully resolved if the ratio J_2/J_1 is greater than 3–5.

As mentioned in the previous section, the corrected DD phase pseudorange $\hat{\phi}_{ij}^{kl}(t)$ can be modeled as a linear function of time. So, in the positioning analysis of GWSN, a linear function is approximately estimated by applying least squares method to the time series of corrected DD phase pseudorange. And the values corresponding to the first and last epochs are calculated using the estimated linear function. The observation equations only at the first and last epochs are solved together with least squares method to estimate the position and DD phase ambiguities. This method may slightly improve the quality of coefficient matrix of normal equation and suppress the effect of noises on the solution.

The success rate of determining the DD phase ambiguities can be improved by using the linear approximation method, especially in case of short data length. Table 1 shows the comparison between the results of the conventional approach (static positioning) and the described method. The observation data used in the analysis was collected under ideal condition in which no obstacles were over there. In this experiment, GPS antennas for survey work were fixed on tripods, and GPS data was logged over 24 h with a 1 Hz sampling rate. In the analysis, data segment with a specified data length was selected

Global Navigation Satellite System (GNSS) in Earthquake Engineering, Usage of, Table 1 Comparison between success rates of the static positioning and the linear approximation method (Saeki et al. 2006)

Data length (s)	Success rate (%)	
	Static positioning	Linear approximation
30	71.1	87.3
60	89.8	96.0
180	98.7	99.1
300	99.5	99.6
600	100.0	100.0

and analyzed using both the conventional and the linear approximation approaches. This estimation was performed 86,400 times by shifting the selected data by 1 s.

Multi-hop Positioning

The observation equation of DD phase pseudorange is solved with the assumption of white noise. This assumption may be valid for some pairs of sensor nodes existing under similar surrounding condition because such condition gives similar multipath noises and the noises are canceled out in the double-difference operation. However, for the pairs existing under different condition, multipath noises are different and not canceled out in double-difference operation. In this case, the accuracy of float solution becomes worse, and the DD phase ambiguities are not successfully resolved as integer value. Therefore, it is meaningful to select a good pair of sensor nodes whose noises are similar to each other.

Figure 5 shows a simple example of relative position determination. Suppose that three sensor nodes are there and the noises are different from each other but the noise of ID3 has some similarities to those of ID1 and ID2. This situation often happens in actual observations when many sensor nodes are deployed densely. In this case, the relative position from ID1 to ID2 tends to be estimated incorrectly, and the ratio J_2/J_1 becomes small (as described in Fig. 5a) because the noises are not canceled out in the double-difference operation. On the other hand, the vector ID1 → ID3, and also the vector ID3 → ID2, may be estimated better and the ratio becomes larger

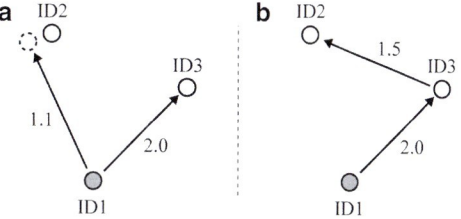

Relative position vectors from a single reference

Estimated position vectors by using multi-hop positioning

Global Navigation Satellite System (GNSS) in Earthquake Engineering, Usage of, Fig. 5 Simple example of (**a**) relative position vectors from a single reference point and (**b**) estimated position vectors by using the multi-hop positioning

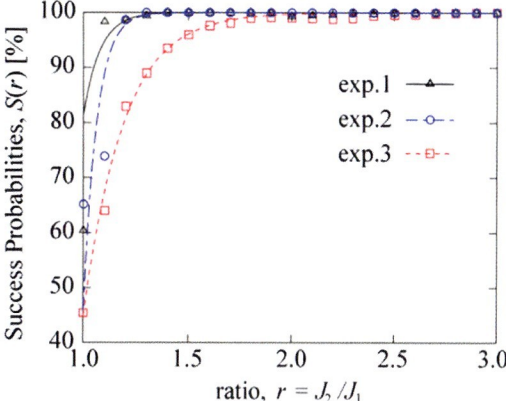

Global Navigation Satellite System (GNSS) in Earthquake Engineering, Usage of, Fig. 6 Function of success probability with respect to the value of ratio test (Saeki 2012)

(Fig. 5b). In this case, the sum of vectors ID1 → ID3 and ID3 → ID2 is more likely to give better solution. This is a basic idea of multi-hop positioning (Saeki and Oguni 2012).

To select an optimum path quantitatively among the candidates (possible sum of vectors), success probability function $P(r)$ is introduced. It is empirically known that higher ratio $J_2/J_1 (= r)$ gives higher success probability. Figure 6 shows the experimental results of the success probability versus the ratio J_2/J_1. The marks represent the results obtained from an experiment, and the lines are the fitted curves using the function $P(r) = 1.0 - be^{-ar}$ where a and b are the fitting parameters. The experiment was conducted

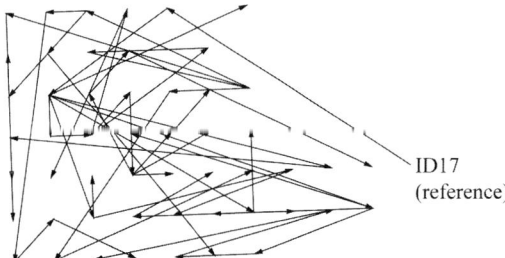

Global Navigation Satellite System (GNSS) in Earthquake Engineering, Usage of, Fig. 7 Relative position vectors determined by Dijkstra's algorithm

under an ideal condition (no obstacles over the site), but the GPS antennas were differently fixed. In the first pair (exp.1) the antennas were both fixed on a flat concrete floor. In the second pair (exp.2) one antenna was fixed on a flat concrete floor, but another one was attached on a concrete block with a height of 15 cm. This difference gave slightly different multipath noises. In the third pair (exp.3) one antenna was fixed on a flat concrete floor, but another one was fixed on a tripod. In this case, the multipath noises were very different to each other.

The success probability of path can be evaluated by multiplying the success probabilities of the corresponding vectors.

$$P(r) = P(r_1)P(r_2)\cdots P(r_n)$$

The optimal path is searched from numerous candidates so that the success probability is maximized. In the multi-hop positioning, Dijkstra's algorithm is used as a search algorithm. This algorithm is widely used in many applications such as network routing protocols or mobile navigation systems to find the shortest (or lowest const) path efficiently.

Figure 7 shows an experimental result of the optimum path estimated using the multi-hop positioning method. In the experiment, 53 sensor nodes were arranged on a rooftop of a building in a grid form at 2 m spaces. The raw GPS data was saved for 4 min with 1 Hz sampling rate and transmitted to the central server through wireless communication. The multi-hop positioning was applied in postprocessing manner to estimate the best relative positions of sensor nodes which maximized the success probabilities.

Cross-References

▶ Early Earthquake Warning (EEW) System: Overview
▶ Earthquake Mechanisms and Tectonics
▶ Remote Sensing in Seismology: An Overview

References

Allen RM, Ziv A (2011) Application of real-time GPS to earthquake early warning. Geophys Res Lett 38, L16310
Dana PH (1997) Global positioning system (GPS) time, dissemination for real-time applications. Real Time Syst 12:9–40
Dow JM, Neilan RE, Rizos C (2009) The international GNSS service in a changing landscape of global navigation satellite systems, J Geod 83:191–198
Falck C, Ramatschi M, Subarya C, Bartsch M, Merx A, Hoeberechts J, Schmidt G (2010) Near real-time GPS applications for tsunami early warning systems. Nat Hazard Earth Syst Sci 10:181–189
GSI HP (2013) http://www.gsi.go.jp/ENGLISH/
Hirose H, Obara K (2005) Repeating short- and long-term slow slip events with deep tremor activity around the Bungo channel region, southwest Japan, Earth Planets Space 57:961–972
Hofmann-Wellenhof B, Lichtengger H, Collins J (1994) GPS theory and practice, 3rd edn. Springer, New York
Hudnut KW, King NE, Galetzka JE, Stark KF, Behr JA, Aspitoes A, van Wyk S, Moffitt R, Dockter S, Wyatt F (2002) Continuous GPS observations of post-seismic deformation following the 16 October 1999 Hector Mine, California, Earthquake (Mw 7.1). Bull Seismol Soc Am 92(4):1403–1422
IGS HP (2013) http://igscb.jpl.nasa.gov/
Kato T, Terada Y, Kinoshita M, Kakimoto H, Isshiki H, Matsuishi M, Yokoyama A, Tanno T (2000) Real-time observation of tsunami by RTK-GPS, Earth Planets Space 52:841–845
Nikokadis RM, Bock Y, de Jonge PJ, Shearer P, Agnew DC, Domselaar MV (2001) Seismic wave observations with the Global Positioning System. J Geophys Res 106(B10):21897–21916
Oguni K, Miyazaki T, Saeki M, Yurimoto N (2011) Wireless sensor network for post-seismic building-wise damage detection. In: Proceedings of the 20th IEEE international workshops on enabling technologies: infrastructure for collaborative enterprises. Paris, pp 238–243

Okada Y, Kasahara K, Hori S, Obara K, Sekiguchi S, Fujiwara H, Yamamoto A (2004) Recent progress of seismic observation networks in Japan - Hi-net, F-net, K-NET and kiK-net -, Earth Planets Space 56:15–28

Ozawa S, Nishimura T, Suito H, Kobayashi T, Tobita M, Imakiire T (2011) Coseismic and postseismic slip of the 2011 magnitude-9 Tohoku-Oki earthquake. Nature 475:373–376. doi:10.1038/nature10227

Saeki M, Oguni K (2012) Multi-hop positioning, relative positioning method for GPS wireless sensor network. In: PECCS2012 – international conference on pervasive and embedded computing and communication systems, Rome, Italy 24–26

Saeki M, Kosaka T, Kaneko S (2006) Phase ambiguity resolution based on linear modeling of DD carrier phase. In: ION GNSS 19th international technical meeting of the satellite division, Fort Worth, 26–29 Sept 2006

Segall P, Davis JL (1997) GPS applications for geodynamics and earthquake studies. Annu Rev Earth Planet Sci 25:301–336

Teunissen PJG, de Jonge PJ, Tiberius CCJM (1997) The least-squares ambiguity decorrelation adjustment: its performance on short GPS baselines and short observation spans. J Geodesy 71:589–602

Yue H, Lay T (2011) Inversion of high-rate (1 sps) GPS data for rupture process of the 11 March 2011 Tohoku earthquake (Mw 9.1). Geophys Res Lett 38:L00G09. doi:10.1029/2011GL048700

Global View of Seismic Code and Building Practice Factors

James Edward Daniell
Geophysical Institute, Karlsruhe Institute of Technology, Karlsruhe, Germany

Synonyms

Building design codes; Construction practice; Earthquake construction; Seismic-resistant codes; Worldwide earthquake standards comparison

Introduction

This entry looks at a review of seismic-resistant codes for each country in the world by examining

---

The online version includes code changes and code types through time.

the historical changes of codes from 1900 to 2013, the years in which updates were made, the number of buildings these codes influence, and a review of the code quality itself with respect to the hazard of that country. Over 160 countries and nations have some form of seismic code. However, the quality, extent of application, and methodologies between seismic-resistant codes differ around the world. An exploration of the location of each of these code changes, in terms of when they have been implemented, to what extent they have been implemented, and which percentage of buildings they encompass, has been undertaken.

The History of Seismic-Resistant Codes Worldwide and Their Components

Seismic-resistant building codes traditionally started being implemented during reconstruction following major earthquakes of the past, including 1755 Lisbon, 1880 Luzon, and 1908 Messina. Although no country had a formal seismic-resistant code in 1900, some countries such as Italy, Portugal, the USA, and other locations had influences from historical earthquakes, which aided the building styles of the time being more seismic resistant. This was also the case with colonial building styles in Africa. Using the components of lateral earthquake forces, ductility and drift and building height, these were slowly implemented into design, as explained in the previous chapter of Gülkan and Reitherman.

The first seismic-resistant codes simply applied a seismic coefficient and building height limits, such as those implemented within Italy (1910, 1917), Costa Rica (1914), Turkey and Japan (1923), the USA (1906, 1927), and other locations pre-1930s. It should be noted that, pre-1930, some municipalities and cities applied different seismic-resistant building codes or practices, such as building from wood after the 1848 Wellington earthquake (Beattie and Thurston 2006) or after the 1770 Haiti earthquake.

Lateral loads were often included with the wind, snow, and gravity loads in analysis.

Post-1930, the difference of the response spectrum to various structural periods was looked at, with the influence being that the lateral forces would be changed depending on the fundamental period of the building. Elastic analysis was, however, still undertaken, as opposed to elastoplastic analysis (ductile analysis).

Only in the 1960s were ductility considerations taken into account in the form of force-reduction factors, given that ductile structures were able to survive ground shaking and thus inertia forces many times greater than elastic structures. These force-reduction factors were placed slowly into the seismic-resistant codes in different countries from the 1970s to 1990s; however, greater research in the field showed the need for analysis into the effects of damage potential as a result of displacements rather than strength alone.

There were also two types of building analysis undertaken: force-based design and displacement-based design. Force-based design uses a seismic coefficient (C) and a seismic weight (W) calculated to give the design seismic forces (V):

$$V = Cs * W, \quad \text{where}$$
$$Cs = Sa/(\text{Reduction Factor/Importance Factor})$$

The seismic coefficient is made up of the design spectra of the earthquake horizontal and vertical components calculated from various parameters detailed in Table 1 below, in combination with a reduction factor corresponding to the global ductility capacity and the inherent overstrength in the lateral force-resisting system. Using this system generally works; however, it relies on the fact that strength is being used to control damage and that strength and stiffness are independent. In reality, these two are interlinked. Force-reduction factors assume that the ductility demand is the same for each type of structure unless changed; however, each building generally acts differently with respect to ductility. The distribution of strength in a building, including the amount of displacement at different limit states, has been determined to be increasingly important, despite both methods looking at the location of plastic hinges (shear strength of members being greater than that of the shear due to flexural strength) and hence looking at the displacement ductility of the components of the building or structure. The previous chapter of Gülkan and Reitherman provides a useful background to the implementation of such parameters.

These difficulties resulted in displacement-based design, rather than force-based design, being undertaken in order to construct buildings safely (Priestley et al. 2007). Displacement-based design is still generally not used in seismic codes but can be applied as a valid alternative where wanted. A comment to this effect has been incorporated in the NZS code 1170 Part 5 (Standards New Zealand 2004).

Thus, there are now two main types of seismic codes used worldwide. The first, strength-based design, relies on a design base shear with force-based design generally based on historical earthquake hazard maps, without any view as to the performance of the structure in different events. The other, performance-based design, still uses these uniform spectra metrics from seismic zoning and traditional force-based design but adds in performance criteria that need to be satisfied at more than one earthquake level. In this way, performance levels are also defined differently for schools versus normal buildings, as a different return period can be defined for the performance, thus designating different displacements or interstory drifts required. Better levels of performance-based design are creeping slowly into revisions of seismic-resistant codes worldwide, with life safety and damage limitation of buildings at various corresponding levels of earthquake motions being implemented and even displacement-based design (Priestley 1993) being included.

Table 1 represents the key elements of a modern seismic-resistant code, which then will be summarized for the purposes of the seismic code index developed.

Global View of Seismic Code and Building Practice Factors, Table 1 The criteria examined for each of the seismic-resistant codes within the worldwide seismic code index (1900–2012) discussed in this study

Element	Sub-element	Key components	Value given out of 104
Structural design method	Material loads/ strengths, dead and live load	Basic principles of design as denoted by the general building code in the country	30
Seismic actions	Horizontal components	Design response spectra (Sa) generally determined by the ground seismic acceleration, importance factor, at characteristic design periods, damping modification factor. It has different shapes for design periods, depending on these cutoffs	5
Seismic actions	Vertical components	Generally taken as a percentage of the horizontal factor to create Av	3
Seismic actions	Ground seismic acceleration	Acceleration at which the seismic zone map coefficient will occur for a return period (475 years generally) Can be intensity or PGA based	Included as a percentage of actuality over 30
Seismic actions	Near fault	Looks at whether near-fault effects are taken into account	2
Seismic actions	Soil classification	Depending on soil type, extra factors applied to account for the increased shaking	5
Seismic actions	Importance factor	Structures generally split into various factors in order to define different levels of design in terms of life safety, e.g., hospitals given a higher level of safety	5
Seismic actions	Behavior factor (ductility factor)	Reduction of seismic loads based on post-elastic behavior – depending on the material of the structural system	5
Seismic actions	Foundation factor	Interaction between the foundation and the soil is included in some cases	5
	Existing buildings		% of buildings
	1–2-story building factor		% of buildings
Design method	Simplified spectrum method	Uses the fundamental mode of oscillation and looks at equivalent seismic forces. Simplified and does not take into account so many complexities, just using base shear multiplied by the design spectra value and distributing the lateral forces over the height of the building	5
		Also includes eccentricities, other loading details	
Design method	Drift	Drift is used to calculate seismic forces as a boundary condition	2
Design method	Dynamic response method	Uses dynamic response spectrum method, taking into account multiple modes of oscillation, which has higher quality until the sum of effective modal masses reaches a certain percentage of the oscillating mass of the system	5
Design criteria	Nonstructural	Are nonstructural elements designed for?	2
Design criteria	Avoidance of collapse	Special notes with application to the earthquake load combination, avoidance of soft-story mechanisms, plastic hinge locations, and other combinations	5
Design criteria	Damage limitation	Load-bearing systems, pounding taken into account, drift, appendages such as chimney, facades	5

(continued)

Global View of Seismic Code and Building Practice Factors, Table 1 (continued)

Element	Sub-element	Key components	Value given out of 104
Foundation design	Foundations, retaining structures, slopes	Basic checks as to slope stability and other foundation design applications such as liquefaction hazard, shear settlement, differences for shallow and deep foundations, etc.	5
Code quality assessment	Displacement-based design/overall	Displacement-based criteria in terms of the design methodology or overengineered status, additional components and methodologies encompassing many design levels for all types of structures	15

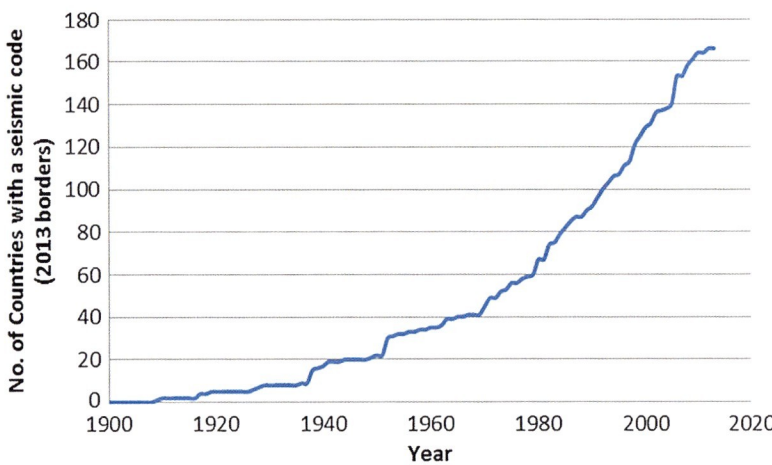

Global View of Seismic Code and Building Practice Factors, Fig. 1 The number of countries with a seismic-resistant code/zonation as of 2013 borders

Which Countries Have Seismic Codes?

Not all countries have seismic codes; however, documentation does not exist on a global scale to see which countries have seismic-resistant codes and whether they are part of law, enforced or ignored. In addition, some countries have differences on the province-level enforcement or adherence to such codes.

In order to determine the factor by which damage is changing and the complexity of codes, a significant review and analysis was undertaken to create a harmonized up-to-date list of locations where seismic-resistant codes have been considered. There were several early initiatives which attempted to undertake this by looking at only a list of some of the countries worldwide with seismic-resistant building codes, including the World List initiative by the IAEE which includes details as to some seismic-resistant codes in various countries – IISEE 1992, 1996, 2000, 2004, 2008, and 2012 (IAEE 1996, 2000, 2013). In addition, there has been a list of countries which have some form of seismic codes in the practice of earthquake hazard assessment (92 countries) (McGuire 1993). The "Seismic Code Evaluation" work in Central America as well as South America has been collected and implemented in the database (Chin and Association of Caribbean States 2003).

One hundred and sixty-six nations out of two hundred and forty-four nations have some form of seismic-resistant code, but they are usually not implemented for all styles of buildings, given the large number of nonengineered building styles worldwide as shown in Fig. 1.

In addition, there are nations which have some form of code; however, this has not been ratified and has not been implemented in practice. For earthquakes, a review of the various seismic-resistant codes around the world has been

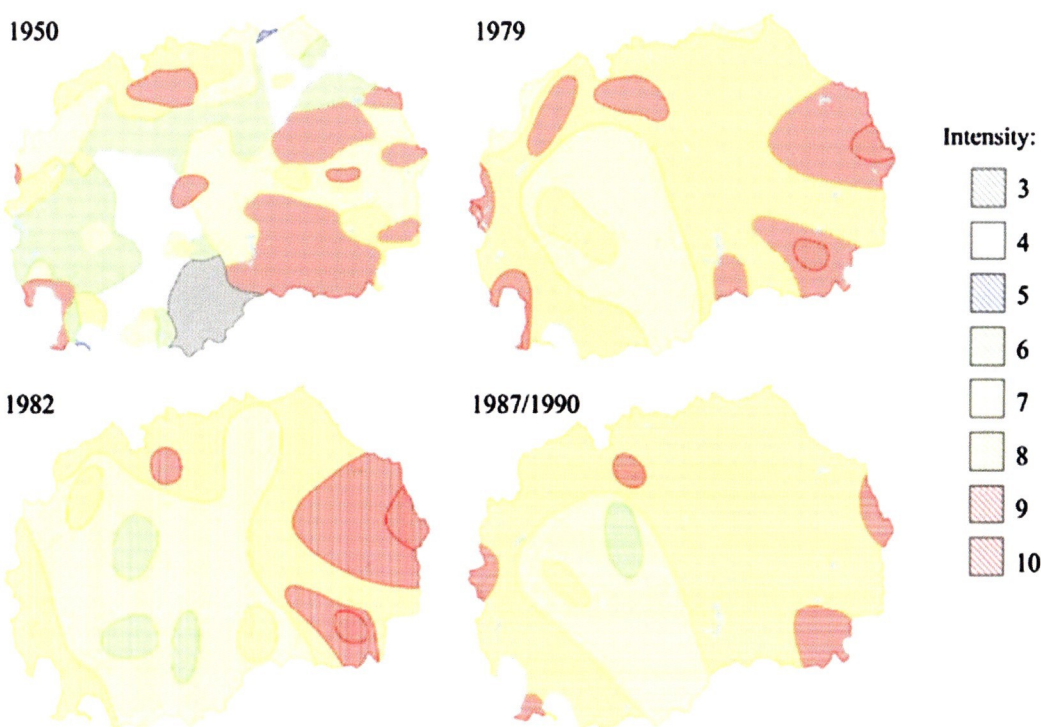

Global View of Seismic Code and Building Practice Factors, Fig. 2 The seismic zonation changes for the country of Macedonia (Salic et al. 2010)

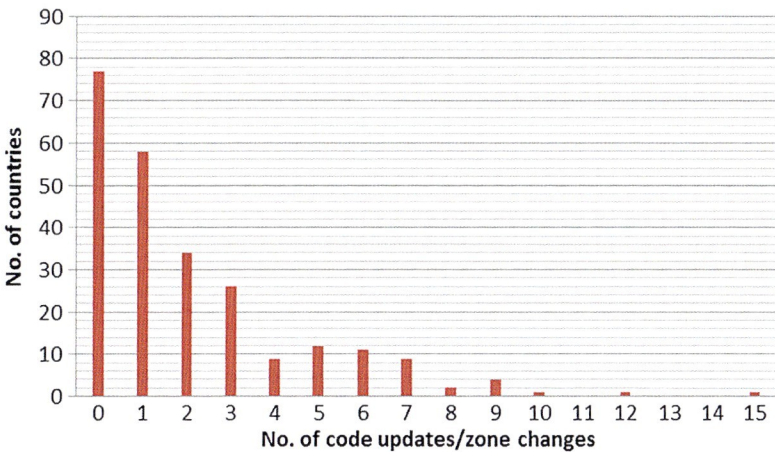

Global View of Seismic Code and Building Practice Factors, Fig. 3 The number of countries with code changes and updates since 1900

undertaken, examining the level of base shear relative to hazard, seismic zoning, and other parameters ranked in comparison to the relative hazard of a particular country versus the seismic hazard actually represented.

Many different countries have had multiple editions, adjustments, and changes to their seismic-resistant codes and zonations in the past century. Studies such as Salic et al. (2010), as depicted in Fig. 2 for Macedonia, or Romeo (2007) for Italy show the move from seismic hazard based on historic earthquakes to probabilistic hazard with a smoothing of hazard zonations used in code through time.

Shown in Fig. 3 are the number of code changes through time for each nation.

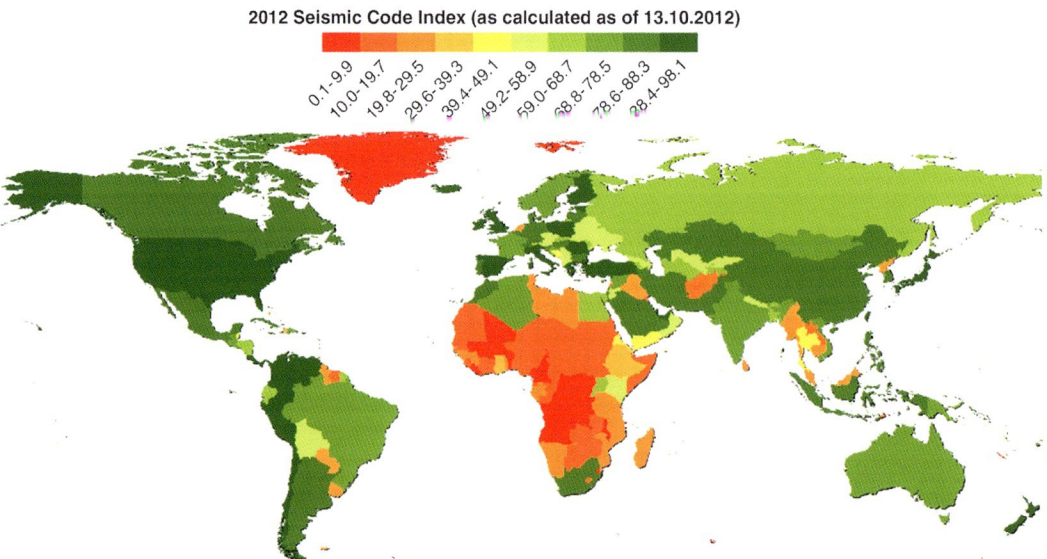

Global View of Seismic Code and Building Practice Factors, Fig. 4 The worldwide seismic code index (level 1) as calculated for 2012, showing the quality of the seismic-resistant and/or building code (or lack thereof) in each country regardless of enforcement and building practice factor

The seismic code index ranks the quality of seismic codes since 1900 in each nation, giving a score between 0 and 104, normalized to 100 as two extra components were added later (Daniell et al. 2011). The criteria was set that a country with a unified national code was placed at a value of 30/100, utilizing the minimum criteria set which is the use of earthquake loading in some form of design (not necessarily for all structures). The quality of this code was then objectively ranked by using the above criteria elements to make an assessment of the code quality at the time of implementation.

It can be seen from the 2012 picture of the seismic code index in Fig. 4 that much work is needed solely on the writing of seismic code indices. Despite the adherence or intention to adhere to the new Eurocodes in Africa and through Asia, many nations still have not signed or enforced codes with respect to seismic resistance.

The current state of the world seismic codes with respect to nonengineered buildings is also a key concern, given the lack of seismic codes for residential buildings and the age of buildings. Many of these buildings are largely vulnerable, use heavy or inappropriate materials, and are not built based on historical earthquake or future earthquake influences. Bilham (2013) estimates that there are 1–3 billion family homes in this category. However, many societies have learned to build with lightweight materials such as thatch, thus protecting themselves from fatal roof and wall collapses. The age distribution of buildings thus needs to be explored.

Age Distribution of Infrastructure and Its Effect on the Number of Buildings Under Seismic-Resistant Codes

A global building inventory including age and number of stories has been built (Daniell et al. 2011) using census information. The number of possible buildings that were built under some form of code was then reduced, as there are codes where not all building typologies are included under the seismic-resistant codes. The age of buildings has also been used to give a comparison as to the influence of a seismic code on the vulnerable percentage of buildings.

By using the same census data and methodology, the age distribution of infrastructure was collected, with the average shown in Fig. 5.

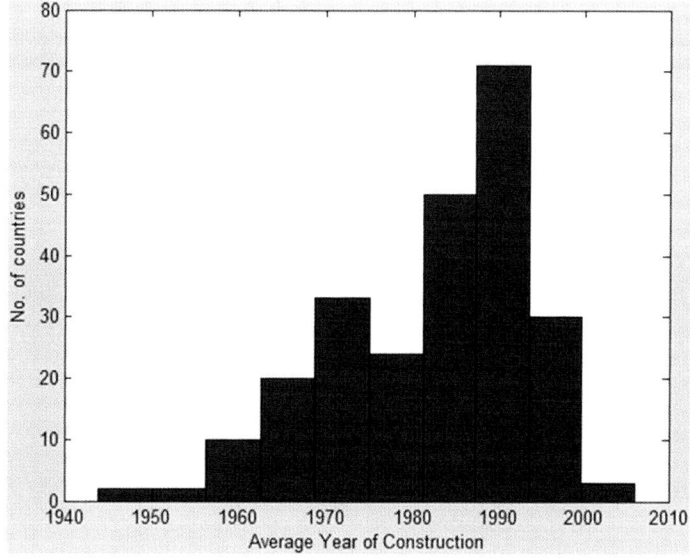

Global View of Seismic Code and Building Practice Factors, Fig. 5 Average year of construction for the 244 nations worldwide using census data

In countries such as Australia, the average age of infrastructure has stayed fairly constant at around 18 years from 1950 to 2005 (ABS 1995–1996), with a dip toward 1971 as building increased, and is now currently at about 18 years. In terms of the average lifetime of a building, this value is quoted as approximately 75 years. By comparison, this value is around 30 years in China but in Great Britain is around 132 years. For European buildings, a study named "Housing Statistics in the European Union" used census data and studies in order to examine the age of buildings (EUROSTAT 2013). The large difference in the design life of structures leads to an interesting problem when filling in the remaining countries.

If there is not enough new investment in buildings, structures, and infrastructure, the average age of buildings will increase, if there is no new population. As an example, Germany has had very low population growth since 1991. The age of their infrastructure increased from 23.7 years to 26.9 years in 2009. Their general government buildings increased from 22.1 to 28.4 years as compared to dwellings from 25.1 to 27.8 years. Figure 6 shows the average building year in each country as a result of the analysis.

Figure 7 depicts the locations in the world that should be under seismic code as per GSHAP (1999)-derived intensity exceedances for 475 years globally (Giardini 1999). Different intensity-PGA relations were used in Fig. 7 to convert PGA to MMI depending on the location worldwide. It can be seen that most nations should thus have some form of seismic code, and when looking at longer return periods, even more of the globe can be exposed to damaging earthquake intensities.

The percentage of structures in each country under seismic code differs greatly when only the age percentage of structures is only looked at; however, there are usually clauses that buildings of one or two stories are not built to the earthquake code. By using household size and building counts globally, a percentage of buildings built under code has been estimated. The age of structures in each country is integrated against the seismic code index in order to create a percentage of structures that are definitely not built under code. This is, of course, assuming that the age percentage of the buildings is the same over the whole country, which is a reasonable assumption as the natural weighting is toward greater building in cities, which is represented by the entire country values.

The locations of seismic code zonation were then split from the non-code zones by calculating the percentage of buildings in these zones. It should be noted that as a proxy, the MMI >6

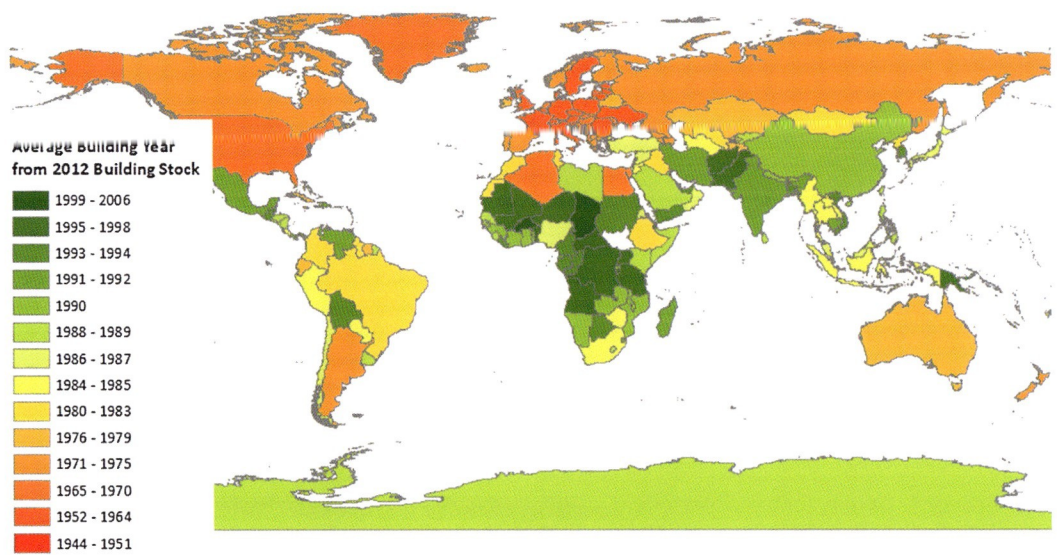

Global View of Seismic Code and Building Practice Factors, Fig. 6 The average building year in each country

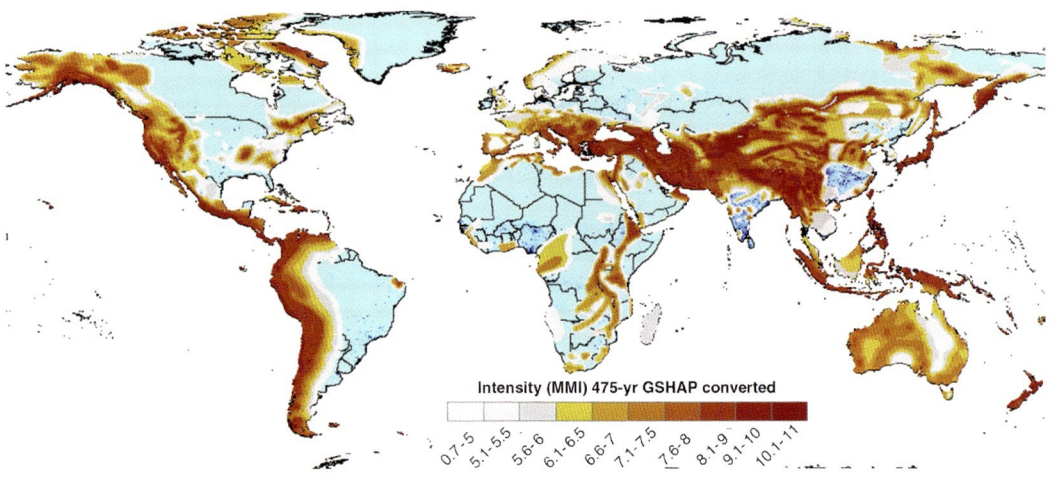

Global View of Seismic Code and Building Practice Factors, Fig. 7 The locations in the world that should be under seismic code as per GSHAP-derived intensity exceedances for 475 years globally. Different intensity-PGA relations were used to convert PGA to MMI, depending on the location worldwide

contour was used from GSHAP, as the seismic zone maps collected have not yet been digitized. However, these maps were looked at to check the assumptions, with major changes being made in Italy (where historical codes have not covered the whole of Italy) and China.

Unfortunately, every building type is not covered by earthquake-resistant codes. The next provision that is examined is that of small buildings or those not covered under code. In order to calculate this value, the individual building codes were looked at. It is difficult to make an assessment in some respects as to small buildings (residential 1–2 stories, or other forms) as, although these are included in some codes, there are many exemptions. The Chinese code (PRC 2001/2008/2010), for instance, uses intensity zones and story heights. Most rural housing is only 1 story and is thus not covered under earthquake-resistant codes. In addition, many

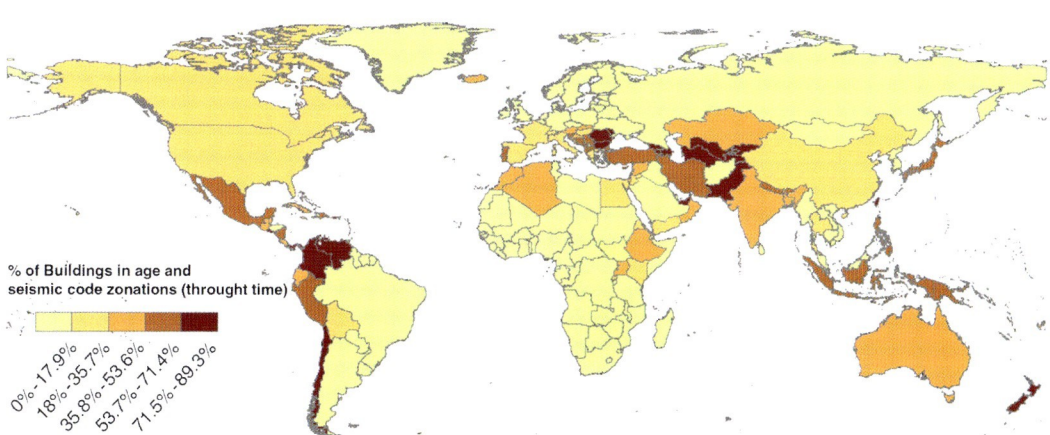

Global View of Seismic Code and Building Practice Factors, Fig. 8 The percentage of buildings in each country that should have been built under a seismic-resistant code (without small building provision)

urban buildings under 4 stories in the intensity zone 6 and 3 stories in the intensity zone 7 do not need tie columns for brick buildings. There is an exception for 1-story public buildings, which need to be built under code (2001 code, Chaps. 7 and 10). In Australia, all structures are included; (Standards Australia (1979, 1993, 2007)) however, for domestic buildings less than 3 stories, which are ductile, no provision is required. This code, being an objective-based code, also provides a good basis for engineers to check that the code is being adhered to. An approximate percentage of buildings coming under the provisions have been recorded. In addition, the number of informal and rural buildings is also taken into account.

Existing buildings are also included in some seismic-resistant codes in terms of requiring retrofitting, such as in Argentina, Italy, EC8 (European Union), the USA, Korea, Mexico, New Zealand, Nicaragua, and Venezuela; however, it is unsure as to what extent this has been enforced currently and thus has not been included.

By combining the age of structures which could have a seismic code and the number of nonengineered structures not built to code structures in those zones, a relative percentage of buildings in earthquake zones coming under code can be established. It should be noted that through Central Asia many buildings, although built to former Soviet codes, have become dilapidated and thus also should be retrofitted or rebuilt.

In general, those countries which have high percentages in Fig. 8 are smaller countries with most of their building stock built in the last 30 years. Marked differences occur across nations with different zonations and building ages across nations not described by such diagrams.

Seismic Code Provisions: Germany

An example is presented below for Germany, in order to show the analysis procedure. The DIN4149 was first implemented in Germany in 1981 after the 1978 Albstadt earthquake. The update to this was given in 2005. Before this, a few other codes had been applied on regional levels, without any federal application (Abrahamczyk et al. 2005). The German code has five zones (no zone, 0, 1, 2, and 3).

Only 31 % of Germany's building stock has been produced after 1981. The code actually applied in zones 1–3 in terms of earthquake loss is for very little of the building stock (15.3 %). This means that under 5 % of possible buildings have been built under some form of earthquake-resistant code. In DIN4149, not every building needs to have had the code change applied, due to the premise of the number of stories (Table 2);

Global View of Seismic Code and Building Practice Factors, Table 2 Various limits on the requirements not to adhere to earthquake codes in zones 1–3

Earthquake zone	Importance class	Maximum story height
1	I–III	4
2	I–II	3
3	I–II	2

thus, even less than 5 % of buildings are applicable to this code. State enforcement ratification of codes is necessary in order for the DIN4149 to be used. The 2005 version has been ratified by most states; however, the new Eurocodes have only currently (as of mid-2013) been ratified by Baden-Württemberg, a state in the southwest of Germany (European Union 2006).

In zone 1, 89 % of buildings do not need to be built to code. In zone 2, around 75 % of buildings do not need to be built to code and in zone 3 around 48 %. This means that less than 0.9 % of buildings in Germany are built to a seismic-resistant code. When focusing on the population and value of the 0.9 % of buildings, this increases to approximately 3 % of capital stock, given the multi-story buildings, infrastructure, and important buildings in importance classes 3 and 4 protected by the earthquake code (hospitals, schools, industrial facilities, etc.). However, a more important problem is posed by the existing hazard, building typology, and standard building code. This formulates a similarly bleak engineered view of the world in terms of loss functions, which presents an additional difficulty in applying a single building approach.

Building Practice Versus Code Compliance

Just because a country has a seismic code in law, this does not mean that the seismic code will be enforced. Thus, building practice plays a major role in earthquake-resistant building, and not only the quality of the seismic code is important but its enforcement, adherence, and lack of corruption in the building. If a building was designed with a high-quality code, yet the builders were not made to adhere to that code and used substandard materials, then the building would likely not stand up to the earthquake forces.

Corruption has been identified as a key component of earthquake losses, first through the Bilham and Hough (2006) paper in India and then Ambraseys and Bilham (2011) and Bilham (2013) exploring the Transparency International corruption index with respect to earthquakes. There is no doubt that engineered constructions have increased the quality of building against earthquake forces, but much of this has to do with the education of the individuals of that country. Education can be used as a proxy for the quality of understanding of building practice as well as the quality of the mathematics, engineering, and drawings that govern building construction. A study looking at the number of universities offering earthquake-related subjects is another proxy that has been previously used (Santa-Ana et al. 2012). Building error and mistakes increase with lack of education.

Summary

Seismic-resistant design codes have been implemented in many nations around the world over the past century. A summary has been provided as to their current application and to the extent of their use. Although many countries have implemented a seismic-resistant building code, the current percentage of buildings covered by these codes is generally less than 50 % of the building stock.

The enforcement and implementation of seismic-resistant codes is important as even though a country may have a very high-quality code, if it is not implemented or enforced by the government, the code is useless. Given the exponential rise in the number of seismic-resistant codes and the quality of these codes, it can be seen that more work is needed in implementing seismic-resistant codes for existing buildings and assessing residual risk, and care must be taken when referring to the use of codes without looking at exclusion clauses.

Cross-References

▸ Building Codes and Standards
▸ Damage to Infrastructure: Modeling
▸ European Structural Design Codes: Seismic Actions
▸ Structural Design Codes of Australia and New Zealand: Seismic Actions

References

Abrahamczyk L, Langhammer T, Schwarz J (2005) Earthquake regions of the Federal Republic of Germany – a statistical analysis (in German). Bautechnik 82(8): 500–507

Ambraseys NN, Bilham R (2011) Corruption kills. Nature 469(7329):153–155

Australian Bureau of Statistics (ABS) (1995–1996) Australian National Accounts: estimates of capital stock, catalogue no 5221, various issues. Retrieved from http://www.abs.gov.au/AUSSTATS/abs@.nsf/ProductsbyCatalogue/8D771B840483335BCA25722E001A7CC8?OpenDocument

Beattie G, Thurston SJ (2006) Changes to the seismic design of houses in New Zealand. In: Proceedings, 2006 New Zealand society for earthquake engineering conference, New Zealand Society for Earthquake Engineering, Wellington

Bilham R (2013) Societal and observational problems in earthquake risk assessments and their delivery to those most at risk. Tectonophysics 584:166–173

Bilham R, Hough S (2006) Future earthquakes on the Indian subcontinent: inevitable hazard, preventable risk. South Asian J 12:1–9

Chin MW, Association of Caribbean States (2003) Model building codes for earthquakes and wind loads. Retrieved from http://www.eird.org/cd/acs/English/enmodel.html

Daniell JE, Wenzel F, Khazai B, Vervaeck A (2011) A Country-by-Country Building Inventory and Vulnerability Index for Earthquakes in comparison to historical CATDAT Damaging Earthquakes Database losses. In: Australian Earthquake Engineering Society 2011 conference, Barossa Valley

European Union (2006) EN1998-1:2004. Eurocode 8: design of structures for earthquake resistance: part 1: general rules, seismic actions and rules for buildings. Brussels: Comite Europeen de Normalisation

Giardini D (1999) The global seismic hazard assessment program (GSHAP)-1992/1999, Annals of Geophysics, 42(6)

International Association of Earthquake Engineering (IAEE) (1996) Regulations for seismic design – a world list. Retrieved from http://www.iaee.or.jp/worldlist.html

International Association of Earthquake Engineering (IAEE) (2000) Regulations for seismic design – a world list, Supplement 2000. Retrieved from http://www.iaee.or.jp/worldlist.html

International Association of Earthquake Engineering (IAEE) (2013) Regulations for seismic design – a world list. Retrieved from http://www.iaee.or.jp/worldlist.html

International Institute of Seismology and Earthquake Engineering (IISEE) (2012) Seismic design code. Retrieved from http://iisee.kenken.go.jp/net/seismic_design_code

McGuire RK (ed) (1993) The practice of earthquake hazard assessment. U.S. Geological Survey, Denver

Ministry of Construction (PRC) (2001/2008/2010) Chinese seismic design code: code for seismic design of buildings (GB50011-2001, 2008) (in Chinese), GB 50011-2010. China Building Industry Press, Beijing

Priestley MJN (1993) Myths and fallacies in earthquake engineering – conflicts between design and reality. Bull New Zeal Natl Soc Earthquake Eng 26(3):329–341

Priestley MJN, Calvi GM, Kowalsky MJ (2007) Displacement-based seismic design of structures. IUSS Press, Pavia; Distributed by Fondazione EUCENTRE

Romeo RW (2007) Italian seismic hazard: experiences and new building code application. In: Geohazards conference and EFG council meeting 2007. Retrieved from http://www.icog.es/_portal/uploads/pub_fedgeol/12.pdf

Salic RB, Garevski MA, Milutinovic ZV (2010) The need for advanced seismic hazard assessment of the Republic of Macedonia. In: Proceedings of 14th European conference of earthquake engineering (ECEE) 2010, Ohrid

Santa-Ana PR, Santa-Ana LG, Baez Garcia J (2012) What we have forgotten about earthquake engineering, paper no 4566. In: Proceedings of the 15th world conference on earthquake engineering, Lisbon

Standards Australia (1979) The design of earthquake resistant buildings. AS2121-1979. Standards Australia, Sydney

Standards Australia (1993) Minimum design loads on structures: part 4 – earthquake loads. AS1170.4-1993. Standards Australia, Sydney

Standards Australia (2007) Structural design actions: part 4: earthquake actions in Australia. AS 1170.4-2007, 2nd edn. Standards Australia, Sydney

Standards New Zealand (2004) NZS 1170.5:2004 - Structural design actions. Earthquake actions., Wellington, New Zealand

Statistical Office of the European Union (EUROSTAT) (2013) Housing statistics: tables and figures. Retrieved from http://epp.eurostat.ec.europa.eu/statistics_explained/images/5/59/Housing_statistics_YB2014.xls

Graphics Processing Unit (GPU) Technology in Earthquake Engineering, Application of

Xinzheng Lu and Zhen Xu
Key Laboratory of Civil Engineering Safety and Durability of China Education Ministry, Department of Civil Engineering, Tsinghua University, Beijing, People's Republic of China

Synonyms

Earthquake engineering; GPU (Graphics Processing Unit); Massive parallel computing; Seismic damages simulation

Overview

The Graphics Processing Unit (GPU) located in the display card of a computer was originally designed for graphics display. Due to advances in massive parallel computing capacity, the current applications of GPUs have grown far beyond the graphics field.

General-Purpose computation on Graphics Processing Units (GPGPU) was proposed as early as the 1990s. However, such technology was not widely applied for general-purpose computation because, at the time, the application of GPGPU required a graphics application programming interface (API) (e.g., DirectX, OpenGL), which became a barrier to effective coding in general-purpose computation. In addition, many other difficulties have existed for GPGPU with respect to hardware, such as the exchange of data between the GPU and CPU. In 2006, the GPU G80 and the corresponding Compute Unified Device Architecture (CUDA) programming model were released by the NVIDIA Corporation (NVIDIA 2013), which was the first unified shader architecture and solved many of the GPGPU software and hardware problems. Since that time, GPU technology has been widely used in many fields, including many applications in earthquake engineering involving seismic wave propagation and seismic damage prediction.

GPU Technology

Hardware Platform for GPU Computing

The term "GPU" was introduced by NVIDIA in 1999 with the release of GeForce 256, which is known as "the world's first 'GPU'". Subsequently, ATI Technologies (acquired by AMD in 2006) introduced the visual processing unit (VPU) in 2002, which is not essentially different from a GPU. In 2006, the GeForce 8 series and Compute Unified Device Architecture (CUDA) were introduced by NVIDIA and constituted the earliest and subsequently most widely adopted programming model for GPU computing. In response, GPU hardware had undergone rapid development. Since 2006, the development history of the architectures for the two major types of GPUs may be depicted as follows:

NVIDIA GPUs: G80 (Nov. 2006) – GT200 (Jun. 2008) – Fermi (GF100/GF110, Mar. 2010/Nov. 2010) – Kepler (Mar. 2012).
ATI/AMD GPUs: R600 (May 2007) – RV770 (Jul. 2008) – RV870 (Sep. 2009) – Cayman (Dec. 2010) – Tahiti (Jan. 2012).

At present, Intel, NVIDIA, and AMD/ATI are the market share leaders. Although over half of the market share is owned by Intel, most of the GPUs from Intel are Intel's integrated graphics solutions, the performance of which is not as high as that of the GPUs introduced by the other two companies. Excluding these integrated GPUs, NVIDIA and AMD/ATI control over 95 % of the market. In addition, there are other companies that produce GPUs, such as VIA Technologies and Matrox.

The reason behind the discrepancy that exists in floating-point capability between the CPU and the GPU is that the GPU is specialized for compute-intensive, highly parallel computation and therefore designed in such a manner that more transistors are devoted to data processing rather than data caching and flow control

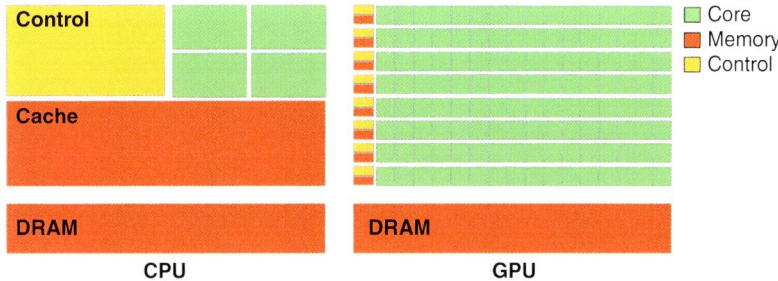

Graphics Processing Unit (GPU) Technology in Earthquake Engineering, Application of, Fig. 1 Comparison between the CPU architecture and the GPU architecture

(NVIDIA 2013), as schematically illustrated in Fig. 1. It is obvious that the GPU has much more cores than the CPU, as shown in Fig. 1, so the GPU is well suited to problems that can be expressed as data-parallel computations with a high level of arithmetic intensity (i.e., the ratio of the demand of arithmetic operations to memory operations).

Software Platform for GPU Computing

Currently, CUDA is the most widely used software platform for GPU computing around the world (NVIDIA 2013). In addition, the Stream platforms of AMD, OpenCL, and DirectCompute are also important GPU computing platforms.

CUDA is a software environment that allows developers to use C as a high-level programming language. Other languages, application programming interfaces or directive-based approaches are also supported, such as FORTRAN, DirectCompute, and OpenACC.

CUDA is carried out by means of a hybrid computing architecture that includes a GPU and CPU. The CPU is responsible for the sequential and logical tasks, and the GPU is used to implement the massive parallel computing tasks. A typical flow chart for a CUDA program is illustrated as follows. First, data are copied from the main memory to the GPU memory. Next, the CPU instructs the main processing and the GPU executes in parallel in each core. Finally, the result is copied from the GPU memory to the main memory.

Two cases of applications of the CUDA program from the CUDA C Programming Guide (NVIDIA 2013) are provided herein to present the CUDA platform in detail.

Example 1: Vector Addition CUDA C extends C by allowing the programmer to define C functions, called *kernels*, that, when called, are executed N times in parallel by N different CUDA threads, as opposed to only once like regular C functions.

A kernel is defined using the __global__ declaration specifier and the number of CUDA threads that execute that kernel for a given kernel call is specified using a new <<<...>>> execution configuration syntax. Each thread that executes the kernel is given a unique thread ID that is accessible within the kernel through the built-in *threadIdx* variable.

As an illustration, the following sample code adds the two vectors A and B of size N and stores the result into vector C:

```
// Kernel definition
__global__ void VecAdd(float* A,
float* B, float* C)
{
int i = threadIdx.x;
C[i] = A[i] + B[i];
}
int main()
{
...
    // Kernel invocation with
    N threads
VecAdd<<<1, N>>>(A, B, C);
}
```

Example 2: Matrix Addition The following code adds two matrices, A and B of size $N \times N$, and stores the result into matrix C:

```
// Kernel definition
```

```
__global__ void MatAdd(float A
[N][N], float B[N][N], float C
[N][N])
{
int i = threadIdx.x;
int j = threadIdx.y;
C[i][j] = A[i][j] + B[i][j];
}
int main()
{
...
  // Kernel invocation with one
  block of N * N * 1 threads
int numBlocks = 1;
dim3 threadsPerBlock(N, N);
       MatAdd<<<numBlocks,
       threadsPerBlock>>>
       (A, B, C);
}
```

For convenience, *threadIdx* is a three-component vector, by which threads can be identified using a one-dimensional, two-dimensional, or three-dimensional thread index, thus forming a one-dimensional, two-dimensional, or three-dimensional thread block. This provides an easy means to invoke computation across the elements in a domain, such as a vector, matrix, or volume.

The index of a thread and its thread ID are related to each other in a straightforward way: For a one-dimensional block, they are the same; for a two-dimensional block of size (Dx, Dy), the thread ID of a thread of index (x, y) is $(x + y\,Dx)$; for a three-dimensional block of size (Dx, Dy, Dz), the thread ID of a thread of index (x, y, z) is $(x + y\,Dx + z\,Dx\,Dy)$.

Earthquake Engineering Applications

Simulation of Ground Motion
The propagation and attenuation of a seismic wave is an important issue in earthquake engineering. Abdelkhalek et al. (2012) adopted a GPU to solve the problem of seismic wave propagation. Considering both the 2D and 3D cases, this group used the GPU Tesla S1070 to perform an analysis of a seismic wave based on a finite difference approach, obtaining an increased speed of ten times for reverse time migration and up to 30 times for a sequential simulation of seismic waves. Komatitsch et al. (2009) implemented a higher-order finite element application of seismic wave propagation based on two types of GPUs, the NVIDIA GeForce 8800 GTX and the GTX 280, and the acceleration ratios reached 15 and 25, respectively.

Okamoto et al. (2011) used the Japanese TSUBAME-1.2, supercomputer to implement a finite-difference-based simulation of seismic wave propagation. In this simulation, the computing efficiency of a single GPU (Tesla S1070s) reached 56 GFLOPS, which is 45 times the efficiency of a CPU (Intel Core i7). Unat et al. (2012) accelerated a 3D finite difference earthquake simulation with a C-to-CUDA Translator. Running on the Tesla C2050, their program was faster than a highly tuned message passing interface (MPI) implementation running on 32 Nehalem cores. Similarly, Zhou et al. (2012) designed 3D finite difference seismic wave propagation code based on the latest GPU Fermi chipset that implemented the asymmetric 3D stencil on Fermi in order to make the best use of the GPU on-chip memory and achieved aggressive parallel efficiency. The benchmark test on an NVIDIA Tesla M2090 demonstrated that a speedup factor of 10 was achieved compared with the original fully optimized AWP-ODC FORTRAN MPI code running on a single Intel Nehalem 2.4 GHz CPU socket (4 cores/CPU), and a speedup factor of 15 was achieved compared with the same MPI code running on a single AMD Istanbul 2.6 GHz CPU socket (6 cores/CPU). The sustained single-GPU performance of 143.8 GFLOPS in single precision mode was benchmarked for the test case of a $128 \times 128 \times 960$ sized mesh.

Structural Numerical Analysis
The GPU has been applied to structural numerical analysis. The finite element method is an important structural numerical analysis approach. In the elastic structural analysis based on finite element method, most of the computing time is spent on solving matrices. Many matrix computations are well suited to GPU computing. For

example, matrix multiplication can be easily processed by a GPU due to its natural parallel characteristics. Ryoo (2008) conducted a similar test for matrix multiplication with a program that attained an efficiency of 91 GFLOPS on the GPU of a GeForce 8800 GTX. Similarly, matrix decomposition, which is the foundation for solving a linear equation, can also be easily accelerated using a GPU. Volkov and Demmel (2008) optimized a matrix decomposition algorithm on the Geforce 8800 platform, and the decomposition efficiencies of the LU, QR and Cholesky methods were all greater than 180 GFLOPS.

Furthermore, Liu et al. (2008) proposed a general finite element method solution that enables dynamically fast simulation of deformation on the newly available GPU hardware with CUDA. Their test results indicate that the GPU with CUDA enables an increase speed of approximately four times for FEM deformation computation on an Intel(R) Core 2 Quad 2.0 GHz machine with GeForce 8800 GTX. Many commercial finite element analysis codes, e.g., ABAQUS, MSC. MARC, and ANSYS, have released their own GPU-based versions. The performance studies demonstrate that the GPU results in speedups in the range of 3–6 times.

In addition to the finite element simulation, other structural numerical simulations also can be greatly accelerated by the use of GPUs. For example, Durand et al. (2012) simulated rock impact on a concrete slab using the discrete element method (DEM), and the use of a GPU (Tesla C2050) reportedly showed a speedup factor of 30.

Earthquake Detection

Meng et al. (2012) proposed a GPU-based computation method to accelerate the detection algorithm for earthquakes during intensive aftershocks or swarm sequences. This group applied parallel code to search the Salton Sea geothermal field for missing earthquakes in a 90–day time window around the occurrence time of the 2010 Mw7.2 El Mayor-Cucapah Earthquake. By dividing the procedure into several routines and processing them in parallel, this group achieved an approximate 40 speedup for one NVIDIA GPU card compared with the sequential CPU code.

Repeating earthquakes are occurring on the similar asperity at the plate boundary. These earthquakes have an important property; the seismic waveforms observed at the identical observation site are very similar regardless of their occurrence time. Kawakami et al. (2013) proposed a high-speed digital signal processing method for the detection of repeating earthquakes using GPGPU-acceleration. They compared the execution time between GPU (NVIDIA GeForce GTX 580) and CPU (Intel Core i7 960) processing. In the case of a band-limited phase only correlation, the obtained results indicate that a single GPU is approximately 8.0 times faster than a 4-core CPU (auto-optimization with OpenMP). In the case of the coherence function using three components, the GPU is 12.7 times as fast as the CPU. This study examines the high-speed signal processing of a huge amount of seismic data using the GPU architecture. It was found that the GPGPU-based acceleration for the temporal signal processing is very useful. In the future, multi-GPU computing will be employed to expand the GPGPU-based high-speed signal processing framework for the detection of repeating earthquakes in the future.

The Visualization of Seismic Data

The GPU can also be used for the high-performance visualization of earthquake data. Hsieh and Yang (2011) presented a method using volume rendering based on a GPU to visualize the spectral information of the 1999 Chi-Chi earthquake and the GPU-based volume-rendering system achieved a rendering speed of 30 frames per second (FPS). Xie et al. (2010) proposed a parallel visualization technique for large seismic datasets based on a GPU and CPU cooperation platform. Their experimental results showed that the method exhibited effective real-time performance for a notably large amount of seismic data (10 GB) on standard PC hardware.

Chen and Hsieh (2012) proposed a GPU-based method using volume rendering to better understand and analyze time-varying earthquake ground-motion data, including acceleration, velocity and displacement. This group used CUDA to implement volume rendering (ray

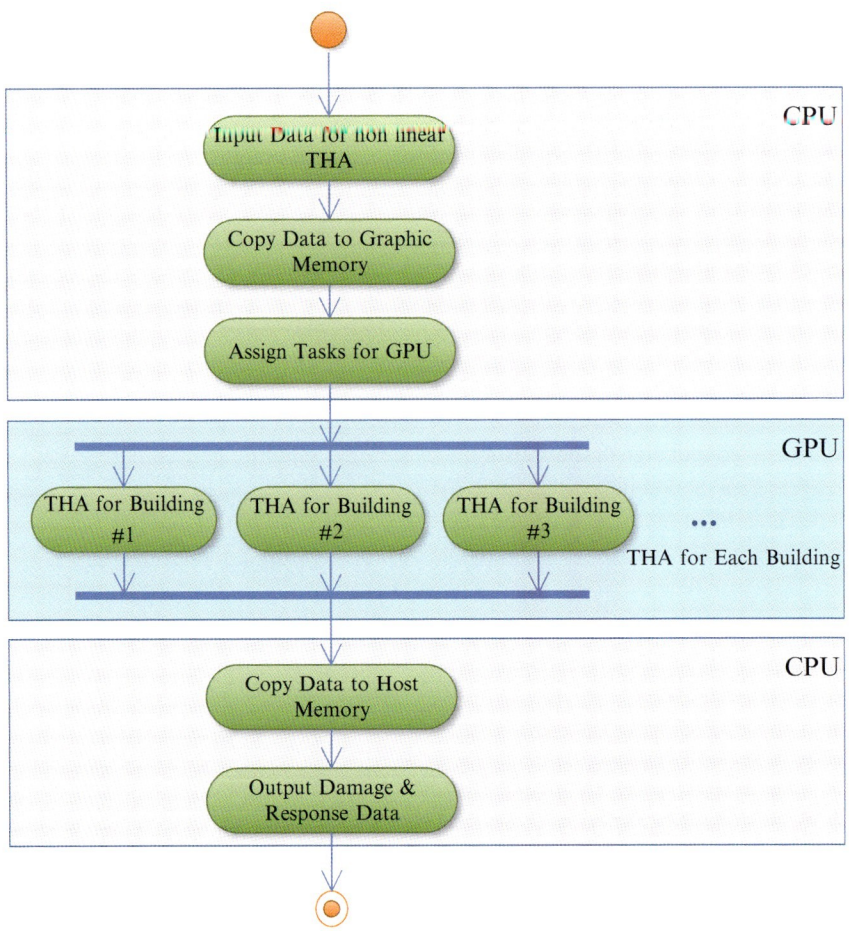

Graphics Processing Unit (GPU) Technology in Earthquake Engineering, Application of, Fig. 2 Flow chart of the seismic damage analysis using GPU

casting), numerical integration (trapezoidal quadrature) and numerical differentiation (centered finite difference), and achieved a speedup of approximately 100× versus a CPU.

Seismic Damage Prediction

Although thousands of buildings may exist in an urban area, if each building is treated as a subtask and a proper computing model is adopted, the computing workload of each subtask is sufficiently small that it can be performed on a single GPU core. Furthermore, there is little interaction between different buildings during an earthquake, which results in there being little data exchange between different GPU cores. Because there are hundreds of cores in a GPU and each GPU core can be used for the simulation of one building, only a few task assignment rounds are required to complete the simulation of a city with thousands of buildings using GPU computing. Thus, the computational efficiency is expected to be notably high.

A novel solution proposed by Han et al. (2012) used a GPU to accelerate urban seismic damage prediction based on refined building models and nonlinear time-history analysis (THA). Figure 2 shows the flow chart of the Han's program in which the CPU reads the data, copies the data into the GPU, and assigns the computational tasks to each GPU core, whereas the GPU implements the nonlinear THA for individual buildings and copies the results back to the host memory for output.

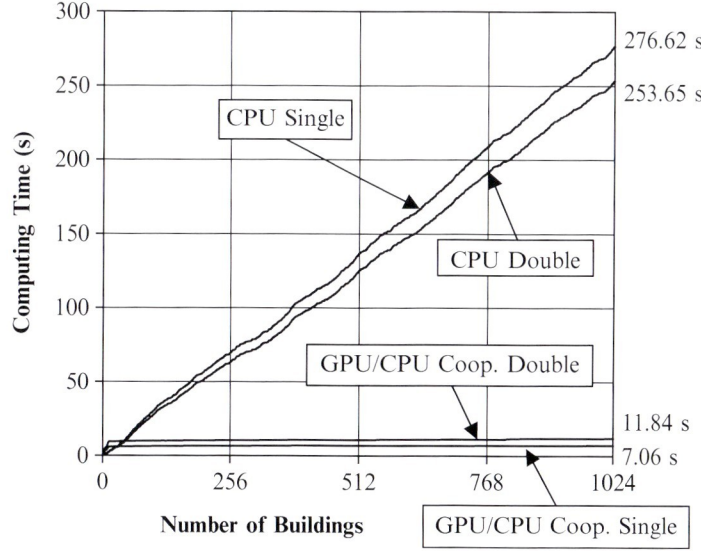

Graphics Processing Unit (GPU) Technology in Earthquake Engineering, Application of,
Fig. 3 Benchmark for the CPU and GPU/CPU cooperative programs

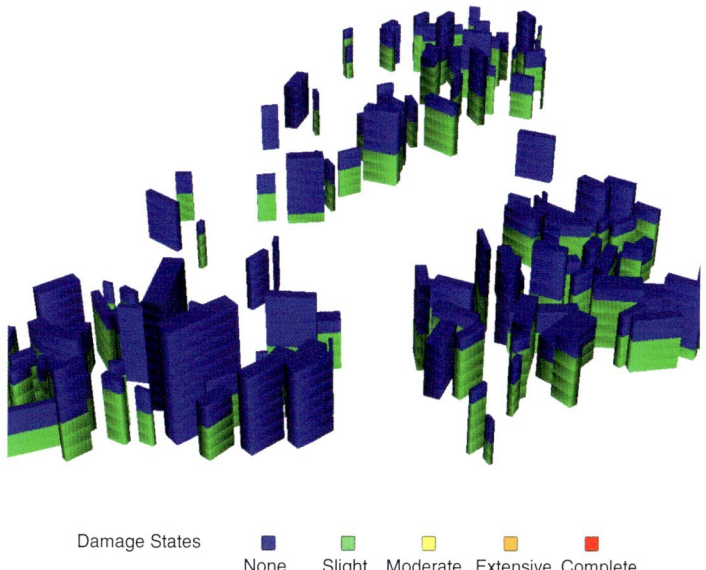

Graphics Processing Unit (GPU) Technology in Earthquake Engineering, Application of,
Fig. 4 Partial view of the damage states of each story

In the work of Han, a benchmark case of 1,024 buildings was investigated to validate the efficiency of seismic damage analysis based on GPU. The well-known El Centro record was selected as the ground motion for the benchmark. The peak ground acceleration (PGA) normalized to 200 cm/s^2.

Two hardware platforms with the same performance-to-price ratio were adopted, as follows:

CPU platform
  Hardware: A 2.93-GHz Intel Core i3 530 processor with 4 GB of 1,333-MHz DDR3 RAM.
  Compiler: Microsoft Visual C++ 2008 SP1.
GPU/CPU cooperative platform
  Hardware: A 2.4-GHz Intel Celeron E3200 CPU and an NVIDIA GeForce GTX 460 with 1 GB of graphics memory.

Compiler: Microsoft Visual C++ 2008 SP1 and CUDA 4.2.

The benchmark cases demonstrated that the GPU/CPU cooperative computation time is approximately 1/39 times the CPU computational time if single precision is used, as shown in Fig. 3. This ratio increases to 1/21 if double precision is used. Seismic damage prediction based on a GPU is shown in Fig. 4, which provides an important reference for earthquake disaster prevention and mitigation.

In summary, existing studies have demonstrated that GPUs provide a highly efficient but low-cost environment for computation and visualization in earthquake engineering. Therefore, in the future, GPU technology will come to be widely used in earthquake engineering.

Acknowledgements The authors are grateful for the financial support received from the National Key Technology R&D Program (No. 2013BAJ08B02), the National Nature Science Foundation of China (Nos. 51222804, 51178249, 51378299, 51308321) and China Postdoctoral Science Foundation (2013M530632).

Cross-References

▶ Earthquake Damage Assessment from VHR Data: Case Studies

References

Abdelkhalek R, Calandra H, Couland O, Roman J et al (2012) Fast seismic modeling and reverse time migration on a GPU cluster. Concurrency Comput Pract Exp 24:739–750

Chen MD, Hsieh TJ (2012) Run-tme GPU computing and rendering of earthquake ground-motion data. In: Proceedings of the IEEE 9th international conference on high performance computing and communication, pp 812–817

Durand M, Marin P, Faure F, Raffin B (2012) DEM-based simulation of concrete structures on GPU. Eur J Environ Civil Eng 16:1102–1114

Han B, Lu XZ, Xu Z (2012) Urban regional seismic damage prediction based on GPU-CPU hybrid computing. In: Proceedings of the 15th world conference in earthquake engineering, Lisbon

Hsieh TJ, Yang YS (2011) Visualizing the seismic spectral response of the 1999 Chi-Chi earthquake using volume rendering technique. J Comput Civil Eng-ASCE 26:225–235

Kawakami T, Okubo K, Uchida N, Takeuchi N, Matsuzawa T (2013) High-speed digital signal processing method for detection of repeating earthquakes using GPGPU-acceleration. EGU General Assembly 2013, Vienna

Komatitsch D, Michéa D, Erlebacher G (2009) Porting a high-order finite-element earthquake modeling application to NVIDIA graphics cards using CUDA. J Parallel Distrib Comput 69:451–460

Liu Y, Jiao S, Wu W, De S (2008) GPU accelerated fast FEM deformation simulation. In: IEEE Asia Pacific conference on circuits and systems, Macao, pp 606–609

Meng XF, Yu X, Peng ZG, Hong B (2012) Detecting earthquakes around Salton Sea following the 2010 Mw7.2 El Mayor-Cucapah earthquake using GPU parallel computing. Procedia Comput Sci 9:937–946

NVIDIA (2013) CUDA programming guide. http://docs.nvidia.com/cuda/pdf/CUDA_C_Programming_Guide.pdf

Okamoto T, Takenaka H, Nakamura T, Aoki T (2011) Accelerating large-scale simulation of seismic wave propagation by multi-GPUs and three-dimensional domain decomposition. Earth Planets Space 62:939–942

Ryoo S (2008) Program optimization strategies for data-parallel many-core processors. University of Illinois at Urbana-Champaign, Urbana

Unat D, Zhou J, Cui Y, Baden SB et al (2012) Accelerating a 3D finite-difference earthquake simulation with a C-to-CUDA translator. Comput Sci Eng 14:48–59

Volkov V, Demmel J (2008) LU, QR and Cholesky factorizations using vector capabilities of GPUs. Technical report UCB/EECS-2008-49. EECS Department, UC Berkeley

Xie K, Wu P, Yang S (2010) GPU and CPU cooperation parallel visualization for large seismic data. Electron Lett 46:1196–1197

Zhou J, Unat D, Choi DJ, Guest CC, Cui Y (2012) Hands-on performance tuning of 3D finite difference earthquake simulation on GPU Fermi chipset. Procedia Comput Sci 9:976–985

H

Historical Seismometer

Josep Batlló
Instituto Dom Luiz (IDL), Faculdade de Ciõncias da Univ. de Lisboa, Lisbon, Portugal

Synonyms

Accelerograph; Early seismic recording; Old seismographs; Seismometer; Seismoscope

Introduction

A seismometer is an instrument that detects ground motion and a seismograph is a seismometer together with a recorder that records it (a seismogram). Historical seismometers or ancient seismographs are those instruments developed previous to the design and development of the World-Wide Standardized Seismographic Network (WWSSN, Oliver and Murphy 1971). Other previous studies on this topic used to define old seismographs as those designed up to the very early years of the twentieth century because around those years the main basic features of such instruments were already the same as what our present knowledge recognizes as seismographs with analog recording. But it should be realized that the younger graduates in seismology and related topics rarely have had contact with seismic analog recording. For this reason, some relevant instruments and topics on seismic recording developed on the first half of the twentieth century will be included here. Seismoscopes (a device that only indicates ground motion) and other devices designed and used previous to the development of seismographs will be also considered in this entry.

The ancient seismoscopes and seismographs will primarily be studied from a chronological point of view, but with specific insight into its relevance from the point of view of innovation, that is, their contribution to the evolution of the technical features and/or capabilities of seismic recording.

It is impossible to describe all the models of instruments developed. Therefore, preference will be given to those instruments showing some new development with respect to previous ones. For the readers more interested in this topic or for more details and primary references on the described instruments, there are the outstanding studies of Dewey and Byerly (1969), Howell (1990), and Ferrari (1992) where more information is available and to which this entry is highly indebted.

Recording Earthquakes

The history of early earthquake recording instruments is not a linear one or of continuous progress. It was clear from old times that earthquakes involved ground motion and, thus, recording the ground motion was an important issue to get some insight in the earthquake phenomena. Also it cannot be ignored that rough devices to detect ground motion were known and used through antiquity and medieval times. It is confirmed that such a device was used on the Exeter siege in 1549. It was used to determine the location of enemies digging underground tunnels. The coeval document says: "hearing a noise underground, he takes a pan of water, and by removing it from place to place, came at length to the very spot where the miners were working; which he knew for certain by the shaking of the water in the pan." But no application of this or other instruments for the recording or study of earthquakes is known from the renaissance up to the eighteenth century.

In fact, theoretical and technical issues were not developed enough at that time to allow the recording of ground motion. For these reasons the history of the development of early instruments for earthquake recording is one of accumulation of empirical knowledge and of trial and error. Only at the end of the nineteenth century were the key issues concerning ground motion recording started to be properly defined and mastered and the technology of instrumental ground motion recording started to progress steadily.

In order to clarify the following discussion, it should be remembered that, from a technical point of view, two main components can be identified in a ground motion recording instrument. The first is the transducer; this is a device that transforms ground shaking into some other variable (e.g., a voltage) that can be recorded. The second one is the recording system itself, a device allowing to visualize and to store the output of the transducer for further inspection and analysis. The third item is a signal-amplifying and preprocessing system or, in other words, a signal-conditioning system. But this was not a key issue in the early development of seismographs.

In the following sections a review will be made on the evolution of seismic recording up to the middle of the twentieth century. As it is impossible to describe all types of seismographs, the description will concentrate on the instruments that defined new characteristics or new evolution in the recording concepts. Instruments not described here will be similar to the ones covered here.

An Overview of Seismic Recording

To better understand the descriptions of seismic instruments, some basic theoretical aspects will be described here. For more details, see e.g., Wielandt 2009 or Havskov and Alguacil 2010.

The Nature of Seismic Ground Motion

Earthquakes release energy as elastic waves. The frequency content of the generated waves depends on the size (total energy) released at the earthquake source. Larger earthquakes generate larger amplitudes at lower frequencies relative to small earthquakes. Attenuation of elastic waves propagating on Earth is larger for higher frequencies. Thus, with equal amplitude, lower-frequency waves propagate to longer distances than higher-frequency waves.

For these reasons seismographs adapted to record lower frequencies (<0.1 Hz) are needed to record large and distant earthquakes while instruments with larger amplification at higher frequencies (>1 Hz) are needed to record near, small earthquakes.

From Ground Motion to Seismic Recording

As it will be seen in the following sections, the most widely used principle for seismic recording is the inertia principle.

A device of such type, a mechanical sensor, has a mass with one degree of freedom coupled to a frame (the ground) through a linear restoring force (see Fig. 1). This can be achieved with a spring with restoring force k. A sensor will also have friction which will damp the swinging system. The equation of motion of this system can then be expressed as (Havskov and Alguacil 2010, p. 15)

$$\ddot{z} + 2h\omega_0 \dot{z} + \omega_0^2 z = -\ddot{u},$$

where u is the ground displacement (the displacement of the frame); z the relative displacement between the frame and the mass and $\omega_0 = \sqrt{k/m}$,

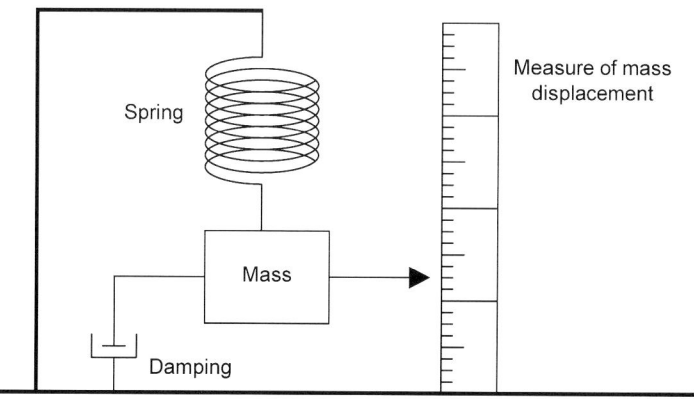

Historical Seismometer, Fig. 1 Schematic illustration of the most simple seismometer, the inertial mechanical seismometer. A mass is held on a spiral spring. The spring connects the mass with the ground and introduces a linear restoring force. The motion of the mass is damped using a "dash pot" so that the mass will not swing excessively near the resonance frequency of the system. A ruler is mounted on the side to measure the motion of the mass relative to the ground (From Havskov and Alguacil 2010)

with $\omega_0 = 2\pi/T_0$; and T_0 is the sensor's natural period. The damping constant is h and it controls how much the system swings. For $h = 1$ (critical damping) there will be no overswing if the mass gets a push (see Fig. 2).

Thus, just two parameters are involved in the solution of the equation, the natural period T_0 and the damping h. If the system is submitted to a sinusoidal ground motion, the equation has an analytical solution, and the ratio between the mass displacement and ground displacement (amplitude response curve for displacement) is (Havskov and Alguacil 2010, p. 16)

$$A(\omega) = \frac{\omega^2}{\sqrt{\left(\omega_0^2 - \omega^2\right)^2 + 4h^2\omega^2\omega_0^2}}$$

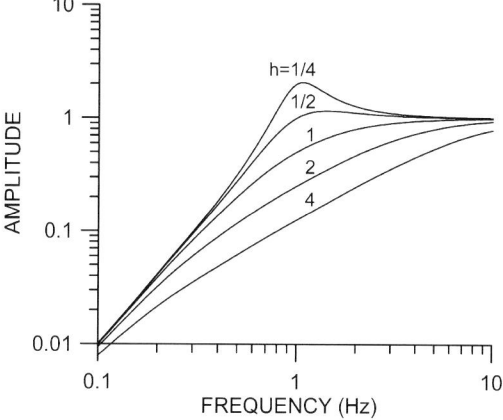

Historical Seismometer, Fig. 2 The amplitude response function $A(\omega)$ for a seismometer with a natural frequency of 1 Hz. Curves for various level of damping h are shown (From Havskov and Alguacil 2010)

From this equation it is seen that for high frequencies $A(\omega) = 1$, so the sensor will measure pure displacement, and $A(\omega)$ diminishes rapidly for lower frequencies, as represented in Fig. 3a. Thus, it is possible to obtain a record (z) reproducing the amplitudes of the ground motion using a sensor with a natural frequency lower than the expected frequencies of the ground motion.

For low frequencies, $A(\omega) = \omega^2/\omega_0^2$ which is proportional to acceleration. Thus, the recording for frequencies lower than the sensor's natural frequency is directly related to the ground motion acceleration (Fig. 3c).

Such principle can be easily applied to the record of vertical and horizontal ground motion. The simplest sensor for vertical motion consists in a mass suspended by a spring (Fig. 1). However, getting a natural frequency of 0.1 Hz would require an unrealistic soft spring combined with a large mass so some tricks are needed; see below. In the case of horizontal motion, it is possible to avoid a spring by using gravity acceleration as the restoring force. The simplest way is

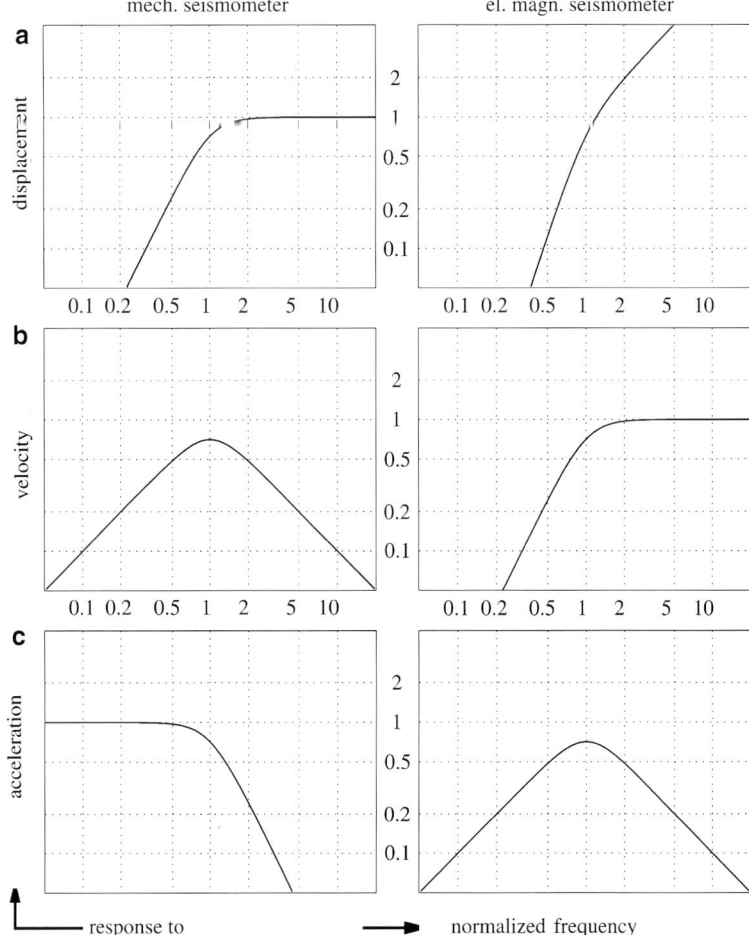

Historical Seismometer, Fig. 3 Response curves of a mechanical seismometer and electromagnetic seismometer with respect to different kinds of input signals: (**a**) displacement, (**b**) velocity, and (**c**) acceleration. The normalized frequency is the signal frequency divided by the natural frequency (corner frequency) of the seismometer. Plots are in bi-logarithmic scale (From Wielandt 2009, Fig. 5.6)

to use a common pendulum (Fig. 4a). In this case $\omega_0 = \sqrt{g/L}$, where g is the gravity acceleration and L the length of the pendulum. A problem arises when long periods are required since this would require a very large L. For a frequency of 0.1 Hz, $L = 25$ m, not very practical. This problem can be solved using the "garden-gate" sensor (Fig. 4b). In this case $\omega_0 = \sqrt{g \sin \alpha / L}$. Theoretically any period can be reached with such a configuration, but in practice, due to friction, it is hard to get a frequency lower than 0.03 Hz.

Electromagnetic sensors were introduced early in the twentieth century (Fig. 5). The output voltage of such a sensor is proportional to ground velocity for frequencies higher than the natural frequency (Fig. 3b). But good voltage recorders for electromagnetic sensors were not available until the second half of the century. Early electromagnetic sensors were connected to galvanometers which record the output current generated by the sensor. Galvanometers have a natural period and damping as sensors so the whole response of the electromagnetic sensor-galvanometer is the result of combining both elements (see Fig. 6).

Seismoscopes: The First Known Instrument

The first known devices used to detect ground motion are not seismographs but seismoscopes. These devices make it possible to observe ground shaking, but they do not give a time history of the

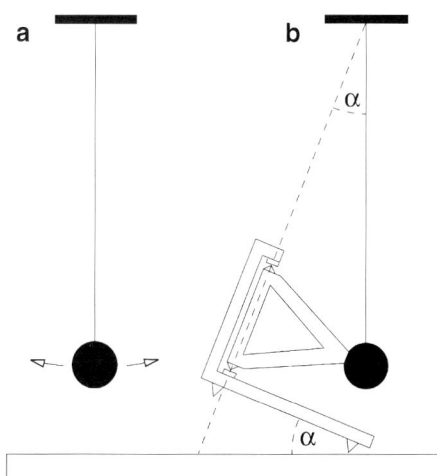

Historical Seismometer, Fig. 4 Schematic illustrations of (**a**) a common pendulum and (**b**) a tilted "garden gate." For a free period of 20 s, the common pendulum must be 100 m long. The tilt angle α of a garden-gate pendulum with the same free period and a length of 30 cm is about 0.2°. The longer the period is made, the less stable it will be under the influence of small tilt changes (Adapted from Wielandt 2009, Fig. 5.8)

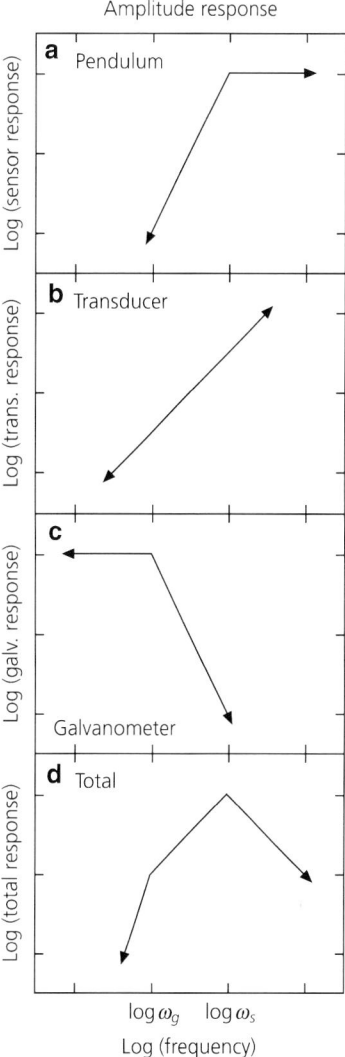

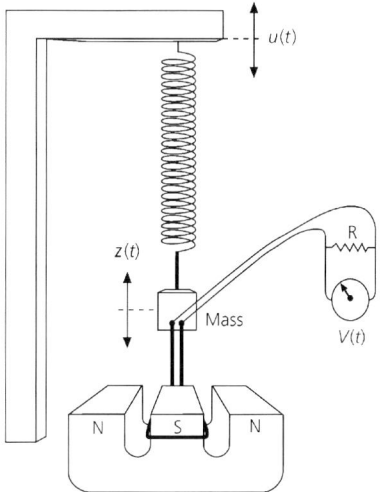

Historical Seismometer, Fig. 5 Schematic illustration of an electromagnetic seismograph, in which the mass is coupled to an electromagnetic transducer. $u(t)$ represents the ground motion and $z(t)$ the relative motion of the mass with respect to the frame. Motion of the coil attached to the mass through the magnetic field generates an electric current (Adapted from Stein and Wysession 2003)

Historical Seismometer, Fig. 6 Schematic illustration of the amplitude responses of the components of an electromagnetic seismograph system using a galvanometer. ω_s represents the sensor's natural frequency and ω_g the galvanometer's natural frequency. The response of the electromagnetic sensor can be decomposed with the response of the mechanical sensor (**a**) and the response of the coil (**b**). The combination of this with the response of the galvanometer (**c**) gives the overall response of the system (**d**). All plots are drawn in bi-logarithmic scale (Adapted from Stein and Wysession 2003)

motion (a seismogram). In its most basic form, they give an indication of ground shaking exceeding a threshold, but they do not give any temporal reference to when it happens or about the real shape of the motion. Others, more sophisticated, may indicate the direction of shaking, time of the event, or maximum amplitude.

The first known device built specifically for the recording of earthquakes is a seismoscope of Chinese origin. It is acknowledged that it was designed by Zhang Heng (78–139) in 132 A.D (Zhang Heng is the name of the Chinese philosopher as written in the Pinyin transcription system. In many old articles, books, etc., it is found as Chang Hêng or also as Choko and Tyoko, Japanese adaptations of his name). It should be noted that this instrument has not been preserved, and what can be seen in books and museums are recreations produced from the descriptions of the instrument given by Fan Ye and others in the fifth century (Hsiao and Yan 2009).

From the translation of these descriptions given by Dewey and Byerly (1969) and Hsiao and Yan (2009), "The instrument was made of bronze and it resembled a wine jar with a diameter of eight chi (a Chinese measure equivalent to 2 m approx.). Inside, there was a pillar in the center and eight transmitting rods near the pillar. It carried out a switch ball and started the mechanism. On the outside of the vessel there were eight dragon-heads, facing the eight principal directions of the compass. Below each of the dragon-heads was a toad, with its mouth opened toward the dragon. The mouth of each dragon held a ball. At the occurrence of an earthquake, one of the eight dragon-mouths would release a ball into the open mouth of the toad situated below. The direction of the shaking determined which of the dragons released its ball and the instrument made a sound (probably the ball beating the toad) to warn the operator." See Fig. 7a for a possible reconstruction. It is known that the instrument was operational because there is record of the detection of, at least, a 400-mile distant earthquake which was not felt at the location of the seismoscope (Hsiao and Yan 2009).

The Heng seismoscope is a noteworthy machine in itself. It has a place in the history of technology and it has been studied by many scholars (notice its place in time with respect to Western history realizing that it was built at the time of Roman emperor Hadrian and just half a century after the Vesuvius eruption that destroyed Pompeii). But it is also noteworthy from the seismological point of view because all studies on its possible detecting mechanism point to the use of the inertia principle, the same that led to the construction of most of the early seismographs. Moreover, it gave an indication not only of the occurrence of an earthquake but also of the direction of the shaking.

Many discussions have taken place about its exact internal mechanism, but a consensus has not been reached. Models with common or inverted pendulums and a model assuming that the oscillating element was the external jar suspended from the pillar have been proposed. Assuming the chronicles are correct, the most puzzling question is that only one ball was delivered and that it pointed in the direction of the epicenter. Thinking on simple mechanisms with pendulums, it looks like more than one ball can be delivered and the pointing direction may be toward or opposite to the epicenter. Anyway, the use of more complex designs, already known at that time, has shown the feasibility of such a mechanism (Fig. 7b–e).

Seismoscopes in Europe: The Eighteenth Century

After Heng's seismoscope, further references to such instruments disappeared from Eastern literature, and it is necessary to jump in time to the eighteenth century in Europe to find new instruments for earthquake recording. The first mention of such an instrument was written by Jean de Hautefeuille (1647–1724) who, in 1703, proposed a device to detect earthquakes. The background idea was very similar to that of Heng. He suggested to fill a bowl to the brim with mercury, so that the shaking produced by an earthquake would cause some of the liquid to spill out. In order to determine the direction of the shock, the mercury spilling out in each of the eight principal directions of the compass was to be collected in cavities (see Fig. 8 for a sketch of the instrument).

Jean de Hautefeuille was not using the inertia principle. His instrument was designed as an inclinometer because he was assuming that, in

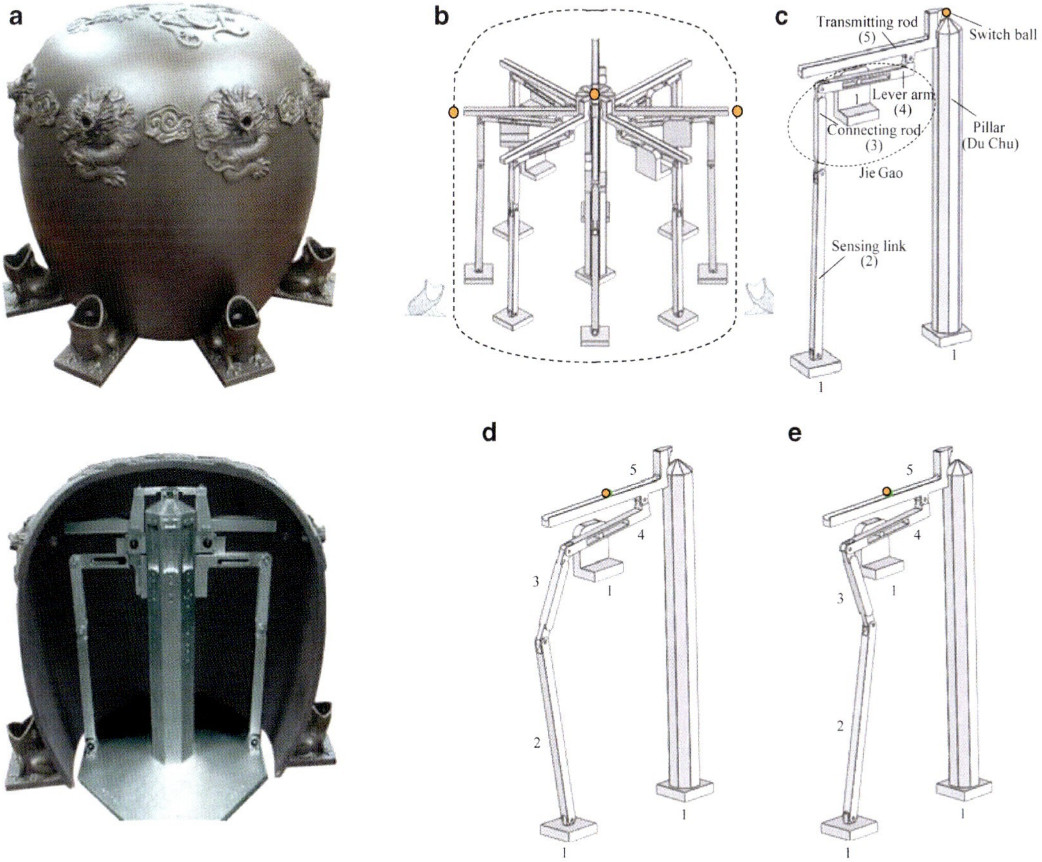

Historical Seismometer, Fig. 7 (a) A possible reconstruction of the outside and inside parts of the Zhang Heng seismoscope. (b) and (c): the position of the eight rods and the detail of one of the rods. (d) and (e): detail showing the feasibility of a mechanism delivering a ball pointing in the direction of the epicenter due to the compression and dilatation of the incident wave (Adapted from Hsiao and Yan (2009))

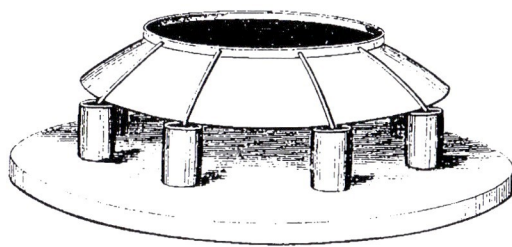

Historical Seismometer, Fig. 8 A simple sketch of the seismoscope designed by de Hautefeuille according to Hörnes (1893). A bowl is filled to the brim with mercury so that the shaking produced by an earthquake would cause some of the liquid to spill out. In order to determine the direction of the shock, the mercury spilling out in each of the eight principal directions of the compass is collected in cavities (just the front four are shown)

an earthquake, the ground would be inclined, and the direction in which the mercury spilled would be away from the origin, "where the ground began to be raised." This is because updated Aristotelian theories about the origin of earthquakes in the eighteenth century suggested underground explosions caused by accumulations of sulfur, bituminous materials, coal, etc., causing dilatation. Thus, this earliest European instrument was expected to detect tilting of the Earth's surface rather than horizontal displacements. Following the author, it could be possible to get an indication of the distance of the epicenter and the size of the disturbance from the amount of mercury which had sloshed out.

But, as far as it is known, this instrument remained as a design and it was never built.

Anyway, earthquake detection was a "hot topic" at that time, and in the following years a lot of references to seismoscopes are encountered in literature. Design of such instruments became popular among Italian scientists on the eighteenth century, and many devices were designed and developed.

Among them, Nicola Cirillo (1671–1735) is the first confirmed constructor and user of seismoscopes. It was the occasion of the Puglia earthquakes of March 1731 that made him construct the instrument. He used common pendulums of small dimensions (the length of the string was a Neapolitan palm, or 26 cm long approx.). With several instruments installed at different places, he was able to deduce that the amplitude of seismic motion attenuates with distance (Ferrari 1992).

Another instrument at that time worth mentioning is the common pendulum designed by Andrea Bina (1724–1792) in 1751. It is almost certainly the first design of a recording instrument in the history of seismology. According to Bina's description, a sphere of lead of considerable weight was to be suspended with a rope from a beam on the ceiling to form a long common pendulum (several meters long). In the lower part of the sphere was a needle one and a half inches long penetrating slightly into a layer of sand placed in a wooden box. In the event of a tremor, the needle would leave marks in the sand that, which depending on their form, width, and depth, would have allowed the observer to draw conclusions as to the intensity, direction, and quality of the seismic movement (Ferrari 1992). As it is seen, this instrument could have given indications on the amplitudes and direction of the ground shaking, but without any timing reference. But, as in the case of de Hautefeuille seismoscope, it looks like this instrument was never built.

In 1796, Ascanio Filomarino (1751–1799) described an instrument which he called a seismograph but in fact was a seismoscope. It consisted of a long pendulum attached to a wall whose spherical mass brought down a pencil that, when shaken, marked a trace on a paper of circular shape (3 in. diameter) placed on a support. In the sphere of iron there were also three bells connected which served as an alarm. Finally, and this is the innovative point, a wire of horsehair blocked the rocker of a clock. When the pendulum swung, the receding hair caused the clock to start, giving an indication on how long ago the earthquake had happened. In this case, the time of the shock is recorded (see a sketch in Fig. 9). It is known that several written records were obtained with this instrument. However, the observations given by Filomarino do not indicate that the timing device was functioning (Dewey and Byerly 1969).

Historical Seismometer, Fig. 9 The seismoscope by Filomarino according to Ferrari (1990). See detailed description on the text

Seismoscopes in the Nineteenth Century

There are references to many more seismoscopes from the middle of the eighteenth century, but it

Historical Seismometer, Fig. 10 The seismoscope of Forbes (1844). See detailed description on the text

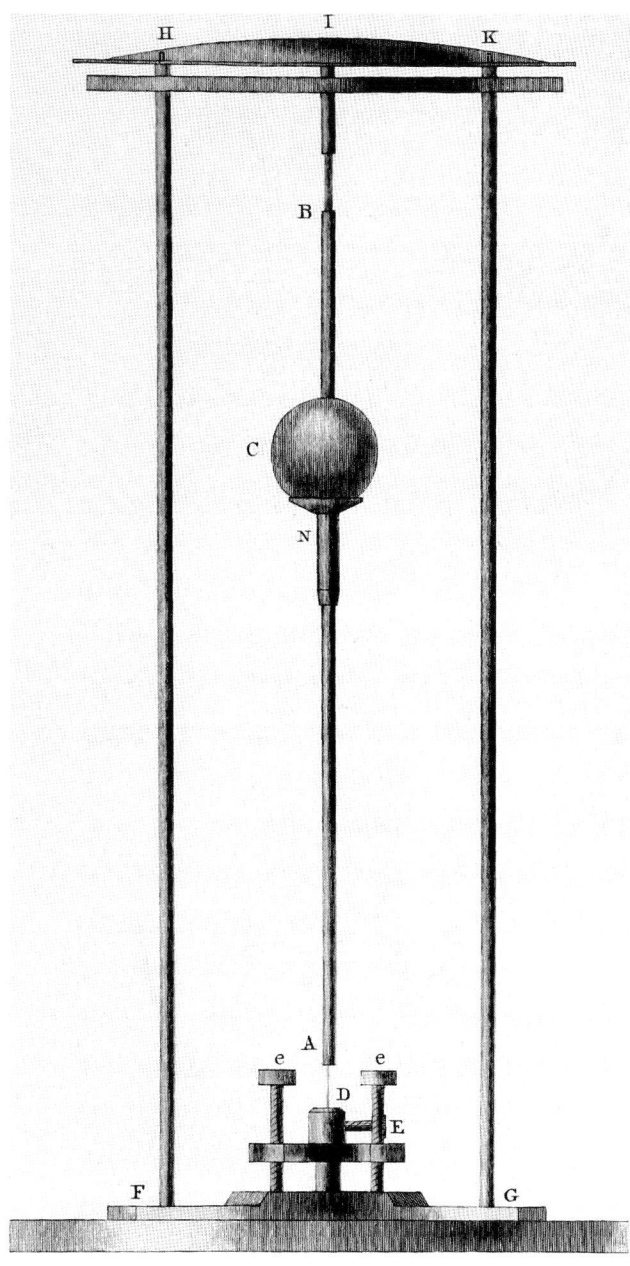

is necessary to wait until the nineteenth century to find the first improved recording device. The most interesting instrument was designed by Scottish James D. Forbes (1809–1868). Due to a series of small earthquakes felt near Comrie, Scotland, a special committee was appointed by the British Association for the Advancement of Science being one of its goals to get instruments to record the shocks. Forbes built an inverted pendulum (see Fig. 10). Using the same words of Forbes (1844), "a vertical metal-rod AB, having a ball of lead C moveable upon it, is supported upon a cylindrical steel-wire D, which is capable of being made more or less stiff by pinching it at a shorter or greater length by means of the screw E [...] The self-registering part of the apparatus [...] consists of a spherical segment HIK of copper lined with paper, against which a pencil L,

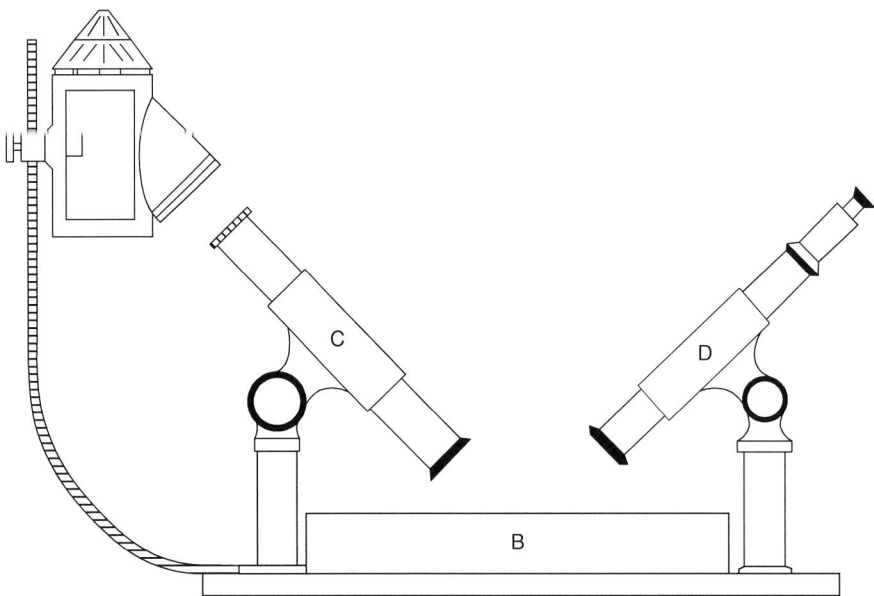

Historical Seismometer, Fig. 11 A sketch of the seismoscope of Mallet. The image of a crosshair in *C* is reflected from the surface of mercury in the basin *B* and viewed through a magnifier, *D* (From Dewey and Byerly 1969)

inserted in the top of the pendulum-rod, is gently pressed by a spiral spring. The marks thus traced on the concave surface indicate at once the direction and maximum extent of the pendulum's vibration."

He used deliberately the inverted pendulum because he knew this was an easy way to obtain long free periods with small dimensions. It is worth pointing out that Forbes was probably the first person to study, from a theoretical point of view and with the aim to calculate its response to ground motion, the effect of an earthquake on a pendulum. But Forbes assumed that an earthquake was just a lateral displacement of the ground with constant speed, and therefore, the results did not give him valuable information about the best settings for his instrument. As Dewey and Byerly (1969) pointed out, Forbes "desired a long period in order that the pendulum remained stationary as the Earth moved beneath it. He clearly wanted to measure ground displacement in an earthquake. However, [...] we can't be sure if Forbes knew that a long-period pendulum would function as a displacement meter for very-short-period oscillations of the ground."

Another interesting point about Forbes pendulum was his deliberate use of "magnification" through the increase of the distance of the writing pencil to the mass. It is the first time a reference to this issue is found. Anyway, the real performance of the instrument was somehow disappointing because it just recorded three of the sixty earthquakes felt. From our present knowledge, it is likely to assume that friction was too high.

During the same years, Robert Mallet (1810–1881), trying to calculate the velocity of seismic waves, used a totally different seismoscope consisting of a mercury container and an eleven-power magnifier. Mallet looked through at the image of a crosshair reflected at the surface of mercury (see Fig. 11). This is a sophisticated version of the "pan" used at the Exeter siege and specifically designed to measure a propagation time, there being no interest in any estimate of the amplitude of the ground motion. In the following years this instrument was to be improved with the addition of photographic recording transforming it into an almost real seismograph without using the inertia principle.

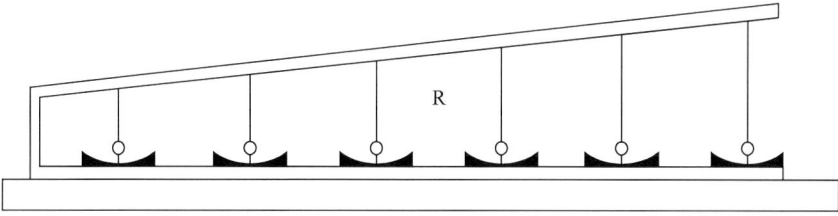

Historical Seismometer, Fig. 12 Seismoscope by Cavalleri. He studied the frequency contents of seismic waves by comparing the amplitudes of oscillation of the different pendulums with different periods (Adapted from Cavalleri 1858)

In Italy, Giovanni M. Cavalleri (1807–1874) designed and built several seismoscopes. Cavalleri constructed an instrument with six short pendulums of different periods, each of which traced the record of its motion in fine powder (Fig. 12). He used them to study the frequency contents of earthquakes and concluded (wrongly) that a range of frequencies from two to four cycles per second would be "sufficient to embrace every undulation occasioned by any earthquake." Anyway, the Cavalleri studies showed that the ground motion during an earthquake is a composition of oscillatory motions with different frequencies, and it is the first attempt to make this kind of analysis.

It is also around these years that the first devices to record vertical ground motion are made and that electrical circuits are used for the detection of ground motion. The most conspicuous of such instruments was the "Sismografo Elettromagnetico" designed by Luigi Palmieri (1807–1896) in 1856. This instrument, really remarkable at that time, was able to assert time, direction, intensity, and duration of an earthquake. It was not yet a seismograph, as its name may suggest, but a collection of seismoscopes, each intended to record particular parameters of an earthquake (Fig. 13). Palmieri's instrument seems to have been an effective earthquake detector for local shocks. From the historical records, it looks like it was more sensitive to earthquakes than humans, being the first instrument to clearly attaint this goal. Copies of this instrument reached Japan and were in use (and are still preserved) in Tokyo, at least from 1875 to 1885.

It is difficult to describe with words all the complexities of the amplitude indicators and recorders of this instrument, not the least because the existing copies are different. Of specific interest, the seismoscope for detecting vertical motion consisted of a conical mass hanging on a spiral spring. The mass was suspended just over a basin of mercury. When a slight motion caused the tip of the cone to touch the mercury, an electric circuit was completed, which caused a clock to stop, indicating the time of the shock. Horizontal motion was detected with common pendulums, whose swinging completed the same circuit as that completed by the mass-spring seismoscope. In this case, a circular narrow channel filled with mercury was placed around the bottom of the oscillating mass. It was possible to reduce the distance between the mass at rest and the mercury to tens of millimeter.

These types of seismoscopes had a high sensitivity, and their design was copied in a large number of further starting/stopping devices for recording apparatus as it will be mentioned below.

Already at the end of the century, in 1895, Giovanni Agamennone (1858–1949) introduced his double-action electric seismoscope (Fig. 14). This instrument can be considered as the final synthesis of the Italian know-how on seismoscopes, closing its evolution. It uses the oscillatory character of the earthquake ground motion in an elegant form. Two inverted pendulums of different periods are set aside each other. Their suspensions bars, connected to the poles of a battery, have an electric contact at the top. When at rest, the contact is open through a gap of few tens of a millimeter. Any slight motion of the ground causes the pendulums to oscillate (at different frequencies) and, thus, to complete the electric circuit. A bell rings and a clock, or

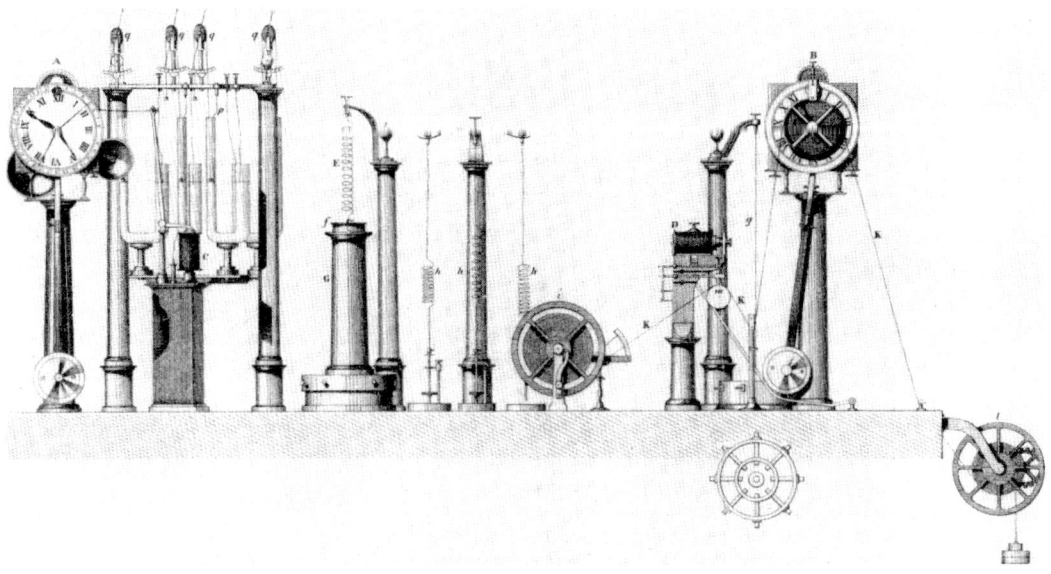

Historical Seismometer, Fig. 13 Palmieri seismograph. The spiral spring (*E*) was used to detect vertical motion. Horizontal motions along different axes were detected and recorded electrically by the movement of mercury in the U-tubes (on the *left* side). The closing of an electric circuit activated electromagnets (*C* and *D*), which caused the clock to be stopped, thus recording the time of the arrival of the seismic waves, and the rotation of the drum (*i*) was started, with paper being unwound and marked at *m* (From Palmieri 1873)

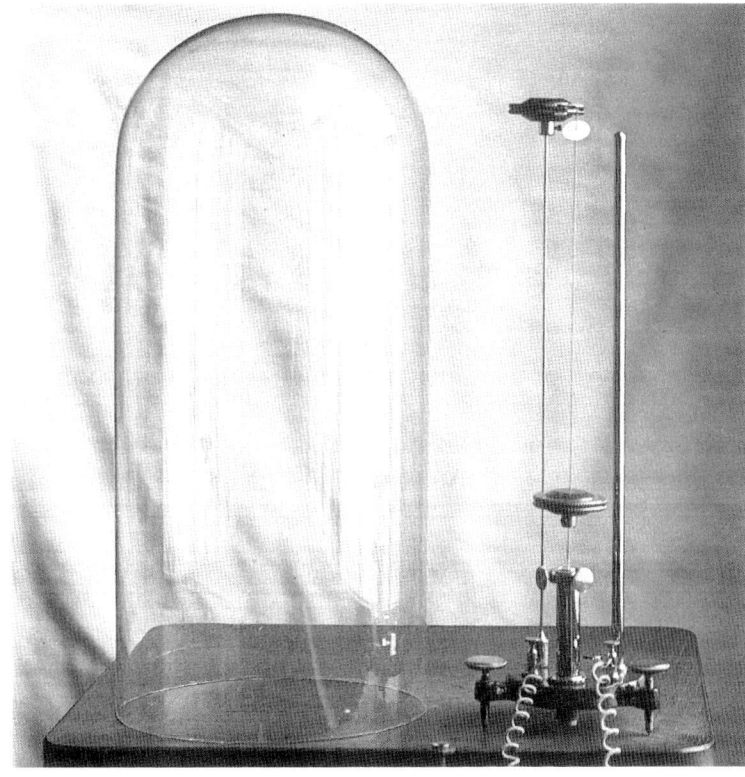

Historical Seismometer, Fig. 14 Agamennone double-action electric seismoscope. Two inverted pendulums, with different free period, are set aside. The close proximity of the two upper extremities of the pendulums had the effect, in the case of a tremor, of making them come into contact with one another, completing an electric circuit (From Ferrari 1992)

The First Seismograph

A fact ignored presently is that by the middle of the nineteenth century, seismoscopes became widely distributed around the world. Its presence and use other than in Italy and the UK is documented in Cincinnati, USA; on the occasion of the New Madrid earthquakes of 1811 and 1812; in Tabriz, Iran, for the October 4, 1856, Persian earthquake; in Manila, Philippines, for the June 20, 1857, event; or in several places in Germany for the June 24, 1877, Herzogenrath earthquake.

But seismoscopes do not give much information about the nature of earthquakes, and this is the most probable cause for the present low relevance of seismoscopic observations. The characteristics of ground shaking during earthquakes continued to be unknown as well as the characteristics of wave propagation at far distances. From our present knowledge about earthquake recording, it is possible to infer that sensors (seismometers) were becoming adequate for their task, but recorders were not. The lack of earthquake recordings (amplitude as a function of time, a seismogram) made it difficult to further study the earthquake phenomena.

But around 1875 Filippo Cecchi (1822–1877) built what it is now considered the first seismograph. However at the time it passed quite unnoticed. Unlike any of the instruments discussed up to this point, the Cecchi seismograph was expected to record the relative motion of a pendulum with respect to the Earth as a function of time. It was based in the inertia principle, and as described by Dewey and Byerly (1969), two common pendulums were used to record horizontal vibrations, each one oscillating in a unique direction and placed perpendicular to each other. Their free period was 1 Hz and the ground motion was magnified three times by a thread-and-pulley apparatus. A mass held on a spiral spring, as shown in Fig. 1, was used for the recording of the vertical motion. A device designed to record rotary motions was also incorporated into the seismograph. It consisted of a crossbar with weights at both ends, much like a dumbbell, which was pivoted at its center of mass so as to rotate in a horizontal plane. Restoring force was applied to the dumbbell by springs so that it oscillated with a period of one second (Fig. 15). A more detailed description can be found in Ferrari (2006).

The instrument was not designed for continuous recording and Cecchi arranged a seismoscope to start a clock and to put into motion the recording surface, a smoked glass plate, at the time of an earthquake. The recording surface would move under the needles at a speed of one centimeter per second for 20 s. From the time on the clock, an observer arriving at the seismograph would determine how long before his arrival the earthquake had occurred.

It looks like several similar instruments were installed at Italian observatories and, certainly, one in Manila, Philippines, but the first confirmed record of this instrument was on February 23, 1887, at the occasion of the large earthquake that occurred near Nice, France (Fig. 16). This points to an insensitive instrument and explains the reason why such an achievement passed almost unnoticed. Our present interpretation is that the friction of the recording system was too high and, most likely, the starting seismoscope was not sensitive enough.

At the same time, other developments important for the future of seismometry were taking place; among them were experimental and theoretical studies about horizontal pendulums. Researchers needing to detect small gravity variations on laboratory and those working in geodesy studied the properties and behavior of horizontal pendulums (its specific characteristic was the way to hold the oscillating mass, as can be seen in Fig. 20-3) and he also developed the theory of such an instrument. A mirror was attached to the pendulum thereby magnifying its motion using a light beam reflected on it.

Another important theoretical development was done by Perry and Ayrton (1879), both British professors working at Japan. They published a paper entitled *On a neglected Principle that*

Historical Seismometer, Fig. 15 Cecchi seismograph. Restoration of the original instrument presented at Genoa in 2000. In the front, the two common pendulums are seen. Among them, the elongated box is the place where recording plates were fixed (four plates for the three translation components and the rotational one). The rotational sensor is seen just behind to the right of the common pendulum (Photograph by the author)

may be employed in Earthquake Measurements. In this paper the authors analyzed the motion of damped and undamped mass-spring oscillators subjected to a periodic force. Now the importance of this contribution is clear, but it was almost ignored at the time of publication.

Developments in Japan

In its efforts to modernize the science and technology of the country, Japan hired, in addition to Perry and Ayrton, many foreign professors in the last quarter of the nineteenth century. Among them were John Milne (1849–1913), Alfred Ewing (1855–1935), and Thomas Gray (1850–1908). By living in one of the most earthquake-prone countries of the world, they were encouraged to study how to record earthquakes. The three, with solid engineering background (Milne served as professor of mining and geology, Ewing of mechanical engineering, and Gray of telegraph Engineering), were in an optimum position to solve the problem of how to record ground motion; however, they were more handicapped by technical problems than by theoretical ones.

Few weeks after the earthquake on February 22, 1880, at Yokohama, and following a proposal by Milne, a seismological society was founded in Japan. Since the three were already involved in earthquake studies and have an engineering background, they gave priority to solving the technical problems of ground motion recording.

They studied different possible types of seismometers and recorders. It is known from their writings that they were looking for instruments with a free period larger than the periods of the signals they wanted to record, taking full advantage of the inertia of pendulums. The first

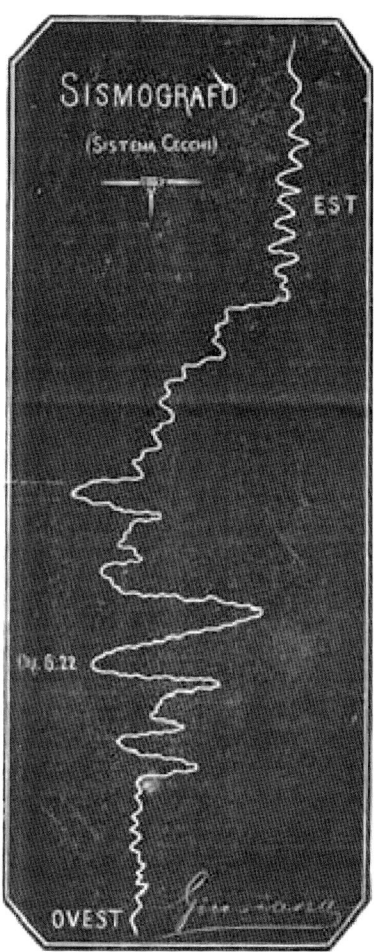

Historical Seismometer, Fig. 16 Cecchi seismograph record of the 1887 Ligurian earthquake. This record was obtained from one of the glass recording plates attached to one side of the elongated box described in Fig. 15 (From Ferrari 2006)

instrument of their long production, built by Ewing, was a common pendulum of 5 s period (approximately 7 m long) because they thought this period was enough for their goal. A simple lever whose fulcrum was attached to the ground gave a six times magnification and recorded on a continuously revolving, circular smoked glass plate. The motion of the pendulum was resolved into two perpendicular components by the recording apparatus (two perpendicular levers were used). Not properly acknowledged up to now, it should be pointed that the great innovation of this instrument was the recording system, giving a long continuous record, and that it was applied successively to other models.

Almost in parallel Ewing tried recording earthquakes with a horizontal pendulum. His design was really successful, being the first to do so in the field of seismology. With small dimensions, the horizontal Ewing pendulum consisted (see Fig. 17) of a light rigid frame c pivoted at points e so as to swing like a garden gate around that axis of rotation. The cylindrical mass a is pivoted at the axis of percussion of the frame b. Again he used the revolving smoked glass and six times magnification. The weight of the oscillating mass was less than one kg.

Both of Ewing's instruments, the common-pendulum and horizontal-pendulum seismographs, recorded a small earthquake on November 3, 1880, giving the first lengthy seismic records of earthquake motion as a function of time. Finally, it was possible to have an idea of

Historical Seismometer, Fig. 17 Drawing of the Ewing horizontal pendulum (From Ewing 1881)

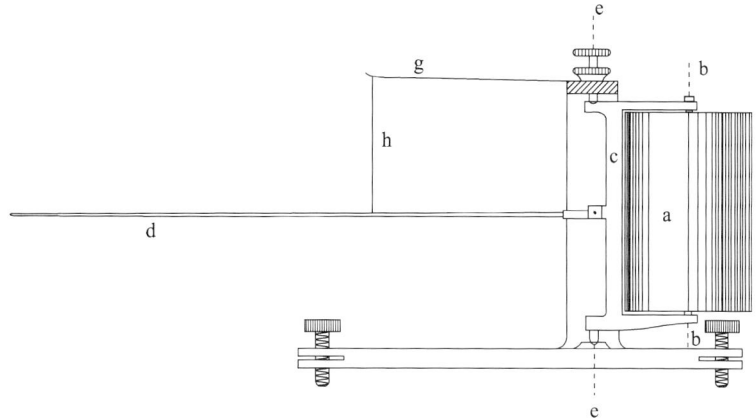

the ground motion during an earthquake, and for the first time, seismologists could design their instruments with some knowledge of the phenomena the instruments were to record.

But an instrument for the vertical motion with a similar long period as those designed for the horizontal ones was still elusive. The solution came when Gray introduced a method of increasing the period of a mass-spring system by attaching the spiral spring to the short arm of a lever and attaching the mass to the long arm of the lever. This increased the period to few seconds by increasing the effective mass of the pendulum bob (see Figs. 20-8 and 9). Thus, even with many drawbacks with respect to our present knowledge (not a damped instrument, large dry frictions), a three-component instrument was now available.

Recording Earthquakes Globally: Teleseismic Recording

Another step was needed for the seismograph to be recognized as the key instrument not only for the study of earthquakes but also for the study of the Earth's interior. This was initiated by Ernst von Rebeur-Paschwitz (1861–1895). His main research area was geodesy and, specifically, the lateral deformations of the Earth's crust. To measure the deformations he built sensitive horizontal pendulums to measure slight changes in the direction of the vertical gravity. On April 17, 1889, two of these instruments, located in Potsdam and Wilhelmshaven, recorded simultaneously an anomalous signal that, from the newspapers information, he related to a large earthquake which had occurred in Japan.

He published a note about this observation in *Nature* that has become one of the milestones of observational seismology (Rebeur-Paschwitz 1889). The accompanying published seismic records have been reproduced many times (here also, see Fig. 18). The main result of his discovery was that, for the first time, the possibility to record distant earthquakes was confirmed. This possibility had already been proposed by others. The observation also confirmed that the whole Earth is elastic and that therefore elastic wave theory can be used to analyze the behavior of seismic waves. Looking at the record, a curvature on the baseline can be seen. This curvature corresponds to the deformation caused by earth tides. The instrument was a tiltmeter but it recorded earthquakes, as gravimeters do.

The instrument consisted of a heavy base containing, inside, a horizontal pendulum sitting on a rigid frame, rotating around two bearings at the top and the bottom, each consisting of a point pressed into a socket (Figs. 19 and 20-2). He used a photographic recording. It was obtained through a mirror attached to the oscillating frame, which reflected light from a lamp through a cylindrical lens to a rotating drum which was covered with photographic paper. The drum turned 11 mm in an hour. As this system reduces friction significantly, the pendulum was only 10 cm long and carried a mass of only 42 g. It was usually used with free periods from 12 to 17 s and a static magnification of 100 (Dewey and Byerly 1969). A time reference was obtained with a second fixed light trace which wrote on the same photographic paper. Every hour, this second trace was eclipsed for 5 min. The magnification and the free period of Rebeur's pendulum were 20 times bigger than Ewing's pendulum. This instrument can be considered as the first approximation to what was needed to record teleseismic events.

The original instruments were, for many years, assumed lost, but recently original frames and other parts have been found by Fréchet and Rivera (2012).

The Seismograph Evolves

It is possible to identify, in the last quarter of the nineteenth century, three main groups or "schools" of seismograph designers and even to assign nationality to each one: the Italians, with their already long tradition of experimentation, in which Palmieri and Cecchi can be their best heralds; the Anglo-Japanese, with the Milne-Ewing-Gray trio, a rapidly progressing group characterized by its good equilibrium between the

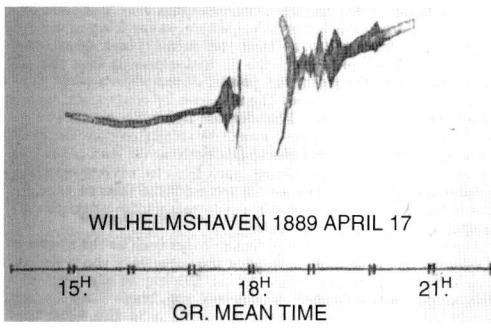

Historical Seismometer, Fig. 18 Rebeur-Paschwitz seismogram of the 1889 Japanese earthquake (From Rebeur-Paschwitz 1889)

Historical Seismometer, Fig. 19 Photography of the original Rebeur-Paschwitz pendulum. The top bearing is clearly seen. Also seen is a hole behind the bearing for the light beam used for photographic recording (From Fréchet and Rivera 2012)

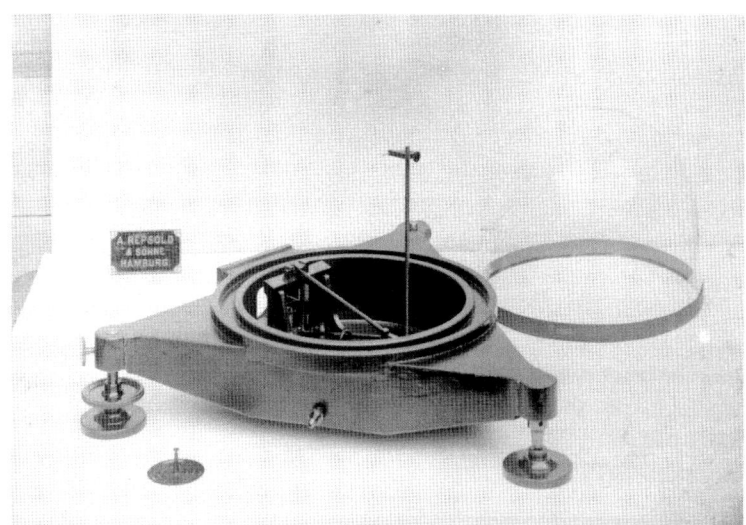

Historical Seismometer, Fig. 20 Different pendulum settings used for earthquake recording. It is possible to establish a seismograph classification according to the mechanical constitution of the transducer: (*1*) conical or bifilar pendulum, (*2*) rigid pendulum, (*3*) Zöllner pendulum, (*4*) vertical pendulum, (*5*) torsion pendulum, (*6*) inverted pendulum, (*7*) flexure pendulum, (*8*) vertical oscillator, (*9*) vertical rotational oscillator (From Batlló 2004)

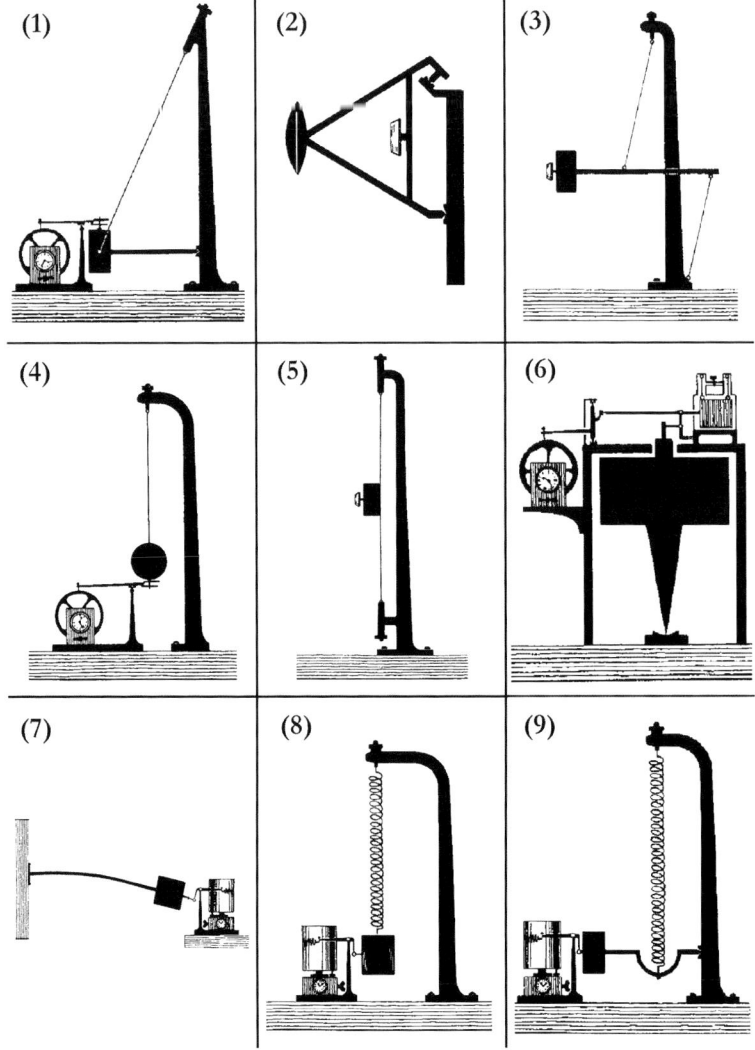

theoretical and experimental approaches to the topic; and finally, with a really poor experience of earthquakes due to the low seismicity of its country, the Germans, Rebeur-Paschwitz at the front, with a geodetic background and a large experience in detecting very small ground motions at low frequencies.

The approach of each group to the seismic recording problem was almost independent and quite different. But when, within few years, the three arrived to obtain their first valuable records, the information among the different groups traveled fast, and the new technical procedures of one group were really soon implemented by the others.

Although other principles were tested, it is clear that all designers adopted the inertia principle for the design of the main instrument. All kinds of pendulums were tested (see Fig. 20).

Around that time (probably of Italian invention) the classical analog continuous seismic record consisting in a spiral line plotted on a paper placed on a drum was introduced. The lateral translation of a drum was obtained by just substituting its rotational axis with a screw. It was now possible to have a 24 h record on a unique sheet of paper with manageable dimensions.

In 1895, Giuseppe Vicentini (1860–1944) and Giulio Pacher (1867–1900) introduced a new

Historical Seismometer, Fig. 21 Vicentini seismograph, with horizontal and vertical components as installed in the Alicante Observatory, Spain (Original photograph from the IGN archive)

common pendulum specially designed to record near earthquakes. Now it was quite clear that signals of near earthquakes contained higher frequencies than those from teleseismic events. For this reason they designed an instrument optimized for near-earthquake recording. A common pendulum of 100 kg with a free period 2.3 s recorded on smoked paper with a speed of 15 mm/min. The motion of the pendulum was amplified and resolved into perpendicular components by levers. Static magnification was set to 80. The recording was on smoked paper, and following a trend common to Italian instruments at that time, they used large masses to overcome the frictions of the amplifying and inscription mechanisms. The good results obtained made them add a vertical component. It was quite original in design and consisted of a flat spring clamped to a wall from one side and with the oscillating mass (50 kg) at the other end (Fig. 20-7). The free period was 1 s approx. and magnification was 130. The resulting instrument was known as Vicentini seismograph and it was used at many places for near-earthquake recording up to the middle of the twentieth century when it was finally superseded by electromagnetic short-period seismographs (see Fig. 21). The three components were written side by side on the same continuous record.

Milne, on coming back to England, centered his interest on global seismicity. For this objective he used the horizontal pendulum he designed in Japan in 1894 (Fig. 22). In this instrument he adopted the photographic record and, thus, he combined it with a light oscillating mass (less than half kilograms). The free period was set near 16–18 s and magnification was 6–9. He adopted also continuous recording, but did not adopt the laterally translating drum (he did in later versions of the instrument). For this reason the early records were long rolls of photographic paper. Paper speed was just 1 mm/min and one roll of 10 m lasted 1 week. The instrument, thus, was not convenient for immediate analysis of seismograms. Even not being an instrument of outstanding performance, it is really important in the history of seismology because, due to a proposal of Milne, it was selected by the British Association for the Advancement of Science to be deployed around the world (and, effectively, it was) in a first attempt of making a worldwide network.

By the end of the nineteenth century, the British professors in Japan returned to Europe and the Japanese took full responsibility for continuing the outstanding seismological research in Japan. Fusakichi Omori (1868–1923), a collaborator of Milne, continued the design of horizontal

Historical Seismometer, Fig. 22 Milne seismograph installed at the Toledo Observatory, Spain. This instrument has a drum recording instead of a roll, which was common in the early instruments (From Batllo 2004)

Historical Seismometer, Fig. 23 Bosch-Omori seismograph single component (the most common set was a pair recording in perpendicular horizontal directions) installed at the Toledo Observatory, Spain (From Batllo 2004)

pendulums. He turned to continuous recording on smoked paper and, as needed in this case, heavier masses (10 or more kg). In its most common form, Omori's pendulum consisted of a mass on a rod, pivoting about a socket, with the mass held up by a flexible wire. With a free period of about 20 s, static magnifications about 10–15, and paper speed of 5/10 mm/min, they were rough instruments of easy use and maintenance. These qualities make them popular and were copied elsewhere. Still in the middle of the twentieth century, they were common in secondary seismic stations. The Bosch factory, in Strasbourg, commercialized a popular version under the name "Bosch-Omori" pendulums (Fig. 23). The worldwide distribution, relatively long period, and adequate paper speed make their records still valuable for the study of large earthquakes which occurred in the early years of the twentieth century (Batlló et al. 2004, 2008).

Wiechert Inverted Pendulum

Emile Wiechert (1861–1928) began his experiments with earthquake recording in 1898, at the just created Institute for Geophysics at the University of Göttingen, with a pendulum with photographic recording and viscous damping. He introduced the damping to lessen the effects of the pendulum's free oscillations and he was the first to introduce such a device in a seismograph. But after a scientific round trip to Italian observatories, he developed already in 1900 a totally new instrument, his worldwide famous astatic pendulum. The new instrument combined characteristics of the instruments already in use in Germany and Italy and some "premiers" to be attributed to Wiechert himself. After some years of experimentation and improvements, Wiechert published it in 1904.

The first "premier" of this instrument was (recovering the idea of Forbes) the use of an inverted pendulum stabilized by springs and free to oscillate in any direction horizontally (Wiechert 1904). In this case the restoring springs were applied to the top of the inertial mass. A joint (of cardan type) at the base of the pendulum permitted motion in any horizontal direction. The 1904 pendulum (Fig. 24) had an oscillating mass M of 1,000 kg. Using such a large mass made it possible to overcome dry frictions. The free period of the pendulum was around 10 s. The relative motion of the mass with respect to the ground was resolved into perpendicular components, magnified 200 times by a mechanical lever system (H, OS, and S), and written on a smoked paper for continuous record P. Paper speed was between 10 and 15 mm/min. Time marks were put on the paper every minute by lifting the writing index off of the paper. The viscous damping was obtained by connecting the motion of the pendulum mass to pistons which fit closely inside cylinders D attached to the seismometer stand. Resistance of the air to the piston motion provided damping for the pendulum; this resistance was controlled by a valve which had the effect of regulating the amount of air space between the piston and the cylinder. The lever system and recording are indebted to Italian technology

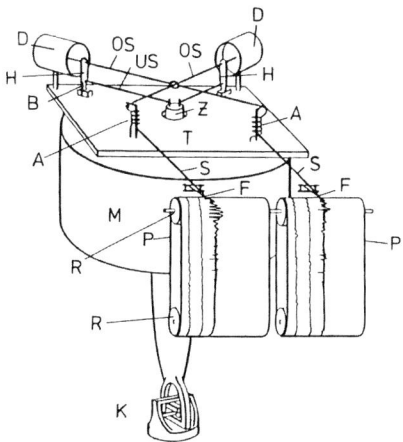

Historical Seismometer, Fig. 24 General sketch of the horizontal Wiechert astatic seismograph. See description in the text (From Ferrari 1992)

while the free mass period and high (at that time) amplification reflected German experience and the air damping device was another premier due to Wiechert. The damping device was the most important innovation because its behavior approaches the expected mathematically modeled viscous damping.

Last but not the least, in parallel with the instrument development, Wiechert studied carefully the theory of the motion of pendulums submitted to ground oscillatory motion and as a result he developed the theory of the response of mechanical seismographs to ground motion. The presentation of this problem in actual textbooks is very similar to the description by Wiechert.

Finally, in 1905 a vertical component complemented the astatic horizontal pendulum. Using as much as possible the sound design of the horizontal instrument, the overall resulting instrument was much more conventional, the large mass being held with springs, and it was impossible to obtain a free period larger than 5 s.

The Wiechert astatic instrument became a reference instrument for many years. Several manufacturers commercialized different versions of the instrument, and by the time of the second World War, more than 100 seismic stations around the world were using these kinds of instruments. Several are still in operation used as

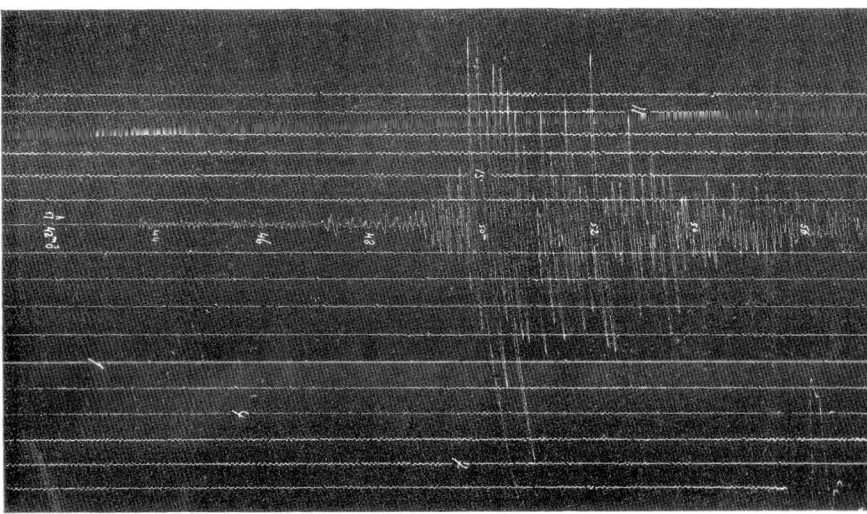

Historical Seismometer, Fig. 25 Record of the earthquake that occurred on April 23, 1909, near Lisbon, Portugal, as recorded in the N-S component of the Wiechert seismograph installed at Munich, Germany, at 2,000 km from the epicenter. P and S waves arrival and surface waves are clearly identified (From Batlló et al. 2008)

reference instruments. Figure 25 shows an example of a typical record obtained with a Wiechert instrument. P and S waves are clearly identified and the record is not different for our present seismograms.

The Electromagnetic Seismograph

Boris B. Galitzin (1862–1916), in charge of the Pulkovo seismic station, was, as Wiechert, deeply interested in the improvement of seismic recording. In 1905 he came with a totally new instrument, the electromagnetic seismograph (Fig. 26). The new seismograph used, instead of mechanical transducers amplifying the ground motion with levers, an electromagnetic transducer. To do so coils attached to the oscillating pendulum were moving through a magnetic field created by permanent magnets. Ground motion was thus converted to an electric current and optically recorded, with the help of a galvanometer, in a photographic paper introducing, thus, a second oscillating element. Viscous damping was obtained magnetically, with a sheet of copper attached to the oscillating mass moving in the magnetic field created by a second pair of magnets. The horizontal pendulums (one for each component) used the Zöllner suspension (Fig. 20-3) and the vertical instrument used the suspension system introduced by Ewing in Japan (Fig. 20-9). Their masses were few kilograms. Its design allowed to separate the "sensing" device from the "recording" device. From then it is called the seismometer-galvanometer system.

The new instrument was the fruit of careful experimentation but also, as in the case of the Wiechert instrument, a result of theoretical study. The horizontal pendulum's free period was set around 24 s for the horizontal and 12 s for the vertical components. The galvanometer's free period was set to the same frequency as the seismometers and both the seismometer and galvanometer had critical damping. It was then possible to get an analytical solution to the equation of motion for the whole coupled system seismometer-galvanometer. The peak magnification was more than one thousand times and a typical paper speed was 30 mm/min. Another advantage was that by using parallel and serial coupled resistances, both the damping and the gain of the instrument could be regulated easily.

The new instrument was much more sensitive than any previous instrument and superior to them in almost all aspects, except complexity of recording and price. At that time photographic

Historical Seismometer, Fig. 26 Group of Galitzin electromagnetic seismometers installed at the Toledo Observatory, Spain. The two perpendicular horizontal components are placed to the left and the vertical component to the *right*. The size of the magnets is notorious (From Batlló 2004)

recording was not affordable to any seismic station. While a standard mechanical seismograph (if properly damped) has a flat response to displacement for frequencies above the natural frequency, the transfer function of the seismometer-galvanometer system is no longer flat (see Fig. 6) making it a bit more difficult to analyze the seismograms. The available magnets suffered from demagnetization, and thus, the overall magnification of the system changed. For these reasons, their use expanded slowly.

From Teleseisms to Near Earthquakes

Wiechert astatic seismograph and Galitzin's electromagnetic instruments gave a good solution to the recording of teleseisms. They were for many years the standard instruments of the most important seismic stations around the world, complemented at secondary stations with a variety of instruments, many of them horizontal pendulums of the Omori type, as the Mainka pendulums (Fig. 27), with nominal characteristics similar to the Wiechert instruments but of easier maintenance. But overall the quality of the mechanical parts of these cheaper instruments was not as good as those of the original Wiechert (and, as pointed by McComb (1936), performance of damping oil systems as viscous damping was not satisfactory) lowering the quality of the obtained seismograms.

Nevertheless, the studies on the structure of the Earth progressed fast. But recording of small earthquakes in the near field was still deficient. To get a high amplification at frequencies near 1 Hz and higher was not an easy task. Some solutions were proposed. To improve the mechanical transducers (the main problem was to overcome the friction), large masses were used. Italian designers recognized this problem already at the end of the nineteenth century. Wiechert himself moved this way constructing a short-period Wiechert-type instrument with 17 t. Later, in 1922, the Quervain-Piccard mechanical seismograph was built in Switzerland. A mass of 21 t was "floating" on springs, and separate recording in three directions was obtained by levers. The static magnification was 1,700 and the free period was 3 s for the horizontal components and 1 s for the vertical component.

At the same time a totally different solution was adopted in California. The result was the famous Wood-Anderson instrument (Fig. 28). It takes advantage of a torsion suspension (Fig. 20-5) for the pendulum (Anderson and Wood 1925). A small mass is enough to obtain

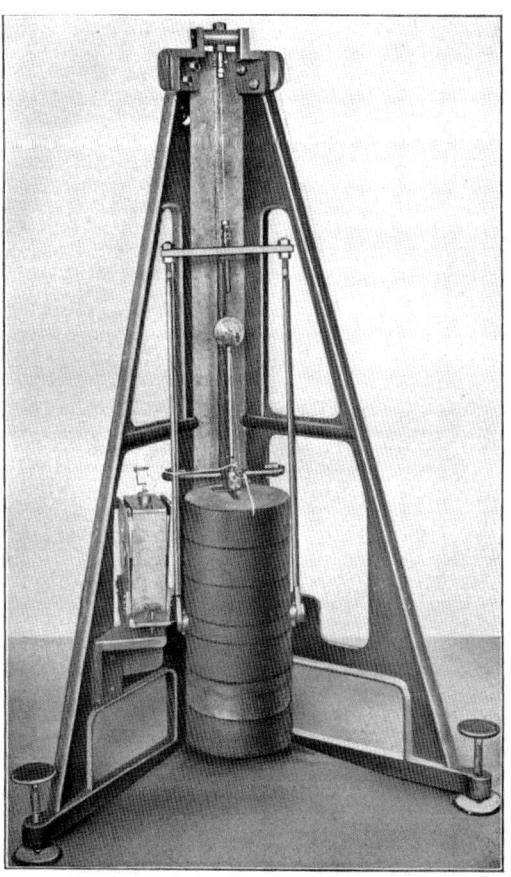

Historical Seismometer, Fig. 27 Front view of the Mainka horizontal pendulum. The recording system has been removed to show the oscillating mass and the structure holding it. Mass is held at the top of the structure and at a second point (not seen), just behind the axis at its center, as in the Bosch-Omori instrument (From Batllo 2004)

high gain when, to avoid friction, it is combined with magnetic damping and photographic recording. In this case, the most cumbersome element is the photographic recording. With a mass of less than 1 g, it has a free period of 0.8 s and magnifications larger than 2,000 (the nominal magnification of the commercial instruments was 2,800 even though further studies by Urhammer and Collins 1990 showed it is smaller). Unfortunately, a vertical component was never produced even though its construction does not offer any further technical complication.

Benioff Reluctance Seismometer

Further research of the ways to increase seismometer sensitivity to be able to detect smaller signals got Hugo Benioff (1899–1968) to introduce, in 1930, the variable reluctance seismometer (Benioff 1955). Its principle is similar to the electromagnetic sensor, but instead of a moving coil in a magnetic field, the transducer is of the same type of those used in telephone receivers. A moving magnet supplies flux across two or more air gaps to an armature around which is wound a coil of wire. Changing air gaps, caused by the motion of the magnet due to ground shaking, generates the current in the coils. The device generates much stronger signals than the classical electromagnetic sensor. Using photographic recording it then becomes easy to get amplifications of 100,000 at 1 Hz. Horizontal and vertical seismometers were produced and they became later, almost unmodified, the short-period equipment of the WWSSN (Fig. 29).

The sensitive transducer permitted to use some recording principles previously abandoned because of the lack of adequate technology. This was the case of the strainmeter developed also by Benioff in the 1930s. The strainmeter dismissed the cherished principle of pendulum inertia and measures the variation of distance between two points at the passing of seismic waves. The new variable reluctance transducer could be used to properly measure this distance and thus to use the strainmeter as seismometer.

Strong Motion Measurement

Earthquake engineering and related fields also use instruments that record ground motion. As it is seen in the previous paragraphs, shortly after the first World War, technological leadership, at least in the seismic recording, transferred to the USA. This is also the case for strong motion recording.

Since early times, there was an interest in measuring vibrations of buildings and other constructions, and attempts were made to use seismic instruments for that purpose. It is possible to give

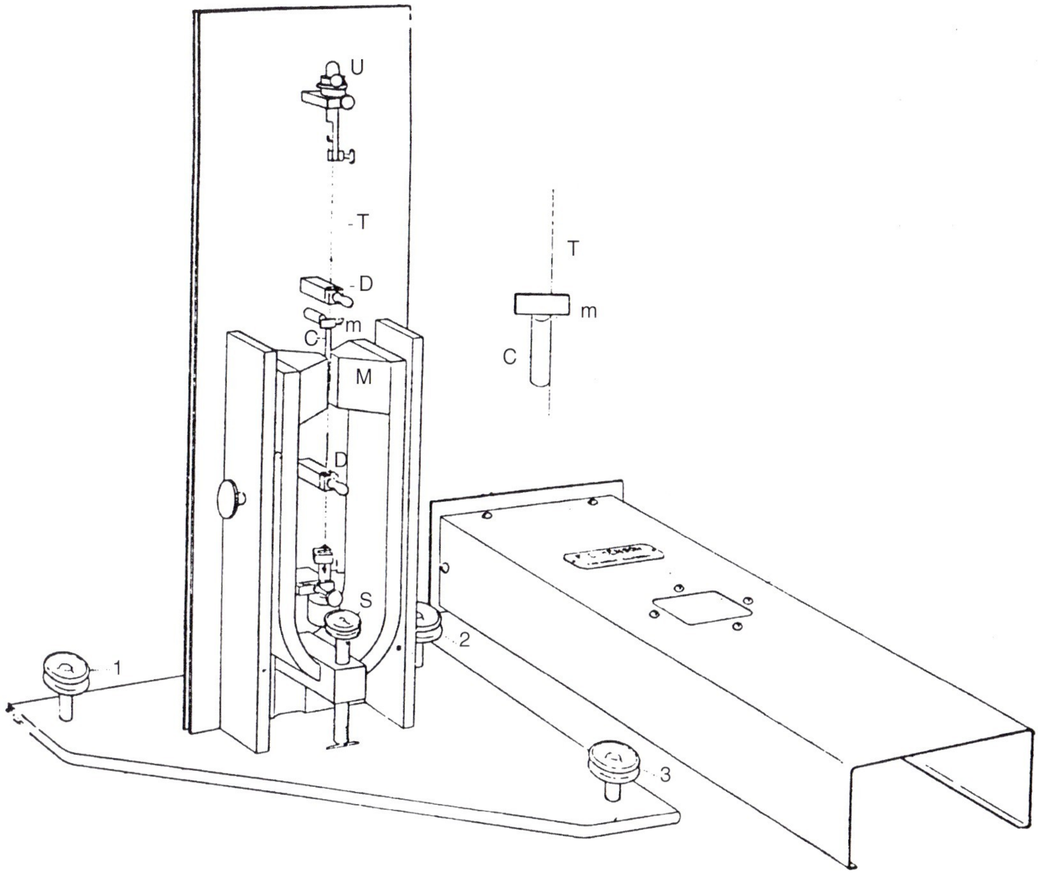

Historical Seismometer, Fig. 28 General sketch of the Wood-Anderson seismometer. A zoom of the oscillating mass is shown in the right part (elements T, C, and m) (From Anderson and Wood 1925)

Historical Seismometer, Fig. 29 WWSSN Benioff variable reluctance seismometer for the horizontal component. The transducer is placed in the central part of the instrument between the two mass disks (Original photograph in the IGN archive)

many examples. Ewing, already in 1888 in the UK, used his double pendulum to measure the vibration of the new Tay Bridge during the passing of railway trains. Guido Alfani (1976–1940), in 1911, with a portable seismograph specially designed for use in buildings, called "trepidometer," studied the vibration of the famous Pisa inclined tower when bells were ringing. Contemporaries Albin Belar (1864–1939) in Ljubljana and Manuel Sánchez Navarro (1867–1941) in Granada, among others, built also "tromometers" to use them in engineering works. But all of these instruments lacked enough sensibility in the needed range of frequencies to properly analyze the obtained signals (even the frequency contents of those signals were not well defined at that time). Even more, most of them were not damped instruments. As it happened in many other occasions related to the seismic recording, proper ideas already existed many years before its proper realization. It is worth to point out that already in 1915, Galitzin studied the possibility of using the piezoelectric effect to build accelerometers (Fremd 1997), but lacking the electronic technology needed to amplify signals, the device was not feasible at that time.

In the early 1930s several damaging earthquakes in the USA started a discussion about seismic safety, and the engineers realized that real progress in earthquake engineering (an engineering branch not yet defined at that time) passes through detailed knowledge of accelerations in buildings and structures, but there was no instrument capable of doing these measurements. However, theoretical and technological studies were properly developed at that time to describe the problem. So just 2 years passed since the starting of the Strong Motion Program, in 1931, up to the recording of the first accelerograms for the Long Beach earthquake of May 10, 1933.

The appointed committee for the Strong Motion Program was very fast in determining that it was important to record acceleration instead of displacement and also in the interesting range of frequencies and the expected amplitudes (f < 10 Hz and amplitude up to 0.5 g) to be recorded. All this knowledge about expected values was already accumulated from the previously available observations as well as how to use mechanical sensors to record ground acceleration.

Once specifications were defined, a review of available technology was made. The choice was to use a Wood-Anderson torsion pendulum-type seismometer (Trifunac 2009). In this case a vertical component was also built. Problems with the proper set of the delicate suspension (in this case with four wires, as explained by Trifunac 2009) required a change for a more robust pivoted and spring-stabilized system already in 1933. Analog recording on photographic paper forced the choice for a triggered instrument. The triggering seismoscope was still of the Palmieri type. The resulting accelerometer (Fig. 30) had three components with a free period of 10 Hz (to properly record accelerations at lower frequencies). The amplitude of the trace was 4 cm for an acceleration of 0.2 g (remember that it was expected for this instrument to have a flat response to acceleration in the range of interest). The recording speed was 1 cm/s and time marks were introduced each 0.5 s.

The instrument worked quite well but was quite expensive for particulars (an indication of that is that accelerometers were not commercialized before 1963). Thus, and mainly for engineering purposes, well-calibrated seismoscopes continued to be on the first line of earthquake engineering up to the advent of electronic systems and digital recording, in the mid-1970s (Fig. 31).

Summary

The development of the early stages of seismometry and the evolution of the early seismographs and of seismic recording have been presented. Evolution of the instruments, from the first known devices of the second century A.D. up to the devices of the nineteenth century, was slow and with many failures, the main issue being the lack of knowledge of the real ground motion during earthquakes and, thus, of a clear goal.

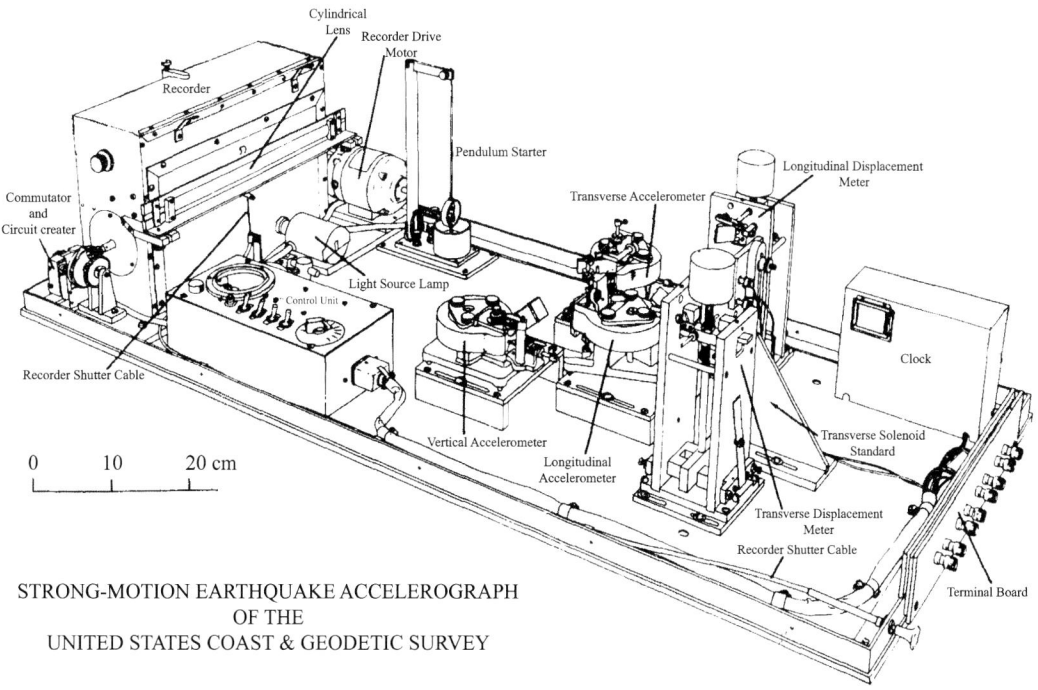

Historical Seismometer, Fig. 30 USCGS accelerograph. The three accelerometers are placed at the center of the instrument. The starter pendulum is clearly seen at the top left of the accelerometers (From Hudson 1963)

Historical Seismometer, Fig. 31 USCGS seismoscope for strong motion recording. This instrument was designed in the 1950s and continued in use up to the time of the digital accelerograph (From Hudson 1963)

This lack of knowledge was also due to the lack of proper recorders able to record continuously. This was a common problem to all sciences and it was not properly solved until the second part of the nineteenth century. Once the first records were obtained, seismometry progressed much faster. The tools for the theoretical analysis of seismic recording (mainly Newtonian mechanics) were already available. Electromagnetic theory, useful to develop electromagnetic transducers, was developed at the same time and was soon incorporated to seismometry.

Humans experience near earthquakes (felt earthquakes) and this led to the early attempts of seismic recording. But the large development of seismic networks at the beginning of the twentieth century was mainly made with the purpose of recording teleseisms. Near-field recording was not properly addressed up to the interwar period of the twentieth century. At first specific instruments for the recording of near earthquakes were developed and later, by the demand of engineers, accelerometers were designed and deployed. In the middle of the twentieth century, with the development of the WWSSN and similar instruments (like the Kirnos seismographs), seismic instruments gave results similar to our present instruments. Nevertheless, sophisticated seismoscopes, cheap and easy to handle, were generally deployed in seismic areas up to the advent of new active seismic sensors and digital recording which solved many of the problems of the older instrumentation.

Cross-References

▶ Passive Seismometers
▶ Recording Seismic Signals
▶ Seismic Instrument Response, Correction for
▶ Sensors, Calibration of

References

Anderson JA, Wood HO (1925) Description and theory of the torsion seismometer. Bull Seism Soc Am 15:1–72

Batlló J (2004) Catálogo Inventario de sismógrafos antiguos españoles. Instituto Geográfico Nacional, Madrid, 413

Batlló J, Clemente C, Pérez-Blanco F, Vidal F (2004) Reconstruction of a Bosch-Omori seismograph. Nuncius 19:701–711

Batlló J, Stich D, Macià R (2008) Quantitative analysis of early seismograph recording. In: Fréchet J, Meghraoui M, Stucchi M (eds) Historical seismology: interdisciplinary studies of past and recent earthquakes. Springer, Berlin, pp 379–396

Benioff H (1955) Earthquake seismographs and associated instruments. Adv Geophys 2:219–275

Cavalleri GM (1858) Di un nuovo sismometro collocate nel Collegio di Monza. Atti dell'I R Istituto Lombardo di Scienze, Lettere ed Arti 1:34–44

Dewey J, Byerly P (1969) The early history of seismometry (to 1900). Bull Seism Soc Am 59:183–227

Ewing JA (1881) On a new seismograph. Proc R Soc 31:440–446

Ferrari G (ed) (1990) Gli strumenti sismici storici: Italia e contesto europeo. SGA Storia- Geofisica-Ambiente, Bologna, p 198

Ferrari G (ed) (1992) Two hundred years of seismic instruments in Italy 1731–1940. SGA Storia- Geofisica-Ambiente, Bologna, p 156

Ferrari G (2006) Note on the historical rotation seismographs. In: Teisseyre R, Takeo M, Majewski E (eds) Earthquake source asymmetry, structural media and rotation effects. Springer, Berlin, pp 367–376

Forbes JD (1844) On the theory and construction of a seismometer, or instrument for measuring earthquake shocks, and other concussions. Trans R Soc Edinb 15:219–228

Fréchet F, Rivera L (2012) Horizontal pendulum development and the legacy of Ernst von Rebeur-Paschwitz. J Seismol 16:315–343

Fremd VM (1997) Historical seismometers as predecessors of modern instruments. Cahiers du Centre Européen de Géodynamique et de Séismologie 13:91–93

Havskov J, Alguacil G (2010) Instrumentation in earthquake seismology. Springer, Dordrecht, xii + 358 pp

Hörnes R (1893) Erdbebenkunde: die Erscheinungen und Ursachen der Erdbeben, die Methoden ihrer Beobachtung. Veit, Leipzig, vi + 452 p

Howell FH (1990) An Introduction to Seismological Research: History and Development (Chapter 4). Cambridge University Press, Cambridge, pp 57–72.

Hsiao KH, Yan HS (2009) The review of reconstruction designs of Zhang Heng's seismoscope. J Jpn Assoc Earthq Eng 9(4):1–10

Hudson DE (1963) The measurement of ground motion of destructive earthquakes. Bull Seism Soc Am 53:419–437

McComb HE (1936) Selection, installation and operation of seismographs. Special publication no. 206. Coast and Geodetic Survey, Washington, iv + 43 p

Oliver J, Murphy L (1971) WWNSS: seismology's global network of observing stations. Science 174:254–261

Palmieri L (1873) The eruption of Vesuvius in 1872. Asher, London, p 148

Perry J, Ayrton WE (1879) On a neglected principle that may be employed in earthquake measurement. Phil Mag Ser 5(8):30–50

Rebeur-Paschwitz E (1889) The earthquake of Tokyo. Nature 40:294–295

Stein S, Wysession M (2003) An introduction to seismology, earthquakes and earth structure. Blackwell Publishing, Oxford, xi + 498 p

Trifunac MD (2009) 75th anniversary of strong motion observation – a historical review. Soil Dyn Earthq Eng 29:591–606

Urhammer RA, Collins ER (1990) Synthesis of Wood-Anderson seismograms from broadband digital records. Bull Seism Soc Am 80:702–716

Wiechert E (1904) Ein astatisches Pendel höher Empfindlichkeit zur mechanischen Registrierung von Erdbeben. Beitr Geophys 6:435–450

Wielandt E (2009) Seismic sensors and their calibration. In: Bormann P (ed) New manual of seismological observatory practice (NMSOP). Deutsches GeoForschungsZentrum GFZ, Potsdam, p 46

Hyperspectral Data in Urban Areas

Giorgio Antonino Licciardi
GIPSA-Lab – INP Grenoble, Grenoble Institute of Technology, Saint Martin d'Hères, France

Synonyms

Change detection; Data fusion; Hyperspectral images; Hyperspectral sensors; Image processing; Material identification; Remote sensing

Introduction

The scope of this entry is to introduce hyperspectral remote sensing data and its applications in urban environment.

In the last decades imaging spectroscopy (Goetz et al. 1985), commonly referred to as hyperspectral remote sensing, has become a widely used method for identification and quantification of surface materials in different kinds of environments, such as urban, rural, and geological areas (Heiden et al. 2012).

A hyperspectral sensor is able to record a high number of spectral bands, corresponding to narrow contiguous wavelength intervals. This characteristic permits the discrimination of different materials based on their unique spectral characteristics, called spectral signature. In general, two materials can be distinguished by broad and narrow spectral reflectance features determined by the chemical composition of the materials. However, some materials may present strong spectral similarities except for a narrow part of the spectrum. From this point of view, hyperspectral sensors have the ability to detect more materials than multispectral sensors.

In the literature, several studies have analyzed the spectral characteristics of material detected by hyperspectral instruments, usually by means of laboratory and field spectroscopic investigations (Ben-Dor 2001).

Hyperspectral remote sensing demonstrated its effectiveness in several fields of application, such as precision agriculture, security surveillance, physics and astronomy, environmental analysis, and mineralogy, as well as in hazard monitoring and management.

In urban environment, hyperspectral remote sensing can be extremely useful, in particular for the management of pre-/post-seismic events. In this optic, the ability of hyperspectral sensors to discriminate several materials can be extremely useful in the characterization of the building type, by identifying the material used for the construction. Another important use of hyperspectral imaging is the ability to produce accurate geological maps of the soil.

Hyperspectral images provide a great amount of spectral information, however, when compared with multispectral images, the main limitation of this kind of data resides in their dimensionality. Since hyperspectral data cubes are large, their dimension may have a negative impact in terms of computational processing load and also on the storage. The dimensionality problem limited also a proper development of satellite-based hyperspectral systems, since transmission and storage of acquired data become extremely difficult. From this point of view, reducing the dimensionality of hyperspectral images is one of the most important processes for the exploitation of the information provided by hyperspectral.

In the following paragraphs, different aspects of hyperspectral imaging in urban area, with a particular attention to those related to seismic events, will be addressed.

Hyperspectral Image Overview

Hyperspectral Sensors

Both scientists and common people are becoming increasingly concerned with environmental phenomena such as the photosynthetic conditions of the vegetation, wide deforestation and fires, desertification, sea pollution, and hazard monitoring, together with the general health of the Earth. The monitoring of these events and the understanding of the impact which they could have on the fragile biophysical mechanisms are becoming more and more important than in the past. For this reason, sensors like Medium-Spectral Resolution Imaging Spectrometer (MERIS), Moderate Resolution Imaging Spectroradiometer (MODIS), and Advanced Very High Resolution Radiometer (AVHRR) have been designed and placed in orbit. These measurements are performed using several spectral bands (up to 36 for MODIS) located into the visible and the infrared range in order to collect a noteworthy dataset for every kind of global investigation (land use, ocean color, snow cover, sea ice observation, etc.). A following step has been the allocation of many contiguous and narrow bands (more than one hundred) available for the measurement. This technological evolution led to the hyperspectral imagery, which has demonstrated very high performance in several cases of material identification and urban mapping, including sub-pixel classification. Managing such dissimilar type of data is not a simple task and requires the adoption of information extraction techniques that are appropriate for each specific sensor data. From this point of view, a hyperspectral sensor can be considered as an extremely precise and sensitive optical sensor. However, due to its sensitivity, a hyperspectral sensor, compared to a multispectral one, is more prone to be affected by imprecise measurements. The precision of a measurement is determined by the instrument response R. The transformation from the input physical quantity I to the measurement O is described mathematically by the convolution:

$$O = I \otimes R$$

The instrument response depends on several factors, such as optical imperfections, image motion, the nonzero spatial area of each detector, as well as the filtering produced by electronic elements. Except for the image motion, all the elements influencing the instrument response can be attributed to the acquisition mode. In particular the type of scanning system and the spectral selection mode can have a strong impact on the precision of the measurements.

In general, an optical sensor can be divided into three different scanning systems for acquiring the image: whiskbroom, pushbroom, and staring imagers. While whiskbroom scanners have a simple optical system and can acquire image in a wide swath width, if compared to the pushbroom scanners, their mechanical system is extremely complex expensive and prone to damages. For these reasons pushbroom scanners are considered as the standard for high-resolution imaging spectrometers. Another characteristic that can influence the hyperspectral measurement is the spectral selection mode. In particular, the spectral selection modes used in hyperspectral sensors can be divided in three main groups: dispersion elements, where the incoming electromagnetic radiation is separated into different angles by means of a prism or grating elements; filter-based elements, based on optical bandpass filters (tunable filters, discrete filters, and linear wedge filters); and Fourier-transform spectrometers (FTS): a Fourier-transform spectrometer is an adaption of the Michelson interferometer where a collimated beam from a light source is divided into two by a beam splitter and sent to two mirrors. These mirrors reflect the beams back along the same paths to the beam splitter, where they interfere. The signal recorded at the output depends on the wavelength of the light and the optical path difference between the beam splitter and each of the two mirrors. If the optical path difference between the two beams is zero or a multiple of the wavelength of the light, then the output will be bright; otherwise if the optical path difference is an odd multiple of half the wavelength of the light, then the output will be dark.

Hyperspectral Sensors
Most of the developed hyperspectral sensors are designed for airborne carriers. One of the first airborne hyperspectral sensors was the AVIRIS (Airborne Visible/Infrared Imaging Spectrometer) developed by National Aeronautics and Space Administration (NASA). AVIRIS contains 242 different detectors, each with a bandwidth of approximately 10 nm, allowing it to cover the entire range between 380 nm and 2,500 nm. Similar sensors were the Compact Airborne Spectrographic Imager (CASI), able to acquire images in 288 bands in the 380 nm–1,050 nm spectral range, and the Hyperspectral Mapper (HyMap) providing 128 bands in the wavelength region of 450–2,500 nm. The Airborne Hyperspectral Scanner (AHS) and the Multispectral Infrared and Visible Imaging Spectrometer (MIVIS) extended the spectral range up to thermal infrared, allowing a wider range of applications.

The development of hyperspectral technology for the space satellites remains difficult and very expensive in terms of payload design, maintenance, and calibration. However, these difficulties have not deterred the space agencies from finding interesting missions carrying on board hyperspectral payloads. This is the case of Hyperion developed by NASA, CHRIS (Compact High Resolution Imaging Spectrometer) PROBA-1 (Project for On-Board Autonomy) developed by a European consortium founded by European Space Agency (ESA), the upcoming PRISMA developed by ASI (Agenzia Spaziale Italiana), and EnMAP (Environmental Mapping and Analysis Program) developed by Deutschen Zentrums für Luft- und Raumfahrt (DLR).

Hyperion instrument, mounted on board of the EO-1 (Earth Observation 1) satellite, provides a high-resolution hyperspectral imager capable of resolving 220 spectral bands (from 400 to 2,500 nm) with a 30 m spatial resolution. The instrument covers a 7.5 km by 100 km land area per image and provides detailed spectral mapping across all 220 channels with high radiometric accuracy.

CHRIS is a high-resolution hyperspectral sensor installed on board the PROBA satellite. Distinctive feature of CHRIS is its ability to observe the same area under five different angles of view in the VIS (visible)/NIR (near infra-red) bands. Another important feature of CHRIS is the ability to be reprogrammable in terms of the number of spectral bands and spectral resolution (up to 150 spectral bands with a spectral resolution of 1.25 nm). This characteristic allows the acquisition of 62 narrow and quasi-contiguous spectral bands with the spatial resolution of 34–40 m or only 18 spectral bands with an enhanced spatial resolution of 18 m.

PRISMA (PRecursore IperSpettrale della Missione Applicativa) is a new earth observation integrating a hyperspectral sensor with a panchromatic camera. The advantage of using both sensors is to integrate the classical geometric feature recognition to the capability offered by the hyperspectral sensor to identify the chemical/physical feature present in the scene. The primary applications will be the environmental monitoring, geological and agricultural mapping, atmosphere monitoring, and homeland security.

The main goal of EnMAP project is to investigate a wide range of ecosystem parameters encompassing agriculture, forestry, soil and geological environments, coastal zones, and inland waters. Thanks to the chosen sun-synchronous orbit, each point on earth can be investigated and revisited within 4 days.

Hyperspectral Data Processing
Since hyperspectral sensors are able to measure the radiance, one of the most important processing steps in hyperspectral image processing is the conversion to reflectance. Usually, the techniques for retrieving surface reflectance can be roughly divided into two major categories, relative reflectance retrieval techniques and absolute reflectance retrieval techniques. Relative reflectance retrieval is based on the knowledge of two or more target reflectances in the image. Knowing the reflectance of

a material in the image permits to define by regression a linear equation between each wavelength and the measured radiance. However, this approach is also based on the assumption that the atmosphere is uniform, resulting in significant artifacts in strong atmospheric bands if ranges of elevations are present in the scene.

Absolute approaches, on the other hand, permit the retrieval of surface reflectance based on the physical principles, in which measured radiance is typically compared to radiance generated by an atmospheric radiative transfer model, such as the MODerate resolution atmospheric TRANsmission (MODTRAN) model. Usually, the relative approach can be useful in processing airborne data, where the influence of the atmosphere is not relevant and can be considered uniform. On the other hand, absolute approaches are dominant in spaceborne images.

Another common preprocessing step is georectification, used to relate the acquired data to the ground.

In order to make maps and relate them to ground reference data, it is critical that aircraft-related and ground-related distortions are removed. The most common approach is to use a "rubber-sheet" stretching approach and numerous tie points between a base map and measured image. Recent improvements in onboard navigation information and recent software development have made it possible to georectify images to within a few pixels (~10 m for a 4 m GIFOV [Ground instantaneous field of view]) in a near-automated fashion.

Aside from the common processing techniques to obtain images more interpretable, there are other techniques that are more related to the dimension of hyperspectral data. If we consider the consistency of the hyperspectral data, we can easily understand the importance of finding a method which can transform the original data cube into one with reduced dimensionality and maintain, at the same time, as much information content as possible. In particular, dataset composed of hundreds of narrowband channels, besides storage and transmission, may cause problems in terms of complexity of processing and inversion phases. Therefore, dimensionality reduction may become a key parameter to obtain a good performance.

These techniques are known under the general name of feature reduction. Besides enabling an easier storage and management of the data, feature-reduction procedures can be crucial for the implementation of optimum inversion algorithms.

Many methods have been developed to tackle the issue of high dimensionality of hyperspectral data (Serpico and Bruzzone 1994). In summary, we may say that feature-reduction methods can be divided into two classes: "feature-selection" algorithms (which suitably select a suboptimal subset of the original set of features while discarding the remaining ones) and "feature extraction" by data transformation which projects the original data space onto a lower-dimensional feature subspace that preserves most of the information, such as nonlinear principal component analysis (NLPCA; Licciardi and Del Frate 2011).

The first analysis suggests that feature selection is a more simple and direct approach compared to feature extraction and that the resulting reduced set of features is easier to interpret. Nevertheless, extraction methods can be expected to be more effective in representing the information content in lower dimensionality domain.

Feature-selection techniques can be generally considered as a combination of both a search algorithm and a criterion function. The solution to feature-selection problem is offered by the search algorithm, which generates a subset of features and compares them on the basis of the criterion function. From a computational viewpoint, an exhaustive search for the optimal solution becomes intractable even for moderate values of features (Siedlecki and Sklansky 1988). Despite these apparent difficulties, many feature-selection approaches have been developed (Serpico and Bruzzone 1994). The sequential forward selection (SFS) and the sequential backward selection (SBS) techniques are the simplest suboptimal search strategies: they can identify the best feature subset achievable by adding (to an empty set in SFS) or removing (from the complete set from SBS) one feature at a time, until the desired number of features is achieved.

The sequential forward floating selection (SFFS) and the sequential backward floating selection (SBFS) methods improve the standard SFS and SBS techniques by dynamically changing the number of features included (SFFS) or removed (SBFS) at each step.

A feature-extraction technique aims at reducing the data dimensionality by mapping the feature space onto a new lower-dimensional space. Both supervised and unsupervised methods have been developed. Unsupervised feature-extraction methods do not require any prior knowledge or training data, even though these are not directly aimed at optimizing the accuracy in a given classification task. The class comprises the "principal component analysis" (PCA, Fukunaga 1990), where a set of uncorrelated transformed features is generated; the "independent component analysis" (ICA) (Jutten and Herault 1991), a computational method for separating a multivariate signal into additive subcomponents supposing the mutual statistical independence of the non-Gaussian source signal; and the "maximum noise fraction" (MNF, Green et al. 1988), where an operator calculates a set of transformed features according to a signal-to-noise ratio optimization criterion.

Challenges in Hyperspectral Data Processing

The analysis of hyperspectral images is not an easy task. Due to the complexity of a hyperspectral sensor, hyperspectral images can be affected by several problems in terms of data uniformity. Any nonuniformity in the system generates degrading artifacts. Since pushbroom scanners mainly characterize hyperspectral sensors, in this entry we will focus on the main problems that afflict this kind of sensors.

In pushbroom imaging spectroscopy, the image is generated in the spectral and spatial dimension simultaneously. This lead to two main kinds of problems:

- Spectral nonuniformity: is the nonuniformity of the spectral response within a sensor's spectral band and can be imaged on a detector row. This nonuniformity is typically represented by the position and shape of the spectral response function. The related artifacts of spectral misregistration are denoted as "smile" or "frown."
- Spatial nonuniformity: is the nonuniformity of the spatial response within an acquired spectrum and is usually imaged on a detector column. This nonuniformity is represented by the position and shape of the spatial response function in both the along-track and across-track dimensions of a spatial pixel. The related artifacts in the across-track dimension are denoted as "keystone."

As for the spatial nonuniformity, the influence of the "keystone" effect results in a black pixel in the image that can be easily replaced by meaning the surrounding pixels with a 3×3 moving window. On the other hand, the removal of the "smile" effect is not an easy task. The consequence of this effect is that the central wavelength of a band varies with spatial position across the width of the image in a smoothly curving pattern. Very often the peak of the smooth curve tends to be in the middle of the image and gives it a shape of "smile" or "frown." The effect of the smile is not obvious in the individual bands. Therefore, an indicator is needed to make evident whether or not a given image suffers from smile effect. A way to check for the smile effect is to look at the band difference images around atmospheric absorption (760 nm). In fact, the region of red-near infrared transition has high information content of vegetation spectra. This region is generally called "red edge" (670–780 nm) and identifies the red-edge position (REP). The smile effect is acute due to sharp absorption at 760 nm, which is within the red-edge region, and for this reason atmospheric correction of smiled data will be incorrect.

The methodologies developed so far can only reduce the intensity of smile effect but cannot remove it entirely, because during its life a detector element can change its response; therefore, the knowledge and the correction of this phenomenon became fundamental in the analysis of hyperspectral images.

Another problem that affects pushbroom scanners is the so-called "striping" noise. The striping effect consists in strong variations in the average value in the column for every band of the image. Theoretically the elements of the detector array are identical. Thus, the detectors in a row should be well calibrated to provide the same output along the spatial domain, if illuminated by the same source. In the real world, a residual nonuniformity of the electro-optical response generates different transfer functions. Thermal fluctuations can cause additive small variations in the alignment of the detector, producing a non-predictable noise not constant during the time. Only the systematic knowledge of the gain function permits an accurate compensation of these fluctuations. The correction of the striping can be carried out by means of filtering approaches. This type of destriping involves two filters on the original image creating two new images, a low-pass filtered image and a high-pass filtered image, and combining the results. If both a low-pass and a high-pass filter of the same size are run on an image and then added together, the result will be the original image. The goal in choosing sizes for the two filters is to create an output for each without the unwanted striping. If the filter sizes are not the same, the result will not be exactly the same as the original. When the filter outputs are added together, the striping will have been removed or subdued. If the sizes of the two filters differ too much, artifacts may be introduced into the result.

From this brief overview of the principal preprocessing techniques, it is possible to understand the amount of efforts that are necessary in order to extract useful and reliable information from hyperspectral data.

Building Vulnerability Assessment

One of the aspects of pre-seismic event management is related to the assessment of the vulnerability of buildings in urban environment. Many features that can influence the seismic performance of a building can be extrapolated from the outside and then be detected by remote sensing instruments. From this point of view, urban areas can be considered as an ensemble of different types of surfaces characterized by a continuous spatial change. This results in small urban structures mostly dominated by artificial (man-made) materials. The identification of these materials requires an accurate analysis of the spectral characteristics of these materials. Information about the material used in the roof of the buildings permits to determinate the construction technique of the building and consequently its level of vulnerability. The identification of the roof spectral characteristics can be useful also for the derivation of the type of urban structure where the building is located. For instance, the historical development of European countries leads the urban structure type to be correlated with a certain time period and construction style. Thus the knowledge of the urban structure type may give information about the type and age of the buildings. In particular, the necessary information for the evaluation of pre-seismic events scenarios can be resumed in the detection of the building type (i.e., the material) and in the detection of the identification of the contextual information, such as the position of each building in relation to the surrounding ones. Analysis of the geological information of the soil can be useful but become extremely difficult if performed by using remote sensing techniques in urban environment, since the soil tends to be covered by man-made structures.

Material Identification

Surface materials can be detected based on their spectral characteristics. In the literature there exist several works on the extraction of information about the material of the roofs by means of multispectral images. However, these techniques permit to identify a limited number of elements, due to the low spectral resolution of multispectral imager. For instance, multispectral images are not sufficient to differentiate between roofs and other surfaces having similar spectral characteristics, such as asphalt and bitumen used to seal rooftops. In this context hyperspectral imaging can be an extremely useful instrument for an effective estimation of the vulnerability, identifying more information about the material used in the construction of the investigated buildings.

In general, several tools have been developed to match reference spectra with those measured by an imaging spectrometer. The most common approach is based on the use of standard supervised classification techniques, where known spectra are used to determine the statistical properties of each class based on spectral characteristics. Examples of supervised classification approaches applied to hyperspectral data are described in McKeown et al. (1999) and Roessner et al. (2001), where the maximum likelihood classifier (MLC) was applied to map urban land cover. Other techniques are based on the use of support vector machines (SVM) (Melgani and Bruzzone 2004) and neural networks (NNs) (Licciardi et al. 2009, 2012). Other approaches have been designed explicitly for the analysis of imaging spectrometry data, such as the Spectral Angle Mapper (SAM; Kruse et al. 1993).

In order to extract as much information content as possible from hyperspectral images in urban environment, specific methods for automated information extraction have been developed (Roessner et al. 2001). In all these methods, the comprehensive mapping of the different surfaces present on urban environment can be obtained through a classification technique (Roessner et al. 2001).

For example, Clark et al. (2001), in their environmental studies on the World Trade Center area after the September 11, 2001, attack, successfully applied hyperspectral remote sensing to map material in the surrounding area. In this study, the authors used high-spatial-resolution AVIRIS data acquired over the World Trade Center to map thermal sources, asbestiform minerals, and dust and debris from the collapse, producing maps of environmental contaminants that are difficult to produce cost-effectively in any other way. McKeown et al. (1999) used data acquired by the Hyperspectral Digital Imagery Collection Experiment (HYDICE) to classify urban and natural materials. The authors then fused the classification results with a 3D model of the buildings. Ben-Dor (2001) developed an urban spectral library using a combination of an existing spectral library developed by Price (1995) and CASI data acquired over the city of Tel Aviv, Israel. Data acquired by the Digital Airborne Imaging Spectrometer (DAIS) have been used by Roessner et al. (2001) to obtain a map of urban materials in the city of Dresden, Germany. In this study, a maximum likelihood classifier has been used to derive a first map of pure spectral features and then used these features to unmix the other spectra.

However, since the spectral variations characterizing a certain material are caused by several factors, such as the age and deterioration of the material, or the color, and also by the illumination angles, several aspects need to be considered. In particular, aside from the spectral heterogeneity of different materials, it is important to take into account the high intra-class variability of spectra corresponding to the same material. Moreover, in urban environment there are many objects having a size smaller than the pixel. This results in mixed pixels, presenting spectra that are a combination of two or more materials that are imaged in the IFOV (instantaneous field of view) of the sensor. Moreover, the modeling of the spectral mixture becomes more complex as the pixel size increases. From this point of view, airborne hyperspectral sensors, characterized by a high spectral and/or spatial resolution, permit to obtain reliable quantitative measurements of specific absorption features of urban materials.

In terms of spectral unmixing, a number of approaches in the literature are based on the assumption that the detected spectra are linear combination of pure spectra. In particular, in linear spectral unmixing the mixed spectrum is modeled as the sum of pure spectra, each weighted by the fraction of the material within the field of view. Roberts et al. (1998) have developed a general approach for spectral unmixing in which the number and types of end-members are allowed to vary per pixel. In Roessner et al. (2001), the authors used an MLC to detect the pure spectra in the image and then use them as end-members to produce the abundance maps. An approach based on nonlinear unmixing has been proposed in Licciardi and Del Frate (2011), where a neural network (NN) has been used to quantify the abundances of previously detected end-members.

Fusion of Hyperspectral and Geometric Information

While the spectral information is extremely useful to identify different materials, it does not provide any information on the use of these materials. This means that spectral information is not sufficient to differentiate, for instance, a roof from other surfaces characterized by the same material. For this reason, additional information is required for the univocal identification of the buildings. One possible solution is to derive the shape characteristics of the buildings present in the image. Usually, when seen from above, most of the buildings present a rectangular shape. However, some exceptions exist in terms of regularity of the shape. For instance, industrial or commercial buildings present differences in terms of size and covering materials, such as bitumen or metal. Residential buildings are, on the other hand, of small size and often covered with tiles or concrete. Roads are characterized by bitumen but present oblong shapes and sometimes interrupted by the tree canopies. Parking lots are spectrally characterized by bitumen but, differently from the roads, are characterized by rectangular shapes. This requires the development of algorithms, which can handle this variety in the shape detection process. The basic approach for the extraction of objects is segmentation, consisting in labeling each pixel in the image and grouping the adjacent ones according to their label. In the literature a variety of methods for segmentation of hyperspectral image exist. Segl and Kaufmann (2001) proposed a method for the identification of buildings based on an iterative segmentation followed by a classification. In Tarabalka et al. (2010), the watershed segmentation algorithm has been applied to hyperspectral images, while in Priego et al. (2013), a segmentation technique based on the application of cellular automata is used in order to produce homogeneous regions. In both cases an support vector machine (SVM) approach is applied to label the obtained objects and classify them. The use of geodesic opening and closing operations of different sizes in order to build a morphological profile and a neural network approach for the classification of features has been introduced in Pesaresi and Benediktsson (2001) and applied successfully to hyperspectral data in Licciardi et al. (2012).

An emerging technique in the last decades is based on the fusion of hyperspectral and Light Detection and Ranging (LiDAR) data, thanks to the increased availability of these data taken from the same area. In particular, in urban environment, the introduction of high-resolution digital surface models derived from LiDAR can effectively improve the classification and consequently the identification of different materials and surfaces.

In Heiden et al. (2012), the height information obtained from LiDAR data has been introduced into orthorectification in order to improve the geometric accuracy of urban objects in the higher-resolution hyperspectral image data. This permitted to eliminate the influence of the facades of the buildings in large field of view (FOV) of such sensors and to improve the accuracy of surface material mapping.

The fusion of LiDAR and hyperspectral data through a physical model has been used in Zhang et al. (2013) to eliminate the direct illumination component in the hyperspectral radiance data and consequently permitted to detect also materials under shadows.

Change Detection

The accurate analysis of the spectral characteristics of the materials present in urban environment covers an important role in the framework of building vulnerability assessment. However, a complete analysis of the vulnerability of buildings in urban environment and more in general environmental monitoring requires also an accurate study of the changes occurred in the time domain. From this point of view, the monitoring of urban areas using multi-temporal images has received increasingly attention in the last decades. In particular, remote sensing technology is able to provide a large amount of images on the same area over a long period of time. Land-cover change detection is a useful instrument for the description of changes in urban environments. Change detection by remote sensing has been widely used in many applications such as land-use/land-cover monitoring, urban development, and ecosystem and disaster monitoring. Conventional change detection methods generally depend on a difference value in each individual band, such as image differencing or the image ratio. However, most of these methods are based on multispectral images and do not take into account the physical meaning of the continuous spectral signatures. In hyperspectral images, the change can be associated to change of a spectral signature from one material to another material. For this reason most of the techniques presented in the literature are based on the difference in the spectral signatures of different materials. However, the detection of changes in hyperspectral images is not a trivial task, mainly because of several factors influencing the final result. Problems that can complicate the detection of changes in urban environment are caused by differences in the illumination level, such as the presence of shadows in two different images. Geometrical differences in terms of parallax and error of misregistration may result in the detection of false changes, as well as differences in atmospheric content. Moreover, since different cities may present different patterns and characteristics, it is difficult to define a general technique that can be applied indifferently to any kind of urban pattern. In the literature the existing change detection techniques can be divided into three main groups.

Post-classification methods are easy to implement and provide "from-to" change maps but may present limits in their validity because of misclassification errors in one of the classification maps. Other methods are based on image transformation techniques such as PCA or multivariate alteration detection (MAD). These techniques project the original hyperspectral data into a feature space in order to label the changes (Nielsen et al. 1998). A third kind of hyperspectral change detection methods is based on the detection of anomalies present in the images. In particular, anomaly detection techniques are able to distinguishing unusual targets from a typical background, and this assumption can be extended to change monitoring (Eismann et al. 2008).

Summary

In this entry we provided a brief description of the different aspects of the use of hyperspectral data in urban environment. In the introduction we presented a brief overview of imaging spectrometry in urban areas. The discussion continued with a description of the characteristics of the hyperspectral sensors and the problems related to the processing of hyperspectral images. We illustrated then the different techniques for the assessment of building vulnerability through the use of hyperspectral data.

The management of pre-seismic event in urban environment can be improved with the use of remote sensing data. In recent years remote sensing becomes more and more a popular instrument in the monitoring of urban areas since many features influencing the seismic performance of a building can be extrapolated from remote sensing instruments. Hyperspectral sensors, able to provide images with unprecedented spectral detail, can have an important role in the assessment of the vulnerability of buildings in urban environment. From this point of view, urban environment is extremely challenging because of the need to discriminate as much materials as possible. Hyperspectral sensors, able to provide images with unprecedented spectral detail, can have an important role in the assessment of the vulnerability of buildings in urban environment. However, this task is often complicated by several factors, such as the different reflectance of the material, different lighting conditions, as well as different atmospheric characteristics. To solve these problems, several techniques have been developed in order to obtain efficient mapping of urban materials with the use of hyperspectral images.

Cross-References

▶ Earthquake Damage Assessment from VHR Data: Case Studies
▶ Features Extraction from Satellite Data
▶ Urban Change Monitoring: Multi-temporal SAR Images

References

Ben-Dor E (2001) Imaging spectrometry for urban applications. In: Van der Meer FD, De Jong SM (eds) Imaging spectrometry. Basic principles and prospective applications. Kluwer, Dordrecht

Clark RN, Green RO, Swayze GA, Meeker G, Sutley S, Hoefen TM, Livo KE, Plumlee G, Pavri B, Sarture C, Wilson S, Hageman P, Lamothe P, Vance JS, Boardman J, Brownfield I, Gent C, Morath LC, Taggart J, Theodorakos PM, Adams M (2001) Environmental studies of the world trade center area after the September 11, 2001 attack. U.S. Geological Survey, Open File Report OFR-01-0429

Eismann MT, Meola J, Hardie RC (2008) Hyperspectral change detection in the presence of diurnal and seasonal variations. IEEE Trans Geosci Remote Sens 46:237–249

Fukunaga K (1990) Introduction to statistical pattern recognition, 2nd edn. Academic, San Diego

Goetz AFH, Vane G, Solomon JE, Barrett NR (1985) Imaging spectrometry for earth remote sensing. Science 228(4704):1147–1153

Green AA, Berman M, Switzer P, Craig MD (1988) A transformation for ordering multispectral data in terms of image quality with implications for noise removal. IEEE Trans Geosci Remote Sens 26(1):65–74

Heiden U, Heldens W, Roessner S, Segl K, Esch T, Mueller A (2012) Urban structure type characterization using hyperspectral remote sensing and height information. Landscape Urban Plann 105(4):361–375

Jutten C, Herault J (1991) Blind separation of sources, part I: an adaptive algorithm based on neuromimetic architecture. Signal Processing 1991(24):1–10

Kruse FA, Lefkoff AB, Boradman JB, Heidebrecht KG, Shapiro AT, Barloon PJ, Goetz AFH (1993) The spectral image processing system (SIPS)- iterative visualization and analysis of imaging spectrometer data. Remote Sens Environ 44:145–163

Licciardi G, Del Frate F (2011) Pixel unmixing in hyperspectral data by means of neural networks. IEEE Trans Geosci Remote Sens 49(11):4163–4172

Licciardi G, Pacifici F, Tuia D, Prasad S, West T, Giacco F, Inglada J, Christophe E, Chanussot J, Gamba P (2009) Decision fusion for the classification of hyperspectral data: outcome of the 2008 GRS-S data fusion contest. IEEE Trans Geosci Remote Sens 47(11):3857–3865

Licciardi G, Marpu PR, Chanussot J, Benediktsson JA (2012) Linear versus nonlinear PCA for the classification of hyperspectral data based on the extended morphological profiles. IEEE Geosci Remote Sens Lett 9(3):447–451

Mckeown DM Jr, Cochran SD, Fored SJ, Mcglone JC, Shufelt JA, Yokum DA (1999) Fusion of HYDICE hyperspectral data with panchromatic imagery for cartographic feature extraction. IEEE Trans Geosci Remote Sens 37(3):1261–1277

Melgani F, Bruzzone L (2004) Classification of hyperspectral remote sensing images with support vector machines. Geosci Remote Sens IEEE Trans 42(8):1778–1790

Nielsen AA, Conradsen K, Simpson JJ (1998) Multivariate alteration detection (MAD) and MAF postprocessing in multispectral, bitemporal image data: new approaches to change detection studies. Remote Sens Environ 64:1–19

Pesaresi M, Benediktsson JA (2001) A new approach for the morphological segmentation of high-resolution satellite imagery. Geosci Remote Sens, IEEE Trans 39(2):309–320

Price JC (1995) Examples of high resolution visible to near-infrared reflectance and a standardize collection for remote sensing studies. Int J Remote Sens 16:993–1000

Priego B, Souto D, Bellas F, Duro RJ (2013) Hyperspectral image segmentation through evolved cellular automata. Pattern Recogn Lett 34(14):1648–1658

Roberts DA, Gardner M, Church R, Ustin S, Scheer G, Green RO (1998) Mapping chaparral in the Santa Monica mountains using multiple end member spectral mixture models. Remote Sens Environ 65:267–279

Roessner S, Segl K, Heiden U, Kaufmann H (2001) Automated differentiation of urban surfaces based on airborne hyperspectral imagery. IEEE Trans Geosci Remote Sens 39:1525–1532

Segl K, Kaufmann H (2001) Detection of small objects from high resolution panchromatic satellite imagery based on supervised image segmentation. IEEE Trans Geosci Remote Sens 39(9):2080–2083

Serpico SB, Bruzzone L (1994) A new search algorithm for feature selection in hyperspectral remote sensing images. IEEE Trans Geosci Remote Sens 39(7):1360–1367

Siedlecki W, Sklansky J (1988) On automatic feature selection. Int J Pattern Recognit Artif Intell 2(2):197–210

Tarabalka Y, Chanussot J, Benediktsson JA (2010) Segmentation and classification of hyperspectral images using watershed transformation. Pattern Recogn 43(7):2367–2379

Zhang Q, Pauca VP, Plemmons RJ, Nikic DD (2013) Detecting objects under shadows by fusion of hyperspectral and lidar data: a physical model approach. In: 5th workshop hyperspectral image and signal processing: evolution in remote sensing, WHISPERS2013, 2013, pp 1–4

Incremental Dynamic Analysis

Dimitrios Vamvatsikos
School of Civil Engineering, National Technical University of Athens (N.T.U.A.), Athens, Greece

Introduction

An important issue in performance-based earthquake engineering is the estimation of structural performance under seismic loads. In particular, one is interested in estimating the probabilistic distribution of structural response in the form of engineering demand parameters (EDPs) such as peak interstory drift, peak floor acceleration, moment, or shear, given the level of seismic intensity represented by a (typically scalar) intensity measure (IM). Incremental dynamic analysis (IDA) has been developed to provide such information by employing nonlinear dynamic analyses of the structural model under a suite of ground motion records, each scaled to several intensity levels designed to force the structure all the way from elasticity to final global dynamic instability.

The use of ground motion scaling for determining the response of a structure at increasing levels of intensity is an old technique. Still, it had not been used systematically to quantify the probabilistic nature of structural response until the start of the SAC/FEMA (2000a, b) project that was conceived in the aftermath of the Northridge 1994 earthquake. Therein, a precursor to IDA was proposed in the form of the "dynamic pushover," a method to determine the global collapse capacity of structures (Luco and Cornell 1998). Subsequently, this was recast and formalized by Vamvatsikos and Cornell (2002) as a comprehensive procedure to assess the statistics of EDP demand given the IM as well as the required ("capacity") IM to achieve given values of EDP at any level of structural behavior. It is this format that is known as IDA and will be described in the following sections, discussing the necessary steps and concepts in executing, postprocessing, and applying IDA to solve problems of engineering significance.

Intensity Measure and Ground Motions

First and foremost for IDA, an efficient and sufficient IM (Luco and Cornell 2007) should be selected. An efficient IM will generally be well correlated with the EDPs of choice, thus showing low dispersion of demand given the IM and subsequently allowing the determination of EDP demand or IM capacity statistics using a relatively low number of ground motion records. In other words, efficiency is synonymous with economy in computational resources, becoming a nontrivial issue when considering that a single nonlinear dynamic analysis of a moderate complexity model can easily last more than 30 min. The second requirement, sufficiency, is defined as the independence of the distribution of EDP given the IM from any other seismological parameters that may characterize the ground motion, e.g., duration, magnitude, spectral shape, or the presence of a pulse indicative of near-source forward directivity. A sufficient IM essentially captures all seismological information needed to determine the effect of a ground motion record on the structure being investigated, thus allowing unrestricted scaling in theory. Naturally, any IM for which a seismic hazard curve can be practically estimated will never be fully sufficient; thus, excessive scaling may introduce biased estimates of response (Luco and Bazzurro 2007).

A standard choice used in the literature is the 5 % damped first-mode (pseudo)spectral acceleration $S_a(T_1)$ (Shome et al 1998; Shome and Cornell 1999). This is generally adequate for first-mode-dominated structures that do not displace far into the nonlinear region, as is the case of most existing brittle or moderately ductile low/mid-rise buildings. For taller or asymmetric structures where higher modes become important or modern buildings that exhibit significant ductility, improved IM alternatives should be sought. One particularly attractive option is the geometric mean of spectral acceleration values at several periods (Cordova et al 2000; Vamvatsikos and Cornell 2005; Bianchini et al 2009) or an inelastic spectral displacement-based IM (Luco and Cornell 2007). Both can largely alleviate the effect of spectral shape, being able to capture the period elongation characterizing ductile structures and the effect of higher modes.

In terms of selecting ground motion records, the general idea is to reduce the bias that a less than ideal IM may introduce. A general rule of thumb is to utilize accelerograms that can significantly damage the investigated structure without much scaling. For modern ductile structures, this generally means records having naturally high $S_a(T_1)$ values. Still, IDA is structured in such a way that the same set of records is used throughout the entire range of structural behavior (elastic–inelastic collapse). Whenever site conditions are deemed to have different impact at different levels of intensity, thus fundamentally changing the nature of records that one would use at low versus high levels of the IM, then methods of analysis other than IDA should probably be employed.

Structural Model

For performing IDA, a (a) realistic, (b) low-to-medium-complexity, (c) robust nonlinear structural model should be formed. Each of these three requirements comes with its own reasons.

For one, realism means incorporating all pertinent sources of material and geometric nonlinearity that are expected to arise. This should include, for example, plastic-hinge formation zones for moment-resisting frames, brace buckling for braced frames, and P-Delta effects. Nonsimulated failure modes, such as the shear failure of members or the brittle failure of beam–column joints, can be incorporated in the analysis a posteriori. Still, they essentially remove the model's ability to track structural behavior beyond their first occurrence. This means that whenever nonsimulated failures are found to have occurred, one cannot trust the model to provide estimates beyond that point.

The requirement of low-to-moderate complexity is necessary for ensuring that the computationally intensive IDA will generally be confined to reasonable amounts of time. Complex nonlinear models can complicate the execution to no end, easily forcing each nonlinear analysis to last several hours or, worse, often forcing the analyst to consider reducing the number of records or dynamic runs per record employed. This can be a tricky situation that may lead to inaccurate or biased results. Thus, it is best to strike a trade-off between model complexity and fidelity, striving to find a good balance that allows using a rich set of at least 20–30 ground motion records with six or more time history analyses for each.

Finally, robustness becomes important in tracking structural behavior in the postelastic region, especially beyond the structure's maximum-strength point, where it starts deteriorating and approaching collapse. To be able to ensure such results, a highly reliable software platform should be used. IDA by nature will drive both the model and the analysis software to their limits, making robustness difficult to achieve. Even then, typical complex finite element models that have been used successfully for elastic or mildly inelastic analysis tend to fare poorly when driven close to global dynamic instability, being consistently less reliable than their simpler versions besides significantly raising the computational cost. For example, it is often the case that distributed plasticity models are numerically less stable and always more expensive than lumped-plasticity ones. Therefore, considerable attention should be paid to the formation and testing of the model before IDA is performed, checking its behavior and stability via nonlinear static pushover analyses.

Execution

Performing IDA is conceptually simple. One only needs to take one record at a time, incrementally scale it at constant or variable IM steps, and perform a nonlinear dynamic analysis each time. Start from a low IM value where the structure behaves elastically, and stop when global collapse is encountered. The latter is defined as the occurrence of a nonsimulated failure mode or the appearance of a global dynamic instability as a collapse mechanism showing "infinite" EDP values at a given IM level. For a well-executed analysis and robust structural model, global dynamic instability manifests itself as numerical nonconvergence.

Selecting constant IM steps is often the simplest but also the most wasteful approach to IDA. As global collapse is encountered at widely varying IM levels for each record, some ground motions will require twice or thrice the dynamic analyses than others. Using variable IM steps estimated via the hunt & fill procedure of Vamvatsikos and Cornell (2002, 2004), offers instead consistent accuracy at a predefined computational cost. Due to the nearly perfect computational independence of each dynamic analysis, IDA execution is an easily parallelizable problem (Vamvatsikos 2011). Hence, one can benefit from using a cluster of N computers to divide the total running time (almost) by N. Such savings make possible the use of Monte Carlo–based algorithms (Vamvatsikos 2014) to economically estimate any bias and additional variability introduced by modeling uncertainty.

Postprocessing

The results of IDA initially appear as distinct points, one per each dynamic analysis, in the EDP versus IM plane, as observed in Fig. 1 for a nine-story steel moment-resisting frame. Linear or spline interpolation (Vamvatsikos and Cornell 2004) is then employed to generate continuous IDA curves, one for each individual ground motion record, shown in Fig. 2. The variability offered by such curves is actually one of the eye-opening features of IDA, visually representing the probabilistic nature of seismic loading and the differences introduced even by (seemingly) similar ground motions. At the highest IM level that each IDA curve can reach, one can identify the characteristic flatline representing global collapse. In the example of Fig. 2, this is the intensity where the maximum (over all stories) peak interstory drift (the EDP of choice) increases without bound for each record, leading to global system collapse.

The complex picture of structural response shown by the IDA curves can be significantly simplified by summarizing them into the 16/50/84 % fractile curves shown in Fig. 3. This is simply the determination of the 16/50/84 % percentile values of EDP at each level of the IM, taking horizontal cross sections of the IDA curves or, equivalently, the 84/50/16 % percentiles of the IM given the EDP when vertical cross sections are assumed instead. The three curves of Fig. 3 thus provide the full characterization of the distribution of structural response via the central

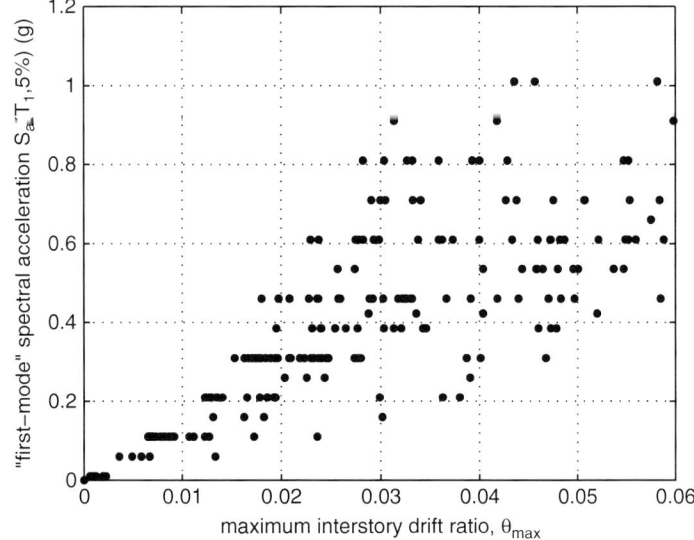

Incremental Dynamic Analysis, Fig. 1 The resulting IM-EDP points for 30 ground motion records when performing IDA for a 9-story steel frame using $S_a(T_1)$ as the IM and the maximum over all stories peak interstory drift θ_{max} as the EDP

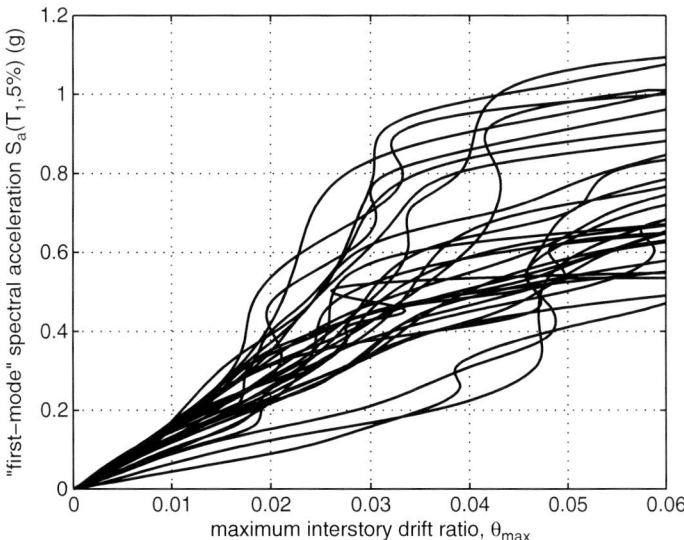

Incremental Dynamic Analysis, Fig. 2 The 30 IDA curves derived by interpolating the IM-EDP points of Fig. 1 for the 9-story steel frame

value (median) and the dispersion of EDP structural demand given the IM. For the example of the nine-story steel frame, one can observe how the median demand obeys the "equal displacement" rule, originally suggested by Veletsos and Newmark (1960) for elastoplastic single-degree-of-freedom systems: The median IDA maintains the same slope in the elastic and the inelastic range, meaning that an elastic version of this nine-story steel frame subject to the same level of ground motion would (in the median sense) experience the same maximum interstory drift as the inelastic one. This only changes when the structure enters into the "negative stiffness" range where its strength starts degrading and the 50 % IDA deviates to the right to merge into the flatline representing the median IM collapse capacity value.

Finally, one can employ the IDA curves to define appropriate limit states and estimate the corresponding IM values of capacity. A limit state is usually tied to specific threshold EDP

Incremental Dynamic Analysis, Fig. 3 The 16/50/84 % fractile IDA curves obtained by summarizing the individual IDA curves of Fig. 2 for the 9-story steel frame

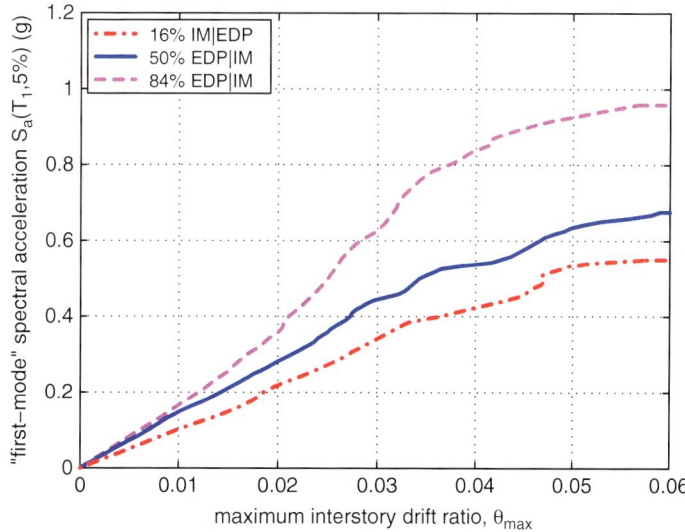

Incremental Dynamic Analysis, Fig. 4 Thirty IDA curves, 30 limit-state capacity points and the corresponding EDP and IM capacity distributions for the 9-story steel frame (From Vamvatsikos 2013)

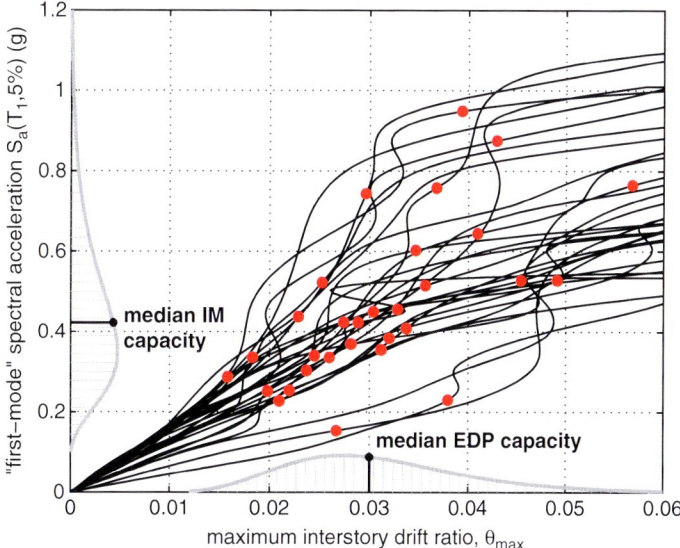

capacity values of one or more EDPs that can be either deterministic, i.e., assumed to be perfectly known, or probabilistic, the latter obviously being the more realistic option. When such threshold values are exceeded, the limit state is deemed to have been violated. The first (in terms of the lowest IM) exceedance of the limit state appears as a single point on each IDA curve (Fig. 4). The resulting distribution of IM coordinates (or IM capacities) of the limit-state points can be used to define the so-called fragility functions, as discussed in the following section.

Use and Applications

IDA has found numerous uses within the framework performance-based earthquake engineering. The three main applications are discussed below.

First, IDA curves can be employed to evaluate fragility curves. A building-level fragility curve is a probability-valued function of the IM. It represents the probability of violating a limit state given the IM level. In terms of IDA, all that is needed is the determination of the limit-state points for each IDA curve. In the case where a single EDP is used to test for limit-state violation, these can be simply defined by simulating equiprobable values of the (random) EDP capacity. Then, one only needs to find the lowest-IM point on a given IDA curve that corresponds to each given EDP value of capacity. For example, when the EDP capacity is deterministic, all such points will align themselves along a single vertical line in the IM-EDP plane that intersects the horizontal axis at that specific value of EDP capacity. In either case, the cumulative distribution function (CDF) of the IM coordinates (or IM capacities) of the limit-state points (Fig. 4) directly provides the corresponding fragility curve. One can either employ the empirical CDF or an analytical approximation (typically via a lognormal assumption) to obtain either a point-by-point accurate estimate of the fragility curve or instead offer a simple (two-parameter) fit.

Second, the results of IDA either in their summarized form (Fig. 3) or as raw IDA curves (Fig. 2) are used within PEER-style frameworks (Cornell and Krawinkler 2000) to determine the distribution of EDP demand given the IM or P (EDP|IM) for any story or component of interest. As such, it is often one of the prominent methods employed for building seismic loss assessment, most notably in FEMA P-58 (FEMA 2012). The results of IDA are essentially convolved with the seismic hazard and appropriate loss/downtime/casualty calculations to derive the mean annual frequency of exceeding any specified level of repair cost, time to repair, or number of people injured or dead.

Finally, IDA has been adopted as the method of choice for determining strength reduction R-factors for USA, as presented in FEMA P-695 (FEMA 2009). R-factors, also known as behavior q-factors, are used in every modern seismic code to appropriately reduce the required design strength of a structure to take advantage of its ductility. Their assessment requires a careful evaluation of the collapse capacity of a comprehensive set of archetypes to ensure that a structure designed according to the specified reduced strength can still offer the required level of safety.

Summary

Incremental dynamic analysis is a method for assessing the distribution of structural response at every level of structural behavior from elasticity to final global dynamic instability. It employs numerous nonlinear dynamic analyses of the structural model subject to a suite of ground motion records that are scaled to multiple levels of intensity. By interpolating such results and appropriately summarizing them, one can estimate any desired statistic of response given the seismic intensity. Additionally, using the IDA results, any number of limit states from minor damage to global collapse may be defined, and the associated fragility functions can be effortlessly estimated. Still, to preserve the validity of such results and ensure unbiased estimation, attention should be paid to the ground motions used and the intensity measure selected to represent them. Then, IDA becomes a powerful tool for performance-based assessment of structures, factually linking seismic hazard and structural behavior to estimate losses, downtime, and casualties.

Cross-References

▶ Analytic Fragility and Limit States [P(EDP|IM)]: Nonlinear Dynamic Procedures
▶ Analytic Fragility and Limit States [P(EDP|IM)]: Nonlinear Static Procedures
▶ Assessment of Existing Structures Using Response History Analysis
▶ Nonlinear Dynamic Seismic Analysis
▶ Performance-Based Design Procedure for Structures with Magneto-Rheological Dampers
▶ Seismic Collapse Assessment
▶ Seismic Fragility Analysis

▶ Seismic Loss Assessment
▶ Time History Seismic Analysis

References

Bianchini M, Diotallevi PP, Baker JW (2009) Predictions of inelastic structural response using an average of spectral accelerations. In: Proceedings of the 10th International Conference on Structural Safety & Reliability (ICOSSAR), Osaka

Cordova PP, Deierlein GG, Mehanny SS, Cornell CA (2000) Development of a two-parameter seismic intensity measure and probabilistic assessment procedure. In: Proceedings of the 2nd U.S.-Japan workshop on performance-based earthquake engineering methodology for reinforced concrete building structures, Sapporo, pp 187–206

Cornell CA, Krawinkler H (2000) Progress and challenges in seismic performance assessment. PEER Center News. http://peer.berkeley.edu/news/2000spring/index.html. Accessed May 2014

FEMA (2009) Quantification of seismic performance factors. FEMA P-695 report, prepared by the Applied Technology Council for the Federal Emergency Management Agency, Washington, DC

FEMA (2012) Seismic performance assessment of buildings. Methodology, vol 1. FEMA P-58-1 report, prepared by the Applied Technology Council for the Federal Emergency Management Agency, Washington, DC

Luco N, Bazzurro P (2007) Does amplitude scaling of ground motion records result in biased nonlinear structural drift responses? Earthq Eng Struct Dyn 36:1813–1835

Luco N, Cornell CA (1998) Effects of random connection fractures on demands and reliability for a 3-storey pre-Northridge SMRF structure. In: Proceedings of the 6th US national conference on earthquake engineering, Seattle

Luco N, Cornell CA (2007) Structure-specific scalar intensity measures for near-source and ordinary earthquake ground motions. Earthq Spectra 23(2):357–392

SAC/FEMA (2000a) Recommended seismic design criteria for new steel moment-frame buildings, FEMA-350, SAC Joint Venture, Federal Emergency Management Agency, Washington, DC

SAC/FEMA (2000b) Recommended seismic evaluation and upgrade criteria for existing welded steel moment-frame buildings, FEMA-351, SAC Joint Venture, Federal Emergency Management Agency, Washington, DC

Shome N, Cornell CA (1999) Probabilistic seismic demand analysis of nonlinear structures. RMS program, report no RMS35, PhD thesis, Stanford University

Shome N, Cornell CA, Bazzurro P, Carballo JE (1998) Earthquakes, records and nonlinear responses. Earthq Spectra 14(3):469–500

Vamvatsikos D (2011) Performing incremental dynamic analysis in parallel. Comput Struct 89(1–2):170–180

Vamvatsikos D (2013) Derivation of new SAC/FEMA performance evaluation solutions with second-order hazard approximation. Earthq Eng Struct Dyn 42:1171–1188

Vamvatsikos D (2014) Seismic performance uncertainty estimation via IDA with progressive accelerogram-wise latin hypercube sampling. ASCE J Struct Eng 140(8):A4014015

Vamvatsikos D, Cornell CA (2002) Incremental dynamic analysis. Earthq Eng Struct Dyn 31(3):491–514

Vamvatsikos D, Cornell CA (2004) Applied incremental dynamic analysis. Earthq Spectra 20(2):523–553

Vamvatsikos D, Cornell CA (2005) Developing efficient scalar and vector intensity measures for IDA capacity estimation by incorporating elastic spectral shape information. Earthq Eng Struct Dyn 34:1573–1600

Veletsos AS, Newmark NM (1960) Effect of inelastic behavior on the response of simple systems to earthquake motions. In: Proceedings of the 2nd world conference on earthquake engineering, vol 2. Japan, pp 895–912

InSAR and A-InSAR: Theory

Andrew Hooper
School of Earth and Environment,
The University of Leeds, Maths/Earth and Environment Building, Leeds, UK

Synonyms

DInSAR; Interferometric synthetic aperture radar; MT-InSAR; Multi-temporal InSAR; Persistent scatterer InSAR; PSI; PS-InSAR; SAR interferometry; SBAS; SB-INSAR; Small baseline InSAR; Time-series InSAR; TS-InSAR

Introduction

The SAR technique allows the formation of high-resolution radar images from the data acquired by side-looking instruments installed on spacecraft, aircraft, or the ground. The fundamentals underlying SAR image processing are presented in Chapter 3 "▶ SAR Images, Interpretation of". Each pixel of an image corresponds to the

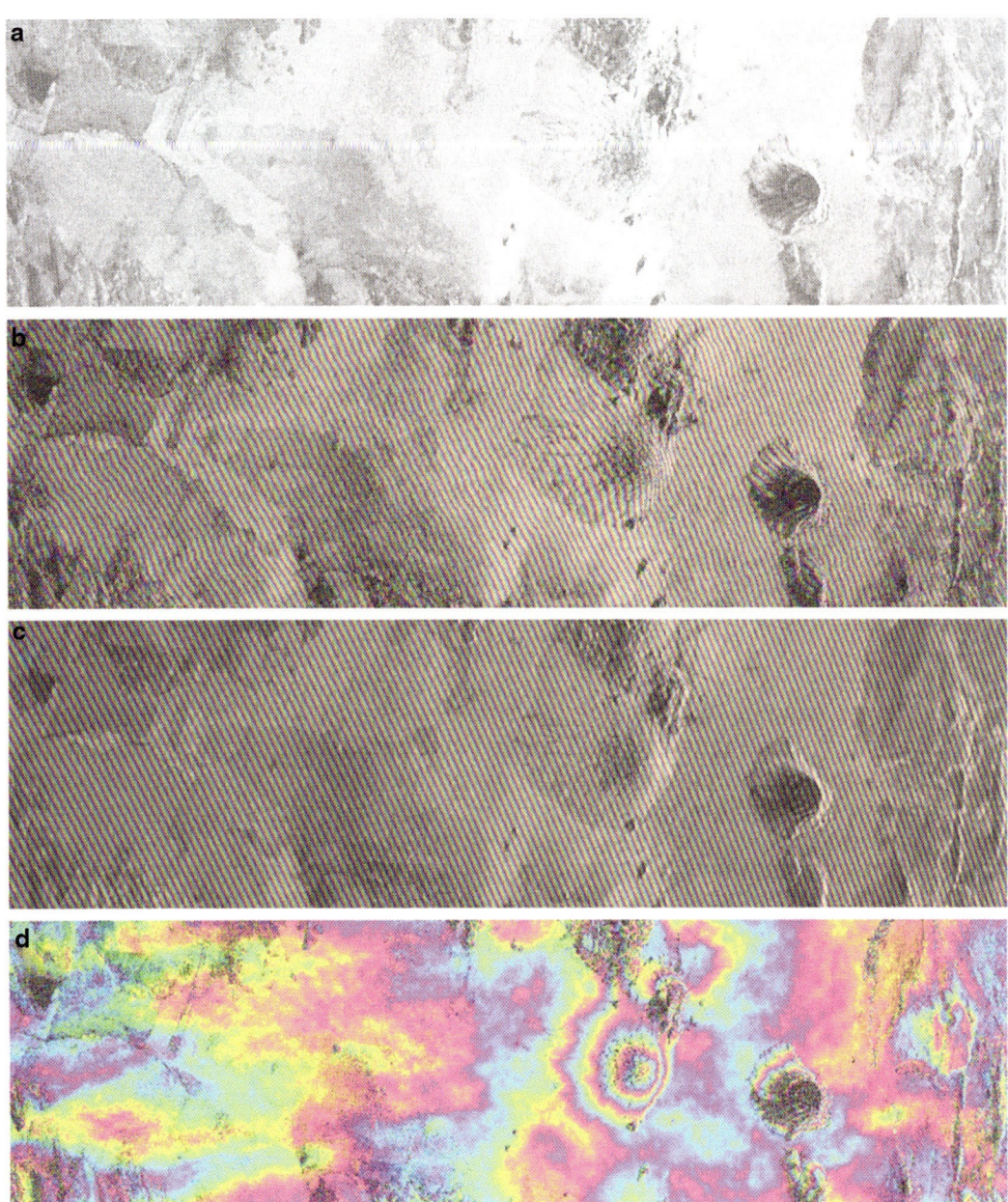

InSAR and A-InSAR: Theory, Fig. 1 Images formed from Envisat data acquired 35 days apart over north Iceland: (**a**) amplitude of master acquisition, (**b**) raw interferogram, (**c**) simulated reference phase projected onto interferometric amplitude, and (**d**) interferogram after reference phase is subtracted from the raw interferogram phase. The perpendicular baseline between the two acquisition geometries is 63 m, giving an altitude of ambiguity of 145 m. In other words each fringe in (**d**) represents another 145 m of elevation, meaning that the a large volcano just east of center (Kollóttadyngja) rises $\sim 3 \times 145 = 435$ m above the surroundings

scattered signal from a resolution element on the ground, which is transmitted and received by the SAR. A pixel is characterized by two values: the amplitude and the phase. While the amplitude of a single image can be interpreted in terms of the backscattering properties of the ground (Fig. 1a), the phase is not very informative because it is a pseudorandom contribution from the configuration of all scatterers within the resolution element. However, provided that the scattering properties of the element are similar for the two acquisitions, the difference in phase between two images can be interpreted in terms of the change in range between the radar instrument and the target, which is the principle of SAR interferometry (InSAR).

For two SAR images acquired at the same moment from slightly different positions, the change in range can be interpreted in terms of the topography, allowing for the production of digital elevation models. For two images acquired at different times, movement of the ground between acquisitions can also contribute to the change in range. Thus, InSAR is also a powerful technique for measuring deformation of the ground, and this is the most common application of InSAR (sometimes referred to as differential InSAR or DInSAR). While the InSAR technique is applied to individual pairs of images, advanced InSAR (A-InSAR) techniques have been developed to analyze multiple SAR images together. These have the advantage of allowing error terms to be better estimated and also provide a framework for looking at the temporal evolution of deformation.

Since the launch of ERS-1 by the European Space Agency (ESA) in 1991, there has always been at least one SAR satellite in operation. Therefore, in addition to the ability to use InSAR to monitor ongoing deformation, there is a long archive of data that allows us to also look at deformation that occurred in the past. In the past two decades, SAR satellites have evolved from mostly C-band sensors with ground-pixel resolutions of tens of meters to C-, X-, and L-band systems that can have spatial resolutions better than 1 m, with some satellites flying in formation or as part of a constellation. Currently, we are entering a golden age for InSAR, with a new generation of satellite missions that have been launched recently or will be launched in the next few years. These include ESA's Sentinel-1 – the first SAR mission that is operational in nature rather than purely scientific, with data being available within minutes of acquisition. A list of all satellites useful for InSAR is provided in Table 1.

InSAR and A-InSAR: Theory, Table 1 Past and present side-looking SAR satellite missions (as of 1 November 2014)

Mission	Period of operation	Wavelength (cm)	Orbit repeat (days)
SEASAT	Jun–Oct 1978	23.5	17
ERS-1	Jul 1991 to Mar 2000	5.66	3 or 35
ERS-2	Apr 1995 to Sep 2011	5.66	3 or 35
JERS-1	Feb 1992 to Oct 1998	23.5	44
SIR-C/X-SAR	9–20 Apr 1994 and 30 Sep to 11 Oct 1994	23.5, 5.8 and 3	N/A
RADARSAT-1	Nov 1995 to present	5.6	24
SRTM	11–22 Feb 2000	5.8 and 3.1	N/A
Envisat	Mar 2002 to Apr 2012	5.63	35[a]
ALOS	Jan 2006 to Apr 2011	23.5	46
COSMO-SkyMed (constellation of four satellites)	Jun 2007 to present	3.1	16
	Dec 2007 to present	3.1	16
	Oct 2008 to present	3.1	16
	Nov 2010 to present	3.1	16
TerraSAR-X	Jun 2007 to present	3.1	11
TanDEM-X	Jun 2010 to present	3.1	11
RADARSAT-2	Dec 2007 to present	5.6	24
Sentinel-1	2014 to present	5.6	12[b]
ALOS-2	2014 to present	23.5	14

[a]From November 2010 to the end of its operation, Envisat operated in a 30 day orbit, which was not optimal for interferometry at high latitudes
[b]In 2016 a second Sentinel-1 satellite will be launched reducing the repeat time to 6 days

In this entry, the theory behind InSAR is described, including all the necessary steps to form and interpret an "interferogram." Subsequently the major A-InSAR algorithms are described.

InSAR

Two SAR images acquired from approximately the same look direction can be combined to form an interferogram. The phase of an interferogram is related to the change in range between the sensor and the ground and forms the basis for interpretation in terms of ground elevation or ground deformation. This section covers the theory behind forming and interpreting a single interferogram (refer to Massonnet and Souyris (2008) for further details on processing).

Coregistration of SAR Images and Resampling

Before an interferogram can be formed from two images, the geometry of one of the images (denoted the "slave") needs to be resampled into the geometry of the other (denoted the "master"). The reason for the difference in geometry is that the sensor acquires the images from slightly different locations. The primary difference is a translation and a linear stretch in range, although the presence of topography introduces higher-order offsets.

To carry out the resampling, the offset for each pixel must first be estimated, in a process known as "coregistration." Coregistration can be achieved using the data themselves by cross-correlation of image amplitude, with satellite orbits being used only to estimate an initial approximate offset for the whole slave image. Range and azimuth offsets are estimated for hundreds to thousands of small subsets of the image. A 2-D polynomial is then fit to both the range and azimuth offsets, using weighting based on correlation. The polynomial, which is typically second or third order, is then used to calculate the offset for each individual pixel.

However, if the satellite orbits and the topography are accurately known, which tends to be the case with modern satellites, it is possible to calculate the coregistration offset for each individual pixel geometrically. Amplitude cross-correlation is then used only to estimate a single offset in azimuth, due to timing errors, and a single offset in range, due to the average difference in atmospheric delay (see section "Atmospheric Phase").

After coregistration, the slave is resampled into the master geometry by interpolation using an appropriate kernel, typically a sinc or raised cosine function, depending on the filtering that has been applied to the image. The output from the coregistration procedure is used to calculate the position of each master pixel within the slave image and the interpolation is used to calculate the value of the slave image at that point.

Raw Interferogram Formation

An interferogram is generated simply by multiplying the master image, pixel by pixel, by the complex conjugate of the resampled slave image. The phase of the interferogram is the phase difference between master and slave images and the amplitude is the product of the amplitude of the two images (Fig. 1b).

The phase of the raw interferogram contains contributions from several sources:

$$\phi = \phi_{\text{ref}} + \phi_{\text{topo}} + \phi_{\text{def}} + \phi_{\text{atmos}} + \phi_{\text{noise}}, \quad (1)$$

where ϕ_{ref} is the reference surface phase due to a difference in look angle of the master and slave (section "Reference Phase"), ϕ_{topo} is phase due to topography (section "Topographic Phase"), ϕ_{def} is phase due to deformation (section "Deformation Phase"), ϕ_{atmos} is phase due to propagation through the atmosphere (section "Atmospheric Phase"), and ϕ_{noise} is phase noise due to decorrelation and other effects (section "Decorrelation").

Once an interferogram is generated, the next steps involve reducing the contributions from sources which are not of interest, in order to isolate those which are.

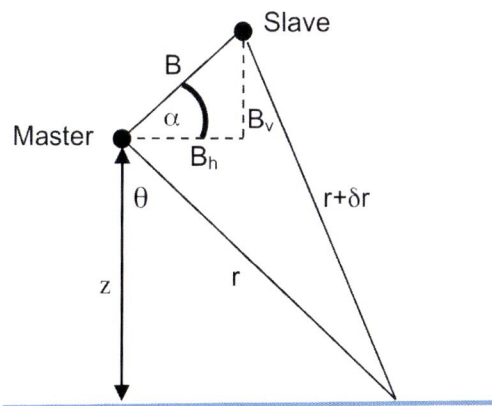

InSAR and A-InSAR: Theory, Fig. 2 Imaging geometry (not to scale; B is typically 10^3–10^4 times smaller than r). Master and slave sensors are flying into the page and looking *down* and to the *right*

Reference Phase

There is a contribution to the interferometric phase from the change in range resulting from the two images being acquired from different points in space. This can be separated into the reference phase expected if the ground surface were at a reference surface (e.g., WGS84) and phase due to modulation of the reference surface by topography. The reference phase can always be estimated and subtracted, although the accuracy of the estimate depends on the accuracy of the orbit data.

The change in range due to geometry for a reference surface can be estimated as follows (Fig. 2). From the law of cosines,

$$\frac{r^2 + B^2 + (r+\delta r)^2}{2rB} = \cos(\alpha + \pi/2 - \theta). \quad (2)$$

where r is the range from the satellite to the ground for the master acquisition, B (baseline) is the distance between the master and slave sensor, α is the angle between the horizontal and the baseline vector, and θ is the look angle. Neglecting second-order terms in δr^2 and B/r, this reduces to

$$\delta r = -B\left(\sqrt{1 - \frac{z_0^2}{r^2}}\cos\alpha - \frac{z_0}{r}\sin\alpha\right), \quad (3)$$

where z_0 is the height of the sensor above the reference surface. Change in range is converted into interferometric phase by dividing by the wavelength, λ, multiplying by -4π (-4π rather than -2π as the signal travels a distance $2\delta r$ farther), and dropping whole phase cycles. The reference phase, also known as "flat-Earth phase," is then given by

$$\phi_{\text{ref}} = W\left\{\frac{4\pi B}{\lambda r}\left(\sqrt{r^2 - z_0^2}\cos\alpha - z_0\sin\alpha\right)\right\}, \quad (4)$$

where $W\{\cdot\}$ is the "wrapping" operator that drops whole phase cycles (Fig. 1c).

Topographic Phase

When generation of a DEM is the goal of the InSAR processing, the remaining phase is interpreted as the topographic contribution. To minimize noise from non-topographic contributions, simultaneous acquisitions therefore provide the best data for this application. In the case where isolation of another signal (usually deformation) is the goal of the processing, a DEM is used to simulate the topographic contribution and subtract it. If topography is considered as a perturbation to the reference surface, the relationship between elevation and phase can be determined by substituting z for z_0 in Eq. 4 and differentiating, giving

$$\frac{d\phi}{dz} = -W\left\{\frac{4\pi B_\perp}{\lambda r \sin\theta}\right\}, \quad (5)$$

where $B_\perp$ is the perpendicular component of the baseline, or "perpendicular baseline."

$\frac{d\phi}{dz}$ gives change in elevation with respect to the vertical vector below the satellite. To put topography in terms of the local vertical vector, the reference frame needs to be rotated, which has the effect that $\sin\theta$ becomes $\sin\theta_i$ where θ_i is the angle of incidence at the reference surface, rather than the look angle.

Considering only the first term in a Taylor expansion, the total geometric phase is given by

$$\phi \approx W\left\{\phi_{\text{ref}} + \frac{4\pi B_\perp}{\lambda r \sin\theta_i}h\right\}, \qquad (6)$$

where $h = -(z - z_0)$, the elevation above the reference surface. Topographic phase is then approximately proportional to elevation, which means that phase contours follow elevation contours, with one color "fringe" representing $\sim \lambda r \sin\theta_i / 2B_\perp$ m (Fig. 1d). This value is known as the *altitude of ambiguity*. Increasing perpendicular baseline therefore provides greater sensitivity to elevation and also leads to greater decorrelation (section "Decorrelation").

Note that for greater accuracy, higher-order terms in the Taylor series must also be taken into account. Considering only the first-order term equates to a 40 m DEM error over 2,000 m of elevation change for ERS or typical Envisat acquisitions.

Given a DEM it is therefore relatively straightforward to project it into the radar coordinate system, calculate the topographic contribution to the phase using the formula above, and subtract it from the interferometric phase. Note that any errors in the DEM will lead to residual phase error that is proportional to the perpendicular baseline. Therefore, when the goal of the processing is to isolate a signal other than that due to topography, a small value for the perpendicular baseline is good for two separate reasons: firstly, the residual phase due to any DEM error is small and secondly, the geometric decorrelation is small (section "Decorrelation").

When the aim is to generate a DEM from the interferometric phase, on the other hand, reversing the operation is not quite so simple, as the wrapping operator must first be inverted in a process known as *phase unwrapping*. This topic is covered in section "Phase Unwrapping." Phase unwrapping provides the total phase difference (including whole phase cycles) between any two pixels, but not the absolute total phase for each pixel, so extra information is required to "tie" the DEM to the ground surface.

Deformation Phase

The most common application of InSAR is to measure deformation of the ground surface. When an interferogram is formed between images acquired at different times, any change in range due to movement of the ground between the acquisitions will contribute to the interferometric phase; each displacement of half the wavelength of the carrier signal away from the satellite will increase the phase delay by one more phase cycle (Fig. 3):

$$\phi_{\text{def}} = -W\left\{\frac{4\pi}{\lambda}\delta r\right\}, \qquad (7)$$

where δr is the change in range. Note that only the component of any displacement in the line-of-sight direction will result in a range change, and any displacement perpendicular to the line of sight will be invisible in the interferogram.

As in the case of DEM generation (section "Topographic Phase"), to interpret the phase as displacement, it must first be unwrapped (section "Phase Unwrapping"). The resulting unwrapped phase can then be used to derive relative line-of-sight displacements between any two points in the interferogram. To derive absolute displacements, the displacement at one or more points in the interferogram must be known a priori.

Atmospheric Phase

Propagation of the signal through the atmosphere influences the phase delay, by an amount that depends on atmospheric conditions, which are spatially variable (Hanssen 2001). Most of the spatial variability in the induced interferometric phase occurs in the ionosphere and the troposphere. The influence of the ionosphere on phase delay tends to be long wavelength (hundreds of km) and, except in the case of L-band data, is commonly ignored. There are, however, several methods for estimating the effect, if necessary (Meyer 2011). Phase delay during propagation through the troposphere, on the other hand, is equally significant at all microwave frequencies and variability is significant over short distances.

Propagation through the troposphere induces an additional phase delay which depends primarily on pressure, temperature, and humidity. These properties vary on two characteristic lateral

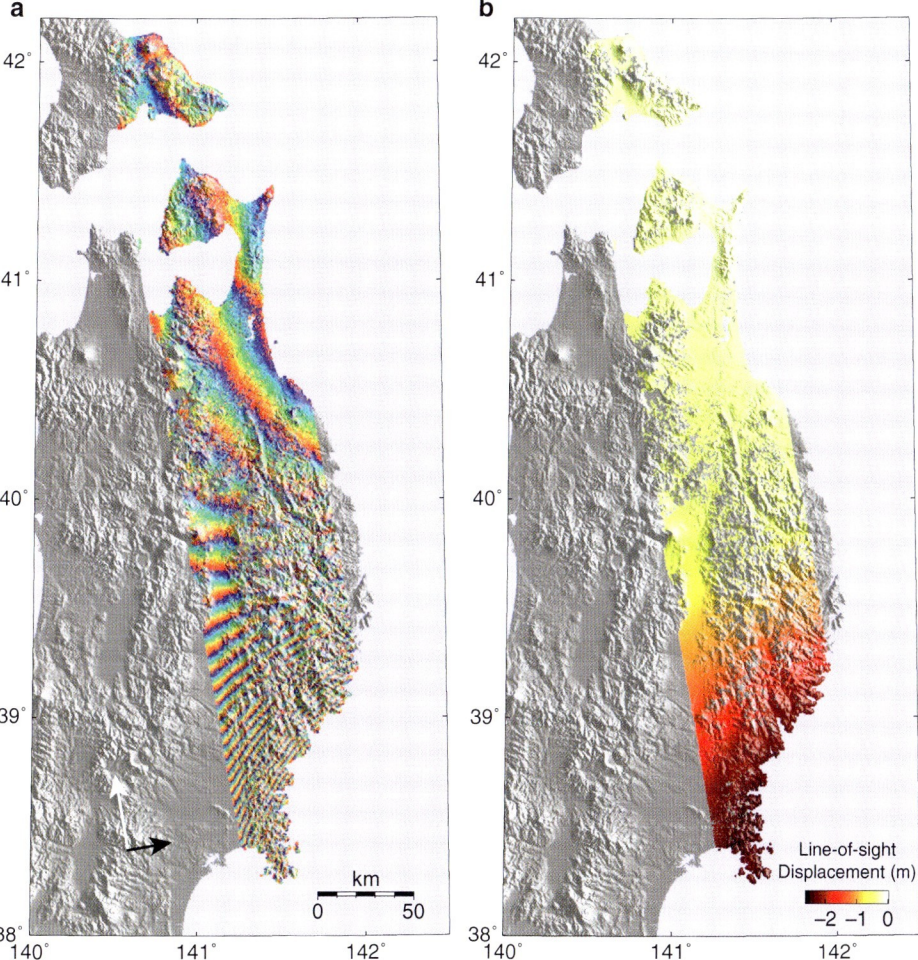

InSAR and A-InSAR: Theory, Fig. 3 An example interferogram displaying coseismic deformation for the Mw = 9.0 Tohoku-Oki earthquake, Japan, which occurred on 11 March 2011. The SAR data were acquired by the L-band ALOS satellite on 28 October 2010 and 15 March 2011. In (**a**) the interferometric phase is displayed with each color cycle representing 11.8 cm of displacement away from the satellite, which was moving in the direction of the *white arrow* and looking in the direction of the *black arrow*. In (**b**) the phase is integrated and converted to line-of-sight displacement, indicating that the southernmost point moved more than 2.5 m away from the satellite, relative to the northern end. The orbit data provided for 15 March were preliminary and so some orbital signal remains. Signal is partially lost in the mountainous regions due to decorrelation, caused chiefly by snow (Modified after Hooper et al. (2012))

length scales: a short scale (a few km) induced by turbulent troposphere dynamics and a long scale (10s of km). The long-scale variation induces not only long-wavelength artifacts but also interferometric fringes that are strongly correlated with topography, as phase delay depends on how far through the troposphere the signal has traveled.

The turbulent variation is difficult to model and currently can only be reduced by filtering in time and space (see section "Phase Unwrapping"). However, temporal filtering methods are less successful for reducing topographically correlated delay, especially when deformation is not linear in time, because SAR data sets typically do not sample seasonal atmospheric fluctuations evenly, resulting in biased estimates.

Alternatively, topographically correlated tropospheric artifacts can be partially corrected using the information contained within the SAR data, based on the correlation between phase and elevation in nondeforming areas. In addition, the long-wavelength artifacts can be estimated over a whole image by consideration of how this correlation varies (Bekaert et al. 2015).

Complementary data sets, such as dense GNSS or meteorological measurements acquired at the same time as the SAR images, or global meteorological models, provide an alternative method for estimating the topographically correlated and long-wavelength tropospheric artifacts (Jolivet et al. 2011).

Decorrelation

A resolution element on the ground usually contains many scatterers, all of which contribute to the echo for that element. If the scatterers move with respect to each other, the echoes from the scatterers sum differently, a phenomenon known as *temporal decorrelation*. This induces a change to the interferometric phase. The echoes also sum differently if viewed from a different angle (*geometric* or *spatial decorrelation*) or probed with a different range of Doppler frequencies (*rotational decorrelation*). There is also *thermal decorrelation* term due to the signal-to-noise properties of the radar.

How closely the phase in one acquisition tracks that of another can be measured in terms of the complex correlation coefficient, or coherence, defined as

$$\gamma = \frac{\mathrm{E}\left[u_1 u_2^*\right]}{\sqrt{\mathrm{E}\left[|u_1|^2\right]\mathrm{E}\left[|u_2|^2\right]}}, \qquad (8)$$

where u_1 and u_2 are the complex signals from each image and $\mathrm{E}[\cdot]$ represents the expected value. In practice, the expected values are usually estimated as the mean values for pixels in a small spatial window. Values of $|\gamma|$ range between 0 (no correlation) and 1 (perfect correlation).

Because the contribution from decorrelation to the interferometric phase is pseudorandom, the effects can be reduced, at the cost of resolution, by averaging the signal of all pixels within a window of some specified size, or *multilooking*, or by spatial filtering (Goldstein and Werner 1998).

Rotational decorrelation can be lessened by only using the Doppler frequencies that were used to sample both images; frequencies outside of the overlapping frequency band are discarded for each image by bandwidth filtering. This has the effect of coarsening resolution, as resolution = 1/bandwidth. In the case of there being no overlapping bandwidth, the result is total decorrelation.

In a similar way, geometric decorrelation can also be reduced by bandwidth filtering, at a cost of resolution (Gatelli et al. 1994). This is possible because changing the look angle slightly is approximately equivalent to shifting the chirp frequency,

$$\Delta f \approx -\frac{c}{\lambda}\frac{\Delta\theta}{\tan(\theta-\alpha)}, \qquad (9)$$

where c is the speed of light, λ is the wavelength, $\Delta\theta$ is the change in look angle, and α is the angle between the horizontal and the baseline vector. $\Delta\theta$ is proportional to the perpendicular baseline; thus, if the perpendicular baseline is too large, there is no overlapping bandwidth, resulting in total decorrelation. The perpendicular baseline at which this occurs is referred to as the *critical baseline*, which is about 1,100 m for ERS and Envisat satellites as an example.

Phase Unwrapping

Phase unwrapping is the process of estimating the difference in the number of whole phase cycles between an arbitrary reference pixel and every other pixel. Pictorially, this can be considered as counting the number of color fringes from the reference pixel (Fig. 3). The basic assumption of phase-unwrapping algorithms is that the phase field is generally sampled at above the Nyquist rate, so that the phase difference between neighboring pixels is generally less than half of a phase cycle. The basis then, for most algorithms, is to calculate the phase difference between neighboring pixels, wrap this into the interval between $-1/2$ and $1/2$ of a phase cycle, and then integrate these phase differences.

However, it is rarely the case across that this assumption holds across a whole interferogram, due to steep gradients and/or discontinuities in the topographic or deformation phase or due to decorrelation noise. In this case, the path taken in the integration step matters – the algorithm needs to avoid integrating between pixels where the phase difference is more than half of a cycle. Mathematically these can be considered as branch cuts. In order to place these branch cuts, an extra constraint is required, and for most algorithms this can be framed in terms of minimization of a weighted function of the residuals between the estimated unwrapped phase differences and the wrapped phase differences:

$$\sum_{i,j} w_{i,j}^{(x)} \left| \Delta\phi_{i,j}^{(x)} - \Delta\psi_{i,j}^{(x)} \right|^p + \sum_{i,j} w_{i,j}^{(y)} \left| \Delta\phi_{i,j}^{(y)} - \Delta\psi_{i,j}^{(y)} \right|^p,$$

(10)

where $\Delta\phi^{(x)}$ and $\Delta\psi^{(x)}$ are the unwrapped and wrapped phase differences, respectively, between pixels in the x direction, $\Delta\phi^{(y)}$ and $\Delta\psi^{(y)}$ are the equivalent in the y direction, and w are user-defined weights (Chen 2001). The summations are carried out in both x and y directions over all i and j, respectively. This is sometimes referred to as an L^p-norm objective function, although strictly speaking, to meet the condition of positive scalability, the sum must be raised to the power of $1/p$, and p must be greater than or equal to 1. However, for $p \geq 1$, the solution that minimizes this objective function also minimizes the L^p-norm.

Two common algorithms that are applied to solve phase-unwrapping problems are the branch-cut algorithm (Goldstein et al. 1988) and the minimum cost flow algorithm (Costantini 1998). In the former case, the algorithm seeks to minimize the total number of nonzero residuals, equivalent to setting $p = 0$ and $w_{i,j}^{(x)} = w_{i,j}^{(y)} = 1$. In the latter case, the algorithm seeks to minimize the weighted sum of the absolute value of the residuals, equivalent to $p = 1$.

The third well-established algorithm implements a statistical cost flow methods and seeks to find the most probable solution overall for any given probability distributions allocated to each phase difference (Chen 2001). The probability distributions are assigned based on the statistics of various measures, such as correlation and amplitude.

A-InSAR

Displacements can be estimated more accurately by processing many images together, rather than the two image approaches described above. While the approach to conventional InSAR is reasonably standardized, there are many variations on A-InSAR algorithms. However, an overview of the main approaches is described in this section.

The simplest approach for combining many images is to sum or "stack" the unwrapped phase of many conventionally formed interferograms. The deformation signal reinforces, whereas other signals typically do not. However, this approach is only appropriate when the deformation is episodic or purely steady state, with no seasonal deformation. Even then it is not optimal, as non-deformation signals are reduced only by unweighted averaging rather than by optimized estimation.

Algorithms for time-series analysis of SAR data have thus been developed to better address the issues facing conventional InSAR; decorrelation is tackled by using phase behavior in time to select pixels for which decorrelation noise is minimized, and non-deformation signals are estimated by a combination of modeling and filtering of the time series. These time-series algorithms fall into two categories, the first being persistent scatterer InSAR, which targets pixels whose scattering properties remain consistent both in time and from variable look directions, and the second being a more general small baseline approach. Because the two approaches are optimized for resolution elements with different scattering characteristics, they are complimentary, and techniques that combine both approaches are able to extract the signal with greater coverage than either method alone (Hooper 2008).

Persistent Scatterer InSAR

Decorrelation is caused by the contributions from scatterers within a resolution element summing differently. This can be due to a relative movement of the scatterers, a change in the looking direction of the radar platform, or the appearance or disappearance of scatterers, as in the case of snow cover. If one scatterer returns significantly more energy than other scatterers within a resolution element, however, decorrelation is reduced. This is the principle behind a "persistent scatterer" (PS) pixel (sometimes referred to as a "permanent scatterer"). Urban environments typically contain many of these dominant scatterers, for instance, roofs oriented such that they reflect energy directly back to the sensor and perpendicular structures that lead to a "double bounce," where energy is reflected once from the structure and once from the ground, returning directly back to the sensor. Dominant scatterers can also occur in areas without man-made structures, e.g., appropriately oriented rocks, but there are fewer of them, and they tend to be less dominant.

PS algorithms operate on a time series of interferograms all formed with respect to a single "master" SAR image (Fig. 4a). The first step in the processing is the identification and selection of the usable PS pixels. There are two approaches to this; the first relies on modeling the deformation in time (Kampes 2005), and the second relies on the spatial correlation of the deformation (Hooper et al. 2007). In the first approach, the phase is unwrapped during the selection process, by fitting a temporal model of evolution to the wrapped phase difference between pairs of nearby PS, although later enhancements to the technique allow for non-model-based improvements to the unwrapping. In the second approach a phase-unwrapping algorithm is applied to the selected pixels without assuming a particular model for the temporal evolution, other than it should be generally smooth.

In both approaches, deformation phase is then separated from atmospheric phase and noise by filtering in time and space, the assumption being that deformation is correlated in time, atmosphere is correlated in space but not in time, and noise is uncorrelated in space and time. Additional approaches to reducing atmospheric phase (section "Atmospheric Phase") can be applied prior to the filtering. In comparative studies between the two PS approaches, estimates for the deformation estimates tend to agree quite well, but the second approach tends to result in better coverage, particularly in rural areas, which is where most volcanoes are sited.

The result of PS processing is a time series of displacement for each PS pixel, with much reduced noise terms (Fig. 5). The technique also has the advantage of being able to associate the deformation with a specific scatterer, rather than

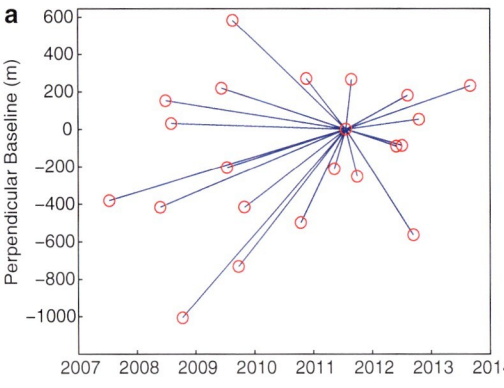
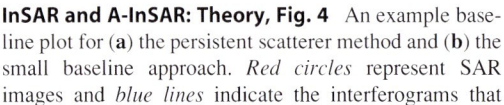
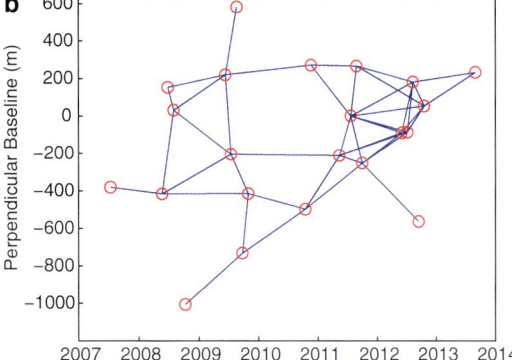

InSAR and A-InSAR: Theory, Fig. 4 An example baseline plot for (**a**) the persistent scatterer method and (**b**) the small baseline approach. *Red circles* represent SAR images and *blue lines* indicate the interferograms that are formed. Perpendicular baseline refers to the component of the satellite separation distance that is perpendicular to the look direction and is proportional to the difference in look angle

InSAR and A-InSAR: Theory, Fig. 5 Average displacement rates (mm/year) for Venice using a persistent scatterer approach applied to COSMO-SkyMed data from 2008 to 2011. The image background is an aerophotograph acquired in 2009. Negative values indicate settlement, positive mean uplift (Modified after Tosi et al. (2013))

a resolution element of dimensions dictated by the radar system, usually on the order of many meters. This allows for very high-resolution monitoring of infrastructure.

Small Baseline InSAR

A drawback of the PS technique for many applications is that the number of PS pixels in rural environments may be limited. For non-PS pixels, containing no dominant scatterer, phase variation due to decorrelation may be large enough to obscure the underlying signal. However, by forming interferograms only between images separated by a short time interval and with a small difference in look direction, decorrelation is minimized, and for some resolution elements can be small enough that the underlying signal is still detectable. Pixels for which the phase decorrelates little over short time intervals are the targets of small baseline methods.

Interferograms are formed between SAR images that are likely to result in low decorrelation noise, in other words, those that tend to minimize the difference in time and look direction (Fig. 4b). As it is not possible to minimize both of these at once, assumptions are made about their relative importance, based on the scattering characteristics of the area of interest.

The interferograms can then be multilooked to further decrease decorrelation noise, either using a standard windowing approach or by using amplitude to identify pixels with similar scattering characteristics (Ferretti et al. 2011). However, there may be isolated ground resolution elements with low decorrelation properties that are surrounded by elements that decorrelate highly, such as a small clearing in a forest, for which multilooking will increase the noise. Therefore, algorithms have been developed that operate at full resolution (Hooper 2008), with the option to reduce resolution later in the processing chain by selective multilooking.

Pixels are selected based on their estimated coherence in each of the interferograms, using either standard ensemble coherence estimation or enhanced techniques in the case of

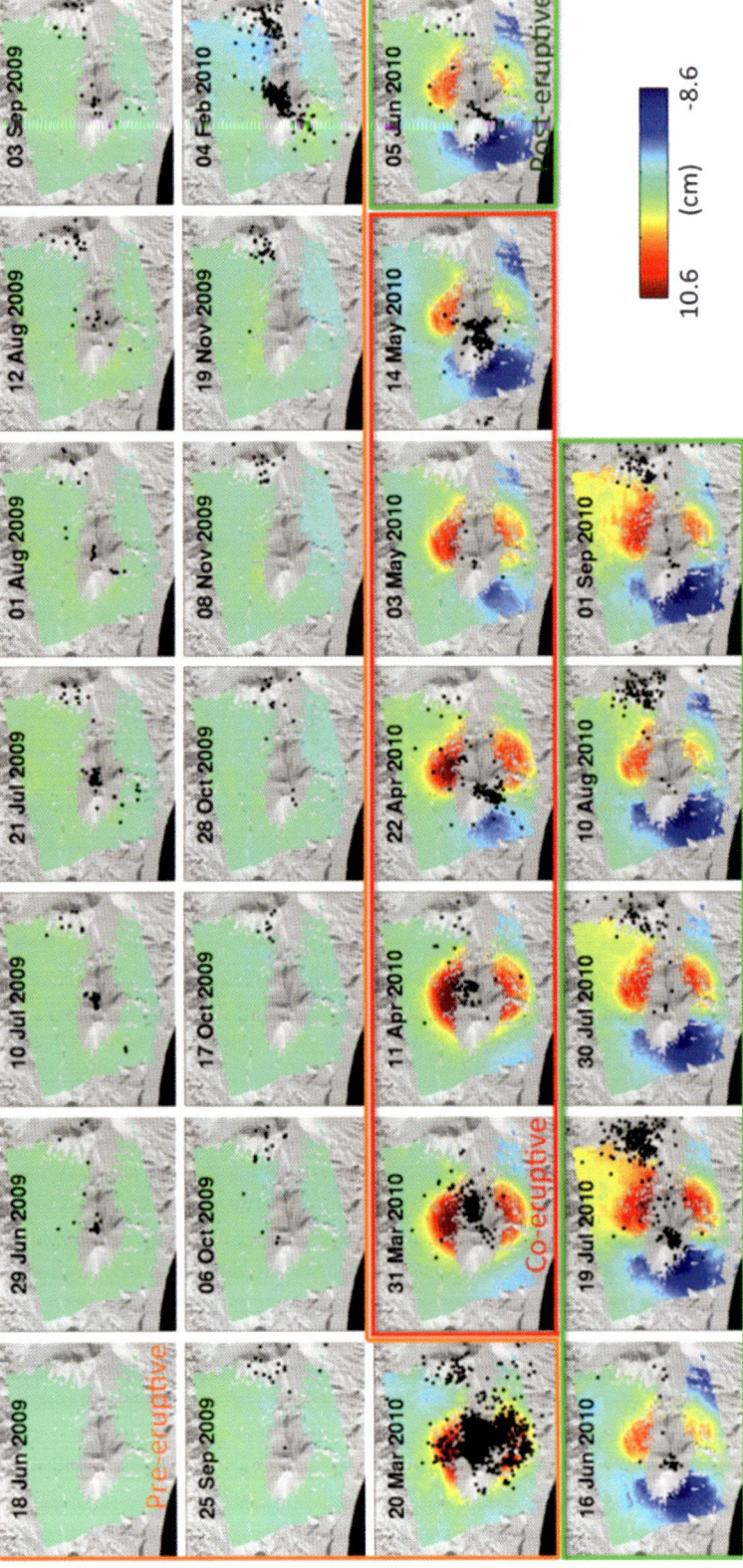

InSAR and A-InSAR: Theory, Fig. 6 Cumulative line-of-sight displacement over Eyjafjallajökull Volcano, Iceland, from 18 June 2009 to 01 September 2010, processed using a small baseline approach (track 132 ascending mode). The co-eruptive interval includes a flank eruption, during which there was very little deformation, and the later summit eruption, during which the volcano deflated. The background is shaded topography. The hole in the data is due to ice and snow cover. *Black dots* are earthquake epicenters for each epoch (Icelandic Meteorological Office) (Modified after Pinel et al. (2014))

full-resolution algorithms. The phase is then unwrapped either spatially in two dimensions using standard approaches or using the additional dimension of time in three-dimensional approaches (Pepe and Lanari 2006; Hooper 2010). At this point the phase is inverted to give the phase at each acquisition time with respect to a single image, using least squares, singular value decomposition, or minimization of the L^1-norm (Fig. 6).

Separation of deformation and atmospheric signals is achieved by filtering the resulting time series in time and space, as in the PS approach. Alternatively, if an appropriate model for the evolution of deformation in time is known, the different components can be directly estimated from the small baseline interferograms (Biggs et al. 2007).

Summary

InSAR is a powerful tool for high-resolution measurement of surface heights with meter scale accuracy, surface displacements with cm accuracy, and surface velocities with mm/year accuracy. Spatial resolution varies from under a meter to tens of meters, depending on the sensor and mode of operation.

Studies of ground deformation, in particular, have benefitted from more than 20 years of continuous satellite SAR observations. This archive is about to be enhanced by new satellite SAR missions, in addition to the growing application of ground-based and airborne SAR systems, which will provide improved temporal and spatial resolution, broader geographic coverage, and better availability to the scientific community.

The main advantage of InSAR remains its ability to collect useful imagery without regard to the time of the day or weather conditions and to record surface deformation on the scale of mm/year over broad regions with a spatial resolution on the order of meters. InSAR applications to seismology can thus be both regional in scope (entire tectonic regions) and highly focused (individual buildings).

Cross-References

▶ Damage Detection in Built-Up Areas Using SAR Images
▶ Earthquake Magnitude Estimation
▶ SAR Images, Interpretation of
▶ SAR Tomography for 3D Reconstruction and Monitoring
▶ Urban Change Monitoring: Multi-temporal SAR Images

References

Bekaert DPS, Hooper A, Wright TJ (2015) A spatially-variable power-law tropospheric correction technique for InSAR data. J Geophys Res Solid Earth, doi:10.1002/2014JB011558

Biggs J, Wright T, Lu Z, Parsons B (2007) Multi-interferogram method for measuring interseismic deformation: Denali fault, Alaska. Geophys J Int 170:1165–1179

Chen CW (2001) Statistical-cost network-flow approaches to two-dimensional phase unwrapping for radar interferometry. PhD thesis, Stanford University

Costantini M (1998) A novel phase unwrapping method based on network programming. IEEE Trans Geosci Remote Sens 36(3):813–821

Ferretti A, Fumagalli A, Novali F, Prati C, Rocca F, Rucci A (2011) A new algorithm for processing interferometric data-stacks: SqueeSAR. IEEE Trans Geosci Remote Sens 49(9):3460–3470

Gatelli F, Guamieri AM, Parizzi F, Pasquali P, Prati C, Rocca F (1994) The wave-number shift in SAR interferometry. IEEE Trans Geosci Remote Sens 32(4):855–865

Goldstein RM, Werner CL (1998) Radar interferogram filtering for geophysical applications. Geophys Res Lett 25(21):4035–4038

Goldstein RM, Zebker HA, Werner CL (1988) Satellite radar interferometry: two-dimensional phase unwrapping. Radio Sci 23(4):713–720

Hanssen RF (2001) Radar interferometry data interpretation and error analysis. Springer, Dordrecht

Hooper A (2008) A multi-temporal InSAR method incorporating both persistent scatterer and small baseline approaches. Geophys Res Lett 35:L16302

Hooper A (2010) A statistical-cost approach to unwrapping the phase of InSAR time series. European Space Agency (Special Publication) ESA, Noordwijk, The Netherlands SP-677

Hooper A, Segall P, Zebker H (2007) Persistent scatterer InSAR for crustal deformation analysis, with application to Volcán Alcedo, Galápagos. J Geophys Res 112: B07407

Hooper A, Bekaert D, Spaans K, Arıkan M (2012) Recent advances in SAR interferometry time series analysis for measuring crustal deformation. Tectonophysics 514:1–13

Jolivet R, Grandin R, Lasserre C, Doin M-P, Peltzer G (2011) Systematic InSAR tropospheric phase delay corrections from global meteorological reanalysis data. Geophys Res Lett 38:l17311. doi:10.1029/2011GL048757

Kampes BM (2005) Displacement parameter estimation using permanent scatterer interferometry. PhD thesis, Delft University of Technology

Massonnet D, Souyris JC (2008) Imaging with synthetic aperture radar. EPFL-CRC Press, Lausanne, 280 p

Meyer FJ (2011) Performance requirements for ionospheric correction of low-frequency SAR data. IEEE Trans Geosci Remote Sens 49(10):3694–3702

Pepe A, Lanari R (2006) On the extension of the minimum cost flow algorithm for phase unwrapping of multitemporal differential SAR interferograms. IEEE Trans Geosci Remote Sens 44(9):2374–2383

Pinel V, Poland M, Hooper A (2014) Volcanology: lessons learned from synthetic aperture radar imagery. J Volcanol Geotherm Res 289:81–113

Tosi L, Teatini P, Strozzi T (2013) Natural versus anthropogenic subsidence of Venice. Sci Rep 3:2710

Insurance and Reinsurance Models for Earthquake

Katsuichiro Goda[1], Friedemann Wenzel[2] and James Edward Daniell[2]

[1]Department of Civil Engineering, University of Bristol, Bristol, UK

[2]Geophysical Institute, Karlsruhe Institute of Technology, Karlsruhe, Germany

List of Acronyms

AEL	Annual expected loss
ART	Alternative risk transfer
CAT	Catastrophe
CEA	California Earthquake Authority
CNSF	Comision Nacional de Seguros y Fianzas
CVaR	Conditional value at risk
DRR	Disaster risk reduction
EP	Exceedance probability
EQC	Earthquake Commission
GDP	Gross domestic product
GP	Generalized Pareto
HNDECI	Hybrid Natural Disaster Economic Conversion Index
ICI	Iceland Catastrophe Insurance
JERC	Japan Earthquake Reinsurance Company
LP-HC	Low-probability high-consequence
PML	Probable maximum loss
POT	Peak over threshold
SEL	Scenario expected loss
SUL	Scenario upper loss
TCIP	Turkish Catastrophe Insurance Pool
TREIF	Taiwan Residential Earthquake Insurance Fund
VaR	Value at risk

Introduction

Building sustainable and resilient communities against extremely large earthquakes is a global and urgent problem in active seismic regions. A catastrophic earthquake and its cascading events, such as tsunami and fire, affect multiple structures and infrastructure simultaneously and have far-reaching economic impact across various sectors in a complex manner. Incurred seismic damage includes loss of life and limb, direct financial loss to building properties and lifeline facilities (capital stock), and indirect loss due to the ripple effects of the direct loss across regional and national economies. All these consequences are revealed in an abrupt reduction of gross domestic product (GDP) for national economy and of stock indices and prices for companies.

One of such devastating events was the 2011 M_w9.0 Tohoku earthquake in Japan among recent earthquake disasters around the world (e.g., 2004 Sumatra, 2008 Wenchuan, and 2010 Haiti earthquakes). For instance, a post-earthquake economic impact analysis of the 2011 Tohoku earthquake by Kajitani et al. (2013) revealed that the economic impact was significant not only in directly affected regions but also in regions outside the damaged areas due to disruptions of supply chains and domestic/international trade networks. The financial consequences of

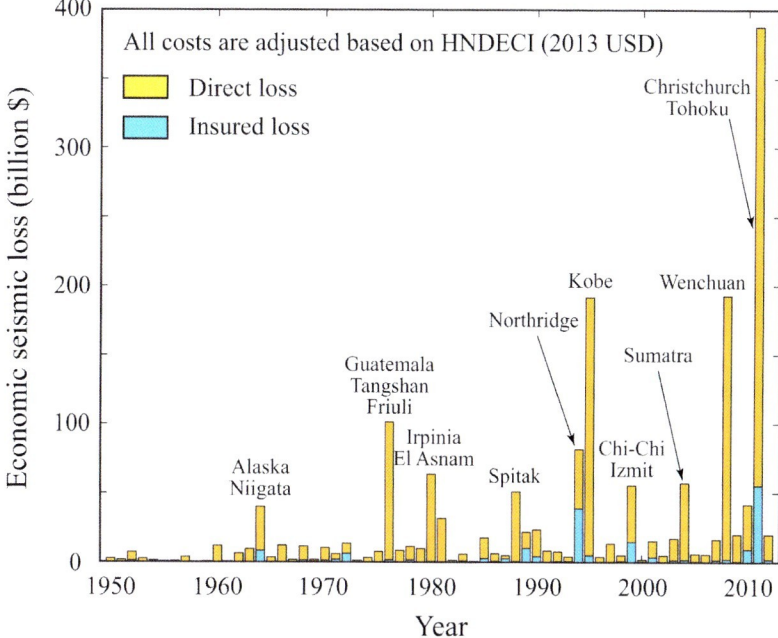

Insurance and Reinsurance Models for Earthquake, Fig. 1 Earthquake-related economic and insurance losses

such major earthquakes have long-term impacts (e.g., stagnant regional economy due to loss of market share). Households in disaster areas suffer from loss of asset and income; recovery from earthquakes can be extremely difficult, and financial public support/aid from governments and municipalities is not sufficient to cover the loss. Moreover, catastrophic earthquakes impose tremendous financial stress on insurers and reinsurers underwriting earthquake insurance policies (Grace et al. 2003; Muir-Wood 2011). The CATDAT damaging earthquakes database (Daniell et al. 2011) reports that the 2011 Tohoku earthquake and tsunami, the 2010–2011 Christchurch (New Zealand) sequences, and the 2010 Maule (Chile) earthquake rank in the top 10 list for the highest insured loss since 1900, exceeding $10 billion. Figure 1 shows a history of estimated economic and insurance losses since 1950 according to the CATDAT database. The CATDAT values are adjusted based on the Hybrid Natural Disaster Economic Conversion Index (HNDECI), which takes into account temporal as well as geographical variation of consumer price indices, wage and low-wage indices, exchange rate, and other factors to facilitate the uniform and consistent comparison of the disaster consequences (Daniell et al. 2011). The figure illustrates the significant magnitude of earthquake catastrophes, which is caused by a combination of physical hazard, exposure, vulnerability, and socioeconomic factors of the interconnected economies around the world.

Recent catastrophes have highlighted that the financial consequences to stakeholders can be significant. It is necessary to enhance the capability of dealing with such low-probability high-consequence (LP-HC) events through integrated seismic risk management by combining hard and soft disaster risk reduction (DRR) measures. Generally, hard and soft measures are complementary, and such integrated DRR solutions are robust against unforeseen and uncertain events. An important difference between hard risk mitigation measure (e.g., seismic retrofitting) and soft risk transfer measure (e.g., insurance) is that the former physically reduces the actual extent of seismic damage, while the latter transfers the incurred loss to a third party based on a pre-agreed risk-sharing scheme. Because the recovery process is (highly) nonlinear and dependent on various post-disaster situations (e.g., prolonged business interruption may result in loss of share in a competitive market), earthquake insurance has the effects of accelerating the recovering process.

This entry is focused upon earthquake risk management using insurance and reinsurance. Earthquake insurance is a form of loss indemnity contract, where an insurer pays policyholders in the event of a major earthquake that causes damage to their properties. Insurance smoothes out the policyholder's wealth at different contingent states (i.e., no earthquake versus major earthquake), while an insurer underwrites policyholders' potential seismic risks collectively in exchange for insurance premium. Because an insurer may face an unacceptable level of insolvency risks due to catastrophic earthquakes, the risks may be transferred to a reinsurer, i.e., reinsurance refers to a risk-sharing contract between insurer and reinsurer. A reinsurer usually has greater capacity to absorb large financial shocks and helps diversify the seismic risks geographically and across different perils. It is also important to recognize the critical role of governments in national earthquake insurance programs, as they back up some layers of potential earthquake loss. An overview of the earthquake insurance and reinsurance system is given in the "Insurance and Reinsurance System" section, while a summary of existing insurance systems in major seismic countries is provided in the "Existing Insurance Systems Around the World" section.

The nature of catastrophic earthquake risks, i.e., correlated loss generation, poses major challenges in achieving the effective diversification of collective seismic risk and may overwhelm the financial capacities of insurers and reinsurers. To avoid the potential insolvency and major constraint in business operation, utilizing alternative risk transfer (ART) tools, such as earthquake catastrophe (CAT) bonds, may be considered (Lalonde 2005; Cummins 2008). Another concern of providing earthquake loss coverage includes adverse selection and moral hazard (Grace et al. 2003), which are influenced by various factors, such as risk perception, asymmetrical information, (cognitive) psychology, framing, personal experience, etc. (Johnson et al. 1993; Palm 1995; Kunreuther 1996; Kahneman 2003; Viscusi and Evans 2006). A summary of these challenges is provided in the "Challenges in Insuring Low-Probability High-Consequence Risks" section.

Approaches for modeling (catastrophe) insurance and reinsurance, adopted in different academic disciplines, are diverse, as their purposes and focuses are often different. Economists formulate insurance purchase problems as rational decision-making under uncertainty and examine arrangements for optimal insurance (Mossin 1968; Ehrlich and Becker 1972; Schlesinger 2000). On the other hand, behavioral economists and psychologists take a flexible view on decision-making problems by incorporating cognitive aspects of human beings (Johnson et al. 1993; Kahneman 2003; Viscusi and Evans 2006). In actuarial sciences, the problems are defined mathematically by capturing the key elements of the stochastic insurance risk processes (Rolski et al. 1999; Cossette et al. 2003; Goda and Yoshikawa 2012), which are not modeled in economic analyses. In earthquake engineering, earthquake CAT models that incorporate the state of the art in engineering seismology and earthquake engineering by taking into account uncertainties are used to assess the financial seismic risk quantitatively (Bommer et al. 2002; Dong and Grossi 2005; Wesson et al. 2009; Asprone et al. 2013; Yucemen 2013). Moreover, an extended approach considers the applications of actuarial/engineering methods in the context of economical decision-making problems (e.g., CAT models are embedded into the expected utility framework for insurance decision-making). Figure 2 depicts an overview of insurance models, involving various disciplines, such as seismology, earthquake engineering, statistics, economics, actuarial sciences, and psychology. Key features of different approaches are described in the "Insurance and Reinsurance System" section. Specifically, the "Earthquake Catastrophe Model" section discusses applications of the CAT models to earthquake insurance; the "Actuarial and Financial Insurance Model" section presents stochastic modeling of insurance

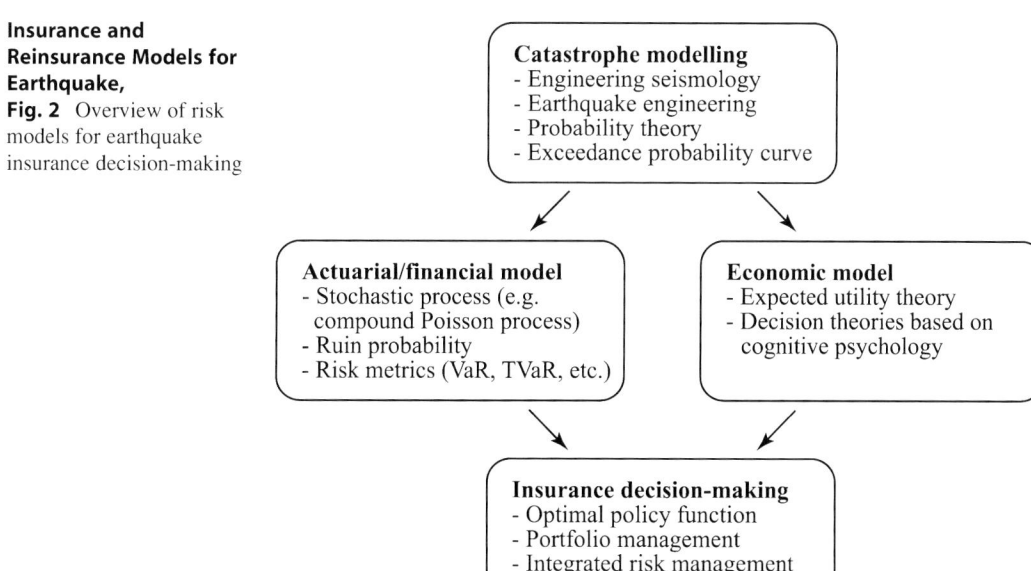

Insurance and Reinsurance Models for Earthquake, Fig. 2 Overview of risk models for earthquake insurance decision-making

risk processes; and the "Economic Insurance Model" section gives a summary of methods based on economics and expected utility decision theories. Finally, in the "Integrated Earthquake Risk Management" section, the role of insurance and reinsurance in the context of integrated earthquake risk management is discussed by broadening the scope of this entry (e.g., earthquake risk premium discount as incentive to promote seismic retrofitting; Kleindorfer et al. 2005; Michel-Kerjan et al. 2013).

Insurance and Reinsurance System

This section gives an overview of current insurance-reinsurance systems for earthquake risk coverage. In the "Insurance and Reinsurance System" section, key stakeholders as well as their interrelationships are defined. Subsequently, main features of the current insurance systems in major seismic regions/countries are described in the "Existing Insurance Systems Around the World" section. Subsequently, issues and challenges in insuring LP-HC earthquake events are discussed in the "Challenges in Insuring Low-Probability High-Consequence Risks" section.

Insurance-Reinsurance System

A generic representation of insurance-reinsurance system for managing earthquake risks is shown in Fig. 3. The identified key stakeholders are policyholders (e.g., owners of properties and enterprises), insurers operating at local/regional/national levels, reinsurers dealing with risk diversification at international scale, insurance commissioners/governments, rating agencies as auditor, and capital markets where investors gather for financial transactions.

Policyholders, Insurers, and Governments: In the event of a major destructive earthquake, policyholders may suffer from seismic damage to their properties and assets and incur financial loss. To mitigate the potential financial impact, they may insure such risks by purchasing earthquake insurance coverage. Alternatively, they have options to mitigate the earthquake damage by adopting physical mitigation measures, such as seismic retrofitting and upgrading. Acceptance of earthquake insurance and/or physical risk mitigation measures largely depends on the risk perception of property owners and on the type of properties (e.g., home versus business). Issues related to risk perception and acceptance of earthquake protection are discussed in the "Challenges

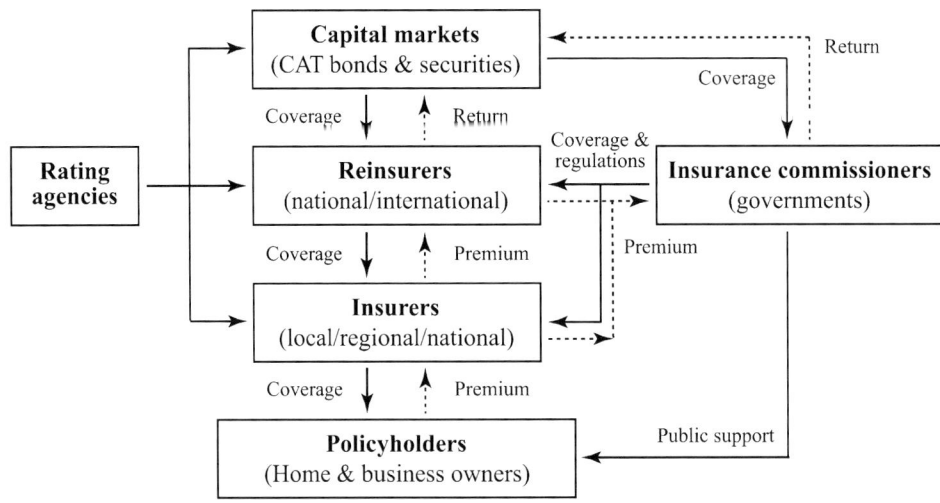

Insurance and Reinsurance Models for Earthquake, Fig. 3 Overview of an insurance-reinsurance system

in Insuring Low-Probability High-Consequence Risks" section.

Insurers offer earthquake risk coverage, usually as part of other insurance (e.g., fire insurance and multi-peril coverage). Depending on country and property type, purchase of earthquake coverage is compulsory or voluntary. A primary insurer collects insurance premiums from policyholders and reimburses incurred seismic loss based on a pre-agreed policy upon the occurrence of a damaging earthquake. The insurance premium P_I typically consists of three components:

$$P_I = P_{\text{pure}} + P_{\text{risk}} + C_{\text{transaction}} \quad (1)$$

where P_{pure} is the pure premium and is the expected cost of the ceded risks (i.e., fair price for the risk coverage), P_{risk} is the risk premium and is determined based on various factors (e.g., insurer's capital reserve, reinsurance price/availability, and regulatory requirements), and $C_{\text{transaction}}$ is the transaction cost related to marketing, monitoring, and claim settlement. As both P_{risk} and $C_{\text{transaction}}$ are greater than zero and not negligible (particularly for earthquake coverage, P_{risk} can be large, in comparison with P_{pure}), risk-neutral owners, who make decisions based on the expected values without considering variability of the consequences, regard insurance coverage as overpriced and unfavorable economically. Thus only risk-averse owners, who take into account uncertainties of the consequences, may seek insurance coverage. In cases owners are affected/biased due to their risk perception (e.g., overestimation of potential consequences), earthquake insurance coverage may or may not be attractive for them. Another important aspect to note is that insurance has additional benefits of providing financial liquidity with policyholders in post-disaster situations. For example, a policyholder who has purchased a property with a mortgage may face liquidity constraints after a major earthquake that damages the property severely; the policyholder may not be able to borrow money to recover the loss immediately after the earthquake, because the collateral (i.e., property) is damaged/lost and there is an outstanding mortgage repayment. In such cases, insurance coverage is beneficial in enhancing financial resilience against major shocks.

The policy (payout) function of an insurance contract F_{payout} for the earthquake damage cost C_{EQ} takes the following form:

$$F_{\text{payout}}(C_{EQ}) = \begin{cases} 0 & C_{EQ} \leq D \\ \gamma(C_{EQ} - D) & D < C_{EQ} < L \\ \gamma(L - D) & C_{EQ} \geq L \end{cases} \quad (2)$$

where D, L, and γ are the deductible, limit (cap), and coinsurance factor, respectively. The deductible is the starting loss value of the insurance payout and is usually specified in terms of total insured value. The limit is the maximum loss value of the insurance payout, expressed in terms of total insured value. From insurers' viewpoint, the deductible is useful for reducing transaction costs related to small claims, while the upper limit is effective in avoiding very large claims (critical for their solvency). The coinsurance factor specifies the proportional share of an incurred seismic loss between a policyholder and an insurer; this helps suppress the problem related to moral hazard. For instance, consider that an owner has a house of total value equal to $200,000 and has purchased earthquake insurance coverage with $D = 10\%$, $L = 80\%$, and $\gamma = 90\%$. In cases where the owner incurs seismic losses of $60,000 and $180,000, he/she will receive $36,000 and $126,000, respectively.

The reinsurance policy has essentially the same structure as primary insurance and is mainly classified into two types: pro rata contract and excess-of-loss contract. In pro rata contracts, an insurer and reinsurer share the risk proportionally (similar to coinsurance), while in excess-of-loss contracts, a reinsurer provides coverage in a specific layer between attachment point and exhaustion point (similar to deductible and limit).

The sales of earthquake risk coverage are often regulated by statutory bodies, such as insurance commissioners. The main reasons are that earthquake insurance, unlike other liability insurance, tends to be *public* (i.e., protection against involuntary risks) and after major earthquake disasters, governments and states may intervene to provide public support to the affected people (e.g., monetary gift and special loan programs). Although, in reality, such disaster relief activities are inevitable (which are funded by tax and borrowing), it is desirable to minimize post-disaster intervention (ex ante subsidies) as this may discourage DRR efforts prior to earthquakes (Kunreuther 1996). This is referred to as charity hazard. It has been demonstrated that DRR measures are often cost-effective, in comparison with situations without DRR (Kleindorfer et al. 2005; Michel-Kerjan et al. 2013).

Insurers, Reinsurers, Auditors, Regulators, and Capital Markets: The basic principle of insurance portfolio management is the law of large numbers; with the increase in the number of policies, variability of aggregate claims of a portfolio becomes small, and thus actual total loss approaches the expected loss value (which increases with the number of policies). In simple terms, the financial risk becomes more predictable. It is important to recognize that this theory is applicable to *independently distributed events* (e.g., car insurance), while the earthquake risks are both spatially and temporally correlated. Insurers may experience a devastating surge of seismic loss claims in a disaster-hit region and face the possibility of insolvency. Therefore, insurers need to diversify their earthquake exposures geographically.

A conventional approach for catastrophe risk transfer is reinsurance by global reinsurers (Munich Re, Swiss Re, etc.) and governments (Fig. 3). A notable functionality of governments as reinsurer is their ability to borrow money by issuing bonds; this achieves temporal diversification of earthquake risks. Although reinsurers have greater risk-bearing capacities than insurers and achieve geographical risk diversification through their national/global portfolios, the potential size of the catastrophic earthquakes can be significant and their default risk is not zero. Because of this, reinsurers charge quite high risk premium to insurers and seek alternative means for risk transfer (Cummins 2008), which in turn affects the risk premiums for primary insurance coverage. Reinsurers also utilize retrocession, i.e., risk transfer from a reinsurer to another reinsurer to reduce the impact of peak risks. Recently, financial market instruments, so-called insurance-linked securities, become available through capital markets (Grace et al. 2003; Lalonde 2005) and have been used more frequently by insurers, reinsurers, and governments/states (Cummins 2008; see also recent market review articles/reports published by Munich Re and Swiss Re). The emerging tools

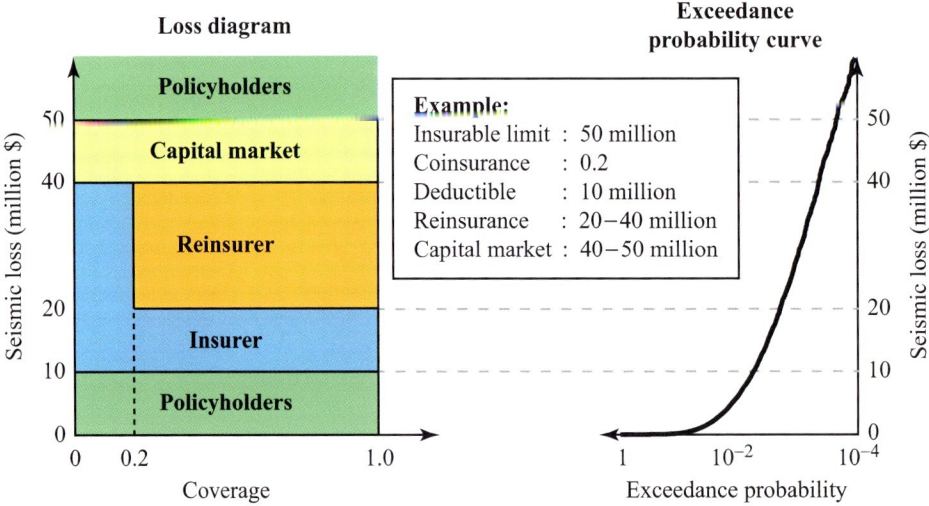

Insurance and Reinsurance Models for Earthquake, Fig. 4 Loss diagram for stakeholders, insurer, reinsurer, and capital market

take advantage of much greater financial capacity of capital markets, covering high-loss layers of insurance-reinsurance portfolios. Since earthquake insurance is an integrated part of catastrophe risk management, insurance commissioners, on behalf of governments, regulate the solvency status and insurance policy/pricing of insurers. A high insolvency risk is not acceptable for public earthquake insurance systems. For monitoring the financial status of the earthquake risk underwriters, credit rating agencies (e.g., Standard & Poor's and Moody's) send useful signals to involved parties on their catastrophe exposures. However, the amount of correlations created by retrocession is not transparent to regulators and auditors.

A useful tool for visualizing the composition of catastrophe risk diversification by different stakeholders is the loss diagram (Dong and Grossi 2005). It displays how the total loss is divided and covered by involved parties. Figure 4 shows a hypothetical loss diagram for aggregate seismic loss; in addition, a so-called exceedance probability (EP) curve of the entire portfolio is included. In the diagram, policyholders retain seismic loss up to $10 million (i.e., deductible for the entire portfolio) and seismic loss exceeding $50 million (i.e., insurable limit). The loss layer between $10 and $40 million is covered by an insurer and a reinsurer; the first $10 million layer is solely covered by the primary insurer, while the next $20 million layer is shared by the excess-of-loss reinsurance with proportional rate of 0.2 (i.e., $20 and $40 million are the attachment and exhaustion points of the reinsurance contract, respectively). The high-loss layer between $40 and $50 million is ceded to capital markets via CAT bonds/securities.

Existing Insurance Systems Around the World

Earthquake risks are prevalent globally. Countries located near the plate boundaries constantly experience major seismic and volcanic activities. One notable example is the Ring of Fire, or the circum-Pacific seismic belt, which hosts the majority of large earthquakes around the world. The circum-Pacific countries include Chile, Peru, Mexico, the USA (California and Alaska), Japan, Taiwan, the Philippines, Indonesia, and New Zealand, while other major seismic countries are Iceland, India, Iran, Italy, Pakistan, and Turkey. In some of these countries, earthquake insurance systems have been established and operated (NLIRO 2008; OECD 2008). This section provides a brief summary of existing insurance systems in California (USA), Iceland, Japan, Mexico, New Zealand, Taiwan, and Turkey. Table 1 compares the main features of different insurance systems. In addition, a brief description of natural catastrophe coverage in the EU is included.

California (USA): In California, earthquake insurance coverage had been offered privately since early twentieth century. Currently, the privately funded, state-run California Earthquake Authority (CEA), which was established in 1996 in the aftermath of the 1994 Northridge earthquake, provides earthquake insurance coverage to homeowners; policies can be obtained incidental to fire insurance (optional). Insurer's CEA membership is voluntary and member companies satisfy the mandatory offer law for selling the CEA policy. It is noteworthy that the CEA is not backed up by the State of California. The payable limit of the CEA is about US$ 100 million; if this amount is exceeded, the claim repayments are proportionally reduced. The policy covers damage and loss due to earthquakes. The premium rates range from 0.036 % to 0.90 %, depending on zip code, structural type, built year, and number of stories. A relatively high deductible of 15 % is specified (10 % can be selected). Policyholders can insure residential properties up to the total insurance value, while they can choose the level of coverage for content (e.g., US$ 5,000–100,000) and incidental expenses (e.g., US$ 1,500–15,000).

Iceland: The state-owned Iceland Catastrophe Insurance (ICI), which was established in 1975 after the 1973 Mt. Heimaey volcanic eruptions, provides catastrophe coverage for natural hazards including volcanic eruptions, earthquakes, landslides, snow avalanches, and floods. The participation is compulsory, incidental to fire insurance, and covers direct loss to buildings, content, and lifelines. The deductible is 5 % for each loss. The premium rate for properties and content is 0.025 %, while that for lifelines is 0.02 %. The ICI is liable up to 1 % of the total insured capital; if this amount is exceeded, the claim repayments are proportionally reduced.

Japan: The government-endorsed Japan Earthquake Reinsurance Company (JERC), which was established in 1966 after the 1964 Niigata earthquake, plays a major role in promoting earthquake insurance for residential buildings. The purchase is incidental to fire insurance and is not mandatory (thus the penetration rate is not so high). Primary insurers sell policies and collect fees from policyholders; the covered loss is retained by the JERC, government, and private insurers. The aggregate limit of the indemnity for earthquake insurance is JPY 6.2 trillion; if this amount is exceeded, the claim repayments are proportionally reduced. The policyholder can obtain earthquake coverage up to about 30–50 % of the total insured value under fire insurance policy; in addition, the maximum insured amount is limited to JPY 50 million for buildings and JPY 10 million for movables, respectively. There is a discount mechanism of premiums to encourage seismic risk mitigation; for instance, properties with base isolation are granted for 30 % discount, and certified properties which meet the current seismic design requirements are granted for 10 % discount. Other cases include "buildings constructed after 1981 (when the new seismic design code was implemented in Japan)," and "certified buildings according to the seismic grading system."

Mexico: Earthquake insurance in Mexico is mainly provided by private insurers, under the regulation by the Comision Nacional de Seguros y Fianzas (CNSF). The major seismic event that led to adoption of earthquake insurance was the 1985 Michoacan earthquake. The purchase is optional and is often accompanied by fire insurance. The insurance covers damage and loss of buildings and movables due to earthquakes and volcanic eruptions. The premium rates vary widely, depending on occupancy type, number of stories, and geographical region. Typical values of deductible are 2–5 %. In addition to private earthquake insurance, the government-supported Natural Disasters' Fund, which was created in 1996, provides financial support to the recovery/restoration of impaired infrastructure and to the disaster victims.

Insurance and Reinsurance Models for Earthquake, Table 1 Comparison of earthquake insurance systems in different countries (NLIRO 2008; OECD 2008)

Country	Insurance commissioner	Participation method	Coverage	Rate	Insured amount and limit	Payable claim amount	Role of primary insurers
California (USA)	California Earthquake Authority (CEA)	Incidental to fire insurance (optional)	Residential building and content; damage due to earthquakes	0.036–0.90 %; 19 classes (zip code); 8 classes (structural type and built year); 2 classes (story number)	Building: total insurance value; content, US$5,000–100,000; incidental expense, US$ 1,500–15,000; deductible, 15 %	US$ 96 million per event	Sales, fee collection, certificate issuance, remittance, and assessment
Iceland	Iceland Catastrophe Insurance (ICI)	Incidental to fire insurance (compulsory)	Building, content, and lifelines; damage due to earthquakes, volcanic eruptions, landslides, avalanches, and floods	0.025 % for buildings and content; 0.02 % for lifelines	Deductible, 5 % of each loss (minimum deductible – ISK 85,000 for building, ISK 20,000 for content, and ISK 750,000 for lifelines)	1 % of the total insured capital	Sales, fee collection, and certificate issuance
Japan	Japan Earthquake Reinsurance Company (JERC)	Incidental to fire insurance (optional)	Residential building and movables for living; damage due to earthquakes, volcanic eruptions, and tsunamis	0.05–0.313 %; 8 classes (structural type and geographical region)	30–50 % of the total insured value under fire insurance policy (maximum insured amount is up to JPY 50 million for buildings and JPY 10 million for movables); deductible, 3 % (building) and 10 % (movables)	JPY 6.2 trillion	Sales, fee collection, certificate issuance, damage assessment, and insurance underwriting

Insurance and Reinsurance Models for Earthquake

Mexico	Private insurers regulated by Comisión Nacional de Seguros y Fianzas (CNSF)	Optional	Residential building, commercial/industrial building, and movables for living; damage due to earthquakes and volcanic eruptions	0.018–0.356 % for residential building; 12 classes (geographical region); 0.028–0.726 % for commercial/industrial building; 12 classes (geographical region) and 2 classes (story number)	Total insurance value (but some upper limits [70–90 %] may be applied in high-seismic regions); deductible, 2–5 % (depending on geographical region)	N/A	Sales, fee collection, certificate issuance, damage assessment, and insurance underwriting
New Zealand	Earthquake Commission (EQC) – EQCover	Incidental to fire insurance (compulsory)	Residential building, content, and land; damage due to earthquakes, landslides, volcanic eruptions, and tsunamis	0.05 % for residential building	Building, NZ$ 100,000; content, NZ$ 20,000; land, total insurance value Deductible, 1 % (building), NZ$ 200 (content), and 10 % (land, maximum deductible is set to NZ$ 5.00)	All claims to be repaid (backed up by the government)	Sales and fee collection
Taiwan	Taiwan Residential Earthquake Insurance Fund (TREIF)	Incidental to fire insurance (compulsory)	Residential building; damage due to earthquakes	0.122 %	Building, TW$ 1.2 million; contingent expenses, TW$ 180,000; no deductible	TW$ 60 billion	Sales
Turkey	Turkish Catastrophe Insurance Pool (TCIP)	Compulsory	Building; damage due to earthquakes	0.044–0.55 %; 15 classes (structural type and geographical region)	TRY110,000 (in Turkish lira [TRY]); deductible, 2 %	EUR 1 billion (from reinsurance)	Sales, fee collection, certificate issuance, and damage assessment

***New Zealand*:** The government-owned Earthquake Commission (EQC) provides natural disaster insurance (EQCover) to homeowners in New Zealand. The EQC was established in 1944 after the devastating 1942 Wairarapa earthquakes. The participation is incidental to fire insurance and is automatic/compulsory. The contract covers damage and loss of residential properties, content, and land due to earthquakes, landslides, volcanic eruptions, and tsunamis. The premium rate is 0.05 %. The insured amount is limited to NZ$ 100,000 for buildings and NZ$20,000 for content, respectively. The deductible is set to NZ$ 200 or 1 % of the loss for residential properties, NZ$200 for content, and the greater of NZ$ 500 or 10 % of the insured loss for land, with the maximum deductible of NZ$ 5,000. In addition to the EQCover, private insurers offer natural disaster damage extension, which can insure commercial buildings (variable rates depending on geographical region, structural type, built year, and number of stories).

***Taiwan*:** The government-endorsed Taiwan Residential Earthquake Insurance Fund (TREIF) operates the earthquake insurance program in Taiwan. TREIF was established in 2002 in the aftermath of the 1999 Chi-Chi earthquake. The policy is incidental to fire insurance and its purchase is compulsory. The insurance covers damage and loss of residential properties due to earthquakes. The upper limit of the repayment per household is TW$ 1.2 million for residential properties and TW$ 180,000 for contingent expenses, respectively. The premium rate is 0.122 %. The total amount of the payable claims is TW$ 60 billion.

***Turkey*:** The state-owned Turkish Catastrophe Insurance Pool (TCIP) was created after the 1999 Izmit earthquake to reduce catastrophe earthquake exposure to the government. The owners of residential properties within municipality boundaries are obliged to take earthquake insurance coverage, which insures damage and loss due to earthquakes. The premium rates range from 0.044 % to 0.55 %, depending on structural type and geographical region. The maximum insured limit is TRY 110,000 and the deductible is 2 %. Currently, as the financial status of the TCIP is not sufficient, it cedes a large amount of its risks to international reinsurers. Note that commercial and industrial properties as well as residential properties in rural areas can be insured on a voluntary basis.

***EU Countries*:** Porrini and Schwarze (2012) classify variable insurance schemes for natural catastrophe coverage (including earthquakes for some countries) in EU member states into five categories: (1) regional public monopoly insurer of natural hazards guided by statutory provisions and public consultation processes; (2) compulsory insurance for all natural hazards; (3) compulsory inclusion of natural hazards into general house ownership insurance; (4) free-market natural hazard insurance with ad hoc governmental relief programs; and (5) taxpayer-financed governmental relief funds. The Netherlands, Denmark, and Switzerland have adopted Schemes 1 and 2; France and Spain have used Schemes 2 and 3; the UK has adopted Schemes 3 and 4; Germany has followed Scheme 4; and Poland, Italy, and Austria have adopted Schemes 4 and 5. These schemes suppress insurance-related problems, such as adverse selection, moral hazard, and charity hazard, differently. For instance, Scheme 1 can be used to mitigate all three issues, whereas Schemes 2 and 3 can deal with adverse selection and charity hazard effectively but may be susceptible to moral hazard. Effectiveness of Schemes 4 and 5 is limited to avoiding charity hazard and adverse selection, respectively. It is noteworthy that the selection of a suitable scheme has important influence on how risk mitigation and (financial) risk management are implemented. Furthermore, insurance penetration rates for earthquake coverage in Europe are unevenly distributed among member states of the EU (Maccaferri et al. 2011). Notably, many earthquake-prone countries (e.g., Italy, Greece, Romania, Spain, and Portugal) show rates below 10 %, although high maximum losses within the past 20 years have been experienced (e.g., Greece, 2.1 % in 2010 GDP value, and Italy, 0.6 % in 2010 GDP value).

Challenges in Insuring Low-Probability High-Consequence Risks

Risk perception has major influence on decision-making processes regarding the adoption of earthquake insurance (Kunreuther 1996). Facing LP-HC risks, home and business owners may have diverse opinions on the potential magnitude of earthquake consequences. Therefore, appreciation of benefits of earthquake insurance coverage varies significantly, depending on personal characteristics (e.g., optimistic or pessimistic), proximity to physical hazards (e.g., living near a major fault), previous experience with earthquakes, and socioeconomic/demographic profiles, such as age, gender, income, and education (Palm 1995). Another important aspect influencing insurance-related decisions is cognitive limitation of human beings (Kahneman 2003). It is well known that people are affected by framing (Johnson et al. 1993) and have difficulties in estimating very low probabilities (Viscusi and Evans 2006). An example of framing is that a factory owner who is considering earthquake insurance for financial protection is likely to take insurance if he/she is told that the chance of experiencing major earthquake damage is 1 in 5 over the entire period (20 years), rather than the chance of such an event is 1 in 100 per year. Therefore, even when the overall premium is reasonably priced, not many stakeholders voluntarily purchase earthquake risk coverage.

Two known issues in offering insurance coverage are moral hazard and adverse selection (note: they are applicable to not only catastrophe earthquake but other more common types). Moral hazard refers to a case where more risky situations (i.e., increase in chances of experiencing loss and/or increase in the extent of loss) are created by the behavior of a stakeholder. An example of moral hazard is that a policyholder, after a damaging earthquake, may cause additional damage to his/her property to receive insurance repayment (when the incurred loss is relatively minor and is near the threshold of deductible). On the other hand, adverse selection refers to a situation where an insurer cannot distinguish between stakeholders having lower and higher earthquake risk potential. In such cases, the insurer cannot charge appropriate insurance premiums that are commensurate with the risks. For instance, if an insurer charges a flat rate for all homeowners in a specific geographical region, earthquake insurance is more attractive for those who live in seismically more vulnerable houses (e.g., non-ductile old construction). This leads to higher-risk exposure for the insurer than it is originally expected and may result in increasing the premium rate (i.e., adverse selection spiral can occur). One of the main causes of moral hazard and adverse selection is the information asymmetry between an insurer and a policyholder. The private information that the stakeholder has on his/her property is not entirely accessible to the insurer; in such cases, the stakeholder may have an incentive to behave inappropriately from the insurer's perspective.

To alleviate the effects due to moral hazard and adverse selection, introducing pro rata share of the risks and differentiating the rates according to physical parameters related to seismic hazard and vulnerability of the covered properties (e.g., geographical region, structural type, built year, and number of stories) are useful (see Table 1). Other proactive ways to avoid adverse selection and encourage physical DRR measures are to implement an effective incentive scheme in determining insurance premium rates (e.g., Japanese insurance system).

Insurance and Reinsurance Models

This section gives an introduction to a wide range of earthquake insurance and reinsurance models that have been developed and applied in different academic fields (Fig. 2). The "Earthquake Catastrophe Model" section provides a brief summary of earthquake CAT modeling and quantitative risk metrics. The key requirements of viable CAT models are mentioned. Next, actuarial/financial insurance models are presented in the "Actuarial and Financial Insurance Model" section. Subsequently, economic insurance models are summarized in the "Economic Insurance

Model" section to give an overview of the decision-making processes for policyholders (demand) and for insurers (supply).

Earthquake Catastrophe Model

Earthquake CAT modeling is an integrated process of conducting numerical simulations of earthquake occurrence, ground motion prediction, damage assessment, and seismic loss calculation. It typically involves: (i) inventory/exposure database, (ii) hazard characterization, (iii) structural vulnerability assessment, and (iv) loss estimation and insurance portfolio analysis. The output for the CAT models is the probability distribution of estimated seismic loss for a building/infrastructure portfolio (i.e., EP curve). There are a variety of different methods and models used for earthquake insurance risk analysis in research and in practice (e.g., Bommer et al. 2002; Asprone et al. 2013; Yucemen 2013). Current challenges of the CAT modeling include the validation of the CAT models using scarce loss data and dealing with rapidly changing exposure data (particularly for developing countries) and the incorporation of multiple concurrent hazards and risks, such as tsunami, aftershock, liquefaction, landslide, and fire, within a unified analytical/computational framework.

A stochastic earthquake catalog method (e.g., Goda and Yoshikawa 2012) starts with a regional probabilistic seismic hazard model and generates a synthetic earthquake catalog using Monte Carlo sampling techniques. For each earthquake in a catalog, seismic vulnerability assessment and loss estimation are carried out using fragility curves and damage-loss functions. A fragility curve evaluates conditional probability of exceeding a certain damage state (e.g., slight, moderate, and extensive), and a damage-loss function transforms damage severity into corresponding seismic loss. It is noted that uncertainty associated with every step of the computation is taken into account and is propagated. The insurance-related risk can be evaluated by applying an insurance policy function to the estimated seismic loss (either at policy level or at aggregate level). The estimated loss results can be sorted in an ascending order to develop an EP curve for the portfolio of interest. An illustrative EP curve for an insurance portfolio, which plots estimated insurance loss as a function of annual exceedance probability, is shown in Fig. 5. The EP curve is often used for insurance portfolio decision-making. Note that there are many variants of the abovementioned method, such as a multiple-event/event-loss-table method (e.g., Dong and Grossi 2005).

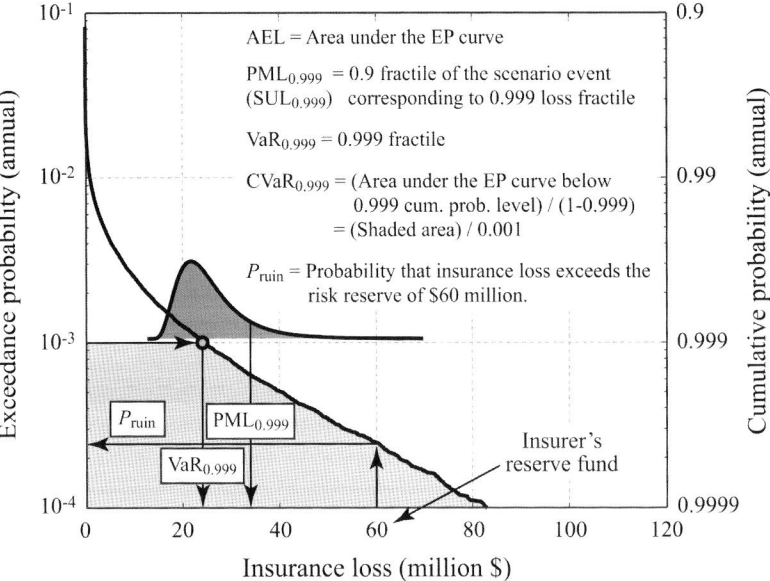

Insurance and Reinsurance Models for Earthquake, Fig. 5 Exceedance probability curve and risk metrics

Insurance and Reinsurance Models for Earthquake, Table 2 Summary of risk metrics

Risk metrics	Equation	Relationship with the EP curve
Annual expected loss: AEL	$AEL = E[L]$ where $E[\bullet]$ is the expectation	Area under the EP curve
Value at risk: VaR_α	$VaR_\alpha = \inf\{l : P(L > l) \leq 1 - \alpha\}$ where $\inf\{\bullet\}$ is the infimum and $P(L > l)$ is the probability of loss L exceeding a certain loss level l	Fractile value at a selected cumulative probability level α
Conditional value at risk: $CVaR_\alpha$	$CVaR_\alpha = \frac{1}{1-\alpha} \int_\alpha^1 VaR_u du$	Area under the EP curve below a selected cumulative probability level α, normalized by the exceedance probability $1-\alpha$
Ruin probability	$P_{\text{ruin}} = P(L > R_{\text{insurer}})$ where R_{insurer} is the insurer's risk reserve	Probability when insurer's reserve fund is depleted

Note: α represents the cumulative probability; in other words, $1-\alpha$ represents the exceedance probability

To facilitate the risk-based decision-making and risk communication among different stakeholders, risk metrics can be derived from the calculated EP curve. Risk metrics that are widely used include: annual expected loss (AEL), probable maximum loss (PML), value at risk (VaR), conditional value at risk (CVaR), and ruin probability. The equations to compute these measures (except for PML) are summarized in Table 2 (see also Fig. 5 for graphical representation of the metrics). The AEL is a useful metric for ordinary risks; however, it is not suitable for catastrophic risks as it fails to capture the extent of devastating consequences due to rare events. The VaR is the fractile value on an EP curve corresponding to a selected probability level, while the CVaR, which is also referred to as expected shortfall in finance, accounts for rare events in terms of their severity and frequency by taking the conditional expectation of the EP curve. One of the crucial factors is "which probability level" to be focused upon in evaluating VaR and CVaR (also applicable to PML). The ruin probability represents the chance of insurer's insolvency (i.e., all reserve funds are depleted) and is often used for insurance regulatory purposes (e.g., an insurer needs to retain a sufficient reserve fund such that annual probability of insolvency is less than 0.005). The PML is one of the most popular metrics in financial risk management, and there are several definitions. Conventionally, it was defined as the fractile of the loss corresponding to the return period of 475 years (i.e., essentially same as VaR).

However, in different industries and countries, variants of the PML had been adopted for use. For example, in Japan, the PML is defined as the (conditional) 0.9-fractile value for a scenario that corresponds to a selected probability level (typically, return period of 475 years is considered, but this can be varied). Currently, specific nomenclatures are in use: scenario expected loss (SEL) corresponds to the original PML definition (i.e., VaR), and scenario upper loss (SUL) corresponds to the PML definition in Japan.

The CAT modeling is useful for quantifying insurers'/reinsurers' earthquake risk exposure under different situations and for deciding risk management strategies. For instance, impact of the policy function parameters (i.e., coinsurance, deductible, and limit) on insurer's earthquake risk exposure is illustrated in Fig. 6. The insurance portfolio is based on 4,000 wooden houses in Vancouver, Canada, and the EP curve is obtained from the stochastic earthquake catalog method (Goda and Yoshikawa 2012). Figure 6 shows that a high deductible and low coinsurance factor reduces earthquake risk exposure significantly at wide probability levels, whereas a low limit curtails the maximum earthquake risk exposure effectively. The limit tapers heavy right tail of the loss distribution and affects the deficit at ruin, which is one of the key decision variables in earthquake insurance management. Importantly, from insurer's viewpoint, flexibility to select a suitable policy arrangement is beneficial to enhance the performance of earthquake portfolio

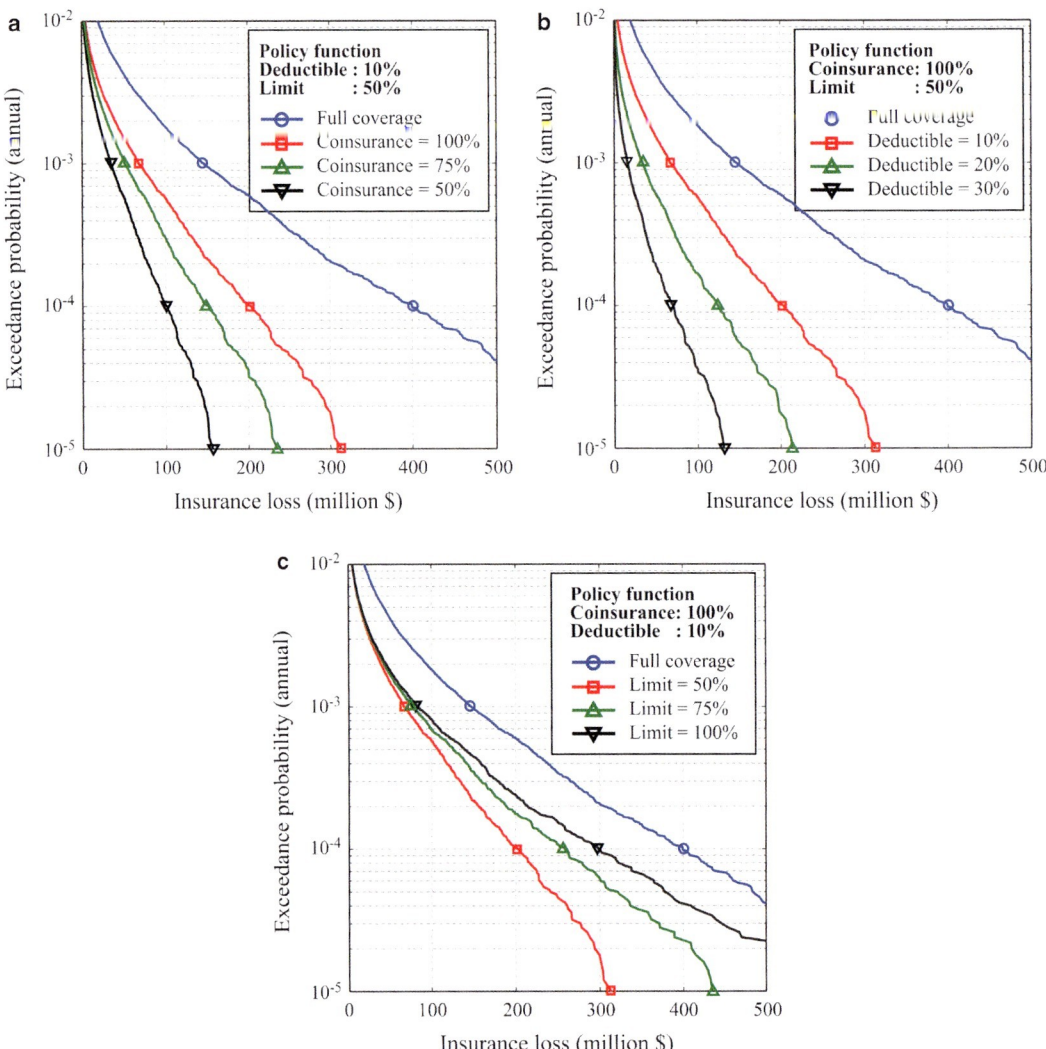

Insurance and Reinsurance Models for Earthquake, Fig. 6 Effects of policy function parameters on exceedance probability curve: (**a**) coinsurance, (**b**) deductible, and (**c**) limit

management and to offer affordable insurance rates to the policyholders. However, such flexibility is not usually possible due to regulatory rules; in such cases, ART tools can be used.

Actuarial and Financial Insurance Model

Insurance models in actuarial sciences and financial engineering have a long history of development and wide applications in assessing underwriting risks. Popular approaches for characterizing catastrophic risk processes include a compound Poisson process (e.g., Rolski et al. 1999) and an extreme value theory (e.g., McNeil et al. 2005). The risk process can be formulated in form of stochastic differential equations. Although analytical solutions for insurer's ruin probability are available for simple cases (e.g., compound Poisson process with exponential insurance loss), for realistic and complex problems, such solutions are not available and numerical simulations of the risk process are often required. On the other hand, statistical modeling of dependent risk processes is useful for assessing insurer's risk exposure due to extreme events. In particular, accurate modeling of exposure dependence among different portfolios and lines of business is important in evaluating the entire risk process. This leads to advancement of dynamic financial analysis (Kaufmann et al. 2001), which integrates large-scale stochastic simulations of all related insurance and reinsurance processes (including fluctuating interest rates and catastrophic as well as non-catastrophic risks). The key challenge for both approaches is how to incorporate spatiotemporally correlated insurance claims in the risk process. In this section, two insurance models based on a compound Poisson (including diffusion-jump) process and based on extreme value theory are briefly mentioned (note: rigorous mathematical formulations are beyond the scope of this entry).

Compound Poisson and Diffusion-Jump Processes: The fluctuation of an insurer's wealth $W(t)$ subjected to both catastrophe and non-catastrophe risks, $W_{\text{noncat}}(t)$ and $W_{\text{cat}}(t)$, can be expressed as

$$W(t) = W_{\text{noncat}}(t) + W_{\text{cat}}(t) \quad (3)$$

In the context of earthquake insurance, $W_{\text{noncat}}(t)$ is related to ordinary risks (e.g., diversifiable liability insurance underwriting) and can be represented by a stochastic growth model, such as Brownian process and geometric Brownian process. The Brownian process assumes that changes of the wealth state over a short period are independent and identically distributed (*i.i.d.*) as normal variate, whereas the geometric Brownian process considers that the logarithmic ratio of two wealth states separated by short time lag is an *i.i.d.* normal variate. For instance, by adopting the geometric Brownian process, $W_{\text{noncat}}(t)$, representing incurred loss process, payout process, expense process, and investment process, can be described as

$$W_{\text{noncat}}(t) = W_0 \exp\left[(\alpha - 0.5\beta^2)t + \beta Z(t)\right] \quad (4)$$

where W_0 is the initial asset; α and β are the instantaneous growth rate and infinitesimal standard deviation (or volatility) of the insurer's wealth related to non-catastrophic risks, respectively; and $Z(t)$ is a standard Brownian motion due to perturbation of non-catastrophic risk processes. In practical applications, the drift and volatility parameters α and β can be estimated from relevant insurance data (see Goda and Yoshikawa 2012). The stochastic process shown in Eq. 4 can be evaluated using Monte Carlo simulation, and thus many realizations can be generated to assess the temporal evolution of the insurer's asset.

The catastrophic jump component of the risk process $W_{\text{cat}}(t)$ occurs infrequently, but when it occurs, a drop of the insurer's net worth can be extremely large. Potentially, such a downward jump causes a negative net worth, forcing an insurer to become insolvent (i.e., $W(t) < 0$). This can be mathematically represented by a compound Poisson jump process:

$$W_{\text{cat}}(t) = -\sum_{i=1}^{N(t)} L_i \quad (5)$$

where jumps (i.e., damaging earthquakes) arrive randomly according to a Poisson counting process $N(t)$ and the random size of the jumps is characterized by a probabilistic loss model. Cossette et al. (2003) presented three different formulations of the catastrophic loss model L in Eq. 5 by considering independent, fully correlated, and partially correlated cases. The degree of loss correlation from multiple policies was defined as a function of seismic intensity, which is realistic. In addition, they derived the risk-ordering relationships of the different types of portfolios with respect to loss correlation for several risk metrics, including VaR and CVaR. It is noteworthy that the jump process shown in Eq. 5 can be substituted by an earthquake CAT model, which is more rigorous and accurate than a compound Poisson process.

The abovementioned stochastic representation of insurer's wealth process is illustrated in Fig. 7. Due to the earthquake's strike, insurer's wealth state is affected by earthquake insurance claims; this may force the insurer to become insolvent or

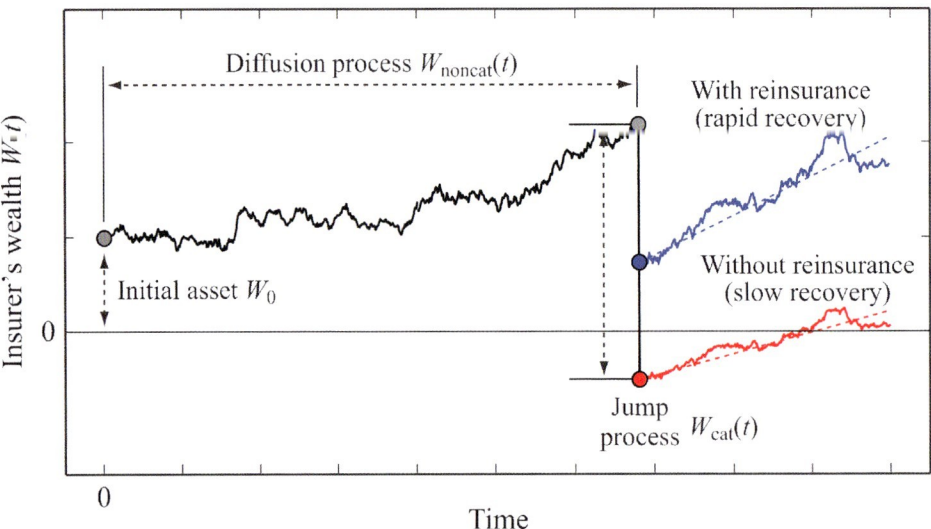

Insurance and Reinsurance Models for Earthquake, Fig. 7 Stochastic wealth process with and without insurance

be in debt. Furthermore, financial constraints caused by the earthquake risk may result in slow recovery from the shock after the earthquake. The figure also includes a hypothetical case where the insurer has arranged reinsurance coverage for the loss; in this case, the wealth state is dropped but not to the level of insolvency, and the post-disaster recovery rate may be more rapid than the situation without reinsurance (e.g., liquidity constraints are avoided). Stochastic simulations of underwriters' risk processes are useful for investigating the effects of insurance policies and reinsurance on the portfolio and for developing effective risk management strategies.

Extreme Value Theory and Copula-Based Dependence Modeling: The statistical modeling of extreme earthquake events is one of the active research areas. Useful statistical models for capturing the key features of a highly correlated loss generation process are the extreme value distributions (McNeil et al. 2005). In particular, the peak-over-threshold (POT) model is suitable as it utilizes all data exceeding a high threshold value more efficiently than the block maxima model and is more applicable to insurance loss processes (focusing on large, rare loss events). The limiting distribution of the POT data Y that exceeds a sufficiently high threshold μ converges to the generalized Pareto (GP) distribution:

$$P(X \leq x | X > \mu) = \begin{cases} 1 - (1 + \xi[x - \mu]/\beta)^{-1/\xi} & \xi \neq 0 \\ 1 - \exp(-[x - \mu]/\beta) & \xi = 0 \end{cases} \quad (6)$$

where ξ is the shape parameter and determines the upper tail behavior of the data and β is the scale parameter. Usually, a suitable value of μ can be determined based on mean excess and shape parameter plots (McNeil et al. 2005), and given data μ, it is straightforward to estimate the parameters of the GP model based on the maximum likelihood method.

In insurance portfolio management, multivariate modeling of insurance data may be required (e.g., portfolio aggregation; if locations of two portfolios are geographically close, they are subjected to similar but not identical earthquake risks). In such a case, a dependence structure of the data needs to be characterized. To deal with insurance data having heavy right-tail and

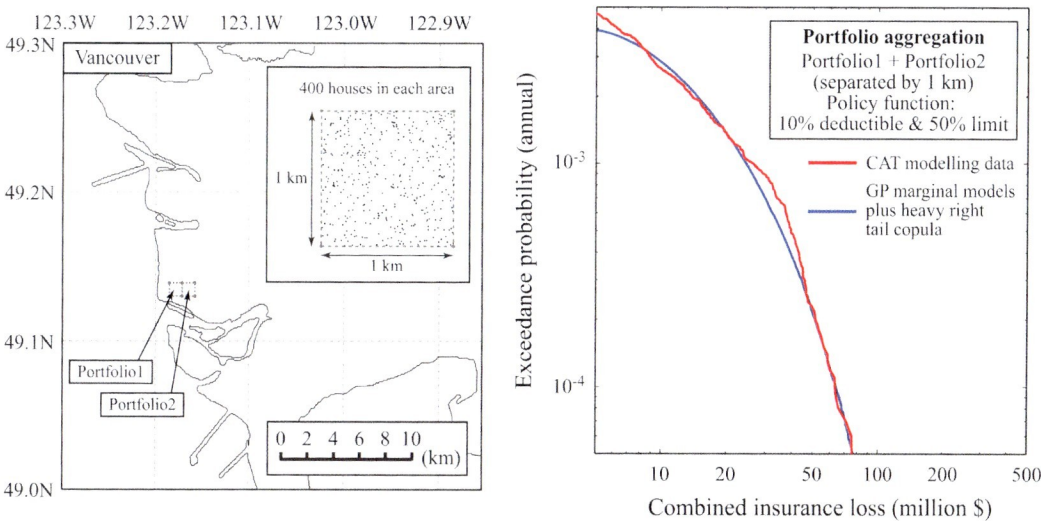

Insurance and Reinsurance Models for Earthquake, Fig. 8 Statistical aggregation of two insurance portfolios

nonlinear dependence, a copula technique can be used (McNeil et al. 2005); the major advantage of the copula is that the dependence modeling can be carried out in isolation from the marginal distribution modeling. The copula model fitting can be performed by finding a parametric copula function that fits best to empirical copula data (transformed data such that their marginal distributions are uniformly distributed) using the maximum likelihood method. There is a wide class of copula functions (e.g., elliptical copula and Archimedean copula).

To illustrate an application of the statistical modeling approach to earthquake insurance loss, results for aggregation of two insurance portfolios at closely located locations are presented in Fig. 8 (Goda and Ren 2013). Two portfolios, each consisting of 400 houses in a 1 km by 1 km area, are considered; the two portfolios are separated by 1 km, and thus high correlation is anticipated, particularly in the upper tail (i.e., large loss in one portfolio is likely to be accompanied by large loss in the other portfolio). The insurance loss samples are generated from the stochastic earthquake catalog approach. To model the joint seismic loss distribution of the two portfolios, GP models are fitted to the marginal distributions of the two portfolios, and a suitable copula function, which is found to be the heavy right-tail copula, is obtained. Then, using the constructed joint probability distribution, insurance loss samples are generated and plotted in Fig. 8. Comparison of the two EP curves indicates that, except for the intermediate loss range, agreement of the combined insurance loss data is good; see Goda and Ren (2013) for more comprehensive results. It is emphasized that the simulation from the statistical model is very quick, in comparison with the CAT model. Therefore, once the accuracy of the statistical method is verified, the proposed method can serve as a viable alternative to conduct various sensitivity analyses related to earthquake insurance portfolios (i.e., comparison of EP curves and risk metrics under different policy functions and reinsurance arrangements).

Economic Insurance Model

In economics, decision problems of providing insurance coverage have been studied extensively from two perspectives: (i) homeowner's demand (Mossin 1968; Schlesinger 2000) and (ii) insurer's supply (Kleindorfer et al. 2005). Other relevant problems include economic valuation and decision-making related to risk mitigation actions, such as self-insurance, self-protection, and market insurance (Ehrlich and Becker 1972). The essential tools that have been used in such analyses are based on the expected utility framework (note: recent studies adopt descriptive decision theories to analyze the abovementioned problems). Typically, a homeowner is assumed to be a risk-averse decision maker, having a concave utility function, whereas an insurer is considered to be a risk-neutral decision maker, having a linear utility function. To simplify situations, two contingent states only, i.e., no earthquake damage and full earthquake damage, are often considered. The conclusions and theorems derived from these idealized analyses provide useful insight on rational behavior of homeowners and insurers toward insurance decision-making under uncertainty. It is important to recognize that these theoretical predictions are obtained based on many (critical) assumptions about risk attitudes of decision makers and market conditions, which may be invalid in reality and inapplicable to catastrophic earthquake risks. In the following, brief summaries of insurance decision-making models for the two cases are given.

Insurance Demand Decisions by Homeowners: A simple form of homeowner's demand problem for earthquake coverage is the following: (i) a homeowner has a utility function U (typically, concave function); (ii) earthquake damage can occur with probability of π (note: there are two contingent states only), and should this happen, he/she will lose a property valued at L; (iii) the owner has an additional asset valued at A; (iv) the owner pays a premium P_I. Under these assumptions, the expected utility without insurance can be expressed as $\pi U(A) + (1-\pi)U(A+L)$, while that with (full) insurance is $U(A + L - P_I)$. The maximum insurance premium that the owner is willing to pay can be obtained by setting the derivative of the objective function:

$$O = \pi U(A) + (1 - \pi)U(A+L) - U(A+L-P_I) \tag{7}$$

with respect to an insurance parameter equal to zero and by finding a solution of the equation. A similar setup can be used for more elaborated cases to analyze the effects due to asset, policy function structure (e.g., coinsurance/deductible/ limit), risk attitude (utility function and degree of risk aversion), and risk premium (Schlesinger 2000). Recently, Asprone et al. (2013) have applied an extended version of the insurance demand model by considering multiple damage/ loss levels to determine insurance premiums of building stocks in Italy.

Insurance Supply Decisions by Insurers: The supply of earthquake coverage is influenced by various factors. Insurer's main concerns in offering earthquake coverage are profitability and fear of insolvency. This insurer's decision problem can be formulated as the profit maximization subjected to survival and stability constraints. The survival criterion requires that ruin probability should be less than a threshold value (which may be given by a regulatory body exogenously), whereas the stability criterion requires that probability of operational costs (i.e., sum of claim payments and expenses) exceeding a certain threshold value should be sufficiently small. Each of these criteria produces the maximum number of policies that can be offered for a given policy-premium structure. If the maximum number of policies based on the two constraints does not exceed the minimum number of policies required to initiate such coverage, the insurer will not offer earthquake insurance. Although the above economic analysis highlights a rational basis for insurer's supply decisions, insurers tend to adopt heuristic criteria based on simpler risk metrics, such as PML and VaR.

Integrated Earthquake Risk Management

Recent catastrophes in Chile and Japan have revealed insufficiency of conventional financial risk management tools and systems, such as self-insurance, public support, and earthquake insurance. Muir-Wood (2011) suggested that for Chile, micro-insurance, targeted for low-income homeowners and small commercial enterprises, should be introduced and a new risk-pooling system should be set up to cover earthquake risks for government-owned buildings. The former issue is compounded by anticipation of ex post subsidies from governments by Chilean people (charity hazard); however, it has been demonstrated that DRR measures are far more cost-effective than no DRR situation (Kleindorfer et al. 2005; Michel-Kerjan et al. 2013). In Japan, relatively low penetration rates of earthquake insurance (about 27 % in 2013) should be increased to achieve a nationwide risk-sharing mechanism with stable incomes; this situation was contrasted by the earthquake insurance coverage in New Zealand after the 2010–2011 Christchurch sequences. It is also important to implement risk-based insurance premiums for different structural types and seismic regions (e.g., California, Japan, and Turkey) and introduce financial incentive schemes to promote DRR investments (e.g., Japan). Such proactive earthquake risk management is the key to achieve sustainable and resilient communities against earthquake disasters.

In the envisaged risk management framework, the role of insurers, reinsurers, and governments is important. Major challenges in the current earthquake risk finance systems are related to securing sufficient funds to cover potential earthquake loss. The development of catastrophe securitization markets enables earthquake risk underwriters to have access to greater financial resources (Grace et al. 2003; Lalonde 2005; Cummins 2008). Broadly, there are three types of catastrophe insurance securities: indemnity-based securitization, index-based securitization, and parametric securitization. The indemnity-based securities are essentially the same as reinsurance contracts (but offered in form of securities), for which the loss repayment is determined based on actual security-issuer's (sponsor's) loss. Thus it suffers from the same issues as reinsurance, such as moral hazard and asymmetrical information. The index-based securities adopt actual industry-wide loss due to a damaging earthquake, rather than actual sponsor's loss. Advantages of this type of securitization include: reduced effects of moral hazard and transparency of loss information (note: individual companies are generally reluctant to reveal detailed information on their risk exposures). Difficulties arise in pricing such securities because industry loss and company's loss are not fully correlated. In other words, the index-based securities are subjected to basis risk.

Recently, parametric securities, such as CAT bonds, have become popular. A typical setup for the CAT bonds is as follows. A single-purpose reinsurer collects funds (principal) from investors in the markets and issues CAT bonds. In cases pre-agreed/specified trigger conditions are met, the principal is paid to the sponsor; otherwise, the principal together with return higher than typical securities is paid back to the investors when the bonds mature. Triggering of the bond payments is determined based on the intensity of physical earthquake parameters, such as a combination of earthquake magnitude and location and instrumentally obtained seismic intensity measures. As estimates of the earthquake intensity parameters are promptly available from reliable third bodies (e.g., US Geological Survey and Japan Meteorological Agency) immediately after an event, bond payments become available soon after the earthquake. This is not the case for the other two types of securities; assessment and settlement of earthquake loss take a long time. Generally, in comparison with reinsurance, the CAT bonds have three main advantages: (i) relatively large total value of bonds can be issued with low default risk, because the capital markets have significantly greater risk-bearing capacity compared with reinsurance industry, (ii) principal is accessible immediately after a disaster (high liquidity), and (3) moral hazard is minimized. A main disadvantage of the CAT bonds is basis risk, i.e., discrepancy between actual loss incurred by the sponsor and payment received by the sponsor (similar to index-based securities). It consists of model risk and trigger risk. Model risk arises because the structure of the CAT bonds is often designed based on numerical CAT models, whereas trigger risk is related to inaccuracy of an adopted trigger mechanism in terms of physical seismic parameters. The basis risk can be reduced by adopting an improved version of the CAT model and by implementing an advanced trigger mechanism that uses more local seismic information (e.g., instrumental measures from nearby recording stations, rather

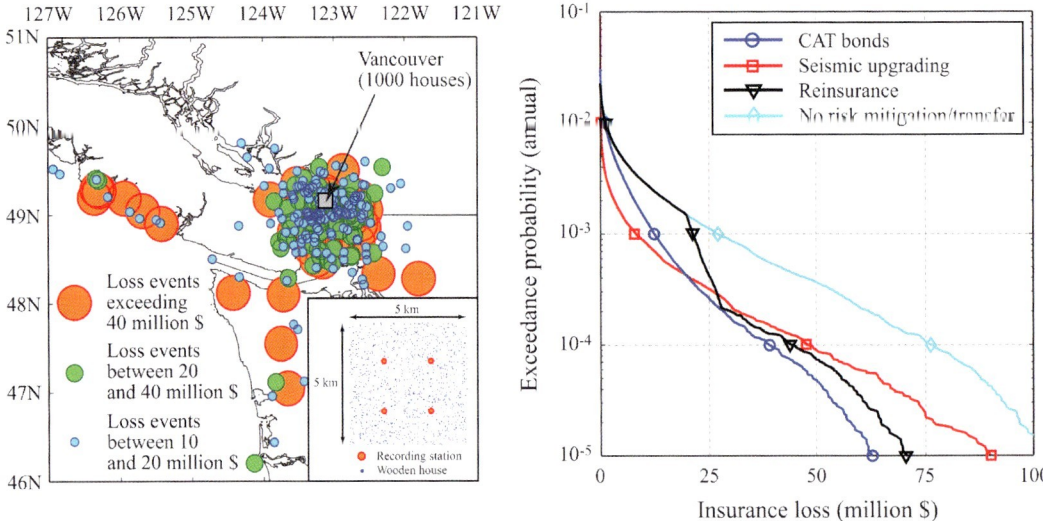

Insurance and Reinsurance Models for Earthquake, Fig. 9 Insurance portfolio management using seismic upgrading, reinsurance, and CAT bonds

than earthquake event characteristics; Goda 2014). In this regard, progress in the CAT modeling, contributed by all involved disciplines, has direct impact on the reliability of the CAT bonds.

Lastly, to illustrate a practical application of CAT bonds in insurance portfolio management, EP curves for four cases are presented in Fig. 9: (i) no risk mitigation/transfer; (ii) CAT bonds with trigger attachment and exhaustion points of $20 and $40 million, respectively; (iii) reinsurance with attachment and exhaustion points of $20 and $60 million and pro rata ratio of 0.2; and (iv) seismic upgrading (implementation of seismic provisions). The base portfolio is comprised of 1,000 wooden houses designed without seismic provisions, and the costs to implement the risk mitigation/transfer measures are included in the EP curves. The details of the analysis setup can be found in Goda (2014). All three risk mitigation/transfer options are effective in reducing insurer's earthquake exposure. The EP curve for the reinsurance contract has clear kinks, which correspond to the policy attachment and exhaustion points. The three options transform the original EP curve differently; reinsurance has clear influence on the specific loss layer between the attachment and exhaustion points, while seismic upgrading essentially shifts the EP curve downward (i.e., chance of experiencing seismic loss is decreased because of enhanced seismic resistance). The influence of the CAT bonds is more widely distributed (note: the EP curve for the CAT-bond case includes the triggering errors). Importantly, the numerical CAT modeling results combined with quantitative risk metrics for different risk management strategies provide valuable insight on the earthquake risk exposure. This facilitates risk-based decision-making and cost-benefit analysis.

Summary

This entry reviewed a broad range of topics related to earthquake insurance and reinsurance. An emphasis was given to the role of insurance and reinsurance in mitigating financial consequences of catastrophic earthquakes as integrated part of disaster risk reduction. Basic terminologies and current insurance-reinsurance systems were introduced. In particular, challenges related to low-probability high-consequence risk characteristics were discussed. Various insurance and reinsurance models that have been developed and applied in different academic fields were described. The fast-evolving earthquake catastrophe modeling techniques were summarized in the context of earthquake insurance portfolio management (including quantitative risk metrics). Subsequently, actuarial and financial insurance models, both statistical and computational, were presented. The homeowner's and insurer's decision models were then summarized. A sequential application of different insurance models (e.g., catastrophe modeling combined with dynamic financial analysis) is the state of the art in practice, requiring broad knowledge and understanding of available techniques. Finally, utilization of emerging alternative risk transfer tools, such as catastrophe bonds, was discussed to promote an integrated earthquake risk management framework.

Cross-References

▶ Performance-Based Design Procedure for Structures with Magneto-Rheological Dampers
▶ Seismic Loss Assessment

References

Asprone D, Jalayer F, Simonelli S, Acconcia A, Prota A, Manfredi G (2013) Seismic insurance model for the Italian residential building stock. Struct Saf 44:70–79

Bommer J, Spence R, Erdik M, Tabuchi S, Aydinoglu N, Booth E, del Re D, Peterken O (2002) Development of an earthquake loss model for Turkish catastrophe insurance. J Seismol 6:431–446

Cossette H, Duchesne T, Marceau E (2003) Modeling catastrophes and their impact on insurance portfolios. N Am Actuar J 7:1–22

Cummins JD (2008) CAT bonds and other risk-linked securities: state of the market and recent developments. Risk Manag Insur Rev 11:23–47

Daniell JE, Khazai B, Wenzel F, Vervaeck A (2011) The CATDAT damaging earthquakes database. Nat Hazards Earth Sys Sci 11:2235–2251

Dong W, Grossi P (2005) Insurance portfolio management. In: Grossi P, Kunreuther H (eds) Catastrophe modeling: a new approach to managing risk. Springer, New York, pp 119–133

Ehrlich I, Becker GS (1972) Market insurance, self-insurance, and self-protection. J Polit Econ 80:623–648

Goda K (2014) Seismic risk management of insurance portfolio using catastrophe bonds. Comput Aided Civ Infrastruct Eng. doi:10.1111/mice.12093

Goda K, Ren J (2013) Seismic risk exposure modelling of insurance portfolio using extreme value theory and copula. In: Tesfamariam S, Goda K (eds) Handbook of seismic risk analysis and management of civil infrastructure systems. Woodhead Publishing, Cambridge, pp 760–786

Goda K, Yoshikawa H (2012) Earthquake insurance portfolio analysis of wood-frame houses in south-western British Columbia, Canada. Bull Earthq Eng 10:615–643

Grace MF, Klein RW, Kleindorfer PR, Murray MR (2003) Catastrophe insurance: consumer demand, markets and regulation. Kluwer, Boston

Johnson EJ, Hershey J, Meszaros J, Kunreuther H (1993) Framing, probability distortions, and insurance decisions. J Risk Uncertainty 7:35–51

Kahneman D (2003) Maps of bounded rationality: psychology for behavioral economics. Am Econ Rev 93:1449–1475

Kajitani Y, Chang SE, Tatano H (2013) Economic impacts of the 2011 Tohoku-oki earthquake and tsunami. Earthq Spectra 29:S457–S478

Kaufmann R, Gadmer A, Klett R (2001) Introduction to dynamic financial analysis. ASTIN Bull 31:213–249

Kleindorfer P, Grossi P, Kunreuther H (2005) The impact of mitigation on homeowners and insurers: an analysis of model cities. In: Grossi P, Kunreuther H (eds) Catastrophe modeling: a new approach to managing risk. Springer, New York, pp 167–188

Kunreuther H (1996) Mitigating disaster losses through insurance. J Risk Uncertainty 12:171–187

Lalonde D (2005) Risk financing. In: Grossi P, Kunreuther H (eds) Catastrophe modeling: a new approach to managing risk. Springer, New York, pp 135–164

Maccaferri S, Cariboni F, Campolongo F (2011) Natural Catastrophes: risk relevance and insurance coverage in the EU. European Commission, Joint Research Centre, Ispra

McNeil AJ, Frey R, Embrechts P (2005) Quantitative risk management: concepts, techniques, and tools. Princeton University Press, Princeton

Michel-Kerjan E, Hochrainer-Stigler S, Kunreuther H, Linnerooth-Bayer J, Melcher R, Muir-Wood R, Ranger N, Vaziri P, Young M (2013) Catastrophe risk models for evaluating disaster risk reduction investments in developing countries. Risk Anal 33:984–999

Mossin J (1968) Aspects of rational insurance purchasing. J Polit Econ 91:304–311

Muir-Wood R (2011) Designing optimal risk mitigation and risk transfer mechanisms to improve the management of earthquake risk in Chile. OECD working papers on Finance, Insurance and Private Pensions (No. 12). OECD, Paris

Non-Life Insurance Rating Organization of Japan (NLIRO) (2008) Earthquake insurance in Japan.

NLIRO, Tokyo. http://www.giroj.or.jp/english/index.html

Organisation for Economic Co-operation and Development (OECD) (2008) Financial management of large scale catastrophes. Policy Issues in Insurance (No. 12). OECD, Paris

Palm R (1995) Earthquake insurance – a longitudinal study of California homeowners. Westview Press, Boulder

Porrini P, Schwarze R (2012) Insurance models and European climate change policies: an assessment. Eur J Law Econ. doi:10.1007/s10657-012-9376-6

Rolski T, Schmidli H, Schmidt V, Teugels J (1999) Stochastic processes for insurance and finance. Wiley, New York

Schlesinger H (2000) The theory of insurance demand. In: Dionne G (ed) Handbook of insurance. Kluwer, Boston, pp 131–151

Viscusi WK, Evans WN (2006) Behavioral probabilities. J Risk Uncertainty 32:5–15

Wesson RL, Perkins DM, Luco N, Karaca E (2009) Direct calculation of the probability distribution for earthquake losses to a portfolio. Earthq Spectra 25:687–706

Yucemen MS (2013) Probabilistic assessment of earthquake insurance rates for buildings. In: Tesfamariam S, Goda K (eds) Handbook of seismic risk analysis and management of civil infrastructure systems. Woodhead Publishing, Cambridge, pp 787–814

Integrated Earthquake Simulation

Lalith Wijerathne
Earthquake Research Institute, University of Tokyo, Tokyo, Japan

Synonyms

End-to-end earthquake simulations; Large-scale earthquake simulations

Introduction

A major earthquake in a commercial metropolis can inflict catastrophic disasters, crippling economy, causing loss of lives, and damaging infrastructure and properties. Even though the earthquake itself cannot be predicted, it is a possibility that the associated catastrophic losses can be prevented by implementing sound disaster mitigation plans; retrofitting the infrastructures, implementing good crisis management and recovery plans, etc. (NEES report 2004). This requires reliable predictions of damages due to anticipating earthquake scenarios. The conventional predictions are based on empirical relations which are obtained by statistical analysis of past earthquake disasters. These empirical predictions are less reliable since the conditions during the recorded major earthquakes, which are rare and have return periods of half a century or more, are significantly different from the conditions at the time of next major earthquake. During the long periods between major earthquakes, built environment and social and economic systems undergo significant changes. As an example, social and economic systems increasingly rely on lifeline networks, like electricity, communication, water, gas, sewer, transportation, etc., making it possible for any malfunction or damage to propagate way beyond what have been experienced during the past earthquake disasters. Even some of these vital lifelines may not have existed at the time of many of the past major events. Therefore, the next major earthquakes can easily outpace the earthquake mitigation efforts based on these empirical predictions.

Predictions, which take the current state of an urban system into account, enable to implement disaster mitigation and recovery plans, which must be significantly successful compared to those made solely based on the past earthquake disaster information. More reliable and detailed prediction of damages to buildings, lifelines, and other structures can be made if the enhanced materials and updated design codes used for the constructions and the current state are taken into account. Such predictions based on up-to-date information can restrain the propagation of damages and malfunctions in lifeline networks, which are becoming increasingly interdependent. With detailed, comprehensive, and reliable predictions, decision makers can prepare a sound recovery plan which minimizes the post disaster economic damages and restores the social system in a shorter time. Thus, what is necessary is detailed and comprehensive disaster prediction which considers the current state of built environment, social system, economy, etc. (Hori 2011).

Integrated earthquake simulation is a new paradigm which enables detailed and comprehensive analysis earthquake disasters in large urban system, taking the current state of built environment, social systems, etc. into account. The rapid progress of earth sciences, earthquake engineering, disaster mitigation sciences, and social sciences has produced well-established numerical tools for simulating many phases involved in an earthquake scenario. These numerical tools can be utilized to seamlessly simulate all the phases involved in an earthquake disaster, in high spatial and temporal resolutions. Opposing to the conventional disaster predictions, the use of such numerical simulations for disaster predictions has two main advantages: current state of built environment and socioeconomic systems can be taken into account; a large ensemble of plausible future earthquake scenarios can be taken into account. These two advantages address the major weaknesses of the conventional predictions, making the predictions based on integrated earthquake simulations more reliable.

Many of the numerical methods and tools necessary for building an integrated earthquake simulator already exist. Hence, the major task involved in developing an integrated earthquake simulator is to build a system which seamlessly integrates the necessary numerical tools for end-to-end earthquake simulations, for suitable computer hardware. There are well-developed numerical tools for earthquake hazard and disaster simulations, while the methodologies and numerical tools for simulating human responses, like evacuation, recovery, etc., are being developed. Despite the availability of some of the required numerical tools, it is not a trivial task to build a seamlessly integrated system. Several problems have to be addressed: lack of consistency between the inputs and outputs of some numerical tools, lack of numerical tools for simulating some phases, automation of analysis model constructions for large metropolis, efficient utilization of high-performance computing resources, etc.

The need for detailed, comprehensive, and reliable earthquake disaster predictions and the wide availability of high-performance computing resources have initiated several research activities on integrated earthquake simulation systems (Hori and Ichimura 2008; Hori 2011; Taborda and Bielak 2011; Muto et al. 2008). Further, the need of an integrated system for more reliable earthquake predictions has been emphasized by the Committee to Develop a Long-Term Research Agenda for the Network for Earthquake Engineering Simulation of the USA (NEES report 2004). Most of the existing integrated systems consider only the integration of earthquake hazard and seismic response of man-made structures. Taborda and Bielak (2011) have demonstrated the application of an integrated system to simulate wave propagation in large basins, amplification in surface layers, and seismic response of a small number of buildings. Muto et al. (2008) have presented a system for simulating rupture process to building damage and estimating repair costs, with small-scale seismic response analysis simulations. Hori et al. (Hori and Ichimura 2008; Hori 2011; Sobhaninejad et al. 2011) have developed a system called integrated earthquake simulator (IES) with the aim of seamlessly simulating all the involved phenomena from earthquake hazard to human responses, in large scale. They have proposed a macro-micro analysis method which not only efficiently simulate wave propagation and amplification in large domains but also treats the uncertainties due to lack of underground structure data. Also, the IES consists of an evacuation simulation module which can simulate evacuation in smaller-scale domains (Hori 2011).

The objective of this entry is to present an overview of integrated earthquake simulations. Being the most complete available integrated earthquake simulation system, most of the contents of this entry are prepared based on Hori et al.'s integrated earthquake simulator (Hori and Ichimura 2008; Hori 2011; Sobhaninejad et al. 2011). In the next section, an overview of integrated earthquake simulations is presented. The remaining sections present some details of earthquake hazard and disaster simulations;

neither tsunami nor human responses are included since these are not included in the available integrated systems. A short description of the numerical methods used for earthquake hazard and disaster simulations and automated analysis model construction based on GIS data are presented in these two sections. The last section presents a summary of the contents and a list of challenging problems which have to be addressed in the future.

Integrated System for End-to-End Earthquake Simulations

An integrated earthquake simulation system consists of a set of numerical tools, which are seamlessly integrated, for simulating the three main phases of an earthquake; earthquake and tsunami hazard, earthquake and tsunami disaster, and human actions against the earthquake disaster (see Fig. 1). Different numerical simulations that are used for simulating each of these three phases are summarized below.

1. Hazard simulations
   (a) Earthquake hazard simulation: For a given fault mechanism, an earthquake wave is synthesized and propagation through the crust is computed. Considering the 3D topographical effects and nonlinear properties of surface soil layers, wave amplification (i.e., strong ground motion) at the target area is computed. The output of the strong ground motion calculation should be consistent with the spatial and temporal resolutions required for seismic response analysis of man-made structures.
   (b) Tsunami hazard simulation: For a given fault mechanism, ocean bottom deformation and tsunami wave propagation are simulated.
2. Disaster simulations
   (a) Earthquake disaster simulation: Seismic responses of all the structures, both on the surface and underground, are computed using the strong ground motion data obtained in the hazard simulations. Depending on the type of each structure, a suitable numerical analysis method has to be used. This simulation should produce the necessary outputs, like the damage state of a building, which are required for human action simulations.
   (b) Tsunami disaster simulation: Simulation of tsunami flow in the target area considering the fluid structure interaction and damages to man-made structures due to

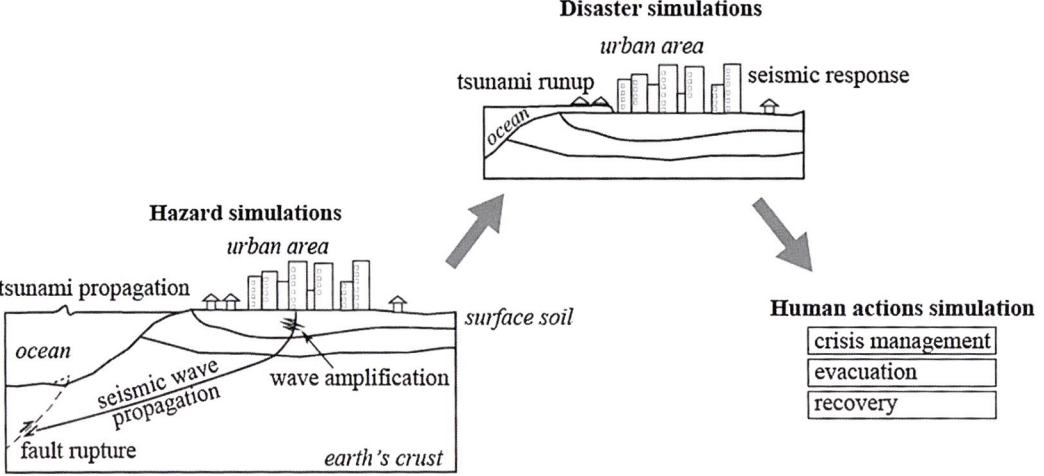

Integrated Earthquake Simulation, Fig. 1 Overview of integrated earthquake simulations

tsunami wave and the debris it carries. Though it is a challenging problem, simulation of damages to man-made structures is essential for the recovery process simulation.
3. Human actions against disasters
   (a) Evacuation: Generally, this involves the simulation of evacuation of buildings and moving to a safe location. In regions prone to tsunami disasters, moving to a safe high ground has to be simulated considering the human behaviors in time-critical events and the time history of inundation. The environment model of evacuation simulation should be modified, according to the results of the earthquake disaster simulation, so that unexpected delays caused by narrowing and blockages of roads are incorporated.
   (b) Crisis management: Simulation of possible emergency situations like fire, flooding, etc.
   (c) Recovery: Simulation of allocation and consumption of resource, like manpower, energy, materials, etc., to recover from the disaster, and estimate the resulting social and economic benefits.

In short, given the input data of a target area and a fault mechanism, an integrated system calculates the damages to the built environment and both the short-term and long-term human actions against the disaster. For the seismic hazard and disaster simulations, the required inputs are the data of underground structures, man-made structures, and a possible fault mechanism. The output of the disaster simulations should produce the necessary data for the simulation of human actions. In addition to the data from the disaster simulation, the human-response simulations require more data like the distribution of people at the time of the disaster and their properties, cost of repairing various damaged components, availability of various resources and manpower, etc. It would be a useful tool for decision makers, if the recovery simulation includes capabilities to find optimal way to allocate resources for a given damage state.

Applications

The main application of an integrated earthquake simulation system is to make detailed and comprehensive earthquake disaster predictions using the current state of built environment, social systems, and economy. This involves many uncertainties like the fault mechanism and the location of the next earthquake, lack of geological and underground structure data, etc. Unless these uncertainties are taken into consideration, the predictions of the integrated system are useless; what the decision makers need is predictions of high reliability so that they can use the predictions to implement disaster mitigation and recovery plans. In order to make reliable predictions, accounting the uncertainties of input data, some kind of stochastic approach like Monte Carlo simulation is necessary; Monte Carlo simulation is the widely used method for simulating this kind of phenomena involving significant uncertainties in input data. The detailed, comprehensive, and reliable predictions using an integrated earthquake simulator will ultimately make it possible to prevent catastrophic losses even though the anticipating earthquakes cannot be predicted.

In addition to the abovementioned main application, an integrated earthquake simulation system can be used for many other applications, some of which are listed below:

– As a tool for land-use planners to decide the required regulations or usages compatible with expected level of risk, when developing regions prone to seismic hazards. An integrated system can be used to assess the hazard level and find the construction standards and response measures which are necessary to prevent disasters in regions with seismic hazards. Then the public and policy makers can be informed of the earthquake risk, necessary planning and construction, and necessary response measures to reduce risks (NEES report 2004).
– For performance-based structural designs.
– Augmented reality and the results from a validated integrated system can be used to convince the political leaders and the public

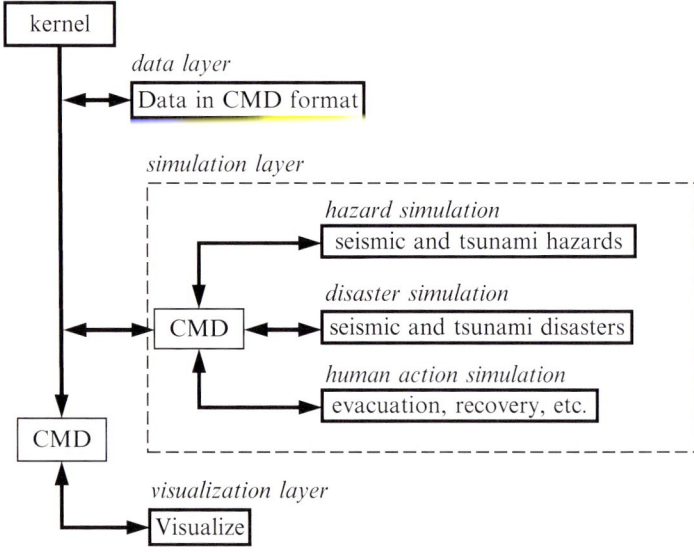

Integrated Earthquake Simulation, Fig. 2 A basic structure of an integrated earthquake simulator. Common modeling data (*CMD*) is the internal data format used by Hori et al.

that new disaster mitigation strategies are effective and economically viable risks (NEES report 2004).
- As a training tool for emergency and relief supply staffs, etc.

Not all the abovementioned numerical tools are included in present integrated earthquake simulators. Most of the existing integrated simulators include only the numerical tools for earthquake hazard and earthquake disaster simulations. The remaining sections provide detailed information of these numerical tools available in present-day integrated earthquake simulators. Though several numerical methods for tsunami hazard simulations are available (Goto et al. 1997; Mader 2004), there is a lack of numerical tools for simulating the tsunami disaster. The existing numerical tools only estimate the inundation and run-up height, and damages to man-made structures in large areas are not considered. Also, there are no well-established techniques for simulating other human responses, except the evacuation simulations. Multi-agent systems have been developed to simulate evacuation triggered by earthquakes or tsunami. However, these tools are not sufficiently mature to simulate large urban area evacuations, which is the target domain of an integrated system.

Therefore, the remaining sections only focus on the earthquake hazard and disaster simulations.

Development of an Integrated System

Figure 2 shows a basic structure for an integrated earthquake simulator, proposed by Sobhaninejad et al. (see also Hori 2011). The kernel is the control center which controls the whole system. It controls inputs and outputs of the system and calls the individual simulation codes in proper order, providing the necessary input data and retrieving output data which are required for subsequent analysis codes. Each item shown in the simulation layer of Fig. 2 consists of a microkernel which controls the flow of each simulated phase. As an example, the seismic response analysis's microkernel controls how a large number of structures in a city is being executed, using suitable numerical tools to simulate different types of structures made of concrete, steel, wood, etc. Also, the microkernel distributes these structures among a large number of CPUs in high-performance computing (HPC) hardware such that all the CPUs have equal work loads. The use of HPC is essential to simulate millions of buildings in large metropolis.

Geographic information system (GIS) data is one of the prominent input data for an integrated earthquake simulator (Hori 2011). GISs are widely used by many governmental and private organizations to store data. As an example, in Japan, GISs of ground elevation, borehole data, shapes of buildings, etc. are available. These data can be utilized to generate analysis model of metropolis. However, these GISs do not contain all the essential information for generating analysis models. As an example, GIS does not contain essential data for seismic response analysis like structural skeleton of buildings, material properties, etc. Although the necessary information is unavailable in current systems, the quantity and quality of GIS data are continuously improved. The next two sections present how to construct analysis models for earthquake hazard and disaster simulations based on GISs containing borehole data and configurations of buildings.

As for the development language, an object-oriented language, like C++ or python, is suitable since object-oriented features are convenient in large and complex system development: the ease of large project management with multiple developers, the ease of long-term software maintenance, ability to develop intuitive user interfaces, availability of generic programming, etc. Further, the ability to call libraries written from any of the standard languages and availability of several inter-process communication methods make it easier to integrate numerical tools developed with any language.

Being a system consisting of a number of different numerical tools, an integrated system requires a suitable internal data format to efficiently manage the flow of data among the numerical tools (Hori 2011; Sobhaninejad et al. 2011). The numerical tools for simulating various phases are usually provided by different research groups and have different data formats. Usage of multiple data formats has several problems: requires many data conversions, readability and maintainability of the source codes are poor, etc. The standard method of solving this problem is maintaining one internal data format with which the kernels interact with each other and the various numerical tools. Microkernels are responsible for translating the internal data format when interacting with the numerical tools. It would be an advantage if the basic elements of the selected internal data format are readily exchangeable among CPUs with Message Passing Interface (MPI), which is the widely used standard library in parallel computing.

Some of the Difficulties Involved in Integrated System Development

Though it appears that integrating existing numerical tool to build an integrated earthquake simulator is simple, it involves several difficulties and challenges like:

- Lack of necessary numerical tools
- Incompatibility of inputs and outputs of some numerical tools
- Need of automated model construction
- Optimal utilization of high-performance computing resources
- Verification and validation of the integrated system

As mentioned above, there is a lack of well-developed numerical tools or methods to simulate certain phases like tsunami disaster and human responses. The incompatibility of inputs and outputs of different numerical tools is also a problem in integrating numerical tools. The outputs of most of the necessary numerical tools, which are developed in different disciplines, may not have consistent spatial and temporal resolutions required for subsequent analysis. To address this problem, close collaborations with the developers of relevant simulation tools are necessary. Being the analysis domains very large, automation of analysis model construction is essential, which is a challenging task; automated construction of suitable analysis model for each of millions of structures in a metropolis, based on available CAD or GIS data, is not trivial.

Enhancing an integrated system to optimally use high-performance computing (HPC) resources is also a major challenge. Obviously, optimal utilization of cutting-edge supercomputing resources is necessary to simulate a large metropolis. While there are HPC-enhanced numerical tools for hazard simulations, HPC-enhanced codes disaster simulation and human response have to be developed.

Further, building a seamlessly integrated system (i.e., a system which will simulate end-to-end process in a single execution) is challenging due to two main problems. The first problem is that even though each numerical tool is enhanced to optimally use supercomputing resources, it does not guarantee that a seamlessly integrated system can optimally utilize the supercomputing hardware. The main reason is that different numerical tools involve different amount of computations; hence, there is a drastic difference in the maximum number of CPUs required for the optimal performance of each numerical tool. As an example, the earthquake hazard simulation requires several hundred thousand CPUs, while evacuation simulation requires only a few thousands of CPUs. This results in waste of CPU time. The second difficulty is that different numerical tools require different parallel computing approaches, which cannot coexist in single execution. As an example, hybrid of shared and distributed memory parallelism is the best for earthquake hazard simulations, while earthquake disaster and evacuation simulations require only the distributed parallelism. These two HPC challenges, involved in end-to-end simulations in a single execution, can be easily avoided by executing each numerical tool independently, which is the same as replacing the main kernel (see Fig. 2) with human control. However, this is not an acceptable solution since simulating a large number of earthquake scenarios is going to be laborious and error prone; as mentioned, Monte Carlo simulations are necessary for improving the reliability of the earthquake disaster simulations.

Hazard Simulation

There are two main hazards involved: earthquake and tsunami. Earthquake hazard involves simulations of wave propagation through crust and amplification in the surface layers. Tsunami hazard involves ocean bottom deformation and tsunami wave propagation from source to shore. There are well-developed numerical methods and large-scale simulation programs for both the methods. However, existing integrated earthquake simulators only include the earthquake hazard, probably because tsunami hazards are rare though cause severe destruction. In this section, the available methods of earthquake hazard simulations and model generation from GIS data is presented.

Earthquake Hazard Simulations

Earthquake hazard simulation involves three main steps: simulation of fault rupture, simulation of seismic wave propagation though the earth's crust, and wave amplification when traveling through surface soil layers (see Fig. 1). Out of these three, fault mechanism is the least understood; rupture process has complex spatial and temporal characteristics. The other two steps are mechanical processes which are well understood, and there are several numerical methods to simulate wave propagation and amplification processes (Bielak et al. 2005; Hori and Ichimura 2008; Cui et al. 2009; Käser et al. 2008). The objective of earthquake hazard simulation is, for a given fault scenario, numerical simulation of wave propagation and amplification process, at sufficiently high spatial and temporal resolutions required for earthquake disaster simulations. Though the two mechanical processes are well understood, these involve two difficulties: required high-resolution simulations involve large-scale computing and lack of accurate geological and underground structure data. This subsection presents brief descriptions of available methods to overcome these problems.

While the dramatically improving supercomputing performance provides the necessary computation power, selection of proper numerical method is essential for the optimal usage of the computing resources. The two main numerical methods used for large-scale earthquake hazard simulations are finite difference method (FDM) and finite element method (FEM) (Moczo et al. 2007). Both methods have advantages and disadvantages (Bielak et al. 2005; Hori 2011; Taborda and Bielak 2011). FDM has been the most popular due to several reasons like the ease of implementation, the wide choice of explicit and implicit time stepping methods, sufficient robustness and accuracy, efficiency, etc.

However, due to the reliance of regular grid, the conventional FDM has several drawbacks, like the involvement of over-refined grid when modeling media with high contrast in material properties, need of smaller time steps, etc. The involvement of over-refined grid and smaller time steps dramatically increases the necessary computer resources and computing time. On the other hand, FEM does not involve these problems; longer time steps can be used and mesh can be adapted to model local features. These advantages become more dramatic with increasing contrast in shear wave velocity in the heterogeneous medium. The two major drawbacks of FEM are that a considerable effort is necessary to generate a refined mesh and storing global stiffness matrix requires considerable memory. Bielak et al. (2005) have proposed an octree-based FEM method to overcome both of these problems. The near-linear scalability up to 98,000 CPUs, when simulating $180 \times 135 \times 62$ km^3 region, shows the potential of octree-based approach (Taborda and Bielak 2011). The macro-micro analysis method, proposed by Hori et al. (Hori and Ichimura 2008; Hori 2011), overcomes the abovementioned two problems using singular perturbation expansion and bounding medium theory. A brief description of macro-micro analysis is given below.

The uncertainty of geological and underground structures is a serious problem which hinders the high-resolution earthquake hazard simulations. The presently used in situ observations and other technologies for mapping the geological and underground structures are inefficient or not sufficiently accurate for generating high-resolution analysis models, which are required for accurate and high-resolution seismic hazard simulations. This hinders accurate disaster simulations; accurate damage predictions require seismic hazard simulations of high spatial and temporal resolutions. As explained below, the macro-micro analysis method uses bounding medium theory to overcome this uncertainty problem.

Macro-Micro Analysis Method
Macro-micro analysis method (Hori and Ichimura 2008; Hori 2011) is a stochastic method which uses singular perturbation expansion and bounding medium theory to overcome the abovementioned two difficulties. In macro-micro analysis, a stochastic variational problem is posed to account the abovementioned uncertainties in geological and underground structures and material properties. In stochastic problems both the input variables with uncertainties and the output variables become random variables, which are defined both in physical space and probabilistic space. The inputs for the stochastic seismic hazard problem are the mean and variance of crust and ground structure configurations and mean and variance of uncertain material properties like density, Young's modulus, etc. Solving such stochastic problems is complex since the response of the system also becomes stochastic and not proportional to the variability of the input data. Micromicro analysis method uses bounding medium theory to find optimistic and pessimistic estimates for the mean response of the stochastic model, without explicitly solving the stochastic problem using laborious or complex methods like Monte Carlo simulations or spectral method. The bounding medium theory determines two deterministic, but fictitious, models which provide optimistic and pessimistic estimates for the mean response.

The upper and lower bound models obtained with the bounding medium theory require large number of discretization according to the geometry and shear wave velocity of materials; hence, solving the resulting analysis models is computationally intensive. In order to efficiently solve this large-scale problem, which has wildly varying parameters, macro-micro analysis takes a multiscale approach with singular perturbation expansion. Singular perturbation expansion is a mathematical technique which leads to multiscale analysis. The resulting analyses in long and short scale are called macroanalysis and microanalysis. Macroanalysis considers the wave propagation in the entire domain in lower spatial resolution, while the microanalysis considers wave propagation in higher spatial resolution for each subdomain of interest. These two analyses are analogous to the wave propagation analysis in seismological scale and strong ground

Integrated Earthquake Simulation,
Fig. 3 Determination of consistent soil layers

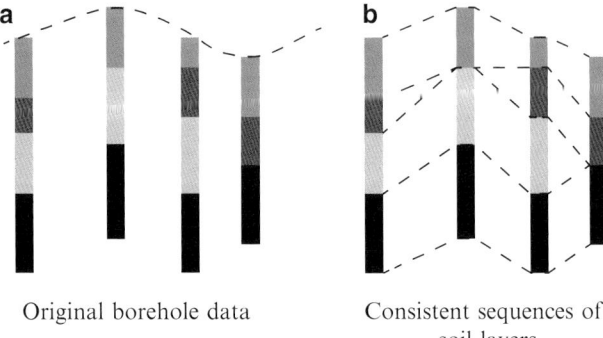

Original borehole data   Consistent sequences of soil layers

motion analysis in surface layers of urban areas. For the sake of reducing memory consumption and efficient computer implementation, macro-micro analysis uses voxel finite element method (VFEM) (Koketsu et al. 2004; Hori 2011). In summary, macro-micro uses bounding medium theory and singular perturbation expansion to numerically efficiently simulate the seismic wave propagation and amplification in large domains, which is a problem involving parameters with uncertainties, in high spatial and temporal resolutions.

Construction of Ground Structure Model from Sparse Borehole Data

As mentioned, accurate models of surface layers are necessary for high-resolution prediction of wave amplification in mediums with low shear wave velocities. However, constructing an accurate model is difficult due to limited availability of borehole data, some of which are poor in quality. Often these borehole data are coming from scattered locations and are inconsistent; borehole data is considered inconsistent if the sequence and the number of distinct layers differ from site to site (see Fig. 3a). To make a model of underground structures, based on borehole data, Hori (Hori 2011) has proposed a method which involves two steps: find a reference sequence of soil layers consistent with the data of each borehole site and interpolate the obtained consistent layers to find soil layer information at any arbitrary point. The proposed scheme, which is used in geo-information geology, for estimating a consistent set of soil layers at each site involves the following four steps:

(i) Choose one borehole and use its layers as the reference sequence.
(ii) Pick another borehole and compare its layers with the reference sequence.
  (a) If the sequences are matching, choose another.
  (b) Else make a new reference sequence consistent with the both by combining those; some layers can be of zero thickness (see Fig. 3b).
(iii) Repeat step ii to obtain a reference layer sequence consistent with all the boreholes.
(iv) Finalize by eliminating any insignificant layers.

Once a sequence of soil layers consistent with each borehole is found, soil layer information at any location can be evaluated using a suitable interpolation scheme. Hori (2011) has proposed the following grid algorithm to interpolate and obtain soil layer information in a regular grid. In order to explain this grid algorithm, consider the problem of interpolating the elevation of a certain soil layer interface. Assume that consistent layer information at K number of borehole sites of the domain D are available. Let the spatial location boreholes and the elevation of the specific layer interface be $\{x^k\}$ and $\{f^k\}$, where $k = 1, 2,..., K$. In order to interpolate $\{f^k\}$, a coarse fictitious grid is introduced. Let the grid points denote by $(x^{n,m})$ and the fictitious elevation of the specific layer interface at corresponding grid locations be $f^{n,m}$. The elevation of the specific layer $f(x)$, at borehole location x, can be expressed by using the data of the four neighboring fictitious grid points as follows (see Fig. 4):

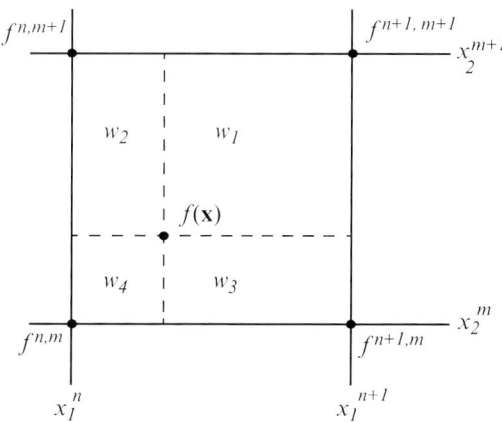

Integrated Earthquake Simulation, Fig. 4 Fictitious grid and the weights used for the bilinear interpolation

$$f(\mathbf{x}) = w_1(\mathbf{x})f^{n,m} + w_2(\mathbf{x})f^{n+1,m} + w_3(\mathbf{x})f^{n,m+1} + w_4(\mathbf{x})f^{n+1,m+1},$$

where $w_1, w_2, \ldots, w_4$ are bilinear weight functions. The weights are the area fractions of the rectangles formed by $\mathbf{x}$ and the four grid points. As an example

$$w_1 = \frac{\left(x_1^{n+1} - x_1\right)\left(x_2^{m+1} - x_2\right)}{\left(x_1^{n+1} - x_1^{n}\right)\left(x_2^{m+1} - x_2^{m}\right)}.$$

Once a coarse grid is made, the above bilinear interpolation can be repeatedly applied to obtain soil layer data in a grid, which is sufficiently fine for setting material properties of VFEM model for the microanalysis. As explained above, macro-micro analysis uses bounding medium theory to handle the uncertainties in the layer configurations and material properties.

Disaster Simulation

Out of the two main disasters involved, seismic disaster has received more attention (Hori and Ichimura 2008; Muto et al. 2008; Hori 2011; Taborda et al. 2011). There are a number of numerical methods developed for seismic response analysis of buildings and other structures. Some of these methods are validated and have been extensively used for the purposes like structural designs, performance evaluations, structural health monitoring, etc. On the contrary, there are no well-established numerical methods for simulating large urban area tsunami disasters, incorporating the damages to structures inflicted by tsunami wave and debris. Development, verification, and validation of numerical methods capable of simulating large area tsunami inundation, including the destruction it brings, are challenging research areas which are essential for integrated earthquake simulators. In this section, the available seismic response analysis methods for buildings and approximate analysis model construction based on GIS data are briefly explained.

Available Numerical Methods for Seismic Response Analysis of Buildings

There are several numerical methods for the seismic response analysis, ranging from simple single degree of freedom (SDOF) systems to sophisticated nonlinear FEM models which consider large deformation and damage. Some well-established numerical techniques used in structural response analysis are SDOF, multi-degree of freedom systems (MDOF), one-component model (OCM), discrete element method (DEM), fiber element model, and nonlinear FEM. Though simple, nonlinear MDOF systems are capable of reproducing the complicated nonlinear structural responses (Pampanin et al. 2003). OCM models the nonlinear behavior of structural members by concentrating inelastic deformations at the ends of elements, with inelastic springs with hysteretic properties. The fiber element method (Spacone et al. 1996a, b; Haselton et al. 2008) is a moderately complex method widely used for simulating nonlinear seismic response of medium rise concrete structures. It is best to include numerical tools for all these methods in an integrated simulator. Then a user has a high flexibility in choosing a suitable analysis method depending on the characteristics and importance of each structure and available computing resources.

Automated Analysis Model Generation

It is a challenging task to automate the analysis model construction for millions of buildings in a metropolis, based on GIS data. Although GIS is a prominent source of input data for an integrated earthquake simulator, the current GIS data does not contain the necessary data like structural skeleton and material properties; only the information like location, height, shape, etc. are included in the current GIS. The difficulties in making analysis model significantly increase with the complexity of the analysis method. As an example, analysis model construction for SDOF or MDOF is way easier than analysis model construction for OCM or fiber element method. OCM and fiber element methods require detailed information of the structural skeleton, which is difficult to be generated automatically from GIS data.

Even though the present GIS does not include the necessary data for structural response analysis, it is possible to generate an approximate analysis model based on the design procedures and recommendations provided in relevant design codes and handbooks. In the absence of necessary data in GIS or other sources, such approximate models, which comply with design manuals, are useful in obtaining a probable seismic disaster of a large metropolis. An approximate method of constructing MDOF models of residential buildings using the available GIS data, which is proposed by Hori (2011), is given below.

Approximate MDOF Model Generation Based on GIS Data

The governing equation of a building, which is subjected to seismic acceleration $\ddot{z}(t)$, modeled as an MDOF system is given by

$$M\ddot{U}(t) + C\dot{U}(t) + KU(t) = \ddot{z}(t)M1, \quad (1)$$

where the vector $1 = [1, 1, \ldots, 1]^T$. The data provided by the current GIS is insufficient to define the mass, damping, and stiffness matrices, which are represented by M, C, and K. However, if rewritten in the form used for modal analysis, the recommendations given in design manual can be utilized to solve the above problem. To that end, the natural frequencies and corresponding modes of the undamped system is found by solving the following eigenvalue problem:

$$(K - \omega_n M)\Phi_n = 0$$

The above eigenvalue problem can be easily obtained by setting $U = \Phi \exp(i\omega t)$, $C = 0$ and $\ddot{z}(t) = 0$ in Eq. 1. Solving this eigenvalue problem, the nth natural frequency, ω_n, and the corresponding mode shape, Φ_n, can be obtained; $n = 1, 2, \ldots, N$, where N is the dimension of the matrices. The N mode shapes or eigenvectors form a basis; in the presence of any repeated frequencies, the associated modes should be orthogonalized. The unknown U can be expressed using these N basis functions as

$$U = \sum q_n \Phi_n. \quad (2)$$

Substituting Eq. 2 to Eq. 1 and following the standard modal analysis procedures, the following N set of differential equations can be obtained (Datta 2010):

$$\ddot{q}_n(t) + 2\xi_n \omega_n \dot{q}_n(t) + \omega_n^2 q_n(t) = \ddot{z}_n(t), \quad (3)$$

where ξ_n is the damping ratio and $\ddot{z}_n(t)$ is the external inertial force corresponding to the nth mode. The damping ratio and inertial force due to seismic loading are given by

$$\xi_n = \frac{\Phi_n^T C \Phi_n}{2\Phi_n^T M \Phi_n} \text{ and } \ddot{z}_n = \frac{\Phi_n^T M 1}{2\Phi_n^T M \Phi_n}.$$

According to the standard procedure of MDOF, the N equations given by Eq. 3 are solved for q_n's with proper boundary conditions and the unknown U is found using Eq. 2. It is not necessary to use all the N modes, and the solution can be approximated with a selected set of modes.

Integrated Earthquake Simulation, Table 1 Period and damping ratio of the fundamental mode for different types of structures. *H* is the height of a building

Building type	T_1/(s)	ξ_1
Wooden house	0.2–0.7	0.02
Reinforced concrete	0.02 H	0.03
Steel reinforced concrete	0.03 H	0.02

Often, only the first few modes are used for short buildings. As an example, the first three modes are used for buildings with the period of fundamental mode shorter than 0.5 s (Hori 2011).

In the absence of necessary data to define M, C, and K of a particular building, approximate analysis model can be constructed based on the standard dynamic properties given in building design codes. As an example, the standard values of the period T_1 and the damping ratio ξ_1 of the fundamental mode residential buildings recommended by the Japanese design code are given in Table 1.

Further, the periods and damping ratios of second and third modes can be obtained from the following two empirical relations:

$$\begin{bmatrix} T_2 \\ T_3 \end{bmatrix} = \begin{bmatrix} \frac{1}{3} \\ \frac{1}{5} \end{bmatrix} T_1$$

$$\xi_{n+1} = \begin{cases} 1.4\,\xi_n \text{ for reinforced buildings} \\ 1.3\,\xi_n \text{ for steel reinforced buildings} \end{cases}$$

The above empirical relations make it possible to estimate the natural frequencies, $\omega_n = 2\pi/T_i$, and the damping ratios of the first three modes, which are sufficient for solving Eq. 3 for the amplitudes of the first three modes. Finally, the response can be estimated with Eq. 2. The required modal shapes can be easily estimated since the vibration modes of the dominant modes are known. The dominant modal shapes of buildings are given in standard textbooks, and even the modes of a cantilever beam are a good approximation (Hans and Boutin 2008).

Similar to the above given procedure, approximate analysis models for more complex analysis methods like OCM and fiber element model can be generated based on recommendations given in design manuals and handbooks. Unlike in the case of MDOF, automated model constructions for these methods are more challenging since these methods require structural skeleton details. The generation of structural skeletons from GIS data is straightforward when floor plan of a structure is made of multiple rectangles. However, if a building's shape changes with the floor number or has complex outer shape, the process of structural skeleton generation is complex and requires expertise in architecture and computational geometry. The spacing and the dimensions of the structural elements can be estimated based on the recommendations given in design manuals. As an example, 1/5 of floor height is often recommended as the height of beams. Obviously, automated FEM model construction is the most challenging.

Figure 5 shows the maximums of the interstory drift angle compiled out of 400 simulations with different observed earthquake observations. The input strong ground motion for each building is computed using the 1D wave theory, using the details of underground structures and seismic waves recorded during 400 different earthquakes. The analysis models for buildings are automatically generated using the abovementioned approximate method for MDOF model generation from GIS data. Such Monte Carlo simulations can be used for more reliable earthquake disaster predictions.

Summary

In this chapter, the need of integrated earthquake simulation, a short overview of existing integrated earthquake simulators, and some remaining challenges are presented. The available integrated earthquake simulators have included only the earthquake hazard simulation

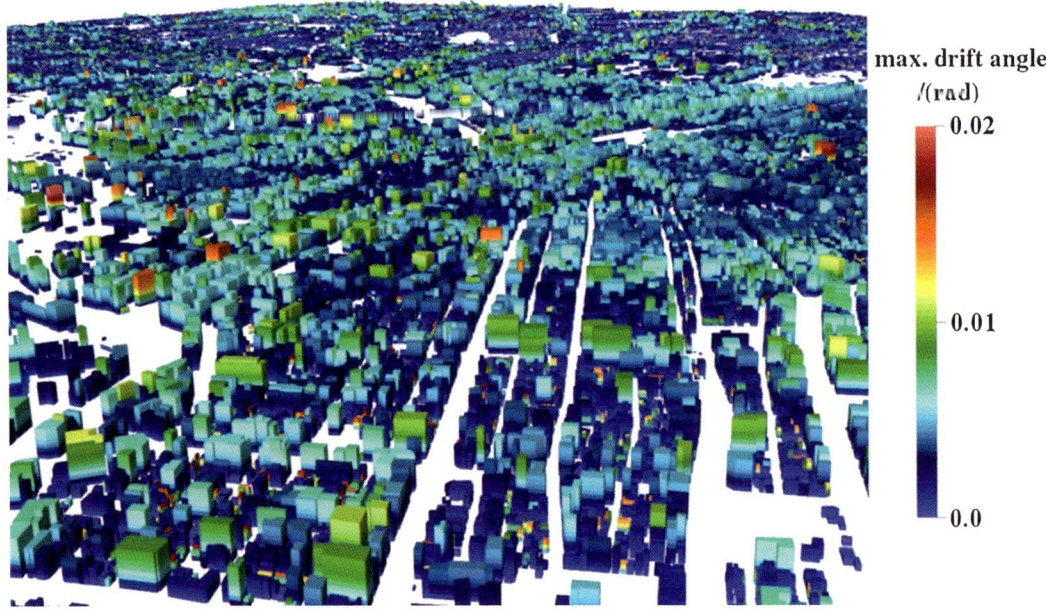

Integrated Earthquake Simulation, Fig. 5 Maximum inter-story drift angle obtained from 400 seismic response analysis simulations (Provided by Kohei Fujita, Department of Civil Engineering, University of Tokyo)

and earthquake disaster simulations. One of the remaining challenges in earthquake disaster simulation is automation of analysis model construction for complex analysis methods like fiber element model or FEM. Even though tsunami disasters are catastrophic, the present integrated systems have not included tsunami disaster simulations. It surely is challenging to develop numerical tools for tsunami disaster simulations including damages to structures. Even though tsunami disasters are rare, it is an essential item for an integrated system. The least developed is the simulations of human actions against earthquake disasters. The available numerical tools for evacuation simulations have to be further improved and enhanced with high-performance computing for simulating large domains. Out of all the numerical tools required for an integrated earthquake simulator, simulation of the recovery process is the least developed. It should be an invaluable tool for the decision makers, if a numerical method is developed to find the optimal way to allocate resources for recovering from a given earthquake disaster.

Cross-References

▶ Performance-Based Design Procedure for Structures with Magneto-Rheological Dampers
▶ Physics-Based Ground-Motion Simulation
▶ Site Response: 1-D Time Domain Analyses
▶ Stochastic Ground Motion Simulation

References

Bielak J, Ghattas O, Kim EJ (2005) Parallel octree-based finite element method for large-scale earthquake ground motion simulation. Comput Model Eng Sci 10–2:99–112

Committee to develop a long-term research agenda for the network for earthquake engineering simulation (NEES), National research council (2004) Preventing earthquake disasters: the grand challenge in earthquake engineering: a research agenda for the network for earthquake engineering simulation. The National Academic Press. http://www.nap.edu/catalog/10799.html. Accessed 15 July 2013

Cui Y, Moore R, Olsen K, Chourasia A, Maechling P, Minster B, Day S, Hu Y, Zhu J, Jordan T (2009) Toward petascale earthquake simulations. Acta Geotech 4–2:79–93

Datta TK (2010) Seismic analysis of structures. Wiley, Hoboken

Goto C, Ogawa Y, Shuto N, Imamura F (1997) IUGG/IOC time project, numerical method of tsunami simulation with the leap-frog scheme, IOC manuals and guides. UNESCO, Paris

Hans S, Boutin C (2008) Dynamics of discrete framed structures: a unified homogenized description. J Mech Mater Struct 3–9:1709–1739

Haselton CB, Liel AB, Lange ST, Deierlein GG (2008) Beam-column element model calibrated for predicting flexural response leading to global collapse of RC frame buildings. PEER report 2007/03, Pacific Earthquake Engineering Research Center, College of Engineering, University of California, Berkeley

Hori M (2011) Introduction to computational earthquake engineering. Imperial College Press, London

Hori M, Ichimura T (2008) Current state of integrated earthquake simulation for earthquake hazard and disaster. J Seismol 12:307–321

Käser M, Hermann V, Puente J (2008) Quantitative accuracy analysis of the discontinuous Galerkin method for seismic wave propagation. Geophys J Int 173(3):990–999

Koketsu K, Fujiwara H, Ikegami Y (2004) Finite-element Simulation of Seismic Ground Motion with a Voxel Mesh. Pure Appl Geophys 161(11–12):2183–2198

Mader CL (2004) Numerical modeling of water waves. CRC press, Boca Raton

Moczo P, Kristeka J, Galisb M, Pazaka P, Balazovjech M (2007) The finite-difference and finite-element modeling of seismic wave propagation and earthquake motion. Acta Physica Slovaca 57–2:177–406

Muto M, Krishnan S, Beck JL, Mitrani-Reiser J (2008) Seismic loss estimation based on end-to-end simulation, life-cycle civil engineering. In: Proceedings of the first international symposium on life-cycle civil engineering. Varenna/Lake Como, CRC Press, Boca Raton, 10–14 June 2008. ISBN 0415468574

Pampanin S, Christopoulos C, Priestley JN (2003) Performance-based seismic response of frame structures including residual deformations part II: multi-degree of freedom systems. J Earthq Eng 7(1):119–147

Sobhaninejad G, Hori M, Kabeyasawa T (2011) Enhancing integrated earthquake simulation with high performance computing. Adv Eng Softw 42:286–292

Spacone E, Filippou FC, Taucer FF (1996a) Fibre beam-column model for non-linear analysis of R/C frames: part I. Formulation. Earthq Eng Struct Dyn 25–7:711–725

Spacone E, Filippou FC, Taucer FF (1996b) Fibre beam–column model for non-linear analysis of R/C frames: part II. Applications. Earthq Eng Struct Dyn 25–7:727–742

Taborda R, Bielak J (2011) Large-scale earthquake simulation: computational seismology and complex engineering systems. Comput Sci Eng 13–4:14–27

Intensity Scale ESI 2007 for Assessing Earthquake Intensities

Pablo G. Silva[1], Alessandro Maria Michetti[2] and Luca Guerrieri[3]
[1]Departamento de Geología, Universidad de Salamanca, Escuela Politécnica Superior de Ávila, Avila, Spain
[2]Dipartimento di Scienza e Alta Tecnologia, Università dell'Insubria, Como, Italy
[3]Department of Geological Survey, ISPRA, Istituto Superiore per la Protezione e la Ricerca Ambientale, Rome, Italy

Synonyms

Earthquake geological effects; Environmental seismic intensity; Macroseismic scale

Introduction

Earthquake intensity scales were introduced at the end of the nineteenth century (e.g., Rossi-Forel, Cancani, Mercalli) in order to characterize source parameters, damage distribution, and environmental impact of relevant seismic events. These intensity scales were based on a classification of earthquake effects on humans, on buildings, and on the natural environment.

Intensity provides a measure of earthquake-induced damage both at a site (local intensity) and at the epicenter (epicentral intensity). It is important to note that intensity evaluations consider the coseismic effects in the whole range of frequencies of vibratory ground motion, together with those resulting from static, finite deformations (fault ground ruptures). It is common practice to use the term "macroseismicity" to describe effects that are measurable with the intensity scales, because very small earthquakes ("microseismicity"), commonly considered intensity degree one (I), are beyond the sensitivity level of these scales.

Most common intensity scales are divided in to twelve degrees using roman numbers (I to XII).

The diagnostic effects for low intensity degrees (I to V) are those on humans, then (VI to IX) those on buildings, and for the highest degrees (X to XII) those on the natural environment. In fact, the earthquake environmental effects (EEEs) start to be noticeable from intensity $\geq$IV as indicated in the Environmental Seismic Intensity scale (ESI-07 scale; Michetti et al. 2007), which is the subject of this chapter.

In spite of the increasing development of the instrumental record of earthquakes from the second half of the twentieth century, macroseismic intensity analyses are still the usual practice when analyzing both historical and recent earthquakes. Recent earthquakes provide the opportunity to compare earthquake damage levels and instrumental records of horizontal ground acceleration to classify and properly quantify the recorded effects on buildings and environment and consequently to refine intensity scales.

By contrast, historical earthquakes only can be explored by macroseismic analyses, and intensity is the only parameter that allows an estimate of the earthquake size to be made. Experience gained from the macroseismic analyses of instrumental events led to the development of empirical relationships in order to approximately estimate the moment magnitude (Mw) of historical events from earthquake intensity distribution. As a result, instrumental records are in the best case available from the first half of the twentieth century; however, historical intensity analyses (extended in some regions back to the sixth to seventh century B.C.) are behind most of the seismic hazard analyses, seismic-building codes, and nuclear regulatory guides worldwide. Therefore, the use and refinement of intensity scales in earthquake analysis is still, and will continue to be, imperative to the society.

ESI-07 scale (Michetti et al. 2007) classifies earthquake environmental effects (EEEs) qualitatively as described in previous macroseismic intensity scales (e.g., MCS, MSK, MM). The ESI-07 scale has been developed primarily to properly describe and quantify EEEs for different intensity levels, based on the extensive amount of observations and research accumulated in the past decades in this field (e.g., Reicherter et al. 2009; Audemard and Michetti 2011). Moreover, the ESI-07 scale has been introduced to fill the gaps of some modern macroseismic intensity scales, such as the European Macroseismic Scale (EMS-98; Grunthal 1998; Musson et al. 2010), which basically excludes EEEs from intensity assessment. Finally, the ESI-07 scale has also the objective of incorporating paleoseismic analysis into the macroseismic studies, extending the information on earthquake intensity data to a geologic time window in the order of some tens of kyrs.

The ESI-07 scale draws upon the expertise of a group of geologists, seismologists, and engineers under the support of the International Union for Quaternary Research (INQUA). The definition of the intensity degrees has been the result of a revision of the EEEs caused by a large number of earthquakes occurring in different parts of the world, conducted by the "INQUA International Focus Group on ▶ Paleoseismology and Active Tectonics" since the year 2003. The scale was formally approved during the XVII INQUA Congress, held in Cairns (Australia) on August 2007. The ensuing literature includes applications of the scale with a global coverage (e.g., Ali et al. 2009; Berzhinskii et al. 2010; Di Manna et al. (2013); Fountoulis and Mavroulis 2013; Gojsar 2012; Guerrieri et al. 2008; Lalinde and Sanchez 2007; Mavroulis et al. 2013; Ota et al. 2009; Papanikolaou 2011; Papathanassiou and Pavlides 2007; Serva et al. 2007; Tatevossian 2007). More recently, Silva et al. (2014) implemented the first official Catalogue on Earthquake Environmental Effects published by the Geological Survey of Spain (IGME) based on the On-line EEE Catalogue hosted at the Geological Survey of Italy (ISPRA) (http://www.eeecatalog.sinanet.apat.it/terremoti/index.php).

In the following sections, we will analyze the role of earthquake environmental effects in the existing macroseismic scales, how EEEs are classified (primary and secondary effects) and incorporated to the ESI-07 scale, the guidelines for the application of the scale, and finally the use of this intensity scale in paleoseismic research. The original text of the scale after Michetti et al. (2007) can be found in Appendix 1.

Earthquake Environmental Effects in the Macroseismic Scales

As mentioned in the introduction, intensity is a description of an earthquake based on the effect to humans, to buildings, and to the natural environment. Macroseismic scales classify these effects for each intensity level. A classification is not a measurement, and there is no way to define the "energetic distance" between different degrees of an intensity scale. In fact, macroseismic classifications are based on empirical qualitative scales, which by definition include significant subjectivity or, better, expert judgment. For this reason, all the scales use integer values from I to XII. Therefore, the only mathematical property that we can use is the order (and this is why the roman numerals are used for defining the intensity degrees).

The scientific worth of intensity scales, however, is to order the coseismic effects in degrees that try to include phenomena indicative of large differences in the strength of the earthquake at a site. Intensity IX in all scales of the "Mercalli family" (or, for the sake of precision, "Cancani family," since Cancani was the first to introduce a 12° intensity scale) describes a scenario of effects, or a level of shaking, that is completely different from (much higher than) intensity VIII and completely different from (much lower than) intensity X. This is the "power of resolution" of the scales. Even with quite a "naive" description of the earthquake, intensity is of paramount importance in seismology and geology and engineering, primarily because intensity allows us to retrieve invaluable information from the historical record of seismicity. The evolution of the scales in the past century has been an effort to keep intensity in line with new knowledge in earthquake engineering and geology and to provide classifications of the effects with increasing objectivity and repeatability but within a substantial continuity with the original intensity scales developed by Rossi-Forel in 1883, Mercalli in 1889, and Mercalli-Cancani-Sieberg (MCS; Sieberg 1930).

All the intensity scales (Rossi-Forel, Mercalli, MCS, MSK, Mercalli Modified) consider the earthquake environmental effects as diagnostic elements for intensity evaluation. This is especially relevant from intensity degrees ≥IX, where intensity evaluation based on building damage nearly saturates (Michetti et al. 2007). However, some modern practices of macroseismic investigation (e.g., Grunthal 1998) tend to only focus on the effects on humans and the anthropic environment, overlooking the earthquake environmental effects, based on the opinion that these effects are too variable and aleatory. As aforementioned, this is especially the case for the new European Macroseismic Scales (EMS-98; Grunthal 1998; Musson et al. 2010) and its impact in European building codes.

Nevertheless, other authors (e.g., Dengler and McPherson 1993; Serva 1994; Serva et al. 2007; Dowrick 1996; Hancox et al. 2002; Michetti et al. 2004; McCalpin 2009) have provided clear evidence that the characteristics of geological and environmental effects are an essential piece of information in order to understand how intensity grows with the earthquake size (for seismic events with the same hypocentral depth). In this sense, intensity combines the effects of seismic shacking with local topographic and geological conditions (topographic and site-effect amplifications or attenuations) triggering temporal or permanent environmental effects in the landscape, which nowadays are widely retrievable from historical sources and/or paleoseismic research. Intensity is therefore a way to classify the interaction between the earthquake and the environment (people, objects, buildings, lifelines, trees, soils, slopes, rivers, aquifers, coastlines, etc.).

The record of earthquake environmental effects extends the set of information even far from urban zones, offering a better and more complete picture of the seismic scenario. Additionally, environmental effects are in many cases a direct source of damage in urban areas (e.g., liquefaction processes, large slope movements), sometimes even exceeding the damage produced by seismic shaking itself (e.g., tsunamis). This constitutes a keystone for intensity assessment in some damaged zones, as it is difficult to differentiate seismic-shaking damage from that produced by environmental effects through the application of traditional intensity scales. Most of these traditional scales consider the occurrence of observable earthquake environmental effects from intensities ≥VI, noticeable from intensity VIII, and destructive or devastating for intensities ≥IX, but with descriptions somewhat qualitative and not properly classified.

The ESI-07 scale therefore provides a classification and quantification of earthquake environmental effects for the different intensity degrees, offering quantitative numerical data for the size of primary effects (surface faulting and coseismic uplift/subsidence) and secondary effects (slope movements, ground cracks, liquefaction processes, tsunamis). Other environmental effects, with poor to no potential of preservation in the geological or geomorphological records, such as hydrological anomalies in springs and wells, tree shacking, jumping stones, and dust clouds, are also included but providing qualitative descriptions. The scale also highlights the occurrence of frequent, characteristic, and diagnostic features on the assemblage of earthquake environmental effects for the different intensity degrees (see Appendix 1).

Consequently, ESI-07 scale allows an assessment of seismic intensity based only on the earthquake environmental effects. This can be regarded as is an update, reclassification, and quantification of environmental effects included in the traditional macroseismic scales (MCS, MSK, MM). Its use, alone (in unpopulated regions) or, as a rule, integrated with the macroseismic traditional scales (for instance, intensity degree IX on the ESI-07 is equivalent to degree IX on the MCS, MM, and MSK scales; Michetti et al. 2004), will offer a more comprehensive picture of the earthquake's scenario and impact and a more consolidated estimate of intensity, allowing, therefore, a better comparability among different earthquakes.

Primary and Secondary Earthquake Environmental Effects in the ESI-07 scale

Earthquake environmental effects (EEEs) are all the phenomena generated in the natural environment by a seismic event. These can be temporal or permanent, holding a different potential of preservation in the geological record. Typically, from intensities >VIII, EEEs can be permanently printed in the geology and landscape of a zone, being susceptible of paleoseismic analysis (Fig. 1). The ESI-07 scale classifies the EEEs in two main types: (a) primary and (b) secondary effects. While primary effects (e.g., surface faulting) are useful to directly assess the earthquake intensity, the nature, dimensions, and areal extent of secondary effects (mainly landslides and liquefaction processes) allow the complementary estimation of the epicentral intensity of an earthquake.

Primary Effects

Primary effects include any surface expression of the seismogenic tectonic source, including surface faulting, surface tectonic uplift, and subsidence. These effects are typically observed for shallow crustal earthquakes over a certain threshold value of magnitude, commonly considered to be between 6.0 and 6.5 Mw, being directly linked to the size, hence the energy of the earthquake. Primary effects in principle do not suffer saturation, although there is a physical limit to their dimensions (included in the XII degree). The size of primary effects is typically expressed in terms of two parameters: (i) total surface rupture length (SRL) and (ii) maximum displacement (MD). The amount of tectonic surface deformation (uplift, subsidence) is also considered in the assessment.

The focal depth and the stress environment of an earthquake obviously control the occurrence and the size of the observed effects. Two crustal earthquakes with the same energy but very different focal depths and stress environment can produce a different range of environmental effects, and, therefore, their respective local intensity values can significantly differ. Especially in volcanic areas, earthquakes of very shallow focus (c. 3–4 km) and low magnitude can be associated with the manifestation of primary effects. Taking these considerations into account, the ESI-07 scale considers the lower cutoff for the occurrence of primary effects

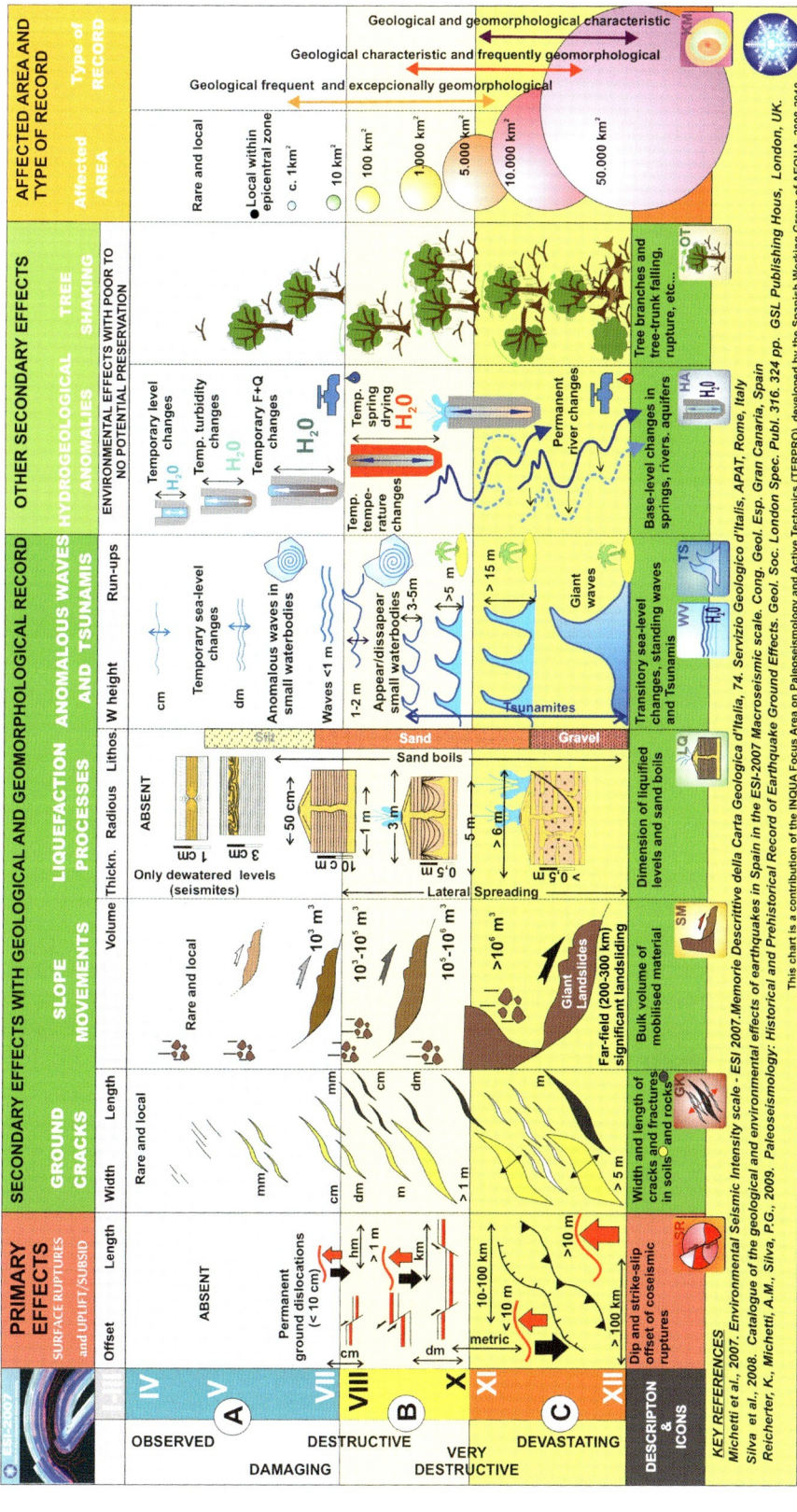

Intensity Scale ESI 2007 for Assessing Earthquake Intensities, Fig. 1 ESI-07 chart illustrating the main features and size parameters of the macroseismic scale for the different intensity degrees described in Appendix 1 (Modified from Reicherter et al. (2009))

(i.e., surface faulting) in intensity VII, whereas for typical crustal earthquakes (focal depth 5–15 km), in non-volcanic zones, primary effects start from intensity VIII.

Secondary Effects

Secondary effects include a wide range of phenomena affecting the natural environment, generally induced by the ground shaking. Their occurrence is typically observed in a specific range of intensities. The ESI-07 scale describes the characteristics and size of each type of secondary effect as a diagnostic feature in a range of intensity degrees. In some cases, it is only possible to establish a minimum intensity value. The total areal distribution of secondary effects grows with the size of the event, without saturation. Therefore, the ESI-07 scale uses this criterion as an independent tool for the assessment of the epicentral intensity I_0.

EEEs are more commonly observed and characterized from intensity IV, which is the lower bound of the ESI-07 scale. Some types of environmental effects concerning hydrological anomalies may be observed even at lower degrees, but they cannot be considered as diagnostic for intensity assessments. The evaluation accuracy increases toward the highest degrees, particularly in the range where primary effects occur (typically for intensity VIII and above), preserving its resolution up to intensity XII. Indeed, effects on man and man-made structures saturate (i.e., all buildings in the epicentral area are completely destroyed) starting from intensity X of traditional scales, so that a reliable intensity becomes difficult to assess. Instead, in the same range, EEEs still provide reliable evidence for an intensity appraisal.

In the ESI-07 scale, EEEs are classified into eight main categories: (1) hydrological anomalies, (2) anomalous waves/tsunamis, (3) ground cracks, (4) slope movements, (5) tree shaking, (6) liquefaction, (7) dust clouds, and (8) jumping stones.

Hydrological anomalies (HA). Hydrological phenomena show up from intensity $\geq$IV and saturate (i.e., their size does not increase) at intensity X. They can be divided in two groups: (a) surface water effects including overflow, discharge variations, and increasing turbidity of rivers and lakes and (b) groundwater effects including drying up of springs, appearance of new springs, temperature changes, anomalies in chemical components, and turbidity of springs. Further useful information might be the amount and rates of variation in temperature and discharge, the presence of anomalous chemical element, the duration of the anomaly, and the time delay (the time interval between the seismic shock and the observation of the specific effect).

Anomalous waves/Tsunamis (AW). In this category are included seiches in closed basins, outpouring of water from pools and basins, and tsunami waves. In the case of tsunamis, more than the size of the tsunami wave itself, the effects on the shores (especially run-up, beach erosion, change of coastal morphology), without neglecting those on humans and man-made structures, are taken as diagnostic of the suffered intensity. These effects may already occur from intensity IV but are more diagnostic from intensities IX to XII. The definition of intensity degrees has taken advantage from previous tsunami intensity scales (e.g., Papadopoulos and Imamura 2001) and from many recent descriptions of the aforementioned effects worldwide (e.g., Lander et al. 2003).

Ground cracks (GK). Ground cracks appear from intensity IV and saturate (i.e., their size does not increase) at intensity X. Diagnostic parameters are lithology, strike and dip, maximum width, and areal density. Different dimensions of fractures are considered for soft materials (loose alluvial deposits and/or saturated soils) and hard rocks. This category also includes the occurrence of cracking, undulations, and/or pressure ridges in paved (asphalt or stone) roads.

Slope movements (SM). Slope movements, including underwater landslides, consider a wide variety of gravitational slope processes such as rockfalls, debris falls, and toppling; rock or debris slides, avalanches, and mudslides; debris-, earth-, and mudflows; and sackungen, complex landslides, and eventually lateral spreads. This category may show up at intensity IV and saturate (i.e., their size does not increase) at intensity X. The total volume of mobilized material is diagnostic for intensity assessment. It can be roughly estimated on the basis of the landslide area when the depth of sliding mass can be reasonably estimated. The uncertainties introduced with this procedure do not appear to significantly influence the intensity evaluation. Additional useful information is maximum dimension of individual blocks (rockfalls), area affected by slope movements, density of slope movements, amount of slip, humidity, and time delay.

Liquefaction processes (LQ). Liquefactions generally materialize at intensity V or greater. The diagnostic features for liquefactions are the diameter of sand volcanoes and their lithology (silt, sands, and gravel). Saturation (i.e., their size does not increase) occurs at intensity X. Other useful characteristics are shape, the time delay, the depth of water table, and the occurrence of water and sand ejection.

Other environmental effects (OT) considered in the scale have a scarce to null preservation potential and a low diagnostic relevance, but of use of well-documented historical and recent events. Among them are the following:

- **Tree shaking (TR)**. This phenomenon is reported to occur from a minimum intensity IV but noticeable for intensity VII and diagnostic from intensity VIII. It is important to record the occurrence of broken branches and the morphologic characteristics of the area (flat, slope). The definition of intensity degrees basically follows that provided by Dengler and McPherson (1993).
- **Dust clouds (DC)** are reported from intensity VIII, typically in dry areas.
- **Jumping stones (JS)** have been reported from minimum intensity IX. The size of stones and their imprint in soft soil are considered as diagnostic parameters for intensity evaluation.

Structure of the ESI-07 Scale

The ESI-07 scale (Fig. 1; Appendix 1) is structured in 12° and has been developed to be consistent with the modified Mercalli macroseismic scale (MM-31; Wood and Neumann 1931; MM-56, Richter, 1958) and the MSK-64 (Medvedev-Sponheuer-Karnik Scale), since these are the most applied scales worldwide and include explicit references to environmental effects (especially in the highest degrees, X to XII). Comparing the ESI-07 with these other twelve-degree scales, we can identify four main intensity subsets as illustrated in Fig. 1:

From I to III: There are no environmental effects that can be used as diagnostic.

From IV to VII: Secondary environmental effects are easily observable starting from intensity IV, but only often permanent and diagnostic starting from intensity VII. Except for volcanic areas, this range of intensity degrees does not record primary effects. Within this range, effects on natural environment are less suitable for intensity assessment than effects on humans and man-made structures. Their use is therefore recommended especially in sparsely populated areas or in combination with damage-based intensity scales. Their use in field surveys for recent events will help to improve future versions of the scale. Environmental damage affects an area normally ≤ 10 km^2.

From VIII to X: In this range of intensities, primary and secondary environmental effects become characteristic and normally permanent and diagnostic. Thus, they are suitable for intensity assessment in combination with damage-based intensity scales but can be used as an independent tool, especially in uninhabited areas and for paleoseismic events. This range of intensity degrees is also a valuable tool in the analysis of well-documented historical earthquakes, refining intensity values in urban areas but mainly filling intensity data gaps among populated zones. Environmental damage typically affects areas ranging from 100 km^2 (VIII) to about 5,000 km^2(X).

From X to XII: Effects on humans and anthropic structures saturate, while effects on natural environment, especially primary, become dominant and diagnostic for this range of intensity degrees; in fact, several types of environmental effects do not suffer saturation in this range. Consequently earthquake environmental effects (EEEs) are the most effective tool to evaluate the intensity. As in the previous range of intensities, this one is fully applicable to well-documented historical events and also to paleoevents. Environmental damage can extend over areas ranging from 5,000 to 50,000 km^2.

Guidelines for the Application of the ESI-07 Scale

The ESI-07 scale might also provide a better estimation for intensity assessments than the traditional intensity scales, because earthquake environmental effects are independent from time (modern, historical, or paleo) and do not depend from the socioeconomic development of the region affected by the seismic event. This is true both:

(A) *In time*, earthquake environmental effects are comparable for a time window much larger than the period of instrumental record (recent, historic, and paleoseismic events). The instrumental records cover about one century, actually much less, if we consider that the introduction of modern global seismic networks started about 50 years ago. As the impact of an earthquake on the artificial environment depends on the distribution and quality of urban buildings and infrastructures, today the comparison of a contemporary event with a historical one (i.e., seventeenth century A.D.) based only on damage indicators might reflect the level of economy in the study area more than the size of the earthquake. Furthermore, the ESI-07 scale allows the expansion of the temporal extent of earthquake catalogues (Silva et al. 2014). Evidence of surface faulting and secondary effects (i.e., Liquefaction, Landslides, Tsunamis) generated by prehistoric events can be identified via detailed paleoseismic investigations, and pertinent ESI-07 values can be derived and used for a seismic hazard assessment over a geologic time interval.

(B) *In different geographic areas*, earthquake environmental effects do not depend on particular socioeconomic conditions or different building practices of the damaged zones. In other words, intensity assessments based on building-damage macroseismic scales are commonly influenced by cultural and technological aspects, which may differ significantly from region to region. The ESI-07 scale is independent from these limitations. Moreover, earthquake-prone areas can be located completely or partially in sparsely populated regions, where the effects on the natural environment might be the only evidence available to estimate intensity.

In particular, the ESI-07 scale complements and/or replaces traditional macroseismic scales in the following aspects:

(a) For earthquake intensity degrees $\geq$X, when damage-based assessments are extremely difficult because the built environment is virtually totally destroyed (so that any assessment based on it is not significant), while environmental effects are still diagnostic.
(b) In sparsely populated areas, where man-made structures are absent or rare, so that only the environmental effects allow intensity estimates.
(c) In paleoseismic analyses, where traditional macroseismic scales are not applicable. Additionally, the ESI-07 scale has an advantage over conventional paleoseismological research (i.e., fault-trenching), since the analysis of secondary environmental effects allows the investigation of earthquake cases where no surface faulting occurs, normally in the lower bound of the scale ($\leq$VIII). These kinds of events are commonly "lost earthquakes" for conventional fault-trenching surveys but source of significant building and environmental damage as illustrated by recent moderate seismic events such as the Lorca 2011 (Spain, 5.2 Mw) or the Emilia-Romagna 2012 (Italy, 6.1 Mw) earthquakes.

Intensity Scale ESI 2007 for Assessing Earthquake Intensities, Table 1 Ranges of surface faulting parameters (primary effects) and typical extent of total area affected by secondary effects for each intensity degree

I_0	Primary effects		Secondary effects
	Surface rupture length	Max surface displacement/ deformation	Total area
IV	–	–	–
V	–	–	–
VI		–	≤ 5 km^2
VII	a	a	10 km^2
VIII	<1 km	<5 cm	100 km^2
IX	1–10 km	5–40 cm	1,000 km^2
X	10–60 km	40–300 cm	5,000 km^2
XI	60–150 km	300–700 cm	10,000 km^2
XII	>150 km	>700 cm	>50,000 km^2

[a]Limited surface fault ruptures, 10–100 m long with centimeter-wide offset may occur in volcanic areas, generally associated to very shallow earthquakes (From Michetti et al. (2007))

The ESI-07 scale is an independent tool for intensity assessment. So, it can be used alone when only environmental effects are the diagnostic features present; this is the case when effects on man and man-made structures are too scarce or have saturated an area. Other than these cases, it is advisable to estimate two independent intensity fields, one for damage-based scales and one for ESI-07. Their subsequent comparison and integration, based on an expert's judgment, will lead to the best intensity scenario and the most reliable intensity estimate.

Generally, the epicentral region includes the area where the highest values of intensity are recorded. The epicentral intensity (I_0) is commonly close to, and usually corresponds with, the highest estimated value. Surface faulting parameters and the total spatial distribution of secondary effects (Landslides and/or Liquefactions) are two independent tools for assessing an ESI-07 I_0, starting from intensity VII (Table 1). Of course, specific care has to be paid when surface faulting parameters are at the boundaries between two different degrees. In this case, the intensity value more consistent with the characteristics and areal distribution of the secondary effects should be selected. Moreover, in the evaluation of the total area, it is recommended that isolated effects that occurred in the far field are not included. This evaluation also requires expert judgment.

Local intensity is generally evaluated through the description of secondary effects (EEEs) that have occurred in different "sites" included within a specific "locality." In this way, a locality (i.e., a village, a cliff zone, a particular beach, etc.) may record EEE intensity data from different sites, and the eventual assessment of the local intensity is based on the EEE mean and maximum dimensions. This type of intensity has to be comparable with the corresponding traditional local intensity based on damage. It must be noted that a "locality" can be referred to an inhabited area (a village, a town) but also to natural areas without human settlements. When only primary effects are present, the local evidence of surface faulting in terms of maximum observed displacement may be used. Table 2 provides the diagnostic range of intensity in which primary and secondary EEEs can be recorded. The specific descriptions of EEEs for each intensity degree are given in Appendix 1.

Application of the ESI-07 Scale to Paleoseismic Data

Conventional paleoseismic research is focused on fault-trenching analyses, which involve the excavation of trenches of different dimensions across fault traces evidencing past coseismic surface ruptures. These kinds of analyses are focused on the identification, measurement, quantification, and geochronological dating of key stratigraphic features, indicating past surface faulting events (earthquake horizons). The final purpose is to establish the earthquake history of a fault, the seismic potential or capability to generate strong seismic events, and the evaluation of the recurrence periods of earthquakes along the fault (e.g., McCalpin 2009). Widely used empirical relations (e.g., Wells and Coppersmith 1994) among surface rupture length (SRL), maximum fault displacement (MD), and earthquake

Intensity Scale ESI 2007 for Assessing Earthquake Intensities, Table 2 Diagnostic ranges of intensity degrees for each category of environmental effects

	Earthquake Environmental effects	IV	V	VI	VII	VIII	IX	X	XI	XII
	Surface Faulting and uplift/subsidence	Absent	Absent	Absent	VII[a]	VIII				XII
A	(HA) Hydrological Anomalies	IV			VII			X		
B	(AW) Anomalous Waves/Tsunamis	IV					IX			XII
C	(GK) Ground Cracks	IV						X		
D	(SM) Slope Movements	IV						X		
E	(TR) Tree Shaking	IV							XI	
F	(LQ) Liquefactions		V					X		
G	(DC) Dust Clouds	Absent	Absent	Absent		VIII				XII
H	(JS) Jumping Stones						IX			XII

[a]For intensity degree VII, limited surface fault ruptures, tens to hundreds meters long with several centimeter-wide offset may occur essentially associated to very shallow earthquakes in volcanic areas. Green-shaded intensity zones represent "rarely to commonly observed" EEEs at indicated intensity levels. Orange-shaded zones represent intensity levels at which individual EEE types are "characteristic and diagnostic" to assess the intensity (Modified from Michetti et al. (2007))

moment magnitude (Mw) allow the estimation of this seismic size parameter. The application of the ESI-07 scale permits estimating the maximum epicentral intensity (I_0) from this kind of paleoseismic approach (see Table 1).

In zones where surface faulting is absent, the most helpful source of information is multiarchive paleoseismic records provided by a variety of secondary earthquake environmental effects, such as paleoliquefaction features and slope movements considered in the ESI-07 scale, but also lake seismites and speleoseismites. For example, the work of Becker et al. (2005) in the Swiss Alps provides good guidelines in the use of secondary EEEs in paleoseismic research. Accordingly, in a case where multiple paleo-earthquake evidence is available in a zone, integrated paleoseismological research is feasible from different geological archives (slope movements, liquefaction, lake, and cave geological records) to identify the seismic history of the zone. Depending on the size of the earthquake (magnitude), attenuation/amplification effects (intensities), and geographical distribution with respect to the suspect macroseismic epicenter, a same geological archive may record evidence of a given earthquake but also of repeated events in the zone (e.g., Becker et al. 2005). The mapping and dating of paleoseismic features in different geological archives (EEEs) can provide sufficient information, making possible to apply the ESI-07 scale. In zones where surface faulting occurs, the combination of primary and secondary EEEs may offer the best scenario in order to reconstruct the seismic history of the zone, providing a reliable image of the potential seismic hazard.

Figure 2 illustrates a theoretical scenario for the paleoseismic record of EEEs in which it is possible to identify several paleo-earthquakes by means of the recognition and quantification of primary and secondary effects, allowing one to trace paleoseismic ESI-07 zones. Of course, the recognition and attribution of the observed EEEs in each site to a particular paleo-earthquake should be made on the basis of the geochronological dating of the observed features. For Holocene events, the most suitable dating technique when possible is radiocarbon dating (^{14}C). For lake seismites,

Intensity Scale ESI 2007 for Assessing Earthquake Intensities, Fig 2 Sketch illustrating the concept of integrated paleoseismology using the slope, liquefaction, ground cracks, lake, cave (other effects), and active fault geological archives. The chance to achieve a complete record of prehistoric strong earthquakes in a region is highest if the evidence from different geological archives and different sites is combined. In the absence of fault surface ruptures, the illustrated archives of secondary earthquake environmental effects can be used in a similar way (based on Becker et al. 2005)

paleoliquefaction features, slope movements, and faulted stratigraphic horizons older than the Holocene or in the limit of the radiocarbon dating technique (c. 45–50 ky), other dating techniques such as thermoluminescence (TL) or derived methods such as OSL or IRSL can offer good results. For speleoseismites, normally Th/U series dating techniques are the more suitable ones. In summary, multi-approach paleoseismic analyses based on the ESI-07 scale are useful to unravel past seismic events and estimate their ▶ earthquake location and earthquake size parameters. Further developments in the study of EEEs will help to understand the contribution of environmental damage on the overall damage of a zone leading the upgrading of earthquake cascading effects in seismic hazard and seismic risk analyses. Recently, the emergence of ▶ archeoseismology leads to the ability to perform combined archeoseismological and paleoseismological analyses of ancient and historical earthquakes (Reicherter et al. 2009; Silva et al. 2011).

Summary

The Environmental Seismic Intensity scale (ESI-07) is a twenty-first-century macroseismic scale based on the analysis of geological and environmental effects caused by the earthquakes. The ESI-07 scale represents a quantification of the natural effects considered in traditional macroseismic scales such as Mercalli, MCS, MM, and MSK. It can be used alone but preferentially combined with the aforementioned macroseismic scales based on building damage. Some recent macroseismic practices in Europe (EMS-98) do not consider natural effects for intensity assessments, and consequently the application of the ESI-07 scale constitutes a relevant complementary approach. Research on earthquake environmental damage (Earthquake Environmental Effects: EEEs) is of use in populated areas and inhabited zones but also at different time windows, including those covering, instrumental, historical, ancient, and paleoseismic events. These latter effects are beyond range of the traditional macroseismic scales, and thus the ESI-07 constitutes one of the most useful tools to expand conventional seismic catalogues to prehistoric and geologic time scales, improving the knowledge on the seismic history of a region or particular tectonic structure and refine evaluations on recurrence periods. This new macroseismic scale considers both on-fault primary effects caused by surface faulting and uplift and a wide range of off-fault secondary ground effects triggered by seismic ground shaking (ground cracks, slope movements, liquefaction processes, hydrological anomalies, and tsunamis). All these primary and secondary EEEs are able to imprint permanent traces in the geological and geomorphological records, especially from intensities above VII–VIII. The comprehensive analysis of EEEs will allow a parameterization of past earthquakes as well as definition of the overall distribution of associated intensity zones, to be applied to future seismic building codes and seismic hazard analyses.

Appendix 1: Definition of Intensity Degrees of the Environmental Seismic Intensity Scale ESI-2007 (*In "italics" highlighted the diagnostic descriptions for different intensity degrees*)

From Intensities I to III

There are no environmental effects that can be used as diagnostic.

Intensity IV

Largely observed/first unequivocal effects in the environment
   Primary effects: absent
   Secondary effects

(a) Rare small variations of the water level in wells and/or of the flow rate of springs are locally recorded, as well as extremely rare small variations of chemical-physical properties of water and turbidity in springs and wells, especially within large karstic spring systems, which appear to be most prone to this phenomenon.
(b) In closed basins (lakes, even seas), seiches with height not exceeding a few centimeters may develop, commonly observed only by tidal gauges, exceptionally even by naked eye, typically in the far field of strong earthquakes. Anomalous waves are perceived by all people on small boats, few people on larger boats, and most people on the coast. Water in swimming pools swings and may sometimes overflow.
(c) Hair-thin cracks (millimeter wide) might be occasionally seen where lithology (e.g., loose alluvial deposits, saturated soils) and/or morphology (slopes or ridge crests) are most prone to this phenomenon.
(d) Exceptionally, rocks may fall and small landslides may be (re)activated, along slopes where the equilibrium is already near the limit state, e.g., steep slopes and cuts, with loose and generally saturated soil.
(e) Tree limbs shake feebly.

Intensity V

Strong/marginal effects in the environment
   Primary effects: absent
   Secondary effects

(a) Rare variations of the water level in wells and/or of the flow rate of springs are locally recorded, as well as small variations of chemical-physical properties of water and turbidity in lakes, springs, and wells.
(b) In closed basins (lakes, even seas), seiches with height of decimeters may develop,

sometimes noted also by naked eye, typically in the far field of strong earthquakes. Anomalous waves up to several tens of cm high are perceived by all people on boats and on the coast. Water in swimming pools overflows.
(c) Thin cracks (millimeter wide and several cm up to one meter long) are locally seen where lithology (e.g., loose alluvial deposits, saturated soils) and/or morphology (slopes or ridge crests) are most prone to this phenomenon.
(d) Rare small rockfalls, rotational landslides, and slump earth flows may take place, along often but not necessarily steep slopes where equilibrium is near the limit state, mainly loose deposits and saturated soil. Underwater landslides may be triggered, which can induce small anomalous waves in coastal areas of sea and lakes.
(e) Tree limbs and bushes shake slightly, very rare cases of fallen dead limbs and ripe fruit.
(f) Extremely rare cases are reported of liquefaction (sand boil), small in size and in areas most prone to this phenomenon (highly susceptible, recent, alluvial and coastal deposits, near-surface water table).

Intensity VI
Slightly damaging/modest effects in the environment
Primary effects: absent
Secondary effects

(a) Significant variations of the water level in wells and/or of the flow rate of springs are locally recorded, as well as small variations of chemical-physical properties of water and turbidity in lakes, springs, and wells.
(b) Anomalous waves up to many tens of cm high flood in very limited areas nearshore. Water in swimming pools and small ponds and basins overflows.
(c) *Occasionally, millimeter-centimeter wide and up to several meters long fractures are observed in loose alluvial deposits and/or saturated soils; along steep slopes or riverbanks, they can be 1–2 cm wide. A few minor cracks develop in paved (either asphalt or stone) roads.*

(d) Rockfalls and landslides with volume reaching ca. 10^3 m^3 can take place, especially where equilibrium is near the limit state, e.g., steep slopes and cuts, with loose saturated soil or highly weathered/fractured rocks. Underwater landslides can be triggered, occasionally provoking small anomalous waves in coastal areas of sea and lakes, commonly seen by instrumental records.
(e) *Trees and bushes shake moderately to strongly; a very few treetops and unstable-dead limbs may break and fall, also depending on species, fruit load, and state of health.*
(f) *Rare cases are reported of liquefaction (sand boil), small in size and in areas most prone to this phenomenon (highly susceptible, recent, alluvial and coastal deposits, near-surface water table).*

Intensity VII
Damaging/appreciable effects in the environment
Primary effects observed very rarely and almost exclusively in volcanic areas. Limited surface fault ruptures, tens to hundreds of meters long and with centimetric offset, may occur, essentially associated to very shallow earthquakes.
Secondary effects: *The total affected area is in the order of 10 km^2*.

(a) Significant temporary variations of the water level in wells and/or of the flow rate of springs are locally recorded. Seldom, small springs may temporarily run dry or appear. Weak variations of chemical-physical properties of water and turbidity in lakes, springs, and wells are locally observed.
(b) Anomalous waves even higher than a meter may flood limited nearshore areas and damage or wash away objects of variable size. Water overflows from small basins and watercourses.
(c) *Fractures up to 5–10 cm wide and up to hundred meters long are observed, commonly in loose alluvial deposits and/or saturated soils, rarely in dry sand, sand clay, and clay soil fractures, up to 1 cm wide.*

Centimeter-wide cracks are common in paved (asphalt or stone) roads.

(d) Scattered landslides occur in prone areas, where equilibrium is unstable (steep slopes of loose/saturated soils), while modest rockfalls are common on steep gorges, cliffs. Their size is sometimes significant (10^3–10^5 m^3); in dry sand, sand clay, and clay soil, the volumes are usually up to 100 m^3. Ruptures, slides, and falls may affect riverbanks and artificial embankments and excavations (e.g., road cuts, quarries) in loose sediment or weathered/fractured rock. Significant underwater landslides can be triggered, provoking anomalous waves in coastal areas of sea and lakes, directly felt by people on boats and ports.

(e) Trees and bushes shake vigorously, especially in densely forested areas, many limbs, and tops break and fall.

(f) *Rare cases are reported of liquefaction, with sand boils up to 50 cm in diameter, in areas most prone to this phenomenon (highly susceptible, recent, alluvial and coastal deposits, near-surface water table).*

Intensity VIII

Heavily damaging/extensive effects in the environment

Primary effects: observed rarely

Ground ruptures (surface faulting) may develop, up to several hundred meters long, with offsets not exceeding a few cm, particularly for very shallow focus earthquakes such as those common in volcanic areas. Tectonic subsidence or uplift of the ground surface with maximum values on the order of a few centimeters may occur.

Secondary effects: The total affected area is in the order of 100 km^2.

(a) Springs may change, generally temporarily, their flow rate, and/or elevation of outcrop. Some small springs may even run dry. Variations in water level are observed in wells. Weak variations of chemical-physical properties of water, most commonly temperature, may be observed in springs and/or wells. Water turbidity may appear in closed basins, rivers, wells, and springs. Gas emissions, often sulfureous, are locally observed.

(b) Anomalous waves up to 1–2 m high flood nearshore areas and may damage or wash away objects of variable size. Erosion and dumping of waste is observed along the beaches, where some bushes and even small weak-rooted trees can be eradicated and drifted away. Water violently overflows from small basins and watercourses.

(c) *Fractures up to 50 cm wide and up to 100 m long are commonly observed in loose alluvial deposits and/or saturated soils; in rare cases, fractures up to 1 cm can be observed in competent dry rocks. Decimetric cracks common in paved (asphalt or stone) roads as well as small pressure undulations.*

(d) Small to moderate (10^3–10^5 m^3) landslides widespread in prone areas; rarely they can occur also on gentle slopes, where equilibrium is unstable (steep slopes of loose/saturated soils, rockfalls on steep gorges, coastal cliffs); their size is sometimes large (10^5–10^6 m^3). Landslides can occasionally dam narrow valleys causing temporary or even permanent lakes. Ruptures, slides, and falls affect riverbanks and artificial embankments and excavations (e.g., road cuts, quarries) in loose sediment or weathered/fractured rock. Frequent occurrence of landslides under the sea level in coastal areas.

(e) *Trees shake vigorously; branches may break and fall, even uprooted trees, especially along steep slopes.*

(f) *Liquefaction may be frequent in the epicentral area, depending on local conditions; sand boils up to ca. 1 m in diameter; apparent water fountains in still waters; localized lateral spreading and settlements (subsidence up to ca. 30 cm), with fissuring parallel to waterfront areas (riverbanks, lakes, canals, seashores).*

(g) *In dry areas, dust clouds may rise from the ground in the epicentral area.*

(h) Stones and even small boulders and tree trunks may be thrown in the air, leaving typical imprints in soft soil.

Intensity IX

Destructive/effects in the environment are a widespread source of considerable hazard and become important for intensity assessment.

Primary effects: observed commonly

Ground ruptures (surface faulting) develop, up to a few km long, with offsets generally in the order of several cm. Tectonic subsidence or uplift of the ground surface with maximum values in the order of a few decimeters may occur.

Secondary effects: The total affected area is in the order of 1,000 km^2.

(a) *Springs can change, generally temporarily, their flow rate and/or location to a considerable extent. Some modest springs may even run dry. Temporary variations of water level are commonly observed in wells. Water temperature often changes in springs and/or wells. Variations of chemical-physical properties of water, most commonly temperature, are observed in springs and/or wells. Water turbidity is common in closed basins, rivers, wells, and springs. Gas emissions, often sulfureous, are observed, and bushes and grass near emission zones may burn.*
(b) *Meters high waves develop in still and running waters. In flood plains, water streams may even change their course, also because of land subsidence. Small basins may appear or be emptied. Depending on shape of sea bottom and coastline, dangerous tsunamis may reach the shores with run-ups of up to several meters flooding wide areas. Widespread erosion and dumping of waste is observed along the beaches, where bushes and trees can be eradicated and drifted away.*
(c) *Fractures up to 100 cm wide and up to 100 m long are commonly observed in loose alluvial deposits and/or saturated soils; in competent rocks, they can reach up to 10 cm. Significant cracks common in paved (asphalt or stone) roads, as well as small pressure undulations.*
(d) *Landslides are widespread in prone areas* and *also on gentle slopes; where equilibrium is unstable (steep slopes of loose/saturated soils; rockfalls on steep gorges, coastal cliffs), their size is frequently large (10^5 m^3), sometimes very large (10^6 m^3). Landslides can dam narrow valleys causing temporary or even permanent lakes. Riverbanks, artificial embankments, and excavations (e.g., road cuts, quarries) frequently collapse. Frequent large landslides under the sea level in coastal areas.*
(e) *Trees shake vigorously; branches and thin tree trunks frequently break and fall. Some trees might be uprooted and fall, especially along steep slopes.*
(f) *Liquefaction and water upsurge are frequent; sand boils up to 3 m in diameter; apparent water fountains in still waters; frequent lateral spreading and settlements (subsidence of more than ca. 30 cm), with fissuring parallel to waterfront areas (riverbanks, lakes, canals, seashores).*
(g) *In dry areas, dust clouds commonly rise from the ground.*
(h) *Small boulders and tree trunks may be thrown in the air and move away from their site for meters, also depending on slope angle and roundness, leaving typical imprints in soft soil.*

Intensity X

Very destructive/effects on the environment become a leading source of hazard and are critical for intensity assessment.

Primary effects become leading.

Surface faulting can extend for few tens of km, with offsets from tens of cm up to a few meters. Gravity grabens and elongated depressions develop; for very shallow focus, earthquakes in volcanic areas rupture lengths might be much lower. Tectonic subsidence or uplift of the ground surface with maximum values in the order of few meters may occur.

Secondary effects: The total affected area is in the order of 5,000 km^2.

(a) Many springs significantly change their flow rate and/or elevation of outcrop. Some springs may run temporarily or even permanently dry. Temporary variations of water level are commonly observed in wells. Even strong variations of chemical-physical properties of water, most commonly temperature, are observed in springs and/or wells. Often water becomes very muddy in even large basins, rivers, wells, and springs. Gas emissions, often sulfureous, are observed, and bushes and grass near emission zones may burn.

(b) *Meters high waves develop in even big lakes and rivers, which overflow from their beds. In flood plains, rivers may change their course, temporarily or even permanently, also because of widespread land subsidence. Basins may appear or be emptied. Depending on the shape of sea bottom and coastline, tsunamis may reach the shores with run-ups exceeding 5 m flooding flat areas for thousands of meters inland. Small boulders can be dragged for many meters. Widespread deep erosion is observed along the shores, with noteworthy changes of the coastline profile. Trees nearshore are eradicated and drifted away.*

(c) *Open ground cracks up to more than 1 m wide and up to 100 m long are frequent, mainly in loose alluvial deposits and/or saturated soils; in competent rocks they can reach several decimeters. Wide cracks develop in paved (asphalt or stone) roads, as well as pressure undulations.*

(d) *Large landslides and rockfalls ($>10^5–10^6$ m^3) are frequent, practically regardless of equilibrium state of the slopes, causing temporary or permanent barrier lakes. Riverbanks, artificial embankments, and sides of excavations typically collapse. Levees and earth dams may even incur serious damage. Frequent large landslides under the sea level in coastal areas.*

(e) *Trees shake vigorously; many branches and tree trunks break and fall. Some trees might be uprooted and fall.*

(f) *Liquefaction, with water upsurge and soil compaction, may change the aspect of wide zones; sand volcanoes even more than 6 m in diameter; vertical subsidence even >1 m; large and long fissures due to lateral spreading are common.*

(g) In dry areas, dust clouds may rise from the ground.

(h) *Boulders (diameter in excess of 2–3 m) can be thrown in the air and move away from their site for hundreds of meters down even gentle slopes, leaving typical imprints in soil.*

Intensity XI

Devastating/effects on the environment become decisive for intensity assessment, due to saturation of structural damage.

Primary effects: dominant

Surface faulting extends from several tens of km up to more than 100 km, accompanied by offsets reaching several meters. Gravity graben, elongated depressions, and pressure ridges develop. Drainage lines can be seriously offset. Tectonic subsidence or uplift of the ground surface with maximum values in the order of numerous meters may occur.

Secondary effects: The total affected area is in the order of 10,000 km^2.

(a) Many springs significantly change their flow rate and/or elevation of outcrop. Many springs may run temporarily or even permanently dry. Temporary or permanent variations of water level are generally observed in wells. Even strong variations of chemical-physical properties of water, most commonly temperature, are observed in springs and/or wells. Often water becomes very muddy in even large basins, rivers, wells, and springs. Gas emissions, often sulfureous, are observed, and bushes and grass near emission zones may burn.

(b) *Large waves develop in big lakes and rivers, which overflow from their beds. In flood plains, rivers can change their course, temporarily or even permanently, also because of widespread land subsidence and landsliding. Basins may appear or be emptied. Depending on shape of sea bottom and coastline, tsunamis may reach the shores with run-ups reaching 15 m and more devastating flat areas for kilometers inland. Even meter-sized boulders can be dragged for long distances. Widespread deep erosion is observed along the shores, with noteworthy changes of the coastal morphology. Trees nearshore are eradicated and drifted away.*

(c) Open ground cracks up to several meters wide are very frequent, mainly in loose alluvial deposits and/or saturated soils. In competent rocks, they can reach 1 m. Very wide cracks develop in paved (asphalt or stone) roads, as well as large pressure undulations.
(d) *Large landslides and rockfalls ($>10^5$–10^6 m^3) are frequent, practically regardless to equilibrium state of the slopes, causing many temporary or permanent barrier lakes. Riverbanks, artificial embankments, and sides of excavations typically collapse. Levees and earth dams incur serious damage. Significant landslides can occur at 200–300 km distance from the epicenter. Frequent large landslides under the sea level in coastal areas.*
(e) *Trees shake vigorously; many branches and tree trunks break and fall. Many trees are uprooted and fall.*
(f) *Liquefaction changes the aspect of extensive zones of lowland, determining vertical subsidence possibly exceeding several meters, numerous large sand volcanoes, and severe lateral spreading features.*
(g) In dry areas, dust clouds arise from the ground.
(h) *Big boulders (diameter of several meters) can be thrown in the air and move away from their site for long distances down even gentle slopes, leaving typical imprints in soil.*

Intensity XII

Completely devastating/effects in the environment are the only tool for intensity assessment.

Primary effects: dominant

Surface faulting is at least few hundreds of km long, accompanied by offsets reaching several tens of meters. Gravity graben, elongated depressions, and pressure ridges develop. Drainage lines can be seriously offset. Landscape and geomorphological changes induced by primary effects can attain extraordinary extent and size (typical examples are the uplift or subsidence of coastlines by several meters, appearance or disappearance from sight of significant landscape elements, rivers changing course, origination of waterfalls, formation or disappearance of lakes).

Secondary effects: *The total affected area is in the order of 50,000 km^2 and more.*

(a) Many springs significantly change their flow rate and/or elevation of outcrop. Temporary or permanent variations of water level are generally observed in wells. Many springs and wells may run temporarily or even permanently dry. Strong variations of chemical-physical properties of water, most commonly temperature, are observed in springs and/or wells. Water becomes very muddy in even large basins, rivers, wells, and springs. Gas emissions, often sulfureous, are observed, and bushes and grass near emission zones may burn.
(b) *Giant waves develop in lakes and rivers, which overflow from their beds. In flood plains, rivers change their course and even their flow direction, temporarily or even permanently, also because of widespread land subsidence and landsliding. Large basins may appear or be emptied. Depending on the shape of sea bottom and coastline, tsunamis may reach the shores with run-ups of several tens of meters devastating flat areas for many kilometers inland. Big boulders can be dragged for long distances. Widespread deep erosion is observed along the shores, with outstanding changes of the coastal morphology. Many trees are eradicated and drifted away. All boats are tore from their moorings and swept away or carried onshore even for long distances. All people outdoor are swept away.*
(c) Ground open cracks are very frequent, up to one meter or more wide in the bedrock, up to more than 10 m wide in loose alluvial deposits and/or saturated soils. These may extend up to several kilometers in length.
(d) *Large landslides and rockfalls ($>10^5$–10^6 m^3) are frequent, practically regardless to equilibrium state of the slopes, causing many temporary or permanent barrier lakes. Riverbanks, artificial embankments, and sides of excavations typically collapse. Levees and earth dams incur serious damage. Significant landslides can occur at more than*

200–300 km distance from the epicenter. Frequent very large landslides under the sea level in coastal areas
(e) Trees shake vigorously; many branches and tree trunks break and fall. Many trees are uprooted and fall.
(f) *Liquefaction occurs over large areas and changes the morphology of extensive flat zones, determining vertical subsidence exceeding several meters, widespread large sand volcanoes, and extensive severe lateral spreading features.*
(g) In dry areas, dust clouds arise from the ground.
(h) *Also very big boulders can be thrown in the air and move for long distances even down very gentle slopes, leaving typical imprints in soil.*

Cross-References

▶ Archeoseismology
▶ Earthquake Location
▶ European Structural Design Codes: Seismic Actions
▶ Luminescence Dating in Paleoseismology
▶ Paleoseismology and Landslides
▶ Paleoseismology: Integration with Seismic Hazard
▶ Paleoseismic Trenching
▶ Paleoseismology
▶ Probabilistic Seismic Hazard Models
▶ Radiocarbon Dating in Paleoseismology
▶ Seismic Actions Due to Near-Fault Ground Motion
▶ Seismic Risk Assessment, Cascading Effects
▶ Time History Seismic Analysis
▶ Tsunamis as Paleoseismic Indicators

References

Ali Z, Qaisar M, Mahmood T, Shah MA, Iqbal T, Serva L, Michetti AM, Burton PW (2009) The Muzaffarabad, Pakistan, earthquake of 8 October 2005: surface faulting, environmental effects and macroseismic intensity. Geol Soc London Spec Publ 316:155–172

Audemard FA, Michetti AM (2011) Geological criteria for evaluating seismicity revisited: forty Years of paleoseismic investigations and the natural record of past earthquakes. Geological Society of America special paper, Washington, USA, 479. pp 1–22

Becker A, Ferry M, Monecke K, Schnellmann M, Giardini D (2005) Multiarchive paleoseismic record of late Pleistocene and Holocene strong earthquakes in Switzerland. Tectonophysics 400(1–4):153–177

Berzhinskii YA, Ordynskaya AP, Gladkov AS, Lunina OV, Berzhinskaya LP, Radziminovich NA, Radziminovich YB, Imayev VS, Chipizubov AV, Smekalin OP (2010) Application of the ESI_2007 scale for estimating the intensity of the Kultuk earthquake, August 27, 2008 (South Baikal). Seismic Instruments 46(4):307–324

Dengler L, McPherson R (1993) The 17 August 1991 Honeydew earthquake north coast California: a case for revising the Modified Mercalli Scale in sparsely populated areas. Bull Seismol Soc Am 83(4):1081–1094

Di Manna P, Guerrieri L, Piccardi L, Vittori E, Castaldini D, Berlusconi A, Michetti AM (2013) Ground effects induced by the 2012 seismic sequence in Emilia: implications for seismic hazard assessment in the Po Plain. Ann Geophys 55(4):727–733

Dowrick DJ (1996) The modified Mercalli earthquake intensity scale- revisions arising from recent studies of New Zealand earthquakes. Bull New Zealand Nat Soc Earthq Eng 29(2):92–106

Fountoulis IG, Mavroulis SD (2013) Application of the environmental seismic intensity scale (ESI 2007) and the European Macroseismic Scale (EMS-98) to the Kalamata (SW Peloponnese, Greece) earthquake (Ms = 6.2, September 13, 1986) and correlation with neotectonic structures and active faults. Ann Geophys 56(6):S0675

Gojsar A (2012) Application of environmental seismic intensity scale (ESI 2007) to Krn Mountains 1998 Mw = 5.6 earthquake (NW Slovenia) with emphasis on rockfalls. Nat Hazards Earth Syst Sci 12:1659–1670

Grunthal G (1998) European macroseismic scale 1998 (EMS-98). European Seismological Commission, Subcommission on Engineering Seismology, Working Group Macroseismic Scales, Conseil de l'Europe, Cahiers du Centre Européen de Géodynamique et de Séismologie, 15, Luxembourg, 99 pp

Guerrieri L, Blumetti AM, Esposito E, Michetti AM, Porfido S, Serva L, Tondi E, Vittori E (2008) Capable faulting, environmental effects and seismic landscape in the area affected by the 1997 Umbria-Marche (Central Italy) seismic sequence. Tectonophysics 476(1–2):269–281

Hancox GT, Perrin ND, Dellow GD (2002) Recent studies of historical earthquake-induced landsliding, ground damage, and MM intensity in New Zealand. Bull New Zealand Nat Soc Earthq Eng 35(2):59–95

Lalinde CP, Sanchez JA (2007) Earthquake and environmental effects in Colombia in the last 35 years. INQUA Scale Project. Bull Seism Soc Am 97(2):646–654

Lander JF, Whiteside LS, Lockridge PA (2003) Two decades of global Tsunamis 1982–2002. The International Journal of the Tsunami Society (21–1), NOAA – NGDC, Boulder, 73 pp

Mavroulis SD, Fountoulis IG, Skourtsos EN, Lekkas EL, Papanikolaou I (2013) Seismic intensity assignments for the 2008 Andravida (NW Peloponnese, Greece) strike-slip event (June 8, Mw 6.4) based on the application of the Environmental Seismic Intensity scale (ESI 2007) and the European Macroseismic scale (EMS-98). Geological structure, active tectonics, earthquake environmental effects and damage pattern. Ann Geophys 56(6):S0681. doi:10.4401/ag-6239

McCalpin JP (2009) Paleoseismology, 2nd edn. Elsevier, Amsterdam, 613 pp

Michetti AM, Esposito E, Gürpinar A, Mohammadioun B, Mohammadioun J, Porfido S, Roghozin E, Serva L, Tatevossian R, Vittori E, Audemard F, Comerci V, Marco S, McCalpin J, Mörner NA (2004) The INQUA scale. An innovative approach for assessing earthquake intensities based on seismically-induced ground effects in natural environment. In: Vittori E, Comerci V (ed) Memorie Descrittive della Carta Geologicad' Italia, APAT, Roma, LXVII-116

Michetti AM, Esposito E, Guerrieri L, Porfido S, Serva L, Tatevossian R, Vittori E, Audemard F, Azuma T, Clague J, Comerci V, Gürpinar A, Mc Calpin J, Mohammadioun B, Mörner NA, Ota Y, Roghozin E (2007) Intensity scale ESI 2007. In: Guerrieri L, Vittori E (ed) Mem. Descr. Carta Geologica d'Italia, Servizio Geologico d'Italia, Dipartimento Difesa del Suolo, APAT, Rome, 74p

Musson RMW, Grunthal G, Stucchi M (2010) The comparison of macroseismic intensity scales. J Seismol 14:413–428

Ota Y, Azuma T, Lin N (2009) Application of INQUA environmental seismic intensity scale to recent earthquakes in Japan and Taiwan. Geol Soc London Spec Publ 316:55–71

Papadopoulos G, Imamura F (2001) A proposal for a new tsunami intensity scale, ITS 2001 Proceedings 20th International Tsunami Conference, Seattle, USA. Session 5–1. pp. 569–577

Papanikolaou ID (2011) Uncertainty in intensity assignment and attenuation relationships: how seismic hazard maps can benefit from the implementation of the environmental seismic intensity scale (ESI 2007). Quatern Int 242:42–51

Papathanassiou G, Pavlides S (2007) Using the INQUA scale for the assessment of intensity: case study of the 2003 Lefkada (Ionian Islands), Greece earthquake. Quatern Int 173–174:4–14

Reicherter K, Michetti AM, Silva PG (eds) (2009) Palaeoseismology: historical and prehistorical records of earthquake ground effects for seismic hazard assessment. Geological Society of London, Special Publications, 316. London, 160 pp

Richter CF (1958) Elementary Seismology. Freeman & Co. San Francisco, USA. 768p

Serva L (1994) The effects on the ground in the intensity scales. Terra Nova 6:414–416

Serva L, Esposito E, Guerrieri L, Porfido S, Vittori E, Comerci V (2007) Environmental effects from some historical earthquakes in Southern Apennines (Italy) and macroseismic intensity assessment. Contribution to INQUA EEE scale project. Quatern Int 173–174:30–44

Sieberg A (1930) Geologie der Erdbeben. Handboch der Geophysic 4:552–554

Silva PG, Sintubin M, Reicherter K (eds) (2011) Earthquake archaeology and palaeoseismology. Quaternary International, vol 241(1). Elsevier, Amsterdam, 258 pp

Silva PG, Rodríguez-Pascua MA, Giner JL, Pérez-López R, Lario J, Perucha MA, Bardají T, Huerta P, Roquero E, Bautista B (2014) Catálogo de los efectos geológicos de los terremotos en España. Publications Geological Survey of Spain (IGME) Serie Riesgos Geológicos/Geotécnia, 4. IGME, Madrid, 358 pp

Tatevossian RE (2007) The Verny, 1887, Earthquake in Central Asia: application of the INQUA scale based on coseismic environmental effects. Dark nature: rapid environmental change and human response. Quatern Int 173–174:23–29

Wells LD, Coppersmith JK (1994) New empirical relationships among magnitude, rupture length, rupture width, rupture area, and surface displacement. Bull Seismol Soc Am 84:974–1002

Wood HO, Neumann F (1931) Modified Mercalli intensity scale of 1931. Bull Seismol Soc Am 21(4):277–283

Interim Housing Provision Following Earthquake Disaster

Regan Potangaroa
Unitec School of Architecture, Auckland, New Zealand

Synonyms

Core housing; Housing; Recovery phase; Shelter; Transitional housing

Introduction

The provision of interim housing following an earthquake disaster is important because it is the first substantial community step away from the effects of a sudden and unexpected seismic event and back towards some "normality."

It is defined as housing that is "short term that can be in place for 1 month to 2 years – preferably on sites close to the damaged housing. (To the extent that displaced residents can remain near their previous homes means that economic impacts are reduced.) Interim housing will be needed because those organizations "donating" temporary shelter space need to reclaim that space for its original uses as schools – community centers – churches – that serve the community." (ABAG 2008).

The idea of "shelter in place" is important not only from a psychosocial perspective of those affected but moreover as part of the subsequent recovery and reconstruction. Shelter in place will encourage local businesses to stay; allow families to return to local schools, friends, social networks, and places; and allow for the restoration of infrastructure based on known numbers rather than estimates or "best guesses." Thus, it can be the trigger for an early resilient community and city wide recovery.

Interim housing will involve strategic planning and the allocation and use of significant resources, but it is the processes and priorities to be used that have been identified as "problematic" (EERI 2004; Davidson et al. 2007; Johnson 2007; Giovinazzi and Stevenson 2012).

Some of these processes and priorities are evidenced by the "philosophical-one-liners" associated with the operational side of interim housing such as "community engagement" (discussed by Davidson et al. above), "build back better" (Lyons et al. 2010), the notion of disasters as "opportunities," and of a "one size fits all" or "cookie cutter" housing solutions (Jha et al. 2010). There are several rights-based issues relating to housing land and property that emerge post disaster (IASC 2011), and there are philosophic positions such as the "do no harm principle" (Shelter Cluster 2014) and the "low-hanging fruit" approach where you assist those that are easiest first rather than those based on need. All balanced against the SPHERE standards in humanitarian assistance (SPHERE 2011) or National codes and the political positions of Government departments in developed economies. These all serve to underline its importance but also suggest a complexity that goes beyond "business as usual."

The provision of interim housing following an earthquake has several similarities to other fast onset disasters such as typhoons and flooding (Johnson 2002). However, the seismic design and detailing of buildings seems to demand greater technical skills, and as noted by da Silva following the Asian Tsunami disaster in Banda Aceh, Indonesia (da Silva 2010), "... a key issue was confusion as to what codes and standards should be followed. . . . . . Several agencies complied with the Building Code for Aceh assuming that it was sufficient or that local designers and contractors knew what they were doing without realising that safe construction practices were not common practice. Local engineering consultants employed by implementing agencies to develop structural designs generally had limited experience of seismic design, which typically requires an additional post-graduate qualification. This resulted in poor design solutions which were not compliant with the Indonesian code. Recognising this, some agencies employed specialist International consultants or firms to develop or check designs, or sought advice from local and national universities. International engineers were also employed as consultants in-house. However, many of these engineers did not previously have seismic design experience and so were ascending a learning curve, trying to follow available guidance and incorporate it into the construction drawings."

This still remains an area of confusion for aid agencies (Potangaroa 2010a); and is highlighted in the later case studies.

So how does a building professional survive, let alone remain operationally focused in such an environment? This section considers some of the theoretical context to interim housing and then discusses several case studies from recent earthquakes through the following "lens" (adapted from Davis 2010):

- Planning strategy
- Cost
- Connection to subsequent stages
- The meaning of place

Interim Housing Provision Following Earthquake Disaster, Fig. 1 Examples of interim housing after an earthquake disaster

- Seismic design
- Climatic design
- Community engagement

It then comes back to consider the potential role of building professionals in the provision of interim housing.

Background

The provision of interim housing following an earthquake seems to have the following patterns (Quarantelli 1995) or phases (MCDEM 2004) as shown below in Table 1.

The above patterns are perhaps indicative of the operational level of shelter/housing provision, while the phases are more policy and strategic. And, for example, the response phase also includes rapid shelter assessments to ascertain the "what," the "where," and the "who" (the so-called 3W analysis) so that any later housing assistance (including interim housing) can be more effective, can be better targeted, and have a level of accountability; all of which must be updated as the earthquake disaster response and reconstruction develops.

The recovery phase also prioritizes the economic impacts and the apparent need to get businesses back up and running. But as noted earlier, the decisions around housing and where people will stay should preempt such a priority; otherwise, there may not be an available "local" market or the required work force, whether a business has local or national markets.

The word coordination is increasingly used, but the dilemma of the recovery phase (and possibly the provision of interim housing) has been well framed by the Natural Hazards Center comments "How does one make a holistic disaster

Interim Housing Provision Following Earthquake Disaster, Table 1 The patterns and phases of interim housing following an earthquake disaster

Pattern	Measured in...	Objectives and locations	Phase
Spontaneous shelter[a]	Days	To provide an interim, safe haven while the situation stabilizes. This could be a school, church/mosque, stadium, park, community hall, or other public facility. Sometimes referred to as evacuation centers. It could also be the home of a friend or family member. It is often on the site of the original house such as in a self-built shelter made from debris material, a camping or basic tent, the garage, or in an undamaged part of the house. There is no outside aid such as food or shelter materials at this point	Response
Emergency or temporary shelter[a]	Months	To provide emergency shelter and contact points for Government and aid agencies supplying food, shelter materials, water, medical/health, and basic supplies such as soaps, blankets, etc. It can be a tent, a larger self-built shelter than before, a public facility, the home of family or friends, hotel, camping ground, park, or a second home. The length varies but will determine the scale of assistance and services required	Response
Interim housing[a]	Years	To provide safe and secure shelter/housing, water, and electricity so that some resumption of family life can be achieved, while efforts are under way to determine and construct permanent, sustainable housing options. This may involve repairs to original housing, the construction of new dwellings, insurance determinations, or relocation options or to find other suitable permanent housing. This can a prefabricated temporary house, specifically built barracks, camp or motor camp area, a winterized tent, a self-built shelter, a mobile home, an apartment, or the home of family member or friend: using either a transitional or core design approach	Recovery
Permanent housing[a]	Generations	To provide long-term, permanent housing solutions for those affected. This is usually on the original house site but can be within or nearby the community. A final option would be to relocate out of the original community, and this is particularly so where other long-term services such as education/schools are required	Reconstruction

[a]The resumption of family activities is noted as the difference between shelter and housing, and where possible, the difference is noted in the text

recovery happen? How can a decision maker reshape a process that operates within an emotional, reactionary, time-sensitive, expensive, and politically charged atmosphere that is based upon incomplete information, disproportionate needs, and the worst working conditions imaginable?" (Natural Hazards Center 2001).

Davis is more specific with his comments "Finally, a question that arises in the early stages of almost every recovery programme is whether reconstruction and the wider aspects of community rehabilitation should be conducted in the same, original, disaster-prone location, or whether the population should be encouraged or required to relocate to a new and possibly less vulnerable location."(Davis 2007).

And it is against this background that the provision of interim housing occurs.

But it gets worse. Strangely, it is not clear who is responsible for interim housing. In a developed economy, the response phase shelter is usually taken up by the Red Cross (ARC 2012), civil society, NGOs, and faith-based organizations, but it is not completely clear who is then responsible for any subsequent interim housing in the recovery phase. Governments do set up specific commissions, but the basis and objectives of them have been historically questioned (Hewitt 2013). This disconnection was evident in case studies 2 and 3.

A similar situation exists in developing counties and with humanitarian assistance under a cluster approach (OCHA 2012) where the response phase (tents and tarpaulins) would be taken up by the Emergency Shelter and Non-Food Items (NFI) Cluster led by either the International Federation of Red Cross and Red Crescent Societies (IFRC) for natural disasters or United Nations High Commissioner for Refugees (UNHCR) in conflict-based situations. However, the interim housing is subsequently spread across the two clusters of early recovery and camp management and is potentially involved in the WASH (water, sanitation, and hygiene promotion), Protection, and Early Education clusters. Interim housing is essentially an "orphan."

Timelines for Interim Housing

The time it takes to provide interim housing is often/always underestimated, but that should not be the case. A model developed in the 1970s by Haas suggests the following (Haas et al. 1977):

- If the time it takes to complete the response phase is 1, then:
- The recovery phase will be 10 times.
- The reconstruction phase will be 100 times.

These are approximations, and there are certainly difficulties in ascertaining when the response phase is actually complete. But as reported by Zuo, the approach is widely accepted and quoted which he details (Zuo 2010). It does provide a useful operational and planning base that is firstly "sobering" but also one that secondly better reflects the actual times (rather than the planned ones) allowed for earthquake disaster responses.

For example, the 22 Feb 2011 Christchurch Earthquake based on a response time of around 3 months would suggest the following timeline:

- Response phase of 3 months: 22 February 2011 to 22 May 2011

- Recovery phase of 30 months: May 2011 to November 2013
- Reconstruction phase of 300 months: November 2013 to November 2037.

This timeline would be difficult for any public authority or the Government to acknowledge and especially present to its electorate. The result is the conviction that all must be OK and a filtering of information to make it fit. Consequently, the Government is communicating exactly that message, that all is well and under control. On one hand, a local respected NGO called CanCERNs has initiated the 900 Project to identify what they suspect are the remaining 900 uninhabitable houses in Christchurch 3+ years after the earthquake and hence possibly out of control (CanCERN 2014).

Such timelines suggested by Haas are unfortunately ignored by Governments and aid agencies for various reasons (Lloyd-Jones 2006). However, on the flip side, it does suggest that any reduction of the response phase will have huge reductions for the subsequent recovery and reconstruction phases. Therefore, the planning and implementation of interim housing should start immediately and not wait till after the response phase has been completed. This "lesson learnt" seems to be repeated in subsequent seismic disasters.

Costs of Interim Housing

Typically, the costs for interim housing can be comparable or at least significant against the cost of permanent housing. For this reason, attempts have been made to miss out the interim housing and go more directly to permanent housing (as in the Banda Aceh case study). Alternatively, humanitarian and aid agencies have developed transitional and core housing approaches that uses any interim house as part of a permanent housing development thus "saving" on overall cost. There are additional political and planning reasons for this approach as it allows for more permanent outcomes for those affected based on an apparent interim solution which is politically expedient when those being assisted may be settled informally and do not have any formal land title and have been that way for some time before the earthquake.

Transitional and Core Approaches to Interim Housing

Transitional housing is where an interim house is designed to merge into a permanent one. This usually means that the structural system (and in particular the lateral load resisting part) is retained and the cladding usually changed over time. This in part seeks to save overall housing costs but more importantly to minimize the emotional impacts on families having to relocate, yet again. It can also make the shelter-in-place option more viable though it has been used in off-site as a prefabricated kitset that goes with the occupants. However, its big advance is that the occupants can upgrade as money and resources become available, and perhaps other family needs arise that require more space such as births and marriages. Typically, the transitional house could be supplied clad in plastic tarpaulins, which can be later replaced with mud bricks, mud-plastered stone, or cement-plastered concrete blocks. The design trick is to structurally ensure that the lateral load resisting as supplied has sufficient capacity for this anticipated loading of a heavier cladding.

A core house on the other hand is where some part of a permanent house is supplied which is then added and extended over time as money and resources become available.

The objective of both approaches is to "transition" the family/occupants into a more permanent or "durable" housing solution based on their priorities (rather than the aid agency) under their own direction. Such approaches are more readily found with aid agencies and in developing economies rather than Governments and developed economies despite their appropriateness for both.

Codes and Planning for Interim Housing

The level and criteria for code compliance for interim housing in post-earthquake situations are contentious. On one hand is whether there should be any/minimal code and planning compliance given the urgency of the need, the temporaryness of the building, and the often dire living conditions of those affected, while on the other hand, the ongoing seismic activity, the relative expense of the interim house, and its possibly semi-permanence suggest that it fully complies with the stipulated country and possibly international codes. Moreover, if these are being done in large numbers (rather than as a one off), it seems to support full code compliance. There is also the issue of their disposal once they have done their job. This can be more difficult than expected (Arslan and Coşgun 2007).

Most developed economies do not have specific codes or planning procedures for interim housing, and because it usually occurs outside, any declared state of emergency will often require some type of executive order. In Christchurch, for example, specific guidance notes were compiled to facilitate the compliance and provision of post-disaster worker accommodation (MBIE 2012). FEMA on the other hand has a suite of interim house types that it can use (DHS 2013) which have had issues in the past and would be pushed to respond in a timely manner to a disaster (McIntosh 2013). There are various guidelines for short-term/interim housing for the military, for farm workers, and for the homeless (USAF 2001; QG 2010; SAMB 2009;Wilson 2013). Strangely, there is minimal design code or guidance post disaster and seemingly nothing that specifies the roles and responsibilities of local authorities or Government in terms of interim housing of affected people.

Moreover, the issue for the worker accommodation built in Christchurch will be what happens to them when they reach their use by date of 31 December 2022. Will the worker's accommodation be demolished or simply on sold possibly creating substandard long-term housing somewhere else? Or will the date simply be extended?

In developing economies and the humanitarian aid sector, there are two "codes of compliance." In a refugee situation, it is the UNHCR's Handbook (UNHCR 2007), while in natural disasters, it is SPHERE (2011).

The UNHCR Handbook was developed from experience in the field and was widely accepted as the de facto standard and code even outside the refugee situation. However, its "camp" focus meant that it increasingly did not translate well into non-refugee situations. And hence the SPHERE Project was initiated to set up universal minimum standards in core areas of humanitarian response. A lack of such standards has in the past resulted in "competition" between shelter agencies. Though it was started in 1997 (with the first trial edition in 1998), the shelter response in Banda Aceh after the 2004 Asian Tsunami is often remembered because of the competition that developed. Shelter options were seemingly racketed up with one agency offering a house, then another offering a house and $2,000USD, and the next a house fully furnished and later with an opening party for friends.

The size and specifications of temporary shelter also generated unintentional competition between shelter agencies, and, for example, the "China tent" in Pakistan following the 2005 Kashmir Earthquake was the "tent of choice" firstly because it had walls which meant that one could stand up anywhere inside it but secondly because it had a steel frame which performed better under snow loading. SPHERE has been revised in 2000, 2004, and in the current green-colored version of 2011. The key quantitative shelter points are the following:

- Site slope should neither exceed 5 %, unless extensive drainage and erosion control measures are taken, nor be less than 1 % to provide for adequate drainage. The lowest point of the site should be not less than 3 m above the estimated maximum level of the water table.
- There should be a minimum usable surface area of 45 m^2 for each person when including household plots/gardens or 30 m^2 otherwise. And where this cannot be provided, the consequences of higher-density occupation should be mitigated, for example, through ensuring adequate separation and privacy between individual households, space for the required facilities, etc.
- There should be a 30 m firebreak for every 300 m of built-up area and a minimum of 2 m (but preferably twice the overall height of any structure) between individual buildings or shelters to prevent collapsing structures from touching adjacent buildings.
- In cold climates, household activities typically take place within the covered area, and affected populations may spend substantial time inside to ensure adequate thermal comfort. In urban settings, household activities typically occur within the covered area as there is usually less adjacent external space that can be used.
- A covered floor area in excess of 3.5 m^2 per person will often be required to meet these considerations.
- The internal floor-to-ceiling height should be a minimum of 2 m at the highest point. In warmer climates, the adjacent shaded external space can be used for food preparation and cooking. Where materials for a complete shelter cannot be provided, roofing materials to provide the minimum covered area should be prioritized. The resulting enclosure may not provide the necessary protection from the climate nor security, privacy, and dignity, so steps should be taken to meet these needs as soon as possible.
- Standards and guidelines on construction should be agreed with the relevant authorities to ensure that key safety and performance requirements are met. Where applicable local or national building codes have not been customarily adhered to or enforced, incremental compliance should be agreed, reflecting local housing culture, climatic conditions, resources, building and maintenance capacities, accessibility, and affordability.

These have been included because of an otherwise scarcity of guidelines.

Interim Housing and Diversity

The standardized nature of interim housing can miss out the special needs of sections of an affected population. These can include pregnant and lactating mothers, the elderly, the blind, the wheel chair bound, the bed ridden, single parent families, and those with medical or mental challenges. It can also happen in unexpected areas. For example, it was unfortunate that a Maori family with previous gang ties was evicted from an evacuation center following the September 2010 earthquake in Christchurch (Kipa et al. 2013). This exclusion leads to the setting up of a specific Maori response when the February 2011 earthquake occurred 5 months later. The notion that New Zealand society could be fractured like this still remains deep, as it no doubt does for the Lower 9th Wards of New Orleans.

Such people are essentially invisible in a disaster, and there is a reliance on self-identifying. The sudden nature of earthquakes means that self-identification before the event is at best limited. As a result, such issues only surface (unfortunately) around the provision of interim housing, and their special needs were compounded by that time by the loss of vital connections with personal care providers and sources for medication, service animals, community liaisons, public transportation, neighbors, and their network of contacts. And the diversity challenge is whether the provision of interim housing addresses a diverse community in an inclusive manner. This is not easy or straightforward, and failures are not recorded because they are seen as failures of the wider community or society such as the above example.

Case Study 1: Sichuan, China, 2008

Seismic event: 12 May 2008, magnitude 8.0.

Deaths: 69,195 deaths and 18,392 missing (Wikipedia 2008).

People affected: the earthquake left about 4.8 million people homeless, though the number could be as high as 11 million.

Total amount of direct damage: 845.1 billion RMD (137 billion USD) (UNCRD 2008, p. 5).

Disasters of this scale are not uncommon in China, and interestingly, this was the first time that China allowed foreigners into a disaster zone (perhaps because it was hosting the Olympics in Beijing later that same year in August), and hence for many, it was a unique opportunity to observe how the Chinese disaster response works. The two key aspects relevant to interim housing seem to be their community (or village)-based approach and partnering. This allowed the construction of around 500,000 temporary houses in 3 months following the earthquake and was one of the best disaster responses measured by the well-being of affected families (Potangaroa et al. 2009). The temporary homes or shelters had a floor area of 20 m^2 and were designed for three people (EERI 2008).

On 6 June 2008 (3 weeks after the disaster) the State Council published the Wenchuan Earthquake Restoration and Reconstruction Regulations and provided legal guidance for the formulation of the overall reconstruction plan (Kristin et al. 2012). The plan emphasized restoring residential housing and in particular the need for addressing the differences of multistory urban apartment blocks and low-rise rural housing. The plan also laid out the process to be adopted which consisted of the three following actions:

- Survey and assessment
- Plan drafting
- Policy measures

In response to this, the Sichuan Provincial Government published in that same month of June the "Sichuan Rural Housing Reconstruction Plan" and then in September the "Urban Housing Reconstruction Plan," while the Central Government published the "Sichuan Rural Housing Repair and Reinforcement Plan" in December 2008.

These plans laid out six methods for reconstructing or replacing houses that were damaged in the earthquake, namely:

- Reinforcing the house
- Rebuilding the house on the original land
- Rebuilding the house on replacement land
- Renting a low-cost house
- Buying a low-cost house
- Rebuilding the house organized by "workplaces"

There were additional options in urban areas of buying a low-cost house, renting public housing, or receiving a cash subsidy.

Houses were assessed and classified according to the extent of damage into the following five categories:

- No damage
- Minor damage
- Medium damage
- Serious damage
- Collapse

Apart from houses that had "collapsed" or had sustained "no damage," categories would receive a cash subsidy of 1,000–2,000 ($163–$325), 2,000–4,000 ($325–$650), or 4,000–5,000 ($650–$813) RMD ($USD), respectively. Urban households whose houses had collapsed or were inhabitable could receive a cash subsidy, a housing tax exemption or reduction, and preferential low-cost housing. The average cash subsidy was 25,000 RMD($4,066USD) depending on the household's size and economic status.

In rural areas, many houses were rebuilt on different land, and in these cases, the town and village authorities could choose to rebuild the houses collectively, with each household contributing a share. The point to note is that the whole village was relocated and not individual families. Households were entitled to a cash subsidy of RMD 20,000 ($3,250USD) (on average) to rebuild with the exact amount determined by the number of people and their overall economic situation. Households that decided on an alternative living arrangement such as living with other family received a reduced subsidy of RMD 2,000 ($325USD).

In October 2008, the Sichuan Government announced various preferential loan policies for households to rebuild their house or buy a new house. For households taking out a bank loan to purchase a new house, the minimum interest rate was adjusted to 0.6 of the benchmark lending interest rate, and the minimum down payment was reduced to 10 %. And for rebuilding, buying or strengthening a house received a 1 % point discount on their interest rate.

Where ever possible, communities are relocated as a community or village, and not individually (UNCRD 2008, pp. 22–24). Each village has its own organizational structure that interfaces between the people of the village and the city authority. This provided a ready two-way information conduit with assessment and damage data being fed up and strategy and priorities being fed down. Thus, the authorities had rapid and relatively accurate data very early and hence could plan an appropriate response. This included a partnering or "twinning" arrangement whereby affected cities and towns are linked to other cities and towns in China who are then expected to assist. The partnering arrangement can be competitive as results from it are published by the Central Government. But this could also be because the workers from coastal cities working in the mountainous and colder regions of Sichuan and staying in tents and their own cars perhaps feel more motivated to get the work completed quickly.

However, despite its important role in terms of financial and resource allocation, it does seem to have three areas of concern (UNCRD 2008, p. 5):

- Variation in support from provinces.
- Inappropriate support that may not necessarily reflect the needs of a different climate, culture, and social background of the partner region. For example, facilities provided from Northern Provinces had double-panel windows and contain heat-insulating materials, while those provided from Southern Provinces have no such equipment. This is not a big problem in the phases and rescue and rehabilitation because short-term responses or restoration is required in the phases. In the reconstruction phase, however, supporting materials need to be accepted by the supported region.
- Supporting provinces tend to have a larger voice and decision-making power than supported areas.

Summary

It has been historically difficult getting access to disasters in China, and certainly while access was granted to foreigners following the earthquake in Sichuan, it was nonetheless limited and understandably constrained. Thus, reports and feedback from the field are scarce. This is a pity as the scale and response to seismic disasters has much that other countries could learn from and adapt for their own situations. The partnering approach and the speed at which interim housing was achieved are two clear examples. While it was conceivably possible for the Government to impose price controls, this did not seem to happen during the four field trips made by the author and several others made by PhD students studying aspects outside the provision of interim housing (Chang et al. 2011). Moreover, the use of loans and other seemingly market instruments to manage the response perhaps suggested a sophistication beyond simply Government structures.

Case Study 2: Port-au-Prince, Haiti, 2010

Seismic Event: Magnitude 7.0, 12 January, 2010.

Deaths: 220,000 deaths with some variance between different estimates.

People affected: an estimated three million people were affected by the earthquake together with the collapse of 250,000 residences and 30,000 commercial buildings (Ferris 2010) (Wikipedia 2010).

This was the first large-scale humanitarian response involving a "crowded" urban context. The rural response seems to lead the urban one, and the reasons why become clear when people see photographs of the affected area in the capital of Port-au-Prince. The poorer areas effectively are 100 % covered in corrugated iron roofing, and hence, the immediate question was: Where would any interim housing be established? The planned shelter solution was to build on the site of the former house which required demolition and clearing of the site before the interim housing could be erected. Needless to say, this was problematic in itself; the "owners" had to be found, land tenure had to be identified, the rubble had to be typically carried out by hand out to the road, and the interim house had to be built. However, when this was completed, where was any permanent housing to be erected? Consequently, the interim house or T-Shelter took on a more permanent role that it was not intended, and the role of what was intended as a "transitional" house became increasingly questioned as transition to what?

Nonetheless, providing emergency shelter support to over 1.5 million people in less than 4 months was an outstanding achievement (Rees-Gildea and Moles 2012). The Shelter Cluster's "harmonized" emergency shelter and NFI response seemed to be a coordination success (the United Nations emergency response consists of 11 sector clusters and one of those is the Shelter Cluster. It is headed by the IFRC for natural disasters and UNHCR in conflict situations). However, the emergency shelter (E-Shelters) response was essentially based on a one-time massive delivery of tarpaulins and tents, there was some variation in quality, but more importantly, there was little or no control over the shelter outcomes that those affected were selecting. Camps both small and large blossomed throughout Port-au-Prince, and at one time, there were over 1,300 camps with 100 people or more. Camps were established in parks, on golf courses, around churches, and on street corners. The shelter agencies did not have the means to guide or direct the final outcome of the E-Shelters. The large numbers and the tight urban situation made it impossible to meet the SPHERE standards such as a minimum of 3.5 m^2/person in covered area (such as a tent or tarpaulin), 30 m^2/person in camp land area, and the prescribed numbers of toilets and showers. It was hoped that these perhaps could be progressively meet as camp numbers decreased. This did not happen as people seemed to "settle" into camp life, and 1 year later, nearly one million people were still in camps.

Replacement of such shelters that have a service life of around 6 months was not initially planned with the intent was to move into T-Shelters by the end of 2010. That did not happen for many reasons, and consequently, there was a shelter gap as the hurricane season rolled in, and another call for replacement tarps and tents was made. This unplanned double effort put more pressure on the overall T-Shelter program and meant a higher workload for shelter agencies together with an unexpected additional budget line. This vulnerability became evident when a relatively small depression Ivan hit Port-au-Prince destroying 9,000 E-Shelters.

The Shelter Cluster was also seeking to harmonize the T-Shelter in much the same way it had for the E-Shelter. There were several different types developed by aid agencies based around the following standardized criteria (IASC 2010):

- An area of 12–18 m^2 for a single-story structure
- Cost between $1,000 and $1,500USD (including transport and labor) later increased to $1,800
- Designed for a 100 mph wind load
- Designed for seismic loads
- A 3-year life span but later increased to 15 depending on possible building up grading

This harmonization proved problematic for several reasons, and as a result, shelter numbers had to be lowered to fit existing budgets. For example, there were no seismic codes and, hence, no seismic zones for Haiti. In addition, the nature of the wind loading was not clear but also not questioned. Was it a 3 s gust or a sustained wind load; what was the specified terrain and at what height was this measured (Potangaroa 2010b). Consequently, agencies and their respective designers in other parts of the world using at times non-hurricane wind loading codes seemed to make their own assumptions of what this wind speed meant. For example, the ASCE 7 code produced basic wind loads of around 1.2 kPa, while the same wind speed in the British code produced a 2.5 kPa loading (it should also be noted that hurricanes are not featured in the British code). This was not apparently picked up by agencies or their designers who invariably were not based in Haiti but were working from information provided by aid agencies in the field. This resulted in significant cost over runs particularly for foundations and costs quickly blew out with the largest being in the order of $4,500USD, nearly three times over the capped value.

In addition, there were other aspects of concern with the seismic and wind design of the T-Shelter design that included the following:

- The lack of a diaphragm transfer system. This structural system transfers lateral loads (such as wind and seismic) through the building to vertical structure which takes it to the foundations. It is usually located in a horizontal plane such as the ceiling or roof.
- Foundations and their spatial relationship to the above wall bracing. It was noted that diagonal bracing in walls were not located above foundations and as such would be unintentionally bending and potential breaking the wall bottom plates.
- A lack of gable end bracing. This is where braces are required to prop gable ends back to a horizontal structure.
- Roof slope: pitched roofs were advised to be +30° and mono-sloped roofs 12–14°. This was done to minimize wind up lift loads.

The 11 exemplar T-Shelters on the Shelter Cluster web page were examined, and these are tabulated below (Table 2).

A lack of a horizontal structure or suitable foundations would constitute a systemic issue, while the gable aspect could/would be more localized. Thus, at least six had systemic issues and nine localized issues, and given that these 11 are probably better examples underlines an issue that aid agencies need to address. This same issue of design is also raised in the Aceh case study that follows. However, aid agencies are usually reluctant to discuss these issues given the large resources and costs involved in the provision of interim housing.

Perhaps this was partly why many agencies "were inclined to work outside Port-Au-Prince, mainly funding the direct delivery of transitional shelter [T-Shelters] and not integrating a housing repair approach into their programmes...... This widely supported strategy [T-Shelters] was unrealistic in timing and cost-wise and was not flexible enough. Relevant progress in T-Shelter delivery started by May 2010, and the sum of the agencies capacities (125,000 units) proved to be insufficient to meet total transitional shelter need (over 180,000) within the foreseen timeline (January 2011), even after maximum delivery capacity was reached (7,300 units per month, as

Interim Housing Provision Following Earthquake Disaster, Table 2 T-shelter design aspects

	Horizontal structure	Foundations	Gable brace	Roof slope
Agency 1	No apparent structure	Do not appear to relate to diagonal braces	Partial separation for side gables?	OK
Agency 2	OK diagonal steel ties	OK concrete slab	OK diagonal timber brace	OK
Agency 3	No apparent structure	Do not appear to relate to diagonal braces	OK diagonal timber ties	OK
Agency 4	OK diagonal timber ties	Probably OK concrete block and slab foundations	No apparent brace	OK
Agency 5	OK diagonal timber ties	Probably OK because of external plywood	No apparent brace	OK
Agency 6	No apparent structure	Do not appear to relate to diagonal braces	Full height member but no apparent bracing at roof level	OK
Agency 7	No apparent structure	No apparent bracing?	No apparent brace	Not clear
Agency 8	OK diagonal steel ties	Probably OK because of external plywood	No apparent brace	OK
Agency 9	No apparent structure	Do not appear to relate to diagonal braces	No apparent brace	Less than 30°
Agency 10	OK diagonal timber ties	Probably OK because of concrete ring beam	No apparent brace	Probably OK
Agency 11	OK diagonal steel ties	Do not appear to relate across the building	No apparent brace	OK
Totals	5 with potential seismic/wind issues	6 with potential seismic/wind issues	9 with potential seismic/wind issues	1 with potential wind issue

of November 2010). All of which led to an accumulated delay, to bottlenecks in the implementation process, and to unforeseen relief responses as new emergencies arose. In addition, the approach of faster delivery of simpler transitional shelter could not be always achieved, and the shelter agencies had serious difficulties in meeting the agreed upon standards and average costs, leading upwards of a 200 % increase in the anticipated costs..." (EPYPSA 2011, p. 5).

That report goes on to outline the complexities of supplying transitional shelter with its comments that "The shelter sector in Haiti promoted a return to a neighbourhood strategy aimed at drawing the displaced population out of the camps through the provision of shelter (transitional, permanent, home repairs), but the shelter agencies procrastinated their engagement in housing repair." This low engagement in housing repair and seismic strengthening may have been caused by the following (EPYPSA 2011, p. 46):

- Lack of experience and perhaps a fear of the complexity involved (first "urban"-based seismic disaster for the humanitarian agencies with a lack of physical access, emphasis on rubble, problems of logistics, and few trained staff and labor)
- Difficulty to design big-scale funding proposals (each situation was different).
- Budget over runs (unforeseen costs that can only become clear during construction).
- Lack of funding (HQ and/or donors not willing to go into complex construction projects).
- Beneficiary selection and involvement (who and how should assistance be provided?).
- Repair or retrofit? (Repairs would fix the damage but not it's lack of earthquake capacity. Retrofitting would address both issues but

could be expensive and would require specialist skills)
- Should this be seen as an emergency rather than a recovery/reconstruction response?
- Camp-driven response easier to manage and control (less constraints and higher visibility).
- Overall perceived as "easy to get into, but hard to get out."

Some agencies such as the IFRC did manage to develop a seismic retrofitting program (Potangaroa et al. 2011b). This approach had the distinct advantage of being well under the $2,200–2,400USD/family cost to relocate families out of the camps. However, as indicated earlier, this option was developed late in the response in Port-au-Prince because there was the sense that any seismic retrofitting would have to be prohibitively expensive. The approach was based on the following two factors (Potangaroa et al. 2011):

1. An acceptance of higher level of building damage in an earthquake; up to what is often termed "near collapse." Most seismic codes are based around a "safe exit" strategy which requires a lower level of damage.
2. A realistic determination of the life span of the building. Again most codes are based around approximately 100 years, but the low building quality suggested a 25–30-year life span.

And these two factors allowed for a code compliant design to levels that were appropriate to the building culture of Port-au-Prince rather than a developed one in San Francisco or Wellington.

The seemingly final part of the interim housing in Port-au-Prince was a Government initiative to set a repair program for 16 neighborhoods and to decongest six camps associated with those neighborhoods under a proposed 16-6 Program (IOM 2012). Camp dwellers and residents in damaged neighborhoods could choose between either having their homes fixed or taking a 1-year rental subsidy of $500USD. About 10,000 families took this option. However, what was going to happen beyond that year was not clear; again it was the "transition to what"?

Summary

The timelines required for interim housing were seemingly underestimated in Port-au-Prince. This caused several issues and, for example, the relevance of the T-Shelter decreased, "while other approaches like repairs, rental subsidies or even permanent housing became relevant and even more cost efficient as time elapsed" (EPYPSA 2011). The affected population also started to look at the T-Shelter as permanent solutions, both because they were of a higher standard than their original housing and because they were unaware of any planned transition. Shelter agencies planned to build 125,000 T-Shelters in 10 months (March 2010 to January 2011) but were unable possibly because the numbers were too ambitious to start off with. With time, material costs escalated, the T-Shelter increased its life span beyond the specified 3 years, and issues such as land tenure, rubble clearance, and urban setting resulted in lengthy delays. Hence, by November 2010, only 19,000 T-Shelters (15 % of the total goal) had been completed and delivered.

Secondly, the role of appropriate structural design was missed by aid agencies. This will be highlighted in the next case study on Banda Aceh. It would seem that this lesson has not been learnt. And while aid agencies may be unnecessarily concerned about design risks and responsibilities, the apparent situation of six systemic and nine localized failures from 11 exemplar T-Shelters in Port-au-Prince would seem to deserve decisive attention from the aid community beyond a concern with design risk and responsibilities. Moreover, there seems to be opportunities to better mainstream the social needs and concerns of communities by a reflective design approach suggested by the IFRC retrofitting approach.

Finally, it would also seem that there is currently no urban game plan for provision of interim housing. While the rural response seemed to be managed in Haiti, the urban one in Port-au-Prince was not. It would seem that the provision of interim shelter in urban areas needs more discussion and research, hopefully before the next earthquake disaster. This is supported by the case study 1 on Christchurch.

Case Study 3: Banda Aceh 2004 and Yogyakarta 2006

Seismic event: Magnitude 9.0, 26 December 2004	Seismic event: Magnitude 5.9, 27 May 2006
Deaths: 130,000 deaths	Deaths: 5,716 deaths
People affected: million houses were destroyed or made uninhabitable	People affected: 240,396 houses were destroyed (IRC 2009)

These two responses were connected firstly because of their geography and timing but also because the Banda Aceh experience is now seen as a kind of watershed for humanitarian agencies working in the interim shelter sector that flowed over into Yogyakarta. The sense at the time of Banda Aceh was to miss out any interim housing and instead go directly to permanent housing (Potangaroa 2006). The Government had indicated that they would build barrack type accommodation and that combined with people returning to their original tsunami affected sites and staying with other family members seemed to be the basis of the Government's strategy. Land tenure was an issue as much of it was oral and informally agreed between male community members. Thus, the loss of land marks and people meant that ownership was no longer certain. In response, agencies put together a multi-program such as the one proposed by UNHCR (Potangaroa 2005):

- New Housing on new or former home sites. (Four sizes of housing proposed depending on family size).
- A "Granny" flat option addition to an existing house where those affected wish to stay long term with other "host" family members (the size was based on the living/lounge area and toilet bathroom areas from a spatial survey of non affected houses in Banda Aceh).
- Shelter Kits for repair of an existing house.

However, in March 2005, the Government suggested that agencies such as UNHCR (and also IOM) should not be in the once military zone of Aceh; thus plans such as the above were never realized at the scales planned (TIME 2005). This created a strategy and agency vacuum and consequently it was not till September 2005 (9 months after the disaster) that the IFRC launched its Temporary Shelter Plan of Action. This consisted of a 25 m^2 lightweight bolted steel frame, with timber cladding and a metal roof. The foundations were supposed to be anchored down but usually were leveled and left. Problems with appropriate sustainable timber supplies delayed its release but by December 2005 a prototype was in place. Consequently, many families did not receive these interim/transitional shelter kits until 2006. This interim kit was valuable and on the informal market could be purchased for $5,000–$6,000USD. This was also before the advent of the Shelter Cluster approach, and so agencies tended to operate independently and without any coordination.

Da Silva explains that the (da Silva 2010) "Shelter policy was not developed within the overall context of the journey from emergency shelter to durable solutions, and over the first six months there was considerable confusion and no clear policy as to the type of shelter assistance required. In June 2005, BRR [the Government's Disaster Response Agency] announced that families should be encouraged to return to their own land or to voluntarily resettle on land purchased by communities themselves or by BRR. They also stated that each affected household would be eligible for a permanent 36 m^2 house, with an expectation that this could be realised within a year, in the belief that in the interim affected communities were adequately housed in barracks, transitional shelter or with host families. Agencies who had already constructed 'semi-permanent' houses were faced with having to upgrade or replace housing, as these were no longer deemed adequate. This timeline proved unrealistic and did not reflect the realities on the ground, led to false expectations by the media, donors, government and beneficiaries and placed considerable pressure on implementing agencies."

Consequently, Da Silva's conclusion was that the notion that everyone who was affected should get a house led to a focus on "reconstruction rather than recovery" and an "emphasis on providing houses rather than assistance to reconstruct."

One further "lesson" was that agencies needed to have experienced seismic design engineers involved; otherwise, a less than adequate seismic solution could be used. And that having solely civil or structural engineers or local engineering input was no guarantee of finding an appropriate or even adequate seismic solution. Some agencies had again to come back and seismically strengthen the housing they had already constructed. Invariably this relates to "simple" structural detailing such as the spacing of column ties or beam stirrups/links. This was not the common practice amongst local consultants and builders in Aceh. In addition, the Indonesian Seismic Code excluded single-story dwellings, and the building code for Aceh issued by BRR (UNHIC 2005) did not include basic seismic design principles. This resulted in widespread confusion that compounded the view that seismic design would be considerably more expensive. BRR did not enforce the need for seismic design and themselves built houses and schools which did not conform to basic seismic guidelines. Da Silva goes on to report that "... ...most DEC [Disasters Emergency Committee (UK)] Member Agencies ultimately did adopt seismic design approaches but this was not always matched by quality construction and the design intent was compromised by poor quality materials and workmanship on site. In some cases remedial works and retro-fitting was needed to create a safe structure."

Agencies tried to dodge these issues with the response in Yogyakarta; but unfortunately repeated many of the same problems.

The Indonesian Government was reported as not being able to cover the need for shelter in Yogyakarta, not simply because of a lack of finances but also due to its scale and that it was much greater than initially believed (Roychansyah 2009). To meet the need for shelter, the Indonesian Government, United Nations agencies, and other humanitarian partners developed a "roof first" interim housing strategy to be implemented from the time aid is first provided up to the stage of community rehabilitation.

The Emergency Shelter Cluster was coordinated by the IFRC who calculated that with an average household size of 4.3 that at least 1,173,742 people, and perhaps as many as 1,542,380 were left homeless by the earthquake. However, the provision of tarpaulins and tents was limited, and by the end of July 2006 (2 months after the earthquake) at the end of the Response Phase, almost 30 % of these numbers were still homeless.

Despite this short fall interim housing was achieved using emergency bamboo and tarpaulins and recycling materials (such as bricks, timber, and window, and door frames), and finally villages that had sustained less than 20 % damage accommodated those made homeless using a kinship family model within the village.

The goal during the response phase was in addition to the usual emergency tents to also provide interim housing or T-Shelter (Setiawan 2007). Most were constructed from bamboo which was selected because of its low cost, availability, and strength. Sizes varied between 18 and 36 m^2. There were three different levels of community participation with regard to shelters: participation in the design process, participation in contributing materials for transitional shelters, and participation in the construction process (IRP 2009; JRF 2008). Somewhere in these phases, there were two significant structural changes; the first was the inclusion of a heavy clay tile roof for cultural reasons. It seemed that no normal Javanese man would sleep under anything other than a clay tile roof. And secondly, the bamboo wall braces were lifted so that people could readily enter in the side of the interim houses. The clay tile roof rendered the house as earthquake prone, and the movement of the wall braces minimized any remaining seismic resistance. This clash between community engagement and the confusion around seismic standards that was so problematic in Aceh had not been learnt for Yogyakarta or for Haiti.

Summary

The provision of interim shelter is important because it is often the first step towards normalizing a community following an earthquake. It is perhaps more complicated and certainly takes longer than expected, and the key appears to be the engagement with the affected people. This seems to apply whether it occurs in a developed or developing economy and whether it is provided as aid assistance or not.

The definitions for interim housing given in the literature need to be reviewed. And, for example, the one provided at the start of this chapter of between 1 month and 2 years is probably too soon on both the 1 month and the 2 years for a large-scale disaster or catastrophe.

While affected communities are the key focus for what is needed, Governments are the key influence for how that could be provided. Strangely Government departments (representing the Government) seem to have problems communicating with their affected communities when there is not an existing communication structure prior to the earthquake. Where it does not exist, departments must ensure that what is happening in the field is directly measured and observed rather than remotely from a headquarters/office location.

Interim housing is expensive and can unintentionally become more permanent housing. Its design can be problematic again in both developing and developed economies, and there are gaps for interim housing strategies for urban contexts. It is perhaps sobering to reflect upon Jean Jacques Rousseau's letter to Voltaire after the 1755 Lisbon Earthquake who is reported to have written that "...... if this tragedy happened, it is not the fault of mother nature. It was not nature who gathered together twenty thousand buildings in this place." This must be a concern with over half of the world now living in cities and how the provision of interim housing after an earthquake in an urban context will be achieved.

The ability of interim housing to merge into or be the basis of any following housing strategy certainly can be an advantage. It is commonly used in the humanitarian and development sector as transitional or core shelter/housing. However, in developed economies, more reliance is placed on insurance and that those affected will seek their own solutions, and hence emergency shelters and evacuation centers seem to be the basis for response.

Finally, it seems that shelter in place is the preferred option for those affected by earthquake, and that individual relocation is the least preferred. Yet as noted above with 50 % + of the worlds people being urban based suggest that some relocation whether temporary or permanent may need to happen. Consequently, the provision of interim shelter will increasingly become more complex and involved but will still remain as one of the important normalizing steps for communities after an earthquake disaster.

Cross-References

▶ Building Codes and Standards
▶ Community Recovery Following Earthquake Disasters
▶ Earthquake Disaster Recovery: Leadership and Governance
▶ Learning from Earthquake Disasters
▶ Resilience to Earthquake Disasters

References

ABAG (2008) Long term housing recovery Pub. by Earthquake and hazards program local and regional disaster recovery planning issues Association of Bay Area Governments Issues paper, 8 July 2008

ARC (2012) Preferred sheltering practices for emergency sheltering in Australia. The application of international humanitarian best practice. Pub. by Australian Red Cross, Victoria

Arslan H, Coşgun N (2007) The evaluation of temporary earthquake houses dismantling process in the context of building waste management. In: International earthquake symposium Koceau, Turkey 22–26 Oct 2007

Cancern (2014) Newsletter # 113, 31 January 2014. http://www.rebuildchristchurch.co.n2/blog/2014/1/cancern-newsletter-113-31-January-2014. Accessed Feb 2015

Chang Y, Wilkinson S, Potangaroa R, Seville E (2011) Identifying factors affecting resource availability for post-disaster reconstruction: a case study in China. Construct Manage Econ 29(1):37–48

da Silva J (2010) Lessons from Aceh Key considerations in post-disaster reconstruction. Practical Action Publishing, Rugby/Warwickshire, p 63. ISBN 978 1 85339 700 4. www.practicalactionpublishing.org

Davidson C, Johnson C, Lizarralde G, Sliwinski A, Dikmen N (2007) Truths and myths about community participation in post-disaster housing projects. Habitat Int 31:100–115

Davis I (2007) Learning from disaster recovery guidance for decision makers. International Recovery Platform (IRP), Geneva and Koba, p 24

Davis I (2010) What is the vision for sheltering and housing in Haiti? Summary observations of reconstruction progress following the Haiti earthquake of January 12th 2010 online at www.onuhabitat.org/haiti, accessed Feb 2015

DHS (2013) Unless modified, FEMA's temporary housing plans will increase costs by an estimated $76 million annually. Department of Homeland Security, Washinton, DC, p 3

EERI (2004) Learning from earthquakes. Social and public policy issues following the Bam, Iran, Earthquake. EERI special earthquake report, California, Aug 2001

EERI (2008) Reconnaissance report on the China Wenchuan Earthquake May 12, 2008. Chu-Chieh J. Lin, Associate Research Fellow of NCREE Juin-Fu Chai, Research Fellow of NCREE, Berkeley, CA

EPYPSA (2011) Estudios Proyectos y Planificación S.A. An evaluation of the Haiti earthquake 2010 meeting shelter needs: issues, achievements and constraints. International Federation of Red Cross and Red Crescent Societies (IFRC), Geneva, p 5

Ferris E (2010) Earthquakes and floods: comparing Haiti and Pakistan. The Brookings Institution

Giovinazzi S, Stevenson J (2012) Temporary housing issues following the 22nd Christchurch earthquake, NZ. In: 2012 NZSEE conference University of Canterbury, Christchurch

Haas J, Kates R, Bowden M (1977) Reconstruction following disaster. The MIT Press, Cambridge, MA/London

Hewitt J (2013). Errors of commission. In: AUBEA 2013 conference, Auckland, 20–22 Nov 2013

IASC (2010) Shelter Cluster Haiti transitional shelter technical guidance 19/02/10. Shelter Cluster technical working group, Port au Princes

IASC (2011) United Nations Inter Agency Standing Committee operational guidelines on the protection of persons in situations of natural disasters. The Brookings – Bern Project on Internal Displacement, IASC, Geneva, Jan 2011

IOM (2012) Haiti: from emergency to sustainable recovery IOM Haiti two year report (2010–2011), pp 9–11, 37-3B. http://www.com.int/Jah 19/webdav/shared/main site/activities/countries/docs/haiti/10M-Haiti-Two-year-Report-2010-2011-From-Emergency-to-sustainable-Recovery.pdf. Accessed Feb 2015

IRP (2009) International Recovery Platform, recovery status report. The Yogyakarta and Central Java Earthquake 2006 with Gadjah mada uni. pub. IRP Kobe Dec 2009

Jha A, Barenstein J, Phelps P, Pittet D, Sena S (2010) The social dimension of housing reconstruction. In: Safer homes, stronger communities: a handbook for reconstructing after natural disasters. World Bank, Washinton, DC, pp 59–76

Johnson C (2002) What's the big deal about temporary housing? Planning considerations for temporary accommodation after disasters: example of the 1999 Turkish earthquakes. In: TIEMS disaster management conference 2002, Waterloo

Johnson C (2007) Impacts of prefabricated temporary housing after disasters: 1999 earthquakes in Turkey. Habitat 31(1):36–52

JRF (2008) Two years after the Java earthquake and tsunami: implementing community based reconstruction, increasing transparency. Java Reconstruction Fund progress report 2008, World bank, Washington, DC. http://documents.worldbank.org/wrote/en/2008/01/97921 accessed Feb 2015

Kipa M, Potangaroa R, Wilkinson S (2013) Iwi resilience? The Maori response following the February 22, 2011 Christchurch earthquake. In: AUBEA 2013 conference, Auckland, 20–22 Nov 2013

Kristin D, Hedda F, Liu J, Zhang H (2012) Recovering from the Wenchuan earthquake. FAFO-Report 2012:39, Oslo. ISBN 978-82-7422-917-4, ISSN 0801-6143. www.fafo.no/english

Lloyd-Jones T (2006) Mind the gap! Post-disaster reconstruction and the transition from humanitarian relief. RICS, London

Lyons M, Schilderman T, Boano C (2010) Building back better: delivering people-centred housing reconstruction at scale. Practical Action, Rugby; South Bank University/IFRC, London

MBIE (2012) Guidance to assist the development of temporary accommodation to house workers in greater Christchurch. NZ Government; Ministry of Business, Innovation and Employment, Wellington, Sept 2012

MCDEM (2004) National civil defence and emergency management CDEM strategy 2003–2006. Ministry of Civil Defence Emergency Management, NZ Govt, Wellington, pp 14–16

McIntosh J (2013) Chapter10: The implications of post disaster recovery for affordable housing. In: Approaches to disaster management – examining the implications of hazards, emergencies and disasters. Intech, Rijeka

Natural Hazards Center (2001) Holistic disaster recovery ideas for building local sustainability after a natural disaster. Natural Hazards Research and Applications Information Center, University of Colorado 482 UCB, Boulder

OCHA (2012) What is the cluster approach? OCHA on message: the cluster approach. UN office for the coordination of humanitarian affairs. https://docs.unocha.org/sites/dms/Documents/120320_OOM-ClusterApproach_eng.pdf

Potangaroa R (2005) The strategy for a shelter program. Report for UNHCR Banda Aceh, Indonesia. Technical Coordinator UNHCR Bandar Aceh, 24 Jan 2005

Potangaroa R (2006) The development of a permanent shelter program for Aceh, North Sumatra.

In: Proceedings of the scientific forum on the tsunami, its impact and recovery regional symposium, AIT, Thailand, 6–7th June 2005
Potangaroa R (2010a) The seismic gap: issues of seismic design in post disaster reconstruction. i-Rec2010 participatory design and appropriate technology for post-disaster reconstruction, Ahmadabad
Potangaroa R (2010b) Lessons learnt as part of the terms of reference deliverables. Shelter cluster technical coordinator available via the Shelter Cluster Web site. Haiti, Nov 2010
Potangaroa R, Wang M, Chang Y (2009) Identifying resilience in those affected by the 2008 Sichuan earthquake. J Eng Sci (in Chinese)
Potangaroa R, Wilkinson S, Zare M, Steinfort P (2011a) The management of portable toilets in the eastern suburbs of Christchurch after the february 22, 2011 earthquake. Australas J Disaster Trauma Stud 2011(2):35–48
Potangaroa R, Mattar Neri R, Brown D (2011b)The design development and costs of several key housing issues for Delmas 19, Port Au Prince. Report for the British Red Cross Haiti, Dec 2011
QG (2010) Form23 temporary accommodation buildings checklist.(version 1) The State of Queensland (Department of Infrastructure and Planning). Queensland Government, Brisbane, May 2010
Quarantelli EL (1995) Patterns of shelter and housing in US disasters. Disaster Prev Manage 4(3):43–53
Rees-Gildea P, Moles O (2012) Lessons learned & best practices. In: The international federation of Red Cross and Red Crescent Societies Shelter Programme in Haiti 2010–2012, IFRC, Geneva
Roychansyah M (2009) Chapter 3: sector-specific recovery and Case Studies: 3.1 Shelter international Recovery Platform recovery status report. The Yogyakarta and Central Java Earthquake 2006, IRP Washington, DC, p 9
SAMB (2009) Regulating temporary farm worker housing in the ALR. Discussion paper and standards. The Sustainable Agriculture Management Branch Ministry of Agriculture and Lands, Victoria
Setiawan B (2007) Lessons from Aceh and Jogya. Towards better disaster management. Partnership for Governance Reform Yogyakarta, Dec 2007 ISBN 978-979-26-9629-9
Shelter Cluster Philippines (2014) Shelter and environment – an overview Typhoon Yolanda response. Philippines, Haiyan Shelter Cluster, 12 Jan 2014 download from https://www.sheltercluster.org/
SPHERE (2011) The Sphere project, the humanitarian charter and minimum standards in humanitarian response. Practical Action Publishing, Rugby. ISBN 978-1-908176-00-4
TIME (2005) After the tsunami: a time to build. Special issue, 4 Apr 2005, 165(14)
UNCRD (2008) The disaster management planning Hyogo Office of the United Nations Centre for Regional Development (UNCRD). Report on the 2008 Great Sichuan earthquake, Mar 2009. Available at http://www.preventionweb.net/files/13114_UNCRDSichuanReport200903EN.pdf
UNHCR (2007) Handbook for emergencies, 3rd edn. United Nations High Commissioner for Refugees, Geneva
UNHIC (2005) Guidelines on housing repair and construction. United Nations Humanitarian Information Centre for Sumatra, Book 3 SWG, pp 1–5
USAF (2001) Temporary lodging design guide. United States Air Force, Brooks AFD, Texas, Rec 2001
Wikipedia (2008) 2008 Sichuan earthquake. http://en.wikipedia.org/wiki/2008_Sichuan_earthquake
Wikipedia (2010) http://en.wikipedia.org/wiki/2010_Haiti_earthquake. Accessed Feb 2014
Wilson W (2013) Homelessness in England – commons library standard note. Published 12 Dec 2013, Standard notes SN01164
Zuo K (2010) Procurement and contractual arrangements for post-disaster reconstruction. PhD thesis, Auckland University School of Engineering, p 60

Land Use Planning Following an Earthquake Disaster

Wendy Saunders[1], Suzanne Vallance[2] and Ljubica Mamula-Seadon[3]
[1]GNS Science, Lower Hutt, New Zealand
[2]Lincoln University, Christchurch, New Zealand
[3]Centre for Infrastructure Research at the Department of Civil and Environmental Engineering, Faculty of Engineering, The University of Auckland, Auckland, New Zealand

Synonyms

Canterbury; Governance; Post-earthquake; Pre-event; Recovery

Introduction

The term "land use planning" covers many aspects of planning: spatial, urban design, heritage, transportation, recreation, access to education, health facilities, natural hazards, landscape, biodiversity, air and water quality, affordable housing, cultural values, sustainability, etc., and this is not a complete list of the many aspects and parts to land use planning. A planner may specialize in one of these aspects, e.g., a transportation planner, or be a generalist – knowing a little about many aspects of planning.

There are many definitions of land use planning; however, March and Henry (2007, p. 17) provide a concise discussion on what planning, and more specifically land use planning, is. They state:

> Land-use planning ... is focussed upon establishing the best spatial arrangements of land use, development, and management ... To do this, land-use planning is confronted with the task of establishing which of the many potential patterns and organisations of land use are likely to be the most advantageous in the future, in a particular place, and for a particular community.

Therefore, land use planning is an important mechanism for the reduction of risks in the built and natural environments (March and Henry 2007). Although planning cannot avoid all natural hazards, the thoughtfully planned location and design of structures and land uses can reduce the risks, particularly when combined with emergency management systems, i.e., early warning and response systems.

So how does land use planning contribute to recovery following an earthquake? Often touted as the key tool available for sustainable hazard mitigation, land use planning has a key role in determining how a city recovers, its long-term sustainability, and resilience. The following sections will provide an overview of the role of land use planning during recovery following a major earthquake using the 2010–2011 Canterbury earthquake sequence as a case study. This will provide a brief overview of the Canterbury earthquake sequence and outline the role of governance in planning outcomes, the value of pre-event recovery planning for land use, and the complex post-earthquake realities of planning within the context of the Canterbury earthquakes.

The Canterbury Earthquakes

On September 4, 2010, at 4.36 am, the Canterbury (New Zealand) region experienced a 7.1 magnitude quake. Although there were no casualties, the regional center of Christchurch and the township of Kaiapoi in the Waimakariri District (to the north) suffered extensive land damage in the form of liquefaction, lateral spread (surface rupture and slippage), and rockfall. Six months later, a 6.3 magnitude earthquake occurred almost directly under the city of Christchurch. Though "smaller" in magnitude, the peak ground acceleration of 2.2 (over twice that of gravity) was one of the highest ever recorded and would have "flattened" most world cities (Anderson 2011). This time, 185 people lost their lives and thousands were injured. Almost 80% (Markham 2013) of core inner city buildings were (or will be) demolished; many schools and other major services and facilities closed either permanently or temporarily; there was significant, widespread infrastructural damage; and residents in some hillside and eastern suburbs were left homeless or living in substandard housing. Following this, a 6.3 magnitude quake on June 13, 2011, caused further damage, as did the 50 other quakes with a magnitude of 5 or more. As of March 2013 (Markham 2013):

- With 91 % of the 191,000 homes in greater Christchurch damaged, more than 450,000 residential land, building, and contents damage claims had been made to the Earthquake Commission across 200,000 South Island properties.
- Approximately 7,800 residential homes have been "Red Zoned," i.e., tagged for demolition or removal with the land under them deemed too expensive to remediate at this time.
- Over 1,500 central business district (80 % of CBD building stock) and 300 suburban commercial buildings have been, or will be, demolished with over 4,100 businesses (and 30,000 employees) affected by the demolition exclusion zone (initially 390 ha).
- Loss of, or disruption to, 75% of hotel accommodation and key convention, cultural, and recreation facilities, i.e., 1,600 community facilities and more than 250 heritage buildings.
- At least 528 km of the wastewater network, 1,021 km of roads, 22 freshwater wells, and 30 bridges need to be replaced or repaired.
- An estimated rebuild cost of over $40 billion.

These events have been referred to as the Canterbury Quakes as their effects have cut across district/city council administrative boundaries and have prompted regional migration. The extended sequence has put pressure on local authorities trying to balance "business as usual" with earthquake recovery. New Zealand also has some of the highest levels of insurance in the world; consequently, the region is undergoing extensive "demolish and rebuild" rather than "repair," the scale of which is causing widespread disruption to residents and debris disposal issues (discussed in section What is Governance and Why is It Important?). Further, the Red Zone decision has meant the displacement and relocation of almost 8,000 households causing significant land use, infrastructure, housing, and investment changes.

It is within this context that the following sections outline the role of land use planning both pre and post event.

The Role of Land Use Planning in Earthquake Recovery

To ensure consistent outcomes are achieved during recovery, land use planning and emergency management objectives need to be integrated (Saunders et al. 2007). In New Zealand, emergency management is generally

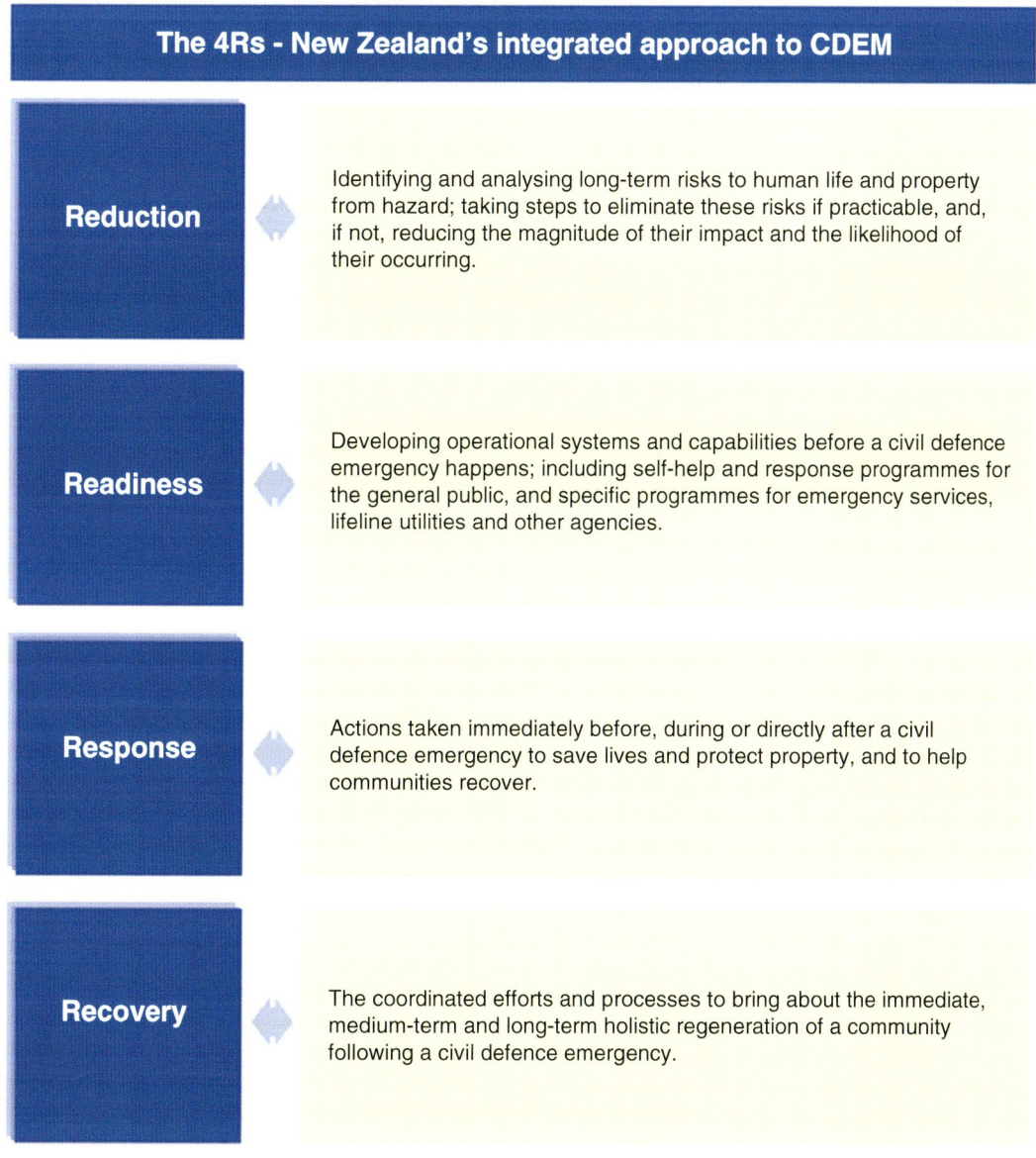

Land Use Planning Following an Earthquake Disaster, Fig. 1 The 4R's of integrated civil defense emergency management in New Zealand (MCDEM 2008, p. 5)

divided into four complementary phases, often referred to as the "4R's": reduction (i.e., prevention, mitigation), readiness (i.e., preparation), response, and recovery (see Fig. 1). Land use planning has a key role in the reduction and recovery aspects: reducing risks through avoidance and mitigation of natural hazards, both pre event and during the recovery phase.

How planning and planners respond to recovery is very much dependent on the governance arrangements in place – both pre and post event – as the following discussion outlines.

Role of Governance in Pre- and Post-event Land Use Planning

Disaster management systems around the world have evolved gradually, largely in response to disastrous events. Typically, the management of disasters, especially large-scale disasters, is a responsibility of governments, as societal institutions and life-sustaining systems are in crisis. During the last few decades, the disaster management practice has been characterized by frameworks and processes based upon a shared system of governance and policy making, with common and overlapping responsibilities apportioned among layers of government (May and Williams 1986). The approach is *ideally* structured to engage from the "bottom-up," with local governments having primary responsibility for supplying disaster-related resources and regional, subnational, and national agencies providing support as requested. By design, the system requires extensive coordination and cooperation among all levels of government, as well as with the many NGOs involved in disaster management, often in a tiered governance structure (national, state/regional, local) (Johnson and Mamula-Seadon 2014). The related legislation, policies, guidelines, and plans together help to promote a more integrated natural hazards risk management framework and are commonly based upon sustainable development principles (Mamula-Seadon 2009; Saunders and Beban 2012).

This "ideal" approach presumes integrated, decentralized, and deliberative planning and decision-making processes for land use, for development, and for comprehensive risk management, with the central government role defined as setting the direction and providing national policies, while the implementation matters were left to local government (Britton and Clark 2000; Mamula-Seadon 2009). However, recent societal and political changes, such as the emergence of public-private partnerships, outsourcing, and contracting, have meant that functions that may formerly have been carried out by public entities are now frequently dispersed among diverse sets of actors that include not only governmental institutions but also private sector and civil society entities.

What Is Governance and Why Is It Important?

Disaster governance takes a form of collaborative activities through "processes and structures of public policy decision making and management that engage people constructively across the boundaries of public agencies, levels of government, and/or the public, private and civic spheres in order to carry out a public purpose that could not otherwise be accomplished" (Emerson et al. 2012, p. 2). Implications for land use planning have yet to be fully understood though there is no doubt that the implications are important. Recent literature has started to demonstrate important connections between "procedural" (the process) and "substantial" (the plan) issues that plague peacetime planning, but which are exacerbated by the conditions of uncertainty and urgency that accompany disasters. For example, Kweit and Kweit (2004) compared recovery processes in Grand Forks (North Dakota, USA) and East Grand Forks following severe floods in 1997. After the disaster, East Grand Forks engaged in extensive citizen participation initiatives and subsequently reported on high levels of political stability and citizen satisfaction. In contrast, Grand Forks instigated a more top-down, bureaucratic approach and has since experienced changes to their government structure, a high turnover of elected and appointed officials, and more negative citizens' evaluations. Similarly, the faltering recovery of New Orleans post Katrina has been attributed, at least in part, to recovery planning processes that failed to engage local communities.

There is a lack of research detailing communities' planning efforts and aspirations, or their engagement with formal state representatives *post disaster*; however, a great deal of work has been conducted on orthodox "peacetime" models based on, for example, the International Association of Public Participation 2's spectrum (2014, see Fig. 2). This has a continuum of engagement/participatory practices that range from "token" or "passive" *informing* to consulting, involving, and collaborating, to "meaningful" or "active" *empowering* forms of participation where the agency agrees to implement the community's decisions. Though these provide useful guidelines, the post-disaster context does add a layer of complexity to these typologies, largely because normal state processes of engagement

	INFORM	**CONSULT**	**INVOLVE**	**COLLABORATE**	**EMPOWER**
PUBLIC PARTICIPATION GOAL	To provide the public with balanced and objective information to assist them in understanding the problems, alternatives and/or solutions.	To obtain public feedback on analysis, alternatives and/or decision.	To work directly with the public throughout the process to ensure that public issues and concerns are consistently understood and considered.	To partner with the public in each aspect of the decision including the development of alternatives and the identification of the preferred solution.	To place final decision-making in the hands of the public.
PROMISE TO THE PUBLIC	We will keep you informed.	We will keep you informed, listen to and acknowledge concerns and provide feedback on how public input influenced the decision.	We will work with you to ensure that your concerns and issues are directly reflected in the alternatives developed and provide feedback on how public input influenced the decision.	We will look to you for direct advice and innovation in formulating solutions and incorporate your advise and recommendations into the decisions to the maximum extent possible.	We will implement what you decide.
EXAMPLE TOOLS	• Fact sheets • Websites • Open houses	• Public comment • Focus groups • Surveys • Public meetings	• Workshops • Deliberate polling	• Citizen Advisory committees • Consensus-building • Participatory decision-making	• Citizen juries • Ballots • Delegated decisions

Increasing Level of Public Impact →

Land Use Planning Following an Earthquake Disaster, Fig. 2 IAP2's public participation spectrum (IAP2) (c) *International Association for Public Participation* www.iap2.org

may be suspended (formally under a state of national emergency or informally due to dysfunction).

Large-scale disasters deplete capital stock and services to a degree where complex rebuilding and societal activities (involving changes to governance) must happen in a compressed period of time. Governments often struggle with these and other time-compressed demands of post-disaster recovery, and as a result, preexisting governance structures – including those around the type and extent of public involvement – are modified or new structures are introduced to help provide more capacity, information, money, and other resources (Johnson and Olshansky 2013).

While these observations have been made, as yet, there has been little detailed study as to how governance structures and processes are often modified post disaster, the conditions by which governance transformations and new or modified institutions for managing recovery are likely to occur, or what the new structures are able to achieve compared to preexisting structures (Johnson and Mamula-Seadon 2014). Unlike top-down or bottom-up approaches to disaster management, the need for flexibility calls on governments and communities to adapt to, and negotiate, different roles, responsibilities, and opportunities seeking innovative strategies for state-civil society engagement. These offer dramatically improved approaches to building better relations, stimulating more effective activities, and sustaining a flexible and adaptable organizational response (Multi-national Resilience Policy Group 2013).

It is also important to note the disaster recovery planning is embedded in more general governance structures, playing out within ideological, socioeconomic, cultural, and environmental contexts. Examination of the Canterbury response and recovery process shows how post-disaster "time compression" can spotlight the strengths and weaknesses of disaster management processes and governance frameworks, including land use planning. For governance in general, the Canterbury transformations could be viewed as a rapid governance adaptation embedded within an overall trend in governance reforms underway in New Zealand (Johnson and Mamula-Seadon 2014). Before the earthquakes, the government had already instigated reforms to several key government policy frameworks and accompanying legislation, particularly the Resource Management Act 1991 and Local Government Act 2002, which represent two significant pieces of legislation that govern land use planning in New Zealand.

Ideally governance arrangements should allow for pre-event land use recovery planning to be undertaken. This type of planning requires political commitment and resources; however, the process can prove invaluable post event, as discussed in the following section.

Pre-event Planning for Post-disaster Recovery

While land use planners have a critical role to play in the disaster recovery process, they are often overlooked when preplanning for disaster recovery (Smith 2011). How land use planning proceeds following an earthquake is dependent on what level of pre-event planning has been undertaken before the disaster occurs. The recovery phase can be both prolonged and political; while the response phase might demand rapid technical assessments of the structural safety of a building, the recovery phase associated with that same building may involve interwoven and subjective issues of security, weather-tightness, aesthetics, or broader contextual factors based on infrastructure provision, out-migration of residents, traffic, and access to services and facilities, including schools. Consequently, it may take years, or even decades, for something like "recovery" to occur.

Once a disaster occurs, land use planning efforts for recovery should not be something new – ideally, some planning should begin prior to an event occurring. Planning for land use recovery before an event occurs is different from emergency preparedness and response planning. However, those involved in preparedness and response planning need to be aware of any pre-event recovery planning, and vice versa, to ensure consistency in decision making (Johnson and Olshansky 2013). Where post-disaster recovery is undertaken in an environment that can be detrimental to successful outcomes (e.g., the meaningful involvement of key stakeholders, including communities), pre-event recovery planning allows key stakeholders to invest time and resources to support and encourage cooperative behavior (Smith 2011). Planning for post-disaster recovery and reconstruction requires a collaborative approach to undertake hazard identification, risk assessment, public education, consensus and partnership building, visioning, and the identification of activities that reduce risks (Johnson and Olshansky 2013). A poorly planned or inefficient response to a disaster can impose additional social, economic, and environmental burdens on a community already impacted by the disaster (Johnston et al. 2009).

An example of a framework for pre-event planning is provided in Fig. 3, from New Zealand. This framework has been based on the Standards Australia/New Zealand 31000:2009 Risk Management standard.

There are many opportunities for pre-event planning prior to an event. Questions that can be addressed include:

- Are current land use activities appropriate for their locations? What land is available for relocating particular land use activities?
- What are the social, cultural, economic, and other implications of the land use?
- What aspects of "place" do the community highly value, e.g., historic and cultural character?
- Are there any restrictions (e.g., other natural hazards, sites of cultural significance, building requirements) on the land?

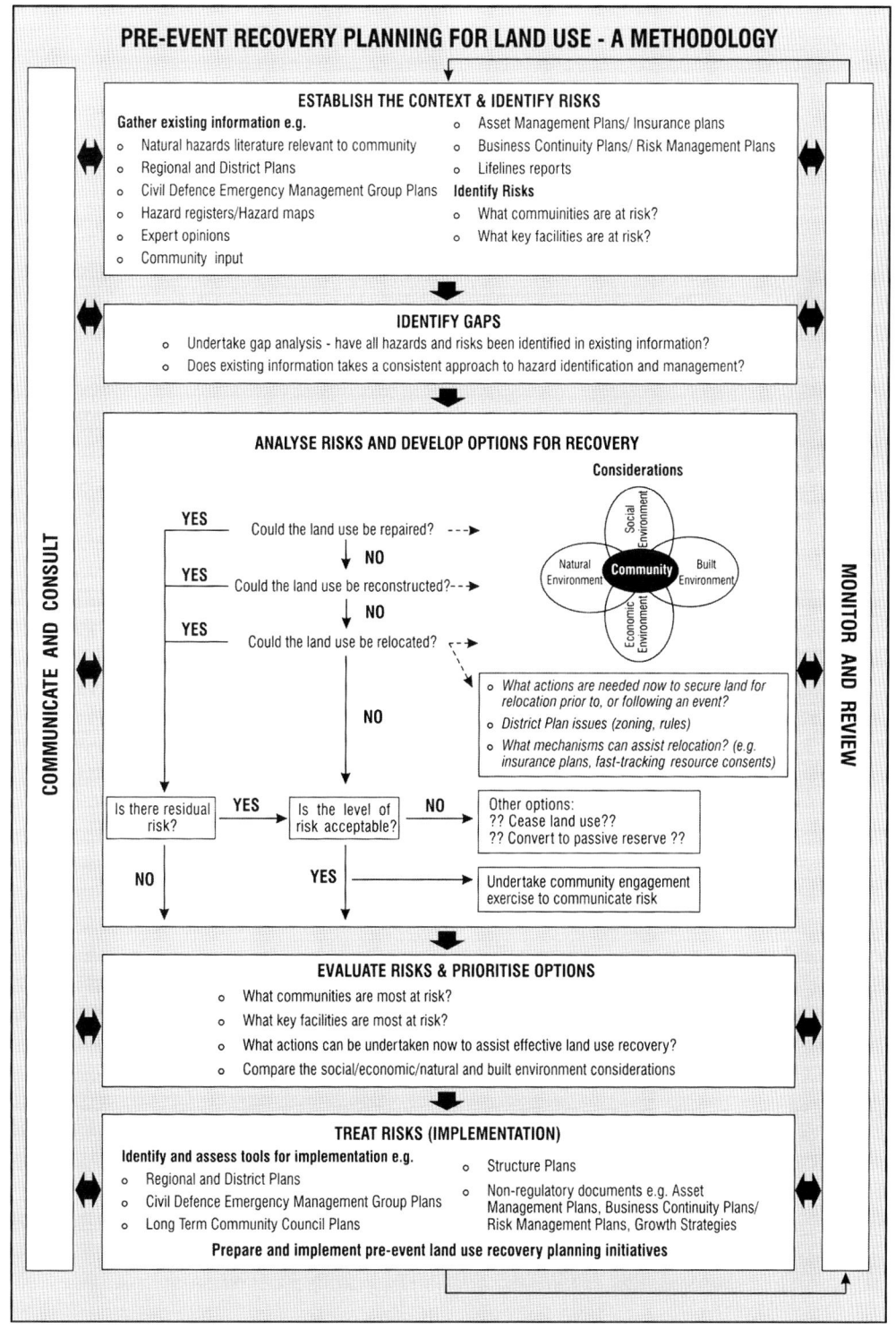

Land Use Planning Following an Earthquake Disaster, Fig. 3 A framework for pre-event planning for post-disaster recovery from New Zealand (Becker et al. 2011, p. 538)

- Do current land use planning policies adequately address the potential for relocating land uses?
- Can the regulatory planning process for (re)development applications be streamlined?
- Where will debris from damaged/demolished buildings, landslides, liquefaction, etc., be disposed of? Consideration needs to be given to:
  – Contaminated material, green material, wood versus concrete, etc.
  – Overflow sites, or prearranging for debris to be transported to another location outside of disaster area
  – The location of staging and sorting sites
- Who are the key stakeholders and community members that need to be involved in preplanning and do they have a recognized role in, for example, civil defense and emergency management programs?
- How healthy are democratic and participatory processes and to what extent is decision-making distributed among different stakeholders?
- What are the short- and long-term planning requirements?
- Can any land use activities be relocated before an event occurs in order to reduce the risk?

There is a significant amount of international research and reporting on the value of pre-event planning for a range of natural hazards, including earthquake. It has also been recognized that preplanning for disaster recovery is important. An example of preplanning for debris disposal following an earthquake is that from New Zealand, where Johnston et al. (2009) have outlined 10 key steps that can be planned by both emergency managers and land use planners, ahead of time. These steps, with those key positions involved in each step, are shown in Table 1.

As shown in the example in Table 1, preplanning for a disaster requires many different parties to be involved in the development of the plan: in this case, emergency managers, planners, communication, and enforcement officers. Undertaking a multidisciplinary approach to pre-event planning, it results in a more comprehensive and robust planning process and outcome.

Pre-event planning can be valuable in building relationships between all of the involved parties, to produce a shared vision and responsibilities for completing pre-event planning tasks, rather than producing the actual plan. As shown in Table 1, building the relationships between the emergency managers, planners, and communication team pre event will mean valuable time is not required to develop these relationships post event. While pre-event planning for recovery should be undertaken before a hazard event occurs, planning can also occur after the event, either before or even while major reconstruction begins. Examples of this type of planning following earthquakes in China, Alaska, and Japan are outlined in Becker et al (2011).

Best practice around both pre-event and post-disaster land use planning shares some similarities, particularly in terms of emphasizing participatory processes. However, much of the theory around disaster-recovery land use planning either belies or underestimates the complexities involved. Recovery planning is unlike "peacetime" planning in that the landscape – geographical, social, economic, political, and conceptual – is likely to be changing rapidly even as the plans are developing. If a disaster is defined as an event which exceeds the ability of a system to cope with the disturbance, causing a loss of functionality, then "chaos," "confusion," and "disorder" are trademarks of the response phase. It is this dysfunction that arguably separates disaster from mere misfortune.

In restoring order – or functionality – and transitioning to the recovery phase, it can be helpful to draw upon pre-event comprehensive land use plans and strategies. This ensures that key principles and concerns embedded in "peacetime" (i.e., business as usual) documents continue to guide recovery planning, even if those plans and strategies need extensive modification.

If pre-event plans and strategies do require significant changes, recovery planners will be faced with some key challenges. The obvious need to act with urgency must be balanced against deliberative, inclusive, and participatory processes that are often seen as costly in terms of

Land Use Planning Following an Earthquake Disaster, Table 1 Ten steps for preplanning for debris disposal (Adapted from Johnston et al. 2009)

Key step	Positions involved
1. Develop debris disposal plans at a regional and local level, or make appropriate provisions in existing solid waste management plans. Plans should include a detailed strategy for debris collection, temporary storage and staging areas, recycling, disposal, hazardous waste identification and handling, administration, and communication with the public	Emergency managers
	Planners (i.e., making provisions in existing plans)
	Communication team
2. Identify potential locations or sites for the temporary and/or permanent disposal of debris. Outline a process for final site selection/confirmation in response to post-event requirements in both regional and local plans	Planners
	Emergency managers
3. Prepare a communication strategy ahead of time. You will need to tell your community when, where, and how normal rubbish collection will resume and give special instructions for reporting and sorting disaster debris	Communication team
4. Prepare for increased demands on council staff in terms of public information, operation, and enforcement of debris management systems. This may require additional staff resources to manage the increased workload	Communication team
	Enforcement officers
	Planners
	Emergency managers
5. Make arrangements for additional equipment and supplies to deal with disposal ahead of time. Identify the types of equipment and supplies that staff and contractors will need to carry out operations	Emergency managers
6. Develop mutual aid agreements with neighboring councils (and/or CDEM groups) for plant, equipment, and expertise	Emergency managers
7. Establish pre-event debris management contracts with private contractors	Emergency managers
8. Establish possible organizational structures, roles and responsibilities, and authority between various stakeholders, regulatory authorities, and decision-makers including Ministry for the Environment, Department of Labour, Ministry of Health, and within local authorities	Emergency managers
	Planners
9. Determine financing strategies at local, regional, and national level for differing sizes and types of disaster events	Emergency managers
10. Assess and develop an approach to the potential impact of disaster debris management on normal environmental processes and standards. For example, undertake a Strategic Environmental Assessment identifying practical strategies to minimize environmental and social impacts	Planners

time and other resources. An example concerns the demolition of damaged, and therefore hazardous, buildings versus the preservation of valued and meaningful heritage sites. "Participation" speaks to the notion that engineers, while experts in determining structural safety, may not be best placed to identify a community's iconic buildings or to assess a building's importance. There is now enormous debate about the extent to which seemingly apolitical structural or geotechnical matters impact on communities and how both urgent safety concerns and high-level strategic documents can be addressed collaboratively given the degree of trauma experienced by the affected population, the extreme need for timeliness, and, often, a lack of information and other resources.

In summary, pre-event planning for post-disaster recovery can be undertaken before an event occurs, or post-event before reconstruction begins. Pre-event planning should address both short- and long-term needs and involve key stakeholders, including community members. It should be adaptable to post-event realities – the importance of the preplanning is in relationship building and the active participation of key stakeholders.

The Reality of Planning Post-earthquake: The Canterbury Experience

Following the February 2011 earthquake, it was seen as necessary to create an administrative body to oversee recovery in the post-quake Canterbury region. There were four key reasons for this decision:

1. The earthquake sequence was prolonged with new fault lines identified and significant aftershocks still occurring 2 years after the first event
2. The effects of the sequence were geographically widespread, cutting across four administrative boundaries: the Canterbury Regional Council (ECan), the Christchurch City Council (CCC), the Waimakariri District Council (WMK), and, to a lesser extent, Selwyn District Council (SDC).
3. With a rebuild cost of over $40 billion, an external coordinating body was seen as necessary to manage the "the political and fiscal risks involved" (Brookie 2012, p. 28).
4. There were concerns that the Christchurch City Council was not able to cope adequately with the scale of the disaster. Councillors were publicly describing the council as "dysfunctional" and "lacking leadership," and a divide between some of the elected representatives, the Mayor, and the bureaucracy was becoming apparent (Sachdeva 2011).

Consequently, in April 2011 the Canterbury Earthquake Recovery Authority (CERA) was created by an Act of Parliament. Essentially a government department, CERA is responsible to the Minister for Earthquake Recovery, Hon. Gerry Brownlee. CERA is a controversial entity because, through the Canterbury Earthquake Recovery (CER) Act of 2011, the Minister has extraordinary powers to expedite rapid decision-making. While this suits those who want to "get on with it," others are concerned about the suspension of citizens' rights and the lack of a layer of governance between CERA and the central government National Party. The CER Act (2011) gives the Minister the ability to suspend, amend, cancel, (or) delay ECan, CCC, or WMK plans or policies, and further, these local government recovery plans must be consistent with CERA's recovery strategy and signed off by the Minister (Brookie 2012). Among other rights, CERA's chief executive has the authority to do the following:

- Require councils to act as directed and or to provide information on request
- Amend or revoke Resource Management Act documents and city plans (i.e., peacetime land use legislation)
- Close or otherwise restrict access to roads and other geographical areas
- Demolish buildings, or otherwise enter and deal with people's land and property (with notice, in the case of Marae and dwelling houses)
- Require compliance of any person with a direction made under the Act

The CER Act thus forms a controversial legislative and political space where the ends justify whatever means – within which more detailed plans and strategies have been formed. We turn now to a more detailed description of five of these including:

1. The Greater Christchurch Recovery Strategy (GCRS) which is a high-level document providing *strategic direction and principles*
2. The pre-quake Greater Christchurch Urban Development Strategy (UDS) and the post-quake Land Use Recovery Plan (LURP) which has a *regional* focus
3. The Christchurch City Plan covering *metropolitan* Christchurch;
4. The Christchurch Central Recovery Plan (CCRP) which covers the Christchurch *CBD*;
5. The Waimakariri District Council's Integrated and Community-based Recovery Framework which guided the infrastructure rebuild program in Kaiapoi

The Greater Christchurch Recovery Strategy (GCRS)

The GCRS was approved by the Canterbury Earthquake (Recovery Strategy Approval) Order 2012 made by the Governor-General. The strategy (Canterbury Earthquake Recovery Authority 2012):

IMMEDIATE

Repair, patch and plan
- Provide basic human needs and support services.
- Address health and safety issues.
- Make safe or demolish unsafe and damaged buildings and structures.
- Investigate, scope and initiate recovery programmes and initiatives.
- Plan for land use and settlement patterns so land can be made available for displaced residents.
- Conduct ongoing programme of investigation and research to understand the geotechnical issues and seismic conditions. Use this information to guide recovery activities and decisions on land suitability for rebuilding.

SHORT TERM

Begin to rebuild, replace and reconstruct
- Engage both established and new communities and inform them about rebuilding and future planning.
- Establish new social and health support and service delivery models.
- Continue demolition of damaged buildings.
- Continue repair and rebuild.
- Deliver early projects to instil confidence.
- Planning and supporting community resilience.
- Begin replacement activity.
- Begin restoration and adaptive reuse of heritage features.
- Continue, monitor and review recovery.

MEDIUM TO LONGER TERM

Construct, restore and improve
- Continue to build resilient communities.
- Continue reconstruction.
- Major construction projects are underway.
- Complete restoration and adaptive reuse of heritage features.
- Phase out recovery organisations.
- Economy is growing and businesses are sustainable.
- Labour market is active and attracting employees.

Land Use Planning Following an Earthquake Disaster, Fig. 4 Typical phases of recovery (Canterbury Earthquake Recovery Authority 2012, p. 14)

- Is a high-level document which defines what "recovery" means for greater Christchurch
- Establishes principles to guide how CERA and other agencies will work together towards recovery
- Describes in broad terms the pace and phases of recovery
- Identifies work programs and which organizations will lead specific projects
- Identifies priorities for recovery efforts
- Sets up governance structures to oversee and coordinate the work programs and links them to wider initiatives
- Commits to measuring and reporting on progress towards recovery

In particular, the plan sets out a framework and provides delivery mechanisms necessary to rebuild existing communities, develop new communities, meet the land use needs of commercial and industrial businesses, rebuild and develop the infrastructure needed to support these activities, and take account of natural hazards and environmental constraints that may affect rebuilding and recovery.

The strategy identifies six components of recovery – cultural, social, built, natural (including geotechnical considerations), and economic. These components link together with the sixth component – community – at the center. Each requires leadership and integration.

The "recovery" process is then divided into three phases (see Fig. 4) – immediate, short-term, and long-term – with each phase assigned milestones and timelines. Immediate actions, like "repairing, patching, and planning," for example, would ideally have been completed within a year of the first earthquake. Long-term "constructing, restoring, and improving," on the other hand, may take until 2020 or beyond.

As an example of a recovery strategy, this document does give a good overview of six components that contribute to recovery (i.e., cultural, social, built, natural, economic, and community), and these components should be considered by others in post-disaster recovery planning. It also provides a good indication of the sorts of recovery milestones associated with different stages of recovery that may usefully serve as a guide for other recovery agencies. On the other hand, implementation is key, and one of the main challenges has been to build both the capability and capacity necessary to deliver on the strategy. Much of the strategy's implementation has been overseen by CERA, but implemented by other local authorities (see Fig. 5) using a blend of both new and existing plans and processes.

The Pre-quake Urban Development Strategy and the Land Use Recovery Plan

In a pre-quake process initiated in 2004 by the Canterbury Regional Council (ECan), the Greater Christchurch Urban Development Strategy (UDS) (Greater Christchurch 2013) was the result of 3 years of collaboration between stakeholders and communities. Based on over 3,250 submissions, the UDS sought to provide a framework for future growth management taking into account risk assessments, transportation, housing demand, and so on. Four options were presented that essentially represented unfettered sprawl at one end, to extreme containment and intensification around core suburban centers and townships in the districts at the other. Discussion around these four options resulted in a vision, guiding principles, strategic directions, and a framework for implementation that was given statutory weight in chapter 12 "Development of Greater Christchurch" of ECan's Regional Policy Statement. This, in turn, guided land use patterns – including areas that could be zoned for residential, commercial, and retail activities – in city- and district-level plans developed by the Christchurch City (CCC), Waimakariri District (WMK), and Selwyn District (SDC) Councils.

The UDS had already signaled areas in the region suitable for future growth when the earthquakes and subsequent Red Zoning decision displaced almost 8,000 households. The comprehensive planning behind the UDS allowed for a fairly rapid appraisal of land that could be made available to accommodate the demand for new housing. However, the UDS was designed to be implemented over a period of 30 years and considered inadequate for the post-quake context. Consequently, in November 2012, the Minister for Canterbury Earthquake Recovery directed ECan (at their request) to prepare a Draft Land Use Recovery Plan (LURP) through a collaborative multiagency approach involving all of the earthquake recovery strategic partners (including the three local councils, Ngāi Tahu, and the New Zealand Transport Agency), with input from key stakeholders and the wider community.

Some of the key issues included tensions between market-led development and regulation, leadership, certainty, council processes, brownfield development, urban design, housing affordability, infrastructure, public transport, and suburban centers (AERU, 2012). Further, although the LURP had been through several consultative rounds, as the agency responsible for the city plan that will give effect to the LURP, the CCC was particularly concerned that residents had not been sufficiently informed about, for example, the new "design-led" Comprehensive Development Mechanisms (CDMs). These would see high-density housing (defined within the Canterbury cultural context as between 30 and 65 households per hectare) throughout the city. Given that even medium-density housing has a controversial history in Christchurch, the CCC argued for CDMs to be limited to Living 2 and 3 Zones (with minimum site sizes of 330 m^2 and 300 m^2, respectively). As so much of the city's public (social) housing stock has been damaged or demolished, a new Community Housing Development Mechanism (CHDM) has also been introduced to facilitate the urgent rebuilding of social housing stock in the city. These are likely to be located in areas that already accommodate such housing. This, too, is likely to be controversial, though criticism may be mitigated somewhat by the review of both mechanisms that is to take place after 5 years of implementation. Nonetheless, tension between the CCC and CERA over CDMs and CHDMs has climbed to very high levels.

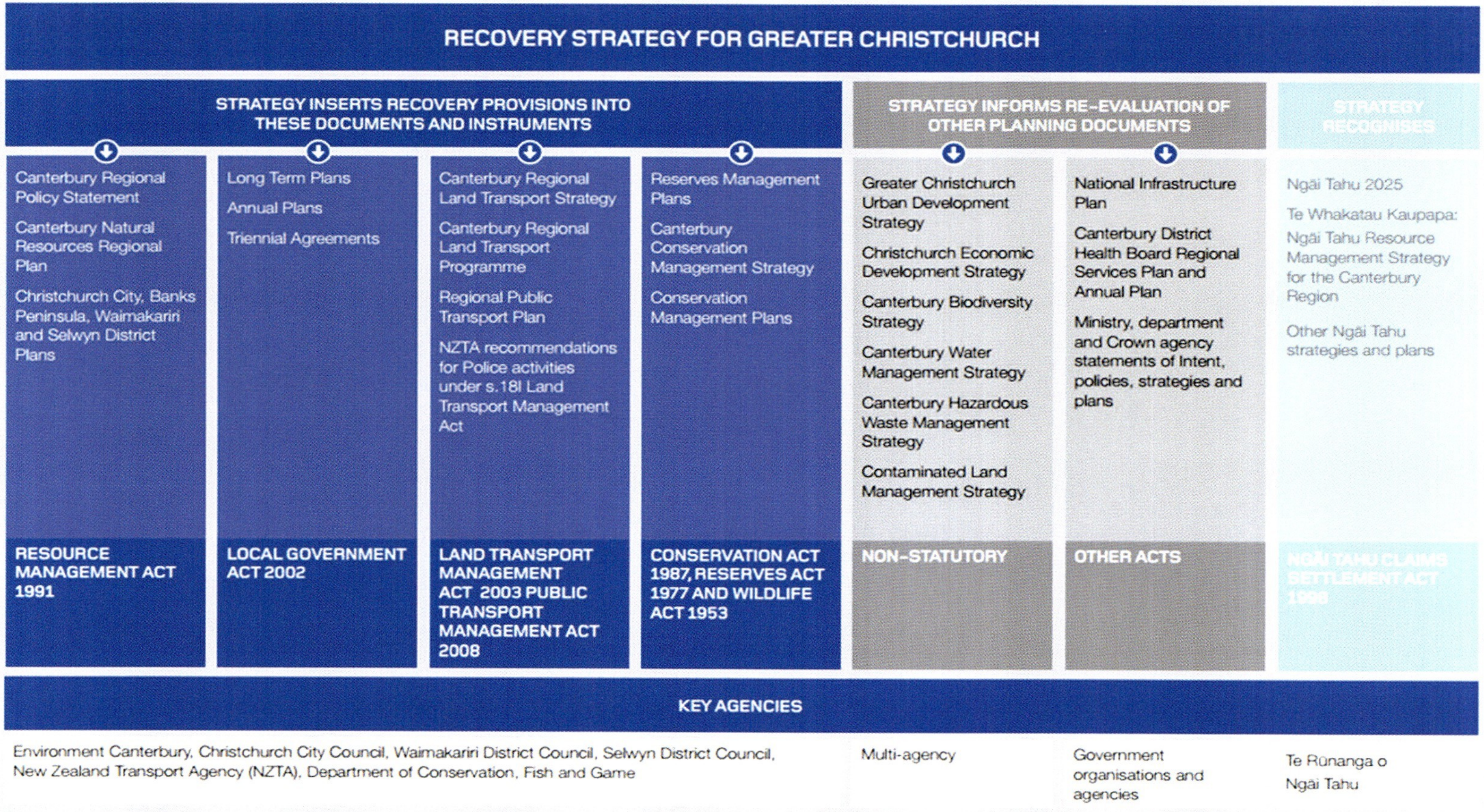

Land Use Planning Following an Earthquake Disaster, Fig. 5 The relationship between the recovery strategy and other strategies, policies, and plans for greater Christchurch (Canterbury Earthquake Recovery Authority 2012, p. 22)

The LURP also identifies areas deemed suitable for greenfield development, and with the Minister's approval, statutory effect will be achieved through the insertion of chapter 6 "Recovery and Rebuilding of Greater Christchurch" into the Canterbury Regional Policy Statement. Chapter 6 will then inform district and city plans that more specifically determine the location, type, and mix of business and residential activities within specific geographical areas based on hazard identification, transportation routes, and migration patterns. Through this blend of regional, district, and city plans, the LURP is a significant document that has adopted much of the pre-quake UDS vision; however, it has been described as a "developer-friendly" document that gives the market considerable scope. It has also been criticized for promoting – through peri-urban greenfield and suburban mall development – a further hollowing out of the inner city, the rebuild of which falls under the Central City Recovery Plan (discussed further in section The (Christchurch) Central City Recovery Plan).

The Christchurch City Plan

City and district plans are a pre-quake requirement of all territorial local authorities in New Zealand under the Resource Management Act (1991). These local council plans provide a framework for land use and subdivision in a manner that is consistent with the Regional Council's Plans and Policy Statements (e.g., the UDS/LURP) and are reviewed every 10 years. Christchurch's City Plan was due for review when the quake sequence began and was subsequently delayed until 2013.

Given the extent of the changes that will need to be made to the plan, the council has decided to undertake the review in two stages with the first notification due in 2014 and set to address overall strategic direction; residential, commercial, and industrial land use; subdivisions (e.g., minimum site size); natural hazards; transport; and contaminated land. The second round (in 2015) will notify issues around heritage and the natural environment, the coast, public open space, and so on. Submissions can be made as the review takes place or once the draft is notified.

While the council has signaled the process may take 3 years, some of the proposed changes – such as the CDMs, CHDMs, and greenfield development mentioned above – are likely to be controversial as are other issues around, for example, maximum height and parking restrictions, "permitted" activities such as light retail, and so on.

The (Christchurch) Central City Recovery Plan

Following the February 22, 2011, earthquake – which led to the demolition of almost 80 % of building stock in the CBD core (Markham 2013) – the Christchurch City Council began the task of developing a recovery plan for the central city. In May of 2011, the CCC initiated the 6-week *Share an Idea* campaign which, according to the CCC website (Christchurch City Council 2013a), allowed "the public to tell us their ideas about how the Central City should be redeveloped to be a great place again...." Combining a weekend of on-site opportunities and an online crowd-sourcing tool, the campaign generated more than 106,000 ideas from over 60,000 participants. These ideas were analyzed so as to inform the draft Central City Plan and guide inner city development for the next 10–20 years.

Share an Idea was followed by another participatory tool – the *48 h Design Challenge* – run by the CCC at Lincoln University. This provided an opportunity for the council to work with the design and architecture industry and practically test the draft Central City Plan over seven preselected sites. A total of 15 teams comprising engineers, planners, urban designers, architects and landscape architects, and students took part in the *Challenge*.

Following these exercises and other, smaller workshops, the Draft CCP was delivered to the Mayor to sign off in August of 2011 before being sent to the Minister for Earthquake Recovery for his approval. Submissions were invited and 79 comments were received. As outlined on the council's website (Christchurch City Council 2013b), the Minister's view was that "taking into account its impact, effect and funding implications, [the Plan] could not be approved without amendment...given insufficient information...on how the Recovery Plan would be implemented and changes to the District Plan that [are] unnecessarily complex."

Consequently, the Minister established a special unit within CERA, the Christchurch Central Development Unit (CCDU) which was to work with a "professional consortium" to deliver, within 100 days, the Central City Plan and take a lead role in its implementation alongside the CCC, Te Rūnanga o Ngāi Tahu, and other key stakeholders. The CCRP is based on a master-planned blueprint for the CBD providing street layout, the location of "anchor" projects (including a convention centre and sports facility), and various precincts. These they hope will stimulate further investment and development (though this has been hamstrung by the need to meet new building standards, current requirements for parking spaces, and so on, often with insurance pay outs that only covered the replacement value of old buildings).

Given the responsibility for the 100 Day Plan was taken from the CCC and given to a nonelected, newly created entity – the CCDU – amidst controversy around compulsory land acquisition to enable the anchor projects to go ahead, the level of public involvement may have been limited. Though the plan's executive summary states that "The city centre [will be]...a reflection of where we have come from, and a vision of what we want to become" (Central City Development Unit 2012, p. 4), it has been suggested that, in practice, the process has been "top-down, centralised, and highly bureaucratic" (in Brookie 2012, p. 29). Drawing on the IAP2's spectrum of participation (see Fig. 2 above), it could be argued that much of the "involvement" behind the plan's development is limited to "informative and consultative" rather than "collaborative and empowered."

Underlying the attempt to design a Central City Plan within 100 days is a sense that urgency both demands and justifies the suspension of peacetime processes around public engagement, transparency, and accountability. It is claimed, for example, that "The experience of other cities after a natural disaster shows that substantial redevelopment must start within 3 years if recovery is to be successful. One year has passed. Speed is of the essence..." (Central City Development Unit 2012, p. 4). Consequently, the policies and procedures for land acquisition have been both praised as providing swifter certainty for homeowners and businesses to relocate and recover and criticized as having evolved over time with often reactive and unclear reasoning. Similarly, while the CCDU has been credited with accelerating the design and implementation of the CBD rebuild, concerns have also been raised about the acceptance of development applications without the establishment of clear criteria for screening and selecting among competing applications (McDonald 2013).

This raises questions about whether it is possible to effectively and adequately reconcile the conflict between the perceived need for urgent decision-making and action, on one hand, and robust process on the other. In exploring possible answers to this question, we turn now to an example of recovery best practice: The Waimakariri District Council's Integrated and Community-based Recovery Framework.

The Waimakariri District Council's Integrated and Community-Based Recovery Framework and the Kaiapoi Infrastructure Rebuild

The Waimakariri District is adjacent to the city of Christchurch with a population of about 50,000 people. It has traditionally been described as agricultural with large farms, lifestyle blocks, and smallholdings devoted to horticulture; however, a few settlements have grown rapidly and now form a commuter corridor of townships, with many residents traveling to work in Christchurch each day.

After the September 2010 earthquake, parts of the Kaiapoi township suffered extensive land damage from liquefaction and lateral spread, and over 1,000 households and the town center were severely affected. Besides serious damage to a quarter of its housing stock, Kaiapoi also lost its library, aquatic center, community halls, churches, the movie theater, bars, and cafes. Over 25% of businesses were immediately affected and there have been significant relocations and a number of permanent closures. Much of the infrastructure was affected leaving 5,000 people without water and sewer. The damage, *proportionally*, was similar to that experienced in Christchurch.

Despite it being a Saturday, many WMK staff members came to work either to assume their post in the Emergency Operations Centre or to volunteer to help in Kaiapoi. By late afternoon, council staff members were distributing information on foot to affected residents. Other departures from "business as usual" started taking place, some of which meant literally and figuratively "stepping across a boundary." As reported by a WMK senior manager (see Vallance 2013 for the full document):

> Traditionally councils do not step across the homeowner's boundary and any infrastructure issues between the house and the front boundary is the home-owner's problem. But [due to demand] post-earthquake it would have been impossible to just call a plumber to get the issue fixed. So we [Waimakariri District Council] made a decision fairly early on to liaise with [and secure funding from] the Earthquake Commission and coordinate repairs *across the boundary* because there's no point us fixing our side of the sewer and people still not being able to use [the toilet] because the pipe between the house and the boundary is broken (italics added).

Other councils did not take on this coordination role as quickly and, as a result, some of their residents suffered for months with poor sanitation.

Other initiatives included three new Recovery Manager positions. The Social Recovery Manager (previously the council's peacetime Community Team leader) was moved from the Rangiora offices to Kaiapoi to "colocate" with engineering and utilities council staff and contractors at the newly established Recovery Assistance Centre. She began acting as a liaison between engineers and contractors and Kaiapoi residents, local NGOs, and community groups. This communication between, and early involvement of, locals significantly shaped subsequent decisions to undertake the project management of what was thought to be (at the time) a neighborhood infrastructure rebuild program.

The neighborhood infrastructure rebuild program – coordinated by the newly appointed Infrastructure Recovery Manager – served about 1,200 households, broken down into smaller clusters for rebuild in different stages. It was argued that 1,200 was about the "right amount" and that if the project had been much larger, a different approach may have been required.

Importantly, the communication with residents around this staged rebuild was both deliberative and inclusive and began with numerous public meetings. These included "Q and A sessions" where, often, residents were seated according to the street in which they lived so they could get to know each other (for support) and triangulate information at a very local level. Elected councillors and senior managers from the council (often including the CEO and/or the Mayor) along with representatives from the insurance companies and geotechnical experts were also present. These meetings were well documented and the notes turned into a booklet that was then put on the council's New Foundations website that had been established solely to disseminate earthquake-related information.

The council then published an order of works with timeframes (see Fig. 6). This was critical as it allowed affected residents to plan ahead for their temporary relocation. However, it is noted that events overtook the feasibility of that particular plan, so it was not implemented. The council also helped coordinate the provision of temporary housing located on council reserve land to accommodate residents displaced by the rebuild. Residents who were "last in the queue" of the staged rebuild were supported by the volunteer-led pastoral care team who liaised weekly with the Social Recovery Manager and who now shared a port-a-com with the Infrastructure Recovery Manager at "the Hub" in Kaiapoi.

The Hub was designed to be a "one-stop recovery shop" and was located in the council-owned community hall in Kaiapoi. It also housed the council's pre-quake Kaiapoi Service Centre staff, some of the council's Building Unit and Community Team usually located in Rangiora, Work and Income New Zealand, the Inland Revenue Department, local NGOs providing business

Land Use Planning Following an Earthquake Disaster, Fig. 6 The Waimakariri District Council's Schedule of Works for Kaiapoi (www.newfoundations.org.nz)

and whānau support, psychosocial and pastoral care teams, a tenancy service, Fletchers (one of the approved building repair companies), and the 15 Waimakariri Earthquake Support Service (WESS) coordinators who, in the first 2 years, assisted between 400 and 600 cases with accommodation, help with earthquake claims, legal aid, and counseling for earthquake stress. A dedicated earthquake communications manager was also appointed and colocated in the Hub. This colocation, right in the heart of Kaiapoi, enabled engineering expertise to reach residents and community intelligence to inform infrastructural recovery. The communication between the Social Recovery and Infrastructure Recovery Managers helped the engineering and utilities teams to appreciate the positive and negative impacts of their work and for residents to better understand what engineers and other experts were doing, and why.

While it is difficult to make direct comparisons between Christchurch and Waimakariri given the different contexts, types of plans, and extent of damage, satisfaction with the Waimakariri District Council's recovery framework for the first 3 years is relatively high. Indeed, as reported in Vallance (2012), a Residents' Association representative said:

> There is a very good understanding on behalf of the majority of the community that the Waimakariri District Council understood the community's needs, that recovery was shared with the community so the community were engaged in different ways. If you want an example of recovery best practice, you've got it right there.

Wider support for their framework may also be reflected in the local government elections of 2013 where 9 of 10 councillors were reelected, with the Mayor unchallenged. In Christchurch, the results were quite different with only 4 of 13 representatives reelected (with the Mayor and two others resigning). There are, therefore, some interesting parallels between these results and those documented by Kweit and Kweit (2004), reported above.

Summary

This chapter has outlined the role of land use planning during the complex and demanding recovery phase, with examples from the Canterbury region in New Zealand. Land use planning has a key role to play in the recovery phase; however, to strengthen this role, planning needs to be undertaken before an event occurs.

While any pre-event plan will need to be adapted or even rewritten post an event, the value is in the process of bringing key stakeholders together to discuss issues, agree on common goals, and form relationships that will be key during a recovery phase. Pre-event recovery planning can be useful as a means of rapidly appraising available land, as the Urban Development Strategy showed; however, such plans need to be flexible and adaptable. The Greater Christchurch Recovery Strategy highlights the value of developing an overall guiding strategy, but building capability and capacity to deliver on the strategy can be difficult and lead to high levels of tension between different recovery agencies. To support pre-event recovery planning for land use, governance arrangements need to be in place that enable pre-event planning and support the planning process post event when time compression factors influence the "business as usual" community consultation and engagement process.

As the five case studies from Canterbury have shown, the reality of recovery is complex, involving political intervention from a national level into local decision-making and governance arrangements. This in turn has directed how recovery plans have been produced, the involvement of the community in their development, and their outcomes. The Canterbury earthquake case study highlights some important lessons for those involved in recovery planning and the relationship between process and reality.

Cross-References

▶ Building Disaster Resiliency Through Disaster Risk Management Master Planning
▶ Community Recovery Following Earthquake Disasters
▶ Earthquake Disaster Recovery: Leadership and Governance
▶ Learning from Earthquake Disasters
▶ Legislation Changes Following Earthquake Disasters

References

Anderson V (2011) Tuesday quake 'no aftershock'. The Press, Christchurch

Becker JS, Saunders WSA et al (2011) Preplanning for recovery. In: Miller DS, Rivera JD (eds) Community disaster recovery and resiliency: exploring global opportunities and challenges. Taylor & Francis, Boca Raton, pp 525–550

Britton NR, Clark GJ (2000). Non regulatory approaches to earthquake risk reduction: the New Zealand experience. In: 12th world conference on earthquake engineering, Auckland

Brookie R (2012) Governing the recovery from the Canterbury Earthquakes 2010–11: the debate over institutional design. Institute for Governance and Policy Studies, Victoria University, Wellington

Canterbury Earthquake Recovery Authority (2012) Recovery strategy for Greater Christchurch Mahere Haumanutanga o Waitaha. Canterbury Earthquake Recovery Authority, Christchurch

Central City Development Unit (2012) The plan. http://ccdu.govt.nz/the-plan. Retrieved Dec 2013

Christchurch City Council (2013a) Share an idea. http://www.ccc.govt.nz/homeliving/civildefence/chchearthquake/ShareAnIdea.aspx. Retrieved 15 Dec 2013

Christchurch City Council (2013b) Central city. http://www.ccc.govt.nz/homeliving/civildefence/chchearthquake/centralcityplan.aspx. Retrieved 15 Dec 2013

Emerson K, Nabatchi T et al (2012) An integrated framework for collaborative governance. J Public Adm Res Theory 22(1):1–29

Greater Christchurch (2013) The greater Christchurch urban development strategy. http://www.greaterchristchurch.org.nz/. Retrieved 1 Dec 2013

IAP2 (2014) Public participation spectrum. http://www.iap2.org.au/documents/item/84. Retrieved 7 Jan 2014

Johnson L, Mamula-Seadon L (2014) Transforming governance: organizing for and managing disaster recovery following the September 4, 2010 and February 22, 2011 Canterbury earthquakes. Earthquake Spectra

(Special Edition – Canterbury Earthquakes) Vol 30, No 1, pp 577–605
Johnson LA, Olshansky RB (2013) The road to recovery: governing post-disaster reconstruction. Land Lines (July), The Lincoln Institute of Land Policy, https://www.lincolninst.edu/pubs/2259_The-Road-to-Recovery–Governing-PostDisaster-Reconstruction, p 8
Johnston D, Dolan L et al (2009) Disposal of debris following urban earthquakes: guiding the development of comprehensive pre-event plans, GNS science report 2009/33. GNS Science, Lower Hutt, p 27
Kweit M, Kweit R (2004) Citizen participation and citizen evaluation in disaster recovery. Am Rev Public Adm 34(4):354–373
Mamula-Seadon L (2009) CDEM, integrated planning and resilience: what is the connection? Tephra Commun Resil: Res, Plan Civ Def Emerg Manag 22:3–8
March A, Henry S (2007) A better future from imagining the worst: land use planning and training responses to natural disaster. Aust J Emerg Manag 22(3):17–22
Markham S (2013) The Waimakariri District Council's recover approach and programme. Otago CDEM Recovery Workshop, Alexandra
May PJ, Williams W (1986) Disaster policy implementation: managing programs under shared governance. Plenum, New York
MCDEM (2008) National civil defence emergency management strategy 2007. Department of Internal Affairs, Wellington
McDonald L (2013) Call for action on mall impasse. The Press, Christchurch
Multi-national Resilience Policy Group (2013) The New Zealand discussion – notes from meeting, Wellington/Christchurch
Sachdeva S (2011) 'I'm off', says council identity Williams. The Press, Christchurch
Saunders WSA, Beban JG (2012) Putting R(isk) in the RMA: technical advisory group recommendations on the resource management act 1991 and implications for natural hazards planning, GNS science miscellaneous series 48. GNS Science, Lower Hutt, p 52
Saunders WSA, Forsyth J et al (2007) Strengthening linkages between land-use planning and emergency management in New Zealand. Aust J Emerg Manag 22(1):36–43
Smith G (2011) Planning for post-disaster recovery: a review of the United States disaster assistance framework. Public Entity Risk Institute, Fairfax
Vallance S (2012) Urban resilience: Bouncing back, coping, thriving. A paper presented at the Australian & New Zealand Disaster and Emergency Management Conference, Brisbane – 16 – 18, April 2012
Vallance S (2013) The Waimakariri District Council's Integrated Community-based Recovery Framework. Department of Environmental Management, PO Box 84, Lincoln University, 85084

Laser-Based Structural Health Monitoring

Hoon Sohn and Byeongjin Park
Department of Civil and Environmental Engineering, Korea Advanced Institute of Science and Technology (KAIST), Yuseong-Gu, Daejeon, Republic of Korea

Synonyms

Damage detection; Laser Doppler vibrometer; Laser scanning; Laser thermography; Laser ultrasonics; Light detection and ranging; Nondestructive testing; Structural health monitoring

Introduction

With an average occurrence rate of 20,000 events worldwide every year, earthquakes have produced innumerable fatalities and economic loss (Fig. 1). For example, the recent Haiti earthquake (2010) was responsible for at least 300,000 injuries, 316,000 deaths, and 300,000 destroyed houses. There is a high demand for rapid and real-time health evaluation of post-earthquake structures as it is critical to distinguish which structures are inhabitable and serviceable and which are damaged and unavailable anymore. However, current evaluation methods are labor intensive, time consuming, and not cost effective.

Structural health monitoring (SHM) refers to the process of determining and tracking structural integrity and assessing the nature of damage in a structure (Chang et al. 2003). The process of SHM usually consists of four stages: (1) damage definition, (2) damage identification, (3) damage localization, and (4) damage quantification. SHM has been studied since the 1970s and primarily studied within the context of contact-type sensors. These contact sensors include accelerometers, strain gauges, acoustic emission transducers, and fiber-optic sensors, to name a few. However, it is challenging to utilize these sensors for

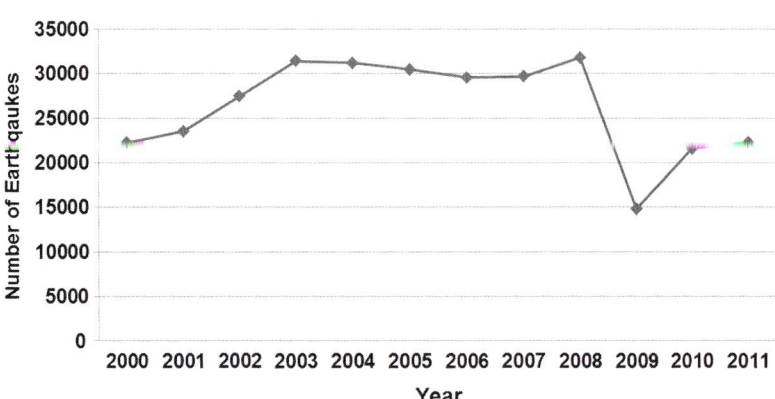

Laser-Based Structural Health Monitoring, **Fig. 1** Number of worldwide earthquakes during 2000–2011 (Source: United States Geological Survey)

post-earthquake monitoring as most of the installed sensors would be destroyed during the earthquake, and there would be no enough power supplies to operate these sensor systems.

In this entry, noncontact laser sensing is introduced as an alternative means for SHM. Light detection and ranging (LiDAR) and laser Doppler vibrometer (LDV) are two of the most common laser-sensing systems. A LiDAR estimates a distance from the device to a target point by measuring the time of flight (TOF) of the incident laser pulse reflected off from the target point or by measuring the phase shift of the reflected laser beam with respect to the incident continuous wave laser beam (Wehr and Lohr 1999). On the other hand, an LDV estimates the out-of-plane velocity of a vibrating target by relating the velocity to the frequency shift of the reflected laser beam with respect to the incident laser beam (Castellini et al. 2006). Furthermore, these devices can be easily deployed for long-range sensing and large area scanning. Figure 2 shows typical LiDAR and LDV devices. Riegl VZ-6000 LiDAR (Fig. 2a) can measure up to 6 km with an accuracy of 15 mm and a measurement speed of 37,000 points/s. Its measurement speed can be increased up to 222,000 points/s, but the maximum measurement range is shortened to 3.3 km. Polytec PSV-500 LDV (Fig. 2b) can measure up to 50 m/s vibration with a resolution of 2 $\mu m/s/\sqrt{Hz}$ and a maximum vibration frequency of 2.5 MHz.

This entry is organized as follows: In Section "Working Principles of LiDAR and LDV," airborne laser scanning techniques for regional damage assessment are summarized. Section "Airborne Laser Scanning for Regional Monitoring" presents ground-based laser scanning techniques for global and local damage detection, and Section "Ground-Based Laser Scanning Techniques" discusses laser ultrasonics and thermography for local damage detection. Finally, the conclusions and discussions are summarized in Section "Other Laser-Based Techniques for Local Damage Detection."

Working Principles of LiDAR and LDV

In this entry, the working principles of LiDAR and LDV are discussed in more detail. LiDAR can work in two manners, (1) TOF measurement and (2) phase-shift measurement. The first one is more commonly used for commercial LiDAR, which is rather simple to understand (Fig. 3a). A high-power laser pulse is emitted from the LiDAR, and the laser pulse reflects back from the target object. When the reflected laser pulse is detected by the LiDAR, the distance to the target D is determined as

$$D = \frac{T}{2}c \qquad (1)$$

where T and c denote the TOF, which is the time difference between laser pulse emission and reception, and the speed of light (3×10^8 m/s), respectively. As the distance measurement via LiDAR

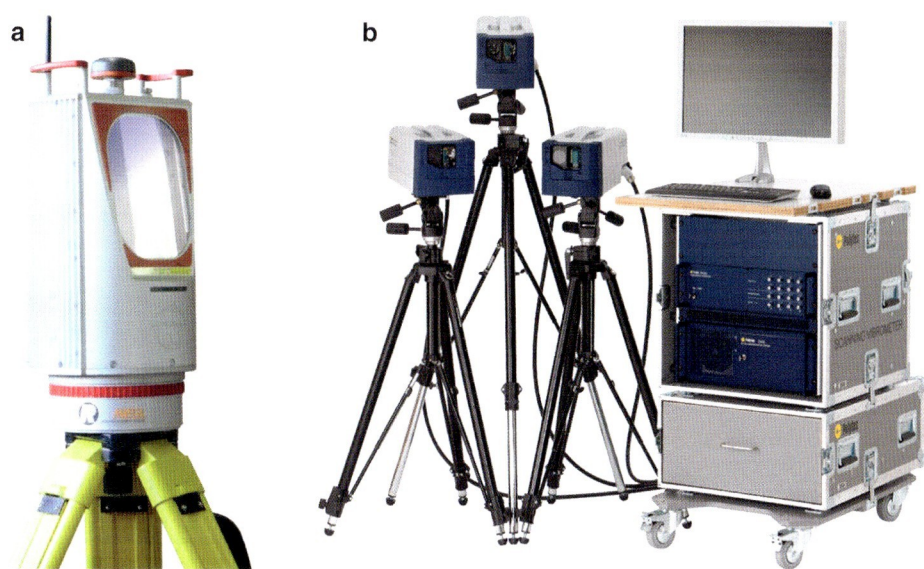

Riegl VZ-6000 LiDAR
(© RIEGL LMS, Austria, www.riegl.com)

Polytec PSV-500 LDV (© Polytec GmbH)

Laser-Based Structural Health Monitoring, Fig. 2 Representative laser scanning devices: (**a**) light detection and ranging (LiDAR) and (**b**) laser Doppler vibrometer (LDV)

depends on the reception of the reflected laser pulse, the sensing ranging of LiDAR is mainly limited by the laser pulse power. Some commercial LDV devices with a very high-power pulse can measure distance up to several kilometers.

Displacement measurement based on phase shift estimates the TOF information from the phase shift of the reflected laser beam (Fig. 3b) with respect to the incident laser beam. It emits an amplitude modulated laser beam and compares it with the reference laser beam. Then the phase difference φ between two laser beams can be expressed as follows:

$$\varphi = 2\pi T f_m \quad (2)$$

where f_m refers to the modulation frequency of the laser beam. By substituting Eq. 2 into Eq. 1, the distance is estimated as

$$D = \frac{T}{2}c = \frac{\varphi}{4\pi f_m}c = \frac{\varphi}{4\pi}\lambda_m \quad (3)$$

where λ_m refers to the wavelength of the amplitude modulation which is typically ranging from 100 to 500 m for common LiDARs. The biggest drawback of the phase-shift-based displacement measurement is that its ranging capability is usually limited to several hundred meters, as $0 \leq D \leq \frac{\lambda_m}{2}$ in Eq. 3. However, its measurement speed can be up to 1 million points/s, which is much faster than the TOF-based measurement technique (usually ∼10,000 points/s).

Figure 4 represents a schematic diagram of a common LDV. Assuming that a target is vibrating with a velocity of v, the frequency of the laser beam reflected from the moving target changes from its initial value of f_0 to the following value based on the Doppler effect:

$$f = \left(1 - \frac{v}{c}\right)^2 f_0 \quad (4)$$

where f denotes the frequency of the reflected laser beam. Note that v becomes negative when the target is vibrating toward the LDV. The intensity of the reflected laser beam can be expressed as

$$I_r = I\cos(2\pi f t) \quad (5)$$

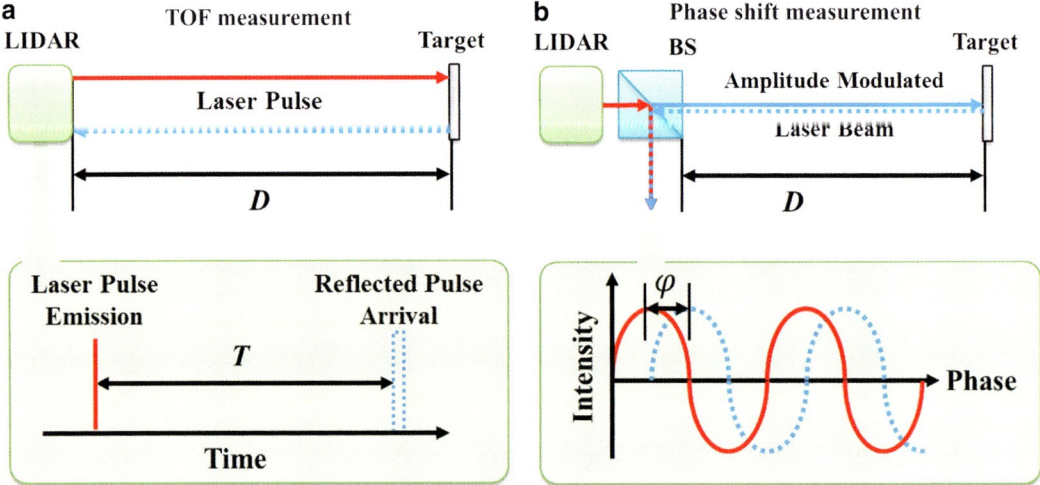

Laser-Based Structural Health Monitoring, Fig. 3 Schematic diagram of LiDAR measurement working principle. *Red solid line* and *blue dotted line* denote the paths of the reference and reflected laser beam, respectively. BS, D, T, and φ refer to a beam splitter, distance to the target, time of flight of the reflected laser pulse, and phase difference between two laser beams, respectively

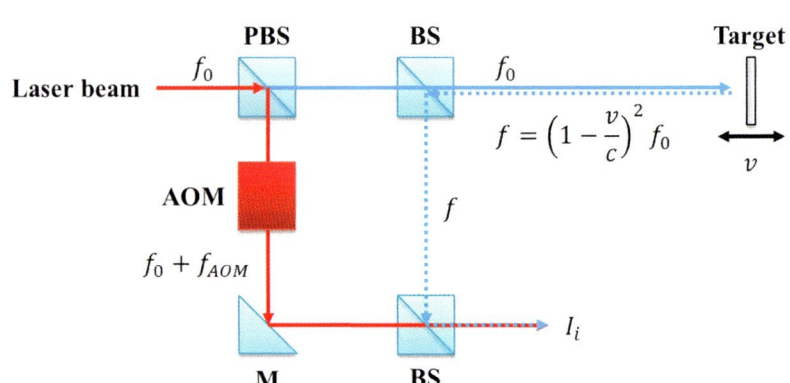

Laser-Based Structural Health Monitoring, Fig. 4 Schematic diagram of a common LDV. *Red solid* and *blue dotted lines* denote the paths of the reference and reflected laser beams, respectively. PBS, BS, AOM, and M refer to a partial beam splitter, a beam splitter, an acousto-optic modulator, and a mirror, respectively

where I and t refer to the maximum intensity of the laser beam and time, respectively. This reflected laser beam interferes with the reference laser beam:

$$I_0 = I \cos(2\pi(f_0 + f_{AOM})t) \quad (6)$$

which originally has a frequency of f_0 but shifted to $f_0 + f_{AOM}$ by an acousto-optic modulator (AOM). f_{AOM} is usually set to 40 MHz for many commercial LDVs. Then the intensity of the interfered laser beam is

$$\begin{aligned} I_i &= I\cos(2\pi(f_0 + f_{AOM})t) + I\cos(2\pi ft) \\ &= \frac{I}{2}\cos(2\pi(f_0 + f_{AOM} + f)t) \\ &\quad + \frac{I}{2}\cos(2\pi(f_0 - f + f_{AOM})t) \end{aligned} \quad (7)$$

By applying a low pass filter to I_i,

$$I_i^{LPF} = \frac{I}{2}\cos(2\pi(f_0 - f + f_{AOM})t) \quad (8)$$

where $f_0 - f$ can be expressed as

$$f_0 - f = f_0 - \left(1 - \frac{v}{c}\right)^2 f_0 = f_0\left(1 - \left(1 - \frac{2v}{c} + \frac{v^2}{c^2}\right)\right)$$
$$= f_0\left(\frac{2v}{c} + \frac{v^2}{c^2}\right) \qquad (9)$$

Since v is very small compared to c, we have

$$f_0 - f \simeq f_0\left(\frac{2v}{c}\right) = \frac{2v}{\lambda_0} \qquad (10)$$

$$I_i^{\text{LPF}} = \frac{I}{2}\cos\left(2\pi\left(\frac{2v}{\lambda_0} + f_{\text{AOM}}\right)t\right) \qquad (11)$$

where λ_0 is the wavelength of the incident laser beam. Then, by measuring the frequency of the interfered laser beam, the velocity of the target v can be estimated. Note that f_{AOM} is introduced to distinguish positive and negative frequency shift from the interfered laser beam frequency. Because LDV measures velocity from the Doppler frequency shift of the reflected laser beam, the signal-to-noise ratio of LDV heavily depends on the reflectivity of the target surface. Therefore, special surface treatments such as a retroreflective tape are often required for the improved signal-to-noise ratio, but it can be rather challenging to treat surfaces of target structures in the field.

Airborne Laser Scanning for Regional Monitoring

Airborne laser scanning (ALS) techniques utilize a laser device mounted onto aircraft to monitor earthquake-affected areas. Aerial techniques with fast and extensive data acquisition availability, such as aerial and satellite imaging, have been extensively used for large area scanning. The ALS techniques have a big advantage over the 2D imaging techniques because they can gather and reconstruct height/depth information for 3D model construction (Fig. 5). Two ALS schemes are introduced in this section. In the first scheme, ALS is used to detect changes in urban features and discern which structures are collapsed or damaged. Then, the structural damage type is identified and classified through the second ALS scheme.

Change Detection from Urban Features

Once a 3D model of the region is constructed by ALS scanning, the regional changes can be detected by comparing models before and after any event occurrence. This scheme has been applied for various purposes including forest observation and coastline monitoring. This ALS has also been applied to extracting urban features including land uses and structural identification (Priestnall et al. 2000).

Vu et al. (2004) scanned over Roppongi, Tokyo, Japan in 1999 and 2004. They compared two models and calculated the height difference at each 2D grid. Via the use of statistical analysis and filtering processes, they were able to identify outliers, which have distinctive height changes compared to the average height change of the whole scanned region. These outlier points are considered possible new constructions (height increased) and demolitions (height decreased). This concept can be applied to rapid inspection of earthquake-affected regions, but requires continuous update of the urban ALS models for comparison.

Shen et al. (2010) proposed another monitoring technique to identify demolished buildings by measuring buildings' inclinations from ALS data. They extracted a vertical axis vector of a building from its roof geometry and compared it with the ground normal vector to determine the buildings' inclination. As these two vectors can be gathered from a single ALS image, there is no requirement for previous or baseline model scanned for this technique. They applied this technique to a building affected by the Haiti earthquake and identified its inclination angle.

Classification of Structural Damage

During the rescue operation for earthquake victims, rapid assessment and classification of structural damage types is very critical to set the rescue strategy, allocate recourse, and save as many lives as possible. Schweier and Markus (2006) classified types of building damage after

Laser-Based Structural Health Monitoring, Fig. 5 (a) Conceptual figure for the airborne laser scanning (ALS) technique and (b) an example of scanned model

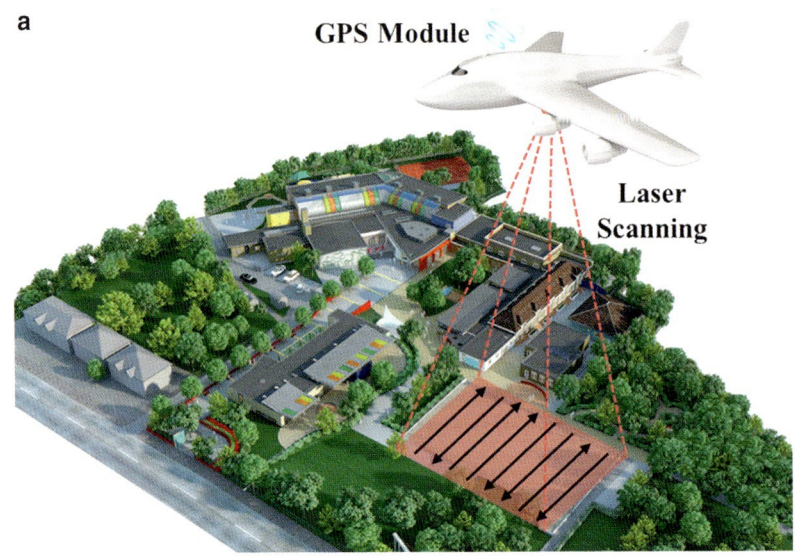

Airborne Laser Scanning

Scanned 3D Model (© Merret Surveys)

earthquakes into 10 categories including inclination, pancake collapse, and heap of debris (Fig. 6). However, it turned out to be vexing and challenging to classify them using 2D imaging data as it does not contain height information. For example, a pancake-collapsed building cannot be easily recognized with 2D imaging techniques as its identical top-view image remains intact.

Rehor (2007) scanned a military training area in Swiss and compared the model with its original CAD blueprint to classify structural damage types of each building. Changes in building features including area size, inclination angle, volume, and height are identified from the comparison.

The damage classification process is automated by introducing fuzzy logic membership functions for each feature. Fuzzy logic is one of the soft computing techniques, which enables the use of probability for logical analysis. For damage classification, the probabilities of all damage categories associated with each specific feature are computed using Fuzzy logic membership functions. For example, the membership functions of the feature "change of inclination" are shown for "the pancake collapse" and "inclined

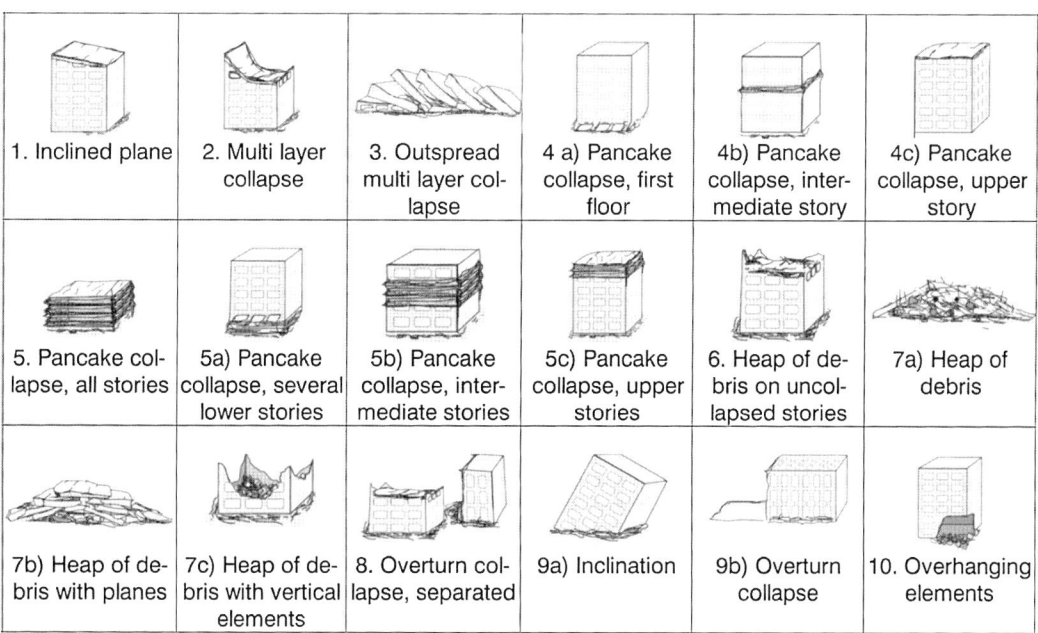

Laser-Based Structural Health Monitoring, Fig. 6 Building damage types after an earthquake (Rehor 2007)

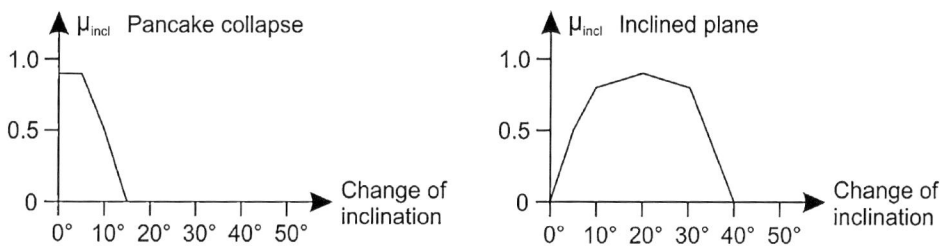

Laser-Based Structural Health Monitoring, Fig. 7 Fuzzy logic membership function example for pancake collapse and inclined plane (Rehor 2007)

plane" damage categories in Fig. 7. If a building had an inclination angle of 5°, this building would have a 0.9 probability for the pancake collapse and 0.25 for the inclined plane. On the other hand, if the inclination angle of another building were 20°, the probabilities for the pancake collapse and the inclined plane would be 0 and 0.9, respectively. The probabilities for other feature values are estimated in a similar manner, and the damage category with maximum probability is classified as the most probable damage category. As a result, damage types of 13 out of 17 (77 %) buildings are correctly classified. Rehor and Voegtle (2008) improved the classification accuracy to 88 % by integrating multispectral scanner data and aerial images with the above ALS model.

Ground-Based Laser Scanning Techniques

While ALS techniques are able to inspect large areas with a high speed, its damage detectability is rather limited to the detection and classification of large-scale defects. Further investigation of the structures, which are classified as intact with the previous ALS techniques, is often required as there still is a high possibility of internal damage

Laser-Based Structural Health Monitoring, Fig. 8 (**a**) Scanning of the Ashdown House in UK using a ground-based laser scanner and (**b**) the resultant 3D model (© Halcrow Group)

Ground-based Laser Scanning

Scanned 3D Model

in these structures. In particular, this detailed inspection could be critical for post-earthquake re-occupation of buildings as the undetected internal damage may lead to the collapse of the structures and additional casualties. A number of contact-type sensors have been suggested for the detailed inspections, and recently LiDAR and LDV are gaining prominence due to their noncontact nature. In this section, ground-based laser scanning techniques for global and local damage diagnosis are presented (Fig. 8).

Serviceability Monitoring Through Deflection/Deformation Measurement

Serviceability refers to the conditions under which a building is still considered useful. The serviceability is typically evaluated by measuring a local deflection and deformation, and the structure is considered in a proper serviceable condition when the deflection remains within a certain limit. However, deflection measurement on large-scale structure has always been a daunting task.

This problem can be readily solved with a LiDAR. As previously mentioned, LiDAR can measure 3D displacement of a structure as well as its static and dynamic deformed shapes. One of the biggest problems is that the maximum displacement measurement error of LiDAR is up to 10 mm, which may be insufficient for SHM purposes. Park et al. (2007) overcame this problem by introducing a displacement measurement model to calibrate the error and reduced the measurement error to less than 1 mm compared to the one measured by contact-type sensors such as LVDT, strain gauge, and fiber-optic sensor.

LiDAR has been applied not only in laboratory settings but also in the field. Watson et al. (2012) scanned a highway bridge for underclearance measurement with traffic and temperature variations. They performed correlation analysis between the measured underclearance, traffic, and temperature variations and concluded that LiDAR can consistently measure the underclearance value without environmental dependence. Liu et al. (2010) carried out a static loading test on a steel girder bridge while measuring its deformation with a LiDAR and reported its behavior under truck loading.

Vibration-Based Damage Detection

Vibration-based damage detection is one of the most widely adopted SHM methods for field applications. This technique attempts to evaluate the global integrity of a structure by relating the change of modal parameters such as natural frequencies, mode shapes, and damping to the physical properties of the structure such as stiffness and mass. Conventionally, these modal parameters are measured by contact-type accelerometers, but they can be substituted with a noncontact laser scanner. Nassif et al. (2005) reported that the bridge girder vibration and displacement measured by an LDV match very well with the ones obtained by geophone sensors and LVDTs attached to the girder.

When it comes to vibration measurements, LDV is a preferred noncontact solution rather than LiDAR because modal analysis requires measurement of vibration signals with a high sampling rate. LDV can achieve a sampling rate up to several MHz, while it is usually limited up to 1 kHz in LiDAR. Siringoringo and Fujino (2009) detected damage in a structural member by performing modal analysis using an LDV. They scanned a bolted joint plate and compared modal parameters measured from different bolt-loosening conditions. Stiffness and damping changes due to the loosened bolt torques were observed, and damage size was quantified. Khan et al. (2000) detected various cracks with an LDV, including through cracks in thin plates, narrow slots in a solid cantilever beam, and cracks in a reinforced concrete beam by comparing mode shapes and frequency response functions of the specimens before/after introducing damages.

Laser Model-Based Damage Quantification

The scope of the previously overviewed vibration-based damage detection is typically limited to damage identification and localization. For damage quantification, additional steps are warranted. The principle of the damage quantification techniques presented here is quite similar to the aforementioned ALS techniques. One of the quantification algorithms is volumetric change analysis (Olsen et al. 2010). By comparing the original and the posttest 3D models obtained from laser scanning, volumetric changes, or defects of the concrete structure, can be detected. The scanning process in Olsen et al.'s study took about 6 h to scan a concrete structure of 6.706 m^3, while a conventional photographic technique required 20 h of manual photo-taking. Liu et al. (2011) suggested two other damage quantification methods, which do not require a baseline laser-scanned model. In these methods, damage is identified by measuring surface flatness and smoothness. When a local defect like spalling appears on a concrete surface, the depth and normal direction of the defect area will deviate from a reference (undamaged) plane. The depth and angle deviation of the damaged concrete surface from the intact surface are computed using the full 3D coordinate information of the target concrete structure obtained by LiDAR. While the depth information is good for detecting the deepest portion of the damage area, the angle variation is more sensitive to the edge of the defect, where a sudden gradient change of the concrete surface is expected (Fig. 9). They inspected a real concrete bridge by integrating both methods and identified mass loss in concrete members (Fig. 10).

Laser-Based Structural Health Monitoring, Fig. 9 Schematic diagram for surface damage quantification by measuring the depth and angle deviation

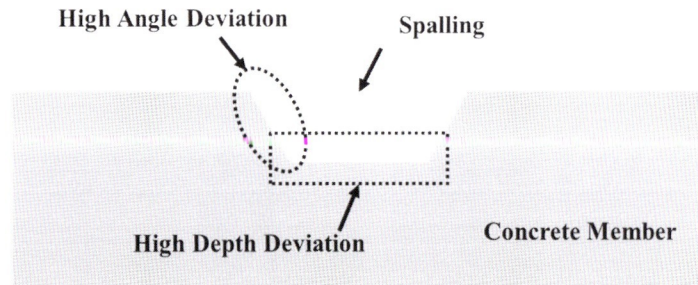

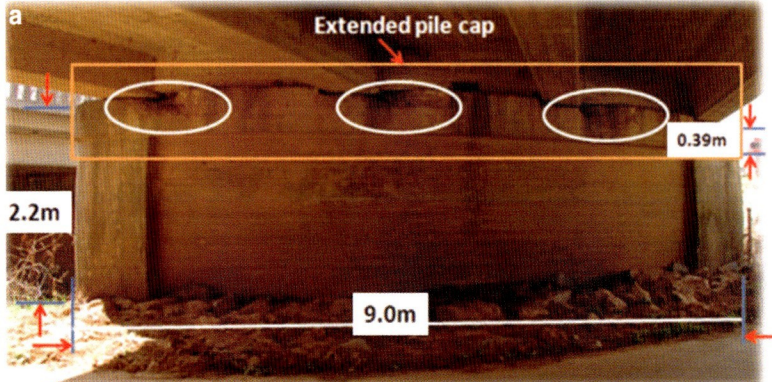

Laser-Based Structural Health Monitoring, Fig. 10 Defect detection for a concrete bridge member using LiDAR scanning, presented by Liu et al. (2011)

Tested bridge showing distress in pile cap

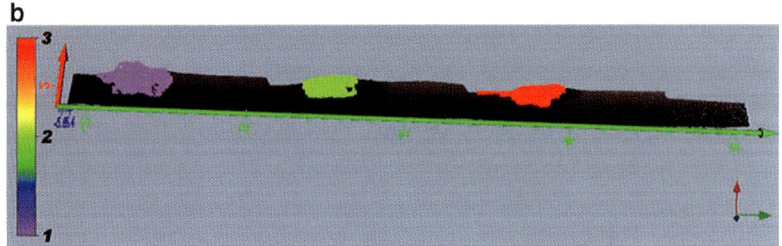

Defect detection result for the scanned region (Orange rectangle in (a))

Other Laser-Based Techniques for Local Damage Detection

After earthquakes, small defects may not pose an immediate threat to the overall structural integrity or cause a sudden collapse of the structure. However, hidden and untreated defects will continuously grow and eventually compromise the safety of the system. Therefore, there is an increasing demand to further improve damage detectability and identify defects as early as possible. Furthermore, as sensing techniques advances, the detection of small-scale defects, which was not feasible using conventional techniques, is becoming feasible. Here, two specific state-of-the-art techniques for such microscale damage detection, laser ultrasonic imaging and laser thermography, are presented. Although they have been mostly applied for inspection of mechanical systems so far, their brief introduction is included here because of their great potential for SHM of earthquake-affected structures (Fig. 11).

Laser Ultrasonic Imaging

Ultrasonic techniques identify a defect by measuring interaction between ultrasonic waves and the defect. Conventionally ultrasonic waves are

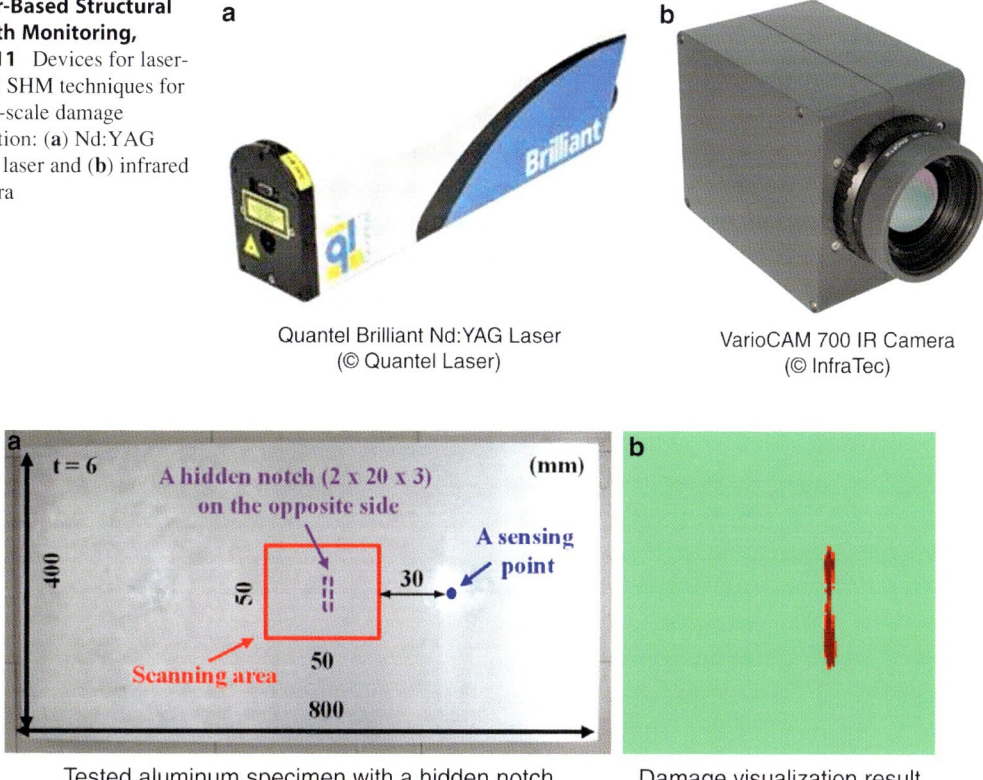

Laser-Based Structural Health Monitoring, Fig. 11 Devices for laser-based SHM techniques for small-scale damage detection: (**a**) Nd:YAG pulse laser and (**b**) infrared camera

Laser-Based Structural Health Monitoring, Fig. 12 Hidden notch detection in an aluminum plate using laser ultrasonic imaging, presented by An et al. (2013). The notch location and its size are visualized with *red* color in (**b**)

generated and measured using contact-type transducers such as piezoelectric transducers (Kim and Sohn 2007), but nowadays noncontact laser solutions are available. When a high-power pulse laser beam is exerted onto a solid surface, a localized heating and corresponding elastic expansion of the specimen produce ultrasonic waves (Scruby and Drain 1990). The generated ultrasonic waves can be measured using an LDV. With its scanning capability, ultrasonic waves generated from a single point can be measured from multiple points with high spatial resolution in the order of mm. Once the ultrasonic responses are measured from multiple scan points, the wave propagation in a spatial domain and its interaction with a defect can be visualized. An et al. (2012, 2013) apply laser ultrasonic imaging techniques to the detection of a notch in a steel bridge girder and a notch in a metal plate (Fig. 12). In their study, a high-power pulse laser beam is scanned over the inspection region for ultrasonic wave generation, and the corresponding responses are measured at a single point using an LDV. In theory, the ultrasonic image obtained by scanning the excitation laser beam with a fixed sensing point is identical to the one obtained by scanning the sensing laser beam with a fixed excitation point based on linear reciprocity. In practice, scanning the excitation laser beam is more advantageous as the performance of LDV usually varies with the surface condition and the incident angle of the sensing laser beam.

However, there are several hurdles that need to be overcome for field applications. The laser ultrasonic generation requires the use of a class IV high-power laser over 1 MW, which may harm human eyes. The scanning time for large area inspection with high spatial resolution can be prohibitively long. Currently laser ultrasonic techniques are best suited for inspection of

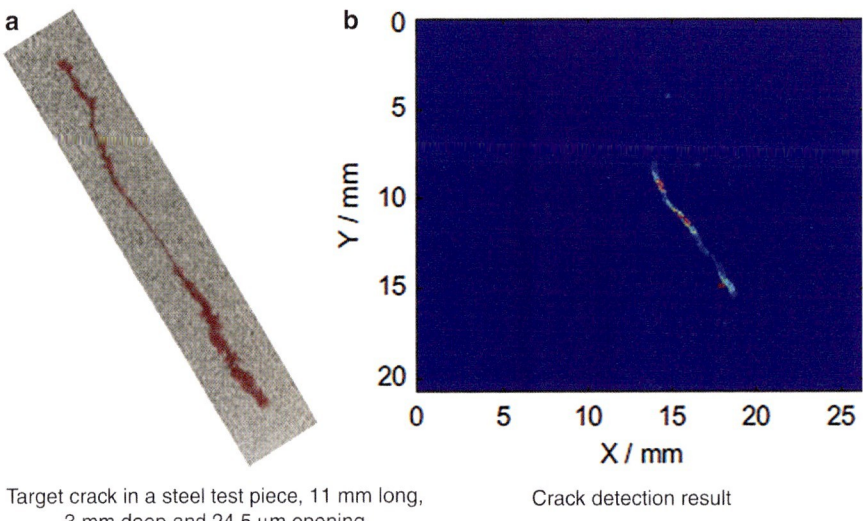

Laser-Based Structural Health Monitoring, Fig. 13 Crack detection in a steel test specimen using laser thermography, presented by Li et al. (2011)

metallic and composite materials, and the applicability to porous materials including concrete is limited due to rough surface conditions.

Laser Thermography

Infrared (IR) thermography techniques visualize and quantify damage by detecting changes in heat transfer characteristics near the defect. As thermal images can be taken over a large area rather quickly using the IR camera, this technique has been popular for nondestructive testing of mechanical systems. Clark et al. (2003) detected delamination in a concrete slab of a real bridge, by using natural sunlight as a passive heat source and taking thermal images. By detecting regions with abnormal temperature in comparison with the rest of the areas, possible damage locations were recognized. However, the maximum temperature differences remained below 1 °C, making it challenging to distinguish damage regions from the rest.

This leads to the development of active thermography techniques, where a controlled heat source such as a halogen lamp and acoustic transducer is used (Chatterjee et al. 2011). A laser thermography technique is relatively new in the field of active thermography and uses a high-power laser as a heat source. Compared to a traditional heat source such as a halogen lamp, the biggest advantage of the laser is its highly localized nature of heating spot. This helps to improve the damage detection resolution and defect micro-cracks. Li et al. (2011) visualized a 50 μm wide and 5 mm long fatigue crack in a mild steel test piece (Fig. 13). However, it should be noted that the application of the laser thermography technique is usually limited to the detection of surface damages in thin structures, and it requires history baseline data collected from the intact condition to identify damage, which may show false alarms in the presence of operational and environmental variations.

Summary

Laser-based structural health monitoring is a promising technology for the monitoring of earthquake-affected structures as it does not require any sensor nor power/data cable installation. Light detection and ranging (LiDAR) and laser Doppler vibrometer (LDV) are two of the most widely adopted laser-sensing devices for civil applications due to their noncontact nature and high measurement accuracy. LiDAR estimates the distance to the target by measuring time-of-flight or phase-shift information of the laser beam reflected off the target, and LDV measures the vibration velocity of the target by relating the frequency shift of the reflected laser beam.

Various laser-based structural health monitoring techniques applicable to earthquake-affected structures have been proposed and briefly discussed in this entry. By means of airborne laser scanning (ALS), earthquake-affected regions may be rapidly monitored for distinguishing damaged structures and further identifying their damage types. Even for intact structures, LiDAR-based displacement/deformation measurement can be used to check the serviceability of civil infrastructure. The damage location and size can be quantified with LDV-based vibration analysis and LiDAR-based 3D structural model. Additional laser techniques with potential application to post-earthquake inspection include laser ultrasonic techniques and laser thermography techniques, which are able to detect small-scale damages.

Still there are several challenges associated with field deployment of these laser devices: (1) the time-of-flight-based LiDAR cannot collect data with a high sampling rate; (2) the applicability of the phase-shift-based LiDAR is limited to a short range below several hundred meters; (3) special surface treatment is often necessary for LDV; and (4) the use of high-power pulse laser may have implications associated with eye safety. However, the developments and improvements of not only hardware devices but also damage detection algorithms will inevitably contribute to the fast and accurate health assessment of earthquake-affected structures and prevent further catastrophic failures.

Cross-References

▶ Post-Earthquake Diagnosis of Partially Instrumented Building Structures
▶ Seismic Behavior of Ancient Monuments: From Collapse Observation to Permanent Monitoring
▶ System and Damage Identification of Civil Structures

References

An YK, Song HM, Park HJ, Sohn H, Yun CB (2012) Remote guided wave imaging using wireless PZT excitation and laser vibrometer scanning for local bridge monitoring. In: Bridge Maintenance, Safety, Management, Resilience and Sustainability: Proceedings of the Sixth International IABMAS Conference: 173, http://koasas.kaist.ac.kr/handle/10203/171481

An YK, Park B, Sohn H (2013) Complete noncontact laser ultrasonic imaging for automated crack visualization in a plate. Smart Mater Struct 22(2):025022

Castellini P, Maratarelli M, Tomasini EP (2006) Laser Doppler Vibrometry: development of advanced solutions answering to technology's needs. Mech Syst Signal Process 20:1265–1285

Chang PC, Flatau A, Liu SC (2003) Review paper: health monitoring of civil infrastructure. Struct Health Monit Int J 2(3):257–267

Chatterjee K, Tuli S, Pickering SG, Almond DP (2011) A comparison of the pulsed, lock-in and frequency modulated thermography nondestructive evaluation techniques. NDT & E Int 44(7):655–667

Clark MR, McCann DM, Forde MC (2003) Application of infrared thermography to the non-destructive testing of concrete and masonry bridges. NDT&E International 36(4):265–275

Khan AZ, Stanbridge AB, Ewins DJ (2000) Detecting damage in vibrating structures with a scanning LDV. Opt Lasers Eng 32(6):583–592

Kim SB, Sohn H (2007) Instantaneous reference-free crack detection based on polarization characteristics of piezoelectric materials. Smart Mater Struct 16(6):2375

Li T, Almond DP, Rees DAS (2011) Crack imaging by scanning pulsed laser spot thermography. NDT & E Int 44(2):216–225

Liu W, Chen S, Boyajian D, Hauser E (2010) Application of 3D LiDAR scan of bridge under static loading testing. Mater Eval 68(12):1359–1367

Liu W, Chen S, Hauser E (2011) LiDAR-based bridge structure defect detection. Exp Tech 35(6):27–34

Nassif HH, Gindy M, Davis J (2005) Comparison of laser Doppler vibrometer with contact sensors for monitoring bridge deflection and vibration. NDT & E Int 38(3):213–218

Olsen MJ, Kuester F, Chang BJ, Hutchinson TC (2010) Terrestrial laser scanning-based structural damage assessment. J Comput Civ Eng 24(3):264–272

Park HS, Lee HM, Adeli H, Lee I (2007) A new approach for health monitoring of structures: terrestrial laser scanning. Comput Aided Civ Infrastruct Eng 22:19–30

Priestnall G, Jaafar J, Duncan A (2000) Extracting urban features from LiDAR digital surface models. Comput Environ Urban Syst 24(2):65–78

Rehor M (2007) Classification of building damage based on laser scanning data. Photogramm J Finl 20(2):54–63

Rehor M, Voegtle T (2008) Improvement of building damage detection and classification based on laser scanning data by integrating spectral information. Int Arch Photogramm Remote Sens Spat Inf Sci XXXVII (Part B7). Beijing, pp 1599–1606

Schweier C, Markus M (2006) Classification of collapsed buildings for fast damage and loss assessment. Bull Earthq Eng 4(2):177–192

Scruby CB, Drain LE (1990) Laser ultrasonics: techniques and applications. Adam Hilgher, Norfolk

Shen Y, Lixin W, Zhi W (2010) Identification of inclined buildings from aerial LIDAR data for disaster management. In: Proceedings of 18th international conference on geoinformatics, pp 1–5, http://ieeexplore.ieee.org/xpls/abs_all.jsp?arnumber=5567852&tag=1

Siringoringo DM, Fujino Y (2009) Noncontact operational modal analysis of structural members by laser Doppler vibrometer. Comput Aided Civ Infrastruct Eng 24:249–265

Vu TT, Matsuoka M, Yamazaki F (2004) LIDAR-based change detection of buildings in dense urban areas. In: Proceedings of IEEE international geoscience and remote sensing symposium, vol 5, pp 3413–3416, http://ieeexplore.ieee.org/xpls/abs_all.jsp?arnumber=1370438

Watson C, Chen S, Bian H, Hauser E (2012) Three-dimensional terrestrial LIDAR for operational bridge clearance measurements. J Perform Constr Facil 26(6):803–811

Wehr A, Lohr U (1999) Airborne laser scanning-an introduction and overview. ISPRS J Photogramm Remote Sens 54:68–82

Lead-Rubber Bearings with Emphasis on Their Implementation to Structural Design

Ioannis Kalpakidis
KBR, Houston, TX, USA

Synonyms

Base isolation; Bridge bearings; Elastomeric bearings; Lead-plug bearings; Lead-rubber bearings; Rubber bearings; Seismic isolation

Introduction

The seismic design of conventionally framed bridges and buildings relies on the dissipation of earthquake-induced energy through inelastic (nonlinear) response in selected components of the structural frame. Such response is associated with structural damage that produces direct (capital) loss repair cost, indirect loss (possible closure, rerouting, business interruption), and perhaps casualties (injuries, loss of life) (Constantinou et al. 2007b).

There are various mechanisms/devices that have been invented in the past for the protection of structures from the damaging effects of earthquakes. A lot of these devices are based on the idea of uncoupling the structures from the ground. Rollers, layers of sand, or similar materials are among the ideas proposed in the past to allow a building to slide. Such mechanisms are early versions of the now widely accepted and applied concept of base/seismic isolation (Kelly 2004).

Lead-rubber bearings are one of various existing types of seismic isolation devices and have been used for years for the protection of structures from the damaging effect of earthquakes. This entry presents summarized information regarding the use, benefits, properties, and applications of lead-rubber bearings along with literature sources for further reading on the subject.

Seismic Isolation

The concept of base isolation is quite simple. The seismic isolation system is a group of low-stiffness elements that is placed between the foundation and the structure and reduces the effects of the earthquake on the latter. Use of these elements causes a decrease of the fundamental frequency of the structure shifting it away from both its fixed-base frequency and the predominant frequencies of the ground motion (Kelly 2004).

Seismic isolators are typically installed between the girders and bent caps (abutments) in bridges and the foundation and first suspended level in a building. Fig. 1 (Constantinou et al. 2007b) shows a typical bridge installation of seismic isolators; the isolators were installed at the top of the pier as shown in Fig. 1b (the steel plate on the side of the isolator is part of the lateral wind-restraint system). For bridge

Lead-Rubber Bearings with Emphasis on Their Implementation to Structural Design, Fig. 1 Typical installation of a seismic isolator in a bridge (Constantinou et al. 2007b). (**a**) Bridge elevation. (**b**) Friction pendulum seismic isolator

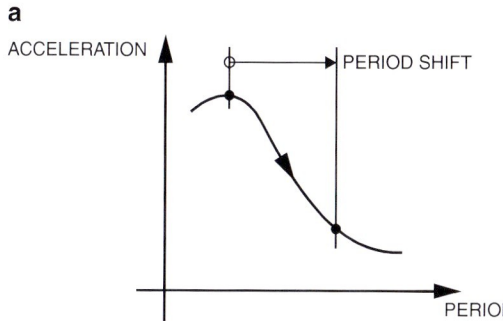
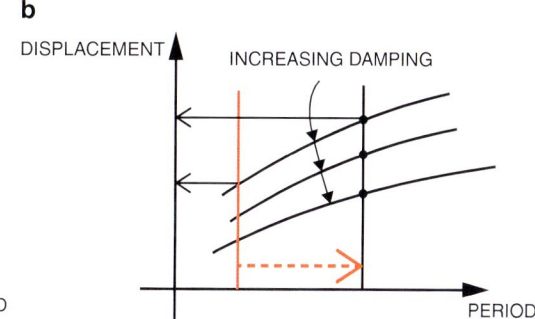

Lead-Rubber Bearings with Emphasis on Their Implementation to Structural Design, Fig. 2 Concept of seismic isolation (Constantinou et al. 2007b). (**a**) Decrease in spectral accelerations. (**b**) ZIncrease in spectral displacements

construction, the typical design goals associated with the use of seismic isolation are (a) reduction of forces (accelerations) in the superstructure and substructure and (b) force redistribution between the piers and the abutments (Constantinou et al. 2007b). Reduction of forces is achieved through the aforementioned structural frequency shift as shown in Fig. 2a (Constantinou et al. 2007b). Control of the increase in displacements that results from the use of seismic isolation is achieved through introducing damping (energy dissipation) in the isolator as shown in Fig. 2b (Constantinou et al. 2007b).

There are various types of seismic isolation devices, including but not limited to: (a) elastomeric-based systems, (b) low-damping natural and synthetic rubber bearings, (c) high-damping natural rubber (HDNR) systems, (d) isolation systems based on sliding, (e) friction pendulum system, (f) double and triple concave friction pendulum sliding bearings, (g) spring-type systems, and (h) lead-rubber bearings (Kelly 2004, Constantinou et al. 2007b). Elastomeric bearings (low- and high-damping rubber, lead rubber) and sliding bearings (spherical sliding or friction pendulum bearings, flat sliding or EradiQuake bearings) are the two common types of seismic isolation bearings used in the United States (Constantinou et al. 2007b).

Lead-Rubber Bearings: Short Description and Historical Background

The lead-rubber bearing (LRB) was invented in 1975 by W. H. Robinson in New Zealand (Skinner et al. 1993; Kelly 2004; Robinson and Tucker 1977, 1983). It has been extensively used

Lead-Rubber Bearings with Emphasis on Their Implementation to Structural Design, Fig. 3 Lead-rubber bearing construction details (Constantinou et al. 2007b)

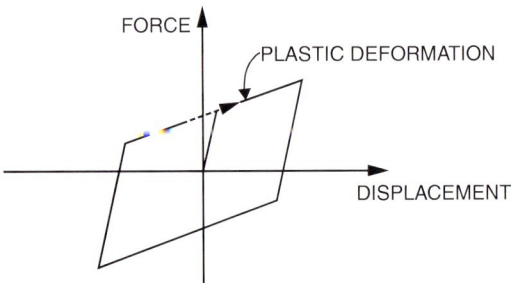

Lead-Rubber Bearings with Emphasis on Their Implementation to Structural Design, Fig. 4 Hysteretic energy dissipation in lead-rubber bearings (Constantinou et al. 2007b)

in New Zealand, Japan, and the United States (Kelly 2004). Such bearings have been used to isolate many buildings, and buildings using them performed well during the 1994 Northridge and 1995 Kobe earthquakes (Kelly 2004). A short summary of the response of one of these buildings during the 1994 Northridge earthquake is provided in the next section.

Generally, a lead-rubber bearing is composed of a low-damping (unfilled) elastomer and one (typically) or more (less common) lead cores. A cutaway view of a lead-rubber bearing is illustrated in Fig. 3 (Constantinou et al. 2007b). In this illustration, the separate components of the bearing can be seen clearly: (a) the lead plug in the core of the bearing, (b) the end and flange plates on top and bottom of the bearing, and (c) the low-damping elastomer with alternating rubber layers and steel shim plates.

The elastomer rubber layers serve as the "isolation" (i.e., stiffness and fundamental structural frequency reduction) component; lead also contributes to the stiffness of the bearing, but this contribution is relatively minor. The steel shim plates assist with vertical load support. The lead core mainly serves as the "damping" (i.e., energy dissipation) component. A typical hysteretic (force-displacement) loop for a lead-rubber bearing is shown in Fig. 4 (Constantinou et al. 2007b).

The lead core provides excellent energy dissipation capability under confined conditions. Such conditions can be achieved by constructing the plug slightly (less than 5 %) longer than the height of the rubber bearing and connecting the top and bottom flange plates to the rubber bearing end plates through countersunk bolts. Then the plug is compressed upon bolting the flange plates to the end plates; it expands laterally and wedges into the rubber layers between the shim plates. The magnitude of energy dissipation capacity depends on the lead plug diameter (Constantinou et al. 2007b).

Example: Performance of a LRB-Isolated Building During the 1994 Northridge Earthquake

As an example of the benefits from the use of lead-rubber bearings, a brief summary of the behavior of the LRB-isolated University of Southern California Teaching Hospital during the 1994 Northridge earthquake is provided in this section. A very detailed description has been outlined by Kelly (2004). Design and construction details of this building have been described by Asher and Van Volkinburg (1989), Asher et al. (1990), and KPFF Consulting Engineers (1991).

The hospital is an eight-story concentrically braced steel frame with significant irregularities in its plan and elevation. The isolation system consists of 68 lead-rubber isolators and 81 elastomeric isolators. The lead-rubber bearings are distributed in the perimeter of the building under the braced frames to minimize torsional response in the isolated structure. Elastomeric bearings are used in the interior.

The fixed-base period of the superstructure is approximately 0.7 s. The design base shear of the hospital is 0.15 g, while the design corner bearing displacement including torsion is 26.0 cm. The effective period of the isolation system at this displacement is approximately 2.3 s.

The strongest motions recorded at the site were in the north–south direction. The pseudo-acceleration response spectrum of the north–south foundation motion (Clark et al. 1996) shows significant short-period energy in the range of 0.2–0.6 s, but relatively little energy above a period of approximately 1.3 s. At periods greater than 2 s, the corresponding 10 % damped spectral displacement is less than 3 cm.

The isolation frequency due to the north–south acceleration was approximately 0.7 Hz and the damping ratio approximately 14 %. In the east–west direction, the isolation frequency was 0.75 Hz and the damping ratio was 13 %. The isolation system performed well, its damping capacity was good, and the structure was effectively isolated from the earthquake ground motions. It is noted that the Northridge earthquake caused significant damage to other buildings in the medical center (Kelly 2004).

Entry Roadmap

Section Introduction of this entry has presented a brief introduction, description, historical background, and an application of lead-rubber isolators. The following sections of this entry are outlined below.

Section Lead-Rubber Bearing Properties and Structural Design provides a more detailed description of the properties of lead-rubber bearings along with their significance in structural design.

Section Applications provides a few more examples of lead-rubber bearing applications in buildings, bridges, and structure retrofit.

Section Summary is a summary of this entry.

Section Cross-References provides a list of related entries from this Encyclopedia.

Section References provides a list of references used in this entry.

Lead-Rubber Bearing Properties and Structural Design

A brief description of lead-rubber bearings and their components has been provided in section Introduction. This section provides further details regarding the properties of lead-rubber bearings.

An idealized force-displacement loop for a lead-rubber bearing is shown in Fig. 5 (Constantinou et al. 2007b) with its significant parameters/mechanical properties being the characteristic strength, Q_d, and the post-elastic stiffness, K_d, defined as follows:

$$Q_d = A_L \sigma_L \qquad (1)$$

$$K_d = \frac{f_L G A_r}{T_r} \qquad (2)$$

where f_L is a parameter that represents the effect of the lead core on the post-yield stiffness (typically close to unity), A_L is the cross-sectional area of the lead core, T_r is the total thickness of the rubber layers, G is the shear modulus of

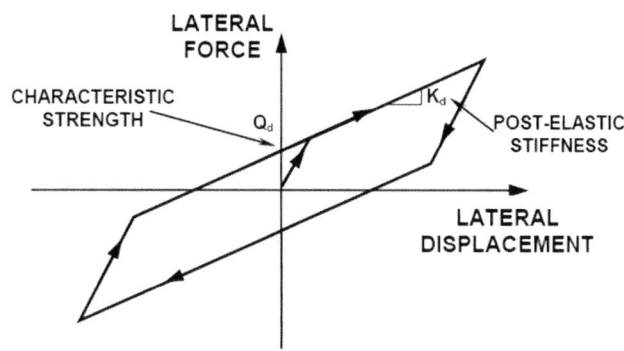

Lead-Rubber Bearings with Emphasis on Their Implementation to Structural Design, Fig. 5 Idealized force-displacement relation of a typical seismic isolation system (Constantinou et al. 2007b)

Lead-Rubber Bearings with Emphasis on Their Implementation to Structural Design, Table 1 Lead-rubber bearing components/properties and environmental/other effects

Property	Controlling component	Component significance
Post-elastic stiffness, K_d	Elastomer (typically low-damping rubber)	Structure stiffness reduction Structure fundamental frequency reduction
Characteristic strength, Q_d	Lead core	Energy dissipation

Environmental or other factor	Component/property	Effect
Aging	Elastomer/post-elastic stiffness, K_d	Minor increase
	Lead/characteristic strength, Q_d	Insignificant (lead yield strength not affected by aging); confinement changes may affect strength slightly
Temperature	Elastomer/post-elastic stiffness, K_d	Increase for low temperatures; effect depends on exposure duration; elastomer becomes brittle below glass transition temperature (compound dependent, $-55\,°C$ for natural rubber)
	Lead/characteristic strength, Q_d	Increase/decrease for temperature decrease/increase (the same happens to lead yield strength)
Heating	Elastomer/post-elastic stiffness, K_d	Insignificant
	Lead/characteristic strength, Q_d	Strength drops significantly, especially during the first few cycles and for high-frequency and large-amplitude motion due to lead core temperature increase; strength recovers its original value after lead core is allowed to cool down
Cumulative travel	Elastomer/post-elastic stiffness, K_d	None
	Lead/characteristic strength, Q_d	Low speed strength increase after high-amplitude cumulative travel; none in all other cases

rubber, A_r is the bonded rubber area (i.e., excluding the outer rubber cover beyond the steel shims), and σ_L is the effective yield stress of lead (Constantinou et al. 2007b).

The elastomer has a shear modulus between 65 and 100 psi at 100 % rubber shear strain. The lead core has a diameter ranging between 15 % and 33 % of the bearing bonded (i.e., distance between the outer edges of the shim plates) diameter. The maximum shear strain range for LR bearings varies depending on the manufacturer but is generally between 125 % and 200 % (Constantinou et al. 2007b). The effective yield stress of lead typically ranges between 10 and 20 MPa (Skinner et al. 1993; Constantinou et al. 2007b).

Numerous tests and studies have been conducted to investigate and determine how these properties are affected by environmental and other factors such as aging, temperature, heating, history of loading, etc., and modification factors for these properties have been proposed (Constantinou et al. 1999, 2007a, b; Kalpakidis and Constantinou 2009a, b; Kalpakidis et al. 2010). Some of these effects and a summary of lead-rubber bearing properties and components along with the significance of these components are outlined in Table 1.

A theoretical interpretation of the effects of seismic isolation on the dynamic properties and, consequently, on the dynamic behavior of a structure has been performed by Kelly (1990; 2004) through the use of a simplified two-degree-of-freedom (2-DOF) model. This simplified model treats the superstructure as a single-degree-of-freedom (SDOF) system that has appropriate period, stiffness, damping, and mass values. This SDOF system representing the superstructure is on top of a mass (equal to the base mass) which in turn is on top of the isolation system (defined by appropriate stiffness and damping values).

Assuming a very wide separation (order of 10^{-2} to 10^{-1}) between the square of the isolation frequency (i.e., frequency of the combined isolation-superstructure system when the superstructure is rigid) and the square of the fixed-base frequency of the superstructure, as well as a value with a 10^{-2} to 10^{-1} order of magnitude for both of the respective damping ratios, the resulting two modes for the combined superstructure-isolation system (simplified 2-DOF model) have the following characteristics (Kelly 1990, 2004):

- Isolation frequency/mode 1: The frequency is affected very slightly by the flexibility of the superstructure. The mode shape involves very little structural deformation (the structure is nearly rigid), while the isolation system picks up much more deformation.
- Structural frequency/mode 2: The mode shape involves much more structural deformation than the first mode, but its effect is insignificant due to the very small participation factor associated with this mode.

In short, as mentioned earlier in this entry, the presence of a seismic isolation system (with either lead-rubber bearings or other types of isolators) is beneficial because it increases the fundamental period of the superstructure (i.e., decreases the corresponding spectral acceleration) while at the same time it adds an energy dissipation mechanism to the system.

Applications

This section contains a brief summary of example applications of lead-rubber bearings (a building application has been briefly described in section Example: Performance of a LRB-Isolated Building during the 1994 Northridge Earthquake). It is noted that each example herein is merely a high-level summary of an application. For further details, the reader is referred to the respective sources provided for each example.

Richmond–San Rafael Bridge (Retrofit)

In general, use of seismic isolation can reduce forces in bridge foundations by up to 70 %. Isolated bridges are superior in performance, and their cost is lower than that of non-isolated bridges (DIS, Inc.).

The Richmond–San Rafael Bridge is the northernmost east–west crossing of the San Francisco Bay in California. It connects Richmond (east end) to San Rafael (west end). It opened in 1956, at which time it was one of the longest bridges in the world. Its length is 5.5 miles and it spans two principal ship channels. In the fall of 2001, extensive seismic retrofit work was performed along with necessary age-related maintenance. The retrofit was completed in 2005.

The bridge structure has a characteristic shape shown in Fig. 6 that has been described as "roller coaster span" and "bent coat hanger." Without the use of seismic isolators, shorter and stiffer piers would attract most of the lateral load. 55-in. diameter lead-rubber isolators were used for the retrofit. The benefit from using isolators was lateral load redistribution throughout the structure (DIS, Inc.).

San Francisco City Hall (Retrofit)

The City Hall was built in the late 1800s and was destroyed in the 1906 San Francisco earthquake. It was subsequently rebuilt at a nearby location and reopened in 1915. The City Hall was again damaged in the 1989 Loma Prieta earthquake. Subsequently, seismic retrofit work was performed and the building was reopened in 1999.

For the seismic retrofit, the two-block-long building was cut from its foundation and made to "float" on more than 500 lead-rubber isolators, designed to dissipate seismic energy and to allow the building to sway horizontally up to 26 in. without shaking apart (MCEER). A photo of the building can be seen in Fig. 7.

Oakland City Hall (Retrofit)

Construction of the Oakland City Hall was completed in 1914. A photo of the building can be seen in Fig. 8. The structure was the first high-rise government office building in the United States (Walters 2003). It is an 18-story building with full basement. Its podium contains a central rotunda, council chambers, and administration offices. An office tower is above the podium, and a clock tower is above the office tower (MCEER; Walters 2003). The structure suffered extensive damage in

Lead-Rubber Bearings with Emphasis on Their Implementation to Structural Design, Fig. 6 Richmond-San Rafael Bridge (MTC CA) Photo: William Hall, Caltrans

Lead-Rubber Bearings with Emphasis on Their Implementation to Structural Design, Fig. 7 San Francisco City Hall (Wikimedia-a) Photo by Andreas Praefcke is licensed under CC BY 3.0

the 1989 Loma Prieta earthquake, losing about 20 % of its strength in the north–south direction and 30 % in the east–west direction (Walters 2003).

Subsequently, several studies were performed resulting in seismic isolation retrofit work that was completed in 1995 (Walters 2003). 42 lead-rubber bearings and 69 natural rubber bearings were utilized for the retrofit (MCEER).

The main reason for selecting seismic isolation as the retrofit method (instead of other conventional strengthening methods) was the fact that the existing structural system of the City Hall (riveted steel frames with infill masonry) could provide sufficient lateral strength and stiffness only until the development of cracks. After cracks develop, the response would degrade under repeated cyclic loading (Benjamin and Williams 1958, Walters 2003). The result would be lack of ductility and of reliable post-cracking strength, which would require

Lead-Rubber Bearings with Emphasis on Their Implementation to Structural Design, Fig. 8 Oakland City Hall (Wikimedia-b) Photo by Sanfranman59 is licensed under CC BY-SA 3.0

a nearly elastic resistance of maximum earthquake forces (Walters 2003). The most effective way to satisfy this demand would be the use of seismic isolation that would significantly reduce the seismic forces acting on the structure.

Emergency Headquarters at Fukushima Daiichi Nuclear Power Plant (New Construction)

During the July 2007 Niigataken Chuetsu-oki earthquake (M6.8), there was no damage to the safety of nuclear power plant facilities at TEPCO Kashiwazaki-Kariwa Nuclear Power Plant. The reactors were shut down safely. In addition, the Emergency Headquarters located on the first floor of the office building suffered little structural damage. However, minor damage (such as door lock failure, overturning of racks, sanitary system trouble, etc.) disrupted the emergency response (Hijikata et al. 2012; TEPCO 2007).

Subsequently, TEPCO decided to use base isolation in "Emergency Headquarters" buildings at every nuclear power plant site. One such building was constructed at the Fukushima Daiichi site in March 2010. The isolation system consisted of four lead-rubber bearings at the corners, 10 natural rubber bearings at the perimeter, 31 sliding bearings, and 16 oil dampers (Hijikata et al. 2012).

The building suffered no damage from the Great East Japan earthquake of March 2011 (Hijikata et al. 2012).

Summary

This entry has presented an overview of the behavior, properties, and applications of lead-rubber seismic isolation bearings.

Seismic hazard mitigation resulting from the use of lead-rubber bearings is twofold. First and foremost, these devices lengthen the structural period and thus decrease the seismic demand on the structure. In addition, they absorb a portion of the seismic energy.

Lead-rubber bearings have been used to isolate many buildings, and buildings using them performed well during strong earthquakes including the 1994 Northridge and 1995 Kobe earthquakes (Kelly 2004) and the 2011 Great East Japan earthquake (Hijikata et al. 2012). Benefits from their use are numerous. Applications include newly constructed buildings and bridges as well as retrofit of existing buildings and bridges. A very significant benefit from the use of lead-rubber bearings for seismic retrofit, other than the (obvious) mitigation of seismic hazard, is the cost-effective preservation and seismic upgrade of historic buildings and other old architecturally and culturally significant structures.

Additionally, benefits from using lead-rubber bearings in high-importance structures are tremendous. Such structures (e.g., bridges, hospitals, government buildings) are expected to resume their normal operation immediately after an extremely strong earthquake. Seismic isolation serves as an additional means to that scope. This benefit is even more evident in high-importance structures the destruction of which would mean not only disruption of crucial service (e.g., transportation, health care, energy production) but also potential health hazard for whole communities (nuclear power plants, chemical plants, etc.).

Cross-References

▶ Code-Based Design: Seismic Isolation of Buildings
▶ Earthquake Protection of Essential Facilities
▶ Friction Dampers for Seismic Protections of Steel Buildings Subjected to Earthquakes: Emphasis on Structural Design
▶ Passive Control Techniques for Retrofitting of Existing Structures
▶ Performance-Based Design Procedure for Structures with Magneto-Rheological Dampers

References

Asher JW, Van Volkinburg DR (1989) Seismic isolation of the USC University Hospital. In: Proceedings of ASCE structures congress, vol 1, San Francisco, pp 605–614
Asher JW et al (1990) Seismic isolation design of the USC University Hospital. In: Proceedings of 4th U.S. National conference on earthquake engineering, Earthquake Engineering Research Institute, vol 3, Palm Springs, pp 529–538
Benjamin JR, Williams HA (1958) The behavior of one-story brick shear walls. American Society of Civil Engineers, Journal of the Structural Engineering Division, Proceedings of the American Society of Civil Engineers, Vol. 84, No. ST4, Paper 1723
Clark PW, Highashino M, Kelly JM (1996) Performance of seismically isolated structures in the January 17, 1994 Northridge Earthquake. In: Proceedings of sixth U.S.-Japan workshop on the improvement of building structural design and construction practices in the United States and Japan. Applied Technology Council and Japan Structural Consultants Association, ATC-15–5, Victoria
Constantinou MC, Tsopelas P, Kasalanati A, Wolff ED (1999) Property modification factors for seismic isolation bearings. Technical report MCEER-99-0012, Multidisciplinary center for earthquake engineering research, University at Buffalo, State University of New York, Department of Civil, Structural and Environmental Engineering, Ketter Hall, Buffalo, NY 14260
Constantinou MC, Whittaker AS, Fenz DM, Apostolakis G (2007a) Seismic isolation of bridges. Department of Civil, Structural and Environmental Engineering, University at Buffalo, State University of New York, Buffalo, NY 14260
Constantinou MC, Whittaker AS, Kalpakidis Y, Fenz DM, Warn GP (2007b) Performance of seismic isolation hardware under service and seismic loading. Technical Report MCEER-07-0012, Multidisciplinary Center for Earthquake Engineering Research, University at Buffalo, State University of New York, Department of Civil, Structural and Environmental Engineering, Ketter Hall, Buffalo, NY 14260
KPFF Consulting Engineers (1991) USC University Hospital, Project summary. Santa Monica
Dynamic Isolation Systems, Incorporated (DIS, Inc.). Portfolio. Online resource. http://www.dis-inc.com/portfolio.html. Accessed 1 June 2014
Hijikata K, Takahashi M, Aoyagi T, Mashimo M (2012) Behavior of a base-isolated building at Fukushima Dai-Ichi nuclear power plant during the Great East Japan Earthquake. In: Proceedings of the international symposium on engineering lessons learned from the 2011 Great East Japan Earthquake, 1–4 Mar 2012, Tokyo
Kalpakidis IV, Constantinou MC (2009a) Effects of heating on the behavior of lead-rubber bearings. I: theory. ASCE J Struct Eng 135(12):1440–1449
Kalpakidis IV, Constantinou MC (2009b) Effects of heating on the behavior of lead-rubber bearings. II: verification of theory. ASCE J Struct Eng 135(12):1450–1461
Kalpakidis IV, Constantinou MC, Whittaker AS (2010) Effects of large cumulative travel on the behavior of lead-rubber seismic isolation bearings. ASCE J Struct Eng 136(5):491–501
Kelly JM (1990) Base isolation: linear theory and design. J Earthq Spectra 6:223–244

Kelly JM (2004) Seismic isolation. In: Bozorgnia Y, Bertero VV (eds) Earthquake engineering from engineering seismology to performance-based engineering. CRC Press, Boca Raton, pp 676–715

Multidisciplinary Center for Earthquake Engineering Research (MCEER) Online resource. http://mceer.buffalo.edu/infoservice/reference_services/baseIsolation2.asp. Accessed 1 June 2014

Robinson WH, Tucker AG (1977) A lead-rubber shear damper. Bull NZ Nat Soc Earthq Eng 10:151–153

Robinson WH, Tucker AG (1983) Test results for lead-rubber bearings for the William M. Clayton building, Toe Toe bridge, and Waiotukupuna Bridge. Bull NZ Nat Soc Earthq Eng 14:21–33

Skinner RI, Robinson WH, McVerry GH (1993) An introduction to seismic isolation. Wiley, Chichester

TEPCO (2007) Damage of Kashiwazaki-Kariwa nuclear power plant site in the Niigataken Chuetsu-oki Earthquake in 2007. Resource document. http://www.tepco.co.jp/nu/kk-np/chuetsu/another-j.html. Accessed 1 June 2014

Walters M (2003) The seismic retrofit of the Oakland City Hall. In: Huang M (ed) Proceedings of SMIP03 seminar on utilization of strong-motion data, Oakland, 22 May 2003, pp 149–164

Learning from Earthquake Disasters

Sarah Beaven[1,3], Thomas Wilson[1,3], Lucy Johnston[2], David Johnston[4,5] and Richard Smith[6]

[1]Department of Geological Sciences, University of Canterbury, Christchurch, New Zealand
[2]University of Canterbury, Christchurch, New Zealand
[3]Natural Hazards Research Group, University of Canterbury, Christchurch, New Zealand
[4]Joint Centre for Disaster Research, GNS Science/Massey University, Wellington, New Zealand
[5]Risk and Society, GNS Science, Lower Hutt, New Zealand
[6]Science and Education, Earthquake Commission, Wellington, New Zealand

Synonyms

All-hazards risk mitigation; Co-creation of knowledge; Disaster data management; Earthquake risk reduction; Global disaster rates; Policy learning

Introduction

The United Nations International Strategy for Disaster Reduction (UNISDR) defines disaster as the "serious disruption of the functioning of a community or society involving widespread human, material economic or environmental losses and impacts, which exceeds the ability of the affected community or society to cope using its own resources" (UNISDR 2014). The idea of learning, from earthquake disasters, to reduce the risks associated with future earthquakes is widely held around the world and across a range of sectors. The United Nations defines disaster risk reduction as the "concept and practice of reducing disaster risks through systematic efforts to analyse and manage the causal factors of disasters, including through reduced exposure to hazards, lessened vulnerability of people and property, wise management of land and the environment, and improved preparedness for adverse events" (UNISDR 2014). The research sector has traditionally contributed to these efforts according to discipline. The physical sciences identify the causes and probability of earthquake hazards and relevant branches of engineering focus on mitigating the impact of earthquakes on the built environment, while the medical and social sciences research a range of human impacts, in socioeconomic, political, and cultural spheres and measures to address them.

The growing emphasis on risk reduction, however, is creating a demand for integrated research approaches that focus directly on contributing to disaster mitigation initiatives. The US Earthquake Engineering Research Institute (EERI), for example, aims to integrate engineering with physical and social science initiatives and so reduce earthquake risk by (1) advancing the science and practice of earthquake engineering; (2) improving understanding of the impact of earthquakes on the physical, social, economic, political, and cultural environment; and (3) advocating comprehensive and realistic measures for reducing the harmful effects of earthquakes (EERI 2014).

UNISDR has also stressed the importance of more multidisciplinary, problem-solving approaches to disaster risk reduction, warning that such initiatives will require better mechanisms "for integrating science and technology into policy processes" and "for bridging the gaps between environmental, humanitarian, development and governmental actors" (UNISDR 2014).

This emphasis on increased coordination between disciplines, scientific and policy domains, and between global sectors is informed, in turn, by a growing perception that existing knowledge is *not* being translated into effective disaster risk reduction measures. The 2009 Science Plan for Integrated Research into Disaster Risk (IRDR), for example, is a response to concern that "the widening gap between advancing science and technology and society's ability to capture and use them" is a major factor in rising global disaster losses (ICSU 2008). Jointly sponsored by UNISDR, the International Council for Science (ICSU), and the International Social Science Council (ISSC), this plan aims to foster a better understanding of the processes involved in the translation of "research findings about natural hazards and human behavior" into practices and policies that are "effective in minimizing the human and economic costs of hazards" (ICSU 2008, p. 5, 7).

This Science Plan draws on White et al.'s seminal 2001 review of trends in the hazard and disaster field. Explicitly couched as an investigation of the relationship between increasing disaster losses and the science that could have been expected to slow, arrest, and reverse that rate of increase, this review began with a series of questions: "Is human knowledge and understanding of the causes of the losses inadequate despite the research effort, or is it that existing knowledge is not applied, or not used in an effective fashion? Could there be other explanations that lie outside our scholastic assumption that knowledge has a key role to play?" (White et al. 2001, p. 81).

Finding that the preceding decades had featured increasing convergence in the literature concerning broader causal factors, White et al. concluded that the lack of knowledge explanation was unlikely. Instead the evidence they draw from this field suggested that existing scientific knowledge had either not been applied at all or had not been applied effectively. In addition, they found that global patterns of unsustainable development had also contributed to rising disaster losses (White et al. 2001).

This encyclopedia chapter is largely concerned with research into the broader, *collective* learning processes and pathways found to have been involved in the translation of scientific knowledge into earthquake disaster risk reduction policy, as well as into challenges to such learning. The first section begins with examples taken from US disaster policy research that are directly concerned with the translation of research findings into policy and practice after disaster events. Using social learning policy approaches, this work has confirmed that earthquake and other disasters have had a focusing effect on policy agendas in the US. Higher-level support, policy frameworks that facilitate risk reduction, increasingly networked global communities, and multilateral learning networks have been found to be key factors in learning disaster risk reduction lessons after major events. These US examples are followed by an example from New Zealand, in which learning from an earthquake disaster has informed an all-hazards risk reduction approach and a policy framework that has devolved responsibility for risk reduction to local levels.

Where the first section is concerned with research focused on the opportunities that contemporary developments offer for disaster risk reduction, the second section turns to consider the inhibiting effects the same developments have had on disaster risk reduction. Outlining evidence of barriers to consensus building both within and across disciplines, sectors, and domains, it also touches on data management issues associated with recent technological and social developments. The article concludes with a brief discussion of the larger global issues that have been found to be associated with the distribution of vulnerability to disasters, unsustainable development, the emergence of subgovernment policy communities, and the emerging focus on major disasters and risk.

Largely limited to considerations of *broader collective learning processes*, the focus is schematic, rather than comprehensive, and so leaves significant research outside the immediate frame. For a comprehensive review of the current wider hazard and disaster research landscape, readers are referred to White et al.'s elegant 2001 account. The rich body of work concerned with historical and other understandings of lesson learning after earthquake disasters also falls largely outside scope, as does extensive recent progress in understanding the psychological and other processes involved in decision-making by individuals.

Finally, a brief clarification concerning the tone of this entry, and expectations as to audience. The expectation for the encyclopedia entry is that the majority of its readers are unlikely to be familiar with the field. Accordingly current findings in the relevant topic area are conveyed as clearly as possible for a wider disciplinary audience. This particular field is often divided, particularly over the significance of the extension of governance to include social actors and current emphases on consensus and risk reduction. Such debates are understood to be integral to the production of knowledge and, where relevant to the focus of this entry, are conveyed as evenhandedly as possible. The field is also characterized by densely logical argument, the comprehensive referencing of influences, carefully defined terminology, and painstaking attention to the detail, diversity, and complexity of subject matter. For reasons of scope, and in the interests of clarity, the following comparatively brusque reduction of this diverse wider body of work to a few broad findings runs counter to these hallmarks. References are provided for those who wish to pursue the arguments in more depth.

Translating Science into Policy and Practice

It is widely acknowledged that disaster risk reduction is the responsibility of national governments and needs to be effected at local levels (UNISDR 2014; Global Network 2013). Recent global developments, however, have impacted on political authority at both national and local levels. Economic integration and market liberalism have limited the authority of governments over multinational economic actors, global bodies are able to impose political requirements on nation states that often conflict with the demands of their constituents, and information and communication technologies have facilitated both global economic transactions and transnational networks of civil society actors (Skogstad 2003). This has contributed to an increase in expert and market-based authority, at the expense of democratic process, as governments respond to an increasingly complex and fragmented policy-making environment by turning to these and other sectors for resources and cooperation (Skogstad 2003, 2005). But at the same time, increased appetite for public participation – fuelled by transnational networks of civil society actors – has also contributed to an international trend of greater public involvement in policy formation, so offering the potential, at least, for new forms of democratic process (Rowe and Frewer 2005; Skogstad 2003).

As a result, a large part of policy making in industrial societies is now carried out collaboratively by government officials, scientific and other experts, and public representatives, through "decentralized, and more or less regularized and coordinated, interactions between state and societal actors" (Skogstad 2005). As a means "to coordinate resources of information, support and authority across state and non-state actors," this subgovernment sphere has been recognized to have the potential to provide "more effective government, and arguably... more legitimate government" (Skogstad 2005). This section is concerned with research that has focused on the *opportunities* for earthquake risk reduction in the United States and New Zealand that have been provided by these changes.

Higher-Level Coordination in the US Earthquake Policy Domain

US policy theorists interested in the role of technical information in policy making have responded to these changes in the political environment and to emerging complexity theory, by focusing on the wider influences on the policy process over time. In particular, this expanded focus includes all actors who regularly and actively attempt to influence public policy concerning a particular issue or problem. Collectively described as policy subsystems, communities, networks, or domains, these larger groupings have been found to typically manifest (one or more) advocacy coalitions (Skogstad 2005; Birkland 1998; Sabatier 1998). Usually including actors from a range of government and other organizations and interest groups, this kind of coalition has been found to be held together by shared beliefs about the relevant policy issue. By engaging "in coordinated activity over time," advocacy coalitions have been found to influence policy outcomes (Sabatier 1998). It has been widely recognized, however, that major policy changes have not been caused solely by the activity of such coalitions. Instead, these policy changes only happen when external factors create a much wider perception that the existing policy has *failed*. Equally, although necessary, this wider consensus has not been found sufficient in itself to trigger major shifts in policy (Birkland 1998; Sabatier 1998).

The Focusing Effect of Earthquake Disasters

The only significant application of this theory in the earthquake risk reduction context is Birkland's (1998) influential comparative study of four US disaster policy domains between 1960 and 1990. This project compares all congressional hearing testimony concerning earthquake, hurricane, oil spill, and nuclear accident disasters over this 30-year period. It was designed to test the idea that "sudden, attention-grabbing events, known as focusing events," could advance issues on the relevant policy agenda and – by contributing to the perception that existing policy had failed – potentially trigger major changes in risk reduction policy (Birkland 1998). Birkland established that major US earthquake events between 1960 and 1990 had had a significant focusing effect on the congressional earthquake policy agenda. After each major event, the dominant topic on the policy agenda changed, the volume of earthquake testimony increased, and much more of this post-event testimony was critical of existing policy. Testimony that referred to specific earthquake events was also much more likely to be critical of existing policy than other earthquake testimonies (Birkland 1998). This contributed to the much wider consensus concerning the need for major change to earthquake risk reduction policy that led to the 1977 National Earthquake Hazard Reduction (NEHR) Act (Birkland 1998).

Birkland's (1998) research also established that the focusing effect of major events was much greater in the earthquake domain than in the other three disaster domains. The comparison between congressional testimony in the oil spill and nuclear power plant accident policy domains indicated that such differences were likely to be related to disaster coverage (Birkland 1998). Major nuclear accidents had had little impact on the volume or tone of congressional testimony, which continued to feature a deadlock between the powerful industry lobby and those opposed to it. By contrast, the high-profile Exxon Valdez disaster triggered the resolution of a similar 14-year deadlock between the oil industry and environmental advocacy groups, by way of the major policy change enacted through the 1990 Oil Pollution Act (Birkland 1998). Birkland (1998) found that the ambiguity of the harm caused by nuclear accidents contributed to their lack of focusing effect. Unable to be visually conveyed to the public, this harm was open to scientific dispute, unlike the dramatic footage of oil-soaked beaches and struggling wildlife following the Exxon Valdez disaster.

Coordinated Scientific/Political Advocacy

Testimony in the earthquake policy domain revealed that the 1977 NEHR Act was drafted and enacted by a very active coalition of senior scientists, academics, and government officials, who had come together over the preceding 15 years to lobby congress for changes to the existing earthquake policy (Birkland 1998). Coalescing around a shared commitment to science-based earthquake risk reduction, this group dominated the wider policy community involved in congressional earthquake hearings and policy formation between 1960 and 1990. After the 1964 9.2 Alaskan and the 1971 6.6 San Fernando events in particular, this coalition was responsible for authoritative critiques of existing policy in congressional science committee hearings (Birkland 1998; Hamilton 2003).

This distinguished the earthquake policy domain from the hurricane domain over the same period, which continued to be dominated by local governmental interests pushing for disaster relief and engineered solutions to hurricanes. Although prominent researchers and officials had warned of the need for hurricane mitigation over this period, they had not succeeded in activating a multi-sectoral coalition that lobbied "for improved federal policy to deal with hurricanes in *testimony before the Congress*" (Birkland 1998; emphasis original). Where the earthquake agenda featured a mix of mitigation and disaster relief policy and was usually heard in congressional science committees, the majority of hurricane testimony concerned disaster relief and was heard in congressional public works committees, where mitigation was subordinate to local distributive spending on disaster relief and public works projects (Birkland 1998).

Higher-Level Coordination of Disciplinary and Agency Fragmentation and Competition

The translation of science into policy in this US earthquake disaster domain thus required that the scientific community became politically active and forged alliances within the government sector, first in pushing for changes to policy and then in drafting and jointly enacting these changes. This was a complex and extensive process, however, which involved as much competition as it did consensus building, both within and between scientific and government spheres.

After the 9.2 1964 Alaskan earthquake drew attention to potential earthquake risk in the US, for example, several major earthquake research centers were established which then "competed vigorously to stake out a lead-agency position for earthquake research" (Hamilton 2003). Over the same period three high-level reports proposing competing national earthquake research programs were presented to congressional committees in 1965, 1968, and 1969 by agencies representing seismological, geological, and engineering interests, respectively. By 1970, "competition among the disciplines and between the agencies, combined with the waning concern after the Alaskan earthquake, contributed to a lack of budgetary attention to the earthquake threat." (Hamilton 2003).

Then on 9 February 1971, the 6.6 San Fernando earthquake caused 65 fatalities and spectacular footage of collapsed freeway overpasses and hospitals. Renewing both public and high-level political interest in the issue, this event led to federal reforms merging relevant agencies, at the same time as renewed congressional activity led to the 1977 NEHR Act (Hamilton 2003; Birkland 1998). A sequence of further changes to the Act established the NEHR Program (NEHRP), formalizing the role of the advocacy coalition and addressing the fragmentation and competition that had characterized this domain by integrating all earthquake-related disciplines, policy, and practice into a single coordinated program (Hamilton 2003). Responsible for policy formation and for funding and coordinating a range of research programs, this high-level multilateral initiative continues to include US federal agencies responsible for emergency response (Federal Emergency Management Association [FEMA]), standards and technology (National Institute of Standards and Technology [NIST]), and for both academic (National Academy of Sciences [NAS]) and government research (United States Geological Survey [USGS]). The directors of these agencies, together with those of the White House Office of Science and Technology Policy (OSTP), and the Office of Management and Budget (OMB) make up the NEHRP Interagency Coordinating Committee (ICC) (NEHRP 2014).

This research into the US earthquake policy domain established that scientific findings concerning earthquake risk reduction have been translated into policies and practices through complex, high-level processes that have extended well beyond decision-making interactions between individual scientists and policy makers. Key factors included:

- A political environment in which governance structures were being extended to include non-state actors, through emerging subgovernment networks responsible for drafting and implementing policy
- High-profile earthquake disasters, which triggered a widespread consensus that policy change was required – lessons really have been learned *from* earthquake events
- A politically active scientific community, able and open to forming the ongoing coalitions required in the formation of such networks and committed to using the focusing effect of disasters to push for policy change
- Ongoing high-level political intervention to reduce disciplinary and government sector fragmentation and competition, by creating an integrated national earthquake risk reduction coordinating program.

Disaster Risk Reduction Policy and Practice: Coordination at Local Levels

Research into the role of policy subsystems in disaster risk reduction has not been restricted to higher-level changes to policy. Others have focused on such activity at the local level. Busenberg (2000), for example, has studied the local policy subsystem that emerged in Alaska after the 1989 Exxon Valdez disaster. The 1990 Oil Pollution Act (OPA) provided for a Regional Citizens Advisory Council (RCAC) to oversee environmental risk management in the region, legislating a permanent role for the advocacy coalition involved in its design and implementation (as did the 1977 NEHR Act) (Birkland 1998). Guaranteeing the independence of the RCAC, the 1990 OPA requires the oil industry to fund the operations of this body for as long as the Alaskan Pipeline remains in operation. The RCAC has 19 member organizations. These include a range of local interest and local government groups, each with a representative on the Board of Directors; five issue-specific committees include board members, experts, and local citizens.

Busenberg focused on the innovative learning arrangements that brought the RCAC into the wider, multilateral policy network involved in significant policy decisions concerning the environmental management of the industry in the region (Busenberg 2000). Between 1990 and 2000, these decisions resulted in policy changes that significantly reduced the risk of future oil spill disasters. This included legislation requiring a sophisticated local weather forecasting system, an integrated regional marine firefighting service, and supported tanker navigation (including redundant tug capacity) within Prince William Sound. The first of these initiatives was immediately enacted, while the latter two required supporting evidence established by research programs partially or completely funded by the RCAC (Busenberg 2000).

Social Learning in Relation to Other Theories of Learning and Participation

Busenberg and Birkland are both working within the social learning policy theory tradition. For both, the term "learning" does not simply refer to an increase in knowledge but is rather defined in relation to policy change. Busenberg has distinguished the incremental, ongoing local-level policy learning that he traces over 10 years in Prince William Sound from the larger changes in policy after major events, which interests Birkland. The coordination and consensus building process is, however, very similar, and both researchers define the social learning process in relation to policy outcome. Busenberg finds that local, incremental learning within the relevant networks of organizations and individuals occurred only when initiatives gained enough support from key stakeholders to be *enacted as risk-reducing policy change*. In the case of supported tanker navigation, this learning had required a higher-level intervention from the state governor. By contrast, initiatives that were researched and promoted but were not enacted in policy are interpreted as evidence that, in these cases, social learning did *not* occur. Similarly, Birkland defines social learning with reference *to* evidence of policy outcome: "Policy learning can be identified if there is prima facie evidence of policy changes that are reasonably linked to the causal factors that connected the event under consideration to its harms, and if addressing these factors would be likely to mitigate the problem" (Birkland 2009).

The social learning process explored by these and other policy theorists is described as social because it is not restricted to individuals, or organizations, or even sectors, but relies on a *much wider social consensus* concerning the failure of existing policy. It is understood to be a learning process in part because it draws from other learning theories that also feature this negatively driven sequence, in which previously unexamined assumptions are understood to have failed, and so triggered a new set of assumptions. Informed by both Kuhn's scientific theory of paradigm shifts and so-called double-loop organizational learning theory, for example, social learning theory also relies on the punctuated equilibrium pattern identified by complexity theory as that in which complex human, social, ecological, and biological systems learn by adapting to disruption. Equally, for most of these theorists, the activities of policy subsystems are described as learning because they *increase knowledge* concerning both the policy problem and likely outcomes of policy alternatives (Sabatier 1998). The emphasis on improved policy outcomes, however, is distinct to Busenberg and Birkland's application of this approach to disaster risk reduction policy.

It is useful to clarify the relationship between this social or policy learning framework and similar concepts current in the hazard and disaster context. A variety of closely related learning arrangements have been recently reviewed by Voorberg et al. (2013). Covering a variety of disciplines, this review finds the terms social innovation, knowledge co-creation, or knowledge coproduction are all used to describe what appears to be the same broad learning process (note that their review does *not* include constructivist uses of the term knowledge coproduction, which refer to the mutually constitutive relationship between science and society). This embedded process takes place in a specific local and institutional context, in which relevant stakeholders participate and collaborate across organizational boundaries and jurisdictions, engaging in interactive exchanges of knowledge, information, and experiences through intra- and interorganizational networks (Voorberg et al. 2013). To this extent, the social learning process constitutes a particular variant of this collaborative learning process. However, Voorberg et al. (2013) found that irrespective of discipline, the focus is almost entirely restricted to considering the process, and value, of such participation, rather than its outcomes, if any. By contrast, the social learning process researched by Birkland and Busenberg is defined in relation to policy outcome.

Secondly, it is also useful to relate the social learning policy process to the international trend of greater public and community involvement in policy formation. Such participation is recognized to be an essential component of any strategy designed to increase community resilience to disasters. In a review of relevant literature, Rowe and Frewer (2005) found more than 120 types of public participation described. They distinguish between communication, consultation, and participation, in order to develop a useful typology (Table 1).

According to this typology the RCAC studied by Busenberg can be seen as a *participatory mechanism* of public engagement – although established as a council, the focus on oil spill disaster risk reduction puts it in the task force category: the use of "a small group of participants (public representatives) with ready access to all pertinent information, to solve specific problems" (Rowe and Frewer 2005). Congressional hearings, however, largely constitute a public *communication mechanism*. The social learning that concerns Busenberg and Birkland does thus rely on forms of public engagement. Their focus, however, is not on public engagement per se, but rather on the larger policy process, including all events and factors that influence policy design and enactment; again, their emphasis is on the policy outcome.

Learning from Earthquake Disasters, Table 1 Mechanisms of public engagement (Adapted from Rowe and Frewer 2005)

Engagement mechanism	Examples	Characteristics
Communication		**Information to > public**
Communication type 1	Traditional "publicity"	Controlled selection; targets particular population with set information
Communication type 2	Public hearings, seminars, etc.	Rely on public initiative, provide flexible information
Communication type 3	Drop-in centers, website information	Rely on public initiative; provide set information (even if FTF)
Communication type 4	Hotlines	Rely on public initiative; uncontrolled selection/flexible information
Consultation		**Information from <- public**
Consultation type 1	Opinion poll, survey, referendum, etc.	Large samples – specific questions; controlled selection/closed response
Consultation type 2	Consultation document	Open responses on specific issues – controlled selection
Consultation type 3	Electronic consultation – interactive website	Open responses on specific issue – uncontrolled selection
Consultation type 4	Focus group	Emphasis on information quality; open responses, controlled selection
Consultation type 5	Study circle	Emphasis on information quality; open responses, uncontrolled selection
Consultation type 6	Citizen panel – e.g., health panel	Ongoing debate of topics; controlled selection, open responses
Participation		**Information to/from <-> public**
Participation type 1	Planning workshop, citizens jury	Controlled selection, facilitated expert and other inputs, open responses
Participation type 2	Negotiated rule making; task force	Groups access pertinent information to solve problems; open responses controlled selection, inputs unfacilitated
Participation type 3	Deliberative polling or planning cell	Polled before/after deliberation; controlled selection, open responses
Participation type 4	Town meeting with voting	Voting after debate; uncontrolled selection; open response mode

All-Hazard, Risk-Based, Problem-Solving Approaches to Disaster Mitigation

The third example of this kind of social learning from earthquake events is taken from New Zealand and occurred in the aftermath of the 1987 M_w6.3 Edgecumbe earthquake. An earlier history of seismic activity, including the destructive M_w 7.8 Napier earthquake in 1931, had contributed to well-enforced New Zealand building codes. These codes, and the fact that the Edgecumbe event occurred in a largely rural area, contributed to a lack of fatalities. These factors had not, however, been able to prevent total losses of NZ$430 million. Drawing political and international attention to the potential for much greater earthquake losses in New Zealand and the Pacific region, the Edgecumbe event also highlighted the potential for losses associated with damage to critical facilities, including supply of water, electricity, and communications, as well as sewerage disposal and transport (Shephard 1997).

Largely in response to this event, major national policy changes "designed to encourage more effective safety and loss prevention strategies" were enacted in a series of subsequent government reforms, including the 1987 Recovery Plan for Natural Disasters, and reforms to the Local Government Act 1989, the Resource Management Act 1991 and the Building Act 1991

(Shephard 1997). The "essential idea" of these reforms: "was that central government would accept shared responsibility for the restoration of damage from natural disasters *only if the local authority concerned had done its part to minimize, mitigate and manage the risk to its assets.... they were expected to... put in place protection and damage limitation measures that would reduce the consequences.*" (Helm 1996; cited in CAE 1997, emphasis added).

Where the focusing effect of US earthquake and oil spill disaster events had led to *hazard-specific policy reforms*, the New Zealand government emphasis on risk and decentralized responsibility effectively meant that the lesson learned from the Edgecumbe earthquake was *to prepare locally for all hazards*.

This political climate facilitated the collaborative Christchurch Engineering Lifelines Project, conducted in 1993 and 1994 by task groups that included physical scientists, academic and professional engineers, local and national government agencies, and representatives from relevant private sectors, including insurance and utility providers. The project was run out of the University of Canterbury's Center for Advanced Engineering (CAE), which was founded in May 1987, 2 months after the Edgecumbe earthquake. The Christchurch Engineering Lifelines Project aimed to identify the vulnerability of critical facilities to damage from a *range of natural hazards* and to arrive at practical strategies to mitigate potential damage; it also aimed to communicate relevant issues to the people involved in managing the services and the public (CAE 1997). Creating detailed hazard and critical facility maps, multi-sector task groups came together to superimpose these maps, going over critical facilities in detail, network by network, to assess the likely vulnerability and importance of elements, discuss often inexpensive mitigation measures, and assess risks associated with network interdependence.

Rather than simply translating scientific findings into risk reduction policy and practice, this project involved the collective, cogeneration of risk reduction knowledge through the interactions of scientists, engineers, local government, and those with day-to-day experience of the relevant lifelines. In addition to being deeply local and enabled by a strong national policy framework, this project was also informed by globally networked international research and industry communities.

The initial assessment phase was conducted during 1993 and 1994. In collaboration with the New Zealand Society for Earthquake Engineering (NZSEE) and with financial support from New Zealand's Earthquake Commission (EQC), project members also visited Los Angeles in 1994 following the M_w 7.2 Northridge earthquake and Kobe after the 1995 Great Hanshin earthquake. The Christchurch project was detailed and disseminated in a 1997 report (CAE 1997). In addition to summarizing the mitigation measures already carried out and ongoing project work, this report noted that "a very significant benefit" had been the formation of the multilateral working group, which had brought together those with "dissimilar working interests," and so established useful ongoing relationships "in preparation, as it were, for a major emergency requiring the various utilities to work together" (CAE 1997). The subsequent Civil Defence and Emergency Management Act (2002) built on this legacy. Devolving hazard and emergency management to the local level and supporting the integration of regional lifelines groups in emergency management training programs, this act reinforced the culture of hazard mitigation, contributing to a community of practice involving regional providers of critical facilities.

In 2010 the M_w 7.2 Greendale earthquake began the destructive "Canterbury earthquake sequence" that included the M_w 6.2 Christchurch earthquake on 22 February 2011. Leading to 185 deaths in Christchurch, widespread liquefaction and lateral spreading in the central and eastern suburbs of the city, this event caused extensive damage to buildings and critical facilities. In a recent study, the Lifelines project and its outcomes have been estimated to have significantly reduced the impacts of the earthquake sequence on the city (Fenwick 2012).

Risk reduction initiatives developed through the project are estimated to have greatly reduced the overall community losses associated with critical facility outages. Investment in alternative, redundant supply routes for electricity and telecommunications allowed for rapid restoration of supply in all but the most badly damaged regions. Telecommunications companies with "strong node" and alternative "hot spot" buildings were able to maintain coverage in the city. Seismic strengthening in the city's port, directly over the epicenter of the damaging 22 February 2011 event, made it possible for a New Zealand naval frigate to dock and unload aid and supplies within 4 h of this event; the port was operational again within days (Fenwick 2012).

Utility companies that had adopted a consistent focus on risk mitigation also saved significantly in direct asset replacement and repair costs. Electricity provider Orion Energy, for example, spent $NZ6 million on seismic strengthening and is estimated to have saved $NZ60-65 million, while the total repair costs after the February event paid by another provider, Transpower, amounted to only $NZ150,000 (Fenwick 2012). By contrast, much of the aging water and wastewater network the project had identified as a seismic risk had proved too expensive and difficult to upgrade. The 2014 estimate of the cost of repairing these seismically damaged networks is NZ$2 billion (SCIRT 2014).

At least as valuable as the financial saving was the networked community of professional and academic engineers, infrastructure companies, and local and national authorities created by the project and further developed by EQC, NZSEE, and MCDEM initiatives. This collaborative critical facilities culture greatly facilitated the restoration of lifeline function after each of the four major events in the 2010–2011 Christchurch earthquake sequence. It has also informed the "Stronger Christchurch Infrastructure Rebuild Team" (SCIRT), a formal alliance of national and local government agencies and private organizations tasked with rebuilding the city's horizontal infrastructure (SCIRT 2014). Designed to "harness the expertise of public and private sectors, promote information sharing and collaboration, and manage risks and opportunities" this work program includes seismic strengthening (SCIRT 2014).

Learning from Earthquake Disasters, Table 2 Comparative policy learning from earthquake disasters in the United States and New Zealand

In the United States and New Zealand:	
Major earthquakes caused extensive damage	
At a time when governance structures were extending to include non-state actors	
Creating a widespread perception that existing legislation had failed	
Leading to a range of legislation designed to address issues exposed by the event(s)	
Generating increased research activity	
Motivating ongoing coalitions between state and non-state actors in learning networks	
Differences:	
United States	New Zealand
Hazard-specific legislation	All-hazard legislative focus
Hazard-specific research and coalitions	All-hazard research and coalitions
Higher-level initiatives and organizational coalitions	Devolution risk management to local-level initiatives and organizational coalitions

As in the US examples, it is clear that the 1987 Edgecumbe earthquake occurred at a time in which governance structures were being extended; the earthquake thus contributed to a widespread consensus that policy change was required, a range of legislation devolving responsibility for risk reduction to the local level, increased research activity, and the motivation to form the ongoing coalitions holding the relevant learning network together. The 2010–2011 Canterbury earthquake sequence confirmed the value of this learning, while also focusing attention again on the need for seismic risk reduction and so informing the Stronger Christchurch Infrastructure Rebuild Team initiative.

However, the multi-hazard approach and the devolution of responsibility for all hazards to the local level differ significantly from the *hazard-specific* US policy responses to earthquake disasters (see Table 2).

To the extent that it has required that local authorities, communities, researchers, and the private sector work together to reduce the risks of a wide range of hazards in their own local region, the focusing effect of an earthquake disaster has been used to reduce the risks associated with a much wider range of disaster types. This can be seen as an example, then, of the kind of "all-hazard, risk-based, problem-solving approach" to disaster risk reduction recommended by UNISDR, which requires "collaboration and communication across the scientific disciplines and with all stakeholders, including representatives of governmental institutions, scientific and technical specialists and members of the communities at risk to guide scientific research, set research agendas, [and] bridge the various gaps between risks and between stakeholders" (UNISDR 2014). On the other hand, however, this process remained limited in focus to the vulnerability of city infrastructure – it did not include social vulnerability or involve the significant public participation constituted by the RCAC; so to this extent, it did not involve the collaboration with "all stakeholders" specified by the UNISDR.

Challenges to Learning from Earthquake Disasters

Considering the challenges to this kind of learning after earthquakes means turning from work focused on the opportunities offered by the focusing effects of earthquake events, increasingly networked global communities, and the extension of governance to include social actors to work that is focused on the *risks* posed by the same developments. This section starts with challenges to the formation and operation of collaborative, "innovative" learning or policy communities, and a brief discussion of identified barriers to higher-level responses to earthquake disasters. Data management issues and wider implications of the role of focusing disaster events and unsustainable global development conclude the section.

Challenges to Multidisciplinary, Multi-sector Learning After Earthquakes

The processes that bring actors together from across sectors and disciplines to coalesce around the shared goal of risk reduction are also at work around the sets of ideas or paradigms that are shared within individual disciplines, organizations, and sectors. At every level, incompatibility between these idea sets, or shared norms and values, have been found to have the potential to inhibit knowledge sharing and so impede the translation of scientific knowledge into policy and practice.

This is apparent in Birkland's (1998) comparative study, for example, which revealed that a relative lack of focusing effect effectively left the hurricane policy domain dominated by local government interests, who prioritized policies that were *relevant to their constituents.* This meant they preferred engineered solutions like levees nonstructural mitigation techniques, such as stricter enforcement of flood plain regulation or coastal development, considered by Birkland and other researchers to be a more *scientifically credible* mitigation approach. Similarly, the "ambiguity" of the harms caused by nuclear accidents had left *commercial* industrial interests dominating the nuclear disaster policy domain. Focusing events had been required in the oil spill and earthquake domains to generate enough social and political pressure to trigger higher-level commitment to risk-reducing policy change. Without this shift in the political environment, the natural disaster domain remained dominated by the drive for *political relevance*, while the industrial disaster domain remained dominated by the commercial drive for *financial profit;* neither of these drivers was compatible with the drive for *scientific credibility* of concern to Birkland and Busenberg.

Incompatibility between these drivers has also been found to significantly inhibit the translation of scientific findings into policy and practice at the level of individual decision-making. In a recent study, for example, policy makers and practitioners working in disaster risk reduction and environmental management domains, in both developed and developing countries, did not access relevant academic publications unless they had been directly involved in their production (Weichselgartner and Kasperson 2010). They were in any case reluctant to use this research – assessing the vulnerability and resilience of populations in their own geographic jurisdictions – to inform decisions about disaster risk reduction and environmental policy and practice. Although conceding broad relevance, policy makers and practitioners did not consider these research findings to be relevant to their specific knowledge needs (Weichselgartner and Kasperson 2010). At the same time, even in this domain, where researchers are typically highly motivated to contribute to disaster risk mitigation, the authors of the relevant publications acknowledged that their work was largely driven by considerations of *scientific credibility* and so designed, conducted, and disseminated with an academic audience in mind, rather than the needs of potential users (Weichselgartner and Kasperson 2010).

By creating incompatible cultures and values, these drivers have also been found to contribute to value clashes between academic and other spheres that inhibit cross-sector collaborative activity (Cash et al. 2003). Academic incentive regimes are based on publications and peer recognition, which have been described as academic "status competitions," sometimes taking "the form of winner takes all" when it comes to publishing first or winning research grants (Bruneel et al. 2010). Contesting the credibility of findings is an essential part of the peer review process, informing a value set in which contest is more usual than that in private and government sectors, where individuals are required to subordinate their own interests to those of their organization and where commercial and political incentive regimes can require that knowledge is kept private, rather than contested in the public domain (Bruneel et al. 2010).

In a range of collaborations involving research and private sectors, these differences in cultures, norms, and values have been found to inhibit collaborative knowledge production. Conversely, ongoing cross-sector engagement over time has been found to *significantly reduce* this inhibiting effect (Bruneel et al. 2010; Cash et al. 2003). The balance of contestation and consensus building required to initiate and continue such cross-sector engagements, however, has been found to be delicate; major value clashes have been found to be particularly compromising.

Challenges to Higher-Level Learning from Earthquake Disasters

The tensions between potentially incompatible scientific, economic, and political drivers that have worked against coordinated disaster risk reduction initiatives at local and national levels are also in evidence in the recent global developments that have reduced political authority at these levels. Economic integration and market liberalism have limited the political authority of governments over multinational economic actors; bodies like the WTO are able to impose global political requirements on nation states that often conflict with the demands of their constituents, and information and communication technologies have facilitated both global economic transactions and transnational networks of civil society actors (Skogstad 2003). These limitations on political authority at the national level, moreover, have not been evenly distributed. Less affluent, developing nations have been much less well positioned than developed nations when it comes to asserting political sovereignty over the activities of multinational companies and against the political strictures of global organizations. Consequently the nations *most exposed* to disaster risk have also been those nations least able to shoulder the responsibility for disaster risk reduction that continues to fall to national government level.

This has meant that developed countries that experience earthquake disasters are more likely to benefit from the distribution of political, public, and international attention and have more to gain from the resultant focusing effect than developing countries. A sequence of Global Network reports has found that networked disaster risk reduction activities involving "state and non-state actors" working "together to ensure the safety and well-being of their communities" have been much more effective "in countries where capable, accountable and responsive local governments worked collaboratively with civil society and at-risk communities" (Global Network 2013). However, there has been very little progress in developing countries across a range of risk governance indicators, including "lack of political authority; inadequate capacities and financial resources; and minimal support from central government," which have all been identified as "significant barriers" to the implementation of disaster risk reduction policies at the local level (Global Network 2013).

In many countries, this can mean that rather than triggering risk-reducing policy changes, the focusing effect of a major earthquake disaster can instead generate a tidal wave of international aid and more or less fragmented and competitive activity among aid organizations. Dombrowsky, for example, has found US$4 billion in aid donated worldwide after the 2004 Sumatran earthquake and subsequent tsunami provided an "excessive" volume of food and clothing that "wrecked local agricultural, market and production systems," while the "accelerated" building of new housing created "extreme shortages of materials, harmful deforestation, profiteering and, more recently, inappropriate construction methods" (Dombrowsky 2007). The "tens of thousands of projects" carried out by competing international organizations in Southeast Asia after the Sumatran earthquake and tsunami "were not networked and there was no co-ordination aimed at developing a long-term, preventive protection and sustainability strategy" (Dombrowsky 2007).

Conversely, Christoplos (2006) has argued that although scientifically credible, disaster risk reduction initiatives can be incompatible with local and global *political* imperatives after disasters. Informing the humanitarian focus on "addressing acute human suffering," these political drivers also inform the immediate priorities for local "politicians, donors and disaster affected people," who need to get "people out of tents and into houses and off food aid and into jobs" (Christoplos 2006). In addition, he noted that the 2004 tsunami decimated many of the local governments and communities required to develop networked community disaster risk reduction initiatives. He has called for compromise, rather than blame casting and identifying "villains," from those working toward disaster risk reduction, in order to achieve a consensus concerning larger, shared goals (Christoplos 2006).

Data Management Challenges to Learning from Earthquake Disasters

The focusing effect of disasters has also been found to contribute to the enormous volume of post-disaster reviews, investigations, and evaluations, most entitled "lessons learned," which are conducted and documented at different levels, in a range of government, academic, NGO, and private sectors. Again, this material is largely produced in response to *political* pressures and so conducted to reassure relevant authorities and the public that an attempt to learn lessons has been made, rather than to generate scientifically credible evidence for policy change (Birkland 2009). International aid agencies are also under intense political pressure to prove that their money is being well spent. This pressure can lead NGOs to compete for prestige projects and the associated positive publicity. It also contributes to considerable information gathering, largely for internal consumption, within organizations (Dombrowsky 2007). Much of this material may contain risk reduction lessons of scientific and practical value. It cannot, however, be easily accessed, since it is not coordinated or collated, but remains fragmented, dispersed within agencies and organizations, and across a diverse range of academic journals (Birkland 2009).

The same issue occurs at the global scale. As the pace with which information technology evolves outstrips the capacity to *coordinate* data management across local, national, and global levels, it becomes increasingly difficult to establish global disaster data. Organizations like Munich Re and the Center for Research on the Epidemiology of Disasters (CRED) have been set up to address this issue by gathering credible, comparable global data and providing it in an accessible form. Even these organizations, however, are struggling with inconsistencies created by "the use of different data sources, different frameworks, different metrics, or different scales" and because of "a lack of consistency in the management of the data on which the assessments are based" (ICSU 2009).

The management and accessibility of *research data* in this area has also been identified as an urgent issue. The 2013 OECD Global Science Forum report on Data and Research Infrastructure for the Social Sciences, for example, addressed "rapid growth in new forms of data collected in conjunction with commercial transactions, internet searches, social networking, and the like, and by technological advances in the capacity to access and link existing survey, census, and administrative data sets." It was tasked with weighing the opportunities offered by this growth for evidence-based research concerning disasters and other global challenges related to the health and well-being of populations around the world (OECD 2013). Issues identified in this report as imperative included the difficulty of accessing the large volumes of data collected and held by the government and private sectors; ethical issues associated with the use and sharing of data concerning human subjects; the need to standardize, manage, and archive data in accessible formats; and the urgent need for more international comparability and collaboration. All these issues reflected a larger problem: the need for a more integrated, coordinated approach to the collection, management, and archiving of data, at every level. The report concluded that the "the need for global coordination of activities designed to address the challenges identified in this report is paramount" (OECD 2013).

This has been identified as the paradox of the twenty-first century (Dombrowsky 2007). More data is available than ever before, in a huge range of areas. Existing national structures and systems, however, are no longer able to manage the complexity and volume of multinational data flows, and there are as yet no global structures and guidelines through which such data can be managed, coordinated, and applied for disaster risk reduction purposes.

Challenges to Learning from Disasters Arising from Focusing Effects

The lack of coordinating structures at the global level is also an issue when it comes to coordinating disaster risk reduction efforts. Incompatible economic, political, and scientific drivers have been found to be best coordinated, at the national level, through policy frameworks that facilitate an integrated focus on the issue of disaster risk reduction. At the global level, however, where similar pressures are contributing to rising disaster losses, there are as yet no structures with the authority or means to establish equivalent global frameworks.

Disaster sociologists point to the extent to which the social issues created by global tensions, rather than being foregrounded by rising disaster rates, have instead been *obscured* by the focusing effect of major natural disasters like earthquakes and tsunami. By drawing attention to the drama associated with such natural forces, in other words, earthquakes and other natural disasters have distracted attention from the extent to which the capacity to cope with and recover from such extreme events can be reduced by lack of resources, including access to information, financial credit, social support, and services. The focus on major natural disasters can thus obscure the fact that populations constrained in this way are more vulnerable to death, injury, and loss during disasters and less able to recover (Gaillard 2010).

On the larger scale, the focusing effect of dramatic earthquake disasters is understood to work in a similar way, grabbing attention from the impacts of climate-related disasters, which are particularly common in less developed global regions. Climate-related events have the most destructive global impact; cause the greatest losses, the most deaths; and are also the hazard event category that is increasing in frequency (Global Network 2013). Moreover, this class of disaster includes the less dramatic, ongoing "everyday disasters" such as droughts and floods, which are collectively responsible for more disaster losses than the more dramatic major events. The hazards associated with extreme poverty have also been found to fall into the category of "everyday disaster" – responsible for the vast majority of harms and deaths affecting the global human population, but obscured by the focus on so-called "natural" disasters (Global Network 2013).

These issues have had a fragmenting effect within the scientific community. It has been argued, for example, that hazard and vulnerability disaster risk reduction traditions are informed by inverse and fundamentally incompatible paradigms. Such arguments have contrasted a hazard mitigation emphasis on infrequent, major natural disasters, and reliance on expensive technical solutions, designed and implemented at high levels by expert groups and delivered through top-down mechanisms, with the emphasis of vulnerability-based approaches on the social causes of disasters, and on bottom-up, participatory approaches designed to counteract these social constraints (Gaillard 2010). Related constructivist critiques have emphasized the risks to democratic process associated with the delegation of policy formation to subgovernment networks and have been similarly critical of disaster risk reduction policy and initiatives developed by experts that fund technical risk reduction measures, rather than public engagement mechanisms, and measures to address the social issues that contribute to disaster impacts (Hayward 2013).

Summary

There is general agreement within research and policy communities when it comes to some broad principles concerning learning from disasters. Governments are responsible for putting policy frameworks in place to support risk reduction initiatives. Implemented at both national and local levels, the most effective initiatives have reduced fragmentation and competition by adopting collaborative, cross-sector participatory approaches in an attempt to include as many stakeholders as possible, including those most exposed to disaster. Earthquake disasters have had a focusing effect on policy agendas in the US. Higher-level support, policy frameworks that facilitate risk reduction, increasingly networked global communities, and multilateral learning networks have been found to be key factors in the translation of science into disaster risk reduction policy after major events. In New Zealand, learning from earthquake disasters has informed an all-hazards risk reduction approach and a policy framework that has devolved responsibility for risk reduction to local levels. There is broad agreement that this kind of current extension of the government sphere to include both state and non-state actors in policy formation and implementation can facilitate such approaches and can also facilitate the translation of science into policy and practice, when researchers take the opportunity to engage with all end-users, at both national and local levels.

White et al. have argued that this kind of consensus reflects an increasing convergence in the hazard and disaster research community around the common goal of disaster risk reduction, notwithstanding cross-disciplinary tensions concerning disaster risk reduction priorities and the risks and opportunities associated with the extension of the government sphere. This convergence is also indicated in a cross-disciplinary agreement concerning the importance of working through such tensions collaboratively, both within the scientific community and with other sectors involved in disaster risk reduction.

Fragmentation, competition, and lack of coordination, however, have been found to inhibit the development of multilateral disaster risk reduction initiatives at local, national, and global levels and within and across disciplines, sectors, and domains. At both local and national levels, *coordinating measures* introduced after disasters have been found to reduce fragmentation and balance competition, both within and between sectors. Such measures have facilitated the compromises between economic, political, and scientific drivers required by collaborative risk reduction initiatives. Global disaster risk reduction is challenged by the absence of similarly authoritative coordinating mechanisms at this level and related issues arising from the rapid and uncoordinated development of information technology, the distribution of vulnerability to disasters, the emergence of subgovernment policy communities, and aspects of current emphasis on risk and disasters. Unsustainable global development, widely acknowledged to be a causal factor in rising disaster losses, also continues to significantly inhibit disaster risk reduction at both national and local levels and so the wider global disaster risk reduction effort.

Cross-References

▶ "Build Back Better" Principles for Reconstruction
▶ Community Recovery Following Earthquake Disasters
▶ Earthquake Disaster Recovery: Leadership and Governance
▶ Earthquake Risk Mitigation of Lifelines and Critical Facilities
▶ Legislation Changes Following Earthquake Disasters
▶ Resilience to Earthquake Disasters
▶ Resiliency of Water, Wastewater, and Inundation Protection Systems
▶ Seismic Resilience
▶ Structural Design Codes of Australia and New Zealand: Seismic Actions

References

Birkland T (1998) Focusing events, mobilisation, and agenda setting. J Public Policy 18(1):53–74
Birkland T (2009) Disasters, lessons learned, and fantasy documents. J Conting Crisis Manag 17(3):146–156
Bruneel J, D'Este P, Salter A (2010) Investigating the factors that diminish the barriers to university-industry collaboration. Res Policy 39:858–868
Busenberg G (2000) Resources, political support, and citizen participation in environmental policy: a reexamination of conventional wisdom. Soc Nat Res 13:579–587
Cash DW, Clark WC, Alcock F, Dickson NM, Eckley N, Guston DH, Mitchell RB (2003) Knowledge systems for sustainable development. Proc Natl Acad Sci 100(14):8086–8091
Center for Advanced Engineering (1997) Risks and realities: a multi-disciplinary approach to the vulnerability of lifelines to natural hazards. Center for Advanced Engineering, Christchurch
Christoplos I (2006) The elusive "window of opportunity" for risk reduction in post-disaster recovery. ProVention Consortium Forum, 2–3 Feb 2006, Bangkok
Dombrowsky WR (2007) Lessons learned? Disasters, rapid change and globalization. International Review of the Red Cross, 89(866)
EERI (2014) Earthquake engineering research institute. https://www.eeri.org/. Retrieved 20 Jan 2014
Fenwick T (2012) The value of lifeline seismic risk mitigation in Christchurch. NZ Earthquake Commission: Wellington, New Zealand
Gaillard JC (2010) Vulnerability, capacity and resilience: perspectives for climate and development policy. J Int Dev 22:218–232
Global network of Civil Society Organisations For Disaster Reduction (2013) Views from the frontline: beyond 2015, http://www.globalnetwork-dr.org/views-from-the-frontline/vfl-2013.html. Retrieved 2 January 2014
Hamilton RM (2003) Milestones in earthquake research. Geotimes, March. www.agiweb.org/geotimes/mar03/comment.hetml. Retrieved 08 Jan 2014
Hayward BM (2013) Rethinking resilience: reflections on the earthquakes in Christchurch, New Zealand, 2010 and 2011. Ecol Soc 18(4):37, http://www.ecologyandsociety.org/vol18/iss4/art37/. Retrieved 01 Jan 2014
ICSU (2008) A science plan for integrated research on disaster risk. International Council for Science, Paris, http://www.icsu.org/publications/reports-and-reviews/IRDR-science-plan. Retrieved 24 Oct 2013
NEHRP (2014) US National earthquake hazard reduction program. www.nehrp.gov. Retrieved 09 Jan 2014
OECD Global Science Forum (2013) New data for understanding the human condition: international perspectives. http://www.oecd.org/sti/sci-tech/new-data-for-understanding-the-human-condition.htm. Retrieved 10 December 2013
Rowe G, Frewer LJ (2005) A typology of public engagement mechanisms. Sci Technol Hum Values 30(2):252–290
Sabatier P (1998) The advocacy coalition framework: revisions and relevance for Europe. J Eur Public Policy 5(1):98–130
SCIRT (2014) Stronger Christchurch infrastructure rebuild team
Shephard RB (1997) A decade of progress since the Edgecumbe earthquake: risk and insurance. Bull N Z Soc Earthquake Eng 30(2):185–193
Skogstad G (2003) Who governs? Who should govern?: political authority and legitimacy in Canada in the twenty first century. Can J Polit Sci 36(5):955–973
Skogstad G (2005) Policy networks and policy communities: conceptual evolution and governing realities. In: Annual meeting of the Canadian political science association. London, pp 1–19
UNISDR (2014) United Nations international strategy for disaster risk reduction. www.unisdr.org. Retrieved 09 Jan 2014

Voorberg W, Bekkers V, Tummers L (2013) Co-creation and co-production in social innovation: a systematic review and future research agenda. In: European group for public administration conference 11–13 Sept. Edinburgh

Weichselgartner J, Kasperson R (2010) Barriers in the science-policy-practice interface: toward a knowledge-action-system in global environmental change research. Glob Environ Chang 20:266–277

White GF, Kates RW, Burton I (2001) Knowing better and losing even more: the use of knowledge in hazard management. Glob Environ Chang 3(3–4):81–92

Legislation Changes Following Earthquake Disasters

Suzanne Wilkinson[1], James Olabode Bamidele Rotimi[2] and Sandeeka Mannakarra[1]
[1]Department of Civil and Environmental Engineering, The University of Auckland, Auckland, New Zealand
[2]Auckland University of Technology, Auckland, New Zealand

Synonyms

Disaster legislation; Legislation changes post-disaster

Introduction

Most disasters are followed by legislative changes, emergency legislation, and new disaster legislative measures. Some of the changes are a reaction to the need to build back safer and to be seen to be facilitating a better future environment. For instance, the Royal Commission into the Canterbury Earthquakes in New Zealand recommended changes to the Building Act, a rewrite of building codes and building standards. Other disasters, such as the Wenchuan earthquake in China, Black Saturday bushfires in Australia, and Hurricane Katrina and Northridge earthquake both in the USA, had all occasioned changes in legislative requirements for reconstruction and code changes for buildings. The impact of legislative changes tends to slow recovery, but often they facilitate in building a better, more resilient, post-disaster environment. In this entry, the focus will be on the common post-disaster legislative changes encountered for reconstruction, such as those affecting building acts and building codes. Examples and effects of the changes are discussed using a variety of cases from earthquakes and other disasters. This entry provides an account of the ways in which legislation changes post-disaster can help and/or hinder reconstruction programs.

Legislation Challenges Post-disaster

The organization and coordination of recovery is usually complex because a wide range of activities occur simultaneously after a significant disaster. There is an equally wide range of needs that would have to be met, some of which may be conflicting and adds to the complexity of post-disaster recovery. Experiences from past disaster recovery arrangements show a struggle to meet recovery needs and changes to the legislative, and regulatory recovery environments seem common place. Several contributory factors account for the success or otherwise of disaster management goals and objectives, irrespective of whether they are in legislation or in recovery management plans. These include:

– Pre-disaster trends and levels of preparedness which are linked to vulnerability
– The extent of damage resulting from the disaster
– Availability and accessibility to the required resources for both response and recovery
– The prevailing political will and governmental interests in disaster management activities

In terms of legislation and regulation, there is often little provision in legislation to facilitate large-scale reconstruction programs. Prior to the Canterbury Earthquakes in New Zealand, Feast (1995) had identified several issues in relation to planning and construction legislation that could impede reconstruction of Wellington, following a major earthquake. Feast's study suggested that much of the legislation (in particular the Resource Management Act (RMA) and the Building Act (BA) that existed during the period was neither drafted to cope with an emergency situation nor developed to operate under the conditions that would prevail in the aftermath of a severe seismic event. Specifically commenting on the RMA, Feast had explained that its consultation procedure could prevent meeting the reconstruction requirements of a devastated city within a reasonable time period (Feast 1995). Decades after in 2012 (following the Canterbury earthquakes), some of Feast's suggestions were relived, and changes to both the RMA and BA were, and are, still being made based on recommendations by the Royal Commission set up to review the Canterbury recovery program. Changes around seismic codes, strengthening of existing buildings, changes to land use, and planning parameters are being made as a response to the Canterbury disaster.

Rolfe and Britton (1995) are of the opinion that the pace of reconstruction is severely impacted by political and cultural conflicts over recovery plans, which means the successful achievement of disaster management goals will depend on the political environment. Once a disaster happens, there is a continuous tension between strictly applying reconstruction regulations which aim at preventing a recurrence of the previous community vulnerabilities and allowing an affected community to move back to their former habitation quickly. Clearly, the quicker communities return to habitability of as many of their homes as possible, the better it will be for restoring a sense of normality. However, to reduce future vulnerabilities, it is rational to approach rebuilding programs cautiously to build in resilience and build back better than before the disaster. In New Zealand, the rebuilding of Napier following the 1931 earthquake did not reduce vulnerabilities to future disasters as the city was rebuilt in the same location mainly producing unreinforced masonry buildings. Retrofitting of Napier buildings was subsequently required under new seismic codes because, as has been repeatedly demonstrated, unreinforced masonry buildings are vulnerable to earthquakes.

When considering the effects of legislation changes after disasters, it can be seen that legislation changes post-disaster can be broadly classified into two categories: (1) legislation for compliance and (2) legislation for facilitation. *Legislation for compliance* entails using legislation to enforce recovery initiatives, such as those that would require buildings to be upgraded or rebuilt to different standards and codes. The lack of enforcement of hazard-related laws and adequate risk-based building controls contributed to the large-scale devastation caused by the 2004 Indian Ocean tsunami (Mulligan and Shaw 2007). The same was witnessed in countries such as Pakistan, Turkey, Samoa, and Haiti. In reconstruction programs, there is a need to enforce the new building design codes which inevitably appear after a disaster. Inspection of the reconstruction, including buildings using the new codes, requires clear guidelines to maintain the required standards.

Legislation for Facilitation denotes legislation being used to simplify and assist recovery activities to speed up the recovery process. These changes can usually be seen in changed processes, such as in building and development consent processing. However, legislation that is customarily used to impose security and safety controls (such as building consents) can become an obstacle to rebuilding programs. Time-consuming procedures, insufficient resources to process permits, and lack of fast-tracked methods could delay reconstruction. For example, delays in the issuance of permits were a major reason for the holdup in housing repair and rebuilding following the 2005 Bay of Plenty storm in New Zealand (Middleton 2008). Fast-tracked consenting procedures, collaboration between councils, open access to information between stakeholders, and using legislation to remove bureaucracy are viable options to speed up recovery. During the 1994 Northridge earthquake, legislation was suspended, and emergency powers were used with the consequence of reducing highway reconstruction time. Similarly, during recovery from the Australian bushfires, planning and building permits were exempted for temporary accommodation so they could be put up quickly. For permanent dwellings, planning permits were exempted, and only building permits were needed after the bushfire (DPCD 2013). Also property falling under the Wildlife Management Overlay which would normally be subject to more rigorous planning and building permit requirements had to be relaxed to only require a simplified planning consent and building permit (DPCD 2013).

From the forgoing, it is clear that legislation and regulatory requirements can have significant influence on the rate of recovery after a disaster event. The overall desire is for legislation to enhance the recovery and reconstruction process so that it improves the functioning of an affected community and reduces risks from future events. Any legislation changes that need to be made post-disaster may be better considered before a disaster so that its implementation could be facilitated early on during recovery. Often the opportunities to introduce mitigating measures become limited over the course of recovery because of the desire to return to normalcy and thus rebuild quickly after disasters. Menoni (2001) notes that "market forces put pressures to reconstruct as quickly as possible ... hampering efforts to implement lessons learnt from the disaster in the attempt to reduce pre-earthquake vulnerability."

Pressures to rebuild critical infrastructure quickly are borne by national and local administration with the possible implication of reduced quality of delivery. Rushed rebuilding programs have led to increased vulnerability of poorly planned and designed built environments to future disasters (Jigyasu 2010). Ingram et al. (2006) also explain that the clamor to rebuild quickly amplifies the underlying social, economic, and environmental weaknesses that result in large-scale disasters (Ingram et al. 2006).

Improving Recovery Through Legislation

Well-articulated and implemented legislation should not only provide an effective means of reducing and containing vulnerabilities (disaster mitigation) but also become a means of facilitating better thought out and designed reconstruction programs.

After damage assessments and evaluations, building and environmental legislation should not present impediments to reconstruction and rebuilding programs. Martín (2005) suggests that there is a relationship between building/environmental regulations and rehabilitation works (Martín 2005) and that regulations could become burdensome in rehabilitation and reconstruction projects and are worthy of considerations (Martín 2005). Martín (2005) describes burdensome regulations as those which incorporate excessive rules and regulations and red tapes (statutory procedures) that add unnecessarily to costs. Listokin and Hattis (2004) provide useful analysis on two kinds of barriers that building codes could pose to rehabilitation works. They are "hard" and "soft" barriers to rehabilitation. The hard barriers are impediments to rehabilitation as a result of overregulation, which would not add appreciably to building value or public safety (Burby et al. 2006) and could discourage housing development or rehabilitation because they are added burdens (May 2004). For instance, building and environmental regulations that do not reduce the vulnerability of built assets to a hazard event are unnecessary. Also to insist on expensive structural solutions in a highly hazardous zone, where a simple alternative will be to restrict development in that zone, is another example of regulation that could fall under Listokin and Hattis' hard category.

Soft impediments, on the other hand, are administrative requirements that require extra time, money, and effort to accomplish rehabilitation and reconstruction projects (Listokin and Hattis 2004). These are red tapes (bureaucratic procedures) that could delay new construction and rehabilitation/reconstruction of physical facilities (May 2004). Such soft impediments are the focus in this entry.

Bureaucratic procedures must be supportive of emergency management under different emergency scenarios whether routine or chaotic. However, research suggests that bureaucracies have been less supportive of the expediency that is desired in disaster response and recovery. May (2004) suggests three sources of impediments by regulatory processes:

- Regulatory approvals. These are delays associated with consent processes and approvals that arise from cumbersome decision-making processes and duplication of regulations. These types of delays are inherent in building and environmental legislation.
- Regulatory enforcement strategies and practices. These are overly rigid practices that foster an unsupportive regulatory environment for the development and rehabilitation of the built stock. In post-disaster situations, rigid enforcement strategies discourage genuine recovery efforts.
- Patchwork of administrative arrangements. This could result from duplication of administrative structures (as in layers or hierarchies of control) and gaps in regulatory decision processes. May (2004) explains that patchwork frustrates regulatory implementation and adds to complexities in regulatory processes.

Regulatory process barriers could also result from *administrative conflicts* in and among disaster agencies (Listokin and Hattis 2004). For example, rivalry between responding agencies is not foreign to emergency services and is an obstacle to effective emergency management McEntire 2002; Quarantelli 1998). Rotimi (2010) explains that rivalry may result from existing silos, from the absence of a coordinating agency, and from the ability of a coordinating agency incorporating other agencies perceived responsibilities. Hence, a broad cooperative effort is needed for the success of post-disaster reconstruction activities. Rotimi (2010) therefore contends that organizations have to coalesce to plan for resource utilization in the restoration of physical assets. Coordination is therefore central to multi-organizational response and recovery programs (McEntire 2002). This multi-organizational coordination is often embedded into emergency legislation post-disaster and usually involves the creation of new recovery coordinating agencies such as Victorian Bushfire Reconstruction and Recovery Authority (VBRRA) in Australia or Canterbury Earthquake Recovery Authority (CERA) in New Zealand.

Another useful dimension to the problems with burdensome regulations is provided by Listokin and Hattis (2004). It is that regulatory procedures could become too rigid forcing implementers to "go by the book" even though variations may be warranted. This places implementers in a state of continuous fear of liability should things go wrong. Rotimi (2010) explains that some latitude of control and discretion is often required to aid decision-making as long as such decisions are pragmatic. Commenting on the rebuilding program after the flooding incident in New Orleans, Stackhouse (2006) says "removing democratic processes from the rebuilding process has the advantage of expediting decision making by allowing politically dangerous but practical outcomes." This statement suggests that greater freedom in decision-making by officials of coordinating agencies could increase the speed of rebuilding programs after significant disaster events. Evidence from literature on recovery often shows slow recovery. For example, on the Bay of Plenty storm in New Zealand in 2005, Middleton (2008) showed that at 300 days after the event, 35 households still required permanent rehousing out of a total of 300 compulsory evacuations. At the same time, nine households were still occupying temporary accommodation. Middleton (2008) suggests that the slow pace of recovery was attributable to the inadequacy of personnel to carry out the stipulated building safety evaluations and to process consents for reconstruction work. Both of these factors have legislative connotations.

Considering New Zealand before the Canterbury earthquakes, there was an emphasis on readiness and response activities, with little consideration given to planning for sustained recovery activities. Where recovery was considered, it was for the short term, as evident in emergency awareness campaigns that encouraged communities to prepare for 3–7 days after an event. Recent emergency events clearly show that longer-term recovery plans and more robust legislation were required given the complexities associated with the rebuilding of damaged built assets. Warnings about the inadequacies of routine construction processes being modified on an ad hoc basis during the small disaster recovery phases in previous hazard events were unheeded in New Zealand (Rotimi et al. 2006). While such an approach works reasonably well for small-scale emergencies, the effectiveness of reconstruction in a large disaster required large-scale legislative changes (Rotimi et al. 2009). Post-Canterbury, an extensive review of legislation to cope with future disasters is now underway.

Post-disaster Legislation Changes in Action: Three Case Studies

Despite the type of disaster, there are often commonalities in the post-disaster legislative changes enacted and the impacts of these changes. Starting with an earthquake, and ending with a fire, this section demonstrates through three cases, the impact of legislation on recovery and some good and bad practices.

The Northridge Earthquake, USA, 1994

The Northridge earthquake provides an example of a disaster situation where legislative changes helped to facilitate reconstruction projects. The moderate earthquake struck Southern California in the early hours of 17 January 1994 with a magnitude of 6.8 on the Richter scale, small compared to other earthquakes, but causing significant damage.

Comerio (1996) gave an insight into the extent of damage. There were damage to 27 bridges and a collapse of sections of six freeways. Four hundred and fifty public buildings suffered significant damage, 6,000 commercial buildings, 49,000 housing units in 10,200 buildings had serious structural damages, while 388,000 housing units in 85,000 buildings experienced minor damages. The total value of damage to houses in Los Angeles was estimated to be about $1.5 billion.

The Northridge earthquake caused a shift in emphasis from disaster preparation and relief to recovery (Comerio 1996), and this shift largely resulted in the success of emergency management programs for the restoration of the affected areas. Reconstruction contributed to the economic revitalization of the affected area, and bureaucratic requirements were suspended to encourage rapid rebuilding of damaged infrastructure. Marano and Fraser (2006) conclude that "identifying and easing regulations and statutes that inhibit reconstruction can mean a dramatically faster and less costly recovery."

Wu and Lindell (2004) provide an insight into some of the actions that were taken to increase the speed of housing reconstruction in Los Angeles after the earthquake showing that it is possible to expedite procedural requirements by establishing fast-tracked processes that would operate after a disaster would benefit recovery.

While the rapid recovery experienced after the Northridge earthquake can be attributed to other factors, such as political will, public policy changes and enabling emergency management legislation played a substantial role in the rebuilding programs after the earthquake (Comerio 2004).

Hurricane Katrina, New Orleans, USA, 2005

The disaster that followed Hurricane Katrina provides lessons on legislative changes either proposed or already implemented to enable its recovery from the event. The Hurricane was a category 3 storm when it struck New Orleans and the Gulf Coast in the morning of August 29, 2005. The storm surge caused severe destruction along the Gulf Coast from central Florida to Texas in the USA. The most severe damage occurred in New Orleans, Louisiana, because of the failure of the levee system that was designed to contain the resulting storm surges. The worst damage was caused by floods with an estimated 1.2 million people evacuated before the incident and another 100–120 thousand afterward. About 350,000 houses were destroyed and over 200,000 persons required temporary shelters scattered around 16 states in the USA. By all accounts, Hurricane Katrina was a catastrophic event with economic loss estimates of about $200 billion (Burby et al. 2006).

A brief description of some of the policy changes and legislative reviews that occurred as a result of Katrina is provided by Rotimi (2010) including:

1. Changes in building codes and standards. There were changes made to the building codes in New Orleans with a view in improving the resilience of built spaces in New Orleans. For example, there were revisions made to the base flood elevation levels for new construction to 3 ft or higher (Colton et al. 2008). This is a risk mitigation strategy which has been tied to flood insurance cover so that only buildings that meet these new guidelines can qualify for flood insurance and subsequent compensations. Overall, funding sources and budget priorities have been developed for reconstructing flood protection in New Orleans (Colton et al. 2008).
2. Changes in emergency management regulations and guidelines. Colten et al. (2008) explain that the Katrina event necessitated the review and updating of Louisiana and New Orleans' response strategies and their emergency operations plans. The legislative reviews included the adoption of an all-hazards approach thus expanding the scope and magnitude of anticipated hazards and allowing greater involvement of non-agency actors who proved crucial to response and recovery after the event. Colton et al. (2008) noted that partnering with nongovernmental stakeholders was a paradigm shift that emerged out of the Katrina experience.
3. Changes in land development regulations. Changes to land and development regulations are largely seen as a veritable tool for mitigating disaster risk in disaster management (Ingram et al. 2006). After Hurricane Katrina, changes in land use planning and zoning systems have been proposed to reduce the vulnerability of the New Orleans region from future flooding disasters (Burby et al. 2006).

The legislative changes in New Orleans show the importance placed on built asset reinstatements as a major input to holistic recovery. Similar to Northridge, building reconstruction in New Orleans aimed to stimulate development and growth. The rate at which recovery is achieved is therefore tied to the speed of reconstruction, and any recovery is underpinned by well-thought-out and implemented legislative and regulatory changes.

Victorian Bushfires, Australia, 2009

The "Black Saturday" disaster was one of the most damaging disasters in Australian history, leaving 173 people dead and many seriously injured, affecting 6,000 households, destroying more than 2,000 homes, and damaging around 430,000 ha of land. Despite the institutions and procedures set up for expediting community recovery, reconstruction proceeded slowly. Shortly after the bushfires in March 2009, the Victorian Government introduced a new residential bushfire construction building standard AS3959-2009 in response to the need to better protect the bushfire-affected communities from future fire events. Under the new codes, there are increased construction requirements depending where a building is situated, ranging from ember protection to direct flame contact protection.

Changes have been documented by Mannakkara and Wilkinson (2013) which showed that one of the first steps taken in Australia was to publish a revised edition of the Australian building code for bushfire-prone areas (AS 3959) on March 11, 2009 (VBBRA 2009). The revisions introduced Bushfire Attack Levels (BAL) to identify the bushfire risk of properties. Stringent design and construction requirements were specified for each BAL to provide greater fire protection. Another key change in legislation was regarding land use. Soon after the fires, the entire state of Victoria was declared bushfire prone and placed under the Wildfire Management Overlay (WMO) which imposed stricter planning regulations (VBBRA 2009). The introduction of the buy-back scheme posed a solution for people on high-risk lands (Victorian Government 2012).

As with earthquake recovery, constructing bushfire-resistant buildings contributes to saving lives and properties in the event of a bushfire. Mannakkara and Wilkinson (2013) report on the bushfire legislative and code change impacts. The more stringent construction requirements stipulated in the new Australian building standards are mainly concerned with the use of noncombustible materials for housing reconstruction. Bushfire housing reconstruction proceeded slowly because the advanced resources for direct flame contact protection such as the window systems, roof systems, shutters, and external cladding materials as required in the new AS3959-2009 were not available in the market. The main reason for the unavailability of these resources was that it required a considerable amount of time for manufacturers to undertake the research and development needed to test and release these new materials onto the market. Only in March 2010, a year after the bushfires, for instance, a combined window and screen system manufactured for use in direct flame contact protection zones was ready for release onto the market. The delays with production of compliant materials for the bushfire recovery, combined with growing demand on building services in the local construction market, created a series of scarcities, which greatly hindered housing recovery in the fire-affected areas. During the bushfire recovery, there were extra costs for the construction requirements of bushfire houses; these extra costs were significantly underrepresented (i.e., officially given assessments were from AUD 10,000 to 40,000 depending on the protection levels required, whereas the extra real cost of a house to be rebuilt to new codes was somewhere up to AUD 100,000). The increase in costs caused financial pressure for the affected house owners who already struggled to procure suitable resources for rebuilding their houses. The uncertainty about the number of houses in the direct flame contact protection zone was another major concern for the building product manufacturers.

Given few incentives from the government and the low likelihood of profitability, material producers were reluctant to put effort into developing materials for houses in the direct flame zone, which they believed would account for only a small fraction of their market. Lack of training and understanding of the new standards introduced in Australia slowed recovery; even one and a half years after the bushfires, designers and builders were still trying to come to terms with the application of the standard.

General Implications of Legislation on Recovery

Having highlighted some of the issues that are connected to the appropriateness of legislation and regulatory provisions in the previous sections, this section presents a summary of their implication on recovery. The summary is in line with Rotimi (2010) who describes the effect that poor legislative provisions could have on post-disaster reconstruction activities, thus:

1. Loss of vital momentum of action. The efficiency of post-disaster reconstruction activities is impacted as a result of delays caused by poor planning and implementation, restrictive legislation and regulatory provisions, and lack of government commitment in reconstruction programs.
2. Loss of commitment to the reconstruction process. There is a tendency for poor commitment to recovery programs by responsible authority because disaster practitioners are unable to apply pragmatic solutions to real-time reconstruction problems for fear of being held liable for their decisions.
3. Difficulties in achieving reconstruction deliverables and inability to accelerate the process of reinstatement. Introduce measures for risk and vulnerability reduction and aid planning for sustainable developments.
4. Impairment of overall community recovery and quality of life. Of essence, reconstruction should become a tool for empowerment till a level of functioning is reached where communities are self-sustaining and require no external interventions and also a therapeutic process for overall community recovery.

Summary

This entry has shown common practices relating to changes in legislation for disaster recovery. Changing legislation after a disaster is a common response to the disasters but takes on different forms. Most disasters are followed by legislative changes, emergency legislation, and new disaster legislative, and these measures aim to encourage better building practices and future resilience. Putting in place legislation which enforces revised building codes to increase disaster resilience in the built environment is recommended as is that which speeds up slow processes or offers flexibility. Putting in place rigid rules and bureaucracies slows recovery and creates confusion. As cautioned by Ingram et al. (2006), legislative and regulatory changes need scrutiny to avoid issues such as resource constraints, high costs, and impacts on livelihood that can unnecessarily hinder recovery progress.

Cross-References

- "Build Back Better" Principles for Reconstruction
- Economic Recovery Following Earthquakes Disasters
- Reconstruction Following Earthquake Disasters
- Reconstruction in Indonesia Post-2004 Tsunami: Lessons Learnt
- Resilience to Earthquake Disasters
- Resourcing Issues Following Earthquake Disaster

References

Burby RJ, Salvesen D, Creed M (2006) Encouraging residential rehabilitation with building codes: New Jersey experience. J Am Plann Assoc 72(2):183–196

Colton CE, Kates RW, Laska SB (2008) Community resilience: lessons from New Orleans and Hurricane Katrina, CARRI research report 3, Community and Regional Resilience Initiative of the Oak Ridge National Laboratory. www.resilientUS.org

Comerio MC (2004) Public policy for reducing earthquake risks: a US perspective. Build Res Inform 32(5):403–413

Comerio MC (1996) The impact of housing losses in the northridge earthquake: Recovery and Reconstruction issues, Center for environmental design research, University of California, US

DPCD (2013) List of amendments to the victoria planning provisions. In: Department of Planning and Community Development (ed). A Guide to the Planning System Victoria State Government, Victoria

Feast J (1995) Current planning and construction law: the practical consequences for rebuilding Wellington after the quake. Paper presented at the Wellington after the Quake: the challenges of rebuilding cities, Wellington

Ingram JC, Franco G, del Rio CR, Khazai B (2006) Post-disaster recovery dilemmas: challenges in balancing short-term and long-term needs for vulnerability reduction. Environ Sci Policy 9(7–8):607–613

Jigyasu R (2010) Appropriate technology for reconstruction. In: Lizarralde G, Davidson C, Johnson C (eds) Rebuilding after disasters: from emergency to sustainability. Taylor and Francis, London

Listokin D, Hattis D (2004) Building codes and housing. Paper presented at the workshop on regulatory barriers to affordable housing, US Department of Housing and Urban Development, Washington, DC

Mannakkara S, Wilkinson S (2013) The impact of post-disaster legislative and regulatory changes on the recovery of the built environment. The Society of Construction Law Australia, Sydney

Marano N, Fraser AA (2006) Speeding reconstruction by cutting red tape. Retrieved 20 Oct 2006 from www.heritage.org/

Martín C (2005) Response to "building codes and housing" by David Listokin and David B. Hattis. Cityscape J Policy Devel Res 8(1):253–259

May PJ (2004) Regulatory implementation: examining barriers from regulatory processes. Paper presented at the workshop on regulatory barriers to affordable housing, US Department of Housing and Urban Development from www.2004nationalconference.com

McEntire DA (2002) Coordinating multi-organisational responses to disaster: lessons from the March 28, 2000, Fort Worth tonardo. Disaster Prev Manag 11(5):369–379

Menoni S (2001) Chains of damages and failures in a metropolitan environment: some observations on the Kobe earthquake in 1995. J Hazard Mater 86(1–3):101–119

Middleton D (2008) Habitability of homes after a disaster. Paper presented at the 4th international i-REC conference on building resilience: achieving effective post-disaster reconstruction, Christchurch New Zealand

Mulligan M, Shaw J (2007) What the world can learn from Sri Lanka's Post-Tsunami experiences. Int J Asia Pac Stud 3(2):65–91

Quarantelli EL (1998) what is a disaster? Perspectives on the question. London: Routledge

Rolfe J, Britton NR (1995) Organisation, government and legislation: who coordinates recovery? Paper presented at the Wellington after the quake: the challenge of rebuilding cities, Wellington

Rotimi JOB (2010) An examination of improvements required to legislative provisions for post disaster reconstruction in New Zealand. Doctoral thesis, University of Canterbury, Christchurch

Rotimi JOB, Le Masurier J, Wilkinson S (2006) The regulatory framework for effective post-disaster reconstruction in New Zealand. Paper presented at the 3rd international conference on post-disaster reconstruction: meeting stakeholder interests, Florence

Rotimi JO, Myburgh D, Wilkinson S, Zuo K (2009) Legislation for effective post-disaster reconstruction. Int J Strateg Prop Manag 13(2):143

VBBRA (2009) 100 day report. Victorian Bushfire Reconstruction and Recovery Authority, Victorian Bushfire Reconstruction and Recovery Authority, Melbourne

Victorian Government (2012) Victorian bushfire recovery three year report. Victorian Government, Melbourne

Wu J, Lindell MK (2004) Housing reconstruction after two major earthquakes: the 1994 Northridge earthquakes in the U.S. and the Chi-Chi earthquake in Taiwan. Disasters 28(1):63–81

Liquefaction: Countermeasures to Mitigate Risk

Rolando P. Orense
Department of Civil and Environmental Engineering, University of Auckland, Auckland, New Zealand

Synonyms

Ground improvement; Liquefaction mitigation; Liquefaction prevention; Liquefaction remediation

Introduction

Past several earthquakes have vividly demonstrated the impact of soil liquefaction and the associated ground deformations to buildings, bridges, buried lifelines, and other civil infrastructure. To reduce the risk of damage to the structures, remediation measures are usually employed. In order to implement a successful remediation work for a target structure, a thorough understanding of the following are required: the liquefaction hazard at the site, the potential consequences of liquefaction for the structure, the performance requirements of the work, and the available construction materials and methods. Several alternative approaches can be taken if liquefaction poses as threat to existing or proposed structures. For existing structures, the choices are:

(1) Retrofitting the structure and/or site to reduce the potential failure
(2) Abandoning the structure if the retrofit costs exceed the potential benefits derived from maintaining the structure
(3) Accepting the risk and maintaining the existing use

Additional options would be to continue the use of the structure but to change the operation or to alter the use of the structure such that the hazard becomes tolerable even if failure occurs.

For new construction, moving the project assumes that there are alternative sites, where liquefaction hazard is not a concern, but is equally appropriate for the proposed project. The costs of necessary mitigation actions may, in some cases, make selection of an alternative site a more cost-effective alternative than the use of the primary site. The choice of accepting the risk associated with the hazard, and the potential loss of life, social disruption, economic loss, and political ramifications that might result from structural failure, must be based on engineering evaluation of the site.

Thus, the overall goal of any remediation work is to investigate alternatives to mitigate the liquefaction hazard so that relative benefits versus costs for different levels of mitigation efforts can be assessed. In investigating potential remedial measures, the following factors should be considered:

- Technical adequacy
- Field verifiability
- Costs

- Maintenance
- Long-term performance
- Environmental impact

In this section, general descriptions of the principles involved in ground improvement methods, i.e., densification, solidification, drainage, reinforcement, containment, and soil replacement, as well as some representative methods, are presented. Verification techniques and design issues associated with general ground improvement methods are also explained.

General Description of Countermeasures

Countermeasures to mitigate the damaging effects of liquefaction can be classified into two categories: (1) ground improvement (to remediate liquefiable soils) and (2) structural enhancement (to relieve the effects of liquefaction). Some projects may require the combination of the two methods.

Ground improvement, whose goal is to limit soil displacements and settlements to acceptable levels, is being used increasingly for remediating liquefiable soils due to the wide variety of methods that are available. These methods can be modified and customized to specific site settings, including space/operation restrictions and the presence of existing structures. In addition, one or more methods can often provide economical solutions for liquefaction remediation problems.

Ground improvement methods are generally based on one or more of the following principles:

- Densification
- Solidification
- Drainage
- Reinforcement and containment
- Soil replacement

Descriptions of the principles behind the improvement mechanism associated with each category are provided below, along with some particular treatment techniques in the category that can potentially be used at existing structures and for future sites. In addition, some design considerations are also briefly described. Details concerning the specific treatment methods can be found in references on ground improvement (e.g., Cooke and Mitchell 1999; Andrus and Chung 1995; Xanthakos et al. 1994; Hausmann 1990; JGS 1998; PHRI 1997). Note that there is no attempt herein to provide a comprehensive discussion of all available liquefaction countermeasures; rather, only representative methods are discussed for each improvement mechanism. Moreover, new techniques are being developed when difficult circumstances are encountered in the field, while available techniques are being refined to suit the requirement of the problem. Thus, more and more ground improvement techniques are becoming part of the expanding set of available mitigation measures.

Structural enhancement can reduce the damage to the structure while allowing liquefaction to occur. Examples include strengthening the foundations of the structures and the ground supporting the structures to avoid reduction of bearing capacity or possible uplift due to soil liquefaction. The use of flexible structures, which can deform together with the ground deformation, is a viable option for underground structures.

It should be mentioned that mitigation of liquefaction risk is an area subject to considerable controversy and that the current understanding of the efficacy of some of these methods is still evolving (Seed et al. 2003). Moreover, while the case histories clearly indicate that ground improvement leads to a significant decrease and/or elimination of large liquefaction-induced ground deformations, the data is not yet sufficient to be able to predict the ground deformations for a given set of site conditions and earthquake motion (Hausler and Sitar 2001). Furthermore, large earthquakes which can verify the effectiveness of these techniques are infrequent and current database of observed field performance has marginal quality. Fortunately, this is being addressed by more detailed studies of the field case histories (e.g., Yasuda et al. 1996; Ishii et al. 2013) coupled with carefully designed experimental testing and numerical analyses.

Densification Methods

Densification or compaction methods involve rearranging the soil particles into tighter configuration, resulting in increased density. This increases the shear strength and liquefaction resistance of the soil and encourages a dilative instead of a contractive dynamic soil response. Densifying loose sandy deposit with vibration and/or impact has been used extensively, making it the most popular liquefaction countermeasure.

An increase in soil density can be achieved through a variety of means (JGS 1998). These include (1) compaction by penetration (penetration of sandy material into the liquefiable deposit will push the surrounding soil and result in reduced void ratio and therefore increase in soil strength), (2) compaction by vibration (subjecting the loose sandy deposit to vibration energy will compact the soil and increase its strength), and (3) compaction by impact energy (impact energy can densify loose granular deposit).

The suite of methods which involve compaction by displacement and/or vibration involves a vibrating probe repetitively penetrating the ground to densify the liquefiable soils and is generally referred to as *vibromethod*. These methods involve various kinds of equipment and procedures and are known by some generic or proprietary names. The simplest categorization of vibromethods is as follows:

- Vibro-replacement: the cavity formed by probe penetration is filled with imported materials, such as crushed stone gravel and sand.
- Vibro-flotation: the surrounding soil is compacted with a horizontally vibrating motor attached to the end of the probe, sometimes aided by water sprayed from the nozzle tip end.
- Vibro-rod: compaction is achieved using vertical vibratory penetration to the top of the probe or rod.
- Vibratory tamper method: a rigid steel plate (tamper) attached to a vibrator at the surface of the ground to compact the surface soil.

Vibro-replacement methods can improve liquefiable soil deposits not only by densification but also by increasing the in situ lateral stress, replacing the liquefiable in situ soil with non-liquefiable material, reinforcing the original ground with stiffer columns of fill materials, and providing drainage paths for excess pore water pressure. Two of the most widely implemented methods are the sand compaction pile (SCP) method (popular in Japan) and stone column (SC) method (common in the United States). Sand compaction piles are constructed by driving a sand-filled steel casing into the ground after which the casing is then slowly withdrawn and the deposited sand is compacted to form a high density sand column (Schaefer 1997; JGS 1998). The sand backfill must have specific properties and sometimes may be difficult to source. Recent case histories indicate that metal slag, crushed oyster shell, granulated fly ash, or recycled crushed concrete and asphalt are feasible alternatives to sand (Kitazume 2005). A variation on the sand compaction pile technique is the use of gravel backfill, to create stone columns (Munfakh et al. 1987; Sondermann and Wehr 2004). Generally the fill material consists of crushed coarse aggregates of various sizes, with the ratio in which the stones of different sizes will be mixed decided by design criteria.

Vibro-flotation method compacts loose soil by inserting a horizontally vibrating vibroflot into the ground while jetting water from the bottom end of the vibroflot, followed by gradually extracting it while forcing sand or gravel from the ground surface into the voids created adjacent to the vibroflot. On the other hand, vibro-rod method involves driving a vibrating rod or pile into the liquefiable soil, thereby causing it to locally and temporarily liquefy, after which new sand is placed along the rod or pile to densify the in situ soil. In vibro-flotation and vibro-probe methods, sand or gravel fill is injected from the ground surface into the voids created in the original ground during vibration.

Dynamic compaction involves repetitively dropping a large weight from a significant height onto the ground causing the soil grains to rearrange and form a denser arrangement. Additionally, the impact of the dropped weight on the ground surface produces dynamic stress waves, which can be large enough to generate significant excess pore water pressure in the soils beneath the point of impact (Idriss and Boulanger 2008). The dissipation of these excess pore water pressures results in densification, accompanied by surface settlement. The drop height, weight, and spacing vary, depending on ground and groundwater conditions. Dynamic compaction is known to be fast and economic, especially in treating large areas. However, it has obvious disadvantages due to the noise and vibration that are produced. Moreover, it is often effective only in the upper 10 m of the deposit and it is less effective for soils with high fines content.

In compaction grouting, a very stiff grout is injected into the soil such that it does not permeate the native soil, but results in coordinated growth of the bulb-shaped grout that pushes and displaces the surrounding soil. Typically the grout consists of soil–cement–water mixture with sufficient silt sizes to provide plasticity, together with sand and gravel sizes to develop internal friction (Welsh 1992). Note that the strength of the grout is unimportant because the purpose of the technique is to densify the surrounding soil by displacement. Since the technique involves the pressurized injection of grout into the soil deposit using small-scale, maneuverable, and vibration-free equipment, there is minimum disturbance to the structure and surrounding ground during implementation. Additional advantages include relatively little site disruption and the ability to work in constricted space, resulting in greater economy. However, it has some disadvantages; for example, stabilization of near-surface soils is generally ineffective due to the fact that the overlying restraint is small and the grouting pressures can heave the ground surface rather than densify the soil.

Compaction by man-made explosions can also be employed to densify loose liquefiable deposits. By detonating explosive charges placed at various depths in boreholes across the site in a controlled manner, the explosions propagate dynamic shear stresses through the ground which can induce liquefaction. The post-liquefaction consolidation following the dissipation of excess pore water pressure results in densification of the soil, together with ground settlement (Narin van Court and Mitchell 1994; Maeda et al. 2006). This method is effective in sites where cleaner loose sands, which will undergo relatively large and rapid reconsolidation, are present at greater depths. Conversely, it may be less effective in densifying shallow deposits.

In general, densification or compaction methods are more suitable for use in saturated, cohesionless soils with a limited percentage of fines. They cause noise and vibration during installation and also increase horizontal earth pressures against adjacent structures. This increase in pressure is the major disadvantage of using compaction methods in close proximity to retaining walls and pile foundations. The major advantage of compaction methods is the relatively low cost/benefit ratio. The necessary degree of compaction can be evaluated using penetration resistances that have been back-calculated from an acceptable factor of safety against liquefaction.

Solidification Methods

Increase in liquefaction resistance can also be obtained through a more stable skeleton of soil particles. This can be achieved by solidification methods, which involve filling the voids with cementitious materials resulting in the soil particles being bound together. This will prevent the development of excess pore water pressure, preventing the occurrence of liquefaction. Typical methods include deep mixing method, premixing method, and grouting techniques.

The deep mixing method achieves liquefaction remediation by agitating and mixing stabilizing material such as cement in sandy soil and solidifying the soil. The soil–cement materials in the mixed columns can have a wide range of unconfined compressive strengths, depending on the amount of cementitious material used and the in situ soil characteristics (Kitazume and Terashi 2013). This technique can be applied over a whole area of liquefiable soil (block type) or over partial sections (wall type, pile type, or lattice type). In block type, the mix columns overlap each other, and the whole layer is improved due to the increased stiffness. In partial improvement, unimproved sections remain and may liquefy; however, depending on the shape of installation, the mix columns are effective in preventing liquefaction by restraining the shear deformation of the ground through reinforcement and confinement, as described below.

The premixing method, on the other hand, is done by adding stabilizing material to the soil in advance and placing the treated soil at the site (JGS 1998). This is recommended for constructing new reclaimed land or in filling behind caissons, allowing both reclamation work and anti-liquefaction measures to be implemented at the same time to develop stable ground.

Chemical grouting, sometimes called permeation grouting, is a technique that transforms granular soils into hardened soil mass by injecting cement or other grouting materials that permeate and fill the pore space. The hardened grout improves the native soil by cementing the soil particles together and filling the voids in between (minimizing the tendency of the soil to contract during shearing). The treated soil would have increased stiffness and strength and decreased permeability. Because of its minimal disturbance to the in situ soil, it is an effective method in treating liquefiable deposits adjacent to existing foundations or buried structures.

In jet grouting, high-pressure jets of air, water, and/or grout are injected into the native soil in order to break up and loosen the ground and mix it with thin slurry of cementitious materials. In essence, it is not truly grouting but rather a mix-in-place technique to produce a soil–cement material. Depending on the application and soils to be improved, different kinds of jet are combined by using single fluid system (slurry grout jet), double fluid system (slurry grout jet surrounded by an air jet), and triple fluid system (water jet surrounded by an air jet, with a lower grout jet). The process can construct grout panels, full columns, or anything in between (partial columns) with a specified strength and permeability.

Solidification techniques can generally be used in a wide range of soil types, including those with high fines content. They are advantageous because the installation methods are relatively quiet and induce relatively small vibrations as compared to compaction methods. The induced horizontal earth pressures are smaller than with compaction methods and are larger than with drainage methods. Their disadvantage is the relatively high cost/benefit ratio as compared to compaction and drainage methods.

Drainage Methods

Excess pore water pressure generated by cyclic loading can be reduced and/or dissipated by installing permeable drain within the deposit. The methods employed can be divided into two categories depending on the material used as drain: grave drain and artificial drain methods. These methods rely on two mechanisms to reduce damage due to liquefaction: (1) delaying the development of excess pore water pressure due to earthquake shaking and (2) preventing the migration of high excess pore water pressure from untreated liquefied zones into non-liquefied areas (say underneath the structure) to prevent secondary liquefaction caused by pore water pressure redistribution.

Gravel drains are typically installed either as column-like drains in a closely spaced grid pattern or as backfill around underground structures to control the levels of maximum excess pore water pressure ratio during earthquake shaking. They can also be installed as wall-like or column-like perimeter drains at both sides of densified (treated) zones or to isolate the high excess pore water pressure from liquefied areas. In the installation of gravel drains, casing with an auger inside is drilled into the ground down to the specified depth. Crushed stone is then discharged into the casing and the gravel drain is formed by lifting the casing pipe.

Artificial drains can be made of geosynthetic composites or piles with drainage functions. They have not only drainage effect but also reinforcing effect. For example, plastic drain consists of a plastic perforated pipe drainage skeleton wrapped in geofabric to prevent clogging from soil particles. These can be easily installed; however, close spacing is usually required due to the limited capacity of each drain.

Design charts for drains were initially developed by Seed and Booker (1977) to control the maximum excess pore water pressure levels, but more recent design charts and analytical methods (e.g., Iai and Koizumi 1986; Pestana et al. 1997) provide better methods of taking into account various factors affecting the drain performance, such as the hydraulic properties of the drain and permeability and volumetric compressibility of the native soil.

Drainage remediation methods are suitable for use in sands, silts, or clays. One of the greatest advantages of drains is that they induce relatively small horizontal earth pressures during installation. Therefore, they are suitable for use adjacent to sensitive structures. In the design of drains, it is necessary to select a suitable drain material that has a coefficient of permeability substantially larger than the in situ soils. Installation can be carried out with low noised and vibration. However, since the in situ soils improved by this method remain in a loose condition, the method has obvious disadvantages when compared to compacted deposits, such as negligible ductility and significant residual settlement of the treated soils. It is effective only if it successfully promotes sufficiently rapid dissipation of pore pressures as to prevent the occurrence of liquefaction; if pore pressure dissipation is not sufficiently rapid during the relatively few critical seconds of the earthquake, this method does relatively little to improve post-liquefaction performance (Seed at al. 2003). Thus, the method is usually combined with densification method, i.e., the surrounding ground is compacted to some extent during the drain installation.

Replacement Methods

The remove and replace method is another liquefaction mitigation technique. This involves the removal of the in situ liquefiable material and replacement with a non-liquefiable material such as clay, gravel, or dense sand. Its advantage is that it provides high degree of confidence in the final product and uses construction equipment and practices that are widely available and easily tested (Idriss and Boulanger 2008). However, this method is only feasible when relatively small volumes of liquefiable material are involved at shallow depths.

Reinforcement and Containment Methods

When saturated sand deposits are sheared during seismic loading, excess pore water pressure is generated. To prevent the occurrence of liquefaction, the soil should undergo smaller shear deformation during earthquake. This can be achieved by underground diaphragm walls, sheet piles, or lattice-shaped walls using deep mixing techniques. This method simply surrounds the liquefiable soil with continuous underground wall, and thus, it can be used for the soil at the bottom of existing structures where liquefaction remediation is difficult with other methods. The shear deformation in the ground is reduced during an earthquake to mitigate liquefaction and at the same time provide support to the overlying structure. It can also be used to prevent excessive deformation of structures by restraining the lateral flow of ground after liquefaction.

The presence of stone columns, timber piles, and other stiff elements provide additional shear resistance against earthquake shaking. These reinforcements would take much of the earthquake load, thereby minimizing the shear deformation of the soil.

The key advantages of the method include (1) less vibration and noise, (2) applicable to existing structures, (3) no restriction to soil type and depth, and (4) effective use of land due to smaller treatment area. On the other hand, disadvantages of the method are the (1) lack of simplified methods to evaluate effectiveness and (2) generally higher construction costs.

Increasing In Situ Stress

In situ effective stresses within the soil mass can be increased, resulting in an increase in shear resistance. Pre-loading involves overconsolidating the soil with surcharge embankment by initially inducing increased vertical and horizontal effective stresses, and when the surcharge is removed, the resulting overconsolidation leaves the in situ soil somewhat more resistant against liquefaction. This method is applicable to soils with large fines content and can also be applied even below existing structures since the soil can be overconsolidated.

Lowering the groundwater table makes the soil above the water table unsaturated, resulting in increased effective stress for soils below the water table. In addition, if the water table is reduced to a level below the liquefiable soil layer, liquefaction is prevented because the absence of water makes the buildup of excess pore water pressure impossible (Cox and Griffiths 2010). For this method, it is always necessary to maintain the lower groundwater level, and this may result in higher operating cost and maintenance cost associated with continuous pumping. Thus, this technique is typically limited to short-term applications.

Structural Strengthening Techniques

There are many types of measures which can be adopted to minimize damage to structures while allowing the occurrence of liquefaction itself. The selection of the appropriate remedial measure depends on the type of structure and the results of assessment of the required stability of the structure considering the effect of liquefaction, e.g., loss of bearing capacity, increase in earth pressure, etc. Several structural strengthening techniques implemented to various structures are described below.

Deep foundations, such as piles or piers whose lengths extend below the occurrence of liquefaction (or significant cyclic softening due to partial liquefaction), can provide reliable vertical support and so can reduce or eliminate the risk of unacceptable liquefaction-induced settlements. However, pile or pier foundations do not necessarily prevent damages that may occur as a result of differential lateral structural displacements, so piles and/or piers must be coupled with sufficient lateral structural connectivity at the foundation as to safely resist unacceptable differential lateral displacements. Moreover, pile foundations are likely to be damaged by lateral forces caused by a decrease in subgrade reaction accompanying liquefaction and associated lateral spreading of liquefied soil. To reduce the damage from such ground deformations, inclined piles or large-diameter vertical piles have frequently been used, or the number of piles has been increased. Moreover, significant research efforts over the past 20 years have led to the development of a range of methods to address the effect of laterally spreading ground on pile foundations, ranging from simplified pseudo-dynamic approach (e.g., Hamada et al. 1986; Cubrinovski et al. 2012) to fully nonlinear, time-domain, fully coupled effective stress analysis of soil/pile/superstructure interaction analyses (e.g., Iai et al. 1998; Bowen and Cubrinovski 2008). These types of methods, complemented with appropriate conservatism, can provide a suitable basis for analysis of this issue and for the design and detailing of piles (or piers) and pile/cap connections.

In structures with spread foundations, differential settlement and cracks in foundation can occur as a result of loss in bearing capacity of foundation ground following liquefaction. To address this, spread foundations can be supported by piles. In addition, stiff, reinforced shallow foundations are used to resist both differential lateral and vertical displacements. Japanese practice has increasingly employed both grade beams and continuous reinforced foundations for low to moderate height structures, and performance of these types of systems in earthquakes has been good. Note that although stiff, shallow foundations can undergo differential settlements resulting in rotational tilting of the structure; these can be easily re-leveled after earthquake-induced settlements through several techniques, including jacking, under-excavation, or compaction grouting.

Quay walls can become unstable because of increase in active earth pressure and reduction in passive earth pressure. To prevent damage to such waterfront structures, strong quay walls such as pile-supported quay walls have been employed.

Buried pipelines may be bent, compressed, or extended as a result of large-ground displacement induced by liquefaction. Underground tanks are likely to be uplifted as a result of high excess pore water pressure developing underneath. For buried pipes, flexible joints have been used to absorb the large-ground deformation while some pipes have been supported by pile foundations. To prevent floating of underground structures, counterweights are usually placed on top of the structure or the structure is supported by piles.

When the foundation ground liquefies or softens, embankments built on top may slide or settle due to the decrease in bearing capacity. Possible countermeasures that can be used include installation of sheet piles on both edges of the embankment connected by tie rod or placement of berms and counterweight fills to minimize damage to the embankments.

Verification of Effectiveness

After the appropriate remediation technique has been selected and applied to the target site, attempts should be made to verify that the desired level of improvement has been achieved. It is no longer an acceptable practice to simply implement mitigation; the adequacy of the mitigation must also be evaluated.

The most direct way of effectiveness verification is to compare the soil characteristic which was deemed deficient before and after the improvement. For example, if the remediation was performed to increase the strength of the soil, then strength measurements before and after the improvement would be necessary to confirm the effectiveness of the chosen technique. In some cases, direct verification of the target characteristic may not be feasible and checking of related characteristics can be done more easily.

Laboratory Testing Techniques

Verification of the effectiveness of improvement technique through laboratory tests, such as triaxial tests, simple shear tests, and torsional shear tests, offers several advantages and disadvantages. One of the merits of these element tests is that by obtaining soil samples before and after improvement, a more direct inspection can be made. In addition, more accurate measurement of stress, strain, and other parameters can be performed, and therefore, a more accurate characterization of the improved soil is possible.

On the other hand, since soil sampling has to be done only at several representative locations, verification of effectiveness only at these discrete points can be made and not throughout the whole target area. Moreover, the effect of sample disturbance during sampling, handling, and transportation can have adverse effects on the characteristics of the sample which could affect the verification process.

In Situ Testing Techniques

By far, the most popular approach of checking the level of improvement is by performing in situ tests before and after the treatment process. Procedures like the standard penetration tests (SPT), cone penetration tests (CPT), flat plate dilatometer tests, and pressure meter tests are quite common. This approach has the advantage that the in situ tests can be done quickly, especially the SPT and CPT, and therefore, the level of improvement can be immediately assessed.

However, the results of these in situ tests should be interpreted carefully. For example, the SPT N-value is affected not only by density and confining pressure but also by lateral stress. Therefore, evaluating liquefaction potential using simplified approaches based on SPT N-values alone should be done with caution. These semiempirical methodologies are typically defined for normally consolidated soils, while ground improvement techniques can undoubtedly alter the soil's density, lateral stress, stress and strain history, fabric, and prior effect of ageing. Moreover, the time-dependent nature of the improvement should be considered. Results obtained immediately after the implementation of ground improvement technique may not represent the long-term effect of the technique due to factors like stress redistribution, creep-induced relaxation in lateral stress, etc. In addition, the location of in situ tests and location of treatment points should be considered. For example, to evaluate the effectiveness of sand compaction pile method, it has been customary to perform the penetration test at the center of the grid of sand piles, deemed as the point where the effectiveness of sand compaction is weakest.

Geophysical Testing Techniques

Another approach to verification is through direct transmission seismic testing, such as downhole and cross-hole seismic testing as well as spectral analysis of surface waves (SASW), which provides calculation of improvement values with some accuracy. In this procedure, the velocity of seismic waves, which is a function of basic soil properties such as modulus of elasticity, density, and Poisson's ratio, is measured between two points and this can give a more accurate picture of the level of improvement as compared to traditional penetration tests. Seismic reflection and refraction tests can be performed to verify the effectiveness of the technique for a wide area.

Design Issues in Ground Improvement

Several factors influence the stability and deformation of improved ground zones and supported structures during and after an earthquake (PHRI 1997; JGS 1998; Mitchell et al. 1998). Some of the more important factors are discussed below.

Size, Location, and Type of Treated Zone

Even when a wide area of soil undergoes liquefaction, the zone requiring soil improvement is generally limited to the area which directly affects the stability of the structure. For example, the portion of the subsoil which provides stability to structures on spread foundation is that directly underneath and on the perimeter of the structure; the portion of the subsoil at considerable distance from the structure has negligible effect on the stability of the said structure. The question then boils down to how wide the soil improvement should be.

In practice, the region to be remediated should be determined in both horizontal and vertical directions. For the vertical direction, it is usual practice to improve the ground to the deepest part of the liquefiable layer. If only the shallow portion of the liquefiable layer is improved, the excess pore water pressure generated at the bottom of the unimproved soil, if it liquefies, may induce upward seepage towards the surface and may result in potential instability at the upper improved zone. As for the lateral extent of treatment necessary, this is dependent on the type of remediation measure employed and the allowable deformation the structure can undergo. This can be investigated by performing detailed analysis and, if possible, supplemented by model tests. In many cases, the lateral extent of treatment beyond the edge of the structure is a distance equal to the depth of treatment beneath the structure.

Pore Pressure Migration

The difference in stiffness between a compacted zone of material and the surrounding loose liquefiable material could result in seepage between these two materials during and after earthquakes. Higher excess pore water pressure is developed in the untreated material, resulting in propagation into the adjacent improved soil. Although the effect of pore pressure migration into a dense material is very complex, it may nevertheless lead to strength loss of the densified soil.

Ground Motion Amplification

Due to the increase in its stiffness, remediated ground may induce an increase in ground motion, and this may be disadvantageous to the structure on top of it. Although little information is available regarding the influence of size of improved ground and its stiffness on ground motions, engineers should consider a balance in design such that the size and stiffness of the remediated ground would result in acceptable deformations and ground accelerations of the structure during an earthquake.

Dynamic Fluid Pressure

At the boundary between the improved and unimproved ground, both dynamic fluid pressure and static pressure corresponding to earth pressure coefficient equal to 1.0 are applied due to the liquefaction of the untreated ground. These forces must be taken into account when investigating the stability and deformation of structure supported in the improved zone.

Inertia Force

The dynamic response of the improved ground may change as a result of the liquefaction of the unimproved soil and the migration of excess pore water pressure, leading to a change in the inertia force on the improved ground. This inertia force should be included when evaluating the stability of an improved ground zone supporting a structure. Note however, that the phasing between inertia forces acting on the improved ground and supported structure may be different and this can have significant impact on the results of simplified stability and deformation analyses.

Influence of Structure

The existence of a structure either on top of or adjacent to an improved ground will change the stress state within the said zone and can affect the dynamic response and stress–strain behavior of the improved soil. The effects would depend on the soil and structure conditions, as well as on the installation process.

Forces Exerted by Laterally Spreading Soil

Liquefaction-induced lateral ground movements have been observed to occur at gently sloping sites, with gradients less than 1 %. Therefore, the improved ground should be sufficiently designed to resist the forces induced by the laterally moving ground. There are two force components that should be considered: first, the load exerted by the moving liquefied ground itself, and second, the force exerted by the unliquefied surficial crust riding on top of the liquefied ground.

Summary

Various remediation techniques are available to mitigate the liquefaction risk to structures constructed on loose saturated sandy deposits. These techniques can be roughly classified into two groups: the first is to improve the soil so that liquefaction may not occur, while the second is to provide measures for structures to prevent or minimize damage even if liquefaction occurs. Remedial treatment of soils to prevent liquefaction is based either on the principle of improving the properties of soils or the principle of changing stress or deformation conditions. In some cases, liquefaction is allowed to occur, but the damage is prevented or minimized by strengthening the structure or foundations.

Ground improvement is being used increasingly for remediating liquefiable soils due to the wide variety of methods that are available. Backed by several decades of research and experience, these methods have been utilized in many soil conditions and have been developed by both the construction industry and research institutions. These techniques can be adapted to site-specific conditions and, in some cases, one or more methods being combined to provide economical solutions for liquefaction remediation problems. Often, new ground improvement techniques are developed when difficult circumstances are encountered and these new techniques are then refined and become part of the expanding set of available mitigation measures.

A few of the remediation methods which have been implemented at many sites all over the world have been tested by few large-scale earthquakes, and most of them performed well, with negligible or minor ground damage. Thus, these remediation techniques will continue to play an important role in the mitigation of seismic risk to existing and future structures.

Cross-References

▶ Liquefaction: Performance of Building Foundation Systems

References

Andrus RD, Chung RM (1995) Ground improvement techniques for liquefaction remediation near existing lifelines. National Institute of Standards and Technology report NISTIR 5714, Gaithersburg, p 74

Bowen HJ, Cubrinovski M (2008) Effective stress analysis of piles in liquefiable soil: a case study of a bridge foundation. Bull NZ Soc Earthq Eng 41(4):247–262

Cooke HG, Mitchell JK(1999) Guide to remedial measures for liquefaction mitigation at existing highway bridge sites. Technical report MCEER-99-015, Multidisciplinary Center for Earthquake Engineering Research, Buffalo, p 176

Cox BR, Griffiths SC(2010) Practical recommendations for evaluation and mitigation of soil liquefaction in Arkansas. Project report MBTC 3017, p 175

Cubrinovski M, Haskell JJM, Bradley BA (2012) Analysis of piles in liquefying soils by the pseudo-static approach. In: Sakr MA, Ansal A (eds) Special topics in earthquake geotechnical engineering, vol 16, Geotechnical, geological and earthquake engineering. Springer, Dordrecht, pp 147–174

Hamada M, Yasuda S, Isoyama R, Emoto K (1986) Study on liquefaction-induced permanent ground displacements. Association for the Development of Earthquake Prediction, Tokai University, Shimizu, Japan

Hausler EA, Sitar N (2001) Performance of soil improvement techniques in earthquakes. In: Fourth international conference on recent advances in geotechnical earthquake engineering and soil dynamics, San Diego, CA, Paper 10.15

Hausmann MR (1990) Engineering principles of ground modification. McGraw-Hill, New York, p 632

Iai S, Koizumi K (1986) Estimation of earthquake induced excess pore pressure for gravel drains. In: Proceedings, 7th Japan earthquake engineering symposium, Tokyo, Japan, pp 679–684

Iai S, Ichii K, Li H, Morita T (1998) Effective stress analyses of port structures. Soils Found Special issue on Geotechnical Aspects of the January 17, 1995 Hyogoken-Nambu earthquake 2:97–114

Idriss IM, Boulanger RW (2008) Soil liquefaction during earthquakes, vol MNO-12, Monograph series. Earthquake Engineering Research Institute, Oakland

Ishii H, Funahara H, Matsui H, Horikoshi K (2013) Effectiveness of liquefaction countermeasures in the 2011 M = 9 gigantic earthquake, and an innovative soil improvement method. Indian Geotech J 43(2):153–160

Japanese Geotechnical Society, JGS (1998) Remedial measures against soil liquefaction. A.A. Balkema, Rotterdam

Kitazume M (2005) The sand compaction pile method. A. A. Balkema, London

Kitazume M, Terashi M (2013) The deep mixing method. CRC Press/Balkema, Leiden

Maeda S, Nagao K, Tsujino S (2006) Shock wave densification method – countermeasures against soil liquefaction by controlled blasting. Report of Satokogyo Engineering Research Institute, no. 31, pp 21–27 (in Japanese)

Mitchell JK, Cooke HG, Schaeffer J (1998) Design considerations in ground improvement for seismic risk mitigation. In: Proceedings, geotechnical earthquake engineering and soil dynamics III, ASCE geotechnical publication no. 75, vol 1, American Society of Civil Engineers, Reston, VA, pp 580–613

Munfakh GA, AbramsonLW, Barksdale RD, Juran I (1987) In situ ground reinforcement. In: Soil improvement – a ten year update. ASCE geotechnical special publication, vol 12. American Society of Civil Engineers, Reston, VA, pp 1–17

Narin Court W, Mitchell JK (1994) Explosive compaction: densification of loose, saturated cohesionless soils by blasting, Geotechnical engineering report UCB/GT/94-03. University of California, Berkeley

Pestana JM, Hunt CE, Goughnour RR (1997) FEQDrain: a finite element computer program for the analysis of the earthquake generation and dissipation of pore water pressure in layered sand deposits with vertical drains. Report no. EERC 97–17. Earthquake Engineering Research Center, UC-Berkeley

Port and Harbour Research Institute, PHRI (1997) Handbook on liquefaction remediation of reclaimed land. A.A. Balkema, Rotterdam

Schaefer V (ed) (1997) Ground improvement, ground reinforcement, ground treatment: developments 1987–1997. ASCE geotechnical special publication, vol 69. American Society of Civil Engineers, Reston, VA, p 618

Seed HB, Booker JR (1977) Stabilization of potentially liquefiable sand deposits using gravel drains. J Geotech Eng Div ASCE 103(GT7):757–768

Seed RB, Cetin KO, Moss RES, Kammerer A, Wu J, Pestana J, Riemer M, Sancio RB, Bray JD, Kayen RE, Faris A (2003) Recent advances in soil liquefaction engineering: a unified and consistent framework. Geotechnical report no. UCB/EERC-2003/06. Earthquake Engineering Research Center, University of California, Berkeley

Sondermann W, Wehr W (2004) Chapter 2: deep vibro techniques. In: Moseley MP, Kirsch K (eds) Ground improvement 2nd edn. Spon Press, Taylor & Francis Group, London, UK

Welsh JP (1992) Grouting techniques for excavation support. Excavation Support for the Urban Infrastructure. ASCE, New York, pp 240–261

Xanthakos PP, Abramson LW, Bruce DA (1994) Ground control and improvement. Wiley, New York, p 910

Yasuda S, Ishihara K, Harada K, Shinkawa N (1996) Effect of soil improvement on ground subsidence due to liquefaction. Spec Issue Soils Found 99–107

Liquefaction: Performance of Building Foundation Systems

Shideh Dashti
Department of Civil, Environmental, and Architectural Engineering, University of Colorado at Boulder, Boulder, CO, USA

Synonyms

Building settlement on softened ground; Shallow-founded structures on liquefiable soils; Soil-structure-interaction on softened ground

Definition

When founded on liquefiable ground, building response is commonly evaluated by the geotechnical engineer and structural engineer in a decoupled manner. The geotechnical engineer evaluates the potential for liquefaction triggering under a likely earthquake scenario (in most cases ignoring the presence of the structure), assesses the resulting building settlements, and typically designs a mitigation strategy to prevent liquefaction from occurring or to reduce settlements. On the other side, the structural engineer typically evaluates the seismic performance of the structure under fixed-base conditions (no soil-structure interaction considered), assuming that the liquefaction hazard is mitigated or bypassed. In performance-based structural design, the building *performance* is evaluated in terms of critical engineering demand parameters that correlate well with its damage potential (e.g., inter-story drift). In this article, the *performance* of a building on liquefiable ground is mainly discussed from the geotechnical engineer's viewpoint: the settlement potential.

Introduction

Liquefaction continues to pose a significant risk of damage to the built environment, as observed in both New Zealand and Japan in 2011 (e.g., Fig. 1). More than 50 % of the Christchurch area in New Zealand was affected by soil liquefaction during the 2011 Christchurch earthquake (Fig. 2). Damage caused by settlement, tilt, and lateral sliding, which mostly affected buildings on shallow foundations and their surrounding lifelines, caused severe economic losses (Green et al. 2011; Cubrinovski and McCahon 2012). In many cases, the structures were uneconomical to repair and were demolished. Future earthquakes in major cities around the world are expected to continue causing liquefaction-related damage.

Effective mitigation of the soil liquefaction hazard requires a thorough understanding of the potential consequences of liquefaction and the building performance objectives. The consequences of liquefaction in terms of ground displacement, in turn, depend on site conditions, earthquake loading characteristics, and building properties. This article provides an overview of

Liquefaction: Performance of Building Foundation Systems, Fig. 1 Settlement and tilting of residential buildings in Kamisu City, Japan, following the 2011 Tohoku-Kanto earthquake (Ashford et al. 2011)

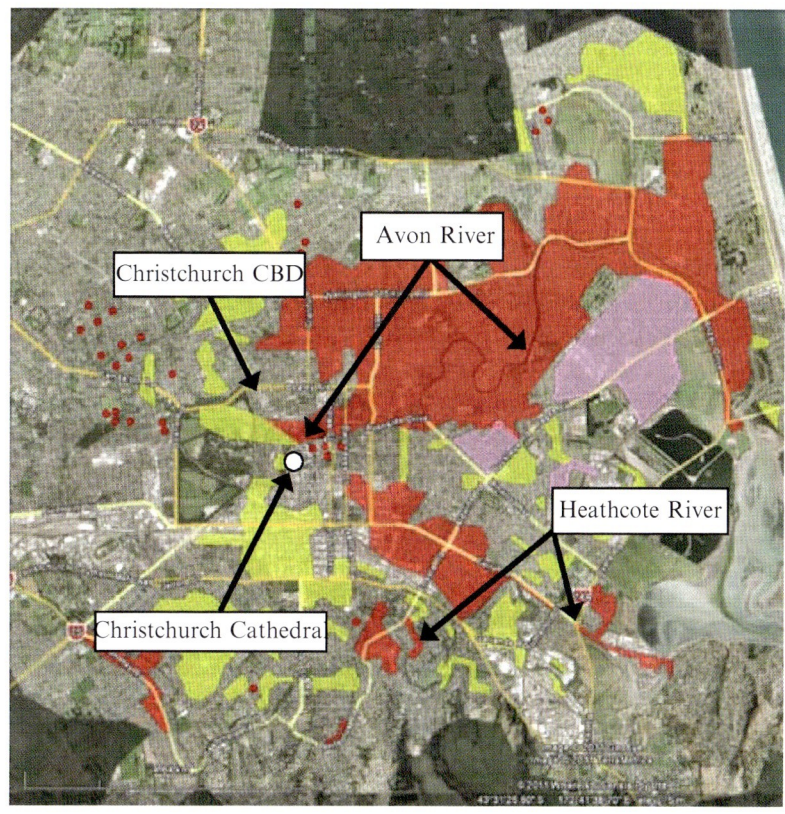

Liquefaction: Performance of Building Foundation Systems, Fig. 2 Liquefaction map of eastern Christchurch during the 2011 Christchurch, New Zealand, earthquake (*red*, moderate to severe; *yellow*, low to moderate; *pink*, liquefaction of roads – Green et al. 2011)

the state of practice in evaluating liquefaction-induced building settlements in geotechnical engineering. A summary of observations from recent case histories, physical model studies, and numerical analyses is presented to provide recommendations for practice and future research.

Background and Developments in Liquefaction Engineering

State of Practice in Evaluating Liquefaction-Induced Building Displacements

Figure 3 provides an overview of the various steps used to analyze the likelihood and impact of liquefaction in design and building assessment (Seed et al. 2003). The process starts by assessing the likelihood of liquefaction triggering (e.g., Youd et al. 2001; Seed et al. 2003). If the soil is judged to liquefy during the expected earthquake scenario, the overall stability of the foundation is evaluated using the available post-liquefaction,

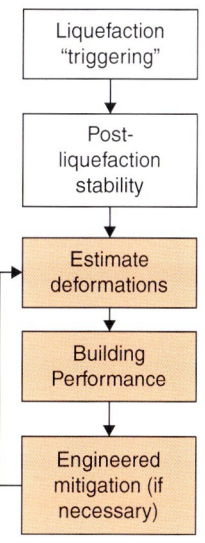

Liquefaction: Performance of Building Foundation Systems, Fig. 3 Steps in liquefaction engineering (Adapted from Seed et al. 2003)

residual soil strength to avoid failure and large deformations. If overall stability is not a concern, the consequences of liquefaction need to be evaluated in terms of building displacement. The evaluation of liquefaction-induced building settlements and the impact of those settlements on building performance (highlighted boxes in Fig. 3, which are discussed in this article) are currently steeped in empiricism with significant uncertainties (Seed et al. 2003).

Liquefaction-induced settlements are commonly estimated using simplified empirical procedures (e.g., Tokimatsu and Seed 1987; Ishihara and Yoshimine 1992), which were based on undrained cyclic laboratory tests on samples of saturated, clean sand. These empirical procedures evaluate post-liquefaction, volumetric reconsolidation settlements due to the dissipation of excess pore water pressures after cyclic loading. These methods have generally compared well with previous case histories and physical model studies in the free field or away from buildings. They, however, ignore the presence of the structure and the displacement mechanisms that are active under its foundation. Therefore, these procedures may be misleading in assessing the extent of building settlement and subsequent damage.

To bring in the influence of a structure, practicing engineers often use a combination of (1) empirical methods (e.g., Tokimatsu and Seed 1987; Ishihara and Yoshimine 1992) with the added confining pressure of the structure and (2) case history observations of the effect of foundation width on foundation settlement both normalized by liquefiable layer thickness (Liu and Dobry 1997). In current practice, if the estimated settlements obtained in this manner are judged to be excessive, mitigation techniques are considered to reduce the settlements to an acceptable level or to avoid soil liquefaction altogether. The same empirical procedures as above are then used to evaluate the effectiveness of the proposed mitigation techniques (e.g., ground improvement) in terms of reduction in settlement.

There are presently no well-calibrated, simplified design procedures for estimating the combined and complex effects of deviatoric and volumetric settlements due to cyclic softening under the static and dynamic loads of structures. This is in contrast with those procedures available for evaluating liquefaction triggering and post-liquefaction reconsolidation settlements in the free field. The lack of a proper understanding of the dominant displacement mechanisms near a building and a reliable analytical tool for estimating liquefaction-induced building settlements has sometimes resulted in the implementation of inadequate hazard mitigation measures (as observed during the 1999 Kocaeli earthquake in Turkey).

Case Histories of Liquefaction-Induced Building Damage

Observations of building performance on liquefied sites during previous earthquakes showed punching settlement, bearing failure, tilt, and lateral shifting of buildings. In the 1964 Niigata (Japan) and the 1990 Luzon (Philippines) earthquakes, most of the damaged buildings were two to four stories, founded on shallow foundations and relatively thick and uniform deposits of clean sand. The confining pressure and shear stress imposed by the buildings and their adjacent structures on the soil affected building movements (e.g., Tokimatsu et al. 1994). Contrary to previous case histories, in the 1999 Kocaeli (Turkey) earthquake, many of the damaged structures were influenced by the liquefaction of thin deposits of silt and silty sand (Sancio et al. 2004; Bray et al. 2000; Bird and Bommer 2004). The building settlement was directly proportional to its contact pressure, and the building's height/width (H/B) aspect ratio greatly affected the degree of tilt (Sancio et al. 2004), showing the importance of the building's dynamic properties. Some buildings translated laterally up to 1 m.

More recently, liquefaction-induced settlements of 1–2 m and tilts exceeding 2° were observed in low- to mid-rise structures in the M_w 6.1, 2011 Christchurch, New Zealand, earthquake. Ground motions recorded on liquefiable sites showed amplified spectral content for periods greater than 2 s. The uplift forces from groundwater pressures caused floors to bulge upward and the foundations to damage and tilt due to lateral spreading (Cubrinovski and McCahon 2012). Similarly, during the M_w 9, 2011 Tohoku-Kanto earthquake in Japan, the building damage in the Kanto region was dominated by liquefaction, not ground shaking alone (Ashford et al. 2011).

Despite these well-documented case histories, the relation between key ground motion characteristics, soil properties, and the response of building (in terms of settlement, tilt, lateral displacement, and subsequently building performance and damage potential) associated with soil liquefaction, with or without remediation, is not well understood. Buildings that are significantly tilted may need to be demolished and rebuilt, representing a complete loss, despite having a rigid body rotation and intact structural components. On the other hand, there may also be cases in which the liquefied soil, serving as a base isolator, reduces shaking-induced damage to structures by dissipating energy and altering motion frequency content.

Physical Model Studies of Shallow-Founded Structures on Liquefiable Sand

Because of the uncertainties involved in interpreting case histories and limited instrumental recordings at key locations, physical modeling under controlled and simplified conditions provides valuable insights to improve the understanding of liquefaction and its effects. Several researchers have used reduced-scale shaking table and centrifuge tests in the past to study the response of rigid, shallow model foundations situated atop deposits of saturated, loose to medium dense, clean sand (e.g., Yoshimi and Tokimatsu 1977; Liu and Dobry 1997; Hausler 2002; Madabhushi and Haigh 2010).

In a recent study, centrifuge experiments were performed to identify the dominant mechanisms of building settlement on relatively thin and shallow deposits of liquefiable, clean sand (Dashti 2009; Dashti et al. 2010a). These tests employed elastic, single-degree-of-freedom (SDOF) structural models with realistic fundamental frequencies (as opposed to a rigid foundation) on liquefiable ground (as shown in Fig. 4). Structures and the soil adjacent to the structures settled significantly more than the soil in the free field (soil away from the structures) in most cases, as shown in Fig. 5 for a representative test. These comparisons demonstrate the inadequacy of using empirical, free-field procedures to evaluate building settlements.

A strong tendency for flow away from the structures toward the free field was observed during stronger shaking events, which likely led to localized volumetric strains due to partially drained cyclic loading (Dashti et al. 2010a). The relative influence of various testing parameters and structural properties was investigated. Conceptually, the study classified the primary settlement mechanisms as (**a**) volumetric types, rapid drainage during cyclic loading ($\varepsilon_{p\text{-}DR}$), sedimentation ($\varepsilon_{p\text{-}SED}$), and consolidation due to excess pore pressure dissipation ($\varepsilon_{p\text{-}CON}$) and (**b**) deviatoric types, partial bearing capacity loss under the static load of the structure ($\varepsilon_{q\text{-}BC}$) and soil-structure interaction (SSI)-induced building ratcheting under the dynamic load of the building ($\varepsilon_{q\text{-}SSI}$).

Dashti et al. (2010a) showed that building settlement is not proportional to the thickness of the liquefiable material, as volumetric settlements may not be dominant underneath a structure. As long as there is sufficient thickness of liquefiable soils present under the building, significant settlements may be observed due to deviatoric or shear-induced strains. Hence, the available empirical charts of building settlement that are normalized by the thickness of liquefiable layer may be misleading and should not be used in the engineering practice.

The relative importance of the identified displacement mechanisms, which affects the choice of an effective remediation strategy, depended strongly on several parameters: the liquefiable soil's initial relative density (D_r) and thickness, flow boundary conditions, and ground shaking intensity and the building's fundamental frequency, foundation geometry, and contact pressure. Further, centrifuge experiments revealed that the settlement time history of buildings during each earthquake followed the shape of the Arias Intensity time history of the motion, as shown by the representative results in Fig. 6.

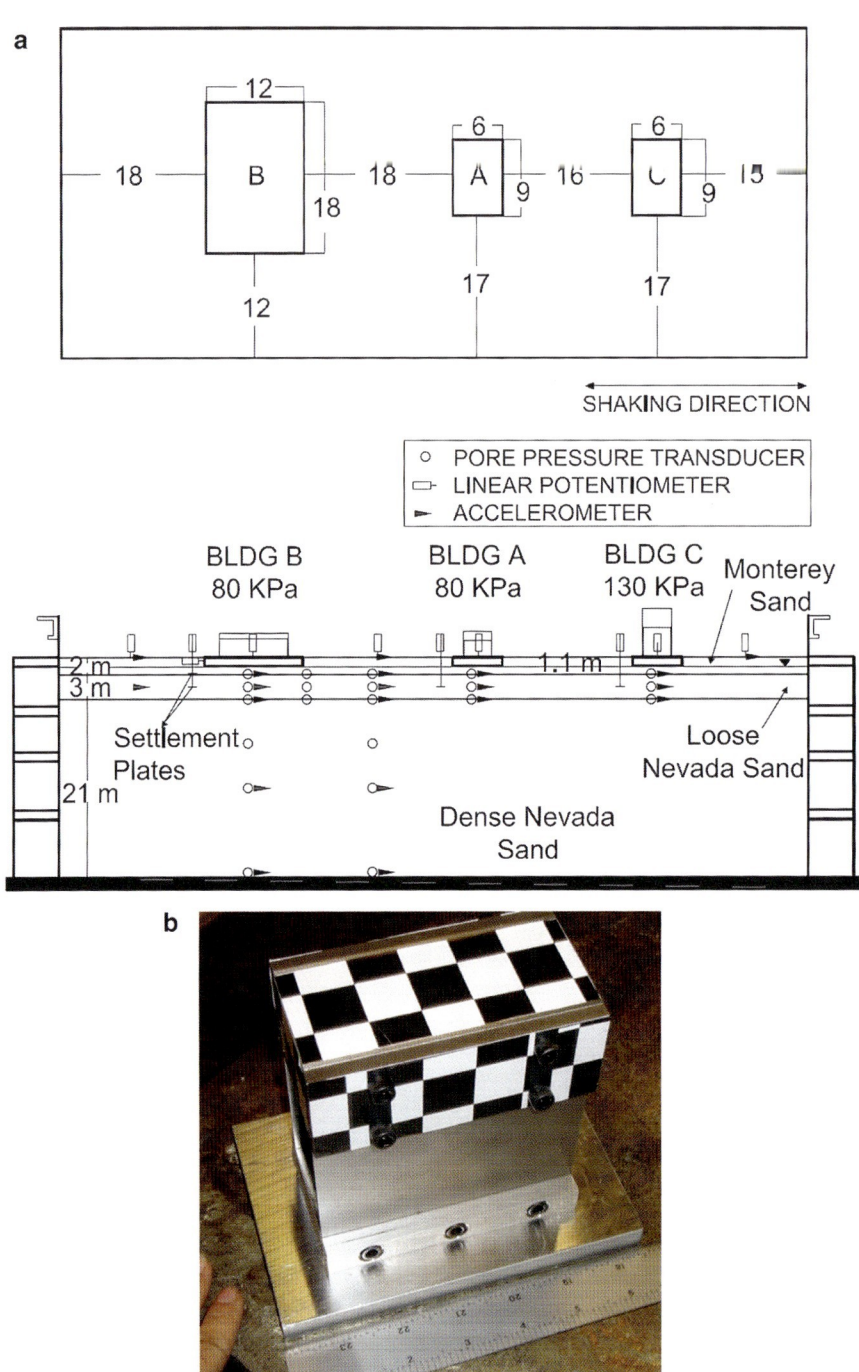

Liquefaction: Performance of Building Foundation Systems, Fig. 4 Centrifuge experiment by Dashti (2009): (**a**) schematic drawings (plan and elevation views) of a representative test with three SDOF structures with different properties and (**b**) picture of model building C used in centrifuge testing. Dimensions shown in prototype scale meters (when the model was spun to 55 g of centrifugal acceleration)

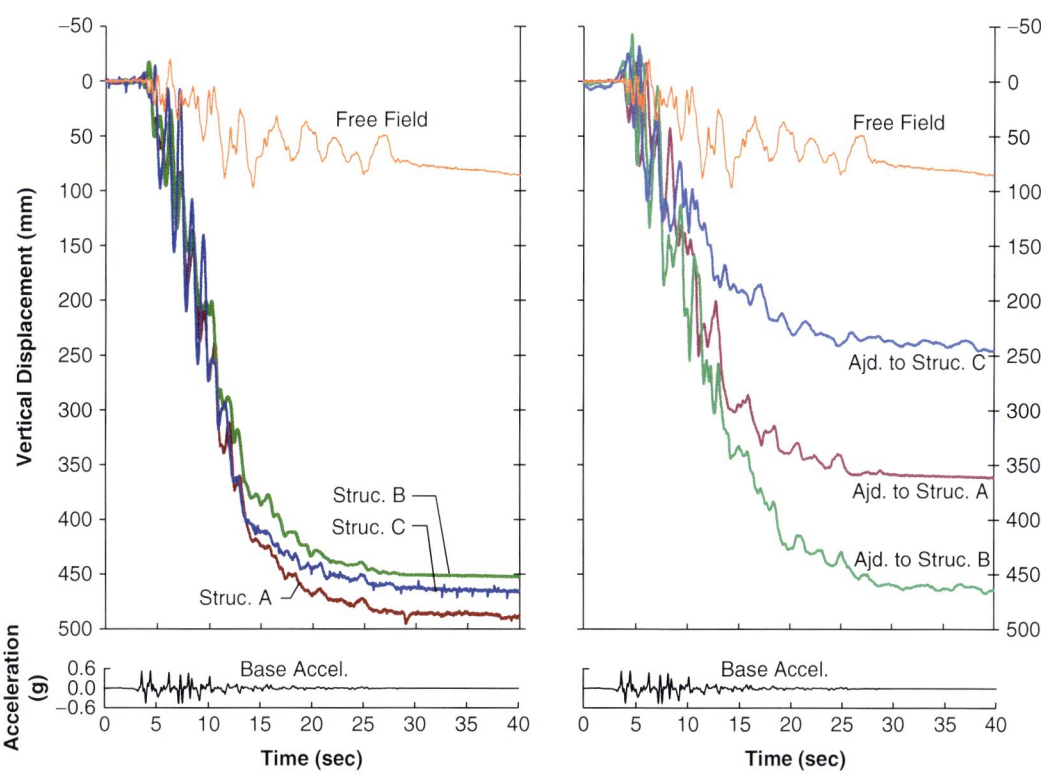

Liquefaction: Performance of Building Foundation Systems, Fig. 5 Settlement of different structures compared to the free field and the adjacent soil during a representative centrifuge experiment T3-30, with liquefiable layer thickness of 3 m and relative density of 30 % (Adapted from Dashti 2009)

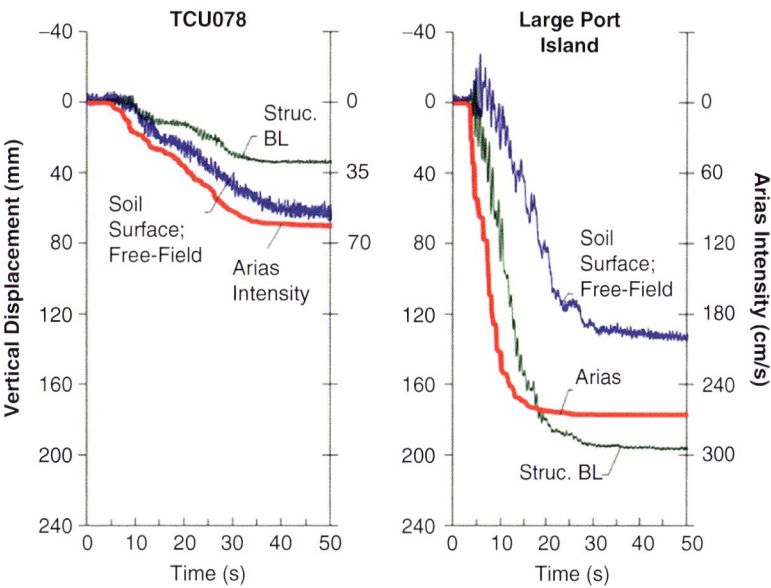

Liquefaction: Performance of Building Foundation Systems, Fig. 6 Settlement time history of the baseline structure (BL) and free field soil surface compared to the corresponding Arias Intensity time history of the base motion during a representative centrifuge experiment T3-50, with liquefiable layer thickness of 3 m and relative density of 50 % (Adapted from Dashti et al. 2010b)

Arias Intensity (I_a) is an index representing the energy of the ground motion in units of L/T (Arias 1970) and defined as

$$I_a(T) = \frac{\pi}{2 \cdot g} \int_0^T a^2(t) \cdot dt \qquad (1)$$

over the time period from 0 to T, where a = the measured acceleration value. The Arias Intensity of an earthquake motion depends on the intensity, frequency content, and duration of the ground motion. Its rate represents roughly the rate of earthquake energy buildup. These preliminary experimental results show the importance of taking into account holistic ground motion intensity parameters (e.g., I_a or its slope) in the selection of ground motions for design.

Numerical Studies of Liquefaction Effects

In analytical studies, Soil-Foundation-Structure Interaction (SFSI) effects are commonly evaluated employing equivalent linear, visco-elastic soil models. In reality, the soil response is nonlinear, particularly when it undergoes a significant strength loss and softening as in liquefaction. The insight gained from the previous case histories and physical model studies (summarized in the previous sections) points to the need for performing fully coupled, effective stress, nonlinear dynamic analysis of the soil-structure system, to capture the interacting influences of the key factors on soil and structural response. Popescu and Prevost (1993), Elgamal et al. (2005), Popescu et al. (2006), Andrianopoulos et al. (2006), Lopez-Caballero and Modaressi Farahmand-Razavi (2008), Andrianopoulos et al. (2010), Shahir and Pak (2010), Dashti and Bray (2013), Lopez-Caballero and Modaressi Farahmand-Razavi (2013), and Karamitros et al. (2013) conducted 2-D (plane strain) and 3-D, fully coupled, nonlinear, finite element and finite difference analyses to study the dynamic interaction between homogeneous and layered liquefiable sand and a structure. In these analyses, the structure was simulated either as a surface load, rigid structure, or an elastic SDOF model. The foundation was mostly fixed to the soil mesh, assuming no relative movement at the interface.

Numerical simulations of SDOF structures on liquefiable ground have shown that liquefaction increases the building's settlement potential, but reduces the seismic demand on the structure and the peak inter-story drift ratio (Lopez-Caballero and Farahmand-Razavi 2013). This observation agrees with previous case histories and may influence the choice and design of remediation techniques in the future. Further, the characteristics of input ground motion used in numerical simulations have been shown to govern the effectiveness of the mitigation technique in terms of settlement and inter-story drift.

Structural engineers employ high-fidelity analytical models of structures, often with no consideration of soil response or soil-structure interaction. In addition, most work in numerical modeling of earthquake effects on structures has focused on the effects of ground shaking alone. Only a few studies have addressed the combined influence of shaking and liquefaction-induced permanent ground movements on the performance of a structure (e.g., Fotopoulou and Pitilakis 2012; Negulescu and Foerster 2010; Aygun et al. 2011). At this time, the influence of ground settlement on the building's engineering demand parameters and damage potential is not clearly understood for a wide range of structures, ground motions, and soil conditions. Complex, nonlinear, dynamic simulations of the soil-structure system are expensive and time consuming and often not justifiable in most projects.

In summary, most analyses of soil liquefaction have ignored structural performance and nonlinearities. On the other hand, nonlinear analyses of structures typically ignore soil nonlinearities or bring in ground shaking and permanent soil displacements in a decoupled manner. It is advantageous, if the project resources allow, that the triggering of liquefaction, post-liquefaction instability, the resulting displacements and ground shaking, and the nonlinear response of the structure be considered simultaneously in a coupled analysis. This approach brings in the complexities of both soil and structure, in order to accurately evaluate the performance of structures on liquefied ground and the effectiveness of mitigation methods in terms of improved performance. At this time, the cost and expertise required for reliably running such analyses are often prohibitive for most projects.

Summary

The profession has largely addressed the problem of liquefaction triggering assessment. However, there is still much uncertainty in evaluating the consequences of liquefaction in terms of building settlements and the effect of those settlements on building performance. Reliable and simplified procedures for estimating liquefaction-induced building movements that take into account the key displacement mechanisms are currently lacking. As such, observations of ground and building response in the previous case histories, physical model studies, and numerical analyses are combined in this article to provide an overview of the complexities of the problem and guidance for practicing engineers.

The governing displacement mechanisms near a structure are primarily deviatoric induced when the liquefiable soil layer is sufficiently thick and shallow. Methods that estimate free-field, post-liquefaction, volumetric reconsolidation settlement (e.g., Tokimatsu and Seed 1987; Ishihara and Yoshimine 1992) cannot be used to estimate building settlement, because they cannot estimate deviatoric displacement. These procedures either completely ignore the presence of the structure or simplistically bring in its influence through an added foundation load.

It is not appropriate to normalize building settlements by the thickness of the liquefied soil layer. This type of normalization implies that volumetric-induced displacement mechanisms govern building settlement. This is not the case for shallow liquefiable layers under a building, where deviatoric displacement mechanisms dominate ($\varepsilon_{q\text{-BC}}$ and $\varepsilon_{q\text{-SSI}}$). Localized volumetric strains resulting from partial drainage in response to intense transient hydraulic gradients ($\varepsilon_{p\text{-DR}}$) are also important.

If liquefaction is triggered in the free field, it is likely that it will also occur under the edges of a building's shallow foundation and in the adjacent soil (Travasarou et al. 2006; Dashti et al. 2010a, b). If significant excess pore water pressures are expected to be generated, liquefaction-induced building movements should be evaluated. A well-calibrated, fully coupled, nonlinear, effective stress dynamic analysis may be performed to provide insight into the consequences of liquefaction in terms of displacements and building performance. This type of advanced analysis may be justifiable for sensitive projects and should only be performed by well-trained and experienced engineers with calibrated and validated constitutive models.

For many regular projects, performing sophisticated, nonlinear, effective stress, dynamic analyses of the soil-structure system may not be practical. First, as an initial check, the seismic bearing capacity of the building must be evaluated taking into account the available post-liquefaction, residual strength of the foundation soil, and the inertial loading of the building. If bearing capacity failure is a concern (factor of safety of near 1), large differential building settlements are expected with a possibility of failure. Hence, remediation is necessary. If the building passes this first instability check, small to moderate settlements are expected typically, for which there are currently no reliable, simplified analytical methods available. In these cases, the available empirical methods for estimating volumetric settlement in the free field must be used with caution and engineering judgment. Considerable localized volumetric strains due to partial drainage as well as static and dynamic deviatoric strains may produce large building settlements and tilt, none of which are taken into account by the available procedures. Special attention should be given to taller buildings with higher aspect ratios (i.e., $H/B > 1.5$), which are more prone to tilting or extreme differential settlements. Previous case histories and physical model studies have shown that a building's tilting potential and overall settlements increase proportionally with its aspect ratio (i.e., H/B) and contact pressure, respectively (Sancio et al. 2004; Dashti et al. 2010a). In designing a mitigation measure, the dominating mechanisms of displacement should be considered and minimized.

The liquefaction hazard facing existing structures in many parts of the world demands more reliable, performance-based engineering procedures to design remediation techniques. These methods should focus on reducing displacements under and around the foundation as well as structural nonlinearities. Soil, foundation, and building response need to be explicitly linked to damage in developing effective and economic remediation strategies in the future.

Cross-References

▶ Dynamic Soil Properties: In Situ Characterization Using Penetration Tests
▶ Geotechnical Earthquake Engineering: Damage Mechanism Observed
▶ Liquefaction: Countermeasures to Mitigate Risk
▶ Seismic Analysis of Masonry Buildings: Numerical Modeling

References

Andrianopoulos KI, Bouckovalas GD, Karamitros DK, Papadimitriou AG (2006) Effective stress analyses for the seismic response of shallow foundations on liquefiable sand. In: Proceedings of the 6th European conference on numerical methods in geotechnical engineering, Graz

Andrianopoulos KI, Papadimitriou AG, Bouckovalas GD (2010) Bounding surface plasticity model for the seismic liquefaction analysis of geostructures. J Soil Dyn EQ Eng 30:895–911

Arias A (1970) A measure of earthquake intensity. In: Hansen RJ (ed) Seismic design for nuclear power plants. MIT Press, Cambridge, Mass

Ashford SA, Boulanger RW, Donahue JL, Stewart JP (2011) Geotechnical quick report on the Kanto plain region during the March 11, 2011, Off pacific coast of Tohoku earthquake, Japan. In: NSF supported geotechnical extreme events reconnaissance (GEER) report, 5 April 2011

Aygun B, Dueas-Osorio L, Padgett JE, Desroches R (2011) Efficient longitudinal seismic fragility assessment of a multispan continuous steel bridge on liquefiable soils. J Bridge Eng 16(1):93–107

Bird JF, Bommer JJ (2004) Earthquake losses due to ground failure. J Eng Geol 74(2):147–179

Bray JD, Stewart JP, Baturay MB, Durgunoglu T, Onalp A, Sancio RB, Ural D, Ansal A, Bardet JB, Barka A, Boulanger R, Cetin O, Erten D (2000) Damage patterns and foundation performance in Adapazari. Earthq Spectra 16(S1):163–189

Cubrinovski M, McCahon I (2012) Short term recovery project 7: CBD foundation damage. Natural hazards research platform. University of Canterbury, Christchurch

Dashti S (2009) Toward evaluating building performance on softened ground. PhD dissertation, University of California, Berkeley

Dashti S, Bray JD (2013) Numerical simulation of building response on liquefiable sand. J Geotechnic Geoenviron Eng, ASCE 139(8):1235–1249

Dashti S, Bray JD, Pestana JM, Riemer MR, Wilson D (2010a) Mechanisms of seismically-induced settlement of buildings with shallow foundations on liquefiable soil. J Geotechnic Geoenviron Eng, ASCE 136(1):151–164

Dashti S, Bray JD, Pestana J, Riemer MR, Wilson D (2010b) Centrifuge testing to evaluate and mitigate liquefaction-induced building settlement mechanisms. J Geotechnic Geoenviron Eng, ASCE 136(7):918–929

Elgamal A, Lu J, Yang Z (2005) Liquefaction-induced settlement of shallow foundations and remediation: 3D numerical simulation. J Earthq Eng 9(1):17–45

Fotopoulou SD, Pitilakis KD (2012) Vulnerability assessment of reinforced concrete buildings subjected to seismically triggered slow-moving earth slides. Landslides (published online)

Green RA, Cubrinovski M, Wotherspoon L, Allen J, Bradley B, Bradshaw A, Bray J, De Pascale G, Orense R, Orense R, O'Rourke T, Pender M, Rix G, Wells D, Wood C, Wotherspoon L, Cox B, Henderson D, Hogan L, Kailey P, Lasley S, Robinson K, Taylor M, Winkley A, and Zupan J (2011) Geotechnical reconnaissance of the 2011 Christchurch, New Zealand earthquake. NSF supported geotechnical extreme events reconnaissance (GEER) report, vol 1, 8 Nov 2001

Hausler EA (2002) Influence of ground improvement on settlement and liquefaction: a study based on field case history evidence and dynamic geotechnical centrifuge tests. PhD dissertation, University of California, Berkeley

Ishihara K, Yoshimine M (1992) Evaluation of settlements in sand deposits following liquefaction during earthquakes. J Soils Found 32(1):173–188

Karamitros DK, Bouckovalas GD, Chaloulos YK (2013) Seismic settlement of shallow foundations on liquefiable soil with a clay crust. Soil Dyn Earthq Eng 46:64–76

Liu L, Dobry R (1997) Seismic response of shallow foundation on liquefiable sand. J Geotech Geoenviron Eng 123(6):557–567

Lopez-Caballero F, Modaressi Farahmand-Razavi A (2008) Numerical simulation of liquefaction effects on seismic SSI. Soil Dyn Earthq Eng 28:85–98

Lopez-Caballero F, Modaressi Farahmand-Razavi A (2013) Numerical simulation of mitigation of liquefaction seismic risk by preloading and its effects on the performance of structures. Soil Dyn Earthq Eng 49:27–38

Madabhushi SPG, Haigh SK (2010) Effect of superstructure stiffness on liquefaction-induced failure mechanisms. Int J Geotech Earthq Eng 1:71–87, ISSN 1947-8488

Negulescu C, Foerster E (2010) Parametric studies and quantitative assessment of the vulnerability of an RC frame building exposed to differential settlements. Nat Hazards Earth Syst Sci 10: 1781–1792

Popescu R, Prevost JH (1993) Centrifuge validation of a numerical model for dynamic soil liquefaction. Soil Dyn Earthq Eng 12:73–90

Popescu R, Prevost JH, Deodatis G, Chakrabortty P (2006) Dynamics of nonlinear porous media with applications to soil liquefaction. Soil Dyn Earthq Eng 26:648–65

Sancio R, Bray JD, Durgunoglu T, Onalp A (2004) Performance of buildings over liquefiable ground in Adapazari, Turkey. In: Proceedings, 13th world conference on earthquake engineering, Vancouver, Aug 2004, No. 935

Seed RB, Cetin KO, Moss RES, Kammerer AM, Wu J, Pestana JM, Riemer MF, Sancio RB, Bray JD, Kayen RE, Faris A (2003) Recent advances in soil liquefaction engineering: a unified and consistent framework. In: Proceedings, 26th annual ASCE Los Angeles geotechnical spring seminar, Keynote Presentation, H.M.S. Queen Mary, Long Beach

Shahir H, Pak A (2010) Estimating liquefaction-induced settlement of shallow foundations by numerical approach. J Comput Geotech 37:267–279

Tokimatsu K, Seed HB (1987) Evaluation of settlements in sands due to earthquake shaking. J Geotech Eng, ASCE 113(8):861–878

Tokimatsu K, Kojima J, Kuwayama AA, Midorikawa S (1994) Liquefaction-induced damage to buildings I 1990 Luzon earthquake. J Geotech Eng, ASCE 120(2):290–307

Travasarou T, Bray JD, Sancio RB (2006) Soil-Structure Interaction Analyses of Building Responses During the 1999 Kocaeli Earthquake. In: Proceedings, 8th US Nat. Conf. EQ Engrg., 100th Anniv. EQ Conf. Comm. the 1906 San Francisco Earthquake, EERI, April, Paper 1877

Yoshimi Y, Tokimatsu K (1977) Settlement of buildings on saturated sand during earthquakes. Soils Found 17(1):23–38

Youd TL, Idriss IM, Andrus RD, Arango I, Castro G, Christian JT, Dobry R, Finn WDL, Harder LF, Hynes ME, Ishihara K, Koester JP, Liao SSC, Marcuson WF, Martin GR, Mitchell JK, Moriwaki Y, Power MS, Robertson PK, Seed RB, Stokoe KH (2001) Liquefaction resistance of soils: summary report from the 1996 NCEER and 1998 NCEER/NSF workshops on evaluation of liquefaction resistance of soils. J Geotech Geoenv Eng, ASCE 127(10):817–833

Liquid Storage Tanks: Seismic Analysis

Maria Vathi and Spyros A. Karamanos
Department of Mechanical Engineering, University of Thessaly, Volos, Greece

Synonyms

Dynamic analysis of liquid storage tanks; Earthquake analysis of liquid containers

Introduction

Liquid-containing tanks are used in water distribution systems and in industrial plants for the storage and/or processing of a variety of liquids and liquid-like materials, including oil, liquefied natural gas, chemical fluids, and wastes of different forms. These tanks are mainly categorized in two types: ground-supported tanks and elevated tanks. In the present chapter, ground-supported tanks of upright cylindrical shape are presented.

As the necessity and importance of liquid storage tanks has increased over the years, mainly because of their association with the need for continuous supply of energy and water resources, so has the need to understand better their structural behavior and performance. In particular, the analysis and design of liquid storage tanks against earthquake-induced actions has been recognized as an important issue towards safeguarding their structural integrity and has been the subject of numerous analytical, numerical, and experimental works. A pioneering work of Housner (1957), motivated by the need for accurate determination of seismic actions in tanks, presented a solution for the hydrodynamic effects in non-deformable vertical cylinders and rectangles. For the first time, the total seismic action was split in two parts, namely, the "impulsive" part and the "convective" part. This concept constitutes the basis for the American Petroleum Institute (2003, 2007) Standard provisions (Appendix E) for vertical cylindrical tanks. Veletsos and Yang (1977) and Haroun (1983) have extended this formulation to include the effects of shell deformation and its interaction with hydrodynamic effects. More recently, the case of uplifting of unanchored tanks as well as soil-structure interaction effects have been studied extensively, in the papers by Peek (1988), Natsiavas (1988), Veletsos and Tang (1990), and Malhotra (1995). Notable contributions on the seismic response of anchored and unanchored liquid storage tanks have been presented by Fischer (1979) and Rammerstorfer et al. (1988), with particular emphasis on design implications. Apparently, those papers constitute the basis for the seismic design provisions concerning vertical cylindrical tanks in Part 4 of Eurocode 8, Annex A (European Committee for Standardization 2006b). In addition to the numerous analytical/numerical works, notable experimental contributions on this subject have been reported (Niwa and Clough 1982; Manos and Clough 1982). The reader is referred to the review paper of Rammerstorfer et al. (1990) for a literature review of liquid storage tank response under seismic loads, including fluid-structure and soil-structure interaction effects. Recently, the European research project INDUSE (Pappa et al. 2012) has been completed on the seismic analysis and design of industrial facilities, with particular emphasis on liquid storage tanks. Results from this study have been incorporated in specific design guidelines.

During a strong seismic event, liquid storage tanks may exhibit significant damages. Typical tank damage can be in the following form: (a) buckling of the shell, precipitated by excessive axial compression due to overall bending or beamlike action of the structure; (b) damage to the roof, caused by sloshing of the upper part of the contained liquid when there is insufficient freeboard between the liquid surface and the roof; and (c) failure of attached piping and other accessories, due to the inability of these elements to accommodate the deformations of the flexible shell. Shell buckling usually takes the form of "diamond shaped" or "elephant's foot" buckling that appears at a short distance above the base and usually extends around most or all of the circumferences. In previous earthquake events, damages of various severity have been reported, e.g., from fracture of a pipe that created only slight leaks on minor repairs (damage to an overflow pipe) to complete loss of tank content.

Seismic analysis of liquid-containing tanks differs from typical civil engineering structures (i.e., buildings and bridges) in two ways: First, during seismic excitation, liquid inside the tank exerts hydrodynamic force on tank walls and base due to liquid-tank interaction. Second, liquid storage tanks are generally very thin walled and this results in relatively low ductility and low redundancy.

The present chapter outlines principles and methodologies for the seismic analysis of aboveground ground-supported liquid storage tanks of upright cylindrical shape without restrictions on their size and their use. The contents of this chapter are applicable to tanks for both hydrocarbon (oil) and water storage. However, they do not cover containers of different shape, such as horizontal-cylindrical and spherical vessels. For the seismic design of such vessels, the reader is referred to the publication of Karamanos et al. (2006). In addition, they do not refer to elevated tanks, where the tank is supported on braced or unbraced legs, on skirt, or on a more complex structural system. Moreover, the present chapter refers mainly to steel tanks, which are widely used in industrial applications. However, the basic principles of this chapter can be applicable to tanks made of different material.

In the present chapter, following a short presentation of current design standards in "Relevant Standards and Guidelines", and a short reference to the seismic input in the course of spectral analysis in "Seismic Input", seismic actions in liquid storage tanks are discussed in "Seismic Loading on Liquid Storage Tanks" in a concise manner, with special reference to hydrodynamic loading. In "Special Issues on Seismic Action", some special issues including the response of unanchored tanks are discussed, whereas in "Seismic Resistance of Liquid Storage Tanks", the resistance of liquid storage tanks is outlined. Finally, in "Summary", some concluding remarks are stated.

Relevant Standards and Guidelines

There exist several relevant standards and guidelines for the seismic analysis of liquid storage tanks. In this paragraph, the most important European and American Standards as well as the New Zealand Recommendations (Priestley et al. 1986) are briefly presented.

A basic standard on tank seismic design is the 4th part of Eurocode 8 (European Committee for Standardization 2006b). It is a standard that concerns the seismic design of silos, tanks, and pipelines for earthquake resistance and is part of the CEN/TC250 standard group, often referred to as "Structural Eurocodes." It specifies principles and application rules for the seismic design of structural components including aboveground and buried pipeline systems, silos, and storage tanks of different types and uses. This standard contains significant information on the calculation of seismic action based mainly on the work of Scharf (1990). Furthermore, seismic resistance in terms of shell buckling is based mainly on the work of Rotter (1990). It should be noted that EN 1998-4 rules are quite general and may refer to tanks of different material, i.e., steel, concrete, or plastic. It should be noted that this standard should be used together with the other relevant Structural Eurocodes. More specifically, Part 1 of Eurocode 8 (European Committee for Standardization 2004a) should be used for the seismic input in terms of the design spectrum. In addition, Part 1–6 of Eurocode 3 (European Committee for Standardization 2007) gives basic design rules for the strength and structural stability of steel tanks, covering the basic failure modes of steel shells, namely, plasticity, cyclic plasticity, buckling, and fatigue, with emphasis on buckling. Finally, Part 4.2 of Eurocode 3 (European Committee for Standardization 2006a) provides principles and application rules for the general structural design of vertical cylindrical aboveground steel tanks for the storage of liquid products that should be taken into account.

API 650 (American Petroleum Institute 2007) is a standard dedicated to the general design of liquid storage tanks, developed by the American Petroleum Institute and widely used by the petrochemical industry for the design and construction of tanks in petrochemical facilities. In particular, Appendix E of API 650 refers exclusively to seismic design and contains provisions for both determining seismic actions on tanks and calculating the strength of the tank. Appendix E of the new 11th edition of API 650 (American Petroleum Institute 2007) has significant revisions with respect to the previous 2003 edition (American Petroleum Institute 2003) and includes provisions for site-specific seismic input, calculation of hoop hydrodynamic stresses, distinction between ringwall and slab overturning moments, freeboard requirements, and consideration of vertical excitation effects. It should be noted that the seismic provisions of the updated Appendix E are in accordance with the provisions of ASCE 7 Standard (American Society of Civil Engineers 2006), which refers to structural loading.

EN 14015 (European Committee for Standardization 2004b) is a European standard outside the Structural Eurocode framework, i.e., independent of EN 199x Standards, which specifies the requirements for the materials, design, fabrication, erection, testing, and inspection of site built, vertical, cylindrical, flat-bottomed, aboveground, welded, steel tanks for the storage of liquids at ambient temperatures and above. It constitutes an updated version of the classical British Standard BS 2654 (British Standards Institution 1989), which is a specification for the manufacture of vertical steel welded non-refrigerated storage tanks with butt-welded shells for the petroleum industry. It is a complete standard for tank design analogous to API 650 (American Petroleum Institute 2007). Recommendations for seismic design and analysis of storage tanks are stated in Annex G of EN 14015, which are quite similar with the requirements of Annex E of API 650 (American Petroleum Institute 2003).

The New Zealand Seismic Tank Design Recommendations (Priestley et al. 1986) is a document that, when published in 1986, contained pioneering recommendations for the seismic design of storage tanks, developed by a study group of the New Zealand National Society for Earthquake Engineering. The intention of this study group was to collate existing information, available in research papers and codes, and to produce uniform recommendations that would cover a wide range of tank configurations and contained materials. These recommendations attempted to produce a unified approach for the seismic design of storage tanks, regardless of material or function, and to provide additional information to that already available in alternative sources. Several decades after their publication, these recommendations have been widely referenced and adopted internationally, including EN 1998-4 (European Committee for Standardization 2006b).

Seismic Input

In the framework of a pseudo-dynamic spectral analysis, the earthquake motion is represented by an appropriate design spectrum, $S_a(T)$. As an example, EN 1998-1 (European Committee for Standardization 2004a) provides a design spectrum for general use. Similar design spectra are provided by all seismic design codes. The design spectrum for horizontal earthquake excitation is usually expressed through an analytical equation for the spectral acceleration S_a in terms of the natural period T of the structural system, the design ground acceleration a_g, the damping coefficient ξ, the behavior factor q, the importance factor γ_I, and the soil factor S.

In addition, provisions for the vertical component of the seismic action are given in terms of the peak vertical ground acceleration a_{gv} usually as a percentage of a_g. In this section, two main issues, the behavior factor and the importance factor, are discussed in more detail.

Behavior Factor

The behavior factor, q, accounts for the capacity of the structural system to dissipate seismic energy. This results in a reduction of the forces obtained from a linear elastic analysis, in order to account for the nonlinear response of the structure, associated with the material, the structural system, and the design procedure. However, the structural system should be able to sustain such nonelastic behavior and dissipate seismic energy.

Upright cylindrical steel tanks (non-base-isolated) supported directly on the ground or on the foundation may be designed with a behavior factor q that represents a quasi-elastic behavior, i.e., $q \leq 1.5$. More specifically, a value of $q = 1$ is recommended for the convective motion where negligible dissipation of energy occurs, whereas a value of $q = 1.5$ (quasi-elastic behavior) is recommended for the impulsive motion.

The relatively low values of q for the impulsive motion are justified by the unstable nature of shell buckling primary mode of failure. This philosophy is adopted by European Standard EN 1998-4 (European Committee for Standardization 2006b). According to this standard, values of behavior factor greater than 1.5 for the impulsive motion can be used (up to 2.5); in special cases when the tank or its foundation is appropriately designed to allow uplift and the localization of deformations in the shell wall, the bottom plate or their intersection does not cause failure.

American Standard API 650 (American Petroleum Institute 2007) uses a similar concept to account for the capacity of the tank to dissipate seismic energy through the so-called response modification factor or simply reduction factor, denoted by "R." The response modification factor for ground-supported, liquid storage tanks designed and detailed to these provisions shall be less than or equal to the values shown in Table E-4 of API 650.

Nevertheless, there exist significant differences between API 650 (American Petroleum Institute 2007) and EN 1998-4 (European Committee for Standardization 2006b) for the value of the behavior factor. In particular, API

650 proposes larger values for the behavior factor for both the impulsive and convective actions than EN 1998-4. It is important to note that in both standards those values are significantly higher than unity which implies the presence of a dissipating mechanism.

In conclusion, the behavior (reduction) factor for the seismic design of liquid storage tanks is a controversial issue that has not been examined thoroughly so far.

Importance Factor

Reliability differentiation in liquid storage tank design is implemented by classifying tanks into different "importance classes." The importance classes depend on the potential loss of life due to the failure of the particular tank and on the economic and social consequences of failure. Clearly, the importance of a specific tank should be assessed considering the facility or plant it belongs to and its proximity to urban areas. In EN 1998-4 (European Committee for Standardization 2006b), the following classes for liquid storage tanks are specified:

- *Class I* refers to situations where the risk to life is low and the economic and social consequences of failure are small or negligible.
- Situations with medium risk to life and local economic or social consequences of failure belong to *Class II*.
- *Class III* refers to situations with a high risk to life and large economic and social consequences of failure.
- *Class IV* refers to situations with exceptional risk to life and extreme economic and social consequences of failure.

An importance factor γ_I is assigned to each importance class, ranging from 0.8 to 1.6 for the four importance classes unless otherwise specified in National Annexes, depending on the local seismic hazard conditions or by the owner. An analogous classification (with three important classes) is also specified in API 650 (American Petroleum Institute 2007), Appendix E, Section E.5.1.2, Table E-5.

Seismic Loading on Liquid Storage Tanks

Horizontal Seismic Action

General
A cylindrical coordinate system is usually used to describe the tank geometry with orthogonal coordinates: r, φ, z where the origin at the centre of the tank bottom and the z axis is vertical (Fig. 1). The height of the tank to the original of the free surface of the fluid and its radius are denoted by H and R, respectively; ρ is the mass density of the fluid, while the nondimensional coordinates $\xi = r/R$ and $\zeta = z/H$ in the radial direction and along the height of the tank, respectively, are usually considered.

The liquid contained in an upright cylinder subjected to horizontal seismic excitation $\ddot{X}(t)$ may be considered as ideal fluid (potential flow) resulting in a hydrodynamic problem (Ibrahim 2005) associated with the motion of the liquid free surface and the development of standing waves. The solution of the hydrodynamic problem shows that the liquid motion can be expressed as the sum of two separate contributions, called impulsive and convective, respectively. The impulsive component of the motion satisfies exactly the boundary conditions at the tank wall and the bottom of the tank and corresponds to zero pressure at the free surface of the fluid. It represents the motion of the fluid part that "follows" the motion of the container. The convective term is associated with liquid sloshing of

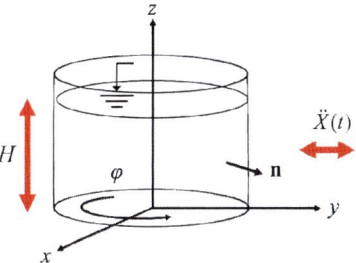

Liquid Storage Tanks: Seismic Analysis, Fig. 1 Cylindrical coordinate system describing tank geometry

its free surface and satisfies the kinematic and dynamic conditions at the free surface.

As a result, the hydrodynamic pressure is decomposed into an impulsive part and a convective part. In an analogous manner, the total mass of the liquid, m_L, can be considered as the sum of two parts, an impulsive part, m_I, and a convective part, m_C, so that $m_L = m_I + m_C$.

The convective motion is associated with an infinite number of modes and the corresponding masses are computed as follows:

$$\frac{m_{Cn}}{m_L} = \frac{2\tanh(k_n R\gamma)}{k_n R\gamma (k_n^2 R^2 - 1)}, \quad n = 1, 2, 3, \ldots \quad (1)$$

where n is the number of sloshing mode, R is the radius of the tank, γ is the tank aspect ratio defined as the ratio of the liquid height over the tank radius ($= H/R$), and $k_n R = \lambda_n$ are the roots of equation $J_1'(\cdot) = 0$. More specifically the first three roots are $k_1 R = \lambda_1 = 1.841$, $k_2 R = \lambda_2 = 5.331$, and $k_3 R = \lambda_3 = 8.536$ corresponding to the first three sloshing modes. Note that the total convective mass is the sum of the individual convective masses, $m_C = \sum_{n=1}^{\infty} m_{Cn}$. Usually, consideration of only the first one ($n = 1$), or the first two modes ($n = 1, 2$) in some special cases, is adequate for design purposes.

Masses m_I and m_{Cn} for $n = 1, 2$ are plotted in Fig. 2 as functions of the tank aspect ratio $\gamma = H/R$. The impulsive mass is expressed as the ratio m_I/m_L of the impulsive mass over the total liquid mass. This ratio increases with increasing values of γ, tending asymptotically to unity. On the other hand, the convective mass, expressed in normalized form as m_{C1}/m_L, decreases with increasing values of γ. The value of m_{C2} is quite small but may have some effect on the dynamic response in broad tanks where $\gamma \to 0$.

In the course of a design procedure, the impulsive mass m_I should consist of the above impulsive mass of the liquid, plus the mass of the moving tank wall, m_{SH}, and the mass of the tank roof, m_r.

Vibration Periods and Damping Coefficients

In case of rigid (non-deformable) tank walls, the impulsive motion follows exactly the ground motion. In such a case the impulsive circular frequency has an infinite value ($\omega_I \to \infty$) and the impulsive period is zero ($T_I = 0$). This is not applicable in the case of flexible (deformable) tank walls. To simplify the analysis, it is reasonable to assume that the flexibility of tank wall affects almost exclusively the impulsive component of the response, and therefore, the impulsive hydrodynamic effects for flexible tanks are generally significantly larger than those for rigid tanks. On the other hand, the convective components of the response can be considered insensitive to variations in wall flexibility, because they are associated with natural periods of vibration that are significantly longer than the dominant periods of the ground motion or pipe wall vibration motion. Therefore, the convective components of seismic action for tank proportions normally encountered in practice can be

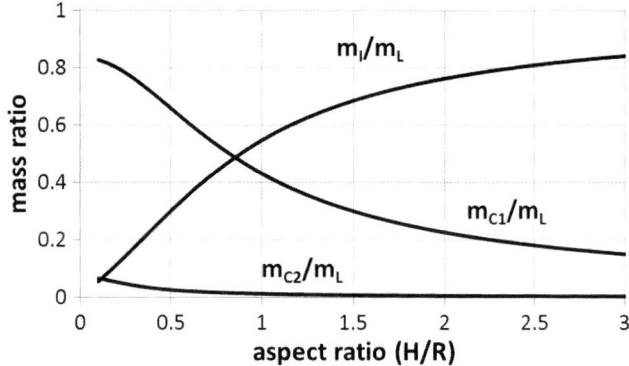

Liquid Storage Tanks: Seismic Analysis, Fig. 2 Mass ratio as a function of the aspect ratio

considered the same as those obtained for rigid tanks.

Analytical expressions are available for the impulsive vibration period and frequency accounting for tank flexibility. From hydrodynamics, the natural circular frequency and natural period of the impulsive response of flexible tanks are calculated by the expressions:

$$\omega_I = \frac{2\pi}{Y_I H \sqrt{\frac{\rho R}{t_{eq} E}}} \quad (2)$$

$$T_I = Y_I H \sqrt{\frac{\rho R}{t_{eq} E}} \quad (3)$$

where ρ is the mass density of the fluid, E is Young's modulus, t_{eq} is the equivalent uniform thickness of the tank wall (can be taken as the weighted average over the wetted height of the tank wall, the weight may be taken proportional to the strain in the wall of the tank, which is maximum at the base of the tank), and coefficient Y_I is given in Table 1 for specific values of the aspect ratio γ of the tank ($\gamma = H/R$). For intermediate values of γ, a simple interpolation may be used.

Furthermore, the circular frequency (Fig. 3) and natural period of the convective response, denoted as ω_{Cn} and T_{Cn}, respectively, for the n-th sloshing mode can be calculated by the following expressions:

$$\omega_{Cn} = \sqrt{g \frac{\lambda_n}{R} \tanh(\lambda_n \gamma)}, \quad n = 1, 2, 3, \ldots \quad (4)$$

$$T_{Cn} = \frac{2\pi}{\omega_{Cn}}, \quad n = 1, 2, 3, \ldots \quad (5)$$

where λ_n are the roots of $J_1'(\cdot) = 0$, $\lambda_1 = 1.841$, $\lambda_2 = 5.331$, and $\lambda_3 = 8.536$ for the first three sloshing modes).

The impulsive and convective damping coefficients can be taken equal to $\xi_I = 5\%$ and $\xi_C = 0.5\%$ for the impulsive and the convective motion, respectively.

Impulsive Motion and Pressure

The impulsive motion can be expressed in terms of the impulsive acceleration of the liquid storage tank, $\ddot{u}_I(t)$, which can be calculated from the following linear oscillator equation:

$$\ddot{u}_I + 2\xi_I \omega_I (\dot{u}_I - \dot{X}) + \omega_I^2 (u_I - X) = 0 \quad (6)$$

where $u_I(t)$ is the generalized coordinate representing the total impulsive displacement

Liquid Storage Tanks: Seismic Analysis, Table 1 Coefficient Y_I in terms of the aspect ratio

H/R	0.3	0.5	0.7	1.0	1.5	2.0	2.5	3.0
Y_I	9.28	7.74	6.97	6.36	6.06	6.21	6.56	7.03

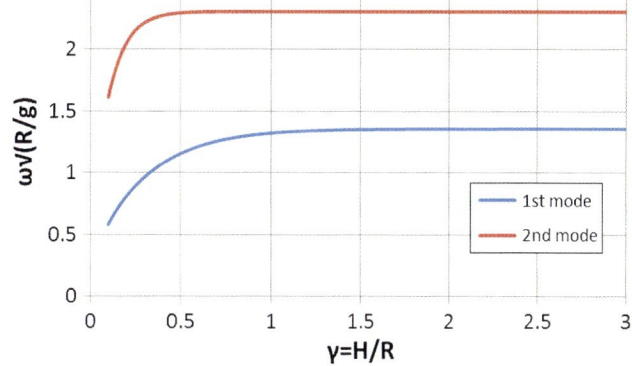

Liquid Storage Tanks: Seismic Analysis, Fig. 3 Values of the first two sloshing frequencies as functions of γ

motion of the tank, ξ_I the viscous damping ratio of the impulsive motion, and ω_I the impulsive frequency defined in Eq. 2. Setting

$$a_I(t) = u_I(t) - X \quad (7)$$

one can obtain

$$\ddot{a}_I + 2\xi_I \omega_I \dot{a}_I + \omega_I^2 a_I = -\ddot{X}(t) \quad (8)$$

In the above expressions, function $a_I(t)$ can be considered as the generalized coordinate representing the relative impulsive displacement of the tank with respect to its base.

The spatial-time variation of the *impulsive* component of hydrodynamic pressure within the liquid domain is given by the following expression:

$$p_I(\xi, \zeta, \varphi, t) = C_I(\xi, \zeta) \rho H \cos \varphi \, \ddot{u}_I(t) \quad (9)$$

where the spatial coefficient C_I is defined:

$$C_I(\xi, \zeta) = 2 \sum_{n=0}^{\infty} \frac{(-1)^n}{I_1'(v_n/\gamma) v_n^2} \cos(v_n \zeta) I_1\left(\frac{v_n}{\gamma} \xi\right) \quad (10)$$

in which $v_n = \frac{2n+1}{2}\pi$, $\gamma = H/R$ is the aspect ratio of the tank, whereas $I_1(\cdot)$ and $I_1'(\cdot)$ denote the modified Bessel function of order 1 and its derivative, respectively.

For seismic design purposes, the distribution of the impulsive hydrodynamic pressure on the pipe walls is assumed as follows:

$$p_I(\xi, \zeta, \varphi) = C_I(\xi, \zeta) \rho H \cos \varphi \, S_a(T_I) \quad (11)$$

where ξ is taken equal to 1 for computing the pressure on the lateral surface of tank shell, whereas ζ should be considered equal to zero for calculating the pressure on the tank bottom plate (Fig. 4).

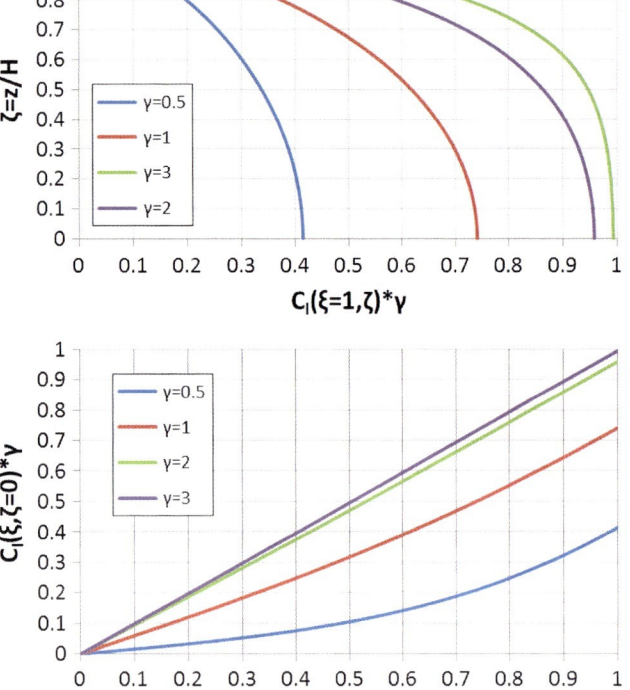

Liquid Storage Tanks: Seismic Analysis, Fig. 4 Variation of the impulsive pressure (normalized to $\rho R a_g$) for three values of $\gamma = H/R$, (**a**) variation along the height, (**b**) radial variation on the tank bottom. a_g is the maximum ground acceleration

Convective Motion and Pressure

The convective acceleration, $\ddot{u}_{Cn}(t)$, is calculated from the following linear oscillator equation:

$$\ddot{u}_{Cn} + 2\xi_{Cn}\omega_{Cn}(\dot{u}_{Cn} - \dot{X}) + \omega_{Cn}^2(u_{Cn} - X) = 0 \quad (12)$$

where $u_{Cn}(t)$ is the generalized coordinate representing the total convective motion of the liquid associated with the n-th mode, ξ_{Cn} the viscous damping ratio of the convective motion for the n-th sloshing mode, and ω_{Cn} the convective frequency. Setting

$$a_{Cn}(t) = u_{Cn}(t) - X \quad (13)$$

one can write

$$\ddot{a}_{Cn} + 2\xi_{Cn}\omega_{Cn}\dot{a}_{Cn} + \omega_{Cn}^2 a_{Cn} = -\ddot{X}(t) \quad (14)$$

where $a_{Cn}(t)$ is the generalized coordinate representing the relative convective motion of the liquid associated with the n-th sloshing mode.

The spatial distribution within the liquid domain and time variation of the "convective" component of hydrodynamic pressure associated with the n-th convective mode is given by the following expression:

$$p_{Cn}(\xi,\zeta,\varphi,t) = \rho\psi_n \cosh(\lambda_n\gamma\zeta)J_1(\lambda_n\xi)\cos\varphi\, \ddot{u}_{Cn}(t) \quad (15)$$

where

$$\psi_n = \frac{2R}{(\lambda_n^2 - 1)J_1(\lambda_n)\cosh(\lambda_n\gamma)} \quad (16)$$

J_1 is the Bessel function of the first order and $\lambda_1 = 1.841$, $\lambda_2 = 5.331$, and $\lambda_3 = 8.536$ are the roots of $J_1'(\lambda) = 0$. In Eq. 15, r is taken equal to R ($\xi = 1$) for the pressure on the lateral surface, whereas z is taken equal to zero ($\zeta = 0$) for the pressure on the bottom plate. In the majority of practical applications, only the first convective mode ($n = 1$) of the fluid is necessary (Pappa et al. 2011) and the distribution of the pressure is as follows:

$$p_C(\xi,\zeta,\varphi) = 1.146 R\rho \frac{\cosh(1.841\gamma\zeta)J_1(1.841\xi)}{\cosh(1.841\gamma)} \cos\varphi\, S_a(T_{C1}) \quad (17)$$

where J_1 is the Bessel function of first order. In the above expressions, r is taken equal to R ($\xi=1$) for determining the hydrodynamic pressure on the lateral tank (shell) surface (Fig. 5), whereas z is taken equal to zero ($\zeta = 0$) for the hydrodynamic pressure on the bottom plate. In addition, 1.841 is the first root of $J_1'(0) = 0$.

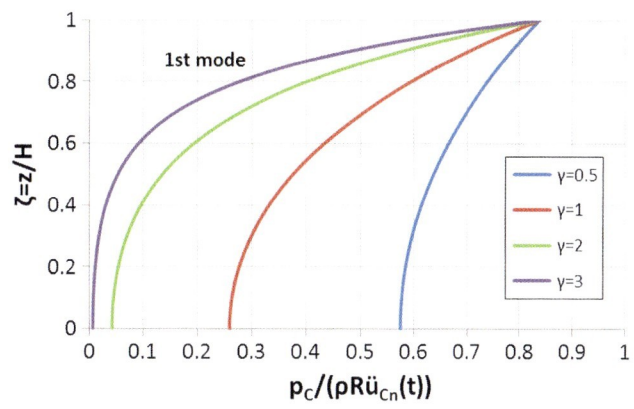

Liquid Storage Tanks: Seismic Analysis, Fig. 5 Variation of first-mode sloshing pressures on tank wall along the tank height

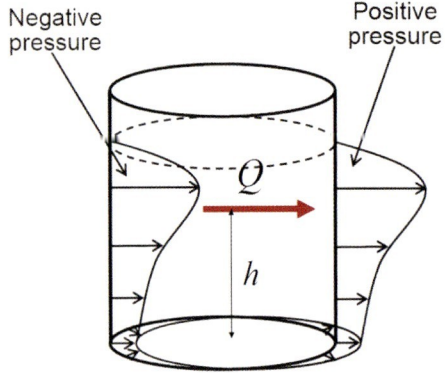

Liquid Storage Tanks: Seismic Analysis, Fig. 6 Horizontal seismic force Q acting at a height h

In broad tanks (i.e., tanks with large values of H/R ratio), the sloshing pressures may be significant near the bottom, while in slender (tall) tanks, the sloshing effect is limited to the vicinity of liquid surface.

Horizontal Seismic Forces

Integrating the pressure distribution on the tank wall, the impulsive base shear force at the base of the wall can be calculated as follows:

$$Q_I(t) = \int_{B_q} p_I(\mathbf{n} \cdot \mathbf{e}_x) dB_q \quad (18)$$

where $\mathbf{e}_x$ is the unit vector in the direction of the earthquake, $\mathbf{n}$ is the unit normal vector to the liquid surface, and B_q is the boundary surface of the liquid (Fig. 6).

The result of this integration can be readily written as follows:

$$Q_I(t) = m_I \ddot{u}_I(t) \quad (19)$$

where m_I is the impulsive mass defined in Fig. 2 and $\ddot{u}_I(t)$ is the impulsive acceleration parameter, defined in Eq. 6.

The convective base shear can be obtained by integrating the convective pressure (p_C) components on the lateral shell of the tank:

$$Q_C(t) = \int_{B_q} p_C(\mathbf{n} \cdot \mathbf{e}_x) dB_q \quad (20)$$

The result of this integration can be readily written as follows:

$$Q_C(t) = \sum_{n=1}^{\infty} m_{Cn} \ddot{u}_{Cn}(t) \quad (21)$$

where m_{Cn} is the convective mass for the n-th mode defined in Fig. 2 and $\ddot{u}_{Cn}(t)$ is the convective acceleration parameter for the n-th sloshing mode, defined in Eq. 12.

In the course of a spectral seismic analysis of the tank, the impulsive base shear Q_I and convective base shear Q_{Cn} for the n-th mode at the base of the wall can be computed in terms of the corresponding spectral acceleration as follows:

$$Q_I = m_I S_a(T_I) \quad (22)$$

$$Q_{Cn} = m_{Cn} S_a(T_{Cn}), \quad n = 1, 2, 3, \ldots \quad (23)$$

Ringwall and Slab Overturning Moment/Heights

The ringwall moment, M, is the overturning moment that acts immediately above the bottom plate of the tank. It includes only the contribution of pressure on the wall and should be used for the calculation of stresses and stress resultants in the tank wall and at its connection to the base, for tank structural verification. On the other hand, the slab moment, M', is the overturning moment that acts immediately below the bottom plate of the tank. It includes the contribution of pressure on the wall together with the pressure on the tank bottom and should be used for the verification of its support structure, base anchors, or foundation.

The impulsive ringwall base moment is calculated through the following integration:

$$M_I(t) = \int_{B_q} p_I z(\mathbf{n} \cdot \mathbf{e}_x) dB_q \quad (24)$$

The result of this integration can be readily written as follows:

$$M_I(t) = m_I h_I \ddot{u}_I(t) \quad (25)$$

or

$$M_I(t) = Q_I(t) h_I \quad (26)$$

where m_I is the impulsive mass defined in Fig. 2; h_I is the height at which acts the horizontal seismic force, $Q_I(t)$; and $\ddot{u}_I(t)$ is the impulsive acceleration parameter, defined in Eq. 6.

The convective ringwall base moment is calculated in an analogous manner:

$$M_C(t) = \int_{B_q} p_C z(\mathbf{n} \cdot \mathbf{e}_x) dB_q \quad (27)$$

The result of this integration can be readily written as follows:

$$M_C(t) = \sum_{n=1}^{\infty} m_{Cn} \ddot{u}_{Cn}(t) h_{Cn}$$
$$= \sum_{n=1}^{\infty} Q_{Cn}(t) h_{Cn} \quad (28)$$

where m_{Cn} is the convective mass for the n-th mode defined in Fig. 2; h_{Cn} is the height at which acts the horizontal seismic force for the n-th sloshing mode, $Q_{Cn}(t)$; and $\ddot{u}_{Cn}(t)$ is the convective acceleration parameter for the n-th sloshing mode, defined in Eq. 12.

Analytical expressions for the heights at which impulsive and convective forces are acting are available from hydrodynamics, represented also in graphical form in Fig. 7.

For the impulsive moment:

$$h_I = H \frac{\sum_{n=0}^{\infty} \frac{(-1)^n I_1(v_n/\gamma)}{v_n^4 I_1'(v_n/\gamma)} (v_n(-1)^n - 1)}{\sum_{n=0}^{\infty} \frac{I_1(v_n/\gamma)}{v_n^3 I_1'(v_n/\gamma)}} \quad (29)$$

$$h_I' = H \frac{\frac{1}{2} + 2\gamma \sum_{n=0}^{\infty} \frac{v_n + 2(-1)^{n+1} I_1(v_n/\gamma)}{v_n^4 I_1'(v_n/\gamma)}}{2\gamma \sum_{n=0}^{\infty} \frac{I_1(v_n/\gamma)}{v_n^3 I_1'(v_n/\gamma)}} \quad (30)$$

For the convective moment:

$$h_{Cn} = H \left(1 + \frac{1 - \cosh(\lambda_n \gamma)}{\lambda_n \gamma \sinh(\lambda_n \gamma)} \right) \quad (31)$$

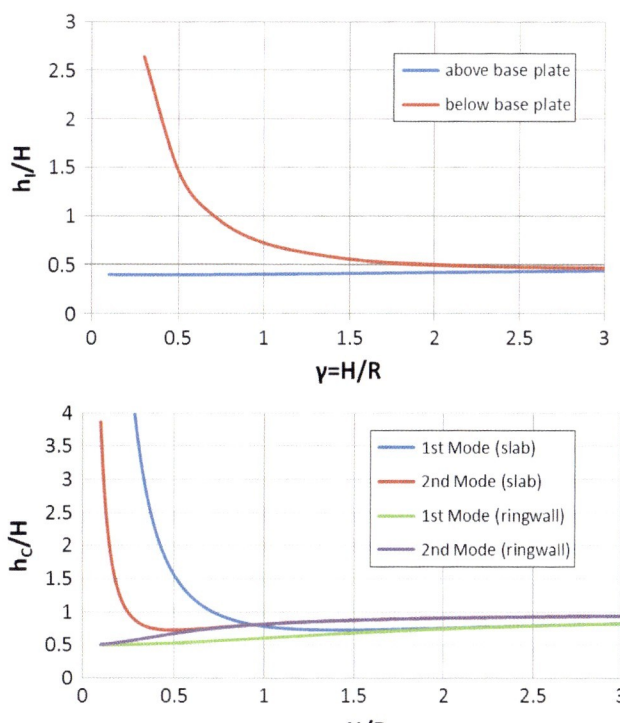

Liquid Storage Tanks: Seismic Analysis, Fig. 7 Ratios h_I/H, h_I'/H, h_C/H, and h_C'/H as functions of the aspect ratio

$$h'_{Cn} = H\left(1 + \frac{2 - \cosh(\lambda_n \gamma)}{\lambda_n \gamma \sinh(\lambda_n \gamma)}\right) \quad (32)$$

In the course of a spectral design and analysis procedure, the impulsive and convective ringwall moments are given by the following expressions:

$$M_I = m_I h_I S_a(T_I) \quad (33)$$

$$M_C = m_{Cn} h_C S_a(T_{Cn}) \quad (34)$$

Furthermore, simplified equations for the height h_I are stated in Section 6.1.2 of API 650 (American Petroleum Institute 2007) as alternative to Eq. 29:

when $D/H \geq 1.333$:

$$h_I = 0.375H \quad (35)$$

when $D/H < 1.333$:

$$h_I = \left[0.5 - 0.094\frac{D}{H}\right]H \quad (36)$$

and

$$h_C = \left[1 - \frac{\cosh\left(\frac{3.67H}{D}\right) - 1}{\frac{3.67H}{D}\sinh\left(\frac{3.67H}{D}\right)}\right]H \quad (37)$$

Similarly, the impulsive and convective slab moments are given by the following expressions:

$$M'_I = m_I h'_I S_a(T_I) \quad (38)$$

$$M'_C = m_{Cn} h'_C S_a(T_{Cn}) \quad (39)$$

and an alternative equation for the height h'_I can be found in Section 6.1.2 of API 650 (American Petroleum Institute 2007):

when $D/H \geq 1.333$:

$$h'_I = 0.375\left[1 + 1.333\left(\frac{0.866\frac{D}{H}}{\tanh\left(0.866\frac{D}{H}\right)} - 1\right)\right]H \quad (40)$$

when $D/H < 1.333$:

$$h'_I = \left[0.5 - 0.06\frac{D}{H}\right]H \quad (41)$$

Combination of Impulsive/Convective Action

In the course of a spectral analysis and design procedure, the impulsive and convective action should be combined in order to compute the total horizontal hydrodynamic force. In particular it is necessary to combine (a) the sloshing modes to obtain the resultant convective action and (b) the impulsive and convective actions for the total seismic force.

Both combinations should be conducted according to the SRSS rule, rather than the addition of the maximum convective and maximum impulsive forces as noted in a recent publication (Pappa et al. 2011).

More specifically, the total horizontal seismic force is

$$Q_T = \sqrt{[m_I S_a(T_I)]^2 + \sum_{n=1}^{N_1} [m_{Cn} S_a(T_{Cn})]^2} \quad (42)$$

where N_1 is the number of sloshing modes considered.

The above SRSS combination rule applies also to the total "ringwall" and "slab" moments:

$$M_T = \sqrt{[m_I h_I S_a(T_I)]^2 + \sum_{n=1}^{\infty} [m_{Cn} h_C S_a(T_{Cn})]^2} \quad (43)$$

$$M'_T = \sqrt{[m_I h'_I S_a(T_I)]^2 + \sum_{n=1}^{\infty} [m_{Cn} h'_C S_a(T_{Cn})]^2} \quad (44)$$

One should note that the impulsive force is significantly higher than the convective force. Therefore, in most of the cases, refinements on the value of the convective force and the combination of the corresponding modes are of little importance on the calculation of the total seismic force (Pappa et al. 2011).

Meridional Stresses

The meridional stress, σ_z, in mechanically anchored tanks is related to meridional membrane force, N, per unit circumferential length:

$$\sigma_z = N/t_i \quad (45)$$

where t_i is the thickness of the tank shell course under consideration and the axial force N per unit circumferential length is given by the following equations:

- On the compression side of the tank (Fig. 8 – location 1):

$$N = -1.273\frac{M_T}{D^2} - w_t(1 + cA_v) \quad (46)$$

- On the tension side of the tank (Fig. 8 – location 2):

$$N = 1.273\frac{M_T}{D^2} - w_t(1 - cA_v) \quad (47)$$

where A_v is the spectral acceleration coefficient for the vertical motion in [% g] ($A_v = 100a_{gv}/g$), w_t is the load per unit circumferential length because of shell and roof weight acting at the base of shell [N/m] calculated by

$$w_t = \frac{W_s}{\pi D} + w_{rs} \quad (48)$$

and c is a factor that reflects the contribution of the vertical component. In Eq. 48 W_s is the total weight of the tank shell and appurtenances and w_{rs} is the load acting on the shell roof. Note that the first term in Eqs. 46 and 47 is directly obtained from simple mechanics of materials considering bending stresses of a beam of circular cross section.

In API 650 (American Petroleum Institute 2007), the value of c is taken equal to 0.40, whereas EN 1998-4 (European Committee for Standardization 2006b) does not give an explicit expression for the effect of the vertical ground acceleration component on the meridional stresses.

Hoop Hydrodynamic Stresses

In the course of seismic loading, the hydrodynamic pressure loading on the tank wall, apart from the development of significant meridional stresses, results also in the development of additional stresses in the circumferential direction, denoted as σ_h, called hoop hydrodynamic stresses (Fig. 9).

Explicit equations for computing hoop hydrodynamic stresses for the convective and impulsive component of liquid motion (denoted as σ_{hC} and σ_{hI} respectively), in terms of the z coordinate of the tank and the D/H (aspect) ratio of the tank, are reported by Wozniak and Mitchell (1978), also adopted by API 650 (American Petroleum Institute 2007), Appendix E, Section E.6.1.4, as follows:

For tanks with $D/H \geq 1.333$:

$$\sigma_{hI} = \frac{8.48A_I GDH}{t_i}\left[\frac{z_1}{H} - 0.5\left(\frac{z_1}{H}\right)^2\right]\tanh\left(0.866\frac{D}{H}\right) \quad (49)$$

For tanks with $D/H < 1.33$ and $z_1 < 0.75D$:

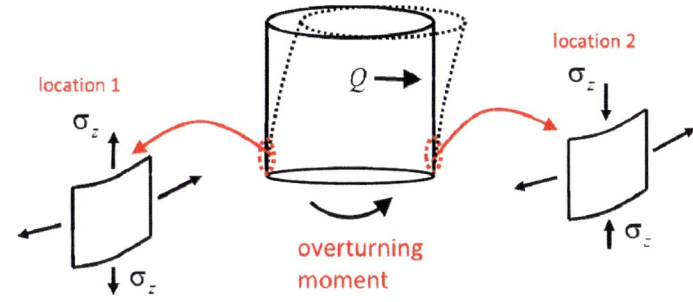

Liquid Storage Tanks: Seismic Analysis, Fig. 8 Meridional stresses σ_z on the tension side (*location 1*) and the compression side (*location 2*) of the tank under seismic loading

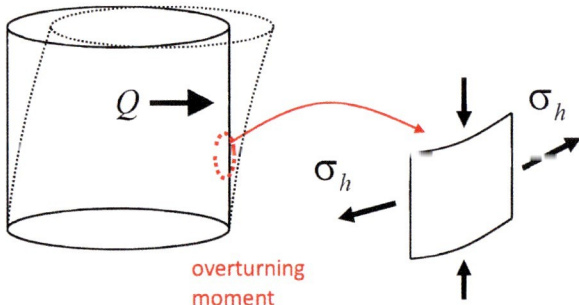

Liquid Storage Tanks: Seismic Analysis, Fig. 9 Hoop hydrodynamic stresses σ_h

$$\sigma_{hI} = \frac{5.22 A_I G D^2}{t_i} \left[\frac{z_1}{0.75D} - 0.5 \left(\frac{z_1}{0.75D} \right)^2 \right] \quad (50)$$

For tanks with $D/H < 1.33$ and $z_1 \geq 0.75D$:

$$\sigma_{hI} = \frac{2.6 A_I G D^2}{t_i} \quad (51)$$

For all proportions of D/H:

$$\sigma_{hC} = \frac{1.85 A_C G D^2 \cosh\left[\frac{3.68(H - z_1)}{D}\right]}{t_i \cosh\left[\frac{3.68H}{D}\right]} \quad (52)$$

In the above equations, stresses are expressed in [MPa], D is the nominal tank diameter in [m], H is the maximum design product level in [m], t_i is the thickness of the shell ring under consideration in [mm], $z_1 = H - z$ is the distance from the liquid surface in [m], G is the specific gravity of tank contents, and A_I and A_C are the impulsive and convective spectral acceleration coefficients in [% g] defined as $A_I = 100 S_a(T_I)/g$ and $A_C = 100 S_a(T_C)/g$.

The hoop stresses due to impulsive and convective motion should be combined, using the SRSS rule, and superimposed to the hydrostatic stress σ_{hs} to obtain the total hoop stress as follows:

$$\sigma_h = \sigma_{hs} + \sqrt{\sigma_{hI}^2 + \sigma_{hC}^2} \quad (53)$$

where σ_h is the total hoop stress in the shell, to be used for hoop strength design.

When vertical acceleration is taken into account, the above equation is written as follows:

$$\sigma_h = \sigma_{hs} + \sqrt{\sigma_{hI}^2 + \sigma_{hC}^2 + (A_v \sigma_{hs})^2} \quad (54)$$

where A_v is the spectral acceleration coefficient for the vertical motion in [% g].

Sloshing Wave Height

From hydrodynamics, the elevation of the liquid free surface (sloshing wave height) is given by the following expression:

$$\eta(r, \theta, t) = \sum_{n=1,2,3\ldots}^{\infty} \frac{2\lambda_n R J_1(\lambda_n \xi) \tanh(\lambda_n \gamma) \cos\theta \; a_{Cn}}{J_1(\lambda_n)(\lambda_n^2 - 1)} \quad (55)$$

In this expression, $\lambda_1 = 1.841$, $\lambda_2 = 5.331$, and $\lambda_3 = 8.536$ are the roots of $J_1'(\lambda) = 0$, and a_{Cn} are the generalized coordinates of the convective motion with respect to the tank base given by Eq. 14. Considering only 1 term ($n = 1$), $r = R$, one obtains Eq. 56 for the maximum elevation $\eta_{\max}$:

$$\eta_{\max} = 0.84 R S_a(T_C)/g \quad (56)$$

that can be used for design purposes.

Vertical Seismic Action

The hydrodynamic pressure on the walls of a rigid tank due to vertical ground acceleration $A_v(t)$ can be computed by the following equation:

$$p_v(\zeta, t) = \rho H (1 - \zeta) \ddot{X}_v(t) \quad (57)$$

where $\ddot{X}_v(t)$ is the vertical ground acceleration input.

The corresponding hydrodynamic pressure distribution is axisymmetric and does not produce a shear force or moment resultant at any horizontal level of the tank or immediately above or below the base.

In the course of a spectral dynamic analysis, the peak vertical ground acceleration a_{gv} can be used for the calculation of p_v in the above equation replacing $\ddot{X}_v(t)$.

If one assumes that the tank is not moving rigidly with vertical acceleration $\ddot{X}_v(t)$, it may be reasonable to assume a contribution to the hydrodynamic pressure, $p_{vf}(\zeta,t)$, due to the deformability (radial "breathing") of the shell as suggested by Fischer and Seeber (1988). This additional term may be calculated as

$$p_{vf}(\zeta,t) = 0.815 f(\gamma) \rho H \cos\left(\frac{\pi}{2}\zeta\right) a_{vf}(t) \quad (58)$$

where

$$f(\gamma) = 1.078 + 0.274 \ln\gamma \quad \text{for } 0.8 \leq \gamma < 4 \quad (59)$$

$$f(\gamma) = 1.0 \quad \text{for } \gamma < 0.8 \quad (60)$$

and $a_{vf}(t)$ is the acceleration response of a simple oscillator having a frequency equal to the fundamental frequency ω_{vd} of the axisymmetric vibration of the tank containing with the liquid. This fundamental frequency ω_{vd} can be calculated by the corresponding fundamental period, estimated by the expression

$$T_{vd} = 4R\left[\frac{\pi\rho H(1-v^2)I_0(\gamma_1)}{2EI_1(\gamma_1)t_{eq}}\right]^{1/2} \quad (\text{for } \zeta = 1/3) \quad (61)$$

where $\gamma_1 = \frac{\pi}{2\gamma}$ and $I_0(\cdot), I_1(\cdot)$ denote the modified Bessel function of order 0 and 1, respectively. In the course of a spectral dynamic analysis, the maximum value of $a_{vf}(t)$ is obtained from the vertical acceleration response spectrum for the appropriate values of period and damping. If soil flexibility is neglected, the applicable damping values are those of the material of the shell. Furthermore, the behavior factor value, q, adopted for the response due to the impulsive component of the pressure and the tank wall inertia may be used for the response to the vertical component of the seismic action. The maximum value of the pressure due to the combined effect of p_v and p_{vf} may be obtained by applying the SRSS combination rule to the individual maxima.

Special Issues on Seismic Action

Unanchored Tanks

Numerous aboveground tanks are not anchored to the foundation and, therefore, in the course of a strong earthquake event, uplift of tank bottom from the ground may occur due to seismic overturning moment. Uplift may cause significant plastic deformations in the tank, especially in its base plate or in the area of nozzles and piping attachments, leading to shell wall fracture and leakage of the liquid.

Base plate uplifting in those tanks referred to as unanchored or self-anchored, apart from its effect on the base plate connection may lead to an increase of the compressive vertical stress in the shell, which is critical for buckling-related modes of failure. At the wall side opposite to the uplifted area, vertical compression is maximum and hoop compressive stresses are generated in the shell, due to the membrane action of the base plate. The uplifting parameters of the tank can be seen in (Fig. 10).

The vertical uplift of two tanks with aspect ratios $\gamma = 1.131$ and 0.783 at the edge of the tank base, w, derived from a nonlinear finite element model, is given in Fig. 11 as a function of the normalized overturning moment M/WH (Vathi and Karamanos 2014).

API 650 (American Petroleum Institute 2007) provisions for tank uplifting effects are based on the value of the anchorage ratio J, defined as follows:

$$J = \frac{M_T}{D^2\left[w_t(1 - 0.4a_{gv}) + w_a - 0.4w_{\text{int}}\right]} \quad (62)$$

where w_a is the force per unit circumferential length resisting uplift in the annular region and w_{int} the uplift load per unit circumferential length due to product pressure. The values of w_a and w_{int} can be computed by appropriate formulae in API 650.

Tanks with $J < 0.785$ do not uplift, whereas if $J > 1.54$, the tank is unstable and mechanical anchorage is necessary.

An estimate of the membrane stress σ_{rb} in the base plate due to uplift can be obtained by Eq. 63:

$$\sigma_{rb} = 0.568 E \left(\frac{pL(w)}{Et}\right)^{2/3} \quad (63)$$

proposed in EN 1998-4 (European Committee for Standardization 2006b) and written herein in more convenient terms. In this equation, L is the uplifted length of the tank which is a function of the uplifting size w, so that $L = L(w)$. The corresponding strain can be computed as follows:

$$\varepsilon_{rb} = \frac{\sigma_{rb}}{E} = 0.568 \left(\frac{pL(w)}{Et}\right)^{2/3} \quad (64)$$

When significant uplift takes place in large-diameter tanks, the state of stresses in the uplifted part of the base plate at the ultimate limit state is dominated by plate bending (including the effect of the pressure acting on the tank base), not by membrane stresses. In such cases, a finite element analysis method should be used for the calculation of the state of stresses.

As far as the plastic rotation of the base plate connection is concerned, it is recommended to design the bottom annular ring with a thickness less than the wall thickness, so as to avoid flexural yielding at the base of the wall.

Finally, the plastic rotation associated to uplift w at the edge of the tank and a base separation of L can be computed as follows:

$$\theta_p = \left(\frac{2w}{L} - \frac{w}{2R}\right) \quad (65)$$

which should be less than the rotation capacity of the welded connection.

Forces on Anchors

In the case of mechanically anchored tanks, the seismic design load, P_{AB}, at each anchor

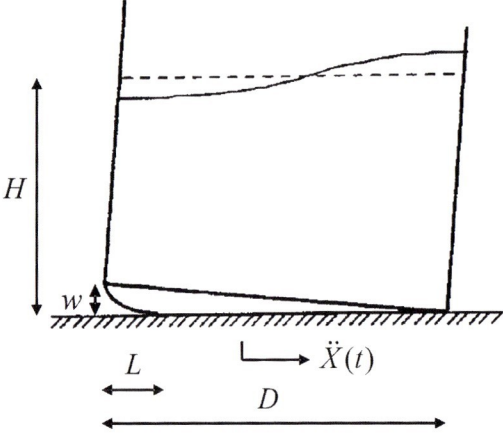

Liquid Storage Tanks: Seismic Analysis, Fig. 10 Uplifting parameters of an unanchored tank

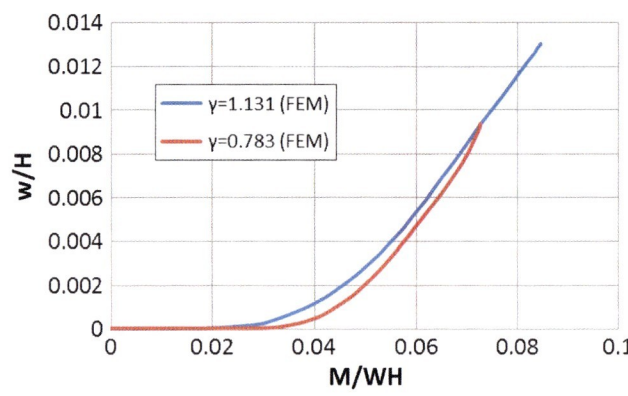

Liquid Storage Tanks: Seismic Analysis, Fig. 11 Maximum vertical uplift of the tank w versus overturning moment M of two fixed-roof unanchored cylindrical tanks; finite element results (Vathi and Karamanos 2014)

necessary to design the anchor can be calculated by the following simple equation:

$$P_{AB} = N\left(\frac{\pi D}{n_A}\right) \quad (66)$$

where N is the maximum tensile force per unit circumferential length given in Eq. 47 and n_A is the number of anchors around the tank circumference.

Soil-Tank Interaction

For tanks on relatively deformable ground, the base motion can be significantly different from the free-field motion; in general the translational component is modified and there is also a rocking component. Moreover, for the same input motion, as the flexibility of the ground increases, both the fundamental period of the tank-fluid system and the total damping increase, reducing the peak seismic force. The increase in the fundamental period is more pronounced for tall, slender tanks, because the contribution of the rocking component is greater. The reduction of the peak seismic force, however, is less pronounced for tall tanks, since the damping associated with rocking is smaller than that associated with horizontal translation.

The convective periods and pressures are assumed not to be affected by soil-structure interaction. A simple procedure, initially proposed for buildings and consisting of an increase of the fundamental period and of the damping of the structure on a rigid soil, has been extended to the impulsive (rigid and flexible) components of the response of tanks (Habenberger and Schwarz 2002). A good approximation can be obtained through the use of an equivalent simple oscillator with parameters adjusted to match frequency and peak response of the actual system. The properties of this "substitute oscillator" are given in the form of graphs, as functions of the ratio H/R described in the above publications.

An alternative procedure to account for soil flexibility has been suggested by Priestley et al. (1986). The procedure suggests the modification of the frequency and the damping of the impulsive motion, based on the work of Gazetas (1983), and the corresponding equations for the modified natural periods and damping values can be found in Annex A, Section A.7.2.2 of EN 1998-4 (European Committee for Standardization 2006b).

Seismic Action on Nozzles (Pipe Attachments)

The region of the tank shell where piping is attached to should be designed to resist the forces transmitted by the piping amplified by an appropriate load factor. If reliable information is not available for the attached piping system or an accurate analysis cannot be performed, a relative displacement Δ between the first anchoring point of the piping and the tank should be assumed to take place in the most adverse direction and used for the analysis of the nozzle, with a value, as suggested in EN 1998-4 (European Committee for Standardization 2006b):

$$\Delta = \frac{x}{x_0} d_g \quad (67)$$

In the above expression, x_0 is a reference length equal to 500 m, x the distance between the anchoring point of the piping and the point of connection with the tank, and d_g the design ground displacement.

An alternative methodology for calculating actions on piping attachments is stated by API 650 (American Petroleum Institute 2007). Piping systems shall provide a minimum imposed displacement in the piping, supports, and tank (Table E-8 of API 650) in lieu of a more detailed analysis of the piping system and its connection.

Seismic Resistance of Liquid Storage Tanks

Limit States (General)

The failures against which liquid storage tanks have to be verified when subjected to seismic loading are:

- Development of elephant's foot buckling at the lower course of the tank

- Buckling at the top of the tank due to the alternating sign of the hydrostatic and hydrodynamic pressures
- Rupture or leakage because of the increased induced hoop hydrodynamic stresses
- Plastification of the base plate in unanchored tanks due to uplifting
- Failure of the anchor bolts
- Failure at the nozzles and piping system, especially in the case of unanchored tanks
- Sliding of the tank

The above modes are examined below in more detail. It is noted that a thorough presentation of each failure mode and its modeling in the course of a seismic design and analysis procedure is out of the scope of the present chapter.

Shell Buckling Mode

The maximum actions (e.g., membrane forces and bending moments, circumferential or meridional) induced in the seismic design shall be less or equal to the resistance of the shell evaluated as in the persistent or transient design situations. For the cylindrical steel shell buckling by vertical (meridional) compression with simultaneous action of hoop tension due to hydrostatic loading ("elephant's foot mode of failure") is a critical situation.

Steel tanks with very thin shells have also displayed another type of shell buckling mode involving diamond-shaped buckles at a distance above the base of the tank, usually with low internal pressure.

Observations from previous seismic events have indicated that buckling of the lower courses has occasionally resulted in the loss of tank contents due to weld or attached piping fracture and, in some cases, total collapse of the tank. Therefore, in the course of a seismic design procedure, it should be verified that the tank is safe against local buckling. Thus, to ensure "integrity" of the tank under the seismic action against elephant's foot buckling:

- The geometry (thickness) of the tank should be such that it can withstand significant compressive stresses without buckling in the case of an earthquake.

- The connection area of hydraulic/piping systems with the tank (nozzles) shall be verified to accommodate stresses and distortions due to relative displacements between tanks or between tanks and soil, without their functions being impaired (see also provisions for attached piping).

Section 8.5 and Annex D of EN 1993-1-6 (European Committee for Standardization 2007), paragraphs D.1.2 and D.1.5, give a systematic methodology for the calculation of cylindrical shell resistance against meridional axial compression in the presence of hoop tension due to internal pressure. Moreover, paragraph A.10 of EN 1998-4 (European Committee for Standardization 2006b) provides alternative rules for internally pressurized shell buckling, based on the work by Rotter (1990).

Roof Damage

Sloshing motion of tank liquid contents occurs during earthquake motion. Tank inspections following strong earthquakes indicated that the actual amplitude of liquid surface elevation may be of several meters in some cases. For full or near-full tanks, resistance of the roof to the impact of sloshing waves results in an upward (internal) loading on the roof which may cause buckling of the upper courses of the shell walls. Damage may also occur at the frangible joints between the walls and the ribs of the roof, with accompanying spillage of tank contents over the top of the tank wall.

In practice, a freeboard shall be provided having a height not less than the calculated height of the slosh waves. Freeboard at least equal to the calculated height of slosh waves shall be provided, if the contents are toxic or if spilling could cause damage to piping or scouring of the foundation. Freeboard less than the calculated height of the slosh waves may be sufficient, if the roof is designed for the associated upward pressure or if an overflow spillway is provided to control spilling of contents. Damping devices against sloshing, e.g., grillages or vertical partitions, can be used to reduce sloshing wave height. If the tank has no rigid roof, adequate freeboard is

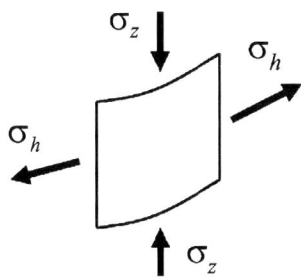

Liquid Storage Tanks: Seismic Analysis, Fig. 12 Hoop hydrodynamic stresses, σ_h, and meridional stresses, σ_z, on the tank wall

needed to prevent undesirable effects of liquid spilling.

Hoop Stresses (Hydrostatic and Hydrodynamic)

Tensile hoop stresses, σ_h, may increase due to hydrodynamically induced pressures between the fluid and the tank and exceeding allowable hoop stress, leading to fracture and leakage, as shown in Fig. 12. This phenomenon has occurred extensively in old riveted tanks, where leakage at the riveted joints resulted from seismic pressure-induced yielding. Welded steel tanks with butt-welded joints are less vulnerable in such a rupture because of better ductility of the welded joints; however, large tensile hoop stresses, σ_h, can contribute to the premature development of elephant foot buckling near the tank base due to excessive meridional compression, σ_z, and its interaction with the hoop stress σ_h.

Relevant shell design provisions for tensile stress failure exist in all specifications, where allowable stresses and efficiency factors for welds are specified. If overmatched butt-welded joints are used, efficiency factor is usually equal to unity.

Base Plate Connection Failure

In the case of unanchored tanks, uplifting may occur and, instantaneously, the welded connection between the tank bottom plate and the tank shell at the uplifting side of the tank is subjected to significant tensile stresses that may lead to weld failure. This may result in fracture at the base plate welded connection and immediate loss of tank content. To avoid this failure the plastic rotation at the welded connection, calculated through Eq. 65, should be limited to a certain value. Such limit values are stated in Section A.9 of EN 1998-4 (European Committee for Standardization 2006b). In addition, strong cyclic loading at the weld, associated with plastic rotation less than the allowable plastic rotation of the connection, may result in fracture due to low-cycle fatigue (Prinz and Nussbaumer 2012).

Resistance of Anchor Bolts

Numerous steel tanks, especially tall (slender) tanks, are anchored with bolts or straps. However, these anchors may be insufficient to withstand the total imposed load in strong earthquake events and can be damaged, in the form of anchor pull-out, stretching, or rupture. One should consider that failure of an anchor does not necessarily lead to tank collapse or loss of tank contents. On the other hand, failure of a significant number of bolts may lead to failure of the tank supporting system and eventually to tank overturning and loss of contents.

The resistance of an anchor bolt subjected to tension can be considered as a fraction of the ultimate force, defined as the product of the bolt net area with the ultimate stress of the bolt material. Bolts should also be designed so that their attachment plate to the shell tank does not fail in shear. Finally, the bolts should be adequately anchored and designed against pullout or punching shear failure.

Failure at Attached Piping Locations

Piping systems connected to tanks should be designed in a way that during earthquakes the potential deformation of the connection location with piping is accommodated. If not, fracture may occur at this area, at the vicinity of the weld leading to loss of content (Wieschollek, Hoffmeister, and Feldmann 2013). To minimize the risk of such a failure, piping systems should provide sufficient flexibility to avoid release of the product by failure of the piping system. To achieve this, the piping system and supports shall be designed so as to not induce significant mechanical loading on the tank shell attachment.

Local loads at piping connections shall be considered in the design of the tank shell. Moreover, the piping shall be designed to minimize unfavorable effects of interaction between tanks and other structures. Mechanical devices which add flexibility to the piping connection, such as bellows, expansion joints, and other flexible components, may be used when they are designed for seismic loads and displacements.

If there is no flexibility loop between the point where the pipe is independently (rigidly) supported and the nozzle, and there are relative anchor motions, the tank nozzle and tank shell should be appropriately checked. Construction of a flexibility loop in the pipe between the tank nozzle and the independent piping supports is always beneficial.

Stability Against Sliding

In general, sliding of the tank should be prevented during a seismic event. The resisting shear force at the interface of the base of the structure and the foundation shall be evaluated taking into account the effects of the vertical component of the seismic action.

For the evaluation of the sliding resistance, the transfer of the total lateral shear force between the tank and the subgrade shall be considered. For self-anchored (unanchored) flat-bottom steel tanks, the overall horizontal seismic shear force shall be resisted by the friction between the tank bottom and the foundation or subgrade. Self-anchored storage tanks shall be proportioned such that the seismic base shear, Q, does not exceed a limit value Q_s, defined as follows:

$$Q_s = \mu(W_s + W_r + W_f + W)(1 - 0.4a_{gv}) \tag{68}$$

where μ is an appropriate friction coefficient.

Limited sliding in unanchored tanks may be acceptable, only if the implications of sliding for the connections between the various parts of the structure and between the structure and any piping are taken into account in the tank analysis and the corresponding verifications.

Summary

In this chapter, important information concerning the seismic analysis of circular cylindrical liquid storage tanks is stated. This information includes a short presentation of the current design standards, a short reference to the seismic input in the course of spectral analysis, the seismic actions that occur during a seismic event, and the resistance of the tank to these actions. Both rigid and flexible tanks that are anchored or unanchored may be considered. It is important to remember that liquid-tank interaction is a key issue for tank seismic response. Furthermore, the total motion can be decomposed in an impulsive part that follows the motion of the container and a convective part, associated with free surface liquid sloshing.

Using the analysis described in the present chapter, for determining the seismic action and considering all possible failure modes, the seismic response of a liquid storage tank can be evaluated rationally and effectively towards safeguarding its structural integrity.

Cross-References

▶ Earthquake Response Spectra and Design Spectra
▶ European Structural Design Codes: Seismic Actions
▶ Modal Analysis
▶ Performance-Based Design Procedure for Structures with Magneto-Rheological Dampers

References

American Petroleum Institute (2003) Seismic design of storage tanks – Appendix E, welded steel tanks for oil storage, API 650, 10th edn. American Petroleum Institute, Washington, DC

American Petroleum Institute (2007) Seismic design of storage tanks – Appendix E, welded steel tanks for oil storage, API 650, 11th edn. American Petroleum Institute, Washington, DC

American Society of Civil Engineers (2006) Minimum design loads for buildings and other structures, ASCE 7-05. American Society of Civil Engineers, Reston

British Standards Institution (1989) Specification for manufacture of vertical steel welded non-refrigerated storage tanks with butt-welded shells for the petroleum industry, BS 2654. British Standards Institution, London, UK

European Committee for Standardization (2004a) General rules, seismic actions and rules for buildings. Eurocode 8. Part 1, CEN/TC 25., EN 1998-1. European Committee for Standardization, Brussels

European Committee for Standardization (2004b) Specification for the design and manufacture of site built, vertical, cylindrical, flat-bottomed, above ground, welded, steel tanks for the storage of liquids at ambient temperature and above, EN 14015. European Committee for Standardization, Brussels

European Committee for Standardization (2006a) Tanks. Eurocode 3. Part 4-2. CEN/TC 250, EN 1993-4-2. European Committee for Standardization, Brussels

European Committee for Standardization (2006b) Silos, tanks and pipelines. Eurocode 8. Part 4. CEN/TC 250, EN 1998-4. European Committee for Standardization, Brussels

European Committee for Standardization (2007) Strength and stability of shell structures. Eurocode 3. Part 1-6. CEN/TC 250, EN 1993-1-6. European Committee for Standardization, Brussels

Fischer FD (1979) Dynamic fluid effects in liquid-filled flexible cylindrical tanks. Earthq Eng Struct Dyn 7:587–601

Fischer FD, Seeber R (1988) Dynamic response of vertically excited liquid storage tanks considering liquid-soil-interaction. Earthq Eng Struct Dyn 16:329–342

Gazetas G (1983) Analysis of machine foundation vibrations: state-of-the-art. Soil Dyn Earthq Eng 2(1):2–43

Habenberger J, Schwarz J (2002) Seismic response of flexibly supported anchored liquid storage tanks. In: Proceedings of the 13th European conference on earthquake engineering, London

Haroun MA (1983) Vibration studies and tests of liquid storage tanks. Earthq Eng Struct Dyn 11:179–206

Housner GW (1957) Dynamic pressures on accelerated fluid containers. Bull Seismol Soc Am 47:15–35

Ibrahim RA (2005) Liquid sloshing dynamics: theory and applications. Cambridge University Press, Cambridge, UK

Karamanos SA, Patkas LA, Platyrrachos MA (2006) Sloshing effects on the seismic design of horizontal-cylindrical and spherical industrial vessels. J Press Vessel Technol ASME 128(3):328–340

Malhotra PK (1995) Base uplifting analysis of flexibly supported liquid-storage tanks. Earthq Eng Struct Dyn 24(12):1591–1607

Manos GC, Clough RW (1982) Further study of the earthquake response of a broad cylindrical liquid-storage tank model, Report No. UCB/EERC-82/7. University of California, Berkeley

Natsiavas S (1988) An analytical model for unanchored fluid-filled tanks under base excitation. J Appl Mech ASME 55:648–653

Niwa A, Clough RW (1982) Buckling of cylindrical liquid-storage tanks under earthquake excitation. Earthq Eng Struct Dyn 10:107–122

Pappa P, Vasilikis D, Vazouras P, Karamanos SA (2011) On the seismic behaviour and design of liquid storage tanks. In: III ECCOMAS thematic conference on computational methods in structural dynamics and earthquake engineering, COMPDYN 2011, Corfu

Pappa P et al (2012) Structural safety of industrial steel tanks, pressure vessels and piping systems under seismic loading (INDUSE). RFCS project final report, Brussels. www.mie.uth.gr/induse

Peek R (1988) Analysis of unanchored liquid storage tanks under lateral loads. Earthq Eng Struct Dyn 16:1087–1100

Priestley MJN, Davidson BJ, Honey GD, Hopkins DC, Martin RJ, Ramsey G, Vessey JV, Wood JH (1986) Seismic design of storage tanks, recommendations of a study group of the New Zealand National Society for Earthquake Engineering, Wellington, New Zealand

Prinz GS, Nussbaumer A (2012) Fatigue analysis of liquid-storage tank shell-to-base connections under multi-axial loading. Eng Struct 40:77–82

Rammerstorfer FG, Fischer FD, Scharf K (1988) A proposal for the earthquake resistant design of tanks – results from the Austrian project. In: Proceedings of the 9th world conference on earthquake engineering, vol 6. Tokyo, pp 715–720

Rammerstorfer FG, Fischer FD, Scharf K (1990) Storage tanks under earthquake loading. Appl Mech Rev ASME 43(11):261–283

Rotter JM (1990) Local inelastic collapse of pressurized thin cylindrical steel shells under axial compression. ASCE J Struc Eng 116(7):1955–1970

Scharf K (1990) Beiträge zur Erfassung des Verhaltens von erdbebenerregten, ober-irdischen Tankbauwerken. Fortschritt-Berichte VDI. Reihe 4. Bauingenieurwesen. Nr. 97. VDI Verlag, Düsseldorf

Vathi M, Karamanos SA (2014) Modelling of uplifting mechanism in unanchored liquid storage tanks subjected to seismic loading. In: Second European conference on earthquake engineering and seismology (2ECEES), Istanbul

Veletsos AS, Tang Y (1990) Soil-structure interaction effects for laterally excited liquid storage tanks. Earthq Eng Struct Dyn 19:473–496

Veletsos AS, Yang JY (1977) Earthquake response of liquid storage tanks. In: 2nd engineering mechanics conference, ASCE, Raleigh, pp 1–24

Wieschollek M, Hoffmeister B, Feldmann M (2013) Experimental and numerical investigations on nozzle reinforcements. In: Proceedings of the ASME 2013 pressure vessels & piping division conference (PVP2013), Paris, 14–18 July 2013

Wozniak RS, Mitchell WW (1978) Basis of seismic design provisions for welded steel oil storage tanks. Advances in storage tank design, API 43rd mid-year meeting, Toronto

Long-Period Moment-Tensor Inversion: The Global CMT Project

Goran Ekstrom and Meredith Nettles
Department of Earth and Environmental Sciences, Lamont-Doherty Earth Observatory, Columbia University, Palisades, NY, USA

Synonyms

Global centroid-moment-tensor inversion; Long-period earthquake analysis

Introduction

Long-period moment-tensor inversion is a methodology to estimate first-order earthquake parameters by inverse modeling of long-period seismograms, typically recorded at teleseismic distances. Data from large numbers of globally distributed digital broadband seismographs and knowledge of elastic wave propagation through the heterogeneous Earth currently make possible the uniform determination of moment tensors of earthquakes with $M \geq 5.0$. The Global CMT Project is an ongoing systematic research effort to study all significant earthquakes occurring worldwide using long-period moment-tensor inversion and to maintain and curate a catalog of the resulting earthquake parameters.

Earthquakes are complex geophysical events with the potential to cause vast destruction. When and where earthquakes occur and a characterization of their size and effects form knowledge of scientific and societal value. For example, systematic analysis of earthquakes led to the mapping of seismic belts and, subsequently, the understanding that most earthquakes occur at the boundaries of tectonic plates. The monitoring and cataloging of earthquake characteristics and their geographical distribution were key ingredients for the development and acceptance of the theory of plate tectonics.

The size of an earthquake is a fundamental descriptive parameter. It was traditionally assessed using the maximum intensity of ground shaking experienced near the earthquake or, following the development and global deployment of seismographs, by measurement of the logarithmic amplitude of weak ground motion recorded at large distances, corrected for distance and attenuation effects – the quantity known as the earthquake magnitude. Since most earthquakes are not felt and the determination of a magnitude is an analysis step that often accompanies earthquake detection and location, the seismic magnitude is the most commonly reported measure of earthquake size. While useful in determining the relative sizes of earthquakes and for estimating and predicting the frequency of earthquakes of a given magnitude using observed frequency–magnitude relationships, traditional magnitude scales are phenomenological, empirical, and imprecise and cannot be directly related to physical properties of the earthquake.

The fundamental insight that earthquakes reflect the sudden relaxation of tectonic stresses across faults dates back more than a century. A deeper physical and quantitative understanding of the relationship between fault motion and seismic-wave generation developed in the 1960s and 1970s. In particular, the seismic moment M_0 was identified as the central physical parameter that describes the size of an earthquake. An earthquake that involves average relative motion $\bar{d}$ of the two sides of a fault over an area A has a seismic moment $M_0 = \mu A \bar{d}$, where μ is the rigidity of the rock surrounding the fault. The seismic moment carries units of force × distance and represents the strength of two perpendicular force couples with no net rotation acting at the hypocenter. The double-couple model of forces is consistent with the notion that the elastic rebound in an earthquake is the response to sudden unrestrained stress at the earthquake fault. This "stress glut" is a tensor quantity closely related to the moment tensor of the earthquake. It reflects the strength and three-dimensional geometry of fault motion. From the seismic moment given in Nm, a moment magnitude M_W (sometimes simply M) can be calculated using a standard formula,

$$M_W = \frac{2}{3}(\log_{10} M_0 - 9.1), \quad (1)$$

designed to agree, on average, with traditional earthquake magnitudes for moderate earthquakes (Kanamori 1977). The seismic moment and the moment magnitude are the preferred modern measures of earthquake size. The moment-tensor elements are directly related to the strike and dip angles of the earthquake fault and the rake angle of fault motion in the earthquake.

The development of detailed models of the layered elastic and density structure of the Earth in the 1970s and 1980s made possible the accurate forward calculation of the seismic wavefield caused by forces acting in the Earth. In particular, in the formalism of the Earth's long-period elastic normal modes, the excitation of seismic waves by body forces and moment-tensor sources is straightforward (Gilbert 1971). Linear elastic wave theory thus provides the framework in which earthquake source characteristics can be deduced from observed seismic waves. At long periods, the seismic wavefield is simple and can be explained by a small number of fundamental earthquake parameters.

Improvements in the digital recording of broadband ground motion on a global scale during the 1980s provided the observations needed for systematic application of inverse modeling methods to characterize global earthquake activity. At present, several groups routinely determine moment tensors and seismic moments for earthquakes worldwide. Knowledge about global seismic activity collected and derived from these efforts is important for improved understanding of earthquake sources, active tectonics, and global seismic hazard.

Long-Period Earthquake Analysis

Earthquake fault rupture is complex. The slip on the fault is distributed heterogeneously over a surface, and the seismic rupture and slip take time to complete. Seismic waves excited by the earthquake reflect the full complexity of the earthquake forces, but the observed vibrations average out the details in a way that depends on period and wavelength. At long periods and long wavelengths, the effect of the entire earthquake is well represented by a small number of point-source parameters. Analogously to distributions of other physical quantities, such as masses or forces, it is useful to expand the description of the earthquake source using spatial and temporal moments. Only the lowest moments of the source influence the long-period source radiation. At the same time, the lowest moments carry fundamental information about the earthquake source including the seismic moment and the fault geometry.

The Moment Tensor and Its Centroid

The spatial and temporal complexity of an earthquake can be described by a distribution of stress glut in space and time, $\dot{\Gamma}_{ij}(\mathbf{r}, t)$. The zeroth-order moment tensor is the time and space integral of $\dot{\Gamma}_{ij}(\mathbf{r}, t)$,

$$M_{ij} = M_{ij}^{(0,0)} = \int_{t_1}^{t_2} \int_V \dot{\Gamma}_{ij}(\mathbf{r}, t) dV\, dt, \quad (2)$$

where the zeros (0,0) in the superscript refer to the zeroth spatial and temporal moments, respectively. In spherical coordinates, the elements of the moment tensor are

$$\mathbf{M} = \begin{bmatrix} M_{rr} & M_{r\theta} & M_{r\varphi} \\ M_{\theta r} & M_{\theta\theta} & M_{\theta\varphi} \\ M_{\varphi r} & M_{\varphi\theta} & M_{\varphi\varphi} \end{bmatrix}, \quad (3)$$

with (r, θ, φ) denoting (up, south, east). Just as the stress-glut-rate tensor, the moment tensor is symmetric:

$$M_{r\theta} = M_{\theta r},\ M_{r\varphi} = M_{\varphi r},\ M_{\theta\varphi} = M_{\varphi\theta}.$$

For a planar fault with a strike angle ϕ and dip angle δ and with earthquake fault motion described by the rake angle λ, the moment-tensor elements are

$$M_{rr} = M_0 \sin 2\delta \sin \lambda$$
$$M_{\theta\theta} = M_0(\sin\delta\cos\lambda\sin 2\phi + \sin 2\delta\sin\lambda\sin^2\phi)$$
$$M_{\varphi\varphi} = M_0(\sin\delta\cos\lambda\sin 2\phi - \sin 2\delta\sin\lambda\cos^2\phi)$$
$$M_{r\theta} = -M_0(\cos\delta\cos\lambda\cos\phi + \cos 2\delta\sin\lambda\sin\phi)$$
$$M_{r\varphi} = M_0(\cos\delta\cos\lambda\sin\phi - \cos 2\delta\sin\lambda\cos\phi)$$
$$M_{\theta\varphi} = -M_0\left(\sin\delta\cos\lambda\cos 2\phi + \frac{1}{2}\sin 2\delta\sin\lambda\sin 2\phi\right), \quad (4)$$

where M_0 denotes the seismic moment of the earthquake.

A point-source representation of a seismic source can be expressed as

$$\dot{\Gamma}_{ij}(\mathbf{r},t) = M_{ij}\delta^3(\mathbf{r}-\mathbf{r}_s)\delta(t-t_s), \quad (5)$$

where $(\mathbf{r}_s, t_s)$ is the location and time of the instantaneous earthquake. The location of the point source that is most representative of a given stress-glut distribution can be defined by considering the first moments of the stress-glut-rate distribution. The first spatial moment of $\dot{\Gamma}_{ij}(\mathbf{r},t)$ around some reference location $\mathbf{r}_h$ is defined as

$$M_{ijp}^{(1,0)} = \int_{t_1}^{t_2}\int_V \dot{\Gamma}_{ij}(\mathbf{r},t)(r_p - r_{hp})dV\,dt, \quad (6)$$

where the subscript p represents one of the three spatial coordinates (r, θ, φ). The first temporal moment around some reference time t_h is

$$M_{ij}^{(0,1)} = \int_{t_1}^{t_2}\int_V \dot{\Gamma}_{ij}(\mathbf{r},t)(t - t_h)dV\,dt. \quad (7)$$

The time and location for which the sum of the squares of these four moments is minimum can be considered the center of the stress-glut-rate distribution and defines the moment-tensor centroid. A point-source representation of an earthquake is therefore given by the following ten parameters: the six independent elements of M_{ij}, the centroid location $\mathbf{r}_c$ (latitude, longitude, depth), and the centroid time t_c.

A common extension to the point-source representation in Eq. 5 is to incorporate a duration of faulting or, more generally, source activity. In a moment expansion of the stress glut, this corresponds to a nonzero second temporal moment of the stress glut, $M_{ij}^{(0,2)}$. A convenient way to incorporate this complexity is to introduce a source-time function $S(t)$ and write the source representation as

$$\dot{\Gamma}_{ij}(\mathbf{r},t) = M_{ij}\delta^3(\mathbf{r}-\mathbf{r}_s)\delta(t-t_s) * S(t;\tau_h), \quad (8)$$

where $*$ indicates convolution and $S(t;\tau_h)$ is a function of unit area, symmetric around $t = 0$, with a total duration $2\tau_h$.

Synthetic Seismograms and Inversion

The relationship between an instantaneous moment-tensor point source and the ground motion $u(\mathbf{r}, t)$ at an arbitrary point on the Earth is illustrated by the expression describing the excitation and summation of the Earth's normal modes,

$$u(\mathbf{r},t) = \sum_k \left[1 - \cos\omega_k(t-t_s)e^{-\alpha_k(t-t_s)}\right] \quad (9)$$
$$\mathbf{M} : \mathbf{e}_k(\mathbf{r}_s)\mathbf{s}_k(\mathbf{r}),$$

where the summation is over modes k, ω_k and α_k are the normal-mode frequency and decay constants, $\mathbf{e}_k(\mathbf{r}_s)$ is the strain of the kth mode at the source location, and $\mathbf{s}_k(\mathbf{r})$ is the displacement of the kth mode at the receiver location. From this expression, it is clear that there is a linear relationship between the moment-tensor elements and the resulting ground motion. The linear relationship is not limited to the normal-mode formulation but is a consequence of linear elastic theory.

For convenience, the six independent elements of the moment tensor can be written as a vector f_i, with $f_1 = M_{rr}$, $f_2 = M_{\theta\theta}$, $f_3 = M_{\varphi\varphi}$, $f_4 = M_{r\theta}$, $f_5 = M_{r\varphi}$, and $f_6 = M_{\theta\varphi}$. Ground motion u_k at a location $\mathbf{r}$ in a direction given by the subscript k and caused by a moment-tensor

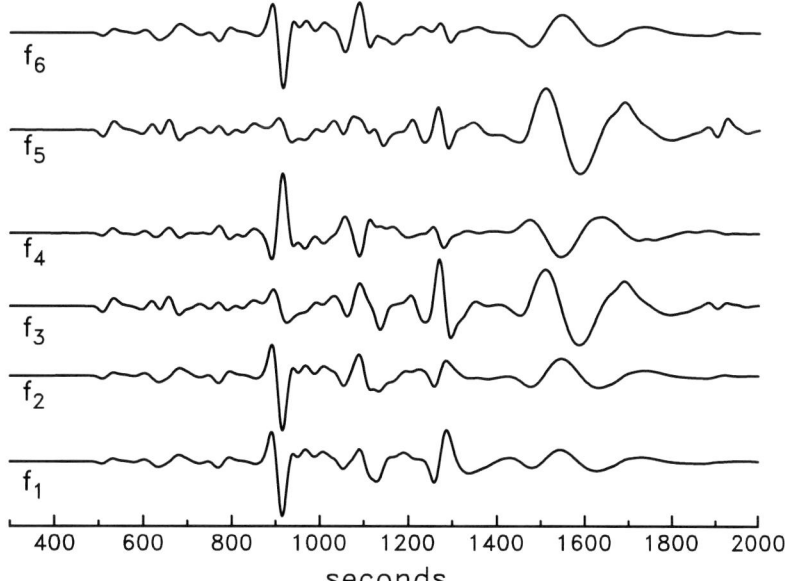

Long-Period Moment-Tensor Inversion: The Global CMT Project, Fig. 1 Kernel east–west seismograms for a hypothetical deep earthquake ($h = 500$ km) in Bolivia recorded at a station in Los Angeles, California. Each trace corresponds to the motion associated with a single moment-tensor element f_i. The ground motion has been filtered between 30 and 300 s

point source at $\mathbf{r}_s$ with a time history $S(t; \tau_h)$ can then generally be written as

$$u_k(\mathbf{r},t) = \sum_{i=1}^{6} \psi_{ik}(\mathbf{r},\mathbf{r}_s,t) * S(t;\tau_h) \cdot f_i, \quad (10)$$

where the kernel seismograms $\psi_{ik}(\mathbf{r}, \mathbf{r}_s, t)$ represent the individual contributions to the seismogram from each moment-tensor component i. Figure 1 shows the six moment-tensor kernels for a hypothetical deep earthquake in Bolivia recorded on the east–west component of a seismometer in Los Angeles, California. The different shapes of the kernels illustrate the linear independence of the different moment-tensor contributions to any recorded ground motion.

Equation 10 and Fig. 1 suggest the least-squares minimization problem

$$\min \left(\int_{t_1}^{t_2} [o_k(t) - u_k(t)]^2 dt \right) \quad (11)$$

to estimate the six moment-tensor elements f_i, where $o_k(t)$ represents the observed seismogram, t_1 and t_2 define a signal time window, and it is understood that the observed and synthetic waveforms have been filtered in the same way. Expressed as in Eqs. 10 and 11, the inverse problem for f_i is strictly linear and can be solved in a single step. However, this formulation is based on a fixed source location ($\mathbf{r}_s$, t_s). In practice, this location is a reported hypocenter that may not correspond closely to the optimal moment-tensor point-source location and time, the centroid ($\mathbf{r}_c$, t_c). To estimate the centroid as well as the moment-tensor elements, additional steps are needed.

When an initial estimate of the moment-tensor elements has been obtained by solving the inverse problem in Eq. 11, the change in the synthetic seismograms with respect to small changes in source location and the moment-tensor elements can be written as

$$\begin{aligned}\delta u_k(t) &= u_k(t) - u_k^{(0)}(t) = b_k(t)\delta r_s \\ &\quad + c_k(t)\delta\theta_s + d_k(t)\delta\varphi_s + e_k(t)\delta t_s \\ &\quad + \sum_{i=1}^{6} \psi_{ik}^{(0)}(t) * S(t) \cdot \delta f_i,\end{aligned}$$
$$(12)$$

where the superscript (0) refers to the synthetic seismogram and moment-tensor kernels calculated for the initial source location. The quantities b_k, c_k, d_k, and e_k represent linear-perturbation kernels,

$$b_k(t) = \sum_{i=1}^{6} \frac{\partial \psi_{ki}^{(0)}}{\partial r_s}(t) * S(t) \cdot f_i^{(0)}$$
$$c_k(t) = \sum_{i=1}^{6} \frac{\partial \varphi_{ki}^{(0)}}{\partial \theta_s}(t) * S(t) \cdot f_i^{(0)}$$
$$d_k(t) = \sum_{i=1}^{6} \frac{\partial \varphi_{ki}^{(0)}}{\partial \varphi_s}(t) * S(t) \cdot f_i^{(0)} \qquad (13)$$
$$e_k(t) = \sum_{i=1}^{6} \frac{\partial \varphi_{ki}^{(0)}}{\partial t_s}(t) * S(t) \cdot f_i^{(0)},$$

reflecting changes in the seismogram due to a small perturbation in the source location. A second step in the inversion can then be formalized by minimizing the remaining misfit between observed and synthetic waveforms with respect to perturbations in the moment-tensor elements as well as the centroid parameters,

$$\min \left(\int_{t_1}^{t_2} \left[\left(o_k(t) - u_k^{(0)}(t) \right) - \delta u_k(t) \right]^2 dt \right). \qquad (14)$$

By iteratively updating the moment-tensor and centroid estimates, as well as the corresponding synthetic seismograms and kernels, and inverting the residual signal for additional perturbations in the source parameters, an optimal estimate of the point-source parameters is obtained.

The preceding paragraphs summarize the theoretical developments and technical approaches described by Gilbert (1971) and Dziewonski et al. (1981). Dziewonski et al. (1981) introduced the full formalism for the simultaneous estimation of the moment tensor and the source centroid, giving rise to the concept of the centroid-moment tensor, or the CMT. The CMT concept is general and can in principle be used with any recorded seismograms, provided that the point-source approximation remains valid. Conditions for success are that high-quality, well-calibrated seismograms are available and that synthetic waveforms corresponding to the observed seismograms can be generated with sufficient fidelity to allow a quantitative comparison.

The Global CMT Project

A systematic effort to use long-period centroid-moment-tensor inversion for the study of global seismicity was initiated by A. M. Dziewoński and J. H. Woodhouse at Harvard University in the early 1980s. This sustained analysis effort was known as "the Harvard CMT Project" and the catalog of earthquake parameters that resulted as "the Harvard CMT Catalog." In 2006, the research effort moved with G. Ekström and M. Nettles to Lamont–Doherty Earth Observatory of Columbia University, where it continues under the name "the Global CMT Project," with the primary research goal of curating, improving, and extending the CMT catalog of global earthquake focal mechanisms (Ekström et al. 2012).

Proof of Concept

In the first demonstration of the CMT approach (Dziewonski et al. 1981), it was shown that CMTs could be calculated for globally distributed earthquakes as small as $M_W = 5.5$ by least-squares inversion of long-period ($T > 45$ s) body-wave seismograms recorded on a sparse network of stations. The synthetic seismograms were calculated by summation of normal modes for the spherically symmetric Preliminary Reference Earth Model (PREM) (Dziewonski and Anderson 1981). Three specific circumstances explain the success of this approach. First, the general characteristics of the seismic source are such that a point-source approximation is appropriate at the periods considered. For example, a typical duration of an $M = 6.0$ earthquake is 5 s, and a typical fault dimension is 10 km, both of which are small compared with a wave period of ~50 s and a wavelength of ~200 km. Second, the background seismic noise in the Earth is low at periods of 30–200 s. Even relatively small earthquakes generate ground motion in this period range that is observable at large distances with modern digital instrumentation. Third, while the three-dimensional (3D) elastic heterogeneity of the Earth is significant, the selected part of the long-period seismic wavefield can be simulated with high accuracy using a one-dimensional (1D) Earth model. In particular, seismic body-wave phases diving deep into the Earth's mantle and core typically do not exhibit travel-time residuals greater than several seconds, which is small with respect to the periods used in the CMT waveform inversion.

Project Evolution

Early applications of the CMT approach were constrained by the very limited availability of high-quality digital seismograms. Only 10–20 stations provided digital data on a global scale in the early 1980s. In addition, although the importance of the Earth's shallow elastic heterogeneity was appreciated, the 3D elastic structure of the Earth was unknown. Methods for synthesizing seismograms in heterogeneous 3D Earth structure were also limited and not appropriate for routine calculations on then-available computers. Rapid progress occurred during the 1980s and 1990s. New networks of globally distributed seismographic stations were deployed, soon providing broadband data in near-real time that could be incorporated in CMT analysis. Low-degree tomographic models of the Earth's upper mantle were developed, and new approximate methods for synthesizing normal-mode seismograms for these models in a computationally efficient manner made possible the inclusion of very long-period ($T > 135$ s) surface waves in the CMT analysis. These waves, often called mantle waves because their primary sensitivity is to elastic structure in the mantle, are particularly useful for the analysis of large earthquakes since their long period and wavelength make the point-source approximation valid even for very large events.

Since 1984, the methodology used in the CMT project for synthesizing normal-mode seismograms in an elliptical and heterogeneous 3D Earth has been based on the path-average approximation. In this approach, the average 1D structure for the minor arc and the great circle connecting the earthquake and each recording station are calculated from an existing 3D Earth model. Small adjustments in the normal-mode frequencies corresponding to deviations from the global-average 1D Earth structure are calculated for each path and applied in the mode sum. To account for the different average structures along the great-circle and minor-arc paths, a fictitious shift in the geographical location of the earthquake is also applied for each mode, using an approach introduced by Woodhouse and Dziewonski (1984). This method for seismogram synthesis corrects the phase of long-period seismic waves for the effect of large-wavelength 3D heterogeneity to first order and retains the computational efficiency of normal-mode summation.

New generations of tomographic models of the Earth's mantle have led to an improved ability to match long-period body waves and very long-period surface waves (mantle waves). These 3D models have not, however, been adequate for matching the large propagation delays observed for fundamental-mode Love and Rayleigh surface waves at periods shorter than 100 s. In addition, the linear-perturbation approach to correcting the mode sum is inappropriate for heterogeneity that causes travel-time delays of the same order as, or longer than, the period of the wave. The value of including intermediate-period surface waves in the CMT analysis is that these waves frequently are the largest phases in long-period seismograms. Following the development in the late 1990s of high-resolution global dispersion models for Love and Rayleigh waves, the possibility arose to incorporate propagation delays from such maps in the calculation of the fundamental-mode contribution to the synthetic seismograms. In 2004, a hybrid approach to the calculation of synthetic seismograms was introduced in the CMT analysis. The body-wave portion and very long-period mantle-wave portion of the wavefield are synthesized using the path-average normal-mode approach, while intermediate-period minor-arc fundamental-mode surface waves are calculated using a surface-wave approach, with phase delays derived from surface-wave phase-velocity maps (Ekström et al. 1997). The consideration of intermediate-period surface waves allows for the analysis of smaller earthquakes than was previously possible and offers improved constraints on the moment tensors for shallow earthquakes.

Current Practice

Analysis of seismicity in the Global CMT Project is divided into calendar months. The analysis is initiated 2–3 months after real time and is completed before the end of 4 months. Earthquakes that have the potential to yield robust CMT results are selected for analysis, primarily from the hypocenter catalog of the National Earthquake Information Center (NEIC) of the United States Geological Survey (USGS). The current objective is to attempt analysis for all earthquakes $M_W \geq 5.0$. Because of the large uncertainties in traditional magnitudes, all events with m_b or $M_S \geq 4.8$ are selected for analysis. This typically leads to a monthly list of ~300 candidate earthquakes, of which more than half are successfully analyzed.

Seismograms are collected from the IRIS-USGS Global Seismographic Network (seismic network codes II and IU) with additional data from stations of the China Digital Seismograph Network (IC), the Geoscope (G), Mednet (MN), Geofon (GE), and Caribbean (CU) networks. Normally, more than 150 stations are available for any earthquake. Figure 2 shows a map with the stations contributing to CMT analysis for earthquakes in 2012.

Three-component data from all stations are extracted for each event, and the responses are equalized by deconvolution of the instrument response and convolution with standard band-pass filters that depend on the data type. Rotated horizontal components (longitudinal and transverse) are constructed and considered in the analysis. Three types of data are used: (1) body waves, which are the signals that arrive in the time window before the arrival of the minor-arc surface waves; (2) mantle waves, which consist of up to 4.5 h of data, including several very long-period Love and Rayleigh wave arrivals (e.g., G1–G4 and R1–R4); and (3) intermediate-period surface waves, which consist of a time window centered on the minor-arc arrival times for surface waves.

Filtering strategies and data selection depend on the earthquake magnitude. Body-wave data for earthquakes $5.5 \leq M_W \leq 6.5$ are filtered to displacement between 150 and 40 s. For earthquakes $M_W < 5.5$, the seismograms are filtered to velocity. For earthquakes $M_W \geq 6.5$, the waves are filtered to displacement between 150 and 50 s. Mantle-wave data are considered only for earthquakes $M_W \geq 5.5$, since smaller earthquakes do not produce observable signals in this band.

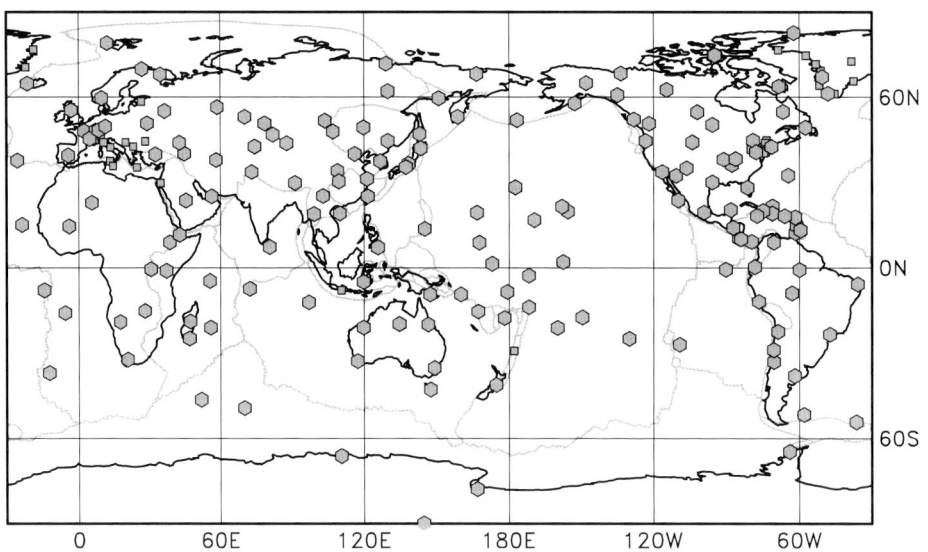

Long-Period Moment-Tensor Inversion: The Global CMT Project, Fig. 2 Map showing the locations of 206 stations that contributed seismograms to the GCMT analyses in 2012. Stations that contributed data for more than 200 earthquakes are shown with *hexagons*; other stations are shown with *squares*

For earthquakes $M_W \leq 7.0$, the waves are filtered to displacement between 350 and 125 s. For earthquakes $M_W > 7.0$, the filter is shifted in period with earthquake size so that for earthquakes $M_W > 8.0$, the range is 450–200 s. Intermediate-period surface waves for earthquakes $M_W \geq 5.5$ are filtered to displacement between 150 and 50 s. For smaller earthquakes, the response is modified to velocity, to emphasize the shorter-period signals, which are less obscured by noise. Intermediate-period surface waves are not considered for earthquakes deeper than 300 km.

The three data types provide complementary constraints on the seismic source. Body waves sample the focal sphere of the earthquake in a relatively uniform fashion. Intermediate-period surface waves provide good constraints for smaller earthquakes. The very long-period mantle waves are less sensitive to the spatial extent and temporal history of the seismic source, an advantage for large earthquakes. Seismograms of the three wave types are weighted in the CMT inversion in such a way that the mantle waves do not contribute for earthquakes $M_W < 5.5$, have a relative weight that increases linearly with magnitude for $5.5 \leq M_W \leq 6.5$, and are weighted fully for earthquakes $M_W > 6.5$. Body and surface waves are weighted fully for $M_W < 6.5$, have a decreasing relative weight for $6.5 \geq M_W \geq 7.5$, and do not contribute for earthquakes with $M_W > 7.5$ since, for earthquakes of this size, the signal at shorter periods is dominated by spatial and temporal source complexities that are not accounted for in the CMT inversion.

Data selection is necessary to eliminate erroneous seismograms and to exclude traces that are dominated by noise or signals not associated with the earthquake. An automated editing program is used to assist with the data-selection tasks, including assessing observed signal levels and waveform fits in predicted time windows across the network and eliminating faulty and outlying traces. The final inversion results and corresponding fit to the data undergo human review. The validity and quality of the result is assessed based on the total number of seismograms that were fit in the inversion and the reduction in data variance obtained. The convergence and stability of the results are used as additional criteria. For small earthquakes and occurrences of overlapping signals from two or more earthquakes, careful human editing of the waveforms is required for the generation of stable and reliable results.

An example of the quality of fit achieved between observed and model waveforms is shown in Fig. 3, with surface-wave seismograms observed at a number of stations at different distances and azimuths from the $M_W = 6.1$ Emilia-Romagna, Italy, earthquake of May 20, 2012.

While in theory all six elements of the moment tensor are resolvable from the data, in practice, trade-offs among the diagonal elements of the moment tensor make the isotropic component of the moment tensor, $M_{rr} + M_{\theta\theta} + M_{\varphi\varphi}$, poorly constrained, except for large and deep earthquakes. However, because the stress glut associated with tectonic faulting is not expected to have a significant isotropic component, there is no expectation that the isotropic component for standard earthquakes should be nonzero. In the CMT inversions, the trace of the moment tensor is therefore constrained to equal zero, $M_{rr} + M_{\theta\theta} + M_{\varphi\varphi} = 0$. In addition, for some earthquakes, the focal depth of the source centroid is held fixed in the inversion. A minimum source depth of 12 km is used to reduce instabilities that can occur in long-period moment-tensor inversions for sources located at or near the Earth's surface. When the depth exhibits persistent instability in the inversions, the depth is fixed at the reported hypocentral depth or at a depth deemed appropriate for the source region.

The source half duration is a parameter assumed in the inversion based on the scalar moment of the event, with an initial estimate derived from the reported magnitude. The empirical relationship $\tau_h = 2.26 \times 10^{-6} M_0^{1/3}$ is used to estimate source half duration, where τ_h is the half duration measured in seconds and M_0 is the scalar moment measured in Nm. Before 2004, the moment-rate function used in the CMT inversions was modeled as a boxcar. Since 2004, the moment-rate function has been modeled as an isosceles triangle with half duration τ_h.

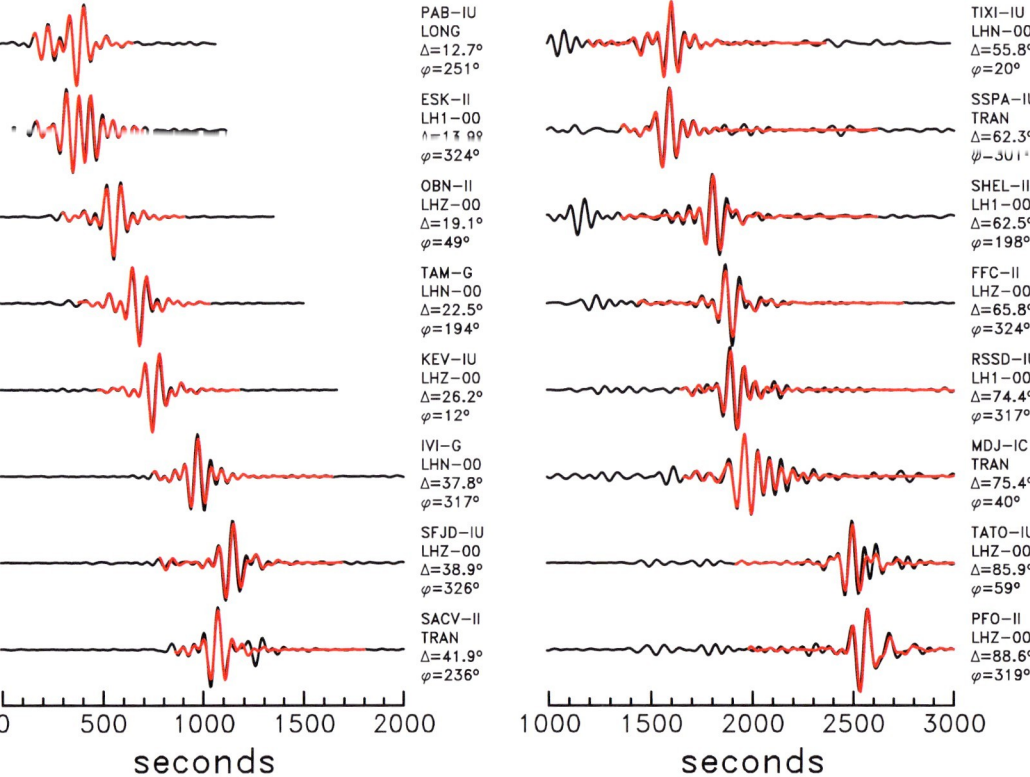

Long-Period Moment-Tensor Inversion: The Global CMT Project, Fig. 3 A selection of observed (*black*) and modeled (*red*) waveforms for the May 20, 2012, $M_W = 6.1$ Emilia-Romagna, Italy, earthquake (875 waveforms were used in the CMT inversion). The station, network, and channel of motion are given to the right of each pair of waveforms, along with the distance Δ and azimuth φ from the earthquake epicenter. In the channel designation, LHZ refers to the vertical component, LHN, LHE, and LH1 refer to recorded horizontal components, and TRANS and LONG refer to components rotated into the directions of transverse and longitudinal motion at the station

The Global CMT Catalog

The catalog of earthquake parameters that has resulted from the research effort described in the preceding paragraphs now contains more than 40,000 earthquakes for the period 1976–2013. It is the most comprehensive and internally consistent global collection of moment tensors. Typical (one-sigma) uncertainties in the estimates of seismic moment are probably no larger than 20 %, implying uncertainties in M_W of approximately 0.05 magnitude units. The catalog is uniform in its methodology, and for large earthquakes, it can be considered uniform and complete, with exception for a small number of large earthquakes that have occurred in the coda of even larger mainshocks. The cumulative moment of earthquakes occurring globally since 1976 is shown in Fig. 4. The figure illustrates the well-known observation that most of the moment is associated with large earthquakes. The largest steps in the graph correspond to the 2004 $M_W = 9.3$ Sumatra earthquake, the 2010 $M_W = 8.8$ Chile earthquake, and the 2011 $M_W = 9.1$ Tohoku earthquake.

Figure 5 shows focal mechanisms from the Global CMT catalog for Eastern Africa, illustrating the use of the catalog for characterizing seismicity and the mode of seismic deformation in areas with limited regional earthquake-monitoring capabilities. The dominant normal-faulting focal mechanisms reflect the process of continental rifting and extension in the area.

For moderate and small earthquakes, general improvements in station coverage and the 2004

Long-Period Moment-Tensor Inversion: The Global CMT Project, Fig. 4 Cumulative moment of all earthquakes in the GCMT catalog through the end of 2013. *Red stars* indicate times of earthquakes of $M_W \geq 8.0$. The area shaded *green* reflects the moment of earthquakes with $M_W \leq 6.5$

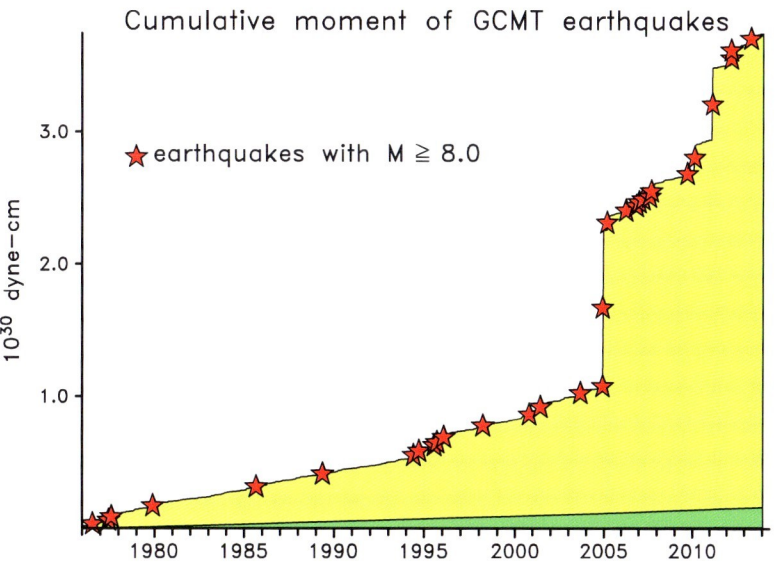

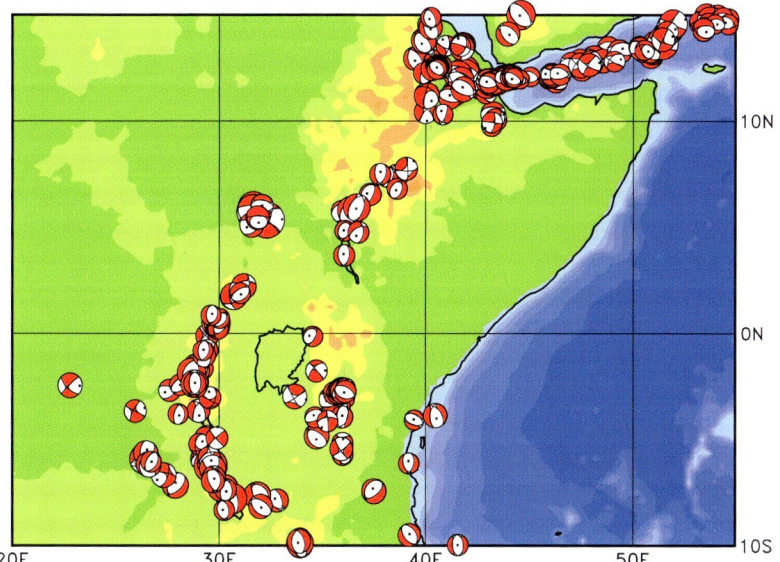

Long-Period Moment-Tensor Inversion: The Global CMT Project, Fig. 5 Map showing the Horn of Africa and surrounding area, with focal mechanisms for earthquakes from the GCMT catalog for the period 1976–2013. Background map shows topography. The quadrants associated with compressional P-wave motion in the lower-hemisphere focal-mechanism plots are shaded *red*. The radii of the focal mechanisms increase linearly with M_W. The magnitudes of the earthquakes shown range from 4.7 to 7.1

incorporation of intermediate-period surface waves in the CMT analysis have extended the completeness of the catalog to smaller magnitudes. Figure 6 shows a magnitude–frequency diagram for the catalog divided in two parts: 1976–2003 and 2004–2013. The data for 2004–2013 indicate linearity of the slope of the magnitude–frequency relation (and thus completeness) to $M_W = 5.0$, while the earlier data show completeness to about $M_W = 5.4$.

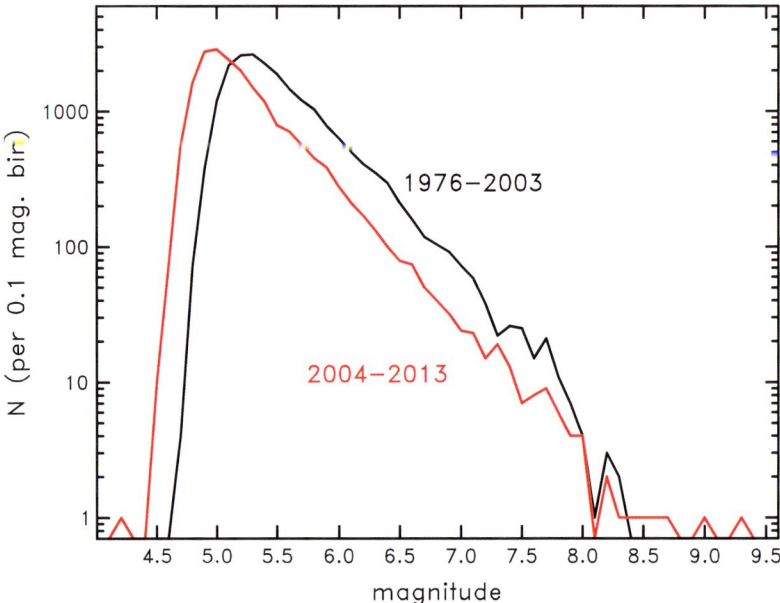

Long-Period Moment-Tensor Inversion: The Global CMT Project, Fig. 6 Magnitude–frequency diagram for earthquakes in the GCMT catalog for two time periods: (*black line*) 1976–2003, (*red line*) 2004–2013. The number of earthquakes is counted in bins of 0.1-magnitude-unit width

The Global CMT catalog is updated routinely and is available from a number of sources. The primary distribution point is the Global CMT website www.globalcmt.org. The full catalog is also available via the website of the Incorporated Research Institutions for Seismology (IRIS), as well as from the International Seismological Centre (ISC) and the NEIC.

Summary

Long-period moment-tensor inversion is a robust method to characterize the geometry and size of earthquakes worldwide. The linear relationship between the amplitude of long-period seismograms and the seismic moment leads to precise and accurate measurements of earthquake size. Catalogs of moment tensors provide primary data on active faults and the geometry of regional seismotectonic deformation.

Global moment-tensor determination is a mature field. Further advances are, however, anticipated as improved 3D models of the Earth's elastic structure are developed and as methods for calculating synthetic seismograms in a heterogeneous Earth in a routine manner become available.

Cross-References

- ▶ Earthquake Magnitude Estimation
- ▶ Earthquake Mechanism Description and Inversion
- ▶ Earthquake Mechanisms and Tectonics
- ▶ Moment Tensors: Decomposition and Visualization
- ▶ Regional Moment Tensor Review: An Example from the European–Mediterranean Region
- ▶ Reliable Moment Tensor Inversion for Regional- to Local-Distance Earthquakes

References

Dziewonski AM, Anderson DL (1981) Preliminary reference Earth model. Phys Earth Planet Int 25:297–356

Dziewonski AM, Chou T-A, Woodhouse JH (1981) Determination of earthquake source parameters from waveform data for studies of global and regional seismicity. J Geophys Res 86:2825–2852

Ekström G, Tromp J, Larson EWF (1997) Measurements and global models of surface wave propagation. J Geophys Res 102:8137–8157

Ekström G, Nettles M, Dziewonski AM (2012) The global CMT project 2004–2010: centroid-moment tensors for 13,017 earthquakes. Phys Earth Planet Int 200–201:1–9

Gilbert F (1971) Excitation of the normal modes of the Earth by earthquake sources. Geophys J Roy Astron Soc 22(2):223–226

Kanamori H (1977) The energy release in great earthquakes. J Geophys Res 82:2981–2987

Woodhouse JH, Dziewonski AM (1984) Mapping the upper mantle: three-dimensional modeling of Earth structure by inversion of seismic waveforms. J Geophys Res 89:5953–5986

Luminescence Dating in Paleoseismology

Steven L. Forman
Department of Geology, Baylor University, Waco, TX, USA

Synonyms

Optical dating; Optically stimulated dating; OSL; Photostimulation

Introduction

Optically stimulated luminescence (OSL) dating or optical dating provides a measure of time since sediment grains were deposited and shielded from further light or heat exposure, which often effectively resets the luminescence signal (Fig. 1). This technique, as thermoluminescence, was originally developed in the 1950s and 1960s to date fired archeological materials, like ceramics (Aitken 1985). Ensuing research in the 1970s documented that marine and other sediments with a prior sunlight exposure of hours to days were suitable for thermoluminescence dating (Wintle and Huntley 1980). Discoveries in the 1980s and 1990s that exposure of quartz and feldspar grains to a tunable light source, initially with lasers and later by light-emitting diodes, yields luminescence components that are solar reset within seconds to minutes expanded greatly the utility of the method (Huntley et al. 1985; Hutt et al. 1988; Aitken 1998). In the past 15 years, there have been significant advances in luminescence dating with the advent of single aliquot and grain analysis, and associated protocols with blue/green diodes that can effectively compensate for laboratory-induced sensitivity changes (Murray and Wintle 2003; Wintle and Murray 2006; Duller 2012) and render accurate ages for the past ca. 100,000 years. Most recently, the development of protocols for inducing the thermal transfer of deeply trapped electrons has extended potentially OSL dating to the 10^6 year timescale for well solar-reset quartz and potassium feldspar grains from eolian and littoral environments (Duller and Wintle 2012).

Common silicate minerals like quartz and potassium feldspar contain lattice-charge defects formed during crystallization and from subsequent exposure to ionizing radiation. These charge defects are potential sites of electron storage with a variety of trap-depth energies. A subpopulation of stored electrons with trap depths of ~1.3–3 meV is a subsequent source for time-diagnostic luminescence emissions. Free electrons are generated within the mineral matrix by exposure to ionizing radiation from the radioactive decay of daughter isotopes in the ^{235}U, ^{238}U, and ^{232}Th decay series and a radioactive isotope of potassium, ^{40}K, with lesser contributions from the decay of ^{85}Rb and cosmic sources. The radioactive decay of ^{40}K releases beta and gamma radiation, whereas the decay in the U and Th series generates mostly alpha particles and some beta and gamma radiation. Alpha particles are about 90–95 % less effective in inducing luminescence compared to beta and gamma radiation. Thus, the population of stored electrons in lattice-charge defects increases with prolonged exposure to ionizing radiation, and the resolved luminescence emission increases with time. Exposure of mineral grains to light or heat (at least 300 °C) reduces the luminescence to a low and definable residual level. Often this

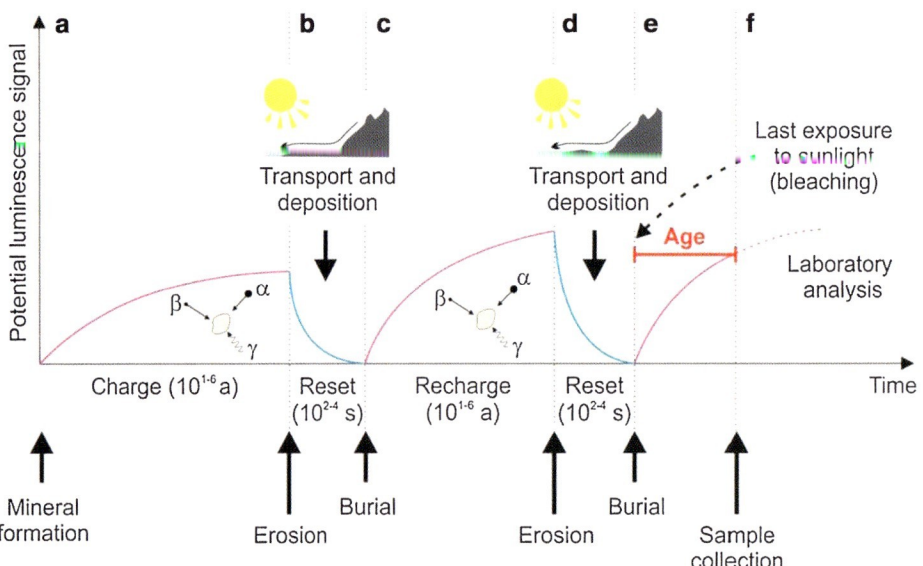

Luminescence Dating in Paleoseismology, Fig. 1 Processes associated with OSL dating. (*a*) Luminescence is acquired in mineral grains with exposure to ionizing radiation and trapping of electrons. (*b*) The luminescence for grains is zeroed by exposure to sunlight with erosion and transport. (*c*) With burial and exposure to ionizing radiation, free electrons are stored in charge defects within grains crystal lattice. (*d*) Further light exposure of grains with erosion and transport zeros the luminescence. (*e*) The grains are buried again and luminescence is acquired with exposure to ionizing radiation. (*f*) Careful sampling without light exposure and measuring of the natural luminescence, followed by a normalizing test dose (L_n/T_n) compared to the regenerative dose to yield an equivalent dose (D_e) (From Mellett 2013)

luminescence "cycle" occurs repeatedly in many depositional environments with signal acquisition of mineral grains by exposure to ionizing radiation during the burial period and signal resetting ("zeroing") with light exposure concurrent to sediment erosion and transportation. Often mineral grains that are fresh from bedrock sources have significantly lower luminescence emissions per radiation dose in comparison to grains that have cycled repeatedly. OSL dating provides an estimate of the time elapsed with latest period of burial and, thus, yields a depositional age (Fig. 1).

Principles of Luminescence Dating

The exposure of quartz and feldspar grains to sunlight for >60 s effectively diminishes the time-stored OSL signal to a low definable level. This residual level is the point from which the geological OSL signal accumulates post burial (Fig. 2). Many types of sediment receive prolonged (>1 h) light exposure with transport and deposition, particularly in eolian, littoral, and sublittoral sedimentary environments. In addition, the inherent residual level is influenced by the susceptibility of the luminescence signal of a specific mineral to solar resetting. The OSL signal of potassium feldspar is usually more resistant to solar resetting than most quartz. There is significant variability in the luminescence properties of quartz and potassium feldspar grains related to crystalline structure, minor and rare-earth impurities, solid-solution relations, number of luminescence cycles (Fig. 1), and radiation history (Mejdahl 1986). Thus, because of this inherent variability in dose sensitivity of quartz and feldspar, analytical procedures for dating often need to be tailored for a specific geologic provenance. The advent of single aliquot regenerative (SAR) dose procedures for quartz (Murray and Wintle 2003; Wintle and Murray 2006) has provided the needed analytical flexibility to compensate for

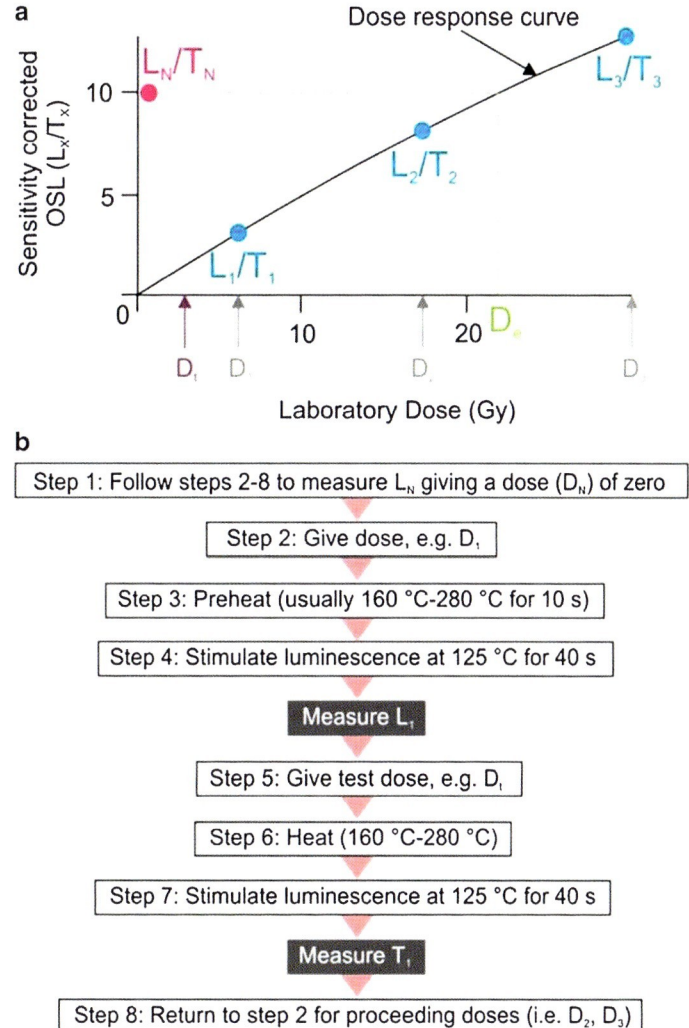

Luminescence Dating in Paleoseismology, Fig. 2 (**a**) Determination of equivalent dose (in grays) using the single aliquot regenerative (SAR) protocols, where the natural luminescence emission is L_n/T_n and the regenerative dose is L_x/T_x; sensitivity changes are corrected by the administration of a small text dose (e.g., 5 grays). (**b**) Generalized SAR protocol (From Mellett 2013)

variable luminescence properties of quartz and feldspar grains and laboratory-induced sensitivity changes, particularly associated with preheat treatments and with laboratory beta irradiation (Fig. 2).

OSL dating is predicated on measurements of a specific mineral and particle size, usually quartz or potassium feldspar. Mineral separations are performed by standard techniques using heavy liquids and hydrofluoric acid (HF) to digest non-quartz minerals and etch the outer 10–20 μm of quartz grains, which are affected by alpha radiation. The purity of the quartz extract is primal for effective dating because a small amount contamination (1 %) by potassium feldspar and other minerals can dominate the luminescence emissions. Multiple soaks in HF may be needed to obtain a pure quartz separate. Purity of the separate is accessed through microscopic inspection and point counting of grain mineralogy. Spectral purity of quartz is often determined by excitation by infrared light from a diode array with subsequent light emissions associated presumably with feldspar contaminants. However, some quartz grains yield considerable emissions with infrared excitation and may host feldspathic or other mineral inclusions; such grains should be analyzed as feldspar grains.

Sediment grains act as long-term radiation dosimeters when shielded from further light exposure with the luminescence signal a measure of radiation exposure during the burial period. The radiation dose that

$$\text{OSL age (year)} = \frac{\text{Equivalent dose}(D_e, \text{grays})}{\text{Dose rate (Dr; grays/year)}}$$
$$\alpha D_\alpha w + D_\beta w + D_\gamma w + D_c \quad (1)$$

$\alpha = $ *alpha efficiency* (0.03–0.15) *for grains* $>50~\mu m = 0$
$D_\alpha = $ alpha dose
$D_\beta = $ beta dose
$D_\gamma = $ gamma dose
$D_c = $ cosmic dose
w = attenuation by water

is equivalent to the natural luminescence emission of isolated quartz and feldspar grains is referred to as the equivalent dose (D_e, measured in grays: 100 rads = 1 gray) and is one half of the OSL age equation (Eq. 1). In most dating applications, quartz is often the favored mineral because of its abundance in sediments, ease of physical separation, and known stability of luminescence emissions. In contrast, feldspar minerals are often less abundant and have a troubling signal instability (anomalous fading), though yield considerably brighter OSL emissions. The recent development of charge transfer techniques for potassium feldspar (e.g., post IR290) that use elevated preheats (~290 °C) to transfer electrons from stable deeper to shallower traps for ease of measurement has extended dating possibilities to 10^5–10^6 timescales for well solar-reset grains (Duller and Wintle 2012). Similar protocols have been also developed for quartz that has been particularly useful for dating Pleistocene loess deposits (e.g., Brown and Forman 2012).

A common approach in OSL dating is to use SAR protocols on quartz aliquots with the protocols customized for a specific sample, study site, or area (Fig. 2). The SAR approach is predicated on a number of assumptions. First, the fast component of luminescence emissions, light released within the first 4 s, is the dominant signal, usually >90 % of the total light output above background levels. Second, the zero dose ratio is <5 % of the natural ratio. Furthermore, the natural ratio is not at dose saturation that errors in calculating aliquot or grain equivalent doses are ≤10 % and the recycling ratio is between 1.1 and 0.9, a measure of the coincident of the first and the last same regenerative dose and indicative that the test dose compensates well for sensitivity changes (Murray and Wintle 2003). A critical practice in luminescence dating is the application of preheats (160–260 °C) to isolate a stable and a time-indicative luminescence emission (Fig. 2). The accuracy of the SAR protocols is evaluated by applying a known dose and testing for specific heat treatments within a sequence of analysis (Fig. 2) if the known dose can be recovered. Often such tests indicate that the known dose can be recovered using a variety of preheat treatments. In practice, using the SAR protocols to determine an equivalent dose involves calculation of a ratio between the natural luminescence and the luminescence from a known test dose (L_n/T_n ratio), which is compared to the luminescence emissions for regenerative doses (L_x/T_x) and also divided by the luminescence from the same test dose (Fig. 2). Applying a known test dose with each measurement cycle provides a metric to correct for changes in the sensitivity of the dated quartz grain(s) to acquire luminescence. Often with each successive measurement cycle, the luminescence output increases by 5–15 % or more from the same radiation dose.

The equivalent dose using the SAR protocols is determined often on >30 aliquots of quartz or feldspar grains. Each aliquot often contains 10–100s of quartz grains, the total number dependent on grain size (e.g., 100–150 μm), plate area, and luminescence yield. Statistical analyses of equivalent dose distributions are critical to render accurate OSL ages with specific age models (Galbraith and Roberts 2012). A common metric used is an overdispersion percentage of a D_e distribution and is an estimate of the relative standard deviation from a central D_e value in context of a statistical estimate of errors. A zero overdispersion percentage indicates high internal consistency in D_e values with 95 % of the D_e values within 2σ errors, though rarely, if ever is a naught value calculated with equivalent dose data. Overdispersion values $\sim \leq 20$ % (at two sigma errors) are routinely assessed for quartz grains that are well solar reset, like eolian sands (e.g., Olley et al. 2004; Wright et al. 2011), and this value is considered a threshold metric for calculation of a D_e value using the central age model of Galbraith and Roberts (2012). Overdispersion values >20 % may indicate mixing of grains of various ages or the partial solar resetting of grains, though overdispersion values up to 32 % have been associated with a single equivalent dose population, particularly if the equivalent dose distribution is symmetrical (Arnold and Roberts 2009). The minimum or maximum age model may be an appropriate statistical treatment for equivalent dose data that is positively or negatively skewed (Galbraith and Roberts 2012), depending on the pedologic and the sedimentologic context. The net effect of pedogenesis and bioturbation is often the mixing in of younger grains, and thus, a maximum age model may be appropriate (Ahr et al. 2013). In contrast, high-energy fluvial and colluvial depositional environments often incorporate partially solar-reset grains; thus, the minimum age model may be appropriate. Other numeric treatments such as finite mixture modeling may also be applicable for separating multiple populations of equivalent dose (Galbraith and Green 1990). The efficacy of these numeric analyses for isolating grain populations should be questioned, and independent tests of accuracy are advised.

Another advantageous approach in OSL dating is single-grain dating using the SAR protocols (e.g., Duller 2008, 2012). Dating single grains of quartz and feldspar is particularly suitable with a mixture of grain populations in sediments, a common occurrence in some fluvial and colluvial sedimentary environments. A significant challenge with single-grain dating is that often 1000s of quartz grains need to be analyzed because 20–90 % of the grains yield little or no luminescence emissions. Thus, a single age determination can occupy a luminescence reader for weeks to months to render sufficient number of analyses to statistically separate equivalent dose populations. Though this approach is time consuming, highly credible data is generated for identifying different equivalent dose grain populations and ultimately accurate ages.

A determination of the environmental dose rate (grays/ka) is needed to render an OSL age (Eq. 1), which is an estimate of the exposure of mineral grains to ionizing radiation from the decay of U and Th series, ^{40}K, and cosmic sources during the burial period. The U, Th, and ^{40}K content of the sediments can be determined by elemental analyses by neutron activation analyses or inductively coupled plasma-mass spectrometry, though such analyses assume secular equilibrium in the U and Th decay series. A suitable alternative is either gamma spectrometry in the field or in laboratory; the laboratory variant (germanium gamma spectrometry) can detect isotopic disequilibrium in the U and Th decay series. If disequilibrium is detected, the dose rate should be suitably modified. The beta and gamma doses should be adjusted according to grain diameter to compensate for mass attenuation (Fain et al. 1999). There is no appreciable alpha dose for coarse grains (>50 μm) with the outer 10–20 μm of grains etched by soaking in HF during preparation. Fine grain sediments (<40 μm) are fully affected by alpha radiation. A cosmic ray component must be included in dose rate calculations which compensates for geographic position, elevation, depth of burial, and density of burial material (Prescott and Hutton 1994). An estimate of the moisture content (by weight) during the burial period is also needed for age calculation. Often estimating moisture is difficult because there are few sedimentologic or diagentic criteria to quantify changes in moisture content during the burial period. Thus, realistic moisture contents are usually derived from particle size considerations, measured moisture contents in an undisturbed setting, evaluation of sediment compaction, level of water table to sampling site, and climatic conditions; and appropriate uncertainty (10–20 %) should be included.

Sedimentologic Context for OSL Dating

There are a number of sedimentary facies in tectonic settings that can be or have the potential to be dated by luminescence. A luminescence age for clastic sediment is a measure of the time since the last sunlight exposure. The zeroing of luminescence in mineral grains, usually by sunlight, must be related to a significant tectonic event, like burial of soil with sand-blow emplacement or colluviation post faulting, for the OSL age to be meaningful. In turn, dating sediments that were displaced or deformed provides maximum limiting ages on tectonic activity (e.g., Chen et al. 2013). Eolian sediments, like loess, sand sheet deposits, dune sands, and cover loams, are the most preferred sediments for OSL dating because of the long (hours) light exposure prior to deposition. Also, littoral and sublittoral sediments often receive long light exposure within the swash zone (Argyilan et al. 2005), though storm beach deposits may be variably solar reset.

Mineral grains in water-lain environments such as glacial-marine, certain lacustrine, and fluvial can be incompletely solar reset reflecting attenuation of spectra as light penetrates a turbid water column. Low-energy fluvial sediment, like shallow channel fills with millimeter-scale, horizontally bedded, medium sands, is the preferred facies for OSL dating (Schaetzl and Forman 2008). Higher-energy facies with erosion of previous deposited sediments and formation of new grains with clast percussion during saltation and creep at the base of the water column often yield a mixture of apparent grain ages. Colluvial and alluvial sediment, often associated with normal and reverse faulting, may be amenable for OSL dating, though care is needed to avoid sampling intraclasts in proximal colluvium, closest to the fault zone. Proximal colluvium is difficult to date by OSL and yields often highly dispersed equivalent dose values by aliquots and single grains, necessitating the use of finite mixture models (e.g., Cortes et al. 2012). Distal colluvium, deposited by grain transport with overland flow downslope of the fault offset, can be well solar reset and is more suitable for OSL dating (Forman et al. 1989; Sohbati et al. 2012). There is some promise in OSL dating of minerals from fault gouge to constrain the timing of tectonic activity, though luminescence characteristics are highly variable (Spencer et al. 2012; Banerjee et al. 1999).

Weathered horizons exhibiting pedogenic accumulations of clay, silt, carbonate, or silica should be avoided scrupulously for OSL dating. Postdepositional pedogenesis can affect detrimentally the luminescence time signal by altering the radionuclide concentration, disrupting crystal structure, and mixing in fully or partially solar-reset grains and thus increasing overdispersion values (Ahr et al. 2013).

A detailed study of luminescence characteristics of quartz grains in soils indicates that the upper (<5 cm) part of A horizons in well-developed soils (e.g., alfisols) is well solar reset and is another target for OSL dating, particularly in the buried context (Ahr et al. 2013).

Sampling of Sediment for OSL Dating

Sampling sediment for luminescence dating is relatively straightforward, though care should be exercised that the appropriate sedimentary facies is sampled. It is critical that the sedimentologic, pedologic, stratigraphic, and tectonic context is well known prior to sampling, so a judicious decision is made on sample selection that weighs the likelihood of solar resetting with original deposition, minimizes pedogenic alterations, and maximizes that the sediment-depositional (OSL) age closely constrains tectonic activity. Ideally, at the sampling site, at least 20 cm of homogenous sediment should surround the collected sediment to maximize uniformity in the dose rate environment during the burial period. Sampling within 20 cm of boulders or major lithologic contacts should be avoided to obviate potential inhomogeneities in radioactivity. If such inhomogeneities cannot be avoided, field gamma spectrometry or collection of sediment from the disparate units is advised to document the in situ dose rate. Also, care should be exercised in collection samples within 1–1.5 m of the surface, with likely pedogenic alterations. Approximately 30 g of sediment should be collected, though more or less sediment may be adequate depending on the concentration of the chosen particle size for dating. The geological luminescence signal of sediment is reduced rapidly with exposure to sunlight; thus, care must be taken not to expose the sediment to any light during sampling. The OSL signal of quartz is significantly reduced with a few seconds of light exposure. Prior to sampling, the section should be excavated back at least 20 cm to expose a fresh face. Immediately prior to sampling, the face should be scraped free of light-exposed grains on the surface. It is best to take the sediment intact; though the outer mineral grains may have been light exposed, the internal grains have been shielded from sunlight. The most efficacious approach for sampling for OSL dating is to use light-tight sediment coring tubes, composed of aluminum, copper, steel, or black, stiff plastic, like ABS. The tube should be 10–20 cm long, 2–4 cm diameter, and with an end cap, the tube can be hammered or pushed into the desired sediment for dating. For clay-rich or indurated sediments, blocks can be carved from the section and placed in Kubiena tins or other metallic containers. A secondary bag sample of sediment (~50 g) that can be light exposed should also be taken from a 20 cm radius from the tube/block collection point for elemental (dose rate) and particle size analyses.

aliquots (10–100s of grains) or single grains. In many sedimentary environments, like eolian, sublittoral, and littoral, mineral grains are well solar reset and are most amenable for OSL dating. However, in certain fluvial, colluvial, and alluvial environments, solar resetting of mineral grains is partial, or older grains are incorporated, and thus, there are often multiple populations of grain ages. The collection of unsuitable samples for OSL dating can be obviated by a careful sedimentary facies analysis that maximizes for light exposure of grains with transportation and deposition. In turn, there are a variety of computational solutions for single aliquot data to isolate grain populations that may be temporally significant, though the efficaciousness of such solutions should be questioned. Single-grain analyses, though laborious requiring the analyses of 100–1000s of grains, provide the needed analytical data to define grain populations and accurate ages.

The effective age range of OSL dating of quartz is ca. 100,000 years, though this limit will vary by ca. ±50 ka with environmental dose rate and electron trap, stability density, and energy. Thermal transfer techniques may extend the dating of quartz and feldspar grains from well solar-reset sediments to 10^5–10^6 timescales. OSL has wide application in tectonic settings in dating fault offset or uplifted sediments or sediment deposition that closely precede or succeed a faulting event. Judicious application of OSL dating is hinged on documenting the complex sedimentologic association with fault movement and understanding equally if not more complex the luminescence emissions from grains of quartz and feldspar which yield a time signature.

Summary

Optically stimulating luminescence is a technique that dates the burial time of quartz and feldspar grains. The zeroing mechanism is the exposure of grains to >1 min of sunlight, and there is subsequent signal acquisition with burial, shielding from sunlight, and exposure to ionizing radiation, mostly from radioactive decay during the burial period. Specific minerals, like quartz and potassium feldspar, are dated, with a known particle size range (e.g., 100–150 µm) either as

Cross-References

▶ Archeoseismology
▶ Paleoseismic Trenching
▶ Paleoseismology
▶ Radiocarbon Dating in Paleoseismology
▶ Tsunamis as Paleoseismic Indicators

References

Ahr SW, Nordt LC, Forman SL (2013) Soil genesis, optical dating, and geoarchaeological evaluation of two upland Alfisol pedons within the Tertiary Gulf Coastal Plain. Geoderma 192:211–226

Aitken MJ (1985) Thermoluminescence dating. Academic, New York

Aitken MJ (1998) An introduction to optical dating: the dating of quaternary sediments by the use of photon-stimulated luminescence. Oxford University Press, New York

Argyilan EP, Forman SL, Johnston JW, Wilcox DA (2005) Optically stimulated luminescence dating of late holocene raised strandplain sequences adjacent to Lakes Michigan and Superior, Upper Peninsula, Michigan, USA. Quatern Res 63(2):122–135

Arnold LJ, Roberts RG (2009) Stochastic modelling of multi-grain equivalent dose (D-e) distributions: implications for OSL dating of sediment mixtures. Quat Geochronol 4(3):204–230

Banerjee D, Singhvi AK, Pande K, Gogte VD, Chandra BP (1999) Towards a direct dating of fault gouges using luminescence dating techniques – methodological aspects. Curr Sci 77(2):256–268

Brown ND, Forman SL (2012) Evaluating a SAR TT-OSL protocol for dating fine-grained quartz within Late Pleistocene loess deposits in the Missouri and Mississippi river valleys, United States. Quat Geochronol 12:87–97

Chen Y, Li S-H, Sun J, Fu B (2013) OSL dating of offset streams across the Altyn Tagh Fault: channel deflection, loess deposition and implication for the slip rate. Tectonophysics 594:182–194

Cortes AJ, Gonzalez LG, Binnie SA, Robinson R, Freeman SPHT, Vargas EG (2012) Paleoseismology of the Mejillones Fault, northern Chile: insights from cosmogenic Be-10 and optically stimulated luminescence determinations. Tectonics 31, TC2017

Duller G (2008) Single-grain optical dating of quaternary sediments: why aliquot size matters in luminescence dating. Boreas 37(4):589–612

Duller G (2012) Improving the accuracy and precision of equivalent doses determined using the optically stimulated luminescence signal from single grains of quartz. Radiat Meas 47:770–777

Duller G, Wintle AG (2012) A review of the thermally transferred optically stimulated luminescence signal from quartz for dating sediments. Quat Geochronol 7:6–20

Fain J, Soumana S, Montret M, Miallier D, Pilleyre T, Sanzelle S (1999) Luminescence and ESR dating-Beta-dose attenuation for various grain shapes calculated by a Monte-Carlo method. Quat Sci Rev 18:231–234

Forman SL, Machette MN, Jackson ME, Matt P (1989) Evaluation of thermoluminescence dating of paleoearthquakes on the American Fork segment, Wasatch fault zone, Utah. J Geophys Res 94:1622–1630

Galbraith RF, Green PF (1990) Estimating the component ages in a finite mixture. Nucl Tracks Radiat Meas 17(3):197–206

Galbraith RF, Roberts RG (2012) Statistical aspects of equivalent dose and error calculation and display in OSL dating: an overview and some recommendations. Quat Geochronol 11:1–27

Huntley DW, Godfrey-Smith DI, Thewalt MLW (1985) Optical dating of sediments. Nature 313:105–107

Hutt G, Jaek I, Tchonka J (1988) Optical dating: K-feldspars optical response stimulation spectra. Quat Sci Rev 7:381–385

Mejdahl V (1986) Thermoluminescene dating of sediments. Radiat Prot Dosimetry 17:219–227

Mellett CL (2013) Luminescence dating. In: Clarke LE (ed) Geomorphical techniques (online edition). British Society for Geomorphology, London, pp 1–11. http://www.geomorphology.org.uk/assets/publications/subsections/pdfs/OnsitePublicationSubsection/90/4.2.6_luminescencedating.pdf

Murray AS, Wintle AG (2003) The single aliquot regenerative dose protocol: potential for improvements in reliability. Radiat Meas 37(4–5):377–381

Olley JM, Pietsch T, Roberts RG (2004) Optical dating of Holocene sediments from a variety of geomorphic settings using single grains of quartz. Geomorphology 60(3–4):337–358

Prescott JR, Hutton JT (1994) Cosmic ray contributions to dose rates for luminescence and ESR dating: large depths and long-term time variations. Radiat Meas 23:497–500

Schaetzl RJ, Forman SL (2008) OSL ages on glaciofluvial sediment in northern Lower Michigan constrain expansion of the Laurentide ice sheet. Quatern Res 70(1):81–90

Sohbati R, Murray AS, Buylaert J-P, Ortuno M, Cunha PP, Masana E (2012) Luminescence dating of Pleistocene alluvial sediments affected by the Alhama de Murcia fault (eastern Betics, Spain) – a comparison between OSL, IRSL and post-IR IRSL ages. Boreas 41(2):250–262

Spencer JQG, Hadizadeh J, Gratier J-P, Doan M-L (2012) Dating deep? Luminescence studies of fault gouge from the San Andreas Fault zone 2.6 km beneath Earth's surface. Quat Geochronol 10:280–284

Wintle AG, Huntley DJ (1980) Thermoluminescence dating of ocean sediments. Can J Earth Sci 17:348–360

Wintle AG, Murray AS (2006) A review of quartz optically stimulated luminescence characteristics and their relevance in single-aliquot regeneration dating protocols. Radiat Meas 41(4):369–391

Wright DK, Forman SL, Waters MR, Raveslot JC (2011) Holocene eolian activation as a proxy for broad-scale landscape change on the Gila River Indian Community, Arizona. Quatern Res 76(1):10–21

M

Masonry Box Behavior

Rui Marques
Engineering Department, Civil Engineering Section, Pontifical Catholic University of Peru, Lima, Peru

Synonyms

Box action; Building global behavior; Diaphragm effect; Structural connectivity

Introduction

This entry deals with the box behavior of masonry structures, which is a major hypothesis for the application of seismic assessment procedures based on the in-plane response of the walls, when using performance-based approaches for seismic safety. A general description is also made of the unreinforced and confined masonry building typologies, regarding the constructive technique, industrial development, and applicability. Methods and procedures are presented for the seismic assessment and safety verification of masonry buildings, which are applied to the case of a dwelling.

Masonry has been continually used for building, employing very different materials and bond patterns, but the main distinction between masonry constructions probably is the presence or absence of rigid floor diaphragms well connected to the walls. According to this aspect, masonry constructions can be divided in two categories: structures with and without box behaviour, which present a very dissimilar response when subjected to seismic actions. For the case without box behavior, the walls of the building behave independently and out of phase, with combined in-plane and out-of-plane deformations, and presenting mainly out-of-plane damage (Fig. 1a). On the other hand, when with box behavior the building acts as a jointly assemblage of walls and roofs, with mainly in-plane response of the walls (Fig. 1b).

The concept of box behavior was a key assumption of methods developed for the seismic analysis of masonry structures, e.g., the POR (Tomaževič 1978). In this case, a macroelement discretization has been adopted for the walls through a frame-type assemblage of pier, spandrel, and connection elements (Marques and Lourenço 2011). For the POR, it was initially considered that the walls deform jointly and without plane rotation, thus the global response of a masonry structure being computed as the sum of the in-plane response of individual walls. Within this hypothesis, even if the masonry is a brittle material, the variety of wall sections (deformation capacities) in a building allows a significant structural ductility.

These methods consider generally an elastic–perfectly plastic load–displacement response for the walls according to given strength

© Springer-Verlag Berlin Heidelberg 2015
M. Beer et al. (eds.), *Encyclopedia of Earthquake Engineering*,
DOI 10.1007/978-3-642-35344-4

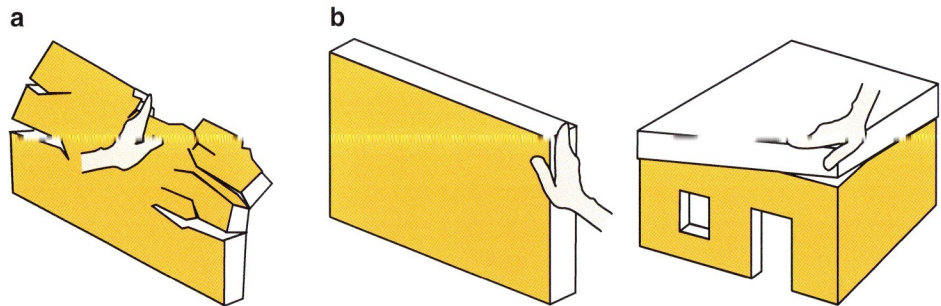

Masonry Box Behavior, Fig. 1 Illustration of box behavior concepts (Adapted from Touliatos 1996): (**a**) out-of-plane failure, (**b**) in-plane response due to box behavior

criterion and ultimate drift, depending on the formed collapse mechanism by shear or flexure. The response of the building is then evaluated by a pushover loading analysis, which allows identifying the sequence of damage events on panels, reflecting in a graphical representation of the base shear force against the displacement of a control point, e.g., the centroid of the roof slab, which is called the capacity curve. This curve is an intrinsic feature of the building that reflects, beyond the base shear strength and deformation capacity, the dissipated energy.

The concept of box behavior and the current performance-based approaches for seismic assessment specified in codes (e.g., EN 1998-1 2004; Sullivan et al. 2012) are explained here referring to cases of unreinforced and confined masonry building structures. In the next two sections, these typologies are presented in general terms.

Unreinforced Masonry

Unreinforced masonry is an ancient, traditional, and ongoing construction typology, whose system consists in the juxtaposition and superposition of masonry pieces through joints. In the antiquity, stone pieces and mud bricks were used for the construction of unreinforced masonry monumental buildings, but the great revolution of this construction technique was with the production of burnt clay bricks. In recent years, a large development was verified in the industry of ceramic bricks allowing for high quality and efficiency, regarding mechanical properties and functional aspects of the brick itself and of the masonry components as an assemblage. At the same time, concrete and calcium silicate blocks were also developed, which are an alternative to ceramic bricks. Masonry requires also for a good quality of remaining constituents, particularly the joint mortar. Block masonry systems have been also developed with mortar pocket and/or tongue-and-groove head joints, e.g., in Fig. 2. Aiming to control cracking due to creep and shrinkage movements, bed-joint steel truss reinforcement is in many cases recommended.

Unreinforced masonry is widely used in Germany (mainly ceramic brick masonry, e.g., Jäger et al. 2010) and in Brazil (mainly concrete block masonry, e.g., Parsekian et al. 2012), which demonstrates the wide applicability of unreinforced masonry around the world, both in cases of strong-economy nations and developing countries. In these countries, great development was observed with the industrialization of, more than a masonry unit, a masonry system allowing for a modular design with use of complementary pieces, such as half bricks, corner blocks, lintel bricks, etc. Masonry systems need to consider also for functionality regarding the installation of water supplies and cable networks, namely, through driving of shafts into the walls. Another aspect to consider is the adequacy of used solutions for slabs, particularly concerning the condition of box behavior. Regarding this point, solutions need to be considered that allow an effective connection of the

 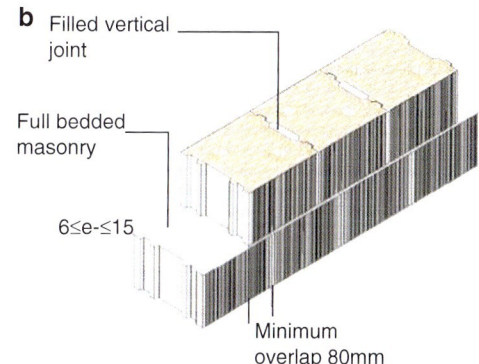

Masonry Box Behavior, Fig. 2 Example of an unreinforced masonry system: (**a**) building view, (**b**) scheme of the bond

walls to the slab system, namely, through the use of a bond beam.

Even if the constructive process of masonry is very simple, skilled labor is a major aspect for unreinforced masonry construction, since the practices for infill walls have caused a great degradation of the masonry works, which requires for a re-specialization. Once more, the architectural trends and requested esthetics for buildings are a great challenge for masonry as a material and for engineers, when requiring for asymmetric, complex, and irregular configurations. For this reason, more than other construction typology, unreinforced masonry requires a much synchronized planning involving both architects and engineers to allow the feasibility of the construction. However, unreinforced masonry presents a restricted applicability in high seismicity regions due to limited earthquake resistance, in which cases an improvement to the masonry system is necessary, namely, by using the confined masonry typology presented in the next section. Reinforced masonry is also an alternative, which is largely disseminated in the United States and Canada (Parsekian et al. 2012), but which implementation seems not easily and directly adaptable to other countries.

Confined Masonry

Confined masonry was historically introduced as a reaction to destructive earthquakes in Italy and Chile (1908 Messina and 1928 Talca quakes, respectively), which almost completely destroyed traditional unreinforced masonry buildings. The improvement of this technique was reflected on the good performance of confined masonry buildings subjected to the 1939 Chilean earthquake, which was probably the main reason for the large dissemination of this technique in all Latin America (Brzev 2007). The construction with confined masonry spread widely to all continents, in countries with moderate-to-high seismicity such as Slovenia, India, New Zealand, Japan, and Canada.

Conceptually, the confined masonry system is based on embracing masonry panels with frame elements, similarly to reinforced concrete frames, but with the difference that in the confined masonry the reinforced concrete elements are cast only after the masonry construction (Fig. 3a). For this reason, contrarily to reinforced concrete structures where infill masonry is built after concrete hardening, in the case of confined masonry most of the building weight rests on the masonry panels. In addition, due to concrete shrinkage, the connection between masonry and concrete is very effective (Fig. 3b). The interaction between the confining elements and the masonry panel allows a confined masonry wall under lateral loading to behave as a whole up to large deformation levels, allowing improved strength and ductility (e.g., Gouveia and Lourenço 2007). A typical phenomenon in confined masonry is the dowel

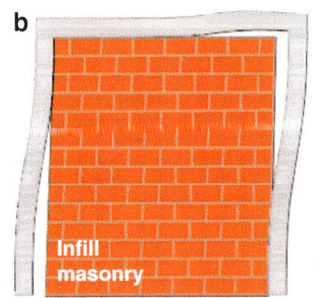

Masonry Box Behavior, Fig. 3 Illustrations of confined masonry typology: (**a**) construction example (Courtesy of Paulo B. Lourenço), (**b**) difference between infill and confined masonry regarding the frame-to-masonry connection

action of the longitudinal reinforcement bars of the concrete columns, which balances the loss of shear strength in the masonry panel, allowing to keep the strength necessary for a ductile response.

In practice, and relatively to the reinforced concrete system, confined masonry allows normally for economical savings due to the required smaller sections of concrete and steel. However, even if extensive experimental studies have been performed to evaluate the shear response of confined masonry structures (e.g., San Bartolomé 1994), the practice of self-construction and the use of prescriptive rules have led to poor earthquake resistance. Additionally, the use in many cases of poor-quality units leads to its break and to an inadequate behavior of the confined walls. The observed seismic damage in confined masonry buildings occurs in some cases at upper stories of the buildings, with associated out-of-plane damage, and is mostly due to the absence of box behavior in the affected stories. Brzev et al. (2010) attribute the seismic damage on confined masonry buildings mainly to inadequate wall density, poor quality of masonry and construction, deficiencies in detailing of reinforced concrete confining elements, absence of confinements at openings, and geotechnical issues.

In Europe, industrialized systems for confined masonry have been developed, namely, in Spain and Portugal, which allow for a modular design (e.g., Lourenço et al. 2008). An important aspect in the case of confined masonry is the formation of thermal bridges due to masonry discontinuity, whose problem can normally be solved by coating the columns with thermal insulation plates or by casting the piers within special blocks. This construction typology is also a potential solution for the problem of housing in developing countries due to its quick construction and economy. Confined masonry was, for example, used in the reconstruction of Haiti after the 2010 earthquake.

Macroelement Models

Masonry structures present specific and diverse bond typologies, for which several modeling approaches have been adopted. In the academic-research field, the modeling of masonry buildings has been applied using two different scales, namely, the finite element and macroelement approaches (Lourenço 2002). The concept of using structural component models for masonry structures, designated by macroelement modeling, was introduced by Tomaževič (1978) and applied to perform seismic assessment. This concept is the one addressed next, given the easy implementation of material laws and the formulation of structural equilibrium. In addition, the adopted structural component discretization largely reduces the number of degrees of freedom in relation to the traditional finite element modeling approaches, allowing for more resource- and time-efficient computations and making them attractive to practitioners. In the following, the available models are briefly described for unreinforced and confined masonry.

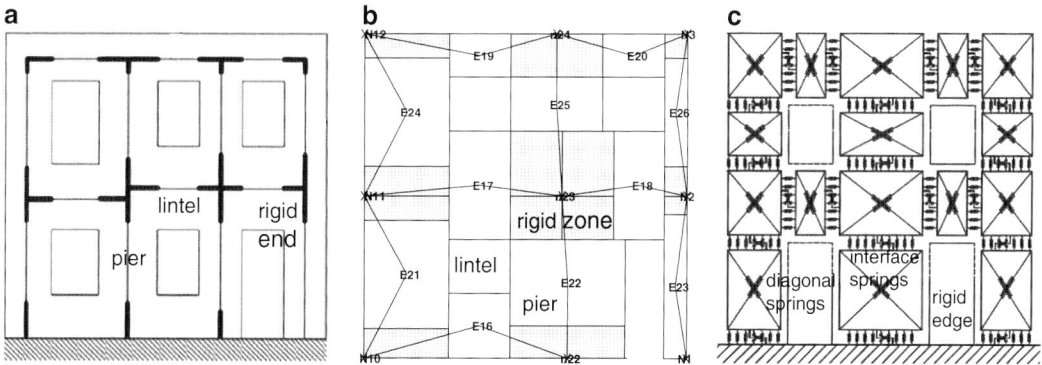

Masonry Box Behavior, Fig. 4 Schematization of wall models in (a) SAM II, (b) TreMuri, and (c) 3DMacro

Recently, and mainly in Italy, several user-friendly computer codes based on macroelements have been developed for assessing the seismic safety of unreinforced masonry buildings. Marques and Lourenço (2011) benchmarked the ANDILWall/SAM II (Calliari et al. 2010), the TreMuri (Lagomarsino et al. 2009), and the 3DMacro (Gruppo Sismica 2013) software codes and provided the basic description of the macroelement formulation and assemblage used in these methods. Briefly, SAM II and TreMuri are based on frame-type modeling by using one-dimensional macroelements, while the 3DMacro is based on a discretization with two-dimensional discrete elements, as shown in Fig. 4. The seismic assessment is made through performance-based approaches, i.e., procedures for nonlinear static (pushover) analysis, according to recent design codes (EN 1998-1 2004; Sullivan et al. 2012).

The methods above were developed throughout a sequential process, including first the idealization of the macroelement, then the definition of the wall assemblage, and finally the simulation of the full building model. The validation process of these methods was made by referring to experimental testing and more advanced computations, for individual masonry panels, full plane masonry walls, and three-dimensional structures, e.g., for the building tested by Magenes et al. (1995). This validation process is not easy due to the large variability of masonry materials and due to the importance that the complex structural organization in full buildings might assume.

On the other hand, rather few quasi-static tests on unreinforced masonry buildings have been carried out allowing this validation. Many unreinforced masonry building models have been tested in shaking tables, but, in general, the dynamic experimental behavior is difficult to compare with the pushover response.

Confined masonry is a particular case of masonry structures, even if it presents some similarity with reinforced concrete structures because of the presence of a frame. Confined masonry is characterized by casting of the reinforced concrete elements only after the masonry work, which provides a good connection between the confining elements and the masonry panels due to the combination of bond effects, shrinkage of the reinforced concrete elements, and the fact that the vertical dead load is transferred to the walls. The interaction behavior between the confining elements and masonry through the existing interface is a specific aspect that needs to be considered in the response of confined masonry walls under lateral loading. Some models have been implemented for confined masonry structures based on a wide-column approach (e.g., (Marques and Lourenço 2013), considering the interaction behavior between the confining elements and the masonry implicitly in the wall shear response.

Micro-modeling strategies can also be used for confined masonry, namely, based on the finite element method, to model explicitly the concrete–masonry interface (e.g., Calderini et al. 2008). Alternatively, a discrete element

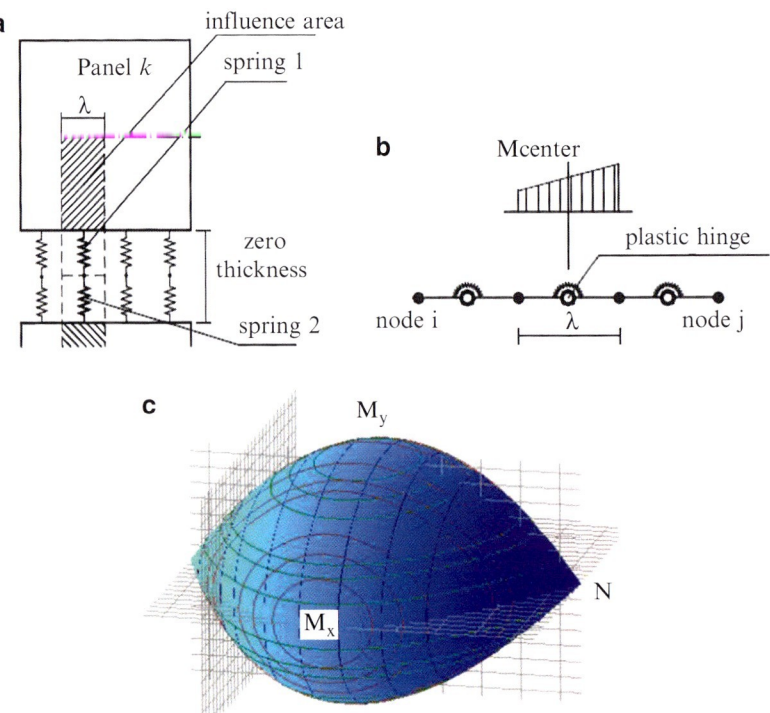

Masonry Box Behavior, Fig. 5 Discrete macroelement: (**a**) interface model, (**b**) frame element, and (**c**) example of interaction domain N–Mx–My

approach is also applicable, such as that idealized by Caliò et al. (2012) originally for unreinforced masonry and which has been extended in the 3DMacro software (Gruppo Sismica 2013) to model reinforced concrete/steel/masonry mixed structures. This last approach uses an interface connection (constituted by nonlinear springs) between the masonry panels (Fig. 5a), which in the case of two neighboring confined masonry panels is interposed by a frame modeled through nonlinear beam finite elements with concentrated plasticity (Fig. 5b). For the beam elements and in agreement to a given type of interaction (axial, flexural, or axial-flexural), the corresponding hinges are considered according to the respective N–Mx–My domain (such as in Fig. 5c).

Nonlinear Static (Pushover) Analysis

The first approaches for seismic analysis were based on the concept of seismic coefficient, which represents the proportion of the building weight that is applied as a lateral force to the structure. Later, approaches were introduced considering the response in displacements of the masonry walls, accounting for the greater sensitivity of the damage to the imposed displacements (e.g., Turnšek and Čačovič 1970).

The masonry structures can present significant ductility when subjected to lateral actions, which is the greater the higher the ductility of sections and the larger the variety of wall geometries (deformation capacities). This concept is early evidenced from damage patterns observed in masonry buildings subjected to earthquakes (e.g., 1963 Skopje, 1965 Oaxaca and 1976 Friuli earthquakes) and also based on experimental evidence (Tomaževič 1999). The masonry walls show a great capacity of inelastic cyclic deformation and energy dissipation due to progressive activation of damage mechanisms. Furthermore, the walls are capable to support the vertical loading even if significantly damaged (e.g., with diagonal cracks).

The study of the nonlinear behavior of structures was started in the 1960s for reinforced concrete, particularly based on the observation of earthquake-damaged buildings. Later, numerical models were developed for the nonlinear static

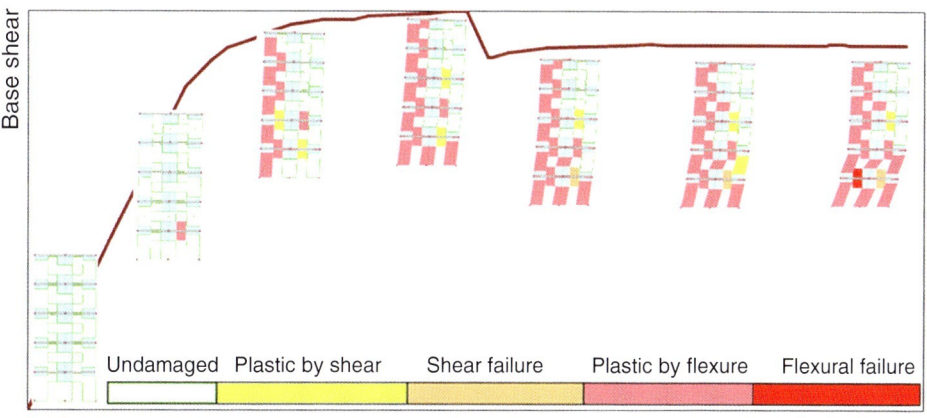

Masonry Box Behavior, Fig. 6 Illustration of the pushover analysis on a masonry building

analysis of reinforced concrete structures (Saiidi and Sozen 1981). A similar investigation sequence and line time was observed for masonry structures (Tomaževič 1978). However, if for the case of reinforced concrete structures advanced models were proposed in the background of extensive resources and studies, the proposed methods for masonry were simpler. This was due to the secondary role of masonry structures in the construction sector and particularly to the very complex behavior of masonry.

The nonlinear static analysis, or pushover analysis as popularized, is focused on the evaluation of the seismic performance of structures based on the control of damage mechanisms and deformation. Several procedures for pushover analysis have been developed, which allow considering the inelastic reserve of capacity in the design and safety verification of structures as a simplified alternative to the nonlinear dynamic time history analysis, e.g., the N2 method (Fajfar and Fischinger 1988) specified in seismic codes (EN 1998-1 2004; Sullivan et al. 2012). The procedures for pushover analysis are based on the control of deformation, where the displacement demands due to the design earthquakes are obtained through a spectral analysis of the response of an equivalent single-degree-of-freedom system, which are after compared with the displacement capacities corresponding to given performance levels.

The displacement capacities are computed from the capacity curve obtained for the building, which presents the evolution of the base shear force (horizontal force representative of the seismic action) against the displacement of a control point (significant point of the structure, usually the centroid of the roof slab), e.g., in Fig. 6. This curve is computed by applying an incremental lateral static loading on the structure, which is processed with force and/or displacement control, and for which two load distribution patterns are usually assumed: proportional to the inertial masses multiplied by the displacements of the first vibrational mode of the structure (modal distribution) and proportional to the inertial masses (uniform distribution).

Seismic Safety Assessment

Traditionally, the seismic design of buildings is based on computing a solicitation for the structure in terms of a base shear force through the consideration of a seismic coefficient or spectrum, which is reduced by a behavior factor and distributed after to the structure. However, the behavior factor values recommended in Eurocode 8 (EN 1998-1 2004) for masonry structures (1.5–2.0) turned out to be very conservative, and often the buildings are considered unsafe. In this case, the consideration of an overstrength ratio that multiplies the basic value of the behavior factor is mandatory, which reflects the possibility of force redistribution between elements beyond the elastic limit (Magenes 2006). The values for the overstrength ratio are specified in the Italian code (IBC 2008) based on experimental and analytical evidences.

Currently, a transition stage from force-based to performance-based design is verified through the use of approaches that consider the ductility or displacement capacity of the structure. The displacement-based approach, which in effect provides a safety verification rather than an explicit design, is presently the more developed level of seismic safety assessment. In this case, performance is related to acceptable damage and damage to displacement. The N2 method (Fajfar and Fischinger 1988) specified in Eurocode 8 (EN 1998-1 2004) is the method considered here.

For the safety evaluation, the N2 method defines a bilinear representation of the capacity curve corresponding to an equivalent single-degree-of-freedom system. According to the Italian code (IBC 2008), this representation includes a first straight line that passes by the origin and intercepts the capacity curve of the actual system for a base shear force of 70 % of the maximum and a second line, horizontal and defined in mode that the area below the envelope of the actual system equalizes the area under the bilinear idealized response. The bilinear representation allows to obtain the reference period of vibration for the computation of the target displacement.

The procedure for safety verification is based on the evaluation of the seismic performance of the building, in terms of deformation, verifying that the target displacement is lower than the displacement capacity of the building, obtained on the capacity curve in correspondence with the established limit states. Commonly, damage and ultimate limit states are considered, which are, respectively, associated with the peak base shear and with a post-peak remaining base shear force of 80 %. A factor is also considered as the ratio between the elastic response force and the yield force of the equivalent system, which value (imposed to be less than 3.0) represents a limitation to the ductility of the structural system as a whole.

The out-of-plane assessment is an issue not clearly evidenced in seismic codes, which implies a personal interpretation of safety from the engineer. However, in general, it can be considered that the control of slenderness limitations, minimum thickness requirements, and appropriate structural conception and detailing (rigid diaphragms and efficient floor-to-wall connection) avoid out-of-plane driven failures. A possibility to check the out-of-plane safety is the application of a kinematic limit analysis, which in effect has been implemented in several commercial software codes, additionally to the global displacement-based safety verification. In this case, the user needs to define the potential kinematic blocks (portions of connected walls), the kind of mechanisms, and respective hinges and constraints.

Application to Dwelling House

Dwelling houses up to two stories are the majority of the building stock in Europe, in terms of both existing and newly constructed buildings. The loss of masonry as a structural solution, due to an ungrounded perception of its lack of capacity to resist earthquakes and also to the dissemination of reinforced concrete structures, caused a strong reduction of the use of structural masonry in new buildings. On the other hand, a large development occurred in the masonry industry, namely, with the introduction of high-quality masonry systems regarding functional and mechanical features. In the academic field important efforts have been also made to develop adequate tools to account for the intrinsic nonlinear capacity of masonry structures when subjected to earthquakes. A first comparison of these tools was made by Marques and Lourenço (2011) referring to a simple building configuration. In this work the comparison is extended, concerning a real and more complex structure.

The studied two-story semidetached dwelling is representative of typical housing in southern

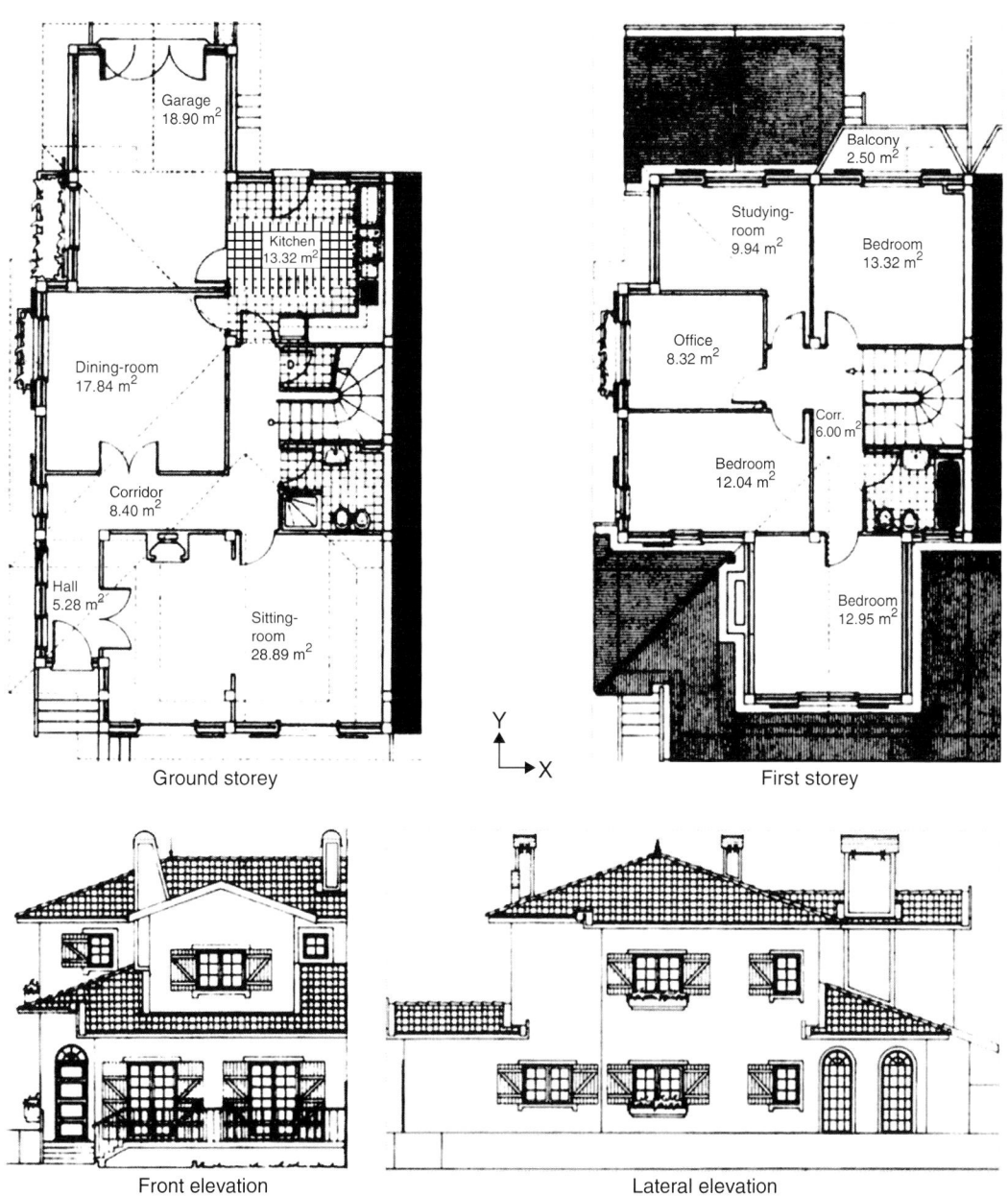

Masonry Box Behavior, Fig. 7 Architectural drawings of the dwelling: plans and elevation views

Europe, with rooms and kitchen in the ground story and bedrooms and an office in the first story. The house presents also a garage and a multilevel roof, as shown in Fig. 7. The structure of the building was originally designed in reinforced concrete. In the following, alternative to the original structure, solutions of unreinforced masonry and confined masonry structures are presented and compared. The building is assumed to be constructed on type B ground (deposits of very dense sand, gravel, or very stiff clay).

Unreinforced masonry, considering that is a simple construction technique which allows an energy-efficient enclosure with no thermal bridges concern, is the first adopted option. In this case,

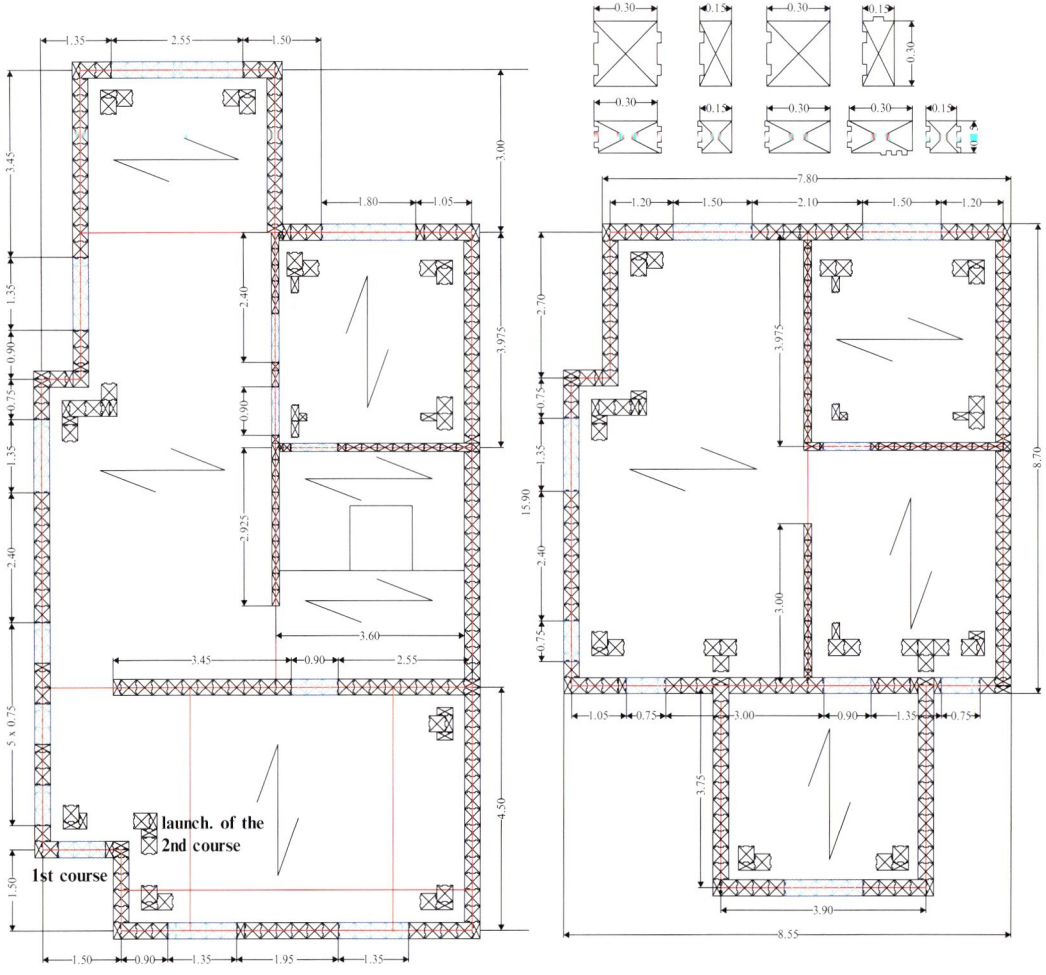

Masonry Box Behavior, Fig. 8 Structural plans of the constructive solution in unreinforced masonry

a clay block masonry system is used with tongue-and-groove head joints and with shell-bedded horizontal joints. The system uses a vertically perforated block of 0.30 m × 0.30 m × 0.19 m and complementary pieces (half block, end piece, corner block, adjusting piece, and lintel block) allowing a geometry in plan with a module of 0.15 m. A M10 mortar pre-batched, according to Eurocode 6 (EN 1996-1-1 2005), is used for the joints. The plans for the proposed structure are presented in Fig. 8, which show the bond in the first course and the second course arrangement around the wall crossings.

The floors of the building are made with prestressed ribbed slabs, which have a thickness of 0.19 and 0.15 m in the accessible zones and for the ceiling/roof, respectively. The slabs are similar to those used in the original reinforced concrete structure and are subjected to dead loads of 4.5 and 3.0 kN/m^2 and to live loads of 2.0 and 0.4 kN/m^2 for the same areas. A bond beam is made with the lintel block that serves as a formwork, with a concrete section of 0.20 × 0.35 m^2 and reinforcement of 4ϕ10 mm with stirrups ϕ6 mm@0.20 m (0.10 m at the ends). The properties for the masonry were considered according to Eurocode 6 and the producer-declared values and selected materials, resulting in a compressive strength of 3.25 MPa, a pure shear strength of 0.1 MPa, an elastic modulus of 3,250 MPa, and a shear modulus of 1,300 MPa. Additionally, the in-plane ultimate

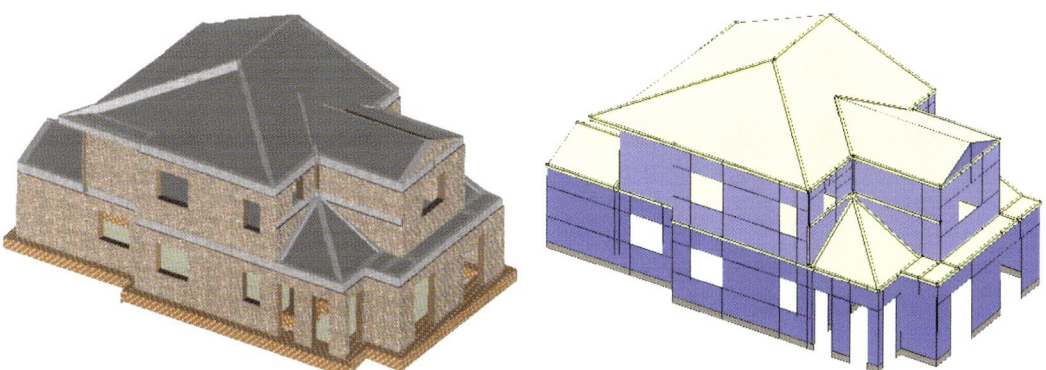

Masonry Box Behavior, Fig. 9 Geometrical and computational models of an unreinforced masonry building

drifts for the masonry panels were assumed, according to Eurocode 8 (EN 1998-1 2004) and the Italian code (IBC 2008) with values of 0.4 % and 0.8 %, respectively, for the shear and flexural failure modes. The weight for the masonry with plastering is 12.0 kN/m^3.

The building was modeled in the macroelement software code developed by Gruppo Sismica (2013), according to the geometrical and computational models presented in Fig. 9. The floors are simulated as polygonal diaphragms elastically deformable, considering an orthotropic slab element. The model was analyzed under pushover loading in the principal directions, and considering "mass" and "first-mode" load distributions. Concerning the +X and +Y analysis with first-mode load distribution, the damage sequence for the building is shown in Fig. 10. For +X and regarding the front façade, a trend is captured in the ground story with shear failure of the central pier and rocking of the extreme piers and with concentration of displacements at this story. A similar deformation mechanism and substantial shear damage are observed in Fig. 10a for the other wall alignments. In +Y, rocking and shear failures are predicted for slender and squat panels, respectively, and identifying a first story mechanism.

A confined masonry solution was also considered following the prescriptive rules from the European regulations. For the confining columns, according to Eurocode 6 (EN 1996-1-1 2005), a longitudinal steel reinforcement must be included equivalent to 0.8 % of the column's cross-section area, with a minimum of 4ϕ8 mm. The transversal reinforcement consists of 6 mm diameter stirrups spaced in general of 0.20 m and of 0.10 m at the column ends. The solution for the horizontal confining element is the same adopted for the unreinforced masonry structure. The structural plans for the confined masonry solution are presented in Fig. 11. The confined masonry structure was also modeled in the software by Gruppo Sismica (2013) with the geometrical and computational models shown in Fig. 12. Note that the main change with respect to the unreinforced masonry model is the inclusion of the reinforced concrete confining columns, which are simulated as beam finite elements with concentrated plasticity and feature a tridimensional constitutive behavior N–Mx–My. The model was then analyzed under pushover loading in the main directions and assuming load distributions proportional to the mass and to the first vibration mode.

The damage sequence for the confined masonry model is illustrated in Figs. 13 and 14 for the +X and +Y analysis with "first-mode" load distribution, corresponding to different displacement levels. For both analyses, damage in the building starts to be relevant at 1 mm displacement, when several masonry panels develop tensile cracks in the interfaces and also diagonal shear cracking. For a displacement of 4 mm, damage spreads widely to all walls of the building. Then, damage evolves with increasing deformation due to the progressive formation of flexural plastic hinges at the ends of the confining elements.

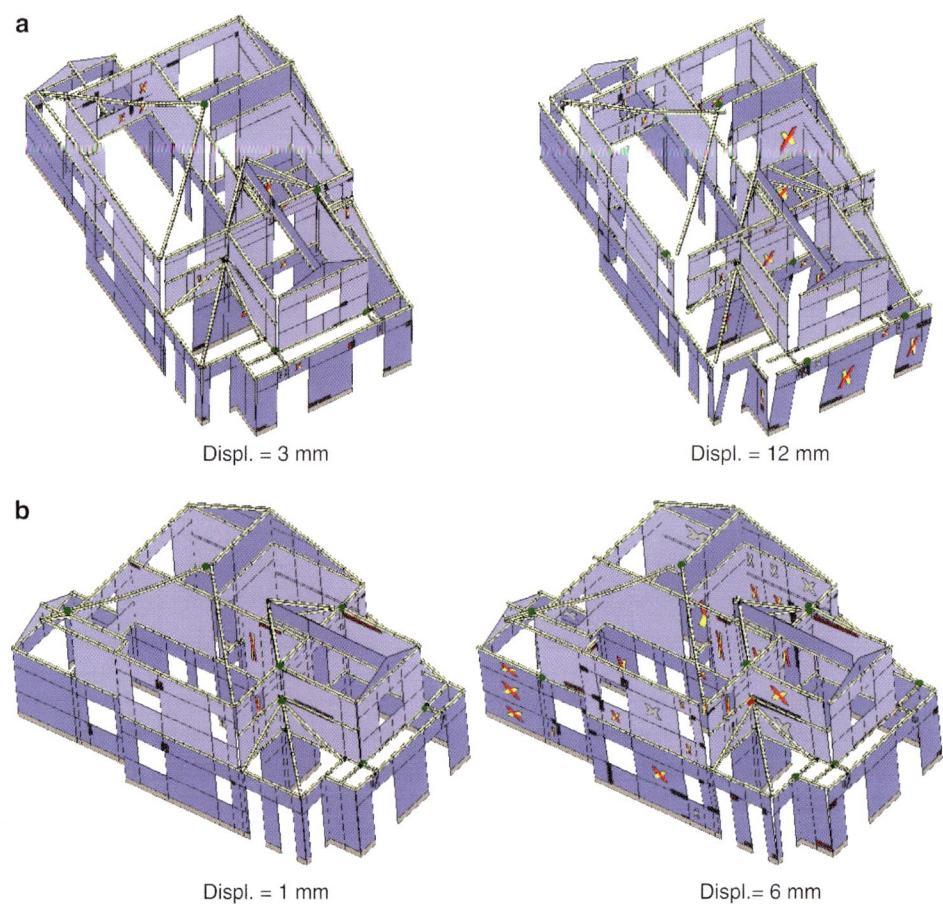

Masonry Box Behavior, Fig. 10 Predicted damage sequence corresponding to the (**a**) +X and (**b**) +Y analysis (X diagonal shear, = tension cracks, • flexural plastic hinge)

The capacity curves for both unreinforced masonry and confined masonry structural solutions are compared in Figs. 15 and 16, where the base shear is expressed as a percentage of the building weight (%g). In X, the weaker direction, the unreinforced masonry building presents a base shear strength of about 23 % of the weight and a deformation capacity higher than 10 mm. Regarding the confined masonry solution, an increase of the base shear capacity is clearly observed for the X direction, comparatively to the unreinforced masonry solution. Still, a sharp decrease of capacity is found after peak due to a small number of confined panels and the absence of walls at the first story in the left part of the building. The high degradation of the confined walls at peak load generates a ground story mechanism. Concerning the Y direction, a remarkable improvement of the base shear strength and ductility is observed.

Finally, considering that the building is subjected to a type 2 spectrum ("near-field earthquake") in Eurocode 8 (EN 1998-1 2004), and from using the N2 method (Fajfar and Fischinger 1988) in the software code by Gruppo Sismica (2013), an evaluation is made of the values of the reference peak ground acceleration on type A ground (rock formations), $a_{g,u}$, corresponding to the ultimate displacement allowed for the building. These values are presented in Fig. 17, which can be considered as an indicator of the seismic performance of the structure. The $a_{g,u}$

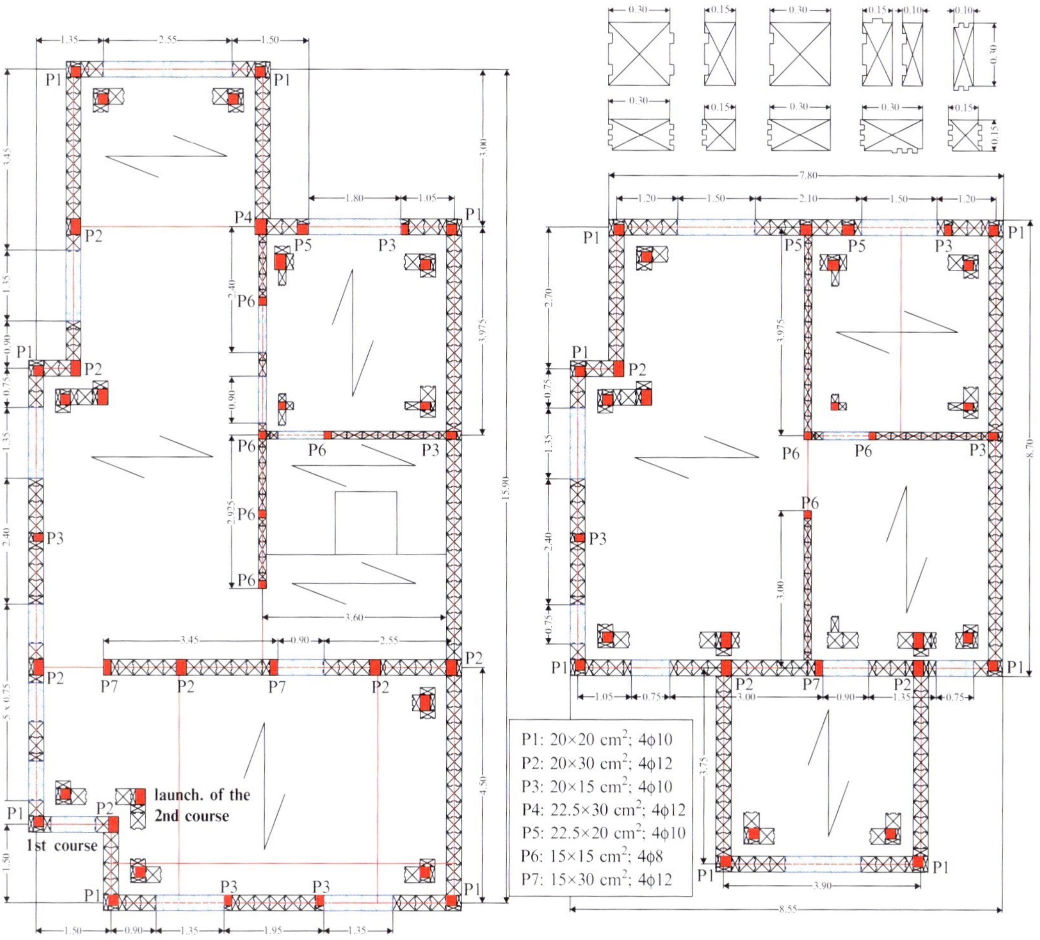

Masonry Box Behavior, Fig. 11 Structural plans of the constructive solution in confined masonry

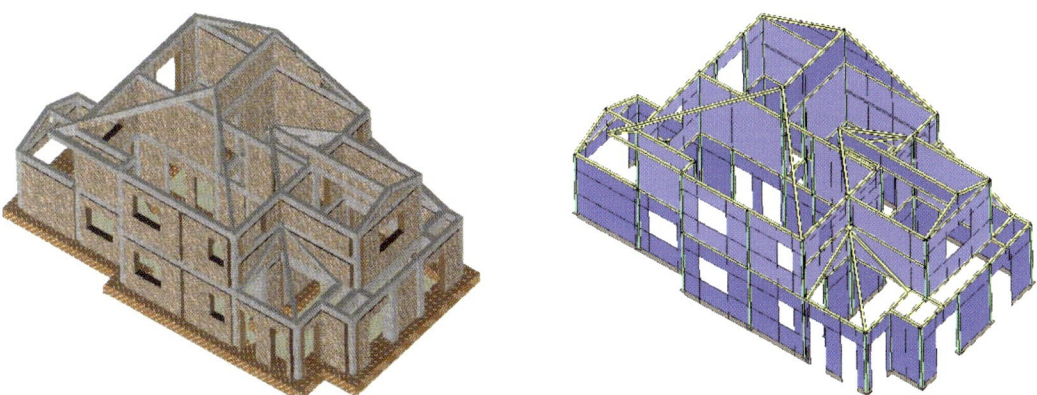

Masonry Box Behavior, Fig. 12 Geometrical and computational models of a confined masonry building

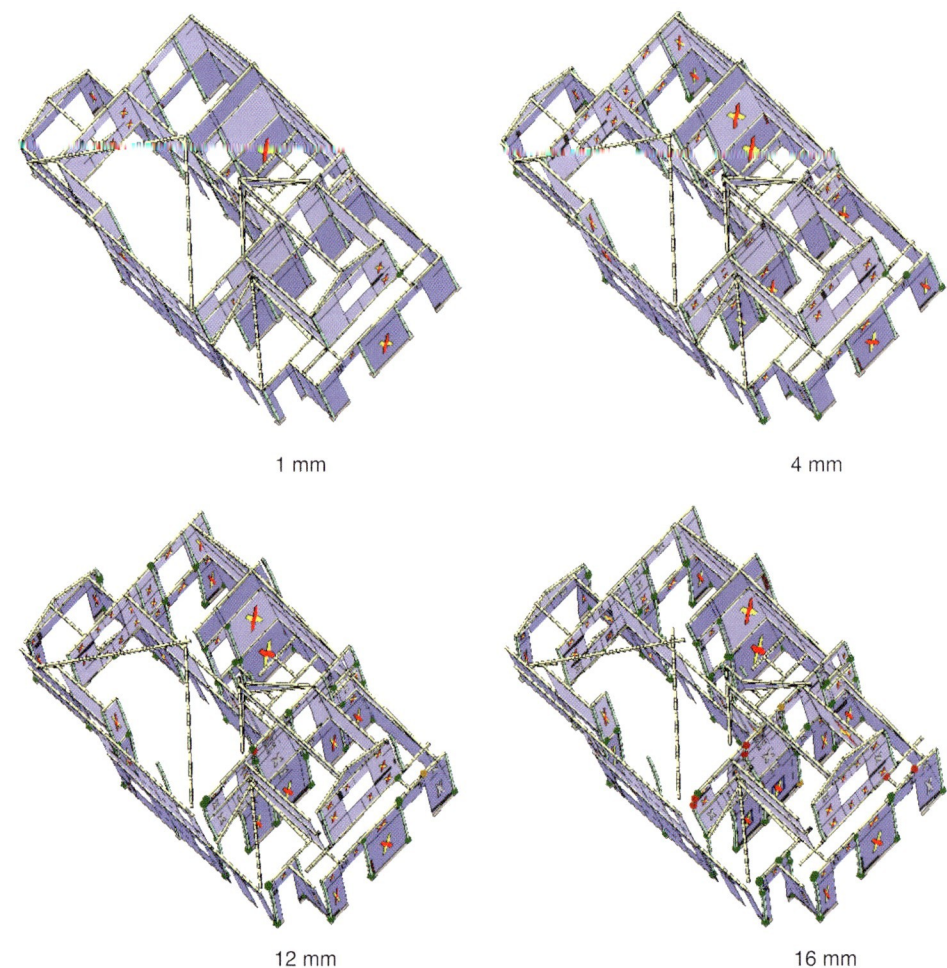

Masonry Box Behavior, Fig. 13 Damage sequence for the +X analysis with first-mode load distribution (X diagonal shear, = tension cracks, ● flexural plastic hinge)

can be interpreted as the maximum ground acceleration supported by the building for a condition of near collapse or ultimate state, which can be directly compared with the design ground acceleration for a global safety verification.

Conclusions and Future Directions

Masonry constructions are, more than a legacy from the past, a solution for the future, given the existing development in the sector. The box behavior is a fundamental feature of masonry buildings, which allows the structures to present a significant ductility when subjected to earthquakes. This ductility is reflected in the capacity of inelastic displacements for the masonry buildings, allowing them to support significant ground accelerations. Using a typical house in southern Europe as a case study, structural masonry typologies featuring box behavior allow ensuring seismic safety up to large ground acceleration levels, namely, 0.15 and 0.20 g for unreinforced masonry and confined masonry solutions, respectively.

The box behavior is then a request for earthquake-resistant masonry construction, which can be obtained through the adoption of slab solutions allowing for an enough diaphragm effect and good floor-to-wall connection.

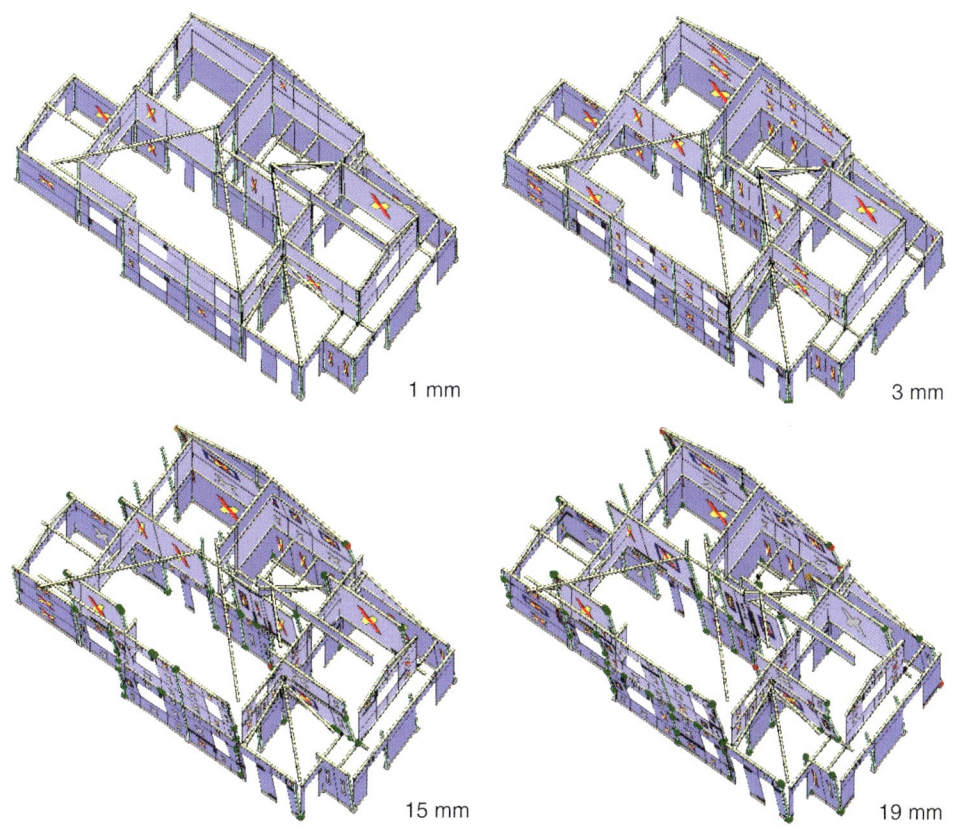

Masonry Box Behavior, Fig. 14 Damage sequence for the +Y analysis with first-mode load distribution (X diagonal shear, = tension cracks, • flexural plastic hinge)

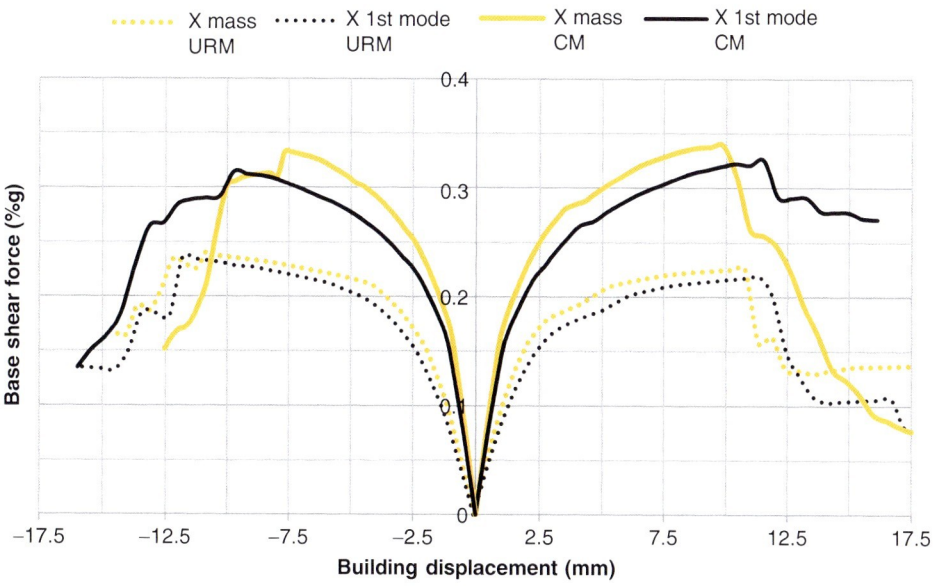

Masonry Box Behavior, Fig. 15 Capacity curves in X direction for the unreinforced masonry and confined masonry models

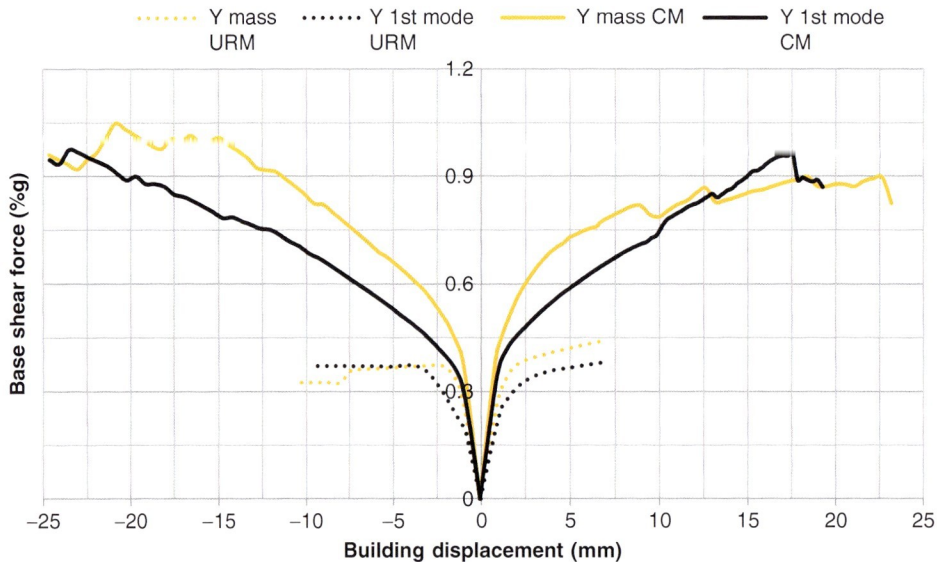

Masonry Box Behavior, Fig. 16 Capacity curves in Y direction for the unreinforced masonry and confined masonry models

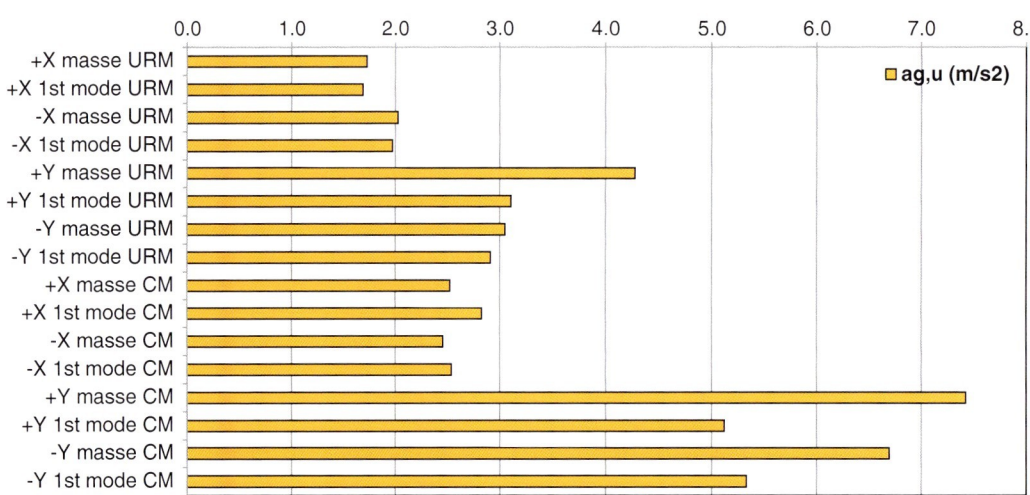

Masonry Box Behavior, Fig. 17 Evaluation of the maximum acceleration capacity of the building

The supply of box behavior needs also to be considered when retrofitting existing masonry buildings with flexible timber floors, e.g., through the adoption of a multi-plank system for the floors. The development of solutions to provide box behavior to social housing buildings in developing countries is also a problem that requires study. This is, for example, the case of traditional adobe masonry constructions in mountainous regions of Peru, which are usually covered with lightweight wooden roofs poorly connected to the adobe walls.

Summary

Masonry box behavior is a feature of masonry structures that present floors significantly rigid and well connected to the walls. This is the case of new masonry buildings constructed with a floor system of ribbed or reinforced concrete slab bonded to the walls by a ring beam, unlike existing buildings with timber floors. However, for this last case, a retrofit solution by stiffening the floor system and providing adequate wall-to-floor connections allows also a box behavior.

Nevertheless, the concept of box behavior is mainly applicable to current unreinforced and confined masonry buildings. The box action is a key assumption of macroelement methods developed for the seismic assessment of masonry buildings, when considering that the walls in a given direction respond jointly under lateral loading. Under this hypothesis, the seismic response of a building can be obtained through nonlinear static (pushover) analysis, i.e., submitting the building to an incremental static loading and considering the inelastic stage of behavior.

The pushover analysis provides the capacity curve (base shear force vs. displacement of the centroid of the roof slab) of the building, whose evolution depends on the sequence of cracking and failure events verified for the structure. The capacity curve is the reference to proceed with a performance-based seismic assessment, which is commonly based on the comparison of the displacement capacity and demand. Under this approach, the masonry box behavior normally allows the building to perform better when subjected to earthquakes.

This entry provides a review of current masonry typologies featuring a box behavior, and is explained with application to the seismic assessment of a dwelling house.

Cross-References

▶ Behavior Factor and Ductility
▶ European Structural Design Codes: Seismic Actions
▶ Masonry Components
▶ Masonry Modeling
▶ Masonry Structures: Overview
▶ Performance-Based Design Procedure for Structures with Magneto-Rheological Dampers
▶ Response Spectrum Analysis of Structures Subjected to Seismic Actions

References

Brzev S (2007) Earthquake-resistant confined masonry construction. National Information Center of Earthquake Engineering (NICEE), Kanpur

Brzev S, Astroza M, Moroni MO (2010) Performance of confined masonry buildings in the February 27, 2010 Chile earthquake. EERI report, Earthquake Engineering Research Institute, Oakland

Calderini C, Cattari S, Lagomarsino S (2008) Numerical investigations on the seismic behaviour of confined masonry walls. In: Proceedings of the 2008 seismic engineering conference commemorating the 1908 Messina and Reggio Calabria earthquake. AIP Conference Proceedings 1020, New York, pp 816–823

Caliò I, Marletta M, Pantò B (2012) A new discrete element model for the evaluation of the seismic behaviour of unreinforced masonry buildings. Eng Struct 40:327–338

Calliari R, Manzini CF, Morandi P, Magenes G, Remino M (2010) User manual of ANDILWall, version 2.5. ANDIL Assolaterizi, Rome

EN 1996-1-1 (2005) Eurocode 6: design of masonry structures – part 1–1: general rules for reinforced and unreinforced masonry structures. European Committee for Standardization, Brussels

EN 1998-1 (2004) Eurocode 8: design of structures for earthquake resistance – part 1: general rules, seismic actions and rules for buildings. European Committee for Standardization, Brussels

Fajfar P, Fischinger M (1988) N2 – a method for nonlinear seismic analysis of regular buildings. In: Proceedings of the 9th world conference on earthquake engineering, vol 5, Tokyo-Kyoto, pp 111–116

Gouveia JP, Lourenço PB (2007) Masonry shear walls subjected to cyclic loading: influence of confinement and horizontal reinforcement. In: Proceedings of the 10th North American Masonry conference (paper 042), St. Louis

Gruppo Sismica (2013) Theoretical manual of the 3DMacro software, beta version. Gruppo Sismica, Catania

IBC (2008) Italian building code, Ministerial Decree dated of 14-01-2008. Ministero delle Infrastrutture e dei Trasporti, Rome

Jäger W, Hirsch R, Masou R (2010) Product and system development in masonry construction under requirements of sustainable construction. In: Proceedings of the 8th international Masonry conference (keynote speech), Dresden (CD-ROM)

Lagomarsino S, Penna A, Galasco A, Cattari S (2009) User guide of TreMuri (Seismic analysis program for 3D masonry buildings), version 1.7.34. University of Genoa

Lourenço PB (2002) Computations of historical masonry constructions. Prog Struct Eng Mater 4(3):301–319

Lourenço PB, Vasconcelos G, Gouveia JP, Medeiros P, Marques R (2008) CBloco: handbook of structural design. Cerâmica Vale da Gândara SA, Viseu

Magenes G (2006) Masonry building design in seismic areas: recent experiences and prospects from a European standpoint. In: Proceedings of the 1st European conference on earthquake engineering and seismology (keynote k9), Geneva

Magenes G, Calvi GM, Kingsley R (1995) Seismic testing of a full-scale, two-story masonry building: test procedure and measured experimental response. In: Experimental and numerical investigation on a brick masonry building prototype: numerical prediction of the experiment. Report 3.0. Gruppo Nazionale per la Difesa dai Terremoti, Pavia

Marques R, Lourenço PB (2011) Possibilities and comparison of structural component models for the seismic assessment of modern unreinforced masonry buildings. Comput Struct 89(21–22):2079–2091

Marques R, Lourenço PB (2013) A model for pushover analysis of confined masonry structures: implementation and validation. Bull Earthquake Eng. 11(6): 2133–2150 doi:10.1007/s10518-013-9497-5

Parsekian GA, Hamid AA, Drysdale RG (2012) Behavior and design of structural masonry. EdUFSCar-Editora da Universidade Federal de São Carlos, São Paulo

Saiidi M, Sozen MA (1981) Simple nonlinear seismic analysis of R/C structures. ASCE J Struct Div 107(5):937–953

San Bartolomé A (1994) Masonry buildings: seismic behavior and structural design. Fondo Editorial, Pontificia Universedad Católica del Perú, Lima

Sullivan TJ, Priestley MJN, Calvi GM (eds) (2012) A model code for the displacement-based seismic design of structures, DBD12. IUSS Press, Pavia

Tomaževič M (1978) The computer program POR. Report ZRMK. Institute for Testing and Research in Materials and Structures, Ljubljana

Tomaževič M (1999) Earthquake-resistant design of masonry buildings, vol 1, Series on innovation in structures and construction. Imperial College Press, London

Touliatos PG (1996) Seismic behaviour of traditionally-built constructions: repair and strengthening. In: Courses and lectures-international centre for mechanical sciences. Springer, New York, pp 57–70

Turnšek V, Čačovič F (1970) Some experimental results on the strength of brick masonry walls. In: Proceedings of the 2nd international brick Masonry conference. British Masonry Society, Stoke-on-Trent, pp 149–156

Masonry Components

Graça Vasconcelos
Department of Civil Engineering, ISISE, University of Minho, Guimarães, Portugal

Synonyms

Masonry materials; Mechanical behavior

Introduction

Masonry is a nonhomogeneous material, composed of units and mortar, which can be of different types, with distinct mechanical properties. The design of both masonry units and mortar is based on the role of the walls in the building. Load-bearing walls relate to structural elements that bear mainly vertical loads but can serve also to resist to horizontal loads. When a structural masonry building is submitted to in-plane and out-of-plane loadings induced by an earthquake, for example, the masonry walls are the structural elements that ensure the global stability of the building. This means that the walls should have adequate mechanical properties that enable them to resist to different combinations of compressive, shear, and tensile stresses. The boundary conditions influence the resisting mechanisms of the structural walls under in-plane loading, and in a building, the connection at the intersection walls is of paramount importance for the out-of-plane resisting mechanism. However, it is well established that the masonry mechanical properties are also relevant for the global mechanical performance of the structural masonry walls. Masonry units for load-bearing walls are usually laid so that their perforations are vertically oriented, whereas for partition walls, brick units with horizontal perforation are mostly adopted.

As a composite material, the mechanical properties of masonry under different loading configurations depend on the properties of masonry components. The unit has an important contribution for the resisting mechanisms of masonry particularly in case of resisting mechanisms that are associated to crushing and tensile cracking of masonry. On the other hand, mortar joints acting as a plane of weakness on the composite behavior of masonry can control the shear behavior, which is particularly relevant in case of strong unit–weak mortar joint combinations. The mortar joints have also a central role on the flexural behavior in the direction perpendicular to the bed joints, as it is tightly controlled by the tensile bond adherence. In case of experimental characterization, flexural testing of masonry in the direction perpendicular to bed joints can even

Masonry Components, Fig. 1 Example of a masonry unit: concrete block

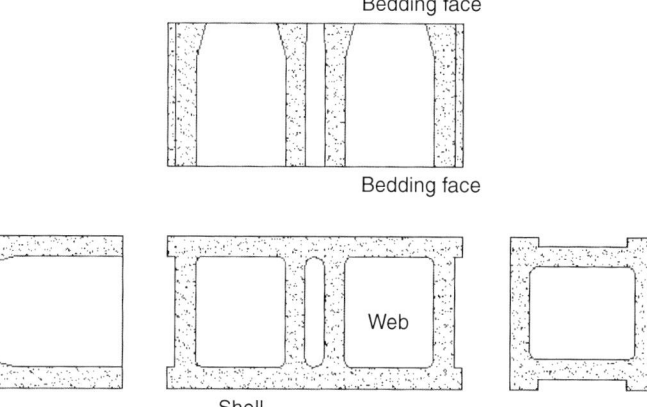

be used as a means of obtaining the tensile bond strength.

Besides the review on masonry material components, it is important to present and discuss the most important mechanical properties of masonry assemblages, together with a discussion of the main parameters affecting the seismic performance of masonry under distinct loading configurations. Understanding the response of the basic masonry materials is essential for interpreting the seismic response of masonry structural systems.

Masonry Units

Masonry is a composite material composed of masonry units with a regular arrangement that are connected with mortar commonly at horizontal bed and vertical head joints. The interface between units and mortar represents in general an important role on the mechanical behavior of the composite material submitted to distinct types of loading.

The masonry units represent the fundamental material for the formation of the main body of the masonry structural element and can be made of distinct raw materials, namely, clay, mud, concrete, calcium silicate, and stone. However, the clay and concrete are far the most common raw materials used in structural masonry units.

The masonry units have commonly a rough rectangular shape, and the dimensions are defined generally by the (length) × (height) × (width) and are laid usually according to the larger dimension (length). The length and height of the masonry units are usually multiples of 200 mm (nominal dimensions), including the 10 mm for the mortar thickness so that modularity of the structural elements can be achieved. The modularity is an important characteristic of masonry to make the construction technology and geometrical implementation of the structural elements (walls) with openings easier. The external vertical surface of the masonry units is known as the shell of the unit, and the walls perpendicular to the face are the webs of the units (Fig. 1). The top and bottom faces of the masonry units are known as the bedding areas.

The masonry units can be solid or have vertical and horizontal voids/perforations known as cores with smaller dimension and cells. Generally, solid units can have up to 25 % of perforations in relation to their gross area. For structural purposes, it is more common that concrete or clay bricks should have vertical perforations along their full height. The horizontally perforated brick units are more common for nonstructural purposes, namely, for masonry infill walls typically used in some south European countries. Examples of common masonry units are shown in Fig. 2. Clay units have generally a set of vertical cores with reduced area, whereas the concrete blocks commonly have large hollow cells, as seen in Fig. 2a. In vertical perforated units, the faces are called as face shells connected

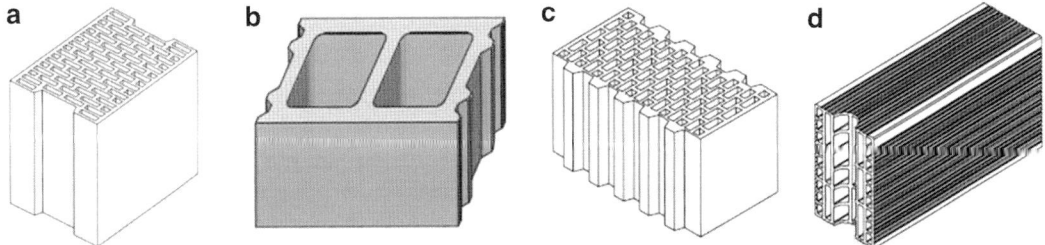

Masonry Components, Fig. 2 Examples on different types of masonry units: (**a**) clay brick unit, (**b**) concrete block unit, (**c**) masonry unit with tongue and groove vertical joints, and (**d**) masonry unit with mortar pocket

by the internal solid parts called as webs. When the perforation does not go through the entire height of the unit, it is called a frog. This has usually a conical shape. Sometimes, the units have indented ends over the full height and are called as frogged ends. When it is intended to have dry vertical joints, without the addition of mortar, tongue and groove or interlocking systems are designed (see Fig. 2c). In this case, the out-of-plane resistance should rely on the combination of the bed joint resistance and on the tongue and groove system. Other times, the geometry of the units foresees the addition of mortar into vertical pockets (see Fig. 2d), where, depending on the geometry, different reinforcing systems can be added to improve the resistance of masonry to in-plane and out-of-plane loads.

The raw materials used in the manufacturing process of clay units are commonly surface clays (recent sedimentary formations) but shales formed from clays under pressure or fire clay, mined at deeper levels. All of these clays are equivalent in terms of silica and alumina compounds with different types of metallic oxides. The surface clays present a great variability, and in some cases, a mixture of clay of distinct locations can be used to reduce the variability. The material used in the concrete units is a dry concrete composed of Portland cement, stone aggregates, and water. It is also common to use other blended cements including blast-furnace cement and fly ash and inert fillers, considered commonly as by-products, aiming at reducing the percentage of Portland cement. In other instances, expanded clay aggregates are used to reduce the weight of the concrete units. Additive such as pozzolanic materials and other workability agents can be also used. The calcium silicate units are composed of sand and hydrated lime.

Besides the raw materials, the production technologies used to produce the clay, concrete masonry, and calcium silicate units are very different. The clay units are normally extruded and fired at different temperatures, whereas the concrete units are produced in molds with a required geometry through a vibration and pressing process. The calcium silicate units are manufactured by pressing the mixture of sand and hydrated lime and then autoclaving them in order to produce a tightly grained unit.

Mechanical Properties

The most relevant mechanical properties of masonry units consist of the compressive strength, elastic modulus, and tensile strength. The mechanical behavior of the masonry assemblages depends greatly on the mechanical properties of the masonry units.

The compressive strength of the unit can be seen as a measure of its quality, and it is important for predicting the compressive strength of masonry assemblages. The compressive strength and elastic modulus can be obtained experimentally from uniaxial compressive tests according to the European standards. The compressive strength is calculated from the loaded area, which is the gross area (length) × (width) of the unit when the units are oriented in the same way as they are intended to be laid in a bed of mortar. In general, an average value is obtained from the experimental results, being possible to calculate

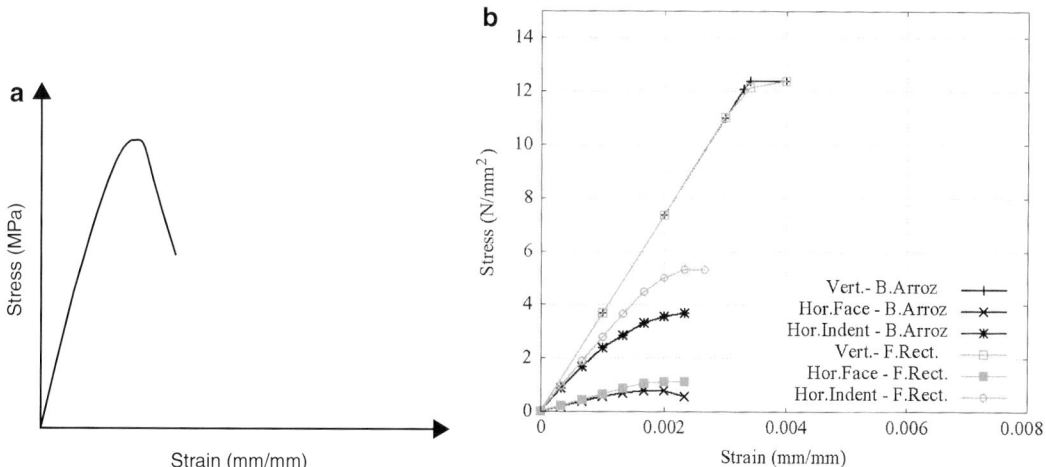

Masonry Components, Fig. 3 Details about compression behavior of masonry units: (**a**) typical stress–strain diagrams and (**b**) effect of loading direction on the compressive strength of units

the characteristic value and the corresponding normalized compressive strength of the masonry unit, f_b, by multiplying the average values by a coefficient taking into account the moisture environment of the curing conditions (oven dry as a reference) and also by the shape factor, accounting for the dimensions of the width and the length. The modulus of elasticity can be calculated as a secant modulus of elasticity between 0 % and 33 % of the compressive strength of masonry unit in the stress–strain diagram obtained from the uniaxial compressive tests. This property can be important if advanced modeling of masonry is required.

In case of modern clay and concrete masonry units, considerably high values of compressive strength can be achieved, being generally higher than the strength requirements for units to be used in seismic zones. It is common to have an average compressive strength higher than 10 MPa. It should be noted that the compressive strength of masonry units is different from the compressive strength of the raw material due to the effect of the shape and geometry of the units. In spite of attempts that have been made to obtain the complete stress–strain diagram of masonry units in compression, it is hard to obtain the post-peak branch of the stress–strain diagrams describing the high-rate crack damage progress of the units after the maximum load is attained (see Fig. 3a).

It should be also noticed that the compressive strength of the masonry units can differ significantly according to the loading direction, namely, in the directions perpendicular and parallel to the laying and in the direction perpendicular to the face. According to the work carried out by Lourenço et al. (2010), it was observed that compressive strength is considerably higher in the direction perpendicular to the bed joints, due to the orientation of vertical perforations, in comparison with when the direction is parallel to the bed joints. A reduction of more than 30 % in the normalized compressive strength obtained in the parallel direction to the bed joints was also found experimentally for concrete units, as seen in Fig. 3b. This difference is attributed mainly to the geometry of the masonry units with distinct arrangements of the internal perforations and cells. This results naturally in the different compressive behavior of masonry under compression for the different loading directions. The failure modes recorded in clay and concrete masonry units confirm its brittle character. In clay units with vertical perforation, it is common to observe cracking and splitting of the internal webs and shells. In case of concrete masonry units, the failure mode has a commonly pyramidal trunk (Gihad et al. 2007; Haach 2009) (see Fig. 4). The first cracks appear vertically in corners of the units.

Masonry Components, Fig. 4 Crack patterns of concrete blocks under uniaxial compression

With increase of the load, there was a tendency of the connection of vertical cracks by a horizontal crack in the upper region of the unit. This behavior can be explained by the lateral restrictions caused by the steel plates in top and bottom of the specimen, generating friction forces. This horizontal crack occurs because the upper part of the units slides over the pyramidal-trunk surface of rupture. In some specimens near the collapse limit, a vertical crack also appeared in the central region of the unit. The brittle failure mode of the masonry units can influence the seismic performance of the masonry under shear walls due to local failures determining the failure mode of the walls (Tomažević et al. 2006).

Mortars for Masonry

Mortar is a component of masonry, and it is used to bond individual masonry units into a composite assemblage. It has a central role in the stress transfer among units when masonry is loaded by promoting and more uniform bearing and avoiding stress concentrations that can result in the premature collapse of masonry. The mortar has also the role of smoothing the irregularities of blocks and accommodating deformations associated with thermal expansion and shrinkage. As pointed out by Vasconcelos and Lourenço (2009), the deformability of masonry is clearly influenced by the material at the bed joints. Very distinct pre-peak behavior was found by considering dry saw unit–mortar interfaces, rough dry joints, lime mortar, or dry clay resulting from sieving granitic soil. The mortar also influences the bond strength (tensile and shear) of the joints.

The mortars for laying masonry units and filling the vertical joints are commonly a combination of Portland cement, lime, sand, and water in specified proportions. The strength of mortars is controlled by the cement, and the workability is controlled mainly by lime and by the grading of the sand, as shown in Fig. 5a. The sand can be natural or artificial resulting from crushing stone. The size and grading of the sand particles influence both the plastic and hardened properties. More graded sand (increased variation of the size and distribution of particles) contributes to improve workability of mortars, which play a major role on the laying process of masonry units. The workability can also be improved through the use of additives like clay fillers and air entrainment. Mortar mixes can be defined by specific proportions of the compounds in volume or in weight of the cement or binder (cement and lime) content. For example, the mortar mix defined by the trace 1:2:9 (cement:lime:sand) by volume means that it has double the volume of lime in relation to cement and has sand with a volume nine times the volume of cement. It can be considered also that the mortar mix has three times more sand in volume that the total binder of the mixture (cement and lime).

Properties of Fresh Mortar

The knowledge about the fresh and hardened properties of mortar is fundamental in ensuring a good performance of masonry walls. The most important properties in the fresh state of mortars are the workability, air content, and setting time (rate of hardening).

The workability of mortars plays an important role on the construction process of masonry

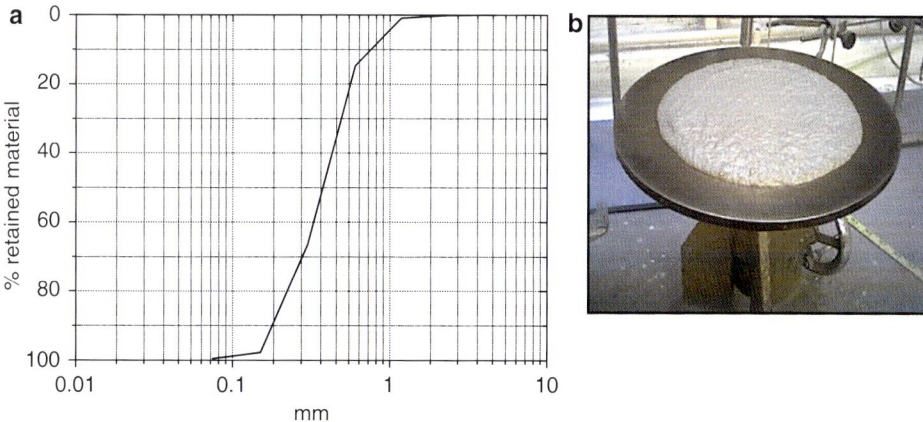

Masonry Components, Fig. 5 (a) Grading curve for sand and (b) measurement of mortar flow

structures. Workability may be considered as one of the most important properties of mortar, and it is related to the process of laying masonry units, and, thus, it influences directly the bricklayer's work. On the other hand, it is important to mention that the quality of the workmanship can influence considerably the mechanical properties of masonry. A workable mortar is easy to adhere to the surface of the trowel, slides off easily, spreads readily, and adheres easily to vertical surfaces. The workability can be improved by the addition of air entrainment agents to the cementitious materials, enhancing also the durability. The addition of lime and the use of an appropriate curve grade for the sand influence also positively the workability of mortar (Haach et al. 2011). The workability is an outcome of several properties such as consistency, plasticity, and cohesion. Given that plasticity and cohesion are difficult to measure, consistency is frequently used as the measure of workability. The consistency is obtained experimentally by the flow table test, as shown in Fig. 5b. An acceptable value for workability for masonry construction ranges from 150 to 180 mm.

The water retention is the property of the mortar that avoids the rapid loss of the mixing water in the masonry units and to the air, and it plays a major role on the bond adherence of the mortar to the masonry units. The ability of the mortar to retain water is important to prevent the excessive stiffening of the mortar before it is used in the laying of masonry units to ensure an adequate hydration of cement and to prevent the water from bleeding out of the mortar. The ability of the mortar to retain water is related to the masonry unit's absorption and should be higher for higher absorption units.

The setting time of fresh mortar relates to the hardening process of mortar. If the setting time is low, the mortar can extrude out of the joints as laying is carried out. If the setting time is high, the mortar placing on the joints can be difficult. The proper hardening rate of the mortar contributes for the adequate bond to the masonry units.

Properties of Hardened Mortar

The most important mechanical properties of hardened mortar are compressive strength and bond. The bond presents a central role, not only in the mechanical performance of masonry under different loading configurations (shear, tension), but it is also important for the durability of masonry.

The bond between mortar and masonry units develops through mechanical interlocking resulting from its adherence. The bond can in certain extent result also from chemical adhesion. Several factors influence the bond between mortar and masonry units, such as properties of masonry units, type of mortar, water–cement ratio, air content, workmanship, workability, and curing conditions. The initial water absorption of the masonry units should be compatible

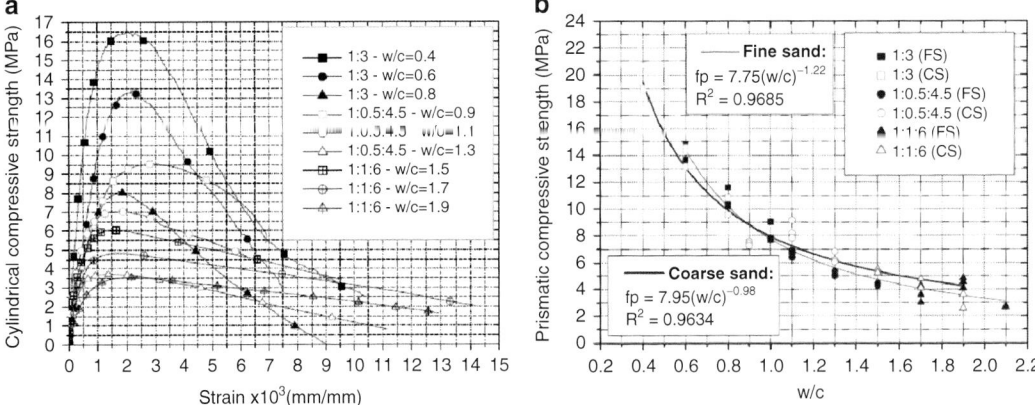

Masonry Components, Fig. 6 Behavior of hardened mortar: (**a**) stress–strain diagrams and influence of the addition of lime in the compressive strength and (**b**) influence of the w/c ratio in the compressive strength

with a good workability and appropriate water retention of the mortar to avoid rapid absorption of the water by the units from the mortar, reducing considerably the water availability for the hardening. The roughness of the masonry units is also important, being enhanced by rougher surfaces of the masonry units. More workable mortars result in better penetration through the voids of masonry units, improving the mechanical interlocking between mortar and masonry units. The additives that are placed in the mortar mix to enhance the workability also contribute to the enhancement of the bond.

Although it is known that the mortar plays a major role in the deformation of masonry under compression, it has also some influence on its compressive strength. More deformable mortar, with lower modulus of elasticity, increases the deformability of masonry under compression. The compressive strength of mortar is also used as an indicator of the workmanship quality, being common to take some mortar specimens during the construction for posterior testing and comparison with the compressive strength required. The compressive strength of mortar is affected by several factors, such as the cement content, the addition of lime, and water–cement ratio. The cement gives the strength to mortar, and if it is replaced by a certain quantity of lime, the compressive strength reduces, as shown in Fig. 6a. Additionally, the compressive strength of mortar is also strongly reduced by the increase in the water–cement ratio (w/c) (Fig. 6b).

Masonry as a Composite Material

The masonry is considered as a composite material composed of units and mortar and unit–mortar interfaces, and its mechanical behavior depends on the mechanical characteristics of the elements and also on its arrangements. The loading configurations to which masonry is subjected depend on the structural element to which it belongs.

Compressive Behavior

The compressive strength of masonry is the primary mechanical property characterizing its structural quality and is fundamental for structural stability in case of load-bearing masonry walls. Compressive behavior is also important when masonry is subjected to lateral loading because the in-plane behavior depends on the compressive properties of masonry, especially if flexural resistance mechanisms predominate (Haach et al. 2011). The finite element numerical analysis of masonry walls based on macro-modeling also requires the data regarding the mechanical behavior of masonry under compression and the key mechanical properties, namely, the compressive strength, elastic modulus, and fracture energy. Masonry is a composite material made of units and mortar, so it has been largely accepted that its failure mechanism and resistance is governed by the interaction between the different components.

In case of hollow or solid units with full mortar bedding and when the mortar has lower compressive strength than the masonry units, the cracking paths and overall behavior are considerably controlled by mechanical properties of mortar and masonry units. As the mortar has generally lower modulus of elasticity than the masonry unit, it exhibits a trend to expand laterally within the mortar joints more than the masonry, being restrained by the masonry units. This interaction results in a triaxial compression state of the mortar and on a compression-lateral tensile state on the masonry units. This stress state results in the vertical cracking of the units, as seen commonly in the experimental testing of masonry. In case of masonry units with face shell mortar bedding, it is common to find vertical cracking along the webs of the units. This is related to the nonuniform stress distribution of stresses along the height of the unit and along the thickness of the masonry and to the principal tensile stresses mainly at the top and bottom of the units (Lourenço et al. 2010).

The typical stress–strain diagram describing the compressive behavior of masonry is shown in Fig. 7. The pre-peak behavior is characterized by nonlinearity beyond approximately 60 % of the peak stress, particularly in concrete masonry. The clay brick masonry tends to present a more linear elastic behavior with nonlinearly close to the peak load, achieving also higher values of the compressive strength. Almost no post-peak is usually recorded, which is associated with the brittle character of unreinforced masonry under compression. However, the post-peak behavior of masonry is dependent on the type of units and also dependent on the mortar used. Concrete masonry specimens built with lime-based mortar present slight lower strength than masonry prisms built with cement-based mortar. Additionally, higher deformations characterized the compressive behavior of masonry built with lower strength mortar. In this case, the ability

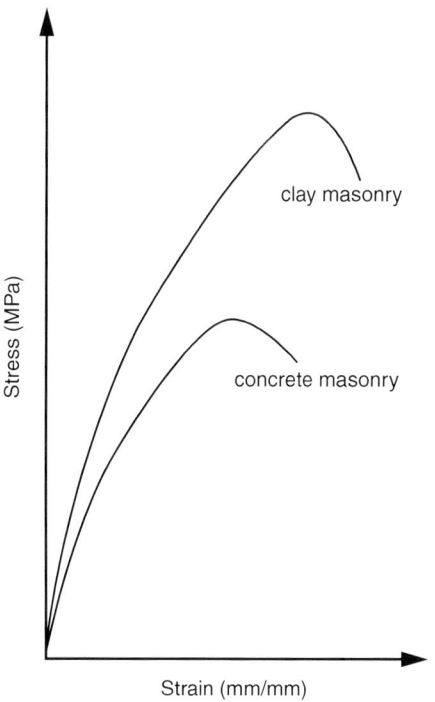

Masonry Components, Fig. 7 Typical stress–strain diagram of concrete masonry under compression

deformation after peak load is higher, enabling more ductile response of masonry under compression (Haach et al. 2014). The compressive strength of masonry is always higher than the compressive strength of mortar and lower than the compressive strength of masonry unit. However, the compressive strength of masonry units takes a central role on the compressive strength of masonry. Higher strength masonry units lead to higher strength of masonry. The relation appears to be linear in concrete block masonry (Drysdale and Hamid 2005).

The direction of compression of masonry units is also an important factor to take into account. Even if, in general, masonry is load in the direction normal to bed joints, in case of masonry beams and flexural walls (out-of-plane loading), the compression in the parallel direction to the bed joints is relevant for their mechanical behavior. The compressive strength in the parallel direction to the bed joints reduces considerably in relation to the compressive strength in the normal direction to the bed joints, particularly in

Masonry Components, Fig. 8 Typical failure patterns that can develop in masonry walls under shear

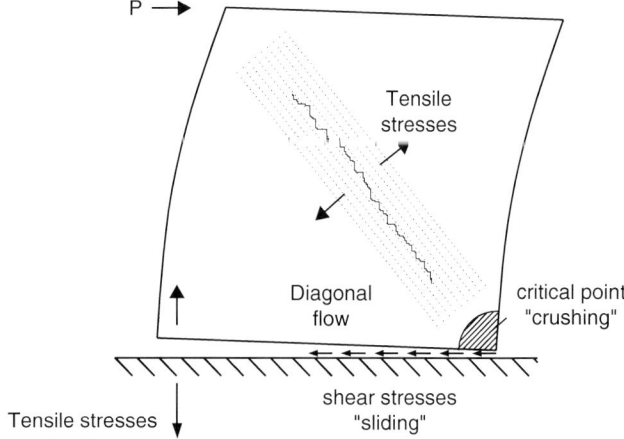

case of hollow units as it depends on the geometry of perforations. In case of clay brick units with vertical perforation, the compressive strength in the parallel direction to the bed joints is about 30 % of the compressive strength of the units in the normal direction to the bed joints, resulting in the lowering of the compressive strength of masonry in this direction. The failure modes are also distinct, being the failure in the parallel direction more ductile.

Masonry Under Shear

The main resisting mechanisms that are characteristic of the response of the masonry walls submitted to combined in-plane loading are shear and flexure, which results in distinct failure modes (Fig. 8). In general, in squat walls, shear resisting mechanism predominates, and in slender walls, the flexural resistance mechanism plays the major role. Low pre-compression load levels are associated to flexural resisting mechanisms, and high pre-compression load levels are in general associated with the development of more dominant shear resisting mechanism. The shear resisting mechanism is associated with diagonal cracks in the alignment of the compressive strut related to the tensile stresses developed in the perpendicular direction of the strut. Diagonal shear cracking can occur with distinct patterns, namely, cracks developing along the unit mortar interfaces or through both unit and mortar interfaces and through masonry units as a combination of joint failure or brick shear–tension splitting. In diagonal cracking along the unit–mortar interfaces, the shear behavior of the bed joints plays an important role on the response of the walls.

Shear Behavior Along the Bed Joints

The influence of mortar joints acting as a plane of weakness on the composite behavior of masonry is particularly relevant in case of strong unit–weak mortar joint combinations. Two basic failure modes can occur at the level of the unit–mortar interface: tensile failure (mode I) associated with stresses acting normal to joints and leading to the separation of the interface and shear failure (mode II) corresponding to a sliding mechanism of the units or shear failure of the mortar joint. The preponderance of one failure mode over another or the combination of various failure modes is essentially related to the orientation of the bed joints with respect to the principal stresses and to the ratio between the principal stresses.

The shear behavior of mortar masonry joints is characterized experimentally based on direct shear tests by following the typical shear test configuration as shown in Fig. 9a. The typical behavior of mortar masonry joints under increasing shear load and constant pre-compression load is presented in Fig. 9b. The general shape of the shear stress–shear displacement is characterized by a sharp initial linear stretch. The peak load is rapidly attained for very small shear displacements. Nonlinear deformations develop in the

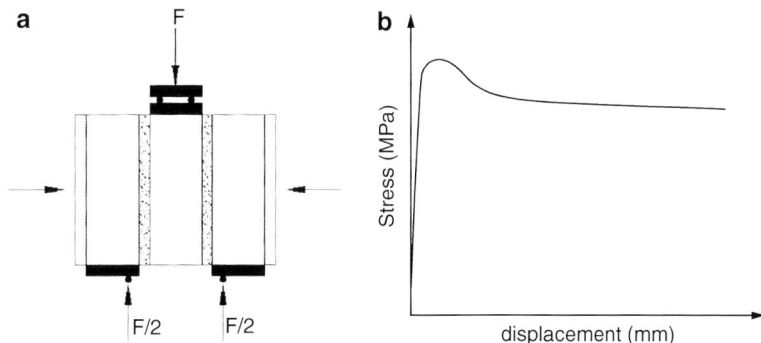

Masonry Components, Fig. 9 Typical shear stress–slipping diagram of mortar masonry joints under shear

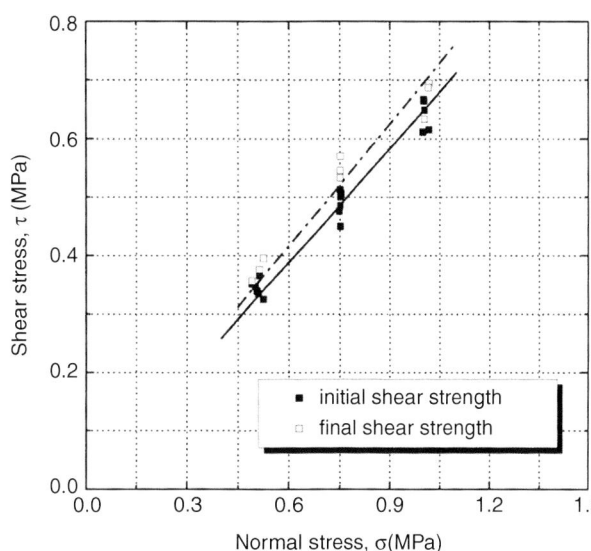

Masonry Components, Fig. 10 Relation between shear stress and normal stress

pre-peak regime only very close to the peak stress. After peak load is attained, there is a softening branch corresponding to progressive reduction of the cohesion of the joint, until reaching a constant dry-friction value. This stabilization is followed by the development of large plastic deformations.

After peak load is attained, there is a softening branch corresponding to progressive reduction of the cohesion of the joint, until reaching a constant dry-friction value. This stabilization is followed by the development of large plastic deformations.

In case of moderate pre-compression stresses, for which the nonlinear behavior of mortar is negligible and the friction resistance takes the central role, the shear resistance of masonry bed joints is linearly dependent on the compressive stress (see Fig. 10) and is given by the Coulomb friction criterion:

$$\tau = c + \mu\sigma \quad (1)$$

where c is the shear strength at zero compressive stress (usually denoted by cohesion) and μ is the friction coefficient or tangent of the friction angle. For dry joints, the cohesion is obviously zero. It should be kept in mind that the failure envelope given by Eq. 1 describes only a local failure of masonry joints and cannot be directly assumed as the shear strength of the walls submitted to in-plane horizontal loads (Mann and Müller 1982; Calvi et al. 1996). In any case, the shear bond strength of masonry joints assumes a major role on the shear resistance when it can be

Masonry Components, Fig. 11 Dilation of shear mortar joints: (**a**) definition and (**b**) effect of vertical pre-compression on the dilation

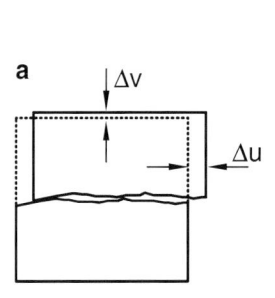
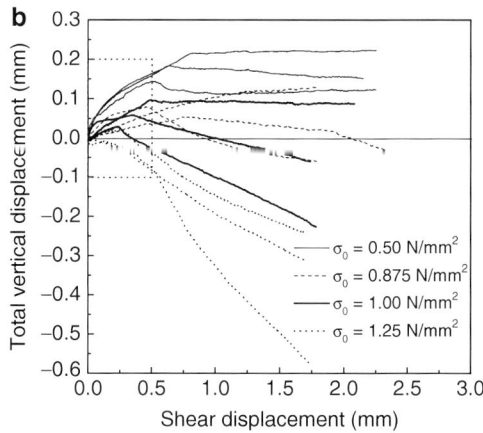

described by the Coulomb friction criterion (diagonal cracking along the unit–mortar interfaces). The shear bond strength of masonry units can be seen as the initial shear strength used to calculate the shear strength according to the Coulomb friction criterion, as suggested by Eurocode 6 (2005).

The strength values, particularly the cohesion, are greatly dependent on the moisture content, porosity of the units, initial rate of absorption of the units, the strength and composition of mortar, as well as the nature of the interface (Amadio and Rajgelj 1991). More plastic mortar enhances shear behavior of joints by promoting better adherence. Binda et al. (1994) pointed out that when strong mortar is considered, the strength of the units can also determine the shear behavior of the joints. In case of hollow concrete masonry, the mortar placed on the internal webs contributes considerably to the increase of the shear strength as it increases the mechanical interlocking. This implies that a wide range of shear strength values may be pointed out for various combinations of units and mortar. Mann and Müller (1982) indicated a mean friction coefficient of approximately 0.65 on brick–mortar assemblages and a cohesion ranging from 0.15 MPa up to 0.25 MPa, depending on the mortar grade. From the results of direct shear tests carried out by Pluijm (1999), the coefficient of internal friction ranges between 0.61 and 1.17, whereas cohesion varies from 0.28 MPa up to 4.76 MPa, depending on different types of units and mortar.

Another important issue regarding shear tests is the dilatant behavior of masonry joints. The dilatancy represents the difference between the variation on the normal displacements of the upper and the lower unit, Δv, as a result of the variation of the shear displacement Δu (see Fig. 11). The opening of the joint is associated with positive dilation, whereas negative values of dilatancy represent the compaction of the joint. The dilatancy of rock joints is mostly controlled by the joint roughness. In conjunction with the cohesion and the friction angle, the dilatancy is also required as a parameter for micro-modeling of masonry. As pointed out by Lourenço (1996), dilatancy in masonry wall structures leads to a significant increase of the shear strength in case of confinement.

In-Plane Tensile Strength

The other approach for the shear resistance of masonry shear walls is based on the Turnšek and Sheppard (1980) criterion, which is based on the assumption that diagonal cracking occurs when the maximum principal stress at the center of the wall reaches the tensile strength of the masonry.

The stress state is calculated by assuming that masonry is an isotropic and homogeneous material, which does not correspond to its actual behavior, since tensile strength is dependent on the orientation of the principal stress regarding the mortar bed joints. For height-to-width ratios (h/l) higher than 1.5, from which walls can be considered as a solid in the Saint-Venant sense, the tensile principal stress can be calculated by Eq. 2:

$$\sigma_t = \sqrt{\left(\frac{\sigma_0}{2}\right)^2 + \tau^2_{max}} - \frac{\sigma_0}{2} \quad (2)$$

being the vertical stress considered as the average stress, σ_0, calculated as the ratio between the compressive load (N) and the area (txl) of the walls $N/(txl)$, and the horizontal stress is negligible. This assumption was confirmed by using photoelastic analysis as reported by Turnšek and Čačovič (1971).

Considering that the maximum shear stress, τ_{max}, assumes a parabolic distribution, the horizontal shear force corresponding to the opening of shear cracks, H_s, is derived by Eq. 2, presenting the following expression:

$$H_s = \frac{f_t \, lt}{b}\sqrt{1 + \frac{\sigma_0}{f_t}} \quad (3)$$

where f_t is taken as the tensile strength of masonry. The variable b takes the value of 1.5 for walls with height-to-length ratios larger than 1.5. In case of height-to-width ratios (h/l) ranging between 1.0 and 1.5, the shear stress distribution deviates from the parabolic shape, and the horizontal normal stress becomes different from zero. In case of unreinforced masonry, this force is considered as the shear resistance when the failure mode corresponds to the cracking involving the cracking of units along the diagonal compression strut.

The in-plane tensile strength of masonry, f_t, can be obtained experimentally through diagonal compression tests by following the recommendation of standard ASTM E519 (2002). The tensile strength of masonry is calculated through Eq. 4, assuming that in the center of the panel, a pure shear stress state develops corresponding to the tensile strength to the maximum principal stress given by Eq. 4:

$$f_t = 0.707 \times \frac{P}{A} \quad (4)$$

where P is the vertical load applied and A is the horizontal cross section of the specimens. The shear deformation is calculated based on Eq. 5, where ΔH and ΔV are the deformation measured along the compression and tension diagonals and g is the width of the diagonal of the panel. The shear modulus is calculated by the ratio between the shear stress and the shear deformation (Eq. 6):

$$\gamma = \frac{\Delta V + \Delta H}{g} \quad (5)$$

$$G = \frac{\tau}{\gamma} \quad (6)$$

The typical failure mode found in current modern unreinforced masonry composed of regular units and submitted to diagonal compression load results from the opening of a stair-stepped crack along the unit–mortar interface developing in the direction of load. The crack is developed in the perpendicular direction to the tensile stresses, which means that it appears when the tensile stress in masonry is reached. The failure of unreinforced masonry occurs suddenly in very brittle style (see Fig. 12a).

According to Haach (2009), the non-filling of vertical joints appears not to significantly influence the crack patterns and failure modes of unreinforced masonry, even if it can clearly influence the shear strength of masonry. The mortar type also influences the tensile strength as it influences the tensile and shear bond strengths of masonry joints, particularly in case of cracks that develop along the unit–mortar interfaces. The tense strength of units influences the values of the tensile strength when the crack passes through the units.

Flexural Strength

The flexural strength of masonry assemblages subjected to out-of-plane bending relates to the resistance of walls submitted to lateral loads from wind, earthquakes, or earth pressures. Depending on the boundary conditions and wall geometry, the bending can develop about vertical axis, about the horizontal axis, or about both directions. Thus, the tensile strength is referred to the

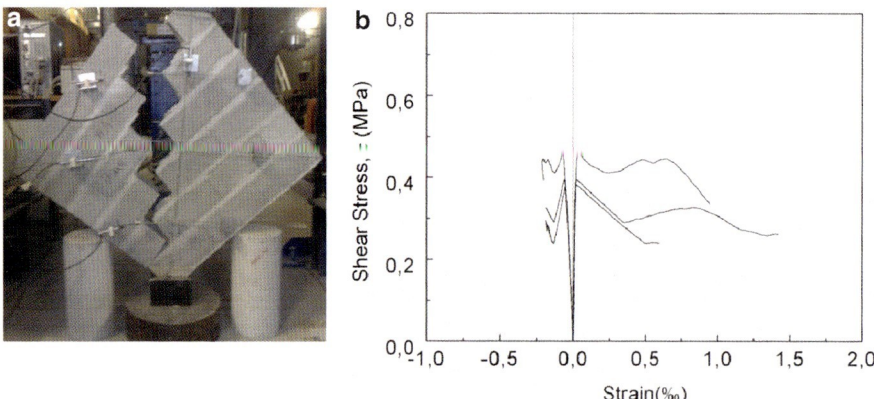

Masonry Components, Fig. 12 Details of diagonal compression tests on unreinforced masonry: (**a**) failure patterns and (**b**) typical shear stress–strain diagrams (negative values correspond to vertical strains)

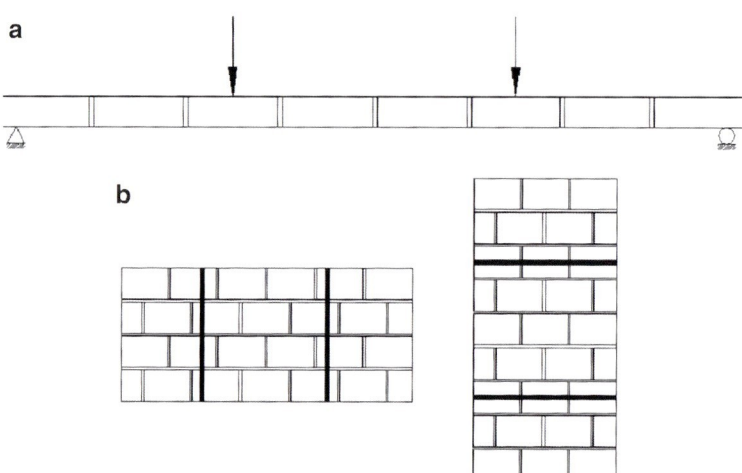

Masonry Components, Fig. 13 Loading configuration for the experimental determination of the flexural strength of masonry: (**a**) geral scheme and (**b**) bending parallel and in the normal direction to the bed joints

direction of the tension that can develop in the direction normal to the bed joints, f_{tn}, or in the direction parallel to the bed joints, f_{tp}.

The flexural strength of masonry units can be obtained experimentally according to EN 1052-2 (1999), by considering a four-point load testing configuration (see Fig. 13a) being the load applied typically according to the scheme as shown in Fig. 13b to obtain the flexural strength in the parallel and perpendicular direction to the bed joints.

The unreinforced masonry under flexure is characterized by a very brittle behavior, which is associated with the localized central crack involving the failure of the unit–mortar interface and the units (see Fig. 14a). When the flexure develops in the normal direction to the bed joints, usually the crack patterns develop along a bed joint (de-bonding of the mortar from the masonry unit). In flexure parallel to the bed joints, the usually observed crack patterns are (1) stepped cracks along the unit–mortar interface, when masonry is made with strong units and weak mortar joints, and (2) cracks passing through head joints and masonry units. The flexural strength in the perpendicular direction to the bed joints can be also taken as the tensile bond strength of mortar joints. The typical force–displacement diagram relating to the vertical load applied and the maximum displacement measured at midspan, presented in Fig. 14b, confirms the brittle nature of masonry under flexure.

 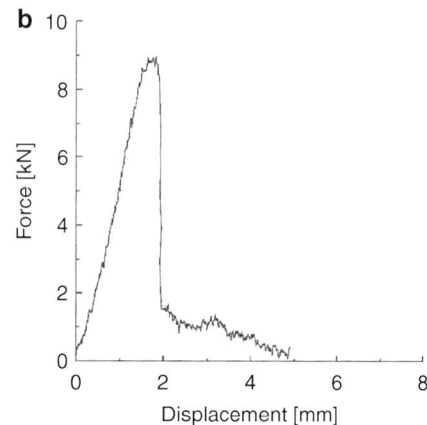

Masonry Components, Fig. 14 Details about the flexure behavior of unreinforced masonry: (**a**) crack pattern in flexure in the normal direction to the bed joints and (**b**) typical force–displacement diagram (direction parallel to bed joints)

After peak load has been attained, there is an abrupt reduction of the bearing capacity, meaning that the masonry loses almost all resistance with no increment of displacement. The pre-peak regime is characterized by an elastic range with only a small nonlinearity very close to the peak load.

The flexural strength depends on the type of mortar, especially on the resistance and tensile bond strength. Also here the workability of mortar plays a major role as the tensile bond strength depends on the adequate adherence of the mortar to the unit. The flexural strength depends also on the tensile strength of the masonry units, particularly when flexure develops in the parallel direction to the bed joints.

Brief Code Considerations

The compressive strength of masonry can be estimated through empirical formulas generally based on the results of experimental tests. European masonry code (Eurocode 6 2005) proposes Eq. 7 to estimate the compressive strength of masonry:

$$f_k = k f_b^{0.7} f_m^{0.3} \qquad (7)$$

where k depends on the type and shape of units and mortar at bed joints, f_b is the normalized compressive strength of the unit, and f_m is the characteristic compressive strength of mortar. For hollow clay units of group 2 and general-purpose mortar, k is 0.45.

The modulus of elasticity of masonry can be determined based on the experimental results, generally by taking the tangent value at 1/3 of the compressive strength of masonry in the stress–strain diagrams or by considering the secant values in a range between 0.1 and 0.4 of the compressive strength. It can be also estimated from the compressive strength of masonry. According to Eurocode 6 (2005), the elastic modulus can be obtained from Eq. 8:

$$E = k_E f_k \qquad (8)$$

where k_E is recommended to be 1,000.

On the other hand, the values of shear modulus, G, used, for example, in the calculation of the lateral stiffness of masonry walls, can be estimated by multiplying the modulus of elasticity by 0.4 (Eurocode 6 2005).

In terms of shear, the Eurocode 6 (2005) suggests that the shear strength of masonry is calculated through Eq. 9:

$$f_{vk} = f_{vk0} + 0.4 \sigma_d \qquad (9)$$

where f_{vk0} is the characteristic initial strength of masonry, obtained for zero compressive stress, and σ_d is the average normal stress.

The value of the characteristic shear stress should not be higher than $0.065 f_b$ (f_b is the normalized compressive strength of masonry units) neither exceed f_{vlt}, which should be defined in the National Annex.

The values of the initial shear strength given in Eurocode 6 (2005) depend on the type of unit (clay, calcium silicate, concrete, stone) and on the type of mortar (general-purpose mortar, lightweight mortar, or thin-layer mortar).

The characteristic flexural strength of masonry can be obtained by experimental tests or alternatively through the values suggested in Eurocode 6 (2005), depending on the type of unit and type of mortar. Typically, the characteristic flexural strength in the direction normal to the bed joints ranges from 0.1 to 0.2 MPa, and the characteristic flexural strength in the direction parallel to the bed joints ranges from 0.2 to 0.4 MPa.

There are requirements for the masonry units and mortar to be used in earthquake-prone regions. In case of masonry units, the normalized compressive strength should be higher than 5 MPa in the normal direction to the bed joints and 2.0 MPa in the parallel direction to the bed joints. The recommended values for the compressive strength of mortar are 5.0 MPa for unreinforced masonry and 10.0 MPa for reinforced masonry (Eurocode 8 2004).

Summary

In this section, a review on the masonry components and on mechanical properties of masonry under distinct loading configurations has been made. Additionally, some code considerations about the design mechanical properties of masonry are provided.

The masonry units have a wide range of possibilities either from the viewpoint of raw materials or from geometrical configurations. The geometrical configuration should comply with thermal and mechanical requirements to optimize the performance both from the mechanical and physical point of view. Besides the strength, it is required that the mortar should present an adequate workability and water retention ability so that an adequate mechanical behavior of masonry is attained. These properties play a major role on the bond strength of masonry (tensile and shear bond strength), which in turn have an important contribution on the shear and flexural resistance of masonry.

The compressive strength of masonry is clearly dependent on the mechanical properties of the components, the masonry unit being more important than mortar, which contributes mainly to the deformability of masonry. Depending on the failure patterns, which are dependent on the level of vertical load applied and boundary conditions, the shear response of masonry walls under in-plane loading can be largely dependent on the shear bond strength (failure load described by the Coulomb friction criterion) or on the in-plane tensile strength of masonry (failure load described by the Turnšek and Čačovič criterion). The bond strength of the masonry joints and the in-plane tensile strength of masonry play an important role on the flexural resistance of masonry. The tensile strength is more important than the flexural strength in the direction parallel to the bed joints and the tensile bond strength of primary importance in case of the flexural strength in the normal direction to the bed joints.

Cross-References

▶ Masonry Structures: Overview

References

Amadio C, Rajgelj S (1991) *Shear behavior of brick-mortar joints*. Masonry Int 5(1):19–22

ASTM E 519–02 (2002) Standard test method for diagonal tension (shear) in masonry assemblages. Annual book of ASTM standards, American Society for Testing and Materials, West Conshohocken, USA

Binda L, Fontana A, Mirabella G (1994) Mechanical behavior and stress distribution in multiple-leaf stone walls. In: Proceedings of 10th international brick block masonry conference, Calgary, pp 51–59

Calvi GM, Kingsley GR, Magenes G (1996) Testing masonry structures for seismic assessment,

Earthquake Spectra. J Earthquake Eng Res Inst 12(1): 145–162
Drysdale R, Hamid AA (2005) Masonry structures: behavior and design, Canadian edition. Canada Masonry Design Centre, Mississauga. ISBN 0-9737209-0-5
EN 1052-2 (1999) Methods of test for masonry – Part 2: determination of flexural strength, CEN - European Committee for Standardization, Brussels
EN 1996-1-1 (2005) Eurocode 6: design of masonry structures – Part 1–1: general rules for reinforced and unreinforced masonry structures, CEN - European Committee for Standardization, Brussels
EN 1998-1-1 (2004) Eurocode 8: design of structures for earthquake resistance – Part 1: general rules, seismic actions and rules for buildings, CEN - European Committee for Standardization, Brussels
Gihad M, Lourenco PB, Roman HR (2007) Mechanics of hollow concrete block masonry prisms under compression: review and prospects. Cem Concr Compos 29(3):181–192
Haach VG (2009) Development of a design method for reinforced masonry subjected to in-plane loading based on experimental and numerical analysis. PhD thesis, University of Minho, Guimarães, 367 pp
Haach VG, Vasconcelos G, Lourenço PB (2011) Parametric study of masonry walls subjected to in-plane loading through numerical modeling. Eng Struct 33(4):1377–1389
Haach VG, Vasconcelos G, Lourenço PB (2014) Assessment of compressive behavior of concrete masonry prisms partially filled by general purpose mortar. J Mater Civ Eng 26(10):04014068
Lourenço PB (1996) Computational strategies for masonry structures. PhD thesis, Delft University of technology, Delft. ISBN 90-407-1221-2
Lourenço PB, Vasconcelos G, Medeiros P, Gouveia J (2010) Vertically perforated clay brick masonry for load-bearing and non-load-bearing masonry walls. Construct Build Mater 24(11):2317–2330
Mann W, Müller H (1982) Failure shear-stressed masonry – an enlarged theory, tests and application to shear walls, Proc Br Ceram Soc 30:223–235
Pluijm RVD (1999) Out-of-Plane bending of masonry, behavior and strength. PhD thesis, Eindhoven University of Technology
Tomaževič M, Lutman M, Bosiljkov V (2006) Robustness of hollow clay masonry units and seismic behavior of masonry walls. Construct Build Mater 20(10):1028–1039
Turnšek V, Čačovič F (1971) Some experimental results on the strength of brick masonry walls. In: SIBMAC proceedings, London, pp 149–156
Turnšek V, Sheppard P (1980) The shear and flexural resistance of masonry walls. In: International research conference on earthquake engineering, Sophie
Vasconcelos G, Lourenço PB (2009) Experimental characterization of stone masonry in shear and compression. Construct Build Mater 23(11):3337–3345

Masonry Macro-block Analysis

Nuno Mendes
ISISE, University of Minho, Guimarães, Portugal

Synonyms

Macro-block; Macro-element

Introduction

The structural analysis involves the definition of the model and selection of the analysis type. The model should represent the stiffness, the mass, and the loads of the structure. The structures can be represented using simplified models, such as the lumped mass models, and advanced models resorting the finite element method (FEM) and discrete element method (DEM). Depending on the characteristics of the structure, different types of analysis can be used such as limit analysis, linear and nonlinear static analysis, and linear and nonlinear dynamic analysis.

Unreinforced masonry structures present low tensile strength, and the linear analyses seem to not be adequate for assessing their structural behavior. On the other hand, the static and dynamic nonlinear analyses are complex, since they involve large time computational requirements and advanced knowledge of the practitioner. The nonlinear analysis requires advanced knowledge on the material properties, analysis tools, and interpretation of results. The limit analysis with macro-blocks can be assumed as a more practical method in the estimation of maximum load capacity of structure. Furthermore, the limit analysis requires a reduced number of parameters, which is an advantage for the assessment of ancient and historical masonry structures, due to the difficulty in obtaining reliable data.

The observation of the damage in masonry buildings caused by earthquakes in the past has been shown that the masonry structures can be discretized into several macro-blocks and interfaces.

The macro-blocks are portions of a structure with similar material properties and structural behavior, to which the mechanical properties of the material can be assigned or, by simplification, they can be assumed to be infinitely rigid. The interfaces in a macro-modelling represent, in general, the cracks associated to the failure mechanisms. However, in a micro-modelling strategy applied to masonry, in which the units and the joints are individually considered, the interfaces simulate the behavior of the joints. Furthermore, different criteria for the strength parameters of macro-blocks and interfaces can be considered.

The limit analysis (Fig. 1) with macro-blocks is a simplified and powerful structural analysis tool to evaluate the ultimate capacity of masonry structures by static models, involving the equilibrium of the macro-blocks through the limit analysis basic concepts.

Classic Limit Analysis Concepts

In the classic limit analysis using rigid-plastic material behavior, a yield function τ based on stresses is defined, and three states are considered (Fig. 2): (i) if $\tau < 0$, the material presents rigid behavior and the material remains undamaged; (ii) if $\tau = 0$, the material becomes plastic and corresponds to the yield surface; (iii) if $\tau > 0$, the stress state is inadmissible. The points inside or on the yield function correspond to the states in which the stress can be sustained by the material, and the points outside correspond to stress states inadmissible for the material. The classic limit analysis considers that when the stress states are on the yield surface, the material becomes plastic and the flow direction is normal to the yield surface. This assumption is called as associated flow or normality condition. The associated flow implies that the yield surface must be convex and provides the maximum energy dissipation (Nielsen 1999).

A general load on a structure can be multiplied by a load factor δ and increased from zero to a limit (failure load factor δ_F) for which the structure remains the equilibrium condition (equilibrium between internal and external forces) and the yield condition (the stresses must be less than or equal to the material strength),

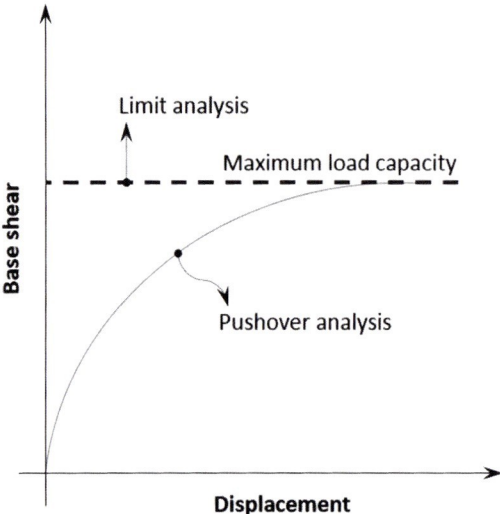

Masonry Macro-block Analysis, Fig. 1 Maximum load capacity obtained from the pushover analysis and limit analysis

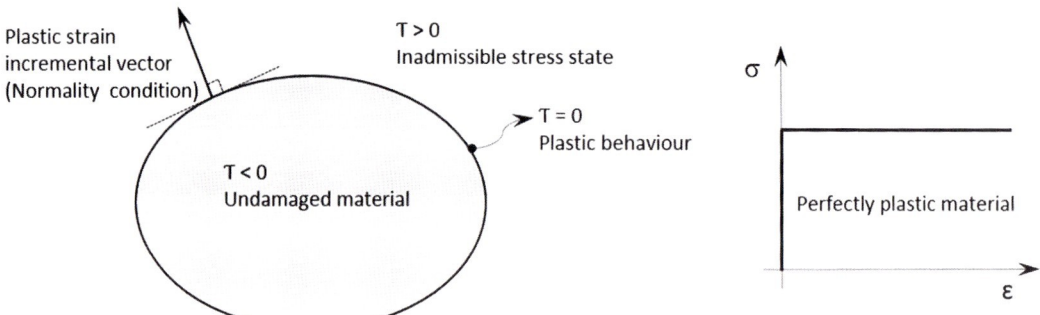

Masonry Macro-block Analysis, Fig. 2 Classic limit analysis theory

Masonry Macro-block Analysis,
Fig. 3 Graphical representation of the limit analysis theorems

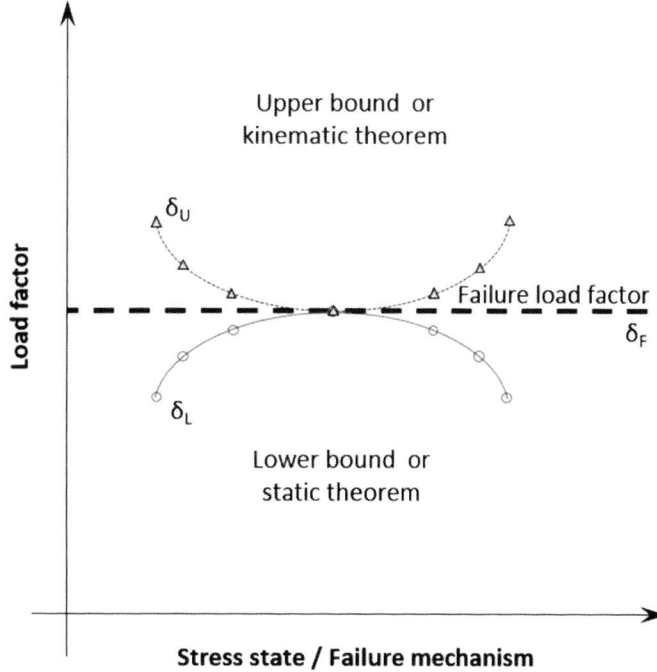

i.e., the structure remains safe. This approach is associated to the static or lower bound theorem of the classic limit analysis, in which the safety factor is the largest of all the statically admissible load factors (δ_L). Furthermore, the classic limit analysis presents the kinematic approach, in which a sufficient number of releases (e.g., plastic hinges) have to occur to transform the structure into a mechanism. Each kinematically admissible mechanism presents an associated load factor, which is, in general, larger or equal to the safety factor. This approach corresponds to the kinematic or upper bound theorem of the classic limit analysis, in which the safety factor is the smallest of all kinematically admissible load factors (δ_U). The largest load factor obtained from the static theorem and the smallest load factor obtained from the kinematic theorem are the same, stating that a structure can present a statically admissible state and be unsafe if a sufficient number of sections reach the yield surface to transform the structure into a mechanism. This particular case corresponds to the uniqueness theorem of the classic limit analysis. The graphical representation of the theorems of classic limit analysis is presented in Fig. 3 and is summarized as follows (Kamenjarzh 1996; Nielsen 1999):

- Static or lower bound theorem: If a load presents a magnitude such that the stress state satisfies the equilibrium and yield conditions, then the structure will not collapse and the load factor (δ_L) is less than or equal to failure load factor (δ_F). In this static approach the failure load factor is determined by searching for the maximum load factor.
- Kinematic or upper bound theorem: If a kinematic admissible mechanism can be found for which the work of the external loads exceeds the internal work, then the structure will collapse and the load factor (δ_U) is greater than or equal to the failure load factor (δ_F). In the kinematic approach the failure load factor is determined by searching for the minimum load factor.
- Uniqueness theorem: If the equilibrium, mechanism, and yield conditions are satisfied, then the load factor obtained from the static and kinematic approach is the same and is equal to the failure load factor (δ_F). Thus, the

failure load factor is determined by equating the load factors of both approaches and using, for example, optimization techniques.

Application of Limit Analysis in Masonry Structures

Heyman (1966) was among the first and the most famous promoters to use the limit analysis based on plasticity theory on arches and other masonry structures. Although the masonry is a quasi-brittle material, the plastic limit analysis was used to assess the structural behavior of masonry structures, assuming the following assumptions:

- Masonry has no tensile strength: Masonry presents low tensile strength and quasi-brittle tensile behavior, mainly in ancient masonry structures built with weak mortar, justifying this assumption.
- Masonry has infinite compressive strength: In general, masonry presents a compressive strength higher than the compressive stresses present in the structure. For this reason, the most cases of collapse of masonry buildings are related to tensile and shear failures. Thus, this assumption is in general accepted and the crushing failure of masonry is not considered.
- Sliding failure is not permitted: Although the sliding between units can sometimes occur at the joints, in general the sliding failure is not the most relevant failure mechanism and this assumption can be accepted.

The failure load assessment of masonry structures based on the limit analysis and the assumptions defined by Heyman (1966) can be carried out using two approaches: (a) static approach and (b) kinematic approach.

In the static approach the resultant of the compressive stresses is plotted at each cross section of the structure. The line containing all the positions of the compressive stress resultants is called by thrust line (Fig. 4a). The thrust line can be determined by analytical methods or graphical methods, such as funicular polygon (Kooharian 1952). The structure is safe when the thrust line is located totally inside of the geometry of structure and the equilibrium with the external loads can be found. In this conditions and according to the lower bound theorem, the applied load is less

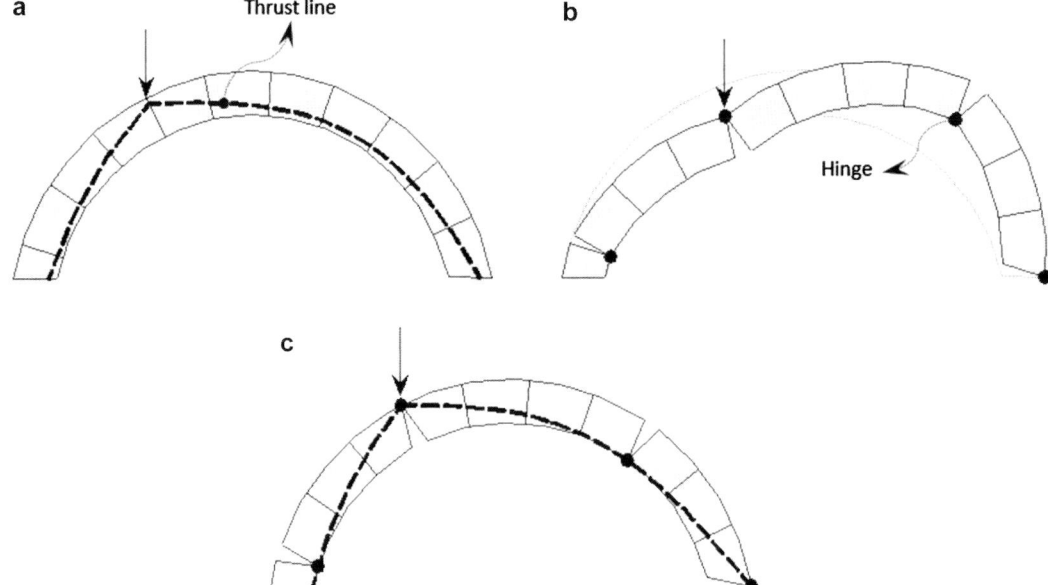

Masonry Macro-block Analysis, Fig. 4 Application of the limit analysis to a masonry arch: (**a**) static approach (lower bound theorem); (*b*) kinematic approach (upper bound theorem); (**c**) uniqueness theorem

than or equal to the failure load. Nevertheless, several thrust lines contained in the geometry of the structure can be found, meaning that the thrust line definition is an indeterminate problem (hyperstatic structure), where a solution can be obtained by using an iterative method. Thus, if the thrust line is found to be totally inside of the geometry of structure that was found, the structure is safe, which does not mean that the structure necessarily works according to this solution. However, nothing can be concluded with respect to the stability if the thrust line is outside of the geometry of structure, unless the infinite possibilities were tried.

With regard to kinematic approach, when the structure presents an enough number of hinges to transform it into a mechanism (Fig. 4b), the load factor obtained by equating the work of the external loads to zero corresponds to an upper bound limit (upper bound theorem). The load obtained from the kinematic approach is greater than or equal to the failure load. Different positions for the hinges on the structure can be adopted, and several mechanisms, and respective load factors, can be found. The mechanism that presents the lowest load factor corresponds to the collapse mechanism. Thus, the kinematic approach presents also several solutions.

Taking into account the results obtained from both approaches, the static approach presents a safe solution of the maximum load capacity of masonry structures. However, for complex masonry structures it can be difficult to predict the position of the thrust line. On the other hand, the kinematic approach can lead to an unsafe estimation of the ultimate load. This problem can be solved by implementing computational tools of optimization and searching by the failure load (minimum load factor).

A particular case can occur when the thrust line becomes tangent to the boundaries of the geometry at an enough number of cross sections to create a mechanism (Fig. 4c). In this condition, the mechanism corresponds to the collapse mechanism, the thrust line is unique, and the load corresponds to the failure load (uniqueness theorem).

The assumptions adopted for the limit analysis of masonry structures are simplifications of its real and complex structural behavior. In fact, the compressive strength of masonry is not infinite, and the structures can present high concentrations of compressive stress. The use of finite compressive strength for masonry is more realistic and can provide a small reduction of the load factor at the collapse. Furthermore and as previously mentioned, the sliding failure can occur, which can be a nonconservative assumption for some particular construction techniques of masonry (e.g., sliding at the interface between the rings of a multi-ring masonry bridge). The consideration of the sliding failure implies the lack of the normality condition (nonassociated flow). Thus, several researchers have been conducting efforts to develop limit analysis considering the crushing, the sliding, and the twisting failures of masonry. For more information on innovate limit analyses of masonry structures, see, e.g., Orduña and Lourenço (2003), Gilbert et al. (2006), and Portioli et al. (2013).

Macro-block Approach

The limit analysis with macro-blocks allows to determine the maximum load capacity of structures by using simple procedures based on kinematic approach and involving the equilibrium of the macro-blocks. A macro-block corresponds to a portion of structures with similar material properties and structural behavior, which can represent the structural element (e.g., piers) or a set of structural elements (e.g., a façade). In this approach, the structure is discretized in several macro-blocks with independent structural behaviors. This type of analysis is a practical tool for assessing of the structural behavior of structures and does not require a high number of parameters for the material properties.

The existing masonry structures present in general local modes under an earthquake, due to the loss of equilibrium of parts of the structure. Thus, the limit analysis using macro-blocks is particularly suitable for evaluating the seismic performance of existing masonry buildings, considering even the in-plane or the out-of plane collapse mechanisms (Fig. 5). It should be noted that the out-of-plane behavior of masonry

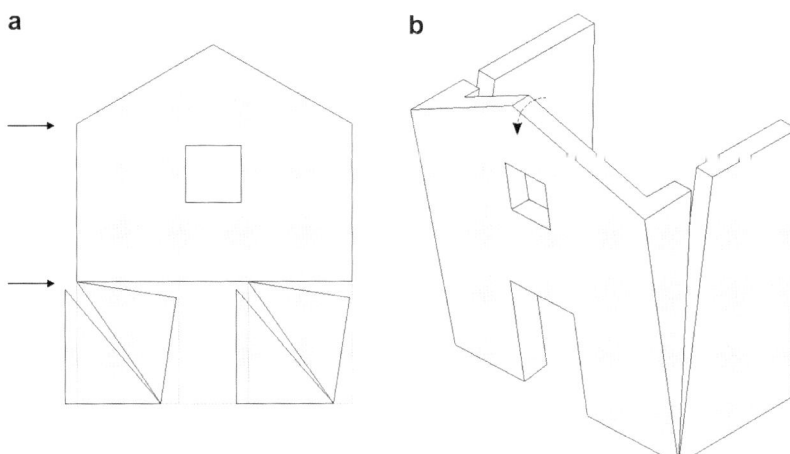

Masonry Macro-block Analysis, Fig. 5 Examples of collapse mechanisms using macro-blocks: (**a**) in-plane collapse of façade; (**b**) out-of-plane collapse of façade

structures is complex and that in general this typology of buildings does not satisfy the assumptions of the typical static nonlinear analysis imposing lateral forces proposed for reinforced concrete buildings (rigid floors, regularity in plan and in elevation, and response governed by the first mode). Furthermore, this type of analysis is also a useful tool for evaluating the efficiency of strengthening techniques applied to existing masonry buildings.

In general, the application of limit analysis with rigid macro-blocks for masonry structures is based on the assumptions proposed by Heyman (1966) (no tension strength, infinite compressive strength, and the sliding cannot occur). However, more realist assumptions can be considered, such as a finite compressive strength.

The macro-block approach involves a first step for selection of mechanisms. The mechanisms can be proposed on the basis of the knowledge obtained from the post-earthquake survey of similar buildings, using the crack patterns obtained from the experimental research, and on the basis of the practitioner experience. The selection of the mechanisms to consider is a fundamental step in this type of analysis and requires a detailed knowledge of the features of the structure, such as the quality of the connections between masonry and the connections between masonry walls and floors, the presence of cracks, and the interaction with adjacent buildings. In fact, a bad evaluation of the possible mechanisms can lead to the non-consideration of the mechanism with the lowest load factor and, consequently, can lead to a failure load higher than the real maximum capacity of the structure. However, the main local collapse mechanisms of masonry building typologies have been compiled in abacus on the form of graphical interpretations schemes, such as the abacus for religious and urban masonry buildings (D'Ayala and Speranza 2002; Lagomarsino and Podesta 2004). Next, the load factor of each considered mechanism is determined by applying the principle of virtual works, and the mechanism that presents the lowest load factor is assumed as the collapse mechanism. Finally, the stability of the structure is analyzed, taking into account the load factor of the collapse mechanism and the requirements defined in the codes (e.g., by comparing the maximum capacity in terms of horizontal acceleration on the structural and the demands in terms of peak ground acceleration (PGA)).

Subsequently, an example of application of the limit analysis with rigid macro-blocks for masonry structures is presented.

Example of Application

Figure 6a presents a possible out-of-plane mechanism of a façade from masonry buildings with

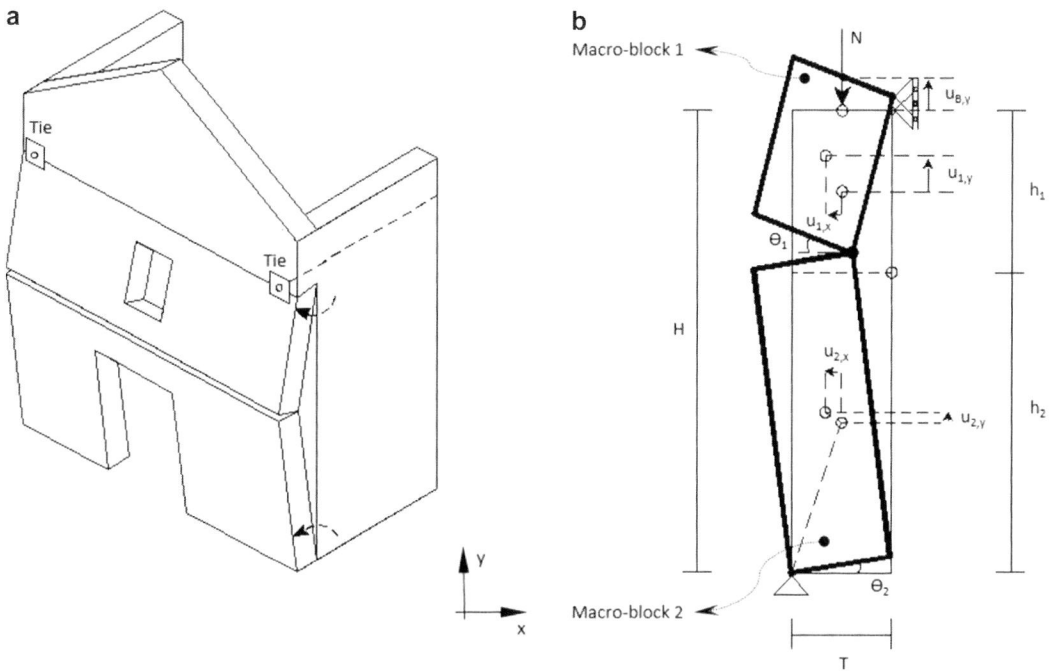

Masonry Macro-block Analysis, Fig. 6 Out-of-plane collapse of tied masonry wall: (**a**) general view; (**b**) scheme of the mechanism

ties at the top and subjected to a seismic action with uniform acceleration acting in the orthogonal direction to the plan of the façade. The structure presents three horizontal cracks at the façade and two cracks at the corners. This local mechanism is composed by three hinges, in which the hinge associated to the crack within the wall is not initially known. Figure 6b shows the scheme of the mechanism in analysis. The height H and thickness T of the masonry wall are equal to 3.50 m and 0.30 m, respectively. The specific mass of the masonry Υ is equal to 20 kN/m³. Furthermore, it assumed that the vertical force at the top of the masonry wall N is equal to 10 kN/m.

The first step corresponds to the determination of the variables of rotation (θ) and displacements of the macro-blocks at the center of mass (u) in the vertical and horizontal directions. Since the position of the crack within the wall is not known, the heights of the macro-blocks are defined as a function of the variable X:

$$h_1 = XH = 3.5H \quad (1)$$

$$h_2 = (1-X)H = (1-3.5)H \quad (2)$$

If a rotation $\theta_1 = 1$ is given to macro-block 1, then the macro-block 2 rotation is:

$$\theta_1 h_1 = \theta_2 h_2$$
$$\theta_2 = \frac{h_1}{h_2}\theta_1$$
$$\theta_2 = \left(\frac{X}{1-X}\right)\theta_1 = \frac{X}{1-X} \quad (3)$$

and the displacements of the macro-blocks and of application of the vertical force N are:

$$u_{1,x} = \frac{h_1}{2}\theta_1 = \frac{XH}{2} = \frac{3.5}{2}X \quad (4)$$

$$u_{1,y} = \frac{T}{2}\theta_1 + T\theta_2 = \frac{T}{2} + T\frac{X}{1-X}$$
$$= 0.15 + 0.30\frac{X}{1-X} \quad (5)$$

$$u_{2,x} = \frac{h_2}{2}\theta_2 = \frac{(1-X)}{2}H\frac{X}{(1-X)} = \frac{XH}{2} = \frac{3.5}{2}X \quad (6)$$

$$u_{2,y} = \frac{T}{2}\theta_2 = \frac{T}{2}\frac{X}{(1-X)} = 0.15\frac{X}{(1-X)} \quad (7)$$

$$u_{B,y} = u_{1,y} = \frac{T}{2} = 0.15 + 0.30\frac{X}{1-X} \quad (8)$$

The forces applied at the macro-blocks due to the self-weight (P) and the inertial forces proportional to mass (F) are:

$$P = \gamma TH = 21 \text{ kN/m} \quad (9)$$

$$P_1 = PX = 21X \quad (10)$$

$$P_2 = P(1-X) = 21(1-X) \quad (11)$$

$$F = \delta P = 21\delta \quad (12)$$

$$F_1 = \delta P_1 = 21X\delta \quad (13)$$

$$F_2 = \delta P_2 = 21(1-X)\delta \quad (14)$$

where δ is the load factor. P and F are forces by meter of wall width.

The application of the principal of virtual works provides:

$$-P_1 u_{1,y} - P_2 u_{2,y} - Nu_{B,y} + F_1 u_{1,x} + F_2 u_{2,x} = 0$$
$$-\left(3.15X + \frac{6.3X^2}{1-X}\right) - (3.15X) - \left(1.5 + \frac{3X}{1-X}\right)$$
$$+ (36.75X^2\delta) + (36.75X\delta - 36.75X^2\delta)$$
$$= 0 - 6.3X - \frac{(3X + 6.3X^2)}{1-X} - 1.5 + 36.75X\delta = 0 \quad (15)$$

Thus, it is possible to obtain the load factor as a function of X:

$$\delta = \frac{6.3X + \frac{(3X + 6.3X^2)}{1-X} + 1.5}{36.75X} \quad (16)$$

The minimum load factor δ can be computed by equating the:

$$\frac{d\delta}{dx} = 0 \quad (17)$$

and a value of X equal to 0.287 is obtained. Replacing the value of X in the Eq. 16, the minimum load factor is equal to 0.50, leading to the conclusion that position of the crack within of façade that cause the minimum load factor is equal to 2.50 m with respect to the base of structure ($h_2 = 2.50$ m).

Summary

The limit analysis with macro-blocks is a simple tool for evaluating the maximum load capacity of masonry structures. The possible mechanisms are proposed and then the respective load factors are determined. The mechanism that presents the minimum load factor corresponds to the collapse, and its load factor is assumed as the failure load. Thus, a careful evaluation of the possible mechanisms is needed, aiming at not excluding any important mechanism and predicting correctly the maximum load capacity of the structure.

The macro-block analysis applied to masonry structures is based on the first assumptions used for the classic limit analysis of arches (no tension strength, infinite compressive strength, the sliding cannot occur), which is acceptable for the majority of the masonry buildings. However, several approaches considering more realistic assumptions have been proposed and implemented in computational tools, aiming at improving the capabilities and the applicability of this type of analysis.

As final conclusion, the limit analysis with macro-blocks is a powerful tool for the seismic assessment of masonry structures, including the historical buildings and the out-of-plane behavior. Furthermore, it allows evaluating the efficiency of strengthening techniques.

Cross-References

▶ Masonry Structures: Overview
▶ Nonlinear Dynamic Seismic Analysis

References

D'Ayala D, Speranza E (2002) An integrated procedure for the assessment of the seismic vulnerability of historic buildings. In: 12th European conference on earthquake engineering, paper no. 561, London, UK

Gilbert M, Casapulla C, Ahmed HM (2006) Limit analysis of masonry block structures with non-associative frictional joints using linear programming. Comput Struct 84:873–887. doi:10.1016/j.compstruc.2006.02.005

Heyman J (1966) The stone skeleton. Int J Solids Struct 2(2):249–279

Kamenjarzh J (1996) Limit analysis of solids and structures. CRC Press, Boca Raton

Kooharian A (1952) Limit analysis of voussoir (segmental) and concrete arches. J Am Concr Inst 24(4):317–328

Lagomarsino S, Podesta S (2004) Seismic vulnerability of ancient churches: I. Damage assessment and emergency planning. Earthquake Spectra 20(2):377–394, http://dx.doi.org/10.1193/1.1737735

Nielsen M (1999) Limit analysis and concrete plasticity, 2nd edn. CRC Press, Boca Raton

Orduña A, Lourenço PB (2003) Cap model for limit analysis and strengthening of masonry structures. J Struct Eng 129(10):1367–1375, http://dx.doi.org/10.1061/(ASCE)0733-9445(2003)129:10(1367)

Portioli F, Casapulla C, Cascini L, D'Aniello M, Landolfo R (2013) Limit analysis by linear programming of 3D masonry structures with associative friction laws and torsion interaction effects. Arch Appl Mech 83(10):1415–1438. doi:10.1007/s00419-013-0755-4

Masonry Modeling

Paulo B. Lourenço
Department of Civil Engineering, University of Minho, ISISE, Guimarães, Portugal

Synonyms

Computational strategies; Finite element modeling; Numerical modeling; Structural analysis

Introduction

This entry deals with modeling of masonry structures. In general, the approach towards the numerical representation of masonry can focus on the micro-modeling of the individual components, units and mortar, or the macro-modeling of masonry as a composite (Rots 1991). Much effort is made today in the link between the micro- and macro-modeling approaches using homogenization techniques.

Masonry joints act as planes of weakness, and the explicit representation of the joints and units in a numerical model seems a logical step towards a rigorous analysis tool (Fig. 1). This kind of analysis is particularly adequate for small structures, subjected to states of stress and strain strongly heterogeneous, and demands the knowledge of each of the constituents of masonry (unit and mortar) as well as the interface. In terms of modeling, all the nonlinear behavior can be concentrated in the joints and in straight potential vertical cracks in the centerline of all units. In general, a higher computational effort ensues, so this approach still has a wider application in research and in small models for localized analysis.

The approach based on the use of averaged constitutive equations seems to be the only one suitable to be employed a large-scale finite element analyses. Two different approaches are illustrated in Fig. 2, one collating experimental data at average level and another from homogenization techniques. A major difference is that homogenization techniques provide continuum average results as a mathematical process that includes the information on the microstructure.

These three different approaches, namely, micro-modeling, macro-modeling, and homogenization, are reviewed next.

Micro-Modeling

Different approaches are possible to represent heterogeneous media, namely, the discrete element method, the discontinuous finite element method (FEM), and limit analysis. The finite element method remains the most used tool for numerical analysis in solid mechanics, and an extension from standard continuum finite elements to represent discrete joints was developed in the early days of nonlinear mechanics.

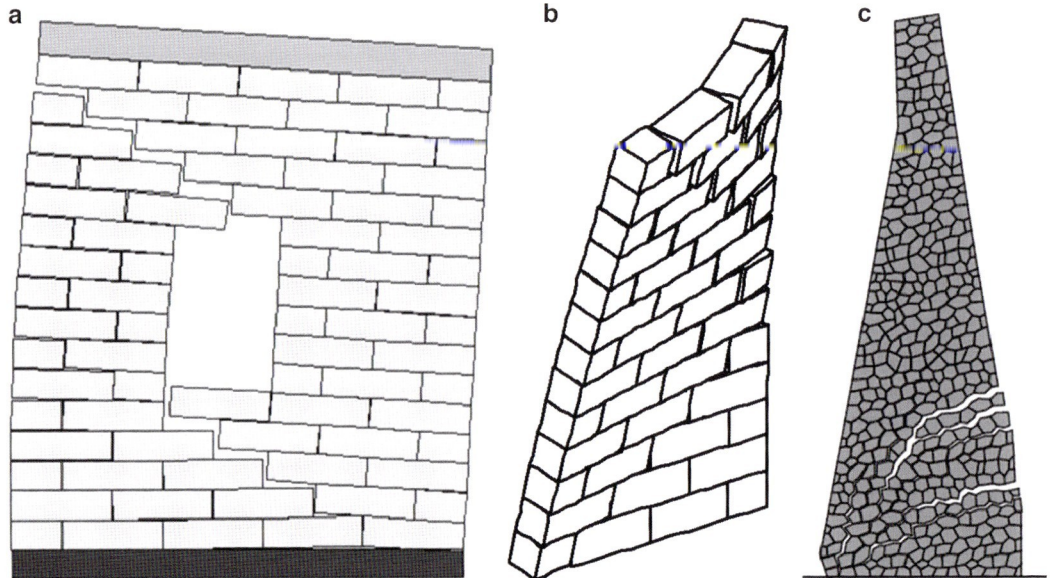

Masonry Modeling, Fig. 1 Examples of cracking at failure using structural micro-modeling: (**a**) shear wall with opening; (**b**) wall subjected to out-of-plane loading; (**c**) retaining wall

Interface elements were initially employed in concrete and rock mechanics, being used since then in a great variety of structural problems. A complete micro-model must include all the failure mechanisms of masonry, namely, cracking of joints, sliding over one head or bed joint, cracking of the units, and crushing of masonry, as done in (Lourenço and Rots 1997) for monotonic loading and in (Oliveira and Lourenço 2004) for cyclic loading.

The typical characteristics of discrete element methods are (a) consideration of rigid or FEM deformable blocks, (b) connection between vertices and sides/faces, (c) interpenetration usually possible, and (d) explicit integration of the equations of motion for the blocks using the real damping coefficient (dynamics) or artificially large (statics). The main advantages are the formulation for large displacements, including contact update, and an independent mesh for each block, in case of deformable blocks. The main disadvantages are the large number of contact points required for accurate representation of interface stresses and a time-consuming analysis, especially for 3D problems. Masonry applications can be found in (Lemos 2007).

Computational limit analysis received far less attention from the technical and scientific community for blocky structures. Still, limit analysis has the advantage of being a simple tool, while having the disadvantages that only collapse load and collapse mechanism can be obtained and loading history can hardly be included. A limit analysis constitutive model for masonry that incorporates nonassociated flow at the joints, tensile, shear, and compressive failure and a novel formulation for torsion is given (Orduña and Lourenço 2005a, b). The salient characteristics of limit analysis are (a) rigid blocks, (b) interpenetration not allowed, and (c) mathematical formulation that leads to an optimization problem (linear or nonlinear). The main advantages of the technique are adequate formulation for design problems (requires a low number of parameters) and fast analysis. The main advantages are that only the collapse load and mechanism can be obtained, tensile strength cannot be included in the analysis, and the introduction of the loading history remains a challenge.

As an example of the possibilities that can be achieved with micro-modeling, a powerful

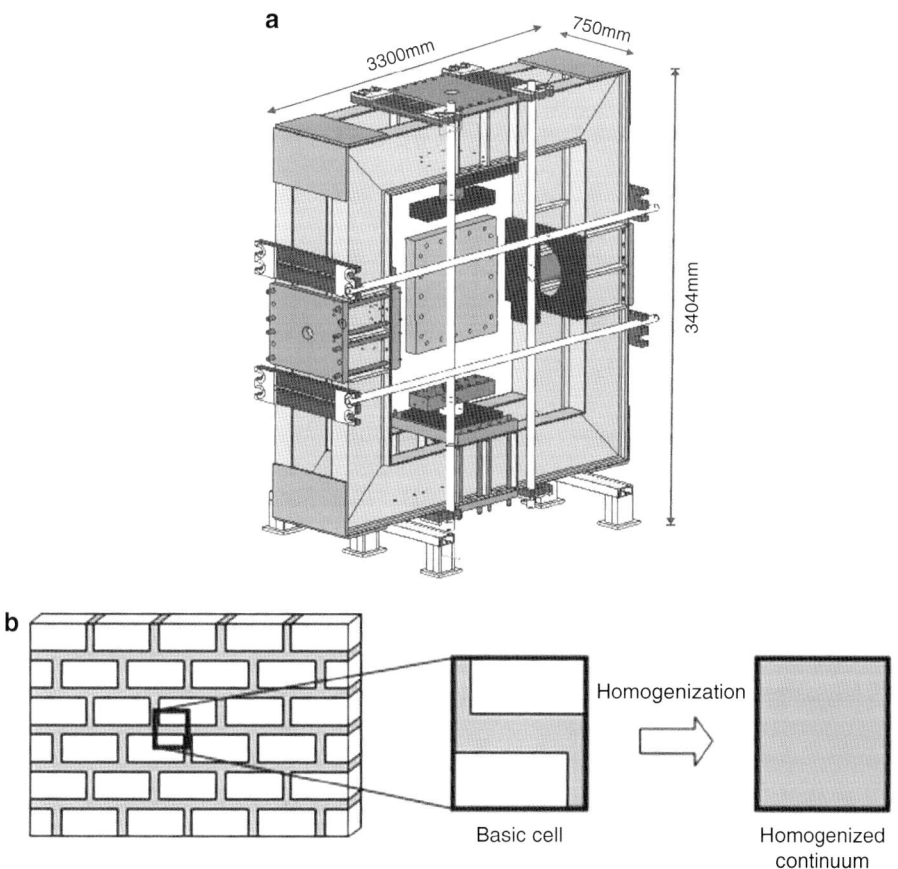

Masonry Modeling, Fig. 2 Constitutive behavior of materials with microstructure: (**a**) collating experimental data; (**b**) mathematical process using geometry and mechanics of components

Masonry Modeling, Fig. 3 Multi-surface interface constitutive model

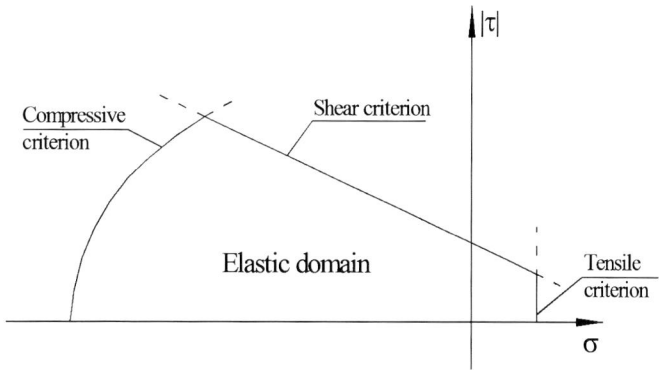

interface model is applied to illustrative examples. A constitutive interface model can be defined by a convex composite yield criterion (see Fig. 3), composed by three individual yield functions, usually with softening included for all modes so that experimental observations can be replicated. The three functions include the usual Coulomb friction criterion for shear, combined with tension and compression cutoffs. The adoption of appropriate inelastic constitutive relations

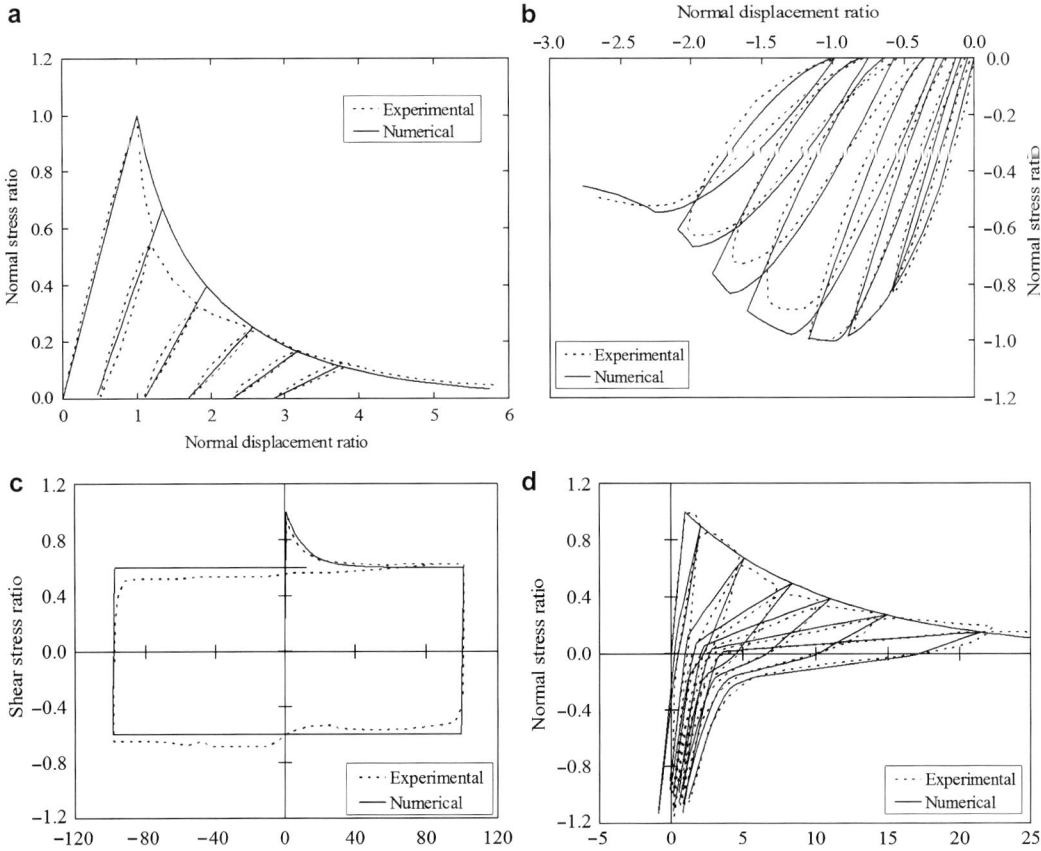

Masonry Modeling, Fig. 4 Comparison between experimental and numerical results for uniaxial testing: (**a**) tension; (**b**) compression; (**c**) shear; (**d**) tension–compression

and evolution rules makes possible to reproduce nonlinear behavior during loading, unloading, and reloading, consisting of exponential softening for tension and shear, and parabolic hardening followed by exponential softening in compression (see Fig. 4). In this case, isotropic and kinematic hardening laws are used, and aspects related to the algorithm can be found in (Oliveira and Lourenço 2004).

The ability of the model to reproduce the main features of structural masonry elements is shown in Fig. 5. Shear walls have been used traditionally to validate constitutive models and to characterize masonry structural solutions, due to the fact that the most demanding demand for a masonry building with box behavior usually leads to a combination of vertical and horizontal loading.

For the first wall considered, it was found that the geometric asymmetry in the microstructure (arrangement of the units) influenced significantly its structural behavior. Note that, depending in the loading direction, the masonry course starts either with a full unit or only with half unit. Figure 5a shows that the monotonic collapse load is 112.0 kN in the LR direction and 90.8 kN in the RL direction, where L indicates left and R indicates right. The cyclic collapse load is 78.7 kN, which represents a loss of about 13 % with respect to the minimum monotonic value but a loss of about 30 % with respect to the maximum monotonic value. This demonstrates the importance of cyclic loading but also the importance of taking into account the microstructure. It is also clear from these analyses that masonry shear walls with diagonal zigzag cracks

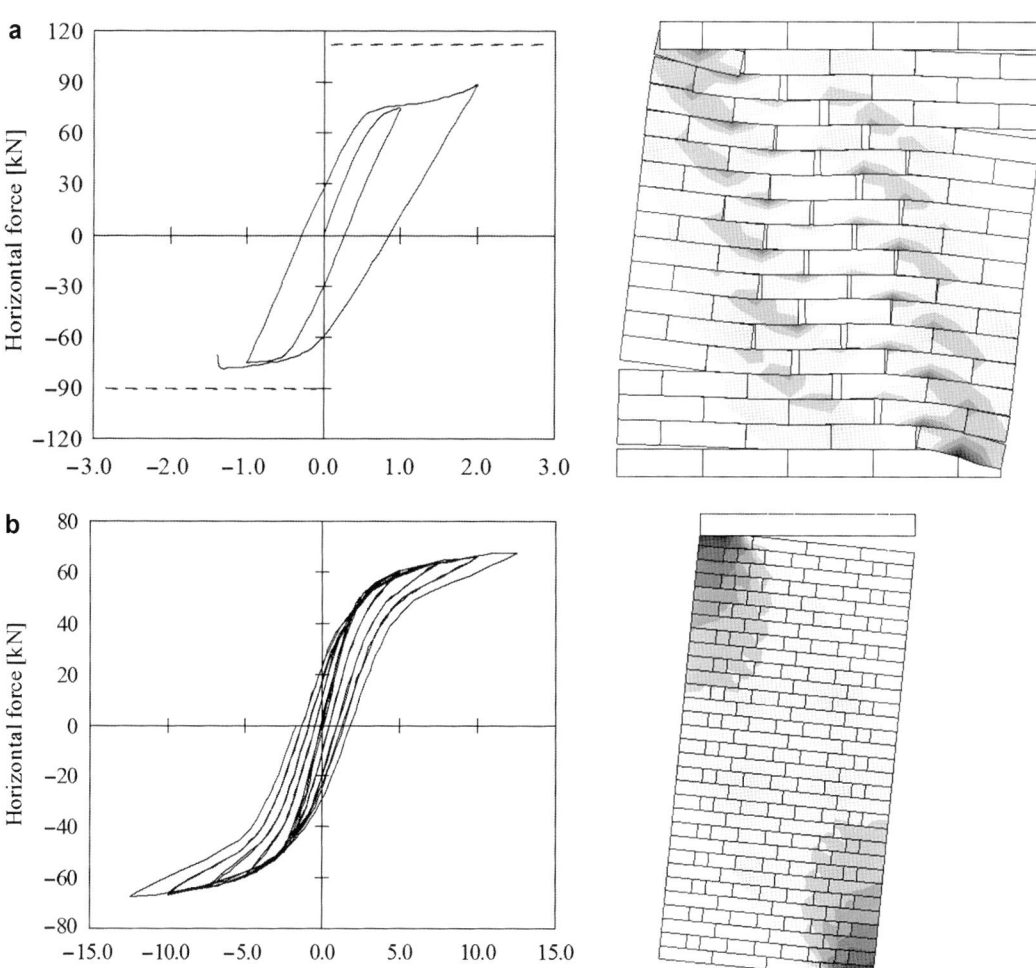

Masonry Modeling, Fig. 5 Results obtained with interface cyclic loading model for shear walls, in terms of force–displacement diagram and failure mode: (**a**) wall failing in shear; (**b**) wall failing in bending

possess an appropriate seismic behavior with respect to energy dissipation.

The second wall considered (Fig. 5b) is a tall wall that simply rocks in both ways. The highly nonlinear shape of the load–displacement curve is essentially due to the opening and subsequent closing, under load reversal, of the top and bottom bed joints. Similar deformed patterns, involving the opening of extreme bed joints, were observed during the experimental test. Numerical results show that the cyclic behavior of the wall is controlled by the opening and closing of the top and bottom bed joints, where damage is mainly concentrated. The model also shows low (static) energy dissipation, which is a consequence of the activated nonlinear mechanism (opening–closing of joints and the absence of a diagonal shear crack). As shown, the failure of this wall is much different from the previous wall, stressing the relevance of the internal structure.

Macro-Modeling

Difficulties of conceiving and implementing macro-models for the analysis of masonry structures arise especially due to the fact that few

Masonry Modeling, Fig. 6 Continuum failure surface for masonry (plane stress representation)

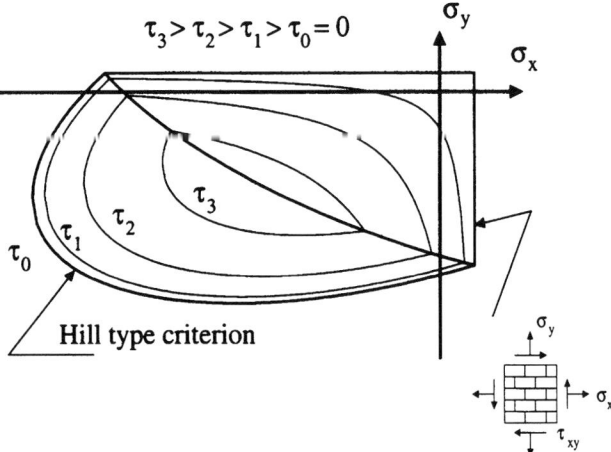

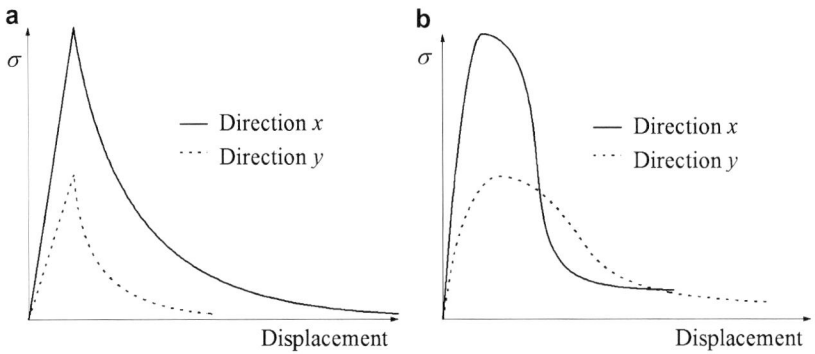

Masonry Modeling, Fig. 7 Behavior of the model for (**a**) tension and (**b**) compression, along two orthogonal directions

experimental results are available (either for pre- and post-peak behavior), but also due to the intrinsic complexity of formulating anisotropic inelastic behavior. Only a reduced number of authors tried to develop specific models for the analysis of masonry structures (Dhanasekar et al. 1985; Seim 1994; Lourenço et al. 1998; Berto et al. 2002; Pelà et al. 2011), always using the finite element method.

Formulations of isotropic quasi-brittle materials behavior consider, generally, different inelastic criteria for tension and compression. The new model introduced in (Lourenço et al. 1997), extended to accommodate shell masonry behavior (Lourenço 2000), combines the advantages of modern plasticity concepts with a powerful representation of anisotropic material behavior, which includes different hardening/softening behaviors along each material axis. The model includes the combination of a Rankine-like yield surface for tension and a Hill-like yield surface in compression (see Fig. 6). The behavior of the model in uniaxial tension and uniaxial compression, along two orthogonal directions, is given in Fig. 7. This composite surface permits to reproduce the results obtained in uniaxial tests, in which different behaviors are obtained along different directions. The application of the model in structural modeling of masonry structures leads to excellent results, both in terms of collapse loads and in terms of reproduced behavior (see Lourenço et al. 1998; Lourenço 2000).

Figure 8 shows the results of modeling another shear wall with an initial vertical pre-compression pressure. The horizontal force

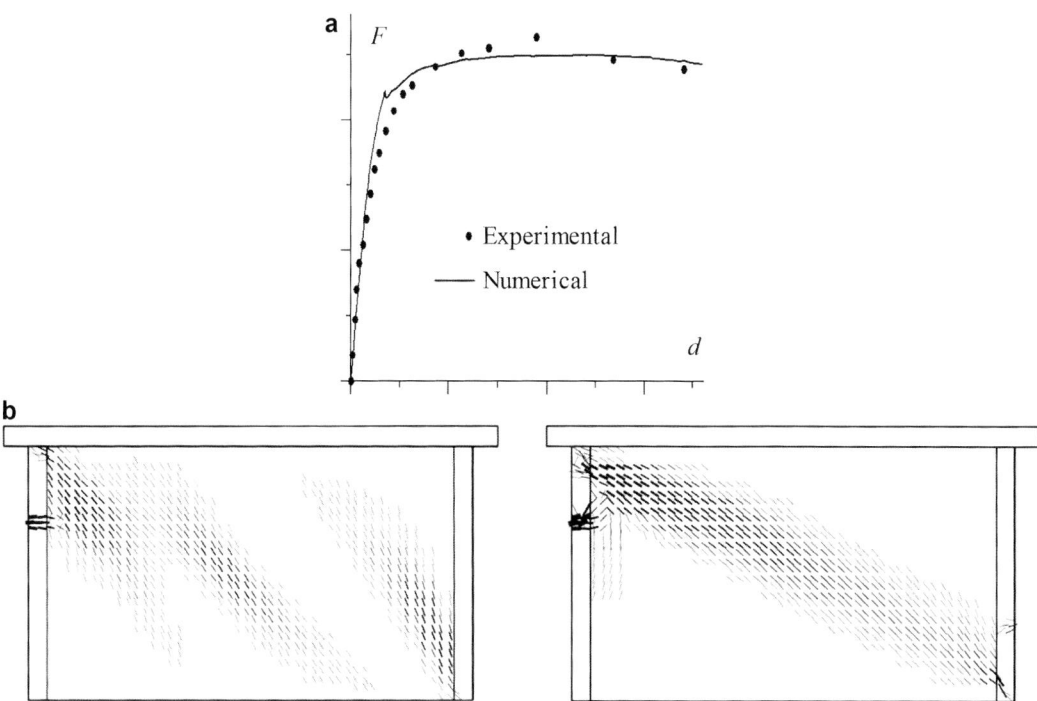

Masonry Modeling, Fig. 8 Results for masonry shear wall: (**a**) load–displacement diagram; (**b**) predicted cracking pattern at peak and ultimate load

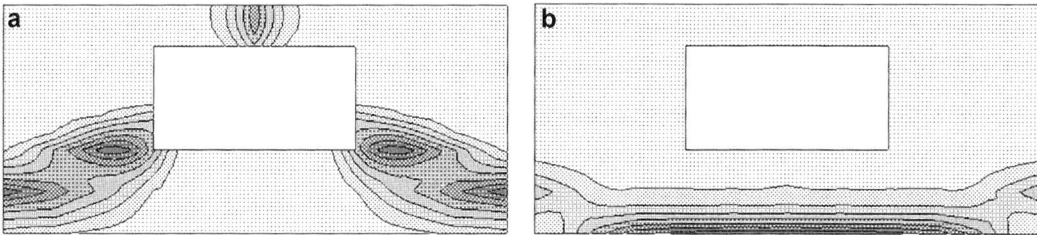

Masonry Modeling, Fig. 9 Results for a panel subjected to uniform out-of-plane loading: predicted cracking pattern at (**a**) *bottom* and (**b**) *top* face of the panel

F drives the wall to failure and produces a horizontal displacement d at top. The wall is confined by two concrete slabs (top and bottom) and two masonry flanges (left and right). This confinement and the large size of the wall make it appropriate for continuum modeling. Initially, cracking occurs well distributed in the panel and finally concentrates in a single shear band from one corner of the panel to the other. The compressive stresses are well below the crushing strength of masonry, i.e., failure is dominated by tension. A complete discussion of the numerical results has been given in (Lourenço et al. 1998). Figure 9 shows the results of modeling a panel with out-of-plane pressure. The panel is simply supported on two sides (left and right), fully clamped on one side (bottom), and free on the other (top). The central opening simulates a window, and the panel was loaded with an air bag with a uniformly distributed load. The predicted form of collapse includes diagonal cracks from each lower corner of the panel up to the opening, which were also observed in the experiments. This form of yield line collapse does not mean that yield line design is safe due to the quasi-brittle behavior of the material. A complete discussion of the numerical results has been given in Lourenço (2000). A discussion on the application of these models to large historic constructions is given in Lourenço (2002).

Homogenization Approaches

Homogenization techniques permit the establishment of constitutive relations in terms of average stresses and strains from the geometry and constitutive relations of the individual components. This can represent a step forward in masonry modeling because of the possibility to use standard isotropic material models and data for masonry components, instead of the rather expensive approach of testing large masonry specimens under homogenous loading conditions.

The complex geometry of the masonry representative volume, i.e., the geometrical pattern that repeats periodically in space, means that no closed-form solution of the problems exists for running bond masonry. One of the first ideas presented was to substitute the complex geometry of the basic cell with a simplified geometry, so that a closed-form solution for the homogenization problem was possible. This approach, rooted in geotechnical engineering applications, assumed masonry as a layered material (Lourenço 1996). This simplification does not allow including information on the arrangement of the masonry units and provides significant errors in the case of nonlinear analysis. Moreover, the results depend on the sequence of homogenization steps. To overcome the limitation, micro-mechanical homogenization approaches that consider additional internal deformation mechanisms have been derived (Zucchini and Lourenço 2002, 2009). The implementation of these approaches in standard macroscopic finite element nonlinear codes is relatively simple, and the approaches can compete with macroscopic approaches (see Lourenço et al. (2007) for a detailed review on homogenization approaches).

A powerful micro-mechanical model for the limit analysis for masonry has been recently developed (Milani et al. 2006a, b). In this model, the elementary cell is subdivided along its thickness in several layers. For each layer, fully equilibrated stress fields are assumed, adopting polynomial expressions for the stress tensor components in a finite number of sub-domains. The continuity of the stress vector on the interfaces between adjacent sub-domains and suitable anti-periodicity conditions on the boundary surface are further imposed. In this way, linearized homogenized surfaces in six dimensions for masonry in- and out-of-plane loaded are obtained. Such surfaces are then implemented in a finite element limit analysis code for simulation of 3D structures and including, as recent advances, blast analysis, quasiperiodic masonry internal structure, and FRP strengthening. Figure 10 shows a masonry wall constituted by a periodic arrangement of bricks and mortar arranged in running bond. For a general rigid-plastic heterogeneous material, homogenization techniques combined with limit analysis can be applied for the evaluation of the homogenized in- and out-of-plane strength domain.

The homogenized failure surface obtained has been coupled with finite element limit analysis. Both upper- and lower-bound approaches have been developed, with the aim to provide a complete set of numerical data for the design and/or the structural assessment of complex structures. The finite element lower-bound analysis is based on an equilibrated triangular element, while the upper bound is based on a triangular element with discontinuities of the velocity field in the interfaces. Recent developments include the extension of the model to blast analysis (Milani et al. 2009), to quasiperiodic masonry (Milani et al. 2010), and to FRP strengthening (Milani and Lourenço 2013).

An enclosure running bond masonry wall subjected to a distributed blast pressure is considered first (Milani et al. 2010). The wall is supposed simply supported at the base and on vertical edges and free on top due to the typical imperfect connection between infill wall and RC

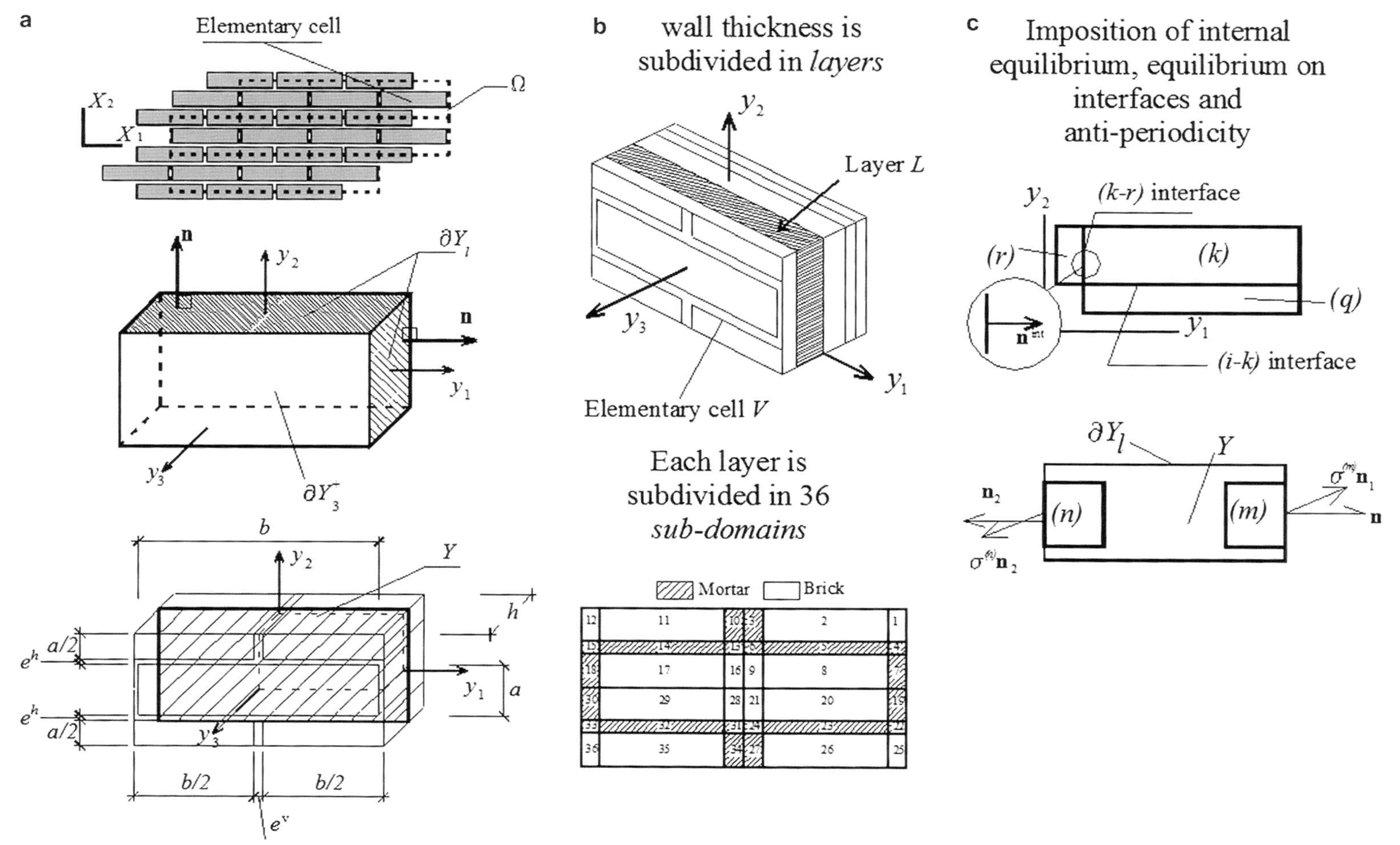

Masonry Modeling, Fig. 10 Proposed micro-mechanical model: (**a**) elementary cell; (**b**) subdivision in layers along thickness and subdivision of each layer in sub-domains; (**c**) imposition of internal equilibrium, equilibrium on interfaces and anti-periodicity

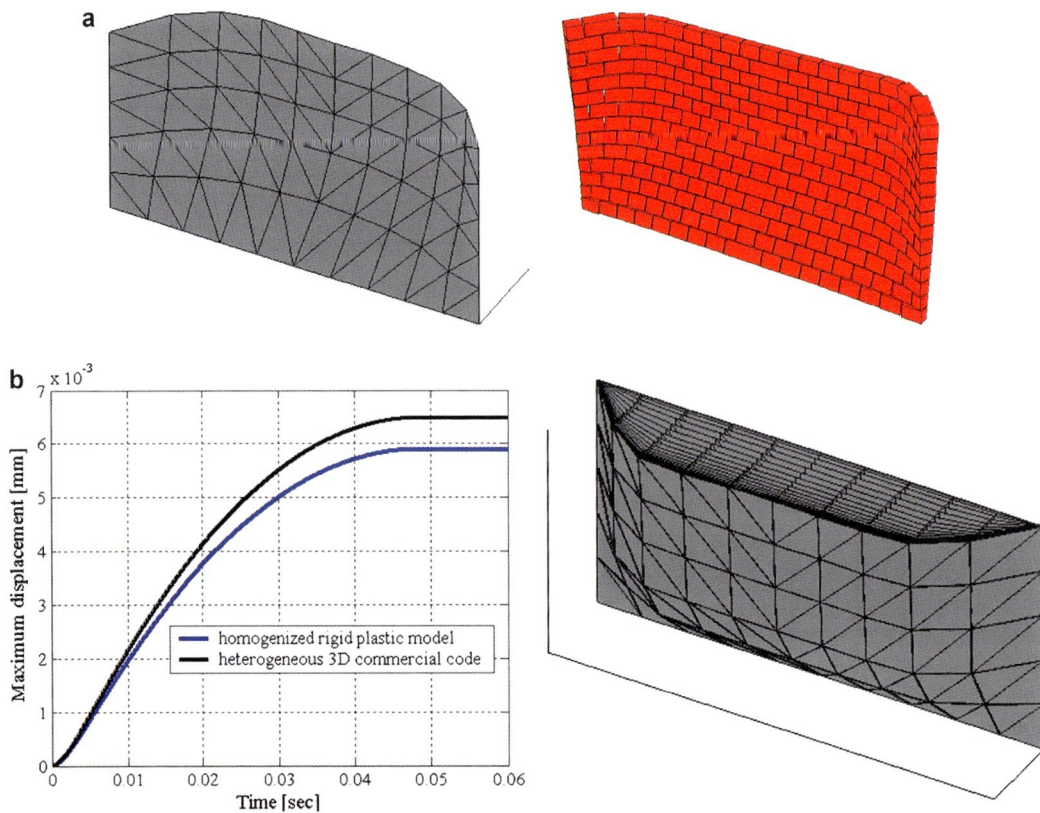

Masonry Modeling, Fig. 11 Masonry infill wall subjected to blast pressure: comparison among deformed shapes at t = 400 μs for (**a**) homogenized limit analysis and heterogeneous 3D elastic–plastic FE approach; (**b**) comparison between maximum out of plane displacements and limit analysis failure mode

beam. A full 3D FE heterogeneous elastic–plastic dynamic analysis has been also conducted, in order to have a deep insight into the problem and to collect alternative data to compare with. For the 3D model, a rigid infinitely resistant behavior for bricks was assumed, whereas for joints, a Mohr–Coulomb failure criterion with the same tensile strength and friction angle used in the homogenized approach for joints was adopted. Eight-noded brick elements were utilized both for joints and bricks, with a double row of elements along wall thickness. A comparison between the deformed shapes at t = 400 μs obtained with the present model and the commercial software is schematically depicted in Fig. 11a. As it is possible to notice, the models give almost the same response in terms of deformed shape for the particular instant time inspected, confirming that reliable results may be obtained with the model proposed. On the other hand, it is worth underlining that the homogenized rigid-plastic model required only 101 s to be performed on a standard PC Intel Celeron 1.40 GHz equipped with 1Gb RAM, a processing time around 10–3 lower than the 3D case. Comparisons of time-maximum displacement diagrams provided by the two models analyzed are reported in Fig. 11b, together with the evolution of the deformation provided by the homogenized model proposed.

Recently, two different classes of problems have been investigated (Milani et al. 2010), the first consisting of full stochastic representative element of volume (REV) assemblages without horizontal and vertical alignment of joints and the second assuming the presence of a horizontal

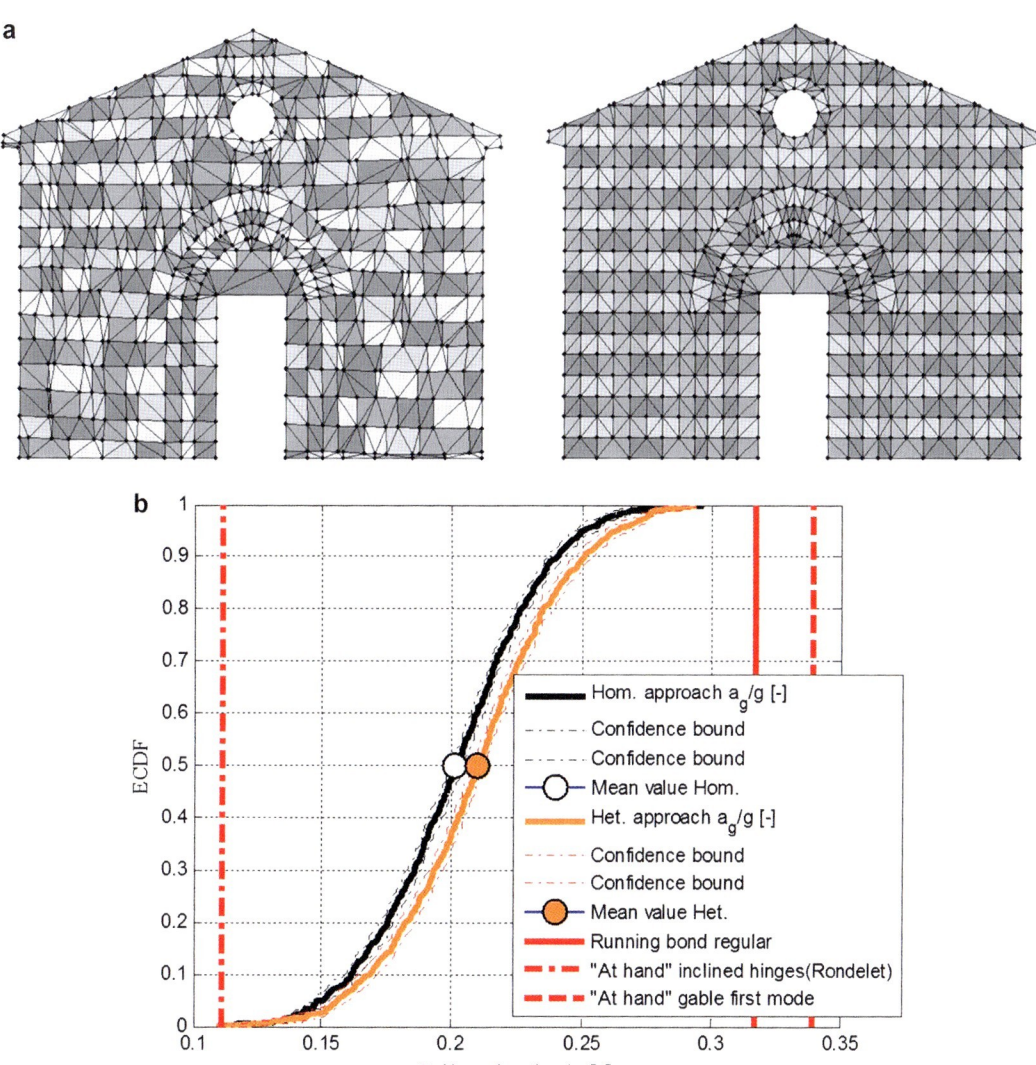

Masonry Modeling, Fig. 12 Church of Gondar (Portugal), FE discretization adopted: (**a**) heterogeneous random mesh vs. mesh for running bond regular heterogeneous and homogenized random analysis; (**b**) ECDF of the failure load provide through a direct heterogeneous approach and homogenized limit analysis simulations

alignment along bed joints, i.e., allowing block height variability only row by row. The model is characterized by a few material parameters, and it is therefore particularly suited to perform large-scale Monte Carlo simulations. Masonry strength domains are obtained equating the power dissipated in the heterogeneous model with the power dissipated by a fictitious homogeneous macroscopic plate. A stochastic estimation of out-of-plane masonry strength domains (both bending moments and torsion are considered) accounting for the geometrical statistical variability of blocks dimensions is obtained with the proposed model. The case of deterministic block height (quasiperiodic texture) can be obtained as a subclass of this latter case. As an important benchmark, the case in which joints obey a Mohr–Coulomb failure criterion is also tested and compared with results obtained assuming a more complex interfacial behavior for mortar. In order to show the capabilities of the approach proposed when dealing with large-scale structures, the ultimate behavior prediction of a Romanesque masonry church facade located in Portugal are presented. Comparisons with finite element heterogeneous approaches and "at hand" calculations show that reliable predictions of the load bearing capacity of real large-scale structures may be obtained with a very limited computational effort (see Fig. 12).

Conclusions

Constraints to be considered in the use of advanced modeling are the cost, the need of an experienced user/ engineer, the level of accuracy required, the availability of input data, the need for validation, and the use of the results. As a rule, advanced modeling is a necessary means for understanding the behavior and damage of masonry constructions. For the study of isolated structural elements and problems controlled by the geometry of a few units, micro-modeling strategies are available. For large-scale applications, average continuum mechanics is usually adopted, and homogenization techniques represent a popular and active field in masonry research. The assessment and design of unreinforced masonry structures subjected to seismic loading is particularly challenging when masonry box behavior is not present. In this case, macro-block analysis seems adequate for simplified assessment and design of strengthening measures.

Summary

Masonry joints act as planes of weakness, and the explicit representation of the joints and units in a numerical model seems a logical step towards a rigorous analysis tool (Fig. 1). This kind of analysis is particularly adequate for small structures, subjected to states of stress and strain strongly heterogeneous. In general, a higher computational effort ensues, so this approach still has a wider application in research and in small models for localized analysis. The approach based on the use of averaged constitutive equations seems to be the most suitable to be employed a large-scale finite element analyses, either by collating experimental data at average level (macro-modeling) or from homogenization techniques.

A micro-model for masonry joints is usually composed by three individual yield functions, consisting of the usual Coulomb friction criterion for shear, combined with tension and compression cutoffs. A macro-model for the masonry composite must permit to reproduce the results obtained in uniaxial tests, with distinct behavior obtained along different directions. Finally, homogenization techniques require as input the geometry of the units and joints, as well as their mechanical properties. The most successful applications are based on micromechanical models that perform a simplified single-step homogenization because a close form solution is not available.

This entry provides a review of different modeling approaches for masonry structures with illustrative applications.

References

Berto L, Saetta A, Scotta R, Vitaliani R (2002) Orthotropic damage model for masonry structures. Int J Numer Methods Eng 55(2):127–157

Dhanasekar M, Page AW, Kleeman PW (1985) The failure of brick masonry under biaxial stresses. Proc Inst Civ Eng, Part 2 79:295–313

Lemos JV (2007) Discrete element modeling of masonry structures. Int J Archit Herit 1(2):190–213

Lourenço PB (1996) A matrix formulation for the elastoplastic homogenisation of layered materials. Mech Cohes-Frict Mater 1:273–294

Lourenço PB (2000) Anisotropic softening model for masonry plates and shells. J Struct Eng 126(9):1008–1016

Lourenço PB (2002) Computations of historical masonry constructions. Prog Struct Eng Mater 4(3):301–319

Lourenço PB, Rots JG (1997) Multisurface interface model for the analysis of masonry structures. J Eng Mech ASCE 123(7):660–668

Lourenço PB, de Borst R, Rots JG (1997) A plane stress softening plasticity model for orthotropic materials. Int J Numer Methods Eng 40:4033–4057

Lourenço PB, Rots JG, Blaauwendraad J (1998) Continuum model for masonry: parameter estimation and validation. J Struct Eng 124(6):642–652

Lourenço PB, Milani G, Tralli A, Zucchini A (2007) Analysis of masonry structures: review of and recent trends of homogenization techniques. Can J Civ Eng 34(11):1443–1457

Milani G, Lourenço PB (2013) Simple homogenized model for the non-linear analysis of FRP strengthened masonry structures. Part I: theory. J Comput Const ASCE 139(1):59–76

Milani G, Lourenço PB, Tralli A (2006a) Homogenised limit analysis of masonry walls. Part I: failure surfaces. Comput Struct 84(3–4):166–180

Milani G, Lourenço PB, Tralli A (2006b) A homogenization approach for the limit analysis of out-of-plane loaded masonry walls. J Struct Eng ASCE 132(10):1650–1663

Milani G, Lourenço PB, Tralli A (2009) Homogenized rigid-plastic model for masonry walls subjected to impact. Int J Solids Struct 46(22–23):4133–4149

Milani G, Lourenço PB, Tralli A (2010) A simplified homogenized limit analysis model for randomly assembled blocks out-of-plane loaded. Comput Struct 88(11–12):690–717

Oliveira DV, Lourenço PB (2004) Implementation and validation of a constitutive model for the cyclic behaviour of interface elements. Comput Struct 82(17–19):1451–1461

Orduña A, Lourenço PB (2005a) Three-dimensional limit analysis of rigid blocks assemblages. Part I: torsion failure on frictional joints and limit analysis formulation. Int J Solids Struct 42(18–19):5140–5160

Orduña A, Lourenço PB (2005b) Three-dimensional limit analysis of rigid blocks assemblages. Part II: load-path following solution procedure and validation. Int J Solids Struct 42(18–19):5161–5180

Pelà L, Cervera M, Roca P (2011) Continuum damage model for orthotropic materials: application to masonry. Comput Methods Appl Mech Eng 200(9–12):917–930

Rots JG (1991) Numerical simulation of cracking in structural masonry. Heron 36(2):49–63

Seim W (1994) Numerical modeling of the failure of biaxially loaded masonry walls with consideration of anisotropy (in German). PhD thesis, University of Karlsruhe

Zucchini A, Lourenço PB (2002) A micromechanical model for the homogenisation of masonry. Int J Solids Struct 39(12):3233–3255

Zucchini A, Lourenço PB (2009) Validation of a micromechanical homogenisation model: application to shear walls. Int J Solids Struct 46(3–4):871–886

Masonry Structures: Overview

Paulo B. Lourenço
Department of Civil Engineering, University of Minho, ISISE, Guimarães, Portugal

Synonyms

Brick; Building; Concrete; Housing; Mortar; Stone; Structural analysis

Introduction

Masonry is the oldest building material that still finds wide use in today's building industries. The most important characteristic of masonry construction is its simplicity. Laying pieces of stone, bricks, or blocks on top of each other, either with or without cohesion via mortar, is a simple, though adequate, technique that has been successfully used ever since remote ages. Naturally, innumerable variations of masonry materials, techniques, and applications occurred during the course of time. The influence factors were mainly the local culture and wealth, the knowledge of materials and tools, the availability of material, and architectural reasons.

The primitive savage endeavors of mankind to secure protection against the elements and from attack included seeking shelter in rock caves, learning how to build tents of bark, skins, turfs, or brushwood and huts of wattle and daub. Some of such types crystallized into houses of stone, clay, or timber. The evolution of mankind is thus linked to the history of building materials.

The first masonry material to be used was probably stone. In the ancient Near East, evolution of housing was from huts, to apsidal houses (Fig. 1a), and finally to rectangular houses (Fig. 1b). The earliest examples of the first permanent houses can be found near Lake Hullen, Israel (c. 9000–8000 BC), where dry-stone huts, circular and semisubterranean, were found. Several other legacies survived until present

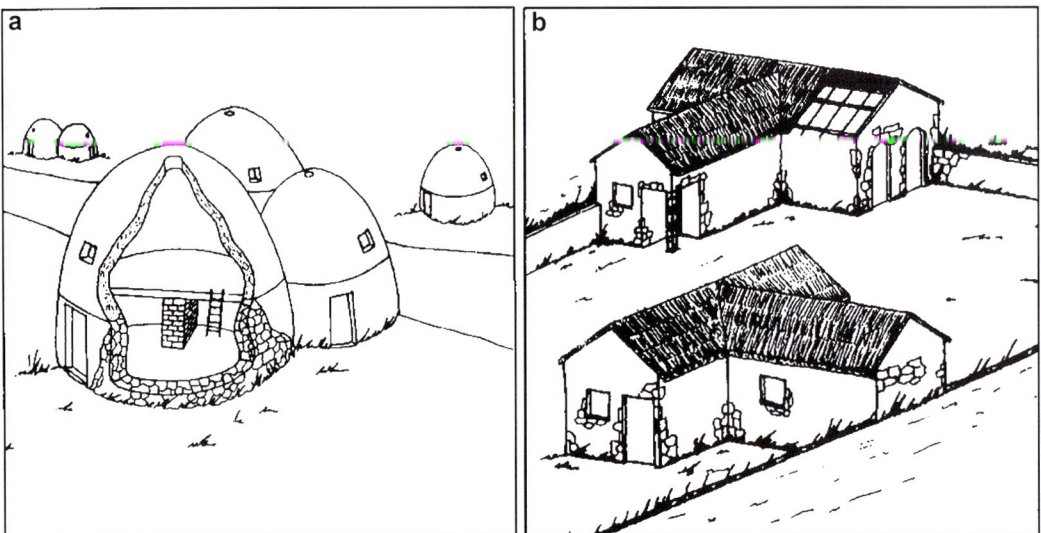

Masonry Structures: Overview, Fig. 1 Examples of prehistoric architecture of masonry in the ancient Near East: (**a**) beehive houses from a village in Cyprus (*c*. 5650 BC); (**b**) rectangular dwellings from a village in Iraq

as testimonies of ancient and medieval cultures (Fletcher and Cruikshank 1996), often as a stone skeleton (Heyman 1997).

The first assault to the use of structural masonry happened at the middle of the nineteenth century, when cast-iron beams and columns started to be produced. By the end of the century, skyscraper constructions methods had eliminated the necessity of massive ground-level piers of masonry. Nevertheless, the collapse of masonry as a structural material started in the beginning of the twentieth century, with the introduction of German, French, and British regulations for design of reinforced concrete structures. Concrete was used in constructions of walls as early as the fourth-century BC around Rome. But, only in 1854, a system for reinforced concrete was patented in Britain by W.B. Wilkinson. By the beginning of the twentieth century, it was clear that reinforced concrete was a durable, strong, moldable, and inexpensive material, and masonry was practically forgotten as a structural material in several developed countries.

In Europe, the building solutions using unreinforced structural masonry represent about 15 % to more than 50 % of the new housing construction, taking as reference countries with low seismicity (e.g., Germany, Netherlands, or Norway) but also countries with high seismicity (e.g., Italy). A usual solution is the adoption of masonry units with large thickness in the building envelope to fulfill thermal requirements. It is stressed that an integrated and complete building technology is needed, including units with different shapes and solutions for floors. Reinforced masonry was developed in different countries as a response to the lower performance of unreinforced masonry buildings under large horizontal loading, but no unified solution was found. Diverse solutions with different levels of success coexist, together with recent innovative solutions.

It is common practice to combine prefabricated slabs with load-resisting walls so that formwork, scaffolding, and execution times can be significantly reduced. In the USA, in the last 30–40 years, reinforced masonry became an attractive and efficient solution from a perspective of cost-benefit analysis for buildings in regions of low to high seismicity, e.g., hotels, residential buildings, office buildings, schools, commercial buildings, or warehouses. The American standard solution includes reinforced concrete horizontal bond beams, two-cell blocks filled with grout, and vertical reinforcement. Besides other reinforced masonry approaches, confined masonry is popular in developing countries, as the changes with respect to unreinforced masonry construction are small. In this system vertical and horizontal reinforced concrete elements of small section are included in the masonry. These elements aim at providing an increase of shear and flexural strength, together with a larger energy dissipation capacity and larger ductility with respect to horizontal actions.

The last decades have witnessed an enormous development in the characterization of masonry materials and in numerical methods for structural analysis. Today, with the help of a computer, it is possible to analyze structures with a high level of accuracy. The material science (directly dependent on the observation and analysis of the experimental behavior) has suffered a slower evolution, but important advances in constitutive models and structural component models have occurred in various fields. For a long time, it was believed that the decline of masonry as a structural material was not only due to economic reasons but also to underdeveloped masonry codes and lack of insight in the behavior of this type of structures. This is not presently the case, as modern codes are available, e.g., Eurocode 6 or EN 1996-1-1:2005 (CEN 2005), but also modern engineering books, e.g., Drysdale and and Hamid (2008), Hendry (1998), Pfefferman (1999), or Sahlin (1971), which provide general overviews of design. More specific books, with a focus on earthquake performance are also available, e.g., Tassios (1988) or Tomaževič (1999). It is stressed that the design of unreinforced masonry structures under seismic loading has not yet received general consensus.

Description

Masonry is a heterogeneous material that consists of units and joints. Units are such as bricks, blocks, ashlars, adobes, irregular stones, and others. Mortar can be clay, bitumen, chalk, lime-/cement-based mortar, glue, or others. A (very) simple classification of stone masonry is shown in Fig. 2. The huge number of possible combinations generated by the geometry, nature, and arrangement of units as well as the characteristics of mortars raises doubts about the accuracy of the term "masonry." Just for brick masonry, some usual combinations are shown in Fig. 3.

When the walls of ancient constructions were of small width, stone units could be of the full width (bond stone or through stone). If the walls were very thick, ashlars would only be used for the outer leafs and the inside would be filled with irregular stones or rubble, or more than one leaf of masonry would be used. Indeed, physical evidence shows us that ancient masonry is a very complex material with three-dimensional internal arrangement, usually unreinforced, but which can include some form of traditional reinforcement, see Fig. 4. Moreover, these materials are associated with complex structural systems, where the separation between architectural features and structural elements is not always clear.

Modern masonry can also exhibit significant variations, not only of materials but also of building technology, see Fig. 5. The choice of materials and the thermal solution, particularly for the enclosure walls, which is a matter of growing concern, is mostly due to tradition and local availability of the materials. Also, the use of reinforcement is associated with tradition and local technological developments, with different approaches from one country to the other.

Nevertheless, the mechanical behavior of the different types of masonry has generally a common feature: a very low tensile strength. This property is so important that it has determined the shape of ancient constructions. The difficulties in performing advanced testing of ancient structures are quite large due to the innumerable variations of masonry, the variability of the masonry itself in a specific structure, and the impossibility of reproducing it all in a specimen. Therefore, most of the advanced experimental research carried out in the last decade has concentrated in brick/block masonry and its relevance for design. The masonry components are detailed in another essay.

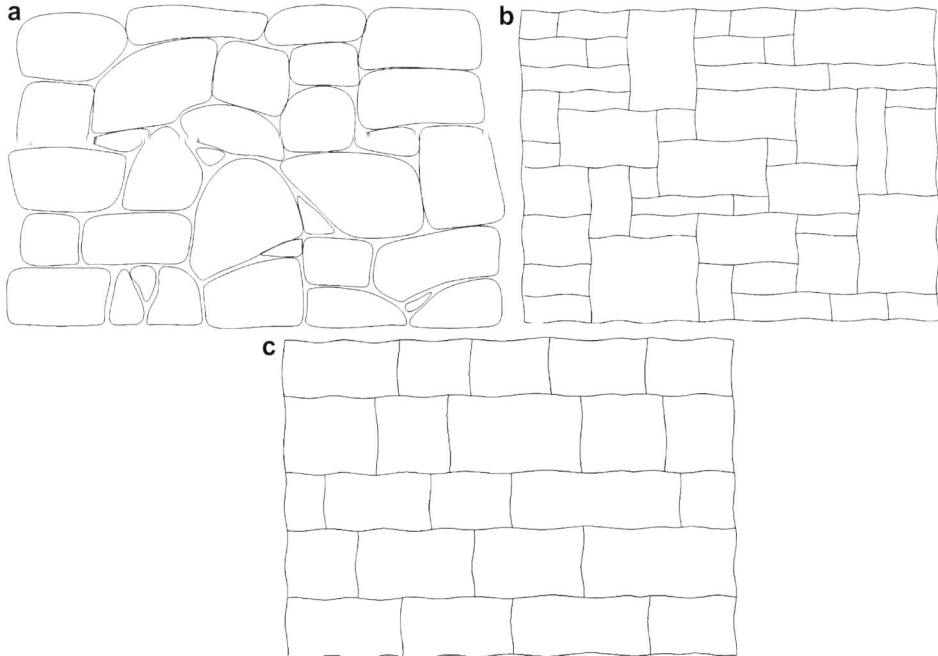

Masonry Structures: Overview, Fig. 2 Different kinds of stone masonry: (**a**) rubble masonry; (**b**) ashlar masonry; (**c**) coursed ashlar masonry

Mechanics

The fact that ancient and modern masonry have so much variability in materials and technology make the task of structural analysis of these structures particularly complex. From a very simplified perspective, it is possible to distinguish masonry as reinforced and unreinforced. The presence of (distributed) reinforcement provides masonry with tensile strength and renders masonry closer to reinforced concrete. In such a case, the orthotropic behavior of masonry and the nonlinear constitutive behavior become less relevant, and the techniques normally used for the design and analysis of reinforced concrete structures can possibly be used. Conversely, in the case of unreinforced masonry structures, the very low tensile strength of the material renders the use of nonlinear constitutive behavior more obvious. This is particularly true in the assessment of existing structures and in seismic analysis.

Masonry is usually described as a material that exhibits distinct directional properties due to the mortar joints, which act as planes of weakness. This description is associated mostly with the material, whereas a different description can be given at structural level.

Description at material level: In general, the approach toward the numerical representation of masonry, i.e., masonry modeling, can address the micro-modeling of the individual components, viz., unit (brick, block, etc.) and mortar, or the macro-modeling of masonry as a composite (CUR 1997). Depending on the level of accuracy and the simplicity desired, it is possible to use the following modeling strategies (see Fig. 6):

1. Detailed micro-modeling – units and mortar in the joints are represented by continuum elements, whereas the unit-mortar interface is represented by discontinuum elements.
2. Simplified micro-modeling – expanded units are represented by continuum elements, whereas the behavior of the mortar joints and unit-mortar interface is lumped in discontinuum elements.

Masonry Structures: Overview, Fig. 3 Different arrangements for brick masonry: (**a**) English (or cross) bond; (**b**) Flemish bond; (**c**) stretcher bond

3. Macro-modeling – units, mortar, and unit-mortar interface are smeared out in a homogeneous continuum.

In fact, the term "micro-modeling" is probably not the most adequate and the term "meso-modeling" would be more reasonable, leaving the former designation for approaches at a lower scale. But the terms macro- and micro-modeling are now widely accepted by the masonry community. A major step in the computational representation using modern analysis techniques is provided in Lourenço (1996).

Description at structural level: The simplest approach related to the modeling of masonry buildings is given by the application of different structural elements resorting to, e.g., truss, beam, panel, plate, or shell elements to represent columns, piers, arches, and vaults, with the assumption of homogeneous (macro) material behavior. Fig. 7 illustrates various possibilities. The lumped approach or mass-spring-dashpot model of Fig. 7a is at best a crude approximation of the actual geometry of the structure, using floor levels and lumped parameters as structural components. The simplicity of the geometric model allows increased complexity on the loading side and in the nonlinear dynamic response. The structural component model in Fig. 7b approximates the actual structural geometry more accurately by using beams and joints as structural components. This approach allows the assessment of the system behavior in more detail. In particular, it is possible to determine the sequential formation of local, predefined failure mechanisms and overall collapse, both statically and dynamically. Finally, the structural model in Fig. 7c approximates the actual structural geometry using macro-blocks with a discrete set of failure lines. Most of these efforts address seismic design and assessment. Models such as the ones shown in Fig. 7a, b are adequate in case that masonry box

Masonry Structures: Overview, Fig. 4 Examples of different masonry types: (**a**) irregular stone wall with a complex transverse cross section, from the eighteenth century in Northern Portugal; (**b**) timber-braced "Pombalino" system emerging after the 1755 earthquake in Lisbon

behavior is exhibited by the building, whereas in Fig. 7c, the so-called macro-block analysis is adequate in case that the masonry building does not exhibit box behavior.

Accurate modeling of masonry structures requires a thorough experimental description of the material. Obtaining experimental data, which is reliable and useful for numerical models, has been hindered by the lack of communication between analysts and experimentalists. The use of different testing methods, test parameters, and materials preclude comparisons and conclusions between most experimental results. It is also current practice to report and measure only strength values and to disregard deformation characteristics. Only recently, information in the post-peak or softening regime became more widely available.

Masonry Behavior

Softening is a gradual decrease of mechanical resistance under a continuous increase of deformation forced upon a material specimen or structure. It is a salient feature of quasi-brittle materials like clay brick, mortar, ceramics, rock, or concrete, which fail due to a process of progressive internal crack growth. Such mechanical behavior is commonly attributed to the heterogeneity of the material, due to the presence of different phases and material defects, like flaws and voids. Even prior to loading, mortar contains micro-cracks due to the shrinkage during curing and the presence of the aggregate. The clay brick contains inclusions and micro-cracks due to the shrinkage during the burning process. Stone also, usually, contains inclusions and micro-cracks. The initial stresses and cracks as well as variations of internal stiffness and strength cause progressive crack growth when the material is subjected to progressive deformation. Initially, the micro-cracks are stable which means that they grow only when the load is increased. Around peak load an acceleration of crack formation takes place and the formation of macro-cracks starts. The macro-cracks are unstable,

Masonry Structures: Overview, Fig. 5 Examples of modern masonry: (**a**) confined masonry in areas of moderate to high seismicity, with thick blocks; (**b**, **c**) different reinforced masonry solutions, adopted in Switzerland (left) and the USA (right)

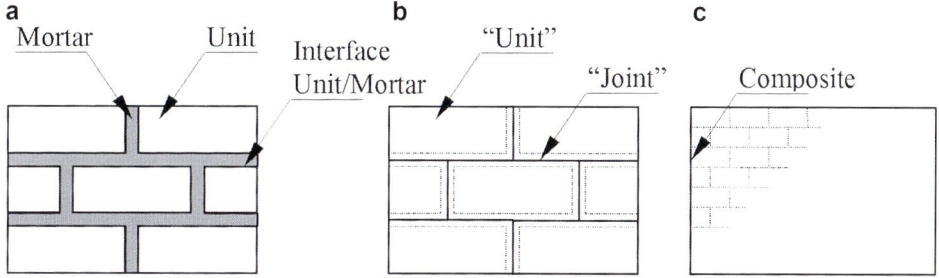

Masonry Structures: Overview, Fig. 6 Modeling strategies for masonry structures: (**a**) detailed micro-modeling; (**b**) simplified micro-modeling; (**c**) macro-modeling

which means that the load has to decrease to avoid an uncontrolled growth. In a deformation-controlled test, the macro-crack growth results in softening and localization of cracking in a small zone while the rest of the specimen unloads.

The properties of masonry are strongly dependent upon the properties of its constituents. Compressive strength tests are easy to perform and give a good indication of the general quality of the materials used. It is difficult to relate the tensile strength to the compressive strength due to the different shapes, materials, manufacture processes, and volume of perforations in the units. The bond between the unit and mortar is often the weakest link in masonry assemblages.

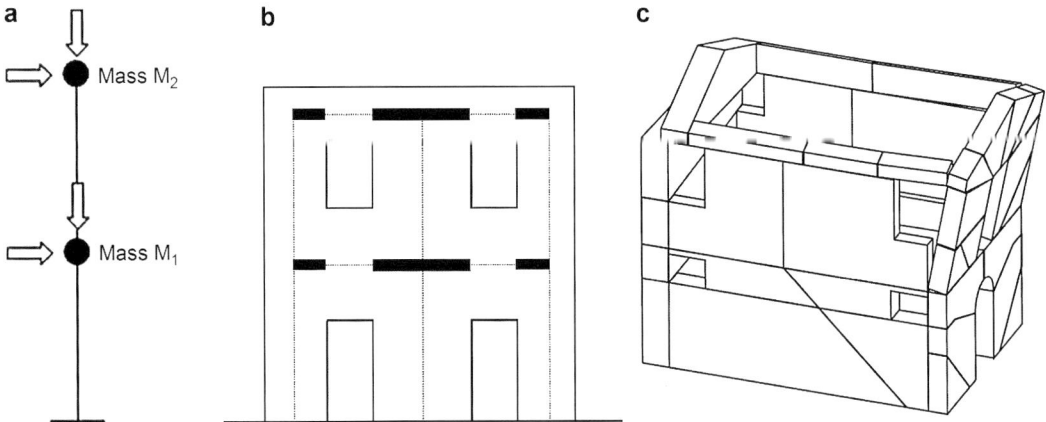

Masonry Structures: Overview, Fig. 7 Examples of structural component models: (**a**) lumped parameters for a complete building with 3 degrees of freedom per story; (**b**) beam elements for wall with openings; (**c**) macro-elements for seismic assessment

The nonlinear response of the joints, which is then controlled by the unit-mortar interface, is one of the most relevant features of masonry behavior. Two different phenomena occur in the unit-mortar interface, one associated with tensile failure (mode I) and the other associated with shear failure (mode II).

Different test setups have been used for the characterization of the tensile behavior of the unit-mortar interface. These include (three-point, four-point, bond-wrench) flexural testing, diametral compression (splitting test), and direct tension testing. An important aspect in the determination of the shear response of masonry joints is the ability of the test setup to generate a uniform state of stress in the joints. This objective is difficult because the equilibrium constraints introduce nonuniform normal stresses in the joint.

Finally, the uniaxial behavior of the composite material is anisotropic, i.e., depending on the loading direction with respect to the material axes, namely, the directions parallel and normal to the bed joints. The compressive strength of masonry in the direction normal to the bed joints has been traditionally regarded as the sole relevant structural material property, at least until the recent introduction of numerical methods for masonry structures. A test frequently used to obtain this uniaxial compressive strength is the stacked bond prism, but it is still somewhat unclear what the consequences are of using this type of specimens in the masonry strength. It has been accepted by the masonry community that the difference in elastic properties of the unit and mortar is the precursor of failure in compression, with uniaxial compression of masonry leading to a state of triaxial compression in the mortar and of compression/biaxial tension in the unit. Uniaxial compression tests in the direction parallel to the bed joints have received substantially less attention from the masonry community. However, regular masonry is an anisotropic material and, particularly in the case of low longitudinal compressive strength of the units due to high or unfavorable perforation, the resistance to compressive loads parallel to the bed joints can have a decisive effect on the load-bearing capacity.

The constitutive behavior of masonry under biaxial states of stress cannot be completely described from the constitutive behavior under uniaxial loading conditions. The influence of the biaxial stress state has been investigated up to peak stress to provide a biaxial strength envelope, which cannot be described solely in terms of principal stresses because masonry is an anisotropic material. Therefore, the biaxial strength envelope of masonry must be described either in terms of the full stress vector in a fixed set of material axes or in terms of principal stresses and the rotation angle between the principal stresses and the material axes.

Summary

Masonry is a building material with many variations and different levels of success in modern construction worldwide. Masonry is also widely available in the built heritage, as most of the pre-twentieth-century buildings are made of masonry. Masonry consists of units laid with a certain bond, usually including mortar in the joints. Masonry can be typically unreinforced or reinforced and an understanding of this composite material requires the characterization of its components. Because masonry is a composite material, different modeling strategies are available either with the discretization of the bond or not. These approaches are denoted by micro- and macro-modeling, usually combined with advanced nonlinear analysis tools. Adequate seismic design and simple assessment of unreinforced masonry buildings are a challenging task and the applicability of the methods depends on the existence of box behavior for the building. In the absence of box behavior, macro-block analysis is a popular and powerful approach. Finally, it is noted that adequate characterization of the experimental behavior is needed for an effective analysis and design of masonry structures, and this is currently available in the literature.

References

CEN (2005) EN 1996-1-1:2005, General rules for reinforced and unreinforced masonry structures. Committee for Standardization, European
CUR (1997) Structural masonry: an experimental/numerical basis for practical design rules (CUR report 171). A.A. Balkema. Rotterdam. ISBN 9789054106807
Drysdale R, Hamid A (2008) Masonry structures: behavior and design, 3rd edn. Boulder, Colorado. ISBN 1929081332
Fletcher B, Cruikshank D (1996) Sir Banister Fletcher's history of architecture: centenary edition. Butterworth-Heinemann, Oxford. ISBN 9780750622677
Hendry AW (1998) Structural masonry. Macmillan Press, Houndmills. ISBN 0333733096
Heyman J (1997) The stone skeleton: structural engineering of masonry architecture. Cambridge University Press, Cambridge. ISBN 0521629632
Lourenço PB (1996) Computational strategies for masonry structures. PhD dissertation, Delft University of Technology. ISBN 9040712212
Pfefferman O (1999) Maçonnerie portante. Kluwer Academic
Sahlin S (1971) Structural masonry. Prentice Hall, Englewood. ISBN 0138539375
Tassios TP (1988) Meccanica delle murature. Liguori Editore. ISBN 8820716380
Tomaževič M (1999) Earthquake-resistant design of masonry buildings. Imperial College Press, London. ISBN 1860940668

Mechanisms of Earthquakes in Aegean

Anastasia A. Kiratzi
Department of Geophysics, Aristotle University of Thessaloniki, Thessaloniki, Greece

Synonyms

Fault-plane solutions; Focal mechanisms; Moment-tensor solutions

Introduction

Definition of Basic Concepts

The **focal mechanism** illustrates the geometry of faulting during an earthquake. For a fault-related earthquake, the focal mechanism describes the orientation of the causative fault (its strike and dip angle) and the direction of the slip on its plane (rake angle). The pattern of the radiated seismic waves depends on the fault geometry, thus the focal mechanism is determined from seismograms recorded at various distances and azimuths from the seismic source. The simplest method relies on the first-motion polarity of body waves, and the more sophisticated methods are based on waveform modeling of body and surface waves. Fault-plane solutions are described by two

Mechanisms of Earthquakes in Aegean, Fig. 1 (a) Fault geometry used in earthquake studies, characterized by the strike, dip, and rake angles of the fault plane; (b) beach-ball presentation of focal mechanisms for the three major fault types

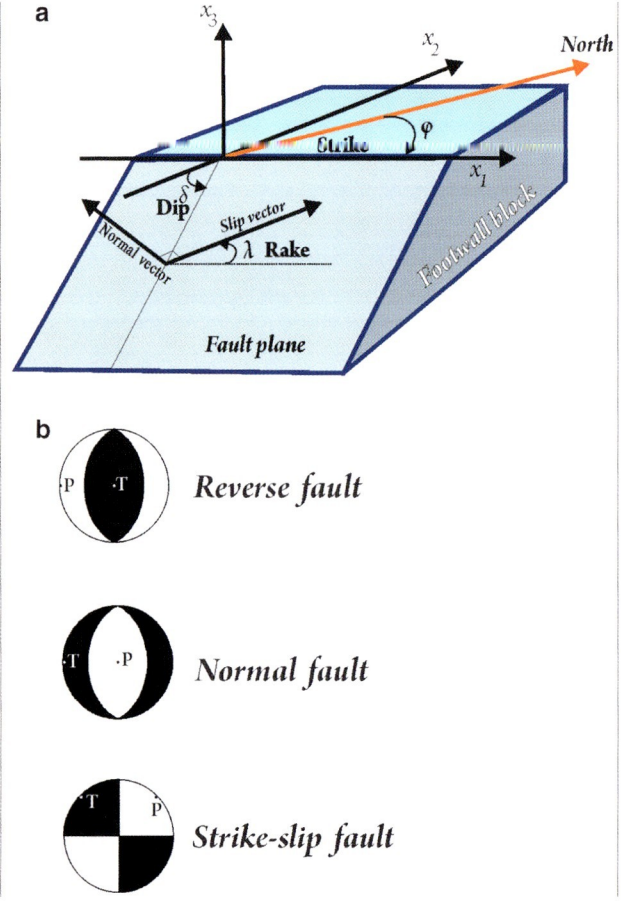

mutually orthogonal planes in space, one of which is the fault plane and the other is the auxiliary plane, to which the slip vector is normal. There is an ambiguity in identifying the fault plane on which the slip occurred, from the orthogonal, mathematically equivalent, auxiliary plane. The ambiguity can be resolved using additional information, for example, if the fault reaches the surface, if the aftershocks define a clear pattern which is aligned with one of the two nodal planes, and if directivity effects could be identified.

The fault geometry is shown in Fig. 1a. The fault plane is characterized by its normal vector. The direction of motion is given by the slip vector which is supposed to lie in the fault plane, so it is perpendicular to the normal vector. The slip vector indicates the direction in which the upper block of the fault, e.g., the hanging wall block, moved with respect to the lower block, e.g., the footwall block. In the coordinate system adopted (Fig. 1a), the parameters used to describe the focal mechanism are: (a) the strike, φ, which is the angle between the trace of the fault (the intersection of the fault plane with the horizontal) and North ($0° \leq \varphi \leq 360°$) (the strike angle is measured in such a manner so that the fault plane dips to the right-hand side); (b) the dip, δ, which is the angle between the fault plane and the horizontal at a right angle to the trace ($0° \leq \delta \leq 90°$); and (c) the rake, λ, which is the angle between the direction of relative displacement or slip and the horizontal measured on the fault plane ($0° \leq \lambda \leq 360°$).

The three major fault types e.g., reverse dip slip (reverse faults that dip at shallow angles $\sim <45°$ are called thrusts), normal dip slip, and strike-slip are described by their corresponding focal mechanisms, providing the values of these three angles, e.g., the strike, the dip, and the rake. The focal mechanism solution is typically displayed graphically using the so-called beach-ball diagram (Fig. 1b). The beach ball hints the stress-field orientation at the time of rupture, and the basic fault types can be related to the orientation of the principal stress directions. The tension axis (T) and the pressure axis (P) tend to approach the minimum and the maximum principal compressive stress directions, respectively, of the stress tensor. They coincide in the case the fault surface corresponds to the plane of maximum shear. The principal stress directions can be obtained by the inversion of fault-plane solutions. The T- and P-axes are determined from an eigenvalue and eigenvector analysis of the moment tensor. Since this tensor is symmetric, its eigenvalues are real, and its eigenvectors are mutually orthogonal, forming the set of T, P, and N (null) axes.

The Aegean

The Aegean Sea and the surrounding lands, mainland of Greece, southern Balkans, and western Anatolia, hereafter referred to as the Aegean, exhibit active deformation as manifested by the record of intense earthquake activity (the area hosts more than 60 % of the total European seismicity). Active tectonics apart from forming magnificent landscapes also provide a wealth of data to the geoscientists. To this end, the Aegean, even though it covers a small area in eastern Mediterranean, roughly 700 km × 700 km, is listed among the best studied regions of continental deformation and is often called a *natural geophysical laboratory*.

The major kinematic boundary conditions within the eastern Mediterranean are sketched in Fig. 2 and outline the following:

(a) The northward movement of the African (Nubia) plate relative to the Eurasian plate

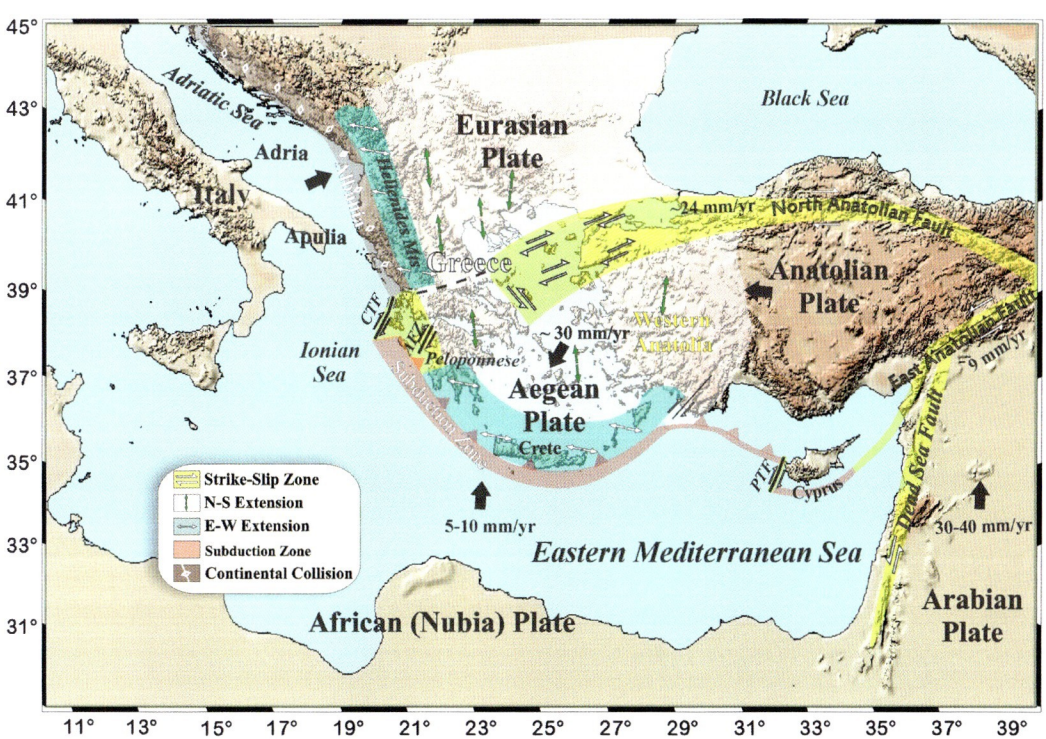

Mechanisms of Earthquakes in Aegean, Fig. 2 A sketch of the major kinematic boundaries within eastern Mediterranean, with the rates of motion relative to Eurasia. *CTF* Cephalonia transform fault zone, *AFZ* Achaia fault zone, *PTF* Pafos transform fault zone (Figure modified from Papazachos et al. 1998)

(convergence rate between 5 and 10 mm/year) and the subduction of its frontal part beneath the overriding Aegean lithosphere. The subduction rate exceeds the convergence rate, because of the rapid SW motion of the southern Aegean relative to Eurasia which leads to a subduction rate of ~35 mm/year; in this setting, the overriding Aegean plate must extend to compensate for the difference between the subduction rate and the convergence rate.

(b) The westward motion of Anatolia relative to Eurasia, toward the Aegean, driven by the N-S convergence (at a rate of 30–40 mm/year), between the Arabian plate and Eurasia. This lateral extrusion of Anatolia, as a coherent rigid block, is facilitated by the operation of two strike-slip faults: the right-lateral North Anatolian Fault (NAF), with a slip rate ~24 mm/year and the more diffuse left-lateral East Anatolian Fault (EAF), with a slip rate ~9 mm/year (McClusky et al. 2000). The presence of the Hellenic subduction allows the Aegean plate to easily override the retreating subduction boundary.

(c) The collision of the Adria–Apulia block with Eurasia along the western coastal region of Albania and western Greece.

In this geodynamic context, the deformation is distributed and is taken up by a number of faults, of variable length and orientation, on land and offshore, as illustrated in Fig. 3. Major fault segments are populated by families of minor faults forming wide zones of distributed faulting.

The Hellenic Subduction Zone

The Hellenic subduction zone (Jolivet et al. 2013; Ring et al. 2010) extends (Fig. 3) from approximately south of Zakynthos in the west up to Rhodes Island in the east. South of Crete, there is the *Mediterranean Ridge* which is the accretionary prism formed by the pile of sediments, up to 10 km thick that overlies the oceanic crust. The *Hellenic Trench* is a scarp in the bathymetry 2–3 km high, which is more clearly developed in the western part of the subduction zone, between Peloponnese and Crete, while it splits into branches toward the east, the best developed of which are known as the *Ptolemy*, *Pliny*, and *Strabo Trenches*. After the recognition that the Mediterranean Ridge is an accretionary complex, it became apparent that the Hellenic Trench can be better understood as sediment-loaded fore-arc basin or else as a backstop to the deformed accretionary complex. The Hellenic Trench does not define the plate boundary, as it is common in other ocean trenches; it is not the surface projection of the subduction interface. The African–Aegean plate boundary itself lies south of the Mediterranean Ridge or very likely is buried beneath it. This can be easily understood from an assumption and simple geometry. Assuming that the low-angle thrusts, underlying the Hellenic Trench at depths up to 40–50 km, represent the interface between the subducting African lithosphere and the overriding Aegean plate, then their surface projection should be sought at least 150 km south of Crete at the Mediterranean Ridge. The degree of coupling on the Hellenic subduction interface, i.e., the fraction of the motion across the plate boundary accommodated by elastic strain accumulation, is estimated to be low (10–20 % of the full African–Aegean convergence rate).

The *Aegean Volcanic Arc* (Fig. 3) extends from Methana on the eastern coast of Peloponnese to the Nisyros Island in the east, close to coast of western Anatolia. The Hellenic Trench, the south Aegean Volcanic Arc, and the back-arc sea are the tectonic elements of the *Hellenic Arc*, which exhibits a strong curvature. It must be noted that in the present work, the back-arc area refers to the region behind the south Aegean Volcanic Arc and the fore-arc area refers to the region in front of the south Aegean Volcanic Arc (toward Crete). The identification of the Wadati–Benioff zone in the Hellenic Arc has been accomplished since the early 1970s (Papazachos and Comninakis 1971) and was subsequently better defined (Papazachos et al. 2000). This zone is relatively flat (~16°) beneath Peloponnese, has an average dip of 30° beneath Crete, and is steeper (50°) in the eastern part, near Rhodes Island.

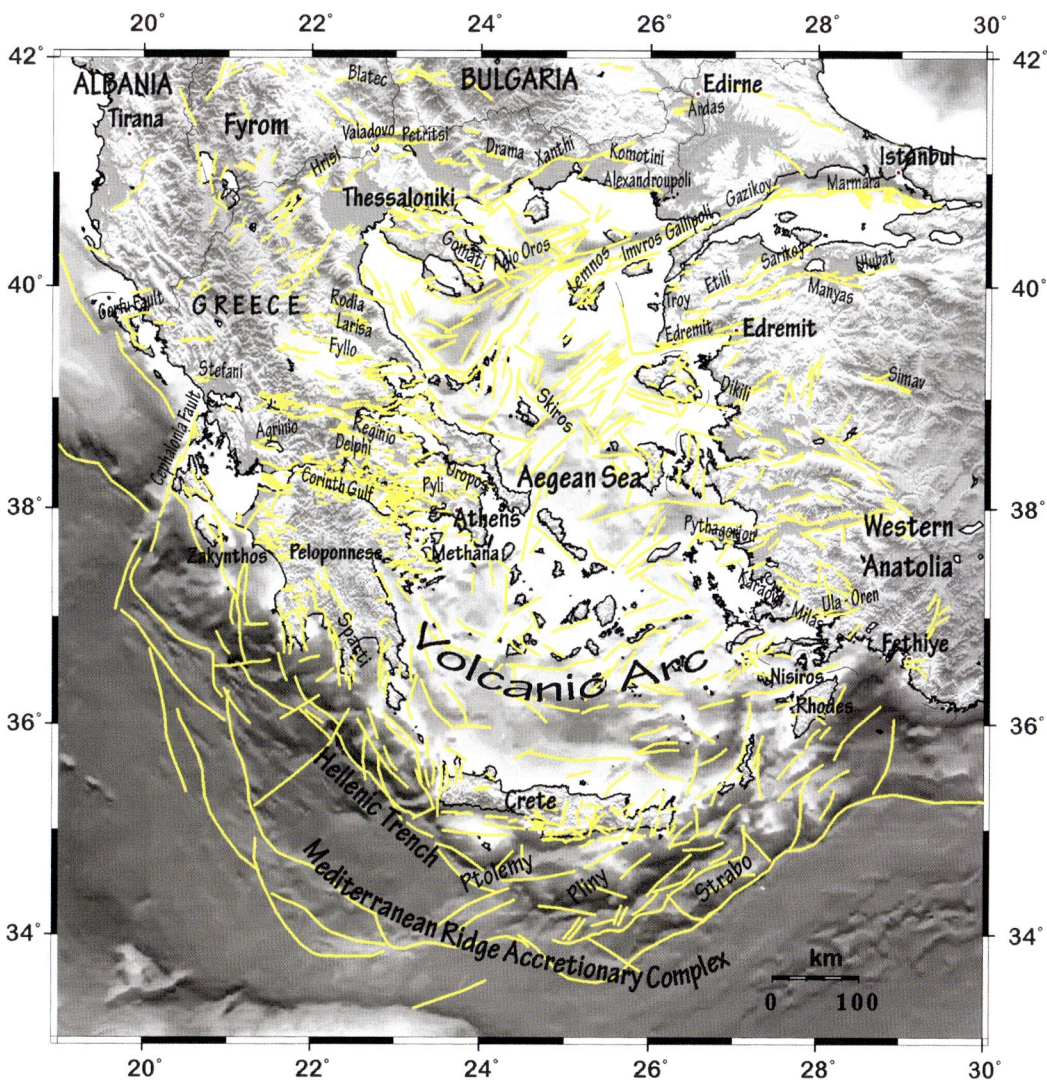

Mechanisms of Earthquakes in Aegean, Fig. 3 Major tectonic structures in the Aegean, which are discussed in the text. The fault lines (available from the database of Basili et al. 2013) are plotted aiming to show that in this tectonic setting, the deformation is distributed along wide zones including major and minor fault segments

The present work is focused on the earthquake focal mechanisms of the Aegean and the information they provide on the faulting characteristics and the regional stress field. Thus, in the sections that follow, a brief background on the sources of focal mechanisms for earthquakes in the Aegean is presented. Then, the available focal mechanisms are classified depending on the faulting type, and the general pattern of the spatial distribution of each category is presented. A number of features of specific interest are discussed in more detail, for the region between the meridians 19–28°E and the parallels 34–41°N, even though in a number of cases broader regions are included in the figures.

Published Earthquake Focal Mechanisms in the Aegean: Historical Background

The published focal mechanisms for the Aegean follow the worldwide advances in instrumentation and computational methods. In a global scale, the period 1962–1976 marks the era of analogue records, and from 1976 onward, the digital era takes place. In a regional scale, the first seismograph in Greece was installed in Athens in 1911 and marks the era of instrumental seismology for the Aegean. In mid-1960s, the World-Wide Standardized Seismographic Network (WWSSN) was established which provided high-quality analogue seismograms to the scientific community. These data were collected at Lamont–Doherty Observatory and at the United States Geological Survey (USGS) in Denver and Golden, Colorado.

By the 1980s, WWSSN was upgraded to form the 150+ station Global Seismographic Network (GSN), which currently has substantial capability in recording a broad spectrum of motions from Earth tides to local ground noise. Nowadays, free access to three-component broadband time series of ground motion is provided by the International Federation of Digital Seismograph Networks (FDSN), through the IRIS Data Management System (iris.edu/data). The advent of digital seismology together with the subsequent theoretical and computational advances in calculating synthetic seismograms increased the number of the published earthquake focal mechanisms worldwide as well as for the Aegean area. Thus, since 1976, nearly all earthquakes with Mw > 5.5 have a published solution based on waveform modeling.

Over the years, the published earthquake mechanisms for the Aegean have been determined using different methods and waveforms, which are briefly presented in the following section.

Focal Mechanisms Determined from First-Motion Polarities

These are focal mechanisms of Aegean earthquakes determined from the first-motion polarity of P-waves. From the published first-motion solutions, more reliable are those for which (a) the first P-wave polarity was read on long-period instruments, preferably on the WWSSN 15–100 s instruments, and (b) the observed polarity distributions are also published. Long-period records are generally less noisy, so the polarity is clearer to read, while the publication of the observed polarities provides the means to evaluate the solution, judging from the number of the polarities, the azimuthal distribution of the stations used, and the number of nodal readings that constrain the orientation of the nodal planes, to mention a few criteria. Depending on the distribution and quality of first-motion data, more than one focal mechanism solution may fit the data equally well (see, e.g., the compilation by Udias et al. 1989). From the first-motion solutions which are available for Greece, a significant number include focal mechanisms for small and moderate magnitude ($\sim M < 5$) earthquakes and aftershock sequences, recorded by temporary networks. These were installed in mainland Greece and the islands, in the framework of a series of projects, mainly in the decades from 1980 to 2000 (e.g., Hatzfeld et al. 1990). First-motion polarity solutions for earthquakes in the Aegean remain an important database even though their significance was later surpassed by the solutions obtained using waveform modeling.

Focal Mechanisms Determined from Waveform Modeling at Teleseismic Distances

These are focal mechanisms of Aegean earthquakes determined by the computation of synthetic seismograms that best fit the observed seismograms at teleseismic distances in the range of 30°–90°. The technique is known as waveform modeling, and synthetics are calculated using a ray theory approach, a well-known code is MT5 (Zwick et al. 1994), usually referred to as *body-wave analysis*, or on a normal-mode approach commonly referred to as *centroid moment tensor* or CMT analysis. The main tenet in both approaches is to use wavelengths of seismic waves that are longer compared to the earthquake source dimensions (i.e., the fault length). If this is fulfilled and if the source-station separation is larger than the fault length, then the source appears as a point in space (the so-called centroid) even though it may have finite time duration. In a number of cases, the point-source approximation is not valid. For example, when the observed waveforms support the existence of discrete sub-events, then the contribution of these sub-events can be modeled introducing a number of separate sources appropriately lagged in space and time. Well-known source for CMT solutions is the Global CMT Project (globalCMT.org), formerly known as Harvard CMT catalog (Ekström et al. 2012). Other sources include the body-wave moment-tensor solutions, which have been reported from the United States Geological Survey since 1981 (Sipkin 1994).

There are differences in the previous inversion schemes. In the body-wave analysis, the focal mechanisms are constrained to be double-couple solutions (i.e., to result from slip on a fault). The window of data examined in the inversion is short; it includes the first part of the P and SH (horizontal component of S-wave) waveforms. For example, for shallow earthquake sources, the data window examined usually ranges from ~20 to 40 s and allows the inspection of relatively high frequencies. Focal depths constrained by the so-called depth phases, pP and sP, which leave the source upward, are reflected at the free surface and continue further as P-waves. To identify the separation between P and pP or sP is more difficult the shallower the source. For example, for depths in the range 10–20 km, the time separation between P and pP and between P and sP is of the order of 4–10 s. Commonly, during the processing stage of digital waveforms from the modern global networks, after removing the instrument response, the waveforms are convolved with the response of the old long-period WWSSN 15–100 instruments (with a peak response at ~15 s), because it provides a frequency range suitable for the inversion.

The CMT method is based on the linear relationship that exists between the six independent elements of a zeroth-order moment-tensor representation of an earthquake and the ground motion that the earthquake generates. The "centroid" of the CMT method refers to the center of the earthquake moment distribution in time and space, defined by four additional parameters. Ten parameters (six moment-tensor elements and the centroid latitude, longitude, depth, and centroid time) thus provide the point-source CMT representation of an earthquake. No volumetric component is allowed in the moment tensor, and the published focal mechanisms are constrained to be purely deviatoric. Longer time windows are used compared to the time windows used in the classic body-wave analysis. For the typical $5.5 \leq Mw \leq 6.5$ Aegean earthquakes analyzed, the body waves are filtered to displacement between 150 and 40 s, with a flat response between 100 and 50 s. In a number of cases, the CMT inversions cannot sufficiently resolve the spatial centroid of the source, in particular the focal depth for shallow sources. Thus, in the GCMT catalog, sometimes the focal depth is fixed, when the inversion attempts to place the source above the minimum allowed depth of 12 km or when the depth exhibits persistent instability in the inversions.

Focal Mechanisms Determined from Waveform Modeling at Regional Distances

These are focal mechanisms of Aegean earthquakes determined by waveform analysis, moment-tensor inversion from body waves (Dreger 2003; Sokos and Zahradnik 2008), and centroid moment-tensor inversion, using ground motion records from broadband sensors at regional distances. The trick in this methodology is to use long-period regional-distance seismic waves. The source process can be reduced to a simple delta function "spike" in space and time. The wave propagation is also simplified because filtering regional seismograms to long periods (or low frequencies) results in waves that have only propagated in a few wave length cycles and can be predicted using relatively simple one-dimensional layered Earth models. Canceling out all of these complex effects leaves the robust extraction of the radiation pattern. A more generalized moment tensor can be solved with simple constraints added e.g., that the source has no isotropic component, although this constraint is not valid when estimating the moment tensor of volcanic sources.

The Era of Real-Time Focal Mechanisms in the Aegean

Seismic instrumentation in Greece was gradually upgraded to obtain its present-day configuration which counts 200+ broadband sensors in operation. The funding to the Greek seismological

Mechanisms of Earthquakes in Aegean, Table 1 Classification of the Aegean focal mechanisms based on the plunge of P, T, and N stress axes (After Zoback 1992)

Plunge (pl) of axes			
P-axis	T-axis	N-axis	Faulting type
pl ≥ 52°	pl ≤ 35°		Normal faulting [NF]
40° ≤ pl < 52°	pl ≤ 20°		Normal oblique faulting [NS]
pl ≤ 35°	pl ≥ 52°		Thrust faulting [TF]
pl ≤ 20°	40° ≤ pl < 52°		Thrust oblique faulting [TS]
pl < 40°	pl < 40°	pl ≥ 45°	Strike-slip faulting [SS]

community usually increases in the aftermath of a strong event, thus the seismological networks were mainly upgraded after the Athens 1999 earthquake. The establishment of the Hellenic Unified Seismological Network (HUSN) marks the beginning of a new era for instrumentation coverage in Greece (see http://bbnet.gein.noa.gr/HL/ for network geometry). Four centers constitute the core members of HUSN: (a) the Geodynamic Institute of the National Observatory of Athens (NOA), (b) the Seismological Laboratory of the Aristotle University of Thessaloniki (AUTH), (c) the Seismological Laboratory of Patras University (UPSL), and (d) the Seismological Laboratory of Athens University (UOA).

In late 2005, these Greek seismological agencies began a systematic determination of moment-tensor solutions for strong earthquakes, and these solutions are transmitted to the European-Mediterranean Seismological Centre (EMSC-CSEM/). They can be retrieved from the following website using the code of each institute (http://emsc-csem.org).

Gross Features of the Spatial Organization of the Aegean Mechanisms

In the following sections, the spatial organization of earthquake focal mechanisms in the Aegean is discussed, using data from published sources and for earthquakes after mid-1950s mainly with Mw ≥ 5.5. For the period 1950–1959, the focal mechanisms are mainly from first-motion polarities. Even though smaller magnitude events are usually not a good proxy for regional tectonics, in a number of cases, they are included in the plots for emphasis to some patterns. To classify the focal mechanisms, the criteria listed in Table 1 were used. Focal mechanisms from shallow-focus (h ≤ 50 km) earthquakes and deeper than 50 km are examined separately. This cutoff depth, even though arbitrarily chosen, provides a useful depth range, because the deeper than 50 km earthquake mechanisms are all confined within the subducting African lithosphere.

It should be noted that the maximum depth of earthquakes within the lithosphere, which defines the thickness of the seismogenic layer, is of the order of 10–20 km in the Aegean. This thickness provides a scale for the dimensions of active structures, as well. Earthquakes which are strong enough to rupture the entire thickness of the seismogenic layer are produced from large faults. Moderate-size events with ruptures confined within the seismogenic layer are produced from smaller faults (Jackson 2001).

Figure 4a–d summarize the classification of the shallow (h ≤ 50 km) focus and intermediate (h > 50 km) focus earthquake focal mechanisms, respectively, into three major categories: reverse/thrust faulting, normal faulting, and strike-slip faulting. Some general features are summarized below, while selective features of the spatial organization of the focal mechanisms, in all cases for shallow-focus events, are discussed in more detail in the following sections.

The Zone of Thrust/Reverse Faulting

The Thrust Zone Along the Adria: Eurasia Convergence
Thrust faulting (Fig. 4a) dominates the foreland and hinterland flanks of Albania, where the

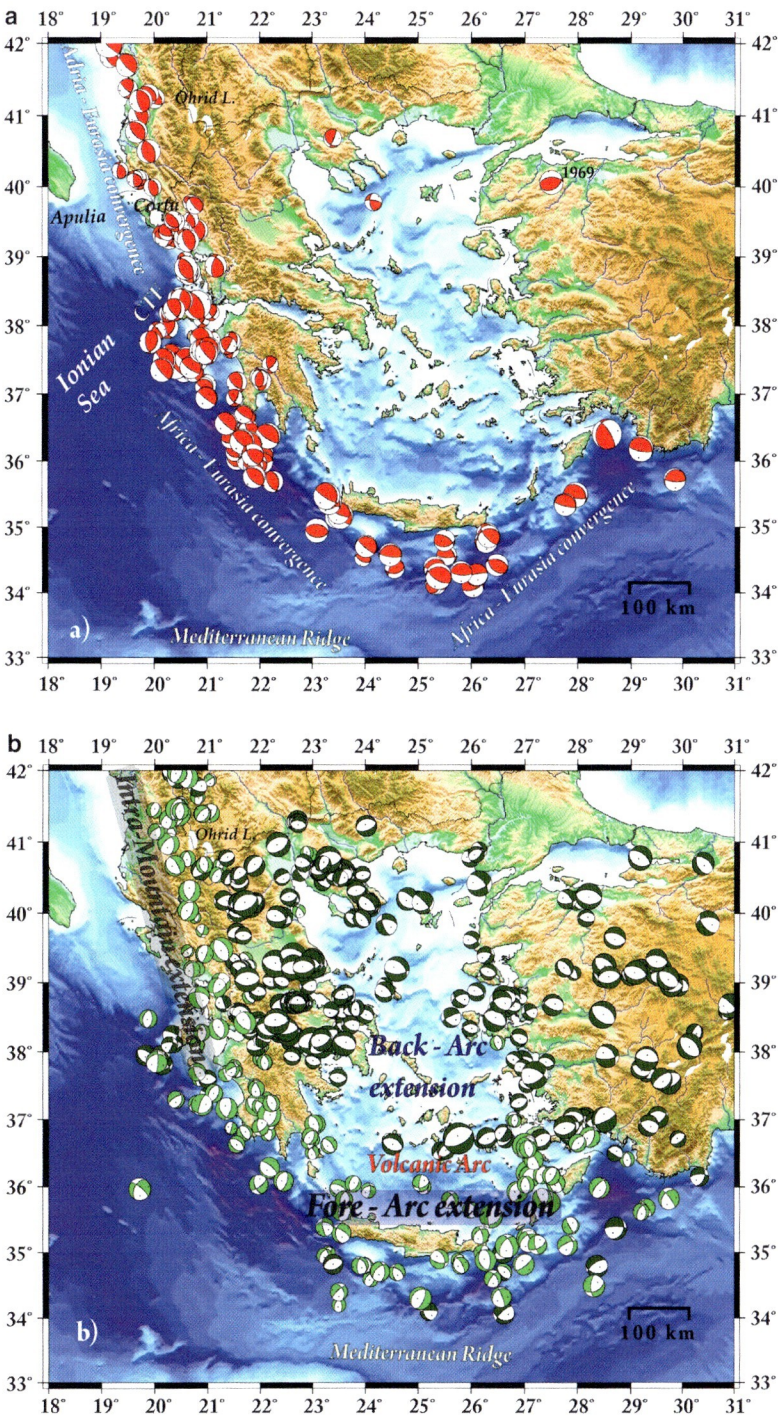

Mechanisms of Earthquakes in Aegean, Fig. 4 (continued)

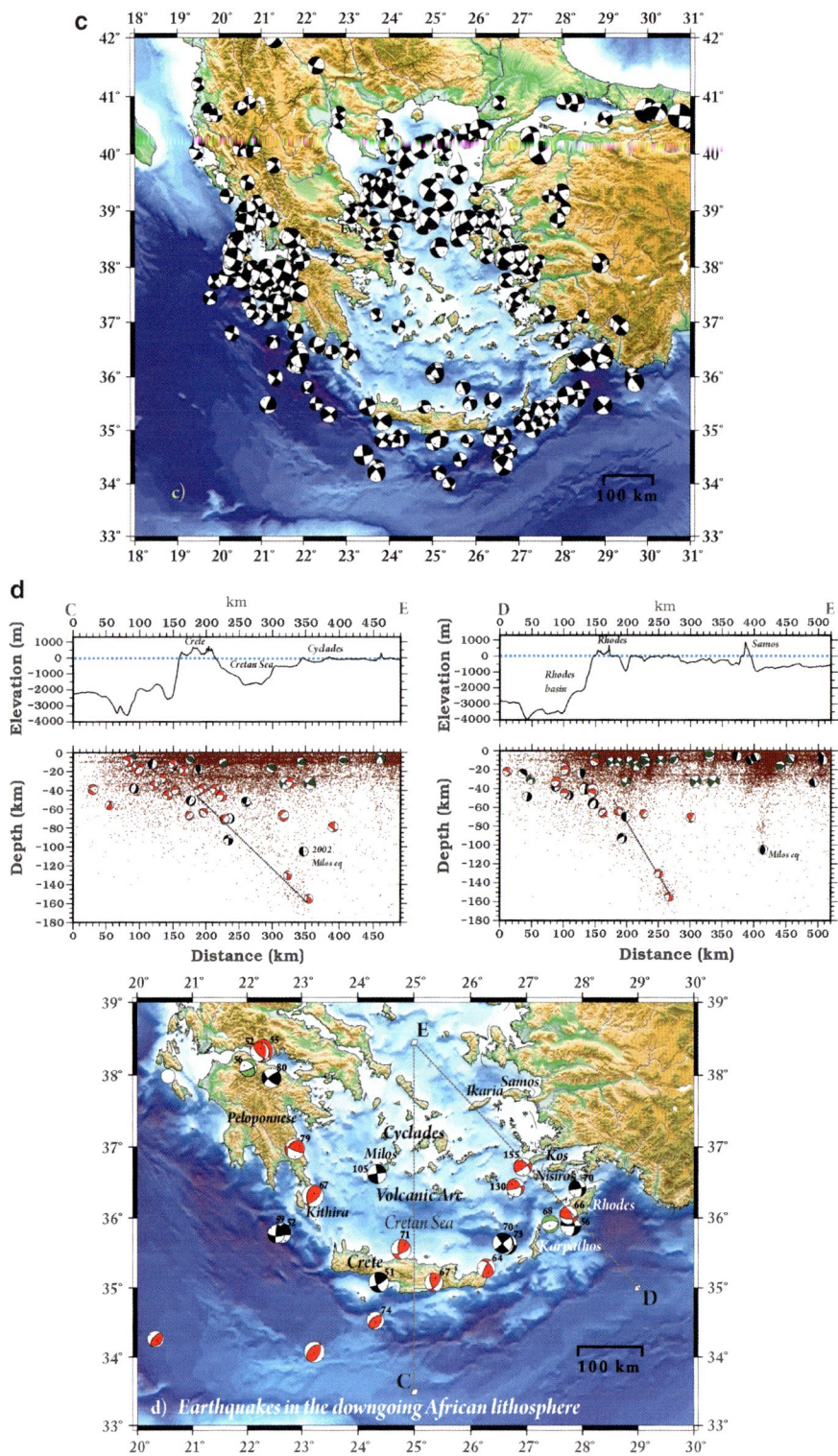

Mechanisms of Earthquakes in Aegean, Fig. 4 (continued)

collision of Adria–Eurasia occurs. The active thrust front is related with (a) low-angle thrusting (<35°) toward the coastal region, which extends to depths up to 30 km, and (b) steeply dipping (>35°) reverse type faulting, probably blind, which is overlain by anticlines, clearly visible in the topography. In the anticline geometry, the large thickness of salt formations plays a significant role (Copley et al. 2009). The majority of earthquakes are connected with the low-angle thrust faults, and fewer are connected with the steeper reverse faults (Kiratzi and Louvari 2003; Tselentis et al. 2006).

The zone of thrusting from Albania continues southward (Fig. 4a), through the coast of western Greece, along the inner east coast of the Ionian Islands and connects south of Zakynthos Island with the western termination of the Hellenic subduction zone. This thrusting is mainly along N-S trending planes, with slip vectors trending E-W. High-resolution seismic images of P- and S-wave velocity perturbations, across a line that starts from Corfu and crosses western and central Greece, provide new evidence that continental crust is subducting below northern Greece (Pearce et al. 2012) in contrast to the oceanic crust subducting between Peloponnese in the south. The images show a ~20 km thick low velocity layer (LVL) dipping at ~17° within the upper mantle beneath northern Greece. It is interpreted as subducted continental crust. This interpretation requires that ~8 km of sediments have been scraped from the subducting slab and have been accreted into the thrust belt of the overriding plate, in accordance with previous results (Royden and Papanikolaou 2011). This is required because marine seismic data indicate that the thickness of the continental crust within the foreland is ~28 km, consisting of ~20 km of crystalline crust overlain by ~8 km of sediments. The low seismicity level between Africa and Apulia and the very similar pattern of the subducted slab north and south of the Cephalonia transform fault support the model of a continuous slab from north to south (Pearce et al. 2012). This continuity and the offset induced by the Cephalonia transform fault would provide a lateral ramp geometry for this fault (Shaw 2012) instead of a slab tear.

The Thrust Zone Along the Africa–Eurasia Convergence Zone

South of Zakynthos Island and up to Rhodes Island and the southern coast of western Anatolia, the thrust and reverse faulting reflects the African–Aegean collision (Fig. 4a). The

Mechanisms of Earthquakes in Aegean, Fig. 4 Spatial organization of earthquake focal mechanisms in the Aegean (the size of the beach-ball scales with magnitude). Data were classified into the three major fault types, based on the plunges of the principal axes, and are presented for shallow (h ≤ 50 km) and deeper sources, separately. The data included in the figures are for earthquakes with Mw > 4 which occurred in the period 1950–2013, with the smaller magnitude events to span the most recent years. From 1960 onward, most of the mechanisms have been determined with waveform modeling. (**a**) Reverse/thrust faulting in the Aegean along the subduction zones. In western Albania and Greece, the compression reflects the Adria–Eurasia collision zone; south of Zakynthos Island, the compression reflects the African-Aegean convergence. Sparse thrusting in northern Greece is also evident together with the focal mechanism for the event of 3 March 1969 Mw 5.8, south of the North Anatolian Fault, which probably reflects contractional irregularities on strike-slip faults. (**b**) Normal faulting in the Aegean. Dark-shaded beach balls denote the N-S extension in the back-arc region (as defined here), whereas the light-shaded beach balls denote the ~E-W intra-mountain extension (Hellenides Mts) which connects with the along-arc extension that the overriding Aegean plate undergoes, south of the volcanic arc. (**c**) Strike-slip faulting in the Aegean discussed in more detail in the text. (**d**) Distribution of intermediate-depth (h > 50 km) earthquake focal mechanisms (Mw > 5.5), whose depths are also plotted. They are all confined within the subducting African slab which undergoes along-arc compression. The majority of the data show strike-slip and reverse/thrust faulting, with the deepest events in the eastern part of the subduction. The two cross sections depict the steeper dip angle of the subduction in the eastern edge compared to the central part. The seismicity plotted (*dots*) is from HypoDD relocated events and from field experiments

earthquakes along this zone follow the bathymetry of the Hellenic Trench and show a tendency, at least during the years of observation, to form dense discrete clusters, separated by areas with no strong earthquakes, as, for example, south of Peloponnese and Crete. The low-angle landward dipping planes are the fault planes, and the P-axis is orthogonal to the subduction interface in its western margin and parallel to it, due to arc-oblique convergence in its eastern termination.

The Zone of Normal Faulting

E-W Extension Along the Albanides: Hellenides Mountainous Chains

There is a distinct zone of ~N-S normal faulting (Fig. 4b – light-shaded beach balls) that runs along the mountainous area of eastern Albania and continues in the mainland of Greece, along the backbone of Hellenides Mountains. The ~E-W extension along this zone is perpendicular to the long axis of the mountains. There is evidence that this N-S normal faulting reaches the surface. This can be deduced from the shape of a number of intra-mountain basins, as, for example, Lake Ohrid, which has the characteristic shape of a normal fault-bounded basin, and from the surface expression of the faults at its southern end in NW Greece (Goldsworthy et al. 2002).

E-W Extension Along the Southern Part of the Overriding Aegean Plate (Fore-Arc)

South of the Volcanic Arc in southern Aegean, the distribution of the focal mechanisms reveals a broad zone of N-S normal faulting (Fig. 4b – light-shaded beach balls). It was first identified in western Crete and Peloponnese (Lyon-Caen et al. 1988), but as the focal mechanism database enriches, it is evident that this E-W extension is the dominant mode of deformation within the entire southern part of the overriding Aegean plate.

N-S Extension Along Mainland Greece and Western Anatolia (Back-Arc)

The back-arc Aegean and western Anatolia are dominated by normal faulting along roughly E-W trending planes. This is the dominant mode of faulting (Fig. 4b – dark-shaded beach balls), which is frequently combined with a considerable strike-slip component. The dip angle of the nodal planes (both planes considered) is in the range of 30–65° with an average value of ~50°. The age of the onset of the back-arc Aegean extension is from 15 to 45 Ma, depending on the author and the data used to define it (as, e.g., normal faulting bounding sedimentary basins, metamorphic core complexes in central Cyclades islands). For ~30 Ma, only back-arc extension driven by slab rollback was dominant. The interaction between the back-arc Aegean extension and the westward extrusion of Anatolia initiated around 13 Ma ago (middle Miocene) and dominates the Aegean tectonics ever since. It is worth mentioning that since the 1970s, the extension in the Aegean has been connected to the sedimentary basin formation (McKenzie 1978) and to the subduction process itself (Le Pichon and Angelier 1979).

The Zone of Strike-Slip Faulting

The Shallow-Depth Strike-Slip Fault Zones

Figure 4c shows all earthquakes with strike-slip focal mechanisms. It is evident that there are several narrow zones which accommodate shear motions, and in general, the NE-SW trending structures are associated with dextral strike-slip motions, whereas the NW-SE trending ones are associated with left-lateral strike-slip motions. In eastern Aegean Sea, the strike-slip zones reflect the shear transmitted from the east, and the major and minor segments of the North Anatolian Fault are well outlined. The strike-slip faulting reaches Evia Island but it does not cross central Greece. In western Greece, the most striking feature is the wide strike-slip zone which covers the Ionian Islands and western Peloponnese. North of ~40° parallel, the sparse strike-slip faulting is mainly confined at the edges of major normal faults. It is worth noting here the abundant strike-slip faulting along the Hellenic arc, which follows bathymetric lineaments. It is more evident at its eastern termination from the occurrence of a few strong events.

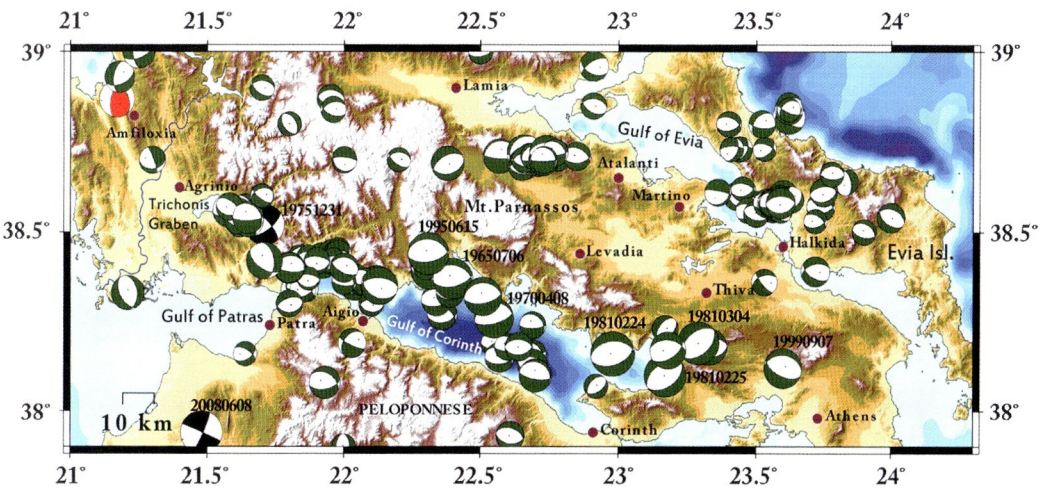

Mechanisms of Earthquakes in Aegean, Fig. 5 Development and continuation of ~E-W normal faulting bounding graben structures and basins in the region extending from the Gulf of Corinth to the Gulf of Evia, as discussed in the text. In this and the following figures, the dates for all earthquakes with Mw ≥ 5.9 are shown next to the beach balls

The Intermediate-Depth Mechanisms Within the Subducting African Lithosphere

Figure 4d summarizes the focal mechanisms for the intermediate-depth (h > 50 km) earthquakes (Mw > 5.5), which are all confined within the subducting African plate. These focal mechanisms indicate reverse faulting with considerable strike-slip component or pure strike-slip faulting. Evidently, the subducting African slab undergoes along-arc compression. The downdip length of the slab is approximately 300 km and reaches to a depth of about 160 km, as deduced from the results of field experiments (Hatzfeld and Martin 1992). There is a pronounced consistency regarding the P-axes, which, in general, are parallel to the local curvature of the subduction zone, whereas the T-axes are aligned downdip within the subducting slab (Benetatos et al. 2004). The deepest events are concentrated at the eastern part of the subducting slab, NW of Rhodes, where the slab has a steeper dip (see the cross sections in Fig. 4d), as was mentioned in the introduction.

In the following sections, special features of the earthquake focal mechanisms in the Aegean are discussed in more detail.

Characteristics of Normal Faulting in Mainland Greece

The prevalence of normal faulting in the mainland of Greece is manifested in a number of grabens and basins in central Greece, whose edges are bounded by large segmented normal faults, which in many cases are associated with impressive topographic escarpments. The maximum typical lengths of these fault segments are of the order of 15–25 km. In the following section, starting from south toward north, a number of these regions of significant normal faulting concentration are discussed.

The best studied region is the Gulf of Corinth (Fig. 5), an asymmetric graben (or rift) 100 km long, which ranks first among the rapidly extending regions in Greece. The geodetic measurements indicate a fast rate of 12–14 mm/year in the west part of the Gulf of Corinth and a slower rate of 6–8 mm/year in the eastern part. The present-day structure and kinematics of the Corinth Rift can be explained with a series of décollements relayed by steeper ramps that altogether formed a mechanically weak, crustal-scale detachment (Jolivet et al. 2010). The E-W normal

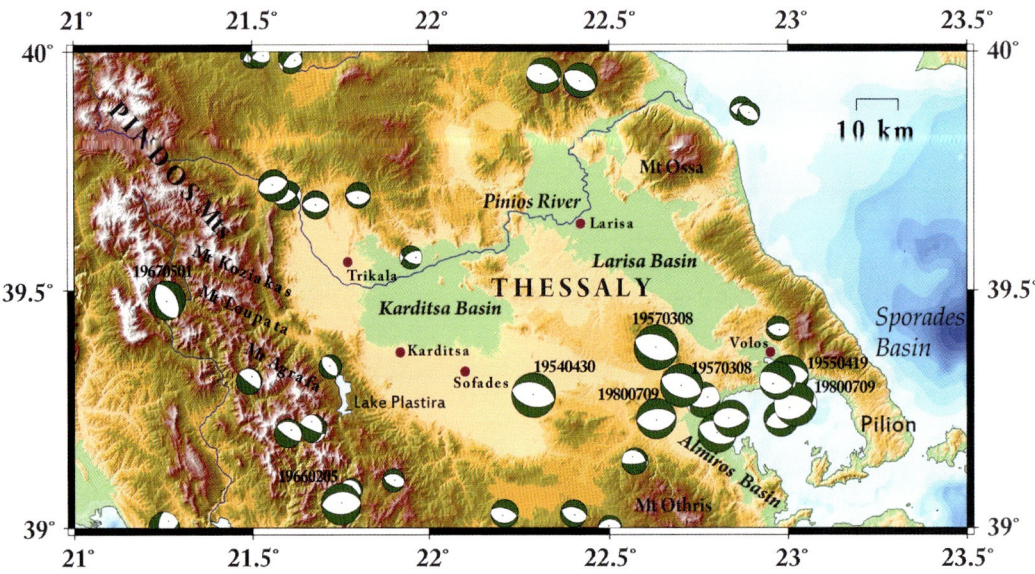

Mechanisms of Earthquakes in Aegean, Fig. 6 Development of WNW-ESE and E-W normal faulting in Thessaly, as deduced from focal mechanisms, mainly along the southern margins of the basins, extending from Sofades to the Pelion Peninsula. Along this zone, a series of strong events occurred in the 1950s and 1980s (dates designated in the figure)

faulting observed in the Gulf of Corinth continues eastward through the Thiva basin and into the Gulf of Evia. In the family of faults within the broader region of this basin, the most recent earthquake activity occurred during February–March 1981 and on September 1999 earthquake sequence. Few mechanisms are available for strong earthquakes on Evia Island itself; however from a number of moderate-size events, there are indications of normal faulting along mainly E-W trending planes.

Further to the north in central Greece (Fig. 6), most of the available focal mechanisms are from earthquakes with epicenters along the southern side of the Karditsa and Larisa basins, between Sofades and Volos. These basins are bounded by normal faulting striking NW-SE or E-W.

In northern Greece, a region of pronounced normal faulting is the Mygdonian basin (Fig. 7) close to Thessaloniki, the second metropolitan city in Greece, after the capital, Athens. The basin between the Lagkada and Volvi Lakes shows the operation of nearly E-W trending normal faulting (Tranos and Lacombe 2014), which is considered more active and more recent and is optimally oriented to the present-day stress field. One of these faults moved during the 1978 Mw 6.5 earthquake which produced significant losses to the city of Thessaloniki. The zone of E-W normal faulting close to eastern Thessaloniki is not clear if it continues westward, crossing the Gulf of Thermaikos. In this area, there is only moderate-size recent seismicity, with focal mechanisms predominantly showing strike-slip motions; however it is doubtful whether they reflect the regional tectonics. The epicenters of this modern seismicity are in the flood plain of the Axios and Aliakmon rivers, and the thick pile of sediments obscures field observations.

Further to the west, (Fig. 7) along the Ptolemais plains, the focal mechanisms indicate the operation of normal faulting along NE-SW trending planes, and the most well-studied event in this area is the 15, 1995, sequence between the cities of Grevena and Kozani. There are indications, from field observations and moderate-size seismicity, on the existence of gradually evolving network of distributed active normal faulting, between the Kastoria Lake and Vegoritis Lake (see also Fig. 3).

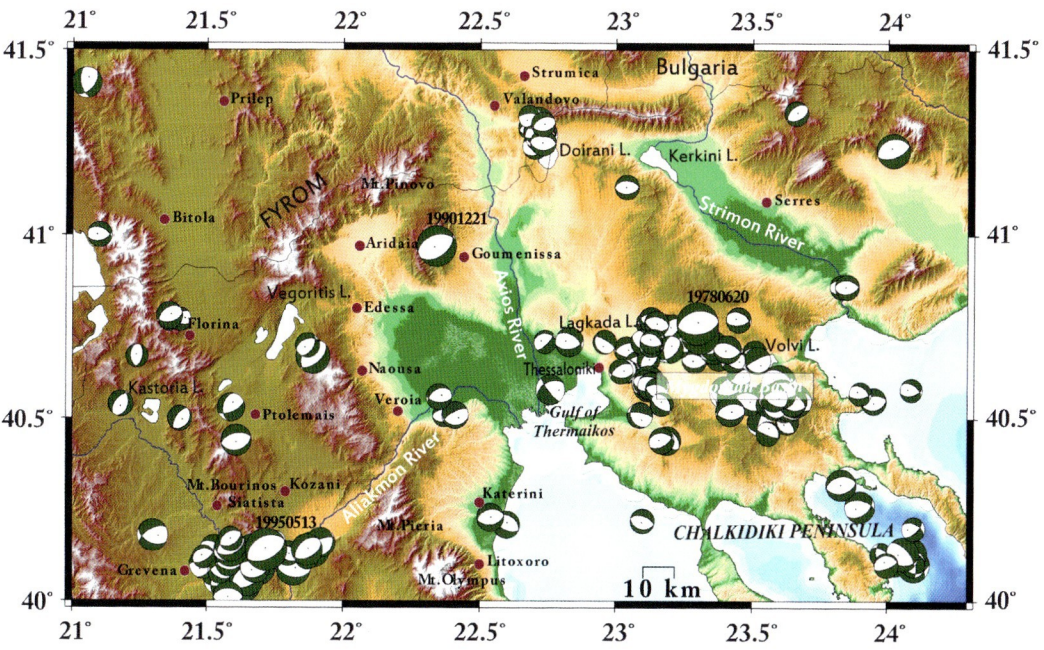

Mechanisms of Earthquakes in Aegean, Fig. 7 Development and continuation of normal faulting in northern Greece. Along the Mygdonian basin, there are abundant focal mechanisms that show development of mainly E-W normal faulting, whereas in other parts of this region, the number of the available mechanisms is limited. Nevertheless, they do depict the acting stress field and the large-scale topography

Characteristics of Strike-Slip Motions in the Aegean

Strike-Slip Motions in Central and Eastern Aegean Sea

The main characteristics of the strike-slip faulting in the eastern Aegean Sea and Thrace are (Fig. 8) (a) its connection with topographic and bathymetric escarpments that run for several hundreds of kilometers and (b) its concurrent operation with normal faulting, which differs in orientation from the classic E-W trending faults characterizing the Greek mainland. The most prominent example for the latter is the normal faulting event of 19670304 Mw 6.6 in Skiros basin; this ENE-WSW extension orientation characterizes most of the normal faulting events in central Aegean Sea (Chatzipetros et al. 2013). It can be traced up to the three-leg Chalkidiki peninsula as deduced from a few moderate earthquake focal mechanisms.

In northern-eastern Greece, the Kavala-Xanthi-Komotini fault zone has a prominent escarpment in the morphology, which suggests normal oblique faulting as the prevailing mechanism. Unfortunately, the instrumental seismicity and the available focal mechanisms are not sufficient to describe the type of faulting.

Along the eastern Aegean Sea, strike-slip faulting forms distinctive stands, which bound a number of basins/troughs, the most prominent of which are: (a) The North Aegean Basin (NAB) which consists a number of fault-bounded basins which extend from Saros bay in the east to the Sporades islands in the west. Moreover, along the NAB which accommodates a large part of the westward motion of Anatolia relative to Eurasia, the occurrence of strong events is abundant. (b) The Skiros- Edremit Trough and the Lesvos-Psara Trough, which are both the sites of several strong events. There is evidence from the bathymetry and the focal mechanisms of small magnitude events that the strike-slip faulting reaches the Cyclades Islands in the west.

The dextral strike-slip motions in the north and central Aegean Sea extend westward and

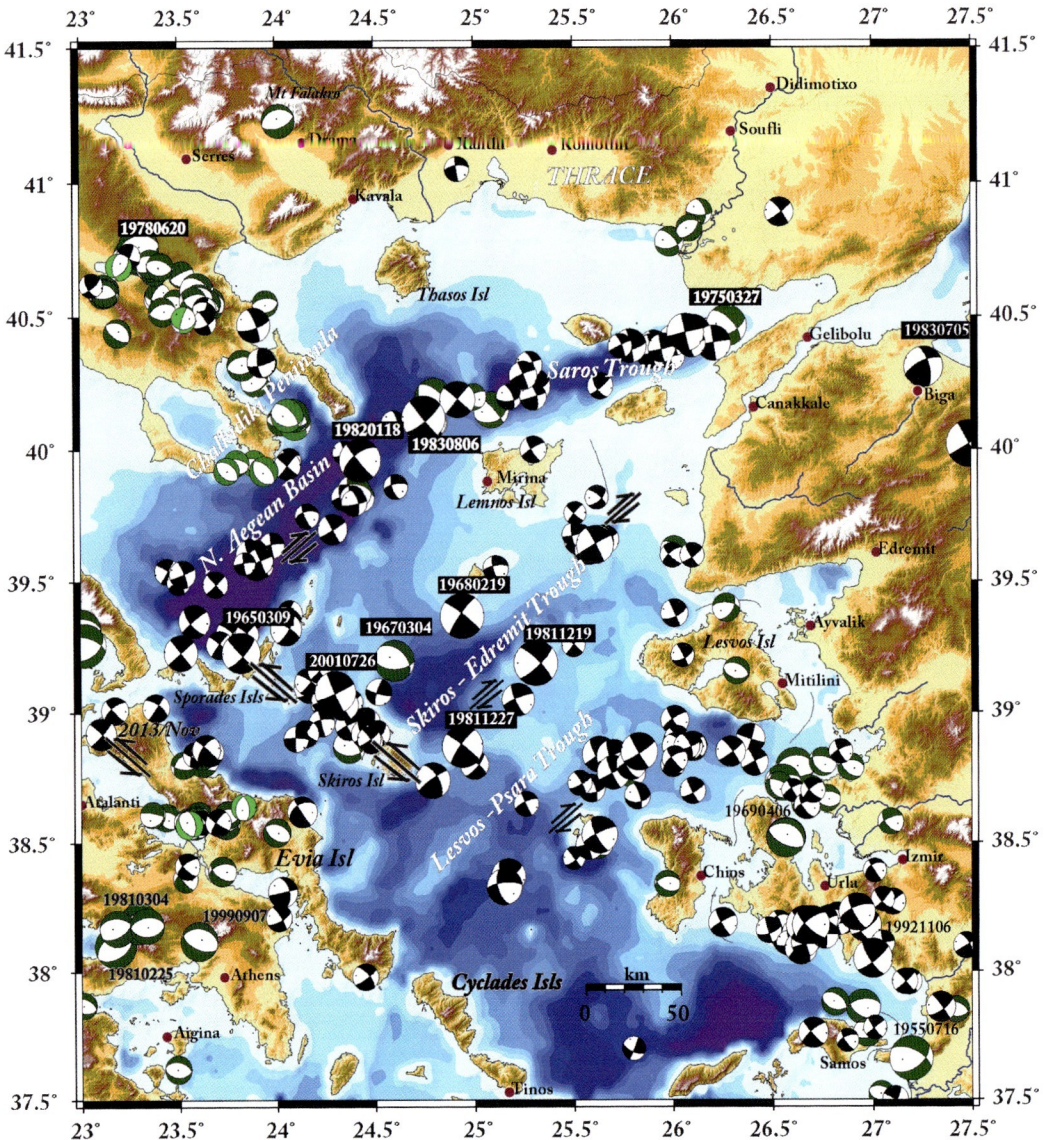

Mechanisms of Earthquakes in Aegean, Fig. 8 Distribution of strike-slip faulting in central and eastern Aegean Sea and regions of localized normal faulting within it, as discussed in the text. The *arrows* indicate the polarity of shear motions, right-lateral in the central and eastern part and left-lateral near the interface with the mainland of Greece

terminate close to the mainly E-W striking normal faulting system that controls the eastern coastline of the Greek mainland. This transition from dextral NE-SW shear motion to E-W normal faulting is reflected in the azimuth of the slip vectors, which changes from NE-SW to N-S.

Strike-Slip Motions Near the Intersection with the Eastern Greek Mainland

The linkage of the strike-slip motions in eastern Aegean Sea with the normal faulting in mainland Greece has been an interesting field of research and debate. In this transition zone, the Sporades Islands and Evia Island have a key location (see Fig. 8).

One point that is worth mentioning is the presence of NW-SE trending structures, close to this transition zone, that are activated as strike-slip faults with left-lateral component and not dextral as in the eastern Aegean Sea.

The first well-recorded information for left-lateral strike-slip motions was from the 20010726 Skiros Island earthquake sequence (Fig. 8). A number of previous strong earthquakes, for example, the event of 19650309 in Sporades Islands, could be also connected with left-lateral strike-slip motions. A well-located seismic sequence in northern Evia, which burst in November 2013, provided additional evidence for left-lateral strike-slip motions.

In view of the above, the operation of these NNW-SSE trending structures as left-lateral strike-slip faults should be also taken into account in the discussions regarding the change in the direction of the slip vectors. It is also worth mentioning that this left-lateral strike-slip component along the western shoreline of the North Aegean Sea is well resolved in the GPS-derived strain field (Müller et al. 2013). The change of strike of the active structures from east to west, e.g., from NE-SW to east Aegean to NW-SE at the edge of the Greek mainland to E-W on the mainland itself, and the subsequent change in the direction of the slip vectors can be accommodated by clockwise rotations of the fault-bounded blocks in central Greece relative to Eurasia (broken-slats model of Taymaz et al. 1991).

Strike-Slip Motions in Western Greece

As previously mentioned, western Greece lies in the crossroads of continent–continent collision in the north and ocean–continent subduction in the south (Fig. 9). These two regimes are offset by ~100 km (Pearce et al. 2012) across the Cephalonia–Lefkada Transform Fault, whose existence has been postulated by McKenzie (1978). It is expressed as an abrupt lateral ramp in the bathymetry (Shaw 2012) approximately 4,000 m deep, between the Adriatic Sea and the Ionian Islands. Two segments have been identified along this fault zone, the 40 km long Lefkada segment and the 90 km long Cephalonia segment (Louvari et al. 1999). This fault zone is very active, and this is the reason why the Hellenic Seismic Code is designed for the highest resistant constructions. Recently, on 26 January 2014, an Mw 6.1 seismic sequence burst on Cephalonia Island, followed by another Mw 6.0 event on 3 February 2014. The first event was connected with a N24°E and steeply dipping strike-slip fault, whereas the second event was connected with a N297°E and shallow dipping thrust fault. These thrust faulting events are not unusual in the Ionian Islands and are usually associated with severe ground shaking. This zone of thrusting follows the Ionian Thrust which is a major structural feature, a crustal–scale thrust fault (Kokkalas et al. 2013). It is worth noting that the 1953 earthquakes that destroyed the islands of Cephalonia and Zakynthos are associated with coastal uplift and thrusting mechanisms.

The strike-slip motions in western Greece cover a wide zone between the Ionian Islands and western Peloponnese. Strike-slip motions along the northwestern coast of Peloponnese have been already noticed in Louvari et al. (1999) and were confirmed by the occurrence of the 8 June 2008 Mw 6.3 Achaia earthquake. This event, which had a source depth of ~22 km, was the first strong event which provided evidence for a gradually developing right-lateral strike-slip zone, the Achaia strike-slip fault zone.

From a number of moderate-size earthquake focal mechanisms, there is evidence for left-lateral strike-slip motions, oblique to the main branches of the right-lateral shear. For example, such a left-lateral shear zone extends from Lake Trichonis toward Amfiloxia and the Gulf of Amvrakikos (Fig. 9), and its operation is supported by the morphology and a characteristic drainage pattern (Vassilakis et al. 2011).

This tectonic setting with distinct parallel branches of NE-SW right-lateral strike-slip faulting and oblique to it NW-SE left-lateral strike-slip faulting resembles the tectonic setting in central and eastern Aegean Sea, as it was previously discussed.

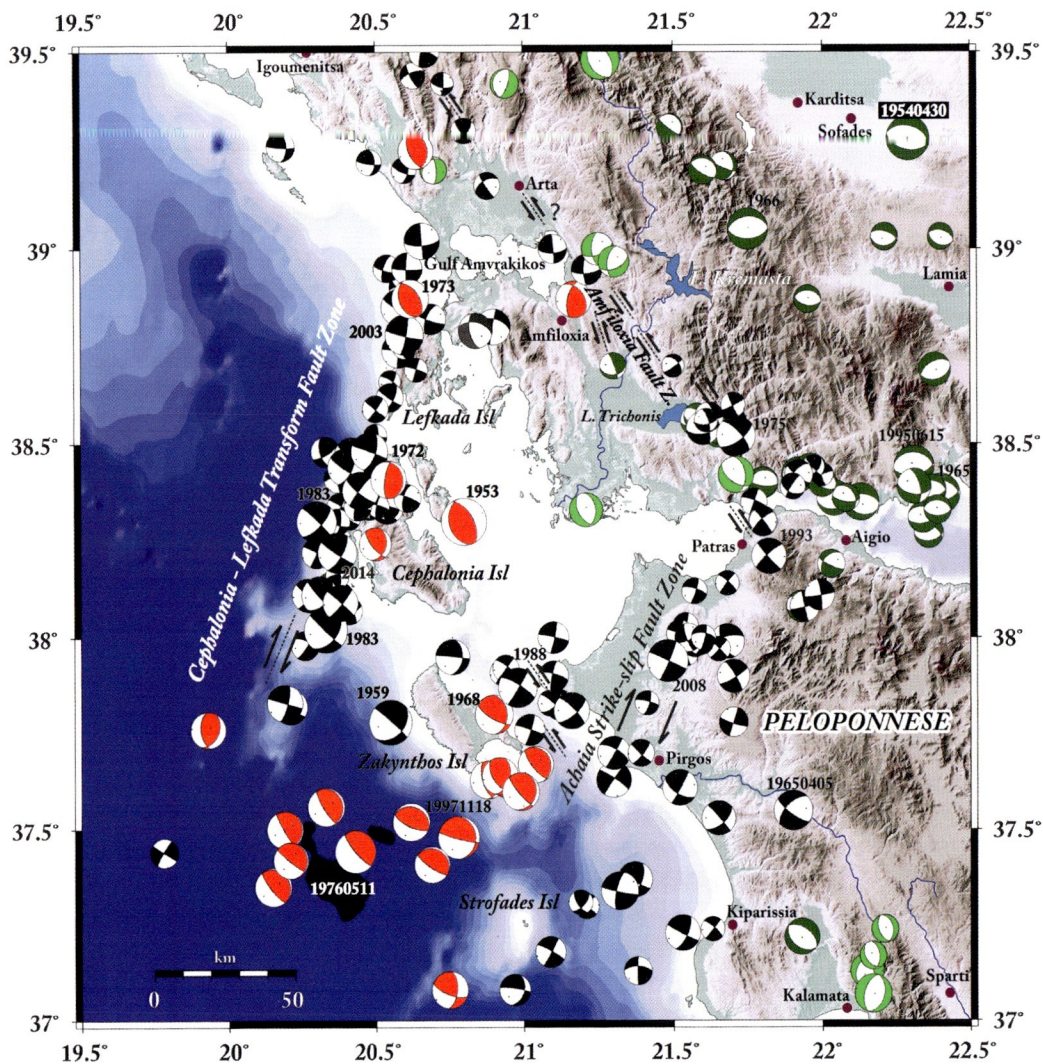

Mechanisms of Earthquakes in Aegean, Fig. 9 Distribution of strike-slip faulting in the western margin of the Aegean. The two parallel fault zones (e.g., the Cephalonia–Lefkada in the Ionian Islands and the Achaia in western Peloponnese) are associated with well-documented right-lateral shear motions. It is worth noting here the thrust faulting which develops onshore the islands of Lefkada, Cephalonia, and Zakynthos which constitutes a significant threat for the populated regions. Regions of left-lateral strike-slip motions evidenced from moderate-size earthquake mechanisms are discussed in the text

Summary

A wealth of earthquake focal mechanisms is now available for the broader Aegean which provides the tools to study regional tectonics and elucidate the presently acting stress field. From the early 1960s onward, nearly a thousand of focal mechanisms determined from synthetic seismograms can be traced in the published literature. These mechanisms, if classified into the three major fault types (reverse/thrust, normal, and strike-slip), exhibit uniform spatial organization in localized regions, especially in central Greece, and more complex organization at the western and eastern borders of the Aegean.

Mechanisms of Earthquakes in Aegean, Table 2 Typical focal mechanisms for normal/reverse/strike-slip type faulting in the Aegean, derived by summing magnitude-normalized moment-tensor components (for Mw > 4 and h ≤ 50 km) of each category. The average depths provide a scale of the thickness of the seismogenic layer

Typical focal mechanisms for shallow earthquakes in the Aegean											
Nodal plane 1			Nodal plane 2			P-axis az°/pl°	T-axis az°/pl°	N-axis az°/pl°	Depth (km)	Fault type	Number of events
Strike°	Dip°	Rake°	Strike°	Dip°	Rake°						
89	45	−94	275	45	−86	276/87	182/0	92/3	13 ± 9	Normal	422
331	33	102	136	58	82	232/13	23/76	141/7	16 ± 11	Reverse	182
45	84	177	135	87	6	270/2	0/6	166/83	15 ± 9	Strike-slip	400

The available focal mechanisms for earthquakes with M > 4 and for depths less than 50 km are not balanced in terms of the faulting type. The majority is for normal (40 %) and strike-slip (40 %) faulting and the remaining (20 %) is for reverse/thrust faulting. This crude quantification also reflects the dominant modes of deformation within the Aegean area. Taking this gross statistics a step further, Table 2 summarizes the typical focal mechanisms for each fault type, deduced by summing the moment-tensor components and normalizing for the magnitude.

Summarizing, the compression revealed by the distribution of the reverse/thrust faulting (Fig. 4a) is well confined in two regimes: (a) in the north, the collision of Adria and Eurasia takes place, and the active thrust front is confined in a narrow belt in the lowlands of Albania and western Greece; new evidence suggest that continental crust is subducting below northwestern Greece; (b) in the south, at the Ionian Islands, and southward the zone of low-angle thrusting follows the curvature of the Hellenic Trench and continues up to western Anatolia; this low-angle land dipping thrusting reflects the African–Aegean convergence; it does not mark the subduction interface itself, as this is projected south of the Mediterranean Ridge (very likely below the thick pile of sediments). These two subduction zones are separated by the Cephalonia–Lefkada Transform Fault.

~N-S extension (Fig. 4b) is confined in continental Greece and western Anatolia. Within the Aegean Sea itself, the extension is localized, and its orientation is mainly ENE-WSW. In central Greece, the extension is accommodated by slip along families of normal fault segments. The major and most prominent faults bound the southern part of a series of grabens. The Gulf of Corinth and the Gulf of Evia are good examples of this geometry.

~E-W extension (Fig. 4b) occurs along a distinct belt which can be traced all the way from the mountainous inner Albania, through the backbone of the Hellenides Mountains, spreading along the southern Aegean and ending in western Anatolia. In Albania and northwestern Greece, the zone of E-W extension is subparallel to the outer zone of thrusting and is attributed to contrasts of the gravitational potential energy, between the high mountains in the west and the lowlands of western Albania and the Adriatic Sea. E-W extension occurs south of the Volcanic Arc, in the southern Aegean Sea and Crete itself, and reflects the deformation within the overriding Aegean plate. The boundary, if any, between these two extensional regimes of similar orientation should be sought somewhere south of the 38° parallel where the signature of the subduction process begins to emerge.

Strike-slip faulting (Fig. 4c) is abundant in the Aegean especially in its eastern and western edge. Shear motions tend to develop in parallel strands. For example, in eastern and central Aegean Sea (Fig. 8), the best developed and more clearly related to the prolongation of the North Anatolian Fault into the Aegean is the strike-slip faulting along the North Aegean Basin (or else North Aegean Trough). The right-lateral shear ends abruptly at the western edge of the Greek mainland, near the Sporades Islands and Evia Island. New evidence from modern strong events suggests that along this interface, left-lateral strike-slip faulting is taking place, which is further supported by updated versions of the GPS-derived strain field (Müller et al. 2013). In brief, as previously suggested depending on the fault strike, the present deformation field allows for both right-lateral and left lateral shear motions (Kiratzi 2002).

The shear imposed from the Anatolia extrusion passes central Greece as it is depicted by the orientation of the earthquake slip vectors, which alter from being parallel to the shear motions to become approximately N-S. Clockwise rotations of the fault-bounded blocks in central Greece relative to Eurasia (broken-slats model of Taymaz et al. 1991) provide a suitable kinematic mechanism to explain this change in the orientation of the slip vectors.

The shear motions which pass the Greek mainland are accommodated in the western Aegean margin in a similar pattern as in central and eastern Aegean Sea (Fig. 9). This is evidenced by the wide zones of strike-slip faulting which develop, again in parallel strands, along the western coasts of the Ionian Islands and western Peloponnese. These strike-slip faults are mechanically necessary in order to transport the strain and act as planes of weakness. The right-lateral Cephalonia-Lefkada and the more immature Achaia strike-slip fault zones are well resolved from modern strong events.

There are indications, from moderate-size events, for left-lateral strike-slip faulting along two zones: (a) one extending from the Gulf of Patras and Lake Trichonis up to Amfiloxia, following the NNE-trending Amfiloxia fault zone (Kiratzi et al. 2008) and (b) one extending from Pirgos in western Peloponnese up to Zakynthos Island (see Fig. 9 for locations). These two strands are not yet well documented from the occurrence of strong events.

Regarding the intermediate-depth (h > 50 km) earthquake focal mechanisms, these are all confined within the subducting Africa slab (see Fig. 4d). The azimuths of the P-axes are consistently parallel to the local trend of the Hellenic arc showing along-arc shortening. The azimuths of the T-axes are radial to the arc and are aligned with the dipping slab (Benetatos et al. 2004).

In conclusion, the earthquake focal mechanisms in the Aegean most of the times reveal remarkable uniformity, order, and kinematic links. They provide the most significant tool to study present day tectonics and the key to study past tectonics.

Data Sources

The data used in this work have been all retrieved from published sources, especially for the stronger earthquakes. A number of focal mechanisms from smaller magnitude events are from the author's unpublished work. Most of the figures were produced using the GMT software (Wessel and Smith 1998).

Cross-References

▶ Earthquake Mechanisms and Stress Field
▶ Earthquake Mechanism Description and Inversion
▶ Earthquake Mechanisms and Tectonics
▶ Long-Period Moment-Tensor Inversion: The Global CMT Project
▶ Moment Tensors: Decomposition and Visualization
▶ Non-Double-Couple Earthquakes

References

Basili R, Kastelic V, Demircioglu MB, Garcia Moreno D, Nemser ES, Petricca P, Sboras SP, Besana-Ostman GM, Cabral J, Camelbeeck T, Caputo R, Danciu L, Domac H, Fonseca J, García-Mayordomo J, Giardini D, Glavatovic B, Gulen L, Ince Y, Pavlides S, Sesetyan K, Tarabusi G, Tiberti MM, Utkucu M, Valensise G, Vanneste K, Vilanova S, Wössner J (2013) The European database of seismogenic faults (EDSF) compiled in the framework of the project SHARE. http://diss.rm.ingv.it/share-edsf/. doi:10.6092/INGV.IT-SHARE-EDSF

Benetatos C, Kiratzi A, Papazachos C, Karakaisis G (2004) Focal mechanisms of shallow and intermediate depth earthquakes along the Hellenic arc. J Geophys Res 37(253–296):2004

Chatzipetros A, Kiratzi A, Sboras S, Pavlides S (2013) Active faulting in the north-eastern Aegean Sea

Islands. Tectonophysics 597:106–122. doi:10.1016/j.tecto.2012.11.026

Copley A, Boait F, Hollingsworth J, Jackson J, McKenzie D (2009) Subparallel thrust and normal faulting in Albania and the roles of gravitational potential energy and rheology contrasts in mountain belts. J Geophys Res 114, B05407. doi:10.1029/2008JB005931

Dreger DS (2003) TDMT_INV: time domain seismic moment tensor INVersion. Int Handb Earthq Eng Seismol 81B:1627

Ekström G, Nettles M, Dziewonski AM (2012) The global CMT project 2004–2010: centroid-moment tensors for 13,017 earthquakes. Phys Earth Planet Inter 200–201:1–9. doi:10.1016/j.pepi.2012.04.002

Goldsworthy M, Jackson J, Haines J (2002) The continuity of active fault systems in Greece. Geophys J Int 148(3):596–618. doi:10.1046/j.1365-246X.2002.01609.x

Hatzfeld D, Martin C (1992) Intermediate depth seismicity in the Aegean defined by teleseismic data. Earth Planet Sci Lett 113:267–275

Hatzfeld D, Pedotti G, Hatzidimitriou P, Makropoulos K (1990) The strain pattern in the western Hellenic arc deduced from a microearthquake survey. Geophys J Int 101:181–202

Jackson J (2001) Living with earthquakes: know your faults. J Earthq Eng 5:5–123

Jolivet L, Labrousse L, Agard P, Lacombe O, Bailly V, Lecomte E, Mouthereau F, Mehl C (2010) Rifting and shallow-dipping detachments, clues from the Corinth Rift and the Aegean. Tectonophysics 483:287–304

Jolivet L, Faccenna C, Huet B, Labrousse L, Le Pourhiet L, Lacombe O, Lecomte E, Burov E, Denèle Y, Brun J-P, Philippon M, Paul A, Salaün G, Karabulut H, Piromallo C, Monié P, Gueydan F, Okay A, Oberhänsli R, Pourteau A, Augier R, Gadenne L, Driussi O (2013) Aegean tectonics: strain localisation, slab tearing and trench retreat. Tectonophysics 597–598:1–33

Kiratzi A (2002) Stress tensor inversions along the westernmost north and central Aegean Sea. Geophys J Int 106:433–490

Kiratzi A, Louvari E (2003) Focal mechanisms of shallow earthquakes in the Aegean Sea and the surrounding lands determined by waveform modelling: a new database. J Geophys Res 36:251–274

Kiratzi A, Sokos E, Ganas A, Tselentis A, Benetatos C, Roumelioti Z, Serpetsidaki A, Andriopoulos G, Galanis O, Petrou P (2008) The April 2007 earthquake swarm near Lake Trichonis and implications for active tectonics in western Greece. Tectonophysics 452:51–65

Kokkalas S, Kamberis E, Xypolias P, Sotiropoulos S, Koukouvelas I (2013) Coexistence of thin- and thick-skinned tectonics in Zakynthos area (western Greece): insights from seismic sections and regional seismicity. Tectonophysics 597–598:73–84

Le Pichon X, Angelier J (1979) The Hellenic arc and trench system: a key to the neotectonic evolution of the eastern Mediterranean area. Tectonophys 60(1–2):1–42. doi:10.1016/0040-1951(79)90131-8

Louvari E, Kiratzi A and Papazachos B (1999) The Cephalonia Transform Fault and its continuation to western Lefkada Island. Tectonophysics 308:223–236

Lyon-Caen H, Armijo R, Drakopoulos J, Baskoutas J, Delibasis N, Gaulon R, Kouskouna V, Latoussakis J, Makropoulos N, Papadimitriou P, Papanastasiou D, Pedotti G (1988) The 1986 Kalamata (South Peloponnesus) earthquake: detailed study of a normal fault, evidences for east–west extension in the Hellenic arc. J Geophys Res 93:14967–15000

McClusky S, Balassanian S, Barka A, Demir C, Georgiev I, Hamburg M, Hurst K, Kahle H, Kastens K, Kekelidze G, King R, Kotzev V, Lenk O, Mahmoud S, Mishin A, Nadariya M, Ouzounis A, Paradissis D, Peter Y, Prilepin M, Reilinger R, Sanli I, Seeger H, Tealeb A, Toksoz MN, Veis G (2000) Global positioning system constraints on plate kinematics and dynamics in the eastern Mediterranean and Caucasus. J Geophys Res 105:5695–5720

McKenzie D (1978) Active tectonics of the Alpine-Himalayan belt: the Aegean Sea and surrounding regions. Geophys J Roy Astron Soc 55:217–254

Müller M, Geiger A, Kahle H-G, Veis G, Billiris H, Paradissis D, Felekis S (2013) Velocity and deformation fields in the North Aegean domain, Greece, and implications for fault kinematics, derived from GPS data 1993–2009. Tectonophysics 597–598:34–49

Papazachos BC, Comninakis PE (1971) Geophysical and tectonic features of the Aegean arc. J Geophys Res 76:8517–8533

Papazachos BC, Papadimitriou EE, Kiratzi AA, Papazachos CB, Louvari E (1998) Fault plane solutions in the Aegean Sea and the surrounding area and their tectonic implication. Boll Geof Teor App 39:199–218

Papazachos BC, Karakostas VG, Papazachos CB, Scordilis EM (2000) The geometry of the Wadati-Benioff zone and lithospheric kinematics in the Hellenic arc. Tectonophysics 319:275–300

Pearce FD, Rondenay S, Sachpazi M, Charalampakis M, Royden LH (2012) Seismic investigation of the transition from continental to oceanic subduction along the western Hellenic subduction zone. J Geophys Res 117, B07306. doi:10.1029/2011JB009023

Ring U, Glodny J, Will T, Thomson S (2010) The Hellenic subduction system: high-pressure metamorphism, exhumation, normal faulting, and large-scale extension. Annu Rev Earth Planet Sci 38:45–76. doi:10.1146/annurev.earth.050708.170910

Royden LH, Papanikolaou DJ (2011) Slab segmentation and late Cenozoic disruption of the Hellenic arc. Geochem Geophys Geosyst 12:Q03010. doi:10.1029/2010GC003280

Shaw B (2012) Active ectonics of the Hellenic subduction zone. Springer theses, vol 1. doi:10.1007/978-3-642-20804-1_1. © Springer, Berlin

Sipkin SA (1994) Rapid determination of global moment-tensor solutions. Geophys Res Lett 21:1667–1670

Sokos E, Zahradnik J (2008) ISOLA a Fortran code and a Matlab GUI to perform multiple-point source inversion of seismic data. Computers & Geosciences 34:967–977. doi: 10.1016/j.cageo.2007.07.005

Taymaz T, Jackson J, McKenzie D (1991) Active tectonics of the north and central Aegean Sea. Geophys J Int 106(2):433–490. doi:10.1111/j.1365-246X.1991.tb03906.x

Tranos MD, Lacombe O (2014) Late Cenozoic faulting in SW Bulgaria: fault geometry, kinematics and driving stress regimes. Implications for late orogenic processes in the Hellenic hinterland. J Geodyn 74:32–55

Tselentis G-A, Sokos E, Martakis N, Serpetsidaki A (2006) Seismicity and seismotectonics in Epirus, western Greece: results from a microearthquake survey. Bull Seismol Soc Am 96(5):1706–1717. doi:10.1785/0120020086

Udias A, Buforn E, de Gauna R (1989) Catalogue of focal mechanisms of European earthquakes. Universidad Complutense de Madrid, Madrid, p 274, ISBN 84-600-721903

Vassilakis E, Royden L, Papanikolaou D (2011) Kinematic links between subduction along the Hellenic trench and extension in the Gulf of Corinth, Greece: a multidisciplinary analysis. Earth Planet Sci Lett 303:108–120

Wessel P, Smith WHF (1998) New improved version of the generic mapping tools released. Eos Trans AGU 79:579

Zoback ML (1992) First- and second-order patterns of stress in the lithosphere: the world stress map project. J Geophys Res 97:11703–11728

Zwick P, McCaffrey R, Abers G (1994) MT5 Program, IASPEI Software Library, 4

Mechanisms of Earthquakes in Iceland

Páll Einarsson
Institute of Earth Sciences, University of Iceland, Reykjavík, Iceland

Synonyms

Bookshelf faulting; Divergent plate boundaries; Hreppar microplate; Seismicity of Iceland; South Iceland Seismic Zone; Tjörnes fracture zone; Transform zones; Volcanic earthquakes

Introduction

Most of the seismicity of Iceland is due to processes related to the mid-Atlantic plate boundary that crosses the country (Fig. 1). These processes include transform faulting, rifting, dike injection, inflation and deflation of magma chambers, cooling of magma bodies, and volume changes of geothermal fluids (Einarsson 1986, 1991, 2008; Sigmundsson 2006).

South of Iceland the seismically active belt follows the axis of the Reykjanes Ridge, and north of the country it follows the Kolbeinsey Ridge. On both ridges the seismicity decreases toward Iceland. This may be due to the existence of the Iceland hotspot, an area of high topography and heat flow, a result of excessive volcanism. The excessive volcanism produces a relatively thick crust of elevated temperature gradient. Combined with the spreading of the plates away from the plate boundary, the thick crust forms a complex of submarine highs extending between Scotland and Greenland (Fig. 1). The spreading velocity along the plate boundary is relatively low, less than 2 cm/year. Yet the topography of the ridge segments near Iceland, Reykjanes and Kolbeinsey Ridges, shows characteristics of fast-spreading ridges. Near its southern end, 53–59°N, on the other hand, the Reykjanes Ridge has the characteristics of a slow-spreading ridge, rough topography, deep rift valley, and relatively high seismicity. North of 59°, the fast-spreading characteristics become more prominent (Einarsson 2008). There is no direct evidence for faster spreading, however. Magnetic anomalies show spreading rates of 18–20 mm per year in a direction of 105°, corresponding to the distance to the Euler pole between the North America and Eurasia plates that is located in Siberia (Einarsson 2008; Sigmundsson 2006).

The plate boundary on Iceland is expressed by a series of seismic and volcanic zones. Two transform zones connect the presently active Northern and Eastern Volcanic Zones to the ridges offshore (Fig. 2). The transform zones host the largest earthquakes in the area, as large as magnitude 7, whereas the volcanic rift zones produce only smaller earthquakes, rarely exceeding magnitude 5.5.

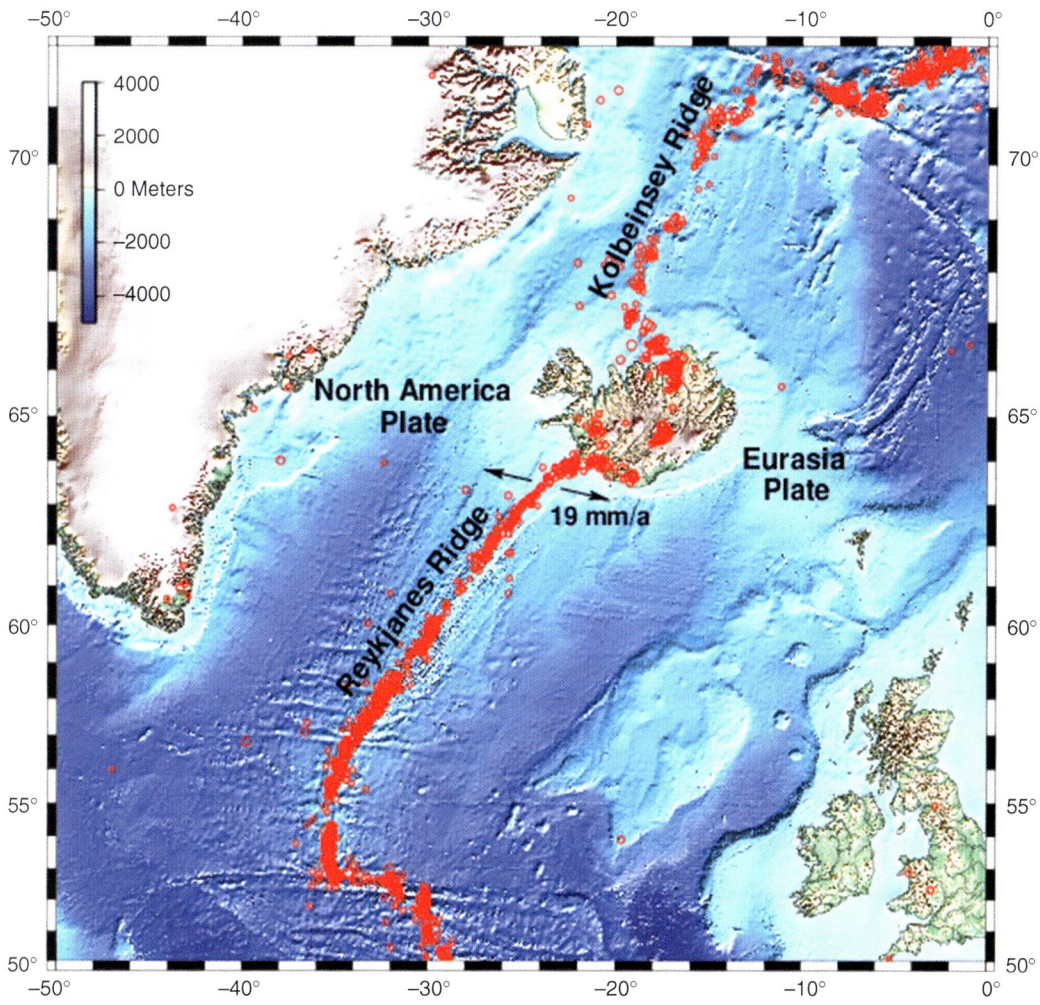

Mechanisms of Earthquakes in Iceland, Fig. 1 Epicenters in Iceland and along the Mid-Atlantic Ridge system 1964–2004. Data are from the catalogs of the International Seismological Centre. Bathymetry is from General Bathymetric Chart of the Oceans (Modified from Sólnes et al. (2013) with help from Gunnar B. Guðmundsson at the Icelandic Meteorological Office)

The structure of the plate boundary is strongly influenced by the Iceland mantle plume presumed to be centered beneath central Iceland. The plume is thought to have a deep root in the mantle, and its upwelling leads to the production of the magma that feeds the volcanoes of the hotspot. The relative motion of the Mid-Atlantic Ridge with respect to the plume leads to ridge jumps, propagating rifts and other complexities (e.g., Einarsson 1991, 2008). A new rift zone is initiated when the old rift has migrated some critical distance off the plume center. The transform zones are also unstable and respond to changes in the configuration of the rifts.

Seismic and Tectonic Characteristics of Seismic Areas

Plates, Microplates, and Boundaries

The mid-Atlantic plate boundary is relatively simple in most parts of the NE Atlantic consisting of rifting and transform segments separating the two major plates, Eurasia and North America.

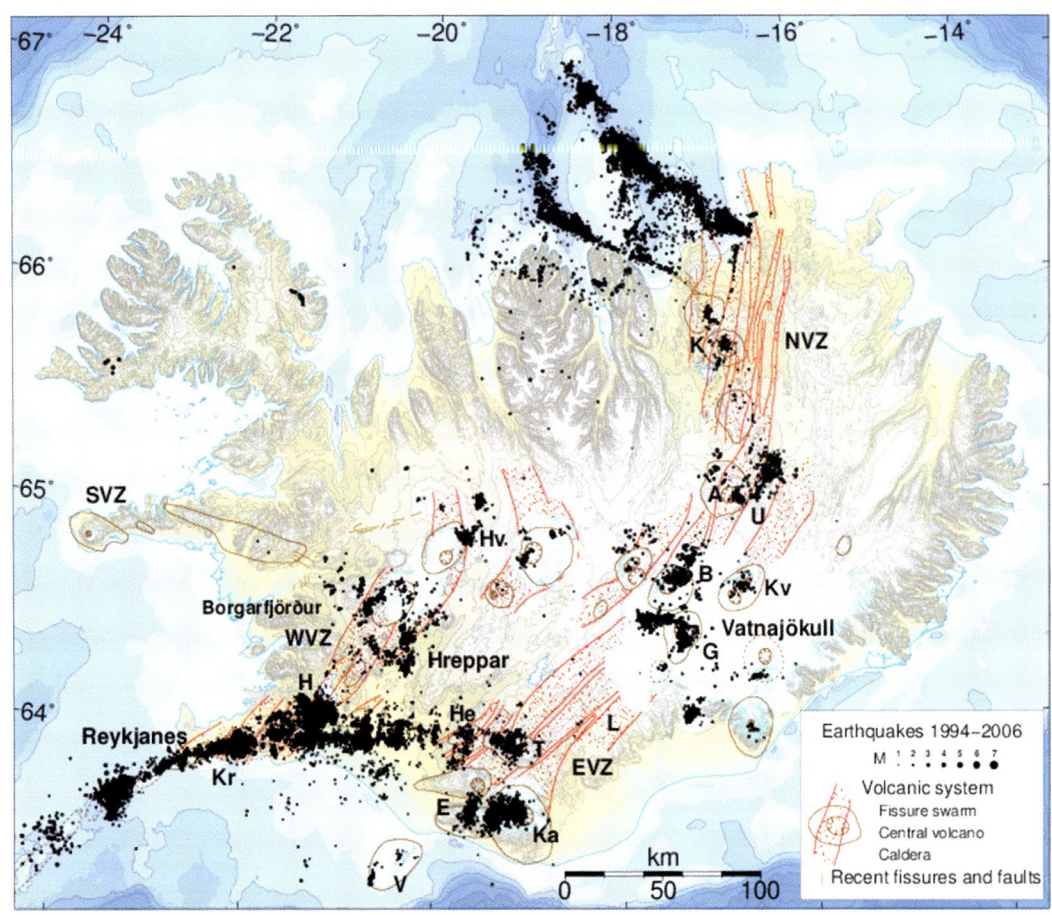

Mechanisms of Earthquakes in Iceland, Fig. 2 Volcanic systems of Iceland and epicenters located by the local seismic network in the period 1994–2006. The volcanic systems are from Einarsson and Sæmundsson (1987) and the seismic data from the Icelandic Meteorological Office. The fissure swarms of the volcanic systems delineate the divergent segments of the plate boundary, the Northern, Western, and Eastern Volcanic Zones (NVZ, WVZ, and EVZ), the last two of which outline the Hreppar Microplate. SVZ marks the Snæfellsnes Volcanic Zone and K, A, B, Kv, G, T, He, Ka, E, V, H, and Kr mark the Krafla, Askja, Bárðarbunga, Kverkfjöll, Grímsvötn, Torfajökull, Hekla, Katla, Eyjafjallajökull, Vestmannaeyjar, Hengill, and Krísuvík volcanic systems. Hveravellir, Laki, and Upptyppingar are shown with Hv, L, and U (Modified from Sólnes et al. (2013) with the help of GBG at IMO)

The boundary is clearly defined by the epicenters of earthquakes that show a narrow zone of deformation (Fig. 1). As it crosses the Iceland platform, however, the deformation zone becomes wider, as shown by the distribution of earthquakes and volcanism in Fig. 2. The boundary breaks up into a series of more or less oblique segments. Sometimes the boundary branches out, and small microplates or blocks are formed that may move independently of the large plates. The largest of these is the Hreppar Microplate that is bounded by the two subparallel volcanic rift zones in South Iceland (Einarsson 2008; Sigmundsson et al. 1995; La Femina et al. 2005), the Western and Eastern Volcanic Zones (Fig. 2). Each branch or segment of the plate boundary has its own characteristics, both regarding tectonic style and seismicity.

South Iceland Seismic Zone

The southern boundary of the Hreppar Microplate, near 64°N, is a zone of high seismic

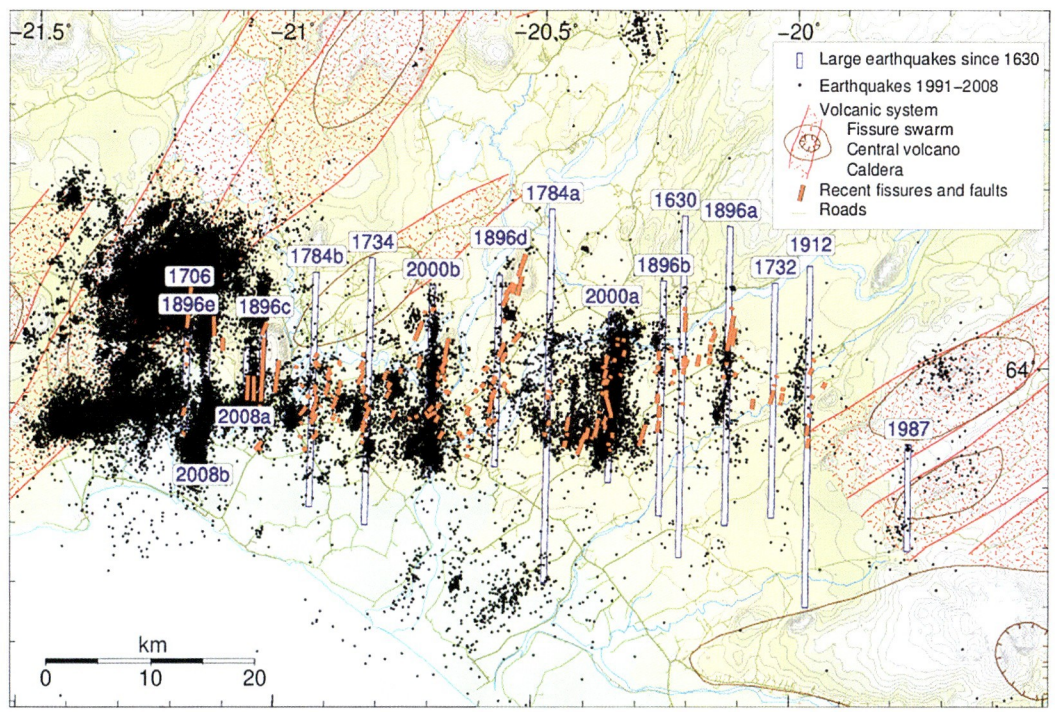

Mechanisms of Earthquakes in Iceland, Fig. 3 South Iceland Seismic Zone, epicenters, and suggested source faults of historical earthquakes. Epicenters for the time period 1991–2006 are from the Icelandic Meteorological Office, volcanic systems are from Einarsson and Sæmundsson (1987), recent fissures and faults are from Einarsson (2010), and source faults of historical earthquakes (1630–1912) are based on fault mapping and investigation of historical documents (Modified from Sólnes et al. (2013) with the help of GBG at IMO)

activity, the South Iceland Seismic Zone (SISZ), which takes up most of the transform motion between the Reykjanes Ridge and the Eastern Volcanic Zone (Fig. 3). It has been argued that rifting is dying out in the Western Volcanic Zone and is being taken over by the Eastern Volcanic Zone (Einarsson 2008).

The majority of destructive, historical earthquakes in Iceland have taken place within the South Iceland Seismic Zone. The zone has been defined by destruction areas of historical earthquakes, Holocene surface ruptures, and instrumentally determined epicenters (Figs. 2 and 3). It is oriented E-W and is 10–15 km wide. Destruction areas of individual earthquakes and surface faulting show, however, that each event is associated with faulting on N-S striking planes, perpendicular to the main zone. The overall left-lateral transform motion across the zone thus appears to be accommodated by right-lateral faulting on many parallel, transverse faults and counterclockwise rotation of the blocks between them, "bookshelf faulting" (Einarsson 1991, 2010).

Earthquakes in South Iceland tend to occur in major sequences in which most of the zone is affected (Einarsson et al. 1981). These sequences tend to last from a few days to about 3 years or even more. Each sequence typically begins with a magnitude 7 event in the eastern part of the zone, followed by smaller events farther west. Sequences of this type occurred in 1896, 1784, 1732–1734, 1630–1633, 1389–1391, 1339, and 1294. The sequences thus occur at intervals that range between 45 and 112 years, and it has been argued that a complete strain release of the whole zone is accomplished in about 140 years (Stefánsson et al. 1993). In addition to the sequences, single events occasionally occur near the ends of the zone, e.g., in 1912 at the eastern end.

The 1896 and 1912 earthquakes were followed by a long quiet interval, which in 1985 led to a long-term forecast of a major earthquake sequence within the next decades (Stefánsson et al. 1993). The forecast was fulfilled in June 2000 when two magnitude 6.5 events occurred in the central part of the zone. The sequence started on June 17 with a magnitude 6.5 event in the eastern part of the zone which was immediately followed by triggered activity along at least a 80 km long stretch of the plate boundary to the west (Árnadóttir et al. 2001, 2004). The second mainshock of about the same magnitude occurred 21 km west of the first one on June 21. It was clearly preceded by clustering of microearthquakes near the coming hypocenter (Stefánsson 2011). The mainshocks of the sequence occurred on N-S striking faults, transverse to the zone itself. The sense of faulting was right-lateral strike-slip consistent with the model of bookshelf faulting for the SISZ. The two mainshocks occurred on preexisting faults and were accompanied by surface ruptures expressed by en echelon tension gashes and push-up structures typical for strike-slip faults (Clifton and Einarsson 2005). The main zones of rupture were 15 km long, had a N-S trend, and coincided with the epicentral distributions. Faulting along conjugate faults was also observed, but was less pronounced than the main rupture zones. Faulting extended to a depth of about 10 km as shown by aftershock distribution and modeling of displacement fields at the surface determined by GPS and InSAR measurements (Árnadóttir et al. 2001; Pedersen et al. 2003). The maximum fault displacements according to these models were 2.5 m for both earthquakes.

The observed faulting structures were clearly smaller than those observed in association with earlier historical earthquakes such as those of 1630, 1784, 1896, and 1912. These earlier earthquakes were therefore larger than the earthquakes of June 2000, which is consistent with the measured magnitude of 7 for the 1912 earthquake.

High acceleration was recorded by a network of strong-motion accelerographs. The highest recorded acceleration was 0.83 g, recorded by an instrument in a bridge abutment within 3 km of the source fault of the June 21 event. Indications of acceleration in excess of 1 g, such as overturned stones, were abundant in the source areas. In light of the high accelerations, the damage to man-made structures was surprisingly low. Only 20 houses were deemed unusable, none of them collapsed, however. Damage was strongly correlated with age of the buildings. There were no casualties and very little injury to humans or animals (Sigbjörnsson and Ólafsson 2004).

The earthquakes of 2000 only released about one fourth of the strain accumulated across the SISZ since the major sequence of 1896–1912. Further activity was therefore anticipated. A strong earthquake occurred near the western end of the zone, in the Ölfus district, on May 29, 2008, between the towns of Selfoss and Hveragerði (Fig. 3). The moment magnitude of the event was about 6.3, but it turned out to be a double event (Decriem et al 2010; Sigbjörnsson et al. 2009). The first event took place on a fault at the W-foot of Mt. Ingólfsfjall. It immediately triggered another slip event on a parallel fault, 4 km to the west. Both sub-events were associated with right-lateral strike-slip on N-S striking faults and had moment magnitudes (M_w) of 5.8 and 5.9, respectively. The surface effects of the earthquakes were locally very strong. High acceleration was seen in the epicentral areas, rock shattering occurred along many topographic protrusions, rockfalls were common on steep slopes, and sandblows were observed in the alluvium of Ölfusá river that crosses the affected area. Damage to houses was widespread even though no house collapsed. Injuries to people were minor.

Combining field knowledge and mapping of surface ruptures, historical documents, and experience from the 2000 and 2008 earthquakes, it is possible to identify with some certainty the source faults of some of the historical earthquakes since 1630 (Fig. 3, Table 1). Magnitudes of events prior to 1912 are estimated by Stefánsson and Halldórsson (1988). It is noteworthy that each earthquake appears to have occurred on separate fault, except possibly those of 1706 and 1896 in the western part of the zone. The repeat time of events on a particular fault therefore is generally longer than 400 years.

Mechanisms of Earthquakes in Iceland, Table 1 Earthquakes in the South Iceland Seismic Zone

Year	Day	Latitude	Longitude	Magnitude	Area
1630	02 21	64.0	−20.21	~7	Land, Minnivellir
1633					Ölfus
1706	04 20	63.9	−21.20	6.0	Ölfus, Hveragerði
1732	09 07	64.0	−20.04	6.7	Land, Leirubakki ?
1734	03 21	63.9	−20.83	6.8	Flói, Litlureykir ?
1766	09 09	63.9	−21.16	6.0	Ölfus, Gljúfur – Kross – Kaldaðarnes
1784	08 14	63.9	−20.47	7.1	Holt – Gíslholtsvatn
1784	08 16	63.9	−20.95	6.7	Flói – Laugardælir
1829	02 21	63.9	−20.0	6.0	Rangárvellir – Hekla
1896	08 26	64.0	−20.13	6.9	Skarðsfjall – Fellsmúli
1896	08 27	64.0	−20.26	6.7	Flagbjarnarholt -Lækjarbotnar
1896	09 05	63.9	−21.04	6.0	Selfoss – Ingólfsfjall
1896	09 05	64.0	−20.57	6.5	Skeið – Arakot -Borgarkot
1896	09 06	63.9	−21.20	6.0	Ölfus – Hveragerði
1912	05 06	63.9	−19.95	7.0 M_S	Selsund – Galtalækur
1935	10 09	64.0	−21.37	6.0 M_L	Hellisheiði
1987	05 25	63.9	−19.78	5.9 M_w	Vatnafjöll
2000	06 17	63.97	−20.35	6.5 M_w	Holt, Skammbeinsstaðir
2000	06 21	63.97	−20.70	6.4 M_w	Flói, Grímsnes, Hestvatn
2008	05 29	63.9	−21.09	5.8 M_w	Ölfus, Ingólfsfjall
2008	0529	63.9	−21.16	5.9 M_w	Ölfus, Kross

Reykjanes Peninsula Oblique Rift

The mid-Atlantic plate boundary goes onshore at the SW tip of the Reykjanes Peninsula and extends from there with a trend of 70° (azimuth) along the whole peninsula (Figs. 2 and 4). The rift zone is highly oblique and is relatively homogeneous until it joins with the Western Volcanic Zone and the South Iceland Seismic Zone at the Hengill triple junction near 64°N and 21.4°W. The most prominent structural elements are extensional, i.e., hyaloclastite ridges (tindar) formed by fissure eruptions beneath the Pleistocene glacier, postglacial eruptive fissures, normal faults, and open fissures. These structures trend mostly highly obliquely to both the plate boundary and the plate velocity vector (Fig. 4). The average trend of several of the most prominent features is 35° (Einarsson 2008). The fissures and normal faults are grouped into swarms, 4–5 of which can be identified and are shown in Fig. 4. Less conspicuous, but probably equally important, are N-S strike-slip faults that cut across the plate boundary at a high angle. They are expressed as N-S trending arrays of left-stepping, en echelon fissures with push-up hillocks between them. Typical spacing between the faults is 1 km but in some areas it may be as small as 0.4 km. The N-S strike-slip faults may act to accommodate the transcurrent component along the plate boundary in the same way as is explained for the South Iceland Seismic Zone by the bookshelf faulting model.

Seismicity on the Reykjanes Peninsula is high and appears to be episodic in nature with active episodes occurring every 30 years or so. Recent active periods were in the beginning of the 20th century, 1929–1935, 1967–1973, and in 2000. The largest earthquakes in the latest two episodes (1967–1973 and since 2000) are known to have been associated with strike-slip faulting (e.g., Einarsson 1991; Árnadóttir et al. 2004). At the present time, therefore, deformation along the plate boundary appears to be accommodated by strike-slip faulting, possibly with some contribution of crustal stretching. No evidence of magmatic contribution has been detected in the last decades, such as inflation sources, volcanic tremor, or earthquake swarms propagating along

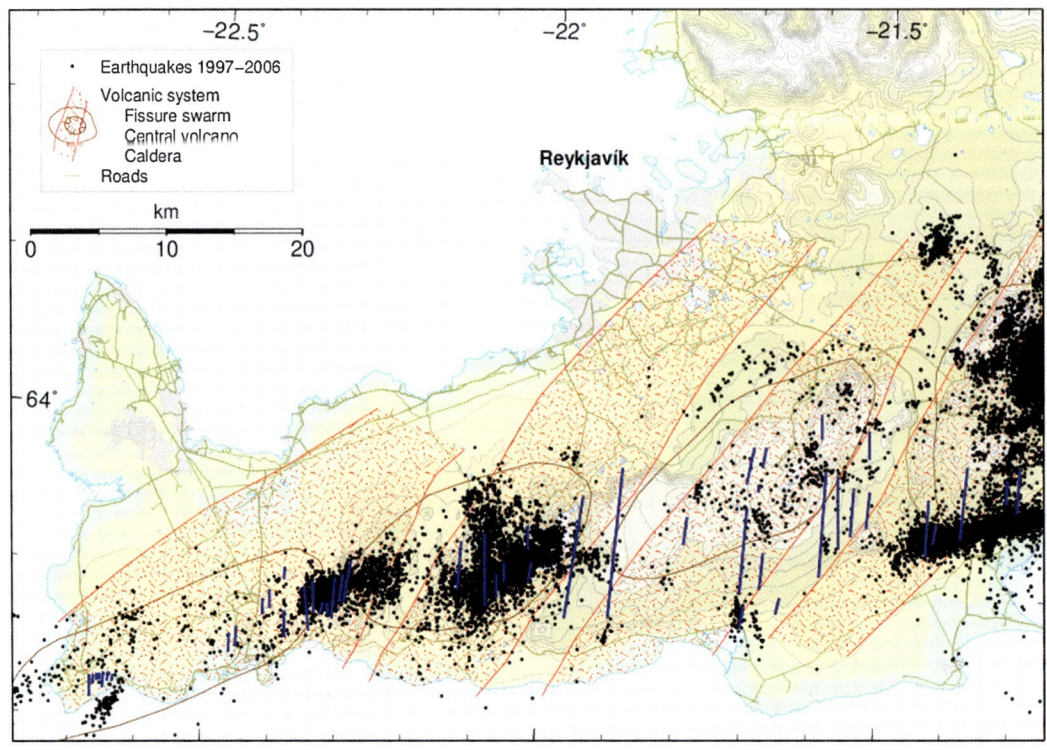

Mechanisms of Earthquakes in Iceland, Fig. 4 Reykjanes Peninsula Oblique Rift, fissure swarms, epicenters of 1997–2006, and identified strike-slip faults shown in blue. Fissure swarms are from Einarsson and Sæmundsson (1987), seismic data are from the Icelandic Meteorological Office, and the strike-slip faults are based on Clifton and Kattenhorn (2006) to a large extent. The four volcanic systems shown are Reykjanes, Krísuvík, Brennisteinsfjöll, and Hengill (Modified from Sólnes et al. (2013) with the help of GBG at IMO)

the fissure swarms. Magmatic activity within this segment also appears to be highly episodic, but with a much longer period, of the order of 1000 years. The latest eruptive period ended in 1240 AD. During this most recent episode, all the volcanic systems of the peninsula were active with several lava eruptions issuing from their fissure swarms (see, e.g., Sólnes et al. 2013, pp. 379–401).

Crustal deformation along the plate boundary on the Reykjanes Peninsula appears to occur in two different modes: (1) dry or seismic mode and (2) wet or magmatic mode (Hreinsdóttir et al. 2001). Deformation in the dry mode occurs mostly by strike-slip faulting during periods when magma is not available to the crust in any appreciable quantity. The mode of deformation changes when magma enters the crust. Dikes are opened up against the local minimum compressive stress, and they propagate into the plates on both sides of the plate boundary and the fissure swarms are activated.

Known significant earthquakes of the Reykjanes Peninsula Oblique Rift are summarized in Table 2. It is noteworthy that following the eruptive period that ended in 1240 AD, no earthquakes are mentioned until 1724. This could be the result of poor documentation, but it is also possible that diking during the magmatic episode left this plate boundary segment completely de-stressed. It then took 500 years of plate movements to build up sufficient stress to produce earthquakes again.

Tjörnes Fracture Zone

The transform motion between the offshore Kolbeinsey Ridge and the Northern Volcanic Rift Zone of Iceland is taken up by the Tjörnes

Mechanisms of Earthquakes in Iceland, Table 2 Earthquakes in the Reykjanes Peninsula Oblique Rift

Year	Day	Latitude	Longitude	Magnitude	Area
1211		63.8	−22.8		Reykjanes
1724	08	63.9	−22.0		Krísuvík – Geitafell ?
1929	07 23	63.9	−21.70	6.3 M_S	Heiðin há, Hvalhnúkur
1933	06 10	63.9	−22.2	6.0 M_S	Núpshlíðarháls
1935	10 09	64.0	−21.37	6.0 M_S	Hellisheiði
1968	12 05	63.9	−21.70	6.0 M_S	Heiðin há, Hvalhnúkur
1973	09 15	63.9	−22.1	5.4 M_S	Móhálsadalur
1973	09 16	63.9	−22.3	5.2 m_b	Fagradalsfjall
1998	06 04	64.0	−21.28	5.4 M_w	Hellisheiði
1998	11 13	63.95	−21.34	5.1 M_w	Ölfus, Hjalli
1998	11 14	63.96	−21.24	4.5 M_S	Ölfus, Hjalli
2000	06 17	63.96	−21.70	5.5 M_w	Hvalhnúkur
2000	06 17	63.95	−21.95	5.9 M_w	Kleifarvatn
2000	06 17	63.90	−22.12	5.3 M_w	Núpshlíðarháls
2003	08 23	63.9	−22.06	5.2 M_w	Krísuvík
2013	10 13	63.81	−22.66	5.2 M_w	Reykjanes

Fracture Zone, a broad zone of seismicity, strike-slip faulting, and crustal extension (Einarsson 1991; Stefánsson et al. 2008). The zone consists of several subparallel NW-trending seismic zones, termed the Grímsey Oblique Rift, the Húsavík-Flatey Faults, and the Dalvík Seismic Zone (Einarsson 2008), and a few northerly trending rift or graben structures (Fig. 5). Recent GPS measurements indicate that about ¾ of the plate transform motion is taken up by the Grímsey Oblique Rift, ¼ by the Húsavík-Flatey Faults, and an insignificant part by the Dalvík Zone (Metzger et al. 2013). Yet, all the zones have produced significant and damaging earthquakes in the last 150 years.

The Grímsey Oblique Rift is the northernmost seismic zone and is entirely offshore. It has an overall NW-SE trend but is composed of several volcanic systems with N-S trend. Evidence for Holocene volcanism is abundant on the seafloor. At its SE end, the zone connects to the Krafla fissure swarm, but the NW end joins the Kolbeinsey Ridge just south of the rapidly disappearing Kolbeinsey Island. The larger earthquakes of the Grímsey Zone seem to be associated with left-lateral strike-slip faulting on NNE-striking faults (Rögnvaldsson et al. 1998). The zone has the characteristics of an oblique rift. There is a striking similarity between the Reykjanes Peninsula Oblique Rift and the Grímsey Zone. The two zones are symmetrical with respect to the plate separation vector. They are nearly perfect mirror images of each other, with the axis of symmetry parallel to the plate separation vector. The similarity is seen in the overall trend, the en echelon fissure swarms, the transverse bookshelf faults, and the occurrence of geothermal areas. Earthquake swarms are common in both zones.

The second seismic zone, the **Húsavík-Flatey Faults**, is about 40 km south of the Grímsey Zone and is well defined by the seismicity near its western end, where it joins with the Eyjafjarðaráll Rift (E in Fig. 5). This transform zone can be traced on the ocean bottom to the coast in the Húsavík town, continuing on land into the volcanic zone, where it merges into one of its fissure swarms (Fig. 5). This is a highly active seismic zone. The last major earthquakes occurred in 1872 when the town of Húsavík suffered heavy damage during an earthquake sequence, including at least two M 6.5 events. Well-developed faults can be seen along the whole length of this zone. They show evidence of right-lateral faulting and are subparallel to the zone itself. No evidence of bookshelf faulting has been found so far. Volcanism only plays a subordinate role if any, in this zone.

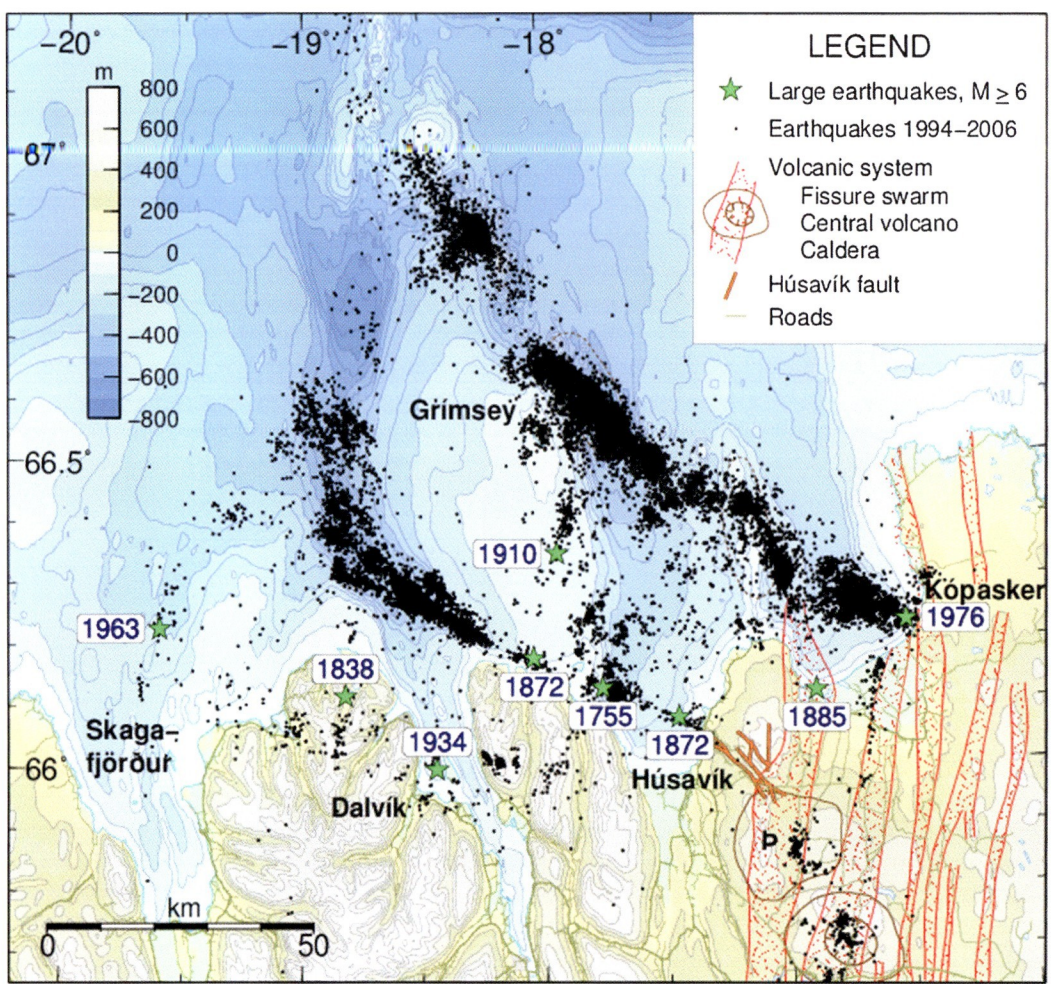

Mechanisms of Earthquakes in Iceland, Fig. 5 Tjörnes Fracture Zone, epicenters 1994–2006, and suggested sources of historical earthquakes since 1755. Fissure swarms are from Einarsson and Sæmundsson (1987), seismic data are from the Icelandic Meteorological Office, and the sources of the historical events are from Stefánsson et al. (2008). E marks the Eyjafjarðaráll Graben and Þ the Þeistareykir volcanic system (Modified from Sólnes et al. (2013) with the help of GBG at IMO)

A third zone, the **Dalvík Zone**, is indicated by scattered seismicity about 30 km south of the Húsavík-Flatey Faults (Fig. 5). Earthquakes as large as M_S 7 have occurred in this zone, but it lacks clear topographic expression. In spite of rather clear alignment of epicenters, the Dalvík Zone is not seen as a throughgoing fault on the surface. A common feature of all three seismic zones is the occurrence of earthquakes on transverse structures, similar to that observed in the South Iceland Seismic Zone and the Reykjanes Peninsula. This is clearly seen in the Dalvík Zone (Fig. 5). In spite of apparently slow strain accumulation in this zone, it has produced two significant earthquakes in the last century, the Dalvík earthquake of 1934 (M 6.3) that caused considerable damage in the town of Dalvík and the 1963 Skagafjörður earthquake (M 7.0) offshore in the mouth of Skagafjörður, one of the largest earthquakes in Iceland in the last century.

Large earthquakes of the Tjörnes Fracture Zone during the last centuries are summarized

Mechanisms of Earthquakes in Iceland, Table 3 Earthquakes in the Tjörnes Fracture Zone

Year	Day	Latitude	Longitude	Magnitude	Area
1755	09 11	66.13	−17.76	~7	Húsavík-Flatey
1838	06 12	66.12	−18.82	6.5	Fljót, Dalvík Zone
1872	04 18	66.08	−17.45	6.5	Húsavík
1872	04 18	66.18	−18.04	6.5	Flatey
1885	01 25	66.12	−16.88	6.3	Kelduhverfi
1910	01 22	66.35	−17.94	7.0	Grímsey
1921	08 23	67	−18	6.3	
1934	06 02	66.00	−18.44	6.2	Dalvík
1963	03 27	66.23	−19.60	7.0 M_S	Skagafjörður, Dalvík Zone
1976	01 13	66.23	−16.50	6.4 M_S	Kópasker, Grímsey Oblique Rift

in Table 3, based mainly on Stefánsson et al. (2008). For the pre-instrumental era, i.e., prior to 1960, the location of the epicenters is uncertain. Most of the zone is offshore and the felt areas are therefore poorly constrained. In addition, some of the events may belong to a series of triggered events, similar to recent events in South Iceland. The felt reports may therefore belong to different events within the sequence that will further confuse the issue. In addition to the large historical earthquakes in Table 3, several sequences and swarms containing events as large as magnitude 5.6 have occurred in recent decades, along the GOR in 1967–1969 and 2013, and in Eyjafjarðaráll-HFF in 2012.

The focal mechanisms have only been instrumentally determined for the 1963 and 1976 events. Both are strike-slip mechanisms on ESE-striking (right-lateral) or NNE-striking (left-lateral) fault planes. The aftershock distributions (see Fig. 5) would favor the ESE plane as the fault plane for the 1976 event, but the NNE plane for the 1963 event. It may be inferred that the other earthquakes are also caused by strike-slip faulting, except possibly the 1885 earthquake. It occurred within the fissure swarm of the Þeistareykir volcanic system and may have been related to rifting activity. From the distribution of background seismicity, it may be inferred that the 1755 and 1872 earthquakes took place on the main Húsavík-Flatey Faults, whereas the 1838, 1910, 1934, and 1963 earthquakes occurred on bookshelf-type faults, transverse to the seismic zones.

Volcanic Zones

Earthquakes occur in the volcanic zones of Iceland but are rarely large (e.g., Pedersen et al. 2007). Events larger than magnitude 5 are only known in connection with unusually large events such as during the Krafla rifting episode in 1975–1984 and at Bárðarbunga volcano prior to the Gjálp eruption of 1996. Events of magnitude 6 and larger have so far not been recorded. It should be noted, however, that large earthquakes have sometimes occurred in the transform zones of South and North Iceland following or in connection with major rifting events in the adjacent divergent parts of the plate boundary. The best documented case is the Kópasker earthquake of TFZ in 1976 during the first rifting event of Krafla (Einarsson 1991). Other cases are the SISZ earthquakes of 1784 following the Laki eruption of 1783, the 1872 earthquakes of the Húsavík-Flatey Faults in connection with an eruption 1867 offshore and the Askja rifting episode of 1874–1875, and the SISZ earthquakes of 1732–1734 following an unusually active period in the rift zones.

Inspection of Fig. 2 reveals a low-magnitude background seismicity of several of the central volcanoes. Apparent spatial correlation with high-temperature geothermal areas suggests that this activity is the expression of heat mining of magmatic sources within these volcanoes. The cooling of hot rock bodies leads to contraction, stress changes, and fracturing. The resulting earthquakes are expected to have a large non-double-couple component in their source mechanism, which has been confirmed in the case of the Hengill and Krafla volcanoes (Miller et al. 1998). Other low-level seismicity sources that may have similar origin include Þeistareykir, Askja (SE caldera part), Kverkfjöll, Grímsvötn, Hveravellir, Torfajökull, Katla (caldera part), Krísuvík, and Reykjanes.

Some of the volcanoes have been shown to go through inflation-deflation cycles during limited times, i.e., a magma chamber is inflated by inflowing magma and subsequently deflates, either by eruption to the surface or by intrusion of a dike. The best documented case is that of Krafla during the rifting episode of 1975–1984, but also Grímsvötn has shown this behavior since 1997 and Hekla during the last decades (Sturkell et al. 2006). Inflation and deflation of a magma chamber leads to stress changes in the chamber roof and eventually to earthquakes if the stress change is large enough. Krafla and Grímsvötn have shown seismicity that conforms with this mechanism, but the Hekla chamber is deep-seated. It is apparently below the brittle-ductile transition and has not produced stress changes large enough to cause brittle failure during recent cycles. The largest earthquakes of this origin have reached magnitude 5, during deflation of Krafla in 1976 (Einarsson 1991).

The brittle-ductile transition is shown to be at about 8–9 km depth at the constructive plate boundaries in Iceland but becomes deeper with increasing age (e.g., Stefánsson et al. 1993). Occasionally earthquakes have been detected at considerably larger depth in the volcanic zones, indicating brittle failure of crustal material that would be ductile at normal strain rates, i.e., strain rates in the crust associated with plate movements. In the case of the 1973 Heimaey (Vestmannaeyjar) and the 2010 Eyjafjallajökull eruptions, these "deep" events occurred in connection with obvious magmatic movements and high strain rates. In other cases the magmatic connection is inferred, mainly because high strain rate is the only known mechanism to cause brittle failure at abnormal depth. A dike intrusion at 15–30 km depth at Upptyppingar in the Northern Volcanic Zone in 2007–2008 was accompanied by an earthquake swarm and verified by surface deformation measurements (e.g., Hooper et al. 2011). Persistent deep earthquake activity in the Askja volcano area at depths of 15–34 km has been interpreted as the expression of magma movements in the lower crust (Soosalu et al. 2009), similarly deep events in the Katla area (Jakobsdóttir 2008).

A large sequence of magmatic rifting events occurred in the Northern Volcanic Zone, beginning in 1974 and lasting until 1989, with diking and eruptions in the interval 1975–1984. The activity was mainly concentrated on the Krafla volcanic system but also extended into the adjacent Grímsey Oblique Rift, a part of the Tjörnes Fracture Zone. Most of this time, magma apparently ascended from depth and accumulated in a magma chamber at about 3 km depth beneath the Krafla volcano. The inflation periods were punctuated by sudden deflation events lasting from several hours to 3 months when the walls of the chamber were breeched and magma was injected into the adjacent fissure swarm accompanied by propagating earthquake swarms (Wright et al. 2012). Twenty deflation events were documented. Large-scale rifting was observed in the fissure swarm during these deflation events, and in nine of them, magma reached the surface in basaltic fissure eruptions lasting from a half hour to 14 days. The earthquakes accompanying the diking were smaller than magnitude 4.5, except the triggered events in the adjacent Tjörnes Fracture Zone (Einarsson 1986, 1991).

A unique earthquake sequence occurred at the Bárðarbunga volcano in Central Iceland during the period 1974–1996. Fifteen earthquakes of magnitude 5 and larger occurred in the caldera region at fairly regular intervals during this period. The focal mechanisms were characterized by reverse faulting but with a significant non-double-couple component (Einarsson 1991; Ekström 1994). Each one of the large events was preceded by a few months of increasing seismic activity, but followed by quiescence or only very few earthquakes. This behavior is highly anomalous. Most other large earthquakes are followed by months or years of high aftershock activity that gradually decays with time. The last large event of the sequence occurred on September

29, 1996 ($M_w = 5.6$). In contrast to previous events, it was immediately followed by an intense earthquake swarm in the summit region of Bárðarbunga. The earthquakes migrated to the SE and S, and in the evening of September 30, they were replaced on the seismic records by continuous volcanic tremor signifying the beginning of an eruption. The eruption is termed the Gjálp eruption and was one of the larger eruptions of the twentieth century in Iceland (Einarsson et al. 1997; Guðmundsson et al. 1997). It occurred on a 7 km long fissure SE of the caldera and beneath the Vatnajökull glacier. It resulted in a catastrophic flood of about 4 km^3 of meltwater.

Several interpretations of the cause of the Bárðarbunga earthquakes have been proposed. The one favored here assumes that they are the result of pressure drop in a magma chamber beneath the caldera of Bárðarbunga (Einarsson 1991). This drop began prior to 1974 and continued until the plumbing of the volcano changed in 1996. The earthquakes were, according to this interpretation, generated by piecemeal collapse of the caldera, by reverse faulting along the curved caldera main fault. The curvature produced the non-double-couple component of the focal mechanism. In the final event in 1996, a pocket of magma was cut by the fault, leading to the formation of a ring dike and triggering a dike propagating out of the caldera to feed the Gjálp eruption.

Intraplate Earthquakes

Even though earthquakes are rare outside the plate boundary areas of Iceland, the class of intraplate earthquakes is of considerable importance for the assessment of seismic risk. The main reason is that the population of Iceland is highly clustered, with half the population concentrated in the Reykjavík metropolitan area, 10–30 km off the plate boundary. Estimating correctly the probability of a large earthquake within the city is a fundamental issue.

Considering Fig. 2 and an earlier analysis of intraplate earthquakes in Iceland (Einarsson 1989), three classes of intraplate earthquakes can be identified:

1. *Earthquakes in West Iceland.* Earthquakes occur west of the main rift zone through Iceland, within the tongue of lithosphere that is bounded by the two transform zones in North and South Iceland. This seismicity is associated with internal deformation of the North American Plate. Best known are the Borgarfjörður events of 1974 in Central West Iceland (Fig. 2), an extended sequence that culminated with an earthquake of m_b 5.5. They were associated with normal faulting on NE- or E-striking faults. Crustal extension in this area may be linked with the center of the Iceland hotspot and its track into the North America Plate. The Snæfellsnes Volcanic Zone may be another expression of this extension (Fig. 2). It is a zone of recent volcanic activity laying unconformably on basalts of 6 Ma age. Distributed, low-level seismicity is seen in Fig. 2 connecting the Snæfellsnes Volcanic Zone to the Western Volcanic Zone.

2. *Earthquakes at the insular shelf edge off Eastern and SE Iceland.* Iceland sits on a platform bounded by a relatively sharp edge that resembles a continental shelf edge (see Fig. 1). It is particularly sharp off the E and SE coast, and there it is seismically active. The edge in this area represents a discontinuity in crustal and mantle structure related to eastward jumps in the plate boundary. There is probably an age jump of 15–25 m.y. and a resulting thermal discontinuity across the shelf edge. The present seismicity is thus interpreted as the result of differential lithospheric response to loading or different cooling rate on either side of the edge.

3. *Earthquakes under and around the Vatnajökull glacier.* Small earthquakes are occasionally seen in the area of Vatnajökull glacier, the largest ice mass of Iceland. The ice is shrinking due to the global warming of the climate. This leads to crustal uplift to restore isostatic equilibrium. The area around the glacier has been shown to rise 10–25 mm/year in recent years (Árnadóttir et al. 2009), and it is a probable cause of this low-level seismic activity.

Summary

Most earthquakes in Iceland are causally linked with the mid-Atlantic plate boundary that crosses the country. The largest earthquakes occur in the two major transform zones, one in South Iceland and the other in North Iceland, that link the subaerial volcanic rift zones with the spreading ridge segments offshore. These earthquakes are caused by strike-slip faulting and may reach magnitude of 7. The transform zones are rather complicated structures, and in many cases the faulting takes place on several parallel strike-slip faults that strike almost perpendicularly to the plate boundary, a faulting mode termed "bookshelf faulting." Triggering of earthquakes has been observed, both by other earthquakes and by rifting in an adjacent zone. Earthquakes in the volcanic and divergent zones of Iceland are generally much smaller. Events of magnitude 5–6 have been found to occur within volcanic calderas. In the case of Krafla in 1975–1976, they were associated with deflation of the volcano's magma chamber. Deflation is also suspected to be the cause of a persistent series of events in the Bárðarbunga caldera in 1974–1996. The focal mechanisms were of the reverse faulting type with a significant non-double-couple component. Dike intrusions and other magmatic processes are responsible for some earthquake activity, mostly at lower magnitudes still. Intraplate earthquakes have been documented in several cases. In West Iceland normal faulting was found to be responsible for a sequence of events in 1974, culminating in an earthquake of magnitude 5.5. The shelf edge off the SE, E, and NE coast is found to be seismically active, and the present isostatic uplift around the shrinking Icelandic glaciers is probably responsible for occasional events there.

Cross-References

▶ Earthquake Magnitude Estimation
▶ Earthquake Mechanisms and Stress Field
▶ Earthquake Mechanisms and Tectonics
▶ Earthquake Swarms
▶ Frequency-Magnitude Distribution of Seismicity in Volcanic Regions
▶ InSAR and A-InSAR: Theory
▶ Non-Double-Couple Earthquakes
▶ Paleoseismology
▶ Seismic Monitoring of Volcanoes

References

Árnadóttir Þ, Hreinsdóttir S, Guðmundsson G, Einarsson P, Heinert M, Völksen C (2001) Crustal deformation measured by GPS in the South Iceland Seismic Zone due to two large earthquakes in June 2000. Geophys Res Lett 28:4031–4033

Árnadóttir Þ, Geirsson H, Einarsson P (2004) Coseismic stress changes and crustal deformation on the Reykjanes Peninsula due to triggered earthquakes on June 17, 2000. J Geophys Res 109, B09307. doi:10.1029/2004JB003130

Árnadóttir Þ, Lund B, Jiang W, Geirsson H, Björnsson H, Einarsson P, Sigurdsson Þ (2009) Glacial rebound and plate spreading: results from the first countrywide GPS observations in Iceland. Geophys J Int 177(2):691–716. doi:10.1111/j.1365-246X. 2008.04059.x

Clifton A, Einarsson P (2005) Styles of surface rupture accompanying the June 17 and 21, 2000 earthquakes in the South Iceland Seismic Zone. Tectonophysics 396:141–159

Clifton A, Kattenhorn S (2006) Structural architecture of a highly oblique divergent plate boundary segment. Tectonophysics 419:27–40

Decriem J, Árnadóttir Þ, Hooper A, Geirsson H, Sigmundsson F, Keiding M, Ófeigsson BG, Hreinsdóttir S, Einarsson P, LaFemina P, Bennett RA (2010) The 2008 May 29 earthquake doublet in SW Iceland. Geophys J Int. doi:10.1111/j.1365-246X.2010.04565.x

Einarsson P (1986) Seismicity along the eastern margin of the North American Plate. In: Vogt PR, Tucholke BE (eds) The geology of North America, vol M. The Western North Atlantic Region. Geological Society of America, Boulder, Colorado, pp 99–116

Einarsson P (1989) Intraplate earthquakes in Iceland. In: Gregersen S, Basham PW (eds) Earthquakes at North-Atlantic passive margins: neotectonics and postglacial rebound. Kluwer Academic, Dordrecht, pp 329–341

Einarsson P (1991) Earthquakes and present-day tectonism in Iceland. Tectonophysics 189:261–279

Einarsson P (2008) Plate boundaries, rifts and transforms in Iceland. Jökull 58:35–58

Einarsson P (2010) Mapping of Holocene surface ruptures in the South Iceland Seismic Zone. Jökull 60:121–138

Einarsson P, Björnsson S, Foulger GR, Stefánsson R, Skaftadóttir Þ (1981) Seismicity pattern in the South Iceland seismic zone. In: Simpson D, Richards P (eds) Earthquake prediction – an international review. American Geophysical Union, Maurice Ewing Series, vol 4, pp 141–151

Einarsson P, Sæmundsson K (1987) Earthquake epicenters 1982–1985 and volcanic systems in Iceland. In: Sigfússon ÞI (ed) Í hlutarins eðli, Festschrift for Þorbjörn Sigurgeirsson (Map). Menningarsjóður, Reykjavík

Einarsson P, Brandsdóttir B, Guðmundsson MT, Björnsson H, Grönvold K, Sigmundsson F (1997) Center of the Iceland hotspot experiences volcanic unrest. Eos 78:369–375

Ekström G (1994) Anomalous earthquakes on volcano ring-fault structures. Earth Planet Sci Lett 128:707–712

Guðmundsson MT, Sigmundsson F, Björnsson H (1997) Ice-volcano interaction of the 1996 Gjálp subglacial eruption, Vatnajökull, Iceland. Nature 389:954–957

Hooper A, Ofeigsson BG, Sigmundsson F, Lund B, Geirsson H, Einarsson P, Sturkell E (2011) Increased capture of magma in the crust promoted by ice-cap retreat in Iceland. Nat Geosci. doi:10.1038/NGEO1269

Hreinsdóttir S, Einarsson P, Sigmundsson F (2001) Crustal deformation at the oblique spreading Reykjanes Peninsula, SW Iceland: GPS measurements from 1993 to 1998. J Geophys Res 106:13803–13816

Jakobsdóttir S (2008) Seismicity in Iceland: 1994–2007. Jökull 58:75–100

La Femina PC, Dixon TH, Malservisi R, Árnadóttir Þ, Sturkell E, Sigmundsson F, Einarsson P (2005) Geodetic GPS measurements in south Iceland: strain accumulation and partitioning in a propagating ridge system. J Geophys Res 110, B11405. doi:10.1029/2005JB003675

Metzger S, Jónsson S, Danielsen G, Hreinsdóttir S, Jouanne F, Giardini D, Villemin T (2013) Present kinematics of the Tjörnes Fracture Zone, North Iceland, from campaign and continuous GPS measurements. Geophys J Int 192:441–455

Miller A, Foulger GR, Julian B (1998) Non-double-couple earthquakes. 2. Observations. Rev Geophys 36:551–568

Pedersen R, Jónsson S, Árnadóttir Þ, Sigmundsson F, Feigl K (2003) Fault slip distribution of two June 2000 M_w 6.4 earthquakes in South Iceland estimated from joint inversion of InSAR and GPS measurements. Earth Planet Sci Lett 213:487–502

Pedersen R, Sigmundsson F, Einarsson P (2007) Controlling factors on earthquake swarms associated with magmatic intrusions; constraints from Iceland. J Volcanol Geotherm Res 162:73–80. doi:10.1016/j.jvolgeores.2006.12.010

Rögnvaldsson S, Gudmundsson A, Slunga R (1998) Seismotectonic analysis of the Tjörnes Fracture Zone, an active transform fault in north Iceland. J Geophys Res 103:30117–30129

Sigbjörnsson R, Ólafsson S (2004) On the South Iceland earthquakes in June 2000: strong-motion effects and damage. Boll Geof Teor Appl 45:131–152

Sigbjörnsson R, Snæbjörnsson JÞ, Higgins SM, Halldórsson B, Olafsson S (2009) A note on the M_w 6.3 earthquake in Iceland on 29 May 2008 at 15:45 UTC. Bull Earthq Eng 7:113–126. doi:10.1007/s10518-008-9087-0

Sigmundsson F (2006) Iceland geodynamics, crustal deformation and divergent plate tectonics. Springer – Praxis, Chichester

Sigmundsson F, Einarsson P, Bilham R, Sturkell R (1995) Rift-transform kinematics in south Iceland: deformation from Global Positioning System measurements, 1986 to 1992. J Geophys Res 100:6235–6248

Sólnes J, Sigmundsson F, Bessason B (eds) (2013) Natural hazards in Iceland: volcanic and seismic hazard (in Icelandic). University of Iceland Press, Reykjavík

Soosalu H, Key AJ, White RS, Knox C, Einarsson P, Jakobsdóttir SS (2009) Lower-crustal earthquakes caused by magma movement beneath Askja volcano on the north Iceland rift. Bull Volcanol 72:55–62. doi:10.1007/s00445-009-0297-3

Stefánsson R (2011) Advances in earthquake prediction, research and risk mitigation. Springer, Berlin/Heidelberg

Stefánsson R, Böðvarsson R, Slunga R, Einarsson P, Jakobsdóttir S, Bungum H, Gregersen S, Havskov J, Hjelme J, Korhonen H (1993) Earthquake prediction research in the South Iceland seismic zone and the SIL project. Bull Seismol Soc Am 83:696–716

Stefánsson R, Gudmundsson GB, Halldorsson P (2008) Tjörnes fracture zone, new and old seismic evidences for the link between the North Iceland rift zone and the mid-Atlantic ridge. Tectonophysics 447:117–126. doi:10.1016/j.tecto.2006.09.019

Stefánsson R, Halldórsson P (1988) Strain release and strain build-up in the South Iceland Seismic Zone. Tectonophysics 152: 267–276

Sturkell E, Einarsson P, Sigmundsson F, Geirsson H, Pedersen R, Van Dalfsen E, Linde A, Sacks S, Stefánsson R (2006) Volcano geodesy and magma dynamics in Iceland. J Volcanol Geotherm Res 150:14–34

Wright TJ, Sigmundsson F, Ayele A, Belachew M, Brandsdóttir B, Calais E, Ebinger C, Einarsson P, Hamling I, Keir D, Lewi E, Pagli C, Pedersen R (2012) Geophysical constraints on the dynamics of spreading centres from rifting episodes on land. Nat Geosci. doi:10.1038/NGEO1428

Mechanisms of Earthquakes in Vrancea

Mircea Radulian
National Institute for Earth Physics, Măgurele, Romania

Synonyms

Earthquake; Focal mechanism; Vrancea

Introduction

The Vrancea seismic region is defined as an earthquake nest, meaning a particular kind of intermediate-depth earthquake concentration or clustering. First, an earthquake nest implies a volume of intense activity that is isolated from nearby activity. Second, this activity is persistent over time (at least for decades). From this point of view, nests are distinct from aftershock sequences or earthquake swarms. The definition of what represents an earthquake nest is somewhat arbitrary, and there are many regions in the world where clustering of seismic activity can be categorized as earthquake nests. Nevertheless, earthquake nests are associated to three most famous and intensively studied areas, Bucaramanga (Colombia), Hindu Kush (Afghanistan), and Vrancea (Romania).

Earthquake nests provide perhaps the best setting for understanding the physical mechanism responsible for intermediate-depth earthquakes (waveform data related to a well-defined seismogenic zone are practically available continuously). Mechanisms involving dehydration embrittlement or thermal shear runaway are mostly invoked to explain brittle or brittle-like failure at intermediate depths.

The Vrancea nest occurred in the last stage of the Carpathians tectonic history. Formation of Carpathians comprises first an extension phase between the European and Apulian Plates in Triassic to Early Cretaceous, then a continental collision phase of the European Eurasian and African Plates and finally a post-collision phase (11–0 Ma), when a set of mechanisms were developed such as slab detachment, oceanic or continental lithosphere delamination, thermal or lithospheric instabilities, or combination of them. The collision led to the closure of the Alpine Tethys and to the lateral extrusion of Intra-Alpine plate. The extrusion led to a fan-shaped migration of the Carpathian collision front, accommodated by subduction of the last remnant of the Tethys Ocean. The Vrancea, located at the southeastern bending of Carpathians, represents the only segment of the mountain belt where significant seismicity is recorded at present, concentrated in a finger-shaped pattern descending almost vertically in the mantle.

Several geodynamic models have been proposed to explain the nature of the seismogenic body beneath the Vrancea region (Ismail-Zadeh et al. 2012). Basically, they belong to two categories: (1) a relic oceanic lithosphere, either attached or already detached from the continental crust, and (2) continental lithosphere that has been delaminated, after continental collision and orogenic thickening. There are arguments in favor and against these two types of models. Oceanic-type models involve break-off process (Fuchs et al. 1979; Wortel and Spakman 2000; Sperner et al. 2001), lateral migration of an oceanic slab (Girbacea and Frisch 1998; Gvirtzman 2002), or subduction and lateral tearing of a slab (Martin and Rietbrock 2006; Wenzel et al. 1998; Wortel and Spakman 2000). Continental-type models involve active delamination (Knapp et al. 2005; Koulakov et al. 2010) or gravitational instability (Lorinczi and Houseman 2009; Ren et al. 2012). The recent results of Bokelman et al. (2014) regarding the dispersion of P waves and the hypothesis of a "double seismic zone" (Bonjer et al. 2005; Radulian et al. 2007; Cărbunar and Radulian 2011) bring evidence in favor of an oceanic nature of the subducting lithosphere.

Tectonic Setting

The earthquake-prone Vrancea region is situated at the bend of the Southeast Carpathians at the contact of East European Platform (EEP) to the north and northeast, Scythian platform to the east, North Dobrogea orogeny to the southeast, Moesian Platform to the south and southwest, and Carpathian orogen and Transylvanian basin (Intra-Alpine plate) to the west and northwest (Fig. 1).

The region is characterized by the frequent occurrence of strong intermediate-depth earthquakes (M $\sim$ 7), confined in a narrow zone ($\sim 60 \times 20$ km^2 on a horizontal section) within a cold vertically descending lithosphere. The part of the descending lithosphere which is

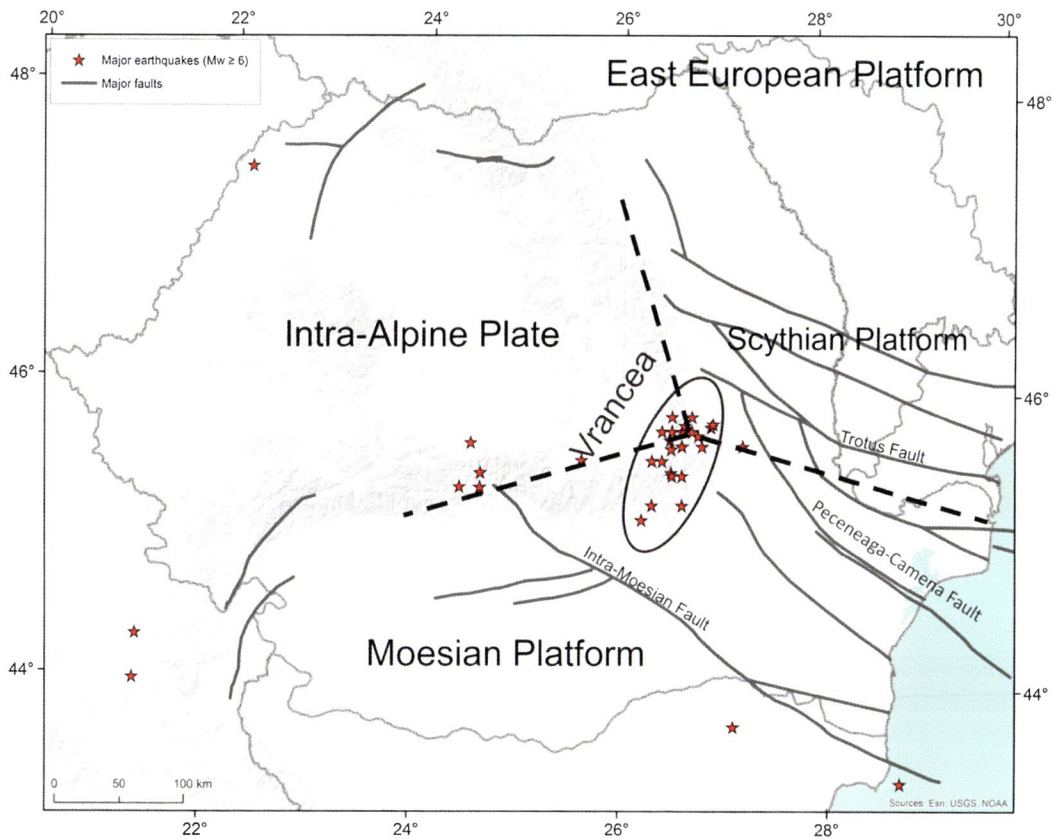

Mechanisms of Earthquakes in Vrancea, Fig. 1 Location of the Vrancea region. The North Dobrogea zone is located between the Scythian and Moesian Platforms, bounded by Trotuş and Peceneaga-Camena Faults

seismically active is confined in the depths between 70 and 170 km, with no seismicity practically detected below this range. This volume fits well within a high seismic wave velocity body, evidenced by all tomographic studies (e.g., Oncescu 1984; Koch 1985; Wortel and Spakman 2000; Sperner et al. 2001; Martin and Rietbrock 2006; Koulakov et al. 2010). The mantle seismogenic volume is separated from the sub-Carpathian crust by a relatively aseismic zone between 40 and 70 km depth.

The seismicity in the Vrancea overlying crust, spread mostly to the southeast between Trotuş and Intra-Moesian faults, is significantly less important as magnitude (below 5.5) and energy release rate (about four orders of magnitude smaller than in the mantle; Ismail-Zadeh et al. 2012). There was debate on a possible causative relation between mantle and shallow events, as was suggested by the simple geometry of seismicity alignments (Răileanu et al. 2009; Bonjer et al. 2008; Cărbunar and Radulian 2011). Mitrofan et al. (2014) showed that the triggering effect of the Vrancea intermediate-depth events upon the shallow events in different segments of the Carpathians foredeep region is not noticeable, except one lineament located to the east, close to Danube River, in the Mărăşti-Galaţi-Brăila area. They explain this triggering mechanism by a process of exhumation of the material originating in a slab slicing progression that is consistent with the reverse-faulting mechanism predominating in the Vrancea intermediate-depth domain (see text below).

The intermediate-depth earthquakes along Vrancea strongly affect extended areas in the Southeast and Central Europe. For this reason, the Vrancea source is controlling seismic hazard level in almost half of the territory of Romania and in cross-border areas as well (in Bulgaria, Republic of Moldova, Serbia, Ukraine). At the same time, the particular pattern of the macroseismic effects provides an efficient clue to identify a major earthquake that occurred in the Vrancea zone in historical times (24 events with $M_w > 7$ since 1,000 according to SHEEC catalogue; Stucchi et al. 2013). The four large events recorded during the last century (10 November 1940, 4 March 1977, 30 August 1986, 30 May 1990) had a significant impact on a large geographical area (Kronrod et al. 2013). The event of 4 March 1977 was the most disastrous, especially toward SW of the epicenter: 1,570 people died, 11,300 were injured, and 32,500 residential and 763 industrial units were destroyed or were seriously damaged in Romania (Sandi 2001).

Analysis of the macroseismic and instrumental data from the intermediate-depth Vrancea earthquakes revealed several peculiarities of the earthquake effects (e.g., Mândrescu and Radulian 1999; Ismail-Zadeh et al. 2012; Kronrod et al. 2013) that can be summarized as follows:

- Extended areas, N-W elongated, are strongly affected.
- NE-SW enhancement of effects coincides with the geometry of seismicity and of the fault-plane solutions.
- Source directivity caused by particular rupture orientation can play a significant role in distributing asymmetrically the intensity effects at regional scale.
- Local and regional geological conditions can control the amplitudes of earthquake ground motion to a larger degree than magnitude or distance.
- Apparently the focal depth can shape to some extent specific patterns in the ground motion distribution.

Focal Mechanisms in the Vrancea Source

All the previous studies on the subcrustal seismic activity in the Vrancea region outline a predominant dip-slip, reverse faulting, characterizing both the moderate and strong earthquakes (Enescu 1980; Oncescu 1987; Oncescu and Trifu 1987). In most of the cases, the principal T axis tends to be vertical, the principal P axis tends to be horizontal and oriented perpendicular to the Carpathians arc, and one nodal plane is dipping about 70° toward NW, while the other nodal plane is dipping SE.

The focal mechanisms of the largest Vrancea earthquakes that occurred since 1977 up to the present ($M_w \geq 5.0$) from the CMT catalogue (Dziewonski et al. 1981; http://www.globalcmt.org/, accessed on February 2014), represented in Figs. 2 and 3, indicate reverse faulting, and in a number of cases, the focal mechanisms are connected with a strong strike-slip component. The parameters of the nodal planes and P and T principal axes are given in Table 1. These mechanisms are connected with the sinking of the lithosphere in the mantle beneath the Vrancea region. The rate of seismic moment per volume, $\sim 0.8 \times 10^{19}$ Nm/year, in Vrancea is comparable to southern California (Wenzel et al. 1998) since the seismic active volume is extremely concentrated in space.

The fault-plane solutions of the instrumentally recorded large earthquakes are remarkably similar (Fig. 4). The individual solutions for the strong earthquakes of 10 November 1940 ($M_w = 7.7$), 4 March 1977 ($M_w = 7.4$), 30 August 1986 ($M_w = 7.1$), and 30 May 1990 ($M_w = 6.9$) show one another only slight variations. These results suggest that the large earthquakes in the Vrancea subcrustal region are governed by the same geodynamic process.

The nodal plane typically striking NE-SW ($\sim 220°$) and dipping 60–70° to the NW is ascribed to be the rupture plane (Oncescu and Bonjer 1997; Sperner et al. 2001; Wenzel et al. 2002). This plane coincides with the location of the aftershocks of the major events (Fuchs et al. 1979; Oncescu and Trifu 1987; Trifu et al. 1992). For some of the moderate-size events ($M_w < 7$), the rupture plane is rotated about 90^0 and dips to the SE (events of 02 October 1978 M_w 5.2, 31 May 1990 M_w 6.4, 28 April 1999 M_w 5.3, and 18 June 2005 M_w 5.2 in Fig. 2). In all cases, the rake angle is close to 90° (i.e., up dip).

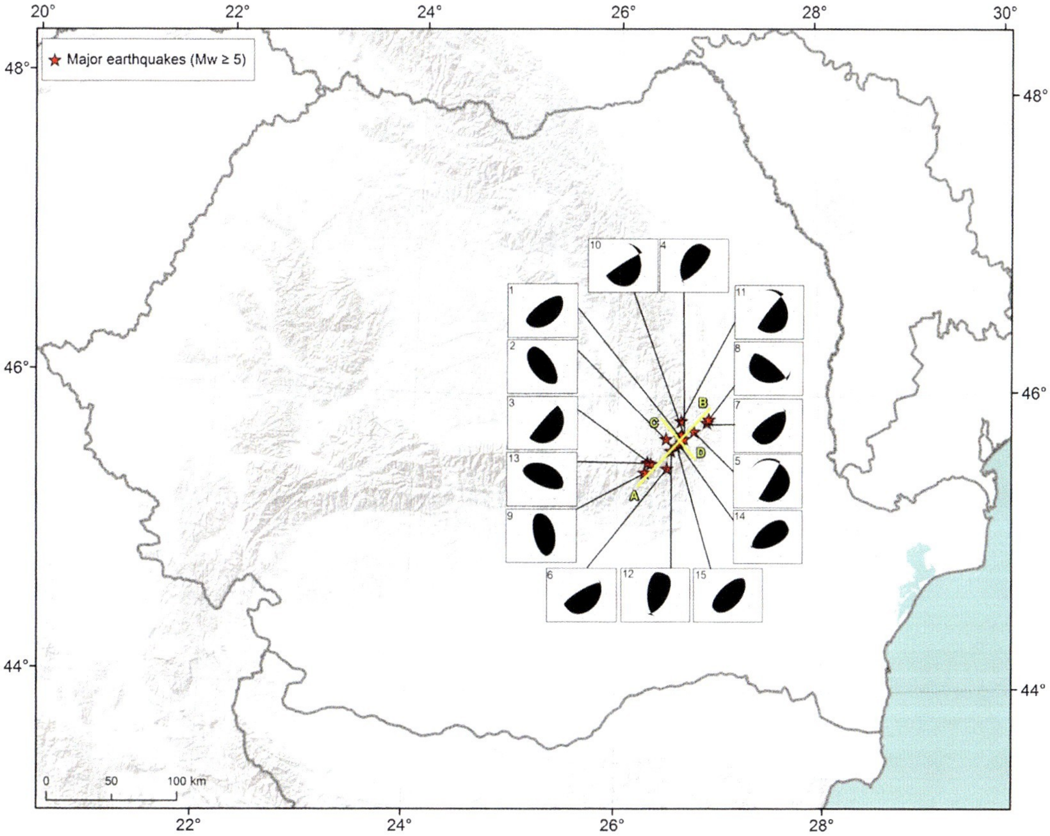

Mechanisms of Earthquakes in Vrancea

Mechanisms of Earthquakes in Vrancea, Fig. 2 Focal mechanisms of the Vrancea largest earthquakes ($M_w \geq 5.0$) recorded since 1977 from the CMT catalogue (Harvard, USA) at surface. The events are numbered according to Table 1. Profiles A–B and C–D point the position of the vertical cross sections in Fig. 3

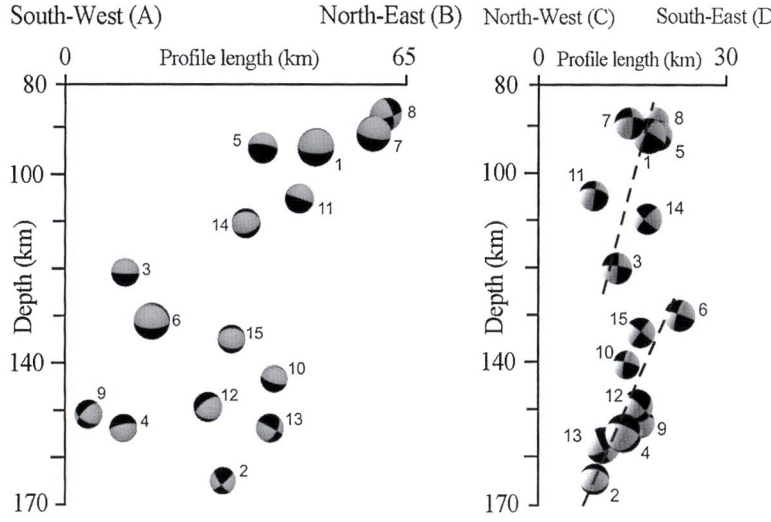

Mechanisms of Earthquakes in Vrancea, Fig. 3 Representation of the focal mechanisms of the earthquakes in Fig. 2 on two vertical cross sections NE-SW (A–B) and NW-SE (C–D). All the events are located at intermediate depths beneath the bend joining Eastern and Southern Carpathians

Mechanisms of Earthquakes in Vrancea, Table 1 CMT fault-plane solutions for Vrancea earthquakes with $M_w \geq 5.0$ recorded since 1977 to the present

No.	Date day/month/year	Time UTC	Latitude °N	Longitude °E	h km	M_w	Plane 1 az.	Plane 1 dip	Plane 1 slip	Plane 2 az.	Plane 2 dip	Plane 2 slip	P axis az.	P axis pl.	T axis az.	T axis pl.
1.	04/03/1977	19:21	45.77	26.76	94	7.4	50	28	86	235	62	92	323	17	151	73
2.	02/10/1978	20:28	45.72	26.48	164	5.2	140	39	85	326	51	94	53	6	263	83
3.	31/05/1979	07:20	45.55	26.33	120	5.3	23	6	72	221	85	92	309	40	133	50
4.	11/09/1979	15:36	45.56	26.30	154	5.3	202	29	70	45	63	101	127	17	338	70
5.	01/08/1985	14:35	45.73	26.62	94	5.8	288	9	−14	32	88	−98	293	46	130	42
6.	30/08/1986	21:28	45.52	26.49	131	7.0	39	19	70	240	72	97	324	27	160	63
7.	30/05/1990	10:40	45.83	26.89	91	6.9	33	29	70	236	63	101	318	17	168	70
8.	31/05/1990	00:17	45.85	26.91	87	6.3	90	26	54	309	69	106	27	22	244	63
9.	28/04/1999	08:47	45.49	26.27	151	5.3	350	36	93	166	54	88	258	9	66	80
10.	06/04/2000	00:10	45.75	26.64	143	5.0	356	18	29	238	81	106	315	34	167	51
11.	27/10/2004	20:34	45.84	26.63	105	6.0	335	19	27	219	81	107	294	34	149	51
12.	14/05/2005	01:53	45.64	26.53	149	5.5	31	44	111	183	50	71	286	3	29	76
13.	18/06/2005	15:16	45.72	26.66	154	5.2	293	32	90	112	58	90	203	13	22	77
14.	25/04/2009	17:18	45.68	26.62	110	5.4	230	46	77	68	46	102	149	0	58	81
15.	06/10/2013	01:37	45.67	26.58	135	5.2	40	37	84	228	53	95	315	8	160	81

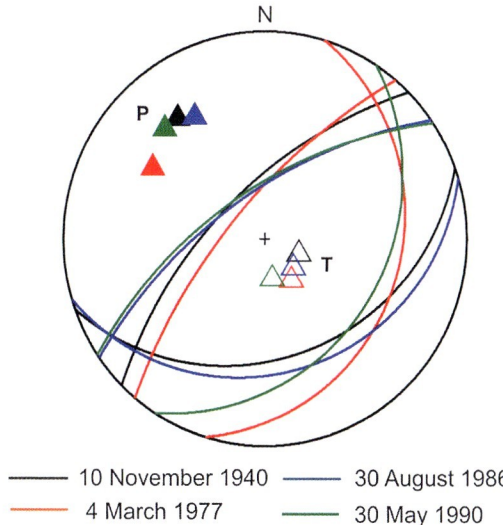

Mechanisms of Earthquakes in Vrancea, Fig. 4 Fault-plane solutions for the major Vrancea earthquakes instrumentally recorded (equal-area lower-hemisphere projection)

Reverse faulting with down-dip extension is predominant along the entire depth range of the seismogenic volume except some variations noticed around 100 km depth (Oncescu and Trifu 1987; Oncescu and Bonjer 1997; Radulian et al. 2002).

The projection of the earthquake foci on a vertical plane crossing the Vrancea area perpendicularly to the Carpathians bend (projection C-D in Fig. 2) suggests the presence of a double seismic zone (Bonjer et al. 2008; Cărbunar and Radulian 2011), which is a characteristic of many subduction zones. Only that in this case the shift between the two nearly parallel "layers" is of the order of location errors (10 km). According to the earthquake triggering scenario suggested by Ganas et al. (2010), there are two NW-dipping nearly parallel planes responsible for triggering successive major events. They are optimally oriented planes of weakness inside the slab to drive the release of the accumulated strain. The foci of the large Vrancea shocks instrumentally recorded are clustered on these two faults accommodating the deformation of the slab (down-dip extension) by shear failure. One cluster is located around 90–100 km depth and the other around 130–140 km depth. A third, subparallel, major, NW-dipping plane of weakness at depth of about 150–160 km cannot be excluded if the hypothesis of seismic gap at this depth range (Oncescu and Bonjer 1997 and Hurukawa et al. 2008) is valid.

Summary

The Vrancea seismic region is a particularly active source in Europe, embedded in the mantle beneath the Southeastern Carpathians arc bend in a continental tectonic setting. Despite the extreme concentration of the focal volume, frequent earthquakes of magnitude above 7 are generated here in an isolated high-velocity lithospheric body. The focal mechanisms show predominant reverse faulting with the T axis along down dip. The distribution of the strong ground motion is characterized by specific asymmetric patterns, affecting extended areas elongated along NE-SW direction. The elongation coincides with the pattern of the seismicity and the geometry of the rupture plane dipping toward NW. Recent higher-resolution investigations reveal the existence of possible parallel alignments of weakness crossing the seismogenic volume along planes oriented NE-SW and dipping nearly vertically toward NW. These weakness planes are developed separately in two or eventually three segments on depth (centered at 90–100, 130–140, and 150–160 km depth, respectively) where the seismicity is clustered and major shocks nucleate alternatively in time. Repeated generation of large events along parallel sub-faults separated in depth suggests a combination of dehydration process that preferentially weakens preexisting sub-faults in the subducting crust (Kiser et al. 2011) and thermal shear runaway process (Wiens and Sniders 2001).

Cross-References

▶ Earthquake Mechanism and Seafloor Deformation for Tsunami Generation
▶ Earthquake Mechanisms and Stress Field

▶ Earthquake Mechanism Description and Inversion
▶ Earthquake Mechanisms and Tectonics
▶ Long-Period Moment-Tensor Inversion: The Global CMT Project
▶ Mechanisms of Earthquakes in Aegean
▶ Mechanisms of Earthquakes in Iceland
▶ Regional Moment Tensor Review: An Example from the European–Mediterranean Region

References

Bokelmann G, Rodler FA (2014) Nature of the Vrancea seismic zone (Eastern Carpathians) – New constraints from dispersion of the first-arriving P-waves. Earth and Planetary Science Letters 390:59–68

Bonjer KP, Ionescu C, Sokolov V, Radulian M, Grecu B, Popa M, Popescu E (2005) Source parameters and ground motion pattern of the October 27, 2004 intermediate depth Vrancea earthquake. In: EGU General Assembly 2005, Vienna (abstracts)

Bonjer KP, Ionescu C, Sokolov V, Radulian M, Grecu B, Popa M, Popescu E (2008) Ground motion patterns of intermediate-depth vrancea earthquakes: the October 27, 2004 event. In: Zaicenco A, Craifaleanu I, Paskaleva I (eds) Harmonization of seismic hazard in Vrancea zone, NATO Science for Peace and Security Series – C. Springer, Dordrecht, pp 47–62

Cărbunar OF, Radulian M (2011) Geometrical constraints for the configuration of the Vrancea (Romania) intermediate-depth seismicity nest. J Seismol 15:579–598

Dziewonski AM, Chou TA, Woodhouse JH (1981) Determination of earthquake source parameters from waveform data for studies of global and regional seismicity. J Geophys Res 86(2825–2852):1981

Enescu D (1980) Contributions to the knowledge of the focal mechanism of the Vrancea strong earthquakes of March 4, 1977. Rev Roum Géol Géophys Géogr Ser Géophys 24(1):3–18

Fuchs K, Bonjer K-P, Bock G, Cornea I, Radu C, Enescu D, Jianu D, Nourescu A, Merkler G, Moldoveanu T, Tudorache G (1979) The Romanian earthquake of March 4, 1977. II. Aftershocks and migration of seismic activity. Tectonophysics 53:225–247

Ganas A, Grecu B, Batsi E, Radulian M (2010) Vrancea slab earthquakes triggered by static stress transfer. Nat Hazards Earth Syst Sci 10:2565–2577

Girbacea R, Frisch W (1998) Slab in the wrong place: lower lithospheric mantle delamination in the last stage of the Eastern Carpathian subduction retreat. Geology 26:611–614

Gvirtzman Z (2002) Partial detachment of a lithospheric root under the southeast Carpathians: toward a better definition of the detachment concept. Geology 30(1):51–54

Hurukawa N, Popa M, Radulian M (2008) Relocation of large intermediate-depth earthquakes in the Vrancea region, Romania, since 1934 and a seismic gap. Earth Planets Space 60:565–572

Ismail-Zadeh A, Matenco L, Radulian M, Cloetingh S, Panza GF (2012) Geodynamics and intermediate-depth seismicity in Vrancea (The South-Eastern Carpathians): current state-of-the art. Tectonophyiscs 530–531:50–79

Kiser E, Ishii M, Langmuir CH, Shearer PM, Hirose H (2011) Insights into the mechanism of intermediate-depth earthquakes from source properties as imaged by back projection of multiple seismic phases. J Geophys Res 116, B06310. doi:10.1029/2010JB007831

Knapp JH, Knapp CC, Raileanu V, Matenco L, Mocanu V, Dinu C (2005) Crustal constraints on the origin of mantle seismicity in the Vrancea Zone, Romania: the case for active continental lithospheric delamination. Tectonophysics 410:311–323

Koch M (1985) Nonlinear inversion of local seismic travel times for the simultaneous determination of the 3D-velocity structure and hypocentres – application to the seismic zone Vrancea. J Geophys 56:160–173

Koulakov I, Zaharia B, Enescu B, Radulian M, Popa M, Parolai S, Zschau J (2010) Delamination or slab detachment beneath Vrancea? New arguments from local earthquake tomography. Geochem Geophys Geosyst 11(G3):Q03002. doi:10.1029/2009GC002811

Kronrod T, Radulian M, Panza GF, Popa M, Paskaleva I, Radovanovich S, Grjibovski K, Sandu I, Pekevski L (2013) Integrated transnational macroseismic data set for the strongest earthquakes of Vrancea (Romania). Tectonophysics 590:1–23

Lorinczi P, Houseman GA (2009) Lithospheric gravitational instability beneath the Southeast Carpathians. Tectonophysics 474:322–336

Mândrescu N, Radulian M (1999) Macroseismic field of the Romanian intermediate-depth earthquakes. In: Wenzel F, Lungu D, Novak O (eds) Vrancea earthquakes: tectonics, hazard and risk mitigation. Kluwer, Dordrecht, pp 163–174

Martin S, Rietbrock A (2006) Guided waves at subduction zones: dependencies on slab geometry, receiver locations and earthquake sources. Geophys J Int 167:693–704

Mitrofan H, Chitea F, Anghelache M-A, Visan M (2014) Possible triggered seismicity signatures associated with the Vrancea intermediate-depth strong earthquakes (Southeast Carpathians, Romania). Seismol Res Lett 85(2):314–323

Oncescu MC (1984) Deep structure of Vrancea region, Romania, inferred from simultaneous inversion for hypocenters and 3-D velocity structure. Ann Geophys 2:23–28

Oncescu MC (1987) On the stress tensor in Vrancea region. J Geophys Res 62:62–65

Oncescu MC, Bonjer KP (1997) A note on the depth recurrence and strain release of large Vrancea earthquakes. Tectonophysics 272:291–302

Oncescu MC, Trifu CI (1987) Depth variation of the moment tensor principal axes in Vrancea (Romania) seismic region. Ann Geophysicae 5B:149–154

Radulian M, Bălă A, Popescu E (2002) Earthquakes distribution and their focal mechanism in seismogenic zones of Romania. Roman Report Phys 47:945–963

Radulian M, Bonjer K-P, Popescu E, Popa M, Ionescu C, Grecu B (2007) The October 27th, 2004 Vrancea (Romania) earthquake. ORFEUS Newsletter 1(7), Jan 2007

Răileanu V, Dinu C, Ardeleanu L, Diaconescu V, Popescu E, Bălă A (2009) Crustal seismicity and associated fault systems in Romania. In: Proceedings of the 27th ECGS workshop: seismicity patterns in the euro-med region, Luxembourg, 17–19 Nov.2008, in "Cahiers du Centre Europeen de Geodynamique et de Seismologie", pp 153–159

Ren Y, Stuart GW, Houseman GA, Dando B, Ionescu C, Hegedüs E, Radovanović S, Shen Y, South Carpathian Project Working Group (2012) Upper mantle structures beneath the Carpathian–Pannonian region: implications for the geodynamics of continental collision. Earth Planet Sci Lett 349–350:139–152

Sandi H (2001) Obstacles to earthquake risk reduction encountered in Romania. In: Lungu D, Saito T (eds) Earthquake hazard and countermeasures for existing fragile buildings. Independent Film, Bucharest, pp 261–266

Sperner B, Lorenz F, Bonjer K, Hettel S, Muller B, Wenzel F (2001) Slab break-off – abrupt cut or gradual detachment? New insights from the Vrancea Region (SE Carpathians, Romania). Terra Nova 13:172–179

Stucchi M, Rovida A, Gomez Capera AA, Alexandre P, Camelbeeck T, Demircioglu MB, Kouskouna V, Gasperini P, Musson RMW, Radulian M, Sesetyan K, Vilanova S, Baumont D, Faeh D, Lenhardt W, Martinez Solares JM, Scotti O, Zivcic M, Albini P, Batllo J, Papaioannou C, Tatevossian R, Locati M, Meletti C, Viganò D, Giardini D (2013) The SHARE European Earthquake Catalogue (SHEEC) 1000–1899. J Seismol 17:523–544

Trifu CI, Deschamps A, Radulian M, Lyon-Caen H (1992) The Vrancea earthquake of May 30, 1990: an estimate of the source parameters. In: Proceedings of the XXIInd ESC Gen. Ass., Barcelona, 1990, vol 1, pp 449–454

Wenzel F, Achauer U, Enescu D, Kissling E, Mocanu V (1998) Detailed look at final stage of plate break-off is target of study in Romania. EOS Trans Am Geophys Union 79(48):589–594

Wenzel F, Sperner B, Lorenz F, Mocanu V (2002) Geodynamics, tomographic images and seismicity of the Vrancea region (SE-Carpathians, Romania), vol 3, EGU Stephan Mueller Special Publication Series. Katlenburg-Lindau, Germany, pp 95–104

Wiens DA, Snider NO (2001) Repeating deep earthquakes: evidence for fault reactivation at great depth. Science 293:1463–1466

Wortel MJR, Spakman W (2000) Subduction and slab detachment in the Mediterranean-Carpathian region. Science 290:1910–1917

MEMS Sensors for Measurement of Structure Seismic Response and Their Application

Kenji Oguni
Department of System Design Engineering,
Keio University, Yokohama, Japan

Synonyms

Accelerometer; Displacement sensor; MEMS (microelectromechanical systems); Restoring force characteristics; SDOF (single degree of freedom) mass-spring system; Sensor network; Velocimeter

Introduction

MEMS is an abbreviation for microelectromechanical systems. MEMS sensor is a measurement device consisting of mechanical parts for picking up events or environmental changes (i.e., sensors) and electric circuit for amplifying and filtering the signals from sensors. MEMS sensor is usually integrated on a small circuit board.

The major advantages of MEMS sensors for measurement of structure seismic response are their small size, low power consumption, and low price in contrast with their high accuracy in measurement. For example, a typical price of MEMS accelerometers is less than 10 USD. In spite of this low cost, a MEMS accelerometer with 3 axes, ± 2 g full scale, 5 mm $\times$ 5 mm $\times$ 2 mm size, 1.1 mA current consumption at the input voltage of 2.7 V, and noise density of 175 $\mu g/\sqrt{Hz}$ is available off the shelf.

MEMS sensors are basically implemented as fine patterns printed on silicon wafers. These patterns are the combination of the mechanical parts corresponding to the sensor pickup and the electric circuit to convert, amplify, and filter the signal from sensor pickup. The size and the energy consumption of the MEMS can be suppressed by implementing the patterns on the silicon wafers as small as possible. The micro-fabrication technology made mass production and distribution of low-cost MEMS sensors possible.

A typical micro-fabrication process of the MEMS sensors is a photolithograph method. In photolithography, photosensitive material is painted on a silicon wafer. Then, this layer is masked by the patterns of the electric circuit. After the exposure of this wafer to the light, unmasked area is scraped off by reactive gas or fluid (gas, dry etching; fluid, wet etching). Finally, the masked area remains as fine patterns of the electric circuit. This photolithographic method enables us to automatically produce more than 1,000 circuit boards (each of them corresponds to a MEMS sensor) from one silicon wafer plate (30 cm in diameter). This is the key production technology for low-cost MEMS sensors.

Among many types of MEMS sensors, MEMS accelerometers are mainly used in the field of earthquake engineering. The measurement principle of most MEMS accelerometers is that of SDOF (single degree of freedom) mass-spring system. If some condition (mentioned below in this manuscript) is satisfied, the relative displacement between the mass and the support is proportional to the absolute acceleration of the support.

This is the overview of the MEMS technology and the MEMS sensors related to the field of earthquake engineering. This manuscript is organized as follows. First, the difference between a displacement sensor, a velocimeter, and an accelerometer is illustrated by the analysis of SDOF mass-spring system. Through this analysis, the reason for (i) many MEMS accelerometers are available with low cost, and (ii) no MEMS velocimeters and MEMS displacement sensors are available. Also, the limitation of the MEMS accelerometers and the point to be noted for usage of the MEMS accelerometers are discussed. This discussion leads us to a possible application of the MEMS accelerometers in the field of earthquake engineering.

Vibration Measurement Device

Consider an SDOF (single degree of freedom) mass-spring system shown in Fig. 1. The mass is m, spring constant is k, and the dashpot is c. The governing equation of this system is

$$m(\ddot{y}(t) + \ddot{z}(t)) + c\dot{z}(t) + kz(t) = 0 \quad (1)$$

where t is time. Suppose that the displacement of the support is $y(t) = Y_0 \sin \omega t$. Then Eq. 1 becomes

$$\ddot{z}(t) + 2\xi\omega_0\dot{z}(t) + \omega_0^2 z(t) = \omega^2 Y_0 \sin \omega t \quad (2)$$

where $\omega_0^2 = k/m$ and $\xi = c/\sqrt{4mk}$. The following expression for $z(t)$ can be obtained by solving Eq. 2.

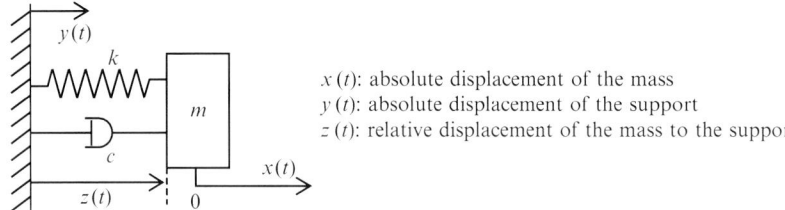

$x(t)$: absolute displacement of the mass
$y(t)$: absolute displacement of the support
$z(t)$: relative displacement of the mass to the support

MEMS Sensors for Measurement of Structure Seismic Response and Their Application, Fig. 1 SDOF mass-spring system (simplified model for vibration measurement device)

$$z(t) = \frac{(\omega/\omega_0)^2}{\sqrt{\left[1-(\omega/\omega_0)^2\right]^2 + (2\xi\omega/\omega_0)^2}} Y_0 \sin(\omega t - \phi),$$

$$\phi = \tan^{-1}\left(\frac{2\xi\omega/\omega_0}{1-(\omega/\omega_0)^2}\right) \quad (3)$$

Consider the following three cases for the angular frequency of the absolute displacement of the support ω and the natural angular frequency of the mass-spring system ω_0:

Case 1: $\omega_0 \ll \omega$
Case 2: $\omega_0 \cong \omega$
Case 3: $\omega_0 \gg \omega$

In Case 1, since $\left[1-(\omega/\omega_0)^2\right]^2 \cong (\omega/\omega_0)^4 \gg (2\xi\omega/\omega_0)^2$ and $\phi = \pi$,

$$z(t) \cong -Y_0 \sin \omega t = -y(t). \quad (4)$$

Equation 4 means that for $\omega_0 \ll \omega$, the absolute displacement of the support $y(t)$ is almost equal to $-z(t)$, where $z(t)$, the relative displacement of the mass to the support, is the measurable quantity. Therefore, for the case of $\omega_0 \ll \omega$, this SDOF mass-spring system can be used as a displacement sensor by converting the relative displacement between the mass and the support to some electric signal.

In Case 2, $\omega_0 \cong \omega$ and $\phi = \pi/2$ result in

$$z(t) \cong \frac{1}{2\xi\omega_0} Y_0 \cos \omega t = \frac{1}{2\xi\omega_0} \dot{y}(t). \quad (5)$$

Equation 5 means that for $\omega_0 \cong \omega$, the relative displacement of the mass to the support, $z(t)$, is almost proportional to the absolute velocity of the support, $\dot{y}(t)$. Therefore, when the condition for Case 2 is satisfied, this SDOF mass-spring system can be used as a velocimeter by converting the relative displacement between the mass and the support to some electric signal.

In Case 3, since $\omega_0 \gg \omega$, $\omega/\omega_0 \cong 0$ and $\phi = 0$. Therefore,

$$z(t) \cong \frac{1}{\omega_0^2} \omega^2 Y_0 \sin \omega t = -\frac{1}{\omega_0^2} \ddot{y}(t). \quad (6)$$

Equation 6 means that in Case 3, the relative displacement of the mass to the support, $z(t)$, is almost proportional to the absolute acceleration of the support, $\ddot{y}(t)$. Therefore, for the case of $\omega_0 \gg \omega$, this SDOF mass-spring system can be used as an accelerometer by converting the relative displacement between the mass and the support to some electric signal.

Here, the physical meaning of each case for the SDOF mass-spring system shown in Fig. 1 should be discussed. In Case 1, this SDOF mass-spring system works as a displacement sensor. The condition $\omega_0 \ll \omega$ requires relatively small ω_0. Since $\omega_0 = \sqrt{k/m}$, small ω_0 means large m and small k. Thus, an SDOF displacement sensor consists of a very heavy mass supported by an extremely soft spring which results in a bulky and fragile device. This is the reason for nonexistence of a MEMS displacement meter based on the SDOF mass-spring system.

In Case 2, $\omega_0 \cong \omega$ requires a subtle tuning of the device. Thus, a velocimeter becomes a precision instrument and this is not suitable for the mass production on a silicon wafer (typical production method for MEMS devices). This is why there exists no MEMS velocimeter based on the SDOF mass-spring system. One comment should be added for an SDOF velocimeter. As shown in Eq. 5, the proportional constant between the absolute velocity of the support $\dot{y}(t)$ and the measured relative displacement of the mass $z(t)$ is $1/(2\xi\omega_0)$. Ordinary materials have small ξ in the order of 0.05. Therefore, $\dot{y}(t)$ is amplified a lot. The gain of the SDOF velocimeter is large enough for using this as a precise seismometer.

In Case 3, $\omega_0 \gg \omega$ means relatively large ω_0. Since $\omega_0 = \sqrt{k/m}$, large ω_0 means small m and large k. Thus, an SDOF accelerometer consists of a light mass supported by an extremely hard spring which results in a small and tough device. The smaller and the tougher, the device works as a better accelerometer. This nature of the SDOF accelerometer is suitable for the mass production on a silicon wafer. This is why MEMS accelerometers with reasonable cost are widely supplied in the market.

There is an additional explanation for MEMS accelerometers. As shown in Eq. 6, the proportional constant between the absolute acceleration of the support $\ddot{y}(t)$ and the measured relative displacement of the mass $z(t)$ is $-1/\omega_0^2$. Since large ω_0 is the requirement for an SDOF accelerometer, the amplification factor $1/\omega_0^2$ becomes very small. Therefore, a MEMS accelerometer should have an internal electric circuit for amplifying the signal. The measuring accuracy of the MEMS accelerometer is limited by this amplifier circuit. The accuracy of this amplifier circuit is the noise density mentioned at the beginning of this manuscript. The lowest noise density of currently available MEMS accelerometers is in the order of several tens of $\mu g/\sqrt{Hz}$. Thus, acceleration measurement using a MEMS accelerometer with the bandwidth of 100 Hz results in the noise in the order of 1 gal. Compared with non-MEMS precise accelerometers whose measurement noise is in the order of a few milligal, the measurement accuracy of the MEMS accelerometer is not high. This implies limitation of the application of the MEMS accelerometer in the field of earthquake engineering.

Application of MEMS Accelerometers for Earthquake Engineering

Sensor network has been started by Smart Dust project (Kahn et al. 1999) in the late 1990s. Ever since, sensor network technology has been rapidly advanced. This is mainly because of the advancement of MEMS technology. This sensor network technology has been also imported to the field of earthquake engineering. Measurements of the ambient vibration of the structures and the structure seismic response to weak ground motion are performed by using MEMS accelerometer sensor network. These measurements are intended for structural health monitoring. However, no successful example has been observed in these structural health monitoring applications by densely deployed MEMS accelerometers (Farrar et al. 2001; Farrar and Worden 2007). The major reason for these unsuccessful attempts for MEMS accelerometer-based structural health monitoring is the insufficient accuracy of MEMS accelerometers. Structural health monitoring using the ambient vibration of the structures and the structure seismic response to weak ground motion requires the change in the structural response to weak input (less than 10 gal). Therefore, the accuracy of MEMS accelerometers mentioned before (1 gal for 100 Hz measurement bandwidth) is obviously insufficient. MEMS accelerometer-based application for earthquake engineering has to be the measurement of structure seismic response to strong (at least 100 gal) ground motion.

An example application of a sensor network system equipped with MEMS accelerometers is the measurement of restoring force characteristics of structures subjected to strong ground motion. Right after a severe earthquake, the damage level of the structures in disaster-stricken region has to be quickly identified. For this

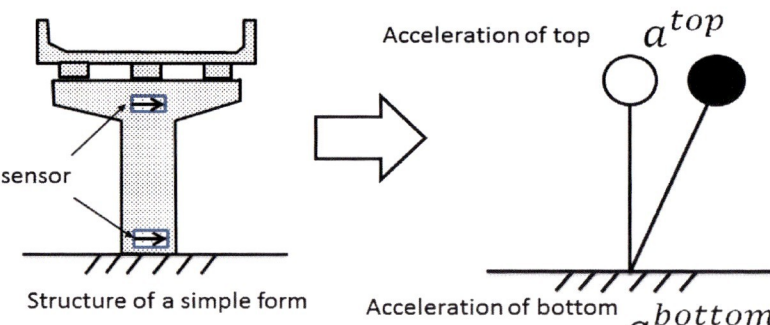

MEMS Sensors for Measurement of Structure Seismic Response and Their Application, Fig. 2 Restoring force characteristics measurement using MEMS accelerometer sensor network

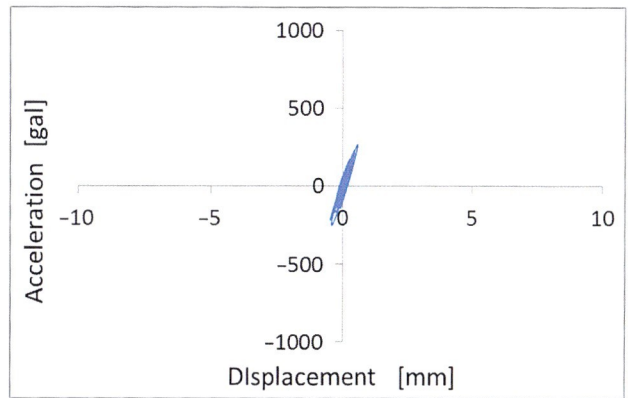

MEMS Sensors for Measurement of Structure Seismic Response and Their Application, Fig. 3 Typical restoring force characteristics under weak seismic ground motion

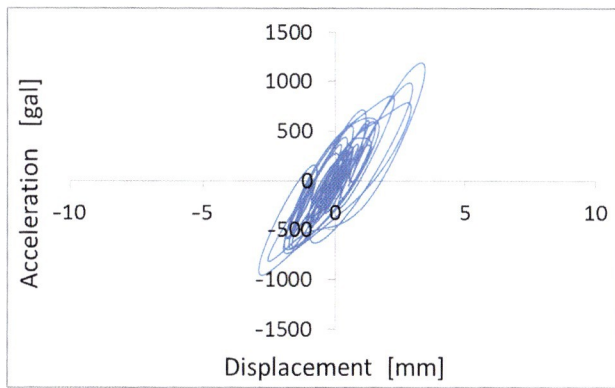

MEMS Sensors for Measurement of Structure Seismic Response and Their Application, Fig. 4 Typical restoring force characteristics under strong seismic ground motion

purpose, MEMS accelerometer sensor network system can be used.

Structures of a simple shape such as the standard viaducts for railways and highways can be safely approximated as an SDOF mass-spring system as shown in Fig. 2. Then, the acceleration at the top of the structure is proportional to the inertia force acting on the mass and the relative displacement between the top and the bottom of the structure corresponds to the relative displacement between the mass and the support. Therefore, if the acceleration at the top of the structure and the relative displacement between the top and the bottom of the structure subjected to ground motion can be observed, the property of the spring, i.e., the restoring force characteristics of the structure, can be obtained.

At the top and the bottom of the structure shown in Fig. 2, sensor nodes equipped with a MEMS accelerometer are installed. These sensor nodes are also equipped with wireless communication device and a synchronized measurement can be performed between these sensor nodes (Ikeda et al. 2012).

Typical restoring force characteristics are shown in Figs. 3 and 4. The vertical axis is the acceleration at the top of the structure and the horizontal axis is the relative displacement between the top and the bottom of the structure. Subjected to weak ground motion shown in Fig. 3, the structure remains elastic and the restoring force characteristics are linear. On the other hand, Figure 4 corresponding to the case of strong ground motion shows hysteresis curves which indicate the plastic response of the structure.

These restoring force characteristic can be obtained by a synchronized acceleration measurement using a pair of sensor nodes installed at the top and the bottom of the structure. The relative displacement can be obtained through time integration of the acceleration data using a band-pass filter which includes the lowest natural frequency of the structure and excludes the higher natural frequencies.

The advantage of this application over unsuccessful attempts for MEMS accelerometer-based structural health monitoring is the required accuracy in the acceleration measurement. The measurement accuracy of the commercially available MEMS accelerometers is good enough for constructing restoring force characteristics.

Summary

MEMS sensors for measurement of structure seismic response and their application are discussed. The principle of the measurement of displacement, velocity, and acceleration is illustrated through an analysis of an SDOF mass-spring system. Based on this analysis, the reason for the popularity of the MEMS accelerometers and their limitations are discussed. An example of a possible application of MEMS accelerometer sensor network is presented.

References

Farrar CR, Worden K (2007) An introduction to structural health monitoring. Phil Trans R Soc Lond A 365(1851):303–315

Farrar CR, Doebling SW, Nix DA (2001) Vibration-based structural damage identification. Phil Trans R Soc Lond A 359(1778):131–149

Ikeda H, Kawaguchi T, Oguni K (2012) Automatic detection of damage level of structures under the severe earthquake using sensor network. Appl Mech Mater 166–169:2216–2220

Kahn JM, Katz RH, Pister KSJ (1999) Mobile networking for smart dust. In: Proceedings of MobiCom 99, online publication

Mixed In-Height Concrete-Steel Buildings Under Seismic Actions: Modeling and Analysis

Thanasis Papageorgiou[1] and Charis J. Gantes[2]
[1]Department of Structural Engineering, School of Civil Engineering, National Technical University of Athens (N.T.U.A.), Athens, Greece
[2]Institute of Steel Structures, School of Civil Engineering, National Technical University of Athens, Zografou Campus, Athens, Greece

Synonyms

Complex response; Decoupling; Equivalent damping; Irregular structures; Primary-secondary systems

Introduction

Issues pertaining to the dynamic behavior of mixed in-height structures, and more specifically concrete-steel buildings, are presented. The term "mixed structures" denotes here buildings that consist of a lower and an upper part with different dynamic response characteristics. When such structures are subjected to a dynamic excitation, e.g., an earthquake, they respond in a different way than more uniform structures; thus, their dynamic analysis and seismic design must account for these peculiarities.

The dissimilarity in the dynamic response characteristics of the two parts can be due to several factors. It may occur due to different lateral force resisting systems in the upper and lower part, or it might be the result of the action of seismic isolation or energy dissipation devices. One of the most common causes, which is the coexistence of different construction materials leading to nonuniform damping and elastoplastic behavior over the height of the building, is studied here.

More specifically, this section is devoted to modeling and analysis issues of structures consisting of a lower part made of concrete that

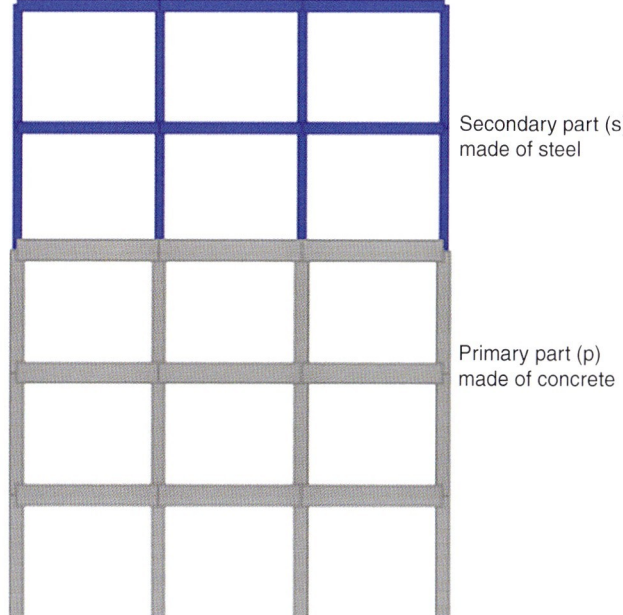

Mixed In-Height Concrete-Steel Buildings Under Seismic Actions: Modeling and Analysis, Fig. 1 Schematic representation of mixed in-height structure

Mixed In-Height Concrete-Steel Buildings Under Seismic Actions: Modeling and Analysis, Fig. 2 Mixed in-height concrete-steel building

rests on the ground and supports an upper part made of steel. A schematic representation of such a structure is shown in Fig. 1. The lower part in the pertinent bibliography is usually referred to as primary structure and is denoted with the letter p, while the upper part is called secondary structure and is denoted with the letter s. Emphasis is directed here towards the different damping characteristics of the two parts, as reinforced concrete structures are known to exhibit higher damping than steel structures. Without loss of generality, it is assumed that the concrete and steel parts have damping ratios equal to 5 % and 2 %, respectively, which are values commonly encountered in design codes for these two materials.

This configuration is very common in the case that an addition of floors occurs sometime in the life span of a building originally made of concrete, where, for reasons of speed of construction and reduction of inertia forces and floor mass or merely for architectural reasons, the added floors are made of steel (Fig. 2). Other typical examples

Mixed In-Height Concrete-Steel Buildings Under Seismic Actions: Modeling and Analysis,
Fig. 3 Telecommunications tower made of steel mounted on concrete building

Mixed In-Height Concrete-Steel Buildings Under Seismic Actions: Modeling and Analysis, Fig. 4 Typical stadium configuration with concrete seats and steel cover

of mixed in-height structures are the cases of concrete buildings with telecommunication towers mounted on top of them (Fig. 3) and stadiums with the structures supporting the spectator's seats made of concrete and covered by a steel truss canopy (Fig. 4). Thus, mixed concrete-steel structural systems are widespread and offer in many occasions desirable solutions due to architectural, structural, and functional reasons.

As a result, there are numerous structures with mixed concrete-steel configuration, having parts

Mixed In-Height Concrete-Steel Buildings Under Seismic Actions: Modeling and Analysis, Fig. 5 Typical MDoF structure configured in one material

that, during a seismic event, exhibit different damping and dissipate energy in different ways. Given that the pertinent seismic design regulations refer to structures consisting of only one material and specify accordingly analysis and design rules, the modeling, calculation of seismic actions, analysis, and design of mixed in-height concrete-steel buildings require a different treatment. The aim of this section is to investigate analysis possibilities of such mixed structures in ways similar to the ones followed for typical homogeneous structures, e.g., with methods that engineers are well acquainted with.

Fundamental Seismic Engineering Concepts

A typical multi-degree of freedom (MDoF) structure consisting of only one structural material, like the one shown in Fig. 5, is the case usually encountered in engineering practice and fully covered by pertinent seismic codes in terms of analysis and design specifications. The seismic oscillation of a structure with n degrees of freedom is described by Eq. 1, where the ground motion is denoted with $\ddot{x}_g$; the displacement with x; M, C, K are the mass, damping, and stiffness matrices, respectively; and r is a unit vector. The damping matrix here represents a typical Rayleigh type damping and can be easily produced using the mass and stiffness matrices.

$$M\{\ddot{x}\} + C\{\dot{x}\} + K\{x\} = -M\{r\}\ddot{x}_g \quad (1)$$

This equation is easy to handle and may be solved to obtain the response of the structure in a manner of ways, widely implemented and incorporated in commercial software. For example, the equation itself can be solved in the time or in the frequency domain, yielding thus the time history of the response in terms of total accelerations or displacements at each degree of freedom. Furthermore, a choice closer to engineering practice is to perform an eigenvalue analysis using the mass and stiffness matrices of Eq. 1 and thus obtain the modal quantities of the structure. Next, each mode is analyzed separately as a single degree of freedom (SDoF) structure as in Eq. 2, where q, ω, Γ are the response, eigenfrequency, and participation factor of each mode.

$$\ddot{q}_i + 2\zeta_i\omega_i\dot{q}_i + \omega_i^2 q_i = -\Gamma_i\ddot{x}_g, \quad i = 1, 2 \ldots n \quad (2)$$

After formulating Eq. 2, it is possible either to analyze it also in the time or frequency domain and thus obtain the time history of the response of each mode or to estimate the maxima of each modal response via the use of a response spectrum, like the ones defined by EC8 (European Committee of Standardization 2004) and shown in Fig. 6. Then, by means of superimposing the modal maxima, it is possible to obtain the maxima of the overall response.

Peculiarities in the Seismic Response of Mixed Structures

In contrast to the case of typical structures consisting of only one material, the dynamic analysis of irregular structures made of concrete

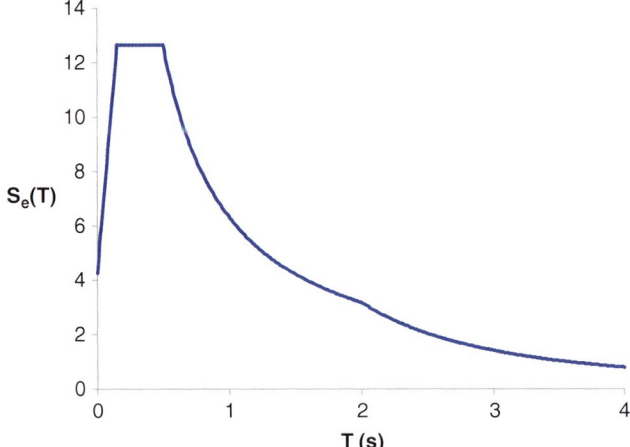

Fig. 6 EC8 elastic response spectrum

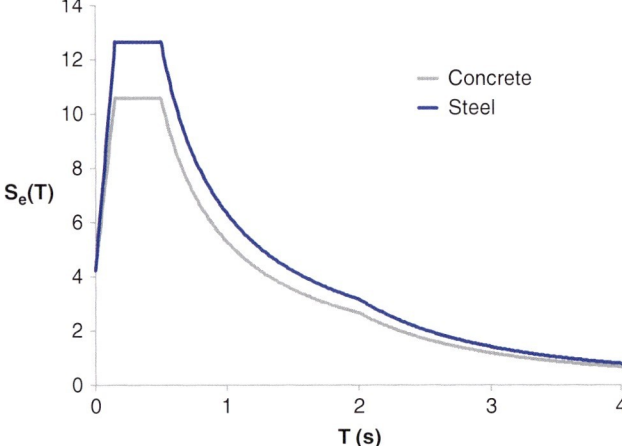

Fig. 7 Comparison between the concrete and steel, compatible to EC8, elastic spectra

and steel as the ones schematically shown in Fig. 1 is not straightforward. The difficulties occur already from the beginning, when forming the equation of motion. Unlike the case of uniform structures, where the damping matrix of Eq. 1 can be easily produced by the mass and stiffness matrices, now a rather cumbersome procedure must be followed in order to produce a damping matrix that manages to represent the actual damping distribution of the structure. During this approach, which is not supported by readily available commercial software, two separate damping matrices of Rayleigh type are formed, one for each part of the structure, which are then merged into a single matrix to be used in Eq. 1.

Then, the equation of motion itself cannot be decoupled in real-valued modal equations of motion like in Eq. 2. Instead, complex-valued modes occur, leading thus to complex-valued eigenfrequencies and modal damping ratios as well as modal participation factors. Moreover, should someone neglect the above and try to implement a typical spectral analysis using code compatible or even seismic spectra, the dilemma of choosing a single universal damping ratio comes up, which may alter significantly the response, as shown in the comparison of the EC8 compatible elastic spectra for damping values of 5 % and 2 %, shown in Fig. 7.

Thus, a common engineering analysis problem turns into a very complex one. As a result,

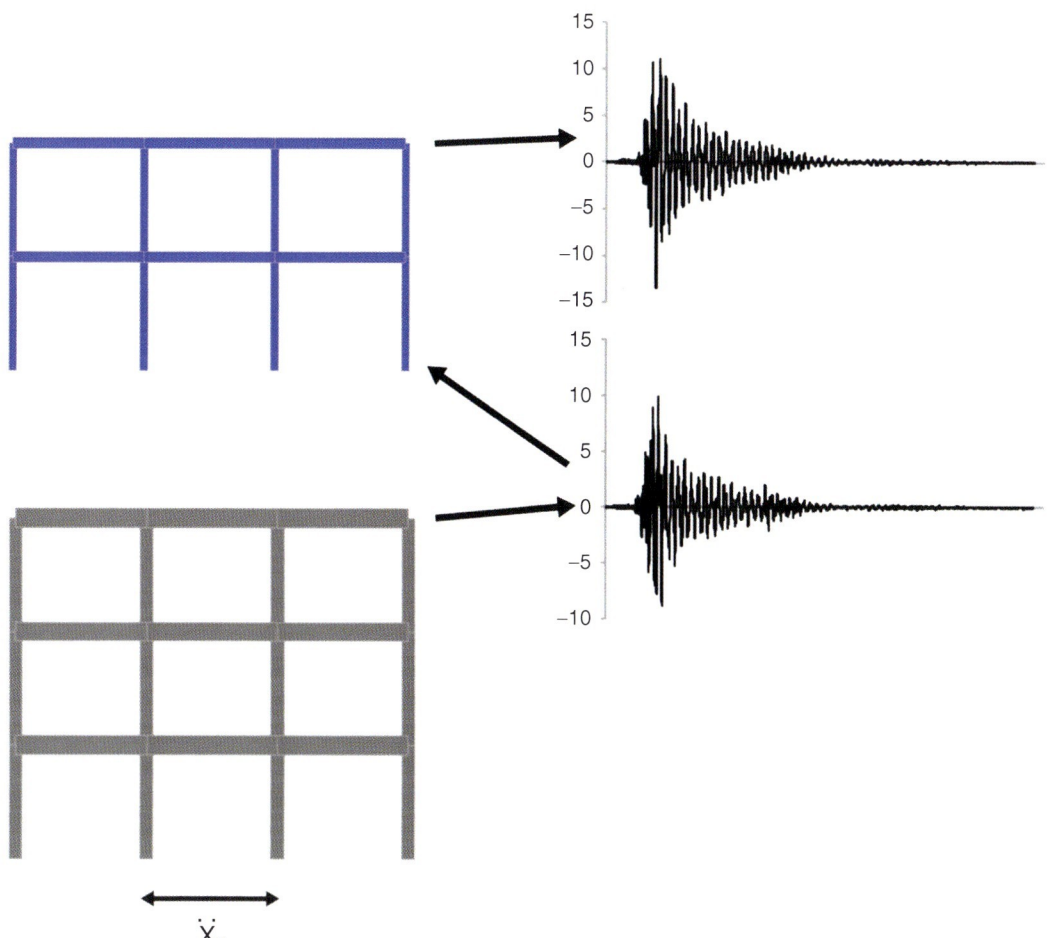

Mixed In-Height Concrete-Steel Buildings Under Seismic Actions: Modeling and Analysis, Fig. 8 Decoupled analysis procedure

and in order to avoid out of the ordinary analysis procedures, the common practice for such cases of irregularity is to make conservative assumptions for the damping ratios, which then leads to overestimation of the response. Due to the frequency with which such irregular structures are encountered, a significant amount of research has been carried out in order to provide analysis solutions that are compatible with common engineering practice.

An interesting approach that has been developed in that direction is the decoupled analysis of such structures. In this procedure the irregular structure is actually divided in two homogeneous parts, and each of them is analyzed separately, as shown in Fig. 8, in contrast to the typical coupled analysis procedure shown in Fig. 9. In the decoupled analysis, the ground acceleration is induced in the concrete part, and its response is then used as a fictitious excitation for the steel part of the structure. The advantage of this method is that in each step the structural part analyzed is homogeneous and therefore typical analysis techniques are applicable. The disadvantage is that the so-called decoupling error is introduced, due to the fact that the analysis is not carried out in the actual structure.

An analytical prediction of this decoupling error for the case of harmonic excitations has been attempted (Chen and Wu 1999), by

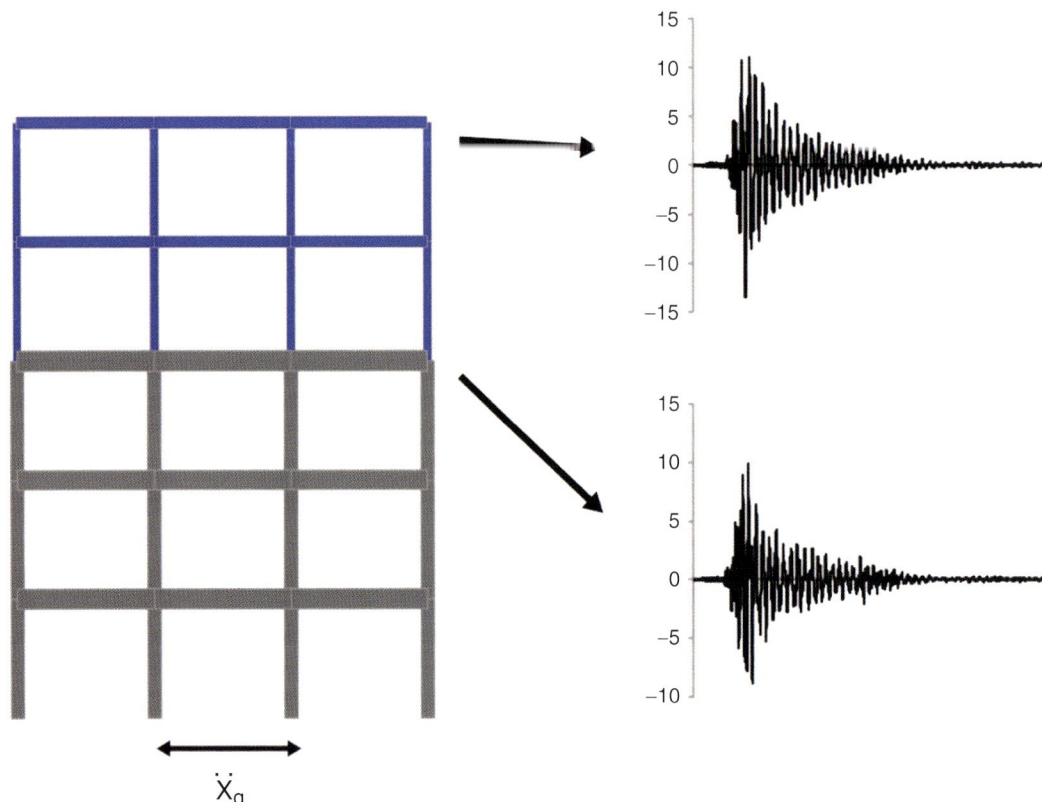

Mixed In-Height Concrete-Steel Buildings Under Seismic Actions: Modeling and Analysis, Fig. 9 Coupled analysis procedure

introducing for that purpose a 2-DoF oscillator and thus managing a limited connection to MDoF structures, while analytical solutions for the decoupling error based on a complex-valued process are also provided (Gupta and Tembulkar 1984a, b).

In a modification of these approaches (Papageorgiou 2010a; Papageorgiou and Gantes 2005), the decoupling error is investigated numerically, by performing time history analyses of the 2-DoF model, so that other issues can be taken into account, such as inelastic material behavior as well as nonharmonic excitations including earthquake accelerograms. The analyses are also based on 2-DoF oscillators, but the results are efficiently correlated to MDoF structures by means of assigning the first mode characteristics of each part, i.e., modal mass and eigenfrequency, to the dynamic characteristics of each degree of freedom of the 2-DoF model, as shown in Figs. 10 and 11. Moreover, graphical tools are produced allowing the prediction of the decoupling error at a preliminary design level, as will be shown next.

In the case that the errors arising from the decoupling procedure are above acceptable limits, a coupled analysis of the irregular structure is the proper solution. In this case and in order to avoid the adoption of a conservative damping ratio, the possibility of using equivalent damping ratios has been thoroughly investigated. By means of such ratios, the exact analysis of the structure, which is difficult to handle with readily available commercial software, is substituted with approximate analyses (Bilbao et al. 2006; Papagiannopoulos and Beskos 2009, 2006), even for the specific case of composite concrete-steel structures (Huang et al. 1996). More specifically, for the case of irregular concrete-steel structures, the results of numerical time history analyses

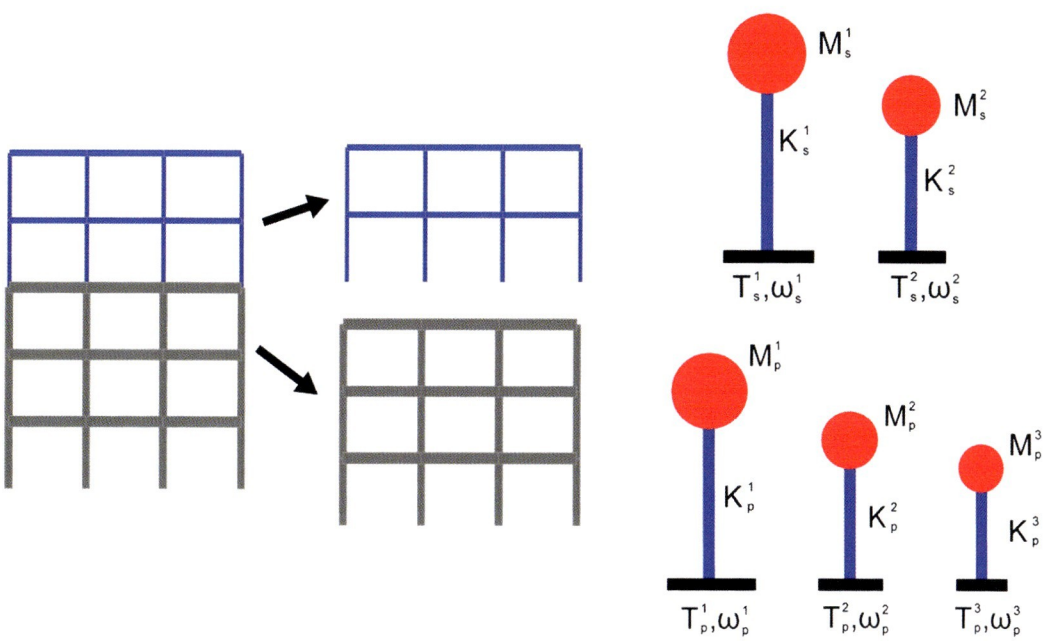

Mixed In-Height Concrete-Steel Buildings Under Seismic Actions: Modeling and Analysis, Fig. 10 Partition of the mixed structure and modal analysis of each part

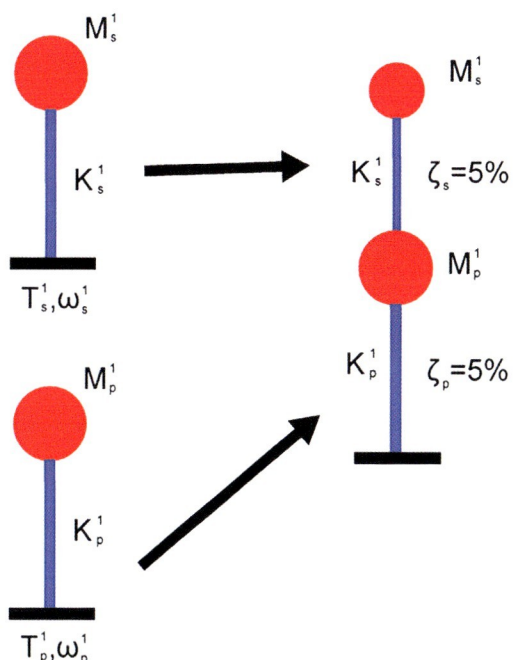

Mixed In-Height Concrete-Steel Buildings Under Seismic Actions: Modeling and Analysis, Fig. 11 Formation of the equivalent 2-DoF oscillator

over a wide range of dynamic characteristics of the two parts permitted the development of graphical tools proposing equivalent damping ratios, both modal and uniform for the entire structure (Papageorgiou and Gantes 2008a, b, 2010b, 2011), as will be presented in the following.

Decoupling

As explained above, a common way to overcome the difficulties in the analysis and design of irregular concrete-steel structures is to divide them in two homogeneous parts and analyze each part separately. The concrete primary structure is analyzed first and its response at the support level of the secondary structure is then used as a fictitious excitation for the steel secondary structure. This decoupled procedure resembles the provisions of current seismic codes relevant to appendages, with the drawback that these provisions cannot be implemented in the case of superstructures with significant masses like the added steel floors

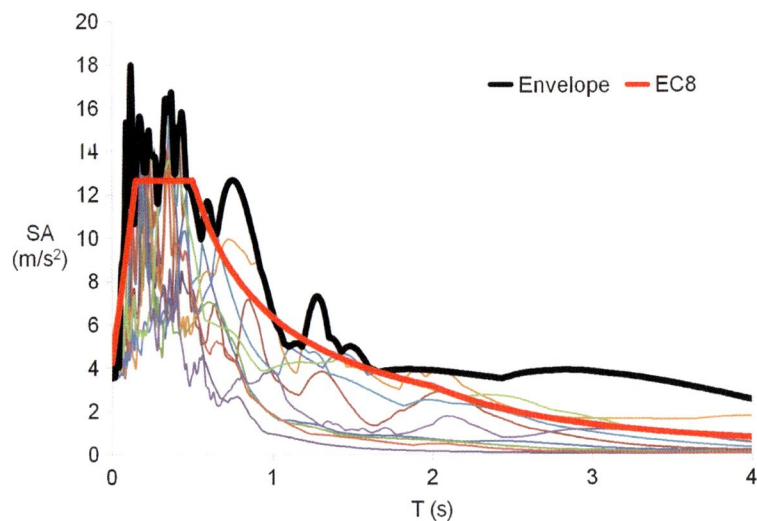

Mixed In-Height Concrete-Steel Buildings Under Seismic Actions: Modeling and Analysis, Fig. 12 Spectra of the seismic records used and comparison of their envelope to EC8 compatible spectrum

of structures schematically represented in Figs. 1, 2, 3, and 4. This division also allows different engineering teams to undertake the design of each part, thus allowing offices specialized in constructions of each material to undertake the analysis and design of each relevant part of the structure.

The disadvantage of the decoupled procedure is that in either of its two analysis stages, the structure analyzed is a part of the real one and not the actual structure with its full mass, damping, and stiffness matrices. Thus, the interaction of the two parts is neglected and an error is introduced, in the prediction of the response of both the primary and the secondary structure. In order to have a reliable and safe, i.e., conservative, prediction of the decoupling error, a thorough investigation has been carried out in order to obtain its distribution for a wide range of dynamic characteristics of the two parts of the structure.

For that purpose, the 2DoF representation illustrated in Fig. 11 has been implemented. From the fundamental eigenfrequencies and the corresponding modal masses of the two parts, two ratios are defined, namely, the eigenfrequency and the mass ratios of secondary to primary structure, R_ω and R_m, respectively, as in Eq. 3. Thus, an (R_ω, R_m) plane can be defined containing a wide range of ratio pairs correlated through the first mode dynamic quantities to a wide range of irregular structures.

$$R_m = \frac{M_s}{M_p}, \quad R_\omega = \frac{\omega_s}{\omega_p} \qquad (3)$$

For the specific case where earthquake excitations are considered, a range of seismic records have been chosen based on the dispersion of the frequencies they resonate the most, and they have been used in the analyses. More specifically, the records were chosen from seismic events in Greece as well as other seismic active areas of the world, and they were selected so as to produce a group of spectra enveloping the code prescribed ones, as shown in Fig. 12.

For each point in the (R_ω, R_m) plane, the corresponding structure is analyzed both approximately, with a decoupled analysis, and "exactly," with a coupled analysis. Thus, the decoupling error is calculated for all points of the (R_ω, R_m) plane, i.e., for the chosen wide range of dynamic characteristics of the irregular structures, as described in Eq. 4 for all levels of the structure. Three reference error levels are chosen, namely, 10 %, 20 %, and 30 %, and the points of the (R_ω, R_m) plane at which these error levels appear are noted. Finally, the envelope of these points is plotted in a set of curves named decoupling curves, one for each error level, as

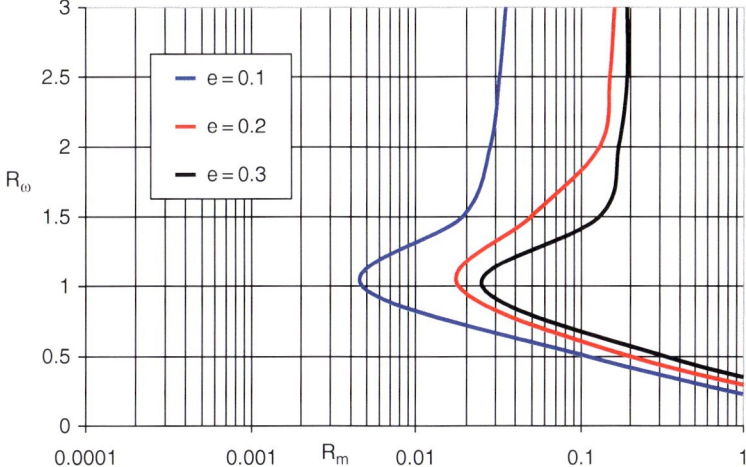

Mixed In-Height Concrete-Steel Buildings Under Seismic Actions: Modeling and Analysis, Fig. 13 Seismic decoupling curves for the primary structure

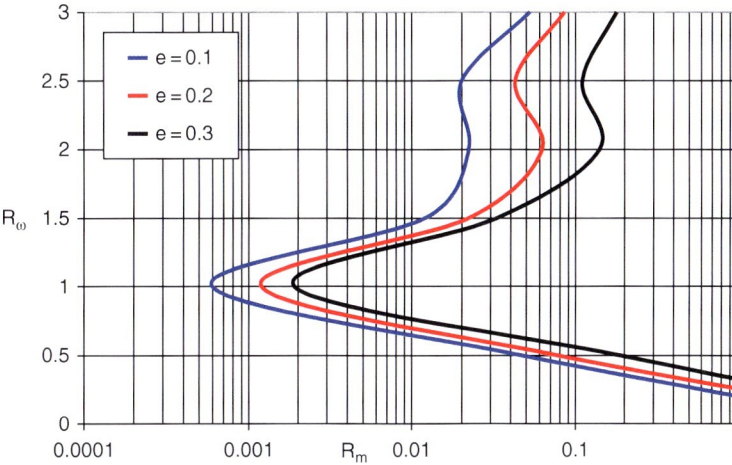

Mixed In-Height Concrete-Steel Buildings Under Seismic Actions: Modeling and Analysis, Fig. 14 Seismic decoupling curves for the secondary structure

shown in Fig. 13 for the concrete part and in Fig. 14 for the steel part of the irregular structure.

$$e = \left| \frac{\max(|\ddot{x}^{dec}|) - \max(|\ddot{x}^{coup}|)}{\max(|\ddot{x}^{coup}|)} \right| \quad (4)$$

These decoupling curves may be of use at a preliminary design stage, when the decision has to be taken whether to proceed with the decoupled or the coupled analysis procedure. At this stage and after the conceptual design of the structure has been concluded, modal analyses of each part of the structure can be carried out. Thus, the R_ω and R_m ratios can be defined so that the structure can be positioned as a point in the (R_ω, R_m) plane. If the point representing the structure lies to the left of a curve, it means that the expected decoupling error is smaller than the one predicted by this specific curve, while if it is on the right, it means larger expected error. Hence, with a maximum error tolerance defined and with the decoupling error prediction by the curves of Figs. 13 and 14, the decision whether to proceed with decoupling can be taken.

The fact that the decoupling curves of Figs. 13 and 14 have been constructed as envelopes of the points arising from each seismic record makes them a conservative, and thus safe, prediction of the decoupling error. The veracity of this claim is

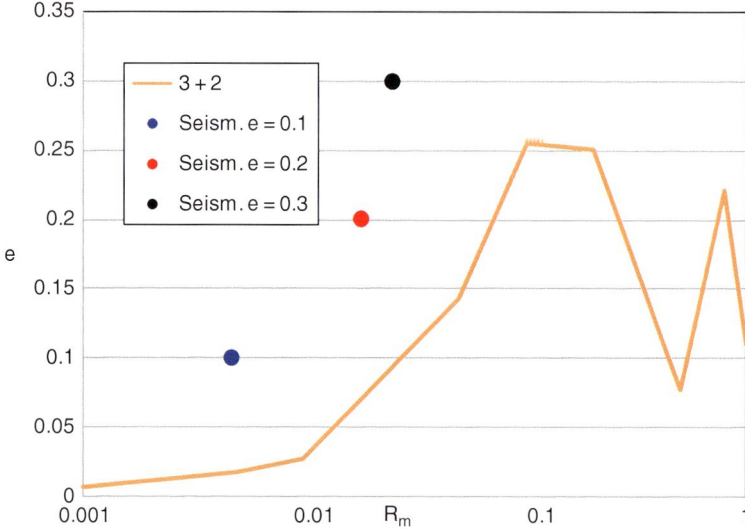

Mixed In-Height Concrete-Steel Buildings Under Seismic Actions: Modeling and Analysis, Fig. 15 Comparison between actual error and prediction from seismic decoupling curves for the primary structure

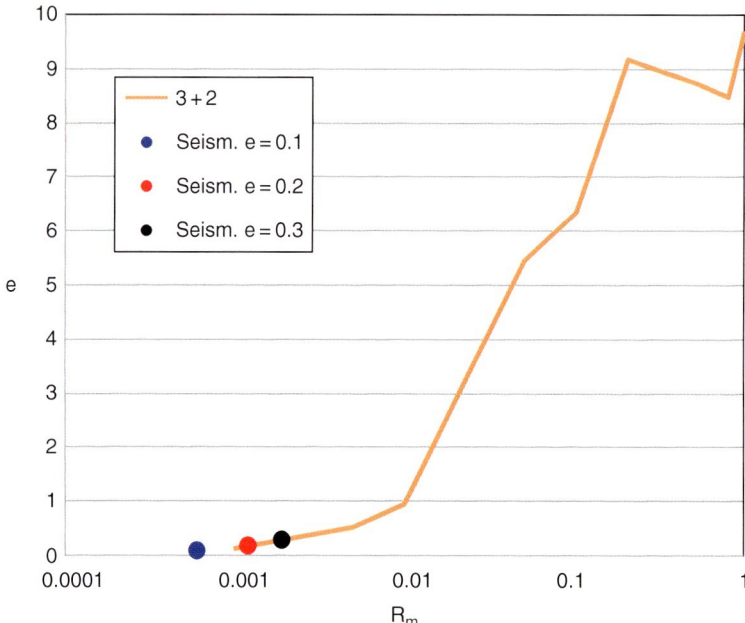

Mixed In-Height Concrete-Steel Buildings Under Seismic Actions: Modeling and Analysis, Fig. 16 Comparison between actual error and prediction from seismic decoupling curves for the secondary structure

tested over an irregular structure comprised by a two-floor steel superstructure mounted atop of a three-story concrete substructure, as the example shown in Fig. 1. In order to test the efficiency of the decoupling curves towards a conservative prediction of the decoupling error, a series of analyses is performed with the first modes of the two parts being in resonance. This is the worst-case scenario for the decoupling error, and the structure of Fig. 1 is modified accordingly in order to be tested over a range of mass ratios. The decoupling error is monitored and shown in Fig. 15 for the primary structure and in Fig. 16 for the secondary structure. In order to provide an assessment of the efficacy of the decoupling curves, the points of the decoupling curves of Figs. 13 and 14 for $R_\omega = 1$ are shown in the respective colors.

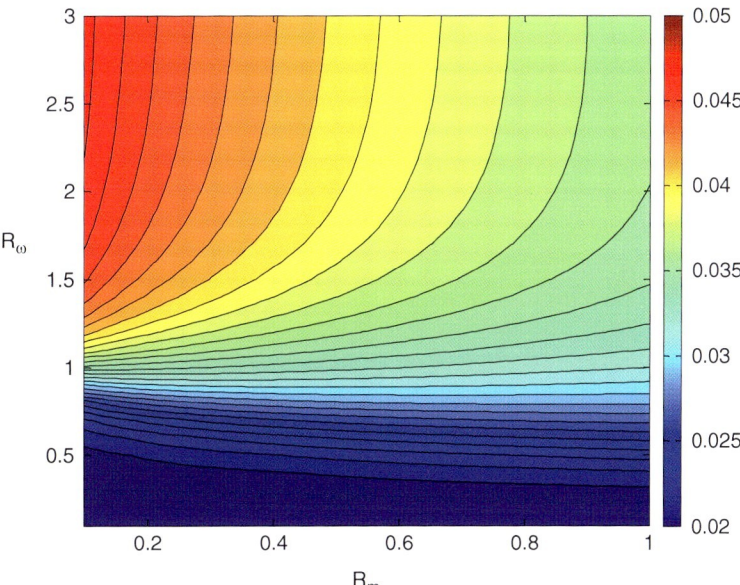

Mixed In-Height Concrete-Steel Buildings Under Seismic Actions: Modeling and Analysis, **Fig. 17** First mode damping ratio

As seen from Figs. 15 and 16, the decoupling curves manage a conservative prediction of the decoupling error for the specific mass ratios. Even considering the worst-case scenario exhibited above, i.e., that of resonance between primary and secondary structures, the error prediction actually coincides with the actual error of the superstructure (Fig. 16), while it is very conservative as far as the primary structure is concerned (Fig. 15). Therefore, the developed decoupling curves can be a useful and at the same time safe tool at a preliminary design stage of analysis in order to enable decision taking whether to proceed with coupled or decoupled analysis.

Equivalent Modal Damping Ratios

It is evident from Figs. 13 and 14 that there are areas of the (R_ω, R_m) plane, i.e., irregular structures with certain combination of dynamic characteristics of the two parts, for which the decoupling error is larger than 30 %. Furthermore, as shown in Fig. 16, this error can be disproportionately large for $R_\omega = 1$ and large R_m values. In such cases, the decoupled analysis procedure is not acceptable. Thus, in order to overcome the coupled analysis difficulties of an irregular structure consisting of two parts with different damping ratios, in this section equivalent damping ratios are proposed, suitably defined so as to effectively approximate the actual response by means of typical analysis that can be performed using commercially available software.

Initially, for each point in the (R_ω, R_m) plane, i.e., for each corresponding irregular structure, the complex damping ratios that normally occur from the complex-valued modal analysis are modified to real-valued ones, by simply adopting their absolute value. The distribution of the modal damping ratios for the first two modes of the structure over the (R_ω, R_m) plane is shown in Figs. 17 and 18.

As in the case of the decoupled analysis, the use of the absolute value of the complex damping ratios in a common real-valued modal analysis described in Eq. 2 also introduces an error, since the correct approach requires an analysis procedure that is complex valued in all steps. This error is again quantified for structures covering all points in the (R_ω, R_m) plane, as shown in Eq. 5 in terms of displacements, for all structure levels. For every point of the (R_ω, R_m) plane, i.e., for every structure investigated, the absolute value of

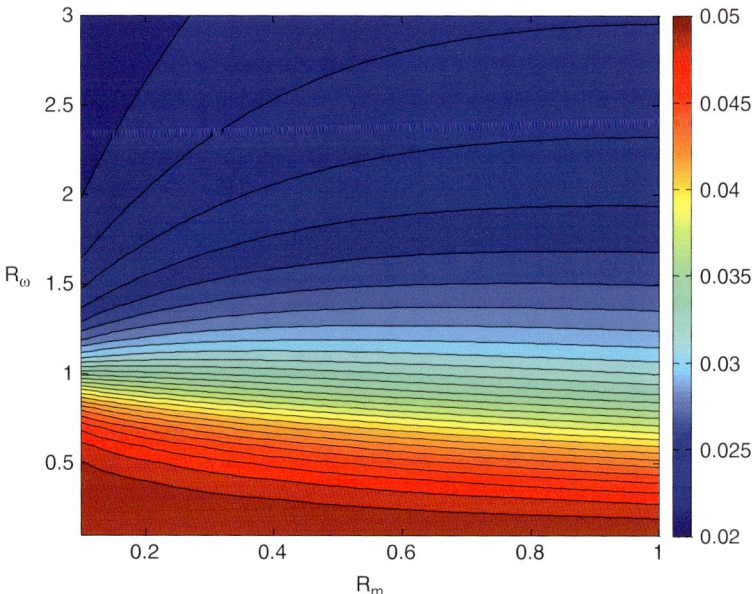

Mixed In-Height Concrete-Steel Buildings Under Seismic Actions: Modeling and Analysis, Fig. 18 Second mode damping ratio

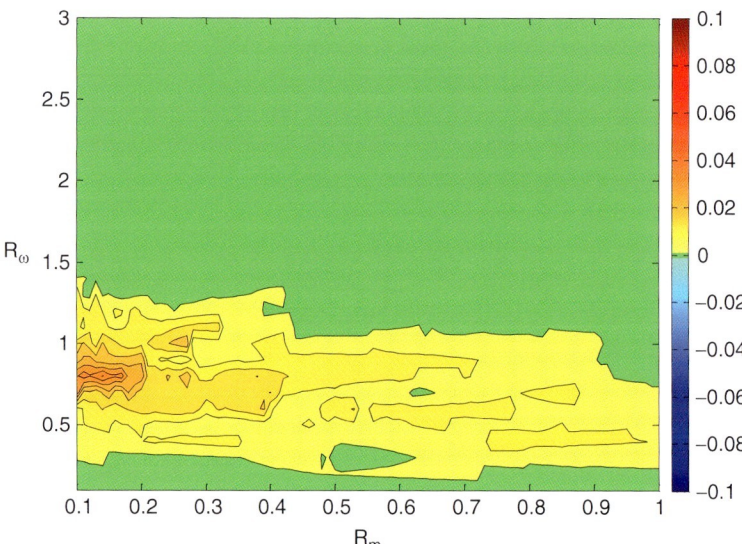

Mixed In-Height Concrete-Steel Buildings Under Seismic Actions: Modeling and Analysis, Fig. 19 Error envelopes for the primary structure

these errors is obtained for each seismic record used, and the envelopes of the maxima are presented in Figs. 19 and 20.

$$e = \frac{\max(x^{appr}) - \max(x^{exact})}{\max(x^{exact})} \quad (5)$$

The fact that the error distribution both for the primary structure and for the secondary one is sufficiently small, at least for a preliminary design level, encourages the adoption of the equivalent damping ratios in the analysis of actual irregular structures. For the structure with two steel floors mounted atop of three concrete levels, schematically shown in Fig. 1, with such dynamic characteristics of the two parts that lead to $R_\omega = 1$ and $R_m = 0.5$, i.e., a case that the two parts are in resonance, the maximum total

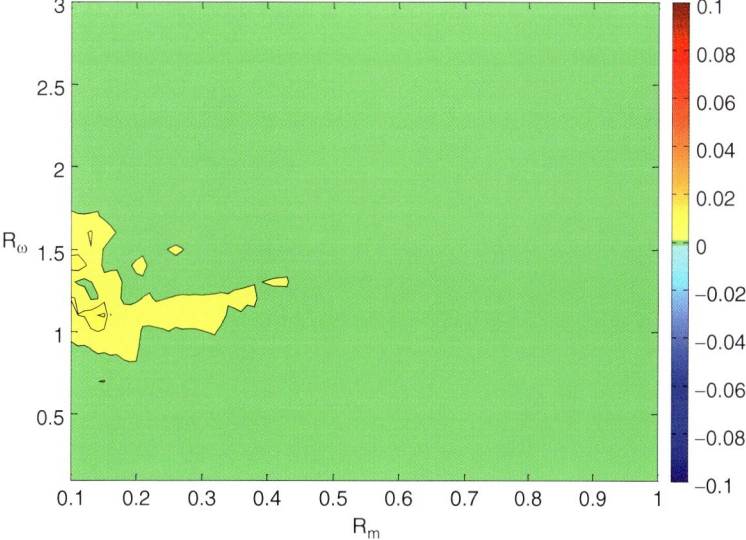

Mixed In-Height Concrete-Steel Buildings Under Seismic Actions: Modeling and Analysis, **Fig. 20** Error envelopes for the secondary structure

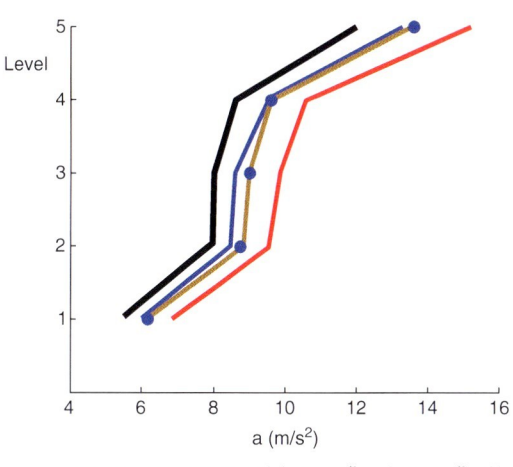

Mixed In-Height Concrete-Steel Buildings Under Seismic Actions: Modeling and Analysis, **Fig. 21** Total acceleration profile for a five-story irregular structure

Mixed In-Height Concrete-Steel Buildings Under Seismic Actions: Modeling and Analysis, **Fig. 22** Error profile for a five-story irregular structure

acceleration distribution has been compared between the actual response and the one arising from the adoption of the equivalent modal damping ratios, and the results are shown in Fig. 21. Moreover, in order to allow comparison with common engineering practice, the distribution obtained by the adoption of overall damping ratios equal to 2 % and 5 % is presented as well. The good performance of the proposed equivalent damping ratios is evident.

Furthermore, in order to demonstrate the effectiveness of the proposed equivalent damping ratios, the error distribution as defined in Eq. 6, in absolute values, over the profile of the structure, is presented in Fig. 22. Along with the error occurring from the modification of the complex-valued damping ratios in order to be used in a real-valued modal analysis, the error that would occur should the damping ratio be assumed to have overall values equal to 2 % or 5 %, again in absolute values, is presented also. It is made clear that the complex-valued modal

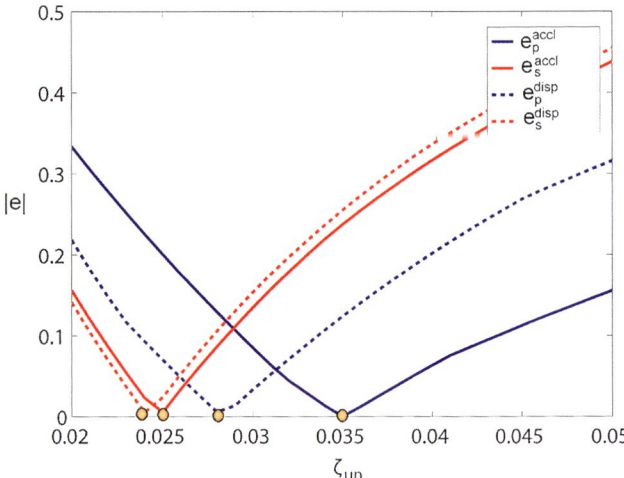

Mixed In-Height Concrete-Steel Buildings Under Seismic Actions: Modeling and Analysis, Fig. 23 Error variation with respect to the uniform damping ratio

damping ratios, suitably modified in order to be used in a common, real-valued, modal analysis procedure, give totally acceptable results, and thus, the whole process is deemed suitable to be adopted in engineering practice. Moreover, the fact that there is readily available commercial software capable of performing structural analysis allowing the specification of modal damping ratios enhances the application prospects of the above-described procedure.

$$e = \left| \frac{\max(\ddot{x}^{appr}) - \max(\ddot{x}^{exact})}{\max(\ddot{x}^{exact})} \right| \quad (6)$$

Equivalent Uniform Damping Ratios

In an effort to provide engineers with equivalent damping ratios that are even closer to engineering practice compared to the modal ones, uniform damping ratios that apply to the whole structure and not just to certain modes are proposed next. In order to come up with these uniform equivalent damping ratios for each structure of the (R_ω, R_m) plane, a trial and error procedure is followed during which the set of possible uniform damping ratios between 0.02 and 0.05 with a step of 0.001 is tested and the response at each level of the structure is compared to the actual one.

In order to increase the accuracy of this procedure, the efficacy of the proposed uniform damping ratios is tested independently for the primary and the secondary structure, in terms of total accelerations and displacements, and the dependence of the error on the damping ratios is shown in Fig. 23. The value of equivalent damping ratio within the above range that minimizes the error in the prediction of the response of the structure, whether in terms of accelerations or in terms of displacements, is chosen as optimum and used as equivalent uniform damping ratio for obtaining the corresponding response quantity.

This procedure is carried out for all points of the (R_ω, R_m) plane, i.e., for irregular structures with a wide range of dynamic characteristics, and for each point the optimum damping ratio in terms of total accelerations and displacements is noted for the concrete and the steel part of the structure. Thus, a distribution of proposed equivalent damping ratios is obtained in terms of total accelerations and displacements in the primary and the secondary part of the structure, as shown in Figs. 24, 25, 26, 27, 28, 29, 30, and 31, along with the corresponding errors.

The error distributions are sufficiently small in this case as well, and thus, their adoption instead of the actual damping distribution is recommended. For the irregular structure of Fig. 1, and for the specific case $R_\omega = 1$, $R_m = 0.5$, i.e., a case that the two parts are in resonance, the response of the structure using the proposed equivalent damping ratios is tested against the actual response.

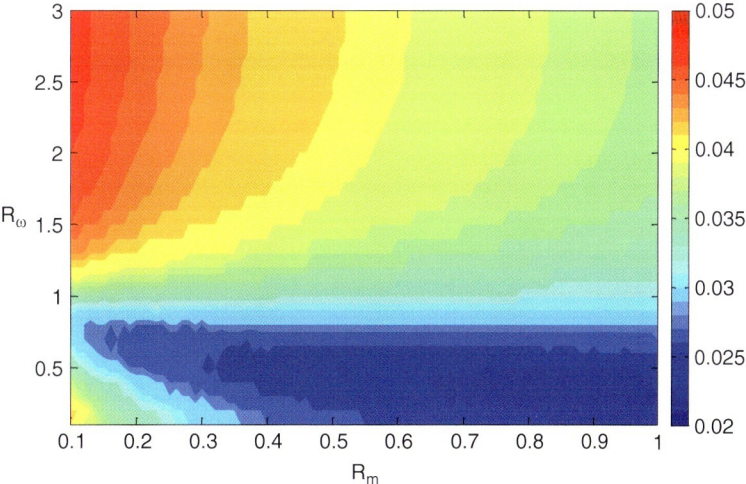

Mixed In-Height Concrete-Steel Buildings Under Seismic Actions: Modeling and Analysis, Fig. 24 Equivalent damping ratio minimizing the displacement error in the primary structure

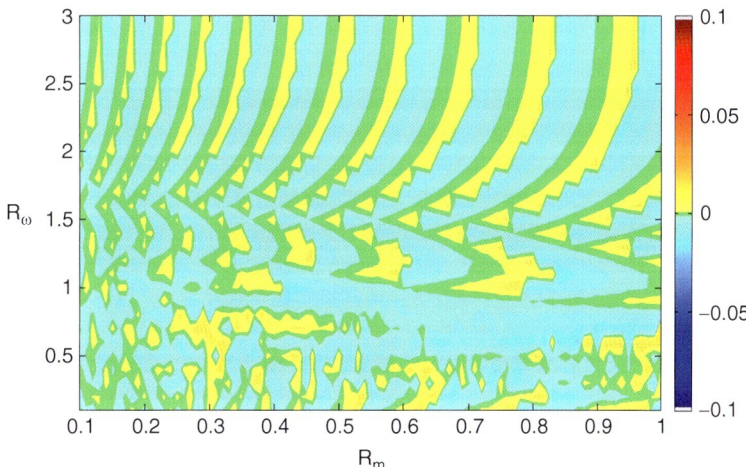

Mixed In-Height Concrete-Steel Buildings Under Seismic Actions: Modeling and Analysis, Fig. 25 Minimized displacement error in the primary structure

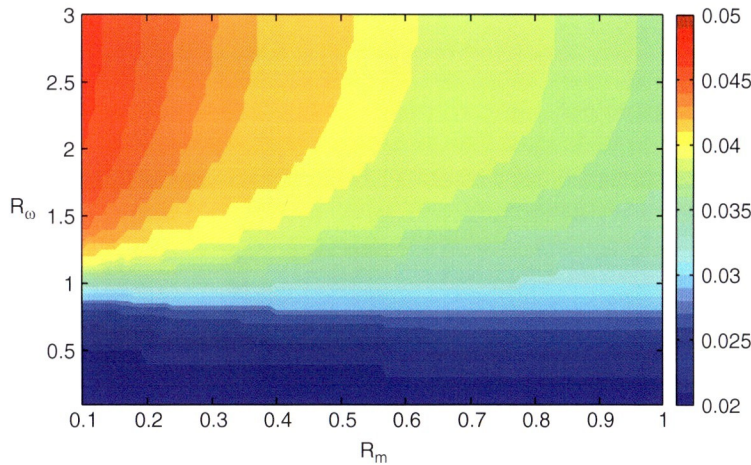

Mixed In-Height Concrete-Steel Buildings Under Seismic Actions: Modeling and Analysis, Fig. 26 Equivalent damping ratio minimizing the displacement error in the secondary structure

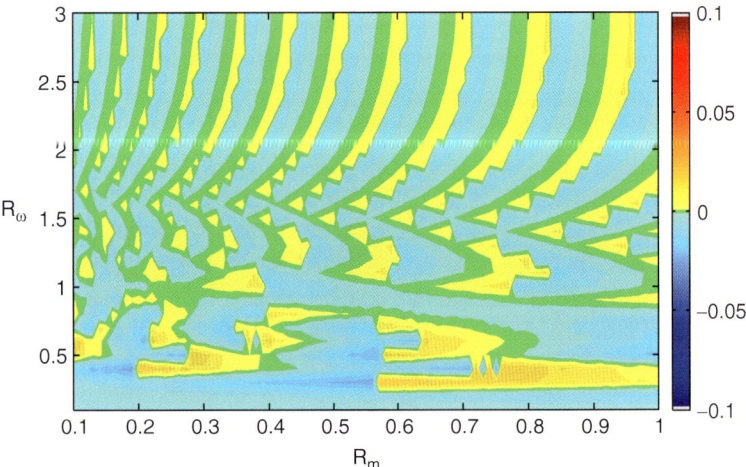

Mixed In-Height Concrete-Steel Buildings Under Seismic Actions: Modeling and Analysis, Fig. 27 Minimized displacement error in the secondary structure

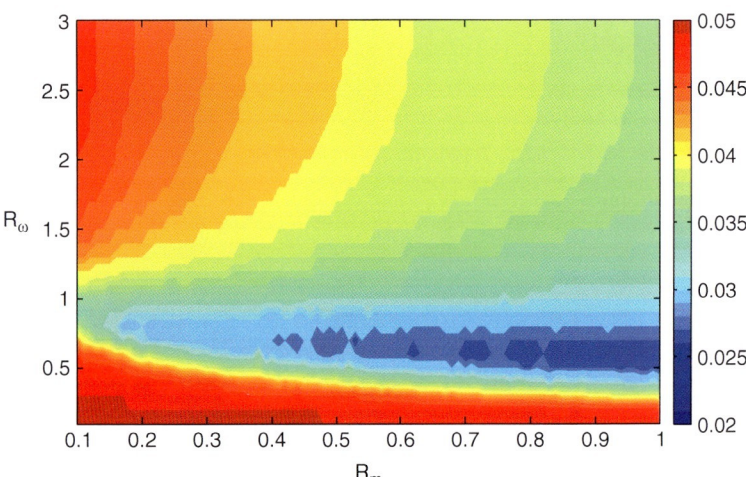

Mixed In-Height Concrete-Steel Buildings Under Seismic Actions: Modeling and Analysis, Fig. 28 Equivalent damping ratio minimizing the acceleration error in the primary structure

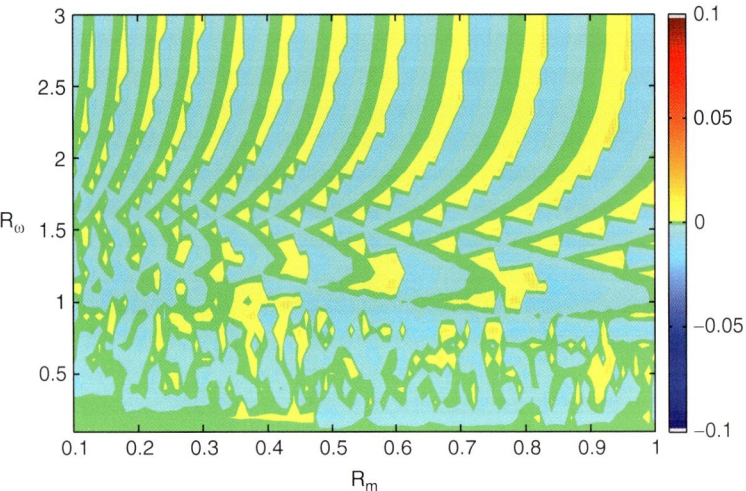

Mixed In-Height Concrete-Steel Buildings Under Seismic Actions: Modeling and Analysis, Fig. 29 Minimized acceleration error in the primary structure

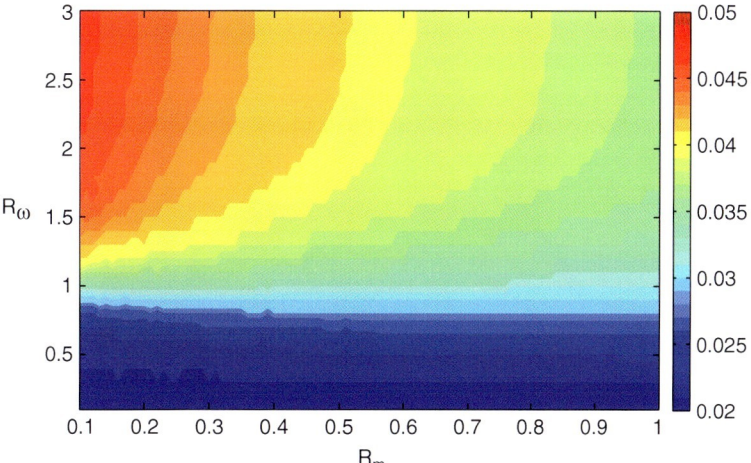

Mixed In-Height Concrete-Steel Buildings Under Seismic Actions: Modeling and Analysis, Fig. 30 Equivalent damping ratio minimizing the acceleration error in the secondary structure

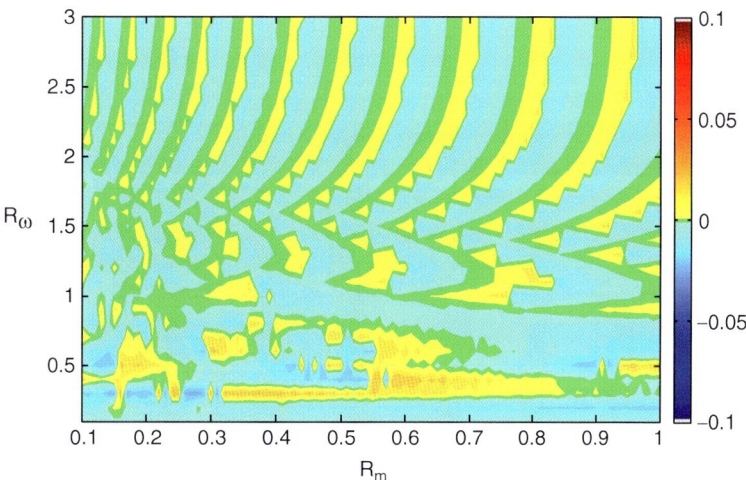

Mixed In-Height Concrete-Steel Buildings Under Seismic Actions: Modeling and Analysis, Fig. 31 Minimized acceleration error in the secondary structure

More specifically, the structure is analyzed two times, each time with the proposed damping ratios for each part, and the displacement profile is constructed as a synthesis of the occurring profiles from each corresponding part. Thus, the total response is obtained and also compared with the response that would occur should overall damping ratios equal to 2 % and 5 % be selected in Fig. 32, while in Fig. 33 the absolute error profiles are compared. A similar approach would be needed to compute accelerations.

Thus, it is evident that even the uniform damping ratios occurring from a trial and error procedure can be applied instead of the adoption of the actual damping distribution. Furthermore, the uniform damping ratios are even closer to engineering practice and may be applied in a response spectrum analysis procedure, where the response occurs by using accordingly produced spectra taking into consideration the proposed damping ratios.

Summary

Despite the frequency with which irregular concrete-steel structures are encountered in engineering practice, there are still no code provisions regarding their seismic design. As a result, intense research efforts have been directed towards

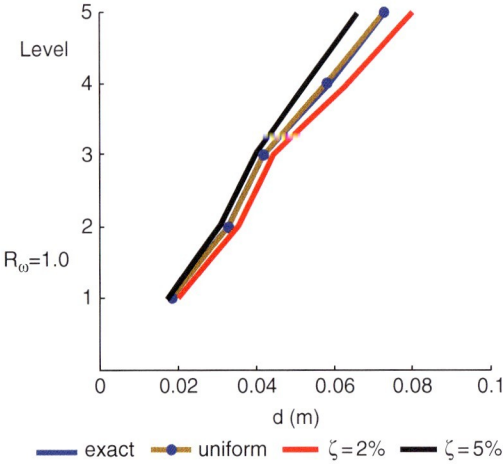

Mixed In-Height Concrete-Steel Buildings Under Seismic Actions: Modeling and Analysis, Fig. 32 Displacements profile for a five-story irregular structure

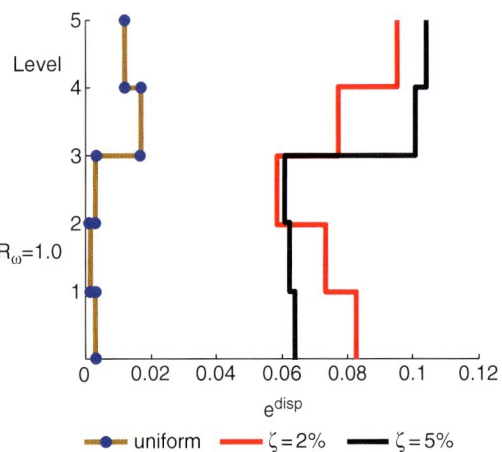

Mixed In-Height Concrete-Steel Buildings Under Seismic Actions: Modeling and Analysis, Fig. 33 Error profile for a five-story irregular structure

supplying answers to the problems arising in such cases. These efforts, as presented here, have started providing results that are very useful for engineering practice, aiming at developing tools that can be used by structural analysis and design teams using available commercial software.

These results come in the form of decoupling criteria, aiming at helping engineers to decide whether they can proceed with a convenient decoupled analysis or not, while in the case that the decoupling error is intolerable, equivalent damping ratios are proposed, modal as well as uniform ones. Still, further research is necessary in order to cover various issues, regarding, for example, structures that are torsionally sensitive or have nonuniform concrete and steel parts, so that higher vibration modes are significant to the overall response, or aiming to address the differences in the elastoplastic behavior of the two parts.

Cross-References

▶ Classically and Nonclassically Damped Multi-degree of Freedom (MDOF) Structural Systems, Dynamic Response Characterization of
▶ European Structural Design Codes: Seismic Actions
▶ Incremental Dynamic Analysis
▶ Modal Analysis
▶ Nonlinear Dynamic Seismic Analysis
▶ Response Spectrum Analysis of Structures Subjected to Seismic Actions
▶ Seismic Analysis of Steel Buildings: Numerical Modeling
▶ Seismic Analysis of Steel–Concrete Composite Buildings: Numerical Modeling
▶ Structures with Nonviscous Damping, Modeling, and Analysis
▶ Time History Seismic Analysis

References

Bilbao A, Avilés R, Agirrebeitia J, Ajuria G (2006) Proportional damping approximation for structures with added viscoelastic dampers. Finite Elem Anal Des 42:492–502

Chen G, Wu J (1999) Transfer-function-based criteria for decoupling of secondary systems. J Eng Mech 125(3):340–346

Eurocode 8 (EC8) (2004) Design of structures for earthquake resistance

Gupta AK, Tembulkar JM (1984a) Dynamic decoupling of secondary systems. Nucl Eng Des 81:359–373

Gupta AK, Tembulkar JM (1984b) Dynamic decoupling of multiply connected MDOF secondary systems. Nucl Eng Des 81:375–383

Huang BC, Leung AYT, Lam KL, Cheung VK (1996) Analytical determination of equivalent modal damping ratios of a composite tower in wind-induced vibrations. Comput Struct 59(2):311–316

Papageorgiou AV, Gantes CJ (2005) Decoupling criteria for the dynamic response of primary/secondary structural systems. In: 4th European workshop on irregular and complex structures (4EWICS), Thessaloniki

Papageorgiou AV, Gantes CJ (2008) Equivalent uniform damping ratios for irregular in height concrete/steel structural systems. In: 5th European conference on steel and composite structures (Eurosteel 2008), Graz, 3–5 Sept 2008, pp 1485–1490

Papageorgiou AV, Gantes CJ (2008) Modal damping ratios for irregular in height concrete/steel structures. In: 5th European workshop on the seismic behavior of irregular and complex structures (5EWICS), Catania, 16–17 Sept 2008, pp 157–168

Papageorgiou AV, Gantes CJ (2010a) Decoupling criteria for inelastic irregular primary/secondary structural systems subject to seismic excitation. J Eng Mech 136(10):1234–1247

Papageorgiou AV, Gantes CJ (2010b) Equivalent modal damping ratios for concrete/steel mixed structures. Comput Struct 88(19–20):1124–1136

Papageorgiou AV, Gantes CJ (2011) Equivalent uniform damping ratios for linear irregularly damped concrete/steel mixed structures. Soil Dyn Earthquake Eng 31(3):418–430

Papagiannopoulos GA, Beskos DE (2006) On a modal damping identification model of building structures. Arch Appl Mech 76:443–463

Papagiannopoulos GA, Beskos DE (2009) On a modal damping identification modal for non-classically damped linear building structures subjected to earthquakes. Soil Dyn Earthquake Eng 29:583–589

Modal Analysis

Nikolaos Nikitas[1], John Hugh George Macdonald[2] and Konstantinos Daniel Tsavdaridis[1]
[1]School of Civil Engineering, University of Leeds, Leeds, UK
[2]Civil Engineering Department, University of Bristol, Bristol, UK

Synonyms

Generalized coordinate analysis; Modal expansion; Mode acceleration method; Mode displacement method; Mode superposition method; Multi-degree-of-freedom (MDOF) system decoupling; Order reduction; Response series truncation

Introduction

The current entry attempts to synoptically guide the reader through the merits and critical details of modal analysis having a particular focus on earthquake engineering, which is the main subject of this encyclopedia. Tracking the history of modal analysis, the conception of vibration modes dates back to the eighteenth century and the pioneering studies and debates of Daniel and John Bernoulli, Euler, and d'Alembert (see Kline 1990) who while studying the problem of the vibration of a taut string came up with the notion of the modal shape contributions that build up the total observed oscillations. Such revolutionary for the time vibration knowledge along with a systematic treatment and practical extensions of modal analysis topics, more focused on continuous systems, appeared first in a very concerted way in the historical "Theory of Sound" by Lord Rayleigh (1877). Much later, today, the term modal analysis is used with different context in engineering language depending on the engineering stream and application ranging from the simple eigenvalue analysis, which recovers the normal mode frequencies and shapes, to the complex experimental dynamic properties identification. Its typical and closest to earthquake engineering definition refers to the ensemble of theoretical tools and operations, which are employed in simplifying the dynamic response calculation of a very large extended structural system under generic time-dependent loading. A much wider picture of modal analysis encompassing apart from the previous side a vast experimental background that has an invaluable contribution in understanding, enhancing, and controlling structures pertains primarily in mechanical engineering. This latter broad view was captured for the first time in a very holistic manner by a seminal book publication by Ewins (1984), which was later succeeded by similar influential titles like the ones by his students Maia et al. (1997) and He and Fu (2001).

The description that follows does not intend to be an alike broad exhaustive review of the so-called *experimental* or *operational modal analysis*, and the reader is suggested to refer to the quoted titles for this purpose. The definition assumed herein is closer to the first solely analytical approach, and as such it was previously very effectively included in the fundamental dynamics handbooks by Biggs (1964), Clough and Penzien (1993), and Chopra (1995) or in latest educational dynamics literature like Rajasekaran (2009). There the main attributes, limitations, and benefits of expressing a discretized structure's response to some random or deterministic dynamic excitation (earthquakes being an ideal generic example) as the superposition of modal responses are discussed and exemplified through practical exercises. The influence of damping in the modal decomposition process along with the crucial question of how many modes are needed in accurately reproducing the full dynamic information hidden in response metrics are meticulously addressed for becoming a tool in the hands of practicing structural engineers. The presentation that follows collects and reproduces a synopsis of all this essential digested knowledge, which is of high priority for seismic analysis and design. This information directly links to other entries of the encyclopedia and allows their better understanding and connection. Although it aims to be a self-contained entry, it is founded on some previous knowledge of structural modeling and the dynamics of single-degree-of-freedom (SDOF) and multi-degree-of-freedom (MDOF) systems, which are presented assuming the reader was previously exposed to basic vibration textbooks.

The entry begins with the study of a typical linear undamped n-degree-of-freedom system (with the application of multistory buildings always in mind) under a common form of dynamic loading. This system, in a straightforward manner, is transformed to n independent/uncoupled generalized SDOF systems acted upon by certain fractions of the initially defined loading, and two different approaches are introduced for performing any further practical analysis. Subsequently, findings are extrapolated to a variously damped viscous system (the proportional or classical and the generic or nonclassical), which is a very useful and effective approximation of the energy dissipation mechanism even in complex structures consisting of many interconnecting nonuniform parts. The alternatives of working either in the time or frequency domain are both discussed in brief. For the subsequent main topic of modal truncation that emerges, where one has to decide the minimum number of modes r ($r < n =$ number of degrees of freedom) that suffice for a sufficiently accurate reduced representation of the dynamic solutions (which may be displacements, rotations, cumulative shear forces, moments, etc.), all main different approaches and practices are presented. Miscellaneous details on various alterations of the analysis (e.g., mode acceleration or mode displacement form) practiced for increasing its numerical efficacy together with connections to aseismic structural design and brief essential reference to valuable extensions (e.g., nonlinear systems, health monitoring, and control) are provided throughout.

Modal Analysis Basics

The Undamped/Conservative System

The topic of dynamic-degree-of-freedom (DOF) selection, which is typical when transforming continuous systems to discrete models, along with reference to *nonholonomic* systems (i.e., where independent motion coordinates are essentially accompanied by additional constraints for fully determining motion) is not pursued herein. For the presentation in hand, the simple form of the already discrete engineering models assumed yield such discussions unnecessary. Thus, for this point, imagine a simplified n-degree-of-freedom discrete undamped structural model relieved by any *gyroscopic effects*. Such can be the translational toy mechanical system in Fig. 1a or the sway multistory shear frames in Fig. 1b, c.

In all cases depicted in Fig. 1, the envisaged finite degrees of freedom (DOFs) necessary are clearly designated. For all systems additional assumptions (e.g., rigid beams, massless columns, and springs) were employed for reducing the required DOFs. The set of governing equations of motion may be put, by either following the force equilibrium approach on n free-body

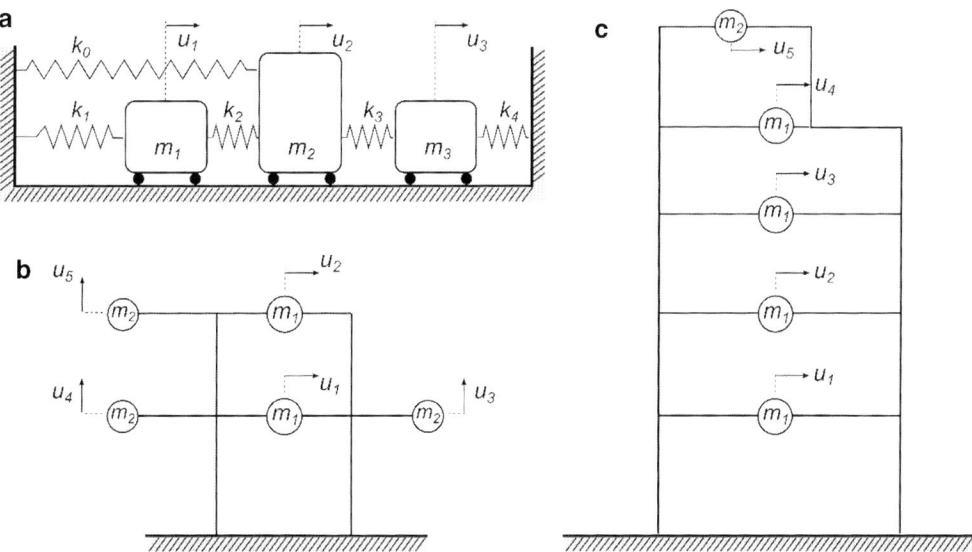

Modal Analysis, Fig. 1 MDOF structural models (**a**) toy mechanical discrete system (**b**, **c**) multistorey sway shear frames

diagrams or the energy approach, in the matrix form

$$\mathbf{M}\ddot{\mathbf{u}}(t) + \mathbf{K}\mathbf{u}(t) = \mathbf{F}(t), \quad (1)$$

where **M** and **K** are the constant symmetric mass and stiffness matrices ($n \times n$) $\ddot{\mathbf{u}}(t)$ and $\mathbf{u}(t)$ are the column matrices or vectors ($n \times 1$) of accelerations and displacements expressed in physical coordinates varying with time t and $\mathbf{F}(t)$ the vector ($n \times 1$) of the externally applied (not motion-dependent) dynamic loads. As a convention, all matrices are herein denoted by bold letters, while common letters are used to denote scalar magnitudes. Regarding the stiffness matrix **K**, we must notice that it encompasses both the elastic and geometric stiffness information of the structural model, e.g., forces causing buckling are assigned to geometric stiffness. Also the constant nature of **K** translates to a linear elastic operational regime. As it can be easily verified through all Fig. 1 examples, for **M** it is straightforward to make it diagonal using the so-called *lumped mass* form, where translational DOFs are defined on each discrete mass. Yet this, which is not always possible or practical, will generally result in a non-diagonal **K** and, thus, in statically coupled Eq. 1, with the term *statically coupled* meaning that each of the constituting linear equations of motion includes components of the displacement vector $\mathbf{u}(t)$ corresponding to various DOFs and not only the DOF referred to the existing component of the acceleration vector $\ddot{\mathbf{u}}(t)$. This coupling is in fact a coordinate coupling that perplexes the solution of Eq. 1 especially in the common case of large n. Alternatively, a different formulation practiced in finite element (FE) derivations for either **M** or **K**, entitled *consistent form*, leads in general to non-diagonal forms for both these matrices, thereby leading to *statically and dynamically* coupled equations of motion. This means that each one of these equations includes components of the vectors $\mathbf{u}(t)$ and $\ddot{\mathbf{u}}(t)$ corresponding to various DOFs and not only to one DOF. And exactly, modal analysis aims at decoupling these equations. If Eq. 1 gets uncoupled, this will yield n independent equations of motion, i.e., each one of which refers to only one DOF, with the result that the MDOF system described by Eq. 1 can be treated as n equivalent uncoupled SDOF systems. Each of them corresponds to a so-called natural mode of vibration, which is characterized by a particular frequency and shape (modal/eigen frequency ω_i and modal/eigen vector $\boldsymbol{\varphi}_i$, ($n \times 1$)). Such an advantageous simplification can be established by transforming the vector $\mathbf{u}(t)$ into a new vector $\mathbf{q}(t)$, $(n \times 1)$, whose components are called normal or principal coordinates, as well as generalized displacements. This transformation is equivalent to the separation of variables technique, exercised in partial differential equations (PDEs), and is described as below

$$\mathbf{u}(t) = \mathbf{\Phi}\mathbf{q}(t) = \sum_{i=1}^{n} \boldsymbol{\varphi}_i q_i(t), \qquad (2)$$

or alternatively for the displacement vector $\mathbf{u}_i(t)$ corresponding to each individual mode i

$$\mathbf{u}_i(t) = \boldsymbol{\varphi}_i q_i(t). \qquad (3)$$

In Eqs. 2 and 3, $\mathbf{\Phi}$ is the modal matrix $(n \times n)$ consisting of all n mode shapes $\boldsymbol{\varphi}_i$ arranged as its columns, which includes the entire modal shape information. And $\mathbf{q}(t)$ is a vector (i.e., column matrix) of n components, with the arbitrarily chosen ith component $q_i(t)$ representing the unknown time-functioning generalized displacement for the ith mode.

The modal parameters of frequency ω_i and shape $\boldsymbol{\varphi}_i$ are deduced by solving the so-called eigenvalue problem. This comes up when replacing Eq. 3 inside the free/unforced counterpart of Eq. 1 resulting from Eq. 1 after setting $\mathbf{F}(t) = 0$ and recognizing the pure harmonic nature of the associated $\mathbf{q}(t)$. Specifically speaking, considering the free/unforced counterpart of Eq. 1, i.e., the free-vibration equations

$$\mathbf{M}\ddot{\mathbf{u}}(t) + \mathbf{K}\mathbf{u}(t) = 0, \qquad (4)$$

and assuming that each modal response $\mathbf{u}_i(t) = \boldsymbol{\varphi}_i q_i(t)$ is a solution to Eq. 4, it follows

$$\mathbf{M}\boldsymbol{\varphi}_i \ddot{q}_i(t) + \mathbf{K}\boldsymbol{\varphi}_i q_i(t) = 0, \qquad (5)$$

which after premultiplying with the transposed mode shape vector $\boldsymbol{\varphi}_i^T$, i.e., a row matrix with the same elements as the column matrix $\boldsymbol{\varphi}_i$, gives

$$\boldsymbol{\varphi}_i^T \mathbf{M}\boldsymbol{\varphi}_i \ddot{q}_i(t) + \boldsymbol{\varphi}_i^T \mathbf{K}\boldsymbol{\varphi}_i q_i(t) = 0. \qquad (6)$$

Now, defining the frequency ω_i as equal to

$$\omega_i^2 = \frac{\boldsymbol{\varphi}_i^T \mathbf{K}\boldsymbol{\varphi}_i}{\boldsymbol{\varphi}_i^T \mathbf{M}\boldsymbol{\varphi}_i}, \qquad (7)$$

Equation 6 becomes equal to the harmonic vibration in $q_i(t)$

$$\ddot{q}_i(t) + \omega_i^2 q_i(t) = 0, \qquad (8)$$

and combining with Eq. 5 results in

$$\left(\mathbf{M} - \omega_i^2 \mathbf{K}\right) \boldsymbol{\varphi}_i = 0, \qquad (9)$$

which has a nonzero solution $\boldsymbol{\varphi}_i$ only for the zero determinant

$$\left|\mathbf{M} - \omega_i^2 \mathbf{K}\right| = 0. \qquad (10)$$

Equation 10 is also termed the characteristic or frequency equation of the system. As expected due to the problem's indeterminate nature, the resulting vectors $\boldsymbol{\varphi}_i$ are not unique solutions but represent families of solutions of the form $\alpha \boldsymbol{\varphi}_i$, with α being an arbitrary scalar constant. The computational effort/time required for a solution is proportional to n^3. It is noticed that deriving such sets of ω_i, $\boldsymbol{\varphi}_i$ solutions for large structural systems have introduced a whole new field for devising efficient solvers of Eqs. 9 and 10 (i.e., eigensolvers); the curious reader may consult §8 (Rades 2010). One of the most renown equivalent numerical techniques of great application belongs to Rayleigh and Ritz (for some historical notes and disputes on its origins, one can read Leissa (2005)). Further, it is vital to refer to the important orthogonality properties, viz.,

$$\boldsymbol{\varphi}_i^T \mathbf{K}\boldsymbol{\varphi}_j = \boldsymbol{\varphi}_i^T \mathbf{M}\boldsymbol{\varphi}_j = 0 \text{ for all } i \neq j, \qquad (11)$$

which result from the symmetrical form of the mass and stiffness matrices $\mathbf{M}$ and $\mathbf{K}$, respectively, and allow Eq. 1 to be uncoupled into n SDOF equations. Indeed, premultiplying Eq. 1 with $\boldsymbol{\varphi}_i^T$ and replacing $\mathbf{u}(t)$ with $\mathbf{\Phi}\mathbf{q}(t)$ according to Eq. 2, one gets

$$\boldsymbol{\varphi}_i^T \mathbf{M} \boldsymbol{\Phi} \ddot{\mathbf{q}}(t) + \boldsymbol{\varphi}_i^T \mathbf{K} \boldsymbol{\Phi} \mathbf{q}(t) = \boldsymbol{\varphi}_i^T \mathbf{F}(t). \quad (12)$$

Owning to the orthogonality conditions, Eq. 12 becomes

$$\boldsymbol{\varphi}_i^T \mathbf{M} \boldsymbol{\varphi}_i \ddot{q}_i(t) + \boldsymbol{\varphi}_i^T \mathbf{K} \boldsymbol{\varphi}_i q_i(t) = \boldsymbol{\varphi}_i^T \mathbf{F}(t), \quad (12a)$$

or equally

$$M_i \ddot{q}_i(t) + K_i q_i(t) = P_i(t), \quad (12b)$$

which after division with M_i becomes

$$\ddot{q}_i(t) + \omega_i q_i(t) = \frac{P_i(t)}{M_i}, \quad (12c)$$

where

$$\begin{aligned} K_i &= \boldsymbol{\varphi}_i^T \mathbf{K} \boldsymbol{\varphi}_i, \; M_i = \boldsymbol{\varphi}_i^T \mathbf{M} \boldsymbol{\varphi}_i, \\ \omega_i &= K_i/M_i \text{ and } P_i(t) = \boldsymbol{\varphi}_i^T \mathbf{F}(t) \end{aligned} \quad (13)$$

with K_i, M_i, ω_i and $P_i(t)$ being the so-called scalar modal stiffness, mass, frequency, and force, respectively. Note that combining orthogonality with Eq. 9 the typical SDOF relation, $K_i = \omega_i^2 M_i$ comes up. The values of K_i and M_i are subject to the scaling opted for the modal vectors $\boldsymbol{\varphi}_i$ and as such on their own are not very informative of the whole system. A common scaling, or else referred to as normalization, used for $\boldsymbol{\varphi}_i$ is the one that results in unit modal masses. Namely, if each $\boldsymbol{\varphi}_i$ is divided by $1/\sqrt{M_i}$, then evidently the second of Eq. 13 yields

$$\frac{\boldsymbol{\varphi}_i^T}{\sqrt{M_i}} \mathbf{M} \frac{\boldsymbol{\varphi}_i}{\sqrt{M_i}} = 1. \quad (14)$$

With all the previous data in hand, one can now proceed with solving the initial problem of determining the response of the undamped system to some generic forcing $\mathbf{F}(t)$. The complete solution to Eq. 12b, which is a generalized SDOF equation, provides the part of the total response that owes exclusively to the ith mode and can be solved independently of all other modes. As is well known from the solution to an SDOF equation, the modal solution $q_i(t)$ of Eq. 12c consists of a free-vibration solution $q_i^c(t)$ due to exclusively the initial conditions $q_i(0)$ and $\dot{q}_i(0)$ and a particular forced-vibration solution $q_i^p(t)$ due to exclusively the imposed modal force $P_i(t)$. Namely, the resultant solution $q_i(t)$ in the time domain equals

$$q_i(t) = q_i^c(t) + q_i^p(t), \quad (15)$$

with

$$q_i^c(t) = q_i(0)\cos \omega_i t + \frac{\dot{q}_i(0)}{\omega_i} \sin \omega_i t, \quad (16)$$

and

$$q_i^p(t) = \frac{1}{\omega_i} \int_0^t \frac{P_i(\tau)}{M_i} \sin \omega_i(t - \tau) d\tau. \quad (17)$$

The forced-vibration solution $q_i^p(t)$ described in the time domain by Eq. 17 is the Duhamel integral expression, for the calculation of which a number of numerical techniques (e.g., Newton-Cotes formulas, Simpson's rule, Romberg integration, etc.) may be employed. It is worth pointing out that recovering the modal initial conditions $q_i(0)$ and $\dot{q}_i(0)$ from the initial conditions $\mathbf{u}(0)$ and $\dot{\mathbf{u}}(0)$ can be a straightforward result of Eq. 2 and the orthogonality properties (11). Namely, by multiplying Eq. 2 with $\boldsymbol{\varphi}_i^T \mathbf{M}$ and taking into account the orthogonality properties (11), it follows that

$$\begin{aligned} \boldsymbol{\varphi}_i^T \mathbf{M} \mathbf{u}(0) &= \boldsymbol{\varphi}_i^T \mathbf{M} \boldsymbol{\Phi} \mathbf{q}(0) \\ &= \boldsymbol{\varphi}_i^T \mathbf{M} \boldsymbol{\varphi}_i q_i(0) = M_i q_i(0), \end{aligned} \quad (18a)$$

$$\boldsymbol{\varphi}_i^T \mathbf{M} \dot{\mathbf{u}}(0) = \boldsymbol{\varphi}_i^T \mathbf{M} \boldsymbol{\Phi} \dot{\mathbf{q}}(0) = \boldsymbol{\varphi}_i^T \mathbf{M} \boldsymbol{\varphi}_i \dot{q}_i(0) = M_i \dot{q}_i(0), \quad (18b)$$

from which the magnitudes $q_i(0) = \boldsymbol{\varphi}_i^T \mathbf{M} \mathbf{u}(0)/M_i$ and $\dot{q}_i(0) = \boldsymbol{\varphi}_i^T \mathbf{M} \dot{\mathbf{u}}(0)/M_i$ are derived.

Finally, after solving Eqs. 16 and 17 for the unknowns $q_i^c(t)$ and $q_i^p(t)$, and then Eq. 15 for the unknown $q_i(t)$, Equation 2 will be used for synthesizing the total response time histories $\mathbf{u}(t)$ from the individual synchronous modal contributions $\mathbf{u}_i(t)$, which have been determined as the products $\boldsymbol{\varphi}_i q_i(t)$ of the evaluated modal shapes $\boldsymbol{\varphi}_i$ and generalized displacements $q_i(t)$. This writes as

$$\mathbf{u}(t) = \sum_{i=1}^{n} \mathbf{u}_i(t) = \sum_{i=1}^{n} \boldsymbol{\varphi}_i q_i(t). \qquad (19)$$

The most valuable feature of Eq. 19 is that not all n terms of the sum are required for accurately representing the total response. A large-scale structure like a tall building has hundreds of DOFs, which will produce hundreds of natural modes. Yet, it is common that for typical operational loading conditions, the lower (i.e., of lower frequencies) few modes, whose number r is far smaller than the number n of all the DOFs of the structure, are enough to compute the vector $\mathbf{u}(t)$ by using Eq. 19, for the higher (i.e., of higher frequencies) modes of a structure are not excited considerably. The value of r, as well as to what extent it can be reduced with respect to n, is a function not only of the initial loading $\mathbf{F}(t)$'s characteristics (i.e., shape distribution and frequency content) and their match to specific natural modes but also of the response metric that needs to be considered. For instance, representation of dynamic moment, shear, and stress variables in general requires more natural modes than representing the displacement solutions. This finding that was meticulously exemplified in Chopra (1995, 1996) and Clough and Penzien (1993) owes primarily to the fact that higher-order response attributes need additional information for their description (analogous to the added information embedded in derivatives of a function compared to the actual function).

The Classical/Proportional Damping Addition

The conservative (i.e., undamped, i.e., without energy losses) system treated earlier is fundamental for any dynamic study. Yet, it is not completely accurate in representing reality. All structures have the inherent ability of damping with which they can dissipate energy inputs. As a matter of fact, apart from being a structural property, effective damping can also surface from the interaction of the structure with environmental or other types of loads. Distinctive examples of such cases, also described as *self-excitation*, are the Tacoma Narrows Bridge collapse (with the associated pure torsional flutter) and the large lateral vibrations of London Millennium Bridge (with pedestrians acting as negative dampers). The damping property is almost impossible to faithfully model ab initio. This is a result of a multitude of reasons. There is little information on inherent material damping, and further energy dissipation can originate from sources like connections, crack openings, or inter-element friction that are really hard to capture entirely. The practical and plausible solution is to model damping as a linear viscous term $\mathbf{C}\dot{\mathbf{u}}(t)$, i.e., function of velocity, which efficiently provides a route toward changing the energy balance in the equations of motion. This will lead to modifying the set of Eq. 1 as follows:

$$\mathbf{M}\ddot{\mathbf{u}}(t) + \mathbf{C}\dot{\mathbf{u}}(t) + \mathbf{K}\mathbf{u}(t) = \mathbf{F}(t), \qquad (20)$$

where $\mathbf{C}$ stands for the damping matrix ($n \times n$). This is in general non-diagonal. A work-around toward approximating the many unknown elements of $\mathbf{C}$ is to connect it to the straightforwardly derived structural matrices $\mathbf{K}$ and $\mathbf{M}$. The approximation of stiffness-proportional damping, which is actually appealing to intuition, can easily be shown to result in damping increasing with frequency. This is contrary to many experimental findings, which in general refute such a pattern. Mass-proportional damping on the other hand, which is simplistically linked to air/medium damping, also contradicts experience. However, combining the two proportionality approaches yields the so-called Rayleigh damping, which seems to constitute a good and much-used idealization for the energy dissipation mechanism of evenly distributed structures. A typical example of the latter is a multistory building of similar material and structural system throughout its height. According to the Rayleigh approach, $\mathbf{C}$ is given the form

$$\mathbf{C} = a_0 \mathbf{M} + a_1 \mathbf{K}, \qquad (21)$$

where a_0 and a_1 are unknown scalar constants. Employing Rayleigh damping allows the classical modal decomposition (uncoupling) process earlier performed for the undamped case to apply to the damped case as well. Accordingly, one can firstly insert Eq. 2 in Eq. 20 and then multiply with the transposed mode shape vector $\boldsymbol{\varphi}_i^T$ derived from the undamped free-vibration Eq. 4 to get

$$\boldsymbol{\varphi}_i^T \mathbf{M} \boldsymbol{\Phi} \ddot{\mathbf{q}}(t) + \boldsymbol{\varphi}_i^T (a_0 \mathbf{M} + a_1 \mathbf{K}) \boldsymbol{\Phi} \dot{\mathbf{q}}(t) \\ + \boldsymbol{\varphi}_i^T \mathbf{K} \boldsymbol{\Phi} \mathbf{q}(t) = \boldsymbol{\varphi}_i^T \mathbf{F}(t). \quad (22)$$

Equation 22 due to Eqs. 11 and 13 becomes alike to Eq. 12

$$\boldsymbol{\varphi}_i^T \mathbf{M} \boldsymbol{\varphi}_i \ddot{q}_i(t) + \boldsymbol{\varphi}_i^T (a_0 \mathbf{M} + a_1 \mathbf{K}) \boldsymbol{\varphi}_i \dot{q}_i(t) \\ + \boldsymbol{\varphi}_i^T \mathbf{K} \boldsymbol{\varphi}_i q_i(t) = \boldsymbol{\varphi}_i^T \mathbf{F}(t). \quad (22a)$$

Or alternatively

$$M_i \ddot{q}_i(t) + (a_0 M_i + a_1 K_i) \dot{q}_i(t) + K_i q_i(t) = P_i(t). \quad (22b)$$

In the last pair of equations, it should be noted that exclusively due to the fact that $\mathbf{C}$ was assumed to be a weighted sum of $\mathbf{M}$ and $\mathbf{K}$, orthogonality still applies and results in uncoupled modal damping expressions C_i for each mode i. Namely,

$$\boldsymbol{\varphi}_i^T \mathbf{C} \boldsymbol{\varphi}_j = 0 \text{ for all } i \neq j \text{ and} \\ C_i = a_0 M_i + a_1 K_i = \boldsymbol{\varphi}_i^T \mathbf{C} \boldsymbol{\varphi}_i. \quad (23)$$

Equation 23 allows Eq. 22b to be rewritten as

$$M_i \ddot{q}_i(t) + C_i \dot{q}_i(t) + K_i q_i(t) = P_i(t), \quad (22c)$$

which after division with M_i becomes equal to

$$\ddot{q}_i(t) + 2\xi_i \omega_i \dot{q}_i(t) + \omega_i q_i(t) = \frac{P_i(t)}{M_i}, \quad (22d)$$

with the modal mass and stiffness M_i and K_i, the modal frequency ω_i, and the modal force $P_i(t)$ defined by Eq. 13. As for the so-called modal damping ratio (or fraction of the critical damping) ξ_i, it stands for the ratio $C_i/2\omega_i M_i$, i.e.,

$$\xi_i = C_i / 2\omega_i M_i. \quad (24)$$

It is worth noting that Eq. 23 characterizes not only the Rayleigh damping model but also any damping model that offers the possibility for the classical modes of the undamped free vibration to be used in order to uncouple the equations of damped motion of the structure under consideration. Any such damping model is grouped as classical damping and substantially simplifies all subsequent dynamic calculations. Obviously, the subcases of mass-proportional and stiffness-proportional damping brought up earlier must also be categorized as classical damping.

Subsequently the process of calculating modal responses is identical to the previous description, where someone solves independently each modal equation and then synthesizes the total response by summing up the few lower modal contributions up to a certain order. The modal solution $q_i(t)$ of the Eq. 22 is now also affected by the modal damping as expressed by the modal damping ratio ξ_i. In analogy with Eqs. 15 up to 17, now the modal solution $q_i(t)$ generalizes to

$$\begin{aligned} q_i(t) &= q_i^c(t) + q_i^p(t) = \\ &= e^{-\xi_i \omega_i t} \left(q_i(0) \cos \widetilde{\omega}_i t + \frac{\dot{q}_i(0) + \xi_i \omega_i q_i(0)}{\widetilde{\omega}_i} \sin \widetilde{\omega}_i t \right) \\ &+ \int_0^t \frac{P_i(\tau)}{M_i} h_i(t-\tau) d\tau, \end{aligned}$$

$$(25)$$

where $\widetilde{\omega}_i$ is the so-called damped modal frequency, which stands for the multiple $\omega_i \sqrt{1 - \xi_i^2}$ of the undamped modal frequency ω_i, i.e.,

$$\widetilde{\omega}_i = \omega_i \sqrt{1 - \xi_i^2}, \quad (26)$$

and $h_i(t-\tau)$ is the modal function known as unit-impulse response function given by

Modal Analysis, Table 1 Recommended damping values from Chopra (1995)

Type of structure	Damping ratio % (stress below ½ of yield)	Damping ratio % (stress at or just below yield)
Welded steel, prestressed concrete	2–3	5–7
Reinforced concrete (with considerable cracking)	3–5	7–10
Bolted steel, timber structures with bolted joints	5–7	10–15

$$h_i(t-\tau) = \frac{1}{\widetilde{\omega}_i} e^{-\xi_i \omega_i (t-\tau)} \sin \widetilde{\omega}_i(t-\tau). \quad (27)$$

In the above analysis, one step was tacitly skipped. When adopting the Rayleigh damping expression of Eq. 21, the unknown constants a_0 and a_1 need to be determined for the damping matrix to be fully described. Indicatively this can be done as follows:

The modal damping can be determined experimentally through analyzing response measurements. This is the exact inverse problem of what is dealt in this entry. In short an answer is enabled through different techniques, e.g., the half-power bandwidth method, the free decays of isolated modes, or some other more elaborate modal identification technique. In the absence of such experimental data, figures that have developed from long structural experience can be used. Such information is given in Table 1 noting the stress amplitude dependence.

With values for the modal damping ratios in hand, one can write Eq. 23 for two modes, customarily the first and the last one considered (i.e., rth), viz.,

$$\left. \begin{array}{l} C_1 = a_0 M_1 + a_1 K_1 = 2\omega_1 M_1 \xi_1 \\ C_r = a_0 M_r + a_1 K_r = 2\omega_r M_r \xi_r \end{array} \right\}. \quad (28)$$

The latter system of two algebraic equations is then to be solved simultaneously for a_0 and a_1. A further generalization of classical damping can also be considered when more than two modal damping values are taken into account in capturing the damping distribution. Then we have the Caughey damping approach according to which C is expressed as

$$\mathbf{C} = \mathbf{M} \sum_{j=0}^{l-1} a_j (\mathbf{M}^{-1}\mathbf{K})^j, \quad (29)$$

where l is the number of modal damping ratios to be used. Evidently for $l = 2$ this reduces to Rayleigh damping, while for $l = 3$ Eq. 29 results in

$$\mathbf{C} = a_0 \mathbf{M} + a_1 \mathbf{K} + a_2 \mathbf{K}\mathbf{M}^{-1}\mathbf{K}. \quad (30)$$

The l unknown constants are found again in a straightforward manner by solving a $l \times l$ system of algebraic equations.

The Nonclassical Damping Case

The classical damping approach although not physically realizable has found extensive use in engineering practice and particularly in standard earthquake engineering design. However, for structures that are heavily damped or unevenly distributed or even when they possess closely spaced modes, the couplings induced by the damping matrix may produce significant shortcomings and complications in the earlier presented modal analysis. An example of such a case, where nonclassical damping is to be expected, would appear in a nuclear reactor containment vessel founded on a soft soil. The quite different damping properties between the stiff structure and the foundation will induce a highly damped, very complex dynamic soil-structure interaction that establishes itself through a very generically shaped damping matrix (Clough and Mojtahedi 1976). For making things clear, in these instances the solution again starts from writing the damped equation of motion

$$\mathbf{M}\ddot{\mathbf{u}}(t) + \mathbf{C}\dot{\mathbf{u}}(t) + \mathbf{K}\mathbf{u}(t) = \mathbf{F}(t). \quad (20)$$

Still, what changes is that after substituting in the latter the modal decomposition indicated in Eq. 2

and premultiplying with the undamped mode vector $\boldsymbol{\varphi}_i^T$, one gets

$$\boldsymbol{\varphi}_i^T \mathbf{M} \boldsymbol{\varphi}_i \ddot{q}_i(t) + \boldsymbol{\varphi}_i^T \mathbf{C} \boldsymbol{\Phi} \dot{\mathbf{q}}(t) + \boldsymbol{\varphi}_i^T \mathbf{K} \boldsymbol{\varphi}_i q_i(t) = \boldsymbol{\varphi}_i^T \mathbf{F}(t) \quad (31a)$$

or alternatively

$$M_i \ddot{q}_i(t) + \boldsymbol{\varphi}_i^T \mathbf{C} \boldsymbol{\Phi} \dot{\mathbf{q}}(t) + K_i q_i(t) = P_i(t). \quad (31b)$$

Equation 31b obviously is a system of n-coupled equations, since the row vector $\tilde{\mathbf{C}} = \boldsymbol{\varphi}_i^T \mathbf{C} \boldsymbol{\Phi}$, $(1 \times n)$, has in general nonzero elements that introduce in Eq. 31b apart from $\dot{q}_i(t)$ also terms $\dot{q}_j(t)$ with $j \neq i$. Thus, contrary to the principle and target of modal analysis, it is not possible to deal with n-independent generalized SDOF modal systems. It actually seems that there is no benefit in transforming the initial coupled equation of motion to its modal form using the undamped modes. There are a number of approaches to tackle this problem:

The first more simplistic approach disregards all the coupled contributions of Eq. 31b by essentially zeroing in the row vector $\tilde{\mathbf{C}}$ all elements apart from C_i, i.e., $\tilde{\mathbf{C}} = [0 \ldots C_i \ldots 0]$. Subsequently, the remaining analysis becomes identical to the above described for the classically damped case.

A different also approximate approach would be for one to keep all terms inside Eq. 31b and attempt to solve the system of coupled equations using numerical techniques, that is, direct integration (e.g., Newmark's method). More on the latter will appear in the entry of nonlinear systems. The relative benefit of employing Eq. 31b against applying direct integration to the initial n equations of damped motion described by the matrix Eq. 20 lies in the fact that due to generally few modes dominating response, a much reduced number of r equations can be considered. This is of great importance in large structural systems with thousands of DOFs.

Yet, both of the above analytical approaches may introduce errors of undefined magnitude. The most rigorous treatment of the problem is to uncouple the equations of motion using damped modes instead of the undamped ones. Their derivation, indicatively provided herein, proceeds by first writing the free-vibration (unforced) counterpart of the forced-vibration matrix Eq. 20

$$\mathbf{M}\ddot{\mathbf{u}}(t) + \mathbf{C}\dot{\mathbf{u}}(t) + \mathbf{K}\mathbf{u}(t) = \mathbf{0}. \quad (32)$$

And then, assuming a solution of the form

$$\mathbf{u}_i(t) = \boldsymbol{\phi}_i e^{\lambda_i t}, \quad (33)$$

Equation 32 leads to

$$\left(\lambda_i^2 \mathbf{M} + \lambda_i \mathbf{C} + \mathbf{K}\right) \boldsymbol{\phi}_i = \mathbf{0}, \quad (34)$$

where λ_i and $\boldsymbol{\phi}_i$ are the damped eigenvalues and eigenvectors, respectively.

The methodology to solve Eq. 34 relies on the so-called state-space formulation. According to it, Eq. 34 may be rewritten as

$$\begin{bmatrix} -\mathbf{K} & \mathbf{0} \\ \mathbf{0} & \mathbf{M} \end{bmatrix} \begin{bmatrix} \lambda_i \boldsymbol{\phi}_i \\ \lambda_i^2 \boldsymbol{\phi}_i \end{bmatrix} + \begin{bmatrix} \mathbf{0} & \mathbf{K} \\ \mathbf{K} & \mathbf{C} \end{bmatrix} \begin{bmatrix} \boldsymbol{\phi}_i \\ \lambda_i \boldsymbol{\phi}_i \end{bmatrix} = \begin{bmatrix} \mathbf{0} \\ \mathbf{0} \end{bmatrix}, \quad (35)$$

and introducing the block matrices $\mathbf{A}$ and $\mathbf{B}$, $(2n \times 2n)$, and column block vector $\mathbf{v}_i$, $(2n \times 1)$, i.e.,

$$\mathbf{A} = \begin{bmatrix} -\mathbf{K} & \mathbf{0} \\ \mathbf{0} & \mathbf{M} \end{bmatrix}, \mathbf{B} = \begin{bmatrix} \mathbf{0} & \mathbf{K} \\ \mathbf{K} & \mathbf{C} \end{bmatrix} \text{ and } \mathbf{v}_i = \begin{bmatrix} \boldsymbol{\phi}_i \\ \lambda_i \boldsymbol{\phi}_i \end{bmatrix}, \quad (36)$$

reshapes Eq. 35 into

$$(\lambda_i \mathbf{A} + \mathbf{B}) \mathbf{v}_i = \mathbf{0}, \quad (37)$$

which has a nonzero solution $\mathbf{v}_i$ only for the zero determinant

$$|\lambda_i \mathbf{A} + \mathbf{B}| = 0. \quad (38)$$

The latter is an eigenvalue problem of order $2n$, which is known as the complex eigenvalue problem owing to the fact that both λ_i and $\boldsymbol{\phi}_i$ are in general complex valued. Before proceeding any further, it is worth noticing that the increased order $2n$ of the complex eigenvalue problem in comparison with the order n of the classical eigenvalue problem requires an increased computational effort/time proportional to $(2n)^3 = 8n^3$ for a solution. This is the reason for this approach to be outside the main pursuits of applied structural dynamics. Now, returning to Eq. 38, if λ_i is a solution, then its conjugate is also one. Typically, λ_i is expressed as

$$\lambda_i = \omega_i \left(-\xi_i + j\sqrt{1-\xi_i^2}\right), \quad (39)$$

where $j = \sqrt{-1}$ is the imaginary unit and ω_i and ξ_i are the modal frequency and modal damping ratio, respectively. The complex nature of the eigenvectors $\boldsymbol{\phi}_i$ accounts for phase differences other than $0°$ and $180°$ between their components. Orthogonality relations now become

$$\mathbf{v}_i^T \mathbf{A} \mathbf{v}_j = \mathbf{v}_i^T \mathbf{B} \mathbf{v}_j = 0 \text{ for all } i \neq j, \quad (40)$$

which allow the uncoupling of Eq. 35 or of its forced counterpart. Having laid the ground for subsequent analysis (e.g., time response solutions), no additional details are presented herein, and the reader is suggested to consult relevant fundamental literature on this special topic for additional resources and worked examples, e.g., (Crandall and McCalley 2002; Itoh 1973; Hansen et al. 2012).

Member Forces

The procedures described above yielded the displacement time response, both modal and total, of a linear elastic structure. Yet, it is of great relevance for design purposes to derive also the force and stress developing on individual members/elements. There are two distinct ways to achieve this:

1. Having in hand the modal displacements $\mathbf{u}_i(t)$ along the structure, one can use the member stiffness properties in order to derive the modal force $R_i(t)$ on the member and subsequently the required stresses. Considering all the modal contributions, one can write

$$R(t) = \sum_{i=1}^{r} R_i(t), \quad (41)$$

which apart from forces also applies to the resultant stresses.

2. Alternatively the modal displacements $\mathbf{u}_i(t)$ can be used to define equivalent modal static forces

$$\mathbf{F}_i^s(t) = \mathbf{K} u_i(t) = \omega_i^2 \mathbf{M} u_i(t). \quad (42)$$

Solving the structure statically for the obtained $\mathbf{F}_i^s(t)$, which are now applied as external forces at each time instant, will give again the modal member force $R_i(t)$. All individual modal contributions again have to be added together in accordance with Eq. 41.

The Earthquake Problem

The general modal analysis background presented above will now shift the focus to the earthquake-loading scenario of a classically damped n-DOF structure. Such a structure could be a typical multistory frame, e.g., Fig. 1c, which is customarily used as a benchmark in similar studies. Earthquake loading is the result of seismic action of the ground on the supports of the structure, that is, the result of a vibration motion of the supports of the structure caused by an earthquake. Specifically speaking, the seismic action translates to a ground acceleration input $\ddot{u}_g(t)$ at the foundation level (this is in most instances horizontal but can also be vertical), whose transmission to the upper structure creates an equivalent dynamic force distribution that constitutes what is called earthquake loading. Disregarding the special topic of multiple support excitation (for such an extension, consult Clough and Penzien (1993) or Chopra (1995)), the earthquake (or seismic) load vector that enters the matrix equation of motion, Eq. 20, can be expressed as

$$\mathbf{F}(t) = -\mathbf{M} \boldsymbol{\iota} \ddot{u}_g(t). \quad (43)$$

The latter implies a synchronous though spatially varying loading of the different lumped masses. The vector $\boldsymbol{\iota}$, ($n \times 1$), is an influence vector of the ground motion on the structure and represents the displacements that will result for each DOF when a static unit ground displacement $u_g(t) = 1$, of the same direction as $\ddot{u}_g(t)$, is applied to an undeformed (absolutely rigid) model of the structure. For example, in the frame of Fig. 1b, considering a horizontal earthquake input, $\boldsymbol{\iota}$ becomes $[1\ 1\ 0\ 0\ 0]^T$, while for the frame of Fig. 1c, it turns to $[1\ 1\ 1\ 1\ 1]^T$. For proof purposes, Eq. 43 and the definition of $\boldsymbol{\iota}$ can naturally be derived when expressing the equilibrium at each mass of the MDOF structure in terms of the total displacement $\mathbf{u}_t(t)$, which evidently is the vector sum of the relative-to-base displacement $\mathbf{u}(t)$ and the ground/base displacement $\boldsymbol{\iota}\, u_g(t)$, i.e.,

$$\mathbf{u}_t(t) = \mathbf{u}(t) + \boldsymbol{\iota} u_g(t). \quad (44)$$

Indeed, the equilibrium of the masses of a seismically excited MDOF structure is expressed by

$$\mathbf{M}\ddot{\mathbf{u}}_t(t) + \mathbf{C}\dot{\mathbf{u}}(t) + \mathbf{K}\mathbf{u}(t) = 0, \quad (45)$$

and substituting $\ddot{\mathbf{u}}_t(t)$ in accordance with Eq. 24, it follows

$$\mathbf{M}\ddot{\mathbf{u}}(t) + \mathbf{C}\dot{\mathbf{u}}(t) + \mathbf{K}\mathbf{u}(t) = -\mathbf{M}\boldsymbol{\iota}\ddot{u}_g(t), \quad (46)$$

which proves that the seismic load vector $\mathbf{F}(t)$ equals $-\mathbf{M}\boldsymbol{\iota}\ddot{u}_g(t)$, thereby proving Eq. 32.

In a broader perspective, the expression for the seismic load can be generalized to describe a wider family of loading cases for which

$$\mathbf{F}(t) = \mathbf{s}f(t), \quad (47)$$

with the vector $\mathbf{s}$, ($n \times 1$), embodying all spatial variations and the scalar function $f(t)$ embodying all synchronous time variations. Trivially for $\mathbf{s} = -\mathbf{M}\boldsymbol{\iota}$ and $f(t) = \ddot{u}_g(t)$, one reverts to the earthquake-related problem.

This is actually the starting point for any subsequent analysis intending to obtain response histories. Yet, it must be noted that for most usual practices in seismic calculations, a different characterization for the earthquake motion is opted. This is the instrumental approach of the so-called response spectra and attempts to remove the tedious numerical burdens that are related to obtaining all the history of the response of a specific structure to the long time sequence of the ground acceleration of a specific earthquake excitation in order that the maximum value of the response is evaluated. Namely, with response spectra, one only has to consider the maximum response output (this may be in terms of displacement, S_d, or velocity, S_v, or acceleration, S_a) that an SDOF system produces under a specific earthquake excitation. Thus, an earthquake is no more an explicit time function $\ddot{u}_g(t)$. It is described through its maximum effect on SDOF systems with specific natural frequency and damping. This approach best serves the engineering need for evaluating the strength of a structure against the maximum value of its response to a given seismic excitation.

Whatever the earthquake definition assumed and the analysis opted (time histories or response maxima), there are substantial benefits associated with the use of modal analysis. Yet, the inherent in response spectra need for a decomposition of an initial MDOF structure to equivalent SDOF ones makes modal analysis in that instance more than a beneficial feature an irreplaceable prerequisite.

Excitation and Participation Factors

In what follows, some essential notions relevant to the earthquake analysis of structures are developed through exemplifying the modal analysis application. Combining Eqs. 13 and 47 yields

$$P_i(t) = \boldsymbol{\varphi}_i^T \mathbf{F}(t) = \boldsymbol{\varphi}_i^T \mathbf{s} f(t)$$
$$= L_i f(t) \quad \text{with } L_i = \boldsymbol{\varphi}_i^T \mathbf{s}, \quad (48)$$

as well as

$$\frac{P_i(t)}{M_i} = \frac{\boldsymbol{\varphi}_i^T \mathbf{s}}{M_i} f(t) = \frac{L_i}{M_i} f(t)$$
$$= \Gamma_i f(t) \quad \text{with } \Gamma_i = \frac{L_i}{M_i}, \quad (49)$$

and, hence,

$$\Gamma_i = \frac{L_i}{M_i} = \frac{\boldsymbol{\varphi}_i^T \mathbf{s}}{M_i}. \quad (50)$$

The modal factors L_i and Γ_i are termed the *modal excitation factor* and the *modal participation factor*, respectively. It is easy to verify that the values of both the factors Γ_i and L_i depend on the normalization used for $\boldsymbol{\varphi}_i$ (since the ratio L_i/Γ_i equals the modal mass $M_i = \boldsymbol{\varphi}_i^T \mathbf{M} \boldsymbol{\varphi}_i$).

In view of Eq. 50 and the orthogonality conditions, it holds true

$$\boldsymbol{\varphi}_i^T \mathbf{s} = \Gamma_i M_i = \Gamma_i \boldsymbol{\varphi}_i^T \mathbf{M} \boldsymbol{\varphi}_i = \boldsymbol{\varphi}_i^T \sum_{j=1}^n \Gamma_j \mathbf{M} \boldsymbol{\varphi}_j, \quad (51)$$

which allows expanding of the vector $\mathbf{s}$ in terms of the modal contributions $\mathbf{s}_i$ as follows:

$$\mathbf{s} = \sum_{i=1}^n \mathbf{s}_i = \sum_{i=1}^n \Gamma_i \mathbf{M} \boldsymbol{\varphi}_i. \quad (52)$$

Evidently, the vector $\mathbf{s}$ and its modal components $\mathbf{s}_i$ are independent of the absolute magnitudes of $\boldsymbol{\varphi}_i$. The modal factors L_i and Γ_i naturally emerge when performing the modal analysis steps indicated in Eqs. 22c and 22d. Namely, when substituting for the load form adopted in Eq. 47 and writing the i modal matrix equation of motion, one gets

$$M_i \ddot{q}_i(t) + C_i \dot{q}_i(t) + K_i q_i(t) = L_i f(t), \quad (53a)$$

or equally

$$\ddot{q}_i(t) + 2\omega_i \xi_i \dot{q}_i(t) + \omega_i^2 q_i(t) = \Gamma_i f(t). \quad (53b)$$

The rationale behind the modal decomposition of the load distribution $\mathbf{s}$ lies in introducing inertial modal load contributions. To explain this, the inertial force in the ith mode is given as $\mathbf{M}\ddot{\mathbf{u}}_i$ or, taking into account Eq. 3, as $\mathbf{M}\boldsymbol{\varphi}_i \ddot{q}_i(t)$. The $\mathbf{M}\boldsymbol{\varphi}_i$ pre-factor, thus, makes apparent the underlying inertial concept linking to Eq. 52. Γ_i evidently from Eq. 53b is a weighting factor that scales the load fraction to be assigned to each specific mode. The obscurity of the normalization dependence for Γ_i hinders it, becoming a direct clear indicator of the relative significance of the ith mode.

Relevant to this, the most important and useful feature of the expansion employed in Eq. 52 is that the load distribution $\mathbf{s}_i$ produces response exclusive to mode i or else that the response at each mode i owes solely to $\mathbf{s}_i$. This can be easily derived by observation of Eq. 52. When multiplying it with $\boldsymbol{\varphi}_j^T$ and $f(t)$ to create what is defined as the modal load $P_j(t)$ one gets

$$P_j(t) = \boldsymbol{\varphi}_j^T \mathbf{s} f(t)$$
$$= \sum_{i=1}^n \boldsymbol{\varphi}_j^T \mathbf{s}_i f(t) = \sum_{i=1}^n \Gamma_i \boldsymbol{\varphi}_j^T \mathbf{M} \boldsymbol{\varphi}_i f(t). \quad (54)$$

This expression due to orthogonality yields a nonzero value only for $j = i$.

Contribution Factors (Chopra's Physical Interpretation)

Subsequently, if one follows the above procedures, the ensuing earthquake response problem can be solved. Yet, as first developed by Chopra (1995), a transformation is put forward that can further assist in a physical interpretation of modal analysis relevant to earthquake engineering. Namely, due to the linearity of the structural system, if one assumes

$$q_i(t) = \Gamma_i D_i(t), \quad (55)$$

then the modal Eq. 53b can be replaced by the equivalent

$$\ddot{D}_i(t) + 2\omega_i \xi_i \dot{D}_i(t) + \omega_i^2 D_i(t) = f(t), \quad (56)$$

which describes a unit mass SDOF system, with frequency ω_i and damping ξ_i loaded by $f(t)$; the latter can be either a force or a ground acceleration without any distinction. Accordingly, $D_i(t)$ is but the ith modal response of the system to the loading $f(t)$. The contribution of the ith mode to response can then be given by Eq. 3 as

$$\mathbf{u}_i(t) = \boldsymbol{\varphi}_i q_i(t) = \Gamma_i \boldsymbol{\varphi}_i D_i(t). \quad (57)$$

Defining equivalent modal static forces $\mathbf{F}_i^s(t)$ as in Eq. 42

$$\begin{aligned}\mathbf{F}_i^s(t) &= \mathbf{K}\mathbf{u}_i(t) = \omega_i^2 \mathbf{M}\mathbf{u}_i(t)\\ &= \omega_i^2(\mathbf{M}\Gamma_i \boldsymbol{\varphi}_i)D_i(t) = \mathbf{s}_i \omega_i^2 D_i(t),\end{aligned} \quad (58)$$

one can write for any dynamic modal response quantity $R_i(t)$

$$R_i(t) = R_i^s \omega_i^2 D_i(t). \quad (59)$$

In the latter R_i^s is the modal response quantity that results from static analysis of the structure under the influence of the vector $\mathbf{s}_i$ of static external forces. This clearly shows that within the realm of modal decomposition, any complicated dynamic analysis of a n-DOF structure reduces to the product of two simple steps. In the first one performs static calculations for deriving all the n different R_i^s and then multiplies with the response solutions $D_i(t)$ of the n different SDOF systems loaded by $f(t)$. The total response is evidently given by the modal summation illustrated in Eq. 41.

The above modal analysis allows the introduction of another set of quantities, named the *contribution factors* $\tilde{R}_i$. These similarly to the participation factors indicate the significance of each mode in the total response. Still, their added merit lies in the fact that they are dimensionless and relieved by the abstractness of normalization. To produce them, one needs to write Eq. 59 as

$$R_i(t) = \tilde{R}_i R^s \omega_i^2 D_i(t), \quad (60)$$

in which R^s is the total response to the vector $\mathbf{s}$ of static external forces, and $\tilde{R}_i$ is the contribution factor of the ith mode defined as

$$\tilde{R}_i = \frac{R_i^s}{R^s}. \quad (61)$$

It can easily be proved that

$$\sum_{i=1}^{n} \tilde{R}_i = 1, \quad (62)$$

and, hence, if all modes beyond the r first modes are truncated, the corresponding error e_r is

$$e_r = 1 - \sum_{j=1}^{r} \tilde{R}_j, \quad (63)$$

which is usually very small for the truncation beyond the first three or four modes.

Extensions and Miscellanea

There are a number of additional topics that were intentionally excluded from the above described backbone of modal analysis. A handful of them is further elaborated in what follows. The particular intention is not to fully capture the extensive available knowledge but rather motivate the reader to further pursue the vast capabilities and multifaceted merits associated with this unique dynamics tool.

Mode Acceleration Method

The entirety of the above approach is based on estimating modal displacements, which subsequently will combine to form the total response. This displacement-based approach that naturally acquires the characterization mode displacement method was physically shown to extend into the calculation of any response variable.

Although the mode displacement method was developed typically first, it is not the only option available. One of the most useful alternatives is the so-called mode acceleration method along with its implementation variations. In identical form, though founded on a different basis, this may also be found under the name of static correction method, and the origin of this naming will below become clear. It was first devised for an undamped system, and its main attribute is that instead of modal displacements, one can seek solutions in terms of only modal accelerations for an undamped system or in terms of modal accelerations and velocities for a damped system. A brief presentation is herein put forward following the derivations and critique proposed by Cornwell et al. (1983).

Starting from the classically damped modal equation Eq. 22d, one can rewrite it as

$$q_i(t) = \frac{1}{\omega_i^2}\left(\frac{P_i(t)}{M_i} - \ddot{q}_i(t) - 2\xi_i\omega_i\dot{q}_i(t)\right)$$
$$= \frac{P_i(t)}{K_i} - \frac{1}{\omega_i^2}\ddot{q}_i(t) - \frac{2\xi_i}{\omega_i}\dot{q}_i(t), \quad (64)$$

which substituting in the modal superposition Eq. 19 yields

$$\mathbf{u}(t) = \sum_{i=1}^{n} \boldsymbol{\varphi}_i \left(\frac{P_i(t)}{K_i} - \frac{1}{\omega_i^2}\ddot{q}_i(t) - \frac{2\xi_i}{\omega_i}\dot{q}_i(t)\right) \quad (65a)$$

or equally

$$\mathbf{u}(t) = \sum_{i=1}^{n} \boldsymbol{\varphi}_i \frac{\boldsymbol{\varphi}_i^T F(t)}{K_i}$$
$$- \sum_{i=1}^{n} \boldsymbol{\varphi}_i \left(\frac{1}{\omega_i^2}\ddot{q}_i(t) + \frac{2\xi_i}{\omega_i}\dot{q}_i(t)\right). \quad (65b)$$

A plausible replacement is now sought for the first term in Eq. 65b. Namely, taking into account the definition of the equivalent static displacement of the structure $\mathbf{u}^s(t)$ as $\mathbf{u}^s(t) = \mathbf{K}^{-1}\mathbf{F}(t)$, as well as the definition of the equivalent static generalized displacements $q_i^s(t)$ as $q_i^s(t) = P_i(t)/K_i$, and applying Eq. 19, one gets

$$\mathbf{K}^{-1}\mathbf{F}(t) = \mathbf{u}^s(t) = \sum_{i=1}^{n} \mathbf{u}_i^s(t)$$
$$= \sum_{i=1}^{n} \boldsymbol{\varphi}_i \frac{P_i(t)}{K_i} = \sum_{i=1}^{n} \boldsymbol{\varphi}_i \frac{\boldsymbol{\varphi}_i^T F(t)}{K_i}. \quad (66)$$

Thus, the first term of the right-hand member of Eq. 65b only accounts for the total equivalent static displacement, as shown by Eq. 66. This fact evidently transforms the total displacement solution expressed by Eq. 65b into

$$\mathbf{u}(t) = \mathbf{K}^{-1}\mathbf{F}(t)$$
$$- \sum_{i=1}^{n} \boldsymbol{\varphi}_i \left(\frac{1}{\omega_i^2}\ddot{q}_i(t) + \frac{2\xi_i}{\omega_i}\dot{q}_i(t)\right). \quad (67)$$

The latter relation gives this whole alternative solution process a unique physical meaning. To obtain the dynamic response of a structure, one needs to consider first the easy to derive equivalent static (or pseudostatic) displacement and, subsequently, fine-tune it by adding the dynamic effects as introduced by the modal acceleration and velocity responses of the n individual modes. The higher mode, which trivially translates to higher frequency, would impose a relatively faster quadratic convergence for the second term in the right-hand side of Eq. 67 (see ω_i^2 denominator). Thus, this would allow less modes to be considered when truncating for economy and practicality the full series expansion. The method actually even when modes higher than r ($\ll n$) are discarded clearly considers partly their influence through their static function.

Evidently, for an undamped system, Eq. 67 reduces to

$$\mathbf{u}(t) = \mathbf{K}^{-1}\mathbf{F}(t) - \sum_{i=1}^{n} \boldsymbol{\varphi}_i \frac{1}{\omega_i^2}\ddot{q}_i(t) \text{ with } \xi_i = 0, \quad (67a)$$

where the dynamic contribution of the modal velocities has been canceled out and only remains the dynamic effect of modal accelerations, which justifies the name mode acceleration method.

It is worth noting that even for a damped system, what matters in structural analysis is the maximum displacement max $\mathbf{u}$, which causes the maximum strain and, hence, the maximum stress, required for proportioning the structural members. As a rule of thumb, this max $\mathbf{u}$ is taken for equal to a combination of the maximum displacements max $\mathbf{u}_i$ of the three or four first modes of the structure, which necessitates that the corresponding synchronous modal velocities become zero, $\dot{\mathbf{u}}_i = 0$. Thus, the dynamic contribution of the modal accelerations to account for the maximum value of the dynamic response of the damped system only remains.

The method is greatly superior in terms of numerical efficacy outperforming the conventional mode displacement method in all cases by consistently requiring reduced number of modes for accurate, similar minimal error, solutions. Further the mode acceleration solution was found to be more sensitive to damping by means that increasing uniformly the damping of all modes would lead to faster convergence rates of associated dynamic solutions on reduced modal information.

Nonlinearity

The most critical assumption for applying modal analysis lies in the validity of the superposition of modes adopted in Eq. 2. This in simple terms translates to linearity of the structural system. With nonlinear behavior, developing the pursuit for a response-independent stationary modal reference system is not straightforward. Thus, the economy in the analysis of linear structures achieved through the modal order reduction cannot be realized in the analysis of nonlinear structures. Interestingly, in most real-life cases, structures move substantially, yield, or just interact with their loading environment. In all such instances, nonlinear descriptions are needed to capture the underlying phenomena waiving the validity of most of the above analysis.

A generalized manifold-type similar concept of nonlinear normal modes (NNM) has recently developed within the nonlinear dynamics field. Yet, this should be seen by engineers as more of a mathematical intricacy rather than a practical tool credible to be used in ordinary analysis of typical high-order systems. In generic nonlinear structures (this is to distinguish from cases of linear subsystems connecting through nonlinear links), the strict approach consists of numerically integrating the n-coupled equations of motion concurrently. To this purpose the time-continuous matrix Eq. 20 is written in a variational form to enable a time-stepping approach to be used for calculating purposes. Namely, Eq. 20 changes to

$$\mathbf{M}\Delta\ddot{\mathbf{u}}_i + \mathbf{C}_i\Delta\dot{\mathbf{u}}_i + \mathbf{K}_i\Delta\mathbf{u}_i = \Delta\mathbf{F}_i, \quad (68)$$

where

$$\Delta\ddot{\mathbf{u}}_i = \ddot{\mathbf{u}}_{i+1} - \ddot{\mathbf{u}}_i, \ \Delta\dot{\mathbf{u}}_i = \dot{\mathbf{u}}_i + 1 - \dot{\mathbf{u}}_i,$$
$$\Delta\mathbf{u}_i = \mathbf{u}_{i+1} - \mathbf{u}_i, \ \Delta\mathbf{F}_i = \mathbf{F}_{i+1} - \mathbf{F}_i, \quad (69)$$

with i representing the time-step index contrary to its earlier modal meaning, while $\mathbf{C}_i$ and $\mathbf{K}_i$ are nonlinear damping and stiffness matrices, respectively, whose elements are functions of the displacement $\mathbf{u}(t)$ and the velocity $\dot{\mathbf{u}}(t)$ and take on values depending on the considered time step. The index i would translate in terms of absolute time to $i\Delta t$, where Δt denotes the time step. Alternatively, the time index $i+1$ can be used (i.e., this is the implicit vs the explicit method when i is used). The overall response calculation problem includes three unknowns $\ddot{\mathbf{u}}_i$, $\dot{\mathbf{u}}_i$, and $\mathbf{u}_i$ for which apart from Eq. 68 two more equations could be derived from different approximations for $\ddot{\mathbf{u}}_i$ and $\dot{\mathbf{u}}_i$. As a matter of fact, depending on these approximations, which link to an assumed variance of the $\ddot{\mathbf{u}}_i$ and/or $\dot{\mathbf{u}}_i$ variables during the time-step duration, a number of different methods have developed. The most broadly used for earthquake-related purposes is Newmark's β-method.

It is noteworthy that although methods like the latter would easily produce bounded solutions (i.e., stable, meaning results will not "blow up"), whilst the time-step Δt is accurately prescribed. As a matter of fact, this is the most important parameter for any numerical integration scheme and must be sufficiently small. A first approximation for the Δt value can come from the associated modal analysis pertaining in the linear operation regime of the structure. Namely, the highest linear mode would correspond to the lowest natural period T_n. Δt should be sufficiently smaller than this T_n value. It should be reminded that this imposes large numerical limitations and costs and that the inherent softening associated with yielding would increase T_n (on the other hand, hardening would have an adverse effect by further reducing T_n).

Frequency Domain Solutions

After bringing the structural system to its modal description equivalent, the solutions pursued whether in terms of modal displacements or in terms of modal accelerations and velocities were always expressed in the time domain. Considering the case of the classically damped system with periodic loading and focusing on the probably most significant part of the response, the forced or else for this case steady state, one may suggest some alternatives to Eqs. 17 and 25. The reason is that the Duhamel's integral that provide the steady-state time response involves the convolution operation between the applied load and the unit-impulse response function. This term tends to perplex calculations.

Without detailing proofs, the alternative representation is in the frequency domain and can easily surface when one assigns inside the SDOF a periodic loading of the form $P_i(t) = P_i^o e^{j\omega t}$, where P_i^o is a scalar amplitude and again $j = \sqrt{-1}$. Then by rearranging the generalized SDOF response becomes

$$q_i(t) = \frac{P_i^o e^{j\omega_i t}}{(K_i - \omega_i^2 M_i) + j\omega_i C_i}. \qquad (70)$$

Dividing Eq. 70 with $P_i(t)$ in order to create an output over input ratio, one gets

$$H(\omega) = \frac{q_i(t)}{P_i(t)} = \frac{\tilde{q}_i(\omega)}{\tilde{P}_i(\omega)}$$
$$= \frac{1}{(K_i - \omega^2 M_i) + j\omega C_i}, \qquad (71)$$

where over dash stands for the Fourier transform. The $H(\omega)$ function depending only on ω is called the receptance frequency response function. If, instead of displacement response, acceleration or velocity were used for the numerator, one would have the acceleration and mobility frequency response function, respectively. $H(\omega)$ is the Fourier transform of the unit-impulse response function given in Eq. 27. Generalizing the concept for an MDOF system, this instead of a scalar becomes a matrix $\mathbf{H}(\omega)$ with components $h_{kl}(\omega)$, denoting the ratio of displacement at position k and force at position l

$$h_{kl}(\omega) = \sum_{i=1}^{n} \frac{\varphi_i(k)\varphi_i(l)}{M_i(\omega_i^2 - \omega^2) + j\omega C_i}. \qquad (72)$$

Subsequently the convolution in the time domain is replaced in the frequency domain from a simple product of the frequency response function with the input force.

Operational Modal Analysis

To this point the presentation was focused on the so-called direct problem, which is concerned with obtaining the dynamic response of a fully known system to a given loading. By fully known it is meant that all the necessary properties to describe its behavior through a model, mainly an FE model, are in hand. Such can be the modal properties instead of the full stiffness, damping, and mass matrices. This in fact can bring a vast simplification and data economy having merits even more exaggerated than the ones modal truncation introduced before. Indicatively, for an n-DOF system, one, instead of determining $3n^2$ matrix terms, could obtain similar results by employing only a reduced number of modal frequency, damping, and shape sets. Still, this a priori assumed knowledge is not always the case. There are many uncertainties assigned to modeling (e.g., links' behavior, foundations, damping, etc.) which can bring up substantial errors in any associated model parameters and subsequently response predictions, used either for design, assessment, or control of the structure. To deal with this significant engineering issue, a major part of modal analysis nowadays has been devoted to the so-called inverse analysis. This translates to a mapping of the measured response to the system modal characteristics, and it first appeared with the development of the space program of the USA back in the 1950s. Interestingly, although the direct problem is a single-valued one, the inverse is not. The latter brings up the need for optimizing any of the prediction/analysis routines that will eventually produce modal parameter outputs from measured response data.

The equation-free description that follows leaves aside the very interesting and demanding experimental (including signal preprocessing) techniques devised through the years to obtain the measured response from a structure under different forms of loading. The main scope of this entry is to only refer briefly to some of the most typical available tools able to obtain modal estimates once measured data are available. The mathematical background necessary to fully explain these methods are superseding the space limitations of this entry. Thus, the reader is motivated to also direct to the very insightful and broad §4 of Maia et al. (1997) as well as to a true encyclopedia on the topic by Ljung (1999) for further studying. Excellent software can be found to practice inverse analysis, some even in freeware form, which encompasses the full information provided herein and beyond. Such availability is a true late year's addition.

Practically all useful inverse analysis methods refer to MDOF systems. For them one may have information on their structural excitation or not. Their difficulty ranges from the simple peak picking method, where one observes peaks in the frequency response functions, to much more demanding alternatives employing complex algebra (Hankel matrices, Markov parameters, etc.) together with the earlier introduced state-space formulation. Most of the MDOF methods break down the structure to a superposition of independent-uncoupled SDOF systems, making even more profound the linearity and modal decomposition demands. In general, the strict target is to fit an analytical function of the form introduced in Eq. 72 to the measured data, which are customarily transformed to frequency response functions. This fitting operation inherently involved in the process lends the name curve fitting to the concept. Depending on the specific algorithm opted for the fitting, one may get the circle fitting, rational function polynomial, or many other refined approaches along the same lines.

Alternatively, one, instead of working in the frequency domain, may work employing the time counterpart of frequency response functions, the impulse response functions. Typically the complex exponential method is used for such fittings. A similar very widespread approach fitting free decays instead of impulse response functions is named after Ibrahim, i.e., Ibrahim time domain method. Alternatively time-based methods like the option of the eigen-realization algorithm which will recover families of systems with identical eigenvalues to the measured one or generic time series tools like the auto-regressive moving average can further be employed.

In all cases there is a major parameter affecting results. This is the order of the assumed model in the identification analysis. Namely, earlier treating the direct problem, one had to decide the order to which the modal solutions were to be truncated. The order that should be decided for the number of modes to be identified here is the direct equivalent.

Summary

Although every effort was made to synoptically present all the critical points that could not be missed of any educational work addressing modal analysis, the reader may still find omissions. To this reason, the wealth of references provided throughout is an excellent resource to complement the current entry as a further reading that could cover practically every aspect relating to the subject.

To recapitulate the main attributes presented herein, a short mind map is put forward that can illustrate the simple steps involved in modal analysis when employed in practical earthquake engineering purposes. This consists of:

– Determining the structural matrices **K**, **M**, and **C**
– Determining the modal natural frequencies ω_i and shapes φ_i
– Computing each modal-generalized response q_i (or $\ddot{q}_i$ and $\dot{q}_i$) and turning it to modal displacement $\mathbf{u}_i$

- Calculating the total displacement vector **u** after deciding the minimum number of modes r needed for an accurate solution

References

Biggs JM (1964) Introduction to structural dynamics. McGraw Hill, New York

Chopra AK (1995) Dynamics of structures: theory and applications to earthquake engineering. Prentice Hall, Englewood Cliffs

Chopra AK (1996) Modal analysis of linear dynamics systems: physical interpretation. J Struct Eng 122(5):517–527

Clough RW, Mojtahedi S (1976) Earthquake response analysis considering non-proportional damping. Earthq Eng Struct Dyn 4:489–496

Clough RW, Penzien J (1993) Dynamics of structures, 2nd edn. McGraw Hill, New York

Cornwell RE, Craig RR Jr, Johnson CP (1983) On the application of the mode-acceleration method to structural engineering problems. Earthq Eng Struct Dyn 11:679–688

Crandall SH, McCalley RB Jr (2002) Chapter 28, Part I, matrix methods of analysis. In: Harris CM, Piersol AG (eds) Harris' shock and vibration handbook. McGraw-Hill, New York

Ewins DJ (1984) Modal testing: theory and practice. Research Studies Press, Somerset

Hansen C, Snyder S, Qiu X, Brooks L, Moreau D (2012) Active control of noise and vibration. CRC Press, Boca Raton

He J, Fu ZF (2001) Modal analysis. Butterworth Heinemann, Oxford

Itoh T (1973) Damped vibration mode superposition method for dynamic response analysis. Earthq Eng Struct Dyn 2:47–57

Kline M (1990) Mathematical though from ancient to modern times, vol II. Oxford University Press, Oxford, UK

Leissa AW (2005) The historical bases of the Rayleigh and Ritz methods. J Sound Vib 287:961–978

Ljung L (1999) System identification – theory for the user, 2nd edn. Prentice Hall, New Jersey

Lord R (1877) The theory of sound, vol I. Macmillan, UK (reprinted 1945 by Dover Publications, New York)

Maia NMM, Silva JMM, He J, Lieven NAJ, Lin RM, Skingle GW, To WM, Urgueira APV (1997) Theoretical and experimental modal analysis. Research Studies Press, Somerset

Rades M (2010) Mechanical vibrations II, structural dynamic modelling. Editura Printech, Bucharest

Rajasekaran S (2009) Structural dynamics of earthquake engineering: theory and application using MATHEMATICA and MATLAB. Woodhead, Cambridge, UK

Model Class Selection for Prediction Error Estimation

E. Simoen[1], C. Papadimitriou[2] and Geert Lombaert[1]
[1]Department of Civil Engineering, KU Leuven, Leuven, Belgium
[2]Department of Mechanical Engineering, University of Thessaly, Volos, Greece

Synonyms

Bayesian error estimation; Bayesian model class updating; Likelihood function

Definitions

Model class selection Determining the most suitable model class based on the available observed data, by applying Bayesian inference at the model class level.

Prediction error The discrepancy between the observed data and the predictions made by a numerical model. In Bayesian model updating, the likelihood function is constructed as the probability density function of the prediction error.

Introduction

In Bayesian model updating, probability density functions (PDFs) of model parameters are updated based on prior knowledge and information contained in experimental data (Beck and Katafygiotis 1998). This requires the construction of the *likelihood function*, i.e., the PDF of the experimental data, or, equivalently, the PDF of the error between model predictions and observed data. Most often, a fixed zero-mean uncorrelated Gaussian prediction error (i.e., "white noise") is assumed or presumed given; however, in many engineering applications, the magnitude of the true prediction error is unknown. Moreover, a white noise model may not always be appropriate, for instance, when the errors show spatial or temporal correlations or a systematic component. In such cases, one can make use of the available observed data to try and estimate (characteristics of) the prediction error, provided sufficient data are at hand. Bayesian model class selection (MCS) (Beck and Yuen 2004) constitutes one of the most effective tools to this end.

Bayesian Model Updating and Prediction Error

The objective of model updating (often also referred to as *parameter estimation*) is to calibrate unknown system properties which appear as parameters in numerical models, based on actually observed behavior of the system of interest. In Bayesian model updating, this is performed in a probabilistic uncertainty quantification framework: PDFs representing the uncertainty on the model parameters are updated through the experimental data; this procedure is described briefly below.

Uncertainty in Model Updating

In general terms, a model $\mathcal{M}_M(\boldsymbol{\theta}_M)$ belonging to the model class $\mathcal{M}_M$ provides a mapping from the parameters $\boldsymbol{\theta}_M \in \mathbb{R}^{N_M}$ to an output vector $\mathbf{G}_M(\boldsymbol{\theta}_M) \in \mathbb{R}^N$ through the transfer operator $\mathbf{G}_M$:

$$\mathcal{M}_M(\boldsymbol{\theta}_M) : \mathbb{R}^{N_M} \to \mathbb{R}^N : \boldsymbol{\theta}_M \mapsto \mathbf{G}_M(\boldsymbol{\theta}_M) \quad (1)$$

In structural engineering, the prediction model class usually involves a mechanical model such as a finite element model. The model parameters $\boldsymbol{\theta}_M$ are typically properties directly connected to (sub)structure stiffness or mass (e.g., Young's modulus, moment of inertia, mass density), material properties, or parameters describing connections and boundary conditions. The model outputs $\mathbf{G}_M(\boldsymbol{\theta}_M)$ should correspond to the observed system behavior, which can consist of directly measured structural responses (displacements, accelerations) or derived system properties such as modal data (natural frequencies, mode shapes, modal flexibilities, modal curvatures, modal strain energies). Most often, the latter are preferred: modal data are rich in information content and can be obtained in an *operational* state of the structure, meaning that no external excitation needs to be applied or known. This also explains why, for the cases considered in this text, no input vector $\mathbf{x}$ is present in Eq. 1.

In the ideal case, with a perfect model and perfect measurements, the model output $\mathbf{G}_M(\boldsymbol{\theta}_M)$ corresponds perfectly to the true system output $\mathbf{d}$, so that $\mathbf{G}_M(\boldsymbol{\theta}_M) = \mathbf{d}$. This is the underlying assumption in deterministic parameter identification, where the objective is to determine the model parameters $\boldsymbol{\theta}_M$ for a given set of observed system outputs $\mathbf{d}$. In real applications, however, the numerical model is never capable of perfectly representing the behavior of the true physical system; in other words, a modeling error is always present. This error can be described as the discrepancy between the model predictions $\mathbf{G}_M(\boldsymbol{\theta}_M)$ and the true system output $\mathbf{d}$, i.e., $\boldsymbol{\eta}_G = \mathbf{d} - \mathbf{G}_M(\boldsymbol{\theta}_M)$.

As the true system output has to be measured and processed experimentally, the data $\mathbf{d}$ are also imperfect: they are always subject to measurement error, resulting in a discrepancy between the true system output $\mathbf{d}$ and the actually observed data $\bar{\mathbf{d}}$. This difference is defined as the measurement error $\boldsymbol{\eta}_D = \bar{\mathbf{d}} = \mathbf{d}$. Eliminating the unknown true system output $\mathbf{d}$ from the error equations and collecting both errors on the right-hand side of the equation yield

$$\bar{\mathbf{d}} - \mathbf{G}_M(\boldsymbol{\theta}_M) = \boldsymbol{\eta}_G + \boldsymbol{\eta}_D = \boldsymbol{\eta} \quad (2)$$

The sum of both errors is the difference between the model predictions and the observed quantities and is defined as the total observed prediction error $\boldsymbol{\eta}$. The above expression serves as a starting point for the Bayesian uncertainty quantification method.

Bayesian Model Updating Methodology

The general principle of Bayesian parameter estimation (Beck and Katafygiotis 1998) is that uncertainties in the model parameters

$\boldsymbol{\theta}_M \in \mathbb{R}^{N_M}$ that parameterize the model class $\mathcal{M}_M$ are quantified by probability density functions (PDFs), which are updated in an inference scheme based on the available information. Measurement and modeling uncertainty are taken into account by modeling the respective errors as random variables: PDFs are appointed to $\boldsymbol{\eta}_G$ and $\boldsymbol{\eta}_D$, which are parameterized by parameters $\boldsymbol{\theta}_G \in \mathbb{R}^{N_G}$ and $\boldsymbol{\theta}_D \in \mathbb{R}^{N_D}$. These parameters are added to the structural model parameters $\boldsymbol{\theta}_M$ to form the general model parameter set $\boldsymbol{\theta} = \{\boldsymbol{\theta}_M, \boldsymbol{\theta}_G, \boldsymbol{\theta}_D\}^T \in \mathbb{R}^{N_\theta}$. This in fact corresponds to adding two probabilistic model classes to the structural model class $\mathcal{M}_M$ to form a *joint* model class $\mathcal{M} = \mathcal{M}_M \times \mathcal{M}_G \times \mathcal{M}_D$, parameterized by $\boldsymbol{\theta}$. This procedure of probabilistic modeling is commonly referred to as *stochastic embedding* (Beck 2010).

It is important to note that in Bayesian statistics, the *Bayesian* interpretation of probability is pertained as opposed to the classical *frequentist* interpretation. In the frequentist interpretation, the probability attributed to a random variable is seen as a measure for the long-term frequency of occurrence of that variable. The Bayesian interpretation is a subjective interpretation where probability reflects a measure of plausibility or degree of belief attributed to a variable, given the current state of information. It is clear that only the Bayesian interpretation is meaningful in the context of forming inferences on model parameters using observed data.

To express the updated probabilities of the unknown parameters $\boldsymbol{\theta}$, given some observations $\bar{\mathbf{d}}$ and a certain joint model class $\mathcal{M}$, Bayes' theorem is used:

$$p(\boldsymbol{\theta}|\bar{\mathbf{d}}, \mathcal{M}) = c p(\bar{\mathbf{d}}|\boldsymbol{\theta}, \mathcal{M}) p(\boldsymbol{\theta}|\mathcal{M}) \qquad (3)$$

where $p(\boldsymbol{\theta}|\bar{\mathbf{d}}, \mathcal{M})$ is the updated or posterior PDF of the model parameters given the measured data $\bar{\mathbf{d}}$ and the assumed model class $\mathcal{M}$; c is a normalizing constant that ensures the posterior PDF integrates to one; $p(\bar{\mathbf{d}}|\boldsymbol{\theta}, \mathcal{M})$ is the PDF of the observed data given the parameters $\boldsymbol{\theta}$; and $p(\boldsymbol{\theta}|\mathcal{M})$ is the initial or prior PDF of the parameters. In the following, the explicit dependence on the model class $\mathcal{M}$ is omitted in order to simplify the notations.

The prior PDF $p(\boldsymbol{\theta})$ quantifies the uncertainty associated with model parameters $\boldsymbol{\theta}$ in the absence of measurement results. In many cases, the prior PDF is chosen based on engineering judgment or on computational tractability (e.g., conjugate priors (Diaconis and Ylvisaker 1979)). Except in cases where a large amount of data is at hand, the prior PDF has a significant influence on the Bayesian updating results. A wide range of methods has been developed to obtain "objective" prior PDFs based on the given prior information. One of the most commonly used approaches in this respect is the method based on the maximum entropy (ME) principle (Jaynes 1957; Soize 2008), which determines the prior PDF that, given the current state of prior information, results in maximum entropy.

The PDF of the experimental data $p(\bar{\mathbf{d}}|\boldsymbol{\theta})$ can be interpreted as a measure of how good a model succeeds in explaining the observations $\bar{\mathbf{d}}$. As this PDF reflects the likelihood of observing the data $\bar{\mathbf{d}}$ when the model is parameterized by $\boldsymbol{\theta}$, it is also referred to as the *likelihood* function $L(\boldsymbol{\theta}|\bar{\mathbf{d}})$. Since the data set $\bar{\mathbf{d}}$ is fixed, this function in fact no longer represents a conditional PDF and can be denoted as $L(\boldsymbol{\theta}; \bar{\mathbf{d}})$; in the following, however, the common notation of $L(\boldsymbol{\theta}|\bar{\mathbf{d}})$ is pertained. The likelihood function is determined according to the total probability theorem in terms of the probabilistic models of the measurement and modeling errors:

$$p(\bar{\mathbf{d}}|\boldsymbol{\theta}) \equiv L(\boldsymbol{\theta}|\bar{\mathbf{d}}) = \int_{\mathbb{R}^N} p_{\bar{\mathbf{d}}}(\bar{\mathbf{d}}|\boldsymbol{\theta}, \mathbf{d}) p_{\mathbf{d}}(\mathbf{d}|\boldsymbol{\theta}) d\mathbf{d} \qquad (4)$$

$$= \int_{\mathbb{R}^N} p_{\eta_D}(\bar{\mathbf{d}} - \mathbf{d}|\boldsymbol{\theta}_D) p_{\mathbf{d}}(\mathbf{d}|\boldsymbol{\theta}) d\mathbf{d} \qquad (5)$$

$$= \int_{\mathbb{R}^N} p_{\eta_D}(\bar{\mathbf{d}} - \mathbf{d}|\boldsymbol{\theta}_D) p_{\eta_G}(\mathbf{d} - \mathbf{G}_M(\boldsymbol{\theta}_M)|\boldsymbol{\theta}_G) d\mathbf{d} \qquad (6)$$

where $p_{\eta_D}(\bar{\mathbf{d}} - \mathbf{d}|\boldsymbol{\theta}_D)$ corresponds to the probability of obtaining a measurement error $\boldsymbol{\eta}_D$, given the PDF of $\boldsymbol{\eta}_D$ parameterized by $\boldsymbol{\theta}_D$, and where $p_{\eta_G}(\mathbf{d} - \mathbf{G}_M(\boldsymbol{\theta}_M)|\boldsymbol{\theta}_G)$ represents the probability of obtaining a modeling error $\boldsymbol{\eta}_G$ when the PDF of $\boldsymbol{\eta}_G$ is known and parameterized by $\boldsymbol{\theta}_G$. Here, it is implicitly assumed that the modeling error and measurement error are independent variables.

The above equations show that the likelihood function can be computed as the convolution of the PDFs of the measurement and modeling error. In most realistic applications, however, no distinction can be made between measurement and modeling error. In those cases, the likelihood function can be constructed using the probabilistic model of the total prediction error $\boldsymbol{\eta}$, parameterized by $\boldsymbol{\theta}_\eta$:

$$p(\bar{\mathbf{d}}|\boldsymbol{\theta}) \equiv L(\boldsymbol{\theta}|\bar{\mathbf{d}}) = p(\boldsymbol{\eta}|\boldsymbol{\theta}_\eta) \qquad (7)$$

Even though the likelihood function has a significant influence on the model updating results, its construction has received much less attention than the construction of the prior PDF. This is mainly due to the fact that most often very little or no information is at hand regarding the characteristics of the error(s); only in selected cases, a realistic estimate can be made concerning the probabilistic model representing the prediction error, for instance, based on the analysis of measurement results. Most often, it is simply assumed that the probabilistic model of the prediction error is known and fixed, so that the parameter set reduces to $\boldsymbol{\theta} = \{\boldsymbol{\theta}_M\} \in \mathbb{R}^{N_M}$.

In structural mechanics applications, an uncorrelated zero-mean Gaussian model (i.e., "white noise") with a fixed and constant variance is typically selected to represent the prediction error. However, this common assumption is not necessarily justified or realistic. For instance, in most practical applications, the magnitude (or variance) of the errors is not (precisely) known. Moreover, a white noise model might not be appropriate to represent the true prediction error, as significant spatial or temporal correlations between the modeling error components are often likely to be present. Consider, for instance, the case where mode shapes need to be predicted along a densely populated sensor grid. Then, it can be expected that the modeling errors for two nearby mode shape components are related. In the case where time-domain data are employed, it can similarly be suspected that modeling errors at consecutive time steps are correlated at high sampling frequencies. Besides correlations, the prediction error can show a systematic component, e.g., due to incorrect modeling assumptions or inexact measurement equipment or setup, which means assuming a zero-mean error is also not always realistic.

In order to avoid incorrect or unsuitable assumptions and thereby influencing the Bayesian updating results in an unfounded way, one can make use of the available observed data to try and estimate (characteristics of) the prediction error, for instance, using Bayesian model class selection.

Bayesian Model Class Selection

Bayesian inference can be applied at model class level to assess the plausibility of several alternative model classes based on the available observations $\bar{\mathbf{d}}$; this is referred to as Bayesian model class selection or MCS (Beck and Yuen 2004; Yuen 2010). The set of alternative model classes commonly concern (mechanical) prediction model classes $\mathcal{M}_M$, but here the method will be used to distinguish between alternative probabilistic prediction error models $\mathcal{M}_\eta$. The following, however, is elaborated for a set of general model classes $\mathcal{M}_i$.

Suppose there is a set $\boldsymbol{\mathcal{M}}$ of N_C candidate model classes $\mathcal{M}_i$:

$$\boldsymbol{\mathcal{M}} = \{\mathcal{M}_i\} \quad \text{where} \quad i = 1, \ldots, N_C \qquad (8)$$

Then, the posterior probability of each model class $\mathcal{M}_i$ is given by Bayes' theorem as

$$P(\mathcal{M}_i|\bar{\mathbf{d}}, \boldsymbol{\mathcal{M}}) = \frac{p(\bar{\mathbf{d}}|\mathcal{M}_i)P(\mathcal{M}_i|\boldsymbol{\mathcal{M}})}{p(\bar{\mathbf{d}}|\boldsymbol{\mathcal{M}})} \qquad (9)$$

where $P(\mathcal{M}_i|\boldsymbol{\mathcal{M}})$ is the prior probability of each model class $\mathcal{M}_i$ (e.g., taken equal to $1/N_C$ when the model classes are considered equally plausible a priori), where usually the sum of all prior probabilities over all model classes is taken to be equal to 1.

The term $p(\bar{\mathbf{d}}|\mathcal{M}_i)$ in the numerator on the right-hand side of Eq. 9 is the *evidence* (sometimes also referred to as the *model class likelihood*) for the model class $\mathcal{M}_i$ provided by the data $\bar{\mathbf{d}}$. The evidence, hereafter denoted with ϵ, is a very important quantity in Bayesian model class selection and can be determined based on the law of total probability as

$$\epsilon_i = p(\bar{\mathbf{d}}|\mathcal{M}_i) = \int_{D_i} p(\bar{\mathbf{d}}|\boldsymbol{\theta}_i, \mathcal{M}_i) p(\boldsymbol{\theta}_i|\mathcal{M}_i) d\boldsymbol{\theta}_i \quad (10)$$

where $\boldsymbol{\theta}_i$ is the parameter vector in a parameter space D_i that defines each model in $\mathcal{M}_i$. By definition, the evidence ϵ_i gives the probability of the data according to model class $\mathcal{M}_i$. Following Eq. 10, it is determined as the weighted average of the probability of the data according to each model in the model class $\mathcal{M}_i$, where the weights are equal to the prior probability of the parameter values corresponding to each model. In fact, the evidence is equal to the a priori expected value of the likelihood. Note that it is also equal to the reciprocal of the normalizing constant c in the Bayes' theorem described by Eq. 3.

The denominator in Eq. 9, $p(\bar{\mathbf{d}}|\mathcal{M})$, is the probability of the data according to all model classes in $\mathcal{M}$ and is determined by applying the law of total probability as

$$p(\bar{\mathbf{d}}|\mathcal{M}) = \sum_{i=1}^{N_C} p(\bar{\mathbf{d}}|\mathcal{M}_i) P(\mathcal{M}_i|\mathcal{M}) \quad (11)$$

As this value is a constant, it is clear that in order to establish the most probable model class, the numerator term $p(\bar{\mathbf{d}}|\mathcal{M}_i)P(\mathcal{M}_i|\mathcal{M}) = \epsilon_i P(\mathcal{M}_i|\mathcal{M})$ in Eq. 9 has to be maximized with respect to i. Usually, the process of model class selection consists of determining the posterior probabilities of all model classes in the set $\mathcal{M}$ and ranking them accordingly, where the best model class is the one that results in the highest value of the quantity $\epsilon_i P(\mathcal{M}_i|\mathcal{M})$. Since most often equal prior model class probabilities are adopted, it suffices to compute the evidence values ϵ_i for all model classes and to rank them accordingly. Then, the most probable model class – according to the available data – corresponds to the model class with the highest evidence value.

The actual computation of the evidence values according to Eq. 10 poses a challenging problem in most practical applications, as it entails computing complex and usually high-dimensional integrals. To overcome this problem, asymptotic approximations can be applied (Papadimitriou et al. 1997; Beck 2010), or, alternatively, methods based on stochastic simulation such as MCMC sampling (Gelfand and Dey 1994; Cheung and Beck 2010) may be used.

Model Parsimony Through Bayesian MCS

In the context of model class selection, the principle of model parsimony expresses that simpler models that are reasonably consistent with the observed data are preferable to more complicated models (e.g., with more modeling parameters), even though the more complex models might result in a better data fit. In other words, the selected class of models should be able to reproduce the observed system behavior closely, but should otherwise be as simple as possible. This principle of model simplicity is also often referred to as "Occam's razor" and is pursued in MCS for two reasons: firstly, the achieved data fit improvement is in most cases only marginal, and secondly, an over-parameterized model may lead to an over-fit of the data. This often results in relatively poor predictions made by the model, due to the high sensitivity of the model to small changes in the data.

Many authors have suggested adding penalty terms to the log likelihood to discourage highly parameterized model classes. However, it can be shown that applying a Bayesian inference scheme at the model class level automatically enforces model parsimony, thereby avoiding the need for ad hoc penalty terms (Beck 2009). Using Bayes' theorem and the fact that the posterior PDF integrates to one, the logarithm of the evidence in Eq. 10 can be reformulated and expanded for a general model class $\mathcal{M}$ as

$$\log \epsilon = \log p(\bar{\mathbf{d}}|\mathcal{M}) = \mathbb{E}\{\log p(\bar{\mathbf{d}}|\boldsymbol{\theta}, \mathcal{M})\}_{\text{po}}$$
$$- \mathbb{E}\left\{\log \frac{p(\boldsymbol{\theta}|\bar{\mathbf{d}}, \mathcal{M})}{p(\boldsymbol{\theta}|\mathcal{M})}\right\}_{\text{po}} \quad (12)$$

The first term in the above log evidence expression is the posterior mean value of the log likelihood function, termed log ϵ_{data}, which gives a measure for the average data fit for the model class $\mathcal{M}$. It is easily verified that the second term equals the Kullback–Leibler divergence D_{KL} between the prior and the posterior PDF. This term gives a measure for the difference between the prior and posterior PDF: it can be interpreted as a measure for the information that is gained from the observations $\bar{\mathbf{d}}$. It can be shown that this term is nonnegative and increases with the number of model parameters in the model class (Beck and Yuen 2004; Beck 2007); therefore, this term penalizes more complex models that attempt to extract more information from the data, thus eliminating the need for ad hoc penalty terms. In the following, the negative of this term is denoted as the Occam term log ϵ_{occam}, as it enforces the Occam simplicity principle. As such, the log evidence can be rewritten as

$$\log \epsilon = \log \epsilon_{\text{data}} + \log \epsilon_{\text{occam}} \quad (13)$$

where

$$\log \epsilon_{\text{data}} = \mathbb{E}\{\log p(\bar{\mathbf{d}}|\boldsymbol{\theta}, \mathcal{M})\}_{\text{po}} \quad (14)$$

$$\log \epsilon_{\text{occam}} = -\mathbb{E}\left\{\log \frac{p(\boldsymbol{\theta}|\bar{\mathbf{d}}, \mathcal{M})}{p(\boldsymbol{\theta}|\mathcal{M})}\right\}_{\text{po}} \quad (15)$$

Thereby, it is shown that the log evidence consists of a data fit term and a term which provides a penalty for more complex models that attempt to extract more information from the data. In other words, by applying the Bayesian inference scheme for model class selection, an explicit trade-off is made between the data fit of the model class and its complexity, without introducing ad hoc concepts.

Application: Estimation of Prediction Error Correlation Model Class

This example treats a common application of model updating in structural mechanics, namely, vibration-based structural health monitoring (SHM). In this technique, modal data extracted from vibration experiments are used to update stiffness values along a structure; as such, structural damage (i.e., loss in stiffness) can be identified, located, and quantified. In this context, Bayesian model updating provides an effective tool to assess the influence of measurement and modeling error on the SHM results.

The practical application of the prediction error estimation procedure explained above is demonstrated in a small-scale example, where a vibration test on a damaged reinforced concrete (RC) beam is simulated. This is done by modeling the beam using a detailed 3D FE model consisting of 5250 volume elements (Fig. 1) where the steel reinforcement is taken into account through an adapted material density. Structural damage is simulated by reducing the Young's modulus of a small area around one third of the length of the beam to 18.75 GPa; the rest of the beam is appointed a stiffness modulus of 37.5 GPa.

The simulated modal data consists of the first four bending modes $\bar{\boldsymbol{\phi}}_r \in \mathbb{R}^{N_s}$ at N_s degrees of freedom (DOFs) or sensors and their associated natural frequencies $\bar{f}_r$ with eigenvalues

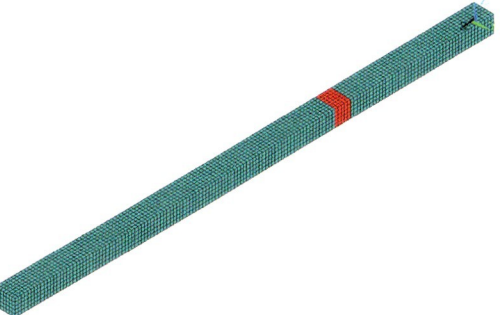

Model Class Selection for Prediction Error Estimation, Fig. 1 3D solid FE model of the RC beam; the area with reduced stiffness is shown in *red*

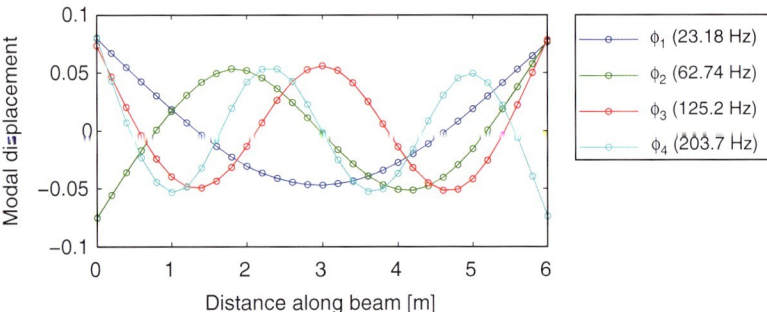

Model Class Selection for Prediction Error Estimation, Fig. 2 Simulated mode shapes $\bar{\phi}_r$ at $N_s = 31$ sensors and corresponding natural frequencies $\bar{f}_r = \sqrt{\bar{\lambda}}/2\pi$

$\bar{\lambda}_r = (2\pi \bar{f}_r)^2$, where free–free boundary conditions are adopted in the 3D model. To simulate a measurement error, an uncorrelated zero-mean Gaussian error with a standard deviation of 0.1 % is superimposed on these modal data. In order to assess the influence of the sensor density, three sensor configurations are considered, with $N_s = 6$, 31, and 151 equidistant sensors, and where a denser configuration includes the previous one. The final simulated experimental data set $\bar{\mathbf{d}} = \left\{ \bar{\lambda}_1, \ldots, \bar{\lambda}_4, \bar{\boldsymbol{\phi}}_1^\mathrm{T}, \ldots, \bar{\boldsymbol{\phi}}_4^\mathrm{T} \right\}^\mathrm{T}$ for 31 sensors is shown in Fig. 2.

The model of the RC beam that will be updated using a Bayesian inference approach is a 2D finite element model which consists of 150 beam elements and 151 nodes, with two DOFs per node, resulting in a total of $N_d = 302$ DOFs in the FE model. This prediction model is parameterized by a single parameter $\boldsymbol{\theta}_\mathrm{M}$ representing the Young's modulus of the damaged region of elements at around one third of the length of the beam. It is expected that this model combined with the simulated data will lead to prediction errors that are correlated in space; the objective is therefore to effectively account for these correlations through the proposed Bayesian MCS approach.

A zero-mean Gaussian prediction error is adopted so that $\boldsymbol{\eta} \sim \mathcal{N}(0, \boldsymbol{\Sigma}_\eta)$, where it is assumed that the eigenvalue errors $\eta_{\lambda,r} = \bar{\lambda}_r - \lambda_r(\boldsymbol{\theta}_\mathrm{M})$ are independent from the mode shape errors $\boldsymbol{\eta}_{\phi,r} = \bar{\boldsymbol{\phi}}_r - \boldsymbol{\phi}_r(\boldsymbol{\theta}_\mathrm{M})$, meaning that the covariance matrix $\boldsymbol{\Sigma}_\eta$ can be constructed as $\boldsymbol{\Sigma}_\eta = \mathrm{blkdiag}(\boldsymbol{\Sigma}_\lambda, \boldsymbol{\Sigma}_\phi)$. The eigenvalue covariance matrix is assumed to be diagonal and parameterized as $\boldsymbol{\Sigma}_\lambda = \theta_\lambda^2 \mathrm{diag}\left(\bar{\lambda}_1^2, \ldots, \bar{\lambda}_4^2\right)$. For the mode shapes, it is suspected that the correlation length scale depends on the considered mode; therefore, it is assumed that the covariance matrix for the mode shape components can be constructed as the block diagonal matrix of 4 individual covariance matrices: $\boldsymbol{\Sigma}_\phi = \mathrm{blkdiag}\left(\boldsymbol{\Sigma}_\phi^1, \ldots, \boldsymbol{\Sigma}_\phi^4\right)$.

Because the spatial correlation structure is unknown, a set of three alternative prediction error model classes is determined for $\boldsymbol{\Sigma}_\phi$: an uncorrelated model class A, a model class B with an exponential correlation function, and a model class C with an exponentially damped cosine correlation function. Each of these model classes is parameterized by a number of prediction error parameters as follows:

$$\boldsymbol{\Sigma}_\phi^{A,r} = \left(\theta_{\phi,1}^{A,r}\right)^2 \mathbf{I}_{N_s} \quad (16)$$

$$\left[\boldsymbol{\Sigma}_\phi^{B,r}\right]_{ij} = \left(\theta_{\phi,1}^{B,r}\right)^2 \exp\left(-\frac{\tau_{ij}}{\theta_{\phi,2}^{B,r}}\right) \quad (17)$$

$$\left[\boldsymbol{\Sigma}_\phi^{C,r}\right]_{ij} = \left(\theta_{\phi,1}^{C,r}\right)^2 \exp\left(-\frac{\tau_{ij}}{\theta_{\phi,2}^{C,r}}\right) \cos\left(\theta_{\phi,3}^{C,r} \tau_{ij}\right) \quad (18)$$

where τ_{ij} represents the distance between sensors i and j. These correlation models of increasing

Model Class Selection for Prediction Error Estimation, Fig. 3 Illustration of the three employed correlation models

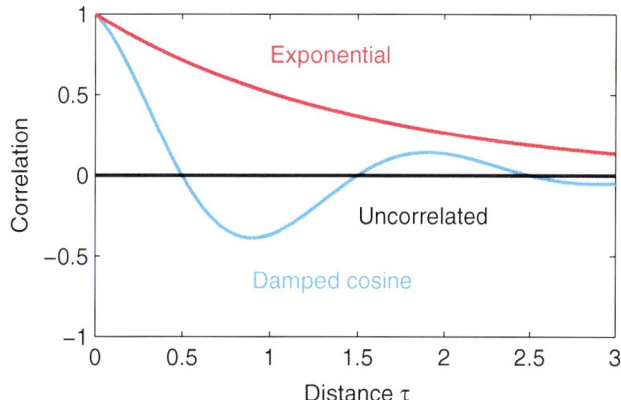

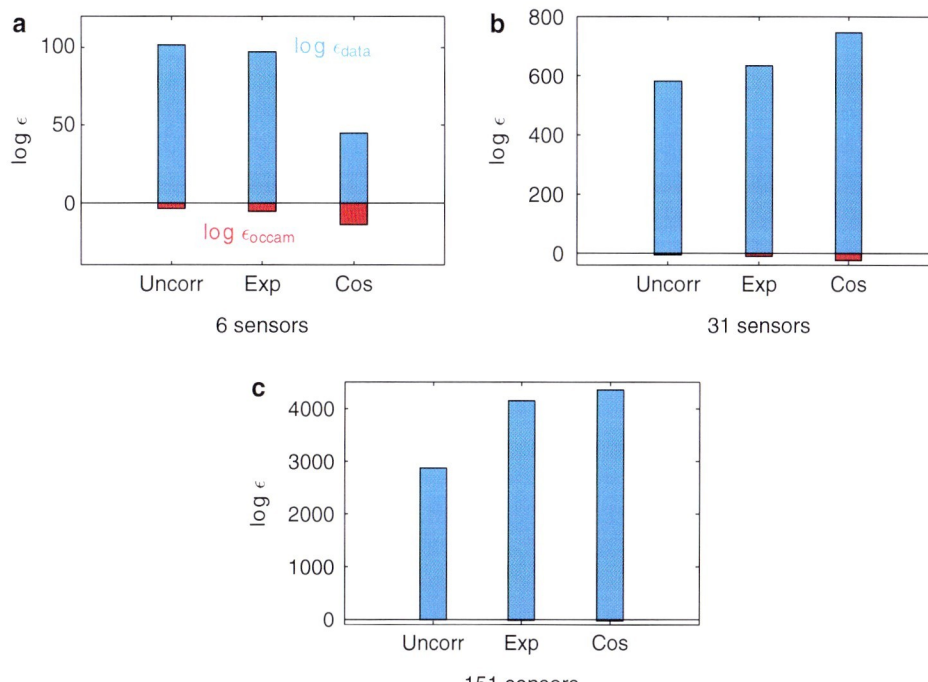

Model Class Selection for Prediction Error Estimation, Fig. 4 Log data fit values log ϵ_{data} and log Occam values log ϵ_{occam} for 6, 31, and 151 sensors

complexity (i.e., with more parameters) are illustrated in Fig. 3.

Based on the ME principle, gamma-distributed prior PDFs are assumed for all parameters (Soize 2003), and the Bayesian updating scheme is performed for all three alternative model classes using Markov Chain Monte Carlo (MCMC) sampling where a Metropolis–Hastings sampling algorithm is used to obtain 50,000 samples of the posterior PDF. The evidence values are computed using asymptotic expressions (Simoen et al. 2013) and are shown in Fig. 4 for the three sensor configurations. Several observations can be made. Firstly, the log Occam value clearly increases in absolute value as the model class complexity increases, confirming the statements made above regarding model parsimony. Secondly, for a low number of sensors, the uncorrelated model class is selected, indicating that correlation only becomes important for relatively dense sensor grids. However, as the number of sensors increases, the uncorrelated model class is clearly rejected, and the cosine correlation model class is eventually distinctly preferred.

The PDFs of the parameters characterizing this correlation model are estimated as well as the PDF of the stiffness parameter $\boldsymbol{\theta}_M$. For 151 sensors (and the selected cosine prediction error model class), a posterior mean value of $\mu(\boldsymbol{\theta}_M) = 18.95$ GPa is found, with a standard deviation of $\sigma(\boldsymbol{\theta}_M) = 0.80$ GPa.

This application shows that the Bayesian MCS approach can effectively be used to estimate a suitable correlation structure. Moreover, the MCS approach can also be applied to distinguish between different types of prediction error models (e.g., Gaussian vs. non-Gaussian) or between alternative descriptions for systematic prediction errors.

Summary

This reference entry describes the use of Bayesian model class selection to determine, among a set of possible model classes, the prediction error model class most suited for Bayesian model updating, according to the available experimental data. It is demonstrated that, provided sufficient information is available, the Bayesian MCS approach is an effective tool to this end, ensuring a more realistic joint structural-probabilistic model and corresponding Bayesian model updating results.

Cross-References

▶ Bayesian Statistics: Applications to Earthquake Engineering

▶ Uncertainty Quantification in Structural Health Monitoring

▶ Vibration-Based Damage Identification: The Z24 Bridge Benchmark

References

Beck J (2007) Bayesian updating and model class selection of deteriorating hysteretic structural models using seismic response data. In: Proceedings of the ECCOMAS thematic conference on computational methods in structural dynamics and earthquake engineering, Rethymno

Beck J (2009) Using model classes in system identification for robust response predictions. In: Proceedings of COMPDYN 2009, ECCOMAS thematic conference on computational methods in structural dynamics and earthquake engineering, Rhodos

Beck JL (2010) Bayesian system identification based on probability logic. Struct Control Health Monit 17(7):825–847

Beck J, Katafygiotis L (1998) Updating models and their uncertainties. I: Bayesian statistical framework. ASCE J Eng Mech 124(4):455–461

Beck J, Yuen K-V (2004) Model selection using response measurements: Bayesian probabilistic approach. ASCE J Eng Mech 130(2):192–203

Cheung SH, Beck JL (2010) Calculation of posterior probabilities for bayesian model class assessment and averaging from posterior samples based on dynamic system data. Comput-Aided Civ Infrastruct Eng 25(5):304–321

Diaconis P, Ylvisaker D (1979) Conjugate priors for exponential families. Ann Stat 7(2):269–281

Gelfand A, Dey D (1994) Bayesian model choice: asymptotics and exact calculations. J R Stat Soc 56(3):501–514

Jaynes E (1957) Information theory and statistical mechanics. Phys Rev 106(4):620–630

Papadimitriou C, Beck J, Katafygiotis L (1997) Asymptotic expansions for reliability and moments of uncertain systems. ASCE J Eng Mech 123(12):1219–1229

Simoen E, Papadimitriou C, Lombaert G (2013) On prediction error correlation in Bayesian model updating. J Sound Vib 332(18):4136–4152

Soize C (2003) Probabilités et modélisation des incertitudes: éléments de base et concepts fondamentaux. Handed out at the séminaire de formation de l'école doctorale MODES, Paris

Soize C (2008) Construction of probability distributions in high dimensions using the maximum entropy principle: applications to stochastic processes, random fields and random matrices. Int J Numer Methods Eng 75:1583–1611

Yuen K-V (2010) Recent developments of Bayesian model class selection and applications in civil engineering. Struct Saf 32(5):338–346

Model-Form Uncertainty Quantification for Structural Design

Ramana V. Grandhi[1] and Christopher Corey Fischer[2]
[1]210 Russ Engineering Center, Wright State University, Dayton, OH, USA
[2]Wright State University, Dayton, OH, USA

Synonyms

Additive Adjustment Factor; Adjustment Factor Approach; Bayesian Model Averaging; Bayes' Theorem; Model Combination; Model-Form Uncertainty; Multiplicative Adjustment Factor; Probabilistic Adjustment Factor; Uncertainty; Uncertainty Quantification

Introduction

Multiple forms of uncertainty exist in the prediction of a system response through the use of any type of modeling process. These uncertainties can be thought of presenting in one of three forms: parametric, predictive, and model-form uncertainty (Kennedy and O'Hagan 2001). The first of these three forms, parametric uncertainty, refers to the natural variability present within any input parameter, parameter in which a given model is reliant for predicting the response of interest. The latter two forms of uncertainty, predictive and model-form uncertainty, refer to the natural variability present within the modeling process itself. Uncertainty quantification work in the literature has primarily focused on the quantification of parametric uncertainty through the exploration, adaptation, and application of multiple approaches and methodologies. However, the majority of these approaches and methodologies fail to account for the presence of uncertainties that arise as a result of the modeling process.

Multiple models often arise as possibilities for representing the same physical situation when the solution approach for solving an engineering problem utilizes physics-based simulations. This is often a result of differing assumptions on governing physics, physical representation of boundary or loading conditions, or solution approaches. The availability of multiple models in simulating a given physical scenario can also result from the use of various modeling packages utilizing various fidelities or even operating on differing assumptions and/or mesh sizes within the same fidelity level. In very rare situations, the phenomena being modeled have been explored extensively and are well enough understood that a "best" model can emerge from the set of possible models, where in this work the term "best" model refers to the model that most accurately represents the true physical scenario being modeled. However, in most physical scenarios in which multiple sets of differing, complex physics can couple in various ways, there exists uncertainty in the selection of this "best" model as a result of a lack of knowledge of the true physics governing the physical scenario. As a result of this uncertainty, a single "best" model will not emerge from the model set, and thus, multiple models operating on assumptions of differing physics can produce inconsistent results for the prediction of the same physical problem. Therefore, in order to completely quantify the uncertainty present in the modeling process, it is necessary to not only quantify parametric uncertainties, but also the uncertainty inherent in selecting the "best" model – the model-form uncertainty.

Multiple approaches for the quantification of model-form uncertainties are presented, discussed, and applied to a nonlinear transient concrete creep engineering problem. This entry presents the application bounds of each approach such as requirements on the availability of experimental data and whether or not the approach can account for probabilistic model predictions – predictions including the quantification of parametric and/or predictive uncertainties. These approaches include the traditional adjustment factor approaches, probabilistic adjustment factor approach (an adaptation of the prior), and the Bayesian model averaging (BMA) approach.

Uncertainty in Modeling Process

The process involved in modeling a physical scenario requires the construction of a mathematical model which entails discretization of the physical scenario as well as making assumptions on the physics governing the true scenario. It is however, often beyond the capability of the designer to fully understand the true engineering problem at hand and thus capture the full complexity and all aspects of the physical scenario. Therefore, as a result of the assumptions made during development of physics-based models, discrepancy inevitably exists between the physical scenario being modeled and the prediction of the response by the computational model. This discrepancy is referred to as the predictive uncertainty of a particular model (Droguett and Mosleh 2008), and is unique to each individual model within a set of models considered for predicting a system response.

As mentioned previously, in the solution of an engineering problem, it is common practice for multiple models to be constructed for the purpose of representing the same physical scenario. Such examples include models of varying fidelities, such as assumptions on linear, quasilinear, or nonlinearity within a model, or models that account for complex phenomenon or boundary conditions in different manners, using different simplification techniques. As a result, the predictions of each of these models, for the particular output of interest, can be different from one another. Guedes-Soares states that in situations where multiple models yield different responses, that there can exist only one correct model (Soares 1997). However, as stated earlier, it is often beyond the capability of the designer to select the "best" model within the design space, or even a subset of the design space, and arises as a result of the inability to completely understand the full complexity of the physics governing the problem. As is the case in many engineering problems, due to the physical complexity and multidisciplinary interactions, a correct model will often not exist, instead one merely seeks to identify the "best" model among the model set. For reasons mentioned above, there exists uncertainty in the selection of this "best" model.

This uncertainty in the identification of the model most accurately predicting the correct outcome is referred to as model-form uncertainty (Zhang and Mahadevan 2000).

The definition of input parameters to each model within a model set, such as dimensions, material properties, environmental conditions, or modeling constants, are often defined as being deterministic. However, in nature, the parameters are rarely deterministic, or seldom able to be accurately represented as deterministic values, within the true physical scenario. As a result of this fact, there exists a third type of uncertainty in the modeling process – parametric uncertainty – which refers to the natural variability present in the values of parameters that a mathematical model is dependent upon (Droguett and Mosleh 2008). Parametric uncertainties are often classified as either aleatory or epistemic uncertainty, dependent upon the degree of information known in regards to the form of their uncertainty (Paté-Cornell 1996). Aleatory uncertainty refers to the form of uncertainty that arises from the natural unpredictable variation in the performance of a system (Hacking 1984), and can often be represented through distribution functions of a parameter's variability. Epistemic uncertainty, on the other hand, is defined as uncertainty due to a lack of knowledge regarding the performance of a system that can, in theory, be reduced through the introduction of additional data or information (Chernoff and Moses 2012).

All of the aforementioned uncertainties are present in nearly every physics-based modeling problem. A general representation for the modeling of an output of interest, y, is given in Eq. 1. As can be seen, this output of interest is a function of three terms: $\tilde{f}_k$, $\bar{x}_i$, and $\hat{\varepsilon}_k$.

$$y = \tilde{f}_k(\bar{x}_i) + \hat{\varepsilon}_k \qquad (1)$$

Here, $\tilde{f}_k$ represents the result/prediction obtained using model k, to the set of input parameters, $\bar{x}_i$. The second term, $\hat{\varepsilon}_k$, represents the discrepancy between the result/prediction obtained from model k, and the true physical value of the output of interest. The selection of the appropriate model, $\tilde{f}_k(\bar{x}_i)$, given this

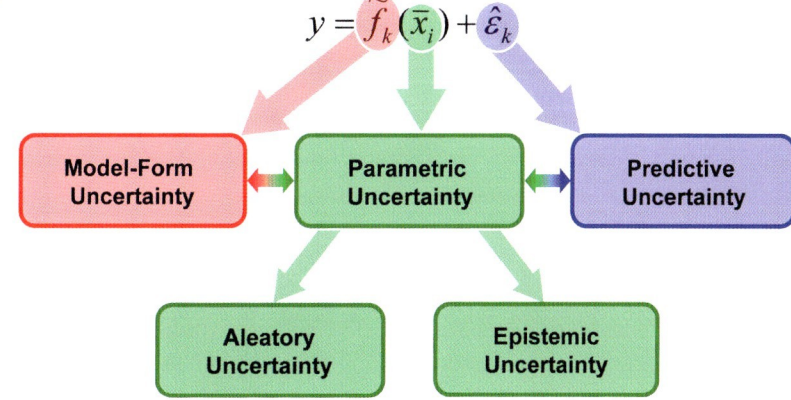

Model-Form Uncertainty Quantification for Structural Design, **Fig. 1** Modeling uncertainty breakdown

representation of a modeling problem, is representative of the inherent model-form uncertainty whereas, the variation in $\tilde{f}_k(\bar{x}_i)$ due to the uncertainty present in the set of input parameters, $\bar{x}_i$, is representative of parametric uncertainty. Finally, the determination of a discrepancy between response/prediction of model k and the true physical value of the output of interest, $\hat{\varepsilon}_k$, is representative of the predictive uncertainty present in model k. The breakdown of these three distinct types of uncertainty is demonstrated in Fig. 1.

Bayes' Theorem for Quantifying Model Probability

The first step for quantifying model-form uncertainty is to define model probabilities. Model probability is defined as the degree of belief that a model is the best approximating model from within a model set (Park et al. 2010). Original model probabilities can be assigned on one of two different concepts. Either expert opinion, if available, can be used to assign model probabilities based upon some prior knowledge of each model within the model set, or the more likely basis is to assign these probabilities based upon a uniform distribution of the probabilities assigned to each model within the model set. In the case of the uniformly distributed model probabilities where there exists K models within the model set, each model is assigned a probability of $\Pr(M_k) = 1/K$. Note that the sum of model probabilities, as shown in Eq. 2, for a model set, must be equal to 1.

$$\sum_{k=1}^{K} \Pr(M_k) = 1, \quad k = 1, 2, \ldots, K \quad (2)$$

In the event that experimental data is available, Bayes' theorem can then be used to update these original (prior) probabilities into posterior probabilities. Applying Eq. 3 updates the model probability of model k.

$$\Pr(M_k|D) = \frac{\Pr(D|M_k)\Pr(M_k)}{\sum_{k=1}^{K}\Pr(D|M_k)\Pr(M_k)} \quad (3)$$

In Eq. 3, $\Pr(D|M_k)$, which is broadly known as $L(M_k|D)$, is called the likelihood of model M_k given experimental data D, or the likeliness that model M_k will predict the known data D, because the first argument D is known, while the second argument M_k is not held constant (Robert 2007). The numerical value of the likelihood, $\Pr(D|M_k)$, is not needed in the uncertainty quantification process; however, it is of importance in the evaluation of model probabilities given experimental data. Note that the denominator in Eq. 3 is common to all models, and this model likelihood value is calculated using Eq. 4 as shown by Park et al. in (Park et al. 2010).

$$\Pr(D|M_k) = \left(\frac{1}{2\pi(\hat{\sigma})_{mle}^2}\right)^{N/2} \exp\left(-\frac{N}{2}\right) \quad (4)$$

Model likelihood then becomes a function of the maximum likelihood estimator, $(\hat{\sigma})_{mle}^2$, a measure of the variation within model k, as calculated in Eq. 5, where N in Eqs. 4 and 5 represents the number of experimental data points available for calculating the error term, ϵ. This maximum likelihood estimator provides an estimate for the model's variance based upon given experimental data in which the error term is the difference between the given experimental data point i and the prediction for that data point from model k. In Eq. 5 this error term is squared so as to eliminate the possibility of a positive and negative error canceling each other out.

$$(\hat{\sigma})^2_{mle} = \frac{\sum_{i=1}^{N} \epsilon^2_{k,i}}{N} \qquad (5)$$

Model-Form Uncertainty Quantification Techniques

Many methods for the quantification of model-form uncertainty have recently arisen within the literature. Each one of these methods exhibit unique advantages and disadvantages. Original work in model-form uncertainty quantification gave birth to methods requiring the presence of experimental data, such as Bayesian model averaging which requires the presence of experimental data in order to develop a maximum likelihood estimator of variance within a model prediction as is used in updating model probabilities through Bayes' theorem. Riley and Grandhi quantified the model-form and predictive uncertainty in the calculation of the flutter velocity of the AGARD 445.6 wing using BMA in (Riley and Grandhi 2011b). Other methods requiring the availability of experimental data include the methods of Continuous Model Expansion explored by Drapert, which showed a reliance on experimental data points, as well as difficulty in handling asymmetric distributions of parametric uncertainties (Draper 1995), and work done recently by Allaire and Willcox, which require a maximum entropy representation of the modeling uncertainty, an extra step that could be cost intensive for a high simulation-cost model (Allaire and Willcox 2010). As previously mentioned, the common thread among these three methods is the dependence on the availability of experimental data. The adjustment factor approach, however, was demonstrated by Mosleh and Apostolakis as a possible method for model-form uncertainty quantification which operates on the utilization of expert opinion to assign model probabilities in the absence of empirical data (Mosleh and Apostolakis 1986). Riley and Grandhi present an adaptation of this approach for use with nondeterministic model predictions, called the probabilistic adjustment actor approach (Riley and Grandhi 2011a). This work will explore the adjustment factor approaches, probabilistic adjustment factor approach, and the Bayesian Model Averaging approach and apply these methods to a multiphysics problem.

Additive Adjustment Factor Approach

As mentioned previously, the adjustment factor approach was first demonstrated by Mosleh and Apostolakis as a method for quantifying model-form uncertainty in the absence of experimental data by using an adaptation of Bayes' Theorem. This approach modifies the result of the "best" model–model within the model set being considered for use in solving the problem that obtains the highest model probability–by applying an adjustment factor to account for the uncertainty that exist in selection of the "best" model. The applicability of this approach to engineering problems has been well demonstrated in the literature. Zio and Apostolakis utilized an adjustment factor approach to quantify the uncertainty present in the selection of a "best" radioactive waste repository model in (Zio and Apostolakis 1996). Reinert and Apostolakis also used this approach in the assessment of risk for decision-making processes in (Reinert and Apostolakis 2006).

Multiple derivations of the adjustment factor approach exist in the literature. These derivatives all employ a similar technique of quantifying the model-form uncertainty through the use of expert opinion regarding a model's accuracy with respect to other models in the model set by assigning model probabilities, and updating those probabilities through the use of Bayes' theorem upon the availability of experimental data. In this approach, $\Pr(M_k)$ represents the model probability assigned to model k. Recall that this model probability is the probability that model k is the "best" model among model set M being considered, where Eq. 6 defines the model set M.

$$M = \{M_1, M_2, \ldots, M_K\} \quad (6)$$

Note that model probabilities for each of the K individual models within the model set M remain bounded by the laws of probability theory. Thus, constraints are applied to the model probability values, as shown in Eq. 7.

$$\sum_{k=1}^{K} \Pr(M_k) = 1 \quad \text{such that} \quad 0 \leq \Pr(M_k) \leq 1 \quad (7)$$

The various derivations of the adjustment factor approach differ here, in the form of the adjustment factor being applied–used to adjust–the "best" model. In the additive adjustment factor approach, the adjusted model, y, is formed by adding an additive adjustment factor, E_a^*, to the "best" model from the model set being considered, as shown in Eq. 8.

$$y = y^* + E_a^* \quad (8)$$

This additive adjustment factor, E_a^*, is assumed to be a normally distributed factor representing the uncertainty present in the selection of the "best" model as being most accurate at predicting the true physical response. In assuming a Gaussian form for this factor, the first and second moments–expected value and variance–of the adjustment factor are calculated as shown in Eqs. 9 and 10.

$$E[E_a^*] = \sum_{k=1}^{K} \Pr(M_k)(y_k - y^*) \quad (9)$$

$$\mathrm{Var}[E_a^*] = \sum_{k=1}^{K} \Pr(M_k)(y_k - E[y])^2 \quad (10)$$

It is important to note that use of the adjustment factor approach is reliant on models within the model set to be deterministic; this is to say that the models cannot incorporate parametric uncertainty. Therefore, the expected value of the adjusted model, $E[y]$, is calculated as shown in Eq. 11, as the sum of the prediction of the "best" model and the expected value of the additive adjustment factor, $E[E_a^*]$, found using Eq. 9.

$$E[y] = y^* + E[E_a^*] \quad (11)$$

Similarly, due to each individual model being deterministic, the variance of the adjusted model, $\mathrm{Var}[y]$, assumes only the variance of the additive adjustment factor, $\mathrm{Var}[E_a^*]$, as shown in Eq. 12.

$$\mathrm{Var}[y] = \mathrm{Var}[E_a^*] \quad (12)$$

The first and second moments–expected value and variance–of the adjusted prediction, y, are calculated by Eqs. 11 and 12 which fully define a normal distribution representing the uncertainty inherent in the prediction of a system response as a result of model-form uncertainty.

Multiplicative Adjustment Factor

Another derivative of the adjustment factor approach is that of a multiplicative adjustment factor. In the multiplicative adjustment factor approach, the adjusted model, y, is formed by multiplying the prediction of the "best" model, y^* from the model set being considered by a multiplicative adjustment factor, E_m^*, as shown in Eq. 13.

$$y = y^* * E_m^* \quad (13)$$

This multiplicative adjustment factor, E_m^*, is assumed to be a log normally distributed factor representing the uncertainty present in the selection of the "best" model as being most accurate at predicting the true physical response. In assuming a log normally distributed form for this factor, the first and second moments–expected value and variance–of the adjustment factor are calculated as shown in Eqs. 14 and 15.

$$\mathrm{E}\left[\ln\left(E_m^*\right)\right] = \sum_{k=1}^{K} \Pr(M_k)(\ln(y_k) - \ln(y^*)) \quad (14)$$

$$\mathrm{Var}\left[\ln\left(E_m^*\right)\right] = \sum_{k=1}^{K} \Pr(M_k)(\ln(y_k) - \mathrm{E}[\ln(y)])^2 \quad (15)$$

As was the case with the additive adjustment factor approach, it is important to note that use of the adjustment factor approach is reliant on models within the model set to be deterministic. Therefore, the expected value of the adjusted model, E[ln(y)], is calculated as shown in Eq. 16, as the sum of the natural logarithm of the prediction of the "best" model and the expected value of the multiplicative adjustment factor, $\mathrm{E}[\ln(E_m^*)]$, from Eq. 14.

$$\mathrm{E}[\ln(y)] = \ln(y^*) + \mathrm{E}\left[\ln\left(E_m^*\right)\right] \quad (16)$$

Much like the additive adjustment factor approach, due to each individual model being deterministic, the variance of the adjusted model, Var[ln(y)], assumes only the variance of the multiplicative adjustment factor, $\mathrm{Var}[\ln(E_m^*)]$, as shown in Eq. 17.

$$\mathrm{Var}[\ln(y)] = \mathrm{Var}\left[\ln\left(E_m^*\right)\right] \quad (17)$$

While these adjustment factor approaches have the benefit of being able to quantify model-form uncertainty in the absence of experimental data, they lack the ability to handle probabilistic model predictions, this is to say that the adjustment factor approach cannot quantify the model-form uncertainty present in the selection of a "best" model, from a model set, when the models within said model set contain quantified parametric uncertainty. This is a result of the assumption made during derivation of the approach that individual models are deterministic. Therefore, a rederivation of the approach, working on the assumption that individual models are probabilistic/stochastic in nature, would allow for the approach to be adapted to handle parametric uncertainty.

Probabilistic Adjustment Factor Approach

The probabilistic adjustment factor approach is an adaptation of the traditional adjustment factor approach in that it is capable of handling stochastic models, and thus parametric uncertainty. Note that this approach does not quantify the parametric uncertainty within each model of a model set, but it can handle parametrically uncertain models in its analysis (Riley and Grandhi 2011a). As with the traditional adjustment factor approach, a distribution must first be assumed for each of the individual models within the model set for derivation of the probabilistic adjustment factor approach. In general, there is no restriction on the form of this distribution; however, as was the case with the additive adjustment factor approach, this work will focus on the approach derived from assumption of a normal distribution.

The first step to quantifying model-form uncertainty through the use of the probabilistic adjustment factor approach is the same as that of the traditional adjustment factor approach. Model probabilities are applied to each individual model within the model set using either expert opinion or uniformly distributed probabilities, and then updating through the application of Bayes' theorem given the availability of experimental data, such that constraints of Eq. 7 are satisfied. The adjusted model for the probabilistic adjustment factor approach can then be computed as shown in Eq. 18.

$$y = \mathrm{E}[y^*] + E_{pafa}^* \quad (18)$$

Equation 18 is similar to Eq. 8 from the additive adjustment factor approach with the slight difference in the utilization of the expected value of the "best" model, $\mathrm{E}[y^*]$, rather than the deterministic "best" model prediction. Calculating the first and second moments–expected value and variance–of the probabilistic adjustment factor is also different for this new approach, as the approach had to be rederived to handle the stochastic model set. The calculation of these two moments can be performed using Eqs. 19 and 20.

$$E\left[E^*_{pafa}\right] = \sum_{k=1}^{K} \Pr(M_k)(E[y_k] - E[y^*]) \quad (19)$$

$$\mathrm{Var}\left[E^*_{pafa}\right] = \sum_{k=1}^{K} \Pr(M_k)(E[y_k] - E[y])^2 \quad (20)$$

After calculating the first and second moments of the probabilistic adjustment factor, the expected value and variance of the adjusted model can then be calculated using Eqs. 21 and 22.

$$E[y] = E[y^*] + E\left[E^*_{pafa}\right] \quad (21)$$

$$\mathrm{Var}[y] = \mathrm{Var}\left[E^*_{pafa}\right] + \sum_{k=1}^{K} \Pr(M_k)(\mathrm{Var}[y_k])^2 \quad (22)$$

As can be seen, there is only a slight difference in the formulations of the expected value equations for the traditional adjustment factor approach and the probabilistic adjustment factor approach, only in that of operating on the expected value of the "best" model prediction rather than the deterministic model prediction, as is the case for the traditional approach. The key difference in the two derivations comes in the formulation of the variance equation. The variance, shown in Eq. 22, includes the addition of the summation of weighted individual model variances–weighted by model predictions–to the variance of the adjustment factor calculated using Eq. 20, known as the between-model variance. The second term in the equation represents the variance in the adjusted model due to variances within each of the individual models–the within-model variance. Therefore, the first term in Eq. 22 can be thought of as representing the model-form uncertainty within the problem, whereas the second term represents the parametric uncertainty inherent to each model within the model set, and quantified using parametric uncertainty quantification techniques. In the event that individual models within the model set are deterministic, do not account for parametric uncertainties, and experimental data is available, the probabilistic adjustment factor approach can still be utilized by assuming the maximum likelihood estimator of variance for each individual model to be said model's variance in the evaluation of Eq. 22.

Bayesian Model Averaging

The Bayesian model averaging (BMA) technique is one that requires the availability of experimental data as mentioned previously; however, this technique can handle both stochastic and deterministic models. BMA is a technique for quantifying model-form uncertainty by averaging the predictions of each individual model within a model set using each model's corresponding model probability as a weighting factor. As is the case with the previous approaches, a distribution must be assumed for the averaged model, and the individual models, if stochastic, must follow this same distribution. For the purpose of this work, each model, if stochastic, must follow a normal distribution–defined by an expected value and variance. Thus, the averaged model assumes the form of a Gaussian distribution. The expected value of the averaged model, $E[y|D]$, is calculated by Eq. 23 as the summation of the model probabilities, $\Pr(M_k|D)$, multiplied by the corresponding model's expected value, $E[y|M_k, D]$.

$$E[y|D] = \sum_{k=1}^{K} \Pr(M_k|D) E[y|M_k, D] \quad (23)$$

Similarly, the equation for calculating the variance of the averaged model is shown in Eq. 24. This equation states that the variance of the averaged model, $\mathrm{Var}[y|D]$, is the summation of the model probabilities, $\Pr(M_k|D)$, multiplied by the corresponding model's variance, $\mathrm{Var}[y|M_k, D]$, plus the summation of the model probabilities multiplied by the squared difference between the corresponding model's expected value, $E[y|M_k,D]$, and the averaged model's expected value, $E[y|D]$.

$$\text{Var}[y|D] = \sum_{k=1}^{K} \Pr(M_k|D)\text{Var}[y|M_k,D]$$
$$+ \sum_{k=1}^{K} \Pr(M_k|D)(E[y|M_k,D] - E[y|D])^2 \quad (24)$$

In the event that BMA is being applied to a deterministic model set, case in which uncertainties due to input parameters are not included, Eqs. 25 and 26 are used to calculate the averaged model expected value and variance. The difference in the formulation between Eqs. 23 and 25 is that the latter used the individual model deterministic predictions, y_k, rather than their expected values. The key difference in these two derivations is seen in the formulation of the equation for calculation of averaged model variance, where the first term of Eq. 24 changes by using the maximum likelihood estimator of variance, $(\hat{\sigma})^2_{mle}$ calculated using Eq. 5, in place of the individual stochastic model variances as well as substituting the individual deterministic model predictions, y_k, for the stochastic model expected values in the first term of Eq. 26, corresponding to the second term of Eq. 24 (Park and Grandhi 2011).

$$E[y|D] = \sum_{k=1}^{K} \Pr(M_k|D)y_k \quad (25)$$

$$\text{Var}[y|D] = \sum_{k=1}^{K} \Pr(M_k|D)(y_k - E[y|D])^2$$
$$+ \sum_{k=1}^{K} \Pr(M_k|D)(\hat{\sigma}_k)^2_{mle} \quad (26)$$

Application

The outlined methodology is demonstrated through application to a concrete creep engineering problem. Creep is defined as the slow deformation a solid material undergoes under the influence of sustained stresses generated by external loads over an extended period of time. The creep deformation experienced by concrete may be three or four times as large as the initial elastic deformation caused by externally applied loads. It is widely accepted that creep deformation is largely attributed to shearing forces acting on material particles which cause them to slip against each other. Water within concrete has a large influence on the amount of slip occurring between these particles as a result of the action of weakening attractive forces binding those particles. Creep in concrete is not fully understood because it involves many factors that influence the total amount of creep a specimen experiences. The factors that considerably influence concrete creep are the aggregate, cement type, water-cement ratio, member size, curing condition, temperature and relative humidity, age of loading, and stress intensity.

The aspects of the aggregate that influence creep are the volume and the material properties. The aggregate serves to restrain creep from occurring; therefore, the magnitude of creep largely depends on the quantity and properties of the aggregate added into the mixture. Creep is inversely related to the aggregate. The content and type of cement paste are also key affecters on concrete creep. This is because creep mainly occurs in the hydrated cement paste that surrounds the aggregate. The amount of cement and amount of creep are directly correlated. Type of cement affects the strength of the concrete mixture at the time of loading. Therefore, when rapid hardening cements are used, the concrete matrix stiffness is increased at the time of loading allowing for the concrete to be more resistant to creep. Water-cement ratio is also directly related to creep. Lower water content results in higher concrete strength and thus fewer pores in the mature cement. This leads to a decrease in the amount of creep. It is thought that creep deformation decreases with an increase in member size or concrete specimen which is highly related to mobility of moisture in the concrete mixture. It has been shown that curing conditions have a substantial effect on the maturity of concrete, and thus concrete strength. Longer curing durations tend to lead to an increase in strength and consequently decrease in creep. Ambient conditions such as temperature and relative humidity are also sources that influence creep. Higher temperatures traditionally result in higher creep until a maximum is reached in the vicinity of 71 °C at which point further rise in temperature results in a decrease in creep. Humidity results in moisture within the concrete to be diffused to the ambient, thus allowing for more creep. Due to the phenomenon that concrete strength increases as it matures, age of loading has an inverse correlation with creep. Finally, it is general acceptance that the amount of creep is approximately proportional to the applied stress; however, this proportionality is only valid up to 0.2 to 0.5 times the ultimate strength.

Several mathematical models have been suggested to estimate the amount of creep in concrete. Most of the creep-prediction models developed were empirically derived based on the outcomes of experiments carried out on concrete specimens. Here, four empirical models recommended by different committees are used to predict the amount of creep that a concrete specimen experiences under a particular environment. The four creep-prediction models are the equations contained in the ACI 209 (1997) (209 1997), the AASHTO (2007) (AASHTO 2007), the CEB-FIP (1990) (CEB 1990), and the JSCE (1996) (Sakata and Shimomura 2004) design codes. For a given condition, the empirical models for predicting creep strain, given by these four codes as a function of time (number of days after onset of loading), are given in Eqs. (27–30) respectively.

$$M_1 = 1297^*10^{-1} \frac{t^{0.6}}{10 + t^{0.6}} \quad (27)$$

$$M_2 = 671^*10^{-6} \frac{t}{24.7 + t} \quad (28)$$

$$M_3 = 1527^*10^{-6} \left(\frac{t}{307.16 + t}\right)^{0.3} \quad (29)$$

$$M_4 = 1039^*10^{-6}\left(1 - \exp\left(-0.09(1.14(t+28) - 31.8)^{0.6}\right)\right) \quad (30)$$

The transient prediction of creep strain as predicted by each of the four models is plotted in Fig. 2. As can be seen, there is some discrepancy among the four different models. Consequently, this discrepancy is the source of model-form uncertainty. The uncertainty that arises as a result of not knowing with perfect certainty the true system response and thus which model among those available is most accurate at predicting said result.

Utilizing the additive adjustment factor approach to quantify the aforementioned model-form uncertainty, model probabilities must first be assigned using expert opinion in the absence of experimental data. The model probabilities were taken to be 0.5, 0.15, 0.05, and 0.3 for models M_1, M_2, M_3, and M_4 respectively. Note that this approach is only applicable to deterministic model predictions. Application of this technique at each point in the time domain is plotted in Fig. 3, where the upper and lower 95 % confidence bounds are determined using Eqs. 31 and 32 respectively.

$$95\% \text{ Upper Bound} = E(y|D) - 1.96\sqrt{\sum_{k=1}^{K} \Pr(M_k|D)\sigma_k^2 + \sum_{k=1}^{K} \Pr(M_k|D)(f_k - E(y|D))^2} \quad (31)$$

$$95\% \text{ Lower Bound} = E(y|D) - 1.96\sqrt{\sum_{k=1}^{K} \Pr(M_k|D)\sigma_k^2 + \sum_{k=1}^{K} \Pr(M_k|D)(f_k - E(y|D))^2} \quad (32)$$

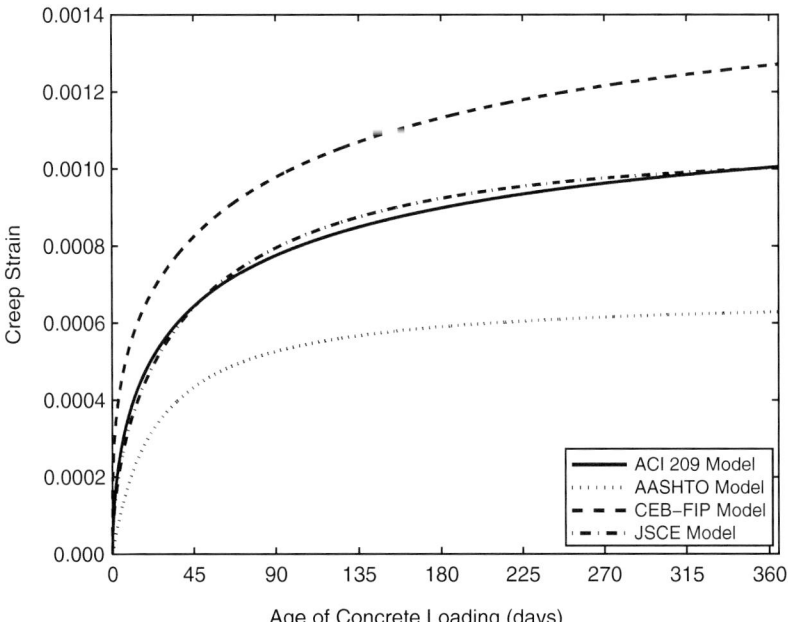

Model-Form Uncertainty Quantification for Structural Design, Fig. 2 Plot of each of the four models used for prediction of creep strain as a function of time

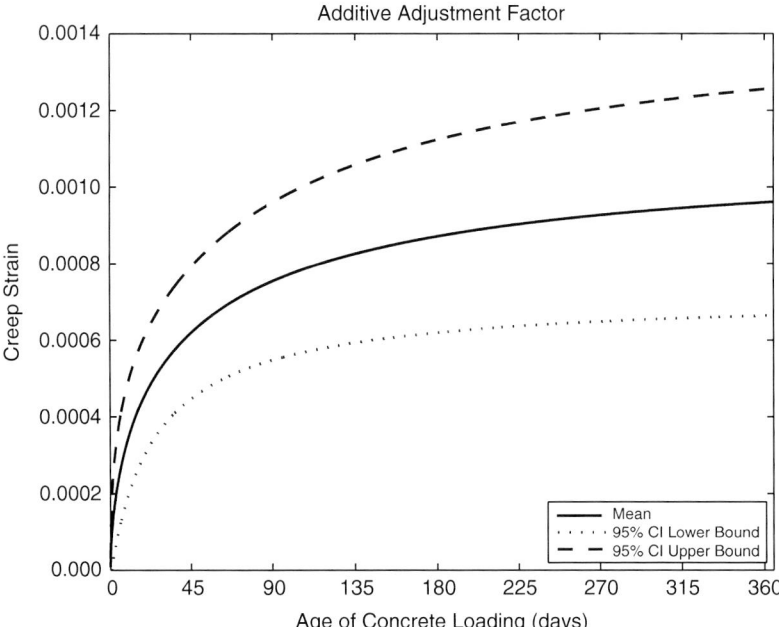

Model-Form Uncertainty Quantification for Structural Design, Fig. 3 Transient response obtained utilizing an additive adjustment factor approach

Utilizing the multiplicative adjustment factor approach to quantify the aforementioned model-form uncertainty, model probabilities in the same manner as with the additive adjustment factor approach. Note that this approach is also only applicable to deterministic model predictions. Application of this technique at each point in the time domain is plotted in Fig. 4. It can be seen that the latter two model-form uncertainty quantification techniques yield nearly identical results.

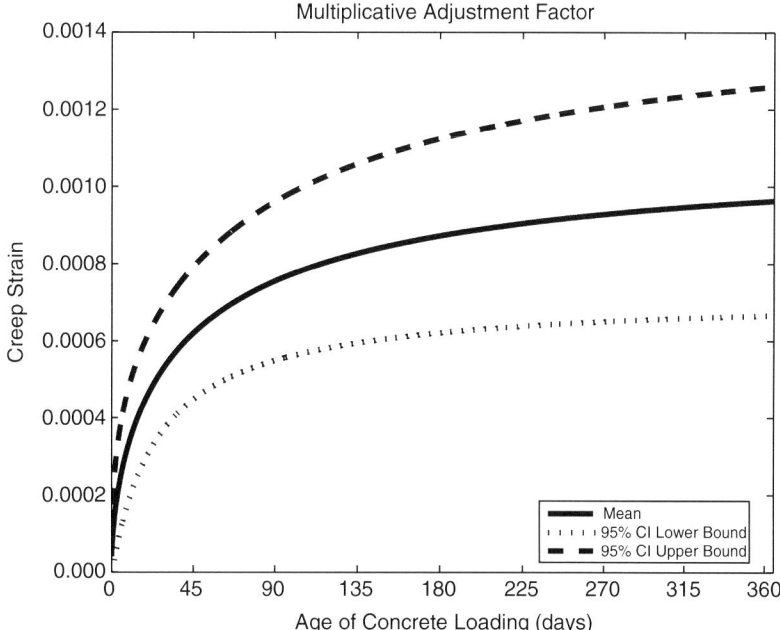

Model-Form Uncertainty Quantification for Structural Design, Fig. 4 Transient response obtained utilizing a multiplicative adjustment factor approach

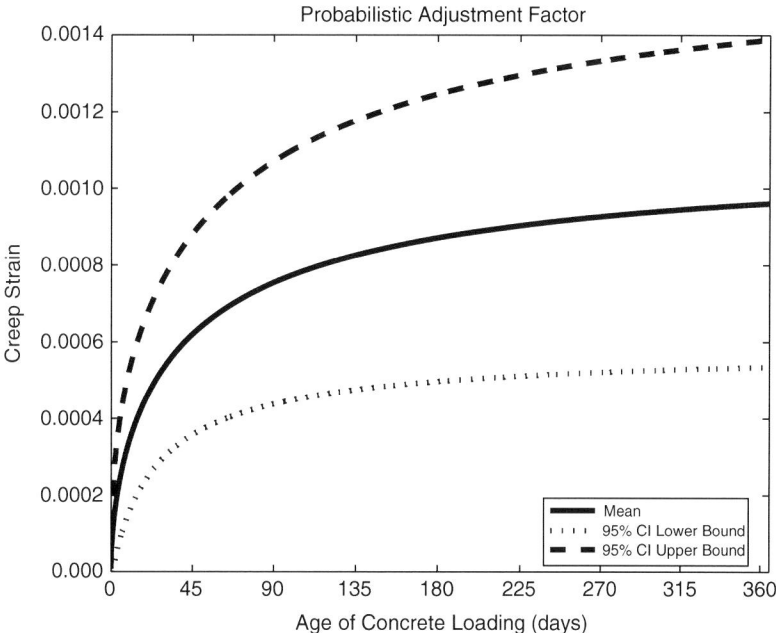

Model-Form Uncertainty Quantification for Structural Design, Fig. 5 Transient response obtained utilizing a probabilistic adjustment factor approach

Utilization of the probabilistic adjustment factor approach to quantify the aforementioned model-form uncertainty is performed in a nearly identical manner as the previous two approaches. Model probabilities defined for the previous two techniques were also applied in this approach. Note that this approach is applicable to probabilistic model predictions, thus allowing for the

Model-Form Uncertainty Quantification for Structural Design, Table 1 Measured creep strains of tested concreted specimen

Age of Concrete after loading	1	7	14	21	28	90	180	270	360
Measured Creep Strain (10^{-6})	175	367	431	490	537	711	802	875	951

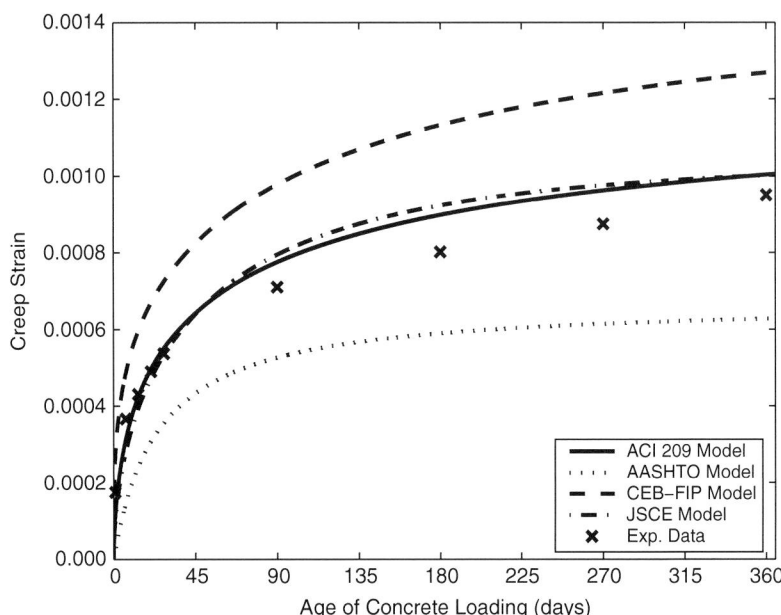

Model-Form Uncertainty Quantification for Structural Design, Fig. 6 Plot of each of the four models used for prediction of creep strain as a function of time with measured creep strains

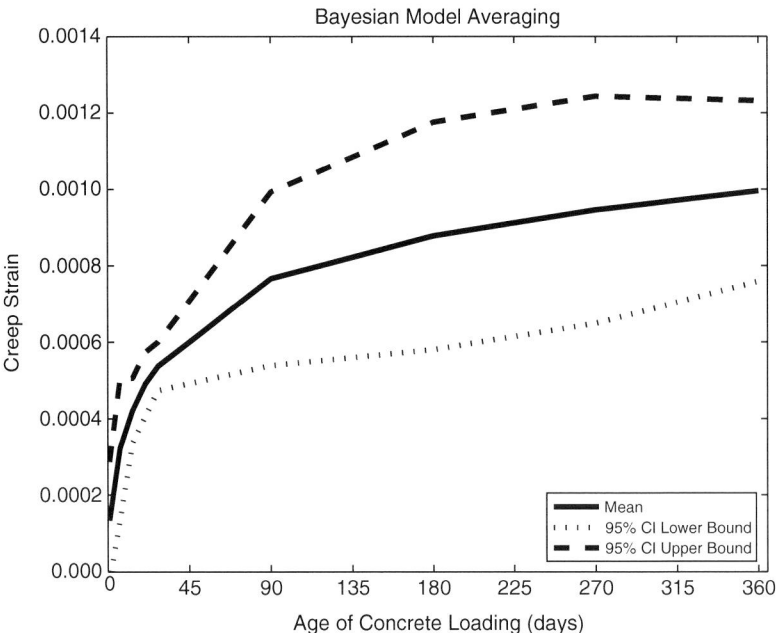

Model-Form Uncertainty Quantification for Structural Design, Fig. 7 Transient response obtained utilizing Bayesian model averaging approach

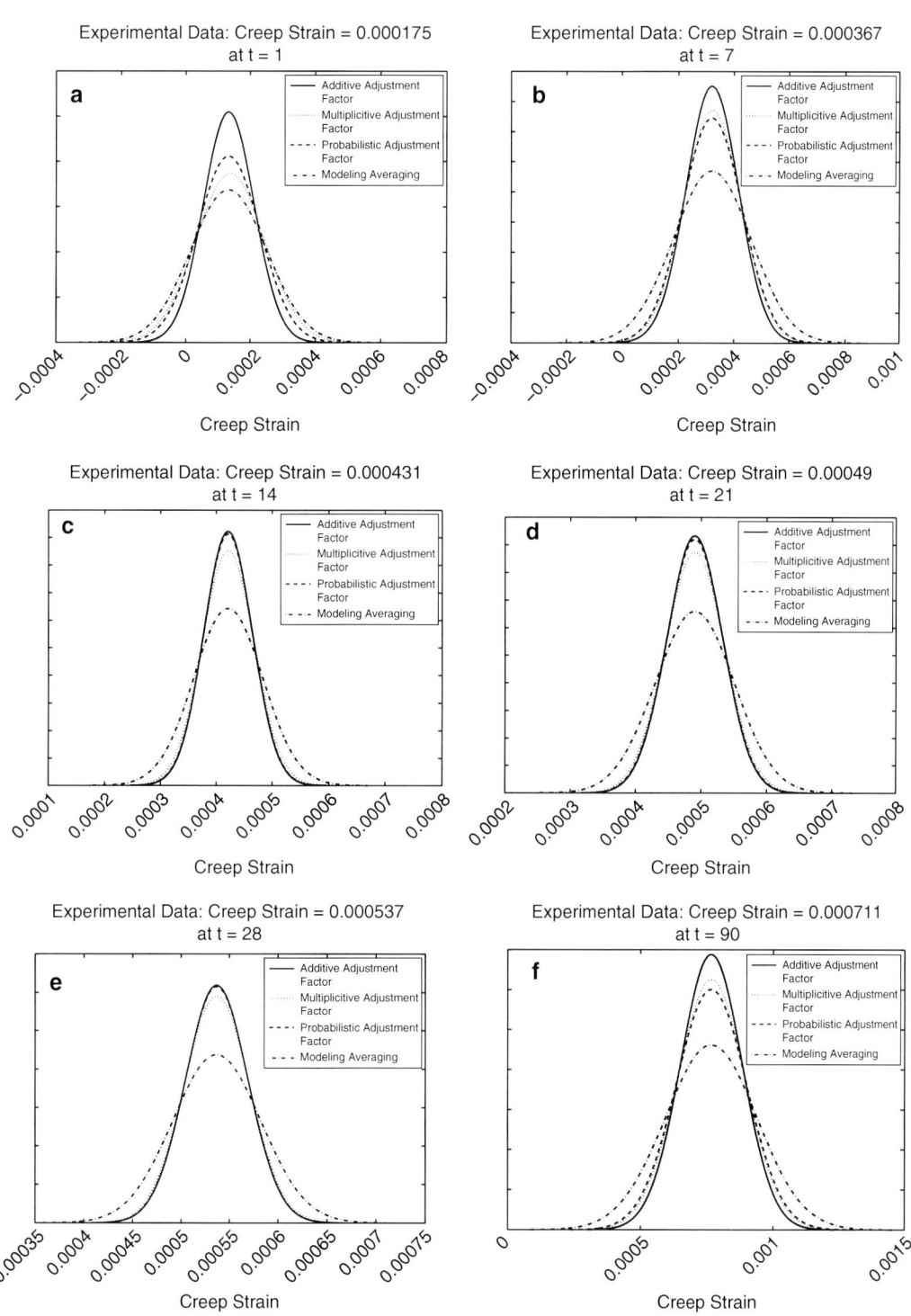

Model-Form Uncertainty Quantification for Structural Design, Fig. 8 (continued)

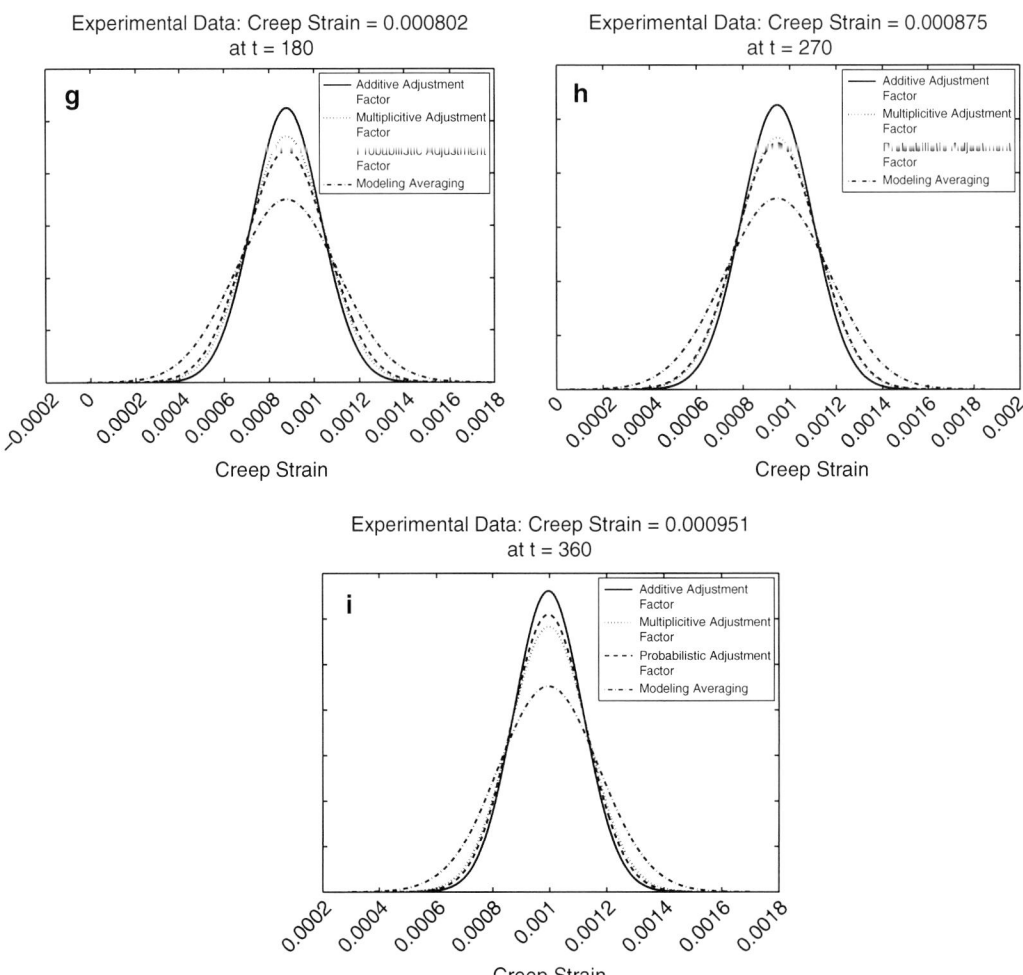

Model-Form Uncertainty Quantification for Structural Design, Fig. 8 Model-form uncertainty quantification results

incorporation of parametric or predictive uncertainties. Here, predictive uncertainties of 15, 19, 20, and 16 % of the model prediction are accounted for in models M_1, M_2, M_3, and M_4 respectively. Application of this technique at each point in the time domain is plotted in Fig. 5. It can be seen that this model-form uncertainty quantification technique yields wider confidence intervals than the previous two approaches due to incorporation of predictive uncertainties.

As mentioned previously, the Bayesian model averaging technique for quantifying model-form uncertainty requires the availability of experimental data for proper application. The data given in Table 1 contains experimental data of measured creep strain to be used in applying the Bayesian model averaging approach. This data is plotted along with the transient responses of Fig. 2 in Fig. 6. Bayes' theorem is applied using this measured data and uniformly distributed prior model probabilities (therefore each model assumes a probability of 0.25). Upon updating prior probabilities using Bayes' theorem and obtaining maximum likelihood estimators, at which point Bayesian model averaging is carried out. The transient response is plotted in Fig. 7. It can be seen that this response is not as seamless as the adjustment factor approaches. This is a result of the fact that this approach can only be implemented at the points that experimental data is available.

Figure 8 shows the normal distribution generated by each model-form uncertainty quantification technique, at each point an experimental data point is available. The results in this figure were all applied on model probabilities updated from uniformly distributed model probabilities using Bayes' theorem. The model variances required for applying the probabilistic adjustment factor approach are assumed to be the maximum likelihood estimators found in performing Bayes' theorem. It can be seen that the additive and multiplicative adjustment factor approaches yield nearly identical results; whereas the probabilistic adjustment factor approach yields slightly different results. The Bayesian model averaging approach on the other hand yields the most conservative distribution. Note that the adjustment factor approaches can be applied under conditions necessary for the BMA approach, as shown here, but the opposite is not possible.

Summary

Model-form uncertainty arises as a result of lack of knowledge or certainty about the models available for predicting a given response. Four methodologies were presented for the quantification of model-form uncertainty in structural design. These methodologies were the additive adjustment factor, multiplicative adjustment factor, probabilistic adjustment factor, and Bayesian model averaging approaches. Limitations on these techniques include a constraint that experimental data points be available for Bayesian model averaging and that model predictions be deterministic for the additive and multiplicative adjustment factor approaches. The probabilistic adjustment factor approach, a novel derivation of the additive adjustment factor approach, can handle stochastic model predictions in contrast to that from which it was derived.

Cross-References

▶ Bayesian Statistics: Applications to Earthquake Engineering
▶ Model Class Selection for Prediction Error Estimation
▶ Physics-Based Ground-Motion Simulation
▶ Probabilistic Seismic Hazard Models
▶ Reliability Estimation and Analysis
▶ Site Response for Seismic Hazard Assessment
▶ Structural Reliability Estimation for Seismic Loading
▶ Uncertainty Quantification in Structural Health Monitoring
▶ Uncertainty Theories: Overview

References

209 AC (1997) Aci 209r-92: prediction of creep, shrinkage, and temperature effects in concrete structures (reapproved 1997). Technical report, American Concrete Institute, Farmington Hills

AASHTO (2007) Aashto lrfd bridge design specifications: customary u.s. units, 3rd edn. Technical report, American Association of State Highway and Transportation Officials (AASHTO), Washington, DC

Allaire D, Willcox K (2010) Surrogate modeling for uncertainty assessment with application to aviation environmental system models. AIAA J 48(8): 1791–1803

CEB (1990) Creep and shrinkage, in ceb-fip model code 1990. Technical report, Comite Euro-International du Beton (CEB), Lausanne

Chernoff H, Moses LE (2012) Elementary decision theory. Courier Dover Publications, New York

Draper D (1995) Assessment and propagation of model uncertainty. J R Stat Soc Ser B (Methodol) 57:45–97

Droguett EL, Mosleh A (2008) Bayesian methodology for model uncertainty using model performance data. Risk Anal 28(5):1457–1476

Hacking I (1984) The emergence of probability: a philosophical study of early ideas about probability, induction and statistical inference. Cambridge University Press, London

Kennedy MC, O'Hagan A (2001) Bayesian calibration of computer models. J R Stat Soc Ser B (Stat Method) 63(3):425–464

Mosleh A, Apostolakis G (1986) The assessment of probability distributions from expert opinions with an application to seismic fragility curves. Risk Anal 6(4):447–461

Park I, Amarchinta HK, Grandhi RV (2010) A Bayesian approach for quantification of model uncertainty. Reliab Eng Syst Saf 95(7):777–785

Park I, Grandhi RV (2011) Quantifying multiple types of uncertainty in physics-based simulation using Bayesian model averaging. AIAA J 49(3): 1038–1045

Paté-Cornell ME (1996) Uncertainties in risk analysis: six levels of treatment. Reliab Eng Syst Saf 54(2): 95–111

Reinert JM, Apostolakis GE (2006) Including model uncertainty in risk-informed decision making. Ann Nucl Energy 33(4):354–369

Riley ME, Grandhi RV (2011a) A method for the quantification of model-form and parametric uncertainties in physics-based simulations. In: 52nd AIAA/ASME/ASCE/AHS/ASC structures, structural dynamics and materials conference, Denver. AIAA

Riley ME, Grandhi RV (2011b) Quantification of model-form and predictive uncertainty for multi-physics simulation. Comput Struct 89(11):1206–1213

Robert C (2007) The Bayesian choice: from decision-theoretic foundations to computational implementation. Springer, New York

Sakata K, Shimomura T (2004) Recent progress in research on and code evaluation of concrete creep and shrink-age in Japan. J Adv Concr Technol 2:133–140

Soares CG (1997) Quantification of model uncertainty in structural reliability. In: Probabilistic methods for structural design. Springer, Norwell, MA, pp 17–37

Zhang R, Mahadevan S (2000) Model uncertainty and Bayesian updating in reliability-based inspection. Struct Saf 22(2):145–160

Zio E, Apostolakis G (1996) Two methods for the structured assessment of model uncertainty by experts in performance assessments of radioactive waste repositories. Reliab Eng Syst Saf 54(2):225–241

Moment Tensors: Decomposition and Visualization

Václav Vavryčuk
Institute of Geophysics, Czech Academy of Sciences, Prague, Czech Republic

Synonyms

Dynamics and mechanics of faulting; Earthquake source observations; Focal mechanism; Seismic anisotropy; Theoretical seismology

Introduction

Elastic waves are generated by forces acting at the source and affected by the response of a medium to these forces. Mathematically, they are expressed using the representation theorem (Aki and Richards 2002, Eq. 2.41):

$$u_i(\mathbf{x},t) = \int_{-\infty}^{\infty} \iiint_V f_k(\boldsymbol{\xi},\tau)\, G_{ik}(\mathbf{x},t;\boldsymbol{\xi},\tau)\, d\tau\, dV(\boldsymbol{\xi}), \tag{1}$$

where $u_i = u_i(\mathbf{x},t)$ is the observed displacement field, $\mathbf{x}$ is the position vector of an observer, t is time, and Green's tensor $G_{ik} = G_{ik}(\mathbf{x},t;\boldsymbol{\xi},\tau)$ is the solution of the equation of motion for a point single force with time dependence of the Dirac delta function. The Green's tensor is defined as the ith component of the displacement produced by a force in the x_k direction and describes propagation effects on waves along a path from the source to a receiver. Vector $f_k = f_k(\boldsymbol{\xi},\tau)$ is the density of the body forces acting at the source being a function of the position vector $\boldsymbol{\xi}$ and time τ at the source. The integration in Eq. 1 is over focal volume V and time τ. For simplicity, an infinite medium is assumed in Eq. 1.

Seismic waves generated at an earthquake source and propagating in the Earth have some specific properties. Firstly, the body forces associated with the earthquake source are not distributed in a volume but along a fault. Secondly, the earthquake source is not represented by single forces but rather by dipole forces. The dipole forces cause the two blocks at opposite sides of the fault to mutually move (Fig. 1a). They are described by moment tensor density $m_{kl} = m_{kl}(\boldsymbol{\xi},\tau)$ defined along fault Σ (Fig. 1b). A substitution of single forces by dipole forces leads to modifying Eq. 1 as follows (Burridge and Knopoff 1964; Aki and Richards 2002, Eq. 3.20):

$$u_i(\mathbf{x},t) = \int_{-\infty}^{\infty} \iint_\Sigma m_{kl}(\boldsymbol{\xi},\tau)\, \frac{\partial}{\partial \xi_l} G_{ik}(\mathbf{x},t;\boldsymbol{\xi},\tau)\, d\tau\, d\Sigma(\boldsymbol{\xi}). \tag{2}$$

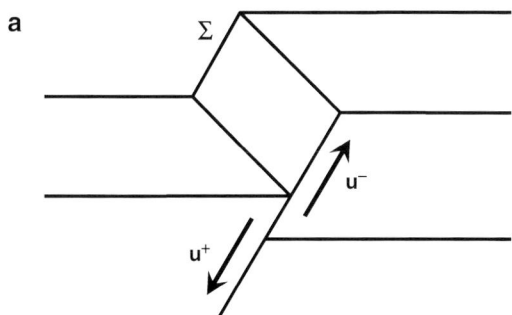

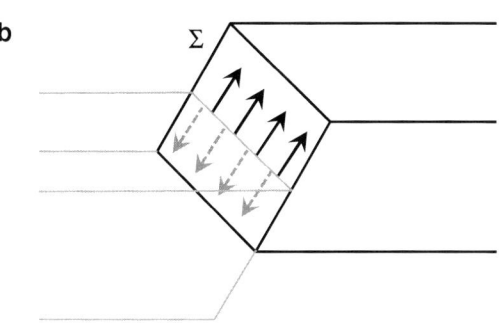

Moment Tensors: Decomposition and Visualization, Fig. 1 (a) Example of motion (*normal faulting*) and (b) distribution of equivalent dipole forces along fault Σ

If the size of the fault is small with respect to distance between the source and the receiver, the representation theorem can be simplified by introducing the point source approximation:

$$u_i(\mathbf{x},t) = \int_{-\infty}^{\infty} M_{kl}(\tau) \frac{\partial}{\partial \xi_l} G_{ik}(\mathbf{x},t;\boldsymbol{\xi},\tau)\, d\tau, \quad (3)$$

or simply

$$u_i = M_{kl} * G_{ik,l}, \quad (4)$$

where $M_{kl} = M_{kl}(t)$ is the seismic moment tensor

$$M_{kl} = \iint_\Sigma m_{kl}\, d\Sigma, \quad (5)$$

and symbol "*" means the time convolution. Integration in Eq. 5 is performed over fault Σ. Moment tensor **M** is a symmetric tensor describing nine couples of equivalent dipole forces which can act at the earthquakes source (Fig. 2).

It is a basic quantity evaluated for earthquakes on all scales from acoustic emissions to large devastating earthquakes (see entries "▶ Long-Period Moment-Tensor Inversion: The Global CMT Project;" "▶ Reliable Moment Tensor Inversion for Regional- to Local-Distance Earthquakes;" and "▶ Regional Moment Tensor Review: An Example from the European–Mediterranean Region").

The most common type of the moment tensor is the double-couple (DC) source which represents the force equivalent of shear faulting on a planar fault in isotropic media. However, many studies reveal that seismic sources often display more general moment tensors with significant non-double-couple (non-DC) components (Julian et al. 1998; Miller et al. 1998; see entry "▶ Non-Double-Couple Earthquakes"). An explosion is an obvious example of a non-DC source, but non-DC components can also be produced by a collapse of a cavity in mines (Rudajev and Šílený 1985), by inflation or deflation of magma chambers in volcanic areas (Mori and McKee 1987), by shear faulting on a nonplanar (curved or irregular) fault, by tensile faulting induced by fluid injection when the slip vector is inclined from the fault and causes its opening (Vavryčuk 2001, 2011), or by shear faulting in anisotropic media (Kawasaki and Tanimoto 1981; Vavryčuk 2005).

Moment Tensor Decomposition

In order to identify which type of seismic source is physically represented by a retrieved moment tensor **M**, the moment tensor is usually diagonalized and decomposed into some elementary parts. The first step is the decomposition into its isotropic (ISO) and deviatoric (DEV) parts:

$$\mathbf{M} = \mathbf{M}_{\text{ISO}} + \mathbf{M}_{\text{DEV}}, \quad (6)$$

where

$$\mathbf{M}_{\text{ISO}} = \frac{1}{3}\text{Tr}(\mathbf{M}) \cdot \mathbf{I}, \quad (7)$$

Moment Tensors: Decomposition and Visualization, Fig. 2 Set of nine couples of equivalent dipole forces forming the moment tensor

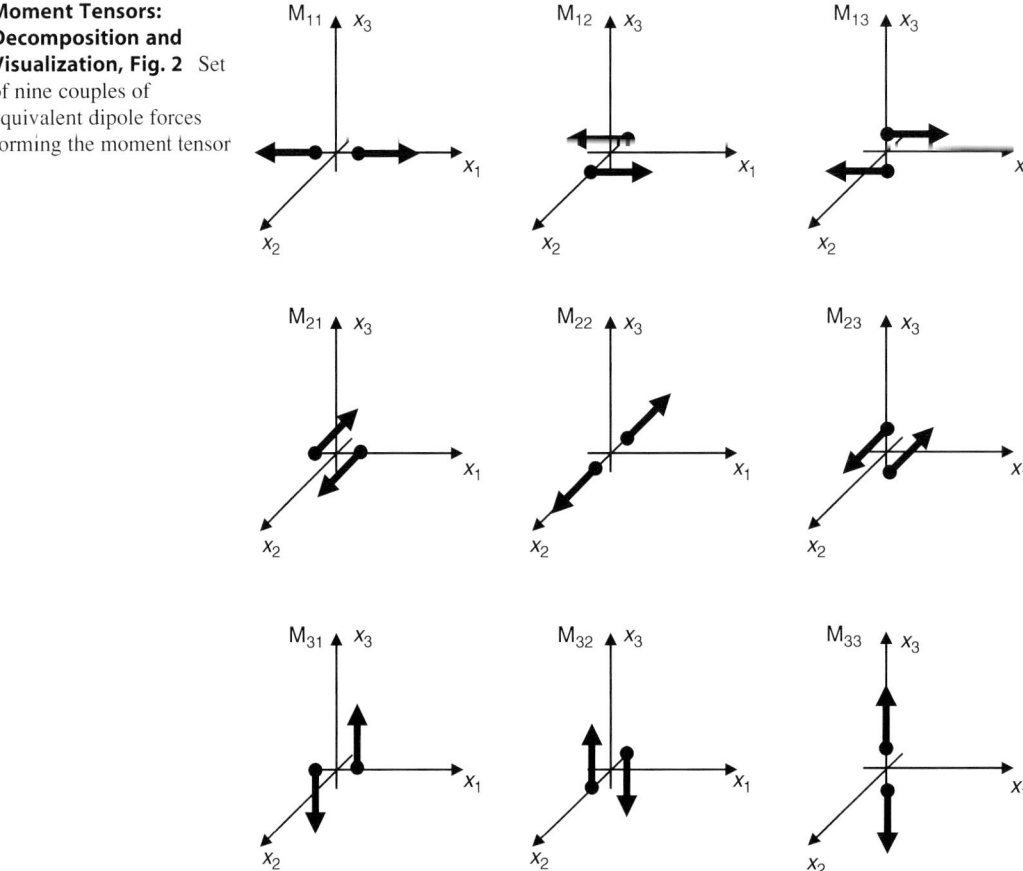

and matrix **I** is the identity matrix. The second step is the decomposition of the deviatoric part of **M**. This step is more ambiguous and can be performed in many alternative ways. The deviatoric part can be decomposed into three double couples (Jost and Herrmann 1989), into major and minor double couples (Kanamori and Given 1981; Wallace 1985), into double couples with the same T axis (Wallace 1985), or into a double couple and a compensated linear vector dipole (CLVD) component (Knopoff and Randall 1970). The last decomposition into the DC and CLVD components proved to be useful for physical interpretations and became widely accepted. This decomposition was further developed and applied by Sipkin (1986), Jost and Herrmann (1989), Kuge and Lay (1994), Vavryčuk (2015), and others, and it will be treated here in detail.

Definition of ISO, CLVD, and DC

The seismic moment tensor **M** can be decomposed using eigenvalues and an orthonormal basis of eigenvectors in the following way:

$$\mathbf{M} = M_1\, \mathbf{e}_1 \otimes \mathbf{e}_1 + M_2\, \mathbf{e}_2 \otimes \mathbf{e}_2 + M_3\, \mathbf{e}_3 \otimes \mathbf{e}_3, \tag{8}$$

where

$$M_1 \geq M_2 \geq M_3, \tag{9}$$

and vectors $\mathbf{e}_1$, $\mathbf{e}_2$, and $\mathbf{e}_3$ define the T (tension), N (intermediate or neutral), and P (pressure) axes, respectively. Symbol "$\otimes$" in Eq. 8 means the dyadic product of two vectors. Two basic properties of the moment tensor are separated in Eq. 8: (1) the orientation of the source defined by three eigenvectors and (2) the type and size of the source defined by three eigenvalues of **M**.

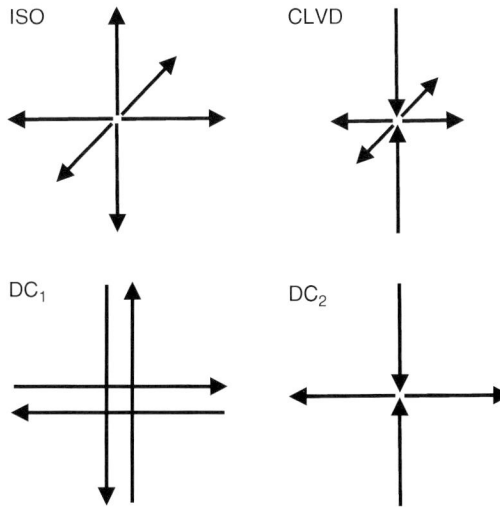

Moment Tensors: Decomposition and Visualization, Fig. 3 The ISO, DC, and CLVD⁻ base tensors of the moment tensor. The DC part is plotted in the original coordinate system associated with the fault (DC$_1$) and after its diagonalization (DC$_2$)

The eigenvalues can be represented as a point in three-dimensional (3-D) space (Riedesel and Jordan 1989):

$$\mathbf{m} = M_1 \hat{\mathbf{e}}_1 + M_2 \hat{\mathbf{e}}_2 + M_3 \hat{\mathbf{e}}_3, \quad (10)$$

where vectors $\hat{\mathbf{e}}_1$, $\hat{\mathbf{e}}_2$, and $\hat{\mathbf{e}}_3$ define the coordinate system in this space. In order to get a unique representation, the eigenvalues must be ordered according to Eq. 9. Consequently, the points representing the source type cannot cover the whole 3-D space but only its wedge called the "source-type space." The choice of the coordinate system and the metric for parameterizing the source-type space differ for individual moment tensor decompositions.

For physical reasons, the three terms in Eq. 8 are further restructured to form isotropic (ISO), double-couple (DC), and compensated linear vector dipole (CLVD) parts (Fig. 3) in the following way (Knopoff and Randall 1970; Dziewonski et al. 1981; Sipkin 1986; Jost and Herrmann 1989):

$$\mathbf{M} = \mathbf{M}_{\text{ISO}} + \mathbf{M}_{\text{DC}} + \mathbf{M}_{\text{CLVD}}$$
$$= M_{\text{ISO}} \mathbf{E}_{\text{ISO}} + M_{\text{DC}} \mathbf{E}_{\text{DC}} + M_{\text{CLVD}} \mathbf{E}_{\text{CLVD}}, \quad (11)$$

where $\mathbf{E}_{\text{ISO}}$, $\mathbf{E}_{\text{DC}}$, and $\mathbf{E}_{\text{CLVD}}$ are the ISO, DC, and CLVD elementary (base) tensors and M_{ISO}, M_{DC}, and M_{CLVD} are the ISO, DC, and CLVD moments. The base tensors read

$$\mathbf{E}_{\text{ISO}} = \begin{bmatrix} 1 & 0 & 0 \\ 0 & 1 & 0 \\ 0 & 0 & 1 \end{bmatrix}, \quad \mathbf{E}_{\text{DC}} = \begin{bmatrix} 1 & 0 & 0 \\ 0 & 0 & 0 \\ 0 & 0 & -1 \end{bmatrix},$$

$$\mathbf{E}_{\text{CLVD}}^+ = \frac{1}{2} \begin{bmatrix} 2 & 0 & 0 \\ 0 & -1 & 0 \\ 0 & 0 & -1 \end{bmatrix}, \quad \mathbf{E}_{\text{CLVD}}^- = \frac{1}{2} \begin{bmatrix} 1 & 0 & 0 \\ 0 & 1 & 0 \\ 0 & 0 & -2 \end{bmatrix}, \quad (12)$$

where $\mathbf{E}_{\text{CLVD}}^+$ or $\mathbf{E}_{\text{CLVD}}^-$ is used if $M_1 + M_3 - 2M_2 \geq 0$ or $M_1 + M_3 - 2M_2 < 0$, respectively. Hence, the CLVD tensor is aligned along the axis with the largest magnitude deviatoric eigenvalue. The eigenvalues of the base tensors are ordered according to Eq. 9 in order to lie in the source-type space. The norm of all base tensors, calculated as the largest magnitude eigenvalue (i.e., the maximum of $|M_i|$, $i = 1,2,3$), is equal to 1. This condition is called the unit "spectral norm" and physically means that the maximum dipole force of the base tensors is unity (Fig. 3). Alternative alignments of the CLVD tensor and other normalizations of the base tensors are also admissible (Chapman and Leaney 2012; Tape and Tape 2012) but have less straightforward physical interpretations.

Equations 11 and 12 uniquely define values M_{ISO}, M_{DC}, and M_{CLVD} expressed as follows:

$$M_{\text{ISO}} = \frac{1}{3}(M_1 + M_2 + M_3), \quad (13)$$

$$M_{\text{CLVD}} = \frac{2}{3}(M_1 + M_3 - 2M_2), \quad (14)$$

$$M_{\text{DC}} = \frac{1}{2}(M_1 - M_3 - |M_1 + M_3 - 2M_2|), \quad (15)$$

where M_{CLVD} includes also the sign of the elementary CLVD tensor. If the elementary CLVD tensor $\mathbf{E}_{\text{CLVD}}$ is considered with its sign as in Eq. 11, then M_{CLVD} should be calculated as the absolute value of Eq. 14. Values M_{ISO}, M_{DC}, and

M_{CLVD} in Eqs. 13, 14, and 15 are usually further normalized and expressed using scalar seismic moment M and relative scale factors C_{ISO}, C_{DC}, and C_{CLVD}:

$$\begin{bmatrix} C_{\text{ISO}} \\ C_{\text{CLVD}} \\ C_{\text{DC}} \end{bmatrix} = \frac{1}{M} \begin{bmatrix} M_{\text{ISO}} \\ M_{\text{CLVD}} \\ M_{\text{DC}} \end{bmatrix}, \quad (16)$$

where M reads

$$M = |M_{\text{ISO}}| + |M_{\text{CLVD}}| + M_{\text{DC}}, \quad (17)$$

or equivalently (Bowers and Hudson 1999)

$$M = ||\mathbf{M}_{\text{ISO}}|| + ||\mathbf{M}_{\text{DEV}}||, \quad (18)$$

where $||\mathbf{M}_{\text{ISO}}||$ and $||\mathbf{M}_{\text{DEV}}||$ are the spectral norms of the isotropic and deviatoric parts of moment tensor $\mathbf{M}$, respectively. Scale factors C_{ISO}, C_{DC}, and C_{CLVD} satisfy the following equation:

$$|C_{\text{ISO}}| + |C_{\text{CLVD}}| + C_{\text{DC}} = 1. \quad (19)$$

Equations 13, 14, 15, 16, and 17 imply that C_{DC} is always positive and in the range from 0 to 1; C_{CLVD} and C_{ISO} are in the range from -1 to 1. Consequently, the decomposition of $\mathbf{M}$ can be expressed as

$$\mathbf{M} = M(C_{\text{ISO}} \mathbf{E}_{\text{ISO}} + C_{\text{DC}} \mathbf{E}_{\text{DC}} + |C_{\text{CLVD}}| \mathbf{E}_{\text{CLVD}}), \quad (20)$$

where M is the norm of $\mathbf{M}$ calculated using Eq. 17 and represents a scalar seismic moment for a general seismic source. The absolute value of the CLVD term in Eq. 20 is used because the sign of CLVD is included in the elementary tensor $\mathbf{E}_{\text{CLVD}}$.

Physical Properties of the Decomposition

The above decomposition of the moment tensor is performed in order to physically interpret a set of nine dipole forces representing a general point seismic source and to identify easily some basic types of the source in isotropic media:

- The explosion/implosion is an isotropic source, and thus, it is characterized by $C_{\text{ISO}} = \pm 1$ and by zero C_{CLVD} and C_{DC}.
- Shear faulting is represented by the double-couple force and characterized by $C_{\text{DC}} = 1$ and by zero C_{ISO} and C_{CLVD}.
- Pure tensile or compressive faulting is free of shearing and thus characterized by zero C_{DC}. However, the non-DC components contain both ISO and CLVD. The ISO and CLVD components are of the same sign: they are positive for tensile faulting but negative for compressive faulting,
- Shear-tensile (dislocation) source defined as the source, which combines both shear and tensile faulting (Vavryčuk 2001, 2011), is characterized by nonzero ISO, CLVD, and DC components. The positive values of C_{ISO} and C_{CLVD} correspond to tensile mechanisms when fault is opening during rupturing. The negative values of C_{ISO} and C_{CLVD} correspond to compressive mechanisms when a fault is closing during rupturing. The ratio between non-DC and DC components defines the angle between the slip and the fault.
- Shear faulting on a nonplanar fault is characterized generally by a nonzero C_{DC} and C_{CLVD}. The C_{ISO} is zero, because no volumetric changes are associated with this type of source.

Source-Type Plots

For physical interpretations, it is advantageous to visualize the retrieved moment tensors graphically. Double-couple components of moment tensors are displayed using the well-known "beach balls" which show orientations of the fault together with the slip vector defining the shear motion along the fault (see entry "▶ Earthquake Mechanism Description and Inversion"). The non-double-couple components of moment tensors are displayed in the so-called source-type plots.

All moment tensors fill a source-type space which is a wedge in the full 3-D space. The magnitude of the vector in this space is the scalar

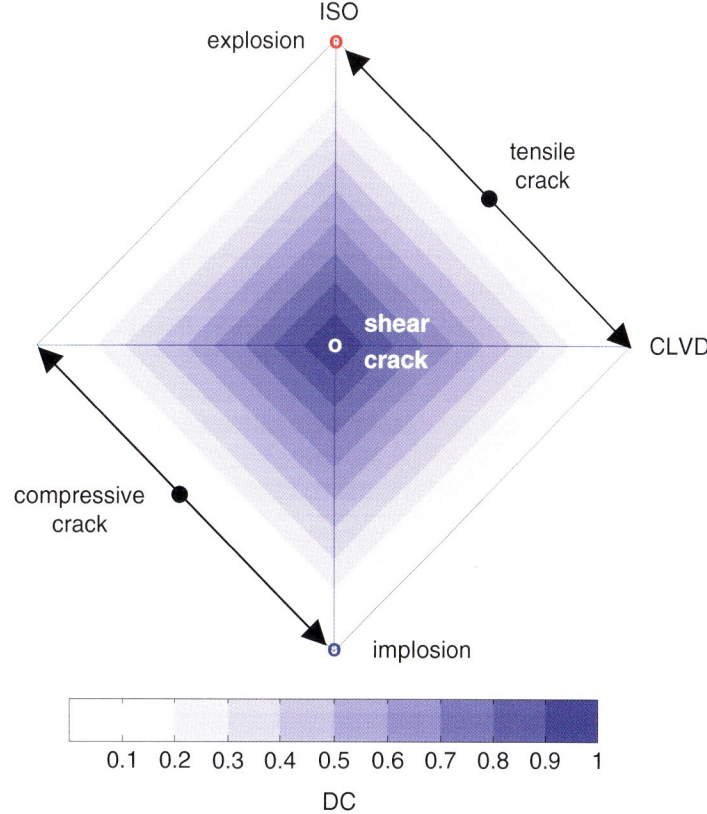

Moment Tensors: Decomposition and Visualization, Fig. 4 Diamond CLVD-ISO source-type plot with positions of basic types of seismic sources. The *arrows* indicate the range of possible positions of moment tensors for pure tensile or compressive cracks

moment, and its direction defines the type of the source. In order to identify the type of the source visually, it is convenient to plot all unit vectors of the source-type space in a 2-D figure using some projections. Here, three basic plots are mentioned: diamond CLVD-ISO plot (Vavryčuk 2015), Hudson's skewed diamond plot (Hudson et al. 1989), and the Riedesel-Jordan lune plot (Riedesel and Jordan 1989).

Diamond CLVD-ISO Plot

The diamond CLVD-ISO plot shows the position of the source in the CLVD-ISO coordinate system in which the DC component is represented by the color intensity (Fig. 4). Since the sum of absolute values of the CLVD and ISO cannot exceed 1, moment tensors must lie inside a diamond. A source with pure or predominant shear faulting is located at the origin of coordinates or close to it. An explosion or implosion source is located at the top or bottom vertex of the diamond, respectively. Motion on a pure tensile or compressive crack is plotted at the margin of the diamond. Points along the CLVD axis correspond to faulting on nonplanar faults, and points in the first and third quadrants of the diamond correspond to shear-tensile sources.

For pure tensile and shear-tensile sources, the ISO/CLVD ratio depends on the elastic properties of the medium surrounding the source. In isotropic media, this ratio is (Vavryčuk 2001, 2011)

$$\frac{C_{ISO}}{C_{CLVD}} = \frac{3}{4}\left(\frac{v_P}{v_S}\right)^2 - 1. \qquad (21)$$

Hence, the point representing the pure tensile faulting in Fig. 4 (black dot) can be close to $C_{ISO} = 1$ (corresponding to an explosion) for high values of v_P/v_S but also close to $C_{CLVD} = 1$ for low values of v_P/v_S. The limiting cases are

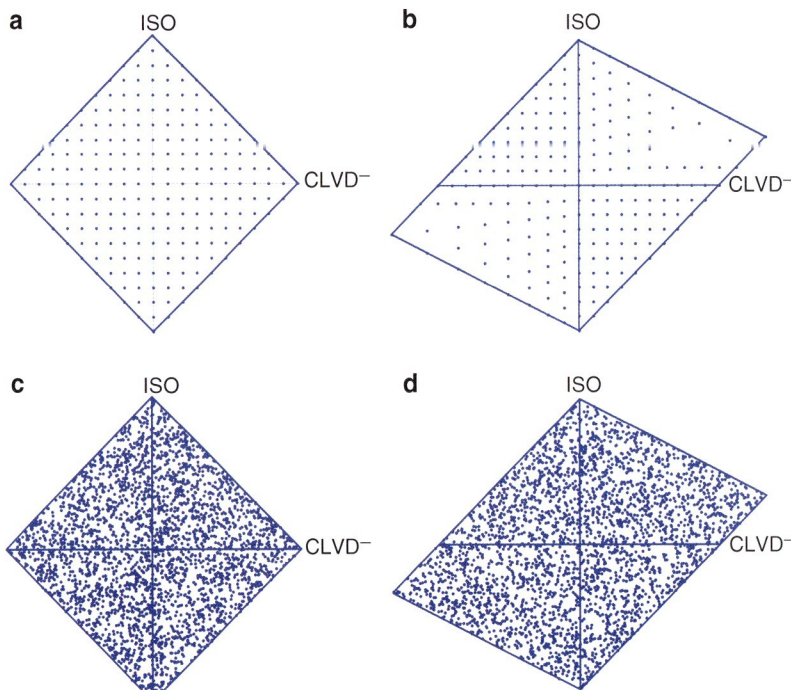

Moment Tensors: Decomposition and Visualization, Fig. 5 The Hudson's diamond τ-k plot (a, c) and the skewed diamond u-v plot (b, d). The CLVD$^-$ means the reversed CLVD axis. The *dots* in (a, b) show a regular grid in C_{ISO} and C_{CLVD} from -1 to $+1$ with step of 0.1. The *dots* in (c, d) show 3,000 sources defined by moment tensors with randomly generated eigenvalues. For scaling of the eigenvalues, see the text. Plots (c) and (d) indicate that the distribution of random sources is uniform in both projections

$$\frac{v_P}{v_S} \to \infty \quad \text{and} \quad \frac{v_P}{v_S} = \frac{2}{\sqrt{3}}, \quad (22)$$

describing fluids and the lower limit of stable solids ($\lambda = -2/3\,\mu$), respectively. Similar conclusions can be drawn for pure compressive faulting (see Fig. 4).

Note that the abovementioned basic types of sources cannot be located in the second or fourth quadrants of the diamond source-type plot in Fig. 4. Moment tensors located in this area indicate numerical errors of the moment tensor inversion, a more complicated source model or faulting in anisotropic media.

As mentioned above, values $C_{ISO} = \pm 1$ and $C_{DC} = 1$ correspond to an explosion/implosion and to shear faulting in isotropic media, respectively. Their physical meaning is thus straightforward. However, the moment tensor with $C_{CLVD} = 1$ does not correspond to any simple seismic source, and the presence of CLVD in moment tensors often causes confusions and poses questions whether it is necessary to introduce the CLVD. The decomposition described above indicates that the CLVD component is required to render the decomposition mathematically complete, and the CLVD component cannot thus be avoided. Although, it has no physical meaning itself, it can be interpreted physically in combination with the ISO component as a product of tensile faulting. In the case of a pure tensile crack, the CLVD component's major dipole is aligned with the normal to the crack surface and the volume change associated with the opening crack is described by the ISO component.

Hudson's Skewed Diamond Plot

Hudson et al. (1989) introduced two source-type plots: a diamond τ-k plot which is the diamond CLVD-ISO plot described in the previous section but with the opposite direction of the CLVD axis (Fig. 5a) and a skewed diamond u-v plot (Fig. 5b). The latter plot is introduced in order to conserve the uniform probability of moment tensor eigenvalues. If eigenvalues M_1, M_2, and M_3 have a uniform probability distribution between -1 and $+1$ and satisfy the ordering condition (9), then all points fill uniformly the skewed diamond plot. Axis u defines the deviatoric sources and axis v connects the pure explosive and implosive sources.

The moment tensor with arbitrary (but ordered) eigenvalues M_1, M_2, and M_3 is projected into the u-v plot using the following equations:

$$u = -\frac{2}{3M}(M_1 + M_3 - 2M_2),$$
$$v = \frac{1}{3M}(M_1 + M_2 + M_3), \quad (23)$$

where M is the scalar seismic moment calculated as the spectral norm of complete moment tensor $\mathbf{M}$

$$M = \max(|M_1|, |M_2|, |M_3|). \quad (24)$$

Equation 23 is similar to Eqs. 13 and 14 in the ISO-CLVD-DC decomposition except for scaling.

Figure 5a, b show mapping of a regular grid in C_{ISO} and C_{CLVD} calculated using Eqs. 13, 14, 15, 16, and 17 into the diamond CLVD-ISO plot and into the skewed diamond u-v plot. Figure 5b shows that the CLVD-ISO grid is deformed in the first and third quadrants of the u-v plot. Figure 5c, d demonstrate that sources with randomly generated eigenvalues cover uniformly the source-type plots. The uniform probability distribution function (PDF) is produced by the Hudson's skewed diamond plot (Fig. 5d) but also for the diamond CLVD-ISO plot. In this respect, the Hudson's skewed diamond plot does not provide any particular advantage compared to the standard CLVD-ISO plot (for details, see Tape and Tape 2012; Vavryčuk 2015).

Riedesel-Jordan Plot

A completely different approach is suggested by Riedesel and Jordan (1989) who introduce a compact plot displaying both the orientation and type of source on the focal sphere. Apparently, this plot looks simple and mathematically elegant but introduces difficulties. The moment tensor is represented by a vector defined in Eq. 10, and the coordinate axes $\hat{\mathbf{e}}_1$, $\hat{\mathbf{e}}_2$, and $\hat{\mathbf{e}}_3$ are identified with the T, N, and P axes of $\mathbf{M}$: $\mathbf{e}_1$, $\mathbf{e}_2$, and $\mathbf{e}_3$ defined in Eq. 8. The vector is normalized using the Euclidean norm

$$M = \sqrt{\frac{1}{2}(M_1^2 + M_2^2 + M_3^2)} \quad (25)$$

and projected on the sphere using a lower-hemisphere equal-area projection (see Fig. 6a).

Chapman and Leaney (2012) pointed out, however, that this representation is not optimum for several reasons. Firstly, vector $\mathbf{m}$ cannot lie everywhere on the focal sphere but inside its small part called the "lune" (Tape and Tape 2012). The lune covers only one sixth of the whole sphere (see Fig. 6b). Secondly, vectors characterizing positive and negative isotropic sources (explosion and implosion) are physically quite different, but they are displayed in the same area on the focal sphere in this projection. Thirdly, analysis of uncertainties of a moment tensor solution by plotting a cluster of vectors $\mathbf{m}$ includes both effects – uncertainties in the orientation and in the source type. This is fine if the moment tensor is nondegenerate. However, difficulties arise when the moment tensor is degenerate or nearly degenerate, because small perturbations cause significant changes of eigenvectors.

Some of the mentioned difficulties can be avoided by fixing the eigenvectors and analyzing the size of clusters produced by a varying source type only. If we fix the eigenvectors in the form

$$\mathbf{e}_1 = \left(\frac{1}{\sqrt{3}}, \frac{1}{\sqrt{6}}, \frac{1}{\sqrt{2}}\right)^T,$$
$$\mathbf{e}_2 = \left(\frac{1}{\sqrt{3}}, -\frac{2}{\sqrt{6}}, 0\right)^T, \quad (26)$$
$$\mathbf{e}_3 = \left(\frac{1}{\sqrt{3}}, \frac{1}{\sqrt{6}}, -\frac{1}{\sqrt{2}}\right)^T,$$

in the north-east-down coordinate system, we obtain a plot shown in Fig. 6b. This plot resembles the diamond CLVD-ISO plot (Fig. 4) but adapted to a spherical metric. The basic source types are characterized by the following unit vectors:

$$\mathbf{e}_{\text{ISO}} = \frac{1}{\sqrt{3}}(\mathbf{e}_1 + \mathbf{e}_2 + \mathbf{e}_3) = (1, 0, 0)^T, \quad (27)$$

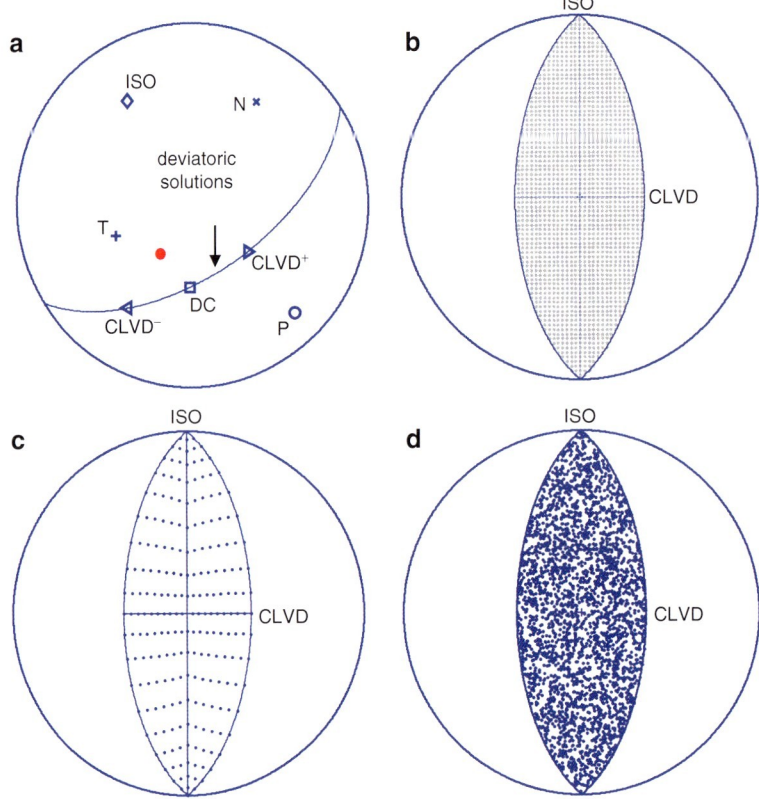

Moment Tensors: Decomposition and Visualization, Fig. 6 Riedesel-Jordan source-type plot. (**a**) The original compact plot proposed by Riedesel and Jordan (1989) displaying the orientation of the moment tensor eigenvectors (*P*, *T*, and *N* axes), basic source types (ISO, CLVD, DC), and the position of the studied moment tensor (*red dot*). (**b, c, d**) A modified Riedesel-Jordan plot proposed by Chapman and Leaney (2012). The *dashed area* in (**b**) shows the area of admissible positions of sources. The *dots* in (**c**) show a regular grid in C_{ISO} and C_{CLVD} from −1 to +1 with step of 0.1. The *dots* in (**d**) show 3,000 sources defined by moment tensors with randomly generated eigenvalues. Plot (**d**) indicates that the distribution of random sources is uniform in this projection

$$\mathbf{e}_{DC} = \frac{1}{\sqrt{2}}(\mathbf{e}_1 - \mathbf{e}_3) = (0, 0, 1)^T, \quad (28)$$

$$\mathbf{e}_{CLVD}^+ = \sqrt{\frac{2}{3}}\left(\mathbf{e}_1 - \frac{1}{2}\mathbf{e}_2 - \frac{1}{2}\mathbf{e}_3\right)$$

$$= \left(0, \frac{1}{2}, \frac{\sqrt{3}}{2}\right)^T, \quad (29)$$

$$\mathbf{e}_{CLVD}^- = \sqrt{\frac{2}{3}}\left(\frac{1}{2}\mathbf{e}_1 + \frac{1}{2}\mathbf{e}_2 - \mathbf{e}_3\right)$$

$$= \left(0, -\frac{1}{2}, \frac{\sqrt{3}}{2}\right)^T. \quad (30)$$

Basic properties of the Riedesel-Jordan projection are exemplified in Fig. 6c, d. Figure 6c shows mapping of a regular grid in C_{ISO} and C_{CLVD} calculated using Eqs. 13, 14, 15, 16, and 17, and Fig. 6d indicates that the PDF of sources with randomly distributed eigenvalues M_1, M_2, and M_3 is uniform. For a detailed analysis on the probability of eigenvalues in the spherical projection, see Tape and Tape (2012).

Analysis of Moment Tensor Uncertainties Using Source-Type Plots

The source-type plots are often used for assessing uncertainties of the ISO, CLVD, and DC components of moment tensors. The reason for using the source-type plots for assessing the errors is

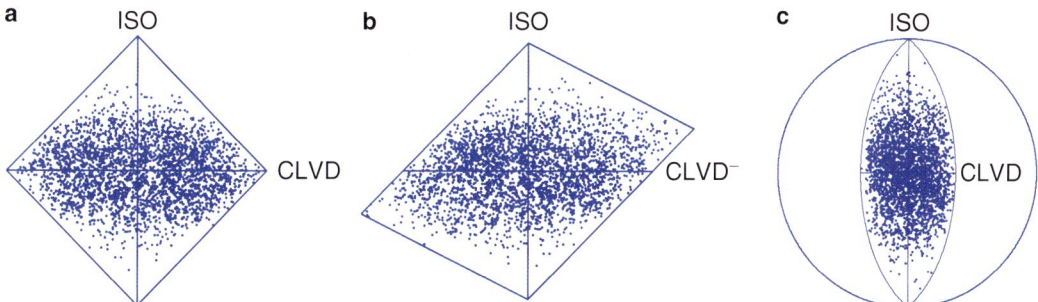

Moment Tensors: Decomposition and Visualization, Fig. 7 Distribution of random sources displayed in three different source-type plots. (**a**) The diamond CLVD-ISO plot, (**b**) the Hudson's skewed diamond plot, and (**c**) the Riedesel-Jordan plot. The *dots* show 3,000 sources defined by moment tensors with randomly generated components in the interval from −1 to +1. The distribution of sources is nonuniform for all three source-type plots

simple. The moment tensor is usually plotted as a cluster of acceptable solutions, and the size of the cluster reflects uncertainties of the solution. Such approach is, however, simplistic and rough because the same uncertainties produce differently large clusters in dependence of the position of the cluster. Although the source-type plots display a uniform PDF for randomly generated eigenvalues (see section "Source-Type Plots"), the behavior of moment tensor uncertainties is not simple. When uncertainties of moment tensor components are analyzed, the moment tensor is not in the diagonal form. After diagonalizing the moment tensor, the errors are projected into the errors of eigenvalues in a rather complicated way. This is demonstrated in Fig. 7. Moment tensors in this figure have all components random and distributed with a uniform probability in the interval from −1 to 1. Nevertheless, some source types are quite rare. In particular, sources with a high explosive or implosive component are almost missing. This observation is common for all source-type plots.

More realistic sources are modeled in Fig. 8: the pure DC and ISO sources defined by tensors $\mathbf{E}_{DC}$ and $\mathbf{E}_{ISO}$ from Eq. 12 are contaminated by random noise with a uniform distribution in the interval from −0.25 to 0.25. The noise is superimposed to all tensor components and 1,000 random moment tensors are generated. As expected, the randomly generated source tensors form clusters, but their shape is different for different projections and their size depends also on the type of the source. For the DC source (Fig. 8, left-hand plots), the maximum PDF is in the center of the cluster which coincides with the position of the uncontaminated source. In the diamond CLVD-ISO plot and in the skewed diamond plot, the cluster is asymmetric being stretched along the CLVD axis. A more symmetric shape of the cluster is produced in the Riedesel-Jordan plot. However, the symmetry of the cluster is apparent because the CLVD and ISO axes are of different lengths. A significantly higher scatter of the CLVD components compared to the ISO components in moment tensor inversions has been observed and discussed also in Vavryčuk (2011). For the pure ISO source (Fig. 8, right-hand plots), the clusters are smaller than for the DC source, and the maximum PDF is out of the position of the uncontaminated source. This means that the ISO percentage is systematically underestimated due to errors of the inversion for highly explosive or implosive sources.

Source Tensor Decomposition

A simple classification of sources based on the moment tensor decomposition is possible in isotropic media only. In anisotropic media, the problem is more complicated. The moment tensor is affected not only by the geometry of faulting but also by the elastic properties of the focal zone. Depending on these properties, the moment tensors can take a general form with nonzero DC,

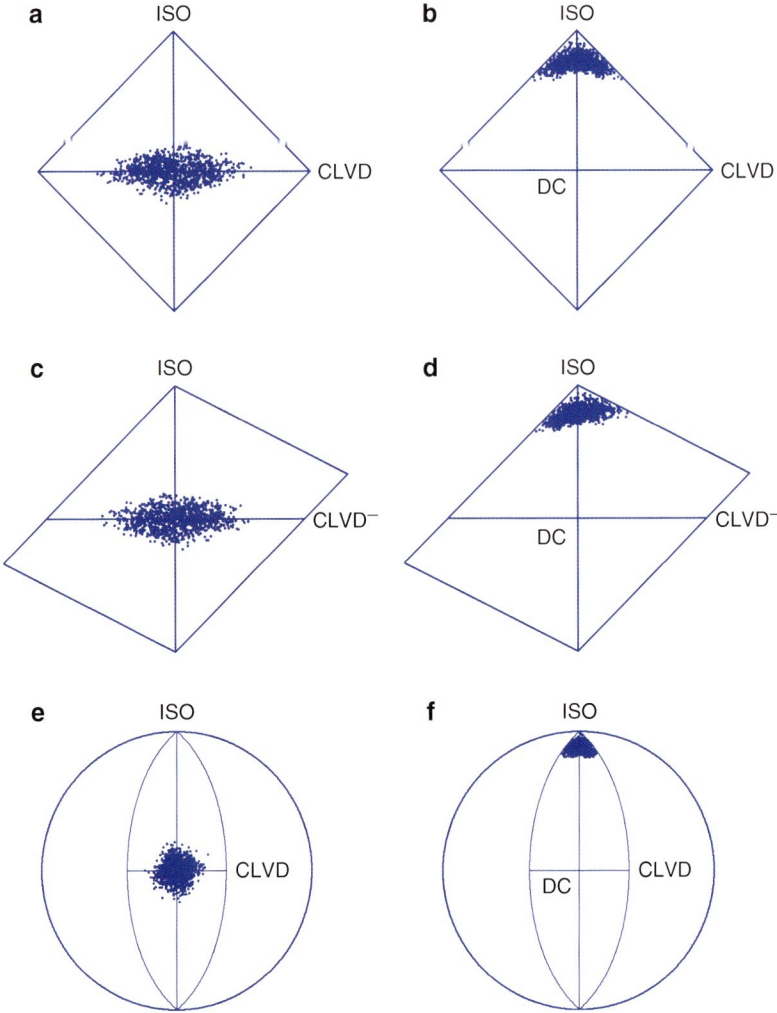

Moment Tensors: Decomposition and Visualization, Fig. 8 Distribution of pure DC (*left-hand plots*) and pure explosive (*right-hand plots*) sources contaminated by random noise and displayed in three different source-type plots. (**a, b**) The diamond CLVD-ISO plot, (**c, d**) the Hudson's skewed diamond plot, and (**e, f**) the Riedesel-Jordan plot. The *dots* show 1,000 DC sources defined by the elementary tensor $\mathbf{E}_{DC}$ (see Eq. (12)) and contaminated by noise with a uniform distribution from −0.25 to +0.25. The noise is superimposed to all tensor components

CLVD, and ISO components even for simple shear faulting on a planar fault (Vavryčuk 2005). For this reason, physical interpretations of shear or tensile dislocation sources in anisotropic media should be based on the decomposition of the source tensor, which is directly related to geometry of faulting.

The source tensor **D** (also called the potency tensor) is a symmetric dyadic tensor defined as (Ben-Zion 2003; Vavryčuk 2005)

$$D_{kl} = \frac{uS}{2}(s_k n_l + s_l n_k), \quad (31)$$

where vectors **n** and **s** denote the fault normal and the direction of the slip vector, respectively, u is the slip and S is the fault size. The relation between the source and moment tensors reads in anisotropic media (Vavryčuk 2005, his Eq. 4)

$$M_{ij} = c_{ijkl} D_{kl}, \quad (32)$$

and in isotropic media

$$M_{ij} = \lambda D_{kk} \delta_{ij} + 2\mu D_{ij}, \quad (33)$$

where c_{ijkl} is the tensor of elastic parameters and λ and μ are the Lamé's parameters. While the moment and source tensors diagonalize in anisotropic media in different systems of eigenvectors and thus their relation is complicated, the eigenvectors of the moment and source tensors are the

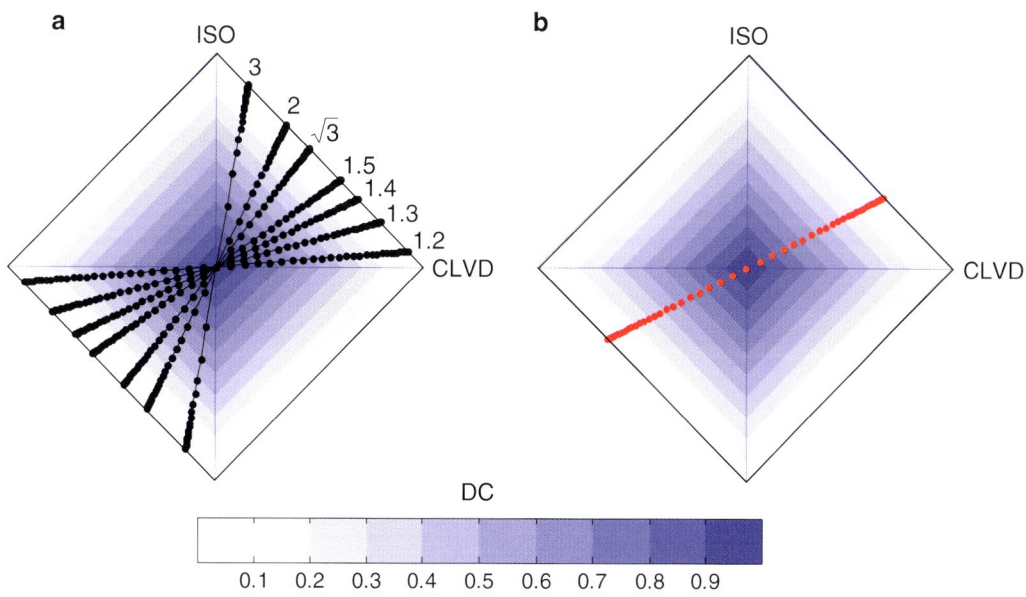

Moment Tensors: Decomposition and Visualization, Fig. 9 Diamond source-type plots for the shear-tensile source model in an isotropic medium characterized by various values of the v_P/v_S ratios (the values are indicated in the plot). *Red dots*, source tensors; *black dots*, moment tensors. The *dots* correspond to the individual sources. The slope angle (i.e., the deviation of the slip vector from the fault) ranges from $-90°$ (pure compressive crack) to $+90°$ (pure tensile crack) in steps of $3°$

same in isotropic media and their decomposition according to formulas in section "Definition of ISO, CLVD, and DC" yields similar results.

Properties of the moment and source tensor decompositions for shear and tensile sources in isotropic and anisotropic media are illustrated in Figs. 9 and 10. Figure 9 shows the source-type plots for tensile sources with a variable slope angle (i.e., the deviation of the slip vector from the fault) situated in an isotropic medium. The plot shows that the ISO and CLVD components are linearly dependent for both moment and source tensors. For the moment tensors, the line direction depends on the v_P/v_S ratio (Fig. 9a). For the source tensors, the line is independent of the properties of the elastic medium, and the C_{ISO}/C_{CLVD} ratio is always 1/2 (Fig. 9b). The differences between the behavior of the source and moment tensors are even more visible in anisotropic media. Figure 10 indicates that the ISO and CLVD components of the moment tensors of shear faulting (Fig. 10a, black dots) or tensile faulting (Fig. 10b, black dots) may behave in a complicated way. For example, shear faulting in anisotropic media can produce strongly non-DC moment tensors (Vavryčuk 2005). This prevents a straightforward interpretation of moment tensors in terms of the physical faulting parameters. Therefore, first, the source tensors must be calculated from moment tensors and then interpreted (Fig. 10, red dots). If elastic properties of the medium in the focal zone needed for calculating the source tensors are not known, they can be inverted from the non-DC components of the moment tensors (Vavryčuk 2004, 2011; Vavryčuk et al. 2008). Note that the retrieved medium parameters do not refer to local material properties of the fault, but to the medium surrounding the fault.

Summary

The moment tensor represents equivalent body forces of a seismic source. The forces described by the moment tensor are not the actual forces

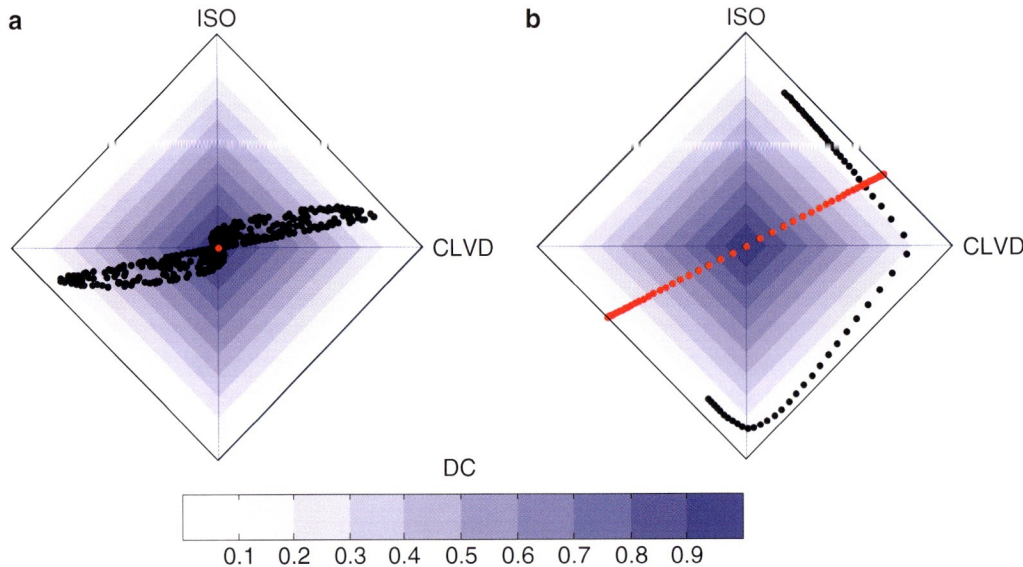

Moment Tensors: Decomposition and Visualization, Fig. 10 Diamond source-type plots for the shear (**a**) and tensile (**b**) source models in an anisotropic medium. The *black dots* in (**a**) correspond to 500 moment tensors of shear sources with randomly oriented fault and slip. The *black dots* in (**b**) correspond to moment tensors of tensile sources with strike = 0°, dip = 20°, and rake = −90° (normal faulting). The slope angle ranges from −90° (pure compressive crack) to +90° (pure tensile crack) in steps of 3°. The *red dots* in (**a**) and (**b**) show the corresponding source tensors. The medium is transversely isotropic with the following elastic parameters (in 10^9 kg m^{-1} s^{-2}): $c_{11} = 58.81$, $c_{33} = 27.23$, $c_{44} = 13.23$, $c_{66} = 23.54$, and $c_{13} = 23.64$. The medium density is 2,500 kg/m^3. The parameters are taken from Vernik and Liu (1997) and describe the Bazhenov shale (depth of 12.507 ft)

acting at the source because the moment tensor description assumes elastic behavior of the medium and ignores nonlinear rheology at the focal area. Nevertheless, the moment tensor proved to be a useful quantity and became widely accepted in seismological practice for studying seismic sources. The moment tensor is evaluated for earthquakes on all scales from acoustic emissions to large devastating earthquakes.

In order to understand physical processes at the earthquake source, the moment tensor is commonly decomposed into double-couple (DC), isotropic (ISO), and compensated linear vector dipole (CLVD) components. High percentage of DC indicates a source with shear faulting in isotropic media, and high percentage of ISO indicates an explosive or implosive source. A combination of positive (negative) ISO and CLVD is produced by tensile (compressive) faulting. The type of the source can be visualized using the so-called source-type plots; among them, the diamond CLVD-ISO plot, the Hudson's skewed diamond plot, and the Riedesel-Jordan lune plot are in common use. In anisotropic media, the physical interpretation of the DC, ISO, and CLVD percentages is not straightforward, and the decomposition of the moment tensor must be substituted by that of the source tensor.

Cross-References

▶ Earthquake Mechanism Description and Inversion
▶ Long-Period Moment-Tensor Inversion: The Global CMT Project
▶ Non-Double-Couple Earthquakes
▶ Reliable Moment Tensor Inversion for Regional- to Local-Distance Earthquakes
▶ Regional Moment Tensor Review: An Example from the European–Mediterranean Region

References

Aki K, Richards PG (2002) Quantitative seismology. University Science, Sausalito

Ben-Zion Y (2003) Appendix 2: key formulas in earthquake seismology. In: Lee WHK, Kanamori H, Jennings PC, Kisslinger C (eds) International handbook of earthquake and engineering seismology, Part B. Academic, London, p 1857

Bowers D, Hudson JA (1999) Defining the scalar moment of a seismic source with a general moment tensor. Bull Seismol Soc Am 89(5):1390–1394

Burridge R, Knopoff L (1964) Body force equivalents for seismic dislocations. Bull Seismol Soc Am 54:1875–1888

Chapman CH, Leaney WS (2012) A new moment-tensor decomposition for seismic events in anisotropic media. Geophys J Int 188:343–370

Dziewonski AM, Chou T-A, Woodhouse JH (1981) Determination of earthquake source parameters from waveform data for studies of global and regional seismicity. J Geophys Res 86:2825–2852

Hudson JA, Pearce RG, Rogers RM (1989) Source type plot for inversion of the moment tensor. J Geophys Res 94:765–774

Jost ML, Herrmann RB (1989) A student's guide to and review of moment tensors. Seismol Res Lett 60:37–57

Julian BR, Miller AD, Foulger GR (1998) Non-double-couple earthquakes 1. Theory Rev Geophy 36:525–549

Kanamori H, Given JW (1981) Use of long-period surface waves for rapid determination of earthquake-source parameters. Phys Earth Planet In 27:8–31

Kawasaki I, Tanimoto T (1981) Radiation patterns of body waves due to the seismic dislocation occurring in an anisotropic source medium. Bull Seismol Soc Am 71:37–50

Knopoff L, Randall MJ (1970) The compensated linear-vector dipole: a possible mechanism for deep earthquakes. J Geophys Res 75:4957–4963

Kuge K, Lay T (1994) Data-dependent non-double-couple components of shallow earthquake source mechanisms: effects of waveform inversion instability. Geophys Res Lett 21:9–12

Miller AD, Foulger GR, Julian BR (1998) Non-double-couple earthquakes 2. Observations. Rev Geophys 36:551–568

Mori J, McKee C (1987) Outward-dipping ring-fault structure at Rabaul caldera as shown by earthquake locations. Science 235:193–195

Riedesel MA, Jordan TH (1989) Display and assessment of seismic moment tensors. Bull Seismol Soc Am 79:85–100

Rudajev V, Šílený J (1985) Seismic events with non-shear components. II. Rockbursts with implosive source component. Pure Appl Geophys 123:17–25

Sipkin SA (1986) Interpretation of non-double-couple earthquake mechanisms derived from moment tensor inversion. J Geophys Res 91:531–547

Tape W, Tape C (2012) A geometric comparison of source-type plots for moment tensors. Geophys J Int 190:499–510

Vavryčuk V (2001) Inversion for parameters of tensile earthquakes. J Geophys Res 106(B8):16.339–16.355. doi:10.1029/2001JB000372

Vavryčuk V (2004) Inversion for anisotropy from non-double-couple components of moment tensors. J Geophys Res 109:B07306. doi:10.1029/2003JB002926

Vavryčuk V (2005) Focal mechanisms in anisotropic media. Geophys J Int 161:334–346. doi:10.1111/j.1365-246X.2005.02585.x

Vavryčuk V (2011) Tensile earthquakes: theory, modeling, and inversion. J Geophys Res 116:B12320. doi:10.1029/2011JB008770

Vavryčuk V (2015) Moment tensor decompositions revisited. J Seismol 19:231–252. doi:10.1007/s10950-014-9463-y

Vavryčuk V, Bohnhoff M, Jechumtálová Z, Kolář P, Šílený J (2008) Non-double-couple mechanisms of micro-earthquakes induced during the 2000 injection experiment at the KTB site, Germany: a result of tensile faulting or anisotropy of a rock? Tectonophysics 456:74–93. doi:10.1016/j.tecto.2007.08.019

Vernik L, Liu X (1997) Velocity anisotropy in shales: a petrological study. Geophysics 62:521–532

Wallace TC (1985) A reexamination of the moment tensor solutions of the 1980 Mammoth Lakes earthquakes. J Geophys Res 90:11,171–11,176

Noise-Based Seismic Imaging and Monitoring of Volcanoes

Florent Brenguier
Institut des Sciences de la Terre, University of Grenoble, Grenoble, France

Synonyms

Eruption forecasting; Piton de la Fournaise Volcano; Seismic imaging; Seismic noise; Volcano monitoring

Introduction

Volcanic eruptions represent a great hazard to population leaving nearby volcanoes. In order to better understand volcanic activity and thus to improve eruption forecasting, it is of primary importance to better image and monitor the structure of volcanic edifices.

In the field of ultrasonics, it has been shown both theoretically and experimentally that a random wavefield has correlations which, on average, take the form of the Green's function of the media (Weaver and Lobkis 2001). In seismology, recent studies also showed that the Green's function between pairs of seismographs can be extracted from cross-correlations of coda waves and ambient noise (Campillo 2006). Moreover, some authors used correlations of long ambient seismic noise sequences from regional networks in order to extract and to invert Rayleigh waves to produce high-resolution seismic images of the shallow crustal layers under these networks (Shapiro et al. 2005). The first part of this entry will show how noise-based seismic imaging can be employed to image volcano interiors.

The early detection of volcanic unrest mainly relies on the monitoring of volcanic seismicity and ground deformation. Those methods provide insights into the dynamics of magma pressurization and transport. However, despite considerable effort, the precise forecasting of eruptions and their intensity has proven to be difficult. Therefore, there is a constant need for novel observational methods to obtain information about ongoing volcanic processes. Pressurized volcanic fluids (magma, water) or gases induce deformation and thus perturbations of the elastic properties of volcanic edifices. These small perturbations can be detected as changes of seismic wave properties using repetitive seismic sources (Ratdomopurbo and Poupinet 1995; Grêt et al. 2005; Wegler et al. 2006). However, none of these approaches were apt to provide a continuous monitoring of volcano elastic properties. In a pioneering work, Sens-Schoenfelder and Wegler (2006) proposed to use the repetitive waveforms of seismic noise cross-correlations to track for subsurface volcanic edifice velocity changes. In this manner, the continuous recording of ambient seismic noise allows continuous monitoring of volcano interiors. The second part of this paper will show how

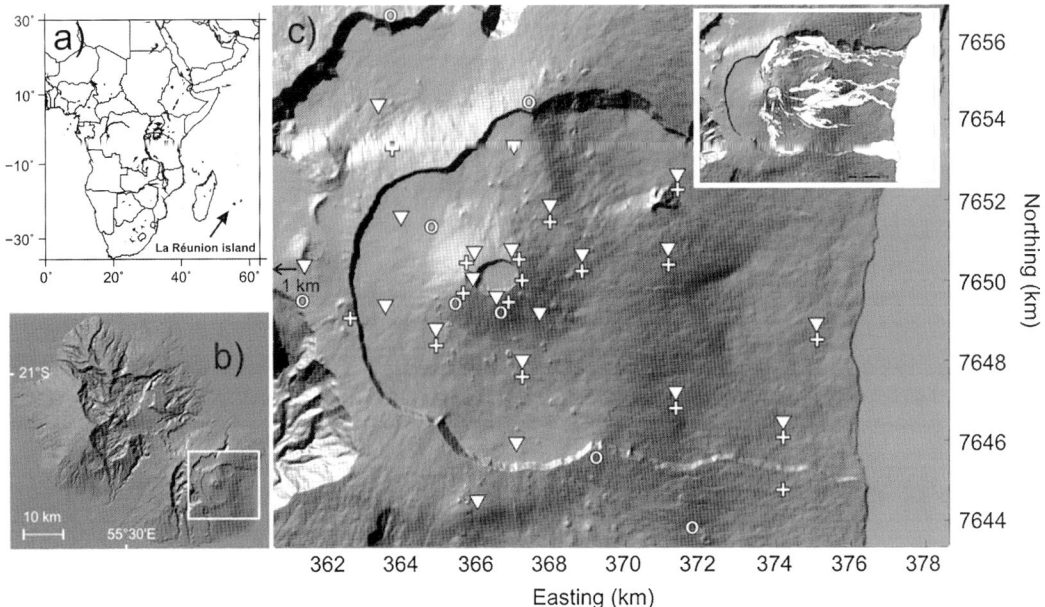

Noise-Based Seismic Imaging and Monitoring of Volcanoes, Fig. 1 Location of (**a**) La Réunion island, (**b**) Piton de la Fournaise Volcano, and (**c**) the UnderVolc and observatory seismic and GPS networks. UnderVolc broadband seismic stations are shown as inverted triangles, GPS as crosses, and observatory short-period seismic stations as circles. The *inset* map shows lava flows from 2000 to 2010 (Courtesy of OVPF, T. Staudacher, Z. Servadio)

noise-based seismic monitoring can be employed to monitor volcano interiors.

Noise-Based Seismic Imaging of Piton de la Fournaise Volcano

Piton de la Fournaise Volcano (PdF) is a hot spot, shield volcano located on La Réunion island in the Indian Ocean (Fig. 1). It erupted more than 30 times between 2000 and 2010. These eruptions lasted from a few hours to a few months and were associated with the emission of mainly basaltic lava with volume ranging from less than one to tens of million cubic meters (Peltier et al. 2009). The time period that is considered here (1999–2011) started and ended with two major eruptions, namely, the March 1998 eruption (60 million of cubic meter of lava emitted) and the April 2007 eruption associated to the 300 m high collapse of the main Dolomieu crater (130 million of cubic meter of lava emitted) (Staudacher et al. 2009).

Piton de la Fournaise Volcano is monitored by OVPF/IPGP volcano observatory. In 2009, in the framework of UnderVolc project (Brenguier et al. 2012), 15 new broadband seismic stations complemented the observatory seismic network that was composed of about 20 seismic stations. The intense eruptive activity together with a weak tectonic activity makes Piton de la Fournaise Volcano well suited for studies focused on the processes of magma pressurization and injection and for the development of innovative imaging and monitoring methods.

By using continuous records of ambient seismic noise from the PdF Volcano observatory short-period seismic network, Brenguier et al. (2007) were able to measure Rayleigh wave velocity dispersion curves obtained from the cross-correlations of 18 months of ambient seismic noise. The tomography of these measurements as well as the depth inversion led to a high-resolution image of the shallow structure of PdF Volcano (Fig. 2). The results show a high seismic velocity body located in the central part of the

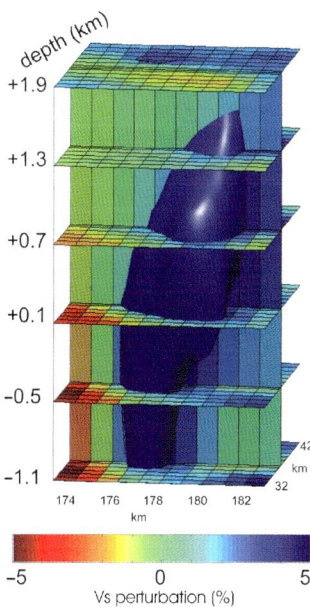

Noise-Based Seismic Imaging and Monitoring of Volcanoes, Fig. 2 3D tomographic image of the shallow central part of Piton de la Fournaise Volcano using cross-correlations of ambient seismic noise (Brenguier et al. 2007)

volcano. This body is being interpreted as a preferential zone of magma injection within the edifice. The high velocities are associated with intrusive magma slowly cooling within the surrounding effusive low velocity material. This information is crucial in order to define a more precise structural model for the volcano edifice and thus to better model and then understand the response of the volcano to magmatic activity.

Noise-Based Seismic Monitoring of Piton de la Fournaise Volcano

The approach consists in measuring very small waveform time delays in the coda of noise cross-correlations. This method is described as the so-called Moving Window Cross Spectrum (Ratdomopurbo and Poupinet 1995; Clarke et al. 2011) or Coda Wave Interferometry (Snieder et al. 2002) techniques. Coda waves (late part of seismograms) are scattered waves that travel long distances and thus accumulate time delays as a consequence of a seismic velocity change in the propagating medium. Measuring travel time perturbations in the coda thus allows detecting very small velocity changes that would not be detectable by a classical measure of first arrival time delays. The drawback of that approach is that it is difficult to estimate the travel paths of the scattered waves constituting the coda. However, recent promising results suggest it may be possible to produce refined 3D maps of small changes in a near future (Larose et al. 2010).

In a previous work Brenguier et al. (2012) measured seismic velocity variations within the Piton de la Fournaise edifice from June 2010 to April 2011 following Brenguier et al. (2008). Continuous seismic velocity changes as well as the daily seismicity rate are shown on Figs. 3 and 4 for two different time periods at PdF volcano. Figure 3 shows small (0.1 %) seismic velocity drops preceding eruptions at PdF volcano. On Fig. 4, the authors observe a velocity drop starting about 1 month before the 2010 October 14th eruption. The velocity decrease correlates in time with an increase of seismicity and has been interpreted as the deformation of the edifice induced by magma pressure buildup and injection. These velocity drops are interpreted as being associated with the deformation of the edifice induced by magma pressure buildup. Interestingly, seismic velocity increases during the eruption of October 2010 and after the short eruption of December 2010 indicating a possible mechanism of stress relaxation induced by the emptying of the magmatic reservoir. Furthermore, to improve the accuracy and thus time resolution of velocity measurements, Baig et al. (2009) developed a cross-correlation filtering method based on time-frequency transforms and phase coherence filtering.

Noise-based seismic velocity monitoring is thus a unique method that allows to precisely monitor volcanic activity. It must however be mentioned that other phenomena perturb seismic velocities of the rock mass such as the presence of water in the medium, temperature and barometric atmospheric pressure changes, and solid and oceanic tides. It is thus important

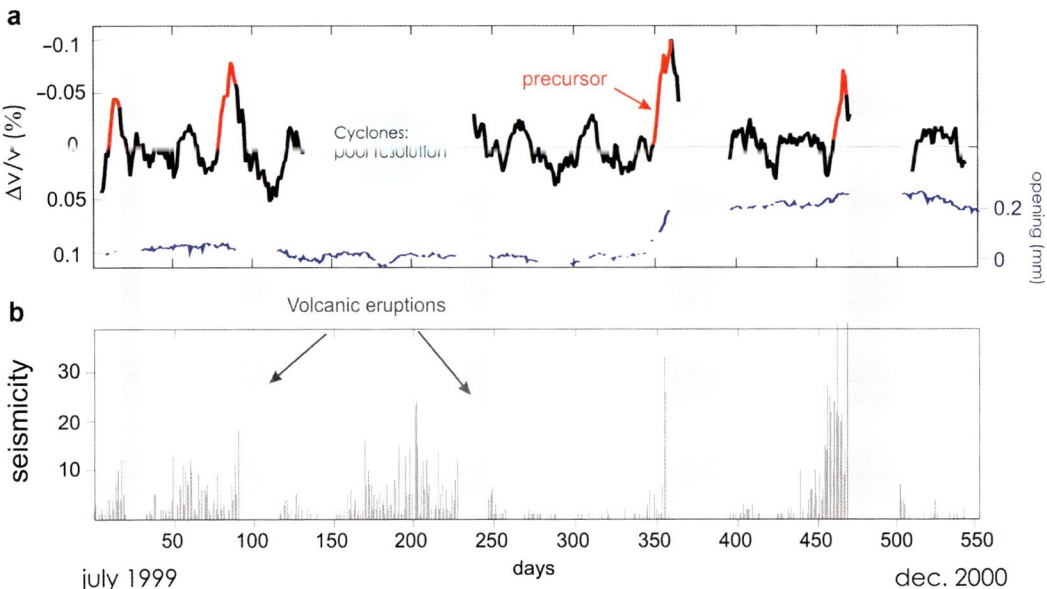

Noise-Based Seismic Imaging and Monitoring of Volcanoes, Fig. 3 (a) Relative velocity changes compared to extensometer (FORX) located on PdF volcano. The travel time shifts are measured in the frequency range 0.1–0.9 Hz. For details, see Brenguier et al. (2008) (b) Inter-eruptive seismicity (pre-eruptive swarms are excluded)

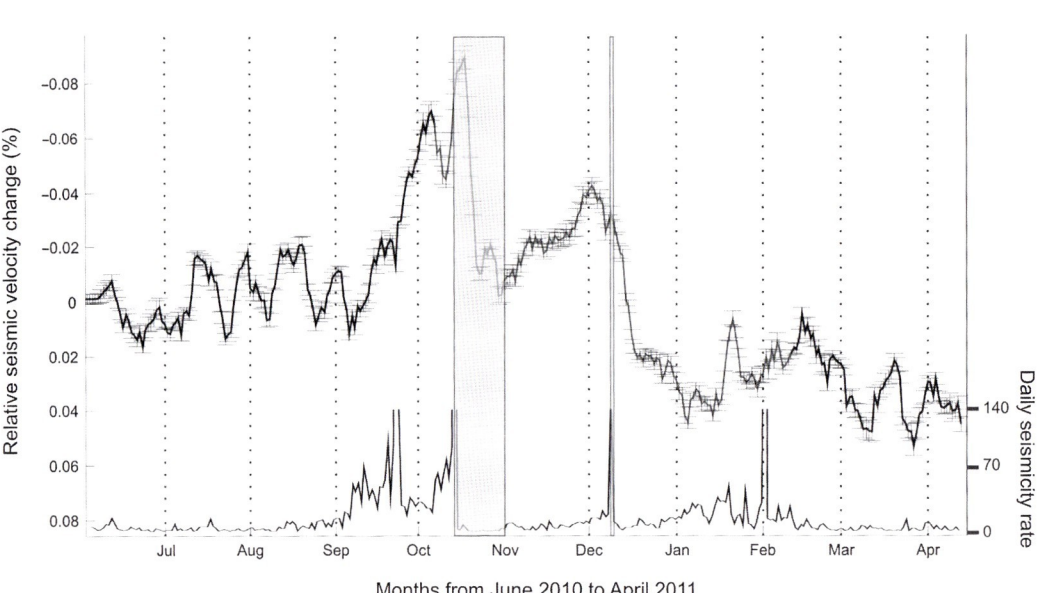

Noise-Based Seismic Imaging and Monitoring of Volcanoes, Fig. 4 Relative seismic velocity changes with error bars and daily seismicity rate between June 2010 and April 2011. Eruption periods are shown as gray rectangles

to correctly model these effects in order to correct them from the observed seismic velocity changes in order to extract the volcanic-related signal.

In April 2007, a major eruption occurred at Piton de la Fournaise Volcano ejecting more than 250 million cubic meters of lava and located a few kilometers east of the main active central part of the volcano. Clarke et al. (2013) observed an unusual high seismic velocity drop associated with this volcanic episode that could not be explained by the pre-eruptive edifice inflation due to magma pressure buildup as described by (Brenguier et al. 2008). Also, (Clarke et al. 2013) showed surface displacements images obtained from InSAR data inversion by J-L. Froger (OPGC, France). These results show that between March and May 2007, a widespread volcanic edifice flank movement occurred with a maximum of 1.4 m of eastward displacement. However, the timing of this movement and, in particular, its relation with the occurrence of the April 2007 eruption could not be identified by a lack of temporal resolution of the InSAR images. Clarke et al. (2013) proved using high temporal resolution seismic velocity change measurements that the strong drop of seismic velocities started at the time of a small eruption preceding the main April eruption by a few days. Inferring the association between this high seismic velocity drop and the widespread volcanic flank movement, the authors concluded that the volcanic flank movement started at the time of the small eruption preceding the main April 2007 eruption by a few days. Considering new simulation of volcanic edifice deformation from Got et al. (2013), it is likely that the small eruption preceding the main one triggered a large elastoplastic deformation of the edifice flank that released stresses and favored the horizontal migration of magma to a long distance through the April 2007 eruptive fissure. This study showed how noise-based monitoring has been used to explain the origin of the unusual April 2007 eruption at Piton de la Fournaise Volcano and that this method can also be used to detect volcanic flank movements and possible instabilities on volcanoes.

This method is now used routinely at the Piton de la Fournaise Observatory in order to improve eruption forecasting together with earthquake and deformation observations. Lecocq et al. (2013) developed an integrated package that includes all processing steps required to obtain continuous seismic velocity changes from continuous seismic records. This package is available for free at www.msnoise.org.

Summary

Ambient seismic noise continuously travels along the surface and the interiors of the Earth. These seismic waves thus carry crucial information about the medium they propagate through. When crossing volcanoes, these waves are affected by the mechanical heterogeneities of the volcanic edifice and are also perturbed during unrest periods by temporal changes of volcano interiors. Noise-based seismic imaging and monitoring allows imaging the structure and temporal evolution of volcanic edifices. In particular, it proved success in detecting very small changes of seismic velocities of volcanic edifices (0.1 %) that have been shown, on Piton de la Fournaise Volcano (La Réunion island), to be precursors of volcanic eruptions.

Cross-References

▶ Seismic Monitoring of Volcanoes
▶ Seismic Noise
▶ Seismic Tomography of Volcanoes
▶ Tracking Changes in Volcanic Systems with Seismic Interferometry
▶ Volcanic Eruptions, Real-Time Forecasting of

References

Baig A, Campillo M, Brenguier F (2009) Denoising seismic noise cross correlations. J Geophys Res Solid Earth (1978–2012), 114(B8)
Brenguier F, Shapiro N, Campillo M, Nercessian A, Ferrazzini V (2007) 3-D surface wave tomography of

the Piton de la Fournaise volcano using seismic noise correlations. Geophys Res Lett 34(2):2305

Brenguier F, Shapiro N, Campillo M, Ferrazzini V, Duputel Z, Coutant O, Nercessian A (2008) Towards forecasting volcanic eruptions using seismic noise. Nat Geosci 1(2):126–130

Brenguier F, Kowalski P, Staudacher T, Ferrazzini V, Lauret F, Boissier P, Di Muro A (2012) First results from the UnderVolc high resolution seismic and GPS network deployed on Piton de la Fournaise Volcano. Seismological Research Letters 83(1):97–102

Campillo M (2006) Phase and correlation in random seismic fields and the reconstruction of the green function. Pure Appl Geophys 163(2):475–502

Clarke D, Zaccarelli L, Shapiro NM, Brenguier F (2011) Assessment of resolution and accuracy of the Moving Window Cross Spectral technique for monitoring crustal temporal variations using ambient seismic noise. Geophys J Int 186(2):867–882

Clarke D, Brenguier F, Froger J-L, Shapiro N, Peltier A, Staudacher T (2013) Timing of a large volcanic flank movement at Piton de la Fournaise volcano using noise-based seismic monitoring and ground deformation measurements. Geophys J Int 195(2): 1132–1140

Got JL, Peltier A, Staudacher T, Kowalski P, Boissier P (2013) Edifice strength and magma transfer modulation at Piton de la Fournaise volcano. J Geophys Res Solid Earth 118:1–18. doi:10.1002/jgrb.50350

Grêt A, Snieder R, Aster R, Kyle P (2005) Monitoring rapid temporal changes in a volcano with coda wave interferometry. Geophys Res Lett 32:1–4

Larose E, Planes T, Rossetto V, Margerin L (2010) Locating a small change in a multiple scattering environment. Appl Phys Lett 96:204101

Lecocq T, Caudron C, Brenguier F (2013) MSNoise, a Python package for monitoring seismic velocity changes using ambient seismic noise. Seismol Res Lett 85(3), 715–726.

Peltier A, Bachèlery P, Staudacher T (2009) Magma transport and storage at Piton de La Fournaise (La Réunion) between 1972 and 2007: a review of geophysical and geochemical data. J Volcanol Geotherm Res 184(1–2):93–108

Ratdomopurbo A, Poupinet G (1995) Monitoring a temporal change of seismic velocity in a volcano: application to the 1992 eruption of Mt. Merapi (Indonesia). Geophys Res Lett 22(7):775–778

Sens-Schoenfelder C, Wegler U (2006) Passive image interferometry and seasonal variations of seismic velocities at Merapi Volcano. Indones Geophys Res Lett 33:1–5

Shapiro N, Campillo M, Stehly L, Ritzwoller M (2005) High-resolution surface-wave tomography from ambient seismic noise. Science 307(5715):1615

Snieder R, Grêt A, Douma H, Scales J (2002) Coda wave interferometry for estimating nonlinear behavior in seismic velocity. Science 295(5563):2253

Staudacher T, Ferrazzini V, Peltier A, Kowalski P, Boissier P, Catherine P, Lauret F, Massin F (2009) The April 2007 eruption and the Dolomieu crater collapse, two major events at Piton de la Fournaise (La Réunion Island, Indian Ocean). J Volcanol Geotherm Res 184(1–2):126–137

Weaver RL, Lobkis OI (2001) Ultrasonics without a source: thermal fluctuation correlations at MHz frequencies. Phys Rev Lett 87(13). doi:10.1103/PhysRevLett.87.134301

Wegler U, Lühr B, Snieder R, Ratdomopurbo A (2006) Increase of shear wave velocity before the 1998 eruption of Merapi volcano (Indonesia). Geophys Res Lett 33:1–4

Non-Double-Couple Earthquakes

G. R. Foulger and Bruce R. Julian
Department of Geological Sciences, Durham University, Durham, UK

Introduction

Non-double-couple ("non-DC") earthquake mechanisms differ from what is expected for pure shear faulting in a homogeneous, isotropic, elastic medium. Until the mid-1980s, the DC assumption underlay nearly all seismological analysis, and was highly successful in advancing our understanding of tectonic processes and of seismology in general. In recent years, though, many earthquakes have been found that do not fit the DC model. Earthquakes that depart strongly from DC theory range in size over many orders of magnitude and occur in many environments, but are particularly common in volcanic and geothermal areas. Moreover, minor departures from the DC model are detected increasingly frequently in studies using high-quality data. These observations probably reflect departures from idealized models, caused by effects such as rock anisotropy or fault curvature.

At the same time, it has become clear that industrial activities such as oil and gas production and storage, hydrofracturing, geothermal energy exploitation, CO_2 sequestration, and waste disposal can induce earthquakes, and that these events often have non-DC mechanisms. Induced earthquakes pose legal hazards because of their destructive potential, but they can also be beneficial, because they can be used to monitoring physical processes that accompany industrial activity.

Non-DC earthquakes are therefore important for improving our understanding of how faults and volcanoes work, perhaps leading some day to an ability to predict earthquakes and eruptions, for avoiding nuisance seismicity in connection with industrial activity, and for monitoring such activity in detail.

Non-DC earthquake mechanisms, by definition, require for their description a more general mathematical formalism than the DC model. The most widely used such formalism is the expansion of the elastodynamic field in terms of the spatial moments of the equivalent-force system (Gilbert 1970). Usually, attention is restricted to second moments, and thus to second-rank moment tensors, and moreover these tensors are usually assumed to be symmetric, so that they have six independent components (A DC has four independent components.). This restriction is not always justified, however. Sources involving net forces or torques are theoretically possible, and phenomena such as landslides and volcanic eruptions provide clear examples of them.

Because it has two extra adjustable parameters, a moment tensor can describe volume changes and general kinds of shear deformation. This increased generality allows the moment tensor to encompass physical processes such as geometrically complex faulting, tensile faulting, faulting in anisotropic media, faulting in heterogeneous media (e.g., near an interface), and polymorphic phase transformations.

The moment tensor also has an important computational property: Mathematical expressions for static and dynamic displacement fields are linear in the moment-tensor components. This property facilitates the evaluation of these fields and enormously simplifies the inverse problem of deducing source mechanisms from observations.

Representation of Earthquake Mechanisms

The Equivalent Force System

Physically, an earthquake involves a nonlinear failure process occurring within a limited region. The equivalent force system, acting in an intact (unfaulted) "model" medium, would produce the same displacement field outside the source region. Any physical source has a unique equivalent force system in a given model medium, but the converse statement is not true. Different physical processes can have identical force systems and therefore identical static and dynamic displacement fields.

The equivalent force system is all that can be deduced from observations of displacement fields, so it constitutes a phenomenological description of the source. There is a one-to-one correspondence between equivalent force systems and elastodynamic fields outside the source region, so we can, in principle at least, determine force systems from observations. On the other hand, the correspondence between force systems and physical source processes is one-to-many, so equivalent force systems (and therefore seismic and geodetic observations) cannot uniquely diagnose physical source processes.

Failure can be regarded as a sudden localized change in the constitutive relation (stress–strain law) in the Earth (Backus 1977). Before an earthquake, the stress field satisfies the equations of equilibrium. At the time of failure, a rapid change in the constitutive relation causes the stress field to change. The resulting disequilibrium causes dynamic motions that radiate elastic waves. We disregard the effect of gravity in the following discussion. In the absence of external forces, the equation of motion is

$$\rho \ddot{u}_i = \sigma_{ij,j}, \qquad (1)$$

where $\rho(\mathbf{x})$ is density, $\mathbf{u}(\mathbf{x}, t)$ is the particle displacement vector, $\sigma(\mathbf{x}, t)$ is the physical stress tensor, $\mathbf{x}$ is position, and t is time. Dots indicate differentiation with respect to t, ordinary subscripts indicate Cartesian components of vectors or tensors, the subscript, j indicates differentiation with respect to the jth Cartesian spatial coordinate x_j, and duplicated indices indicate summation. The true stress is unknown, however, so in theoretical calculations we use the stress $\mathbf{s}(\mathbf{x}, t)$, given by the constitutive law of the model medium (usually Hooke's law). If we replace σ_{ij} by s_{ij} in the equations of motion, though, we must also introduce a correction term, $\mathbf{f}(\mathbf{x}, t)$:

$$\rho \ddot{u}_i = s_{ij,j} + f_i, \qquad (2)$$

$$f_i \stackrel{\text{def}}{=} \left(\sigma_{ij} - s_{ij}\right)_{,j}. \qquad (3)$$

This term has the form of a body-force density and is the equivalent force system of the earthquake. It differs from zero only within the source region.

Net Forces and Torques

Nearly all analyses of earthquake source mechanisms explicitly exclude net forces and torques. The equivalent forces given by Eq. 3, which arise from the imbalance between true physical stresses and those in the model, are consistent with these restrictions. The stress glut $\sigma_{ij} - s_{ij}$ is symmetric, so $\mathbf{f}$ exerts no net torque at any point. Furthermore, $\sigma_{ij} - s_{ij}$ vanishes outside the source region, so Gauss's theorem implies that the total force vanishes at each instant.

More complete analysis, however, including the effects of gravitation and mass advection, shows that Eq. 3 is based on overly restrictive assumptions and that net force and torque components are possible for realistic sources within the Earth (Takei and Kumazawa 1994). These forces and torques transfer linear and angular momentum between the source region and the rest of the Earth, with both types of momentum conserved for the entire Earth. An easily understood example is the collapse of a cavity, in which rocks fall from the ceiling to the floor. While the rocks are falling, the Earth outside the cavity experiences a net upward force, relative to the state before and after the event.

The net force component in any source is constrained by the principle of conservation of momentum; if the source region is at rest before and after the earthquake, the total impulse of the equivalent force (its time integral) must vanish. No such requirement holds for the torque. The total torque exerted by gravitational forces need not vanish even after the earthquake. Horizontal displacement of the center of mass of the source region leads to a gravitational torque, which must be balanced by stresses on the boundary of the source and causes the radiation of elastic waves. Because gravity acts vertically, there can be no net torque about a vertical axis.

The Moment Tensor

We cannot use the equivalent force system $\mathbf{f}(\mathbf{x}, t)$ and the elastodynamic Eq. 2 to determine the displacement field for a hypothetical source. The equivalent force system itself depends on the displacement field that is being sought. Two different approaches are commonly used: (i) In the kinematic approach we assume some mathematically tractable displacement field in the source region (e.g., suddenly imposed slip, constant over a rectangular fault plane), derive the equivalent force system from Eq. 3, and solve Eq. 2 for the resulting displacement field outside the source region; and (ii) in the inverse approach, we use Eq. 2 to determine the force system $\mathbf{f}(\mathbf{x}, t)$ from the observed displacement field and compare the result with force systems predicted theoretically for hypothesized source processes. The most useful way to parameterize the force system in this approach is to use its spatial moments.

The Moment-Tensor Expansion for the Response
Given the equivalent force system $\mathbf{f}(\mathbf{x}, t)$, computing the response of the Earth is a linear problem, and its solution can be expressed as an integral over the source volume V (Aki and Richards 2002, Eq. 3.1, omitting displacement and traction discontinuities for simplicity):

$$u_i(\mathbf{x}, t) = \iiint_V G_{ij}(\mathbf{x}, \xi, t) * f_j(\xi, t) d^3\xi, \qquad (4)$$

where $G_{ij}(\mathbf{x}, \xi, t)$ is the Green's function, which gives the ith component of displacement at position **x** and time t caused by an impulsive force in the j direction applied at position ξ and time 0, and the symbol * indicates temporal convolution. If we expand the Green's function in a Taylor series in the source position ξ,

$$G_{ij}(\mathbf{x}, \xi, t) = G_{ij}(\mathbf{x}, \mathbf{0}, t) + G_{ij,k}(\mathbf{x}, \mathbf{0}, t)\xi_k + \ldots , \quad (5)$$

Eq. 4 for the response becomes

$$u_i(\mathbf{x}, t) = G_{ij}(\mathbf{x}, \mathbf{0}, t) * F_j(t) + G_{ij,k}(\mathbf{x}, \mathbf{0}, t) * M_{jk}(t) + \ldots , \quad (6)$$

where

$$F_j(t) \stackrel{\text{def}}{=} \iiint_V f_j(\xi, t) d^3\xi \quad (7)$$

is the total force exerted by the source and

$$M_{jk}(t) \stackrel{\text{def}}{=} \iiint_V \xi_k f_j(\xi, t) d^3\xi \quad (8)$$

is the moment tensor. If the equivalent force is derivable from a stress glut via Eq. 3, then the moment tensor is the negative of the volume integral of the stress glut:

$$M_{jk}(t) = -\iiint_V (\sigma_{ij} - s_{ij}) d^3\xi \quad (9)$$

The moment tensor is a second-rank tensor, which describes a superposition of nine elementary force systems, with each component of the tensor giving the strength (moment) of one force system. The diagonal components M_{11}, M_{22}, and M_{33} correspond to linear dipoles that exert no torque, and the off-diagonal elements M_{12}, M_{13}, M_{21}, M_{23}, M_{31}, and M_{32} correspond to force couples. It is usually assumed that the moment tensor is symmetric, ($M_{12} = M_{21}$, $M_{13} = M_{31}$, $M_{23} = M_{32}$), so that the force couples exert no net torque (see above), in which case only six moment-tensor components are independent. In this case, the off-diagonal components correspond to three pairs of force couples, each exerting no net torque ("double couples").

The magnitudes of the six (or nine) elementary force systems (the moment-tensor components) transform according to standard tensor laws under rotations of the coordinate system, so there exist many different combinations of elementary forces that are equivalent. In particular, for a symmetric (six-element) moment tensor one can always choose a coordinate system in which the force system consists of three orthogonal linear dipoles, so that the moment tensor is diagonal. In other words, a general point source can be described by three values (the principal moments) that describe its physics and three values that specify its orientation.

The moment tensor has three important properties that make it useful for representing seismic sources. (1) It makes the "forward problem" of computing theoretical seismic-wave excitation linear. A general source is represented as a weighted sum of elementary force systems, so any seismic wave is just the same weighted sum of the waves excited by the elementary sources. The linearity of the forward problem in turn makes much more tractable the inverse problem of determining source mechanisms from observations. (2) It simplifies the computation of wave excitation. By transforming the moment tensor into an appropriately oriented coordinate system, the angles defining the observation direction can be made to take on special values such as 0 and $\pi/2$. Thus radiation by the elementary sources must be computed not for a general direction, but for only a few directions for which the computation is easier. In a laterally homogeneous medium, for example, radiation must be computed for only a single azimuth. (3) The moment tensor is more general than the DC representation. It includes DCs as special cases, but has two more free parameters than a DC (six vs. four), which enable it to represent sources involving volume changes and more general types of shear.

Higher-Rank Moment Tensors

As Eq. 6 shows, "the" moment tensor described above is only one of an infinite sequence of spatial moments that appear in the expansion of the Earth's response to an earthquake. The later terms involve higher-rank moment tensors, which contain information about the spatial and temporal distribution of failure in an earthquake, and have great potential value for studying source finiteness and rupture propagation.

Surface Sources (Faults)

A fault is a surface across which there is a discontinuity in displacement. The equivalent force distribution, **f**, for a generally oriented fault in an elastic medium is, from Eq. 2,

$$f_k(\eta,t) = -\iint_A [u_i(\xi,t)]c_{ijkl}v_j \frac{\partial}{\partial \eta_l}\delta(\eta-\xi)dA, \tag{10}$$

where η is the position where the force is evaluated, ξ is the position of the element of area dA, and the integration extends over the fault surface (Aki and Richards 2002, Eq. 3.5). The unit vector normal to the fault surface is $v(\xi)$ and $[\mathbf{u}(\xi,t)]$ is the displacement discontinuity across the fault in the direction of v. The components of the elastic modulus tensor are c_{ijkl} and $\delta(\mathbf{x})$ is the three-dimensional Dirac delta function.

Substituting the force distribution from Eq. 10 into Eq. 8 gives the moment tensor of a general fault,

$$M_{ij} = -c_{ijkl}A\overline{v_k[u_l]}, \tag{11}$$

where A is the total fault area and the overbar indicates the average value over the fault.

Shear Faults

For a planar shear fault (with normal v in the x_3 direction and displacement discontinuity $[\mathbf{u}]$ in the x_1 direction, say, so that $v_1 = v_2 = 0$ and $[u_2] = [u_3] = 0$) in a homogeneous isotropic medium ($c_{ijkl} = \lambda\delta_{ij}\delta_{km}\delta_{lm} + 2\mu\delta_{ik}\delta_{jl}$), Eq. 8 gives a moment tensor with two nonzero components $M_{13} = M_{31} = \mu A\bar{u}$, where $\bar{u}$ is the average slip. This corresponds to a pair of force couples, one with forces in the x_1 direction and moment arm in the x_3 direction, and the other with these directions interchanged.

We get the same DC moment tensor for a fault with normal in the x_1 direction and displacement discontinuity in the x_3 direction. This ambiguity between "conjugate" faults is an example of the fundamental limitations on the information that can be deduced from equivalent force systems.

Tensile Faults

For a tensile fault in a homogeneous isotropic medium, with the fault lying in the x_1-x_2 plane and opening in the x_3 direction, $v_1 = v_2 = 0$ and $[u_1] = [u_2] = 0$. Equation 8 gives a moment tensor with three non-zero components: $M_{11} = M_{22} = \lambda A\bar{u}$ and $M_{33} = (\lambda + 2\mu)A\bar{u}$, corresponding to two dipoles in the fault plane with moments of $\lambda A\bar{u}$, and a third dipole oriented normal to the fault and with a moment of $(\lambda + 2\mu)A\bar{u}$.

Volume Sources

Some possible non-DC source processes, such as polymorphic phase transformation, occur throughout a finite volume rather than on a surface. The equivalent force system often can be expressed in terms of the "stress-free strain" $\Delta_\sigma e_{ij}$, which is the strain that would occur in the source volume if the tractions on its boundary were held constant by externally applied artificial forces (It might more accurately be called the "fixed-stress strain."). By reasoning that involves a sequence of imaginary cutting, straining, and welding operations (e.g., Aki and Richards 2002, Sect. 3.4), the moment tensor of such a volume source is found to be

$$M_{ij} = \iiint_V c_{ijkl}\Delta_\sigma e_{kl}dV. \tag{12}$$

For a stress-free volume change $\Delta_\sigma V$ in an isotropic medium, for example,

$$M_{ij} = K\Delta_\sigma V \delta_{ij}, \tag{13}$$

where $K = \lambda + (2/3)\mu$ is the bulk modulus.

The stress-free strain is *not*, in general, the strain that actually occurs in a seismic event (Richards and Kim 2005). Because the source is imbedded in the Earth, its deformation is resisted by the stiffness of the surrounding medium, making the true strain changes smaller than the stress-free values. For example, in terms of the true volume change ΔV_t, the moment tensor of an isotropic source in an infinite isotropic elastic medium is

$$M_{ij} = (\lambda + 2\mu)\Delta_t V \delta_{ij}, \quad (14)$$

so

$$\Delta_t V = \frac{\lambda + (2/3)\mu}{\lambda + 2\mu} \Delta_\sigma V. \quad (15)$$

The distinction is quantitatively significant; for an infinite homogeneous Poisson solid, $\Delta_t V = (5/9) \Delta_\sigma V$. Furthermore, the stiffness seen by the source, and thus the true strain, depends on the elasticity structure outside the source volume.

Decomposing Moment Tensors

To make a general moment tensor easier to understand, it helps to decompose it into simpler force systems. First, we express the moment tensor in its principal axis coordinate system. Three values (the Euler angles, for example) are needed to specify the orientation of this system, and three other values specify the moments of three orthogonal dipoles oriented parallel to the coordinate axes. Writing these three principal moments as a column vector, we first decompose the moment tensor into an isotropic force system and a deviatoric remainder,

$$\begin{bmatrix} M_1 \\ M_2 \\ M_3 \end{bmatrix} = M^{(V)} \begin{bmatrix} 1 \\ 1 \\ 1 \end{bmatrix} + \begin{bmatrix} M'_1 \\ M'_2 \\ M'_3 \end{bmatrix}, \quad (16)$$

with $M^{(V)} \stackrel{\text{def}}{=} (M_1 + M_2 + M_3)/3$, and then decompose the deviatoric part into a DC (principal moments in the ratio $1:-1:0$) and a "compensated linear vector dipole" (CLVD), a source with principal moments in the ratio $1:-1/2:-1/2$ (Knopoff and Randall 1970). Figure 1 illustrates the three elementary force systems used in this decomposition, showing their compressional-wave radiation patterns and the distributions of compressional-wave polarities on the focal sphere (an imaginary sphere surrounding the earthquake hypocenter, to which observations are often referred). Many other decompositions of the deviatoric part are possible, including many decompositions into two DCs or into two CLVDs. The method of Knopoff and Randall (1970), which makes the largest axis of the CLVD coincide with the corresponding axis of the DC, avoids pathological behavior and is the method most widely used in seismological research:

$$\begin{bmatrix} M'_1 \\ M'_2 \\ M'_3 \end{bmatrix} = M^{(DC)} \begin{bmatrix} 0 \\ -1 \\ 1 \end{bmatrix} + M^{(CLVD)} \begin{bmatrix} -1/2 \\ -1/2 \\ 1 \end{bmatrix}, \quad (17)$$

with $M^{(DC)} = M'_1 - M'_2$, and $M^{(CLVD)} = -2M'_1$. The moment-tensor elements are arranged so that $|M'_1| \leq |M'_2| \leq |M'_3|$.

The quantity

$$\varepsilon \stackrel{\text{def}}{=} \frac{-M'_1}{|M'_3|} \equiv \frac{1}{2} \frac{M^{(CLVD)}}{|M^{(DC)} + M^{(CLVD)}|} \quad (18)$$

is sometimes used as a measure of the departure of the deviatoric part of a moment tensor from a pure DC. It ranges in value from zero for a pure DC to $\pm 1/2$ for a pure CLVD. Positive ε corresponds to extensional polarity for the major dipole of the CLVD component.

Displaying Focal Mechanisms

Focal-Sphere Polarity Plots

Since the 1960s, common practice has been to display DC mechanisms by showing the compressional-wave polarity fields on maps of a focal hemisphere, and this method is now also widely used for general symmetric moment tensors. Both azimuthal-equal-area (Lambert) and stereographic projections are used, with teleseismic data usually plotted on lower hemispheres and local data on upper hemispheres. Shaded fields usually represent positive polarities (outward first motions). Figures 2, 4, 6, 8, 9, 11, 12, and 13 below contain examples of focal-sphere polarity plots.

A disadvantage of focal-sphere polarity plots is that different mechanisms can lead to

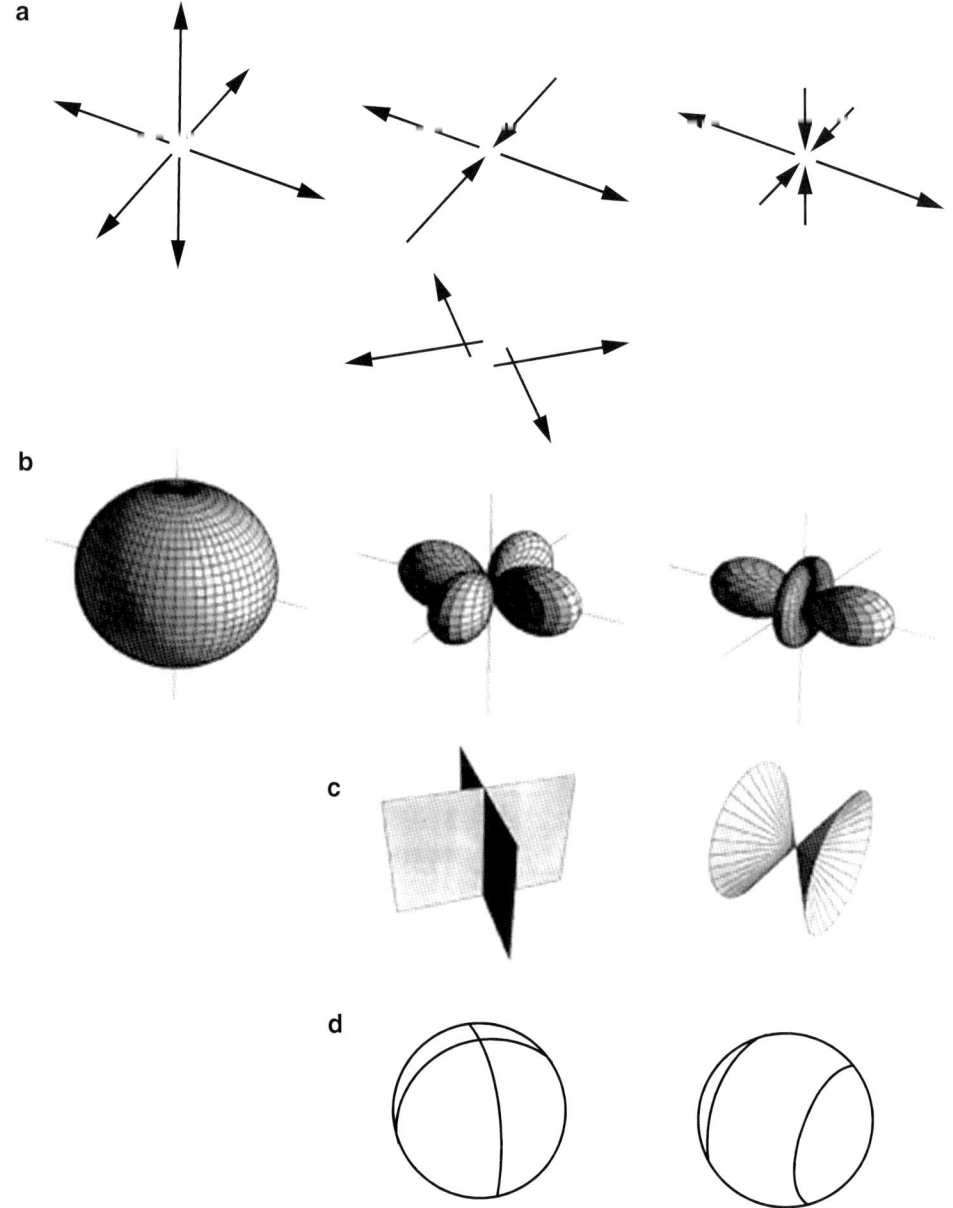

Non-Double-Couple Earthquakes, Fig. 1 Three source types commonly used in decomposing moment tensors: (from *left* to *right*) isotropic, double couple (DC), and compensated linear-vector dipole (CLVD). *Top row*: equivalent force systems, in principal-axis coordinates. For the DC, the force system in a fault-oriented coordinate system is shown underneath. *Second row*: compressional-wave radiation patterns. *Third row*: compressional-wave nodal surfaces. *Fourth row*: curves of intersection of nodal surfaces with the focal sphere

identical plots. Figure 2 shows, for example, that a rather wide range of mechanisms can have positive (or negative) compressional-wave polarities over the entire focal sphere.

Source-Type Plots

Focal-sphere polarity plots display information about source orientation as well as "source type" (the relative values of the principal

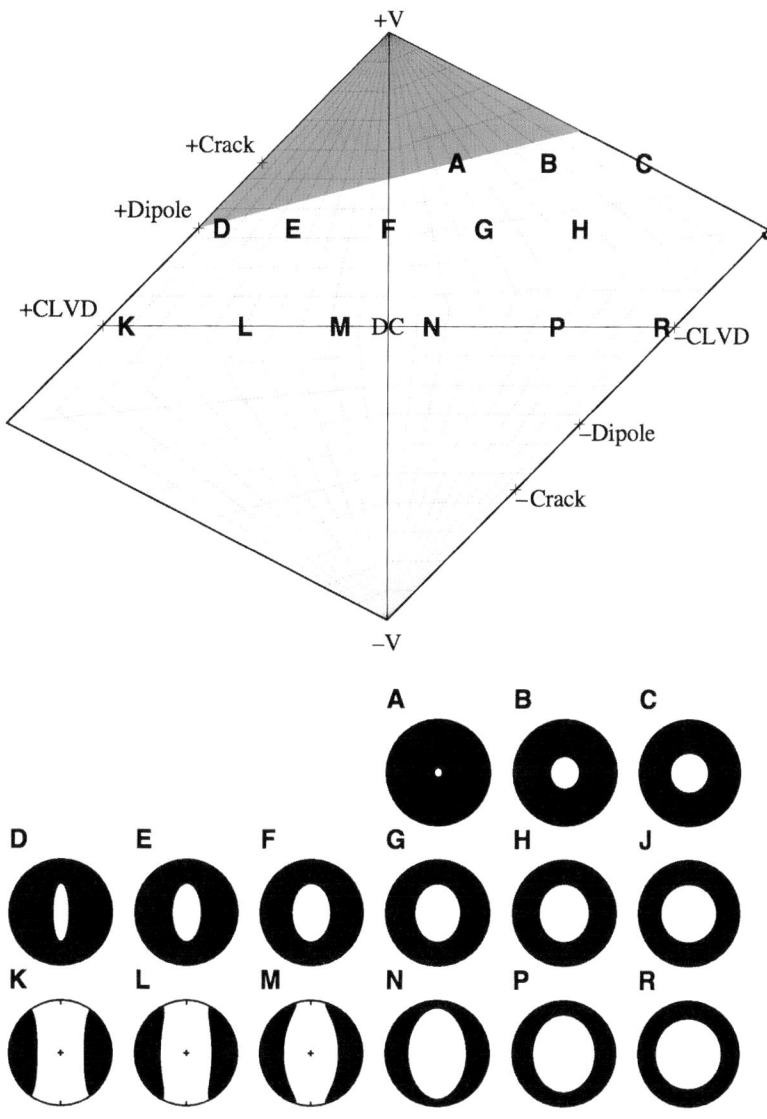

Non-Double-Couple Earthquakes, Fig. 2 *Top*: "Source-type plot" of Hudson et al. (1989), which displays earthquake mechanisms (symmetric moment tensors) without regard to their orientations. The quantity k (Eq. 19), which measures volume change, ranges from -1 at the *bottom* of the plot to $+1$ at the *top*, and is constant along the sub-horizontal grid lines. The quantity -2ε (Eq. 18), which describes the deviatoric part of the moment tensor, ranges from -1 on the *left-hand side* of the plot to $+1$ on the *right-hand side*, and is constant along the grid lines that run from *top* to *bottom*. DC: double-couple mechanisms; +Crack: Opening tensile faults; +Dipole: Force dipoles with forces directed outward; +CLVD: "Compensated linear-vector dipoles," with dominant dipoles directed outward. −Crack, −Dipole, and −CLVD: The same mechanisms with opposite polarities. *Shaded area*: Region in which all compressional waves have outward polarities. A similar region of inward polarities occurs at the *bottom* of the plot. A–R: Some representative mechanisms. *Bottom*: Conventional equal-area focal-hemisphere plots of compressional-wave polarities for 15 representative mechanisms shown on the source-type plot

moments). In fact, conventional DC polarity plots contain information about *only* source orientation. When considering non-DC mechanisms, however, it is useful to display the source type without regard to orientation. The normalized principal moments contain two independent degrees of freedom, and there are many possible ways to display such information in two dimensions. Figure 2 shows the "source-type plot" (Hudson et al. 1989), which gives $T \stackrel{\text{def}}{=} -2\varepsilon$ (Eq. 18) vs.

$$k \stackrel{\text{def}}{=} \frac{M^{(V)}}{|M^{(V)}| + |M'_3|}, \quad (19)$$

a measure of the volume change.

Determining Earthquake Mechanisms

Many types of seismic and geodetic data can be used to determine earthquake focal mechanisms. These range from simple polarities (signs) of observable quantities, through measurements of their amplitudes, to complete time histories of their evolution.

Wave Polarities

First-Motion Methods
P-phase polarities are the most commonly used data in focal-mechanism studies, because they can be determined easily, even using recordings of low dynamic range and accuracy from single-component seismometers. Such polarities alone are of very limited use, however, for studying (or even for identifying) non-DC earthquakes. Even with observations well distributed on the focal sphere, it usually is difficult to rule out DC mechanisms in practice. This difficulty is especially severe if the earthquake lacks a large isotropic component.

In analyzing P-wave polarities, seismologists usually constrain earthquake mechanism to be DCs. Finding a DC mechanism amounts to finding two orthogonal nodal planes (great circles on the focal sphere) that separate the compressional and dilatational polarities into four equal quadrants. Three independent quantities (fault-plane strike and dip angles and the rake angle of the slip vector, say) are needed to specify the orientations of the nodal planes. Nodal planes can be sought either manually, by plotting data on maps of the focal sphere and using graphical methods to find suitable nodal planes, or by using computerized methods, most of which systematically search through the space of possible solutions.

Moment-Tensor Methods
For general moment-tensor mechanisms, hand-fitting mechanisms to observed polarities is impractical. The number of unknown parameters increases from three to five (the six moment tensor components, normalized in some arbitrary way), and furthermore the theoretical nodal surfaces become general ellipses, rather than great circles, on the focal sphere. Searching methods still work, but the addition of two more unknown parameters makes them costly in terms of computer time. Linear programming (sections "Wave Amplitudes" and "Amplitude Ratios," below), an analytical method that treats linear inequalities, is well suited to determining moment tensors from observed wave polarities (Julian 1986).

Near-Field Polarities
Near-field observations (made within a few wavelengths of the source) provide a simple and elegant method of detecting an isotropic source component using a single radial-component seismogram (McGarr 1992). The method is not guaranteed to detect isotropic components whenever they exist, but it involves few assumptions, so detected isotropic components are comparatively reliable.

Assume that all the moment-tensor components are proportional to the unit step function, $U(t)$ (any monotonic function of time will work). Place the origin of the coordinate system at the source, with the x_1 axis directed toward the observer. Then from Eq. 4.29 of Aki and Richards (2002), in an infinite homogeneous medium the radial displacement (the x_1 component observed on the x_1 axis) is

$$u_1(t) = \frac{3}{4\pi\rho r^4}(3M_{11} - Tr\mathbf{M})w(t) \quad (20a)$$

$$+ \frac{1}{4\pi\rho V_P^2 r^2}(4M_{11} - Tr\mathbf{M})U(t - r/V_P) \quad (20b)$$

$$- \frac{1}{4\pi\rho V_S^2 r^2}(3M_{11} - Tr\mathbf{M})U(t - r/V_S) \quad (20c)$$

$$+ \frac{1}{4\pi\rho V_P^3 r}M_{11}\delta(t - r/V_P), \quad (20d)$$

where

$$w(t) \stackrel{\text{def}}{=} \int_{r/V_P}^{r/V_S} \tau U(t - \tau)d\tau \quad (21)$$

vanishes for $t \leq r/V_P$, increases monotonically for $r/V_P \leq t \leq r/V_S$, and is constant for $r/V_S \leq t$. Here $r \equiv x_1$ is the source-observer distance and V_P and V_S are the compressional- and shear-wave speeds. The near-field term Eq. 20a produces a monotonic trend on the seismogram between the compressional and shear waves. For a purely deviatoric source ($Tr\,\mathbf{M} = 0$), the polarity of this near-field term must be the same as that of the compressional wave Eqs. 20b and 20d. Opposite polarities on any radial seismogram imply that the source mechanism has an isotropic component. Similarly, the near-field shear wave Eq. 20c must have the opposite polarity to the compressional wave, and any observations to the contrary indicate an isotropic source component.

Polarities of Other Seismic Waves

The polarities of shear waves can furnish valuable information to complement compressional-wave first motions, but because shear seismic phases arrive later in the seismogram, where scattered energy ("signal-generated noise") is also arriving, their polarities often are difficult to measure reliably. An additional severe difficulty arises for vertically polarized shear (SV) waves incident at the surface beyond the critical angle $\sin^{-1}(V_S/V_P)$ (at epicentral distances between about 0.7 times the focal depth and several thousand kilometers). Outside the "shear-wave windows" at small and large distances, incident shear waves excite evanescent compressional waves that complicate signals and render them practically useless for analysis (Booth and Crampin 1985). Still another problem arises because wave scattering by heterogeneities near the surface excites compressional waves that arrive slightly before the direct shear wave, making the true S onset difficult to identify. These latter two difficulties are caused by conditions at or near the free surface and can be partially alleviated by installing seismometers in deep boreholes.

None of these problems affect horizontally polarized shear waves in spherically symmetric media, and in practice useful SH polarities can usually be determined reliably on transverse-component seismograms obtained by numerical rotation.

Wave Amplitudes

The amplitude of a radiated seismic wave contains far more information about the earthquake mechanism than does its polarity alone, so amplitude data can be valuable in studies of non-DC earthquakes. Moreover, because seismic-wave amplitudes are linear functions of the moment-tensor components, determining moment tensors from observed amplitudes is a linear inverse problem, which can be solved by standard methods such as least squares. Conventional least-squares methods, however, cannot invert polarity observations such as first motions, which typically are the most abundant data available. Linear programming methods, which can treat linear inequalities, are well suited to inverting observations that include both amplitudes and polarities (Julian 1986). In this approach, bounds are placed on observed amplitudes, so that they can be expressed as linear inequality constraints. Polarities are already in the form of linear inequality constraints if the moment-tensor representation is used.

Linear programming methods seek solutions by attempting to minimize the L1 norm (the sum of the absolute values) of the residuals between the constraints and the theoretical predictions.

Amplitude Ratios

Seismic-wave amplitudes are subject to distortion during propagation, particularly because of focusing and de-focusing by structural heterogeneities. A simple way to reduce the effect of this distortion when deriving earthquake mechanisms is to use as data the ratios of amplitudes of waves that have followed similar paths, such as $P:SV$, $P:SH$, or $SH:SV$. If the ratios of the wave speeds is constant in the Earth, then the amplitudes of the waves are affected similarly and the ratio is relatively unaffected.

Using amplitude ratios makes inverting for source mechanisms more difficult, however, because a ratio is a nonlinear function of its denominator. Systematic searching methods still work, but because the dimensionality of the model space is increased by two over that for a DC mechanism, the computational labor is greatly increased (typically by a factor of more than 100).

The efficient linear programming method described above is easily extended to treat amplitude-ratio data in addition to polarities and amplitudes (Julian and Foulger 1996). An observed ratio is expressed as a pair of bounding values, each of which gives a linear inequality that is mathematically equivalent to a polarity observation with a suitably modified Green's function.

Waveforms

A digitized waveform is just a series of amplitude measurements, so waveform inversion may be regarded as an extension of amplitude inversion.

Mathematical Formulation

A theoretical seismogram can be written as a sum of terms, each of which is the temporal convolution of a Green's function and the time function of one source component (Eq. 6). The source components can include components of the equivalent force and moment-tensor components of any order. If there are n such source components (six, in the usual case of a symmetric second-rank moment tensor), which we arrange in a column vector $\phi(t)$, then a set of m seismograms corresponds to the system of simultaneous equations

$$\begin{bmatrix} u_1(t) \\ u_2(t) \\ \vdots \\ u_m(t) \end{bmatrix} = \begin{bmatrix} A_{11}(t) & \cdots & A_{1n}(t) \\ A_{12}(t) & \cdots & A_{2n}(t) \\ \vdots & \ddots & \vdots \\ A_{m1}(t) & \cdots & A_{mn}(t) \end{bmatrix} * \begin{bmatrix} \phi_1(t) \\ \phi_2(t) \\ \vdots \\ \phi_n(t) \end{bmatrix}, \quad (22)$$

where each matrix element A_{ij} is a Green's function giving the ith seismogram generated by a source whose jth component is the impulse $\delta(t)$, and whose other components are zero. The symbol $*$ indicates temporal convolution. The Green's functions can be thought of as a multichannel filter, which takes the n source time functions $\phi_j(t)$ as inputs and generates as output the m synthetic seismograms $u_i(t)$.

Estimating the source time functions from a set of observed seismograms and assumed Green's functions is thus a multichannel inverse filtering, or deconvolution, problem. It can be solved, for example, by transforming to the frequency domain (so that the convolutions become multiplications) and then applying standard least-squares methods to solve for each spectral component. Alternatively, the output seismograms in Eq. 22 can be concatenated into a single time series, and the Green functions in each column of the matrix **A** similarly concatenated. Then, when each time series is expressed explicitly in terms of its samples, the least-squares normal equations turn out to have a "block-Toeplitz" structure, so that they can be solved efficiently by Levinson Recursion (Sipkin 1982).

Limitations

Imperfect Earth Models

If the Earth model used to analyze the radiation from an earthquake differs from the true structure of the Earth, systematic errors will be introduced into the Green functions and thus into inferred source mechanisms. Near-source anisotropy (section "Shear Faulting in an Anisotropic Medium") is one cause of such errors.

Deficient Combinations of Modes

Even if the Green's functions are correct, it may be impossible to determine the source mechanism completely in some circumstances. For particular types of seismic waves, there may be certain source characteristics that cannot be determined. For example, shear waves alone cannot detect isotropic source components, because purely isotropic sources do not excite shear waves. Similarly, sources with vertical symmetry axes (those whose only nonzero components are $M_{11} = M_{22}$ and M_{33}, with the x_3 axis vertical) excite no horizontally polarized shear (SH) or Love waves, and cannot be detected using such waves alone. For any Rayleigh mode, the component M_{33} occurs only in the combination $M_{11} + M_{22} + f(\omega)M_{33}$, where the frequency function $f(\omega)$ depends on the mode and the Earth model. The M_{33} component can be traded off against $M_{11} + M_{22}$ without changing this combination, so isotropic sources are unresolvable by Rayleigh waves of a single frequency and mode (Mendiguren 1977), and even with multimode observations, determining M_{33} requires a priori assumptions about its spectrum. In most studies, enough different modes and/or frequencies are used so that none of these degenerate situations arises. Furthermore, all general inversion methods in widespread use provide objective information about uncertainty and nonuniqueness in derived values, so degeneracies can be detected if they happen to occur.

Shallow Earthquakes

If an earthquake is effectively at the free surface (shallow compared to the seismic wavelengths used), then it is impossible to determine its full moment tensor. Only three moment-tensor components can be determined, and these are not enough even under the a priori assumption of a DC mechanism (which requires four parameters). This degeneracy follows from the proportionality between the coefficient C_{ij} giving the amplitude of a seismic mode excited by the moment-tensor component M_{ij} and the displacement derivative $u_{i,j}$ for the mode. (This proportionality follows from the principle of reciprocity.) Vanishing of the traction on the free surface,

$$\sigma_{13} = \mu(u_{1,3} + u_{3,1}) = 0$$
$$\sigma_{23} = \mu(u_{2,3} + u_{3,2}) = 0$$
$$\sigma_{33} = \lambda(u_{1,1} + u_{2,2}) + (\lambda + 2\mu)u_{3,3} = 0, \quad (23)$$

implies that three linear combinations of the excitation coefficients vanish:

$$C_{13} = 0$$
$$C_{23} = 0 \quad (24)$$
$$\lambda C_{11} + \lambda C_{22} + (\lambda + 2\mu)C_{33} = 0.$$

(Here we restrict ourselves to symmetric moment tensors.) Therefore a moment tensor whose only nonzero components are M_{13} and M_{23} radiates no seismic waves (to this order of approximation), and moreover a diagonal moment tensor with elements in the ratio $\lambda : \lambda : (\lambda + 2\mu)$ likewise radiates no seismic waves. The first case corresponds to a vertical dip-slip shear fault or a horizontal shear fault, and the second corresponds to a horizontal tensile fault. This second undeterminable source type has an isotropic component, so isotropic components cannot be determined for shallow sources.

This degeneracy is most often important in studies using surface waves or normal modes, because these are usually observed at frequencies below 0.05 Hz, for which the wavelengths are greater than the focal depths of many earthquakes.

Non-DC Earthquake Processes

Processes Involving Net Forces

Most experimental investigations of earthquake source mechanisms have excluded net forces and torques from consideration a priori. As discussed above in section "Net Forces and Torques," the laws of physics do not require such restrictions. Net forces are possible for an internal source because momentum can be transferred between the source region and the rest of the Earth. Momentum conservation does, however, require that the impulse (time integral) of the net force component must vanish if the source is at rest before and after the event.

Landslides

Among sources that involve net forces, landslides have received the most attention. Modeling a landslide as a block of mass M sliding down a ramp gives an equivalent force of $-M\mathbf{a}$, where $\mathbf{a}$ is the acceleration of the block.

The gravitational forces on a landslide also produce a torque of magnitude $Mg\Delta x$, where g is the acceleration of gravity and Δx is the horizontal distance the slide travels.

Volcanic Eruptions

The eruption of material by a volcano applies a net force to the Earth, much as an upward-directed rocket exhaust would. Of course, the total impulse imparted to the Earth-atmosphere system is zero, as with any internal source, but the spatially and temporally concentrated force at the volcanic vent can generate observable seismic waves, whereas the balancing forces transmitted from the ejected material through the atmosphere back to the Earth's surface excite waves that probably are unobservable in practice. Therefore a volcanic eruption may be modeled as a point force $S\Delta P$, where S is the area of the vent and ΔP is the pressure difference between the source reservoir within the volcano and the atmosphere (Kanamori et al. 1984).

Other processes accompanying volcanic eruptions might act as seismic-wave sources. A change in pressure in a deep spherically symmetric reservoir acts as an isotropic source with a moment tensor given by Eq. 14 (section "Volume Sources"). For a tabular or crack-shaped reservoir, the force system is the same as that for a tensile fault, discussed in sections "Tensile Faults" and "Tensile Faulting."

Unsteady Fluid Flow

If the speed, and thus the momentum, of fluid flowing in a conduit varies with time, a time-varying net force,

$$\mathbf{F} = -\iiint_V \rho \mathbf{a} dV, \qquad (25)$$

is exerted on the surrounding rocks, where ρ is the density of the fluid and $\mathbf{a}$ is its acceleration. This process may cause "long-period" volcanic earthquakes and the closely related phenomenon of volcanic tremor. Time variations in the flow speed might be caused by the breaking of barriers to flow, or be self-excited by nonlinear interaction between the flowing fluid and deformable channel walls (Julian 1994).

Complex Shear Faulting

Multiple Shear Events

If earthquakes occur close together in space and time, observed seismic waves may not be able to resolve them, and they may be misinterpreted as a single event. The apparent moment tensor of the composite event is then the sum of the true moment tensors of the earthquakes, and because the sum of two DCs is not, in general, a DC, shear faulting can produce non-DC mechanisms in this way. Combining DCs cannot ever produce mechanisms with isotropic (volume change) components, because the trace of the moment tensor is a linear function of its components, so multiple shear-faulting mechanisms lie on the horizontal axis of the source-type plot.

By analyzing complete seismic waveforms, it is often possible to resolve a multiple event into subevents with different mechanisms. Doing this requires use of algorithms that allow the moment tensor to vary with time in a general way. Algorithms that assume that all the moment-tensor components have identical time functions can produce spurious non-DC mechanisms, even if the subevents have identical DC mechanisms.

Volcanic Ring Faults

Dikes intruded along conical surfaces with both outward and inward dips are often found in

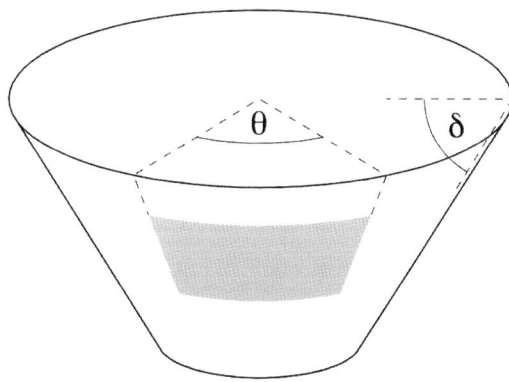

Non-Double-Couple Earthquakes, Fig. 3 Geometry of volcanic ring faulting used in computing mechanisms shown in Fig. 4. Dip-slip motion on fault of dip δ is uniformly distributed over azimuth range θ

exhumed extinct volcanoes, and are expected consequences of the stresses caused by inflation and deflation of magma chambers (Anderson 1936). These dikes are of two types: "cone sheets," which dip inward at about 30–70°, form as tensile faults during inflation, and nearly vertical or steeply outward-dipping "ring dikes" form through shear failure accommodating subsidence following deflation or eruption of magma. For both types, the axes of the cones are approximately vertical. If dip-slip shear faulting occurs on a conical fault, and the rupture in an earthquake spans a significant azimuth range (Fig. 3), the resulting mechanism, considered as a point source, can have a non-DC component (Ekström 1994). (Strike-slip motion on such a surface always gives pure DC mechanisms.) Figure 4 shows a suite of theoretical source mechanisms corresponding to dip-slip ruptures spanning various azimuth ranges on conical faults of different dips. For steeply dipping faults the non-DC components are small (for vertical faults they vanish), so cone sheets are more efficient than ring dikes as a non-DC process.

Tensile Faulting

Opening Tensile Faults in the Earth

An obvious candidate mechanism for geothermal and volcanic earthquakes is tensile faulting, in which the displacement discontinuity is normal, rather than parallel, to a fault surface. The equivalent force system of a tensile fault consists of three orthogonal linear dipoles with moments in the ratio $(\lambda + 2\mu):\lambda:\lambda$. It is equivalent to an isotropic source of moment $(\lambda + 2\mu/3)A\bar{u}$ plus a CLVD of moment $(4\mu/3)$ $A\bar{u}$ (section "Tensile Faults"). The far-field compressional waves have all first motions outward, with amplitudes largest (by a factor of $1 + 2\mu/\lambda$) in the direction normal to the fault. Figure 2 shows the position of tensile faults on a source-type plot.

Compressive stress tends to prevent voids from forming at depth in the Earth, but high fluid pressure can overcome this effect and allow tensile failure to occur. The situation is conveniently analyzed using Mohr's circle diagrams (Fig. 5). The effect of interstitial fluid at pressure p in a polycrystalline medium such as a rock is to lower the effective principal stresses by the amount p. Thus fluid pressure, if high enough, can cancel out much of the compressive stress caused by the overburden. Fluid pressures comparable to the lithostatic load are found surprisingly often in deep boreholes.

A second prerequisite for tensile failure is that the shear stresses be small, or equivalently that the principal stresses be nearly equal. The diameter of the Mohr's circle in Fig. 5 is equal to the maximum shear stress (difference between the extreme principal stresses). If this diameter is too large, the circle can touch the failure envelope only along its straight portion, which corresponds to shear failure. Only if the shear stress, and thus the diameter, is small will the circle first touch the failure envelope in the tensile field to the left of the τ axis.

Combined Tensile and Shear Faulting

Although tensile faults could cause earthquakes that involve volume increases, they cannot explain non-DC earthquakes whose isotropic components indicate volume decreases. Tensile faults can open suddenly for a variety of reasons, but they would be expected to close gradually, and not to radiate elastic waves. If a tensile fault and a shear fault intersect, however, then stick–slip instability could cause sudden episodes of either opening or closing, with volume increases or decreases. The stresses around the

Non-Double-Couple Earthquakes, Fig. 4 Non-DC mechanisms for volcanic ring faulting (Fig. 3). Theoretical compressional-wave nodal surfaces for an arcuate dip-slip fault whose strike spans a range θ and averages north–south. Each focal sphere corresponds to two situations: a fault dipping to the west by the smaller angle δ given or to the east by the larger angle. All mechanisms are purely deviatoric. Numbers below each mechanism give values of ε (Eq. 18), which describe the deviatoric parts of the moment tensors. Upper focal hemispheres are shown in equal-area projection. Lower hemisphere plots are *left-right* mirror images. For normal faulting, central fields have dilatational polarity

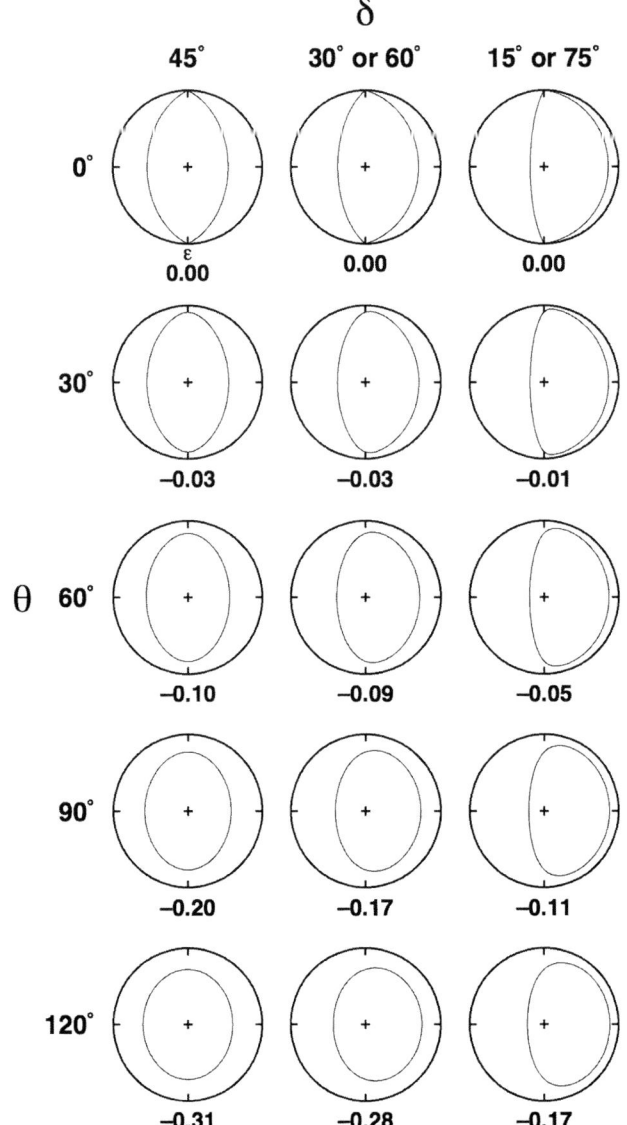

tips of both shear and tensile faults favor this type of fault pairing, and faulting of this kind occurs in rocks in the laboratory (Brace et al. 1966). A similar situation can occur in the case of shear faulting near mines, with the tunnel playing the role of the tensile fault.

Figure 6 shows the theoretical source mechanisms for combined tensile and shear faults of different geometries and relative seismic moments. When the tensile fault opens or closes in the direction normal to its face, the mechanisms have large isotropic components, with most of the focal sphere having the same polarity as the tensile fault and two unconnected fields having the opposite polarity. The symmetry of the moment tensors makes it impossible to determine the angle between the two faults. An angle of $45° - \alpha$ is equivalent to an angle of $45° + \alpha$. When the tensile fault opens or closes obliquely, in the direction parallel to the shear fault, then the mechanisms are closer to DCs, and less sensitive to the relative seismic moments.

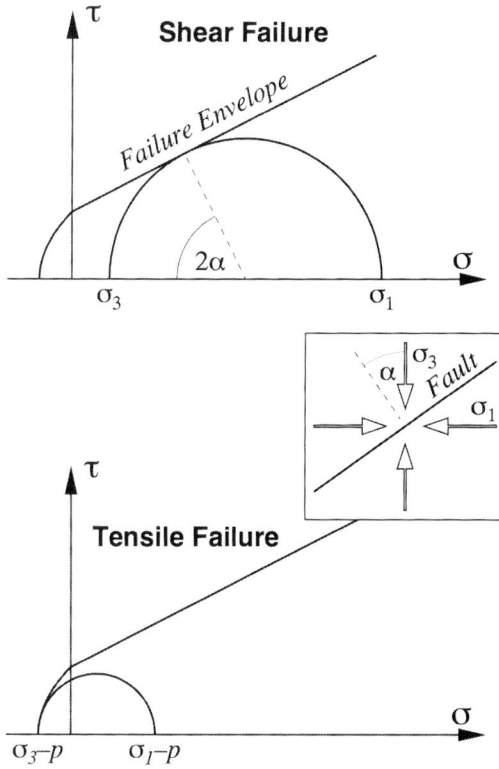

Non-Double-Couple Earthquakes, Fig. 5 Conditions for shear and tensile failure. Mohr's circle diagram shows the relationship between shear traction τ and normal traction σ across a plane in a stressed medium. Locus of (σ, τ) points for different orientations of the plane is a circle of diameter $\sigma_1 - \sigma_3$, centered at $((\sigma_1 + \sigma_3)/2)$, where σ_1 and σ_3 are the extreme principal stresses. Failure occurs when circle touches the "failure envelope," and the point of tangency determines the orientation of the resulting fault (see *inset*). (Theoretical failure envelope shown corresponds to Griffith theory of failure, as modified by F. A. McClintock and J. B. Walsh (Price 1966).) Straight portion of failure envelope in compressional field ($\sigma > 0$) represents the Navier-Coulomb criterion for shear failure. (*Top*) At high confining stress with no fluid pressure, only shear failure occurs. (*Bottom*) High fluid pressure lowers the effective confining stress, and tensile failure occurs for low stress differences

If the dominant principal axis of the tensile fault lies in the same plane as the *P* and *T* axes of the shear fault (i.e., if the null axis of the shear fault lies in the tensile-fault plane), then the composite mechanism lies on the line between the DC and Crack positions on a focal-sphere plot. For more general (and less physically plausible) geometrical arrangements, the composite mechanism lies within a region consisting of two triangles (Fig. 7).

Shear Faulting in an Anisotropic Medium

The equivalent force system of an earthquake depends on the constitutive law used to compute the model stress s_{ij} in Eq. 2. This means that a fault in an anisotropic elastic medium has a different equivalent force system than it would if the medium were isotropic, and in particular that a shear fault in an anisotropic medium generally has a non-DC moment tensor, which can be determined, for example, from Eq. 11 (Vavryčuk 2005). Most rocks are anisotropic, because of layering on a scale smaller than seismic wavelengths, preferential orientation of crystals, and the presence of cracks and inclusions, so most earthquakes are affected by anisotropy.

Elastic wave propagation in an anisotropic medium is more complicated in several ways than in an isotropic medium. The particle motion in body waves is no longer either longitudinal or transverse to the direction of propagation, but is generally oblique, so body-wave modes are referred to as "quasi-compressional" and "quasi-shear." The "direction of propagation," in fact is no longer a single direction, but rather two directions for each mode: the normal to the wavefront (the "phase velocity" direction) and the direction of energy transport (the "group velocity" direction, from the source to the observer).

Green's functions for the anisotropic medium must be used to compute the radiated seismic waves for the force system given by Eq. 11 (Gajewski 1993) and to solve the inverse problem of determining the force system from observed seismograms. If enough information is available about source-region anisotropy to determine such Green's functions, then it will be possible to recognize when non-DC force systems are consistent with shear faulting. In practice, however, information about anisotropy in earthquake focal regions is seldom available, so Green's functions appropriate for isotropic constitutive laws are used instead. In this case, the non-DC force system of most interest is not the one given by Eq. 11, but rather the one that would be derived

Non-Double-Couple Earthquakes, Fig. 6 Non-DC mechanisms for combined tensile and shear faulting with different geometries. Both faults are vertical, with the shear fault striking north–south and the tensile fault striking west of north at the angle α, indicated by bold ticks. *Left column*: tensile fault opening normal to its face. *Right column*: tensile fault opening obliquely, parallel to the shear fault. Compressional-wave nodal surfaces are shown for different relative moments of the tensile (M^T) and shear (M^S) components. *Solid curves*: $M^T = 0.5M^S$. *Dashed curves*: $M^T = 0.2M^S$. *Dotted curves*: $M^T = 0.1M^S$. Numbers to the *right* of each plot give values of k and ε (Eqs. 18 and 19), for each mechanism. Focal hemispheres (either upper or lower, because of symmetry) are shown in equal-area projection

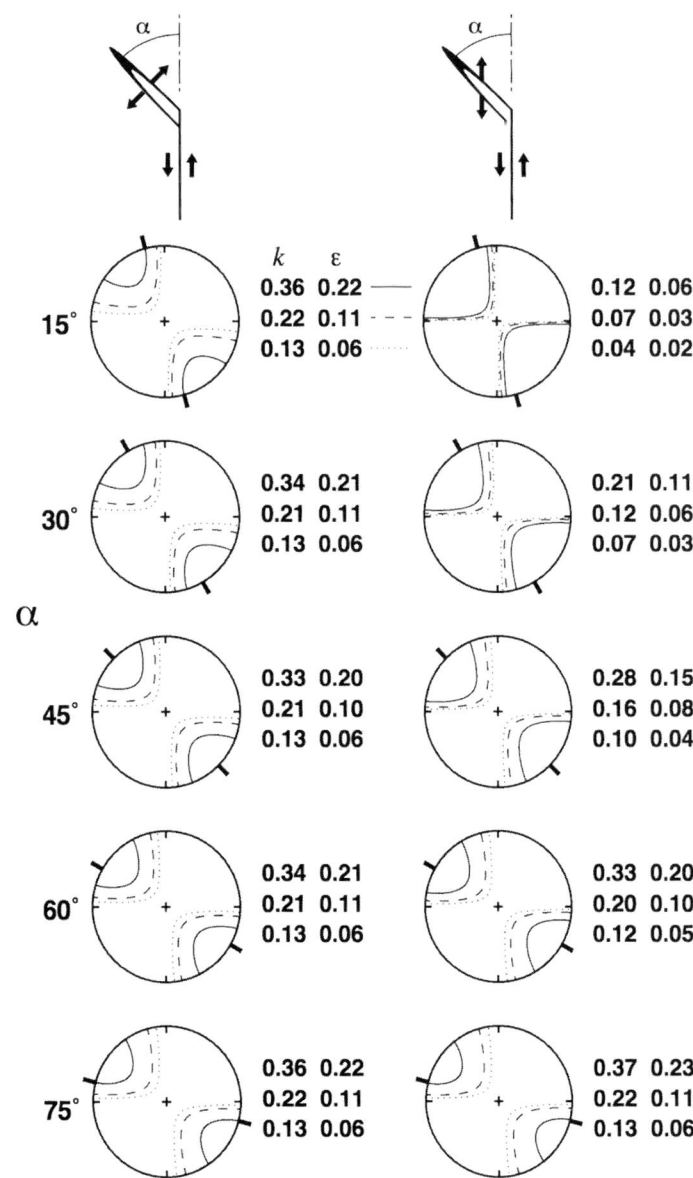

from seismic waves using an isotropic constitutive law when the focal region is actually anisotropic.

Shear Faulting in a Heterogeneous Medium

If an earthquake occurs in a place where the elastic moduli vary spatially, its apparent mechanism will be distorted, and a DC earthquake may appear to have non-DC components. This occurs when the spatial derivatives of Green's functions (strains) appearing in the second term on the right side of the moment expansion Eq. 6 vary significantly over the source region, so that the values at $\xi = 0$ are inappropriate for a portion of the moment release. In effect, neglected higher moments are contaminating estimates of the lower moments.

This effect is *not* a consequence of using an incorrect Earth model to compute the Green function. In this discussion, we assume that the Earth model and Green's function are exact. The distortion of the focal mechanism is caused by the finiteness of the source region. Errors in the

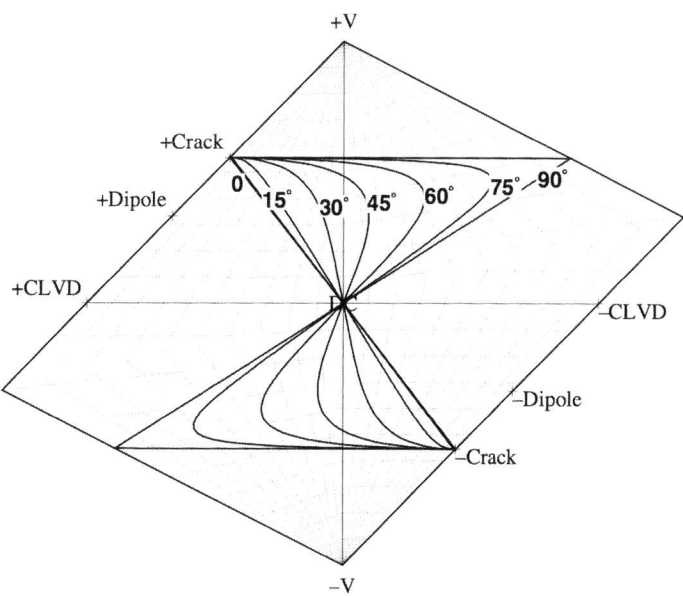

Non-Double-Couple Earthquakes, Fig. 7 Source types for combined tensile and shear faulting. Numbers give angles between the tensile-fault planes and the intermediate principal (null) axes of the shear faults. Small angles are physically most plausible (For the mechanisms shown in Fig. 6 this angle is zero.). For all possible relative orientations and moments, the source type lies between the corresponding curve and the *straight line* from +Crack to −Crack. The upper half of the plot corresponds to opening faults and the lower half to closing faults. For an explanation of the plotting method, see Fig. 2

Green function due to our incomplete knowledge of Earth structure and to the mathematical complexity of elastodynamics can cause severe errors in derived earthquake mechanisms, but that is a different phenomenon.

Consider an earthquake near an interface across which the elastic moduli change discontinuously (Woodhouse 1981), so that the truncated Taylor series Eq. 5 is a particularly poor approximation. Then the inferred mechanism of an earthquake that is assumed to occur on one side of the interface will be distorted if the earthquake actually is on the other side. If the source region includes both sides of the interface, it is unavoidable that a portion of the moment release will be distorted in this manner.

If we arrange the independent components of the moment tensor in a column vector,

$$\mathbf{m} \stackrel{\text{def}}{=} [M_{11}\ M_{12}\ M_{22}\ M_{13}\ M_{23}\ M_{33}]^T, \quad (26)$$

then the seismic waves excited can be written $\mathbf{g}^T \mathbf{m}$, where $\mathbf{g}$ is a column vector whose components are spatial derivatives of Green's functions appearing in the second term on the right side of Eq. 6. Because displacement and stress are continuous at the interface, the elements of $\mathbf{g}$ on one side of the interface are related to those on the other side by a relation that can be written

$$\mathbf{A}^+ \mathbf{g}^+ = \mathbf{A}^- \mathbf{g}^-, \quad (27)$$

where $\mathbf{A}$ is a matrix that depends on the orientation of the interface and the elastic moduli adjacent to it, and the superscripts $^+$ and $^-$ indicate values on the two sides of the interface. It follows that

$$\mathbf{g}^{+T}\mathbf{m} = \mathbf{g}^{-T}[\mathbf{A}^{+-1}\mathbf{A}^-]^T\mathbf{m}, \quad (28)$$

or in other words that an earthquake with moment tensor $\mathbf{m}$ occurring on the + side of the interface excites the same waves as an earthquake with moment tensor $[\mathbf{A}^{+-1}\mathbf{A}^-]^T \mathbf{m}$ occurring on the opposite (−) side. In a coordinate system with the x_3 axis normal to the interface, the matrix connecting the true and apparent moment tensors is

$$[\mathbf{A}^{+-1}\mathbf{A}^-]^T = \begin{vmatrix} 1 & 0 & 0 & 0 & 0 & R_1 \\ 0 & 1 & 0 & 0 & 0 & 0 \\ 0 & 0 & 1 & 0 & 0 & 0 \\ 0 & 0 & 0 & R_2 & 0 & 0 \\ 0 & 0 & 0 & 0 & R_2 & 0 \\ 0 & 0 & 0 & 0 & 0 & R_3 \end{vmatrix},$$

(29)

where

$$R_1 \stackrel{\text{def}}{=} \frac{\lambda^- - \lambda^+}{\lambda^+ + 2\mu^+},$$

(30)

$$R_2 \stackrel{\text{def}}{=} \frac{\mu^-}{\mu^+},$$

(31)

and

$$R_3 \stackrel{\text{def}}{=} \frac{\lambda^- + 2\mu^-}{\lambda^+ + 2\mu^+}.$$

(32)

Figure 8 shows the distortion of the apparent mechanisms of DCs of various orientations occurring adjacent to a horizontal interface across which the elastic-wave speeds change by 20 %. If the fault plane is parallel to the interface (or if the interface *is* the fault plane), then shear faulting does not lead to apparent non-DC mechanisms, although the scalar seismic moment is distorted. This case is not illustrated in the figure, but is clear from the structure of the matrix in Eq. 29. Only M_{13} and M_{23} are nonzero and matrix multiplication merely multiplies these elements by R_2, producing a DC of the same orientation, but with its moment multiplied by the contrast in rigidity modulus. If the fault is perpendicular to the interface, then the apparent mechanism is still a DC for all slip directions, but its orientation and seismic moment are distorted, as the first column of the figure illustrates. For general fault orientations, the apparent mechanism has artificial isotropic and CLVD components.

Rapid Polymorphic Phase Changes

Except in the shallow crust, compressional stresses in the Earth greatly exceed shear stresses. Therefore, earthquake processes that involve even relatively small volume changes could release large amounts of energy. For this reason, it has long been speculated that polymorphic phase transformations in minerals might cause deep earthquakes. Such speculation has been stimulated also by consideration of the simple theory of frictional sliding, which seems to require impossibly large shear tractions when the confining pressure is high, and by the theory of plate tectonics, which involves large-scale vertical motions in the upper mantle. Many common minerals undergo polymorphic changes in crystal structure in response to changes in pressure and temperature, and some of the major structural features in the Earth, most notably the "transition region" at depths between about 400 and 800 km in the upper mantle, are attributed to such phase changes (in this case, involving the mineral olivine, $(Fe, Mg)_2SiO_4$, transforming to the spinel and then perovskite crystal structures).

As slabs of lithosphere subduct into the mantle, olivine and other minerals are carried out of their stability fields and into the stability fields of denser phases, into which they transform. If these changes occur rapidly enough to radiate seismic waves, they constitute earthquakes, and their mechanisms will have isotropic components. They probably will also have deviatoric components, because the process of phase transformation will release shear strain, much as explosions are often observed to release tectonic shear strain. There seems to be no reason, however, why such a deviatoric component should be a DC rather than a CLVD, and in fact the CLVD force system was first invented as a possible mechanism for deep earthquakes caused by phase transformations (Knopoff and Randall 1970).

Observations

Non-DC earthquakes are observed in many environments, including volcanic and geothermal areas, mines, and deep subduction zones.

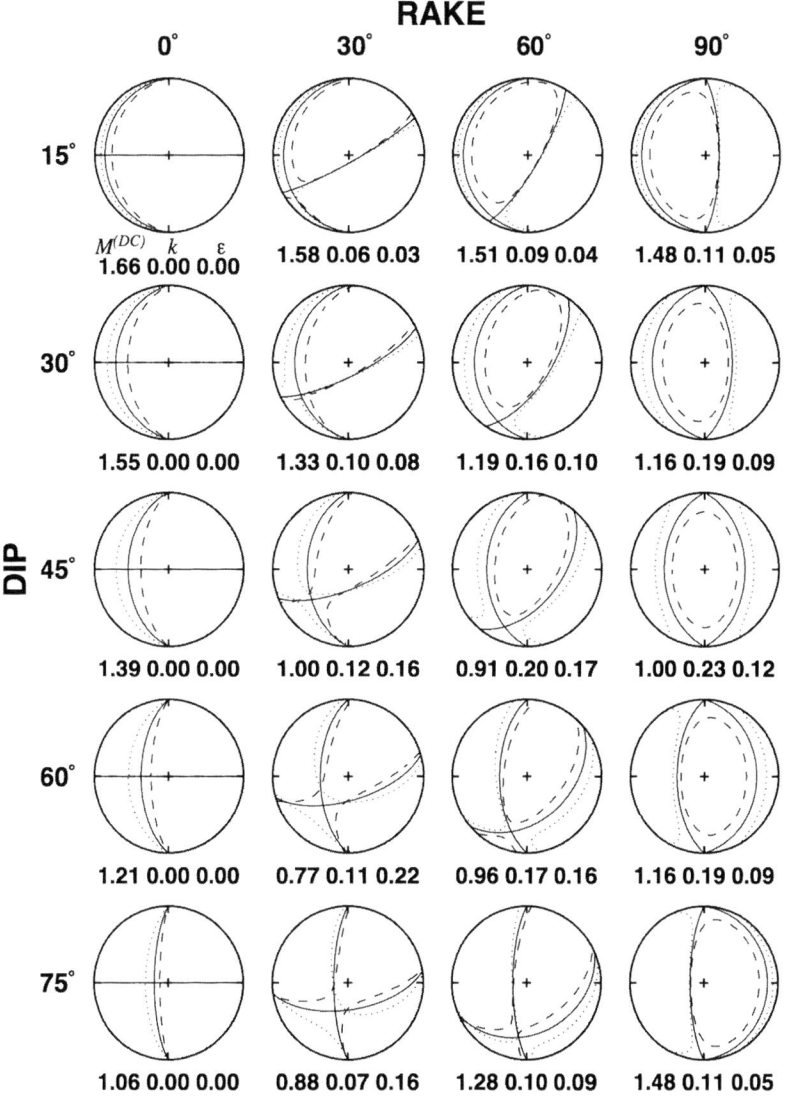

Non-Double-Couple Earthquakes, Fig. 8 Apparent non-DC mechanisms caused by shear faulting with unit moment near a horizontal interface, for various dip and rake angles. *Solid curves*: compressional-wave nodal planes corresponding to true (DC) mechanisms. *Dotted curves*: nodal surfaces for apparent mechanisms obtained if DC moment release on the low-speed side of interface is assumed to be on the high-speed side. *Dashed curves*: same, with sides interchanged. Both media are Poisson solids, and the ratio of elastic moduli across the interface is 1.7:1, corresponding to a wave-speed contrast of about 20 % if density is proportional to wave speed. Numbers below each mechanism give the DC moment and the values of k and ε (Eqs. 18 and 19) corresponding to the *dotted curves*. Dips of 0 are not shown because the apparent mechanisms are DCs (although with moments distorted by the ratio of the rigidity moduli). Dips of 90° correspond to cases with rake of 0 (first column). Focal hemispheres (which may be considered either upper or lower) are shown in equal-area projection

They include both natural earthquakes and events induced by human activity. An in-depth review of many observed examples is given by Miller et al. (1998b).

Landslides and Volcanic Eruptions

Landslides and volcanic eruptions have equivalent force systems that include net forces (section "Landslides"). An example is the Mount

St. Helens, Washington, eruption of May 18, 1980, which began with a landslide with a mass of about 5×10^{13} kg that traveled about 10 km from the volcano. The seismic observations are well explained by two forces, a near-horizontal, southward directed force representing the landslide, and a vertical force representing the eruption (Kanamori et al. 1984).

The gravitational forces on a landslide also exert a torque, which can be substantial, on the source volume (section "Landslides"). The dimensions of the landslide that accompanied the May 18, 1980, eruption of Mt. St. Helens, for example, correspond to a net torque of about 5×10^{18} N m. By comparison, the two largest earthquakes accompanying the eruption had surface-wave magnitudes of about 5.3, which correspond to seismic moments of about 2.6×10^{17} N m. Apparently, no seismological analyses of landslides to date have included torques in the source mechanism.

Long-Period Volcanic Earthquakes

Long-period volcanic earthquakes have dominant seismic frequencies an order of magnitude lower than those of most earthquakes with comparable magnitudes. They are caused by unsteady flow of underground magmatic fluids and are expected to have mechanisms involving net forces (section "Unsteady Fluid Flow"). An example is a 33-km-deep, long-period earthquake that occurred a year before the 1986 eruption of Izu Ōshima volcano, Japan. The seismograms are characterized by nearly monochromatic shear-wave trains that have dominant frequencies of about 1 Hz and last for more than a minute. The shear-wave polarization directions are consistent with a net force mechanism.

Several types of non-DC earthquakes occur at the intensively monitored active Sakurajima volcano in southern Kyushu, Japan. It is a rich source of low-frequency earthquakes and explosion earthquakes, which accompany crater eruptions that radiate spectacular visible shock waves into the atmosphere. The moment tensors of the explosion earthquakes have been explained as either deflation of cracks or other cavities that might rapidly expel gas into the atmosphere (Uhira and Takeo 1994), or as sources dominated by vertical dipoles involving, for example, the opening of horizontal cracks (Iguchi 1994). The discrepancies between the results of these two studies reflect the difficulty of distinguishing between vertical forces and vertical dipoles.

Short-Period Volcanic and Geothermal Earthquakes

Observations from dense local seismic networks show that earthquakes in volcanic areas commonly have non-DC mechanisms. In many cases, the data have been subjected to careful analysis, including waveform modeling (e.g., Dreger et al. 2000; Julian and Sipkin 1985), corrections for wave propagation through three-dimensional crustal structure determined from tomography (e.g., Miller et al. 1998a; Ross et al. 1996), and the use of multiple seismic phases. The clearest cases are from Iceland, California, and Japan.

Three volcano-geothermal areas on the spreading plate boundary in Iceland are rich sources of short-period, non-DC earthquakes. These are the Reykjanes, Hengill-Grensdalur, and Krafla volcanic-geothermal systems. At all three areas, small earthquakes occur that have unequal dilatational and compressional areas on the focal sphere, suggestive of isotropic components in their mechanisms. The non-DC earthquakes are intermingled spatially with DC events, indicating that the non-DC mechanisms are not artifacts of instrumental errors or propagation through heterogeneous structures.

The Hengill-Grensdalur triple junction is the best studied of these areas. It was there that experiments were first performed in sufficient detail to demonstrate beyond reasonable doubt that non-DC earthquakes with net volume changes occur in nature (Foulger 1988; Miller 1998a) (Fig. 9). The area is currently the richest source of extreme non-DC earthquakes known. Approximately 70 % of the earthquakes involve volume increases, i.e., their radiation patterns are dominated by compressions, and they are thought to be caused by thermal contraction in the heat source of the geothermal area (Fig. 10).

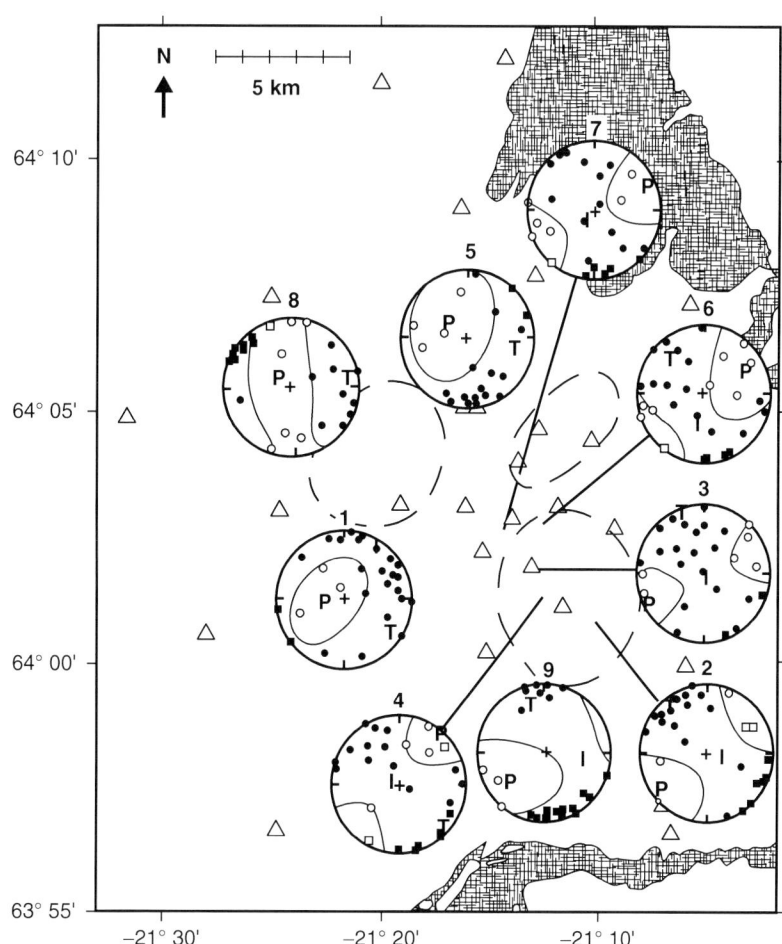

Non-Double-Couple Earthquakes, Fig. 9 Map of the Hengill-Grensdalur volcanic complex, Iceland, showing focal mechanisms of nine representative earthquakes. Events 5, 6, 8, and 9 are indistinguishable from DCs. Compressional-wave polarities (*solid circles* indicate compressions; *open circles* indicate dilatations) and P-wave nodes are shown in upper focal hemisphere equal area projection. *Squares* denote downgoing rays plotted on the upper hemisphere. T, I, and P are positions of principal axes. *Lines* indicate epicentral locations. Where *lines* are absent, mechanisms are centered on epicenters. *Triangles* indicate seismometer locations; *dashed lines* show outlines of the main volcanic centers

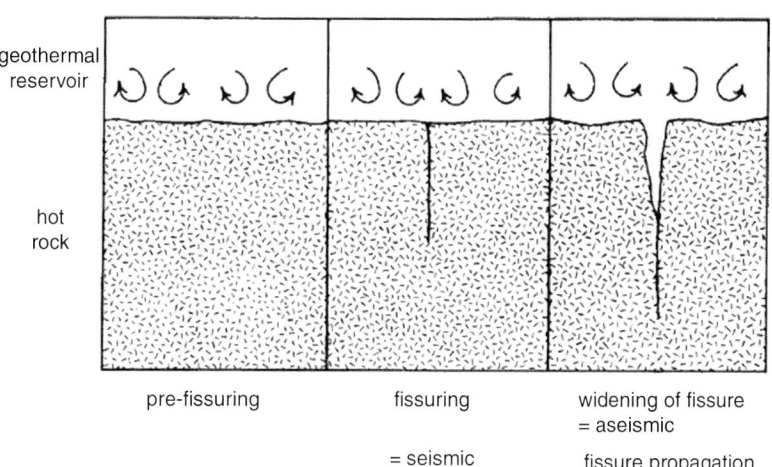

Non-Double-Couple Earthquakes, Fig. 10 Schematic illustration of seismogenic tensile cracking by thermal stresses caused by convective cooling of rocks at the heat source of a geothermal system (From Foulger 1988)

Non-Double-Couple Earthquakes, Fig. 11 Non-shear fault plane delineated by earthquake hypocenters. Views from three orthogonal directions showing the locations of 314 earthquakes (*black dots*) that lie between 2 and 5 km depth. *Upper left*: map view; *upper right*: SSW-NNE vertical cross-section; *lower left*: WNW-ESE vertical cross-section. *Star*: earthquake whose focal mechanism is shown at *lower right* as an upper-hemisphere plot of compressional-wave polarities. The mechanism for this earthquake is compatible with compensated tensile failure

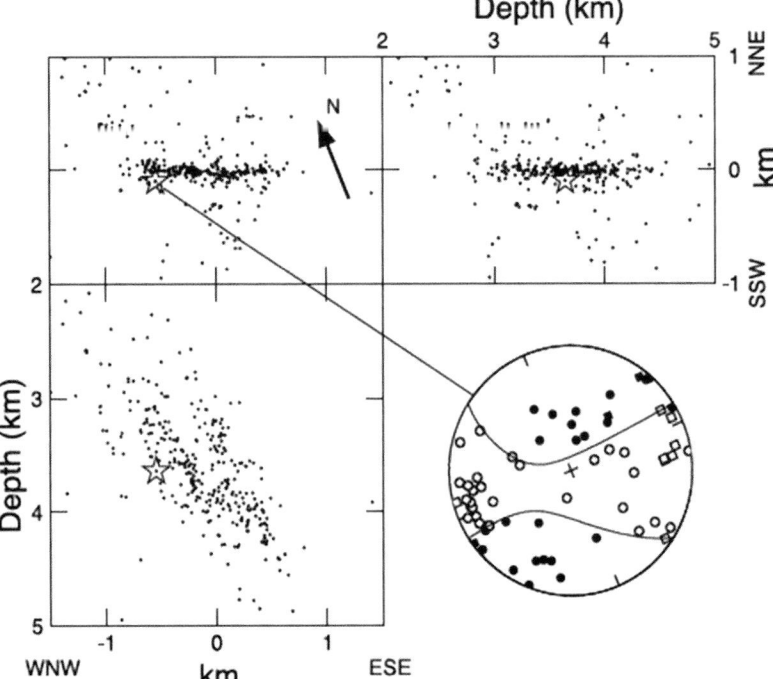

The source mechanisms require more than simple tensile cracking, because the radiation patterns include both compressions and dilatations. Fluid flow and/or shear faulting probably also contribute components.

Non-DC geothermal earthquakes with volumetric mechanisms occur also at several volcanic-geothermal areas in California, including the Geysers steam field (Ross et al. 1996), the Coso Hot Springs (Julian et al. 2010), and Long Valley caldera (Foulger et al. 2004). Industrial steam extraction and water reinjection at the Geysers induces thousands of small earthquakes per month in the reservoir. The mechanisms are typically closer to DCs than those of Icelandic geothermal earthquakes, and the volumetric non-DC components range from explosive to implosive, with approximately equal numbers of each type.

Non-DC earthquake focal mechanisms were observed at the Coso Hot Springs geothermal field, California, in March 2005. There, a fluid injection activated a fault plane on which earthquakes with non-DC mechanisms of uniform type occurred. These involved volume increases. The orientation of the mechanisms relative to the fault plane implied a process dominated by tensile failure with subsidiary shear faulting.

Abundant non-DC earthquakes were observed in a detailed experiment at Long Valley caldera in 1997. More than 80 % of the mechanisms had explosive volumetric components and most had compensated linear-vector dipole components with outward-directed major dipoles. The simplest interpretation of these mechanisms is combined shear and extensional faulting accompanied by a volume-compensating process such as the rapid flow of water, steam, or CO_2 into opening tensile cracks A particularly well-recorded swarm was consistent with extensional faulting on an ESE-striking subvertical plane, an orientation consistent with the activated zone defined by the earthquake hypocenters. The orientations of non-DC earthquakes beneath Mammoth Mountain, a volcanic cone on the southwestern caldera rim, varied systematically with location, reflecting a variable local stress field. Events in a spasmodic burst indicated nearly pure compensated tensile failure and high fluid mobility (Foulger et al. 2004) (Fig. 11). Earthquakes with M_W 4.6–4.9 observed at

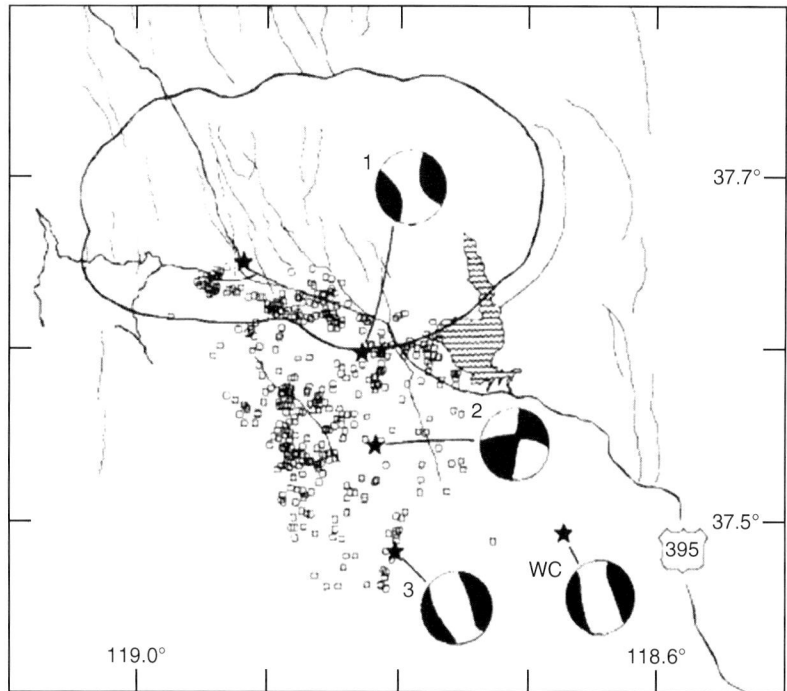

Non-Double-Couple Earthquakes, Fig. 12 Long Valley caldera, California, and vicinity, showing the best located earthquakes in 1980 and mechanisms for the largest earthquakes of 1978 and 1980 as lower-hemisphere, equal-area projections, with fields of compressional-wave polarity in *black*

regional distances on broadband sensors also had non-DC moment tensors with significant volumetric components and were probably caused by hydrothermal or magmatic processes (Dreger et al. 2000).

Earthquakes with non-DC radiation patterns, some of them moderately large, accompany magmatic and volcanic activity. Many non-DC earthquakes with P-wave polarities that are either all dilatational or all compressional accompanied the 1983 eruption of Miyakejima volcano, Japan. The earthquakes radiated significant shear waves, indicating that their mechanisms were not purely isotropic. The P-wave polarities and P- and SV-wave amplitudes are compatible with combined tensile and shear faulting (section "Combined Tensile and Shear Faulting"), an interpretation supported by the numerous open cracks that formed prior to the eruption. A similar but substantially larger earthquake, of magnitude 3.2, occurred 10 km beneath the Unzen volcanic region, western Kyushu, in 1987 (Shimizu et al. 1987).

Long Valley caldera, in eastern California, has experienced some of the largest clearly non-DC volcanic earthquakes. Four events with M_L greater than six occurred there in May 1980 (Fig. 12), and at least two had non-DC mechanisms. These events followed 2 years of volcanic unrest characterized by escalating seismic activity and surface deformation indicating magma-chamber inflation. Moment tensor inversions of different subsets of the available data conducted independently and using different methods all gave mechanisms that were similar and close to CLVDs. Any volumetric components were unresolvably small. The source processes probably involved tensile faulting at high fluid pressure, perhaps associated with dike intrusion. A M_S 5.6 earthquake similar in mechanism to the Long Valley earthquakes, though differently orientated, occurred near Tori Shima island, in the Izu-Bonin arc, on June 13, 1984. Because the earthquake was shallow, its full moment tensor could not be determined well, but the data were consistent with a mechanism close to a CLVD with a vertical symmetry axis. The seismic moment and source duration required intrusion of approximately 0.02 km^3 within 10–40 s (Kanamori et al. 1993).

Non-Double-Couple Earthquakes,
Fig. 13 Harvard centroid moment tensor mechanisms of earthquakes at Bardarbunga volcano, southeast Iceland, from Ekström (1994)

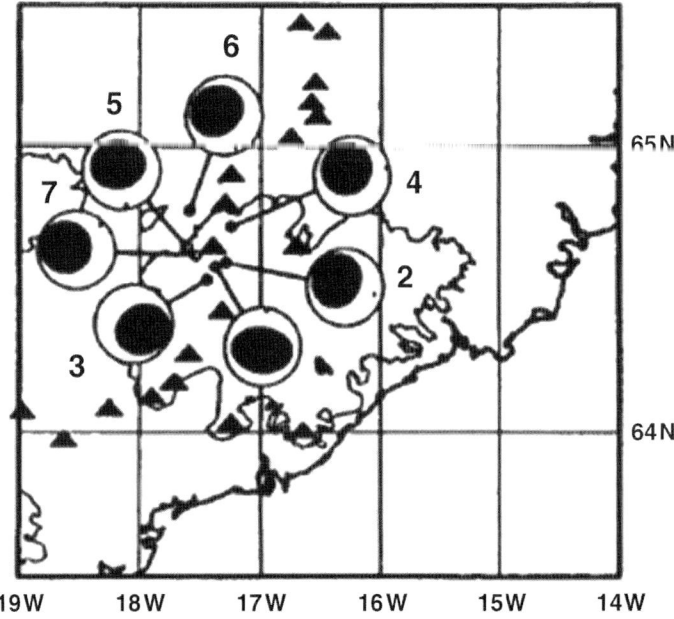

Bardarbunga volcano, beneath the Vatnajökull ice cap in Iceland, regularly generates non-DC earthquakes with M_W 5.2–5.6 ($M_0 = 8$ to 30×10^{16} Nm) (Ekström 1994). They typically have nearly vertical CLVD-like mechanisms. A particularly well-studied example that occurred in 1996 yielded a non-DC mechanism with a 67 % vertically oriented compensated linear vector dipole component, a 32 % DC component, and a statistically insignificant (2 %) volumetric contraction. They may result from slip on a ring fault and can be explained by activation of less than 55° of such a fault. Cone sheets, which dip inward, have the most favorable geometry (Fig. 13).

Earthquakes at Mines

Deep mine excavations perturb the stresses in the surrounding rocks, reducing some components from values initially of the order of 100 MPa practically to zero. The resulting stress differences can exceed the strength of competent rock and cause earthquakes (often called "rock bursts," "coal bumps," etc.). Seismic data show clearly that many earthquakes at mines have non-DC mechanisms, usually with predominantly dilatational radiation patterns, that are incompatible with shear faulting.

An M_L 3.5 earthquake on May 14, 1981, at the Gentry Mountain mine, Utah, coincided with the collapse of a large cavity at 200-m depth in the mine.

Well-recorded earthquakes occurring at the Coeur d'Alene mining district, Idaho, had dilatational P-wave polarities, consistent with cavity collapse.

A M_W 3.9 event occurred in the Crandall Canyon coal mine, Utah, August 6, 2007, in association with a gallery collapse that killed six miners and, later, three rescuers. Its focal mechanism is dominated by an implosive component, consistent with a shallow underground collapse (Dreger et al. 2008).

Many non-DC earthquakes occurred in association with the excavation of a 3.5-m-wide tunnel in unfractured, homogeneous granite at the Underground Research Laboratory in Manitoba, Canada (Feignier and Young 1992). Moment tensors computed for 33 earthquakes with M_W -2 to -4 include tensile, implosive, and shear mechanisms. Some were associated with breakout on the tunnel roof, and others occurred ahead of the active face.

Earthquakes located within 150 m of deep gold mines in the Witwatersrand, South Africa, had large volume decreases consistent with the partial collapse of mine galleries (McGarr 1992).

Fluid Injection

Fluids are injected into the shallow crust increasingly often for reasons such as increasing permeability within geothermal and hydrocarbon reservoirs by hydrofracturing, sequestration of CO_2, storing gas, and disposing of waste. Such injection often induces earthquakes, but they are not always monitored in detail. Nevertheless, in a number of well-studied cases non-DC source mechanisms have been observed (e.g., Baig and Urbancic 2010; Julian et al. 2010). Source types typically range from explosive to implosive dipoles. As underground fluid injections increase in importance over the coming decades, more detailed studies of the source mechanisms of the earthquakes induced may prove to be important for understanding the response of the Earth to such operations, project optimization and hazard mitigation.

Other Shallow Earthquakes

Real fault surfaces are not perfectly planar, and earthquakes on curved faults can, for some geometries, have non-DC mechanisms (Frohlich et al. 1989). Effects of this sort may account for the non-DC components observed in some large earthquakes, such as the 28-km deep M_S 7.8 event that occurred near Taiwan on November 14, 1986 (Zheng et al. 1995). The focal mechanism is inconsistent with shear faulting, but requires an implosive component. A similar, M 4.6 earthquake that occurred February 10, 1987, beneath the Kanto district, Japan. It had primarily dilatational compressional-wave polarities and was consistent with a conical nodal plane.

Deep-Focus Earthquakes

Earthquakes up to 700 hundred km deep occur in the mantle beneath subduction zones. The physical causes of deep earthquakes are not well understood. This is because minerals are expected to flow plastically at depths greater than about 30 km in normal areas. Processes that have been suggested to explain deep earthquakes include plastic instabilities, shear-induced melting, polymorphic phase transformations, and transformational faulting. Focal mechanisms can potentially shed light on this problem.

Volume changes in deep earthquakes are statistically unresolvable (less than 10 % of the seismic moment) (Kawakatsu 1991), indicating that polymorphic phase changes (section "Rapid Polymorphic Phase Changes"), play an insignificant role in deep earthquakes. Deep earthquakes do, however, have larger CLVD components than shallow earthquakes, and in some cases the sizes of the CLVD components increases systematically with depth and magnitude. The lack of large volume changes in deep earthquakes is compatible with complex shear faulting, and detailed waveform analysis resolves some deep non-DC earthquakes into DC subevents.

Glacial Earthquakes

Glaciers are subject to several kinds of sudden transient phenomena, apparently caused by diverse physical processes. Application of seismological analysis techniques to these events promises to provide fundamentally new types of data bearing on glacier dynamics and on understanding glaciers' response to influences such as climate change (Ekström et al. 2006).

Earthquakes apparently caused by glacier surges or calving, even though they can be large ($M \geq 4.5$), were recognized on seismograms only recently, because they have source durations of the order of a minute or more and excite high-frequency seismic waves only weakly. The application of array-processing techniques to digital seismic data from the global network now enables us to detect dozens of large glacial earthquakes per year that previously went unnoticed. A particularly well-recorded event, of magnitude about 5, occurred in 1999 near the Dall glacier in the Denali range of Alaska, within a regional seismometer network and near a second, temporarily deployed, network. The low-frequency (~0.01–0.2 Hz) seismic waves are inconsistent with any moment-tensor mechanism, but agree well with those predicted for an approximately horizontal force directed in the direction of the glacier flow (Ekström et al. 2003). Seismic data from numerous other events, most of them near the margins of the Greenland icecap, are consistent with similar single-force mechanisms.

Most small ($M \geq -2.5$) icequakes in mountain glaciers have mechanisms describable by moment tensors, but they often differ greatly from DCs. Walter et al. (2009, 2010) recorded tens of thousands of events on Gornergletscher, in the Swiss Alps, on dense seismometer networks, and extended Dreger's waveform-inversion method, originally developed for earthquakes above magnitude 4, to frequencies approaching 1 KHz, in order to study the mechanisms of these microearthquakes. Most events had large explosive volumetric components, with source types close to those for opening tensile cracks. These included shallow events, probably representing crevasse opening, and events 100 m below the surface. Many shallow events were of a different type, equivalent to a DC combined with a volume decrease. A later study (Walter et al. 2010) of events near the base of the glacier found opening-crack mechanisms, but with inferred crack planes oriented horizontally.

Summary

A wide variety of processes can cause earthquake mechanisms to depart from the ideal DC force system that characterizes planar shear faulting in a homogeneous isotropic medium. These departures can be as extreme as unbalanced forces or torques associated with major landslides, glacial surges, or volcanic eruptions, or can be minor anomalies that are barely resolvable with the best data currently available. Studying non-DC earthquake mechanisms can be valuable for refining our knowledge of how faults work, for learning to understand and predict volcanic activity, for prospecting and exploiting geothermal energy and hydrocarbons, and for avoiding damage to civil and industrial infrastructure. Moreover, non-DC mechanisms are of direct value for monitoring mining and other industrial activities, particularly ones that involve fluid injection.

References

Aki K, Richards PG (2002) Quantitative seismology, 2nd edn. University Science, Sausalito

Anderson EM (1936) The dynamics of the formation of cone-sheets, ring-dykes, and cauldron-subsidences. Proc R Soc Edinb 56:128–156

Backus G (1977) Interpreting the seismic glut moments of total degree two or less. Geophys J R Astron Soc 51:1–25

Baig A, Urbancic T (2010) Microseismic moment tensors: a path to understanding frac growth. Lead Edge 29:320–324

Booth DC, Crampin S (1985) Shear-wave polarizations on a curved wavefront at an isotropic free surface. Geophys J R Astron Soc 83:31–45

Brace WF, Paulding BW Jr, Scholz CH (1966) Dilatancy in the fracture of crystalline rocks. J Geophys Res 71:3939–3953

Dreger DS, Tkalčić H, Johnston M (2000) Dilational processes accompanying earthquakes in the Long Valley caldera. Science 288:122–125

Dreger D, Ford SR, Walter WR (2008) Source analysis of the Crandall Canyon, Utah mine collapse. Science 321:217

Ekström G (1994) Anomalous earthquakes on volcano ring-fault structures. Earth Planet Sci Lett 128:707–712

Ekström G, Nettles M, Abers GA (2003) Glacial earthquakes. Science 302:622–624

Ekström G, Nettles M, Tsai VC (2006) Seasonality and increasing frequency of Greenland Glacial earthquakes. Science 311:1756–1758

Feignier B, Young RP (1992) Moment tensor inversion of induced microseismic events: evidence of non-shear failures in the $-4 < M < -2$ moment magnitude range. Geophys Res Lett 19:1503–1506

Foulger GR (1988) Hengill triple junction, SW Iceland; 2. Anomalous earthquake focal mechanisms and implications for process within the geothermal reservoir and at accretionary plate boundaries. J Geophys Res 93:13507–13523

Foulger GR, Julian BR, Hill DP, Pitt AM, Malin P, Shalev E (2004) Non-double-couple microearthquakes at Long Valley caldera, California, provide evidence for hydraulic fracturing. J Volcanol Geotherm Res 132:45–71

Frohlich C, Riedesel MA, Apperson KD (1989) Note concerning possible mechanisms for non-double-couple earthquake sources. Geophys Res Lett 16:523–526

Gajewski D (1993) Radiation from point sources in general anisotropic media. Geophys J Int 113:299–317

Gilbert F (1970) Excitation of the normal modes of the Earth by earthquake sources. Geophys J R Astron Soc 22:223–226

Hudson JA, Pearce RG, Rogers RM (1989) Source type plot for inversion of the moment tensor. J Geophys Res 94:765–774

Iguchi M (1994) A vertical expansion source model for the mechanisms of earthquakes originated in the magma conduit of an andesitic volcano: Sakurajima, Japan. Bull Volcanol Soc Jpn 39:49–67

Julian BR (1986) Analysing seismic-source mechanisms by linear-programming methods. Geophys J R Astron Soc 84:431–443

Julian BR (1994) Volcanic tremor: nonlinear excitation by fluid flow. J Geophys Res 99:11859–11877

Julian BR, Foulger GR (1996) Earthquake mechanisms from linear-programming inversion of seismic-wave amplitude ratios. Bull Seismol Soc Am 86: 972–980

Julian BR, Sipkin SA (1985) Earthquake processes in the Long Valley caldera area, California. J Geophys Res 90:11155–11169

Julian BR, Foulger GR, Monastero FC, Bjornstad S (2010) Imaging hydraulic fractures in a geothermal reservoir. Geophys Res Lett 37:L07305. doi:10.1029/2009GL040933

Kanamori H, Given JW, Lay T (1984) Analysis of seismic body waves excited by the Mount St. Helens eruption of May 18, 1980. J Geophys Res 89:1856–1866

Kanamori H, Ekström G, Dziewonski A, Barker JS, Sipkin SA (1993) Seismic radiation by magma injection – an anomalous seismic event near Tori Shima, Japan. J Geophys Res 98:6511–6522

Kawakatsu H (1991) Insignificant isotropic component in the moment tensor of deep earthquakes. Nature 351:50–53

Knopoff L, Randall MJ (1970) The compensated linear vector dipole: a possible mechanism for deep earthquakes. J Geophys Res 75:4957–4963

McGarr A (1992) Moment tensors of ten Witwatersrand mine tremors. Pure Appl Geophys 139:781–800

Mendiguren JA (1977) Inversion of surface wave data in source mechanism studies. J Geophys Res 82:889–894

Miller AD, Julian BR, Foulger GR (1998a) Three-dimensional seismic structure and moment tensors of non-double-couple earthquakes at the Hengill-Grensdalur volcanic complex, Iceland. Geophys J Int 133:309–325

Miller AD, Foulger GR, Julian BR (1998b) Non-double-couple earthquakes II. Observations. Rev Geophys 36:551–568

Price NJ (1966) Fault and joint development in brittle and semi-brittle rock. Pergammon Press, New York

Richards PG, Kim WY, (2005) Equivalent volume sources for explosions at depth: Theory and observations. Bull Seismol Soc Am 95:401–407

Ross A, Foulger GR, Julian BR (1996) Non-double-couple earthquake mechanisms at The Geysers geothermal area, California. Geophys Res Lett 23:877–880

Shimizu H, Ueki S, Koyama J (1987) A tensile-shear crack model for the mechanism of volcanic earthquakes. Tectonophysics 144:287–300

Sipkin SA (1982) Estimation of earthquake source parameters by the inversion of waveform data: synthetic waveforms. Phys Earth Planet Inter 30:242–259

Takei Y, Kumazawa M (1994) Why have the single force and torque been excluded from seismic source models? Geophys J Int 118:20–30

Uhira K, Takeo M (1994) The source of explosive eruptions of Sakurajima volcano, Japan. J Geophys Res 99:17775–17789

Vavryčuk V (2005) Focal mechanisms in anisotropic media. Geophys J Int 161:334–346

Walter F, Clinton JF, Deichmann N, Dreger DS, Minson SE, Funk M (2009) Moment tensor inversions of icequakes on Gornergletscher, Switzerland. Bull Seismol Soc Am 99:852–870

Walter F, Dreger DS, Clinton JF, Deichmann N, Funk M (2010) Evidence for near-horizontal tensile faulting at the base of Gornergletscher, a Swiss Alpine glacier. Bull Seismol Soc Am 100:458–472

Woodhouse JH (1981) The excitation of long period seismic waves by a source spanning a structural discontinuity. Geophys Res Lett 8:1129–1131

Zheng T, Yao Z, Liu P (1995) The 14 November 1986 Taiwan earthquake – an event with isotropic component. Phys Earth Planet Inter 91:285–298

Nonlinear Analysis and Collapse Simulation Using Serial Computation

Xinzheng Lu[1] and Xiao Lu[2]
[1]Key Laboratory of Civil Engineering Safety and Durability of China Education Ministry, Department of Civil Engineering, Tsinghua University, Beijing, People's Republic of China
[2]School of Civil Engineering, Beijing Jiaotong University, Beijing, People's Republic of China

Synonyms

Collapse simulation; Elemental nonlinearity; Geometric nonlinearity; Material nonlinearity; Nonlinear analysis

Introduction

Due to the extensive use of performance-based seismic design in earthquake engineering, nonlinear analysis has become an important method for evaluating the seismic performances of structures. In comparison with elastic analysis, nonlinear analysis is usually more complex and expensive. Nonlinear analysis generally involves nonlinear static analysis (pushover) and nonlinear dynamic analysis in earthquake engineering. The nonlinearities mainly come from material nonlinearity, geometric nonlinearity, elemental nonlinearity, and contact nonlinearity. Only when these nonlinearities are considered correctly in the analysis models can the structural responses be reasonably predicted.

Material Nonlinearity

Most materials used in structural engineering exhibit nonlinear behavior, such as concrete, steel, and soil. For example, the deformation is not proportional to the load, and permanent plastic deformation remains after being unloaded. In addition to the elastic material constants (Young's modulus and Poisson's ratio), the yield criterion, the work hardening, and the flow rules are very essential to describe the nonlinear material behavior (Chen 2005).

The yield stress of a material is a measured stress level that separates the elastic and inelastic behavior of the material. The magnitude of the yield stress is generally obtained from the results of a uniaxial test. However, the stresses in a structure are usually multiaxial. A measurement of yielding for the multiaxial state of stress is called the yield criterion. For example, the Mohr-Coulomb criterion, the Buyukozturk criterion, and the Drucker-Prager criterion are common yield criteria for concrete, and the von Mises criterion is the widely used yield criterion for steel.

Isotropic hardening, kinematic hardening, and combined hardening are three typical work hardening rule types in civil engineering (Chen 2005). The isotropic work hardening rule assumes that, due to work hardening, the center of the yield surface remains stationary in the stress space and the size (radius) of the yield surface expands. In contrast, for the kinematic hardening rule, the yield surface does not change in size or shape, and the center of the yield surface can move in the stress space.

Yield stress and work hardening rules are two experimentally related phenomena that characterize plastic material behavior. The flow rule is also essential in establishing the incremental stress-strain relationships for plastic material. The flow rule describes the differential changes in the plastic strain components as a function of the current stress state (Chen 2005). In most nonlinear analyses, the associated flow rule in which the direction of inelastic straining is normal to the yield surface is adopted.

Geometric Nonlinearity

There are two natural classes of geometric nonlinearities: the large displacement, small strain problem and the large displacement, large strain problem. For the large displacement, small strain problem, changes in the stress-strain law can be neglected, but the contributions from the nonlinear terms in the strain-displacement relations cannot be neglected. For the large displacement, large strain problem, the constitutive relation, which must be defined in the correct frame of reference, is transformed from this frame of reference to the one in which the equilibrium equations are written. Most of the geometric nonlinearities in structural engineering correspond to the large displacement, small strain problem; two typical examples are the P-Δ effect of a structure and the P-δ effect of a component. The total Lagrangian procedure can be used to describe the geometric nonlinearity in both the static and dynamic analyses. The equilibrium is expressed with the original undeformed state as the reference. The nonlinear analysis method should have the ability to address this geometric nonlinearity; if it does not have the ability, the structural capacity will be overestimated, leading to an unsafe design.

Elemental Nonlinearity

During the nonlinear analysis, the structural components may reach their load-carrying capacities and be out of work, at which point, the elemental nonlinearity occurs. To handle

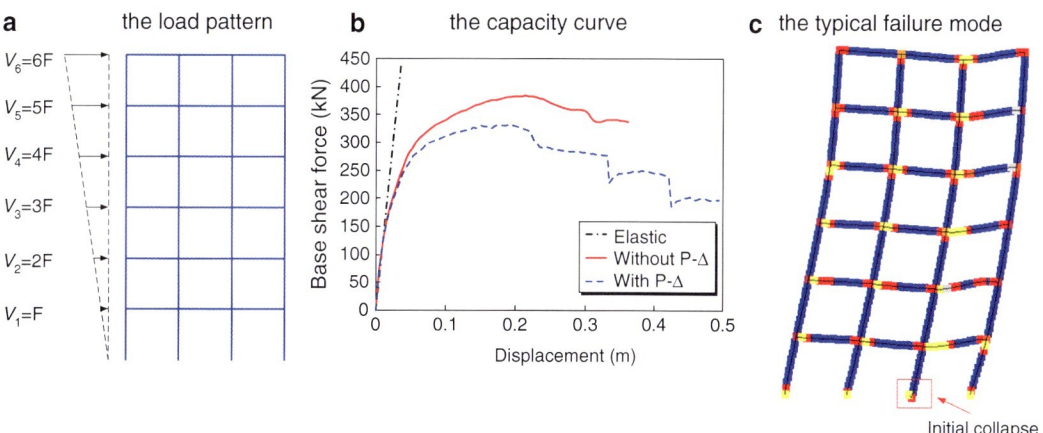

Nonlinear Analysis and Collapse Simulation Using Serial Computation, Fig. 1 Example of a pushover analysis: (**a**) the load pattern, (**b**) the capacity curve, and (**c**) the typical failure mode

this problem, the elemental deactivation technology, in which the failed elements are deactivated when a specified elemental-failure criterion is satisfied, can be used (Lu et al. 2011, 2013). For example, the material-related failure criterion should be used to monitor the failure of structural elements. If the strain at any integration point exceeds the material failure criterion, the stress and the stiffness of this element are deactivated, which means that this element no longer contributes to the stiffness computation of the entire structure. The nodes that are connected to the deactivated element are simultaneously checked. A node is considered "isolated" when all its constituent elements are deactivated, which implies that this node contains zero stiffness in relation to the degrees of freedom (DOFs). Because the "isolated" nodes are more likely to cause convergence problems, all of the DOFs of the "isolated" nodes are removed from the global stiffness matrix.

Nonlinear Static Analysis

Nonlinear static analysis, which is commonly referred to as pushover analysis in earthquake engineering, enables the simple evaluation of the bearing of structural loads and the deformation capacities, as well as the evaluation of the internal forces and deformations of components. This analysis method was first proposed by Freeman et al. (1975) to approximately assess the structural seismic performance using the seismic demand spectrum. During the pushover analysis, a given lateral load pattern is applied to the building and increases gradually to obtain the structural internal forces and displacements. Combined with the use of the seismic demand spectrum or the target displacement performance, the approximate seismic performance under the expected earthquake can be evaluated. As a simplified performance evaluation method, the following assumption should be complied: the deformation of the building is controlled by the first vibration mode, and this deformation pattern remains unchanged along the building height during the analysis process. Correspondingly, four types of load patterns are widely used in the pushover analysis, including the height-based load pattern, the fundamental mode-dominated load pattern, the SRSS-based load pattern, and the uniform load pattern.

Taking a simple frame structure (shown in Fig. 1a) as an example, the fiber-beam element (Lu et al. 2013) or plastic hinge model (Ibarra et al. 2005) can be used to simulate the nonlinear behavior of beams and columns. The inverted-triangle load pattern, which is a particular

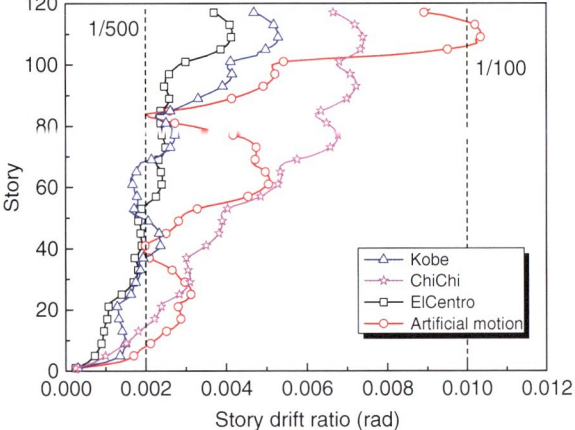

Nonlinear Analysis and Collapse Simulation Using Serial Computation, Fig. 2 Displacement responses at the MCE level

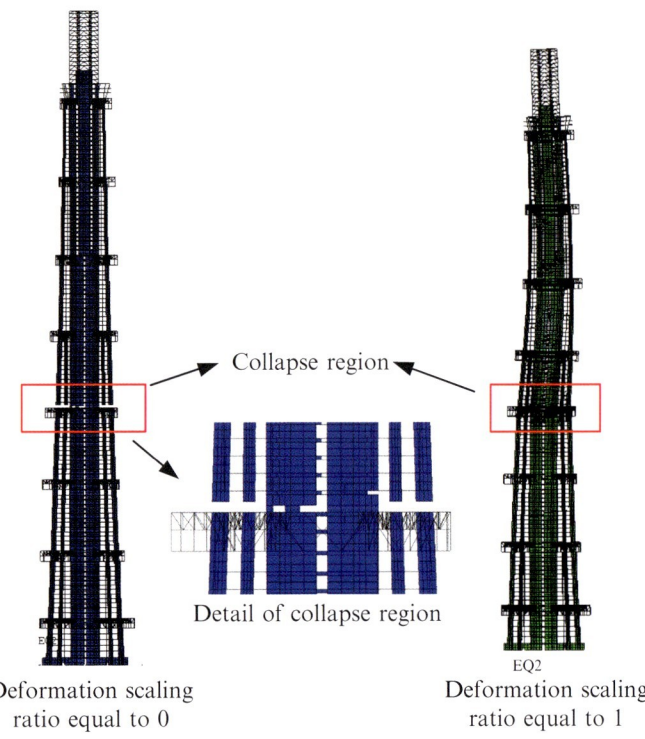

Nonlinear Analysis and Collapse Simulation Using Serial Computation, Fig. 3 Typical collapse mode subjected to the El-Centro ground motion

pattern of the height-based load pattern, is applied to this building, resulting in the capacity curve. If the linear analysis is conducted, the capacity curve is the straight line in Fig. 1b. In addition, if only material nonlinearity is considered, the capacity curve is the solid line in Fig. 1b, which overestimates the load-carrying capacity of the building. Correspondingly, if both the material and geometric nonlinearities are considered, the capacity curve is the dashed line in Fig. 1b, which better reflects the real seismic performance of the building. In addition, the initial collapse is predicted to occur in the middle column (shown in Fig. 1c).

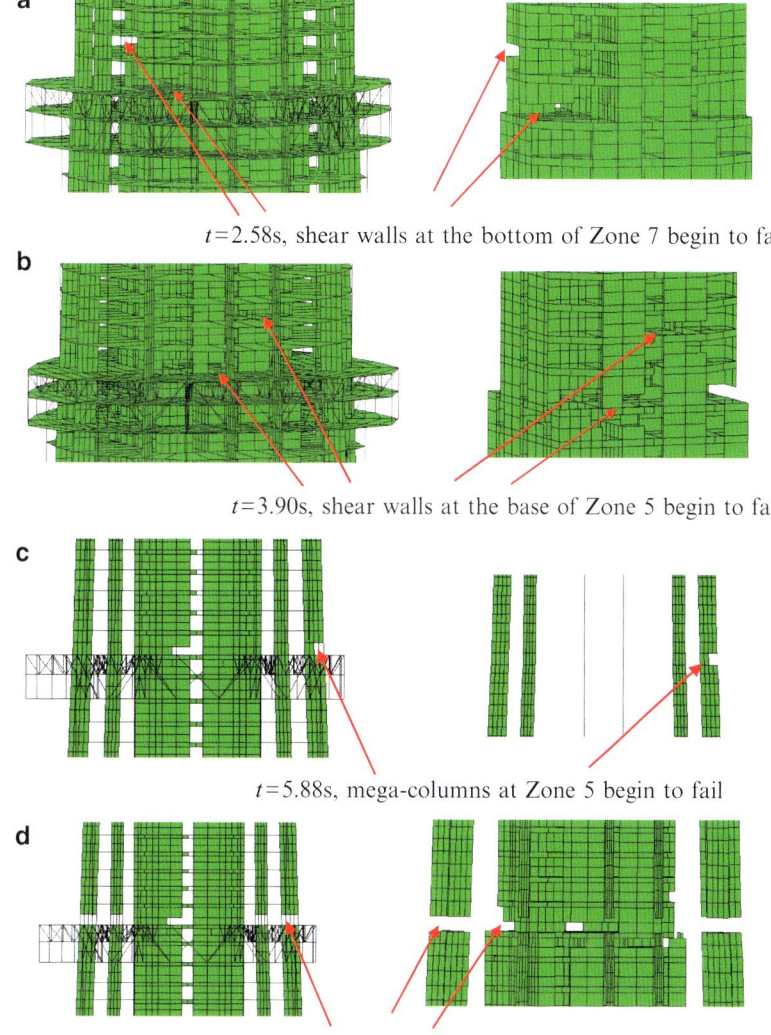

Nonlinear Analysis and Collapse Simulation Using Serial Computation, Fig. 4 Details of the collapse process

(a) $t=2.58$s, shear walls at the bottom of Zone 7 begin to fail

(b) $t=3.90$s, shear walls at the base of Zone 5 begin to fail

(c) $t=5.88$s, mega-columns at Zone 5 begin to fail

(d) $t=6.18$s, more than 50% shear walls and all mega-columns destroyed at Zone 5 the whole structure begins to collapse

Nonlinear Dynamic Analysis

In comparison with the nonlinear static analysis, the nonlinear dynamic analysis, which has attracted more attention in recent years, exhibits a broader range of applications. Because nonlinear dynamic analysis is directly based on the structural dynamic equation, the displacement, velocity, and acceleration of each story of a building and the internal forces, cracking, and yielding of the individual structural components of a building can be obtained at any moment. The analysis also reflects the characteristics of ground motion, including amplitude and duration. Therefore, this method has been widely used in many building codes for the seismic evaluation of buildings with unusual configurations or of special importance. However, the structural response may be very sensitive to the characteristics of the individual ground motion used as the seismic input; therefore, seven pairs of ground motion records, at least three pairs, are suggested in most building codes as the basic seismic input to achieve a reliable estimation of the mean structural responses. In addition, the damping effect must be considered in the dynamic analysis. Rayleigh damping is one of the most effective means of deriving a suitable damping matrix, which can be expressed in proportion to the structural mass matrix and the initial or current tangent stiffness matrix (Clough and Penzien 1975). Subsequently, based on the nonlinear detailed finite element model and appropriate ground motion records, the seismic performance of the building at various seismic hazard levels can be predicted and evaluated, as well as the potential collapse modes.

Taking an actual super-tall building, for example, the nonlinear dynamic analysis of the Shanghai Tower, with the total height of 632 m, is described in the following. The beams and columns are simulated by fiber-beam elements, the shear walls are simulated by multilayer shell elements, and elemental deactivation technology is used to simulate the components failure (Lu et al. 2011). The fundamental period is approximately 9.8 s. The displacement responses under four widely used ground motion records at the maximum considered earthquake (MCE) level and the typical potential collapse mode under the extreme earthquake are shown in Figs. 2 and 3, respectively.

The details of the collapse process are illustrated in Fig. 4. When $t = 2.58$ s, some coupling beams in the core tube begin to fail, and the flange wall of the core tube at the bottom of Zone 7 is crushed. This crushing is caused by the changes in the layout of the openings in the core tube between Zones 6 and 7, which results in sudden changes in the stiffness and the stress concentration. When $t = 3.90$ s, the shear wall at the bottom of Zone 5 begins to fail because the cross section of the core tube changes from Zone 4 to Zone 5, as shown in Fig. 4b. When $t = 5.88$ s, over 50 % of the shear walls at the bottom of Zone 5 fail, and the internal forces are redistributed to the remaining components. The vertical and horizontal loads in the mega-columns gradually increase and reach their load capacities. Subsequently, the mega-columns begin to fail. When $t = 6.18$ s, the core tube and mega-columns in Zone 5 are completely destroyed, and the collapse begins to propagate throughout the entire structure. Obviously, Zone 5 is a typical potentially weak part.

Summary

To improve the structural seismic safety of new and existing buildings is always the most important objective in earthquake engineering. Based on the appropriate numerical models, including the models of the material nonlinearity, geometric nonlinearity, and elemental nonlinearity, the nonlinear analysis, especially the nonlinear dynamic analysis, is currently one of the best tools for the prediction of structural responses at varying seismic hazard levels. In addition, the potential collapse modes and the corresponding weak parts can be predicted using a collapse simulation, which provides a better understanding of the collapse mechanism of buildings and contributes to the further development of both the optimum design and the structural health monitoring methodologies.

Cross-References

▶ Nonlinear Dynamic Seismic Analysis
▶ Seismic Collapse Assessment

References

Chen HF (2005) Elasticity and plasticity. China Architecture & Building Press, Beijing

Clough RW, Penzien J (1975) Dynamics of structures. McGraw Hill, New York

Freeman SA, Nicoletti JP, Tyrell JV (1975) Evaluation of existing buildings for seismic risk-a case study of Puget Sound Naval Shipyard, Bremerton, Washington. In: Proceedings of the 1st U.S. national conference earthquake engineering, EERI, Berkeley, pp 113–122

Ibarra LF, Medina RA, Krawinkler H (2005) Hysteretic models that incorporate strength and stiffness deterioration. Earthq Eng Struct Dyn 34(12):1489–1511

Lu X, Lu XZ, Zhang WK, Ye LP (2011) Collapse simulation of a super high-rise building subjected to extremely strong earthquakes. Sci China Technol Sci 54(10):2549–2560

Lu X, Lu XZ, Guan H, Ye LP (2013) Collapse simulation of reinforced concrete high-rise building induced by extreme earthquakes. Earthq Eng Struct Dyn 42(5):705–723

Nonlinear Dynamic Seismic Analysis

Evangelos Sapountzakis
Department of Structural Engineering, School of Civil Engineering, National Technical University of Athens (N.T.U.A.), Athens, Greece

Synonyms

Flexural–torsional analysis; Geometrically nonlinear dynamic analysis; Secondary torsional moment; Shear deformation warping

Introduction

In engineering practice beam-like continuous systems are encountered in the vast majority of structural engineering projects (Fig. 1). The term "beam-like" refers to these structural members, one dimension of which is significantly larger than the other two. As far as the analysis of such members is concerned, many modeling strategies can be applied (based mainly on finite elements); however, it is common knowledge that, especially in cases of complicated beam assemblages, beam theories are a very attractive modeling approach. Beam elements are practical and computationally efficient and offer better insight of the structural phenomena since they permit their isolation and their independent investigation. Furthermore, due to easy parameterization of all necessary data, beam elements are more convenient for parametric analyses than more refined models which, in most cases, require the construction of multiple models.

Apart from static external loading, structures are very often subjected to external loading, the magnitude, the direction, and the position of which vary with time, i.e., dynamic external loading. Such cases of loading include seismic excitations, excitations due to wind conditions, vibrations induced by large mechanical devices, impact loading due to explosions, etc. The aforementioned arbitrary external dynamic loading leads to the formulation of the flexural–torsional vibration problem. The complexity of this problem increases significantly in the case the cross section's centroid does not coincide with its shear center (asymmetric beams). Furthermore, when arbitrary torsional boundary conditions are applied either at the edges or at any other interior point of a beam due to construction requirements, the beam under the action of general twisting loading is leaded to nonuniform torsion. Moreover, since requirement of weight saving is a major aspect in the design of structures, thin-walled elements of arbitrary cross section and low flexural and/or torsional stiffness are extensively used. Treating displacements and angles of rotation of these elements as being small leads in many cases to inadequate prediction of the dynamic response; hence, the occurring nonlinear effects should be taken into account. This can be achieved by retaining the nonlinear terms in the strain–displacement relations (finite displacement–small strain theory).

When the displacement components of a member are small, a wide range of linear analysis tools, such as modal analysis, can be used, and some analytical results are possible. As these components become larger, the induced geometrical nonlinearities result in effects that are not observed in linear systems. When finite displacements are considered, the flexural–torsional dynamic analysis of beams becomes much more complicated, leading to the formulation of coupled and nonlinear flexural, torsional, and axial equations of motion. The analysis of these systems becomes even more complicated when shear deformation effect in flexure and secondary torsional moment deformation effect (STMDE) in torsion, which are significant in many cases (e.g., short beams, beams of box-shaped cross sections, folded structural members, beams made of materials weak in shear, etc.), are taken into account.

In section "Nonlinear Flexural Dynamic Analysis of Beams with Shear Deformation Effect" of this chapter, the geometrically nonlinear dynamic flexural analysis of homogeneous prismatic beam members taking into account shear deformation and rotary inertia effects (Timoshenko beam theory) is presented. The differential equations of

Nonlinear Dynamic Seismic Analysis, Fig. 1 Structural engineering projects composed by beam structural members

motion governing free or forced vibrations of Timoshenko beams are formulated employing the principle of virtual work. In order to be able to examine the flexural problem independently (i.e., excluding torsional response) and without loss of generality, the case of beams of arbitrary doubly symmetric cross sections subjected to torsionless bending (external transverse loading passes through the shear center of the cross section which coincides with its centroid) is examined. For a detailed review of the relevant literature, the reader is referred to the study of Sapountzakis and Dourakopoulos (2009a, b).

Since very often beam members under arbitrary external dynamic loading develop torsional response, the rigorous dynamic analysis of nonuniform torsion of beams in the nonlinear range is necessary. In section "Nonlinear Torsional Dynamic Analysis of Beams with Secondary Torsional Moment Deformation Effect (STMDE)" of this chapter, the geometrically nonlinear dynamic torsional analysis of homogeneous prismatic beam members taking into account STMDE (analogy with Timoshenko beam theory) is presented. The differential equations of motion governing free or forced vibrations of beams under nonuniform torsion are formulated employing the principle of virtual work. Similarly with section "Nonlinear Flexural Dynamic Analysis of Beams with Shear Deformation Effect," the case of beams of arbitrary doubly symmetric cross sections subjected to twisting and/or warping moments is examined.

For a detailed review of the relevant literature, the reader is referred to the studies of Sapountzakis and Tsipiras (2010a, b, c).

Finally, in order to highlight the significant coupling between flexure and torsion when displacements and rotations become large, in section "Nonlinear Flexural–Torsional Dynamic Analysis of Beams of Arbitrary Cross Section," the formulation for the geometrically nonlinear flexural–torsional dynamic analysis of prismatic beams is analyzed. In this section, the most general case is examined by considering arbitrary thin- or thick-walled cross sections possessing no axis of symmetry. The beam may undergo moderately large deflections and twisting rotations, under the most general boundary conditions. The beam is subjected to arbitrarily distributed or concentrated transverse loading, which can be applied on any arbitrary point of the lateral surface of the beam, to bending moments, as well as to twisting and/or axial loading. For a detailed review of the relevant literature, the reader is referred to the study of Sapountzakis and Dikaros (2011).

Nonlinear Flexural Dynamic Analysis of Beams with Shear Deformation Effect

Statement of the Problem

Let us consider a prismatic beam of length L of doubly symmetric cross section of area A, occupying the multiply connected region Ω of the y, z

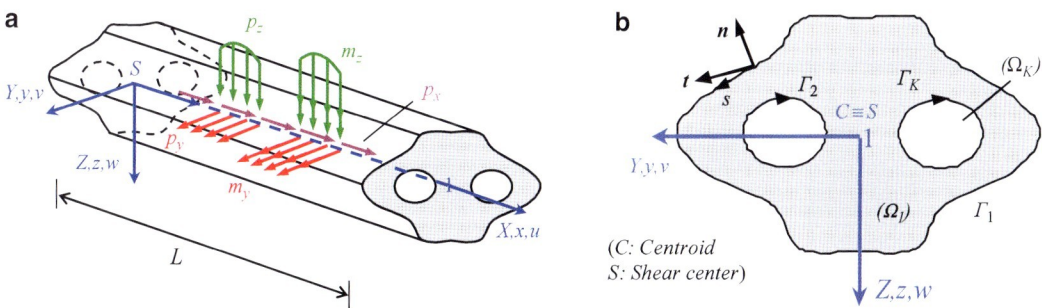

Nonlinear Dynamic Seismic Analysis, Fig. 2 Prismatic beam in axial–flexural loading (**a**) with an arbitrary doubly symmetric cross section occupying the two-dimensional region Ω (**b**)

plane and is bounded by the curve $\Gamma = \cup_{j=1}^{K} \Gamma_j$, which is piecewise smooth, i.e., it may have a finite number of corners (Fig. 2). The material of the beam is considered to be homogeneous, isotropic, and linearly elastic with modulus of elasticity E, shear modulus G, Poisson ratio ν, and mass density ρ. In Fig. 2b, $Cxyz \equiv CXYZ$ is the principal bending system of axes passing through the cross section's centroid C (coinciding with the shear center S). The beam is subjected to the combined action of the arbitrarily distributed or concentrated time-dependent axial loading $p_x = p_x(x, t)$; transverse loading $p_y = p_y(x, t)$, $p_z = p_z(x, t)$ acting in the y, z directions, respectively; and bending moments $m_y = m_y(x, t)$, $m_z = m_z(x, t)$ along y, z axes, respectively.

Under the action of the aforementioned loading, the beam does not develop twisting rotation, and the displacement field taking into account shear deformation effect can be written as

$$\bar{u}(x, y, z, t) = u(x, t) - y\theta_z(x, t) + z\theta_y(x, t) \quad (1a)$$

$$\bar{v}(x, t) = v(x, t) \quad (1b)$$

$$\bar{w}(x, t) = w(x, t) \quad (1c)$$

where $\bar{u}, \bar{v}, \bar{w}$ are the displacement components of an arbitrary point of the cross section with respect to $Cxyz$; $u(x, t)$, $v(x, t)$, $w(x, t)$ are the corresponding components of the centroid, and $\theta_y(x, t)$, $\theta_z(x, t)$ are the angles of rotation due to bending of the cross section which do not coincide with the deflection slopes (w', v'). Employing the strain–displacement relations of the Green–Lagrange strain tensor (Rothert and Gensichen 1987; Ramm and Hofmann 1995), the strain components can be obtained as

$$\begin{aligned}\varepsilon_{xx} &= \frac{\partial \bar{u}}{\partial x} + \frac{1}{2}\left[\left(\frac{\partial \bar{u}}{\partial x}\right)^2 + \left(\frac{\partial \bar{v}}{\partial x}\right)^2 + \left(\frac{\partial \bar{w}}{\partial x}\right)^2\right] \\ &\approx \frac{\partial \bar{u}}{\partial x} + \frac{1}{2}\left[\left(\frac{\partial \bar{v}}{\partial x}\right)^2 + \left(\frac{\partial \bar{w}}{\partial x}\right)^2\right]\end{aligned} \quad (2a)$$

$$\begin{aligned}\gamma_{xz} &= \frac{\partial \bar{w}}{\partial x} + \frac{\partial \bar{u}}{\partial z} + \left(\frac{\partial \bar{u}}{\partial x}\frac{\partial \bar{u}}{\partial z} + \frac{\partial \bar{v}}{\partial x}\frac{\partial \bar{v}}{\partial z} + \frac{\partial \bar{w}}{\partial x}\frac{\partial \bar{w}}{\partial z}\right) \\ &\approx \frac{\partial \bar{w}}{\partial x} + \frac{\partial \bar{u}}{\partial z} + \left(\frac{\partial \bar{v}}{\partial x}\frac{\partial \bar{v}}{\partial z} + \frac{\partial \bar{w}}{\partial x}\frac{\partial \bar{w}}{\partial z}\right)\end{aligned} \quad (2b)$$

$$\begin{aligned}\gamma_{xy} &= \frac{\partial \bar{v}}{\partial x} + \frac{\partial \bar{u}}{\partial y} + \left(\frac{\partial \bar{u}}{\partial x}\frac{\partial \bar{u}}{\partial y} + \frac{\partial \bar{v}}{\partial x}\frac{\partial \bar{v}}{\partial y} + \frac{\partial \bar{w}}{\partial x}\frac{\partial \bar{w}}{\partial y}\right) \\ &\approx \frac{\partial \bar{v}}{\partial x} + \frac{\partial \bar{u}}{\partial y} + \left(\frac{\partial \bar{v}}{\partial x}\frac{\partial \bar{v}}{\partial y} + \frac{\partial \bar{w}}{\partial x}\frac{\partial \bar{w}}{\partial y}\right)\end{aligned} \quad (2c)$$

$$\varepsilon_{yy} = \varepsilon_{zz} = \gamma_{yz} = 0 \quad (2d)$$

where it has been assumed that for moderate displacements $(\partial \bar{u}/\partial x)^2 \ll (\partial \bar{u}/\partial x)$, $(\partial \bar{u}/\partial x)(\partial \bar{u}/\partial z) \ll (\partial \bar{u}/\partial x) + (\partial \bar{u}/\partial z)$, $(\partial \bar{u}/\partial x)(\partial \bar{u}/\partial y) \ll (\partial \bar{u}/\partial x) + (\partial \bar{u}/\partial y)$. Substituting the displacement components Eq. 1 to the strain–displacement relations Eq. 2, the strain components are written as

$$\varepsilon_{xx} = u' + z\theta'_y - y\theta'_z + \frac{1}{2}\left(v'^2 + w'^2\right) \quad (3a)$$

$$\gamma_{xy} = v' - \theta_z \quad (3b)$$

$$\gamma_{xz} = w' + \theta_y \quad (3c)$$

It is noted that within the Euler–Bernoulli theory context, it holds that $\theta_z = v'$, $\theta_y = -w'$ (the cross section remains perpendicular to the deformed axis); hence, shear strain components Eqs. 3b and 3c vanish. Considering strains to be small and employing the second Piola–Kirchhoff stress tensor, the nonvanishing stress components are defined in terms of the strain ones as

$$\begin{Bmatrix} S_{xx} \\ S_{xy} \\ S_{xz} \end{Bmatrix} = \begin{bmatrix} E^* & 0 & 0 \\ 0 & G & 0 \\ 0 & 0 & G \end{bmatrix} \begin{Bmatrix} \varepsilon_{xx} \\ \gamma_{xy} \\ \gamma_{xz} \end{Bmatrix} \quad (4)$$

where E^* is obtained from Hooke's stress–strain law as $E^* = E(1 - v)/(1 + v)(1 - 2v)$. If the assumption of plane stress condition is made, the above expression is reduced in $E^* = E/(1 - v^2)$ (Vlasov 1963), while in beam formulations, E is frequently considered instead of E^* ($E^* \approx E$). This last consideration has been followed throughout this study, while any other reasonable value of E^* could be used without any difficulty. Substituting Eq. 3 into Eq. 4, the nonvanishing stress components are obtained as

$$S_{xx} = E\left[u' + z\theta'_y - y\theta'_z + \frac{1}{2}\left(v'^2 + w'^2\right)\right] \quad (5a)$$

$$S_{xy} = G \cdot (v' - \theta_z) \quad (5b)$$

$$S_{xz} = G \cdot (w' + \theta_y) \quad (5c)$$

In order to establish the nonlinear equations of motion, the principle of virtual work

$$\delta W_{\text{int}} + \delta W_{\text{mass}} = \delta W_{\text{ext}} \quad (6)$$

where

$$\delta W_{\text{int}} = \int_V \left(S_{xx}\delta\varepsilon_{xx} + S_{xy}\delta\gamma_{xy} + S_{xz}\delta\gamma_{xz}\right)dV \quad (7a)$$

$$\delta W_{\text{mass}} = \int_V \rho\left(\ddot{u}\delta u + \ddot{v}\delta v + \ddot{w}\delta w\right)dV \quad (7b)$$

$$\delta W_{\text{ext}} = \int_F \left(t_x\delta\bar{u} + t_y\delta\bar{v} + t_z\delta\bar{w}\right)dF \quad (7c)$$

under a total Lagrangian formulation is employed. In the above equations, $\delta(\cdot)$ and $(\dot{\cdot})$ denote virtual quantities and derivatives with respect to time, respectively, V is the volume of the beam in the initial (undeformed) configuration, F is the lateral surface of the beam in the initial configuration including the end cross sections, and t_x, t_y, t_z are the components of the traction vector with respect to the undeformed surface of the beam. Moreover, the stress resultants of the beam are given as

$$N = \int_\Omega S_{xx}dx \quad (8a)$$

$$M_y = \int_\Omega S_{xx}zdx \quad (8b)$$

$$M_z = -\int_\Omega S_{xx}ydx \quad (8c)$$

$$Q_y = \int_\Omega S_{xy}dx \quad (8d)$$

$$Q_z = \int_\Omega S_{xz}dx \quad (8e)$$

Substituting the expressions of the stress components Eq. 5 into Eq. 8, the stress resultants are obtained as

$$N = EA\left[u' + \frac{1}{2}\left(v'^2 + w'^2\right)\right] \quad (9a)$$

$$M_y = EI_y \theta_y' \tag{9b}$$

$$M_z = EI_z \theta_z' \tag{9c}$$

$$Q_y = GA_y \gamma_{xy} = GA_y(v' - \theta_z) \tag{9d}$$

$$Q_z = GA_z \gamma_{xz} = GA_z(w' + \theta_y) \tag{9e}$$

where A, I_y, I_z are the cross-sectional area and the principal moments of inertia of the cross section given as

$$A = \int_\Omega dx \tag{10a}$$

$$I_y = \int_\Omega z^2 dx \tag{10b}$$

$$I_z = \int_\Omega y^2 dx \tag{10c}$$

and GA_y, GA_z are the shear rigidities of the Timoshenko theory, where

$$A_y = \kappa_y A = \frac{1}{a_y} A \tag{11a}$$

$$A_z = \kappa_z A = \frac{1}{a_z} A \tag{11b}$$

are the shear areas with respect to y, z axes, respectively, with κ_y, κ_z being the shear correction factors and a_y, a_z being the shear deformation coefficients.

Employing Eq. 8 and substituting Eqs. 1, 3, and 9 to 6, the differential equations of motion are derived as

$$- EA(u'' + w'w'' + v'v'') + \rho \varsigma \ddot{u} = p_x \tag{12a}$$

$$- (Nv')' + \rho \varsigma \ddot{v} - GA_y(v'' - \theta_z') = p_y \tag{12b}$$

$$- EI_z \theta_z'' + \rho I_z \ddot{\theta}_z - GA_y(v' - \theta_z) = m_z \tag{12c}$$

$$- (Nw')' + \rho \varsigma \ddot{w} - GA_z(w'' + \theta_y') = p_z \tag{12d}$$

$$- EI_y \theta_y'' + \rho I_y \ddot{\theta}_y + GA_z(w' + \theta_y) = m_y \tag{12e}$$

Combining Eqs. 12b and 12c with 12d and 12e in order to eliminate θ_y, θ_z, the following differential equations are derived

$$EI_z v'''' - \left(\frac{EI_z}{G\varsigma_y}\rho\varsigma + \rho I_z\right)\ddot{v}'' + \rho\varsigma\ddot{v} + \frac{EI_z}{GA_y}(Nv')''' - (Nv')'$$
$$- \frac{\rho I_z}{GA_y}\left(\frac{\partial^2(Nv')'}{\partial t^2} - \rho A \ddot{v}''\right) = p_y - \frac{EI_z}{GA_y}p_y'' + \frac{\rho I_z}{GA_y}\ddot{p}_y - m_z' \tag{13a}$$

$$EI_y w'''' - \left(\frac{EI_y}{GA_z}\rho A + \rho I_y\right)\ddot{w}'' + \rho\varsigma\ddot{w} + \frac{EI_y}{GA_z}(Nw')''' - (Nw')'$$
$$- \frac{\rho I_y}{GA_z}\left(\frac{\partial^2(Nw')'}{\partial t^2} - \rho A\ddot{w}''\right) = p_z - \frac{EI_y}{GA_z}p_z'' + \frac{\rho I_y}{GA_z}\ddot{p}_z + m_y' \tag{13b}$$

In Eqs. 12a and 13, the quantities p_x, p_y, p_z, m_z, m_y (external axial loading, transverse loading, and bending moments along each direction, respectively) constitute the externally applied time-dependent loading along the beam (Fig. 2a) and arise (through the principle of virtual work) from the externally applied traction vectors t_x, t_y, t_z. Ignoring time derivatives of fourth order from the above equations (Thomson 1981), the following governing differential equations are derived as

$$- EA(u'' + w'w'' + v'v'') + \rho\varsigma\ddot{u} = p_x \tag{14a}$$

$$EI_z v'''' - \rho I_z\left(\frac{Ea_y}{G} + 1\right)\ddot{v}'' + \rho\varsigma\ddot{v} + \frac{EI_z}{GA_y}(Nv')''' - (Nv')'$$
$$- \frac{\rho I_z}{GA_y}\frac{\partial^2(Nv')'}{\partial t^2} = p_y - \frac{EI_z}{GA_y}p_y'' + \frac{\rho I_z}{GA_y}\ddot{p}_y - m_z' \tag{14b}$$

$$EI_y w'''' - \rho I_y\left(\frac{Ea_z}{G} + 1\right)\ddot{w}'' + \rho\varsigma\ddot{w} + \frac{EI_y}{GA_z}(Nw')''' - (Nw')'$$
$$- \frac{\rho I_y}{GA_z}\frac{\partial^2(Nw')'}{\partial t^2} = p_z - \frac{EI_y}{GA_z}p_z'' + \frac{\rho I_y}{GA_z}\ddot{p}_z + m_y' \tag{14c}$$

The stress resultants and bending rotations are expressed as

$$M_z = EI_z v'' + \frac{EI_z}{GA_y}\left[p_y - \rho A \ddot{v} + (Nv')'\right] \tag{15a}$$

$$M_y = -EI_y w'' + \frac{EI_y}{GA_z}\left[\rho A \ddot{w} - p_z - (Nw')'\right] \tag{15b}$$

$$Q_y = -EI_z v''' - \frac{EI_z}{GA_y}\left[p'_y - \rho A \ddot{v}' + (Nv')''\right]$$
$$+ \rho I_z \ddot{\theta}_z - m_z \qquad (15c)$$

$$Q_z = -EI_y w''' + \frac{EI_y}{GA_z}\left[\rho A \ddot{w}' - p'_z - (Nw')''\right]$$
$$- \rho I_y \ddot{\theta}_y + m_y \qquad (15d)$$

$$\theta_z = v' - \frac{Q_y}{GA_y} \qquad (15e)$$

$$\theta_y = -w' + \frac{Q_z}{GA_z} \qquad (15f)$$

The above equations are also subjected to the pertinent boundary conditions of the problem which are given as

$$\alpha_1 u(x,t) + \alpha_2 N(x) = \alpha_3 \qquad (16a)$$

$$\beta_1 v(x,t) + \beta_2 R_y^b(x) = \beta_3 \qquad (16b)$$

$$\overline{\beta}_1 \theta_z(x,t) + \overline{\beta}_2 M_z^b(x) = \overline{\beta}_3 \qquad (16c)$$

$$\gamma_1 w(x,t) + \gamma_2 R_z^b(x) = \gamma_3 \qquad (16d)$$

$$\overline{\gamma}_1 \theta_y(x,t) + \overline{\gamma}_2 M_y^b(x) = \overline{\gamma}_3 \qquad (16e)$$

at the beam ends $x = 0, L$, where $\alpha_i, \beta_i, \gamma_i, \overline{\beta}_i, \overline{\gamma}_i$ $i = 1, 2, 3$ are known time-dependent functions. The boundary conditions Eq. 16 are the most general boundary conditions for the problem at hand, including also the elastic support. It is apparent that all types of the conventional boundary conditions (clamped, simply supported, free, or guided edge) can be derived from Eq. 16 by specifying appropriately the functions $\alpha_i, \beta_i, \gamma_i, \overline{\beta}_i, \overline{\gamma}_i$ (e.g., for a clamped edge, it is $\alpha_1 = \beta_1 = \overline{\beta}_1 = \gamma_1 = \overline{\gamma}_1 = 1, \alpha_2 = \alpha_3 = \beta_2 = \beta_3 =, \gamma_2 = \gamma_3 = \overline{\beta}_2 = \overline{\beta}_3 = \overline{\gamma}_2 = \overline{\gamma}_3 = 0$). In Eq. 16, the reaction forces R_y^b, R_z^b; bending moments M_z^b, M_y^b; and the angles of bending rotations θ_y, θ_z at the beam ends ignoring the inertia terms are written as

$$R_y^b = Nv' - EI_z v''' - \frac{EI_z}{GA_y} Nv''' \qquad (17a)$$

$$R_z^b = Nw' - EI_y w''' - \frac{EI_y}{GA_z} Nw''' \qquad (17b)$$

$$M_z^b = EI_z v'' + \frac{EI_z}{GA_y} Nv'' \qquad (17c)$$

$$M_y^b = -EI_y w'' - \frac{EI_y}{GA_z} Nw'' \qquad (17d)$$

$$\theta_y^b = -w' - \frac{EI_y}{G^2 A_z^2} Nw''' - \frac{EI_y}{GA_z} w''' \qquad (17e)$$

$$\theta_z^b = v' + \frac{EI_z}{G^2 A_y^2} Nv''' + \frac{EI_z}{GA_y} v''' \qquad (17f)$$

In case of nonzero initial conditions, the following relations are valid for $t = 0$

$$u(x,0) = \overline{u}_0(x) \qquad (18a)$$

$$\dot{u}(x,0) = \dot{\overline{u}}_0(x) \qquad (18b)$$

$$v(x,0) = \overline{v}_0(x) \qquad (18c)$$

$$\dot{v}(x,0) = \dot{\overline{v}}_0(x) \qquad (18d)$$

$$w(x,0) = \overline{w}_0(x) \qquad (18e)$$

$$\dot{w}(x,0) = \dot{\overline{w}}_0(x) \qquad (18f)$$

$$\theta_z(x,0) = \overline{\theta}_{z0}(x) \qquad (18g)$$

$$\dot{\theta}_z(x,0) = \dot{\overline{\theta}}_{z0}(x) \qquad (18h)$$

$$\theta_y(x,0) = \overline{\theta}_{y0}(x) \qquad (18i)$$

$$\dot{\theta}_y(x,0) = \dot{\overline{\theta}}_{y0}(x) \qquad (18j)$$

Applications

Free Vibrations of Hinged Beam of Rectangular Cross Section

In order to examine the influence of shear deformation on the free vibrations of beams in the

Nonlinear Dynamic Seismic Analysis, Table 1 Material properties, geometric, inertia constants, and shear deformation coefficients of the rectangular cross section of application section "Free Vibrations of Hinged Beam of Rectangular Cross Section"

$E = 70$ GPa	$G = 27$ GPa
$\rho = 2.7$ tn/m^3	$A = 150 \times 10^{-6}$ m^2
$I_y = 7,812.5 \times 10^{-12}$ m^4	$I_z = 450 \times 10^{-12}$ m^4
$a_y = 1.74$	$a_z = 1.20$

geometrically nonlinear range, a hinged beam of a 25×6 mm rectangular cross section subjected to free vibrations is studied. It is worth here mentioning that the axially immovable ends lead to the development of axial loading along the beam as deflections increase. The employed initial conditions are represented by the following expression $w(x, 0) = w_{max} \sin(\pi x/L)$ where w_{max} is the central beam deflection. In Table 1, the material properties, the geometric and inertia constants, and the shear deformation coefficients of the examined cross section are presented.

In Fig. 3, the obtained results of the period ratios T/T_0, at various amplitude ratios w_{max}/r for three different cases of beam slenderness L/r, as obtained from the numerical solution of the initial-boundary value problem Eqs. 14, 16, and 18 using analog equation method (AEM) and boundary element method (BEM)-based techniques (Sapountzakis and Dourakopoulos 2009a, b), are presented taking into account or ignoring shear deformation effect. T denotes the period of free vibration obtained by nonlinear analysis including rotary inertia and taking into account or ignoring shear deformation effect, T_0 is the corresponding period of the linear vibration ignoring both rotary inertia and shear deformation effect, and $r = \sqrt{I_y/A}$ is the radius of gyration about y axis. The obtained results are compared with the corresponding ones available in the study of Foda (1999) obtained by a multiple scales procedure. In Table 2 the ratios of the nonlinear frequency ω_{NL} over the linear one ω_L at various amplitude ratios w_{max}/r and slenderness L/r, obtained by AEM, are presented as compared with those obtained from a finite element solution employing polynomial expressions for the displacement components (Rao et al. 1976).

From this table the convergence of both methods can be observed. Moreover, from the above figure and table, the increment of vibration frequency with the increment of deflections can be observed due to the fact that axially immovable ends induce tensile forces to the beam, thus increasing its stiffness.

Free Vibrations or Beam with Very Flexible Boundary Conditions

In this application the free vibrations of Timoshenko beams with very flexible boundary conditions is examined, since in this case the natural frequencies and the corresponding modeshapes are highly sensitive to the effects of shear deformation and rotary inertia (Aristizabal-Ochoa 2004). Thus, the linear free vibration analysis of pinned-free beams of length $L = 5.0$ m of various cross sections is examined ($\rho = 2.40$ tn/m^3, $E = 25.42$ GPa, $G = 11.05$ GPa). In Table 3 the geometric and inertia constants as well as the shear correction factors of the examined cross sections are presented ($\kappa = 1/\alpha$). The differential equations for the linear free vibrations of this special case can be obtained by Eq. 14 for zero axial stress resultant and external force as

$$EI_z v'''' - \rho I_z \left(\frac{Ea_y}{G} + 1\right) \ddot{v}'' + \rho_\varsigma \ddot{v} = 0 \quad (19a)$$

$$EI_y w'''' - \rho I_y \left(\frac{Ea_z}{G} + 1\right) \ddot{w}'' + \rho A \ddot{w} = 0 \quad (19b)$$

The boundary conditions for the examined case can be derived by Eq. 16 as

$$v(0,t) = w(0,t) = 0 \quad (20a)$$

$$M_y(0,t) = M_z(0,t) = 0 \quad (20b)$$

$$V_y(L,t) = V_z(L,t) = 0 \quad (20c)$$

$$M_y(L,t) = M_z(L,t) = 0 \quad (20d)$$

In Table 4 the first three natural frequencies of the pinned-free Timoshenko beams of various cross sections and in Fig. 4 the corresponding modeshapes for the cross section with web

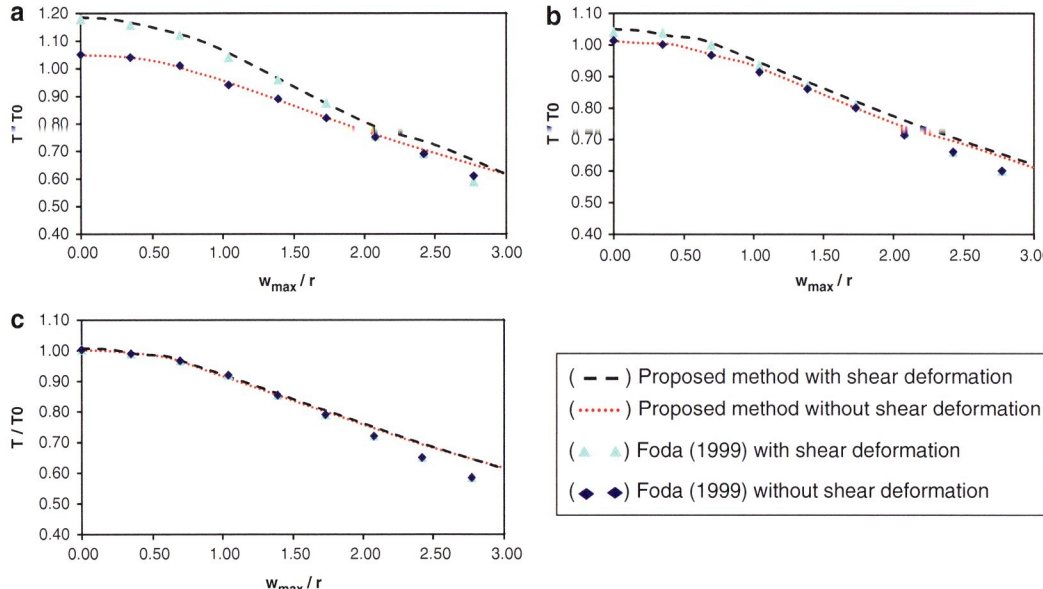

Nonlinear Dynamic Seismic Analysis, Fig. 3 Free vibration period ratios T/T_0 at various amplitude ratios w_{max}/r for beam slenderness $l/r = 50$ of the simply supported beam of application section "Free Vibrations of Hinged Beam of Rectangular Cross Section"

Nonlinear Dynamic Seismic Analysis, Table 2 Frequency ratios ω_{NL}/ω_L at various amplitude ratios w_{max}/r of the simply supported beam of application section "Free Vibrations of Hinged Beam of Rectangular Cross Section"

	$L/r = 10$		$L/r = 50$		$L/r = 100$	
w_{max}/r	AEM	Rao et al. (1976)	AEM	Rao et al. (1976)	AEM	Rao et al. (1976)
0.4	1.0214	1.0193	1.0198	1.0154	1.0103	1.0149
0.8	1.0694	1.0737	1.0587	1.0585	1.0551	1.0581
1.0	1.1116	1.1116	1.0928	1.0897	1.0920	1.0890
2.0	1.4296	1.3701	1.3230	1.3143	1.3227	1.3125
3.0	1.9268	1.6884	1.6432	1.6052	1.6410	1.6030

thickness $t_w = 0.254$ mm are presented as obtained from an AEM solution (Sapountzakis and Dourakopoulos 2009a, b) and semi-analytical methods employing sinusoidal functions (Aristizabal-Ochoa 2004; Kausel 2002).

Forced Vibrations or Beam of Hollow Rectangular Cross Section Under Biaxial Bending

In order to examine the influence of shear deformation on the free vibrations of beams in the geometrically nonlinear range, a clamped beam

Nonlinear Dynamic Seismic Analysis, Table 3 Geometric constants of the pinned-free beam of application section "Free Vibrations or Beam with Very Flexible Boundary Conditions"

Web thickness t_w (mm)	Cross-sectional area A(m^2)	Moment of inertia I (m^4)	Shear correction factor $\kappa = 1/a$
101.6	0.359	0.193	0.54
25.4	0.229	0.162	0.21
7.62	0.199	0.154	0.07
0.254	0.186	0.151	0.002

Nonlinear Dynamic Seismic Analysis, Table 4 Natural frequencies $f = \omega/2\pi$ (Hz) of the beam of application section "Free Vibrations or Beam with Very Flexible Boundary Conditions," for various cross sections

	f_1			f_2	f_3
t_w (mm)	Aristizabal-Ochoa (2004)	Kausel (2002)	AEM	AEM	AEM
101.6	148.2	218	153.9	315.0	467.6
25.4	116.8	133.4	120.6	222.9	120.6
7.62	73.4	76.4	77.0	135.8	193.0
0.254	12.8	12.8	13.7	23.6	33.3

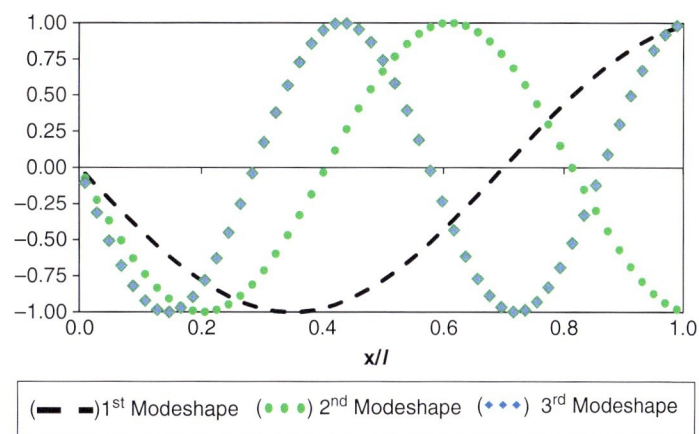

Nonlinear Dynamic Seismic Analysis, Fig. 4 Three first vibration modeshapes of the beam of application section "Free Vibrations or Beam with Very Flexible Boundary Conditions," for $t_w = 0.254$ mm

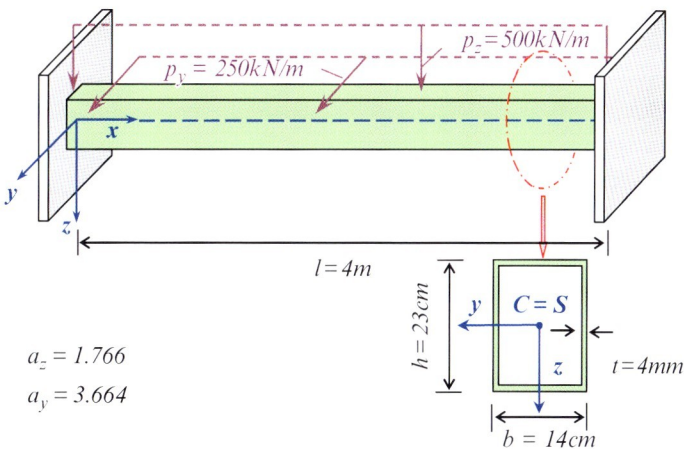

Nonlinear Dynamic Seismic Analysis, Fig. 5 Clamped beam of hollow rectangular cross section of application section "Forced Vibrations or Beam of Hollow Rectangular Cross Section Under Biaxial Bending"

($E = 210$ GPa, $v = 0.3$, $\rho = 7.85$ tn/m^3) of length $L = 4.0$ m and of a hollow rectangular cross section, subjected to the suddenly applied uniformly distributed transverse loadings $p_y(t) = 250$ kN/m, $p_z(t) = 500$ kN/m, $t \geq 0$, (Fig. 5) is examined. In Fig. 6, the time histories of central displacements $v_{L/2}$, $w_{L/2}$, taking into account or ignoring shear deformation effect, are presented. Moreover, in Table 5, the maximum values of the central deflections $v_{l/2}$, $w_{l/2}$

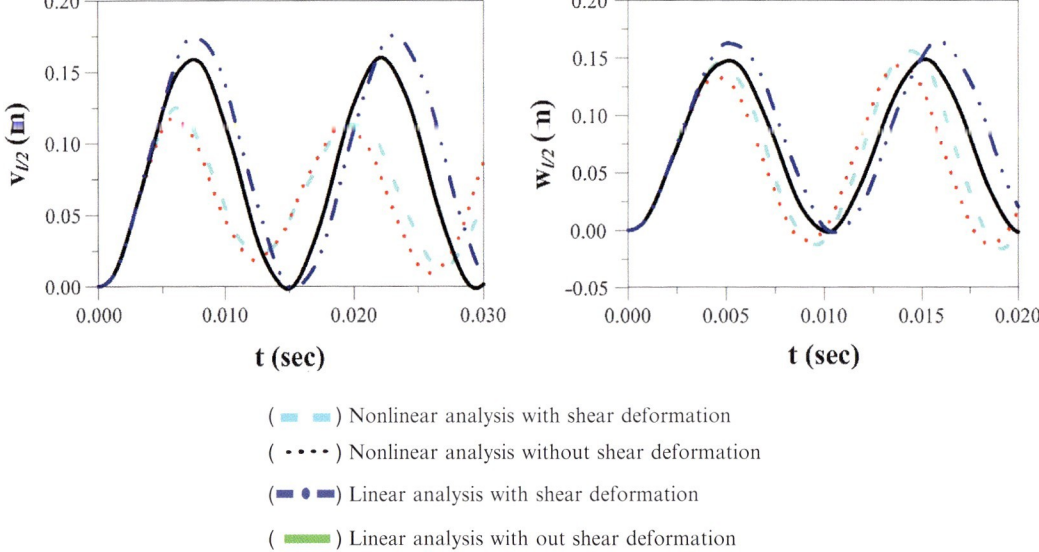

(— —) Nonlinear analysis with shear deformation
(· · · ·) Nonlinear analysis without shear deformation
(— • —) Linear analysis with shear deformation
(▬▬▬) Linear analysis with out shear deformation

Nonlinear Dynamic Seismic Analysis, Fig. 6 Time histories of central deflections along directions $y(v_{L/2})$, $z(w_{L/2})$ of the beam of application section "Forced Vibrations or Beam of Hollow Rectangular Cross Section Under Biaxial Bending"

Nonlinear Dynamic Seismic Analysis, Table 5 Maximum central deflections v_{max}(m), w_{max}(m) and periods T_y, T_z(s) of the first cycle of the clamped beam of application section "Forced Vibrations or Beam of Hollow Rectangular Cross Section Under Biaxial Bending"

	Without shear deformation		With shear deformation	
	Linear analysis	Nonlinear analysis	Linear analysis	Nonlinear analysis
v_{max}	0.1588	0.1180	0.1739	0.1252
w_{max}	0.1476	0.1330	0.1626	0.1460
T_y	0.01483	0.01229	0.01532	0.01240
T_z	0.01019	0.00911	0.01051	0.00958

and the periods T_y, T_z of the first cycle are presented for the aforementioned cases of analysis, taking into account or ignoring shear deformation effect. From this figure and table, the influence of geometrical nonlinearity can be observed reducing the magnitude of deflections and the period of vibrations due to axially immovable ends which lead to the development of axial forces, thus increasing the stiffness of the beam. Shear deformation effect is also apparent increasing both the maximum central transverse displacements (especially in z direction) and the calculated periods of the first cycle of motion.

Forced Vibrations or Beam of Hollow Rectangular Cross Section Subjected to Axial Compressive Load

In order to further investigate the influence of geometrical nonlinearity on the response of Timoshenko beams, the nonlinear dynamic analysis of a clamped-rolled beam of length $l = 3.0$ m, having a hollow rectangular cross section ($E = 210$ GPa, $v = 0.3$, $\rho = 7.85$ tn/m^3) subjected to the suddenly applied uniformly distributed loads $p_x(t) = -250$ kN/m, $p_y(t) = -150$ kN/m, $p_z(t) = -150$ kN/m and concentrated axial load $P_x(t) = -500$ kN at the rolled end ($t \geq 0$), as is shown in Fig. 7, has been studied.

In Fig. 8, the time histories of the central transverse displacements $v_{l/2}$, $w_{l/2}$ of the beam performing either a linear or a nonlinear analysis are presented taking into account or ignoring shear deformation effect. Moreover, in Table 6 the maximum values of the central deflections

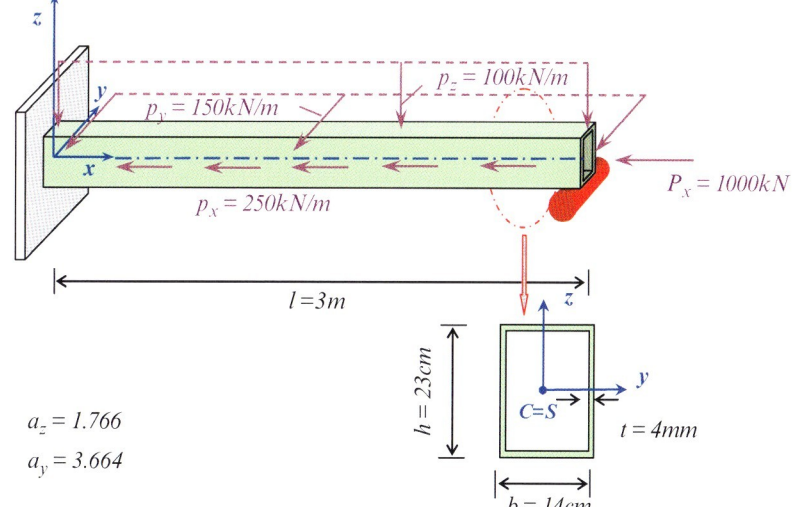

Nonlinear Dynamic Seismic Analysis, Fig. 7 Clamped-rolled beam of hollow rectangular cross section of application section "Forced Vibrations or Beam of Hollow Rectangular Cross Section Subjected to Axial Compressive Load"

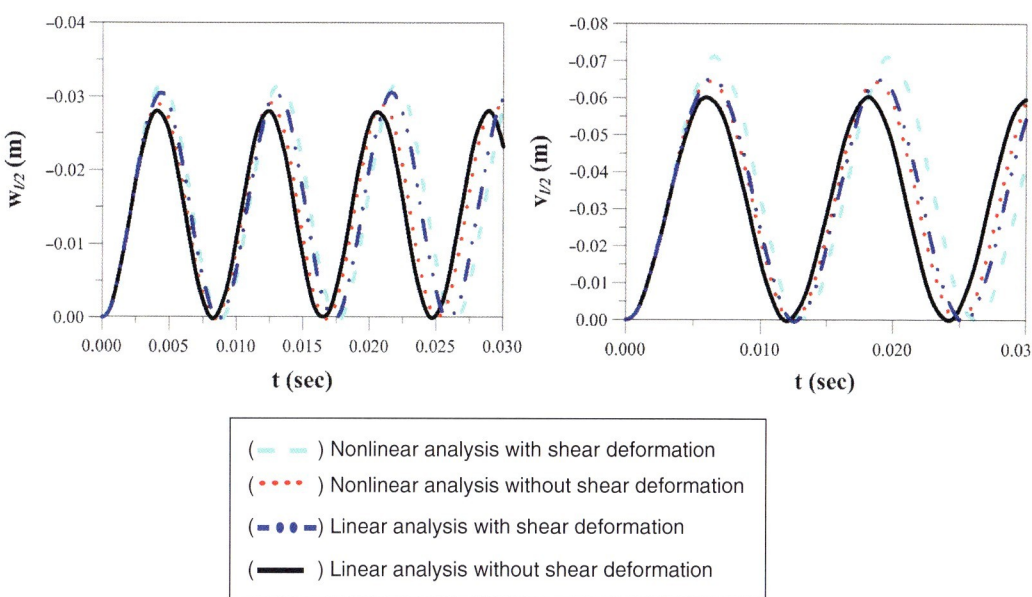

Nonlinear Dynamic Seismic Analysis, Fig. 8 Time histories of central deflections along directions $y(v_{L/2})$, $z(w_{L/2})$ of the beam of application section "Forced Vibrations or Beam of Hollow Rectangular Cross Section Subjected to Axial Compressive Load"

v_{max}, w_{max} and the periods T_y, T_z of the first cycle are presented for the aforementioned cases of analysis, including or ignoring shear deformation effect. In all of the aforementioned cases, rotary inertia has been taken into account. From the obtained results, the influence of geometrical nonlinearity is apparent, increasing the magnitude of kinematical components due to compressive loading, while the influence of shear deformation effect is also intense leading to further increment of the kinematical components.

Nonlinear Torsional Dynamic Analysis of Beams with Secondary Torsional Moment Deformation Effect (STMDE)

Statement of the Problem

The prismatic beam of doubly symmetric cross section described in section "Statement of the Problem" is examined and is subjected to the combined action of the arbitrarily distributed or concentrated time-dependent conservative axial load $p_x(x, t)$, twisting $m_t = m_t(x, t)$, and warping $m_w = m_w(x, t)$ moments acting in the x direction (Fig. 9a), while its ends are subjected to the most general boundary conditions with respect to the axial displacement, the angle of twist, and warping intensity. The beam can be twisted freely about the longitudinal axis and its flexural behavior is not examined. The analysis is carried out with respect to the coordinate system $Sxyz$, the longitudinal axis Sx of which passes through the cross sections' shear center. It is noted that due to the fact that the cross section is a doubly symmetric one, the shear center of the cross section coincides with the centroid (Fig. 9b).

Assuming that the cross section maintains its shape during deformation (no distortional effects), considering that the points of the cross section follow a circular trajectory around axis Sx, and considering twisting rotations to be large and taking into account STMD effect, the displacement field of the beam is written as

$$\bar{u}(x,y,z,t) = u(x,t) + \left(\theta_x^P(x,t)\right)' \phi_S^P(y,z)$$
$$- \left[\left(\theta_x^P(x,t)\right)' - \theta_x'(x,t)\right]\left[\phi_S^P(y,z) + \phi_S^S(x,y,z,t)\right] \quad (21a)$$

$$\bar{v}(x,y,z,t) = -z \sin \theta_x(x,t) - y(1 - \cos \theta_x(x,t)) \quad (21b)$$

$$\bar{w}(x,y,z,t) = y \sin \theta_x(x,t) - z(1 - \cos \theta_x(x,t)) \quad (21c)$$

In Eq. 21, apart from the "average" axial displacement $u(x, t)$ of the cross section defined in section "Nonlinear Flexural Dynamic Analysis of Beams with Shear Deformation Effect," $\theta_x^P(x, t)$, $\theta_x(x, t)$ denote the primary part and the total angle

Nonlinear Dynamic Seismic Analysis, Table 6 Periods T_y, T_z(s) of the first cycle and maximum central displacements v_{max}(m), w_{max}(m) of the beam of application section "Forced Vibrations or Beam of Hollow Rectangular Cross Section Subjected to Axial Compressive Load"

	Without shear deformation		With shear deformation	
	Linear analysis	Nonlinear analysis	Linear analysis	Nonlinear analysis
v_{max}	-2.90×10^{-2}	-3.23×10^{-2}	-3.25×10^{-2}	-3.68×10^{-2}
w_{max}	-1.35×10^{-2}	-1.41×10^{-2}	-1.51×10^{-2}	-1.60×10^{-2}
T_y	8.17×10^{-3}	8.75×10^{-3}	9.03×10^{-3}	9.46×10^{-3}
T_z	5.74×10^{-3}	5.86×10^{-3}	6.20×10^{-3}	6.36×10^{-2}

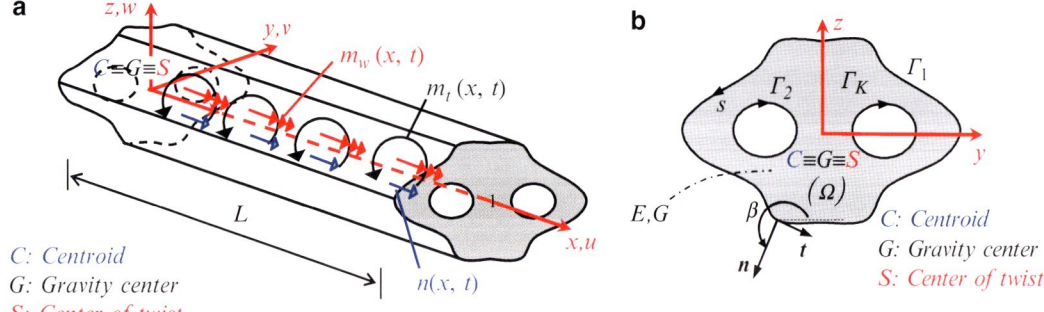

Nonlinear Dynamic Seismic Analysis, Fig. 9 Prismatic beam of an arbitrarily shaped doubly symmetric constant cross section occupying region Ω (a) subjected to axial and torsional loading (b)

of twist, respectively; ϕ_S^P, ϕ_S^S are the primary and secondary warping functions with respect to the shear center S. Considering the "average" axial displacement and the warping displacements of the cross section to be small, strain components Eq. 2 and stress–strain relationship Eq. 4 are once again employed to derive the nonvanishing stress components of the second Piola–Kirchhoff stress tensor as

$$S_{xx} = E\left[u' + \left(\theta_x^P\right)''\phi_S^P + \frac{1}{2}\left(y^2 + z^2\right)\left(\theta_x'\right)^2\right] \quad (22a)$$

$$S_{xy} = \underbrace{G\theta_x'\left(\frac{\partial \phi_S^P}{\partial y} - z\right)}_{\text{primary}} + \underbrace{G\left(\left(\theta_x^P\right)' - \theta_x'\right)\left(-\frac{\partial \phi_S^S}{\partial y}\right)}_{\text{secondary}} \quad (22b)$$

$$S_{xz} = \underbrace{G\theta_x'\left(\frac{\partial \phi_S^P}{\partial z} + y\right)}_{\text{primary}} + \underbrace{G\left(\left(\theta_x^P\right)' - \theta_x'\right)\left(-\frac{\partial \phi_S^S}{\partial z}\right)}_{\text{secondary}} \quad (22c)$$

In Eq. 22a, the term $(y^2 + z^2) \cdot (\theta_x')^2/2$ is often described as the "Wagner strain" (Sapountzakis and Tsipiras 2010a).

Local Equilibrium Equation, Primary Torsional Warping Function

In order to establish the local equilibrium equation in the longitudinal direction, the calculus of variations is applied. The principle of virtual work under a total Lagrangian formulation is applied (Eq. 6). Taking into account the primary warping function's ϕ_S^P virtual components and applying the procedure analyzed in the study of Sapountzakis and Tsipiras (2010a), the following local equilibrium equation:

$$\frac{\partial S_{xx}}{\partial x} + \frac{\partial S_{xy}}{\partial y} + \frac{\partial S_{xz}}{\partial z} - \rho \ddot{u} = 0 \quad \text{στο } \Omega, \ \forall x \in (0, L) \quad (23)$$

along with its corresponding boundary condition

$$S_{xy}n_y + S_{xz}n_z = t_x \quad \text{στο } \Gamma_j, \ \forall x \in (0, L) \quad (24)$$

Subsequently, taking into account the secondary warping function's ϕ_S^S virtual components and applying the same procedure, Eq. 23 can be once again derived. Hence, the two warping functions introduced in the displacement field of the beam should fulfill the local equilibrium Eq. 23 and the corresponding boundary condition Eq. 24. Employing the stress expressions (Eq. 22) and after some algebraic manipulations, it can be easily concluded that the warping functions can be established by the following boundary value problems as

$$\nabla^2 \phi_S^P = 0 \quad \text{στο } \Omega \quad (25a)$$

$$\frac{\partial \phi_S^P}{\partial n} = (zn_y - yn_z) \quad \text{στο } \Gamma_j \quad (25b)$$

$$G\left(\left(\theta_x^P\right)' - \theta_x'\right)\nabla^2 \phi_S^S$$
$$- E\left[u'' + \left(\theta_x^P\right)'''\phi_S^P + \left(y^2 + z^2\right)\theta_x'\theta_x''\right]$$
$$+ \rho\left[\ddot{u} + \left(\ddot{\theta}_x^P\right)'\phi_S^P$$
$$- \left(\left(\ddot{\theta}_x^P\right)' - \ddot{\theta}_x'\right)\left(\phi_S^P + \phi_S^S\right) - \left(\left(\theta_x^P\right)' - \theta_x'\right)\ddot{\phi}_S^S\right]$$
$$= 0 \quad \text{στο } \Omega, \ \forall x \in [0, L] \quad (26a)$$

$$-G\left(\left(\theta_x^P\right)' - \theta_x'\right)\frac{\partial \phi_S^S}{\partial n} = t_x \quad \text{στο } \Gamma_j, \ \forall x \in [0, L] \quad (26b)$$

which are considered to be valid at the beam ends as well. In the above equations, $\nabla^2 = \partial^2/\partial y^2 + \partial^2/\partial z^2$ is the Laplace operator, while $\partial/\partial n = (\partial/\partial y)n_y + (\partial/\partial z)n_z$ denotes the derivative with respect to the outward normal vector $\mathbf{n}$ to the boundary

Γ_j (directional derivative). In order to simplify the boundary value problem yielding ϕ_S^S, the usual assumption of ignoring the axial traction component ($t_x \approx 0$) is adopted; hence, relation Eq. 26b is modified as

$$\frac{\partial \phi_S^S}{\partial n} = 0 \quad \sigma\tau o \; \Gamma_j, \; \forall x \in [0, L] \quad (27)$$

Boundary value problem Eq. 26 can be further simplified, by ignoring the influence of geometrical nonlinearity terms and inertia terms of the secondary warping function and taking advantage of the global equation of motion for the beam which will be formulated in the next section. It can be also proved that ϕ_S^S is a two-dimensional function. Finally, in order to compute unique solutions of Eqs. 25 and 26, ϕ_S^P, ϕ_S^S are considered to fulfill the following orthogonality conditions:

$$\int_\Omega \phi_S^P d\Omega = 0 \quad (28a)$$

$$\int_\Omega \phi_S^P y d\Omega = 0 \quad (28b)$$

$$\int_\Omega \phi_S^P z d\Omega = 0 \quad (28c)$$

$$\int_\Omega \phi_S^S d\Omega = 0 \quad (28d)$$

Stress Resultants, Global Equations of Motion
The kinematical unknowns of the problem depend only on the variable x (u, θ_x, $(\theta_x^P)'$) and will be established from the solution of global differential equations of motion, which will be formulated employing the calculus of variations. In order to facilitate the analysis, the following stress resultants are defined as

$$N = \int_\Omega S_{xx} d\Omega \quad (29a)$$

$$M_t^P = \int_\Omega \left[S_{xy}\left(\frac{\partial \phi_S^P}{\partial y} - z\right) + S_{xz}\left(\frac{\partial \phi_S^P}{\partial z} + y\right) \right] d\Omega \quad (29b)$$

$$M_t^S = \int_\Omega \left(S_{xy}\frac{\partial \phi_S^S}{\partial y} + S_{xz}\frac{\partial \phi_S^S}{\partial z} \right) d\Omega \quad (29c)$$

$$M_w = \int_\Omega S_{xx} \phi_S^P d\Omega \quad (29d)$$

where N is the axial force, M_t^P, M_t^S are the primary and secondary twisting moment (Tsipiras and Sapountzakis 2012) resulting from the primary (S_{xy}^P, S_{xz}^P) and secondary (S_{xy}^S, S_{xz}^S) shear stress distribution, respectively, and M_w is the warping moment due to the torsional curvature. Substituting Eqs. 22 into Eq. 29, the stress resultants are obtained as

$$N = EA\left[u' + \frac{1}{2}\frac{I_P}{A}\left(\theta_x'\right)^2\right] \quad (30a)$$

$$M_t^P = GI_t^P \theta_x' \quad (30b)$$

$$M_t^S = -GI_t^S\left(\left(\theta_x^P\right)' - \theta_x'\right) \quad (30c)$$

$$M_w = EC_S \left(\theta_x^P\right)'' \quad (30d)$$

where $A = \int_\Omega d\Omega$ is the area of the cross section. The geometric constants I_P (polar moment of inertia with respect to the shear center S), I_t^P (primary torsion constant), I_t^S (secondary torsion constant), and C_S (warping constant) are given by the following definitions:

$$I_P = \int_\Omega (y^2 + z^2) d\Omega \quad (31a)$$

$$I_t^P = \int_\Omega \left(y^2 + z^2 + y\frac{\partial \phi_S^P}{\partial z} - z\frac{\partial \phi_S^P}{\partial y} \right) d\Omega \quad (31b)$$

$$I_t^S = \int_\Omega \left[\left(\frac{\partial \phi_S^S}{\partial y}\right)^2 + \left(\frac{\partial \phi_S^S}{\partial z}\right)^2 \right] d\Omega \quad (31c)$$

$$C_S = \int_\Omega \left(\phi_S^P\right)^2 d\Omega \quad (31d)$$

Applying the principle of virtual work under a total Lagrangian formulation (Eq. 6), taking into account the virtual quantities of u, θ_x, $(\theta_x^P)'$ (and their derivatives) and the stress components given by Eq. 22, the Euler–Lagrange equations governing the global dynamic equilibrium of the beam, together with the pertinent boundary conditions at the beam ends $x=0,L$, are expressed as

$$-\rho A \ddot{u} + \frac{dN}{dx} = -p_x(x,t) \qquad (32a)$$

$$-\rho I_P \ddot{\theta}_x + \frac{dM_t^P}{dx} + \frac{dM_t^S}{dx} + \frac{d}{dx}\left[\frac{1}{2}EI_n(\theta_x')^3 + N\frac{I_P}{A}\theta_x'\right]$$
$$= -m_t(x,t) \qquad (32b)$$

$$-\rho C_S \left(\ddot{\theta}_x^P\right)' + \frac{dM_w}{dx} + M_t^S = -m_w(x,t) \qquad (32c)$$

$$(N + \overline{N}_0)\delta u(0) = 0 \qquad (33a)$$

$$(N - \overline{N}_L)\delta u(L) = 0 \qquad (33b)$$

$$\left[M_t^P + M_t^S + \frac{1}{2}EI_n(\theta_x')^3 + N\frac{I_P}{A}\theta_x' + \overline{M}_{t0}\right]\delta\theta_x(0) = 0 \qquad (33c)$$

$$\left[M_t^P + M_t^S + \frac{1}{2}EI_n(\theta_x')^3 + N\frac{I_P}{A}\theta_x' - \overline{M}_{tL}\right]\delta\theta_x(L) = 0 \qquad (33d)$$

$$(M_w + \overline{M}_{w0})\delta(\theta_x^P)'(0) = 0 \qquad (33e)$$

$$(M_w - \overline{M}_{wL})\delta(\theta_x^P)'(L) = 0 \qquad (33f)$$

where $I_n = I_{PP} - I_P^2/A$ and $I_{PP} = \int_\Omega (y^2 + z^2)^2 d\Omega$ is the fourth moment of inertia with respect to the shear center S (coinciding with the centroid). In all expressions of Eq. 33, the quantities p_x, m_t, m_w (external axial loading, twisting, and warping moment, respectively) constitute the externally applied time-dependent loading along the beam (Fig. 9a) and arise (through the principle of virtual work) from the externally applied traction vectors t_x, t_y, t_z. The quantities $\overline{N}_0, \overline{N}_L, \overline{M}_{t0}, \overline{M}_{tL}, \overline{M}_{w0}, \overline{M}_{wL}$ (appearing in boundary conditions Eqs. 33a and 33b) constitute externally applied reactions (axial force, twisting moment, and warping moment, respectively) at the beam end cross sections and are given as

$$\overline{N}_0 = \int_{\Omega_0} t_x d\Omega \qquad (34a)$$

$$\overline{N}_L = \int_{\Omega_L} t_x d\Omega \qquad (34b)$$

$$\overline{M}_{t0} = \int_{\Omega_0} [t_y(-z\cos\theta_x - y\sin\theta_x) + t_z(y\cos\theta_x - z\sin\theta_x)]d\Omega \qquad (34c)$$

$$\overline{M}_{tL} = \int_{\Omega_L} [t_y(-z\cos\theta_x - y\sin\theta_x) + t_z(y\cos\theta_x - z\sin\theta_x)]d\Omega \qquad (34d)$$

$$\overline{M}_{w0} = \int_{\Omega_0} t_x \phi_S^P d\Omega \qquad (34e)$$

$$\overline{M}_{wL} = \int_{\Omega_L} t_x \phi_S^P d\Omega \qquad (34f)$$

From Eqs. 32b, 33c, and 33d, it can be concluded that axial and torsional stress resultants are coupled and cannot be studied independently, which is the case in linear analysis. Introducing the expressions of stress resultants Eq. 30 into Eqs. 32 and 33, the following governing differential equations of motion are obtained as

$$-\rho A \ddot{u} + EAu'' + EI_P \theta_x' \theta_x'' = -p_x(x,t) \qquad (35a)$$

$$-\rho I_P \ddot{\theta}_x + G(I_t^P + I_t^S)\theta_x'' - GI_t^S (\theta_x^P)''$$
$$+ \frac{3}{2}EI_{PP}(\theta_x')^2 \theta_x'' + EI_P u' \theta_x'' \qquad (35b)$$
$$+ EI_P u'' \theta_x' = -m_t(x,t)$$

$$-\rho C_S \left(\ddot{\theta}_x^P\right)' + EC_S (\theta_x^P)''' - GI_t^S \left((\theta_x^P)' - \theta_x'\right)$$
$$= -m_w(x,t) \qquad (35c)$$

subjected to the initial conditions ($x \in (0,L)$)

$$u(x,0) = \overline{u}_0(x) \qquad (36a)$$

$$\dot{u}(x,0) = \dot{\bar{u}}_0(x) \quad (36b)$$

$$\theta_x(x,0) = \bar{\theta}_{x0}(x) \quad (36c)$$

$$\dot{\theta}_x(x,0) = \dot{\bar{\theta}}_{x0}(x) \quad (36d)$$

$$\left(\theta_x^P\right)'(x,0) = \left(\bar{\theta}_{x0}^P\right)'(x) \quad (36e)$$

$$\left(\dot{\theta}_x^P\right)'(x,0) = \left(\dot{\bar{\theta}}_{x0}^P\right)'(x) \quad (36f)$$

and to the most general boundary conditions

$$\alpha_1 N + \alpha_2 u = \alpha_3 \quad (37a)$$

$$\beta_1 M_t + \beta_2 \theta_x = \beta_3 \quad (37b)$$

$$\bar{\beta}_1 M_w + \bar{\beta}_2 \left(\theta_x^P\right)' = \bar{\beta}_3 \quad (37c)$$

where N, M_w are the axial force and the warping moment at the beam ends given by relations Eqs. 30a and 30d, respectively, while M_t is the twisting moment at the beam ends given as

$$M_t = G\left(I_t^P + I_t^S\right)\theta_x' - GI_t^S\left(\theta_x^P\right)'$$
$$+ \frac{1}{2}EI_{PP}\left(\theta_x'\right)^3 + EI_P u'\theta_x' \quad (38)$$

and α_i, β_i, $\bar{\beta}_i$ ($i = 1, 2, 3$) are time-dependent functions specified at the boundary of the beam defining any possible support condition as described in section "Statement of the Problem." In the above differential equations, viscous damping can be also added without any difficulty. Finally, it is noted that in Eq. 35b, the derivatives of secondary torsion constant with respect to x have been ignored since it can be proved that it is independent of the axial coordinate.

In Eq. 35c, the third derivative of the primary angle of twisting rotation can be observed, thus making the solution of the system more complicated. In order to remove this term, the independent warping parameter $\eta_x = (\theta_x^P)'$ is introduced, and Eqs. 35, 36, 37, and 38 are expressed as

$$-\rho A\ddot{u} + EAu'' + EI_P\theta_x'\theta_x'' = -p_x(x,t) \quad (39a)$$

$$-\rho I_P\ddot{\theta}_x + G\left(I_t^P + I_t^S\right)\theta_x'' - GI_t^S\eta_x'$$
$$+ \frac{3}{2}EI_{PP}\left(\theta_x'\right)^2\theta_x'' + EI_P u'\theta_x'' + EI_P u''\theta_x' \quad (39b)$$
$$= -m_t(x,t)$$

$$-\rho C_S\ddot{\eta}_x + EC_S\eta_x'' - GI_t^S\left(\eta_x - \theta_x'\right) = -m_w(x,t) \quad (39c)$$

$$u(x,0) = \bar{u}_0(x) \quad (40a)$$

$$\dot{u}(x,0) = \dot{\bar{u}}_0(x) \quad (40b)$$

$$\theta_x(x,0) = \bar{\theta}_{x0}(x) \quad (40c)$$

$$\dot{\theta}_x(x,0) = \dot{\bar{\theta}}_{x0}(x) \quad (40d)$$

$$\eta_x(x,0) = \bar{\eta}_{x0}(x) \quad (40e)$$

$$\dot{\eta}_x(x,0) = \dot{\bar{\eta}}_{x0}(x) \quad (40f)$$

$$\alpha_1 N + \alpha_2 u = \alpha_3 \quad (41a)$$

$$\beta_1 M_t + \beta_2 \theta_x = \beta_3 \quad (41b)$$

$$\bar{\beta}_1 M_w + \bar{\beta}_2 \eta_x = \bar{\beta}_3 \quad (41c)$$

Ignoring all the nonlinear terms of the stress resultants, the initial-boundary value problem of nonuniform torsion problem with STMDE within the range of linear elastic dynamic analysis can be derived. It can be easily concluded that in this simplified problem, the axial and torsional deformation are uncoupled; hence, they can be investigated independently. Moreover, the geometrically nonlinear dynamic problem of nonuniform torsion without STMDE can be described by setting $(\theta_x^P)' = \theta_x'$ in Eq. 21 and following the procedure presented in this section. The differential equations and the boundary conditions of the corresponding simplified initial-boundary value problem are given as

$$-\rho A\ddot{u} + EAu'' + EI_P\theta_x'\theta_x'' = -p_x \quad (42a)$$

$$\rho I_P\ddot{\theta}_x - \rho C_S\ddot{\theta}_x'' + EC_S\theta_x'''' - GI_t^P\theta_x''$$
$$- \frac{3}{2}EI_{PP}\left(\theta_x'\right)^2\theta_x'' - EI_P u'\theta_x'' - EI_P u''\theta_x' = m_t - m_w' \quad (42b)$$

$$u(x,0) = \bar{u}_0(x) \tag{43a}$$

$$\dot{u}(x,0) = \dot{\bar{u}}_0(x) \tag{43b}$$

$$\theta_x(x,0) = \bar{\theta}_{x0}(x) \tag{43c}$$

$$\dot{\theta}_x(x,0) = \dot{\bar{\theta}}_{x0}(x) \tag{43d}$$

$$\alpha_1 N + \alpha_2 u = \alpha_3 \tag{44a}$$

$$\beta_1 M_t + \beta_2 \theta_x = \beta_3 \tag{44b}$$

$$\bar{\beta}_1 M_w + \bar{\beta}_2 \theta'_x = \bar{\beta}_3 \tag{44c}$$

where M_t, M_w at the beam ends are given as

$$M_t = \rho C_S \ddot{\theta}'_x + GI_t^P \theta'_x - EC_S \theta'''_x$$
$$+ \frac{1}{2} EI_{PP}(\theta'_x)^3 + EI_P u' \theta'_x - m_w \tag{45a}$$

$$M_w = EC_S \theta''_x \tag{45b}$$

Finally, in order to neglect the influence of nonuniform torsion, it is considered that $C_S = 0$, $m_w = 0$ in Eq. 42b, 45a and in the corresponding simplified initial-boundary value problem, the boundary condition Eq. 44c with respect to warping is neglected.

Secondary Warping Function

The boundary value problem for the evaluation of ϕ_S^S has been defined in section "Local Equilibrium Equation, Primary Torsional Warping Function" (Eqs. 26a and 27). With the aid of the global equilibrium equation of motion Eq. 39c and ignoring the influence of m_w, Eq. 26a is written as

$$(-\rho \ddot{\eta}_x + E\eta''_x)\left(\frac{C_S}{I_t^S}\nabla^2 \phi_S^S - \rho \frac{C_S}{GI_t^S}\ddot{\phi}_S^S - \phi_S^P\right)$$
$$- E\left[u'' + (y^2 + z^2)\theta'_x \theta''_x\right]$$
$$+ \rho\left[\ddot{u} - (\ddot{\eta}_x - \ddot{\theta}'_x)(\phi_S^P + \phi_S^S)\right]$$
$$= 0 \quad \text{στο } \Omega, \quad \forall x \in [0, L] \tag{46}$$

Removing m_w constitutes an assumption similar to $t_x \approx 0$ which has been adopted in section "Local Equilibrium Equation, Primary Torsional Warping Function." In order to further simplify the computation of secondary warping function, nonlinear terms (Eu'', $E(y^2 + z^2)\theta'_x \theta''_x$, $\rho \ddot{u}$) as well as inertia terms ($-\ddot{\phi}_S^S \rho C_S/(GI_t^S)$, $\rho(\ddot{\eta}_x - \ddot{\theta}'_x)(\phi_S^P + \phi_S^S)$) appearing in Eq. 26a are neglected. As a consequence, the longitudinal local equilibrium in geometrically nonlinear analysis is partially violated; however, in the corresponding problem in the linear range, local equilibrium is fulfilled, if the inertia terms of the secondary warping function are neglected. Since $\eta_x \neq \theta'_x$ along the beam length, it holds that $(-\rho \ddot{\eta}_x + E\eta''_x) \neq 0$. Consequently, Eq. 46 is written as

$$\nabla^2 \phi_S^S = \frac{I_t^S}{C_S}\phi_S^P \quad \text{στο } \Omega, \quad \forall x \in [0, L], \tag{47}$$

from which it can be concluded that ϕ_S^S is a two-dimensional time-independent warping function.

Applications

Free Vibrations of Beam of Thin-Walled I-Section Without STMDE

In order to investigate the influence of geometrical nonlinearity on the free vibrations of beams under nonuniform torsion, a beam of an I-shaped cross section ($E = 2.1 \times 10^8$ kN/m^2, $G = 8.1 \times 10^7$ kN/m^2, $\rho = 8.002$ kNs^2/m^4) of length $L = 4.0$ m, having flange and web width $t_f = t_w = 0.01$ m and total height and total width $h = b = 0.20$ m, has been studied. Two cases of boundary conditions are studied, while the influence of STMDE is neglected. The ends of the beam are simply supported (according to the torsional boundary conditions), the left end is axially immovable, while the right one is subjected to a constant force $\bar{N}(L, t)$. The initial-boundary value problems described by Eqs. 42, 43, and 44 are solved, in order to obtain the response of the beam in the time domain taking into account or ignoring the influence of axial inertia.

The linear fundamental modeshape of the angle of twist is firstly considered as initial

Nonlinear Dynamic Seismic Analysis, Fig. 10 Time histories of the angle of twist at the midpoint (**a**) and axial displacement at the right end point (**b**) of the beam of application section "Free Vibrations of Beam of Thin-Walled I-Section Without STMDE," for the case of initial twisting rotations $\bar{\theta}_{x0}(x)$

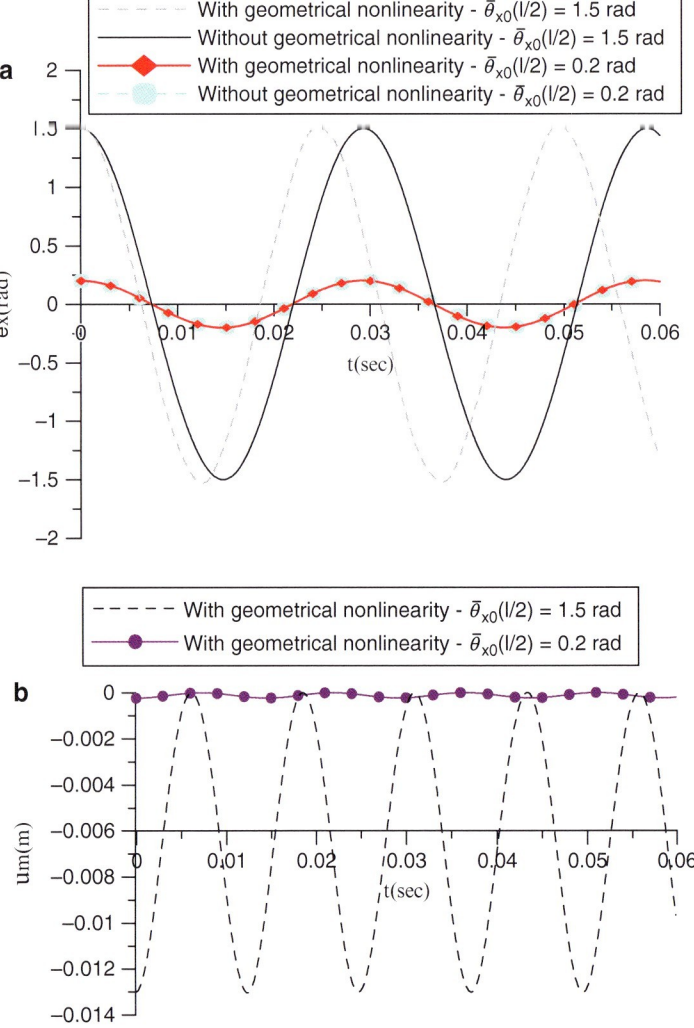

twisting rotation $\bar{\theta}_{x0}(x)$ along with zero initial twisting velocity $\bar{\dot{\theta}}_{x0}(x)$, while the axial force at the right end is considered to be zero. The normalization of the modeshape is performed by specifying the amplitude of the angle of twist at the midpoint of the beam $\bar{\theta}_{x0}(L/2)$. In Figs. 10 and 11, the time histories of characteristic kinematical components (angle of twist $\theta_x(l/2, t)$ at the midpoint of the beam, axial displacement $u(l, t)$ at the beam's right end), and the twisting moment at the beam's left end $M_t(0, t)$, respectively, are presented for two values of the initial midpoint angle of twist amplitude ($\bar{\theta}_{x0}(l/2) = 1.5$ rad, $\bar{\theta}_{x0}(l/2) = 0.2$ rad) performing two cases of analysis (linear analysis and nonlinear analysis ignoring axial inertia term). From these figures, the effects of both the nonlinear terms and the initial midpoint angle of twist amplitude are observed on both the kinematical components and stress resultants. It is worth here noting that the pronounced alteration of the twisting moment's time history pattern and a slight change of the amplitude of the angle of twist (which is not clearly shown in Fig. 10a) are attributed to the fact that the fundamental nonlinear modeshape does not coincide with the linear one. Moreover, as it is easily verified from Fig. 10, the frequency of the axial displacement's response is twice as much as the one of the twisting rotation. Moreover, the case of constant along the beam initial

Nonlinear Dynamic Seismic Analysis, Fig. 11 Time history of the twisting moment at the left end of the beam of application section "Free Vibrations of Beam of Thin-Walled I-Section Without STMDE," for the case of initial twisting rotations $\bar{\theta}_{x0}(x)$

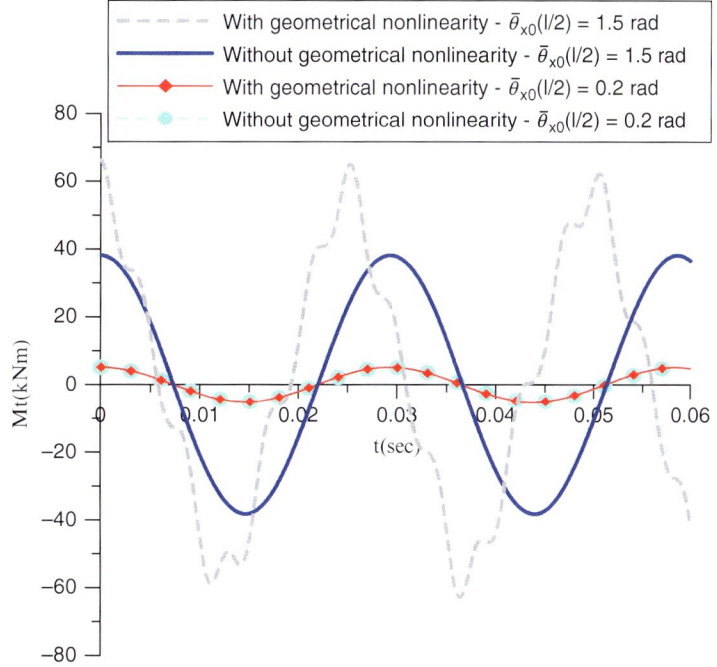

twisting velocities $\dot{\bar{\theta}}_{x0}(x) = 50$ rad/s, variable along the beam initial axial displacements given from $\bar{u}_0(x) = [\overline{N}(l,0)/(EA)]x$, along with vanishing initial twisting rotations $\bar{\theta}_{x0}(x) = 0$ and initial axial velocities $\dot{\bar{u}}_0(x) = 0$, applied to the beam with immovable left and free right end subjected to constant axial load $\overline{N}(l,t) = -3,000$ kN according to the axial boundary conditions, has been also studied. In Figs. 12 and 13, the time history of the angle of twist $\theta_x(l/2, t)$ at the midpoint of the beam and $u(l, t)$ at the beam end, respectively, is presented for three cases of analysis, namely, a linear one, a nonlinear complete, and a nonlinear one ignoring the influence of axial inertia. From these figures the influence of geometrical nonlinearity can be once again observed, while the influence of axial inertia proves to be negligible.

Forced Vibrations of Beam of Thin-Walled I-Section Without STMDE

In order to investigate the influence of geometrical nonlinearity on the forced vibrations of beams under nonuniform torsion, the beam of application section "Free Vibrations of Beam of Thin-Walled I-Section Without STMDE" has been studied ignoring the influence of STMDE. The studied beam is considered to be simply supported (according to the torsional boundary conditions) and subjected to a moving concentrated twisting moment. All of the initial conditions are zero except from the initial axial displacements, which are given from $\bar{u}_0(x) = [\overline{N}(l,0)/(EA)]x$, while the beam has an immovable left and free right end subjected to constant axial load $\overline{N}(l,t) = -2,500$ kN according to the axial boundary conditions. The concentrated twisting moment has a constant numerical value $M_t = 20.0$ kNm and "travels" with a constant velocity $v = 40$ m/s; thus, the beam is subjected to free vibrations after $t = 0.1$ s. In Fig. 14, the time history of the angle of twist $\theta_x(l/2, t)$ at the midpoint of the beam is presented for the aforementioned cases of analysis (linear, nonlinear) demonstrating the discrepancy of the nonlinear response of the beam compared with that of the linear one in both its forced and free vibrating phase. In Fig. 15, the time history of the axial displacement at the beam's right end $u(l, t)$ (which is constant in the linear case) is presented for whichever of the aforementioned cases of analysis, the results are not trivial.

Nonlinear Dynamic Seismic Analysis, Fig. 12 Time history of the angle of twist at the midpoint of the beam of application section "Free Vibrations of Beam of Thin-walled I-Section Without STMDE," for the case of initial twisting velocities $\dot{\bar{\theta}}_{x0}(x)$ and initial axial displacements $\bar{u}_0(x)$

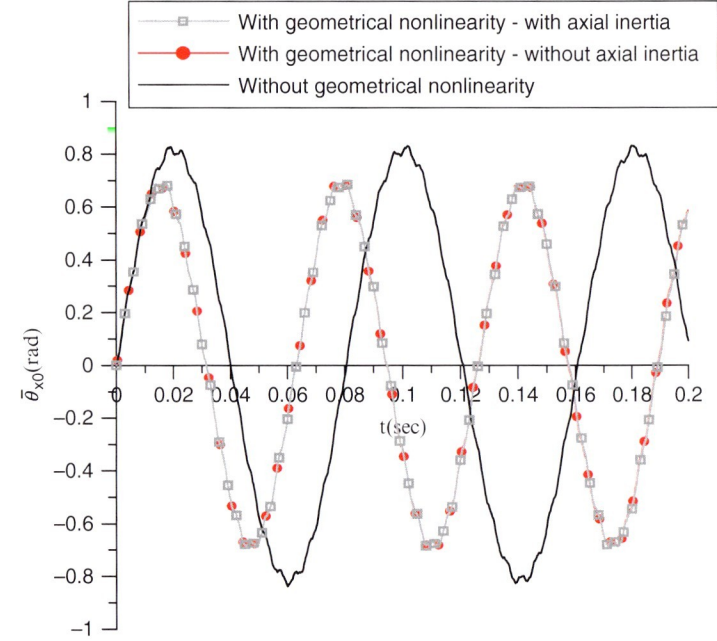

Nonlinear Dynamic Seismic Analysis, Fig. 13 Time histories of the axial stress resultant at the left end (**a**) and the axial displacement at the right end (**b**) of the beam of application section "Free Vibrations of Beam of Thin-walled I-Section Without STMDE," for the case of initial twisting velocities $\dot{\bar{\theta}}_{x0}(x)$ and initial axial displacements $\bar{u}_0(x)$

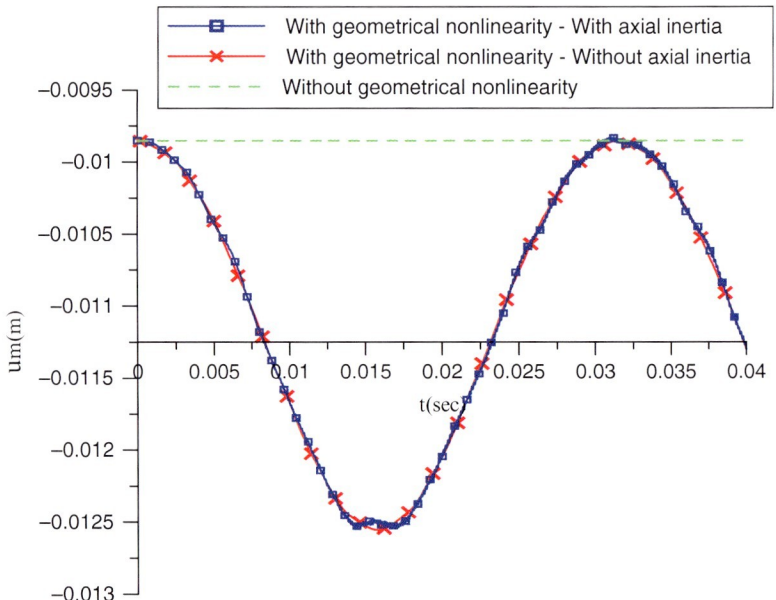

Free Vibrations of Beam of Thin-Walled I-Section with STMDE

In order to investigate the influence of STMDE in linear free vibrations of a beam of an open cross section under nonuniform torsion, the beam of application section "Free Vibrations of Beam of Thin-Walled I-Section Without STMDE" ($I_t^S = 3.048 \times 10^{-4} \text{m}^4$) is examined. The beam's ends are simply supported according to its torsional boundary conditions, while the left end is immovable and the right end is subjected to a compressive axial load according to its axial

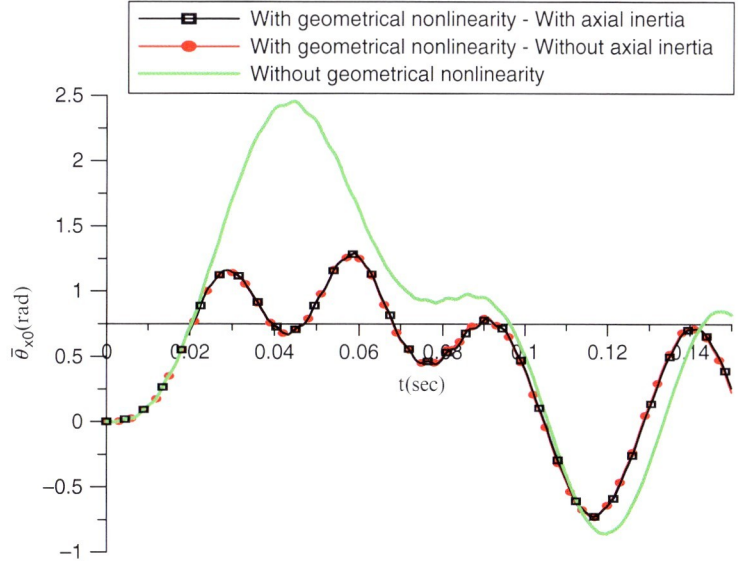

Nonlinear Dynamic Seismic Analysis, Fig. 14 Time history of the angle of twist at the midpoint of the beam of application section "Forced Vibrations of Beam of Thin-Walled I-Section Without STMDE"

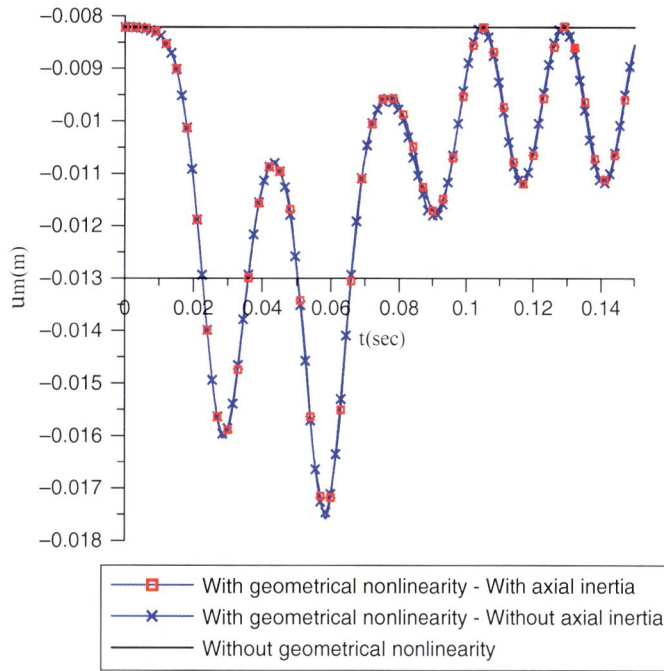

Nonlinear Dynamic Seismic Analysis, Fig. 15 Time history of the axial displacement at the right end of the beam of application section "Forced Vibrations of Beam of Thin-Walled I-Section Without STMDE"

boundary conditions. For comparison reasons, the axial load–torsional fundamental natural frequency relation of the aforementioned beam has been investigated. Mohri et al. (2004), without considering the secondary twisting moment deformation effect, proposed analytical load–frequency relations by dropping the higher-order warping inertia term and assuming that (i) the fundamental modeshape of vibration follows a sinusoidal form $\sin(\pi x/l)$ both in the pre- and post-buckling region, (ii) the beam vibrates harmonically, and (iii) vibrations of small amplitude around a static equilibrium state are performed. These load ($\overline{N}$)–frequency

(ω_f) relations for the pre- and post-buckling regions are given, respectively, as

$$\omega_f^2 = \omega_0^2 \left(1 - \frac{\overline{N}}{\overline{N}_{cr,\theta}}\right) \quad (48a)$$

$$\omega_f^2 = 2\omega_0^2 \left(\frac{\overline{N}}{\overline{N}_{cr,\theta}} - 1\right) \quad (48b)$$

where ω_θ is the natural frequency for vanishing axial load and $\overline{N}_{cr,\theta}$ is the torsional buckling load of the beam, respectively, given as (Mohri et al. 2004)

$$\omega_\theta^2 = \frac{\pi^2}{\rho l^2 I_P} \left(\frac{\pi^2 E C_S}{l^2} + G I_t\right) \quad (49a)$$

$$\overline{N}_{cr,\theta} = \frac{A}{I_P} \left(\frac{\pi^2 E C_S}{l^2} + G I_t\right) \quad (49b)$$

According to the results of the presented methodology, the nonlinear initial-boundary value problem (Eqs. 39, 40, and 41) has been solved for several values of constant axial load $\overline{N}$ to obtain the response of the beam in the time domain, taking into account or neglecting the STMDE. The fundamental natural frequency ω_f is then computed by exploiting the first few full cycles of vibration. The computation is based on the fast Fourier transform (FFT) of the time history, while a Hanning data window is employed (Brigham 1988). The free vibrations are initiated by dropping all the torsional loading terms and by employing the linear fundamental modeshape of the angle of twist as initial twisting rotations $\overline{\theta}_{x0}(x)$ along with zero initial twisting velocities $\dot{\overline{\theta}}_{x0}(x)$. In all cases, the STMDE is neglected in the evaluation of $\overline{\theta}_{x0}(x)$. $\overline{\theta}_{x0}(x)$ is computed by employing the methodology presented in the study of Sapountzakis and Tsipiras (2010a) and corresponds to the sinusoidal form assumed by Mohri et al. (2004). In order to perform vibrations of small amplitude, the initial midpoint angle of twist amplitude $\overline{\theta}_{x0}(l/2)$ is chosen closely to the midpoint angle of twist amplitude $\theta_0(l/2)$ corresponding to the static equilibrium state. Mohri et al. (2004) assuming that the twisting deformation mode follows a sinusoidal form $\sin(\pi x/l)$ proposed the following analytical expression to evaluate $\theta_0(l/2)$

$$\theta_0(l/2) = \begin{cases} 0, & \text{prebuckling region} \\ \pm\sqrt{\frac{8l^2}{3\pi^2}\frac{I_P}{EAI_n}(\overline{N}-\overline{N}_{cr,\theta})}, & \text{postbuckling region} \end{cases} \quad (50)$$

In Fig. 16, the load–frequency relations of Eq. 50 are presented along with pairs of values (ω_f, $\overline{N}$) obtained from the proposed method, ignoring STMDE. The pairs (ω_f, $\overline{N}$) are computed by using the values of initial midpoint angle of twist amplitude $\overline{\theta}_{x0}(l/2)$, which are given in Table 7 along with the corresponding values of $\theta_0(l/2)$ computed from Eq. 50. From Fig. 16, the validity of the developed method in predicting $\overline{N} - \omega_f$ pairs of values of beams undergoing small-amplitude torsional vibrations is concluded. The higher discrepancies between the two sets of solutions in the post-buckling region as compared to the ones in the pre-buckling region are attributed to the fact that the previously mentioned assumptions (i), (ii) are valid only in the pre-buckling region. To illustrate this point better, in Fig. 17, the obtained time histories of the angle of twist $\theta_x(l/2, t)$ at the midpoint of the beam for a pre-buckling ($\overline{N} = -1,000$ kN) and a post-buckling ($\overline{N} = -5,000$ kN) axial loading are presented, respectively, demonstrating that only the buckled beam undergoes multifrequency vibrations. For comparison purposes, in Fig. 17b, the time history of $\theta_x(l/2, t)$ employing the nonlinear fundamental modeshape of the angle of twist as initial twisting rotations $\overline{\theta}_{x0}(x)$ (along with zero initial twisting velocities $\dot{\overline{\theta}}_{x0}(x)$) is also included ($\overline{N} = -5,000$ kN), showing that the initiation of free vibrations with the nonlinear modeshape does not induce higher harmonics in the response of the beam. Finally, in Table 8, the obtained results of pairs of values (ω_f, $\overline{N}$) are presented taking into account or ignoring STMDE and employing either the linear or the nonlinear fundamental modeshape to initiate the free vibrations, showing that in both pre- and post-buckling regions, STMDE does not affect the fundamental natural frequency of beams of open thin-walled cross sections undergoing

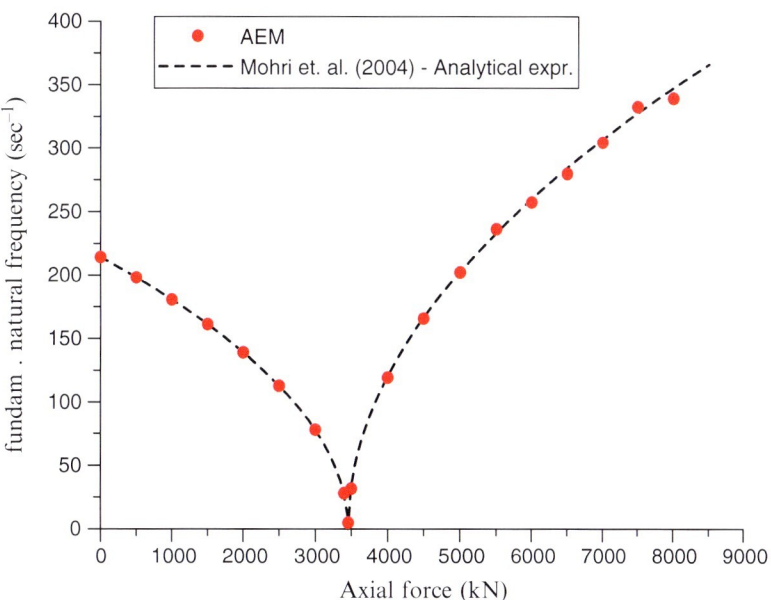

Nonlinear Dynamic Seismic Analysis, Fig. 16 Axial load $\overline{N}$ –torsional fundamental natural frequency ω_f relation of Eq. 50 (Mohri et al. 2004) along with pairs of values $(\omega_f, \overline{N})$ obtained for application section "Free Vibrations of Beam of Thin-walled I-Section with STMDE" (AEM)

Nonlinear Dynamic Seismic Analysis, Table 7 Axial load $\overline{N}$, midpoint angle of twist amplitude $\theta_0(l/2)$ corresponding to the static equilibrium state (Mohri et al. 2004), along with values of initial midpoint angle of twist amplitude $\overline{\theta}_{x0}(l/2)$ used to initiate free vibrations of application section "Free Vibrations of Beam of Thin-Walled I-Section with STMDE"

$|\overline{N}|$(kN)	$\theta_0(L/2)$(rad)	$\overline{\theta}_{x0}(L/2)$
0–3,458	0.00	0.05
3,500	0.22	0.27
4,000	0.80	0.85
4,500	1.11	1.16
5,000	1.35	1.40
5,500	1.55	1.60
6,000	1.73	1.85
6,500	1.90	2.00
7,000	2.05	2.20
7,500	2.19	2.35
8,000	2.32	2.50

small-amplitude vibrations. The insignificance of STMDE for linear static loading conditions of beams of such cross sections has already been reported in the literature (Murín and Kutiš 2008).

Primary Resonance of Beam of Thin-Walled I-Section with STMDE

In this application, in order to investigate the influence of geometrical nonlinearity and STMDE on the response of beams of open cross section under the action of external twisting loading (forced vibrations), the beam of application section "Free Vibrations of Beam of Thin-walled I-Section Without STMDE" is examined under nonuniform torsion in its pre-buckled state. More specifically, the primary resonance of the beam is studied by applying a concentrated twisting moment $M_{t,\text{ext}}$ at its midpoint given as $M_{t,\text{ext}}(t) = M_{t0} \sin(\omega_{f,\text{lin}}t)$, where $M_{t0} = 5$ kNm and $\omega_{f,\text{lin}} = 214.23$ s^{-1} (initial conditions $\overline{\theta}_{x0}(x) = 0$, $\dot{\overline{\theta}}_{x0}(x) = 0$). $\omega_{f,\text{lin}}$ is the fundamental natural frequency of the beam undergoing linear torsional vibrations, ignoring STMDE, and is numerically evaluated by performing modal analysis employing Eqs. 39, 40, and 41 and ignoring geometrical nonlinearity. In Figs. 18 and 19, the time histories of the angle of twist $\theta_x(l/2, t)$ at the midpoint of the beam are presented (with or without STMDE) for two cases of analysis, namely, ignoring or considering geometrical nonlinearity, respectively. The linear analyses are carried out by dropping all the nonlinear terms of Eqs. 39, 40, and 41. As expected, only in the linear cases (Fig. 18), deformations continue to increase with time. The beating phenomenon observed in the nonlinear response

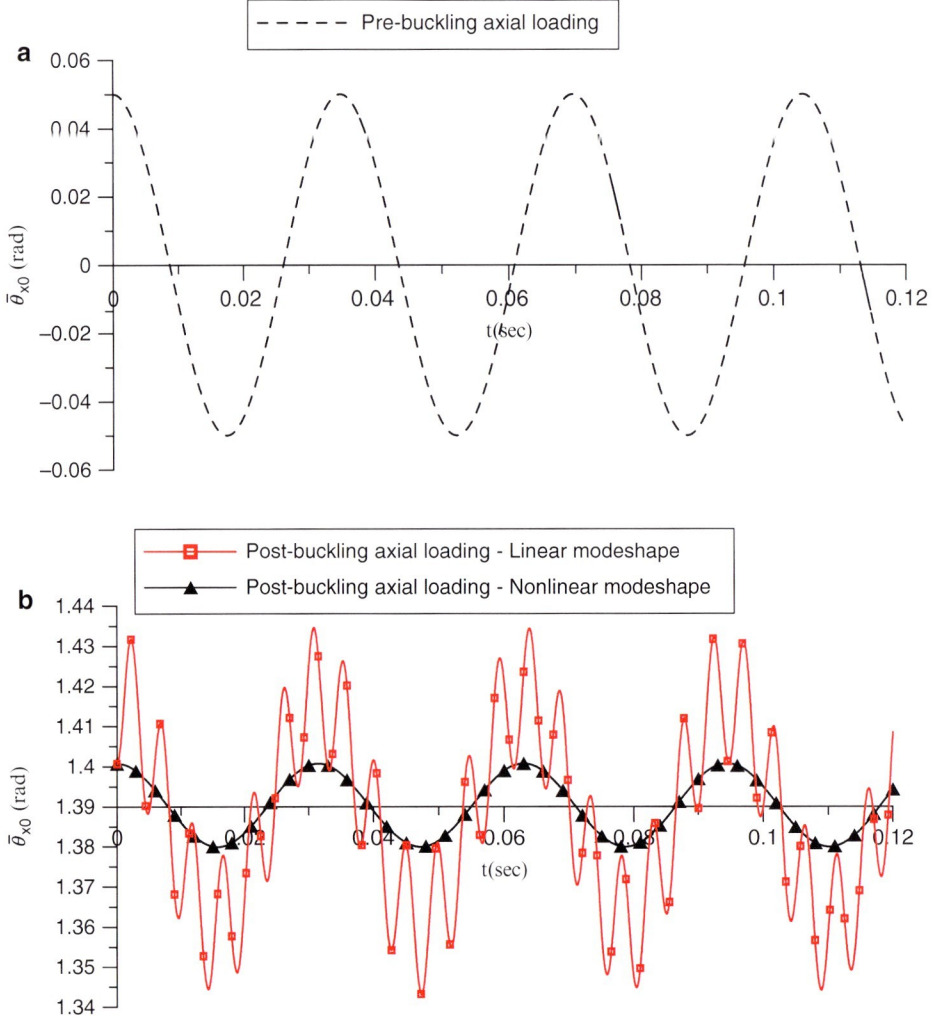

Nonlinear Dynamic Seismic Analysis, Fig. 17 Time histories of the angle of twist at the midpoint of the beam of application "Free Vibrations of Beam of Thin-walled I-Section with STMDE," for a pre-buckling ($\overline{N} = -1,000$ kN) (**a**) and a post-buckling ($\overline{N} = -5,000$ kN) (**b**) axial loading

Nonlinear Dynamic Seismic Analysis, Table 8 Pairs of values $\overline{N}$(kN), ω_f(s^{-1}) of the beam of application section "Free Vibrations of Beam of Thin-Walled I-Section with STMDE"

	ω_f(s^{-1})			
	$\overline{\theta}_{x0}(x)$ from linear analysis		$\overline{\theta}_{x0}(x)$ from nonlinear analysis	
$|\overline{N}|$(kN)	With STMDE	Without STMDE	With STMDE	Without STMDE
1,000	180.48	180.70	–	–
5,000	199.77	199.46	199.98	199.67

(Fig. 19) is explained from the fact that large twisting rotations increase the beam's fundamental natural frequency ω_f (by increasing the stiffness of the beam), thereby causing a detuning of ω_f with the frequency of the external loading ($\omega_{f,\text{lin}}$). After the angle of twist reaches its maximum value, the amplitude of twisting deformations decreases leading to the reversal of the previously mentioned effects.

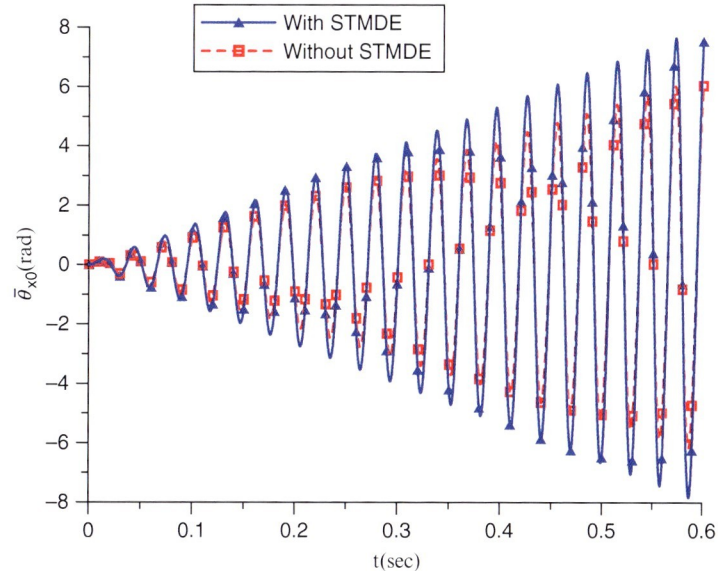

Nonlinear Dynamic Seismic Analysis, Fig. 18 Time history of the angle of twist at the midpoint of the beam of application section "Primary Resonance of Beam of Thin-Walled I-Section with STMDE" taking into account or ignoring secondary twisting moment deformation effect – geometrically linear case

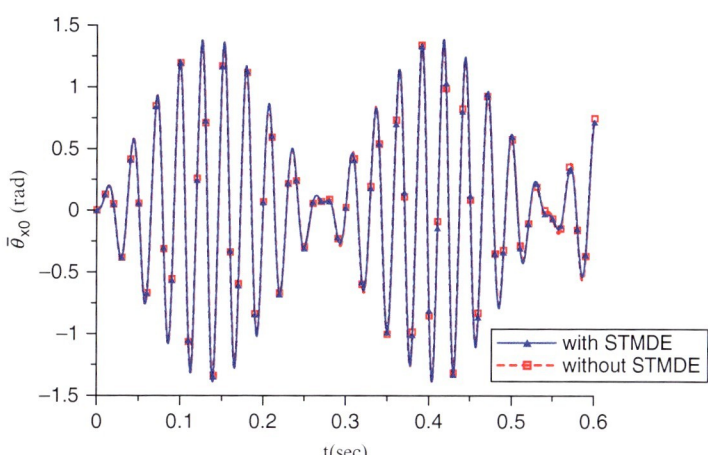

Nonlinear Dynamic Seismic Analysis, Fig. 19 Time history of the angle of twist at the midpoint of the beam of application section "Primary Resonance of Beam of Thin-Walled I-Section with STMDE" taking into account or ignoring secondary twisting moment deformation effect – geometrically nonlinear case

Moreover, in Fig. 20, the time histories of the primary twisting moment M_t^P, the linear secondary twisting moment $M_{t,\text{lin}}^S$, the nonlinear twisting moment $M_{t,nl}$, and the total twisting moment $M_t = M_t^P + M_{t,\text{lin}}^S + M_{t,nl}$ at the beam's left end are presented taking into account STMDE and performing geometrically nonlinear analysis. It is deduced that the STMDE influences negligibly both kinematical and stress components of beams of open thin-walled cross sections undergoing forced vibrations.

Primary Resonance of Beam of Closed Cross Section with STMDE

In this application, in order to investigate the influence of STMDE on the forced linear vibrations of a beam of closed cross section under nonuniform torsion, a steel beam ($E = 2.1 \times 10^8 \text{kN/m}^2$, $G = 8.0769 \times 10^7 \text{kN/m}^2$) of length $L = 5.0$m and of a hollow rectangular cross section (total height $H = 1.64$m, total width $B = 1.05$m, web thickness $t_w = 0.05$m, and flange thickness $t_f = 0.04$m) is examined.

Nonlinear Dynamic Seismic Analysis, Fig. 20 Time histories of various twisting moment components at the left end of the beam of application section "Primary Resonance of Beam of Thin-Walled I-Section with STMDE" taking into account the secondary twisting moment deformation effect (geometrically nonlinear analysis)

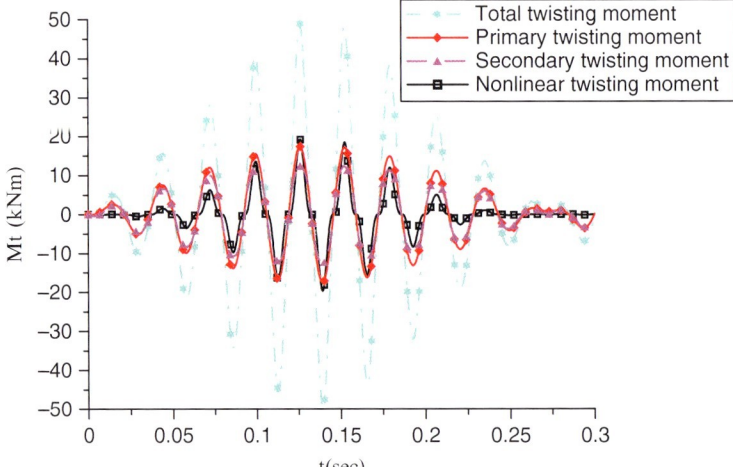

The geometric constants of the beam are assumed to take the values presented in the study of Murín and Kutiš (2008) ($A = 0.240\text{m}^2$, $I_t = 0.089824\text{m}^4$, $I_t^S = 0.001107\text{m}^4$, $C_S = 0.000193\text{m}^6$). The beam's left end is clamped, and its resonance is studied by applying a distributed twisting moment $m_{t,\text{ext}}$ at $0 < x < 5$ (m) given as $m_{t,\text{ext}}(x,t) = m_{t0} \sin(\omega_{f,\text{lin}} t)$, where $m_{t0} = 5$ kNm/m and $\omega_{f,\text{lin}} = 835, 793\text{s}^{-1}$ ($M_t(l,t) = 0$, vanishing initial conditions $\overline{\theta}_{x0}(x) = 0$, $\dot{\overline{\theta}}_{x0}(x) = 0$). $\omega_{f,\text{lin}}$ is the fundamental natural frequency of the beam undergoing linear torsional vibrations, ignoring STMDE, and is numerically evaluated by performing modal analysis similarly with applications sections "Free Vibrations of Beam of Thin-Walled I-Section Without STMDE" and "Free Vibrations of Beam of Thin-Walled I-Section with STMDE."

In Fig. 21, the time histories of the angle of twist $\theta_x(L,t)$, the secondary twisting moment $M_t^S(0,t)$ (scaled quantity $M_t^S(0,t) \times 0,1$), and the warping moment $M_w(0,t)$, respectively, are presented taking into account or ignoring STMDE. It can be observed that the influence of STMDE on the magnitude of the angle of twist and the torsional stiffness of the beam is negligible, while its influence on the stress resultants is significant and cannot be neglected.

Nonlinear Flexural–Torsional Dynamic Analysis of Beams of Arbitrary Cross Section

Statement of the Problem

Let us consider a prismatic beam of length L (Fig. 22), of constant arbitrary cross section of area A. The homogeneous isotropic and linearly elastic material of the beam's cross section, with modulus of elasticity E, shear modulus G, and Poisson's ratio v, occupies the two-dimensional multiply connected region Ω of the y, z plane and is bounded by the $\Gamma_j (j = 1, 2, \ldots, K)$ boundary curves, which are piecewise smooth, i.e., they may have a finite number of corners. In Fig. 22, CYZ is the principal bending coordinate system through the cross section's centroid C, while y_C, z_C are its coordinates with respect to the Syz shear system of axes through the cross section's shear center S, with axes parallel to those of the CYZ system. The beam is subjected to the combined action of the arbitrarily distributed or concentrated, time-dependent, and conservative axial loading $p_X = p_X(X, t)$ along the X direction, twisting moment $m_x = m_x(x, t)$ along the x direction, and transverse loading $p_y = p_y(x, t)$, $p_z = p_z(x, t)$ acting along the y and z directions, applied at distances y_{p_y}, z_{p_y} and y_{p_z}, z_{p_z} with respect to the Syz shear system of axes, respectively (Fig. 22b).

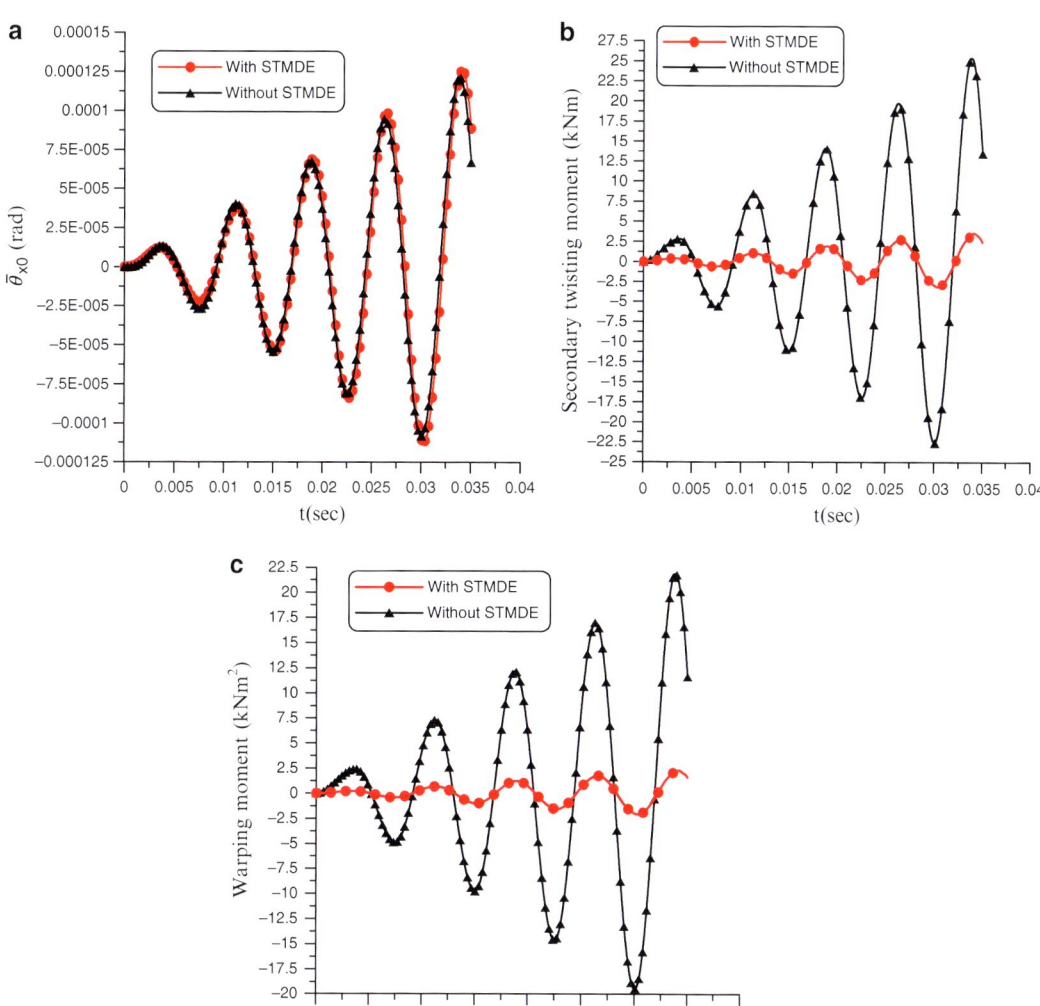

Nonlinear Dynamic Seismic Analysis, Fig. 21 Time history of the angle of twist $\theta_x(L, t)$ (**a**), the secondary twisting moment $M_t^S(0, t)$ (**b**), the warping moment $M_w(0, t)$ (**c**) of the beam of application section "Primary Resonance of Beam of Closed Cross Section with STMDE"

Under the action of the aforementioned loading, the displacement field of an arbitrary point of the cross section can be derived with respect to those of the shear center as

$$\bar{u}(x, y, z, t) = u(x, t) - (y - y_C)\theta_Z(x, t) \\ + (z - z_C)\theta_Y(x, t) \\ + \theta'_x(x, t)\varphi_S^P(y, z) \quad (51a)$$

$$\bar{v}(x, y, z, t) = v(x, t) - z \sin(\theta_x(x, t)) \\ - y[1 - \cos(\theta_x(x, t))] \quad (51b)$$

$$\bar{w}(x, y, z, t) = w(x, t) + y \sin(\theta_x(x, t)) \\ - z[1 - \cos(\theta_x(x, t))] \quad (51c)$$

$$\theta_Y(x, t) = v'(x, t) \sin(\theta_x(x, t)) \\ - w'(x, t) \cos(\theta_x(x, t)) \quad (51d)$$

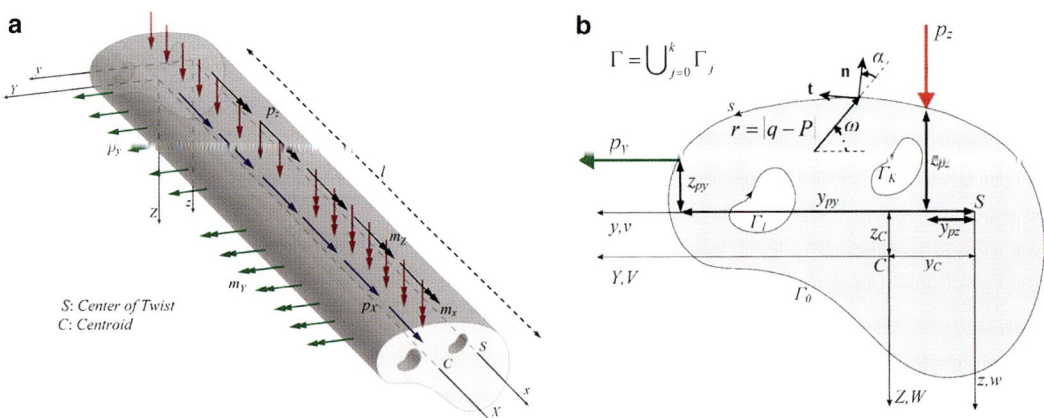

Nonlinear Dynamic Seismic Analysis, Fig. 22 Prismatic beam in axial–flexural–torsional loading (**a**) of an arbitrary cross section occupying the two-dimensional region Ω (**b**)

$$\theta_Z(x,t) = v'(x,t)\cos(\theta_x(x,t)) + w'(x,t)\sin(\theta_x(x,t)) \quad (51e)$$

Similarly with the previous sections, $\bar{u}, \bar{v}, \bar{w}$ are the axial and transverse beam displacement components with respect to the Syz shear system of axes; $u(x, t)$ denotes the "average" axial displacement of the cross section, and $v(x, t)$, $w(x, t)$ are the corresponding components of the shear center S; $\theta_Y(x,t)$, $\theta_Z(x,t)$ are the angles of rotation of the cross section due to bending, with respect to its centroid; $\theta_x'(x,t)$ denotes the rate of change of the angle of twist $\theta_x(x,t)$ regarded as the torsional curvature; and φ_S^P is the primary warping function with respect to the shear center S.

Employing the strain–displacement relations Eq. 2 and stress–strain relations Eq. 4, the non-vanishing stress components are obtained as

$$S_{xx} = E\Big[u' + (z - z_C)(v''\sin\theta_x - w''\cos\theta_x) - (y - y_C)(v''\cos\theta_x + w''\sin\theta_x) \\ + \theta_x''\varphi_S^P - \theta_x'(z_C(v'\cos\theta_x + w'\sin\theta_x) + y_C(v'\sin\theta_x - w'\cos\theta_x)) \\ + \frac{1}{2}\Big(v'^2 + w'^2 + (y^2 + z^2)(\theta_x')^2\Big)\Big] \quad (52a)$$

$$S_{xy} = G \cdot \theta_x' \cdot \left(\frac{\partial \varphi_S^P}{\partial y} - z\right) \quad (52b)$$

$$S_{xz} = G \cdot \theta_x' \cdot \left(\frac{\partial \varphi_S^P}{\partial z} + y\right) \quad (52c)$$

In order to establish the nonlinear equations of motion, the principle of virtual work Eq. 6 under a total Lagrangian formulation is employed. It is worth here noting that in the examined general case, the expression of the external work (Eq. 7c) takes into account the change of the eccentricity of the external conservative transverse loading, arising from the cross-section torsional rotation, inducing additional (positive or negative) torsional moment. This effect may influence substantially the torsional response of the beam. Moreover, the corresponding stress resultants of the beam are defined by relations Eqs. 8, 29b, and 29d. It is noted that in the examined problem, the secondary twisting moment M_t^S is not defined, and it holds that $M_t = M_t^P$. Due to arbitrary shape of the cross section, in the geometrically nonlinear analysis, the following stress resultant is also defined as

$$M_R = \int_\Omega S_{xx}(y^2 + z^2)d\Omega \qquad (53)$$

which is a higher-order stress resultant. Using the expressions of the stress components (Eq. 52), the stress resultants are obtained as

$$N = EA\left\{u' + \frac{1}{2}\left(v'^2 + w'^2 + \frac{I_p}{A}\theta'^2_x\right) - \theta'_x[z_C(v'\cos\theta_x + w'\sin\theta_x) + y_C(v'\sin\theta_x - w'\cos\theta_x)]\right\} \qquad (54a)$$

$$M_Y = -EI_Y\left(w''\cos\theta_x - v''\sin\theta_x - \beta_Z\theta'^2_x\right) \qquad (54b)$$

$$M_Z = EI_Z\left(v''\cos\theta_x + w''\sin\theta_x - \beta_Y\theta'^2_x\right) \qquad (54c)$$

$$M_t = GI_t\theta'_x \qquad (54d)$$

$$M_w = -EC_S\left(\theta''_x + \beta_\omega\theta'^2_x\right) \qquad (54e)$$

$$M_R = N\frac{I_p}{A} - 2EI_Z\beta_Y(v''\cos\theta_x + w''\sin\theta_x)$$
$$- 2EI_Y\beta_Z(w''\cos\theta_x - v''\sin\theta_x)$$
$$+ 2EC_S\beta_\omega\theta''_x + \frac{1}{2}E\left(I_{pp} - \frac{I_p^2}{A}\right)\theta'^2_x \qquad (54f, g)$$

where the area A, the principal moments of inertia I_Y, I_Z with respect to the cross section's centroid, the polar moment of inertia I_p, the torsion constant I_t, and the warping constant C_S with respect to the shear center S are given by relations Eqs. 10, 31a, 31b, and 31d, respectively, while $I_{pp} = \int_\Omega (y^2 + z^2)^2 d\Omega$ is the fourth moment of inertia with respect to the shear center S defined in section "Stress Resultants, Global Equations of Motion" as well. Finally, the Wagner's coefficients β_Z, β_Y, and β_ω are given as

$$\beta_Z = \frac{1}{2I_Y}\int_\Omega (z - z_C)(y^2 + z^2)d\Omega \qquad (55a)$$

$$\beta_Y = \frac{1}{2I_Z}\int_\Omega (y - y_C)(y^2 + z^2)d\Omega \qquad (55b)$$

$$\beta_\omega = \frac{1}{2C_S}\int_\Omega (y^2 + z^2)\phi^P_S d\Omega \qquad (55c)$$

Similarly with the previous sections, applying the principle of virtual work (Eq. 6), the equations of motion of the beam can be derived as

$$-N' + \rho A\ddot{u} = p_X \qquad (56a)$$

$$[-N(v' - y_C\theta'_x\sin\theta_x - z_C\theta'_x\cos\theta_x)]'$$
$$+ (M_Z\cos\theta_x)'' + (M_Y\sin\theta_x)''$$
$$- \rho A\ddot{v} - \rho A(y_C\sin\theta_x + z_C\cos\theta_x)\ddot{\theta}_x$$
$$+ \rho A(z_C\sin\theta_x - y_C\cos\theta_x)\dot{\theta}^2_x$$
$$- (\rho\{(I_Y - I_Z)\sin\theta_x\cos\theta_x v'$$
$$- [(I_Y - I_Z)\cos^2\theta_x - I_Y]w'\}\ddot{\theta}_x)'$$
$$- (\rho\{(I_Y - I_Z)\sin\theta_x\cos\theta_x w'$$
$$+ [(I_Y - I_Z)\cos^2\theta_x - I_Y]v'\}\dot{\theta}^2_x)'$$
$$- \left[\rho(I_Y - I_Z)\cos\theta_x\sin\theta_x\left(2\dot{\theta}_x\dot{v}' - \ddot{w}'\right)\right]'$$
$$+ \left\{\rho[(I_Y - I_Z)\cos^2\theta_x - I_Y]\left(2\dot{\theta}_x\dot{w}' + \ddot{v}'\right)\right\}' = p_y \qquad (56b)$$

$$[-N(w' + y_C\theta'_x\cos\theta_x - z_C\theta'_x\sin\theta_x)]'$$
$$- (M_Y\cos\theta_x)'' + (M_Z\sin\theta_x)''$$
$$+ \rho A\ddot{w} - \rho A(-y_C\cos\theta_x + z_C\sin\theta_x)\ddot{\theta}_x$$
$$- \rho A(z_C\cos\theta_x + y_C\sin\theta_x)\dot{\theta}^2_x$$
$$- (\rho\{(I_Z - I_Y)\cos\theta_x\sin\theta_x w'$$
$$+ [(I_Z - I_Y)\cos^2\theta_x - I_Z]v'\}\ddot{\theta}_x)'$$
$$+ (\rho\{(I_Z - I_Y)\cos\theta_x\sin\theta_x v'$$
$$- [(I_Z - I_Y)\cos^2\theta_x - I_Z]w'\}\dot{\theta}^2_x)'$$
$$- \left[\rho(I_Z - I_Y)\cos\theta_x\sin\theta_x\left(2\dot{\theta}_x\dot{w}' + \ddot{v}'\right)\right]'$$
$$+ \left\{\rho[(I_Z - I_Y)\cos^2\theta_x - I_Z]\left(\ddot{w}' - 2\dot{\theta}_x\dot{v}'\right)\right\}' = p_z \qquad (56c)$$

$$N(-y_C v'\theta_x' \cos\theta_x - y_C w'\theta_x' \sin\theta_x - z_C w'\theta_x' \cos\theta_x + z_C v'\theta_x' \sin\theta_x)+$$

$$[N(z_C(v'\cos\theta_x + w'\sin\theta_x))]'$$
$$+ [N(y_C(v'\sin\theta_x - w'\cos\theta_x))]'$$
$$M_Y(w''\sin\theta_x + v''\cos\theta_x)$$
$$+ M_Z(w''\cos\theta_x - v''\sin\theta_x) - M_t' - M_w''$$
$$- (M_R\theta_x')' - \rho A[(z_C\cos\theta_x + y_C\sin\theta_x)\ddot{v}$$
$$+ (-z_C\sin\theta_x + y_C\cos\theta_x)\ddot{w}]$$
$$+ \rho\{[(I_Y - I_Z)\cos^2\theta_x + I_Z]v'^2$$
$$- 2(I_Z - I_Y)v'w'\sin\theta_x \cos\theta_x + I_s$$
$$+ [(I_Z - I_Y)\cos^2\theta_x + I_Y]w'^2\}\ddot{\theta}_x$$
$$+ \rho\{[(I_Z - I_Y)\cos^2\theta_x + I_Y]w'$$
$$+ (I_Y - I_Z)\sin\theta_x \cos\theta_x v'\}(\ddot{v}' + 2\dot{\theta}_x\dot{w}')$$
$$+ \rho\{[(I_Z - I_Y)\cos^2\theta_x - I_Z]v'$$
$$+ (I_Z - I_Y)\sin\theta_x \cos\theta_x w'\}(\ddot{w}' - 2\dot{\theta}_x\dot{v}')$$
$$- \rho\{(I_Y - I_Z)\sin\theta_x \cos\theta_x v'^2$$
$$+ (I_Z - I_Y)\sin\theta_x \cos\theta_x w'^2$$
$$+ [2(I_Z - I_Y)\cos^2\theta_x - I_Z + I_Y]v'w'\}\dot{\theta}_x^2$$
$$- (\rho C_S \ddot{\theta}_x')' = m_x + p_z y_{p_z}\cos\theta_x$$

$$- p_y z_{p_y}\cos\theta_x - p_z z_{p_z}\sin\theta_x$$
$$- p_y y_{p_y}\sin\theta_x \quad (56d)$$

where the expressions of the stress resultants are given from Eq. 54. Equation 56 is coupled and highly complicated. This set of equations can be simplified if the approximate expressions (Sapountzakis and Dikaros 2011)

$$\cos\theta_x = 1 - \frac{\theta_x^2}{2!} = 1 - \frac{\theta_x^2}{2} \quad (57a)$$

$$\sin\theta_x = \theta_x - \frac{\theta_x^3}{3!} = \theta_x - \frac{\theta_x^3}{6} \quad (57b)$$

are employed. Thus, using the aforementioned approximations, neglecting the term $\rho A \ddot{u}$ of Eq. 56a denoting the axial inertia of the beam, employing the expressions of the stress resultants (Eq. 54), and ignoring the nonlinear terms of the fourth or greater order (Sapountzakis and Dikaros 2011), the governing partial differential equations of motion for the beam at hand can be written as

$$- EA\left[u'' + w'w'' + v'v'' + \frac{I_p}{A}\theta_x'\theta_x'' - z_C(\theta_x\theta_x'w'' + \theta_x\theta_x''w' + \theta_x''v' + \theta_x'v'' + \theta_x'^2 w') \right.$$
$$\left. - y_C(\theta_x\theta_x''v' + \theta_x\theta_x'v'' - \theta_x''w' - \theta_x'w'' + \theta_x'^2 v')\right] = p_X \quad (58a)$$

$$EI_Z v'''' - N\left[-z_C\theta_x'' - y_C(\theta_x'^2 + \theta_x''\theta_x) + v''\right]$$
$$+ (EI_Z - EI_Y)(w''''\theta_x + 2w'''\theta_x' + w''\theta_x''$$
$$- v''''\theta_x^2 - 4v'''\theta_x\theta_x' - 2v''\theta_x\theta_x'' - 2v''\theta_x'^2)$$
$$+ EI_Z\beta_Y(-2\theta_x'\theta_x''' - 2\theta_x''^2)$$
$$+ EI_Y\beta_Z(2\theta_x'\theta_x'''\theta_x + 2\theta_x''^2\theta_x + 5\theta_x'^2\theta_x'') + \rho A\ddot{v} +$$
$$+ \rho[(I_Z - I_Y)(\theta_x v'' + \theta_x'v') - I_Z w''$$
$$- A\left(y_C\theta_x + z_C - \frac{1}{2}z_C\theta_x^2\right)]\ddot{\theta}_x + \rho(I_Z - I_Y)$$
$$\times \left[2\theta_x\theta_x'\ddot{v}' + \theta_x^2\ddot{v}'' + 2(\theta_x'\dot{\theta}_x + \theta_x\dot{\theta}_x')\dot{v}' + 2\theta_x\dot{\theta}_x\dot{v}''\right.$$
$$\left. - \theta_x'\ddot{w}' - \theta_x\ddot{w}'' + v'\theta_x\ddot{\theta}_x'\right] + \rho I_Z$$
$$\times \left[-w'\ddot{\theta}_x' - \ddot{v}'' - 2\dot{\theta}_x\dot{w}'' + v''\dot{\theta}_x^2 + 2v'\dot{\theta}_x\dot{\theta}_x' - 2\dot{\theta}_x\dot{w}'\right]$$
$$- \rho A(y_C - z_C\theta_x)\dot{\theta}_x^2 = p_y$$
$$- p_X(v' - y_C\theta_x'\theta_x - z_C\theta_x') \quad (58b)$$

$$EI_Y w'''' - N\left[w'' + y_C\theta_x'' - z_C(\theta_x'^2 + \theta_x\theta_x'')\right]$$

$$
\begin{aligned}
&+ (EI_Z - EI_Y)(v''''\theta_x + 2v'''\theta_x' + v''\theta_x'' \\
&+ w''''\theta_x^2 + 4w'''\theta_x\theta_x' + 2w''\theta_x\theta_x'' + 2w''\theta_x'^2) \\
&+ EI_Z\beta_Y\left(-2\theta_x'\theta_x'''\theta_x - 5\theta_x'^2\theta_x'' - 2\theta_x''^2\theta_x\right) \\
&+ EI_Y\beta_Z\left(-2\theta_x'\theta_x''' - 2\theta_x''^2\right) + \rho A \ddot{w} \\
&+ \rho\left[(I_Z - I_Y)(-\theta_x w'' - \theta_x' w') + I_Y v'' \right. \\
&\left. + A\left(-z_C\theta_x + y_C - \tfrac{1}{2}y_C\theta_x^2\right)\right]\ddot{\theta}_x - \rho(I_Z - I_Y) \\
&\cdot \left[2\theta_x\theta_x'\ddot{w}' + \theta_x^2\ddot{w}'' + 2\left(\theta_x'\dot{\theta}_x + \theta_x\dot{\theta}_x'\right)\dot{w}' \right. \\
&\left. + 2\theta_x\dot{\theta}_x\dot{w}'' + \theta_x'\ddot{v}'' + \theta_x\ddot{v}''' + w'\theta_x\ddot{\theta}_x'\right] - \rho I_Y \\
&\cdot \left[-v'\ddot{\theta}_x' + \ddot{w}'' - 2\dot{\theta}_x\dot{v}'' - w''\dot{\theta}_x^2 - 2w'\dot{\theta}_x\dot{\theta}_x' - 2\theta_x'\dot{v}'\right] \\
&- \rho A(z_C + y_C\theta_x)\dot{\theta}_x^2 = p_z \\
&\quad - p_X(w' + y_C\theta_x' - z_C\theta_x'\theta_x) \qquad (58c)
\end{aligned}
$$

$$
\begin{aligned}
&EC_S\theta_x'''' - GI_t\theta_x'' - \tfrac{3}{2}E\left(I_{pp} - \tfrac{I_p^2}{A}\right)\theta_x'^2\theta_x'' \\
&- N\left[\tfrac{I_p}{A}\theta_x'' + y_C(w'' - v''\theta_x) - z_C(v'' + w''\theta_x)\right] \\
&+ (EI_Z - EI_Y)\left(v''w'' - v''^2\theta_x + w''^2\theta_x\right) \\
&+ EI_Z\beta_Y\left(2\theta_x''w''\theta_x + w''\theta_x'^2 \right.\\
&\left. + 2\theta_x'v'''' + 2\theta_x''v''' + 2\theta_x'w'''\theta_x\right) \\
&+ EI_Y\beta_Z(-2\theta_x''v''\theta_x + 2\theta_x''w'' + 2\theta_x'w'''' \\
&- v''\theta_x'^2 - 2\theta_x'v'''\theta_x) + \rho\left(v'^2 I_Y + w'^2 I_Z + I_S\right)\ddot{\theta}_x \\
&- \rho A\left(z_C + y_C\theta_x - \tfrac{1}{2}z_C\theta_x^2\right)\ddot{v} \\
&+ \rho A\left(y_C - z_C\theta_x - \tfrac{1}{2}y_C\theta_x^2\right)\ddot{w} \\
&+ \rho(I_Z - I_Y)\theta_x(w'\ddot{w}' - v'\ddot{v}') \\
&+ \rho I_Z\left(w'\ddot{v}' + 2\dot{\theta}_x\dot{w}'w'\right) \\
&+ \rho I_Y\left(-v'\ddot{w}' + 2\dot{\theta}_x\dot{v}'v'\right) - \rho C_S\ddot{\theta}_x'' \\
&= m_x + p_z y_{p_z} - p_y z_{p_z}
\end{aligned}
$$

$$
\begin{aligned}
&+ \left(\tfrac{1}{2}\theta_x^2 z_{p_y} + \tfrac{1}{6}\theta_x^3 y_{p_y} - \theta_x y_{p_y}\right) p_y \\
&+ \left(\tfrac{1}{6}\theta_x^3 z_{p_z} - \tfrac{1}{2}\theta_x^2 y_{p_z} - z_{p_z}\theta_x\right) p_z \\
&- p_X\left(\tfrac{I_p}{A}\theta_x' - y_C v'\theta_x - z_C w'\theta_x - z_C v' + y_C w'\right)
\end{aligned}
\qquad (58d)
$$

while the expression of the axial stress resultant N is given as

$$
N = EA\left[u' + \tfrac{1}{2}\left(v'^2 + w'^2 + \tfrac{I_p}{A}\theta_x'^2\right) \right.\\
\left. - \theta_x'(z_C(w'\theta_x + v') + y_C(v'\theta_x - w'))\right] \qquad (59)
$$

The above governing differential equations (Eq. 58) are also subjected to the initial conditions ($x \in (0, l)$)

$$u(x,0) = u_0(x) \qquad (60a)$$
$$u_{,t}(x,0) = u_{0,t}(x) \qquad (60b)$$
$$v(x,0) = v_0(x) \qquad (61a)$$
$$v_{,t}(x,0) = v_{0,t}(x) \qquad (61b)$$
$$w(x,0) = w_0(x) \qquad (62a)$$
$$w_{,t}(x,0) = w_{0,t}(x) \qquad (62b)$$
$$\theta_x(x,0) = \theta_{x0}(x) \qquad (63a)$$
$$\theta_{x,t}(x,0) = \theta_{x0,t}(x) \qquad (63b)$$

together with the corresponding boundary conditions of the problem at hand, which are given as

$$\alpha_1 u(x,t) + \alpha_2 N_b(x,t) = \alpha_3 \qquad (64)$$
$$\beta_1 v(x,t) + \beta_2 R_{by}(x,t) = \beta_3 \qquad (65a)$$
$$\overline{\beta}_1 \theta_Z(x,t) + \overline{\beta}_2 M_{bZ}(x,t) = \overline{\beta}_3 \qquad (65b)$$

$$\gamma_1 w(x,t) + \gamma_2 R_{bz}(x,t) = \gamma_3 \quad (66a)$$

$$\overline{\gamma}_1 \theta_Y(x,t) + \overline{\gamma}_2 M_{bY}(x,t) = \overline{\gamma}_3 \quad (66b)$$

$$\delta_1 \theta_x(x,t) + \delta_2 M_{bt}(x,t) = \delta_3 \quad (67a)$$

$$\overline{\delta}_1 \theta_{x,x}(x,t) + \overline{\delta}_2 M_{bw}(x,t) = \overline{\delta}_3 \quad (67b)$$

at the beam ends $x = 0, l$, where R_{by}, R_{bz} and M_{bZ}, M_{bY} are the reactions and bending moments with respect to y, z or to Y, Z axes, respectively, given by the following relations (ignoring again the nonlinear terms of the fourth or greater order)

$$R_{by} = N(v' - z_C \theta'_x - y_C \theta_x \theta'_x)$$
$$+ EI_Z \left(v'''' \theta_x^2 - w''' \theta'_x - w'''' \theta_x - v'''' + 2v'' \theta_x \theta'_x + 2\beta_Y \theta'_x \theta''_x \right)$$
$$+ EI_Y \left(w''' \theta_x + w'''' \theta'_x - 2v'' \theta_x \theta'_x - v''' \theta_x'^2 - \beta_Z \theta'^3_x - 2\beta_Z \theta_x \theta'_x \theta''_x \right) \quad (68a)$$

$$R_{bz} = N(w' + y_C \theta'_x - z_C \theta'_x \theta_x)$$
$$+ EI_Y \left(w''' \theta_x^2 + v'''' \theta_x - w'''' + v'' \theta'_x + 2w'' \theta_x \theta'_x + 2\beta_Z \theta'_x \theta''_x \right)$$
$$- EI_Z \left(v'' \theta'_x + w''' \theta_x^2 + v''' \theta_x + 2w'' \theta_x \theta'_x - 2\beta_Y \theta_x \theta'_x \theta''_x - \beta_Y \theta'^3_x \right) \quad (68b)$$

$$M_{bZ} = EI_Z \left(w'' \theta_x - \beta_Y \theta'^2_x + v'' - v'' \theta_x^2 \right)$$
$$+ EI_Y \left(-w'' \theta_x + v'' \theta_x^2 + \beta_Z \theta'^2_x \theta_x \right) \quad (68c)$$

$$M_{bY} = EI_Z \left(-w'' \theta_x^2 + \beta_Y \theta'^2_x \theta_x - v'' \theta_x \right)$$
$$+ EI_Y \left(\beta_Z \theta'^2_x - w'' + w'' \theta_x^2 + v'' \theta_x \right) \quad (68d)$$

while M_t and M_w are the torsional and warping moments at the boundaries of the beam, respectively, given as

$$M_{bt} = GI_t \theta'_x - EC_S \theta'''_x$$
$$+ N \left(-z_C w' \theta_x + y_C w' - y_C v' \theta_x - z_C v' + \frac{I_p}{A} \theta'_x \right)$$
$$+ EI_Z \beta_Y \cdot (-2\theta'_x v'' - 2\theta'_x w'' \theta_x)$$
$$+ EI_Y \beta_Z (-2\theta'_x w'' + 2\theta'_x v'' \theta_x) + \frac{1}{2} E \left(I_{pp} - \frac{I_p^2}{A} \right) \theta'^3_x \quad (69a)$$

$$M_{bw} = -EC_S \left(\theta''_x + \beta_\omega \theta'^2_x \right) \quad (69b)$$

Finally, α_k, β_k, $\overline{\beta}_k$, γ_k, $\overline{\gamma}_k$, δ_k, $\overline{\delta}_k$ ($k = 1, 2, 3$) are time-dependent functions specified at the end cross sections of the beam ($x = 0, l$) as described in section "Statement of the Problem."

Applications

Forced Vibrations or Beam of Hollow Rectangular Cross Section Under Biaxial Bending

In the first application, in order to demonstrate the strong coupling between flexure and torsion, the beam of application section "Forced Vibrations or Beam of Hollow Rectangular Cross Section Under Biaxial Bending" ($I_p = 3.16407 \times 10^{-5} \mathrm{m}^4$, $I_{pp} = 3.94264 \times 10^{-7} \mathrm{m}^6$, $I_t = 2.10844 \times 10^{-5} \mathrm{m}^4$, $C_S = 3.59634 \times 10^{-9} \mathrm{m}^6$) is reexamined employing the governing differential equations derived in this section for the most general case (Eqs. 58, 59, 60, 61, 62, 63, 64, 65, 66, and 67). The time histories of $v(l/2, t)$, $w(l/2, t)$ evaluated employing AEM to solve the differential equations of motion (Sapountzakis and Dikaros 2011) occurred identical with the ones presented in Fig. 6 (case without shear deformation). However, what is of interest in this case

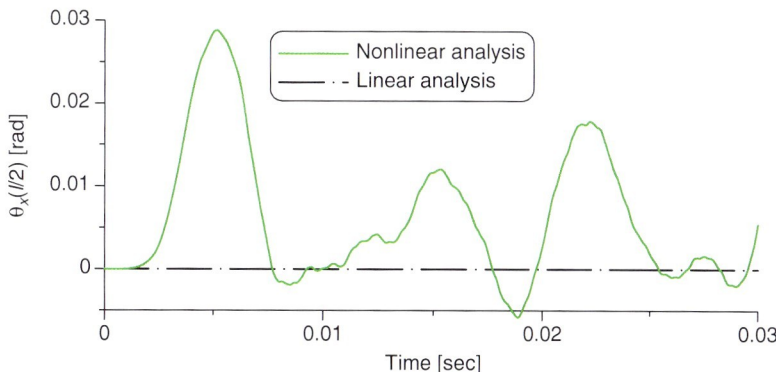

Nonlinear Dynamic Seismic Analysis, Fig. 23 Time history of the angle of twist θ_x at the midpoint of the clamped beam of application section "Forced Vibrations or Beam of Hollow Rectangular Cross Section Under Biaxial Bending"

is the fact that even though the beam is of a doubly symmetric cross section and loaded on its centroid (i.e., no twisting loading is applied), torsional deformation is developed. In Fig. 23, the time history of the angle of twist $\theta_x(l/2, t)$ is presented. As it can be observed, this phenomenon cannot be predicted from linear analysis.

Forced Vibrations or Beam of Monosymmetric Cross Section Under Eccentric Transverse Loading

In this application, in order to investigate the response of a monosymmetric beam and the influence of the loading point upon the cross section, in nonlinear flexural–torsional vibrations, the forced vibrations of a cantilever beam ($E = 2.164 \times 10^8$ kN/m², $G = 8.0148 \times 10^7$ kN/m², $\rho = 7.85$ tn/m³, $l = 1$ m) of a thin-walled open-shaped cross section (Fig. 24), under two load cases, have been studied (its geometric constants are given in Table 9). More specifically, the beam is subjected to a suddenly applied concentrated force $P_Y(t) = 5$ kN either on the right (load case (i), Fig. 24b) or on the left (load case (ii), Fig. 24c) flange. In Fig. 25, the time histories of the axial displacement $u(l, t)$, the transverse displacements $v(l, t)$, $w(l, t)$, and the angle of twist $\theta_x(l, t)$ of the cantilever beam, respectively, and in Table 10 the maximum values of these kinematical components are presented. From the aforementioned figures and table, it can easily be observed that geometrical nonlinearity affects substantially the dynamic response of the beam inducing non-vanishing axial displacement and displacement with respect to the z axis, while in linear analysis, these kinematical components vanish. From the obtained results, it can also be verified that the loading position has significant influence on the response, altering substantially the magnitude of the kinematical components. This discrepancy can be explained by the fact that in load case (ii), the change of eccentricity of the transverse load during torsional rotation increases the magnitude of the twisting moment, acting adversely compared with load case (i), where the applied twisting moment is reduced during torsional rotation.

Prismatic Beam of Asymmetric Cross Section Under Harmonic Excitation

In this application, the forced vibrations of a hinged steel L-shaped beam of unequal legs (asymmetric cross section) (Fig. 26a) ($E = 2.1 \times 10^8$ kN/m², $\rho = 7.85$ tn/m³, $G = 8.0769 \times 10^7$ kN/m², $l = 1.0$ m), having the geometric constants presented in Table 11, are studied. The beam is subjected to a uniformly distributed harmonic loading $p_{\bar{z}}(x, t) = p_0(x) \sin(\omega_{f\text{lin}} t)$, $p_0(x) = 10$ kN/m as this is shown in Fig. 26. The frequency $\omega_{f\text{lin}}$ is considered as $\omega_{f\text{lin}} = 2\pi f_{1\text{lin}}$, where $f_{1\text{lin}} = 118.2$ Hz is the first natural frequency of the examined beam, performing a linear analysis (Sapountzakis and Dourakopoulos 2009c). Due to lack of

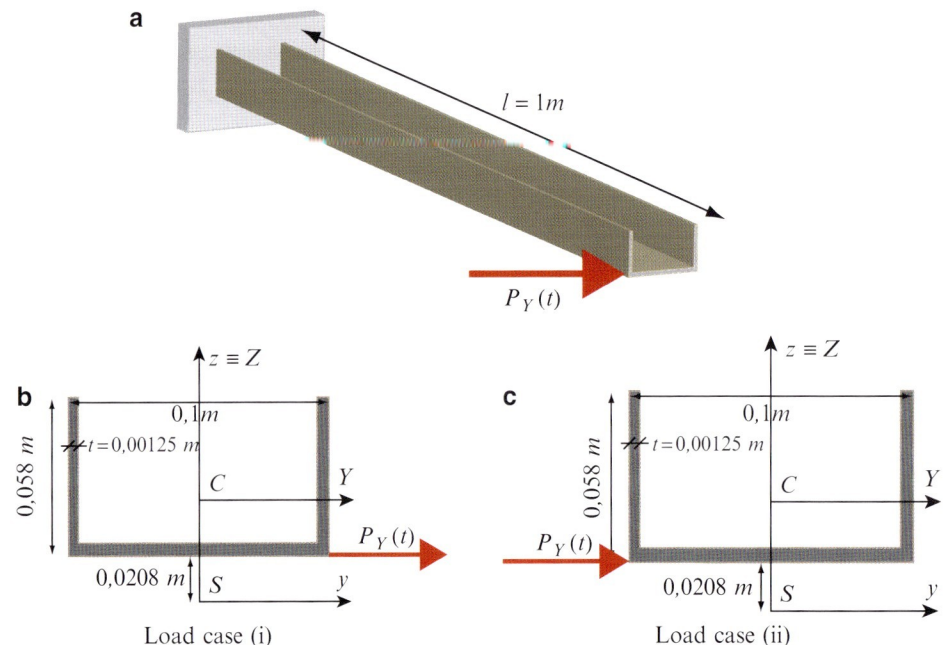

Nonlinear Dynamic Seismic Analysis, Fig. 24 Cantilever beam of application section "Forced Vibrations or Beam of Monosymmetric Cross Section Under Eccentric Transverse Loading" (**a**). Transverse force applied on the *right* (**b**) or on the *left* (**c**) flange

Nonlinear Dynamic Seismic Analysis, Table 9 Geometric constants of the beam of example 2

$A = 2.66875 \times 10^{-4}$ m^2	$I_Y = 9.39789 \times 10^{-8}$ m^4	$I_Z = 4.50061 \times 10^{-7}$ m^4
$I_p = 9.06833 \times 10^{-7}$ m^4	$I_t = 9.10243 \times 10^{-9}$ m^4	$C_s = 1.31047 \times 10^{-10}$ m^6
$I_{pp} = 4.58807 \times 10^{-9}$ m^6	$\beta_z = 6.10287 \times 10^{-2}$ m	$z_c = 3.687 \times 10^{-2}$ m

symmetry, the beam is expected to develop transverse displacements as well as angle of twist. In Fig. 27, the time histories of the displacements $\tilde{v}(l/2,t)$, $\tilde{w}(l/2,t)$, with respect to the $S\tilde{x}\tilde{y}\tilde{z}$ system of axes and the angle of twist $\theta_x(l/2,t)$, are presented, noting the significant difference in response between linear and nonlinear analyses. More specifically, it is observed that only in the linear response deformations continue to increase with time, while the beating phenomenon observed in the nonlinear one is explained from the fact that large kinematical components increase the beam's fundamental natural frequency ω_f (by increasing the stiffness of the beam due to the tensile axial force induced by the axially immovable ends), thereby causing a detuning of ω_f with the frequency of the external loading (ω_{flin}). After the kinematical components reach their maximum values, the amplitude of deformations decreases, leading to the reversal of the previously mentioned effects.

Summary

The main conclusions that can be drawn from this investigation are:

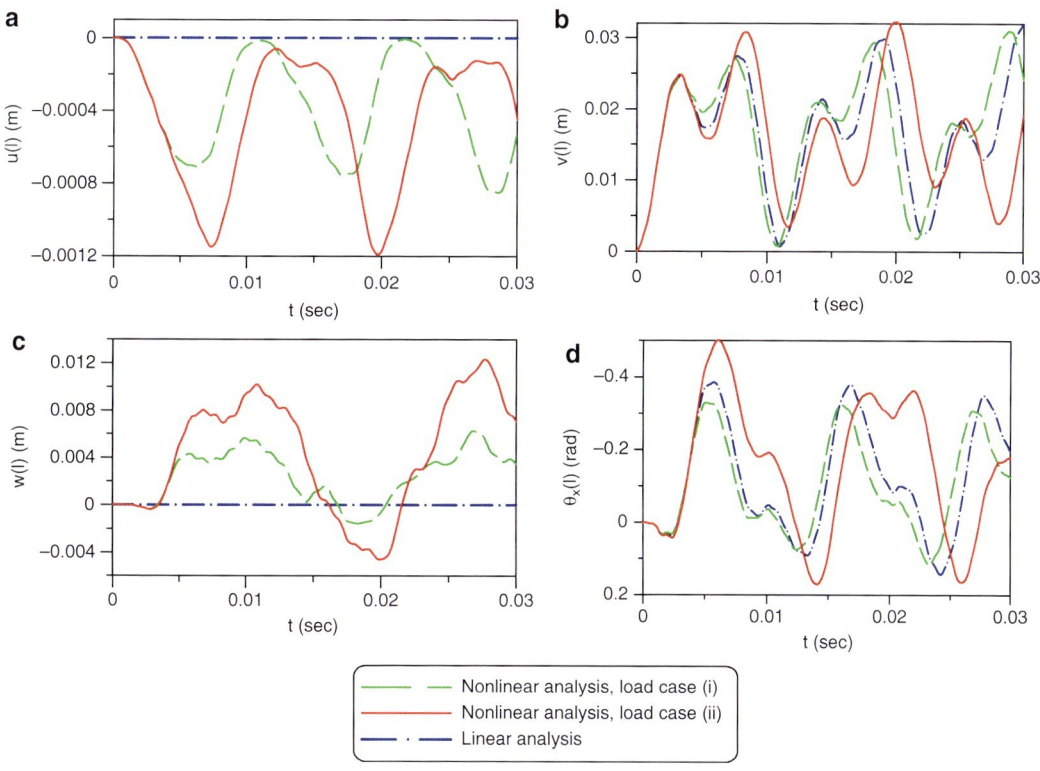

Nonlinear Dynamic Seismic Analysis, Fig. 25 Time history of the axial u (**a**), transverse v (**b**), w (**c**), displacements and angle of twist θ_x (**d**) at the tip of the cantilever beam of application section "Forced Vibrations or Beam of Monosymmetric Cross Section Under Eccentric Transverse Loading"

Nonlinear Dynamic Seismic Analysis, Table 10 Maximum values of the kinematical components $u(l, t)$ (m), $v(l, t)$ (m), $w(l, t)$ (m), and $\theta_x(l, t)$ (rad) of the cantilever beam of application section "Forced Vibrations or Beam of Monosymmetric Cross Section Under Eccentric Transverse Loading" for load cases (i), (ii)

	Linear analysis	Nonlinear analysis	
		Load case (i)	Load case (ii)
$u(l)_{max}$	0.00000	−0.00085	−0.00119
$v(l)_{max}$	0.03190	0.03091	0.03217
$w(l)_{max}$	0.00000	0.00626	0.01230
$\theta_x(l)_{max}$	−0.38479	−0.32897	−0.50086

a. In some cases, the effect of shear deformation is significant, especially for low beam slenderness values, increasing both the maximum transverse displacements and the calculated periods of the first cycle of motion.
b. The discrepancy between the results of the linear and the nonlinear analysis is remarkable. The coupling effect of the transverse displacements in both directions in the nonlinear analysis influences these displacements.
c. The geometrical nonlinearity leads to coupling between the torsional and axial equilibrium equations and alters the modeshapes of vibration. Consequently, the initiation of

Nonlinear Dynamic Seismic Analysis, Fig. 26 L-shaped cross section of unequal legs of application section "Prismatic Beam of Asymmetric Cross Section Under Harmonic Excitation" (**a**) and transverse harmonic excitation (**b**)

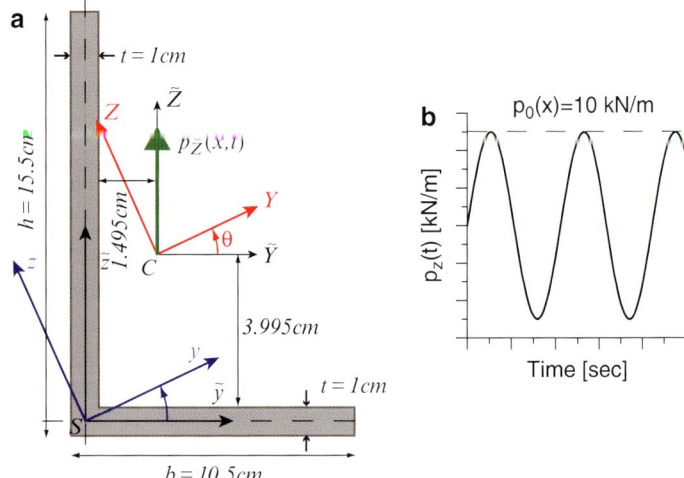

Nonlinear Dynamic Seismic Analysis, Table 11 Geometric constants of the asymmetric beam of application section "Prismatic Beam of Asymmetric Cross Section Under Harmonic Excitation"

$A = 2.500\text{E} - 3 \text{ m}^2$	$I_t = 8.3903\text{E} - 8 \text{ m}^4$	$y_C = 3.6620\text{E} - 2 \text{ m}$
$I_Y = 7.2359\text{E} - 6 \text{ m}^4$	$C_S = 1.1937\text{E} - 10 \text{ m}^6$	$z_C = 3.1900\text{E} - 2 \text{ m}$
$I_Z = 1.3220\text{E} - 6 \text{ m}^4$	$\beta_Y = 8.1915\text{E} - 2 \text{ m}$	$\theta = 4.3000\text{E} - 1 \text{ rad}$
$I_p = 1.4552\text{E} - 5 \text{ m}^4$	$\beta_Z = 3.8866\text{E} - 2 \text{ m}$	
$I_{pp} = 1.6873\text{E} - 7 \text{ m}^6$	$\beta_\omega = -1.6050\text{E} - 1$	

small-amplitude free vibrations of buckled beams with the linear fundamental modeshape as initial twisting rotations induces higher harmonics in the beam's response, but its fundamental natural frequency is only slightly affected.

d. Large twisting rotations have a profound effect on both the positions around which vibrations are performed and the fundamental natural frequency of buckled beams undergoing large-amplitude free vibrations. The computation of dynamic characteristics of buckled beams differs from the one of beams at a pre-buckled state where the increase of initial amplitude of vibration always leads to an increase of the fundamental natural frequency.

e. The natural frequency of the axial displacement's response of buckled beams undergoing large-amplitude free vibrations is twice as much as the one of the twisting rotation, when vibrations are performed around the static equilibrium position of the pre-buckling configuration.

f. As expected, geometrical nonlinearity bounds the (twisting) deformations of beams at a pre-buckled state subjected to primary resonance excitations. A beating phenomenon is observed in the time histories of kinematical and stress components.

g. STMDE affects negligibly the kinematical and stress components of beams of open-shaped thin-walled cross sections undergoing free or forced vibrations of small or large amplitude.

h. STMDE affects the kinematical components of beams of closed-shaped cross sections

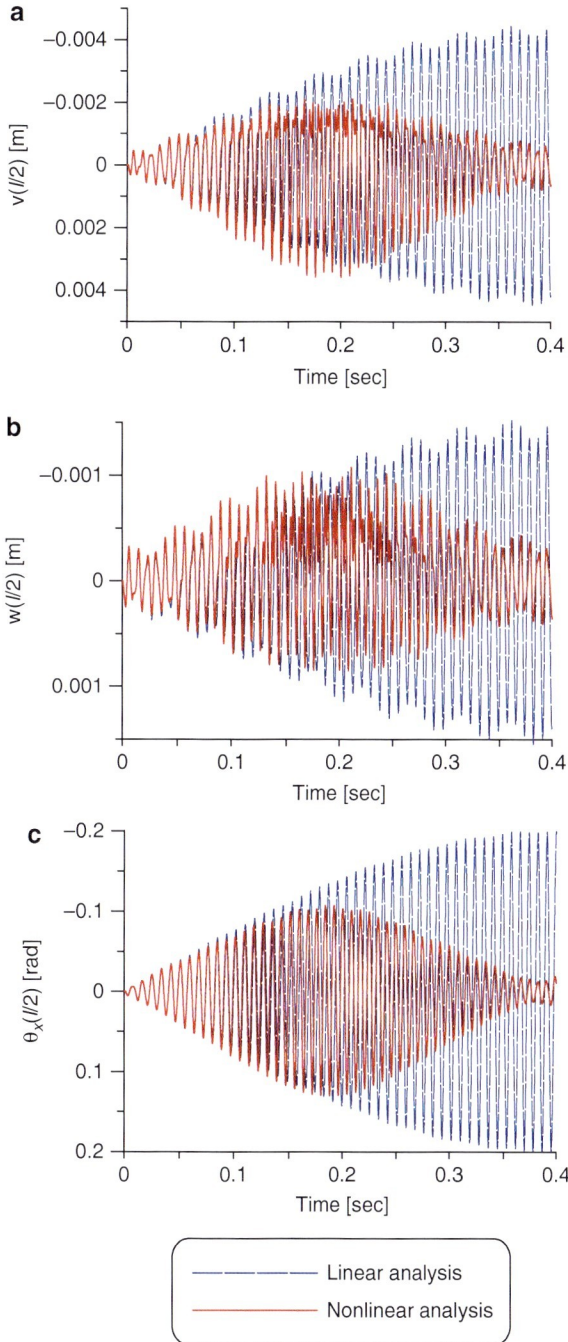

Nonlinear Dynamic Seismic Analysis, Fig. 27 Time history of the displacement $\tilde{v}$ (**a**), $\tilde{w}$ (**b**) and of the angle of twist θ_x (**c**) at the midpoint of the hinged–hinged beam of application section "Prismatic Beam of Asymmetric Cross Section Under Harmonic Excitation"

undergoing linear vibrations. Its effect is much more pronounced on stress components, concluding that it cannot be neglected in linear dynamic analysis of beams of such cross sections.

i. The strong coupling between flexure and torsion may lead beams of doubly symmetric cross sections, undergoing biaxial transverse loading applied on their centroid, to develop torsional rotation.

j. The eccentricity change of the transverse loading during the torsional beam motion, resulting in additional torsional moment influences the beam response.

Cross-References

▶ Nonlinear Finite Element Analysis
▶ Seismic Analysis of Concrete Bridges: Numerical Modeling
▶ Seismic Analysis of Steel and Composite Bridges: Numerical Modeling
▶ Seismic Analysis of Steel Buildings: Numerical Modeling
▶ Seismic Analysis of Steel–Concrete Composite Buildings: Numerical Modeling

References

Aristizabal-Ochoa JD (2004) Timoshenko beam-column with generalized end conditions and nonclassical modes of vibration of shear beams. ASCE J Eng Mech 130(10):1151–1159
Brigham E (1988) Fast Fourier transform and its applications. Prentice Hall, Englewood Cliffs
Foda MA (1999) Influence of shear deformation and rotary inertia on nonlinear free vibration of a beam with pinned ends. Comput Struct 71:663–670
Kausel E (2002) Nonclassical modes of unrestrained shear beams. ASCE J Eng Mech 128(6):663–667
Mohri F, Azrar L, Potier-Ferry M (2004) Vibration analysis of buckled thin-walled beams with open sections. J Sound Vib 275:434–446
Murín J, Kutis V (2008) An effective finite element for torsion of constant cross-sections including warping with secondary torsion moment deformation effect. Eng Struct 30:2716–2723
Ramm E, Hofmann TJ (1995) Stabtragwerke, Der Ingenieurbau. In: Mehlhorn G (ed) Band Baustatik/Baudynamik. Ernst & Sohn, Berlin
Rao GV, Raju IS, Kanaka Raju K (1976) Nonlinear vibrations of beams considering shear deformation and rotary inertia. AIAA J 14(5):685–687
Rothert H, Gensichen V (1987) Nichtlineare Stabstatik. Springer, Berlin
Sapountzakis EJ, Dikaros IC (2011) Non-linear flexural-torsional dynamic analysis of beams of arbitrary cross section by BEM. Int J Non-Linear Mech 46:782–794
Sapountzakis EJ, Dourakopoulos JA (2009a) Nonlinear dynamic analysis of Timoshenko beams by BEM. Part I: Theory and numerical implementation. Nonlinear Dyn 58:295–306
Sapountzakis EJ, Dourakopoulos JA (2009b) Nonlinear dynamic analysis of Timoshenko beams by BEM. Part II: Applications and validation. Nonlinear Dyn 58:307–318
Sapountzakis EJ, Dourakopoulos JA (2009c) Shear deformation effect in flexural torsional vibrations of beams by BEM. Acta Mech 203:197–221
Sapountzakis EJ, Tsipiras VJ (2010a) Nonlinear nonuniform torsional vibrations of bars by the boundary element method. J Sound Vib 329:1853–1874
Sapountzakis EJ, Tsipiras VJ (2010b) Warping shear stresses in nonlinear no uniform torsional vibrations of bars by BEM. Eng Struct 32:741–752
Sapountzakis EJ, Tsipiras VJ (2010c) Shear deformable bars of doubly symmetrical cross section under nonlinear nonuniform torsional vibrations—application to torsional postbuckling configurations and primary resonance excitations. Nonlinear Dyn 62:967–987
Thomson WT (1981) Theory of vibration with applications. Prentice Hall, Englewood Cliffs
Tsipiras VJ, Sapountzakis EJ (2012) Secondary torsional moment deformation effect in inelastic nonuniform torsion of bars of doubly symmetric cross section by BEM. Int J Non-Linear Mech 47(4):68–84
Vlasov V (1963) Thin-walled elastic beams. Israel Program for Scientific Translations, Jerusalem

Nonlinear Finite Element Analysis

Charis J. Gantes
Institute of Steel Structures, School of Civil Engineering, National Technical University of Athens, Zografou Campus, Athens, Greece

Synonyms

Advanced numerical analysis; Geometrically and material nonlinear analysis with imperfections; Numerical collapse prediction

Introduction into Basic Concepts of Linear and Nonlinear Behavior

The aim of this contribution is to present the state of the art of nonlinear finite element analysis tools for predicting the collapse load of structures, hence their strength. Reliable evaluation

Nonlinear Finite Element Analysis

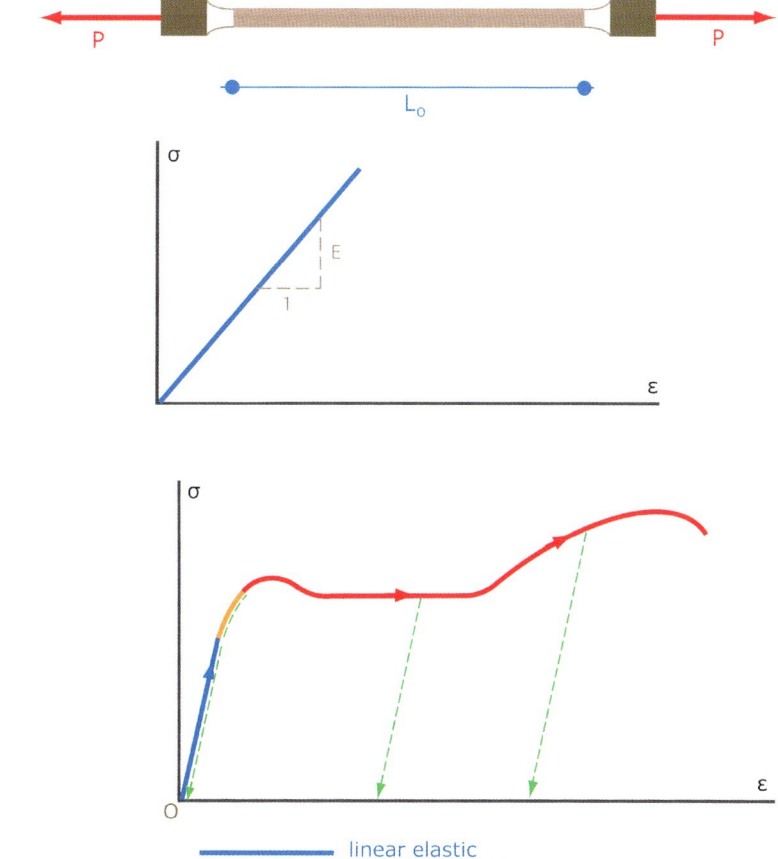

Nonlinear Finite Element Analysis, Fig. 1 Tension test and linear material behavior

Nonlinear Finite Element Analysis, Fig. 2 Nonlinear material behavior

of collapse load is a powerful tool in structural design in general and in earthquake-resistant design in particular. Collapse may occur due to material nonlinearity, geometric nonlinearity, or a combination of both types of nonlinearity. Some basic concepts of linear and nonlinear structural behavior are therefore briefly reviewed first. For a more in-depth understanding of the associated concepts, the reader should consult pertinent text books (e.g., Bazant and Cedolin 1991).

In classical structural analysis materials are assumed to behave in a linear manner. This behavior is commonly described by a linear stress-strain relation obtained by means of a tension test, as shown in Fig. 1. Stress σ, defined as the applied tensile force P divided by the cross-sectional area A, is then proportional to strain ε, which is equal to the elongation ΔL of the bar divided by its undeformed length L_o. This is commonly known as Hooke's law and the proportionality ratio is the modulus of elasticity E of the material:

$$\sigma = E \cdot \varepsilon \qquad (1)$$

In case the applied load is increased beyond a certain level, the stress-strain relation is no longer proportional. The material behavior then becomes nonlinear, as illustrated in Fig. 2, which provides an out-of-scale description of a tension test for a steel specimen by means of the stress-strain curve, frequently referred also as the material's constitutive law. The green lines denote unloading and may coincide with the corresponding loading curve, in which case the response is called elastic, or not, and then the response is inelastic. Thus, the behavior is linear in the blue part of the curve in Fig. 2, nonlinear elastic in the orange part, and nonlinear inelastic in the red part. The nonlinear response of common structural materials is characterized by a sharp reduction of stiffness, corresponding to a decrease of the slope of the stress-strain curve.

The ability of a material to deform well into the inelastic range without rupture is called ductility. It is directly related to the material's ability to absorb energy when it is subjected to extreme actions and is a very desirable feature for good seismic behavior. Steel is such a material; thus, a key objective of structural design in seismic regions is to configure layout and detailing of steel and reinforced concrete structures in such a way that material failure rather than other undesirable failure modes is critical, so that proper advantage of its ductility is taken.

In engineering practice, the nonlinear material behavior is commonly described in a simplified manner. Two common such idealizations are the bilinear ones, either elastic-perfectly plastic or elastic-plastic with linear hardening, shown in Fig. 3. The stress corresponding to the end of the linear region is called yield stress of the material and is denoted by f_y, while the maximum stress, also called ultimate stress, is denoted by f_u. The corresponding strains are ε_y and ε_u, and the difference between them is a measure of the material's ductility. In the elastic-plastic idealization, the higher value of f_u compared to f_y is conservatively ignored.

Thus, by material nonlinearity, we refer to cases where a structure is loaded to such levels that at certain cross sections of certain members, the strains exceed ε_y and Hooke's law no longer applies. Material nonlinearity is commonly a critical failure mode for "stocky" structures (Fig. 4), consisting of short members with respect to their cross section and thick sections compared to their overall dimensions.

In order to explain the concept of geometric nonlinearity, it is reminded that in classical structural analysis it is commonly assumed that equilibrium equations can be written in the structure's undeformed configuration. This is theoretically wrong, as equilibrium of a loaded structure is actually achieved at its deformed configuration. However, in most cases engineering structures possess sufficient stiffness, so that the difference between undeformed and deformed configuration is small and can be safely ignored. As a result, action (usually external loads) and effect (e.g., deformations or internal forces and moments) are proportionally related, as shown in Fig. 5, for a simply supported beam. This type of response is known as geometrically linear. The above assumption simplifies structural analysis significantly and is a prerequisite for the applicability of the principle of superposition, which is at the core of most widely used analysis methods for everyday structures.

Some structures, however, are more flexible, so that the difference between their undeformed and deformed configuration is no longer negligible. One such example is the shallow, two-member truss of Fig. 6, known in the literature as von Mises truss, which exhibits gradual stiffness reduction as applied load increases, eventually leading to a so-called limit point or critical point corresponding to maximum load and zero stiffness.

This type of structural behavior is called geometrically nonlinear and invalidates the use of the principle of superposition, thus of all commonly used structural analysis methods. The so-called nonlinear analysis methods are required, often involving iterations to achieve convergence, as equilibrium equations must now be formulated in the deformed geometry, which however is unknown. Thus, nonlinear analysis is a lot more time and computationally demanding than linear ones. Typical examples of structures that are prone to geometric nonlinearity are cables subjected to transverse loads, shallow arches and domes, slender bars that buckle under axial compression, and thin-walled shells (Fig. 7).

It is noted here that almost all structures behave linearly for low load levels and nonlinearly above a certain load level. Moreover, collapse of most structures is associated with combined material and geometric nonlinearity. As noted above, stocky structures (structures of small slenderness) are more susceptible to

Nonlinear Finite Element Analysis, Fig. 3 Common idealizations of nonlinear material behavior

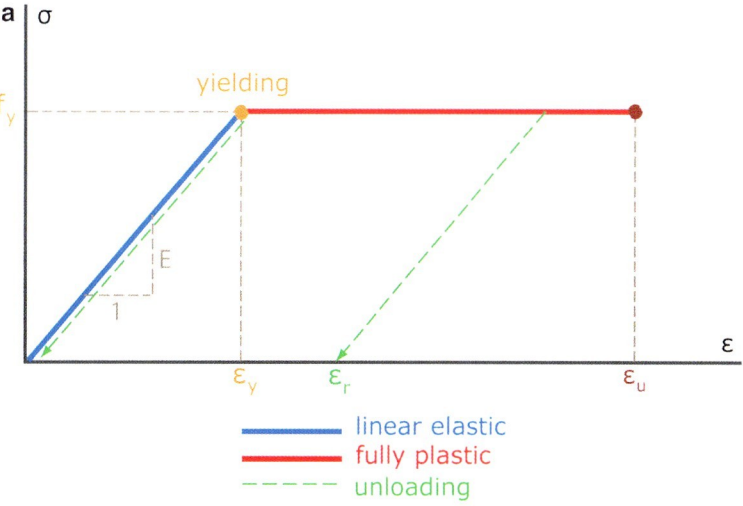

Elastic-perfectly plastic behavior

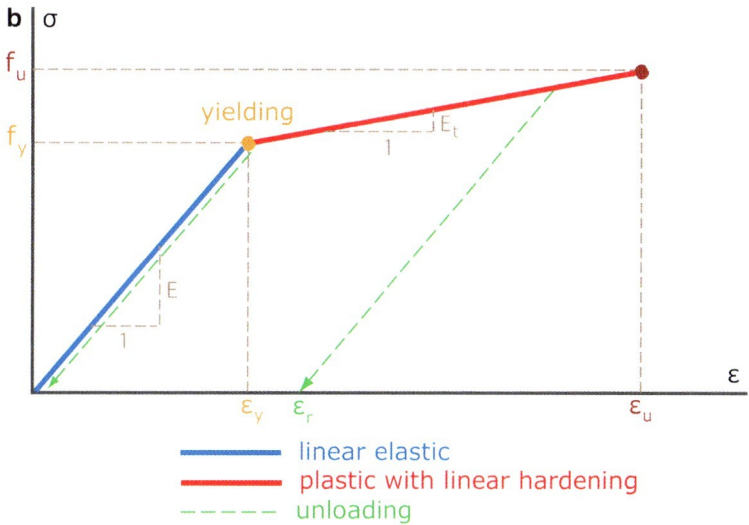

Elastic-plastic behavior with linear hardening

material nonlinearity; therefore, collapse initiation is governed by material nonlinearity. Then, as collapse progresses, stiffness decreases very rapidly, large deformations and strains occur, and therefore geometric nonlinearity develops as well, as illustrated in Fig. 8, where linear behavior is denoted with blue color, material nonlinearity with orange, and combined material and geometric nonlinearity with red.

Slender structures on the other hand are flexible and exhibit large deformations; thus, they are more susceptible to geometric nonlinearity; therefore, collapse initiation is due to geometric nonlinearity. As collapse progresses, stiffness decreases further very rapidly, even larger deformations and strains occur, and therefore material nonlinearity develops as well. This type of behavior is described in Fig. 9, where linear behavior is denoted with blue color, geometric nonlinearity with green, and combined material and geometric nonlinearity with red.

Moreover, as structures enter into their nonlinear range, their behavior is affected by the unavoidable presence of initial imperfections,

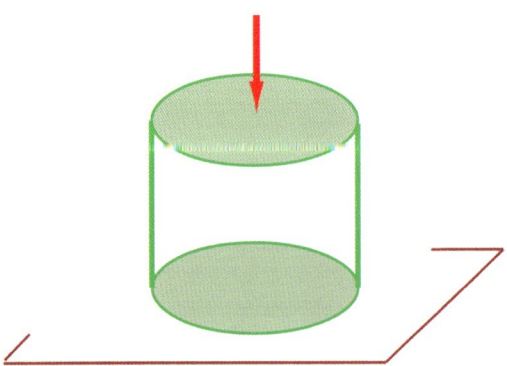

Nonlinear Finite Element Analysis, Fig. 4 Stocky structure that is prone to material nonlinearity

including geometric irregularities, material and cross section inhomogeneities, load eccentricities, etc. This is illustrated in Fig. 10 for the example of a shallow arch subjected to a vertical concentrated load at its crown, where the presence of imperfections drastically reduces the arch's strength, depending also on their magnitude w_0.

It is thus concluded that the capacity of a structure depends on its sensitivity to material nonlinearity, geometric nonlinearity, and initial imperfections. Therefore, all these factors should be taken into account in order to predict the structure's collapse, thus evaluate its strength.

Treatment of Nonlinear Behavior in Engineering Practice

In recent years structural design codes have adopted limit state design (LSD), replacing the older concept of allowable stress design (ASD). Limit state design requires the structure to satisfy two principal criteria: the ultimate limit state (ULS) and the serviceability limit state (SLS). A limit state is a set of performance criteria that must be met when the structure is subject to loads. To satisfy the ultimate limit state, the structure must not collapse when subjected to any of the pertinent combinations of its design loads. To satisfy the serviceability limit state criteria, a structure must remain functional for its intended use subject to routine, everyday loading.

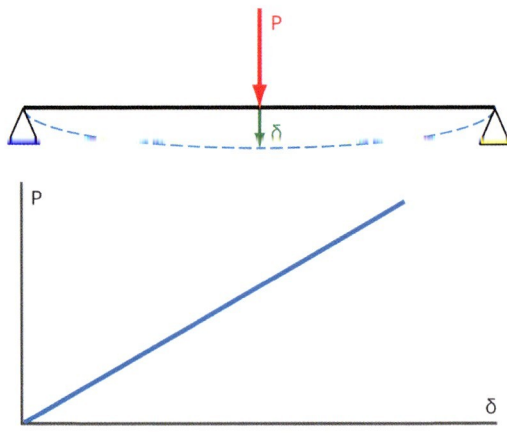

Nonlinear Finite Element Analysis, Fig. 5 Linear response of a simply supported beam

The prevailing approach proposed in modern codes for carrying out ultimate limit state checks consists of obtaining action effects by means of linear elastic analysis of the structure subjected to design loads and comparing them to resistances that account for both types of eventual nonlinearity, material, and geometric. Thus, prediction of collapse, a highly nonlinear phenomenon, is accomplished by means of linear elastic analysis.

In order to evaluate strength, modern codes assume full exploitation of material capacity and appropriate reduction factors, by means of which geometric nonlinearities and imperfections are taken into account. For example, the strength of a steel bar subjected to compression is given by Eurocode 3 (European Committee for Standardisation 2004) as:

$$N_{b.Rd} = \frac{\chi \cdot \beta \cdot A \cdot f_y}{\gamma_{M1}} \quad (2)$$

In this relation, A is the cross-sectional area and f_y the material's yield strength; thus, the product $A \cdot f_y$ represents the cross section's axial strength and addresses the issue of material nonlinearity, adopting an elastic-perfectly plastic idealization of the constitutive law. On the other hand, χ is a reduction factor accounting for global buckling, obtained from the so-called buckling curves, which have been calibrated for

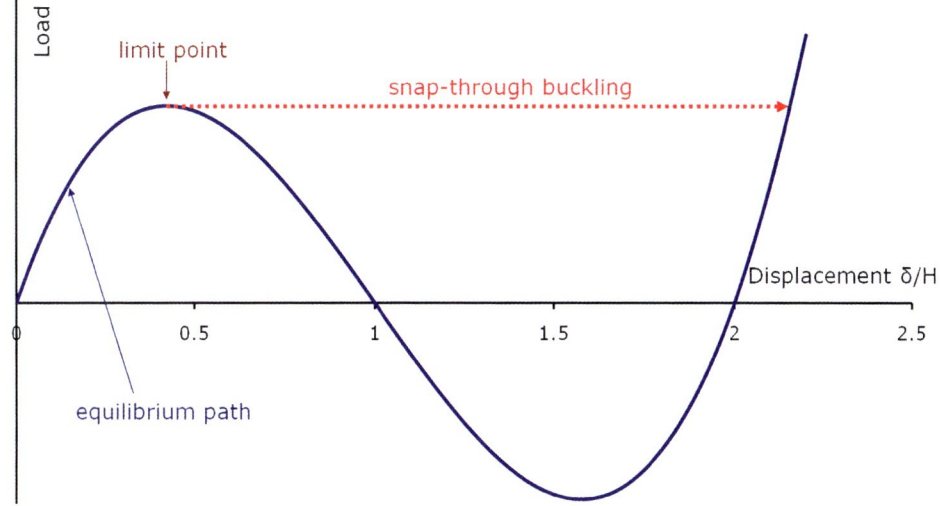

Nonlinear Finite Element Analysis, Fig. 6 Shallow two-bar truss and its geometrically nonlinear response

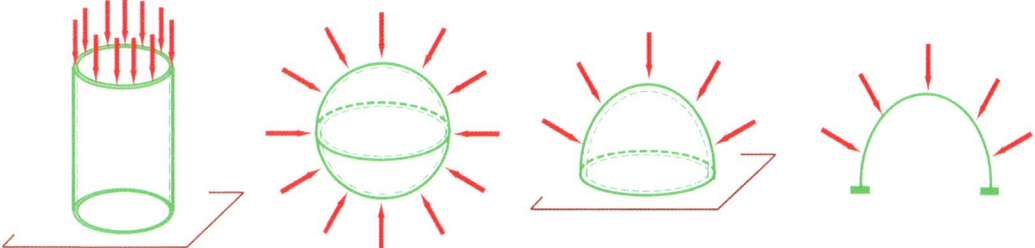

Nonlinear Finite Element Analysis, Fig. 7 Slender structures that are prone to geometric nonlinearity

specific types of cross sections by means of a wide range of experimental results, and incorporates the effects of initial imperfections and residual stresses, while β is a reduction factor accounting for local buckling, depending on the classification of the cross section into appropriate slenderness categories. Thus, the product $\chi \cdot \beta$ addresses the issues of geometric nonlinearity and imperfections.

The above approach offers significant advantages, as it is straightforward and the prediction of collapse is accomplished by means of linear elastic analysis. This means that it can be carried out using widely available commercial software, is computationally inexpensive, and does not require special expertise of the engineer in setting up the structural model, selecting the analysis parameters, and interpreting the results.

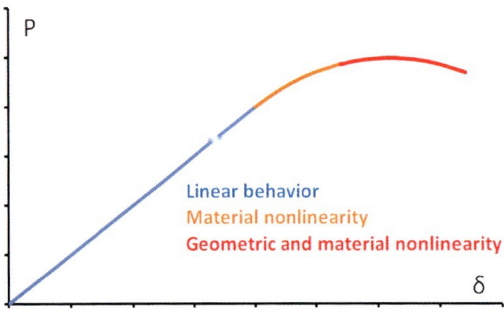

Nonlinear Finite Element Analysis, Fig. 8 Typical load-displacement curve of stocky structures

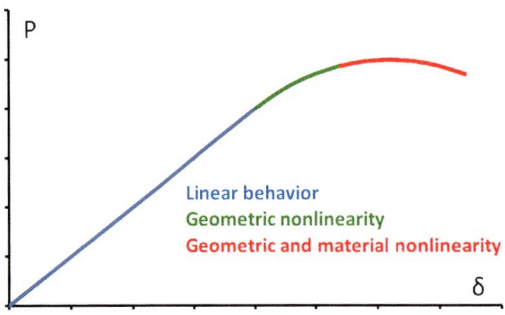

Nonlinear Finite Element Analysis, Fig. 9 Typical load-displacement curve of slender structures

Nonlinear Finite Element Analysis, Fig. 10 Influence of imperfections on strength of shallow arch

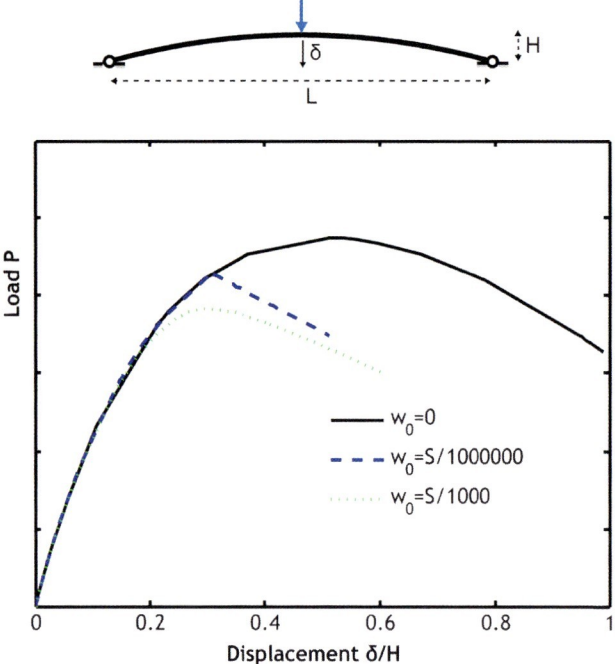

Furthermore, the results are quite reliable for ordinary structural systems.

Although this approach is appropriate for the design of ordinary structural systems, more complex structures with unusual geometry or cross sections require more accurate methods of their strength evaluation, as such cases are not directly covered by buckling curves and interaction equations between different action components, while simplifying assumptions that are made in an effort to approximate their behavior may lead to significant inaccuracies. To address such cases, modern structural design codes allow predicting collapse by means of more elaborate, nonlinear numerical analysis, taking also initial imperfections into account.

Prediction of collapse of structures can thus be accomplished by means of such nonlinear numerical analysis, making use of commercially available finite element software, setting up an appropriate finite element model, obtaining critical buckling modes from Linearized Buckling Analysis (LBA), and then using a linear combination of these modes as imperfection pattern for a Geometrically and Material Nonlinear

Imperfection Analysis (GMNIA). Equilibrium paths accompanied by snapshots of deformation and stress distribution at characteristic points are then used for evaluating the results of the GMNI analysis. This approach is particularly applicable to seismic design, as all major seismic design codes allow structures to behave inelastically during major earthquakes, defining "no collapse" as the design objective.

In the following, practical details of the implementation of this approach are discussed, and pertinent case studies are presented. It is thus aimed to contribute towards making advanced numerical analysis methods accessible to practicing structural engineers, by providing guidelines for using such methods for structural design and by demonstrating some of the capabilities afforded by such methods to the structural engineering community.

Overview of Linearized Buckling and Nonlinear Numerical Algorithms

The geometric and material nonlinear phenomena characterizing structures near collapse are very complicated in nature and require, in most cases, highly sophisticated numerical tools for their simulation. A common feature of most nonlinear analysis algorithms is that the load is applied in successive steps and iterations are performed within each step to achieve convergence (Stricklin et al. 1973).

The simplest numerical method for a detailed geometrically and material nonlinear (GMN) analysis is the Newton-Raphson scheme (Crisfield 1979; Bathe 1995), which can be found in three forms: (i) the full Newton-Raphson, which is the most accurate, but also the most time consuming, since the tangent stiffness of the structure has to be calculated and factorized within each iteration in the solution procedure; (ii) the modified Newton-Raphson, which differs from the full Newton-Raphson in that the calculation and the factorization of the tangent stiffness matrix take place only in some iterations within each step, thus requiring in most cases a larger number of iterations per step but a smaller computational effort per iteration; and (iii) the "initial stress" Newton-Raphson procedure, where the calculation of the tangent stiffness matrix is performed once, at the beginning of the analysis, and then kept the same throughout the analysis (Fig. 11).

Applying the Newton-Raphson method in a nonlinear finite element system will yield results only in the pre-collapse range, but it will fail to give information about the post-collapse response. To circumvent this limitation, a constraint can be added into the finite element system, which relates the load increment and the incremental displacements within each iteration (Fig. 12). This technique allows the calculation of the whole equilibrium path, even beyond the critical limit points. A number of different solution algorithms have been proposed in the literature (Riks 1979; Crisfield 1981; Ramm 1981; Bathe and Dvorkin 1983).

In case initial imperfections are also included in the numerical model, the above GMN analysis is denoted as GMNIA. Such analyses are often preceded by linearized buckling analyses (LBA) (Bathe 1995), which seeks deformed configurations of the structure that correspond to zero determinant of a characteristic stiffness matrix, assuming linear response beyond an initial configuration. Such deformed configurations are known as buckling modes and the corresponding loads as linear critical buckling loads. Critical buckling loads obtained by means of LBA are, in most cases, not safe predictions of strength. However, it is advisable that this type of analysis precedes the subsequent, more exact, nonlinear analyses, for several reasons (Dimopoulos and Gantes 2012):

(i) Critical buckling loads obtained by means of linearized buckling analyses are an initial indication, and in most cases an upper bound, of actual strength. Taking also into account the fact that this is a very fast and inexpensive type of analysis, it may be very useful as a tool for evaluating alternative solutions during preliminary design, before resorting to the much more time-consuming and expensive nonlinear analyses.

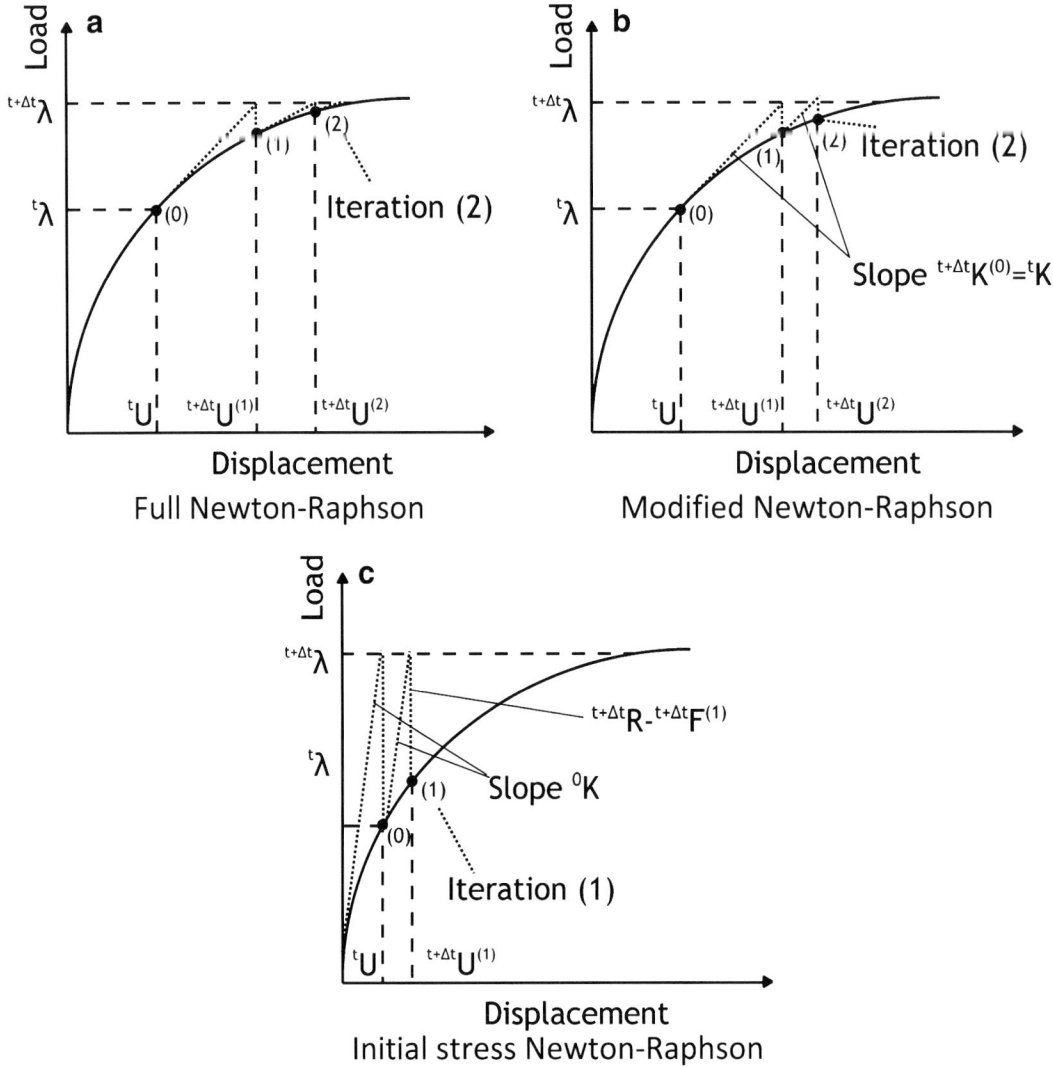

Nonlinear Finite Element Analysis, Fig. 11 Variations of the Newton-Raphson method

(ii) Critical buckling loads are often needed for calculating nondimensional slenderness ratios to be used with buckling curves in the framework of code-based design procedures. Even though analytical expressions for these buckling loads are available for simple cases of geometry, loading patterns, and boundary conditions, in the majority of other cases, buckling loads must be obtained numerically.

(iii) It has been shown that buckling analysis can be used in place of a geometrically nonlinear analysis for the calculation of the nonlinear stability limit of a number of structures by performing a number of successive buckling analyses on the unstressed and some stressed configurations of the structures (Chang and Chen 1986; Dimopoulos and Gantes 2012).

(iv) Buckling modes obtained by means of linearized buckling analyses are commonly used as initial imperfections for geometrically and material nonlinear imperfection analyses. Such imperfections are necessary

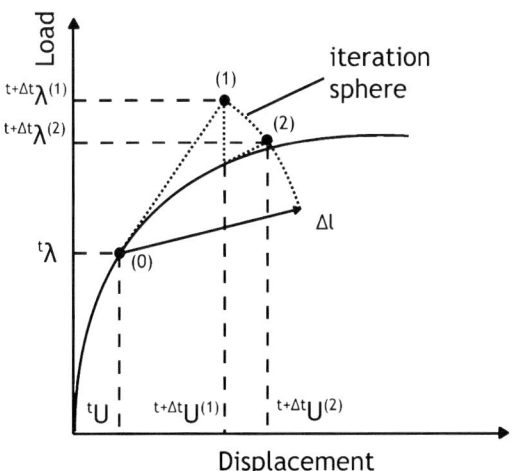

Nonlinear Finite Element Analysis, Fig. 12 Arc-length method by Crisfield

in order to trigger all possible failure mechanisms and to make sure that the critical failure mechanism is captured by the nonlinear numerical analysis algorithm.

It should be noted that the choice of buckling modes as imperfection patterns is not unique and has in many cases been found to lead to lower compliance to experimental results than other shapes of initial imperfections. For the case of cylindrical shell structures, a type of structures that is particularly sensitive to imperfections, imperfections affine to the collapse mode of the perfect shell were found to be more unfavorable than ideal buckling modes (Schneider and Brede 2005; Schneider et al. 2005). Furthermore, fabrication processes are responsible for imperfection patterns, creating residual stress profiles in the body of the cross sections, which can have significant consequences on the ultimate bearing capacity of steel structures (Berry et al. 2000; Herynk et al. 2007). However, the choice of manufacturing-related imperfection patterns is usually dependent upon the structural type as well as the specific manufacturing method, cannot be easily generalized, and often lacks adequate data to be properly implemented. In case such information is available, it is, in general, preferable to use these imperfection patterns, as they more accurately represent real strength.

In the absence of such data, however, the use of buckling modes as imperfection shapes is considered to be the next best available choice.

Details of the formulation and implementation of these algorithms are beyond the scope of this presentation, and the reader should consult other specialized texts for that purpose. The engineers applying this methodology should have good knowledge of theoretical and practical aspects of the finite element method for linear and nonlinear structural applications (as outlined, e.g., by Crisfield et al. 1997; Bathe and Cimento 1980; Bathe et al. 1980; Bathe 1995; Belytschko et al. 2000; Kojic and Bathe 2004). Emphasis here is placed on the application of such methods to determine structural collapse, thus strength. For that purpose, the successive steps of a proposed numerical methodology are presented, followed by two case studies, referring to steel members with unconventional geometry that are part of two long-span roofs, which have been recently designed in Greece.

Numerical Methodology for Collapse Prediction

A state-of-the-art strategy for understanding the behavior, predicting all possible failure mechanisms, evaluating the strength, and assessing the vulnerability of structures based on advanced nonlinear numerical analyses consists of the following steps (Gantes and Fragkopoulos 2010; Gantes 2011; Gantes and Koulatsou 2013):

Step 1. Setting up an appropriate finite element model

In this step special attention should be paid to the following:
(i) The finite element software, the specific elements used to model the structure, as well as the numerical solution algorithm employed for numerical analysis should be verified, by means of comparisons to analytical solutions and/or numerical solutions from the literature.
(ii) The model should be able to predict all anticipated failure mechanisms.

Nonlinear Finite Element Analysis, Fig. 13 Architectural proposal for the steel roof over Aristotle's Lyceum

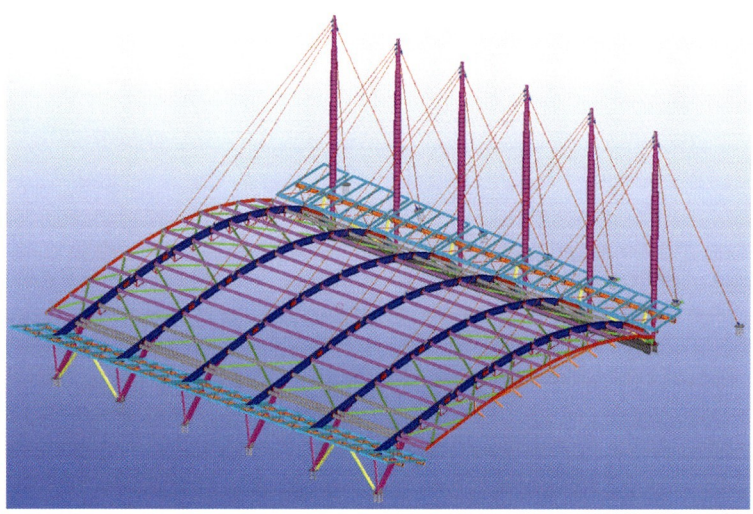

Nonlinear Finite Element Analysis, Fig. 14 Perspective view of the suspended steel roof over Aristotle's Lyceum

For example, modeling a steel column with beam elements is not appropriate for predicting failure where local or lateral buckling may be predominant. Shell or plate elements should be used instead, and the mesh should be fine enough to capture the curvature part of the cross section where local or lateral buckling occurs.

(iii) Basic features of the model, such as element connectivity, boundary conditions, and loads, should be checked by means of simple, linear, static, and modal analyses. Sum of support reactions can be used to check load application. Mode shapes and periods of vibration can be compared to expected values, in order to identify possible errors in mass or stiffness.

(iv) The mesh density should be verified by performing a convergence study. Element size should be successively divided in half, and representative quantities of the response should be plotted against mesh density until sufficient convergence is achieved. It should be noted that a finer mesh may be required for subsequent nonlinear analyses than the one obtained from linear static

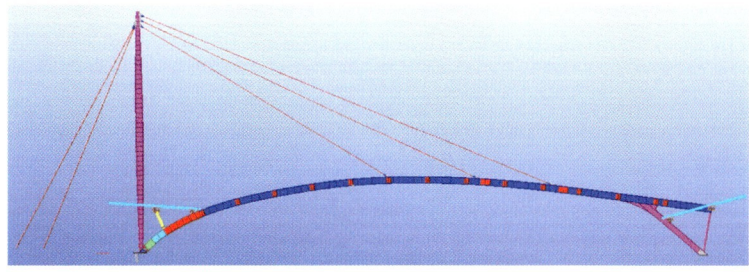

Nonlinear Finite Element Analysis, Fig. 15 View of a suspended main arch and the corresponding pylon of the steel roof over Aristotle's Lyceum

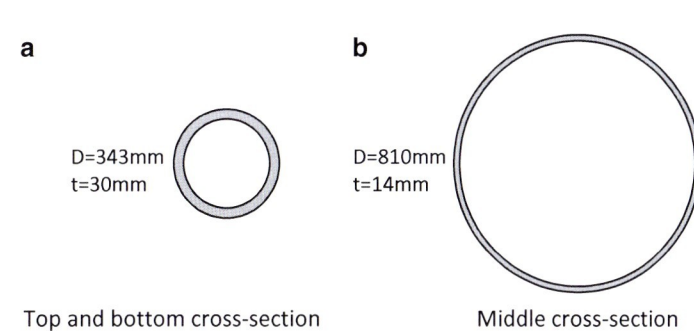

Nonlinear Finite Element Analysis, Fig. 16 Cross sections of pylon

convergence study. However, the latter one should be considered as a minimum.

Step 2. Carrying out linearized buckling analysis

One question arising when carrying out such an analysis concerns the number of modes that should be evaluated. Even though the first buckling mode is the one corresponding to the lowest critical buckling load, it is, in general, not sufficient, for several reasons. A higher mode associated with an unstable post-buckling equilibrium path may be more critical than a lower one with a stable post-buckling equilibrium path, due to its sensitivity to imperfections. Even two modes with stable post-buckling equilibrium paths, but with nearly coinciding critical buckling loads, may interact in the presence of imperfections and lead to unstable post-buckling behavior (e.g., Bazant and Cedolin 1991; Livanou et al. 2013). Therefore, several buckling modes should be obtained. There is no clear-cut criterion as to their number; however, it is recommended to obtain sufficient number of modes, so that the critical buckling load of the last mode is sufficiently larger than that of the first mode, in order to eliminate any possibility of mode interactions and the associated detrimental effects for the bearing capacity of the structure.

Step 3. Carrying out nonlinear analysis with imperfections

Taking all above discussion into account, a geometrically and material nonlinear imperfection analysis (GMNIA) is recommended for reliably understanding the behavior, predicting all possible failure mechanisms, evaluating the strength, and assessing the vulnerability of structures. In many cases it is useful to carry out first separate analyses accounting only for material nonlinearity (MNIA) and geometric nonlinearity (GNIA), which help identify the prevailing failure mechanism of the structure in question. But the ultimate strength evaluation should always be based on GMNI analyses, accounting for both types of nonlinearity. Several decisions, related to the details of this type of analysis, arise and must be considered by the engineer:

(i) *Shape of initial imperfections*: As stated above, in the absence of experimental data on manufacturing-related imperfection patterns, it is widely recognized as appropriate to use a linear combination of the critical buckling modes obtained from

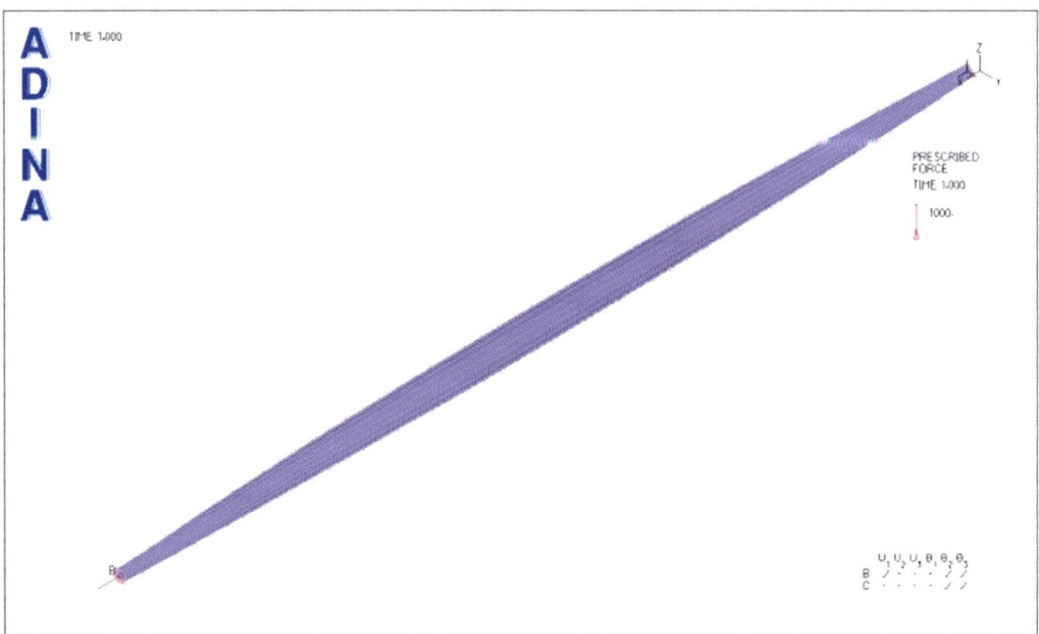

Nonlinear Finite Element Analysis, Fig. 17 Pylon shell element model of steel roof over Aristotle's Lyceum

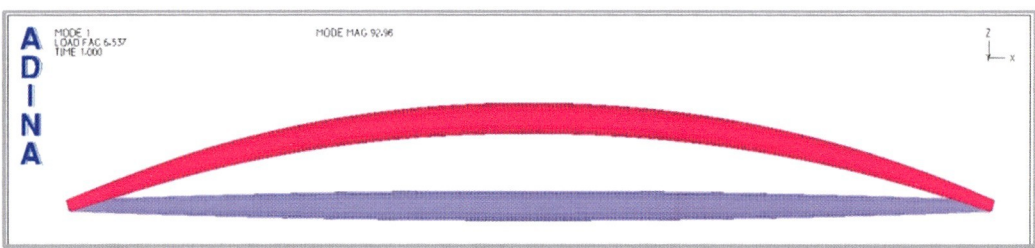

Nonlinear Finite Element Analysis, Fig. 18 First global flexural buckling mode of pylon of the steel roof over Aristotle's Lyceum

Nonlinear Finite Element Analysis, Fig. 19 Equilibrium paths of pylon of the steel roof over Aristotle's Lyceum

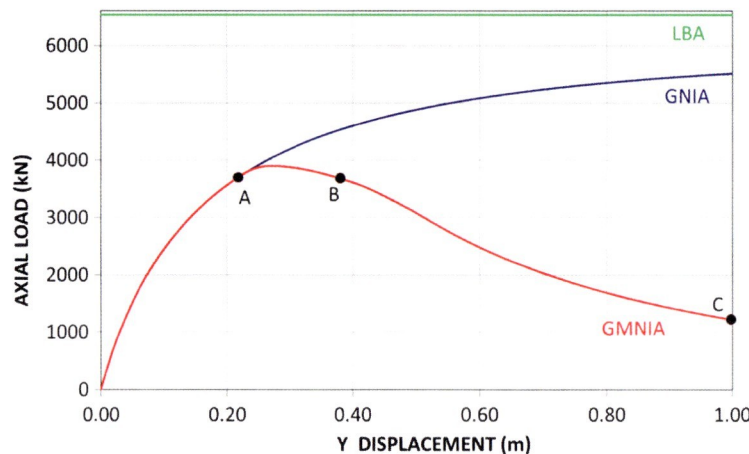

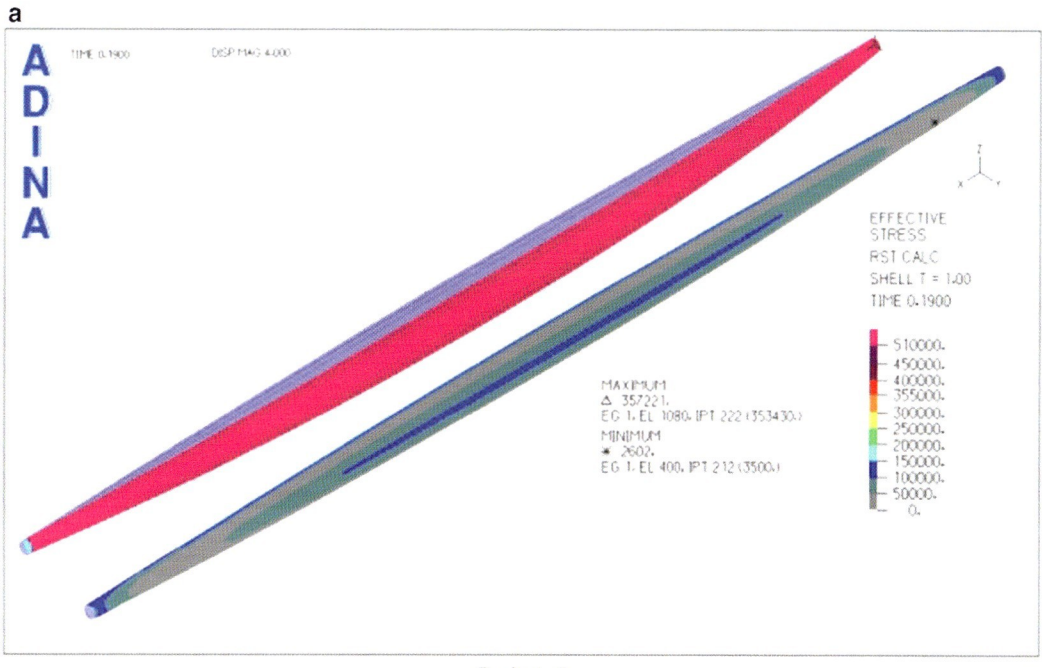

Point A

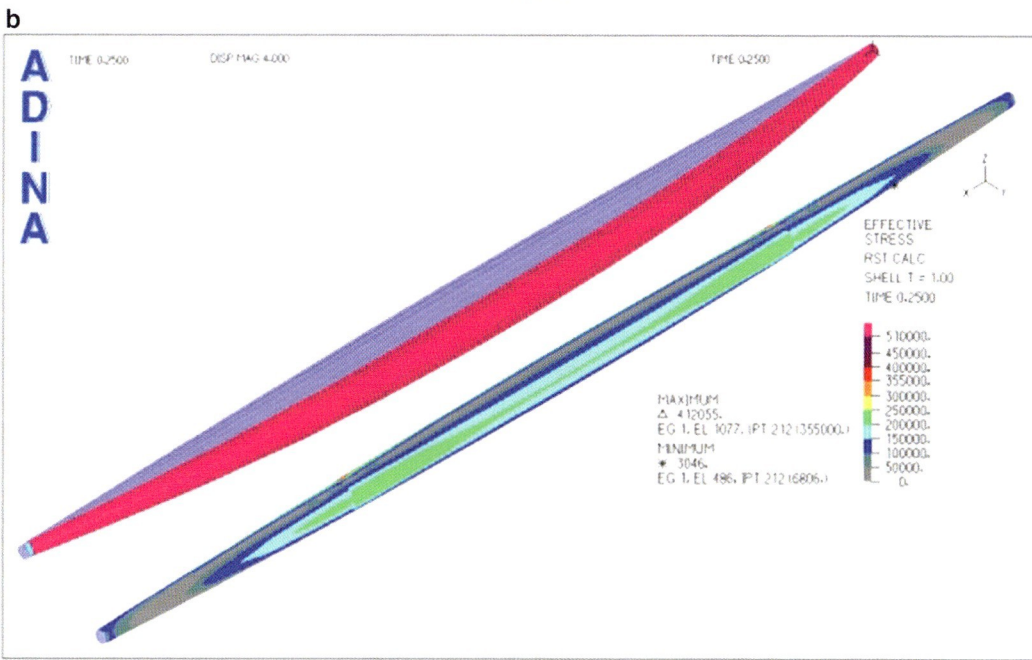

Point B

Nonlinear Finite Element Analysis, Fig. 20 (continued)

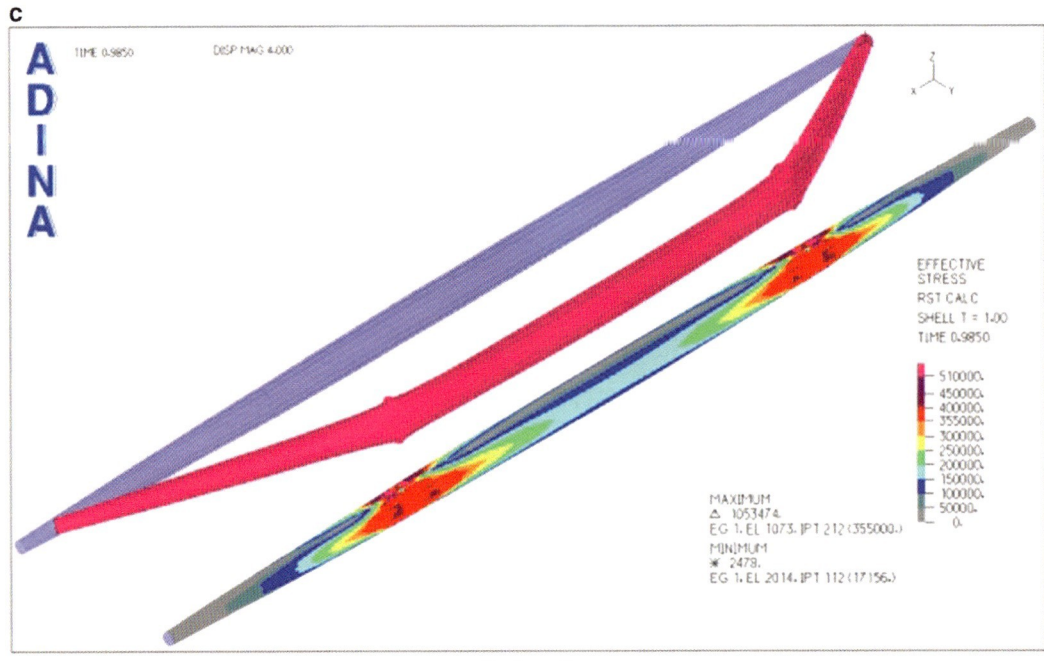

Nonlinear Finite Element Analysis, Fig. 20 Deformation and von Mises stress distribution at characteristic points A, B, and C along the GMNIA equilibrium path of pylon of steel roof over Aristotle's Lyceum (kN/m^2)

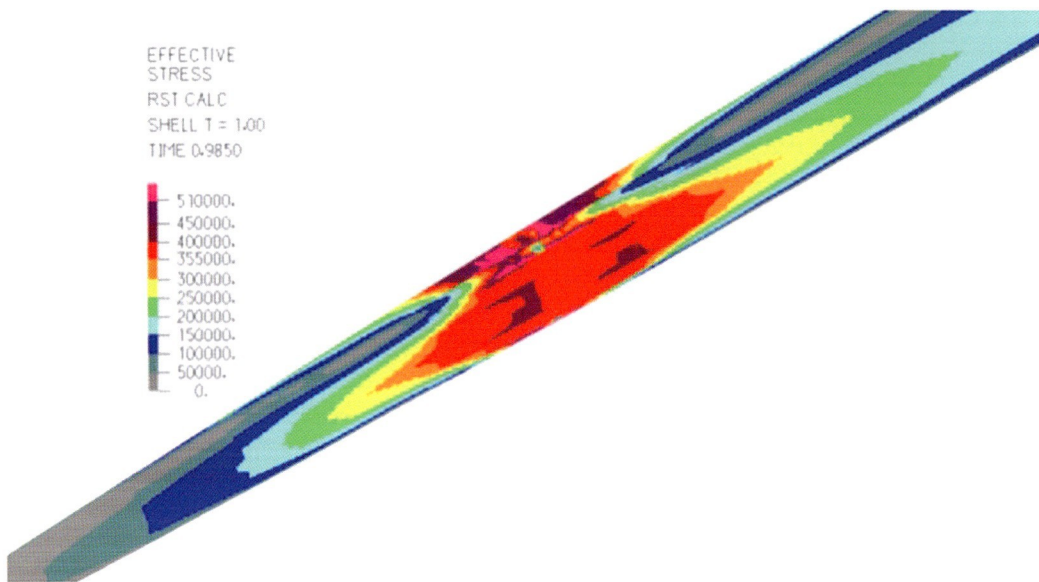

Nonlinear Finite Element Analysis, Fig. 21 Detail of von Mises stress distribution at characteristic point C along the GMNIA equilibrium path of pylon of steel roof over Aristotle's Lyceum (kN/m^2)

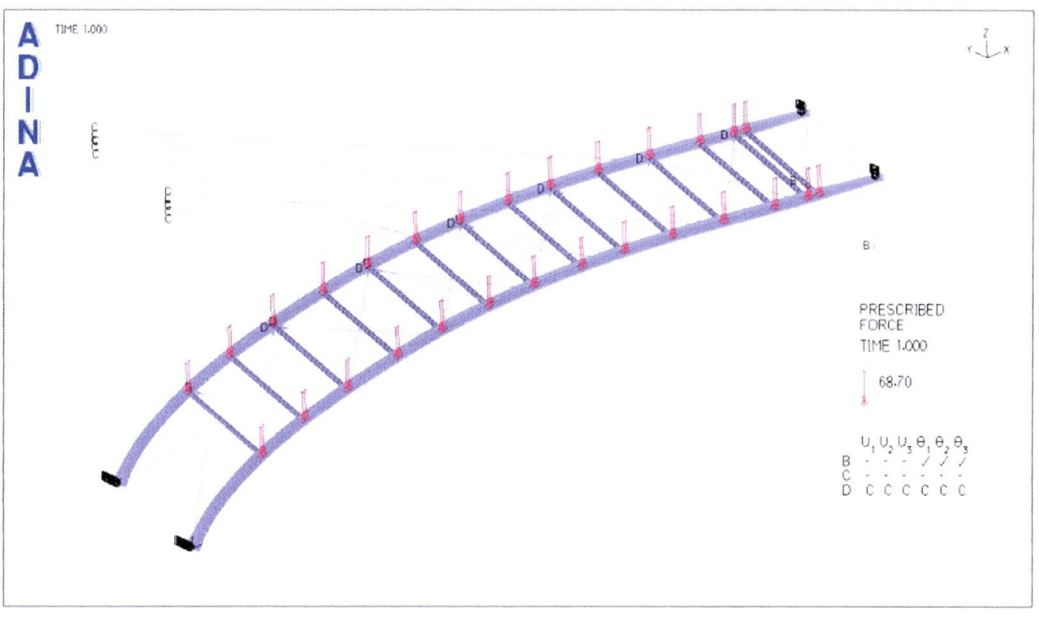

Nonlinear Finite Element Analysis, Fig. 22 Finite element model of two adjacent main arches of steel roof over Aristotle's Lyceum

linearized buckling analysis as the shape of initial imperfections. Regarding the number of critical buckling modes, reference is made to the discussion in step 2 above. In the absence of any available information about the relative importance of individual modes, equal weights may be assumed for their linear combination.

(ii) *Amplitude of initial imperfections*: With the exception of cases with steep unstable post-buckling equilibrium paths, for example, frequently encountered for thin-walled shells, the size of initial imperfections is not very important for the type of response and the values of critical response quantities obtained from the analysis. For certain types of structures, pertinent codes specify recommended magnitudes of imperfections or manufacturing and erection tolerances that can be regarded as upper bounds of imperfection magnitudes (European Committee for Standardisation 2004). In some cases, a parametric study for different sizes within the range of realistic construction tolerances may provide valuable insight into the effect of imperfection magnitude on ultimate strength.

(iii) *Details of numerical algorithm*: As mentioned above, all numerical algorithms for nonlinear structural analysis rely on applying the loads in small steps, performing iterations of linear analyses within each step until a predefined convergence criterion is met, and using the final configuration of the structure at each step as a starting configuration for the next one. The choice of an arc-length algorithm with either MNR (as a first alternative) or NR (in cases of convergence difficulties) iterations is recommended for GMNI analyses for practical applications. If convergence difficulties are still encountered, the engineer may have to increase the default number of maximum iterations per step or relax the convergence tolerances, but in the second case, special attention is recommended, as erroneous

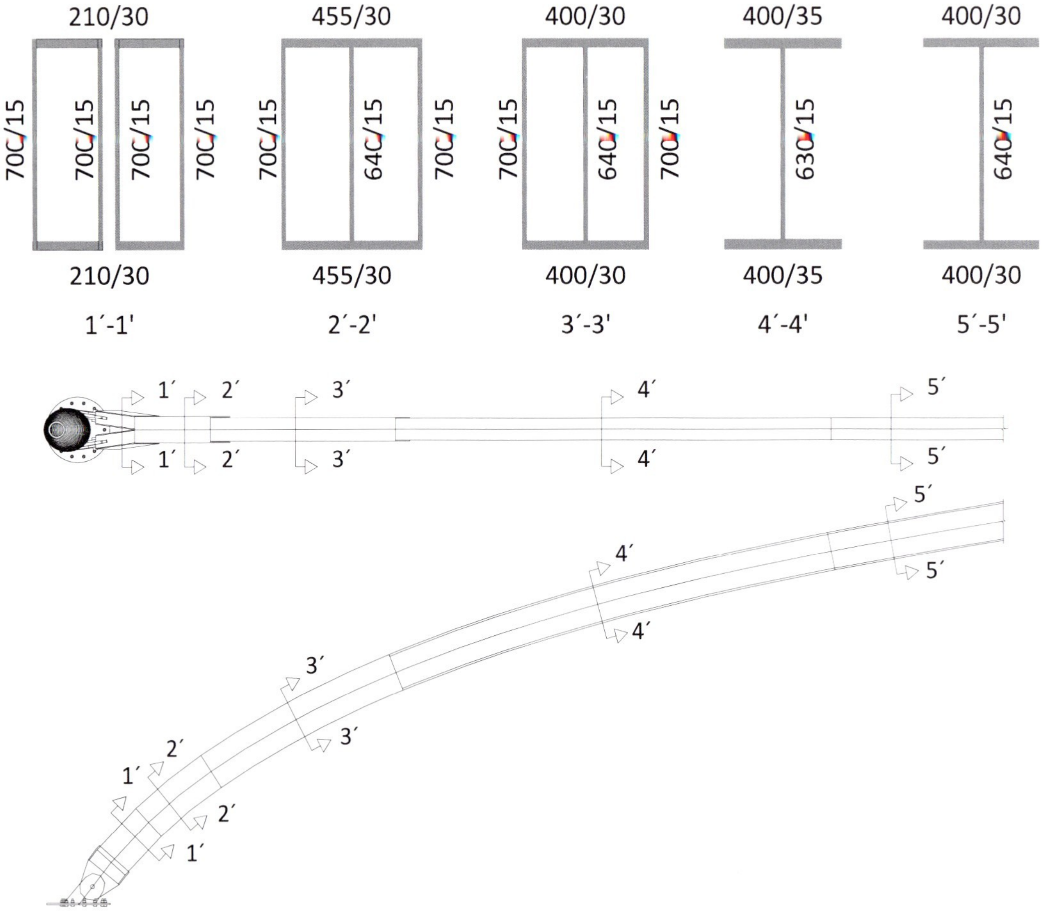

Nonlinear Finite Element Analysis, Fig. 23 Cross sections of main arch

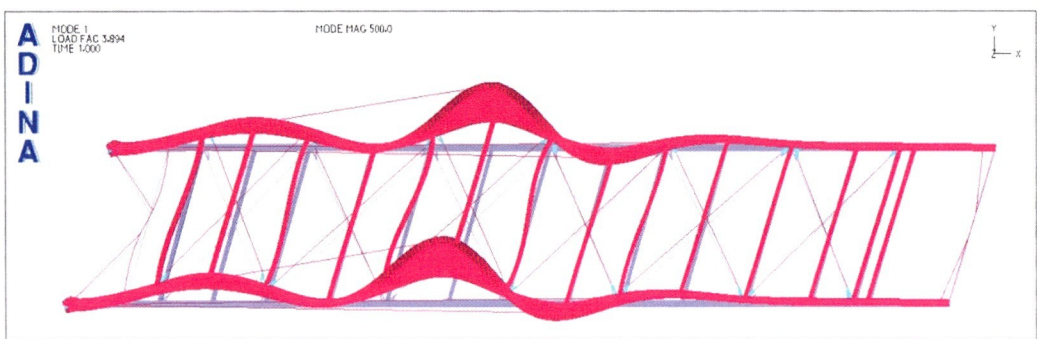

Nonlinear Finite Element Analysis, Fig. 24 First buckling mode of two adjacent main arches of steel roof over Aristotle's Lyceum

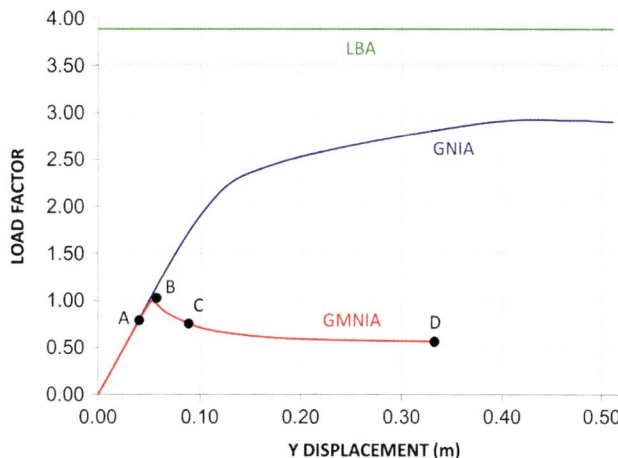

Nonlinear Finite Element Analysis, Fig. 25 Equilibrium paths of two adjacent main arches of steel roof over Aristotle's Lyceum

convergence to configurations without physical meaning may be obtained.

(iv) *Evaluation of results*: It is recommended to evaluate the results of GMNI analyses by using a plot of the equilibrium path, accompanied by characteristic snapshots of the structure's deformation and stress distribution along this path. The equilibrium path is a plot of a characteristic response quantity (usually on the horizontal axis) with respect to the external action (usually on the vertical axis). Very often the external action is represented by a load multiplier, λ, which acts upon the required design loads. Particular attention is needed for the selection of the characteristic response quantity of the horizontal axis, which should be representative of the overall structural response and not of some local phenomena, depending also on the critical failure mode. Multiple equilibrium paths for more than one response quantity may be useful in some cases.

Snapshots of the structure's deformation and stress distribution at characteristic points (e.g., before and after change of slope or change of curvature, at the ultimate limit point, at the end of the curve) are very useful in appreciating the physical phenomena occurring with progressing external loading.

Important quantities of an equilibrium path include: (a) the slope of its initial, usually straight, part, representing the structure's initial stiffness; (b) the maximum value of the external action, representing the structure's strength; (c) the amplitude of displacements corresponding to the maximum value of the external action, pointing out whether serviceability problems may be encountered before failure; and (d) the maximum deformation at failure, as well as the area between the equilibrium path and the horizontal axis, particularly in the post-linear range, representing the structure's ductility.

Comparison of representative numerical results with experimental measurements, if available, is strongly encouraged as a means for calibrating and verifying the numerical results. Numerical analysis can then, with increased reliability, be extended for other cases of geometry, cross sections, material, boundary conditions, or loads, thus performing "numerical experiments" that may help optimize the design.

Case Studies

The abovementioned approach, making use of advanced nonlinear numerical analyses, is

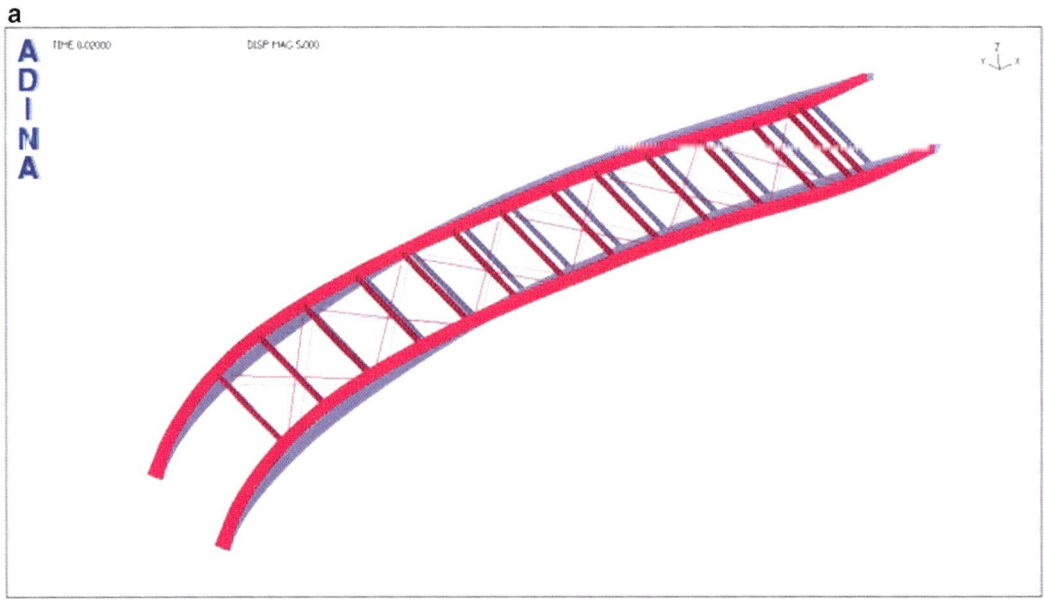

Point A

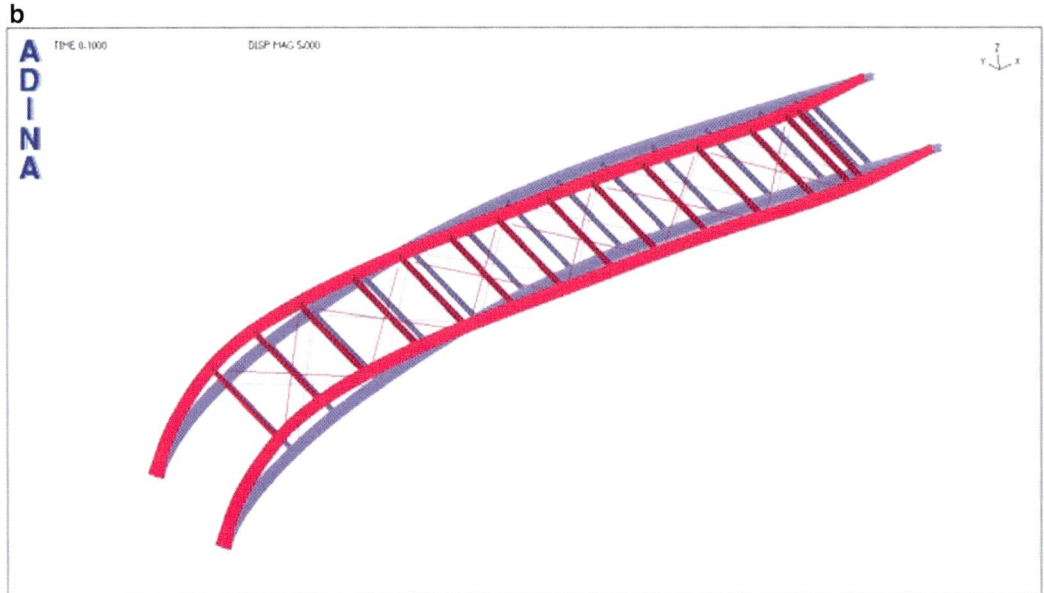

Point B

Nonlinear Finite Element Analysis, Fig. 26 (continued)

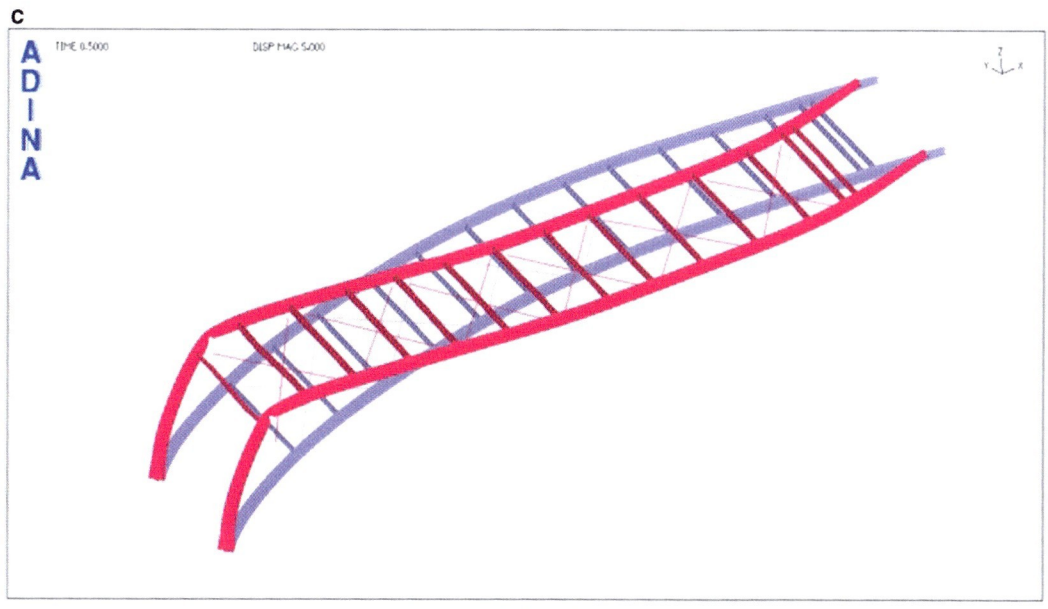

Point C

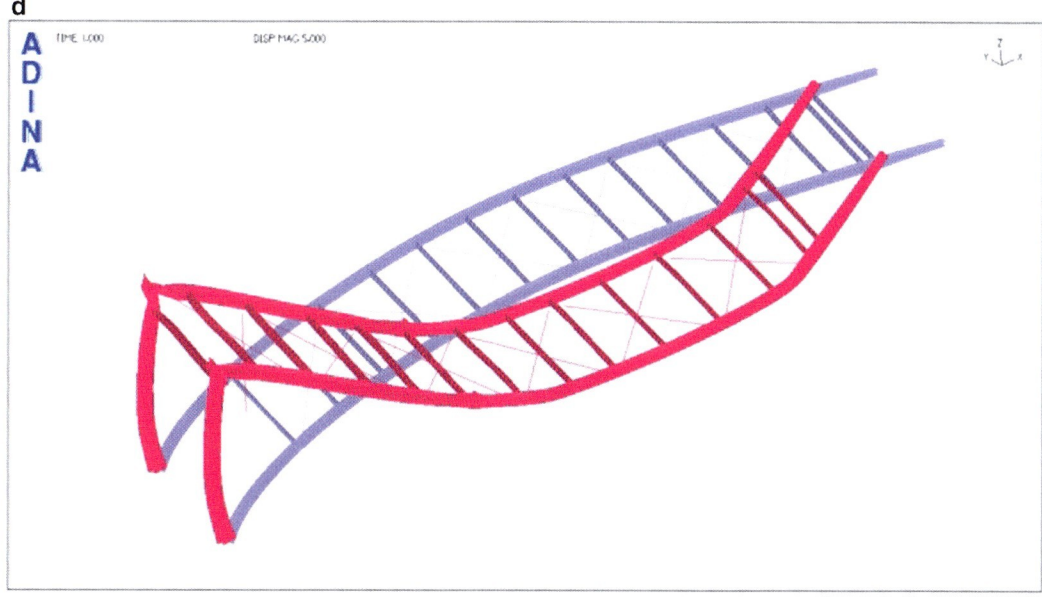

Point D

Nonlinear Finite Element Analysis, Fig. 26 Deformation at characteristic points along the GMNIA equilibrium path of two adjacent main arches of steel roof over Aristotle's Lyceum

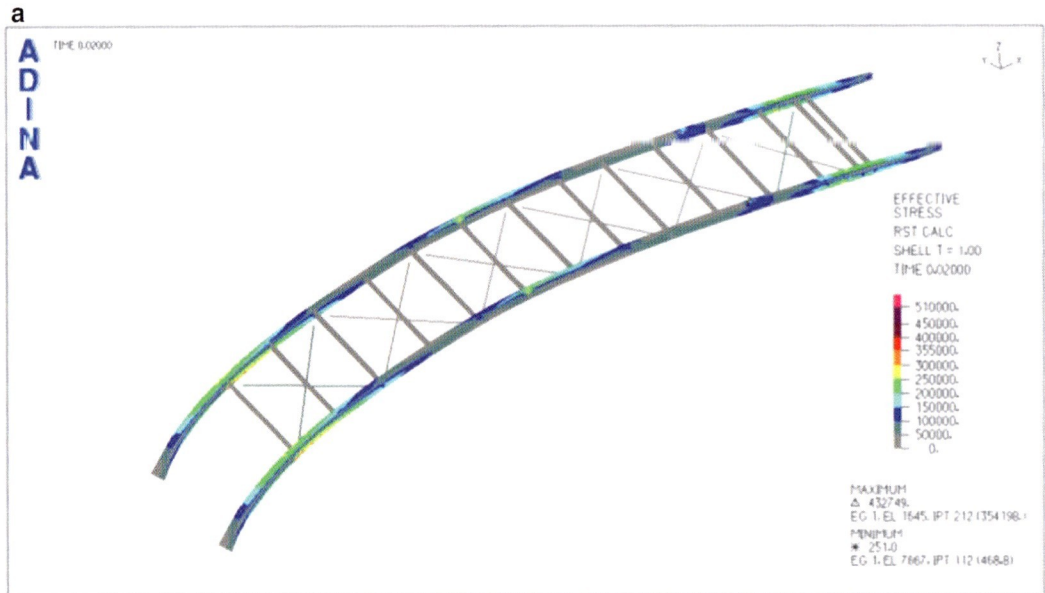

Point A

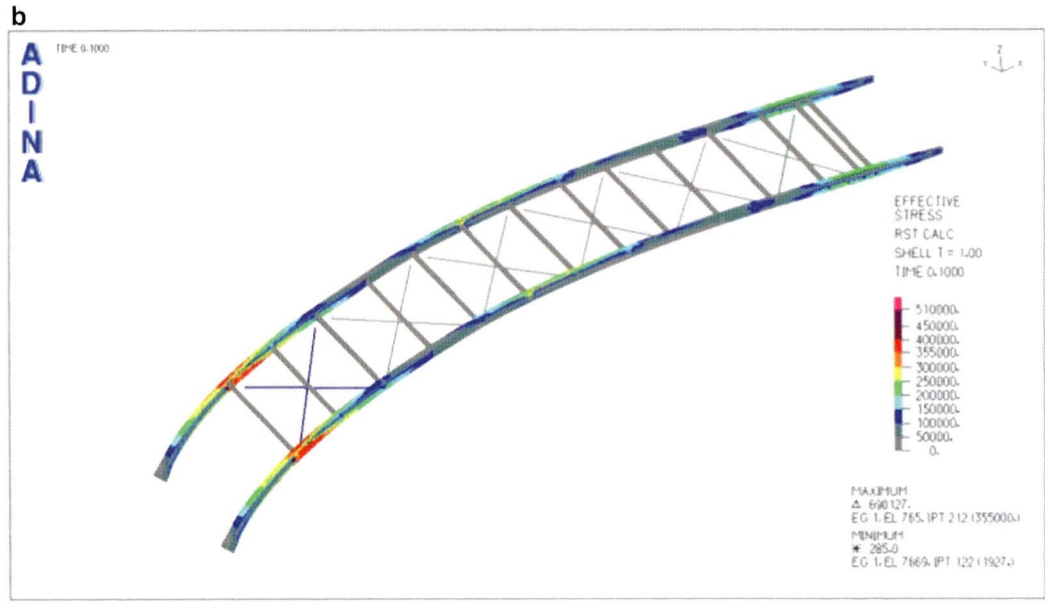

Point B

Nonlinear Finite Element Analysis, Fig. 27 (continued)

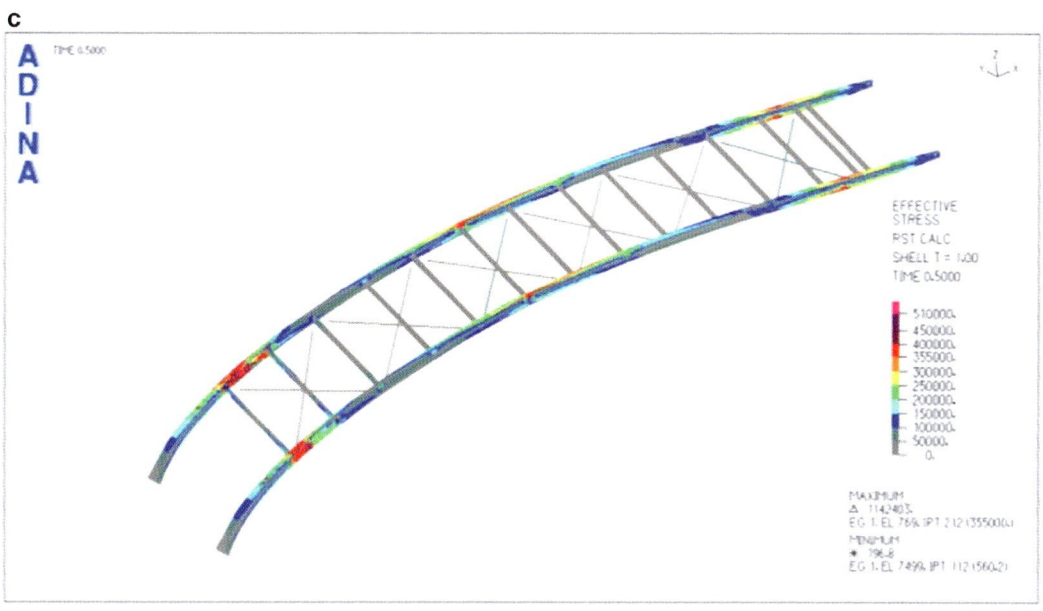

Point C

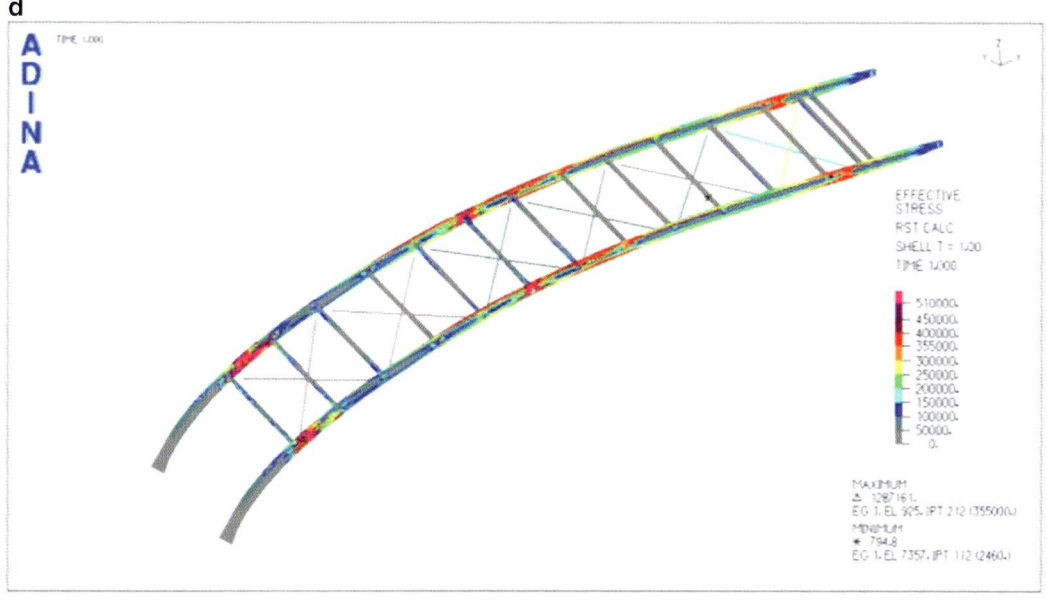

Point D

Nonlinear Finite Element Analysis, Fig. 27 Von Mises stress distribution at characteristic points along the GMNIA equilibrium path of two adjacent main arches of steel roof over Aristotle's Lyceum (kN/m^2)

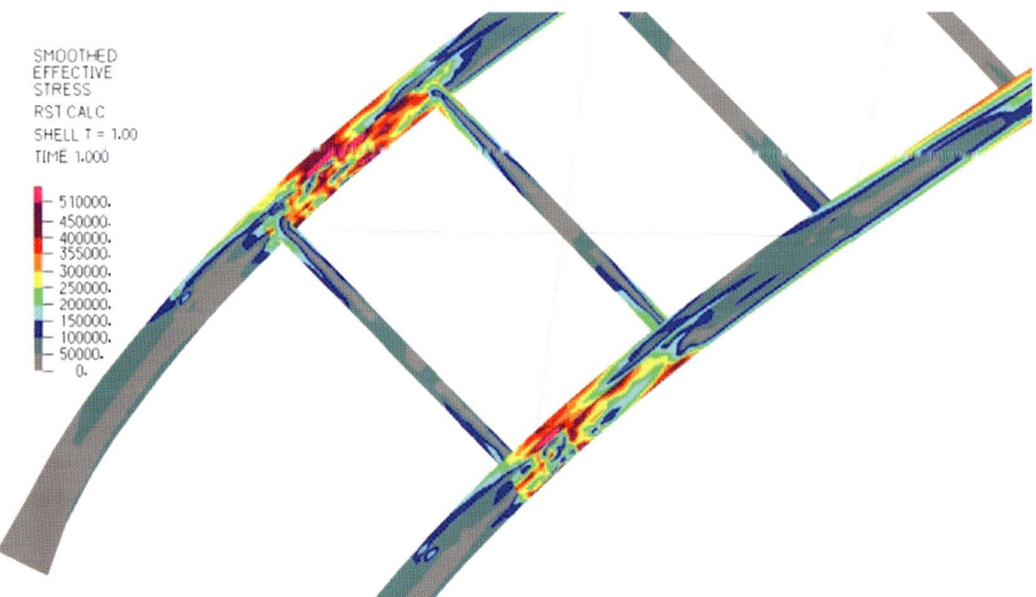

Nonlinear Finite Element Analysis, Fig. 28 Detail of von Mises stresses distribution of characteristic point D along the GMNIA equilibrium path of two adjacent main arches of steel roof over Aristotle's Lyceum (kN/m^2)

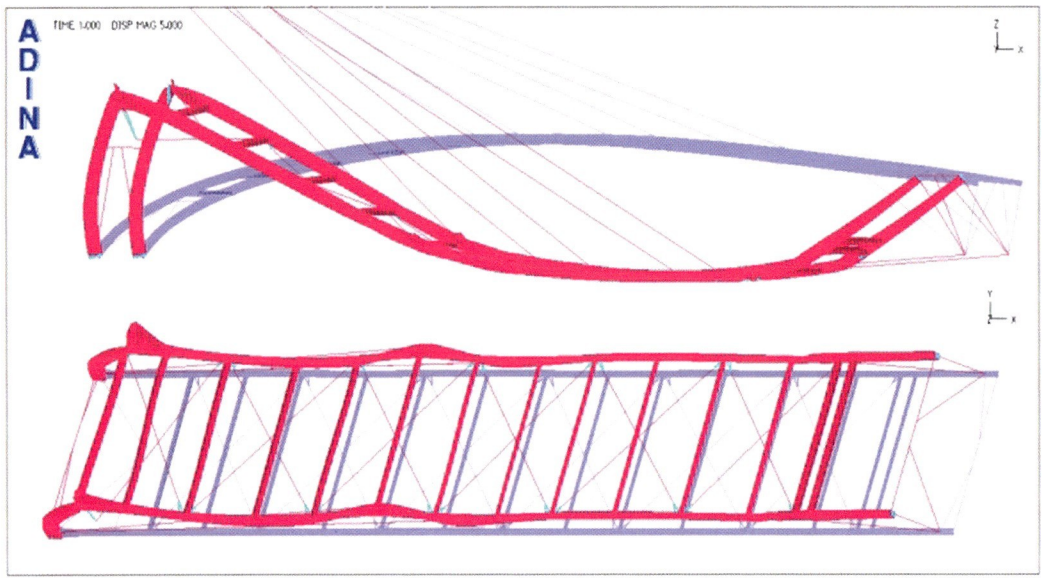

Nonlinear Finite Element Analysis, Fig. 29 Elevation and plan view of main arches at characteristic point D along the GMNIA equilibrium path (same scale of deformations in both views)

next illustrated by presenting the results of evaluation of the ultimate strength of members with unconventional geometry and/or cross section belonging to two long-span roofs that have been recently designed in Greece (Gantes and Koulatsou 2013). Due to the unconventional nature of these members, their design according to design codes had to rely upon simplifying

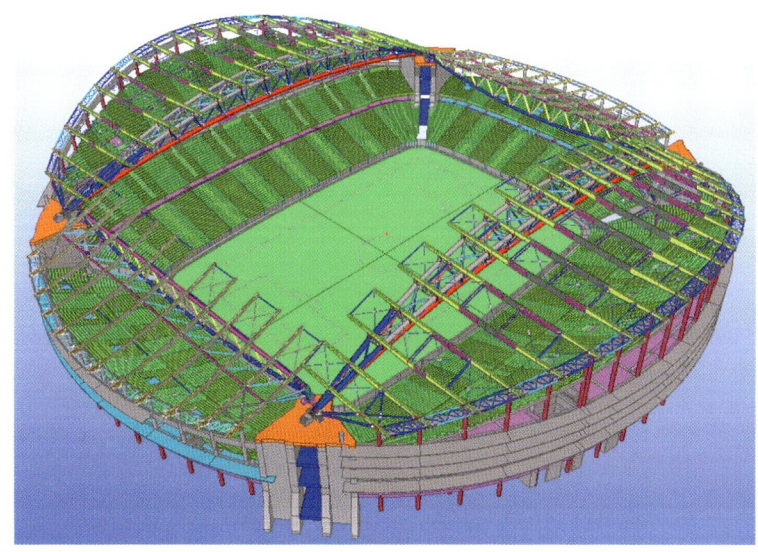

Nonlinear Finite Element Analysis, Fig. 30 Perspective view of football stadium of Panathinaikos FC

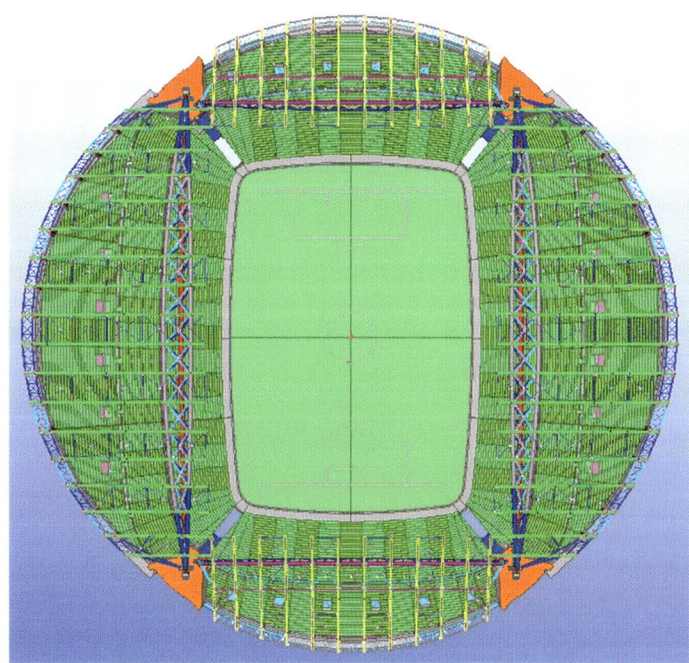

Nonlinear Finite Element Analysis, Fig. 31 Plan view of football stadium of Panathinaikos FC

assumptions of questionable validity, thus application of this approach was meaningful and appropriate. The analyses have been carried out using the well-known and extensively tested finite element software Adina [ADINA R&D 2012].

Suspended Steel Roof Over Aristotle's Lyceum

The archeological findings of the School of Aristotle are located in the center of Athens, Greece. For the protection of the archeological site, an architectural competition was held by the Greek

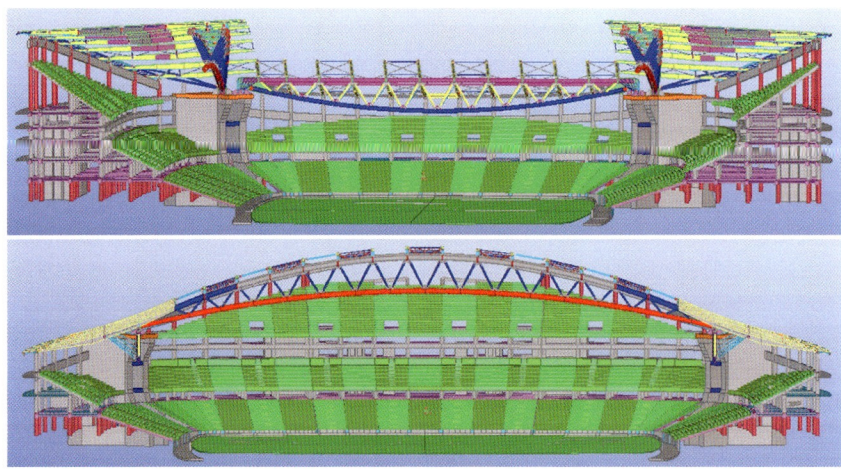

Nonlinear Finite Element Analysis, Fig. 32 Sections of the stadium and main truss girders

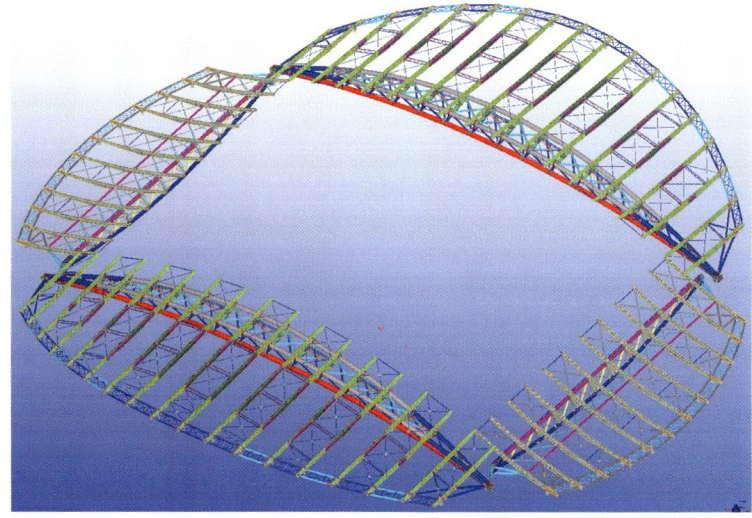

Nonlinear Finite Element Analysis, Fig. 33 Secondary beams and bracings of stadium's roof

Nonlinear Finite Element Analysis, Fig. 34 Support of secondary beams on main truss girder

Nonlinear Finite Element Analysis, Fig. 35 Cross sections of secondary beams

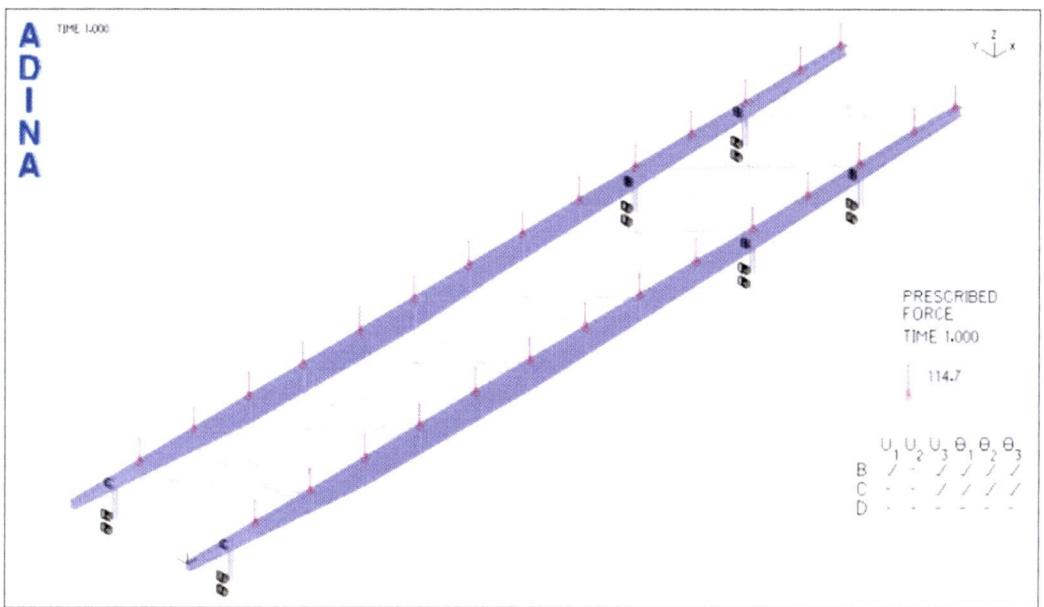

Nonlinear Finite Element Analysis, Fig. 36 Finite element model of two secondary beams and their cross-bracing in the steel roof of Panathinaikos stadium

Ministry of Culture, and the preliminary architectural proposal (Fig. 13) of the present suspended steel roof was awarded the first prize.

The roof steel structure (Fig. 14) consists of six parallel main arches having a span of 60 m and spaced at a distance of 10.5 m that are interconnected by means of purlins and horizontal bracings. Each main arch is pinned on the ground on one edge, while the other edge is supported by a V-shaped column. The arches are suspended by prestressed cables from a 25 m tall pylon, which has a slight inclination with respect to the vertical plane. The stability in and out of plane of the pylons is ensured by two back-stayed prestressed cables.

On the exterior of the pylons and V-shaped columns, independent steel structures for the extension of the cladding as well as eccentric vertical bracings for contributing to the resistance against lateral loads transversely to the main arches are provided.

The main arches (Fig. 15) consist of bending and compression members, which are made of built-up cross sections, varying along the length, having the shape of a rectangular box (green, cyan, and red parts in Fig. 15) and an I shape in the remaining parts (blue parts in Fig. 15). The 25 m tall pylons (Fig. 15) are compression members with hollow circular cross section, varying over the height in both diameter and wall thickness. For the design of these two members, main arches and pylons, numerical models have been created and analyzed linearly and nonlinearly.

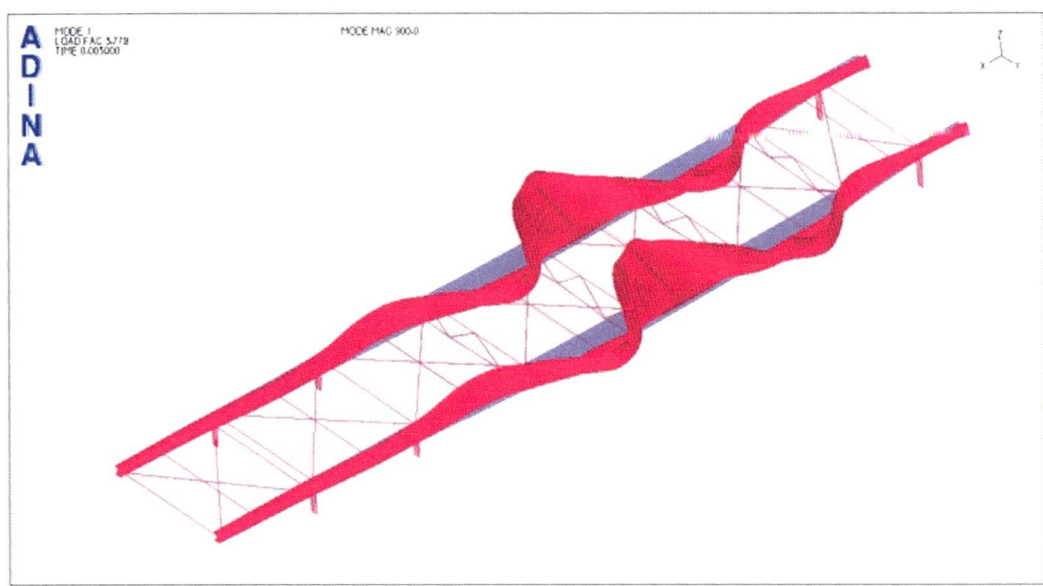

Nonlinear Finite Element Analysis, Fig. 37 First buckling mode (lateral buckling) of two secondary beams and their cross-bracing in the steel roof of Panathinaikos stadium

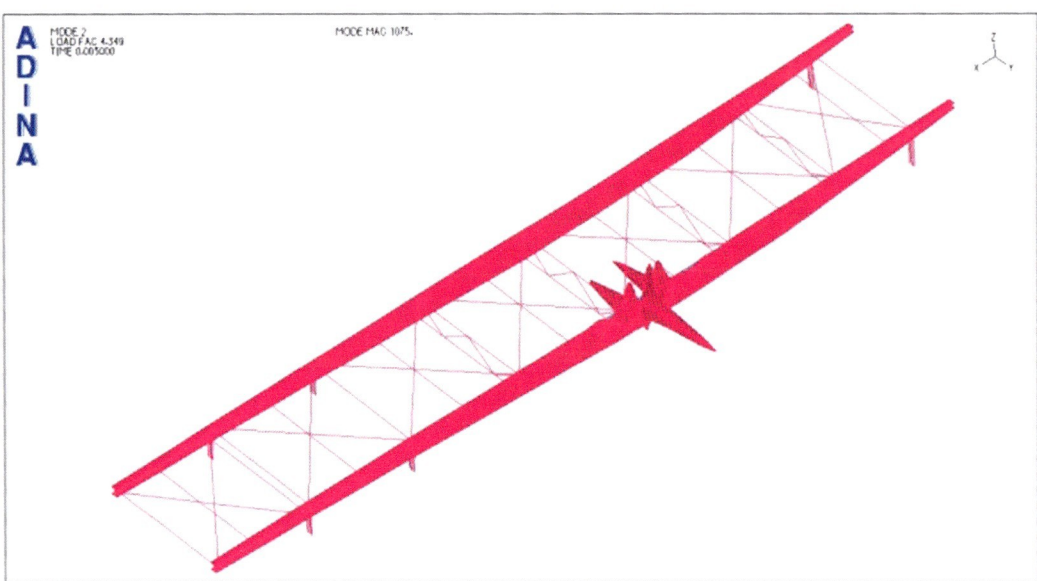

Nonlinear Finite Element Analysis, Fig. 38 Second buckling mode (local buckling) of two secondary beams and their cross-bracing in the steel roof of Panathinaikos stadium

With respect to the resistance of the pylon in compression, the applicability of buckling curves and associated reduction factors is questionable due to the varying cross section (Fig. 16). In order to obtain the pylon's strength against compression, a shell finite element model has been created, shown in Fig. 17. A conceptual design decision was taken, to use cross sections of class

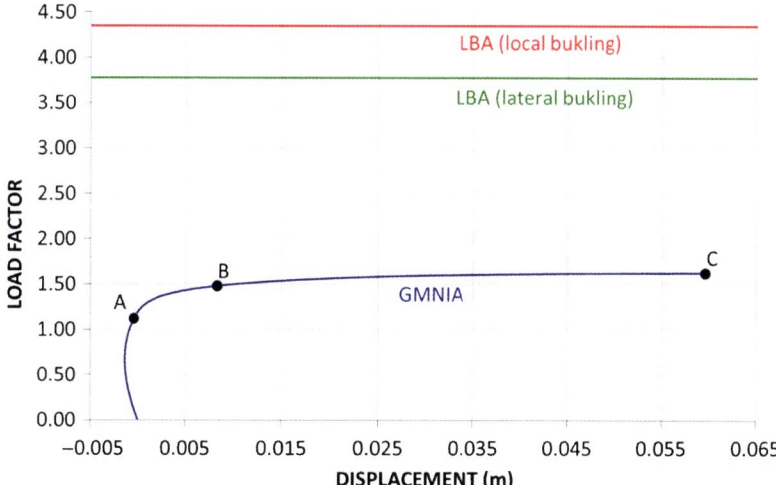

Nonlinear Finite Element Analysis, Fig. 39 Equilibrium paths of secondary beams of Panathinaikos stadium

1 or 2 for the pylon, to avoid local buckling as a possible failure mechanism. This was confirmed by the linearized buckling analysis, which showed no local buckling modes among the first ten, and thus, only the shape of the first global flexural buckling mode, presented in Fig. 18, was used as initial imperfection for nonlinear analyses.

Then, GNI and GMNI analyses were carried out for this imperfection shape having the magnitude prescribed by Eurocode 3 specifications, and the corresponding equilibrium paths are shown in Fig. 19. It is observed that the presence of imperfections reduces significantly the elastic critical load with respect to the linear one, as expected. Failure is encountered on the ascending branch of the equilibrium path, and it is mainly due to material yielding, accompanied by moderate global flexural buckling. This is verified by the deformation and stress snapshots on three characteristic points A, B, and C, before, on, and beyond the critical point, shown in Figs. 20 and 21. The clear formation of plastic hinges at two locations, highlighted in Fig. 21, helps identify these cross sections as weak. These locations are characterized by a decrease of cross-sectional thickness, so that increased thickness at these areas would be the most appropriate way for improving overall pylon strength.

With respect to calculation of the ultimate strength of the main arch, the use of code provisions is also of questionable validity due to its unusual shape and cross sections, thus rendering evaluation of the resistance to lateral buckling with analytical tools unreliable. For that purpose, a finite element model of two arches with their interconnecting purlins and bracings has been created, illustrated in Fig. 22. Similarly to the pylon, cross sections (Fig. 23) of class 1 or 2 were used for the main arches, so that local buckling was not critical. This model is able to simulate the behavior of the system regarding lateral buckling in a reliable manner, since bracings provide realistic lateral support to the system. Main arches and purlins are modeled with shell elements, whereas horizontal bracings are modeled with beam elements and cables with nonlinear, tension-only prestressed truss elements.

First, linearized buckling analysis was performed in order to obtain the first critical buckling mode (Fig. 24), which was used as shape of initial imperfection for the subsequent nonlinear analyses. GNI and GMNI analyses were then carried out, and their equilibrium paths are displayed in Fig. 25, accompanied by corresponding snapshots of deformed geometry and stresses for characteristic points A, B, C, and D along the GMNIA paths, illustrated in Figs. 26

and 27, respectively. The reference loads applied for the nonlinear analyses were chosen equal to the design loads corresponding to the most onerous load combinations. The multiplying load factor reported on the vertical axis of the equilibrium path (Fig. 25) denotes the percentage of reference loads acting on the structure at each load step. As observed, the maximum load factor for the main arches is approximately equal to 1.1, which represents a rather low factor of safety against collapse and signifies a possible need for strengthening.

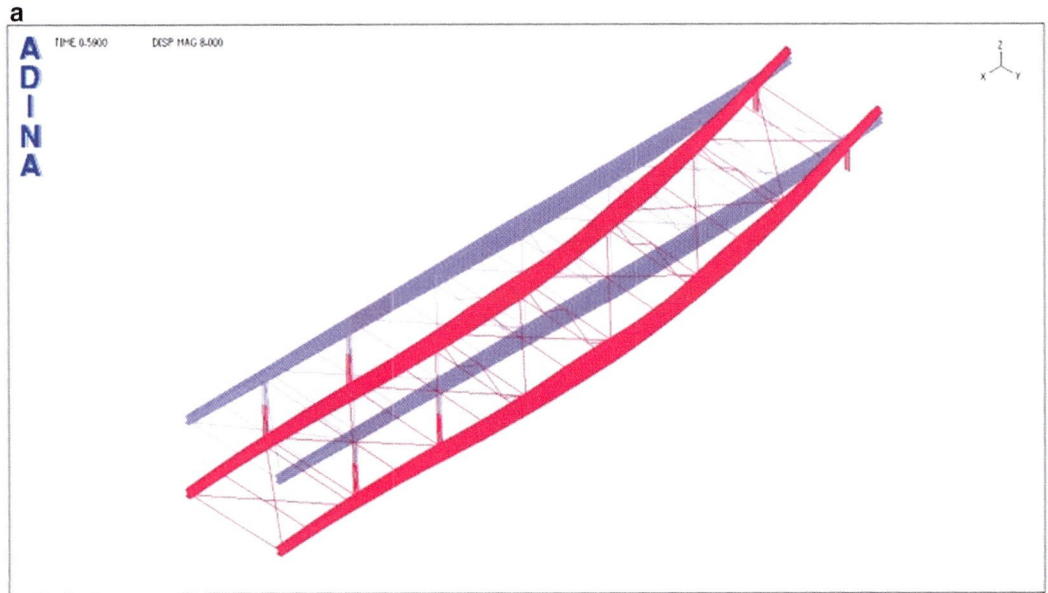

Point A

Point B

Nonlinear Finite Element Analysis, Fig. 40 (continued)

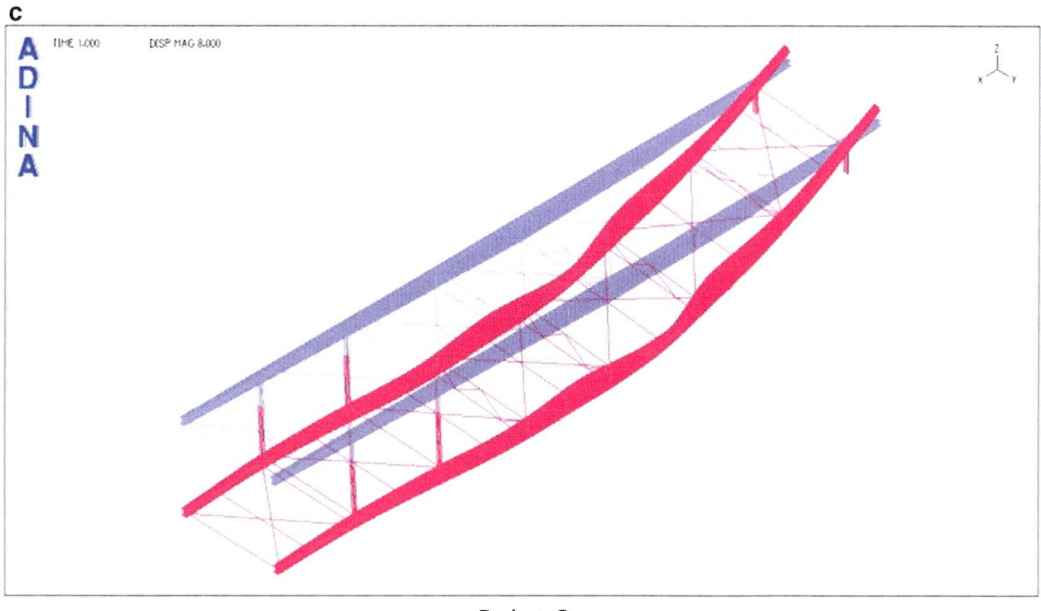

Nonlinear Finite Element Analysis, Fig. 40 Deformation at characteristic points along the GMNIA equilibrium path of the secondary beams of Panathinaikos stadium

From the above discussion and the deformed shapes and stress contours of Figs. 26 and 27, it is concluded that the box-shaped cross sections at the lower end, combined with the lateral support provided by the bracing system, are sufficient for preventing lateral buckling; thus, failure is primarily due to material yielding, with the formation of plastic hinges becoming particularly evident from the deformation and stress pictures on the descending branch of the equilibrium path. This is also verified by the stress contours detail shown in Fig. 28, corresponding to the equilibrium path of GMNI analysis point D. From the deformed shape of main arches (Fig. 29), it is also observed that the system has little lateral deformation, while vertical deflections dominate.

Figures 28 and 29 are also very useful in identifying as weak the locations of cross section change from double box section to I section (sections 3'-3' and 4'-4' in Fig. 23). An extension of the box section or a more gradual transition to the I section would thus be an appropriate intervention for improving the collapse behavior of the arch and enhancing its ultimate strength.

Moreover, it is verified for this example also that the elastic critical load derived from linearized buckling analysis is reduced significantly due to geometric imperfections and material yielding.

Roof of Panathinaikos Football Stadium

The second case study refers to the secondary beams of the steel roof in the new football stadium of Panathinaikos FC, which will be constructed in Votanikos, Athens, Greece, with a maximum capacity of approximately 40,000 spectators. A perspective view of the structural parts of the stadium is shown in Fig. 30 and its plan view in Fig. 31.

The roof has the shape of a cylindrical surface, and it consists of four structurally independent parts. The basic element of each part is a main truss girder, simply supported on reinforced concrete pylons, arranged at the four corners of the stadium, as shown also in the sections of Fig. 32. Secondary beams are supported on these main girders and on the exterior peripheral reinforced concrete columns of the grandstands (Figs. 33

and 34). These beams carry the roof cladding. Appropriately arranged cross-bracings contribute to the overall stability.

Aim of the numerical investigation presented here is the evaluation of the ultimate capacity of secondary beams. These beams have different lengths, depending on their location, so that they adjust to the overall roof geometry. In the initial structural design, which is described here, they were made of welded I-shaped cross sections with variable height over their length, as illustrated in Fig. 35. The unusual geometry and cross

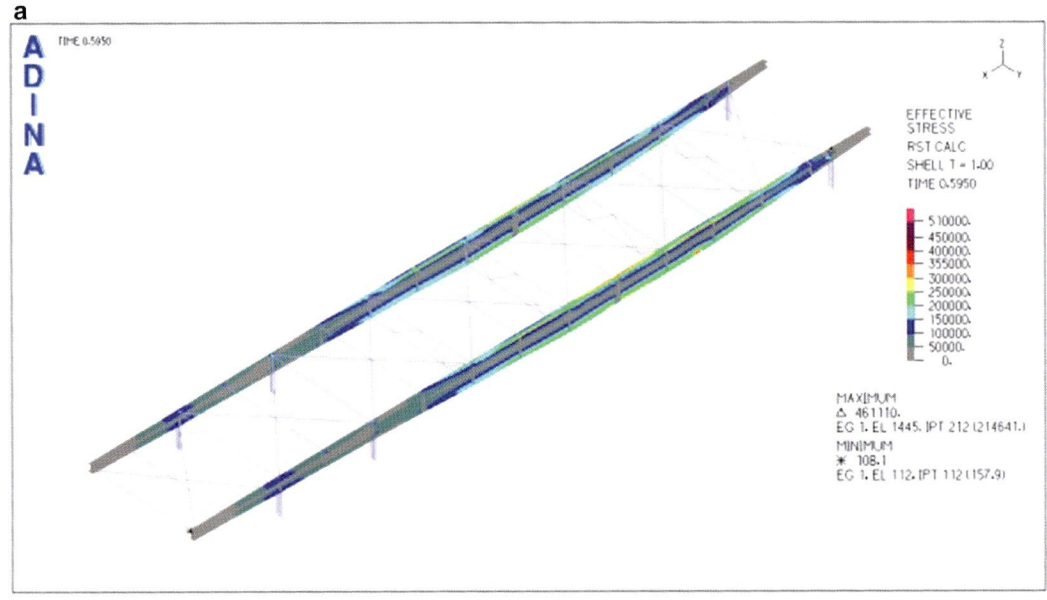

Point A

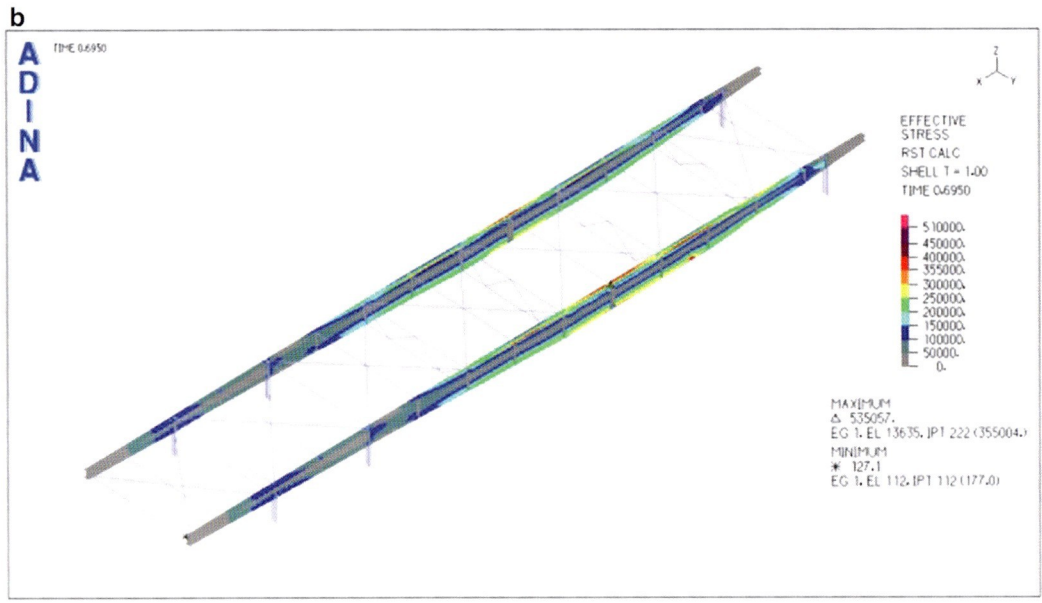

Point B

Nonlinear Finite Element Analysis, Fig. 41 (continued)

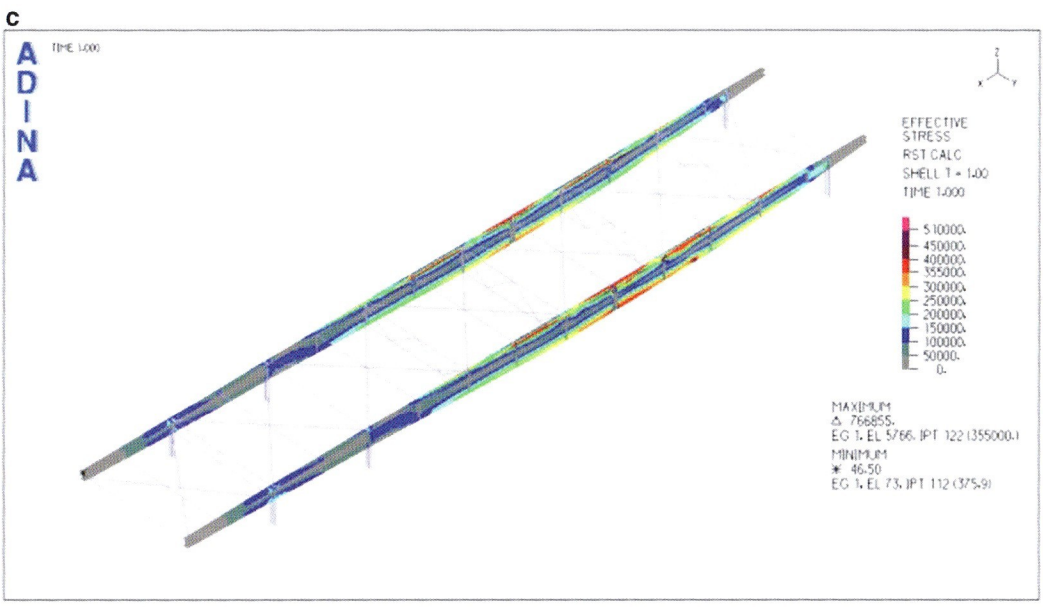

Point C

Nonlinear Finite Element Analysis, Fig. 41 Von Mises stresses at characteristic points along the GMNIA equilibrium path of secondary beams of Panathinaikos stadium (kN/m^2)

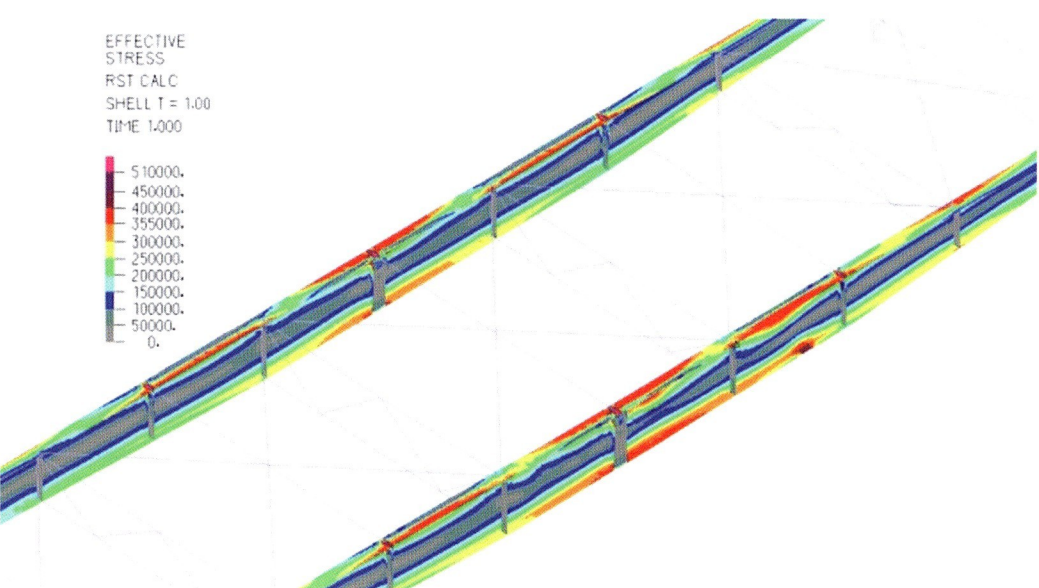

Nonlinear Finite Element Analysis, Fig. 42 Detail of von Mises stresses distribution of characteristic point C along the GMNIA equilibrium path of two adjacent secondary beams of Panathinaikos stadium (kN/m^2)

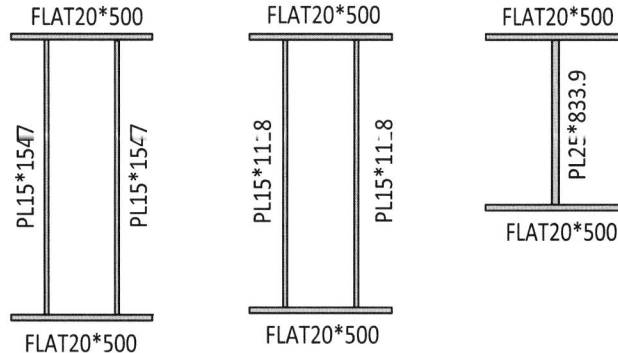

Nonlinear Finite Element Analysis, Fig. 43 Final cross sections of secondary beams of Panathinaikos stadium

sections of the beams highlighted the need for advanced numerical modeling and analyses, in order to address in a reliable manner the lateral and local buckling issues that were encountered.

In order to evaluate the behavior of secondary beams against lateral and local buckling and their strength, a finite element model consisting of two adjacent secondary beams and their cross-bracing members was created, shown in Fig. 36. Secondary beams were modeled with shell elements and cross-bracing members with beam elements. Secondary beam support on the main truss girders was modeled with spring elements.

First, linearized buckling analysis was performed in order to estimate the buckling modes and corresponding critical loads for both lateral and local buckling. The first two buckling modes are presented in Figs. 37 and 38, which illustrate the first lateral buckling mode and the first local buckling mode, respectively. A linear combination of these two first buckling modes was used as initial imperfection for the GMNI analysis.

GNI and GMNI analysis were performed, and the results are displayed by means of the equilibrium paths, shown in Fig. 39, and the snapshots of deformation and stress distribution for characteristic points A, B, and C along the GMNIA equilibrium path (Figs. 40 and 41). As in the previous example, the multiplying load factor reported on the vertical axis of the equilibrium path (Fig. 39) denotes the percentage of reference loads acting on the structure at each load step. In this case the maximum load factor for the main arches is approximately equal to 1.5, however, a significant reduction of stiffness is observed for much lower loads.

It is observed that failure was due to combined lateral buckling and material yielding, with small contribution from local buckling, as indicated by the equilibrium path and the corresponding deformation and stress distribution (Fig. 42). As in the first case study, here also the elastic critical buckling load derived by linearized buckling analysis decreases due to geometric imperfections and material yielding.

Due to the stiffness reduction occurring at low load levels, as mentioned above, associated with onset of lateral buckling, it was decided to change the beams' cross section from I-shaped to narrow box cross sections with protruding flanges that vary over their length (Fig. 43), in order to improve their torsional inertia, thus increasing their stiffness against lateral buckling.

Summary

For structures with unusual shape and variable cross sections, for which the use of pertinent codes and recommendations requires simplifying assumptions of questionable validity, an alternative approach has been presented for understanding the behavior, predicting all possible failure mechanisms, and evaluating the ultimate strength by means of linearized buckling and geometrically and material nonlinear imperfection analyses with commercially available finite element software. Steps include setting up an appropriate finite element model, obtaining critical buckling modes from linearized buckling analysis (LBA), and then using a linear combination of these modes as imperfection pattern for a geometrically and material nonlinear imperfection analysis (GMNIA).

Failure dominated by either material yielding or instability can thus be addressed, as well as interaction of failure modes. Equilibrium paths accompanied by snapshots of deformation and stress distribution at characteristic points are used for evaluating the results of the GMNI analysis, identifying the dominant failure modes as well as proposing and evaluating alternative strengthening measures.

Results of two case studies have been used to demonstrate this methodology, involving tall compression members and long-span beams with varying cross section, encountered in two steel roofs that have been recently designed in Greece.

Cross-References

▶ Assessment of Existing Structures Using Inelastic Static Analysis
▶ Behavior Factor and Ductility
▶ Nonlinear Analysis and Collapse Simulation Using Serial Computation
▶ Plastic Hinge and Plastic Zone Seismic Analysis of Frames
▶ Seismic Collapse Assessment

References

ADINA R&D (2012) Report ARD 12-8: Theory and Modeling Guide – volume I: ADINA Solids & Structures
Bathe KJ (1995) Finite element procedures. Prentice-Hall, Englewood Cliffs
Bathe KJ, Cimento AP (1980) Some practical procedures for the solution of nonlinear finite element equations. Comput Methods Appl Mech Eng 22(1):59–85
Bathe KJ, Dvorkin EN (1983) On the automatic solution of nonlinear finite element equations. Comput Struct 17(5–6):871–879
Bathe KJ, Snyder MD, Cimento AP, Rolph WD (1980) On some current procedures and difficulties in finite element analysis of elastic-plastic response. Comput Struct 12(4):607–624
Bazant ZP, Cedolin L (1991) Stability of structures: elastic, inelastic, fracture, and damage theories. Oxford University Press, Oxford
Belytschko T, Liu WK, Moran B (2000) Nonlinear finite elements for continua and structures. Wiley, New Jersey
Berry PA, Rotter JM, Bridge RQ (2000) Compression tests on cylinders with circumferential weld depressions. J Eng Mech (ASCE) 126(4):405–413
Chang SC, Chen JJ (1986) Effectiveness of linear bifurcation analysis for predicting the nonlinear stability limits of structures. Int J Numer Methods Eng 23:831–846
Crisfield MA (1979) A faster modified Newton-Raphson iteration. Comput Methods Appl Mech Eng 20(3):267–278
Crisfield MA (1981) A fast incremental/iterative solution procedure that handles snap-through. Comput Struct 13(1–3):55–62
Crisfield MA, Jelenic G, Mi Y, Zhong H-G, Fan Z (1997) Some aspects of the non-linear finite element method. Finite Elem Anal Des 27(1):19–40
Dimopoulos CA, Gantes CJ (2012) Comparison of alternative algorithms for buckling analysis of slender steel structures. Struct Eng Mech 44(2):219–238
European Committee for Standardisation (2005) Eurocode 3: Design of steel structures, Part 1-1: General rules and rules for buildings
Gantes CJ (2011) Numerical evaluation of ultimate bearing capacity of steel structures. In: Topping BHV, Tsompanakis Y (eds) Civil and structural engineering computational technology. Saxe-Coburg Publications, Stirlingshire
Gantes CJ, Fragkopoulos Kς (2010) Strategy for numerical verification of steel structures at the ultimate limit state. Struct Infrastruct Eng 6(1–2):225–255
Gantes CJ, Koulatsou K (2013) Methodology for nonlinear finite element analyses to evaluate strength of steel structures. In: Obrębski JB, Tarczewski R (eds) Beyond the limits of man. Paper presented at International Association for Shell and Spatial Structures (IASS), headquartered in Madrid, Spain, symposium 2013, Wroclaw, 23–27 Sept 2013
Herynk MD, Kyriakides S, Onoufriou A, Yun HD (2007) Effects of the UOE/UOC pipe manufacturing processes on pipe collapse pressure. Int J Mech Sci 49:533–553
Kojic M, Bathe KJ (2004) Inelastic analysis of solids and structures, Series in computational fluid and solid mechanics. Springer, Berlin
Livanou MA, Gantes CJ, Avraam TP (2013) Revisiting the problem of buckling mode interaction in 2-DOF models and built-up columns. In: Obrębski JB, Tarczewski R (eds) Beyond the limits of man. Paper presented at International Association for Shell and Spatial Structures (IASS), headquartered in Madrid, Spain, symposium 2013, Wroclaw, 23–27 Sept 2013
Ramm E (1981) Strategies for tracing nonlinear responses near limit points. In: Wunderlich W, Stein E, Bathe KJ

(eds) Nonlinear finite element analysis in structural mechanics. Springer, New York, pp 63–89

Riks E (1979) An incremental approach to the solution of snapping and buckling problems. Int J Solids Struct 15(7):529–551

Schneider W, Brede A (2005) Consistent equivalent geometric imperfections for the numerical buckling strength verification of cylindrical shells under uniform external pressure. Thin-Walled Struct 43(2):175–188

Schneider W, Timmel I, Höhn K (2005) The conception of quasi-collapse-affine imperfections: a new approach to unfavourable imperfections of thin-walled shell structures. Thin-Walled Struct 43(8):1202–1224

Stricklin JA, Haisler WE, von Riesemann WA (1973) Evaluation of solution procedures for material and/or geometrically nonlinear structural analysis. AIAA J 11(3):292–299

Nonlinear Seismic Ground Response Analysis of Local Site Effects with Three-Dimensional High-Fidelity Model

Tsuyoshi Ichimura[1], Kohei Fujita[2,3] and Muneo Hori[1]
[1]Earthquake Research Institute, University of Tokyo, Bunkyo, Tokyo, Japan
[2]Department of Architecture and Architectural Engineering, Graduate School of Engineering, Kyoto University, Kyoto, Japan
[3]RIKEN Advanced Institute for Computational Science, Kobe, Japan

Synonyms

Characterization and spatial variability of earthquake ground motions; Nonlinear finite element analysis; Site response

Introduction

An accurate estimation of seismic ground motion plays an important role, in order to realize rational countermeasures against an earthquake disaster. This review focuses on seismic ground response analysis above engineering bedrock. As has been pointed out, in a large earthquake, a local concentration of seismic ground motion due to complex ground structure often results in inducing severe damage to structures (e.g., see Liang and Sun (2000) for a review of relationships between complex ground structures and damage to structures). Although many countermeasures have been implemented, there remain sites in which local concentration of ground motion takes place; for example, a concentration of earthquake ground motion due to the complex ground structure caused severe damage to a buried gas pipeline network in the 2011 Tohoku Earthquake (Advisory Committee for Natural Resources and Energy et al. 2012). Because the impedance contrast between the engineering bedrock and the sedimentary layer is sharp, seismic ground motion is greatly amplified in sedimentary layers of complicated configuration showing nonlinear behavior. As a result, remarkably different seismic ground motion may be observed at even very close points. To prevent damage to structures due to locally concentrated seismic ground motion, a method is needed which can estimate ground motion amplification due to complex ground structure with high resolution and high accuracy.

A candidate for such an estimation method is observation that uses a dense network of seismographs. For example, a superdense seismograph network is operated in Tokyo (Shimizu et al. 2000), where the network of average distance between seismographs being 3 km has been installed for the real-time monitoring of earthquakes. However, it is difficult to estimate a local concentration of earthquake ground motion even with such a superdense seismograph network because the complex ground structure varies over distances of 10^{1-2} m. Also, it is difficult to estimate strain distribution because a seismograph can only observe point-wise wave profiles.

Another candidate is numerical simulation. The increase in ground structure data now enables the construction of a high-resolution three-dimensional (3D) numerical model that considers surface geometries and heterogeneities of the ground structure as faithfully as possible; from now on, this model is referred to as a "high-fidelity model." Although 3D nonlinear seismic ground response analysis with the high-fidelity model can be expected to accurately estimate 3D local site effects, it is not an easy task to conduct such numerical analyses because of the huge computation and modeling costs. As a result, numerical simulation with a simplified ground model is conventionally used. A representative analysis is a one-dimensional (1D) amplification analysis (e.g., SHAKE (Schnabel et al. 1972)). Because the 3D ground structure is approximated to a layered medium in the 1D amplification analysis, the 3D problem is reduced to the 1D problem. Stiffness and damping change momentarily, depending on shear strain due to 1D wave propagation. Although the 1D amplification analysis can estimate nonlinear seismic ground responses with less computational cost, it is not possible to estimate the local concentration which is caused by a 3D complex ground structure, by using the stratified-layer approximation. Recently, a two-dimensional (2D) nonlinear seismic ground response analysis has often been conducted for complex ground structure. It is quite natural that the 2D analysis gives more accurate estimations than the 1D analysis. However, a serious problem remains for the 2D analysis, with regard to selection of a suitable 2D cross section of the ground structure. Thus, 3D nonlinear seismic ground response analysis with a high-fidelity model is desirable.

Because numerical simulation performance and ground structure data have increased with recent advances in Information and Communications Technology (ICT), nonlinear seismic ground response analysis which computes a 3D high-fidelity model is becoming possible (e.g., Ichimura et al. (2014a, b)). Such an analysis enables estimation of the local concentration and strain distribution, which are difficult to observe. This review explains the detail and utility of the nonlinear seismic ground response analysis with a 3D high-fidelity model (Ichimura et al. 2014a).

Three-Dimensional Nonlinear Seismic Ground Response Analysis Method with a High-Fidelity Model

Because the finite element method (FEM) can accurately generate a 3D high-fidelity numerical model for complex ground structures and accurately satisfy traction-free boundary conditions on the surface, the following nonlinear dynamic FEM with unstructured finite elements is used for the nonlinear seismic ground response analysis:

$$\left(\frac{4}{dt^2}\mathbf{M} + \frac{2}{dt}\mathbf{C}^n + \mathbf{K}^n\right)\delta\mathbf{u}^n = \mathbf{F}^n - \mathbf{Q}^{n-1} \\ + \mathbf{C}^n\mathbf{v}^{n-1} + \mathbf{M}\left(\mathbf{a}^{n-1} + \frac{4}{dt}\mathbf{v}^{n-1}\right), \quad (1)$$

with

$$\begin{cases} \mathbf{Q}^n = \mathbf{Q}^{n-1} + \mathbf{K}^n\delta\mathbf{u}^n, \\ \mathbf{u}^n = \mathbf{u}^{n-1} + \delta\mathbf{u}^n, \\ \mathbf{v}^n = -\mathbf{v}^{n-1} + \frac{2}{dt}\delta\mathbf{u}^n, \\ \mathbf{a}^n = -\mathbf{a}^{n-1} - \frac{4}{dt}\mathbf{v}^{n-1} + \frac{4}{dt^2}\delta\mathbf{u}^n. \end{cases} \quad (2)$$

Here, $\delta\mathbf{u}$, $\mathbf{u}$, $\mathbf{v}$, $\mathbf{a}$, and $\mathbf{F}$ indicate the incremental displacement, displacement, velocity, acceleration, and external force vectors, respectively. $\mathbf{M}$, $\mathbf{C}$, and $\mathbf{K}$ indicate mass, damping, and stiffness matrices, respectively. dt and n indicate time increment and time step number, respectively. The Rayleigh damping matrix is used for $\mathbf{C}$; the element damping matrix $\mathbf{C}_e^n$ is calculated using the element mass matrix $\mathbf{M}_e$ and the element stiffness matrix $\mathbf{K}_e^n$ as

$$\mathbf{C}_e^n = \alpha\mathbf{M}_e + \beta\mathbf{K}_e^n.$$

Here, α and β are determined by solving the least squares equation,

$$\text{minimize}\left[\int_{f_{\min}}^{f_{\max}}\left(h^n - \frac{1}{2}\left(\frac{\alpha}{2\pi f} + 2\pi f\beta\right)\right)^2 df\right],$$

where $f_{\max}$, $f_{\min}$, and h^n indicate the maximum and minimum target frequencies and the damping ratio at time step n, respectively. Note that the damping matrix is calculated for every time step because the stiffness and damping ratio change by time step number, according to the constitutive relation of the ground material. Small elements are generated locally when modeling complex geometry with solid elements. Satisfying the

Courant condition when using an explicit time integration method (e.g., central difference method) for such small elements leads to small time increments and a huge computational cost. To resolve this problem, the Newmark-β method ($\beta = 1/4, \delta = 1/2$) is used for the time integration. Note that an explicit scheme is selected to compute the response robustly; an implicit scheme sometimes does not reach an accurate solution because of complex nonlinear constitutive relation. Thus, to verify the convergence of a solution, attention must be paid to the spatial and temporal discretization.

The most significant difference between the ordinary FEM analysis and the present nonlinear seismic ground response FEM analysis is the treatment of boundary conditions. Because the target domain is cut off from a half-infinite ground structure, half-infinite and absorbing conditions must be added on the bottom and side boundaries of the domain that describe the half-infinite continuity of the ground structure in the nonlinear seismic ground response FEM analysis. Several methods have been proposed, and one simple and effective method is based on damper and free motion ($\mathbf{u}^f$) on the side and bottom boundaries; $\mathbf{u}^f$ is computed based on the 1D wave theory, by applying the stress caused by $\mathbf{u}^f$ onto the side and bottom boundaries and by absorbing the scattered wave $(\dot{\mathbf{u}} - \dot{\mathbf{u}}^f)$ on the side and bottom boundaries by the damper. This method can describe the half-infinite continuity with the assumption that the ground structure on the side boundaries is extended infinitely.

The computation is performed by repeating the following steps:

1. Compute stiffness and damping ratio for n-th time step with nonlinear constitutive relation using $\mathbf{u}^{n-1}$.
2. Compute $\mathbf{K}^n$ and $\mathbf{C}^n$ using stiffness and damping ratio for nth time step.
3. Compute $\delta \mathbf{u}^n$ by solving Eq. 1, and update values in Eq. 2 using $\delta \mathbf{u}^n$.

The bottlenecks of the nonlinear seismic ground response FEM analysis with a 3D high-fidelity model are the modeling cost and the computation cost. For example, with the size of the target region of the order of $10^3 \times 10^3 \times 10^2$ m and a spatial resolution in the FE model of the order of $10^{-1 \sim 0}$ m to satisfy the temporal resolution of a few Hz, the degrees of freedom (DOF) of the FE model become of the order of $10^{7 \sim 9}$ and the time steps needed in the analysis would be of the order of $10^{3 \sim 4}$. Because constructing such a large 3D FE model is difficult and the computational cost becomes enormous, it is still difficult to make a nonlinear, dynamic analysis for such a large problem. A fully automated modeling method that can generate elements of high quality is needed for constructing a 3D FE model of ground structures with large DOF. Here, the elements of high quality mean that the aspect ratio of the elements is small; an element with large aspect ratio behaves more stiffly, leading to a significant loss of accuracy in the analysis results. Although it is common to perform manual tuning to avoid elements of low quality, manual tuning of a huge model is extremely costly. Thus, a suitable method for the automated generation of an FE model is desirable. The remaining bottleneck is the large computational cost. The unknown in Eq. 1 is $\delta \mathbf{u}^n$; this is solved for each time step to obtain the time-history response of $\mathbf{u}$. Most of the computation time is spent solving the matrix equation of Eq. 1. Thus, a fast solver is also desirable.

Seismic Ground Motion Computed with a 3D High-Fidelity Model

Seismic ground motion is computed by nonlinear seismic ground response FEM analysis with a 3D high-fidelity model (see Ichimura et al. (2014a for details)). This example was an effort to reproduce the earthquake ground motion distribution in the 2011 Tohoku-Oki earthquake. The size of the target region modeled was 1,696 m in the east to west direction and 1,920 m in the north to south direction, where the soft sedimentary layers form a complex soil structure (see Fig. 1 for the 3D high-fidelity model of the target ground structures). The site is known to have had significantly larger

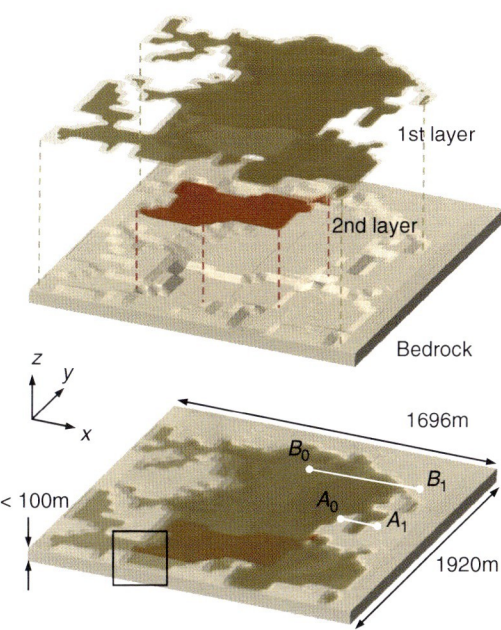

Nonlinear Seismic Ground Response Analysis of Local Site Effects with Three-Dimensional High-Fidelity Model, Fig. 1 Overall view of 3D ground structure model

observed seismic ground motion compared with that of nearby observation sites in past earthquake events. The modified Ramberg-Osgood model (Idriss et al. 1978) and the Masing rule (Masing 1926) were used as the nonlinear constitutive relation of soil material, and semi-infinite domain absorbing boundary conditions for the bottom and side boundaries of the domain were implemented, as explained above.

Here, a fully automated modeling method of Ichimura et al. (2009) with modifications was applied to constructing a 3D FE model, with the underground 3D soil structure being defined as a layered medium. A structured grid was used as background cells covering the target domain, and elements were made for each background cell. Here, the elements were generated in each cell so that they have low aspect ratios. This leads to a fully automated, robust method for making elements of high quality. Because the generation of elements in each cell is performed individually, it is suitable for parallelization, leading to fast element generation using parallel computation. First, elements with linear-tetrahedral elements and cubic elements are generated, and the both types of elements are converted to second-order tetrahedral elements, leading to a 3D FE soil structure model consisting of only second-order tetrahedral elements. Also, a fast solver was implemented with the Block-Jacobi method, mixed-precision arithmetic, a geometric multigrid method, a predictor-corrector method, and parallel computing using hybrid parallelization MPI-OpenMP to reduce computational cost and shorten computation time.

Figure 2 shows a close-up view of the 3D FE soil structure model. It can be seen that the complex 3D geometry consisting of three soil layers is modeled faithfully. The minimum size of the second-order tetrahedral elements is 4 m. With 32,509,107 DOF, the number of tetrahedral elements and nodes are 7,779,048 and 10,836,369, respectively. Using this 3D FE model, a seismic response analysis with nonlinear dynamic FE analysis of 60,000 time steps with 0.005 s time increments was conducted. Here, Eq. 1 is solved for each time step with relative error less than 10×10^{-6}. A PC cluster with eight computing nodes was used, each computing node with dual hexa-core Intel Xeon X5680 CPUs, connected with the InfiniBand QDR network. Using this small system, an analysis can be conducted in only 1,107,495 s. On average, one step is solved in 18.46 s, which is fast enough, considering that a 32 million DOF matrix equation is solved. This analysis time would be expected to become much shorter as a faster computer or a larger computer system is used. Due to recent technological progress, a dynamic nonlinear FE analysis is becoming feasible for application to nonlinear soil response analyses.

The analysis results and observed data are compared at observation points shown in Fig. 3. The computed and observed waves matched well, although no tuning was performed on the input wave, soil parameters, or soil model. It is also possible to obtain better conformity using dense observation networks to tune models based on data obtained from such dense networks. The seismic ground motion distribution is evaluated at the surface. Figure 4a shows the maximum

Nonlinear Seismic Ground Response Analysis of Local Site Effects with Three-Dimensional High-Fidelity Model, Fig. 2 3D ground structure FE model. The close-up view of the region in the rectangle in Fig. 1

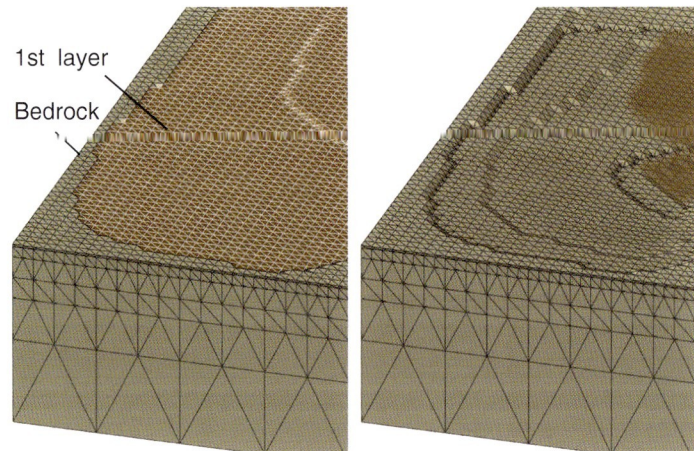

Nonlinear Seismic Ground Response Analysis of Local Site Effects with Three-Dimensional High-Fidelity Model, Fig. 3 Comparison of observed and computed waves

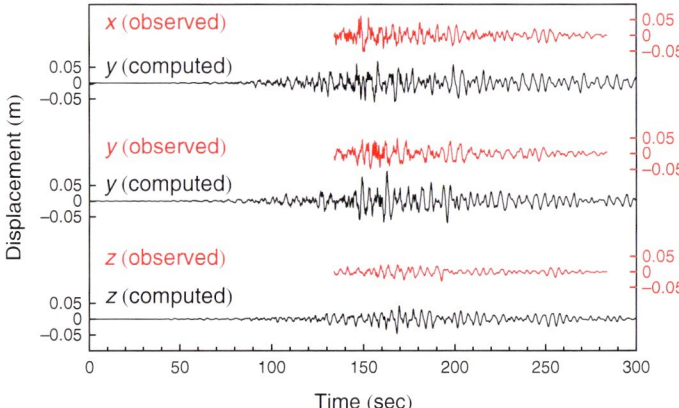

values of magnitude of displacement. It can be seen that the maximum displacement distribution has a high correlation with the depth of layer 1. To evaluate the strain in the axial direction (one of the damage indices for pipeline structure), the maximum values of principal strain in the whole time history are shown in Fig. 4b. Compared with the distribution of maximum displacement, it can be seen that estimating the distribution of strain from the ground structure is difficult; using numerical results of the seismic ground response analysis is effective for evaluating the distribution of strain. Two observation lines (A_0A_1 and B_0B_1) are set that imitate pipelines in parts of the model with nonuniform ground, and the axial strain is analyzed (see Fig. 1). Line A_0A_1 has a V-shaped depth distribution in layer 1, which is similar to that reported in Tsukamoto et al. (1984), and Line B_0B_1 goes through the ground structure with a constantly increasing depth of layer 1, followed by a region with constant depth of layer 1. Figures 5 and 6 show the maximum axial strain and the soil structure under each line. The axial strain along line A_0A_1 is similar to the observed strain on buried pipelines in V-shaped surface soil layers reported in Figs. 1 and 2 of Tsukamoto et al. (1984). The response in the bedrock was small but became larger for the soft layer, with a nearly constant response in the center of the V-shaped valley. This leads to small axial strain in the center of the V-shaped valley and large strain between the bedrock and the soft

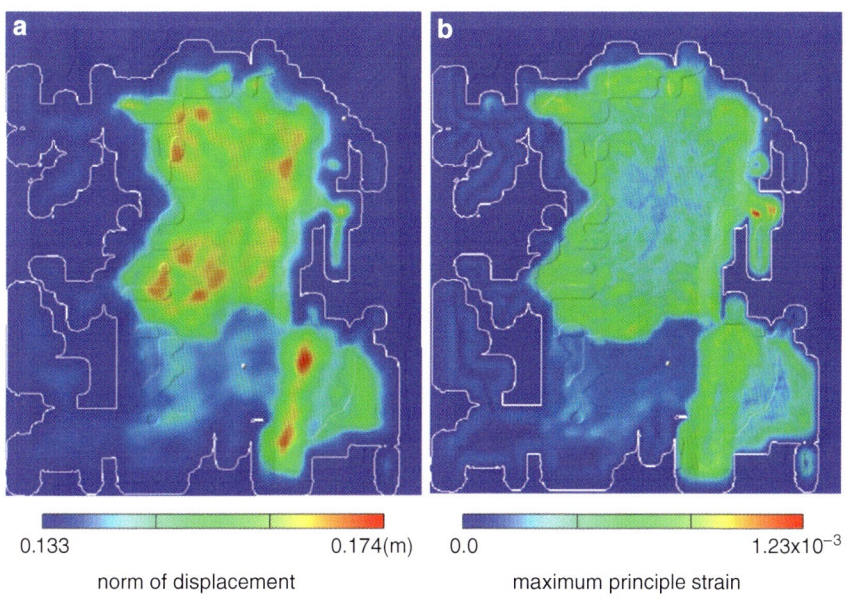

Nonlinear Seismic Ground Response Analysis of Local Site Effects with Three-Dimensional High-Fidelity Model, Fig. 4 Distribution of time-history maximum values of norm of displacement and maximum principal strain

Nonlinear Seismic Ground Response Analysis of Local Site Effects with Three-Dimensional High-Fidelity Model, Fig. 5 Maximum axial displacement, maximum axial strain, and underground structure of line A_0A_1

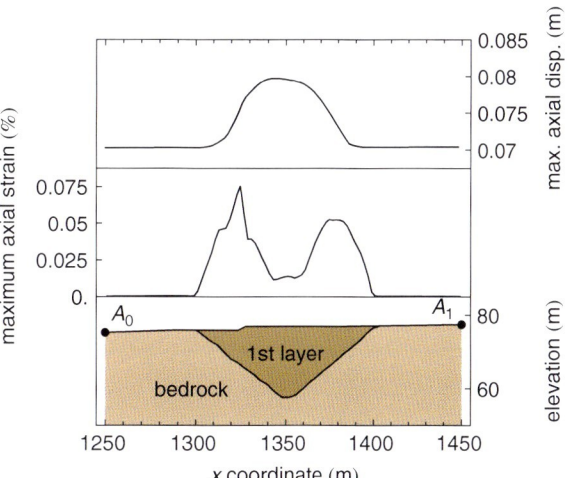

soil layer. This phenomenon is the cause of the pipeline damage reported in Liang and Sun (2000) and Tsukamoto et al. (1984). The response of line B_0B_1 is more complex than that of A_0A_1. Similar to line A_0A_1, high axial strain occurs between the bedrock and the soft layers. On the other hand, a complex distribution of axial strain can be observed from x=900–1,200 m, with a horizontally layered structure. This is because the soil structure around line B_0B_1 is complex and cannot be approximated by a two-dimensional structure, as shown in Fig. 6; the bedrock leans out from the northern side (see Fig. 1). From these observations, it can be seen

Nonlinear Seismic Ground Response Analysis of Local Site Effects with Three-Dimensional High-Fidelity Model,
Fig. 6 Maximum axial displacement, maximum axial strain, and underground structure of line B_0B_1

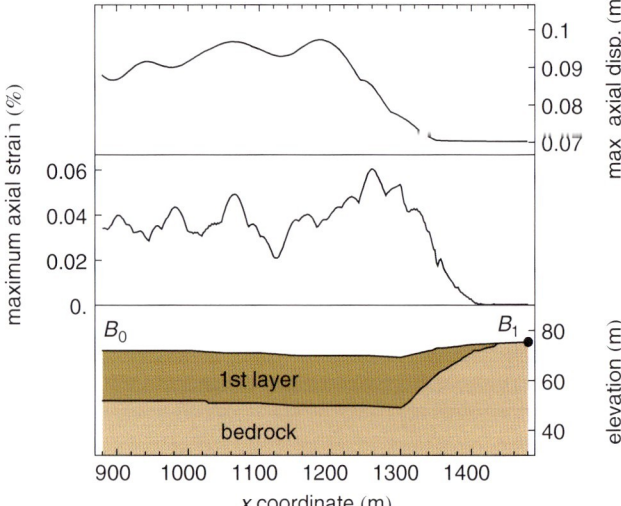

that nonuniform ground causes a complex ground motion distribution, suggesting that even 2D analyses could not be suitable to evaluating ground motion in some cases.

Summary

Increased ground structure data, computer performance, and numerical simulation abilities with advances in ICT now enable the execution of nonlinear seismic ground response analysis with a 3D high-fidelity model. Such analysis can accurately estimate earthquake ground motion in a complex ground structure model which is constructed from observed data. Even today, strain distribution can be estimated in a small computer environment; in the near future, such analyses are expected to become commonplace.

Cross-References

▶ Engineering Characterization of Earthquake Ground Motions
▶ Geotechnical Earthquake Engineering: Damage Mechanism Observed
▶ Nonlinear Dynamic Seismic Analysis
▶ Nonlinear Finite Element Analysis
▶ Physics-Based Ground-Motion Simulation
▶ Seismic Actions Due to Near-Fault Ground Motion
▶ Seismic Design of Pipelines
▶ Site Response: 1-D Time Domain Analyses
▶ Site Response: Comparison Between Theory and Observation
▶ Spatial Variability of Ground Motion: Seismic Analysis

References

Advisory Committee for Natural Resources and Energy, Urban Area Thermal Energy Committee Gas Safety Subcommittee, Working Group for Earthquake Disaster Prevention (2012) Report on disaster mitigation for gas supply in view of great east Japan earthquake 2011 (in Japanese)

Ichimura T, Hori M, Bielak J (2009) A hybrid multiresolution meshing technique for finite element three-dimensional earthquake ground motion modeling in basins including topography. Geophys J Int 177:1221–1232

Ichimura T, Fujita K, Hori M, Sakanoue T, Hamanaka R (2014a) Three-dimensional nonlinear seismic ground response analysis of local site effects for estimating seismic behavior of buried pipelines. J Press Vessel Technol,ASME 136:041702

Ichimura T, Fujita K, Tanaka S, Hori M, Lalith M, Shizawa Y, Kobayashi H (2014b) Physics-based urban earthquake simulation enhanced by 10.7 BlnDOF $\times$ 30 K time-step unstructured FE Non-linear seismic wave simulation. In: Proceedings of the international conference for high performance

computing, networking, storage and analysis, New Orleans, Nov 2014
Idriss IM, Singh RD, Dobry R (1978) Nonlinear behavior of soft clays during cyclic loading. J Geotech Eng 104:1427–1447
Liang J, Sun S (2000) Site effects on seismic behavior of pipelines: a review. J Press Vessel Technol 122:469–475
Masing G (1926) Eigenspannungen und verfestigung beim messing. In: Proceedings of the 2nd international congress of applied mechanics, pp 332–335 (in German)
Schnabel PB, Lysmer J, Seed HB (1972) SHAKE: a computer program for earthquake response analysis of horizontally layered sites, UCB/EERC-72/12. University of California, Berkeley
Shimizu Y, Watanabe A, Koganemaru K, Nakayama W, Yamazaki F (2000) Super high-density realtime disaster mitigation system. In: 12th world conference on earthquake engineering, New Zealand
Tsukamoto K, Nishio N, Satake M, Asano T (1984) Observation of pipeline behavior at geographically complex site during earthquake. In: 8th world conference on earthquake engineering, San Francisco, vol 7, pp 247–254

Nonlinear System Identification: Particle-Based Methods

Eleni N. Chatzi[1] and Andrew W. Smyth[2]
[1]Department of Civil, Environmental and Geomatic Engineering, ETH Zurich, Institute of Structural Engineering, Zurich, CH, Switzerland
[2]Department of Civil Engineering and Engineering Mechanics, Columbia University, New York, NY, USA

Synonyms

Joint state and parameter estimation; Nonlinear systems; Online identification; Particle filter; Structural identification; Unscented Kalman filter

Introduction

The presence of nonlinearity in structural systems is inevitably linked to the dissipation of energy occurring either within particular components of the structure, such as joints or damper elements, or as a result of exceeding material capacity which in turn leads to response that lies outside the usual assumption of the elastic range. Therefore, nonlinearity in the form of hysteretic behavior is commonly observed in structures subjected to significant levels of excitation such as earthquake, wind, or sea waves. Typically, the description of the restoring force that is associated with the behavior of such systems involves the use of nonlinear laws which render the governing system of equations a highly nonlinear one.

The task of joint parameter and state estimation for such systems is nontrivial as it requires the implementation of methods able to account for nonlinearity. The complexity of the problem is further increased when a real-time estimation is sought. The latter is quite commonly the case in structural health monitoring (SHM) implementations and in vibration control applications where the real-time feedback of the restoring force is necessary. In serving such a purpose, time domain identification methods constitute a class of methods particularly well suited for the online monitoring of dynamic systems. Numerous techniques have been proposed for time domain identification in civil engineering, including the more simplified autoregressive-moving average (ARMA) models, the least squares estimation (LSE) method, the Eigensystem realization algorithm (ERA), as well as the Bayesian approximation techniques, such as the observer/Kalman filter identification (OKID), the extended Kalman filter (EKF), the unscented Kalman filter (UKF), and the sequential Monte Carlo methods. Some of these techniques lend themselves more readily than others to problems where the system dynamics or the measurement equations are expressed as nonlinear functions of the states.

The EKF has by far been the most extensively used identification algorithm, for the case of nonlinear systems, over the past 30 years, and has been applied for a number of civil engineering applications, such as structural damage identification, parameter identification of inelastic structures, and so forth. It is based on the propagation of a Gaussian random variable (GRV) through the first-order linearization of the state-space model of the system. Despite its wide use, the EKF is only reliable for systems that are almost linear on the time scale of the updating intervals. In what is presented herein, alternative techniques are presented that prove more efficient in handling higher-order nonlinearities. The UKF is an extension of the Kalman filter that does not require the calculation of Jacobians in order to linearize the state equations. Instead, the state is approximated by a GRV which is now represented by a set of carefully chosen points. These sample points completely capture the true mean and covariance of the GRV when propagated through the actual nonlinear system.

The sequential Monte Carlo method, also known as the particle filter (PF), is especially useful since it is capable of handling nonlinear systems described by non-Gaussian posterior probability of the state, as well as systems with non-Gaussian noise. This attribute makes the particular method quite attractive for SHM applications. The implementation concept is the approximation of the posterior probability of the state through the generation of a large number of samples that are appropriately weighted. Particle filters are essentially an extension to point-mass filters with the difference that the particles are no longer uniformly distributed over the state but instead concentrated in regions of high probability. A significant drawback is the fact that depending on the problem, a large number of samples may be required, thus rendering the PF analysis computationally expensive. Additionally, this method is faced with what is referred to as the sample impoverishment problem, which essentially refers to the loss of diversity among particles as the analysis progresses and can be quite pronounced when dealing with parameter identification problems, as is typically the case in SHM. In what follows, the workings of the methods are described and enhancements are presented to effectively treat issues arising in SHM applications.

As a literature note, a number of authors have implemented time domain modeling approaches within a monitoring framework (Juang and Pappa 1985; Juang et al. 1993; Smyth et al. 2002; Zhang et al. 2002; Nagarajaiah and Li 2004; Yang et al. 2006; Fraraccio et al. 2008). Additionally, several authors have experimented with the use of the EKF on nonlinear identification problems (Corigliano and Mariani 2004; Mariani and Corigliano 2005; Yun and Shinozuka 1980). The alternative nonlinear KF extension, i.e., the unscented Kalman filter, was first introduced in 1994 when J. Uhlmann suggested that rather than approximating the known system function through linearization, it might be beneficial to apply the exact nonlinear function to an approximating probability distribution (Julier and Uhlmann 1997). Wan and Van der Merwe (2000) further verified the validity of this proposal. In the case of SHM applications, Wu and Smyth have shown that the UKF is superior to the EKF for joint state and parameter estimation problems while proving more robust for increased measurement noise levels for high-dimensional systems (Wu and Smyth 2007). Chatzi and Smyth further demonstrated the potential of the UKF in balancing rapid estimation with high precision (Chatzi et al. 2010; Chatzi and Smyth 2009). The particle filter methods have also been widely applied for the case of nonlinear identification problems (Arulampalam et al. 2002; Chen et al. 2005; Ching et al. 2006; Nasrellah and Manohar 2011; Eftekhar et al. 2012).

The Dynamic Problem Formulation and the Optimal Bayesian Solution

Within a structural health monitoring framework, the target commonly is the tracking of dynamic response, described by the general continuous state-space formulation

$$\dot{\mathbf{x}}(t) = f(\mathbf{x}(t)) + \mathbf{v}(t) \quad (1)$$

or by its discretized equivalent which is better suited for SHM implementations where information is provided via sampling at an interval Δt:

$$\mathbf{x}_k = F(\mathbf{x}_{k-1}) + \mathbf{v}_{k-1} \quad (2)$$

The information on the system is provided through noisy observations of the form

$$\mathbf{y}_k = H(\mathbf{x}_k) + \boldsymbol{\eta}_k \qquad (3)$$

where $\mathbf{x}_k \in \mathbb{R}^n$ is the **state** variable **vector** at time $t = k\Delta t$, $\boldsymbol{v}_k$ is the zero mean **process noise** vector with covariance matrix $\mathbf{Q}_k$, $\mathbf{y}_k \in \mathbb{R}^m$ is the current **observation vector**, and $\boldsymbol{\eta}_k$ is the **observation noise** vector with corresponding covariance matrix $\mathbf{R}_k$. It is worth noting that the regimes described herein do not require feedback from all of the system's degrees of freedom (dof's). Instead, we might have limited information drawn from few sensors on selected locations of the structure/system.

From a Bayesian perspective, the problem of determining the estimate of $\mathbf{x}_k$ based on the sequence of all available measurements up to time step k, $\mathbf{y}_k$, is a recursive one, involving the construction of a posterior PDF $p(\mathbf{x}_k|\mathbf{y}_{1:k})$. Assuming the prior distribution, $\mathbf{y}_0$, is known and that the required PDF $p(\mathbf{x}_{k-1}|\mathbf{y}_{1:k-1})$ at time $k-1$ is available, the prior probability $p(\mathbf{x}_k|\mathbf{y}_{1:k-1})$ can be obtained sequentially through:

Step 1 Prediction (Chapman-Kolmogorov Equation)

$$p(\mathbf{x}_k|\mathbf{y}_{1:k-1}) = \int p(\mathbf{x}_k|\mathbf{x}_{k-1})p(\mathbf{x}_{k-1}|\mathbf{y}_{1:k-1})d\mathbf{x}_{k-1} \qquad (4)$$

The probabilistic model of the state evolution $p(\mathbf{x}_k|\mathbf{x}_{k-1})$, also referred to as transitional density, is defined by the process Eq. 2. Consequently, the measurement $\mathbf{y}_k$ is taken into account in formulating a posterior estimate through:

Step 2 Update (Bayesian Theorem)

$$p(\mathbf{x}_k|\mathbf{y}_{1:k}) = p(\mathbf{x}_k|\mathbf{y}_k, \mathbf{y}_{1:k-1}) = \frac{p(\mathbf{y}_k|\mathbf{x}_k)p(\mathbf{x}_k|\mathbf{y}_{1:k-1})}{p(\mathbf{y}_k|\mathbf{y}_{1:k-1})} \qquad (5)$$

where the likelihood function $p(\mathbf{y}_k|\mathbf{x}_k)$ is completely defined by the observation Eq. 3.

Once the posterior PDF is known, the optimal estimate can be computed using different criteria, one of which is the minimum mean square error (MMSE) estimate. Although, the Bayesian solution is hard to compute analytically, the particle-based methods described herein in fact are approximations of this concept. For further details, the interested reader is pointed to the work of Ristic and Arulampalam (2004).

The Joint State and Parameter Estimation Problem

Within the context of SHM implementations, it is often the case that the parameters describing the model of the system are either unknown or known with some uncertainty. To this end, it is often desirable to additionally identify the time-invariant parameters of the system model. In a joint state and parameter estimation problem, the state vector of Eq. 1 is augmented in order to include the parameters to be identified, i.e.,

$$\tilde{\mathbf{x}} = \begin{bmatrix} \mathbf{x} \\ \boldsymbol{\theta} \end{bmatrix}, \; \mathbf{x} \in \mathbb{R}^n, \; \boldsymbol{\theta} \in \mathbb{R}^l \qquad (6)$$

where $\boldsymbol{\theta}$ stands for the unknown parameter vector of dimension l. Given that this component is time invariant, Eq. 2 is modified as follows:

$$\tilde{\mathbf{x}}_k = \begin{bmatrix} \mathbf{x}_k \\ \boldsymbol{\theta}_k \end{bmatrix} = \begin{bmatrix} F(\mathbf{x}_{k-1}) \\ \boldsymbol{\theta}_{k-1} \end{bmatrix} + \begin{bmatrix} \boldsymbol{v}_{k-1} \\ \boldsymbol{v}^\theta_{k-1} \end{bmatrix} \qquad (7)$$

where $\boldsymbol{v}^\theta_{k-1}$ is an added process noise vector, relating to the time-invariant parameter components.

The Unscented Kalman Filter Method

The UKF approximates the state as a GRV, and therefore, assuming $N(\mathbf{x}; \boldsymbol{\mu}, \mathbf{P})$ is a Gaussian density with argument $\mathbf{x}$, mean $\boldsymbol{\mu}$, and covariance $\mathbf{P}$, the UKF relates to the two steps of the Bayesian approach Eqs. 4 and 5 through the following relationships:

$$\begin{aligned} p(\mathbf{x}_{k-1}|\mathbf{y}_{1:k-1}) &= N\left(\mathbf{x}_{k-1}; \hat{\mathbf{x}}_{k-1|k-1}, \mathbf{P}_{k-1|k-1}\right) \\ p(\mathbf{x}_k|\mathbf{y}_{1:k-1}) &= N\left(\mathbf{x}_k; \hat{\mathbf{x}}_{k|k-1}, \mathbf{P}_{k|k-1}\right) \\ p(\mathbf{x}_k|\mathbf{y}_{1:k}) &= N\left(\mathbf{x}_k; \hat{\mathbf{x}}_{k|k}, \mathbf{P}_{k|k}\right) \end{aligned} \qquad (8)$$

In the classic Kalman filter formulation, outlined in the seminal paper by Kalman (1960), the prediction and update steps are formalized in a set of equations derived on the basis of linearity. In the case examined herein, the governing equations are nonlinear and a discretized equivalent is utilized for the propagation of the statistics of $\mathbf{x}_k \in \mathbb{R}^n$ through the system. The unscented transformation is used for this purpose which involves the computation of a set of discrete points, termed the sigma points, $\boldsymbol{\chi}_k^i$, with corresponding weights w_i. In order to do so, the original state vector is redefined as the concatenation of the original state vector and noise variables as $\mathbf{x}_{k-1}^a = [\mathbf{x}_{k-1}^T \ \boldsymbol{v}_{k-1}^T \ \boldsymbol{\eta}_{k-1}^T]^T$. Then, the sigma points are distributed symmetrically around the mean $\hat{\mathbf{x}}_{k-1}^\alpha \in \mathbb{R}^L$ as follows:

$$\boldsymbol{\chi}_{k-1} = \begin{bmatrix} \hat{\mathbf{x}}_{k-1}^\alpha & \hat{\mathbf{x}}_{k-1}^\alpha + \sqrt{(L+\lambda)\mathbf{P}_k^\alpha} \\ & \hat{\mathbf{x}}_{k-1}^\alpha - \sqrt{(L+\lambda)\mathbf{P}_k^\alpha} \end{bmatrix} \quad (9)$$

Each of the columns of the above matrix corresponds to a sigma point vector $\boldsymbol{\chi}_{k-1}^i \in \mathbb{R}^L$ with components corresponding to the state, process, and measurement noise, respectively, and $\mathbf{P}_k^\alpha$ is the corresponding augmented covariance vector, incorporating the process and observation noise components:

$$\boldsymbol{\chi}_{k-1}^i = \begin{bmatrix} \boldsymbol{\chi}_{k-1}^{x,i} \\ \boldsymbol{\chi}_{k-1}^{v,i} \\ \boldsymbol{\chi}_{k-1}^{\eta,i} \end{bmatrix}, \quad \mathbf{P}_{k-1}^a = \begin{bmatrix} \mathbf{P}_{k-1} & 0 & 0 \\ 0 & \mathbf{Q}_{k-1} & 0 \\ 0 & 0 & \mathbf{R}_{k-1} \end{bmatrix} \quad (10)$$

The weights for the state and covariance components of the sigma point vector are calculated as

$$w_0^x = \frac{\lambda}{L+\lambda}, w_0^P = \left[\frac{\lambda}{L+\lambda} + (1 - \alpha^2 + \beta)\right]$$
$$w_i^x = w_i^P = \frac{1}{2(L+\lambda)}, \quad i = 1, \ldots, 2L \quad (11)$$

where α, κ, β are the KF parameters, $\lambda = \alpha^2(L + \kappa) - L$ is a scaling parameter. α and κ are scaling factors (for a two-dimensional state vector, $\kappa = 0$ is optimal), and β is related to the distribution (equals to 2 for Gaussian priors). Finally, L is the dimension of the augmented state vector.

Step 1 UKF Prediction

The state and process noise components of the sigma point vectors are propagated through the nonlinear function $F(\mathbf{x}_k)$ without any need for linearization:

$$\boldsymbol{\chi}_{k|k-1}^{x,i} = F\left(\boldsymbol{\chi}_{k-1}^{x,i}, \boldsymbol{\chi}_{k-1}^{v,i}\right), \quad i = 0, \ldots, 2L \quad (12)$$

The discrete set of the sample points, $\boldsymbol{\chi}_{k|k-1}^{x,i} \in \mathbb{R}^n$, is utilized to derive a representation of the prior density $p(\mathbf{x}_k|\mathbf{y}_{1:k-1})$. The prior mean and covariance are approximated as the weighted sample mean and covariance of the prior sigma points and the time update step is continued as follows:

$$\hat{\mathbf{x}}_{k|k-1} = \sum_{i=0}^{2L} w_i^x \boldsymbol{\chi}_{k|k-1}^{x,i} \quad (13)$$

$$\mathbf{P}_{k|k-1} = \sum_{i=0}^{2L} w_i^P \left[\boldsymbol{\chi}_{k|k-1}^{x,i} - \hat{\mathbf{x}}_{k|k-1}\right] \left[\boldsymbol{\chi}_{k|k-1}^{x,i} - \hat{\mathbf{x}}_{k|k-1}\right]^T \quad (14)$$

Step 2 UKF Update

An estimate of the observation vector (reflecting the measured quantities) may then be approximated as

$$\hat{\mathbf{y}}_{k|k-1} = \sum_{i=0}^{2L} w_i^x H\left(\boldsymbol{\chi}_{k|k-1}^{x,i}, \boldsymbol{\chi}_{k-1}^{\eta,i}\right) \quad (15)$$

where the state and observation noise components of the sigma point vectors are propagated through the nonlinear observation function $H(\mathbf{x}_k)$. Finally, the posterior estimate $\hat{\mathbf{x}}_{k|k}$ is obtained as the weighted sum of the prior $\hat{\mathbf{x}}_{k|k-1}$ and the measurement residual, $\mathbf{y}_k - \hat{\mathbf{y}}_{k|k-1}$, which reflects the discrepancy between the predicted measurement and the actual measurement. The Kalman gain matrix at time step k, $\mathbf{K}_k$, acts as the weighing factor:

$$\hat{\mathbf{x}}_k = \hat{\mathbf{x}}_{k|k} = \hat{\mathbf{x}}_{k|k-1} + \mathbf{K}_k \left(\mathbf{y}_k - \hat{\mathbf{y}}_{k|k-1} \right) \quad \text{where} \quad \mathbf{K}_k = \mathbf{P}_k^{xy} \left(\mathbf{P}_k^{yy} \right)^{-1}$$

$$\text{with} \quad \begin{cases} \mathbf{P}_k^{yy} = \sum_{i=0}^{2L} w_i^P \left[H \left(\boldsymbol{\chi}_{k|k-1}^{x,i} \right) - \hat{\mathbf{y}}_{k|k-1} \right] \left[H \left(\boldsymbol{\chi}_{k|k-1}^{x,i} \right) - \hat{\mathbf{y}}_{k|k-1} \right]^T \\ \mathbf{P}_k^{xy} = \sum_{i=0}^{2L} w_i^P \left[\boldsymbol{\chi}_{k|k-1}^{x,i} - \hat{\mathbf{x}}_{k|k-1} \right] \left[H \left(\boldsymbol{\chi}_{k|k-1}^{x,i} \right) - \hat{\mathbf{y}}_{k|k-1} \right]^T \end{cases} \quad (16)$$

The posterior state covariance is also calculated as

$$\mathbf{P}_k = \mathbf{P}_{k|k} = \mathbf{P}_{k|k-1} - \mathbf{K}_k \mathbf{P}_k^{yy} \mathbf{K}_k^T \quad (17)$$

Figure 1 illustrates the workings of the unscented Kalman filter.

The unscented Kalman filter, as its linear equivalent, is quite sensitive to the selection of the initial state covariance $\mathbf{P}_0$ as well as the noise covariance matrices $\mathbf{Q}_k$, $\mathbf{R}_k$. As in the linear KF case, the autocovariance least-squares (ALS) technique (Rajamani and Rawlings 2009) can be utilized which employs autocovariances of routinely operating data for estimating the noise covariance matrices.

The Particle Filter Method

The unscented Kalman filter is in fact a particular case of the particle filter where the assumption of Gaussianity is made for the state and noise components. The general particle filter, however, does not make any prior assumption on the state distribution. Instead, the posterior probability density function (PDF), $p(\mathbf{x}_k|\mathbf{y}_{1:k})$, is approximated via a set of random samples, also known as support points $\mathbf{x}_k^i$, $i = 1, \ldots, N$, with associated weights w_k^i. This means that the probability density function at time k can be approximated as follows:

$$p(\mathbf{x}_k|\mathbf{y}_{1:k}) = \sum_{i=1}^{N} w_k^i \delta \left(\mathbf{x}_k - \mathbf{x}_k^i \right) \quad (18)$$

where

$$w_k^i \propto \frac{p \left(\mathbf{x}_k^i | \mathbf{y}_{1:k} \right)}{q \left(\mathbf{x}_k^i | \mathbf{y}_{1:k} \right)} \quad (19)$$

where $\mathbf{x}_k^i$ are the N samples drawn at time step k from the **importance density function** $q(\mathbf{x}_k^i|\mathbf{y}_{1:k})$. The appropriate selection of this

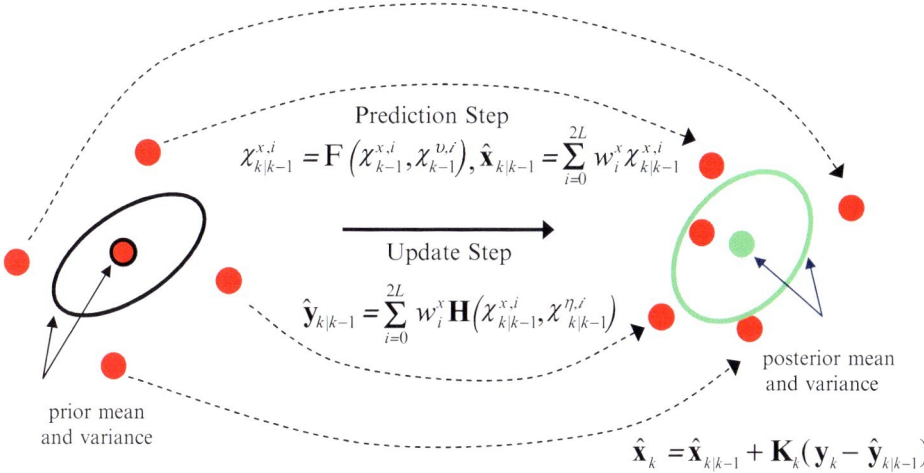

Nonlinear System Identification: Particle-Based Methods, Fig. 1 The unscented Kalman filter process for a two-dimensional state

function constitutes a fundamental consideration in this class of methods as explained next.

Using the state-space assumptions (1st-order Markov process/observational independence for a given state), the importance weights can be estimated recursively by

$$w_k^i \propto w_{k-1}^i \frac{p(\mathbf{y}_k|\mathbf{x}_k^i)p(\mathbf{x}_k^i|\mathbf{x}_{k-1}^i)}{q(\mathbf{x}_k^i|\mathbf{x}_{k-1}^i,\mathbf{y}_k)} \quad (20)$$

where $p(\mathbf{x}_k^i|\mathbf{x}_{k-1}^i)$ is the transitional density, defined by the process Eq. 2, and $p(\mathbf{y}_k|\mathbf{x}_k)$ is the likelihood function, defined by the observation Eq. 3. As the number of samples increases, this sequential Monte Carlo approach becomes an equivalent representation to the function description of the PDF and the solution approaches the optimal Bayesian estimate.

The Importance Density Function

An important issue in the implementation of particle filters is the selection of the importance density. It has been proved that the optimal importance density function that minimizes the variance of the true weights is given by

$$q(\mathbf{x}_k|\mathbf{x}_{k-1}^i,\mathbf{y}_{1:k})_{\text{opt}} = p(\mathbf{x}_k|\mathbf{x}_{k-1}^i,\mathbf{y}_k)$$
$$= \frac{p(\mathbf{y}_k|\mathbf{x}_k,\mathbf{x}_{k-1}^i)p(\mathbf{x}_k|\mathbf{x}_{k-1}^i)}{p(\mathbf{y}_k|\mathbf{x}_{k-1}^i)} \quad (21)$$

However, sampling from $p(\mathbf{x}_k|\mathbf{x}_{k-1}^i,\mathbf{y}_k)$ might not be straightforward, leading to the use of the transitional prior as the importance density function which greatly simplifies the analysis process:

$$q(\mathbf{x}_k|\mathbf{x}_{k-1}^i,\mathbf{y}_{1:k}) = p(\mathbf{x}_k|\mathbf{x}_{k-1}^i) \quad (22)$$

This essentially means that at time step k, the samples $\mathbf{x}_k^i$ are drawn from the transitional density, which is entirely defined by the process equation. Additionally, Eq. 20 yields

$$w_k^i = w_{k-1}^i p(\mathbf{y}_k|\mathbf{x}_k^i) \quad (23)$$

Therefore, the selection of the importance weights is essentially dependent on the likelihood of the error between the estimate and the actual measurement, as this is defined by the observation function H and the assumed properties of the observation noise $\mathbf{\eta}_k$.

The weights are then normalized so that their sum equals unity. In relating this process to the prediction and update steps referenced earlier, a two-stage methodology is once again derived where:

Step 1 PF Prediction

Given the value of the N discrete state vectors (particles), at the previous time step, $\mathbf{x}_{k-1}^i$, these are propagated through the process equation of the dynamic system in order to yield a prior estimate:

$$\mathbf{x}_{k|k-1}^i = F(\mathbf{x}_{k|k-1}^i) + \mathbf{v}_{k-1}, \ i=1,\ldots,N \quad (24)$$

Step 2 PF Update

The evaluation of the importance weights through the use of the likelihood function essentially constitutes the measurement update step, leading to the calculation of the posterior estimate through the weighted mean of the sample points:

$$\hat{\mathbf{x}}_k = \sum_{i=1}^{N} w_k^i \mathbf{x}_{k|k-1}^i \quad (25)$$

The Problem of Degeneracy and the Resampling Process

A second issue pertaining to the implementation of particle filters (PFs) is that of **degeneracy**, meaning that after some time steps, importance weights are unevenly distributed; thus, considerable computational effort is spent on updating particles with "trivial" contribution to the approximation of $p(\mathbf{x}_k|\mathbf{y}_{1:k})$. This may create numerical instabilities and calls for appropriate treatment. In quantifying this divergence, a measure of degeneracy has been specified, also known as the effective sample size:

$$N_{\text{eff}} = \frac{1}{\sum_{i=1}^{N}(w_k^i)^2} \quad (26)$$

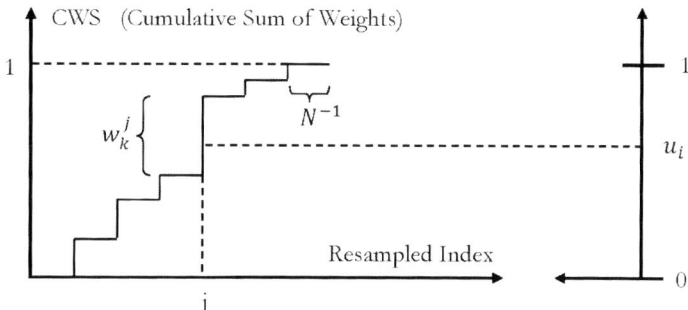

Nonlinear System Identification: Particle-Based Methods, Fig. 2 The process of resampling: the uniformly distributed random variable u_i maps into index j; thus, the corresponding particle $\mathbf{x}_k^j$ is likely to be selected due to its considerable weight w_k^j

Resampling is a technique developed for the tackling of degeneracy. It discards those particles with negligible weights and enhances the ones with more significant weights. Resampling takes place when N_{eff} falls below some user-defined threshold N_T. A particle is more likely to be selected as a member of the remaining set, if it corresponds to a higher weight as schematically shown in Fig. 2.

The workings of the various stages of the particle filter algorithm are illustrated in Fig. 3.

The Problem of Sample Impoverishment

The use of the resampling technique however is known to lead to other problems, relating to the loss of diversity of particles. As the higher-weight particles are duplicated, diversity is lost leading to the **sample impoverishment (or particle depletion) phenomenon**. Such a phenomenon is more likely to occur when the process noise levels are low and is particularly problematic in the case of joint state and parameter estimation. As demonstrated in section "The Joint State and Parameter Estimation Problem" and in the example that follows, in such a case the state vector is augmented to include the unknown, but time-invariant (constant) system parameters yielding state components with minimal variability throughout the analysis process. Since the addition of a significant amount of process noise would lead to algorithm convergence problems and instability issues, one typically needs to resort to the use of a large number of particles. Although, both the UKF and PF present the advantage of parallel implementation, the use of an excessive number of particles inevitably incurs a significant computational cost which can be a major disadvantage. The UKF is free of such a problem as the conditioning of the state via the Kalman gain indices a variability in the evolution of the parameter components $\boldsymbol{\theta}$ of the augmented state vector $\tilde{\mathbf{x}}$ of Eq. 7.

More information on the selection of the importance weights based on importance sampling can be found in Doucet et al. (2001), Bergman (1999), and Bergman et al. (2001).

The Particle Filter with Mutation

In order to tackle the sample impoverishment problem, Chatzi and Smyth (2013) proposed an enhancement of the PF, termed the particle filter with mutation (MPF), which incorporates a mutation operator in the resampling process. **Mutation** is a process typically used in genetic algorithms (GAs) where it serves as a means of maintaining diversity among the members of a population. Within the framework of GAs, mutation is typically enforced under two regimes, the *creep mutation* and the *jump mutation*. Jump mutations involve random modifications in the binary encoding of the system's variables, whereas creep mutation takes place in the real number representation of the variables (i.e., the phenotype). The operation implemented in the MPA scheme resembles the creep mutation process.

In SHM implementations or system identification problems where the estimation of system parameters is required, the state vector is commonly augmented in order to include these

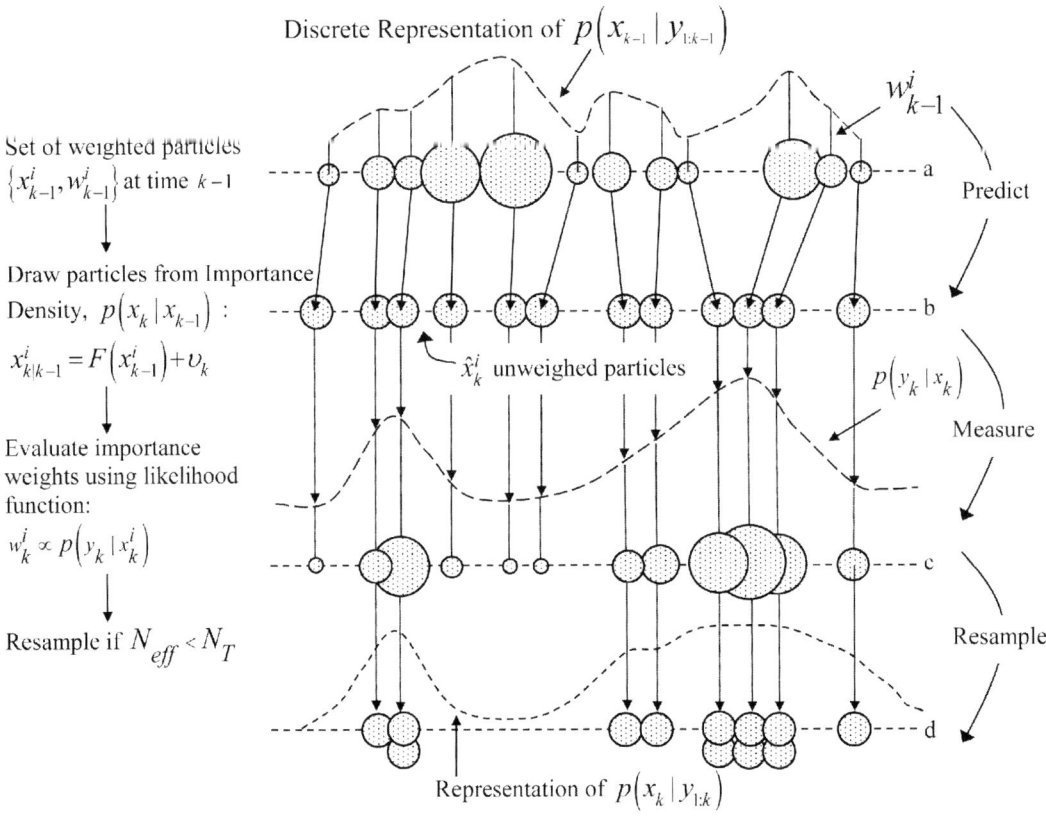

Nonlinear System Identification: Particle-Based Methods, Fig. 3 The particle filter implementation steps

parameters as in Eq. 6. Consequently, the state-space equations are also expanded to include the zero*th* derivative equations that enforce the time invariance of those parameters, which are considered as constants. Equation 7 reflects with the discrete time equivalent, indicating the time invariance of the parameter components. Hence, especially in the case of higher complexity problems, the initial particle selection space needs to be appropriately spanned and quite densely sampled in order to achieve an efficient parameter estimate. In other words the initial sample space needs to include seeds sufficiently close to the true parameter values for a successful prediction. In order to cope with this restriction, the MPF algorithm features a twofold innovation, enforced during the resampling step, i.e., when the effective sample size, N_{eff}, drops below a certain threshold.

Propagating the Weighted Estimate of the State

Firstly, part of the formerly unfit particles is replaced in step k by the prior estimate of the weighted mean of the state, obtained as $\hat{\mathbf{x}}_{k|k-1} = F(\hat{\mathbf{x}}_{k-1})$. Replacement is performed under a uniform probability, p_e, i.e., if rand $< p_e$, where rand is a uniformly distributed random number, then replacement takes place; otherwise, the particles remain unchanged. Function F is the state-space function of the process Eq. 2. In this manner, the actual value of the weighted estimate is incorporated in the particles and propagated through the nonlinear system. Formerly, the optimal estimate would only appear as the weighted result and the actual fitness for that particle was not explicitly evaluated at any step of the algorithm. For strongly nonlinear systems, this may lead to imprecise estimations.

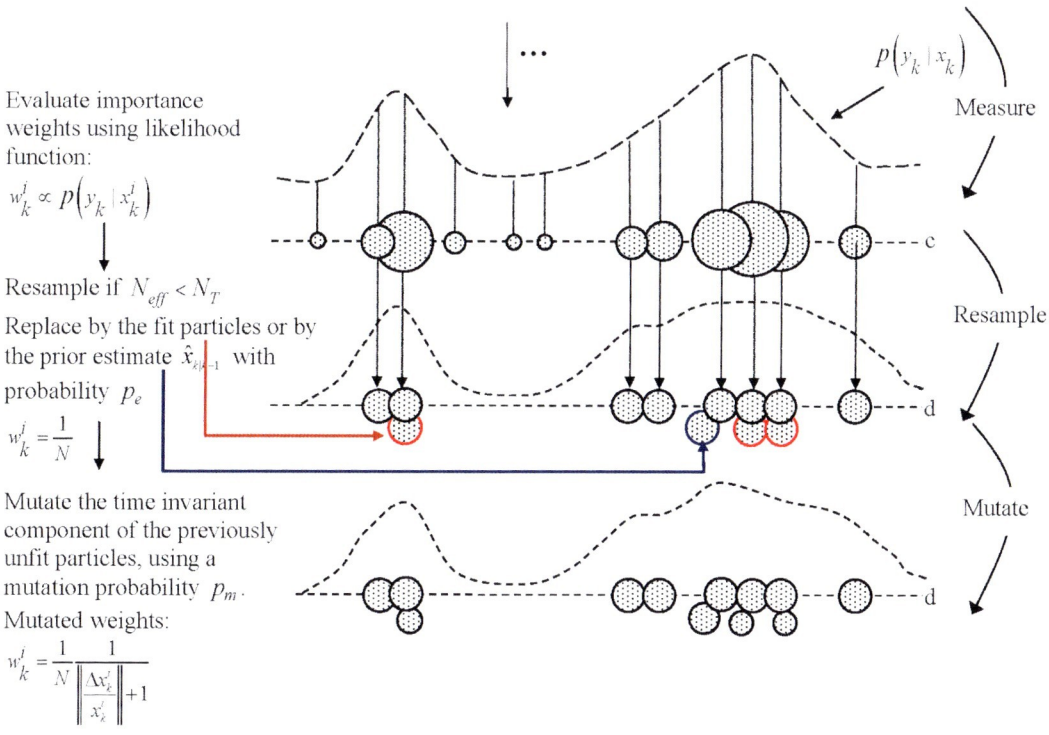

Nonlinear System Identification: Particle-Based Methods, Fig. 4 The particle filter with mutation implementation steps

Mutation of the Time Invariant Components
Secondly, after the formerly unfit particles have been replaced by either particles of more significant weight (standard resampling) or by the propagated value of $\hat{x}_{k-1}$, their time-invariant components are mutated. Mutation takes place by shifting the parameter components by a random amount using a mutation probability p_m. Both the mutation probability and the extent of the shifting interval can be user specified. The mutated particles are assigned a weight that is inversely proportional to the relative difference between the mutated vector and the original one ("parent" vector) according to the following relationship:

$$w_k^i = \frac{1}{N} \frac{1}{\left\|\frac{\Delta x_k^i}{x_k^i}\right\| + 1} \qquad (27)$$

whereas the weights of the non-mutated resampled particles remain equal to $w_k^i = \frac{1}{N}$.

In expression Eq. 27, Δx_k^i is the difference between the mutated vector and the parent one and $\|\cdot\|$ is the L_2 norm. The full set of weights is then once again normalized before proceeding to the next time step. A graphical representation of the MPF, relating it to the stages previously outlined for the standard PF, is displayed in Fig. 4.

As a literature note, the blending of the standard PF with evolutionary concepts has only very recently been explored in the literature. Akhtar et al. (2011) propose a Particle Swarm Optimization accelerated Immune Particle Filter (PSO-acc-IPF). Park et al. (2007) propose the so-called genetic filter which involves a standard genetic algorithm step (with all three GA operators, i.e., crossover, mutation, and selection) in place of standard resampling. Similarly, Kwok et al. (2005) employ the crossover operator to what they call the evolutionary particle filter.

The drawback of the previous approaches is that they more or less incorporate a GA or PSO

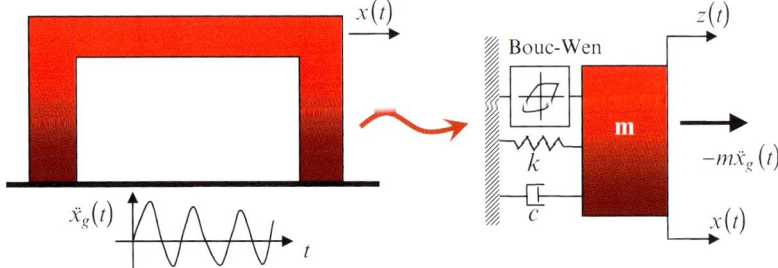

Nonlinear System Identification: Particle-Based Methods, Fig. 5 One floor shear-frame structure with a Bouc-Wen-type material nonlinearity

(Particle Swarm Optimization) step in the process, often requiring several loops that might significantly delay what is already a relatively lengthy estimation procedure. More recently, Yu et al. (2010) have suggested the use of simply the mutation operator for an adaptive mutation PF. The scheme proposed by Yu et al. quite resembles the earlier-mentioned notion of adding artificial process noise to the state vector.

Application: Joint State and Parameter Estimation for a Nonlinear Shear-Frame Structure

The approaches outlined above are tested on the real-time system identification of a nonlinear single degree of freedom (sdof) system. The system is assumed to describe the hysteretic response of a one storey shear-type frame building. This type of structure is described by a single translational degree of freedom in the elastic case. In the present example, the structure exhibits material nonlinearity when subjected to a ground motion, simulating earthquake excitation. As noted in Fig. 5, this type of system is equivalent to a mass-spring-dashpot-hysteretic element system. The Bouc-Wen model is used in order to account for nonlinear hysteretic response, which involves an additional hysteretic degree of freedom, henceforth denoted as $z(t)$.

Using the Bouc-Wen model, it is quite straightforward to compile the system's governing equations of motion. In fact the Bouc-Wen enables an elegant and concise formulation of the nonlinear phenomenon of hysteresis. For reasons of simplicity, dependence on time is not explicitly denoted in the expressions that follow:

$$m\ddot{x} + c\dot{x} + \alpha k x + (1-\alpha)kz = -m\ddot{x}_g$$
$$\dot{z} = \dot{x} - \beta|\dot{u}||z|^{n-1}z - \gamma\dot{u}|z|^n \quad (28)$$

Additionally, we assume that we have measurements of the system's absolute acceleration, via the use of appropriate accelerometer sensors. Since the aim is the joint state and parameter identification, as the nonlinear dof $z(t)$ cannot be measured, the system is cast into the following state-space form:

$$\dot{\mathbf{x}}(t) = \begin{bmatrix} \dot{x}(t) \\ -\dfrac{c}{m}\dot{x}(t) - \alpha\dfrac{k}{m}x(t) - (1-\alpha)\dfrac{k}{m}z(t) - \ddot{x}_g(t) \\ \dot{x}(t) - \beta|\dot{u}(t)||z(t)|^{n-1}z(t) - \gamma\dot{u}(t)|z(t)|^n \\ 0 \\ 0 \\ 0 \\ 0 \\ 0 \end{bmatrix} + \mathbf{v}(t) \quad (29)$$

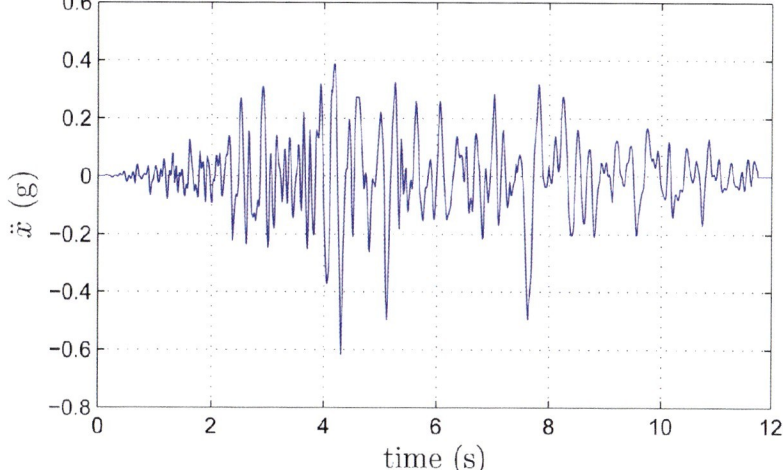

Nonlinear System Identification: Particle-Based Methods, Fig. 6 The input ground acceleration: Northridge earthquake

where $\boldsymbol{v} \in \mathbb{R}^8$ is the process noise vector, and $\tilde{\mathbf{x}} \in \mathbb{R}^8$ is the augmented vector that includes the displacement x, velocity $\dot{x}$, hysteretic dof z, as well as the stiffness k, damping c, and Bouc-Wen parameter constants α, β, γ. Hence, the parameter vector $\boldsymbol{\theta} \in \mathbb{R}^5$ is defined as $\boldsymbol{\theta} = [k \ c \ \alpha \ \beta \ \gamma \ n]^T$ and

$$\tilde{\mathbf{x}} = [x \ \dot{x} \ z \ k \ c \ \beta \ \gamma \ n]^T \quad (30)$$

The state-space (process) equation is complimented by the accompanying measurement (observation) equation:

$$y(t) = \ddot{x}_{tot}(t) = -\frac{c}{m}\dot{x}(t) - \alpha\frac{k}{m}x(t) \\ - (1-\alpha)\frac{k}{m}z(t) + \boldsymbol{\eta}(t) \quad (31)$$

where $\boldsymbol{\eta} \in \mathbb{R}$ is the measurement noise vector. The process equation can be brought into a discrete form by implementing a simple integration scheme. In order to keep the process online, a simple Euler scheme is implemented herein and proves to be sufficient for our purposes.

The actual response of the system is numerically generated using $k = 9$, $m = 1, c = .25, \alpha = 0.2, \beta = 2, \gamma = 1$ & $n = 2$ and the implemented ground motion is a scaled record of the Northridge earthquake (1994) shown in Fig. 6. Furthermore, the discretization is performed using a sampling frequency of 100 Hz which is also the sampling frequency of the input ground motion. The analysis is performed for a total of 20 s.

The particle-based algorithms outlined in the previous sections are implemented, namely, the UKF, the PF, and the MPF algorithm. The UKF employs 21 particles ($=2^*10 + 1$), the PF requires 5,000 particles for achieving admissible accuracy, and the MPF is run using 800 particles. The tuning parameters of the mutation operator for the PF are selected as $p_e = 0.2, p_m = 0.1$.

A Gaussian, process noise level of 0.1 % RMS noise-to-signal ratio is assumed, while the observation noise corresponds to approximately 10 % RMS. No process noise is added for the time-invariant parameter components. The addition of some minor noise could improve parameter estimation for the PF; however, it might also lead to instabilities and non-converging behavior. The observation noise is chosen so as to reflect a realistic instrumentation noise level, whereas the process noise is kept to a low level indicating confidence that the observed system can be described by this type of a model formulation.

Figure 7 illustrates the estimated state evolution for the three filters. It is already obvious that the standard particle filter (PF) underperforms

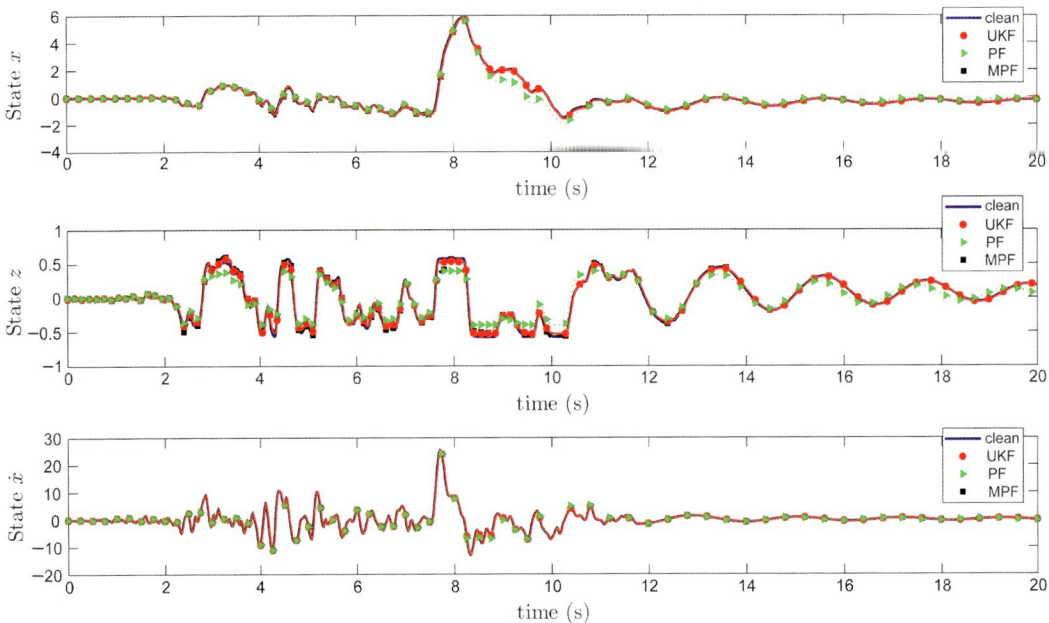

Nonlinear System Identification: Particle-Based Methods, Fig. 7 Estimated versus actual state evolution

and is not able to accurately track the system's states with a relative error of ∼20 % for the displacement estimate. The accuracy of the other two filters is almost identical and very close to the actual time history, with a relative error of ∼3.5 % for the UKF and ∼3 % for the MPF. As verified by the parameter estimation results, summarized in Figs. 8 and 9, the PF fails to accurately identify the true parameter values, as a result of the sample impoverishment problem outlined in section "The Problem of Sample Impoverishment." The use of the mutation operator for the particle filter with mutation (MPF) on the other hand alleviates this problem. The latter is schematically presented in Figs. 10 and 11 where it is obvious that, for the PF, the particle group eventually degenerates to a single parameter set for the time-invariant parameters. The use of the mutation operator for the MPF succeeds in maintaining the diversity of the population with a small scatter around the finally identified value, which prevents the algorithm from reaching a premature convergence.

From a computational cost perspective, in this quite simple sdof problem, all of the suggested methods can be implemented in real time, i.e., on the fly, as data is acquired. The computational time required for the whole analysis on a 4 core CPU is of the order of 1.0 s for the UKF, 5.7 s for the PF (5,000 particles), and 1.5 s for the MPF. The required time would of course increase for a higher dimensionality; nonetheless, one of the benefits of particle-based regimes is the fact that they can be implemented in parallel, as the particle evaluations are mutually independent.

Summary

This reference entry describes the use of particle-based methods for joint state and parameter identification of a structural system for SHM purposes. This class of methods is adopted in order to tackle the difficulties arising due to the nonlinear nature of the physical system and the uncertainty related to our knowledge of the system characteristics. The workings of each method are described, and the advantages, limitations, and enhancements of the presented approaches

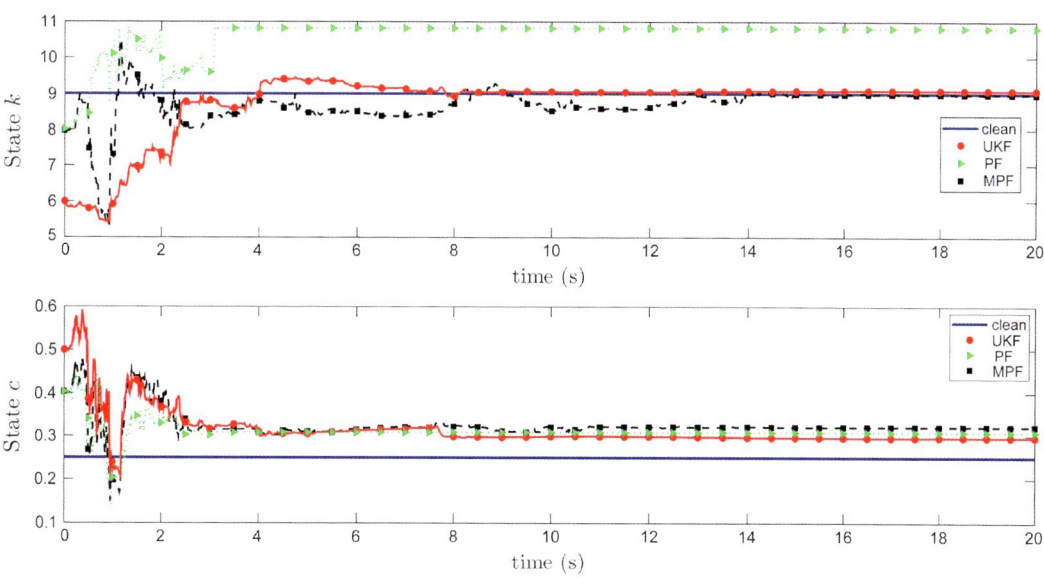

Nonlinear System Identification: Particle-Based Methods, Fig. 8 Estimated versus Actual Material Parameter evolution

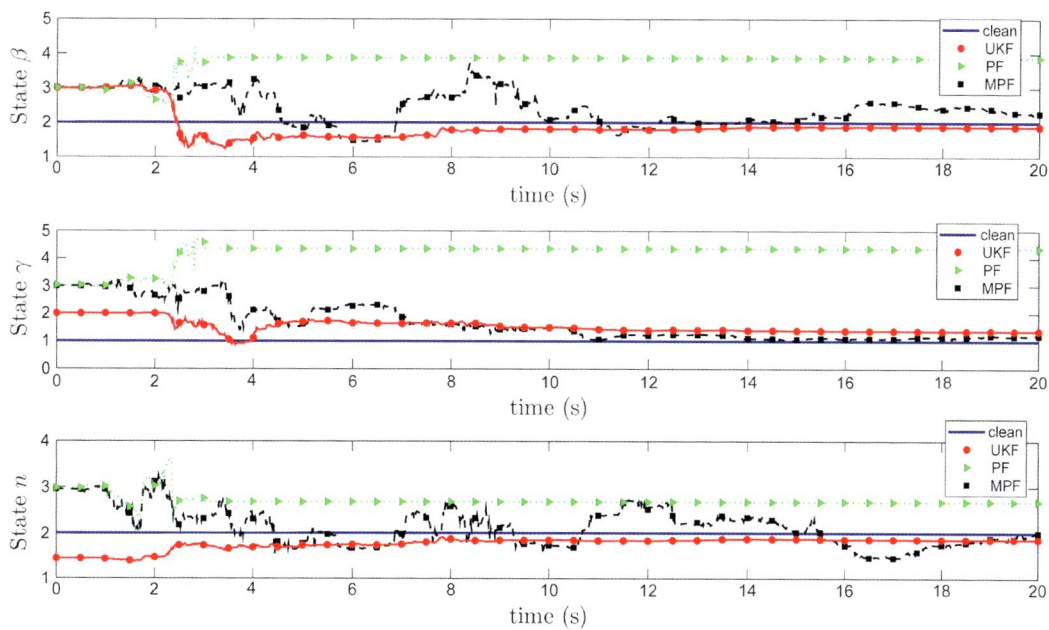

Nonlinear System Identification: Particle-Based Methods, Fig. 9 Estimated versus actual Bouc-Wen parameter evolution

are presented and discussed. As demonstrated, the use of particle-based techniques enables the real-time tracking of state evolution and the accurate identification of unknown system parameters in a robust and reliable manner. The benefits would be even more pronounced if a parallel processing regime were adopted, drastically cutting down the required computational time.

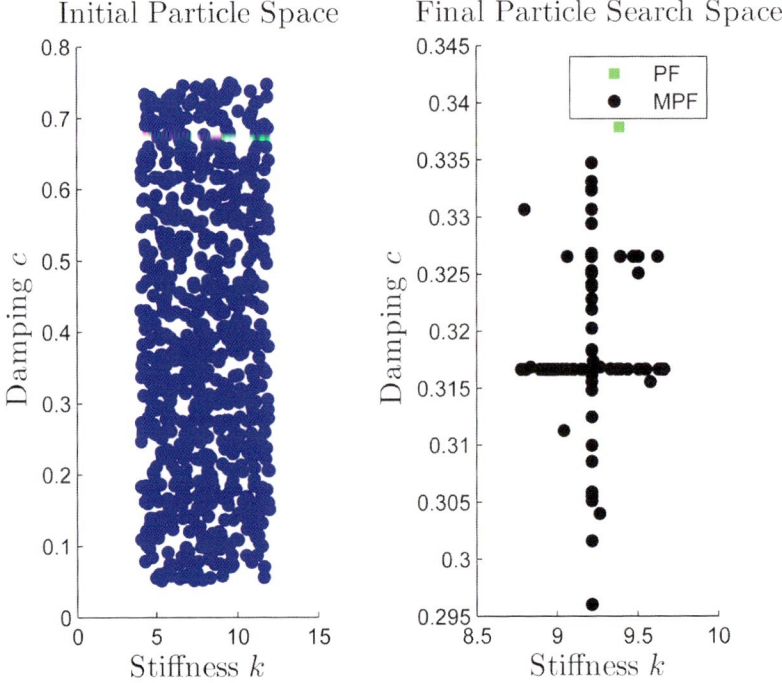

Nonlinear System Identification: Particle-Based Methods, Fig. 10 Initial versus final particle space for the material parameters

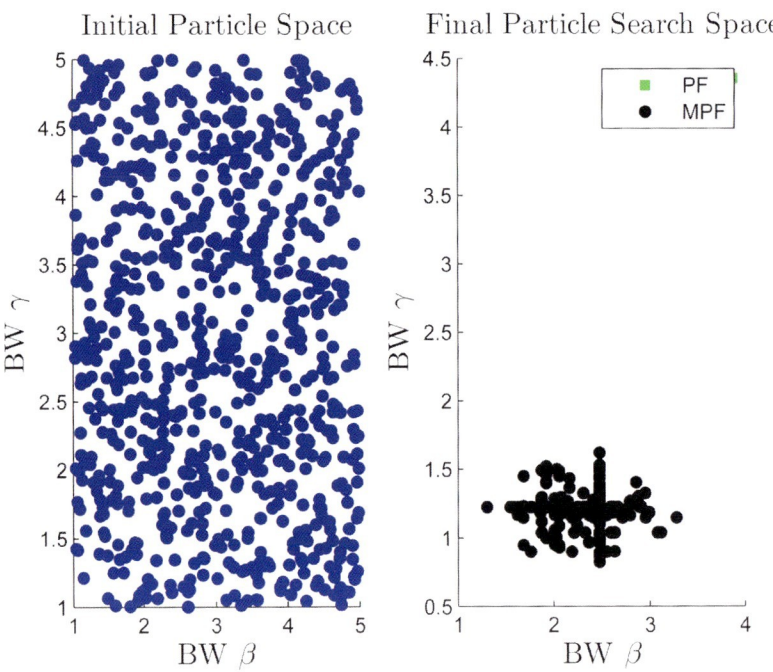

Nonlinear System Identification: Particle-Based Methods, Fig. 11 Initial versus final particle space for the BW parameters

Cross-References

▶ Advances in Online Structural Identification
▶ Parametric Nonstationary Random Vibration Modeling with SHM Applications
▶ System and Damage Identification of Civil Structures
▶ Vibration-Based Damage Identification: The Z24 Bridge Benchmark

References

Akhtar S, Ahmad AR, Abdel-Rahman EM, Naqvi T (2011) A PSO accelerated immune particle filter for dynamic state estimation. In: Canadian Conference on Computer and Robot Vision, St. John's, Newfoundland, 25–27 May 2011. pp 72–79

Arulampalam S, Maskell S, Gordon N, Clapp T (2002) A tutorial on particle filters for on-line non-linear/non-Gaussian Bayesian tracking. IEEE T Signal Process 50(2):174–188

Bergman N (1999) Recursive Bayesian estimation: navigation and tracking applications. PhD thesis, Linkoping University

Bergman N, Doucet A, Gordon N (2001) Optimal estimation and Cramer Rao bounds for partial non-Gaussian state space models. Ann I Stat Math 53(1):97–112

Chatzi EN, Smyth AW (2009) The unscented Kalman filter and particle filter methods for nonlinear structural system identification with non-collocated heterogeneous sensing. Struct Contr Health Monit 16–1:99–123

Chatzi EN, Smyth AW (2013) Particle filter scheme with mutation for the estimation of time-invariant parameters in structural health monitoring applications. J Struct Contr Health Monit 20:1081–1095

Chatzi EN, Smyth AW, Masri SF (2010) Experimental application of on-line parametric identification for nonlinear hysteretic systems with model uncertainty. J Struct Safety 32(5):326–337

Chen T, Morris J, Martin E (2005) Particle filters for state and parameter estimation in batch processes. J Process Contr 15(6):665–673

Ching J, Beck JL, Porter KA, Shaikhutdinov R (2006) Bayesian state estimation method for nonlinear systems and its application to recorded seismic response. J Eng Mech 132(4):396–410

Corigliano A, Mariani S (2004) Parameter identification in explicit structural dynamics: performance of the extended Kalman filter. Comput Method Appl Mech Eng 193(36–38):3807–3835

Doucet A, De Freitas JFG, Gordon NJ (eds) (2001) Sequential Monte Carlo methods in practice, Springer series in statistics for engineering and information science. Springer, New York

Eftekhar AS, Ghisi A, Mariani S (2012) Parallelized sigma-point Kalman filtering for structural dynamics. Comput Struct 92–93:193–205

Fraraccio G, Brügger A, Betti R (2008) Identification and damage detection in structures subjected to base excitation. Exp Mech 48(4):521–528

Juang JN, Pappa RS (1985) An eigensystem realization algorithm for modal parameter identification and model reduction. J Guid Control Dynam 8(5):620–627

Juang JN, Phan MQ, Horta LG, Longman RW (1993) Identification of observer/Kalman filter Markov parameters: theory and experiments. J Guid Control Dynam 16(2):320–329

Julier SJ, Uhlmann JK (1997) A new extension of the Kalman filter to nonlinear systems. In: Proceedings of AeroSense: the 11th international symposium on aerospace/defense sensing, simulation and controls, Orlando

Kalman RE (1960) A new approach to linear filtering and prediction problems. Trans ASME J Basic Eng 82-(Series D):35–45

Kwok N, Fang G, Zhou W (2005) Evolutionary particle filter: re-sampling from the genetic algorithm perspective. In: IEEE/RSJ international conference on intelligent robots and systems (IROS 2005), Edmonton, pp 2935–2940

Mariani S, Corigliano A (2005) Impact induced composite delamination: state and parameter identification via joint and dual extended Kalman filters. Comput Method Appl Mech Eng 194(50–52):5242–5272

Nagarajaiah S, Li Z (2004) Time segmented least squares identification of base isolated buildings. Soil Dyn Earthq Eng J 24(8):577–586

Nasrellah HA, Manohar CS (2011) Particle filters for structural system identification using multiple test and sensor data: a combined computational and experimental study. J Struct Contr Health Monit 18(1):99–120

Park S, Hwang J, Rou K, Kim E (2007) A new particle filter inspired by biological evolution: genetic filter. Proc World Acad Sci Eng Technol 21:459

Rajamani MR, Rawlings JB (2009) Estimation of the disturbance structure from data using semidefinite programming and optimal weighting. Automatica 45:142–148

Ristic NGB, Arulampalam S (2004) Beyond the Kalman filter, particle filters for tracking applications. Artech House Publishers, Boston

Smyth AW, Masri SF, Kosmatopoulos EB, Chassiakos AG, Caughey TK (2002) Development of adaptive modeling techniques for non-linear hysteretic systems. Int J Non-Linear Mech 37(8):1435–1451

Wan E, Van Der Merwe R (2000) The unscented Kalman filter for nonlinear estimation. In: Adaptive systems for signal processing, communications, and control symposium 2000, AS-SPCC. The IEEE 2000, Lake Louise, pp 153–158

Wu M, Smyth AW (2007) Application of the unscented Kalman filter for real-time nonlinear structural system identification. J Struct Contr Health Monit 14(7): 971–990

Yang JN, Lin S, Huang H, Zhou L (2006) An adaptive extended Kalman filter for structural damage identification. J Struct Contr Health Monit 13(4):849–867

Yu JX, Tang YL, Liu WJ (2010) Adaptive mutation particle filter based on diversity guidance. In: 2010 international conference on machine learning and cybernetics (ICMLC), TBD Qingdao, China. vol 1, 11–14 July 2010, pp 369–374

Yun CB, Shinozuka M (1980) Identification of nonlinear structural dynamics systems. J Struct Mech 8(2):187–203

Zhang H, Foliente GC, Yang Y, Ma F (2002) Parameter identification of inelastic structures under dynamic loads. Earthq Eng Struct Dynam 31(5):1113–1130

Non-Poisson Impulse Processes

Radoslaw Iwankiewicz
Institute of Mechanics and Ocean Engineering, Hamburg University of Technology, Hamburg, Germany

Synonyms

Counting process; Equations for response statistical moments; Jump processes; Markov processes; Non-Poisson processes; Point process; Probability density; Random impulses; Random vibrations; Renewal processes

Introduction

Dynamic loads are adequately idealized as discontinuous stochastic processes, or random trains of loading events, in the following problems:

- Behavior of a vehicle traveling over a very rough road: impacts or shocks, due to sudden humps or holes
- Impact loads on booms of bucket-wheel excavators used in opencast mining engineering
- Moving loads on a bridge due to highway traffic
- Dynamic loading due to wind gusts
- Any kind of irregular train of impacts or shocks

As the occurrence (arrival) times of the loads are random, their mathematically sound characterization is in terms of stochastic point (random counting) processes. The most fundamental and the simplest of such processes is a Poisson process. However, it is known that the Poisson process is an adequate model of a random train of events if the events are rare. Otherwise, other models, for example, the renewal processes, are more suitable. The impulse process excitation (or the random train of impulses) may be driven by different processes. If it is driven by a Poisson process, the state vector of the dynamic system is a nondiffusive, so-called Poisson-driven, Markov process. Pertinent mathematical tools such as generalized Itô's differential rule or Kolmogorov–Feller integrodifferential equation may be used. However, if the impulse process is non-Poisson (i.e., it is driven by a counting process other than Poisson), the state vector of the dynamic system is not a Markov process, and no mathematical tools may be directly applied. Then the problem has to be converted into a Markov one. Herewith it is explained how it may be achieved with the aid of two exact methods. The first one is the method of augmentation of state vector by additional state variables, which are pure-jump stochastic processes. In this approach the augmented state vector becomes a nondiffusive, Poisson-driven, Markov process, and the differential equations for response moments may be derived. In the second method, the state vector of the dynamic system is augmented by Markov states of an auxiliary, pure-jump stochastic process. This approach allows to obtain the set of integrodifferential equations governing the joint probability density of the state vector and also the differential equations for response moments. Two example non-Poisson impulse processes are considered: an Erlang renewal impulse process and the process driven by two independent Poisson processes.

Stochastic Point Processes

Specification of a Random Counting Process

Let $N(t)$ denote a random counting process, or a random variable, specifying the number of events (or time points) in an interval $[0, t]$, i.e., excluding the event that possibly occurs at the time instant t. Consequently the sample paths of $N(t)$ are ever-increasing step (pure-jump) functions, which are left continuous with right limits. Strictly speaking, an additional assumption $\Pr\{N(0) = 0\} = 1$ should also be imposed. The occurrences of events are not, in general, assumed to be independent. The increment $dN(t)$ of the counting process during an infinitesimal time interval $[t, t + dt]$, denoted conventionally as $dN(t)$, is defined as $dN(t) = N(t + dt) - N(t)$.

The point process is *regular* or *orderly*, if the probability governing the counting measure satisfies the following condition:

$$\sum_{k>1} \Pr\{dN(t) = k\} = \mathrm{O}(dt^2), \quad (1)$$

which means that in the infinitesimal time interval, there can only occur, with nonzero probability, one event or no event.

Let the disjoint infinitesimal time intervals $[t_i, t_i + dt_i]$, $i = 1, 2, \ldots, n$ be chosen from the interval $[0, t]$. Product density functions are defined as follows (Srinivasan 1974):

$$f_n(t_1, t_2, \ldots, t_n) dt_1 \ldots dt_n = E\left[\prod_{i=1}^{n} dN(t_i)\right]. \quad (2)$$

Equivalently, if the point process is regular, the nth degree product density function $f_n(t_1, \ldots, t_n)$ represents the probability that one event occurs in each of disjoint intervals $[t_i, t_i + dt_i]$, irrespective of other events in the interval $[0, t]$; thus,

$$f_n(t_1, t_2, \ldots, t_n) dt_1 \cdots dt_n = \Pr\left\{\bigwedge_{i=1}^{n} dN(t_i) = 1\right\}.$$
$$t_1 \neq t_2 \neq \cdots \neq t_n. \quad (3)$$

In what follows the attention is confined to regular or orderly point processes. In particular,

$$\Pr\{dN(t) = 1\} = f_1(t) dt, \quad (4)$$

where $f_1(t)$ is the product density of degree one. The regularity assumption Eq. 1 implies that

$$\Pr\{dN(t) = 0\} = 1 - f_1(t) dt + \mathrm{O}(dt^2), \quad (5)$$

$$E[dN(t)] = f_1(t) dt + \mathrm{O}(dt^2), \quad (6)$$

and

$$E[\{dN(t)\}^n] = f_1(t) dt + \mathrm{O}(dt^2), \quad (7)$$

for arbitrary n.

Product density of degree one $f_1(t)$ represents the mean rate of occurrence of events (mean arrival rate). It should be noted that $f_1(t)$ is not a probability density; its integration over the whole time interval $[0,t]$ yields an expected number of events in this interval, which usually is not equal to one:

$$\int_0^t f_1(\tau) d\tau = \int_0^t E[dN(\tau)] = E\left[\int_0^t dN(\tau)\right]$$
$$= E[N(t)]. \quad (8)$$

Product density of degree two, satisfying the relationship

$$f_2(t_1, t_2) dt_1 dt_2 = E[dN(t_1) dN(t_2)], \quad t_1 \neq t_2, \quad (9)$$

specifies the correlation between arrival rates at two different time instants t_1, t_2 (or the correlation of increments of the counting measure $N(t)$ on disjoint infinitesimal time intervals).

If k out of n time instants are set equal, i.e., $t_{j_1} = t_{j_2} = \cdots = t_{j_k}$, or k out of n infinitesimal intervals all overlap, the product density of degree n degenerates to $(n - k + 1)$th degree product density; thus,

$$E\left[\prod_{i=1}^{n} dN(t_i)\right]\Big|_{t_{j_1}=\cdots=t_{j_k}} = E\left[\prod_{i=1}^{n} dN(t_i)\{dN(t_{j_1})\}^k\right]$$
$$i \neq j_r, \quad r = 1, \ldots, k$$
$$= f_{n-k+1}(t_1, \ldots, t_n, t_{j_1}) dt_1 \cdots dt_n dt_{j_1}. \tag{10}$$

For example,

$$E[dN(t_1)dN(t_2)]|_{t_1=t_2} = E\left[\{dN(t_1)\}^2\right]$$
$$= f_1(t)dt. \tag{11}$$

Joint density function, defined as

$$\pi_n(t_1, t_2, \ldots, t_n) dt_1 dt_2 \cdots dt_n$$
$$= \Pr\left\{\bigwedge_{i=1}^{n} dN(t_i) = 1 \wedge N(t) = n\right\}, \tag{12}$$

specifies the probability that one event occurs in each of disjoint intervals $[t_i, t_i + dt_i]$ and there are no other events in the whole time interval $[0, t]$, i.e., that there are exactly $N(t) = n$ events.

The following relationships between the product density and the joint density functions hold (Srinivasan 1974):

$$f_k(t_1, \ldots, t_k) = \sum_{n=k}^{\infty} \frac{1}{(n-k)!} \underbrace{\int_0^t \cdots \int_0^t}_{(n-k)\text{-fold}} \pi_n(t_1, \ldots, t_k, t_{k+1}, \ldots, t_n) dt_{k+1} \cdots dt_n \tag{13}$$

$$\pi_k(t_1, \ldots, t_k) = \sum_{n=k}^{\infty} \frac{(-1)^{n-k}}{(n-k)!} \underbrace{\int_0^t \cdots \int_0^t}_{(n-k)\text{-fold}} f_n(t_1, \ldots, t_k, t_{k+1}, \ldots, t_n) dt_{k+1} \cdots dt_n. \tag{14}$$

The probability that exactly n events occur in the time interval $[0, t]$ is evaluated as (Srinivasan 1974)

$$\Pr\{N(t) = n\} = \frac{1}{n!} \underbrace{\int_0^t \cdots \int_0^t}_{n\text{-fold}} \pi_n(t_1, t_2, \ldots, t_n) dt_1 dt_2 \ldots dt_n. \tag{15}$$

Moreover, the correlation functions of the nth degree are defined in terms of product densities as (Stratonovich 1963)

$$g_1(t) = f_1(t),$$
$$g_2(t_1, t_2) = f_2(t_1, t_2) - f_1(t_1) f_1(t_2),$$
$$g_3(t_1, t_2, t_3) = f_3(t_1, t_2, t_3) - 3\{f_1(t_1) f_2(t_2, t_3)\}_s$$
$$+ 2f_1(t_1) f_1(t_2) f_1(t_3), \tag{16}$$

where $\{\cdots\}_s$ denotes the symmetrizing operation, i.e., the arithmetic mean of all terms similar to the one in brackets and obtained by all possible permutations of t_1, t_2, t_3. For example,

$$\{f_1(t_1) f_2(t_2, t_3)\}_s = \frac{1}{3} (f_1(t_1) f_2(t_2, t_3) + f_1(t_2) f_2(t_1, t_3) + f_1(t_3) f_2(t_1, t_2)). \tag{17}$$

Poisson Process

Poisson process is a special case of a point process, whose increments $dN(t)$ defined on disjoint time intervals dt are independent. The nonhomogeneous Poisson process is completely characterized by its first-order product density:

$$f_1(t) = \nu(t), \tag{18}$$

which is the *intensity* of the Poisson process.

Higher-order correlation functions are equal to zero:

$$g_n(t_1, \ldots, t_n) = 0 \quad (19)$$

for $n > 1$.

Due to the independence of events, Eq. 2 becomes

$$f_n(t_1, \ldots, t_n) = \prod_{i=1}^{n} v(t_i). \quad (20)$$

Substitution of Eq. 20 into Eq. 14 and into Eq. 15 yields, respectively (Snyder and Miller 1991),

$$\pi_n(t_1, \ldots, t_n) = \prod_{i=1}^{n} v(t_i) \exp\left(-\int_0^t v(\tau) d\tau\right), \quad (21)$$

$$\Pr\{N(t) = n\} = \frac{1}{n!} \left(\int_0^t v(\tau) d\tau\right)^n \exp\left(-\int_0^t v(\tau) d\tau\right). \quad (22)$$

For a homogeneous Poisson process ($v(t) = v = $ const.), the following expressions are obtained:

$$f_n(t_1, \ldots, t_n) = v^n, \quad (23)$$

$$\pi_n(t_1, \ldots, t_n) = v^n \exp(-vt), \quad (24)$$

$$\Pr\{N(t) = n\} = \frac{(vt)^n}{n!} \exp(-vt). \quad (25)$$

Renewal Processes

The renewal process can be defined as a sequence of random time points $t_1, t_2, \ldots, t_n$ on the positive real line, such that

$$t_i - t_{i-1} = T_i, \quad i = 2, 3, \ldots, t_1 = T_1, \quad (26)$$

where the time intervals $\{T_i, i = 2, 3, \ldots\}$ between the successive points, called *inter-arrival times*, are the positive, independent, and identically distributed random variables. The point process is called an *ordinary renewal process* if the time T_1 measured from the origin to the first event has the same distribution as other time intervals T_i. This means that the time origin is placed at the instant of 0th, or initial, event which is not counted. If T_1 has another distribution than other time intervals T_i, the point process is called a *general* or *delayed renewal process*. In that case the time origin is placed arbitrarily.

An ordinary renewal process can be defined equivalently as the sequence of positive, independent, and identically distributed random variables $\{T_i, i = 1, 2, \ldots\}$.

Consider an interval $[0, t]$ of the time axis. An *ordinary renewal density* $h_o(t)$ (Cox 1962; Cox and Isham 1980) represents the probability that a random point (not necessarily the first) occurs in $[t, t + dt]$, given that a random point occurs at the origin. A *modified renewal density* $h_m(t)$ (Cox 1962) represents the probability that a random point (not necessarily the first) occurs in $[t, t + dt]$, with arbitrarily chosen time origin. A modified renewal density is the first-order product density of the renewal point process:

$$h_m(t) dt = \Pr\{dN(t) = 1\} = f_1(t) dt. \quad (27)$$

If this probability is irrespective of the position of the interval $[t, t + dt]$ on the time axis, the renewal process is stationary.

Product densities of higher degrees of a renewal process appear to split into a product form or get factorized (Srinivasan 1974):

$$f_n(t_1, \ldots, t_n) dt_1 \cdots dt_n = E[dN(t_1) \cdots dN(t_n)]$$
$$= h_m(t_1) h_o(t_2 - t_1) h_o(t_3 - t_2) \cdots$$
$$h_o(t_n - t_{n-1}) dt_1 dt_2 \cdots dt_n,$$
$$(t_1 < t_2 < \ldots < t_n), \quad (28)$$

hence

$$f_n(t_1, \ldots, t_n) = h_m(t_1) h_o(t_2 - t_1) h_o(t_3 - t_2) \cdots h_o(t_n - t_{n-1})$$
$$(t_1 < t_2 < \cdots < t_n). \quad (29)$$

Let the probability density of the random variable X_1 be denoted as $g_1(t)$ and the probability density of each of the variables $\{T_i, i = 2, 3, \ldots\}$ as $g(t)$. It can be shown that the renewal densities $h_m(t)$ and $h_o(t)$ satisfy, respectively, inhomogeneous Volterra integral equations of the second kind, called *renewal equations*. These equations are derived (Srinivasan 1974) by considering the fact that the occurrence of the point in $[t, t+dt]$ is due to two mutually exclusive events: either it is the first point or it is the subsequent point. If it is the first point, the probability of its occurrence is just $g_1(t)dt$ (in the case of a delayed renewal process) or $g(t)dt$ (in the case of an ordinary renewal process). If it is the subsequent point, the preceding one has occurred at an arbitrary $t - u, u \in [0, t]$, u being the time interval between those two points. This leads to the following integral equations:

$$h_m(t) = g_1(t) + \int_0^t h_m(t-u)g(u)du, \quad (30)$$

$$h_o(t) = g(t) + \int_0^t h_o(t-u)g(u)du. \quad (31)$$

The renewal densities can be evaluated by taking the Laplace transforms of the equations Eqs. 30 and 31, which finally yields (Cox 1962; Cox and Isham 1980)

$$h_m(t) = \mathcal{L}^{-1}\left\{\frac{g_1^*(s)}{1-g^*(s)}\right\}, \quad (32)$$

$$h_o(t) = \mathcal{L}^{-1}\left\{\frac{g^*(s)}{1-g^*(s)}\right\}, \quad (33)$$

where $\mathcal{L}^{-1}\{\ldots\}$ denotes an inverse Laplace transform.

A class of the renewal processes which is important in applications are Erlang processes, where the time intervals between events have gamma (or Pearson type III) probability distribution with the density function

$$g(t) = v^k t^{k-1} \exp(-vt)/(k-1)!, \quad t > 0, \quad (34)$$

where $k = 1, 2, 3, \ldots$. A homogeneous Poisson process is a special case of such a process, specified by letting $k = 1$, in which case the time intervals have the negative exponential distribution characterized by the density function:

$$g(t) = v \exp(-vt), \quad t > 0. \quad (35)$$

An important property of the gamma distribution with density function given by Eq. 34 is that it is the distribution of the sum of k independent random variables, each of whose distribution is negative exponential with parameter v. Hence, the events driven by an Erlang process with parameter k can be regarded as every kth event of the generating Poisson process with the mean arrival rate v.

The renewal densities of the Erlang process are

$$h_o(t) = \frac{v}{2}(1 - \exp(-2vt)), \quad k = 2, \quad (36)$$

$$h_o(t) = \frac{v}{3}\left[1 - \left(\sqrt{3}\sin\frac{\sqrt{3}}{2}vt + \cos\frac{\sqrt{3}}{2}vt\right)\exp\left(-\frac{3}{2}vt\right)\right], \quad k = 3, \quad (37)$$

$$h_o(t) = \frac{v}{4}\left[1 - 2\sin(vt)\exp(-vt) - \exp(-2vt)\right], \quad k = 4. \quad (38)$$

It is interesting to note that, although the Erlang events are every kth Poisson events, the renewal densities, which are the mean arrival rates of Erlang events, only asymptotically (as $t \to \infty$) tend to v/k.

Random Pulse Trains Driven by Different Stochastic Point Processes

Random Trains of Overlapping Pulses: Filtered Stochastic Point Processes

General Case
A filtered stochastic point process $\{X(t), t \in [0, t]\}$ is defined as

$$X(t) = \sum_{i=1}^{N(t)} s(t, t_i, \mathbf{P}_i), \quad (39)$$

where $\{N(t), t \in [0, \infty]\}$ is a general counting process and $\mathbf{P}_i$ is a vector random variable

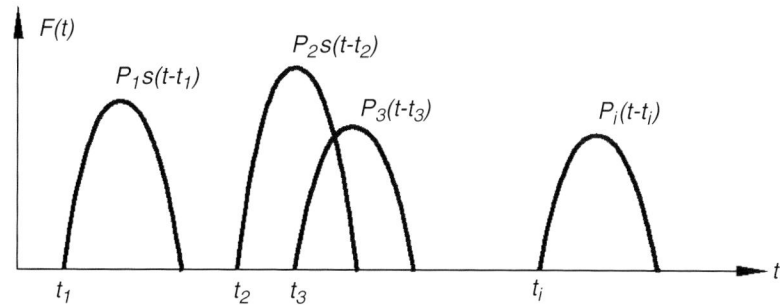

Non-Poisson Impulse Processes, Fig. 1 Train of overlapping pulses

attributed to a random point t_i. In a general case, the random variables which are the components of the vector $\mathbf{P}_i$ do not have to be independent, they can be correlated, and they can be characterized by different probability distributions, nor have the vector random variables $\mathbf{P}_i$ attributed to different points to be mutually independent or identically distributed. The only assumption made at present is that these random variables are statistically independent of the counting process $N(t)$. The nonrandom function $s\,(t, t_i, \cdot)$, called the *filter function*, represents the effect at the time t of an event occurring at the random instant t_i, the event being characterized by a vector random variable $\mathbf{P}_i$. For causality reasons, it is assumed that $s\,(t, \tau, \mathbf{P}_i) = 0$ for $\tau > t$. Hence, the process of Eq. 39 represents the cumulative effect of a train of point events occurring at random instants t_i belonging to the interval [0, t], described by a general stochastic point process. The process $X(t)$, $t \in [0, \infty]\}$ defined by the formula Eq. 39 can be interpreted as a random train of general pulses $F(t)$ (Fig. 1), or signals, with origins at the random times t_i; $s\,(t, t_i, \mathbf{P}_i)$ being the *pulse shape function*.

After the division of the interval [0, t] onto disjoint, contiguous subintervals, $X(t)$ can be written down as the Riemann–Stieltjes sum. The limit, in the mean-square sense, of the sequence of such sums, is the mean-square Riemann–Stieltjes integral with respect to the counting process $N(t)$ or the stochastic integral:

$$X(t) = \int_0^t s(t, \tau, \mathbf{P}(\tau))\,dN\,(\tau), \quad (40)$$

where $\mathbf{P}(\tau)$ is the vector random variable assigned to the point occurring in the interval $[\tau, \tau + d\tau]$.

The expected value of the process $X(t)$ is obtained just by averaging the expression Eq. 40, which yields (Iwankiewicz 1995)

$$E[X(t)] = \int_0^t E[s(t, \tau, \mathbf{P}(\tau))]f_1(\tau)d\tau$$

$$= \int_0^t \int_{\mathcal{P}_\tau} s(t, \tau, \mathbf{p})f_\mathbf{P}(\mathbf{p}, \tau)f_1(\tau)d\mathbf{p}d\tau, \quad (41)$$

where $f_\mathbf{P}\,(\mathbf{p},\tau)$ is the joint probability density of the vector random variable $\mathbf{P}\,(\tau)$, which may be time variant, and $\mathcal{P}_\tau$ is the sample space of this vector random variable.

Subsequent moments of the process $X\,(t)$ are evaluated by averaging of the pertinent multifold integrals obtained based on Eq. 40.

For example, the second-order moment (the mean square) is obtained as

$$E\left[X^2(t)\right] = \int_0^t \int_0^t E[s(t,\tau_1, \mathbf{P}(\tau_1))s(t,\tau_2, \mathbf{P}(\tau_2))]$$
$$E\left[dN\,(\tau_1)dN\,(\tau_2)\right]. \quad (42)$$

In order to evaluate this integral, the degeneracy property of the second-degree product density must be taken into account, which takes place within the integration domain, for $\tau_1 = \tau_2$. This yields

$$E[X^2(t)] = \int_0^t E\left[s^2(t,\tau,\mathbf{P}_\tau)\right] f_1(\tau) d\tau$$

$$+ \int_0^t \int_0^t E\left[s(t,\tau_1,\mathbf{P}_{\tau_1}) s(t,\tau_2,\mathbf{P}_{\tau_2})\right] f_2(\tau_1,\tau_2) d\tau_1 d\tau_2. \quad (43)$$

$$E[X(t_1)X(t_2)] = \int_0^{\min(t_1,t_2)} E[s(t_1,\tau,\mathbf{P}(\tau))s(t_2,\tau,\mathbf{P}(\tau))] f_1(\tau) d\tau$$

$$+ \int_0^{t_1}\int_0^{t_2} E[s(t_1,\tau_1,\mathbf{P}(\tau_1))s(t_2,\tau_2,\mathbf{P}(\tau_2))] f_2(\tau_1,\tau_2) d\tau_1 d\tau_2. \quad (44)$$

Likewise the correlation function of the process $X(t)$ is obtained as

The general expression for the nth order moment is

$$E[X^n(t)] = \underbrace{\int_0^t \cdots \int_0^t}_{n-\text{fold}} E\left[\prod_{k=1}^n s(t,\tau_k,\mathbf{P}(\tau_k))\right] E\left[\prod_{k=1}^n dN(\tau_k)\right]. \quad (45)$$

Of course, as $\tau_1 \neq \tau_2 \neq \ldots \neq \tau_n$, then

$$E\left[\prod_{k=1}^n dN(\tau_k)\right] = f_n(\tau_1,\ldots,\tau_n) d\tau_1 \cdots d\tau_n. \quad (46)$$

In the multidimensional integration domain, any possible equations of the arguments τ_k take place. Therefore, in order to evaluate the integral Eq. 45, all possible degeneracies of nth degree product density $f_n(\tau_1,\ldots,\tau_n)$ must be taken into account. Moreover, in general case, the integration is performed with respect to the joint probability density of n vector random variables $\mathbf{P}_k$, $k = 1, 2, \ldots, n$.

In particular the expression for the third-order moment is obtained as

$$E[X^3(t)] = \int_0^t E\left[s^3(t,\tau,\mathbf{P}_\tau)\right] f_1(\tau) d\tau$$

$$+ 3\int_0^t\int_0^t E\left[s^2(t,\tau_1,\mathbf{P}(\tau_1))s(t,\tau_2,\mathbf{P}(\tau_2))\right] f_2(\tau_1,\tau_2) d\tau_1 d\tau_2 \quad (47)$$

$$+ \int_0^t\int_0^t\int_0^t E\left[s(t,\tau_1,\mathbf{P}(\tau_1))s(t,\tau_2,\mathbf{P}(\tau_2))s(t,\tau_3,\mathbf{P}(\tau_3))\right] f_3(\tau_1,\tau_2,\tau_3) d\tau_1 d\tau_2 d\tau_3.$$

However, in general, the evaluation of the above integrals becomes cumbersome, especially in the case of higher-order moments. Then, in the case of a Poisson process, it is much easier to handle the cumulants, which can be obtained directly from the log-characteristic function, called also a cumulant-generating function. In the case of a filtered renewal process, the recursive expressions for the moments can be obtained from the integral equations governing the characteristic function.

Filtered Renewal Process
Consider a filtered process $\{X(t), t \in [0, \infty]\}$, driven by an ordinary renewal counting process $\{X(t), t \in [0, \infty]\}$, in the form of

$$X(t) = \sum_{j=1}^{N(t)} s(t - t_j, P_j), \quad (48)$$

where the filter function is assumed to be causal, i.e., $s(\tau, P_j) = 0$ for $\tau < 0$. The random variables P_j are assumed to be independent and to have identical probability distributions characterized by the common density function $f_P(p)$. The probability distributions of the inter-arrival times are characterized by the probability distribution function $G(t)$ and by the probability density function $g(t)$.

The characteristic function $\Phi_X(\theta, t)$ of the filtered renewal process $X(t)$ defined by Eq. 48 appears to satisfy the inhomogeneous Volterra integral equation of the second kind. The following derivation is due to Takacs (1956).

The filtered process $X(t)$ given by Eq. 48 may be regarded as a sum of the first pulse occurring after first inter-arrival time T_1 and the filtered process with the origin shifted by T_i, i.e., $X(t - T_1)$; thus,

$$X(t) = s(t - T_1, P_1) + X(t - T_1). \quad (49)$$

Suppose that $T_1 = \tau$. In view of independence of random variables P_j, the conditional characteristic function, given that the first pulse occurs at the time $T_1 = \tau$, is expressed as

$$\Phi_X(\theta, t | T_1 = \tau) = E\left[\exp(i\theta X(t)) | T_1 = \tau\right]$$
$$= E\left[\exp(is(t - \tau, P_1)) \exp(i\theta X(t - \tau))\right]$$
$$= \Gamma(\theta, t - \tau) \Phi_X(\theta, t - \tau), \quad (50)$$

where

$$\Gamma(\theta, t - \tau) = E\left[\exp\{i\theta s(t - \tau, P)\}\right]$$
$$= \int_{-\infty}^{\infty} \exp[i\theta s(t - \tau, p)] f_P(p) dp \quad (51)$$

is the characteristic function of a single general pulse $s(t-\tau, P)$.

By unconditioning one obtains

$$\Phi_X(\theta, t) = \int_0^t \Phi_X(\theta, t - \tau) \Gamma(\theta, t - \tau) g(\tau) d\tau + C, \quad (52)$$

where the integration constant C is evaluated from the obvious condition for $\theta = 0$

$$\Phi_X(0, t) = 1 = \int_0^t g(\tau) d\tau + C, \quad (53)$$

thus $C = 1 - G(t)$.
Hence, the final result is

$$\Phi_X(\theta, t) = \int_0^t \Phi_X(\theta, t - \tau) \Gamma(\theta, t - \tau) g(\tau) d\tau + 1 - G(t). \quad (54)$$

After differentiating the Eq. 54 r times with respect to θ and after substituting $\theta = 0$, the following integral equation governing the moment $E[X^r(t)]$ is arrived at

$$E[X^r(t)] = \sum_{i=0}^{r} \binom{r}{i} \int_0^t \phi_{r-i}(t - \tau) E[X^i(t - \tau)] g(\tau) d\tau, \quad (55)$$

where

$$\phi_k(t - \tau) = E\left[s^k(t - \tau, P)\right]$$
$$= \int_{-\infty}^{\infty} s^k(t - \tau, p) f_P(p) dp. \quad (56)$$

Taking Laplace transforms of the both sides of Eq. 55, solving the obtained algebraic equation for the transform of the rth order moment, and taking the inverse transform, one obtains the following recursive expression (Takacs 1956):

$$E[X^r(t)] = \sum_{i=0}^{r-1} \binom{r}{i} \int_0^t \phi_{r-i}(t - \tau) E[X^i(t - \tau)] h_o(\tau) d\tau, \quad (57)$$

where $h_o(\tau)$ is the ordinary renewal density.

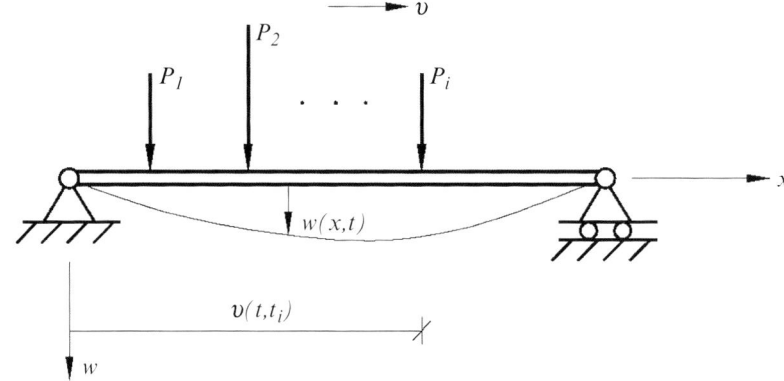

Non-Poisson Impulse Processes, Fig. 2 Beam under a random train of moving loads

The specific formulae for the mean value function $E[X(t)]$ and the mean-square function $E[X^2(t)]$ are obtained as

$$E[X(t)] = \int_0^t \phi_1(t-\tau)h_o(\tau)d\tau, \qquad (58)$$

$$E[X^2(t)] = \int_0^t \phi_2(t-\tau)h_o(\tau)d\tau$$

$$+ 2\int_0^t\int_\tau^t \phi_1(t-\tau)\phi_1(t-u)h_o(u-\tau)h_o(\tau)dud\tau, \qquad (59)$$

where

$$\phi_1(t-\tau) = E[s(t-\tau, P)], \qquad (60)$$

$$\phi_2(t-\tau) = E\left[s^2(t-\tau, P)\right]. \qquad (61)$$

The same result for $E[X^2(t)]$ as Eq. 59 is obtained if the expression Eq. 29 for the second-order product density for the ordinary renewal process is inserted into the general expression Eq. 43. In the integration domain $\tau_1 \in (0, t)$, $\tau_2 \in (0, t)$, it must be assumed

$$f_2(\tau_1, \tau_2) = \begin{cases} h_o(\tau_1)h_o(\tau_2 - \tau_1), & \tau_2 > \tau_1, \\ h_o(\tau_1)h_o(\tau_2 - \tau_1), & \tau_2 > \tau_1 \end{cases}. \qquad (62)$$

More results for the moments of the filtered renewal process (e.g., the response of a linear dynamic system to a renewal pulse train) may be found in Iwankiewicz (1995).

Example Problem: Dynamic Response of a Bridge (Beam) to a Random Train of Moving Loads

Consider the well known in structural engineering problem of the dynamic response of a highway bridge to the moving load due to the vehicular traffic. It is known that if the bridge has a long span, the coupling of the motion of the bridge and of the vehicle as well as the inertia of the vehicle may be neglected and the vehicles may be adequately idealized by point forces. The times of occurrence of vehicles at the bridge are random, and in the simplest case, all the vehicles may be assumed to travel with the same, constant velocity. The bridge, idealized as a beam, is then subjected to a random train of moving forces with random magnitudes P_j and at a given time instant t randomly located at the beam, as shown in Fig. 2.

The equation governing the transverse motion of the beam has the form

$$EI\frac{\partial^4 w(x,t)}{\partial x^4} + c\frac{\partial w(x,t)}{\partial t} + \mu\frac{\partial^2 w(x,t)}{\partial t^2}$$

$$= \sum_{j=1}^{\mathcal{N}(t)} P_j \delta(x - v(t - t_j)). \qquad (63)$$

Application of the normal mode approach

$$w(x,t) = \sum_{i=1}^n q_i(t)\phi_i(x), \qquad (64)$$

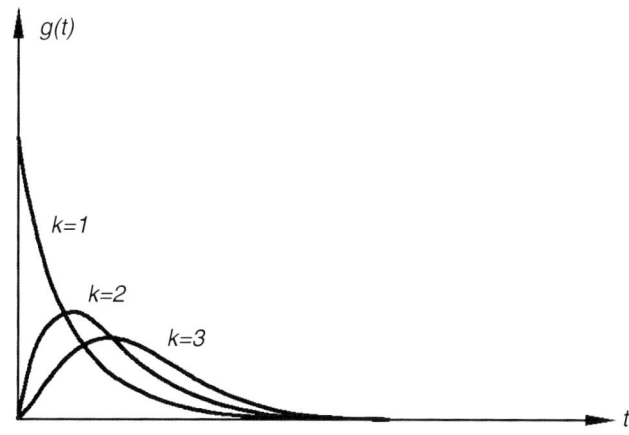

Non-Poisson Impulse Processes, **Fig. 3** Different gamma probability density plots

where $\varphi_i(x)$ are the normal modes, yields the equations

$$\ddot{q}_i(t) + 2\zeta_i\omega_i\dot{q}_i(t) + \omega_i^2 q_i(t)$$
$$= \beta_i \sum_{j=1}^{\mathcal{N}(t)} P_j \phi_i\left(v(t - t_j)\right), \ i = 1, 2, \ldots, n \quad (65)$$

where $\phi_i(v(t - t_j)) = s(t - t_j)$ plays the role of the pulse shape.

Thus, the problem is converted into the problem of the random train of general pulses. The underlying counting process $\mathcal{N}(t)$ may be any counting process. It is known from the highway traffic engineering that the inter-arrival times between the vehicles are adequately idealized as positive random variables with a unimodal probability density function. If the inter-arrival times are independent and identically distributed, the underlying counting process is a renewal process.

For example, if the renewal process is an Erlang process, the probability density function of the inter-arrival times is

$$g(t) = \frac{v^k t^{k-1}}{(k-1)!} \exp(-vt), \ t > 0. \quad (66)$$

Some example probability density plots are shown in Fig. 3.

Example Problem: Dynamic Response of Linear System to a Random Train of Impulses

A random train of impulses, or an impulse process excitation, $F(t)$ is shown in Fig. 4.

Vibrations of a linear oscillator under a random train of impulses are governed by

$$\ddot{Y}(t) + 2\zeta\omega\dot{Y}(t) + \omega^2 Y(t) = F(t)$$
$$= \sum_{i=1}^{\mathcal{N}(t)} P_i \delta(t - t_i), \quad (67)$$

where $\mathcal{N}(t)$ is a stochastic point (random counting) process, giving the random number of time points t_i in the time interval $[0, t)$, i.e., excluding the one that may occur at t. The impulse magnitudes P_i are independent random variables, identically distributed as a random variable P. The variables P_i are also statistically independent of the random times t_i or of the counting process $\mathcal{N}(t)$. The counting process may be, e.g., a Poisson or a renewal process.

From the impulse-momentum principle, it follows that

$$\dot{Y}(t_i)_+ = \dot{Y}(t_i)_- + P_i.$$

Hence, the velocity response process $\dot{Y}(t)$ changes by jumps; it is piecewise continuous.

Consequently the displacement response process $Y(t)$ is continuous, but it is only piecewise continuously differentiable.

Non-Poisson Impulse Processes, Fig. 4 Random train of impulses

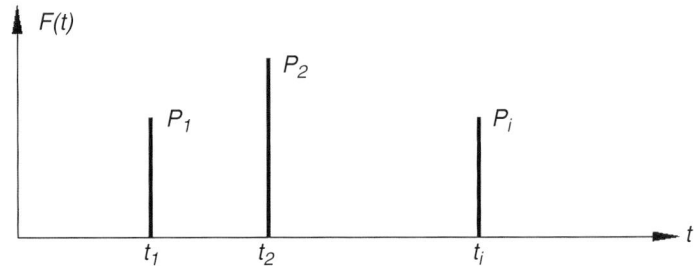

The jump change in the velocity may occur at any point; hence, the usual rules of calculus do not apply.

The usual notation of the differential equation of motion is **not mathematically meaningful**.

Stochastic counterparts of the usual differential equations are the **stochastic differential equations**:

$$dY(t) = \dot{Y}(t)\,dt, \quad d\dot{Y}(t) = -2\zeta\omega\dot{Y}(t)\,dt - \omega^2 Y(t)\,dt + P(t)d\mathcal{N}(t), \quad (68)$$

where $P(t)$ is the magnitude of the impulse which occurs in the time interval $[t, t + dt]$.

Nevertheless, the response of **linear dynamical systems** may be directly analyzed in **time domain**. The explicit expression for the response to a random train of impulses, based on the **linear superposition principle**, is obtained as

$$Y(t) = \sum_{i=1}^{\mathcal{N}(t)} P_i h(t-t_i) = \int_0^t P(\tau) h(t-\tau) d\mathcal{N}(\tau), \quad (69)$$

where $h(t) = \frac{1}{\omega_d}\exp(-\zeta\omega t)\sin\omega_d t$ is the **impulse response function** and $\omega_d = \omega\sqrt{1-\zeta^2}$. The integral with respect to the increments of the stochastic point process is the stochastic counterpart of the usual Duhamel convolution integral.

Statistical moments of the response process may be evaluated by averaging of the pertinent multifold integrals Eq. 45.

Response of Dynamic Systems to Non-Poisson Impulse Processes. State-Space Formulation

Itô's differential rule for nondiffusive, Poisson-driven Markov processes

The stochastic equations of motion of the dynamic system subjected to a Poisson impulse process excitation are

$$d\mathbf{Y}(t) = \mathbf{c}(\mathbf{Y}(t), t)\,dt + \mathbf{b}(\mathbf{Y}(t), t)P(t)d\mathcal{N}(t), \quad \mathbf{Y}(0) = y_0. \quad (70)$$

If the excitation process is statistically independent of the initial conditions, the vector process $\mathbf{Y}(t) = [Y_1(t), Y_2(t), \ldots, Y_n(t)]^T$ is a **nondiffusive, Poisson-driven, Markov process**.

If the function $V(\mathbf{Y}(t), t)$ is bounded for t and $\mathbf{Y}(t)$ finite and is once continuously differentiable with respect to all its arguments, the differential of a compound function $V(t, \mathbf{Y}(t))$ of the state variables $\mathbf{Y}(t)$ is expressed by the following **generalized Itô's differential rule** (Snyder and Miller 1991; Iwankiewicz and Nielsen 1999):

$$dV(t, \mathbf{Y}(t)) = \frac{\partial V(t, \mathbf{Y}(t))}{\partial t}dt + \sum_{i=1}^{n} \frac{\partial V(t, \mathbf{Y}(t))}{\partial Y_i} c_i(\mathbf{Y}(t), t)\,dt + [V(t, \mathbf{Y}(t) + \mathbf{b}(\mathbf{Y}(t), t, P(t))) - V(t, \mathbf{Y}(t))]d\mathcal{N}(t). \quad (71)$$

Differential equations governing the response joint statistical moments $\mu_{ij}(t) = E[Y_i(t)Y_j(t)]$, $\mu_{ijk}(t) = E[Y_i(t)Y_j(t)Y_k(t)]$, $\mu_{ijkl}(t) = E[Y_i(t)Y_j(t)Y_k(t)Y_l(t)]$, etc. are derived from this rule by assuming the function $V(t, \mathbf{Y}(t))$ in form of different products of the state variables, for example, $V(t, \mathbf{Y}(t)) = Y_i(t)Y_j(t)$, $V(t, \mathbf{Y}(t)) = Y_i(t)Y_j(t)Y_k(t)$, and $V(t, \mathbf{Y}(t)) = Y_i(t)Y_j(t)Y_k(t)Y_l(t)$.

Next, the expectation of both sides of the above differential rule must be taken and

$$E[dV(t, \mathbf{Y}(t))] = dE[V(t, \mathbf{Y}(t))].$$

The effective averaging is possible if the driving process is a Poisson process, because then the increment $dN(t) = N(t + dt) - N(t)$ is statistically independent of $V(t, \mathbf{Y}(t))$ and

$$E[V(t, \mathbf{Y}(t))dN(t)] = E[V(t, \mathbf{Y}(t))]E[dN(t)]$$
$$= E[V(t, \mathbf{Y}(t))]v(t)dt.$$

For a general Poisson-driven pulse problem, the equations for the mean values, the second-, third-, and fourth-order joint central moments of the response, are obtained as

$$\dot{\mu}_i(t) = E[c_i(\mathbf{Y}(t), t)] + v(t)E[P]E[b_i(\mathbf{Y}(t), t)] \quad (72)$$

$$\dot{\kappa}_{ij}(t) = 2\left\{E\left[Y_i^0\left(c_j^0(\mathbf{Y}^0(t), t) + v(t)b_j(\mathbf{Y}(t), t)P\right)\right]\right\}_s + v(t)E[P^2]E[b_i(\mathbf{Y}(t), t)b_j(\mathbf{Y}(t), t)] \quad (73)$$

$$\dot{\kappa}_{ijk}(t) = 3\left\{E\left[Y_i^0 Y_j^0\left(c_k^0(\mathbf{Y}^0(t), t) + v(t)b_k(\mathbf{Y}(t), t)P\right)\right]\right\}_s + 3v(t)E[P^2]\left\{E\left[Y_i^0 b_j(\mathbf{Y}(t), t)b_k(\mathbf{Y}(t), t)\right]\right\}_s + v(t)E[P^3]E[b_i(\mathbf{Y}(t), t)b_j(\mathbf{Y}(t), t)b_k(\mathbf{Y}(t), t)] \quad (74)$$

$$\begin{aligned}\dot{\kappa}_{ijkl}(t) &= 4\left\{E\left[Y_i^0 Y_j^0 Y_k^0\left(c_l^0(\mathbf{Y}^0(t), t) + v(t)b_l(\mathbf{Y}(t), t)P\right)\right]\right\}_s \\ &+ 6v(t)E[P^2]\left\{E\left[Y_i^0 Y_j^0 b_k(\mathbf{Y}(t), t)b_l(\mathbf{Y}(t), t)\right]\right\}_s \\ &+ 4v(t)E[P^3]\left\{E\left[Y_i^0 b_j(\mathbf{Y}(t), t)b_k(\mathbf{Y}(t), t)b_l(\mathbf{Y}(t), t)\right]\right\}_s \\ &+ v(t)E[P^4]E[b_i(\mathbf{Y}(t), t)b_j(\mathbf{Y}(t), t)b_k(\mathbf{Y}(t), t)b_l(\mathbf{Y}(t), t)]\end{aligned} \quad (75)$$

where $Y_i^0(t) = Y_i(t) - \mu_i(t)$ and $c_j^0(\mathbf{Y}^0(t), t) = c_j(\mathbf{Y}(t), t) - E[c_j(\mathbf{Y}(t), t)]$ denote the components of the zero-mean (centralized) state vector and drift vector, respectively; $E[P^r]$ denotes the rth moment of the random variable P, i.e.,

$$E[P^r] = \int_P p^r f_P(p)dp \; ; \text{ and } \{\ldots\}_s \text{ denotes the}$$

Stratonovich symmetrizing operation, e.g.,

$$\{Y_i Y_j c_k\}_s = \frac{1}{3}(Y_i Y_j c_k + Y_i Y_k c_j + Y_j Y_k c_i). \quad (76)$$

Equations for response moments of a linear system form always a closed set and can be directly solved numerically. Equations for response moments of a nonlinear system with polynomial nonlinearity form an infinite hierarchy. Then special closure approximations (truncation procedures) must be used. For other types of nonlinearity, the equations involve unknown expectations of nonlinear functions of state variables, and some tentative forms of the joint probability density must be used (Iwankiewicz et al. 1990).

Conversion of Non-Markov Impulse Problems to Markov Ones

Non-Markov Nature of the State Vector of the Dynamic System

A renewal impulse process is

$$F(t) = \sum_{i, R=1}^{R(t)} P_{i,R}\delta(t - t_{i,R}) \quad (77)$$

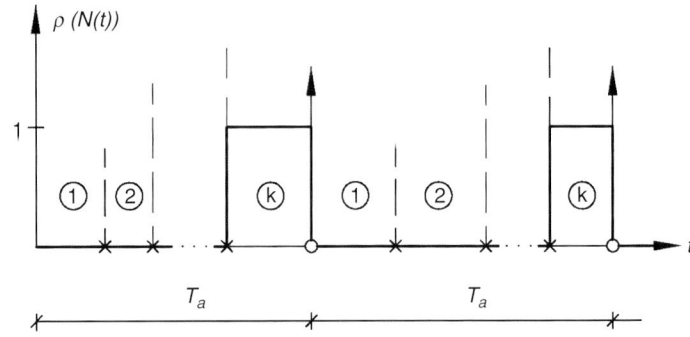

Non-Poisson Impulse Processes, Fig. 5 Train of Erlang impulses. ×: Poisson-driven points, ○: Erlang-driven points

where the occurrence times $t_{i,R}$ of impulses are driven by a renewal process. $R(t)$ and $P_{i,R}$ are statistically independent and identically distributed random magnitudes of the impulses.

The stochastic equations of motion in a general case of a nonlinear oscillator and a parametric excitation are

$$dY(t) = \dot{Y}(t)\,dt, d\dot{Y}(t)$$
$$= f(Y(t), \dot{Y}(t))\,dt + b(Y(t), \dot{Y}(t))\,P(t)\,dR(t), \quad (78)$$

where $f(Y(t), \dot{Y}(t))$ is a nonlinear function of the instantaneous values of $Y(t)$ and $\dot{Y}(t)$ which represents all restoring force together with damping terms of the equation of motion and $b(Y(t), \dot{Y}(t))$ is the parametric excitation term. As the increments $dR(t)$ of the renewal counting process $R(t)$ are not independent, the state vector $\mathbf{Y}(t) = [Y(t), \dot{Y}(t)]^T$ is **not a Markov process**.

Augmentation of a State Vector by Additional Variables

A basic idea of conversion of a non-Markov impulse problem to a Markov one is to replace the original train of impulses by an equivalent one for which the response process becomes a Markov process.

If the impulse process is driven by an Erlang renewal process, the following exact replacement is valid with probability 1 (Iwankiewicz and Nielsen 1999):

$$\sum_{i,R=1}^{R(t)} P_{i,R}\delta(t - t_{i,R}) = \sum_{i=1}^{N(t)} \rho(N(t_i))P_i\delta(t - t_i). \quad (79)$$

Based on the fact that the events of the Erlang renewal process are every second, or every third, or every fourth, etc. Poisson events, a zero-memory transformation $\rho(N(t))$ of the homogeneous Poisson process $N(t)$ is introduced which assumes values $\rho(N(t_i)) = 1$ for every kth Poisson event and $\rho(N(t_i)) = 0$ for all other Poisson events. The sample paths of $\rho(N(t))$ are assumed to be left continuous with right limits. The arrival times t_i are driven by a homogeneous Poisson process $N(t)$ with a mean arrival rate ν, and the impulse magnitudes P_i are compounded with $\rho(N(t_i))$. Thus, the actual impulse process is obtained by selecting, with the aid of an auxiliary stochastic variable $p(N(t))$, every kth impulse from the train driven by a Poisson process $N(t)$. The replacement Eq. 79 is illustrated in Fig. 5, where the dashed-line spikes represent these of the impulses driven by $N(t)$, whose magnitudes are multiplied by $\rho(N(t_i)) = 0$, hence are excluded, and the solid-line spikes represent the remaining $N(t)$-driven impulses, whose magnitudes are multiplied by $\rho(N(t_i)) = 1$; hence, these are the impulses driven by the underlying Erlang renewal process $R(t)$.

Consequently the increments of the Erlang renewal process can be expressed in terms of the Poisson counting process:

$$dR(t) = \rho(N(t))\,dN(t). \quad (80)$$

The transformation $\rho(N(t))$ is expressed in terms of the auxiliary variables $C_j(t)$ and $S_j(t)$ (Iwankiewicz and Nielsen 1999):

$$\rho(N(t)) = \rho(C_j(N(t)), S_j(N(t))) \quad (81)$$

where

$$C_j(N(t)) = \cos\left(\frac{j\pi N(t)}{k}\right), \quad S_j(N(t))$$
$$= \sin\left(\frac{j\pi N(t)}{k}\right) \quad (82)$$

hence

$$dR(t) = \rho\big(C_j(N(t)), S_j(N(t))\big)\, dN(t). \quad (83)$$

For example, if $k = 2$, $k = 3$, and $k = 4$, the required transformations of the Poisson counting process $N(t)$, such that $\rho(N(t)) = 1$ for every second, third, and fourth Poisson event and $\rho(N(t)) = 0$ for all other Poisson events, are, respectively,

$$\rho(N(t)) = \frac{1}{2}(1 - \cos(\pi N(t)))$$
$$= \frac{1}{2}\left(1 - (-1)^{N(t)}\right), \quad (84)$$

$$\rho(N(t)) = \frac{1}{3}\left(1 - \sqrt{3}\sin\left(\frac{2}{3}\pi N(t)\right) - \cos\left(\frac{2}{3}\pi N(t)\right)\right), \quad (85)$$

$$\rho(N(t)) = \frac{1}{4}\left(1 - 2\sin\left(\frac{1}{2}\pi N(t)\right) - \cos(\pi N(t))\right). \quad (86)$$

For all auxiliary variables $C_j(t)$ and $S_j(t)$, the stochastic differential equations are written down, which are all driven by the Poisson process $N(t)$.

It should be noted that as the increment $dN(t)$ is independent of $N(t)$, then $dN(t)$ is independent of $\rho(N(t))$.

The original state vector $\mathbf{Y}(t) = [Y(t), \dot{Y}(t)]^T$ is augmented by the auxiliary variables $C_j(t)$ and $S_j(t)$ to make up a vector $\mathbf{Z}(t)$:

$$d\mathbf{Z}(t) = \mathbf{c}(\mathbf{Z}(t), t)\, dt + \mathbf{b}(\mathbf{Z}(t), t,)P(t)\, dN(t), \quad \mathbf{Z}(0) = \mathbf{z}_0. \quad (87)$$

The augmented state vector $\mathbf{Z}(t)$ is driven by a Poisson process; hence, it is a **nondiffusive (Poisson-driven) Markov process**. Equations for moments may be derived from the Itô's differential rule for Poisson-driven Markov processes.

Now the renewal, impulse process excitation is considered where the inter-arrival times T_a are the sum of two independent, negative exponential distributed variables T_r and T_d, with probability density functions given, respectively, by (for t > 0)

$$g_{T_r}(t) = \nu\exp(-\nu t), \quad g_{T_d}(t)$$
$$= \mu\exp(-\mu t). \quad (88)$$

The following replacement holds with probability 1 for the renewal driven train of impulses (Iwankiewicz 2003):

$$\sum_{i,R=1}^{R(t)} P_{i,R}\delta(t - t_{i,R}) = \sum_{i=1}^{N_\mu(t)} Z(t_i)P_i\delta(t - t_i), \quad (89)$$

where the arrival times $t_{i,R}$ are driven by the underlying renewal process $R(t)$, the arrival times t_i are driven by a homogeneous Poisson process $N_\mu(t)$ with a mean arrival rate μ, and $Z(t_i)$ is a value at t_{i-} of the zero–one auxiliary stochastic jump process $Z(t)$ governed by (Iwankiewicz 2002)

$$dZ(t) = (1 - Z)\, dN_\nu(t) - Z\, dN_\mu(t). \quad (90)$$

The processes $N_\nu(t)$ and $N_\mu(t)$ are independent Poisson processes, with parameters ν and μ, respectively, and $dN_\nu(t)$ and $dN_\mu(t)$ are the increments during the infinitesimal time interval $[t, t+dt)$. The process $Z(t)$ is zero–one valued, and $dZ(t) = Z(t+dt) - Z(t)$ denotes the jump increment, which may be the jump from 0 to 1, i.e., $dZ(t) = 1$, when $Z(t) = Z(t_-) = 0$ and $dN_\nu(t) = 1$, or the jump from 1 to 0, i.e., $dZ(t) = -1$, when $Z(t) = Z(t_-) = 1$ and $dN_\mu(t) = 1$. The sample paths of $N(t)$ and of $Z(t)$ are assumed to be left continuous with right limits. Thus, the actual impulse process is obtained by selecting, with the aid of an auxiliary stochastic variable $Z(t)$, some impulses from the train driven by a Poisson process $N_\mu(t)$. The replacement Eq. 89 is illustrated in Fig. 6, where the dashed-line spikes represent these of

Non-Poisson Impulse Processes, Fig. 6 Train of impulses expressed in terms of the jump process $Z(t)$ driven by two Poisson processes, ×: N_μ-driven points, ○: N_ν-driven points

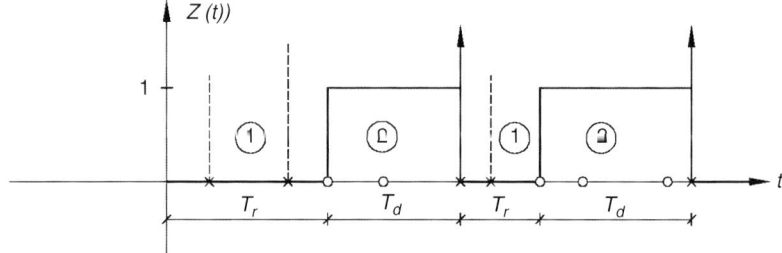

the impulses driven by $N_\mu(t)$, whose magnitudes are multiplied by $Z(t_i) = 0$, hence are excluded, and the solid-line spikes represent the remaining $N\mu(t)$-driven impulses, whose magnitudes are multiplied by $Z(t_i) = Z(t_{i-}) = 1$; hence, these are the impulses driven by the underlying counting process $R(t)$. The inter-arrival time between the solid-line spikes (actual impulses) is exactly equal to the inter-arrival time $T_a = T_r + T_d$, as defined above. The replacement implies the equivalence of the increments $dR(t) = Z(t) dN_\mu(t)$, which holds with probability 1.

The augmented state vector $\mathbf{Z}(t) = [Y(t), \dot{Y}(t), Z(t)]^T = [Z_1(t), Z_2(t), Z_3(t)]^T$ is governed by stochastic equations:

$$dZ_1 = Z_2 dt,$$
$$dZ_2 = f(Z_1, Z_2) dt + b(Z_1, Z_2) P(t) Z_3 dN_\mu(t),$$
$$dZ_3 = (1 - Z_3) dN_\nu(t) - Z_3 dN_\mu(t).$$

(91)

The state vector $\mathbf{Z}(t) = [Z_1(t), Z_2(t), Z_3(t)]^T$ is driven by two independent Poisson processes, and hence, it **is a nondiffusive Markov process**. Equations for moments may be derived from the generalized Itô's differential rule for Poisson-driven Markov processes.

Augmentation of a State Vector by Auxiliary Markov States

General Integrodifferential Equations for the Joint Probability Density of the State Vector

As the explicitly introduced, pure-jump stochastic processes $\rho(N(t))$ and $Z(t)$ are Poisson driven, they are characterized by negative exponential distributed phases, hence by a chain of Markov states. Consequently the original state variables and the states of the auxiliary pure-jump stochastic process are jointly Markovian. The jumps have to be defined in such a way that the actual impulse (i.e., the jump in the velocity response $Z_2(t)$) only occurs if there is a jump between some particular Markov states.

The problem is characterized by the set of joint probability density – discrete distribution functions $q_j(z_1, z_2, t)$ of the response state variables – the displacement $Z_1(t)$ and the velocity $Z_2(t)$, and the m states $S(t)$ of a pertinent Markov chain, defined as

$$q_j(z_1, z_2, t) dz_1 dz_2 = \Pr\{Z_1(t) \in (z_1, z_1 + dz_1) \\ \wedge Z_2(t) \in (z_2, z_2 + dz_2) \wedge S(t) = j\},$$

(92)

where $j = 1, 2, \ldots, m$. The fundamental equation for such a continuous-jump Markov process is the general forward integrodifferential Chapman–Kolmogorov equation (Gardiner 1985; Iwankiewicz and Nielsen 1999):

$$\frac{\partial}{\partial t} q_j(\mathbf{z}, t) = -\sum_{r=1}^{2} \frac{\partial}{\partial z_r} [c_r(\mathbf{z}, t) q_j(\mathbf{z}, t)] \\ + \sum_{i=1}^{m} \int_{-\infty}^{\infty} [J_{\{\mathbf{z}\}}(\mathbf{z}, j | \mathbf{x}, i, t) q_i(\mathbf{x}, t) \\ - J_{\{\mathbf{z}\}}(\mathbf{x}, i | \mathbf{z}, j, t) q_j(\mathbf{z}, t)] d\mathbf{x}$$

(93)

where in the present problem $\mathbf{q}(\mathbf{z}, t) = [q_1(\mathbf{z}, t), q_2(\mathbf{z}, t), \ldots, q_m(\mathbf{z}, t)]$, $c_r(\mathbf{z}, t)$ are the drift terms of the equation of motion written down in the state space form, i.e., $c_1(\mathbf{z}, t) = z_2$, $c_2(\mathbf{z}, t) = f(\mathbf{z}, t)$, $j = 1, 2, \ldots, m$ and $J_{\{\mathbf{z}\}}(z_1, z_2, j | x_1, x_2, i, t) = J_{\{\mathbf{Z}\}}(\mathbf{z}, j | \mathbf{x}, i, t)$ is the **jump probability intensity function** defined as

$$J_{\{\mathbf{Z}\}}(\mathbf{z},j|\mathbf{x},i,t) =$$
$$\lim_{\Delta t \to 0} \frac{\Pr\{Z_1(t+\Delta t)=z_1, Z_2(t+\Delta t)=z_2, S(t+\Delta t)=j|Z_1(t)=x_1, Z_2(t)=x_2, S(t)=i\}}{\Delta t} \quad (94)$$

which is determined from the pertinent chain of Markov states as follows. When $i=j$, the Markov chain remains in the same state, and no actual impulse occurs; hence, both the displacement and the velocity state variables are continuous. The nonzero jump probability intensity functions are only defined for $i \neq j$, such that there is a transition in a Markov chain (jump in the auxiliary process). Only some of those transitions are associated with the occurrence of the actual impulse, or the jump in the velocity process $Z_2(t)$. Hence, if there is a transition from $S(t)=i$ to $S(t+\Delta t)=j$, but no actual impulse occurs (no jump in the velocity process), the jump probability intensity function is

$$J_{\{\mathbf{Z}\}}(z_1,z_2,j|x_1,x_2,i,t) = \pi(j|i)\delta(z_1-x_1)\delta(z_2-x_2), \quad (95)$$

where

$$\pi(j|i) = \frac{\Pr\{S(t+\Delta t)=j|S(t)=i\}}{\Delta t} \quad (96)$$

is determined from the pertinent chain of Markov states. If the transition from $S(t)=i$ to $S(t+\Delta t)=j$ is associated with the actual impulse (the jump in the velocity process $Z_2(t)$), the jump probability intensity function is expressed as

$$J_{\{\mathbf{Z}\}}(z_1,z_2,j|x_1,x_2,i,t)$$
$$= \pi(j|i)\delta(z_1-x_1)\int_{\mathbf{P}}\delta(z_2-(x_2+b(z_1,z_2)p))f_P(p)dp$$
$$(97)$$

where $\mathbf{P}$ denotes the sample space and $f_P(p)$ the probability density function of the random impulse magnitude P.

Summation of the joint probability density-distribution functions $q_{\mathbf{Z}}(z_1, z_2, j, t)$ over all Markov states yields the joint probability density of the original state variables:

$$q_{\mathbf{Z}}(z_1,z_2,t) = \sum_{j=1}^{m} q_j(z_1,z_2,t). \quad (98)$$

Detailed Integrodifferential Equations for the Joint Probability Density of the State Vector for Renewal Impulse Process Driven by Two Independent Poisson Processes The jump process $Z(t)$ driven by two independent Poisson processes Eq. 90 is tantamount to a two-state Markov chain $S(t)$, such that $S(t)=1$ when $Z(t)=0$ and $S(t)=2$ when $Z(t)=1$ (Figs. 6 and 7).

If the excitation is multiplicative to the displacement process, z_1, i.e., $b(Z_1, Z_2)=b(Z_1)$, or if it is external (additive), i.e., $b(Z_1,Z_2)=$ const. $=b$, then when there is an impulse of magnitude p, according to the equation of motion, there is a jump in the velocity by $b(z_1)p$, i.e., a jump from x_2 to $z_2 = x_2 + b(z_1)p$.

The jump probability intensity function $J_{\mathbf{Z}}(z_1, z_2, j | x_1, x_2, i, t)$ determined with the aid of the Markov chain shown in Fig. 7 equals (Iwankiewicz 2008)

$$J_{\mathbf{Z}}(z_1,z_2,j|x_1,x_2,i,t) = \begin{cases} 0, & j=i, \\ \mu\delta(z_1-x_1)\int_{\mathbf{P}}\delta(z_2-(x_2+b(z_1)p))f_P(p)\,dp, & \\ & j=1,\ i=2, \\ \nu\delta(z_1-x_1)\delta(z_2-x_2), & j=2,\ i=1, \end{cases} \quad (99)$$

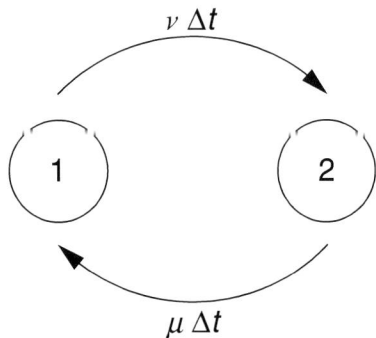

Non-Poisson Impulse Processes, Fig. 7 Markov chain for a two-state jump process driven by two independent Poisson processes

$$\frac{\partial}{\partial t}q_1(z_1,z_2,t) = -\sum_{r=1}^{2}\frac{\partial}{\partial z_r}[c_r(z_1,z_2,t)q_1(z_1,z_2,t)]$$
$$+\mu\int_P q_2(z_1,z_2-b(z_1)p,t)f_P(p)dp - vq_1(z_1,z_2,t) \tag{100}$$

$$\frac{\partial}{\partial t}q_2(z_1,z_2,t) = -\sum_{r=1}^{2}\frac{\partial}{\partial z_r}[c_r(z_1,z_2,t)q_2(z_1,z_2,t)]$$
$$+vq_1(z_1,z_2,t) - \mu q_2(z_1,z_2,t). \tag{101}$$

where $f_P(p)$ is the probability density function of the random impulse magnitude and P denotes the sample space of the impulse magnitude.

The explicit equations for $j = 1$ and $j = 2$ are obtained after the insertion of the jump probability intensity function Eq. 99 into Eq. 93 and integration with respect to x_1, x_2, respectively, as (Iwankiewicz 2008)

At the initial time instant $t = 0$, the continuous processes $Z_1(t) = Y(t)$, $Z_2(t) = \dot{Y}(t)$ are statistically independent of the jump process, or of the Markov states. It may be assumed that the jump process starts with probability 1 from the first ("off") state, i.e., $\Pr\{S(0) = 1\} = P_1(0) = 1$ and $\Pr\{S(0) = 2\} = P_2(0) = 0$. Consequently the random initial conditions are written as

$$\begin{aligned}q_{\mathbf{Z}}(z_1,z_2,1,0) &= q_1(z_1,z_2,0) = p(z_1,z_2)\cdot P_1(0) = p(z_1,z_2)\cdot 1 = p(z_1,z_2)\\ q_{\mathbf{Z}}(z_1,z_2,2,0) &= q_2(z_1,z_2,0) = p(z_1,z_2)\cdot P_2(0) = p(z_1,z_2)\cdot 0 = 0,\end{aligned} \tag{102}$$

where $p(z_1,z_2)$ is the joint probability density of $Z_1(0) = Y(0), Z_2(0) = \dot{Y}(0)$.

If the system starts from rest, i.e., $Z_1(0) = Y(0) = 0, Z_2(0) = \dot{Y}(0) = 0$, then

$$\begin{aligned}q_{\mathbf{Z}}(z_1,z_2,1,0) &= q_1(z_1,z_2,0) = \delta(z_1)\delta(z_2)P_1(0) = \delta(z_1)\delta(z_2)\cdot 1 = \delta(z_1)\delta(z_2)\\ q_{\mathbf{Z}}(z_1,z_2,2,0) &= q_2(z_1,z_2,0) = \delta(z_1)\delta(z_2)P_2(0) = \delta(z_1)\delta(z_2)\cdot 0 = 0.\end{aligned} \tag{103}$$

Detailed Integrodifferential Equations for the Joint Probability Density of the State Vector for Renewal Impulse Process Driven by an Erlang Renewal Process The jump process $\rho(N(t))$ is tantamount to a k-state Markov chain $S(t)$ shown in Fig. 8.

In this model, the actual impulse, and hence the jump in the velocity variable, occurs when the jump is from the state k to 1 (cf. Fig. 5). All other jumps in the auxiliary process occur from $j-1$ to j, for $j = 2, \ldots, k$, but there are no corresponding impulses.

The jump probability intensity function is expressed as (Iwankiewicz 2006)

$$J_{\mathbf{Z}}(z_1,z_2,j|x_1,x_2,i,t)$$
$$= \begin{cases} v\delta(z_1-x_1)\int_P \delta(z_2-(x_2+b(z_1)p))f_P(p)\,dp, \\ \qquad\qquad\qquad j=1,\ i=k. \\ v\delta(z_1-x_1)\delta(z_2-x_2), \\ \qquad\qquad\qquad j=2,3,\ldots,k,\ i=j-1 \end{cases} \tag{104}$$

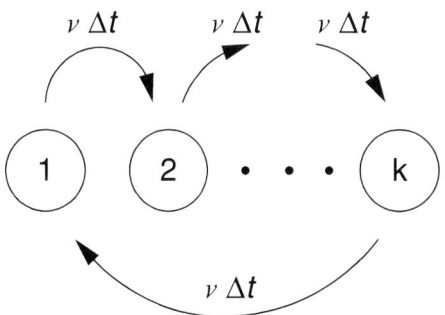

Non-Poisson Impulse Processes, Fig. 8 Markov chain for a jump process driven by an Erlang renewal process

The governing equations for $j = 1, 2, \ldots, k$ are obtained after the insertion of the jump probability intensity function Eq. 104 into Eq. 93 and integration, as (Iwankiewicz 2006)

$$\frac{\partial}{\partial t} q_1(z_1, z_2, t) = -\sum_{r=1}^{n} \frac{\partial}{\partial z_r} \left[c_r(z_1, z_2, t) q_1(z_1, z_2, t) \right]$$
$$+ \nu \int_P q_k(z_1, z_2 - b(z_1)p, t) f_P(p) dp$$
$$- \nu q_1(z_1, z_2, t)$$
$$\vdots$$
$$\frac{\partial}{\partial t} q_j(z_1, z_2, t) = -\sum_{r=1}^{n} \frac{\partial}{\partial z_r} \left[c_r(z_1, z_2, t) q_j(z_1, z_2, t) \right]$$
$$+ \nu q_{j-1}(z_1, z_2, t) - \nu q_j(z_1, z_2, t),$$
$$j = 2, \ldots k.$$
(105)

Generating Equation for Moments

General Equation The original state vector of the dynamic system $\mathbf{Z}(t)$ is not a Markov process, but the original state variables together with the states of the auxiliary jump process are jointly Markovian. Therefore, the generating equation for moments must be derived for the expectations:

$$E_j[V(\mathbf{Z}(t), t] = \int_{-\infty}^{\infty} V(\mathbf{z}(t), t) q_j(\mathbf{z}, t) d\mathbf{z}, \quad j = 1, 2, \ldots, m.$$
(106)

$$\frac{d}{dt} E_j[V(\mathbf{Z}(t), t] = \frac{\partial}{\partial t} \int_{-\infty}^{\infty} V(\mathbf{z}(t), t) q_j(\mathbf{z}, t) d\mathbf{z}$$
$$= E_j \left[\frac{\partial}{\partial t} V(\mathbf{Z}(t), t) \right] + \int_{-\infty}^{\infty} V(\mathbf{z}(t), t) \mathcal{K}_j[\mathrm{q}(\mathbf{z}, t)] d\mathbf{z},$$
(107)

where $\mathcal{K}_j[\ldots]$ is the forward integrodifferential Chapman–Kolmogorov operator (Gardiner 1985; Iwankiewicz and Nielsen 1999):

$$\mathcal{K}_j[\mathrm{q}(\mathbf{z}, t)] = -\sum_{r=1}^{2} \frac{\partial}{\partial z_r} \left[c_r(\mathbf{z}, t) q_j(\mathbf{z}, t) \right]$$
$$+ \sum_{i=1}^{m} \int_{-\infty}^{\infty} \left[J_{\{\mathbf{Z}\}}(\mathbf{z}, j | \mathbf{x}, i, t) q_i(\mathbf{x}, t) \right.$$
$$\left. - J_{\{\mathbf{Z}\}}(\mathbf{x}, j | \mathbf{z}, j, t) q_j(\mathbf{z}, t) \right] d\mathbf{x}.$$
(108)

After the integration by parts of the last term and some rearrangements, the **generating equation for moments** is arrived at (Iwankiewicz 2014)

$$\frac{d}{dt} E_j[V(\mathbf{Z}(t), t)] = E_j \left[\frac{\partial}{\partial t} V(\mathbf{Z}(t), t) \right] + \sum_{r=1}^{2} E_j \left[\frac{\partial V(\mathbf{Z}(t), t)}{\partial Z_r} c_r(\mathbf{Z}(t), t) \right] +$$
$$\sum_{i=1}^{m} \int_{-\infty}^{\infty} \int_{-\infty}^{\infty} \left[V(\mathbf{y}(t), t) J_{\{\mathbf{Z}\}}(\mathbf{y}, j | \mathbf{z}, i, t) q_i(\mathbf{z}, t) - V(\mathbf{z}(t), t) J_{\{\mathbf{Z}\}}(\mathbf{y}, i | \mathbf{z}, j, t) q_j(\mathbf{z}, t) \right] d\mathbf{y} d\mathbf{z},$$
$$j = 1, 2, \ldots, m.$$
(109)

As the usual-sense marginal joint probability density function $q(\mathbf{z}, t)$ of the state variables is obtained by summation, so is the usual-sense expectation:

$$q(\mathbf{z},t) = \sum_{j=1}^{m} q_j(\mathbf{z},t), \implies E[V(\mathbf{Z}(t),t)] \\ = \sum_{j=1}^{m} E_j[V(\mathbf{Z}(t),t)]. \tag{110}$$

Detailed Equations for the Renewal Impulse Process Driven by Two Poisson Processes The insertion of the jump probability intensity function Eq. 99 into the generating equation for moments Eq. 109 followed by the integration with respect to y yields the problem-specific generating equations for moments (Iwankiewicz 2014):

$$\begin{aligned}
\frac{d}{dt} E_1[V(\mathbf{Z}(t),t)] &= E_1\left[\frac{\partial}{\partial t} V(\mathbf{Z}(t),t)\right] + \sum_{r=1}^{2} E_1\left[\frac{\partial V(\mathbf{Z}(t),t)}{\partial Z_r} c_r(\mathbf{Z}(t),t)\right] \\
&\quad + \mu E_2\left[\int_{\mathbf{P}} V(\mathbf{Z}(t) + \mathbf{b}(\mathbf{Z})p,t) f_P(p) dp\right] - \nu E_1[V(\mathbf{Z}(t),t)], \\
\frac{d}{dt} E_2[V(\mathbf{Z}(t),t)] &= E_2\left[\frac{\partial}{\partial t} V(\mathbf{Z}(t),t)\right] + \sum_{r=1}^{2} E_2\left[\frac{\partial V(\mathbf{Z}(t),t)}{\partial Z_r} c_r(\mathbf{Z}(t),t)\right] \\
&\quad + \nu E_1[V(\mathbf{Z}(t),t)] - \mu E_2[V(\mathbf{Z}(t),t)],
\end{aligned} \tag{111}$$

where $\mathbf{b}(\mathbf{Z}) = [0, b(Z_1, Z_2)]^T$. As a result of integration with respect to $\mathbf{z}$, the Markov state probabilities may appear

$$\int_{-\infty}^{\infty} q_j(\mathbf{z},t) d\mathbf{z} = \mathcal{P}_j(t) = \Pr\{S(t) = j\}. \tag{112}$$

Differential equations governing the time evolution of Markov states probabilities $\mathcal{P}_j(t)$ are derived from the general expression for m-state Markov chain:

$$\mathcal{P}_j(t + \Delta t) = \sum_{i=1}^{m} \mathcal{P}_{j|i}(\Delta t) \mathcal{P}_i(t), \quad j = 1, 2, \ldots m. \tag{113}$$

Detailed Equations for the Impulse Process Driven an Erlang Renewal Process The insertion of the jump probability intensity function Eq. 104 into the generating equation for moments Eq. 109 followed by the integration with respect to y yields the problem-specific generating equations for moments (Iwankiewicz 2014):

$$\begin{aligned}
\frac{d}{dt} E_1[V(\mathbf{Z}(t),t)] &= E_1\left[\frac{\partial}{\partial t} V(\mathbf{Z}(t),t)\right] + \sum_{r=1}^{2} E_1\left[\frac{\partial V(\mathbf{Z}(t),t)}{\partial Z_r} c_r(\mathbf{Z}(t),t)\right] \\
&\quad + \nu E_k\left[\int_{\mathbf{P}} V(\mathbf{Z}(t) + \mathbf{b}(\mathbf{Z})p,t) f_P(p) dp\right] - \nu E_1[V(\mathbf{Z}(t),t)], \\
\frac{d}{dt} E_j[V(\mathbf{Z}(t),t)] &= E_j\left[\frac{\partial}{\partial t} V(\mathbf{Z}(t),t)\right] + \sum_{r=1}^{2} E_j\left[\frac{\partial V(\mathbf{Z}(t),t)}{\partial Z_r} c_r(\mathbf{Z}(t),t)\right] \\
&\quad + \nu E_{j-1}[V(\mathbf{Z}(t),t)] - \nu E_j[V(\mathbf{Z}(t),t)], \quad j = 2, 3, \ldots, k.
\end{aligned} \tag{114}$$

Cross-References

▶ Probability Density Evolution Method in Stochastic Dynamics
▶ Stochastic Analysis of Linear Systems
▶ Stochastic Analysis of Nonlinear Systems

References

Cox DR (1962) Renewal theory. Methuen, London
Cox DR, Isham V (1980) Point processes. Chapman and Hall, London
Gardiner CW (1985) Handbook of stochastic methods for physics, chemistry and the natural sciences. Springer, New York
Iwankiewicz R (1995) Dynamical mechanical systems under random impulses. World Scientific, series on advances in mathematics for applied sciences, vol 36. World Scientific, Singapore, New Jersey, London, Hong Kong
Iwankiewicz R (2002) Dynamic response of non-linear systems to random trains of non-overlapping pulses. Meccanica 37:167–178
Iwankiewicz R. (2003). Dynamic systems under random impulses driven by a generalized Erlang renewal process. In: Furuta H, Dogaki M, Sakano M (eds) Proceedings of the 10th IFIP WG 7.5 working conference on reliability and optimization of structural systems, 25–27 March 2002, Kansai University/Balkema, Osaka, pp 103–110
Iwankiewicz R (2006) Equation for probability density of the response of a dynamic system to Erlang renewal random impulse processes. In: Sørensen JD, Frangopol DM (eds) Proceedings of the 12th IFIP WG 7.5 working conference on reliability and optimization of structural systems, 22–25 May 2005, Taylor and Francis, Aalborg, pp 107–113
Iwankiewicz R (2008) Equations for probability density of response of dynamic systems to a class of non-Poisson random impulse process excitations. Probab Eng Mech 23:198–207
Iwankiewicz R (2014) Response of dynamic systems to renewal impulse processes: generating equation for moments based on the integro-differential Chapman-Kolmogorov equations. Probab Eng Mech 35:52–66
Iwankiewicz R, Nielsen SRK (1999) Advanced methods in stochastic dynamics of non-linear systems. Aalborg University Press, Denmark
Iwankiewicz R, Nielsen SRK, Thoft-Christensen P (1990) Dynamic response of non-linear systems to Poisson-distributed pulse trains: Markov approach. Struct Saf 8:223–238
Snyder DL, Miller MI (1991) Random point processes in time and space. Springer, New York
Srinivasan SK (1974) Stochastic point processes and their applications. Griffin, London
Stratonovich RL (1963) Topics in the theory of random noise. Gordon and Breach, New York/London
Takacs L (1956) On secondary stochastic processes generated by recurrent processes. Acta Math Acad Sci Hung 7:17–29

Novel Bio-Inspired Sensor Network for Condition Assessment

Simon Laflamme
Department of Civil, Construction, and Environmental Engineering, Iowa State University, Ames, IA, USA

Synonyms

Bio-inspired sensor; Condition assessment; Flexible strain gauge; Sensing skin; Sensor network; Shape reconstruction; Soft elastomeric capacitor; Structural health monitoring

Introduction

Condition assessment of civil structures is a task dedicated to forecasting future structural performances based on current states and past performances and events. The concept of condition assessment is often integrated within a closed-loop decision, where structural conditions can be adapted based on system prognosis. Figure 1 illustrates a particular way to conduct condition assessment. In the process, various structural states are measured, which may include excitations (e.g., wind, vehicles) and responses (e.g., strain, acceleration). These measurements are processed to extract indicators (e.g., maximum strain, fundamental frequencies) of current structural performance. These indicators are stored in a database, and also used within a forecast model (e.g., time-dependent reliability, Markov decision process) that will lead to a prognosis on the structural system, enabling optimization of

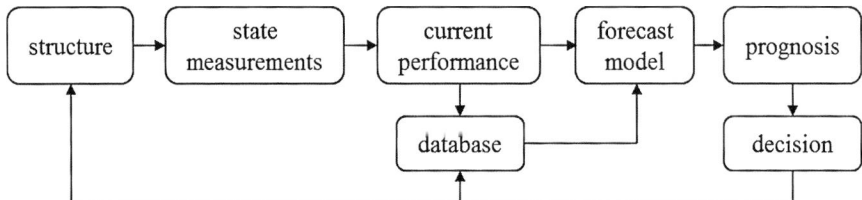

Novel Bio-Inspired Sensor Network for Condition Assessment, Fig. 1 Condition assessment process

structural management decisions (e.g., inspection, repairs, maintenance). The forecast model itself may query information from the database that include past performance indicators and events. One of the fundamental benefits of condition assessment is the availability of condition-based maintenance decisions (CBM), which enables maintenance in function of current and expected conditions rather than based on usage rate (preventive-based maintenance) or breakdown (breakdown-based maintenance). CBM has the potential to significantly improve life-cycle costs and structural resiliency by optimizing maintenance and inspection schedules and forecasting structural behavior (Jardine et al. 2006). It could also be used for automatic assessment of structural conditions following natural hazards (e.g., earthquakes).

The literature on structural prognosis and condition assessment has fundamentally focused on the specialized task of damage diagnosis, including sensing hardware and signal processing methods, but much remains to be done on developing integrated sensing solution that could lead to condition assessment (Jardine et al. 2006). The vast majority of cited work in condition assessment is in the field of machinery, and applications in civil engineering are mostly limited to bridges. See References (Frangopol et al. 2004; Teughels and De Roeck 2004; Perera and Ruiz 2008) for examples. Other specialized civil engineering applications include buildings (Savadkoohi et al. 2011), pipes (Dilena et al. 2011), and wind turbine blades (Abouhnik and Albarbar 2012). While a literature survey indicates a growth in research on condition assessment methods for civil structures, the full potential of these methods is yet to be realized (Farrar and Lieven 2007). Some challenges impeding broad applicability include the needs to develop (1) algorithms enabling real-time decision making; (2) robust sensing systems for online data acquisition; (3) methods for collecting event information; and (4) more accurate forecasting models (Frangopol et al. 2004; Jardine et al. 2006; Farrar and Lieven 2007).

The author has developed a novel sensing method designed for condition assessment of civil structures. The method consists of an array of soft elastomeric capacitors (SECs) acting as flexible strain gages. Arranged in a network, SECs are capable of covering very large areas at low costs. It was demonstrated that a network was capable of covering an area of 70×280 mm^2 with only four sensors (Laflamme et al. 2013a). Analogous to biological skin, the SEC network can localize strain over a global area. The technology is an alternative to fiber optics technologies. With both fiber optic sensors and the SEC technology, strain data can be measured over large systems. Others have proposed alternatives to conventional strain sensing, including conducting cement mixes (Materazzi et al. 2013) and piezoelectric networks (Giurgiutiu 2009). Conducting polymers, such as soft resistors and capacitors, have also gained popularity for structural health monitoring applications (Tata et al. 2009; Loh et al. 2009; Gao et al. 2010). The proposed SEC differs from existing literature in that it combines both a large physical size and high initial capacitance, resulting in a larger surface coverage and higher sensitivity. Also, it combines the advantages of being cost-effective, easy to install, robust with respect to mechanical tampering, and customizable in shapes and sizes, and low powered.

Deployment of the sensor network over large areas allows the measurement of strains over large surfaces. These measurements can be used to reconstruct physics-based features associated with the structural behavior. For instance, the analysis of deflection shapes can give insights on structural performance, whether it is by studying changes in curvature through time, or simply by counting the number of cycles and/or overstresses. These physics-based features can be integrated into a forecast model to establish the severity of the problem and enable decision making.

In this article, the promise of the SEC network for condition assessment applications is presented. The next section describes the SEC used in the network setup. The description includes a discussion of the fabrication process, the electromechanical model used for converting signal into strain and shows a comparison with off-the-shelf resistance-based strain gauges (RSGs). The subsequent section discusses the application of the sensor in a sensor network for conducting condition assessment. It also describes an algorithm used for extracting physics-based features from the network signal and demonstrates the application. The last section concludes the article.

Soft Elastomeric Capacitors

The proposed sensor network for condition assessment applications has been developed by the author (Laflamme et al. 2012, 2013a). It consists of an array of SECs, a type of conducting polymers. The field of conducting polymers has been pioneered in the 1970s, when it was discovered that polymers can not only be used as insulators but also as conducting mediums (Shirakawa et al. 1977). They have since then been used for various purposes, including flexible sensors and actuators (Osada and De Rossi 2000). These synthetic metals typically originate from the constitution of a nanocomposite mix of organic and inorganic particles, which can be obtained via chemical and electromechanical preparations, as discussed in Reference (Gangopadhyay and De 2000). Figure 2 shows the principle using scanning electron microscope (SEM) photos, in which an organic material (poly-styrene-co-ethylene-co-butylene-co-styrene (SEBS), Fig. 2a) is mixed with inorganic particles (titanium dioxide (TiO_2), Fig. 2b) to form a nanocomposite mix SEBS + TiO_2 (Fig. 2c).

The SEC is fabricated using the principles of conducting polymers. Figure 3a shows the schematic of a capacitor, constituted from a dielectric sandwiched between two conducting plates that can be connected to an electric charge. It follows that a soft capacitor can be built from an elastomeric dielectric layer sandwiched between compliant electrodes. Here, the dielectric is a nanocomposite mix of SEBS doped with TiO_2, the same materials showed in Fig. 2. The compliant electrodes are fabricated from a nanocomposite of SEBS and carbon black (CB). Figure 3b shows a picture of a single SEC. In this section, the fabrication process of the SEC is described, its electromechanical model derived, and its performance versus off-the-shelf RSGs compared.

Fabrication

Figure 4 illustrates the fabrication process of a SEC. First, a solution of SEBS dissolved in toluene (solvent) is created. Then, TiO_2 particles are dispersed in part of this solution using a sonication process. The resulting SEBS + TiO_2 mix is spread over a glass slide and allowed to dry, during which phase the solution becomes solid and the toluene evaporates. Meanwhile, the CB particles are dispersed in the remaining SEBS-toluene solution, also using sonication. Finally, this electrode mix is painted or sprayed onto the top and bottom surfaces of the dielectric and allowed to dry.

The SEC's dielectric layer is doped with TiO_2 to improve the sensor's mechanical properties, specifically its dielectric properties. The capacitance value C of a SEC is written

$$C = e_0 e_r \frac{A}{h}$$

Novel Bio-Inspired Sensor Network for Condition Assessment, Fig. 2 SEM photos of (**a**) SEBS, (**b**) TiO$_2$, and (**c**) SEBS + TiO$_2$

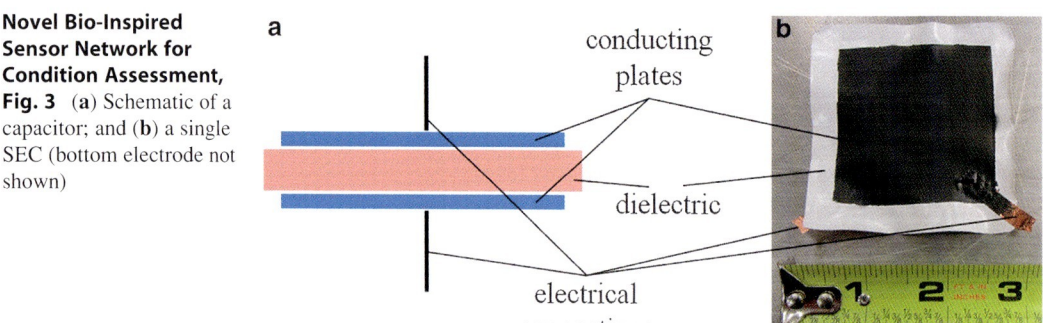

Novel Bio-Inspired Sensor Network for Condition Assessment, Fig. 3 (**a**) Schematic of a capacitor; and (**b**) a single SEC (bottom electrode not shown)

where $e_0 = 8.854$ pF/m is the vacuum permittivity, e_r the dimensionless polymer relative permittivity, $A = wl$ the sensor area with width w and length l, and h the height of the dielectric. Altering the nanocomposition of the dielectric enables a customization of e_r. For instance, the author has shown that it was possible to dramatically increase the relative

Novel Bio-Inspired Sensor Network for Condition Assessment, Fig. 4 Fabrication process

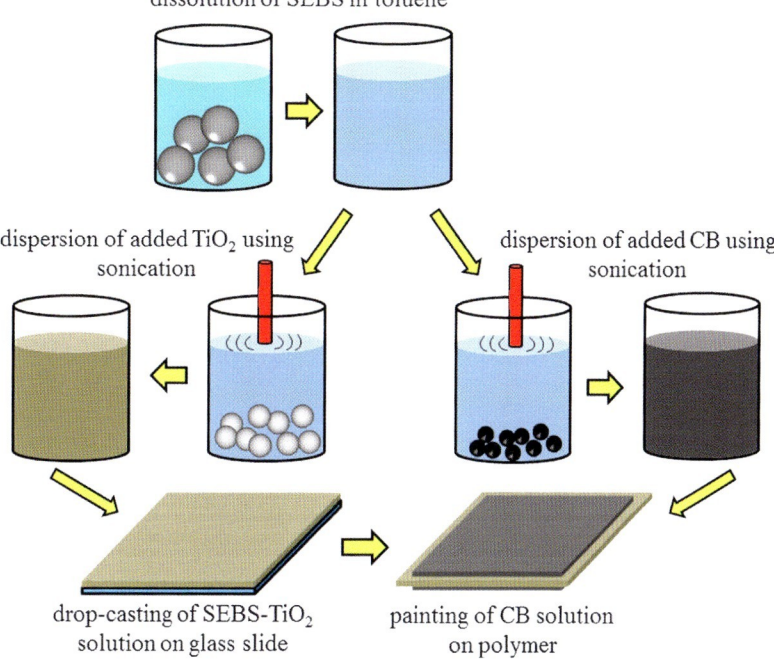

Novel Bio-Inspired Sensor Network for Condition Assessment, Fig. 5 Sensing principle (layers not scaled)

permittivity of the SEBS by grafting polyaniline (PANI) on the polymer backbone (Kollosche et al. 2011). Here, TiO_2 is used due to the low cost and high stability of the particles.

Electromechanical Model

The sensing principle of the electromechanical sensor consists in measuring strain via changes in capacitance. Figure 5 illustrates the sensing principle for a SEC glued onto a monitored surface using an epoxy. In the example, a strain in the monitored surface provokes a change in the sensor geometry Î"l. This strain is linearly transduced by a change in the capacitance Î"C of the sensor. This change is measured by a data acquisition system (DAQ).

The SEC materials can be considered as incompressible (the Poisson ratio of SEBS $v \approx 0.49$ (Wilkinson et al. 2004)). It follows that the sensor volume V is preserved ($\Delta V = 0$):

$$V = V + \Delta V$$
$$w \cdot l \cdot h = (w + \Delta w)\,(w + \Delta l)\,(h + \Delta h)$$

Ignoring higher order terms, the last equation can be written:

$$-\frac{\Delta h}{h} \approx \frac{\Delta l}{l} + \frac{\Delta w}{w}$$
$$-\varepsilon_z \approx \varepsilon_x + \varepsilon_y$$

Also, for small changes in C, the differential of the equation governing capacitance is taken as:

$$\Delta C = \left(\frac{\Delta l}{l} + \frac{\Delta w}{w} - \frac{\Delta h}{h}\right) C$$

$$\frac{\Delta C}{C} = \varepsilon_x + \varepsilon_y - \varepsilon_z$$

Substituting ε_z, the last equation becomes:

$$\frac{\Delta C}{C} = 2(\varepsilon_x + \varepsilon_y)$$

which results in a gauge factor $\lambda = 2$. The equation above shows that the sensor measures additive strain. The principal strain components and magnitudes can be decomposed by leveraging network applications of the SEC. This is out-of-the-scope of this chapter.

While the nanocomposite mix does not influence the gauge factor, it plays an important role in the sensor sensitivity $\frac{\Delta C}{\varepsilon_x + \varepsilon_y} = 2C$. It results that the sensitivity can be improved by increasing the materials permittivity e_r, resulting in a better resolution. The sensitivity can also be increased by altering the sensor geometry.

Comparison Versus Off-the-Shelf Strain Gauge

In this subsection, a performance comparison between an SEC and an off-the-shelf RSG is presented. Additional details on this comparison can be found in Reference (Laflamme et al. 2013a). The test setup consists of a three-point load setup on a simply supported aluminum beam of support-to-support dimensions 406.4 × 101.6 × 6.35 mm³ (16 × 4 × 0.25 in³). A SEC and a RSG (Vishay Micro-Measurements, CEA-06-500UW-120, resolution of 1 $\mu\varepsilon$) are installed centered onto the bottom surface of the beam. The setup is similar to Fig. 8, except with one sensor of each type, centered. Both sensors are installed following a similar procedure. The monitored surface is sanded, painted with a primer, and a thin layer of an off-the-shelf epoxy (JB Kwik) is applied on which the sensors are adhered. Data from the SECs are acquired using an inexpensive off-the-shelf data acquisition system (ACAM PCap01) sampled at 48 Hz. RSG data are acquired using a Hewlett-Packard 3852 data acquisition system, and data sampled at 55 Hz. The excitation history consists of a displacement-based triangular wave loads with increasing frequencies from 0.0167 to 0.40 Hz to remain in a quasi-static range.

Figure 6 shows the results from the test. The time series responses from the SEC and RSG are shown in Fig. 6a, along with the loading history in Fig. 6b. Results from the SEC compares well against readings from the off-the-shelf RSG. Figure 7a shows the absolute error between the SEC and RSG readings. The SEC can track the time history with a resolution of 25–30 $\mu\varepsilon$. This resolution could be improved with the fabrication of a dedicated DAQ system (Laflamme et al. 2012). Figure 7b studies the sensitivity of the SEC obtained experimentally versus the theoretical value. The experimental sensitivity of 1190 pF/ε is close to the theoretical value of 2 × 600 = 1,200 pF/ε, a 0.84 % difference. Also, results from Fig. 7b exhibit linearity, consistent with theory.

Sensor Network for Condition Assessment

The previous section described the theory for SECs and showed that it can be used effectively as a large-scale strain gauges. In this section, the SEC concept is extended to multiple sensors installed in an array form, which allows monitoring of large surface areas. In the context of condition assessment, spatial and temporal surface strain data can be assembled, from which physics-based features can be extracted. As discussed in the introduction, these features can be stored, compared, and analyzed to evaluate structural usage, or to detect changes and/or anomalies in the structural behavior. This study of structural behavior can be incorporated in a forecast model to enable prognosis

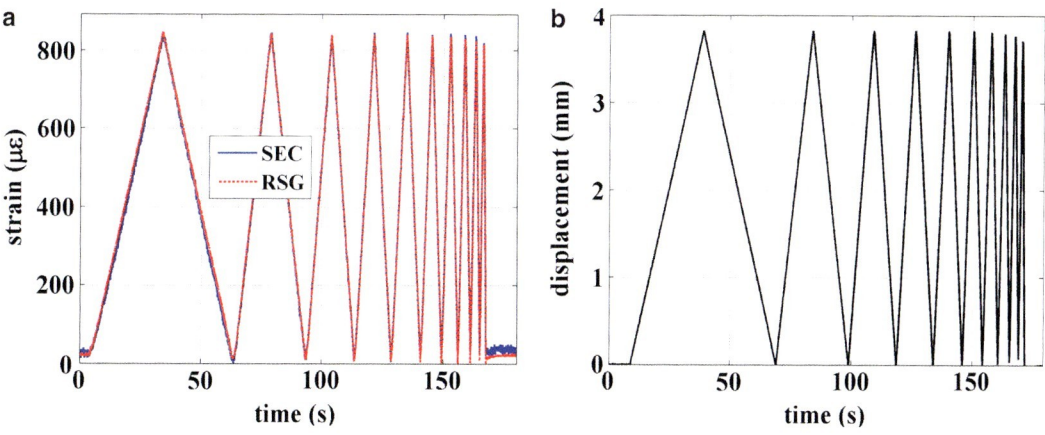

Novel Bio-Inspired Sensor Network for Condition Assessment, Fig. 6 Strain gauges comparison: (**a**) time series response; and (**b**) loading history

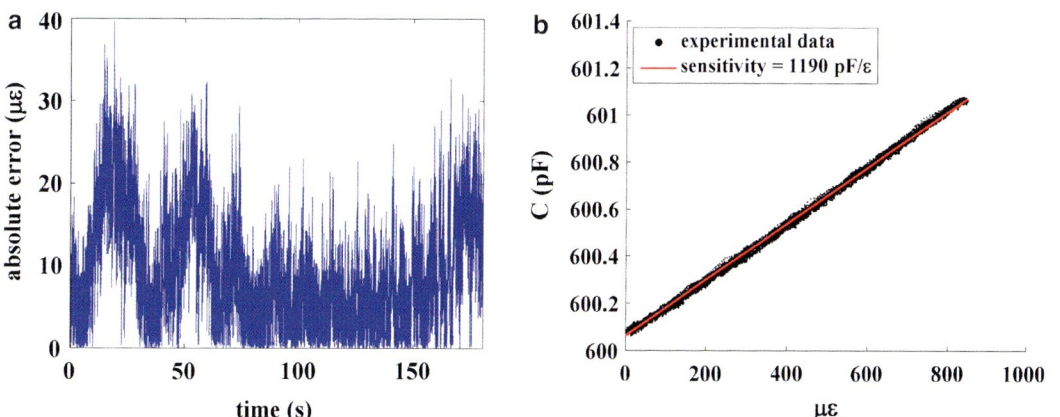

Novel Bio-Inspired Sensor Network for Condition Assessment, Fig. 7 (**a**) Absolute error between the SEC and RSG; and (**b**) capacitance versus strain for the SEC

of the structural system, with the overarching objective to optimize structural management decisions.

Here, the promise of SEC networks at conducting condition assessment is shown by extracting deflection shapes from the measurements. Remark that the design of accurate forecast models based on signal features is a challenging problem and constitutes an active field of research. Reference (Frangopol et al. 2004) reviews some fundamental forecast models, and constitutes a good introduction to the research problem for the interested reader.

Feature Extraction

Spatial and temporal strain data contain rich information about the monitored system. However, the task of condition assessment cannot be successfully conducted without the extraction of meaningful features from data. The problem is analogous to signals from accelerometers from which, in most cases, frequency features (e.g., fundamental frequencies, mode shapes) need to be extracted to enable the analysis of the vibration signature. Several types of features can be extracted from an array of strain gauges, including local plastic deformations, overstrains, cycles, and deflection shapes. Here, the concept of the sensor network for condition assessment is demonstrated by extracting deflection shapes from a monitored surface.

The problem of real-time reconstruction of deflection shapes from position and curvature measurements from sensor networks has been widely studied, with applications to condition assessment, structural health monitoring, and shape control. See references (Jones et al. 1999; Glaser et al. 2012): for instance. Here, the algorithm consists of fitting the curvature data using a polynomial interpolation and double-integrating to obtain the deflection shape. In the case of a two-dimensional beam equipped with four sensors, the fitting function is taken as a third degree polynomial to avoid possible over-fitting and also allows some additional filtering on the measured strain data (Jones et al. 1999; Glaser et al. 2012):

$$\hat{\varepsilon}_{m,i} = a_0 + a_1 x_i + a_2 x_i^2 + a_3 x_i^3$$

where the *hat* denotes an estimation for the *i*th sensor, a_1 to a_4 are constants, and x is the Cartesian location $0 \leq x \leq L$ along the beam of length L. Minimizing the error J for n sensors:

$$J = \sum_i^n \left(\varepsilon_{m,i} - \hat{\varepsilon}_{m,i}\right)^2$$

leads to the expression:

$$\mathbf{A} = \left(\mathbf{X}^T \mathbf{X}\right)^{-1} \mathbf{X}^T \Xi_\mathbf{m}$$

with:

$$\mathbf{A} = \begin{bmatrix} a_0 \\ a_1 \\ a_2 \\ a_3 \end{bmatrix} \quad \Xi_m = \begin{bmatrix} \varepsilon_{m,1} \\ \varepsilon_{m,2} \\ \dots \\ \varepsilon_{m,n} \end{bmatrix}$$

$$\mathbf{X} = \begin{bmatrix} 1 & x_1 & x_1^2 & x_1^3 \\ 1 & x_2 & x_2^2 & x_2^3 \\ \vdots & \vdots & \vdots & \vdots \\ 1 & x_n & x_n^2 & x_n^3 \end{bmatrix}$$

The deflection shape $y(x)$ is obtained by integrating the curvature twice:

$$\begin{aligned} y(x) &= \int_0^L \int_0^L \frac{\delta^2 y}{\delta x^2} dx^2 \\ &= \int_0^L \int_0^L -\frac{\varepsilon_m}{c} dx^2 \\ &= \int_0^L \int_0^L -\frac{1}{c}\left(a_0 + a_1 x_j + a_2 x_j^2 + a_3 x_j^3\right) dx^2 \\ &= -\frac{1}{c}\left(a_0 \frac{x^2}{2} + a_1 \frac{x^3}{6} + a_2 \frac{x^4}{12} + a_3 \frac{x^5}{20}\right) + b_1 x + b_2 \end{aligned}$$

where c is the distance from the surface to the centroid of the beam, and b_1 and b_2, are constants that can be determined by enforcing boundary conditions. For example, in the case of a simply-supported beam ($y(0) = y(L) = 0$):

$$\begin{aligned} b_1 &= \frac{1}{c}\left(a_0 \frac{L}{2} + a_1 \frac{L^2}{6} + a_2 \frac{L^3}{12} + a_3 \frac{L^4}{20}\right) \\ b_2 &= 0 \end{aligned}$$

Laboratory Demonstration

The method for extracting the deflection shape feature explained in the previous subsection is demonstrated in what follows. Additional details on the test can be found in Reference (Laflamme et al. 2013a). The test setup, shown in Fig. 8, consists of the same aluminum specimen used for the comparison against the RSG, with the number of SECs extended to four in order to create a sensor network. The SECs and RSGs are located under the beam at $x = \{0.20, 0.40, 0.60, 0.80\}L$, and a three-point load setup is used. A similar experimental procedure as in the previous section is used. The objective of this laboratory verification is to extract deflection shapes from the sensor network.

Figure 9 plots the sensors signals during the first triangular load for the SECs (Fig. 9a) and for the RSGs (Fig. 9b). SEC_1 and SEC_4 have closely spaced signals, as it would be expected for strain gauges symmetrically placed, while SEC_3 shows a substantial difference with respect to SEC_2. The

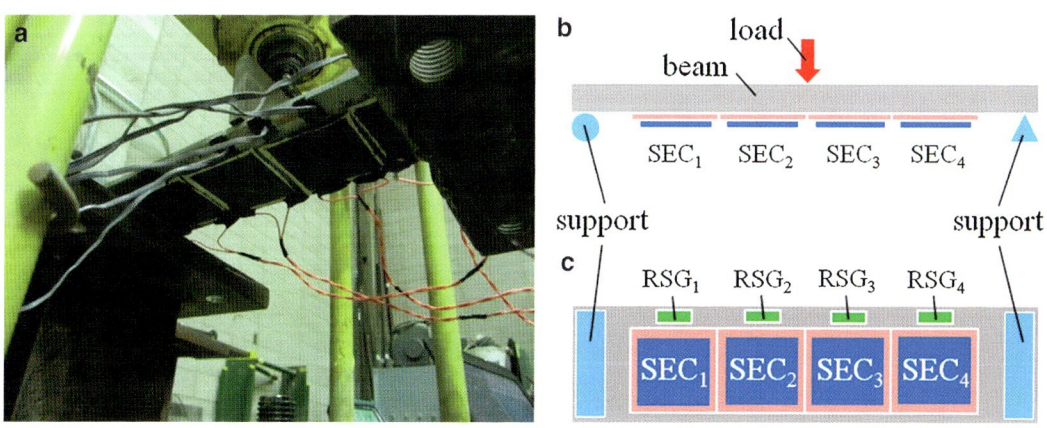

Novel Bio-Inspired Sensor Network for Condition Assessment, Fig. 8 Laboratory setup. (**a**) Picture of the setup; (**b**) setup schematic, elevation view; and (**c**) setup schematic, bottom plan view

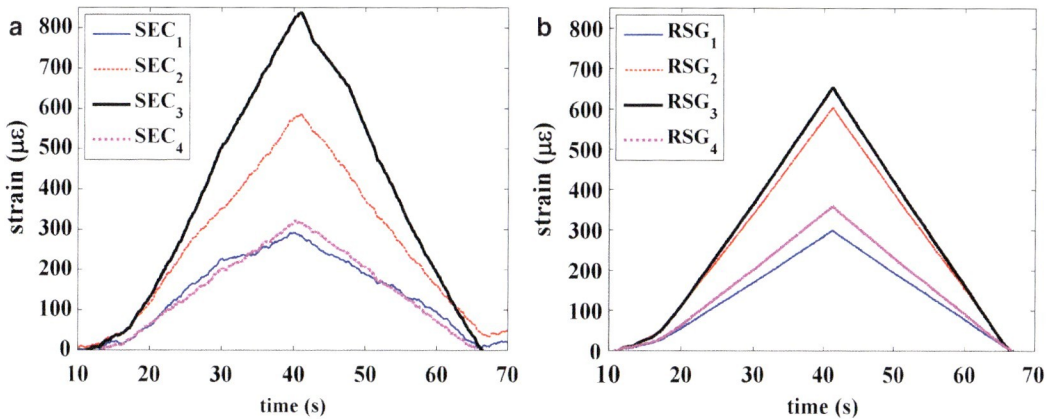

Novel Bio-Inspired Sensor Network for Condition Assessment, Fig. 9 Sensors signals for the first triangular load. (**a**) SECs; and (**b**) RSGs

signals for the RSGs are coupled, with a small difference that can be explained by small asymmetries in the sensors placement. All SECs, except for SEC_3, underestimate strain with respect to RSGs.

Figure 10a shows the deflection shapes taken at time $t = 40$ s extracted using the methodology previously described. Results are benchmarked against the analytical solution obtained from the Euler-Bernoulli beam theory. The SECs underestimate the deflection shape as a result of the underestimation of strain from SEC_1 SEC_2, and SEC_4. The deflection shape from the RSG is closer to the analytical result, but with a shift of the maximum deflection to the right. Figure 10b shows the deflection shapes normalized to their maximum unit deflection. Results show that the SECs give a better estimate of the normalized deflection shapes compared to RSGs. The shift of the maximum deflection point can be explained by the slightly higher strain readings obtained with RSG_3 and RSG_4, both located to the right-hand-side of the beam.

The promise of the SEC network for extracting deflection shapes is further investigated by comparing the root mean square (RMS) error of the normalized deflection shapes with respect to the analytical solution. Results

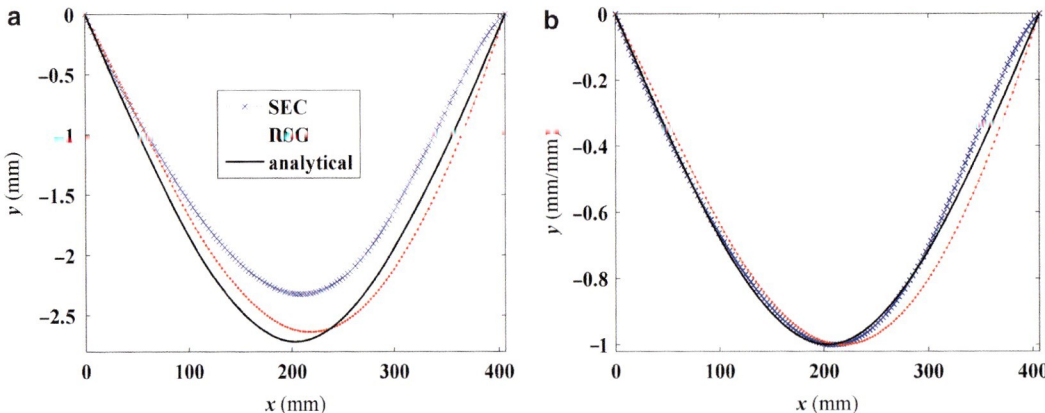

Novel Bio-Inspired Sensor Network for Condition Assessment, Fig. 10 Deflection shapes extracted from the sensors signals. (**a**) Non-normalized; and (**b**) normalized

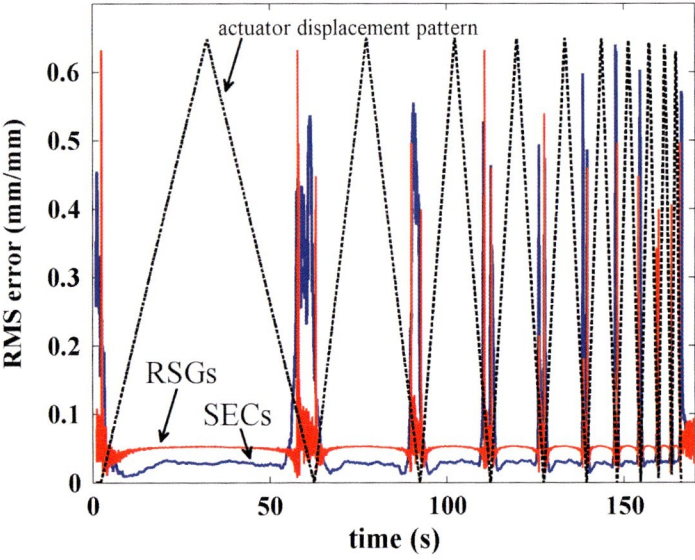

Novel Bio-Inspired Sensor Network for Condition Assessment, Fig. 11 RMS error of the normalized deflection shapes with respect to the analytical solution

are shown in Fig.11 for the entire loading history. The SEC network obtains a more accurate shape than the RSG network beyond an initial level of loading. When the load is around zero, the noise in the sensors signals results in highly inaccurate deflection shapes. The significant difference in performance between both sensors can be attributed to the SECs averaging strain over a large area, while the RSGs measure a localized strain. SECs are less sensitive to placement errors.

Summary

A novel bio-inspired sensing solution has been presented for condition assessment of civil structures. The technology consists of a SEC transducing changes in strain into changes in capacitance. Arranged in a network setup, the technology can be used to extract physics-based features for condition assessment. Comparisons against an off-the-shelf RSG showed that the SEC is capable of tracking a quasi-static strain history at a resolution in the range of 25–30 $\mu\varepsilon$. This resolution is limited by existing off-the-shelf DAQ systems dedicated to capacitance measurements. Laboratory verifications demonstrated the performance of the SEC network at extracting deflection shapes. The study of the RMS error showed that the SEC network provided accurate normalized deflection shapes, with performance levels beyond the RSG network. This performance can be explained partly by the capacity of the SEC to average strain over a large area, unlike RSGs that measure localized strain. Thus, slight misplacement of SECs has minimum consequences on the shape extraction. Given the results from this laboratory demonstration, the SEC network offers great promise as a sensing method for condition assessment of civil structures.

References

Abouhnik A, Albarbar A (2012) Wind turbine blades condition assessment based on vibration measurements and the level of an empirically decomposed feature. Energy Conversion and Management 64(2012):606–613

Dilena M, DellOste M, Morassi A (2011) Detecting cracks in pipes filled with fluid from changes in natural frequencies. Mech Syst Signal Process 25(8):3186–3197

Farrar C, Lieven N (2007) Damage prognosis: the future of structural health monitoring. Philos Trans R Soc A Math Phys Eng Sci 365:623–632

Frangopol D, Kallen M, Noortwijk J (2004) Probabilistic models for life-cycle performance of deteriorating structures: review and future directions. Prog Struct Eng Mater 6(4):197–212

Gangopadhyay R, De A (2000) Conducting polymer nanocomposites: a brief overview. Chem Mater 12(3):608–622

Gao L, Thostenson E, Zhang Z, Byun J, Chou T (2010) Damage monitoring in fiber-reinforced composites under fatigue loading using carbon nanotube networks. Philos Mag 90(31–32):4085–4099

Giurgiutiu V (2009) Piezoelectricity principles and materials. In: Encyclopedia of structural health monitoring. Wiley, pp 981–991

Glaser R, Caccese V, Shahinpoor M (2012) Shape monitoring of a beam structure from measured strain or curvature. Exp Mech 52(6):591–606

Jardine A, Lin D, Banjevic D (2006) A review on machinery diagnostics and prognostics implementing condition-based maintenance. Mech Syst Signal Process 20(7):1483–1510

Jones R, Bellemore D, Berko T, Sirkis J, Davis M, Putnam M, Friebele E, Kersey A (1999) Determination of cantilever plate shapes using wavelength division multiplexed fiber bragg grating sensors and a least-squares strain-fitting algorithm. Smart Mater Struct 7(2):178

Kollosche M, Stoyanov H, Laflamme S, Kofod G (2011) Strongly enhanced sensitivity in elastic capacitive strain sensors. J Mater Chem 21:8292–8294

Laflamme S, Kollosche M, Connor J, Kofod G (2012) Soft capacitive sensor for structural health monitoring of large-scale systems. Struct Control Health Monitor 19.1 (2012):70–81

Laflamme S, Kollosche M, Kollipara VD, Saleem HS, Kofod G (2012) Large-scale surface strain gauge for health monitoring of civil structures. In: SPIE smart structures and materials/nondestructive evaluation and health monitoring. International Society for Optics and Photonics, pp 83471P-83471P

Laflamme S, Saleem HS, Vasan BK, Geiger RL, Chen D, Kessler MR, Rajan K (2013a) Soft elastomeric capacitor network for strain sensing over large surfaces. IEEE/ASME Trans Mech

Laflamme S, Kollosche M, Connor J, Kofod G (2013b) Robust flexible capacitive surface sensor for structural health monitoring applications. ASCE J Eng Mech 139(7):879–885

Loh K, Hou T, Lynch J, Kotov N (2009) Carbon nanotube sensing skins for spatial strain and impact damage identification. J Nondestruct Eval 28(1):9–25

Materazzi A, Ubertini F, D'Alessandro A (2013) Carbon nanotube cement-based transducers for dynamic sensing of strain. Cem Concr Compos 37:2–11

Osada Y, De Rossi DE (2000) Polymer sensors and actuators. Springer, Germany

Perera R, Ruiz A (2008) A multistage fe updating procedure for damage identification in large-scale structures based on multiobjective evolutionary optimization. Mech Syst Signal Process 22(4):970–991

Savadkoohi A, Molinari M, Bursi O, Friswell M (2011) Finite element model updating of a semi-rigid moment resisting structure. Struct Control Health Monit 18(2):149–168

Shirakawa H, Louis EJ, MacDiarmid AG, Chiang CK, Heeger AJ (1977) Synthesis of electrically conducting organic polymers: halogen derivatives of polyacetylene, (CH) x. J Chem Soc Chem Commun 16:578–580

Tata U, Deshmukh S, Chiao J, Carter R, Huang H (2009) Bio-inspired sensor skins for structural health monitoring. Smart Mater Struct 18:104026

Teughels A, De Roeck G (2004) Structural damage identification of the highway bridge Zz24 by FE model updating. J Sound Vib 278(3):589–610

Wilkinson A, Clemens M, Harding V (2004) The effects of sebs-g-maleic anhydride reaction on the morphology and properties of polypropylene/PA6/sebs ternary blends. Polymer 45(15):5239–5249

Numerical Modeling of Masonry Infilled Reinforced Concrete Frame Buildings

Ioannis Koutromanos[1] and P. Benson Shing[2]
[1]Department of Civil and Environmental Engineering, Virginia Polytechnic Institute and State University, Blacksburg, VA, USA
[2]Department of Structural Engineering, University of California, San Diego, La Jolla, CA, USA

Synonyms

Creep; Equivalent strut model; Finite element model; Infilled frame; Masonry; Nonlinear analysis; Reinforced concrete

Introduction

Masonry-infilled reinforced concrete frames constitute a significant portion of the building inventory in seismically active regions around the world. Even though early efforts to analyze the behavior of infilled frames (e.g., Polyakov 1960) date back to more than half a century ago, the modeling of the interaction between the frame members and the masonry infill walls is still an active research area.

It is well known that the resistance of an infilled frame is not a simple sum of the resistance of a bare frame and that of the infill wall due to the fact that the load-resistance mechanism of a frame can change as a result of its interaction with the infill. As described in ASCE/SEI 41 (ASCE/SEI 2007) and shown in Fig. 1, when the system is subjected to a horizontal force acting toward the right, the frame tends to separate from the infill wall at the bottom left and top right corners. Compressive contact stresses develop at the other two corners.

A numerical model for the analysis of infilled frames must be able to capture the effect of the frame–infill separation and the development of compressive contact stresses at two of the four corners of the infill wall. The damage and nonlinear response of the RC frame members and of the infill walls also need to be accounted for. A masonry infill wall can fail by corner crushing, as shown in Fig. 2a, or shear sliding along bed joints, as shown in Fig. 2b. Shear sliding is expected to be important for cases when the mortar bed joints are relatively weak as compared to the masonry units. Another damage mode, which is common for older structures with relatively weak frame members, has diagonal/sliding cracks developed in the infill wall and shear cracks in the reinforced concrete columns, as shown in Fig. 2c.

This entry is aimed to provide an overview of analytical tools that can be used to study the behavior of masonry-infilled RC frames under earthquake loading. It is not intended to be an exhaustive summary of the literature; rather, it will focus on some of the most common and representative analysis methods developed for such structures including their advantages and limitations. In this respect, two types of analysis methods will be considered: (i) simplified, design-oriented analysis tools, and (ii) refined tools based on the finite element method.

Simplified Modeling Using Equivalent Strut Concept

The vast majority of the simplified analysis methods proposed for infilled frames are based on the *equivalent strut* concept, with which the effect of the infill wall is modeled with diagonal struts. An example of this is shown in Fig. 3. The use of diagonal struts can approximately reproduce the frame–infill contact condition and the stress field in the infill walls. Since earthquake ground motions introduce cyclic lateral loading to infilled frames, at least two diagonal truss elements (one for each loading direction) are required to model an infill wall.

Strut Calibration

The use of a single strut for each direction of loading is expected to be adequate when the infill wall is relatively weak as compared to the frame so that failure is expected to occur in the infill

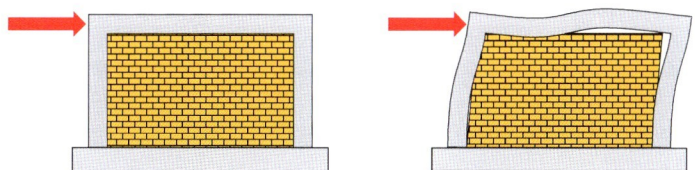

Numerical Modeling of Masonry Infilled Reinforced Concrete Frame Buildings, Fig. 1 Deformation of infilled frame subjected to a horizontal force

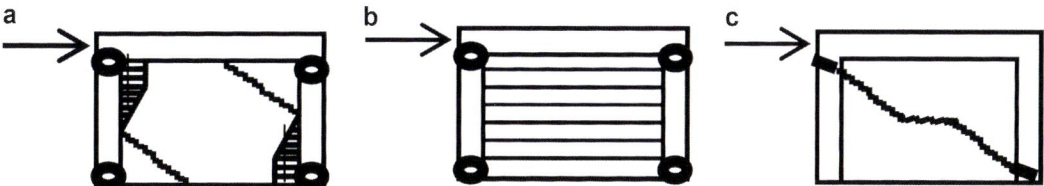

Numerical Modeling of Masonry Infilled Reinforced Concrete Frame Buildings, Fig. 2 Typical damage patterns for infilled frames (Mehrabi et al. 1994). (**a**) Corner crushing (with flexural hinges in columns). (**b**) Sliding along bed joints (with flexural hinges in columns). (**c**) Diagonal/sliding cracks (with shear cracks in columns)

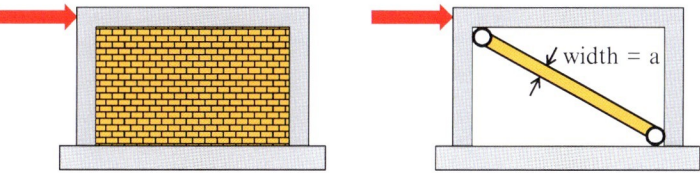

Numerical Modeling of Masonry Infilled Reinforced Concrete Frame Buildings, Fig. 3 Substitution of infill wall with a diagonal strut member for monotonic loading

wall. The width of the equivalent strut depends on the contact length between the infill wall and the adjacent columns, which, in turn, depends on the stiffness of the infill as compared to that of the columns (Stafford Smith 1967; Mainstone 1972). However, the contact length between the infill and the columns and, thereby, the width of the equivalent strut are expected to change as the load experienced by an infilled frame increases and inelastic behavior develops in the masonry wall. Hence, strictly speaking, the effective strut width determined to represent the stiffness of an infilled frame will not be the same as that required to calculate the strength. For determining the lateral stiffness of an infilled frame, ASCE/SEI 41 (ASCE/SEI 2007) recommends the following expression based on the work of Mainstone (1972):

$$a = 0.175(\lambda_1 \cdot h)^{-0.4} \cdot r_{inf} \qquad (1)$$

where h is the height of the column in a centerline representation of the frame geometry, as shown in Fig. 4a, r_{inf} is the length of the diagonal of the infill wall, and ($\lambda_1 \cdot h$) is a dimensionless parameter representing a relative stiffness coefficient for the masonry infill and the frame, with λ_1 given by the following expression:

$$\lambda_1 = \left[\frac{E_{me} \cdot t_{inf} \cdot \sin(2\theta)}{4 E_{fe} \cdot I_{col} \cdot h_{inf}}\right]^{\frac{1}{4}} \qquad (2)$$

In Eq. 2, E_{me} is the modulus of elasticity of the masonry, t_{inf} is the thickness of the infill wall, E_{fe} is the modulus of elasticity of the concrete in the frame, I_{col} is the moment of inertia of the cross

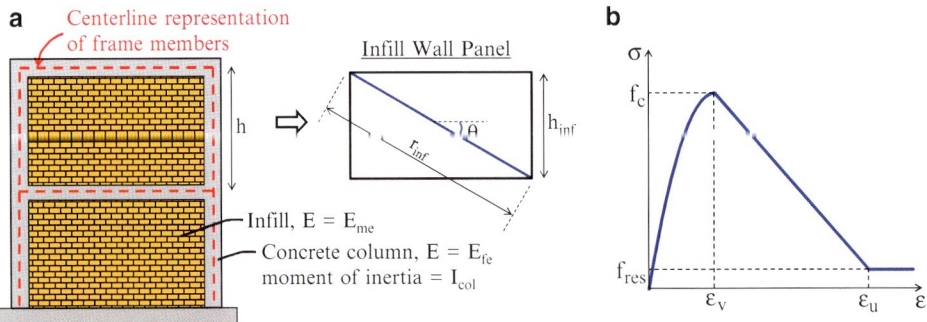

Numerical Modeling of Masonry Infilled Reinforced Concrete Frame Buildings, Fig. 4 Material model and parameters for an equivalent strut. (**a**) Infilled frame properties and dimensions. (**b**) Uniaxial model for masonry

section of the frame columns, and h_{inf} is the height of the infill wall. A number of other expressions have been proposed in the literature to determine the strut width, as summarized in Asteris et al. (2011). The cross-sectional area of the diagonal strut is the product of the width, a, and the thickness of the infill, t_{inf}.

Mainstone's work suggested different effective strut widths for strength and stiffness for the reason mentioned above. However, for computer-based models, it is convenient to have a constant strut width. To this end, one can determine the effective compressive strength of a diagonal strut rather than the actual compressive strength of the masonry. Unfortunately, no general guidelines are available for this purpose. It is not difficult to convince oneself that this strength depends on the failure mechanism of the infill wall, which could be governed by the sliding of the bed joints, corner crushing, or diagonal/sliding failure. The guidelines in ASCE 41 recommend that the resistance of an infill wall be equal to the product of the shear strength of the masonry bed joints and the cross-sectional area of the wall. A different set of guidelines on determining the effective strut width and the shear strength of masonry infill is given in Appendix B of the MSJC code (2011). However, it should be mentioned that the latter is more for design and for performance assessment.

Klingner and Bertero (1976) used strut-based models to simulate the behavior of infilled frames under cyclic loading. They considered infill walls constructed of reinforced masonry. They established a nonlinear force–deformation law for the struts to capture the strength degradation due to masonry crushing, the tensile resistance contributed by reinforcing steel in the wall, and the stiffness degradation of the infill wall due to damage. They used the prism compressive strength for the determination of the strut peak resistance, assuming that the area of the diagonal struts remains constant. Their analytical models provided satisfactory estimates of the cyclic response of experimentally tested infilled frames.

Ideally, two struts should be used to represent an unreinforced masonry infill wall, with each strut carrying only compression. A material model that can be used for this purpose is the Concrete01 material in OpenSees (McKenna et al. 2000), whose compressive stress–strain law is shown in Fig. 4b. The peak strength of the material model can be calibrated such that the horizontal component of the strut force is equal to a target resistance (e.g., the value stipulated in ASCE 41), $V_{u,\,inf}$, using the following equation:

$$f'_m = \frac{V_{u,\,inf}}{a \cdot t_{inf} \cdot \cos\theta} \qquad (3)$$

Other stress–strain laws for infills that have been proposed in the literature (e.g., Crisafulli and Carr 2007) can also be used. A study by El-Dakhakni et al. (2003) has suggested that the

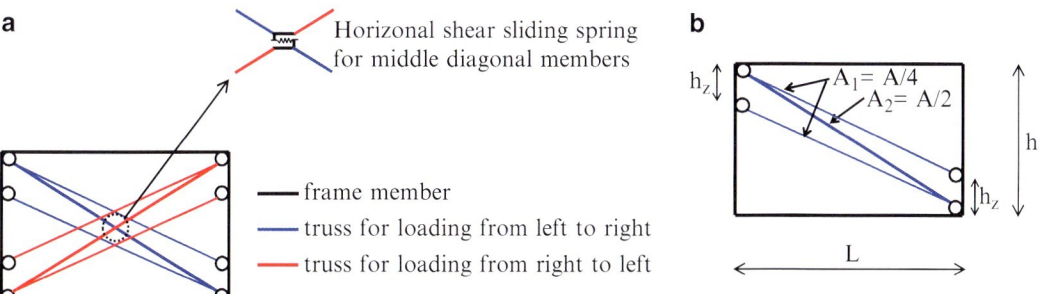

Numerical Modeling of Masonry Infilled Reinforced Concrete Frame Buildings, Fig. 5 Multiple diagonal strut model of an infilled frame under lateral loading. (**a**) Modeling approach. (**b**) Distribution of total area into multiple struts

material anisotropy of masonry should be accounted for when determining the stress–strain law for the equivalent strut. However, there has been no systematic study to determine whether accounting for the anisotropy will significantly increase the accuracy of an equivalent strut analogy.

The equivalent strut method is an oversimplification of the actual behavior of an infill wall and fails to capture some key failure mechanisms, such as the one depicted in Fig. 2c. A strut model will not account for the possible shear failure of a column that could be induced by the frame–wall interaction. There is no simple solution to overcome this problem. A study by Stavridis (2009) based on detailed nonlinear finite element models has demonstrated that the compressive stress field in a masonry infill wall may not be accurately represented by a single diagonal strut and that a strut model ignores the shear transfer between the beam and the infill. Hence, replacing a wall by a diagonal strut will not lead to a realistic representation of the load transfer from the frame to the wall. Moreover, as mentioned previously, it is not possible to have a single strut width to capture both the initial stiffness and load capacity of an infilled frame.

To represent the load transfer mechanism in a more accurate manner, multi-strut approaches have been proposed in a number of studies (El-Dakhakni et al. 2003; Crisafulli and Carr 2007). Crisafulli and Carr (2007) have proposed a multi-strut approach as shown in Fig. 5. The figure shows that several of the struts are connected to the columns with an eccentricity, h_z, which is to be calculated using the following equation:

$$h_z = \kappa \cdot \frac{\pi}{2\lambda_1} \qquad (4)$$

where κ is a constant which can take a value between 0.33 and 0.5.

The total strut area, A, needs to be distributed among the various truss elements as shown in Fig. 5b. The middle truss element is assigned half of the cross-sectional area of the equivalent strut, and each of the remaining two struts are assigned one fourth of the cross-sectional area. To account for the possibility for sliding along the bed joints, the middle strut is subdivided into two members, which are connected by a horizontal sliding "spring" and also by a vertical spring with large stiffness to enforce the continuity of vertical displacement at the location of the sliding connections. The sliding spring must have a large, penalty stiffness, and the frictional resistance can be estimated based on the amount of normal (gravity) stresses carried by the infill wall before the application of lateral loads.

The multi-strut model described above has been implemented by Crisafulli and Carr (2007) in a panel element, which only has four corner nodes and can be connected to the frame only at the points representing the beam–column joints. Thus, the effect of the frame–infill contact forces on the bending moments of the frame members cannot be captured by a panel element. For this reason, it seems preferable to avoid the use of a panel element and simply connect the multiple truss elements to a model representing the RC frame. However, this will necessarily require that each column be represented by two beam–column elements.

While a multi-strut approach may better represent the load transfer mechanism in an infilled frame, it is not certain that it will lead to a significant improvement. A validation analysis by Crisafulli (1997) has shown that excellent results can be obtained by the proposed multi-strut; however, the same reference reported that a very careful adjustment of the properties of the model was required to achieve a good agreement. Details of this adjustment were not discussed. In view of the approximate nature of the equivalent strut approach, the additional complications introduced by a multi-strut model may not warrant such efforts.

In view of the aforementioned issues, a reasonable approach is to treat diagonal struts as purely phenomenological models. They can be calibrated in such a way that they represent not only the behavior of infill walls but that of an infilled frame as a whole. Such calibration can rely on experimental data, or in the absence of experimental data, on refined finite element models. Stavridis (2009) has used experimental data and finite element analysis results to derive a set of simple rules to determine ASCE 41-type pushover curves for infilled frames. Such a curve can be used to determine the load–displacement relation for an equivalent diagonal strut so that the overall load–displacement relation of an infilled frame can be captured. However, that study focused on non-ductile RC frames with relatively strong solid brick infill. More studies are needed to develop pushover curves for other infilled frame configurations.

Modeling of RC Frame Members

The RC frame members in a simplified analysis can be modeled using nonlinear beam elements. A variety of force-based and displacement-based beam elements, with lumped plasticity (having inelastic deformations only at end plastic hinges) or with distributed plasticity, are available in analysis programs (Filippou and Fenves 2004). Many options also exist for modeling the cross-sectional behavior of the beam elements, which relates the stress resultants, namely, the axial forces and bending moments, to the corresponding generalized strains, i.e., the axial strain along the reference axis and the curvature. The most accurate and efficient formulation is the one based on the discretization of the cross section into fibers of concrete and steel reinforcement, with each fiber having an appropriate uniaxial constitutive law.

Shear failure can occur in the columns due to the forces developed from the frame–infill interaction, especially for non-ductile frames with strong masonry infill. If column shear failure is deemed probable, it can be accounted for in the analytical models through, e.g., nonlinear springs representing the shear force - shear deformation relation for the columns. However, this approach must be used with caution, because a strut model neglects many important aspects of the frame–infill interaction, such as the frictional shear transfer along the beam–infill interface and the variation in the axial forces of the columns due to the friction along the column–infill interfaces (Shing and Stavridis 2014). The introduction of shear springs in the columns may lead to unexpected results.

Infilled Frames Under Combined In- and Out-of-Plane Loading

The simplified strut-modeling concept can be extended to the analysis of infilled frames subjected to combined in- and out-of-plane loading. The simplest possible approach is to reduce the in-plane resistance of the strut elements using a reduction coefficient accounting for the out-of-plane force using the following expression (Al-Chaar 2002):

$$R_{i\text{-}o} = 1 + \frac{1}{4}\frac{OP_{demand}}{OP_{capacity}} - \frac{5}{4}\left(\frac{OP_{demand}}{OP_{capacity}}\right)^2 \quad (4)$$

where OP_{demand} is the applied out-of-plane pressure and $OP_{capacity}$ is the out-of-plane capacity of

Numerical Modeling of Masonry Infilled Reinforced Concrete Frame Buildings, Table 1 Values of coefficient λ_2 for determination of out-of-plane capacity of infill walls according to ASCE 41

h_{inf}/t_{inf}	5	10	15	20
λ_2	0.129	0.060	0.034	0.013

the wall, which can be estimated using the following relation proposed in ASCE 41:

$$OP_{capacity} = \frac{0.7 f'_m \lambda_2}{h_{inf}/t_{inf}} \quad (5)$$

where λ_2 depends on the height-to-thickness ratio of the infill wall in accordance with Table 1. Equation (5) can only be used if several criteria, which allow the development of arching action in the infill walls, are satisfied. More specifically, the infill wall must be in full contact with the surrounding frame, the ratio h_{inf}/t_{inf} must not exceed 25 and the frame members must be sufficiently stiff and strong to allow the development of the thrusts from arching action. While this approach is conceptually simple, it requires the estimation of the out-of-plane load demand prior to the analysis.

The interaction of in-plane and out-of-plane loading can also be captured by accounting for the out-of-plane flexure of the masonry infill walls in strut-based models. A method proposed by Kadysiewksi and Mosalam (2009) uses beam elements to account for the interaction of the axial force and out-of-plane bending. As shown in Fig. 6, each infill wall is represented with a diagonal beam member whose material can develop both tensile and compressive resistance. The member is subdivided into two beam elements which use a fiber section model. Each fiber in the section has the same elastic modulus, but the strength, sectional area, and distance of each fiber from the reference axis of the beam need to be determined so that the simplified model provides the in- and out-of-plane strength values established in ASCE 41 and can also reproduce a target interaction relation between in- and out-of plane resistance.

While the method by Kadysiewksi and Mosalam (2009) provides a reasonable generalization of the ASCE 41 recommendations, it still has some issues. First of all, the determination of

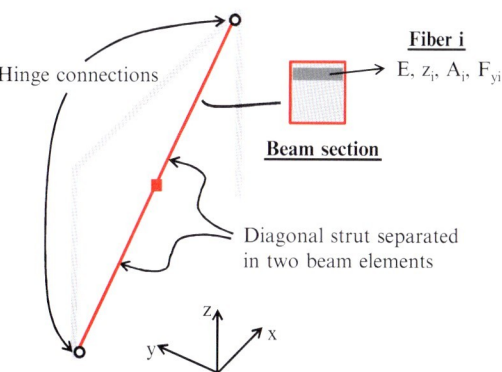

Numerical Modeling of Masonry Infilled Reinforced Concrete Frame Buildings, Fig. 6 Modeling approach for infilled frame under combined in- and out-of-plane loading (Kadysiewksi and Mosalam 2009)

the properties of each fiber is an underconstrained problem and has no unique solution. While Kadysiewksi and Mossalam have proposed a procedure to circumvent this difficulty, there is no sufficient justification for the method they are proposing. Furthermore, other issues exist with the specific model pertaining to the behavior after yielding and also when collapse is expected to occur, because the model cannot capture the strength degradation occurring at the collapse limit state. Thus, further research is required to ensure a more sound calibration process and also capture the strength degradation effect.

Infilled Frames with Wall Openings

The simplified strut approach based on the representation of infill walls with truss elements can also be used for the analysis of structures where the infill walls have openings. A relatively small opening will not have a significant effect on the behavior of an infilled frame, and for this reason strut-based models can still be used with minor modifications. Al-Chaar (2002) has recommended that the cross-sectional area of the diagonal struts be multiplied by a reduction factor, R_i, which accounts for the existence of openings and is given by:

$$R_i = 0.6 \left(\frac{A_{op}}{A_{inf}}\right)^2 - 1.6 \frac{A_{op}}{A_{inf}} + 1 \quad (5)$$

where A_{op} is the area of the openings in an infill wall and A_{inf} is the area of the infill wall assuming that it has no openings. The above expression can be used for $A_{op} < 0.60 A_{inf}$. If the area of the openings in a wall is greater than 60 % of A_{inf}, then the effect of the infill wall can be neglected. Stavridis (2009) has also proposed formulas to account for the effects of an opening on the initial stiffness and strength of an infilled frame.

The current state of knowledge does not allow the establishment of general guidelines for the strut-based modeling of infill walls with openings. Several efforts have been made, especially for infilled steel frames (e.g., Mosalam et al. 1998), but they are far from conclusive. This is the reason why modern evaluation documents such as ASCE/SEI 41 (2007) state that the use of strut-based models for the analysis of infill walls with openings requires judgment and should be conducted on a case-by-case basis. Additional studies are needed to establish appropriate guidelines for truss models that can be applied to perforated infill walls.

Refined Finite Element Models

More refined models, based on the nonlinear finite element method, can be employed for the simulation of infilled frames (Mehrabi and Shing 1997; Hashemi and Mosalam 2007; Stavridis and Shing 2010; Koutromanos et al. 2011). The main advantage of nonlinear finite element analyses is that they can provide detailed information on the initial stiffness, stiffness and strength degradation, and damage pattern of a structure. The analysis of frames with openings in the infill walls poses no additional difficulty, and it can be conducted using the same types of elements and material laws as for solid infill walls.

Analysis for In-Plane Loading

For refined finite element analysis, the fracture behavior of concrete and masonry can be simulated with continuum elements based on the nonlinear fracture mechanics concept. In these elements, cracks are modeled in a smeared fashion, i.e., with a material stress–strain law representing distributed crack development in a continuum rather than a traction–separation law for individual cracks. The constitutive models for these elements must account for the effect of cracking-induced damage and compressive crushing in concrete and masonry. Various formulations, based on plasticity, damage mechanics, or simplified nonlinear orthotropic laws, are available in a number of analysis programs. The reinforcing steel can be represented with truss elements using uniaxial constitutive laws. Appropriate interface or spring elements can be added in a model to capture the bond–slip behavior of the reinforcing steel; however, Mehrabi et al. (1994) have shown that the influence of the bond–slip effect is normally insignificant for infilled frames.

The use of continuum elements alone to model the behavior of the concrete and masonry materials is expected to provide accurate estimates of the response when the damage is dominated by the crushing of the infill. Special care is required for cases where cracks are dominated by mode II fracture or when cracks are not aligned with the element boundaries, which can be the case for diagonal/sliding cracks in infill walls or diagonal shear cracks in RC frame members. In such a case, cohesive crack interface elements must be added in the model to represent cracks in a discrete manner. These elements use "traction–separation" (stress–displacement) laws capable of describing the mixed-mode fracture behavior of cracks and mortar joints. Different elastic–plastic cohesive crack interface constitutive laws have been formulated and are available in analysis programs. A cohesive crack model formulated by Koutromanos and Shing is presented in Fig. 7. The displacement and stress vectors of the model include a normal and tangential (shear) component, as shown in Fig. 7a. The failure surface of the model, shown in Fig. 7b, is characterized by three key strength quantities, namely, the tensile strength, s; the cohesive strength, c (which is the sliding resistance of an interface when there is a zero normal compressive stress); and the asymptotic frictional coefficient, μ. It represents a generalized Mohr–Coulomb law. The failure surface translates

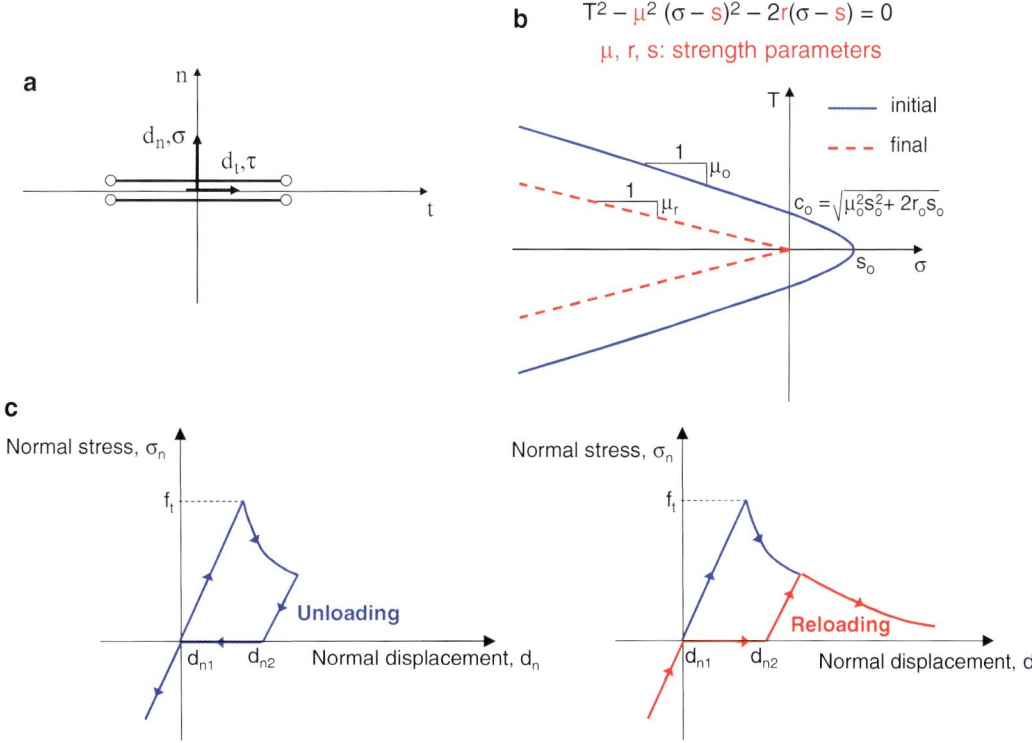

Numerical Modeling of Masonry Infilled Reinforced Concrete Frame Buildings, Fig. 7 Formulation of discrete-cohesive crack interface element for finite element analysis of infilled frames (Koutromanos and Shing 2012). (**a**) Interface element. (**b**) Failure surface. (**c**) Normal tensile unloading–reloading law

and shrinks when fracture occurs, so that the effect of strength degradation is captured. The hysteretic behavior of the cohesive element captures the effect of crack opening and closing as shown in Fig. 7c.

The contact condition between the frame members and the infill walls can also be captured in an analysis using interface elements. Alternatively, if the cohesive strength of the frame–infill interfaces is relatively small and if the frictional resistance along the interface can be entirely attributed to Coulomb friction, standard contact formulations included in many commercial programs can be used instead.

Stavridis and Shing (2010) have used the aforementioned scheme to model masonry-infilled RC frames. They have used both triangular smeared-crack elements and interface elements to simulate the behavior of concrete frame members, as shown in Fig. 8a, and quadrilateral smeared-crack elements and interface elements to model the unreinforced masonry walls, as shown in Fig. 8b. The zero-thickness interface elements used for the mortar joints are only meant to capture the localized mixed-mode fracture along the brick–mortar interface and cannot account for compressive crushing in the mortar. If the compressive crushing of the mortar were to be accounted for, the interaction between the mortar joints and brick units would need to be explicitly accounted for in the analysis. This interaction can strengthen the mortar layer and weaken the brick units and result in a masonry assembly whose compressive strength is between those of the two constituents. Since this interaction is not captured with the use of the zero-thickness interface, it should be accounted for indirectly by adjusting the properties of the continuum elements in compression to represent the compressive behavior of the

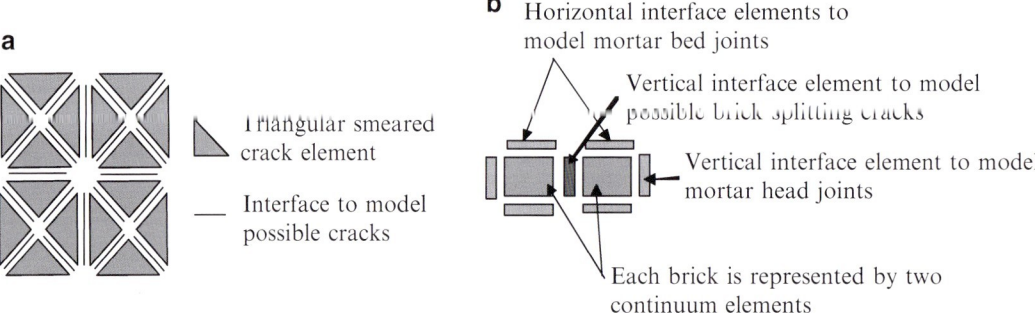

Numerical Modeling of Masonry Infilled Reinforced Concrete Frame Buildings, Fig. 8 Discretization scheme to capture strongly localized cracks in the refined finite element analysis of infilled RC frames. (**a**) Reinforced concrete members. (**b**) Unreinforced masonry panels

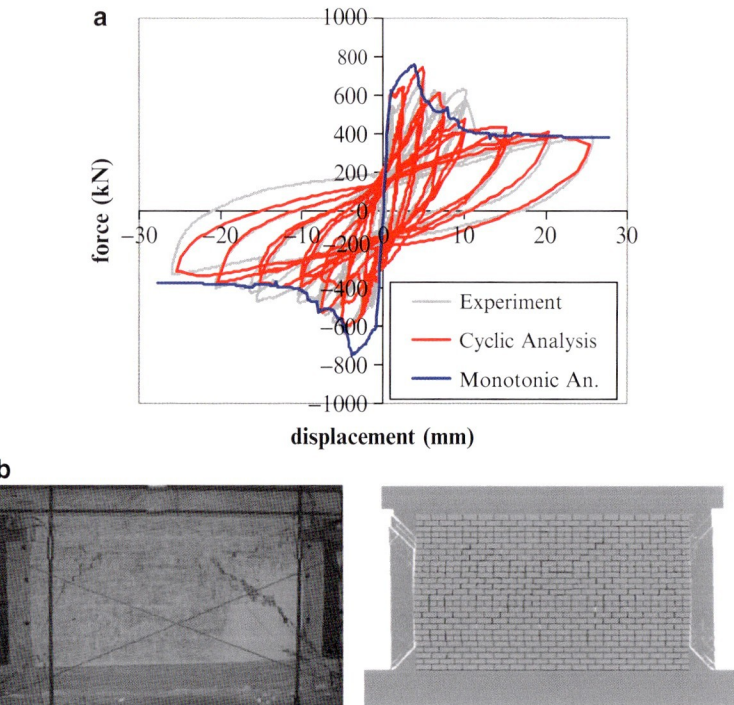

Numerical Modeling of Masonry Infilled Reinforced Concrete Frame Buildings, Fig. 9 Comparison of refined finite element analysis with experimental tests on a single-story, single-bay infilled RC frame. (**a**) Load–displacement curves. (**b**) Crack patterns

masonry assembly. For the RC columns, each reinforcing bar in the columns has been divided into multiple truss elements so that each discrete crack will cross the right quantity of reinforcement.

Koutromanos et al. (2011) have used aforementioned meshing scheme with novel constitutive models to successfully capture the global and local response and damage pattern of infilled frames under static and dynamic loads. One example is shown in Fig. 9.

Analysis for Combined In- and Out-of-Plane Loading

Refined finite element analysis can also consider combined in- and out-of-plane loading. For this purpose, three-dimensional interface elements need to be used with three-dimensional

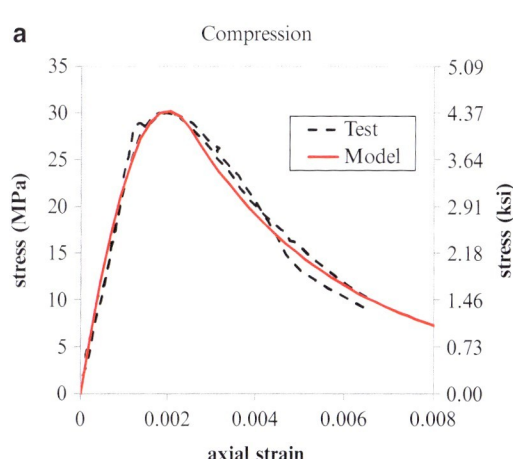
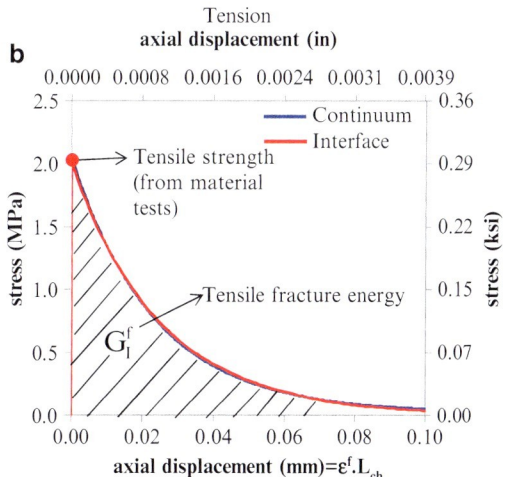
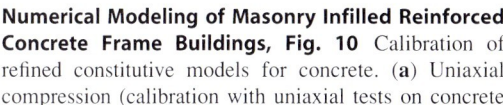

Numerical Modeling of Masonry Infilled Reinforced Concrete Frame Buildings, Fig. 10 Calibration of refined constitutive models for concrete. (**a**) Uniaxial compression (calibration with uniaxial tests on concrete cylinders or masonry prisms). (**b**) Uniaxial tension (calibration with splitting tension tests on concrete cylinders or brick units)

continuum elements or shell elements with smeared-crack formulations. While the current state of the knowledge in material models and element formulations allows the use of three-dimensional finite element analysis for infilled frames under combined in- and out-of-plane loads, there have been relatively few such studies. Hashemi and Mosalam (2007) have used shell elements to model the infill walls subjected to combined in- and out-of-plane loading.

Calibration of Constitutive Models for Nonlinear Finite Element Analysis

Refined constitutive models for continuum elements and cohesive crack interface elements typically include many parameters, which require calibration with data from material tests. Data from uniaxial compression tests on concrete cylinders can be used for the calibration of the continuum material models for the concrete, as shown in Fig. 10a, while data from masonry prism tests (uniaxial compression and bond wrench) can be used for the calibration of the constitutive models for the masonry. Interface elements typically include parameters pertaining to mixed-mode fracture, and data from mixed-mode fracture tests on concrete or masonry mortar joints need to be used for the calibration (e.g.,

Hassanzadeh 1990). For mortar joints in masonry infill walls, the tensile (bond) strength typically ranges between 275 kPa (40 psi) and 690 kPa (100 psi), while the frictional coefficient ranges between 0.65 and 0.90.

An important aspect regarding the calibration of continuum elements that have material laws with strain softening is to avoid the spurious mesh-size sensitivity, which will lead to the loss of objectivity of the numerical results. This is caused by strain localization, as explained, e.g., in Bazant and Planas (1998). To remedy this problem, the softening portions of the stress–strain laws will require regularization, i.e., adjustment to account for the element size. In addition, one needs to consider the fact that in the meshing scheme shown in Fig. 8a, tensile cracking can occur in both the continuum elements and the discrete-cohesive crack interface elements. Obviously, the two types of elements should be calibrated to give identical tensile stress-versus-fracturing displacement behavior, as shown in Fig. 10b. For the continuum elements, the fracturing displacement is equal to the product of the inelastic strain, ε^f, times the characteristic element length, L_{ch}. The area under the stress–fracturing displacement curve is a material constant called the mode I fracture

Numerical Modeling of Masonry Infilled Reinforced Concrete Frame Buildings, Table 2 Values of constant G_{fo}^I for determination of tensile fracture energy of concrete (Fib 1999)

D_{max} mm (in)	8 (0.31)	16 (0.63)	32 (1.26)
G_{fo}^I, N/mm (lb/in)	0.025 (0.14)	0.030 (0.17)	0.058 (0.33)

energy or tensile fracture energy, G_f^I. The following expression, which is proposed in FIB (1999), can be used for the determination of G_f^I:

$$G_f^I = G_{fo}^I \left(\frac{f_{cm}}{f_{cmo}}\right)^{0.7} \quad \text{for } f_{cm} \leq 80 \text{ MPa}$$
$$G_f^I = 4.3 G_{fo}^I \quad \text{for } f_{cm} > 80 \text{ MPa} \quad (6)$$

where f_{cm} is the mean compressive strength of the concrete, f_{cmo} is equal to 10 MPa (1.46 ksi), and G_{fo}^I is a reference value of the fracture energy, which represents the fracture energy for concrete with a compressive strength equal to 10 MPa, and it depends on the maximum aggregate size, d_{max}, as shown in Table 2. The fracture energy for the masonry units and for the mortar joints can be calibrated using data from experimental tests (e.g., van De Pluijm 1997).

Often, there may not be sufficient material test data to calibrate all the material parameters in a finite element model. In such cases, a sensitivity analysis is required to determine the sensitivity of the numerical results to the values of parameters that cannot be determined from material test data. A parametric study by Stavridis and Shing (2010) with finite element models has indicated that the shear strength parameters for mortar joints are most influential on the load–displacement response of an infilled frame.

Determination of Gravity Load Distribution Between Frame Columns and Infill Walls

The behavior of an infill wall strongly depends on the compressive stress in the bed joints. An increased gravity load will increase compressive stress and, thereby, the resistance of the bed joints, thus increasing the stiffness and strength of a wall. An accurate estimate of the gravity load distribution between the infill wall and the surrounding frame is necessary to ensure that the analysis provides accurate estimates of the strength and stiffness of the system. The most straightforward approach to estimate the gravity load distribution for existing structures is to use in situ tests with flat-jacks. Since it may not always be feasible to conduct such tests, analytical models may be used to estimate the fraction of gravity loads carried by the frame columns and infill walls, respectively.

The analytical determination of the gravity load distribution between the frame and the infill wall requires several considerations. First, a part of the gravity loads may be applied onto the RC columns before the construction of the infill walls because these walls could be constructed after the frame has been completed. Second, long-term effects such as concrete and masonry creep, concrete shrinkage, and brick masonry expansion with time due to water absorption can significantly affect the gravity load distribution. While refined finite element models with viscoelastic material properties can be employed for the determination of the gravity load distribution, the increased computational burden of such analyses may not necessarily produce results of increased accuracy due to lack of experimental data to allow the calibration of multiaxial viscoelastic constitutive models.

Based on the above consideration, simplified models are deemed preferable for the determination of the gravity load distribution, since they are easy to calibrate and their reliability is not necessarily inferior to that of more refined models. Such a simplified physical model is shown in Fig. 11a. The model consists of springs and dashpots to represent the instantaneous axial stiffness and the viscoelastic (creep) properties of the RC columns and the masonry infill. The "gap" u_o, shown in Fig. 11a, corresponds to the short-term deformation of the columns due to the gravity loads that are applied before the construction of the infill walls. The spring-and-dashpot assemblages representing the concrete columns and masonry walls can be calibrated with data from

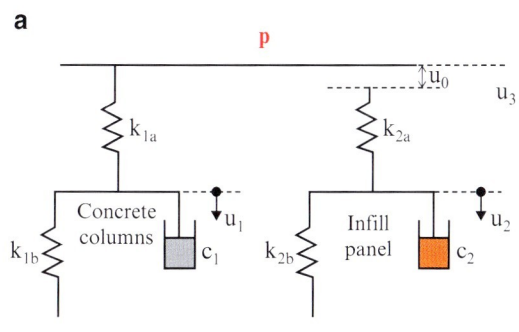

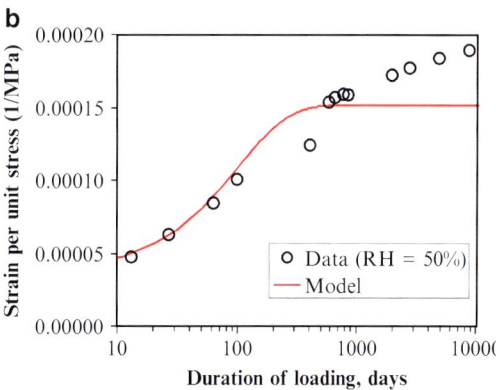

Numerical Modeling of Masonry Infilled Reinforced Concrete Frame Buildings, Fig. 11 Simplified physical model for the estimation of the gravity load distribution in the frame columns and in the infill walls. (**a**) Rheological model for creep. (**b**) Calibration of rheological model for concrete

creep tests on concrete and masonry; Fig. 11b shows an example of the calibration of a creep model for concrete. The effect of the brick expansion can also be easily added in the model. A detailed explanation of the simplified modeling approach and of how the gravity load distribution can be correctly modeled in a finite element model is given in Koutromanos (2011).

Summary: Concluding Remarks

The modeling of infilled frames is a challenging task because a variety of failure mechanisms may affect the load-resistance properties of the system. The analysis can be conducted using simplified models based on the equivalent strut concept or refined finite element models using appropriate constitutive models for the materials. Strut-based models are conceptually simple and easy to calibrate and implement, but their accuracy is inherently limited, especially if walls with openings are to be analyzed. It is not certain whether the use of complicated multi-strut models is meaningful, because such models are bound to misrepresent some important mechanisms that can develop in infilled frames. On the other hand, refined finite element models are more general and realistic, but the efforts on the calibration of the constitutive models and preparation of such analyses are much more significant. Furthermore, refined finite element analyses are computationally demanding. Thus, the selection of the modeling approach depends on the desired level of accuracy, the expertise of the analyst, and the time and computational resources available for the analysis.

References

Al-Chaar G (2002) Evaluating strength and stiffness of unreinforced masonry infill structures. Research report, US Army Corps of Engineering

ASCE/SEI (2007) Seismic rehabilitation of existing buildings, ASCE/SEI 41–06. American Society of Civil Engineers

Asteris PG, Antoniou ST, Sofianopoulos DS, Chrysostomou CZ (2011) Mathematical macro-modeling of infilled frames: state-of-the-art. ASCE J Struct Eng 137(12):1508–1517

Bazant Z, Planas H (1998) Fracture and size effect in concrete and other quasibrittle materials. CRC Press, Boca Raton, 640 p

Crisafulli FJ (1997) Seismic behaviour of reinforced concrete structures with masonry infills. PhD dissertation, University of Canterbury, Cristchurch, New Zealand. Available online at http://ir.canterbury.ac.nz/handle/10092/1221

Crisafulli FJ, Carr AJ (2007) Proposed macro-model for the analysis of infilled frame structures. Bull NZ Soc Earthq Eng 40(2):69–77

El-Dakhakni WW, Elgaaly M, Hamid AA (2003) Three-strut model for concrete masonry-infilled steel frames. ASCE J Struct Eng 129(2):177–185

Federation Internationale Beton du (1999) Structural concrete: textbook on behaviour, design and performance, vol. 1: Introduction – design process – materials. International Federation for Structural Concrete, Lausanne, 224 p

Filippou FC, Fenves GL (2004) Methods of analysis for earthquake-resistant structures. In: Bozorgnia Y, Bertero VV (eds) Earthquake engineering – from engineering seismology to performance-based earthquake engineering. CRC Press, Boca Raton

Hashemi A, Mosalam KM (2007) Seismic evaluation of reinforced concrete buildings including effects of masonry infill walls. In: Report PEER 2007/100. Pacific Earthquake Engineering Research Center, Berkeley

Hassanzadeh M (1990) Determination of fracture zone properties in mixed mode I and II. Eng Fract Mech 35(4/5):845–853

Kadysiewksi S, Mosalam KM (2009) Modeling of unreinforced masonry infill walls. Considering in-plane and out-of-plane interaction. In: Report PEER 2008/102. Pacific Earthquake Engineering Research Center, Berkeley

Klingner RE, Bertero VV (1976) Infilled frames in earthquake-resistant construction. Report UCB/EERC-76/32, Earthquake Engineering Research Center, University of California, Berkeley. Available online at http://nisee.berkeley.edu/elibrary/Text/61000515

Koutromanos I (2011) Numerical analysis of masonry-infilled reinforced concrete frames subjected to seismic loads and experimental evaluation of retrofit techniques. PhD dissertation, University of California, San Diego, La Jolla. Available online at http://nees.org/resources/3579/download/Ioannis_Koutromanos-PhD_Dissertation.pdf

Koutromanos I, Shing PB (2012) A cohesive crack model to simulate cyclic response of concrete and masonry structures. ACI Structural Journal 109: 349-358

Koutromanos I, Stavridis A, Shing PB, Willam K (2011) Numerical modeling of masonry-infilled RC frames subjected to seismic loads. Comput Struct 89(3–4): 1026–1037

Mainstone B (1972) On the stiffnesses and strengths of infilled frames. CP 2/72, Garston [Eng.], Gt. Brit. Building Research [Establishment] Current paper, 90 p

Masonry Standards Joint Committee (2011) TMS 402-11/ACI 530-11/ASCE 5–11: building code requirements and specification for masonry structures, The Masonry Society, American Concrete Institute, and American Society of Civil Engineers, Boulder, CO, Farmington Hills, MI, and Reston, VA

McKenna F, Fenves GL, Scott MH, Jeremic B (2000) Open system for earthquake engineering simulation. http://opensees.berkeley.edu

Mehrabi AB, Shing PB (1997) Finite element modeling of masonry-infilled RC frames. ASCE J Struct Eng 123(5):604–613

Mehrabi AB, Shing PB, Schuller MP, Noland JL (1994) Performance of masonry-infilled R/C frames under in-plane lateral loads. In: Report CU/SR-94/6. Department of civil, environmental & architectural engineering, University of Colorado at Boulder

Mosalam KM, White RN, Ayala G (1998) Response of infilled frames using pseudodynamic experimentation. Earthq Eng Struct Dyn 27(6):589–608

Polyakov SV (1960) On the interaction between masonry filler walls and enclosing frame when loaded in the plane of the wall. In: Polyakov SV (ed) Translation in earthquake engineering. EERI, San Francisco, pp 36–42

Shing PB, Stavridis A (2014) Analysis of seismic response masonry-infilled RC frames through collapse. ACI Special Publication 297

Stafford Smith B (1967) Methods for predicting the lateral stiffness and strength of multi-storey infilled frames. Build Sci 2:247–257

Stavridis A (2009) Analytical and experimental study of seismic performance of reinforced concrete frames infilled with masonry walls. PhD dissertation, University of California, San Diego

Stavridis A, Shing PB (2010) Finite element modeling of nonlinear behavior of masonry-infilled RC frames. ASCE J Struct Eng 136(3):285–296

Van der Pluijm R (1997) Non-linear behaviour of masonry under tension. Heron 42(1):25–54

Ocean-Bottom Seismometer

Mechita C. Schmidt-Aursch[1] and
Wayne C. Crawford[2]
[1]Alfred-Wegener-Institut, Helmholtz-Zentrum für Polar- und Meeresforschung, Bremerhaven, Germany
[2]Institut de Physique du Globe de Paris, Sorbonne Paris Cité, Univ Paris Diderot, UMR 7154 CNRS, Paris, France

Synonyms

OBS

Introduction

About 70 % of the world is covered by oceans. Because of the difficulty of accessing the ocean floor, most of the seafloor and the crust below was unexplored for a long time. In the early 1930s, the seismic refraction method was developed and geoscientists tried to develop techniques to use this method offshore. They experimented with cabled sources and geophones but also with free-fall instruments. The first layout of a stand-alone ocean-bottom seismometer (OBS) was published in 1938 (Ewing and Vine 1938) and tested in the years 1939–1940. This OBS used a gasoline-filled rubber balloon for buoyancy, which floats approx. 3 m above the seafloor. An aluminum housing containing an automatic oscillograph was mounted below the balloon. The iron ballast and the external geophone for recording man-made explosive seismic signals were located on the ocean bottom (Ewing et al. 1946).

After these first experiments, there were only intermittent OBS deployments until the late 1950s and early 1960s, where seismic monitoring of nuclear explosions became suddenly important. For this purpose, a uniform monitoring station distribution around the globe was desired, so seismic stations in the oceans were necessary. Programs like the "Vela Uniform project" promoted advancements in OBS technology (VESIAC 1965). Since then, the usage of OBS for passive earthquake recording and active-source experiments has become more and more common, as many scientific targets are located offshore, including continental margins, mid-ocean ridges, gas and oil reservoirs, and potential tsunami-generating areas.

Today, a great variety of OBS exist for different scientific purposes: short-period instruments mainly for active-source experiments, broadband seismometers for passive earthquake recording, small-sized OBS for short-term deployments, and larger platforms with independent operating times of 1 year or more. Technology developed rapidly since the first OBS deployments, but the principle remains the same: an instrument carrier, equipped with one or more sensors, a data logger with batteries, and a release unit, is weighted with some ballast anchor and sinks freely down to the

Ocean-Bottom Seismometer, Fig. 1 Stacked OBS frames with floatation consisting of glass spheres (*left*) and syntactic foam (*right*) (Photographs W. Crawford, Paris (*left*) and M. Schmidt-Aursch, Bremerhaven (*right*))

seafloor to record autonomously natural or man-made seismic signals. After a certain time or after receiving a hydroacoustic signal, a release unit detaches the anchor and the OBS ascends to the sea surface using some kind of buoyancy. The instrument can then be recovered to retrieve the data for further processing and interpretation.

Besides these free-fall and pop-up ocean-bottom seismometers, there also exist systems with real-time data transfer using moorings or cables. These are sometimes also called ocean-bottom seismometers, although they belong to the category of ocean-bottom observatories. This entry gives an overview of the principles of self-contained OBS without cable connection or any other telemetry, introduces the state of the art of the main component technologies, and explains the special demands of OBS data processing.

General Hardware Requirements

Ocean-bottom seismometers are deployed in a rather hostile environment: the seafloor. Ambient temperatures are low, the ambient pressure is high, and water salinity causes metal parts to corrode. The seismometer should be well coupled to the seafloor and withstand water currents, and the instrument should cause as little noise on the seismometer as possible. The stations might face high accelerations and shocks during handling on the ship's deck, during deployment and recovery, and on touchdown at the seafloor. An OBS must therefore be mechanically robust and solidly built. Pressure casings for water depths of several thousand meters – a water pressure of tens of MPa – must have thick and homogenous walls, and the floatation must be highly incompressible. To resist shocks, the frame, buoyancy, and all other parts must be of rugged design.

On the other hand, the instruments should be as compact as possible, because transport costs are high and deck space is limited on most vessels. Most OBS designs allow stacking of the frames and the floatation (Fig. 1) and storing the other parts in standard transport boxes. As ship time is limited and expensive, the instruments should be easy to mount so that a large number of units can be prepared in a short time. The mounted devices must be able to be handled on deck and by crane during bad weather conditions. For recovery, easy sighting of the instruments in the waves and a safe pick up from the sea surface is desirable. There is no ideal OBS design to

Ocean-Bottom Seismometer, Fig. 2 Examples of ocean-bottom seismometers. *Top*: OBS with a mechanically isolated broadband seismometer and differential pressure gauge for long-term deployments. Floatation: glass spheres; frame: high-density polypropylene (*PEHD*); pressure tubes: anodized aluminum (Photograph W. Crawford, Paris). *Bottom left*: OBS with integrated wideband seismometer and hydrophone for long-term deployments. Floatation: syntactic foam; frame and pressure tubes: titanium alloy (Photograph M. Schmidt-Aursch, Bremerhaven). *Bottom right*: compact OBS with integrated short-period seismometer and hydrophone for short-term deployments. Floatation and housing for seismometer; data logger; batteries; strobe and radio beacon: glass sphere; frame: aluminum (Photograph courtesy of F. Klingelhöfer, IFREMER, Brest)

satisfy these specifications; therefore, various designs have been developed that are optimized for specific purposes, like especially compact and light-weighted OBS for active-source experiments, using large numbers of instruments or particularly solid units with larger internal volumes for long-term deployments (Fig. 2).

The harsh conditions offshore also require enhanced specifications for the electronics. All electronic parts must be shock resistant: this is a challenge for sensitive instruments like broadband seismometers or the most common data storage devices. Cables and connectors outside of the pressure tubes must be watertight, saltwater resistant, and designed for high pressures.

Placed on deck of a ship, OBS are exposed to a wide temperature range as the vessel might operate in the sunlit tropics or in polar regions where even the water temperatures are below zero. Therefore, the electronics must be able to operate over a large temperature range. Batteries are so far the only power supply at the seafloor for untethered systems; hence, all components must have low power consumption. In contrast to onshore stations, limited or no maintenance is possible on the ocean bottom. Most OBS can receive some remote-control commands, but for the acoustic link, the presence of a vessel is necessary. The instruments must therefore be highly reliable during long-term deployments.

Instrument Carrier

Frame and Floatation

The base of every OBS is some kind of frame, on which all the other components are mounted. The frame must resist high pressures at the deep sea and the chemically aggressive subsea environment. It must be rigid and stable enough to carry all attaching parts, but on the other hand, it should be as compact and easy to handle as possible. These structures are mostly built of aluminum or titanium alloy or some synthetic material like, e.g., polypropylene. Aluminum alloy is lightweight and cheap, but not as resistant against corrosion as the more expensive titanium alloy. Synthetic materials are rust-free but the frames need larger diameters or thicknesses to reach the same mechanical strength as the slim metal carriers. Figures 2 and 3 present three different OBS types with structures built of aluminum, high-density polypropylene, and titanium alloy.

The frame needs some buoyancy to ascend to the surface after termination of the seismic measurements. The floatation must withstand very high pressures up to tens of MPa in a couple of thousand-meter water depth. The first OBS prototypes used oil- or gasoline-filled rubber balloons for floatation. The densities of oil and gasoline are lower than the density of water, and in addition, oil and gasoline are nearly incompressible. Therefore, the balloons kept their volume and hence their static buoyancy also in deep water. Soon the application of evacuated glass spheres as floatation was developed; their usage is still very common today. They are lightweight and can simultaneously be used as housing for seismometers and other electronic parts. Figures 2 and 3 (right) show a very compact short-period OBS mainly designed for active-source experiments, where the seismometer, data logger, batteries and recovery strobe light, and radio beacon are integrated into the buoyancy glass sphere. Larger OBS for long-term deployments need a couple of spheres to achieve enough buoyancy to carry two or more pressure cylinders full of electronics and batteries back to the surface (Figs. 2 and 3, top). Glass spheres are very shock sensitive; therefore, an additional plastic casing (e.g., polyethylene) is necessary. If a glass sphere contains electronics, the casing must be perforated to allow access to connectors or to provide visibility of internal flash lights. Nevertheless, they risk implosion at the seafloor and hence the loss of the instrument.

Since the early 1960s, syntactic foam has also been used for floatation. Syntactic foam is a synthetic compound of hollow particles in a background matrix of ceramic, metal, or polymer. An epoxy matrix with hollow glass microballoons with diameters less than 1/100 mm is widely used for deep-sea floatation. This foam is highly incompressible, showing a small reduction in buoyancy only after several years of deep-sea deployments. Syntactic foam can easily be customized to various shapes like plates, blocks, or barrels, so customized OBS designs are possible (Figs. 2 and 3, bottom left). Floatation units made of syntactic foam are robust and easy to handle but much heavier than the glass spheres. The majority of OBS floatation is painted in signal colors to help in spotting the instrument after emerging at the sea surface.

Anchor Weight and Release Unit

An anchor weight is necessary for the OBS to sink to the seafloor. Most anchors are made of untreated metals like iron or steel; some models also use concrete blocks. The entire system of the frame, sensors and batteries, floatation, and anchor must be carefully balanced in air as well as in water. The weight or buoyancy is different in salt water, freshwater, and air, so a good calculation of buoyancy is necessary. The OBS should sit solidly on the deck, hang well balanced beneath a crane, and should sink and rise in the ocean without swinging or rotating. It must have enough weight and a low enough center of gravity in water to be stable at the seafloor, resisting bottom currents and providing a good coupling of the seismometer to the ground. On the other hand, the impact of the OBS on the sea bottom at deployment should be smooth to avoid damage, especially to the sensitive seismometer. If the instrument weighs too much in water or sinks too fast in the water column, it might sink deep enough into soft sediments to be stuck.

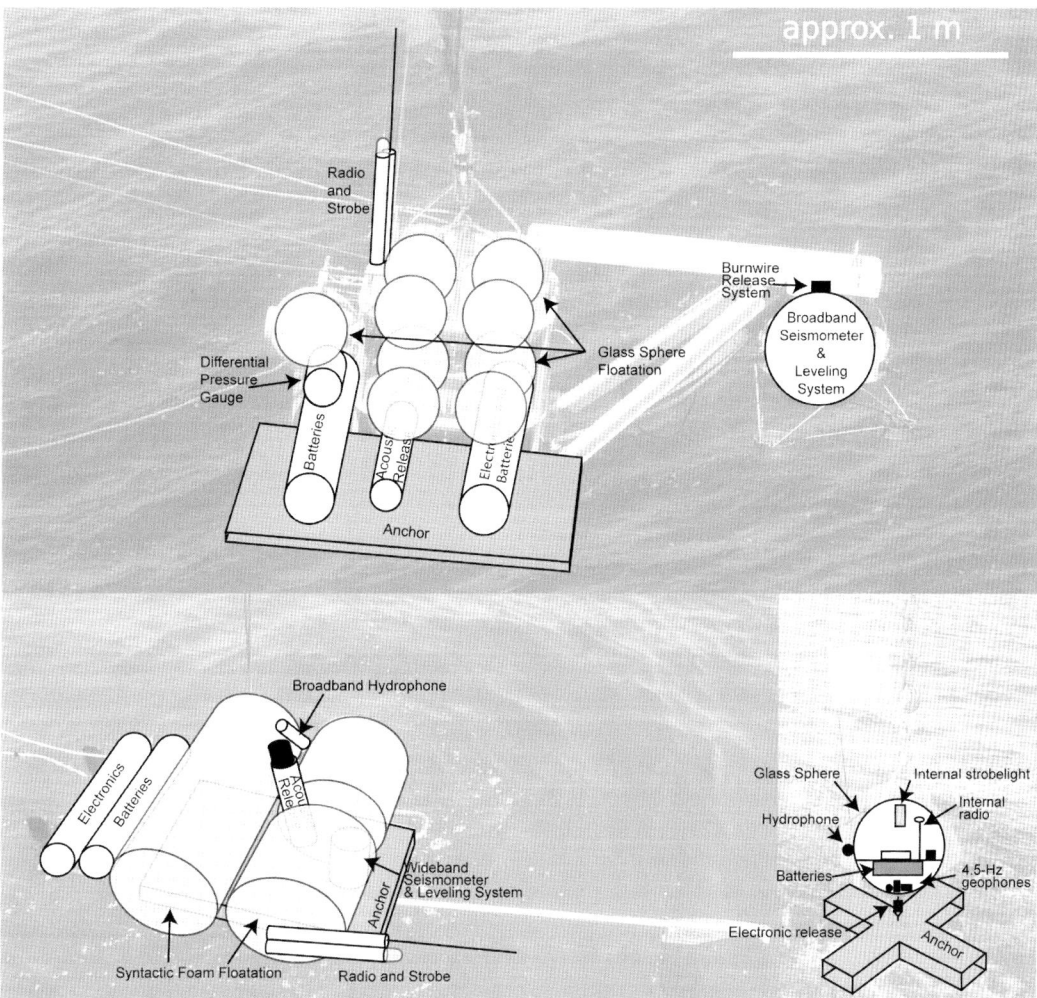

Ocean-Bottom Seismometer, Fig. 3 Sketch of ocean-bottom seismometers. *Top*: OBS with a mechanically isolated broadband seismometer and differential pressure gauge for long-term deployments. *Left*: OBS with integrated wideband seismometer and hydrophone for long-term deployments. *Right*: Compact OBS with integrated short-period seismometer and hydrophone for short-term deployments (Drawing adapted from Auffret et al. (2004))

A central element of an OBS is the release unit, which connects the frame with the anchor weight. After a certain time (timed release) or on receiving an acoustic command (acoustic release), the release unit disconnects the anchor from the OBS and the OBS can rise freely to the sea surface. The ballast remains at the bottom where it degrades naturally. There are two main release mechanisms: a burn wire or a motor-driven turning hook. A motor-driven releaser clamps the anchor with a hook that is held by a bolt (Fig. 4, right). The bolt turns mechanically, the hook is released, and finally the anchor is decoupled. A burn wire release holds the anchor using a wire that is partially exposed to seawater. A second larger wire or metal post is also exposed to seawater. When a positive voltage is applied between the wire and the post, electrolysis corrodes the wire, which then breaks, releasing the anchor (Fig. 4, left). The "burn" time depends on the length and thickness of the exposed wire: 10–15 min is typical for many systems.

Ocean-Bottom Seismometer, Fig. 4 Example of release units. *Left*: motor-driven release unit with integrated transducer (Photograph M. Schmidt-Aursch, Bremerhaven). *Right*: burn wire mechanism (Photograph courtesy of A. Ndiaye, INSU, Paris)

Most release units can also send back a response, allowing a two-way hydroacoustic link to the OBS, e.g., to determine the distance to the system, to ask for state of health, or to send simple commands. Some research vessels are equipped with permanently installed transducers, but more often, mobile transducer systems are used on board. For the mobile units, the ship has to stop and the transducer lowered into the water to achieve a coupling of the acoustic signal to the water column. The hydroacoustic signal uses frequencies between 5 kHz and 20 kHz with different modulation techniques. Normally, the signal is coded; each unit uses unique codes for release and other commands. Because the release unit is the central component to recall the OBS, its power supply is usually completely independent from the main system. Some models even use independent batteries for communication and for activating the release.

Flag, Radio Beacon, and Flasher

After an OBS reaches the surface, the next challenge is to recover it. The OBS is small compared to the area of the ocean in which it may surface and local currents might make the OBS drift away from the deployment position on its way down and up through the water column. After emerging at the surface, currents and wind may also move the OBS away. Waves and decreased visibility due to rain or fog can hinder the search. Painting the OBS with a signal color (yellow, orange, or red) helps observers to spot the instruments. A flag can also help sighting OBS at large distances or in tall waves, but high-profile flags can also generate noise on the data, so not all OBS have flags.

However, almost all OBS are equipped with a VHF radio beacon and a Xenon flasher (Fig. 5, middle and left). The casing of the beacons is usually made of aluminum and the radio antenna is made of steel, so sacrificial anodes are necessary to avoid electrolytic corrosion of the aluminum. The radio beacon and strobe both contain a pressure switch that turns the units off at 1–10-m water depth to save battery power while the instruments are at the seafloor. The strobe light additionally uses a light sensor that only allows flashing in the dark, also for battery saving purposes. The flashers are especially useful during night, the light can be spotted over larger distances than the flags, but the estimation of the station distance is more difficult than during daylight.

Radio beacons are available with different frequencies or different transmitting cycles to distinguish individual OBS in case that several instruments are "on the air" at the same time. The majority of VHF beacons operate in nautical frequency bands, so besides handheld direction finders, also shipborne cross bearing receivers can be used to determine the direction of the radio signals. The signal range at the sea surface is quasi-optical, which means approximately to the optical horizon, from elevated locations or helicopters, the OBS can be located over

Ocean-Bottom Seismometer, Fig. 5 Example of a Xenon flasher (*left*), VHF radio beacon (*middle*), and a satellite transmitter (*right*) (Photograph M. Schmidt-Aursch, Bremerhaven)

distances of several tens of kilometers. Some models integrate a GPS system and send position data using the "Automatic Identification System" (AIS) that is mandatory for many vessel classes. There are also beacons that send an identifying signal to a satellite after emerging at the sea surface (Fig. 5, right). The operator then receives an email containing the position of the OBS. This remote surveillance technique is especially useful, if the station rises ahead of time or is caught by a trawler, but the transponders are bulky and costly, require additional license fees, and are therefore not widespread.

Sensors and Data Logger

Seismometer

The main purpose of an OBS is to record seismic signals on the sea bottom; therefore, a seismometer specially adapted to the marine environment is necessary. For active-source experiments (using man-made sources) or local seismicity studies, a short-period seismometer, sometimes also called geophone (Greek for "earth sound"), is sufficient. The most common are the three-component seismometers with a corner frequency of 4.5 Hz. These passive analog sensors consume no power and are relatively compact. Either they are mounted in a small pressure casing on the frame of the instrument or they can even be integrated into the main casing with other electronics (Fig. 3, bottom right). Some models are equipped with detachable seismometers, which are normally released by a burn wire or corrosion mechanism after arriving at the seafloor. High-frequency geophones are often leveled, usually with a passive, gravity-based system, a so-called gimbaled system.

To study lower-frequency signals, a wideband or broadband seismometer is required. For offshore deployments, broadband seismometers with a pass band out to 120–240 s are available, but wideband seismometers with a pass band out to 30–120 s are more common because of their smaller size and lesser power consumption. These active wide- or broadband seismometers output one vertical and two horizontal components. Similar to onshore stations, the seismometers must be perfectly leveled in order to work properly: in fact, it is even more important at the seafloor because seafloor currents can create noise on the horizontal component that rotates on to the vertical component if the seismometer is not perfectly leveled (Crawford and Webb 2000). Because the tilt of the OBS on the seafloor is unknown, all wideband and broadband seismometers contain an automatic electromechanical leveling system (Fig. 6). Leveling is activated some hours after the OBS arrives at the seafloor and is usually repeated from time to time for long-term deployments. Care must be taken not to level (or even check the level) too frequently, as the switching on of the leveling circuitry can introduce a noise spike in the data. There are several ways to level either the entire seismometer package or each component separately. Some models are gimbal mounted in two directions (Fig. 6); other models contain only one gimbal but additionally turn the seismometer until all three components are leveled.

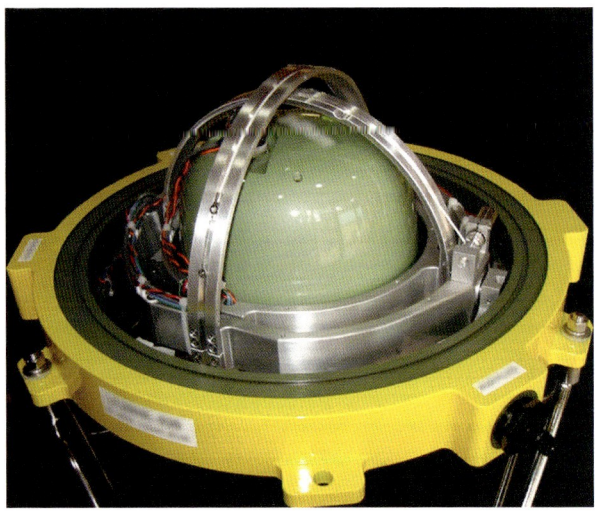

Ocean-Bottom Seismometer, Fig. 6 Example of a broadband seismometer (*green sphere*) mounted in a two-axis gimballing system. The pressure case was removed (Photograph W. Crawford, Paris)

For all seismometers, a good coupling to the seafloor is essential. Two main approaches are in use today: integrated and mechanically isolated seismometers. Integrated seismometers are fixed directly to the instrument carrier and coupling is provided by the entire system including the anchor weight (Figs. 2 and 3, bottom). These systems are tightly arranged and easy to handle during deployment and recovery. Mechanically isolated seismometers have their own pressure case and are normally mounted on a deployment arm (Figs. 2 and 3, top). The sensor package is automatically detached when the station reaches the ocean bottom. These models are more complex, but movements or tilting of the instrument carrier will not be transmitted to the seismometer. The best coupling would be achieved by burying the seismometer into the sediment (e.g., Duennebier and Sutton 2007), but unfortunately all these instruments still need support from remote-operating vehicles (ROV) for deployment and recovery, which makes such experiments very expensive and time-consuming.

Pressure Sensor

Besides the seismic channels, the majority of OBS types provide a fourth channel recording the signals of a pressure sensor. In water, pressure is much easier to measure than displacement or acceleration. Pressure is omnidirectional, so a single channel is sufficient and no special coupling to the surrounding water is necessary.

For applications only interpreting compressional waves (P-waves), like some active-source experiments, the so-called ocean-bottom hydrophones (OBH) can be used. This kind of instrument is equipped solely with a hydrophone. When combined with three-component seismological data, the additional pressure channel can be used to remove water-bounce phases from seismic signals (the pressure and seismic signals from water bounces have different polarities; see Blackman et al. 1995) and at low frequencies (less than approximately 0.1 Hz) remove low-frequency noise on the vertical seismometer channel (Webb and Crawford 1999) and determine the physical properties of the seafloor ("seafloor compliance," e.g., Crawford et al. 1991).

The overall bandwidth of hydrostatic pressure in the oceans is large; it starts at 100 KPa at the sea surface and increases by 10 Mpa per thousand-meter water depth. Compared to this, pressure changes caused by earthquakes are very small (generally less than 1 Pa). An instrument that is able to adapt to the large pressure range as well as to detect small pressure changes over the entire seismic band would be very complex and power consuming. Therefore, various sensors exist, which are optimized for specific purposes. There are absolute gauges quantifying the entire hydrostatic pressure, or differential units, which adapt to the mean surrounding pressure and record only the small changes.

Ocean-Bottom Seismometer, Fig. 7 Examples of pressure sensors. *Left*: instrument carrier with a differential pressure gauge (*DPG*) and an absolute pressure gauge (*APG*). *Right*: APG (*top*), hydrophone (*middle*), and DPG (*bottom*) (Photographs M. Schmidt-Aursch, Bremerhaven and W. Crawford, Paris (*top right*))

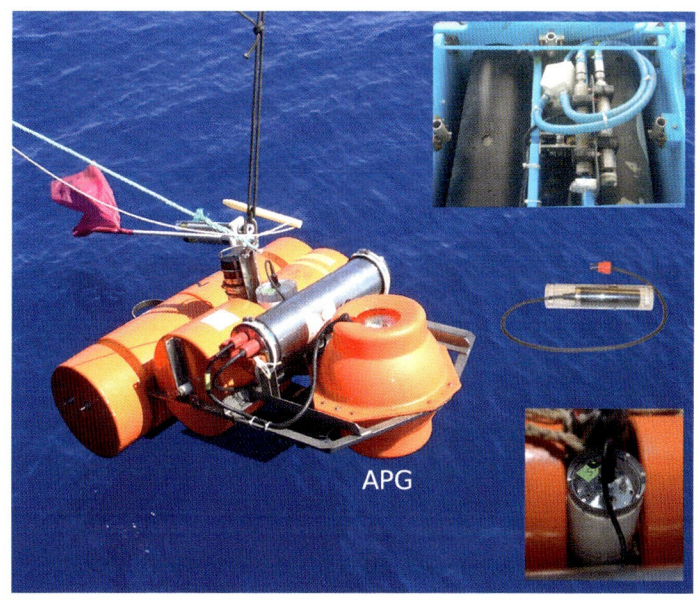

The simplest differential pressure sensor is an analog hydrophone (Fig. 7, middle). The first hydrophones were utilized during World War I, mainly to detect submarines. There are several ways to convert pressure changes into electricity, including moving coils in solenoids actuated by membranes and interferometric fiber-optic coils, but most OBS are equipped with a pressure-compensated hydrophone using the piezoelectric effect. It consists mainly of a piezoelectric ceramic cylinder, which linearly generates a voltage when opposed to mechanical stress. The entire system is encapsulated in an elastomer, e.g., polyurethane. Hydrophones are compact, passive sensors that need no additional power supply unless they integrate a preamplifier. Similar to seismometers, hydrophones are available in various short- and long-period versions with frequencies from approx. 50 kHz down to 0.01 Hz. The hydrophone acts electronically as a capacitor; therefore, the high-impedance input of the data logger must carefully be adapted to achieve both short settling times at the seafloor and linear sensitivity down to very low frequencies.

Differential pressure gauges (DPGs, Cox et al. 1984) can measure pressure signals from approximately 0.0005 Hz to 40 Hz and are popular on broadband OBS. Differential pressure is measured between a reference chamber and the ocean using a strain gauge with small dynamic range (Fig. 7, left and bottom, and Fig. 8). The reference chamber and the gauge around it are filled with silicon oil, and the gauge is left open to ocean pressures by a soft membrane. The pressure in the reference chamber is kept close to that of the ocean using a capillary leak and, for deployment and recovery, overpressure relief valves. The capillary leak creates a high-pass filter whose corner frequency depends on the tube's diameter and length as well as the viscosity of the oil within the gauge; typically, a value of 0.002–0.004 Hz is sought. These gauges have proven difficult to calibrate accurately. In principle, changes in the viscosity of the oil with temperature and pressure should change the high-pass corner frequency, but in practice the biggest uncertainty has proven to be the absolute gain which appears to be 15–20% lower than that measured by an absolute gauge, even for DPGs that were carefully calibrated in the laboratory.

Absolute pressure gauges (APGs, Fig. 7, left and top) can be used to measure pressures from approximately 1 Hz to DC. The most commonly available absolute pressure gauges for ocean floor experiments had, until recently, a noise floor of

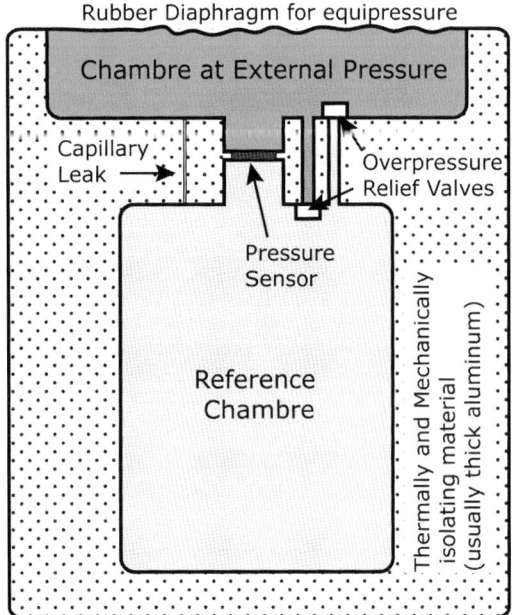

Ocean-Bottom Seismometer, Fig. 8 Sketch of a differential pressure gauge (*DPG*)

about 10 Pa, but sensitivities are now down to about 0.1 Pa. APGs are expensive and power hungry, but they are usually very well calibrated and they can also measure tsunamis and seafloor vertical motions. The pressure is measured by counting the oscillations of a quartz crystal resonator, which is different from the voltage-based signals from hydrophones, DPG, and seismometers. Several commercial companies sell absolute pressure gauges designed for full ocean depth: one possible configuration would be to deploy an APG sampling at a relatively low rate strapped on to an OBS containing a hydrophone.

Data Logger

The analog signals of the sensors are recorded by a data logger. The technology has evolved a long way from the early photographic oscillographs to the present-day 24-bit digitizers with solid-state-disk storage media, and development continues rapidly. Many efforts have been made to provide the same technical specifications – like dynamic range or size of data storage – for OBS as for onshore stations. Nevertheless, this has not been fully possible because of the added requirements of seafloor stations: besides stringent requirements for mechanical and electric robustness, seafloor operations demand small dimensions (to fit into the pressure cases), low power consumption (to enable fully autonomous long-term deployments), and a high-precision clock. Size and power consumption are decreasing and data quality and storage are increasing continually as technology advances, but timing is still an issue.

Onshore stations can receive wireless time signals (e.g., DCF77 or GPS) to synchronize the internal clock. On the seafloor, this is unfortunately not possible, so great efforts have been made to develop high-precision clocks. Temperature-stabilized clocks using miniature ovens have been used but they are generally power hungry. Most loggers currently use as their time base temperature-compensated, microprocessor-controlled crystal oscillators (MCXO) with a frequency deviation in order of 10^{-7} to 10^{-8}. Chip-scale atomic clocks (CSAC) with much higher accuracy up to 10^{-10} may be used in the future, but they consume currently more power than crystal oscillators and there are some reliability/aging issues. The internal clock of the data logger is synchronized to an external time signal (e.g., GPS) before deployment and after recovery to determine the drift ("skew") of the clock. Modern standard clocks show drift rates less than 500 ms/year. The clocks should not only feature a small absolute drift, drift rates should be constant over time to enable a linear time correction of the data afterward. Constant seafloor temperatures contribute to this, but large temperature contrasts and mechanical stress during deployment and recovery must be balanced by the system. For safety reasons, a backup battery independent from the main power supply is desirable for the internal clock.

Specific Requirements of Data Processing

Location of OBS

In contrast to onshore stations, for which the exact location can easily be determined, only the deployment position of the OBS is known.

Most OBS sink relatively slowly to the seafloor (less than 1 m/s) to prevent a hard impact on the ground, so, especially for large water depths, there is much time for currents to push the station away from the direct path. Depending on instrument design, current speed, and sinking time, an OBS might drift several hundreds of meters before reaching the bottom. This discrepancy is too high for experiments mapping small-scale subsurface structures with active sources and specially designed passive arrays that need exact station spacing.

Deployed OBS can be located with the help of the acoustic release unit. Most release units can respond to the onboard unit and the onboard unit can calculate the two-way travel time. Using the sound speed in water enables a calculation of the distance to the station, and ranging from different positions allows a triangulation of the OBS location at known water depth. For good constraints, the instrument should be ranged at from at least three points at approximately equal distance on a circle around the OBS-presumed position. The farther the interrogation, the better the horizontal resolution, but the instrument stops responding beyond some distance, often similar to the water depth. If the water depth is not exactly known, a ranging directly over the instrument is required. Some onboard units are equipped with more than one transducer, and small travel time differences of the response at the transducers can be used to determine the bearing to the station, allowing positioning (in theory) from a single point. Together with the ranged distance and known water depth, the location of the instrument can be determined.

If the OBS experiment is combined with multichannel or wide-angle seismics, the recorded airgun shots of the seismic profiles can be used to determine the position of the instrument on the seafloor. The calculation will be done after recovery; no additional ranging with the onboard unit is necessary. The procedure is similar to the direct ranging via the acoustic release unit; at least three shots evenly distributed around the OBS are necessary. Knowing the exact time and position of the shots, the sound velocity of water, and the water depth, the distances between shots and OBS and hence the location of the instrument can be computed from the first arrivals of the water wave in the seismograms.

Orienting Horizontal Components

In a free-fall deployment, no method is known to assure that the OBS seismometers' north axis is aligned with geographic north. For active-source experiments, the orientation of the horizontal components is not always important, but it is essential for many passive seismological methods (e.g., receiver functions, shear wave splitting). There are two main approaches to identify the OBS orientation: direct determination by an additional sensor or indirect estimation by analyzing the seismological data.

The simplest direct determination of the orientation would be an electronic compass measuring the three components of the Earth's magnetic field and subsequent calculation of the seismometer's orientation with respect to magnetic north. Unfortunately, magnetic compasses are strongly affected by large metal parts (e.g., anchors) and local variations of the declination, which can be large, especially in volcanic areas. Fiber-optic gyros could be an alternative as they compute the true north direction by measuring the rotation of the Earth, but these sensors are still too expensive and power consuming to be installed on OBS as a standard. Microelectromechanical systems (MEMS) could offer an efficient solution in the future. MEMS accelerometers and gyroscopes, which are widespread in inertial navigation and portable electronic devices like mobile phones, would record accelerations and rotations of the instrument on its way through the water column. The spatial orientation of the instrument could then be calculated for each moment, from the initial point on deck, where the orientation of the OBS is well known, to the final arrival at the seafloor. With this method, the known orientation of the station on deck could be transferred to the sea bottom, but the MEMS systems are not yet stable and precise enough to cope with the intense movements of an OBS during deployment.

The horizontal orientation of the seismometer can also be estimated by analyzing the recorded data. The polarity of P-wave arrivals from explosives or airgun shots can be used to determine the orientation of the sensor (e.g., Anderson et al. 1987), but in most cases no sources with equally distributed azimuths are available. Instead of man-made signals, the polarization of P- and Rayleigh waves from teleseismic events can be analyzed (Stachnik et al. 2012), but this also requires a good azimuthal distribution of the signals in order to average out anomalies due to structure. Another problem is the naturally high noise level on the horizontal OBS components in combination with the small amplitudes of teleseismic events. A new approach is to use ambient noise instead of teleseismic events (Zha et al. 2013). For this method, virtual Rayleigh waves are calculated from the cross-correlations of vertical and horizontal components of various station pairs. These virtual Rayleigh waves can be used for a similar polarization analysis as that applied to teleseismic data.

Timing

Although the internal clocks in OBS data loggers are synchronized to an external time signal (e.g., GPS) before launching and after retrieving the instrument, the clock can stop before the final synchronization or an incorrect reference time can be recorded during one of the synchronizations. The offset between the internal clock and the external signal after recovery ("skew") must be corrected, as many applications (e.g., active seismics and event localization) rely on a precise common time base of all stations.

There are two main methods used to correct/verify drift rates in the seismological data, and both procedures assume a constant linear clock drift over the entire recording time. Using the first method, earthquakes or shots are located using data from the OBS, and then the time residuals for each station are plotted as a function of time. A bad timing or non-drift-corrected instrument can be identified by a linear trend in the time residuals. A bad clock drift is often caused by misreading the synchronization time by a multiple of 1 s, so the correction is quite simple to calculate. A non-drift-corrected instrument will have a nonintegral seconds offset at the end of the experiment, allowing a less precise correction.

If several instruments are badly corrected, it can be challenging to determine which one is faulty as several time residuals will exhibit interrelated drift. Ambient noise analysis can also be used to determine the clock drift (Hannemann et al. 2013). Cross-correlations of the vertical seismometer components of station pairs show time shifts if the stations are not properly time corrected. This method also tests the synchronizations and the linearity of the clock drift.

Passive seismological data are normally divided into daily files to keep the huge data amounts manageable. The start time for each daily file is corrected to account for the time drift. Most seismological data formats do not have an accurate enough sample rate specification to account for the drift, which is generally smaller than 1 part in 10^8 (e.g., the true sampling rate for nominally 100 sps data could be 100.000001 sps), so the time at the end of the daily file will be off by the drift rate times the 86,400 s in 1 day. Normally, this offset is smaller than one sample, but this can produce gaps in the data stream. To avoid these gaps, the entire data set can be resampled to a new, skew-corrected sampling rate. Now, the data will be continuous, but the shape of the signal may be slightly changed. Whether resampling is used or not depends on the further usage of the data and analyses to be applied.

Alternatively, some data formats (e.g., miniSEED) allow the time to be specified several times throughout the day. In the miniSEED format, the start time, the time correction, and whether this time correction was applied to the start time can be specified in every record header, which typically occurs every 500–2,000 samples (for a 4,096-byte record length and depending on compression). Although this allows an accurate specification of the clock drift, the software reading the data must still be able to stitch the data together properly when assembling multi-day records. How to do this does not yet appear to have been defined.

Noise Reduction

Noise levels of OBS data are generally higher than of onshore data, because ocean-bottom seismometers are less coupled to the ground and the oceans are a very noisy environment. At low frequencies (<0.1 Hz), the two major sources of noise are seafloor currents and ocean waves, which have the strongest effect on, respectively, the horizontal channels and the pressure channel. Much of this noise can be removed from the vertical channel by removing the correlated signal on the pressure and horizontal seismometer channels (Webb and Crawford 1999; Crawford and Webb 2000).

The current noise is removed by calculating the transfer functions (the amplitude ratio between the two in the frequency domain) between the horizontal channels and the vertical channel in the absence of earthquakes and then subtracting the horizontal channels times this transfer function from the vertical channel. Next, the transfer function between the pressure channel and the vertical channel is calculated, and the pressure data times this transfer function is subtracted from the vertical channel. If $G_{AA}(f)$ is the power spectral density (mean squared fast Fourier transform (FFT) over several windows) of the channel we want to remove noise from, $G_{SS}(f)$ is the power spectral density of the channel containing the noise source, and $C_{AS}(f)$ is the coherency between them, then the transfer function is defined as

$$T_{AS}(f) = C_{AS}(f)\sqrt{\frac{G_{AA}(f)}{G_{SS}(f)}}$$

The transfer function is then a measure of how much of the noise on the "source" channel is induced into the vertical channel. Channel A is then "cleaned" by calculating its FFT, $A(f)$, and that of the source channel, $S(f)$, applying the formula

$$A'(f) = A(f) - T^*_{AS}(f)S(f)$$

and taking the inverse FFT of $A'(f)$ to get the time domain signal. Note that the power spectral densities and coherence should be calculated on an earthquake-free section of data, so that the only relation between them is the noise to be removed.

If removing the noise recorded on multiple channels, either a multi-coherence method must be used or each channel must be cleaned of the other before being sequentially used to clean the vertical channel. Calling the vertical channel "Z," the horizontal channels "X" and "Y," and the pressure channel "P," the standard sequence is:

1. Calculate Z', Y' and P' by subtracting the coherent part of X.
2. Calculate Z" and P" by subtracting the coherent part of Y'.
3. Calculate Z''' by subtracting the coherent part of P".

This correction does not distort seismological signals because $T_{ZS}(f) \ll 1$ for the noise sources, whereas $T_{ZS}(f) = O(1)$ for seismological signals. This method cannot be used to improve the horizontal channels, for two reasons. First, the horizontal channel noise is only reduced by a few dB when the correlated noise from the other horizontal channel is removed (compared to 20+ dB for the vertical channel). Second, $T_{XY}(f) = O(1)$ for both the noise and seismological signals, so applying this correction would greatly distort the seismological signal.

These techniques should only be applied at frequencies below the microseism band. In the microseism band, the dominant "noise" signals are seismic surface waves, so removing the correlated pressure or horizontal channel signal would also eliminate, or strongly distort, seismic signals. Removing the horizontal signal times a transfer function from the vertical channel does not significantly alter the shape of any seismic waves that are also recorded on the horizontal, as the seafloor current transfer function is typically on the order of 10^{-3} or less, much smaller than the transfer function for seismic waves.

Summary

Ocean-bottom seismometers are indispensable for exploring offshore structures like mid-ocean ridges, continental margins, hydrocarbon reservoirs, or potential tsunami-generating areas, as well as "mixed" areas such as hot-spot volcanoes, island chains, and many plate interfaces. Since the first OBS deployments in 1939, technology has advanced rapidly and modern autonomous OBS contain state-of-the art mechanics and electronics. All elements are of a rugged design to cope with the harsh conditions on board and on the sea bottom like high pressure and saltwater environment. An OBS consists of an instrument carrier with floatation made of one or more evacuated glass spheres or syntactic foam and an anchor weight coupled to a release unit. After sinking freely to the seafloor, a seismometer, a hydrophone or an absolute/differential pressure gauge, and a data logger record seismic signals continuously. Receiving a hydroacoustic signal or waiting a prespecified time, the anchor is detached from the instrument and the OBS rises back to the sea surface. Signal color painting, radio beacon, flasher, and a flag facilitate the recovery of the instrument. In contrast to onshore stations, ocean-bottom stations need some additional data processing to enhance the data quality like to orient the seismometer, correction of the drift of the internal clock, and reduction of noise on the vertical component.

Cross-References

▶ MEMS Sensors for Measurement of Structure Seismic Response and Their Application
▶ Principles of Broadband Seismometry
▶ Recording Seismic Signals
▶ Seismometer Arrays
▶ Seismic Instrument Response, Correction for
▶ Seismic Network and Data Quality
▶ Seismic Noise

References

Anderson PN, Duennebier FK, Cessaro RK (1987) Ocean bore-hole horizontal seismic sensor orientation determined from explosive charges. J Geophys Res Solid Earth 92(B5):3573–3579
Auffret Y, Pelleau P, Klingelhoefer F et al (2004) MicrOBS: a new generation of ocean bottom seismometer. First Break 22:41–47
Blackman DK, Orcutt JA, Forsyth DW (1995) Recording teleseismic earthquakes using ocean-bottom seismographs at mid-ocean ridges. Bull Seis Soc Am 85(6):1648–1664
Cox CS, Deaton T, Webb SC (1984) A deep sea differential pressure gauge. J Atmos Oceanic Tech 1:237–246
Crawford WC, Webb SC, Hildebrand JA (1991) Seafloor compliance observed by long-period pressure and displacement measurements. J Geophys Res 96(10):16151–16160
Crawford W, Webb S (2000) Identifying and removing tilt noise from low-frequency (<0.1 Hz) seafloor vertical seismic data. Bull Seismol Soc Am 90(4):952–963
Duennebier FK, Sutton GH (2007) Why bury ocean bottom seismometers? Geochem Geophys Geosyst 8: Q02010. doi:10.1029/2006GC001428
Ewing M, Vine AC (1938) Deep sea measurements without wires and cables. Transactions of American Geophysical Union, 19th annual meeting, pp 248–251
Ewing M, Woollard GP, Vine AC, Worzel JL (1946) Recent results in submarine geophysics. Geol Soc Am Bull 57(10):909–934
Hannemann K, Krüger F, Dahm T (2013) Measuring of clock drift rates and static time offsets of ocean bottom stations by means of ambient noise. Geophys J Int. doi:10.1093/gji/ggt434
Stachnik JC, Sheehan AF, Zietlow DW, Yang Z, Collins J, Ferris A (2012) Determination of New Zealand ocean bottom seismometer orientation via Rayleigh-wave polarization. Seismol Res Lett 83(4):704–713. doi:10.1785/0220110128
VESIAC (VELA Seisonic Information Analysis Center, University of Michigan) (1965) A bibliography of seismology for the VELA uniform program: report of VESIAC. Geophysics Laboratory, Institute for Science and Technology, University of Michigan
Webb SC, Crawford WC (1999) Long period seafloor seismology and deformation under ocean waves. Bull Seis Soc Am 89(6):1535–1542
Zha Y, Webb SC, Menke W (2013) Determining the orientations of ocean bottom seismometers using ambient noise correlation. Geophys Res Lett 40. doi:10.1002/grl.50698

Online Response Estimation in Structural Dynamics

Kristof Maes and Geert Lombaert
Department of Civil Engineering, KU Leuven, Leuven, Belgium

Synonyms

Real-time response extrapolation

Introduction

The dynamic forces acting on a structure and the corresponding system response are of great importance to many engineering applications. Often, however, the dynamic forces can hardly be obtained directly by measurements, e.g., for wind loads, and the number of response measurements during operation is in most cases limited due to practical and economical considerations. Therefore, the dynamic forces and corresponding system response have to be determined indirectly from measurements using system inversion techniques. These techniques allow estimating the dynamic forces acting on a structure from the response in a limited number of sensors, while at the same time the system response can be extrapolated to unmeasured locations from the identified forces and system states.

Originally, force identification and state estimation problems were treated separately. Force identification problems were solved offline in a deterministic setting. Many methods were proposed, most of them based on the inversion of the frequency response function (Guillaume et al. 2002; Parloo et al. 2003) or making use of a time-domain approach (Klinkov and Fritzen 2007; Nordström and Nordberg 2002). Several state estimation algorithms have been proposed for linear as well as for nonlinear systems (Wu and Smyth 2007; Hernandez 2011). Currently, the attention is shifted to joint input-state estimation, resulting in numerous recursive combined deterministic-stochastic approaches (Hwang et al. 2009; Ma et al. 2003; Gillijns and De Moor 2007). These methods allow for online response estimation in many engineering fields. This contribution focuses on recursive state estimation and joint input-state estimation algorithms applied for response estimation in structural dynamics.

System Model

In structural dynamics, first principle models, e.g., finite element (FE) models, are widely used. In many cases, modally reduced-order models are applied, constructed from a limited number of structural modes. When proportional damping is assumed, the continuous-time decoupled equations of motion for modally reduced-order models are given by

$$\ddot{\mathbf{z}}(t) + \boldsymbol{\Gamma}\dot{\mathbf{z}}(t) + \boldsymbol{\Omega}^2 \mathbf{z}(t) = \boldsymbol{\Phi}^T \mathbf{S}_p(t) \mathbf{p}(t) \quad (1)$$

where $\mathbf{z}(t) \in \mathbb{R}^{n_m}$ is the vector of modal coordinates, with n_m the number of modes taken into account in the model. The excitation force is written as the product of a selection matrix $\mathbf{S}_p(t) \in \mathbb{R}^{n_{dof} \times n_p}$, specifying the force locations, and a time history vector $\mathbf{p}(t) \in \mathbb{R}^{n_p}$, with n_{dof} the number of degrees of freedom of the FE model and n_p the number of forces. $\boldsymbol{\Gamma} \in \mathbb{R}^{n_m \times n_m}$ is a diagonal matrix containing the terms $2\xi_j \omega_j$ on its diagonal, where ω_j and ξ_j are the natural frequency and modal damping ratio corresponding to mode j, respectively. $\boldsymbol{\Omega} \in \mathbb{R}^{n_m \times n_m}$ is a diagonal matrix as well, containing the natural frequencies ω_j on its diagonal, and $\boldsymbol{\Phi} \in \mathbb{R}^{n_{dof} \times n_m}$ is a matrix containing the mass normalized mode shapes $\boldsymbol{\Phi}_j$ as columns.

The output vector is generally written as

$$\mathbf{d}(t) = \mathbf{S}_{d,a} \boldsymbol{\Phi} \ddot{\mathbf{z}}(t) + \mathbf{S}_{d,v} \boldsymbol{\Phi} \dot{\mathbf{z}}(t) + \mathbf{S}_{d,d} \boldsymbol{\Phi} \mathbf{z}(t) \quad (2)$$

where $\mathbf{S}_{d,a}$, $\mathbf{S}_{d,v}$, and $\mathbf{S}_{d,d} \in \mathbb{R}^{n_d \times n_{dof}}$ are selection matrices indicating the degrees of freedom corresponding to the acceleration, velocity, and displacement or strain measurements, respectively. The output vector is composed of $n_{d,d}$ displacement or strain measurements, $n_{d,v}$ velocity measurements, and $n_{d,a}$ acceleration measurements, where n_d is the sum of $n_{d,d}$, $n_{d,v}$, and $n_{d,a}$.

Equations 1 and 2 can be written into state-space form. After time discretization and adding noise, the following discrete-time combined deterministic-stochastic state-space description of the system is obtained:

$$\mathbf{x}_{[k+1]} = \mathbf{A}\mathbf{x}_{[k]} + \mathbf{B}\mathbf{p}_{[k]} + \mathbf{w}_{[k]} \quad (3)$$

$$\mathbf{d}_{[k]} = \mathbf{G}\mathbf{x}_{[k]} = \mathbf{J}\mathbf{p}_{[k]} + \mathbf{v}_{[k]} \quad (4)$$

where $\mathbf{x}_{[k]} = \mathbf{x}(k\Delta t)$, $\mathbf{p}_{[k]} = \mathbf{p}(k\Delta t)$, and $\mathbf{d}_{[k]} = \mathbf{d}(k\Delta t)$, $k = 1, \ldots, N$, Δt is the sampling time step, and N is the total number of samples. $\mathbf{d}_{[k]} \in \mathbb{R}^{n_\mathrm{d}}$ is the output vector, assumed to be measured, and $\mathbf{p}_{[k]} \in \mathbb{R}^{n_\mathrm{p}}$ is the input vector. The state vector $\mathbf{x}_{[k]} \in \mathbb{R}^{2n_\mathrm{m}}$ consists of the modal displacements and velocities:

$$\mathbf{x}_{[k]} = \begin{bmatrix} \mathbf{z}_{[k]} \\ \dot{\mathbf{z}}_{[k]} \end{bmatrix} \quad (5)$$

The deterministic system behavior is described by the four system matrices $\mathbf{A}$, $\mathbf{B}$, $\mathbf{G}$, and $\mathbf{J}$. The process noise vector $\mathbf{w}_{[k]} \in \mathbb{R}^{2n_\mathrm{m}}$ and measurement noise vector $\mathbf{v}_{[k]} \in \mathbb{R}^{n_\mathrm{d}}$ account for unknown excitation sources and modeling errors. In addition, the measurement noise vector $\mathbf{v}_{[k]}$ accounts for measurement errors. As an alternative to models based on first principles, models can be directly identified from experimental vibration data using system identification techniques; see, e.g., Van Overschee and De Moor (1993) and Reynders and De Roeck (2008).

State Estimation

The objective of state estimation is to estimate the system states from a limited number of response measurements (e.g., acceleration measurements) and a system model. The state vector can be determined using a deterministic approach (e.g., least squares estimation of the state vector in the frequency domain) or using a combined deterministic-stochastic approach, which mostly results in recursive time-domain algorithms which can be applied for online state estimation. The best-known recursive state estimation algorithm for linear systems is the Kalman filter algorithm (Kalman 1960), which is outlined next.

In the derivation of the Kalman filter, the unknown system input is assumed to be zero-mean white noise and is included in the vectors $\mathbf{w}_{[k]}$ and $\mathbf{v}_{[k]}$. Under this assumption, the system described by Eqs. 3 and 4 becomes

$$\mathbf{x}_{[k+1]} = \mathbf{A}\mathbf{x}_{[k]} + \mathbf{w}_{[k]} \quad (6)$$

$$\mathbf{d}_{[k]} = \mathbf{G}\mathbf{x}_{[k]} + \mathbf{v}_{[k]} \quad (7)$$

The noise processes $\mathbf{w}_{[k]}$ and $\mathbf{v}_{[k]}$, which now account for all excitation sources, are assumed to be zero mean and white, with known covariance matrices $\mathbf{Q}$, $\mathbf{R}$, and $\mathbf{S}$:

$$\mathbb{E}\left[\begin{pmatrix} \mathbf{w}_{[k]} \\ \mathbf{v}_{[k]} \end{pmatrix} \begin{pmatrix} \mathbf{w}_{[l]}^\mathrm{T} & \mathbf{v}_{[l]}^\mathrm{T} \end{pmatrix}\right] = \begin{bmatrix} \mathbf{Q} & \mathbf{S} \\ \mathbf{S}^\mathrm{T} & \mathbf{R} \end{bmatrix} \delta_{[k-l]} \quad (8)$$

with $\mathbf{R} > 0$, $\begin{bmatrix} \mathbf{Q} & \mathbf{S} \\ \mathbf{S}^\mathrm{T} & \mathbf{R} \end{bmatrix} \geq 0$, and $\delta_{[k]} = 1$ for $k = 0$ and 0 otherwise.

If the covariance matrix of the unknown system input $\mathbf{C}_\mathrm{p} \in \mathbb{R}^{n_\mathrm{dof} \times n_\mathrm{dof}}$ and the measurement error covariance matrix $\mathbf{R}_\mathrm{m} \in \mathbb{R}^{n_\mathrm{d} \times n_\mathrm{d}}$ are known, the noise covariance matrices are calculated as

$$\begin{bmatrix} \mathbf{Q} & \mathbf{S} \\ \mathbf{S}^\mathrm{T} & \mathbf{R} \end{bmatrix} = \begin{bmatrix} \mathbf{B}_\mathrm{dof} \\ \mathbf{J}_\mathrm{dof} \end{bmatrix} \mathbf{C}_\mathrm{p} \begin{bmatrix} \mathbf{B}_\mathrm{dof}^\mathrm{T} & \mathbf{J}_\mathrm{dof}^\mathrm{T} \end{bmatrix} + \begin{bmatrix} 0 & 0 \\ 0 & \mathbf{R}_\mathrm{m} \end{bmatrix} \quad (9)$$

where the system matrices $\mathbf{B}_\mathrm{dof} \in \mathbb{R}^{2n_\mathrm{m} \times n_\mathrm{dof}}$ and $\mathbf{J}_\mathrm{dof} \in \mathbb{R}^{n_\mathrm{d} \times n_\mathrm{dof}}$ are obtained by assuming $n_\mathrm{p} = n_\mathrm{dof}$ in Eqs. 3 and 4, respectively, i.e., excitation acting at every degree of freedom in the model.

State estimation consists of estimating the system states $\mathbf{x}_{[k]}$ from a set of response measurements $\mathbf{d}_{[k]}$. A state estimate $\hat{\mathbf{x}}_{[k,l]}$ is defined as an estimate of $\mathbf{x}_{[k]}$, given the output sequence $\mathbf{d}_{[k]}$, with $n = 0, 1, \ldots, l$. The corresponding error covariance matrix, denoted as $\mathbf{P}_{[k|l]}$, is defined as

$$\mathbf{P}_{[k|l]} = \mathbb{E}\left\{ \left(\mathbf{x}_{[k]} - \hat{\mathbf{x}}_{[k|k-l]}\right) \left(\mathbf{x}_{[k]} - \hat{\mathbf{x}}_{[k|k-l]}\right)^\mathrm{T} \right\} \quad (10)$$

The Kalman filter algorithm is initialized using an initial state estimate vector $\hat{\mathbf{x}}_{[0|-1]}$ and its error covariance matrix $\mathbf{P}_{[0|-1]}$, both assumed known. Hereafter, it propagates by computing state estimates $\hat{\mathbf{x}}_{[k+1|k]}$ recursively from the following equation:

$$\hat{\mathbf{x}}_{[k+1|k]} = \mathbf{A}\hat{\mathbf{x}}_{[k|k-1]} + \mathbf{K}_{[k]}\left(\mathbf{d}_{[k]} - \mathbf{G}\hat{\mathbf{x}}_{[k|k-1]}\right) \quad (11)$$

The gain matrix $\mathbf{K}_{[k]}$ is determined such that the state estimates are minimum variance and unbiased (i.e., $\mathbf{K}_{[k]} = \mathrm{argmin}_{\mathbf{K}_{[k]}} \mathrm{tr}\{\mathbf{P}_{[k|k]}\}$, $\mathbb{E}\{\mathbf{x}_{[k]} - \hat{\mathbf{x}}_{[k|k-1]}\} = 0$). The recursive state estimator of the form (11) can be split in two steps, i.e., the measurement update and the time update:

Measurement update

$$\mathbf{L}_{[k]} = \mathbf{P}_{[k|k-1]}\mathbf{G}^{\mathrm{T}}\left(\mathbf{G}\mathbf{P}_{[k|k-1]}\mathbf{G}^{\mathrm{T}} + \mathbf{R}\right)^{-1} \quad (12)$$

$$\hat{\mathbf{x}}_{[k|k]} = \hat{\mathbf{x}}_{[k|k-1]} + \mathbf{L}_{[k]}\left(\mathbf{d}_{[k]} - \mathbf{G}\hat{\mathbf{x}}_{[k|k-1]}\right) \quad (13)$$

$$\mathbf{P}_{[k|k]} = \mathbf{P}_{[k|k-1]} - \mathbf{P}_{[k|k-1]}\mathbf{G}^{\mathrm{T}}\left(\mathbf{G}\mathbf{P}_{[k|k-1]}\mathbf{G}^{\mathrm{T}} + \mathbf{R}\right)^{-1}\mathbf{G}\mathbf{P}_{[k|k-1]} \quad (14)$$

with $\mathbf{K}_{[k]} = \mathbf{A}\mathbf{L}_{[k]}$

Time update

$$\hat{\mathbf{x}}_{[k+1|k]} = \mathbf{A}\hat{\mathbf{x}}_{[k|k]} \quad (15)$$

$$\mathbf{P}_{[k+1|k]} = \mathbf{A}\mathbf{P}_{[k|k]}\mathbf{A}^{\mathrm{T}} + \mathbf{Q} - \mathbf{A}\mathbf{L}_{[k]}\mathbf{S}^{\mathrm{T}} - \mathbf{S}\mathbf{L}_{[k]}^{\mathrm{T}}\mathbf{A}^{\mathrm{T}} \quad (16)$$

The algorithm given in Eqs. 12, 13, 14, 15, and 16 distinguishes itself from the classical Kalman filter by including the correlation between the process noise vector $\mathbf{w}_{[k]}$ and the measurement noise vector $\mathbf{v}_{[k]}$, i.e., $\mathbf{S} \neq \mathbf{0}$. In the equations above, the system is assumed to be time invariant. The algorithm can, however, also be applied to time-variant systems, resulting in system matrices $\mathbf{A}_{[k]}$ and $\mathbf{G}_{[k]}$ depending on the time step k.

In general, the system input does not satisfy the white noise assumption. The Kalman filter, however, yields good results as long as the applied forces are broadband and approximately white. If the forces acting to the structure do not meet this assumption, the quality of the estimated states can be improved by estimating both the system input and states, i.e., joint input-state estimation.

Joint Input-State Estimation

The objective of joint input-state estimation is to jointly estimate the forces applied to the system and the corresponding states from a limited number of response measurements and a system model. Similar to state estimation, the applied forces can be determined using a deterministic approach, e.g., Guillaume et al. (2002), or alternatively using a combined deterministic-stochastic approach, which mostly results in recursive time-domain algorithms which can be applied for online joint input-state estimation. The algorithm outlined next is a joint input-state estimation algorithm which was originally proposed by Gillijns and De Moor (2007). The algorithm was applied for force identification and response estimation in structural dynamics by Lourens et al. (2012) and was extended for applications in structural dynamics where ambient excitation becomes important by Maes et al. (2013).

The system under consideration is described by Eqs. 3 and 4. The noise processes $\mathbf{w}_{[k]}$ and $\mathbf{v}_{[k]}$ are assumed to be zero mean and white, with known covariance matrices $\mathbf{Q}$, $\mathbf{R}$, and $\mathbf{S}$, defined by Eq. 8. Joint input-state estimation consists of estimating the forces $\mathbf{p}_{[k]}$ and states $\mathbf{x}_{[k]}$ from a set of response measurements $\mathbf{d}_{[k]}$. A state estimate $\hat{\mathbf{x}}_{[k|l]}$ is defined as an estimate of $\mathbf{x}_{[k]}$, given the output sequence $\mathbf{d}_{[n]}$, with $n = 0,1,\ldots,l$. The corresponding error covariance matrix, denoted as $\mathbf{P}_{[k|l]}$, is defined in Eq. 10. An input estimate $\hat{\mathbf{p}}_{[k|l]}$ and its error covariance matrix $\mathbf{P}_{\mathrm{p}[k|l]}$ are defined similarly. The filtering algorithm is initialized using an initial state estimate vector $\hat{\mathbf{x}}_{[0|-1]}$ and its error covariance matrix $\mathbf{P}_{[0|-1]}$, both assumed known. Hereafter, it propagates by computing the force and state estimates recursively in three steps, i.e., the input estimation step, the measurement update, and the time update:

Input estimation

$$\tilde{\mathbf{R}}_{[k]} = \mathbf{G}\mathbf{P}_{[k|k-1]}\mathbf{G}^{\mathrm{T}} + \mathbf{R} \quad (17)$$

$$\mathbf{M}_{[k]} = \left(\mathbf{J}^{\mathrm{T}}\tilde{\mathbf{R}}_{[k]}^{-1}\mathbf{J}\right)^{-1}\mathbf{J}^{\mathrm{T}}\tilde{\mathbf{R}}_{[k]}^{-1} \quad (18)$$

$$\hat{\mathbf{p}}_{[k|k]} = \mathbf{M}_{[k]}\left(\mathbf{d}_{[k]} - \mathbf{G}\hat{\mathbf{x}}_{[k|k-1]}\right) \quad (19)$$

$$\mathbf{P}_{\mathrm{p}[k|k]} = \left(\mathbf{J}^{\mathrm{T}}\tilde{\mathbf{R}}_{[k]}^{-1}\mathbf{J}\right)^{-1} \quad (20)$$

Measurement update

$$\mathbf{L}_{[k]} = \mathbf{P}_{[k|k-1]}\mathbf{G}^{\mathrm{T}}\tilde{\mathbf{R}}_{[k]}^{-1} \quad (21)$$

$$\hat{\mathbf{x}}_{[k|k]} = \hat{\mathbf{x}}_{[k|k-1]} + \mathbf{L}_{[k]}\left(\mathbf{d}_{[k]} - \mathbf{G}\hat{\mathbf{x}}_{[k|k-1]} - \mathbf{J}\hat{\mathbf{p}}_{[k|k]}\right) \quad (22)$$

$$\mathbf{P}_{[k|k]} = \mathbf{P}_{[k|k-1]} + \mathbf{L}_{[k]}\left(\tilde{\mathbf{R}}_{[k]} - \mathbf{J}\mathbf{P}_{\mathrm{p}[k|k]}\mathbf{J}^{\mathrm{T}}\right)\mathbf{L}_{[k]}^{\mathrm{T}} \quad (23)$$

$$\mathbf{P}_{\mathrm{xp}[k|k]} = \mathbf{P}_{\mathrm{px}[k|K]}^{\mathrm{T}} - \mathbf{L}_{[k]}\mathbf{J}\mathbf{P}_{\mathrm{p}[k|k]} \quad (24)$$

Time update

$$\hat{\mathbf{x}}_{[k+1|k]} = \mathbf{A}\hat{\mathbf{x}}_{[k|k]} + \mathbf{B}\hat{\mathbf{p}}_{[k|k]} \quad (25)$$

$$\mathbf{N}_{[k]} = \mathbf{A}\mathbf{L}_{[k]}\left(\mathbf{I} - \mathbf{J}\mathbf{M}_{[k]}\right) + \mathbf{B}\mathbf{M}_{[k]} \quad (26)$$

$$\mathbf{P}_{[k+1|k]} = \begin{bmatrix}\mathbf{A} & \mathbf{B}\end{bmatrix}\begin{bmatrix}\mathbf{P}_{[k|k]} & \mathbf{P}_{\mathrm{xp}[k|k]} \\ \mathbf{P}_{\mathrm{px}[k|k]} & \mathbf{P}_{\mathrm{p}[k|k]}\end{bmatrix}\begin{bmatrix}\mathbf{A}^{\mathrm{T}} \\ \mathbf{B}^{\mathrm{T}}\end{bmatrix} + \mathbf{Q} - \mathbf{N}_{[k]}\mathbf{S}^{\mathrm{T}} - \mathbf{S}\mathbf{N}_{[k]}^{\mathrm{T}} \quad (27)$$

The gain matrices $\mathbf{M}_{[k]}$ and $\mathbf{L}_{[k]}$ are determined such that both the input estimates $\hat{\mathbf{p}}_{[k|k]}$ and the state estimates $\hat{\mathbf{x}}_{[k|k-1]}$ are minimum variance and unbiased (i.e., $\mathbf{M}_{[k]} = \mathrm{argmin}_{\mathbf{M}_{[k]}}\mathrm{tr}\{\mathbf{P}_{\mathbf{p}_{[k|k]}}\}$, $\mathbb{E}\{\mathbf{p}_{[k]} - \hat{\mathbf{p}}_{[k|k]}\} = 0$, $\mathbf{L}_{[k]} = \mathrm{argmin}_{\mathbf{L}_{[k]}}\mathrm{tr}\{\mathbf{P}_{[k|k]}\}$, and $\mathbb{E}\{\mathbf{x}_{[k]} - \hat{\mathbf{x}}_{[k|k]}\} = 0$). In the equations above, the system is assumed to be time invariant. The algorithm can, however, also be applied to time-variant systems, resulting in system matrices $\mathbf{A}_{[k]}$, $\mathbf{B}_{[k]}$, $\mathbf{G}_{[k]}$, and $\mathbf{J}_{[k]}$ depending on the time step k.

Very often unknown ambient forces such as wind loads are acting on the structure. For these loads, the force locations or spatial distributions of the forces are not well known. In this case, the estimated forces $\mathbf{p}_{[k]}$ compensate for any unknown source of vibration. They are not the true forces acting on the structure but equivalent forces that act at predefined locations (Lourens et al. 2012).

Response Estimation

After applying the Kalman filter algorithm or the joint input-state estimation algorithm, the estimated state vector $\hat{\mathbf{x}}_{[k|k]}$ and force vector $\hat{\mathbf{p}}_{[k|k]}$ (in the case of joint input-state estimation) can be used to estimate the output at any arbitrary location in the structure, using the following modified output equation:

$$\hat{\mathbf{d}}_{\mathrm{e}[k|k]} = \mathbf{G}_{\mathrm{e}}\hat{\mathbf{x}}_{[k|k]} + \mathbf{J}_{\mathrm{e}}\hat{\mathbf{p}}_{[k|k]} \quad (28)$$

where $\hat{\mathbf{d}}_{\mathrm{e}[k|k]}$ is the estimated response. The matrices $\mathbf{G}_{\mathrm{e}}$ and $\mathbf{J}_{\mathrm{e}}$ are related to the extrapolated output quantities $\hat{\mathbf{d}}_{\mathrm{e}[k|k]}$ and, therefore, do not equal the original matrices $\mathbf{G}$ and $\mathbf{J}$. Note that when Eq. 28 is used after applying the Kalman filter algorithm, the excitation $\hat{\mathbf{p}}_{[k|k]}$ is contained in the noise vectors $\mathbf{w}_{[k]}$ and $\mathbf{v}_{[k]}$ (cf. Eq. 7) and therefore disregarded. In this way, an error is introduced for acceleration estimation. For displacement, velocity, or strain estimation, this is not the case, as there is no direct feedthrough from the ambient forces $\hat{\mathbf{p}}_{[k|k]}$ to the output $\hat{\mathbf{d}}_{\mathrm{e}[k|k]}$ which has to be accounted for (i.e., the rows of the matrix $\mathbf{J}_{\mathrm{e}}$ corresponding to the displacement, velocity, or strain measurements are all zero).

Application: Response Estimation in a Three-Story Shear Building Subject to an Earthquake

The response estimation procedure is illustrated using numerical simulations for a three-story building subjected to horizontal earthquake excitation. The building is modeled as an idealized shear building (Fig. 1). The story masses are lumped to the floor levels, the floors are presumed infinitely rigid, and axial member deformations are neglected. This system is described by three degrees of freedom (DOFs), u_1, u_2, and u_3. Under the assumption that the structure is subjected to a uniform horizontal base motion $\ddot{u}_{\mathrm{g}}(t)$, the system is governed by the following set of equations of motion:

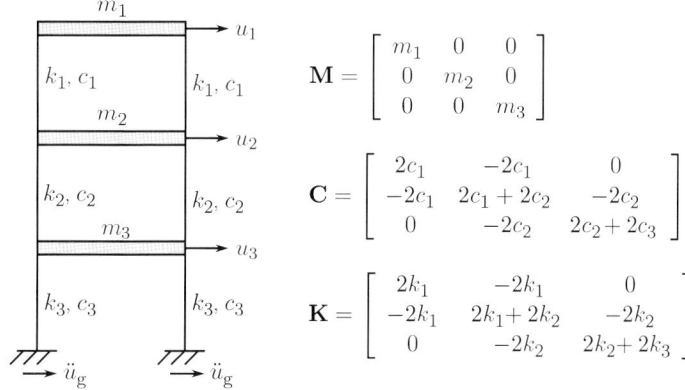

Online Response Estimation in Structural Dynamics, Fig. 1 A three-story shear building model and its mass, damping, and stiffness matrices

$$\mathbf{M}\ddot{\mathbf{u}}(t) + \mathbf{C}\dot{\mathbf{u}}(t) + \mathbf{K}\mathbf{u}(t) = \mathbf{S}_\mathrm{p}(t)\mathbf{p}(t) \quad (29)$$

where $\mathbf{u}(t) = [u_1(t), u_2(t), u_3(t)]^\mathrm{T}$ is the vector of displacements relative to the displacements of the building basement and $\mathbf{M}$, $\mathbf{C}$, and $\mathbf{K}$ are the mass, damping, and stiffness matrix of the structure, respectively, defined in Fig. 1. $\mathbf{S}_\mathrm{p}(t) = \mathbf{I}_3$ and $\mathbf{p}(t) = [-m_1, -m_2, -m_3]^\mathrm{T}\ddot{u}_\mathrm{g}(t)$, where $\mathbf{I}_3 \in \mathbb{R}^{3\times3}$ is an identity matrix.

The story masses m_i, damping values c_i, and stiffnesses k_i are assumed known and are given by

$$\begin{array}{lll} m_1 = 10 \times 10^3 \text{ kg} & c_1 = 6 \times 10^3 \text{ Ns/m} & k_1 = 6 \times 10^6 \text{ N/m} \\ m_2 = 10 \times 10^3 \text{ kg} & c_2 = 7 \times 10^3 \text{ Ns/m} & k_2 = 7 \times 10^6 \text{ N/m} \\ m_3 = 10 \times 10^3 \text{ kg} & c_3 = 8 \times 10^3 \text{ Ns/m} & k_3 = 8 \times 10^6 \text{ N/m} \end{array} \quad (30)$$

The modal properties of the system are found as the solution of the undamped eigenvalue equation $\mathbf{K}\mathbf{\Phi} = \mathbf{M}\mathbf{\Phi}\mathbf{\Lambda}$, where $\mathbf{\Phi} \in \mathbb{R}^{3\times3}$ collects the three eigenvectors $\boldsymbol{\phi}_j \in \mathbb{R}^3$ that correspond to the eigenvalues $\lambda_j = \omega_j^2 = (2\pi f_j)^2$ located on the diagonal of $\mathbf{\Lambda}$. The first three natural frequencies f_j of the shear building are found as 2.72, 7.27, and 10.59 Hz. The corresponding mode shapes are shown in Fig. 2. The modal damping ratios ξ_j are calculated as $\gamma_j/(2\omega_j)$, where γ_j is a diagonal element of the matrix $\mathbf{\Gamma}$, with $\mathbf{\Gamma} = \mathbf{\Phi}^\mathrm{T}\mathbf{C}\mathbf{\Phi}$. Note that $\mathbf{C} = 10^{-3} \text{ s} \times \mathbf{K}$, so proportional damping was implicitly assumed. The first three modal damping ratios of the shear building are 0.85 %, 2.28 %, and 3.33 %.

The building is subject to the horizontal ground motion recorded during the Northridge earthquake that occurred on January 17, 1994. The epicenter of the earthquake was located in Northridge, California. The ground motion considered was recorded in the northwest direction in Montebello, California (http://peer.berkeley.edu/peer_ground_motion_database). The time history and narrowband frequency spectrum of the recorded ground acceleration $\ddot{u}_\mathrm{g}$ are shown in Fig. 3.

A reduced-order state-space model is constructed from the first three eigenmodes of the shear building, applying a zero-order hold assumption on the force. The data assumed for the response estimation consists of the ground acceleration time history $\ddot{u}_\mathrm{g}$, three absolute horizontal acceleration time histories, $\ddot{u}_{\mathrm{a},1}$, $\ddot{u}_{\mathrm{a},2}$, and $\ddot{u}_{\mathrm{a},3}$, and one relative horizontal displacement time history, u_3. The response is calculated at a sampling rate of 100 Hz. The initial state $\mathbf{x}_{[0]}$ is assumed zero. The output vector $\mathbf{d}_{[k]}$ considered for response estimation is composed of the relative acceleration time histories $\ddot{u}_1$, $\ddot{u}_2$, and $\ddot{u}_3$ and the relative displacement time history u_3, where $\ddot{u}_j = \ddot{u}_{\mathrm{a},j} - \ddot{u}_\mathrm{g}$. Note that the ground acceleration time history $\ddot{u}_\mathrm{g}$ can also be used directly as input

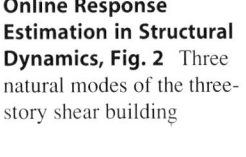

Online Response Estimation in Structural Dynamics, Fig. 2 Three natural modes of the three-story shear building

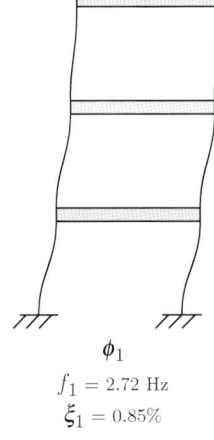

ϕ_1
$f_1 = 2.72$ Hz
$\xi_1 = 0.85\%$

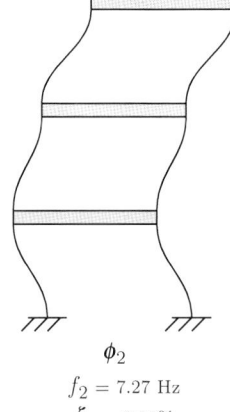

ϕ_2
$f_2 = 7.27$ Hz
$\xi_2 = 2.28\%$

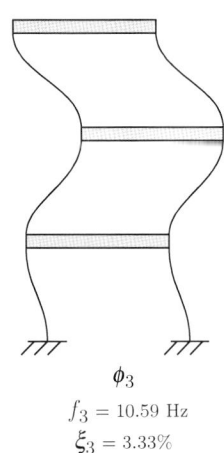

ϕ_3
$f_3 = 10.59$ Hz
$\xi_3 = 3.33\%$

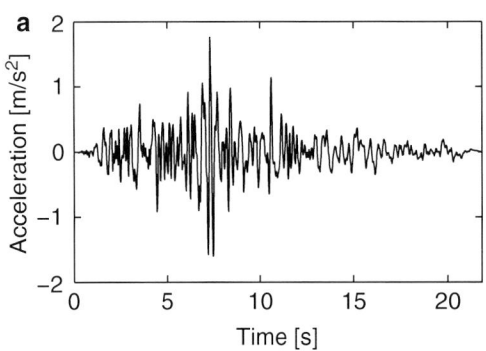
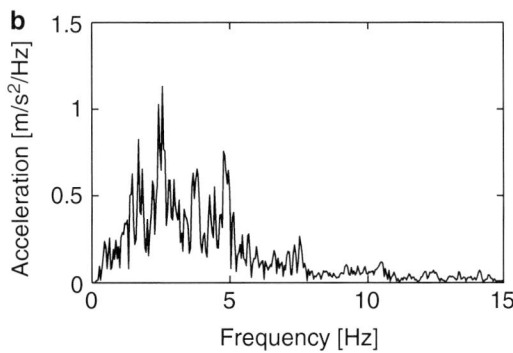

Online Response Estimation in Structural Dynamics, Fig. 3 (a) Time history and (b) narrowband frequency spectrum up to 15 Hz of the recorded ground accelerations in northwest direction due to the Northridge earthquake, January 17, 1994, 12:31 UTC, Montebello, California

to a forward calculation, hereby obtaining the response in the structure at predefined locations. The proposed filtering algorithms, however, have the advantage of canceling out measurement errors and modeling errors. A small error on the natural frequencies of the system, for example, hardly influences the quality of the predicted response obtained by applying the Kalman filter or the joint input-state estimation algorithm, whereas for the forward calculation, this is mostly not the case.

Gaussian white noise is added to the ground acceleration time history $\ddot{u}_g$ and to the calculated time histories ($\ddot{u}_{a,1}$, $\ddot{u}_{g,2}$, $\ddot{u}_{a,3}$, and u_3), in order to represent the measurement error on the data. A choice is made to assume the same noise level γ_a for all acceleration signals and a noise level γ_d for the displacement signal. The noise level γ_a is 5 % of the maximal acceleration $\ddot{u}_{a,3}$, and the noise level γ_d is 5 % of the maximal displacement u_3 ($\gamma_a = 0.17$ m/s^2, $\gamma_d = 0.54$ mm). The polluted output time histories ($\widetilde{\ddot{u}}_{1[k]}, \widetilde{\ddot{u}}_{2[k]}, \widetilde{\ddot{u}}_{3[k]}$, and $\tilde{u}_{3[k]}$) are calculated according to Eq. 31:

$$\begin{aligned}
\widetilde{\ddot{u}}_{1[k]} &= \ddot{u}_{a,1[k]} - \ddot{u}_{g[k]} + \gamma_a\left(r_{a,1[k]} + r_{g[k]}\right) \\
\widetilde{\ddot{u}}_{2[k]} &= \ddot{u}_{a,2[k]} - \ddot{u}_{g[k]} + \gamma_a\left(r_{a,2[k]} + r_{g[k]}\right) \\
\widetilde{\ddot{u}}_{3[k]} &= \ddot{u}_{a,3[k]} - \ddot{u}_{g[k]} + \gamma_a\left(r_{a,3[k]} + r_{g[k]}\right) \\
\tilde{u}_{3[k]} &= u_{3[k]} + \gamma_d r_{d,3[k]}
\end{aligned} \quad (31)$$

where $r_{a,1[k]}$, $r_{a,2[k]}$, $r_{a,3[k]}$, $r_{g[k]}$, and $r_{d,3[k]}$ are random numbers drawn independently from a normal distribution with zero mean and unit

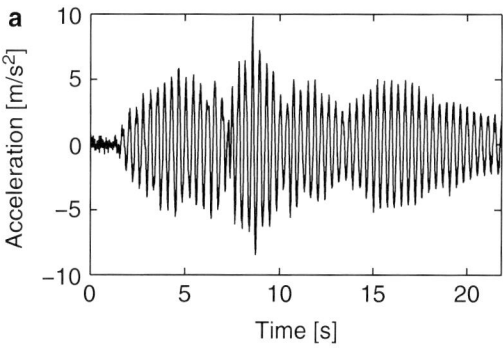
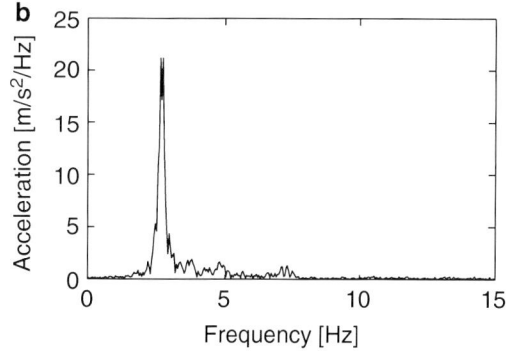

Online Response Estimation in Structural Dynamics, Fig. 4 (a) Time history and (b) narrowband frequency spectrum up to 15 Hz of the simulated acceleration $\tilde{\ddot{u}}_1$, measurement noise included

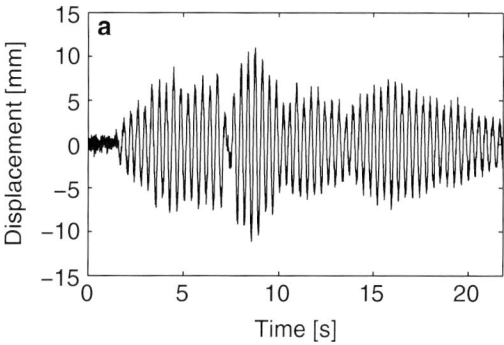
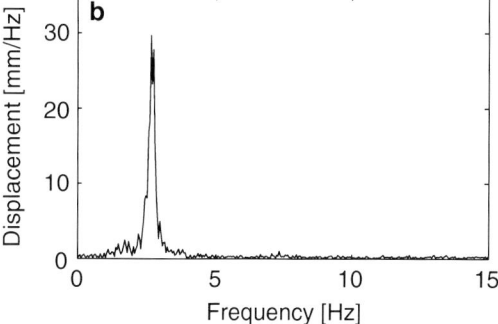

Online Response Estimation in Structural Dynamics, Fig. 5 (a) Time history and (b) narrowband frequency spectrum up to 15 Hz of the simulated displacement $\tilde{u}_3$, measurement noise included

standard deviation. The time history and narrowband frequency spectrum of the simulated acceleration $\tilde{\ddot{u}}_1$ and the displacement $\tilde{u}_3$ are shown in Figs. 4 and 5, respectively.

The response signals are now used as input to the response estimation procedure, hereby comparing the Kalman filter and the joint input-state estimation algorithm presented in sections "State Estimation" and "Joint Input-State Estimation," respectively. The response to be estimated consists of the horizontal displacement u_1 of the top floor.

The Kalman filter is applied to estimate the system states $\hat{\mathbf{x}}_{[k|k]}$, and, subsequently, the response of the top floor $\hat{\mathbf{d}}_{e[k|k]}$ is estimated according to Eq. 28, where the term $\mathbf{J}_e \hat{\mathbf{p}}_{[k|k]}$ is disregarded. The noise covariance matrices $\mathbf{Q}$, $\mathbf{R}$, and $\mathbf{S}$ are calculated from Eq. 9, for $\mathbf{C}_p$ and $\mathbf{R}_m$ given by the following equations:

$$\mathbf{C}_p = \begin{bmatrix} m_1^2 & m_1 m_2 & m_1 m_3 \\ m_2 m_1 & m_2^2 & m_2 m_3 \\ m_3 m_1 & m_3 m_2 & m_3^2 \end{bmatrix} \sigma_{\ddot{u}_g}^2 \quad (32)$$

$$\mathbf{R}_m = \begin{bmatrix} 2\gamma_a^2 & \gamma_a^2 & \gamma_a^2 & 0 \\ \gamma_a^2 & 2\gamma_a^2 & \gamma_a^2 & 0 \\ \gamma_a^2 & \gamma_a^2 & 2\gamma_a^2 & 0 \\ 0 & 0 & 0 & \gamma_d^2 \end{bmatrix} \quad (33)$$

where $\sigma_{\ddot{u}_g}$ is the standard deviation of the ground acceleration $\ddot{u}_g$ imposed to the building, assumed 0.2 m/s^2 for this example. The initial state estimate vector $\hat{\mathbf{x}}_{[0|-1]}$ and its error covariance matrix $\mathbf{P}_{[0|-1]}$ are both assumed zero.

The joint input-state estimation algorithm is applied to estimate the system states $\hat{\mathbf{x}}_{[k|k]}$ and the applied forces $\hat{\mathbf{p}}_{[k|k]}$, hereafter estimating the response of the top floor $\hat{\mathbf{d}}_{e[k|k]}$, according to Eq. 28. A distribution of the equivalent force is

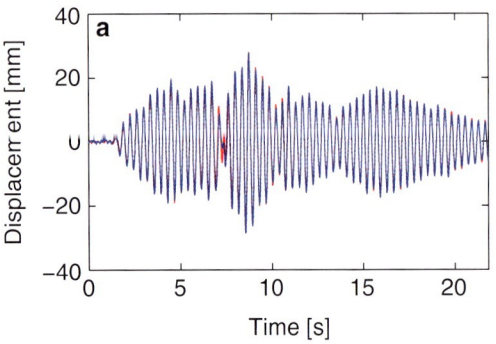 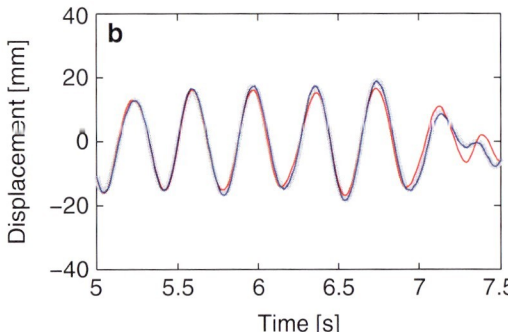

Online Response Estimation in Structural Dynamics, Fig. 6 (**a**) Complete and (**b**) detail of the simulated (*gray*) and estimated displacement time history u_1 (*red*, Kalman filter; *blue*, joint input-state estimation algorithm)

assumed by putting $\mathbf{S}_p(t) = [1, m_2/m_1, m_3/m_1]^T$ in Eq. 29. The estimated force $\hat{\mathbf{p}}_{[k|k]} \in \mathbb{R}$ has a physical meaning, i.e., the inertial force resulting from the ground motion applied at the top of the shear building. The noise covariance matrices $\mathbf{Q}$, $\mathbf{R}$, and $\mathbf{S}$ are calculated from Eq. 9, for $\mathbf{C}_p = \mathbf{0}$ and $\mathbf{R}_m$ given by Eq. 33. The initial state estimate vector $\hat{\mathbf{x}}_{[0|-1]}$ and its error covariance matrix $\mathbf{P}_{[0|-1]}$ are both assumed zero.

The results of the response estimation procedure are shown in Fig. 6. The detail in Fig. 6b reveals that the results obtained by applying the joint input-state estimation algorithm are more accurate than the results from the Kalman filter. The inertial forces applied to the structure, which result from ground motion, do not perfectly fulfill the white noise assumption, which explains the errors on the results from the Kalman filter. The equivalent forces assumed when applying the joint input-state estimation algorithm successfully compensate for the excitation applied to the structure. As a result, only very small errors on the estimated displacement due to measurement errors remain.

Summary

This reference entry focuses on two recursive time-domain algorithms that can be applied for response estimation in structural dynamics. The first algorithm is the Kalman filter, a well-known state estimation algorithm that has been used for many applications in various engineering disciplines. The second algorithm is a recursive joint input-state estimation algorithm. Both algorithms are briefly outlined and thereafter compared using simulations for a three-story building subject to an earthquake.

Cross-References

▶ Site Response: Comparison Between Theory and Observation

References

Gillijns S, De Moor B (2007) Unbiased minimum-variance input and state estimation for linear discrete-time systems with direct feedthrough. Automatica 43(5):934–937

Guillaume P, Parloo E, De Sitter G (2002) Source identification from noisy response measurements using an iterative weighted pseudo-inverse approach. In: Proceedings of ISMA2002 international conference on noise and vibration engineering, Leuven, pp 1817–1824

Hernandez E (2011) A natural observer for optimal state estimation in second order linear structural systems. Mech Syst Signal Process 25(8):2938–2947

Hwang J, Kareem A, Kim W (2009) Estimation of modal loads using structural response. J Sound Vib 326:522–539

Kalman RE (1960) On the general theory of control systems. In: Proceedings of the first international congress of IFAC, Moscow

Klinkov M, Fritzen C (2007) An updated comparison of the force reconstruction methods. Key Eng Mater 347:461–466

Lourens E, Papadimitriou C, Gillijns S, Reynders E, De Roeck G, Lombaert G (2012) Joint input-response estimation for structural systems based on reduced-order models and vibration data from a limited number of sensors. Mech Syst Signal Process 29:310–327

Ma C, Chang J, Lin D (2003) Input forces estimation of beam structures by an inverse method. J Sound Vib 259(2):387–407

Maes K, Lourens E, Van Nimmen K, Reynders E, Van den Broeck P, Guillaume P, De Roeck G, Lombaert G (2013) Verification of joint input-state estimation by in situ measurements on a footbridge. In: Chang F-K (ed) Proceedings of the 9th international workshop on structural health monitoring. Stanford, vol 1, pp 343–350

Nordström L, Nordberg T (2002) A critical comparison of time domain load identification methods. In: Proceedings of the 6th international conference on motion and vibration control, Saitama, pp 1151–1156

Parloo E, Verboven P, Guillaume P, Van Overmeire M (2003) Force identification by means of in-operational modal models. J Sound Vib 262(1):161–173

Reynders E, De Roeck G (2008) Reference-based combined deterministic-stochastic subspace identification for experimental and operational modal analysis. Mech Syst Signal Process 22(3):617–637

Van Overschee P, De Moor B (1993) Subspace algorithm for the stochastic identification problem. Automatica 29(3):649–660

Wu M, Smyth A (2007) Application of the unscented Kalman filter for real-time nonlinear structural system identification. Struct Control Health Monit 14:971–990

Operational Modal Analysis in Civil Engineering: An Overview

Edwin Reynders and Guido De Roeck
Department of Civil Engineering, KU Leuven, Leuven, Belgium

Synonyms

Ambient vibration testing; System identification

Introduction

The susceptibility of civil structures and buildings to vibrations is the major concern when designing earthquake-resistant structures. The experimental verification of the dynamic design values, in particular the relevant standing wave or modal characteristics (natural frequencies, damping ratios, and mode shapes), is essential for validating the dynamic structural design, hence for guaranteeing the safety of the structure under earthquake loading. Furthermore, the modal characteristics depend on the stiffness of the structure, so monitoring their evolution over time allows, in principle, to identify structural damage. For example, comparing modal characteristics that have been measured before and after the occurrence of an earthquake can help in detecting structural damage that is not directly visible. However, modal characteristics may also be sensitive to environmental conditions, such as temperature or relative humidity, and to loading conditions such as excitation amplitude or added mass. These influences need to be properly taken into account when employing the results of a modal test for design validation, model calibration, structural health monitoring, etc.

Depending on the type of excitation that is employed to perform a modal test, a distinction is made between forced vibration testing, ambient vibration testing, and combined or hybrid vibration testing, also termed experimental modal analysis (EMA), operational modal analysis (OMA), and operational modal analysis with exogenous inputs (OMAX), respectively. In EMA, which was historically the first testing approach that was developed, the structure is excited by one or several measured dynamic forces, the response of the structure to these forces is recorded, and the modal characteristics in the frequency range of interest are identified from the measured data. The first EMA techniques were single degree of freedom methods such as peak picking and circle fitting, in which it is assumed that at resonance, only a single mode dominates. Consequently, these techniques are not useful when some modes of interest have close natural frequencies. This disadvantage was later removed with the introduction of multiple degree of freedom methods for EMA. By now, EMA is a well-established and often-used approach in mechanical and aerospace engineering, as documented by, e.g., Maia and Silva (1997), Heylen et al. (1997), Allemang (1999), and Ewins (2000).

EMA methods are in general not well suited for the modal testing of civil structures and buildings, because these structures cannot be tested in laboratory conditions. When employing compact and practical actuators, the contribution of the measured excitation to the total structural response is rather low. This implies that the ever-present ambient or operational excitation – due to wind or traffic loading, for example – can most often not be neglected. OMA techniques have therefore been developed. They extract the modal characteristics from the measured operational response only. The unmeasured, ambient forces are usually modeled as white noise sequences, i.e., as zero-mean time sequences of which all samples are statistically uncorrelated and have the same variance. In civil engineering, OMA has become the primary modal testing method, especially for structures whose performance under earthquake loading is critical, such as bridges, high-rise buildings, and dams. Peeters and De Roeck (2001) review some established OMA techniques, while Magalhaes and Cunha (2011) provide an extensive tutorial introduction to OMA using bridge vibration data.

Nevertheless, the OMA approach has two disadvantages: the mode shapes cannot be scaled in an absolute way (e.g., mass normalized), and there is no control over the ambient excitation signal, which may be narrow banded in the frequency domain. For these reasons, there has been an increasing interest during the last few years toward combined modal testing or OMAX techniques. The main difference between OMAX and the traditional EMA approach is that the operational loads are included in the identified system model: they are not considered as noise but as useful excitation. Consequently, the amplitude of the artificial forces can be equal to, or even lower than, the amplitude of the operational forces. This is of crucial importance for the modal testing of large structures, since it allows using actuators that are small and practical when compared to the ones needed for EMA testing of such structures. Reynders (2012) provides an in-depth review of the most relevant and powerful EMA, OMA, and OMAX techniques and compares their performance in an extensive simulation study.

Whether an EMA, OMA, or OMAX approach is followed, all modal tests consist of three distinct stages:

1. Designing the experiment and collecting the vibration response data (for OMA) or the force and vibration response data (for EMA or OMAX)
2. Identifying a set of dynamic structural models belonging to a given model class (e.g., linear time-invariant (LTI)) from the measured data
3. Extracting and validating a set of modal characteristics based on the identified models

Each of these stages is discussed in more detail in the following sections.

Experiment Design and Data Collection

The first stage of an operational modal analysis consists of designing the experiment and gathering the necessary data. For a pure OMA test, only response data are measured. For an OMAX test, the artificial excitation is also measured.

Modes are standing waves. The largest frequency and the shortest wavelength of all standing waves of interest have a prime influence on the experiment design. The largest frequency of interest determines the sampling rate and the necessary synchronization accuracy among the different sensors, which is especially important for wireless sensor networks. Components of the signal with a frequency that is larger than half of the sampling frequency should be filtered out before sampling in order to avoid aliasing. The shortest wavelength of interest determines the spatial resolution of the test: the shorter the wavelength, the closer the response sensors need to be in order to capture the short-wavelength mode shapes in sufficient detail.

For measuring the response of the structure, many types of sensor are available. Figure 1 provides a few examples. Sensor types that are often used include accelerometers which measure uniaxial or triaxial acceleration, velocity or

Operational Modal Analysis in Civil Engineering: An Overview, Fig. 1 Some sensors that are employed for operational modal analysis: two uniaxial wired accelerometers (*left*), a triaxial wireless accelerometer (*middle*), and two optical fiber sensors (*right*) (Right photo reproduced with permission from: Reynders et al., Damage identification on the Tilff bridge by vibration monitoring using optical fibre strain sensors, ASCE Journal of Engineering Mechanics, 133(2):185–193, 2007)

displacement sensors such as laser vibrometers, LVDTs (linear variable differential transformers) and capacitive sensors, and strain sensors such as strain gages or fiber optic sensors. Most sensors only measure at a single location. Some quasi-distributed sensing techniques, based on, e.g., laser scanning or photogrammetry, are available, but their use on large and/or complex structures can be challenging. The development of wireless sensing technology has greatly improved the ease and speed of occasional modal testing.

The number of sensors employed in a test can be smaller than the total number of locations at which the response needs to be measured. In such case, the test is performed in different setups, each of which yields partial mode shape estimates. They can be assembled when sensor locations overlap between different setups in such a way that there is at least one common nonzero modal displacement among both setups for every mode of interest.

For OMAX testing (and also for traditional EMA testing), actuators are also needed. Figure 2 provides a few examples. Whenever the cost of testing and the ease of installation are a concern, the use of shakers can be excluded since they are not very cost-effective. Due to the low maximum force, the classic impact hammer is only feasible for the testing of relatively small structures. A larger and more controllable version consists of a drop weight system. When specific excitation signals are needed (multisine, swept sine, etc.), pneumatic artificial muscles can in some cases provide a good alternative for shakers (Deckers et al. 2008).

When an a priori estimate of the modal characteristics that one wishes to measure is available, e.g., from a preliminary finite element model, it is possible to estimate the optimal sensor (and actuator) configuration according to a given performance measure. As shown recently by Papadimitriou and Lombaert (2012), it is

Operational Modal Analysis in Civil Engineering: An Overview, Fig. 2 Some actuators that are employed for OMAX testing: drop weight system (*left*), impact hammer (*center*), and pneumatic artificial muscle (*right*) (Reproduced with permission from: Reynders et al., combined experimental-operational modal testing of footbridges, ASCE Journal of Engineering Mechanics, 136(6):687–696, 2010)

important to take the effect of spatially correlated prediction errors into account when computing the optimal sensor placement.

System Identification

Once the data have been gathered, a mathematical model that compactly describes the measured signals can be identified. Since the modal characteristics are classically defined as the eigenparameters of a dynamic, linear time-invariant (LTI) system description with general viscous damping, it is such a description that is identified from the OMA or OMAX data (Fig. 3). A very large number of system identification algorithms are available from the literature, but, as shown by Ljung (1999), they can be considered as particular implementations of just a few general ideas. This section therefore provides a concise overview of the different approaches rather than providing a complete list of algorithms. Most identification methods can be applied to time-domain as well as to frequency-domain data. This means that most time-domain system identification algorithms have a frequency-domain counterpart which often differs from it only by Fourier transforms, and vice versa.

First, a distinction can be made between nonparametric and parametric identification. Nonparametric system identification involves the estimation of an impulse response function, frequency response function (FRF), correlation function, or power spectral density (PSD), not as a mathematical function depending on a few parameters, but as a set of tabulated values for each considered time lag or frequency. Although nonparametric models are sometimes directly used for modal analysis, they are most often used as preprocessed data for parametric identification since the estimation accuracy of parametric approaches is much higher than that of nonparametric approaches (Peeters and De Roeck 2001; Reynders 2012).

Parametric linear system identification as a research discipline originated in the late 1960s and developed along two lines that are still dominant today. The first line is the prediction error framework, where a system model is identified by minimizing the difference between the measured system response and the response predicted by

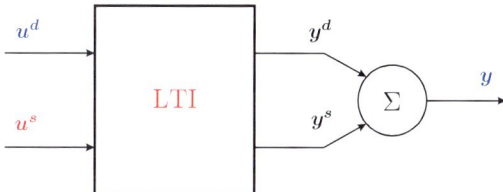

Operational Modal Analysis in Civil Engineering: An Overview, Fig. 3 A linear time-invariant (*LTI*) system model is identified from measured data. The total response y is measured, and part of the excitation u^d may also be measured. The operational excitation u^s is not measured. The subscript d stands for deterministic, while the subscript s stands for stochastic

the model, most often using the maximum likelihood (ML) principle (Ljung 1999; Pintelon and Schoukens 2001). The second line seeks at identifying the system model by exploiting the correlation structure (over time) of the measured response signals (and, if relevant, the measured excitation signals). This line started with the seminal work of Ho and Kalman (1966) and Akaike (1974) on system realization, which was later generalized into the so-called subspace approach (Van Overschee and De Moor 1996; Katayama 2005). Both approaches have distinct advantages: maximum likelihood methods have optimal asymptotic statistical properties under fairly general assumptions, while subspace methods are very robust, statistically accurate, and computationally much less demanding.

Estimating and Validating a Set of Modal Characteristics

The modal characteristics corresponding to an identified system description can be obtained by exploiting its eigenstructure, that is, by decomposing the original description in terms of its free vibration solutions.

When the identified system description is nonparametric and consists of, e.g., tabulated frequency response data, rough estimates of the modal characteristics can be obtained when the damping is low. One of the most popular techniques, because of its simplicity and intuitiveness, is peak picking. Under the assumptions that the modes have low damping and well-separated natural frequencies and that the estimated frequency response functions (in case of input-output measurements) or power spectral densities (in case of output-only measurements) are noiseless and have infinite frequency resolution, the natural frequencies can be obtained as the peaks of the frequency response function or power spectral density. The deflection shapes at those frequencies are estimates of the corresponding mode shapes, which converge to the true mode shapes when the damping goes to zero or the natural frequency spacing goes to infinity. The assumption that the natural frequency spacing is large can be relaxed by extending the peak picking method with singular value decomposition (Shih et al. 1988). The corresponding approach is called complex mode indicator function when applied to FRF data and frequency domain decomposition when applied to PSD data. However, peak picking or its extensions often lead to inaccurate estimates because most of its assumptions are not tenable in practical situations, especially in operational conditions. They are therefore mainly used for the estimation of modal characteristics in the preliminary phase of a modal test, even in laboratory conditions (Ewins 2000, p. 304).

When the identified system description is parametric, e.g., in state-space form, no assumptions – other than that the structure's response can be described with the identified description – are needed for estimating the modal characteristics. The assumed damping model is most often general viscous damping, which contains proportional damping as a special case. This means that the approach can also be used when localized dampers are present and the mode shapes are complex.

When compared to nonparametric methods, parametric system identification methods need at least one additional user-defined integer: the model order, which equals the number of eigenvalues present in the model, hence, in theory, twice the number of natural frequencies. From control theory, several model validation techniques are available for choosing the model order in an automated way, so that the prediction

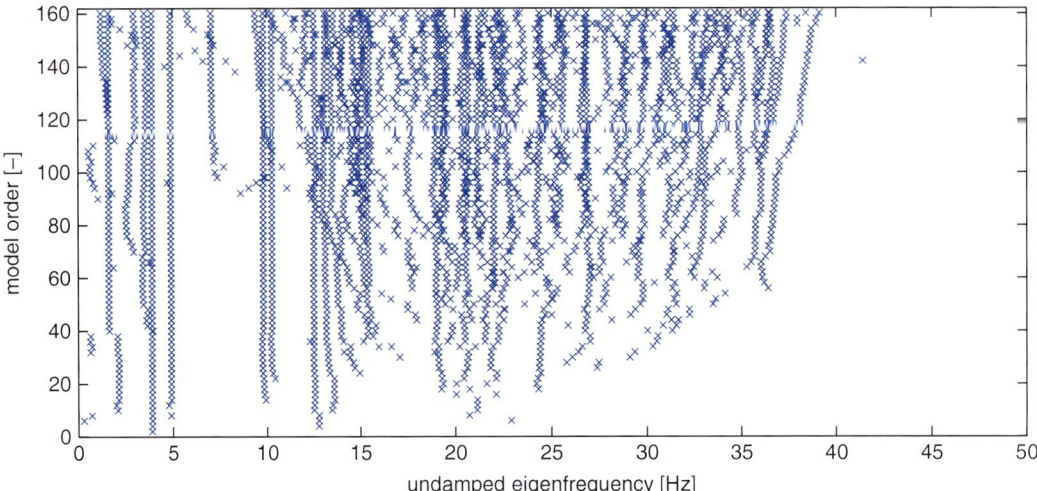

Operational Modal Analysis in Civil Engineering: An Overview, Fig. 4 Stabilization diagram for one setup of a modal test on a three-span prestressed concrete bridge. The modes that are stable in all setups are shown in Fig. 5 (Reproduced with permission from: Reynders et al., Fully automated (operational) modal analysis, Mechanical Systems and Signal Processing, 29:228–250, 2012)

capacity of the identified models is maximized (Ljung 1999). However, in modal testing applications, one is not primarily interested in the prediction capacity of an identified model as such, but rather in the physical relevance of the individual modes that constitute the model. An alternative approach has therefore been developed, based on the empirical observation that in a very large number of modal identification problems, the physical modes of the structure appear at nearly the same eigenfrequency when the model order is over-specified, while other spurious modes do not. In this approach, parametric models are estimated for a wide range of model orders, and the modes corresponding to all these models are plotted in a model order versus natural frequency diagram, called stabilization diagram (Heylen et al. 1997; Allemang 1999). The physical modes then show up as vertical lines in the diagram, and one of the modes of each line can be chosen as a representative of the corresponding physical mode. Figure 4 shows an example of a stabilization diagram, and Fig. 5 presents the corresponding stable modes.

Finally, the obtained set of modes can be validated according to relevant validation criteria. A criterion that is often used is the so-called modal assurance criterion (MAC) which is the dimensionless correlation coefficient between two mode shapes ϕ_j and ϕ_l:

$$MAC\left(\phi_j, \phi_l\right) = \frac{\left|\phi_j^H \phi_l\right|^2}{\left\|\phi_j\right\|^2 \left\|\phi_l\right\|^2},$$

where the subscript H denotes complex conjugate (or Hermitian) transpose, || denotes amplitude, and |||| denotes the Euclidian norm. When the mass of the structure is approximately uniformly distributed and the spatial resolution of the identified mode shapes is sufficiently large, the MAC value between different mode shapes should be zero (Allemang and Brown 1982). When real normal modes are expected, i.e., when proportional damping can be assumed, the modal phase collinearity (MPC) can be used for assessing the quality of the identified mode shapes (Pappa et al. 1992). The MPC value of a real normal mode shape, which is collinear in the complex plane, equals unity, while the MPC value of a complex mode shape is significantly lower than unity. Finally, for some of the most powerful parametric identification methods, the statistical accuracy of the estimated modal

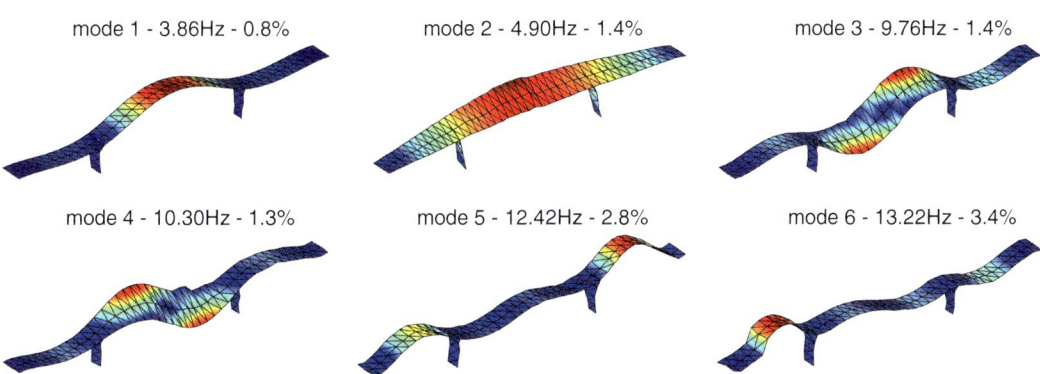

Operational Modal Analysis in Civil Engineering: An Overview, Fig. 5 The final set of validated modal characteristics for an operational modal analysis of a three-span prestressed concrete bridge (Reproduced with permission from: Reynders et al., Fully automated (operational) modal analysis, Mechanical Systems and Signal Processing, 29:228–250, 2012)

characteristics can be assessed. This is, for instance, the case for the stochastic subspace identification method, for which the complete joint probability distribution of the estimated natural frequencies, damping ratios, and mode shape components is available (Reynders et al. 2008; Reynders 2012).

Summary

The identification of modal characteristics from vibration data is very important in earthquake engineering, not only for design validation but also for structural health monitoring. Operational modal analysis, which identifies the modal characteristics from measured response data only, is often the most effective modal testing approach in earthquake engineering. Many system identification algorithms are available for estimating modal characteristics from the measured (force-) response data. Nonparametric algorithms are simple and intuitive, but they often yield inaccurate results. Parametric algorithms lead generally to much more accurate estimates; in particular, subspace identification combines a good statistical accuracy with a low computation cost and a high computational robustness. Since parametric algorithms need the model order as a user-defined integer, they are often combined with the stabilization diagram technique for estimating a final set of modal characteristics. Finally, a range of criteria is available for validating the estimated modal characteristics, from assessing the statistical accuracy of the estimates to assessing the complexity of the mode shapes.

Cross-References

▶ Advances in Online Structural Identification
▶ Ambient Vibration Testing of Cultural Heritage Structures
▶ Bayesian Operational Modal Analysis
▶ Blind Identification of Output-Only Systems and Structural Damage via Sparse Representations
▶ Laser-Based Structural Health Monitoring
▶ Nonlinear System Identification: Particle-Based Methods
▶ Post-Earthquake Diagnosis of Partially Instrumented Building Structures
▶ Seismic Behavior of Ancient Monuments: From Collapse Observation to Permanent Monitoring
▶ Stochastic Structural Identification from Vibrational and Environmental Data
▶ System and Damage Identification of Civil Structures
▶ Vibration-Based Damage Identification: The Z24 Bridge Benchmark

References

Akaike H (1974) Stochastic theory of minimal realization. IEEE Trans Automat Control 19(6):667–674

Allemang RJ (1999) Vibrations: experimental modal analysis, 7th edn. University of Cincinnati, Cincinnati

Allemang RJ, Brown DL (1982) A correlation coefficient for modal vector analysis. In: Proceedings of the 1st international modal analysis conference, Orlando, pp 110–116

Deckers K, De Troyer T, Reynders E, Guillaume P, Lefeber D, De Roeck G (2008) Applicability of low-weight pneumatic artificial muscle actuators in an OMAX framework. In: Sas P, Bergen B (eds) Proceedings of ISMA2008 international conference on noise and vibration engineering, KU Leuven, Leuven, Sept 2008, pp 2445–2456

Ewins DJ (2000) Modal testing, 2nd edn. Research Studies Press, Baldock

Heylen W, Lammens S, Sas P (1997) Modal analysis theory and testing. Department of Mechanical Engineering, KU Leuven, Leuven

Ho BL, Kalman RE (1966) Effective reconstruction of linear state-variable models from input/output functions. Regelungstechnik 14(12):545–548

Katayama T (2005) Subspace methods for system identification. Springer, London

Ljung L (1999) System identification: theory for the user, 2nd edn. Prentice Hall, Upper Saddle River

Magalhaes F, Cunha Á (2011) Explaining operational modal analysis with data from an arch bridge. Mech Syst Signal Process 25(5):1431–1450

Maia NMM, Silva JMM (1997) Theoretical and experimental modal analysis. Research Studies Press, Taunton

Papadimitriou C, Lombaert G (2012) The effect of prediction error correlation on optimal sensor placement in structural dynamics. Mech Syst Signal Process 28:105–127

Pappa RS, Elliott KB, Schenk A (1992) A consistent-mode indicator for the eigensystem realization algorithm. Report NASA TM-107607. National Aeronautics and Space Administration

Peeters B, De Roeck G (2001) Stochastic system identification for operational modal analysis: a review. ASME J Dyn Syst Meas Control 123(4):659–667

Pintelon R, Schoukens J (2001) System identification. IEEE Press, New York

Reynders E (2012) System identification methods for (operational) modal analysis: review and comparison. Arch Comput Method Eng 19(1):51–124

Reynders E, Pintelon R, De Roeck G (2008) Uncertainty bounds on modal parameters obtained from stochastic subspace identification. Mech Syst Signal Process 22(4):948–969

Shih CY, Tsuei YG, Allemang RJ, Brown D (1988) Complex mode indicator function and its applications to spatial domain parameter estimation. Mech Syst Signal Process 2(4):367–377

Van Overschee P, De Moor B (1996) Subspace identification for linear systems. Kluwer, Dordrecht

Paleoseismology of Glaciated Terrain

Björn Lund
Department of Earth Sciences, Uppsala University, Uppsala, Sweden

Synonyms

Deglaciation seismotectonics; Early Holocene faulting; Endglacial faulting; Glacially induced faulting; Glacio-isostatic faulting; Glacio-seismotectonics; Lateglacial faulting; Postglacial faulting

Introduction

Earthquakes generally occur in response to tectonic loading of the crust. However, once the crust has reached a state of stress close to frictional equilibrium other processes that affect the crustal stress field, such as the filling of a reservoir or water injection in a borehole, may trigger seismicity. The discovery of large fault scarps in northern Fennoscandia, and their subsequent dating to the end of the latest glaciation, the Weichselian, showed that also the load of an ice sheet can have a significant impact on earthquake occurrence.

As an ice sheet grows the weight of the ice will make the lithosphere subside into the softer asthenosphere below, a process known as glacial isostatic adjustment (GIA). The load increases both vertical and horizontal stresses in the lithosphere, and additional horizontal stresses develop due to the flexure of the elastic part of the lithosphere. Since variations to the size of the ice sheet takes place on much shorter time scales than asthenospheric flow, the flexural stresses develop at a slower pace. How these glacially-induced stresses affect earthquake occurrence depend on the prevailing stress field prior to the start of the glaciation, but generally in continental areas an ice sheet will tend to decrease differential stress under the load and therefore stabilize faults. As the ice retreats the vertical stress will decrease faster than the horizontal flexural stress due to the delayed response of the asthenosphere. In addition, horizontal tectonic stresses may have accumulated under the stabilizing weight of the ice sheet. The combination of deglaciation stresses, accumulated tectonic stress and the pre-existing stress field can potentially destabilize faults and cause large "postglacial" earthquakes, as was the case in Fennoscandia approximately 10,000 years ago.

Since the discovery of the Fennoscandian faults much effort in searching for and understanding postglacial faulting has been concentrated in the previously glaciated regions of northern Europe and North America. These are generally intraplate, continental regions which are not very seismically active at present. As many of the countries in these regions: Canada, Sweden, Finland and Russia, are investigating the

© Springer-Verlag Berlin Heidelberg 2015
M. Beer et al. (eds.), *Encyclopedia of Earthquake Engineering*,
DOI 10.1007/978-3-642-35344-4

feasibility of long term storage of high-level nuclear waste, which require safety assessments spanning hundreds of thousands to a million years, the possibility of postglacial faulting contributes significantly to the seismic hazard of such repositories. The connection between postglacial faults and nuclear waste repositories have made the identification of these faults a controversial, and highly debated subject.

Over the years, as more workers have been identifying and analyzing faults related to the latest glaciation in Fennoscandia and North America, these faults have become known as *postglacial*, *endglacial* or *lateglacial* faults, with the name indicating both cause and timing. Other terms emphasize either the cause (*glacially induced* or *glacio-isostatic faults*) or occurrence time (*early Holocene faults*). In the broader context of the effect of glaciations on crustal deformation and seismicity, terms such as *deglaciation seismotectonics* and *glacio-seismotectonics* have been used. The general discussion in this article will use the term *glacially induced* fault to emphasize the effect of glaciers on faulting both during their growth and decay, and to avoid the association to the end of the latest glacial period. The term postglacial fault is the one most commonly used and it will be used here to refer to faults that occurred at the end of, or after a glaciation. For specific faults, when the occurrence time has been associated with the last stage of deglaciation, the term endglacial will be used.

This article will review where glacially induced faults have been observed and much of the focus will be on the faults that occurred at the end of the latest glacial period. Identification of the faults will be discussed, their mechanism of faulting and finally a brief look at the current effect of deglaciation on fault stability and seismicity.

Observations of Glacially Induced Faulting

This section briefly reviews the identification and analysis of glacially induced faults in various regions of the world.

Fennoscandia

Northern Fennoscandia's spectacular endglacial faults (EGFs), in a region often referred to as the Lapland fault province (Fig. 1a), have long been known to the local Sami populations as prominent features in the landscape. The name of the 155 km long Pärvie fault scarp translates from Sami along the lines of "Like a breaking wave," an apt description of how the fault disrupts the gentle mountain terrain in northernmost Sweden (Fig. 1b). First described as late- or postglacial in Finland in the 1960s, subsequent investigations in Finland, Sweden and Norway identified approximately a dozen kilometer-scale fault scarps by the end of the 1980s (reviewed in Kuivamäki et al. 1998; Lagerbäck and Sundh 2008; Olesen et al. 2004). Most faults run in a north-northeasterly direction (Fig. 1a) and have reverse motion offsets with downthrow to the northwest, and inferred dips to the southeast. Generally, the faults have reactivated old, Proterozoic or Archean structures, although not necessarily the most prominent of these. The largest known EGF is the 155 km long Pärvie fault, which has a throw of up to 10 m and which runs just east of the Caledonian mountain front in northernmost Sweden. The Lapland fault province extends into Finnmark in northern Norway and into northwestern Finland. To the south, large EGFs had until recently only been detected down to Burträsk, south of Skellefteå in northern Sweden. However, soft-sediment deformation in varved sequences in south and central Sweden has been attributed to strong seismic shaking during end- and postglacial periods, but without identification of a causative fault (Mörner 2004). Recent LiDAR imagery (Fig. 2) have disclosed a number of new, smaller faults in Lapland, more complex fault systems around some of the larger faults, notably on the central Pärvie fault, and a 5 km long scarp in central Sweden of endglacial origin, albeit yet unconfirmed in bedrock (Smith et al. 2014).

The fault scarps are inferred to have been produced by large earthquakes, evidenced by the ubiquitous occurrences of faulting-related phenomena such as low angle landslides and

**Paleoseismology of Glaciated Terrain,
Fig. 1** (*Upper*) Map of northern Fennoscandia showing the Lapland fault province. Faults are: in Norway St: Stuoragurra, N: Nordmannvikdalen; in Sweden P: Pärvie, M: Merasjärv, LS: Lainio-Suijavaara, L: Lansjärv, R: Röjnoret, B: Burträsk; in Finland P: Pasmajärvi, V: Venejärvi, Su: Suasselkä. (*Lower*) The Pärvie fault north of Tjuonajokk, Sweden (Photo: B. Lund)

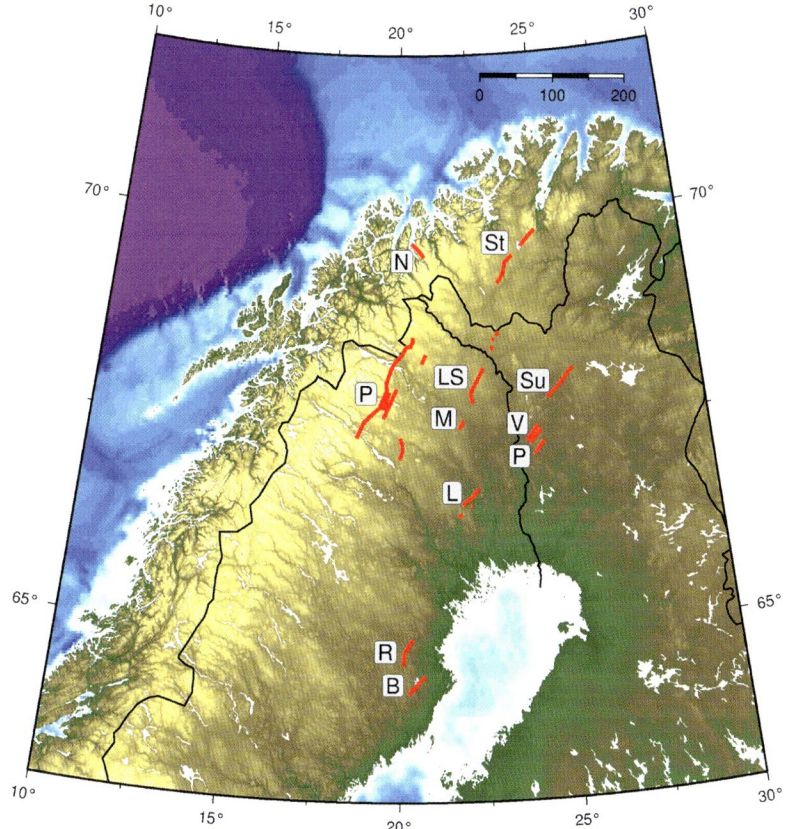

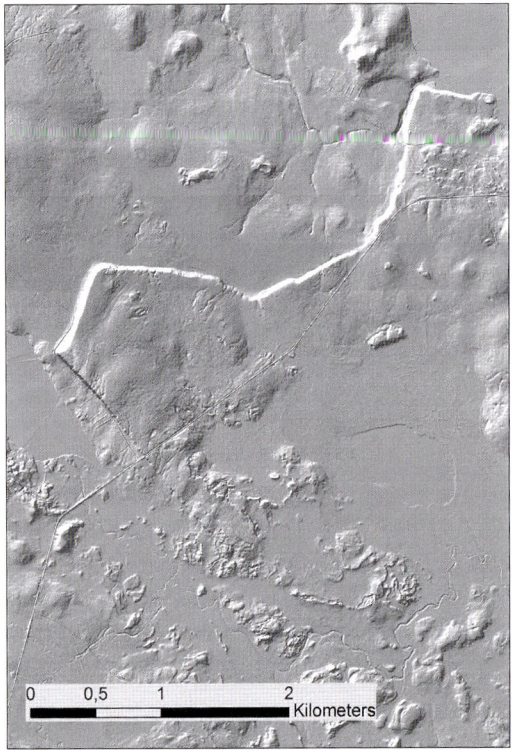

Paleoseismology of Glaciated Terrain, Fig. 2 LiDAR image of the Risträskkölen section of the Lansjärv endglacial fault in northern Fennoscandia. The fault scarp is up to 25 m high in the western corner, where the fault abruptly changes strike and continues to the southeast. Further to the southeast the fault can be seen to displace hummocky moraines just before it disappears (Processed LiDAR image courtesy of Henrik Mikko, Geological Survey of Sweden)

soft-sediment disturbances (seismites). Extensive trenching has been carried out on, and in the vicinity of, some of the more accessible faults and shows both liquefaction phenomena and that faulting occurred as single events rather than repeated movements on the faults. The glacial sediments aid in determining the timing of the earthquakes relative to local deglaciation. Using morphological features associated with ice-flow, glacial melting and marine erosion and deposition (many of the events occurred in areas which at that time were covered by the sea), rupture has been observed to occur just before local deglaciation (the Pärvie fault), during deglaciation, just after local deglaciation (the Lansjärv fault) or later in postglacial times (the Stuoragurra fault). Deglaciation in northern Fennoscandia occurred approximately 10,000 years ago but absolute dating of the ruptures, and also relative dating of the different scarps, has proven difficult. This is due both to the scarcity, and uncertain origins, of organic material in the ruptured sediments and because the course and chronology of the latest Fennoscandian deglaciation is not well known (Lagerbäck and Sundh 2008).

The Lapland fault province occupies a region mostly between the highest mountains in northern Scandinavia and the center of the Weichselian ice sheet, which is inferred to have been located in the Bay of Bothnia where it reached a thickness of approximately 3 km (Fig. 3a). Although the Weichselian glacial period consisted of a number of stadials, the whole region is well within the subsidence bowl of the maximum extent of the Weichselian ice sheet at the last glacial maximum (LGM), as shown clearly even today by the current glacial rebound field. As such, the horizontal flexural stresses in the region have been compressional, and most likely of significant magnitude, during much of the glacial period (see the example of one possible model realization in Fig. 3b).

Earthquake moment magnitude is calculated from the rupture area, the amount of slip and the shear modulus of the rock. First-order estimates of the endglacial earthquake magnitudes can be obtained from empirical relationships between the surface rupture length and magnitude. More accurate estimates need better constraints on the width of the fault plane, on the slip distribution in the earthquake and the shear modulus. Current microearthquake activity on the faults suggests that they are seismogenic down to at least 35 km depth (Arvidsson 1996; Juhlin and Lund 2011; Lindblom et al. 2014). The dip of the fault planes at depth is poorly constrained, but recent reflection seismic and microearthquake investigations indicate that the dip of the Pärvie and Burträsk faults are around 60° to the southeast (Juhlin et al. 2010; Juhlin and Lund 2011; Lindblom et al. 2014). Assuming that the current seismicity reflects the downdip extent of the rupture planes, the width of the fault planes can then be

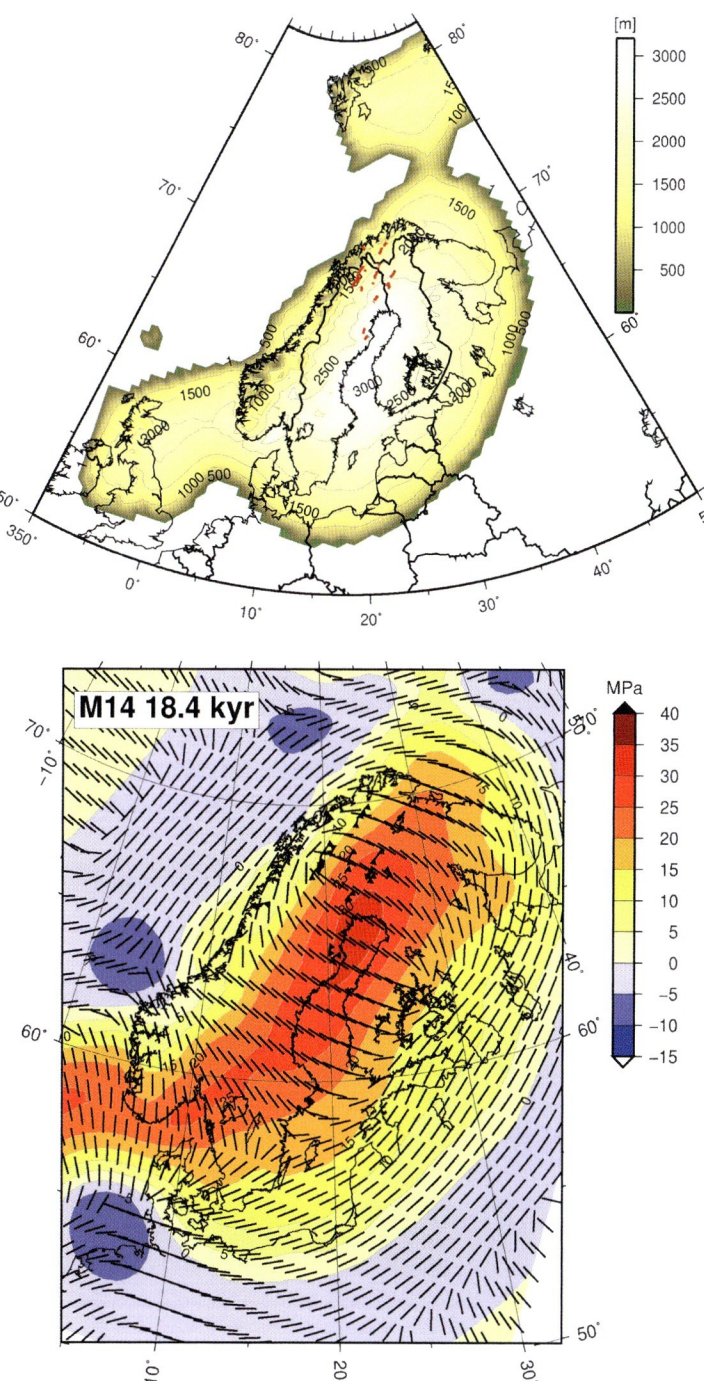

Paleoseismology of Glaciated Terrain, Fig. 3 (*Upper*) Map of the UMISM model of the Weichselian ice sheet at the last glacial maximum (LGM) 18,400 years BP (Schmidt et al. 2014). Endglacial faults indicated with *red lines*. (*Lower*) Modelled glacially induced maximum horizontal stress at LGM (Lund and Schmidt 2011). The model has elastic lithospheric thickness 120 km, upper mantle viscosity $5 \cdot 10^{20}$ Pa s and lower mantle viscosity $3 \cdot 10^{21}$ Pa s. Stress directions indicated by *black bars*, magnitudes with the color scale. Endglacial faults indicated in *black*

calculated. The slip distributions, or even the average slip, on the faults cannot be estimated with any certainty. One available option is to use relations of surface rupture offset to slip distribution estimates for similarly large reverse faulting earthquakes today. It is, however, still unclear if rupture of endglacial faults is similar to that of "regular" tectonic earthquakes, or if the relatively higher stresses produced at shallow depths by the ice produce relatively more shallow slip, thereby making such comparisons less relevant. With these limitations in mind, the large endglacial earthquakes in Lapland have estimated magnitudes between 7 and 8.

As alluded to above, the endglacial faults of northern Scandinavia are still seismically active (Arvidsson 1996; Lindblom et al. 2014). In fact, most of the earthquakes in northern Sweden occur in the vicinity of an endglacial fault, and all the major endglacial faults in northern Sweden and Norway are seismically active. Recent installations of seismic stations in the vicinity of the Suasselkä and Pasmajärvi faults in Finland indicate that these also host microearthquakes, albeit at a lower rate of activity. The magnitudes of the current earthquakes on the faults reach 3.6 for events that have well determined locations. There are indications that historic events of magnitude approximately 4.5 can be associated with some of the endglacial faults. The cause of the current activity is debated and includes hypotheses such as current tectonic strain release in significantly weakened zones or aftershocks to the endglacial events. In spite of the pervasive microseismicity, it has not been possible to detect any fault movement at the surface since the time of rupture, neither using geological indicators, GPS nor InSAR (Mantovani and Scherneck 2013).

Scotland

The ice sheet over Great Britain was significantly smaller and thinner than the Fennoscandian ice sheet, although it was at times connected to the Fennoscandian ice through an ice bridge across the North Sea. Early investigations of postglacial tectonics in Scotland identified displaced late- and postglacial shorelines in the Firth of Forth and the former ice-dammed lake in Glen Roy. Further work in Scotland indicated that a number of fault offsets were spatially and temporally associated with seismically induced features such as landslides and liquefied sediments (Fenton 1992; Ringrose 1989) and reported 10–100 m of horizontal displacements on 1–14 km long faults. Later investigations in the northwest Highlands have questioned some of the earlier work, especially the evidence for significant strike-slip faulting (Firth and Stewart 2000). Although the study still considered the lateglacial and early Holocene period as significantly seismically active, postglacial faulting was found to be limited to meter-scale vertical movements along pre-existing faults. Interestingly, all examples of postglacial seismicity in Scotland have been attributed to the disappearance of the smaller ice cap developed during the Younger Dryas, and not the deglaciation of the main British ice sheet (Firth and Stewart 2000). Current seismicity in Scotland does not seem to concentrate at inferred postglacial faults, nor does it correlate well with the center of uplift. The seismicity is highly variable, but have been suggested to cluster at the limits of the former ice-sheet, or in areas of local uplift anomalies (Firth and Stewart 2000).

Other European Observations

Smaller-scale structures inferred to be glacially induced have been identified in a number of areas around Europe. Close to the Fennoscandian ice sheet there are reports of faults in Russian Karelia with postglacial movement. Initially inferred to have repeated earthquakes, later analysis favor a "one-off" deformation phase approximately 11 kyr BP but found a seismic origin difficult to confirm (Kuivamäki et al. 1998).

In Germany, shallow meter-scale faults and folds were identified in a sand-pit in Upper Pleistocene alluvial-aeolian deposits 1 km away from the Osning Thrust, a major fault system in central Europe (Brandes et al. 2012). The faults have experienced normal faulting motion some 16–13 kyr ago, when the area was in the forebulge of the Fennoscandian ice sheet. Later, during deglaciation reverse motion occurred, inferred to be due to the N-S directed compressional stress field in northern Germany. Glaciotectonics has been ruled out since the faults are south of the maximum extent of the ice sheet.

In Siekierki near the Polish/German border, soft-sediment deformation structures have recently been interpreted as seismites due to glacially induced earthquakes during the Saalian deglaciation 120 kyr ago (Van Loon and Pisarska-Jamroży 2014). However, no causative fault was identified.

In Northern Iceland, the Kerlingar normal fault runs subparallel to the Northern Volcanic Zone (NVZ), but east of the zone itself. It is an unusually long normal fault with throw down to the east and it is not parallel to the fissure swarms in the NVZ (Hjartardóttir et al. 2010). Hjartardóttir et al. (2010) suggests that the Kerlingar fault formed shortly after the late Pleistocene deglaciation due to isostatic rebound with differential movements between two adjacent crustal domains.

North America

Postglacial faults were first described in 1843, in New York State (Mather 1843), although at that time neither the theory of global glacial periods, nor the concept of postglacial faulting were known. The offset grooves and scratches observed by Mather (1843) were in fact glacial striations. Several other observations of small-scale postglacial faults with offsets of a few millimeters were reported from northeastern USA and eastern Canada during the early twentieth century, and many more during the latter part of the century (Fenton 1994). To date, only two large scale phenomena in Canada have been proposed as large post- or endglacial faults: the Aspy fault on Cape Breton Island and the differential uplift at Peel Sound in Arctic Canada. The Aspy Fault offsets an 125,000 year old intertidal rock platform by 15 m, but has no exposed fault scarp, and the surficial deposits show no evidence of Holocene movement (Adams 1996). At Peel Sound, the marine beaches on opposite sides of the sound are tilted and have differential uplift of over 60 m, which is inferred to have occurred during a short period of time 9,000 years ago. The feature controlling this movement is apparently beneath the sound, but it has not been possible to identify a causative fault, or to show that the uplift is related to earthquakes at all (Adams 1996).

In the more tectonically active western parts of North America there have only been a few reports indicative of postglacial faulting: a spectacular fault scarp with suspected postglacial movements in the eastern Sierra Nevada, California, postglacial offsets in Archean basement rock in Montana and indications that glacial loading and unloading in Puget Sound may have controlled the timing of movements on the Seattle fault, above the Cascadia subduction zone (see review in Munier and Fenton 2004).

The Wasatch and Teton faults in the Basin-and-Range province in western USA show an increase in slip rate in the late Pleistocene and in the Holocene (see review in Hampel et al. 2010). This has been interpreted as a response to unloading due to the disappearance of Lake Bonneville and the Yellowstone ice cap, respectively, at the end of the latest glacial period (Hampel et al. 2010).

Grollimund and Zoback (2001) suggested that stresses induced by the Laurentide ice sheet affected the Reelfoot Rift in Missouri, USA, and triggered the New Madrid earthquakes of 1811–1812. Later models indicate, however, that such triggering is unlikely (Wu and Johnston 2000).

South America

Investigations from a bog at Puerto del Hambre in the southern Strait of Magellan indicate that the ground surface was downfaulted by at least 30 m sometime between the occupation of the strait by a proglacial lake (12,640–10,315 ^{14}C years BP) and the marine incursion of the site at 8,265 ^{14}C years BP (Bentley and McCulloch 2005). This inferred postglacial faulting seems not to have been associated with one or more large earthquakes, as sediment cores from the area do not show any indication of soft-sediment deformation, slumping, liquefaction or other disturbances to the overall stratigraphy.

Identification of Glacially Induced Faults

As discussed in the Introduction, the identification of postglacial faults has been an intensively debated issue both in Scandinavia and in Canada. Identifying glacially induced faulting and earthquake activity is non-trivial as the surface expressions produced must be distinguished from similar phenomena created by other processes such as glacial movement, ice-water interactions, gravity and temperature. It is also important, not least from a seismic hazard point of view, to distinguish "one-off" postglacial faulting from other neotectonic faulting which may "just" have long recurrence times.

As the physical characteristics of postglacial faulting, tectonic faulting and glaciotectonic deformation overlap, no single criterion can uniquely distinguish them. Instead, multiple lines of evidence need to be assembled for a correct classification. Several identification and classification schemes have been proposed and discussed over the years (see review in Munier and Fenton 2004), the scheme by Muir-Wood (1993) has for example been widely used in Fennoscandia and Canada to differentiate between postglacial faulting and glaciotectonic deformation. It contains the following key criteria, which should all be fulfilled:

1. The surface or material that appears to be offset has to have originally formed as a continuous, unbroken unit. Can the surface be dated? Is it the same age?
2. Can the apparent evidence of an offset be shown to be related directly to a fault?
3. Is the ratio of displacement to overall length of the feature less that 1:1,000? For most faults this ratio, a function of the strength of the rock prior to fault rupture, is between 1:10,000 and 1:100,000.
4. Is the displacement reasonably consistent along the length of the feature?
5. Can the movement be shown to be synchronous along its entire length?

However, a simple list like the above rarely covers all the variations in scales and postglacial deformation styles in the various glaciated or previously glaciated regions of the world (Munier and Fenton 2004). Detailed investigations at a variety of scales and with multiple techniques are essential in identifying the two over-arching attributes of postglacial faults; their age and their underlying driving mechanism.

The techniques used to investigate postglacial faults are similar to those employed in most regular paleoseismic investigations. In searching for possible candidates in the vast, sparsely inhabited areas of Fennoscandia and Canada, the analysis of stereoscopic areal photographs has been very successful. As referred to above, LiDAR based digital elevation models nowadays provide unprecedented resolution of features on the Earth's surface (Fig. 2), even through dense vegetation. The LiDAR images not only provide a tool to identify faults and landslides, they are also used to to analyze glacial landforms and their spatial relation to the faults. Detailed field mapping is essential to understand regional and local stratigraphy and geomorphic development, in order to aid in determining both the formation process and the age. Geophysical investigations such as magnetics, gravity, ground penetrating radar, sonar profiling (in lakes) and reflection seismic profiling, geodesy and local earthquake seismology reveal various aspects of fault geometry and conditions. Trenching is vital to gain higher confidence in the style of faulting and the age determination, and to investigate fault reactivation patterns. Finally, drilling is used to investigate shallow fault geometry and to determine mechanical, chemical and hydrological properties of the fault.

Mechanics of Glacially Induced Faulting

Although the effects of glaciers and ice sheets on fault stability and seismicity had already received some attention, interest in the topic increased significantly with the observation that there are very few earthquakes below Greenland and Antarctica (Johnston 1987). The analysis indicated that as the ice load increases on the crust, both

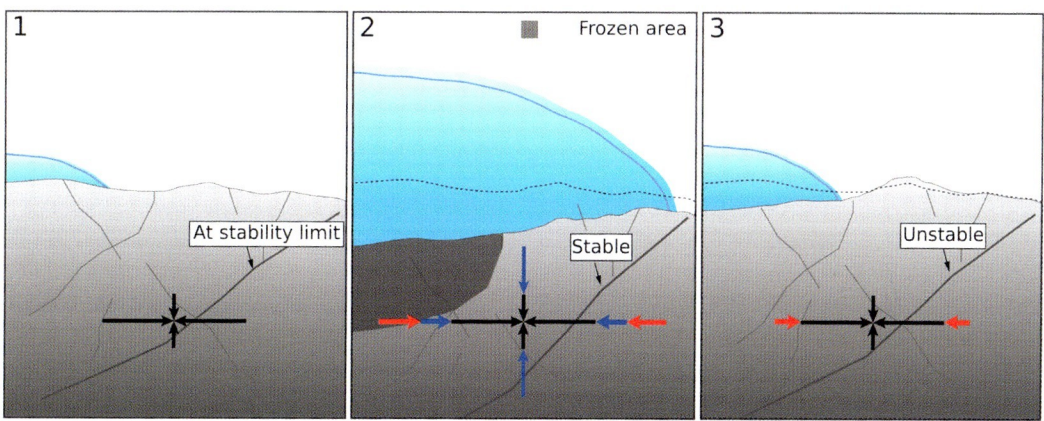

Paleoseismology of Glaciated Terrain, Fig. 4 Schematic stress evolution at a fixed location in the upper crust during a glacial period. Pre-existing (e.g., tectonic) stress field (*black*), glacially induced stresses: direct elastic load (*blue*), flexural (*red*). (**1**) Shortly before glaciation. (**2**) During glaciation. (**3**) Shortly after deglaciation. Note that neither the vertical displacement nor the *arrows* are to scale (Adapted from a figure courtesy of B. Fälth, Uppsala University)

vertical and horizontal stresses increase and thereby stabilize faults (Fig. 4). In Mohr-Coulomb terms, the increase in average stress moves the Mohr-circle to the right, away from the failure envelope. Subsequent progress in modelling glacial isostatic adjustment (GIA) made it possible to refine the analysis of fault stability by including flexural stresses and temporal stress variations due to the viscosity of the mantle (e.g., Wu and Hasegawa 1996). Glacially induced stresses under large ice sheets can amount to 30–40 MPa both horizontally and vertically, depending on the size of the ice sheet and the rheology of the Earth. Although seemingly large, they are only 10–15 % of the weight of the overburden at 10 km depth, appropriate for earthquake nucleation. In addition, there are tectonic, topographic, sediment induced and other stresses that can be significant in a specific area (Fig. 4). It is important that all these stresses, the full stress field, is included in the analysis of fault stability as it is the total stress that determines the stress state, which fault orientations that are most easily activated and the stability of these (e.g., Lund et al. 2009).

Not generally included in the GIA based models is the accumulation of tectonic stress during glaciation. Although much smaller than the glacially induced stress variations, at least in continental interiors, there can be enough stress accumulated during the cause of a glacial period to significantly affect the stability of faults once the stabilizing effect of the ice is disappearing (Johnston 1989). Over even longer time periods, such as the duration of the Greenland or Antarctic ice sheets, tectonic loading may even be able to bring faults back towards frictional equilibrium, and hence destabilize them again in spite of the overriding ice sheet (Stewart et al. 2000).

Modelling Glacially Induced Stresses

GIA models have primarily been concerned with the rates of vertical uplift through time in order to fit observations of Holocene sea-level rise, historical tide-gauge data and recent satellite based data, e.g., GPS, InSAR and GRACE gravity. Such studies have provided a wealth of information on climate, ice sheet configuration and evolution, and the rheology of the Earth. The emergence of suitable finite element models provided a convenient tool to investigate the stress field induced in the crust by a glaciation, and thus the mechanisms of glacially induced faulting (Wu and Hasegawa 1996). The stress field that develops in the Earth due to the ice load depends both on the ice load itself and the rheological and elastic structure of the Earth. Even during glacial periods the large ice sheets evolve constantly, it is nowadays well established that during the latest glaciation, in the last 100,000 years, the ice sheets grew and declined several times and therefore had very varying extent and thickness. Crustal stresses evolve, to first order, with the ice load implying that a good ice model is vital in GIA modelling (e.g., Schmidt et al. 2014).

As outlined above, the growth of an ice sheet and the subsequent down-warping of the Earth's elastic lithosphere induces vertical stress in the Earth proportional to the ice load. It also induces horizontal stresses which are a combination of the Poisson effect of the load and the flexural stresses that develop as the lithosphere bends (Fig. 4). The horizontal stresses will be compressive in the upper part of the lithosphere under the load, and tensional outside the load in the so called fore-bulge. Under the load, the vertical stresses will initially be larger than the horizontal as flexure develops slowly due to the slow flow of the mantle. This produces an induced normal faulting stress regime. As the ice sheet stops growing, horizontal stress magnitudes will generally catch-up, and surpass, the vertical, inducing a strike-slip state of stress. During deglaciation the vertical stress decreases much more rapidly than the horizontal stresses, again due to the slower response of the mantle, such that the stress field will evolve to a reverse state. During and after deglaciation the uplift of the formerly glaciated region will decrease the compression in the upper crust, as the collapse of the fore-bulge will decrease the induced tension in the crust. The inward migration of the fore-bulge may produce induced tensional stresses in areas that were previously subjected to compressional glacial stresses.

Second only to the ice load itself, the rheology and elasticity of the Earth model has a major influence on the glacially induced stresses. For a certain ice load, the elastic structure determines the magnitude of the stresses, which is especially important in the upper crust, where the glacial stresses are the largest. A high resolution layered, or three-dimensional, structure is vital in order to not significantly over-estimate the stress magnitudes in the uppermost crust. A laterally varying thickness of the elastic part of the lithosphere, and lateral variations of the parameters within the lithosphere, also affects the magnitude of the stress field and its spatial variation (Lund et al. 2009). The viscosity of the mantle, including lateral variations and non-linear rheology, strongly affects the temporal evolution of stress magnitudes outside the LGM ice margin, but has little effect inside the margin (e.g., Wu et al. 1999). The interaction of the ice load with Earth structure produces some interesting effects. The horizontal stress magnitudes induced by flexure of the elastic lithosphere depend on the spatial extent of the ice sheet relative to the thickness of the elastic lithosphere (Johnston et al. 1998). Accordingly, for commonly inferred lithospheric thicknesses, a smaller ice sheet, the size of the former British ice sheet, would produce maximum amplification of the horizontal stresses. The Fennoscandian ice sheet would similarly produce higher flexural stresses than the North American ice sheet, consistent with the observation of large endglacial fault in northern Fennoscandia.

Fault Stability

The GIA models discussed above have included neither a pre-existing stress field nor pre-defined faults. Instead, fault stability has been evaluated in post-processing, where the modelled glacially induced stresses have been added to a pre-existing stress field and areas and times susceptible to faulting have been studied. Faulting potential is commonly investigated using Mohr-Coulomb theory, through changes either in the Coulomb Failure Stress (CFS = $\tau - \mu(\sigma_n - P_p)$), where τ is the shear stress on a fault, σ_n the normal stress, μ the coefficient of friction and P_p the pore pressure

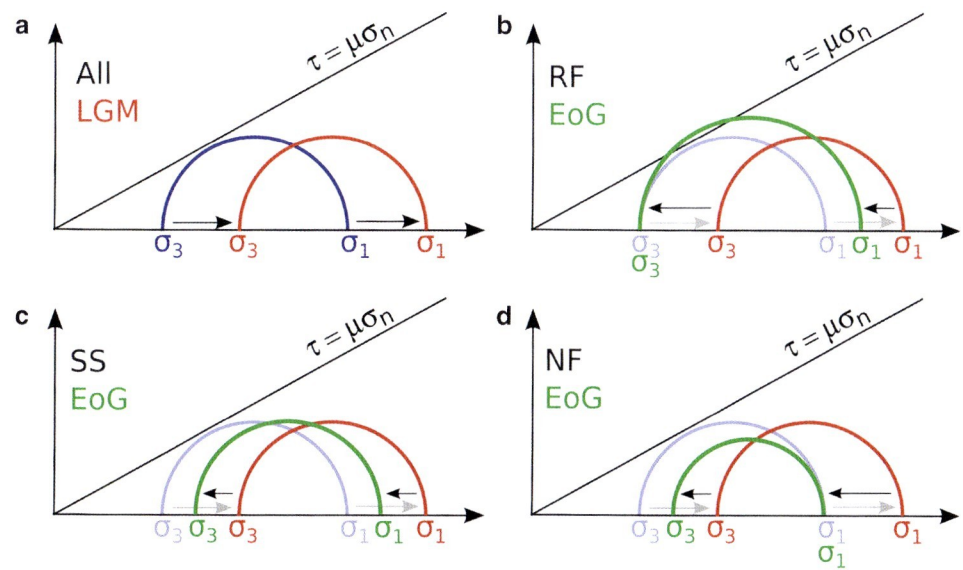

Paleoseismology of Glaciated Terrain, Fig. 5 Variation in fault stability at a location centrally under the ice sheet, from prior to the glaciation (*blue lines*), to LGM (**a**, *red lines*) and to the end of glaciation (*EoG, green lines*) for pre-existing reverse (**b**), strike-slip (**c**) and normal (**d**) stress fields. (**a**) applies to all pre-existing stress fields as it is assumed here that the glacially induced stress field is isotropic ($\sigma_1^G = \sigma_2^G = \sigma_3^G$)

(e.g., Lund 2005), or the similar fault stability margin (Wu and Hasegawa 1996).

The models generally show that for a stable continental interior, with a preexisting stress field in a state of reverse faulting, the region within the LGM ice margin is stabilized by the growth of an ice sheet. Towards the end of deglaciation, as the vertical stress diminishes much faster than the horizontal stresses, optimally oriented faults in the region centrally under the ice sheet tend to become unstable (Fig. 5 and Lund et al. 2009; Wu et al. 1999). This model agrees well with the occurrence of the endglacial earthquakes in northern Fennoscandia, both in timing and strike direction (Lund et al. 2009). For strike-slip or normal faulting stress fields it is not the interior (Fig. 5c, d) but the areas outside the LGM ice margin that are affected by fault instability, not only at the time of deglaciation but also during times of glacier advance and stand still. The stability of the fore-bulge region depends very much on the details of ice and Earth models, and is as such less well constrained.

Fault stability and earthquake occurrence depend on more factors than just the state of stress. Pore pressure is a vital parameter in the analysis of faulting, and pore pressures in the crust during deglaciation are notoriously difficult to estimate due to the lack of data. The Fennoscandian ice sheet was cold based for much of the latest glacial period (Lagerbäck and Sundh 2008) and it is likely that permafrost could have existed in the crust below the ice for long periods of time. Without a hydraulic head, it is uncertain how pore pressure in the crust would be affected by the ice and permafrost. If there was hydraulic contact between water in the ice sheet and crustal pore fluids, high pore pressures would have been able to penetrate deep into the crust during the long duration of the ice sheet (e.g., Lönnqvist and Hökmark 2013). The combination of high pore pressures and permafrost could potentially have maintained higher than hydrostatic pressures in the crust during deglaciation, adding to the instability of faults.

In a further advancement of finite element based GIA models it has become possible to explicitly include faults and pre-existing stresses. Early models imposed velocity boundary conditions to have faults slip at a constant rate prior to glaciation, thereby including tectonic stressing rates, but did not include the deep mantle (see Hampel et al. 2010). The models agree with previous results but more explicitly show the inhibition of slip during glaciation and rapid slip increase at deglaciation. Later models include a proper mantle description and come to similar conclusions (Steffen et al. 2014), although still in a state of early development.

Effects of Current Deglaciation

The current global warming trend increases melting of ice sheets and glaciers worldwide. Does this deglaciation influence seismicity rates in the affected regions? Areas that are glaciated and tectonically active, such as Iceland and Alaska, are good candidates for investigations of the process but few studies have been attempted. Although glacier thinning rates may have increased, the time scales involved, compared to the time of good seismic network coverage, make it difficult to establish solid background rates to compare to. A glacier surge in Alaska affected local seismicity by increasing it in the surge reservoir region and decreasing it in the receiving region, in agreement with predictions for a compressive tectonic stress state (Sauber and Molnia 2004). Also in Alaska, stress changes due to retreating ice in the Icy Bay area in the 80 years between the 1899 Yakataga (M_W 8.1) and Yakutat (M_W 8.1) earthquakes and the 1979 St. Elias (M_S 7.2) earthquake are inferred to have decreased the stability of the St. Elias fault system by 2–3 MPa (Sauber and Molnia 2004). In Iceland, the current thinning of the Vatnajökull glacier has been suggested to contribute to high seismicity and unusual focal mechanisms (Pagli and Sigmundsson 2008). Assessing how the glacially induced stress change affects earthquake activity in Iceland is, however, non-trivial as time varying stresses from volcanic activity and rifting are active in the glaciated region.

In the areas of the two remaining continental ice sheets, Greenland and Antarctica, significantly increased seismic detection capabilities have not seriously challenged the more than 25 year old hypothesis of ice sheets as inhibitors of seismic activity. There is significant earthquake activity along the margins of the Greenland ice sheet but very little in the interior (i.e., Olivieri and Spada 2014). There has been suggestions that glacial isostatic adjustment from the Pleistocene deglaciation still affects the occurrence and mechanisms of earthquakes off the coast of Greenland (Chung 2002), and that the seismicity is affected by elastic rebound stresses from the current thinning. The data is currently to sparse to distinguish between these processes and to validate any of them with numerical models (Olivieri and Spada 2014). What is clear, however, is that the current rapid thinning of outlet glaciers in Greenland cause significant ice-related events, so called glacial earthquakes or ice-quakes (e.g. Ekström et al. 2006). Similarly, in Antarctica there is significant ice-quake activity in ice streams and outlet glaciers, but very few earthquakes detected in the continent itself. Recent investigations have, however, found swarms of seismic activity inferred to be caused by magmatic activity in West Antarctica (Lough et al. 2013).

Summary

Loads on the crust caused by glaciers and ice sheets can have a significant effect on earthquake occurrence, as indicated by the large fault scarps in northern Fennoscandia caused by earthquakes at the end of the latest glaciation, and the current seismic quiescence under the Greenland and Antarctic ice sheets. Therefore, the assessment of long term seismic hazard in regions affected by glaciation need to take this effect into account.

A large number of observations of faulting in North America and Northern Europe indicate that these intraplate, continental regions experienced a substantial increase in seismicity as the ice sheets of the latest glaciation disappeared. In the Lapland fault province in northern Fennoscandia a dozen fault scarps, ranging in length from a few tens of kilometers to 155 km, have been associated with one-step rupture at the end of the deglaciation phase. The association to earthquake rupture comes from detailed investigations of soft-sediment disturbances such as landslides, seismites and stratigraphic offsets in trenches. These investigations also determine the age of faulting relative to the glacial sediments. The magnitudes of the largest endglacial earthquakes have been estimated as M_W 7–8 using fault scarp lengths, fault planes delineated by current microseismicity and empirical scaling relations. Postglacial fault observations in Scotland and Canada generally show shorter lengths and smaller offsets, although there are a few observations of larger scale postglacial phenomena where the cause has not been properly identified. Outside the margin of the former maximum extent of the ice sheets, such as in Germany and USA, glacially induced faulting has been identified or suggested, not necessarily contemporaneous with the latest deglaciation.

Identification of glacially induced faulting is non-trivial as the faults, or faulting related disturbances, first have to be distinguished from structures related to glaciotectonic processes. Then, identified faults have to be shown to have a glacially related origin and not be due to other neotectonic processes. Several classification schemes exist, which all suggest that a wide variety of data should be collected prior to decision making. Areal, satellite, ground mapping, geophysics, trenching and drilling data all provide complementary evidence of fault age and driving mechanisms.

Rapid model development and increase in computing power has enabled more complex models of glacial isostatic adjustment (GIA) and fault stability. A glacier load bends the elastic upper lithosphere down into the softer asthenosphere, inducing both directly load related stresses and flexural stress in the crust. The load tends to stabilize faults inside the load margin, but may destabilize faults outside the margin, depending on the pre-existing stress field. During deglaciation, the vertical stress decreases more rapidly than the horizontal stresses due to the delayed response of the asthenosphere. If the pre-existing stress field inside the former ice margin is suitably oriented, faults may become destabilized. Outside the margin the fore-bulge will collapse and move inward, destabilizing faults if the pre-existing stress conditions combine favorably with the glacial stresses. The latest model generation includes actual faults and can estimate fault offsets in simulated earthquakes. In addition to direct load and flexural stresses, tectonic horizontal stresses may accumulate under the stabilizing weight of the ice sheets. Stress accumulation is generally slow in cratonic interiors and uncertainties in the accumulation rates makes the relative importance of the process debatable. The GIA models can currently to some degree predict location, orientation and onset times of glacially induced fault instability. However, there is still need for considerable progress in understanding the details of endglacial fault mechanisms in terms of location, geometry and size.

Improved models of glacially induced seismicity will allow better prediction of what may happen as glaciers and ice sheets decrease in the current global warming trend. The evidence for current crustal scale deglaciation related seismicity is scarce but it may increase in the next century, emphasizing the need for better seismic hazard estimates in the vicinity of thinning ice sheets.

Cross-References

▸ Earthquake Magnitude Estimation
▸ Earthquake Mechanisms and Stress Field
▸ Earthquake Mechanisms and Tectonics
▸ Earthquake Recurrence

▶ Earthquake Return Period and Its Incorporation into Seismic Actions
▶ Paleoseismology
▶ Paleoseismology and Landslides
▶ Paleoseismology: Integration with Seismic Hazard
▶ Paleoseismic Trenching
▶ Remote Sensing in Seismology: An Overview

References

Adams J (1996) Paleoseismology in Canada: a dozen years of progress. J Geophys Res 101:6193–6207

Arvidsson R (1996) Fennoscandian earthquakes: whole crustal rupturing related to postglacial rebound. Science 274:744–746

Bentley MJ, McCulloch RD (2005) Impact of neotectonics on the record of glacier and sea level fluctuations, Strait of Magellan, southern Chile. Geogr Ann 87:393–402

Brandes C, Winsemann J, Roskosch J, Meinsen J, Tanner DC, Frechen M, ... Wu P (2012) Activity along the Osning Thrust in Central Europe during the Lateglacial: ice-sheet and lithosphere interactions. Quat Sci Rev 38:49–62. doi:10.1016/j.quascirev.2012.01.021

Chung W-Y (2002) Earthquakes along the passive margin of Greenland: evidence for postglacial rebound control. Pure Appl Geophys 159:2567–2584

Ekström G, Nettles M, Tsai VC (2006) Seasonality and increasing frequency of Greenland glacial earthquakes. Science 311:1756–1758

Fenton C (1992) Late quaternary fault activity in north west Scotland. In: Mörner N-A, Owen LA, Stewart I, Vita-Finzi C (eds) Neotectonics – recent advances. Quaternary Research Association, London, pp 62

Fenton C (1994) Postglacial faulting in eastern Canada (Open file no 2774). Geological Survey of Canada, Ottawa

Firth CR, Stewart IS (2000) Postglacial tectonics of the Scottish glacio-isostatic uplift centre. Quat Sci Rev 19:1469–1493

Grollimund B, Zoback MD (2001) Did deglaciation trigger intraplate seismicity in the New Madrid seismic zone? Geology 29:175–178

Hampel A, Hetzel R, Maniatis G (2010) Response of faults to climate-driven changes in ice and water volumes on Earth's surface. Phil Trans R Soc A 368:2501–2517

Hjartardóttir AR, Einarsson P, Brandsdóttir B (2010) The Kerlingar fault, northeast Iceland: a Holocene normal fault east of the divergent plate boundary. Jökull 60:103–116

Johnston AC (1987) Suppression of earthquakes by large continental ice sheets. Nature 330:467–469

Johnston AC (1989) The seismicity of 'stable continental interiors'. In: Earthquakes at North-Atlantic passive margins: neotectonics and postglacial rebound, Kluwer, Dordrecht, pp 299–327

Johnston P, Wu P, Lambeck K (1998) Dependence of horizontal stress magnitude on load dimension in glacial rebound models. Geophys J Int 132:41–60

Juhlin C, Lund B (2011) Refelction seismic studies over the Burträsk fault, Skellefteå, Sweden. Solid Earth 2:9–16. doi:10.5194/se-2-9-2011

Juhlin C, Dehghannejad M, Lund B, Malehmir A, Pratt G (2010) Reflection seismic imaging of the end-glacial Pärvie Fault system, northern Sweden. J Appl Geophys 70:307–316. doi:10.1016/j.jappgeo.2009.06.004

Kuivamäki A, Vuorela P, Paananen M (1998) Indications of postglacial and recent bedrock movements in Finland and Russian Karelia (Technical report no YST-99). Geological Survey of Finland, Helsinki

Lagerbäck R, Sundh M (2008) Early Holocene faulting and paleoseismicity in northern Sweden (Technical report no C 836). Geological Survey of Sweden, Uppsala

Lindblom E, Lund B, Tryggvason A, Uski M, Bödvarsson R, Juhlin C, Roberts R (2014) Microearthquakes illuminate the deep structure of the endglacial Pärvie fault, northern Sweden. Geophys J Int (Submitted)

Lönnqvist M, Hökmark H (2013) Approach to estimating the maximum depth for glacially induced hydraulic jacking in crystalline rock at Forsmark, Sweden. J Geophys Res 118:1–15

Lough AC, Wiens DA, Barcheck CG, Anandakrishnan S, Aster RC, Blankenship DD, ... Wilson TJ (2013) Seismic detection of an active subglacial magmatic complex in Marie Byrd Land, Antarctica. Nat Geosci 6:1031–1035. doi:10.1038/NGEO1992

Lund B (2005) Effects of deglaciation on the crustal stress field and implications for endglacial faulting: a parametric study of simple Earth and ice models (Technical report no TR-05-04). Swedish Nuclear Fuel and Waste Management Co. (SKB), Stockholm. www.skb.se. Accessed 10 Feb 2015

Lund B, Schmidt P (2011) Stress evolution and fault stability at Olikiluoto during the Weichselian glaciation (Technical report no 2011-14). Posiva Oy, Eurajoki. (92 p)

Lund B, Schmidt P, Hieronymus C (2009) Stress evolution and fault stability during the Weichselian glacial cycle (Technical report no TR-09-15). Swedish Nuclear Fuel and Waste Management Co. (SKB), Stockholm. www.skb.se. Accessed 10 Feb 2015

Mantovani M, Scherneck H-G (2013) DInSAR investigation in the Pärvie end-glacial fault region, Lapland,

Sweden. Int J Remote Sens 34:8491–8502. doi:10.1080/01431161.2013.843871

Mather WW (1843) Geology of New York, Part 1, comprising the geology of the first geological district. Carroll and Cook, Albany. (653 p)

Mörner N-A (2004) Active faults and paleoseismicity in Fennoscandia, especially Sweden. Primary structures and secondary effects. Tectonophysics 380:139–157

Muir-Wood R (1993) A review of the seismotectonics of Sweden (Technical report no TR-93-13). Swedish Nuclear Fuel and Waste Managment Co. (SKB), Stockholm

Munier R, Fenton C (2004) Review of postglacial faulting. In: Munier R, Hökmark H (eds) Respect distances (Technical report no TR-04-17). Swedish Nuclear Fuel and Waste Managment Co. (SKB), Stockholm, pp 157–218

Olesen O, Blikra L, Braathen A, Dehls J, Olsen L, Rise L, ... Anda E (2004) Neotectonic deformation in Norway and its implications: a review. Nor J Geol 84:3–34

Olivieri M, Spada G (2014) Ice melting and earthquake suppression in Greenland. Polar Sci. doi:10.1016/j.polar.2014.09.004

Pagli C, Sigmundsson F (2008) Will present day glacier retreat increase volcanic activity? Stress induced by recent glacier retreat and its effect on magmatism at the Vatnajökull ice cap, Iceland. Geophys Res Lett 35. doi:10.1029/2008GL033510

Ringrose PS (1989) Recent fault movement and palaeoseismicity in western Scotland. In: Mörner N-A, Adams J (eds) Paleoseismicity and neotectonics. Elsevier, Amsterdam, pp 305–314

Sauber JM, Molnia BF (2004) Glacier ice mass fluctuations and fault instability in tectonically active southern Alaska. Global Planet Change 42:279–293

Schmidt P, Lund B, Näslund J-O, Fastook J (2014) Comparing a thermomechanical Weichselian Ice Sheet reconstruction to reconstructions based on the sea level equation: aspects of ice configurations and glacial isostatic adjustment. Solid Earth 5:371–388. doi:10.5194/se-5-371-2014

Smith CA, Sundh M, Mikko H (2014) Surficial geology indicates early Holocene faulting and seismicity, central Sweden. Int J Earth Sci 103:1711–1724. doi:10.1007/s00531-014-1025-6

Steffen R, Wu P, Steffen H, Eaton DW (2014) On the implementation of faults in finite-element glacial isostatic adjustment models. Comput Geosci 62:150–159

Stewart IS, Sauber J, Rose J (2000) Glacio-seismotectonics: ice sheets, crustal deformation and seismicity. Quat Sci Rev 19:1367–1389

Van Loon AJ, Pisarska-Jamroży M (2014) Sedimentological evidence of Pleistocene earthquakes in NW Poland induced by glacio-isostatic rebound. Sediment Geol 300:1–10. doi:10.1016/j.sedgeo.2013.11.006

Wu P, Hasegawa HS (1996) Induced stresses and fault potential in eastern Canada due to a disc load: a preliminary analysis. Geophys J Int 125:415–430

Wu P, Johnston P (2000) Can deglaciation trigger earthquakes in N. America? Geophys Res Lett 27:1323–1326

Wu P, Johnston P, Lambeck K (1999) Postglacial rebound and fault instability in Fennoscandia. Geophys J Int 139:657–670

Paleoseismic Trenching

H. Serdar Akyüz[1], Volkan Karabacak[2] and Cengiz Zabcı[1]
[1]İstanbul Teknik Üniversitesi, Maden Fakültesi, Jeoloji Müh. Bölümü, Ayazağa, İstanbul, Turkey
[2]Eskişehir Osmangazi Üniversitesi, Mühendislik-Mimarlık Fakültesi, Jeoloji Müh. Bölümü, Meşelik, Eskişehir, Turkey

Synonyms

Active fault; Earthquake history; Paleoseismology; Trenching

Introduction: The Importance of Trenching in Paleoseismology

Trenching on a fault trace is a direct way to understand the historical (mainly Holocene) evolution of a fault segment. What are signatures of past earthquakes at the Earth's surface? How do we recognize and identify past (paleo-) earthquakes? One of the most famous and fundamental principles of modern geology, "the present is the key to the past," provides the answer, from which historical seismic events are accepted to have the same effect as modern ones do. However, the incompleteness of the historical record makes it hard to determine and identify paleo-earthquake-related structures. The discrimination of paleoseismic evidences from geological and geomorphological processes and the attribution of them to certain historical events make an important contribution to the seismic risk assessment of any region.

Paleoseismology is the art of the identification of past earthquakes in terms of location, timing, and size by using geological and geomorphological evidence (McCalpin 2009). This branch of Earth science is mainly based on the interpretation of features recorded and preserved by depositional and erosional processes. Paleoseismology addresses a specific timescale that connects neotectonics and instrumental seismology, mostly from the Late Pleistocene to the present day. A detailed paleoseismic study can reveal a seismic history even going back several thousand years. Characterizing past earthquake parameters (e.g., fault location, earthquake magnitude, and recurrence interval) helps construct a strong database, which can be used in the modeling of potential seismic hazards.

However, there are no fixed or standard techniques for the study of past earthquake records. Accordingly, different methods and approaches must be specifically applied to each individual case. The most widely used and effective method is the trenching of recent deposits across active faults. The aim of trenching is to expose the stratigraphic and structural relationships of buried modern deposits and to discriminate the signature of paleo-events. A paleoseismologist determines the timing of identified earthquakes by using various Quaternary-dating methods allowing correlation with historical documents. The main target of a paleoseismic study is to understand the history of studied fault segment(s) and to develop a knowledge base for establishing the location and magnitude of future earthquakes for the same region. Paleoseismic results play a vital role for seismic risk assessments. This chapter aims to summarize the main steps undertaken during a paleoseismic trenching study. The most commonly used dating methods in paleoseismology are also briefly outlined.

Flowchart in Trenching Studies

The identification of a paleo-event, its dating, and its correlation with a historical earthquake is the main objective of a paleoseismic study. However, to reach that final goal, there are several steps that need to be undertaken, such as site location, logistic settings, selection of the trench type, taking of safety precautions, preparation of trench walls for logging, sampling for dating, and post-field studies, which will all be briefly mentioned in the following sections.

Site Selection: Geoinformatics, In Situ Observations, Shallow Geophysical Methods, and the Micro-Topographic Surveys

The determination of a suitable site plays a fundamental role in paleoseismology. Not only finding the fault trace but also exposing the suitable sedimentation is crucial for the recovery of a paleo-event in a trench study. Although the techniques may differ from region to region, the general sequence of investigations start with geoinformatics, continue with in situ observations, and usually end with the application of shallow geophysical and micro-topographic surveys.

Geoinformatics: Remote Sensing and Geographic Information Systems (GIS)

Aerial photographs are one of the main sources used in the identification of faults and other tectonic landforms. Analyses of aerial photograph stereopairs allow easily visualization of pseudo-3D reflections of the study region in a 2D environment. Skilled researchers can easily interpret fault scarps and tectonic landforms by using aerial photographs of actively deformed regions. Detailed topographic datasets also assist in the interpretation of physiographic features. Technological progress has greatly advanced remote sensing abilities. For example, one of the latest commercial satellites, Geoeye-2, is planned to start collecting satellite images of the Earth with a ground resolution up to 0.34 m by the second half of 2013 (GeoEye Elevating Insight 2013).

Moreover, advances in LiDAR (Light Detection and Ranging) and high computational technologies make the analysis of high-resolution digital elevation models (DEM) possible. These topographic data with submeter pixel resolution not only provide the easy identification of fault scarps but also yield auto-quantification of displacements (Zielke and Arrowsmith 2012) or precise mapping of seismogenic faults even in highly forested mountainous terrains (e.g., Cunningham et al. 2006). It is possible to merge all these products in a GIS database and to make sophisticated spatial analyses of tectonic landforms in site selection for trenching. These modern tools can focus attention on small study areas; thereby they significantly save time and reduce budgets in paleoseismological projects.

In Situ Observations: Geological and Geomorphological Mapping

To understand geometrical relations among recent (late Quaternary) geological units, and sometimes between older rocks and recent deposits, geological mapping of units and landforms of a trench site should be undertaken. Detailed mapping can also identify piercing points (i.e., offset features) along or across faults. Remote sensing analysis on the region of interest is controlled with field observations during the mapping of geological and geomorphological features. Mapping usually includes fault scarps and detailed subdivisions of Quaternary deposits across the fault. The paleoseismic history is interpreted from the constructed relationship of the mapped Quaternary deposits and the faulting pattern. The displacement measurements are often indicated on these maps. Beside the mapped features, the depositional conditions and sedimentary sources of the site are studied to gather the maximum information for understanding the local stratigraphy before trenching.

Modern depositional and erosional processes at a paleoseismic site supply data so that the trench stratigraphy can be interpreted. Paleoseismologists mostly prefer fine-grained sediments, where slow sedimentation rates allow an older history to be recorded at relatively shallower depths. However, boulder- and cobble-rich sediments can be deposited abruptly and can instantaneously cover paleo-events. Furthermore, it is also difficult to recognize deformational structures within these coarse-grained sediments. As a result, mixed sedimentary environments such as fluvial and alluvial rivers (mostly floodplains and point bars), seasonal swamps, marshes, sag ponds, deltas, alluvial fans (especially the distal part), piedmonts, and deserts are the ideal for paleoseismic trenching.

Colluvium in front of the fault scarp is one of the most important depositional features recording the material shed from the uplifted fault block. The clast type, size, and shape that form colluvial deposits depend on the lithology of uplifted block but may also change laterally from proximal to distal parts.

Shallow Geophysical Methods

Geophysical exploration methods have a critical role for detecting the exact location of main faults and any subsidiary branches. The application of geophysical techniques can be useful in definition of the stratigraphy and the deeper structure of the fault zone, aspects that cannot be determined by trenching or drilling. Geophysical methods are often applied in regions where tectonic landforms are buried and cannot be identified by surficial observations. The most commonly used geophysical methods are seismic reflection/refraction, ground-penetrating radar (GPR), and electric, magnetic (aero and electro), and gravity techniques in paleoseismological studies. For example, along the 1999 İzmit earthquake rupture (North Anatolian Fault, Turkey), GPR profiles clearly show not only the fault zone but also displaced fluvial channels of different ages (Ferry et al. 2004). In addition to 2D representations, for faults with a more complicated tectonic setting, 3D GPR images are needed to constrain fault structure, such as demonstrated on the hidden faults in the transpressional setting of the Alpine Fault (South Island, New Zealand) (Carpentier et al. 2012).

Micro-Topographic Surveys: Construction of Very High-Detailed Morphology

Before the excavation of a natural surface on a trenching site, detailed surveying of the morphology of the field can reveal landscape anomalies, which cannot be easily recognized with the naked eye. Interpretation of high-resolution topographic data can be used to analyze the effect of tectonic activity in the formation of such morphological features. A quantitative assessment of morphological structures provides detailed information especially about the likely characteristics of the faulting. Moreover, well-defined and well-analyzed detailed morphology also decreases the excavation length (and time) at a probable trench site. Toward this goal, micro-topography and the related physiographic features can be measured by using instruments such as the electronic theodolite (total station), d-GPS (or RTK GPS), and the terrestrial LiDAR (TLS) in the field. With the introduction of the high-resolution topographic data collection of the TLS, it is even now possible to constrain the 3D-slip vectors recorded by displaced landforms (Gold et al. 2012). Not only detailed topographic maps but also topographic profiles across fault zones provide invaluable information about the history of deformation.

Final Decision: Site Selection

The best location for a trench is highly site dependent. The site selection is undertaken through the synthesis of all datasets and studies, which are briefly listed above. All combined data are interpreted to identify two (or alternatively three) targets in a trench study in order to expose (a) the full rupture geometry and the style of faulting; (b) the stratigraphy, which reflects the most complete depositional history; and (c) any buried displaced feature, if present. For the first goal, it is important to stay on the main fault strand, which records the most complete seismic history of the deformation zone. However, it is also common to target trench studies on secondary faults. Pantosti et al. (2008) preferred to excavate one of their trenches on an antithetic fault scarp instead of the main displacement zone, because of logistic reasons. Secondly, a paleoseismologist always tries to find the most suitable place with a continuous sedimentation for the most complete stratigraphic record. Even depositional environments with very slow sedimentation rates can be chosen for slow-moving faults, like the Alhama de Murcia Fault (Eastern Betic shear zone, Spain) to expose a longer paleoseismic record (Ortuño et al. 2012). These first two objectives mostly yield the recovery of location and timing of a faulting event by fault-perpendicular trenches. However, on strike-slip faults, the magnitude of displacement can only be obtained with fault-parallel trenches, where buried and laterally displaced features are exposed.

Logistics: Administrative and Environmental Issues

Administrative Issues: Local and Governmental Permissions

Following the site selection, one of the mandatory issues is getting excavation permissions both from local (land owners) and administrative authorities. If the site is located in an archeological site or is close to a military base, then additional permissions may be required.

On cultivated lands, firstly, it is important to get the permission of the landowner but the local authorities must also be informed. The contact also provides logistic information for researchers about the existence of any possible human-made structures, such as waterworks, electricity cables, and oil/gas pipelines, which can easily be destroyed during a trench excavation. Moreover, these authorities, like local municipalities, may help in finding the work machines, which are used in trenching.

Environmental Issues: Water Table, Geography, and Transportation

The high water table in a site can cause the collapse of trench walls or it can make almost impossible for researchers to move inside the trench. The dewatering for a more pleasant working space is done in two general ways: (a) periodically pumping the water out of the trench and releasing it some distance away, and (b) draining the water by digging a shallow pit downslope from the toe of the trench. Also, the drier season is selected to have a lower water table as much possible in most cases.

It ought to be possible to find a proper work machine for excavation (e.g., backhoe, trackhoe, or bulldozer) in most study areas. But the trench location may be located too far from a settlement, or there may be no excavation machine around. In such cases, trenches can be dug by hand. Prentice et al. (2002) dug several hand-made trenches even up to several meters in length and depth in Mongolia, where there are no settlements nearby.

The excavation of a site also means the destruction of the local stratigraphy and structures. Therefore, any trenching should be done in the most preservative way to avoid damage both for the ongoing and any future studies at the same site. Moreover, excavation material should be backfilled carefully, especially on cultivated lands. For example, preservation of the natural soil can be critical and can be done by separating it from other excavated materials and putting it back at the end of the trench study.

Excavation: Choosing the Trench Orientation, Size, and Pattern

The choices for trench excavation depend on several factors such as the kind of material being trenched, the width of the deformation zone, the fault type, the relief of the site, the target depth, the stability of the trench walls, and the style of working.

Trench Orientation: Fault-Perpendicular and Fault-Parallel Trenches

The sense of fault displacement is the major criteria in positioning trenches before starting the excavation. Fault-perpendicular trenches are often used to locate fault zones and identify the recurrence patterns. Dip-slip paleo-events can be characterized in terms of recurrence and displacement by a single fault-perpendicular trench, especially where the deformation is localized in a narrow zone. However, multiple trenches, both fault-perpendicular and fault-parallel, are needed to measure the horizontal slip along strike-slip faults. The concept of 3D trenching was developed for this type of faulting, where both recurrence and displacement history could be inferred. To reveal the slip history, several approaches could be used, such as (a) multiple trenches closely spaced; (b) successive exposures, which are dug orthogonally to the fault zone; or (c) two fault-parallel trenches on each side of the fault, exposing the offset features (piercing lines). In all of these approaches, at least one fault-perpendicular trench is excavated to locate the fault zone. For example, Marco et al. (2005) used buried stream channels as piercing lines to restore the slip history of individual and cumulative events at the Jordan Gorge Segment of the Dead Sea Fault.

Trench Size

The main goal in trench studies is to expose maximum information with the most efficient excavation. Thus, determination of excavation dimensions plays a crucial role in the planning of the study time and budget. Depending on the width of the fault zone, the trench length may vary from a few meters to several hundred meters. Excavations are often started from footwall side and extended toward the hanging wall across dip-slip faults. The trench length is relatively shorter across strike-slip faults. The width is preferred to be large enough for the maintenance of a safe and comfortable working space, especially for taking photos of the trench walls. Depth mostly depends on the stability of walls and water table. Deeper trenches increase the probability of exposing a longer seismic history. Safety precautions are also very important in planning of the trench dimensions. (Please see the relevant section in this article). On the other hand, the geometry of trench walls, either oriented vertically or inclined, should have a planar shape as much as possible for a perfect projection surface to log.

Trench Patterns

Trench arrangement usually differs according to the structural and depositional properties of sites. The most appropriate arrangement is mostly selected according to the water table, stability of trench walls, type of deposits, logistic conditions, and the budget. Figure 1 shows cross sections of

Paleoseismic Trenching, Fig. 1 Cross sections showing different types of trench patterns. Type 1, single slot; Type 2, one-side stepped; Type 3, two-side stepped; Type 4, multistepped; Type 5, one-side sloped; Type 6, two-side sloped (or open pit) (Modified and redrawn after McCalpin 1989)

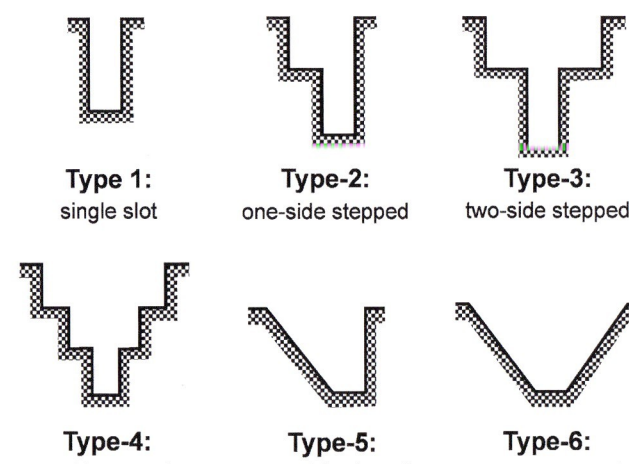

Paleoseismic Trenching, Fig. 2 Sample photos from different trench types: (**a**) single slot (Photo by Dr. Aynur Dikbaş) and (**b**) two-side stepped (Photo by Dr. Taylan Sançar)

different trench types (Type 1 to Type 6). The most often used pattern is single-slot (California-style, Type 1) trench, which is easily shored with a minimum material excavation and is least time consuming (Fig. 2a). This style of trenching is also the cheapest one, but also may acquire hydraulic or wooden shores against collapse of trench walls, especially in loose sediments. On the other hand, it is obvious that deeper sections increase the possibility for the recovery of the longer seismic history. Thus, paleoseismologists tend to excavate deeper trenches in suitable sites (Fig. 2b). Deeper trenches, Type 2 to Type 6, expose more faulting events, but they are expensive and time consuming. Multistepped or sloped patterns are mainly precautions against trench collapse.

Safety Precautions

Trench walls tend to collapse easily under various geological and morphological conditions. Fault zones easily reduce the cohesion of the material by crushing and creating open voids. High water table can evoke the collapse of unconsolidated young sediments. Cohesion differences between stratigraphic units in a section may also cause caving inside the cohesionless unit. Precautions, which are listed under two subtitles below, include only general and most common suggestions. Local regulations must always be consulted and followed. BS 5930 "the code of practice for site investigations" in the UK, for example, gives detailed guidance on legal, environmental, and technical matters relating to site investigation. It should be well understood that trenching is dangerous and loss of life can occur if undertaken unsafely. All available and mandatory risk assessments should be undertaken prior to work commencing and then reviewed once the trench is open to take into account any unexpected ground conditions.

Safety Precautions Inside the Trench

Trenching in loose materials or high water table may cause collapsing or caving of trench walls. Stepped- or sloped-type trenches are relatively safer with respect to single-slot type based on local conditions. For example, according to safety regulations in the USA, the vertical walls exceeding 1.5 m high in Type A or 1.2 m in Type B and Type C soils must contain another 1.5 m-wide horizontal bench on each side (OSHA 1989). In addition to benching, the stability of walls can be supported by hydraulic or wooden shores (Fig. 2a). If trench walls contain blocky materials, fallen boulders or cobbles can easily harm or even kill researches who work inside. Hard hats should always be worn in any trench over 1 m deep, which can be deadly under the right circumstances.

Safety Precautions Outside the Trench

Trenches with considerable depths are dangerous spots not only for domestic animals but also for the people who live nearby. An excavation area can attract especially kids and curious adults, and closing to the trench edges can be dangerous both for them and researchers inside. Paleoseismologists usually use safety tapes or simple fences to avoid such accidents. One member of the study team should also stay outside not only to inform cautious visitors but also to monitor any progressive cracks on the ground following the excavation of the site.

Preparing Trench for the Study

Following the excavation and implementation of safety precautions, the walls must be perfectly cleaned enough to expose all stratigraphic and structural relationships. The shovel of work machines usually leaves many scars, clay packs, or pseudo-corrugations. Tools such as scrapers, masonry towels, hammers, brushes, and water sprays are used for the cleaning of different material types on the trench walls.

After a careful cleaning process, a reference grid is constructed for logging. A typical grid is composed of perfect vertical and horizontal lines, which are generally spaced 1 m (or 0.5 m) apart. Horizontality and verticality can be constructed and checked using a spirit level. Laser levelers, which increase the precision and save lots of time, are getting more common each day in trench studies. At the end of gridding, every horizontal and vertical reference point is labeled successively every 1 or 0.5 m (Fig. 3).

Logging: Recording the Stratigraphic and Structural Relationships

Trench logs are the key records, which contain almost all information regarding the stratigraphy and structures in a trench study. These logs not only document primary and/or secondary earthquake evidence, but they also show the precise coordinates of sampling points. In many trench studies, usually a single wall is logged. However, logging of both walls provides confirmation of structural features and also helps the 3D reconstruction of the distribution of stratigraphic units. There are two common logging methods: manual and photomosaic.

Squared paper is used in the traditional manual method, for which each critical stratigraphic horizon and structure is logged with plotting the measurements of critical points with respect to the reference grid (Fig. 4). An ordinary tape measure or a measuring rod with a spirit level is used in taking these coordinates. In manual logging, there must be at least two people working simultaneously for an efficient study.

Paleoseismic Trenching, Fig. 3 An ideal reference grid (Photo by Dr. Volkan Karabacak)

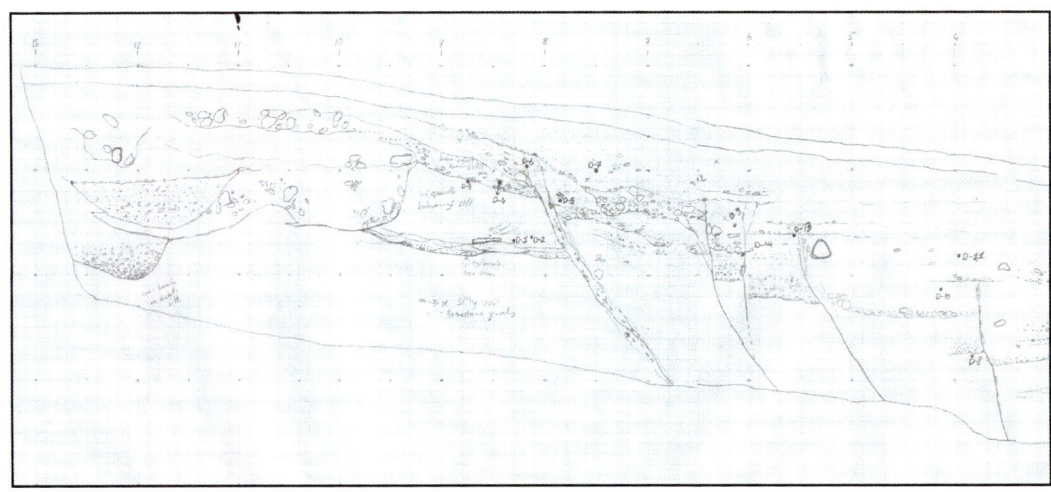

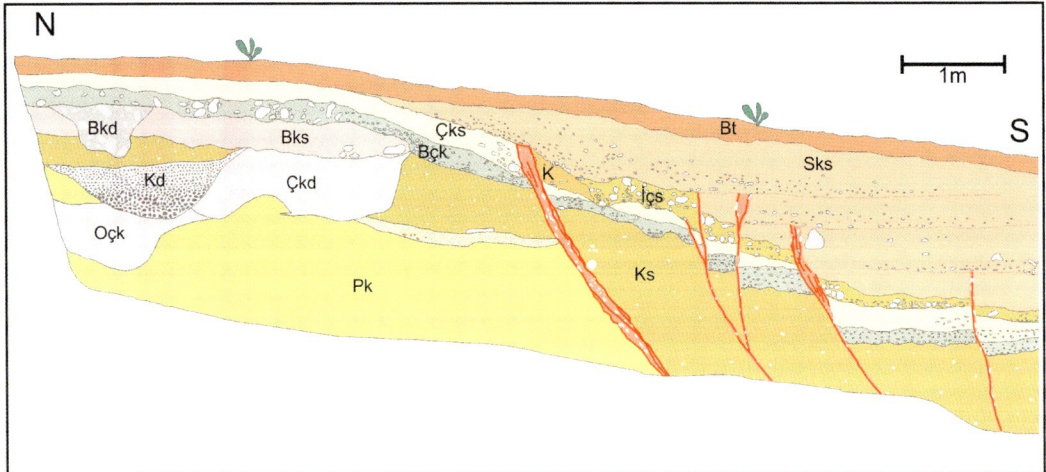

Paleoseismic Trenching, Fig. 4 Traditional manual log of a trench wall (Taken from Dr. Volkan Karabacak)

Paleoseismic Trenching, Fig. 5 Photomosaic of a trench wall (Photo by Dr. Volkan Karabacak)

While one person measures the horizontal and vertical distances from trench features to the nearest grid line, a second person plots the positions of these measurements on the squared paper with a scale.

Especially after the development of the digital photography techniques, logging on the photomosaic of walls has almost become a standard in trench studies. The trench walls are photographed and combined into a mosaic covering the whole wall (Fig. 5). Fast orthorectification of photos provide precise representation of trench exposures.

In addition to these two common techniques, there are different methodological attempts in logging of a trench wall. These relatively new methods aim to provide more quantitative information about structures and the distribution of the depositional units by using such as spectral imaging (Ragona et al. 2006) or the magnetic susceptibility measurements of the trench walls (Fraser et al. 2009).

Irrelevant to which method is chosen during the logging, the final record is digitized by using vector-based drawing software in post-field studies. All stratigraphic units are labeled and shown with proper symbols and colors. In addition to the clear imaging of faults and event horizons, the reference grid is provided to give a true spatial sense (Fig. 6).

Types of Sediment-Structure Relations in the Identification of Paleo-earthquakes

Destructive earthquakes, mainly with magnitude >6, often produce surface deformation, which depends on parameters such as the physical structure of the crust, the type of faulting, and the focal depth. In a properly selected trench site, it is highly probable to see surface faulting within the young deposits. A key objective of paleoseismologists is to identify and define event horizon(s) and recover

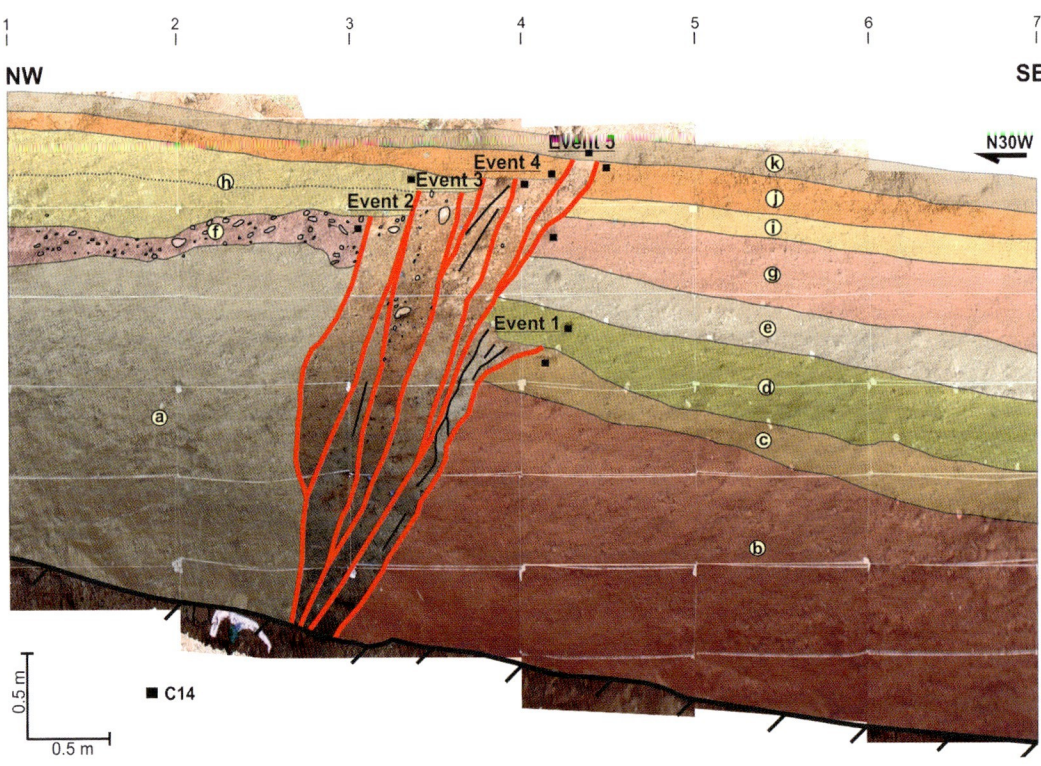

Paleoseismic Trenching, Fig. 6 A final log of a trench study (Taken from Dr. Volkan Karabacak)

the seismic history of the studied fault segment with dating of the relevant stratigraphic units.

Each trench has its own story. The historical evolution of the studied fault segment is hidden on the trench wall exposures. Even though there are some definite event structures for different fault types, all fault-related structures can be classified under two major titles: (a) on-fault (primary) and (b) off-fault (secondary or indirect) structures (Fig. 7). On-fault structures are mostly preferred in trench studies, due to the ambiguous nature of off-fault structures.

On-Fault Structures

The most commonly seen evidence for a past earthquake on the trench wall is upward termination of a fault against overlying strata (Fig. 7a). The thickness change of a layer on both sides of the fault reflects an erosional and depositional stage after an earthquake (Fig. 7b). A colluvial wedge is another clear evidence for a paleo-surface rupture, where a vertical separation or offset occurs between faulted blocks (this is also possible in lateral fault movements in rugged areas) (Fig. 7c). A monoclinal-like structure develops above the faulted zone in environments with high sedimentation rates (Fig. 7d). Open fissures can often be seen both on extensional and strike-slip faults. They can be filled by younger infills, which mark an event on the trench wall (Fig. 7e). Disoriented pebble- or cobble-sized clasts, seen in coarse-grained layers or in shear zones within fine-grained sediments, usually indicate a surface deformation of an earthquake (Fig. 7f, g).

Off-Fault Structures

Where it is not possible to open a trench on the main fault zone, for a variety of reason, paleoseismologists will then look to examine secondary (or off-fault) structures, to reveal the paleoseismic history. Angular unconformities

Paleoseismic Trenching, Fig. 7 Earthquake indicators used in the identification of paleo-events. On-fault structures: (**a**) upward termination of a fault, which is overlain by undeformed stratum; (**b**) difference in thickness of strata at each side of the fault and downward growth of the displacement along the fault trace; (**c**) scarp-derived colluvium; (**d**) formation of monoclinal folding, where fast sedimentation covers the fault scarp; (**e**) open surface cracks, infilled with material of the overlying unit; (**f**) disordered pebbles covered by undeformed layer; and (**g**) sheared layer covered with unsheared one. Off-fault structures: (**h**) angular unconformity within modern sediments; (**i**) sand boil, sand dike, and liquefied sand; (**j**) minor cracks and fissures with no offset; and (**k**) soft-sediment deformation in general that blankets with undeformed one (Modified and redrawn after Allen (1986); reprinted with permission from Active Tectonics (1986) by the National Academy of Sciences, Courtesy of the National Academy Press, Washington, DC)

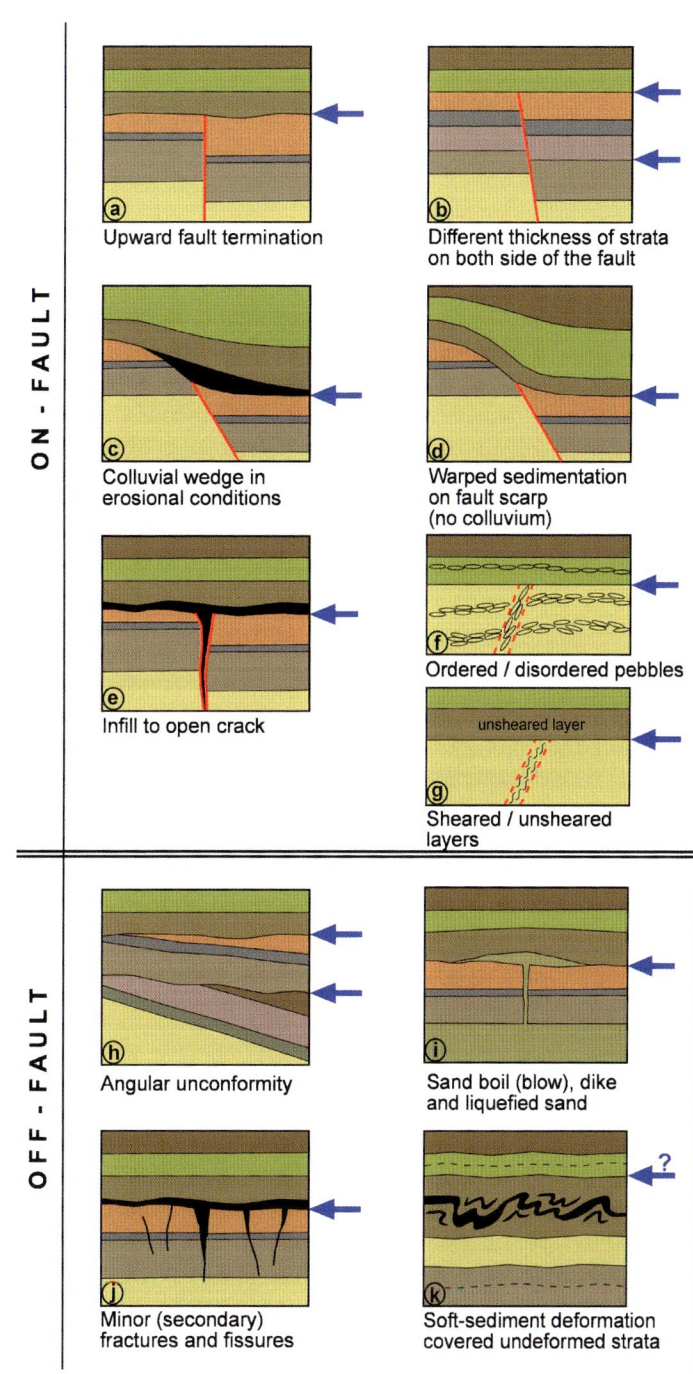

within the modern sediments are important markers, indicating a nearby fault-related deformation (Fig. 7h). Sandblow (or boil), mud diapirism, liquefied sand, minor fractures, and cracks are all soft-sediment deformational structures that are related to ground shaking and can be interpreted as indicators of paleo-earthquakes (Fig. 7i–k).

Dating Paleo-earthquakes

In addition to identification of paleo-earthquakes on trench walls, it is crucial to date important seismic horizons to model paleo-earthquakes and future risk. During or after the logging of stratigraphic horizons and structural features on the trench wall, horizons are sampled for dating to construct the temporal relationships. Especially event horizons, which cover earthquake-related structures, are the key levels in the recovery of the seismic history in a trench study. Late Quaternary deposits can be dated by a wide range of correlated, relative, numerical, and calibrated age methods. Among all, radiocarbon dating is the most common applied method in paleoseismological studies. The main materials are in situ organic compounds, such as bones, teeth, shells, charcoals, peats, chunks, and organic soils for radiocarbon dating. Advances in luminescence dating also introduced the application of this method in trench studies. Other dating methods of young sediments such as varve chronology, electron spin resonance, amino acid racemization, etc. can be used in suitable environments. Prior to sampling for dating, the most proper method should be determined, and the careful sampling techniques should be applied to avoid any contamination.

Advantages and Disadvantages of Trenching Studies

Paleoseismic trenching plays an important role in the reconstruction of a seismic history and makes an invaluable contribution to seismic risk assessments of a region (Gürpınar 2005). However, paleoseismology has advantages and limits like any other applied techniques. It is important to know all these aspects of paleoseismic studies to construct a sound project and solve a scientific problem.

The Advantages of Paleoseismic Trenching

– Paleoseismic trenching can provide the exact location of the fault (or deformational structures), which is very important information in urban or industrial planning.
– The reconstruction of the seismic history from well-chosen trench sites provides reliable information for seismic risk assessments.
– In addition to the location and timing of a paleo-event, well-located fault-perpendicular trenches for dip-slip faults or fault-parallel trenches for strike-slip faults provide the slip history (magnitude of displacement).

The Disadvantages of Paleoseismic Trenching

– Sometimes the destructive earthquakes do not produce any surface rupture along the source fault. These kinds of paleo-earthquakes do not leave their surface expressions on the Earth's surface, which make it impossible to identify anything in a trench study. Thus, paleoseismic trenching may only give information for large earthquakes that produce surface deformations.
– The distributed deformation, especially in thick water-saturated sediments, can be reflected as departed surface faulting after different events. Studies, especially with only a single trench under these circumstances, are resulted with missing recovery of total history.
– The thickness of stratigraphic units, which records the event history in a trench study, is directly affected by the sedimentation rate. Continuous sedimentation is always desired in trench studies; however, very fast sedimentation can cover event horizons with a thick pile of sediments, which prevent reaching the paleo-events.
– High water table usually prevents reaching deeper (older) stratigraphy.
– Logistics (i.e., permissions or environmental issues) may prevent excavating a perfect trench site, which is not ideally located or positioned.
– There may not be enough and suitable sampling material for dating based on environmental and/or sedimentary conditions.
– Measured displacements may also include post-seismic and/or creeping motions in addition to the coseismic slip.
– It is hard to identify all individual earthquakes along fast-moving faults (especially at overlapping sections of neighboring segments) due to limitation of dating methods.

Summary

In the developing world, mankind needs better life conditions, new settlements, and new industrial areas. It is well known that most part of the Earth's crust has a perpetual motion and that earthquakes are one of the most destructive natural hazards. Therefore, scientists have to find safe zones in which to live, especially at actively deformed regions, or provide information to mitigate against the seismic hazard. To achieve this, they have to know the exact location of the deformation zone and understand the characteristic behavior of the fault zone. Then, they can predict the future seismic risk of the area. Paleoseismological trenching is the most important tool to understand the nature of paleo-earthquakes and to predict the present and future behavioral characteristics of studied fault segments. The famous phrase of James Hutton "the present is the key to the past," then, can be improved to "the past is the key to the future" in the paleoseismic point of view. Of course, the term "past" covers mainly Holocene or Late Pleistocene, but not millions of years, while the term "future" embraces tens to thousands of years.

Trenching is a multidisciplinary study, including geology, geophysics, geomorphology, geoinformatics, geochronology, history, even mythology, etc. Pre-field, remote sensing studies provide important information on the location of the deformation zone and possible trench sites, which saves time, manpower, and budget. After the completion of desk and preliminary field studies, the first step is to find a convenient site for trenching. Important issues in proper location are continuous sedimentation, a low water table, presence of dating material, and sufficient logistic conditions (transportation, providing a work machine, etc.). Detailed geomorphological and geological maps are prepared to understand depositional or erosional conditions around trench sites. High-resolution micro-topographic maps give invaluable and precise information about effects of deformation and offset structures. Shallow geophysical methods are also important to define exact deformation zone and the continuation of the fault in depth.

After deciding on the placement of the trench site, the orientation, size, and pattern of the excavation should be chosen based on scientific and logistic conditions. Safety precautions during and after excavation are very important both inside and outside of the trench. These precautions must be strictly applied until the closure of the trench. After excavation, the trench is prepared for a detailed study within the following steps: (1) The trench walls are cleaned and the traces of the work machine are removed; (2) the walls are put in a reference frame with vertical and horizontal lines (mainly by using a plastic string) in 1×1 or 0.5×0.5 m grids; and (3) stratigraphic horizons and critical structures are mapped onto squared paper scaled appropriately. Alternatively, systematic photographs and their photomosaics of trench walls can also be used for logging. Note that the identification of past (paleo-)earthquakes on trench walls needs careful, detailed, and accurate analyses of both exposed trench stratigraphy and structure. On-fault or off-fault deformation and structures (primary and/or secondary) are used to reveal seismic history of the studied fault segment. (4) Sample collection for dating of critical (event) horizons should be undertaken and is important in trench studies to model the timing of events and recurrence intervals. Finally, after the collection of all available data at the trench site, the excavated material is filled back carefully to restore the original environmental conditions as much as possible.

In conclusion, paleoseismological trenching, the direct observation method of historical earthquakes, provides both the exact location and history of the fault. Exposing faulted modern (or Holocene) sediments in suitable environments gives invaluable information about surface-rupturing earthquakes for a few hundreds to thousands of years. These paleoseismic data are very important in the planning of urban and industrial settlements and, most importantly, in providing a safer life for mankind.

Cross-References

▶ Archeoseismology
▶ Earthquake Recurrence
▶ Earthquake Return Period and Its Incorporation into Seismic Actions
▶ Luminescence Dating in Paleoseismology
▶ Paleoseismology
▶ Paleoseismology: Integration with Seismic Hazard
▶ Radiocarbon Dating in Paleoseismology
▶ Seismic Event Detection
▶ Tsunamis as Paleoseismic Indicators

References

Allen CR (1986) Seismological and paleoseismological techniques of research in active tectonics. In: Wallace Chairman RE (ed) Active tectonics: studies in geophysics. National Academy Press, Washington, DC, pp 148–154

Carpentier SFA, Green AG, Langridge R, Boschetti S, Doetsch J, Abächerli AN, Horstmeyer H, Finnemore M (2012) Flower structures and Riedel shears at a step over zone along the Alpine Fault (New Zealand) inferred from 2-D and 3-D GPR images. J Geophys Res Solid Earth 117(B2):B02406

Cunningham D, Grebby S, Tansey K, Gosar A, Kastelic V (2006) Application of airborne LiDAR to mapping seismogenic faults in forested mountainous terrain, southeastern Alps, Slovenia. Geophys Res Lett 33(20), L20308

Ferry M, Meghraoui M, Girard J-F, Rockwell TK, Kozacı Ö, Akyüz S, Barka A (2004) Ground-penetrating radar investigations along the North Anatolian Fault near Izmit, Turkey; constraints on the right-lateral movement and slip history. Geolo Boulder 32(1):85–88

Fraser J, Pigati JS, Hubert-Ferrari A, Vanneste K, Avşar U, Altinok S (2009) A 3000-year record of ground-rupturing earthquakes along the Central North Anatolian Fault near Lake Ladik, Turkey. Bull Seismol Soc Am 99(5):2681–2703

Geo Eye Elevating Insight (2013) About Geoeye-2. http://launch.geoeye.com/LaunchSite/about/Default.aspx. Accessed 03 Aug 2013

Gold PO, Cowgill E, Kreylos O, Gold RD (2012) A terrestrial lidar-based workflow for determining three–dimensional slip vectors and associated uncertainties. Geosphere 8(2):431–442

Gürpınar A (2005) The importance of paleoseismology in seismic hazard studies for critical facilities. Tectonophysics 408(1–4):23–28

Marco S, Rockwell TK, Heimann A, Frieslander U, Agnon A (2005) Late Holocene activity of the Dead Sea Transform revealed in 3D palaeoseismic trenches on the Jordan Gorge segment. Earth Planet Sci Lett 234(1–2):189–205

McCalpin JP (1989) Current investigative techniques and interpretive models for trenching active dip-slip faults. In: Watters RC (ed) Engineering geology and geotechnical engineering. Proceedings of the 25th symposium on engineering geology and geotechnical engineering. A.A. Balkema, Rotterdam, pp 249–258

McCalpin JP (ed) (2009) Paleoseismology, 2nd edn. Academic, Amsterdam

Ortuño M, Masana E, García-Meléndez E, Martínez-Díaz J, Štěpančíková P, Cunha PP, Sohbati R, Canora C, Buylaert J-P, Murray AS (2012) An exceptionally long paleoseismic record of a slow-moving fault: the Alhama de Murcia fault (Eastern Betic shear zone, Spain). Geol Soc Am Bull 124(9–10):1474–1494

Pantosti D, Pucci S, Palyvos N, Martini PMD, D'Addezio G, Collins PEF, Zabcı C (2008) Paleoearthquakes of the Duzce fault (North Anatolian Fault Zone): insights for large surface faulting earthquake recurrence. J Geophys Res 113:B01309

Prentice CS, Kendrick K, Berryman K, Bayasgalan A, Ritz JF, Spencer JQ (2002) Prehistoric ruptures of the Gurvan Bulag fault, Gobi Altay, Mongolia. J Geophys Res Solid Earth 107(B12):2321

Ragona D, Minster B, Rockwell T, Jussila J (2006) Field imaging spectroscopy: a new methodology to assist the description, interpretation, and archiving of paleoseismological information from faulted exposures. J Geophys Res Solid Earth 111(B10):B10309

U.S. Occupational Safety and Health Administration (OSHA) (1989) Occupational safety and health standards – excavations: final rule. Fed Reg 29 CFR 1926, 54(209):45894–45991

Zielke O, Arrowsmith JR (2012) LaDiCaoz and LiDARimager – MATLAB GUIs for LiDAR data handling and lateral displacement measurement. Geosphere 8(1):206–221

Paleoseismology

Sarah J. Boulton
Plymouth University, Centre for Research in Earth Sciences, Plymouth University, Devon, UK

Synonyms

Ancient earthquakes; Earthquake geology; Paleoseismology; Prehistoric earthquakes

Introduction

The aim of paleoseismology is to generate a record of past earthquakes (i.e., magnitude, recurrence interval, timing, etc.) from a range of geological observables preserved within a landscape. This is achieved by identifying features associated with a single paleoearthquake as opposed to the long-term deformation along a fault or within a basin. As such paleoseismology has been defined as "the study of prehistoric earthquakes, their timing, location and size" (McCalpin and Nelson 2009) and as "the study of the ground effects from past earthquakes as preserved in the geologic and geomorphic record" (Michetti et al. 2005). The paleoseismic record provides information on the local and regional deformation accommodated by slip on faults as well as data on patterns of seismicity, which can inform hazard assessments for future earthquakes. In addition to providing information on past earthquakes, paleoseismology can also constrain models of fault behavior and enhance our understanding of the influence of active faulting in the landscape. So although paleoseismology is part of the broader field of earthquake geology, it is distinct and separate from modern instrumental seismology, tectonic and structural geology (including neotectonics), historical seismology, geodesy, and tectonic geomorphology. However, paleoseismology does involve aspects of many of these disciplines as well as elements from stratigraphy, sedimentology, quaternary geology, archeology, remote sensing, and geochronometry.

Paleoseismology began to be recognized as a scientific discipline during the 1970s in the United States and Japan. Although pioneers such as G. K. Gilbert in North America, A. McKay in New Zealand, and B. Koto in Japan made important early descriptions of fault scarps and earthquake ruptures, these studies made little attempt to relate features observed directly along faults to determine specific number or magnitude of earthquakes that had taken place. Prior to the 1970s, historical records represented the sole evidence used in the assessment of past earthquake activity. However, the use of historical records to investigate past earthquakes is problematic in areas where records only exist for a couple of centuries, such as in North America, and where the return period of the fault is much larger than the period of data collection. Even in Europe and Asia where historical records can go back thousands of years, ambiguity can exist on the identity and location of the actual seismogenic fault and/or the earthquake rupture, as fault damage can be spread over large geographic areas. Historical records may also be incomplete due to low population densities in affected areas or political unrest resulting in events going unreported and can suffer a range of biases, such as conflation, amalgamation, or exaggeration of earthquakes, and deliberate or accidental misinformation (Rucker and Niemi 2010).

Paleoseismology, by contrast, is a scientific discipline whose studies identify and document a range of landscape features and excavate paleoseismic trenches, in order to understand and determine reliable earthquake magnitudes and recurrence intervals of prehistoric earthquakes, as well as correlating historical records to geological features along specific structures. It is important to note that the period of time that prehistory (prior to contemporary written accounts) encompasses will vary by location. The associated discipline of archeoseismology that seeks to supplement geological observations with archeological remains and evidence in the geoarcheological record can also complement the understanding of individual earthquakes in regions with long histories of human inhabitancy. However, circular reasoning needs to be avoided in these situations; for example, geological or archeological evidence is used to interpret an earthquake, which is dated with reference to an earthquake catalog and then ends up in such a catalog to be used to date future discoveries (Rucker and Niemi 2010).

Evidence for Paleoearthquakes

Paleoseismic methods (paleoseismology) are used to investigate surface deformation and, thus, independently increase the understanding of historical earthquakes. These methods include the application of the Environmental Seismic Intensity scale (ESI 2007), a scheme developed to quantify the intensity distribution of modern and historic earthquakes using the dimensions of the rupture (length, width, area, volume) and secondary features in the natural landscape (Michetti et al. 2007). However, for an earthquake to generate evidence that can be preserved at the surface, the magnitude of the earthquake must be large, generally considered to be >6.5 M_w (moment magnitude) (McCalpin and Nelson 2009) or intensity VIII on the ESI 2007 scale (Michetti et al. 2007). Smaller earthquakes do not often cause sufficient surface deformation that can subsequently be preserved. This is not always the case though; for example, some large earthquakes that nucleate near the brittle-ductile transition [i.e., 1989 Loma Prieta M 6.9 (California) and 2001 M_w 7.6 Bhuj (India) earthquakes] have little or no surface expression (Yeats 2007). In contrast, some smaller earthquakes occasionally do cause surface rupture, such as in Sicily (Italy) where faults that generate M_w 3.0–5.0 earthquakes can produce surface faulting even during shallow events (Azzaro et al. 2000). Therefore, the local tectonic setting and regional stress regime should always be taken into account when undertaking paleoseismic studies.

The initial identification of potentially seismogenic faults through the identification of geomorphic features that form the "seismic landscape" is the first stage of a paleoseismic investigation (Michetti et al. 2005). A seismic landscape will have its topography, geology, and stratigraphy determined in part by the style of faulting, rate of tectonic activity, thickness, and rheology of the seismogenic layer; the prevailing climate is also critical since this controls rates of deposition and erosion that progressively shape geomorphic landforms over many seismic cycles. Thus, landscapes dominated by extensional faulting will exhibit features such as horsts and grabens and wind gaps on uplifted footwalls with triangular facets along the fault trace, whereas strike-slip faults will be characterized by features such as shutter ridges, sag ponds, and offset or beheaded streams. Such assorted features can aid in the identification of faults both in the field and through the interrogation of satellite imagery, digital elevation models (DEMs), or aerial photographs. In turn, these features are often the focus of subsequent localized investigations such as geophysical surveys, paleoseismic ground investigation, or paleoseismic trenching.

A range of primary and secondary features can be associated with individual earthquakes and are used in paleoseismologic investigation once the active fault has been identified. Primary evidence forms instantaneously at the time of the earthquake due to slip along the fault plane and is therefore termed coseismic deformation. Secondary evidence can be caused by coseismic shaking or by subsequent (postseismic) erosion or deposition taking place at some time after the earthquake. Paleoseismic features can also be considered to be "on-fault" if along or near the fault or "off-fault" if they occur at some distance from the rupture. A further distinction can be made between geomorphic evidence and stratigraphic evidence of fault motion as observed in natural exposures and artificial paleoseismic trenches (McAlpin and Nelson 2009).

Typical examples of primary, on-fault evidence include: fault scarps, fissures, and folded, sheared, or displaced stratigraphic horizons (Table 1). Surface ruptures can be further divided into three categories: primary, subordinate, and sympathetic (Michetti 1994). Primary off-fault evidence can include tilted, uplifted, or submerged surfaces such as terraces (fluvial or coastal) or planation surfaces, drainage anomalies, change in coastal landforms, broken speleothems, and damage to man-made structures (both damage to ancient buildings and modern structures).

Secondary paleoseismic evidence consists of a great range of geomorphic and stratigraphic features including soft-sediment deformation, rockfalls, landslides, mud/sand volcanoes (also known as sand blows or boils), disturbed trees, and turbidite deposits, with little differentiation between the features that form on-fault and off-fault.

Paleoseismology, Table 1 Hierarchical classification of paleoseismic evidence with examples of stratigraphic and geomorphic features, modified from McCalpin and Nelson (2009)

Location	On-fault		Off-fault	
Timing	Coseismic	Postseismic	Coseismic	Postseismic
Primary paleoseismic evidence				
Geomorphic features	Surface ruptures	Colluvial aprons	Titled surfaces	Tectonic alluvial terraces
	1. Primary (i.e., fault scarps on main fault)	Afterslip modification to primary on-fault coseismic features	Uplifted shorelines	Afterslip modification to primary off-fault coseismic features
	2. Subordinate (ruptures branching from main fault)		Subsided shorelines	
	3. Sympathetic (ruptures on isolated faults)			
	Fissures			
	Folds			
	Mole tracks			
	Pressure ridges			
Stratigraphic features	Faulted strata	Scarp-derived colluvial wedges	Tsunami deposits and erosional unconformities caused by tsunamis	Erosional unconformities and deposits induced by uplift, subsidence, and tilting
	Folded strata	Fissure fills		
	Unconformities			
	Disconformities			
Abundance of similar nonseismic features	Few	Few	Some	Common
Secondary paleoseismic evidence				
Geomorphic features	Sand/mud volcanoes	Retrogressive landslides in the fault zone	Sand/mud volcanoes	Retrogressive landslides beyond the fault zone
	Landslides and lateral spreads in the fault zones		Landslides and lateral spreads beyond the fault zone	
	Disturbed trees and tree-throw craters		Disturbed trees and tree-throw craters	
			Fissures and Sackungen	
Stratigraphic features	Sand dykes and sills	Sediments deposited from retrogressive landslides	Sand dykes	Erosion or deposition in response to landscape disturbance (i.e., landslides, fissures, etc.)
	Soft-sediment deformation		Filled craters	
	Landslide toe thrusts		Soft-sediment deformation	
			Turbidites	
			Tsunamiites	
Abundance of similar nonseismic features	Some	Very common	Some	Very common

It should be noted that the distinction between primary and secondary off-fault evidence can be difficult to determine as seismic shaking may generate features that continue to evolve after the cessation of the earthquake. Furthermore, some features could represent both primary and secondary evidence. Tsunami deposits, for example, could be considered as primary evidence where the tsunami is produced directly by primary fault displacement, but could also be considered secondary evidence if say a landslide is triggered by earthquake shaking in a coastal location subsequently resulting in a tsunami. Secondary features are also more ambiguous than primary evidence as other geological mechanisms can result in their formation. McCalpin and Nelson (2009) used these classifications to define 16 categories (Table 1) of paleoseismic evidence that can commonly be distinguished in either geomorphic landforms or in the stratigraphic record. Therefore, using a combination of different lines of paleoseismic evidence tends to produce the most robust results in paleoseismic studies. As on-fault evidence often is clearer but provides little evidence of the strength of shaking or spatial extent of ground deformation, by contrast, geomorphic evidence can be preserved over wide areas, providing this spatial dimension to earthquake studies, yet it can be difficult to find the source of these off-fault features and age control may be lacking due to an absence of datable material.

As a result many paleoseismic studies focus on the excavation of trenches across faults, exposing offset and deformed strata resulting from seismogenic faulting. However, paleoseismic trenching studies commonly only take place at a few sites restricting the lateral applicability of the results. Crucially, these ground investigations can produce critical age control mainly through the use of radiocarbon dating or increasingly the application of optically stimulated luminescence (OSL) dating from material in offset stratigraphic horizons. It must also be noted that geological processes unrelated to earthquake faulting can produce features similar to coseismic and postseismic landforms and deposits and that erosion and sedimentation subsequent to an earthquake will act to modify and conceal the effects of surface faulting.

Seismic Hazard Assessments from Paleoseismology

The main goal of paleoseismic studies is to determine the dimensions, magnitude, and timing of past earthquakes and the slip rate on seismic faults to inform deterministic or probabilistic seismic hazard assessments (SHAs) and seismic risk assessments. The most common way of determining the magnitude of past earthquakes is to use primary and secondary evidence to determine the surface rupture length of the past event and the amount of displacement (maximum or average slip) that occurred. These parameters can then be compared to worldwide catalogs of historical earthquakes and rupture lengths to calculate the probable magnitude of the paleoearthquake (Wells and Coppersmith 1994). Similarly, evidence for the area of displacement on a fault, or the seismic moment, can be used to estimate the magnitude of paleoearthquakes. However, these methods produce uncertain results due to the complex nature of faulting and the widespread use of global relationships that might not hold true on a regional scale. By contrast, spatial distributions of secondary evidence may reveal patterns of ground motion intensity that can be used to infer earthquake magnitudes. These observations can be used in a framework to determine the relative intensity of an earthquake, such as the ESI 2007 scale that combines observations of primary and secondary evidence on a scale of I–XII, with earthquakes intensity IV and above producing observable ground deformation of increasing severity (Michetti et al. 2007).

There are a number of problems with this approach. One is that primary and secondary features of the same age (within error) observed in a stratigraphic or geomorphic context are assumed to be the result of a single earthquake. However, earthquakes are observed to often cluster in time and space, and such earthquake sequences cannot be easily separated in the geological record due to the uncertainties in dating methods resulting in an overestimation of the earthquake magnitude. Displacement is also not constant along the strike of the fault, going from zero displacement at the fault tips to a maximum near the center, which often modeled as a symmetrical profile in seismic hazard assessment but could be an irregular shape – this could lead to bias in displacement reconstructions.

The earthquake recurrence interval (the time elapsed between any two paleoearthquakes) on a fault is also a key component of characterizing the seismic hazard. Faults that rupture over long recurrence intervals in larger events can be less hazardous than faults that rupture more frequently in moderate events, as the likelihood of impacting a critical facility is lower for the long return earthquake. Recurrence intervals can be calculated in one of two ways. The less preferred way is to calculate the average recurrence interval over multiple paleoearthquakes, a method commonly used in neotectonic studies. The preferred alternative is to individually date each paleoearthquake "event horizon" to determine the actual time between each earthquake. It is important to appreciate that the time since the last earthquake on a fault (the elapsed time) does not represent a recurrence interval because it does not bound two earthquakes. Also, when determining recurrence intervals, it is important to assess the degree of underrepresentation or overrepresentation in the seismic record (McCalpin and Nelson 2009). Underrepresentation results from an incomplete preservation record as a result of numerous factors that can affect the potential for preservation in a site as well as subsequent tectonic activity obscuring earlier events. The contrasting issue of overrepresentation is less often considered but can occur when nonseismic features are interpreted as coseismic or postseismic evidence. Overrepresentation can also occur if the paleoseismic record is well preserved during a period of earthquake clustering, which in turn is extrapolated over longer periods and thereby leads to an erroneously large number of past earthquakes being inferred (McCalpin and Nelson 2009). Therefore, it is important to consider the length of the window of observation and compare short-term slip rates with the long-term average to assess the applicability of results.

Fault slip rate is the displacement of the fault over time and, as implied above, can apparently differ when considered at different timescales. The true slip rate is that considered over one or more closed (i.e., complete) seismic cycles, whereas an apparent slip rate is obtained if the slip rate is calculated over a time period containing complete and incomplete seismic cycles. However, due to variations in the displacement and time span between earthquakes, these measurements normally indicate that the slip rate varies between earthquake cycles and can be much higher than the long-term average during earthquake clustering or much lower during periods of fault quiescence.

These various parameters of fault and earthquake behavior can then be used to undertake seismic hazard assessments (SHAs); there are two main types of SHAs, deterministic seismic hazard assessment and probabilistic seismic hazard assessment.

Deterministic Seismic Hazard Assessment

Deterministic seismic hazard assessments (DSHAs) calculate the worst-case scenario for a site using the maximum potential magnitude for the largest active fault (the "controlling fault") and the distance from that fault to the site in question as two parameters that combine (using an attenuation equation) to determine the potential ground motion at any point from the fault. The method also requires knowledge of the rock or soil properties at the site. Yet, DSHAs do not take into account the time probability of an earthquake occurring during the life time of the facility, so data on the recurrence interval and slip rate on the controlling fault is irrelevant.

Probabilistic Seismic Hazard Assessment

Probabilistic seismic hazard assessments (PSHAs), by contrast, calculate the worst seismic effects for a site using both the time-independent factors (magnitude and distance) used in DSHA combined with the time-dependent factors of slip rate and recurrence interval for the controlling fault. Therefore, this method determines both the probability of an earthquake or earthquakes occurring near the critical facility within a defined time period (normally related to the design life of the facility) and also what the effects of these earthquakes would be on the facility. Currently, PSHA is the most commonly used method having generally replaced DSHA for most facilities.

Summary

Paleoseismology seeks to identify and appraise prehistoric or pre-instrumental earthquakes through the observation, documentation, and interpretation of geomorphic and stratigraphic evidence formed on, or near, earthquake ruptures as a result of primary rupture or secondary shaking. Such investigations have greatly improved our understanding of seismogenic fault behavior, revealing the segmentation of major faults (i.e., Schwartz and Coppersmith 1984), the temporal clustering of earthquakes in a nonperiodic fashion, the scaling relations between earthquake recurrence intervals and fault slip rates, and the tendency for faults to have characteristic rupture patterns (i.e., amount of slip: rupture length, magnitude). Of course, paleoseismological studies have also revealed periodic, nonscalar, and noncharacteristic earthquake faults and rigorous field studies are needed to test these apparent conflicting relationships. Variables determined from paleoearthquake studies (magnitude, slip, recurrence intervals) can be used to undertake seismic hazard assessments (SHA), either as a deterministic approach to calculate the worst-case scenario or a probabilistic approach to calculate the probability of an earthquake of a given size occurring within a given time period. However, since many paleoseismic data sets are too incomplete to precisely estimate fault slip rates or recurrence intervals, the contribution of paleoseismology to seismic hazard remains inaccurate (Grant 2002). Refining the record of past earthquakes is crucial for the improvement of earthquake-related hazard (seismic, tsunami, landslide, etc.) assessments for urban planning and critical facilities (i.e., nuclear power stations, dams, chemical/petroleum facilities). Because even in well-constrained areas, lessons can still be learned by refining historical data with geological observations, as witnessed by the 2011 Tohoku (Japan) earthquake, where geological evidence of past megathrust earthquakes was available but not integrated into the existing hazard planning (Silva et al. 2011).

Cross-References

▶ Archeoseismology
▶ Earthquake Magnitude Estimation
▶ Earthquake Mechanisms and Tectonics
▶ Earthquake Recurrence
▶ Intensity Scale ESI 2007 for Assessing Earthquake Intensities
▶ Luminescence Dating in Paleoseismology
▶ Paleoseismic Trenching
▶ Paleoseismology and Landslides
▶ Paleoseismology of Glaciated Terrain
▶ Paleoseismology of Rocky Coasts
▶ Paleoseismology: Integration with Seismic Hazard
▶ Radiocarbon Dating in Paleoseismology
▶ Remote Sensing in Seismology: An Overview
▶ Seismic Risk Assessment, Cascading Effects
▶ Site Response for Seismic Hazard Assessment
▶ Tsunamis as Paleoseismic Indicators

References

Azzaro R, Bella D, Ferreli L, Maria Michetti A, Santagati F, Serva L, & Vittori E (2000) First study of fault trench stratigraphy at Mt. Etna volcano, Southern Italy: understanding Holocene surface faulting along the Moscarello fault. Journal of Geodynamics, 29(3):187–210

Grant LB (2002) Paleoseismology. In: Lee WHK, Kanamori H, Jennings PC, Kisslinger C (eds) International handbook of earthquake and engineering seismology, vol 81. AIASPEI, Elsevier, Amsterdam pp 475–490

McCalpin JP, Nelson AR (2009) Introduction to paleoseismology. In: McCalpin JP (ed) Paleoseismology. vol 95, 2nd edn, International geophysics series. Elsevier, Amsterdam, pp 1–29

Michetti AM (1994) Coseismic surface displacement vs. magnitude: relationships from paleoseismological analyses in the Central Apennines (Italy). In: Special issue 'proceedings of the CRCM'93', Journal Geodetic Society of Japan, Kyoto, pp 375, 380

Michetti AM, Audemard FA, Marco S (2005) Future trends in paleoseismology: integrated study of the seismic landscape as a vital tool in seismic hazard analyses. Tectonophysics 408:3–21

Michetti AM, Esposito E, Guerrieri L, Porfido S, Serva L, Tatevossian R, Vittori E, Audemard F, Azuma T, Clague J, Comerci V, Gurpinar A, McCalpin JP, Mohammadioun B, Morner NA, Ota Y, Roghozin E (2007) Environmental seismic intensity scale-ESI 2007. Memorie Descrittive della Carta Geologica d'Italia 74:7–54

Rucker JD, Niemi TM (2010) Historical earthquake catalogues and archaeological data: achieving synthesis without circular reasoning. In: Sintubin M, Stewart IS, Niemi TM, Altunel E (eds) Ancient earthquakes. Geological Society of America special paper, vol 471. Geological Society of America, Boulder, pp 97–106

Schwartz DP, Coppersmith KJ (1984) Fault behavior and characteristic earthquakes: examples from the Wasatch and San Andreas fault zones. J Geophys Res Solid Earth 89(B7):5681–5698

Silva PG, Sintubin M, Reicherter K (2011) New advances in studies of earthquake archaeology and palaeoseismology. Quat Int 242:1–3

Wells DL, Coppersmith KJ (1994) New empirical relationships among magnitude, rupture length, rupture width, rupture area and surface displacement. Bull Seismol Soc Am 4(84):975–1002

Yeats RS (2007) Paleoseismology: why can't earthquakes keep on schedule? Geology 35:863–864

Paleoseismology and Landslides

Randall W. Jibson
U.S. Geological Survey, Golden, CO, USA

Synonyms

Earthquake-triggered landslides; Seismic ground failure; Slope-stability analysis

Introduction

Most moderate to large earthquakes trigger landslides (Fig. 1). In many environments, landslides preserved in the geologic record can be analyzed

Paleoseismology and Landslides, Fig. 1 Madison Canyon landslide, triggered by the 1959 Hebgen Lake, Montana, earthquake (M_w 7.1). Strong shaking caused 28×10^6 m^3 of rock to slide into the canyon, which dammed the river and created a lake more than 60 m deep. Slide scar at left is 400 m high, debris is as thick as 67 m in valley axis, and slide debris traveled 130 m up the right valley wall. Twenty-eight people were killed by the slide (Photograph courtesy of J.R. Stacy, U.S. Geological Survey Photographic Library, photo no. 209a)

to determine the likelihood of seismic triggering. If evidence indicates that a seismic origin is likely for a landslide or group of landslides, and if the landslides can be dated, then a paleoearthquake can be inferred, and some of its characteristics can be estimated. Such paleoseismic landslide studies thus can help reconstruct the seismic shaking history of a site or region (Jibson 2009).

Paleoseismic landslide studies differ fundamentally from paleoseismic fault studies. Whereas fault studies seek to characterize the movement history of a specific fault, landslide studies characterize the shaking history of a site or region irrespective of the earthquake source. In regions that contain multiple seismic sources and in regions where surface faulting (▶ Seismic Actions due to Near-Fault Ground Motion) is absent, paleoseismic ground-failure studies thus can be valuable tools in hazard and risk studies that are more concerned with shaking hazards than with interpretation of the movement histories of individual faults. In fact, paleoseismic studies in some parts of the world typically rely more on ground failure than on surface fault ruptures.

The practical lower bound earthquake that can be interpreted from paleoseismic landslide investigations is about magnitude 5–6 (▶ Earthquake Magnitude Estimation). This range is comparable or perhaps slightly lower than that for paleoseismic fault studies. Obviously, however, larger earthquakes tend to leave much more abundant and widespread evidence of landsliding than smaller earthquakes; thus, available evidence and confidence in interpretation increase with earthquake size.

Paleoseismic landslide analysis involves three steps: (1) identify a feature as a landslide, (2) date the landslide, and (3) determine if the landslide was triggered by earthquake shaking. This article addresses each of these steps and discusses methods for interpreting the results of such studies. Only subaerial landslides are discussed here; submarine landslides are analyzed using different methods.

In this article, *landslide* is used as a generic term to include all types of downslope movement of earth material, including types of movement that involve little or no true sliding. Thus, rock falls, debris flows, etc., are considered types of landslides. The classification system of Varnes (1978) is used, which categorizes landslides by the type of material involved (soil or rock) and by the type of movement (falls, topples, slides, slumps, flows, or spreads). Other modifiers commonly are used to indicate velocity of movement, degree of internal disruption, state of activity, and moisture content.

Identifying Landslides

Identifying surface features as landslides can be relatively easy for fairly recent, well-developed, simple landslides. Older, more degraded landslides or those having complex or unusual morphologies can be more difficult to identify. In general, landslides are identified by anomalous

Paleoseismology and Landslides, Table 1 Relative abundance of earthquake-induced landslides

Abundance	Landslide type
Very abundant	Rock falls
	Disrupted soil slides
	Rock slides
Abundant	Soil lateral spreads
	Soil slumps
	Soil block slides
	Soil avalanches
Moderately common	Soil falls
	Rapid soil flows
	Rock slumps
Uncommon	Subaqueous landslides
	Slow earth flows
	Rock block slides
	Rock avalanches

Note: Data from Keefer (1984). Landslide types use nomenclature of Varnes (1978) and are listed in decreasing order of abundance

topography, including arcuate or linear scarps, backward-rotated masses, benched or hummocky topography, bulging toes, and ponded or deranged drainage. Abnormal vegetation type and age also are common.

Earthquakes can trigger all types of landslides, and all types of landslides triggered by earthquakes also can occur without seismic triggering. Therefore, an earthquake origin cannot be determined solely on the basis of landslide type. However, some types of landslides tend to be much more abundant in earthquakes than other types. Table 1 shows the relative abundance of various types of earthquake-triggered landslides. Overall, the more disrupted types of landslides are much more abundant than the more coherent types of landslides. Also, most earthquake-induced landslides occur in intact materials rather than in preexisting landslide deposits; thus, the number of reactivated landslides is small compared to the total number of landslides triggered by earthquakes. Earthquake-triggered landslides most commonly occur in materials that are weathered, sheared, intensely fractured or jointed, or saturated.

Sackungen (ridge-crest troughs) are a somewhat controversial type of ground failure that has been related, in some cases, to seismic shaking. Sackungen are identified by one or more of the following: (1) grabens or troughs near and parallel to ridge crests of high mountains, (2) uphill-facing scarps a few meters high that parallel the topography, (3) double-crested ridges, and (4) bulging lower parts of slopes.

Determining Landslide Ages

Paleoseismic interpretation requires establishing the numerical age of a paleoearthquake. In the case of earthquake-triggered landslides, this means that dating landslide movement is required. Several methods for dating landslide movement can be used; some are similar or identical to those used for dating fault scarps, while others are unique to landslides.

Different types of landslides could be datable by different methods, depending on a variety of factors such as distance of movement, degree of internal disruption, landslide geometry, type of landslide material, type and density of vegetation, and local climate. Ideally, multiple, independent dating methods should be used to increase the level of certainty of the age of landslide movement.

Historical Methods
Some old landslides might have been noted by local inhabitants or could have damaged or destroyed human works or natural features. In some parts of the world, potentially useful historical records or human works extend back several hundreds or thousands of years. For fairly recent events, comparing successive generations of topographic maps or aerial photographs can bracket the time period in which mappable landslides first appeared.

Dendrochronology
Dendrochronology can be applied to date landslide movement in several ways. At the simplest level, the oldest undisturbed trees on disrupted or rotated parts of landslides should yield reasonable minimum ages for movement. On rotational slides that remained fairly coherent, preexisting trees that survived the sliding will have been tilted because of headward rotation of the ground surface; if both tilted and straight trees are present on such landslides, the age of slide movement is bracketed between the age of the oldest straight trees and the youngest tilted trees. Using this simple application of dendrochronology to date coherent translational slides is more difficult because trees can remain upright and intact even after landslide movement. On all types of landslides, trees growing from the surface of the scarp will yield minimum ages of scarp formation, from which the age of slide movement can be interpreted. In some cases, trees killed by landslide movement will be preserved and can thus yield the exact date of movement.

A more sophisticated application of dendrochronology involves quantitative analysis of growth rings. For trees that have survived one or more episodes of landslide movement, such analysis can be used to identify and date reaction wood (eccentric growth rings), growth suppression, and corrosion scars, which might be evidence of landslide movement. Some landslides block stream drainages and form dams that impound ponds or lakes. Inundation of areas upstream from landslide dams can drown trees that can be dated dendrochronologically.

Radiometric and Cosmogenic Dating
Radiometric dating (▶ Radiocarbon Dating in Paleoseismology) (most commonly using ^{14}C) can be used in a variety of ways to date organic material buried by landslide movement. Landslide scarps degrade similarly to fault scarps, and so colluvial wedges at the bases of landslide scarps might contain organic material that can be retrieved by trenching or coring and dated radiometrically. Fissures on the body of a landslide, particularly near the head where extension can take place, also can trap and preserve organic matter. If the landslide mass is highly disrupted, as in rock or soil falls or avalanches, then some vegetation from the original ground surface might have become mixed with the slide debris; such organic material excavated from slide debris can be dated radiometrically. At the toes of landslides, slide material commonly is deposited onto undisturbed ground; if this original ground

surface can be excavated beneath the toe of a slide, buried organic material from this surface can be dated to indicate the age of initial movement.

Sag ponds commonly form on landslides, and organic material deposited in such ponds can be dated radiometrically. Organics at the base of the pond deposits should yield reliable dates of pond formation.

Vegetation submerged from inundation of areas upstream from landslide dams also can be dated radiometrically. Similarly, landslides into lakes can submerge and kill vegetation that can be dated.

Rock-fall and rock-avalanche deposits can be dated cosmogenically (▶ Luminescence Dating in Paleoseismology) if the surface of the deposit has been relatively stable since the time of emplacement. This method measures the amount of time that specific types of mineral grains have been exposed to cosmic radiation. The assumption is that a significant proportion of the material on the surface of a rock-fall or rock-avalanche deposit was newly fractured and exposed when the landslide occurred.

Lichenometry

Lichenometry – analysis of the age of lichens based on their size – can be used to date rock-fall and rock-avalanche deposits. By measuring lichen diameters on rock faces freshly exposed at the time of failure, numerical ages can be estimated by assuming that lichens colonized the rock face in the first year after exposure. Because rock-fall and rock-avalanche deposits typically include abundant rocks having freshly exposed faces, numerous samples generally can be taken to create a database for the statistical analysis required by lichenometry. Lichenometric ages must be calibrated at sites of known historical age or by comparison with other numerical dating techniques. Lichenometric dating is subject to considerable uncertainty, however, because several decades can elapse before lichens colonize a fresh rock exposure, and lichens might never colonize unstable landslide deposits on very steep slopes.

Weathering Rinds

For a given climate and rock type, measuring the thickness of weathering rinds can be used to date when rocks were first exposed at the ground surface. For rock falls and rock avalanches and for other landslides whose movement exposed rock fragments at the ground surface, measuring the thickness of weathering rinds can be used to date landslide movement. Determining which rock surfaces were initially exposed at the time of landsliding can be difficult, but if a sufficiently large number of samples can be measured, consistent statistical results of predominant ages that relate to landslide movement can be obtained.

Pollen Analysis

Analysis of pollen in deposits filling depressions on landslides can yield both an estimated age of initial movement and, in some cases, a movement history through time. Such analyses assume that sediment deposition and incorporation of pollen occur immediately following landslide movement and that local climatic and vegetation variations can be accounted for. Pollen samples from the buried ground surface beneath the toes of landslides also have potential for use in dating landslide movement.

Geomorphic Analysis

Landslides are disequilibrium landforms that change through time more rapidly than surrounding terrain. By analyzing the degree of degradation of landslide features such as scarps, ridges, sags, and toes, relative ages can be assigned to various landslides. Criteria for such relative age classification might include degree of definition of landslide features, soil development, tephra cover, stream dissection, preservation of vegetation killed by movement, and drainage integration.

Models of fault-scarp degradation also have potential application in landslide dating because landslide scarps should behave similarly to fault scarps. Several approaches to morphologic fault-scarp dating have been proposed, all of which require calibration for various parameters such as climate and scarp material. Scarp degradation commonly is modeled as a diffusion process, in which degradation rate varies in time and is a function of slope angle, which represents the degree to which the scarp is out of equilibrium with the surrounding landscape.

Analysis of soil-profile development also is a potential tool for dating landslides. New soil profiles will begin to develop on disrupted landslide surfaces. If such surfaces can be identified, dating the newly developed soil profile will indicate the age of movement.

Interpreting an Earthquake Origin for Landslides

Interpreting an earthquake origin for a landslide or group of landslides is by far the most difficult step in the process, and methods and levels of confidence in the resulting interpretation vary widely. This section summarizes several basic approaches that can be used to interpret the seismic origin of landslides.

Regional Analysis of Landslides

Many paleoseismic landslide studies involve analysis of large groups of landslides rather than individual features. The premise of these regional analyses is that a group of landslides of the same age, scattered across a discrete area, probably was triggered by a single event of regional extent. In an active seismic zone, that event commonly is inferred to be an earthquake. Such an interpretation could be justified in areas where landslide types and distributions from historical earthquakes have been documented and can be used as a standard for comparison. In areas where such historical observations are absent, assuming an earthquake origin for landslides of synchronous age is much more tenuous, primarily because large storms also can trigger widespread landslides having identical ages and spatial distributions.

Differentiating between such groups of storm- and earthquake-triggered landslides might be possible using a statistical approach to characterize the distribution of steep slopes in a region. Storm-triggered landslides form most commonly near the bases of slopes and thus tend to form landscapes characterized by steep inner gorges. Earthquake-triggered landslides, on the other hand, tend to form either near ridge crests or more uniformly across the entire reach of slopes; corresponding landscapes generally lack well-developed inner gorges. These landscape patterns provide supportive evidence of the possible seismic origin of landslides in a region, but they cannot be used to definitively determine the origin of any specific landslide.

Some criteria to support a seismic origin for landslides include (1) ongoing seismicity in the region that has triggered landslides; (2) coincidence of landslide distribution with an active fault or seismic zone; (3) geotechnical slope-stability analyses showing that earthquake shaking would have been required to induce slope failure (discussed in detail subsequently); (4) presence of liquefaction features associated with the landslides; (5) correlation between the elongation of a landslide distribution and the location and dimensions of the seismogenic faults in a region; and (6) landslide distribution that cannot be explained solely on the basis of geologic or geomorphic conditions. Obviously, the more of these criteria that are satisfied, the stronger the case for seismic origin.

Landslide Morphology

Some landslides have morphologies that strongly suggest triggering by earthquake shaking. For example, stability analyses of landslides on low-angle basal shear surfaces show that they generally form much more readily under the influence of earthquake shaking than in other conditions. Landslides that formed as a result of liquefaction of subsurface layers also are much more likely to have formed seismically than aseismically. Slides that form as a result of intense rainfall are more fluid and tend to spread out more across a depositional area, whereas seismically induced landslides tend to have a blockier appearance and a more limited depositional extent in some cases. None of these criteria is definitive, but the types and characteristics of landslides described previously do suggest seismic triggering and can be used as corroborative evidence of earthquake triggering.

Landslide size is considered evidence of seismic triggering in some cases. In areas where large landslides have been documented in historical time to occur only during earthquakes, the large size of prehistoric landslides could suggest seismic origin and could even be used to infer the relative size of the triggering earthquake; very large landslides commonly are triggered by longer duration and longer period shaking, which generally relate to larger magnitude earthquakes.

Multiple lines of evidence strengthen an argument for seismic triggering. For example, a large, ancient landslide near an active fault and having a low-angle basal shear surface might be considered a strong candidate for having been seismically triggered, particularly if similar landslides have been documented in recent earthquakes.

Sackungen

Sackungen are geomorphic features in mountainous areas that are characterized by ridge-parallel, uphill-facing scarps; double ridge lines; and troughs or closed depressions along ridge crests. While topography and gravity clearly influence the ridge-parallel geometry of sackungen, several different processes for their origin have been proposed, including gravitational spreading due to long-term creep, stress relief due to deglaciation, faulting, strong shaking, or a combination of factors. It appears that sackungen can form under a variety of conditions or combinations of conditions because sackungen have been documented to have formed in different ways in different tectonic and geologic settings.

Sackungen have been documented in several historical earthquakes, but the specific mechanism by which they form appears to be complex. Strong shaking certainly plays a role, but in many cases sackungen formed in the immediate vicinity of, and parallel to, the seismogenic faults, which suggests that fault-related tectonic deformation as well as strong shaking might contribute to their formation. And some sackungen appear to have multiple episodes of movement, some related to seismic shaking and some to periods of climatically induced increased groundwater levels.

Because sackungen can form in a variety of tectonic and geologic environments and can form by several different processes, paleoseismic interpretation is difficult and commonly tenuous. Criteria for establishing the seismic or nonseismic origin of sackungen have been proposed with the aim of differentiating between features indicating abrupt, episodic movement versus those indicating gradual, continuous movement. McCalpin (1999) proposed seven criteria, including stratigraphic, geomorphic, and structural evidence, to differentiate between seismic and nonseismic movement: (1) evidence of continuous deformation of sediments suggests a nonseismic origin; (2) sackung deformation events that are contemporaneous with other regional paleoseismic features could be coseismic; (3) if sackungen overlie a steeply dipping crustal fault zone that has a net displacement much larger than the scarp height, the fault could be active; (4) gravity-driven sackungen tend to occur in swarms and be shorter, less continuous, and arcuate, whereas tectonic scarps tend to be longer, more continuous, singular, and straighter; (5) height-to-length ratios of gravity-driven sackungen are much greater than those of tectonic faults; (6) an asymmetrical fault zone having a sharp upper boundary and transitional brecciated lower boundary is more likely to be a sackung than a tectonic fault; and (7) subsurface deformation zones of tectonic faults can occur in any spatial relation with the modern topography, whereas subsurface deformation zones of sackungen are closely related to modern topography.

No single criterion is sufficient to unequivocally prove the seismic or aseismic origin of a sackung feature. And a seismic origin could have resulted from strong shaking, primary tectonic faulting, sympathetic faulting on a feature other than the seismogenic fault, or a combination of these factors. Also, a single sackung feature could have had both seismic and aseismic episodes of movement. Therefore, paleoseismic interpretation of sackungen is generally quite challenging and in many cases impossible. In some cases, however, evidence for abrupt episodes of movement that can be linked to seismic event can provide valuable paleoseismic evidence.

Sediment from Earthquake-Triggered Landslides

Earthquake-triggered landslides can profoundly affect alluvial systems by denuding slopes, which generates large amounts of disrupted sediment that will move into the alluvial system and physically disrupt drainage systems. The commonest types of landslides triggered by earthquakes are shallow, highly disrupted slides in unconsolidated surficial material, and the deposits of these types of landslides tend to move quickly into stream drainages. Thus, earthquakes can deposit large pulses of sediment into alluvial systems, which can (1) create new alluvial fans, (2) cause widespread aggradation of channels, (3) provide material for subsequent deposition on fan surfaces by debris flows and hyperconcentrated flows, and (4) affect the overall development of the fan surface on the long term. Thus, landslides triggered by earthquakes can leave evidence in the depositional record of alluvial systems.

Large earthquakes can cause a spectrum of ground-failure effects including abundant landslides, pervasive ground cracking, microfracturing of surficial hillslope materials, collapse of drainage banks over long stretches, widening of hillside rills, and lengthening of first-order tributary channels. Such widespread disruption increases the capacity of channels to carry runoff by enlarging upstream channels and detaches large amounts of loose slope material, which increases the amount of sediment available for transport and deposition.

Comparison of normal debris-flow deposits to those deposited soon after major earthquakes shows that the post-earthquake deposits tend to (1) be abnormally thicker, (2) contain larger clasts, (3) contain a higher percentage of coarse clasts, and (4) have more angular clasts.

Landslides that Straddle Faults

Landslides sometimes occur on slopes immediately above fault traces, and the slide mass can extend across the trace. Subsequent surface movement of such a fault would offset the landslide mass and allow estimation of fault slip rates if the slide could be dated. This approach does not require that the landslide be seismically triggered because the paleoseismic interpretation is based on post-landslide fault offset of the landslide mass. However, landslides triggered in the immediate vicinity of active faults commonly are seismically triggered.

Precariously Balanced Rocks

Precariously balanced rocks have been used as crude paleoseismoscopes. The premise of this approach is that areas containing precariously balanced rocks indicate the absence of strong earthquake motions since the precarious rocks developed; paleoseismic interpretations can be made by estimating the peak accelerations required to cause toppling and the length of time the rocks have been precarious. The shaking required to topple precarious rocks has been estimated using analytical and numerical modeling, physical modeling using shaking-table tests, and field experiments on actual precarious rocks. Precarious rocks have been defined as being capable of being toppled by peak accelerations of 0.1–0.3 g; rocks requiring 0.3–0.5 g are commonly defined as semi-precarious. Precarious rocks can be dated cosmogenically and by analysis of rock-varnish microlaminations.

Results from precarious-rock analyses are not always consistent with other lines of paleoseismic evidence. There are several possible reasons for these inconsistencies, and no consensus currently exists regarding the validity of the results of the precarious-rock studies. Therefore, several caveats regarding precarious-rock studies should be kept in mind: (1) Large uncertainties exist in the required toppling accelerations owing both to the geometric complexity of the rocks and the complexities of 3-D ground motion. (2) Not all precarious rocks in a given area will be toppled by the estimated threshold ground shaking. Studies of overturning of tombstones in Japan have shown that a given threshold acceleration will overturn only a fraction of a group of seemingly identical tombstones. Therefore, finding some precarious rocks in an area does not necessarily mean that the area has not experienced the threshold ground shaking. (3) Establishing the age of precarious rocks does not necessarily determine the minimum time since a certain level of shaking has occurred because the toppling acceleration of a given rock will have been continuously changing as the rock has evolved into a precarious state. For example, a precarious rock with an estimated age of 20 ka and a present-day toppling acceleration of 0.2 g might have had a toppling acceleration of 1.0 g at 20 ka, 0.5 g 10 ka, etc.; therefore, it cannot be concluded that the present-day toppling acceleration has not been exceeded in 20 ka.

Speleoseismology

Speleoseismology is the investigation of earthquake records in caves. Such records can include broken speleothems (stalactites, stalagmites, soda straws, etc.), cave-sediment deformation structures, offset along fractures and bedding planes, simple rock falls (incasion), and coseismic fault displacement. Before an earthquake origin can be inferred, all other possible causes of the disturbance must be ruled out. Such causes include human or animal disturbance, water flow, ice movement, debris flow, and sediment creep.

By measuring and dating the tilting and collapse of many stalagmites in a region, it is possible to differentiate sudden (seismic) versus gradual movements and local versus regional causes. Tilting and collapse events can be dated by analysis of radiometrically determined speleothem growth rates; uranium-series isotopes can be analyzed to date speleothems precisely within the 0–500 ka range. By modeling stalagmites as simple inverted pendulums, it is possible to estimate the minimum ground shaking necessary to cause collapse using pseudostatic engineering analysis.

Numerical and physical models have been developed to examine the ground shaking that would be required to break and topple various types of speleothems. These models have been used to measure the natural frequencies and damping characteristics of speleothems and the peak ground accelerations necessary to break them. The natural frequencies of most speleothems are between 50 and 700 Hz, well above the range of seismically generated ground motion (0.1–30 Hz). The only exceptions are so-called soda straws: long, slender speleothems that can have natural frequencies as low as 20 Hz. Most speleothems would require ground accelerations in excess of 1 g to cause breakage; some very long, thin soda straws, however, could be broken at accelerations as low as 0.1–0.2 g. Thus it appears that only exceptionally long, thin speleothems having weak sections are likely to break during earthquakes, and only about 2 % of such structures have been observed to have broken in recent, well-documented earthquakes.

Summary

Many methods for interpreting the seismic origin of landslides have been developed and, in some cases, successfully applied to paleoseismic analysis. Virtually all of the methods summarized in this section have one aspect in common, which is stated explicitly in most papers: the seismic origin of the features being interpreted remains tentative and cannot be proven, because in each case a nonseismic process could have produced the observed features. Circumstantial evidence for seismic triggering ranges from very strong to extremely tenuous. Indeed, on the latter end of the spectrum, the reasoning can be rather circular: an earthquake origin for a feature is assumed, and then an earthquake origin is interpreted and concluded from analysis of that feature. Any paleoseismic interpretation of a feature is limited primarily by the certainty with which seismic triggering can be established.

Using Stability Analysis to Determine Seismic Landslide Origin

The most direct way to assess the relative likelihood of seismic versus aseismic triggering of an individual landslide is to apply established methods of static and dynamic slope-stability analysis (Jibson and Keefer 1993). Such an analysis involves constructing a detailed slope-stability model of static conditions to determine if failure is likely to have occurred in any reasonable set of groundwater and shear-strength conditions in the absence of earthquake shaking.

All potential nonseismic factors must be considered; these might include processes such as fluvial or coastal erosion that oversteepens the slope or undrained failure resulting from rapid drawdown (for slopes subject to submersion). If aseismic failure can reasonably be excluded even in worst-case conditions (minimum shear strength, maximum piezometric head), then an earthquake origin can be inferred. Dynamic slope-stability analyses can then be used to estimate the minimum shaking conditions that would have been required to cause failure.

This approach is by far the most involved but also yields quantitative results that can be used to assess landslide origins (Jibson and Keefer 1993). Steps involved in a typical stability analysis are summarized in the following sections.

Geotechnical Investigation

Accurately modeling the stability of a slope requires detailed investigation to determine the geotechnical properties of the slope materials. The key properties required for a stability analysis include the material shear strength (friction angle and cohesion), unit weight, and moisture content. Investigating these properties might involve (1) in situ approaches such as cone-penetration testing or (2) acquiring samples that can be tested in the laboratory.

Shear strength can be characterized in different ways to model different types of failure conditions. In aseismic conditions, effective (drained) shear strengths are used because pore-water pressures are assumed to be in static equilibrium. During earthquakes, many soils behave in a so-called undrained manner because excess pore pressures induced by the transient ground deformation cannot dissipate during the brief duration of the shaking; therefore, total (undrained) shear strengths are used to model seismic failure conditions. Effective shear strengths can be measured in the laboratory using various methods: (1) direct shear in which the strain rate is slow enough to allow full drainage and (2) consolidated-undrained triaxial (CUTX) shear in which pore pressure is measured to allow modeling of drained conditions. Total (undrained) shear strength can be measured using CUTX shear tests or simpler methods such as vane shear.

Static (Aseismic) Slope-Stability Analysis

Static slope-stability analysis models the stability of slopes in the absence of earthquake shaking. A stability model using effective shear strengths can be constructed, and the worst possible groundwater conditions can be modeled to determine the likelihood of aseismic failure. If a slope is stable even in worst-case aseismic conditions, then it is likely that seismic shaking was necessary to induce failure.

Slope stability is quantified using the factor of safety (FS), the ratio of the sum of the resisting forces or moments that act to inhibit slope movement to the sum of the driving forces or moments that tend to cause movement. Slopes having factors of safety greater than 1.0 are thus stable; those having factors of safety less than 1.0 should move. Of course, input parameters have uncertainties, and so determining the stability of slopes from the factor of safety requires judgment. A good rule of thumb for interpreting factors of safety for slopes that have well-constrained input parameters is as follows:

FS $<$ 1 is considered unstable.
FS $=$ 1.00–1.25 is considered marginally stable.
FS $=$ 1.25–1.50 is considered stable.
FS $>$ 1.50 is considered very stable.

Dynamic (Seismic) Slope-Stability Analysis

Analysis of slope stability during earthquake shaking is best modeled using sliding-block analysis. This type of analysis was first introduced by Newmark (1965) and is used widely in engineering practice. Newmark's method models a landslide as a rigid-plastic friction block that slides on an inclined plane. The block begins to slide when a given critical (or yield) base acceleration is exceeded; thus, critical acceleration is defined as the base acceleration required to overcome basal shear resistance and initiate sliding. The analysis calculates the cumulative permanent displacement of the block as it is subjected to the effects of an earthquake acceleration-time history, and the user judges the significance of the displacement. Laboratory model tests and analysis of actual earthquake-induced landslides have confirmed that Newmark's method can fairly accurately predict landslide displacements if slope geometry and soil properties are known accurately and if earthquake ground accelerations can be estimated using real or artificial acceleration-time histories. More sophisticated forms of sliding-block analysis have been developed that allow modeling the landslide block as a flexible rather than a rigid mass, which yields more accurate results for deeper, larger landslides.

The critical acceleration is a simple function of the static factor of safety and the landslide geometry; it can be expressed as

$$a_c = (FS - 1)g \sin \alpha, \quad (1)$$

where a_c is the critical acceleration in terms of g, the acceleration of Earth's gravity; FS is the static factor of safety; and α is the thrust angle, the angle from the horizontal that the center of mass of the potential landslide block first moves.

Calculation of the estimated landslide displacement consists of a two-part integration with respect to time: (1) the parts of the selected acceleration-time history that lie above the critical acceleration of the landslide block are integrated to yield the velocity of the block with respect to its base and (2) the velocity curve is then integrated to determine the cumulative permanent displacement of the block.

Conducting a rigorous sliding-block analysis requires knowing the critical acceleration of the landslide and selecting one or more earthquake acceleration-time histories (▶ Selection of Ground Motions for Response History Analysis, ▶ Time History Seismic Analysis) to approximate the earthquake shaking at the site. The critical acceleration of a potential landslide can be determined in two ways: (1) For relatively simple slope models where material properties do not differ significantly between layers, Eq. 1 can be used to estimate the critical acceleration. (2) For more complex slope models that include layers having complex geometries or widely differing material properties, the critical acceleration should be determined using iterative pseudostatic analysis, where different seismic coefficients are used until the static factor of safety reaches 1.0. The seismic coefficient yielding a factor of safety of 1.0 is the yield or critical acceleration.

Selecting an appropriate suite of strong-motion records (▶ Selection of Ground Motions for Response History Analysis, ▶ Time History Seismic Analysis) for the dynamic analysis can be challenging. Some key properties to consider in estimating ground motions and selecting records include earthquake magnitude, source distance, peak ground acceleration, Arias intensity, and shaking duration.

The significance of the Newmark displacements must be judged in terms of the probable effect on the potential landslide mass. For shallower landslides in brittle surficial rock and soil, estimated displacements in the 5–10-cm range commonly correlate with failure. For deeper landslides in more compliant material, estimated displacements in the 10–30-cm range more commonly correlate with landslide initiation. When displacements in this range occur, previously undisturbed materials can lose some of their strength and be in a residual-strength condition. Static factors of safety using residual shear strengths can then be calculated to determine the stability of the landslide after earthquake shaking (and consequent inertial landslide displacement) ceases.

Interpreting Minimum Ground Motions Required to Cause Slope Failure

If static stability analysis clearly indicates that failure of a landslide in aseismic conditions is highly unlikely, then an earthquake origin can be hypothesized. A dynamic analysis can then be used to estimate the minimum shaking necessary to have caused failure. Such an approach requires a general relationship between critical acceleration, shaking intensity (which can be characterized in various ways), and Newmark displacement. Several such relations have been published.

For example, consider a hypothetical landslide that is stable in aseismic conditions and that has a critical acceleration of 0.15 g. The following equation (Jibson 2007) could be used to estimate the minimum peak ground acceleration

(▶ Selection of Ground Motions for Response History Analysis) needed to cause failure:

$$\log D_N = 0.215 + \log \left[\left(1 - \frac{a_c}{a_{max}}\right)^{2.341} \left(\frac{a_c}{a_{max}}\right)^{-1.438} \right] \quad (2)$$

where D_N is Newmark displacement in centimeters, a_c is critical acceleration, and a_{max} is peak ground acceleration; the ratio of a_c to a_{max} is commonly referred to as the critical acceleration ratio. Applying this equation requires judgment regarding the critical amount of Newmark displacement that would reduce shear strength on the failure surface to residual levels and lead to continuing failure. The general guidelines stated previously suggest that a reasonable estimate of critical displacement might be 10 cm. For the hypothetical example under consideration, insertion of a displacement value (D_N) of 10 cm into Eq. 2 yields a critical acceleration ratio of 0.2. For a critical acceleration (a_c) of 0.15 g for the hypothetical landslide, this would yield a peak ground acceleration (a_{max}) of 0.75 g as a minimum ground acceleration required to trigger enough displacement to cause general failure.

The peak ground acceleration from such an analysis could be used by itself as a basis for hazard assessment, or it could be used to estimate various magnitude/distance combinations of possible triggering earthquakes. If more than one landslide of identical age were similarly analyzed in an area, iterative magnitude and distance combinations could be optimized to estimate likely earthquake characteristics.

Equations that use other parameters are available and could be applied similarly. For example, a minimum threshold Arias (1970) intensity leading to slope failure can be estimated using the following equation (Jibson 2007) if a reasonable critical displacement can be specified:

$$\log D_N = 2.401 \log I_A - 3.481 \log a_c - 3.230 \quad (3)$$

where D_N is Newmark displacement in centimeters, I_A is Arias intensity in meters per second, and a_c is critical acceleration in terms of g.

Another approach for estimating ground motions from the results of slope-stability analyses uses a quantity referred to as $(A_c)_{10}$, which is the critical acceleration of a landslide that will yield 10 cm of displacement (the estimated critical displacement leading to catastrophic failure) in a given level of earthquake shaking. The following regression model relates Arias intensity to $(A_c)_{10}$:

$$\log (A_c)_{10} = 0.79 \log I_A - 1.095, \quad (4)$$

where $(A_c)_{10}$ is in g's and I_A is in meters per second (Crozier 1992). If the critical acceleration of a landslide can be determined, then this value can be used as the threshold value of $(A_c)_{10}$ in Eq. 4, and the Arias intensity that would trigger the critical displacement of 10 cm can be calculated.

Interpreting Results of Paleoseismic Landslide Studies

Once a landslide or group of landslides has been identified, dated, and linked to earthquake shaking, interpretations regarding the magnitude and location of the triggering earthquake can be made. The previous section outlined a method for detailed geotechnical analysis to address this issue, but in many cases such an analysis will be impossible owing to lack of data or the unsuitability of the landslide for detailed modeling. Several other approaches to this last level of paleoseismic interpretation are possible; in most cases, multiple lines of evidence will be required to make reasonable estimates of magnitude and location. Perhaps the most important aspect of such interpretation is a thorough understanding of the characteristics of landslides triggered by recent, well-documented earthquakes.

Characteristics of Landslides Triggered by Earthquakes

Comprehensive studies of landslides caused by historical earthquakes have allowed documentation of minimum earthquake magnitudes and intensities that have triggered landslides of

Paleoseismology and Landslides, Fig. 2 Dense concentration of disrupted slides and falls triggered by the 1994 Northridge, California, earthquake (M_w 6.7). Virtually all of the *light-colored* areas in the photo are triggered landslides (Photograph by R.W. Jibson, U.S. Geological Survey)

various types, average and maximum areas affected by landslides as a function of magnitude, and maximum distances of landslides from earthquake sources as a function of magnitude (Keefer 1984, 2002). For these comparisons, landslides were grouped into three categories: disrupted slides and falls, including falls, slides, and avalanches in rock and soil (Fig. 2); coherent slides, including slumps and block slides in rock and soil and slow earth flows (Fig. 3); and lateral spreads and flows, including lateral spreads and rapid flows in soil and subaqueous landslides (Fig. 4).

Minimum Earthquake Magnitudes that Trigger Landslides

Table 2 shows the minimum magnitudes of earthquakes that have triggered various types of landslides. Landslides of various types have threshold magnitudes ranging from 4.0 to 6.5; the more disrupted types of landslides have lower threshold magnitudes than the more coherent types of slides. Although smaller earthquakes could conceivably trigger landslides, such triggering by very weak shaking probably would occur on slopes where failure was imminent before the earthquake.

Minimum Shaking Intensities that Trigger Landslides

Table 3 shows the lowest Modified Mercalli Intensity (MMI) values and the predominant minimum MMI values reported where the three categories of landslides occurred. The data show that landslides of various types are triggered one to five levels lower than indicated in the current language of the MMI scale.

Areas Affected by Earthquake-Triggered Landslides

Drawing boundaries around all reported landslide locations in historical earthquakes and calculating the areas enclosed yields a plot of area versus earthquake magnitude (Fig. 5); a well-defined upper bound curve represents the maximum area that can be affected for a given magnitude (Keefer 1984). Average area affected by landslides as a function of earthquake magnitude

Paleoseismology and Landslides, Fig. 3 Rotational slump that moved as a coherent landslide in the 2004 Niigata Ken Chuetsu, Japan, earthquake (M_w 6.8) (Photograph by D.S. Kieffer, Graz University of Technology, Austria)

Paleoseismology and Landslides, Fig. 4 Lateral-spread landslide triggered by the 1980 Mammoth Lakes, California, earthquake (M_w 6.2) (Photograph by E.L. Harp, U.S. Geological Survey)

can be predicted using the following regression equation (Keefer and Wilson 1989):

$$\log A = M - 3.46 \pm 0.47 \quad (5)$$

where A is area affected by landslides in square kilometers and M is a composite magnitude term, which generally indicates surface-wave magnitudes below 7.5 and moment magnitudes above 7.5. Area affected by landslides also is influenced by the geologic conditions that control the distribution of susceptible slopes and by the focal depth of the earthquake.

Maximum Distance of Landslides from Earthquake Sources

The maximum distance of the three categories of landslides from the earthquake epicenter and from the closest point on the fault-rupture surface relates closely to earthquake magnitude (Fig. 6). Upper bound curves are well defined and are constrained to pass through the minimum threshold magnitudes shown in Table 2 as distance approaches zero. Although the upper bounds shown have been exceeded a few times in subsequent earthquakes, they remain fairly reliable indicators of the maximum possible distances at which the three classes of landslides could be triggered in earthquakes of various magnitudes.

Figure 6 indicates that disrupted slides and falls have the lowest shaking threshold and that lateral spreads and flows have the highest shaking threshold. As with area, earthquakes having focal depths greater than 30 km generally triggered landslides at greater distances than shallower earthquakes of similar magnitude.

Interpreting Earthquake Magnitude and Location

Figures 5 and 6 and Tables 2 and 3 allow interpretation of earthquake magnitude and location in

Paleoseismology and Landslides, Table 2 Minimum earthquake magnitude required to trigger landslides

Earthquake magnitude	Type of landslide
4.0	Rock falls, rock slides, soil falls, disrupted soil slides
4.5	Soil slumps, soil block slides
5.0	Rock slumps, rock block slides, slow earth flows, soil lateral spreads, rapid soil flows, subaqueous landslides
6.0	Rock avalanches
6.5	Soil avalanches

Note: Data from Keefer (1984)

Paleoseismology and Landslides, Table 3 Minimum modified Mercalli intensity required to trigger landslides

Landslide type	Lowest modified Mercalli intensity	Predominant modified Mercalli intensity
Disrupted slides and falls	IV	VI
Coherent slides	V	VII
Lateral spreads and flows	V	VII

Note: Data from Keefer (1984)

Paleoseismology and Landslides, Fig. 5 Area affected by seismically triggered landslides plotted as a function of earthquake magnitude. *Solid line* is upper bound of Keefer (1984); *dashed line* is from Rodriguez et al. (1999); *dotted line* is regression line from Keefer and Wilson (1989)

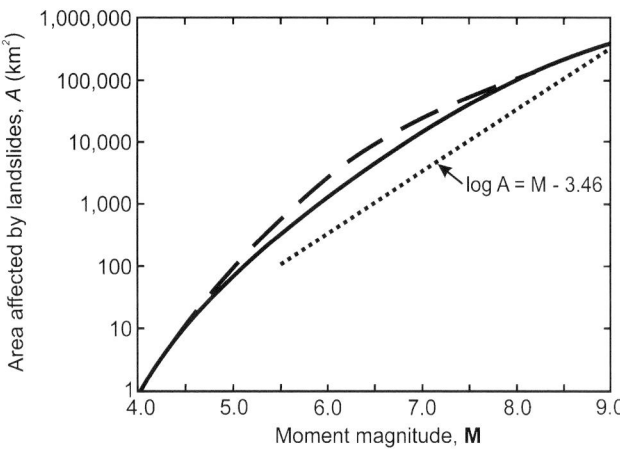

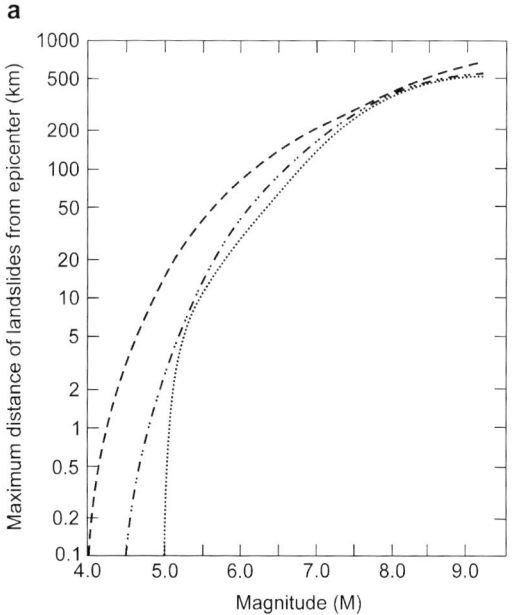 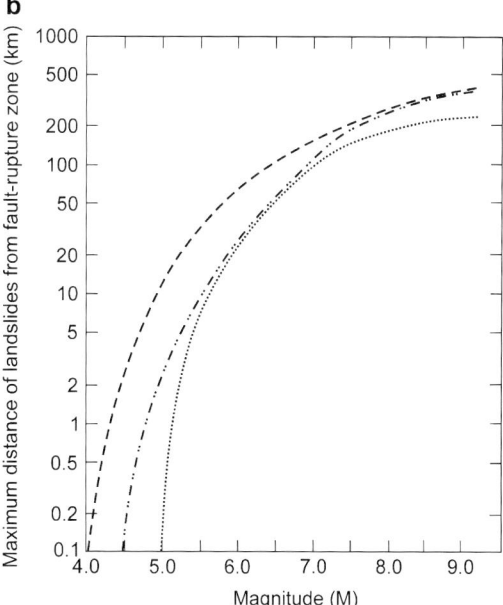

Paleoseismology and Landslides, Fig. 6 Maximum distance to landslides from (**a**) epicenter and (**b**) fault-rupture zone for earthquakes of different magnitudes. *Dashed line* is upper bound for disrupted slides and falls; *dash-double-dot line* is upper bound for coherent slides; and *dotted line* is upper bound for lateral spreads and flows (Modified from Keefer 1984)

a variety of ways. If a single landslide is identified as being seismically triggered, then a minimum magnitude and MMI can be estimated based on the landslide type. If several landslides in an area are identified as being seismically induced, then application of the magnitude-area and magnitude-distance relationships can yield minimum magnitude estimates. As the area in which landslides documented to have been triggered by the same earthquake increases, the estimated magnitude will increase toward the actual magnitude of the triggering earthquake. Therefore, documentation and analysis of landslides over a large area will produce more accurate magnitude estimates. If seismic source zones are well documented, then the distance from the closest source zone to the farthest landslide will yield a reasonable minimum magnitude estimate. The observation that greater source depth relates to greater areas affected and greater source distances for landslides of all types further complicates estimation of earthquake magnitude.

For a specific region, earthquake magnitude can be estimated based on comparison of paleoseismic landslide distribution with landslide distributions from recent, well-documented earthquakes in the region.

Earthquake locations generally are estimated based on the distribution of synchronous landslides attributed to a single seismic event. In a broad area of roughly similar susceptibility to landsliding, the earthquake epicenter probably will coincide fairly closely with the centroid of the landslide distribution. In areas of highly variable or asymmetrical landslide susceptibility, epicentral estimation is much more difficult and subject to error. In areas where seismic source zones are well defined, the epicentral location is best defined as the point in a known seismic source zone (or along a known seismogenic fault) closest to the centroid of the landslide distribution.

Additional Considerations

The primary limitation of paleoseismic analysis of landslides is the inherent uncertainty in interpreting a seismic origin. Unlike liquefaction, which can occur aseismically only in relatively rare conditions, landslides of all types form readily in the absence of earthquake shaking as a result of many different triggering mechanisms. In many cases, ruling out aseismic triggering will be impossible, and the level of confidence in any resulting paleoseismic interpretation will be limited. For this reason, paleoseismic landslide analysis should include, so far as possible, multiple lines of evidence to constrain a seismic origin. In this way, a strong case can be built for seismic triggering of one or more landslides, even if no single line of evidence is unequivocal. Where independent paleoseismic evidence from fault or liquefaction studies is available, paleoseismic landslide evidence can provide useful corroboration.

Detailed slope-stability analyses generally can be performed only on certain types of landslides. Failure conditions of falls, avalanches, and disrupted slides cannot easily be modeled using Newmark's method, and even static stability analyses of these types of slides can be very problematic. Also, the pre-landslide geometry of slides in very steep terrain can be difficult or impossible to reconstruct. Thus, detailed dynamic slope-stability analysis can be applied only to fairly coherent landslides where pre-landslide geometry can be reconstructed with confidence, where groundwater conditions can be modeled reasonably, and where the geotechnical properties of the materials can be accurately measured.

Even allowing for these limitations, paleoseismic landslide studies have been extremely useful where applied successfully, and they hold great potential in the field of paleoseismology. Dating landslide deposits is, in many cases, easier than dating movement along faults because many different dating methods can be used on the same slide to produce redundant results. In addition, landslides have the potential for preserving large amounts of datable material in the various parts of the slide (scarp, body, toe, etc.). In areas containing multiple or poorly defined seismic sources, paleoseismic ground-failure analysis might be preferable to fault studies because landslides preserve a record of the shaking history of a site or region from all seismic sources. Knowing the frequency of strong shaking events could, in many cases, be more critical than knowing the behavior of any individual fault.

Paleoseismic landslide analysis could have greatest utility in assessing earthquake hazards in stable continental interiors where fault exposures are rare or absent but where earthquakes are known to have occurred. In such areas, analysis of earthquake-triggered ground failure, both landslides and liquefaction, might be one of the few paleoseismic tools available.

Another advantage of paleoseismic landslide analysis is that it gets directly at the effects of the earthquakes being studied. Ultimately, most paleoseismic studies are aimed at assessing earthquake hazards. Fault studies can be used to estimate slip rates, recurrence intervals, and, indirectly, magnitudes. From these findings, the effects of a possible earthquake on such a fault are extrapolated. In paleoseismic landslide studies, the effects are observed directly. Thus, if a seismic origin can be established, a landslide shows directly the effects of some previous earthquake. Even if magnitude and location are poorly constrained, at least a partial picture of the actual effects of seismic shaking in a locale or region can be estimated.

In conclusion, paleoseismic landslide analysis can be applied in a variety of ways and can yield many different types of results. Although interpretations are limited by the certainty with which a seismic origin can be established, paleoseismic landslide studies can play a vital role in the paleoseismic interpretation of many areas, particularly those lacking fault exposures.

Summary

In many environments, landslides preserved in the geologic record can be analyzed to determine the likelihood of seismic triggering. If evidence indicates that a seismic origin is likely for a landslide or group of landslides, and if the landslides can be dated, then a paleoearthquake can be inferred, and some of its characteristics can be estimated. Such paleoseismic landslide studies thus can help reconstruct the seismic shaking history of a site or region. In regions that contain multiple seismic sources and in regions where surface faulting is absent, paleoseismic ground-failure studies are valuable tools in hazard and risk studies that are more concerned with shaking hazards than with interpretation of the movement histories of individual faults. Paleoseismic landslide analysis involves three steps: (1) identifying a feature as a landslide, (2) dating the landslide, and (3) showing that the landslide was triggered by earthquake shaking. Showing that a landslide was triggered by seismic shaking can be challenging, but some types of landslides can be analyzed using established static and dynamic methods of slope-stability analysis to determine the likelihood of seismic triggering and to estimate minimum shaking levels required to initiate landslide movement.

Cross-References

▶ Earthquake Magnitude Estimation
▶ Luminescence Dating in Paleoseismology
▶ Radiocarbon Dating in Paleoseismology
▶ Seismic Actions due to Near-Fault Ground Motion
▶ Selection of Ground Motions for Response History Analysis
▶ Time History Seismic Analysis

References

Arias A (1970) A measure of earthquake intensity. In: Hansen RJ (ed) Seismic design for nuclear power plants. MIT Press, Cambridge, MA, pp 438–483
Crozier MJ (1992) Determination of paleoseismicity from landslides. In: Bell DH (ed) Landslides (Glissements de terrain), Proceedings of the 6th international symposium, vol 2. A. A. Balkema, Christchurch/Rotterdam, pp 1173–1180
Jibson RW (2007) Regression models for estimating coseismic landslide displacement. Eng Geol 91:209–281
Jibson RW (2009) Using landslides for paleoseismic analysis. In: McCalpin JP (ed) Paleoseismology, 2nd edn. Academic, New York, pp 565–601
Jibson RW, Keefer DK (1993) Analysis of the seismic origin of landslides – examples from the New Madrid seismic zone. Geol Soc Am Bull 105:521–536
Keefer DK (1984) Landslides caused by earthquakes. Geol Soc Am Bull 95:406–421
Keefer DK (2002) Investigating landslides caused by earthquakes – a historical review. Surv Geophys 23:473–510
Keefer DK, Wilson RC (1989) Predicting earthquake-induced landslides, with emphasis on arid and semi-arid environments. In: Sadler PM, Morton DM (eds) Landslides in a semi-arid environment, vol 2. Inland Geological Society, Riverside, pp 118–149
McCalpin JP (1999) Criteria for determining the seismic significance of sackungen and other scarplike landforms in mountainous regions. In: Techniques for identifying faults and determining their origins. U.S. Nuclear Regulatory Commission, NUREG/CR-5503, Washington, DC pp A-122–A-142
Newmark NM (1965) Effects of earthquakes on dams and embankments. Geotechnique 15:139–160
Rodríguez CE, Bommer JJ, Chandler RJ (1999) Earthquake-induced landslides: 1980–1997. Soil Dyn Earthq Eng 18:325–346
Varnes DJ (1978) Slope movement types and processes. In: Schuster RL, Krizek RJ (eds) Landslides – analysis and control. National Academy of Sciences, Washington, DC, pp 11–33, Transportation Research Board Special report 176

Paleoseismology of Rocky Coasts

Niki Evelpidou[1] and Paolo Pirazzoli[2]
[1]Faculty of Geology and Geoenvironment/ National and Kapodistrian University of Athens, Athens, Greece
[2]Laboratoire de Géographie Physique, Paris, France

Synonyms

Littoral; Sea level; Shoreline; Subsidence; Uplift

Introduction

Important earthquakes are often accompanied by vertical-land displacements. Therefore in coastal areas they may result in rapid changes of the relative sea level. An essential tool for the study of coastal paleoseismicity is the identification of fossil paleoshorelines, paying special attention to sea-level indicators that are consistent or provide evidence of rapid relative sea-level change.

Here, different types of sea-level indicators that are often used in literature in order to determine changes in fossil shorelines are summarized. Information is also provided regarding four case studies of important earthquakes that occurred in Greece (in AD 365, 1953, and 1956) and in Japan (in 1923).

How Can Fossil Paleoshorelines Be Identified?

Fossil paleoshorelines can be identified and traced from geomorphological, biological, sedimentological, stratigraphical, or archeological sea-level indicators.

The coastal geomorphological features that are used as sea-level indicators are the result of either erosional or depositional processes. Erosional features can only be preserved on hard, solid rocks, and in some cases they constitute indicators of sea-level change. Such indicators are marine notches, potholes, abrasion platforms, etc. Among the depositional formations, marine terraces and beachrocks stand out as the most important sea-level indicators.

Erosional Geomorphological Sea-Level Indicators

Tidal notches are well known as precise sea-level indicators that usually undercut limestone cliffs in the midlittoral zone (e.g., Pirazzoli 1986), which renders them the most important erosional geomorphological sea-level indicators. In microtidal areas sheltered from wave action, elevated or submerged notches are used to indicate former sea-level positions, with up to a decimeter confidence.

Bioerosion by endolithic organisms and surface feeders grazing upon epi- and endolithic algae are generally acknowledged to play an important role in tidal-notch development.

The erosion rate is generally highest near mean sea level (MSL) and decreases gradually toward the upper and lower limits of the intertidal range. Accordingly, in places sheltered from continuous wave action, if MSL remains stable, notch profiles will be typically reclined U shaped or V shaped, with a vertex located near MSL, the base of the notch near the lowest-tide, and the top near the highest-tide level. In a moderately exposed site, continuous wave action may splash seawater onto the roof, thus shifting the top of the notch upward, above the highest-tide level.

The rate of maximum undercutting (near MSL) varies with the rock type and the local climate and has been roughly estimated to be of the order of 1 mm/year (Laborel et al. 1999). However this is only a first-order value, while lower rates are generally observed in hard limestones, especially in nontropical areas. More detailed estimations show a range varying from 0.2 to 5 mm/year, depending on lithology, location, and probably duration of bioerosion (for references, see Pirazzoli 1986, Table 1 and Laborel et al. 1999, Table 1).

When a tidal notch is uplifted or submerged, its profile may provide valuable information concerning the type of the paleoseismic event, e.g., whether one or more events took place, whether the movement was coseismic or gradual, etc.

Depositional Geomorphological Sea-Level Indicators

Marine Terraces

Marine terraces are quite common coastal features, owing their presence to a combination of eustatic and crustal processes (Pirazzoli 1994).

The presence of a sequence of uplifted, stepped marine terraces usually corresponds to the superimposition of eustatic change in sea level and of a tectonic uplifting trend that may include sequences of paleoseismic events. The presence of marine terraces several hundred meters above sea level indicates tectonic uplift and allows the estimation of the long-term rate of tectonic deformation.

Marine terraces' age usually increases with altitude, while conservation quality decreases. When uplifted marine terraces are used for the study of sea-level changes, the uplift rate of each section is considered to have remained constant and the eustatic sea-level position corresponding to at least one uplifted terrace has to be known. Marine terraces usually provide evidence of tectonic trends rather than of single paleoseismic events.

Terms often used by geomorphologists for marine terraces are the "beach angle" or "shoreline angle," expected to indicate a frequent sea-level (and wave) position during the period of terrace formation. However, in some cases it can be difficult to precisely determine shoreline angles for Pleistocene marine terraces, during which relative sea-level changes are not known in great detail. This approach can be more useful for Holocene marine terraces and beach ridges formed between two successive coseismic uplifts, like near the Wairarapa Fault in New Zealand where the mean earthquake recurrence time is expected to be of $1,230 \pm 190$ years (Little et al. 2009).

Beachrocks

Beachrock formation is a diachronic and wide-ranging sedimentary process, and since lithification takes place at the coastline, beachrocks have been used as indicators in Quaternary sea-level and neotectonic studies (e.g., Kelletat 2006).

There are, however, several problems in the use of beachrocks as sea-level indicators; the most important are mentioned next. It is difficult to determine the upper level of beachrock cementation, which may correspond to the mean high water springs, and in regions with high tidal range, this limit may be quite large. Another insurmountable problem arises from fluctuations in the tidal range, originating from local changes in the coastal configuration.

Within a sedimentary body, the cementation zone rises and falls following the perpetual tidal cycle of sea-level rise and fall; thus the lower level of cementation zone will correspond to the lowest sea level, at any time during this sedimentary body's existence. Therefore, the cementation of the lower level of an exposed formation may have occurred at an earlier stage, during a rising sea level, more recently during a falling sea level, or during any variation. Consequently, only the upper limit of beachrocks constitutes an accurate indicator of past tidal level.

Additionally, there is the possibility of confusion with other cemented materials in the intertidal zone, although a careful study of the cement, the composition, and the microstructure of the sample can accurately determine the type of formation.

Submerged beachrocks or raised deposits may give evidence of tectonic activity in the area, but it is not possible to provide information on whether this displacement is based on one or more paleoseismic events and whether this movement was rapid or slow.

Biological Sea-Level Indicators

The biological zonation depends essentially on wave exposure. One may distinguish a supralittoral zone, a midlittoral zone, and an infralittoral zone.

The biomass of the supralittoral zone is very low (presence of lichens such as *Verrucaria* or *Lichina* and/or of associated endolithic *Cyanobacteria*) and bioconstruction is uncommon but may occur in the form of incrustations by *Chthamalus* shells in shaded rock crevices exposed to wave splash. Limited erosion may take place, mainly due to boring activity in carbonate substrate. According to Le Campion-Alsumard (1979), endolithic microorganisms may create many microtunnels, with their roots, that are easily preserved and fossilized; these marks may be suitable as paleobathymetric indicators. Other algae, which may occur in the lower supralittoral zone, are the green *Endoderma* and the brown *Chrysophyceae*, which may be outgrown by various other algae. The most widespread animals in the lower supralittoral zone are the gastropods *Littorina* and *Melanerita*, which graze the algae, some crustaceans, and occasionally opportunistic visitors that do not normally reside there, such as hermit crabs, insects, spiders, birds, and small rodents (Pirazzoli 1996).

The midlittoral zone is submerged at close intervals by waves and tide. The upper part of this zone is never rich in species and even barren for most of the year on the Arctic and Antarctic shores, due to the grinding action of ice and may become vegetated by benthic diatoms and ephemeral algae only during the summer. *Fucus* and *Pelvetia* algae are frequent in North Atlantic and North Pacific coasts. Farther south a greater amount of annual species is found, and the importance of *Cyanobacteria* increases at the tropics. Lithothamnium species have been observed on exposed shores of the Indian and Pacific Ocean, as well as *Cladophora*, *Ectocarpus*, and *Ulva*. The lower part of the midlittoral zone is in general densely covered by fucoid and turf algae, which are grazed by several herbivores (littorinoids, limpets).

Barnacles (*Balanus, Elminius, Tetraclita*, and especially *Chthamalus*) live in the upper part of the midlittoral zone, while mussels (*Mytilus*) and oysters tend to occupy lower levels. Barnacles, mussels, oysters, and *Lithophyllum* form a rim just above sea level in the western Mediterranean and may be fossilized in situ after death.

Erosive agents include many *Cyanobacteria* in the upper part of this zone and limpets (*Patella*) and Chitons in the lower part.

The infralittoral (or sublittoral) zone extends from the biological mean sea level (BMSL) (Laborel and Laborel-Deguen 1994), located at the base of the midlittoral zone, to a depth of 25–50 m (depending on water transparency) but may be absent in very turbid coastal waters. The position of the BMSL in relation to the tide-gauge MSL varies with exposure.

This zone is densely populated by brown algae, coralline encrusting algae (*Porolithon, Neogoniolithon, Lithophyllum*), fixed vermetid gastropods (e.g., *Dendropoma, Vermetus, Serpulorbis*), cirripedes like *Balanus*, or coral reefs in warm waters. They may be accompanied by turf-forming algae and fucoid vegetation (*Fucus, Cystoseira, Sargassum*) that are grazed by herbivorous fish and sea urchins; the latter like clionid sponges and other borers (e.g., *Lithophaga*) may also attack the rocky substrate.

All bioconstructions (encrusting algae, vermetids, reef-building corals and associates, barnacles, and oysters) may fossilize in situ in the sublittoral zone. The most accurate altimetric indications for the estimation of past sea levels derive from the comparison of the fossil bioconstructions with their present-day counterpart. The most useful sea-level indicators are those with the narrowest vertical zonation, e.g., the upper level of coral microatolls and rims made by *Dendropoma* or *Neogoniolithon*. Fossil *Lithophaga*, which are frequently used to indicate uplift on carbonate shores, deserve special mention. They are generally a poor sea-level indicator, because they live between the upper limit of the sublittoral zone and depths greater than 30 m. However, when the upper limit of their population is well marked and forms a horizontal line, it corresponds to the BMSL. Nonetheless, as reported by Shaw et al. (2010), *Lithophaga* shells tend to incorporate host-rock carbon, thus giving older apparent radiocarbon ages. More details on the vertical range, altitudinal accuracy, and resistance to erosion of various Mediterranean midlittoral and infralittoral sea-level indicators are given by Laborel and Laborel-Deguen (1994).

Although biological indicators are very useful for dating an uplift, they are unlikely to be preserved when they are submerged because of the subsequent bioerosion. An additional indicator is useful for diagnosing the type of the paleoseismic event, e.g., coseismic or gradual. In any case, if the shoreline has quickly shifted from the tidal zone, the biota will be well preserved, while slow emergence will expose it leading to gradual destruction in the supralittoral wave zone.

Sedimentological/Stratigraphical Sea-Level Indicators

The most common method for locating material that will provide information regarding sea-level change is the use of a vibrating sampler, provided that samples are slightly moderately disturbed. The method of excavation trenches is also often used for smaller depths.

In order to acquire accurate results from coring, it is necessary to obtain a solid and undisturbed sample, so that the conditions are similar to the ones prevailing at the sampling location. There are several biomarkers, present in sampling cores, which can be dated and associated to paleo-sea levels. An indispensable requirement is the recording of the precise location and height of the sampling with differential GPS.

A stratigraphic column may provide evidence of the paleogeographic/paleoenvironmental changes that originated from paleoseismic events, while it may provide clear evidence of paleo-tsunami in the area. In a stratigraphic column history is fossilized making clear whether transition was slow or abrupt. If datable material is available, the event could be dated as well.

Archeological Sea-Level Indicators

Most archeological remains provide no evidence on how far from sea level they were constructed. A human civilization may not be linked to the sea, and in most cases it is impossible to distinguish a house or pottery found near the sea from an equivalent found some kilometers inland. On the other hand, the civilization may have developed specific activities, closely related to the sea, requiring sailors, fishermen, boat builders, and salt gatherers. In this case the settlement was probably built close to the shore.

Apart from the undisputable evidence of minimum relative sea-level rise, provided by submerged structures that had to have their foundations on dry land (houses, tombs, mosaic floors, passageways, storage tanks, tells, middens), many archeological findings refer to structures that were partly in the sea and may be considered as reliable sea-level indicators (slipways, breakwaters, jetties, quays, docks, channels, drains, salt pans, the lowest part of certain flights of steps or of coastal quarries that used a wood splitting wedge for cutting the lowest level of stones, or finally installations for fishing or fish farming). For a reliable estimation of the ancient sea level, a good understanding of the functionality of the structure and the local hydrographic and climatic constraints is always essential.

In any case the ruins may provide valuable information concerning the paleoseismic events in an area. For example, around 3300 BP an earthquake destroyed House B in Grotta (Naxos Island, Cyclades). The archeologists who excavated Grotta and Aplomata on Naxos spoke of two seismic events: one at an early phase of the LH IIIA2 and another one in LH IIIC period, based on the ruins of the archeological sites. The destruction of the site known as "Kolona," literally "Column," at Heraion (Samos Island, Eastern Aegean), has been assigned to seismic damage. The drums of the only surviving column are offset laterally and tilted. This effect was clearly the result of rocking produced by a strong earthquake. The coeval destruction of at least one other temple nearby supports the possibility of a destructive earthquake circa 530 BC.

Structures Partly in the Sea

Most known examples (e.g., Flemming 1979) come from the tideless Mediterranean, where much well-preserved material exists. However, similar methods and techniques have also been occasionally adopted in fully tidal waters.

Gradients for ancient slipways may vary from $4°$ to $15°$, and the original position of the sea level can be calculated approximately by making assumptions about the water depth needed at the foot of the structure to take the bow of the boat and the length of slipway needed to be dry to support the boat while work was done on it. Well-preserved slipways in the Mediterranean can provide an accuracy of sea-level estimate of ± 0.25 m.

Breakwaters, jetties, and moles are generally walls, built to create a barrier between the open sea and a calm sheltered area of water, where ships may safely moor. Mooring quays are built on the inner side. The base of the breakwater was in the water, while the top must have been out of the water, high enough to prevent waves from climbing over. Protection moles have often been constructed around Roman fish tanks. Since the upper part of the majority of these structures has been damaged by countless storms, the identification of the upper level is generally problematic. However, when the upper surface is intact, the ancient sea level may be estimated within ±1 m.

Quays are well-squared structures built to provide mooring for ships, access for men, and storage space for cargoes. They were often associated with steps leading down to the vessels, mooring stones, bollards, and warehouses. Quay surfaces varied generally around 1.0–2.0 m. Through the combination of the water depth at the foot of the quay wall, the level of the mooring stones, surfaces reached by steps, and the main working surfaces of the quay, it may be possible, according to Flemming (1979), to determine the original sea level to about ±0.5 m.

Docks are small rectangular basins in which a ship can be berthed and the basin can be pumped dry. The interpretation of docks is similar to that for quays.

Fish Tanks

Though stone fish traps and turtle pens have been reported in many cultures (e.g., Caribbean Indian and Australia aborigine), Roman fish tanks (*piscinae*) are worth special mention. Such installations were especially frequent along the Tyrrhenian coasts of central Italy between the first century BC and the first century AD, when they came into fashion for wealthy Romans. When certain elements are well preserved and the functioning of the structure is well understood, fish tank remains allow estimating with excellent accuracy (±0.1 m) the sea-level position of about 2,000 year ago. The main sea-level indicators are the top of closing gates (*cataractae*), generally slightly above the high-tide level. In several examples they were located near the top of the walls (*crepidines*) delimiting the basins from where they could be easily worked, by using the *crepidines* as a footwalk. For several other coastal structures, like salt pans, foundations of towers and lighthouses, roads, seaside villas, wells, etc., the exact relation to the sea level may vary between sites.

How Can Fossil Paleoshorelines Be Dated? Radiocarbon, Archeological Remains, Coastal Cores, and Historical Information

Fossil paleoshorelines can be dated through sea-level indicators, using different methods, depending on the type of indicators available. For example, radiocarbon dating of biological indicators is one of the easiest and most trustworthy methods. On the other hand, this is not possible with submerged paleoshorelines since any biological indicator is destroyed by bioerosion occurring after submergence. In this case other, indirect ways of dating fossil paleoshorelines are used. In some cases it is possible to date a relative sea-level change, based on an ancient site in the nearby area.

Emerged or submerged beachrocks may be dated, but their accurate dating is a very difficult task, even though biogenic materials such as shells or corals can easily be dated by radiocarbon. The age acquired corresponds to the time of the organism's death. After death, the organism was transported and deposited on a beach and later was cemented into the beachrock. Although in many cases this sequence may be complete within a few years, it is likely that the time interval between the death of the organism and the cementation of the beachrock may last hundreds or thousands of years. For this reason, the age acquired by dating a constituent organism may be regarded as a maximum for the cementation of beachrock. The minimum age for this process is only attainable by dating the cement. However, several difficulties may arise with extracting the adequate amount of cement, especially in submerged beachrocks. Furthermore, it is possible that after exposure, water passing through beachrock may cause carbonate exchange resulting in continuous renewal of the apparent cementing age.

Dating material provided by coastal cores could often be the only solution, especially in cases where biological, geomorphological, and archeological sea-level indicators are absent from a study area, or their dating is impossible. In relatively recent sediments, which were deposited in various geoenvironments, a variety of information that determines the paleogeography of a region and the various changes in sea level is recorded, through datable material found at a certain depth of the stratigraphic column. In these cases, depth precision is determined by the accuracy of the relationship of the biological marker to mean sea level. However, an uncertainty is added in all estimations, especially if the cores come from marsh deposits where compaction can be significant and the positional uncertainty of the biological markers increases considerably.

How Can Coseismic Sea-Level Changes Be Distinguished from Gradual Changes?

The best indicators to distinguish rapid relative sea-level changes from gradual ones are tidal notches. The shape of the notch profiles may provide qualitative information on the rate of sea-level change and on tectonic movements. In the case of a rapid (e.g., coseismic) emergence greater than the tidal range, the notch that is completely emerged will be preserved from further marine bioerosion, while a new notch will develop in the new lower intertidal zone (Pirazzoli 1986). On the other hand, in the case of rapid submergence greater than the tidal range, the whole notch will be submerged. It will not be completely preserved, due to further bioerosion as in the case of emergence, and its profile will be uniformly deepened by the rate of bioerosion that predominates in the infralittoral zone, while a new tidal notch will develop in the intertidal range. However, the infralittoral bioerosion rate is generally estimated to be one order of magnitude less than the intertidal one.

If the rapid submergence is smaller than the tidal range, the height of the notch will increase, with further deepening in the new intertidal zone, while the part of the notch below the low tide level will continue to deepen at a much slower rate, and the notch profile will be marked by an undulation at the level of the roof of the former notch. Finally, in the case of gradual

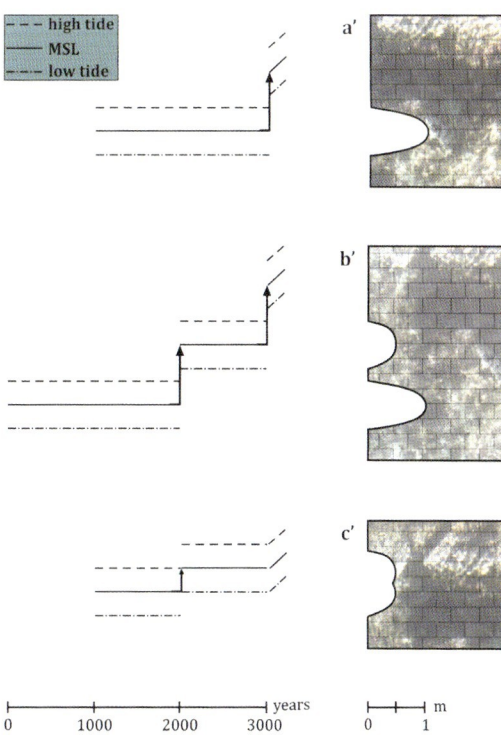

Paleoseismology of Rocky Coasts, Fig. 1 Different profiles of notches that give evidence of a rapid movement caused by earthquakes with magnitude larger than 6.0 Mw, commonly associated to morphogenic faults producing direct surface faulting (Ambraseys and Jackson 1990) and ground deformation (Pavlides and Caputo 2004)

submergence, the height of the notch will gradually increase, the lower part of its profile becoming higher than the upper part, and no marked undulation will appear in the profile. Many combinations of vertical movements leading to emergence or submergence are possible, as well as different erosion rates, tidal ranges, and periods of time, but even so, notch profiles would be similar to those of Fig. 1.

Case Studies

AD 365, Crete Earthquake

On 21 July AD 365, a strong underwater earthquake occurred near the southwestern part of Crete. Nearly all towns in the island were

Paleoseismology of Rocky Coasts, Fig. 2 Remains of at least three fossil ripple notches nave been preserved between +4.00 and +4.75 m below Afrata (1.5 km N of Gonias, near the eastern foot of Rodopos peninsula). These tidal notches have probably developed a few centuries before the general uplift of AD 365 (Photo P.A.P. n° 3904, Sept. 1977)

destroyed, and widespread damage extended to central and southern Greece, northern Libya, Egypt, Cyprus, and Sicily. The earthquake was followed by a tsunami that devastated the southern and eastern coasts of the Mediterranean, particularly Libya, Alexandria, and the Nile Delta, hurling ships nearly two miles inland. The Roman historian Ammianus Marcellinus described in detail the tsunami hitting Alexandria.

In Crete, systematic geomorphological surveys carried out in the late 1970s (Pirazzoli et al. 1982) showed that a block of lithosphere approximately 200 km long, extending from central Crete to the small Antikythera island, was uplifted and inclined northeastward. The uplift reached a maximum of 9 m in southwestern Crete, leaving along the coast a very visible emerged shoreline that has been dated by radiocarbon in many positions and also remains of Roman harbors, like in Kisamos and Falasarna (where the coseismic uplift reached about 6 m).

If the AD 365 tsunami is well dated historically, direct historical evidence confirming that such uplift in Crete was synchronous to the earthquake that produced the tsunami is generally missing, and many radiocarbon dates converge only by a few decades toward the critical period. A closer connection to a major destructive earthquake that occurred shortly after AD 355/361 in Crete was established archeologically by a number of coins found in Kisamos among skeletons of people killed and buried by fallen debris (Stiros and Papageorgiou 2001).

If the type and geometry of the fault, which produced the block movement, are still a matter of debate, some geomorphological elements provide information on the preparatory phase that permitted the necessary strain accumulation of the major earthquake. These elements consist of nine fossil shorelines, preserved as ripple notches in Antikythera and as bioconstructed vermetid rims at Moni Khrisoskalitisas, remnants of which have been reported in several sites (e.g., Plaka, Falasarna, Koutsounari, Afrata, Damnoni, Sougia) (Fig. 2) showing that the western part of the block uplifted in AD 365 underwent between 4,000 and 1,700 year BP, along a length of at least 150 km, ten small coseismic subsidences (from 10 to 25 cm each time), apparently due to gravitational forces, without any noticeable tilting (Pirazzoli et al. 1981, 1982). Shaw et al. (2010) attempted to narrow the chronology of the subsidence movement having preceded the AD

Paleoseismology of Rocky Coasts, Fig. 3 Submerged fossil shoreline in Keros island, which has been also located along the rocky coasts of Sifnos, Antiparos, Paros, Naxos, and Iraklia, reveals a widespread evidence of a recent subsidence, part of which according to Evelpidou et al. (2012b) is due to Amorgos 1956 earthquake

365 uplift event by using *Lithophaga* shells and showing that these shells may incorporate host-rock carbon. If the reality of such incorporation by *Lithophaga* shells has been confirmed by Evelpidou et al. (2012a) for tsunami deposits in the Gulf of Euboea, however the chronological narrowing proposed by Shaw et al. (2010) cannot be applied to Crete before AD 365, because the time of the subsidence events dated by Pirazzoli et al. (1981, 1982) were not deduced from lithophagid dates, but from organic accretions of *Dendropoma, Neogoniolithon, Lithophyllum*, other calcareous algae, and *vermetids*, which are excellent sea-level indicators providing reliable radiocarbon-age estimations.

1956, Cyclades: Amorgos Earthquake

The Amorgos earthquake, on 9 July 1956 in the south central Aegean Sea, was one of the largest ($M_S = 7.4$) and most destructive crustal earthquakes in the twentieth century in the Aegean and was followed by a more destructive aftershock ($M_S = 7.2$) (Makropoulos et al. 1989).

The whole event resulted in 53 deaths and considerable damage notably on the island of Santorini and generated a local tsunami, which affected the shores of the Cyclades and Dodecanese Islands, Crete, and the coast of Asia Minor, with run-up values of 30, 20, and 10 m reported on the southern coast of Amorgos, Astypalaia, and Folegandros, respectively (Papastamatiou et al. 1956; Galanopoulos 1960; Soloviev et al. 2000).

According to Evelpidou et al. (2012b), an extended underwater geomorphological survey located a well-developed submerged notch along the rocky coasts of Sifnos, Antiparos, Paros, Naxos, Iraklia, and Keros (Fig. 3), revealing widespread evidence of a recent 30–40 cm submergence, part of which may have seismic origin.

Comparison with information reported from earthquakes having affected the area suggests that at least part of the recent submergence might be an effect of the 1956 Amorgos earthquake. Modeling of the coseismic and short-term postseismic effects of the earthquake revealed that part of the observed subsidence may be explained in some of the islands by a fast postseismic relaxation of a low-viscosity layer underlying the seismogenic zone. However far-field observations are underestimated by

Evelpidou et al. (2012b) model and may be affected by a wider deformation field induced by the largest aftershock of the Amorgos sequence or by other earthquakes.

1953, Ionion: Cephalonia Earthquake

The region of Cephalonia and Ithaca islands is characterized by the frequent occurrence of shallow seismic events with magnitudes up to 7.2. Based on historical documents, large earthquake events took place in 1469, 1636, 1767, 1867, and 1953, while moderate-to-large ones occurred in 1658, 1723, 1742, 1759, 1766, 1862, 1912, 1915, 1925, 1932, and 1939 (Papagiannopoulos et al. 2012). The most destructive earthquake in the region hit the area in 1953. The 1953 series of earthquakes greatly damaged or destroyed 91 % and 70 % of all houses in Cephalonia and Ithaca, respectively. The first earthquake of the series occurred on 9 August and hit Ithaca and the town of Sami (Cephalonia). The second one occurred on 11 August and was greater than the first, destroying many towns around Cephalonia: Argostoli, Lixouri, Agia Efimia, and Valsamata. The third one on 12 August had a magnitude of 7.2 and completely destroyed Cephalonia, Ithaca, and Zakynthos. The number of human casualties as well as curves that illustrate the distribution of seismic intensity has been given by Grandazzi (1954).

According to the map of seismic intensity by Grandazzi (1954, Fig. 3), in Cephalonia a seismic intensity of IX is observed for most part of the island, with the exception of Fiscardo peninsula, where the intensity was VII. In addition, in Ithaca island an intensity of IX was measured for the southern part, while in the northern one the intensity was smaller (VIII). This is also illustrated on the map of destruction by Grandazzi (1954, Fig. 4) (Poros, Cephalonia), where the percentage of destructions both in Ithaca and Cephalonia are increased toward the south.

The observations of Galanopoulos (1955) and Mueller-Miny (1957) about the uplift of the shorelines of Cephalonia during 12 August 1953 earthquake were also verified by Stiros et al. (1994). In fact, the 1953 earthquake left recognizable uplifted marks in several places around Cephalonia island, which reach a maximum elevation of +0.7 m in Poros area (Fig. 3). In fact, the uplift is ranging from 30 to 70 cm and has been evident in the central part of the island. According to Stiros et al. (1994), the uplifted part is bounded by two subparallel and homothetic major thrusts and can be better explained if assumed that the surface deformation reflects a

Paleoseismology of Rocky Coasts, Fig. 4 Holes made by Lithophaga shells in a tufa layer of Miura Peninsula uplifted by the 1923 earthquake (Photo P.A.P. n° 595, 1974)

Paleoseismology of Rocky Coasts, Fig. 5 Uplifted fossil shorelines in Poros area (Cephalonia). The lower one at +0.7 m is ascribed to the 1953 earthquake

halotectonic deformation, indicating that the seismic deformation of the uppermost crustal strata in the area may mimic the style of the long-term halotectonic deformation.

The study of Stiros et al. (1994) shows that no postseismic displacement occurred after the 1953 coseismic uplift (Fig. 4) (Poros, Cephalonia).

1923, The Great Japan Earthquake

On 1 September 1923, shortly before noon (at 2 h 58' 44" GMT), a violent shock of earthquake occurred, shaking houses and other buildings intensely. A second and a third quake followed shortly after and then many aftershocks, spreading disaster all over. The afflicted zone covered seven prefectures: Tokyo, Kanagawa, Shizuoka, Chiba, Saitama, Yamanashi, and Ibaraki. The fire that followed reduced to ashes a great part of Tokyo and of the port of Yokohama.

According to the Bureau of Social Affairs (1926), the number of houses having been damaged in Tokyo and Kanagawa Prefectures is of:

- 381.090 (54.9 %) entirely burnt
- 517 (0.1 %) partially burnt
- 83.319 (12.1 %) entirely collapsed
- 91.233 (13.1 %) partially collapsed
- 1.390 (0.2 %) swept away
- 136.572 (19.6 %) damaged

This makes a total of 558.049 (80.4 %) houses entirely or partially burnt, entirely or partially collapsed, or swept away due to the earthquake or the fire that followed.

According to http://earthquake.usgs.gov/learn.today/index.php?r, the magnitude M_L of this earthquake was 8.3 and it produced a death toll of 142.800.

The uplift of the ground in the disturbed zone was generally (Fig. 4) (Miura Peninsula) along 15 km of coast, especially in Tokyo Bay and Sagami Bay, reaching a maximum of 1.8 m in

Paleoseismology of Rocky Coasts, Fig. 6 Abrasion coastal platform uplifted in 1923 at Miura Peninsula. Water level (at Aburatsubo tidal station) is at −76 cm. Spring tidal amplitude is ±54 cm (Photo P.A.P. 592, May 1974)

Chiba Prefecture, while more to the south (e.g., OShima (Vries Island)) the ground sunk (Fig. 6).

Summary

Paleo-sea levels estimated through indicators such as geomorphological, biological, sedimentological, stratigraphical, or archeological may give evidence of coastal paleoseismicity. After the indicators' identification comes the important step of interpretation and dating of the event. Different types of indicators provide different amount of confidence and precision in determining the respective tectonic events, mainly due to the variety of their formation process and position as well as their subsequent evolution. Especially tidal notches may additionally provide information about whether a sea-level change has been gradual or coseismic, based on their profiles.

Four case studies of significant paleoseismic events are presented, all of which have greatly affected the coastal zone.

Cross-References

▶ Archeoseismology
▶ Early Earthquake Warning (EEW) System: Overview
▶ EEE Catalogue: A Global Database of Earthquake Environmental Effects
▶ Luminescence Dating in Paleoseismology
▶ Paleoseismic Trenching
▶ Paleoseismology
▶ Radiocarbon Dating in Paleoseismology
▶ Tsunamis as Paleoseismic Indicators

References

Ambraseys NN, Jackson JA (1990) Seismicity and associated strain of central Greece between 1890 and 1988. Geophys J Int 101:663–708

Evelpidou N, Vassilopoulos A, Pirazzoli PA (2012a) Holocene emergence in Euboea island (Greece). Mar Geol 295–298:14–19

Evelpidou N, Melini D, Pirazzoli P, Vassilopoulos A (2012b) Evidence of a recent rapid subsidence in the S–E Cyclades (Greece): an effect of the 1956 Amorgos earthquake? Cont Shelf Res 39–40:27–40

Flemming NC (1979) Archaeological indicators of sea level. In: NIVMER (ed) Les Indicateurs de niveaux marins. pp 149–166. Oceanis, 5, hors-série

Galanopoulos A (1955) Seismic geography of Greece. Ann Géol Pays Hell 6:83–121

Galanopoulos AG (1960) Tsunamis observed on the coasts of Greece from antiquity to present time. Ann Geofisica 13:369–386

Grandazzi M (1954) Le tremblement de terre des Iles Ioniennes (aout 1953). Ann Géogr 63(340):431–453

Kelletat D (2006) Beachrock as sea-level indicator? Remarks from a geomorphological point of view. J Coast Res 22:1558–1564

Laborel J, Laborel-Deguen F (1994) Biological indicators of relative sea level variations and of co-seismic displacements in the Mediterranean region. J Coast Res 10(2):395–415

Laborel J, Morhange C, Collina-Girard J, Laborel-Deguen F (1999) Littoral bioerosion, a tool for the study of sea-level variations during the Holocene. Bull Geol Soc Den 45:164–168

Le Campion-Alsumard T (1979) Les végétaux perforants en tant qu'indicateurs paleobathymétriques. In: NIVMER (ed) Les Indicateurs de niveaux marins. pp 259–264. Oceanis, 5, hors-série

Little TA, Van Dissen R, Schermer E, Came R (2009) Late Holocene surface ruptures on the southern Wairarapa fault, New Zealand: link between earthquakes and the uplifting of beach ridges on a rocky coast. Lithosphere 1(1):4–28

Makropoulos K, Drakopoulos J, Latousakis J (1989) A revised and extended catalogue for Greece since 1900. Geophys J Int 98:391–394

Mueller-Miny H (1957) Beiträge zur Morphologie der mittleren jonischen Inseln. Ann Géol Pays Hell 8:1–28

Papagiannopoulos GA, Hatzigeorgiou GD, Beskos DE (2012) An assessment of seismic hazard and risk in the islands of Cephalonia and Ithaca, Greece. Soil Dyn Earthq Eng 32:15–25

Papastamatiou J, Zachos K, Voutetakis S (1956) The earthquake of Santorini of 9 July 1956. Institute of Geology and Subsurface Research (I.G.S.R.). Athens, Greece

Pavlides S, Caputo R (2004) Magnitude versus faults' surface parameters: quantitative relationships from the Aegean. Tectonophysics 380(3–4):159–188

Pirazzoli PA (1986) Marine notches. In: van de Plassche O (ed) Sea-level research: a manual for the collection and evaluation of data. Geo Books, Norwich, pp 361–400

Pirazzoli PA (1994) Tectonic shorelines. In: Carter RWG, Woodroffe CD (eds) Coastal evolution. University Press, Cambridge, pp 451–476

Pirazzoli PA (1996) Sea-level changes – the last 20000 years. Wiley, Chichester

Pirazzoli PA, Thommeret J, Thommeret Y, Laborel J, Montaggioni LF (1981) Les rivages émergés d'Antikythira (Cerigotto): corrélations avec la Crète Occidentale et implications cinématiques et géodynamiques. In: Actes du Colloque "Niveaux marins et tectonique quaternaires dans l'aire méditerranéenne". CNRS et University Paris I, Paris, pp 49–65

Pirazzoli PA, Thommeret J, Thommeret Y, Laborel J, Montaggioni LF (1982) Crustal block movements from Holocene shorelines: Crete and Antikythira (Greece). Tectonophysics 86:27–43

Shaw B, Jackson JA, Higham TFG, England PC, Thomas AL (2010) Radiometric dates of uplifted marine fauna in Greece: implications for the interpretation of recent earthquake and tectonic histories using lithophagid dates. Earth Planet Sci Lett 297:395–404

Soloviev SL, Solovieva ON, Go CN, Kim KS, Shchetnikov NA (2000) Tsunamis in the Mediterranean sea 2000 B.C.–2000 A.D. Kluwer Academics, Dordrecht

Stiros SC, Papageorgiou S (2001) Seismicity of Western Crete and the destruction of the town of Kisamos at AD 365; archaeological evidence. J Seismol 5:381–397

Stiros SC, Pirazzoli PA, Laborel J, Laborel-Deguen F (1994) The 1953 earthquake in Cephalonia (Western Hellenic Arc): coastal uplift and halotectonic faulting. Geophys J Int 117:834–849

Paleoseismology: Integration with Seismic Hazard

Kuvvet Atakan
Department of Earth Science, University of Bergen, Bergen, Norway

Synonyms

Deterministic seismic hazard analysis; Earthquake geology; Earthquake hazard analysis; Paleoseismology; Probabilistic seismic hazard analysis; Seismic hazard analysis; Seismic hazard assessment

Introduction

Paleoseismology is the branch of science that aims to understand the earthquakes and their effects that have occurred in the geological past – i.e., during the Quaternary Period

(2.588 Ma to present), most commonly during the Holocene. Contributions from these past earthquakes turn out to be very critical not only to the understanding of the deformational processes within the Earth's crust but to a large extent also to develop better models for assessing seismic hazard. The main scope of this entry is to elaborate on the latter, the use of paleoseismological data in seismic hazard assessment.

Paleoseismology, also used as synonymous with "earthquake geology," has become an important and integrated part of the seismic hazard assessment during the late 1970s, following systematic work done on the San Andreas Fault in California, USA. Several studies in California (e.g., Sieh 1978; Wallace 1981, 1990; Sieh and Jahns 1984; Sieh et al. 1989; Grant and Sieh 1994; Grant and Lettis 2002) have been influential in making paleoseismology visible to other branches of Earth science. Researchers working with seismic hazard assessment became especially interested in these studies as they could see the importance of the paleoseismological data in their work. Later in the 1980s and 1990s, paleoseismological studies have spread in many other countries, most notably in Japan, Canada, New Zealand, Italy, Turkey, Greece, and Spain. In the late 1980s several monographs have contributed to a wider recognition of paleoseismology as an important field of geology (e.g., Wallace 1986; Vita-Finzi 1986; Crone and Omdahl 1987). In the late 1990s the two books published on paleoseismology by McCalpin (1996) and Yeats et al. (1997) were influential in introduction paleoseismological studies to a wider community.

Main Elements of a Paleoseismological Study

Large earthquakes occurring on crustal scale faults leave visible traces on the Earth's surface during their rupture process. This evidence on the surface is subject to both erosional and depositional processes controlled by the climatic conditions. As a consequence these evidences may either be destroyed or buried.

Paleoseismological studies are usually conducted on these faults with the aim of identifying individual paleoearthquakes with quantitative constraints on their age and size. Final aim is usually try to establish a maximum earthquake magnitude and a recurrence interval for a given fault.

Paleoseismological studies contain a number of distinct elements. They usually start with a regional analysis on the tectonic setting and structural geology of the area of interest. This is done to find out the regional deformation rates and includes geodetic strain rates obtained through the time-lap analysis of GPS data. Once the regional strain rate is known, it is possible to look for evidence at a local scale that can be expected from that rate of deformation. These regional scale studies are then followed by various types of analyses of the fault segment(s), through geological, geophysical, geodetic techniques, as well as detailed analyses of the local stratigraphy in order to select a suitable site(s) for detailed analysis. Trenching is done at the selected site(s) for identifying the paleoearthquakes. Various types of evidence are then used to identify the individual paleoearthquakes. In order to constrain the ages of these events, various dating techniques are applied on available material on event horizons within the trench stratigraphy. This is followed by establishing the slip per event and the rupture length. Based on these, finally a maximum magnitude and a recurrence interval for the fault are established. These parameters are especially important in probabilistic seismic hazard assessment (PSHA). In the following section how these parameters are integrated in seismic hazard assessment are explained.

Integrating Paleoseismological Data in Seismic Hazard Assessment

Integrating paleoseismological data in Seismic Hazard Assessment (SHA) is dependent on the type of methodology used, i.e., probabilistic (PSHA) or deterministic seismic hazard assessment (DSHA).

Paleoseismic Data and Probabilistic Seismic Hazard Assessment

Standard probabilistic seismic hazard analysis (Cornell 1968; McGuire 1993) is based on statistical treatment of earthquakes using probability density functions assuming different types of earthquake recurrence (Fig. 1). One of the most common applications of PSHA is based on a Poissonian assumption for earthquake recurrence, which assumes that the earthquake occurrence has not have memory, i.e., occurrence of earthquakes in a given area in the future has not have relation to the occurrence of previous earthquakes in the same region. Such a "random" occurrence of earthquakes is applied, not because the earthquakes do not follow a certain pattern of occurrence governed by the physical processes, but because of recognition of the fact that it is simply not always possible to assess these physical parameters that lead to stress accumulation and release process. Therefore the simple assumption of a "random" earthquake occurrence is valid, and hence it is an attempt to cover the large uncertainties associated with the complex physical processes that lead to rupture along a fault. One of the important implications of applying a recurrence relation, let it be Poissonian or other types such as conditional, exponential, Weibull, Brownian, gamma, or log-normal (e.g., Main 1995; Mathews et al. 2002), is that the complete earthquake history is known and that it represents a statistically valid sample to develop a reliable estimate of the frequency of occurrence of various magnitude levels in a given area, the so-called Gutenberg-Richter relation.

The completeness criterion in earthquake catalogs is not always met. This is especially true in areas of low seismicity where the instrumental and historical earthquake catalogs do not cover the occurrence of large and destructive earthquakes when the recurrence time of these is beyond the limits of the total catalog time span. Using such incomplete catalogs in developing the Gutenberg-Richter relation (Fig. 2) may lead to erroneous estimate of the seismic hazard in an area. Important parameters, such as the maximum expected earthquake magnitude and its frequency of occurrence are obtained, based on the maximum observed earthquake. This may be subject to interpretation as, in many cases, simple extrapolation of the frequency-magnitude relation curve will not give realistic estimates of the maximum expected magnitude (m_2 in Fig. 2). What is usually adopted is then a truncation to the curve (m_1 in Fig. 2), whereas in reality paleoseismological data may imply a larger (or lower) level on the maximum expected magnitude (m_3 in Fig. 2). In these cases, it is essential to extend the catalog time span by investigating the earthquake occurrence in the geological past through paleoseismological studies. Extending the time span of the catalogs with the occurrence of large earthquakes in the geological past brings important constraints in establishing the maximum expected magnitude in an area.

Applying paleoseismological data to probabilistic seismic hazard assessment (Fig. 3) using

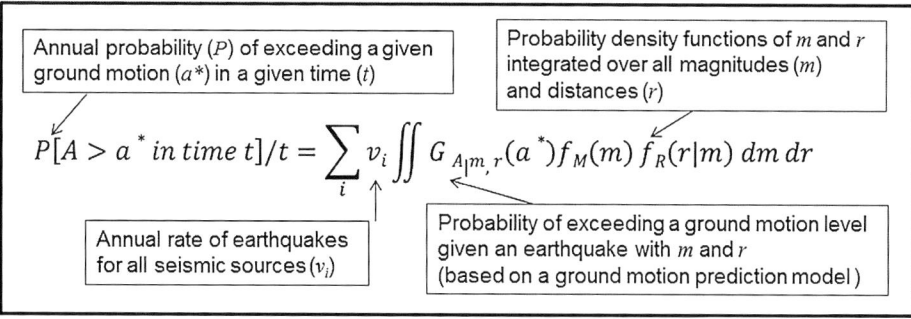

Paleoseismology: Integration with Seismic Hazard, Fig. 1 Standard probabilistic seismic hazard computations (Cornell 1968; McGuire 1993)

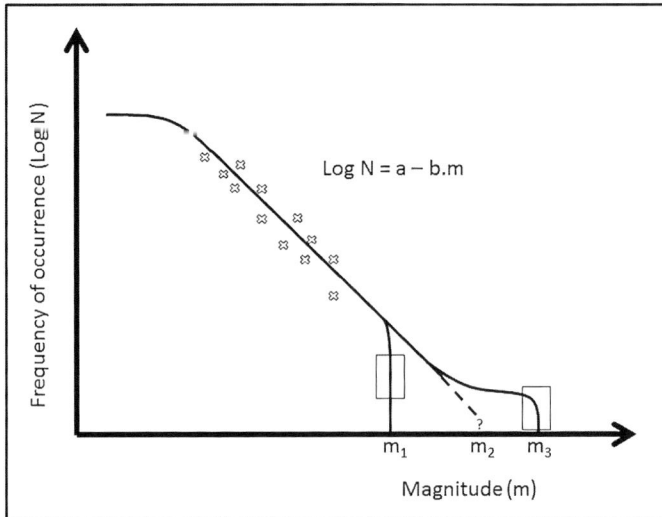

Paleoseismology: Integration with Seismic Hazard, Fig. 2 A hypothetical case of frequency of occurrence of earthquake magnitudes (Gutenberg-Richter relation). Crosses refer to the individual data points of the cumulative number of frequency in the earthquake catalog which is usually limited in incomplete earthquake catalogs. The different possibilities then exist in estimating the maximum expected magnitude (e.g., m_1, m_2, or m_3)

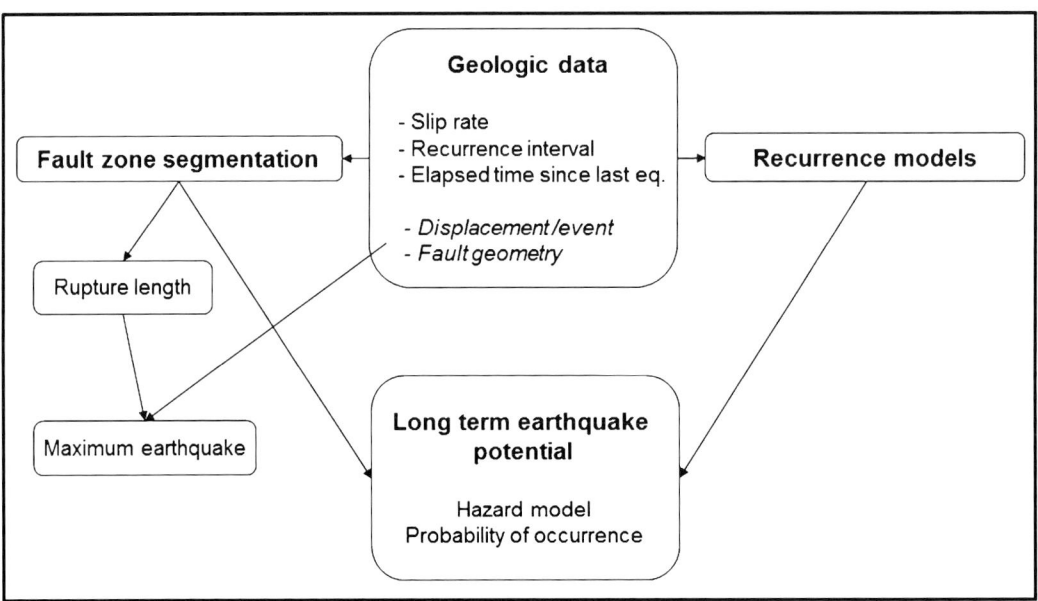

Paleoseismology: Integration with Seismic Hazard, Fig. 3 The use of geologic data in assessing the long-term earthquake potential (Redrawn from Schwartz and Coppersmith 1986)

time-dependent recurrence models on the other hand has typically been related to establishing the last occurrence of a large earthquake along a given fault and its recurrence in time. Together with slip rate this information is then used to develop recurrence models for the probabilistic computations. The maximum expected magnitude of a future earthquake is also used based on the slip per event and the fault geometry together with the study of the fault segmentation and the rupture length (Schwartz and Coppersmith 1986).

Paleoseismic Data and Deterministic Seismic Hazard Assessment

While PSHA provides a useful tool for defining earthquake design loads for noncritical

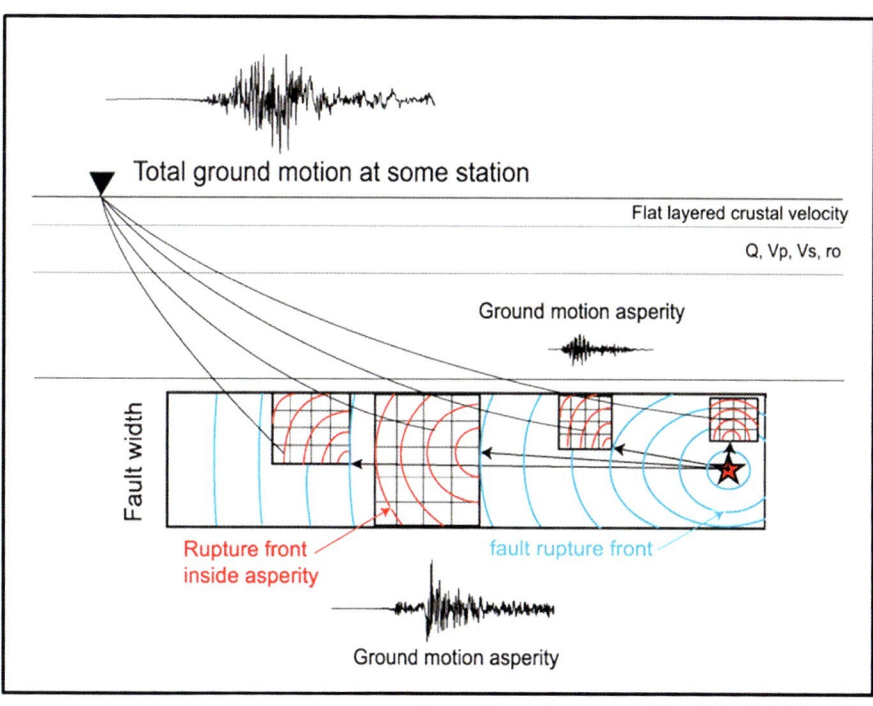

Paleoseismology: Integration with Seismic Hazard, Fig. 4 A simplified sketch showing the main principles of ground motion simulations based on earthquake scenarios including the fault rupture complexity (Courtesy of N. Pulido, 2004)

structures, especially for relatively low hazard levels (Bommer et al. 2000), in cases where the hazard is dominated by a single large earthquake from a nearby fault (e.g., Pulido et al. 2004; Sørensen et al. 2007) or if the engineering design of a critical structure requires realistic ground motions, it is preferable to perform deterministic seismic hazard assessment (DSHA). This is because in DSHA, the earthquake scenarios are defined unambiguously.

During the last decade or so, there has been a growing interest in assessing seismic hazard using ground motion simulations based on deterministic earthquake scenarios. These scenarios are usually built upon a fault rupture model where the rupture complexity is addressed either kinematically or dynamically. In kinematic models (Fig. 4) the variation of slip along the fault is modeled through predefined asperities where dynamic rupture parameters such as the rupture velocity and rise time are kept constant or varied within a given range. The obtained ground motions for each element of the sub-faults are then propagated through a crustal velocity structure either for a simple flat-layered model or more complex 3-D structures to obtain the ground motions on the surface. The simulations can be done either for a broad frequency band using full waveform modeling for the low-frequency part of the ground motion and stochastic simulations for the high-frequency part (e.g., Pulido et al. 2004; Sørensen et al. 2007) or only using stochastic simulations (Boore 2009; Gofrani et al. 2013). Other more complicated uses of dynamic rupture models are only conducted for a few cases where data to constrain the dynamic rupture parameters as well as the three-dimensional velocity structure of the wave propagation path and the local site conditions are available (e.g., Olsen et al. 1997; Olsen 2000).

The use of paleoseismological data in developing better earthquake rupture scenarios in ground motion simulations has significant potential which is not currently well exploited. Although most of the information regarding the fault geometry, its kinematics, rupture length, maximum magnitude, etc., is based on paleoseismological data, other constraints such as the slip distribution from past earthquakes along the fault, the location, and the size of the asperities could be better utilized in the computations. It is therefore important that paleoseismological studies focus not only on the standard parameters relevant for PSHA (e.g., maximum magnitude and recurrence interval) but also focus on other parameters, such as the location of the rupture initiation point and rupture propagation direction with its possible effects on the surface (e.g., Dor et al. 2008), as well as the details of the fault segmentation, asperities, and slip distribution.

Uncertainties in Paleoseismological Studies

Using paleoseismological data in seismic hazard assessment requires that the associated uncertainties are quantified and can be accounted for in the hazard computations. Although there are well-established methodologies that exist accounting for the various types of uncertainties (aleatory or epistemic) in seismic hazard analysis (e.g., Abrahamson and Bommer 2005), quantifying uncertainties in paleoseismic studies are usually only restricted to the analytical uncertainties associated with the age determinations. A few exceptions exist, systematically accounting for uncertainties associated with the paleoseismological (e.g., Atakan et al. 2000) and archeoseismological (Sintubin and Stewart 2008) studies.

Summary

Importance of paleoseismic data in seismic hazard assessment has been recognized since the late 1970s. There are well-established methodologies developed for integrating the paleoseismic data in probabilistic seismic hazard analysis, which usually involve extending the earthquake catalog time span back in the geological past. This is especially true for areas of low deformation rate with sporadic seismic activity, such as plate interiors or stable continental regions, or in areas where the historical earthquake records are incomplete and scarce. The typical application of paleoseismic data in this case is associated with establishing the maximum magnitude based on the paleoearthquakes on a given fault and the recurrence in time. In seismic hazard analysis using renewal models where conditional probabilities are calculated, the occurrence of the last earthquake along a given fault as well as the magnitude and the recurrence interval are parameters that are typically used. There is however an unexploited potential of paleoseismic data in the deterministic seismic hazard studies using ground motion simulations based on earthquake rupture scenarios. The details of the slip distribution, repeated directivity effects, location and size of the asperities, as well as the rupture initiation point can be assessed by paleoseismic investigations. The use of such data in ground motion simulations requires a systematic analysis of the uncertainties associated with the paleoseismological data, not only those related to the analytical uncertainties of the age determinations but also those that are related to the interpretation of the various indices used in identifying the paleoearthquakes and their source parameters.

Cross-References

▶ Archeoseismology
▶ Earthquake Location
▶ Earthquake Magnitude Estimation
▶ Earthquake Mechanisms and Stress Field
▶ Earthquake Mechanisms and Tectonics
▶ Earthquake Recurrence
▶ Earthquake Recurrence Law and the Weibull Distribution
▶ Earthquake Return Period and Its Incorporation into Seismic Actions
▶ Integrated Earthquake Simulation
▶ Luminescence Dating in Paleoseismology
▶ Paleoseismology of Glaciated Terrain
▶ Paleoseismic Trenching

- Paleoseismology
- Paleoseismology and Landslides
- Paleoseismology of Rocky Coasts
- Physics-Based Ground-Motion Simulation
- Probabilistic Seismic Hazard Models
- Radiocarbon Dating in Paleoseismology
- Random Process as Earthquake Motions
- Remote Sensing in Seismology: An Overview
- Seismic Actions Due to Near-Fault Ground Motion
- Seismic Risk Assessment, Cascading Effects
- Site Response for Seismic Hazard Assessment
- Spatial Variability of Ground Motion: Seismic Analysis
- Tsunamis as Paleoseismic Indicators
- Uncertainty Theories: Overview

References

Abrahamson NA, Bommer JJ (2005) Probability and uncertainty in seismic hazard analysis. Earthq Spectra 21(2):603–607. doi:10.1193/1.1899158

Atakan K, Midzi V, Toirán BM, Vanneste K, Camelbeeck T, Meghraoui M (2000) Seismic hazard in regions of present day low seismic activity: uncertainties in paleoseismic investigations in the Bree fault scarp (Roer Graben, Belgium). Soil Dyn Earthq Eng 20(5–8):415–427

Bommer JJ, Scott SG, Sarva SK (2000) Hazard-consistent earthquake scenarios. Soil Dyn Earthq Eng 19:219–231

Boore DM (2009) Comparing stochastic point-source and finite-source ground-motion simulations: SMSIM and EXSIM. Bull Seismol Soc Am 99(6):3202–3216. doi:10.1785/0120090056

Cornell CA (1968) Engineering seismic risk analysis. Bull Seismol Soc Am 58:1583–1906

Crone AJ, Omdahl EM (eds) (1987) Directions in paleoseismology. US Geological survey open file report 87-673, 456 p

Dor O, Yildirim C, Rockwell TK, Ben-Zion Y, Emre O, Sisk M, Duman TY (2008) Geological and geomorphologic asymmetry across the rupture zones of the 1943 and 1944 earthquakes on the North Anatolian Fault: possible signals for preferred earthquake propagation direction. Geophys J Int 173:483–504. doi:10.1111/j.1365-246X.2008.03709.x

Ghofrani H, Atkinson GM, Goda K, Assatourians K (2013) Stochastic finite fault simulations of the 2011 Tohoku, Japan, earthquake. Bull Seismol Soc Am 103(28):1307–1320. doi:10.1785/0120120228

Grant LB, Lettis WR (2002) Introduction to the special issue on paleoseismology of the San Andreas Fault system. Bull Seismol Soc Am 92(7):2551–2554. doi:10.1785/0120000600

Grant LB, Sieh K (1994) Paleoseismic evidence of clustered earthquakes on the San Andreas Fault in the Carrizo Plain, California. J Geophys Res Solid Earth 99(B4):6819–6841. doi:10.1029/94JB00125

Main IG (1995) Earthquakes as critical phenomena: implications for probabilistic seismic hazard analysis. Bull Seismol Soc Am 85(5):1299–1308

Mathews MV, Ellsworth WL, Reasenberg PA (2002) A Brownian model for recurrent earthquakes. Bull Seismol Soc Am 92(6):2233–2250. doi:10.1785/0120010267

McCalpin J (ed) (1996) Paleoseismology. Academic, London, 553 p. ISBN 0-12-481826-9

McGuire R (1993) Computations of seismic hazard. Anali di Geofisica XXXVI(3–4):181–200

Olsen KB (2000) Site amplification in the Los Angeles basin from three-dimensional modeling of ground motion. Bull Seismol Soc Am 90(6B):S77–S94. doi:10.1785/0120000506

Olsen KB, Madariaga R, Archuleta RJ (1997) Three-dimensional dynamic simulation of the 1992 landers earthquake. Science 278(5339):834–838. doi:10.1126/science.278.5339.834

Pulido N, Ojeda A, Atakan K, Kubo T (2004) Strong ground motion estimation in the Marmara Sea region (Turkey) based on a scenario earthquake. Tectonophysics 391:357–374

Schwartz DP, Coppersmith KJ (1986) Seismic hazards: new trends in analysis using geologic data. In: Active tectonics. National Academy Press, Washington, DC, pp 215–230

Sieh KE (1978) Prehistoric large earthquakes produced by slip on the San Andreas Fault at Pallett Creek, California. J Geophys Res Solid Earth 83(B8):3907–3939. doi:10.1029/JB083iB08p03907

Sieh KE, Jahns RH (1984) Holocene activity of the San Andreas Fault at Wallace Creek, California. Geol Soc Am Bull 95(8):883–896. doi:10.1130/0016-7606

Sieh KE, Stuiver M, Brillinger D (1989) A more precise chronology of earthquakes produced by the San Andreas Fault in southern California. J Geophys Res Solid Earth 94(B1):603–623. doi:10.1029/JB094iB01p00603

Sintubin M, Stewart I (2008) A logical methodology for archaeoseismology: a proof of concept at the archaeological site of Sagalassos, Southwest Turkey. Bull Seismol Soc Am 98(5):2209–2230. doi:10.1785/0120070178

Sørensen MB, Pulido N, Atakan K (2007) Sensitivity of ground motion simulations to earthquake source parameters: a case study for Istanbul, Turkey. Bull Seismol Soc Am 97(3):881–900. doi:10.1785/0120060044

Vita-Finzi (1986) Recent earth movements: an introduction to neotectonics. Academic, London, 226p. ISBN 0127223703

Wallace RE (1981) Active faults, paleoseismology, and earthquake hazards in the western United States. In: Simpson DW, Richards PG, Wallace RE (eds) Earthquake prediction. American Geophysical Union (AGU) Publications, Washington, DC, USA, doi:10.1029/ME004p0209

Wallace RE (1986) Active tectonics, Studies in geophysics. National Academy of Sciences, Washington, DC

Wallace RE (ed) (1990) The San Andreas Fault system, California. US Geological survey professional paper 1515

Yeats RS, Sieh KE, Allen CR (1997) The geology of earthquakes. Oxford University Press, New York, 568p. ISBN 0195078276

Parametric Nonstationary Random Vibration Modeling with SHM Applications

Luis David Avendaño-Valencia and
Spilios D. Fassois
Department of Mechanical & Aeronautical Engineering, Stochastic Mechanical Systems and Automation (SMSA) Laboratory, University of Patras, Patras, Greece

Synonyms

Fault diagnosis; Nonstationary random vibration; Signal-based modeling (identification); Structural Health Monitoring; Time-dependent ARMA modeling; Time-frequency analysis

Introduction

Nonstationary random vibration is characterized by *time-dependent* (*evolutionary*) characteristics (Priestley 1988; Roberts and Spanos 1990, Chapter 7; Bendat and Piersol 2000, Chapter 12; Newland 1993, pp. 211–219; Preumont 1994, Chapter 8; Hammond and White 1996; Kitagawa and Gersch 1996). Typical examples of nonstationary random vibration include earthquake ground motion and resulting structural vibration response, as well as the vibration of surface vehicles, flying aircraft, mechanisms, rotating machinery, cranes, bridges with passing vehicles, and so on. Nonstationary random vibration typically originates from *time-varying* dynamics or the linearization of *nonlinear* dynamics.

An example of a system exhibiting nonstationary random vibration is the mechanism of Fig. 1a. It is a 2-DOF pick-and-place mechanism consisting of two coaxially aligned linear motors carrying prismatic links (arms) connected to their ends. The mechanism is clamped on an aluminum base and is excited by a zero-mean Gaussian stationary random excitation force, applied vertically with respect to the base by means of an electromechanical shaker, while the linear motors are following predetermined trajectories. The nonstationary nature of the resulting vibration (see Fig. 1b) is due to the time-varying position of the linear motors, thus controlling the position of the links, and is evident in the time-varying power spectral density (TV-PSD) parametric estimate of Fig. 1c. Further details on this example may be found in (Spiridonakos and Fassois 2013).

From a mathematical point of view, nonstationary random vibration is characterized by time-dependent statistical moments. In the presently assumed *Gaussian* case, this means that the mean is a function of time t and the autocovariance function (ACF) a function of two time arguments t_1 and t_2. That is, for a random vibration signal $x(t)$, one has E$\{\cdot\}$ designates statistical expectation)

$$\text{Mean}: \mu(t) = \mathrm{E}\{x(t)\},$$
$$\text{ACF}: \gamma(t_1,t_2) = \mathrm{E}\{(x(t_1) - \mu(t_1)) \cdot (x(t_2) - \mu(t_2))\}$$
(1)

In many random vibration problems, the mean is constant and is thus easily estimated and removed from the signal (sample-mean adjusted signal). The case of a time-dependent mean (also referred to as deterministic trend) may be treated via proper techniques, such as curve fitting, high-pass filtering, or special parametric models (such as integrated models with a deterministic trend parameter (Box et al. 1994)).

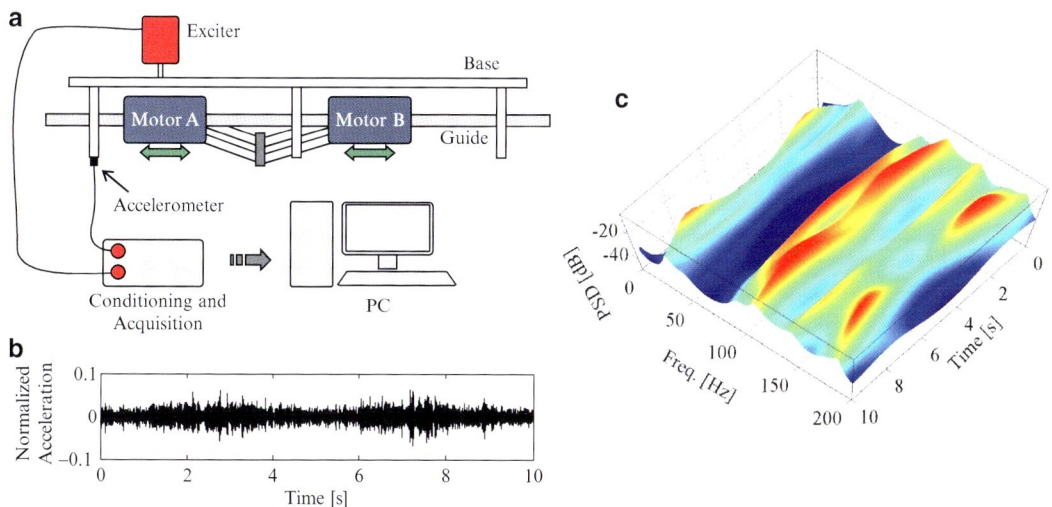

Parametric Nonstationary Random Vibration Modeling with SHM Applications, Fig. 1 Example of a laboratory pick-and-place mechanism exhibiting nonstationary random vibration: (**a**) schematic diagram of the mechanism and the laboratory setup; (**b**) measured vibration acceleration response signal (normalized); (**c**) estimate of the TV-PSD (time-frequency spectrum)

This entry focuses on the following two subjects: (i) The signal-based modeling (identification) of nonstationary random vibration based on a uniformly sampled (with sampling period T) signal realization $x[t]$, for $t = 1, 2, \ldots, N$. Absolute time is $t \cdot T$, N is the signal length in samples, while the use of brackets indicates function of an integer variable. (ii) The use of an identified random vibration model (or in fact a set of such models) for Structural Health Monitoring (SHM), where the objective is the *detection* of potential structural damage and its *characterization* (*identification*) (Fassois and Sakellariou 2009). Other important uses of an identified model – not discussed in this entry – include model-based analysis (like the extraction of time-dependent power spectral density (PSD) and time-dependent vibration modes (Newland 1993, p. 218; Preumont 1994, Chapter 8; Poulimenos and Fassois 2006)), prediction, classification, and control.

Nonstationary random vibration signal-based modeling (identification) has received significant attention in recent years. The available methods may be broadly classified as *parametric or nonparametric* (Poulimenos and Fassois 2006).

Nonparametric methods have received most of the attention and are based upon nonparameterized representations of the vibration energy as a simultaneous function of time and frequency (time-frequency representations). They include the classical spectrogram (based upon the short-time Fourier transform – STFT) and its ramifications (Newland 1993, p. 218; Hammond and White 1996; Bendat and Piersol 2000, p. 504), Mark's physical spectrum (Preumont 1994, Section 8.3), the Cohen class of distributions (Hammond and White 1996), Priestley's evolutionary spectrum (Priestley 1988, Section 6.3; Preumont 1994, Section 8.4), as well as wavelet-based methods (Newland 1993, Chapter 17). On the other hand, *parametric methods* are based upon parameterized representations, usually of the *time-dependent autoregressive moving average (TARMA)* type. These have an apparently similar form to their conventional (stationary) counterparts, but are characterized by *time-dependent* parameters and innovations variance. Thus, a $TARMA(n_a, n_c)$ model, with n_a and n_c designating its AR and MA orders, respectively, is defined as follows (Poulimenos and Fassois 2006):

TARMA(n_a, n_c) model :

$$x[t] = -\sum_{i=1}^{n_a} a_i[t]x[t-i] + \sum_{i=1}^{n_c} c_i[t]w[t-i] + w[t];$$
$$w[t] \sim \text{NID}\left(0, \sigma_W^2[t]\right) \quad (2)$$

where $x[t]$ represents the nonstationary random signal model; $w[t]$ an unobservable *normally and independently distributed* (NID) *(thus white)* nonstationary *innovations* sequence with zero-mean and time-dependent variance $\sigma_W^2[t]$; n_a, n_c the autoregressive (AR) and moving average (MA) orders, respectively; and $a_i[t]$ and $c_i[t]$ the corresponding AR and MA *time-dependent parameters*.

Parametric TARMA models are of three main families, according to the form of "structure" imposed upon the evolution of the time-dependent parameters and innovations variance (Poulimenos and Fassois 2006):

(a) *Unstructured parameter evolution (UPE) models*, in which no particular "structure" is imposed on the parameter evolution. Prime models in this family include short-time ARMA (ST-ARMA) and recursive models (such as recursive ARMA or in short RARMA) models.
(b) *Stochastic parameter evolution (SPE) models*, in which *stochastic* "structure" is imposed on the parameter evolution via stochastic smoothness constraints. Prime models in this family include smoothness-priors ARMA (SP-ARMA) models.
(c) *Deterministic parameter evolution (DPE) models*, in which deterministic "structure" is imposed on the parameter evolution. Prime models in this family include the so-called functional series TARMA (FS-TARMA) models in which the parameters are projected on properly selected functional subspaces.

Parametric models, and their respective signal-based (identification) methods, are known to be characterized by a number of important advantages, such as representation parsimony, improved accuracy and resolution, improved tracking of the TV dynamics, flexibility in analysis, synthesis (simulation), prediction, diagnosis, and control. For instance, once a TARMA model is available, the corresponding "frozen"-type TV-PSD may be readily obtained as

"Frozen" TV-PSD :

$$S_F(\omega, t) = \left|\frac{1 + \sum_{i=1}^{n_c} c_i[t] \cdot e^{-j\omega T_{si}}}{1 + \sum_{i=1}^{n_c} a_i[t] \cdot e^{-j\omega T_{si}}}\right|^2 \cdot \sigma_w^2[t] \quad (3)$$

with ω representing frequency in *rad/s*, j the imaginary unit, and $|\cdot|$ complex magnitude. Notice that this would be the PSD of the vibration signal if the system were "frozen" (made stationary) at each time instant t. This entry focuses on parametric models and methods and in particular on the SPE and DPE families.

The problem of Structural Health Monitoring (SHM) for structures exhibiting nonstationary random vibration responses may be treated in a statistical time series (STS) framework (Fassois and Sakellariou 2009), using either nonparametric or parametric models and statistical decision-making schemes. In the majority of applications thus far, nonparametric models (nonparametric time-frequency-type representations) are used, and damage detection and identification are based on potential discrepancies observed between those obtained in a *baseline* (healthy) phase and an *inspection* (current) phase (Feng et al. 2013). Although the use of parametric models and methods may potentially lead to performance improvements, it has thus far received limited attention (Poulimenos and Fassois 2004; Spiridonakos and Fassois 2013).

Brief Historical Notes. Among parametric models, the unstructured parameter evolution (UPE) family of methods was initially developed. Prime methods in this area include the short-time ARMA (ST-ARMA) method (Niedzwieki 2000, pp. 79–82; Owen et al. 2001) and the class of recursive (or adaptive) methods (Ljung 1999, Chapter 11; Niedzwieki 2000, Chapters 4 and 5).

The stochastic parameter evolution (SPE) family of methods was developed primarily for the modeling of earthquake ground motion signals (Kitagawa and Gersch 1996; Gersch and Akaike 1988). The deterministic parameter evolution (DPE) family was introduced in a broader context in Rao (1970) and later in Kozin (1977). In Kozin (1988) and Fouskitakis and Fassois (2002) it was employed for earthquake ground motion modeling. In Poulimenos and Fassois (2009b) it was applied to the vibration of a bridge-like structure with a moving mass. In a broader context the reader is also referred to (Niedzwieki 2000, Chapter 4).

Article Roadmap. This entry is organized as follows: SPE and DPE nonstationary random vibration modeling is discussed in section "Parametric TARMA Modeling of Nonstationary Random Vibration," where specific model forms and identification schemes are briefly reviewed. Structural Health Monitoring (SHM) based on nonstationary random vibration parametric modeling is presented in section "Structural Health Monitoring (SHM) Based on Nonstationary Random Vibration." The application of these concepts to random vibration modeling and SHM for the pick-and-place mechanism of Fig. 1 is outlined in section "Illustrative Example: Nonstationary Random Vibration Modeling and SHM for a Pick-and-Place Mechanism," while a summary is provided in section "Summary."

Parametric TARMA Modeling of Nonstationary Random Vibration

Stochastic Parameter Evolution (SPE) TARMA Modeling

In the context of *stochastic parameter evolution (SPE) TARMA* models, the parameters are assumed to follow *stochastic smoothness constraints* in the form of linear integrated autoregressive (IAR) models with integration order q (in this context referred to as *smoothness-priors order*). A *smoothness-priors TARMA (SP-TARMA)* model thus has parameters that obey the relations (Kitagawa and Gersch 1996)

$$(1-B)^q \cdot a_i[t] = v_{ai}[t], \quad v_{ai}[t] \sim \text{NID}\left(0, \sigma_v^2\right) \tag{4a}$$

$$(1-B)^q \cdot c_i[t] = v_{ci}[t], \quad v_{ci}[t] \sim \text{NID}\left(0, \sigma_v^2\right) \tag{4b}$$

where the signal innovations $w[t]$ (see Eq. 2) and the parameter innovations $v_{ai}[t]$ and $v_{ci}[t]$ are all mutually independent and normally and identically distributed (NID) sequences, each being zero mean and with variance $\sigma_w^2[t]$ and σ_v^2, respectively. These smoothness constraints are characterized by unit roots that represent integrated stochastic models describing homogeneously nonstationary evolutions (Box et al. 1994, Chapter 4).

Generalization of SP-TARMA models, in the form of *generalized stochastic constraint time-dependent autoregressive moving average TARMA (GSC-TARMA)* models, was recently introduced (Avendaño-Valencia and Fassois 2013). In this, the model parameters are allowed to follow more general autoregressive (AR) models of the forms

$$a_i[t] = -\sum_{k=1}^{q} \mu_k \cdot a_i[t-k] + v_{ai}[t], \tag{5a}$$
$$v_{ai}[t] \sim \text{NID}\left(0, \sigma_v^2\right)$$

$$c_i[t] = -\sum_{k=1}^{q} \mu_k \cdot c_i[t-k] + v_{ci}[t], \tag{5b}$$
$$v_{ci}[t] \sim \text{NID}\left(0, \sigma_v^2\right),$$

where, as in the SP-TARMA case, $w[t]$, $v_{ai}[t]$, and $v_{ci}[t]$ are mutually independent and normally and identically distributed (NID) innovation sequences. The coefficients μ_k are referred to as the *stochastic constraint parameters*. These are collected in the vector $\boldsymbol{\mu} = [\mu_1 \cdots \mu_q]^T$, which is along with the covariance $\boldsymbol{\Sigma}_v = \sigma_v^2 \cdot \boldsymbol{I}_{n_a+n_c}$ (where $\boldsymbol{I}_{n_a+n_c}$ is the identity matrix with the indicated dimensions), and the innovations variance $\sigma_w^2[t]$ defines the model *hyperparameters*. Then the time-dependent AR/MA parameters $\boldsymbol{\theta}[t]$, hyperparameters $\mathcal{P}$, and structural parameters $\mathcal{M}$ of a GSC-TARMA model are

$$\boldsymbol{\theta}[t] = [a_1[t] \cdots a_{n_a}[t] \vdots c_1[t] \cdots c_{n_c}[t]]^T,$$
$$\mathcal{P} = \{\boldsymbol{\mu}, \sigma_w^2[t], \boldsymbol{\Sigma}_v\}, \quad \mathcal{M} = \{n_a, n_c, q\}$$
(6)

It should be noted that the stochastic constraint parameters and parameter innovations variances may be – more generally – different for each AR/MA parameter, but the above simple form is adopted here for purposes of presentation simplicity. Comparing the GSC-TARMA and SP-TARMA model forms, one sees that in the latter case the stochastic constraint parameters are essentially prefixed, limiting the types of trajectories that each AR/MA parameter would be capable of following.

The identification of an SP-TARMA or GSC-TARMA model consists of the selection of the model structure $\mathcal{M}$ and the estimation of the time-dependent parameter vector $\boldsymbol{\theta}[t]$ and hyperparameters $\mathcal{P}$. The estimation of the model parameters and hyperparameters may be posed as the maximization of the *joint a posteriori probability* density function of $\boldsymbol{\theta}_1^N = \{\boldsymbol{\theta}[1], \ldots, \boldsymbol{\theta}[N]\}$ given the available observations $x_1^N = \{x[1], \ldots, x[N]\}$, namely, $p(\boldsymbol{\theta}_1^N, \mathcal{P}|x_1^N)$. These, combined with the Gaussianity assumption for $w[t]$ and $\boldsymbol{v}[t] = [v_{a_1}[t] \, v_{a_2}[t] \cdots v_{c_{n_c}}[t]]^T$, lead to the following cost function:

$$\mathcal{J}(\boldsymbol{\theta}_1^N, \mathcal{P}) = \frac{N \cdot q \cdot (n_a + n_c)}{2} \ln|\sigma_v^2|$$
$$+ \frac{1}{2} \sum_{t=1}^N \left(\ln \sigma_w^2[t] + \frac{w^2[t]}{\sigma_w^2[t]} + \frac{\boldsymbol{v}^T[t] \cdot \boldsymbol{v}[t]}{\sigma_v^2} \right)$$
(7)

which must be minimized in order to provide the optimal *maximum a posteriori* (MAP) estimate of $\boldsymbol{\theta}[t]$ and $\mathcal{P}$.

An estimate of $\boldsymbol{\theta}[t]$ based on fixed values of $\mathcal{P}$ may be obtained recursively using the Kalman filter (or a proper nonlinear approximation filter in the full TARMA case) based on the following state-space representation of the SP/GSC-TARMA model (Poulimenos and Fassois 2006):

$$\boldsymbol{z}[t] = \boldsymbol{F}(\boldsymbol{\mu}) \cdot \boldsymbol{z}[t-1] + \boldsymbol{G} \cdot \boldsymbol{v}[t] \quad (8a)$$
$$x[t] = \boldsymbol{h}^T[t] \cdot \boldsymbol{z}[t] + w[t] \quad (8b)$$

with:

$$\boldsymbol{z}[t-1] = \begin{bmatrix} \boldsymbol{\theta}[t-1] \\ \boldsymbol{\theta}[t-2] \\ \vdots \\ \boldsymbol{\theta}[t-q] \end{bmatrix}, \boldsymbol{F}(\boldsymbol{\mu}) = \begin{bmatrix} \mu_1 & \mu_2 & \cdots & \mu_{q-1} & \mu_q \\ 1 & 0 & \cdots & 0 & 0 \\ 0 & 1 & \cdots & 0 & 0 \\ \vdots & \vdots & \ddots & \vdots & \vdots \\ 0 & 0 & \cdots & 1 & 0 \end{bmatrix} \otimes \boldsymbol{I}_{n_a+n_c},$$

$$\boldsymbol{G} = \begin{bmatrix} 1 \\ 0 \\ \vdots \\ 0 \end{bmatrix} \otimes \boldsymbol{I}_{n_a+n_c}, \boldsymbol{h}[t] = \begin{bmatrix} \boldsymbol{x}[t-1] \\ \boldsymbol{w}[t-1] \\ 0 \\ \vdots \\ 0 \end{bmatrix}$$

where $\boldsymbol{x}[t-1] = [x[t-1] \, x[t-2] \cdots x[t-n_a]]^T$ and $\boldsymbol{w}[t-1] = [w[t-1] \, w[t-2] \cdots w[t-n_c]]^T$. Notice that in estimation and in the full TARMA case, the innovations $w[t]$ may be replaced by their respective a posteriori estimates $\hat{w}[t] = x[t] - \boldsymbol{h}^T[t] \cdot \hat{\boldsymbol{z}}[t|t]$, where $\hat{\boldsymbol{z}}[t|t]$ is the a posteriori state estimate (Niedzwieki 2000, p. 263). After initial estimation with the Kalman filter, refined parameter estimates may be obtained by using the Kalman smoother (Poulimenos and Fassois 2006).

The estimation of an SP-TARMA model is performed by computing $\boldsymbol{\theta}_1^N$ as described above, using fixed values of σ_w^2 and σ_v^2. Since the values of these variances are generally unavailable, a *normalized* form of the Kalman filter may be used, where the prediction/update equations are divided by σ_w^2. In this way, a single parameter $\lambda = \sigma_v^2/\sigma_w^2$ is left as a design (user selected) parameter that adjusts the "tracking speed" versus "smoothness of the estimates" in the algorithm (Poulimenos and Fassois 2006).

In the GSC-TARMA case, an expectation-maximization scheme can be used as follows (Avendaño-Valencia and Fassois 2013): (i) *expectation step*: obtain $\boldsymbol{\theta}[t]$ using the Kalman filter as previously described, based on fixed values of $\mathcal{P}$; (ii) *maximization step*: estimate $\boldsymbol{\mu}$ and $\sigma_w^2[t]$ based on the estimated values of $\boldsymbol{\theta}[t]$. The E and M steps are sequentially repeated until convergence of the cost function $\mathcal{J}(\boldsymbol{\theta}_1^N, \mathcal{P})$ is achieved. The estimation of σ_v^2 is avoided within the M step, as it destabilizes the algorithm. Thus, as in the SP-TARMA case, σ_v^2 is left as a design parameter to adjust the tracking speed and the parameter smoothness.

For both SP and GSC-TARMA models, the selection of the value of σ_v^2 may not be straightforward, since a very low value may over-smooth the estimated parameter trajectories, while a high value may lead to "noisy" trajectories. The selection of σ_v^2 may be guided by comparing the innovations (prediction residuals) with the parameter innovations (prediction error of the parameters) – that is, the *residual sum of squares* (RSS) to the *parameter prediction error sum of squares* (PESS):

$$RSS = \sum_{t=1}^{N} \hat{w}^2[t], \quad PESS = \sum_{t=1}^{N} \hat{\boldsymbol{v}}^T[t] \cdot \hat{\boldsymbol{v}}[t] \quad (9)$$

where $\hat{w}[t] = x[t] - \boldsymbol{h}^T[t] \cdot \hat{\boldsymbol{z}}[t|t-1]$ stands for the one-step-ahead prediction error (residual) at time t and $\hat{\boldsymbol{v}}[t] = \hat{\boldsymbol{\theta}}[t|t] - \hat{\boldsymbol{\theta}}[t|t-1]$ is an estimate of the parameter innovations with $\hat{\boldsymbol{\theta}}[t|t-1]$ being the a priori and $\hat{\boldsymbol{\theta}}[t|t]$ the *a posteriori* Kalman filter estimates of $\boldsymbol{\theta}[t]$. Both RSS and PESS are computed from the Kalman filter predictions of $w[t]$ and $v[t]$ obtained for a specific value of σ_v^2. A high RSS indicates poor modeling accuracy, whereas high PESS indicates noisy parameter estimates. A curve displaying the PESS versus the RSS (parameterized in terms of σ_v^2) may be constructed and used for selecting a good compromise (and hence a proper σ_v^2).

Remarks: (i) In the definition of the GSC-TARMA model, it is also possible to include a stochastically time-dependent innovations variance, as in Kitagawa and Gersch (1996). (ii) Additional definitions are possible for the general SPE-TARMA model class. For instance non-Gaussian signals or nonlinear stochastic parameter evolution dynamics may be included, which may be appropriate for some rapidly evolving processes (Kitagawa and Gersch 1996).

Deterministic Parameter Evolution (DPE) TARMA Modeling

DPE-TARMA models are typically defined in terms of the *functional series TARMA (FS-TARMA)* model form, for which the temporal evolution of the parameters is expressed via projections in proper functional subspaces. Thus, for an *FS-TARMA* $(n_a, n_c)_{[p_a, p_c, p_s]}$ model, with p_a, p_c, p_s designating its AR, MA, and innovations variance functional subspace dimensionalities, the evolution of the parameters is as follows (Poulimenos and Fassois 2006):

$$a_i[t] = \sum_{k=1}^{p_a} a_{i,k} \cdot G_{b_{a(k)}}[t], \quad c_i[t] = \sum_{k=1}^{p_c} c_{i,k} \cdot G_{b_{c(k)}}[t],$$

$$\sigma_w^2[t] = \sum_{k=1}^{p_s} s_k \cdot G_{b_{s(k)}}[t]$$

(10a)

$$\mathcal{F}_{AR} = \left\{ G_{b_{a(1)}}[t], \ldots, G_{b_{a(pa)}}[t] \right\},$$
$$\mathcal{F}_{MA} = \left\{ G_{b_{c(1)}}[t], \ldots, G_{b_{c(pc)}}[t] \right\}, \quad (10b)$$
$$\mathcal{F}_{\sigma_w^2[t]} = \left\{ G_{b_{s(1)}}[t], \ldots, G_{b_{s(ps)}}[t] \right\}$$

with "$\mathcal{F}$" designating the functional subspace of the indicated quantity; $b_{a(k)}$ ($k = 1 \ldots, p_a$), $b_{c(k)}$

($k = 1 \ldots, p_c$), and $b_{s(k)}$ ($k = 1 \ldots, p_s$) indices indicating the specific basis functions included in each subspace; and $a_{i,k}$, $c_{i,k}$ and s_k the AR, MA, and innovations variance, respectively, coefficients of projection. Thus, for an FS-TARMA model the model parameter vector is $\boldsymbol{\vartheta} = [\boldsymbol{\vartheta}_a^T \, \boldsymbol{\vartheta}_c^T \, \boldsymbol{\vartheta}_s^T]^T$, while model structure is specified by the AR and MA orders n_a, n_c and the AR, MA, and innovations variance basis function index vectors $\boldsymbol{b}_a = [b_{a(1)} \ldots b_{a(pa)}]^T$, $\boldsymbol{b}_c = [b_{c(1)} \ldots b_{c(pc)}]^T$, and $\boldsymbol{b}_s = [b_{s(1)} \ldots b_{s(ps)}]^T$, that is,

$$\boldsymbol{\vartheta} = [\boldsymbol{\vartheta}_a^T \, \boldsymbol{\vartheta}_c^T \, \boldsymbol{\vartheta}_s^T]^T$$
$$= [a_{1,1} \ldots a_{n_a, p_c} | c_{1,1} \ldots c_{n_c, p_c} | s_1 \ldots s_{p_s}]^T \quad (11a)$$

$$\mathcal{M} = \{n_a, n_c, \boldsymbol{b}_a, \boldsymbol{b}_c, \boldsymbol{b}_s\} \quad (11b)$$

In a recent extension, adaptable FS-TARMA (AFS-TARMA) models that employ functional subspaces parameterized by a parameter vector $\boldsymbol{\delta}$ were introduced (Spiridonakos and Fassois 2014a). The model definition is as in Eq. 10, but the functional bases are parameterized according to the following forms: $\mathcal{F}_{AR} = \left\{ G_{b_{a(1)}}[t, \boldsymbol{\delta}_a], \ldots, G_{b_{a(p_a)}}[t, \boldsymbol{\delta}_a] \right\}$, $\mathcal{F}_{MA} = \left\{ G_{b_{c(1)}}[t, \boldsymbol{\delta}_c], \ldots, G_{b_{c(p_c)}}[t, \boldsymbol{\delta}_c] \right\}$, and $\mathcal{F}_{\sigma_w^2[t]} = \left\{ G_{b_{s(1)}}[t, \boldsymbol{\delta}_s], \ldots, G_{b_{s(p_s)}}[t, \boldsymbol{\delta}_s] \right\}$, where $\boldsymbol{\delta}_a$, $\boldsymbol{\delta}_c$, and $\boldsymbol{\delta}_s$ indicate the AR, MA, and innovations variance functional subspace parameter vector, respectively. The model structure is in this case defined by just the model orders and functional subspace dimensionalities, while the complete parameter vector includes the functional subspace parameters as well:

$$\boldsymbol{\theta} = \begin{bmatrix} \boldsymbol{\vartheta}^T & \boldsymbol{\delta}^T \end{bmatrix}^T, \quad \boldsymbol{\delta} = \begin{bmatrix} \boldsymbol{\delta}_a^T & \boldsymbol{\delta}_c^T & \boldsymbol{\delta}_s^T \end{bmatrix}^T,$$
$$\mathcal{M} = \{n_a, n_c, p_a, p_c, p_s\} \quad (12)$$

The advantage of the AFS-TARMA model structure is that the selection of the basis functions, which is a structural problem in conventional FS-TARMA models, becomes part of the parameter estimation problem and is thus significantly simplified. The adaptable models may thus better "adapt" to a given nonstationary signal, while the modeling procedure is easier for the user.

The estimation of AFS/FS-TARMA models is typically accomplished within a *maximum likelihood* (ML) framework. In the FS-TARMA case, and under Gaussian innovations, the log-likelihood function is (Spiridonakos and Fassois 2014b)

$$\ln \mathcal{L}(\boldsymbol{\vartheta}|x_1^N) = -\frac{N}{2} \cdot \ln 2\pi$$
$$-\frac{1}{2} \sum_{t=1}^N \left(\ln \sigma_w^2[t] + \frac{w^2[t]}{\sigma_w^2[t]} \right) \quad (13)$$

As the likelihood function in Eq. 13 is non-quadratic in terms of the unknown parameter vector, the maximization is based on iterative schemes and rather accurate initial parameter estimates are required for avoiding potential local extrema.

In the simpler case of FS-TAR models, initial parameter values may be obtained by estimating $\boldsymbol{\vartheta}_a$ via *ordinary least squares* (OLS) and subsequently estimating $\boldsymbol{\vartheta}_s$ via an overdetermined set of equations after estimating $\sigma_w^2[t]$ via a sliding window approach (Poulimenos and Fassois 2006). In the FS-TARMA case more elaborate techniques are necessary, as the prediction error is a nonlinear function of the projection coefficient vector $\boldsymbol{\vartheta}_c$. These include *linear multistage* or *recursive* methods (Poulimenos and Fassois 2006). Linear multistage methods first obtain an initial *estimate* of the prediction error sequence based on high-order TAR models and subsequently employ the obtained values to estimate the FS-TARMA model coefficients of projection (the two-stage least-squares (2SLS) approach). Recursive methods use the recursive extended least squares (RELS) or the recursive maximum likelihood (RML) algorithms to obtain initial coefficient of projection estimates (Poulimenos and Fassois 2006). Several runs over the data are typically recommended to avoid the influence of the unknown initial conditions and ensure convergence of the algorithm. For adaptable

(AFS-TARMA) models, an estimation scheme based on separable nonlinear least squares (SNLS), in which the vectors ϑ and δ are estimated separately in an sequential fashion, has been suggested (Spiridonakos and Fassois 2014a, b). The reader is referred to (Poulimenos and Fassois 2006; Spiridonakos and Fassois 2014b) for further details on AFS/FS-TARMA models and their estimation.

Remarks: (i) For conventional FS-TARMA models, the functional subspaces include linearly independent basis functions selected from an ordered set, such as Chebyshev, trigonometric, b-splines, wavelets, and other functions. For simplicity a functional subspace is often selected to include consecutive basis functions up to a maximum index. Yet, for purposes of model parsimony (economy) and effective estimation, some functions may not be necessary and may be dropped. (ii) An FS-TARMA $(n_a, n_c)_{[p_a, p_c, p_s]}$ model of the form (10) is referred to as a *fully parametric* FS-TARMA model. The term *semi-parametric* FS-TARMA $(n_a, n_c)_{[p_a, p_c]}$ model implies that the innovations variance is not parameterized, that is, it is not projected on a functional subspace.

Model Structure Selection

Model structure selection is the process by which the structural parameters of the model are obtained. This is typically an iterative and tedious procedure, in which models corresponding to various candidate structures are first estimated, and the one providing the *best fitness* is selected. This procedure may be facilitated via integer optimization schemes or backward/forward regression schemes (Spiridonakos and Fassois 2014b). Model fitness may be judged in terms of a number of criteria, which may include the residual sum of squares (RSS) (often normalized by the series sum of squares, SSS), the likelihood function, and the Akaike information criterion (AIC) or the Bayesian information criterion (BIC). The latter two are typically preferred as they maintain a balance between model fit (model accuracy) and model size (thus discouraging overfitting). The RSS/SSS and BIC criteria are defined as

$$RSS/SSS = \sum_{t=1}^{N} \hat{w}^2[t] / \sum_{t=1}^{N} x^2[t],$$

$$BIC = -\ln \mathcal{L}(\cdot) + \frac{\ln N}{2} \cdot d \quad (14)$$

where $\hat{w}[t]$ is the obtained one-step-ahead prediction error at time t, $\mathcal{L}(\cdot)$ is the likelihood of the respective model family, and d is the number of estimated parameters in the model (FS-TARMA, $d = \dim \vartheta$; AFS-TARMA, $d = \dim \theta$; SPE-TARMA, $d = \dim z[t] = (n_a + n_c) \cdot q$) (Poulimenos and Fassois 2006; Kitagawa and Gersch 1996, Chapter 2). Notice that in the DPE-TARMA case, the likelihood is defined by Eq. 13, while in the SPE-TARMA case the log-likelihood function (w.r.t. the hyperparameters) is equivalent to $\mathcal{J}\left(\hat{\theta}_1^N, \hat{\mathcal{P}}\right)$ (Avendaño-Valencia and Fassois 2013).

It should be noted that model estimation always requires attention on part of the user in order to detect numerical problems (for instance, those due to inverting an ill-conditioned matrix) or estimating a number of parameters that is not commensurate with the signal length. As a rough guide, the number of signal samples per estimated parameter (SPP) should be at least 15 (Spiridonakos and Fassois 2014b). Finally formal *model validation* – which examines the validity of the model assumptions (such as innovations whiteness and Gaussianity) – should be performed before final model acceptance (Box et al. 1994, Chapter 8; Poulimenos and Fassois 2006; Spiridonakos and Fassois 2014b).

Structural Health Monitoring (SHM) Based on Nonstationary Random Vibration

Let s_v designate a given structure in one of several potential health states. $v = o$ designates the healthy state, while any other v from the set $V = \{a, b, \ldots\}$ designates the structure in a damaged (faulty) state of a distinct *type a,b*, … and so forth (for instance, damage in a particular region, or of a particular nature).

In general, each damage type may include a continuum of damages, each being characterized by its own damage *magnitude*.

The SHM problem may be then posed as follows: Given the structure in a currently *unknown* state u, first determine whether or not the structure is damaged ($u = o$ or $u \neq o$) (the *damage detection subproblem*). In case the structure is found to be damaged, determine which one of $\{a,b,\ldots\}$ is the current damage type (the *damage identification subproblem*). The *damage magnitude estimation subproblem*, which focuses on estimating the magnitude of the current damage, will not be treated in this entry.

When the main information available for solving this problem is in the form of measured structural vibration response, then the problem is classified as a *vibration-based* SHM problem. If additional information – such as an analytical structural dynamics model – is used for its solution, then the method is classified as *analytical model based*, otherwise as a *data based*. This entry focuses in the latter case where analytical models are not available or are hard to obtain – the reader is referred to Fassois and Sakellariou (2009) and Farrar and Worden (2013) for a broader overview and details. In the context of this entry, the important – but obviously more difficult and scarcely studied – case of *nonstationary* random vibration is considered.

From an operational viewpoint, vibration-based SHM is organized into two distinct phases: First, an initial *baseline phase* in which a set of vibration response signals $x_v[t]$ ($t = 1,\ldots,N$), possibly for all $v \in \{o,a,b,\ldots\}$, are obtained and properly processed (this phase is carried out only once). Second, an *inspection phase* in which (typically) a single vibration response signal $x_u[t]$ is obtained and decisions on the presence and type of damage need to be made (the aforementioned damage detection and identification subproblems). This phase is typically carried out continuously or periodically, each time using a fresh vibration signal.

In this entry two SHM approaches are presented within a nonstationary random vibration context: *a parameter-based* one and a *residual-based* one. Both are based on the modeling of the nonstationary random vibration signals via the earlier discussed FS-TARMA representations. For purposes of simplicity, the damage identification subproblem is treated via successive binary hypothesis testing (instead of a single multiple hypothesis testing). In this (former) context, once the presence of damage is detected, its type is determined via successive (pairwise) comparisons with each potential damage type. An implicit assumption behind both approaches is that the operating and measurement conditions in the baseline and inspection phases are *identical*. Hence, the various signals correspond to each other in a proper way – in particular in their time duration they describe the exact same motion or operational cycle.

A Parameter-Based Approach

The essence of this approach is on the use of the projection coefficient vector $\boldsymbol{\vartheta}$ of an FS-TARMA model of the nonstationary random vibration as the *characteristic quantity* (or *feature*) in the decision-making mechanism. The underlying thesis is that each distinct health state is characterized by its own projection coefficient vector, thus "comparing" that of the current state u to that of the healthy state o leads to damage detection. In the positive (damage) case, "comparing" that of the current state u sequentially (pairwise) to that of each damage state $\{a,b,\ldots\}$ leads to damage identification.

Hence, in the (initial) baseline phase FS-TARMA models corresponding to the healthy o and each damage state $\{a,b,\ldots\}$ are estimated. Then, in the (current) inspection phase, an (identical in structure) FS-TARMA model corresponding to the current structural state u is estimated based on the currently available vibration signal. Then damage detection may be treated via a hypothesis test of the form

$$H_0 : \boldsymbol{\vartheta}_o - \boldsymbol{\vartheta}_u = \mathbf{0} \quad \text{Null hypothesis – healthy state}$$
$$H_1 : \boldsymbol{\vartheta}_o - \boldsymbol{\vartheta}_u \neq \mathbf{0} \quad \text{Alternative hypothesis – damaged state}$$

As the true coefficient of projection vectors are not available, decision making is based on corresponding estimates $\hat{\boldsymbol{\vartheta}}_o$ and $\hat{\boldsymbol{\vartheta}}_u$, and since these are random quantities, on their distributions. In this context, under mild assumptions, the estimators are shown to be asymptotically (for "long" data records, i.e., $N \to \infty$) Gaussian distributed, with mean equal to the true coefficient of projection vector and covariance Σ that may be estimated (Poulimenos and Fassois 2009a; Spiridonakos and Fassois 2013). Under these conditions, the quantity d_M^2, below, follows (assuming negligible variability for the covariance estimator) chi-square distribution with d degrees of freedom ($d = \dim \boldsymbol{\vartheta}$) Hence, decision making may be made as follows at the α risk level (i.e., false alarm probability equal to α):

$$d_M^2 = \left(\hat{\boldsymbol{\vartheta}}_u - \hat{\boldsymbol{\vartheta}}_o\right)^T \cdot \hat{\boldsymbol{\Sigma}}_o^{-1} \cdot \left(\hat{\boldsymbol{\vartheta}}_u - \hat{\boldsymbol{\vartheta}}_o\right) \leq \chi_d^2(1-\alpha) \Rightarrow \text{Accept } H_0 \qquad (15)$$
$$\text{otherwise} \Rightarrow \text{Accept } H_1$$

with $\chi_d^2(1-\alpha)$ designating the $1-\alpha$ critical point of the chi-square distribution (d degrees of freedom) and $\hat{\boldsymbol{\Sigma}}_o$ the estimator covariance in the healthy case. As already indicated, damage identification may be treated via similar pairwise tests in which the current true coefficient of projection vector $\boldsymbol{\vartheta}_u$ is compared to the true vector $\boldsymbol{\vartheta}_v$ corresponding to each damage state (for $v \in \{a, b, \ldots\}$). Of course, ramifications of the method are possible, for instance, by only including specific elements of the coefficient of projection vector, or a properly transformed (for instance, via principal component analysis) version. An obvious disadvantage of the general parameter-based approach is that model estimation needs to be carried out in the inspection phase as well.

A Residual-Based Approach

The essence of this approach is on the use of the FS-TARMA model residual signal $\hat{w}[t]$, and more specifically its time-dependent variance $\sigma_w^2[t]$, as the *characteristic quantity* (or *feature*) in the decision-making mechanism. The underlying thesis is that for each distinct health state (for instance, the healthy state), the model residual signal (then $\hat{w}_o[t]$) is characterized by its own residual variance (then $\sigma_{w_o}^2[t]$) – this may be obtained in the baseline phase. Then, under the hypothesis that the structure is still in the same health state during inspection, the new residual signal ($\hat{w}_u[t]$) obtained by driving the *current* (fresh) vibration signal ($x_u[t]$) through the *same baseline model* (no model reestimation involved) should be characterized by the same time-dependent variance (then $\sigma_{w_o}^2[t]$) *if and only if* the hypothesis of the structure being in the same health state (for instance, healthy state) is correct (as a change would result in increased variance). Then, a decision on the health state of the structure may be made based on comparing the current time-dependent residual variance ($\sigma_{w_u}^2[t]$) to the baseline time-dependent variance (then $\sigma_{w_o}^2[t]$) at each time instant. Of course, this procedure may be repeated for any other health state of the structure in the baseline phase for binary *damage identification*.

Then damage detection may be treated via a hypothesis test of the form (Poulimenos and Fassois 2004):

$$H_0 : \sigma_{w_u}^2[t] \leq \sigma_{w_o}^2[t] \qquad \text{Null hypothesis – healthy state}$$
$$H_1 : \sigma_{w_u}^2[t] > \sigma_{w_o}^2[t] \qquad \text{Alternative hypothesis – damaged state}$$

As the theoretical variances $\sigma_{w_u}^2[t]$ and $\sigma_{w_o}^2[t]$ are unavailable, they need to be estimated from the obtained residual series $\hat{w}_u[t]$ and $\hat{w}_o[t]$ using a moving average filter (sliding window) as follows:

$$\hat{\sigma}_{w_u}^2[t] = \frac{1}{\ell} \sum_{\tau=t-(\ell-1)/2}^{t+(\ell-1)/2} \hat{w}_u^2[\tau],$$
$$\hat{\sigma}_{w_o}^2[t] = \frac{1}{\ell_o} \sum_{\tau=t-(\ell_o-1)/2}^{t+(\ell_o-1)/2} \hat{w}_o^2[\tau] \quad (16)$$

with ℓ, ℓ_o designating the corresponding window lengths. Under the null hypothesis (H_0), given the residual normality and uncorrelatedness, the statistic defined as the ratio of the two variance estimators follows an F distribution with($\ell - 1$, $\ell_o - 1$) degrees of freedom, that is,

$$F[t] = \frac{\hat{\sigma}_{w_u}^2[t]}{\hat{\sigma}_{w_o}^2[t]} \sim F_{\ell-1,\ell_o-1} \quad (17)$$

This leads to the following sequential F-test (at the α risk level, i.e., false alarm probability equal to α):

$$\begin{aligned} F[t] \leq F_{1-\alpha} &\Rightarrow \text{Accept } H_0 \\ \text{otherwise} &\Rightarrow \text{Accept } H_1 \end{aligned} \quad (18)$$

with $F_{1-\alpha} = F_{\ell-1,\ell_o-1}(1-\alpha)$ indicating the distribution's $(1-\alpha)$ critical point. An obvious advantage of this approach is that no model reestimation is required in the inspection phase.

Illustrative Example: Nonstationary Random Vibration Modeling and SHM for a Pick-and-Place Mechanism

The Structure and its Nonstationary Random Vibration Response

The system studied in this example is the 2-DOF pick-and-place mechanism mentioned earlier (Fig. 1a). The random vibration response is measured in the same direction as the excitation, using lightweight piezoelectric accelerometers. During a single experiment a single cycle is performed, in which the linear motors move from their rightmost to their leftmost position and back. The measured vibration response is conditioned and driven into a data acquisition module, which digitizes the signal with a sampling frequency $f_s = 512$ Hz – signal length 10 s ($N = 5{,}120$ samples). Each signal is subsequently sample mean corrected and normalized (scaled). The frequency range of interest is 5–200 Hz, with the lower limit set in order to avoid instrument dynamics and rigid body modes.

For the SHM problem six damage scenarios are considered, which correspond to the loosening or removal of various bolts at different points of the mechanism (damages A to C and E), loosening the slider of motor B (damage D), and adding a mass at the free end of the slider of motor A (damage F). For each health state, a set of 40 random vibration responses are recorded – see (Spiridonakos and Fassois 2013) for details.

Nonstationary Random Vibration Parametric Modeling (Healthy Structure)

The nonstationary random vibration response (see Fig. 1b) of the healthy structure is now modeled via the stochastic parameter evolution (both SP-TAR and GSC-TAR models) and the deterministic parameter evolution (FS-TAR models) methods using a single data record obtained from the healthy structure. The details for each method are summarized as follows:

Stochastic parameter evolution modeling: SP-TAR models are estimated via the Kalman filter – smoother method (Poulimenos and Fassois 2006) – and GSC-TAR models via the expectation-maximization method (Avendaño-Valencia and Fassois 2013). The innovations variance is in each case estimated via a moving rectangular (600 sample long) window. The following structural parameters are considered: AR order $n_a \in \{3,\ldots,32\}$, with $\{q = 1, \sigma_v^2 = \exp(-11)\}$ and $\{q = 2, \sigma_v^2 = \exp(-24)\}$. The value of σ_v^2 is further optimized by estimating SP/GSC-TAR models with the selected model order and $\sigma_v^2 = \exp(v)$, $v = \{-10, -11, \ldots, -30\}$, $q = \{1, 2\}$. The optimal σ_v^2 is determined by considering the RSS and PESS.

Deterministic parameter evolution modeling: Fully parameterized FS-TAR models with trigonometric basis functions of the form

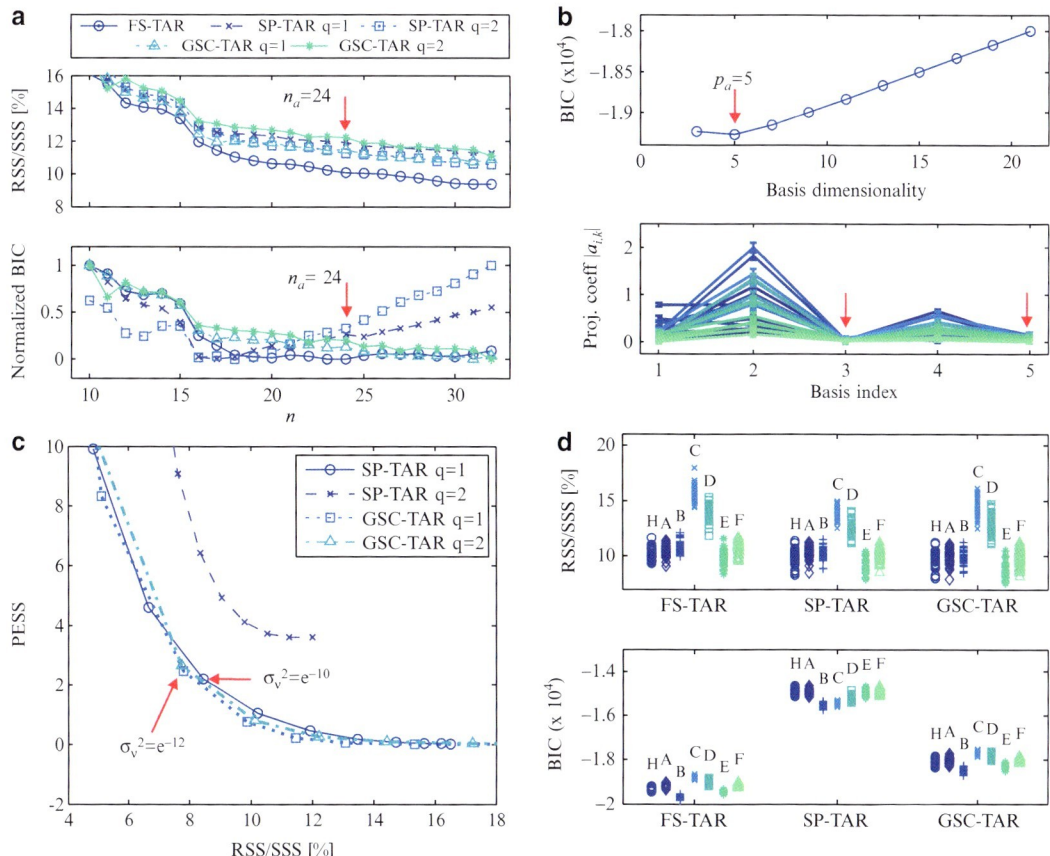

Parametric Nonstationary Random Vibration Modeling with SHM Applications, Fig. 2 Model structure selection for SP-TAR, GSC-TAR, and FS-TAR models: (**a**) model order selection – RSS/SSS and BIC (the latter normalized between 0 and 1) curves for SP/GSC/FS-TAR models with orders $n \in \{10, \ldots, 32\}$. (**b**) Selection of the subspace dimensionality for FS-TAR models: *top*, BIC of FS-TAR(24) models with $p_a = p_s \in \{3, 5, \ldots, 21\}$; *bottom*, estimated projection coefficients for each basis dimensionality (continuous lines connect the point estimates and bars indicate ± one standard deviation). (**c**) Optimization of σ_v^2 for SP/GSC-TAR models – PESS versus RSS/SSS for SP-TAR(24) and GSC-TAR(24) models. (**d**) Sample distribution of RSS/SSS and BIC for the selected model structure reestimated for each healthy (H) and damage (A ... F) data record (40 models per health state)

$$G_0[t] = 1, \quad G_{2k-1}[t] = \sin(2\pi kt/N), \quad G_{2k}[t] = \cos(2\pi kt/N) \quad k = 1, 2, \ldots \quad (19)$$

are considered. The functional subspace dimensionality p_a corresponds to the number of sine and cosine functions used by the FS-TAR model plus one (the constant component), $t = 1, 2, \ldots, N$ is the normalized discrete time, and N the signal length. Parameter estimation is based on ordinary least squares, while the innovations variance is estimated via the instantaneous method (Poulimenos and Fassois 2006). The following structural parameters are considered: AR order $n_a \in \{3, \ldots, 32\}$, $p_a = p_s \in \{3, 5, \ldots, 21\}$.

Modeling results: The results of the model structure selection procedure are depicted in Fig. 2a–c). Figure 2a shows the RSS/SSS and normalized BIC for models from each considered class (FS/SP/GSC-TAR), with the selected order ($n_a = 24$) being indicated by an arrow. Figure 2b shows the functional subspace dimensionality selection for FS models. The top plot shows the BIC of FS-TAR(24) models versus functional basis dimensionality; the selected dimensionality ($p_a = 5$) is indicated by an arrow. The bottom plot shows the absolute value of the estimated coefficients of projection with their corresponding ± 1 standard deviation interval indicated by the bars. It is evident that the standard deviation interval of the estimated coefficients of projection for the basis indices 2 and 4, indicated by the arrows in the plot, consistently contains zero and may be thus removed from the model. An FS-TAR(24)$_{[3,5]}$ model with functional basis indices $b_a = [0,1,3]$ and $b_s = [0,1,2,3,4]$ is finally selected (i.e., including the functions $G_0[t], G_1[t], G_3[t]$ in the AR subspace and the functions $G_0[t], G_1[t], G_3[t], G_4[t]$ in the innovations variance subspace).

Figure 2c shows the PESS versus RSS/SSS plot for the selection of σ_v^2 for SP-TAR and GSC-TAR models. An arrow indicates the model selected, which corresponds to a "good" compromise between low PESS and RSS/SSS. The selected model structures are SP-TAR(24) with $q = 1$, $\sigma_v^2 = e^{-10}$, and GSC-TAR(24) model with $q = 1$, $\sigma_v^2 = e^{-12}$.

Models of the *selected* (under the healthy condition) structure are subsequently fitted (estimated) for each data record corresponding to the healthy and each damaged state of the structure (40 models per health state, each one based on a distinct data record). In Fig. 2d the sample distribution of RSS/SSS and BIC of the above selected models are presented for the various health states. The FS-TAR model structure uniformly (for all health states and data records) achieves the lowest BIC; although its RSS/SSS values are not minimal.

The "frozen"-type TV-PSDs of the healthy structure, as obtained by the aforementioned three model types and a single data record, are presented in Fig. 3. For purposes of comparison, the nonparametric spectrogram (Gaussian window $\sigma = 8$, $N_{fft} = 1,024$ samples, 51 samples advance (~5% of N_{fft})) estimate is also shown. While all TV-PSDs are in rough overall agreement, it is obvious that the parametric model-based ones are much cleaner and informative than their nonparametric (spectrogram) counterpart. This is an important feature of parametric methods.

Nonstationary Vibration Response-Based SHM

As already mentioned the healthy (H) and six damage scenarios (A to F) are considered in SHM, with 40 data records used in each health state.

Two versions of the FS-TAR model parameter-based approach are used: (a) the original version in which the coefficient of projection covariance matrix $\sum_o$ is estimated based on a single data record and (b) an alternative version in which the covariance matrix is estimated based on several (presently 35) data records (using the sample covariance estimator).

Damage detection results using version (a) are depicted in Fig. 4a. All 40 cases corresponding to the healthy structure provide d_M^2 values lying below the selected detection threshold (dashed horizontal line: $\alpha = 10^{-14}$), thus correctly detecting the current healthy state. Also, in all cases corresponding to damages A...D, the obtained d_M^2 values are above the detection threshold, thus correctly detecting damage. Only certain cases corresponding to damages E and F are not properly detected. As indicated by the ROC curves (receiver operating characteristic curves which depict the true positives, TPs, versus false positives, FPs, as the threshold varies) of Fig. 4b, the two versions of the parameter-based approach perform quite adequately and similarly (the performance is almost ideal if damages E and F are excluded).

Next, the FS-TAR residual-based approach is employed using variance estimates obtained via a sliding window of length $\ell = \ell_o = 200$. Figure 4c provides a comparison of the obtained $F[t]$ statistic for a healthy and a damaged (damage D) case. $F[t]$ is, at all times, under the threshold F_u ($\alpha = 10^{-4}$) indicating healthy structure, or above it (for at least some times) indicating damaged structure. As indicated by the ROC curves of Fig. 4d, the residual-based approach

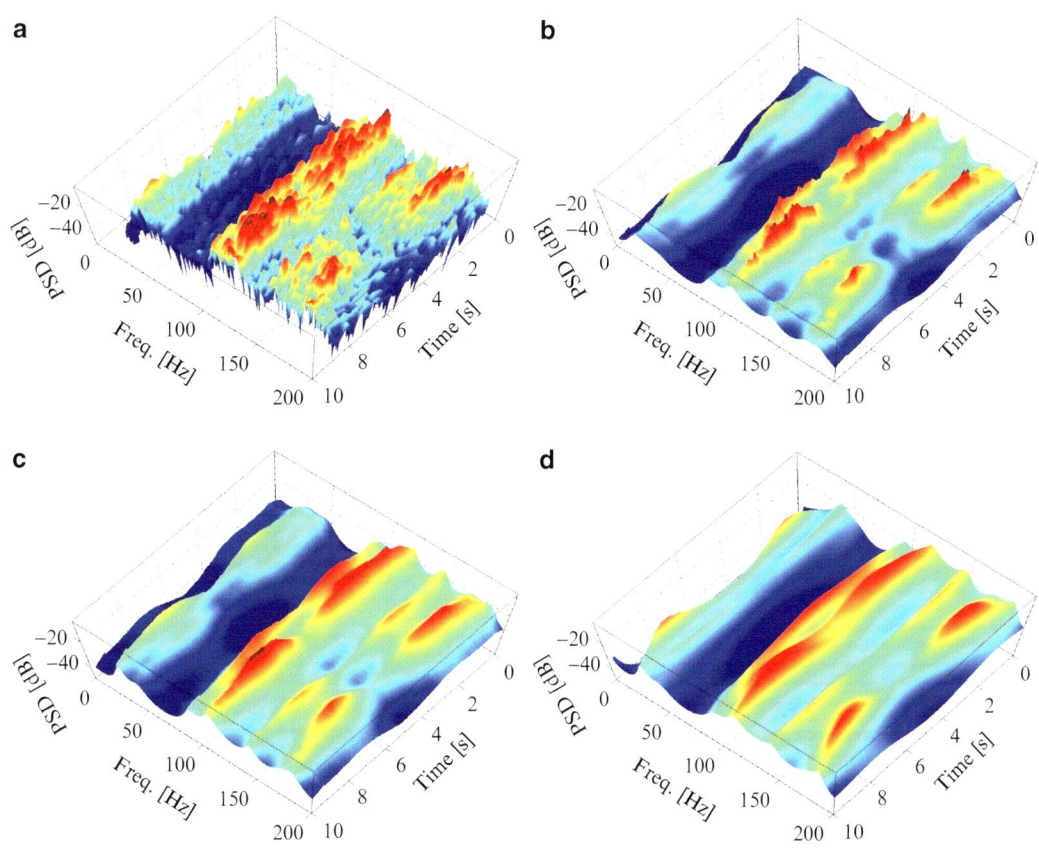

Parametric Nonstationary Random Vibration Modeling with SHM Applications, Fig. 3 Spectrogram and "frozen"-type TV-PSD estimates obtained from estimated TAR models using a single vibration response of the healthy structure: (**a**) spectrogram; (**b**) SP-TAR(24) ($q=1$); (**c**) GSC-TAR(24) ($q=1$); (**d**) FS-TAR(24)$_{[3,5]}$

provides, in this application, inferior performance, which is somewhat improved when only two types of damage, namely, C and D, are considered.

Summary

Models and methods for nonstationary random vibration parametric modeling have been presented, with focus on the stochastic parameter evolution (SP-TAR and GSC-TAR models) and deterministic parameter evolution (FS-TAR models) methods. Approaches for nonstationary random vibration-based Structural Health Monitoring (SHM) employing these models have been also discussed. An illustrative application of the methods to the vibration response modeling and SHM for a laboratory pick-and-place mechanism has been also presented.

Some key points are summarized below:

- Parametric SPE and DPE modelings of nonstationary random vibration are more involved than their nonparametric counterparts but offer unique opportunities for more accurate and compact representations, improved time-frequency resolution, analysis, and SHM. Other areas such as

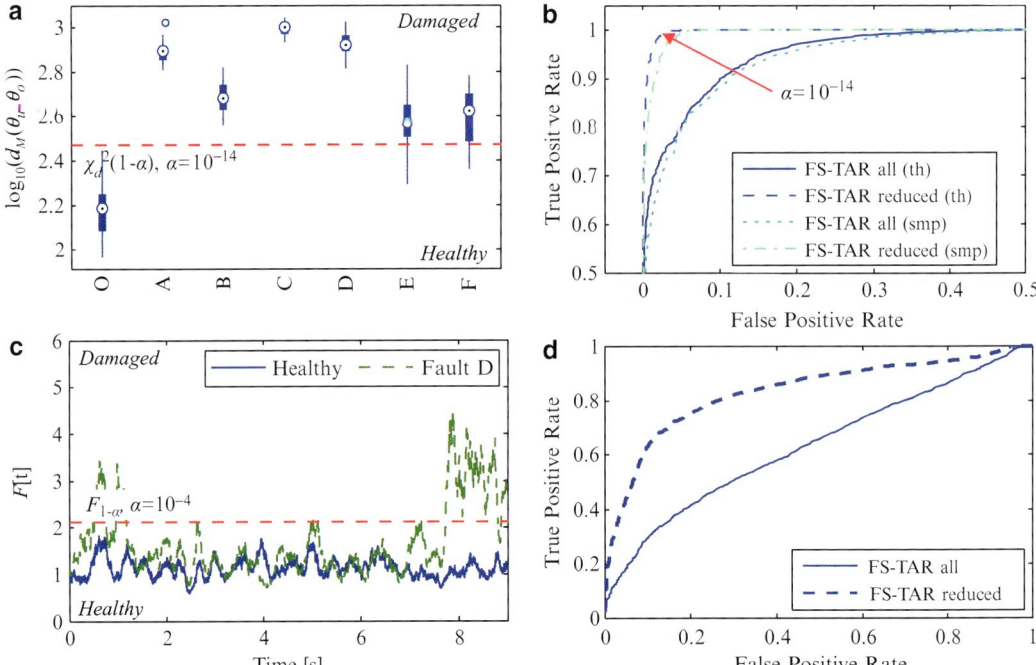

Parametric Nonstationary Random Vibration Modeling with SHM Applications, Fig. 4 Summary of damage detection results: (**a**) boxplots of d_M^2 for the parameter-based approach (original version; 40 experiments per health state) – a damage is detected if d_M^2 exceeds the threshold (*dashed horizontal line*; $\alpha = 10^{-14}$); (**b**) ROC curves for the original ("th") and alternative ("smp") versions of the approach when all damage types are included ("all") and when damage types E and F are excluded ("reduced"); (**c**) performance of the residual-based approach – $F[t]$ versus the threshold ($\ell = \ell_o = 200$, $\alpha = 10^{-4}$) for a single healthy and a damaged state; (**d**) ROC curves for the residual-based approach when all damage types are included ("all") and when only damage types C and D are included ("reduced")

- prediction and control may significantly benefit as well.
- SPE modeling is better suited to "slow" or "medium" variations in the nonstationary dynamics, whereas DPE modeling is also suited (with proper functional subspace selection) for "fast" variations.
- The problem of model structure selection is important for both SPE and DPE modelings. Yet, in the latter case, it may be significantly alleviated via the new class of adaptable models (Spiridonakos and Fassois 2014a).
- The problem of local extrema in the (non-convex) estimation criterion is important for both SPE and DPE modeling. Good initial values and careful use of model validation and diagnostic tools are thus necessary (Poulimenos and Fassois 2009b).
- SHM in nonstationary random vibration environments is a rather recent but important area with many potential applications. The presented approaches should be viewed as initial attempts to address the problem.

Cross-References

▶ Model Class Selection for Prediction Error Estimation
▶ Operational Modal Analysis in Civil Engineering: an Overview
▶ Stochastic Structural Identification from Vibrational and Environmental Data
▶ System and Damage Identification of Civil Structures

References

Avendaño-Valencia L, Fassois S (2013) Generalized stochastic constraint TARMA models for in-operation identification of wind turbine non-stationary dynamics. In: Basu B (ed) Key engineering materials (volumes 569–570) – damage assessment of structures X. Trans Tech Publications, Switzerland, pp 587–594

Bendat J, Piersol A (2000) Random data analysis and measurement procedures. Wiley, New York

Box G, Jenkins G, Reinsel G (1994) Time series analysis, forecasting and control. Prentice-Hall, Englewood Cliffs, Third edition

Farrar C, Worden K (2013) Structural health monitoring, a machine learning perspective. Wiley, Chichester

Fassois S, Sakellariou J (2009) Statistical time series methods for structural health monitoring. In: Encyclopedia of structural health monitoring. Wiley, Chichester, pp 443–472

Feng Z, Liang M, Chu F (2013) Recent advances in time-frequency analysis methods for machinery fault diagnosis: a review with application examples. Mech Syst Signal Pr 38(1):165–205

Fouskitakis G, Fassois S (2002) Functional series TARMA modeling and simulation of earthquake ground motion. Earthq Eng Struct Dyn 31:399–420

Gersch W, Akaike H (1988) Smoothness priors in time series. In: Spall J (ed) Bayesian analysis of time series and dynamic models. Marcel Dekker, New York, pp 431–476

Hammond J, White P (1996) The analysis of non-stationary signals using time-frequency methods. J Sound Vib 190:419–447

Kitagawa G, Gersch W (1996) Smoothness priors analysis of time series. Springer, New York

Kozin F (1977) Estimation and modeling of non-stationary time series. In: Proceedings of the symposium on computational mechanics in engineering, California

Kozin F (1988) Autoregressive moving average models of earthquake records. Probabilist Eng Mech 3(2):58–63

Ljung L (1999) System identification: theory for the user. Prentice Hall PTR, Upper Saddle River

Newland D (1993) An introduction to random vibrations, spectral and wavelet analysis. Dover, New York

Niedzwieki M (2000) Identification of time-varying processes. Wiley, England

Owen J, Eccles B, Choo B, Woodings M (2001) The application of auto-regressive time series modelling for the time-frequency analysis of civil engineering structures. Eng Struct 23:521–536

Poulimenos AG, Fassois SD (2004) Vibration-based on-line fault detection in non-stationary structural systems via statistical model based method. In: Proceedings of the 2nd European workshop on structural health monitoring, Munich

Poulimenos AG, Fassois SD (2006) Parametric time-domain methods for non-stationary random vibration modeling and analysis: a critical survey and comparison. Mech Syst Signal Pr 20(4):763–816

Poulimenos AG, Fassois SD (2009a) Asymptotic analysis of non-stationary functional series TARMA estimators. In: Proceedings of the 15th symposium on system identification, Saint-Malo

Poulimenos AG, Fassois SD (2009b) Output-only stochastic identification of a time-varying structure via functional series TARMA models. Mech Syst Signal Pr 23(4):1180–1204

Preumont A (1994) Random vibration and spectral analysis. Kluwer, Dordrecht

Priestley M (1988) Non-linear and non-stationary time series analysis. Academic Press, London

Rao T (1970) The fitting of non-stationary time-series models with time-dependent parameters. J R Stat Soc B Met 32(2):312–322

Roberts J, Spanos P (1990) Random vibration and statistical linearization. Wiley, Chichester

Spiridonakos M, Fassois S (2013) An FS-TAR based method for vibration-response-based fault diagnosis in stochastic time-varying structures: experimental application to a pick-and-place *mechanism*. Mech Syst Signal Pr 38:206–222

Spiridonakos M, Fassois S (2014a) Adaptable functional series TARMA models for non-stationary signal representation and their application to mechanical random vibration modeling. Signal Process 96:63–79

Spiridonakos M, Fassois S (2014b) Non-stationary random vibration modelling and analysis via functional series time dependent ARMA (FS-TARMA) models – a critical survey. Mech Syst Signal Pr 47(1–3):175–224

Passive Control Techniques for Retrofitting of Existing Structures

Alessandro Palmeri
School of Civil and Building Engineering, Loughborough University, Loughborough, Leicestershire, UK

Introduction

There are many reasons why existing structures may have insufficient capacity to resist the expected seismic events, or their performance may be considered as unsatisfactory by designers and stakeholders. Examples include (but are not limited to):

- Updated hazard assessment for the site, based on new seismological data and/or improved analyses, that increases the intensity of the design earthquake for the various limit states;
- Change of use of an existing structure that increases the overall risk in case of an earthquake (e.g., an office building which now headquarters strategic departments of civil protection or an industrial building which now contains hazardous substances);
- Design and construction of the structure carried out with obsolete methods (e.g., with insufficient ductility of members and joints).

In such cases, two options are available, namely, building a new structure once the existing one has been knocked down or retrofitting the existing structure, in such a way that enhanced/updated performance criteria are satisfied. In many cases, the second option is more advantageous, reducing direct and indirect costs.

Traditional retrofitting of building structures can be achieved by locally increasing stiffness and strength of structural members (e.g., Wu et al. 2006; Hueste and Bai 2007; Ozcan et al. 2008). This can affect both the demand (as the modal frequencies would increase and may have either a beneficial or a detrimental effect) and the capacity (as the overall resistance will be higher). However, traditional retrofitting (e.g., adding steel or FRP jackets and plates to increase the axial and bending performance of RC columns and beams) may be very expensive due to the large number of local interventions and may also prevent the use of the structure for many months (with the associate indirect costs for the loss of use).

A different retrofitting approach consists of equipping the existing structure with passive control devices, i.e., dampers and isolators (e.g., Soong and Spencer 2002; Dargush and Sant 2005; Di Sarno and Elnashai 2005). Such devices are said to be "passive" because they do not require any power supply to work (unlike "active" and "semi-active" devices). Dampers, commonly placed in the superstructure of buildings, allow reducing the seismic demand by increasing the overall energy dissipation capability, although they may also increase the stiffness of the structure. Viscous, viscoelastic and elastoplastic mechanisms of energy dissipation are typically exploited in this type of devices. Conversely, isolators placed between the superstructure and the foundation of a building improve the seismic performance by increasing the fundamental period of vibration of the overall structural system, in such a way that most of the energy of the earthquake is filtered as the seismic motion of the superstructure is decoupled (i.e., isolated) from the ground. High-damping rubber bearings (HDBRs) and friction pendulum bearings (FPBs) are among the most popular types of isolators.

In the following, the effectiveness of different passive control techniques for earthquake protection will be demonstrated by considering the seismic response of SDoF (single-degree-of-freedom) oscillators, with and without dampers, namely:

- Fluid viscous dampers (FVD);
- Elastomeric viscoelastic dampers (EVDs);
- Steel hysteretic dampers (SHDs).

Other control techniques can be adopted in the professional practice, which exploit different dynamic phenomena. This is the case of tuned mass dampers (TMDs), including sloshing liquid dampers (SLDs), in which a secondary mass is tuned to the mode of vibration of the existing structure, allowing part of the seismic energy to be transferred to the TMD, where this energy can be dissipated (e.g., Hoang et al. 2008; Lin et al. 2011). Another possibility is to exploit rocking mechanisms, in which part of the seismic energy is transformed into potential energy, as the center of mass of the structure raises when the structure uplifts, and then this energy is dissipated through impacts (e.g., Marriott et al. 2008; Palmeri and Makris 2008). These alternative control techniques are not discussed in the following, as they are not as effective as dampers and isolators in the seismic protection of building structures (e.g., TMDs are very effective against wild loads, while rocking is particular effective for bridge structures).

Similarly, seismic isolators are not treated in the following, as this technique is more appealing and economically advantageous for new structures rather than for the seismic retrofitting of existing structures. It must be said, however, that in some situations the installation of dampers within an existing building can be very difficult, e.g., because of the large forces that such devices need to transfer to the old structural members, and in this case base isolation system may become the only viable solution.

Governing Equations

The dynamic equilibrium at a generic time t for a SDoF oscillator subjected to seismic input can be written as:

$$m\ddot{u}(t) + f_S(t) + f_D(t) = -m\ddot{u}_g(t), \quad (1)$$

where m is the mass of the oscillator; $u(t)$ is the time history of the relative displacement of the mass with respect to the ground; $\ddot{u}_g(t)$ is the absolute acceleration of the ground; $f_S(t)$ and $f_D(t)$ are the time-varying reaction forces due to stiffness and damping mechanisms. In the above expression, the first term $m\ddot{u}(t) = f_I(t)$ can be interpreted by a ground observer as the inertial force experienced by the mass, while $-m\ddot{u}_g(t) = F(t)$ is the dynamic force induced by the ground shaking. The total inertial force, within a Galilean reference frame, is proportional to the absolute acceleration of the mass m, that is: $f_I^{(G)}(t) = m(\ddot{u}(t) + \ddot{u}_g(t)) = f_I(t) - F(t)$. For a linear system, $f_S(t) = k\,u(t)$ and $f_D(t) = c\dot{u}(t)$ are proportional to the relative displacement and velocity of the mass through the elastic stiffness k and the viscous damping coefficient c.

Multiplying both sides of Eq. 1 by the infinitesimal relative displacement $du(t)$, and integrating from 0 to the generic time instant t, one obtains:

$$\int_0^t m\ddot{u}(\tau)du(\tau) + \int_0^t f_S(\tau)du(\tau) + \int_0^t f_D(\tau)du(\tau) = \int_0^t F(\tau)du(\tau), \quad (2)$$

which gives the energy balance of the dynamic system over the time interval $[0,t]$ (Uang and Bertero 1990).

Assuming that the oscillator is initially at rest (i.e., $u(0) = 0$ and $\dot{u}(0) = 0$), it can be easily shown that the first integral in the left-hand side is the kinetic energy of the system at the generic time t:

$$T(t) = \frac{1}{2}m\dot{u}(t)^2 = \int_0^t m\ddot{u}(\tau)du(\tau), \quad (3)$$

while the integral in the right-hand side gives the cumulative work done by the seismic input:

$$W_{in}(t) = \int_0^t F(\tau)du(\tau). \quad (4)$$

Additionally, the integral of the linear-elastic stiffness term gives the potential energy of the oscillator:

$$V(t) = \frac{1}{2}ku(t)^2 = \int_0^t ku(\tau)du(\tau), \quad (5)$$

and the linear-viscous damping term provides the energy dissipation:

$$W_D(t) = \int_0^t f_D(\tau)du(\tau) = c\int_0^t \dot{u}(\tau)^2 d\tau. \quad (6)$$

Taking into account Eqs. 3, 4, 5, and 6, it follows that for a linear SDoF oscillator the energy balance at a generic time t can be expressed as (Uang and Bertero 1990):

$$E(t) = T(t) + V(t) = W_{in}(t) - W_D(t), \quad (7)$$

where $E(t)$ is the total mechanical energy stored within the system at time t. Since the potential for damaging the structure increases with this quantity, passive control techniques aim to reduce $E(t)$ either by reducing the input energy $W_{in}(t)$ (e.g.,

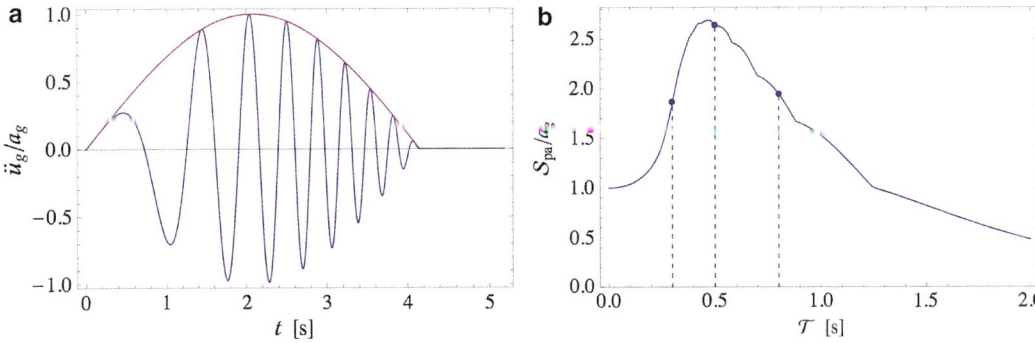

Passive Control Techniques for Retrofitting of Existing Structures, Fig. 1 Seismic input. Time history of ground acceleration (**a**) and response spectrum in terms of pseudo-acceleration (**b**)

changing the stiffness) or increasing the energy dissipation capacities of the structure (e.g., using dampers to increase $W_D(t)$).

For illustration purposes, specific examples will be provided in the next three sections for different types of dampers. The method used for comparing the different systems will be briefly discussed in the next subsection.

Seismic Excitation and Performance Analysis

The seismic performance of any control technique is greatly influenced by the characteristics of the ground motion, e.g., the peak ground acceleration (PGA), the duration of the event, the variation with time of both amplitude and frequency content, as well as by the characteristics of the existing structure, e.g., modal frequencies, available ductility, overstrength, etc. In order to make a fully comprehensive comparison between different techniques, all these aspects should be considered and the results carefully analyzed. This is far beyond the scope and the limits of the present contribution, and for this reason, a very simple signal has been used throughout as seismic input for our numerical investigations, namely, a harmonic function in which both amplitude and frequency vary with time, i.e., an amplitude-modulated cosine-sweep function. The mathematical expression of the envelope of the signal is given by

$$e_g(t) = \begin{cases} a_g \sin\left(\dfrac{\pi t}{t_g}\right), & \text{if } 0 \leq t \leq t_g; \\ 0, & \text{otherwise;} \end{cases} \quad (8)$$

and the ground acceleration is

$$\ddot{u}_g(t) = e_g(t) \cos\left(\dfrac{\omega_g^2 t^2}{2 N_g \pi}\right), \quad (9)$$

where a_g is the PGA and $t_g = 2N_g\pi/\omega_g$ is the duration of the signal, which in turn depends on the mean circular frequency ω_g and number of cycles N_g of the ground motion. As an example, for $\omega_g = 12.5$ rad/s and $N_g = 8.25$, the duration of the signal is $t_g = 4.15$ s. The accelerogram obtained with this set of parameters is shown within Fig. 1a, in which the red half sine gives the envelope.

This signal has then been used in the following to investigate the effects of different protection systems, and Fig. 1b shows the corresponding elastic response spectrum in terms of pseudo-acceleration, $S_{pa}(T_i)$, which is mathematically defined as:

$$S_{pa}(T_i) = \left(\dfrac{2\pi}{T_i}\right)^2 \max_{0 \leq t \leq t_g}\{|u(t)|\}, \quad (10)$$

and represents a measure of the maximum force experienced by the SDoF oscillator of natural period T_i during the seismic event. The three vertical dashed lines denote the three natural periods $T_i = 0.3, 0.5,$ and 0.8 s considered in the analysis, which corresponds to frequency ratios $\beta_i = \omega_g/\omega_i$ equal to 0.60, 0.99, and 1.59, i.e., "stiff," "resonant," and "flexible" test structures.

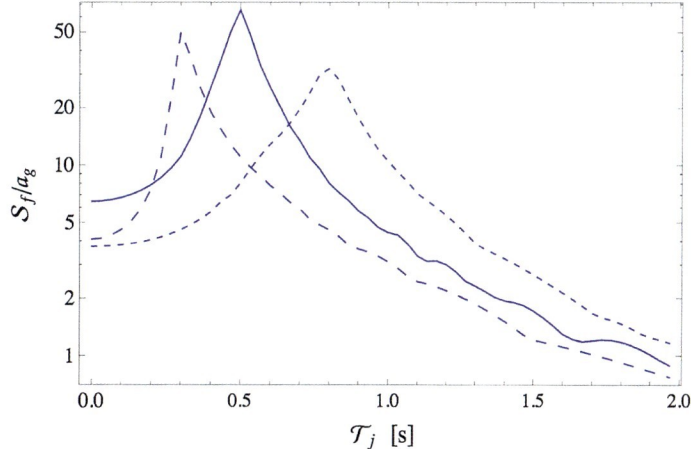

Passive Control Techniques for Retrofitting of Existing Structures, Fig. 2 Floor spectra. Stiff (*dashed line*), Resonant (*solid line*) and Flexible (*dotted line*) existing (primary) structure

The value assumed for the equivalent viscous damping ratio is $\zeta_0 = c/c_{crit} = 0.02$, being $c_{crit} = 2\sqrt{km}$ the viscous damping coefficient for a critically damped system. Even though quite small, this value of ζ_0 is reasonable for an undamaged structure within the linear-elastic range.

The semilogarithmic graph of Fig. 2 compares the so-called floor spectra for the three SDoF oscillators with the selected natural periods T_i. Each floor spectrum is obtained by using the absolute acceleration of the ith (primary) oscillator, $\ddot{u}^{(abs)}(t) = \ddot{u}(t) + \ddot{u}_g(t)$, as seismic input for a (secondary) oscillator with natural period T_j and viscous damping ratio ζ_0. The floor spectrum $S_f(T_i, T_j)$ then gives the maximum pseudo-acceleration for a light attachment (of natural period $T_j = 2\pi/\omega_j$) to the primary oscillator (of natural period $T_i = 2\pi/\omega_i$). The governing equations are:

$$\begin{cases} \ddot{u}(t) + 2\zeta_0\,\omega_i\,\dot{u}(t) + \omega_i^2\,u(t) = -\ddot{u}_g(t); \\ \ddot{y}(t) + 2\zeta_0\,\omega_j\,\dot{y}(t) + \omega_j^2\,y(t) = -\left(\ddot{u}(t) + \ddot{u}_g(t)\right), \end{cases} \quad (11)$$

where $y(t)$ is the relative displacement of the secondary mass with respect to the primary one, and the assumption is made that the secondary mass is light enough for its feedback force on the primary mass to be negligible. The reason why the floor spectra are considered as part of this study is because they allow quantifying the maximum force that a seismic event induces in secondary substructures attached to the primary structural system and then can give an indication about the effectiveness of different retrofitting strategies in terms of building content, including building services (e.g., Sackman and Kelly 1979; Muscolino and Palmeri 2007). As expected, (i) each floor spectrum of Fig. 2 takes the maximum value for $T_j = T_i$, i.e., when primary and secondary oscillators are tuned, and (ii) the maximum amplification is observed for $T_i = 0.5$ s (solid line), i.e., for the resonant primary oscillator.

Fluid Viscous Dampers (FVDs)

Viscous fluids have the potential to dissipate large amount of energy, as friction forces arise when they are forced to flow through orifices (e.g., Lee and Taylor 2001; Martinez-Rodrigo and Romero 2003; Molina et al. 2004; Marano et al. 2013). Typically, a fluid viscous damper (FVD) consists of a stainless steel piston traveling through two adjacent chambers filled with silicone oil (often chosen because it is fire resistant, nontoxic, and stable for long periods of time). The movement of the piston causes the silicone oil to flow through an orifice in the piston head, transforming part of the mechanical energy into heat, which is then dissipated into the atmosphere.

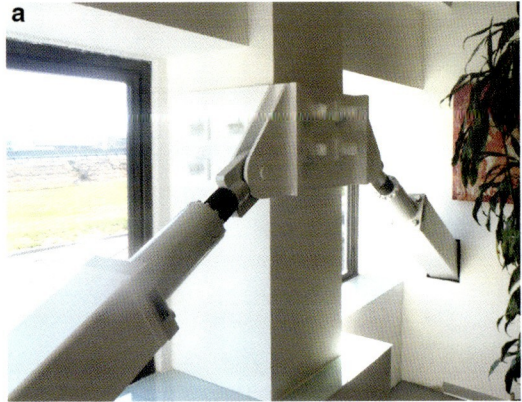

Passive Control Techniques for Retrofitting of Existing Structures, Fig. 3 Fluid viscous dampers. Seismic retrofit of buildings in Imola (Courtesy of FIP Industriale, Franco Baroni and Stefano Silvestri) (**a**) and in L'Aquila (Courtesy of FIP Industriale and Prof Vincenzo Gattulli) (**b**)

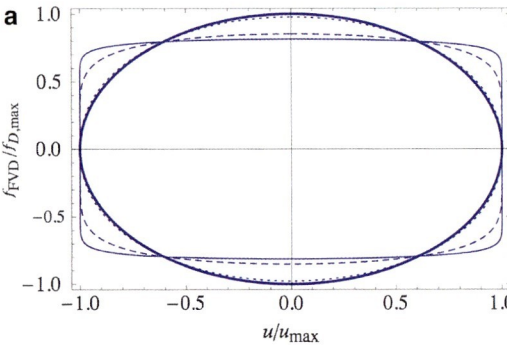
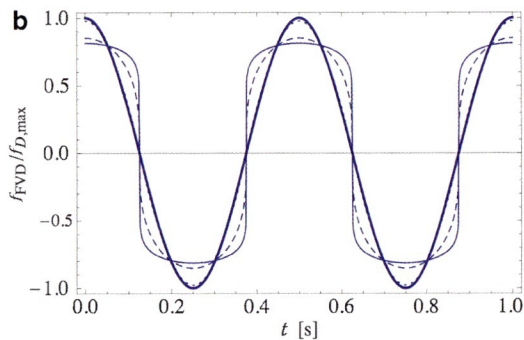

Passive Control Techniques for Retrofitting of Existing Structures, Fig. 4 Fluid viscous dampers. Hysteresis loop (**a**) and time history (**b**) of the reaction force under harmonic cycles for different values of the velocity exponent: $\alpha = 1.00$, *solid thick line*; $\alpha = 0.90$, *dotted line*; $\alpha = 0.30$, *dashed line*; $\alpha = 0.10$, *solid thin line*

Figure 3 shows two recent applications of FVDs for the seismic retrofitting of existing structures in Italy, namely, a prefabricated industrial building in Imola (Fig. 3a), necessary after the 2012 Northern Italy earthquakes, and building "A" of the Faculty of Engineering of the University of L'Aquila (Fig. 3b), required after the 2009 L'Aquila earthquake. In both cases, the presence of the control devices increases the damping capacity of the structure and improves their seismic behavior.

The reaction force $f_{\mathrm{FVD}}(t)$ exerted by FVDs depends on the relative velocity $\dot{u}(t)$ between the opposite ends of the device, and their nonlinear force–velocity relationship is generally expressed as (e.g., Lin and Chopra 2002)

$$f_{\mathrm{FVD}}(t) = c|\dot{u}(t)|^{\alpha}\mathrm{sgn}[\dot{u}(t)], \qquad (12)$$

where c is the damping coefficient (measured in kN s^{α}/m^{α}); α is a velocity exponent, which usually takes a value between 0.3 and 0.9; $|\cdot|$ stands for the absolute value; and $\mathrm{sgn}[\cdot]$ is the signum function, which gives -1 if the argument within square brackets is negative, $+1$ if positive, and 0 if the argument is zero.

Figure 4 illustrates the effect of the velocity exponent α in terms of hysteresis loop $f_{\mathrm{FVD}}(t)$ – $u(t)$ (Fig. 4b) and time history of the reaction force $f_{\mathrm{FVD}}(t)$ (Fig. 4a). Assuming a harmonic displacement $u(t) = u_{\max} \sin(2\pi\, t/T)$ with period $T = 0.5$ s, it is shown that reducing the exponent α changes the hysteresis loop, from the ellipse typical of the linear behavior ($\alpha = 1$, thick solid line) to a rectangle ($\alpha \to 0$). In both graphs, the maximum force $f_{D,max} = 2\pi c\, u_{max}/T$ experienced by the linear device ($\alpha = 1$) is used to make dimensionless the vertical axis. This reveals that, if the damping coefficient c and the period of oscillation T are kept constant, the peak force in the FVD decreases with α; on the other hand, the less α, the larger the reaction force for small velocities.

More precisely, one can prove that linear ($\alpha = 1$) and nonlinear ($0 < \alpha < 1$) devices experience the same reaction force if the velocity takes the value:

$$|\dot{u}(t)| = v_D = \left(\frac{c_{nonlin}}{c_{lin}}\right)^{\frac{1}{1-\alpha}}; \qquad (13)$$

for $|\dot{u}(t)| > v_D$ the linear device reacts with higher forces, and the opposite happens for $|\dot{u}(t)| < v_D$. Given that the maximum velocity tends to increase with the amplitude of the oscillatory motion, it follows that nonlinear VFDs, with a small velocity exponent (say $\alpha < 0.50$), are more efficient in the seismic retrofit of existing structures, as comparatively they require lesser deformations to apply the control forces.

When the nonlinear FVD is added to the SDoF oscillator, the equation of seismic motion becomes:

$$\ddot{u}(t) + 2\zeta_0 \omega_i \dot{u}(t) + \omega_i^2 u(t) + \frac{1}{m} f_{FVD}(t) = -\ddot{u}_g(t), \qquad (14)$$

and any reliable scheme of numerical integration (e.g., the Newmark-β method) can be used to evaluate the dynamic response.

In order to allow assessing the performance of FVDs in improving the seismic behavior of an existing structure, Fig. 5 compares the seismic response of three pairs of SDoF oscillators, with and without the additional FVD, for the three periods of vibration T_i selected in the previous section, namely, $T_i = 0.3$ s (stiff oscillator) for the top row, $T_i = 0.5$ s (resonant oscillator) for the central row, and $T_i = 0.8$ s (flexible oscillator) for the bottom row.

The accelerogram $\ddot{u}_g(t)$ is the sweep function of Eq. 9 (see Fig. 1a), and the PGA is $a_g = 0.25$ g, which is representative of a moderate seismic event. The viscous damping ratio of the existing structure is assumed to be $\zeta_0 = 0.02$, while the additional FVD is characterized by the mechanical parameters $c/m = 1.59$ m$^{0.7}$ s$^{-1.7}$ and $\alpha = 0.3$.

The time histories of the mass displacement, $u(t)$, are plotted in the three graphs in the first column (Fig. 5a–c), using lighter dashed lines for the seismic responses without control devices and thick solid lines when the nonlinear FVD is added. In all the three cases, a similar reduction in the maximum displacements is observed.

The second column from the left (Fig. 5d–f) presents the comparison in terms of force–displacement hysteresis loops: namely, the narrow pseudo-ellipses show the energy dissipated through the inherent viscous damping of the SDoF oscillator, while the pseudo-rectangles are the hysteresis loops associated with the additional nonlinear FVD. In all the cases, the latter source of energy dissipation is larger, which explains why the seismic displacements are reduced. This is confirmed by the graphs in the third column (Fig. 5g–i), in which the comparison is presented in terms of time histories of various forms of energy and the external work. The two dashed lines are used for the oscillator without FVD, and specifically to plot the cumulative work done by the seismic input, $W_{in}(t)$, and the total mechanical energy $E(t)$. Solid lines are used for the oscillator with the FVD: in this case, the $W_{in}(t)$ is the highest among the solid curves, while the total mechanical energy is the envelope of potential energy $V(t)$ and kinetic energy $T(t)$. The other two solid lines are used to plot the energy dissipated by the two damping mechanics, i.e., the additional FVD (black, high curve) and the inherent damping of the existing structure (gray, low curve).

Additionally, the floor spectra $S_f(T_j)$ depicted in the fourth column (Fig. 5j–l) demonstrate that the additional FVD allows reducing the maximum accelerations experienced by any attachment to the building structure (as in previous cases, solid and dashed lines are used for the seismic response with and without additional control device). This is particularly significant if primary and secondary structure are tuned (i.e., $T_i = T_j$).

Figure 6 quantifies the performance of the additional FVD for increasing levels of the seismic input, i.e., PGA $a_g = 0.10\ g$ (low intensity), $a_g = 0.25\ g$ (moderate earthquake), and $a_g = 0.40\ g$ (strong motion). It is evident that in all the three cases, the FVD allows mitigating the seismic response. Due to the highly nonlinear behavior of the selected device, however (i.e., the small velocity exponent $\alpha = 0.3$), this retrofitting technique is more efficient in case of low intensity, that is, the lower the PGA, the higher the percentage reduction in displacements and accelerations.

Elastomeric Viscoelastic Dampers (EVDs)

Viscoelastic damping is another mechanism of energy dissipation which has been successfully used to improve the dynamic behavior of structures (e.g., Zhang and Soong 1992; Chang et al. 1995; Singh and Moreschi 2002; Park et al. 2004). Historically, one of the first and most impressive applications is the use of

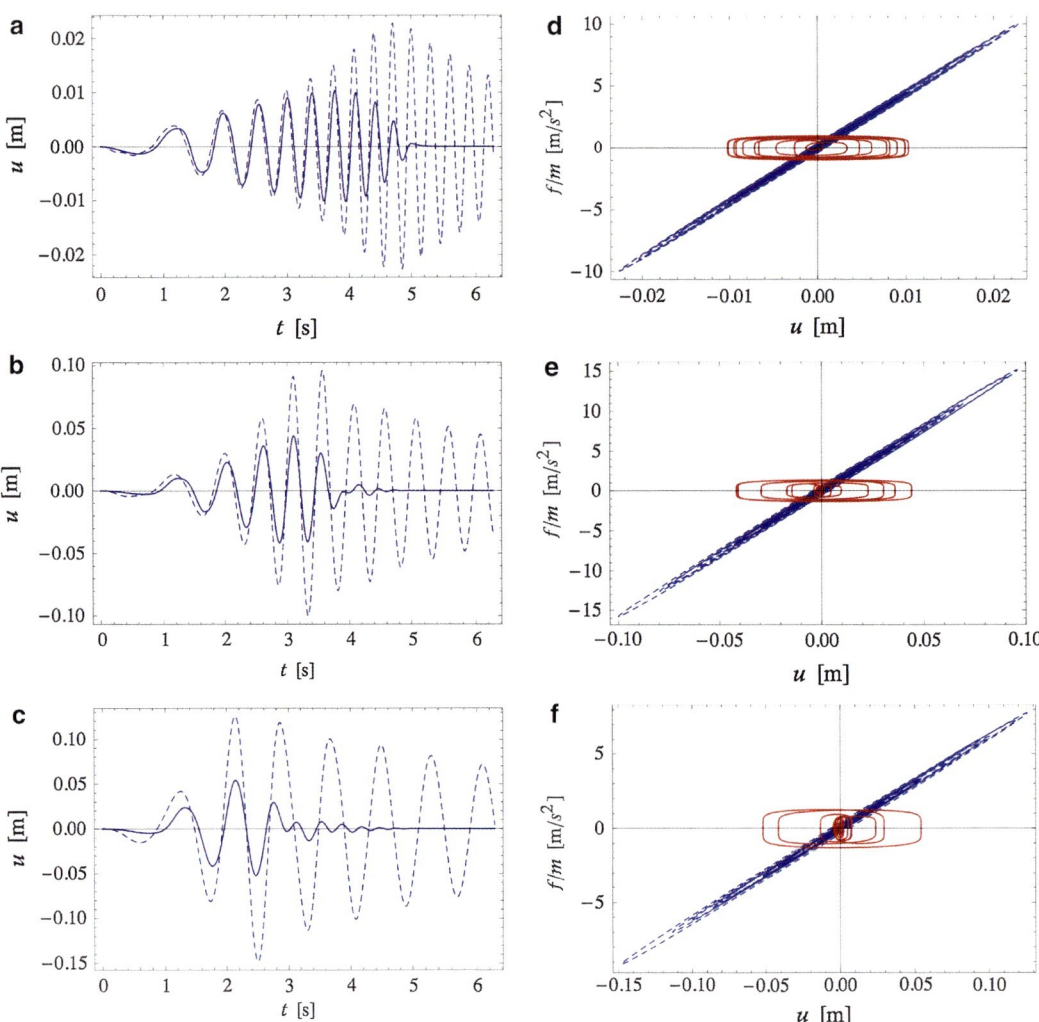

Passive Control Techniques for Retrofitting of Existing Structures, Fig. 5 (continued)

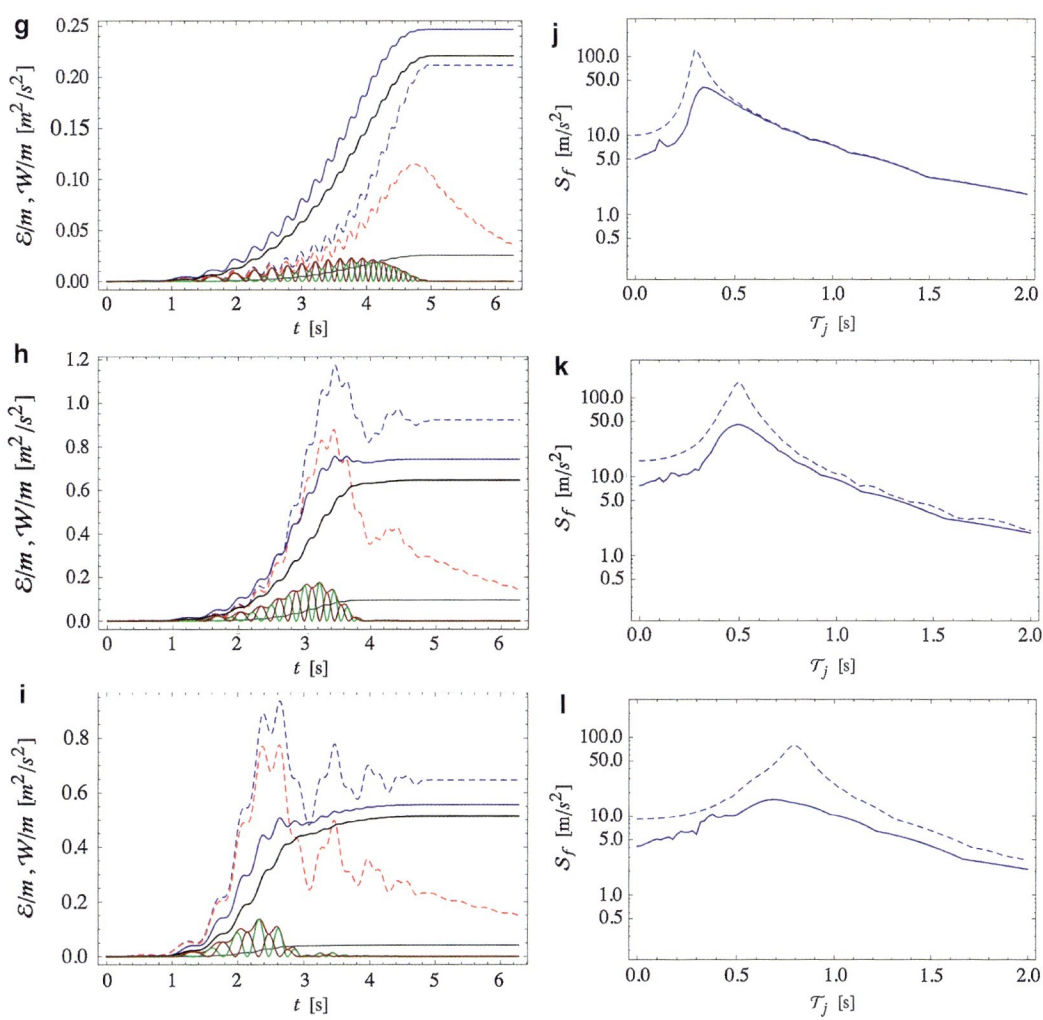

Passive Control Techniques for Retrofitting of Existing Structures, Fig. 5 Effect of the period of vibration. Seismic response of SDoF oscillators with different periods of vibration $T_i = 0.3$ s (*top row*), 0.5 s (*central row*), and 0.8 s (*bottom row*), in terms of time history of displacement (*first column*), hysteresis loop (*second column*), energy dissipated (*third column*) and floor spectra (*fourth column*)

10,000 elastomeric viscoelastic dampers (EVDs) in each of the Twin Towers of the World Trade Center (1968–2001) for mitigating the wind-induced vibrations, which otherwise would have been excessive. Because the amount of energy that can be dissipated through viscoelastic damping is comparatively smaller than for fluid viscous dampers and steel hysteretic dampers, EVDs are mainly used for wind engineering applications, although they can also be effective for the seismic retrofit of existing structures.

Figure 7, for instance, shows an application of EVDs, in which the devices are mounted as part of chevron steel braces externally connected to the main structural frame. In this way, energy is dissipated through the shear deformations within each EVD (Fig. 7a), which in turn are caused by the relative movements of the stories.

From a mathematical point of view, the reaction force experienced by a linear viscoelastic device at rest for $t \leq 0$ can be expressed in the time domain through the following convolution integral:

$$f_{\text{EVD}}(t) = \int_0^t \varphi_{\text{EVD}}(t-s)\dot{u}(s)\mathrm{d}s, \qquad (15)$$

where $\varphi_{\text{EVD}}(t)$ is the relaxation function of the damper, representing the time history of the reaction force due to a unit-step displacement; and $\dot{u}(t)$ is the relative velocity between the two ends of the device. One can easily prove that elastic stiffness and viscous damping are particular cases of the viscoelastic behavior. If the relaxation function for the elastic behavior is a Heaviside's step function of amplitude k

$$\varphi_{\text{EVD}}(t) = k\Theta(t), \qquad (16)$$

where $\Theta(t) = 0$ for $t < 0$; $1/2$ for $t = 0$; 1 for $t > 0$, then substituting Eq. 16 into Eq. 15 gives

$$\begin{aligned} f_{\text{EVD}}(t) &= \int_0^t k\Theta(t-s)\dot{u}(s)\,\mathrm{d}s \\ &= k\int_0^t \dot{u}(s)\mathrm{d}s = ku(t), \end{aligned} \qquad (17)$$

which is the constitutive law of a linear-elastic spring of stiffness k. Similarly, if the relaxation

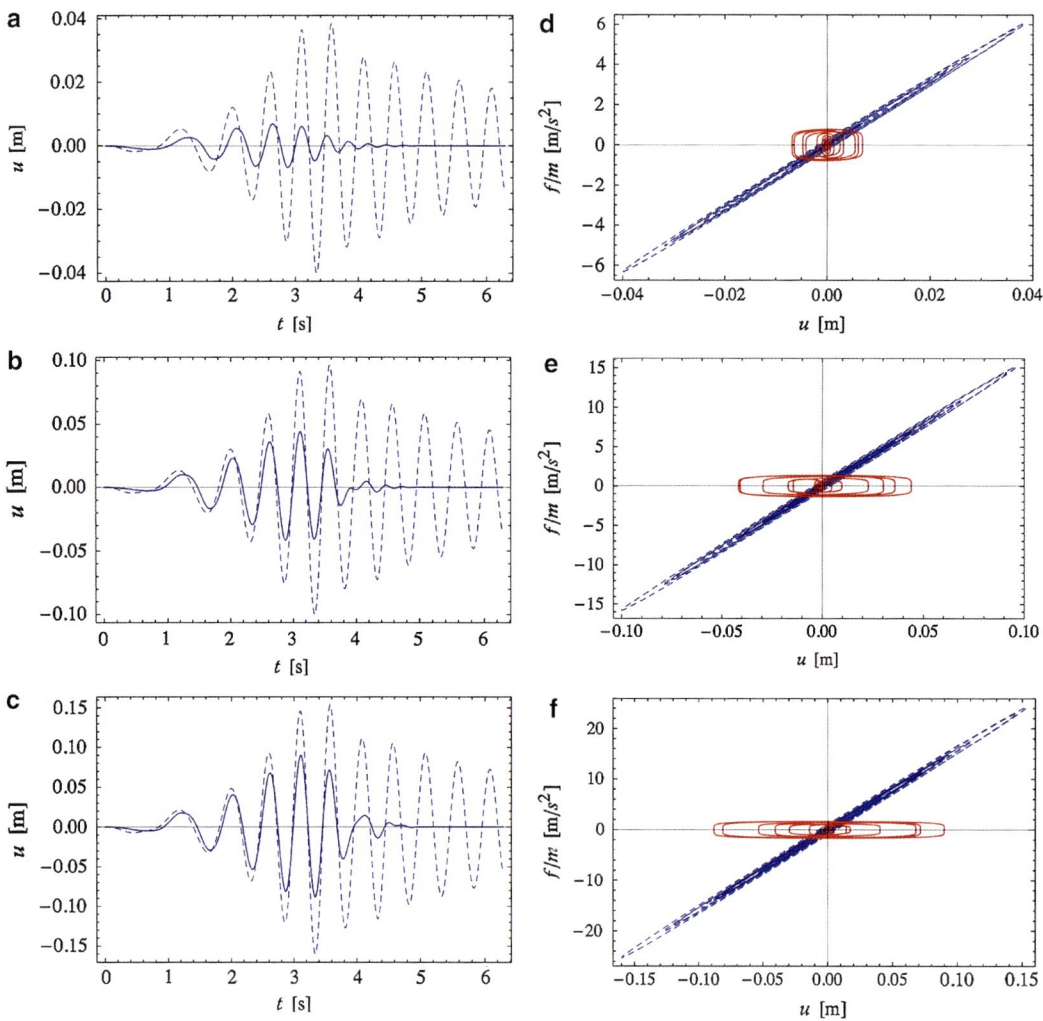

Passive Control Techniques for Retrofitting of Existing Structures, Fig. 6 (continued)

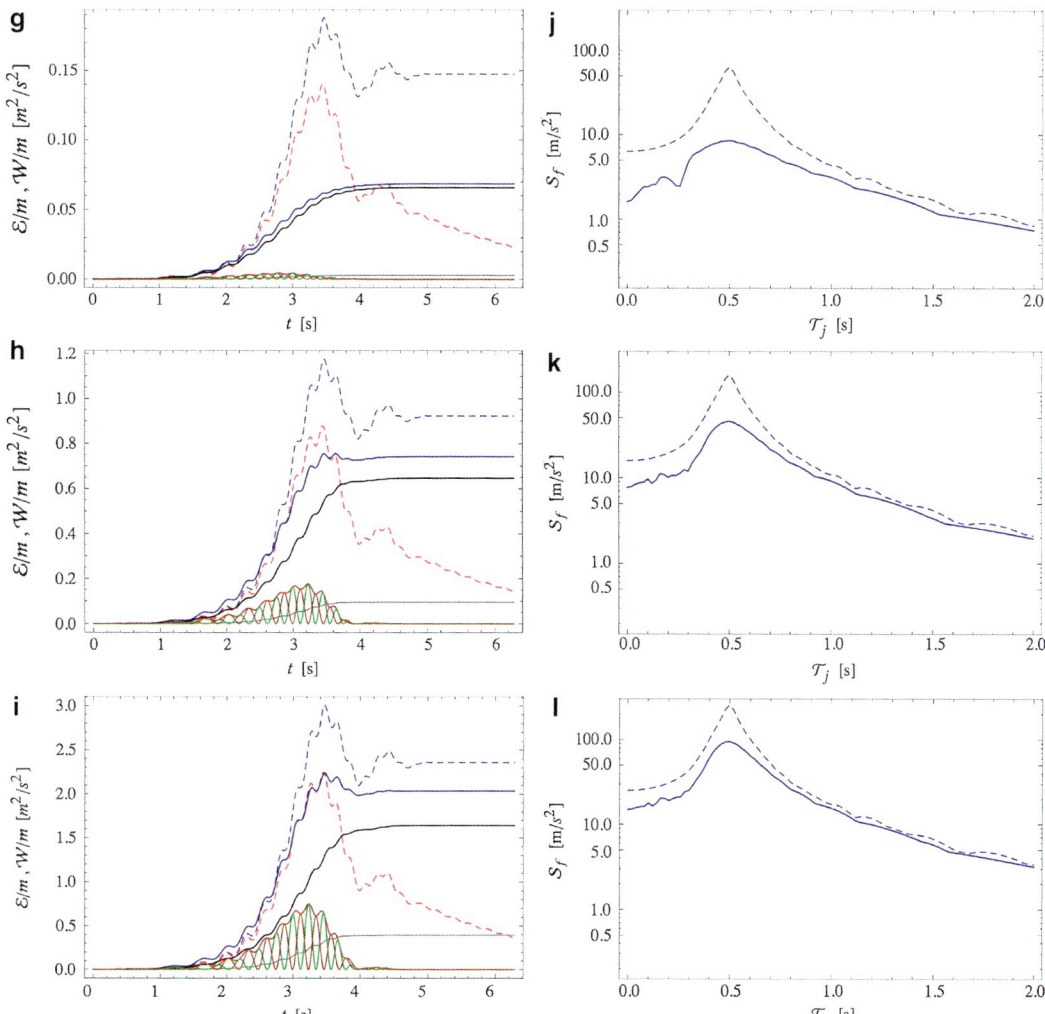

Passive Control Techniques for Retrofitting of Existing Structures, Fig. 6 Effect of the intensity of the ground shaking. Seismic response of SDoF oscillators with different values of PGA $a_g = 0.10$ m/s^2 (*top row*), 0.25 m/s^2 (*central row*), and 0.40 m/s^2 (*bottom row*), in terms of time history of displacement (*first column*), hysteresis loop (*second column*), energy dissipated (*third column*) and floor spectra (*fourth column*)

function is taken as a Dirac's delta function of intensity c

$$\varphi_{\text{EVD}}(t) = c\delta(t), \quad (18)$$

where $\delta(t) = \dot{\Theta}(t)$, Eq. 15 then simplifies as:

$$f_{\text{EVD}}(t) = \int_0^t c\delta(t-s)\dot{u}(s)\mathrm{d}s = c\dot{u}(t), \quad (19)$$

which is the constitutive law of a linear-viscous dashpot of damping coefficient c.

In general, the relaxation function is zero for $t < 0$ (this ensures the causality of the constitutive law) and monotonically decreasing for $t \geq 0$. For structural dampers made of natural rubber or other elastomeric materials, the relaxation function can be accurately represented as the superposition of a certain number N of exponential functions with various rates of decay, that is:

$$\varphi_{\text{EVD}}(t) = \left[\sum_{\ell=1}^{N} R_\ell \exp(-t/\tau_\ell)\right]\Theta(t), \quad (20)$$

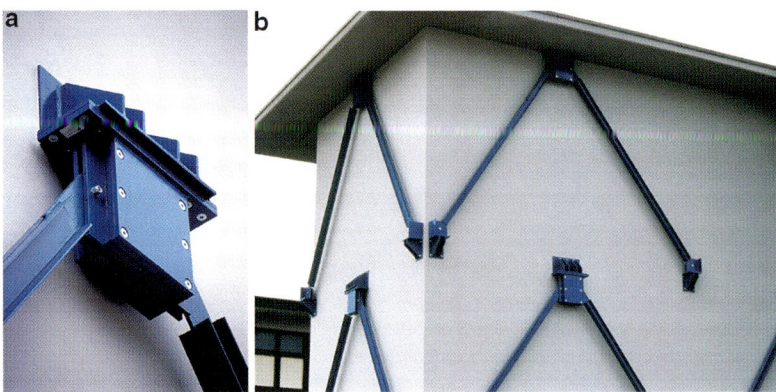

Passive Control Techniques for Retrofitting of Existing Structures, Fig. 7 Elastomeric viscoelastic dampers. Devices as installed in the seismic retrofit of the Gentile-Fermi School in Fabriano, Italy (Courtesy of FIP Industriale, Prof Rodolfo Antonucci and Francesco Balducci); detail of the damper (**a**) and installation with external chevron braces (**b**)

where the N pairs $\{R_\ell, \tau_\ell\}$ define the discrete relaxation spectrum of the EVD, in which the rigidity coefficient R_ℓ gives the amplitude of the ℓth contribution, while the relaxation time τ_ℓ provides a measure of the time required by such term to vanish (i.e., the larger the relaxation time, the slower the relaxation process). It is worth noting here that for $\tau_\ell \to +\infty$, the ℓth contribution in the r.h.s. of Eq. 20 becomes purely elastic; similarly, the pure viscous behavior is recovered for $\tau_\ell \to 0$. Any intermediate value of the relaxation time determines a truly viscoelastic behavior, which combines stiffness and energy dissipation.

Interestingly, one can prove that Eq. 20 is the relaxation function of N Maxwell's elements in parallel, i.e., a set of N parallel elastic springs R_ℓ each one connected in series with a viscous dashpot $D_\ell = R_\ell \tau_\ell$ (see Fig. 8). Accordingly, the reaction force can be expressed as (Palmeri et al. 2003; Adhikari and Wagner 2004)

$$f_{\text{EVD}}(t) = \sum_{\ell=1}^{N} R_\ell \lambda_\ell(t), \qquad (21)$$

where $\lambda_\ell(t)$ is the ℓth additional internal variable associated with the dynamic response of the viscoelastic device, which corresponds to the elastic deformation within the ℓth Maxwell element and is ruled by

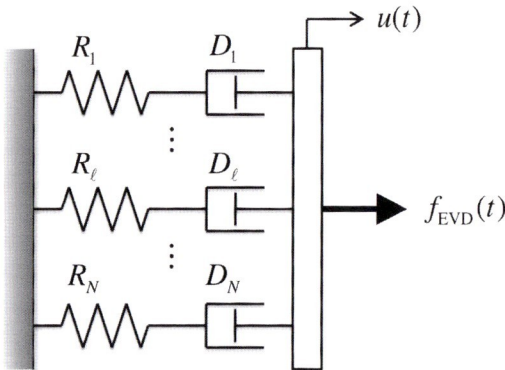

Passive Control Techniques for Retrofitting of Existing Structures, Fig. 8 Generalized Maxwell's model. Spring-dashpot model for EVDs

$$\dot{\lambda}_\ell(t) = \dot{u}(t) - \frac{\lambda_\ell(t)}{\tau_\ell}. \qquad (22)$$

Although EVDs used in the engineering practice usually require two or more Maxwell's elements to capture their dynamic behavior, the simplest case of a single Maxwell's element (i.e., $N = 1$) can be effectively used to assess the effects of the relaxation time τ_1 on the seismic performance of a building structure retrofitted with viscoelastic devices. The resulting governing equations for a SDoF oscillator equipped with additional viscoelastic damping then become

$$\begin{cases} \ddot{u}(t) + 2\zeta_0\omega_i\dot{u}(t) + \omega_i^2 u(t) + \underbrace{\dfrac{R_1}{m}\lambda_1(t)}_{f_{\mathrm{EVD}}(t)/m} = -\ddot{u}_g(t); \\ \dot{\lambda}_1(t) = \dot{u}(t) - \dfrac{\lambda_1(t)}{\tau_1}, \end{cases} \quad (23)$$

in which the two linear differential equations have to be solved simultaneously.

For this parametric study, ground acceleration $\ddot{u}_g(t)$, undamped periods of vibration $T_i = 2\pi/\omega_i$, and viscous damping ratio ζ_0 are the same as in the previous section; the rigidity coefficient is assumed to be $R_1/m = 0.5 \ \mathrm{s}^{-2}$, where m is the mass of the oscillator; three values of the relaxation time have been chosen, namely, $\tau_1 = 0.0136$, 0.136, and 1.36 s.

In a first stage, the intermediate value of the relaxation time has been used, i.e., $\tau_1 = 0.136$ s, and the beneficial effects of the additional viscoelastic damping have been assessed for three periods of vibration, namely, $T_i = 0.3, 0.5$, and 0.8 s (corresponding to "stiff," "resonant," and "flexible" structure, respectively). The comparison between their seismic responses with (solid lines) and without (dashed lines) EVDs is ordinately shown in the three rows of Fig. 9.

Figure 9a–c reveals that in all the three cases, the EVDs are effective in reducing the maximum displacements induced by the ground shaking. Their hysteresis loops (Fig. 9d–f) are ellipses, in which the slope of the major axis increases with the effective stiffness of the device. It is worth stressing here that, while the FVDs considered in the previous section only increase the damping capabilities of the structure, without affecting its stiffness, EVDs have both elastic stiffness and viscous damping. As a result, the retrofitted structure not only is capable of dissipating more energy but is also stiffer. This is confirmed by the floor spectra of Fig. 9j–l, in which the peak of the solid curves (retrofitted system) is always lower (because of the damping) and occurs at a reduced period T_j (because of the increased stiffness).

In a second stage, the effects of the relaxation time τ_1 have been investigated, while keeping constant the undamped period of vibration ($T_i = 0.5$ s). The results of these numerical analyses are offered within Fig. 10, which shows reduced peak displacements in all cases (Fig. 10a–c). Depending on the relaxation time, however, the type of control mechanism provided by the additional EVD can be quite different. In the top row ($\tau_1 = 0.0136$ s, Fig. 10a, d, g, j), for instance, the elastomeric device behaves essentially as a linear-viscous damper, while in the bottom row ($\tau_1 = 1.36$ s, Fig. 10c, f, i, l), the control force is more similar to an elastic force. A better performance is clearly seen in the central row ($\tau_1 = 0.136$ s, Fig. 10b, e, h, k), where the control force is truly viscoelastic: this happens because in this case the timescale for the relaxation of the EVD is comparable to the period of the oscillations, while the relaxation is too fast in the top row ($\tau_1/T_i < 1/20$), too slow in the bottom row ($\tau_1/T_i > 2$).

There are interesting consequences for the different types of control forces exerted by the elastomeric device. In terms of hysteresis loops, for instance, the slope of the major axis of the viscoelastic ellipses in the $f_{\mathrm{EVD}}(t) - u(t)$ diagram increases with the relaxation time τ_1. In the top row (Fig. 10d), in which the viscous damping prevails, the major axis is substantially horizontal (similar to the linear hysteresis loop for $\alpha = 1$ within Fig. 4a); on the contrary, the steepest slope is seen in the bottom row (Fig. 10f), in which the elastic stiffness prevails, and therefore the ellipse is very narrow (meaning that little energy is dissipated in each cycle). In the central row, where the EVD provides both stiffness and damping, the hysteresis loop has the major axis inclined and shows that a significant amount of energy is dissipated.

It is also interesting to note the differences between the floor spectra for the three cases. While the proper viscoelastic behavior in the central row (Fig. 10k) allows the peak of the floor spectrum to reduce (because of the additional damping) and shift to the left (because of the additional stiffness), only the reduction of the peak is seen in the top row (Fig. 10j) and the peak shift in the bottom row (Fig. 10l).

From the above observations, it follows that the relaxation times of the devices (also in relationship to the periods of vibration) are the key parameters to be considered while designing the seismic retrofitting of an existing structure with EVDs.

Steel Hysteretic Dampers (SHDs)

Plastic deformations in metals can be exploited to dissipate energy and therefore mitigate the effects of seismic forces. In this case, the device has to be designed in order to maximize the plastic work and ensure stable hysteretic cycles (e.g., Nakashima et al. 1996).

Low-yield steel plates with hourglass and triangular shapes have been proposed, and they are best known with the acronyms ADAS (added viscous and stiffness) and TADAS (triangular ADAS), respectively. Such devices are usually installed at the top of stiff chevron steel braces, so that the relative lateral movement of two consecutive stories induces bending about the weak

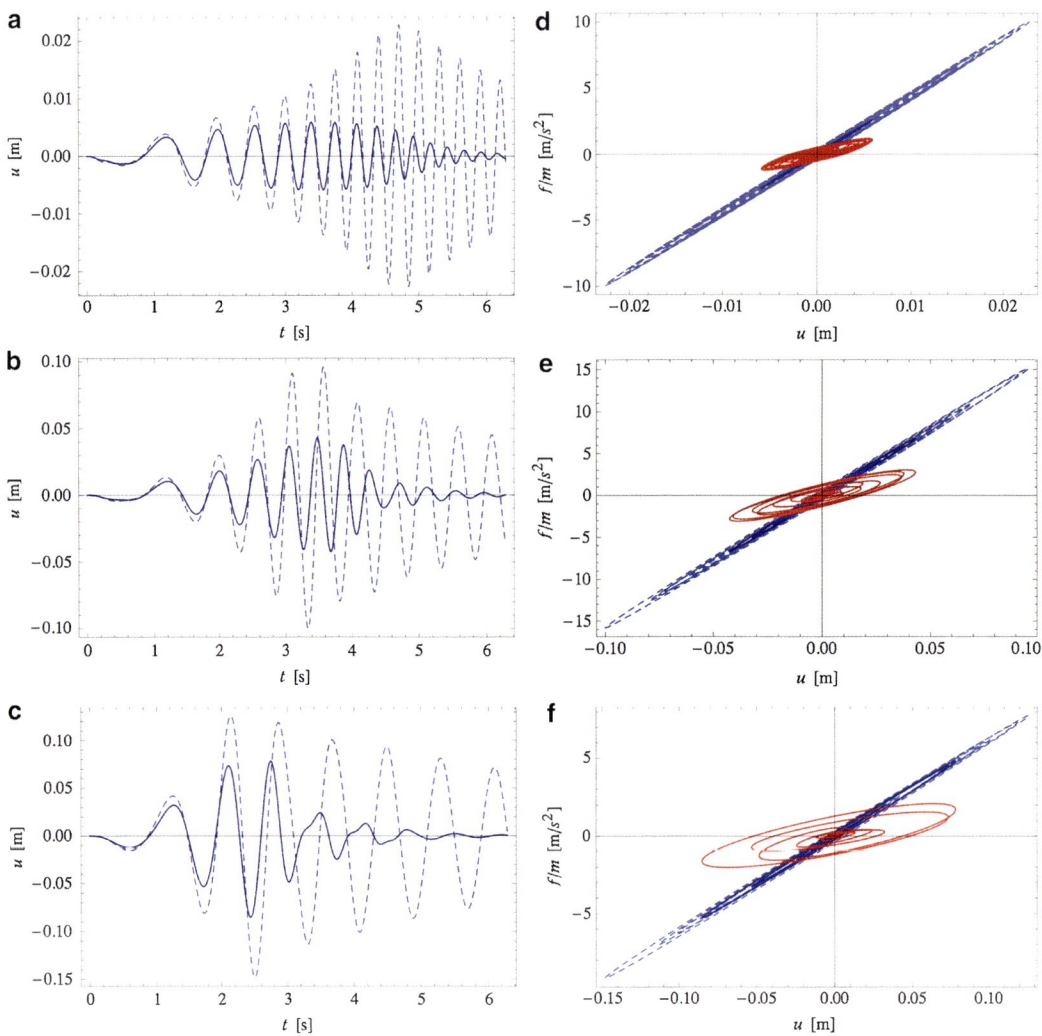

Passive Control Techniques for Retrofitting of Existing Structures, Fig. 9 (continued)

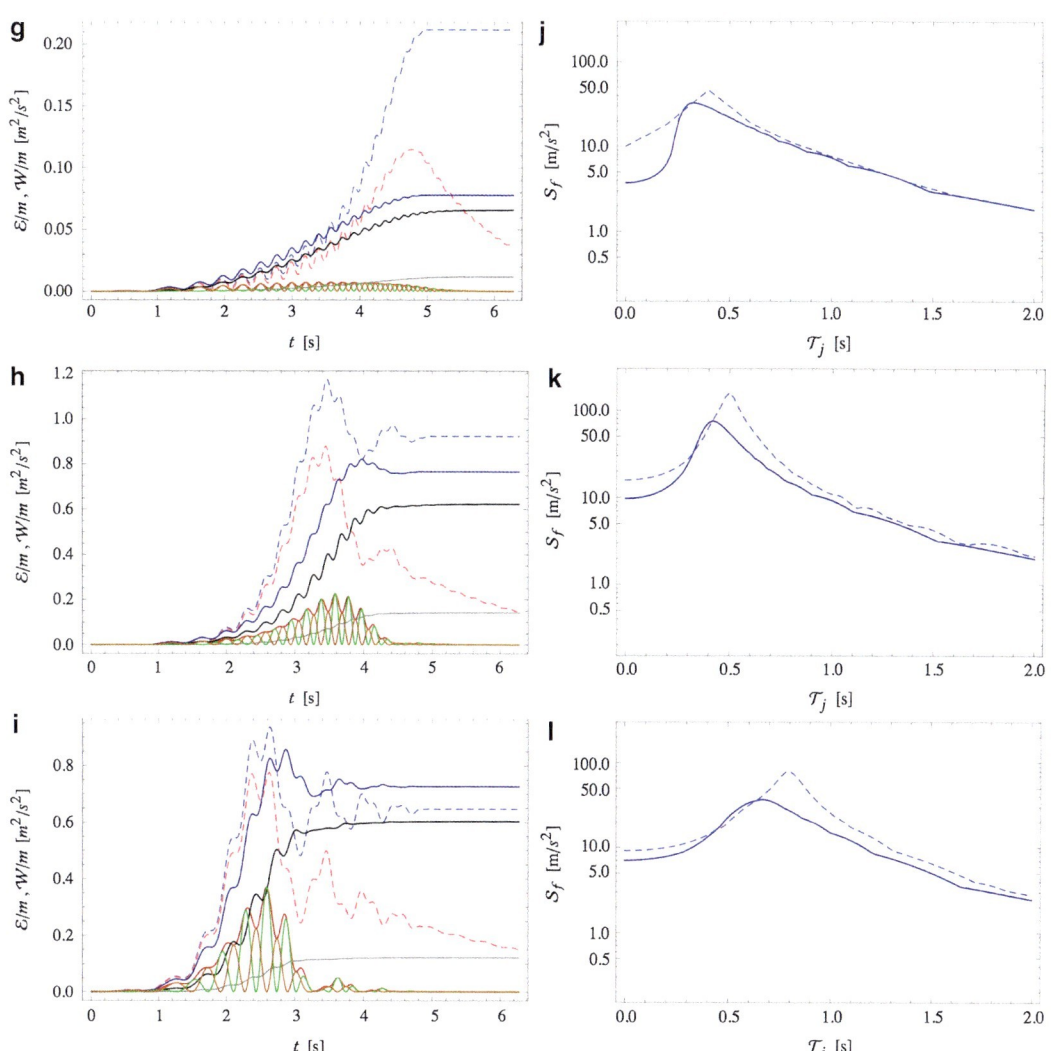

Passive Control Techniques for Retrofitting of Existing Structures, Fig. 9 Effect of the period of vibration. Seismic response of SDoF oscillators with different periods of vibration $T_i = 0.3$ s (*top row*), 0.5 s (*central row*), and 0.8 s (*bottom row*), in terms of time history of displacement (*first column*), hysteresis loop (*second column*), energy dissipated (*third column*) and floor spectra (*bottom column*)

axis of the steel plate (Xia and Hanson 1992; Chou and Tsai 2002; Alehashem et al. 2008; Bayat and Abdollahzadeh 2011).

Figure 11 (reproduced from Alehashem et al. 2008) shows the typical geometry of ADAS and TADAS devices. In the first case (Fig. 11a), the steel plates are clamped, and their hourglass shape allows maximizing the energy dissipation when the plates experience plastic bending. In the second case (Fig. 11b), the steel plates are clamped at the top and pinned at the bottom: the bending moment then increases linearly with the height, and the triangular shape then becomes the most efficient one.

Buckling-restrained braces (BRBs), also known as BRADs (buckling-restrained axial dampers), are another type of metal dampers, in which axial rather than bending stresses in low-yield steel are exploited. The system typically consists of a cruciform low-yield pinned steel brace, whose buckling is prevented by encasing the brace within a concrete-filled steel tube, and a special coat inhibits the bond between inner steel and concrete (so that the brace can freely elongate and shorten in each cycle). This solution is particularly efficient as the brace is subjected to uniaxial stress/strain, and therefore all the volume of material can contribute to energy dissipation (e.g., Sabelli et al. 2003; Black et al. 2004; Di Sarno and Manfredi 2010).

Figure 12 shows the application of this type of devices for the seismic retrofit of two schools with RC framed structure in Italy. In the first case (Fig. 12a), the devices are left exposed (increasing the sense of safety for the occupants), while in the second case (Fig. 12b), the choice has been to cover the devices.

Many mathematical models have been proposed for the representation of the elastoplastic force experienced by SHDs (steel hysteretic dampers) such as ADAS and BRB devices. The Bouc–Wen model is probably the most popular one, as it can be easily implemented to perform time-history analyses (Bouc 1971; Wen 1976; Ismail et al. 2009). The model requires the introduction of a dimensionless hysteretic variable $z(t)$, which is proportional to the elastoplastic force:

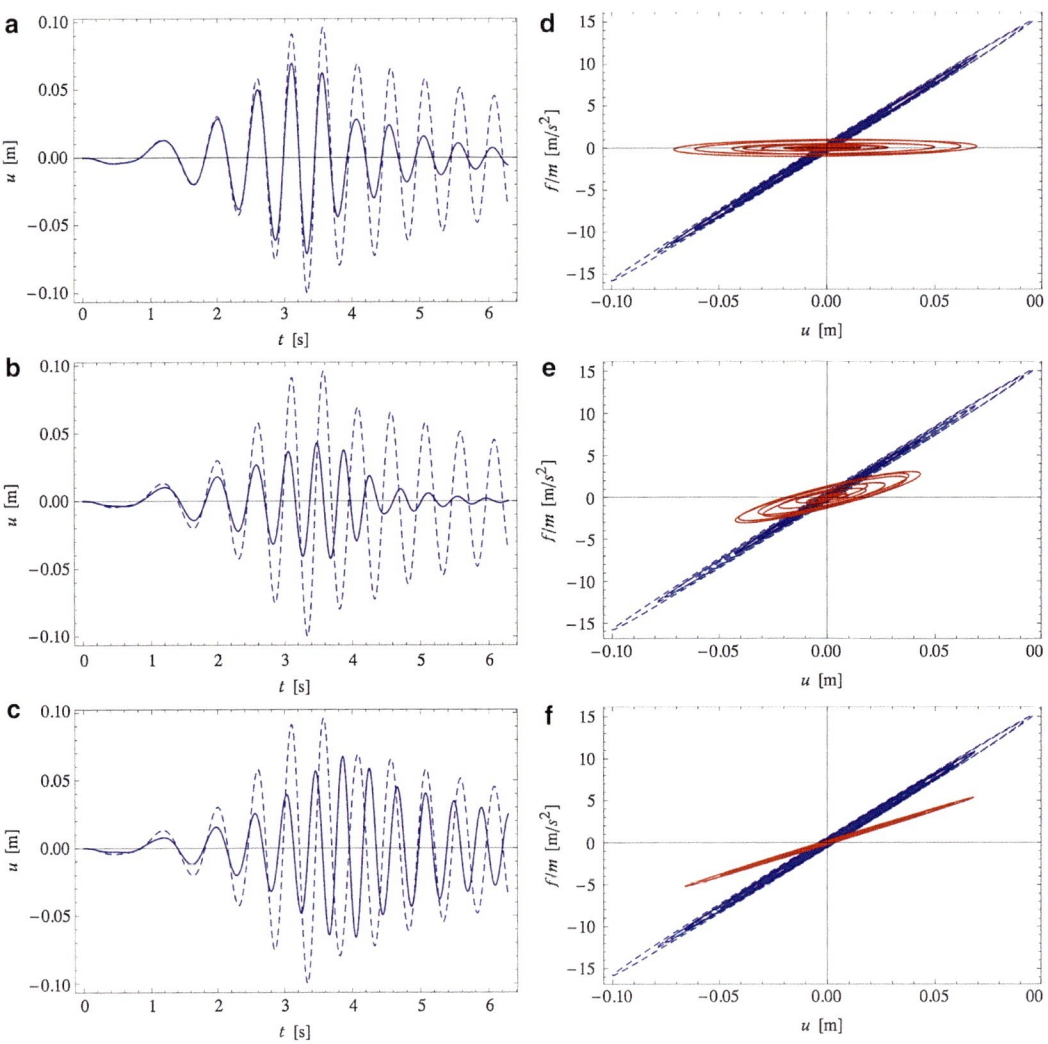

Passive Control Techniques for Retrofitting of Existing Structures, Fig. 10 (continued)

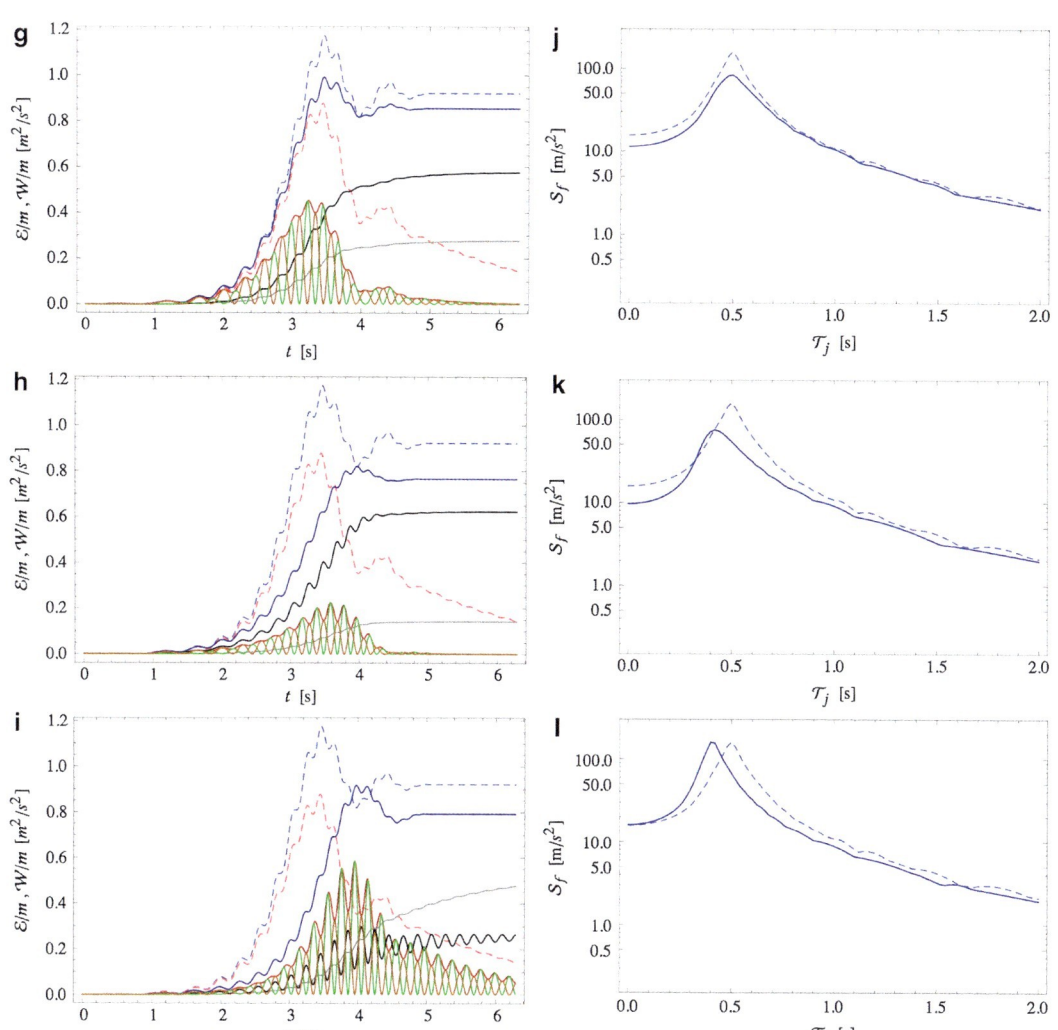

Passive Control Techniques for Retrofitting of Existing Structures, Fig. 10 Effect of the relaxation time. Seismic response of SDoF oscillators with different values of the relaxation time $\tau_i = 0.0136$ s (*top row*), 0.136 s (*central row*), and 1.36 s (*bottom row*), in terms of time history of displacement (*first column*), hysteresis loop (*second column*), energy dissipated (*third column*) and floor spectra (*fourth column*)

$$f_{\text{SHD}}(t) = f_y z(t), \quad (24)$$

and is ruled by the following nonlinear differential equation:

$$\dot{z}(t) = \frac{1}{\delta_y}\left(\dot{u}(t) - \sigma|\dot{u}(t)||z(t)|^{n-1}z(t) - (1-\sigma)\dot{u}(t)|z(t)|^n\right), \quad (25)$$

where f_y and δ_y are the yield force and yield displacement of the device, while σ and n are two dimensionless parameters which control the shape of the hysteretic cycles. That is, the larger $n > 1$, the quicker the transition between elastic and plastic branches, and vice versa (see left column of Fig. 13); the larger $\sigma > 0$, the steeper the slope of the hysteretic loop $f_{\text{SHD}}(t) - u(t)$ when the device reenters the elastic branch (for $\sigma = 0.5$ one obtains the same slope f_y/δ_y as the elastic branch; see right column of Fig. 13).

For the sake of simplicity, the model of Eqs. 24 and 25 assumes that once the plastic

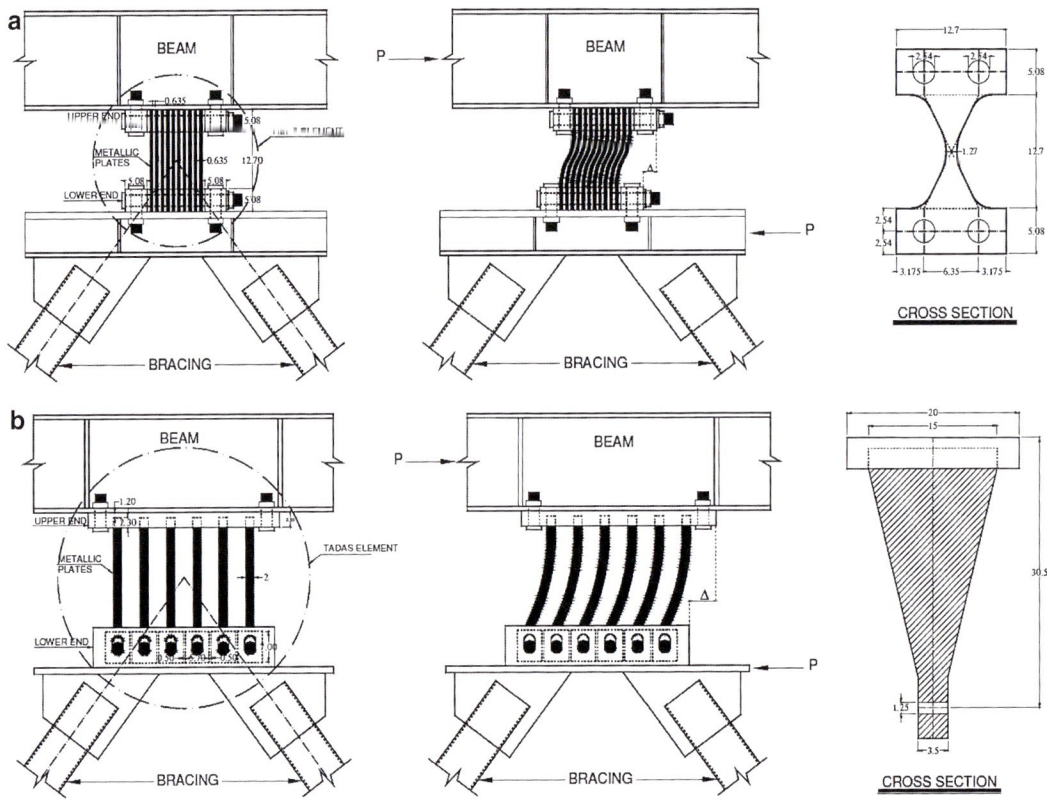

Passive Control Techniques for Retrofitting of Existing Structures, Fig. 11 Steel hysteretic dampers. Schematics of two different devices: ADAS, with hourglass shape (**a**); TABAS, with triangular shape (**b**) (Reproduced from Alehashem et al. 2008)

Passive Control Techniques for Retrofitting of Existing Structures, Fig. 12 Steel hysteretic dampers. Devices as installed in the seismic retrofit of two schools in Italy, the Cappuccini School in Ramacca (**a**) (Courtesy of FIP Industriale and Dr Fabio Neri) and the Giulio Perticari School in Senigallia (**b**) (Courtesy of FIP Industriale and Prof Rodolfo Antonucci)

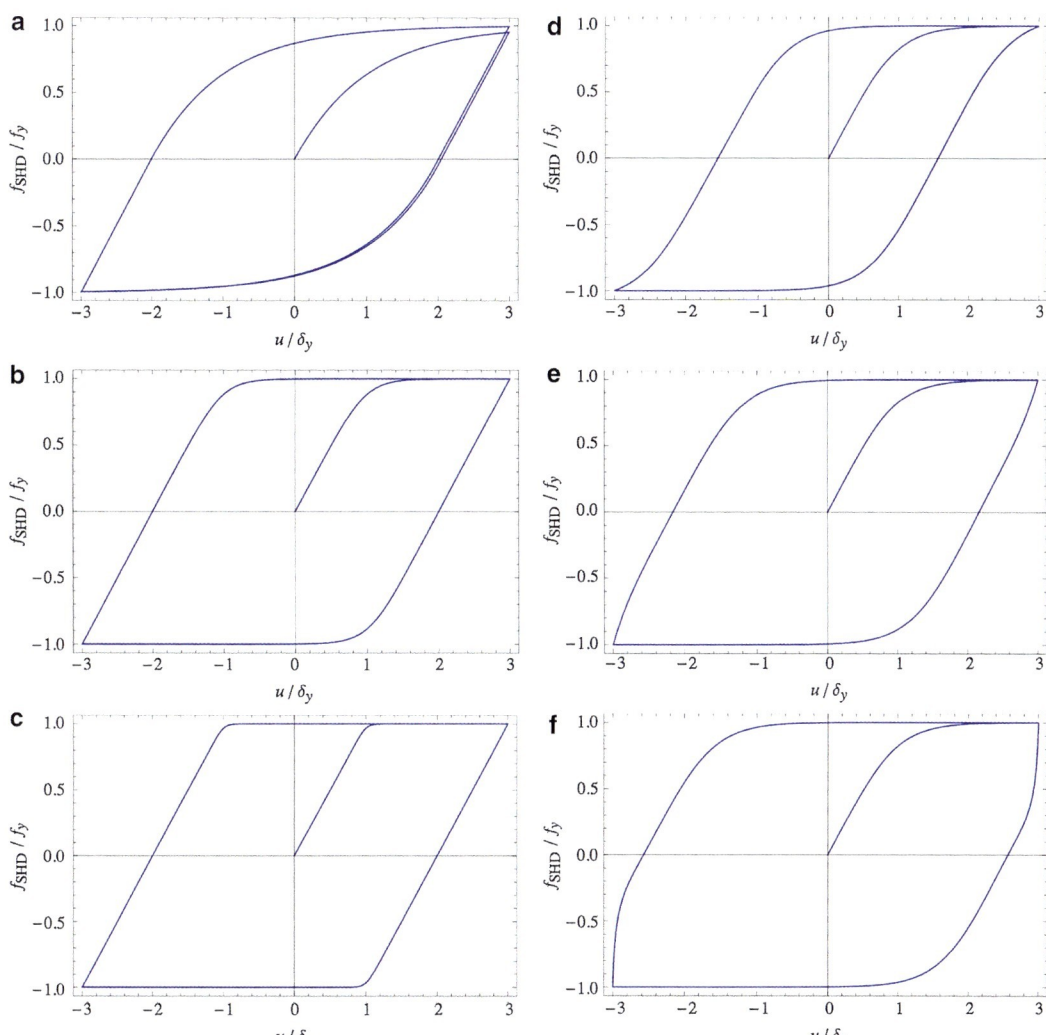

Passive Control Techniques for Retrofitting of Existing Structures, Fig. 13 Bouc–Wen hysteretic loops. Effects of the parameters α and n on the shape of the force–displacement loop: $\sigma = 0.5$, $n = 1.01$ (**a**); $\sigma = 0.5$, $n = 5$ (**b**); $\sigma = 0.5$, $n = 15$ (**c**); $\sigma = 0.1$, $n = 3$ (**d**); $\sigma = 1.0$, $n = 3$ (**e**); $\sigma = 10$, $n = 3$ (**f**)

branch has been reached, there is no further increase in the elastoplastic force, i.e., $|f_{\text{SHD}}(t)| \leq f_y$ (elastic-perfectly plastic behavior). If the device shows a residual post-yielding stiffness $k_{\text{d,res}}$, an elastic term can be added to the r.h.s. of Eq. 24, which then becomes:

$$f_{\text{SHD}}(t) = \left(f_y - k_{\text{d,res}}\delta_y\right)z(t) + k_{\text{d,res}}u(t). \quad (26)$$

Additionally, degradation effects can also be accounted for through a reduction of the yield force f_y and the elastic stiffness f_y/δ_y, which in turn can be expressed as functions of the absorbed hysteretic energy:

$$W_{\text{SHD}}(t) = \int_0^t f_{\text{SHD}}(\tau)\dot{u}(\tau)\mathrm{d}\tau, \quad (27)$$

i.e., the more hysteretic energy is dissipated, the more damage is accumulated in the device, and the larger is the reduction in stiffness and strength. It must be said, however, that the majority of the devices used for the seismic retrofitting of existing structures show stable hysteretic loops for many cycles, without any appreciable degradation.

It is also worth noting here that, while FVDs and EVDs are frequency-dependent, i.e., the rate of change of displacements and forces applied to such devices changes their dynamic response, SHDs are frequency-independent, so their performance is not affected by the energy content of the dynamic excitation. Another peculiar aspect of SHDs is that they require relatively large displacements to dissipate energy, and for this reason they are specifically used for earthquake engineering applications (while both FVDs and EVDs can also be used for other dynamic loads, e.g., to mitigate the effects of wind forces in tall buildings).

In order to perform the dynamic analysis, the equation of motion for a SDoF oscillator equipped with SHD can then be posed in the following state-space form:

$$\dot{\mathbf{y}}(t) = \mathbf{F}(\mathbf{y}(t), t), \qquad (28)$$

where $\mathbf{y}(t) = \{u(t), \dot{u}(t), z(t)\}^{\mathrm{T}}$ is the array of the three state variables of the system (displacement, velocity, and hysteretic variable), and $\mathbf{F}$ collects the differential equation for each state variable:

$$\mathbf{F}(\mathbf{y}(t), t) = \begin{Bmatrix} \dot{u}(t) \\ -\omega_0^2 u(t) - 2\zeta_0 \omega_0 \dot{u}(t) - \beta z(t) - \ddot{u}_{\mathrm{g}}(t) \\ \dfrac{\kappa \omega_0^2}{\beta} \left[\dot{u}(t) - \sigma |\dot{u}(t)| |z(t)|^{n-1} z(t) - (1-\sigma) \dot{u}(t) |z(t)|^n \right] \end{Bmatrix}, \qquad (29)$$

in which $\beta = f_y/m$ is a measure of the ground acceleration needed to reach the yielding point of the SHD and $\kappa = f_y/(k\, \delta y)$ is the dimensionless stiffness of the device, normalized with respect to the elastic stiffness of the structure without damper; the superscripted T stands for the transpose operator. Different schemes of numerical integration can be used for Eqs. 28 and 29, including the 4th-order Runge–Kutta method.

As for the other types of damping mechanics considered in the previous sections, it is interesting to assess the sensitivity of their performance to the two governing parameters for SHDs, i.e., β and κ (or, alternatively, f_y and δ_y). The value of β determines the intensity of the seismic event that mobilizes the devices: if β is too small, even modest earthquakes will result in plastic deformations in the dampers; if β is too large, the dampers would only be effective for the most intense events. The selection of κ is also very important too, as it affects the fundamental period of the retrofitted structure. Indeed, if T_1 is the fundamental period of the existing structure, $T_1^* = T_1/(1+\kappa) < T_1$ is the new period with the SHDs installed. A poor choice of κ may result in an overall increase in the seismic forces (which would then reduce the effectiveness of the SHDs). Additionally, if κ is too small, then the existing structure would still take most of the seismic forces in the retrofitted structure, with the SHDs then being unable to dissipate enough energy during the seismic event.

Figure 14 confirms that a careful selection of the design parameters is required in order to maximize the performance of SHDs. In the top row $\beta = 1.2$ m/s^2, which is about half of the pseudo-spectral acceleration for $T_0 = 0.5$ s (see Fig. 1b), and $\kappa = 1$, meaning that existing structure and retrofitting device will take the same seismic forces in the elastic range. As a result, a significant reduction in the maximum displacement is achieved (Fig. 14a), a large amount of energy is dissipated (Fig. 14d), and the ordinates of the floor spectrum are considerably mitigated (Fig. 14j). The other two rows of Fig. 14 show that lower values of yielding force ($\beta = 0.6$ m/s^2 in the second row) and stiffness ($\kappa = 0.2$ in the third row) diminish the effectiveness of SHDs, as in both cases large displacements are then experienced by the structure.

Conclusions

In this contribution, three among the most popular passive control techniques for the seismic retrofitting of existing building structures have been reviewed. Governing equations and mathematical models have been presented for viscous fluid dampers (VFDs), elastomeric viscoelastic dampers (EVDs), and steel hysteretic dampers (SHDs), along with some examples of recent applications. A simple seismic signal, a sinusoidal function with both amplitude and frequency varying with time, has been used to investigate the effects of the governing parameters for each type of device.

It has been shown that, despite the different damping mechanisms (viscosity of fluids, relaxation of elastomers, plasticity of metals), all these devices can significantly improve the seismic response of an existing structure, in terms of maximum displacements and floor spectra. In all cases, however, the design parameters of the added dampers must be carefully selected in order to maximize the amount of energy

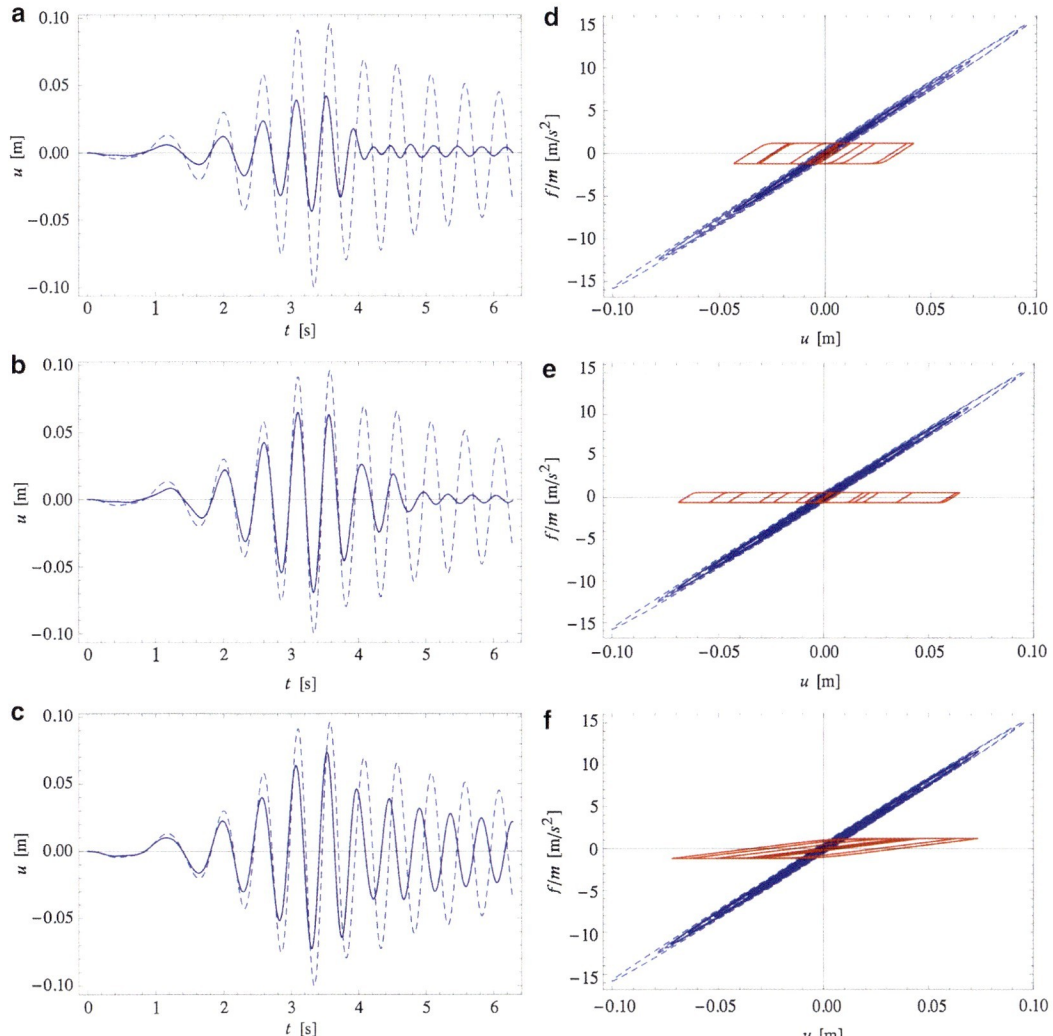

Passive Control Techniques for Retrofitting of Existing Structures, Fig. 14 (continued)

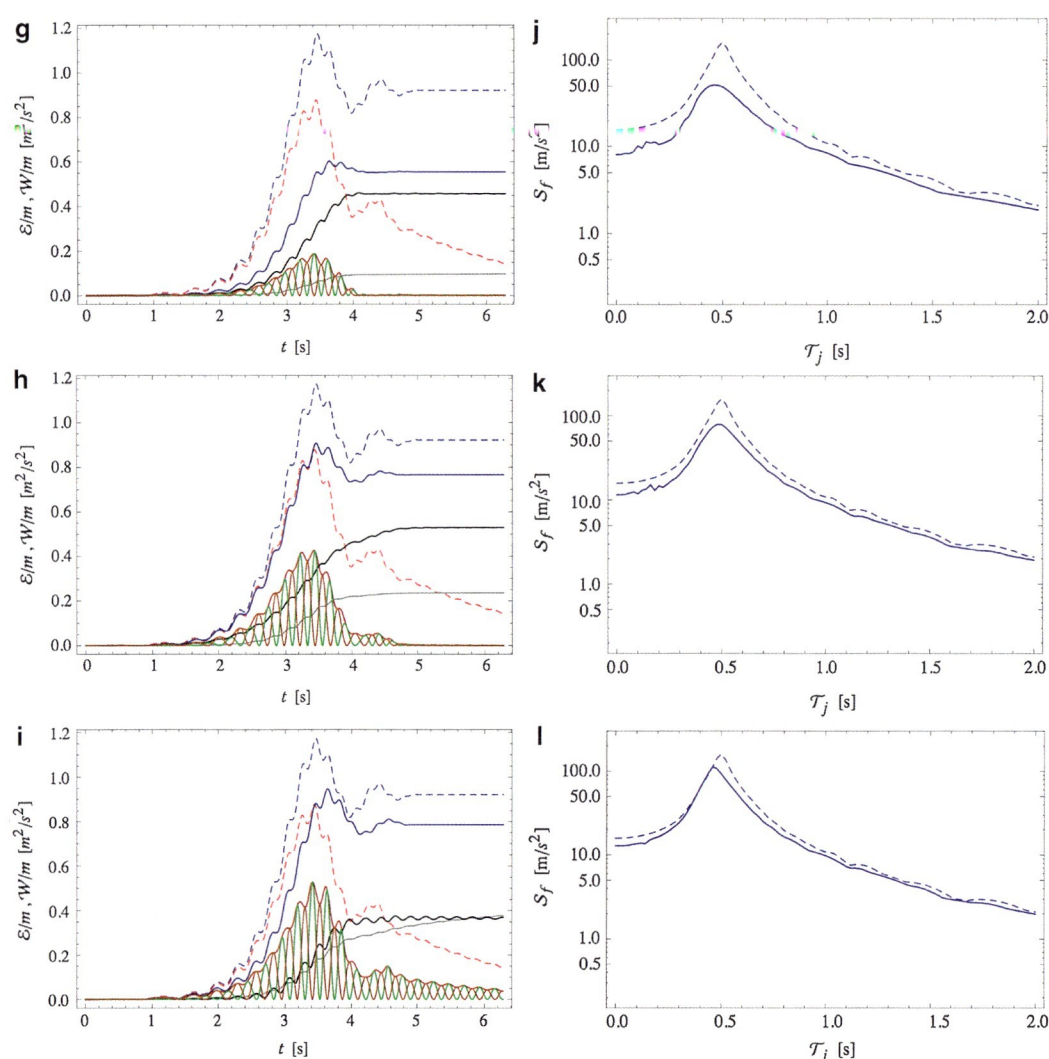

Passive Control Techniques for Retrofitting of Existing Structures, Fig. 14 Effect of the yield force and stiffness. Seismic response of SDoF oscillators with different values of design parameters β and κ, in terms of time history of displacement (*first column*), hysteresis loop (*second column*), energy dissipated (*third column*) and floor spectra (*fourth column*)

dissipation, and nonlinear time-history analyses are instrumental to measure the expected benefit of the control devices.

For the sake of simplicity, all the numerical results presented have been obtained considering a single-degree-of-freedom (SDoF) oscillator. The extension of these results to multi-degree-of-freedom (MDoF) structures is straightforward if the existing building is regular both in plan and in elevation, while further design considerations are needed to handle situations where torsional effects and soft stories are present, as larger inter-story drifts are expected at certain critical locations. Another important design constraint is the available ductility of the existing structural members. Indeed, the additional energy dissipation capacity can only be exploited if the existing structure can accommodate sufficiently large deformations.

References

Adhikari S, Wagner N (2004) Direct time-domain integration method for exponentially damped linear systems. Comput Struct 82:2453–2461

Alehashem SMS, Keyhani A, Pourmohammad H (2008) Behavior and performance of structures equipped with ADAS and TADAS dampers (a comparison with conventional structures). 14th World Conference on Earthquake Engineering, 12–17 Oct 2008, Beijing

Bayat M, Abdollahzadeh G (2011) Analysis of the steel braced frames equipped with ADAS devices under the far field records. Latin Am J Solids Struct 8:163–181

Black C, Makris N, Aiken ID (2004) Component testing, seismic evaluation and characterization of buckling-restrained braces. J Struct Eng ASCE 130:880–894

Bouc R (1971) Modèle mathématique d'hystérésis: application aux systèmes à un degré de liberté (in French). Acustica 24:16–25

Chang KC, Soong TT, Oh S-T, Lai ML (1995) Seismic behavior of steel frame with added viscoelastic dampers. J Struct Eng ASCE 121:1418–1426

Chou C-C, Tsai K-C (2002) Plasticity-fibre model for steel triangular plate energy dissipating devices. Earthq Eng Struct Dyn 31:1643–1655

Dargush GF, Sant RS (2005) Evolutionary aseismic design and retrofit of structures with passive energy dissipation. Earthq Eng Struct Dyn 34:1601–1626

Di Sarno L, Elnashai AS (2005) Innovative strategies for seismic retrofitting of steel and composite structures. Prog Struct Eng Mater 7:115–135

Di Sarno L, Manfredi G (2010) Seismic retrofitting with buckling restrained braces: application to an existing non-ductile RC framed building. Soil Dyn Earthq Eng 30:1279–1297

Hoang N, Fujino Y, Warnitchai P (2008) Optimal tuned mass damper for seismic applications and practical design formulas. Eng Struct 30:707–715

Hueste MBD, Bai J-W (2007) Seismic retrofit of a reinforced concrete flat-slab structure: part I – seismic performance evaluation. Eng Struct 29:1165–1177

Ismail M, Ikhouane F, Rodellar J (2009) The hysteresis Bouc-Wen model, A survey. Arch Comput Meth Eng 16:161–188

Lee D, Taylor DP (2001) Viscous damper development and future trends. Struct Des Tall Build 10:311–320

Lin W-H, Chopra AK (2002) Earthquake response of elastic SDF systems with non-linear fluid viscous dampers. Earthq Eng Struct Dyn 31:1623–1642

Lin J-L, Tsai K-C, Yu Y-J (2011) Bi-directional coupled tuned mass dampers for the seismic response control of two-way asymmetric-plan buildings. Earthq Eng Struct Dyn 40:675–690

Marano GC, Quaranta G, Avakian J, Palmeri A (2013) Identification of passive devices for vibration control by evolutionary algorithms, Ch. 15. In: Gandomi AH, Yang X-S, Marand ST, Alavi AH (eds) Metaheuristic applications in structures and infrastructures. Elsevier, pp 373–387. http://www.sciencedirect.com/science/article/pii/B9780123983640000152

Marriott D, Pampanin S, Bull D, Palermo A (2008) Dynamic testing of precast, post-tensioned rocking wall systems with alternative dissipating solutions. Bull N Z Soc Earthq Eng 41:90–103

Martinez-Rodrigo M, Romero ML (2003) An optimum retrofit strategy for moment resisting frames with nonlinear viscous dampers for seismic applications. Eng Struct 25:913–925

Molina FJ, Sorace S, Terenzi G, Magonette G, Viaccoz B (2004) Seismic tests on reinforced concrete and steel frames retrofitted with dissipative braces. Earthq Eng Struct Dyn 33:1373–1394

Muscolino G, Palmeri A (2007) An earthquake response spectrum method for linear light secondary substructures. ISET J Earthq Technol 44:193–211

Nakashima M, Saburi K, Tsuji B (1996) Energy input and dissipation behaviour of structures with hysteretic dampers. Earthq Eng Struct Dyn 25:483–496

Ozcan O, Binici B, Ozcebe G (2008) Improving seismic performance of deficient reinforced concrete columns using carbon fiber-reinforced polymers. Eng Struct 30:1632–1646

Palmeri A, Makris N (2008) Response analysis of rigid structures rocking on viscoelastic foundation. Earthq Eng Struct Dyn 37:1039–1063

Palmeri A, Ricciardelli F, De Luca A, Muscolino G (2003) State space formulation for linear viscoelastic dynamic systems with memory. J Eng Mech ASCE 129:715–724

Park J-H, Kim J, Min K-W (2004) Optimal design of added viscoelastic dampers and supporting braces. Earthq Eng Struct Dyn 33:465–484

Sabelli R, Mahin S, Chang C (2003) Seismic demands on steel braced frame buildings with buckling-restrained braces. Eng Struct 25:655–666

Sackman JL, Kelly JM (1979) Seismic analysis of internal equipment and components in structures. Eng Struct 1:179–190

Singh MP, Moreschi LM (2002) Optimal placement of dampers for passive response control. Earthq Eng Struct Dyn 31:955–976

Soong TT, Spencer BF Jr (2002) Supplemental energy dissipation: state-of-the-art and state-of-the-practice. Eng Struct 24:243–259

Uang C-M, Bertero VV (1990) Evaluation of seismic energy in structures. Earthq Eng Struct Dyn 19:77–90

Wen YK (1976) Method for random vibration of hysteretic systems. J Eng Mech ASCE 102:246–263

Wu Y-F, Liu T, Oehlers DJ (2006) Fundamental principles that govern retrofitting of reinforced concrete columns by steel and FRP jacketing. Adv Struct Eng 9:507–532

Xia C, Hanson RD (1992) Influence of ADAS element parameters on building seismic response. J Struct Eng ASCE 118:1903–1918

Zhang R-H, Soong TT (1992) Seismic design of viscoelastic dampers for structural applications. J Struct Eng ASCE 118:1375–1392

Passive Seismometers

Gerardo Alguacil[1] and Jens Havskov[2]
[1]Instituto Andaluz de Geofísica, University of Granada, Granada, Spain
[2]Department of Earth Science, University of Bergen, Bergen, Norway

Synonyms

Accelerometer; Amplitude response; Phase response; Response function; Seismometer

Introduction

A seismic sensor measures the ground motion and outputs a voltage, usually proportional with ground velocity. Earlier purely mechanical sensors measured the displacement of a stylus representing the amplified ground motion while newer sensors pick up the motion using a coil moving in a magnetic field. Common for these two types of sensors is that there is no electronics involved and they are therefore called passive sensors in contrast to the many new sensors with active electric circuits as an integrated part of the sensor, the so-called active sensors.

Standard Inertia Seismometer

The objective is to measure the ground motion at a point with respect to this same point undisturbed. The main difficulty is that the measurement is done in moving reference frame. So displacement cannot be measured directly and, according to the inertia principle, an inertial force will appear on a mass only if the reference frame (in this case the ground) has an acceleration, so the seismometer can only measure velocities or displacements associated with nonzero values of ground acceleration.

Since the measurements are done in a moving reference frame (the Earth's surface), almost all seismic sensors are based on the inertia of

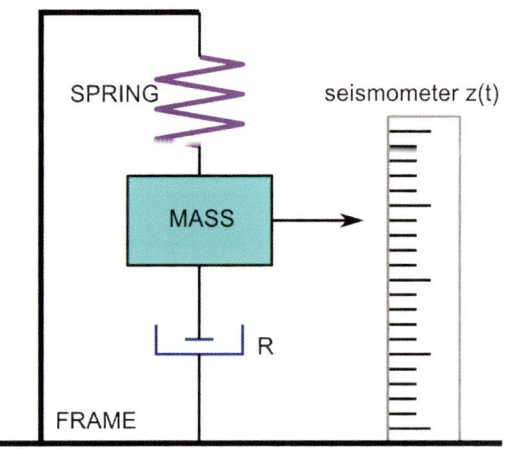

Passive Seismometers, Fig. 1 A mechanical inertial seismometer. R is a dash pot (Figure from Havskov and Alguacil 2010)

a suspended mass, which will tend to remain stationary in response to external motion. The relative motion between the suspended mass and the ground will then be a function of the ground's motion.

Figure 1 shows a simple seismometer that will detect vertical ground motion. It consists of a mass suspended from a spring. The resonance angular frequency of the mass-spring system is $\omega_0 = \sqrt{k/m}$, where $\omega_0 = 2\pi/T_0$ and T_0 is the corresponding natural period (s) of the swinging system. The motion of the mass is damped using a "dash pot" so that the mass will not swing excessively near the resonance frequency of the system. A ruler is mounted on the side to measure the motion of the mass relative to the ground.

If the ground moves with a very fast sinusoidal motion, it would be expected that the mass remains stationary and thus the ground sinusoidal motion can be measured directly as the relative mass-frame motion. The amplitude of the measurement would also be the ground's displacement amplitude and the seismometer would have a gain of 1. It is also seen that if the ground moves up impulsively, the mass moves down relative to the frame, represented by the ruler, so there is a phase shift of π (or 180°) in the measure of ground displacement. In general, a sinusoidal ground motion will produce a sinusoidal motion of the mass with the same frequency but with a frequency-dependent phase shift and amplitude. With the ground moving very slowly, the mass would have time to follow the ground motion; in other words, there would be little relative motion, the gain would be low, and there would be less phase shift. At the resonance frequency, with a low damping, the mass could get a new push at the exact right time, so the mass would move with a larger and larger amplitude, thus the gain would be larger than 1. In order to get the exact motion of the seismometer mass relative the ground motion including the phase shift, the equation for the swinging system must be solved.

If $u(t)$ is the ground's vertical motion and $z(t)$ the displacement of the mass relative to the ground, both positive upwards, there are two real forces acting on the mass m: the force of the deformed spring and the damping.

Spring force. $-kz$, negative since the spring opposes the mass displacement, k is the spring constant. This linear relation is strictly valid for small deformations only.

Damping force. $-d\dot{z}$, where d is the friction constant. Thus the damping force is proportional to the mass times the velocity and is negative since it also opposes the motion.

The acceleration of the mass relative to an inertial reference frame will be the sum of the acceleration $\ddot{z}$ with respect to the frame (or the ground) and the ground acceleration $\ddot{u}$.

Since the sum of forces must be equal to the mass times the acceleration, we have

$$-kz - d\dot{z} = m\ddot{z} + m\ddot{u} \tag{1}$$

For practical reasons, it is convenient to use ω_0 and the seismometer damping constant, $h = \frac{d}{2m\omega_0}$, instead of k and d, since both parameters are directly related to measurable quantities (see Havskov and Alguacil 2010). Equation 1 can then be written

$$\ddot{z} + 2h\omega_0 \dot{z} + \omega_0^2 z = -\ddot{u} \tag{2}$$

This equation shows that the acceleration of the ground can be obtained by measuring the relative displacement of the mass, z, and its time derivatives.

In the general case, there is no simple relationship between the sensor motion and the ground motion, and Eq. 2 will have to be solved so that the input and output signals can be related. This is most simply done assuming a harmonic ground motion

$$u(t) = U(\omega)e^{i\omega t} \tag{3}$$

where $U(\omega)$ is the complex amplitude and ω is the angular frequency. Equation 3 is written in complex form for simplicity of solving the equations and the real part represents the actual ground motion. Since a seismometer is assumed to represent a linear system, the seismometer mass motion is also a harmonic motion with the same frequency, and its amplitude is $Z(\omega)$

$$z(t) = Z(\omega)e^{i\omega t} \tag{4}$$

then

$$\begin{aligned}\ddot{u} &= -\omega^2 U(\omega)e^{i\omega t} \\ \dot{z} &= i\omega Z(\omega)e^{i\omega t} \\ \ddot{z} &= -\omega^2 Z(\omega)e^{i\omega t}\end{aligned} \tag{5}$$

Inserting in Eq. 2 and dividing by the common factor $e^{i\omega t}$, the relationship between the output and input complex amplitudes can be calculated as $T(\omega) = Z(\omega)/U(\omega)$, the so-called displacement frequency response function or transfer function:

$$T_d(\omega) = \frac{Z(\omega)}{U(\omega)} = \frac{\omega^2}{\omega_0^2 - \omega^2 + 2\omega\omega_0 hi} \tag{6}$$

From this expression, the amplitude displacement response $A_d(\omega)$ and phase response $\Phi_d(\omega)$ can be calculated as the modulus and phase of the complex amplitude response:

$$A_d(\omega) = |T_d(\omega)| = \frac{\omega^2}{\sqrt{\left(\omega_0^2 - \omega^2\right)^2 + 4h^2\omega^2\omega_0^2}} \tag{7}$$

$$\Phi_d(\omega) = a\tan\left(\frac{\operatorname{Im}(T_d(\omega))}{\operatorname{Re}(T_d(\omega))}\right) = a\tan\left(\frac{-2h\omega\omega_0}{\omega_0^2 - \omega^2}\right) \quad (8)$$

and $T_d(\omega)$ can be written in polar form as

$$T_d(\omega) = A_d(\omega)e^{i\Phi_d(\omega)} \quad (9)$$

From Eq. 7, it can be seen what happens in the extreme cases. For high frequencies

$$A_d(\omega) \to 1 \quad (10)$$

This is a constant gain of one and the sensor behaves as a pure displacement sensor.

For low frequencies,

$$A_d(\omega) \to \frac{\omega^2}{\omega_0^2} \quad (11)$$

which is proportional to acceleration and the output from the sensor is proportional to acceleration. For a high damping,

$$A_d(\omega) \approx \frac{\omega}{2h\omega_0} \quad (12)$$

and the output is proportional to ground velocity; however, the gain is low since h is high.

Figure 2 shows the amplitude and phase response of a sensor with a natural period of 1 s and damping from 0.1 to 4. As it can be seen, a low damping ($h < 1$) results in a peak in the response function. If $h = 1$, the seismometer mass will return to its rest position in the least possible time without overshooting, and the seismometer is said to be critically damped. From the shape of the curve and Eq. 7, it is seen that the seismometer can be considered a second-order high-pass filter for ground displacement. Seismometers perform optimally at damping close to critical. The most common value to use is $h = 1/\sqrt{2} = 0.707$. Why exactly this value and not 0.6 or 0.8? In practice it does not make much difference, but the value 0.707 is a convenient value to use when describing the amplitude response function. Inserting $h = 0.707$ in Eq. 7, the value for $A_d(\omega_0) = 0.707$. This is the amplitude value used to define the corner frequency of a filter or the −3 dB point. So using $h = 0.707$ means that the response can be described as a second-order high-pass Butterworth filter with a corner frequency of ω_0. This filter has the flattest possible response in its passband.

When the damping increases above 1, the sensitivity decreases, as described in Eq. 7, and the response approaches that of a velocity sensor (mass motion is proportional to ground velocity). From Fig. 2, it can be seen that for $h = 4$, the response approaches a straight line indicating a pure velocity response within a limited frequency band.

The Velocity Transducer

Nearly all traditional seismometers use a velocity transducer to measure the motion of the mass (example in Fig. 3). The principle is to have a moving coil within a magnetic field. This can be implemented by having a fixed coil and a magnet that moves with the mass or a fixed magnet and the coil moving with the mass. The output from the coil is proportional to the velocity of the mass relative to the frame, and this kind of electromagnetic seismometer is therefore called a velocity transducer. Two new constants are brought into the system:

Generator constant G. This constant relates the velocity of the mass to the output of the coil. It has a unit of V/ms^{-1}. Typical values are in the range of 30–500 V/ms^{-1}.

Generator coil resistance R_g. The resistance of the generator coil (also called signal coil) in Ohms (the coil is built as hundreds or thousands of turns of thin wire winding).

The signal coil makes it possible to damp the seismometer in a very simple way by loading the signal coil with a resistor. When a current is generated by the signal coil, it will oppose to the mass motion with a proportional magnetic force (see next section).

The frequency response function for the velocity transducer is different than for the mechanical sensor. With the velocity transducer, the observed output signal is now a voltage proportional to the mass-frame velocity $\dot{Z}(\omega) = i\omega Z(\omega)$ and G,

Passive Seismometers, Fig. 2 The amplitude and phase response functions for a seismometer with a natural frequency of 1 Hz. Curves for various levels of damping h are shown. Note that the phase shift goes toward 0 at low frequencies and toward 180° at high frequencies as qualitatively deduced

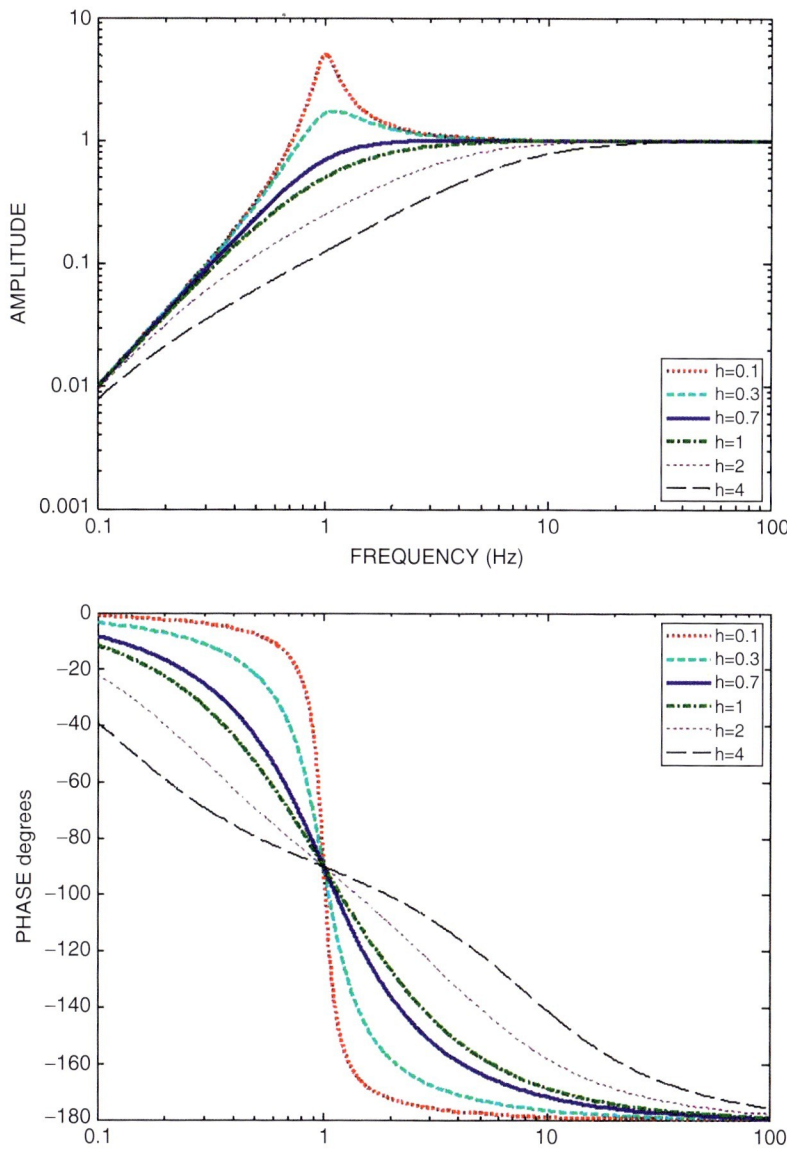

instead of $Z(\omega)$. The displacement response for the velocity sensor is then

$$T_d^v(\omega) = \frac{\dot{Z}(\omega)}{U(\omega)} G = \frac{i\omega\omega^2 G}{\omega_0^2 - \omega^2 + i2\omega\omega_0 h}$$
$$= \frac{i\omega^3 G}{\omega_0^2 - \omega^2 + i2\omega\omega_0 h} \quad (13)$$

and it is seen that the only difference compared to the mechanical sensor is the factors G and $i\omega$.

For the response curve Eq. 13, the unit is (ms^{-1}/m) $(V/ms^{-1}) = V/m$. It is assumed in Eq. 13 that a positive velocity gives a positive voltage. Most often the response for the velocity transducer is shown for input velocity, which is Eq. 13 divided by $i\omega$:

$$T_v^v(\omega) = \frac{G\omega^2}{\omega_0^2 - \omega^2 + i2\omega\omega_0 h} \quad (14)$$

and the response looks like the displacement response for a mechanical sensor (Fig. 2). It is

Passive Seismometers, Fig. 3 A model of an electromagnetic sensor. The coil resistance is R_g, the damping resistor is R, and the voltage output is V_{out}. The dashpot damping has been replaced by the damping from the coil moving in the magnetic field (Figure from Havskov and Alguacil 2010)

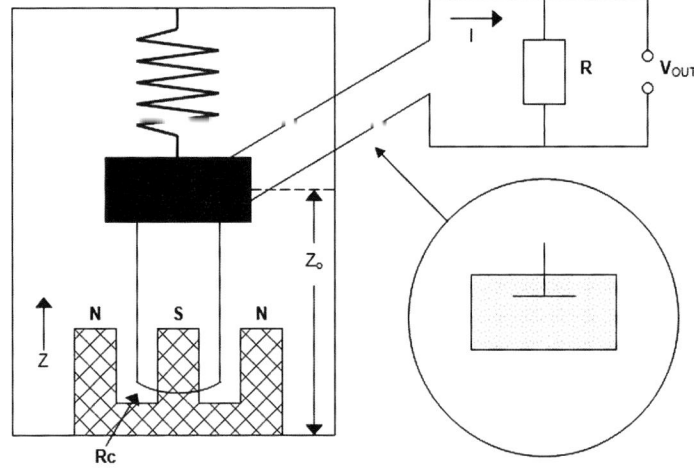

Passive Seismometers, Fig. 4 Recording of a small earthquake from a 1 Hz passive sensor. The first signal recorded is the primary wave (P) and the second is the shear wave (S)

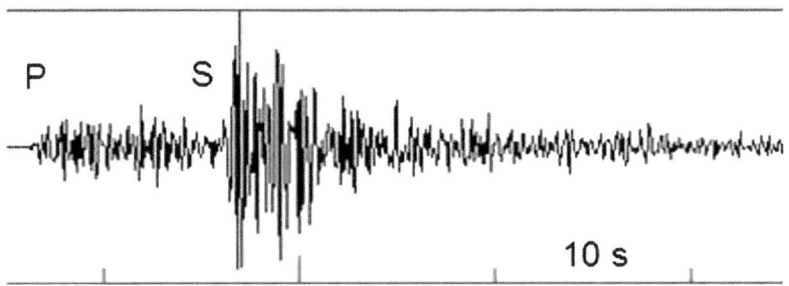

seen that the response to ground velocity is constant for $\omega > \omega_0$ and the sensor is therefore often called a velocity sensor.

In practice, the sensor output is always connected to an external resistor R_e (for damping control and because amplifiers have a finite input impedance). This forms a voltage divider. Thus the effective generator constant (or loaded generator constant) G_e becomes

$$G_e = G \cdot \frac{R_e}{R_e + R_g} \quad (15)$$

and this must be used in Eqs. 13 and 14 instead of G. All available passive seismometers are now using a velocity transducer and the most common way to use the is to digitize it and record it in a computer. The signal can then be plotted to show the so-called seismogram, an example in Fig. 4.

Other passive sensors, seldom used for earthquake recording, but quite common in structural dynamics monitoring, are piezoelectric accelerometers. An inertial mass is fixed on a piezoelectric material (normally in a multilayer arrangement), which has an elastic behavior. Due to the high elastic compliance, the resulting natural frequency of the system is also high and the strain is proportional to ground acceleration up to this frequency. Because of piezoelectricity, the system output is an electric charge proportional to the strain, so a charge amplifier (current integrator) is used to yield a voltage proportional to acceleration. In practice, the charge amplifier cannot perform a perfect integration up to DC ($\omega = 0$), so there is some low-frequency limit for the flat response to acceleration.

The piezoelectric properties of some materials are quite temperature dependent and this has to be somehow compensated for.

Damping

For a purely mechanical seismometer there is a damping h_m due to friction (mainly air friction and spring internal elastic dissipation) of the mechanical motion. In a sensor with electromagnetic transducer, a voltage E is induced in the coil proportional to velocity. If the signal coil is shunted with an external resistance, let R_T be the total circuit resistance, then a current $I = E/R_T$ (neglecting self-induction) will flow through the circuit. This will cause a force on the mass proportional to this current and in the sense opposed to its motion, as is given by Lenz law. This will introduce an additional electrical or electromagnetic damping h_e.

Thus the total damping is the sum of electromagnetic and mechanical contributions:

$$h = h_e + h_m \quad (16)$$

h_m is also called open-circuit damping, since this is the damping of the seismometer with no electrical connection. The mechanical damping, hereafter called open-circuit damping, cannot be changed and range from a very low value of 0.01 to 0.3 while the electrical damping can be regulated with the value of the external resistor to obtain a desired total damping. The electrical damping can be calculated as (Havskov and Alguacil 2010)

$$h_e = \frac{G^2}{2M\omega_0 R_T} \quad (17)$$

where M is the seismometer mass and R_T is the total resistance of the generator coil and the external damping resistor. Seismometer specifications often give the critical damping resistance CDR, which is the total resistance $CDR = R_T$ required to get a damping of 1.0.

From Eqs. 17 and 16, it is seen that if the total damping h_1 is known for one value of R_T, R_{T1}, the required resistance R_{T2} for another required total damping h_2 can be calculated as

$$R_{T2} = R_{T1} \frac{h_1 - h_m}{h_2 - h_m} \quad (18)$$

If the mechanical damping is low ($h_m \approx 0$), Eq. 18 can be written in terms of CDR ($h_2 = 1$, $CDR = R_{T1}$) as

$$R_{T2} = \frac{CDR}{h_2} \quad (19)$$

and thus, the desired total resistance for any required damping can easily be calculated from CDR.

As an example, consider the classical 1 Hz sensor, the Geotech S13 (Fig. 15). The coil resistance is 3,600 Ω and $CDR = 6,300$ Ω. Since the open-circuit damping is low, it can be ignored. The total resistance to get a damping of 0.7 would then, from Eq. 19, be $R = 6,300/0.7 = 9,000$ Ω, and the external damping resistor to connect would have a value of 9,000–3,600 = 5,400 Ω.

Construction of Passive Sensors

The mass-spring system of the vertical seismometer serves as a very useful model for understanding the basics of seismometry. However, in practical design, this system is too simple, since the mass can move in all directions as well as rotate. So, nearly all seismometers have some mechanical device which will restrict the motion to one translational axis. Figure 5 shows how this can be done in principle for a vertical seismometer.

It can be seen that due to the hinged mass, the sensor is restricted to move vertically. The mass does not only move in the z direction but in a circular motion with the tangent to the vertical direction. However, for small displacements, the motion is sufficiently linear. The above pendulum arrangement is in principle the most common way to restrict motion and can also be used for horizontal seismometers. Pendulums are also sensitive to angular motion in seismic waves, which normally is so small that it has no practical importance.

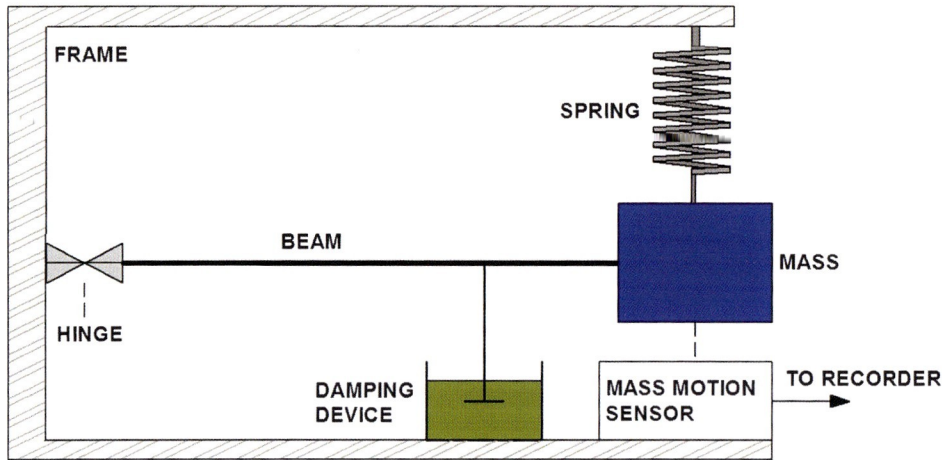

Passive Seismometers, Fig. 5 A vertical mass-spring seismometer where the horizontal motion has been restricted by a horizontal hinged rod. Usually, the hinge is a thin flexible leaf to avoid friction (Figure from Havskov and Alguacil 2010)

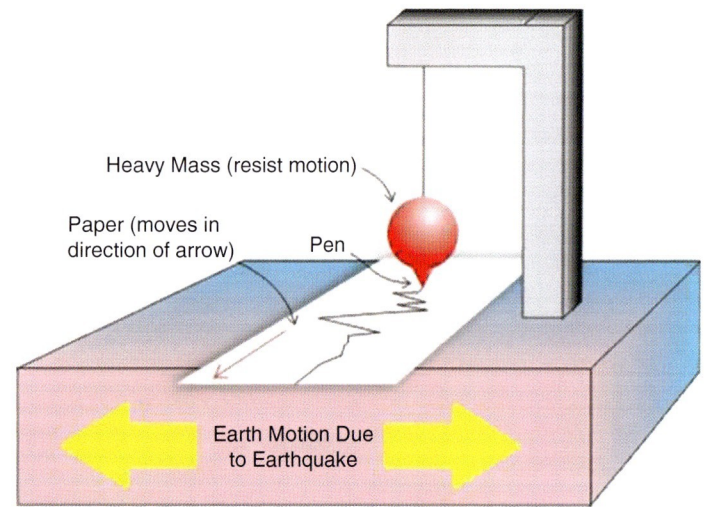

Passive Seismometers, Fig. 6 Horizontal seismometer made with a pendulum (From http://www.azosensors.com/)

So far only a vertical seismometer has been described; however, the ground motion must be measured in all three directions, x, y, and z. Normally z is positive up, x is to the east, and y to the north. A horizontal seismometer is in its simplest form a pendulum (see Fig. 6).

For a small-size mass m compared to the length L of the string, the natural frequency

$$\omega_0 = \sqrt{g/L} \qquad (20)$$

where g is the gravitational constant. For small translational ground motions, the equation of motion is identical to Eq. 1 with z replaced with the angle of rotation. Note that ω_0 is independent of the mass.

It is relatively easy to make sensors with a natural frequency down to 1 Hz (so-called short-period (SP) seismometers), but in seismology it is desirable to measure much lower frequencies using so-called long-period (LP)

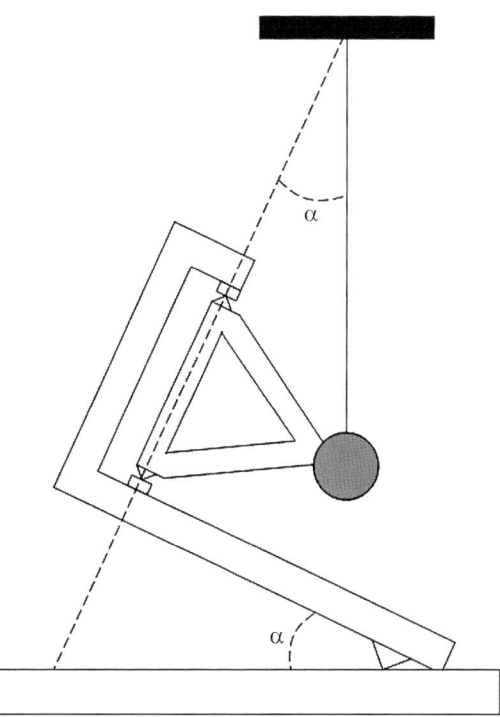

Passive Seismometers, Fig. 7 Principle of the garden-gate pendulum. The tilt angle is exaggerated. A string pendulum must have length L to have the same period, so at a small angle α, L becomes very large (Figure from Havskov and Alguacil 2010)

seismometers, and a good station with a passive sensor should be able to measure down to ideally 0.01 Hz. For the vertical seismometer, this would require a very soft spring combined with a heavy mass, which is not possible in practice. For the pendulum, L would have to be very long. For a 1 Hz seismometer, it would be 9.8 m. So there are various ways of making the natural frequency smaller without using a very large design; however, in practice it is hard to achieve less than 0.03 Hz with a passive seismometer, while active seismometers can go down to 0.003 Hz.

For a horizontal seismometer, the simplest solution is the "garden-gate" pendulum (Fig. 7). The mass moves in a nearly horizontal plane around a nearly vertical axis. The restoring force is now $g \sin(\alpha)$ where α is the angle between the vertical and the rotation axis, so the natural frequency becomes

$$\omega_0 = \sqrt{g \sin(\alpha)/L} \qquad (21)$$

where L is the vertical distance from the mass to the point where the rotation axis intersects the vertical above the mass (see Fig. 7).

To obtain a natural frequency of 0.05 Hz with a pendulum length of 20 cm will require a tilt of $0.1°$. This is close to the lowest stable period that has been obtained in practice with these instruments. Making the angle smaller makes the instrument very sensitive to small tilt changes. The "garden gate" was one of the earliest designs for long-period horizontal seismometers. It is still in use in a few places but no longer produced. Until the new broadband sensors were installed in the GSN (Global Seismographic Network), this kind of sensors was used in the WWSSN (World-Wide Standardized Seismograph Network) (see, e.g., Peterson and Orsini 1976).

The *astatic* spring geometry for vertical seismometers was invented by LaCoste (1934). The principle of the sensor is shown in Fig. 8. The sensor uses a "zero-length spring" which is designed such that the force $F = k \cdot L$, where L is the *total* length of the spring. Normal springs used in seismometers do not behave as zero-length springs, since $F = k \cdot \Delta L$, where ΔL is the change in length relative to the unstressed length of the spring. However, it is possible to construct a "zero-length spring" by, e.g., twisting the wire as it is wound into a spring. The physical setup is to have a mass on a beam and supported by the spring so that the mass is free to pivot around the lower left hand point.

This system can be constructed to have an infinite free period, which means that the vertical restoring force must cancel the gravity force at any mass position. Therefore, if the mass is at equilibrium at one angle, it will also be at equilibrium at another angle, which is similar to what was obtained for the garden-gate horizontal seismometer. Qualitatively, what happens is that if, e.g., the mass moves up, the spring force lessens, but it turns out the force from gravity in direction of mass motion is reduced by the exact same amount, due to the change in angle as will be shown below.

Passive Seismometers, Fig. 8 The principle behind the LaCoste suspension. The mass m is sitting on a hinge, which has an angle α with the horizontal and suspended by a spring of length L (Figure from Havskov and Alguacil 2010)

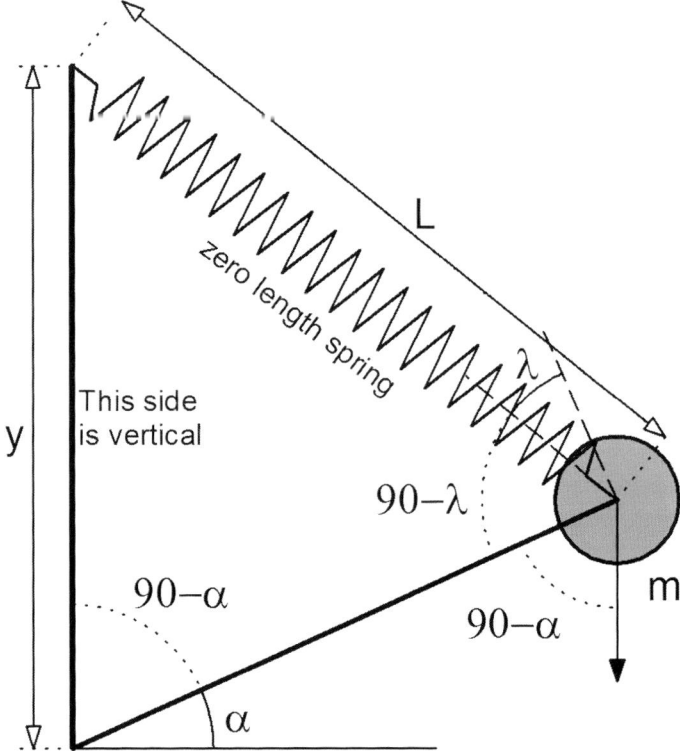

The gravity force F_g acting on the mass in direction of rotation can be written as

$$F_g = mg \cos(\alpha) \quad (22)$$

while the spring restoring force F_s acting in the opposite sense is

$$F_s = kL \cos(\lambda) \quad (23)$$

λ can be replaced by α using the law of sines:

$$\frac{L}{\sin(90-\alpha)} = \frac{y}{\sin(90-\lambda)} \text{ or } \cos(\lambda)$$
$$= \frac{y}{L}\cos(\alpha) \quad (24)$$

Equating F_s and F_g and including the expression for $\cos(\lambda)$ gives

$$ky = mg \text{ or } y = mg/k \quad (25)$$

As long as this condition holds, the total force is zero independent of the angle. As with the garden-gate seismometer, this will not work in practice and, by inclining the vertical axis slightly, any desired period within mechanical stabilization limits can be achieved. In practice, it is difficult to use free periods larger than 20–30 s.

The astatic leaf-spring suspension (Wielandt and Streckeisen 1982) is comparable to the LaCoste suspension, but simpler to make (Fig. 9). The delicate equilibrium of forces in astatic suspensions makes them sensitive to external disturbances, so they are difficult to operate without a stabilizing feedback system.

The LaCoste pendulums can be made to operate as vertical seismometers, as has been the main goal or as sensors with an oblique axis. Sensors with an oblique axis are used to obtain both vertical and horizontal motions. The normal arrangement for a 3-component sensor is to have 3 sensors oriented in the Z, N, and E directions. Since horizontal and vertical seismometers differ in their construction, it is difficult to make them equal. A three-component sensor can be constructed by using three identical sensors whose axes U, V, and W are inclined against the vertical like the edge of

Passive Seismometers, Fig. 9 Leaf-spring astatic suspensions. The figure to the *left* shows a vertical seismometer and to the *right* an oblique axis seismometer (Figure from http://jclahr.com/science/psn/wielandt/node15.html)

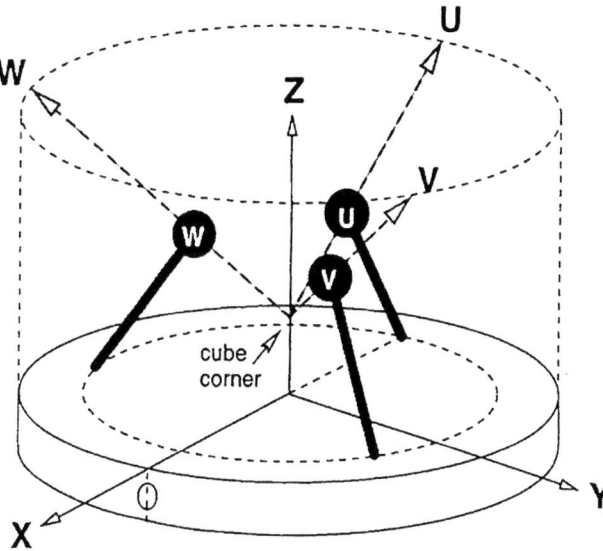

Passive Seismometers, Fig. 10 The triaxial geometry of the STS-2 seismometer. The oblique components are W, V, and U (Figure from Havskov and Alguacil 2010)

a cube standing on its corner (Fig. 10). Each sensor is made with an *astatic* leaf-spring suspension (Fig. 9). The angle of inclination is $\tan^{-1}\sqrt{2} = 54.7$ degrees, which makes it possible to electronically recombine the oblique components to X, Y, and Z simply

$$\begin{pmatrix} X \\ Y \\ Z \end{pmatrix} = \frac{1}{\sqrt{6}} \cdot \begin{pmatrix} -2 & 1 & 1 \\ 0 & \sqrt{3} & -\sqrt{3} \\ \sqrt{2} & \sqrt{2} & \sqrt{2} \end{pmatrix} \begin{pmatrix} U \\ V \\ W \end{pmatrix} \quad (26)$$

This arrangement is now the heart of some modern active seismic sensors with a natural frequency below 0.1 Hz. Figure 11 shows an example of a long-period recording.

For more details on passive seismic sensors, see also the New Manual of Seismological Observatory Practice (NMSOP-2), Bormann (2012), Aki and Richards (1980), and the old Manual of Seismological Observatory Practice, Wilmore (1979).

Sensitivity of Passive Sensors

The output from a passive sensor V_{out} is, above its natural frequency,

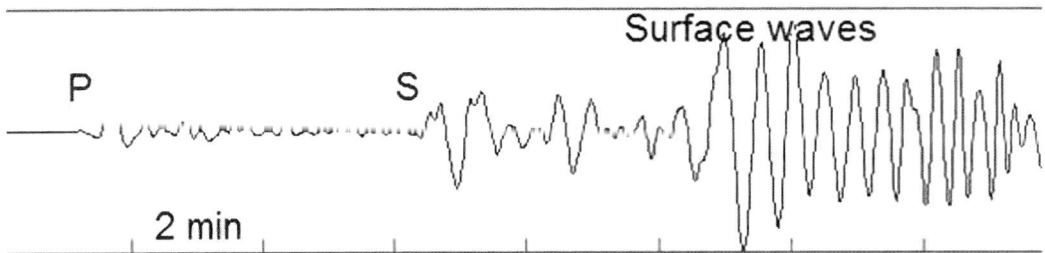

Passive Seismometers, Fig. 11 Example of recording an earthquake using a long-period sensor. Note the low frequency of the signal. The distance from the location of the earthquake (epicenter) to the recording station is 3,500 km. The harmonic waves at the end of the record are surface waves traveling along the Earth's surface

$$V_{\text{out}} = GV \qquad (27)$$

where V is the ground velocity. At a low-noise site, the ground velocity amplitude is typically 10 nm/s and a typical value of the generator constant is 100 V/ms^{-1} so the output would be 1 μV. This is the ground seismic signal. The sensor itself will, since it has no active electronics, usually produce less than 10 times this in electrical noise from the Brownian motion of the mass and the current flowing through the coil. So in general, passive seismic sensors can easily be made sensitive enough for most sites. The low-level output must be amplified to be recorded. For analog recording, this means a sensitive amplifier (usually with a noise level of 0.1 μV), while modern recording systems use digital recording. Many digital recorders will have a noise level also of 0.1 microvolt, meaning the number 1 (one count) corresponds to 0.1 μV and the seismic noise signal would then be recorded with 10 counts of resolution. Active sensors usually have higher electrical outputs, so not all digital recorders have enough sensitivity for passive sensors. For frequencies f below the natural frequency, the response decreases proportional to f^2. So, for a 1 Hz sensor, the sensitivity is down by a factor of 100 at 0.1 Hz. Since the Earth's natural background noise also increases for frequencies below 1 Hz until about 0.2 Hz (Fig. 12), a 1 Hz seismometer will in practice have sufficient sensitivity for most seismological applications down to 0.1 Hz.

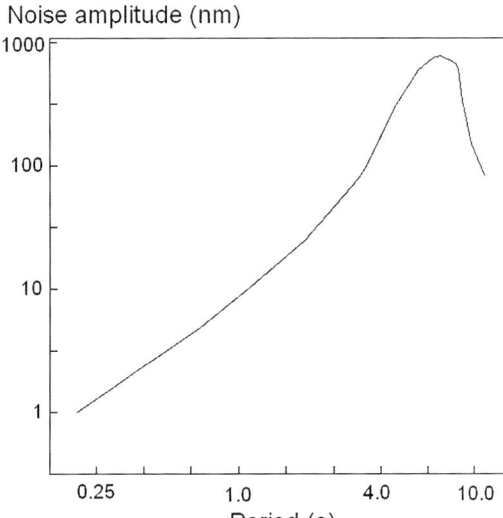

Passive Seismometers, Fig. 12 Typical amplitude of the microseismic background noise as a function of period(s) (The data for the plot are taken from Brune and Oliver 1959)

Examples of Some Passive Sensors

Passive sensors are today only sold for sensors with a natural frequency at 1 Hz and higher. For sensors with a lower natural frequency, active sensors are now both smaller, better, and cheaper than the old passive sensors and therefore no longer constructed. For earthquake seismology, it is particularly sensors with natural frequencies of 1, 2 and 4.5 Hz that are used, with the 1 Hz sensor the natural choice and there are still hundreds of seismic stations with 1 Hz sensors.

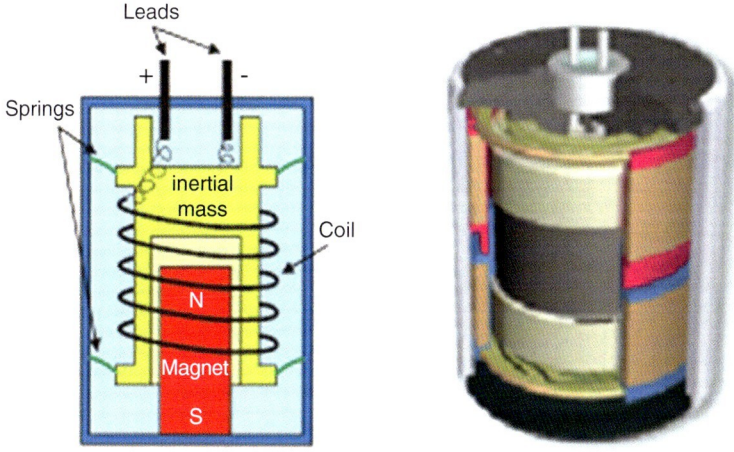

Passive Seismometers, Fig. 13 Typical construction of a geophone. Note the leaf-spring suspension. The magnet is fixed to the case and the coil moves with the mass http://vibration.desy.de/equipment/geophones/

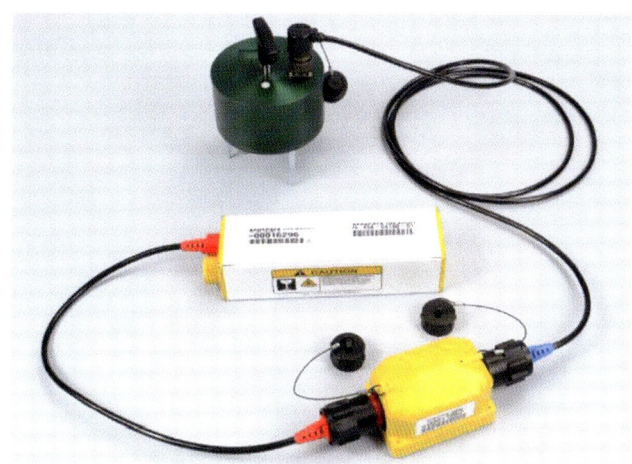

Passive Seismometers, Fig. 14 The Geospace GSX recorder (*box below*) connected to a small battery and a Geospace three-component 2 Hz geophone (*green cylinder*) (Figure from www.geospace.com)

Sensors with a frequency up to 4.5 Hz and above are constructed in the thousands and mostly used for exploration purposes and they are usually called geophones, in contrast to the 1 Hz sensor which is usually called a seismometer.

Geophone: An example is seen in Fig. 13. It is small in size (about 3×5 cm) and only costs about $100, since it is produced in large numbers for seismic exploration. It typically has a mass of 20 g and the generator constant is often around 30 V/ms^{-1}.

The sensor is very simple and robust to use. Due to its low sensitivity, it requires more amplification than the standard sensors. It has traditionally not been used for earthquake recording as much as the 1 Hz sensor; however, with more sensitive and higher dynamic range recorders, it is now possible to use it directly without any special filtering and obtain good recordings down to 0.3 Hz by post processing.

A modern 2 Hz, three-component geophone (Fig. 14). This is a very compact 2 Hz sensor used both for earthquake recording and seismic exploration. The figure shows how compact it is possible to make a complete seismograph.

A classical 1 Hz seismometer, the S13 from Geotech (see Fig. 15): This 1 Hz seismometer is one of the most sensitive passive 1 Hz seismometers produced and it has been the standard by which other 1 Hz seismometers are measured. It is suspended by both leaf and helical springs in such a way that by an internal adjustment, the sensor can be used both as a vertical and horizontal seismometer. It is still produced.

Another classical 1 Hz seismometer from Kinemetrics (Fig. 16). This seismometer was designed for the Moon and then later produced in a terrestrial version and is still produced. It can work both as a horizontal and vertical seismometer by an external adjustment.

A long-period horizontal seismometer (Fig. 17). This seismometer was used by the WWSSN and was thus produced in large numbers. The longest practical free period was 30 s and it was mostly used with a 25 s period. At very stable sites, periods larger than 30 s have been used.

Summary

Passive seismic sensors function without internal electronics, so no power supply is required, and they have been in operation for more than 100 years. All modern passive sensors are of the so-called velocity types, meaning that the voltage output is proportional to the ground velocity for frequencies above the sensor's natural frequency. Passive LP sensors are no longer sold. Compared to active sensors, passive SP sensors are low cost, very stable, and very sensitive but cannot easily resolve signals for frequencies below 1 Hz, although a good 1 Hz sensor can provide useful signals down to 0.1 Hz. Passive sensors (both 1 and

Passive Seismometers, Fig. 15 Geotech S13 seismometer (The picture is from www.bgr.bund.de/EN/Themen/Seismologie/Seismologie/Seismometer_Stationen/stationen_node.html)

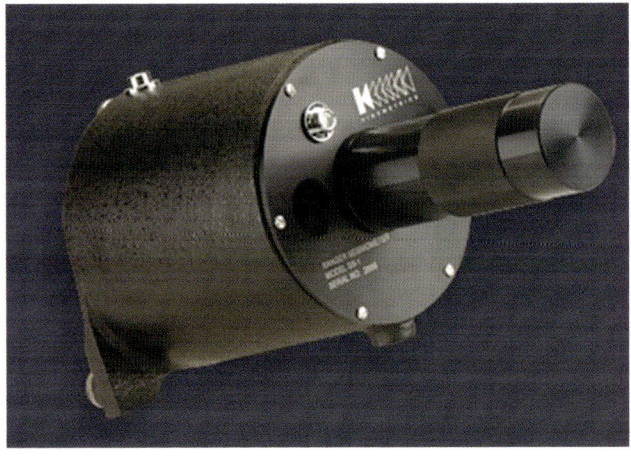

Passive Seismometers, Fig. 16 The ranger 1 Hz seismometer from Kinemetrics (Figure from www.kinemetrics.com)

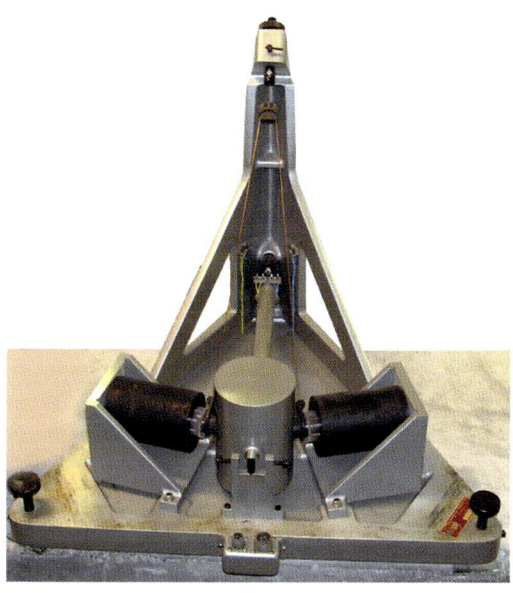

Passive Seismometers, Fig. 17 A typical horizontal long-period seismometer which was made by Sprengnether (no longer exists). The two *black* cylinders seen at each side are the magnets and the coils are moving inside with the mass in the *center*. Leveling is done with the three screws at the corners of the base plate. Once the horizontal leveling is done (front screws), the back screw is used to adjust the period. The base plate side length is 63 cm and the weight 45 kg

4.5 Hz) are still a good choice for recording small earthquakes, while the higher-frequency geophones are mostly used for seismic exploration.

Cross-References

▶ Principles of Broadband Seismometry
▶ Seismic Accelerometers
▶ Seismic Instrument Response, Correction for
▶ Sensors, Calibration of

References

Aki K, Richards PG (1980) Chapter 10: Principles of seismometry. In: Quantitative seismology – theory and methods, vol 1. W. H. Freeman, San Francisco, pp 477–524
Bormann P (ed) (2012) New manual of seismological observatory practice (NMSOP-2). IASPEI, GFZ German Research Centre for Geosciences, Potsdam; nmsop.gfz-potsdam.de
Brune JN, Oliver J (1959) The seismic noise at the earth's surface. Bull Seismol Soc Am 49:349–353
Havskov J, Alguacil G (2010) Instrumentation in earthquake seismology. Springer, Dordrecht, 358 pp
LaCoste LJB (1934) A new type long period seismograph. Physics 5:178–180
Peterson J, Orsini NA (1976) Seismic research observatories: upgrading the world wide seismic data network. Eos Trans Am Geophys Union 57:548–546
Wielandt E, Streckeisen G (1982) The leaf-spring seismometer: design and performance. Bull Seismol Soc Am 72:2349–2367
Wilmore PL (ed) (1979) Manual of seismological observatory practice, report SE-20. World Data Center A for Solid Earth Geophysics/US Department of Commerce/NOAA, Boulder

Performance-Based Design Procedure for Structures with Magneto-Rheological Dampers

Yunbyeong Chae[1], Baiping Dong[2],
James M. Ricles[3] and Richard Sause[4]
[1]Department of Civil and Environmental Engineering, Old Dominion University, Norfolk, VA, USA
[2]Department of Civil and Environmental Engineering, Lehigh University, Bethlehem, PA, USA
[3]Bruce G. Johnston Professor of Structural Engineering, Department of Civil and Environmental Engineering, Lehigh University, Bethlehem, PA, USA
[4]Joseph T. Stuart Professor of Structural Engineering, Department of Civil and Environmental Engineering, Lehigh University, Bethlehem, PA, USA

Synonyms

Magneto-rheological damper; Performance-based seismic design; Simplified design procedure

Introduction

It is well-known that supplemental damping devices increase the energy dissipation capacity of structures, reducing the seismic demand on the primary structure (Constantinou et al. 1998; Soong and Dargush 1997). A structural system with supplemental dampers is often represented by an equivalent linear system. Kwan and Billington (2003) derived optimal equations for the equivalent period and damping ratio of SDOF systems with various nonlinear hysteresis loops based on time-history analysis and regression analysis. Symans and Constantinou (1998) studied the dynamic behavior of SDOF systems with linear or nonlinear viscous fluid dampers and derived an equation for the equivalent damping ratio of the nonlinear viscous fluid damper. Ramirez et al. (2002) proposed a simplified method to estimate displacement, velocity, and acceleration for yielding structures with linear or nonlinear viscous dampers. Lin and Chopra (2003) investigated the behavior of SDOF systems with a diagonal brace in series with a nonlinear viscous damper by transforming the system to an equivalent linear Kelvin model.

Fan (1998) investigated the behavior of nonductile reinforced concrete frame buildings with viscoelastic dampers. He derived an equivalent elastic-viscous model based on the complex stiffness and energy dissipation of the viscoelastic system and proposed a simplified design procedure for a structure with viscoelastic dampers. Lee et al. (2005, 2009) applied this method to structures with elastomeric dampers and validated the simplified design procedure by comparing the design demand with the results from nonlinear time-history analysis.

In this paper, a systematic procedure for the design of structures with MR dampers, referred to as the *Simplified Design Procedure* (SDP), is developed. The procedure is similar to that developed by Lee et al. (2005, 2009), but with modifications to account for the characteristics of the MR dampers. A quasistatic MR damper model for determining the loss factor and the effective stiffness of an MR damper is introduced and incorporated into the procedure to calculate the design demand for the structure with MR dampers. The procedure is evaluated by comparing the predicted design demand to the seismic response determined from nonlinear time-history analysis.

Simplified Design Procedure (SDP)

In the SDP developed by Lee et al. (2005, 2009), the supplemental damper properties are represented by β, which is the ratio of the damper stiffness per story in the global direction to the lateral load resisting frame story stiffness, k_0, without dampers and braces of the structural system. The structural system with dampers is converted into a linear elastic system characterized by the initial stiffness of the structure, α (the ratio of brace stiffness per story in the global direction to the lateral load resisting frame story stiffness k_0), β, and η. α, β, and η may vary among the stories of the structure. By conducting an elastic-static analysis with the RSA method, the design demand for the structure is determined.

Since the loss factor of an MR damper depends on the displacement of the structure, as can be shown below in Eq. 8, the SDP for an elastomeric damper developed by Lee et al. (2005, 2009) needs to be modified for structures with MR dampers. The loss factor η is associated with the energy dissipation of the damper over a cycle. For the purpose of determining the energy dissipation over a cycle of displacement, the properties of the MR damper are assumed to remain constant. The linearization and energy dissipation of an MR damper are discussed later.

Figure 1 summarizes the SDP for structures with MR dampers. In Step 1, the seismic performance objectives and associated design criteria are established for the design of the structure. In Step 2, the structure is designed without MR dampers in accordance with the design code selected in Step 1 to satisfy the strength requirement for the members in the structure. In Step 3, the MR dampers are incorporated into the design of the structure to satisfy the specified performance objectives. The design demand for the structure is estimated for a range of selected values for α, β, and a constant loss factor $\eta = 4/\pi$

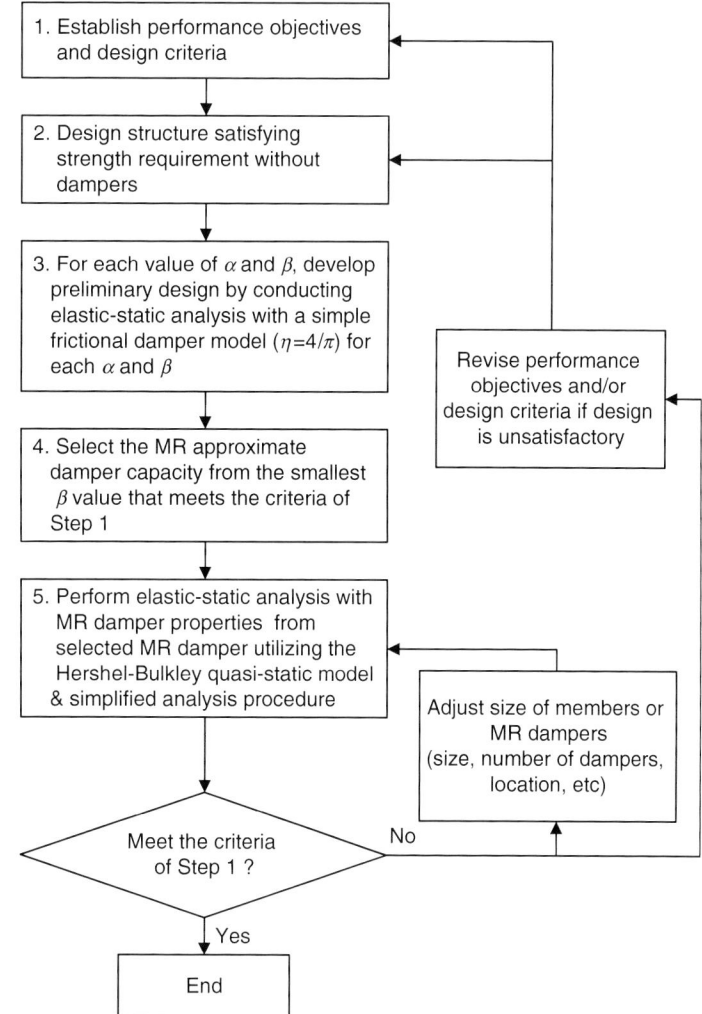

Performance-Based Design Procedure for Structures with Magneto-Rheological Dampers, Fig. 1 Generalized simplified design procedure (SDP) for structures with MR dampers

using a *simplified analysis procedure*, discussed later. In the simplified analysis procedure, the MR damper behavior is based on the simple frictional MR damper model (Chae 2011). The required MR damper sizes are then selected in Step 4 based on the smallest β value that meets the design criteria and performance objectives in Step 1. Since the simple frictional damper model does not account for the velocity dependent behavior of an MR damper, a more accurate determination of the design demand is determined in Step 5 using a more sophisticated MR damper model (i.e., Hershel-Bulkley model) in the simplified analysis procedure. The design is then revised with final member sizes and the MR damper sizes are selected (location, number, force capacity, etc.). If the performance objectives cannot be met in an economical manner, then the performance objectives and/or structural system design need to be revised as indicated in Fig. 1.

Equivalent Linear System for an MR Damper

The SDP requires that the structure with the nonlinear MR dampers be linearized. In order to linearize the system for estimating the response of structures with MR dampers, the

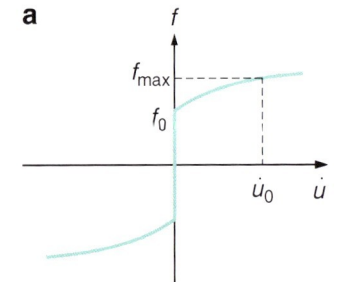
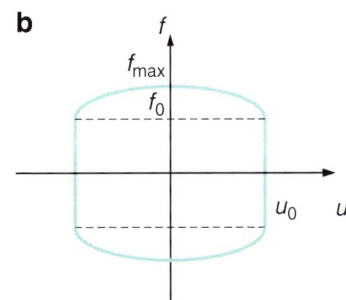

Performance-Based Design Procedure for Structures with Magneto-Rheological Dampers, Fig. 2 Hershel Bulkley visco-plasticity MR damper model: (**a**) force-velocity relationship; (**b**) force-displacement relationship

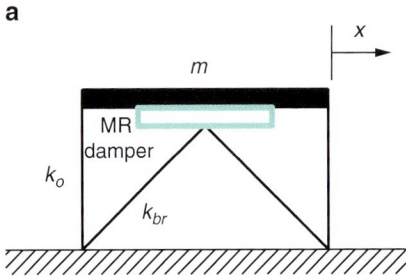
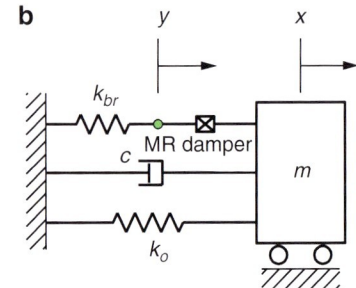

Performance-Based Design Procedure for Structures with Magneto-Rheological Dampers, Fig. 3 SDOF system: (**a**) schematic of equivalent SDOF system with MR damper and brace; (**b**) mechanical model

Hershel-Bulkley quasi-static MR damper model is used. Figure 2 shows the force-velocity and force-displacement relationships for the model where the damper force f is given as

$$f = \text{sign}(\dot{u})\{f_0 + C|\dot{u}|^n\} \quad (1)$$

In Eq. 1 u is the damper displacement relative to the initial position of the damper and $\dot{u}$ is the damper velocity. f_0 is the frictional force. C and n are the coefficients of the nonlinear dashpot. Suppose that the MR damper is subjected to a harmonic displacement motion

$$u(t) = u_0 \sin(\omega t) \quad (2)$$

where u_0 is the amplitude of displacement and ω is the excitation frequency of the damper. The energy dissipated by the damper over one cycle of the harmonic motion is equal to

$$E_{MRD} = \int_0^{\frac{2\pi}{\omega}} f(t)\dot{u}(t)dt$$
$$= 4f_0 u_0 + 2^{n+2} C\gamma(n) u_0^{1+n} \omega^n \quad (3)$$

where

$$\gamma(n) = \frac{\Gamma^2\left(1+\frac{n}{2}\right)}{\Gamma(2+n)} \quad (4)$$

In Eq. 4, $\Gamma()$ is the gamma function (Soong and Dargush 1997). In general, diagonal bracing is installed in the building in series with the dampers. Therefore, the energy dissipation of an MR damper needs to be studied considering the stiffness of the diagonal bracing. Figure 3 shows an SDOF system with an MR damper and diagonal bracing. Under the harmonic motion $x = x_0 \sin(\omega t)$, the maximum damper displacement, u_{d0}, and velocity, $\dot{u}_{d0}$, of the MR damper occurs when the displacement x and velocity $\dot{x}$ are a maximum, respectively, where u_{d0} and $\dot{u}_{d0}$ can be calculated as (Chae 2011)

$$u_{d0} = x_0 - f_0/k_{br} \quad (5a)$$

$$\dot{u}_{d0} = \dot{x}_0 = x_0 \omega \quad (5b)$$

In Eq. 5a k_{br} is the stiffness of the diagonal bracing. Substitution of Eq. 5a into Eq. 3 results in the expression for the energy dissipation of the MR damper for a SDOF system

$$E_{MRD} = 4f_0 u_{d0} + 2^{n+2} C\gamma(n) u_{d0}^{n+1} \omega^n \quad (6)$$

The strain energy of the MR damper, E_S, is calculated from the equivalent stiffness of the damper and the maximum damper displacement:

$$E_S = \frac{1}{2} k_{eq} u_{d0}^2 \quad (7)$$

where k_{eq} is the equivalent stiffness of the MR damper, defined as $k_{eq} = f_0/u_{d0}$ and based on the secant stiffness from the force-displacement relationship of the Hershel-Bulkley model. The loss factor η of the MR damper by definition is

$$\eta = \frac{1}{2\pi} \frac{E_{MRD}}{E_S} = \frac{4\{f_0 + C\gamma(n)(2u_{d0}\omega)^n\}}{\pi k_{eq} u_{d0}} \quad (8)$$

Simplified Analysis Procedure

The simplified analysis procedure provides an elastic-static method for calculation of the design demand of an MDOF system with MR dampers. The simplified analysis procedure utilizing the response spectrum analysis (RSA) method is summarized in Figs. 4 and 5.

In the simplified analysis procedure, the maximum structural displacements are determined by the well-known equal displacement rule. In the equal displacement rule, the maximum displacement of the nonlinear structure, whose lateral stiffness is based on its initial tangent stiffness, is assumed to be equal to that of a linear structure. The equal displacement rule is only applicable to structures that lie in the low-frequency and medium-frequency spectral regions (Newmark and Hall 1973).

In order to obtain the equivalent period for an MDOF structure with MR dampers, the combined stiffness of the MR dampers and diagonal bracing needs to be added to the stiffness of the structure. Thus, the global effective stiffness of the MDOF system is given as

$$\mathbf{K_{eff}} = \mathbf{K_0} + \mathbf{K_{brsystem}} \quad (9)$$

where $\mathbf{K_0}$ is the stiffness of the structure without diagonal braces and MR dampers, and $\mathbf{K_{brsystem}}$ is the stiffness associated with the braces and MR dampers. The structure is assumed to have N DOF, thus the dimension of $\mathbf{K_{eff}}$ is $N \times N$. The combined stiffness $K_{brsystem}^i$ of the diagonal bracing and MR damper at the i-th MR damper location is

$$K_{brsystem}^i = \frac{k_{br}^i k_{eq}^i}{k_{br}^i + k_{eq}^i} \quad (10)$$

where k_{br}^i and k_{eq}^i are the horizontal stiffness of the diagonal bracing and the MR damper associated with the ith MR damper, respectively. k_{eq}^i can be calculated utilizing the secant stiffness method as noted above. The individual combined stiffnesses based on Eq. 10 are appropriately assembled to form $\mathbf{K_{brsystem}}$. The effective periods and mode shapes of the structure are then obtained by performing an eigenvalue analysis of the structure, considering the seismic mass of the structure.

The equivalent damping ratio ξ_{eq} of an MDOF system is determined using the lateral force energy method proposed by Sause et al. (1994), where

$$\xi_{eq} = \frac{1}{2} \frac{\sum_{i=1}^{L} \eta_i F_d^i u_d^i}{\mathbf{F}^T \mathbf{x}_0} + \xi_{in} \quad (11)$$

In Eq. 11 η_i and u_d^i are the loss factor and maximum damper displacement of the ith MR damper, respectively, and L is the number of MR dampers. Since the damper displacement is unique for each MR damper, the loss factor of each MR damper, which is a function of damper displacement, is also unique for each damper. For each damper η_i is obtained from Eq. 8. ξ_{in} in

Given:
- MR damper properties: f_0^i, C_i, n_i (i: index for MR damper location)
- Structural properties: $\mathbf{M}$, $\mathbf{K_0}$, k_{br}^i, ξ_{in}^j (inherent damping ratio of the j-th mode)

Step 1. Assume $\mathbf{x_0}$ and set $\omega_{eff}^1 = \omega_n^1$ (fundamental frequency of structure without MR dampers)

Step 2. Determine maximum damper displacements
$$u_{d0}^i = u_0^i - f_0^i / k_{br}^i$$
u_0^i: maximum deformation of damper and bracing of the ith MR damper

Step 3. Calculate equivalent stiffness of each MR damper
$$k_{eq}^i = f_0^i / u_{d0}^i$$

Step 4. Determine $K_{brsystem}^i = \frac{k_{br}^i k_{eq}^i}{k_{br}^i + k_{eq}^i}$ for each MR damper and update $\mathbf{K_{eff}}$

Step 5. Update modal frequency ω_{eff}^j and modal vector $\boldsymbol{\phi}_j$ ($j = 1,...,N$)
$$\omega_{eff}^j, \boldsymbol{\phi}_j \leftarrow \text{eig}(\mathbf{K_{eff}}, \mathbf{M})$$
where $\mathbf{M}$ is the mass matrix of the structure

Step 6. Calculate loss factor of each MR damper
$$\eta_i = \frac{4\{f_0^i + C_i\gamma(n_i)(2u_{d0}^i \omega_{eff}^1)^{n_i}\}}{\pi k_{eq}^i u_{d0}^i}$$
where ω_{eff}^1 is the fundamental modal frequency

Step 7. Perform modal analysis described in Figure 5

Step 8. Apply the modal combination rule (e.g., SRSS or CQC) to get the final displacement $\mathbf{x_0}$ and velocity of each MR damper $\dot{u}_{d0}^i$
$\mathbf{x_0}$ = function of $(\mathbf{x_0^1},...,\mathbf{x_0^N})$,
$\dot{u}_{d0}^i$ = function of $(\dot{u}_{d0}^{i1},...,\dot{u}_{d0}^{iN})$

Step 9. Repeat **Step 2** ~ **Step 8** until $\mathbf{x_0}$ convergence is achieved.

Step 10. Calculate maximum damper force for each MR damper
$$f_{max}^i = f_0 + C(\dot{u}_{d0}^i)^n$$

Performance-Based Design Procedure for Structures with Magneto-Rheological Dampers, Fig. 4 Simplified analysis procedure used to design MDOF structures with passive MR dampers utilizing response spectrum analysis (RSA) method

Eq. 11 is the inherent damping ratio and $\mathbf{x}_0$ is the vector of the displacements of the structure that develop under the lateral force $\mathbf{F}$. The individual damper force F_d^i and the lateral force vector $\mathbf{F}$ are defined as

$$F_d^i = k_{eq}^i u_d^i, \quad \mathbf{F} = \mathbf{K_{eff}} \mathbf{x}_0 \qquad (12)$$

In the simplified analysis procedure using the RSA, the relationships in Eq. 12 are substituted into Eq. 11, and the inherent damping ξ_{in} and $\mathbf{x}_0$ from each mode are considered, as indicated in Substep 3 in Fig. 5.

Performance-Based Design of a Three-Story Building with MR Dampers

Prototype Building Structure

Based on the proposed SDP, a three-story building with MR dampers is designed. The floor plan and elevation of the prototype structure is shown

Step 7. For $j=1$ to N^{th} mode

    Substep 1. Assume modal displacement vector $\mathbf{x}_0^j$
    Substep 2. Determine maximum displacement for each damper u_{d0}^{ij}
$$u_{d0}^{ij} = u_0^{ij} - f_0^i/k_{br}^i$$
    u_0^{ij}: maximum deformation of damper and bracing at the ith MR damper in mode j

    Substep 3. Calculate equivalent modal damping ratio
$$\xi_{eq}^j = \frac{1}{2}\frac{\sum_{i=1}^{L}\eta_i k_{eq}^i (u_{d0}^{ij})^2}{\mathbf{x}_0^{jT}\mathbf{K}_{eff}\mathbf{x}_0^j} + \xi_{in}^j$$

    Substep 4. Find maximum modal displacement (y_0^j) from response spectrum
$$y_0^j = S_d(T_{eff}^j, \xi_{eq}^j) \text{ where } T_{eff}^j = 2\pi/\omega_{eff}^j$$

    Substep 5. Update modal displacement vector $\mathbf{x}_0^j$
$$\mathbf{x}_0^j = \Gamma_j \boldsymbol{\phi}_j y_0^j$$
    $\boldsymbol{\phi}_j$: mode vector; $\Gamma_j = \boldsymbol{\phi}_j^T \mathbf{M} \mathbf{1}/M_j$; $\mathbf{1}$: unit vector; M_j: modal mass ($=\boldsymbol{\phi}_j^T \mathbf{M}\boldsymbol{\phi}_j$)

    Substep 6. Repeat Substep 2 ~ 5 until convergence in $\mathbf{x}_0^j$ is achieved

    Substep 7. Calculate maximum velocity $\dot{u}_{d0}^{ij}$ for each MR damper, where for the ith MR damper
$$\dot{u}_{d0}^{ij} = u_0^{ij}\omega_{eff}^j$$

Performance-Based Design Procedure for Structures with Magneto-Rheological Dampers, Fig. 5 Modal analysis method for the simplified analysis procedure utilizing response spectrum analysis (RSA) method

in Figs. 6 and 7, respectively. It consists of a three-story, six-bay building and represents a typical office building located in Southern California. Lateral loads are resisted by a four perimeter moment resisting frames (MRFs) and four damped braced frames (DBFs) in the two orthogonal principle directions of the building's floor plan. MR dampers are installed in the DBFs to control the drift of the building, adding supplemental damping to the structure. The DBFs have continuous columns, with pin connections at the beam-to-column connections and at the ends of the diagonal bracing. A rigid diaphragm system is assumed to exist at each floor level and the roof of the building to transfer the floor inertia loads to the MRFs and DBFs. The building has a basement where a point of inflection is assigned at one third of the height of the column from the column base in the analysis model.

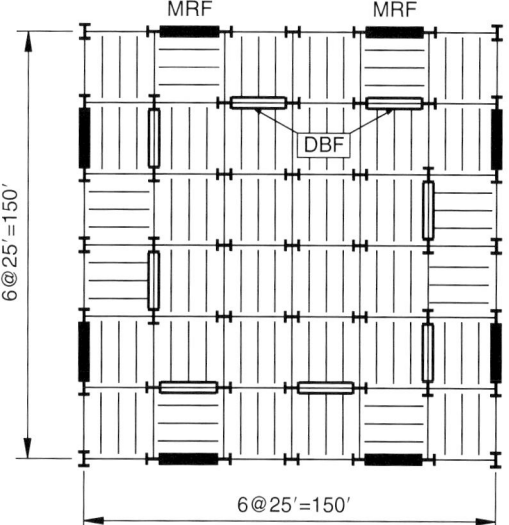

Performance-Based Design Procedure for Structures with Magneto-Rheological Dampers, Fig. 6 Floor plan of prototype building

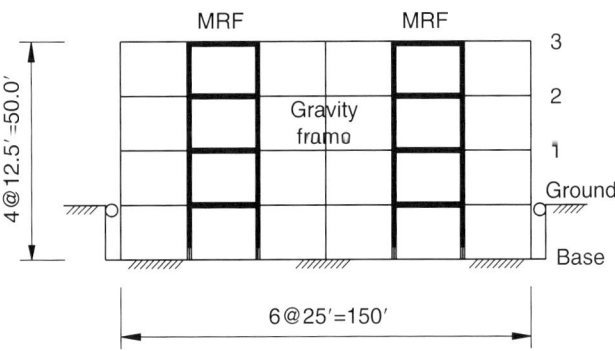

Performance-Based Design Procedure for Structures with Magneto-Rheological Dampers, Fig. 7 Elevation of prototype building

Performance Objectives

In this design, three different performance objectives for the building are considered:

1. Limit the story drift to 1.5 % under the DBE ground motion
2. Limit the story drift to 3.0 % under the MCE ground motion
3. Design strength of members in the DBF shall not be exceeded by the demand imposed by the DBE ground motion

The MCE ground motion is represented by a response spectra that has a 2 % probability of exceedance in 50 years, and the DBE ground motion is two third the intensity of the MCE ground motion (FEMA 2000a). The performance objectives of 1.5 % story drift satisfies the life safety performance level under the DBE, and the 3 % story drift satisfies the collapse prevention level under the MCE. The performance levels are defined in FEMA-356 (2000b). To minimize the damage and repair cost to the DBF structure, the third performance objective is adopted to have the DBF structure remain elastic under the DBE.

Performance-Based Design

The prototype building structure is intended to provide the basis for an MRF and DBF with MR dampers which can be constructed in the laboratory for future tests. Due to laboratory constraints, the prototype building structure and resulting MRF and DBF were designed at 0.6-scale. The MRFs are designed to satisfy the strength requirement of the current building seismic code of ICC (2006); the member design criteria is based on the AISC steel design provisions (2005b). The design response spectrum is based on a site in Southern California where the spectral acceleration for the short period, S_S, and for a 1 s period, S_1, are equal to 1.5 and 0.6, respectively. The strength contribution from the DBFs and MR dampers is not considered when the MRFs are designed since, as noted above, the DBFs and MR dampers are intended only to control the story drift of the building system. More detailed information on the design of the MRFs and gravity frames can be found in Chae (2011).

Once the MRFs and gravity frames are designed for strength, the required capacity of the MR dampers to control the drift is determined. The DBF members are then designed by imposing the displacement and damper force demands on the DBF, which are obtained from the simplified analysis procedure and the required MR damper capacity. The maximum displacements and the maximum MR damper forces are assumed to occur concurrently in the SDP. The design of the three-story building is revised until the performance objectives and strength requirements are satisfied.

Large-scale MR dampers were used for the study which can generate a 200 kN damper force at a velocity of 0.1 m/s (Chae et al. 2010). The parameters for the Hershel-Bulkley model associated with the large-scale MR damper are: $f_0 = 138.5$ kN, $C = 161.8$ kNsec/m, and $n = 0.46$.

The optimal damper location which satisfies the performance objectives is determined by using the simplified analysis procedure, resulting in one large-scale MR damper in the second and third stories, respectively, with $\alpha = 10$ and $\beta = 0.3$ (Chae 2011). Tables 1 and 2 summarize the member sizes for the MRFs, gravity frames, and DBFs. Table 3 summarizes the calculated design demand associated with maximum story drift and maximum damper forces. As can be observed, the design demand for the story drift under the DBE and MCE are less than 1.5 % and 3.0 %, respectively, in order to satisfy the performance objectives. Table 4 shows the DBE demand-to-capacity ratios for the DBF members. The demand-to-capacity ratio for each member is less than 1.0, which means the members are designed to remain elastic under the DBE. The design of the braces was controlled by stiffness, $\alpha = 10$, and not strength, hence, the demand-to-capacity ratios for the braces are small in Table 4.

Assessment of Simplified Design Procedure

The SDP is assessed by comparing the design demand from the SDP with results from a series of nonlinear time-history analyses (NTHA) of the three-story building using the nonlinear finite element program OpenSees (2009).

OpenSees Model

Symmetry in the floor plan and ground motions along only one principal axis of the building were considered in the analysis. Hence, only one-quarter of the building was modeled consisting of one MRF, one DBF, and the gravity frames that are within the tributary area of the MRF and DBF. The OpenSees model is shown in Fig. 8. The beams and columns of the MRF structure are modeled with a nonlinear distributed plasticity force-based beam-column element with five fiber sections along the element length. The cross section of the element is discretized into 18 fibers, including 12 fibers for the web and 3 fibers each for the top and bottom flanges. Each fiber is modeled with a bilinear stress–strain relationship with a post-yielding stiffness that is 0.01 times the elastic stiffness. The beam-column joints in the MRF are modeled using a panel zone element, where shear and symmetric column bending deformations are considered (Seo et al. 2009). The doubler plates in the panel zones of the MRF are included in the model.

Performance-Based Design Procedure for Structures with Magneto-Rheological Dampers, Table 1 Member sizes for MRFs and gravity frames

Story (or floor level)	MRFs		Gravity frames	
	Column	Beam	Column	Beam
1	W8X67	W18X46	W8X48	W10X30
2	W8X67	W14X38	W8X48	W10X30
3	W8X67	W10X17	W8X48	W10X30

Performance-Based Design Procedure for Structures with Magneto-Rheological Dampers, Table 2 Member sizes for DBFs

Story (or floor level)	Column	Beam	Diagonal bracing
1	W10X33	W10X30	–
2	W10X33	W10X30	W6X20
3	W10X33	W10X30	W6X20

Performance-Based Design Procedure for Structures with Magneto-Rheological Dampers, Table 3 Calculated design demand associated with maximum story drift and maximum damper force from SDP

	Maximum story drift (%)		Maximum damper force (kN)	
Story	DBE	MCE	DBE	MCE
1	1.18	1.91	–	–
2	1.35	2.32	222.9	244.4
3	1.41	2.57	233.6	261.6

Performance-Based Design Procedure for Structures with Magneto-Rheological Dampers, Table 4 Demand-to-capacity ratio for DBF frames, DBE

Story (or floor level)	Column (W10X33)	Beam (W10X30)	Brace (W6X20)
1	0.955	0.521	–
2	0.303	0.576	0.270
3	0.079	0.354	0.283

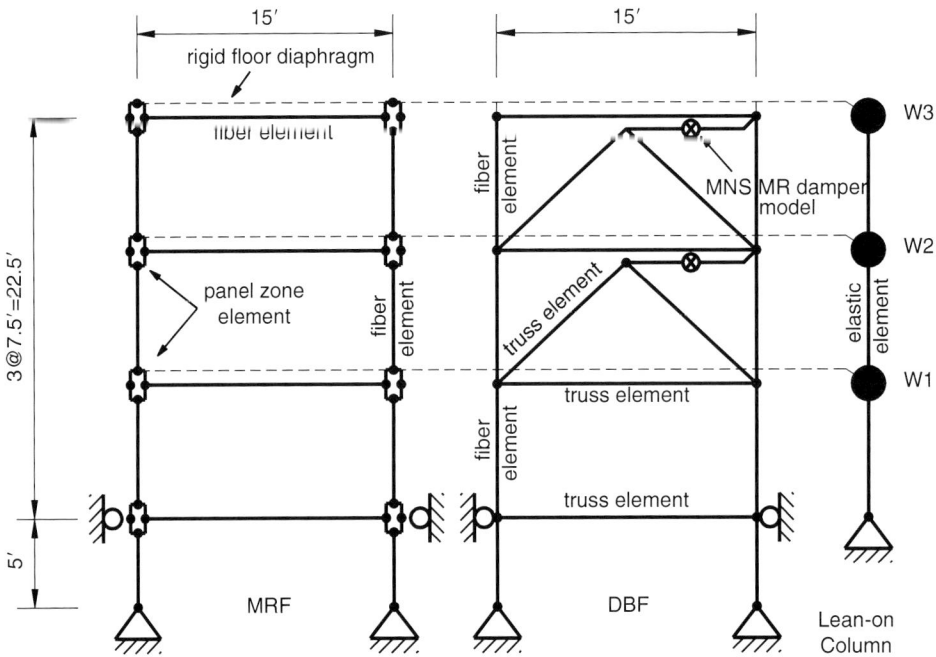

Performance-Based Design Procedure for Structures with Magneto-Rheological Dampers, Fig. 8 OpenSees model for 0.6-scale building

The nonlinear force-based fiber element is also used to model the columns of the DBF. The beams and braces of the DBF are modeled using linear elastic truss elements. The gravity frames are idealized using the concept of a lean-on column, where an elastic beam-column element with geometric stiffness is used to model the lean-on column. The section properties of the lean-on column are obtained by taking the sum of the section properties of each gravity column within the tributary area (i.e., one quarter of the floor plan) of the MRF and the DBF. The MR dampers are modeled using the MNS MR damper model implemented into OpenSees by Chae et al. (2010). The MR damper is assumed to be located between the top of the brace and the adjacent beam-column joint, as shown in Fig. 8. The results reported in this paper are for MR dampers that are passive controlled with a constant current input of 2.5 A. Studies with the MR dampers in semi-active control mode are presented in Chae (2011).

The gravity loads from the tributary gravity frames are applied to the lean-on column to account for the P-Δ effect of the building. To model the effect of the rigid floor diaphragm, the top node of the panel zone element in the MRF and the beam-column joint in the DBF are horizontally constrained to the node of the lean-on column at each floor level, while the vertical and rotational degrees of freedom of these nodes are unconstrained.

Rayleigh damping is used to model the inherent damping of the building with a 5 % damping ratio for the first and second modes.

Comparison of Response

An ensemble of 44 ground motions listed in FEMA P695 (ATC 2009) is scaled to the DBE and MCE levels using the procedure by Somerville et al. (1997) for the NTHA.

A summary of the median and standard deviation of maximum story drift and residual story drift from the NTHA is given in Table 5. Figures 9 and 10 compare the calculated design demand for drift from the SDP with the median values for maximum story drift from the NTHA under the DBE and MCE ground motions. The story drift

Performance-Based Design Procedure for Structures with Magneto-Rheological Dampers, Table 5 Median and standard deviation of maximum and residual story drift from nonlinear time history analysis

Story	DBE level		MCE level	
	Max story drift (%)	Residual drift (%)	Max story drift (%)	Residual drift (%)
1	1.18 (0.35)[a]	0.11 (0.21)	1.86 (0.85)	0.42 (0.62)
2	1.35 (0.36)	0.17 (0.26)	2.10 (0.85)	0.57 (0.66)
3	1.46 (0.33)	0.22 (0.27)	2.32 (0.84)	0.63 (0.69)

[a]Value in () indicates standard deviation response

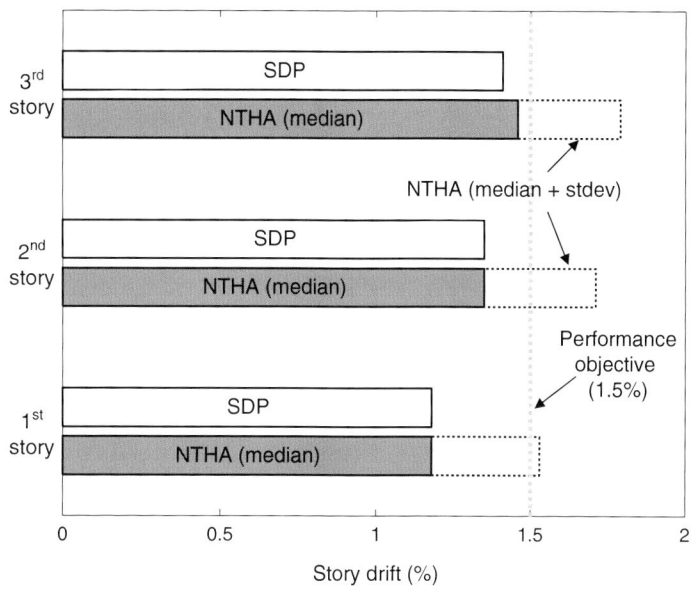

Performance-Based Design Procedure for Structures with Magneto-Rheological Dampers, Fig. 9 Comparison of story drift between SDP and NTHA, DBE ground motions

design demand calculated by the SDP shows good agreement in Fig. 9 with the median maximum story drifts from the NTHA. The calculated design demand for story drift from the SDP also shows good agreement with the median values from the NTHA under the MCE, see Fig. 10. Figures 9 and 10 show that the median values of the maximum story drift from the NTHA satisfies the performance objectives of 1.5 % and 3.0 % under the DBE and MCE levels, respectively. The residual story drift of the building after the DBE has a maximum median value and standard deviation of 0.22 % and 0.27 %, respectively, which occurred in the third story as summarized in Table 5. The residual drift is small.

Table 6 compares the design demand for maximum MR damper forces calculated by the SDP with the median maximum MR damper forces from the NTHA. The MR damper force design demands from the SDP are slightly smaller than the median NTHA results for the DBE. However, the differences between the SDP and the NTHA are only 3.9 % and 3.3 % for the MR dampers in the second and third stories, respectively. For the MCE, the differences between the median NTHA results and the SDP for the MR damper forces in the second and third stories are 1.6 % and 0.5 %, respectively. The design demand calculated by the SDP shows reasonably good agreement with the median results from the NTHA for the maximum MR damper forces.

The linear elastic behavior of the DBF columns under the DBE is confirmed by checking the plastic rotation developed in the columns. Summarized in Table 7 are the median and standard deviation of the DBF maximum magnitude of column plastic rotation from the NTHA for the DBE ground motions. In the first story, some

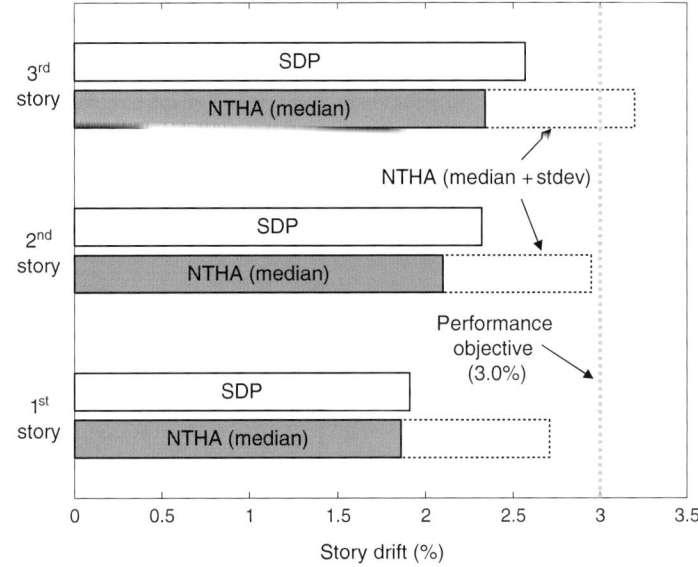

Performance-Based Design Procedure for Structures with Magneto-Rheological Dampers, Fig. 10 Comparison of story drift between SDP and NTHA, MCE ground motions

Performance-Based Design Procedure for Structures with Magneto-Rheological Dampers, Table 6 Comparison of maximum damper forces

Story	DBE level SDP	DBE level NTHA	MCE level SDP	MCE level NTHA
1	–	–	–	–
2	222.9	231.9 (6.4)[a]	244.4	248.4 (7.3)
3	233.6	241.5 (8.3)	261.6	260.2 (9.4)

[a]Value in () indicates standard deviation response

plastic rotation developed at the base of the column under the DBE. However, the median is zero and the standard deviation is 0.0005 rad for the maximum plastic rotation. The median and standard deviation of the maximum plastic rotation in the second and third stories columns are zero under the DBE ground motion, as given in Table 7, which indicates linear elastic behavior of the columns at these stories. The median of the residual plastic rotation at the base of the column is zero and the standard deviation for the residual plastic rotation is 0.0004 rad.

The building's response under the DBE and MCE determined by the NTHA appears to have met the performance objectives for the structure. Moreover, the performance under the DBE meets the immediate occupancy level (FEMA 2000b), where the maximum residual drift is practically equal to the allowable construction tolerance of 0.2 % for steel structures (AISC 2005a), with

Performance-Based Design Procedure for Structures with Magneto-Rheological Dampers, Table 7 Median and standard deviation of DBF maximum magnitude of column plastic rotation from nonlinear time-history analysis and DBE ground motions

Story	Location along column	Max plastic rotation (rad %)	Residual plastic rotation (rad %)
1	Bottom	0.00 (0.05)[a]	0.00 (0.04)
	Top	0.00 (0.00)	0.00 (0.00)
2	Bottom	0.00 (0.00)	0.00 (0.00)
	Top	0.00 (0.00)	0.00 (0.00)
3	Bottom	0.00 (0.00)	0.00 (0.00)
	Top	0.00 (0.00)	0.00 (0.00)

[a]Value in () indicates standard deviation response

minimal damage occurring in the building (the median of the maximum plastic rotations in the beams and columns of the MRF are 0.37 % and 0.07 % radians, respectively, with the MRF having the same residual drift statistics as the DBF due to the rigid diaphragm (Chae 2011)).

Summary

A simplified design procedure was developed to enable the performance-based design of structures with MR dampers. The design procedure utilizes a systematic approach to calculate the design demand of structures with MR dampers.

The simplified analysis procedure enables the design demand to be determined without performing a nonlinear time-history analysis by linearizing the structure and utilizing the response spectrum analysis method.

To linear the nonlinear MR dampers, the simplified analysis procedure is based on an equivalent linear MR damper model using the Hershel-Bulkley quasi-static MR damper model. The energy dissipated by the MR damper over one cycle of a harmonic motion is calculated and the equivalent stiffness determined based on the secant stiffness method from the damper force-displacement relationship. The loss factor of the MR damper is obtained from the energy dissipated by the damper and the strain energy calculated from the equivalent stiffness of the damper. Both the equivalent stiffness and loss factor of the MR damper are dependent on the maximum displacement of the damper. A three-story building was designed using the SDP, where three performance objectives associated with two seismic hazard levels were selected. The SDP was assessed by comparing the design demand calculated by the SDP with the response determined from nonlinear time-history analyses. The MNS MR damper model was implemented into the OpenSees computer program and statistics for the response to DBE, and MCE ground motions were obtained from a series of nonlinear time-history analyses using 44 different ground motions. The performance of the building from the nonlinear time-history analysis indicated that building design satisfies the three performance objectives. The design demand associated with story drift and maximum MR damper forces from the SDP showed good agreement with the median values from the nonlinear time-history analyses, confirming the robustness of the SDP.

References

AISC (2005a) Code of standard practice for steel buildings and bridges. American Institute of Steel Construction, Chicago

AISC (2005b) Seismic provisions for structural steel buildings. American Institute of Steel Construction, Chicago

Applied Technology Council (2009) Quantification of building seismic performance factors. ATC-63 project report (FEMA P695), Redwood City

Chae Y (2011) Seismic Hazard mitigation of building structures using magneto-rheological dampers, PhD dissertation, Lehigh University

Chae Y, Ricles JM, Sause R (2010) Development of a large-scale MR Damper model for seismic hazard mitigation assessment of structures. In: 9th US National and 10th Canadian conference on earthquake engineering, Toronto

Constantinou MC, Soong TT, Dargush GF (1998) Passive energy dissipation systems for structural design and retrofit. Monograph series, MCEER. State University of New York at Buffalo, Buffalo

Fan CP (1998) Seismic analysis, behavior, and retrofit of non-ductile reinforced concrete frame buildings with viscoelastic dampers. PhD dissertation, Lehigh University, Bethlehem

Federal Emergency Management Agency (2000a) Recommended seismic design criteria for new steel moment-frame buildings. Report no. FEMA-350, Washington, DC

Federal Emergency Management Agency (2000b) Prestandard and commentary for the seismic rehabilitation of buildings. Report no. FEMA-356, Washington, DC

International Code Council (2006) International building code. International Code Council, Falls Church

Kwan WP, Billington SL (2003) Influence of hysteretic behavior on equivalent period and damping of structural systems. ASCE J Struct Eng 129(5):576–585

Lee K-S, Fan C-P, Sause R, Ricles J (2005) Simplified design procedure for frame buildings with viscoelastic and elastomeric structural dampers. Earthq Eng Struct Dyn 34:1271–1284

Lee K-S, Ricles J, Sause R (2009) Performance-based seismic design of steel MRFs with elastomeric dampers. J Struct Eng 135(5):489–498

Lin WH, Chopra AK (2003) Earthquake response of elastic Single-degree-of-freedom systems with nonlinear viscoelastic dampers. ASCE J Eng Mech 129(6):597–606

Newmark NM, Hall WJ (1973) Seismic design criteria for nuclear reactor facilities. Report no. 46, Building Practices for Disaster Mitigation, National Bureau of Standards, U.S. Department of Commerce, pp 209–236

OpenSees (2009) Open system for earthquake engineering simulation. Pacific Earthquake Engineering Research Center, University of California, Berkeley

Ramirez OM, Constantinou MC, Gomez JD, Whittaker AS (2002) Evaluation of simplified methods of analysis of yielding structures with damping systems. Earthq Spectra 18(3):501–530

Sause R, Hemingway GJ, Kasai K (1994) Simplified seismic response analysis of viscoelastic-damped frame structures. In: 5th U.S. National conference on earthquake engineering, Vol. I, pp 839–848

Seo CY, Lin YC, Sause R, Ricles JM (2009) Development of analytical models for 0.6 scale self-centering MRF with beam web friction devices. In: 6th International conference for steel structures in seismic area (STESSA), Philadelphia

Somerville P, Smith N, Punyamurthula S, Sun J (1997) Development of ground motion time histories for Phase 2 of the FEMA/SAC steel project. Report no. SAC/BD-97/04, SAC Joint Venture, Sacramento

Soong TT, Dargush GF (1997) Passive energy dissipation systems in structural engineering. Wiley, West Sussex

Symans MD, Constantinou MC (1998) Passive and fluid viscous damping systems for seismic energy dissipation. ISET J Earthq Technol 35(4):185–206

Physics-Based Ground-Motion Simulation

Ricardo Taborda[1] and Daniel Roten[2]
[1]Department of Civil Engineering, and Center for Earthquake Research and Information, University of Memphis, Memphis, TN, USA
[2]San Diego Supercomputer Center, University of California, San Diego, La Jolla, CA, USA

Synonyms

Deterministic earthquake simulation; Deterministic ground-motion simulation; Physics-based earthquake simulation

Introduction

Physics-based earthquake ground-motion simulation, also referred to as deterministic earthquake ground-motion simulation, can be defined as the prediction of the ground motion generated by earthquakes by means of numerical methods and models that incorporate explicitly the physics of the earthquake source and the resulting propagation of seismic waves. Other approaches such as the stochastic ground-motion simulation method or ground-motion prediction equations (i.e., attenuation relationships) integrate the physics and specific characteristics of the earthquake source, directivity, path, attenuation and scattering, basin, and site effects by means of indirect, approximate, or statistical approaches. These methods are valid representations of the physics of earthquakes, but do not necessarily solve the accepted mathematical abstractions that describe the physics of source dynamics and wave propagation. Physics-based ground-motion simulation, on the other hand, seeks to account for most or all of these aspects explicitly, thus the distinction given to its name.

This chapter reviews the background of physics-based ground-motion simulation, describes some of the most popular methods used in this field, and presents application examples. The first section focuses on the background of the method. The second section presents the concept of a physics-based simulation framework. Subsequent sections review the basic methods applied for the solution of seismic waves traveling in solids, including elastic, anelastic, and elastoplastic media, and provide details about the input data used in simulations (source and velocity models) while addressing other important complementary aspects in numerical modeling and computer simulation along the way. The final sections present recent examples and applications and a summary with a perspective on the future use of physics-based simulation in earthquake hazard analysis and engineering.

Background

The foundation for a physics-based approach to earthquake ground-motion simulation was laid down in the late 1960s and early 1970s when the finite-difference (FD) and the finite-element (FE) methods were first introduced in seismology (Alterman and Karal 1968; Lysmer and Drake 1972). Before numerical methods started to be more broadly used, only a few problems under restricted conditions could be solved analytically. Classical examples include the surface response of semi-cylindrical and semi-elliptical canyons and alluvial valleys under incident plane SH waves (e.g., Wong and Trifunac 1974). Analytical approaches were, however, useful only to characterize the ground motion under idealized settings that neglected the geometrical irregularity of the geology and the heterogeneity of the media and were typically considered for sites away from the source, or with simplified source representations.

By contrast, FD and FE approaches offered more flexibility. They could be used to describe the propagation of waves in (irregular) stratified media, alluvial valleys, and sedimentary basins, assess ground-motion amplification, and study source dynamics (e.g., Boore 1972; Smith 1975). Initially, due to computational limitations, numerical methods were used mostly for solving two-dimensional (2D) problems. With time, advances in numerical methods and algorithms and the growth and increased availability of computing power and memory capacity allowed scientists to model larger and more complex 2D and small three-dimensional (3D) problems. Virieux (1984), in particular, introduced the staggered-grid FD scheme in the context of seismology, which would later become a preferred approach for modeling seismic wave propagation problems.

The first 3D simulations done at scales large enough to synthesize source, path, basin, and site effects were published in the early 1990s. Frankel and Vidale (1992), for instance, simulated the ground motion generated by an M_L 4.4 aftershock of the 1989 Loma Prieta, California, earthquake using a point source model and a FD representation of the Santa Clara Valley. The model had about four million nodes in a simulation domain of 30 km by 22 km and 6 km in depth. The synthetic seismograms obtained were valid for a maximum frequency, $f_{max} = 1$ Hz, and the minimum shear-wave velocity ($V_{S_{min}}$) considered was 600 m/s. Although the synthetics lacked the level of fidelity expected from direct comparisons with data, they were, in general, comparable in amplitude and duration. More important, the results of Frankel and Vidale (1992) showed that simulations offered a plausible means to understanding the characteristics of 3D wave propagation in highly heterogeneous media.

Following Frankel and Vidale (1992), FD applications gained significant traction (e.g., Olsen et al. 1995; Graves 1996). Some of these works were done using parallel computers. Olsen et al. (1995), for instance, simulated the ground motion of an M 7.75 scenario earthquake on the San Andreas Fault. The model covered the entire Greater Los Angeles metropolitan region in a simulation domain of 230 km by 140.4 km and 46 km in depth. The simulation was designed for $f_{max} = 0.4$ Hz and $V_{S_{min}} = 1$ km/s, using a grid spacing of 0.4 km. The resulting model had over 23.5 million grid points. This simulation was one of the first to use parallel computers for simulating the ground motion at a regional scale. It required nearly 23 h to complete 4,800 time steps on a 512-processor nCUBE-2 computer. Despite some limitations, Olsen et al. (1995) made important observations about the significance of 3D basin and edge effects that could not have been described at the time using only equivalent 1D or 2D models.

Similarly, FE applications were also developed for both source dynamics and wave propagation problems (e.g., Bao et al. 1998; Aagaard et al. 2001). Bao et al. (1998), for example, used a parallel computer FE application to simulate the ground response of the San Fernando Valley, California, to an aftershock of the 1994 Northridge earthquake. The FE mesh consisted of 76.8 million elements, with parameters $V_{S_{min}} = 220$ m/s and $f_{max} = 1.6$ Hz. The simulation required 7.2 h to execute 16,667 time steps (40 s of ground motion) on 256 processors in a Cray T3D machine. Bao et al. (1998) observed that, besides the larger amplitudes and longer durations associated with the deeper parts of the basin beneath the San Fernando Valley, the ground motion was also significantly amplified by the constructive interference of surface and trapped body waves in the shallower soft-material deposits.

Pseudo-spectral, high-order FE, spectral element (SE), and discontinuous Galerkin (DG) methods have also been used (e.g., Seriani 1998; Komatitsch and Vilotte 1998; Dumbser and Käser 2006). Some SE applications have been particularly successful in simulations at the continental and global scales. Komatitsch et al. (2003), for instance, show results using a SE approach to model seismic waves propagating through the Earth's globe, including the full complexity of a 3D crustal model with a mesh of 5.5 billion grid points, for seismic periods as low as 5 s (i.e., $f_{max} = 0.2$ Hz). By contrast, problems of small and moderate sizes continue to be addressed using analytical or semi-analytical methods with approaches such as the boundary element, coupled boundary-domain element, discrete wave-number, and hybrid methods (e.g., Bouchon 1979; Mossessian and Dravinski 1987; Bielak et al. 1991; Sánchez-Sesma and Luzón 1995).

More recently, the increased availability and power of parallel computers have facilitated enormously the advance of 3D physics-based earthquake ground-motion simulation. Chaljub et al. (2010) and Bielak et al. (2010), for example, show the tremendous advance by both Europe- and US-based research groups to conduct verification of simulation codes used for modeling the wave propagation characteristics of historic and scenario earthquakes. It is now commonplace to see simulation domains ranging in the order of tens to hundreds of kilometers, with total number of elements or cells in the order of tens to hundreds of billions. Maximum frequencies modeled today vary between 2 and 5 Hz for regional-scale ground-motion simulations and are as high as 10 Hz for smaller-scale rupture dynamics and local wave propagation problems (e.g., Cui et al. 2010; Taborda and Bielak 2013; Shi and Day 2013). In addition, the advent of newer technologies such as general purpose graphic processing units (GPGPU), hybrid CPU and GPU systems, and many integrated core (MIC) architectures are further helping to accelerate forward and inverse, regional and global wave propagation simulations (e.g., Komatitsch et al. 2010; Zhou et al. 2012; Rictmann et al. 2012).

Altogether, the progress shown over the past few years has opened the possibility of using physics-based ground-motion simulation in earthquake engineering applications. The framework for using simulations for a physics-based approach to regional seismic hazard mapping, for instance, has already been put in place and is being used at low frequencies (up to 0.5 Hz) in Southern California (Graves et al. 2011). Some obstacles, however, still need to be sorted out. The present knowledge of the earthquake source, crustal structure, material properties, and local site effects is still far from ideal. All these aspects will require much research in the years to come, but the trajectory indicates that physics-based earthquake ground-motion simulation will play a significant role in future seismic hazard estimation, risk assessment, and earthquake engineering analysis and design.

The Physics-Based Simulation Workflow

Physics-based earthquake simulations operate within a basic general workflow that consists of the following elements or steps:

- The selection of a region of interest and simulation domain
- The selection of a source model and a material model
- The definition of the modeling parameters (maximum frequency, minimum velocity, etc.)
- The implementation of solution methods and operation of a simulation engine
- The execution of the simulation and collection of results

This simulation workflow is illustrated in Fig. 1. The top section of the workflow refers to the input data that is required for the simulation, that is, the selection of the simulation domain, the source model, the material model, and the simulation parameters. The source model provides information about the fault rupture characteristics in the form of its location, orientation, and slip history. Most physics-based earthquake simulations use kinematic models to represent the source as will be explained later. However, dynamic rupture simulations that fully solve the rupture evolution on the fault plane and the triggered wave propagation problem can also be combined with the simulation of the ground motion. The material model provides information

Physics-Based Ground-Motion Simulation, Fig. 1 Typical workflow in physics-based earthquake ground-motion simulations. The *top* section (in *yellow*) refers to the input models and parameters; the *middle* section (in *purple*) refers to the solution method and implementation into the simulation engine; and the *bottom* section (in *green*) refers to the simulation execution and results

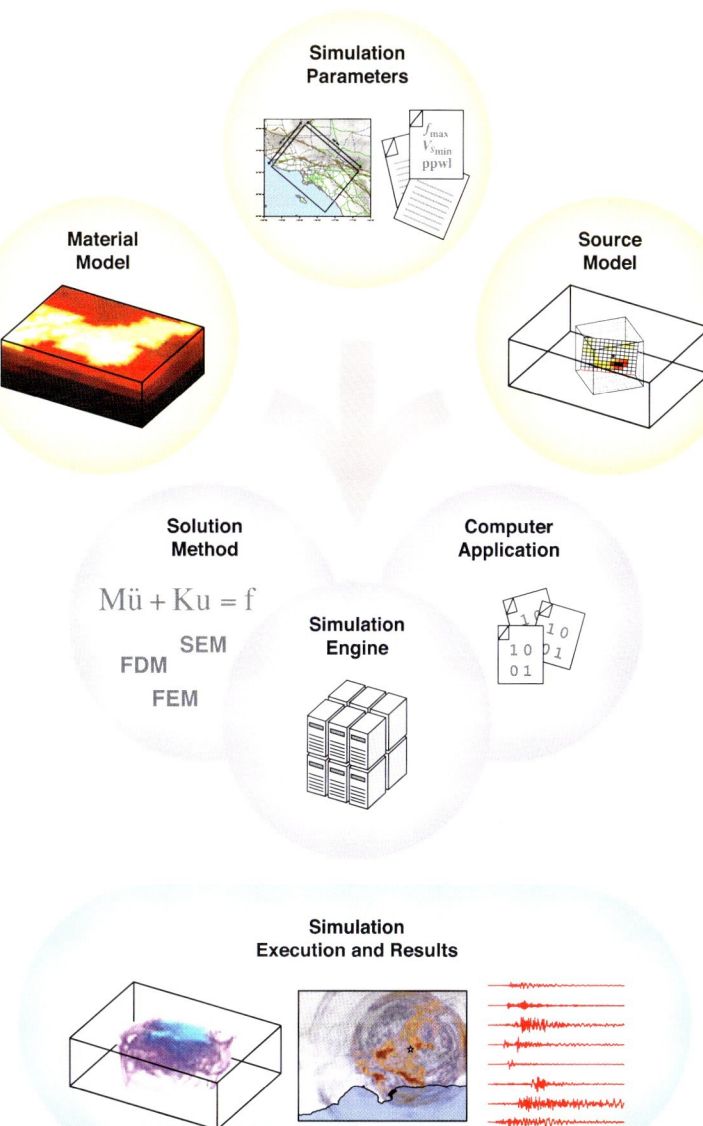

about the properties of the material contained in the chosen simulation domain. Material models are often referred to simply as seismic velocity models, as most of these models only provide information about the P- and S-wave velocities and the density of the medium. However, a complete description of the material also provides information about its dissipation properties and capacity. A later section is dedicated to the description of some publicly accessible velocity models. The last pieces of information at the input data level are the simulation parameters. At the most basic level, these consist of the maximum targeted frequency in the simulation (f_{max}, usually defined in Hz) and the minimum shear-wave propagation velocity ($V_{S_{min}}$, usually defined in m/s). Also relevant at this point is the definition of the number of points per wavelength, which is the integer number of points that will be used to discretize a complete wave cycle. Together, these parameters define the level of refinement or resolution necessary to solve the problem with a certain acceptable level of accuracy to the extent possible.

The second section in the workflow shown in Fig. 1 refers to the solution method and its implementation in a computer code application. Together, they provide a simulation engine that can be used in personal computers, clusters, or supercomputers. There are various simulation computer codes used in research today, some of which provide open-access distribution. A well-documented example of a code available to users is the SPECFEM software family, distributed by the Computational Infrastructure for Geodynamics project (http://www.geodynamics.org). Given the computing resources necessary for the solution of forward wave propagation simulations at regional scales, the implementation of such simulation engines in parallel computer codes and the use of computer clusters and supercomputers have become a commonplace in physics-based earthquake simulation. Until now, this has restricted the use of physics-based earthquake simulations to research-oriented activities. However, as the projected capacity of personal computers over the next 50 years will make parallel computing applications massively accessible to the public, it is expected that the use of these simulation codes and that of physics-based simulations overall will increase within the seismology and earthquake engineering communities in the near future. The last portion of the workflow shown in Fig. 1 makes reference to the execution of the simulation in itself and the gathering of results. Typical simulation output datasets come in three basic flavors: individual station records, plane wave fields, and volumetric wave fields. Station records are usually output in the form of text or binary files with velocities (or displacements) ordered sequentially at every time step. Plane and volumetric wave fields are commonly delivered as binary files indexed in such a way that they can be sliced in 2D and 3D arrays with the ground response for every time step. Simulators usually produced these outputs at a (decimated) time step that is larger than that actually used internally in the computation of the solution. Station records and plane wave field outputs are the most common of the three because their file sizes make them more manageable in terms of memory and disk space capacity.

A station record file is typically of the order of a few hundreds of kilobytes to a couple megabytes, and a plane file ranges between tens and hundreds of gigabytes. Volumetric wave fields, on the other hand, can reach the order of terabytes and thus become onerous to transfer and store.

The following sections expand on some of these aspects, including the most common solution methods, typical representations of source models, and seismic velocity models used in regional simulations, and provide examples about the use given to physics-based earthquake simulations today. Specific computer code applications are not covered here, although some basic principles about their numerical implementation in computer codes are covered in the sections dedicated to the solution methods used in physics-based simulation.

Wave Propagation in Elastic Media

As mentioned in the previous sections, there are various analytical and numerical methods used for solving wave propagation problems. The simulation of the ground motion at scale, that is, the simulation in domains large enough for synthesizing the earthquake source and the local and regional response of the ground, has, however, been primarily approached using FD, FE, SE, and, lately, DG methods. Since the formulation of both the SE and DG methods share some of the basic concepts of FE method, special attention is given here only to the FE and FD methods. This section covers these two methods applied to elastic media first. Subsequent sections address the problem of anelasticity and plasticity separately.

The Finite-Element Method

Earthquake ground-motion simulation entails obtaining the solution of the linear momentum equation, which can be written in Cartesian coordinates and indicial notation as

$$\sigma_{ij,j} + f_i = \rho \ddot{u}_i. \qquad (1)$$

Here, σ_{ij} represents the Cauchy stress tensor, ρ is the mass density, and f_i and u_i are the body forces and displacements in the i direction within a bounded domain Ω. The two dots over the displacements indicate second derivative in time. The indices i and j in the subscripts represent the Cartesian coordinates x, y, and z. When a subscript follows a comma, this indicates a partial derivative in space with respect to the corresponding index. For the special case of elastic isotropic solids, the stress tensor can be expressed in terms of strains following Hooke's law of elasticity, and the strains, in turn, can be expressed in terms of displacements. The resulting expression for the stress tensor is

$$\sigma_{ij} = \lambda u_{k,k} \delta_{ij} + \mu(u_{i,j} + u_{j,i}) \qquad (2)$$

where λ and μ are the Lamé parameters and δ_{ij} is Kronecker's delta. In general, the Lamé parameters in Eq. 2 and the density in Eq. 1 are assumed to be locally constant. Substituting Eq. 2 into Eq. 1 leads to

$$\lambda u_{k,kj} \delta_{ij} + \mu(u_{i,jj} + u_{j,ij}) + f_i = \rho \ddot{u}_i. \qquad (3)$$

This is Navier's equation of elastodynamics. Using the standard Galerkin method, one can obtain the weak form of this equation and then discretize the problem in space. This procedure entails the introduction of set of arbitrary functions v, known as the test functions. The test functions are auxiliary functions which help formulate an approximate solution $\hat{u}$ to the displacements u, called the trial functions. The domain Ω is then discretized in space using a set of global piecewise linear basis functions ϕ, which divide the domain into discrete elements Ω_e. As a result, both the test and trial functions become linear combinations of the global basis functions,

$$v_h(x,y) = \sum_{i=1}^{N} \phi_i(x,y) v_i^h, \text{ and} \qquad (4)$$

$$\hat{u}_h(x,y) = \sum_{i=1}^{N} \phi_i(x,y) \hat{u}_i^h, \qquad (5)$$

respectively. Here, h is used to indicate that the domain Ω has been approximated to a discrete version of it, Ω_h, composed of all the elements with domain Ω_e, where h is a discretization parameter (e.g., element size). These elements Ω_e are connected to each other along their edges and at their vertices. The vertices of the elements are called nodes. Both the nodes and the elements constitute a FE mesh, where N is the total number of nodes. The index i indicates that the values of v and $\hat{u}$ are evaluated at the nodes using the associate global function ϕ_i. It can be shown that substituting Eqs. 4 and 5 into Eq. 3 leads to

$$\mathbf{M}\ddot{u} + \mathbf{K}u = f, \qquad (6)$$

where $\mathbf{M}$ and $\mathbf{K}$ are the assembled global mass and stiffness matrices of the system's discrete FE mesh representation, f is the assembled vector of body forces (which is determined based on the kinematic representation of the source – see the section "Source Models" below), and u is the assembled vector of displacements at the nodes in the FE mesh. Once again, the double dots over the displacement vector mean second derivative in time (i.e., acceleration). Not shown here for brevity is the fact that in following the FE method, the test function terms vanish. The matrices $\mathbf{M}$ and $\mathbf{K}$ and the vector f in Eq. 6 are composed of terms corresponding to the nodes of the elements in the FE mesh. The ith row and jth column terms in these matrices and vector are given by

$$\mathbf{M}_{ij} = \int_{\Omega} \rho \phi_i \phi_j d\Omega, \qquad (7)$$

$$K_{ij} = \int_{\Omega} (\mu + \lambda) \phi_i \phi_j^T d\Omega + \int_{\Omega} \mu \mathbf{I} \phi_i^T \phi_j d\Omega, \qquad (8)$$

$$f_i = \int_{\Omega} \phi_i f d\Omega. \qquad (9)$$

However, the global matrices $\mathbf{M}$ and $\mathbf{K}$ are seldom constructed explicitly for the full simulation domain Ω. A common practice is, instead, to perform the products $\mathbf{M}\ddot{u}$ and $\mathbf{K}u$ at the element level Ω_e using local basis functions ψ_i instead of the global basis functions ϕ_i. This leads to performing the products $\sum_e \mathbf{M}^e \ddot{\mathbf{u}}^e$ and $\sum_e \mathbf{K}^e \mathbf{u}^e$, where $\mathbf{M}^e$ and $\mathbf{K}^e$ are the mass and stiffness matrices of each finite element in the mesh, built using the local basis functions, and $\ddot{\mathbf{u}}^e$ and $\mathbf{u}^e$ are the corresponding acceleration and displacement vectors that have the nodes associated with each element.

When operations are done at the element level using local basis functions as just described, the system of ordinary differential equations in Eq. 6 can be rewritten as

$$\sum_e \mathbf{M}^e \ddot{\mathbf{u}}^e + \sum_e \mathbf{K}^e \mathbf{u}^e = \sum_e \mathbf{f}^e. \qquad (10)$$

Here, the summation symbol means assembling of all elements e in the FE mesh. At any time step n, the acceleration $\ddot{\mathbf{u}}_n$ can then be expressed in terms of the displacements by applying second-order central differences. Then, for the nth time step in the simulation, Eq. 10 becomes

$$\sum_e \mathbf{M}^e \left(\frac{\mathbf{u}^e_{n-1} - 2\mathbf{u}^e_n + \mathbf{u}^e_{n+1}}{\Delta t^2} \right) + \sum_e \mathbf{K}^e \mathbf{u}^e_n = \sum_e \mathbf{f}^e_n, \qquad (11)$$

where Δt is the size of the time step. Furthermore, the system can be uncoupled using a diagonally lumped mass matrix. In that case, the elements of the mass matrix are such that $m_{ij} = m_i$ for $i = j$ and $m_{ij} = 0$ for $i \neq j$. This allows the forward step-by-step explicit solution of the displacements at time step $n + 1$. The solution for each node i in the mesh can be written as

$$u^i_{n+1} = \frac{\Delta t^2}{m_i} f^i_n - \left(u^i_{n-1} - 2u^i_n \right) - \frac{\Delta t^2}{m_i} \left(\sum_e \mathbf{K}^e \mathbf{u}^e_n \right)_i, \qquad (12)$$

where m_i is the mass lumped at node i, u^i_n is the displacement at node i and step n, and the summation within the parenthesis corresponds to the assembling of the stiffness contributions of all the elements that share node i.

Note that in the formulation presented in Eqs. 1 through 9, no mention was made about the boundary conditions. In FE, the traction-free conditions at the free surface are naturally met and no special treatment is needed. For the lateral and the bottom faces of the domain, however, appropriate measures need to be taken to effectively diminish or vanish the occurrence of spurious reflections at the finite boundaries of the simulation domain. There are several alternatives to implement absorbing boundary conditions in FE applications. Perhaps the simplest of them all consists on placing dampers at the nodes on the boundaries designed to absorb compression and shear plane waves locally. This approximation, while far from ideal, has been used in large-scale simulations with minimum reflections and acceptable performance (e.g., Bielak et al. 2010). Other more accurate absorbing boundary conditions can be satisfied with the implementation, for instance, of the perfectly matching layers (PML) method (e.g., Ma and Liu 2006).

Equations 6 through 12 above provide the basic formulation for a forward wave propagation problem in an elastic medium in which the conditions imposed by the geometrical irregularities or the material's heterogeneity are approximated by means of an appropriate discretization of the simulation domain. In earthquake ground-motion modeling, the meshing criteria are determined based on the maximum simulation frequency ($f_{\max}$), the desired number of points per wavelength (p), and the local material properties defined by the shear-wave velocity (V_S). The size e of each element is set so that it satisfies the rule

$$e_{\max} \leq \frac{V_S}{p f_{\max}}. \qquad (13)$$

The adequate numbers of points per wavelength depends on the type of finite element used and the level of accuracy sought in the simulation. Acceptable minimum values for p in the case of first-order (linear) elements vary between 8 and 12, but a minimum of ten points per wavelength is recommended, unless higher-order (quadratic) elements are used.

Adequately setting the size of elements, however, does not necessarily cover all the

Physics-Based Ground-Motion Simulation, Fig. 2 Conventional spatial grid for the velocity–stress FD scheme (Modified from Moczo et al. 2004)

geometrical and material irregularities present in simulation models, such as surface topography, material inhomogeneities, and internal structural interfaces. From a meshing point of view, the best alternative in FE to handle most of these problems is to work with conforming meshes. These are meshes that discretize the simulation domain adjusting both the shape and size of the elements – and thus the location of nodes – to conform to the specific characteristics in the geometry and material properties of the medium. Though powerful in this sense, conforming meshes are more difficult to build and require more memory (if the stiffness matrices of the elements are to be stored) or more computing time (if the stiffness matrices are to be rebuilt for each element at every time step). Nonconforming meshes, on the other hand, rely on the size of the element in order to capture, to a certain extent, the changes in the geometry and material properties of the media. While nonconforming meshes are very efficient because they can be tailored to use template elements whose stiffness matrices need to be computed only once and then scaled up based on material properties (e.g., Tu et al. 2006), this approach requires additional attention when it comes to handling sharp contrasts in material properties, discontinuities, or strongly irregular geometries. Some of these problems can be overcome using hybrid approaches that combine multiple element types or mixed meshing strategies (e.g., Hermann et al. 2011) or by means of special elements with extended or fictitious domains (e.g., Restrepo and Bielak 2014). A more detailed description of these alternatives, however, is out of the scope of this chapter.

Additional aspects pertaining to modeling the earthquake source and the effects of attenuation are discussed in subsequent sections.

The Finite-Difference Method

Attention is now given to the basic concepts and approach used for modeling wave propagation problems using the FD method.

In FD solutions, the derivatives in differential equations are approximated with finite differences computed over a discrete grid. Consider the case of a vertically propagating, planar *SH* wave in a horizontally layered medium. The plane strain approximation reduces the wave Eq. 3 to one dimension:

$$\mu \frac{\partial^2 u_x}{\partial z^2} + f_x = \rho \ddot{u}_x. \tag{14}$$

To eliminate the double derivatives, Eq. 14 is often expressed using the velocity–stress formulation:

$$\rho \frac{\partial v}{\partial t} = \frac{\partial \tau}{\partial z} + f_x, \tag{15}$$

$$\frac{\partial \tau}{\partial t} = \mu \frac{\partial v}{\partial z}, \tag{16}$$

where the velocity in the *x*-direction, $\dot{u}_x$, has been replaced with v and the shear stress component σ_{xz} with τ. FD solutions require discretization of both space and time on a numerical grid (Fig. 2). Partitioning the domain space using a mesh of z_0, $z_1, z_2, \ldots, z_j$ with uniform increment Δz and the time using a mesh of $t_0, t_1, t_2, \ldots, t_n$ with uniform increment Δt, the partial derivatives in Eq. 15 at point j and time step n may be approximated using

$$\frac{\partial v_j^n}{\partial t} \approx \frac{v_j^{n+1} - v_j^n}{\Delta t}, \tag{17}$$

$$\frac{\partial \tau_j^n}{\partial z} \approx \frac{\tau_{j+1}^n - \tau_j^n}{\Delta z}. \quad (18)$$

The approximations in Eqs. 17 and 18 use the forward difference formula. Substituting Eqs. 17 and 18 into Eq. 15 and omitting the body force term yield

$$\rho \frac{v_j^{n+1} - v_j^n}{\Delta t} = \frac{\tau_{j+1}^n - \tau_j^n}{\Delta z}. \quad (19)$$

Using analogous approximations for the temporal and spatial derivatives in Eq. 16 yields

$$\frac{\tau_j^{n+1} - \tau_j^n}{\Delta t} = \mu \frac{v_{j+1}^n - v_j^n}{\Delta z}. \quad (20)$$

By solving Eqs. 19 and 20 for v_j^{n+1} and τ_j^{n+1}, velocities and stresses at time $n+1$ can be determined from the values at time n:

$$v_j^{n+1} = v_j^n + \frac{\Delta t}{\rho} \frac{\tau_{j+1}^n - \tau_j^n}{\Delta z}, \quad (21)$$

$$v_j^{n+1} = v_j^n + \frac{\Delta t}{\rho} \frac{\tau_{j+1}^n - \tau_j^n}{\Delta z}. \quad (22)$$

Solutions of the wave equation require knowledge of the initial velocities and stresses at time $n = 0$, v_j^0 and τ_j^0, respectively. By using Eqs. 21 and 22 for all j, velocities and stresses can iteratively be determined for $n = 1, 2, 3, \ldots$ until the desired time step.

The finite-difference scheme in Eqs. 21 and 22 is *conditionally stable*. The solution converges only if the Courant number

$$C = \frac{\beta \Delta t}{\Delta z} \leq C_{\max}, \quad (23)$$

where $\beta = \sqrt{\mu/\rho}$ is the shear-wave velocity and $C_{\max} = 1$ for the FD scheme in Eqs. 21 and 22. Equation 23 is called the *Courant–Friedrichs–Lewy* condition.

A disadvantage of the forward difference formula is that the approximations to the temporal Eq. 17 and spatial Eq. 18 derivatives are not symmetric with respect to the grid point of interest. A more accurate approximation of the partial derivatives in Eq. 15 can be obtained using the *central difference formula*:

$$\frac{\partial v_j^n}{\partial t} \approx \frac{v_j^{n+1} - v_j^{n-1}}{2\Delta t}, \quad (24)$$

$$\frac{\partial \tau_j^n}{\partial z} \approx \frac{\tau_{j+1}^n - \tau_{j-1}^n}{2\Delta z}. \quad (25)$$

This leads to a numerical scheme that depends both on values from the current time step n and the previous time step $n-1$:

$$v_j^{n+1} = v_j^{n-1} + \frac{2\Delta t}{\rho} \frac{\tau_{j+1}^n - \tau_{j-1}^n}{2\Delta z}, \quad (26)$$

$$\tau_j^{n+1} = \tau_j^{n-1} + 2\Delta t \mu \frac{v_{j+1}^n - v_{j-1}^n}{2\Delta z}. \quad (27)$$

To solve Eqs. 26 and 27, the initial values at $n = 0$ and $n = 1$ must be known. Additionally, storing the velocities and stresses from both the current and the previous time step increases memory requirements. Approximating the spatial derivatives with the central difference formula and the temporal derivatives with the forward difference formula may seem a convenient alternative. However, such a combination leads to a FD scheme that is *unconditionally unstable*, i.e., it will not converge regardless of the value of the Courant number C. The stability of a numerical scheme is often analyzed using the *von Neumann method*, which is based on Fourier decomposition of the numerical solution.

A more efficient numerical scheme is obtained by staggering the position of stresses and velocities on the temporal and spatial grid as shown in Fig. 3 (Virieux 1984). By shifting the grid position for the velocities v by ½ grid point in time and also shifting the grid position of the stresses τ by ½ grid point in space, Eqs. 15 and 16 can be approximated using

$$\rho \frac{v_j^{n+½} - v_j^{n-½}}{\Delta t} = \frac{\tau_{j+½}^n - \tau_{j-½}^n}{\Delta z}, \quad (28)$$

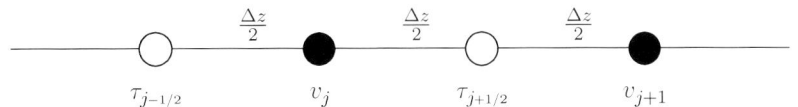

Physics-Based Ground-Motion Simulation, Fig. 3 Staggered spatial grid for the velocity–stress FD scheme (Modified from Moczo et al. 2004)

$$\frac{\tau_{j+\frac{1}{2}}^{n} - \tau_{j+\frac{1}{2}}^{n-1}}{\Delta t} = \mu \frac{v_{j+1}^{n-\frac{1}{2}} - v_{j}^{n-\frac{1}{2}}}{\Delta z}. \quad (29)$$

The required initial conditions for this scheme are the velocities at time $\frac{1}{2}$, $v^{\frac{1}{2}}$, and stresses at time 0, τ_{j}^{0}. Moving forward, the stresses at time n are computed first using

$$\frac{\tau_{j+\frac{1}{2}}^{n} - \tau_{j+\frac{1}{2}}^{n-1}}{\Delta t} = \mu \frac{v_{j+1}^{n-\frac{1}{2}} - v_{j}^{n-\frac{1}{2}}}{\Delta z}, \quad (30)$$

and then the velocities at time $n + \frac{1}{2}$ are determined using

$$v_{j}^{n+\frac{1}{2}} = v_{j}^{n-\frac{1}{2}} + \frac{\Delta t}{\rho} \frac{\tau_{j+\frac{1}{2}}^{n} - \tau_{j-\frac{1}{2}}^{n}}{\Delta z}. \quad (31)$$

The forward difference formula in Eq. 18 represents the first-order approximation to the first spatial derivative, while the central difference formula (25) represents the second-order approximation. Both schemes can be derived by replacing the functional values at $f(z_0 \pm \Delta z)$ with a Taylor expansion:

$$\begin{aligned} f(z_0 \pm \Delta z) = & f(z_0) \pm f'(z_0)\Delta z \\ & \pm f''(z_0)\frac{\Delta z^2}{2} \pm f'''(z_0)\frac{\Delta z^3}{3!} \\ & + O(\Delta z^4). \end{aligned} \quad (32)$$

The order of a numerical scheme is determined by the order of the truncation error, which is defined as the difference between the exact solution and the finite-difference approximation. Higher-order approximations may also be derived. For example, the fourth-order approximation in space combined with the second-order approximation in time has become popular for simulating ground motion (e.g., Olsen 1994; Graves 1996).

All the FD schemes discussed above Eqs. 21, 22, 26, 27, 30, and 31 are explicit, i.e., the velocities (or stresses) at a given grid point and time step are derived only from stresses and velocities of the previous time step(s). In implicit FD schemes, the velocities (stresses) at a given time step depend on velocities and stresses from both the current and previous time step. Implicit FD schemes are more difficult to solve and not frequently used in ground-motion modeling.

FD solutions to the wave Eq. 3 in two and three dimensions approximate the spatial and temporal derivatives in the same way. In computational grids used for 2D and 3D FD solutions, each element in the velocity vector and the stress tensor may be staggered with respect to the other elements (e.g., Graves 1996).

The FD solution derived above discusses only wave propagation in a continuum. In a real-world situation, discontinuities will be encountered at the free surface, at the boundaries of the computational domain, and at the contact between two different media in a heterogeneous medium.

Such internal material discontinuities can be treated using a homogeneous or a heterogeneous approach. In a homogeneous approach, boundary conditions at or near interfaces are explicitly discretized using a separate FD scheme, which is not a trivial problem. Therefore, most ground-motion prediction applications use a heterogeneous approach, where only one FD scheme is used for all internal grid points regardless of their distance to internal interfaces (Moczo et al. 2004). Such heterogeneous approaches typically define effective material parameters to improve accuracy near interfaces (e.g., Zahradnik et al. 1993).

In the above example of a vertically propagating *SH* wave, consider the interface generated by a horizontally layered sedimentary deposit resting on top of a denser, higher-velocity bedrock.

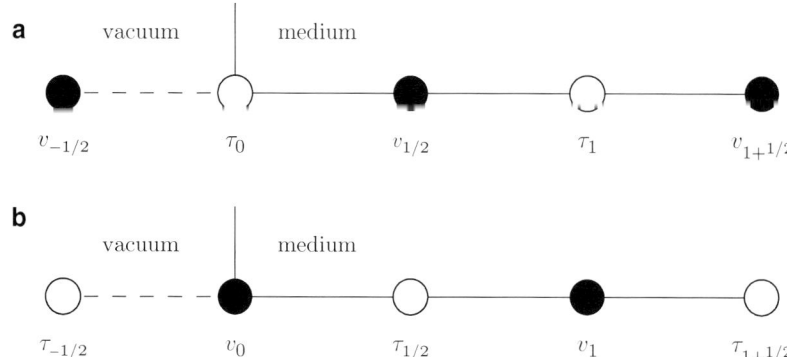

Physics-Based Ground-Motion Simulation, Fig. 4 Free surface defined to coincide with position of (**a**) shear stress and (**b**) velocity (Modified from Moczo et al. 2004)

Assume that the interface coincides with the grid point defining the stress $\tau_{j+1/2}$ in Fig. 3. It can be shown (Moczo et al. 2002) that the boundary conditions at the interface can be fulfilled by defining the shear modulus at the interface as the harmonic average of the shear moduli from the two connected media:

$$\mu_{j+\frac{1}{2}}^{H} = \frac{2}{\frac{1}{\mu_j} + \frac{1}{\mu_{j-1}}}. \tag{33}$$

If the interface coincides with the grid point defining the velocity v_j (Fig. 3), on the other hand, the density ρ at the interface Eq. 31 must be replaced with the arithmetic average of the two materials (Moczo et al. 2002),

$$\rho_j^A = \frac{1}{2}\left(\rho_{j-\frac{1}{2}} + \rho_{j+\frac{1}{2}}\right). \tag{34}$$

Similar averaging methods are implemented in most heterogeneous 2D and 3D FD codes used in research today (e.g., Graves 1996; Cui et al. 2010).

In contrast to FE methods, the free surface requires special attention in FD methods. The boundary conditions at the free surface specify that the shear stress vanishes, i.e., $\tau = 0$. In the simple 1D case described earlier, the location of the free surface can be defined such that it intersects the position of the shear stress (Fig. 4a) or the position of the velocity (Fig. 4b). In the former case, the shear stress at the surface is explicitly set to $\tau_0 = 0$ during each iteration, and the velocity half a grid point below the surface is calculated using Eq. 31. If the free surface coincides with the position of the velocity, v_0 (Fig. 4b), antisymmetry is used to ensure the traction-free boundary condition at the free surface (Levander 1988):

$$\tau_{-\frac{1}{2}} = -\tau_{+\frac{1}{2}}. \tag{35}$$

When staggered finite-difference grids are defined in two and three dimensions, a planar free surface will intersect both stresses and velocities. In that case, both approaches are employed to ensure that stresses vanish at the free surface (e.g., Graves 1996; Gottschämmer and Olsen 2001; Moczo et al. 2004).

Irregular (nonplanar) free-surface boundary conditions are far more difficult to implement in FD methods. Typically, such methods require a much finer sampling of the wave field for accurate results (e.g., Robertsson 1996; Ohminato and Chouet 1997).

Similar to FE methods, absorbing boundary conditions are required at the bottom faces of the computational domain. Both damping zones (e.g., Cerjan et al. 1985) and perfectly matched layers (e.g., Marcinkovich and Olsen 2003) have been implemented in FD codes for simulation of strong ground motion.

Intrinsic Attenuation and Plasticity

The methods for solving wave propagation problems just described apply only to linear elastic conditions. However, the accurate representation of seismic waves requires the consideration of energy losses due to internal friction or intrinsic attenuation and, when earthquake induced deformations are large enough, those due to plastic deformation as well. Both these losses are important because their omission may lead to the overestimation of the amplification and duration of seismic waves in regions with high dissipative properties, for attenuation effects in general, and in regions with soft materials that have low yielding limits or are exposed to large-magnitude earthquakes, in the case of plastic deformation. This section deals with the basic most common approaches used to include realistic attenuation and plasticity in simulations.

Intrinsic Attenuation

The general formulation of the evolution of linear isotropic viscoelastic material in time is governed by the stress–strain relation, in which the stress can be expressed as a convolution of the strain rate with a relaxation function as in

$$\sigma(\mathbf{x},t) = \int_0^t \varphi(\mathbf{x},t-\tau)\dot{\varepsilon}(\mathbf{x},t), \quad (36)$$

which is equivalent to

$$\sigma(\mathbf{x},t) = \varphi(\mathbf{x},t) * \dot{\varepsilon}(\mathbf{x},t) \quad (37)$$

or

$$\sigma(\mathbf{x},t) = \dot{\varphi}(\mathbf{x},t) * \varepsilon(\mathbf{x},t), \quad (38)$$

where the symbol $*$ is used to represent the convolution integral. $\sigma(\mathbf{x},t)$ and $\varepsilon(\mathbf{x},t)$ are the stress and strain states at a point $\mathbf{x}$ in time t, and $\varphi(\mathbf{x},t)$ is the corresponding stress relaxation function. Dots on top indicate derivatives in time. Equations 37 and 38 are the same because of the differentiation properties of convolution.

The solution of the wave propagation problem in viscoelastic media in the time domain which follows from the direct substitution of Eq. 36 in Eq. 1 is, however, inconvenient because that would entail the computation of a convolution term at every time step. Such an approach would require the storage of the complete strain history, making the computational implementation of the viscoelastic problem practically intractable.

It is clear from Eq. 38 that the formulation of the viscoelastic problem in the frequency domain is, on the other hand, straightforward. Applying the Fourier transform, Eq. 38 becomes

$$\sigma(\omega) = M(\omega)\varepsilon(\omega), \quad (39)$$

where

$$M(\omega) = \dot{\varphi}(\omega) \quad (40)$$

is understood as a frequency-dependent viscoelastic modulus. Note that, for simplicity, the spatial variable ($\mathbf{x}$) has been dropped. In general, $M(\omega)$ is defined as a complex quantity and it is such that

$$\lim_{\omega \to 0} M(\omega) = M_R, \quad (41)$$

$$\lim_{\omega \to \infty} M(\omega) = M_U, \quad (42)$$

where M_R and M_U are defined as the relaxed and unrelaxed material viscoelastic moduli. They correspond to the long-term equilibrium and instantaneous elastic response of the material and together define the relaxation modulus:

$$\delta M = M_U - M_R. \quad (43)$$

In practice, the viscoelastic modulus is expressed in terms of the material's quality factor $Q(\omega)$, which is defined as

$$Q(\omega) = \frac{\Re[M(\omega)]}{\Im[M(\omega)]}. \quad (44)$$

Understanding the formulation of anelasticity in the frequency domain facilitates its implementation in the time domain – where the system does not have to be fully assembled as in the frequency domain. The challenges in the formulation of the anelastic wave propagation problem in the time domain are (i) to solve the convolution term efficiently and (ii) to satisfy the behavior of the

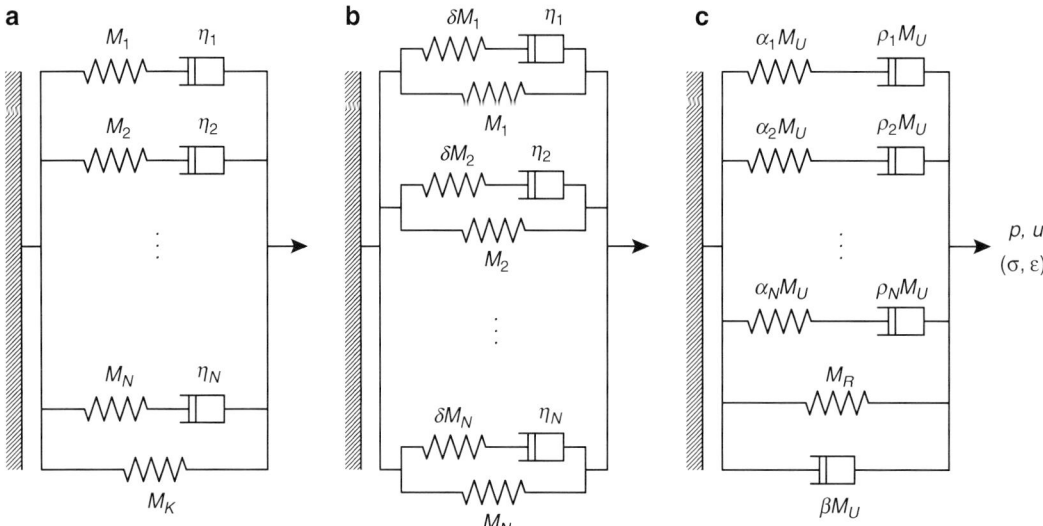

Physics-Based Ground-Motion Simulation, Fig. 5 Examples of rheological models used to incorporate the effect of attenuation. (**a**) Generalized Maxwell model (Emmerich and Korn 1987), (**b**) generalized Zener body (Carcione et al. 1988), and (**c**) a generalized Maxwell model augmented with a viscous damper (Bielak et al. 2011)

quality factor Q in the frequency domain – which for the range of frequencies typically covered by most simulations is considered to be constant in the low frequencies and frequency dependent in the higher frequencies.

Two particular works established the ground for most current approaches used to model the effects of anelasticity in time-stepped solutions (for frequency-independent Q): Liu et al. (1976) and Day and Minster (1984). Liu et al. (1976) were the first to use a rheological model made of a set of mechanical bodies (Zener mechanisms) in the context of seismic problems; and Day and Minster (1984) showed that if $M(\omega)$ is expressed as a rational function, then its inverse form in the time domain can be solved numerically – which then led to the formulation of efficient algorithms to account for anelastic losses.

Following the ideas set forth by Day and Minster (1984), in which a set of internal memory variables are used to represent the relaxation process, others have formulated memory-efficient approaches to anelastic wave propagation simulation using FD and FE approaches (e.g., Day and Bradley 2001; Ma and Liu 2006).

Emmerich and Korn (1987) proposed the use of a rational function $M(\omega)$ corresponding to a rheological model composed of a set of Maxwell bodies in parallel and a Hooke element also in parallel. They defined this model as the generalized Maxwell body (GMB) (Fig. 5a). Similarly, Carcione et al. (1988) employed a generalized Zener body (GZB) (Fig. 5b). Various separate implementations have been inspired in these two models, which were later shown to be equivalent (Moczo and Kristek 2005). Some implementation examples using FD, DG, and SE methods are those described in Chaljub et al. (2010). Most of these are formulated in terms of stresses and strains. More recently, Bielak et al. (2011) introduced a memory-efficient internal friction approach solely based on displacements more suitable for FE. The model proposed by Bielak et al. (2011) uses a set of Maxwell elements in parallel and a Voigt element also in parallel (Fig. 5c).

To correlate the problem of anelastic attenuation with the solution of the wave propagation problem in elastic media shown in the previous section, the model proposed by Bielak et al. (2011) is summarized next for the case of a FE approach. Consider a semidiscretized version of the equations of elastodynamics where a mapping operator T is introduced to modify the displacements associated with the internal body forces as in

$$\mathbf{M}\ddot{\mathbf{u}} + \sum_e [\mathbf{K}^e(T\mathbf{u})^e] = \mathbf{f}. \quad (45)$$

Here, T is such that it represents the convolution term used in the stress–strain relation, which after applying finite elements becomes the stiffness-displacement product $\mathbf{K}^e(T\mathbf{u})^e$ given by

$$\mathbf{K}^e(T\mathbf{u})^e = \mathbf{K}^e \left[\mathbf{u}^e + \beta^e \dot{\mathbf{u}}^e - \sum_j \alpha_j^e \beta_j^e \exp\left(-\gamma_j^e t\right) * \mathbf{u}^e \right]. \quad (46)$$

Note that the convolution term is still present in Eq. 46. This term is replaced by the auxiliary memory variable φ, such that at any time step n with a discrete time-width Δt, the product $\mathbf{K}^e(T\mathbf{u})^e$ is given by

$$\mathbf{K}^e(T\mathbf{u})^e = \mathbf{K}^e \left[\mathbf{u}^e + \beta^e \frac{\mathbf{u}_n^e - \mathbf{u}_{n-1}^e}{\Delta t} - \sum_j \alpha_j^e \beta_j^e \left(\varphi_j^e\right)_n \right], \quad (47)$$

where

$$\left(\varphi_j^e\right)_n = \frac{\Delta t}{2}\left[\left(1 - \Delta t \gamma_j^e\right)\mathbf{u}_n^e + \mathbf{u}_{n-1}^e\right] + \exp\left(-\gamma_j^e \Delta t\right)\left(\varphi_j^e\right)_{n-1}. \quad (48)$$

α, β, and γ are constants associated with the mechanical elements of the adopted model (i.e., spring and dashpot constants) that are determined for each finite element e based on the quality factor Q of the material contained in the element. These constants can be derived separately for the quality factors associated with the propagation of P and S waves, Q_P and Q_S. Details about the particular implementation of this model can be found in Bielak et al. (2011). Information about the derivation of the values of Q_P and Q_S is provided in the section on "Material Models".

Plasticity

The problem of wave propagation in elastoplastic and elasto-visco-plastic media has also been considered in 3D physics-based earthquake simulations, but has not yet reached the same level of maturity as anelastic wave propagation. This is mainly due to the computational complexity involved in incorporating full 3D nonlinear soil behavior in the solution scheme and the entailed computational overhead both in memory and processing time – as well as the difficulty to accurately reproduce ground motions in the frequency ranges where plastic behavior is believed to be more relevant (above 0.5 Hz).

Numerical modeling of the response of nonlinear sedimentary deposits also dates back to the 1960s and 1970s. Initial simulations considered 1D and 2D models and used linear equivalent methods to approximate the stress–strain relationship (e.g., Idriss and Seed 1968). Although the linear equivalent method continues to be used extensively in engineering research and practice, it has long been understood that it does not capture all the characteristics of nonlinear plastic behavior. Alternatively, there exist an abundant number of models based on more rigorous methods (e.g., Prevost 1978) to describe the cyclic stress–strain behavior of geomaterials. However, such models are usually defined in terms of material parameters upon which there is no consensus. In addition, they tend to be computationally expensive. This has made it difficult to apply rigorous plastic models in 3D regional earthquake simulations where the computational aspects are critical and the knowledge about the material properties at regional scales is limited – especially for the near-surface layers, where nonlinear soil effects are more significant.

An approach used to overcome the computational difficulties of incorporating near-surface plastic effects in physics-based simulations has been the combination of 3D linear elastic or anelastic simulation results with 1D nonlinear analyses of the incident motion at the interface between the bedrock and the sedimentary deposits (e.g., Roten et al. 2012). Hybrid simulations, however, cannot completely describe the 3D aspects present when off-fault and near-surface plasticity is combined with the source, path, and basin effects. Alternatively, there have been successful first approximations to obtain full 3D regional-scale simulations that consider plastic deformations using computationally tractable material models such as the classical Drucker–Prager yield criterion (e.g., Dupros et al. 2010; Taborda et al. 2012; Roten et al. 2014). Although the material models used in these simulations do not accurately reproduce the elasto-visco-plastic behavior of most geomaterials, they are useful to understand the 3D nature of nonlinear wave propagation in heterogeneous media.

Following the formulations presented before for the case of the FE method, the solution of the wave propagation problem in plastic media can be described as follows. Consider the semidiscretized version of Navier's Eqs. 1 through 6, but in a manner in which the stresses are preserved explicitly. In that case, Eq. 6 becomes

$$\mathbf{M}\ddot{u} + \sum_e \int_{\Omega_e} \mathbf{B}^T \boldsymbol{\sigma} d\Omega_e = f, \quad (49)$$

where $\mathbf{B}$ is the strain matrix and $\boldsymbol{\sigma}$ is the stress tensor over element Ω_e. The summation, again, means assembling of elements.

Some important points need to be noticed about Eq. 49. See that in order to obtain the contribution of the internal forces given by the product $\mathbf{B}^T \boldsymbol{\sigma}$ in the integral term, one must know the state of stresses. The stresses, however, depend on the state of total strain in the material (ε), and the strains, in turn, need to be compatible with the stresses themselves according to the constitutive model chosen to represent the material's plastic behavior. In the elastic problem, this relationship between the stress and strain tensors is linear. In plasticity, on the contrary and in general, it is not. Therefore, it follows from this that, embedded within the time integration of Eq. 49, there is a nonlinear problem that requires the implementation of an implicit solution scheme, which in most cases requires additional computational effort.

In general, following the classical theory of plasticity, the total strain (ε) can be expressed in terms of the sum of the elastic (ε^e) and plastic (ε^p) deformation components,

$$\varepsilon_{ij} = \varepsilon_{ij}^e + \varepsilon_{ij}^p. \quad (50)$$

This is useful because one can then express the associated admissible stress in terms of the product of the elastic stiffness tensor (D) and the elastic strain,

$$\sigma_{ij} = D_{ijkl}\varepsilon_{kl}^e. \quad (51)$$

Equations 50 and 51 also need to be thought carefully. In solving Eq. 49, at any given time step, one can obtain the total strain from the current state of displacements. The objective then becomes finding its plastic and elastic components, so that the stress given by Eq. 51 remains compatible with the constitutive model and the total strain. Considering that by definition the elastic deformation component is bound to the stresses by the elastic stiffness tensor – as opposed to the plastic deformation which is unbounded in general – the critical point becomes finding the corresponding plastic deformation ε^p.

This is where the choice of the material model plays its role. Constitutive models provide the means for tracking the progress of the plastic deformation through the combination of a yielding potential function (g) and the rate of the plastic strain $\left(\dot{\varepsilon}_{ij}^p\right)$ – that is, a function of how the plastic strain changes in time. Yielding potential functions are of the form

$$g(\sigma_{ij}, k) = F(\sigma_{ij}) - k(\sigma_{ij}, k_n) < 0, \quad (52)$$

where F represents the current state of stresses and k defines the hardening characteristics of the material. At the upper limit of Eq. 52, $F - k = 0$ represents a yielding surface which defines the plastic state of the material. In other words, the yielding surface provides a limiting state on which a material's particle must remain while in a plastic deformation condition. Below this surface, the particle behaves elastically, and on the surface, plastically. Here, the hardening characteristics controlled by k refer to the ability of some materials to regain strength after they have gone over the plastic domain. In perfectly elastoplastic materials, for instance, the yielding surface can be thought of as being "flat" and k is null. In materials with hardening characteristics, on the other hand, the yielding surface changes depending on the state of deformation. That is why k is also a function of σ. k_n is used here generically to represent the material parameters controlling hardening.

As noted in Eq. 52, both F and k are functions of the stresses and therefore depend on the plastic state of deformation. It follows from Eqs. 50, 51, and 52 that the key component toward finding the solution of Eq. 49 is determining the state of plastic deformation, which is given by the plastic potential strain rate:

$$\dot{\varepsilon}_{ij}^{p} = \dot{\lambda} \frac{\partial h(\sigma_{ij}, k)}{\partial \sigma_{ij}} = \dot{\lambda} \frac{\partial g(\sigma_{ij}, k)}{\partial \sigma_{ij}}, \qquad (53)$$

where $\dot{\lambda}$ is a plastic multiplier and h is a function which defines the plastic potential. As noted in the right-hand side of Eq. 53, for simplicity, h is a function similar to and often assumed to be the same as the yielding potential function g explained before. The plastic multiplier $\dot{\lambda}$ sets the magnitude of the plastic deformation in the current state of stresses as a material's particle moves over the yielding surface.

As it can be inferred from this description, the solution of Eq. 49 leads to additional computations to help ensure that the stress–strain relationship is maintained according to the nonlinear constitutive model of choice. This requires the use of implicit solution schemes, as opposed to the explicit approach used in the elastic and viscoelastic problems. This additional effort explains why considering the material's plastic behavior in physics-based ground-motion simulations in 3D has often been ignored or simplified using hybrid and indirect approaches. Future developments, however, are likely to revert this trend, and progress is expected to be done in this area in the near future.

Source Models

As seen in the simulation workflow section, physics-based earthquake simulations depend on two basic models used to represent the earthquake itself and the propagation media, these are the source model and the material model, respectively. They, in turn, define the characteristics of the applied and internal body forces in the formulation of the solution of the wave propagation problem. This section deals with the basic concepts of source models used as input to earthquake simulations.

The fault's rupture and the resulting wave propagation problem are seldom solved together in a single simulation, especially at regional scales. This is mainly due to the complexity of the rupturing process on the fault, which entails a multi-physics problem with plastic deformation and/or the use of dislocation models to represent the loss of friction/contact on the fault's surface. Source models are, instead, resolved prior to performing the simulation of the ground motion and then treated in the simulation as basic input data as seen in the simulation workflow section. Some common alternatives from where source models are derived include:

- Source inversion studies
- Dynamic rupture simulations
- Pseudo-dynamic rupture generators
- Kinematic source model generators
- Basic geologic and seismogenic information about the fault

Regardless of the method employed to obtain the source model, the most common approach used in simulations is to convert the source model into a kinematic source representation, that is, a model that represents the source as a set of equivalent forces (or stresses) that are applied to the forward simulation model to trigger the propagation of seismic waves. These equivalent body forces are such that they produce a displacement field away from the source that is equivalent to that one would have obtained using a more rigorous solution for the source if embedded in the simulation.

In the case of small-magnitude earthquakes ($M < 5$) or sources small enough compared to the wavelength of the radiated energy, the effect of the earthquake rupture and the discontinuity of displacements that occurs at the fault can be modeled using a single set of self-balanced (double-couple) forces acting on a point, that is, a *point source* model. In the case of large-magnitude earthquakes, on the other hand, the rupture occurs over extended areas on the fault's surface and thus cannot be treated as a point source. In such cases, the earthquake is modeled as the sum of many point sources. The approach relies on the idea that it is possible to discretize the fault's ruptured area into a collection of smaller subfaults, each with an assigned point source model. The subfaults are such that they adjust to the geometry of the entire fault and the collective action of the point sources adds up the right amount of energy release, equivalent to that of the complete earthquake model. Extended fault models composed of multiple subfaults are called *finite slip* or *finite source* models.

In both point and extended fault models, the point source at each (sub)fault is defined in terms of its geometry and rupture characteristics. These are given by:

- The location given in latitude, longitude, and depth
- The orientation given by the strike, dip, and rake angles of the (sub)fault
- The (sub)fault's area and average shear modulus of that area
- The evolution of the slip on the (sub)fault with time

Figure 6 shows an example of a kinematic source model. Part (a) shows the total slip distribution on the fault plane of the source model by Graves and Pitarka (2010) for the 1994 M_w 6.7 Northridge, California, earthquake. This model is composed of 140 × 140 subfaults with strike and dip angles of 122° and 40°, respectively, and variable rake angle with an average of 101°. Part (b) shows the concept of the discrete subfault as used in a kinematic finite slip model in which each rectangular (or triangular) patch on the fault plane has an independent geometry (area, strike, dip, rake) and slip (as a function of time). And part (c) shows a typical slip velocity function, which is the function that defines the history of slip associated with the point source in each subfault area.

Point source models used in physics-based simulations can be simply built based on the earthquake focal mechanism and the selection of an appropriate slip function such as that shown in Fig. 6c. Extended fault models, on the other hand, are available from source inversions and rupture generators, and distributed in various formats. The US Geological Survey Earthquake Hazards Program, for instance, offers finite fault models for significant earthquakes in its Historical Earthquake Information database Web portal. Another useful source of fault models is the Finite-Source Rupture Model Database (SRCMOD) maintained by the eQuake-RC project Web site (http://equake-rc.info/). Models distributed by SRCMOD are contributed by researchers from all over the world and distributed in a set of common data files that include basic metadata and simple single-rupture-plane source-model representations. Another popular distribution format among modelers, especially in the USA, is the Standard Rupture Format (SRF) used by Graves and Pitarka (2010). The SRF encapsulates the rupture process in a single (ascii) text file that can have multiple fault planes and subfaults with variable geometry.

One important aspect to note as simulations advance toward higher frequencies is the influence that the source type and source model description have on the characteristics of the ground motion. If seen in the frequency domain, for instance, the slip-rate function of a single point source (as that shown in Fig. 6c) will reveal that the energy of the slip is mostly contained below a certain frequency. This, together with the seismic velocities represented in the model, will influence the energy distribution of the ground motions in the frequency domain,

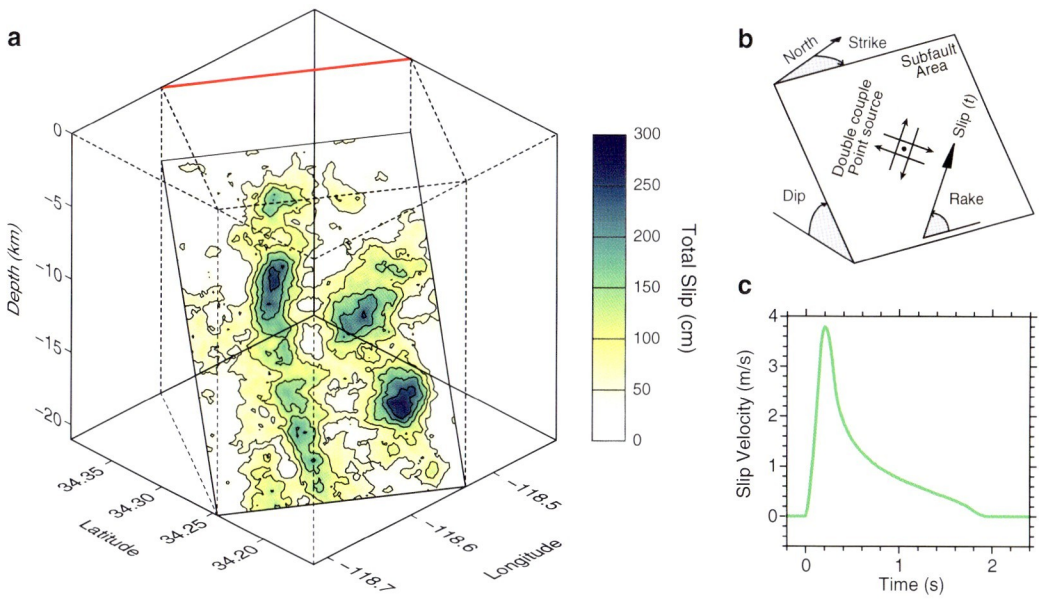

Physics-Based Ground-Motion Simulation, Fig. 6 Example of a kinematic source model: (**a**) finite slip representation of the 1994 *M* 6.7 Northridge earthquake composed of 140 × 140 subfaults (Modified after Graves and Pitarka 2010), (**b**) subfault model concept for a double-couple point source with independent geometry and source-time function, and (**c**) a typical slip-rate function in time

provided models are built with appropriate accuracy. In other words, and it should come as no surprise, the source spectrum is important in determining the frequency content of the ground motion.

Furthermore, in extended source models, the smoothness and homogeneity of the subfault characteristics also influence the outcome. If the distribution of total slip on the subfaults is fairly homogeneous, or if the evolution of the slip through the fault plane follows a certain pattern in space and time, then the wave field will evolve more smoothly as it travels away from the source plane. It will be more coherent. On the other hand, if each subfault slips randomly, the ground motion will be less coherent, and waves will be more likely to interfere with each other as they travel away from the fault. Similarly, the orientation of the slip on the plane, given by the rake angle, will dictate the strength of directivity effects. Additional complexity can be introduced if we also consider the fact that subfaults do not necessarily have to be aligned. Figure 5a depicts extended source models as a collection of point sources with variable rake angle and slip, but constant strike and dip angles, with all the subfaults being part of a single fault plane. In reality, however, faults are not smooth planes, but irregular contact areas. This increases the variability of the ground motion considerably, as shown by recent models developed by Shi and Day (2013) to account for the geometrical heterogeneity (roughness) of the fault.

As mentioned above, these are all important factors that modelers need to consider, especially when simulating ground motions at high frequencies (>1 Hz), because short wavelengths can better capture the variations of the fault structure and rupture characteristics.

Material Models

Provided a source model, the second necessary component for a simulation is the material model, which defines the mechanical properties of the propagating media in the modeling domain. The most basic 3D material model used for elastic wave propagation simulations defines the material density (ρ) and seismic velocities of *P* and *S* waves

(V_P and V_S, respectively) at any arbitrary point within the simulation domain. Anelastic simulations require, in addition, the attenuation properties of the propagating media. The material's attenuation properties are defined in terms of the quality factors Q_P and Q_S associated with the attenuation characteristics of P and S waves, respectively. These quality factors are defined using attenuation rules or attenuation relationships which, in the context of physics-based ground-motion simulation, are empirical functions based on the values of V_P and/or V_S. For elastoplastic simulations, material models must also provide the parameters that define the characteristics of the adopted nonlinear constitutive model. Since elastoplastic material properties are unique to the adopted constitutive model, they are not described here; besides, complete definitions of all the mechanical properties of the propagating media in a single model are uncommon. Instead, for the case of anelastic ground-motion simulation, material models are divided in two parts: (a) seismic velocity models and (b) attenuation relationships. These are described next.

Seismic Velocity Models

Seismic velocity models, also known as community velocity models (CVMs), are datasets or computer programs that define the values of V_P, V_S, and ρ at any point in a particular region of interest. The point's location is usually expressed in terms of latitude, longitude, and depth (with respect to the free surface) or elevation (with respect to the sea level). Internally, CVMs are built using different types of datasets, which may include but are not limited to:

- Source inversion studies
- Surface topography or digital elevation maps
- Subsurface topography or geological horizons
- Gravity observations and refraction surveys
- Teleseismic and 3D tomographic inversions
- Mantle and Moho 1D background models
- Empirical rules correlating V_P, V_S, and ρ
- Shallow and deep boreholes and observations from oil wells
- Geotechnical layer models based on indirect measures (e.g., V_{S30} data)
- Random media representations

Some velocity models are built using voxels (volume elements) that define homogeneous 3D regions. Other models combine different datasets using interpolation rules that assign weights based on the location of the point of interest relative to the data points.

Most CVM computer codes are not necessarily optimized to work seamlessly with earthquake simulation codes. Modelers use additional tools to convert the data retrieved from a given CVM into grids or meshes. One particular tool of the sort is the unified community velocity model (UCVM) software framework (Small et al. 2015). UCVM is a collection of software tools developed and maintained by the Southern California Earthquake Center (SCEC) to provide an efficient and standard access to multiple, alternative velocity models. Although UCVM was primarily built to manage the SCEC community velocity models CVM-S and CVM-H for Southern California (Fig. 7), it supports and can be used to register other models as well.

There exist a good number of velocity models available to the community. It is, however, not possible to cover them all here in detail. A selection of some of the most relevant models used in simulations in the USA, Europe, and Japan is summarized in Table 1.

As simulations aim to produce more realistic ground motions comparable to observations, an important aspect in the construction and use of velocity models is that of the representation of the geotechnical layers and the variability of the material properties at small scales. For the most part, despite their level of detail, seismic velocity models tend to be smooth representations of the crustal structure. This is primarily due to the fact that geologic, exploration, and other survey data are only available at coarse resolutions. Modelers need then to resort to other methods to represent the presence of near-surface soft-soil deposits and the random characteristics of geomaterials. The presence of soil deposits is typically handled by introducing some kind of geotechnical layer

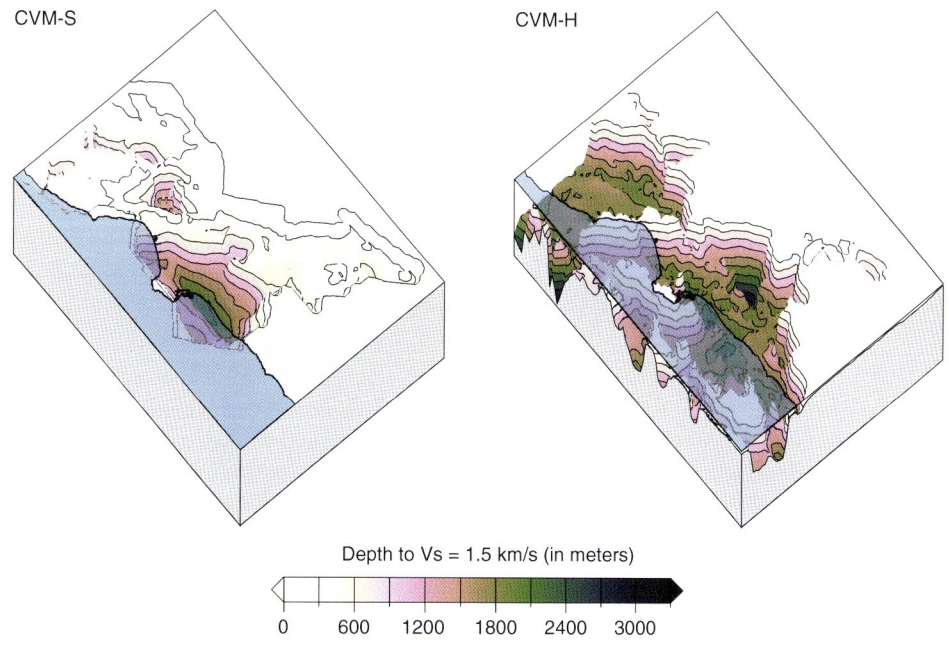

Physics-Based Ground-Motion Simulation, Fig. 7 Comparison of the two Southern California community velocity models CVM-S and CVM-H. Depth of the major basins in the Greater Los Angeles region (in meters) as inferred from the isosurface for $V_S = 1.5$ km/s

Physics-Based Ground-Motion Simulation, Table 1 Summary of velocity models

Model	Region	Available at
CVM-S	Southern California (SCEC Model)	http://scec.usc.edu/scecpedia/
CVM-H	Southern California (Harvard Model)	http://scec.usc.edu/scecpedia/
CenCalVM	Central California (San Francisco Bay Area)	http://earthquake.usgs.gov/research/structure/3dgeologic/
WFCVM	Wasatch Front (Salt Lake City, Utah)	http://geology.utah.gov/ghp/consultants/geophysical_data/cvm.htm
CUSVM	Central U.S. (New Madrid Seismic Zone)	http://earthquake.usgs.gov/research/cus_seisvelmodel/
J-SHIS	Japan Substructure	http://www.j-shis.bosai.go.jp
JIVSM	Japan Integrated Velocity Structure Model	See Koketsu et al. (2009)
Grenoble Basin	Grenoble Basin, France	See Chaljub et al. (2010)
Mygdonian Basin	Northern Greece (Euroseistest)	http://euroseisdb.civil.auth.gr/

model (GTL), which softens the transition from the rock basement to the surface. GTLs use empirical rules to estimate the near-surface (V_{S30}) seismic velocities and then interpolate the material properties from surface to depth. The interpolation is parameterized using existing borehole profiles (Ely et al. 2010). The variability of the medium, on the other hand, is introduced by means of random spatially correlated perturbations to the model at small-scale resolutions (Hartzell et al. 2010; Withers et al. 2013). Both these elements have shown to have a significant effect on the ground motion, especially at higher frequencies, thus their importance in the future of physics-based simulation.

Attenuation (Q) Relationships

As mentioned before, the quality factors associated with P and S waves, Q_P and Q_S, are most commonly defined based on the values of the seismic velocities V_P and V_S. The value of Q_S is usually defined from rules that depend directly on the value of V_S. Typical forms of Q_S–V_S relationships are piecewise linear and continuous polynomial functions. The value of Q_P, on the other hand, is usually defined in terms of Q_S and, in some cases, in terms of the velocity contrast V_P/V_S. Table 2 shows a collection of different Q_S–V_S and Q_P–Q_S relationships used in simulations. Each relationship is listed along with a publication, the earthquake or scenario event for which they were employed, and the minimum shear-wave velocity $V_{S_{min}}$ and maximum frequency (f_{max}) that was used in the associated simulations. Some of the references provided in Table 2 include notes (see superscripts) to clarify the nature, region, or scope within which the relationship was introduced or used. The FD and FE superscripts, in particular, indicate when a reference is being cited specifically in reference to the results published therein using one of these methods. Aagaard et al. (2008) or Bielak et al. (2010), for instance, include simulations using several methods; thus the superscripts indicate which method was used with the relationship associated with these studies.

As it can be seen from this table, there is no consensus on the most appropriate set of relationships between seismic velocities and quality factors. Notice also that the majority of the relationships are independent of depth (z) and all are independent of the frequency, even though it is known that these relationships are both depth and frequency dependent.

Here, the frequency dependence of Q is particularly important as it is understood to be of greater significance at frequencies above 1 Hz. Up until recently, this was not a major factor in physics-based ground-motion simulation because simulations were typically done for maximum frequencies no greater than 1 Hz. However, with the increasing capacity of supercomputers, modelers are now more often being able to simulate ground motions at higher frequencies, thus its relevance in future efforts. In this case, the main ideas of the viscoelastic methods described in the "Intrinsic Attenuation" section will still apply, as they are formulated in frequency. Their current implementation in computer codes used in simulations, however, had been usually calibrated to adjust to constant target values of Q. Therefore, in years to come, the implementation of these viscoelastic models and the relationships shown in Table 2 will need to be revised to offer a consistent approach to modeling high-frequency (0–10 Hz) ground motions. This is a topic of current research and ongoing efforts such as those reported by Withers et al. (2014) offer a glimpse of the future in this regard.

Recent Examples and Applications

There is an ample spectrum of physics-based simulations available in the literature. Most of them have been published over the last two decades after supercomputing centers open for public research became available in the mid- to late 1990s and early 2000s, which boosted the capacity of modelers to conduct regional-scale simulations at resolutions not possible before parallel computer codes were developed and tested. Some of these simulations have already been cited to illustrate the various aspects involved in physics-based simulation. However, they cannot all possibly be covered here, thus only a small selection is addressed next. The selection includes examples of scenario and real earthquake simulations that have been used in verification and validation studies; those are the simulations of the Great Southern California ShakeOut and the 2008 Chino Hills, California, earthquake. Also covered here to a lesser extent is the application of physics-based simulation as a tool to construct a physics-based framework for probabilistic seismic hazard analysis, as it is done in the CyberShake project of the Southern California Earthquake Center.

Physics-Based Ground-Motion Simulation, Table 2 Examples of Q_S–V_S and Q_P–Q_S relationships used in past physics-based simulations. Most of these correspond to the region of Southern California and to past earthquakes, unless noted differently. FD and FE superscripts indicate when a publication is cited specifically in reference to results published therein using a particular method

Publication	Simulation	$V_{S_{min}}$ (m/s)	f_{max} (Hz)	$Q_S = g(V_S)$ (V_S in km/s, z in km)		$Q_P = h(Q_S)$
Olsen et al. (2003)	1994 Northridge	500	0.5	$20V_S$ $100V_S$	$V_S < 1.5$ $V_S \geq 1.5$	$1.5Q_S$
Bielak et al. (2010)[FD]	ShakeOut[a]	500	0.5	$50V_S$		$2Q_S$
Cui et al. (2010)	M8[a]	400	2.0			
Graves et al. (2011)	CyberShake[a]	500	0.5			
Komatitsch et al. (2004)	2001 Hollywood 2002 Yorba Linda	670	0.5	90 ∞	Sediments Bedrock	∞
Taborda et al. (2007)[b]	ShakeOut[a]	200	1.0	$50V_S$		
Bielak et al. (2010)[FE,b]	ShakeOut[a]	500	0.5			
Graves (2008)	2001 Big Bear	250	1.0	$60V_S$ $50V_S$	$V_S < 0.9$	
Aagaard et al. (2008)[FD]	1989 Loma Prieta	330–760	0.5–1.0	$60V_S^{1.5}$ 500 13	$0.9 \leq V_S < 3.4$ $V_S \geq 3.4$ $V_S < 0.3$	$2Q_S$
Brocher (2008)[c]	–	–	–	$-16 + 104.13V_S$ $-25.225V_S^2$ $+8.2184V_S^3$	$0.3 \leq V_S < 5$	
Chaljub et al. (2010)[d]	2003 Lancey, and Event S1[a]	300	2.0	50 ∞	$z < 1$ $z \geq 1$	$3/4(V_P/V_S)^2 Q_S$ ∞
Taborda and Bielak (2013)	2008 Chino Hills	200	4.0	$10.5 - 1.6V_S + 153V_S^2 - 103V_S^3 + 34.7V_S^4 - 5.29V_S^5 + 0.31V_S^6$		$3/4(V_P/V_S)^2 Q_S$

[a]Denotes scenario events
[b]Rayleigh damping instead of a viscoelastic model
[c]Empirical relations (no simulation) for Northern California
[d]Simulations for Grenoble Valley, France
[FE]Finite-element simulation therein
[FD]Finite-difference simulation therein

ShakeOut

The 2008 Great Southern California ShakeOut was a multidisciplinary earthquake preparedness and emergency management exercise involving Earth science, engineering, and social sciences which has now transcended to an annually repeated drill held in many other US and worldwide seismic regions. In its first edition, the ShakeOut included the definition of an M 7.8 scenario earthquake rupturing the southern segment of the San Andreas Fault. The regional ground motion over an area of 600 × 300 km covering all major cities in Southern California for the ShakeOut scenario was computed using physics-based ground-motion simulation. A physics-based approach was chosen over empirical ground-motion prediction equations because the latter were considered poorly constrained for such an event. Details about the scenario and results can be found in Jones et al. (2008). The predicted ground motions helped estimate the dynamic response of buildings, losses, casualties, and the socioeconomic impacts of such an earthquake and are described in a special issue of *Earthquake Spectra* (Porter et al. 2011).

The simulations of ground motions for the ShakeOut were independently carried out by three different groups. Two groups employed FD codes and one group used a FE code. The results were presented in Bielak et al. (2010) and verified using different comparison methods. Ground motions predicted by the three codes (Fig. 8) were found to be in good agreement with each other independently of the differences between the three code implementations. The analysis of the ground motion revealed very strong (>3 m/s) shaking near the fault and damaging ground motions (>0.5 m/s) over large areas of Los Angeles, San Bernardino, and Riverside (Jones et al. 2008). Strong long-period ground motions (>1 m/s) were also anticipated for the Los Angeles Basin, where significant damage and potential collapse was inferred for older high-rise buildings. These strong amplifications were attributed to surface waves channeled along a chain of sedimentary basins acting as a waveguide, which had been previously observed during a similar simulation for the TeraShake scenario earthquake (Olsen et al. 2006).

Also based on ShakeOut simulation results, Graves (2008) reported large variations in Los Angeles Basin ground-motion levels resulting from small (~15 %) adjustments to the rupture velocity. Olsen et al. (2008) had demonstrated that replacing the kinematic rupture models with spontaneous rupture models reduces PGV extremes in the region by a factor of 2–3 and attributed these

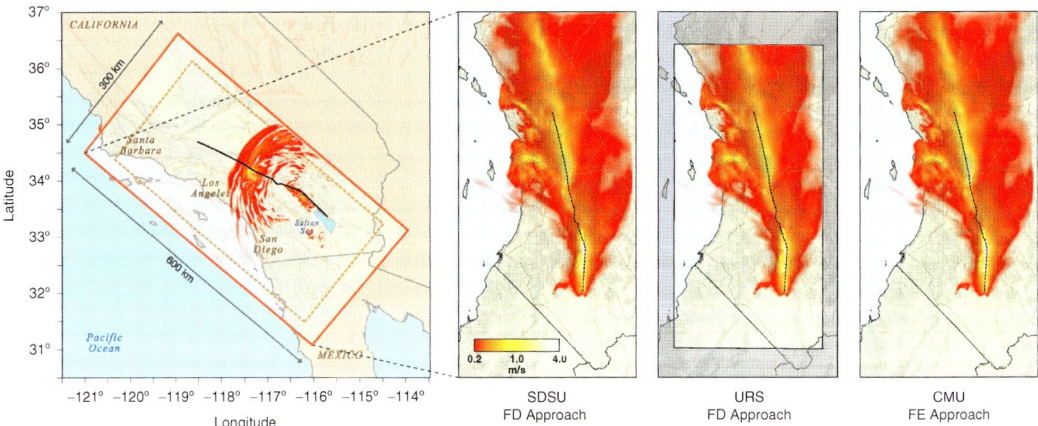

Physics-Based Ground-Motion Simulation, Fig. 8 Region of interest and surface projection of the simulation domain used in the ShakeOut scenario with a superimposed still image of the ground velocity halfway the rupturing segment of the San Andreas Fault (*left*), and comparison of horizontal magnitudes of peak ground velocities obtained from the three simulation sets corresponding to the two finite-difference and one finite-element codes (After Bielak et al. 2010)

reductions to the less coherent wave field excited by dynamic rupture models. Day et al. (2012) identified the segment between the Cajon Pass and the northern Coachella Valley as the main contributor to amplification in the Los Angeles Basin, with the highest excitation resulting from super-shear (or energetically forbidden sub-shear, super-Rayleigh) rupture speeds. These studies have suggested that the level of shaking in the Los Angeles region during a large ShakeOut-type earthquake will depend strongly on the details of the source and warrant more research into the physics of the rupture process and the wave propagation, yet they are excellent examples of the reach of physics-based earthquake simulation as a tool to gain insight about the seismic conditions of regions prone to large-magnitude earthquakes.

2008 M_W 5.4 Chino Hills Earthquake

In contrast to the simulation of scenario earthquakes such as the ShakeOut, the simulation of the wave propagation from past events offers the additional possibility of validating deterministic ground-motion prediction results against recorded data. Various examples of validation of past earthquakes are available for low frequencies ($f < 0.5$ Hz), where physics-based simulations have shown to perform best (e.g., Komatitsch et al. 2004). On the other hand, there are efforts to advance the simulation of earthquakes using fully deterministic simulations alone or in hybrid approaches to predict broadband ground motions at the higher frequencies of engineering interest (up to and above 10 Hz). A series of simulations oriented toward that goal are those done for the 2008 M_w 5.4 Chino Hills, California, earthquake.

Olsen and Mayhew (2010) generated broadband (0–10 Hz) synthetics for the Chino Hills earthquake using a hybrid method based on the combination of low-frequency (<1.6 Hz) FD signals with high-frequency scattering operators. Olsen and Mayhew (2010) modeled the rupture as a point source with a combined strike-slip/thrust fault mechanism and used a minimum S-wave velocity of 500 m/s. Because visual inspection was not deemed suitable for evaluation of the quality of fit between observed and synthetic data at shorter periods, Olsen and Mayhew (2010) used goodness-of-fit criteria to validate the broadband synthetics at 33 selected stations in the Los Angeles Basin. The goodness-of-fit results suggested that the threshold for acceptance was reached at two thirds of all the sites at short periods and at all the analyzed sites at moderate to long periods.

More recently, Taborda and Bielak (2013) used an entirely deterministic FE approach to model the wave propagation in the Greater Los Angeles region during the Chino Hills earthquake for frequencies up to 4 Hz and shear-wave velocities as low as 200 m/s. Their simulations used a finite source model derived from an independent inversion study and performed validations at over 300 stations between simulation synthetics and actual records. Taborda and Bielak (2013) also used a modified version of the goodness-of-fit criterion defined by Anderson (2004). Similar to Olsen and Mayhew (2010), they obtained a good agreement over the entire region at low frequencies (<0.5 Hz), but observed a decay in the goodness of fit with frequency, specially at the higher end (2–4 Hz). Their analysis indicated that this was mainly due to inaccuracies in the velocity model used (CVM-S), and in particular in the near-surface sedimentary layers, which have a stronger influence on the quality of the fit at the higher frequencies. Nonetheless, Taborda and Bielak (2013) found that the synthetics were overall realistic and concluded that future improvements to the material model will eventually render better fits. Figure 9 summarizes the results obtained by Taborda and Bielak (2013) and exemplifies the level of agreement (and lack thereof) that is possible when employing physics-based earthquake simulation to reproduce (well-constrained models of) past earthquakes.

CyberShake

The SCEC CyberShake project is an effort to develop a framework for probabilistic seismic hazard analysis that incorporates explicitly the

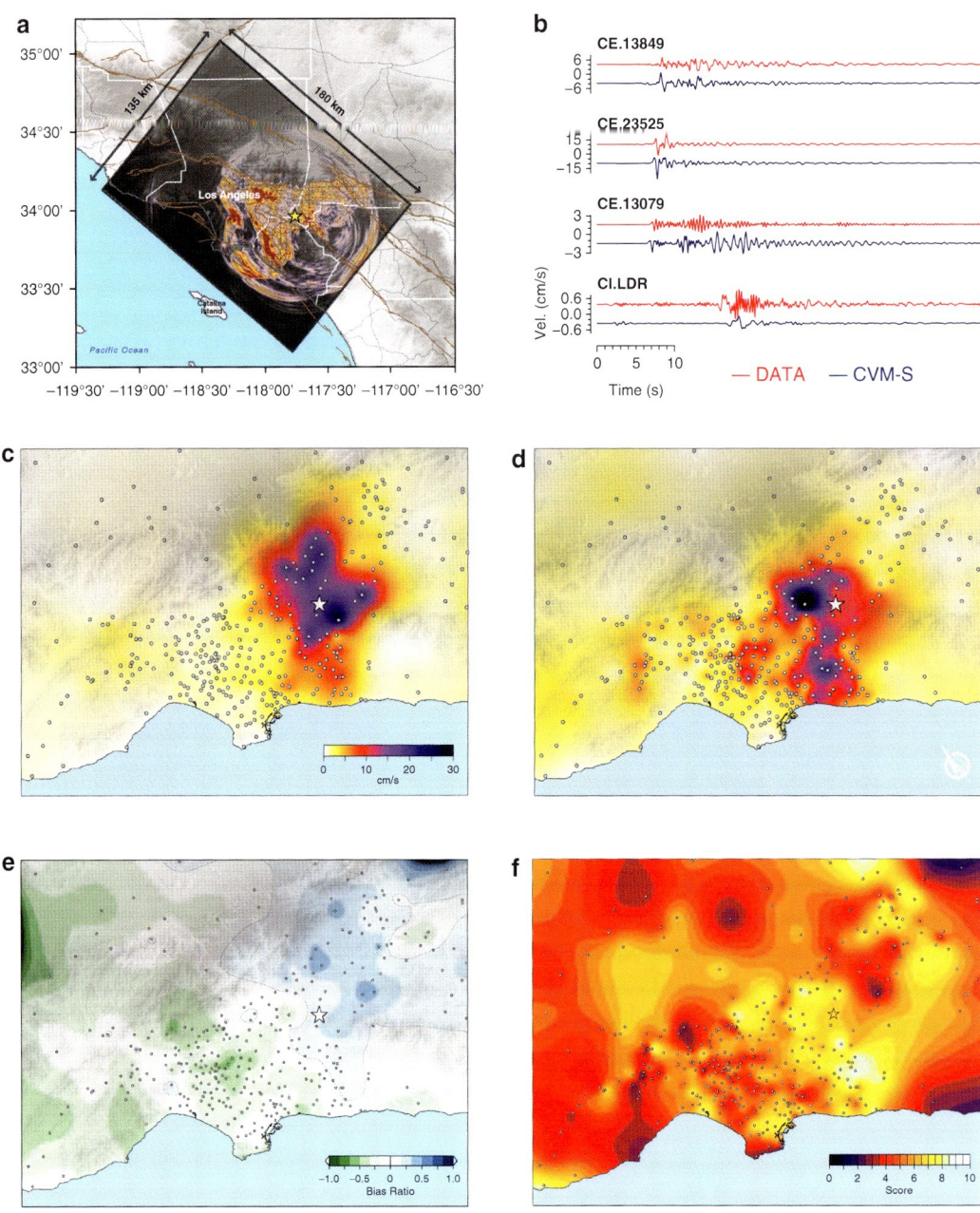

Physics-Based Ground-Motion Simulation, Fig. 9 Summary of results of simulation and validation of the 2008 M_w 5.4 Chino Hills earthquake. (**a**) Region of interest and simulation domain surface projection, (**b**) comparison of data and synthetics at selected locations, (**c**) spatial distribution of PGV interpolated from data at recording stations, (**d**) spatial distribution of PGV interpolated from synthetics at the same stations, (**e**) bias ration between data and synthetic values of PGV, and (**f**) spatial distribution of the goodness-of-fit scores obtained for PGV (After Taborda and Bielak 2013)

source and wave propagation effects by means of physics-based simulations to estimate the ground motions of the expected earthquakes in a region (Graves et al. 2011). CyberShake has been initially tested for the case of the Greater Los Angeles area and will be extended to the Southern and Central California regions in the future. It consists of conducting hundreds of thousands of forward simulations for as many variations of fault ruptures as possible to consider for all active faults in the region of interest. However, in order to avoid the overload of conducting as many forward simulations as possible rupture variations for all the active faults and hypocenter locations, CyberShake precomputes strain Green tensors (SGT) between a collection of receivers (or regular grid of stations), and each of the subfault patches in each fault plane for all the finite fault models in the region. Then, CyberShake uses reciprocity to compute the synthetic seismograms for any given earthquake scenario by combining the corresponding contribution of each subfault's slip to the earthquake's finite fault model. The combined results of all the ground-motion synthetics obtained from the reciprocity computations for all the intended forward simulations offer a unique physics-based approach to the long-term seismic hazard of the region considered because the intensity measures produced by the simulations naturally incorporate 3D source, path, basin, and site effects and spatial variability of the region (as inferred from the employed velocity and source models), thus avoiding other assumptions for these effects implicitly embedded in empirical ground-motion relationships used in traditional probabilistic seismic hazard analysis. While significant further research will be required before CyberShake seismic hazard maps can be adopted in earthquake engineering analysis and design, its operational framework offers a perspective to the potential future use of physics-based earthquake ground-motion simulation.

Summary

This chapter presented the background and basic operational framework of physics-based (deterministic) ground-motion simulation, its most commonly employed methods, and some notable recent application examples. The generation of ground-motion synthetics using a physics-based approach requires a (kinematic) finite source model and a material (seismic velocity and attenuation) model, which are used as input data for solving a forward wave propagation problem. Simulations are resolved using numerical techniques such as the finite-difference, finite-element, spectral element, or discontinuous Galerkin methods. Current uses of physics-based ground-motion simulation include the simulation of large earthquake scenarios for the assessment of the potential impact of seismic events and the evaluation of appropriate emergency management and response strategies and the construction of physics-based probabilistic seismic hazard maps (at long periods) based on local and regional fault models. Physics-based ground-motion simulation methods have been thoroughly tested through verification of scenario events and validation of small- and moderate-magnitude past earthquakes. Validation results, in particular, show that accurate seismograms (comparable to real records) can be obtained through physics-based ground-motion simulation, especially at low frequencies where both the source and the material models are better constrained. It is expected that future advances in all related areas (source models, material models, characterization of site effects, simulation methods) will permit the use of physics-based ground-motion simulations in broadband seismic hazard studies and engineering analysis and design, especially in regions where there is insufficient historical data and where simulations can complement the regional perspective of earthquake hazards.

Cross-References

▶ Earthquake Mechanisms and Tectonics
▶ Earthquake Mechanism Description and Inversion
▶ Engineering Characterization of Earthquake Ground Motions

▶ Integrated Earthquake Simulation
▶ Nonlinear Seismic Ground Response Analysis of Local Site Effects with Three-Dimensional High-Fidelity Model
▶ Probabilistic Seismic Hazard Models
▶ Seismic Actions Due to Near-Fault Ground Motion
▶ Site Response for Seismic Hazard Assessment
▶ Stochastic Ground Motion Simulation

References

Aagaard BT, Hall JF, Heaton TH (2001) Characterization of near source ground motions with earthquake simulations. Earthq Spectra 17(2):177–207

Aagaard BT, Brocher TM, Dolenc D, Dreger D, Graves RW, Harmsen S, Hartzell S, Larsen S, Zoback ML (2008) Ground-motion modeling of the 1906 San Francisco earthquake, part I: validation using the 1989 Loma Prieta earthquake. Bull Seismol Soc Am 98(2):989–1011

Alterman Z, Karal FC (1968) Propagation of elastic waves in layered media by finite difference methods. Bull Seismol Soc Am 58(1):367–398

Anderson JG (2004) Quantitative measure of the goodness-of-fit of synthetic seismograms. In: Proceedings of the 13th world conference on earthquake engineering. International Association for Earthquake Engineering Paper, 243. Vancouver

Bao H, Bielak J, Ghattas O, Kallivokas LF, O'Hallaron DR, Shewchuk JR, Xu J (1998) Large-scale simulation of elastic wave propagation in heterogeneous media on parallel computers. Comput Methods Appl Mech Eng 152(1–2):85–102

Bielak J, MacCamy RC, McGhee DS, Barry A (1991) Unified symmetric BEM-FEM for site effects on ground motion – SH waves. J Eng Mech ASCE 117(10):2265–2285

Bielak J, Graves RW, Olsen KB, Taborda R, Ramírez-Guzmán L, Day SM, Ely GP, Roten D, Jordan TH, Maechling PJ, Urbanic J, Cui Y, Juve G (2010) The ShakeOut earthquake scenario: verification of three simulation sets. Geophys J Int 180(1):375–404

Bielak J, Karaoglu H, Taborda R (2011) Memory-efficient displacement-based internal friction for wave propagation simulation. Geophysics 76(6):T131–T145

Boore DM (1972) Finite difference methods for seismic wave propagation in heterogeneous materials. In: Bolt BA (ed) Methods in computational physics, vol 11. Academic Press, New York

Bouchon M (1979) Discrete wave number representation of elastic wave fields in three-space dimensions. J Geophys Res 84(B7):3609–3614

Brocher TM (2008) Compressional and shear-wave velocity versus depth relations for common rock types in northern California. Bull Seismol Soc Am 98(2):950–968

Carcione JM, Kosloff D, Kosloff R (1988) Wave propagation simulation in a linear viscoelastic medium. Geophys J Int 95(3):597–611

Cerjan C, Kosloff D, Reshef M (1985) A nonreflecting boundary condition for discrete acoustic and elastic wave equations. Geophysics 50:705–708

Chaljub E, Moczo P, Tsuno S, Bard P-Y, Kristek J, Kaser M, Stupazzini M, Kristekova M (2010) Quantitative comparison of four numerical predictions of 3D ground motion in the Grenoble Valley, France. Bull Seismol Soc Am 100(4):1427–1455

Cui Y, Olsen K, Jordan T, Lee K, Zhou J, Small P, Roten D, Ely G, Panda D, Chourasia A, Levesque J, Day S, Maechling P (2010) Scalable earthquake simulation on petascale supercomputers. In: SC'10 Proceedings of the 2010 ACM/IEEE international conference for high performance computing, networking, storage and analysis, New Orleans, LA, November 13-19. pp 1–20

Day SM, Bradley CR (2001) Memory-efficient simulation of anelastic wave propagation. Bull Seismol Soc Am 91(3):520–531

Day SM, Minster JB (1984) Numerical simulation of attenuated wavefields using a Padé approximant method. Geophys J Int 78(1):105–118

Day SM, Roten D, Olsen KB (2012) Adjoint analysis of the source and path sensitivities of basin-guided waves. Geophys J Int 189(2):1103–1124

Dumbser M, Käser M (2006) An arbitrary high-order discontinuous Galerkin method for elastic waves on unstructured meshes – II. The three-dimensional isotropic case. Geophys J Int 167(1):319–336

Dupros F, de Martin F, Foerster E, Komatitsch D, Roman J (2010) High-performance finite-element simulations of seismic wave propagation in three-dimensional nonlinear inelastic geological media. Parallel Comput 36(5–6):308–325

Ely GP, Jordan TH, Small P, Maechling PJ (2010) A Vs30-derived near-surface seismic velocity model. In: Abstract AGU fall meeting, no S51A-1907, San Francisco, 13–17 Dec 2010

Emmerich H, Korn M (1987) Incorporation of attenuation into time-domain computations of seismic wave fields. Geophysics 52(9):1252–1264

Frankel A, Vidale J (1992) A three-dimensional simulation of seismic waves in the Santa Clara Valley, California, from a Loma Prieta aftershock. Bull Seismol Soc Am 82(5):2045–2074

Gottschämmer E, Olsen KB (2001) Accuracy of the explicit planar free-surface boundary condition implemented in a fourth-order staggered-grid velocity-stress finite-difference scheme. Bull Seismol Soc Am 91(3):617–623

Graves RW (1996) Simulating seismic wave propagation in 3D elastic media using staggered-grid finite differences. Bull Seismol Soc Am 86(4):1091–1106

Graves RW (2008) The seismic response of the San Bernardino basin region during the 2001 Big Bear lake earthquake. Bull Seismol Soc Am 98(1):241–252

Graves RW, Pitarka A (2010) Broadband ground-motion simulation using a hybrid approach. Bull Seismol Soc Am 100(5A):2095–2123

Graves R, Jordan T, Callaghan S, Deelman E, Field E, Juve G, Kesselman C, Maechling P, Mehta G, Milner K, Okaya D, Small P, Vahi K (2011) CyberShake: a physics-based seismic hazard model for Southern California. Pure Appl Geophys 168(3–4):367–381

Hartzell S, Harmsen S, Frankel A (2010) Effects of 3D random correlated velocity perturbations on predicted ground motions. Bull Seismol Soc Am 100(4):1415–1426

Hermann V, Käser M, Castro CE (2011) Non-conforming hybrid meshes for efficient 2-D wave propagation using the Discontinuous Galerkin Method. Geophys J Int 184(2):746–758

Idriss IM, Seed HB (1968) Seismic response of horizontal soil layers. J Soil Mech Found Div ASCE 94(SM4):1003–1031

Jones LM, Bernknopf R, Cox D, Goltz J, Hudnut K, Mileti D, Perry S, Ponti D, Porter K, Reichle M, Seligson H, Shoaf K, Treiman J, Wein A (2008) The ShakeOut scenario, Technical report USGS-R1150, CGS-P25. U.S. Geological Survey, Reston, Virginia

Koketsu K, Miyake H, Fujiwara H, Hashimoto T (2009) Progress towards a japan integrated velocity structure model and long-period ground motion hazard map. In: Proceedings of the 14th world conference on earthquake engineering, paper no S10–038, Beijing

Komatitsch D, Vilotte J-P (1998) The spectral element method: an efficient tool to simulate the seismic response of 2D and 3D geological structures. Bull Seismol Soc Am 88(2):368–392

Komatitsch D, Tsuboi S, Ji C, Tromp J (2003) A 14.6 billion degrees of freedom, 5 teraflops, 2.5 terabyte earthquake simulation on the Earth Simulator. In: SC'03 Proceedings of the ACM/IEEE conference for high performance computing and networking. IEEE Computer Society, Phoenix, p 8

Komatitsch D, Liu Q, Tromp J, Suss P, Stidham C, Shaw JH (2004) Simulations of ground motion in the Los Angeles basin based upon the spectral-element method. Bull Seismol Soc Am 94(1):187–206

Komatitsch D, Erlebacher G, Göddeke D, Michéa D (2010) High-order finite-element seismic wave propagation modeling with MPI on a large GPU cluster. J Comput Phys 229(20):7692–7714

Levander AR (1988) Fourth-order finite-difference P-SV seismograms. Geophysics 53(11):1425–1436

Liu H-P, Anderson DL, Kanamori H (1976) Velocity dispersion due to anelasticity; implications for seismology and mantle composition. Geophys J R Astron Soc 47(1):41–58

Lysmer J, Drake LA (1972) A finite element method for seismology, Chapter 6. In: Alder B, Fernbach S, Bolt B (eds) Methods in computational physics, vol 11. Academic, New York

Ma S, Liu P (2006) Modeling of the perfectly matched layer absorbing boundaries and intrinsic attenuation in explicit finite-element methods. Bull Seismol Soc Am 96(5):1779–1794

Marcinkovich C, Olsen K (2003) On the implementation of perfectly matched layers in a three-dimensional fourth-order velocity-stress finite difference scheme. J Geophys Res 108(B5):2276

Moczo P, Kristek J (2005) On the rheological models used for time-domain methods of seismic wave attenuation. Geophys Res Lett 32(L01306):5

Moczo P, Kristek J, Vavryuk V, Archuleta RJ, Halada L (2002) 3D heterogeneous staggered-grid finite-difference modeling of seismic motion with volume harmonic and arithmetic averaging of elastic moduli and densities. Bull Seismol Soc Am 92(8):3042–3066

Moczo P, Kristek J, Halada L (2004) The finite difference method for seismologists – an introduction. Comenius University, Bratislava

Mossessian TK, Dravinski M (1987) Application of a hybrid method for scattering of P, SV, and Rayleigh waves by near-surface irregularities. Bull Seismol Soc Am 77(5):1784–1803

Ohminato T, Chouet BA (1997) A free-surface boundary condition for including 3D topography in the finite-difference method. Bull Seismol Soc Am 87(2):494–515

Olsen KB (1994) Simulation of three-dimensional wave propagation in the Salt Lake basin. PhD thesis, University of Utah, Salt Lake City

Olsen KB, Mayhew JE (2010) Goodness-of-fit criteria for broadband synthetic seismograms, with application to the 2008 M_w 5.4 Chino Hills, California, earthquake. Seismol Res Lett 81(5):715–723

Olsen KB, Pechmann JC, Schuster GT (1995) Simulation of 3D elastic wave propagation in the Salt Lake basin. Bull Seismol Soc Am 85(6):1688–1710

Olsen KB, Day SM, Bradley CR (2003) Estimation of Q for long-period (>2 sec) waves in the Los Angeles basins. Bull Seismol Soc Am 93(2):627–638

Olsen KB, Day SM, Minster JB, Cui Y, Chourasia A, Faerman M, Moore R, Maechling P, Jordan T (2006) Strong shaking in Los Angeles expected from southern San Andreas earthquake. Geophys Res Lett 33(L07305):1–4

Olsen KB, Day SM, Minster JB, Cui Y, Chourasia A, Okaya D, Maechling P, Jordan T (2008) TeraShake2: spontaneous rupture simulations of M_W 7.7 earthquakes on the southern San Andreas fault. Bull Seismol Soc Am 98(3):1162–1185

Porter K, Hudnut K, Perry S, Reichle M, Scawthorn C, Wein A (2011) Foreword. Earthq Spectra 27(2):235–237

Prevost J-H (1978) Plasticity theory for soil stress–strain behavior. J Eng Mech Div ASCE 104(5):1177–1194

Restrepo D, Bielak J (2014) Virtual topography: a fictitious domain approach for analyzing free-surface irregularities in large-scale earthquake ground motion simulation. Int J Numer Methods Eng 100(7):504–533

Rietmann M, Messmer P, Nissen-Meyer T, Peter D, Basini P, Komatitsch D, Schenk O, Tromp J, Dooghi L, Giardini D (2012) Forward and adjoint simulations of seismic wave propagation on emerging large-scale gpu architectures. In: SC'12 Proceedings of the ACM/IEEE international conference on high performance computing, networking, storage and analysis, Salt Lake City, pp 38:1–38:11

Robertsson JO (1996) A numerical free-surface condition for elastic/viscoelastic finite-difference modeling in the presence of topography. Geophysics 61(6):1921–1934

Roten D, Olsen KB, Pechmann JC (2012) 3D simulations of M 7 earthquakes on the Wasatch fault, Utah, part II: broadband (0–10 Hz) ground motions and nonlinear soil behavior. Bull Seismol Soc Am 92(5):2008–2030

Roten D, Olsen KB, Day SM, Cui Y, Fäh D (2014) Expected seismic shaking in Los Angeles reduced by San Andreas fault zone plasticity. Geophys Res Lett 41(8):2769–2777

Sánchez-Sesma FJ, Luzón F (1995) Seismic response of three-dimensional alluvial valleys for incident P, S, and Rayleigh waves. Bull Seismol Soc Am 85(1):269–284

Seriani G (1998) 3-D large-scale wave propagation modeling by spectral element method on Cray T3E multiprocessor. Comput Methods Appl Mech Eng 164(1–2):235–247

Shi Z, Day SM (2013) Rupture dynamics and ground motion from 3-D rough-fault simulations. J Geophys Res 118(3):1122–1141

Small P, Gill D, Maechling PJ, Taborda R, Callagham S, Jordan TH, Olsen KB, Ely G (2015) The unified community velocity model software framework. Comput Geosci (Submitted)

Smith WD (1975) The application of finite element analysis to body wave propagation problems. Geophys J Int 42(2):747–768

Taborda R, Bielak J (2013) Ground-motion simulation and validation of the 2008 Chino Hills, California, earthquake. Bull Seismol Soc Am 103(1):131–156

Taborda R, Ramírez-Guzmán L, López J, Urbanic J, Bielak J, O'Hallaron D (2007) Shake-Out and its effects in Los Angeles and Oxnard areas. Eos Trans AGU 88(52): Fall meeting supplement, abstract IN21B–0477

Taborda R, Bielak J, Restrepo D (2012) Earthquake ground motion simulation including nonlinear soil effects under idealized conditions with application to two case studies. Seismol Res Lett 83(6):1047–1060

Tu T, Yu H, Ramírez-Guzmán L, Bielak J, Ghattas O, Ma K-L, O'Hallaron DR (2006) From mesh generation to scientific visualization: an end-to-end approach to parallel supercomputing. In: SC'06 Proceedings of the ACM/IEEE international conference for high performance computing, networking, storage and analysis. IEEE Computer Society, Tampa, p 15

Virieux J (1984) SH-wave propagation in heterogeneous media: velocity-stress finite-difference method. Geophysics 49(11):1933–1957

Withers KB, Olsen KB, Shi S, Day SM, Takedatsu R (2013) Deterministic high-frequency ground motions from simulations of dynamic rupture along rough faults. In: Abstract SSA annual meeting, Salt Lake City, 17–19 Apr 2013

Withers KB, Olsen KB, Shi Z, Day SM (2014) High-complexity deterministic $Q(f)$ simulation of the 1994 Northridge M_w 6.7 earthquake. In: Proceedings of the SCEC annual meeting, no GMP-066, Palm Springs, 6–10 Sept 2014

Wong HL, Trifunac MD (1974) Surface motion of a semi-elliptical alluvial valley for incident plane SH waves. Bull Seismol Soc Am 64(5):1389–1408

Zahradnik J, Moczo P, Hron F (1993) Testing four elastic finite-difference schemes for behavior at discontinuities. Bull Seismol Soc Am 83(1):107–129

Zhou J, Unat D, Choi DJ, Guest CC, Cui Y (2012) Hands-on performance tuning of 3D finite difference earthquake simulation on GPU fermi chipset. Procedia Comput Sci 9:976–985

Plastic Hinge and Plastic Zone Seismic Analysis of Frames

Vissarion Papadopoulos[1] and Michalis Fragiadakis[2]

[1]Institute of Structural Analysis and Seismic Research, School of Civil Engineering, National Technical University of Athens (N.T.U.A.), Athens, Greece

[2]School of Civil Engineering, Laboratory for Earthquake Engineering, National Technical University of Athens (N.T.U.A.), Athens, Greece

Synonyms

Beam–column elements; Concentrated plasticity; Distributed plasticity; Fiber elements; Lumped plasticity

Introduction

The most common approach to perform seismic design of structures is the traditional one based on a linear elastic analysis assumption. On the other

hand, nonlinear analysis methods due to their higher complexity and computational cost are mainly applied as a verification tool for the assessment of existing structures. The use of nonlinear analysis methods provides a robust and reliable framework for seismic design since they allow the adoption of more elaborate design criteria, based on less simplifying assumptions. For this reason the application of nonlinear analysis approaches for the seismic design of structures as discussed in detail in recent guidelines, e.g., ASCE (2007), is gaining ground among practicing engineers. This development is also attributed to the rapid increase of computing power which makes the implementation of such advance approaches more attractive for real-world applications.

Regarding finite element (FE) modeling assumptions, building structures are primarily modeled with one-dimensional beam–column finite elements (e.g., beams or rods). These models are used to represent the linear skeleton of such structures, while more detailed two- and/or three-dimensional FE models such as the ones proposed by Spiliopoulos and Lykidis 2006 are only rarely utilized. The FE structural model should also be able to adequately capture the response of other structural components that may considerably affect the overall capacity, e.g., infill walls, shear walls, and other nonstructural components. A discussion on these modeling issues can be found in the NIST 2010 document. In addition, soil–foundation–structure interaction may also play a significant role on the overall structural response. A detailed discussion on the effect of soil models of different complexity on the structural response can be found in Assimaki et al. 2012.

Considering the structural nonlinear response, there are two major sources of nonlinearity: material and geometric nonlinearity. Material nonlinearity is considered the primary source of damage for low- and medium-rise building structures, while geometrical nonlinearities should be taken into account in high-rise buildings with small aspect ratios that suffer from large horizontal deflections that introduce P-Δ effects. For the nonlinear material response, the FE simulation with beam–column members falls into two categories: concentrated (or lumped) and distributed plasticity approach. In concentrated plasticity, also called briefly as plastic hinge approach, the plastic deformations are "lumped" at the ends of a linear elastic element and are based on the moment–rotation relationships of the end sections for a given axial force. In distributed plasticity, beam–column elements allow for the formation of plastic zones along the member, while plastic hinges may form anywhere along the member. In the latest case, the constitutive inelastic behavior is monitored at section integration points of the beam–column elements, therefore accounting for the axial–moment interaction in a straightforward manner. Several commercial software packages are available, and usually each follows a different formulation that the user/engineer must be aware in order to obtain reliable predictions of the nonlinear structural response (Fragiadakis and Papadrakakis 2008).

Material Nonlinear Approaches

Concentrated (Lumped) Plasticity Approach

Concentrated plasticity is the approach commonly suggested in most design codes and guidelines (FEMA 2009). The first concentrated plasticity beam elements were the parallel model of Clough and Johnston 1966 and the series model of Giberson (1967). Preceding developments were based on the series model aiming at including the interaction of axial force and bending moment in the form of a nonlinear rotational spring at the element ends (Fig. 1) (Powell and Chen 1986). Further developments on lumped plasticity elements were focused on the implementation of cyclic laws. Models that introduce cyclic stiffness degradation and pinching have been proposed. Such models either modify the path of the reloading branch (Clough and Johnston 1966) or introduce "pinching" (Ibarra et al. 2005). Apart from linear or piecewise linear models, smooth hysteretic models have also been developed (Wen 1976) in order to provide a continuous change of stiffness for the nonlinear springs. Huang and Foutch (2009)

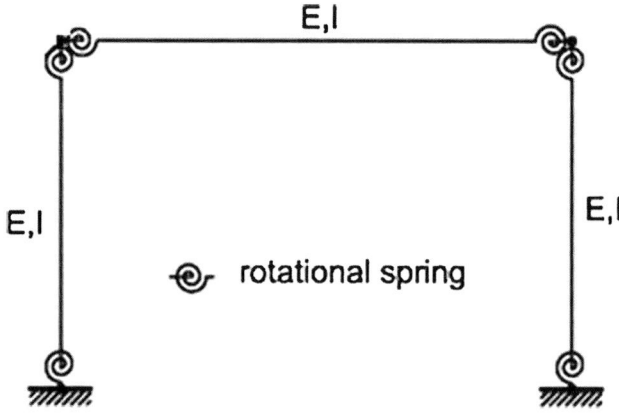

Plastic Hinge and Plastic Zone Seismic Analysis of Frames, Fig. 1 Lumped plasticity model of a portal frame: three beam-column FE and four rotational springs

performed a comparative study investigating the effect of different cyclic laws. An efficient beam–column element is also the force-based lumped plasticity element proposed in Scott and Fenves (2006). This approach combines the benefits of concentrated and distributed plasticity elements since it uses a minimum number of integration sections maintaining all the advantages of force-based distributed plasticity elements which are discussed in the next section.

In lumped plasticity approaches, the constitutive laws are expressed in terms of stress resultants (i.e., moment–rotations). This allows for the adoption of more complex phenomenological relationships with respect to distributed plasticity elements in which direct constitutive stress–strain relationships are used instead. In addition, lumped plasticity beam–column elements are more robust and reduce the computational cost and the memory requirements. This allows simulating complicated responses provided that the springs are appropriately calibrated. However, the concentrated plasticity approach relies on simplifying assumptions, restricting the inelastic deformations to prespecified regions of small length. Furthermore, there is no axial–moment interaction, and thus the springs are calibrated assuming constant axial force during dynamic analysis. Keeping constant the properties of the springs during analysis is another restriction of this modeling approach. Anagnostopoulos (1981) presented a parametric investigation showing that under monotonic loading the response is very sensitive to the parameters of the springs, while Dides and de la Llera (2005) showed that for typical multistory structures, the response is influenced by the ratio of the beam-to-column stiffness. It has to be mentioned that such models usually lead to better response estimates for steel (Papadrakakis and Papadopoulos 1995) rather than for concrete structures. Prior to adopting lumped plasticity modeling, one must also be aware of the inability of lumped plasticity to capture the response under softening behavior (e.g., reinforced concrete structures).

End Section Moment–Rotation
As mentioned above, lumped plasticity elements receive as input the moment–rotation relationships of the cross sections of interest, assuming that it is already known. The section moment–rotation $(M–\theta)$ relationship can be defined assuming a backbone whose parameters are determined combing first principles and default values or performing section (fiber) analysis with appropriate software. This backbone is discussed in the FEMA P695 (2009) guidelines and is shown in Fig. 2. As shown in this figure, the curve essentially consists of three branches, an initial elastic (K_e), a hardening branch (K_h) and a descending branch with slope K_c that is used to introduce degradation. Degradation can be either "in-cycle," i.e., defined with a negative stiffness branch and, or "cyclic." Cyclic degradation usually refers to strength degradation (but not exclusively) at given displacement cycles due to capacity reduction during cyclic loading.

The sensitivity of a building's response to the parameters that define the backbone is demonstrated in Vamvatsikos and Fragiadakis (2010), while a comprehensive discussion on cyclic and in-cycle degradation is also offered in NIST (2010) document.

Distributed Plasticity Approach

Distributed plasticity beam–column elements offer a more accurate description of the inelastic behavior since they allow inelastic deformations to be developed anywhere within the member. These elements are also known as "fiber" elements, since the sections are divided to horizontal and vertical layers forming small areas, known as "fibers," where the strain and stiffness parameters are evaluated (Fig. 3). Fiber elements are based on the classical FE theory with cubic Hermitian shape functions (Hellesland and Scordelis 1981). These elements are known in the literature as "displacement-based" or "stiffness-based" elements, and their shortcoming is that they require a fine mesh of beam–column elements in areas near the sections in which inelastic deformations are expected to be high (Fig. 4a). This is because the cubic shape function assumption leads to a linear distribution of the curvature along the element, which is inaccurate when the element end sections have yielded.

To overcome the problem stemming from the unknown distribution of curvature along the element, flexibility-dependent shape functions can be used instead (Fig. 3b). In this case, it is the distribution of forces (moments) that matters, while the element stiffness matrix can be easily calculated by summing the section flexibility matrices and inverting the resulting element flexibility matrix. This concept was extended by Kaba and Mahin (1984) and later was improved by Zeris and Mahin (1988) and Ciampi and Carlesimo (1986). These elements assume that the element is divided to equally spaced sections that allow the element to produce accurate predictions in the case of softening behavior, or in other words when the post-yield deformation path enters a segment with negative slope (softening). This formulation is also known as

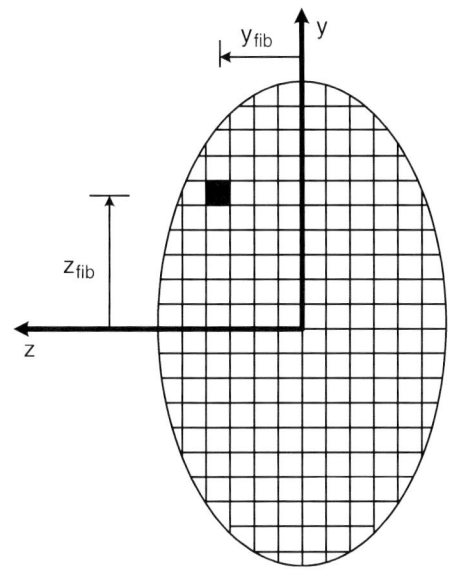

Plastic Hinge and Plastic Zone Seismic Analysis of Frames, Fig. 2 Discretization of an arbitrary cross section to fibers

Plastic Hinge and Plastic Zone Seismic Analysis of Frames, Fig. 3 Typical moment–rotation curve according to FEMA P695 (2009)

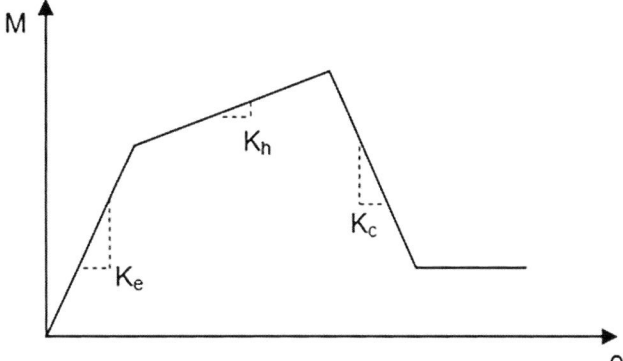

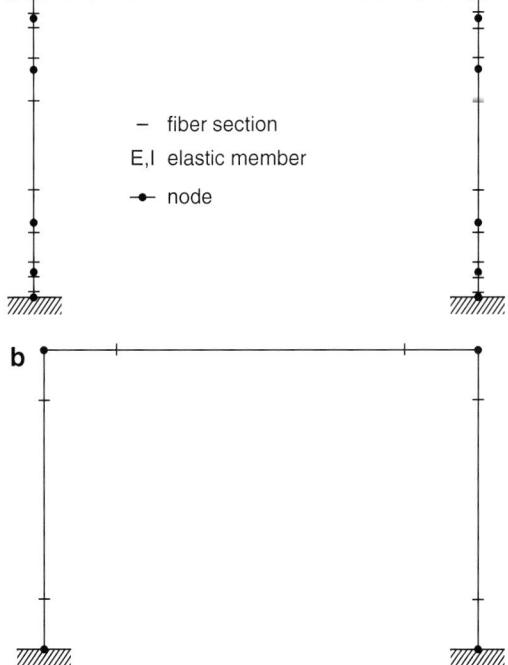

Plastic Hinge and Plastic Zone Seismic Analysis of Frames, Fig. 4 Modeling a portal frame using (**a**) displacement-based and (**b**) force-based elements

"force-based" or "flexibility-based." Following the previous concept, Spacone et al. (1996) introduced a general mixed-type, fiber-based, beam–column element. At the integration sections, this element always maintains equilibrium of both forces and deformations and converges to a state that satisfies the constitutive laws within a specified tolerance.

Compared to lumped plasticity, the main shortcoming of distributed plasticity elements is that they require more computing resources. Numerical instabilities may be encountered if criteria that introduce abrupt loss of capacity are adopted, e.g., when trying to predict collapse. On the other hand, distributed plasticity elements do not require any special calibration and can be easily adopted for sections that consist of different materials since they use stress–strain relationships suitable for each fiber material. A discussion on distributed plasticity elements can be found in Neuenhofer and Filippou (1997), while comparisons between the force and the displacement-based elements are presented in Neuenhofer and Filippou (1997) and Papaioannou et al. (2005). All studies converge that this formulation is accurate and robust and therefore suitable for a wide range of applications. A concise review on concentrated and distributed plasticity approaches can be found in Fragiadakis et al. (2015).

Section/Member Level

Fiber elements differ from lumped plasticity elements in the sense that they perform the integration over the fibers internally and therefore receive as input the parameters that define the stress–strain relationships of their fibers. On the other hand, lumped plasticity elements receive as input the moment–rotation relationships of the cross sections of interest, assuming that it is already known. In both cases, estimating the inelastic response requires integrating the stresses calculated at appropriately selected cross sections. Therefore, it is essential to be able to sufficiently capture the moment–rotation relationship of a cross section of a given geometry and reinforcement. Numerically, moment–rotation relationships are obtained using the fiber approach, i.e., discretizing the cross section to fibers that each follows its own material law.

Example

The simple example of the 1-story, 1-bay moment frame of Fig. 1 is modeled with wide-flange steel sections. The frame's bay width and height are set to 4 m. The beam and columns of the steel frame are standard HEB 220 sections. For the distributed plasticity model, each section is divided into fibers where elastic perfectly plastic material with a yield stress $\sigma_y = 355$ MPa is assigned independently to each fiber. As shown in Fig. 5, the flanges and the web of the HEB 220 cross section are discretized into 64 fibers. For the concentrated plasticity model, the frame is modeled using a beam with hinge elements represented by rotational springs at the element end. The backbone curve describing the nonlinear behavior of the springs is calculated by integrating strains across the section (Fig. 5) and is presented in Fig. 6.

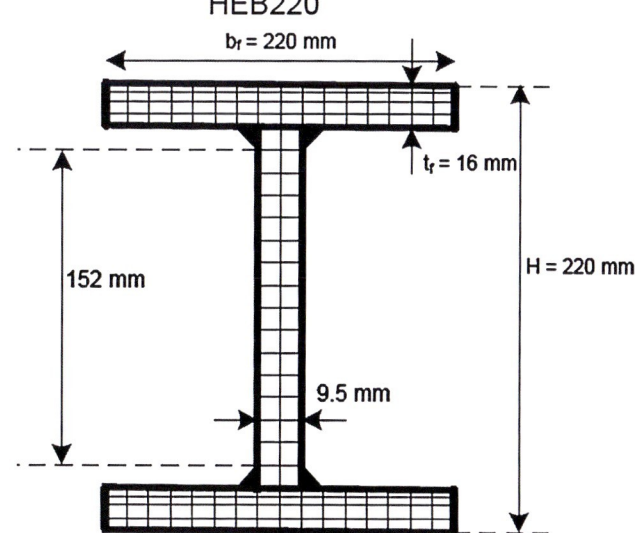

Plastic Hinge and Plastic Zone Seismic Analysis of Frames, Fig. 5 Discretization of HEB 220 section to fibers

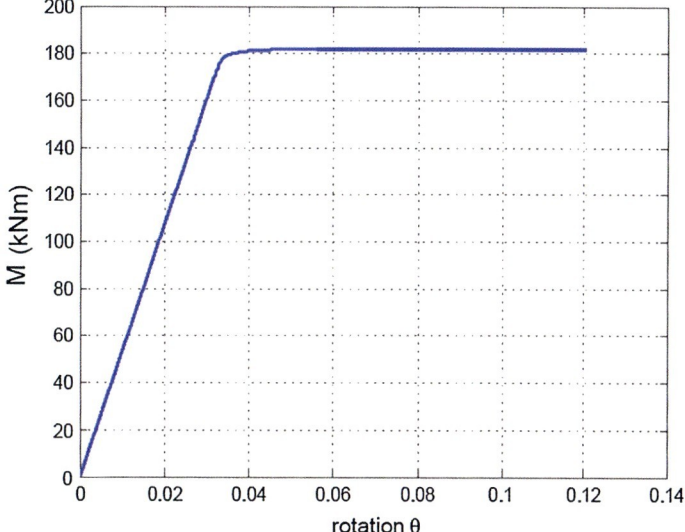

Plastic Hinge and Plastic Zone Seismic Analysis of Frames, Fig. 6 Backbone curve of a HEB 220 steel section

Figure 7a and b presents the pushover curves for both concentrated and distributed plasticity models and the corresponding incremental dynamic analyses (IDA), respectively. For the pushover analysis, the structure is pushed to a maximum 10 % roof drift. For the IDA case, the structural model is subjected to Canoga Park record from the 1994 Northridge earthquake scaled to multiple levels of intensity. The IDA curves of Fig. 7b plot an engineering demand parameter (EDP), in this case the maximum interstory drift θ_{max}, as a function of the earthquake intensity measure (IM) which is selected to be the maximum ground motion acceleration (PGA). From these figures the small difference on the structural response obtained by the two different models may be observed.

Summary

A critical review of the current state of the art of the computing practices adopted by the

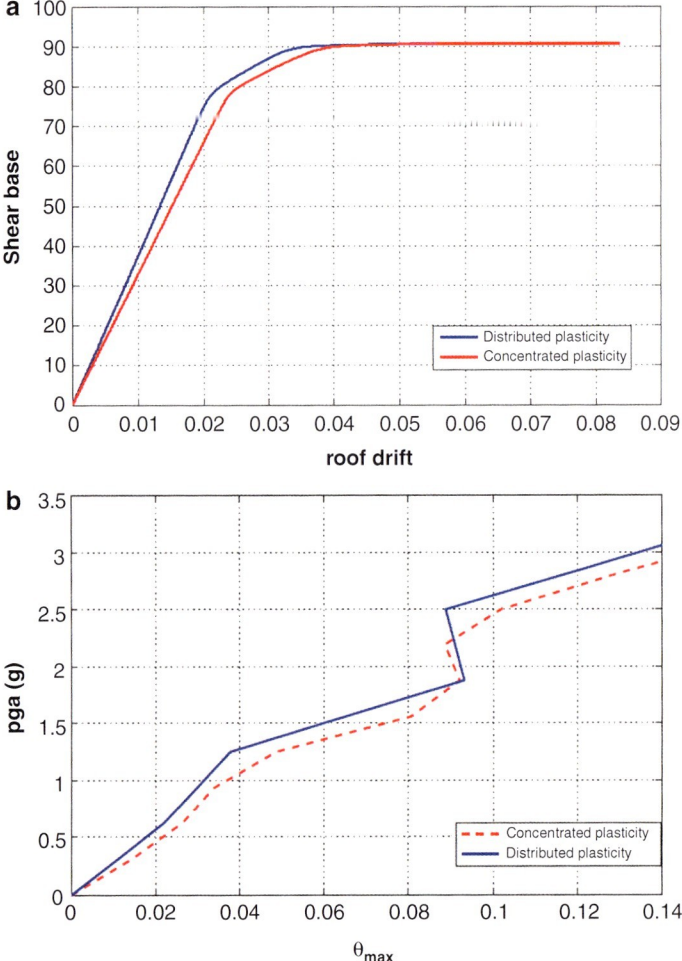

Plastic Hinge and Plastic Zone Seismic Analysis of Frames, Fig. 7 Comparison of (**a**) pushover and (**b**) IDA curves obtained with lumped and distributed plasticity models

earthquake engineering community is presented. The presentation extends from the finite element modeling of earthquake-resistant structures and the analysis procedures currently used to account for material nonlinearity. The two most commonly used approaches, namely, the plastic hinge and plastic zone approach, for estimating the structural capacity of seismically excited structures are discussed, and their relative advantages and disadvantages are presented together with the various aspects, developments, and implementations of these approaches over the past years. A comparison of the two approaches is presented in a simple example for both static pushover and incremental dynamic analysis approaches.

Cross-References

▶ Assessment of Existing Structures Using Inelastic Static Analysis
▶ Incremental Dynamic Analysis
▶ Nonlinear Finite Element Analysis

References

Anagnostopoulos S (1981) Inelastic beams for seismic analysis of structures. J Struct Eng 107(7):1297–1311

ASCE (2007) Seismic rehabilitation of existing buildings, ASCE/SEI Standard 41–06. American Society of Civil Engineers, Reston

Assimaki D, Li W, Fragiadakis M (2012) Site effects in structural performance estimation. Earthq Spectra 28(3):859–88

Ciampi V, Carlesimo L (1986) A nonlinear beam element for seismic analysis of structures. In: 8th European conference in earthquake engineering, Lisbon

Clough RW, Johnston SB (1966) Effect of stiffness degradation on earthquake ductility requirements. In: Proceedings of Japan earthquake engineering symposium, p 232

Dides MA, de la Llera JC (2005) A comparative study of concentrated plasticity models in dynamic analysis of building structures. Earthq Eng Struct Dyn 34(8):1005–1026

FEMA P-695 (2009) Quantification of seismic performance factors, FEMA P-695 Report, prepared by the Applied Technology Council for the Federal Emergency Management Agency, Washington, DC

Fragiadakis M, Papadrakakis M (2008) Modeling, analysis and reliability of seismically excited structures: computational issues. Int J Comput Meth 5(4):483–511

Fragiadakis M, Vamvatsikos D, Karlaftis MG, Lagaros ND, Papadrakakis M (2015) Seismic assessment of structures and lifelines. J Sound Vib 334:29–56

Giberson MF (1967) The response of nonlinear multistorey structures subjected to earthquake excitation. EERL report. Earthquake Engineering Research Laboratory, California Institute of Technology, Pasadena

Hellesland J, Scordelis A (1981) Analysis of RC bridge columns under imposed deformations, In: IABSE colloquium, Delft, Netherlands, pp 545–559

Huang Z, Foutch D (2009) Effect of hysteresis type on drift limit for global collapse of moment frame structures under seismic loads. J Earthq Eng 13(7):939–964

Ibarra LF, Medina RA, Krawinkler H (2005) Hysteretic models that incorporate strength and stiffness deterioration. Earthq Eng Struct Dyn 34(12):1489–1511

Kaba SA, Mahin SA (1984) Refined modelling of reinforced concrete columns for seismic analysis. Report no. UCB/EERC-84/03, College of Engineering, University of California

Neuenhofer A, Filippou FC (1997) Evaluation of nonlinear frame finite-element models. ASCE J Struct Eng 123:958–966

NIST (2010) Applicability of nonlinear multiple-degree-of-freedom modeling for design. Report NIST GCR 10-917-9, prepared for the National Institute of Standards and Technology by the NEHRP Consultants Joint Venture, Gaithersburg

Papadrakakis M, Papadopoulos V (1995) A computationally efficient method for the limit analysis of space frames. J Comput Mech 16(2):132–141

Papaioannou I, Fragiadakis M, Papadrakakis M (2005) Inelastic analysis of framed structures using the fiber approach. In: Proceedings of the 5th international congress on computational mechanics (GRACM 05), Limassol, Cyprus

Powell GH, Chen PF (1986) 3D beam-column element with generalized plastic hinges. J Eng Mech 112(7):627–641

Scott MH, Fenves GL (2006) Plastic hinge integration methods for force-based beam-column elements. J Struct Eng 132(2):244–252

Spacone E, Filippou FC, Taucer FF (1996) Fibre beam-column model for non linear analysis of R/C frames: part I formulation. Earthq Eng Struct Dyn 25(7):711–725

Spiliopoulos KV, Lykidis GC (2006) An efficient three-dimensional solid finite element dynamic analysis of reinforced concrete structures. Earthq Eng Struct Dyn 35(2):137–157

Vamvatsikos D, Fragiadakis M (2010) Incremental dynamic analysis for estimating seismic performance uncertainty and sensitivity. Earthq Eng Struct Dyn 39(2):141–163

Wen YK (1976) Method for random vibration of hysteretic systems. ASCE J Eng Mech Div 102(2):249–263

Zeris C, Mahin SA (1988) Analysis of reinforced concrete beam-columns under uniaxial excitation. J Struct Eng 114(4):804–820

Post-Earthquake Diagnosis of Partially Instrumented Building Structures

Eric M. Hernandez
Department of Civil and Environmental Engineering, College of Engineering and Mathematical Sciences, University of Vermont, Burlington, VT, USA

Synonyms

Damage detection; Instrumented buildings; Post-earthquake diagnosis; Structural dynamics; Wave propagation

Introduction

Post-earthquake diagnosis of building structures is essential in order to determine if a building can be safely re-occupied after a potentially damaging seismic event. The most traditional methodology is the use of visual inspections (ATC-20 2005). These inspections are made by qualified professionals and can take from a few days to possibly months. The effectiveness of the inspection will depend on a large part on the experience of the inspector and accessibility to the structural damage. This last aspect was clearly manifested during the 1994 Northridge earthquake where hidden structural damage in buildings was not detected by qualified inspectors. Recently, researchers have begun to look at methods that rely on computer vision to perform the visual assessment in an automated fashion (German 2013). In principle this approach also relies on the damage being visible.

In the context of post-earthquake damage assessment, the term damage should be interpreted as reduction in stiffness and/or strength of any member in the structural system of the building. The loss of stiffness or strength can occur due to a single occurrence of a maximum relative displacement (drift) or due to short-cycle fatigue induced by alternating displacements that are characteristic of earthquake-induced ground motion. Depending on the type of material used for the building structure, visual symptoms of damage may or may not exist. On the other hand, visual symptoms do not necessarily imply that significant damage has occurred. For example, in reinforced concrete buildings, cracks may appear after a strong ground motion; however, this does not imply a reduction in strength. Due to the shortcomings with visual inspections, vibration-based post-earthquake assessment has been proposed.

Vibration-based post-earthquake damage assessment uses acceleration from the building response measured through accelerometers to detect, localize, and quantify the damage experienced by the building during the seismic event. Ideally, a building would be instrumented with six accelerometers per floor in order to capture the six possible independent motions of the floor (assuming rigid floor diaphragm). However, in most cases instrumentation can only take place at a limited number of locations, typically significantly less than the ideal number. The scientific challenge in vibration-based post-earthquake damage assessment is to develop algorithms that can use all available information (such as structural drawings, previous history of measured seismic events in the building, construction information, etc.) together with limited acceleration measurements during the potentially damaging ground motion in order to perform the correct diagnosis.

Various algorithms have been proposed to perform post-earthquake diagnosis (Moaveni et al. 2011; Todorovska and Trifunac 2010; Bernal and Hernandez 2006; Lynch et al. 2006; Naeim et al. 2005; Revadigar and Mau 1999; Mau and Aruna 1994). Based on the relationship between the window(s) of time used to perform the diagnosis and the time of occurrence of the seismic event, existing methods can be broadly classified into "before-and-after" or "on-line" methods. As the name suggests, in the "before-and-after" strategy, the objective is to make a diagnosis based on measurements of the structural response prior to the potentially damaging event and measurements after the event, typically ambient vibrations. In the "on-line" strategy, the objective is to use the data during the seismic event to make the diagnosis.

Depending on the feature that is selected to characterize damage and on the definition of damage itself, existing methods can be broadly classified into:

- Spectral
- State estimation
- Wave propagation
- Energy methods
- Time series
- Model updating

As the name suggests, spectral methods search for changes in the spectral parameters of the structure (mode shapes and frequencies). Since damage is typically related to the loss of effective stiffness and modal parameters are directly related to the stiffness characteristics of the structure, it is intuitive to use these as a damage indicator. In this approach damage detection involves processing of acceleration records to determine changes in the frequency content of the measured acceleration using sliding windows Fourier analysis or time–frequency methods (Naeim et al. 2005; Todorovska and Trifunac 2010;

Moaveni et al. 2011; Asgarieh et al. 2012). If the effective frequency decreases during the strong motion window of time, then one would tend to infer that the stiffness of the system has decreased and structural damage has occurred.

State estimation methods operate by using partial measurements as feedback in order to estimate the complete dynamic response of the system and possibly track changes in the parameters of the model. Typically, some form of Bayesian filtering is used to perform the estimation (Chatzis et al. 2015). Damage features can be related to estimated functions of the state such as maximum drifts and/or damage indices (Erazo 2015; Hernandez and Erazo 2013). Alternatively, if changes in model parameters are used for damage assessment, then these need to be included in the state vector and estimated jointly with the traditional system state (Ching et al. 2006; Chatzi and Smyth 2009).

Another feature that has been proposed to detect damage in buildings is tracking seismic wave propagation through the building (Safak 1999). Wave propagation parameters include wave travel time between the floors and wave reflection and transmission coefficients of the floors. These have shown to be more sensitive to local damage than natural frequencies and provide yet another set of criteria for damage detection. However, the calculation of wave propagation parameters requires a dense array of sensor and recorders with a high rate of sampling. With the present sensor technology, the associated cost of the required level of instrumentation density may not be viable in all cases.

In energy methods the damage detection feature is the dissipated energy ratio which is a function of the energy dissipated by damping and the total energy dissipated. As mentioned previously and in contrast with existing methods previously described, dissipated energy is not an intrinsic property of linear systems and can be computed independent of whether the system is operating under linear or nonlinear range. In addition, the proposed damage feature has a very clear physical meaning and is by definition intimately related to earthquake-induced damage of engineered structures (Uang and Bertero 1990; Sucuoglu and Erberik 2004; Teran-Gilmore and Jirsa 2007).

Time series methods have also been proposed. This family of methods relies on changes in the time history response or in the identified black-box input–output model coefficients. Lynch et al. (2006) developed a wireless sensor network system and used residuals generated from AR-ARX models as a damage-sensitive feature. The algorithm is based on the work of Sohn and Farrar (2001). Bernal and Hernandez (2006) developed a signal-processing algorithm based on the cumulative integral of residuals from partial ARX models. The partial ARX models are identified from non-damaging events and subsequently used to estimate the expected linear response of the building during the event of interest. The rate of accumulation of the cumulative integral of the residual between the partial ARX prediction and the measured response is used as a damage detection and classification feature.

The time series methods have the advantage (which can also be viewed as a disadvantage in some cases) of working with almost no information from a structural model of the system and relying completely on identified black-box models from measured data. This reduces the sensitivity to structural modeling errors, but the analysis is based entirely on numerical features not easily relatable to physical quantities pertinent to the problem. This makes time series methods less appealing since it is difficult to correlate the results with structural quantities of interest (such as forces, moments, drifts, etc.) related to structural damage.

Finally, model updating methods operate by identifying changes in the parameters of models formulated based on structural mechanics principles. Typically, these parameters are related to stiffness properties of the structural members of the model (beams, columns, walls, etc.). Model updating methods can operate within the "before-and-after" or "on-line" strategy. In the "before-and-after" strategy, the model parameters identified from ambient vibration before the ground motion or simply selected from experience are compared with the model parameters identified from ambient vibrations after the ground motions (Simoen et al. 2015). If a significant change (or probability of change) is detected, then structural damage is deemed to have taken place. In the "on-line" strategy, as the name suggests, the model parameters are identified using the measured response during the ground motion (Ching et al. 2006; Chatzi and Smyth 2009). In this case, a more detailed evaluation can be conducted since in addition to stiffness parameters, hysteretic damage parameters can also be identified (which is not possible in the "before-and-after" strategy). Hysteretic behavior is known to be highly correlated to low-cycle fatigue damage typically experienced by buildings during strong earthquakes.

After this brief description of the various methodologies, a more detailed description of some of the methods is presented in the following sections.

Spectral Methods

Spectral methods operate under the assumption that the dynamic response of a building before and after the strong ground motion is governed by the following linear equation of motion:

$$\mathbf{M}\ddot{x}(t) + \mathbf{C}\dot{x}(t) + \mathbf{K}x(t) = r\ddot{u}_g(t) \qquad (1)$$

where $\mathbf{M}$, $\mathbf{C}$, and $\mathbf{K}$ are the mass, damping, and stiffness matrices of size $n \times n$. The response vector is $x(t)$, $u_g(t)$ is the ground motion, and r is the ground motion influence vector. The corresponding undamped eigenvalue problem is

$$\mathbf{M}\phi_i \omega_i = \mathbf{K}\phi_i \qquad (2)$$

where ω_i is the circular frequency of the ith mode and ϕ_i is the corresponding undamped mode shape. Spectral methods operate by interrogating changes in the mode shapes and frequencies before and after the ground motion, although more recently, data during the ground motion has been also used to identify the "changing" natural frequencies of the system as a sign of damage. The general assumption is that damage induces reduction in stiffness, and reduction in stiffness is reflected in a reduction in the frequencies of the structure.

Challenges with the spectral methods include the following: (i) the intrinsic global nature of the modal properties makes it challenging (although not impossible) to point to the location of the damage; (ii) there are factors other than damage that produce similar effects on the damage-sensitive features, which are not easy to isolate (the effects of soil–structure interaction on the measured frequencies of vibration and environmental influences such as temperature); and (iii) only low-frequency modes can be reliably identified from vibration data in buildings, and these frequencies have low sensitivity to local damage.

State Estimation Methods

In state estimation methods, as the name suggests, the objective is to estimate the complete dynamic response (state) based on a model of the system and noise-contaminated partial measurements. For application in post-earthquake assessment, state estimation methods have been applied in two different avenues: (1) by using acceleration measurements to estimate the complete dynamic response, including displacements, velocities, interstory drifts, interstory shears, and overturning moments (the estimated quantities can be compared with structural capacity values of structural elements) (Mau and Aruna 1994; Hernandez and Bernal 2008), and (2) enlarging the state to include model parameters and identifying the model parameters (Ching et al. 2006). By identifying changes in model parameters, one can infer damage.

The nonlinear response during a strong ground motion can be modeled as

$$\mathbf{M}\ddot{x}(t) + g_R(x,\dot{x},\theta) = r\ddot{u}_g(t) \qquad (3)$$

where g_R is a restoring force function, typically hysteretic. The restoring force function is defined by a set of parameter which we denote as θ.

By defining the state vector as $z^T = \{x^T \dot{x}^T \theta\}$, the equation of motion can be rewritten in first-order (state–space) form as

$$\dot{z}(t) = f(z(t)) + \mathbf{B}\ddot{u}_g \qquad (4)$$

Measurements of the building response can be represented as function of the state as

$$y(t) = h(z(t)) + v(t) \qquad (5)$$

where $v(t)$ is the measurement noise, typically small but not negligible during strong motions. The objective in state estimation methods, as the name suggests, is to estimate the complete state vector from noise-contaminated measurements $y(t)$. State estimation methods constitute an application of Bayes theorem:

$$p(z(t)|Y) = \frac{p(Y|z(t))p(z(t))}{p(Y)} \qquad (6)$$

where $Y(t) = y([0 \ t])$. If the model is linear, then the application of Bayes theorem results in the Kalman filter (Grewal and Andrews 2001). If the model is nonlinear, several extensions of the Kalman filter have been proposed. Among these methods, the most popular ones are the extended Kalman filter, unscented Kalman filter, ensemble Kalman filter, and the particle filter (Sarkka 2013).

Wave Propagation Methods

This method uses data from acceleration sensors to detect changes in structural stiffness based on the analysis of travel times of seismic waves propagating through the structure (Todorovska and Trifunac 2007, 2010). The physical basis for this approach is based on the one-dimensional shear-wave propagation velocity:

$$V_s = \sqrt{\frac{\mu}{\rho}} \qquad (7)$$

where μ is the shear stiffness and ρ is the density. Changes in the shear stiffness (due to damage) will result in a reduction in wave velocity and consequently an increase in travel time between sensors. Wave propagation methods are more sensitive to local damage than the modal methods and should be able to point out to the location of damage with relatively small number of sensors. Additionally, the local changes in travel time should not be sensitive to the effects of soil–structure interaction, which is a major obstacle for the modal methods.

The spatial resolution of the wave methods is limited by the number of sensors. A minimum of two sensors (at the base and at the roof) are required to determine if the structure has been damaged, and additional sensors at the intermediate floors would help point out to the part of the structure that has been damaged. For example, one additional sensor between these two would help identify if the damage has been in the part of the structure above or beyond that sensor.

Energy Methods

Energy methods are based on identifying dissipated energy. Dissipated energy has a very clear physical meaning and is by definition intimately related to earthquake-induced damage of engineered structures (Uang and Bertero 1990; Sucuoglu and Erberik 2004; Teran-Gilmore and Jirsa 2007; Jehel et al. 2008).

To illustrate the concept, consider the governing equation of motion for an SDOF oscillator with viscous damping subject to a base motion time history $x_g(t)$:

$$m\ddot{y} + c\dot{x} + kx = 0 \qquad (8)$$

where m is the mass, c is the viscous damping coefficient, and k is the stiffness coefficient. The absolute acceleration of the mass is $\ddot{y} = \ddot{x} + \ddot{x}_g$, and the relative displacement and velocity are x and $\dot{x}$, respectively. To obtain the energy balance relationship, we multiply both sides of the equation by $\frac{dx}{dt}dt$ and integrate between arbitrary time t_1 and t_2:

$$m\int_{t_1}^{t_2}\ddot{y}\dot{x}\,dt + c\int_{t_1}^{t_2}\dot{x}^2\,dt + k\int_{t_1}^{t_2}x\dot{x}\,dt = 0 \qquad (9)$$

The third term in the left-hand side is the elastic strain energy stored between t_1 and t_2:

$$\int_{t_1}^{t_2} kx\dot{x}\,dt = \int_{t_1}^{t_2} kx\frac{dx}{dt}dt = \int_{x(t_1)}^{x(t_2)} kx\,dx$$
$$= U = \frac{1}{2}kx^2 \Big|_{x(t_1)}^{x(t_2)} \quad (10)$$

Therefore, in the limit, if the system starts from rest at t_1 and we let t_2 go to infinity until the system returns to its static equilibrium point, the strain energy integral U vanishes, and the dissipated energy (E_D) converges to the energy dissipated by viscous damping (E_{damp}), namely,

$$E_D = E_{\text{damp}} = c\int_0^\infty \dot{x}^2\,dt = -m\int_0^\infty \ddot{y}\dot{x}\,dt \quad (11)$$

For a nonlinear system with hysteretic restoring force relationship, the integral in Eq. 2 is modified to incorporate the dissipated energy through hysteresis loops, resulting in

$$m\int_{t_1}^{t_2} \ddot{y}\dot{x}\,dt + c\int_{t_1}^{t_2} \dot{x}^2\,dt + \int_{t_1}^{t_2} f_R\dot{x}\,dt = 0 \quad (12)$$

where f_R is the restoring force function. If the system starts from rest at $t_1 = 0$ and we let t_2 go to infinity until the system returns to one of its equilibrium points, the total energy dissipated by the system is given by

$$E_D = c\int_0^\infty \dot{x}^2\,dt + \int_0^\infty f_R\dot{x}\,dt = E_{\text{damp}} + E_h$$
$$= -m\int_0^\infty \ddot{y}\dot{x}\,dt \quad (13)$$

As can be seen, the expression for total dissipated energy, for linear (Eq. 4) and nonlinear (Eq. 6) hysteretic systems, is exactly the same. This invariance property makes the use of energy as a damage detection feature appealing. The expression for total dissipated energy by hysteresis is found from Eq. 6 as

$$E_h = \int_0^\infty f_R(x,\dot{x})\dot{x}\,dt = -m\int_0^\infty \ddot{y}\dot{x}\,dt - c\int_0^\infty \dot{x}^2\,dt \quad (14)$$

Based on the previous analysis, Hernandez and May (2013) proposed the following damage index (Z), defined for an SDOF as

$$Z = \frac{E_h}{E_D} = 1 - \frac{E_{\text{damp}}}{E_D} = 1 + \left(\frac{c}{m}\right)\frac{\int_0^{t_r} \dot{x}^2\,dt}{\int_0^{t_r} \ddot{y}\dot{x}}\,dt \quad (15)$$

Other energy-based approaches that have been proposed rely on the identification of viscous damping (Satake et al. 2003). In these approaches, the premise is that effective viscous damping increases with damage. This is due to cracking, yielding, and other damage-related phenomena which results in higher levels of dissipated energy. The main drawback of this approach is that damping is difficult to identify accurately and its variability is high.

Time Series

Time series methods operate by identifying ARMA models of the form

$$\sum_{k=0}^{l} A_k y(p-k) = \sum_{k=0}^{n} B_k u(p-k) \quad (16)$$

where $A_k \in R^{m\times m}$ and $B_k \in R^{m\times p}$ are matrices, $y(h)$ is the measurement vector at time $t = h\,\Delta t$, and similarly $u(h)$ is the ground motion vector at time $t = h\,\Delta t$. To detect damage, time series methods search for changes in identified matrices A_k and B_k and(or) by examining the residuals between measurements during the ground motion and predictions provided by an ARMA model identified prior to the occurrence of the ground motion of interest. The residuals are defined as

$$\varepsilon(t) = y_m(t) - y(t) \quad (17)$$

where y_m is the measured acceleration and y is the model prediction. Bernal and Hernandez (2006) proposed two scalars, η and κ, where η measures the total extent of nonlinearity and κ captures the degree to which the structure recuperates the initial stiffness after the strong portion of the earthquake excitation is over. These metrics are computed as follows:

$$\gamma(t) = \frac{\int_0^t \varepsilon^2 dt}{\int_0^{t_{\max}} y_m^2 dt} \quad (18)$$

where

$$\eta = \gamma_{t=t_{\max}}\big|_{\max \text{ over all channels}} \quad (19)$$

and

$$\kappa = \frac{S_e}{S_r}\bigg|_{\max \text{ over all channels}} \quad (20)$$

where $t_{\max}$ = total duration of the earthquake record. It is implicit that Eq. 18 can be computed for every output channel and that the superscript t has been eliminated from the variables to simplify the notation. The numerator in Eq. 20 is the average slope of the curve given by Eq. 18 computed after it reaches 95 % of its final value, and the denominator is the slope of a line joining the 10 % to the 90 % values of the function in Eq. 18 for the earthquake used to identify the ARMA model. Various limits where proposed by Bernal and Hernandez (2006) on the constants γ and κ in order to separate undamaged buildings from potentially damaged ones. One drawback of this approach resides in the fact that residuals between building response and identified model predictions can occur due to many factors not related to damage, such as changes in building boundary conditions.

Model Updating

Model updating methods operate by identifying changes in the parameters (θ) of models formulated based on structural mechanics principles. Typically, these parameters are related to stiffness properties of the structural members of the model (beams, columns, walls, etc.). Model updating methods can operate within the "before-and-after" or "on-line" strategy. In the "before-and-after" strategy, the model parameters identified from ambient vibration before the ground motion or simply selected from experience are compared with the model parameters identified from ambient vibrations after the ground motions (Simoen et al. 2015). If a significant change (or probability of change) is detected, then structural damage is deemed to have taken place. If a Bayesian approach is used, then the joint probability of the parameters given measurements is given by

$$p(\theta|Y_B) \propto p(Y_B|z\theta)p(\theta) \quad (21)$$

$$p(\theta|Y_A) \propto p(Y_A|z\theta)p(\theta) \quad (22)$$

where Y_B and Y_A are sets of ambient response measurements before and after the ground motion, respectively. More details regarding identification and model updating using ambient vibration can be found in Simoen et al. (2015). Damage is typically assessed by contrasting the two probability densities $p(\theta|Y_B)$ and $p(\theta|Y_A)$ or more simply by looking for shifts in the maximum likelihood value of these distributions (if they are unimodal).

In the "on-line" model updating strategy, as the name suggests, the model parameters are identified using the measured response during the ground motion (Ching et al. 2006; Chatzi and Smyth 2009). In this case, a more detailed evaluation can be conducted since in addition to stiffness parameters, hysteretic damage parameters can also be identified (which is not possible in the "before-and-after" strategy). Hysteretic behavior is known to be highly correlated to low-cycle fatigue damage typically experienced by buildings during strong earthquakes. Most of the work found in the literature uses shear building models and Bouc–Wen models (Bouc 1971) to represent the interstory restoring force function. The main drawback of model updating methods is their lack of robustness to model errors (especially model class errors); this has been clearly illustrated in Ching et al. (2006). It is recommended that when using model updating methods, careful selection of the model class be exercised. In addition if the free parameter space is too large, uniqueness problems are to be expected; thus, the user needs to have good a priori knowledge of which elements are likely to exhibit damage and which ones not, thus excluding unnecessary parameters from the updating procedure.

Summary

Post-earthquake diagnosis of building structures is essential in order to determine if a building can be safely re-occupied after a potentially damaging seismic event. The most traditional methodology is the use of visual inspections. Depending on the type of material used for the building structure, visual symptoms of damage may or may not exist. Due to the shortcomings with visual inspections, vibration-based post-earthquake assessment has been proposed, developed, and implemented during the last few decades.

Vibration-based post-earthquake damage assessment uses acceleration from the building response measured through accelerometers to detect, localize, and quantify the damage experienced by the building during the seismic event. Various algorithms have been proposed to perform post-earthquake diagnosis. Existing methods for post-earthquake damage assessment can be broadly classified into spectral, state estimation, wave propagation, time series, energy methods, and model updating. This entry presents a summary of each method, their pros and cons, and range of applicability.

Cross-References

▶ Blind Identification of Output-Only Systems and Structural Damage via Sparse Representations
▶ Damage to Ancient Buildings from Earthquakes
▶ Damage to Buildings: Modeling
▶ Estimation of Potential Seismic Damage in Urban Areas
▶ Operational Modal Analysis in Civil Engineering: An Overview
▶ System and Damage Identification of Civil Structures

References

Asgarieh E, Moaveni B, Stavridis A (2012) Nonlinear structural identification of a three-story infilled frame using instantaneous modal parameters. In: Proceedings of the 30th international conference in modal analysis (IMAC-XXX), Jacksonville

ATC-20 (2005) Field manual: post-earthquake safety evaluation of buildings, 2nd edn. Applied Technology Council, Redwood City, CA

Bernal D, Hernandez E (2006) A data-driven methodology for assessing impact of earthquakes on the health of building structural systems. Struct Des Tall Special Build 15:21–34

Bouc R (1971) Modèle mathématique d'hystérésis. Acustica 21:16–25

Chatzi E, Smyth A (2009) The unscented Kalman filter and particle filter methods for nonlinear structural system identification with non-collocated heterogeneous sensing. Struct Control Health Monit 16(1):99–123

Chatzis MN, Chatzi EN, Smyth AW (2015) An experimental validation of time domain system identification methods with fusion of heterogeneous data. Earthq Eng Struct Dyn 44:523–547

Ching J, Beck J, Porter K, Shaikhutdinov R (2006) Bayesian state estimation method for nonlinear systems and its application to recorded seismic response. ASCE J Eng Mech 132(4):396–410

Erazo K (2015) Non-linear state estimation with application to structural health monitoring. PhD dissertation, University of Vermont

German SA (2013) Automated damage assessment of reinforced concrete columns for post-earthquake evaluations. PhD dissertation, Georgia Institute of Technology

Grewal M, Andrews A (2001) Kalman filtering, theory and practice using MATLAB, 2nd edn. Wiley, New York

Hernandez E, Bernal D (2008) State estimation in structural systems with model uncertainties. J Eng Mech 134(3):252–257

Hernandez E, Erazo K (2013) Non-linear model-data fusión for post-earthquake damage assessment of structures. In: International workshop in structural health monitoring. Stanford University, Stanford, CA

Hernandez EM, May G (2013) The Dissipated energy ratio as tool for earthquake induced damage detection and classification of instrumented structures. ASCE J Eng Mech 139(11):1521–1529

Jehel P, Ibrahimbegovic A, Leger P, Davenne L (2008) On the computation of seismic energy dissipation in reinforced concrete frame elements. In: 8th world congress on computational mechanics, Venice

Lynch JP, Wang Y, Lu KC, Hou TC, Loh CH (2006) Post-seismic damage assessment of steel structures instrumented with self-interrogating wireless sensors. In: Proceedings of the 8th national conference on earthquake engineering, San Francisco

Mau TS, Aruna V (1994) Story-drift, shear and OTM Estimation from building seismic records. ASCE J Struct Eng 120(11):3366–3385

Moaveni B, He X, Conte J, Restrepo J, Panagiotou M (2011) System identification study of a 7-story full-scale building slice tested on the UCSD-NEES shake table. ASCE J Struct Eng 137(6):705–717

Naeim F, Hagie S, Alimoradi A, Miranda E (2005) Automated post-earthquake assessment and safety evaluation of instrumented buildings. JAMA report 2005-10639

Revadigar S, Mau ST (1999) Automated multi-criterion building damage assessment from seismic data. J Struct Eng ASCE 125(2):211–217

Safak E (1999) Wave-propagation formulation of seismic response of multistory buildings. J Struct Eng ASCE 125(4):426–437

Sarkka S (2013) Bayesian filtering and smoothing. Cambridge University Press, Cambridge

Satake N, Suda K, Arakawa T, Sasaki A, Tamura Y (2003) Damping evaluation using full-scale data of buildings in Japan. J Struct Eng 129(4):470–477

Simoen E, DeRoeck G, Lombaert G (2015) Dealing with uncertainty in model updating for damage assessment: a review. Mech Syst Signal Process 56–57:123–149

Sohn H, Farrar C (2001) Damage diagnosis using time-series analysis of vibrating signals. J Smart Mater Struct 10(3):446–451

Sucuoglu H, Erberik A (2004) Energy-based hysteresis and damage models for deteriorating systems. J Earthq Eng Struct Dyn 33:69–88

Teran-Gilmore A, Jirsa J (2007) Energy demands for seismic design against low-cycle fatigue. J Earthq Eng Struct Dyn 36(1):383–404

Todorovska MI, Trifunac MD (2007) Damage detection in the Imperial County Services Building I: the data and time–frequency analysis. Soil Dyn Earthq Eng 27(6):564–76

Todorovska M, Trifunac M (2010) Earthquake damage detection in the Imperial County Services Building II: analysis of novelties via wavelets. J Struct Control Health Monit 17(8):895–917

Uang C, Bertero V (1990) Evaluation of seismic energy in structures. J Earthq Eng Struct Dyn 19(1):77–90

Principles of Broadband Seismometry

Nick Ackerley
Nanometrics, Inc, Kanata, Ottawa, ON, Canada

Synonyms

Weak-motion sensor; Very broadband seismometer

Introduction

There are many different types of instruments which can be used to detect ground motion. Broadband seismometers belong to a class of sensors called inertial sensors. In contrast, methods of sensing ground motion such as strainmeters and Global Positioning Systems (GPS) are not considered inertial sensors.

In this entry the principles of operation of broadband seismometers are discussed and contrasted with those of passive seismometers. The criteria for selecting a seismometer are discussed. Particular attention is paid to noise-generating mechanisms, including self-noise generated within the sensor, environmental sensitivities, and installation-related noise.

Applications

Broadband seismometers are very versatile instruments which have many applications in the field of earthquake engineering, including:

- Event detection and location (see ▶ Seismic Event Detection and ▶ Earthquake Location)
- Earthquake magnitude and moment-tensor estimation (see ▶ Earthquake Magnitude Estimation, ▶ Earthquake Mechanism Description and Inversion and ▶ Long-Period Moment-Tensor Inversion: The Global CMT Project)
- Volcanic eruption early warning (see ▶ Volcanic Eruptions, Real-Time Forecasting of

and ▶ Noise-Based Seismic Imaging and Monitoring of Volcanoes)
- Subsurface imaging
- Site response (see, e.g., ▶ Site Response: Comparison Between Theory and Observation and ▶ Site Response for Seismic Hazard Assessment)
- Restitution of ground displacement (see ▶ Selection of Ground Motions for Response History Analysis and ▶ Seismic Actions due to Near-Fault Ground Motion)

Classes of Inertial Sensors

In the broadest sense, a seismometer is any instrument which responds to ground motion, and a seismograph is any instrument which subsequently makes a recording of that ground motion. In practical usage however, a seismometer refers to a specific subclass of inertial sensors. One way to understand where the broadband seismometer is situated relative to the other classes of instruments, which measure ground motion, is to take a tour of the family tree of inertial sensors.

Inertial sensors consist of a frame and a "proof mass" suspended within the frame. Movement of the frame is sensed by sensing differential motion between the frame and the proof mass. The suspension usually constrains the proof mass to have one or more degrees of freedom, which can be either rotational or translational. If these degrees of freedom are purely rotational, the instrument is called a rotational seismometer or a gyroscope (see Lee et al. 2012).

Translational inertial sensors have many subclasses. One way of understanding their classification is to look primarily at the types of motion each is designed to measure: static or dynamic, strong or weak, horizontal or vertical, and long period or short period (these terms are explained in the paragraphs which follow).

Inertial sensors which are configured to detect static accelerations include gravimeters and tiltmeters, which are arranged to detect vertical and horizontal accelerations, respectively. Neither instrument is required to measure the full acceleration due to gravity. A tiltmeter can be leveled to produce zero output when it is installed; the mainspring of a gravimeter is similarly adjusted to cancel a standard acceleration due to gravity. Thus the sensitivity of a gravimeter or a tiltmeter can be quite high because the range of accelerations to be measured is quite small. Finally, since traveling seismic waves are generally not of interest for gravimeters and tiltmeters, these instruments typically have a low-pass characteristic with an upper corner frequency which is relatively low, typically 1 Hz or lower.

An ▶ accelerometer is another type of inertial sensor which is configured to detect accelerations down to zero frequency. Unlike a gravimeter or a tiltmeter, they are typically configured to read peak accelerations on the order of 1 g or larger without clipping and thus must have relatively low sensitivities. They can be used to detect static accelerations, for the purpose of integrating them to obtain position as in an inertial guidance system, or they can be used to detect strong seismic motions. Like a gravimeter they have a flat response to acceleration down to zero frequency, but the upper corner of their low-pass response will be at a much higher frequency.

A seismometer differs from gravimeters, tiltmeters, and accelerometers in that it is not required to respond to ground motion all the way down to zero frequency; this enables it to have a higher sensitivity and therefore measure much smaller motions than an accelerometer. They typically have transfer functions which are flat to velocity over at least a decade, but many hybrid responses, flat to acceleration over some significant band, are viable alternatives (Wielandt and Streckeisen 1982). "Strong-motion velocity meters" capable of detecting motions as large as those detectable on an accelerometer are also commercially available, but the term seismometer is generally reserved for an instrument which measures weak motion.

Traditionally the basic subgroupings of seismometers were long period or short period, depending on whether the lower corner period is above or below about 7 s period (see section "Ground Motion Spectra" below). Modern broadband seismometers span both frequency bands and have made long-period seismometers obsolete.

Principles of Broadband Seismometry, Fig. 1 Response to acceleration of inertial sensors

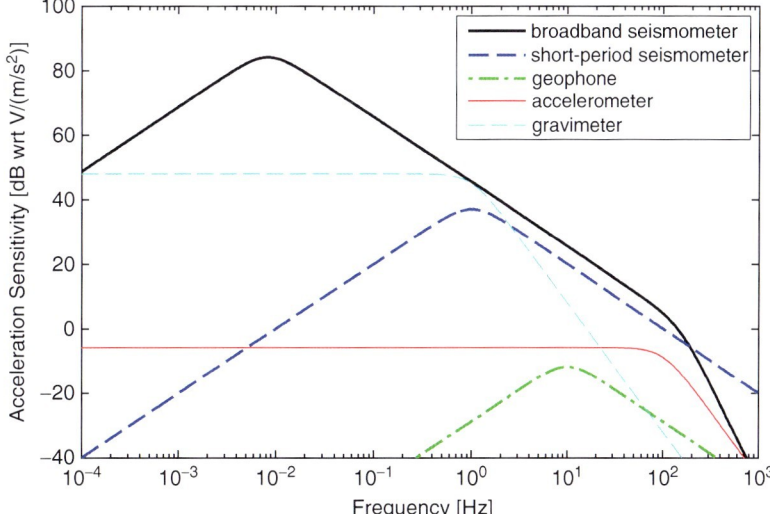

Principles of Broadband Seismometry, Fig. 2 Response to velocity of inertial sensors

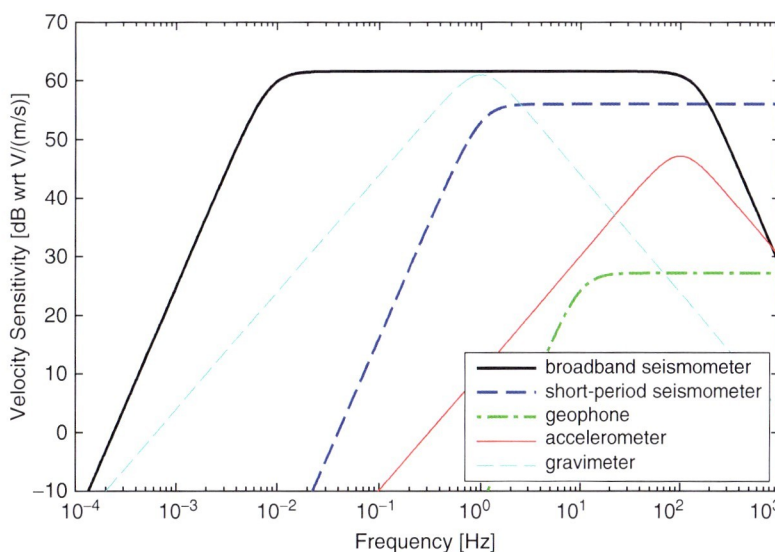

Like a short-period seismometer, a geophone is usually a passive seismometer (see ▶ Historical Seismometer and ▶ Passive Seismometers), but one with a higher corner frequency and lower sensitivity.

The transfer functions of representative examples from the main classes of inertial sensors are plotted in Fig. 1 for ground motion in units of acceleration and in Fig. 2 for units of velocity. Seismometers have the highest sensitivity of all inertial seismometers, and broadband seismometers have the highest sensitivity of all of these classes at long periods, but not at zero frequency.

The following sensors were taken as typical of the classes shown in Figs. 1 and 2:

- **Accelerometer**: a class A accelerometer (Working Group on Instrumentation, Siting, Installation, and Site Metadata 2008), e.g., Nanometrics Titan
- **Gravimeter:** based on the CG-5 (Scintrex Limited 2007)

- **Broadband seismometer:** based on the Trillium 120 (Nanometrics, Inc 2009)
- **Short-period seismometer:** based on the S 13 (Geotech Instruments, LLC 2001)
- **Geophone:** based on the SG-10 (Sercel - France 2012)

For more discussion of the importance of the sensitivity of a seismometer to its performance, see section "Response."

Force Feedback

The principles of operation, benefits, and drawbacks of an active inertial sensor are best understood in relation to the operation of a passive inertial sensor, since the latter is effectively a component in the former.

Modern short-period seismometers and geophones are passive inertial sensors which consist of a pendulum with a velocity transducer with sensitivity G in V·s/m mounted so as to measure the relative velocity $\dot{x}_b$ of a proof mass M and frame of the sensor, as shown in Fig. 3. The spring constant of the mainspring and suspension combined K and the viscous damping B determine the transfer function of the pendulum, required to relate the output voltage v_o to the input motion of the frame, $\dot{x}_i$.

A block diagram of the whole passive seismometer system, from ground motion to the digitizer, is shown in Fig. 4.

The Laplace transform of the transfer function which converts frame velocity to an output voltage is (Aki and Richards 2002; Wielandt 2002)

$$\frac{V_o}{sX_i} = G\frac{s^2}{s^2 + \frac{B}{M}s + \frac{K}{M}}\quad\left[\frac{\text{V·s}}{\text{m}}\right]$$

Here the output voltage V_o and input displacement X_i are written in uppercase to emphasize that this transfer function operates in the frequency domain. It is a property of the Laplace transform that multiplication by the (complex) frequency s in the frequency domain is equivalent to differentiation in the time domain, so sX_i is the input *velocity*. Another property of the Laplace transform is that to evaluate the (complex) transfer function at a given frequency f in Hz or angular frequency $\omega = 2\pi f$ in rad/s, one makes the substitution $s = j\omega$.

For high sensitivity in the passband, the generator constant of the velocity transducer must be large. Typically the velocity transducer is constructed like a voice coil in a speaker: many turns of copper passing through a narrow gap in a magnetic circuit energized by a strong magnet

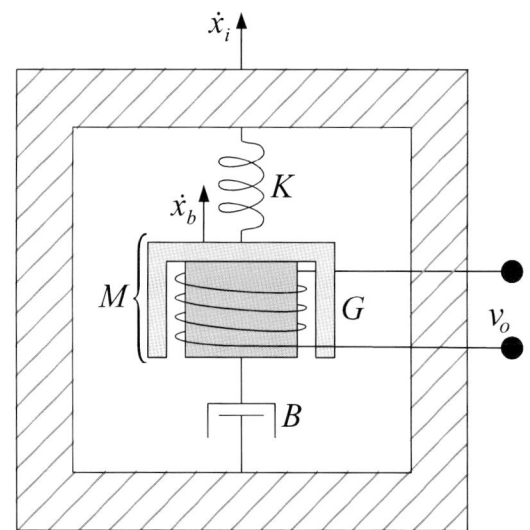

Principles of Broadband Seismometry, Fig. 3 Mechanical schematic of a passive inertial sensor

Principles of Broadband Seismometry, Fig. 4 System block diagram of a passive inertial sensor

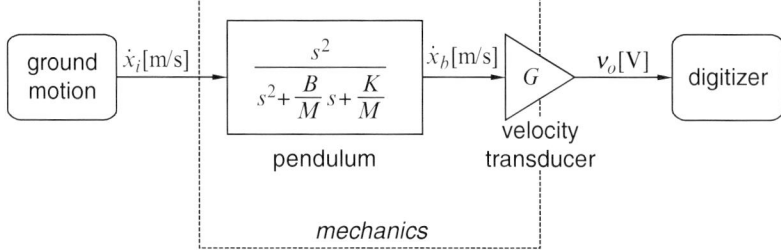

with large diameter are required for a large generator constant. This is shown schematically in Fig. 3.

By comparison with the standard form of a second-order transfer function

$$F(s) = S \frac{s^2}{s^2 + 2\zeta\omega_0 s + \omega_0^2}, \quad (1)$$

we can write expressions for the sensitivity S, the natural frequency f_o in Hz, the angular natural frequency $\omega_0 = 2\pi f_o$, and the (dimensionless) damping constant ζ_0 of the pendulum:

$$S = G$$

$$f_o = \frac{\omega_0}{2\pi} = \frac{1}{2\pi}\sqrt{\frac{K}{M}}$$

$$\zeta_0 = \frac{1}{2\omega_0}\frac{B}{M} = \frac{B}{2\sqrt{KM}}$$

Below the natural frequency, the sensitivity of passive seismometer drops off rapidly, as shown in Figs. 1 and 2. At the resonant frequency, the degree of "peaking" in the transfer function is controlled by the damping constant ζ_0. Low damping corresponds to a strongly peaked response, something which is generally not desirable, because it means the output will be dominated by signal at that frequency, and because the system has a greater tendency to clip at that frequency.

For high sensitivity at low frequencies, the spring stiffness K [N/m] must be low and the mass M [kg] must be large. For maximum bandwidth while maintaining passband flatness, the viscous damping B [N/(m/s)] is generally set – often with the aid of an external damping resistor not modeled here – so that the dimensionless damping constant ζ_0 is approximately $1/\sqrt{2}$.

Passive seismometers have the obvious advantage of requiring no power to operate. Indeed the device is literally a generator, albeit one which produces very little power. However, modern digital seismology requires that some power be expended on digitization and subsequent telemetry or storage, so the seismograph system as a whole always consumes some power.

The chief disadvantages of passive seismometers for recording ground motion at long periods are:

(a) The suspension system and velocity transducer have nonlinearities as a function of the displacement of the proof mass from equilibrium. These nonlinearities can cause excursions in the output signals which cannot be removed by post-processing.
(b) Large masses and compliant suspensions are required to achieve high sensitivity at long periods, making for an instrument, which is physically large, difficult to construct and susceptible to damage when being transported.
(c) Conflicting damping requirements: As we shall see below, the damping must be as low as possible in order to minimize self-noise, but for a flat transfer function the damping must be set to a level much higher than the minimum. A transfer function which is strongly peaked will have a corresponding "notch" in its clip level, limiting its ability to record during large or nearby events.

A force-feedback or "active" seismometer solves these problems by taking a mechanical system like that of the passive seismometer, as shown in Fig. 3, and adding a displacement transducer T, as shown in Fig. 5. In an active seismometer, the voice coil G is used to produce a force on the proof mass by driving it with a feedback current i_f instead of using it to produce a voltage proportional to the velocity of the proof mass.

It is a remarkable fact that every generator is also a motor. The force transducer in the feedback path of an active seismometer has the same physical configuration as the velocity transducer at the output of a passive seismometer. Moreover the electromotive force generated per unit of velocity or "motor constant" of the former in V/(m/s) is identical to the force generated per unit of current or "generator constant" of the latter in N/A:

$$G = \frac{e}{\dot{x}} = \frac{f}{i} \quad \left[\frac{\text{V}\cdot\text{s}}{\text{m}} = \frac{\text{N}}{\text{A}}\right]$$

The displacement transducer shown in Fig. 5 is generally one of two types. A linear-variable differential transformer (LVDT) is effectively a transformer where the coupling to the secondary winding(s) is a function of the position of a movable magnetic core. A capacitive displacement transducer (CDT) is a capacitor where the capacitance is varied by changing the spacing or overlapping area of the capacitor plates. In order to reduce nonlinearity and increase sensitivity, transducers of both types are normally implemented as part of a bridge circuit, requiring three windings in the case of an LVDT or three capacitor plates in the case of a CDT. In both cases the displacement transducers operate at a carrier frequency well above the passband of the instrument, and the outputs require demodulation back down to the passband before they are passed on to the compensator, output, and feedback stages.

The feedback loop is closed using feedback electronics which take the displacement transducer output voltage and produce an appropriate feedback current as shown in the block diagram of Fig. 6. Two main blocks of interest in the feedback electronics are a compensator $C(s)$ in the forward path and the feedback network itself.

Discussions of the components, design, or limitations of the compensator are beyond the scope of this entry, but suffice it to say that the challenge is to simultaneously ensure high loop gain and stability and that the compensator design determines the shape of the roll-off of the closed-loop transfer function at high frequency.

A few of the key components in the feedback network are explicitly indicated in Fig. 6, namely, the "differential" feedback capacitor C_D, a "proportional" feedback resistor R_P, and an "integral" feedback resistor R_I. The integrator time constant

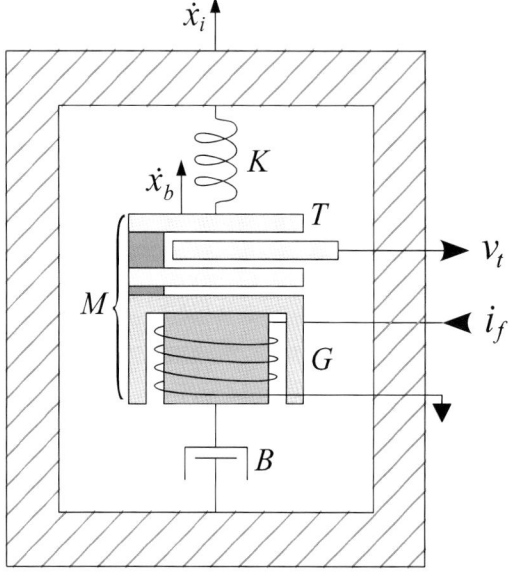

Principles of Broadband Seismometry, Fig. 5 Mechanical schematic of an active inertial sensor

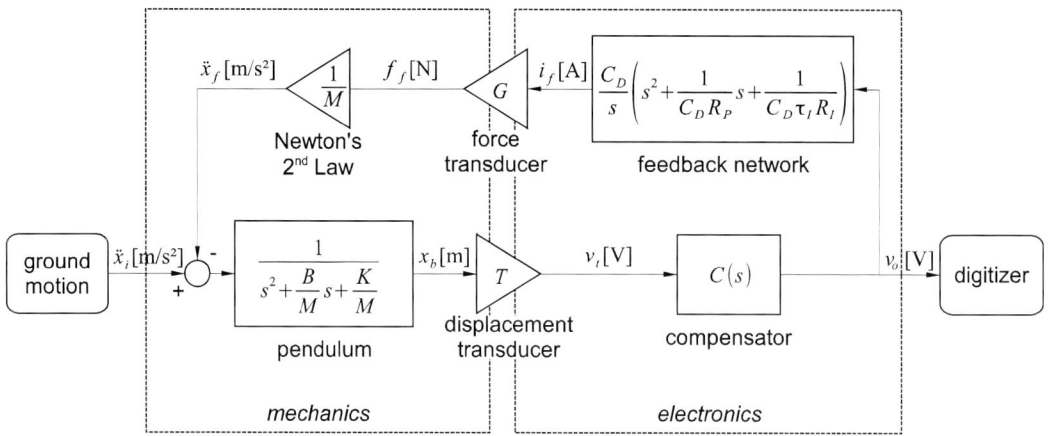

Principles of Broadband Seismometry, Fig. 6 Block diagram of an active inertial sensor using force feedback

τ_I is another key parameter of the feedback circuit.

Feedback control systems have the property that as long as the loop gain is sufficiently high (and the system is stable), the transfer function of the system is the inverse of the feedback transfer function (Phillips and Harbor 1991). In the case of an inertial sensor, this means that the transfer function is determined by electronic components in the feedback network not the physical characteristics of the pendulum. Furthermore this same feedback holds the proof mass substantially at rest with respect to the frame of the sensor, greatly reducing the impact of nonlinearities in the suspension and transducers.

Thus, for high loop gain, a good approximation of the transfer function of the active seismometer of Fig. 6 is

$$\frac{V_o}{sX_i} = \frac{M}{GC_D} \frac{s^2}{s^2 + \frac{1}{C_D R_P}s + \frac{1}{C_D \tau_I R_I}} \left[\frac{\text{V}\cdot\text{s}}{\text{m}}\right]$$

By comparison with the standard form of a second-order transfer function (1), we can write expressions for the sensitivity, corner frequency, and damping of this force-feedback seismometer:

$$S = \frac{M}{GC_D}$$

$$f_o = \frac{\omega_0}{2\pi} = \frac{1}{2\pi}\sqrt{\frac{1}{C_D \tau_I R_I}}$$

$$\zeta_0 = \frac{1}{2C_D R_P \omega_0} = \frac{1}{2R_P}\sqrt{\frac{\tau_I R_I}{C_D}}$$

The properties which make an active force-feedback seismometer a good tool for broadband weak-motion seismology become apparent:

(a) The effects of nonlinearities as a function of displacement of the proof mass from its rest position are greatly reduced, using a displacement transducer and high loop gain.
(b) High sensitivity at long periods is achieved by selecting components in the feedback path. In order to achieve a particular natural frequency, the electronic components required in a force-feedback seismometer are much smaller than the mechanical components required in a passive seismometer with the same response.
(c) The damping of the pendulum can be made low without affecting the closed-loop response of the system, since the damping ζ_0 of the closed-loop system is determined entirely by the components in the feedback network.

The power required to operate a force-feedback seismometer must be considered its chief disadvantage relative to a passive seismometer. Although the power consumption can often be kept to a fraction of the rest of the recording system, the fact that power must be provided at all does increase the size and complexity of cables, connectors, and the digitizer itself. The versatility of the broadband seismometer relative to passive alternatives, however, can more than make up for this.

Ground Motion Spectra

The sensing of ground motion due to earthquakes and other phenomena is fundamentally limited by the background motion of the earth. Efforts to define minimum, maximum, and typical ground motion models go back at least as far as the 1950s (Brune and Oliver 1959).

An important current standard consists of the new high- and low-noise models (NHNM and NHNM) (Peterson 1993) which in practice sets limits on the typical background motion at well-constructed vaults. It represented an improvement over the old noise models (OLNM and ONHM) because it included stations with the then-new STS-1 seismometer, the first high-quality very broadband sensor. These four models are plotted in Fig. 7 below.

In the band 0.05–1 Hz all models of ground motion are dominated by a peak called the microseismic peak. This peak is in fact made up of two peaks. The first peak, typically between 10 and 16 s period, corresponds to the natural period of waves generated by storm winds in mid-ocean. This peak is sometimes called the primary

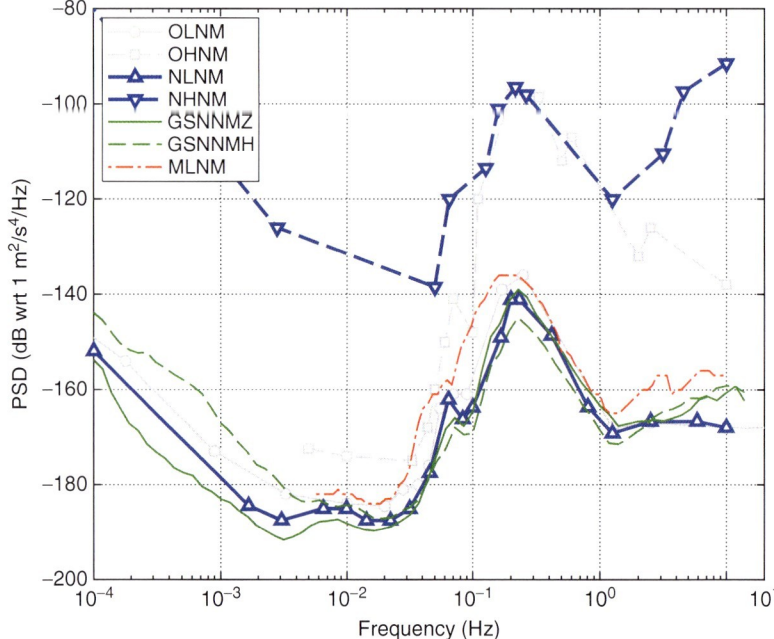

Principles of Broadband Seismometry, Fig. 7 Models of ground motion spectra

microseism. Because the mean pressure at the ocean bottom is proportional to the square of wave height, standing waves at mid-ocean produce a larger, double-frequency peak between 4 and 8 s period (Longuet-Higgins 1950). Ocean microseisms dominate recordings made by sensors with sufficiently low self-noise and divide ground motion into two important bands, commonly called short period (above the microseismic peak) and long period (below the microseismic peak).

Peterson dealt with the non-stationarity of the nonetheless random processes underlying the background motion of the earth by simply discarding time segments containing earthquakes or noise bursts. More recent models of ground motion spectra deal with the problem of non-stationarity in a more sophisticated way, using statistical approaches which evaluate the probability distribution of the power spectral density (PSD) as a function of frequency.

For example, studies of the Global Seismological Network (GSN) resulted in noise models of vertical (Z) and horizontal (H) ground motion (GSNNMZ and GSNNMH, respectively) (Berger et al. 2004), and a study of the continental United States has produced the mode low-noise model (MLNM) (McNamara and Buland 2004): both studies employ the notion of the PSD probability density function (PDF) to produce estimates of ground motion spectra at quiet sites. Figure 8, reprinted from Figure 8 of McNamara and Buland (2004), shows how transient phenomena (e.g., calibration signal injected into the seismometer) are represented alongside stationary phenomena (e.g., the microseismic peak) using this approach.

The PSD PDF allows a more precise statistical definition of ground motion low-noise models. For example, the MLNM was constructed as the per-frequency minimum of the *modes* of the PDF of a set of broadband seismic stations distributed across the continental United States. Similarly, the GSNNM is the per-frequency minimum of the 1st (quietest) percentile of the PDF of the stations in the GSN. As shown in Fig. 7, the GSNNMZ is lower than the NLNM at very long periods (<0.01 Hz); this new lower limit is likely to still be limited by the performance of the STS-1.

Despite the advance in network performance monitoring that the PSD PDF represents, the NLNM remains the standard by which seismometer self-noise floors are measured, and many of

Principles of Broadband Seismometry, Fig. 8 Typical acceleration power spectral density probability density function

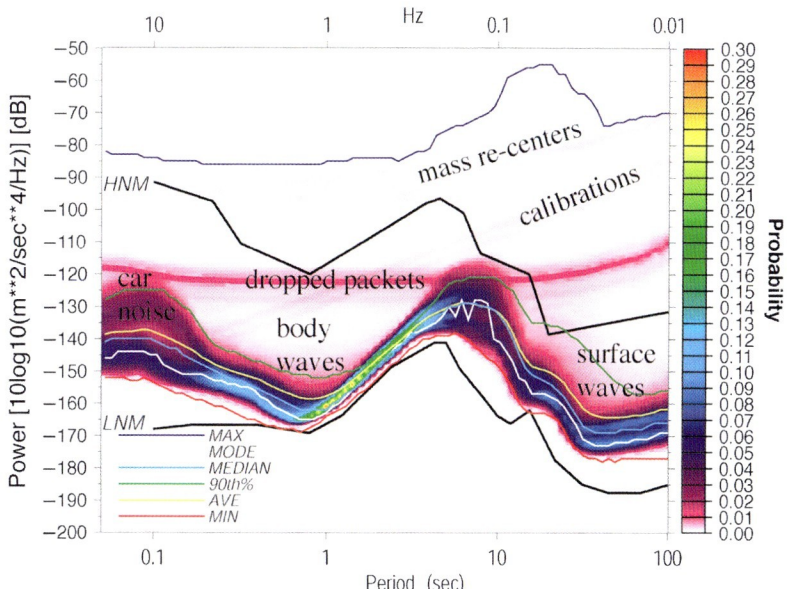

the most common features of background ground motion are represented relative to the NLNM and NHNM.

Some strong generalizations can be made about background motion at long periods:

- Horizontal components are never quieter than vertical components at a given location. See section on "Tilt."
- Sites on alluvium are never quieter than nearby sites on hard rock. See section on "Site Selection."
- Surface sites are never quieter than nearby underground sites. See section "Downhole and Ocean-Bottom."

Indeed, the NHNM at long periods is made up of horizontal ground motion records at surface vaults on alluvium, while the NLNM at long periods is made up of vertical records on hard rock, mainly in subsurface vaults (Peterson 1993).

There are fewer features of ground motion at short periods (above the microseismic peak) which can be generalized. As at long periods, surface sites are never quieter at short periods than nearby underground sites, but horizontals and verticals tend to have similar levels. In both bands "cultural noise" such as that due to traffic can be problematic, as can noise due to wind in trees or running water. Finally, there is always a minimum in the ground noise spectrum at or just above 1 Hz.

Although the NLNM was constructed from the records of land-based seismometers, it may also represent a hypothetical quietest site under the sea. Seismic noise surveys on the ocean bottom have not yet been undertaken on the scale that they have been on land. It is clear from the work which has been done so far, however, that the noise levels are generally significantly higher. Most of the ocean floor is covered with sediment which is more compliant than sediment on land, and seismic velocities are lower. Seafloor compliance means that excess site noise due to deformation and tilt can result from both variations in pressure due to waves at the surface and currents traveling along the bottom of the ocean (Webb and Crawford 2010). When precautions are taken to mitigate these effects, site noise can approach the levels observed on land.

In conclusion, although models based on larger or more regionally focused samples of sites exist, and although statistical methods have improved with time, Peterson's NLNM remains the standard by which the noise performance of broadband seismometers and sites are judged. It represents a hypothetical "quietest site on earth" so a good broadband seismometer should be able to resolve it. The NLNM is therefore commonly used as a reference curve on plots of broadband seismometer self-noise. In turn, at the very longest periods, it may be that the NLNM is limited by the available broadband sensor technology.

Ground Motion Unit Conversion

Ground motion can be expressed in the time domain units of displacement [m], velocity [m/s], or acceleration [m/s^2]. In the frequency domain, a ground motion signal can furthermore be represented either as an amplitude spectrum ["per Hz," e.g., m/s^2/Hz] or as a PSD ["per $\sqrt{\text{Hz}}$," e.g., m/s^2/$\sqrt{\text{Hz}}$]. In assessing the performance of a seismograph station, it is essential to be able to convert between various units fluently.

For example, comparison of earthquake spectra (typically a displacement amplitude spectrum or peak band-passed acceleration spectrum) to the clip level of a station (typically a maximum amplitude of acceleration or velocity) allows determination of the maximum earthquake magnitude which can be detected without clipping. Similarly, comparison of the same event spectra to seismograph self-noise or site noise (typically represented as an acceleration PSD) determines the minimum detectable magnitude. Finally, comparison of seismometer clip and self-noise levels as a function of frequency gives the clearest representation of seismometer dynamic range.

Acceleration-Velocity-Displacement

The simplest kind of conversion is integration and differentiation to obtain different time derivatives of ground motion. For example, to convert an acceleration to a velocity or from velocity to displacement at a given frequency f, one simply divides by the amplitude by the angular frequency $\omega = 2\pi f$. This follows straightforwardly from the equivalence of differentiation in the time domain to multiplication in the frequency domain by the Laplace frequency $s = j\omega$. Thus for sinusoidal amplitudes of acceleration $|a|$, velocity $|v|$, and displacement $|d|$ at frequency f, the following relations hold:

$$|a| = 2\pi f |v| \tag{2}$$

$$|v| = 2\pi f |d| \tag{3}$$

These relations have wide applicability in converting amplitude spectra of finite energy signals of earthquakes between acceleration, velocity, and displacement and in converting PSDs of stationary between acceleration, velocity, and displacement.

Another type of conversion is needed for the comparison of finite energy signals and stationary signals. In particular, conversion from a PSD to the equivalent amplitude requires that a minimum of two factors be specified: a bandwidth and a crest factor.

Relative Bandwidth

Parseval's theorem can be used to compute the root-mean-square amplitude a_{RMS} expected given the one-sided power spectral density as a function of frequency $a_{\text{PSD}}(f)$ and lower and upper frequency band limits $f_H > f_L$:

$$a_{\text{RMS}}^2 = \int_{f_L}^{f_H} a_{\text{PSD}}^2(f) df$$

This equation applies equally well to acceleration, velocity, or displacement, but note that if the RMS is in units, then the PSD is in units/$\sqrt{\text{Hz}}$.

For narrowbands the spectrum can be assumed to be constant within the band so that

$$a_{\text{RMS}}^2 = a_{\text{PSD}}^2(f_H - f_L) = a_{\text{PSD}}^2 k_{\text{RBW}} f$$

$$a_{\text{RMS}} = a_{\text{PSD}} \sqrt{k_{\text{RBW}} f} \tag{4}$$

where the "relative bandwidth factor" is defined as

$$k_{\text{RBW}} \equiv \frac{f_H}{f} - \frac{f_L}{f}$$

Principles of Broadband Seismometry, Table 1 Common choices of relative bandwidth

Bandwidth	b	n	k_{RBW}	$k_{RBW}|_{dB}$
Octave	2	1	0.707	−1.5
1/6 decade	10	6	0.386	−4.1
1/2 octave	2	2	0.348	−4.6
1/3 octave	2	3	0.232	−6.4

It is common to define bands as a fraction 1/n of either an octave (b, 2) or a decade (b, 10), as follows:

$$\frac{f_H}{f_L} = b^{\frac{1}{n}}$$

and furthermore to dispose the band limits symmetrically (in a logarithmic scale) around the band center, so that

$$\frac{f_H}{f} = \frac{f}{f_L} = b^{\frac{1}{2n}}$$

so that the relative bandwidth factor k_{RBW} for various bandwidths can be evaluated using

$$k_{RBW} = b^{\frac{1}{2n}} - b^{-\frac{1}{2n}}$$

Some common choices of bandwidth are listed in Table 1. Two which appear commonly in seismometer specifications, 1/2 octave and 1/6 decade, are close enough that they are often considered interchangeable. The octave bandwidth is considered typical of a passive seismometer and has been used in some important studies of typical earthquake spectra (Clinton and Heaton 2002).

The choice of a relative bandwidth thus permits conversion of a power spectral density to an equivalent RMS amplitude.

Crest Factor

The second factor needed when comparing stationary and transient signals is called a "crest factor," and is defined as the ratio of peak amplitude to RMS amplitude. The correct choice of crest factor depends on the nature of the signal in question. Broadband Gaussian noise will surpass its standard deviation 32 % of the time and twice the standard deviation 5 % of the time, for example. These conditions would correspond to crest factors of 1 and 2, respectively.

The signals which make up background ground motion do often have Gaussian distributions (Peterson 1993), as do the self-noise of seismometers and digitizers. When a Gaussian signal is passed through a narrowband filter, the peak amplitudes of the signal which results have a Rayleigh distribution (Bormann 2002), corresponding to a crest factor:

$$k_{crest} = \sqrt{\frac{\pi}{2}} \cong 1.25$$

Given a crest factor, the equivalent amplitude can be estimated from a PSD using

$$a_{peak} = a_{RMS} k_{crest} \qquad (5)$$

As with the equations accounting for relative bandwidth, this equation applies equally well to acceleration, velocity, or displacement.

Decibels

Ground motions are often expressed in decibels. By convention the reference amplitude for the decibel in this context is the corresponding metric unit. For example, an acceleration PSD is converted to dB using

$$a_{PSD}|_{dB} = 20 \log_{10} \left(\frac{a_{PSD}}{\frac{m}{s^2 \sqrt{Hz}}} \right)$$

while a velocity amplitude is converted using

$$v_{peak}|_{dB} = 20 \log_{10} \left(\frac{v_{peak}}{m/s} \right)$$

Thus, the various unit conversions of Eqs. 2, 3, 4, and 5 become:

$$a|_{dB} = v|_{dB} + 20 \log_{10} \left(\frac{2\pi f}{rad} \right) \qquad (6)$$

Principles of Broadband Seismometry,
Fig. 9 Representative station (noise) and small event (signal) spectra

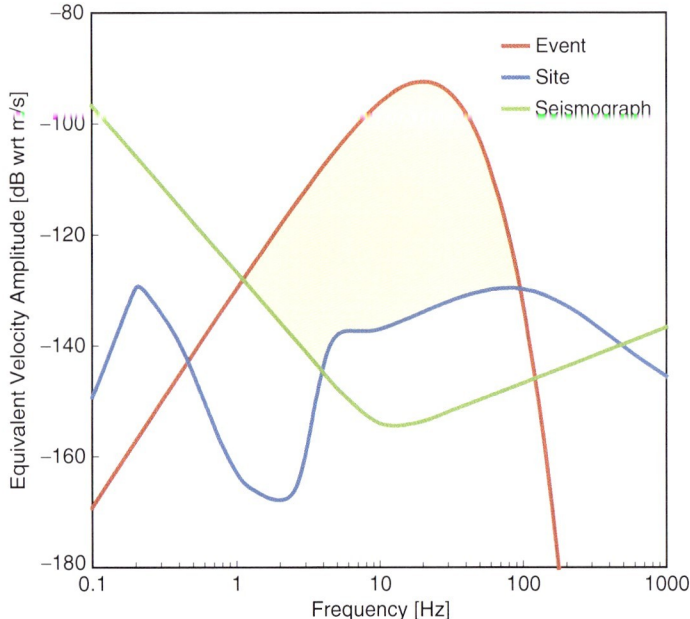

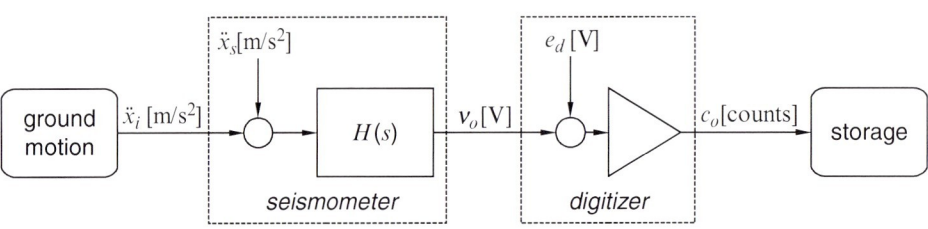

Principles of Broadband Seismometry, Fig. 10 Components of a seismograph station

$$v|_{dB} = d|_{dB} + 20\log_{10}\left(\frac{2\pi f}{\text{rad}}\right) \quad (7)$$

$$a_{RMS}|_{dB} = a_{PSD}|_{dB} + 10\log_{10}(k_{RBW})$$
$$+ 10\log_{10}\left(\frac{f}{\text{Hz}}\right) \quad (8)$$

$$a_{peak} = a_{RMS}|_{dB} + 20\log_{10}(k_{crest}) \quad (9)$$

These relations can be seen in action Fig. 13 below, where the band-passed peak velocity amplitudes of earthquakes are compared to the clip level and noise floor of a seismometer.

Components of Station Noise

The noise power at a seismic station is the sum of the noise powers of the seismograph and the background motion at the site:

$$\left\langle |\ddot{x}_{station}|^2 \right\rangle = \left\langle |\ddot{x}_{seismograph}|^2 \right\rangle + \left\langle |\ddot{x}_{site}|^2 \right\rangle$$

It is important that the station noise be low enough to not obscure the signals of interest when they arrive. This is shown in cartoon form in Fig. 9.

A modern seismograph station consists, in turn, of a seismometer and a digitizer, as shown schematically in Fig. 10. Both elements must be considered in determining the performance of the station as a whole.

Similarly the clip level of a seismograph station is a combination of the clip level of the seismometer and of the digitizer. A difference is that whereas noise powers are summed, the clip level of a system is the minimum of the clip level of all of the components, when referred to common units.

The importance of the sensitivity of a digitizer is now apparent. The sensitivity must be high enough to minimize the digitizer's contribution to the station noise, but at the same time it must be low enough that the clip level of the digitizer does not degrade the station clip level so much that events of interest will be clipped. Note that a system incorporating a seismometer which achieves a high sensitivity to ground motion using a high-gain output stage is equivalent to a system in which the digitizer subsystem includes a preamp with the same gain.

The self-noise of seismometers can be divided roughly into those that are stationary, in the sense that the probability distribution does not change with time, and those that are nonstationary.

Of the sources of stationary noise in a seismometer, there are two main subcategories, called "white" noise and "flicker" noise sources. White noise sources have the same amount of noise power per unit of (absolute) bandwidth in frequency; flicker noise sources have the same amount of noise power per fractional (relative) bandwidth, for example, per octave or per decade. The PSD of a white noise source is independent of frequency; the PSD of a flicker noise source varies as the inverse of frequency (Motchenbacher and Connelly 1993), i.e., proportional to $f^{-\alpha}$ where typically α is close to 1 but can be as high as 1.5.

In the process of referring the noise at various points in the feedback loop to the input of a seismometer, each source must be referred to the sensor input. This is similar to converting ground motion units from velocity to displacement and from displacement to velocity. Thus in principle various components of the noise floor of a seismometer can be proportional to f^n (or equivalently a slope of $10n$ dB/decade) where n is any positive or negative number, usually an integer. In practice, in a well-designed force-feedback seismometer, the equivalent acceleration PSD of the self-noise will be proportional to f^{-3} (-30 dB/decade) at low frequencies and proportional to f^4 (40 dB/decade) at high frequencies (Hutt and Ringler 2011).

For example, the displacement transducer in a feedback seismometer results in a white noise proportional to displacement, but to determine the equivalent acceleration, a double integration is required, so the slope of the acceleration PSD is f^4, and this tends to dominate the self-noise of the sensor at high frequencies. Similarly the digitizer white noise component, when converted from velocity to acceleration PSD, has a f^2 (20 dB/decade) characteristic and will tend to dominate the station noise near 10 Hz.

Two noise sources common to all seismometers, active and passive, are Johnson and Brownian noise. Johnson noise is a noise source due to thermal agitation of the electrons that make up the current flow in a resistor. For a resistance R [Ω] at temperature T [K], the expectation value of the voltage PSD across its terminals will be:

$$\left\langle |e_n|^2 \right\rangle = 4k_b TR$$

where $k_b = 1.38 \times 10^{-23}$ J/K is Boltzmann's constant. For low noise, resistors must have small values, but this generally results in greater power consumption.

Brownian noise is similar to resistor noise, both in terms of the statistical mechanics which explain it, and in the white noise spectrum which results. In particular, the Brownian motion of the boom of a pendulum results in an equivalent acceleration PSD of the frame of that pendulum equal to:

$$\left\langle |\ddot{x}_n|^2 \right\rangle = \frac{4k_b TB}{M^2}$$

Thus it is clear that for low noise, the proof mass must be large while the viscous damping must be small. All known voice-coil and capacitive transducers require narrow gaps for high sensitivity; this is a source of viscous damping and can only be avoided by pumping the pendulum enclosure down to a hard vacuum, something quite difficult in practice. The downside of large proof masses is increased size and fragility; many broadband seismometers required masses to be locked during transportation.

Nonstationary noise sources are also very important to the overall performance of a seismometer. A common example of nonstationary noise is what is sometimes referred to colloquially as a "pop." The characteristic waveform is a randomly occurring disturbance having the shape of the impulse response of the seismometer. When such a signal is referred to the input of the seismometer, it turns out to correspond to a sudden step change in acceleration at the input to the seismometer, equivalent to a sudden step change in tilt of the instrument. Because "pops" occur at random times, they contribute a f^{-2} characteristic to the acceleration PSD noise spectrum.

"Pops" are often associated with temperature transients associated with initial installation and power-on. However, some sensors, which are poorly designed or have been insufficiently tested or damaged during handling, will exhibit an unusual susceptibility to pops after initial installation, and the excess "pop" noise will subsequently never die down to an immeasurable level. "Pop" noise is a low-frequency phenomenon which happens intermittently on a very long time scale. This is why, during testing of broadband seismometers, it is imperative to test sensors for a relatively long period of time, on the order of days or weeks, and include all data collected on a statistical basis, giving the median and maximum noise levels as much attention as the minimum noise level.

The PSD PDF (see section "Ground Motion Spectra") is a particularly useful tool in performing this type of analysis. Figure 8 shows how the PSD PDF represents rare, transient events such as mass-centering operations are captured in the maximum of this distribution, while the typical behavior is captured in the mode. This allows the whole available record to be analyzed, without "window-shopping." This is as it should be: single-occurrence "pops" can be as much of a problem in interpreting seismograms as stationary noise sources and should therefore be considered just as much a part of the overall self-noise performance. See ▶ Seismometer Self-Noise and Measuring Methods.

Selection Criteria

Many factors need to be taken into consideration in choosing a broadband seismometer. Self-noise, clip level, and transfer function are the basic performance criteria, while power and size can crucially determine the cost of the overall installation. These factors and others are treated in this section.

Sensor Self-Noise

Broadband seismometers are required to have a self-noise less than the NLNM, a hypothetical "quietest site on Earth," at least over some wide band. The self-noise cannot be measured deploying the sensor where there is no ground motion: the NLNM shows that there is no such place on Earth. Modern methods require at least three sensors be installed side by side on a rigid pier and a coherence-based analysis technique (Sleeman et al. 2006; Evans et al. 2010). Careful management of thermal, magnetic, and pressure shielding is required (see section on "Installation Procedures"), particularly to observe the performance of the best sensors at very long periods.

One comprehensive self-noise study (Ringler and Hutt 2010) includes all the important broadband seismometers as of its writing. Figures 11 and 12, reprinted from Figures 13 and 14 of that study, show the lowest 5th percentile and median noise of each sensor, respectively.

The abbreviations used in Figs. 11 and 12 are the same as those used in Table 2 below with the following exceptions: T-120, T-240, and T-Compact are the Nanometrics Trillium 120, Trillium 240, and Trillium Compact, respectively, and the 151–120 was the predecessor of the Geotech 151B-120. It is also worth noting that the STS-2 performance shown in these figures is that of a model with higher gain at 20,000 V·s/m and a correspondingly lower clip level, and that the CMG-3TB tested was a borehole variant of the standard CMG-3T.

The median self-noise was for some sensors close to the 5th percentile, while for others it was significantly higher. This can be taken as an indicator of the uniformity and/or the temporal stability of the sensors. The peak in the noise floors

Principles of Broadband Seismometry, Fig. 11 Median self-noise models of weak-motion seismometers

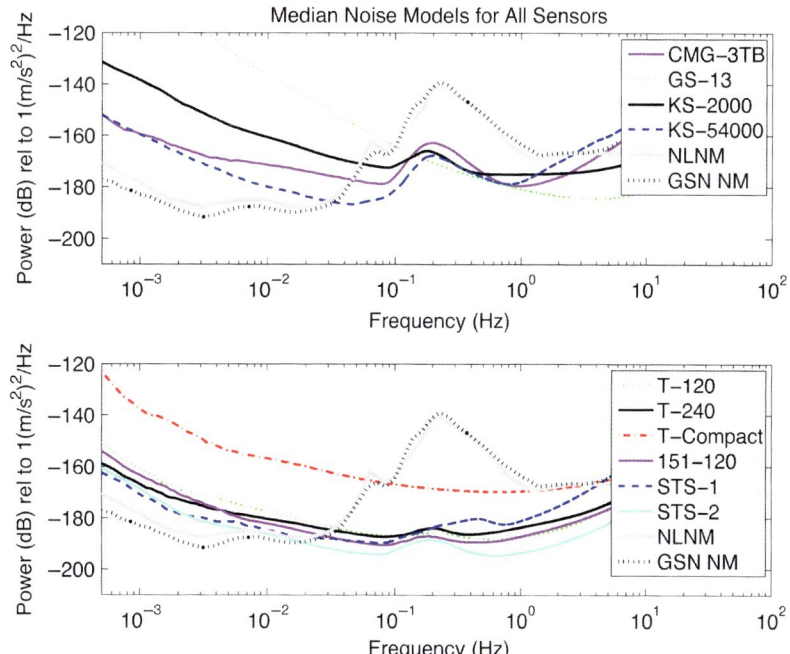

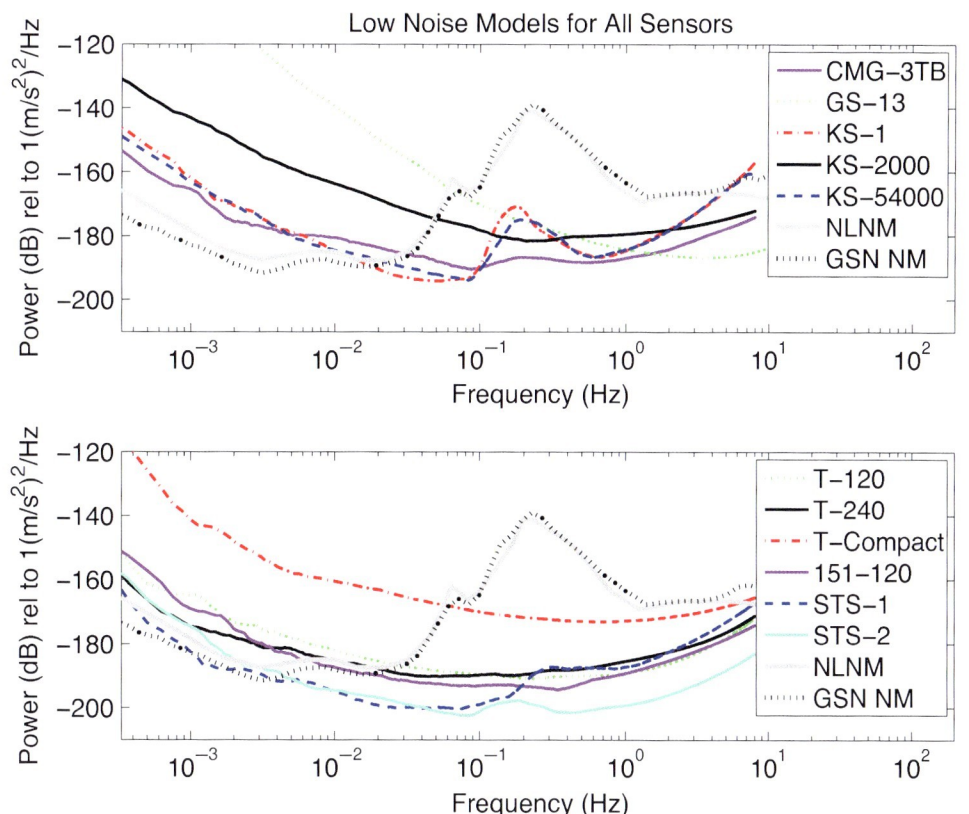

Principles of Broadband Seismometry, Fig. 12 Low self-noise models of weak-motion seismometers

Principles of Broadband Seismometry, Table 2 Specifications of broadband seismometers

Manufacturer	Model	Sensitivity [V·s/m]	Clip [mm/s]	Period [s]	Upper corner [Hz]	Power [W]	Weight [kg]	Sensor volume [L]	Shield volume [L]
Geotech	KS-1	2,400	8	360	5	2.4	45	28	–
Streckeisen	STS-1	2,400	8	360	10	10.5	15	14	450
Nanometrics	Trillium 240	1,200	16	240	200	0.65	14	14	60
Nanometrics	Trillium 120	1,200	16	120	175	0.62	7.2	7.2	32
Guralp	CMG-3 T	1,500	13	120	50	0.75	14	8.4	–
Streckeisen	STS-2	1,500	13	120	50	0.8	13	11	72
Geotech	KS-2000	2,000	10	120	50	0.9	11	8.1	–
REF TEK	151B-120	2,000	10	120	50	1.1	12	12	–
Geodevice	BBVS-120	2,000	10	120	50	1.4	14	8.3	–
Nanometrics	Trillium Compact	750	26	120	100	0.16	1.2	0.8	7.8
Guralp	CMG-3ESP	2,000	10	60	50	0.6	9.5	4.7	–
Guralp	CMG-40 T	800	25	30	50	0.46	2.5	3.9	–

Volumes are for three components and are derived from footprint area and height alone

directly under the microseismic peak is not a feature of the sensor noise floor, but an artifact of imperfect misalignment correction.

Given the importance of the NLNM in setting the target performance of broadband seismometers, it is understandable that self-noise requirements are often specified in terms such as "below the NLNM from 30 s to 30 Hz." Although the intention is clear, this is unfortunately a poor choice of wording. In some bands the NLNM is very steep – which is to say that the typical background acceleration PSD at a quiet site is also very steep – and so a great reduction in self-noise may not result in much improvement in the frequencies at which the self-noise crosses the NLNM. Conversely, where the slope of the NLNM is quite similar to that of the self-noise of a seismometer, a small improvement in self-noise will produce an exaggerated improvement in the NLNM-crossing frequency. Finally, care must be taken at high frequencies because the NLNM is simply not defined above 10 Hz (similarly the MLNM does not extend above 10 Hz and the GSNNM does not extend above 13 Hz).

A better way to set a specification on seismometer self-noise is to pick two key frequencies near the edges of the band of interest and set limits in dB at those frequencies. For example, to require the "median self-noise below −185 dB at 0.01 Hz and below −168 dB at 10 Hz" is equivalent to being "below the NLNM from 100 s to 10 Hz," but for the purposes of comparing one seismometer to another, it is more instructive to compare their median performance in dB at specified frequencies than in the NLNM-crossing frequencies in Hz.

Clip Level

The largest measured ground motions resulting from earthquakes have approached peak accelerations of 40 m/s^2 and peak velocities of 3 m/s, although these limits tend to increase with the densification of networks and the passage of time (Strasser and Bommer 2009). No seismograph can measure both the largest and the smallest ground motions. The dynamic range of the current state of the art in analog-to-digital converters is insufficient to cover the >170 dB range required. Although weak-motion seismometers focus on measuring the smallest motions, they are still expected to measure relatively large motions without clipping, so that weak-motion studies are not "interrupted" by strong motion.

Whether the clip level of a particular seismometer is adequate depends on the expected levels of ground motion at the site of deployment.

Principles of Broadband Seismometry, Fig. 13 Earthquake ground motions and seismograph clip levels and noise floors

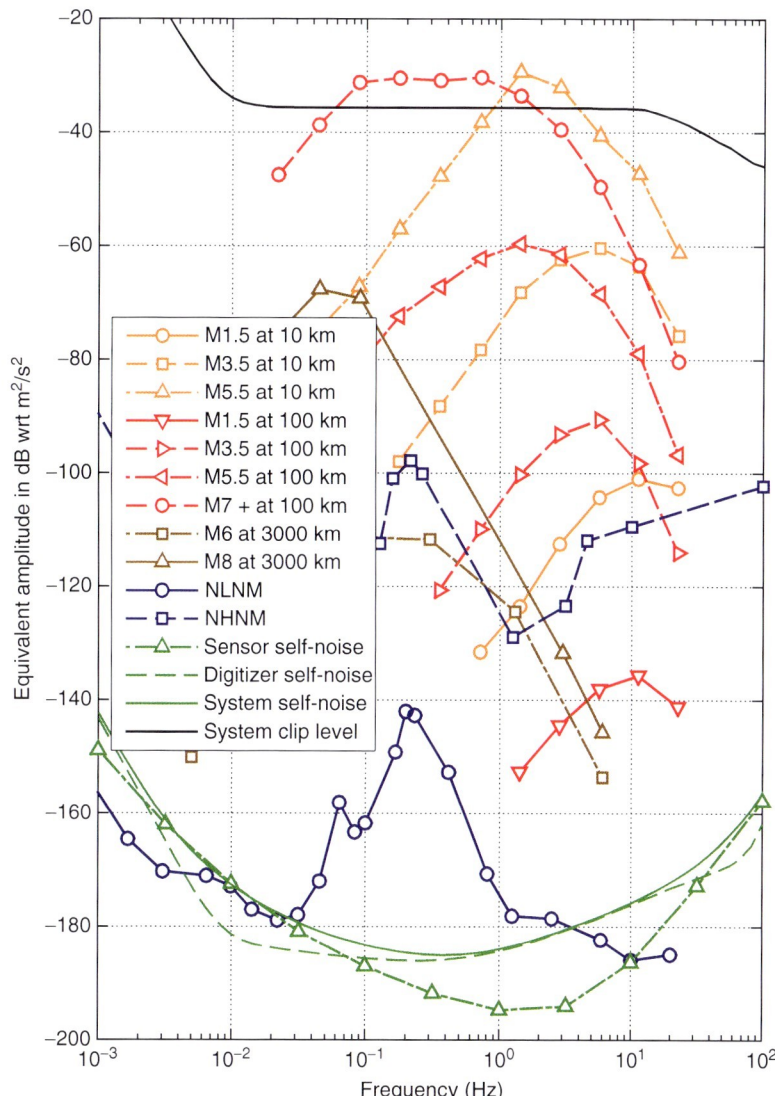

The ground motion resulting from an earthquake varies with the magnitude but also with the distance from the event.

The ground motion resulting from earthquakes is a function of frequency. Figure 13 shows the typical ground motions for large and small earthquakes at local (10 km), regional (100 km) and teleseismic (3,000 km) distances (Clinton and Heaton 2002).

These event spectra are plotted for comparison against the noise floor and clip level of a representative broadband seismograph. It can be seen that the digitizer noise floor is limiting the performance of the seismograph system for frequencies above the microseismic peak. The station operator could choose to increase the digitizer preamp gain or equivalently install a high-gain version of the seismometer. This would result in a reduction of the contribution of the digitizer to the system noise and therefore make it possible to see smaller events, but it would also reduce the clip level of the system, so that large events would be more likely to clip.

The clip level of a seismometer in the middle of its passband depends solely on the output clip level in volts and its sensitivity. Notwithstanding a hypothetical seismometer which runs from very high voltage rails, a requirement for high clip level is low sensitivity, and vice versa. The sensitivity and clip levels of some representative broadband seismometers are given in Table 2, along with other critical performance criteria.

Broadband seismometers generally use displacement transducers to measure the relative displacement of the proof mass and the frame. The proof mass and its suspension together make a pendulum which has a response which is flat to acceleration and independent of its natural frequency above the natural frequency.

There is usually a critical frequency above which the clip level of a broadband seismometer becomes more or less flat to acceleration. For the seismometer depicted in Fig. 6, the acceleration clip level is 0.17 g and the velocity clip level is 16 mm/s, so this critical frequency is near 16 Hz. If this critical frequency is significantly higher than the peak frequency of the spectra of events of interest, it will not affect clipping behavior for real seismic signals.

Response

It is crucial to understand the transfer function of a broadband seismometer when converting a recording in counts or volts back to appropriate units of ground motion. It is a common mistake to think that the signals at frequencies below the lower -3 dB corner or above the upper -3 dB corner are not useful. In fact the only thing which determines whether or not useful signal is present is signal-to-noise ratio, as illustrated by Figs. 9 and 13. With careful application of the inverse response, useful estimates of ground motion can be obtained well outside the -3 dB band.

Seismometers are sometimes available with different options for lower or upper corner frequency and for the mid-band sensitivity. Note that the term "sensitivity" in the context of seismometers is interchangeable with "generator constant" and will have units of voltage per unit of velocity. When selecting a seismometer, it is important to have in mind the largest signal which is likely to be observed, given the seismicity of the nearest seismogenic zone. Seismic risk maps are invaluable for determining the probability of recurrence of a given peak ground acceleration or velocity, which can then be directly compared to the configured seismograph system clip level. High sensitivity means the contribution of the digitizer to the station noise floor is reduced, but it also means the system clip level is reduced. Some users will prefer to use a low-gain seismometer and a digitizer with a built-in variable-gain preamp, so that they can "dial in" the correct station sensitivity after installation.

One feature of a seismometer not normally represented in its nominal transfer functions is the phenomenon of parasitic resonances. Well-characterized seismometers will have a specification for the lowest mechanical resonance; it is important to make sure that this frequency is above the range of frequencies of interest in a particular study.

Power

There are two advantages to low power. The first is that lower power means a physically smaller footprint and a less costly installation for stations which must be located far from main power systems. Such stations are common because the best seismic sites are generally located away from roads and cities and human activity in general, so-called sources of cultural noise. For a temporary deployment, lower power means fewer batteries are needed for a given length of time. For a permanent deployment at a remote site, lower power means less on-site power generation (e.g., fewer solar panels) is needed. A smaller footprint for power generation furthermore generally means less wind-induced seismic noise.

A second advantage to lower power relates specifically to performance at very long periods in vault-type installations. Power dissipation inside the sensor and digitizer means heat generation. This heat causes convection within the vault, and the resulting airflow tends to be turbulent and chaotic, heating and cooling various surfaces around the vault, in particular the floor, causing small but measurable tilts. Thus, for sensors which consume more power, it becomes more difficult to properly thermally shield well enough to drive the resulting apparent horizontal accelerations down below the NLNM at very long periods.

The power consumption of some broadband seismometers is listed in Table 2.

Size and Weight

The physical size of a seismometer has a multiplying effect on the size and thus the cost of deployment of a seismic station. For vault installations, a larger sensor requires a larger volume to be reserved for thermal shielding. Since a good broadband vault must generally be built below ground level, a larger sensor means the minimum volume which must be dug out for the vault is larger.

For temporary deployments, the size and weight of a sensor can significantly affect ease of deployment. When it is a matter of driving to a remote location and hand-carrying the equipment even further away from the road, it can mean that significantly fewer stations can be set up per day, if the sensor is large and heavy.

Aside from the trouble it causes in a temporary deployment, a heavy sensor has an advantage over a light one, in that the associated thermal mass means better temperature stability.

The volume and weight of some broadband seismometers are listed in Table 2.

Enclosure, Leveling, and Topology

The choice of enclosure for broadband seismometers is an important one. Some common options are vault, borehole, posthole, and ocean bottom.

Enclosures designed for deployment in vaults need to be dust- and watertight, but are generally not designed for submersion to significant depths or durations (i.e., ingress protection ratings of IP66 or IP67 are common, but not IP68). There is no particular restriction on the overall diameter of a vault enclosure, but it should be designed for ease of leveling, orientation, and thermal isolation. For example, the connector should be oriented to allow cables to exit the enclosure horizontally near the surface of the pier. This makes it easy to strain-relieve the cable, minimizing the possibility of cable-induced noise, and to place an insulating cover over top of it. The design of broadband seismometer vaults is described in "Installation Procedures" below.

Enclosures designed for deployment in cased boreholes generally need to have smaller diameters and a mechanism to lock the sensor in the hole. Drilling of boreholes is always more economical for smaller diameters than larger ones; a common casing diameter for broadband seismometers is 15 cm. The connector will generally exit at the top of the sensor and the whole assembly should be rated for continuous submersion to a significant depth (i.e., IP 68 to 100 m or more), since flooding of boreholes is common. Boreholes stray from verticality as they are dug deeper; a remote leveling range of up to $\pm 4°$ is thus typically required. See ▸ Downhole Seismometers for a more detailed discussion.

Enclosures designed for deployment in shallow uncased holes called postholes do not need hole locks. Sensors are generally emplaced in backfilled soil or sand and are simply pulled out or dug out at the end of the deployment. As with borehole sensors, the connector should generally exit at the top and must be rated for submersion. There is less control of sensor leveling with deeper holes, and a remote leveling range of $\pm 10°$ may be required. See ▸ Downhole Seismometers for a more detailed discussion.

Enclosures designed for ocean-bottom deployment have several requirements which other sensor types do not. Most of the ocean bottom is near 5 km depth, so in order to be deployable over most of the ocean bottom, a sensor would typically have a continuous submersion rating of 6 km. Most ocean-bottom deployments are done by releasing the sensor at the surface and without controlling exactly where it will come to rest on the ocean floor. The sensors are designed to level themselves, typically at a predetermined time after release, and since the exact resting place is not known in advance, a self-leveling range of $\pm 45°$ or more is required. Prevention of corrosion and biofouling is additional crucial requirement for ocean-bottom enclosures. See ▸ Ocean-Bottom Seismometer for more information.

When an underground vault in bedrock is available, for example, in an inactive mine or in the basement of a building, then a vault-type enclosure is of course the best choice. When there is some significant overburden, so that to reach bedrock a borehole must be dug and cased, then a borehole-type enclosure is required. And of course ocean-bottom deployments require an ocean-bottom enclosure.

It is not uncommon however that a sensor must be deployed in a location where no preexisting vault or borehole is available. In such situations a posthole installation can give performance as good or better than a vault built according to best practices, at significantly less cost for the overall installation.

If leveling motors are included in the sensor, a remote leveling process is initiated via an external electrical signal or at a configurable time after power-on; otherwise manual leveling by adjustment of set screws is sometimes needed. All enclosure types except vault enclosures require remote leveling capability.

Sensor axis topology is a final consideration. Some studies may require only a single axis of seismic sensing, usually vertical. For triaxial seismic sensing, the sensor outputs should be horizontal (X, Y) and vertical (Z). However certain kinds of installation troubleshooting are easier to do if the internal sensing axes are not aligned to horizontal and vertical. See ▶ Symmetric Triaxial Seismometers for more information.

Environmental Sensitivities

Spurious signals due to environmental sensitivities are not normally considered part of the self-noise of a seismometer, but they can deleteriously affect the output signal in many of the same ways. Broadband seismometers are particularly sensitive to changing tilt, temperature, pressure, and magnetic fields.

To understand why, consider that in order to have self-noise just equal to the NLNM at 100 s period, you need to be able to discriminate ground motion from all other effects at a level of

$$a_{\text{PSD}} = 10^{-\frac{185}{20}} \frac{\text{m}}{\text{s}^2 \sqrt{\text{Hz}}} = 0.56 \frac{\text{nm}}{\text{s}^2 \sqrt{\text{Hz}}}$$

And since the self-noise of a seismometer is typically proportional to 1/f in this band, the noise in the decade around $f_{\text{PSD}} = 0.01$ Hz will be

$$a_{\text{NLNM}} = a_{\text{PSD}} \sqrt{f_{\text{PSD}} \ln(10)} \frac{\text{nm}}{\text{s}^2} = 0.08 \frac{\text{nm}}{\text{s}^2}$$

This tiny acceleration, measurable by very broadband seismometers (i.e., a weak-motion inertial sensor with a wide dynamic range over a very broadband), can be overwhelmed by spurious environmental sensitivities, as discussed below.

Tilt

Sensitivity to tilt is an inevitable consequence of inertial sensing, because gravitational equivalence principle tells us that gravity is indistinguishable from accelerations. An inertial sensor tilted from vertical by an angle θ measured in radians experiences an apparent horizontal acceleration of

$$\ddot{x} = g_0 \sin\theta$$

where $g_0 \cong 9.8$ m/s^2 is the standard acceleration due to gravity near the surface of the Earth. For small tilt angles $\sin\theta \cong \theta$, so all inertial seismometers have the same tilt sensitivity α_T, that is, the same apparent horizontal acceleration in response to tilt:

$$\alpha_T \equiv \frac{\ddot{x}}{\theta} = g_0 \cong 9.8 \frac{\text{m/s}^2}{\text{rad}} \cong 0.17 \frac{\text{m/s}^2}{°}$$

Some tilt and rotation is to be expected to accompany the translational motion of a traveling seismic wave, but locally generated non-seismic tilt can prevent critical observations from being made. It is because of their extreme sensitivity to tilt that all inertial translational sensors, including broadband seismometers, record higher levels of apparent horizontal motion than vertical motion at long periods.

In order to resolve the NLNM at 100 s, tilts in the decade band around that frequency would have to be kept smaller than

$$\Delta\theta = \frac{a_{\text{NLNM}}}{g_0} = 5 \times 10^{-10} \, °$$

This is an extremely small angle. It corresponds to lifting one side of a 10 m wide structure pier by just 1 Å, the order of magnitude of atomic radii.

Fortunately, locally generated "excess" tilt is not spontaneous but driven by some other environmental factor and can be greatly reduced with careful vault design. For example, it is common for such tilts to be driven by temperature or pressure sensitivity of the seismic vault or nearby subsurface geology. Tilts can also be driven by changes in insulation or water table, vehicular traffic or other cultural activity, or wind loading on nearby structures. Mitigating these sorts of effects is an overriding concern in designing a seismic vault, as described in the section "Installation Procedures."

Another, more subtle tilt-related effect is that a static tilt will increase off-axis coupling of horizontal motion into vertical. See ▶ Symmetric Triaxial Seismometers for more information. Other than this effect, the actual static tilt of a seismometer is generally not a problem, as long as it is within the operating range of the seismometer.

Most broadband seismometers have an integrator in the feedback circuit, and the operating range of this part of the circuit determines the tilt range of the seismometer. For the lowest possible noise at very long periods, the integrator output resistor must be large, and this restricts the tilt range of the seismometer. Thus very broadband seismometers are typically equipped with centering motors or leveling platforms to extend this tilt range. See ▶ Downhole Seismometers for more information.

Temperature

The operating temperature range of a seismometer is determined by the temperature coefficient of the mechanics and components in the force-feedback circuit. A vertical seismometer involves balancing the acceleration due to the effect of gravity on the proof mass against forces supplied by a suspension. The temperature coefficient relates changes in deflection of the proof mass with temperature and so can be expressed in units of ppm (with respect to g_0 the acceleration due to gravity) per °C. A temperature-compensated axis assembly is one in which changes in forces due to thermoelastic coefficients in the suspension cancel deflections due to coefficients of thermal expansion in the rest of the components (Wielandt 2002), such that a displacement transducer would register no movement of the proof mass.

Some broadband seismometers have very wide temperature ranges, encompassing the full range of possible deployment temperatures. Many of the broadband seismometers with the lowest self-noise, however, have operating temperature ranges of ± 10 °C or less. These seismometers are equipped with a re-centering mechanism which must be activated after the seismometer has been installed in a new vault, ideally after its temperature has stabilized. Just as horizontal sensitivity to tilt determines the tilt range, vertical sensitivity to temperature determines the temperature range.

Even for a sensor operating well within its temperature range, spurious horizontal or vertical output signals can result from tiny changes in temperature. The temperature sensitivity is typically a direct proportionality of equivalent input acceleration to change in temperature.

For a seismometer which is not temperature compensated, the temperature sensitivity is typically dominated by the thermoelastic coefficient of the mainspring, as shown in Fig. 14 (left). The cantilever balances the mass M against the force of gravity g_0 but as the temperature T changes the stiffness of the beam, represented as a spring constant K changes, and the apparent vertical acceleration $\ddot{x}$ changes.

Summing the forces on the mass M in Fig. 14 (left), we find

$$\sum F = -Kx + Mg_0 = M\ddot{x}$$

Principles of Broadband Seismometry, Fig. 14 Schematic representation of vertical (*left*) and horizontal (*right*) temperature sensitivity

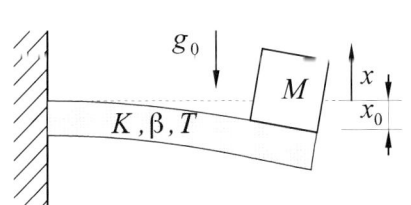
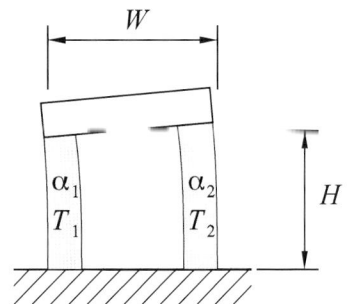

And the temperature sensitivity can be modeled as

$$K = K(\Delta T) = K_0(1 + \beta \Delta T)$$

So if the instrument is designed so that at $T = 0$, there is no apparent acceleration $\ddot{x} = 0$ and the deflection is static at x_0:

$$x_0 = \frac{Mg_0}{K_0}$$

Now if we allow $T \neq 0$, the apparent acceleration is

$$\ddot{x} = g_0 - \frac{x_0}{M} K_0 (1 + \beta \Delta T)$$
$$= g_0 - g_0(1 + \beta \Delta T) = g_0 \beta \Delta T$$

Most copper alloys and steels have a thermoelastic coefficient on the order of $\beta = -300$ ppm/°C, the minus sign indicating that the mainspring relaxes with an increase in temperature.

This translates to a requirement for temperature stability of

$$\Delta T_z = \frac{a_{\text{NLNM}}}{|\beta| g_0} \cong 7 \times 10^{-9} \, °C$$

A related problem is to measure ground motion on the order of the NLNM in the presence of temperature-generated tilts. This problem is significant both at the level of the seismometer and its subassemblies and at the level of the seismic vault and related superstructures. One way to model this effect is to visualize the sensor as a platform with legs which either have different temperature coefficients or which are at different temperatures, as shown in Fig. 14 (right). For such a structure, small differentials produce small tilts:

$$\Delta a = \Delta \theta g_0 = (\alpha_1 T_1 - \alpha_2 T_2) \frac{H}{W} g_0$$

For an enclosure made out of a single material, all that matters is the difference in temperature across the structure. For an enclosure made out of steel or aluminum, the thermal coefficient of expansion is on the order of $\alpha = +20$ ppm/°C, with the positive sign indicating that the material expands with an increase in temperature. For the resulting equivalent horizontal acceleration to be less than the NLNM at a 100 s period, if the height is the same as the width, the temperature difference across the enclosure must be less than

$$\Delta T_x = \frac{a_{\text{NLNM}}}{\alpha g_0} \cong 1 \times 10^{-7} \, °C$$

Obviously the actual dimensions of the structure can result in this effect being significantly amplified or attenuated, as can the geometry and relative stiffness of the members. Furthermore, it is important to note in both cases that static temperatures are not a problem because the seismometer does not respond to static acceleration. What matters is temperature variation with time, in the band of interest, in this case near 100 s period. Although this model is extremely simplistic, the point is that mechanisms for the conversion of changes in temperature into tilt abound, both inside a seismometer and outside.

Principles of Broadband Seismometry

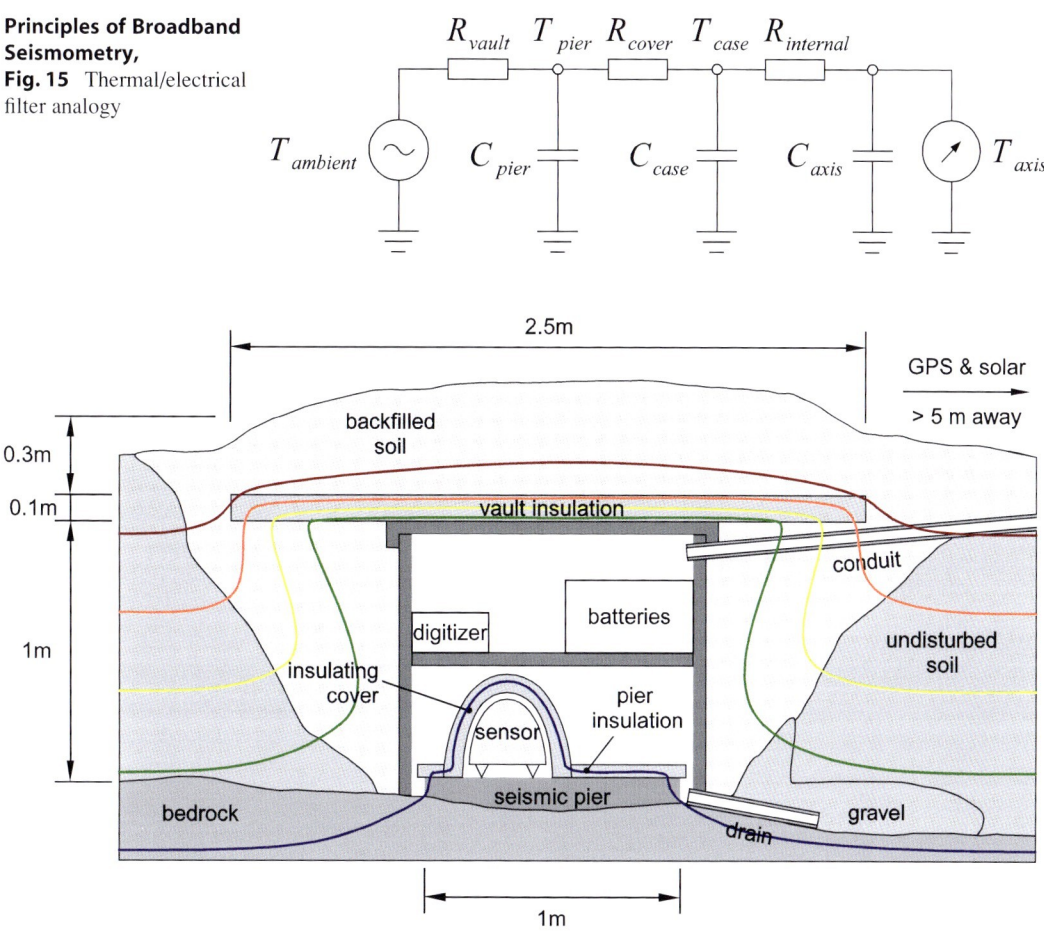

Principles of Broadband Seismometry, Fig. 15 Thermal/electrical filter analogy

Principles of Broadband Seismometry, Fig. 16 Idealized broadband vault

There is a subtle difference between the vertical temperature sensitivity and the horizontal temperature sensitivity of a seismometer. The vertical output of a seismometer is only sensitive to bulk temperature changes; the horizontal outputs are sensitive to differences in temperature across the surface of some part of the enclosure or differences in coefficient of thermal expansion combined with bulk changes in temperature.

The design of a thermal shield against bulk temperature changes is relatively straightforward and can be conceptualized by analogy with electronic circuits as designing a series of cascaded single-time-constant thermal low-pass filters. The design of bulk thermal isolation structures requires a series of concentric shells of thermally resistive elements and thermally massive elements.

In Fig. 15 the first layer of thermal insulation R_{vault} shields the thermal mass C_{pier} of the pier, the sensor insulating cover R_{cover} shields the thermal mass C_{sensor} of the sensor, and the internal insulation of the sensor R_{internal}, if any, shields the thermal mass C_{pier} of the individual axes of the sensor. It is important when designing thermal shields to avoid accidentally including any thermal short circuits which decrease the effectiveness of the shielding. A thermal short circuit or thermal bridge is any path which crosses a thermal insulator and has high thermal conductivity. For example, if the electrical conduit in Fig. 16 was made of metal and therefore thermally conductive, it will act as a thermal short circuit and degrade the performance of the vault insulation. This could be visualized in the electrical analogy of Fig. 15 as a low-value resistance in parallel with R_{vault}. A seismometer which includes some thermal insulation measures within its pressure vessel will require less external thermal shielding.

Bulk thermal isolation of the type described thus far primarily addresses the thermal sensitivity of the vertical output of a seismometer. Inhibiting thermally generated noise on the horizontal outputs of a seismometer is a different problem.

First, a vault must be free of drafts. At the same time, a fully sealed vault can make the vault respond to pressure with tilt, particularly if the vault is not installed on competent rock. The solution in some cases is to design the vault to have a single point at which it vents, so that the internal and external air pressures are equalized without generating drafts across the floor of the vault.

Second, air convection within the vault must be inhibited. This is done by reducing the power dissipated within the vault, by filling airspace within the vault with some material which inhibits airflow. The design of bulk thermal insulation is subtly different from the design of a shield intended to stop convective airflow. See the section on "Installation Procedures" below for more detail.

Pressure

A seismometer not contained within a pressure vessel will exhibit strong pressure sensitivity on the vertical output due to buoyancy of the proof mass. Consider a proof mass with a density $\rho_{proof} = 8$ g/cm^3 at a temperature $T_{air} = 293$ K in dry air with a specific gas constant of $R_{air} = 287$ J/kg · K in standard gravity g_0. For such a seismometer, the vertical sensitivity to air pressure changes due to buoyancy is (Zürn and Wielandt 2007)

$$\alpha_B = \frac{g_0}{P_{air}} \frac{\rho_{air}}{\rho_{proof}} = \frac{g_0}{R_{air} T_{air} \rho_{proof}} = 15 \frac{nm}{s^2 \cdot Pa}$$

In order to be able to measure motions on the order of the NLNM at 100 s period, we need to keep variation in pressure under

$$\Delta P_B = \frac{a_{NLNM}}{g_0} R_{air} T_{air} \rho_{proof} = 0.006 \text{ mPa}$$

This requirement is stringent enough that if the pressure vessel of a seismometer is compromised, the vertical output will be dominated by this buoyancy effect.

A pressure vessel must be well designed in order to ensure that changes in atmospheric pressure do not produce equivalent horizontal or vertical outputs. Three critical specifications, then, are pressure attenuation of the pressure vessel and the pressure sensitivities of the vertical and horizontal outputs.

For the vertical channel of a seismometer in a pressure vessel, the limiting pressure effect is that due to atmospheric gravitation. Using the Bouguer plate model, in which the atmosphere above a station is modeled as a cylindrical plate having constant density, the gravitational pressure sensitivity due to atmospheric gravitation is (Zürn and Wielandt 2007)

$$\alpha_G = 2\pi \frac{G_0}{g_0} = 0.043 \frac{nm}{s^2 \cdot Pa}$$

Where the universal gravitation constant is $G_0 = 6.67 \times 10^{-11}$ N·m^2/kg^2. This, then, places a design constraint on a pressure vessel for a broadband seismometer: the pressure vessel will deform in response to changes in atmospheric pressure and result in a corresponding vertical acceleration, but the resulting pressure sensitivity should be less than α_G.

With this number in hand, we can reconsider the effect of buoyancy on the pressure vessel. An increase in atmospheric pressure will cause the volume of air inside the pressure vessel to become smaller; the rigidity of the vessel determines how much smaller. In order for the buoyancy effect to be much smaller than that due to the unavoidable effect of atmospheric gravitation, the pressure vessel must attenuate pressure changes by a factor much greater than

$$k_P = \frac{\alpha_B}{\alpha_G} = 340$$

For the horizontal channels of a seismometer, pressure sensitivity arises because atmospheric loading deforms the ground near the seismometer and produces measurable tilts. The level of sensitivity depends on geology and on depth, with shallow installations on unconsolidated sediment having the greatest sensitivity. At the Black Forest Observatory, the vault is 150–170 m below the surface in hard rock, and the measured admittance is typically (Zürn et al. 2007)

$$\alpha_T = 0.3 \ \frac{\text{nm}}{\text{s}^2 \cdot \text{Pa}}$$

Thus we can set a reasonable limit on the required horizontal pressure sensitivity of a broadband seismometer. When the pressure vessel deforms in response to pressure and causes the horizontal outputs to tilt and exhibit an apparent horizontal acceleration, the resulting pressure sensitivity should be less than α_T.

With both horizontal and vertical pressure sensitivities, a coherence analysis can be used to find a least-squares best-fit to the relative transfer function. This best-fit pressure sensitivity can then be used to correct seismic records for pressure, if a sufficiently sensitive microbarometer is colocated with the seismometer and recorded.

Magnetic

Temperature-compensated materials suitable for mainsprings tend to be magnetic, so inertial sensors tend to be susceptible to magnetic fields. The susceptibility takes the form of direct proportionality of equivalent output acceleration to magnetic field strength.

The magnetic sensitivity of a very broadband seismometer can vary between 0.05 and 1.4 m/s²/T (Forbriger et al. 2010). In order for the magnetic sensitivity of a seismometer to not interfere with measurement of ground motion down to the NLNM at 100 s during a magnetically quiet period, the vertical magnetic sensitivity must be less than

$$\alpha_M = 0.7 \ \frac{\text{nm}}{\text{s}^2 \cdot \text{T}}$$

For installations which must produce quiet records even during geomagnetic storms, a magnetic shield can be used. Such shields are typically constructed of a high-permeability metal such as permalloy or mu metal.

Geomagnetic storms are not the only source of low-frequency magnetic fields. Hard drives and solar chargers are two examples of common equipment at seismograph stations which tend to generate interfering signals, and should therefore be located as far as away from the seismometer as is practicable.

As with pressure sensitivity, a best-fit magnetic sensitivity can then be used to correct seismic records for pressure, if a sufficiently sensitive magnetometer is colocated with the seismometer and recorded.

Site Selection

There is no substitute for a geological survey when it comes to site selection.

A site survey provides knowledge of the structures over which the seismometer will be installed. Where possible, seismometers should be installed on bedrock and as far away as possible from sources of cultural noise such as roads, dwellings, and tall structures.

The most important factor to consider in terms of geology is the composition of the uppermost stratum. For example, when the boundary of the uppermost layer is clearly defined as roughly horizontal, the S-wave velocity and thickness of that layer will determine the fundamental resonant frequency at that site. Lower velocities and larger drift thicknesses produce greater site amplification at lower frequencies.

Table 3, reprinted with permission from Trnkoczy (2002), grades site quality according to types of sediments or rocks and gives a sense of how the quality of a site relates to the S-wave velocity.

Low porosity is furthermore important as water seepage through the rock can cause tilts which overwhelm the seismic signal at long periods. Clay soils and, to a lesser extent, sand, are especially bad in this sense.

Principles of Broadband Seismometry, Table 3 Classification of rock for site selection

Site quality		S-wave velocity (m/s)	
Grade	Type of sediments/rocks	Min	Max
1	Unconsolidated (alluvial) sediments (clays, sands, mud)	100	600
2	Consolidated clastic sediments (sandstone, marls); schist	500	2,100
3	Less compact carbonatic rocks (limestone, dolomite) Less compact metamorphic rocks; conglomerates, breccia, ophiolite	1,800	3,800
4	Compact metamorphic rocks and carbonatic rocks	2,100	3,800
5	Magmatic rocks (granites, basalts); marble, quartzite	2,500	4,000

Installation Procedures

The question of how to get the best performance out of a seismometer is a very involved topic. In this section the focus will be on broadband seismometers in vault enclosures, and in particular the edges of the usual band of interest, near 100 s period and 10 Hz frequency.

Underground Vaults

The STS-1 defined the limits of ground motion measurement at long periods after it was developed, around 1980. Each component of motion, vertical and horizontals, was detected by a mechanical assembly and controlled by a set of feedback electronics in a separate enclosure.

The advent of the STS-1 was accompanied by the development of systems used to maximize its performance (Holcomb and Hutt 1990). For vertical components, pressure and magnetic shielding was provided for vertical components using a glass bell jar and a permalloy shield, respectively.

Specially designed, so-called "warpless" baseplates prevented pressure-generated tilt noise from contaminating horizontal components. The electronics are housed in a separate enclosure which is not sealed and which can require regular replacement of desiccant to avoid anomalous response characteristics (Hutt and Ringler 2011).

The installation of an STS-1 is a delicate procedure; most of the innovations in the field of broadband seismometry since then have aimed at simplifying this procedure as well as reducing power and overall footprint.

To justify the performance of an STS-1, it is usually necessary to have an underground site in hard rock, something which is expensive to construct and unnecessary for earthquake engineering applications.

Shallow Broadband Vault

An idealized broadband vault design is shown in Fig. 16.

One practical procedure for constructing such a site is as follows. A hole is dug using a backhoe in which a large-diameter plastic tube is placed. A concrete slab is poured at the bottom to serve as a pier for the sensors to rest on. Thermal insulation is added around the sensors, and the digitizer is located in a separate compartment above the sensor. A cover is placed over the tube, and the earth which was dug out to make the hole is backfilled around the tube and tamped down up to the level of the lid. A layer of rigid foam insulation is placed across the lid before piling on the rest of the soil removed in digging the hole for the vault.

This same basic procedure can be tailored to the demands of temporary installations. The seismic vault designed for the "transportable array" of the USArray project (EarthScope 2013), for example, features most of the design elements shown in Fig. 16.

Thermal Insulation

Different thermal insulation components in a broadband vault serve different purposes. The sensor insulating cover serves as bulk insulation and as a breeze cover, and by restricting the airspace around the sensor, it stops convection around the sensor. An insulating layer laid on top of the seismic pier prevents convection-driven air currents from causing the pier to distort as they pass over its surface.

The thick layer of insulation over top of the vault serves to bring the vault closer in temperature to a deeper stratum of the ground. Otherwise, a low thermal-resistance path from the vault to the surface would exist, and much of the benefit of burying a sensor in terms of thermal stability would be lost. Surface air temperature variation does not penetrate very deep into the ground; the effect of the insulation is to drive isotherms of temperature variation deeper into the ground, as shown approximately in Fig. 16. A rule of thumb for good-quality rigid Styrofoam insulation is that 2.5 cm of insulation provides the same thermal insulation as 30 cm of soil.

Pier Construction

The vault is drawn in Fig. 16 to accommodate a seismic pier which is significantly wider than a typical broadband seismometer plus its insulating cover. The reason for this is that some room must be left for the operator to stand beside the sensor and bend over it to orient the sensor to north, level it, and lock its feet. Vaults can be made significantly smaller if the seismometer is self-leveling, such as a ▶ Downhole Seismometer, but of course the problem of sensor orientation still needs to be addressed.

The drain shown schematically in Fig. 16 will only be effective if the water table is at or below the depth of the seismic pier. Broadband vaults such as this one are prone to flooding; the surest remedy to this problem is to make use of a seismometer that is designed for submersion (e.g., ▶ Downhole Seismometers).

Because of the sensitivity of a broadband seismometer to tilt, the seismic pier should be physically decoupled from the vault wall. The soil at the surface will be constantly shifting due to wind and changes in water content or frost heave. Leaving a gap between the vault wall and the pier prevents such soil motion from being transmitted through the vault wall to the pier and producing measurable tilts.

The concrete for the pier should be made from 50 % Portland cement, 50 % sieved sand, and no aggregate. It should be vibrated to eliminate voids and allowed 24 h to harden before use. The pier must not be reinforced with steel; additional strength is not needed, and the different temperature coefficients would result in detectible tilts and cracking with temperature.

All classes of seismometer benefit from being sited on competent rock because levels of high-frequency (>1 Hz) noise of all kinds are lowest when the seismic wave velocities are highest. Broadband seismometers additionally benefit because hard rock sites are less susceptible to tilt, whether driven by pressure, cultural activity, or other phenomena, and horizontal site noise levels will be dominated by tilt at long periods (<0.1 Hz).

Broadband seismometers often must be sited, however, in areas where bedrock does not come near the surface. In these cases the vault design depicted in Fig. 16 is still useful; instead of pouring concrete, it may be convenient to place a stone block on tamped gravel and use that as a seismic pier.

Thermal Shielding

Seismometer insulation is often implemented on a more or less ad hoc basis. Rigid Styrofoam insulation can be glued or taped together to make flat-faced shaped; large-diameter cardboard tubes lined with fiber wool can also be made.

The "Stuttgart shielding" method used in the German Regional Seismic Network (GRSN) is a much more systematic approach, as shown in Fig. 17, reprinted with permission from Hanka (2002). It includes a thick gabbro baseplate, polished on one side, combined with a large stainless steel cooking pot, which provides pressure shielding to improve on the native pressure sensitivity of the STS-2. Both the outside of the pot and the outer surface of the sensor are wrapped with fiber wool to provide thermal insulation. Sometimes the whole assembly is further wrapped in a thermally reflective "space blanket" to provide additional thermal shielding.

A simplified setup was introduced for the GEOFON program. An aluminum enclosure including a "warpless" baseplate provides the needed additional pressure shielding. Thermal shielding is provided by a foam rubber insert inside the aluminum enclosure and polystyrol beads outside the enclosure.

Some manufacturers provide thermal shields as accessories to seismometers. The thermal

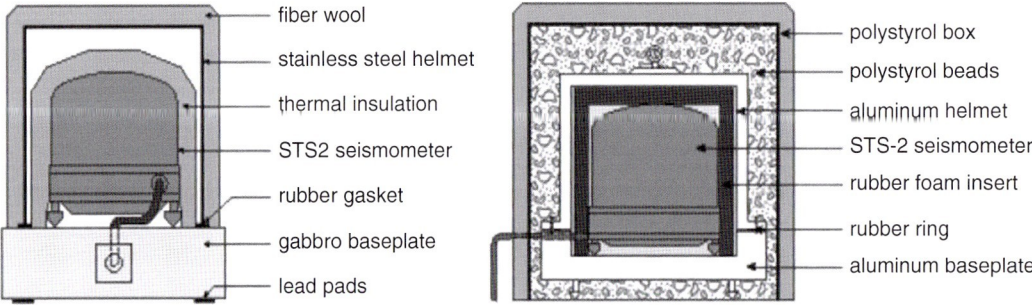

Principles of Broadband Seismometry, Fig. 17 Thermal and pressure shielding for STS-2 (*left*: GRSN, *right*: GEOFON)

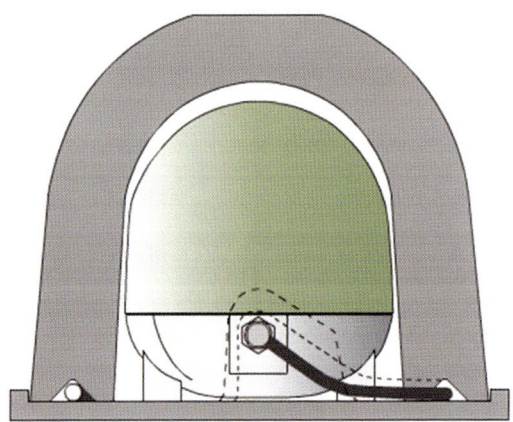

Principles of Broadband Seismometry, Fig. 18 Thermal shield for Trillium 240

shield for a Trillium 240 is shown in Fig. 18. This cover is made of rigid molded plastic filled with insulating foam. The cover is formfitting – without quite touching the sensor – so that it greatly restricts convective airflow around the sensor, without allowing forces to be transferred to the sensor body through the insulation. It includes a race for a turn of the sensor cable, which minimizes heat conduction through the cable. The shield includes a foam base gasket which raises the cover up off the pier and allows the cable to exit the shield, and provides a layer of insulation between the sensor and the surface of the seismic pier. Such shielding systems are rugged and easy to transport and provide excellent shielding performance and repeatability.

Other Installation Details

Many precautions which must be taken in installing a broadband seismometer should go without saying but must be stated anyway:

- Surface of pier must be clear of debris.
- Pier should be free of cracks, particularly beneath sensor.
- Insulating cover must be close to sensor but must not touch it.
- Adjustable feet must be firmly locked.
- Connectors must not touch insulating cover.
- Cable must be strain-relieved close to the sensor.
- Cable must not touch any other structure between connector and strain relief.

Cabling is an underrated source of excess long-period horizontal noise in vault-type installations. It is recommended to "strain-relieve" the cable to the pier, sometimes by placing a heavy weight on it close to the sensor. If this is not done, thermal expansion or other motion induced in the cable can be transferred to the seismometer, generating tilt. Sensitivity to cable-induced noise is particularly acute with stiff cables; cables designed for flexibility make the whole job much easier.

It is always important to follow the manufacturer's instructions for sensor installation. Each sensor, for example, will provide different features for physical alignment of the sensor to north.

Downhole and Ocean Bottom

It is well known that site noise decreases with depth. This is the primary motivation for installations in deep, cased boreholes. Two newer techniques, which offer reduced installation costs and an even smaller surface footprint, are posthole and direct-burial installations (Nanometrics 2013). The design of such installations is however beyond the scope of this entry (see ▶ Downhole Seismometers).

The ocean bottom, in contrast, is an important environment for seismometer deployment not because of reduced levels of site noise, but because most of the earth is covered by ocean. It is similarly beyond the scope of this entry to treat the relevant installation techniques in detail, except to note some parallels with those used on land. Accurate leveling mitigates coupling of horizontal ground motion into the vertical output. Shallow burial can provide significant shielding from the effects of ocean-bottom currents. Colocated pressure sensors can be used to correct excess noise due to infragravity waves on the vertical. See ▶ Ocean-Bottom Seismometers.

Summary

A seismometer is a kind of inertial sensor which can detect the smallest ground motions in some frequency band. Broadband seismometers are those which can detect motions which are as small as the background motion at a hypothetical quiet site (as represented by a model such as the NLNM) over a frequency band which extends both above and below the microseismic peak.

A broadband seismometer has much better performance than a passive seismometer of the same physical size. The use of displacement transducers and force feedback results in an instrument with better linearity, lower self-noise, and higher sensitivity at long periods.

A broadband seismometer must be installed in a carefully designed vault to ensure that spurious signals due to tilt driven by temperature and other environmental factors are minimized. A broadband seismometer capable of resolving the NLNM at 100 s should have a vertical pressure sensitivity less than $\alpha_G = 0.04$ nm/s^2/Pa, a horizontal pressure sensitivity less than $\alpha_T = 0.7$ nm/s^2/Pa, and a magnetic sensitivity less than $\alpha_M = 0.7$ nm/s^2/T.

The construction of an effective broadband seismic vault is described in some detail. Several different kinds of thermal insulation, serving different functions, are required for optimal performance at long periods.

References

Aki K, Richards P (2002) Quantitative seismology, 2nd edn. University Science, Sausalito

Berger J, Davis P, Ekström G (2004) Ambient Earth noise: A survey of the global seismographic network. J Geophys Res 209, B11307

Bormann P (2002) Seismic signals and noise, chapter 4. In: Bormann P (ed) New manual of seismological observatory practice, vol 1. GeoForschungsZentrum, Potsdam

Brune J, Oliver J (1959) The seismic noise of the Earth's surface. Bull Seism Soc Am 49:349–353

Clinton F, Heaton T (2002) Potential advantages of a strong motion velocity meter over a strong motion accelerometer. Seism Res Lett 73(3):332–342

EarthScope (2013) Transportable seismic network: imaging the Earth's interior. Retrieved 29 May 2014, from USArray: http://www.usarray.org/files/docs/pubs/TA_Host-a-Station_Imaging_0411-Final.pdf

Evans JR, Followill F, Hutt CR, Kromer RP, Nigbor RL, Ringler AT, Steim JM, Wielandt E (2010) Method for calculating self-noise spectra and operating ranges for seismographic inertial sensors and recorders. Seism Res Lett 81(4):640–646

Forbriger T, Widmer-Schnidrig R, Wielandt E, Hayman M, Ackerley N (2010) Magnetic field background variations can limit the resolution of seismic broad-band sensors. Geophys J Int 183(1):303–312

Geotech Instruments, LLC (2001) Short-period seismometer model S-13 and GS-13. Retrieved 12 Nov 2013, from http://www.geoinstr.com/ds-s13.pdf

Hanka W (2002) Parameters which influence the very long-period performance of a seismological station: examples from the GEOFON network, Section 7.4.4. In: Bormann P (ed) New manual of seismological observatory practice, vol 1. GeoForschungsZentrum, Potsdam, pp 64–74

Holcomb LG, Hutt CR (1992) An evaluation of installation methods for STS-1 seismometers. Open-file report 92–302. US Geological Survey, Albuquerque

Hutt CR, Ringler AT (2011) Some possible causes of and corrections for STS −1 response changes in the global seismographic network. Seism Res Lett 82(4):560–571

Lee WHK, Evans JR, Huang B-S, Hutt CR, Lin C-J, Liu C-C, Nigbor RL (2012) Measuring rotational ground motions in seismological practice. In: Bormann P (ed) New Manual of Seismological Observatory Practice 2 (NMSOP-2). Deutsches GeoForschungsZentrum GFZ, Potsdam, p. 1–27

Longuet-Higgins MS (1950) A theory of the origin of microseisms. Phil Trans R Soc A 243(857):1–35

McNamara DE, Buland RP (2004) Ambient noise levels in the continental United States. Bull Seism Soc Am 94(4):1517–1527

Motchenbacher CD, Connelly JA (1993) Low-noise electronic system design. Wiley, New York

Nanometrics Inc (2009) Trillium 120P/PA seismometer user guide. (15149R6). Nanometrics, Inc, Kanata

Nanometrics Inc (2013) Trillium posthole user guide. (17217R5). Nanometrics, Inc, Kanata

Peterson J (1993) Observations and modeling of seismic background noise. Open-file report 93–322. US Geological Survey, Albuquerque

Phillips CL, Harbor RD (1991) Feedback control systems, 2nd edn. Prentice-Hall, Englewood Cliffs

Ringler AT, Hutt CR (2010) Self-noise models of seismic instruments. Seism Res Lett 81(6):972–983

Scintrex Limited (2007June 21) CG-5 Scintrex Autograv System operation manual. Retrieved 7 May 2012, from Scintrex web site: http://www.scintrexltd.com/documents/CG5.v2.manual.pdf

Sercel - France (2012) Analog seismic sensors. Retrieved 11 Dec 2013, from Sercel web site: http://www.sercel.com/products/Lists/ProductSpecification/analog-seismic-sensors-specifications-Sercel-Seismometers.pdf

Sleeman R, van Wettum A, Trampert J (2006) Three-channel correlation analysis: a new technique to measure instrumental noise of digitizers and seismic sensors. Bull Seism Soc Am 96(1):258–271

Strasser F, Bommer J (2009) Review: strong ground motions—have we seen the worst? Bull Seism Soc Am 99(5):2613–2637

Trnkoczy A (2002) Factors affecting seismic site quality and site selection procedure, Section 7.1. In: Bormann P (ed) New manual of seismological observatory practice, vol 1. GeoForschungsZentrum, Potsdam, pp 1–14

Webb S, Crawford W (2010) Shallow-water broadband OBS seismology. Bull Seism Soc Am 100(4):1770–1778

Widmer-Schnidrig R, Kurrle D (2006) Evaluation of installation methods for Streckeisen STS-2 seismometers. Retrieved 25 Oct 2013, from http://www.geophys.uni-stuttgart.de/~widmer/ge2.pdf

Wielandt E (2002) Seismic sensors and their calibration, chapter 5. In: Bormann P (ed) New manual of seismological observatory practice, vol 1. GeoForschungsZentrum, Potsdam

Wielandt E, Streckeisen G (1982) The leaf-spring seismometer – design and performance. Bull Seis Soc Am 72(6):2349–2367

Working Group on Instrumentation, Siting, Installation, and Site Metadata (2008) Instrumentation guidelines for the advanced national seismic system. Open file report 2008-1262. US Geological Survey, Reston

Zürn W, Wielandt E (2007) On the minimum of vertical seismic noise near 3 mHz. Geophys J Int 168:647–658

Zürn W, Exß J, Steffen H, Kroner C, Jahr T, Westerhaus M (2007) On reduction of long-period horizontal seismic noise using local barometric pressure. Geophys J Int 171(2):780–796

Probabilistic Seismic Hazard Models

Danielle Hutchings Mieler[1], Tatiana Goded[2] and Mark Stirling[2]
[1]Earthquakes and Hazards Resilience Program, Association of Bay Area Governments, Oakland, CA, USA
[2]GNS Science, Lower Hutt, New Zealand

Synonyms

Deaggregation; Deterministic seismic hazard analysis; Ground motion prediction equations; Hazard curve; Hazard spectra; Probabilistic seismic hazard analysis, PSHA; Seismic hazard

Introduction

This entry provides an overview of probabilistic seismic hazard modeling, which has provided fundamental input to the engineering, planning, insurance sectors, and other fields for over 30 years. In essence probabilistic seismic hazard modeling uses the location, size, and occurrence rate of earthquakes to estimate the frequency or probability of damaging or potentially damaging earthquake motions that may occur at a site. By taking into account the frequency of earthquakes as well as the magnitude, the method captures the contribution to seismic hazard from all relevant earthquakes, from the frequent moderate earthquakes (magnitude $5 \leq M < 7$) to the infrequent large to great earthquakes ($M \geq 7$).

The following sections summarize the history and fundamental steps of probabilistic seismic hazard analysis (PSHA), on which probabilistic seismic hazard modeling is based, providing example applications, discussing strengths and limitations, and describing current research and future needs.

Probabilistic Seismic Hazard Analysis: History, Method, and Outputs

History

Progression from Deterministic to Probabilistic Seismic Hazard Analysis
Prior to development of PSHA, seismic hazard models were primarily based on deterministic methods, in which a limited number of earthquake scenarios (usually just one or two) are selected as most relevant to the seismic hazard of the site. Based on the estimated magnitude of the scenario earthquake, distance from earthquake to site, and other considerations (e.g., slip type and site conditions), a ground motion is estimated for each scenario. While conceptually simple, deterministic methods do not consider the frequency of earthquakes. This can lead to overestimation of the hazard in the case where a source is associated with a very long recurrence interval. Conversely, ignoring sources that produce more frequent earthquakes of lesser size can lead to underestimation of the hazard. For example, a close moderate magnitude earthquake can produce higher ground motions than a more distant, large magnitude earthquake. Recognition of these deficiencies in deterministically based seismic hazard analysis led to the development of the methods of PSHA.

Development of the Probabilistic Hazard Analysis Framework
The framework for PSHA was first developed by Allin Cornell in the late 1960s (Cornell 1968). The motivation for developing PSHA came from the needs of engineers who must make fundamental trade-offs between the initial cost of designing and building a robust structure and the risk of economic loss due to natural disasters. It is important that engineers have the most accurate understanding of the hazard so they can evaluate these trade-offs to achieve the optimal balance of cost, performance, and risk. Cornell's approach was to draw on the analogy of design winds or floods (e.g., 100-year flood) and provide engineers with a single ground motion value expressed in terms of return period. Cornell used the location, recurrence behavior, and predicted ground motion of earthquake sources to estimate the frequency or probability of exceeding a ground motion at a particular site or across a region. In this respect he saw PSHA as a way of providing a standardized method for the engineer or scientist to consider all possible earthquake scenarios and associated ground motions and their uncertainties to gain a clearer picture of the hazard for use in design.

Method

The basic four steps undertaken in a PSHA for a site are:

1. *Source definition.* Use geologic data and historical earthquake record to define the locations and dimensions of earthquake sources.
2. *Magnitude-frequency relationship.* Define the likely magnitudes of earthquakes that may be produced by each source and the associated frequencies of occurrence.
3. *Ground motion prediction.* Estimate the ground motion level that a source will produce at the site.
4. *Probability of exceedance calculation.* Quantify the frequency or probability at which a ground motion level is exceeded at the site, given all sources. Repeating this procedure for all sources and ground motion levels results in a hazard curve.

These steps are represented graphically in Fig. 1 and described in more detail in the following sections.

Source Definition

Seismicity catalogues, active fault data, and increasingly geodetic strain rates derived from global positioning system (GPS) data are used to define the locations and recurrence behavior of earthquake sources relevant to the site of

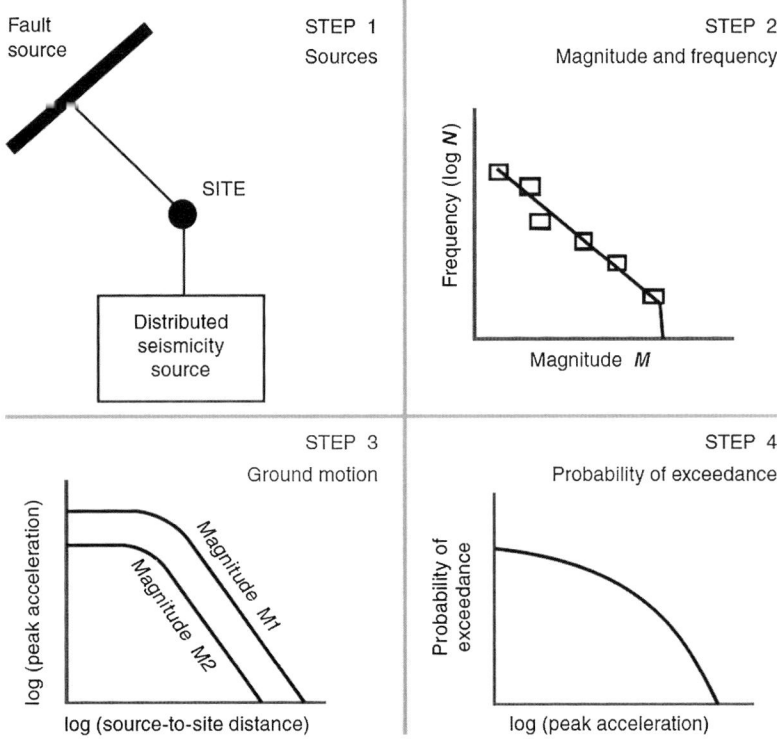

Probabilistic Seismic Hazard Models, Fig. 1 The four steps of probabilistic seismic hazard analysis (After Cornell 1968)

interest. There are two seismic source classes: areal source zone, also referred to as a background or distributed seismicity source (Fig. 1) and fault source. Sources within 100–200 km from the site are usually considered in a PSHA, as sources beyond these distances generally produce ground motions that are too low to be of interest to engineering design.

Areal source zones are used to model the seismicity that is recorded in seismicity catalogues and, in particular, the seismicity that occurs away from known fault sources (e.g., Abrahamson 2011). The most important function of areal sources is therefore to represent sources that lack surface expression and have escaped detection by geologists. Areal source zones actually represent crustal volumes, as they are defined by a geographic area and an assumed thickness based on the maximum depth of seismicity (the seismogenic thickness). Fault sources are commonly modeled as planar features with length and depth that are based on surface fault traces and subsurface information.

Magnitude and Frequency

For fault sources, the dimensions and slip rate and/or paleoseismic (prehistoric) data defining the recurrence behavior of each source is used to develop a magnitude-frequency distribution of earthquakes. The distribution can range from a single magnitude and associated frequency (usually referred to as a characteristic earthquake; Schwartz and Coppersmith 1984) to a range of magnitudes. In the case of the characteristic earthquake, the magnitude is determined by the general equation

$$M_w = \log(A) + b, \quad (1)$$

where A is the fault area and b is a constant that is controlled by factors such as fault type, geology, regional tectonics, and stress drop. Various forms of the equation exist for different fault types and tectonic environments (e.g., Stirling et al. 2013), and there are also relationships that use the length of the fault to determine magnitude.

For areal sources, a range of earthquakes are usually considered for the source, utilizing the well-established Gutenberg-Richter relationship:

$$\text{Log} N(M) = a - bM, \quad (2)$$

where $N(M)$ is the activity rate, M is the magnitude, and a and b are empirical constants. The relationship was found to describe the seismicity of regions across the globe (Gutenberg and Richter 1944). The b-value is typically around 1, which means that there is roughly a tenfold decrease in the number of earthquakes between magnitude M and $M + 1$. The Gutenberg-Richter distributions are typically defined from seismicity catalogue data, and occasionally GPS and geologic data are used as well.

Ground Motion Model

Ground motion prediction equations (GMPEs), formerly called attenuation relations, are used to model the attenuation of ground motions with distance from the source. Various GMPEs have been developed for different tectonic regions (Abrahamson 2011). GMPEs typically account for magnitude, type of faulting, distance from the source, site response, and number of standard deviations. In PSHA the probability that the ground motion S_a exceeds the test value (z) is calculated for each magnitude (M) distance (r), and number of standard deviations (ε).

$$P(S_a > z | M, r, \varepsilon) \quad (3)$$

GMPEs have been developed for shallow crustal earthquakes in active tectonic regions (e.g., California), shallow crustal earthquakes in stable continental regions (e.g., eastern North America), and subduction zone earthquakes (e.g., Japan), rift environments, and others. The most recent suite of GMPEs has been produced by the PEER Next Generation Attenuation project (PEER-NGA; http://peer.berkeley.edu/ngawest). Over time, GMPEs have progressed from being developed for global application to being more focused on tectonic regimes and geographical regions (e.g., New Zealand; McVerry et al. 2006).

Probability of Exceedance Calculation

The final step of PSHA is to develop a hazard curve (Step 4 in Fig. 1), which gives the frequency or probability of exceedance for a suite of ground motion levels. The following equation is the fundamental equation of PSHA that produces the hazard curve:

$$v_i(S_a > z) = N_i(M_{\min}) \int_{r=0}^{\infty} \int_{M_{\min}}^{M_{\max_i}} \int_{\varepsilon_{\min}}^{\varepsilon_{\max}} f_{m_i}(M) f_{r_i}(r) f_{\varepsilon}(\varepsilon)$$
$$P(S_a > z | M, r, \varepsilon) dr dM d\varepsilon$$
$$(4)$$

where $v_i(S_a > z)$ is the annual rate of events (v_i) on a single source that produce a ground motion parameter (S_a) exceeding a specified level (z) at the site of interest. The inverse of v is the return period in years. The hazard is therefore the integration over all possible magnitudes $(f_{m_i}(M))$, distances from the site to the source $(f_{r_i}(r))$, and standard deviations $(f_\varepsilon(\varepsilon))$. The ground motion for each individual scenario is calculated using a GMPE and the probability that the ground motion exceeds the test level is calculated (see "Ground Motion Model").

PSHA is fundamentally a bookkeeping exercise. Instead of developing a small number of deterministic scenarios, a probabilistic seismic hazard model may develop many thousands of scenarios, each with relative contributions to the overall hazard at the site of interest. The rates of ground motions that exceed the specified level z are summed up over all sources to determine how often severe shaking occurs at a site, regardless of the source of the ground motion.

Outputs

Hazard Curve

The hazard curve (Step 4; Fig. 1) gives a suite of ground motion levels and their associated frequencies or return periods. A ground motion level is therefore read off the hazard curve at a user-specified annual frequency, and an important part of the probabilistic seismic hazard modeling process is selecting the appropriate

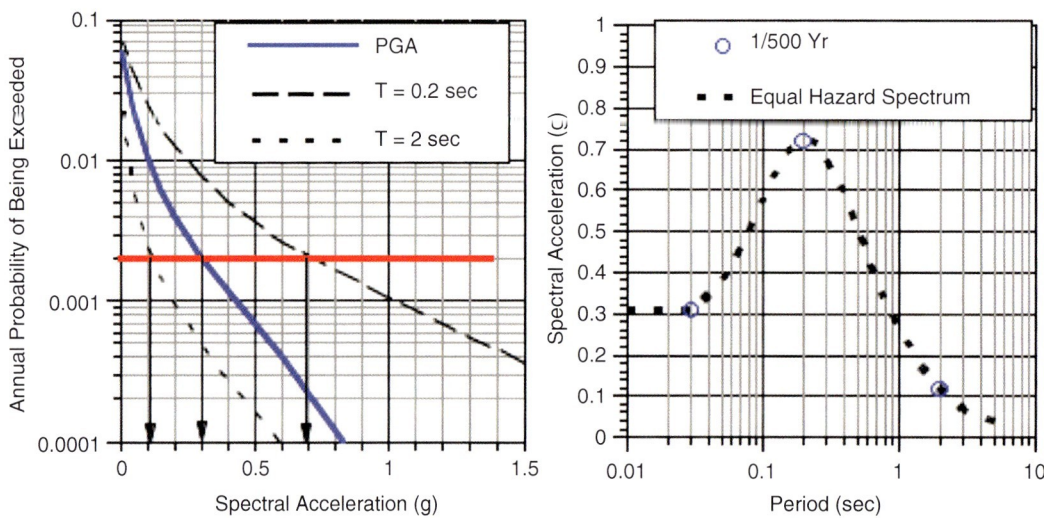

Probabilistic Seismic Hazard Models, Fig. 2 Procedure for developing uniform hazard spectra. In this example a return period of 475 years is used (Abrahamson 2011)

hazard level or return period. The return periods considered for engineering designs typically range from 475 years (often considered for ordinary buildings) to 2,500 years (special buildings, such as hospitals), which are, respectively, equivalent to 10 % and 2 % probability of exceedance in 50 years (e.g., Stirling et al. 2012). In contrast, nuclear facilities and major hydro-dam developments typically consider hazard estimates with 10,000-year return periods or longer. Hazard estimates for these three return periods typically show large quantifiable differences across regions like the USA, Europe, Japan, and New Zealand, reflecting the long-term tectonically driven differences in the expected future activity of earthquake sources across the regions.

Uniform Hazard Spectra

Uniform hazard spectra, or equal hazard spectra (Fig. 2), can be rapidly developed from a probabilistic seismic hazard model to provide seismic design loadings for a range of return periods and spectral periods. The hazard curve (Step 4 in Fig. 1) is plotted for a suite of different spectral periods. At a chosen annual frequency or probability, the spectral acceleration (S_a) for each spectral period is measured from the hazard curve and plotted on a separate graph (Fig. 2). Spectral shapes differ for different sites due to local soil conditions and the different mixes of earthquake magnitudes and distances surrounding the site of interest. The spectra therefore provide meaningful site-specific input to design loadings, including the selection of design earthquake scenarios and associated time histories (actual recordings of earthquakes used in engineering analysis). A response spectrum can also be plotted for a real event (Fig. 3) or a scenario event.

Deaggregation

Because the hazard curve is an ensemble of sources, magnitudes, and distances, it can be difficult to understand the relative contributions to the hazard at a site (Abrahamson 2011). The hazard curve can be broken down, or deaggregated, by magnitude and distance to identify the relative contribution of different earthquake scenarios to the hazard at a site. Similar scenarios are grouped together and the fractional contribution of different scenario groups to the hazard is computed and plotted on a deaggregation graph (Fig. 4). The results of the deaggregation will differ for different return periods and spectral periods. The deaggregation plots are often used to select realistic time histories for input to seismic loading analysis and to design scenario earthquakes for territorial authorities and others to plan for future earthquakes.

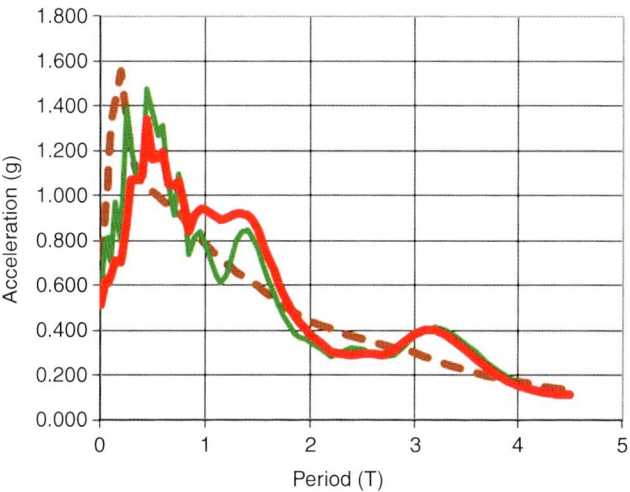

Probabilistic Seismic Hazard Models, Fig. 3 Examples of response spectra for Christchurch, New Zealand, for deep soil site conditions. The *solid lines* are spectral accelerations (SAs) recorded at selected strong motion stations in the city during the M6.2 2011 Christchurch earthquake, and the *dashed line* is a response spectra derived from the New Zealand national seismic hazard model for a 10,000-year return period (Figure courtesy of Graeme McVerry, GNS Science)

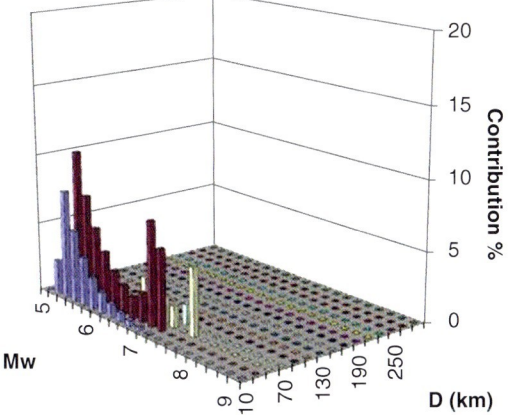

Probabilistic Seismic Hazard Models, Fig. 4 Example of a deaggregation for the city of Christchurch derived from the New Zealand national seismic hazard model (Stirling et al. 2012). The deaggregation plot identifies two relevant classes of earthquakes that dominate the hazard of the city: earthquakes of M5-6.0 at distances of less than 10 km to the city and M6.0-7.5 at distances of 10–50 km. These classes of earthquakes encompass all of the major earthquakes of the Canterbury 2010–2012 earthquake sequence, despite the model being developed prior to initiation of the sequence

Example Applications

Regional, National, and Global Hazard Maps

PSH maps of regions, countries, and the globe are routinely produced by the PSHA process (Fig. 5) at a grid of sites, and then the results are mapped for a given return period. The GSHAP (Global Seismic Hazard Analysis Program, the predecessor of the Global Earthquake Model GEM; globalquakemodel.org) map in Fig. 5a is an example of a global PSH map, which was developed to understand the global distribution of seismic hazard.

Two examples of national-scale PSH maps are from New Zealand (Fig. 5b) and the USA (Fig. 5c). These maps show high hazard along the main plate boundary areas and lower hazard away from the plate boundaries and provide very useful information for engineering and planning, including the development of design standards such as the New Zealand Loadings Standard NZS1170.5 (Standards New Zealand 2004).

At a regional scale, an example from the San Francisco Bay Area is shown in Fig. 6a, b. These

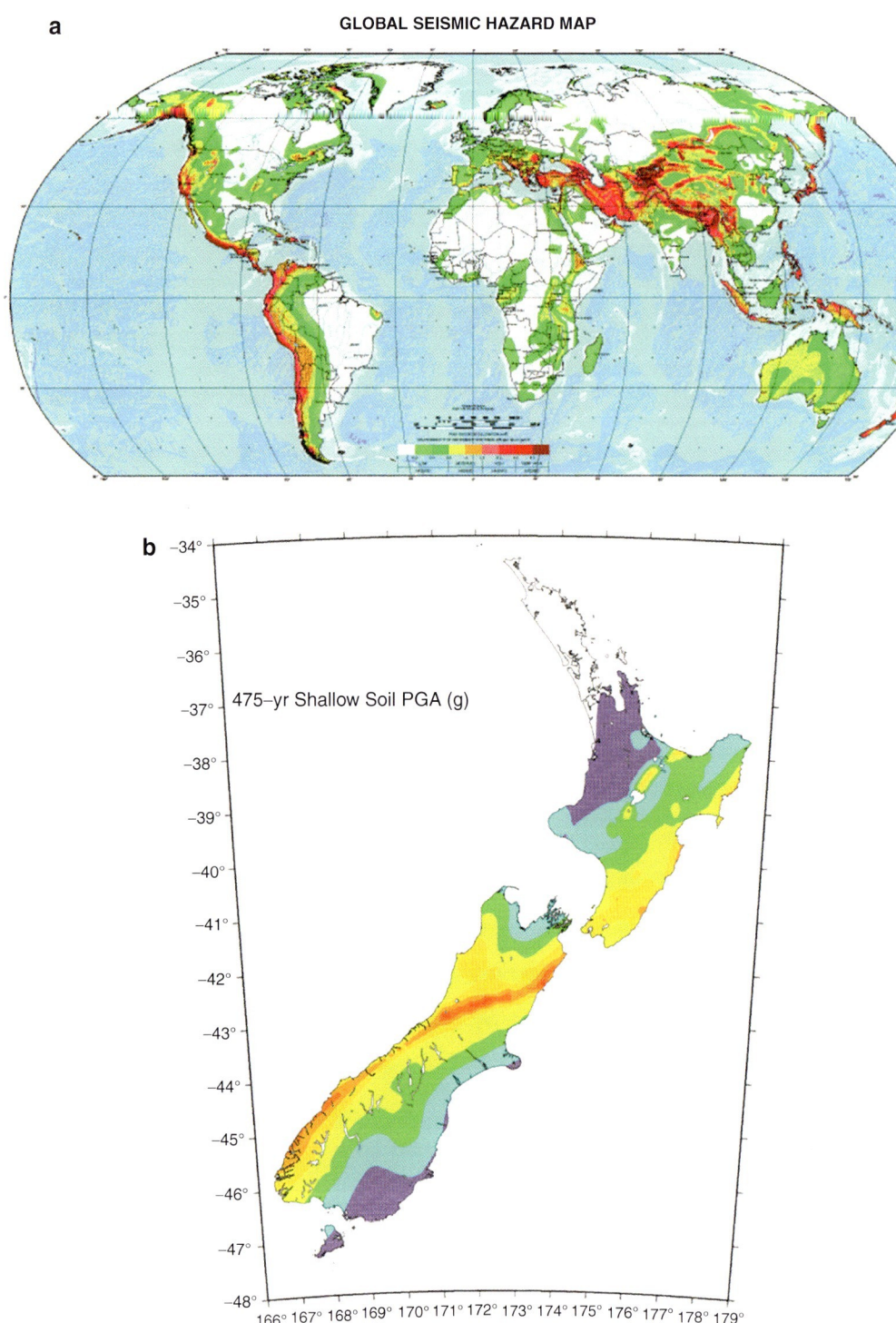

Probabilistic Seismic Hazard Models, Fig. 5 (continued)

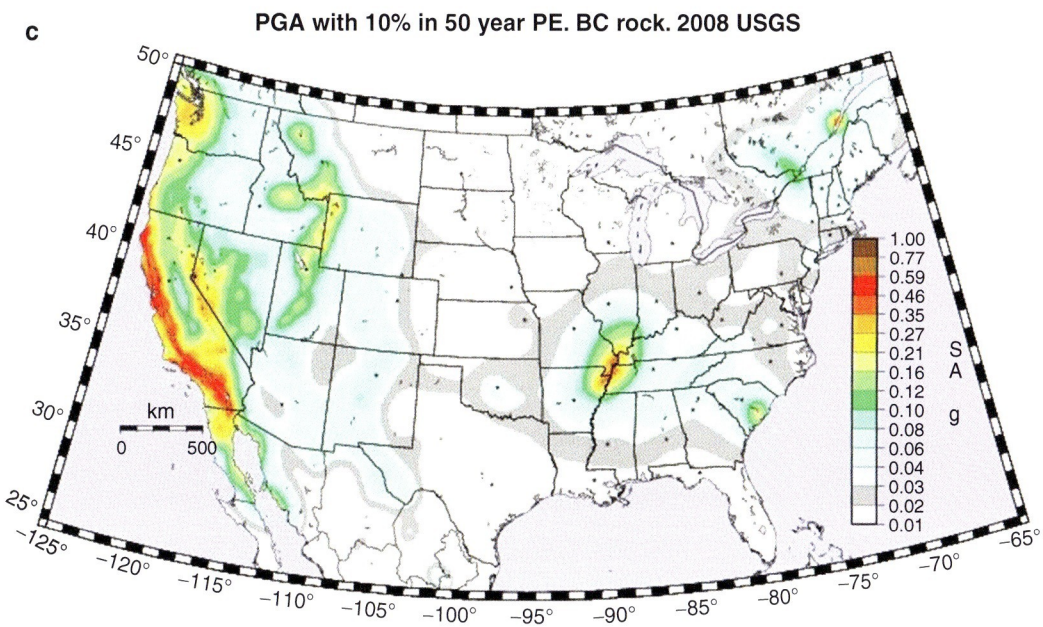

Probabilistic Seismic Hazard Models, Fig. 5 Examples of global and national PSH maps: (**a**) Global seismic hazard analysis program (GSHAP) model (Giardini et al. 1999), (**b**) New Zealand national seismic hazard model (Stirling et al. 2012), and (**c**) US national seismic hazard model (Petersen et al. 2008). Each map shows the peak ground accelerations (*PGA*) expected for a 475-year return period on soft rock sites

PSH maps were developed for use by the public and local governments to help drive hazard mitigation decision-making and policy with appropriate understanding of the likelihood of a significant earthquake event and the expected intensity. The PSHA map was deaggregated across a grid of the region to show the scenario with the highest contribution to the hazard in each location. The deaggregation map provides a guide for selecting the most appropriate earthquake scenarios for a given region or location. Scenario maps are available online for each of the major hazard sources identified in the deaggregation (quake.abag.ca.gov/earthquakes). A legend was developed that links MM (Modified Mercalli) intensity (the earthquake shaking intensity scale that is measured according to its effect on people, objects, and buildings) to expected nonstructural and structural damage of common dwelling types. In this way mitigation decisions by the public are risk informed and appropriate for the expected hazard (Brechwald and Mieler 2013).

Site-Specific Seismic Hazard in Wellington, New Zealand

New Zealand's capital city has long been a focus of site-specific PSHAs in New Zealand. The Wellington region is crossed by a number of major right lateral strike-slip faults and is underlain by the west-dipping subduction interface between the Pacific Plate and overriding Australian Plate (Hikurangi subduction zone) (Holden et al. 2013). In the short historic period of European settlement (ca. 160 years), the region has been shaken by large earthquakes, the largest being the M8.1-8.2 1855 Wairarapa earthquake. This earthquake also stands as the largest historical earthquake to have occurred in New Zealand since European colonization began in 1840. The earthquake was felt over a large part of the North Island and South Island of New Zealand and was severely damaging to settlements in the southern half of the North Island, particularly Wellington and Wanganui (Fig. 7).

Hazard curves for Wellington from the national seismic hazard model (Stirling et al. 2012) are shown for several spectral periods in Fig. 8., The change in hazard as a function of return period is also illustrated by the two peak ground acceleration (PGA) hazard maps in Fig. 9a, b and by graphs of site-specific response spectra for Wellington city (Fig. 10). The highest overall spectrum is associated with the longest return period.

Wellington's 475-year PGA and S_a 1.0 s deaggregation are shown in Fig. 11a, b, respectively. The 475-year PGA hazard is dominantly controlled by fault sources. Peaks on the deaggregation plots show high contributions to overall hazard from the Wellington Fault (M7.5 at less than 1 km; 20 % contribution), Ohariu Fault (M7.6 at 5 km; 20 % contribution), and Wairarapa Fault (M8.1 at 17 km; 13 % contribution). The 475 year S_a 1.0 s graph for Wellington shows an additional contribution to hazard from the local subduction zone (M8.1-9.0 at 23 km; 20 % contribution).

Limitations of Probabilistic Seismic Hazard Models

Recent, devastating earthquakes like the M9.0 2011 Tohoku, Japan, and M6.2 2011

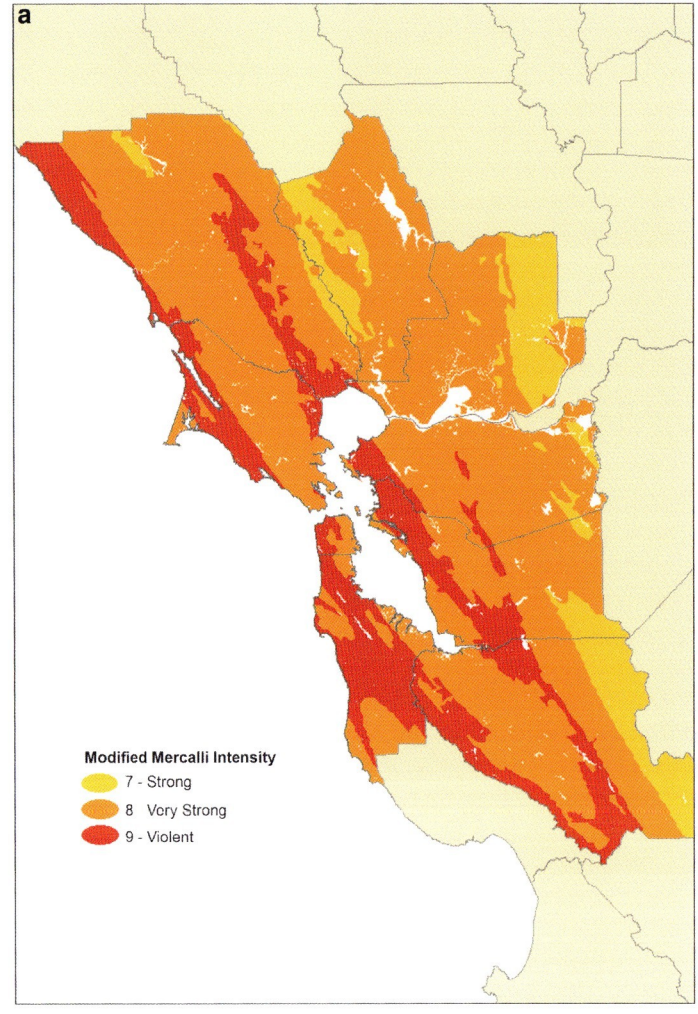

Probabilistic Seismic Hazard Models, Fig. 6 (continued)

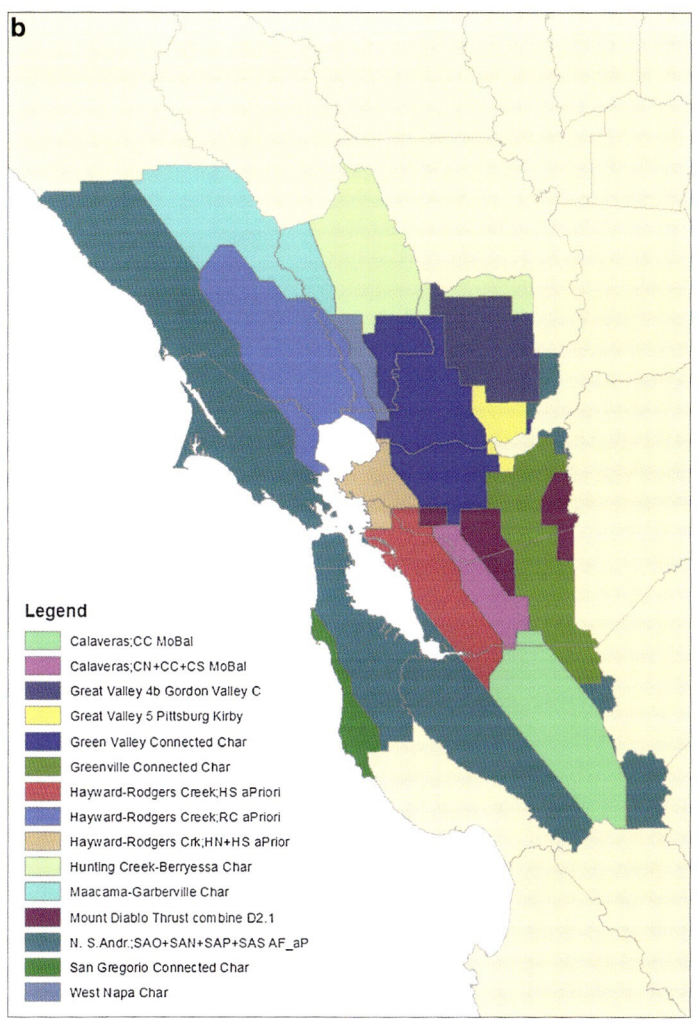

Probabilistic Seismic Hazard Models, Fig. 6 (a) PSH map for the San Francisco Bay Area, California, shown as an example of a regional-scale hazard map. MMI map obtained from 10 % probability in 50-year PGV values (equivalent to 475-year return period) (Brechwald and Mieler 2013). (b) Deaggregation of 10 % in 50-year hazard (a), showing fault scenario with most significant contribution to hazard by location (Brechwald and Mieler 2013)

Christchurch earthquakes have resulted in considerable criticism of PSHA (e.g., Stein et al. 2011). The most frequent criticism is that PSHA did not provide any warning that these events were going to occur in 2011. While this is indeed the case, it is also correct to say that PSH models were never designed to provide short-term earthquake forecasts. The accelerated needs in Japan, New Zealand, and elsewhere to find short-term forecasting solutions are clearly beyond what standard PSHA can provide. Short-term forecasting requires construction of time-dependent or "time-varying" probability models. These models logically require two types of data: (1) detailed knowledge of the earthquake history and prehistory of well-studied faults, so the earthquake recurrence interval and elapsed time since the last earthquake faults can be determined, and (2) high-quality earthquake catalogues, which allow earthquake clustering behavior to be deciphered and modeled with time-varying rate or probability models (e.g., Rhoades et al. 2010).

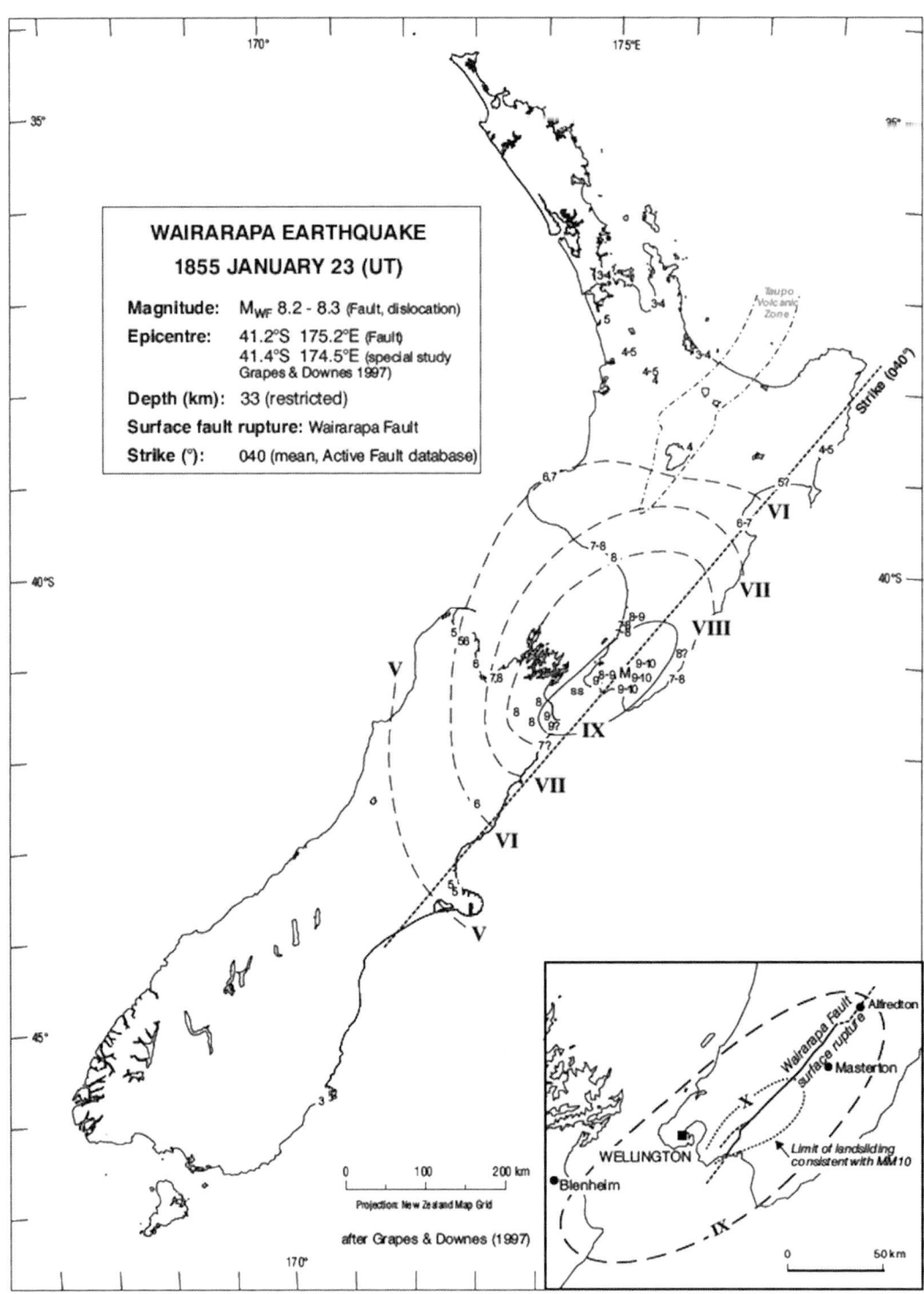

Probabilistic Seismic Hazard Models, Fig. 7 Isoseismal map for the 1855 Wairarapa earthquake (maximum intensity MM9, possibly MM10; Downes and Dowrick 2009)

Probabilistic Seismic Hazard Models, Fig. 8 Example of hazard curves as a result of a PSH analysis for the city of Wellington (data from the Stirling et al. 2012)

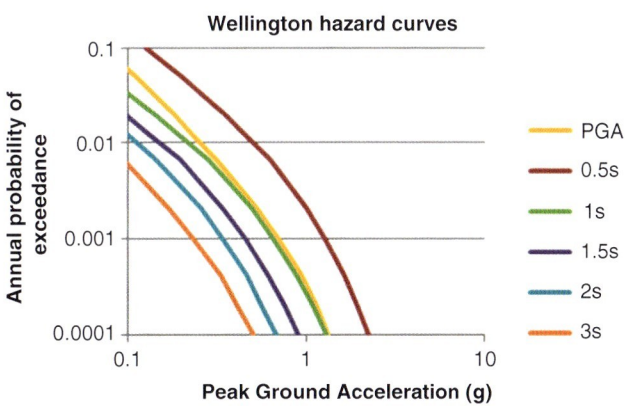

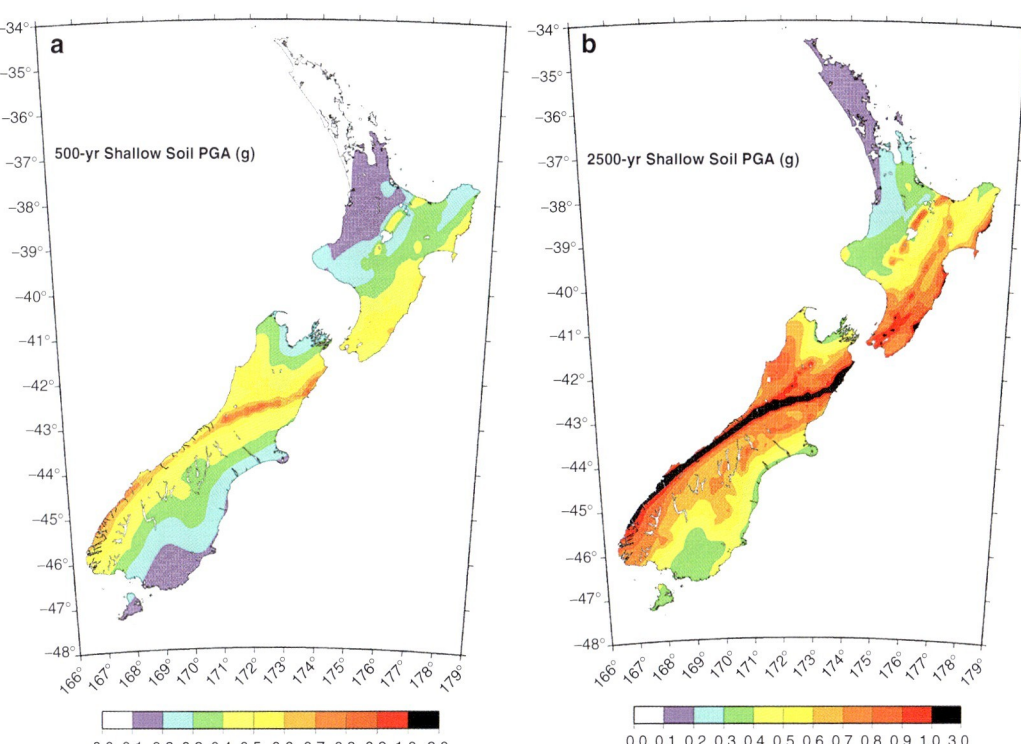

Probabilistic Seismic Hazard Models, Fig. 9 Seismic hazard maps for 475- and 2,500-year return periods (10 % and 2 % probability of exceedance in 50 years) for class C (shallow soil) site conditions: (**a**) peak ground acceleration (PGA) for 475-year return period; (**b**) PGA for 2,500-year return period (Stirling et al. 2012)

However, the resulting models generally differ greatly in terms of the resulting probabilities, and no one model is presently capable of providing a prospective short-term forecast of a large earthquake sequence that suddenly occurs in an area of low seismicity or seismic quiescence (as was the case for the Canterbury earthquake sequence). The ability to provide actual short-term earthquake forecasts in areas of low seismicity still requires some significant advances in relevant scientific research and monitoring/detection.

Ground motion prediction (Step 3 of Fig. 1) is the source of large uncertainties for PSHA.

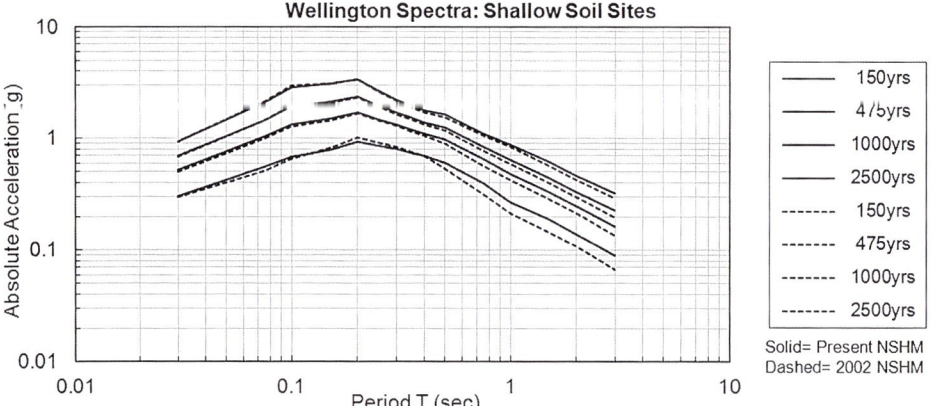

Probabilistic Seismic Hazard Models, Fig. 10 Response spectra for Wellington city for 150, 475, 1,000, and 2,500 years for class C (shallow soil) site conditions (Stirling et al. 2012). *Dashed lines* show the Stirling et al. (2002) spectra for comparison

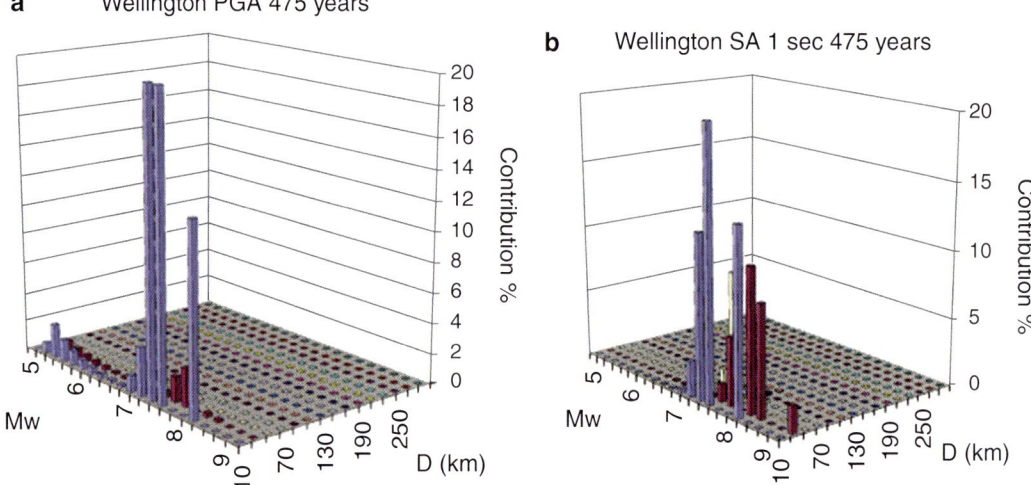

Probabilistic Seismic Hazard Models, Fig. 11 Deaggregation graphs for the city of Wellington for (**a**) 475-year PGA for class C (shallow soil) site conditions and (**b**) 475-year SA 1 s to show the contribution of Hikurangi subduction interface sources at longer spectral periods (Stirling et al. 2012)

GMPEs are typically associated with large standard deviations (about 0.5 in natural log units of ground motion), which represents aleatory (random) uncertainty. Therefore, two earthquakes of the same magnitude and occurring the same distance away from a site can produce hugely different levels of shaking for unknown reason. These ground motion standard deviations do not appear to be reducing despite successive updates of ground motion prediction equations (e.g., Watson-Lamprey 2013). In the Christchurch earthquake, PGAs of over 2 g were produced at some strong motion stations during the earthquake, and this was greatly in excess of what would normally be expected for an earthquake the size of the Christchurch earthquake (M6.2).

Another issue associated with probabilistic seismic hazard models is that earthquakes often occur on sources that were previously unidentified. The causative fault of the main shock of the Canterbury earthquake sequence (the M7.1 2010 Darfield earthquake) was unknown prior to the earthquake on 4 September 2010 due to the long recurrence interval and resulting lack of topographic expression of the fault in the relatively young Pleistocene outwash surface defining the Canterbury Plains. In the national seismic hazard model for New Zealand (Stirling et al. 2012), the earthquake was to some extent accounted for by the areal source model (i.e., consistent in terms of the long-term recurrence interval for Darfield-sized events), but these models do not inform where the earthquake sources and strongest shaking will occur and when the sources will produce the earthquakes. Again, this is an area of science that needs major advances in understanding and monitoring the changes that lead up to the occurrence of a large earthquake.

New Developments

Ground Motion Prediction

Many new GMPEs have been developed in the last decade, capturing the rich strong motion datasets produced by earthquakes in well-instrumented areas (e.g., Taiwan). The PEER-NGA project has involved some of the world's key GMPE developers producing a suite of GMPEs from the same quality-assured strong motion dataset. The models have incorporated more input parameters in an effort to improve ground motion prediction, particularly with respect to source geometry.

Improved Monitoring

Efforts to improve the recording of input data by seismic and GPS networks are of fundamental importance to PSHA. Seismic networks (e.g., Geonet in New Zealand, http://geonet.org.nz) are making large improvements to the detection threshold of earthquakes (the minimum magnitude for a complete record of earthquakes) and the ability to observe temporal and spatial changes in seismicity. GPS is being increasingly used to provide input to source models (e.g., distributed seismicity models and subduction interface models). The generally short temporal coverage of GPS data is compensated for by a large spatial coverage, and as such it can provide a compliment to other source models. Satellite interferometry is another technique that is showing improvements in applicability and resolution over time, and these will allow greater ability to detect the coseismic deformation field (closely related to the source dimension) from earthquakes.

Detection and Characterization of Active Faults

Active fault datasets are the only PSHA input datasets that are able to extend the earthquake record back in time to prehistory. Great improvements in the ability to detect and characterize active faults for input to probabilistic seismic hazard models have been seen in the last 10 years. Fault mapping has improved significantly through accumulated experience and the availability of new tools (e.g., LIDAR). Greater ability to map the surface geometry of faults and distributions of displacement has led to improved characterization of fault sources in PSH models. The use of different disciplines and datasets together for fault characterization, particularly with respect to mapping fault ruptures in three dimensions, has yielded a great deal of understanding of rupture complexity and detail. Furthermore, increased age control on paleoearthquakes has made it possible to establish conditional probabilities of future ruptures and associated uncertainties.

Supercomputing to Consider All Possibilities

PSH models are increasingly drawing on diverse datasets and methods and utilizing high-end computing resources. The UCERF3 model (wgcep.org) incorporates hundreds of thousands of logic tree branches in its comprehensive source model, and access to supercomputers allows the complex model to be run through the four steps of PSHA (Fig. 1). Furthermore, physics-based seismic hazard modeling efforts such as CyberShake (SCEC.org) utilize supercomputers to run multiple realizations of earthquake scenarios from multiple sources, with shaking at the site computed directly from source, path, and site effects for each earthquake. The millions of calculations required would not be possible without access to major computing resources and would not have been possible as recently as a decade ago. Plausible scenarios such as the linking of fault sources to produce extended ruptures and the range of uncertainties in magnitude-frequency statistics for the myriad sources are able to be handled without consideration of CPU demand. Already, exciting scientific results have emerged from the UCERF3 modeling efforts, such as the finding that seismicity on faults may not be well modeled by the Gutenberg-Richter relationship, as this produces a poor fit to the paleoseismic data in California (Ned Field pers. comm. 2013).

Future Needs

Correct Use of PSH Models

Many of the criticisms of PSHA in recent literature have been due to PSH models being used beyond their design capabilities. The models are not designed to be used as short-term forecasting tools, but more appropriately used to estimate seismic hazard for long return periods. The PSH models are also only as good as the input data and generalized methods of source parameterization, ground motion estimation, and probability estimation. However, as long as the associated uncertainties and limitations are fully expressed in the model commentary and appropriate cautionary advice is provided to the end users, the models will continue to provide valuable information. Use of deterministic models and appropriate parameterization of areal source models are two examples of solutions used to compensate for known or suspected deficiencies in PSH models.

Some recent efforts have focused on providing scientific forums to openly debate some of the criticisms leveled at PSHA and the associated input. The American Geophysical Union and Seismological Society of America have held "Earthquake Debates" sessions on several occasions over the last 5 years. Furthermore, the Powell Center for Analysis and Synthesis (http://powellcenter.usgs.gov/) has recently supported a series of workshops that have brought together PSHA experts and critics from around the world, face to face, to address issues associated with maximum magnitude estimation, testability of PSHA, and development of global seismic source models (http://www.nexus.globalquakemodel.org/powell-working-group/). These meetings have been very productive, as people have been working together on common ground rather than talking past each other in the literature.

Earthquake Forecasting Without Earthquake Sequences

Clearly, a major advance in earthquake hazard estimation would be to achieve reliable short-term earthquake forecasts in areas of low seismicity, or in areas experiencing extended periods of seismic quiescence. To this end, future research needs to be focused on improving the ability to monitor and detect microseismicity and crustal deformation and on identifying reliable earthquake precursors. The Canterbury earthquake sequence has resulted in considerable advances being made in the modeling of short-term earthquake probabilities post-mainshock (i.e., aftershocks). These lessons are now being applied to the rebuilding of the city of Christchurch and will be applied to the rest of the country in the coming years. Clearly, these efforts need to be complimented with efforts to identify short-term precursors of future earthquakes in areas that are seismically quiet.

Reduction in Aleatory Uncertainty in Ground Motion Prediction

The aleatory uncertainty in ground motion estimation is very large and does not seem to have been reduced by the increasingly complex GMPEs available today. In other words there is still a very large range in the potential ground motions that could be produced at a single site due to earthquakes of the same magnitude, distance, and slip type. In contrast, the differences between GMPEs (epistemic uncertainties) do appear to have been reduced in recent years, at least within the NGA project. Clearly, effort needs to be focused on better understanding the source, path, and site effects that lead to the large differences in ground motions observed in the strong motion databases.

Testability of Probabilistic Seismic Hazard Models

Finally, efforts need to be supported in the objective testing of PSH models, as to date PSH models have largely been developed in the absence of any form of verification. The Collaboratory for the Study of Earthquake Predictability (CSEP) has been developing testing strategies and methods for a wide variety of applications (SCEC.org), and collaborative work has also been focused on developing ground motion-based tests of the New Zealand and US national seismic hazard models. The Global Earthquake Model (GEM) Foundation (globalquakemodel.org) is including testing and evaluation as an integral part of the overall model development. The Yucca Mountain seismic hazard modeling project developed innovative approaches to consider all viable constraints on ground motions for long return periods for nuclear waste repository storage, prior to cancellation of the project in 2008 (Hanks et al. 2013). The need to verify hazard estimates for return periods of 10^4–10^6 years advanced the use of geomorphic criteria such as fragile geologic features (FGFs) to test the hazard estimates. The rationale is that these FGFs provide evidence for non-exceedance of ground motions for long return periods.

Complex PSHA on Normal Computers

If the future of PSHA is in the development of complex PSH models such as UCERF3, the reliance of these models on supercomputer resources will be a significant barrier to the widespread utility of these models and methods. Significant efforts in the future will therefore need to be focused on making these models usable on standard computers, or uptake will be extremely limited for everyday end-user PSHA applications.

Summary

This entry provides an overview of probabilistic seismic hazard (PSH) models which have provided fundamental input to the engineering, planning, insurance sectors, and other fields for over 30 years. In essence PSH models use the location, size, and occurrence rate of earthquakes to estimate the frequency or probability of damaging or potentially damaging earthquake motions that may occur at a site. By taking into account the frequency of earthquakes as well as the magnitude, the models capture the contribution to seismic hazard from all relevant earthquakes, from the frequent moderate earthquakes (magnitude $5 \leq M < 7$) to the infrequent large to great earthquakes ($M \geq 7$). The entry summarizes the history and fundamental steps of PSHA, provides example applications, discusses strengths and limitations of PSHA, and describes current research and future needs.

Cross-References

▶ Conditional Spectra
▶ Earthquake Recurrence
▶ Earthquake Recurrence Law and the Weibull Distribution
▶ Earthquake Response Spectra and Design Spectra
▶ Earthquake Return Period and Its Incorporation into Seismic Actions
▶ Engineering Characterization of Earthquake Ground Motions
▶ Physics-Based Ground-Motion Simulation
▶ Probability Seismic Hazard Mapping of Taiwan
▶ Review and Implications of Inputs for Seismic Hazard Analysis
▶ Seismic Actions Due to Near-Fault Ground Motion
▶ Seismic Risk Assessment, Cascading Effects
▶ Selection of Ground Motions for Response History Analysis
▶ Site Response for Seismic Hazard Assessment
▶ Spatial Variability of Ground Motion: Seismic Analysis
▶ Spectral Finite Element Approach for Structural Dynamics
▶ Time History Seismic Analysis

References

Abrahamson N (2011) CE 276: seismic hazard analysis and design ground motions [course notes]. Department of Civil Engineering, University of California, Berkeley

Brechwald D, Mieler DH (2013) Sub-regional earthquake hazards and earthquake mapping update. Association of Bay Area Governments, Oakland. http://quake.abag.ca.gov/projects/earthquake-mapping-update/

Cornell CA (1968) Engineering seismic risk analysis. Bull Seismol Soc Am 58(6):1583–1606

Downes GL, Dowrick DJ (2009) Atlas of isoseismal maps of New Zealand earthquakes. GNS science monograph 25. Lower Hutt, New Zealand: GNS Science (Te Pū Ao)

Giardini D, Grunthal G, Shedlock K, Zeng P (1999) The GSHAP global earthquake hazard map. Annali Di Geofisica 42:1225–1230

Gutenberg, Richter (1944) Frequency of earthquakes in California. Bull Seismol Soc Am 34:185–188

Hanks TC, Abrahamson NA, Baker JW, Boore DM, Board M, Brune JN, Cornell CA, Whitney JW (2013) Extreme ground motions and Yucca Mountain: U.S. Geological survey open-file report 2013–1245, 105p. http://dx.doi.org/10.3133/ofr20131245

Holden C, Zhao J, Stirling M (2013) Ground motion modelling of a large subduction interface earthquake in Wellington, New Zealand. In: Proceedings of the New Zealand society of earthquake engineering, annual meeting 2013, Wellington. Paper 7, 8pp

McVerry GH, Zhao JX, Abrahamson NA, Somerville PG (2006) New Zealand acceleration response spectrum attenuation relations for crustal and subduction zone earthquakes. Bull N Z Soc Earthquake Eng 39(1):1–58

Petersen MD, Frankel AD, Harmsen SC, Mueller CS, Haller KM, Wheeler RL, Wesson RL, Zeng Y, Boyd OS, Perkins DM, Luco N, Field EH, Wills CJ, Rukstales KS (2008) Documentation for the 2008 update of the United States national seismic hazard maps: U.S. Geological survey open-file report 2008–1128, 61 pp

Rhoades DA, Van Dissen RJ, Langridge RM, Little TA, Ninis D, Smith EGC, Robinson R (2010) Re-evaluation of the conditional probability of rupture of the Wellington–Hutt valley segment of the Wellington fault. Bull N Z Natl Soc Earthquake Eng 44:77–86

Schwartz DP, Coppersmith KJ (1984) Fault behavior and characteristic earthquakes: examples from the Wasatch and San Andreas Fault Zones. J Geophys Res Solid Earth 89(B7):5681–5698

Standards New Zealand (2004) Structural design actions–Part 5: earthquake actions – New Zealand, New Zealand Standard NZS 1170.5, Department Building and Housing, Wellington

Stein S, Geller R, Liu M (2011) Bad assumptions or bad luck: why earthquake hazard maps need objective testing. Seismol Res Lett 82(5):623–626

Stirling MW, McVerry GH, Berryman KR (2002) A new seismic hazard model for New Zealand. Bull Seismol Soc Am 92:1878–1903

Stirling MW, McVerry GH, Gerstenberger M, Litchfield NJ, Van Dissen R, Berryman KR, Langridge RM, Nicol A, Smith WD, Villamor P, Wallace L, Clark K, Reyners M, Barnes P, Lamarche G, Nodder S, Pettinga J, Bradley B, Rhoades D, Jacobs K (2012) National seismic hazard model for New Zealand: 2010 update. Bull Seismol Soc Am 102(4):1514–1542

Stirling MW, Goded T, Berryman K, Litchfield N (2013) Selection of earthquake scaling relationships for seismic-hazard analysis. Bull Seismol Soc Am 103(6):2993–3011

Watson-Lamprey J (2013) Incorporating the effect of directivity in the intra-event standard deviation of the NGA West 2 ground motion prediction equations. In: Abstracts for annual meeting of the seismological society of America. Seismological Research Letters

Probability Density Evolution Method in Stochastic Dynamics

Jie Li and Jianbing Chen
School of Civil Engineering & State Key Laboratory for Disaster Reduction in Civil Engineering, Tongji University, Shanghai, China

Synonyms

Generalized density evolution equation; Global reliability; Nonlinear stochastic dynamics; PDEM; Stochastic harmonic function; Stochastic response

Introduction

The seismic ground motions are well recognized to be stochastic processes for over 60 years. Under such extreme loadings with large uncertainty, it is almost impossible for engineering structures subjected to earthquakes to avoid nonlinear behaviors during their service life (Roberts and Spanos 1990). Simultaneously, large uncertainties also exist in the models of structures, including the system mechanics parameters and such factors as non-structure effects, boundary conditions, geometric sizes, etc. For instance, the strength of concrete usually has a coefficient of variation (COV) from 10 % to 23 %, whereas even the strength of steel, which is thought to be much more homogeneous, has a COV ranging from 7 % to 9 %. This may induce the fluctuation of static response of structures in the same order of magnitude of COV of the source uncertain parameters. However, the fluctuation in dynamic response may be enlarged greatly. In addition, the coupling of randomness in the system parameters and excitations will make the fluctuation of response much greater than that of response when the randomness is involved only in excitations (Chen and Li 2010). Therefore, to consider the randomness involved in both system parameters and excitations together with their coupling with the development of nonlinearity in structural behaviors is of paramount importance.

Engineering stochastic dynamics has been developed for over half a century, exhibited as two branches, i.e., the random vibration theory and stochastic structural analysis (stochastic finite element method). For linear structures, in both branches the probabilistic information of the second-order statistics could be well obtained (Ghanem and Spanos 1991; Li 1996). In random vibration when the excitations are white noise processes, the joint probability density function (PDF) is governed by the FPK equation, and the solution is well known as a joint Gaussian distribution. But in stochastic structural analysis, where the uncertainty of system parameters is dealt with, no analogous partial differential equation exists in the traditional theory (Li 1996; Ghanem and Spanos 1991). Moreover, in both random-parameter problems and random-excitation problems, huge difficulty exists in dealing with multi-degree-of-freedom (MDOF) nonlinear structures (Goller et al. 2013; Zhu 2006). The coupling of nonlinearity and randomness in MDOF systems is almost unbreakable. This is the common crucial difficulty in both branches.

In the past decade, a family of probability density evolution method (PDEM) was developed. In this method, the thought of physical stochastic systems was advocated (Li and Chen 2009). The principle of preservation of probability was adopted as a unified basis and revisited from the state space description and the random event description (Li and Chen 2008). A decoupled generalized density evolution equation was derived and solved together with the embedded physical equations. By this the instantaneous PDF and reliability of MDOF nonlinear structures with randomness involved in both system parameters and external loadings could be captured (Li and Chen 2003, 2005; Chen and Li 2005; Li et al. 2012a; Goller et al. 2013). This entry will outline its theoretical basis and numerical algorithms and particularly put emphasis on earthquake engineering applications.

Basic Principles of the Probability Density Evolution Method

Without loss of generality, the equation of motion of an n-DOF structure subjected to seismic ground motion is

$$\mathbf{M}(\boldsymbol{\eta})\ddot{\mathbf{X}} + \mathbf{C}(\boldsymbol{\eta})\dot{\mathbf{X}} + \mathbf{f}(\boldsymbol{\eta}, \mathbf{X}) = -\mathbf{M}(\boldsymbol{\eta})\mathbf{I}a_R(\boldsymbol{\xi}, t) \quad (1)$$

where $\ddot{\mathbf{X}}, \dot{\mathbf{X}}, \mathbf{X}$ are the n-dimensional vectors of acceleration, velocity, and displacement relative to ground, respectively; $\mathbf{M}$ and $\mathbf{C}$ are the n by n mass and damping matrices, respectively; $\mathbf{f}$ is the linear or nonlinear restoring forces; $\mathbf{I}$ is the n-dimensional column vector with all components being 1, $a_R(\boldsymbol{\xi}, t)$ is the ground motion accelerogram, which could be specified by the models to be outlined in the later section; $\boldsymbol{\eta} = (\eta_1, \cdots, \eta_{s_1})$ are the basic random parameters in the structural system properties; and $\boldsymbol{\xi} = (\xi_1, \cdots, \xi_{s_2})$ are the basic random parameters in the excitation. For notational convenience, let $\Theta(\varpi) = [\Theta_1(\varpi), \cdots, \Theta_s(\varpi)] = (\boldsymbol{\eta}, \boldsymbol{\xi}) = (\eta_1, \cdots, \eta_{s_1}, \xi_1, \cdots, \xi_{s_2})$, where $s = s_1 + s_2$.

If the state vector $\mathbf{Y} = \left(\mathbf{X}^T, \dot{\mathbf{X}}^T\right)^T = (Y_1, \cdots, Y_{2n})^T$ is introduced, Eq. 1 could be rewritten as a stochastic state equation:

$$\dot{\mathbf{Y}} = \mathbf{A}(\mathbf{Y}, \Theta(\varpi), t) \quad (2)$$

where $\mathbf{A} = (A_1, \cdots, A_{2n})^T = \left(\dot{\mathbf{X}}^T, [-\mathbf{M}^{-1}\mathbf{C}(\boldsymbol{\eta})\dot{\mathbf{X}} - \mathbf{M}^{-1}\mathbf{f}(\boldsymbol{\eta}, \mathbf{X}) - \mathbf{I}a_R(\boldsymbol{\xi}, t)]^T\right)^T$. The initial condition is given by $\mathbf{Y}(0) = \mathbf{Y}_0$.

Let us consider the PDF of $Y_\ell(t)$, the ℓ-th component of $\mathbf{Y}(t)$. Denote the PDF by $p_{Y_\ell}(y,t)$. To understand the evolution of $p_{Y_\ell}(y,t)$, consider the change of probability in an arbitrary interval $[y_L, y_R]$ during the time interval $[t, t + \Delta t]$:

$$\Delta P_D = \int_{y_L}^{y_R} p_{Y_\ell}(y, t+\Delta t)dy - \int_{y_L}^{y_R} p_{Y_\ell}(y,t)dy$$
$$= \left(\int_{y_L}^{y_R} \frac{\partial p_{Y_\ell}(y,t)}{\partial t} dy \right) \Delta t + o(\Delta t) \quad (3)$$

This portion of change of probability is due to the probability transiting through the boundaries y_L and y_R. If during per unit time the probability passing a point z could be denoted by $J(z,t)$, then the change of probability in $[y_L, y_R]$ during $[t, t + \Delta t]$ is given by

$$\Delta P_B = -J(y_R, t)\Delta t + J(y_L, t)\Delta t + o(\Delta t)$$
$$= -\left(\int_{y_L}^{y_R} \frac{\partial J(y,t)}{\partial y} dy \right) \Delta t + o(\Delta t) \quad (4)$$

According to the principle of preservation of probability, $\Delta P_D = \Delta P_B$. Substituting Eqs. 3 and 4 in it and considering the arbitrariness of $[y_L, y_R]$ yield

$$\frac{\partial p_{Y_\ell}(y,t)}{\partial t} = -\frac{\partial J(y,t)}{\partial y} \quad (5)$$

This is nothing but the continuity equation, in which $J(y,t)$ is the flux of probability, i.e., the probability passing a point during per unit time. According to this physical meaning, the flux of probability is $J(y,t) = \lim_{\Delta t \to 0} \frac{\Delta P_y}{\Delta t}$, where ΔP_y is the probability passing the point y during Δt, i.e., $\Delta P_y = p_{Y_\ell}(y,t)\Delta y + o(\Delta y)$, in which Δy is the displacement of the particle during Δt. Note that both $Y_\ell(t)$ and $\dot{Y}_\ell(t)$ depend on $\boldsymbol{\Theta}(\varpi)$. There is

$$\Delta P_y = p_{Y_\ell}(y,t)\Delta y + o(\Delta y)$$
$$= \int_{\Omega_\Theta} \left[\Delta Y_\ell(\boldsymbol{\theta}) p_{Y_\ell|\boldsymbol{\Theta}}(y,t|\boldsymbol{\theta}) \right] p_{\boldsymbol{\Theta}}(\boldsymbol{\theta})d\boldsymbol{\theta} + o(\Delta y)$$
$$= \left(\int_{\Omega_\Theta} \left[\dot{Y}_\ell(\boldsymbol{\theta},t) p_{Y_\ell\boldsymbol{\Theta}}(y,\boldsymbol{\theta},t) \right] d\boldsymbol{\theta} \right) \Delta t + o(\Delta t)$$
$$(6)$$

where $p_{Y_\ell|\boldsymbol{\Theta}}(y,t|\boldsymbol{\theta})$ is the conditional PDF and $p_{Y_\ell\boldsymbol{\Theta}}(y,\boldsymbol{\theta},t)$ is the joint PDF of $(Y_\ell(t), \boldsymbol{\Theta})$. Clearly,

$$p_{Y_\ell}(y,t) = \int_{\Omega_\Theta} p_{Y_\ell\boldsymbol{\Theta}}(y,\boldsymbol{\theta},t)d\boldsymbol{\theta} \quad (7)$$

According to Eq. 6,

$$J(y,t) = \lim_{\Delta t \to 0} \frac{\Delta P_y}{\Delta t} = \int_{\Omega_\Theta} \left[\dot{Y}_\ell(\boldsymbol{\theta},t) p_{Y_\ell\boldsymbol{\Theta}}(y,\boldsymbol{\theta},t) \right] d\boldsymbol{\theta} \quad (8)$$

Substituting Eqs. 7 and 8 in Eq. 5 yields

$$\frac{\partial}{\partial t} \int_{\Omega_\Theta} p_{Y_\ell\boldsymbol{\Theta}}(y,\boldsymbol{\theta},t)d\boldsymbol{\theta} = $$
$$-\frac{\partial}{\partial y} \int_{\Omega_\Theta} \left[\dot{Y}_\ell(\boldsymbol{\theta},t) p_{Y_\ell\boldsymbol{\Theta}}(y,\boldsymbol{\theta},t) \right] d\boldsymbol{\theta} \quad (9)$$

which should hold for any arbitrary domain Ω_Θ, and therefore the integrand should be identical, i.e.,

$$\frac{\partial p_{Y_\ell\boldsymbol{\Theta}}(y,\boldsymbol{\theta},t)}{\partial t} = -\dot{Y}_\ell(\boldsymbol{\theta},t) \frac{\partial p_{Y_\ell\boldsymbol{\Theta}}(y,\boldsymbol{\theta},t)}{\partial y} \quad (10)$$

This is the generalized density evolution equation (GDEE). More rigorous derivation could be found in Li and Chen (2008, 2009) and Chen and Li (2009).

Remark 1 The most important advantage of GDEE compared to the traditional equations, e.g., the FPK equation, is that the dimension is totally untied from the original dynamical system. Actually, although Eq. 10 is in one dimension, from the above heuristic deduction process, it is clear that if any arbitrary number of components are of concern, a corresponding GDEE in appropriate dimensions exists.

Remark 2 In the PDEM there is no need for the stochastic process of concern to be Markovian. Actually, in most cases the process is not Markov. For instance, in engineering practice a complex structure may be modeled by the finite element method where nonlinear constitutive relationship of the material, say, the stochastic damage constitutive law for concrete, is embedded. In this case, usually quite a few of internal variables, say, the damage variables, are involved, and thus the response processes are not Markovian (Li et al. 2014).

Remark 3 In the GDEE, the randomness involved in the system parameters and external loadings is treated simultaneously in a unified way. Traditionally, in random vibration theory and stochastic structural analysis (stochastic finite element method), the methodologies are quite distinct. But as mentioned before in both branches, huge difficulty is encountered for nonlinear MDOF systems.

Stochastic Harmonic Function Representation of Seismic Ground Motions

Random Function Description of Stochastic Processes

Mathematically, a stochastic process $X(t)$ could be regarded as a family of random variables on a parametric set, say $t \in [0, T]$. For a continuous-parameter process, to characterize the probabilistic information of the stochastic process, the finite-dimensional distributions of PDFs, i.e., $p(x_1, t_1), p(x_1, t_1; x_2, t_2), p(x_1, t_1; x_2, t_2; \cdots; x_n, t_n), \cdots$, should be specified. By doing so a stochastic process is regarded as a random function of time, but the dependence of X on t is specified not by an explicit expression of t but in an indirect way of specifying the complete cross-probabilistic information of X at all possible different time instants. This description is complete in mathematics. However, two deficiencies exist for this description at least from the point of view of practical applications (Li et al. 2012b): (i) even if the finite-dimensional distributions are known, how the process X depends on t is still not clear in a sense of physics, but this might be very important for a practical physical problem, and (ii) to capture the high-dimensional distributions of general type other than joint normal distribution by observed information, huge data and prohibitively computational efforts are needed, which is usually impractical either due to lack of data or due to computational difficulty induced by the so-called curse of dimension.

A conceptually more accessible way to a stochastic process is the involvement of an abstract argument representing a sample point, and thus a stochastic process is denoted by $X(\varpi, t)$, where ϖ denotes a sample point in the sample space. By this it is very clear that X is a multi-argument function of ϖ and t. Because X is a function of ϖ, it is "*stochastic*" in nature. Simultaneously because X is also a function of t, it is a "*process*." Thus $X(\varpi, t)$ is a stochastic process. For practical applications the deficiency of this description is that ϖ is a mathematically abstract point in the sample space and its relation to the physical entities is still not exposed. A further step could be made by introducing the embedded basic random variables in the physical problems under consideration, denoted by $\Theta(\varpi) = [\Theta_1(\varpi), \Theta_2(\varpi), \cdots, \Theta_s(\varpi)]$ for convenience, and thus a stochastic process could be represented by $X(\varpi, t) = g(\Theta(\varpi), t)$, where $g(\cdot)$ is an explicit function of $\Theta(\varpi)$ and t. The form of $g(\cdot)$ could be determined by the embedded physical mechanism or by mathematical decomposition if phenomenological statistical models are involved, as shown in the following subsections.

Dynamic excitations encountered in engineering, e.g., earthquakes, wind, and waves, are originated with the embedded physical mechanism (Li et al. 2012b, Lin et al. 2012), although the knowledge on these physical phenomena is still at different levels, some even in very preliminary stage. One of the obstacles to understand these phenomena is the large degree of uncertainty involved and exposed as irregularity in the observed data. However, if the embedded physical mechanism is extracted, then the problem will be understood and captured in a much clearer and easier way. For details, refer to Li et al. (2012b) and Wang and Li (2011).

Representations Based on Mathematical Decomposition

The first two moments, i.e., the mean and the correlation function, could usually capture the major characteristics of a stochastic process. Particularly, these two functions are complete for a Gaussian process. For a zero-mean Gaussian stationary process, the power spectral density function (PSD) is adequate to capture its probabilistic information. These functions essentially belong to phenomenological descriptions although in some cases the physical mechanism is involved when deriving the PSD, e.g., in the Kanai-Tajimi spectrum for ground motions (Tajimi 1960). However, due to its simplicity PSD models are widely employed in most engineering disciplines including earthquake engineering, ocean engineering, wind engineering, etc. Thus, how to represent a stochastic process in time domain by an explicit random function given its PSD is very important. A variety of methods including the Karhunen-Loève decomposition, the spectral representations and improvements were developed (Spanos et al. 2007; Shinozuka and Deodatis 1991; Grigoriu 2002).

The most widely used is the spectral representation method, by which a stochastic process is regarded as the sum of a series of harmonic functions with random phase, i.e.,

$$\hat{X}(t) = \sum_{j=1}^{N} A_j \cos\left(\omega_j t + \phi_j\right) \quad (11)$$

where A_j are deterministic amplitudes; ω_j are deterministic frequencies as inner points, say, uniformly spaced in the interval $[\omega_L, \omega_u]$ over which one-sided PSD is defined; and ϕ_j are independent random variables with identical uniform distribution over $[0, 2\pi]$. Clearly, $\mu_{\hat{X}} = E[\hat{X}(t)] = 0$, and $\sigma_{\hat{X}}^2(t) = E[\hat{X}^2(t)] = \sum_{j=1}^{N} \frac{A_j^2}{2}$. If the target one-sided PSD is $G_X(\omega) = 2S_X(\omega)$ for $\omega \geq 0$ and otherwise $G_X(\omega) = 0$, where $S_X(\omega)$ is the double-sided PSD and symmetric to zero, then

$$\sigma_X^2(t) = \frac{1}{2\pi}\int_{\omega_L}^{\omega_u} G_X(\omega)d\omega \approx \sum_{j=1}^{N} \frac{1}{2\pi} G_X(\omega_j) \Delta\omega_j,$$

where $\Delta\omega_j$ is the length of the j-th frequency subinterval. Letting $\sigma_{\hat{X}}^2(t) = \sigma_X^2(t)$ and comparing the terms one to one lead immediately to $A_j = \sqrt{\frac{1}{\pi} G_X(\omega_j)\Delta\omega_j} = \sqrt{\frac{2}{\pi} S_X(\omega_j)\Delta\omega_j}$.

The properties of the representation in Eq. 11 were elaborately studied in Shinozuka and Deodatis (1991). It should be noted that the expression of the amplitude may vary in different literature due to the different position of the factor $\frac{1}{2\pi}$ in the Fourier transform. The spectral representation method is very simple and straightforward, but usually hundreds or even thousands of terms have to be retained. This leads to a large number of random variables, which induces difficulty in practice. To reduce the number of random variables is of great importance (Spanos et al. 2007).

A modification is to randomize the frequencies, and thus Eq. 11 is modified to

$$\tilde{X}(t) = \sum_{j=1}^{N} A(\tilde{\omega}_j) \cos\left(\tilde{\omega}_j t + \phi_j\right) \quad (12)$$

where $\tilde{\omega}_j$ are now random variables. For simplicity, assume the subintervals over which $\tilde{\omega}_j$ distribute are not overlapping, and construct a partition of $[\omega_L, \omega_u]$, i.e., the support of PDF of $\tilde{\omega}_j$ is $[\omega_{j-1}^{(p)}, \omega_j^{(p)}]$, $\omega_0^{(p)} = \omega_L$, $\omega_N^{(p)} = \omega_u$, $\omega_0^{(p)} < \omega_1^{(p)} \cdots < \omega_{N-1}^{(p)} < \omega_N^{(p)}$, and there are $\cup_{j=1}^{N}[\omega_{j-1}^{(p)}, \omega_j^{(p)}] = [\omega_L, \omega_u]$ and $[\omega_{j-1}^{(p)}, \omega_j^{(p)}) \cap [\omega_{L-1}^{(p)}, \omega_k^{(p)}) = \emptyset$. In this case, there is

$$\sigma_{\tilde{X}}^2(t) = E[\tilde{X}^2(t)] = \sum_{j=1}^{N} \frac{1}{2} E[A^2(\tilde{\omega}_j)] \quad (13)$$

and $\sigma_X^2(t) = \frac{1}{2\pi}\int_{\omega_L}^{\omega_u} G_X(\omega)d\omega = \sum_{j=1}^{N} \frac{1}{2\pi} \int_{\omega_{j-1}^{(p)}}^{\omega_j^{(p)}} G_X(\omega)d\omega$. Letting $\sigma_{\tilde{X}}^2(t) = \sigma_X^2(t)$ and making the terms identical one to one lead to $E[A^2(\tilde{\omega}_j)] = \frac{1}{\pi}\int_{\omega_{j-1}^{(p)}}^{\omega_j^{(p)}} G_X(\omega)d\omega$. If the PDF of $\tilde{\omega}_j$ is $p_{\tilde{\omega}_j}(\omega)$, then it follows that $\int_{\omega_{j-1}^{(p)}}^{\omega_j^{(p)}} A^2(\omega) p_{\tilde{\omega}_j}(\omega)d\omega = \frac{1}{\pi}\int_{\omega_{j-1}^{(p)}}^{\omega_j^{(p)}} G_X(\omega)d\omega$, which leads to

$$A(\tilde{\omega}_j) = \sqrt{\frac{G_X(\tilde{\omega}_j)}{\pi p_{\tilde{\omega}_j}(\tilde{\omega}_j)}} \qquad (14)$$

The representation in Eq. 12 is called the stochastic harmonic function representation (SHF) and is of great flexibility by choosing different PDFs for the randomized frequencies (Chen et al. 2013). Particularly, if the PDF of $\tilde{\omega}_j$ takes the shape of PSD, i.e., $p_{\tilde{\omega}_j}(\omega) = G_X(\omega)/\int_{\omega_{j-1}^{(p)}}^{\omega_j^{(p)}} G_X(\omega)d\omega$, then Eq. 14 becomes $A(\tilde{\omega}_j) = \sqrt{\frac{1}{\pi}\int_{\omega_{j-1}^{(p)}}^{\omega_j^{(p)}} G_X(\omega)d\omega}$. This is called the SHF of the first kind (SHF-I). If $\tilde{\omega}_j$ follows the uniform distribution over $[\omega_{j-1}^{(p)}, \omega_j^{(p)}]$, then from Eq. 14 there is $A(\tilde{\omega}_j) = \sqrt{\frac{1}{\pi}G_X(\tilde{\omega}_j)\left(\omega_j^{(p)} - \omega_{j-1}^{(p)}\right)}$. This is called the SHF of the second kind (SHF-II).

What should be stressed is that the SHF representations are exact in the sense of reproducing the target PSD exactly. This is superior to the spectral representation method which is an approximate approach. Besides, studies show that although any arbitrary number of components could reproduce the target PSD, usually about 7–10 components are adequate considering the shape of the samples and the one-dimensional PDF of the generated process.

Numerical Algorithms

Although for some simple systems the analytical solution of GDEE is possible, for most practical problems, numerical algorithms are needed.

Procedures of Solution

The task is to obtain the instantaneous PDF $p_{Y_\ell}(y, t)$. For this purpose, the GDEE (Eq. 10) should be firstly solved under appropriate initial and boundary conditions which usually take

$$\left.\begin{aligned} p_{Y_\ell\Theta}(y, \boldsymbol{\theta}, t)\big|_{t=t_0} &= \delta(y - y_{\ell,0})p_\Theta(\boldsymbol{\theta}) \\ p_{Y_\ell\Theta}(y, \boldsymbol{\theta}, t)\big|_{y\to\pm\infty} &= 0 \end{aligned}\right\} \qquad (15)$$

if the initial value $y_{\ell,0}$ is deterministic and independent of the random parameters. After Eq. 10 is solved Eq. 7 is employed to yield $p_{Y_\ell}(y, t)$. However, to solve the GDEE (Eq. 10), the coefficient $\dot{Y}_\ell(\boldsymbol{\theta}, t)$ should be determined first. This comes from the embedded physical mechanism and is the solution of the physical equation (Eq. 1, or equivalently Eq. 2). Accordingly, the procedure of solution includes the following steps:

Step 1. Specify a representative point set $Q = \{\boldsymbol{\theta}_q = (\theta_{1,q}, \cdots, \theta_{s,q}) | \boldsymbol{\theta}_q \in \Omega_\Theta, q = 1, 2, \cdots, n_{pt}\}$ and the corresponding assigned probabilities P_q, $q = 1, 2, \cdots, n_{pt}$, where n_{pt} is the total number of selected points and $0 < P_q < 1$, $\sum_{q=1}^{n_{pt}} P_q = 1$. This could be regarded as a first partial discretization of Eq. 10.

Step 2. For each specified $\{\Theta = \boldsymbol{\theta}_q\}$, carry out deterministic analysis, say, the time integration method (Belytschko et al. 2000), to solve the physical equation (Eq. 1, or Eq. 2) to yield $\dot{Y}_\ell(\boldsymbol{\theta}_q, t)$.

Step 3. Substitute $\dot{Y}_\ell(\boldsymbol{\theta}_q, t)$ in the GDEE (Eq. 10) and solve it under the boundary and initial conditions (Eq. 15) by, say, the finite difference method with TVD scheme to yield the numerical result of $p_{Y_\ell\Theta}(y, \boldsymbol{\theta}_q, t)$.

Step 4. Make a summation according to Eq. 7 to yield the numerical results of the instantaneous PDF $p_{Y_\ell}(y, t)$, i.e., $p_{Y_\ell}(y, t) = \sum_{q=1}^{n_{pt}} p_{Y_\ell\Theta}(y, \boldsymbol{\theta}_q, t)$.

Optimal Selection of Representative Points

The point selection in Step 1 is of paramount importance to the efficiency and accuracy of PDEM. A consistent treatment is based on the partition of probability assigned space (Li et al. 2012a).

Let Ω_q denote the representative space of the point $\boldsymbol{\theta}_q$. Ω_qs construct a partition of Ω_Θ, i.e., $\cup_{q=1}^{n_{\text{pt}}} \Omega_q = \Omega_\Theta$ and $\Pr\{\boldsymbol{\varpi} \in \Omega_q \cap \Omega_k\} = 0$ for any $q \neq k$, where $\Pr\{\cdot\}$ denotes the probability of the event. The assigned probability of the point $\boldsymbol{\theta}_q$ is then specified by $P_q = \int_{\Omega_q} p_\Theta(\boldsymbol{\theta})d\boldsymbol{\theta}$, which clearly satisfies $0 < P_q < 1$ and $\sum_{q=1}^{n_{\text{pt}}} P_q = 1$. For problems with $2 \leq s \leq 4$, the method based on tangent spheres performs well, while for problems with $2 \leq s \leq 18$, the number theoretical method together with hyper-ball sieving could be employed (Li and Chen 2009). A versatile approach based on the generalized F-discrepancy (GF-discrepancy) was recently developed for general nonuniform, non-normal distributions.

Let the marginal cumulative distribution function (CDF) of the basic random variable Θ_j, $j = 1, 2, \cdots, s$ be $F_j(\theta) = \int_{-\infty}^{\theta} p_{\Theta_j}(x)dx$. Denote the empirical marginal CDF of Θ_j by $\tilde{F}_j(\theta; Q) = \sum_{q=1}^{n_{\text{pt}}} P_q I\{\theta_{j,q} < \theta\}$, where $I\{\cdot\}$ is the indicator function of which the value is 1 when the bracketed event is true, and otherwise zero, $\theta_{j,q}$ is the j-th component of $\boldsymbol{\theta}_q$. By these a GF-discrepancy could be introduced:

$$D_{\text{GF}}(Q) = \max_{1 \leq j \leq s} \left\{ \sup_{-\infty \leq \theta \leq \infty} |\tilde{F}_j(\theta; Q) - F_j(\theta)| \right\} \quad (16)$$

It is demonstrated that the smaller the GF-discrepancy, the better the performance of the point set in the sense that the error of the involved high-dimensional integral is smaller (Chen and Zhang 2013). Therefore, the task is now to find some point set Q minimizing the GF-discrepancy, i.e.,

$$Q = \arg[\min D_{\text{GF}}(Q)]$$
$$= \arg\left[\min\left(\max_{1 \leq j \leq s}\left\{\sup_{-\infty \leq \theta \leq \infty} |\tilde{F}_j(\theta; Q) - F_j(\theta)|\right\}\right)\right] \quad (17)$$

Plenty of optimization methods could be adopted to complete this task and yield the optimal representative point set.

Finite Difference Method with TVD Scheme

In Step 3 a series of partial differential equations are solved. These equations are of the form

$$\frac{\partial p(y,t)}{\partial t} + a \frac{\partial p(y,t)}{\partial y} = 0 \quad (18)$$

for different values of a in the case of different $\boldsymbol{\theta}_q$. Quite a few difference schemes were developed. According to computational experiences the schemes with TVD (total variation diminishing) property are satisfactory (Li and Chen 2009). One of such schemes that is adopted is as follows:

$$p_j^{(k+1)} = p_j^{(k)} - \frac{1}{2}(\lambda a - |\lambda a|)\Delta p_{j+\frac{1}{2}}^{(k)} - \frac{1}{2}(\lambda a + |\lambda a|)\Delta p_{j-\frac{1}{2}}^{(k)}$$
$$- \frac{1}{2}\left(|\lambda a| - |\lambda a|^2\right)\left(\psi_{j+\frac{1}{2}}\Delta p_{j+\frac{1}{2}}^{(k)} - \psi_{j-\frac{1}{2}}\Delta p_{j-\frac{1}{2}}^{(k)}\right) \quad (19)$$

where $p_j^{(k)}$ denotes $p(x_j, t_k)$, $x_j = j\Delta x$, $j = 0, \pm 1, \pm 2, \cdots$, $t_k = k\Delta t$, $k = 0, 1, 2, \cdots$, and Δx and Δt are the space and time step, respectively; the mesh ratio $\lambda = \Delta t/\Delta x$; $\Delta p_{j+\frac{1}{2}}^{(k)} = p_{j+1}^{(k)} - p_j^{(k)}$; and $\Delta p_{j-\frac{1}{2}}^{(k)} = p_j^{(k)} - p_{j-1}^{(k)}$. The factors $\psi_{j+\frac{1}{2}}, \psi_{j-\frac{1}{2}}$ are called the flux limiters, which are related to the irregularity of the curve of $p(x, t)$ in terms of x. As $\psi_{j+\frac{1}{2}} = \psi_{j-\frac{1}{2}} \equiv 1$ the scheme in Eq. 19 reduces to the well-known Lax-Wendroff scheme, whereas as $\psi_{j+\frac{1}{2}} = \psi_{j-\frac{1}{2}} \equiv 0$ the scheme in Eq. 19 reduces to the one-sided scheme. To achieve the TVD property, $\psi_{j+\frac{1}{2}}, \psi_{j-\frac{1}{2}}$ should depend on the parameters characterizing the irregularity of $p(x, t)$, i.e.,

$$r_{j+\frac{1}{2}}^{+} = \frac{\Delta p_{j+\frac{3}{2}}^{(k)}}{\Delta p_{j+\frac{1}{2}}^{(k)}} = \frac{p_{j+2}^{(k)} - p_{j+1}^{(k)}}{p_{j+1}^{(k)} - p_j^{(k)}},$$
$$r_{j+\frac{1}{2}}^{-} = \frac{\Delta p_{j-\frac{1}{2}}^{(k)}}{\Delta p_{j+\frac{1}{2}}^{(k)}} = \frac{p_j^{(k)} - p_{j-1}^{(k)}}{p_{j+1}^{(k)} - p_j^{(k)}} \quad (20)$$

The flux limiter $\psi_0(r) = \max(0, \min(2r, 1), \min(r, 2))$ could be recommended and then

$$\psi_{j+\frac{1}{2}}\left(r_{j+\frac{1}{2}}^{+}, r_{j+\frac{1}{2}}^{-}\right) = u(-a)\psi_0\left(r_{j+\frac{1}{2}}^{+}\right) + u(a)\psi_0\left(r_{j+\frac{1}{2}}^{-}\right) \quad (21)$$

where $u(\cdot)$ is Heaviside's function, of which the value is 1 if the argument is greater than 0 and is zero otherwise.

First-Passage Reliability and Global Reliability

First-Passage Reliability

The first-passage reliability is of great concern to engineering structures exposed to disastrous dynamic excitations. For instance, if the first-passage failure is defined in terms of $Y_\ell(t)$, the reliability is then

$$R(t) = \Pr\{Y_\ell(\Theta, \tau) \in \Omega_s, \text{ for } 0 \leq \tau \leq t\} \quad (22)$$

Note that the evolution of the probability density of $(Y_\ell(t), \Theta)$ is governed by Eq. 10. Eq. 22 means that once the probability outcrosses the boundary of the safety domain, then it will not return to the safety domain. This could be expressed mathematically as an absorbing boundary condition

$$p_{Y_\ell \Theta}(y, \boldsymbol{\theta}, t)\big|_{y \in \Omega_f} = 0 \quad (23)$$

where Ω_f is the failure domain. This of course is also equivalent to imposing the absorbing boundary condition $p_{Y_\ell}(y,t)\big|_{y \in \Omega_f} = 0$ for Eq. 5. Replacing the boundary condition in Eq. 15 by Eq. 23 and carrying out the procedures in the preceding section will yield the "remaining" PDF $\tilde{p}_{Y_\ell}(y, t)$, and thus the reliability is given by

$$R(t) = \int_{\Omega_s} \tilde{p}_{Y_\ell}(y, t) dy = \int_{-\infty}^{\infty} \tilde{p}_{Y_\ell}(y, t) dy \quad (24)$$

Global Reliability

Traditionally, it is called the system reliability if more than one failure modes are involved. However, due to the fact that the failure of a global structure is usually dependent on the whole process of development of nonlinearity in the system, the reliability should be captured in this whole process involving randomness. In this sense, it is more appropriate to call it the global reliability of a structure (Chen and Li 2007; Li and Chen 2009). For instance, if the failure of the global structure is determined by some combinations or even functional of $Y_\ell(\Theta, t)$ and $Y_k(\Theta, t)$, say, the Park-Ang model (Park and Ang 1985), then a corresponding equivalent process, denoted as $Z(t) = f[Y_k(\Theta, \tau), Y_k(\Theta, \tau); 0 \leq \tau \leq t]$ for convenience, as some kind of compound or even functional of $Y_\ell(\Theta, t)$ and $Y_k(\Theta, t)$, could be constructed such that the global reliability is given by $R_G(t) = \Pr\{Z(\Theta, \tau) \in \Omega_s, \text{ for } 0 \leq \tau \leq t\}$. Then the method similar to that described in the above subsection could be adopted, or the PDF of the equivalent extreme value $\underset{0 \leq \tau \leq t}{\text{ext}} Z(\Theta, \tau)$ could be obtained, and then the global reliability is yielded (Li et al. 2007, Li & Chen 2009).

Numerical Example of Applications

To illustrate the PDEM, a 10-story floor-shear structure with lumped masses on the floors subjected to stochastic ground motion is studied. To be simple, all the system parameters are taken as deterministic variables. The lumped masses from bottom to top are 1.5, 1.5, 1.5, 1.4, 1.4, 1.4, 1.3, 1.3, 1.3, and 0.7 ($\times 10^5$ kg), respectively; the inter-story stiffness from bottom to top are 4.25, 4.25, 4.25, 4.25, 4.0, 4.0, 4.0, 4.0, 3.5, and 3.5 ($\times 10^{10}$ N/m). Rayleigh damping is adopted, i.e., the damping matrix $\mathbf{C} = a\mathbf{M} + b\mathbf{K}$, where $\mathbf{M}$ and $\mathbf{K}$ are the mass and stiffness matrix, and $a = 0.01$ and $b = 0.005$. The Bouc-Wen model is taken to characterize the hysteretic behavior of the restoring force (Ma et al. 2004; Li and Chen 2009), where the basic parameters take $A = 1$, $n = 1$, $q = 0$, $p = 600$, $d_\psi = 0$, $\lambda = 0.5$, $\psi = 0.2$, $\beta = 60$, $\gamma = 10$, $d_\nu = 200$, $d_\eta = 200$, and $\zeta = 0.95$. For the linear system $\alpha = 1$, and for the nonlinear system $\alpha = 0.01$.

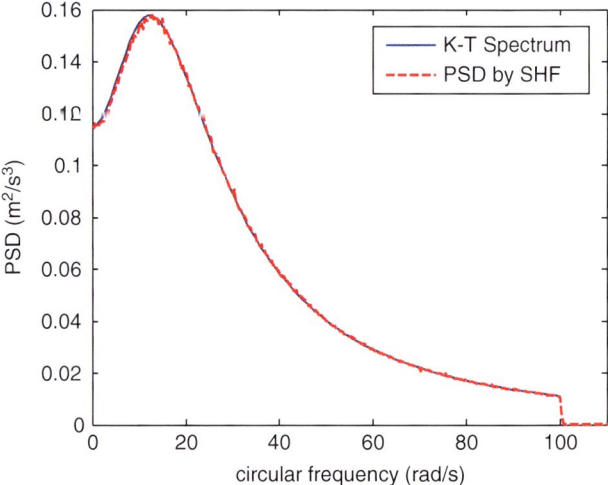

Probability Density Evolution Method in Stochastic Dynamics, Fig. 1 Target and reproduced PSD

The Kanai-Tajimi spectrum is taken as the target spectrum of the stochastic ground motions, i.e.,

$$S(\omega) = \frac{1 + 4\zeta_0^2(\omega/\omega_0)^2}{(1 - (\omega/\omega_0))^2 + 4\zeta_0^2(\omega/\omega_0)^2} S_0 \quad (25)$$

where $S(\omega)$ is the PSD, $\omega_0 = 16.9$, $\zeta_0 = 0.94$, and $\omega_u = 100$ rad/s. S_0 is such determined that the standard deviation of the ground motion acceleration is 1.0 m/s^2.

In Fig. 1 shown is the target PSD in Eq. 25 (labeled as "K-T spectrum") and the PSD reproduced by the SHF-II including 20 terms. It is seen that with only 20 SHF components, the PSD could be reproduced in high accuracy. This will also result with high accuracy in the second-order statistics of response of the linear system, but for the nonlinear structure, it seems that more terms should be retained. Figure 2 shows typical curves of restoring force vs. inter-story drift, demonstrating that in the nonlinear case strong hysteresis is involved. In Fig. 3 pictured are the mean and standard deviation of response (top displacement) of linear and nonlinear structures by PDEM and 20,000 times of Monte Carlo simulations (labeled as "MCS"), respectively. In this paper 800 representative time histories generated by SHF-II are adopted. Computational experiences show that this number could still be reduced. Theoretically, the mean of linear and nonlinear structures should be zero due to the symmetry of distribution of the input ground acceleration process. This is clearly verified from the upper figure of Fig. 3, which shows that the mean by the PDEM is quite small in comparison to the standard deviation (the coordinate in the upper figure of Fig. 3 is in the order of magnitude of the standard deviation). Also it is seen that the standard deviation of the nonlinear response is smaller than that of the linear structure. Pictured in Fig. 4 are the PDF evolution surfaces during an identical period for linear and nonlinear systems. Note that if the input is Gaussian, the output should also be Gaussian for linear systems. Clear differences between the PDFs of linear and nonlinear responses could be observed from the figures.

It is noted that by PDEM the global reliability of the structure could also be obtained by imposing absorbing conditions. In Fig. 5 shown are the time-variant first-passage reliabilities of the nonlinear structure in ordinary and logarithmic coordinates for different threshold values by PDEM and Monte Carlo simulation (20,000 times). The thresholds 0.4, 0.3, 0.25, and 0.2 m correspond to the displacement angle of 1/75, 1/100, 1/120, and

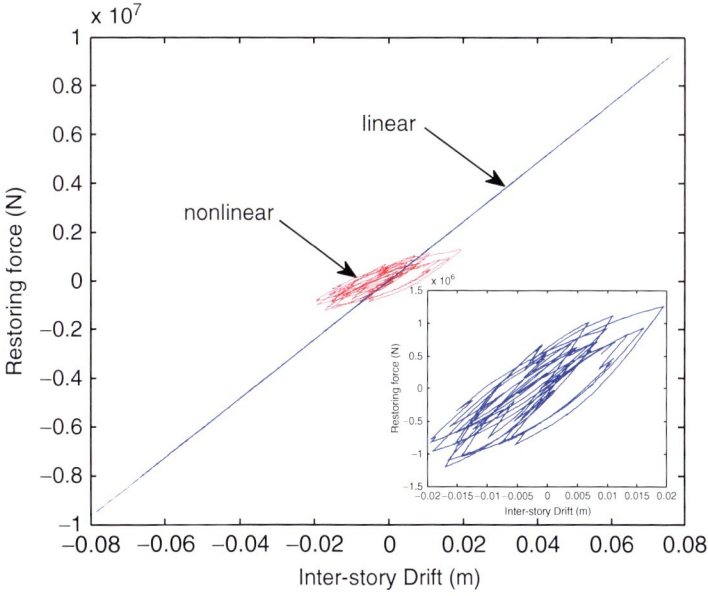

Probability Density Evolution Method in Stochastic Dynamics, Fig. 2 Typical restoring force

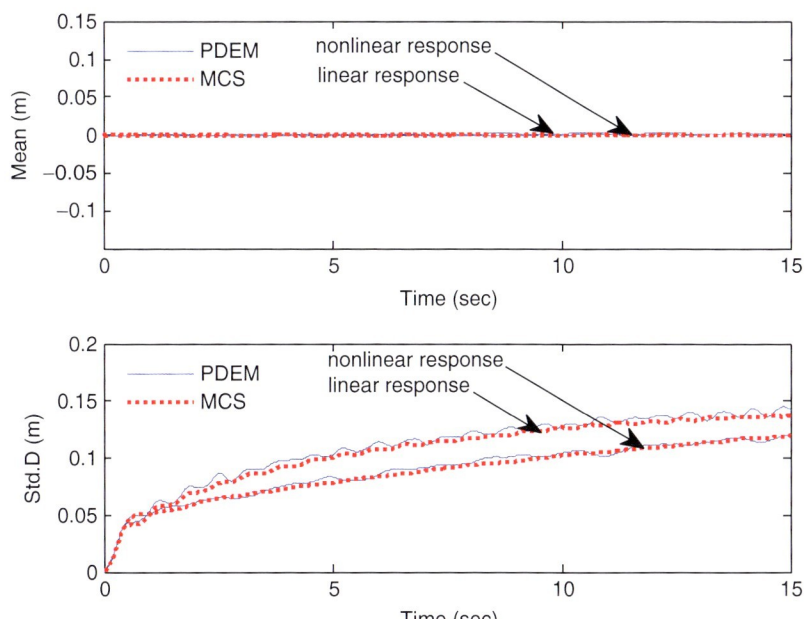

Probability Density Evolution Method in Stochastic Dynamics, Fig. 3 The mean and standard deviation

1/150, respectively. It is shown that the reliability decreases almost continuously from some time monotonically, but noticeably not in an exponential way as predicted by the Poisson assumption. Actually, from Fig. 5b it is seen that the absolute value of slope of the logarithmic reliability is increasing, which means that the hazard rate of the structure exhibiting nonlinear behaviors is increasing as the time elapses.

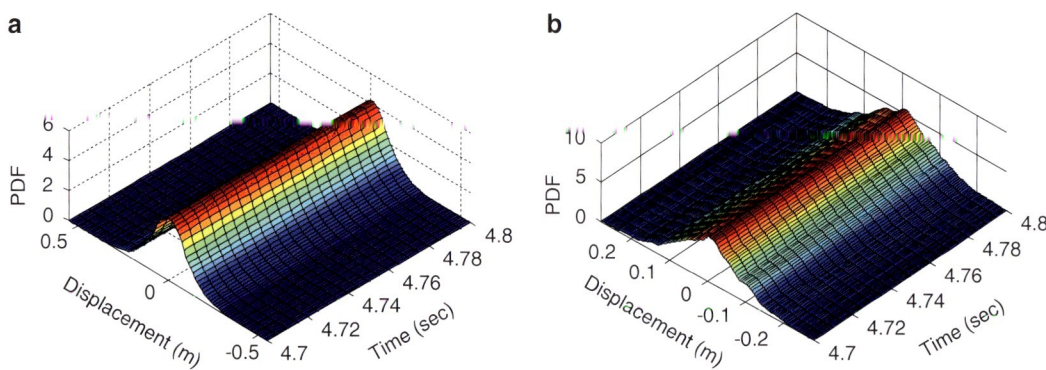

Probability Density Evolution Method in Stochastic Dynamics, Fig. 4 PDF evolution surface. (a) Linear structure; (b) nonlinear structure

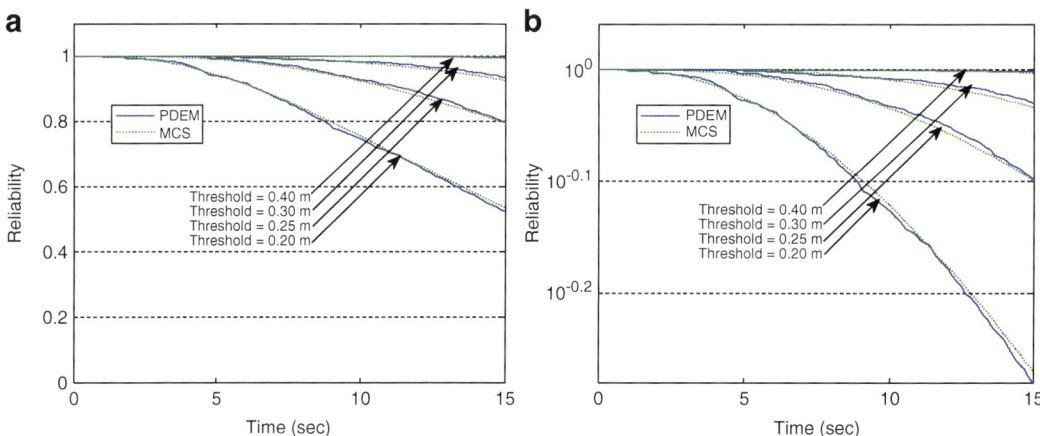

Probability Density Evolution Method in Stochastic Dynamics, Fig. 5 Reliability of the nonlinear structure. (a) Ordinary coordinate; (b) logarithmic coordinate

Summary

The probability density evolution method is outlined and illustrated in this entry. It is concluded that (1) the thought of physical stochastic systems provides a new perspective to stochastic dynamics and (2) the probability density evolution method shows its versatility in stochastic dynamics, particularly for MDOF nonlinear systems subjected to non-white noise excitations. However, improvements and extension of the physical stochastic models of dynamic excitations and more robust and efficient numerical algorithms are still needed.

Cross-References

▶ Reliability Estimation and Analysis
▶ Stochastic Analysis of Nonlinear Systems
▶ Stochastic Ground Motion Simulation
▶ Structural Reliability Estimation for Seismic Loading
▶ Structural Seismic Reliability Analysis

References

Belytschko T, Liu WK, Moran B (2000) Nonlinear finite elements for continua and structures. Wiley, Chichester

Chen JB, Li J (2005) Dynamic response and reliability analysis of nonlinear stochastic structures. Probab Eng Mech 20(1):33–44

Chen JB, Li J (2007) The extreme value distribution and dynamic reliability analysis of nonlinear structures with uncertain parameters. Struct Saf 29:77–93

Chen JB, Li J (2009) A note on the principle of preservation of probability and probability density evolution equation. Probab Eng Mech 24(1):51–59

Chen JB, Li J (2010) Stochastic seismic response analysis of structures exhibiting high nonlinearity. Comput Struct 88(7–8):395–412

Chen JB, Zhang SH (2013) Improving point selection in cubature by a new discrepancy. SIAM J Sci Comput 35(5):A2121–A2149

Chen JB, Sun WL, Li J, Xu J (2013) Stochastic harmonic function representation of stochastic processes. J Appl Mech 80(1):011001-1-11

Ghanem R, Spanos PD (1991) Stochastic finite elements: a spectral approach. Springer, Berlin

Goller B, Pradlwarter HJ, Schuëller GI (2013) Reliability assessment in structural dynamics. J Sound Vib 332:2488–2499

Grigoriu M (2002) Stochastic calculus. Birkhäuser, Boston

Li J, Chen JB, Fan WL (2007) The equivalent extreme-value event and evaluation of the structural system reliability. Struct Saf 29:112–131

Li J (1996) Stochastic structural systems: analysis and modeling. Science Press, Beijing (in Chinese)

Li J, Chen JB (2003) The probability density evolution method for analysis of dynamic nonlinear response of stochastic structures. Acta Mech Sinica 35(6):716–722

Li J, Chen JB (2005) Dynamic response and reliability analysis of structures with uncertain parameters. International Journal for Numerical Methods in Engineering 62:289–315

Li J, Chen JB (2008) The principle of preservation of probability and the generalized density evolution equation. Struct Saf 30:65–77

Li J, Chen JB (2009) Stochastic dynamics of structures. Wiley, Singapore

Li J, Chen JB, Sun W, Peng Y (2012a) Advances of probability density evolution method for nonlinear stochastic systems. Probab Eng Mech 28:132–142

Li J, Yan Q, Chen JB (2012b) Stochastic modeling of engineering dynamic excitations for stochastic dynamics of structures. Probab Eng Mech 27(1):19–28

Li J, Wu JY, Chen JB (2014) Stochastic damage mechanics of concrete structures (in Chinese). Science Press, Beijing

Lin N, Emanuel K, Oppenheimer M, Vanmarcke E (2012) Physically based assessment of hurricane surge threat under climate change. Nat Climate Change 2(6):462–467

Ma F, Zhang H, Bockstedte A, Foliente GC, Paevere P (2004) Parameter analysis of the differential model of hysteresis. J Appl Mech 71:342–349

Shinozuka M, Deodatis G (1991) Simulation of stochastic processes by spectral representation. Appl Mech Rev 44(4):191–204

Spanos PD, Beer M, Red-Horse J (2007) Karhunen–Loéve expansion of stochastic processes with a modified exponential covariance kernel. J Eng Mech 133(7):773–779

Park YJ, Ang AH-S (1985) Mechanistic seismic damage model for reinforced concrete. J Struct Eng 111(4):722–739

Roberts JB, Spanos PD (1990) Random vibration and statistical linearization. Wiley, Chichester

Tajimi H (1960) A statistical method of determining the maximum response of a building structure during an earthquake. In: Proceedings of second world conference on Earthquake engineering, Tokyo, vol 11, pp 781–797

Wang D, Li J (2011) Physical random function model of ground motions for engineering purposes. Sci China Technol Sci 54(1):175–182

Zhu WQ (2006) Nonlinear stochastic dynamics and control in Hamiltonian formulation. Appl Mech Rev 59:230–248

Probability Seismic Hazard Mapping of Taiwan

Chin-Tung Cheng[1], Pao-Shan Hsieh[1], Po-Shen Lin[1], Yin-Tung Yen[1] and Chung-Han Chan[2,3]
[1]Disaster Prevention Technology Research Center, Sinotech Engineering Consultants, Inc., Taipei, Taiwan, ROC
[2]Department of Geosciences, National Taiwan University, Taipei, Taiwan, ROC
[3]Earth Observatory of Singapore, Nanyang Technological University, Singapore, Singapore

Synonyms

Hazard curve; Hazard map; Probabilistic seismic hazard assessment; Seismic hazard mitigation; Taiwan

Introduction

Studies on seismic hazard mitigation are important for seismologists, earthquake engineers, and related scientists. Among studies, probabilistic seismic hazard assessments (PSHAs) provide the probability of exceedance for a specific ground motion level during a time interval (see entry "▶ Site Response for Seismic Hazard Assessment"). PSHA results can provide a key reference for the determination of hazard mitigation policies related to building codes and the site selection of pubic structures. Therefore, multidisciplinary scientists have attempted to build reliable systems for PSHAs.

Due to the plate boundary between the Eurasian and Philippine Sea Plates, Taiwan has a high earthquake activity (Fig. 1). In this region, devastating earthquakes lead to a loss of property and human life. Therefore, it is essential to develop a means of seismic hazard mitigation. One practical approach is to build a seismic hazard assessment system. Over the past few years, several studies have evaluated seismic hazards for Taiwan. For example, the Global Seismic Hazard Assessment Program (GSHAP, http://www.seismo.ethz.ch/static/GSHAP/) obtained a global probabilistic seismic hazard map that included Taiwan. However, this work employed earthquake catalogs obtained from global seismic networks rather than a detailed seismicity catalog from Taiwan. Another application was proposed by Campbell et al. (2002) who utilized seismic catalogs, active fault parameters, and ground motion prediction equations (GMPEs) for the world, the United States, and Taiwan. Cheng et al. (2007) evaluated seismic hazards for Taiwan and proposed a hazard map by integrating a catalog from a local network, active fault parameters, and seismogenic zones in Taiwan. Such studies are crucial for understanding seismic hazards in Taiwan. However, following these studies, many parameters and the database for seismic hazard assessments, such as understanding the tectonic setting, the distribution of active faults, GMPEs, and earthquake catalogs, have been revised and/or updated. By employing state-of-the-art parameters, an evaluation of seismic hazards can be more precise.

Seismic hazards for Taiwan are reevaluated through the PSHA approach as proposed by Cornell (1968). According to this approach, parameters for seismogenic sources, which may result in seismic hazards, are required. Several seismogenic sources were characterized, including shallow regional sources, deep regional sources, crustal active fault sources, subduction intraslab sources, and subduction interface sources. The parameters for each source are discussed according to information from the tectonic setting, geology, geomorphology, geophysics, and earthquake catalog and present the results in the form of hazard maps and hazard curves. The results are compared with those proposed by previous studies and discuss their applicability for the future.

Methodologies and Seismic Activity Models

The PSHA Approach

The applied approach of PSHA was first developed by Cornell (1968) (see entry "Earthquakes and Tectonics: Probabilistic Seismic Hazard Assessment: An Overview"). According to the description of Kramer (1996), seismic hazards based on this approach can be assessed, as follows:

$$P[Y > y^*] = \iint P[Y > y^*|m,r] f_M(m) f_R(r) dm dr, \quad (1)$$

where $P[Y > y^*]$ is the probability, P, for a given ground motion parameter, Y, which exceeds a specific value, y^* $P[Y > y^*|m,r]$ is the probability conditional on an earthquake with magnitude, m, imparted by a seismogenic source with the closest distance, r, between the site of interest and seismogenic source; and $f_M(m)$ and $f_R(r)$ are the probability density functions for magnitude and distance, respectively.

If there are N_S potential seismogenic sources near the site of interest, each of which has an average rate, v_i. The total average exceedance rate, λ_{y^0}, for the region can be presented as follows:

$$\lambda_{y^0} = \sum_{i=1}^{N_S} v_i \iint P[Y > y^*|m,r] f_{M_i}(m) f_{R_i}(r) dm dr, \quad (2)$$

where $\sum_{n=1}^{N_S}$ is the summation of the contribution from N_sth seismogenic sources n.

PSHA uncertainties from different aspects were considered and properly treated. A logic tree approach was introduced to incorporate the uncertainties for seismogenic sources, the corresponding parameters of each source, and the GMPEs. The treatment of the weighting for each parameter is discussed in subsequent sections.

Seismic Activity Models

For the implementation of PSHA, the corresponding seismicity rate as a function of magnitude for each seismogenic source should be introduced (Eq. 2). Generally, there are two models that present the relationships, the truncated exponential model and the characteristic earthquake model. In the following, both of these models are presented and discussed.

The Truncated Exponential Model

The truncated exponential model is based on Gutenberg-Richter's Law (G-R Law) (Gutenberg and Richter 1954), as follows:

$$\log(\dot{N}) = a - bM \quad (3)$$

where $\dot{N}$ is the annual rate for events with magnitudes larger than or equal to M and a and b are constants with values larger than 0. Following on G-R Law, the truncated exponential model represents the rate for a magnitude larger than the maximum magnitude, m_u, as 0. Thus, the cumulative annual rate, $\dot{N}(m)$, for a magnitude larger than or equal to m can be presented, as follows:

$$\dot{N}(m) = \dot{N}(m_0) \frac{\exp(-\beta(m-m_0)) - \exp(-\beta(m_u-m_0))}{1.0 - \exp(-\beta(m_u-m_0))} \quad \text{for} \quad m_u \geq m \geq m_0 \quad (4)$$

where $\dot{N}(m_0)$ represents the cumulative annual rate for a magnitude of a minimum magnitude, m_0, and m_0 and m_u represent the minimum and maximum magnitudes, respectively, of the seismogenic source. β can be represented as follows:

$$\beta = b \cdot \ln(10) \quad (5)$$

where b is the b-value in G-R Law (Eq. 3).

Wesnousky (1994) concluded that this model is suitable for regions with complex tectonic settings or multiple active faults. Thus, it was applied for shallow regional sources, deep regional sources, and subduction intraslab sources.

The Characteristic Earthquake Model

The characteristic earthquake model was first proposed by Youngs and Coppersmith (1985). In addition to earthquake parameters, the model represents the seismicity rate by incorporating geological and geomorphological information. The cumulative annual rate for a magnitude larger or equal to m can be represented as follows:

$$\dot{N}(m) = \dot{N}^e \frac{\exp(-\beta(m-m_0)) - \exp(-\beta(m_u-1/2-m_0))}{1.0 - \exp(-\beta(m_u-1/2-m_0))} + \dot{N}^c \quad \text{for} \quad m_0 \leq m \leq m_u - \frac{1}{2}; \quad \dot{N}(m) = \dot{N}^c \frac{m_u-m}{1/2} \quad \text{for} \quad m_u - \frac{1}{2} \leq m \leq m_u, \quad (6)$$

where $\dot{N}^e$ and $\dot{N}^c$ represent the cumulative annual rates predicted by the truncated exponential and the characteristic earthquake models, respectively. $\dot{N}^e$ can be presented, as follows:

$$\dot{N}^e = \frac{\mu A_f S \cdot (1 - \exp(-\beta(m_u - m_0 - 1/2)))}{\exp(-\beta(m_u - m_0 - 1/2)) \cdot M_0(m_u) \cdot \left[\dfrac{b \cdot 10^{-c/2}}{(c-b)} + \dfrac{b\exp(\beta)\left(1 - 10^{-c/2}\right)}{c}\right]}, \quad (7)$$

where μ is the rigidity shear modulus, generally assumed to be 3×10^{10} Pascal (N/m²); A_f is the fault area; S is the slip rate; and c and d are constants. $\dot{N}^c$ can be represented as follows:

$$\dot{N}^c = \frac{1}{2}\dot{N}^e \frac{b\ln 10 \cdot \exp(-\beta(m_u - 3/2 - m_0))}{(1 - \exp(m_u - 1/2 - m_0))}. \quad (8)$$

In a comparison with the truncated exponential model, the characteristic earthquake model predicted lower rates for smaller magnitudes, whereas higher rates were predicted for larger magnitudes (Youngs and Coppersmith 1985).

To implement the characteristic earthquake model for PSHA, the cumulative annual rate should be in form of the rate $\lambda_n(m_i)$ (Eq. 2) between the magnitude bins of $m_i \pm dm/2$, which can be represented as follows:

$$\lambda_n(m_i) = N(m_i - dm/2) - N(m_i + dm/2), \quad (9)$$

where dm is the magnitude interval for the model and $N(m_i)$ is the rate for a magnitude larger or equal to m_i.

A few previous studies (Youngs and Coppersmith 1985; Wesnousky 1994) have suggested that the behavior of seismic activity along crustal active faults and subduction interfaces follows this model. Therefore, it was applied to the two seismogenic sources.

Seismogenic Tectonics in Taiwan

Tectonic Setting

Taiwan is located within the plate boundary between the Philippine Sea Plate and the Eurasian Plate (Fig. 1). Due to the interaction of the two plates, both subduction and collision take place in this region. Two subduction systems surround this region. In the offshore of northeast Taiwan, the Philippine Sea Plate subducts to the north. As a result of back-arc spreading, the Okinawa Trough and the Ilan Plain formed in the northern section of the Ryukyu Volcanic Arc. In southern Taiwan, the Eurasian Plate subducts to the east. The Longitudinal Valley is the arc-continental collision boundary between the two plates. Collision began in the late Miocene in northern Taiwan. Due to lateral collision between plates, collision activity continues to migrate to the south. Currently, activity takes place in central and southwestern Taiwan. Northern Taiwan, in contrast, is a post-collision region with relatively low seismic activity.

The Earthquake Catalog

The utilized earthquake catalog was collected by the Central Weather Bureau (CWB) and provided earthquake parameters for events from 1900 to 2010 in Taiwan. Prior to 1973, a total of 15 stations equipped with Gray-Milne, Wiechert, or Omori seismographs were maintained. After 1973, the Taiwan Telemetric Seismic Network (TTSN) was established. The TTSN consists of 25 stations within the region of Taiwan. In the TTSN network, real-time signals are transmitted from field stations to a central station via leased telephone lines.

To assess seismic hazards, earthquake parameters should be analyzed using the following procedures: a magnitude harmonization from different scales, an evaluation of the magnitude of completeness, and a declustering process. In the following, each step of the procedure is described in detail.

Magnitude Harmonization

Since the magnitude scales for a catalog during different periods are generally different, it is critical to harmonize the magnitude scales during

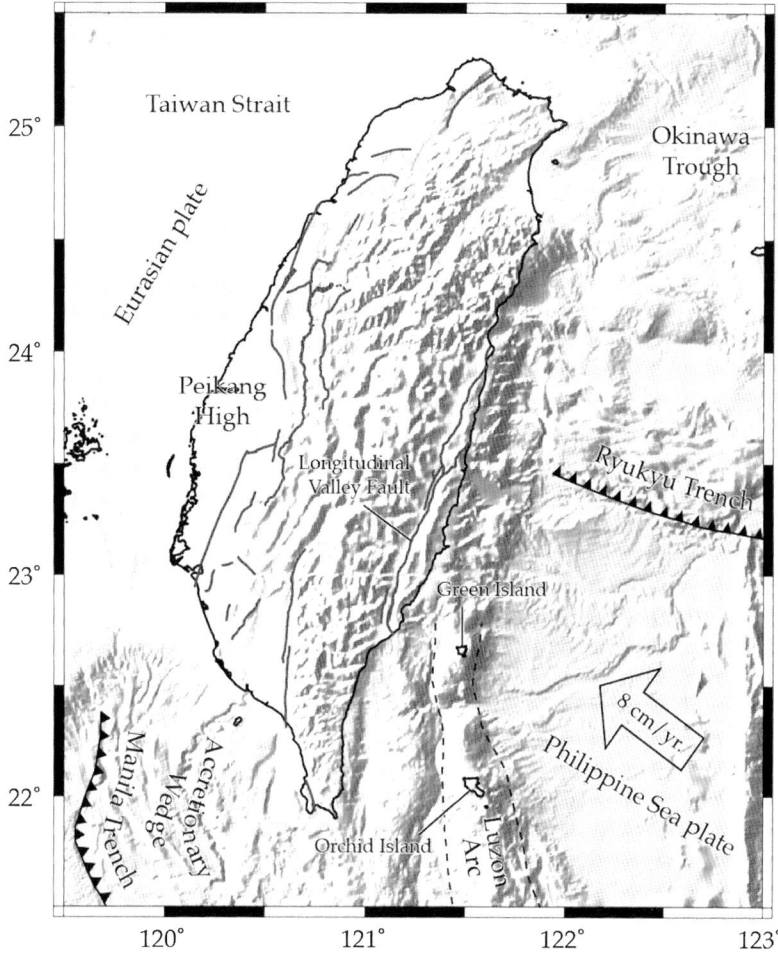

Probability Seismic Hazard Mapping of Taiwan, Fig. 1 The tectonic setting in Taiwan and its vicinity (Modified from Cheng et al. 2007)

different periods. Previous studies (Hanks and Kanamori 1979 and references therein) have suggested a moment magnitude (M_W) for PSHA, since this scale is evaluated based on rupture dimensions and slip magnitudes. Additionally, the M_W scale is not affected by saturation at higher magnitudes. For example, the 1999 Chi-Chi, Taiwan, earthquake was determined to have a M_W of 7.6, whereas its corresponding M_L was 7.3. The discrepancy can be attributed to the saturation of M_L (Cheng et al. 2007). Therefore, the earthquake catalog obtained by Tsai et al. (2000) was considered. The magnitude scales of this catalog have been harmonized as M_W from 1900 to 1999. The magnitude scales of the catalog were harmonized according to the procedure of Tsai et al. (2000).

Magnitude of Completeness

To improve the reliability of the parameters for seismogenic sources, the catalog during the period when the network recorded all earthquakes with a certain magnitude threshold was considered. The threshold is known as the magnitude completeness, M_c. Thus, M_c for catalogs during different periods must be examined in advance. Chen et al. (2012) evaluated the M_c of the CWB catalog using the maximum curvature method. A higher M_c between 4.3 and 4.8 was obtained prior to 1973. Once the TTSN was established, M_c decreased to between 2.0 and 3.0. Based on the temporal distribution of M_c, M ≥ 6.0 earthquakes after 1900 (Fig. 2) and M ≥ 2.0 earthquakes after 1973 (Fig. 3) were considered.

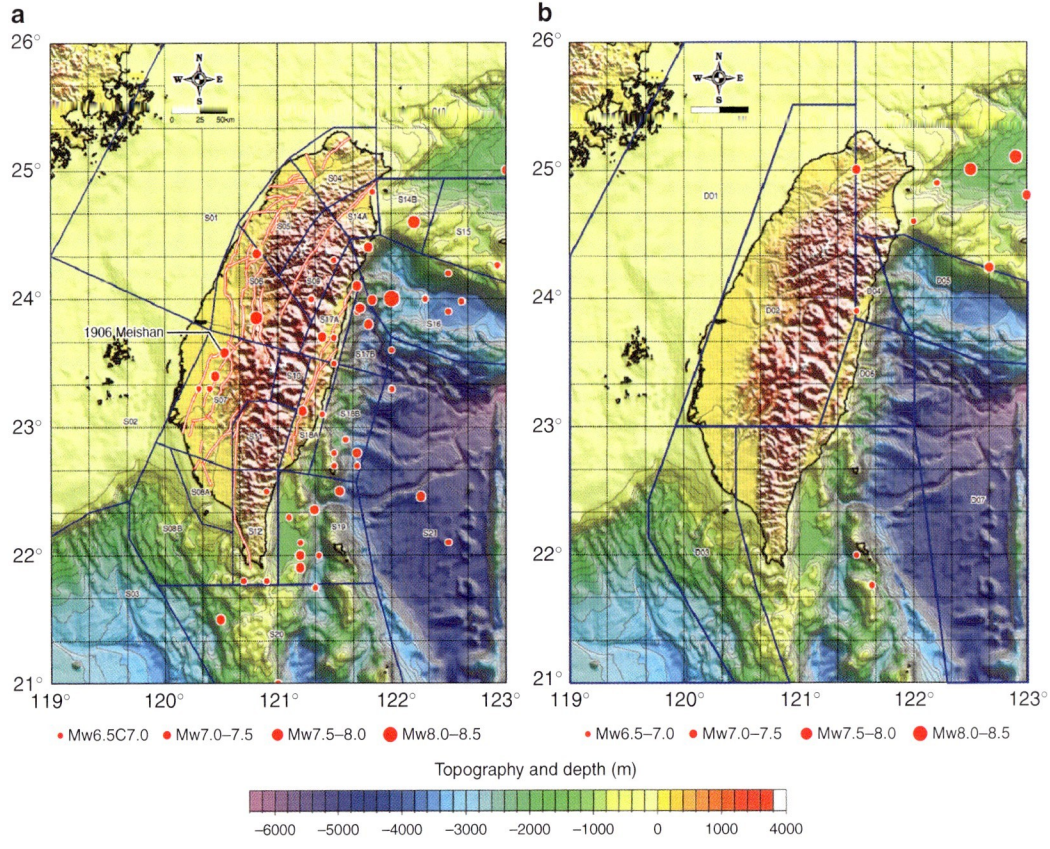

Probability Seismic Hazard Mapping of Taiwan, Fig. 2 The distribution of earthquakes with M ≥ 6.0 at the depth of (**a**) ≤35 km and (**b**) >35 km since 1900. The distribution of shallow and deep regional sources is illustrated by the *blue polygons* in Fig. 2a and b, respectively. In Fig. 2a, crustal active fault sources are presented as *red lines*

The Declustering Process

For application of the PSHA approach by Cornell (1968), it is assumed that the occurrence of earthquakes follows the Poisson procedure. In other words, earthquakes are independent of one another. However, in a catalog, earthquake sequences, which include foreshocks, mainshock, aftershocks, and swarms, can be observed. Therefore, it is critical to obtain a declustered catalog in respect to the PSHA (i.e., to remove foreshocks, aftershocks, and swarms from the catalog). The declustering approaches developed by Wyss (1979), Arabasz and Robinson (1976), Gardner and Knopoff (1974), and Uhrhammer (1986) were implemented. According to these approaches, earthquakes are considered dependent when their distance and time are within the windows (Fig. 4). Earthquakes are regarded as foreshocks or aftershocks if they fulfill at least two of the four declustering approaches.

Focal Mechanisms

Based on the spatial distribution of the focal mechanisms, the seismogenic region for the PSHA can be distinguished. According to the focal mechanisms determined by Wu et al. (2010), the spatial distribution of the crustal stress state in Taiwan can be illustrated. In northern Taiwan, along the Central Range and in the Okinawa Trough, the stress states are normal favorable. In central Taiwan, southwestern Taiwan, and within the interfaces of subduction systems, the stress states are

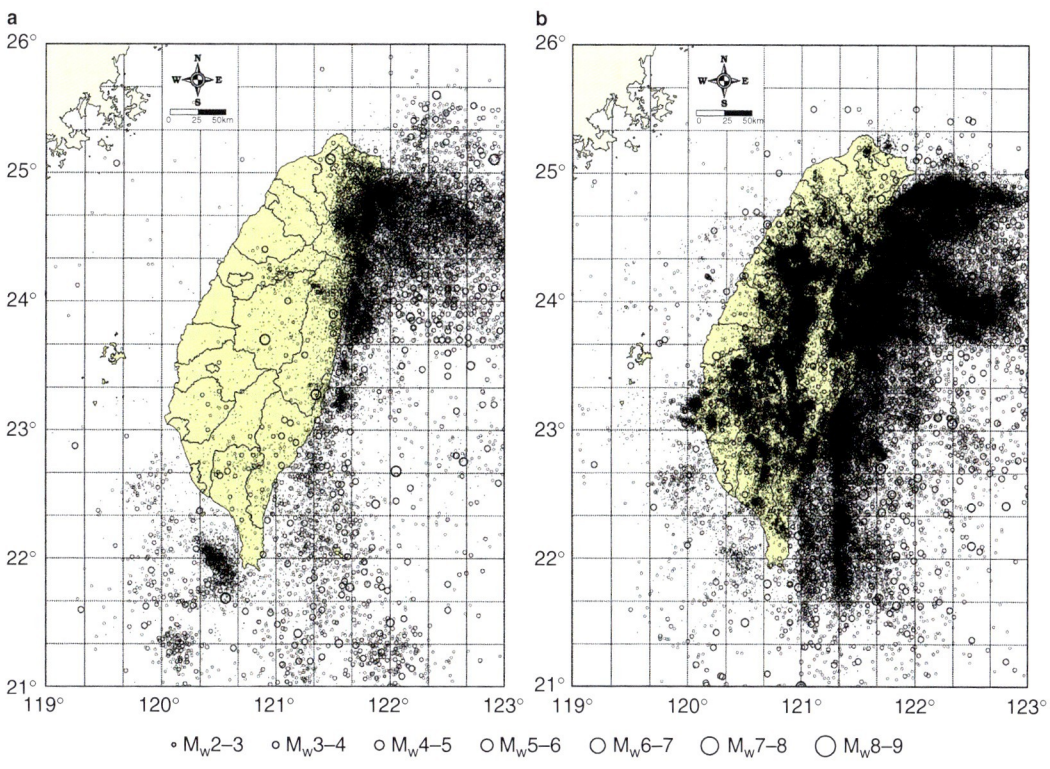

Probability Seismic Hazard Mapping of Taiwan, Fig. 3 The distribution of earthquakes with M ≥ 2.0 at a depth of (**a**) ≤35 km and (**b**) >35 km since 1973. In the column of rake, *N* normal, *RL* right-lateral, *LL* left-lateral, *T* thrust

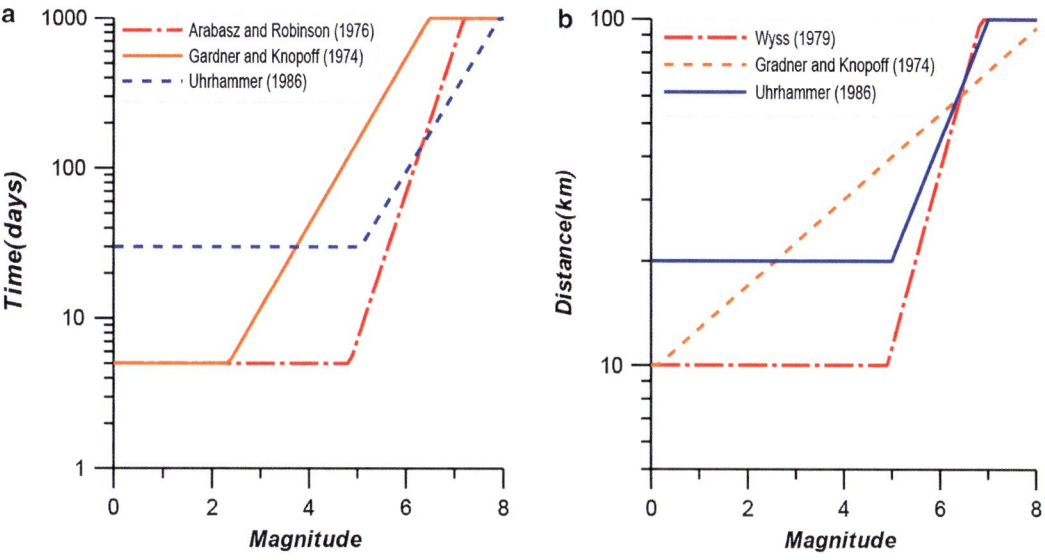

Probability Seismic Hazard Mapping of Taiwan, Fig. 4 The (**a**) distance and (**b**) time windows for each declustering approach. Earthquakes were considered dependent when their distance and time were within the windows

thrust favorable. In northwestern Taiwan, south Taiwan, and along the Longitudinal Valley, the stresses are strike-slip favorable. Additionally, the stress state within the two subduction systems can be comprehended using focal mechanisms determined by Wu et al. (2010).

The distribution of focal mechanisms can be associated with the tectonic setting, as mentioned above (in section "Tectonic Setting"). Representing the spatial distribution of stress states in each region would be of benefit for the PSHA in respect to the determination of seismogenic sources.

Seismogenic Sources

For application of the PSHA approach of Cornell (1968) (Eq. 2), seismogenic sources and corresponding parameters should be defined. Based on the understanding of each source, three types, Type I, Type II, and Type III (Kiureghian and Ang 1977), can be categorized. Type I is a source with a clear fault geometry, Type II is a source with a clear focal mechanism, and Type III is a source with a controversial fault geometry and mechanism. Type II sources were assumed as "regional sources." By further considering the depth boundary of 35 km, a "shallow regional source" and a "deep regional source" were defined. Type I sources were treated as "crustal active fault sources" according to the distribution of active faults obtained by the Central Geological Survey (2010, http://fault.moeacgs.gov.tw/TaiwanFaults_2009/News/NewsView.aspx?id=3). Since two subduction systems exist in Taiwan (Fig. 1) and since the ground motion attenuation behaviors of intraslab and interface events are different, "subduction intraslab sources" and "subduction interface sources" were considered. In the following, each seismogenic source and the corresponding parameters are described in detail. For application of the logic tree in the PSHA, the weights for the parameters of each source are also required. In the following, the treatment of the weighting is described in detail.

Shallow Regional Sources

Using information on geomorphology, seismology, and geophysics, 28 shallow regional sources were defined in Taiwan and its vicinity (Fig. 5). The geometry and the corresponding parameters of each zone are outlined in the following sections.

The Geometry of Each Source

S01, S02, and S03 are located in the stable Eurasian Continental Plate. In these sources, seismicity rates are relatively low in comparison to the region surrounding Taiwan Island. The boundary between S01 and S02 was determined based on the extended alignment of the structure in Taiwan. Additionally, earthquakes for the two sources present different focal mechanisms (Wu et al. 2010). The southern boundary of S02 is defined by the accretionary wedge of the southern subduction zone system. The eastern boundary of S03 is defined by the southern subduction zone system (Fig. 1).

At S04, the focal mechanisms suggest normal favorability, which is significantly different from that in its vicinity (Wu et al. 2010). S05A displays a transient mechanism from the normal mechanism in the northeast (S04) to the strike-slip and one located in the southwest (S05B). The eastern boundary of the two sources is defined based on different mechanisms from S09, where the dipping angles of earthquakes are close to vertical and mechanisms are normal favorable. S06 belongs to the frontal deformation region in the Western Foothills. Both the southern and northern boundaries are defined by fault segmentations and changes in the fault alignments (Fig. 2a). The western boundary is marked by the boarder of the Peikang High (Fig. 1). The eastern boundary is defined due to different mechanisms from S10, where the dipping angles are close to vertical and earthquakes are the normal favorable mechanism.

S07 is also located in the frontal deformation region within the Western Foothills. The eastern boundary is defined according to a significantly different seismicity rate from S11. S08A and S08B are located in the transition region between the frontal deformation region on the north and the subduction system on the south. In comparison with the thrust mechanisms in S07, earthquakes within S08A are strike-slip types with a

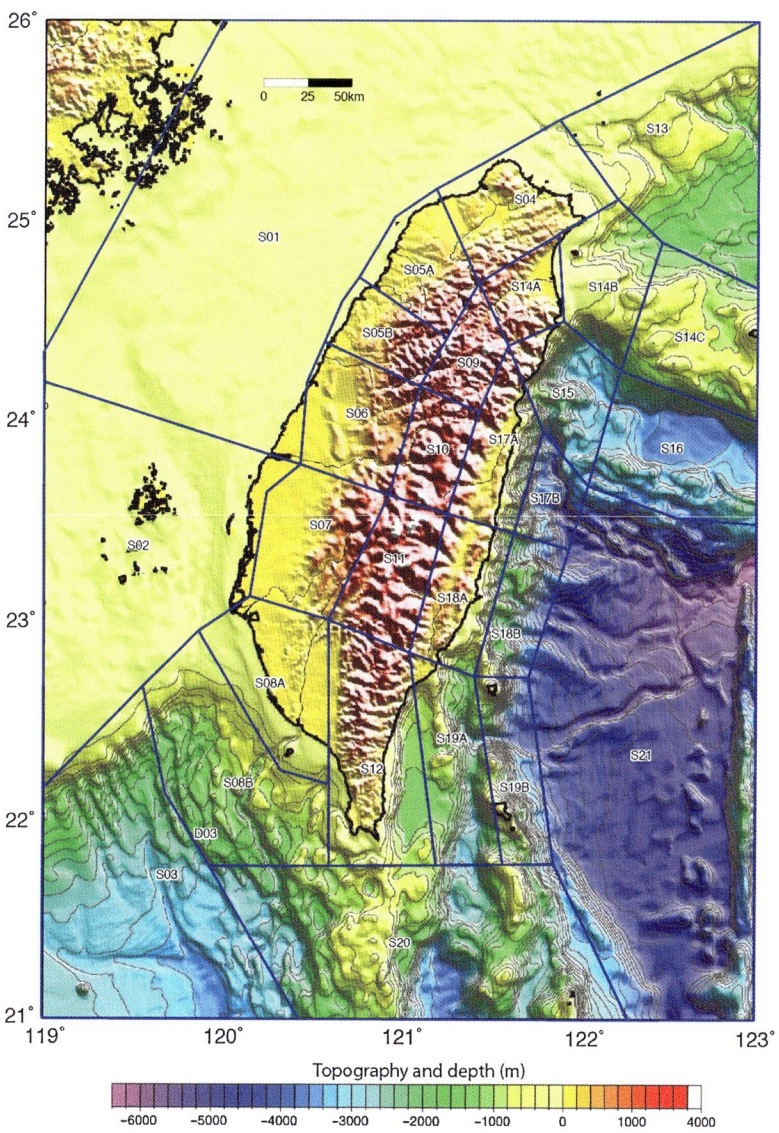

Probability Seismic Hazard Mapping of Taiwan, Fig. 5 The distribution of shallow regional sources

thrust component. The eastern boundary with S12 was determined according to heterogeneous deformation behaviors obtained from GPS observations (Hsu et al. 2009).

S09, S10, S11, and S12 are located within the transition region from the frontal deformation region in the west to the collision boundary between the Eurasian and Philippine Sea Plates in the east. The principal stress axis is vertical. The four sources are distinguished by strike orientations (from a NE-SW orientation in the north to a N-S orientation in the south) and seismicity rates. S13's source is located within the western flank of the Okinawa Trough. Due to back-arc spreading, the mechanism of earthquakes in this source suggests normal favorability. S14A, S14B, and S14C result from back-arc spreading and subduction. In the three sources, the seismicity rates are high and the seismicity behaviors are complex. The southern boundary is defined according to the interface of the subduction system. The three sources are distinguished by their seismicity rates and mechanisms. S15 and S16 are located in the area where the Philippine Sea Plate subducts to the Eurasian Plate. In general, earthquakes have thrust mechanisms with low dipping angles. In comparison, the mechanisms in S15 are complex due to the coexistence of plate collision and subduction. For the same reason, the seismicity rate in S15 is higher than that in S16.

S17A, S17B, S18A, and S18B reside along the eastern coastline and in the offshore region, which is in the collision zone between the two plates. The boundary between S17 and S18 is defined by heterogeneities in respect to active fault activity and deformation behavior according to GPS observations (Hsu et al. 2009). Additionally, seismicity rates in S17A and S18A to the west are higher than those in S17B and S18B to the east. Earthquakes in these sources are mainly thrusts with a strike-slip mechanism with high dipping angles.

S19A and S19B include Green Island and Orchid Island and their vicinity. Since 1900, eight earthquakes with M ≥ 6.0 have taken place in these sources (Fig. 2a). The two sources are distinguished based on differences in the seismicity rate (Fig. 3a). S20 represents the southern offshore region of Taiwan. Additionally, it is above the subduction zone. S21 is located within the Philippine Sea Plate. Since it is located at a distance from the collision zone, the seismicity rate is relatively low (Fig. 1).

The Parameters of Each Source

In the above discussion, the distributions of shallow regional sources were determined. For application of the PSHA, parameters for each source were required. In the following, the acquirement of parameters including the seismicity rate models, maximum magnitudes, and their corresponding weights is described.

Due to insufficient information on fault geometry and the slip rate for regional sources, a truncated exponential model (Eq. 4) is applied for shallow regional sources. The model for each source was obtained through a regression of the declustered catalog using maximum likelihood estimates. For the regression, the magnitude interval for the model (dm in Eq. 7) was assumed to be 0.5. Therefore, the cumulative annual rate ($\dot{N}(m_0)$) and the b-value, as well as the corresponding deviations for each source, were obtained (Table 1).

Another key parameter for the PSHA was the maximum magnitude, m_u. For sources with active faults or subduction systems (i.e., S04, S05A, S05D, S06, S07, S08A, S11, S12, S15, S16, S17A, and S18A) (Fig. 2a), the corresponding m_u's were generally obtained based on the maximum magnitude of active faults within the source (see section "Crustal Active Fault Sources"). Note that a maximum threshold of 6.5 in these sources was assumed since the occurrence of M > 6.5 earthquakes can be attributed to crustal active faults or subduction sources. However, m_u for some sources, S05A, S12, S15, and S16, was determined using other considerations. In S05A, since no earthquake with M ≥ 5.0 has ever been recorded, a smaller m_u of 5.5 was assumed. In S12, S15, and S16, due to the existence of a subduction interface and other offshore active faults, a larger m_u of 7.0 was assumed. On the other hand, for sources without active faults (i.e., S01, S02, S03, S08B, S09, S10, S13, S14A, S14B, S14C, S17B, S18B, S19A, S19B, S20, and S21), the m_u's were assumed based on the maximum observed earthquake plus 0.2 for each source (Table 1).

Deep Regional Sources

According to the distribution of seismicity and the tectonic setting, seven deep regional sources were defined in Taiwan and its vicinity (Fig. 6). Note that subduction interfaces were not included with respect to differences in ground motion attenuation behaviors (Lin and Lee 2008). The geometry and the corresponding parameters of each source are discussed in the following sections.

The Geometry of Each Source

Deep regional sources can be separated into two parts. To the east of longitude 121.5°, they are located within the subduction zone in northeast Taiwan. To the south of latitude 23°, they are regarded as another subduction zone in southern Taiwan. D01, D02, and D03 are located within the Eurasian Plate. The boundary between D02 and D03 is located along the latitude of 23°, which is regarded as the border of the continent to the north and the subduction zone to the south. The accretionary wedge of the southern

Probability Seismic Hazard Mapping of Taiwan, Table 1 The parameters and the corresponding standard deviations for the 28 shallow regional sources; the corresponding weights for the maximum magnitudes, m_u, are denoted in parentheses

NO	m_0	$N(m_0)\ (\pm\sigma_{N(m_0)})$	$b(\pm\sigma_b)$	m_u
S01	3.5	1.108(±0.171)	0.865(±0.114)	6.5(0.2) 6.6(0.6) 6.7(0.2)
S02	3.5	0.968(±0.159)	0.772(±0.114)	6.5(0.2) 6.6(0.6) 6.7(0.2)
S03	3.5	0.806(±0.185)	0.852(±0.162)	6.5(0.2) 6.6(0.6) 6.7(0.2)
S04	2.5	3.578(±0.384)	0.783(±0.066)	6.5(1.0)
S05A	2.5	4.498(±0.349)	1.194(±0.093)	5.5(1.0)
S05B	2.5	5.412(±0.382)	0.811(±0.053)	6.5(1.0)
S06	2.5	11.860(±0.702)	0.816(±0.041)	6.5(1.0)
S07	3	6.107(±0.400)	0.643(±0.034)	6.5(1.0)
S08A	2.5	9.990(±0.649)	0.855(±0.049)	6.5(1.0)
S08B	2.5	14.280(±0.852)	1.028(±0.051)	6.5(0.2) 6.6(0.6) 6.7(0.2)
S09	2.5	1.925(±0.269)	0.0541(±0.062)	6.7(0.2) 6.9(0.6) 7.1(0.2)
S10	3.5	0.935(±0.157)	0.958(±0.136)	6.5(0.2) 6.7(0.6) 6.9(0.2)
S11	3.5	1.543(±0.201)	0.857(±0.094)	6.5(1.0)
S12	2.5	23.510(±0.966)	0.695(±0.024)	7.0(1.0)
S13	3	5.185(±0.497)	0.692(±0.049)	6.3(0.2) 6.5(0.6) 6.7(0.2)
S14A	2.5	3.526(±0.373)	0.676(±0.058)	6.7(0.2) 6.9(0.6) 7.1(0.2)
S14B	2.5	1.523(±0.198)	0.564(±0.061)	7.4(0.2) 7.6(0.6) 7.8(0.2)
S14C	3.5	4.013(±0.326)	0.720(±0.048)	7.4(0.2) 7.6(0.6) 7.8(0.2)
S15	3.5	7.187(±0.528)	0.629(±0.037)	7.0(1.0)
S16	3.5	6.285(±0.0.491)	0.602(±0.036)	7.0(1.0)
S17A	3.5	4.069(±0.405)	0.713(±0.055)	6.5(1.0)
S17B	3.5	1.795(±0.268)	0.685(±0.081)	7.3(0.2) 7.5(0.6) 7.7(0.2)
S18A	3.5	2.979(±0.274)	0.616(±0.048)	6.5(1.0)
S18B	3.5	2.524(±0.253)	0.636(±0.054)	7.3(0.2) 7.5(0.6) 7.7(0.2)
S19A	3.5	4.834(±0.438)	0.678(±0.049)	7.3(0.2) 7.5(0.6) 7.7(0.3)
S19B	3.5	3.017(±0.353)	0.780(±0.071)	7.3(0.2) 7.5(0.6) 7.7(0.3)
S20	3.5	4.866(±0.456)	0.882(±0.064)	7.1(0.2) 7.3(0.6) 7.5(0.2)
S21	3.5	9.581(±0.498)	0.756(±0.033)	7.1(0.2) 7.3(0.6) 7.5(0.2)

subduction zone system defines the western boundary of D03. The eastern boundary of D03 is illustrated by the distribution of the southern subduction zone system. To the east of Taiwan, four deep regional sources, including D04, D05, D06, and D07, were identified. Due to plate collision, higher seismicity rates were observed in D04 and D06 (Fig. 3b). In contrast, since it is a part of the deeper part of the Philippine Sea Plate, D07 has a lower seismicity rate. D05 is located within the region where the Philippine Sea Plate subducts to the Eurasian Plate.

The Parameters of Each Source
As a follow-up to the procedure for shallow regional sources, the acquirement of the corresponding parameters for each deep regional source is described in the following. A truncated exponential model is used to present seismic activity. The cumulative annual rate and the b-value, as well as the corresponding deviations for each source, were obtained (Table 2). The maximum magnitude, m_u, for each source was assumed. Since D01, D02, and D03 are located within the stable Eurasian Continental Plate, a small m_u was assumed. For application of the logic tree, weights of 0.2, 0.6, and 0.2 were assumed for m_u's of 6.5, 6.6, and 6.7, respectively. Due to the plate collision zone, higher m_u's of 7.0, 7.2, and 7.4 were assumed for D04. Since earthquakes with M $\geq$ 6.7 have never been recorded in D05, D06, and D07, m_u's of 6.8, 7.0, and 7.2 were assumed.

Probability Seismic Hazard Mapping of Taiwan, Fig. 6 The distribution of deep regional sources

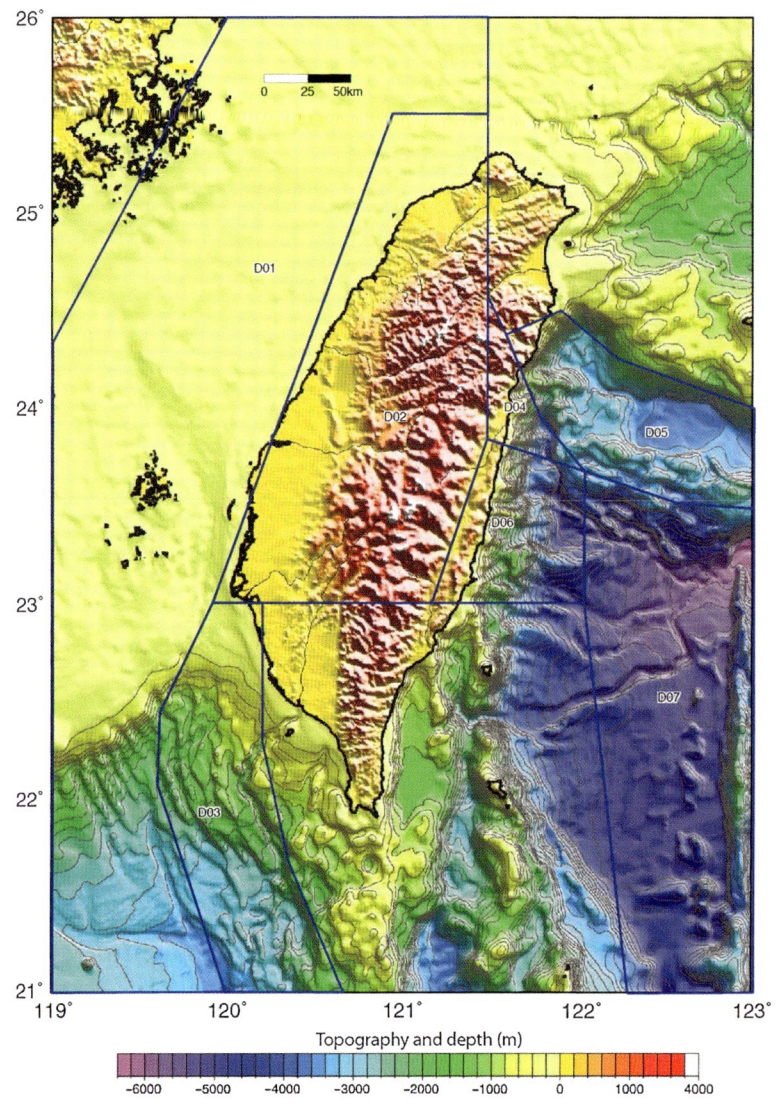

Probability Seismic Hazard Mapping of Taiwan, Table 2 The parameters and the corresponding standard deviations for the seven deep regional sources; the corresponding weights of the maximum magnitudes, m_u, are denoted in parentheses

NO	m_0	$N(m_0) (\pm\sigma_{N(m_0)})$	$b(\pm\sigma_b)$	m_u
D01	3.5	0.673(±0.169)	0.837(±0.162)	6.5(0.2) 6.6(0.6) 6.7(0.2)
D02	3.0	2.142(±0.239)	0.799(±0.078)	6.5(0.2) 6.6(0.6) 6.7(0.2)
D03	3.0	7.341(±0.596)	0.750(±0.055)	6.5(0.2) 6.6(0.6) 6.7(0.2)
D04	3.5	1.340(±0.185)	0.668(±0.077)	7.0(0.2) 7.2(0.6) 7.4(0.2)
D05	3.5	3.394(±0.299)	0.669(±0.052)	6.8(0.2) 7.0(0.6) 7.2(0.2)
D06	3.5	1.585(±0.204)	0.665(±0.072)	6.8(0.2) 7.0(0.6) 7.2(0.2)
D07	3.5	1.374(±0.185)	0.554(±0.070)	6.8(0.2) 7.0(0.6) 7.2(0.2)

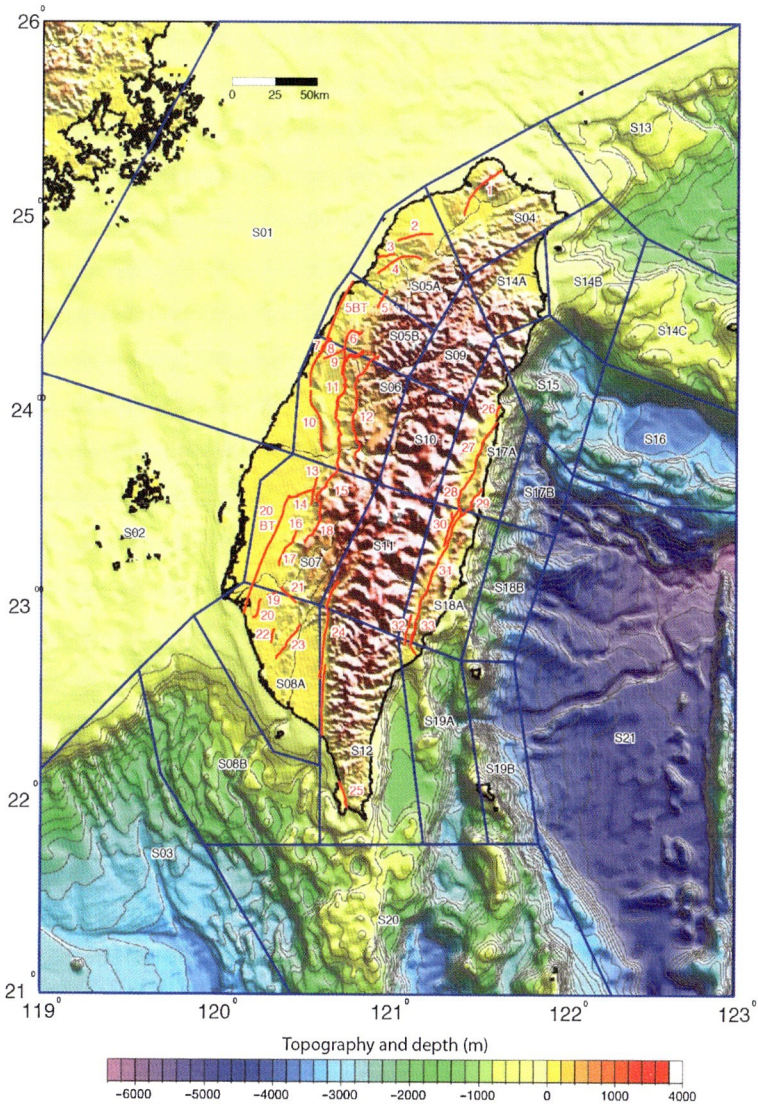

Probability Seismic Hazard Mapping of Taiwan, Fig. 7 The distribution of shallow regional (*blue polygons*) and crustal active fault sources (*red lines*). The corresponding shallow regional source and the *b*-values for each crustal active fault source are denoted in Table 4

Crustal Active Fault Sources

Thirty-three active faults obtained by the Central Geological Survey (2010, http://fault.moeacgs.gov.tw/TaiwanFaults_2009/News/NewsView.aspx?id=3) and three blind faults by Cheng et al. (2007) were considered as crustal active fault sources (Fig. 7). Crustal active fault sources are categorized as Type I sources by Kiureghian and Ang (1977). Sources of this type usually generate earthquakes with a characteristic magnitude repeatedly within a recurrence interval. Such seismic activity does not follow the truncated exponential model, but fulfills the characteristic earthquake model (Wesnousky 1994). For application of this model, some parameters are required (Eqs. 6, 7, and 8). The fault parameters, including segmentation, length, depth, area, the rupture mechanism, dip angle, the possible magnitude of the slip, the slip rate, the recurrence interval, the last slip time, and the magnitude of a characteristic event, were obtained from various references and are listed in Table 3. For treatment of the *b*-value, a constant *b*-value is assumed in each shallow regional source. According to the distribution of shallow

Probability Seismic Hazard Mapping of Taiwan, Table 3 The parameters for crustal active fault sources; the corresponding weights of the slip rate and the maximum magnitude for each source are denoted in parentheses

ID	Length (km)	Depth (km) Top	Depth (km) Bottom	Area (km2)	Rake	Dip angle	Possible slip amount (m)	Slip rate (mm/year)	Recurrence interval (year)	Last occurrence time (year)	Characteristic magnitude (Mw)
1	61	0	15	1,009.6	N	>60°E	1.39 ~ 7.99	0.69(1)	614 ~ 2,511	<11,000	7.20(0.5) 7.33(0.5)
	13	0	15	215.2	N	>60°E	0.23 ~ 1.30	1.2(0.5) 18(0.5)	117 ~ 527	<70,000	6.50(0.5) 6.64(0.5)
2	22	0	12	410.7	T	~40°S	0.46 ~ 3.20	0.9(0.2) 1.7(0.6) 2.5(0.2)	118 ~ 263	Late Pleistocene	5.98(0.5) 6.16(0.5)
3	9	0	12	141	T-RL	~50°S	0.34 ~ 2.33	1.0(0.5) 1.2(0.5)	181 ~ 454	<300 year	6.64(0.5) 6.77(0.5)
4	28	0	12	672	T	~30°S	0.50 ~ 3.48	1.0(0.5) 1.6(0.5)	60 ~ 314	A.D.1935	6.15(0.5) 6.32(0.5)
5	12	0	12	158.8	T	>60°W	0.37 ~ 2.58	1.3(0.2) 2.5(0.6) 3.8(0.2)			
5BT	37	2	15	962	T-RL	30°E	0.56 ~ 3.84	2.5(0.2) 5.0(0.6) 7.5(0.2)	47 ~ 207		6.80(0.5) 6.91(0.5)
6	33	0	15	646.2	T	40–60°E	0.53 ~ 3.69	0.5(0.2) 1.0(0.6) 1.5(0.2)	275 ~ 1,249	Holocene	6.73(0.5) 6.85(0.5)
7	8	2	15	147.1	T	40–50°E	0.46 ~ 3.20	1.7(0.2) 3.6(0.6) 5.5(0.2)	74 ~ 362	Holocene	6.68(0.5) 6.80(0.5)
	22	2	15	404.4		40–50°E	0.32 ~ 2.24				
8	13	2	15	399.9	T	20–30°W	0.38 ~ 2.66	1.2(1)	86 ~ 155	Holocene	6.19(0.5) 6.36(0.5)
9	14	0	15	210	RL	High angle	0.17 ~ 0.79	1.3(0.2) 2.5(0.6) 3.8(0.2)	59 ~ 312	A.D.1935	6.23(0.5) 6.4(0.5)
10	36	2	15	936	T	18–45°E	0.55 ~ 3.81	4.3(0.2) 7.3(0.6) 103(0.2)	33 ~ 119	Holocene	6.78(0.5) 6.9(0.5)
11	36	0	20	1,120.1	T	~40°E	0.55 ~ 3.81	3.5(0.5) 6.9(0.5)	112 ~ 280	A.D.1999	7.36(0.5) 7.43(0.5)
	48	0	20	1,493.5			0.61 ~ 4.22				
	14	0	20	435.6			0.39 ~ 2.73				
12	69	0	15	1,463.7	T	~45°E	0.69 ~ 4.79	0.4(0.2) 0.8(0.6) 1.2(0.2)	670 ~ 2,651	A.D.1999	7.16(0.5) 7.24(0.5)
13	17	0	15	603.4	T	20–30°E	0.42 ~ 2.92	5(0.2) 10(0.6) 15(0.2)	8 ~ 40	<18,540 BP	6.35(0.5) 6.6(0.5)
14	13	0	15	215.2	RL	>60°	0.15 ~ 0.72	3(0.2) 6(0.6) 9(0.2)	22 ~ 117	A.D.1906	6.19(0.2) 6.36(0.5)
14BT	36	2	15	936	T	30°E	0.55 ~ 3.81	6(0.2) 12(0.6) 18(0.2)	19 ~ 87		6.78(0.5) 6.90(0.5)
15	25	0	15	433	T-RL	>60°E	0.48 ~ 3.25	5(0.2) 10(0.6) 15(0.2)	24 ~ 116	A.D.1999	6.57(0.5) 6.71(0.5)

#				Length	Type	Dip	Range		Range2	Age	M1	M2
16	7	0	15	154	T	~30°E	0.31 ~ 2.13	8.2(0.5) 9.3(0.5)	8 ~ 19	Late Pleistocene	5.83(0.5)	6.03(0.5)
17	17	0	15	373.9	T-LL	~30°E	0.42 ~ 2.92	8.2(0.5) 9.3(0.5)	21 ~ 40	<10,000	6.35(0.5)	6.50(0.5)
18	28	0	15	512.7	T	50–60°E	0.50 ~ 3.48	5(0.2) 10(0.6) 15(0.2)	26 ~ 121	<10,000	6.64(0.5)	6.77(0.5)
19	6	0	15	99.3	RL	>60	0.06 ~ 0.27	2.5(0.2) 5(0.6) 7.5(0.2)	12 ~ 72	A.D.1946	5.74(0.5)	5.95(0.5)
20	12	0	15	280	T	>35°W	0.37 ~ 2.58	4.0(0.5) 5.6(0.5)	23 ~ 58	Late Pleistocene	6.15(0.5)	6.32(0.5)
20BT	36	2	15	936	T	30°E	0.55 ~ 3.81	5(0.2) 10(0.6) 15(0.2)	23 ~ 104		6.78(0.5)	6.90(0.5)
21	10	0	15	165.5	LL	>60°N	0.11 ~ 3.81	1.3(0.2) 2.5(0.6) 3.8(0.2)	40 ~ 216	Mid-late Pleistocene	6.04(0.5)	622(0.5)
22	8	0	15	169.7	T	45°E	0.32 ~ 2.24	4.0(0.5) 5.6(0.5)	15 ~ 40	Mid-late Pleistocene	5.91(0.5)	6.10(0.5)
23	30	0	15	587.4	T	High angle	0.52 ~ 3.57	15(0.2) 3(0.6) 4.5(0.2)	85 ~ 385	7,189 BP	6.68(0.5)	6.80(0.5)
24	28	0	20	579.8	T-LL	70–80°E	0.50 ~ 3.48	2(0.2) 4(0.6) 6(0.2)	57 ~ 266	Late Pleistocene	6.64(0.5)	6.77(0.5)
	61			1.263			0.66 ~ 4.59	1.5(0.2) 3(0.6) 4.5(0.2)	26 ~ 122		7.09(05)	7.18(0.5)
25	39	0	20	830.1	T	70°E	0.57 ~ 3.92	4.2(0.5) 7.5(0.5)	61 ~ 159	Late Pleistocene	6.83(0.5)	6.94(0.5)
26	8	0	30	264.8	LL-T	60–70°E	0.32 ~ 2.24	10(0.2) 20(0.6) 30(0.2)	2 ~ 12	A.D.1951	5.91(0.5)	6.10(0.5)
27	30	0	30	1.174.9	LL-T	50°E	0.52 ~ 3.57	10(0.2) 20(0.6) 30(0.2)	6 ~ 29	Late Pleistocene	6.68(0.5)	6.80(0.5)
28	33	0	30	1.292.3	T-LL	40–50°E	0.53 ~ 3.69	12.5(0.5) 16.0(0.5)	13 ~ 25	A.D.1951	6.73(0.5)	6.85(0.5)
29	30	0	30	931.7	T	E	0.52 ~ 3.57	2(0.2) 4(0.6) 6(0.2)	40 ~ 183	Late Pleistocene	6.68(0.5)	680(0.5)
30	23	0	25	750.6	LL-T		0.47 ~ 3.25	8(0.2) 16(0.6) 24(0.2)	7 ~ 35	A.D.1951	6.52(0.5)	6.66(0.5)
31	67	0	30	2.217.8	T-LL	~65°E	0.69 ~ 4.74	26(0.5) 30(0.5)	17 ~ 26	A.D.2003	7.14(0.5)	7.23(0.5)
32	17	0	25	554.8	T	50°E	0.42 ~ 2.92	1.9(0.5) 3.0(0.5) 5.4(0.5) 5.5(0.5)	24 ~ 116	Late Pleistocene	6.35(0.5)	6.50(0.5)
33	20	0	25	551.7	T	65°E	0.45 ~ 3.09	8(0.2) 16(0.6) 24(0.2)	8 ~ 38	Late Pleistocene	6.44(05)	6.59(0.5)

Probability Seismic Hazard Mapping of Taiwan, Table 4 The corresponding shallow regional source and b-values for each crustal active fault source; the spatial distribution of the sources is presented in Fig. 7

Shallow regional sources	b-value	Active fault sources (ID)
S04	0.783	1
S05A	1.194	2, 3, 4,
S05B	0.811	5, 5BT, 6
S06	0.816	7, 8, 9, 10, 11, 12
S07	0.643	13, 14, 14BT, 15, 16, 17, 18, 20BT, 21
S08A	0.855	19, 20, 22, 23
S12	0.695	24, 25
S17A	0.713	26, 27, 28, 29
S18A	0.616	30, 31, 32, 33

regional sources and crustal active fault sources (Fig. 7), the corresponding b-value for each source was obtained. Table 4 summarizes the b-value for crustal active fault sources.

Subduction Interface Sources

Typical interface sources take place along plate boundaries within subduction zones. Two subduction systems are located within the vicinity of Taiwan (Fig. 1). Subduction interface sources exist in both systems (Fig. 8). In the northeast subduction system, an interface earthquake with M8.2 took place in 1920 (Tsai et al. 2000) and is the largest earthquake ever recorded in T01A. m_u's of 8.0, 8.2, and 8.4 with corresponding weights of 0.2, 0.6, and 0.2 were assumed for this source. The interface in the southern subduction system was defined according to the profiles of earthquake distribution. Three interface sources, T02A, T02B, and T02C, were defined (Fig. 8). Segmentations were defined by the bending of the accretionary wedge. Parameters for subduction interface sources are summarized and presented in Table 5. The characteristic earthquake model provides the behaviors of the seismic activity for the sources.

Subduction Intraslab Sources

The geometries of intraslab sources for subduction systems are illustrated based on the profiles of the earthquake distribution. Ten and three interface sources were defined for the northeastern and southern subduction zones, respectively (Fig. 8). Seismic activities of the sources were determined using the truncated exponential model. The corresponding m_u's were inferred from the maximum magnitudes of the intraslab earthquakes surrounding Taiwan and the world. The parameters and the corresponding weights for subduction intraslab sources are summarized and presented in Table 6.

Ground Motion Prediction Equations (GMPEs)

For the application of the PSHA, in addition to reliable seismogenic sources, another key factor is proper GMPEs (Encyclopedia of Earthquake Engineering: Ground Motion Prediction Equations). Lin (2009) pointed out that global GMPEs do not truly represent ground motion attenuation behaviors in Taiwan. Thus, only GMPEs obtained from the regression of strong ground motion observations in Taiwan were considered. Several studies have proposed GMPEs in order to model the attenuation behaviors for different types of earthquakes in Taiwan. For a more complete presence of attenuation behaviors, GMPEs with the form of an acceleration response spectrum (SA) are expected. In the form of SA, attenuation behaviors are presented as a response acceleration as a function of the response period. Lin (2009) considered 5,968 observations of 60 crustal earthquakes recorded by the Taiwan Strong Motion Instrumentation Program (TSMIP). GMPEs as a function of magnitude, m, are determined as follows:

$$\ln y = C_1 + F_1 + C_3(8.5-m)^2 \\ + (C_4 + C_5(m-6.3))\ln\left(\sqrt{R^2 + \exp(H)^2}\right) \\ + C_6 F_{NM} + C_7 F_{RV} + C_8 \ln(Vs_{30}/1,130.0), \quad (10)$$

where y represents the response acceleration for PGA or SA in g; R represents the shortest distance to the rupture surface in kilometers; F_{NM} is 1.0

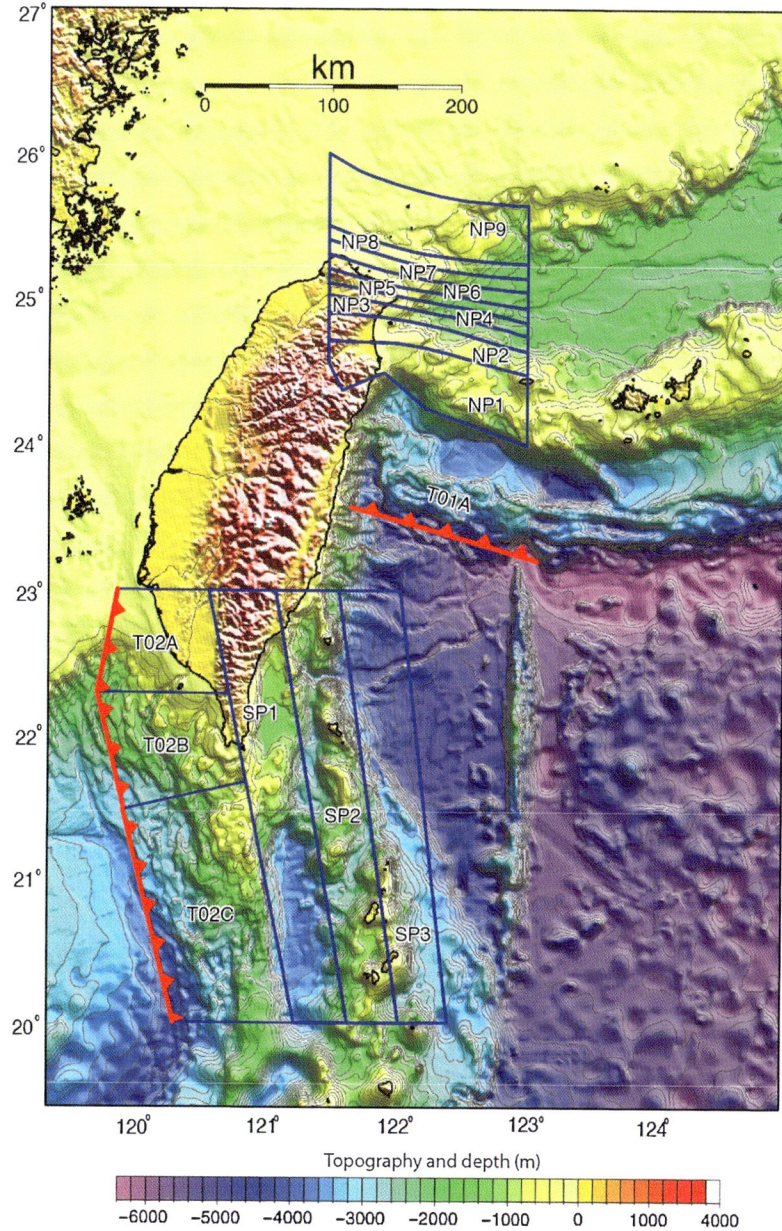

Probability Seismic Hazard Mapping of Taiwan, Fig. 8 The distribution of the subduction interface and the intraslab sources illustrated by *blue polygons*. The surface alignments of the two trenches are presented in *red*

when the sources are normal mechanisms; F_{RV} is 1.0 when the sources are thrust mechanisms; Vs_{30} represents the average shear-wave velocity from the ground surface down to a depth of 30 m; $F_1 = C_{2(M_W - 6.3)}$ for $M_W \leq 6.3$, whereas $F_1 = (-H \cdot C_5)(M_W - 6.3)$ for $M_W > 6.3$; and C_1 to C_8 and H are constants. The corresponding parameters for PGA and SA are presented in Table 7. Note that through these GMPEs, source effects in the form of different focal mechanisms were considered and indicated that the largest amplitudes for thrust events and the smallest for normal ones were obtained based on the same magnitude and distance to the rupture surface. In order to present the reliability of these GMPEs, the corresponding standard deviations were analyzed and are presented as σ_{lny} in Table 8. These GMPEs were applied for crustal sources.

Probability Seismic Hazard Mapping of Taiwan, Table 5 The parameters for five of the subduction interface sources; the corresponding weights for the slip rate, dip angle, and maximum magnitudes, m_u, are denoted in parentheses

NO	m_0	Slip rate (mm/year)	Dip angle (°)	Depth (km)	b-value	m_u
T01A	6.5	20(0.2)	18(0.3)	0–40	0.6	8.0(0.2)
		30(0.6)	20(0.4)			8.2(0.6)
		40(0.2)	22(0.3)			8.4(0.2)
T02A	6.5	5(0.2)	18(0.3)	0–35	0.7	7.6(0.2)
		10(0.6)	20(0.4)			7.8(0.6)
		15(0.2)	22(0.3)			8.0(0.2)
T02B	6.5	5(0.2)	18(0.3)	0–35	0.7	7.7(0.2)
		10(0.6)	20(0.4)			7.9(0.6)
		15(0.2)	22(0.3)			8.1(0.2)
T02C	6.5	5(0.2)	18(0.3)	0–35	0.7	7.9(0.2)
		10(0.6)	20(0.4)			8.1(0.6)
		15(0.2)	22(0.3)			8.3(0.2)

Probability Seismic Hazard Mapping of Taiwan, Table 6 The parameters and the corresponding standard deviations for 13 of the subduction intraslab sources; the corresponding weighs of the maximum magnitudes, m_u, are denoted in parentheses

NO	m_o	$N(m_o)\,(\pm \sigma_{N(m_o)})$	$b(\pm \sigma_b)$	m_u
NP1	4.0	3.735(±0.307)	0.908(±0.063)	7.5(0.2) 7.7(0.6) 7.9(0.2)
NP2	4.0	2.335(±0.247)	0.801(±0.070)	7.5(0.2) 7.7(0.6) 7.9(0.2)
NP3	4.0	1.197(±0.177)	0.866(±0.104)	7.5(0.2) 7.7(0.6) 7.9(0.2)
NP4	4.0	0.580(±0.119)	0.730(±0.124)	7.6(0.2) 7.8(0.6) 8.0(0.2)
NP5	4.0	0.210(±0.074)	0.938(±0.297)	7.6(0.2) 7.8(0.6) 8.0(0.2)
NP6	4.0	0.263(±0.083)	0.959(±0.272)	7.6(0.2) 7.8(0.6) 8.0(0.2)
NP7	4.0	0.344(±0.091)	0.730(±0.173)	7.6(0.2) 7.8(0.6) 8.0(0.2)
NP8	4.0	0.374(±0.098)	1.040(±0.237)	7.6(0.2) 7.8(0.6) 8.0(0.2)
NP9	4.0	0.913(±0.155)	0.913(±0.138)	7.6(0.2) 7.8(0.6) 8.0(0.2)
SP1	4.0	1.041(±0.165)	0.762(±0.106)	7.5(0.2) 7.7(0.6) 7.9(0.2)
SP2	4.0	2.735(±0.268)	0.831(±0.068)	7.6(0.2) 7.8(0.6) 8.0(0.2)
SP3	4.0	1.660(±0.209)	0.876(±0.090)	7.6(0.2) 7.8(0.6) 8.0(0.2)

Many studies (e.g., Chan et al. 2013a) have indicated that ground motion attenuation behaviors for subduction earthquakes are different than those for crustal earthquakes. Since two subduction systems are located within the Taiwan region, as well as its vicinity (Fig. 1), the different GMPEs for both interfaces and intraslab events are considered to apply PSHAs. Lin and Lee (2008) proposed the first GMPEs for subduction events in Taiwan. In order to establish the GMPEs, they analyzed the strong motion records of subduction earthquakes obtained using TSMIP arrays and the Strong Motion Array in Taiwan Phase I (SMARTI). The GMPEs of Lin and Lee (2008) are represented as follows:

$$\ln y = C_1 + C_2 m + C_3 \left(R + C_4 e^{C_5 \cdot m}\right) + C_6 H + C_7 Z_t, \quad (11)$$

where y represents the response acceleration for PGA or SA in g, R represents the hypocentral distance in kilometers, H represents the focal depth in kilometers, Z_t represents the subduction zone earthquake type ($Z_t = 0$ for interface earthquakes, whereas $Z_t = 1$ for intraslab earthquakes), and C_1 to C_7 are constants. The parameters and their corresponding standard

Probability Seismic Hazard Mapping of Taiwan, Table 7 The corresponding parameters and the standard deviations (σ_{lny}) of the GMPEs for the PGA and each response period for crustal events (Eq. 8) as proposed by Lin (2009)

Period (s)	C_1	C_2	C_3	C_4	C_5	H	C_6	C_7	C_8	σ_{lny}
PGA	1.0109	0.3822	0.0000	−1.1634	0.1722	1.5184	−0.1907	0.1322	−0.4741	0.627
0.01	1.0209	0.3822	−0.0003	−1.1633	0.1722	1.5184	−0.1922	0.1314	−0.4738	0.627
0.02	1.0416	0.3822	0.0017	−1.1668	0.1722	1.5184	−0.1942	0.1311	−0.4700	0.627
0.03	1.1961	0.3822	0.0038	−1.2028	0.1722	1.5184	−0.1990	0.1314	−0.4741	0.640
0.04	1.3834	0.3822	0.0087	−1.2499	0.1722	1.5184	−0.1959	0.1362	−0.4806	0.655
0.05	1.5612	0.3822	0.0153	−1.2957	0.1722	1.5184	−0.1922	0.1417	−0.4911	0.670
0.06	1.6907	0.3822	0.0210	−1.3218	0.1722	1.5184	−0.1984	0.1500	−0.4900	0.681
0.07	1.7673	0.3822	0.0261	−1.3336	0.1722	1.5184	−0.2011	0.1557	−0.4920	0.691
0.08	1.8689	0.3822	0.0273	−1.3440	0.1722	1.5184	−0.1947	0.1627	−0.4944	0.699
0.09	1.9430	0.3822	0.0276	−1.3435	0.1722	1.5184	−0.2011	0.1589	−0.4910	0.700
0.10	2.0218	0.3822	0.0254	−1.3409	0.1722	1.5184	−0.1817	0.1607	−0.4825	0.705
0.15	2.0521	0.3822	0.0100	−1.2578	0.1722	1.5184	−0.1851	0.1212	−0.4804	0.691
0.20	2.0333	0.3822	−0.0091	−1.1769	0.1722	1.5184	−0.2265	0.0999	−0.4350	0.676
0.25	1.9887	0.3822	−0.0293	−1.1153	0.1722	1.5184	−0.2355	0.0994	−0.4101	0.679
0.30	1.8827	0.3822	−0.0459	−1.0726	0.1722	1.5184	−0.2163	0.1036	−0.4361	0.686
0.35	1.7459	0.3822	−0.0600	−1.0307	0.1722	1.5184	−0.1949	0.1029	−0.4507	0.692
0.40	1.6821	0.3822	−0.0737	−1.0116	0.1722	1.5184	−0.1955	0.1099	−0.4734	0.695
0.45	1.6139	0.3822	−0.0861	−0.9939	0.1722	1.5184	−0.2011	0.1178	−0.4927	0.699
0.50	1.5288	0.3822	−0.0960	−0.9755	0.1722	1.5184	−0.2089	0.1142	−0.5035	0.699
0.60	1.3081	0.3822	−0.1133	−0.9407	0.1722	1.5184	−0.2212	0.1016	−0.5546	0.704
0.70	1.1383	0.3822	−0.1292	−0.9193	0.1722	1.5184	−0.1900	0.1036	−0.6037	0.710
0.80	1.0757	0.3822	−0.1442	−0.9167	0.1722	1.5184	−0.1865	0.1058	−0.6319	0.718
0.90	0.9935	0.3822	−0.1577	−0.9104	0.1722	1.5184	−0.1643	0.1165	−0.6577	0.723
1.00	0.8642	0.3822	−0.1687	−0.9001	0.1722	1.5184	−0.1505	0.1372	−0.6916	0.728
1.50	0.3150	0.3822	−0.2006	−0.8696	0.1722	1.5184	−0.0377	0.1572	−0.7582	0.738
2.00	−0.1760	0.3822	−0.2190	−0.8328	0.1722	1.5184	0.0780	0.1660	−0.7863	0.726
2.50	−0.4103	0.3822	−0.2319	−0.8415	0.1722	1.5184	0.0907	0.1648	−0.7939	0.709
3.00	−0.5019	0.3822	−0.2431	−0.8684	0.1722	1.5184	0.1195	0.1790	−0.7754	0.707
3.50	−0.7206	0.3822	−0.2479	−0.8689	0.1722	1.5184	0.1206	0.1629	−0.7673	0.708
4.00	−0.9383	0.3822	−0.2493	−0.8618	0.1722	1.5184	0.1267	0.1262	−0.7457	0.707
4.40	−1.0405	0.3822	−0.2559	−0.8472	0.1722	1.5184	0.1655	0.1486	−0.7042	0.717
5.00	−1.3694	0.3822	−0.2535	−0.8287	0.1722	1.5184	0.2208	0.1648	−0.6955	0.715

deviations for PGA and SA are presented in Table 8 and indicate the largest amplitude for intraslab events than interface ones when the magnitude and the hypocentral distance are the same. Since the GMPEs of Lin and Lee (2008) are among the only GMPEs from the regression of ground motion observations in Taiwan, they were applied to this study.

Seismic Hazards in Taiwan

The Hazard Maps

By considering all of the seismogenic sources, the corresponding parameters, and the GMPEs, a PSHA was applied for Taiwan. The spatial distributions of seismic hazards were evaluated in the form of a seismic hazard map. The seismic hazards for 10 % (475-year return period) and 2 % (2,475-year return period) of the probability of exceedance for 50 years are presented (Fig. 9). The results suggest that seismic hazards in Taiwan are mainly contributed by active fault sources. The highest seismic hazard levels were determined from the Western Foothills to the Coastal Plain. A >1.0 g hazard for a 10 % probability of exceedance for 50 years was estimated near the Chiayi blind fault and the Tachienshan Fault (Fig. 9a). A high hazard was also evaluated along the eastern coastline and can be attributed

Probability Seismic Hazard Mapping of Taiwan, Table 8 The corresponding parameters and the standard deviations (σ_{lny}) of the GMPEs for the PGA and each response period for subduction events (Eq. 9) as proposed by Lin and Lee (2008)

Period	C1	C2	C3	C4	C5	C6	C7	σ_{lny}
PGA	−0.9000	1.0000	−1.9000	0.9918	0.5263	0.0040	0.3100	0.6277
0.010	−2.2000	1.0850	−1.7500	0.9918	0.5263	0.0040	0.3100	0.5800
0.020	−2.2900	1.0850	−1.7300	0.9918	0.5263	0.0040	0.3100	0.5730
0.030	−2.3400	1.0950	−1.7200	0.9918	0.5263	0.0040	0.3100	0.5774
0.040	−2.2150	1.0900	−1.7300	0.9918	0.5263	0.0040	0.3100	0.5808
0.050	−1.8950	1.0550	−1.7550	0.9918	0.5263	0.0040	0.3100	0.5937
0.060	−1.1100	1.0100	−1.8360	0.9918	0.5263	0.0040	0.3100	0.6123
0.090	−0.2100	0.9450	−1.8900	0.9918	0.5263	0.0040	0.3100	0.6481
0.100	−0.0500	0.9200	−1.8800	0.9918	0.5263	0.0040	0.3100	0.6535
0.120	0.0550	0.9350	−1.8950	0.9918	0.5263	0.0040	0.3100	0.6585
0.150	−0.0400	0.9550	−1.8800	0.9918	0.5263	0.0040	0.3100	0.6595
0.170	−0.3400	1.0200	−1.8850	0.9918	0.5263	0.0040	0.3100	0.6680
0.200	−0.8000	1.0450	−1.8200	0.9918	0.5263	0.0040	0.3100	0.6565
0.240	−1.5750	1.1200	−1.7550	0.9918	0.5263	0.0040	0.3100	0.6465
0.300	−3.0100	1.3150	−1.6950	0.9918	0.5263	0.0040	0.3100	0.6661
0.360	−3.6800	1.3800	−1.6600	0.9918	0.5263	0.0040	0.3100	0.6876
0.400	−4.2500	1.4150	−1.6000	0.9918	0.5263	0.0040	0.3100	0.7002
0.460	−4.7300	1.4300	−1.5450	0.9918	0.5263	0.0040	0.3100	0.7092
0.500	−5.2200	1.4550	−1.4900	0.9918	0.5263	0.0040	0.3100	0.7122
0.600	−5.7000	1.4700	−1.4450	0.9918	0.5263	0.0040	0.3100	0.7280
0.750	−6.4500	1.5000	−1.3800	0.9918	0.5263	0.0040	0.3100	0.7752
0.850	−7.2500	1.5650	−1.3250	0.9918	0.5263	0.0040	0.3100	0.7931
1.000	−8.1500	1.6050	−1.2350	0.9918	0.5263	0.0040	0.3100	0.8158
1.500	−10.3000	1.8000	−1.1650	0.9918	0.5263	0.0040	0.3100	0.8356
2.000	−11.6200	1.8600	−1.0700	0.9918	0.5263	0.0040	0.3100	0.8474
3.000	−12.6300	1.8900	−1.0600	0.9918	0.5263	0.0040	0.3100	0.8367
4.000	−13.4200	1.8700	−0.9900	0.9918	0.5263	0.0040	0.3100	0.7937
5.000	−13.7500	1.8350	−0.9750	0.9918	0.5263	0.0040	0.3100	0.7468

to crustal active fault sources with short recurrence intervals along the Longitudinal Valley (Fault ID 26–33 in Table 3). In contrast, in northern Taiwan, a low hazard level was obtained. Although the Sanchiao Fault is located in this region, the recurrence intervals of the fault are rather long (in between 614 and 2,511 years).

The Hazard Curves

In order to represent the probability of annual exceedance as a function of ground motion level for different response periods, hazard curves are presented. The hazard curves were evaluated for the six municipalities, i.e., the Taipei, New Taipei, Taoyuan, Taichung, Tainan, and Kaohsiung Cities, respectively (Fig. 10). The hazard curves for PGA and the response period of 0.3 and 1.0 s are obtained. Since some active faults are close to the Tainan and Taichung Cities (Fig. 9), higher hazards were determined. By contrast, the hazards in the Taipei, New Taipei, Taoyuan, and Kaohsiung cities are relatively low because of they are further away from crustal active fault sources with short recurrence intervals.

Discussion and Conclusions

A Comparison to Other PSHA Results

Probabilistic seismic hazards imparted by different seismogenic sources for the Taiwan region

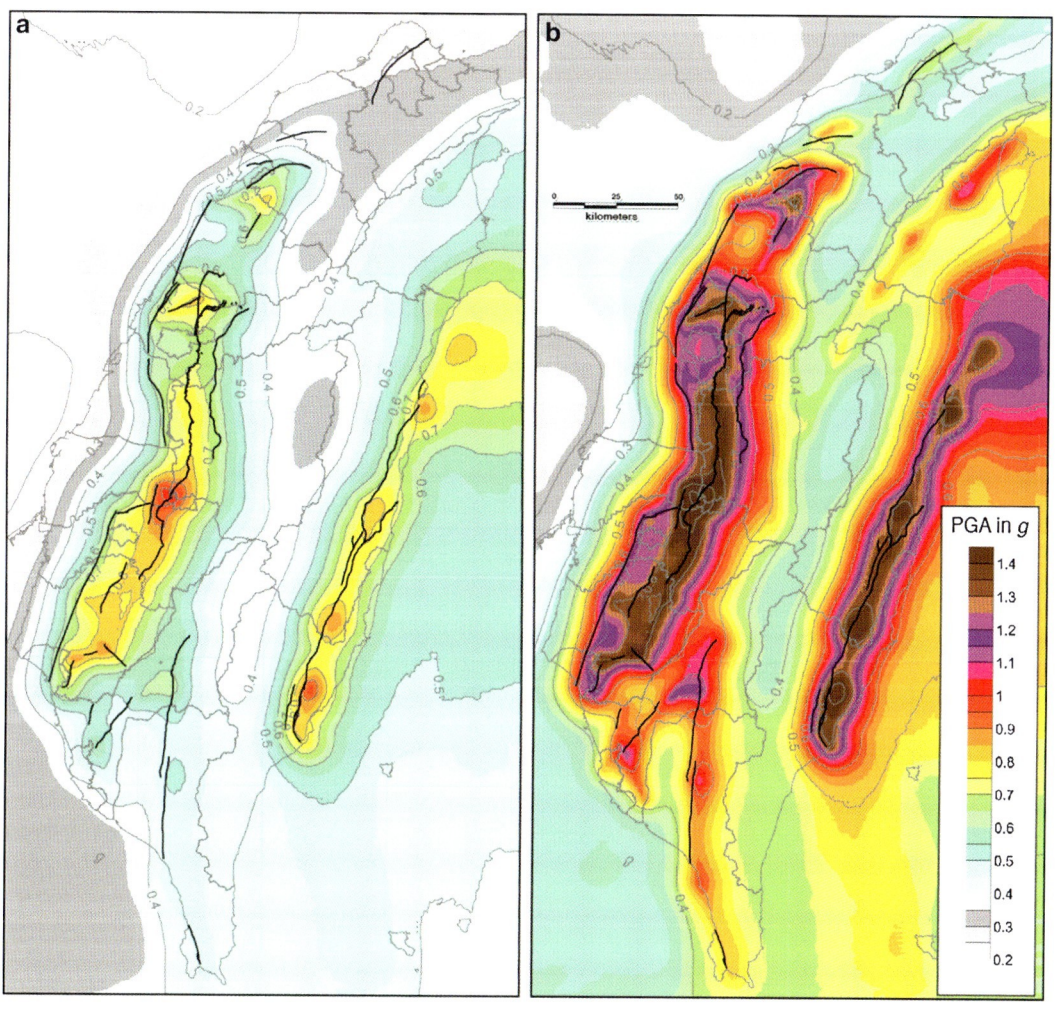

Probability Seismic Hazard Mapping of Taiwan, Fig. 9 The seismic hazard maps in (**a**) a 10 % (a 475-year return period) and (**b**) a 2 % (a 2,475-year return period) probability of exceedance for 50 years

were assessed. Higher hazards were determined along active faults in the Coastal Plain and the Longitudinal Valley. The pattern can be associated with the distribution of seismic activity in Taiwan. The pattern is similar to those obtained from previous studies (e.g., Cheng et al. 2007). In more detail, hazard peaks were evaluated near crustal active fault sources with short recurrence intervals. The results were similar to those obtained from other PSHA studies (e.g., Campbell et al. 2002; Cheng et al. 2007), which also considered the hazards imparted by active faults. The pattern can be attributed to the behaviors of GMPEs (Lin 2009), illustrating the significance of larger ground shakings surrounding regions close to earthquake epicenters or rupture faults. The attenuation behavior is similar to observations of the 1999 Chi-Chi, Taiwan, earthquake. The shake map of this earthquake indicated significant high ground shakings in the form of PGA in the vicinity of the Chelungpu Fault, which ruptured during the coseismic period.

The Importance of the Short-Term Probabilistic Seismic Hazard Assessment

Recently, the concept of a traditional PSHA has been called into question. One of the standard procedures for traditional PSHAs is the

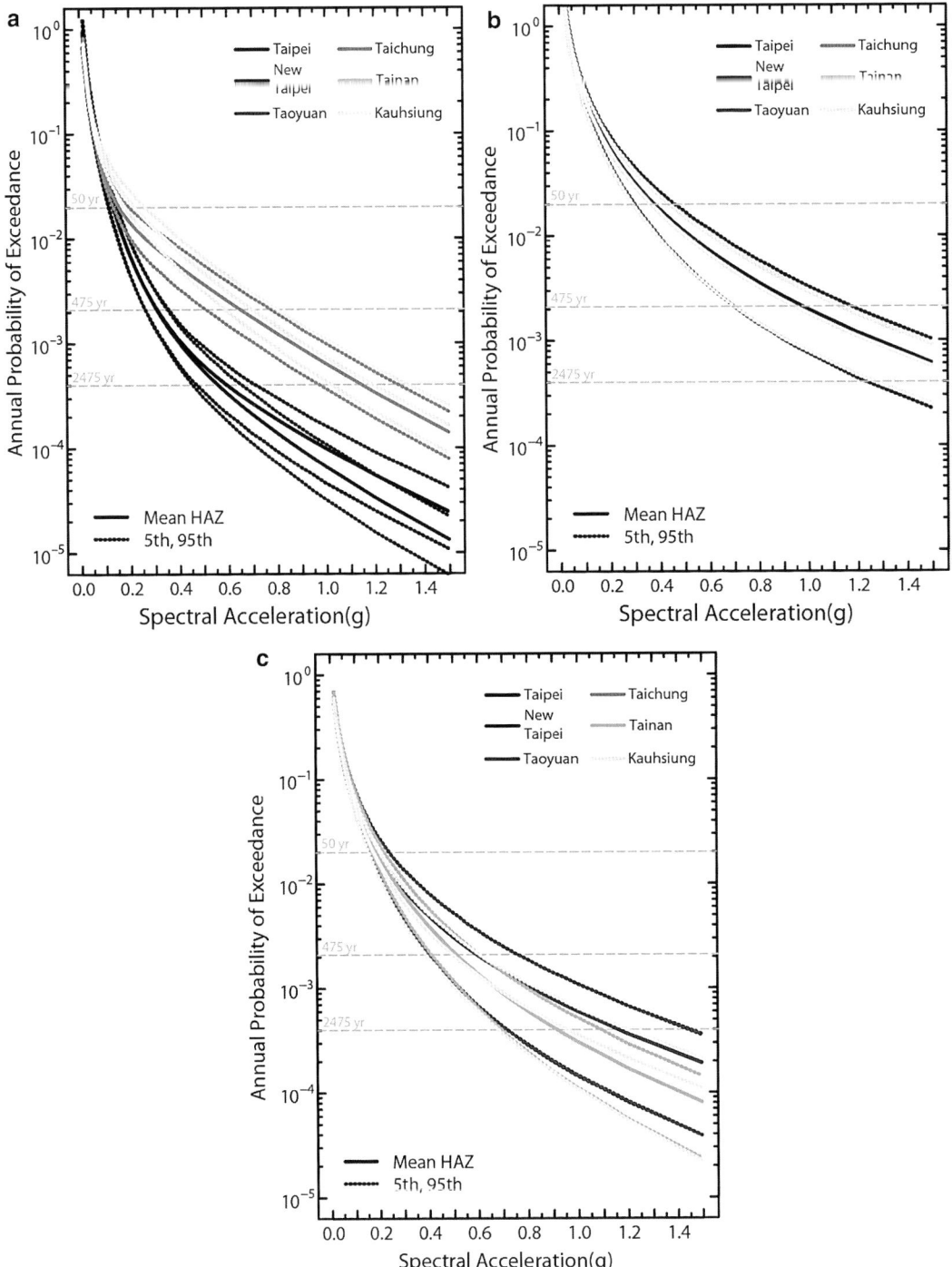

Probability Seismic Hazard Mapping of Taiwan, Fig. 10 The probability of annual exceedance as a function of hazard curves for PGA and the response period of 0.3 and 1.0 s for the six municipalities. The *solid* and *dashed lines* represent mean and two standard deviations of seismic hazards, respectively

construction of return periods for characteristic earthquakes that are independent of one another (see section "The Declustering Process" for details). Thus, an earthquake should be declustered prior to its incorporation for a PSHA. However, some recent cases have determined disadvantages for traditional PSHAs. For example, the M_w 6.3, 21 February 2011, Christchurch earthquake can be regarded as an aftershock of the M_w 7.1 Darfield earthquake that took place on 4 September 2010. The Christchurch earthquake caused severe damage in downtown Christchurch due to a closer epicentral distance. Additionally, not only can aftershocks lead to devastating seismic hazards, but the next earthquake could expand hazards. An earthquake with a M_w 7.4 took place off the Pacific coast of Tohoku, Japan, on 9 March 2011. Due to its distance from urban regions, the resulting damage from this earthquake was negligible. Fifty-one hours after the earthquake (on March 11), a M_w 9.1 earthquake took place in the vicinity and resulted in disasters in Japan. Seismic hazards may also result from several events within an earthquake sequence. Based on spatial and temporal relationships, the 1904 Touliu and the 1906 Yanshuigang earthquakes can be regarded as a foreshock and an aftershock of the 1906 Meishan earthquake (Fig. 2), respectively. However, all three earthquakes caused severe disasters. The case indicates the importance of the seismic hazards imparted by each event in an earthquake sequence and raises the unsuitability of traditional PSHAs (Chan et al. 2013b). Such instances indicate the importance of an earthquake sequence in respect to seismic hazard evaluations and suggest the reevaluation of seismic hazards immediately following large earthquakes.

Several studies have recognized disadvantages in traditional PSHAs and have considered a time dependency for PSHAs. For example, Chan and Wu (2012) and references therein estimated the temporal evolution of the seismicity rate and assessed seismicity hazards as a function of time. The feasibility of this approach has been tested using applications to several cases in Taiwan (e.g., Chan et al. 2013a, b).

The Applicability of PSHA Results and Seismogenic Parameters

An assessment of seismic hazards had been widely applied for each administrative region in Taiwan. In addition, hazard curves for the six municipalities were evaluated. According to the results, the probability of annual exceedance as a function of ground motion level for different response periods was presented. They would be valuable to lawmakers for determining building codes and to decision-makers for the site selection of public structures.

In addition, based on the PSHA results, further applications in respect to seismic hazard mitigation can be assessed (e.g., probabilistic seismic risk assessment). A risk assessment is presented as the probability that a given loss of property and human life exceeds a specific value during a time period. For its application, seismic hazards, exposure (distribution of population or construction), and a corresponding vulnerability are incorporated. The result could be of benefit to decision-makers for territorial planning.

The parameters implemented for the PSHA were acquired through the integration of state-of-the-art information regarding the tectonic setting, geology, geomorphology, earthquake catalog, geophysics, and the ground motion attenuation. The information could be used as a reference for an additional PSHA. For example, these parameters will be regarded as a branch in the logic tree approach for the Committee of Taiwan Earthquake Model (TEM, http://tec.earth.sinica.edu.tw/TEM). TEM is working on a seismic hazard map for the entire region of Taiwan. In the meantime, the work can also be incorporated within the Global Earthquake Model (see entry "The Global Earthquake Model (GEM)"), integrated by several research institutes, insurance, and reinsurance enterprises in order to evaluate seismic hazards around the world. Through collaboration with organizations on different scales and with different individuals, the GEM is expected to establish uniform and open standards for calculating and communicating earthquake hazards and risks worldwide.

Summary

A probabilistic seismic hazard assessment is implemented to the region of Taiwan. Seismogenic sources, which may result in seismic hazards, were divided into the following categories: shallow-crust regional sources, deep-crust regional sources, crustal active fault sources, subduction intraslab sources, and subduction interface sources. By further considering ground motion prediction equations for different types of sources and site conditions, hazard maps in Taiwan and hazard curves in the six municipalities were assessed.

Cross-References

▶ Site Response for Seismic Hazard Assessment

References

Arabasz WJ, Robinson R (1976) Microseismicity and geologic structure in the northern south island, New Zealand. J Geol Geophys 19:569–601

Campbell KW, Thenhaus PC, Barnard TP, Hampson DB (2002) Seismic hazard model for loss estimation and risk management in Taiwan. Soil Dyn Earthq Eng 22:743–754. doi:10.1016/S0267-7261(02)00095-7

Chan CH, Wu YM (2012) The burst of seismicity after the 2010 M6.4 Jiashian earthquake and its implication of short-term seismic hazard in southern Taiwan. J Asian Earth Sci 51:231–239. doi:10.1016/j.jseaes.2012.08.011

Chan CH, Wu YM, Cheng CT, Lin PS, Wu YC (2013a) Time-dependent probabilistic seismic hazard assessment and its application to Hualien City, Taiwan. Nat Hazards Earth Syst Sci 13:1–16. doi:10.5194/nhess-13-1-2013

Chan CH, Wu YM, Cheng CT, Lin PS, Wu YC (2013b) Scenario for a short-term probabilistic seismic hazards assessment (PSHA) in Chiayi, Taiwan. Terr Atmos Ocean Sci 24(4, Part II):671–683. doi:10.3319/TAO.2013.01.22.01(T)

Chen C-H, Wang J-P, Yih-Min W, Chan C-H, Chang C-H (2012) A study of earthquake inter-occurrence times distribution models in Taiwan. Nat Hazards. doi:10.1007/s11069-012-0496-7

Cheng CT, Chiou SJ, Lee CT, Tsai YB (2007) Study on probabilistic seismic hazard maps of Taiwan after Chi-Chi earthquake. J Geo Eng 2:19–28

Cornell CA (1968) Engineering seismic risk analysis. Bull Seismol Soc Am 58(5):1583–1606

Gardner JK, Knopoff L (1974) Is the sequence of earthquakes in southern California, with aftershocks removed, poissonian? Bull Seismol Soc Am 64:1363–1367

Gutenberg B, Richter C (1954) Seismicity of the earth and associated phenomena, 2nd edn. Princeton University Press, Princeton, 310

Hanks TC, Kanamori H (1979) A moment magnitude scale. J Geophys Res 84:2348–2350

Hsu Y-J, Yu S-B, Simons M, Kuo L-C, Chen H-Y (2009) Interseismic crustal deformation in the Taiwan plate boundary zone revealed by GPS observations, seismicity, and earthquake focal mechanisms. Tectonophysics 479:4–18

Kiureghian AD, Ang AH-S (1977) A fault-rupture model for seismic risk analysis. Bull Seismol Soc Am 67:1173–1194

Kramer SL (1996) Geotechnical earthquake engineering. Prentice Hall, Upper Saddle River, 653 pp

Lin PS (2009) Ground-motion attenuation relationship and path-effect study using Taiwan Data set. PhD dissertation, Institute of Geophysics, National Central University, Chung-Li, Taiwan (in Chinese)

Lin PS, Lee CT (2008) Ground-motion attenuation relationships for subduction-zone earthquakes in northeastern Taiwan. Bull Seismol Soc Am 98(1):220–240. doi:10.1785/0120060002

Tsai YB, Wen KL, Chen KB, Kuo JY (2000) Summary of development of the Taiwan earthquake catalog and strong ground motion attenuation equations. Report for the National Project on the Disaster Prevention, 79 pp (in Chinese)

Uhrhammer RA (1986) Characteristic of northern and central California seismicity. Earthquake Notes 57(1):21 (Abstract)

Wesnousky SG (1994) The Gutenberg-Richter or characteristic earthquake distribution, which is it? Bull Seismol Soc Am 84:1940–1959

Wu YM, Hsu YJ, Chang CH, Teng LS, Nakamura M (2010) Temporal and spatial variation of stress field in Taiwan from 1991 to 2007: insights from comprehensive first motion focal mechanism catalog. Earth Planet Sci Lett 298:306–316. doi:10.1016/j.epsl.2010.07.047

Wyss M (1979) Estimating maximum expectable magnitude of earthquakes from fault dimensions. Geology 7:336–340

Youngs RR, Coppersmith KJ (1985) Implications of fault slip rates and earthquake recurrence models to probabilistic seismic hazard estimates. Bull Seismol Soc Am 75:939–964